HANDBOOK OF
ION CHANNELS

HANDBOOK OF ION CHANNELS

EDITED BY

JIE ZHENG
UNIVERSITY OF CALIFORNIA AT DAVIS, USA

MATTHEW C. TRUDEAU
UNIVERSITY OF MARYLAND, BALTIMORE, USA

CRC Press
Taylor & Francis Group
Boca Raton London New York

CRC Press is an imprint of the
Taylor & Francis Group, an **informa** business

MATLAB® is a trademark of The MathWorks, Inc. and is used with permission. The MathWorks does not warrant the accuracy of the text or exercises in this book. This book's use or discussion of MATLAB® software or related products does not constitute endorsement or sponsorship by The MathWorks of a particular pedagogical approach or particular use of the MATLAB® software.

Cover Image. The background is a starburst dendrogram representing the relationship of 143 different ion channels in the human genome. The foreground is the x-ray crystal structure of the sodium channel NavAb. Courtesy of William Catterall.

CRC Press
Taylor & Francis Group
6000 Broken Sound Parkway NW, Suite 300
Boca Raton, FL 33487-2742

Printed on acid-free paper
Version Date: 20141211

International Standard Book Number-13: 978-1-4665-5140-4 (Hardback)

Visit the Taylor & Francis Web site at
http://www.taylorandfrancis.com

and the CRC Press Web site at
http://www.crcpress.com

Contents

Foreword: Early days of ion channels

Throughout their long scientific history, the concepts of cellular electricity and ion-permeable pores have been principally the province of scientists with a physical orientation who often called themselves biophysicists.* Only in the last three decades have ion channels become accessible to a much broader range of scientists. In addition to biophysicists, ion channels are now studied by physiologists, biologists, biochemists, anatomists, molecular biologists, structural biologists, and physicians.

THE DAWN OF MOLECULAR PORES

General physiologists had already postulated molecular pores in membranes several centuries ago. In the late 1700s, osmosis was discovered by Jean-Antoine Nollet. To account for osmosis, Ernst von Brücke (1843) postulated pores (*Kanäle*) no more than a few water molecules wide. He imagined capillary canals in a porous membrane. Soon, Carl Ludwig (1852, 1856) adopted Brücke's pore theory to explain ultrafiltration in the kidney glomerulus, and Adolph Fick (1855) devoted much space to discussing diffusion of salt and water in molecular pores of *membranes* in his paper proposing Fick's first and second laws of diffusion. Such important concepts were discussed by these *physiologists/physicians* before the size or even the clear existence of molecules was known and before the word *membrane* connoted anything but some multicellular sheet, like an epithelium, or even a sheet of parchment. Nevertheless, from 1850 onward, mechanistic books of physiology would discuss the possibility of semipermeability and *membrane pores* in an early section on general physiology as an explanation for osmosis and filtration. German books used the word *Kanäle* and English books said *pore*. In each generation, various new investigators propounded and were given credit for the pore theory, including, for example, in the 1920s and 1930s, Leonor Michaelis of Michaelis–Menten kinetics. From his experiments on artificial collodion membranes, Michaelis (1925) proposed that permeation of small ions might be controlled both by the charge on the pore and by the pore diameter relative to hydrated ions.

ANIMAL ELECTRICITY

Again in the late 1700s, phenomena of animal electricity were first described and debated by Luigi Galvani, Alessandro Volta, Benjamin Franklin, and others. The electric eel soon became a clear exemplar. Starting in 1848, Emil du Bois-Reymond discovered the propagated action potential, and Hermann von Helmholtz measured its conduction velocity in the nerve. For many decades, this velocity measurement would be made with elaborate mechanical devices that could close and open contact switches in the right sequence first to a stimulator and then to a galvanometer recording from extracellular electrodes. Michael Faraday (1834) had postulated that salts contain charged particles and gave the names ion, anion, and cation to them. Eventually, Svante Arrhenius (1887) advocated that these ions dissociate fully in *strong* electrolyte solutions, leading Walther Nernst and Max Planck (1888–1890), while still in their 20s, to formulate diffusion equations for dissolved ions in a thick membrane that included electrical forces as well as concentration forces. Nernst (1888) gave the membrane potential between two salt solutions differing only by the salt concentration. His equation depended on a difference in mobility between the anion and the cation, and if one of them was immobile, resulted in the equilibrium equation we now call the Nernst equation (1889). Nernst and others suggested that ionic gradients might be the source of animal electricity. Using Nernst's theory, Ludwig Bernstein (1902, 1912) (1) hypothesized that muscle and nerve cells are surrounded by a semipermeable membrane that he called the plasma membrane and (2) proposed that there is a negative internal resting potential, caused by potassium selectivity in the membrane. He suggested that the depolarizing action potential is caused by "an increase in the permeability for the impeded ion [chloride] as the result of a chemical change in the plasma membrane." This concept, which he called *membrane theory*, set the stage for understanding the electricity of excitable cells.

THE AXON AS A CABLE

In the mid-1800s, Lord Kelvin developed the equations for decrement of electrical signals in submarine cables. Ludimar Hermann regarded nerve and muscle fibers as *core conductors* similar to cables. He used flow of current down the axis and through the membrane to explain electrotonus in nerve in many publications including his famous *Handbuch der Physiologie* (1879). The cable ideas were given their contemporary form in the later original works of Cole, Hodgkin, Rushton, Lorente de Nó, Rall, and others (1920–1970).

YOUNG HODGKIN

Alan Hodgkin (1914–1998) came to Trinity College, Cambridge, as an undergraduate in 1932. He was extroverted and liberal, and had a brilliant mind. He was persuaded to study physiology. Trinity College had many leading scientists and an extraordinary tradition in mechanistic thinking about the action potential. Keith-Lucas, Edgar Lord Adrian, and William Rushton formed one Trinity lineage. After his studies, Hodgkin took up the problem of how the action potential propagates. One school of thought (Joseph Erlanger), which Hodgkin called the St. Louis school, held that the propagated impulse is a chemical reaction occurring inside the axon, whereas the external electrical action potential is simply an epiphenomenon that reports the fundamental chemical changes going on inside. In contrast, the Cambridge school thought the potentials and currents were essential for propagation and held to Bernstein's membrane theory. Working with frog and crab nerves in 1935 and 1936, Hodgkin (1937a, b) showed that *local circuit currents* can flow forward across a cold-blocked region of nerve into the unexcited

* The ideas of this essay are described and documented much more fully in Hille (2001).

region to lower the threshold for excitation ahead. His two papers, "Evidence for electrical transmission in nerve" stood in contrast to the St. Louis view. Then he took a year abroad, visiting the Herbert Gasser lab in New York. Gasser had introduced improved electronic methods to neurophysiology, the cathode ray oscilloscope, high-impedance cathode followers, and high differential-rejection amplifiers—techniques that Hodgkin absorbed and soon reconstructed in Cambridge. Hodgkin also visited and worked with Kenneth Cole and Howard Curtis in the Woods Hole Oceanographic Institution for a month, learning to use the squid giant axon and having his first experience at experimenting side by side with other scientists.

HUXLEY JOINS HODGKIN

Andrew Huxley (1917–2012) began in Trinity College as an undergraduate in 1935. He was introverted and precise, had exceptional patience, and was already very well versed in machining, designing and building instruments, numerical methods, and optics. In 1939, Hodgkin invited him to go to Plymouth to undertake experiments on the British squid. In this first-ever clear recording of the intracellular action potential, they discovered that the voltage spike had a considerable overshoot beyond 0 mV (Hodgkin and Huxley 1939), unlike the prediction of Bernstein's theory. They were 21 and 25 years old. War in Europe stopped this work within weeks and both investigators joined the war effort, with Hodgkin designing and testing shorter-wavelength radar and Huxley in the gunnery calculating trajectories for ordnance. They acquired a new intensity, urgency, and drive in their work habits. The Plymouth lab was destroyed, and only in 1947 was it rebuilt. Hodgkin wanted to test a Na^+ hypothesis for the overshoot, and since Huxley was not available, Bernard Katz joined him in 1947 for the successful proof (Hodgkin and Katz 1949). The next year, Hodgkin visited Cole again and saw the introduction of an axial wire into the giant axon and the application of feedback to make current clamp and voltage clamp. Immediately, Hodgkin and Huxley joined forces again to do their famous experiments (1948–1950) with voltage clamp, and during 1951 they completed a marathon of analysis, modeling, and calculations—more than 60 years ago.

Their first paper from this work showed that the voltage clamp measures the same currents as the more familiar current clamp, gave the peak and steady-state current-voltage relations, and refined methodological details such as series resistance compensation (Hodgkin et al. 1952). The second paper revealed that time-varying inward and outward currents could be separated into components carried by Na^+ and by K^+ (Hodgkin and Huxley 1952a). The changes of Na^+ current with Na^+ concentration obeyed the independence principle of Hodgkin and Katz (1949). The third paper developed the concept of instantaneous conductance, showing that the current-carrying mechanisms obey Ohm's Law and could be represented as time-varying Na^+ and K conductances and each with Nernstian driving forces (Hodgkin and Huxley 1952b). The fourth paper discovered that the sodium conductance has two kinetic properties that they called activation and inactivation (Hodgkin and Huxley 1952c). And the final paper cast all the kinetic discoveries as a kinetic model and showed that the model predicts all the electrical responses (Hodgkin and Huxley 1952d). Huxley had the idea of representing the gating processes in terms of m^3 h and n^4, and he calculated the predicted action potentials for months on his mechanical hand calculator combining cable theory with the new time-varying conductances.

This final paper introduced the concept of a Boltzmann distribution affecting the disposition of hypothetical regulatory charged particles that we now call gating charges. Presciently, the model corresponded to four sets of charged particles that we now identify as four voltage sensors. It was a disappointment to Hodgkin that the model had too little to say about how the ions crossed the membrane and how this was gated. There was nothing about mechanisms of selectivity and permeation. The words *channel* and *pore* were not used. Only in one later work of Hodgkin and Keynes (1955) on flux ratios did a permeation concept come up. They wrote:

> The very large departures from the independence relation described in this paper can be explained by assuming that K^+ ions tend to move through the membrane in narrow channels, or along chains of sites such as might be provided by the negative charges of a cation exchange resin.... The interaction between potassium ions is of the kind expected in a system in which ions move through the membrane in single file.

Despite the very impressive experiments and theory behind this latter idea (so deferentially and cautiously stated!), it was almost another ten years before other investigators gave the channel concept serious thought and began to exploit the voltage clamp in earnest again. After their brilliant partnership and perfect synergy, Hodgkin and Huxley then independently turned away to other important physiological problems.

THE ION CHANNEL ERA

Ion channels in a more modern sense had their birth in Clay Armstrong's mind, and very quickly I followed, thinking along the same lines. In 1964, Clay had become a postdoctoral fellow (with Cole), working on squid giant axons, and I began as a graduate student to voltage-clamp frog nodes of Ranvier (as taught to me by Fred A. Dodge). Our preconception was that the action potential uses two very distinct gated permeation mechanisms called Na^+ channels and K^+ channels. They are macromolecular aqueous pores made of protein. Charges, potentials, and mechanical interactions at the atomic level govern and catalyze the passage of ions. Gating is a conformational change of the protein. We coined words like inner and outer vestibules, selectivity filter, gating current, and voltage sensor, placing the gate at the cytoplasmic end of the pore. We built our own apparatus and digitized and analyzed ionic currents on the first laboratory computers. There may have been only a dozen people actually interested in these questions. Only gradually did we and many others advance enough evidence that such more molecular channel ideas could be asserted with certainty. The most important technical leaps in the field came when voltage clamp became available to thousands of people by the development and commercialization of the patch clamp in the Neher and Sakmann lab (Hamill et al. 1981) and the ready availability of personal computers with acquisition and analysis programs. This was paralleled by another biomedical revolution, the flowering of molecular biology. Today we have single-channel recording, genes, hundreds of ion channel subtypes, evolutionary trees, crystal structures, 300,000 papers, many targeted drugs, and very many investigators contributing to this vigorous and mainstream field. The chapters in this volume are testimony to that beautiful success, with some bias toward the biomedical side.

Bertil Hille

ACKNOWLEDGMENTS

Our laboratory has been supported for 45 years by the National Institutes of Health (NIH) grant NS08174.

REFERENCES

Arrhenius, S. A. 1887. Über die Dissociation der in Wasser Gelösten Stoffe. *Z. Phys. Chem. (Leipzig)* 1: 631–648.

Bernstein, J. 1902. Untersuchungen zur Thermodynamik der bioelektrischen Ströme. Erster Theil. *Pflügers Arch.* 92: 521–562.

Bernstein, J. 1912. *Elektrobiologie*. Braunschweig, Germany: Vieweg.

Brücke, E. 1843. Beiträge zur Lehre von der Diffusion tropfbarflüssiger Körper durch poröse Scheidenwände. *Ann. Phys. Chem.* 58: 77–94.

Faraday, M. 1834. Experimental researches on electricity. Seventh series. *Phil. Trans. R. Soc. Lond.* 124: 77–122.

Fick, A. 1855. Ueber Diffusion. *Ann. Phys. Chem.* 94: 59–86.

Hamill, O. P., A. Marty, E. Neher, B. Sakmann, and F. J. Sigworth. 1981. Improved patch-clamp techniques for high-resolution current recording from cells and cell-free membrane patches. *Pflügers Arch.* 391: 85–100.

Hermann, L. 1879. *Handbuch der Physiologie*, Vol. 2. Leipzip, Germany: Verlag von F.C.W Vogel.

Hille, B. 2001. *Ion Channels of Excitable Membranes*, 3rd edn. Sunderland, MA: Sinauer Associates.

Hodgkin, A. L. 1937a. Evidence for electrical transmission in nerve. Part I. *J. Physiol. (Lond.)* 90: 183–210.

Hodgkin, A. L. 1937b. Evidence for electrical transmission in nerve. Part II. *J. Physiol. (Lond.)* 90: 211–232.

Hodgkin, A. L. and A. F. Huxley. 1939. Action potentials recorded from inside a nerve fibre. *Nature (Lond.)* 144: 710–711.

Hodgkin, A. L. and A. F. Huxley. 1952a. Currents carried by sodium and potassium ions through the membrane of the giant axon of *Loligo. J. Physiol. (Lond.)* 116: 449–472.

Hodgkin, A. L. and A. F. Huxley. 1952b. The components of membrane conductance in the giant axon of *Loligo. J. Physiol. (Lond.)* 116: 473–496.

Hodgkin, A. L. and A. F. Huxley. 1952c. The dual effect of membrane potential on sodium conductance in the giant axon of *Loligo. J. Physiol. (Lond.)* 116: 497–506.

Hodgkin, A. L. and A. F. Huxley. 1952d. A quantitative description of membrane current and its application to conduction and excitation in nerve. *J. Physiol. (Lond.)* 117: 500–544.

Hodgkin, A. L., A. F. Huxley, and B. Katz. 1949. Ionic currents underlying activity in the giant axon of the squid. *Arch. Sci. Physiol.* 3: 129–150.

Hodgkin, A. L. and B. Katz. 1949. The effect of sodium ions on the electrical activity of the giant axon of the squid. *J. Physiol. (Lond.)* 108: 37–77.

Hodgkin, A. L. and R. D. Keynes. 1955. The potassium permeability of a giant nerve fibre. *J. Physiol. (Lond.)* 128: 61–88.

Ludwig, C. 1852. *Lehrbuch der Physiologie des Menschen*, Vol. 1. Heidelberg, Germany: C.F. Winter'sche Verlagshandlung.

Ludwig, C. 1856. *Lehrbuch der Physiologie des Menschen*, Vol. 2. Heidelberg, Germany: C.F. Winter'sche Verlagshandlung.

Michaelis, L. 1925. Contribution to the theory of permeability of membranes for electrolytes. *J. Gen. Physiol.* 8: 33–59.

Nernst, W. 1888. Zur Kinetik der in Lösung befindlichen Körper: Theorie der Diffusion. *Z. Phys. Chem.* 613–637.

Nernst, W. 1889. Die elektromotorische Wirksamkeit der Ionen. *Z. Phys. Chem.* 4: 129–181.

Preface

This handbook provides a comprehensive reference source on ion channels for students, instructors, and researchers in the field. Ion channels are membrane proteins that control the electrical properties of neurons and cardiac cells; mediate the detection and response to sensory stimuli like light, sound, odor, and taste; and regulate the response to physical stimuli like temperature and pressure, and are therefore fundamentally important for human health and disease. In nonexcitable tissues, ion channels are instrumental for the regulation of basic salt balance that is critical for homeostasis, and because they are often located at the surface of cells, they have the unique ability to interact with their environment. Consequently, ion channels are important targets for pharmaceuticals in mental illness, heart disease, anesthesia, and other clinical applications. The modern methods used in their study are powerful and diverse, ranging from single ion channel measurement techniques to modeling ion channel diseases in animals, and performing human clinical trials for ion channel drugs. This book gives the reader an introduction to the technical aspects of ion channel research as well as offers an inclusive and modern guide to the properties of major ion channels.

Ion channel study has been an active field for decades. Rich knowledge continues to emerge from research, which has direct and significant impact on many aspects of modern biology. We view this effort as a modern extension of the field's most authoritative textbook, which has remained *Ion Channels of Excitable Membranes* by Bertil Hille (also the author of the foreword to this book). We have divided the contents into five major sections. Part I deals with the basic concepts of permeation and gating mechanisms, with a balance of classic theories and latest developments. Part II focuses on various ion channel techniques, covering the basic principle of the method, application to channel research, and practical issues. As a unique feather of this book, the method section covers both classic, well-developed techniques and newly developed powerful techniques. Part III addresses the major channel types, with a combination of well-studied ion channels and *new* channels, and including in-depth, comprehensive coverage of each channel type in classification, properties, gating mechanisms, function, and pharmacology. The organization of this section follows the superfamilies of ion channels. Part IV covers ion channel regulation in addition to trafficking and distribution. Part V examines a few ion channel–related diseases, discussing genetics to mechanism and pharmaceutical developments.

Editors

Jie Zheng, PhD is a professor in the Davis School of Medicine at the University of California, where he serves as a faculty member in the Department of Physiology and Membrane Biology since 2004. He earned his bachelor's degree in physiology and biophysics and a master's degree in biophysics from Peking University, Beijing, China, in 1988. He earned his PhD in physiology in 1998 from Yale University, where he studied with Dr. Fredrick J. Sigworth on patch-clamp recording, single-channel analysis, and voltage-dependent channel activation mechanisms. He received his postdoctoral training at the Howard Hughes Medical Institute (HHMI) and the University of Washington during 1999–2003, working with Dr. William N. Zagotta on the cyclic nucleotide-gated channels activation mechanism and novel ion channel fluorescence techniques. Currently, his research focuses on the activation mechanism of the temperature-dependent thermoTRP channels. He has published over ten book chapters and review articles in addition to a number of original research papers.

Matthew C. Trudeau, PhD is an associate professor in the Department of Physiology at the University of Maryland School of Medicine in Baltimore, Maryland. He earned a bachelor's degree in biochemistry and molecular biology in 1992 and a PhD in physiology in 1998 while working with Gail Robertson, PhD, at the University of Wisconsin-Madison. His thesis work was on the properties of voltage-gated channels in the human gene (hERG) family of potassium channels related to ether-á-go-go and the role of these channels in heart disease. Dr. Trudeau was a postdoctoral fellow with William Zagotta, PhD, at the University of Washington and the Howard Hughes Medical Institute (HHMI) in Seattle from 1998 to 2004, where he focused on the molecular physiology of cyclic nucleotide-gated ion channels and their role in an inherited form of vision loss and the mechanism of their modulation by Ca^{2+}-calmodulin. Currently, Dr. Trudeau's work focuses on hERG potassium channels.

Contributors

Andriy Anishkin
Department of Biology
University of Maryland
College Park, Maryland

Clay M. Armstrong
Department of Physiology
University of Pennsylvania
Philadelphia, Pennsylvania

Donald M. Bers
Department of Pharmacology
University of California, Davis
Davis, California

Rikard Blunck
Departments of Physics and Physiology
Université de Montréal
Montréal, Québec, Canada

Allan H. Bretag
School of Pharmacy and Medical Sciences,
University of South Australia, Adelaide
Adelaide, South Australia, Australia

Cecilia Canessa
Department of Cellular and Molecular Physiology
Yale University
New Haven, Connecticut

William A. Catterall
Department of Pharmacology
University of Washington
Seattle, Washington

Baron Chanda
Department of Neuroscience
School of Medicine and Public Health
University of Wisconsin–Madison
Madison, Wisconsin

Sandipan Chowdhury
Department of Neuroscience
School of Medicine and Public Health
University of Wisconsin–Madison
Madison, Wisconsin

Jianmin Cui
Cardiac Bioelectricity and Arrhythmia Center
and
Center for the Investigation of Membrane Excitability Disorders
and
Department of Biomedical Engineering
Washington University in St. Louis
Saint Louis, Missouri

Gucan Dai
Department of Integrative Physiology and Neuroscience
Washington State University
Pullman, Washington

Declan A. Doyle
Centre for Biological Sciences
University of Southampton
Southampton, United Kingdom

Kenneth S. Eum
Department of Physiology and Membrane Biology
School of Medicine
University of California, Davis
Davis, California

Anthony Fodor
Department of Bioinformatics and Genomics
University of North Carolina at Charlotte
Charlotte, North Carolina

Nikita Gamper
Institute of Membrane and Systems Biology
University of Leeds
Leeds, United Kingdom

Qiong Gao
Department of Molecular, Cellular, and Developmental Biology
University of Michigan
Ann Arbor, Michigan

Steve A.N. Goldstein
Department of Biochemistry
Brandeis University
Waltham, Massachusetts

Sharona E. Gordon
Department of Physiology and Biophysics
University of Washington
Seattle, Washington

Eleonora Grandi
Department of Pharmacology
University of California, Davis
Davis, California

Bertil Hille
Department of Physiology and Biophysics
University of Washington
Seattle, Washington

Frank T. Horrigan
Department of Molecular Physiology and Biophysics
Baylor College of Medicine
Houston, Texas

Toshinori Hoshi
Department of Physiology
University of Pennsylvania
Philadelphia, Pennsylvania

Tzyh-Chang Hwang
John M. Dalton Cardiovascular Research Center
Department of Medical Pharmacology and Physiology
University of Missouri-Columbia
Columbia, Missouri

León D. Islas
Department of Physiology
School of Medicine
National Autonomous University of México
México City, Mexico

Lily Jan
Department of Physiology
University of California, San Francisco
San Francisco, California

Qiu-Xing Jiang
Department of Cell Biology
The University of Texas Southwestern Medical Center
Dallas, Texas

J.P. Johnson, Jr.
Vertex Pharmaceuticals
San Diego, California

Manu Ben-Johny
Departments of Biomedical Engineering and Neuroscience
Johns Hopkins University School of Medicine
Baltimore, Maryland

Dorothy M. Kim
Departments of Physiology and Biophysics
Cornell University
and
Department of Anesthesiology
Weill Cornell Medical College
New York, New York

William R. Kobertz
Department of Biochemistry and Molecular Pharmacology
University of Massachusetts Medical School
Worcester, Massachusetts

Min Li
The Solomon H. Snyder Department of Neuroscience
High Throughput Biology Center
and
Johns Hopkins Ion Channel Center
Johns Hopkins University
Baltimore, Maryland

Diomedes E. Logothetis
Department of Physiology and Biophysics
School of Medicine
Virginia Commonwealth University
Richmond, Virginia

Yungang Lu
Department of Integrative Biology and Pharmacology
The University of Texas Health Science Center at Houston
Houston, Texas

Sarah C.R. Lummis
Department of Biochemistry
University of Cambridge
Cambridge, United Kingdom

Anatoli Lvov
Department of Biochemistry and Molecular Pharmacology
University of Massachusetts Medical School
Worcester, Massachusetts

Linlin Ma
Department of Physiology and Membrane Biology
University of California, Davis
Davis, California

Rahul Mahajan
Department of Physiology and Biophysics
School of Medicine
Virginia Commonwealth University
Richmond, Virginia

Jason G. McCoy
Departments of Physiology and Biophysics
Cornell University
and
Department of Anesthesiology
Weill Cornell Medical College
New York, New York

David D. McKemy
Section of Neurobiology
Department of Biological Sciences
University of Southern California
Los Angeles, California

Andrea L. Meredith
Department of Physiology
University of Maryland School of Medicine
Baltimore, Maryland

Hiroaki Misonou
Graduate School of Brain Science
Doshisha University
Kyoto, Japan

Colin G. Nichols
Department of Cell Biology and Physiology
and
Center for the Investigation of Membrane Excitability Diseases
Washington University
Saint Louis, Missouri

Crina M. Nimigean
Departments of Physiology and Biophysics
Cornell University
and
Department of Anesthesiology
Weill Cornell Medical College
New York, New York

Leigh D. Plant
Department of Biochemistry
Brandeis University
Waltham, Massachusetts

Andrew Plested
Leibniz-Institut für Molekulare Pharmakologie
Berlin, Germany

Murali Prakriya
Department of Pharmacology
Feinberg School of Medicine
Northwestern University
Chicago, Illinois

Michael Pusch
Istituto di Biofisica
Consiglio Nazionale delle Ricerche
Genova, Italy

Tahmina Rahman
Centre for Biological Sciences
University of Southampton
Southampton, United Kingdom

Jon Sack
Department of Physiology and Membrane Biology
School of Medicine
University of California, Davis
Davis, California

Monica Sala-Rabanal
Department of Cell Biology and Physiology
and
Center for the Investigation of Membrane Excitability Diseases
Washington University
Saint Louis, Missouri

Mark S. Shapiro
Department of Physiology
University of Texas Health Science Center at San Antonio
San Antonio, Texas

Liang Shi
Department of Cell Biology
The University of Texas Southwestern Medical Center
Dallas, Texas

Trevor G. Smart
Department of Neuroscience, Physiology and Pharmacology
University College London
London, United Kingdom

Yoshiro Sohma
Department of Pharmacology
School of Medicine
Keio University
Tokyo, Japan

James D. Stockand
Department of Physiology
University of Texas Health Science Center at San Antonio
San Antonio, Texas

Sergei Sukharev
Department of Biology
University of Maryland
College Park, Maryland

Dhananjay Thakur
Department of Integrative Biology and Pharmacology
The University of Texas Health Science Center at Houston
Houston, Texas

Jin-bin Tian
Department of Integrative Biology and Pharmacology
The University of Texas Health Science Center at Houston
Houston, Texas

James S. Trimmer
Department of Neurobiology, Physiology and Behavior
and
Department of Physiology and Membrane Biology
University of California, Davis
Davis, California

Michael D. Varnum
Department of Integrative Physiology and Neuroscience
Washington State University
Pullman, Washington

KeWei Wang
State Key Laboratory of Natural and Biomimetic Drugs
Department of Molecular and Cellular Pharmacology
School of Pharmaceutical Sciences
and
IDG/McGovern Institute for Brain Research
Peking University
Haidian, Beijing, People's Republic of China

Xiuming Wong
Metabolic Health and Diseases Unit
Institute of Molecular and Cell Biology
Singapore, Singapore

Haoxing Xu
Department of Molecular, Cellular, and Developmental Biology
University of Michigan
Ann Arbor, Michigan

Huanghe Yang
Department of Physiology
University of California, San Francisco
San Francisco, California

Vladimir Yarov-Yarovoy
Department of Physiology and Membrane Biology
and
Department of Anesthesiology and Pain Medicine
and
Department of Biochemistry and Molecular Medicine
School of Medicine
University of California, Davis
Davis, California

Haibo Yu
The Solomon H. Snyder Department of Neuroscience
High Throughput Biology Center
and
Johns Hopkins Ion Channel Center
Johns Hopkins University
Baltimore, Maryland

David T. Yue
Calcium Signals Laboratory
Departments of Biomedical Engineering and Neuroscience
School of Medicine
Johns Hopkins University
Baltimore, Maryland

William N. Zagotta
Department of Physiology and Biophysics
University of Washington
Seattle, Washington

Xiaoli Zhang
Department of Molecular, Cellular,
and Developmental Biology
University of Michigan
Ann Arbor, Michigan

Michael X. Zhu
Department of Integrative Biology and Pharmacology
The University of Texas Health Science Center at Houston
Houston, Texas

Giovanni Zifarelli
Istituto di Biofisica
Consiglio Nazionale delle Ricerche
Genova, Italy

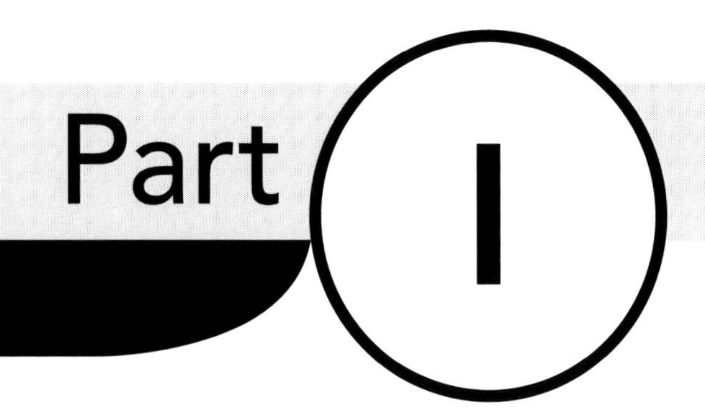

Part I

Basic concepts

1 Electricity, nerves, batteries: A short history

Clay M. Armstrong

Contents

1.1 INTRODUCTION

Beginning with Galvani in the late eighteenth century, many findings and cogent speculations lead to an understanding of the action potential and the role of ion channels in signaling and the very life of a cell. Galvani (1791) was fascinated by lightning and static electricity, which, at the time, were the only known forms of electricity. He had seen that an accidental spark could stimulate a frog sciatic nerve, causing a muscle twitch and wanted also to know if another form of electricity, lightning, could stimulate a nerve–muscle preparation. From his description:

> Therefore having noticed that frog preparations which hung by copper hooks from the iron railings surrounding a balcony of our house contracted not only during thunder storms but also in fine weather, I decided to determine whether or not these contractions were due to the action of atmospheric electricity .... Finally .... I began to scrape and press the hook fastened to the back bone against the iron railing to see whether by such a procedure contractions might be excited, and whether instead of an alteration in the condition of the atmospheric electricity some other changes might be effective. I then noticed frequent contractions, none of which depended on the variations of the weather.

These careful observations regarding copper hooks and iron railings inspired Volta (1800) to devise a battery (Voltaic *pile*): a copper plate separated from a zinc plate by a pad soaked in electrolyte solution, repeated many times in a stack to increase the voltage. The *pile* provided the first steady source of electrical current, and quickly led to enormous advances in science and technology: electromagnetism (Faraday, 1834), the telegraph (Morse), and Maxwell's equations to mention just a few.

The telegraph provided an irresistible analogy to impulse propagation in nerve, which, like a telegraph signal, was thought to be very fast. Helmholtz, however, measured the *speed of thought* in 1850, and found it to be only ~27 m/s, much slower than transmission in a telegraph cable. This slow speed became comprehensible with Kelvin's proposal of the *cable equation* (see Hodgkin and Rushton, 1946), which explained the observed slowness of transmission in underwater telegraph cables.

1.2 CELL MEMBRANES AND MEMBRANE VOLTAGES

The histological substrate of impulse propagation remained unclear for many years. In 1839 Schwann deduced that the nervous system was composed of interconnected cells. The cells did not have microscopically discernible walls but were presumed to be covered by thin membranes, which Quincke (see Hertwig et al., 1895) deduced must be formed by thin layers of lipid. Membrane composition was surmised in more detail (cholesterol and phospholipids) by Overton (1899) from studies on general anesthetics, and the lipid bilayer membrane was proposed by Gorter and Grendel (1925). The bilayer membrane was clearly visualized electron-microscopically still later (Robertson, 1958, 1959; Stoeckenius, 1959). As in a trans-Atlantic cable, the insulator (membrane) surrounds a conductor (the cytoplasm), and the cable is immersed in salt solution.

But how could a cell surrounded by a lipid membrane produce a voltage? In 1889 Nernst published his famous equation relating voltage and ion concentration. The concept that a voltage could result from an ion concentration difference was used by Bernstein in his famous hypothesis of 1902. Muscle cells, and, in fact, all cells have high internal K^+ concentration, and are bathed in extracellular fluid that has a low concentration of K^+ and a high concentration of Na^+. Bernstein postulated that in the resting state, the membrane of a muscle cell is predominantly permeable to K^+, resulting in a negative internal voltage in accordance with Nernst's equation. The selective K^+ permeability, he postulated, was lost when an action potential swept over the fiber, changing the membrane voltage to near zero. (In fact, he recorded an overshoot in the potential change, but seems not to have appreciated its significance.)

1.3 SQUID GIANT AXON AND THE COLE LABORATORY

In the 1930s J.Z. Young brought to the attention of K.S. Cole a very large axon, the giant axon of the squid, the preparation crucial to solving the mechanism of the nerve impulse. This giant axon is large enough (up to 1 mm diameter) that electrodes can be carefully threaded inside, allowing Cole and collaborators to measure directly the axon's membrane voltage (V_m), which they found to be approximately −60 mV inside at rest. Many other important findings and inventions followed from Cole's lab. During an action potential V_m was found to go well beyond 0 mV, that is, it *overshot* the value expected from Bernstein. Slightly before and during the action potential, membrane conductance (or permeability) was found to increase. Marmont (1949), in Cole's lab, introduced the very important *space clamp*: a wire was threaded into the axon, lowering its internal longitudinal resistance to the point that a several centimeter length of axon was made isopotential, and behaved electrically like a single large patch of membrane. Two very important contributions of the Cole lab were the *current clamp* (Marmont) and the *voltage clamp* (Cole, 1949; Marmont, 1949).

1.4 ERA OF HODGKIN AND HUXLEY

The squid giant axon, the space clamp, and an improved voltage clamp were the essential tools of Hodgkin and Huxley (HH) in their astonishing axon work. An important first step was the *sodium theory*: the overshoot of the action potential was explained by Hodgkin and Katz on the basis of an increase in Na^+ permeability, which forced V_m during the action potential to a point close to the equilibrium potential for Na^+. Then a brilliant series of experiments by HH brought a lasting understanding of the nerve impulse, lacking only in the molecular basis. It is worthwhile to describe their analysis and equations in detail, because they are still quite useful for simulations and an excellent learning experience. HH summarized their results in the electrical *equivalent circuit* of a small patch of squid axon membrane, as shown in Figure 1.1.

HH equivalent circuit: The bulk of the *membrane* is a bilipid layer that separates the outside (above) from inside of the membrane patch, and serves as a very high barrier to ion movement. The bilayer and the conducting solutions on either side constitute an electrical capacitance, C_m, in the equivalent circuit. The external plate (or conductor) is the external fluid, which contains a high concentration of NaCl (~0.44 M).

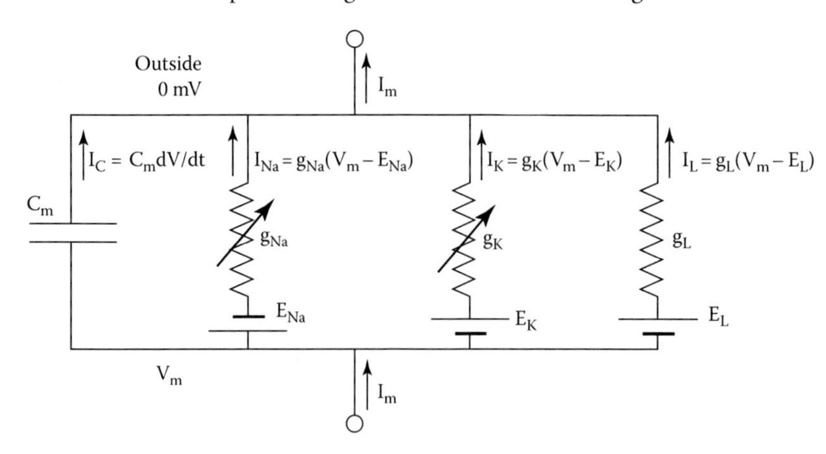

Figure 1.1 HH's equivalent circuit of an axonal membrane, redrawn with modern voltage convention (V_m is the internal voltage). I_C is capacitive current; that is, current flowing not through the membrane, but changing the charge accumulation at the outer and inner membrane surfaces as V_m changes. I_{Na}, I_K, and I_L are the ionic currents passing through membrane pores. I_m is the sum of the four component currents, as measured by a voltage clamp. g_{Na}, g_K, and g_L are the conductances of the ionic currents. g_{Na} and g_K are variables, as indicated by the arrows. E_{Na} and E_K are the equilibrium (Nernst) potentials of Na^+ and K^+, and E_{Leak} is the apparent equilibrium potential of the resting permeability.

The internal conductor is the axoplasm, which contains a high concentration of K^+ (~0.5 M) and a number of unspecified anions (such as phosphates, amino acids, and negatively charged protein). The bilayer is the insulator that separates these conductors. Because the insulating bilayer is very thin, the capacitance is very high, about 1 $\mu F/cm^2$. This high capacitance, together with the relatively high resistance of the axoplasm, is a major factor in the slow *velocity of thought* measured by von Helmholtz (1850).

1.4.1 COMPONENTS OF MEMBRANE CURRENT (I_m)

In parallel with C_m are three pathways for the movement of ions across the membrane. The first, labeled g_K, conducts K^+ ions across the membrane. The arrow through this conductor indicates that g_K is variable; it depends on both V_m and time, as described in the following text. g_K is low at resting V_m and high during the falling phase of the action potential. The battery (E_K) is the Nernst equilibrium voltage for K^+. Movement of K^+ across the membrane is zero when V_m is equal to E_K, about −90 mV in normal solutions.

The second pathway, labeled g_{Na}, conducts Na^+ ions across the membrane. g_{Na} is also variable, depending (like g_K) on voltage and time. It is near zero at resting V_m and very high during the rising phase of the action potential. In normal solution, the equilibrium potential for Na^+ (E_{Na}) is ~+60 mV.

The third pathway is g_{leak}, thought to conduct K^+ predominantly but not exclusively. This conductance (not well known even now) is small and not variable. It serves to set resting V_m.

A voltage clamp experiment imposes a steady voltage across the membrane and measures the resulting current flow (I_m) as a function of time. Thanks to the *space clamp* imposed by the internal wire, the membrane is, ideally, a single giant patch (but aberrations may occur). I_m in the Hodgkin–Huxley model has four components: I_C, I_{Na}, I_K, and I_{leak}. Simplifying slightly, I_c (capacitive current) is nonzero only when V_m is changing, that is, when the charge accumulations at the membrane surfaces are changing. In a normal voltage clamp experiment, V_m changes only during the brief instants when the voltage is stepped to a new value. This is a major advantage of the voltage clamp, as it effectively removes I_C, leaving only the ionic currents I_{Na}, I_K, and I_{leak}. Of these, I_{leak} is usually small enough to ignore. Separation of I_{Na} and I_K requires either removal of Na^+ or K^+ from the internal and external solutions, or an agent, for example, TTX (tetrodotoxin; Narahashi et al., 1964) to block g_{Na}, or TEA^+ to block g_K (Armstrong, 1966).

The equations for I_{Na}, I_K, and I_L in the circuit are approximations, valid only over a limited voltage range, and only in normal solution: The conductance approximation of the Na^+ (and K^+) pathways is sufficient for approximating an action potential in normal solution, but not otherwise.

1.4.2 HH DESCRIPTION OF THE POTASSIUM CONDUCTANCE, g_K

As noted_earlier, g_K can vary with both voltage and time. A family of g_K–time curves is shown in Figure 1.2, as calculated from the HH formula for g_K.

Experimentally, obtaining such traces would require (1) applying voltage-clamp steps to an axon after removing Na^+ inside and out, or applying TTX and (2) dividing the recorded I_K by the *driving force* for K^+, V_m−E_K. The voltage protocol, with steps from −60 mV to various values, is shown below the g_K traces.

The curves closely mirror those seen experimentally (see Figure 1.4). The ordinate gives g_K relative to its maximum value. For negative steps there is no change in g_K. For a step to −40 mV, g_K *activates* after a lag to a small value. Activation is steadily faster and larger as step size increases, but tends to saturate in amplitude above −20 mV. At the end of the voltage step, when the voltage is returned to −60 mV, g_K *deactivates* exponentially. The time constant of the exponential is the same following all of the steps, but is faster if the repolarization voltage is made more negative (not illustrated). Note that deactivation is markedly different from the sigmoid time course of activation at the beginning of the voltage step.

To empirically fit the activation and deactivation curves, HH described g_K with the following equation:

$$g_K = g_{Kmax}\,\mathbf{n}^4 \quad \frac{d\mathbf{n}}{dt} = \alpha_n(1-\mathbf{n}) - \beta_n$$

where g_{Kmax}, the maximum conductance, is the summed conductance of all the many K^+ conducting units in the membrane. HH theorized that each unit behaves in an all or none manner; that is, it is either conducting or not, and has a conductance of γ_K when conducting. Each *unit* is now known to be a transmembrane pore or ion channel, but at the time of HH this was not known. To conform with the present-day terminology (and hopefully minimize confusion) we refer henceforth to HH's *units* as *channels*.

Thus,

$$g_{Kmax} = (\text{total number of channels})^*\gamma_K,$$

Conduction through each individual channel is controlled by four identical n particles. The n particles are charged, and are

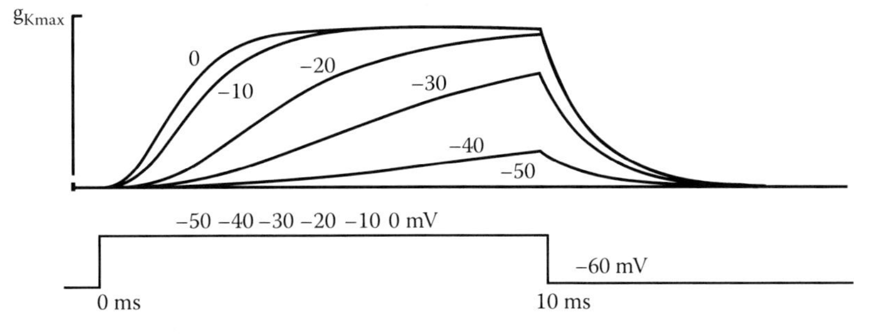

Figure 1.2 The HH predictions for potassium conductance, g_K. The simulations show the change in g_K for step changes of V_m from rest (g_K ~ 0), with return to rest after 10 ms. Experimentally each step would be followed by a rest interval of 1 or 2 s.

confined to the membrane. (Formally, their movement is a slow component of capacitive current.) They migrate outward or inward through the membrane in response to changes of V_m, at rates described by the empirical equations for α_n and β_n. For concreteness, suppose an n particle has a positive charge, and is driven outward (positive V_m; rate α_n) or inward (negative V_m; rate β_n) by the membrane field. Further, suppose all four of a channel's n particles must be at the outer edge of the membrane in order for the channel to conduct. Thus the rate of movement to or from the outer edge depends on voltage, as does the probability (**n**) that a particle will, at equilibrium, be at the outer edge. The probability that a channel is conducting is $\mathbf{n}^4$.

A convenient way of illustrating this model (Fitzhugh, 1965; Armstrong, 1969) is to represent each channel as having five states, N0–N4*, where the asterisk indicates the conducting state:

$$\text{N0} \underset{\beta_n}{\overset{4\alpha_n}{\longleftrightarrow}} \text{N1} \underset{2\beta_n}{\overset{3\alpha_n}{\longleftrightarrow}} \text{N2} \underset{3\beta_n}{\overset{2\alpha_n}{\longleftrightarrow}} \text{N3} \underset{4\beta_n}{\overset{\alpha_n}{\longleftrightarrow}} \text{N4*}$$

At the resting potential, most of a channel's n particles are in state N0 (no n particles at the membrane's outer edge), and essentially none are in state N4*. After a positive voltage step, α_n increases and β_n decreases, driving the channel toward N1. Movement of any one of the four n particles will cause the transition from N0 to N1, so the net rate is $4\alpha_n$. From N1 to N2 any one of the three remaining deactivated particles can move, and so on. Thus, the rate constants for the four steps as a channel moves from N0 to N4* have a ratio of 4:3:2:1 ($4\alpha_n$, $3\alpha_n$, $2\alpha_n$, and α_n). Beginning at N0, there must be four transitions of a channel's n particles before it conducts. This produces a lag, and an overall sigmoid time course for activation.

Once in the conducting state, N4*, inward movement of any one of the four n particles brings the channel back to N3, which is nonconducting. Because only a single step (N4* → N3) is required to turn off the channel, the decay of g_K when V_m returns to rest is exponential (Figure 1.2). The decline in the probability (**n**) that the particles are in the activated state is also exponential, but four times slower than g_K because, after the channel closes in the first step, the remaining three particles must return to the deactivated state. The rate constants for moving through the four steps from fully activated to fully deactivated are 4β, 3β, 2β, 1β.

With splendid candor, Hodgkin and Huxley (1952) said, "the equations governing the potassium conductance do not give as much delay in the conductance rise on depolarization … as was observed in voltage clamps … Better agreement might have been obtained with a fifth or sixth power, but the improvement was not considered to be worth the additional complication."

1.4.3 HH DESCRIPTION OF g_{Na}

The K^+ channel has (in current terminology) a single gate that opens relatively slowly after depolarization, and remains open until V_m is returned to the resting potential.

The Na channel, on the other hand, has two *gates* rather than just one, m^3 and h in HH terminology. The m^3 gate is similar to the n^4 gate of the K^+ channel, but it opens more rapidly after depolarization. The h gate is very different: It is open at rest and closes relatively slowly after depolarization. The result, in normal function, is a transient inward current of Na^+ that depolarizes an axon, followed by the more slowly activating K^+ current that drives V_m back to rest.

The HH equations for g_{Na} are

$$g_{Na} = m^3 h \qquad \frac{dm}{dt} = \alpha_m(1-m) - \beta_m m \qquad \frac{dh}{dt} = -\alpha h + \beta(1-h)$$

Similar to the kinetic scheme for g_K, the two gates for an Na channel can be portrayed in a four-state kinetic scheme for the m^3 gate, and a two-state scheme for h gate:

$$\text{M0} \underset{\beta_m}{\overset{3\alpha_m}{\longleftrightarrow}} \text{M1} \underset{2\beta_m}{\overset{2\alpha_m}{\longleftrightarrow}} \text{M2} \underset{3\beta_m}{\overset{\alpha_m}{\longleftrightarrow}} \text{M3*}$$

$$\text{H1*} \underset{\beta_h}{\overset{\alpha_h}{\longleftrightarrow}} \text{H0}$$

When V_m is stepped from −60 to 0 mV in a voltage-clamped axon, the 3 m particles (assumed positive) of an Na^+ channel move quickly outward to gate-open position, joining the unit's h particle (assumed negative), which is already in its outermost, gate-open position, to form the channel's conducting state M3*H1*. The h particle then shifts inward relatively slowly to gate-closed position, putting the unit in state M3*H0, which is nonconducting and '*refractory*'. On return to the resting potential, the three m particles quickly move inward to the gate-closed position, and the channel remains refractory until the h particle (negative and pushed outward by negative V_m) shifts relatively slowly to gate-open position (state M0H1*). The position of the h particle depends only on V_m and time and is completely independent of the m particles.

1.4.4 SUCCESSES OF THE HH FORMULATION

The HH equivalent circuit and the equations that describe it were a stunning success in describing both the stationary (space-clamped axon) and the propagated action potential, its threshold properties, its refractory period, and the impedance changes that accompany the action potential. HH cautioned, however,

> The agreement must not be taken as evidence that our equations are anything more than an empirical description of the time-course of the changes in permeability to sodium and potassium. An equally satisfactory description of the voltage clamp data could no doubt have been achieved with equations of very different form.

1.4.5 HH FORMULATION APPLIED TO THE HEART

Noble (1966) brilliantly modified the HH equations to apply to the heart. A main change was to slow the rise of g_K and drastically reduce its magnitude, to reproduce the long-duration plateau action. Pacemaking is inherent in the HH equations, and it is simply necessary to apply a steady depolarizing current that makes V_m more positive; or to substantially increase the number of Na^+ channels. It is now clear that several channel types unknown to HH are important in the electrical behavior of the heart.

1.5 AFTER HH

Major questions were raised but left unanswered by HH, and many of them are best expressed in their words:

> At present the thickness and composition of the excitable membrane are unknown. Our experiments are therefore unlikely to give any certain information about the nature of the molecular events underlying changes in permeability.

1.5.1 UNKNOWN MOLECULAR MECHANISM OF THE CONDUCTANCES: HOW IONS GET THROUGH MEMBRANES

For years after the HH papers there was much ferment regarding membranes and the conductances. Relatively little consideration was given to the idea that proteins might be involved in the conductances. This may have been the result of strict adherence to the Davson–Danielli model (Danielli and Davson, 1935), in which the membrane proteins were peripheral; if there were no transmembrane proteins, ions must necessarily go through the lipid. This view was well-exemplified by a quote from David Goldman (1965), "The ions traverse the nonpolar lipid region of the membrane in accordance with the usual electrodiffusion relations."

A related question was whether the conductances involved a pore, which presumably would be formed by a protein penetrating the membrane, or by a lipid-soluble carrier. The *long pore effect* observed by Hodgkin and Keynes (1955) suggested a pore. Arguing against a pore was the discovery of the antibiotic valinomycin, which clearly served as a lipid-soluble selective carrier of K^+ across membranes. To many this seemed the final answer (e.g., Diebler et al., 1969).

Early indications that g_{Na} is, in fact, the result of a transmembrane protein came from the study of tetrodotoxin (TTX), a blocker of g_{Na} (Narahashi et al., 1964). TTX is hydrophilic, highly selective and very potent, with a K_D of ~5 nM. It acts only on g_{Na} and only from the outside. All of these points suggest that it interacts very specifically with a membrane protein, rather than with the relatively amorphous lipids of the bilayer membrane. The conductance of the suspected channel was approximated by dividing the maximum g_{Na} of a myelinated node in frog membrane by the number of conducting units, estimated by TTX labeling of a frog node of Ranvier (Hille, 1984). This yielded an estimate of γ_{Na}, the conductance of a single conducting unit. Dividing experimentally measured I_{Na} by γ_{Na} yielded ~10^7 ions/ms/conducting unit, much too high for a carrier (Parsegian, 1975), and thus supportive of a pore.

Evidence that g_K was composed by transmembrane pores came from studies using the quaternary ammonium ion TEA^+, which has about the same diameter (~8 Å) as a K^+ ion with a single hydration shell. The effects of applying the TEA^+ derivative, nonyl-triethylammonium ion (C_9^+) to a squid axon are shown in Figure 1.3. Part A is a family of normal I_K traces, taken at the voltages indicated, in the presence of TTX to eliminate I_{Na}. The curves show a sigmoid time course as I_K develops after depolarization, and the approximately exponential time course of I_K deactivation, when V_m is returned to −60 mV after 12 ms. In part B, 0.1 mM C_9^+ has been added to the internal perfusion fluid. I_K initially activates normally but after 1 or 2 ms begins to decay as C_9^+ ions diffuse into the channels and block them. I_K decays to a small value, and on return to resting V_m, there is essentially no inward current through the C_9^+-blocked channels.

The picture used to explain the effect of C_9^+ is shown in Figure 1.3c through f. The hypothetical pore has an internally located

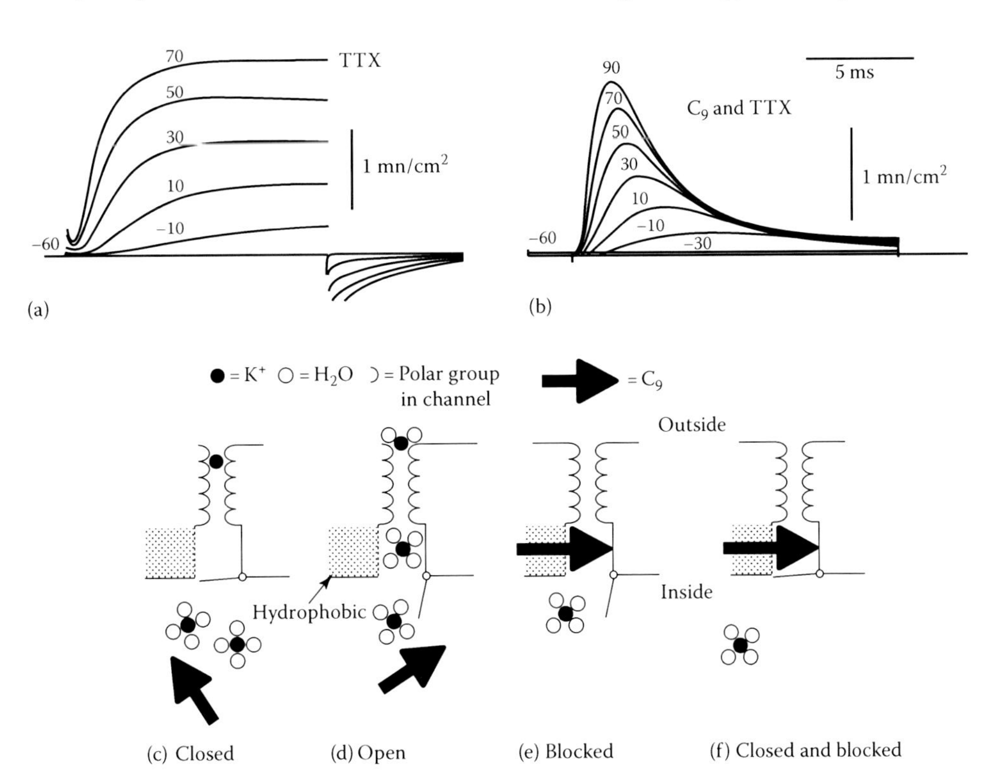

Figure 1.3 The effects of applying the TEA^+ derivative, nonyl-triethylammonium ion (C_9^+) to a squid axon. (a) Normal I_K recorded from an axon with I_{Na} eliminated by TTX (tetrodotoxin), (b) addition of 0.1 mM C_9^+ internally causes I_K to "inactivate", (c) closed, (d) open, (e) blocked, and (f) closed and blocked.

gate, and an internal *vestibule* large enough to accept, normally, a K^+ ion with a hydration shell. C_9^+ or TEA^+ are also of the appropriate size to enter. External to the vestibule is a filter (or tunnel) large enough to admit a dehydrated K^+ ion (2.66 Å) but much too small for a TEA^+ ion. In the resting state (c), the vestibule is protected by the closed gate and is empty. After an activating voltage step, the pore conducts (d) until C_9^+ diffuses in through the open gate (e) to occupy the vestibule and block K^+ access to the vestibule and filter. (This inactivation of the potassium current provided a clear suggestion about the mechanism of sodium channel inactivation, as briefly described later.) A hydrophobic region in the vestibule wall was postulated to bind the long hydrophobic tail of C_9^+. The entry rate of either C_9^+ or TEA^+ was easily measurable and for TEA^+ was much too fast (~1 μs^{-1}; Armstrong, 1966) to be explained by a carrier (Parsegian, 1975).

When V_m was returned to −60 mV after blocking the channels with C_9^+, in-flowing potassium current was near zero as C_9^+ left the channel very slowly. The exit of C_9^+ could be greatly accelerated by increasing the external K^+ concentration: in-moving K^+ ions clearly drove the occupying C_9^+ ions out of the channel vestibules. This *knock-out* phenomenon gave strong support to the idea of a transmembrane pore. Also clear was a *foot in the door* effect: At −60 mV, a channel gate could not close if its vestibule contained C_9^+ and remained open as in Figure 1.3e. At −100 mV, where the likelihood of gate-closing is much higher, a fraction of the channel gates did close, trapping C_9^+ in the vestibule (Figure 1.3f).

Unlike squid axons, for which K^+ channels are very insensitive to external TEA^+, some K^+ channels such as the ones in myelinated frog nerve, are blocked by external TEA^+ with a K_D of ~0.5 mM. The frog fibers are insensitive to external C_9^+, which did, however, block the K^+ channels when applied internally, with the same results as observed in squid.

Final confirmation that ion channels underlie the conductances came with single channel measurements (Neher and Sakman, 1976).

1.5.2 ARE THERE TWO MAJOR CONDUCTANCES OR ONLY ONE?

HH said, "An alternative hypothesis is that only one system is present but that its selectivity changes soon after the membrane is depolarized. A situation of this kind would arise if inactivation of the particles selective for sodium converted them into particles selective for potassium." The differing pharmacological sensitivities of g_{Na} and g_K strongly suggested that the conductances were separate physical entities. A clear proof came with the finding that the enzyme mix pronase destroys inactivation of g_{Na} without disturbing g_K (Armstrong et al., 1973). Some years later came the cloning of the sodium channel protein (Noda et al., 1984), and, a few years later still, the cloning of a quite distinct protein, a potassium channel subunit (see following text).

1.5.3 ION SELECTIVITY

The mechanism of ion selectivity could not be properly addressed until it was clear that ions passed through the membrane in channels. When that question was settled, hypotheses for selectivity of both the Na and the K channel came quickly.

A general principle is that the free energy of the selected ion, be it Na^+ or K^+, must be about the same in the channel as in the external and internal solution (Mullins, 1959; Bezanilla and Armstrong, 1972). Further, to achieve high conductivity, there must be no major energy barriers at entry or exit. If an ion enters a deep energy well, it simply blocks the channel. The free energies in solution are 90 kcal/mol for Na^+ and 73 kcal/mol for K^+ (Robinson and Stokes, 2002). The filter that selects between the ions and allows the passage of one in preference to the other must provide enough binding energy to approximately replace the hydration energy.

For the K channel, the mechanism is relatively simple: The *filter* is lined by carbonyl oxygens (see Figures 1.3 and 1.4) that tightly complex dehydrated K^+ ions, much as they are complexed by water molecules in solution. Because the filter is in a region of much lower dielectric constant than water, it is necessary that the carbonyls bind even more tightly than the oxygens of water. The extra binding energy arises from the higher dielectric constant of carbonyl groups, 2.3 D (ebye) or more, vs. 1.83 D for water.

The Na/K selectivity of sodium channels is 10/1 or less, depending on the method of measurement. Hille (1971) measured the dimensions required for organic cations to permeate, and concluded that the channel had a filter region with a cross section of about 3 × 5 A, large enough to accommodate a Na^+ ion with one water molecule beside it. Also, the filter had a negative charge in it, to help provide the binding energy required.

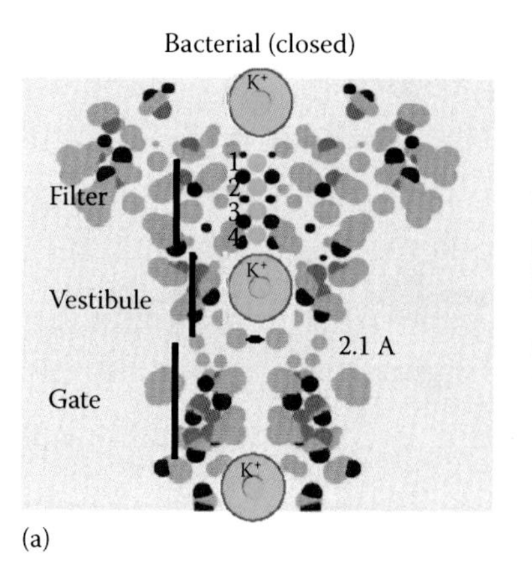

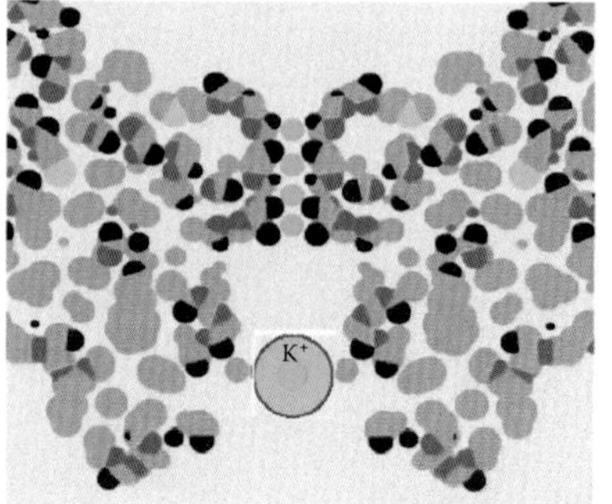

Figure 1.4 Sections of the KcsA bacterial K channel (a; Doyle et al., 1998) and a chimeric channel combining segments from $K_v1.2$ and $K_v2.1$ (b; Long et al., 2007). Carbon atoms are gray, oxygens are red, nitrogens are blue, and K^+ ions are green. Numerals 1–4 are the binding sites in the filter. Hydration shells have been drawn around some K^+ ions.

For neither channel is there yet certainty about the details of ion dehydration as the ions enter the filter.

1.5.4 VOLTAGE DEPENDENCE OF THE CONDUCTANCES

> The dependence of g_{Na} and g_K on membrane potential suggests that the permeability changes arise from the effect of the electric field on the distribution or orientation of molecules with a charge or dipole moment...Details of the mechanism will probably not be settled for some time, but it seems difficult to escape the conclusion that the changes in ionic permeability depend on the movement of some component of the membrane which behaves as though it had a large charge or dipole moment.
>
> **(Hodgkin and Huxley, 1952)**

As explained earlier, the charge in the HH formulation is on the n particles governing g_K, and the m and h particles governing g_{Na}. Movement of this charge or dipole when membrane voltage changed, they predicted, would cause a measurable intramembranous current (*carrier* current, or, later *gating* current), separable from ionic Na^+ or K^+ current, as the conducting units turned on or off.

1.5.5 GATING CURRENTS OF Na^+ CHANNELS

For the HH prediction of intramembranous charge movement there seemed to be no alternative, and gating or carrier currents were widely sought. The first progress in detecting intramembranous current was made not in nerve but muscle. Schneider and Chandler (1973) clearly showed an intramembranous current arising from the transverse tubules of muscle fibers, and proposed that it reflected a voltage-driven molecular rearrangement that led to activation of the muscle cell's contractile machinery.

Measurement of a small *gating current* (I_g) of sodium channels soon followed, and was compared with the predictions of HH for the movement of the m and h particles (Armstrong and Bezanilla, 1974). Although the measurement of I_{gNa} showed that HH were right in principle, the m^3h formulation, unsurprisingly, did not fit in detail. (1) As g_{Na} activates the m^3 formulation predicts a rapid exponential decay of I_g, while the experimental I_g curves have a more complex time course: a rising phase, a peak, and then a two-component decay. I_g can only be measured by an approximate method, as the difference between two intramembranous current records, one in a voltage range where gating occurs, and another in a range where there is capacitive current (assumed linear) but no gating. This leaves some uncertainty, particularly where good time resolution is required, as in the early part of the I_g curve during activation. This technical problem, however, seems unlikely to account for the slow rise of I_g immediately after the activating step. A slow rise in the slower gating current of K^+ channels is also seen, as described later. I_g recorded during a repolarizing step to −70 mV also does not fit the HH formulation, which predicts that I_g should decay 3× more slowly I_{Na}. Experimentally I_g and I_{Na} decay with almost the same time course, clearly not in agreement with the m^3 formulation.

The postulate of a separate, voltage-dependent h particle also has problems. HH predicts (1) a component of I_g with the time course of inactivation and (2) that inactivation does not affect I_g generated by the m^3 particles. Experimentally, however, there is no gating current with the time constant of inactivation. Further, ~2/3 of the gating current inactivates: On return to −70 mV I_g of inactivated channels is ~1/3 as large as after a short pulse in which there is little time for inactivation. This inactivation (or immobilization) of I_g is not seen after inactivation is removed by the enzyme mix pronase applied internally. The pronase results suggest an alternative to h particle inactivation: After opening of the channel gate, an internal, pronase-removable particle diffuses into a site in or near the pore's inner mouth and blocks (1) conduction and (2) the return of about 2/3 of the gating charge to rest position. Functionally, this inactivation mechanism allows the Na channel to close on return to the resting potential without leaking Na^+ ions. Abnormalities of this mechanism, which allow a depolarizing leak of Na^+ after an action potential, underlie the inappropriate tetanic activity that occurs in some paralyses. Details of the inactivation mechanism have been proposed (Armstrong, 2006) but are not yet completely clear.

1.5.6 CLONING

In a remarkable feat, Noda and coworkers cloned the sodium channel of an electric eel (Noda et al., 1984). Cloning of the K_v channel was achieved some years later. Beginning with the simpler K_v channel, it is (or can be) composed of four identical subunits, each of which has six transmembrane crossings. Notable among these is the fourth transmembrane segment (S4), which (in the *Shaker* K channel) contains six periodically spaced arginine residues and one lysine. This positively charged segment is obviously the region sensitive to membrane voltage, and drives the conformational changes of gating. Further description is in the crystallization section.

Bacterial sodium channels are also composed of four identical subunits (Ren et al., 2001), but the Na_v channels of skeletal muscle and brain have a single major peptide with four similar but far from identical domains (Noda et al., 1984). It appears that the four subunits of the more primitive bacterial K or Na channel have been linked together, and then differentiated. One can speculate that differentiation of the domains of the Na channel enhanced Na^+ selectivity and speed of gating, and provided for the complex inactivation process of the channel, in which the differentiated domains play different roles.

1.5.7 MUTATIONS

Mutation of specific residues in ion channels has made major contributions to understanding. Among these are identification of the pore region (Yellen et al., 1991); elucidation of the mechanism of N-type inactivation of *Shaker* K_v channels (Hoshi et al., 1991); pinpointing the region of the Na channel that forms the inactivation gate (Stühmer et al., 1989); and important suggestions regarding motion of the charged S4 segment, which underlies voltage-driven gating (Papazian et al., 1995; Larsson et al., 1996; Smith-Maxwell et al., 1998). Further, a nonconducting mutant of *Shaker* K_v has been found, which is very useful in studying *gating currents* of the K channel (Perozo et al., 1993), as described in the next section.

1.5.8 GATING CURRENTS OF K^+ CHANNELS AND S4 MOTION

K channel gating current (I_{g_K}) has been clearly resolved in a mutant *Shaker* K channel (W434F) that has normal gating properties but no conductance, because of a block in the selectivity filter resulting from C-type inactivation (Hoshi et al., 1991; Yang et al., 1997). Unlike the HH predictions of the n^4 scheme, I_{g_K}, like $I_{g_{Na}}$, has a rising phase as the channels' gates open. This shows that the first step (or steps) of S4 motion is the slowest, unlike the n^4 predictions, where the first step is fastest. Further, on return to resting potential, the initial amplitude of I_{g_K} is near zero: The gating charge movement (S4 motion) is initially paralyzed, then grows in amplitude, and decays as the channels close. This is very different from the n^4 prediction. The paralysis probably results from the need to clear K^+ ions from the vestibule before the channel gate can close (Goodchild et al., 2012). When *Shaker*-IR channels are blocked by internal TEA^+, which is known to prevent gate closing, the duration of the paralysis is profound and long-lasting (Bezanilla et al., 1991).

1.6 CRYSTALLOGRAPHY

The need for detailed structural information became obvious in the 1990s, and was answered by MacKinnon and collaborators. They crystallized and analyzed a potassium channel, quite unexpectedly not from a nerve or muscle cell, but from a bacterium. It took some time to realize that a bacterium, which requires an internal voltage of approximately −180 mV to synthesize ATP, needs to be very careful about what ions it admits from its growth medium. With this membrane voltage, free permeability to Na^+ for a bacterium growing in 0.1 M Na^+ would result in an internal Na^+ concentration of 10 M, obviously impossible. The only acceptable cation would be K^+, which is present in relatively low concentration in ground water, brackish water (and our blood), and the ocean. Hence the bacterium is selectively permeable to K^+, and the negative internal membrane voltage, a by-product of ATP production, combines with selective (and adjustable) K^+ permeability to act as a pump, bringing internal K^+ to an acceptable value.

The crystal structure of the bacterial channel KcsA (Doyle et al., 1998) is shown in sections in Figure 1.4a. Beginning at the outside, one sees a narrow selectivity filter, lined on each side with five red carbonyl oxygens. The filter has four binding sites (1–4) for dehydrated K^+ ions, which are shown as green disks. In operation, it is thought that only two of these sites are occupied at any instant, and that occupancy oscillates rapidly between 2–4 and 1–3 occupancy as ions move outward through the filter (Zhou et al., 2001). Internal to the filter is a vestibule (or cavity) large enough to hold a hydrated K^+ ion, preparatory to its entry into the filter. Details of the steps in dehydration of the ion as it enters the filter have yet to be worked out in detail. Internal to the vestibule the innermost helices (TM2) converge to occlude the channel, forming the channel's closed gate. At its narrowest, the pore diameter in the gate region is ~2.1 A (vs. 2.66 A for a dehydrated K^+ ion) and lined by hydrophobic side chains, creating a formidable barrier to ion movement. A hydrated K^+ is shown just below the gate region, too large to enter the gate. Also shown, at the top, is a K^+ ion that has rehydrated after emerging from the filter (Zhou et al. 2001).

A section through a voltage-gated K channel, K_v1.2–2.1 (Long et al., 2007), is shown in Figure 1.4b. The core of this channel is similar to the KcsA channel, but surrounded by the helices that constitute the voltage-sensor-effector. The selectivity filter and vestibule are similar to the KcsA image, but the gate region at the channel's inner end is wide open, with a diameter (8.1 A), a good fit for a hydrated K^+ ion. The hydrated K^+ can easily pass through the gating region, enter the vestibule, and there shed its coat of water molecules to enter the selectivity filter.

At present there is no crystal structure of a closed K_v channel, and the best available closed model is KcsA. In imagination, the open K_v1.2–2.1 channel can be closed as shown in Figure 1.5. The S4, S4–S5, and S6 helices abstracted from the K_v1.2–2.1 structure are shown at right, with the charged residues of S4 shown in light blue. Corresponding to the open state of the channel, the inner ends of S6 are far apart. At left, two of the TM2 helices (TM2 is the KcsA equivalent of S6) from the KcsA structure are shown. The S4 and S4–S5 segments from K_v1.2–2.1 are (in lighter hue) superimposed, but shifted down in imagination by a negative internal voltage. In this position, the S4–S5 segments lock the gate regions of the TM2 helices in closed position, with the inner ends of the helices relatively close together.

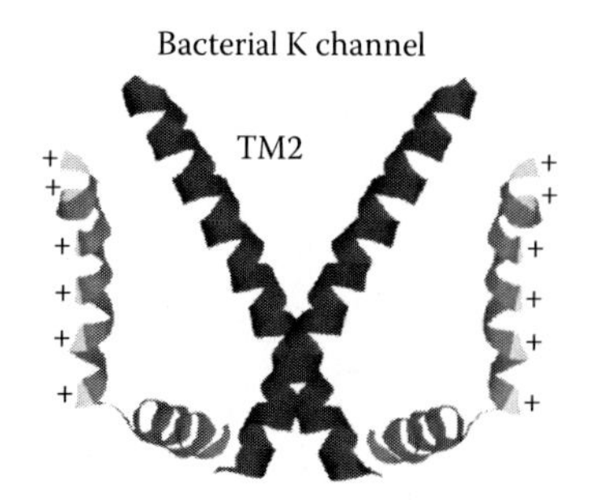

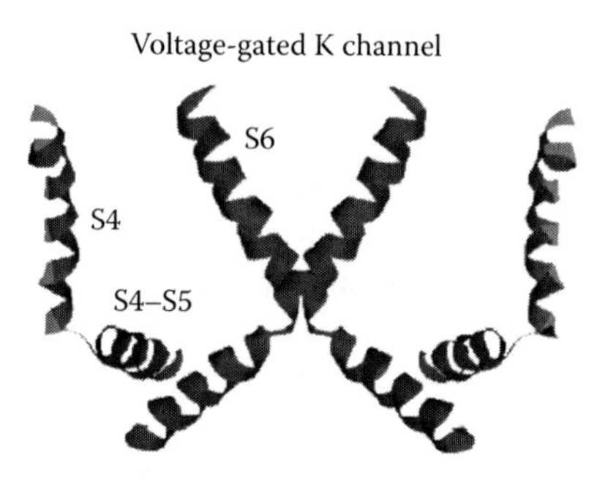

Figure 1.5 At the right is a ribbon diagram of selected parts of the K_v1.2–2.1 crystal structure, with open gate: The inner (lower) part of the S6 helices are far apart, corresponding to the open state shown in Figure 1.4. At left is a partly imaginary structure. The TM2 helices are from the KcsA crystal structure. The inner parts of the TM2 helices are relatively close together, reflecting the closed state shown in Figure 1.4a. The S4 and S4–S5 segments have been transplanted in imagination from K_v1.2–2.1. They have been shifted down as expected at the negative voltage required to close a K_v channel, to a position where they lock the TM2 helices in gate-closed position. The positively charged residues in the S4 segments are labeled with a + sign.

The availability of channel crystal structures will undoubtedly provide the basis for much improved understanding of ion channel function in the coming years.

1.7 CONCLUSION

The mystery of the action potential and the investigations of it by Galvani triggered an avalanche that has transformed science and our lives almost beyond recognition. Science and technology continue to provide useful tools for the study of the nervous system, which is still in its infancy. A major stepping stone on this path was the fundamental and heroic work of HH, which has stimulated many decades of ion-channel exploration. As they recognized, they could not foresee the fascinating complexity since revealed, but even they might have been surprised by the durability and timelessness of their synthesis.

REFERENCES

Armstrong, C.M. 1966. Time course of TEA(+)-induced anomalous rectification in squid giant axons. *The Journal of General Physiology.* 50:491–503.

Armstrong, C.M. 1969. Inactivation of the potassium conductance and related phenomena caused by quaternary ammonium ion injection in squid axons. *The Journal of General Physiology.* 54:553–567.

Armstrong, C.M. 2006. Na channel inactivation from open and closed states. *Proceedings of the National Academy of Sciences of the United States of America.* 103:17991–17996.

Armstrong, C.M. and F. Bezanilla. 1974. Charge movement associated with the opening and closing of the activation gates of the Na channels. *The Journal of General Physiology.* 63:533–552.

Armstrong, C.M., F. Bezanilla, and E. Rojas. 1973. Destruction of sodium conductance inactivation in squid axons perfused with pronase. *The Journal of General Physiology.* 62:375–391.

Bernstein, J. 1902. Untersuchungen zur Thermodynamik der bioelektrischen Ströme. Erster Theil. *Pflügers Archiv.* 92:521–562.

Bezanilla, F. and C.M. Armstrong. 1972. Negative conductance caused by entry of sodium and cesium ions into the potassium channels of squid axons. *The Journal of General Physiology.* 60:588–608.

Bezanilla, F., E. Perozo, D.M. Papazian, and E. Stefani. 1991. Molecular basis of gating charge immobilization in *Shaker* potassium channels. *Science.* 254:679–683.

Cole, K.S. 1949. Dynamic electrical characteristics of the squid axon membrane. *Archives des sciences physiologiques.* 3:253–258.

Danielli, J.F. and H. Davson. 1935. A contribution to the theory of permeability of thin films. *Journal of Cellular and Comparative Physiology.* 5:495–508.

Diebler, H., M. Eigen, G. Ilgenfritz, G. Maass, and R. Winkler. 1969. Kinetics and mechanism of reactions of main group metal ions with biological carriers. *Pure and Applied Chemistry.* 20:93–116.

Doyle, D.A., J. Morais Cabral, R.A. Pfuetzner, A. Kuo, J.M. Gulbis, S.L. Cohen, B.T. Chait, and R. MacKinnon. 1998. The structure of the potassium channel: Molecular basis of K^+ conduction and selectivity. *Science.* 280:69–77.

Faraday, M. 1834. Experimental researches on electricity. Seventh series. *Philosophical Transactions of the Royal Society of London.* 124:77–122.

Fitzhugh, R. 1965. A kinetic model of the conductance changes in nerve membrane. *Journal of Cellular and Comparative Physiology.* 66(Suppl. 2):111.

Galvani, L. 1953. De Viribus Electricitatis in Motu Musculari commentarius (1791). Translated by Robert Montraville Green. Elizabeth Licht Publisher.

Goldman, D.E. 1965. Gate control of ion flux in axons. *The Journal of General Physiology.* 48(Suppl):75–77.

Goodchild, S.J., H. Xu, Z. Es-Salah-Lamoureux, C.A. Ahern, and D. Fedida. 2012. Basis for allosteric open-state stabilization of voltage-gated potassium channels by intracellular cations. *The Journal of General Physiology.* 140:495–511.

Gorter, E. and F. Grendel. 1925. On bimolecular layers of lipoids on the chromocytes of the blood. *The Journal of Experimental Medicine.* 41:439–443.

Hertwig, O., M. Campbell, and H.J. Campbell. 1895. *The Cell: Outlines of General Anatomy and Physiology.* New York: Macmillan and Co.

Hille, B. 1971. The permeability of the sodium channel to organic cations in myelinated nerve. *The Journal of General Physiology.* 58:599–619.

Hille, B. 1984. *Ionic Channels of Excitable Membranes.* Sinauer: Sunderland, MA.

Hodgkin, A.L. and A.F. Huxley. 1952. A quantitative description of membrane current and its application to conduction and excitation in nerve. *The Journal of Physiology.* 117:500–544.

Hodgkin, A.L. and R.D. Keynes. 1955. The potassium permeability of a giant nerve fibre. *The Journal of Physiology.* 128:61–88.

Hodgkin, A.L. and W.A.H. Rushton. 1946. The electrical constants of a crustacean nerve fibre. *Proceedings of the Royal Society B: Biological Sciences.* 133(873):444–479.

Hoshi, T., W.N. Zagotta, and R.W. Aldrich. 1991. Two types of inactivation in *Shaker* K^+ channels: Effects of alterations in the carboxy-terminal region. *Neuron.* 7:547–556.

Larsson, H.P., O.S. Baker, D.S. Dhillon, and E.Y. Isacoff. 1996. Transmembrane movement of the *shaker* K^+ channel S4. *Neuron.* 16:387–397.

Long, S.B., X. Tao, E.B. Campbell, and R. MacKinnon. 2007. Atomic structure of a voltage-dependent K^+ channel in a lipid membrane-like environment. *Nature.* 450:376–382.

Marmont, G. 1949. Studies on the axon membrane; a new method. *Journal of Cellular Physiology.* 34:351–382.

Mullins, L. 1959. The penetration of some cations into muscle. *The Journal of General Physiology.* 42:817–829.

Narahashi, T., J.W. Moore, and W.R. Scott. 1964. Tetrodotoxin blockage of sodium conductance increase in lobster giant axons. *The Journal of General Physiology.* 47:965–974.

Neher, E. and B. Sakmann. 1976. Single-channel currents recorded from membrane of denervated frog muscle fibres. *Nature.* 260:799–802.

Nernst, W. 1889. Die elektromotorische Wirksamkeit der Ionen. *Zeitschrift für Elektrochemie und angewandte physikalische Chemie.* 4:129–181.

Noble, D. 1966. Applications of Hodgkin-Huxley equations to excitable tissues. *Physiological Reviews.* 46:1–50.

Noda, M., S. Shimizu, T. Tanabe, T. Takai, T. Kayano, T. Ikeda, H. Takahashi, H. Nakayama, Y. Kanaoka, N. Minamino et al. 1984. Primary structure of Electrophorus electricus sodium channel deduced from cDNA sequence. *Nature.* 312:121–127.

Overton, E. 1899. Über die allgemeinen osmotischen Eigenschaften der Zelle, ihre vermutlichen Ursachen und ihre Bedeutung für die Physiologie. *Vierteljahrsschr. Naturforsch. Ges. Zurich.* 44:88–114.

Papazian, D.M., X.M. Shao, S.A. Seoh, A.F. Mock, Y. Huang, and D.H. Wainstock. 1995. Electrostatic interactions of S4 voltage sensor in *Shaker* K^+ channel. *Neuron.* 14:1293–1301.

Parsegian, V.A. 1975. Ion-membrane interactions as structural forces. *Annals of the New York Academy of Sciences.* 264:161–171.

Perozo, E., R. MacKinnon, F. Bezanilla, and E. Stefani. 1993. Gating currents from a nonconducting mutant reveal open-closed conformations in *Shaker* K^+ channels. *Neuron.* 11:353–358.

Ren, D., B. Navarro, H. Xu, L. Yue, Q. Shi, and D.E. Clapham. 2001. A prokaryotic voltage-gated sodium channel. *Science.* 294:2372–2375.

Robertson, J.D. 1958. Alterations in nerve fibers produced by hypotonic and hypertonic solutions. *The Journal of Biophysical and Biochemical Cytology.* 4:349–364.

Robertson, J.D. 1959. The ultrastructure of cell membranes and their derivatives. *Biochemical Society Symposium*. 16:3–43.
Robinson, R.A. and R.H. Stokes. 2002. *Electrolyte Solutions*. Dover Publications, Mineola: NY.
Schneider, M.F. and W.K. Chandler. 1973. Voltage dependent charge movement of skeletal muscle: A possible step in excitation-contraction coupling. *Nature*. 242:244–246.
Schwann, T. 1839. *Microscopic Investigations on the Accordance in the Structure and Growth of Plants and Animals*. English translation by the Sydenham Society, 1847: Berlin, Germany.
Smith-Maxwell, C.J., J.L. Ledwell, and R.W. Aldrich. 1998. Uncharged S4 residues and cooperativity in voltage-dependent potassium channel activation. *The Journal of General Physiology*. 111:421–439.
Stoeckenius, W. 1959. An electron microscope study of myelin figures. *The Journal of Biophysical and Biochemical Cytology*. 5:491–500.
Stühmer, W., F. Conti, H. Suzuki, X.D. Wang, M. Noda, N. Yahagi, H. Kubo, and S. Numa. 1989. Structural parts involved in activation and inactivation of the sodium channel. *Nature*. 339:597–603.
Volta, A. 1800. On the electricity excited by the mere contact of conducting substances of different kinds. *Philosophical Transactions of the Royal Society of London*. 90:403–431.
von Helmholtz, H. 1850. Messungen über den zeitlichen Verlauf der Zuckung animalischer Muskeln und die Fortpflanzungsgeschwindigkeit der Reizung in den Nerven. In *Archiv für Anatomie, Physiologie und wissenschaftliche Medicin*. Jg. Veit & Comp: Berlin, Germany, pp. 276–364.
Yang, Y., Y. Yan, and F.J. Sigworth. 1997. How does the W434F mutation block current in *Shaker* potassium channels? *The Journal of General Physiology*. 109:779–789.
Yellen, G., M.E. Jurman, T. Abramson, and R. MacKinnon. 1991. Mutations affecting internal TEA blockade identify the probable pore-forming region of a K^+ channel. *Science*. 251:939–942.
Young, J.Z. 1936. Structure of nerve fibers and synapses in some vertebrates. *Cold Spring Harbor Symposium on Quantitative Biology*. 4:1–6.
Zhou, Y., J.H. Morais-Cabral, A. Kaufman, and R. MacKinnon. 2001. Chemistry of ion coordination and hydration revealed by a K^+ channel-Fab complex at 2.0 A resolution. *Nature*. 414:43–48.

Basic concepts

2 Ion selectivity and conductance

Dorothy M. Kim, Jason G. McCoy, and Crina M. Nimigean

Contents

2.1 INTRODUCTION

Ion channels function to orchestrate an exquisite array of physiological processes, including nerve impulses, muscle contraction, regulation of cell volume, and cell signaling in all organisms. The electric current in a signaling event is generated by ion flux across the cell membrane that is controlled by the opening and closing of ion channels, including those permeable to potassium, sodium, calcium, and chloride ions. The direction of the ion fluxes is determined by the membrane potential and preset transmembrane ionic gradients, which are established and maintained by specific ion channels and the Na^+/K^+ ATPase. For certain signaling modes, such as the generation of the action potential, it is important that the ion channels involved are only permeable to specific ions and impermeable to others. Particularly important to signaling in nerve and muscle are potassium and sodium channels, which display a very high level of selectivity. Therefore, some ion channels have evolved to exhibit high ionic selectivity, which is fundamental to electrical signaling.

As early as 1902, Julius Bernstein predicted that excitation was the result of a change in membrane permeability of excitable cells (Bernstein, 1902), hypothesizing that cells at rest were only permeable to K^+ and that permeability to other ions occurred during excitation; this provided the first suggestion of ion-selective components in the membrane. Breakthrough studies of the squid giant axon in the 1940s and 1950s by Hodgkin, Huxley, Keynes, Goldman, and Katz (Goldman, 1943; Hodgkin and Huxley, 1945, 1946, 1947, 1952a,b,c,d; Hodgkin and Katz, 1949; Hodgkin et al., 1952; Hodgkin and Keynes, 1955) identified the action potential of the axon to be the result of a composite of currents carried by different ions. Using voltage-clamp on the membrane of a squid giant axon bathed in a solution with controlled ion concentrations, they concluded that at rest the membrane was predominantly selective to K^+ resulting in a negative transmembrane potential (according to the electrophysiological convention, the transmembrane potential is measured inside the cell relative to the outside). On the other hand, Na^+ was responsible for the inward current causing the cell membrane to become more positive than at rest, referred to as depolarization. Hodgkin and Huxley identified these two major ionic components responsible for the generation of the action potential in their seminal 1952 papers (Hodgkin and Huxley, 1952a,b,c,d; Hodgkin et al., 1952), and developed a model directly correlating Na^+ and K^+ fluxes with excitation and electric conduction in the squid giant axon, earning them the Nobel Prize in 1963. These studies informed the conclusion that nerve impulse propagation is an electrical process involving a delicate balance of ion fluxes across the cell membrane controlled by the opening and closing (gating) of highly selective ion channels. Ion selectivity and regulated gating are crucial to generating the action potential.

The key to understanding the mechanism of selectivity in channels lies in the aqueous pore, a narrow canal comprising the permeation path for ions. The simplest illustration of the pore is that of a molecular sieve that can only pass ions not exceeding a certain radius. Although pore diameter was initially suggested to be a major determinant in ion permeability, it was insufficient to explain the permeability sequences exhibited by some channels. The region within the channel protein that directly interacts with the conducting ions is called the selectivity filter and both the dimensions and chemical properties of this region influence ion selectivity and conductance. In addition to containing binding sites that can accommodate only certain ions of specific sizes and valences, the pore provides an electrostatic landscape that favors permeation of certain ions over others, thus requiring ions to overcome various energy barriers to permeate the pore. The crystal structure of the bacterial potassium channel KcsA (Doyle et al., 1998; Zhou et al., 2001), for example, reveals that the chemistry within the pore provides binding sites for potassium ions, and may not accommodate sodium ions.

Prior to the elucidation of the atomic structures of these channels, pioneers in the ion channel field predicted some of these pore characteristics for channels of varying ion selectivity. One early hypothesis claimed that in order for channels to discriminate between ions they would need to dehydrate upon entry into the pore (Mullins, 1959; Bezanilla and Armstrong, 1972). The energy required for this step would then need to be balanced by stabilizing interactions between the ion and the walls of the pore (Mullins, 1959; Eisenman, 1962; Hille, 1975). The energetic landscape of a pore consists of specific binding sites for ions representing energetic minima separated by energetic barriers that can also be encountered upon entering and exiting the pore (Hille, 1975; Begenisich and Cahalan, 1980a,b). Ions with lower permeability thus may face a higher energy barrier at external binding sites or alternatively may encounter very deep energy wells within the selectivity filter. This concept can also explain why channels often are permeable to several ions but favor some over others. The determination of permeability ratios led to the conclusion that voltage-gated sodium channels have a large pore that can accommodate many different cations (Hille, 2001). Potassium channels, on the other hand, were imagined to contain a narrower pore than most Na channels, resulting in higher selectivity due to increased interaction and contact between the ion and the pore walls (Hille, 2001). This prediction was confirmed by the determination of crystal structures of voltage-gated Na channels from bacteria, which have a selectivity filter that is ~4.6 Å wide (Payandeh et al., 2011, 2012; Zhang et al., 2012), compared with the ~3 Å wide selectivity filter of the KcsA channel (Doyle et al., 1998; Zhou et al., 2001). Accordingly, potassium channels can conduct other cations such as Tl^+, Rb^+, and NH_4^+ but remain mostly impermeant to other large cations as well as Na^+ and Li^+.

Bezanilla and Armstrong (1972) and Hille (1973) proposed that K^+ ions were stabilized inside the K channel pore by binding sites composed of a bracelet of oxygen dipoles in the selectivity filter region. The selectivity filter would therefore consist of a cylindrical pore with an inner diameter between 3 and 3.4 Å. The K channel selectivity filter was later identified on the pore loop (P-loop), a reentrant loop containing a signature sequence highly conserved among K channels that could contribute these oxygen dipoles (Heginbotham et al., 1992, 1994). These predictions, based entirely on functional data, turned out to be astonishingly accurate and were supported by the crystal structure of the KcsA potassium channel (Doyle et al., 1998; Zhou et al., 2001) with potassium binding sites comprising mostly carbonyl dipoles from amino acid residues in the signature sequence TVGYG (Heginbotham et al., 1994).

At the same time, Bezanilla and Armstrong also hypothesized that ions binding within the pore would slow conduction and thus the payoff for high selectivity should be slow permeation (Bezanilla and Armstrong, 1972). Paradoxically, K channels can have very large conduction rates while also being highly selective. This paradox could be partly addressed by the ability of the pore to accommodate multiple ions at the same time. The multi-ion theory was proposed (Hodgkin and Keynes, 1955; Heckmann, 1965a,b, 1968, 1972; Hille and Schwarz, 1978) in order to explain deviations from the independence principle (Hodgkin and Huxley, 1952b) that states that the probability that an ion crosses the membrane is independent of other ions present. Several lines of evidence support the simultaneous occupancy of more than one ion in the pore. Hodgkin and Keynes hypothesized that the observed anomalous unidirectional flux ratios in squid axons were due to ions passing single-file through long channels that can hold multiple ions at one time (Hodgkin and Keynes, 1955). This was based on the observation that the K channel exhibits flux coupling, which indicates the presence of more than one ion in the pore. The multi-ion channel theory was further supported by the so-called anomalous mole fraction effect (Neher and Sakmann, 1975; Sandblom et al., 1977; Hille and Schwarz, 1978; Hess and Tsien, 1984; Eisenman et al., 1986), which describes changes in permeability ratios in Na (Chandler and Meves, 1965; Cahalan and Begenisich, 1976; Begenisich and Cahalan, 1980a), K (Hagiwara and Takahashi, 1974), and Ca (Almers and McCleskey, 1984; Hess and Tsien, 1984) channels when more than one permeant ion type is present. These studies showed that channel conductance passes through a minimum or maximum when plotted against the mole fraction of two permeant ions and it was understood by allowing the two ions to interact inside the pore and with the pore walls (Nonner et al., 1998; Hille, 2001).

Studies providing strong evidence that supports the multi-ion model were made possible by the advent of recombinant DNA technology, which allowed for the sequencing, cloning, heterologous expression, and purification of ion channel proteins. While sodium and calcium channels had been purified from membrane preparations in the 1970s, it was not until the 1980s that the first ion channel genes, the nicotinic acetylcholine receptor channel nAChR from *Torpedo californica* (Noda et al., 1982, 1983) and the voltage-gated sodium channel from the electric eel *Electrophorus electricus* (Noda et al., 1984), were cloned. The nAChR channel was also the first to be recorded with patch clamp (Neher and Sakmann, 1976) and then eventually purified, reconstituted, and recorded in a lipid bilayer (McCarthy et al., 1986; Montal et al., 1986). Success with the discovery and cloning of potassium channels followed (Papazian et al., 1987). In addition, the discovery of homologous K channels (Milkman, 1994; Schrempf et al., 1995; Derst and Karschin, 1998) and Na channels (Ren et al., 2001; Ito et al., 2004; Koishi et al., 2004; Webster et al., 2004) in prokaryotes paved the way for numerous advancements in structural studies of ion channels. These advances, as well as developments in electrophysiological techniques, enabled the taxonomic grouping of known channels into functional superfamilies as well as the prediction of novel channels by sequence analysis. Site-directed mutagenesis studies revealed the functions of specific residues in channels, while sequence inspection also provided some insight into the structural properties of ion channels.

Ultimately, the rapid accumulation of information in the era of cloning led to a major breakthrough in the first crystal structure of an ion channel in 1998 (Doyle et al., 1998), as described below. This structural data, in conjunction with the wealth of structure-based data that followed, have revolutionized the ion channel field and inspired many new avenues of research. These structures confirmed many of the predictions of early functional data including the physical properties of the pore, the dehydration of ions within the selectivity filter, and the multi-ion nature of ion conduction. In the next section, we describe these structures in detail and explore how they have enriched our knowledge and understanding of selectivity.

2.2 STRUCTURAL BASIS FOR SELECTIVITY IN ION CHANNELS

2.2.1 K CHANNEL SELECTIVITY

Analysis of *Drosophila* mutants resulted in the discovery of a region of genomic DNA encoding a protein involved in the conductance of potassium. The observation of a fly that exhibited uncontrolled shaking led to the identification of the *Shaker* locus. This gene encodes a member of the voltage-gated potassium channel (Kv) family and was the first K channel to be cloned (Kamb et al., 1987; Papazian et al., 1987). This milestone allowed for the heterologous expression of a K channel in *Xenopus* oocytes and functional characterization (Iverson et al., 1988; Timpe et al., 1988). Electrophysiological studies in oocytes revealed that *Shaker* demonstrates a clear selection preference for K^+, Rb^+, and NH_4^+ over Na^+ (Heginbotham and MacKinnon, 1993). In addition to these discoveries, the cloning of additional potassium channels provided a rich trove of amino acid sequence information. Analysis of these sequences identified a stretch of highly conserved residues, and this region is the site of various mutations that lead to distinct changes in ionic selectivity (Heginbotham et al., 1994). These conserved residues ($T_1X_2X_3T_4X_5G_6Y_7G_8$) were coined the signature sequence. Within this sequence, the GYG motif is by far the most conserved region, although some channels such as the EAG-like K channels and the inwardly rectifying Kir6 family of K channels contain a phenylalanine in place of tyrosine.

The significance of these residues was unveiled upon the publication of the first crystal structure of a potassium channel (Doyle et al., 1998; Zhou et al., 2001). The structure of KcsA, from the soil bacterium *Streptomyces lividans*, reveals a tetramer with fourfold symmetry and a clear pore along the fourfold symmetry axis (Figure 2.1a). Each subunit contains two transmembrane domains (T1 and T2). Between the T1 and T2 helices a short helix (the pore helix) is positioned such that its C-terminus points into the center of the channel pore. Immediately following this helix, the selectivity filter loop containing the signature sequence residues, $T_{72}A_{73}T_{74}T_{75}V_{76}G_{77}Y_{78}G_{79}$, extends into the center of the channel. The backbone carbonyl oxygens of Y78, G77, V76, and T75 as well as the side-chain hydroxyl of T75 point directly into the center of the protein, forming the narrowest region of the channel pathway. These oxygens coordinate dehydrated K^+ ions as they permeate the pore, surrounding each K^+ by eight oxygen atoms in a cage-like structure (Figure 2.1b and c). The four potassium binding sites are numbered S1–S4, starting from the most extracellular cage formed by the backbone carbonyls of Y78 and G77 (the S1 site), and ending with the most intracellular cage formed by the side-chain hydroxyl and backbone carbonyl oxygens of T75 (the S4 site). In addition to the four K^+ ions fully coordinated by the protein, a fifth K^+ binding site is observed at the extracellular surface of the channel, coordinated by the backbone carbonyl oxygens of Y78 and four water molecules (the S0 site). The KcsA crystal structure confirmed many of the early predictions about the potassium channel selectivity filter, including its multi-ion nature and the rings of oxygen atoms involved in coordinating K^+ (Bezanilla and Armstrong, 1972; Hille and Schwarz, 1978; Neyton and Miller, 1988a,b).

The structure of the KcsA potassium channel and the three-dimensional architecture of the selectivity filter reveal a detailed picture of the molecular mechanisms by which potassium channels distinguish between different ions. The structures provide not only the identities of functional groups that interact

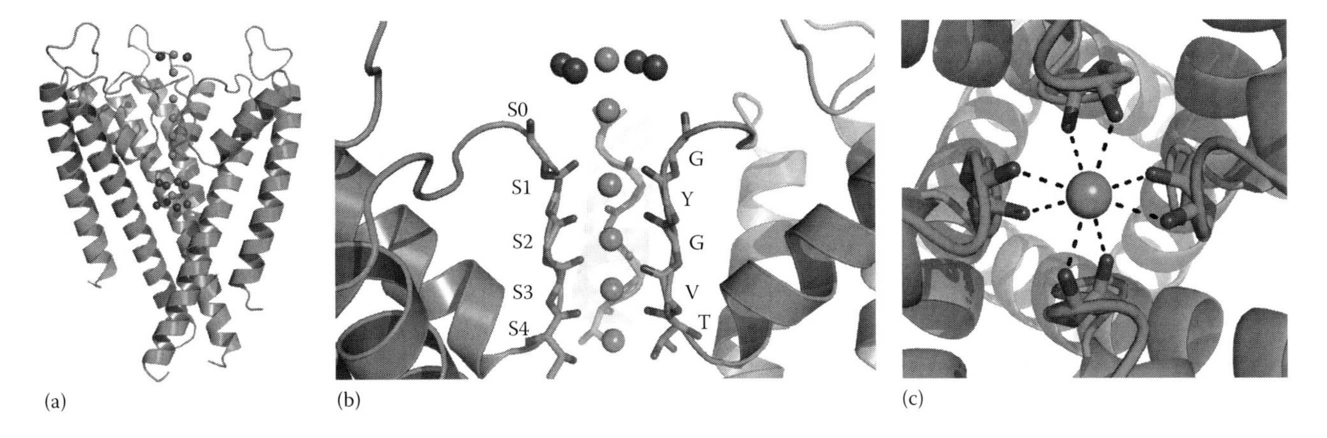

Figure 2.1 Potassium channel selectivity filter. Potassium ions are shown in the selectivity filter in single file. Water molecules are shown in a and b as spheres surrounding the potassium ions at the external and internal sides of the filter. (a) The tetrameric K channel KcsA (PDB ID 1K4C). The front subunit has been removed for clarity. (b) The KcsA selectivity filter. K^+ binding sites are labeled S0 through S4 (from the extracellular to the intracellular surface) and are composed of the contiguous sequence TVGYG. (c) Close-up of the S1 site as observed from the extracellular surface of the channel. The K^+ ion is coordinated in an octahedral fashion by carbonyl oxygens from the selectivity filter residues.

with the K^+ ions and the bond distances, but they also serve as good starting models for molecular dynamic simulations that can also provide a detailed understanding of selectivity. In the following subsections, we describe several models that have been proposed as the basis of selectivity. Many of these models are not necessarily mutually exclusive.

2.2.1.1 Close-fit model of selectivity

The underlying assumption of the close-fit model is that the selectivity filter distinguishes between different ions based on ionic radius. In this model, the distances between the conducting ion and the selectivity filter backbone carbonyl oxygens are tuned specifically to optimize the coordination of K^+ relative to that of smaller or larger ions (Figure 2.2a). This partially alleviates the energetic penalty for dehydrating the K^+, and this is necessary for entry into the selectivity filter. Despite this high affinity, the ions are still rapidly processed through the selectivity filter presumably due to electrostatic repulsions between multiple K^+ ions in close proximity in the filter. In the published 2 Å resolution crystal structure of KcsA, the average coordination distance between the K^+ ions and the selectivity filter oxygen atoms is 2.85 Å, nearly identical to that observed in the potassium-selective antibiotic nonactin (Zhou et al., 2001). Na^+, with an ionic radius approximately 0.4 Å smaller than that of K^+, would require a positional shift of the selectivity filter for optimal binding. It should also be noted that crystal structures depict a positional average over time. Therefore, while the structures show K^+ density in each of the selectivity filter binding sites, the K^+ ions probably occupy only alternate binding sites as they pass through the channel (e.g., S4 and S2 or S3 and S1) (Morais-Cabral et al., 2001).

2.2.1.2 Field-strength model of selectivity

The field-strength model foregoes the idea of rigid cages, and instead treats the selectivity filter as a flexible, liquid-like environment. The dipole moments of the selectivity filter backbone carbonyls generate electrostatic forces that repel each other and attract the cation leading to selectivity for K^+ (Noskov et al., 2004; Noskov and Roux, 2006) (Figure 2.2a). In this model, it is the physical properties of the ligand groups (i.e., the selectivity filter backbone carbonyls) that lead to selection for K^+ over other ions. In simplified simulations, a ligand dipole between 2.5 and 4.5 debye (similar to a carbonyl) selects for K^+, whereas decreasing or increasing the ligand dipole leads to optimal selection for larger or smaller cations, respectively (Noskov et al., 2004).

2.2.1.3 Coordination model of selectivity

The coordination model differs from the close-fit and field strength models in that the number of groups that interact with the ligand also contributes to selectivity. In other words, a protein framework with eight liganding moieties, whether they belong to waters or carbonyls, as in the KcsA channel, will inherently select for K^+ over Na^+ (Bostick and Brooks, 2007) (Figure 2.2a). In another version of the coordination model, the electrical properties of the local environment of the binding sites create strong selectivity by over-coordinating K^+ with eight carbonyl

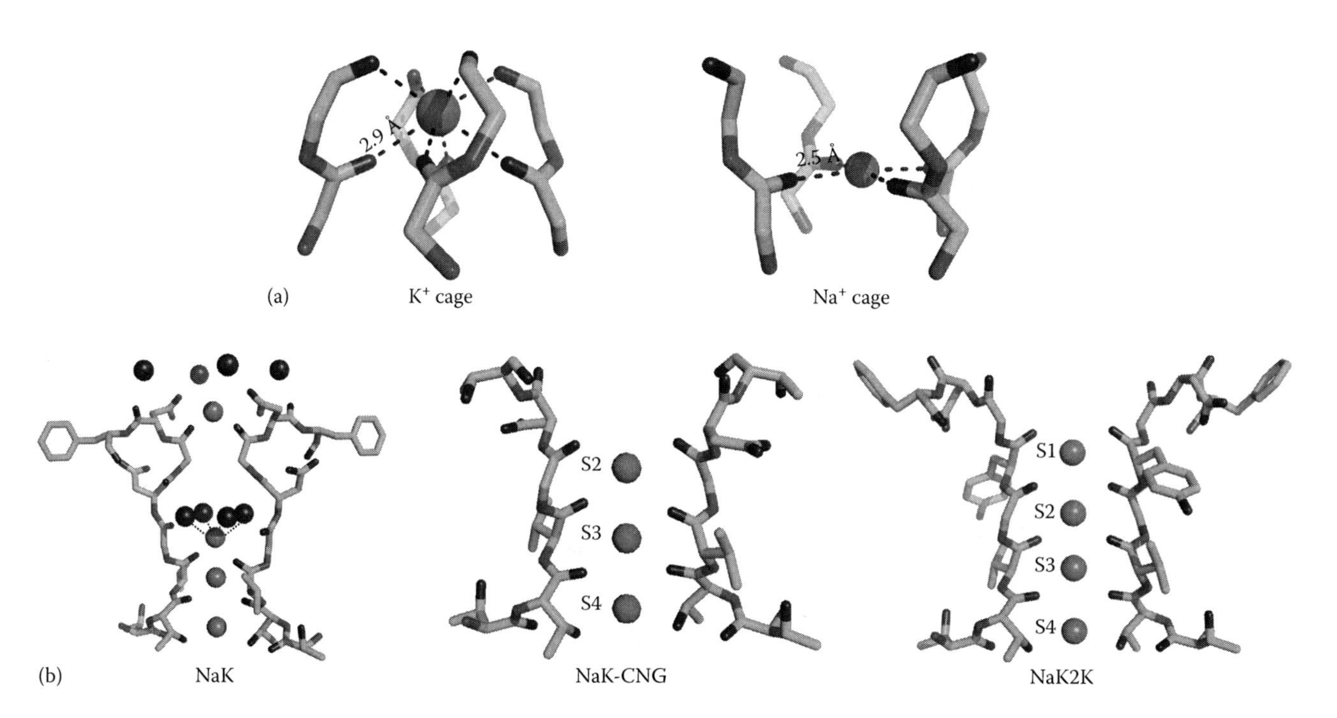

Figure 2.2 Principles of selectivity. (a) Ion coordination in the K channel selectivity filter demonstrates principles of the close-fit, field-strength, coordination, and kinetic models. K^+ binds with octahedral coordination in the center of a cage formed by eight carbonyl oxygens (left). In KcsA this bond distance is 2.9 Å (PDB ID 1K4C). In contrast Na^+ binds with square planar coordination in the plane of the carbonyls (right). In KcsA this bond distance is 2.5 Å (PDB ID 3OGC). (b) The number of contiguous ion binding sites demonstrates principles of the site number model of selectivity. K^+ ions are shown as spheres in single file in the filter. Water molecules are shown only for NaK as spheres coordinating the top two ions. The nonselective NaK channel lacks the S1 and S2 binding sites (left panel). Restoration of the S2 site through mutagenesis of the selectivity filter to TVGDTPP (called NaK-CNG), a sequence very similar to that of nonselective cyclic nucleotide-gated (CNG) channels, fails to increase the K^+ selectivity of the channel (center). Restoration of site S1 through mutagenesis of the NaK selectivity filter to TVGYGDF (called NaK2K), a sequence very similar to that of KcsA, results in a K^+ selective channel (right).

ligands instead of the usual six ligands coordinating K^+ in water (Varma and Rempe, 2006; Varma et al., 2008).

2.2.1.4 Kinetic model of selectivity

In the selectivity mechanisms described above, the origins of selectivity are ultimately derived from a difference in the free energy of binding for different ions within the selectivity filter. An alternative kinetic-based model suggests that an additional layer of K^+ selectivity over other ions may be the result of a high-energy barrier preventing other ions from advancing through the selectivity filter. Molecular dynamics simulations of KcsA predict that the S4 site is slightly selective for both Na^+ and Li^+ over K^+; however, the energy minima for the Na^+ and Li^+ lies in the plane formed by the carbonyl oxygens dividing the S3 and S4 sites (Thompson et al., 2009) (Figure 2.2a). Crystal structures of KcsA in the presence of Na^+ have also shown spherical densities between the carbonyl oxygens instead of centered in the carbonyl cage as observed for K^+ (Lockless et al., 2007; Cheng et al., 2011). The presence of K^+ may then occlude the preferred binding sites of smaller ions such as Na^+, generating a large Coulombic repulsion that prevents Na^+ from entering the selectivity filter (Thompson et al., 2009; Egwolf and Roux, 2010; Kim and Allen, 2011).

2.2.1.5 Site number model for selectivity

Determination of the crystal structure of the NaK channel from *Bacillus cereus* (Shi et al., 2006) led to the hypothesis of the site number model. Unlike KcsA, the NaK channel does not prefer K^+ to Na^+. This difference in selectivity appears to be largely due to a change in the sequence of the selectivity filter. The selectivity filter sequence in NaK is TVGDGNF, in contrast to the KcsA sequence of TVGYGDL. The result of this change is that sites S1 and S2 are disrupted, creating a large cavity in place of the two cage-like binding sites (Figure 2.2b). Modifying the NaK selectivity filter sequence to TVGDTPP, similar to what is observed in nonselective cyclic nucleotide-gated channels, reestablishes the S2 site but this channel still fails to demonstrate selectivity (Derebe et al., 2011). Modifying the NaK selectivity filter to TVGYGDF, similar to that of KcsA, reestablishes both the S1 and S2 sites and results in heightened selectivity for K^+ (Derebe et al., 2011). This suggests that the presence of four contiguous binding sites is a critical component for establishing K^+ selectivity.

2.2.1.6 Other K channel selectivity determinants

While it is clear that the selectivity filter is crucial to mediating channel selectivity, many aspects of selectivity cannot be explained entirely by the filter. For example, hyperpolarization-activated cyclic nucleotide-gated (HCN) channels contain a variant of the signature sequence CIGYG. Replacement of the threonine in K^+-selective channels with cysteine would presumably remove the S4 site and reduce K^+ selectivity. In accordance with this hypothesis, Na^+ permeability in these channels can be up to one-third of that of K^+. However, mutation of cysteine to threonine actually further reduces the K^+ selectivity (D'Avanzo et al., 2009). In addition, channels with identical selectivity filters can exhibit differences in selectivity. Although several channels share the selectivity filter sequence TTVGYG, the voltage-gated channels Kv1.5 and Kv2.1 will conduct Na^+ in the absence of K^+ while KcsA and *Shaker* do not (Heginbotham and MacKinnon, 1993; Korn and Ikeda, 1995; LeMasurier et al., 2001). Addition of negative charges into the cavity of the inwardly rectifying Kir3.2 channel restores K^+ selectivity in a nonselective mutant of the channel, demonstrating that the pore cavity and not just the selectivity filter can influence selectivity (Bichet et al., 2006). C-type inactivation, a process by which the selectivity filter changes conformation to halt K^+ conductance in response to prolonged opening of the channel, has also been shown to influence selectivity (Starkus et al., 1997; Kiss et al., 1999; Cheng et al., 2011). Thus, other channel features in the pore outside of the selectivity filter can clearly modulate selectivity.

2.2.2 Na CHANNEL SELECTIVITY

Channels selective for Na^+ over K^+ were originally observed by Hodgkin and Huxley in studies of the squid giant axon (Hodgkin and Huxley, 1952b). Early work suggested that the Na^+ channel contains a rectangular opening of approximately 3 × 5 Å surrounded by oxygen atoms as well as a negatively charged carboxylate ion (Hille, 1971). The size of this opening suggests that the Na^+ ions conduct in a partially hydrated state. Furthermore, since the selectivity sequence of Na^+ channels ($Na^+ \sim Li^+ > Tl^+ > K^+ >> Ca^{++}$) closely follows the electrostatic model of Eisenman (1962), it was argued that selectivity was dependent on the field strength of the binding site (Hille, 1972). A high field-strength anion, such as a glutamate side chain, would be required to increase the Na^+ selectivity relative to that of K^+.

Cloning and sequencing of mammalian voltage-gated Na channels (Noda and Numa, 1987) and subsequent blocking studies (Terlau et al., 1991) led to the identification of four conserved residues instrumental for maintaining Na^+ selectivity. Unlike K channels, eukaryotic Na channels are monomeric proteins containing four similar repeated domains, thus lacking the fourfold symmetry of K channels. The four residues each reside in a distinct domain but occupy the same location in the primary structure based on sequence alignments of each individual domain. These residues consist of an aspartate in the N-terminal-most repeat followed by a glutamate, a lysine, and an alanine in the following repeats. This is referred to as the DEKA motif. Mutation of these residues has profound effects on selectivity. As Hille predicted, the filter sequence contains two residues with carboxyl-containing side chains (Hille, 1972). Mutation of glutamate to alanine leads to an increase in K^+ permeability (Favre et al., 1996), while mutation of the aspartate to alanine has little effect. Therefore, glutamate appears to play a larger role in selectivity in the Na channel filter. Removal of the lysine side chain leads to an even greater loss in selectivity against K^+ as well as Ca^{++} (Favre et al., 1996). Voltage-gated Ca channels have an EEEE motif in a similar location to that of voltage-gated Na channels. Mutation of the DEKA motif in Na channels to DEEE confers Ca^{++} selection properties on the channel (Favre et al., 1996).

The crystal structures of several bacterial voltage-gated Na channels (Payandeh et al., 2011; McCusker et al., 2012; Zhang et al., 2012) show that in contrast to their eukaryotic counterparts, these channels are homotetramers. The NavAb channel, the first of these to be published, contains a glutamate in the signature position, similar to eukaryotic Ca channels (Payandeh et al., 2011) (Figure 2.3a and b). The four glutamates

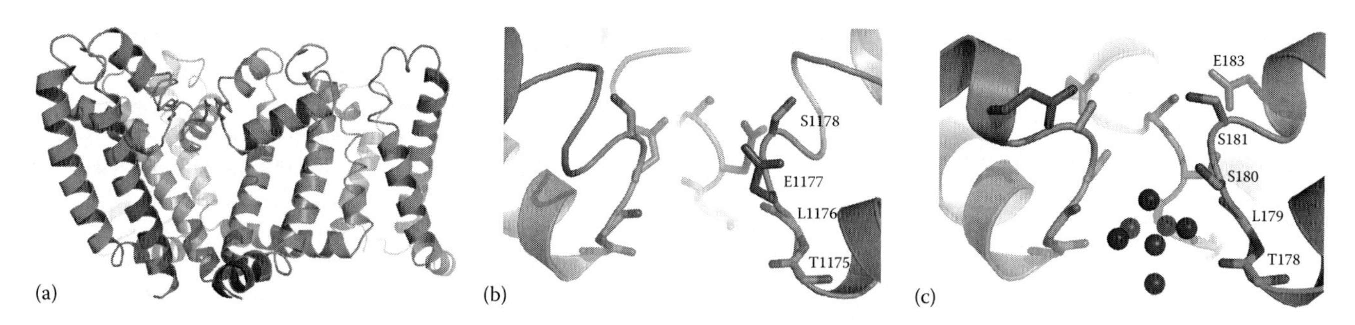

Figure 2.3 Na channel selectivity filter. (a) The crystal structure of the NavAb channel (PDB ID 4EKW). The glutamates of the signature sequence are shown in stick representation. (b) The NavAb channel selectivity filter. The glutamate of the signature sequence and a neighboring serine are proposed to form the extracellular binding site for Na^+. Main-chain carbonyls below the glutamate serve as interaction sites for hydrated Na^+ ions. (c) The NavRh channel selectivity filter (PDB ID 4DXW). The extracellular binding site is also proposed to be formed by a glutamate and a serine; however, in contrast to NavAb, the glutamate is located outside the selectivity filter and a second serine resides in the center of the pore. An ion bound in the NavRh selectivity filter was modeled as a hydrated Ca^{++} (gray sphere) interacting with carbonyls at the bottom of the selectivity filter (waters shown as spheres surrounding the ion.)

are positioned along the pore in the center of the channel near the extracellular opening. The cavity between the four glutamate side chains is the most constricted region within the channel allowing just enough room for the conducting Na^+ to remain partially hydrated while directly interacting with one of the glutamates. Below this glutamate, the backbone carbonyl oxygens of the two following amino acids (Leu and Thr) point into the pore (Figure 2.3b). Solvent molecules coordinated to these carbonyl oxygens presumably help rehydrate Na^+ as it passes through the filter. Molecular dynamics simulations have shown that Na^+ can occupy two sites simultaneously; one site in which it interacts with the side chains of the glutamate, a nearby serine, and multiple solvent molecules, and one site in which the ion is fully hydrated and coordinated by the backbone carbonyl oxygens that point toward the pore of the channel (Corry and Thomas, 2012; Furini and Domene, 2012) (Figure 2.3b). The Na^+ ions in these sites are weakly coupled but do not require the sort of simultaneous ion movement observed in K channels (Corry and Thomas, 2012; Furini and Domene, 2012). The largest energy barrier for Na^+ occurs in between these two sites (Corry and Thomas, 2012; Furini and Domene, 2012). However, as the EEEE sequence of NavAb is more similar to eukaryotic voltage-gated Ca channels than voltage-gated Na channels, it seems likely that there are additional features of the channel that tune its selectivity to Na^+. A similar arrangement is observed in other bacterial Na channels. The NavRh channel contains a serine in the signature position (Figure 2.3c). However, the structure has shown that a glutamate exterior to the selectivity filter points into the pore such that the carboxylate occupies a similar location to that of NavAb, presumably forming a similar Na^+ binding site (Zhang et al., 2012). The interior carbonyl oxygens from a leucine and a threonine form a secondary site that coordinates a hydrated Ca^{++} in the crystal structure.

The structural data described above provide compelling insight into selectivity and conduction mechanisms, and it is evident that the key characteristics of the selectivity filter are observed in both prokaryotic and eukaryotic K channels. Thus, the selectivity filter appears to be highly conserved across the kingdoms for K channels. It remains to be seen whether this universality applies to Na and Ca channels, and future studies of their eukaryotic counterparts are critical for identifying important deviations from the structures of bacterial homologues. These discoveries are paramount for understanding the evolution of selectivity and conduction in ion channels.

2.3 CONDUCTANCE

Ion conductance is a measure of the ease with which ions flux across the membrane. Ion conductance (in picoSiemens) can be calculated from measurements of electrical current generated by the movement of ions through the membrane, normalized to the voltage applied across it (via voltage-clamp), by simply using Ohm's law ($g = I/V$). In electrical terms, conductance is the inverse of resistance ($g = 1/R$). There are two ways to determine the ionic flux rate through a single ion channel (unitary conductance). One approach is to directly measure single-channel currents with patch-clamp and lipid bilayer recordings (Bean et al., 1969; Neher and Sakmann, 1976). Another approach is to indirectly calculate the unitary conductance from a macroscopic current composed of a large number of identical channels by using noise-analysis methods (Neher and Stevens, 1977; Sigworth, 1980a,b; Silberberg and Magleby, 1993). Comparisons of ion conductance among different channels are a useful first metric for assessing the ionic flow within their pores, providing a basis for understanding their respective permeation paths.

Because conductance is given by the ionic flux, it generally increases with increasing permeant ion concentration. However, a linear relationship between conductance and ion concentration is not obeyed at very high ion concentrations due to intrinsic structural properties of the channel or the presence of blockers. Instead, the conductance curves obey simple saturation kinetics that follow a Michaelis–Menten relationship (Michaelis and Menten, 1913; Hille, 2001; Michaelis et al., 2011). Saturation occurs when the steps of ion binding and unbinding are rate limiting; for instance, at high ion concentrations, the rate of ion entry increases and it will approach the maximum rates of unbinding (Hille, 2001). Therefore, as ion concentrations increase further, the rates of unbinding determine the overall rate of conduction and a plateau at maximum conductance is reached. Intrinsic properties of the channels alter this maximum conductance by changing the rate of ion exit from the pore.

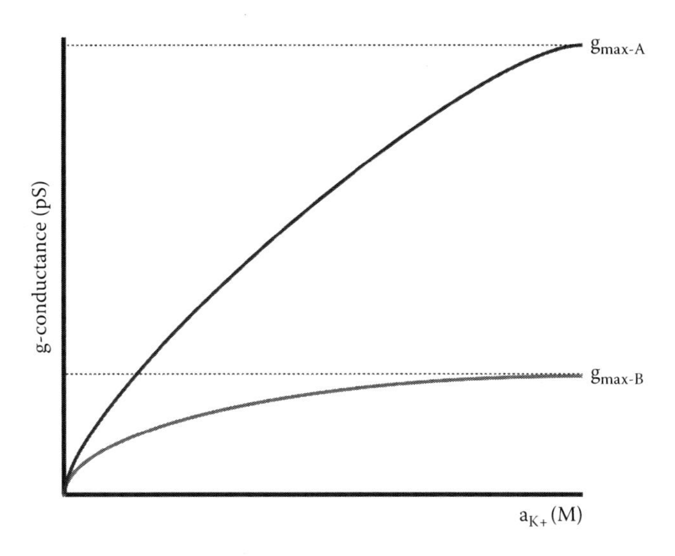

Figure 2.4 Conductance–activity curves. Plot showing conductance as a function of ionic activity for hypothetical channel A with a high maximum conductance ($g_{max\text{-}A}$) such as KcsA, and for hypothetical channel B with a low maximum conductance ($g_{max\text{-}B}$) such as K_{ir}. The g_{max} for each channel is denoted by a dotted black line.

The typical conductance–concentration saturation curve is comprised of two components: an initial quasilinear rise at low ion concentrations and a plateau at higher ion concentrations (Latorre and Miller, 1983) (Figure 2.4). The plateau portion of the curve represents the maximum conductance, or g_{max}, and is a measure of the exit rate of the permeating ion. As described below, this rate is dependent on the length of the pore, the number of binding sites, and the interaction between binding sites and permeating ions. In a multi-ion pore, repulsion between ions increases the rate of ions exiting from the pore, or g_{max}, and therefore, pores that can accommodate multiple ions can achieve a higher maximum conductance. The slope of the initial linear component of the curve measures the rate of ion entry into the pore. This rate is limited by the convergence conductance, described by Latorre and Miller as the absolute upper limit of conductance set by diffusion of ions to the mouth of the pore (Latorre and Miller, 1983). The convergence conductance depends on both the size of the pore (which also plays a role in selectivity) and the radius of the permeant ion. By increasing the capture radius of the pore, the rate of ion entry is also increased. However, widening the pore diameter will consequently decrease the selectivity. To circumvent this issue, many channels (Na and K channels, for instance) widen the pore at the base but contain a narrow constriction at the selectivity filter in order to increase entry rate and conductance but maintain high selectivity (Doyle et al., 1998; Zhou et al., 2001; Long et al., 2005; Payandeh et al., 2011, 2012) (Figures 2.1 and 2.3).

In the context of ion channels, it would be intuitive to assume that high selectivity results in low conductance, due to the fact that selectivity requires intimate contact between the ion and the pore walls, consequently slowing permeation. However, this is not the case. The energetics of ion binding dictates both selectivity and conduction. For example, inter-ion repulsion within the pore results in increased rate of exit of ions from the pore, thus increasing conductance (Hodgkin and Keynes, 1955; Eisenman, 1962; Hille and Schwarz, 1978; Morais-Cabral et al., 2001; Zhou and MacKinnon, 2003). In contrast, attractive forces at ion binding sites within the pore, necessary for high selectivity, can decrease the rate of exit, thus decreasing conductance (Hille, 1975; French, 1976; Hille and Schwarz, 1978). These electrostatic forces contribute to the energetic landscape that the ion must traverse to cross the membrane and can greatly affect the rate of flow.

These principles have been established for many years, but were later illustrated in crystallographic studies of KcsA conducted in the presence of different ions and a range of ion concentrations (Morais-Cabral et al., 2001; Zhou et al., 2001). A model was proposed for the barrierless conduction in K channels based on the observed ionic configurations of the selectivity filter. In high permeant ion concentrations, the selectivity filter is found in a conductive state, where two ions can bind at the same time in either a water-K-water-K (S2, S4) or K-water-K-water (S1, S3) configuration (Figures 2.1 and 2.5). The permeating ions provide counter charges for the 20 electronegative oxygen atoms in the filter, which minimizes the destabilization from same charge repulsion (Figure 2.1b and c). Entry of an additional ion would destabilize this configuration, supporting the knock-on model (Hodgkin and Keynes, 1955) as a mechanism for ion conduction and coupling high conductivity with high selectivity (Figure 2.5).

Interestingly, at low permeant ion concentrations, the selectivity filter of KcsA assumes a collapsed, nonconductive conformation, with presumably only one ion bound at one time (Morais-Cabral et al., 2001; Zhou and MacKinnon, 2003). Binding of the second ion could induce a conformational change in the selectivity

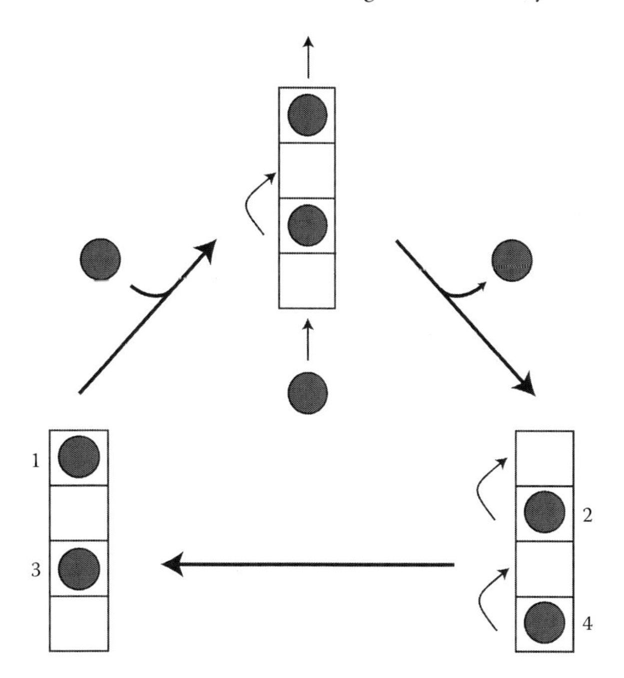

Figure 2.5 Ion conduction pathway in multi-ion pore. (Figure adapted from Zhou, Y. and MacKinnon, R., *J. Mol. Biol.*, 333, 965, 2003; Morais-Cabral, J.H. et al., *Nature*, 414, 37, 2001.) The four potassium binding sites S1–S4 in the KcsA pore are represented by boxes containing either a potassium ion (circle) or water (not shown). In the absence of an electrochemical driving force on the ions, the 1,3 (left) and 2,4 (right) configurations would have similar energies and be in equilibrium. In the presence of an outward driving force (as shown here), an incoming ion (center) enters the pore to occupy site 4, pushing the ion in site 3 to site 2 and also pushing the ion in site 1 out of the pore (top panel), resulting in a net outward K^+ current.

filter, transitioning from the nonconducting to the conductive conformation. These results suggest that conduction requires the presence of two ions in the selectivity filter. Furthermore, because the nonconductive filter contains only one ion, the second ion-binding event is presumed to be the cause of the filter conformational change from collapsed to conductive. The energetic cost of this conformational change can be balanced by a decrease in the energy of ion binding in the conductive filter, thereby increasing the exit rate of ions and allowing for efficient conduction. This study illustrates yet another possibility of how high selectivity could coexist with high conduction; the presence of multiple high affinity binding sites in the pore allows for selection of certain ions over others, while high conduction only occurs in the fully occupied selectivity filter. Thus, high conduction is due to both electrostatic repulsion and, in the case of KcsA, a conformational change that further lowers ion-binding affinity and increases the exit rate of bound ions. Interestingly, other crystallized K channels do not show a collapsed selectivity filter in low concentrations of permeant ions, suggesting that multiple factors play a role in conduction and selectivity and these vary among channels (Shi et al., 2006; Ye et al., 2010; Derebe et al., 2011). Therefore, the multi-ion pore is key to understanding the seemingly paradoxical coupling of high selectivity and high conductance.

While ion selectivity does not vary widely within a channel subfamily, conductance levels can vary dramatically. Channel conductance is dependent on many factors, such as pore chemistry (Hille, 1971, 1975), pore length (Hille and Schwarz, 1978; Latorre and Miller, 1983), the size of the cavity (Jiang et al., 2002a; Brelidze and Magleby, 2005; Geng et al., 2011), rates of dehydration (Mullins, 1959; Bezanilla and Armstrong, 1972), electrostatics (Brelidze et al., 2003; Nimigean et al., 2003; Furini et al., 2007), the energy landscape (Berneche and Roux, 2001; Hille, 2001), rigidity of the selectivity filter (Morais-Cabral et al., 2001; Noskov et al., 2004), pore occupancy (Hille and Schwarz, 1978; Morais-Cabral et al., 2001; Zhou and MacKinnon, 2003; Jensen et al., 2010, 2013; Moscoso et al., 2012), and ionic conditions (Latorre and Miller, 1983). Potassium channels range in conductance from 10 pS (including K_{ir} channels (Latorre and Miller, 1983), and small conductance K (SK) channels (Hille, 1971; Conti et al., 1975; Conti and Neher, 1980)) to 200–400 pS (including BK channels, (Marty, 1981; Pallotta et al., 1981; Yellen, 1984; Latorre et al., 1989; Hille, 2001)) in similar conditions (Latorre and Miller, 1983; Hille, 2001). Since the selectivity filter of K channels is highly conserved (GYG), other factors must account for these varying levels of conductance. One well-studied determinant of conductance in K channels is the size of the entrance to the inner cavity on the intracellular side of the selectivity filter. The maxi-K channels, such as the Ca^{++}-dependent BK channel and its archaeal homolog MthK, exhibit conductance rates close to diffusion limit, ranging from 250 to 300 pS (Eisenman et al., 1986), which is roughly 10–20 times higher than for other K channels (Hille, 2001). Crystal structures of MthK (Jiang et al., 2002a,b) and studies of BK using large quaternary ammonium blockers (Li and Aldrich, 2004) reveal that these channels contain a large aqueous cavity on the inner mouth of the pore that sits directly under the selectivity filter in the plane of the membrane. This cavity is larger than the equivalent in other K channels such as KcsA, K_{ir}, and

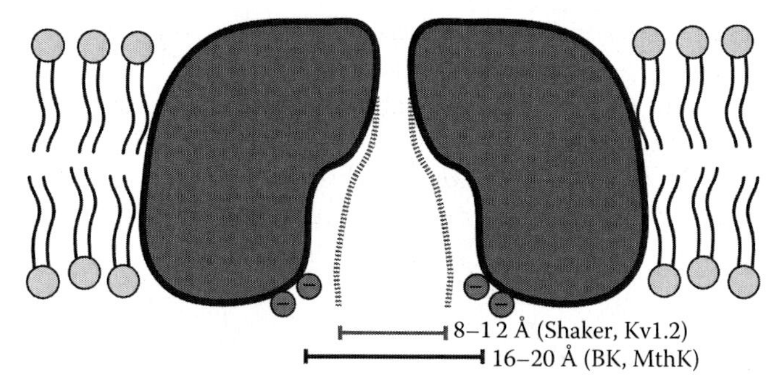

Figure 2.6 Cavity dimensions in different K channels. Variations in cavity width are shown relative to the BK/MthK channel, which is predicted to have a cavity that is 16–20 Å wide (From Brelidze, T.I. and Magleby, K.L., *J. Gen. Physiol.*, 126, 105, 2005; Geng, Y. et al., *J. Gen. Physiol.*, 137, 533, 2011; Jiang et al., 2002). *Shaker*/Kv1.2, which has a more modest cavity compared with BK (8–12 Å, (Li and Aldrich, 2004; Long et al., 2005; Webster et al., 2004)), is shown with dashed lines. Negative charges that ring the mouth of the cavity in BK (Nimigean et al., 2003; Brelidze et al., 2003) are shown in circles at the internal side of the cavity.

Shaker (Doyle et al., 1998; Kuo et al., 2003, 2005; Long et al., 2005) (Figure 2.6). Potassium ions need to pass through this entrance and traverse the cavity before entering the selectivity filter. Therefore, the width of the cavity entrance is key to determining the rate of efflux. This idea was supported by studies showing that an increase in side-chain volume at the cavity entrance results in a reduction in the single-channel conductance in the outward direction (Geng et al., 2011), while having no effect on the inward current. This reduction in outward current is reversed by an increase in intracellular K^+ concentration. This suggests that the ions do not encounter a barrier upon entrance into the cavity, and this contributes to the increase in conduction rate compared with other channels. Additional studies show that the entrance to the inner vestibule of BK is ~16–20 Å in diameter (Brelidze and Magleby, 2005), in contrast to *Shaker*, which contains a more modestly sized vestibule of 8–12 Å (Li and Aldrich, 2004; Webster et al., 2004; Long et al., 2005) and has a correspondingly smaller conductance (30–60 pS). Molecular dynamics simulations provide further support for the importance of the width of the cavity to conductance, demonstrating the increase in conductance of K channels with a widening of the entrance to the inner vestibule (Chung et al., 2002). In addition, eight negative charges ring this entrance, attracting K ions to the cavity (Brelidze et al., 2003; Nimigean et al., 2003) and reducing the diffusion limitation on conductance (Latorre and Miller, 1983), further contributing to the large outward current of BK channels in physiological ion concentrations (Figure 2.6).

In addition to cavity size, pore length has been hypothesized to contribute to the single-channel conductance level characteristic of a particular channel (Latorre and Miller, 1983). This hypothesis is supported by structural and functional studies of the bacterial inward-rectifying K channels KirBac1.1 and KirBac3.1 and the eukaryotic Kir2.2. These each contain a pore that is ~60–85 Å in length extending through the large C-terminal domains (Kuo et al., 2003, 2005; Nishida et al., 2007; Tao et al., 2009; Bavro et al., 2012), about twice the length of the pore of KcsA (Doyle et al., 1998). These channels exhibit a relatively low conductance level of ~10–35 pS at −100 mV (Latorre and Miller, 1983; Cheng et al., 2009). Thus, the long length of the pore of Kir channels may slow the passage of ions through the pore resulting in a decrease in the single-channel conductance compared with channels with shorter pores such as BK and KcsA. However, several other channel families such as the SK channels and *Shaker*/Kv channels also display a low single-channel conductance level (10–20 pS) but do not possess a similarly long pore to Kir. Therefore, while pore length may play a role in conductance levels in K_{ir} channels, additional factors must contribute to the overall single-channel conductance in other types of channels.

The pore dimensions can thus create diversity within ion channel subfamilies by contributing to a wide range of conductance levels while retaining the evolutionarily conserved, highly specialized selectivity filter sequence. With minimal changes to the basic pore architecture, channels selective for a particular ion can exhibit an array of conductance levels that may be appropriate for different physiological situations. Perhaps of even greater significance is the ability of the channels to conduct ions at very high rates of flux without sacrificing the high degree of selectivity that is imperative for electrical signaling. Therefore, ion channels have evolved intrinsic properties critical for achieving the efficiency and specificity necessary for controlling a plethora of biological processes.

REFERENCES

Almers, W. and McCleskey, E. W. 1984. Non-selective conductance in calcium channels of frog muscle: Calcium selectivity in a single-file pore. *J Physiol*, 353, 585–608.

Bavro, V. N., De Zorzi, R., Schmidt, M. R., Muniz, J. R., Zubcevic, L., Sansom, M. S., Venien-Bryan, C., and Tucker, S. J. 2012. Structure of a KirBac potassium channel with an open bundle crossing indicates a mechanism of channel gating. *Nat Struct Mol Biol*, 19, 158–163.

Bean, R. C., Shepherd, W. C., Chan, H., and Eichner, J. 1969. Discrete conductance fluctuations in lipid bilayer protein membranes. *J Gen Physiol*, 53, 741–757.

Begenisich, T. B. and Cahalan, M. D. 1980a. Sodium channel permeation in squid axons. I: Reversal potential experiments. *J Physiol*, 307, 217–242.

Begenisich, T. B. and Cahalan, M. D. 1980b. Sodium channel permeation in squid axons II: Non-independence and current-voltage relations. *J Physiol*, 307, 243–257.

Berneche, S. and Roux, B. 2001. Energetics of ion conduction through the K^+ channel. *Nature*, 414, 73–77.

Bernstein, J. 1902. Untersuchungen zur Thermodynamik der bioelektrischen Ströme. *Pflügers Arch*, 92, 521–562.

Bezanilla, F. and Armstrong, C. M. 1972. Negative conductance caused by entry of sodium and cesium ions into the potassium channels of squid axons. *J Gen Physiol*, 60, 588–608.

Bichet, D., Grabe, M., Jan, Y. N., and Jan, L. Y. 2006. Electrostatic interactions in the channel cavity as an important determinant of potassium channel selectivity. *Proc Natl Acad Sci U S A*, 103, 14355–14360.

Bostick, D. L. and Brooks, C. L., 3RD 2007. Selectivity in K^+ channels is due to topological control of the permeant ion's coordinated state. *Proc Natl Acad Sci U S A*, 104, 9260–9265.

Brelidze, T. I. and Magleby, K. L. 2005. Probing the geometry of the inner vestibule of BK channels with sugars. *J Gen Physiol*, 126, 105–121.

Brelidze, T. I., Niu, X., and Magleby, K. L. 2003. A ring of eight conserved negatively charged amino acids doubles the conductance of BK channels and prevents inward rectification. *Proc Natl Acad Sci U S A*, 100, 9017–9022.

Cahalan, M. and Begenisich, T. 1976. Sodium channel selectivity. Dependence on internal permeant ion concentration. *J Gen Physiol*, 68, 111–125.

Chandler, W. K. and Meves, H. 1965. Voltage clamp experiments on internally perfused giant axons. *J Physiol*, 180, 788–820.

Cheng, W. W., Enkvetchakul, D., and Nichols, C. G. 2009. KirBac1.1: It's an inward rectifying potassium channel. *J Gen Physiol*, 133, 295–305.

Cheng, W. W., Mccoy, J. G., Thompson, A. N., Nichols, C. G., and Nimigean, C. M. 2011. Mechanism for selectivity-inactivation coupling in KcsA potassium channels. *Proc Natl Acad Sci U S A*, 108, 5272–5277.

Chung, S. H., Allen, T. W., and Kuyucak, S. 2002. Conducting-state properties of the KcsA potassium channel from molecular and Brownian dynamics simulations. *Biophys J*, 82, 628–645.

Conti, F., Defelice, L. J., and Wanke, E. 1975. Potassium and sodium ion current noise in the membrane of the squid giant axon. *J Physiol (Lond)*, 248, 45–82.

Conti, F. and Neher, E. 1980. Single channel recordings of K^+ currents in squid axons. *Nature*, 285, 140–143.

Corry, B. and Thomas, M. 2012. Mechanism of ion permeation and selectivity in a voltage gated sodium channel. *J Am Chem Soc*, 134, 1840–1846.

D'avanzo, N., Pekhletski, R., and Backx, P. H. 2009. P-loop residues critical for selectivity in K channels fail to confer selectivity to rabbit HCN4 channels. *PLoS One*, 4, e7712.

Derebe, M. G., Sauer, D. B., Zeng, W., Alam, A., Shi, N., and Jiang, Y. 2011. Tuning the ion selectivity of tetrameric cation channels by changing the number of ion binding sites. *Proc Natl Acad Sci U S A*, 108, 598–602.

Derst, C. and Karschin, A. 1998. Evolutionary link between prokaryotic and eukaryotic K^+ channels. *J Exp Biol*, 201, 2791–2799.

Doyle, D. A., Morais Cabral, J., Pfuetzner, R. A., Kuo, A., Gulbis, J. M., Cohen, S. L., Chait, B. T., and MacKinnon, R. 1998. The structure of the potassium channel: Molecular basis of K^+ conduction and selectivity. *Science*, 280, 69–77.

Egwolf, B. and Roux, B. 2010. Ion selectivity of the KcsA channel: A perspective from multi-ion free energy landscapes. *J Mol Biol*, 401, 831–842.

Eisenman, G., Latorre, R., and Miller, C. 1986. Multi ion conduction and selectivity in the high-conductance Ca^{++}-activated K^+ channel from skeletal muscle. *Biophys J*, 50, 1025–1034.

Eisenman, W. 1962. A two-way affair. *Science*, 136, 182.

Favre, I., Moczydlowski, E., and Schild, L. 1996. On the structural basis for ionic selectivity among Na^+, K^+, and Ca^{2+} in the voltage-gated sodium channel. *Biophys J*, 71, 3110–3125.

French, R. J. and Adelman W. J. Jr. 1976. Competition, saturation, and inhibition-ionic interactions shown by membrane ionic currents in nerve, muscle, and bilayer systems. *Curr Top Membr Transp*, 8, 161–207.

Furini, S. and Domene, C. 2012. On conduction in a bacterial sodium channel. *PLoS Comput Biol*, 8, e1002476.

Furini, S., Zerbetto, F., and Cavalcanti, S. 2007. Role of the intracellular cavity in potassium channel conductivity. *J Phys Chem B*, 111, 13993–14000.

Geng, Y., Niu, X., and Magleby, K. L. 2011. Low resistance, large dimension entrance to the inner cavity of BK channels determined by changing side-chain volume. *J Gen Physiol*, 137, 533–548.

Goldman, D. E. 1943. Potential, impedance, and rectification in membranes. *J Gen Physiol*, 27, 37–60.

Hagiwara, S. and Takahashi, K. 1974. The anomalous rectification and cation selectivity of the membrane of a starfish egg cell. *J Membr Biol*, 18, 61–80.

Heckmann, K. 1965a. Zur Theorie der "Single File"-Diffusion. Part I. *Z Phys Chem*, 44, 184–203.

Heckmann, K. 1965b. Zur Theorie der "Single File"-Diffusion. Part II. *Z Phys Chem*, 46, 1–25.
Heckmann, K. 1968. Zur Theorie der "Single File"-Diffusion. Part III. Sigmoide Konzantratonsabhangigkeit unidirectionaler Flusse bei "single file" Diffusion. *Z Phys Chem*, 58, 210–219.
Heckmann, K. 1972. *Single-File Diffusion*, Plenum: New York.
Heginbotham, L., Abramson, T., and MacKinnon, R. 1992. A functional connection between the pores of distantly related ion channels as revealed by mutant K^+ channels. *Science*, 258, 1152–1155.
Heginbotham, L., Lu, Z., Abramson, T., and MacKinnon, R. 1994. Mutations in the K^+ channel signature sequence. *Biophys J*, 66, 1061–1067.
Heginbotham, L. and MacKinnon, R. 1993. Conduction properties of the cloned *Shaker* K^+ channel. *Biophys J*, 65, 2089–2096.
Hess, P. and Tsien, R. W. 1984. Mechanism of ion permeation through calcium channels. *Nature*, 309, 453–456.
Hille, B. 1971. The permeability of the sodium channel to organic cations in myelinated nerve. *J Gen Physiol*, 58, 599–619.
Hille, B. 1972. The permeability of the sodium channel to metal cations in myelinated nerve. *J Gen Physiol*, 59, 637–658.
Hille, B. 1973. Potassium channels in myelinated nerve. Selective permeability to small cations. *J Gen Physiol*, 61, 669–686.
Hille, B. 1975. Ionic selectivity, saturation, and block in sodium channels. A four-barrier model. *J Gen Physiol*, 66, 535–560.
Hille, B. 2001. *Ion Channels of Excitable Membranes*, 01375, Sinauer Associates, Inc: Sunderland, MA.
Hille, B. and Schwarz, W. 1978. Potassium channels as multi-ion single-file pores. *J Gen Physiol*, 72, 409–442.
Hodgkin, A. L. and Huxley, A. F. 1945. Resting and action potentials in single nerve fibres. *J Physiol*, 104, 176–195.
Hodgkin, A. L. and Huxley, A. F. 1946. Potassium leakage from an active nerve fibre. *Nature*, 158, 376.
Hodgkin, A. L. and Huxley, A. F. 1947. Potassium leakage from an active nerve fibre. *J Physiol*, 106, 341–367.
Hodgkin, A. L. and Huxley, A. F. 1952a. The components of membrane conductance in the giant axon of Loligo. *J Physiol*, 116, 473–496.
Hodgkin, A. L. and Huxley, A. F. 1952b. Currents carried by sodium and potassium ions through the membrane of the giant axon of Loligo. *J Physiol*, 116, 449–472.
Hodgkin, A. L. and Huxley, A. F. 1952c. The dual effect of membrane potential on sodium conductance in the giant axon of Loligo. *J Physiol*, 116, 497–506.
Hodgkin, A. L. and Huxley, A. F. 1952d. A quantitative description of membrane current and its application to conduction and excitation in nerve. *J Physiol*, 117, 500–544.
Hodgkin, A. L., Huxley, A. F., and Katz, B. 1952. Measurement of current-voltage relations in the membrane of the giant axon of Loligo. *J Physiol*, 116, 424–448.
Hodgkin, A. L. and Katz, B. 1949. The effect of sodium ions on the electrical activity of giant axon of the squid. *J Physiol*, 108, 37–77.
Hodgkin, A. L. and Keynes, R. D. 1955. The potassium permeability of a giant nerve fibre. *J Physiol*, 128, 61–88.
Ito, M., Xu, H., Guffanti, A. A., Wei, Y., Zvi, L., Clapham, D. E., and Krulwich, T. A. 2004. The voltage-gated Na^+ channel NaVBP has a role in motility, chemotaxis, and pH homeostasis of an alkaliphilic Bacillus. *Proc Natl Acad Sci U S A*, 101, 10566–10571.
Iverson, L. E., Tanouye, M. A., Lester, H. A., Davidson, N., and Rudy, B. 1988. A-type potassium channels expressed from *Shaker* locus cDNA. *Proc Natl Acad Sci U S A*, 85, 5723–5727.
Jensen, M. O., Borhani, D. W., Lindorff-Larsen, K., Maragakis, P., Jogini, V., Eastwood, M. P., Dror, R. O., and Shaw, D. E. 2010. Principles of conduction and hydrophobic gating in K^+ channels. *Proc Natl Acad Sci U S A*, 107, 5833–5838.
Jensen, M. O., Jogini, V., Eastwood, M. P., and Shaw, D. E. 2013. Atomic-level simulation of current-voltage relationships in single-file ion channels. *J Gen Physiol*, 141, 619–632.
Jiang, Y., Lee, A., Chen, J., Cadene, M., Chait, B. T., and MacKinnon, R. 2002a. Crystal structure and mechanism of a calcium-gated potassium channel. *Nature*, 417, 515–522.
Jiang, Y., Lee, A., Chen, J., Cadene, M., Chait, B. T., and MacKinnon, R. 2002b. The open pore conformation of potassium channels. *Nature*, 417, 523–526.
Kamb, A., Iverson, L. E., and Tanouye, M. A. 1987. Molecular characterization of *Shaker*, a Drosophila gene that encodes a potassium channel. *Cell*, 50, 405–413.
Kim, I. and Allen, T. W. 2011. On the selective ion binding hypothesis for potassium channels. *Proc Natl Acad Sci U S A*, 108, 17963–17968.
Kiss, L., Loturco, J., and Korn, S. J. 1999. Contribution of the selectivity filter to inactivation in potassium channels. *Biophys J*, 76, 253–263.
Koishi, R., Xu, H., Ren, D., Navarro, B., Spiller, B. W., Shi, Q., and Clapham, D. E. 2004. A superfamily of voltage-gated sodium channels in bacteria. *J Biol Chem*, 279, 9532–9538.
Korn, S. J. and Ikeda, S. R. 1995. Permeation selectivity by competition in a delayed rectifier potassium channel. *Science*, 269, 410–412.
Kuo, A., Domene, C., Johnson, L. N., Doyle, D. A., and Venien-Bryan, C. 2005. Two different conformational states of the KirBac3.1 potassium channel revealed by electron crystallography. *Structure*, 13, 1463–1472.
Kuo, A., Gulbis, J. M., Antcliff, J. F., Rahman, T., Lowe, E. D., Zimmer, J., Cuthbertson, J., Ashcroft, F. M., Ezaki, T., and Doyle, D. A. 2003. Crystal structure of the potassium channel KirBac1.1 in the closed state. *Science*, 300, 1922–1926.
Latorre, R. and Miller, C. 1983. Conduction and selectivity in potassium channels. *J Membr Biol*, 71, 11–30.
Latorre, R., Oberhauser, A., Labarca, P., and Alvarez, O. 1989. Varieties of calcium-activated potassium channels. *Annu Rev Physiol*, 51, 385–399.
LeMasurier, M., Heginbotham, L., and Miller, C. 2001. KcsA: it's a potassium channel. *J Gen Physiol*, 118, 303–314.
Li, W. and Aldrich, R. W. 2004. Unique inner pore properties of BK channels revealed by quaternary ammonium block. *J Gen Physiol*, 124, 43–57.
Lockless, S. W., Zhou, M., and MacKinnon, R. 2007. Structural and thermodynamic properties of selective ion binding in a K^+ channel. *PLoS Biol*, 5, e121.
Long, S. B., Campbell, E. B., and MacKinnon, R. 2005. Voltage sensor of Kv1.2: Structural basis of electromechanical coupling. *Science*, 309, 903–908.
Marty, A. 1981. Ca-dependent K channels with large unitary conductance in chromaffin cell membranes. *Nature*, 291, 497–500.
McCarthy, M. P., Earnest, J. P., Young, E. F., Choe, S., and Stroud, R. M. 1986. The molecular neurobiology of the acetylcholine receptor. *Annu Rev Neurosci*, 9, 383–413.
McCusker, E. C., Bagneris, C., Naylor, C. E., Cole, A. R., D'avanzo, N., Nichols, C. G., and Wallace, B. A. 2012. Structure of a bacterial voltage-gated sodium channel pore reveals mechanisms of opening and closing. *Nat Commun*, 3, 1102.
Michaelis, L. and Menten, M. L. 1913. Die Kinetik der Invertinwirkung. *Biochem Z*, 49, 333–369.
Michaelis, L., Menten, M. L., Johnson, K. A., and Goody, R. S. 2011. The original Michaelis constant: Translation of the 1913 Michaelis-Menten paper. *Biochemistry*, 50, 8264–8269.
Milkman, R. 1994. An Escherichia coli homologue of eukaryotic potassium channel proteins. *Proc Natl Acad Sci U S A*, 91, 3510–3514.

Montal, M., Anholt, R., and Labarca, P. 1986. *The Reconstituted Acetylcholine Receptor*, New York: Plenum.
Morais-Cabral, J. H., Zhou, Y., and MacKinnon, R. 2001. Energetic optimization of ion conduction rate by the K^+ selectivity filter. *Nature*, 414, 37–42.
Moscoso, C., Vergara-Jaque, A., Marquez-Miranda, V., Sepulveda, R. V., Valencia, I., Diaz-Franulic, I., Gonzalez-Nilo, F., and Naranjo, D. 2012. K(+) conduction and Mg(2)(+) blockade in a *shaker* Kv-channel single point mutant with an unusually high conductance. *Biophys J*, 103, 1198–1207.
Mullins, L. J. 1959. An analysis of conductance changes in squid axon. *J Gen Physiol*, 42, 1013–1035.
Neher, E. and Sakmann, B. 1975. Voltage-dependence of drug-induced conductance in frog neuromuscular junction. *Proc Natl Acad Sci U S A*, 72, 2140–2144.
Neher, E. and Sakmann, B. 1976. Single-channel currents recorded from membrane of denervated frog muscle fibres. *Nature*, 260, 799–802.
Neher, E. and Stevens, C. F. 1977. Conductance fluctuations and ionic pores in membranes. *Annu Rev Biophys Bioeng*, 6, 345–381.
Neyton, J. and Miller, C. 1988a. Discrete Ba^{2+} block as a probe of ion occupancy and pore structure in the high-conductance Ca^{2+}-activated K^+ channel. *J Gen Physiol*, 92, 569–586.
Neyton, J. and Miller, C. 1988b. Potassium blocks barium permeation through a calcium-activated potassium channel. *J Gen Physiol*, 92, 549–567.
Nimigean, C. M., Chappie, J. S., and Miller, C. 2003. Electrostatic tuning of ion conductance in potassium channels. *Biochemistry*, 42, 9263–9268.
Nishida, M., Cadene, M., Chait, B. T., and MacKinnon, R. 2007. Crystal structure of a Kir3.1-prokaryotic Kir channel chimera. *EMBO J*, 26, 4005–4015.
Noda, M. and Numa, S. 1987. Structure and function of sodium channel. *J Recept Res*, 7, 467–497.
Noda, M., Shimizu, S., Tanabe, T., Takai, T., Kayano, T., Ikeda, T., Takahashi, H., Nakayama, H., Kanaoka, Y., Minamino, N. et al. 1984. Primary structure of *Electrophorus electricus* sodium channel deduced from cDNA sequence. *Nature*, 312, 121–127.
Noda, M., Takahashi, H., Tanabe, T., Toyosato, M., Furutani, Y., Hirose, T., Asai, M., Inayama, S., Miyata, T., and Numa, S. 1982. Primary structure of alpha-subunit precursor of *Torpedo californica* acetylcholine receptor deduced from cDNA sequence. *Nature*, 299, 793–797.
Noda, M., Takahashi, H., Tanabe, T., Toyosato, M., Kikyotani, S., Furutani, Y., Hirose, T., Takashima, H., Inayama, S., Miyata, T., and Numa, S. 1983. Structural homology of *Torpedo californica* acetylcholine receptor subunits. *Nature*, 302, 528–532.
Nonner, W., Chen, D. P., and Eisenberg, B. 1998. Anomalous mole fraction effect, electrostatics, and binding in ionic channels. *Biophys J*, 74, 2327–2334.
Noskov, S. Y., Berneche, S., and Roux, B. 2004. Control of ion selectivity in potassium channels by electrostatic and dynamic properties of carbonyl ligands. *Nature*, 431, 830–834.
Noskov, S. Y. and Roux, B. 2006. Ion selectivity in potassium channels. *Biophys Chem*, 124, 279–291.
Pallotta, B. S., Magleby, K. L., and Barrett, J. N. 1981. Single channel recordings of Ca^{2+}-activated K^+ currents in rat muscle cell culture. *Nature*, 293, 471–474.
Papazian, D. M., Schwarz, T. L., Tempel, B. L., Jan, Y. N., and Jan, L. Y. 1987. Cloning of genomic and complementary DNA from *Shaker*, a putative potassium channel gene from Drosophila. *Science*, 237, 749–753.
Payandeh, J., Gamal El-Din, T. M., Scheuer, T., Zheng, N., and Catterall, W. A. 2012. Crystal structure of a voltage-gated sodium channel in two potentially inactivated states. *Nature*, 486, 135–139.
Payandeh, J., Scheuer, T., Zheng, N., and Catterall, W. A. 2011. The crystal structure of a voltage-gated sodium channel. *Nature*, 475, 353–358.
Ren, D., Navarro, B., Xu, H., Yue, L., Shi, Q., and Clapham, D. E. 2001. A prokaryotic voltage-gated sodium channel. *Science*, 294, 2372–2375.
Sandblom, J., Eisenman, G., and Neher, E. 1977. Ionic selectivity, saturation and block in gramicidin A channels: I. Theory for the electrical properties of ion selective channels having two pairs of binding sites and multiple conductance states. *J Membr Biol*, 31, 383–347.
Schrempf, H., Schmidt, O., Kummerlen, R., Hinnah, S., Muller, D., Betzler, M., Steinkamp, T., and Wagner, R. 1995. A prokaryotic potassium ion channel with two predicted transmembrane segments from *Streptomyces lividans*. *EMBO J*, 14, 5170–5178.
Shi, N., Ye, S., Alam, A., Chen, L., and Jiang, Y. 2006. Atomic structure of a Na^+- and K^+-conducting channel. *Nature*, 440, 570–574.
Sigworth, F. J. 1980a. The conductance of sodium channels under conditions of reduced current at the node of Ranvier. *J Physiol*, 307, 131–142.
Sigworth, F. J. 1980b. The variance of sodium current fluctuations at the node of Ranvier. *J Physiol*, 307, 97–129.
Silberberg, S. D. and Magleby, K. L. 1993. Preventing errors when estimating single channel properties from the analysis of current fluctuations. *Biophys J*, 65, 1570–1584.
Starkus, J. G., Kuschel, L., Rayner, M. D., and Heinemann, S. H. 1997. Ion conduction through C-type inactivated *Shaker* channels. *J Gen Physiol*, 110, 539–550.
Tao, X., Avalos, J. L., Chen, J., and MacKinnon, R. 2009. Crystal structure of the eukaryotic strong inward-rectifier K^+ channel Kir2.2 at 3.1 A resolution. *Science*, 326, 1668–1674.
Terlau, H., Heinemann, S. H., Stuhmer, W., Pusch, M., Conti, F., Imoto, K., and Numa, S. 1991. Mapping the site of block by tetrodotoxin and saxitoxin of sodium channel II. *FEBS Lett*, 293, 93–96.
Thompson, A. N., Kim, I., Panosian, T. D., Iverson, T. M., Allen, T. W., and Nimigean, C. M. 2009. Mechanism of potassium-channel selectivity revealed by Na(+) and Li(+) binding sites within the KcsA pore. *Nat Struct Mol Biol*, 16, 1317–1324.
Timpe, L. C., Schwarz, T. L., Tempel, B. L., Papazian, D. M., Jan, Y. N., and Jan, L. Y. 1988. Expression of functional potassium channels from *Shaker* cDNA in *Xenopus oocytes*. *Nature*, 331, 143–145.
Varma, S. and Rempe, S. B. 2006. Coordination numbers of alkali metal ions in aqueous solutions. *Biophys Chem*, 124, 192–199.
Varma, S., Sabo, D., and Rempe, S. B. 2008. K^+/Na^+ selectivity in K channels and valinomycin: Over-coordination versus cavity-size constraints. *J Mol Biol*, 376, 13–22.
Webster, S. M., Del Camino, D., Dekker, J. P., and Yellen, G. 2004. Intracellular gate opening in *Shaker* K^+ channels defined by high-affinity metal bridges. *Nature*, 428, 864–868.
Ye, S., Li, Y., and Jiang, Y. 2010. Novel insights into K^+ selectivity from high-resolution structures of an open K^+ channel pore. *Nat Struct Mol Biol*, 17, 1019–1023.
Yellen, G. 1984. Ionic permeation and blockade in Ca^{2+}-activated K^+ channels of bovine chromaffin cells. *J Gen Physiol*, 84, 157–186.
Zhang, X., Ren, W., Decaen, P., Yan, C., Tao, X., Tang, L., Wang, J. et al. 2012. Crystal structure of an orthologue of the NaChBac voltage-gated sodium channel. *Nature*, 486, 130–134.
Zhou, Y. and MacKinnon, R. 2003. The occupancy of ions in the K^+ selectivity filter: Charge balance and coupling of ion binding to a protein conformational change underlie high conduction rates. *J Mol Biol*, 333, 965–975.
Zhou, Y., Morais-Cabral, J. H., Kaufman, A., and MacKinnon, R. 2001. Chemistry of ion coordination and hydration revealed by a K^+ channel-Fab complex at 2.0 A resolution. *Nature*, 414, 43–48.

Basic mechanisms of voltage sensing

Sandipan Chowdhury and Baron Chanda

Contents

3.1 INTRODUCTION

Electrical potential across biological membrane serves as a source of energy that drives the transport of molecules and generates electrical signals. Unlike second messenger pathways, electrical signaling is confined to the plane of a cell membrane and can travel long distances extremely rapidly. Thus, electrical impulses are well suited for cell–cell communication and fast processing of complex information in an ever-changing environment. These signaling pathways are ubiquitous throughout the animal kingdom and are also found in plants and prokaryotes. Indeed, the *bashful* response of a mimosa plant has molecular parallels with reflex action in response to noxious stimuli (1).

Since the early pioneering studies by Bernstein, Cole, Hodgkin, Huxley, and others, voltage-sensitive conductances have been extensively studied because of their primary role in electrical excitability in nerves and muscles (2–11). Voltage-dependent K^+, Na^+, and Ca^{++} channels are responsible for generation of an action potential, which is the elementary unit of electrical signals in biology. Calcium influx through calcium channels can also act as a second messenger, which initiates various processes such as muscle contraction (12,13). Isoforms of voltage-dependent potassium channels are involved in diverse physiological processes, which include volume regulation, secretion, proliferation, and migration.

Voltage-dependent ion channels belong to a large superfamily of integral membrane proteins that share a common structural motif known as the voltage-sensing domain (VSD) (Figure 3.1). A majority of the members of this superfamily catalyze transport of ions across the membrane when triggered by changes in membrane potential. Many of them are also polymodal and are activated by other stimuli such as ligands and heat. In this chapter, we will provide an overview of the mechanisms of voltage sensing based primarily on studies of the canonical voltage-dependent ion channels. A number of physiologically important membrane proteins without a putative VSD have also been found to be voltage sensitive, but their mechanisms of voltage sensitivity are poorly understood and will not be considered in this exposition.

3.2 THERMODYNAMIC VIEW OF VOLTAGE SENSING

Statistical thermodynamics lays down general rules to help us understand the molecular forces that determine the function of molecules—simple or complex and of living or inanimate origin. At a fundamental level, interactions determine the structures of macromolecules and changes in interactions underlie conformational rearrangements, which ultimately dictate the function of these molecular machines in our body. In this section, we will consider introductory concepts to deconstruct the energetics that determine the conformation and function of voltage-gated ion channels (VGICs).

For more than half a century, voltage-clamp experiments have been the bedrock of electrophysiological methods to analyze biophysical properties of voltage-gated ion channels. In this technique, we measure the amount of current flowing through the membrane expressing channels in response to step changes in potential. From their time-dependent current responses, we can deduce their rates of opening and closing and the fraction of open channels when a new equilibrium is established. The analysis of voltage dependence of open probabilities provides a measure of the relative stability of the open state vs. closed state of the channel.

From a physical perspective, this voltage-dependent change in channel open probability can be treated as a simple two-state process wherein the channel exists either in the closed or the open state and voltage biases the equilibrium toward one state or the other. For instance, in channels that open in response to depolarization, the equilibrium is biased toward the closed state at resting membrane potentials (approximately −90 mV) and depolarization results in a shift in the bias in favor of the open

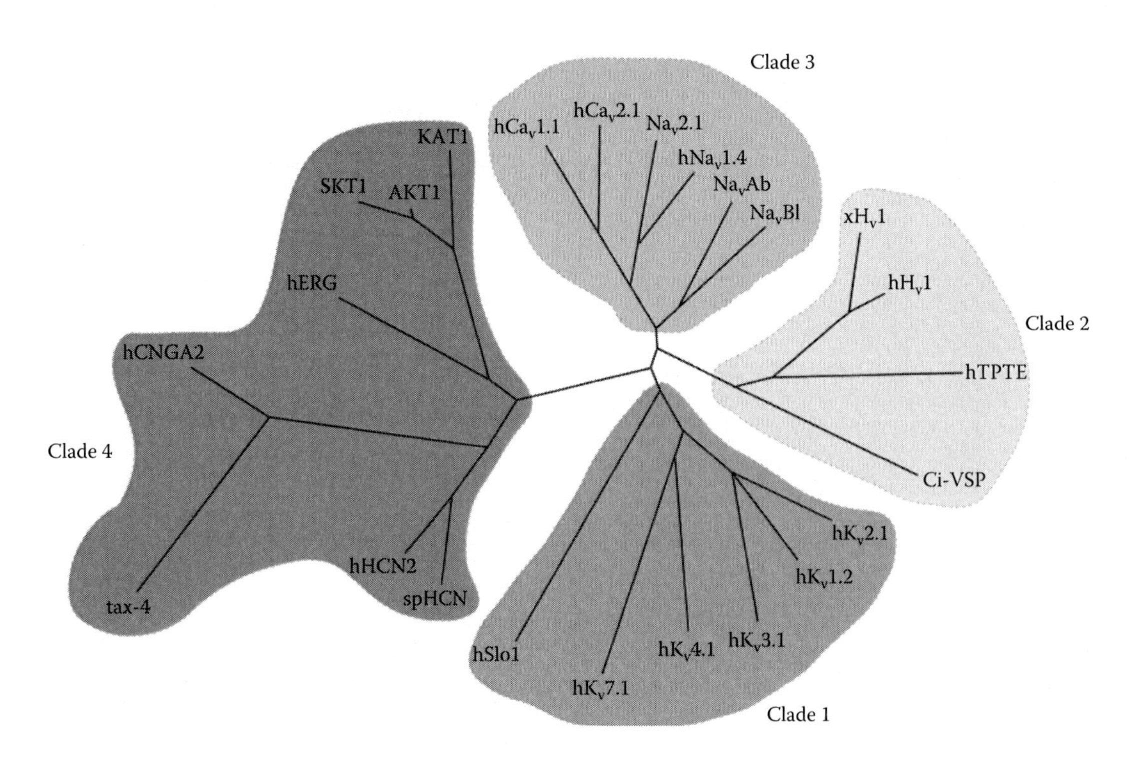

Figure 3.1 Phylogenetic tree of voltage-gated ion channels (VGICs). The cladogram shows the sequence diversity of 24 VSDs from different members of VGIC family, created from a multiple sequence alignment of the VSDs (obtained using MUSCLE from Edgar, R.C., *Nucleic Acids Res.*, 32, 1792, 2004.) The phylogenetic tree calculation was performed using MATLAB® 2012 (Mathworks) and BioNJ. (From Gascuel, O., *Mol. Biol. Evol.*, 14, 685, 1997.) Four distinct clades are directly observable. Clade 1 comprises the different voltage-gated potassium channels members that form homotetrameric channels. Clade 2 comprises the voltage-sensor-only proteins (VSOP)—the proton-selective H$_V$ channel (human and *Xenopus*) and the PTEN phosphatase–domain containing VSOP from *Ciona intestinalis* (Ci-VSP) and its human orthologue (hTPTE). Clade 3 comprises the voltage-gated sodium and calcium channels from eukaryotes (which are heterotetrameric; Na$_V$2.1 represents the sodium channel from *Nematostella vectensis*) and prokaryotes (NavAb and NavBl). Clade 4 comprises members of the VGIC family containing a cyclic nucleotide–binding domain at their C-termini. Members of this clade include the depolarization-activated hERG channels, hyperpolarization-activated channels from metazoans (hHCN2 and spHCN), the inwardly rectifying plant channels (SKT1, AKT1, and KAT1) and the very weakly voltage-sensitive CNG **(cyclic nucleotide gated)** channels (human and *Caenorhabditis elegans* orthologues).

state (Figure 3.2a). Thus, for such a process, the energy difference between the two states at a particular voltage is determined by the chemical energy difference between the two states and electrical work that is done on this system:

$$\Delta G(V) = \Delta G_C - qFV, \tag{3.1}$$

where

ΔG_C is the chemical energy difference
q is the charge that is moved in response to a voltage (V)
F is the faraday constant

In this simple scenario, the energetics of voltage-dependent channel activation can be extracted experimentally in a simple way. The sigmoid P_OV curve (Figure 3.2b) can be fitted to a logistic function (14), commonly known as the Boltzmann equation:

$$P_O = \frac{1}{1+\exp\{zF(V_{1/2}-V)/RT\}}. \tag{3.2}$$

The latter equation has two parameters $-z$ that is a measure of the steepness of the curve and is related to the number of charges transferred during activation (q) and $V_{1/2}$ that is the voltage that elicits half-maximal response (i.e., $P_O = 0.5$). The voltage-independent component of the energy difference between the closed and open states can be estimated from the Boltzmann fit parameters as

$$\Delta G_C = zFV_{1/2}. \tag{3.3}$$

Clearly, a more negative value of $V_{1/2}$ or a steeper P_OV curve (larger z) will imply that at zero membrane potential, the open state is energetically more favorable than the closed state and vice versa.

"…it seems difficult to escape the conclusion that the changes in ionic permeability depend on the movement of some component of the membrane which behaves as though it had a large charge or dipole moment." With these words Hodgkin and Huxley, in their classical descriptions of excitable cells, postulated that channel opening is associated with specific movements of charge (15). The temporal pattern of this charge movement, which is known as gating currents, was first demonstrated independently by two groups in the early 1970s (16,17). Gating currents are observable as (relatively) small transient currents, when all ionic currents are blocked by the application of a channel blocker or depleting permeant ions. Gating charge displaced at a particular voltage is obtained by integrating the gating currents over the time period of depolarization. A plot of the displaced gating charges at various voltages generates a charge–voltage curve (QV) (Figure 3.2c and d). Comparison of the QV and P_OV curves shows that there is a significant displacement of gating charges prior to

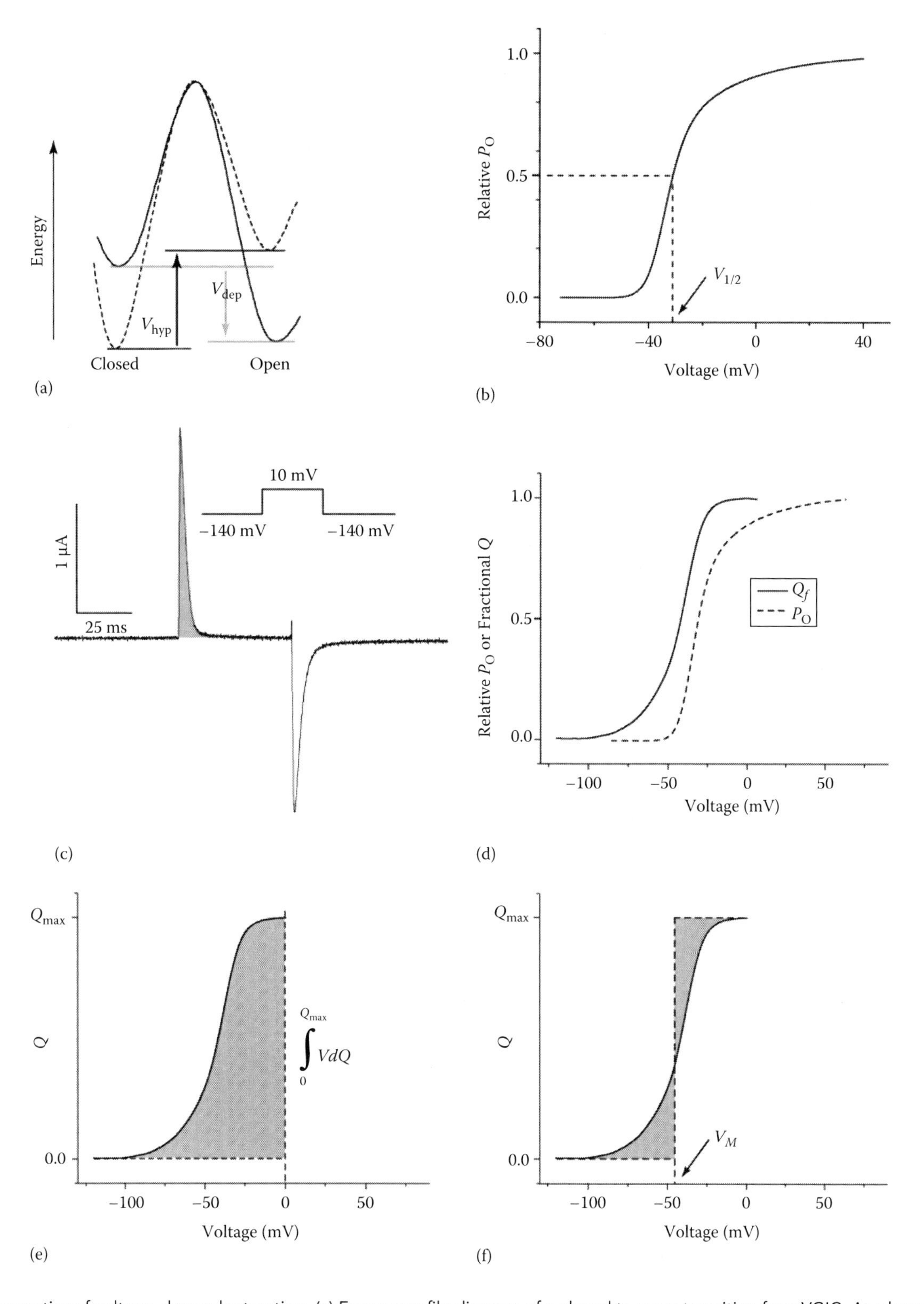

Figure 3.2 Energetics of voltage-dependent gating. (a) Energy profile diagram of a closed to open transition for a VGIC. At a hyperpolarizing voltage, V_{hyp} (dotted profile), the closed state is energetically more favorable than the open state. At a depolarizing voltage, V_{dep} (solid profile), the open state is more stable than the closed state. The net energy difference between the closed and open states at V_{hyp} and V_{dep} is indicated using arrows. (b) The open-probability vs. voltage curve for the *Shaker* potassium channel (with inactivation removed). The voltage at which half the channels are open is indicated by $V_{1/2}$. (c) An example of a gating current trace for the *Shaker* potassium channel (in the background of the W434F and inactivation-removing mutations) obtained in response to a depolarizing pulse. The shaded area represents the gating charge transferred during the depolarization. (d) Comparison of the normalized gating-charge displacement (Q_f) and relative open-probability curves for the *Shaker* potassium channel (inactivation removed mutant). (e) The shaded area represents the area between the *QV* curve (un-normalized) and the ordinate (Q) axis, which represents the chemical components of the total free-energy change associated with voltage-dependent activation of the channel (Equation 3.7). (f) The dashed line represents the median voltage axis (V_M), which is the voltage such that the two shaded areas on either side of the V_M axes are equal.

channel opening, a phenomenon that has been observed in all known VGICs. This is contrary to the predictions of a two-state model where the QV and $P_O V$ curves would be superimposable.

Even before gating currents were obtained, the multistate nature of activation of voltage-dependent ion channels was established based on analysis of channel-opening kinetics. For a two-state process, the time-dependent open probability is

$$P_O(t) = 1 - e^{-kt}, \quad (3.4)$$

where k is the rate constant for channel opening at large depolarization. Hodgkin and Huxley observed that upon depolarization, the sodium and potassium permeabilities arise after a short time delay, incompatible with a two-state process. They proposed that the channel opening requires prior transitions through multiple closed states. More detailed kinetics models of activation constrained by single channel measurements, macroscopic ionic, and gating currents were generated independently by three groups in the 1990s (18–20). The key feature of these models is that the pore opening and voltage-sensor activation are distinct but linked processes. According to the model proposed by Aldrich and colleagues (18), the activation of the voltage-sensor involves two sequential transitions occurring independently in each of the four subunits of the protein. The channel shows no detectable degree of opening until each of these transitions occurs in all four subunits, whereafter a single concerted transition occurring simultaneously in all the four subunits gates the channel open.

In addition to providing a quantitative description of the kinetic features, the well-constrained kinetic models can also provide a robust estimate of the net free-energy associated with activation of ion channels. For instance, let us consider the Zagotta, Hoshi, and Aldrich model:

$$\left(R_1 \overset{K_1}{\rightleftharpoons} R_2 \overset{K_2}{\rightleftharpoons} C\right)_4 \overset{L}{\rightleftharpoons} O$$

K_1, K_2, and L indicate the voltage-independent components of the equilibrium constants for each of the transitions. The net free-energy of activation in this case is

$$\Delta G_C = -RT \ln\{(K_1 K_2)^4 L\}. \quad (3.5)$$

Development of these detailed kinetic models, however, is an extremely resource and time-intensive process, since multiparametric models need to be constrained by orthogonal experimental measurements. This becomes an important consideration from an experimental standpoint when a large number of mutants need to be analyzed or inactivation cannot be removed for developing well-constrained activation gating models. An alternative strategy to obtain ΔG_{net} exploits an important thermodynamic principle that in any reversible process the free-energy change in a system will be equal to the work done on the system by the external force (21,22). The work done, in turn, depends on the external force and its conjugate displacement. For instance, in the case of mechanical work, the external force is surface tension or pressure whose conjugate displacements are surface area or volume expansions. In a similar vein, when the work done is electrical in nature, the external force is voltage and its conjugate displacement is charge. It can be demonstrated that for voltage-dependent ion channels, the free-energy change, ΔG, can be estimated from the QV curve. Specifically, ΔG is given by the integral of the QV curve (Figure 3.2e):

$$\Delta G = \int_{-\infty}^{\infty} Q\, dV. \quad (3.6)$$

Integration of this equation by parts allows us to separate this ΔG into two components:

$$\Delta G = \lim_{V\to\infty} Q_{max} FV - \int_{0}^{Q_{max}} V dQ. \quad (3.7)$$

The first ($Q_{max}FV$) is the electrical component (where Q_{max} is the maximum number of gating charges transferred during channel activation); the second component, which is the area between the QV curve and the Q (ordinate) axis, reflects the chemical component of the free-energy change, ΔG_C, which arises due to the overall conformational change occurring in the entire protein. Using a special parameter, V_M, the median voltage of activation, defined as the voltage axis, which divides the QV curves into two parts such that area on either side of the curve are equal (Figure 3.2f), the ΔG_C can be simply estimated as $Q_{max}FV_M$, that is,

$$\Delta G_C = Q_{max} F V_M. \quad (3.8)$$

This important relationship provides an exact estimate of net free-energy change associated with channel activation despite the presence of intermediate states.

3.3 MOLECULAR MECHANISMS OF VOLTAGE DEPENDENCE

One of the earliest evidence that the sodium and potassium permeabilities of nerve membranes are due to discrete entities came from studies using specific blockers of sodium and potassium conductances (Chapter 1). Narahashi et al. were the first to show that tetrodotoxin can selectively block sodium conductances in lobster giant axons without affecting the potassium or leak conductances (23–25). Around the same time, tetraethylammonium ion was discovered to abolish potassium conductances selectively without affecting the sodium conductances (26,27). Analogous to a standard binding curve, the extent of current block shows a sigmoidal dependence with concentration of the blockers. These plots are typically referred to as dose–response curves by electrophysiologists (28). The block is saturable, demonstrating that there are fixed number of binding sites for these blockers and that the binding is stoichiometric.

Isolation of high-affinity blockers of ion channels also enabled the purification of the molecular components that were responsible for ionic conductances across the membrane.

Availability of purified radiolabeled saxitoxin enabled development of assays to purify voltage-gated sodium channels from the electric organ of electroplax eel, skeletal muscles, and brain (29–31). Reconstitution of the purified proteins into artificial lipid bilayers like membrane vesicles and planer lipid bilayer established that these purified proteins retained biophysical characteristics that were similar to those found in native tissues (32–35). These studies firmly established the notion that discrete cation channels conferred voltage-dependent permeabilities to excitable membranes.

Molecular cloning of voltage-dependent ion channels stimulated mechanistic thinking in terms of chemical structures and their role in determining function. Numa and coworkers, in a series of landmark papers, described the primary structure of voltage-dependent sodium and calcium channels (36–38). Voltage-gated potassium channels were cloned by molecular genetic analysis of a *Drosophila Shaker* mutant (39–41). Expression of these genes in heterologous expression system enabled functional analysis of a relatively homogenous population of these ion channels. The primary structure in combination with biochemical and functional analysis reveals fourfold symmetry with a central ion pore constituted by all four subunits. Instead of four individual subunits, the voltage-gated sodium and calcium channels consist of four homologous domains on a single polypeptide. Each subunit or domain comprises six transmembrane helices (Figure 3.3a), S1 through S6 with a membrane re-entrant P-loop intervening the S5 and S6 helices. The P-loop is

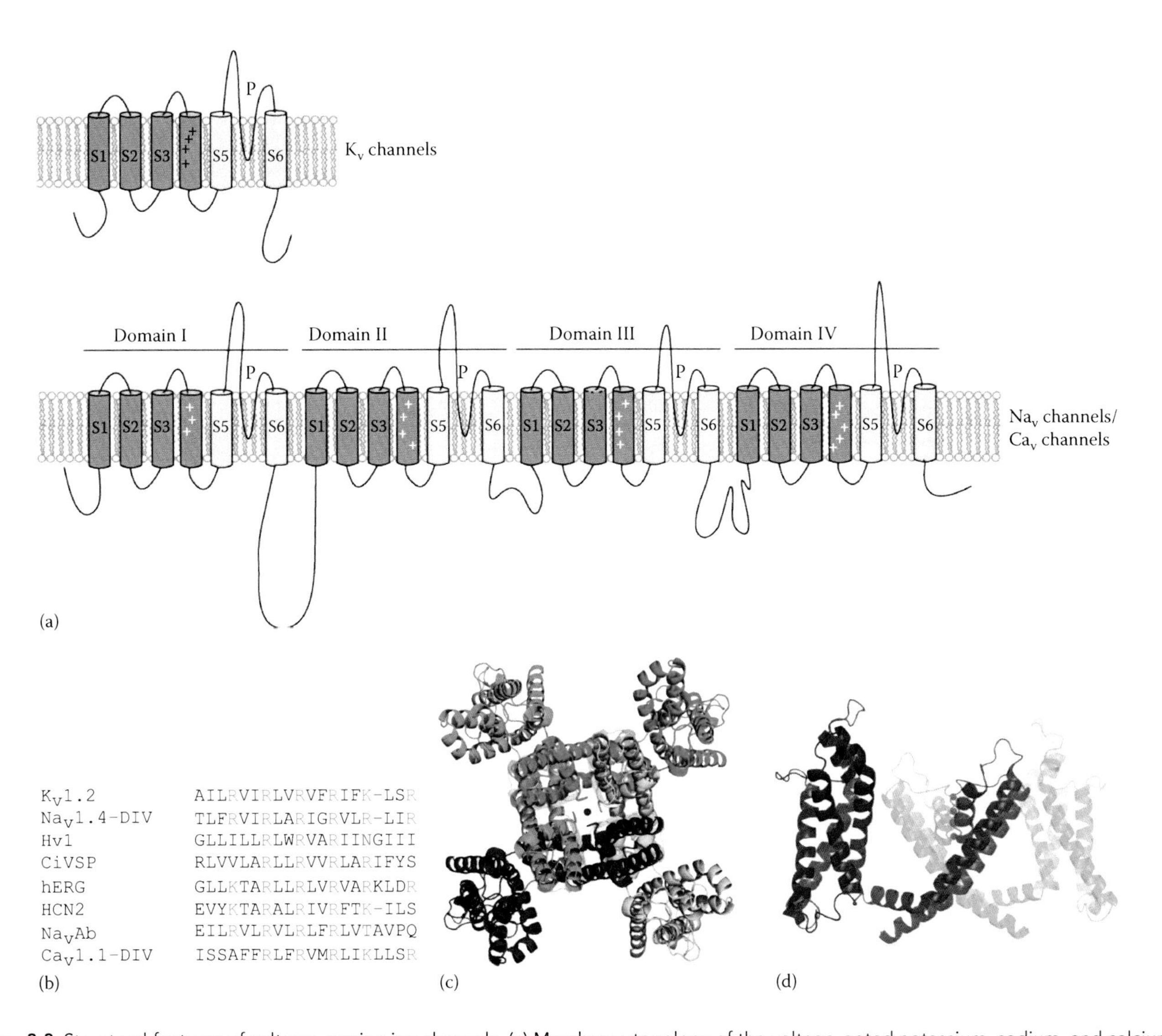

Figure 3.3 Structural features of voltage-sensing ion channels. (a) Membrane topology of the voltage-gated potassium, sodium, and calcium channels. The potassium channels are homotetrameric, and each subunit comprises six transmembrane helices, S1 through S6. The fourth transmembrane segment carries a series of regularly spaced positively charged residues. Between the S5 and S6 segments is the membrane re-entrant loop that houses the residues that confer potassium selectivity to these channels. The eukaryotic Na_V and Ca_V channels are heterotetrameric comprising four homologous domains (Domains I through IV). Each domain shares the same membrane topology as a single subunit of the potassium channels. (b) Sequence alignment of the S4 segments of different members of the voltage-gated ion channels. The conserved positively charged residues are shown in blue. (c) and (d) The high-resolution crystal structures of the mammalian voltage-gated potassium channel (PDB-ID: 2R9R) showing a top view of the tetrameric channel (c) and a side view of two nonadjacent subunits of the channel. The four VSDs of the different subunits are arranged symmetrically around the central pore domain. Potassium ions bound to the selectivity filter are shown as blue circles. The structures represent a putative open-activated state of the channel.

primarily involved in ion selectivity (42,43), but more recently, it has been implicated both in inactivation and activation gating (44–47).

In all voltage-gated ion channels, the fourth transmembrane S4 segment is strikingly conserved and contains multiple repeats of RXXR motif, where X in general represents a hydrophobic residue (Figure 3.3b). Neutralization of these charges reduced the slope of the conductance–voltage curves (48) but so did mutations of hydrophobic residues in the intracellular loops between S4 and S5 (49–51). To unequivocally establish that these residues are gating charges, it is necessary to measure their contributions to the total charge associated with the gating of the channel.

To date, three different approaches have been exploited to measure the total gating charge per channel. Almers showed that for a sequential mode of channel opening, the slope of the ln P_O vs. voltage plot at hyperpolarized voltages becomes equal to the total charge during channel activation (52,53).* Using this limiting slope method, Schoppa and Sigworth calculated that for the *Shaker* potassium channels, 12.8 electronic charges are transferred during activation (54). Mackinnon and Bezanilla groups independently corroborated these values by an alternate approach that involved calculating the total charge by measuring the gating currents and then independently estimating the number of channels in the same cell (55,56). Both groups found that the first four arginines in the S4 segment of the *Shaker* potassium channel were the primary gating charge carrying residues. Seoh et al. also found that a negatively charged residue in the S2 segment also contributes to the gating charge.

The solution of the high-resolution structure of a full-length eukaryotic potassium channel was a remarkable achievement (57,58) and laid a solid foundation for testing the various mechanisms of voltage sensing (Figure 3.3c and d, Box 3.2). In the last several years, multiple structures of the voltage-dependent potassium and prokaryotic sodium channels (59–61) have become available and in all the cases, the quaternary structure is remarkably similar. All of them show that the VSD is an isolated bonafide VSD. The four VSDs are arranged symmetrically around the pore domain and are loosely attached to it, except at an intracellular interface, which may be crucial for coupling VSD to the pore (62).

The structural underpinnings of voltage-sensing mechanisms have been a focus of investigation since the cloning of the first voltage-dependent ion channels. According to the *sliding helix* model (63), voltage stimulus causes a large displacement of the S4 helix perpendicular to the membrane plane. A related structural model is the *helical screw* model (64,65) (Figure 3.4a), in which apart from a sliding motion, the S4 helix also undergoes a rotation about its helical axis. This model facilitates salt-bridge interaction with conserved negative charges in the surrounding S1, S2, and S3 helices (Figure 3.4b, Box 3.1). In functional studies, perturbation of these negative charges caused significant effects on protein expression and function (66). Disulphide and metal bridging assays demonstrate that these ion pairs are in close proximity in functional channels consistent with their position in the crystal structure (67–69).

A central feature in the structures of the VSDs is the presence of water-accessible crevices. In the years leading up to the crystal structure, the presence of these water-filled crevices was proposed based on multiple lines of evidence. Cysteine accessibility studies by Yang and Horn (70,71) on voltage-gated sodium channels show that the third charge on the S4 segment is accessible to a modifying reagent applied on the outside, whereas the fourth gating charge is accessible from the inside. This was the first evidence that there is a septum between the third and fourth charges. From the three-dimensional structure, the *septum* can be deduced to be roughly 7–8 Å in length packed with hydrophobic residues (72) (see Box 3.1). As a consequence, the majority of the electric field drops across a much smaller thickness inside the protein, although the potential difference is applied across the entire membrane width. Thus, voltage-dependent activation can occur with S4 helix moving across a much smaller distance. Evidence of such electric field focusing has emerged through studies that carefully estimated the effect of ionic strength on gating-charge movement (73) and by directly measuring the local electric field gradients by using electrochromic probes (74).

Additionally, in a remarkable set of experiments, Bezanilla and colleagues were first to show that mutating the S4 arginines to histidines allowed protons to go through the VSDs. Furthermore, neutralizing the S4 charges made the VSDs permeable to alkali metal cations in a relatively nonselective manner (75–79) (Figure 3.5a and b). This current, called the gating pore current, further substantiates the idea that there is but a narrow septum that bifurcates the extracellular and intracellular crevices of the VSD and the septum can be *dissolved* through a relatively small number of mutations (80) (Figure 3.5c). Indeed, in the recently identified voltage-gated proton channels (81,82), the protons flux through the VSDs and these channels lack a distinct pore domain.

The gating pore currents have a unique state-dependence—mutation of the R1 charge elicits inwardly rectifying currents (77), whereas R4 neutralization generates outwardly rectifying currents (78). This observation can be reconciled by considering that the R1 is closest to the gating pore septum in the resting configuration, whereas R4 is the most proximal in the activated state and the presence of these charges prevents permeant ions from accessing this pore. Strikingly, mutations of non-S4 residues were also shown to generate state-dependent gating pore currents (80), which would imply that the septum itself undergoes reorganization between the resting and activated conformations. Based on this observation and spectroscopic measurements (83), it has been proposed (Figure 3.5d) that the VSD undergoes a *transporter*-like motion, wherein the septum moves from an outside position to an inside position upon

* We should note that the limiting slope is a reliable estimate of the charge per channel only when the measurements are made in the voltage regimes where the Q–V curves have saturated and the activation process is nonallosteric (see Sigg and Bezanilla (53) for a detailed biophysical analysis).

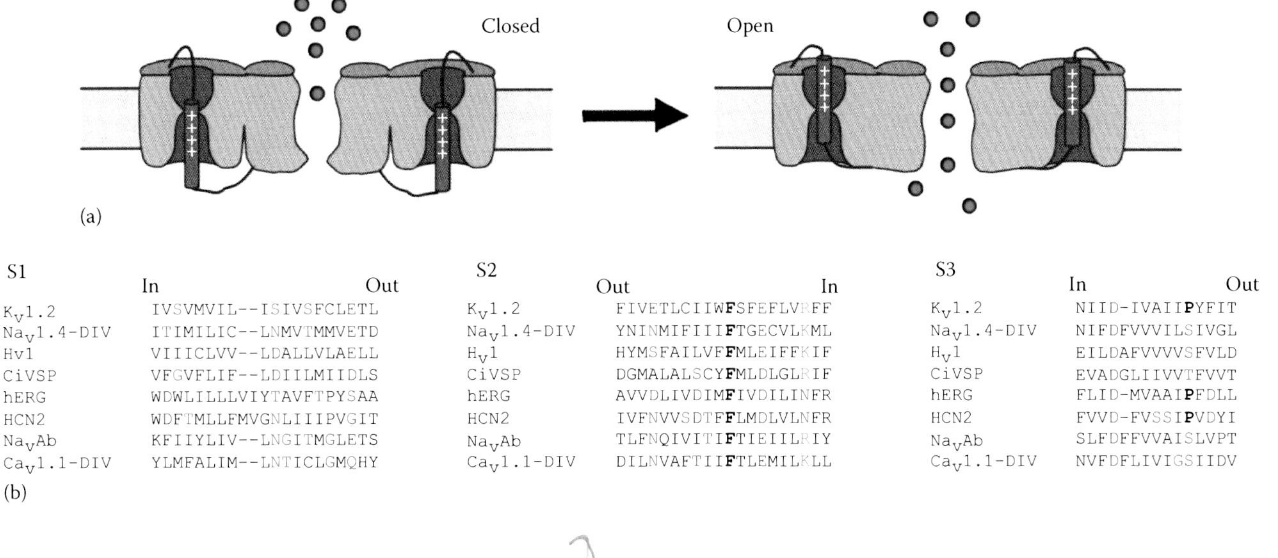

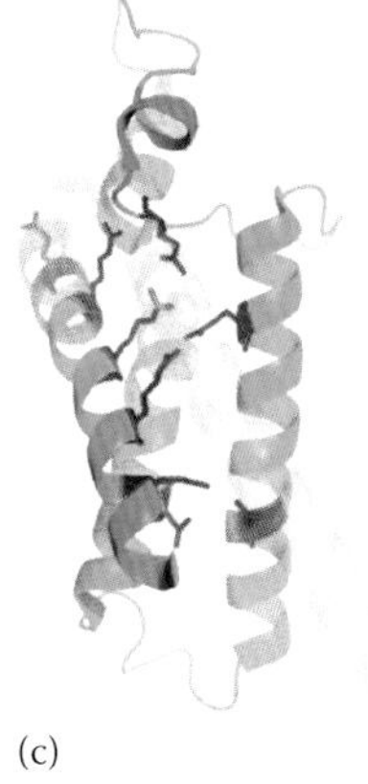

Figure 3.4 Sliding helix and helical screw model of S4 motions. (a) Cartoon depicting the sliding helix/helical screw model of movement of the S4 helix. Voltage-dependent activation, by this model, involves a vertical translation of the S4 helix, accompanied by a rotation of the helix about its axis. (b) Sequence alignment of the S1, S2, and S3 helices of different VGICs. The intracellular and extracellular ends of helical segments are indicated. The positions of the conserved negative charges are shown in red bold characters. Negative charges close to the intracellular ends of the S2 and S3 helices are highly conserved, while those close to the extracellular ends of S1 and S2 segments are more variable. A conserved phenylalanine residue intracellular end of the S2 segment is shown in black bold characters. (c) The VSD of the Kv1.2/2.1 paddle-chimera structure (PDB-ID: 2R9R) showing the positions of the conserved positive charges (on the S4 segment) and the negative countercharges in the S1, S2, and S3 helices in a putatively activated state of the voltage sensor. (For clarity of depiction, the S1 helix is depicted in a transparent mode.) The third and fourth positive charges (from the top of S4) are clearly involved in a salt-bridge interaction, while the fifth positive charge (which is a lysine residue in most Kv channels) is in close proximity to two conserved negative charges on the S2 and S3 segments.

activation. As a result, the region across which the electric field is focused changes between conformations and rather than the charges physically moving a large distance through the electric field, the electric field also moves through the charges. This sort of a movement recapitulates the alternating access model of the transporters in which the opening switches between an outward-facing and inward-facing configuration during the transport cycle (84).

Despite intense focus by many groups, the exact structural details of voltage gating remain unresolved in large part due to the absence of a well-defined structure of a fully resting voltage sensor. In the activated state structure of a voltage-gated potassium channel, the first four charges are in a water-accessible external crevice and the remaining charges are in an internal facing crevice separated by a highly conserved phenylalanine (58). Incisive experiments using nonnatural amino acids revealed that a highly conserved phenylalanine serves as a charge transfer center that separates the two aqueous surfaces and facilitates movement of the charges across this barrier (85). Recent structures of the VSD from Ci-VSP in a putatively resting and activated state show that a single gating charge moves below the conserved charge transfer center that would be consistent with the sliding helix model of charge translocation (86). However, this conserved phenylalanine is only one of the many hydrophobic residues that separate the water-filled crevices in the VSDs. Further work will be required to determine whether this model of voltage gating observed in Ci-VSP structure can be generalized for other VSDs and the role of other residues that constitute the hydrophobic septum.

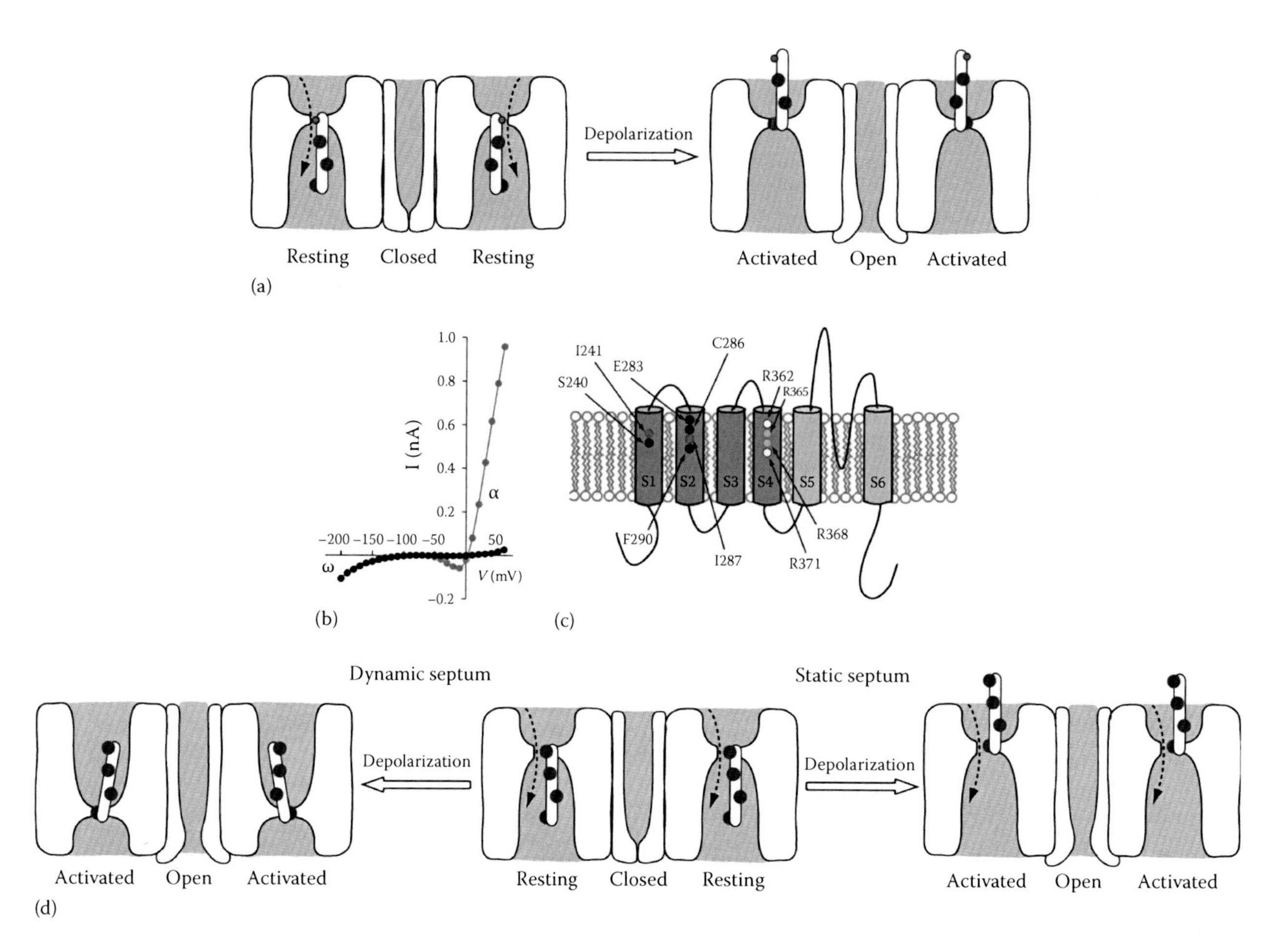

Figure 3.5 Gating pore currents and transporter model of voltage-sensor motion. (a) A cartoon depicting the origin of gating pore currents in the VGICs. At hyperpolarizing voltages, when the channel is closed and the voltage sensor is in a resting conformation, the first arginine (R1) occludes a narrow (gating) pore between the external and internal crevices within the voltage sensor and repels small cations, preventing their passage. Mutation of R1 to small/neutral amino acids removes the pore-occluding moiety, allowing ions to pass through the voltage sensor. (b) Current–voltage plots of the R1 mutant of the *Shaker* potassium channel in the absence and presence of agitoxin. Agitoxin specifically blocks the flux of potassium ions through the principal pore (constituted by S6 helices) but does not inhibit the flux of the ions through the gating pore. (Data from Tombola, F. et al., *Neuron*, 45, 379, 2005.) (c) Positions of the different sites in the *Shaker* potassium channel which when mutated permit ions to flux through the omega pore. (Adapted from Chanda, B. and F. Bezanilla, *Neuron*, 57, 345, 2008). (d) Comparison of the static and dynamic septum models of voltage-sensor activation. Mutations in the non-S4 segments (at sites constituting the hydrophobic septum, referred to as *wall mutants*) elicit gating pore currents only at hyperpolarizing voltages, when the voltage sensor is in a resting conformation. If the septum remains static, gating pore currents should be elicited when the voltage sensor is in resting (hyperpolarizing voltages) as well as activated (depolarizing voltages) conformations. In the dynamic septum model, the hydrophobic septum undergoes reorganization during voltage-dependent activation such that the sites that constitute the primary constriction in the gating pore, in the resting state, are not the same as the sites that make up the septum in the activated state. The result is septum is different in the resting and activated states. The dynamic septum model explains the rectification behavior of the gating pore currents elicited by wall mutants. (From Campos, F.V. et al., *Proc. Natl. Acad. Sci. U.S.A.*, 104, 7904, 2007.) The feature of the dynamic septum is embodied in the transporter model of the voltage-sensor activation.

3.4 TRANSFER OF ACTIVATION ENERGY TO THE PORE

Classic studies by Armstrong and Binstock (27) showed that the pore is accessible from the inside by blockers in a state-dependent manner, implying that the gating machinery is intracellular. High-resolution structures of the pore domain reveal that it is constituted by convergence of S5–P–S6 segments from different subunits (87). The S6 segments line the ion access pathway, and their intracellular portions form a bundle crossing, which may serve as a physical barrier for ion translocation. Substituted cysteine accessibility measurements in the *Shaker* potassium channel clearly demonstrate that channel opening involves splaying of the S6 helices, resulting in dilation of the constriction (88,89). Similar measurements on voltage-gated sodium and calcium channels support the notion that hydrophobic stretch on the S6 segment is the pore gate in canonical voltage-gated ion channels (90,91). Recent structures of the prokaryotic sodium channels in putative open and closed states show that small rearrangement of the intracellular portion of the S6 helices can open the pore sufficiently to allow hydrated ions to permeate (92). Nonetheless, accessibility studies on other members of the VGIC family such as the calcium-activated potassium channel and cyclic nucleotide-gated channel show that intracellular S6

Box 3.1 Electrostatic Considerations for Charge Movement During Gating

I. Principle of Field Focusing

In this section, we briefly describe the principle of electric field focusing and how it is influenced by ionic strength of the surrounding medium (105). We consider a region of low dielectric constant (region 2, $0 < z < L$, where z represents the spatial coordinate with $z = 0$ and $z = L$ defining the boundary planes of the region of the low dielectric constant) sandwiched between aqueous ionic solutions (regions 1 ($z \le 0$) and 3 ($L \le z$)) (Figure 3.6). Regions 1 and 3 have a dielectric constant ε_w and Debye screening lengths, κ_1^{-1} and κ_3^{-1}, respectively. The Debye screening length is proportional to the square root of the ionic strength of electrolytic solution. For region 2, the dielectric constant is ε_m and screening factor is 0 (mobile ions are assumed to be completely excluded from this *septum*). For the simplest case, we assume that each of the regions is isotropic (with respect to the dielectric constants and the screening factors) and extends infinitely in the *x–y* plane (which is the plane normal to the spatial coordinate *z*). When the electromotive forces imposed on the system are small (of the order of k_BT/e), the electrostatic potential in each of the three regions follows the Debye–Huckel equation:

$$\varphi_1''(z) = \kappa_1^2\varphi_1(z) \tag{B.1}$$

$$\varphi_2''(z) = 0 \tag{B.2}$$

$$\varphi_3''(z) = \kappa_3^2\varphi_3(z), \tag{B.3}$$

where $\varphi_i(z)$ indicates the electrostatic potential in the region i (i = 1, 2, or 3) and ″ implies a second order differential. Subjecting these potentials to Dirichlet's boundary conditions at the interfaces between the different regions and the asymptotic boundary conditions, $\varphi_1(-\infty) = 0$ and $\varphi_3(\infty) = V$, analytical solutions for Equations B.1 through B.3 may be obtained:

$$\varphi_1(z) = A\cdot\exp(\kappa_1 z) \tag{B.4}$$

$$\varphi_2(z) = A\left(\frac{\varepsilon_w}{\varepsilon_m}\kappa_1 z + 1\right) \tag{B.5}$$

$$\varphi_3(z) = V - A\frac{\kappa_1}{\kappa_3}\exp\{\kappa_3(L-z)\}, \tag{B.6}$$

where $A = V\left(1+\frac{\kappa_1}{\kappa_3}+\frac{\varepsilon_w}{\varepsilon_m}\kappa_1 L\right)^{-1}$. The fraction of the electric field drop across the septum (region 2), Δf_m, is

$$\Delta f_m = \frac{\varphi_2(L)-\varphi_2(0)}{V} = \frac{1}{1+(1/\alpha\kappa_1 L)+(1/\alpha\kappa_3 L)}, \tag{B.7}$$

where $\alpha = \varepsilon_w/\varepsilon_m$ and ranges between 20 and 40 (since $\varepsilon_w \sim 80$ and $\varepsilon_m \sim 2$–4). In normal physiological solutions that have an ionic strength of ~300 mM the Debye screening lengths (κ^{-1}) are ~7.9 Å. For these values, plotting Δf_m vs. L (Figure 3.6b) shows that even when thickness of the septum is reduced from 30 to 5 Å, more than 80% of the imposed electric field drops across the septum despite the potential difference being physically imposed across relatively larger distances (asymptotic

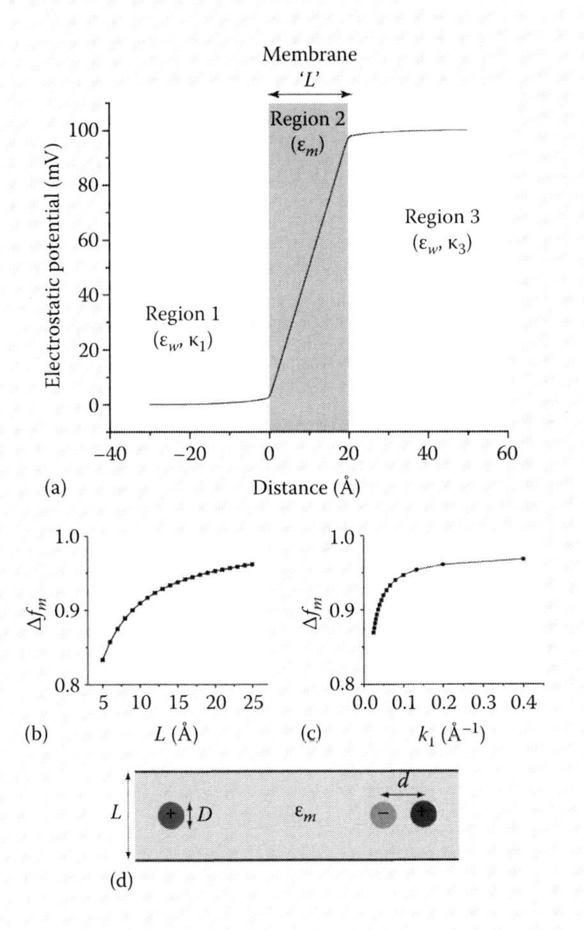

Figure 3.6 Effects of dielectric septum on electric fields and charge transfer energetics. (a) The figure shows the electrostatic potential profile across a model membrane segment (region 2) sandwiched between two regions of high dielectric constant. The electrostatic profile was generated using Equations B4 through B6 (the values of the parameters used for the calculation are as follows: L = 20 Å, κ_1^{-1} = 8Å, κ_3^{-1} = 8Å, ε_m = 5, ε_w = 80, V = 100 mV). (b) The plot shows the fraction of potential difference drop across the membrane, Δf_m, for different values of membrane thickness, L. Even when the thickness is reduced to 5 Å, >80% of the imposed electric field drops across the narrow septum. (c) The plot shows the fraction of potential difference drop across the membrane, Δf_m, for different values of the screening factor of region 1 (κ_1). (d) Configurations of a positive ion of diameter D at the center of a dielectric slab (thickness L and with high dielectric constant media on either side) and a charge–countercharge pair, separated by a distance d, at the center of the slab. Energetic considerations arising out of these configurations are discussed in Box 3.1 Sections II and III, respectively.

boundary conditions). The physical reason behind this focusing effect is that a macroscopic medium with a high dielectric constant is composed of molecules, which can undergo effective reorganization or polarization under the influence of an imposed electric field. Thus, in heterogeneous assemblies of low and high dielectric constant media, the spatial gradients of electrostatic potentials are more focused in regions of low dielectric constant. In addition, the plot of Δf_m vs. κ_1 (or κ_3) (for a fixed value of $L = 8$ Å, Figure 3.6c) shows that increasing the screening factor causes an increase in Δf_m. The screening factors (κ^{-1}) decrease with increase in ionic strength of the corresponding solutions. This implies that increasing concentrations of ions in regions 1 and 3, surrounding the septum, focuses the electric field across region 2 to a greater degree. This occurs because increasing concentrations of free (mobile) ions in regions 1 and 3 effectively increases the capacitance of the corresponding regions (relative to region 2, which has high resistance and low capacitance), which causes lesser voltage drops across them (i.e., between $z \to -\infty$ to $z = 0$ and $z = L$ to $z \to \infty$) and consequently, most of the imposed voltage drops across region 2 (between $z = 0$ and $z = L$). This principle was exploited by Islas and Sigworth to measure the thickness of the septum in a voltage-gated potassium channel (73).

II. Effect of Septum on Gating-Charge Energetics

If the activation of the channel involves a movement of the charged S4 helix across a hydrophobic septum, it is conceivable that some (if not all) of the gating charges undergo a temporary dehydration as they move through the septum. From simple Born energy calculations, the dehydration penalty associated with such a process can be tens of kilocalories, which can make charge movement very slow or energetically unfavorable. For an ion of diameter D placed at the center of a dielectric slab (representing the septum) of finite thickness L (Figure 3.6), the boundaries with surrounding media of high dielectric constant will lower the energy of the ion (106). The fractional reduction in the energy of ion in this situation, relative to its energy when placed in a bulk medium of low dielectric constant, is ~0.7 D/L. This implies that for gating charge movement, a septum thickness of 2–3 times the diameter of a guanidium ion (charged moiety of an arginine residue) can effectively reduce the dehydration penalty by 20%–30%, while a thicker septum (L ~ 10 D) would not offer such an energetic advantage. Thus, a narrow septum, apart from being able to focus most of the imposed electrostatic potential, also reduces the energy barriers for charge movement (107).

Note that this energetic facilitation also depends on the size of the charge that translocates through the septum. The overwhelming majority of the gating charges in VGICs are arginine residues. Lysine residues could also serve as gating charges by virtue of their positive charge. However, due to the smaller size of the positive charge–bearing moiety in Lys (ammonium) relative to Arg (guanidium), the former is likely to experience a much larger dehydration penalty. In addition, the degree to which this penalty can be compensated by the finite (thin) size of the septum will be lesser. This increased energetic cost associated with transferring a Lys might be an important reason for evolutionary selection of arginines as the primary gating-charge-determining amino acid. Additionally, since Arg has a much higher pK_a, it is much less likely to undergo deprotonation (due to local environment effects or when challenged by altered pH conditions) than Lys. Thus, arginines are capable of serving as gating charges under a wider variety of conditions. Some studies have shown that substituting the arginines with lysines at some positions produces large functional effects, while at other positions, the effects of such an isocoulombic substitution appear to be benign (48,56,85,108,109). Our understanding of evolutionary preference of arginine for a lysine remains limited.

III. Countercharge Effect on Gating-Charge Movement

Several members of VGICs, apart from conserved positive charges on the S4 segment, have conserved negative charges in the S1, S2, and S3 segments (110). If we consider a simple physical model where a countercharge is fixed in a low dielectric environment, then the total energy to transfer a positive ion from water into the low dielectric is

$$E_{\pm} = \frac{q^2}{2a_{+}\varepsilon_m}\left(1 - \frac{\varepsilon_m}{\varepsilon_w}\right) - \frac{q^2}{\varepsilon_m d}, \tag{B.8}$$

where

ε_m and ε_w are defined as before

d is the separation between the two charges, q and $-q$ (Figure 3.6d)

The first term is change in the self-energy of the ion during the transfer (comparable to Born solvation energy) and for $\varepsilon_m \ll \varepsilon_w$ is a positive term and the second term reflects coulombic interaction between the two charges. Therefore, if the negative charge is placed permanently in the low dielectric environment, it can reduce the cost of charge translocation through this low dielectric slab.

While the negative charges are conserved in the K_V channel family, in the Na_V/Ca_V (and other) channels, analogous positions do not stringently feature such countercharges. Additionally, while some channels are exquisitely sensitive to perturbations of these negative charges, others can tolerate charge-neutralization or even charge-reversal perturbations at analogous positions (66,111–117). The divergence in the functional consequences of the negative charge perturbations observed in different VGICs would seem to indicate that their importance is fine-tuned differently in different systems. This may be relevant for creating the extraordinary diversity in functional activity of different members of the VGIC family.

Box 3.2 Structural Features of the Voltage-Sensing Apparatus

Before the structures of the VSD became available, scanning mutagenesis studies had suggested that the voltage sensor comprised four α-helical segments (118–120). In particular, these studies had proposed that the S3 helix is kinked at a specific site, which features a proline residue, conserved in the K_V channel family (120). In other VGICs, the analogous position is occupied by serine residue (whose hydroxyl side chain can hydrogen-bond with the backbone), which is known to be a potential *helix breaker*. In the poststructural era, when the existence of such a kink was confirmed in the K_V channels, the S3 helix was further subdivided into two parts—the S3a (below the kink) and the S3b helix (above the kink). Structure–function studies that followed initially suggested that a tight complementarity between the S3b helix and the S4 helix is important for voltage sensing (121). Several studies have also suggested that the S3a helix remains static during voltage-dependent structural changes, but the S3b helix undergoes relatively greater conformational changes (122–124). This helix-turn-helix motif comprises the voltage-sensing charges of the S4 and is referred to as the paddle motif (125) (Figure 3.7). This motif is also an important and selective target for a large number of gating-modifier toxins in different channels (126–128).

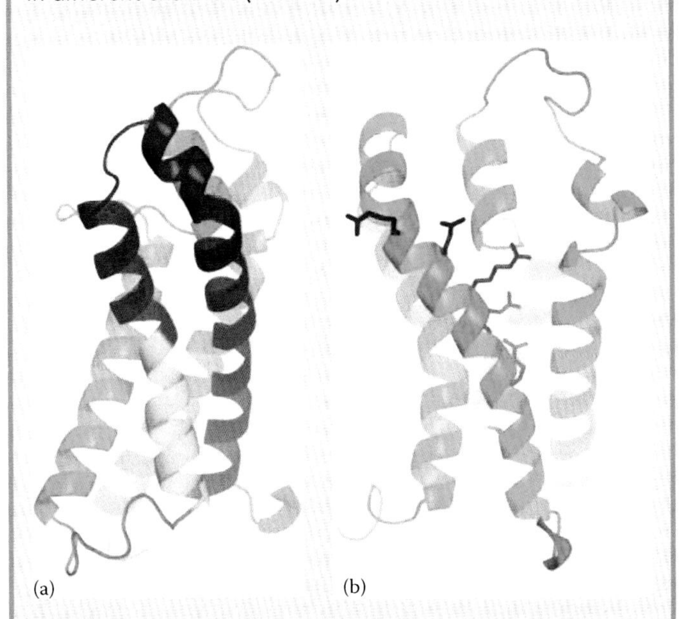

Figure 3.7 Voltage-sensing paddle motifs. (a) The paddle motif of the voltage-gated potassium channel (PDB-ID: 2R9R) is shown in red, while the S1/S2 helices are in green, S3a helix is in yellow, and the intracellular portion of the S4 segment is in blue. (b) The positions of the gating charges on the S4 segment are shown in (blue) stick representations (PDB-ID: 2R9R). The first two side chains are seen to face in the opposite direction from the remaining side chains. The latter point toward the interior of the VSD (crevices), while the former are exposed to the lipid–protein interface. (Note that in this paddle-chimera structure, the first gating charge position is occupied by a glutamine and not an arginine [or lysine], which aligns with the position and orientation of the first arginine residue seen in the crystal structures of K_V1.2 [PDB-ID: 2A79] and the isolated VSD of the prokaryotic channel, K_VAP [PDB-ID: 1ORS]).

In a series of important experiments, Swartz and colleagues showed that the paddle motif was portable across different channels (129,130). By transferring the paddle motif of different (*source*) channels onto a common structural template of K_V 2.1, they showed that the chimeric channels were not only functional but also imbued with some of the macroscopic voltage-dependent properties of the source channels. Additionally, they also became sensitive to gating-modifier toxins that target the source channels (131,132). However, more recent experiments from Lu and colleagues reveal that a tight complementarity between S3b and S4 is not essential for voltage-dependent motions (133,134). They also find that the channels retain their basic voltage dependence even after deletion of nearly the entire paddle motif. Together, these experiments imply that the paddle motif is modular and moves relatively unconstrained in response to depolarization.

Another interesting feature of the structures is that the first charge seems to face away from internal crevices (57). Functional reconstitution of purified proteins in bilayers with different headgroups suggests that the lipid charge may profoundly alter channel function (135). Modification of the lipid headgroups in native membranes by specific lipases also alters the function of voltage-dependent channels (136). Structural studies utilizing nuclear magnetic resonance spectroscopy have shown that the first charge is in close proximity to the lipid headgroups (137). These findings indicate that the arginines may interact with the negative charges on the phospholipid head groups, and this may be an important mechanism for modulation of voltage sensing by lipid-soluble molecules.

segments do not form the primary gate (93–95). Thus, identifying the mechanism of pore gating in members of this superfamily remains an area of active investigation.

If the intracellular portion of the S6 segments serves as ion access gate, how do motions of voltage-sensors control the status of this gate? Strikingly, Lu and colleagues were able to confer voltage dependence to a bacterial pore-only channel by fusing it with a VSD of a eukaryotic potassium channel (96). They were able to generate a functional voltage-dependent chimera as long as the S4–S5 segment and tail end of the S6 belonged to the same channel. This finding clearly demonstrated that this intracellular gating interface is crucial for coupling between voltage sensor and pore. Mutagenesis studies in different VGICs have shown that perturbation of residues in this region seriously compromises the gating properties (97,98). However, clearly assigning the contribution of these residues on interdomain interactions is extremely challenging in multistate processes (see (62) for a detailed discussion).

Recent theoretical developments enable us to estimate the strength of coupling interactions in channels that are activated by voltage sensors in an allosteric fashion (99). In this case, pore opening represents a preexisting equilibrium, initially biased toward the closed state, which is progressively modulated by the activation of each of the four VSDs as observed in the BK channels (100). In most canonical voltage-gated ion channels, like the *Shaker* potassium channel, the mode of coupling appears to be obligatory. Activation of all four VSDs is a prerequisite for the channels to open (101). The difficulties associated with estimating such coupling interactions have been discussed in detail elsewhere (62).

The prevalent notion of electromechanical coupling is that the movement of the S4 helix exerts a force on the S4–S5 linker, which is propagated to the channel gates through the interface between the S4–S5 linker and the S6 helix (102). The intrinsic state preference of the pore domain is an important determinant of the coupling mechanism. If the pore inherently prefers to stay open, then the voltage sensors need to apply a force to close it, while if it prefers to stay closed, then voltage sensors need to do work to pull the gates open. In the *Shaker* potassium channel, the intrinsic preference of the pore is apparently biased toward the closed state and the S4 voltage-sensor activation appears to facilitate the stability of the open pore (103).

3.5 CONCLUDING REMARKS

We should note that, despite some generalizations, our understanding of the mechanisms of voltage sensing remains far from complete. Voltage-gated ion channels constitute one of the largest superfamilies of signaling proteins and exhibit tremendous functional diversity. For instance, some members of the VGIC family are activated upon hyperpolarization. The movements of voltage-sensors in these ion channels resemble those of their outwardly rectifying counterparts, namely, that the positive charges move outward upon depolarization (104). The opposite effect of voltage-sensor movement on pore opening is enabled by inverse coupling between voltage sensor and pore, but the underlying molecular mechanism is poorly understood. Future studies to understand the mechanistic basis for this diversity will be important to obtain a comprehensive understanding of voltage-sensing mechanisms.

REFERENCES

1. Sibaoka, T. 1962. Excitable cells in Mimosa. *Science* 137:226.
2. Bernstein, J. 1902. Untersuchungen zur Thermodynamik der bioelectrischen Ströme. *Erster Theil Pflügers Archiv* 92:521–562.
3. Bernstein, J. 1912. *Electrobiologie.* Vieweg, Braunschweig, Germany.
4. Cole, K. S. and H. J. Curtis. 1939. Electrical impedance of the squid giant axon during activity. *J Gen Physiol* 22:649–670.
5. Hodgkin, A. L. and A. F. Huxley. 1952. A quantitative description of membrane current and its application to conduction and excitation in nerve. *J Physiol* 117:500–544.
6. Hodgkin, A. L. and A. F. Huxley. 1952. The dual effect of membrane potential on sodium conductance in the giant axon of Loligo. *J Physiol* 116:497–506.
7. Hodgkin, A. L. and A. F. Huxley. 1952. The components of membrane conductance in the giant axon of Loligo. *J Physiol* 116:473–496.
8. Hodgkin, A. L. and A. F. Huxley. 1952. Currents carried by sodium and potassium ions through the membrane of the giant axon of Loligo. *J Physiol* 116:449–472.
9. Hodgkin, A. L. and W. A. Rushton. 1946. The electrical constants of a crustacean nerve fibre. *Proc R Soc Med* 134:444–479.
10. Hagiwara, S. and S. Nakajima. 1966. Effects of the intracellular Ca ion concentration upon the excitability of the muscle fiber membrane of a barnacle. *J Gen Physiol* 49:807–818.
11. Hagiwara, S. 1966. Membrane properties of the barnacle muscle fiber. *Ann N Y Acad Sci* 137:1015–1024.
12. Hagiwara, S., M. P. Henkart, and Y. Kidokoro. 1971. Excitation-contraction coupling in amphioxus muscle cells. *J Physiol* 219:233–251.
13. Hagiwara, S. and Y. Kidokoro. 1971. Na and Ca components of action potential in amphioxus muscle cells. *J Physiol* 219:217–232.
14. Ehrenstein, G., R. Blumenthal, R. Latorre, and H. Lecar. 1974. Kinetics of the opening and closing of individual excitability-inducing material channels in a lipid bilayer. *J Gen Physiol* 63:707–721.
15. Hodgkin, A. L. and A. F. Huxley. 1952. Movement of sodium and potassium ions during nervous activity. *Cold Spring Harb Symp Quant Biol* 17:43–52.
16. Armstrong, C. M. and F. Bezanilla. 1973. Currents related to movement of the gating particles of the sodium channels. *Nature* 242:459–461.
17. Schneider, M. F. and W. K. Chandler. 1973. Voltage dependent charge movement of skeletal muscle: A possible step in excitation-contraction coupling. *Nature* 242:244–246.
18. Zagotta, W. N., T. Hoshi, and R. W. Aldrich. 1994. *Shaker* potassium channel gating. III: Evaluation of kinetic models for activation. *J Gen Physiol* 103:321–362.
19. Bezanilla, F., E. Perozo, and E. Stefani. 1994. Gating of *Shaker* K^+ channels: II. The components of gating currents and a model of channel activation. *Biophys J* 66:1011–1021.
20. Schoppa, N. E. and F. J. Sigworth. 1998. Activation of *Shaker* potassium channels. III. An activation gating model for wild-type and V2 mutant channels. *J Gen Physiol* 111:313–342.
21. Chowdhury, S. and B. Chanda. 2012. Estimating the voltage-dependent free energy change of ion channels using the median voltage for activation. *J Gen Physiol* 139:3–17.
22. Chowdhury, S. and B. Chanda. 2013. Free-energy relationships in ion channels activated by voltage and ligand. *J Gen Physiol* 141:11–28.
23. Narahashi, T., J. W. Moore, and W. R. Scott. 1964. Tetrodotoxin blockage of sodium conductance increase in lobster giant axons. *J Gen Physiol* 47:965–974.
24. Nakamura, Y., S. Nakajima, and H. Grundfest. 1965. Analysis of spike electrogenesis and depolarizing K inactivation in electroplaques of *Electrophorus electricus*, L. *J Gen Physiol* 49:321–349.
25. Nakamura, Y., S. Nakajima, and H. Grundfest. 1965. The action of tetrodotoxin on electrogenic components of squid giant axons. *J Gen Physiol* 48:975–996.
26. Hagiwara, S. and N. Saito. 1959. Voltage-current relations in nerve cell membrane of *Onchidium verruculatum*. *J Physiol* 148:161–179.
27. Armstrong, C. M. and L. Binstock. 1965. Anomalous rectification in the squid giant axon injected with tetraethylammonium chloride. *J Gen Physiol* 48:859–872.
28. Hille, B. 1968. Pharmacological modifications of the sodium channels of frog nerve. *J Gen Physiol* 51:199–219.
29. Agnew, W. S., S. R. Levinson, J. S. Brabson, and M. A. Raftery. 1978. Purification of the tetrodotoxin-binding component associated with the voltage-sensitive sodium channel from *Electrophorus electricus* electroplax membranes. *Proc Natl Acad Sci USA* 75:2606–2610.
30. Barchi, R. L. and J. B. Weigele. 1979. Characteristics of saxitoxin binding to the sodium channel of sarcolemma isolated from rat skeletal muscle. *J Physiol* 295:383–396.
31. Hartshorne, R. P. and W. A. Catterall. 1981. Purification of the saxitoxin receptor of the sodium channel from rat brain. *Proc Natl Acad Sci USA* 78:4620–4624.

32. Rosenberg, R. L., S. A. Tomiko, and W. S. Agnew. 1984. Single-channel properties of the reconstituted voltage-regulated Na channel isolated from the electroplax of *Electrophorus electricus*. *Proc Natl Acad Sci USA* 81:5594–5598.
33. Rosenberg, R. L., S. A. Tomiko, and W. S. Agnew. 1984. Reconstitution of neurotoxin-modulated ion transport by the voltage-regulated sodium channel isolated from the electroplax of *Electrophorus electricus*. *Proc Natl Acad Sci USA* 81:1239–1243.
34. Cooper, E. C., S. A. Tomiko, and W. S. Agnew. 1987. Reconstituted voltage-sensitive sodium channel from *Electrophorus electricus*: Chemical modifications that alter regulation of ion permeability. *Proc Natl Acad Sci USA* 84:6282–6286.
35. Keller, B. U., R. P. Hartshorne, J. A. Talvenheimo, W. A. Catterall, and M. Montal. 1986. Sodium channels in planar lipid bilayers. Channel gating kinetics of purified sodium channels modified by batrachotoxin. *J Gen Physiol* 88:1–23.
36. Noda, M., T. Ikeda, H. Suzuki, H. Takeshima, T. Takahashi, M. Kuno, and S. Numa. 1986. Expression of functional sodium channels from cloned cDNA. *Nature* 322:826–828.
37. Suzuki, H., S. Beckh, H. Kubo, N. Yahagi, H. Ishida, T. Kayano, M. Noda, and S. Numa. 1988. Functional expression of cloned cDNA encoding sodium channel III. *FEBS Lett* 228:195–200.
38. Tanabe, T., H. Takeshima, A. Mikami, V. Flockerzi, H. Takahashi, K. Kangawa, M. Kojima, H. Matsuo, T. Hirose, and S. Numa. 1987. Primary structure of the receptor for calcium channel blockers from skeletal muscle. *Nature* 328:313–318.
39. Kamb, A., L. E. Iverson, and M. A. Tanouye. 1987. Molecular charecterization of *Shaker*, a *Drosophila* gene that encodes a potassium channel. *Cell* 50(3):405–413.
40. Papazian, D. M., T. L. Schwarz, B. L. Tempel, Y. N. Jan, and L. Y. Jan. 1987. Cloning of genomic and complementary DNA from *Shaker*, a putative potassium channel gene from *Drosophila*. *Science* 237:749–753.
41. Tempel, B. L., D. M. Papazian, T. L. Schwarz, Y. N. Jan, and L. Y. Jan. 1987. Sequence of a probable potassium channel component encoded at *Shaker* locus of *Drosophila*. *Science* 237:770–775.
42. Heginbotham, L., Z. Lu, T. Abramson, and R. MacKinnon. 1994. Mutations in the K^+ channel signature sequence. *Biophys J* 66:1061–1067.
43. Favre, I., E. Moczydlowski, and L. Schild. 1996. On the structural basis for ionic selectivity among Na^+, K^+, and Ca^{2+} in the voltage-gated sodium channel. *Biophys J* 71:3110–3125.
44. Capes, D. L., M. Arcisio-Miranda, B. W. Jarecki, R. J. French, and B. Chanda. 2012. Gating transitions in the selectivity filter region of a sodium channel are coupled to the domain IV voltage sensor. *Proc Natl Acad Sci USA* 109:2648–2653.
45. Zheng, J. and F. J. Sigworth. 1998. Intermediate conductances during deactivation of heteromultimeric *Shaker* potassium channels. *J Gen Physiol* 112:457–474.
46. Sauer, D. B., W. Zeng, S. Raghunathan, and Y. Jiang. 2011. Protein interactions central to stabilizing the K^+ channel selectivity filter in a four-sited configuration for selective K^+ permeation. *Proc Natl Acad Sci USA* 108:16634–16639.
47. Cuello, L. G., V. Jogini, D. M. Cortes, and E. Perozo. 2010. Structural mechanism of C-type inactivation in K^+ channels. *Nature* 466:203–208.
48. Papazian, D. M., L. C. Timpe, Y. N. Jan, and L. Y. Jan. 1991. Alteration of voltage-dependence of *Shaker* potassium channel by mutations in the S4 sequence. *Nature* 349:305–310.
49. Lopez, G. A., Y. N. Jan, and L. Y. Jan. 1991. Hydrophobic substitution mutations in the S4 sequence alter voltage-dependent gating in *Shaker* K^+ channels. *Neuron* 7:327–336.
50. McCormack, K., M. A. Tanouye, L. E. Iverson, J. W. Lin, M. Ramaswami, T. McCormack, J. T. Campanelli, M. K. Mathew, and B. Rudy. 1991. A role for hydrophobic residues in the voltage-dependent gating of *Shaker* K^+ channels. *Proc Natl Acad Sci USA* 88:2931–2935.
51. Smith-Maxwell, C. J., J. L. Ledwell, and R. W. Aldrich. 1998. Uncharged S4 residues and cooperativity in voltage-dependent potassium channel activation. *J Gen Physiol* 111:421–439.
52. Almers, W. 1978. Gating currents and charge movements in excitable membranes. *Rev Physiol Biochem Pharmacol* 82:96–190.
53. Sigg, D. and F. Bezanilla. 1997. Total charge movement per channel. The relation between gating charge displacement and the voltage sensitivity of activation. *J Gen Physiol* 109:27–39.
54. Schoppa, N. E., K. McCormack, M. A. Tanouye, and F. J. Sigworth. 1992. The size of gating charge in wild-type and mutant *Shaker* potassium channels. *Science* 255:1712–1715.
55. Seoh, S. A., D. Sigg, D. M. Papazian, and F. Bezanilla. 1996. Voltage-sensing residues in the S2 and S4 segments of the *Shaker* K^+ channel. *Neuron* 16:1159–1167.
56. Aggarwal, S. K. and R. MacKinnon. 1996. Contribution of the S4 segment to gating charge in the Shaker K^+ channel. *Neuron* 16:1169–1177.
57. Long, S. B., E. B. Campbell, and R. Mackinnon. 2005. Crystal structure of a mammalian voltage-dependent *Shaker* family K^+ channel. *Science* 309:897–903.
58. Long, S. B., X. Tao, E. B. Campbell, and R. MacKinnon. 2007. Atomic structure of a voltage-dependent K^+ channel in a lipid membrane-like environment. *Nature* 450:376–382.
59. Payandeh, J., T. M. Gamal El-Din, T. Scheuer, N. Zheng, and W. A. Catterall. 2012. Crystal structure of a voltage-gated sodium channel in two potentially inactivated states. *Nature* 486:135–139.
60. Payandeh, J., T. Scheuer, N. Zheng, and W. A. Catterall. 2011. The crystal structure of a voltage-gated sodium channel. *Nature* 475:353–358.
61. Zhang, X. et al. 2012. Crystal structure of an orthologue of the NaChBac voltage-gated sodium channel. *Nature* 486:130–134.
62. Chowdhury, S. and B. Chanda. 2012. Perspectives on: Conformational coupling in ion channels: Thermodynamics of electromechanical coupling in voltage-gated ion channels. *J Gen Physiol* 140:613–623.
63. Catterall, W. A. 2000. From ionic currents to molecular mechanisms: The structure and function of voltage-gated sodium channels. *Neuron* 26:13–25.
64. Guy, H. R. and P. Seetharamulu. 1986. Molecular model of the action potential sodium channel. *Proc Natl Acad Sci USA* 83:508–512.
65. Broomand, A. and F. Elinder. 2008. Large-scale movement within the voltage-sensor paddle of a potassium channel-support for a helical-screw motion. *Neuron* 59:770–777.
66. Tiwari-Woodruff, S. K., C. T. Schulteis, A. F. Mock, and D. M. Papazian. 1997. Electrostatic interactions between transmembrane segments mediate folding of Shaker K^+ channel subunits. *Biophys J* 72:1489–1500.
67. DeCaen, P. G., V. Yarov-Yarovoy, T. Scheuer, and W. A. Catterall. 2011. Gating charge interactions with the S1 segment during activation of a Na^+ channel voltage sensor. *Proc Natl Acad Sci USA* 108:18825–18830.
68. DeCaen, P. G., V. Yarov-Yarovoy, E. M. Sharp, T. Scheuer, and W. A. Catterall. 2009. Sequential formation of ion pairs during activation of a sodium channel voltage sensor. *Proc Natl Acad Sci USA* 106:22498–22503.
69. DeCaen, P. G., V. Yarov-Yarovoy, Y. Zhao, T. Scheuer, and W. A. Catterall. 2008. Disulfide locking a sodium channel voltage sensor reveals ion pair formation during activation. *Proc Natl Acad Sci USA* 105:15142–15147.

70. Yang, N., A. L. George, Jr., and R. Horn. 1996. Molecular basis of charge movement in voltage-gated sodium channels. *Neuron* 16:113–122.
71. Yang, N. and R. Horn. 1995. Evidence for voltage-dependent S4 movement in sodium channels. *Neuron* 15:213–218.
72. Chen, X., Q. Wang, F. Ni, and J. Ma. 2010. Structure of the full-length *Shaker* potassium channel Kv1.2 by normal-mode-based x-ray crystallographic refinement. *Proc Natl Acad Sci USA* 107:11352–11357.
73. Islas, L. D. and F. J. Sigworth. 2001. Electrostatics and the gating pore of *Shaker* potassium channels. *J Gen Physiol* 117:69–89.
74. Asamoah, O. K., J. P. Wuskell, L. M. Loew, and F. Bezanilla. 2003. A fluorometric approach to local electric field measurements in a voltage-gated ion channel. *Neuron* 37:85–97.
75. Sokolov, S., T. Scheuer, and W. A. Catterall. 2007. Gating pore current in an inherited ion channelopathy. *Nature* 446:76–78.
76. Starace, D. M. and F. Bezanilla. 2001. Histidine scanning mutagenesis of basic residues of the S4 segment of the *Shaker* K^+ channel. *J Gen Physiol* 117:469–490.
77. Starace, D. M. and F. Bezanilla. 2004. A proton pore in a potassium channel voltage sensor reveals a focused electric field. *Nature* 427:548–553.
78. Starace, D. M., E. Stefani, and F. Bezanilla. 1997. Voltage-dependent proton transport by the voltage sensor of the *Shaker* K^+ channel. *Neuron* 19:1319–1327.
79. Tombola, F., M. M. Pathak, and E. Y. Isacoff. 2005. Voltage-sensing arginines in a potassium channel permeate and occlude cation-selective pores. *Neuron* 45:379–388.
80. Campos, F. V., B. Chanda, B. Roux, and F. Bezanilla. 2007. Two atomic constraints unambiguously position the S4 segment relative to S1 and S2 segments in the closed state of *Shaker* K^+ channel. *Proc Natl Acad Sci USA* 104:7904–7909.
81. Ramsey, I. S., Y. Mokrab, I. Carvacho, Z. A. Sands, M. S. Sansom, and D. E. Clapham. 2010. An aqueous H^+ permeation pathway in the voltage-gated proton channel Hv1. *Nat Struct Mol Biol* 17:869–875.
82. Ramsey, I. S., M. M. Moran, J. A. Chong, and D. E. Clapham. 2006. A voltage-gated proton-selective channel lacking the pore domain. *Nature* 440:1213–1216.
83. Chanda, B., O. K. Asamoah, R. Blunck, B. Roux, and F. Bezanilla. 2005. Gating charge displacement in voltage-gated ion channels involves limited transmembrane movement. *Nature* 436:852–856.
84. Jardetzky, O. 1966. Simple allosteric model for membrane pumps. *Nature* 211:969–970.
85. Tao, X., A. Lee, W. Limapichat, D. A. Dougherty, and R. MacKinnon. 2010. A gating charge transfer center in voltage sensors. *Science* 328:67–73.
86. Li, Q., Wanderling, S., Paduch, M., Medovoy, D., Singharoy, A., Mcgreevy, R., Villalba-Galea, C. A. et al. 2014. Structural mechanisms of voltage-dependent gating in an isolated voltage-sensing domain. *Nat Struct Mol Biol* 21:244–252.
87. Doyle, D. A., J. Morais Cabral, R. A. Pfuetzner, A. Kuo, J. M. Gulbis, S. L. Cohen, B. T. Chait, and R. MacKinnon. 1998. The structure of the potassium channel: Molecular basis of K^+ conduction and selectivity. *Science* 280:69–77.
88. Liu, Y., M. Holmgren, M. E. Jurman, and G. Yellen. 1997. Gated access to the pore of a voltage-dependent K^+ channel. *Neuron* 19:175–184.
89. del Camino, D., M. Holmgren, Y. Liu, and G. Yellen. 2000. Blocker protection in the pore of a voltage-gated K^+ channel and its structural implications. *Nature* 403:321–325.
90. Oelstrom, K. O., M. P. G. Goldschen-Ohm, M. Holmgren, and B. Chanda. 2014. Evolutionarily conserved intracellular gate for voltage-gated sodium channels. *Nature Commun* 5, Article number: 3420 (doi:10.1038/ncomms4420).
91. Xie, C., X. G. Zhen, and J. Yang. 2005. Localization of the activation gate of a voltage-gated Ca^{2+} channel. *J Gen Physiol* 126:205–212.
92. McCusker, E. C., C. Bagneris, C. E. Naylor, A. R. Cole, N. D'Avanzo, C. G. Nichols, and B. A. Wallace. 2012. Structure of a bacterial voltage-gated sodium channel pore reveals mechanisms of opening and closing. *Nat Commun* 3:1102.
93. Contreras, J. E., D. Srikumar, and M. Holmgren. 2008. Gating at the selectivity filter in cyclic nucleotide-gated channels. *Proc Natl Acad Sci USA* 105:3310–3314.
94. Zhou, Y., X. M. Xia, and C. J. Lingle. 2011. Cysteine scanning and modification reveal major differences between BK channels and Kv channels in the inner pore region. *Proc Natl Acad Sci USA* 108:12161–12166.
95. Flynn, G. E. and W. N. Zagotta. 2001. Conformational changes in S6 coupled to the opening of cyclic nucleotide-gated channels. *Neuron* 30:689–698.
96. Lu, Z., A. M. Klem, and Y. Ramu. 2002. Coupling between voltage sensors and activation gate in voltage-gated K^+ channels. *J Gen Physiol* 120:663–676.
97. Arcisio-Miranda, M., Y. Muroi, S. Chowdhury, and B. Chanda. 2010. Molecular mechanism of allosteric modification of voltage-dependent sodium channels by local anesthetics. *J Gen Physiol* 136:541–554.
98. Muroi, Y., M. Arcisio-Miranda, S. Chowdhury, and B. Chanda. 2010. Molecular determinants of coupling between the domain III voltage sensor and pore of a sodium channel. *Nat Struct Mol Biol* 17:230–237.
99. Chowdhury, S. and B. Chanda. 2010. Deconstructing thermodynamic parameters of a coupled system from site-specific observables. *Proc Natl Acad Sci USA* 107:18856–18861.
100. Horrigan, F. T., J. Cui, and R. W. Aldrich. 1999. Allosteric voltage gating of potassium channels I. Mslo ionic currents in the absence of Ca^{2+}. *J Gen Physiol* 114:277–304.
101. Islas, L. D. and F. J. Sigworth. 1999. Voltage sensitivity and gating charge in *Shaker* and Shab family potassium channels. *J Gen Physiol* 114:723–742.
102. Long, S. B., E. B. Campbell, and R. Mackinnon. 2005. Voltage sensor of Kv1.2: Structural basis of electromechanical coupling. *Science* 309:903–908.
103. Yifrach, O. and R. MacKinnon. 2002. Energetics of pore opening in a voltage-gated K^+ channel. *Cell* 111:231–239.
104. Mannikko, R., F. Elinder, and H. P. Larsson. 2002. Voltage-sensing mechanism is conserved among ion channels gated by opposite voltages. *Nature* 419:837–841.
105. Roux, B. 1997. Influence of the membrane potential on the free energy of an intrinsic protein. *Biophys J* 73:2980–2989.
106. Parsegian, A. 1969. Energy of an ion crossing a low dielectric membrane: Solutions to four relevant electrostatic problems. *Nature* 221:844–846.
107. Chanda, B. and F. Bezanilla. 2008. A common pathway for charge transport through voltage-sensing domains. *Neuron* 57:345–351.
108. Lacroix, J. J. and F. Bezanilla. 2011. Control of a final gating charge transition by a hydrophobic residue in the S2 segment of a K^+ channel voltage sensor. *Proc Natl Acad Sci USA* 108:6444–6449.
109. Cheng, Y. M., C. M. Hull, C. M. Niven, J. Qi., C. R. Allard, and T. W. Claydon. 2013. Functional interactions of voltage sensor charges with an S2 hydrophobic plug in hERG channels. *J Gen Physiol* 142:289–303.
110. Keynes, R. D. and F. Elinder. 1999. The screw-helical voltage gating of ion channels. *Proc Biol Sci* 266:843–852.
111. Ma, Z., X. J. Lou, and F. T. Horrigan. 2006. Role of charged residues in the S1-S4 voltage sensor of BK channels. *J Gen Physiol* 127:309–328.

112. Blanchet, J., S. Pilote, and M. Chahine. 2007. Acidic residues on the voltage-sensor domain determine the activation of the NaChBac sodium channel. *Biophys J* 92:3513–3523.
113. Tiwari-Woodruff, S. K., M. A. Lin, C. T. Schulteis, and D. M. Papazian. 2000. Voltage-dependent structural interactions in the Shaker K^+ channel. *J Gen Physiol* 115:123–138.
114. Pless, S. A., J. D. Galpin, A. P. Niciforovic, and C. A. Ahern. 2011. Contributions of counter-charge in a potassium channel voltage-sensor domain. *Nat Chem Biol* 7:617–623.
115. Chen, J., J. S. Mitcheson, M. Lin, and M. C. Sanguinetti. 2000. Functional roles of charged residues in the putative voltage sensor of the HCN2 pacemaker channel. *J Biol Chem* 275:36465–36471.
116. Pless, S. A., F. D. Elstone, A. P. Niciforovic, J. D. Galpin, R. Yang, H. T. Kurata, and C. A. Ahern. 2014. Asymmetric functional contributions of acidic and aromatic side chains in sodium channel voltage-sensor domains. *J Gen Physiol* 143:645–656.
117. Liu, J., M. Zhang, M. Jiang, and G. N. Tseng. 2003. Negative charges in the transmembrane domains of the HERG K channel are involved in the activation- and deactivation-gating processes. *J Gen Physiol* 121:599–614.
118. Monks, S. A., D. J. Needleman, and C. Miller. 1999. Helical structure and packing orientation of the S2 segment in the *Shaker* K^+ channel. *J Gen Physiol* 113:415–423.
119. Hong, K. H. and C. Miller. 2000. The lipid-protein interface of a Shaker K(+) channel. *J Gen Physiol* 115:51–58.
120. Li-Smerin, Y., D. H. Hackos, and K. J. Swartz. 2000. Alpha-helical structural elements within the voltage-sensing domains of a K^+ channel. *J Gen Physiol* 115:33–50.
121. Ruta, V., J. Chen, and R. MacKinnon. 2005. Calibrated measurement of gating-charge arginine displacement in the KvAP voltage-dependent K^+ channel. *Cell* 123:463–475.
122. Banerjee, A. and R. MacKinnon. 2008. Inferred motions of the S3a helix during voltage-dependent K^+ channel gating. *J Mol Biol* 381:569–580.
123. Nguyen, T. P. and R. Horn. 2002. Movement and crevices around a sodium channel S3 segment. *J Gen Physiol* 120:419–436.
124. Henrion, U., J. Renhorn, S. I. Borjesson, E. M. Nelson, C. S. Schwaiger, P. Bjelkmar, B. Wallner, E. Lindahl, and F. Elinder. 2012. Tracking a complete voltage-sensor cycle with metal-ion bridges. *Proc Natl Acad Sci USA* 109:8552–8557.
125. Swartz, K. J. 2008. Sensing voltage across lipid membranes. *Nature* 456:891–897.
126. Li-Smerin, Y. and K. J. Swartz. 2001. Helical structure of the COOH terminus of S3 and its contribution to the gating modifier toxin receptor in voltage-gated ion channels. *J Gen Physiol* 117:205–218.
127. Li-Smerin, Y. and K. J. Swartz. 1998. Gating modifier toxins reveal a conserved structural motif in voltage-gated Ca^{2+} and K^+ channels. *Proc Natl Acad Sci USA* 95:8585–8589.
128. Winterfield, J. R. and K. J. Swartz. 2000. A hot spot for the interaction of gating modifier toxins with voltage-dependent ion channels. *J Gen Physiol* 116:637–644.
129. Alabi, A. A., M. I. Bahamonde, H. J. Jung, J. I. Kim, and K. J. Swartz. 2007. Portability of paddle motif function and pharmacology in voltage sensors. *Nature* 450:370–375.
130. Kalia, J. and K. J. Swartz. 2013. Exploring structure-function relationships between TRP and Kv channels. *Sci Rep* 3:1523.
131. Bosmans, F., M. F. Martin-Eauclaire, and K. J. Swartz. 2008. Deconstructing voltage sensor function and pharmacology in sodium channels. *Nature* 456:202–208.
132. Milescu, M., F. Bosmans, S. Lee, A. A. Alabi, J. I. Kim, and K. J. Swartz. 2009. Interactions between lipids and voltage sensor paddles detected with tarantula toxins. *Nat Struct Mol Biol* 16:1080–1085.
133. Xu, Y., Y. Ramu, and Z. Lu. 2010. A *Shaker* K^+ channel with a miniature engineered voltage sensor. *Cell* 142:580–589.
134. Xu, Y., Y. Ramu, H. G. Shin, J. Yamakaze, and Z. Lu. 2013. Energetic role of the paddle motif in voltage gating of *Shaker* K^+ channels. *Nat Struct Mol Biol* 20:574–581.
135. Schmidt, D., Q. X. Jiang, and R. MacKinnon. 2006. Phospholipids and the origin of cationic gating charges in voltage sensors. *Nature* 444:775–779.
136. Ramu, Y., Y. Xu, and Z. Lu. 2006. Enzymatic activation of voltage-gated potassium channels. *Nature* 442:696–699.
137. Krepkiy, D., K. Gawrisch, and K. J. Swartz. 2012. Structural interactions between lipids, water and S1-S4 voltage-sensing domains. *J Mol Biol* 423:632–647.
138. Edgar, R. C. 2004. MUSCLE: Multiple sequence alignment with high accuracy and high throughput. *Nucleic Acids Res* 32:1792–1797.
139. Gascuel, O. 1997. BIONJ: An improved version of the NJ algorithm based on a simple model of sequence data. *Mol Biol Evol* 14:685–695.

Ligand-dependent gating mechanism

William N. Zagotta

Contents

4.1 INTRODUCTION

Ligand-gated ion channels are the chemosensors of the brain. Their essential role is to transduce changes in the concentration of a chemical into changes in membrane potential. In this role, they are responsible for the cell's response to neurotransmitters and neuromodulators (e.g., acetylcholine [ACh] and glutamate [Glu]), intracellular second messengers (e.g., Ca^{2+} and cAMP), cellular metabolites (e.g., ATP), and signaling lipids (e.g., PIP_2). The changes in membrane potential they produce may generate (or inhibit) action potentials, release hormone, contract a muscle, or activate a lymphocyte. Ligand-gated ion channels are the most rapid link between a cell and its environment.

Ligand-gated channels are best classified by the structural family in which they belong. Figure 4.1 shows a list of some of the major ligand-gated channel families with some representative members. Later chapters will discuss these channels in more detail. The structural families group by the type of ligand they bind. The 2-TM (transmembrane) P-loop (pore loop) containing channels bind some intracellular ligands such as ATP and G-protein beta–gamma subunits, while the 6-TM P-loop channels (voltage-gated family) bind different intracellular ligands such as Ca^{2+} and cAMP. Both P-loop families contain members that are regulated by PIP_2. Extracellular ligand-gated channels are separated into three families, the Glu receptor family which binds Glu, the cys-loop family which binds ACh, γ-aminobutyric acid (GABA), glycine, and serotonin, and the trimeric ligand-gated channel family that binds ATP (extracellular) and protons. Within each family, the channels can vary in many other properties such as ion selectivity, regulation by other factors (e.g., voltage), and the cell and location on the cell in which they act.

4.2 ENERGETICS OF LIGAND GATING

Despite the differences between the ligand-gated channel families, ligand gating shares several mechanistic features across all channels. The first, and maybe the most important, is that ligands almost always regulate the channels by changing the probability that the channel is open, not by changing the current that flows through an open channel. In other words, ligands affect gating, not permeation. This is illustrated for a single cyclic nucleotide-gated (CNG) channel in Figure 4.2 (Sunderman and Zagotta, 1999). In the absence of ligand, these channels are closed almost all the time (open probability about 1×10^{-6}). However, upon addition of the agonist cGMP to the intracellular face of the channels, their open probability increases. The open probability increases steadily with increasing cGMP concentration until eventually it saturates at a probability near one. However the single-channel current is unaffected by the cGMP concentration as seen in the amplitude histograms in Figure 4.2 (right). This behavior, seen in almost all ligand-gated channels, reflects the fact that the binding of ligand to the channel regulates a conformational change between the closed and open states of the channel, and that the open conformation is the same for different ligand concentrations. For most channels, ligand binding promotes channel opening and the ligand is referred to as an agonist (e.g., cGMP on CNG channels). For some channels, however, the ligand inhibits opening (promotes closing) and is referred to as a reverse agonist (e.g., ATP on K_{ATP} channels). For the rest of this chapter, we will consider only ligand-activated channels, but the same principles apply to ligand-inhibited channels.

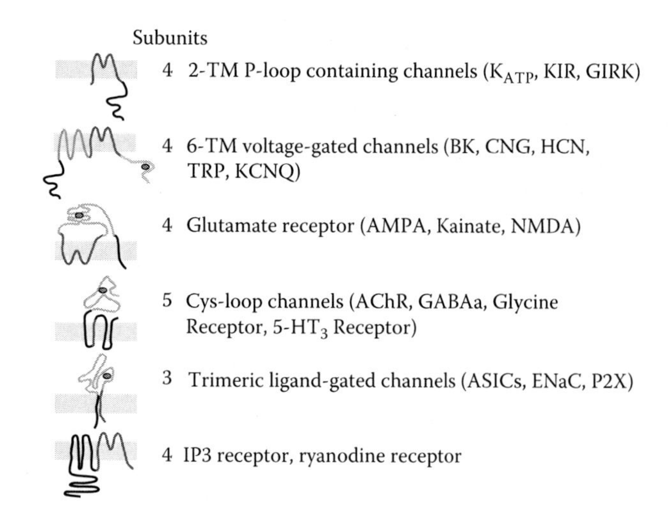

Figure 4.1 A list of some of the major ligand-gated channel families with some representative members.

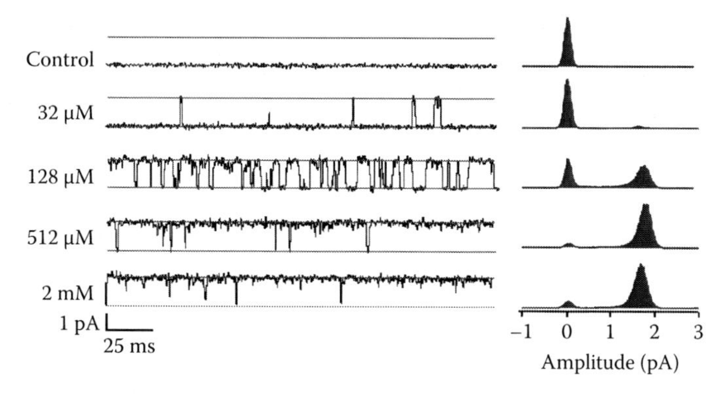

Figure 4.2 Single CNGA1 channels recorded in an inside-out patch in the presence of the indicated concentrations of cAMP. Amplitude histograms are shown on the right. (From Sunderman, E.R. and Zagotta, W.N., *J. Gen. Physiol.*, 113, 601, 1999.)

This effect of ligand binding on the open probability can be viewed from an energy perspective. The progress of the conformational change from closed to open can be described by a plot of the free energy as a function of the reaction coordinate for the closed to open conformational change, as shown in Figure 4.3a. The low points on the free energy profile represent the stable closed and open conformations of the channel, and the high point represents the high energy transition state for the conformational changes between closed and open. According to Boltzmann's Law, the ratio of probability of a channel being in the open state (P_o) to the probability of being in the closed state (P_c) *at equilibrium* is an exponential function of the free energy difference between the states (ΔG):

$$\frac{P_o}{P_c} = L = e^{-\frac{\Delta G}{kT}}, \tag{4.1}$$

where

L is the equilibrium constant for the conformational change
k is Boltzmann's constant
T is absolute temperature

In other words, when the energy difference between the states is on the order of thermal energy (kT), the channel will exist in both conformations at equilibrium. At the single-molecule level, this appears as the channel chattering between closed and open conformations. If ΔG is positive, the channel spends most of its time closed. If ΔG is negative, the channel spends most of its time open. A ligand changes the open probability by simply changing this free energy profile (Figure 4.3b). Most ligands decrease ΔG (stabilizing the open state) and therefore increase the probability of the channel being open.

The rates of opening (γ) and closing (δ) depend on the free energy difference between the closed state and the transition state $\left(\Delta G_f^{\ddagger}\right)$ and the open state and the transition state ($\Delta G_b^{\ddagger}$) respectively, as given by Eyring rate theory:

$$\gamma = Ae^{-\frac{\Delta G_f^{\ddagger}}{kT}}, \tag{4.2}$$

$$\delta = Ae^{-\frac{\Delta G_b^{\ddagger}}{kT}}, \tag{4.3}$$

where A represents the frequency factor (the rate, if there was no transition state) (Figure 4.3a). The frequency factor can vary widely for different reactions, but for a first order reaction, it is typically 10^{-11} to 10^{-12} s^{-1}. Note that by calculating the ratio of the rate constants before and after some perturbation, the frequency factor cancels out and one can calculate the change in $\Delta G^{\ddagger}$ ($\Delta\Delta G^{\ddagger}$) produced by the perturbation. Because the transition state generally has a structure intermediate between the open and closed states, an activating ligand generally partially stabilizes the transition state, speeding channel opening relative to no ligand. If the stabilization of the transition state is less than the stabilization

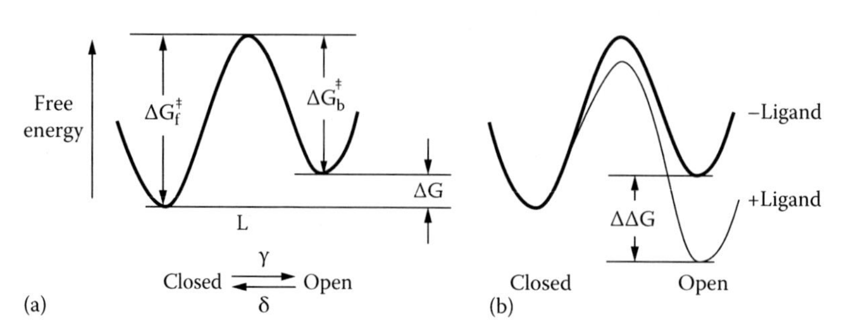

Figure 4.3 Energetics of ligand gating. (a) Hypothetical free energy profile of a closed to open transition. (b) Effect of ligand on the free energy profile.

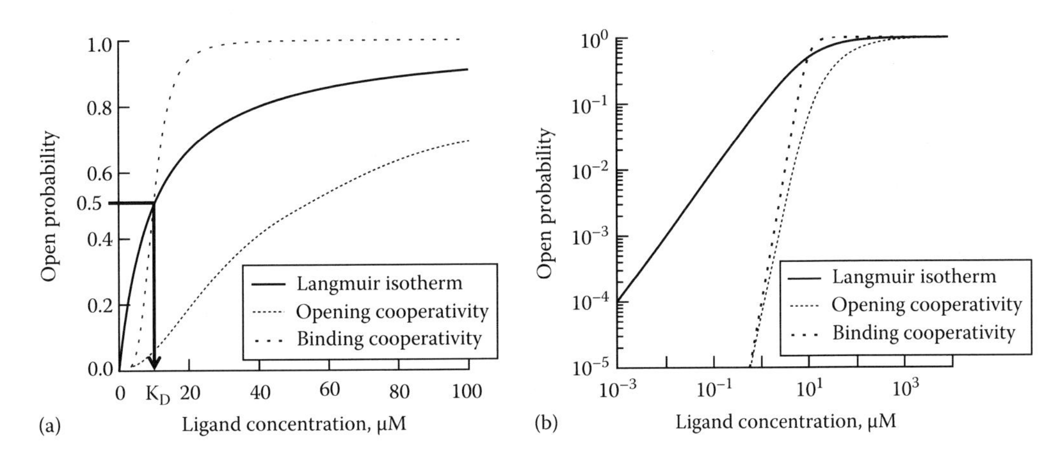

Figure 4.4 Theoretical steady-state dose–response curves for a simple two-state model, a model with opening cooperativity, and a model with binding cooperativity. (a) Linear axes. (b) Log–log axes.

of the open state (relative to the closed state), then the ligand also slows channel closing relative to no ligand. The speeding of channel opening and slowing of channel closing by the agonist increases the open probability of the channel, the hallmark of a ligand-activated channel.

Another general principle of ligand gating (indeed all allosteric modulation) is that ligands exhibit state-dependent binding. For a ligand that promotes channel opening, the ligand binds with higher affinity to the open state than to the closed state. This is a simple consequence of thermodynamics: if ligand binding promotes opening, then opening promotes ligand binding. This means that channels, like all allosteric proteins, actually have two (or more) binding affinities. As discussed later, it is very difficult to measure the affinities of the ligand for a particular state separate from the conformational transitions between the states.

4.3 STEADY-STATE PROPERTIES OF LIGAND GATING

Ion channels behave as molecular switches, opening or closing a channel pore in response to changes in the concentration of a ligand. To understand this behavior, let us start by considering a two-state model with an unbound closed state and an agonist-bound open state:

$$C \underset{}{\overset{A \;\; K_A}{\rightleftharpoons}} O \qquad \text{(Scheme 4.1)}$$

For this model, it can easily be shown that, at equilibrium,

$$P_o = \frac{[A]K_A}{1+[A]K_A} = \frac{[A]}{K_D+[A]}, \qquad (4.4)$$

where

[A] is the free concentration of agonist
K_A is the association equilibrium constant (in units of M^{-1})
K_D is the dissociation equilibrium constant (in units of M)

This equation is the Langmuir isotherm used to describe the saturating concentration dependence of any bimolecular interaction. The equation displays a hyperbolic relationship between P_o (or fraction bound) and the free agonist concentration [A] (Figure 4.4a, solid line). The apparent affinity of the interaction, the concentration that produces half-maximal opening (or binding) is simply the K_D. When plotted on log-log axes, this equation shows a limiting slope at low agonist concentrations of one (x-fold change in [A] produces x-fold change in P_o) (Figure 4.4b).

4.4 COOPERATIVITY

Channels almost always exhibit more complex behavior than predicted by a simple Langmuir isotherm. One of the complexities arises from the fact that most channels are multimers, composed of three to five subunits (Figure 4.1). Because some or all of the subunits usually bind ligand, channels have the potential to exhibit cooperativity. Cooperativity is a loosely used term, which usually means that the dose–response relation or the binding curve has a limiting slope of greater than one. There are two kinds of cooperativity that are found in channels: opening cooperativity and binding cooperativity. Most channels exhibit opening cooperativity, meaning that the binding of more than one agonist is required to open the channel (more precisely, opening of the channel is superlinearly dependent on the number of agonist molecules bound). A simple example is shown in Figure 4.5a where the agonist binds independently and identically to each of four subunits, but the channel only opens when all four agonists are bound. Because the multiple configurations of binding one, two, or three agonists are considered identical, the equilibrium constants in the model are weighted accordingly. For example, the equilibrium constant for binding the first agonist is $4K_A$, because any one of four subunits can bind agonist for the first transition, but only one subunit can unbind to make the reverse transition. For this model, the equilibrium probability of being open by binding n agonists is given simply by the nth power of the probability of binding to a single subunit:

$$P_o = (P_A)^n = \left(\frac{[A]}{K_D+[A]}\right)^n. \qquad (4.5)$$

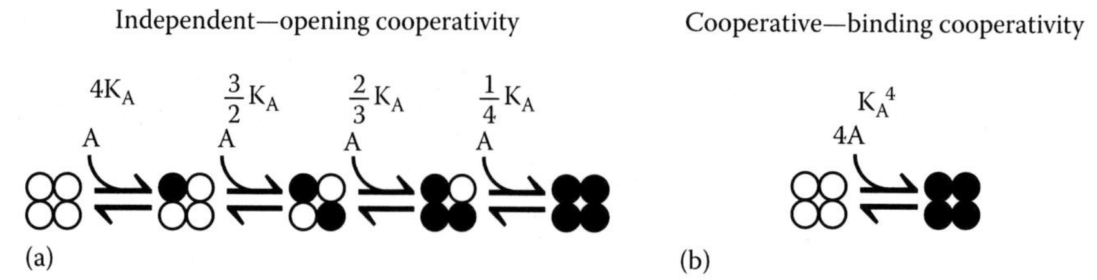

Independent—opening cooperativity

Cooperative—binding cooperativity

(a) (b)

Figure 4.5 Models for gating of tetrameric channels. (a) Model with opening cooperativity. (b) Model with binding cooperativity.

This model for ligand gating is similar to the independent gating particles proposed by Hodgkin and Huxley for voltage-gated channels (Hodgkin and Huxley, 1952). Opening cooperativity makes the opening of the channel more steeply dependent on the agonist concentration, behaving more like a molecular switch (Figure 4.4, dotted trace). This is particularly apparent in log-log dose–response plots where the limiting slope at low agonist concentrations is four, with an x^4-fold change in P_o for an x-fold change in [A] (Figure 4.4b, dotted trace). Note, however, that because the binding is independent, binding curves would still exhibit a limiting slope of one (x-fold change in [A] produces x-fold change in bound A).

The second kind of cooperativity is binding cooperativity. Binding cooperativity means that the binding of one agonist affects the binding of subsequent agonists. Binding cooperativity can arise from two general mechanisms. One way is if the binding of agonist produces a local conformational change (induced fit) that affects the binding of the next agonist, usually to a neighboring subunit. This mechanism was first proposed by Koshland, Nemethy, and Filmer (KNF) in 1966 (Koshland et al., 1966). In the context of the model in Figure 4.5a, the KNF mechanism means that the equilibrium constants no longer conform to the values predicted for independent binding (for a tetramer: $4K_A$, $3K_A/2$, $2K_A/3$, $K_A/4$). If the equilibrium constants get successively bigger, then the binding exhibits positive cooperativity, and if the equilibrium constants get successively smaller, then the binding exhibits negative cooperativity (only the KNF mechanism will produce negative cooperativity). The other general mechanism for binding cooperativity is that the binding of each agonist promotes a quaternary rearrangement in the protein that increases the binding affinity of all of the sites. This mechanism was proposed by Monod, Wyman, and Changeux (MWC) in 1965 (Monod et al., 1965) and will be discussed in more detail later. For both mechanisms, binding cooperativity causes binding curves to exhibit a limiting slope greater than one (assuming positive cooperativity).

The most extreme form of binding cooperativity is produced by the Hill model in Figure 4.5b, where all of the agonists bind at once. This model has both opening and binding cooperativity. The equation that describes the equilibrium open probability for the model in Figure 4.5b is the Hill equation:

$$P_o = \frac{[A]^n}{K_{0.5}^n + [A]^n}, \tag{4.6}$$

where
$K_{0.5}$ is the concentration that produces half-maximal activation
n is the Hill coefficient

Notice that Equation 4.6 is similar to Equation 4.5 except that $K_{0.5}^n + [A]^n$ replaces $(K_D + [A])^n$. While the cooperative opening model and the cooperative binding model predict the same limiting slopes in the dose–response relationship (Figure 4.4b), the slope at intermediate ligand concentrations is steeper for cooperative binding (Figure 4.4a). Furthermore, the binding curve for the Hill model also has a steep slope (for binding, the limiting slope at low agonist concentration is n). This is the hallmark of binding cooperativity.

4.5 SEPARATE LIGAND BINDING AND OPENING TRANSITIONS

Another complexity in the ligand activation of channels is that, unlike in Scheme 4.1, for channels, the ligand-binding step is separate from the channel-opening step. The ligand-binding step involves a docking of the ligand with a binding site on the channel, perhaps with a concurrent induced fit conformational change in the binding site. The channel-opening step, however, involves a conformational change in the protein that opens a pore region located at a distance from the ligand-binding site. This ability of the ligand to produce a change at a distance is the property of allostery, and this conformational change is frequently referred to as the allosteric transition. In fact, ligand-gated channels were one of the first allosteric proteins to be studied. In 1957, working on acetylcholine-gated (AChR) channels, Del Castillo and Katz (1957) were the first to write the binding step and the conformation change as two separate steps:

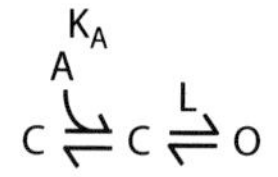

$$C \underset{}{\overset{K_A,\ A}{\rightleftharpoons}} C \overset{L}{\rightleftharpoons} O \qquad \text{(Scheme 4.2)}$$

where
K_A is the association equilibrium constant for binding
L is the equilibrium constant for opening of the fully bound channel

The equilibrium open probability for Scheme 4.2 is given by the following equation:

$$P_o = \frac{[A]K_AL}{1+[A]K_A+[A]K_AL}. \tag{4.7}$$

One of the big differences between Schemes 4.1 and 4.2 is that the maximum open probability (at a saturating agonist

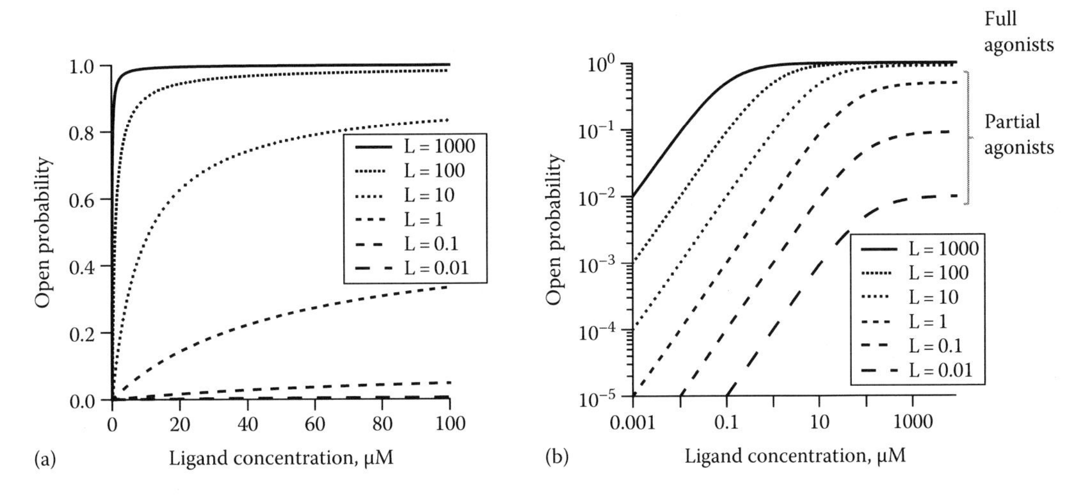

Figure 4.6 Theoretical steady-state dose–response curves for a model with separate binding and opening transitions with different values of L. (a) Linear axes. (b) Log–log axes.

concentration) for Scheme 4.2 is not one but is given by the following equation:

$$P_{o,max} = \frac{L}{1+L}. \quad (4.8)$$

Plots of Equation 4.7 for different values of L are shown in Figure 4.6. When L is large ($\gg$1), $P_{o,max}$ is near one. However, when L is small, $P_{o,max}$ is less than one.

The separation of the binding step from the opening step has two important ramifications. The first is that it is very difficult to differentiate effects of perturbations, such as mutations, on ligand binding vs. channel opening. When $L \gg 1$ (as it frequently is), then changes in L cause a shift in the dose–response relation indistinguishable from changes in K_A (Figure 4.6). This can also be seen by normalizing P_o by $P_{o,max}$, a common practice in ion channels, because P_o is usually not measured directly but measured indirectly from the conductance, which is proportional to P_o:

$$\frac{P_o}{P_{o,max}} = \frac{[A]K_A(1+L)}{1+[A]K_A(1+L)} = \frac{[A]K_{app}}{1+[A]K_{app}}, \quad (4.9)$$

where

$$K_{app}=K_A(1+L). \quad (4.10)$$

Equation 4.9 shows that plots of $P_o/P_{o,max}$ vs. [A] will conform to a Langmuir isotherm (Equation 4.4) with an apparent association equilibrium constant K_{app} (Equation 4.10). Since K_{app} depends on both K_A and L, it is impossible from such a plot to tell if a change in K_{app} resulted from a change in K_A or a change in L. Surprisingly, measuring ligand binding instead of channel opening does not get around this problem. The probability of the agonist being bound for Scheme 4.2 is given by the same equation as for $P_o/P_{o,max}$ (Equation 4.9) with the same apparent association equilibrium constant K_{app} (Equation 4.10). This paradox has been the cause of numerous errors where a mutation of a channel (or other protein) that caused a change in the apparent affinity of a ligand was erroneously thought to be in the binding site for the ligand (for review, see Colquhoun, 1998).

One way to differentiate mutations in the ligand-binding site from mutations that affect channel opening is by measuring the ligand dependence of the alteration. If a mutation has a different energetic effect on two different ligands that bind to the same site, then it is highly likely that the mutated residue resides in the binding site and interacts with the ligand in a portion where the ligands differ. For example, mutation of D604 in the CNGA1 channel causes a decrease in the apparent affinity for cGMP but an increase in the apparent affinity for cAMP (Varnum et al., 1995). From this, it was concluded that D604 interacts directly with the cyclic nucleotide in the portion of the purine ring where cGMP and cAMP differ (Figure 4.7). This conclusion was later confirmed with x-ray crystallography on a related channel (Flynn et al., 2007).

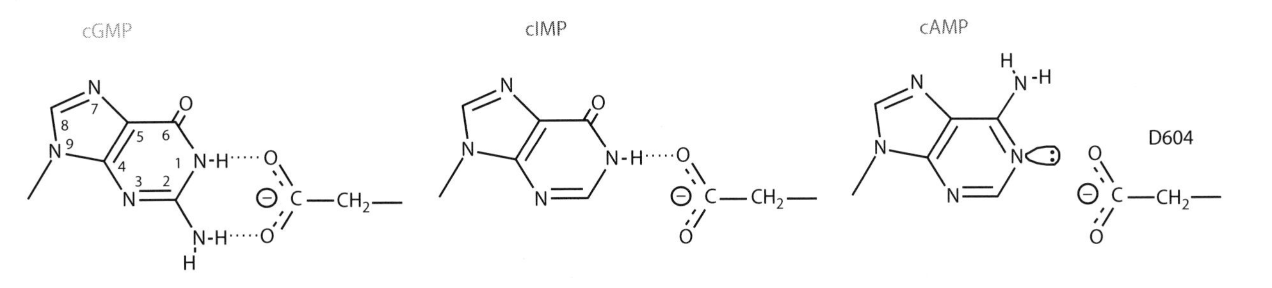

Figure 4.7 Predicted interactions of D604 of CNGA1 with the purine ring of cGMP, cIMP, and cAMP. (Reprinted from *Neuron*, Varnum, MD et al., 15, 619, Copyright 1995, with permission from Elsevier.)

Since the effects of the mutation were primarily on the $P_{o,max}$, it appears that the interaction of the cyclic nucleotide with D604 is one that forms primarily during the allosteric transition (see the following text).

4.6 PARTIAL AGONISTS

The second ramification of the separation of the binding transition from the opening transition (Scheme 4.2) is that sometimes, the ligand does not fully activate the channel even at saturating concentrations. For some channels or ligands or mutants, L is small (<10) so that $P_{o,max}$ is appreciably less than one (Figure 4.6, Equation 4.8). These ligands are referred to as partial agonists to differentiate them from full agonists where $P_{o,max}$ is near one. Desensitization or inactivation can also cause a decreased open probability at saturating ligand concentrations, as discussed later in this chapter.

Partial agonists are a very powerful tool for measuring the properties of the allosteric conformational change. At saturating ligand concentrations, one can isolate just the conformational transitions in the fully liganded channel. Figure 4.8 shows an example of partial agonists for CNG channels. While the open probability in the presence of saturating cGMP is about 0.95, it is about 0.5 for the related agonist cIMP, and only about 0.01 for cAMP. The value of L can then be calculated using Equation 4.8 ($L_{cGMP} = 20$, $L_{cIMP} = 1$, and $L_{cAMP} = 0.01$). Finally, the free energy difference between the closed and open states (ΔG) can be calculated for each of the cyclic nucleotides using Equation 4.1 ($\Delta G_{cGMP} = -1.8$ kcal/mol, $\Delta G_{cIMP} = 0$ kcal/mol, $\Delta G_{cAMP} = 2.7$ kcal/mol). With the assumption that the different ligands induce a similar conformational change in the protein (albeit with different stability), the differences in these ΔG values should reflect structural differences between how the ligands interact with the binding site during the allosteric transition.

Partial agonists offer another way to differentiate alterations in ligand binding from alterations in channel opening. In the context of Scheme 4.2, if a mutation changes $P_{o,max}$, it must reflect changes in L and therefore changes in the allosteric transition. This is the basis for phi analysis for mapping regions of the channel involved in the allosteric transition (Auerbach, 2003). In general, regions that effect L have been identified throughout the structure of channel proteins including the ligand-binding site, the pore, and almost everywhere in between. Conversely, changes in apparent affinity without changes in $P_{o,max}$ for a partial ligand must reflect changes in K_A, presumably due to alterations in the ligand-binding site itself.

4.7 MWC MODEL

A more general model for ligand-dependent activation that incorporates separate binding and opening events is shown in Figure 4.9a. This model explicitly considers the possibility that the channel can open both without ligand bound (the apo state) and with ligand bound. The opening from the apo state has an equilibrium constant of L_0 while the opening from the ligand-bound state has an equilibrium constant of $f\,L_0$. For a ligand that activates the channel, f is greater than one, and opening from the ligand-bound state is more favorable by a factor of f than opening from the apo state. This is equivalent to saying that the binding of the ligand stabilizes the open state relative to the closed state ($\Delta\Delta G$) by $-RT\ln(f)$ kcal/mol. This $\Delta\Delta G$ must come from new or strengthened interactions (or weakened negative interactions) between the ligand and the channel binding site during the allosteric transition.

To preserve microscopic reversibility, more favorable opening with ligand bound also indicates that ligand binding to the open state (fK_A) is more favorable than ligand binding to the closed state (K_A). In other words, the simple cyclic allosteric model in Figure 4.9a illustrates that ligand-gated channels

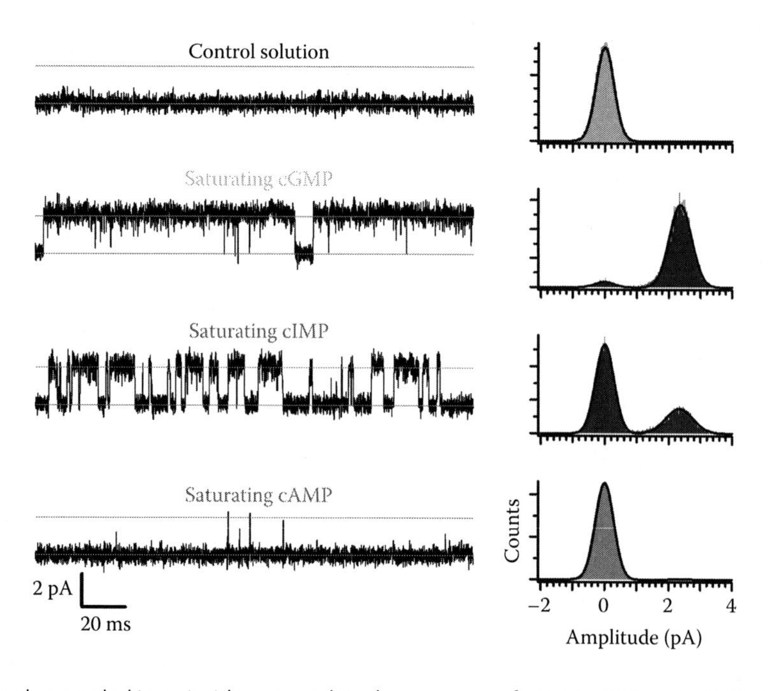

Figure 4.8 Single CNGA1 channels recorded in an inside-out patch in the presence of cGMP, cIMP, and cAMP. Amplitude histograms are shown on the right. (From Sunderman, E.R. and Zagotta, W.N., *J. Gen. Physiol.*, 113, 601, 1999.)

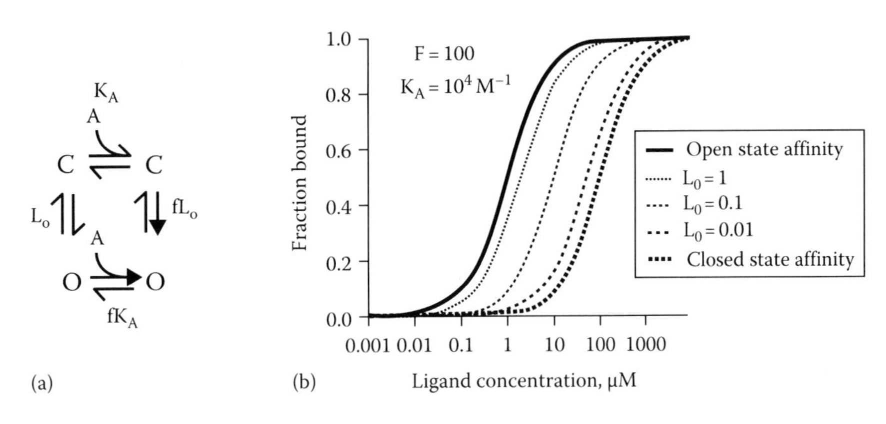

Figure 4.9 Separate but coupled binding and opening transition. (a) Cyclic model. (b) Theoretical steady-state binding curves for different values of L_0 compared to the closed state and open state binding curves.

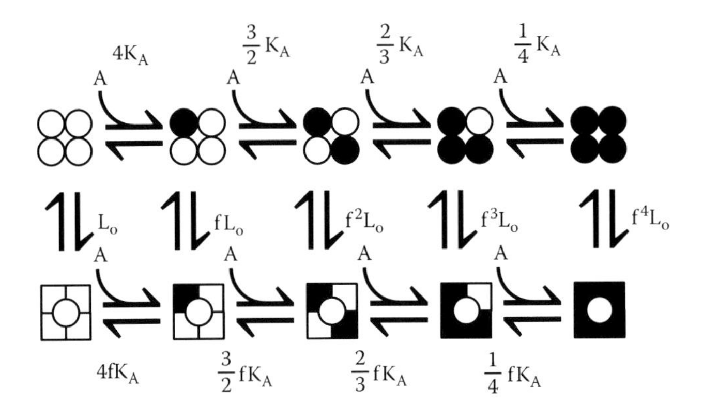

Figure 4.10 MWC model for a tetrameric ion channel.

have at least two different affinities, K_A and fK_A. The ligand dependence of the open probability of this model will still conform to a Langmuir isotherm with an apparent affinity intermediate between K_A and fK_A, depending on the value of L_0 (Figure 4.9b). For low values of L_0 (<1/f), the apparent affinity is similar to K_A, while for large values of L_0 (>1), the apparent affinity is close to fK_A.

Perhaps the most widely used model for allostery in multimeric proteins in general, and ion channels in particular, is the allosteric model of Monod, Wyman, and Changeux (MWC) (Figure 4.10) (Monod et al., 1965). This model is essentially a combination of the independent binding model in Figure 4.5a and the cyclic allosteric model in Figure 4.9a. It supposes that the protein can exist in two different conformational states; for channels, these are closed and open. Within each state, the ligands bind independently and identically. Importantly the binding of each ligand promotes the opening allosteric transition by a factor of f. Therefore, opening with one ligand bound is f-fold more favorable than opening with no ligands, opening with two ligands is f-fold more favorable than opening with one ligand, etc. This produces opening cooperativity—the channel opens more favorably when multiple ligands are bound. Similarly, ligand binding to the open state is f-fold more favorable than ligand binding to the closed state. The binding of ligands to the low affinity closed state promotes a transition to the high affinity open state. Therefore, the MWC model also exhibits binding cooperativity.

4.8 MACROSCOPIC GATING KINETICS

The preceding discussion describes the steady-state properties of ligand-gated ion channels. To understand the molecular mechanisms and physiological role of the channels, however, it is also important to understand their kinetic properties. Kinetics describes the time course of reactions. For ion channels, the time course of the open probability is simply proportional to the time course of the current at a constant voltage (voltage clamp). Electrophysiology is perhaps the most sensitive assay for any protein, able to measure time courses that span eight orders of magnitude (10 μs–1000 s) and channel numbers that range from thousands of channels to just a single channel.

Once again, let us start by considering a two-state model:

$$C \underset{\beta}{\overset{A\alpha}{\rightleftharpoons}} O \qquad \text{(Scheme 4.3)}$$

where

α is the bimolecular rate constant for ligand binding (units of $M^{-1}\,s^{-1}$)

β is the unimolecular rate constant for ligand unbinding (units of s^{-1})

The equilibrium dissociation constant K_D is given by the following equation:

$$K_D = \frac{\beta}{\alpha}. \qquad (4.11)$$

It can be shown that the time course for this two-state model can be described by the following equation:

$$P_o(t) = P_o(\infty) + (P_o(0) - P_o(\infty))e^{-t/\tau}, \qquad (4.12)$$

where

$P_o(0)$ is the initial open probability

$P_o(\infty)$ is the equilibrium open probability

τ is the time constant of relaxation (the time for the open probability to change by e-fold of the total change)

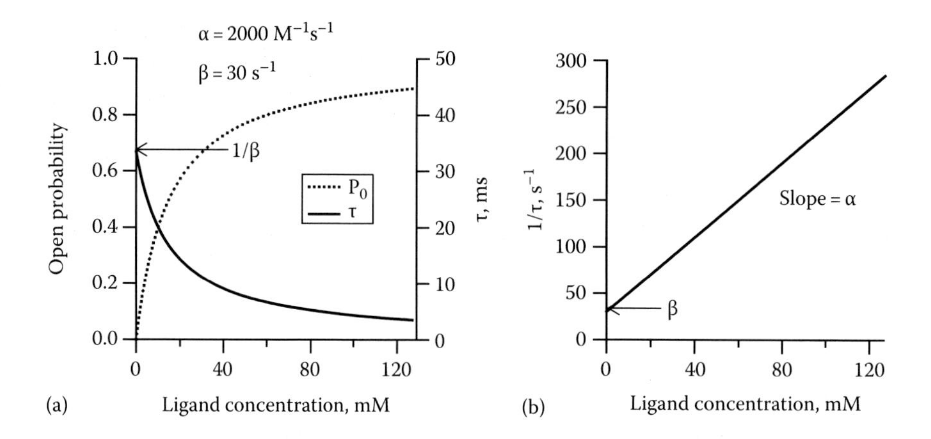

Figure 4.11 Macroscopic kinetics for a two-state model. (a) Concentration dependence of the open probability and the time constant. (b) Linear dependence of 1/τ on the ligand concentration.

For Scheme 4.3,

$$P_o(\infty) = \frac{[A]}{[A] + \beta/\alpha} \quad (4.13)$$

and

$$\tau = \frac{1}{\alpha[A] + \beta}. \quad (4.14)$$

In words, what Equation 4.12 says is that, in a two-state model, the open probability will relax to equilibrium with a single-exponential time course with a time constant that depends equally on both the opening and closing rate constants. For a given [A], the time constant will be the same, whether the channels are opening from a low probability or closing from a high probability. According to Equation 4.14, the time constant will decrease hyperbolically with the increasing agonist concentration, with an affinity of β/α (K_D) (Figure 4.11a). Plots of $1/\tau$ vs. [A] will be linear with a slope of α and a y-intercept of β (Figure 4.11b). In this way, one can get the individual rate constants from the ligand dependence of τ.

The time course of the open probability for a three-state model is described by a double exponential function with two time constants:

$$P_o(t) = P_o(\infty) + F_1 e^{-t/\tau_1} + F_2 e^{-t/\tau_2}, \quad (4.15)$$

where the time constants τ_1 and τ_2 depended on the rate constants between the states. The amplitudes of the exponentials, F_1 and F_2, depend on both the rate constants and the initial conditions (the probability of being in each state at time zero).

In general, any model with N states will be described by N−1 macroscopic time constants. The time constants only depend on the rate constants (and ligand concentration for a ligand-gated channel). Importantly, each time constant, in general, will be a function of some or all of the rate constants. There is not a one-to-one association between the time constants and the transitions. Therefore, it is a mistake to assign any particular time constant to a particular transition except under special circumstances.

4.9 SINGLE-CHANNEL GATING KINETICS

The rate constants for particular transitions can be determined more directly from an analysis of single-channel currents. An example is shown for AChR channels in Figure 4.12a (Purohit et al., 2007). Single-channel currents record the moment a channel makes a conformational change from closed to open and the moment it makes a conformational change from open to closed. Therefore, the duration of the open and closed events reflects the rates of the transitions. Since single-channel behavior (indeed any single-molecule behavior) is stochastic, the rate constants are determined from histograms of the open and closed durations (Figure 4.12b). These duration histograms are exponentially distributed where the time constants are a function of a subset of the opening and closing rate constants. They are frequently plotted on square root of the number of events vs. log duration axes, where each exponential component of the histogram produces a separate peak at the time constant for that component (Figure 4.12b). In general, the open and closed distributions will be multiexponential, where the number of time constants is equal to the number of open and closed states, respectively. Once again, these time constants are generally a function of multiple rate constants, so these cannot be assigned to a particular transition except under special circumstances.

One of those special circumstances is seen in some ligand-gated channels at saturating concentrations of ligand. As predicted by Scheme 4.2, at saturating [A], some ligand-gated channels will transition between just a single closed state and a single open state:

$$C \underset{\delta}{\overset{\gamma}{\rightleftharpoons}} O \quad \text{(Scheme 4.4)}$$

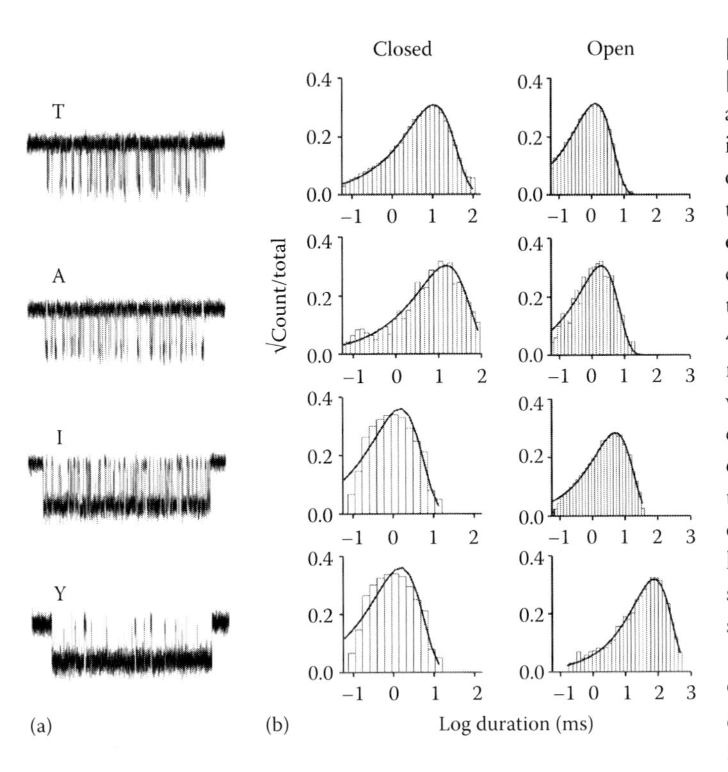

Figure 4.12 Single-channel recordings of AChR channels containing T254, T254A, T254I, or T254Y in the pore region of the α subunit. (a) Single-channel traces recorded with saturating concentrations of the partial agonist choline. (b) Closed and open duration histograms fit with single exponential functions. (From Purohit, P. et al., *Nature*, 446, 930, 2007.)

Under these conditions, the closed durations will be single-exponentially distributed with a time constant (τ_c) dependent on only the opening rate constant (γ):

$$\tau_c = \frac{1}{\gamma}, \qquad (4.16)$$

and the open durations will be single-exponentially distributed with a time constant (τ_o) dependent on only the closing rate constant (δ):

$$\tau_o = \frac{1}{\delta}. \qquad (4.17)$$

Therefore, under this condition, the rate constants can be calculated directly from the time constants. This behavior is observed in AChR channels at saturating concentrations of either the full agonist ACh or the partial agonist choline (Figure 4.12) (Purohit et al., 2007).

4.10 Phi ANALYSIS

Measuring the effects of mutations on the opening and closing rate constants is the basis for a very powerful method for studying the allosteric transition called phi (Φ) analysis (also called Brönsted analysis, linear free energy relationship [LFER] analysis, or rate-equilibrium free energy relationship [REFER] analysis). This method has been most rigorously applied to AChR channels (Purohit et al., 2007). Phi analysis involves measuring the effect of mutations at a particular site on the energy of the transition state relative to the effect on the energy of the open state. From the rate constants of the opening and closing transitions, one can calculate the energy change of the transition state ($\Delta G_f^{\ddagger}$) using Equation 4.2 and the energy change of the open state (ΔG) using Equation 4.1. A plot of $\Delta G_f^{\ddagger}$ vs. ΔG (or simply log γ vs. log L) for a number of mutations at a particular site is frequently linear with a slope Φ (Figure 4.13a). Φ indicates the fraction of the energetic effect of the mutations on the open state that has occurred at the time of the transition state. To the extent that the transition state conformation is intermediate between the open state and the closed state, Φ indicates the open state–like nature of the transition state, with a Φ of one indicating similarity to the open state and a Φ of zero indicating similarity to the closed state.

Different amino acids in the channel will have different Φ values. These Φ values can be mapped onto the three-dimensional structure of the channel (Figure 4.13b). One interpretation of these Φ maps is that they represent the sequence of events that occur during the allosteric transition between binding of the ligand and opening of the channel. For the AChR channel, Φ values near the ACh-binding site are near one while Φ values in the transmembrane segments are near zero (Purohit et al., 2007). This would seem to indicate a conformational wave associated with ligand gating that starts in the ligand-binding site and ends at the gate in the transmembrane segments. Interestingly, the Φ values tend to

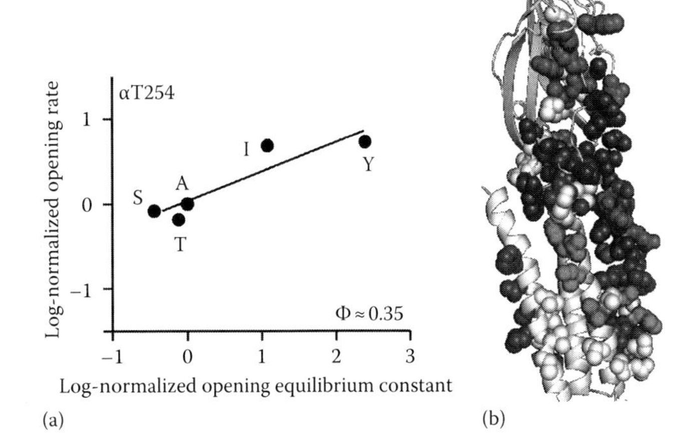

Figure 4.13 Phi analysis of AChR. (a) Plot of log normalized opening rate vs. log normalized equilibrium constant for a number of mutations of T254 in the pore region of the α subunit. The slope of the plot (Φ) is 0.35. (From Purohit, P. et al., *Nature*, 446, 930, 2007. Copyright 2007, reprinted by permission from Macmillan Publishers Ltd.) (b) Map of the Φ values for a number of mutations in the α subunit plotted on the structure of AChR. Similar Φ values are colored as follows: purple (0.85–1), blue (0.7–0.85), green (0.5–0.7), and red (0.25–0.35). (From Auerbach, A., *J. Mol. Biol.*, 425, 1461, 2013.)

cluster, suggesting that some regions of the channel move as a unit, referred to as a gating module or a nanotectonic plate. While other interpretations are possible, it is clear from these maps the broad range of regions of the channel involved in the allosteric transition and their nonequivalency.

4.11 DESENSITIZATION

Our discussion until now has been on the ligand-dependent activation of ion channels. Most ligand-gated channels also undergo a ligand-dependent inactivation or desensitization. Inactivation and desensitization are defined as the processes where, in the maintained presence of an activating stimulus, the channel enters a stable closed state called the inactivated state or desensitized state (Figure 4.14). These processes are not to be confused with deactivation, which is simply the reverse of the process of activation. Classically, inactivation was the term used for voltage-gated channels and desensitization was the term used for ligand-gated channels.

In recent years, however, it has become clear that the mechanisms for inactivation and desensitization are different (Figure 4.15). Inactivation has been shown to involve a separate gate in the channel from the activation gate. The closure of the inactivation gate is generally coupled to the opening of the channel, so that inactivation occurs primarily from the open state. In contrast, desensitization involves reclosure of the same gate that opened during activation due to a decrease in the coupling efficiency between the activation conformational change and channel opening (f in Figure 4.10). One common form of desensitization occurs when the activation process puts a strain on parts of the channel. This strain can be relieved either by channel opening or desensitizing. Desensitization results from a slippage in the channel that reduces the coupling between ligand binding and channel opening, so the channel activation gate recloses. This mechanism requires that desensitization will occur primarily from the closed state before opening.

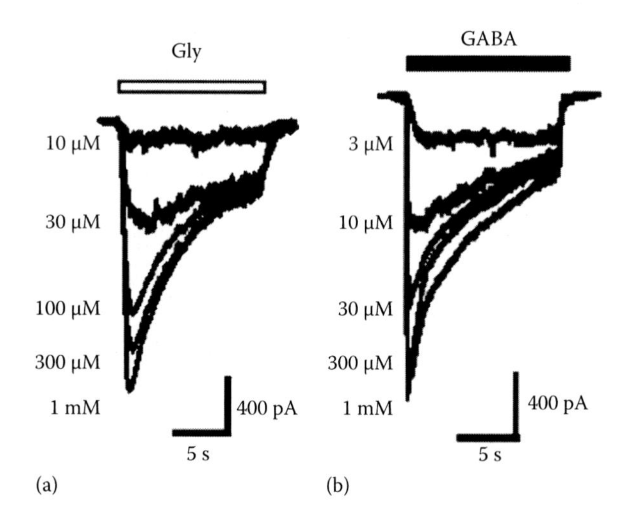

Figure 4.14 Desensitization of glycine receptors (a) and GABA receptors (b) in response to steps in agonist concentration. (From Li, Y. et al., *J. Biol. Chem.*, 278, 3863, 2003.)

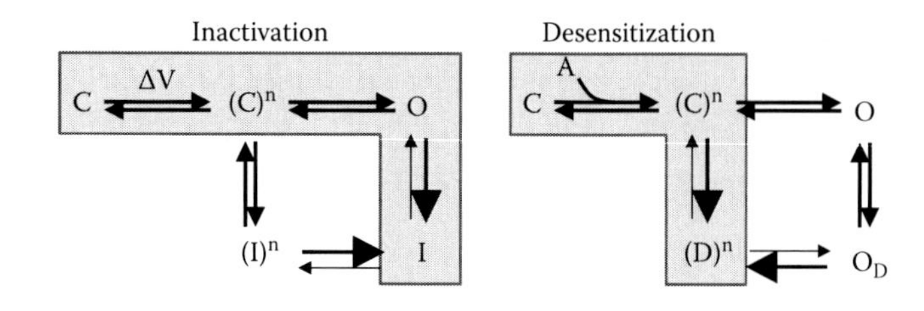

Figure 4.15 Models for inactivation and desensitization.

Desensitization is perhaps best understood in the Glu-activated channels. In response to a step in Glu concentration, GluR2 channels open and then rapidly desensitize (Figure 4.16a) (Sun et al., 2002). This desensitization occurs preferentially from the closed state. Interestingly, two treatments largely eliminate this desensitization: the L483Y mutation and the drug cyclothiazide (CLZ). These same two treatments greatly enhance the dimerization affinity of the isolated Glu-binding domain. In fact, there is a very strong correlation between the energy of desensitization (ΔG_{des}) and the energy of dimerization (ΔG_{dd}) (Figure 4.16b) (Sun et al., 2002). These results suggest that the binding of Glu puts a strain on the intersubunit interaction between the Glu-binding domains (Figure 4.17). This strain either serves to open the channel gate or breaks the subunit interface, causing desensitization. With the subunit interface slipped, the conformational change in the ligand-binding domain is no longer coupled to the opening of the pore, and the activation gate recloses.

A related mechanism is thought to occur in channels involved in sensory transduction. For example, CNG channels (in olfactory receptors) and transient receptor potential (TRP) channels (in nociceptive neurons) are both desensitized by calcium entry into the cell through the channel. Upon activation of CNG channels, the influx of calcium activates calmodulin that directly desensitizes the channel. Similarly, activation of TRP channels (by heat, protons, or capsaicin) causes an influx in calcium, activation of phospholipase C, degradation of PIP_2, and channel desensitization. In both cases, the desensitization involves a reduced coupling between activation and opening, causing activation gate reclosure. This mechanism of desensitization is responsible for a reduced response of the cell to sensory stimulation, a process called adaptation. Adaptation allows the sensory receptors to respond to a much broader range of stimulus intensities.

It now appears that desensitization is not confined to ligand-gated channels. The inactivation found in a class of hyperpolarization-activated cyclic nucleotide-modulated (HCN) channels is thought to involve a *desensitization to voltage* (Shin et al., 2004). It is proposed that activation of the voltage sensor puts strain on parts of the channel that can either open the pore or slip the coupling between the voltage sensor and the gate, producing desensitization (Figure 4.18). Interestingly for these channels, the ligand cAMP acts to prevent the desensitization, not promote desensitization as seen in other ligand-gated channels. This suggests that the mechanism of desensitization occurs in more than just ligand-gated channels.

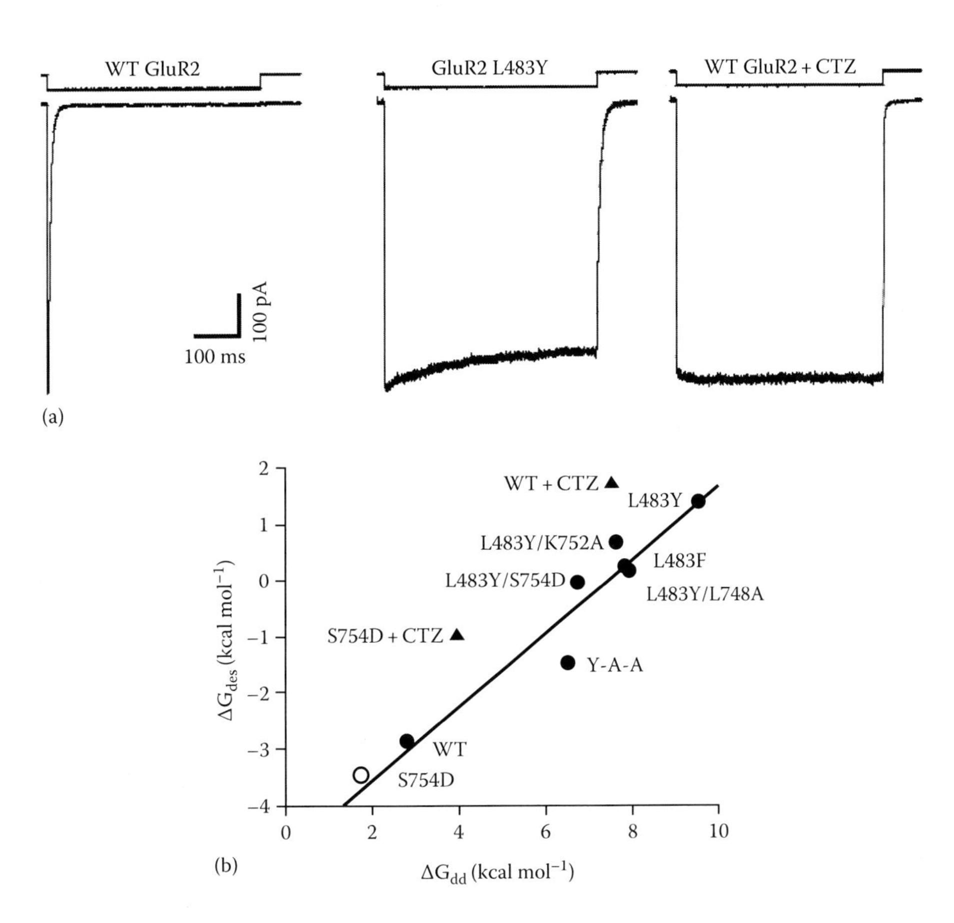

Figure 4.16 Desensitization of GluR2. (a) Desensitization of GluR2 and its removal by the L483Y mutation and CLZ. (b) Correlation between the free energy of desensitization and the free energy of dimerization of the ligand-binding domain. (From Sun, Y. et al., *Nature*, 417, 245, 2002. Copyright 2002, reprinted by permission from Macmillan Publishers Ltd.)

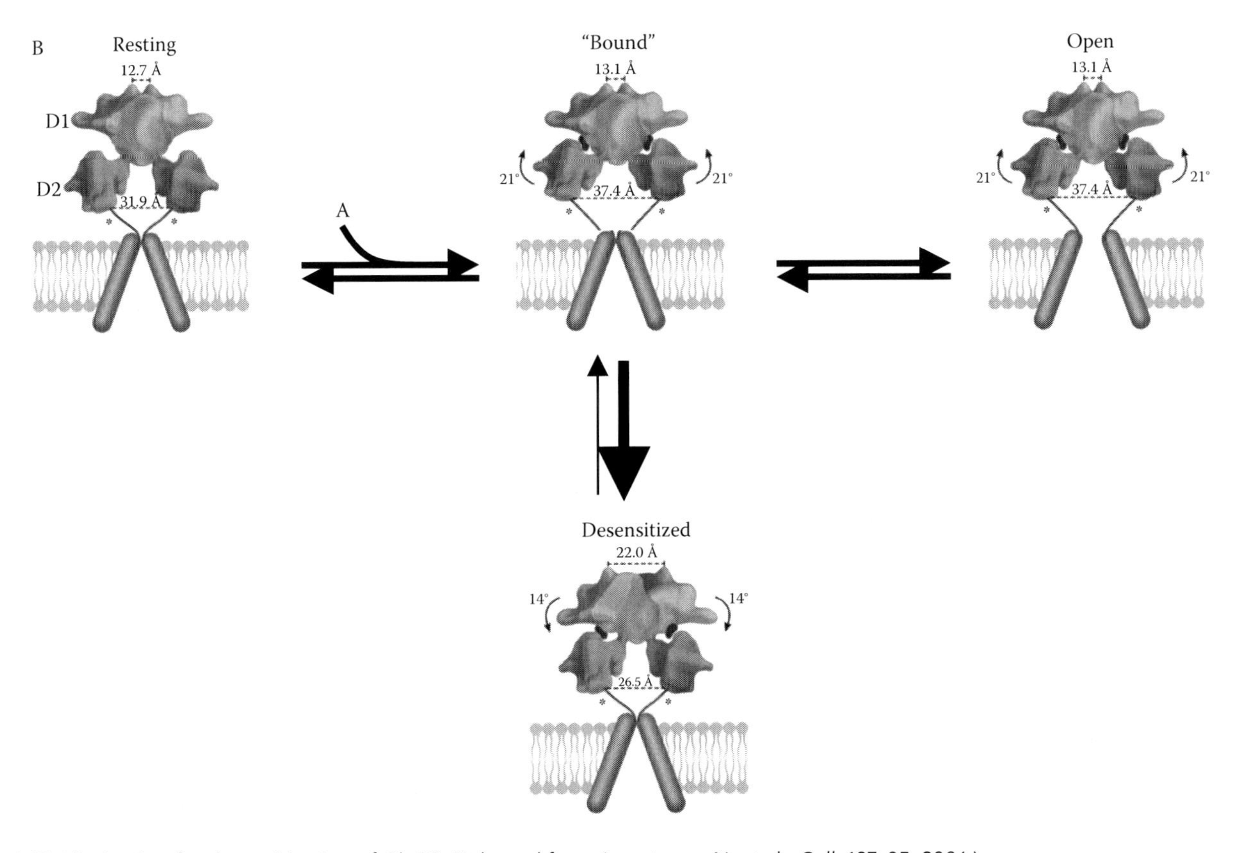

Figure 4.17 Mechanism for desensitization of GluR2. (Adapted from Armstrong, N. et al., *Cell*, 127, 85, 2006.)

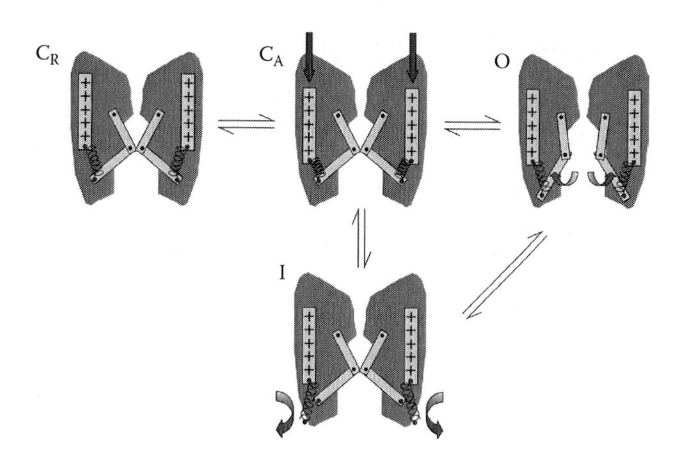

Figure 4.18 Mechanism of voltage-dependent desensitization of HCN channels. (From Shin, K.S. et al., *Neuron*, 41, 737, 2004.)

4.12 SUMMARY

Ligand-gated ion channels are chemosensors that convert chemical signals into electrical signals. They are a diverse set of multimeric allosteric proteins where the binding of ligand energetically regulates the opening and closing of the pore. The mechanism of ligand-dependent gating has been studied in exquisite detail by electrophysiology and single-channel recording. When combined with molecular structures of the channels from x-ray crystallography, these studies make ligand-gated channels one of the best understood of all allosteric proteins.

REFERENCES

Armstrong, N., J. Jasti, M. Beich-Frandsen, and E. Gouaux. 2006. Measurement of conformational changes accompanying desensitization in an ionotropic glutamate receptor. *Cell.* 127:85–97.

Auerbach, A. 2003. Life at the top: The transition state of AChR gating. *Sci STKE.* 2003(188):re11.

Auerbach, A. 2013. The energy and work of a ligand-gated ion channel. *J Mol Biol.* 425:1461–1475.

Colquhoun, D. 1998. Binding, gating, affinity and efficacy: The interpretation of structure-activity relationships for agonists and of the effects of mutating receptors. *Br J Pharmacol.* 125:924–947.

Del Castillo, J. and B. Katz. 1957. Interaction at end-plate receptors between different choline derivatives. *Proc R Soc Lond B Biol Sci.* 146:369–381.

Flynn, G.E., K.D. Black, L.D. Islas, B. Sankaran, and W.N. Zagotta. 2007. Structure and rearrangements in the carboxy-terminal region of SpIH channels. *Structure.* 15:671–682.

Hodgkin, A.L. and A.F. Huxley. 1952. A quantitative description of membrane current and its application to conduction and excitation in nerve. *J Physiol.* 117:500–544.

Koshland, D.E., Jr., G. Nemethy, and D. Filmer. 1966. Comparison of experimental binding data and theoretical models in proteins containing subunits. *Biochemistry.* 5:365–385.

Li, Y., L.J. Wu, P. Legendre, and T.L. Xu. 2003. Asymmetric cross-inhibition between GABAA and glycine receptors in rat spinal dorsal horn neurons. *J Biol Chem.* 278:38637–38645.

Monod, J., J. Wyman, and J.P. Changeux. 1965. On the nature of allosteric transitions: A plausible model. *J Mol Biol.* 12:88–118.

Purohit, P., A. Mitra, and A. Auerbach. 2007. A stepwise mechanism for acetylcholine receptor channel gating. *Nature.* 446:930–933.

Shin, K.S., C. Maertens, C. Proenza, B.S. Rothberg, and G. Yellen. 2004. Inactivation in HCN channels results from reclosure of the activation gate: Desensitization to voltage. *Neuron.* 41:737–744.

Sun, Y., R. Olson, M. Horning, N. Armstrong, M. Mayer, and E. Gouaux. 2002. Mechanism of glutamate receptor desensitization. *Nature.* 417:245–253.

Sunderman, E.R. and W.N. Zagotta. 1999. Mechanism of allosteric modulation of rod cyclic nucleotide-gated channels. *J Gen Physiol.* 113:601–620.

Varnum, M.D., K.D. Black, and W.N. Zagotta. 1995. Molecular mechanism for ligand discrimination of cyclic nucleotide-gated channels. *Neuron.* 15:619–625.

Mechanosensitive channels and their emerging gating mechanisms

Sergei Sukharev and Andriy Anishkin

Contents

5.1 INTRODUCTION

Cells constantly generate mechanical stresses of different scales for reasons of metabolic activity creating an osmotic imbalance, motility, cytoskeletal contractility, or progression through the cell cycle. Cells also perceive external forces characteristic of their specific environments, which are converted into the form of intracellular signals, triggering cell-specific responses. This conversion is implemented by several types of primary mechanotransducers.

In this review, we focus specifically on mechanosensitive (MS) ion channels and their basic mechanisms, although MS channels are not the only type of mechanotransducers in cells. Cytoskeletal filaments, for instance, change the rates of their assembly depending on axial mechanical forces (Abu Shah and Keren 2013). Molecules that connect the cytoskeleton to focal adhesions respond to linear stress by partially unfolding and exposing cryptic binding sites that recruit other proteins, thus leading to remodeling of stress-bearing complexes (Vogel 2006, del Rio et al. 2009). Local stress applied to focal adhesions also leads to accumulation of lipid rafts in that location (Fuentes and Butler 2012). Kinases associated with focal adhesions also respond to forces mediating cell remodeling and differentiation (von Wichert et al. 2008). Bacterial osmosensory kinase EnvZ perceives hydration stresses—apparently through changes in cytoplasmic macromolecular excluded volume (crowding pressure)—forcing a more compact *active* conformation (Wang et al. 2012). G-protein-coupled receptors have also been shown to mediate stretch responses (Sharif-Naeini et al. 2010).

The distinctive traits of MS channels as mechanotransducers are that they reside in membranes separating different compartments and directly respond to stresses in the lipid bilayer or associated intra- or extracellular elements. They use the energy of preexisting ionic or solute gradients to convert mechanical forces into electrical or chemical signals. Direct opening by force and absence of preceding chemical steps makes MS channels the fastest mechanotransducers. Mammalian auditory channels, for instance, respond to hair cell bundle displacement within 10 μs (Vollrath et al. 2007), allowing us to hear a wide range of frequencies. Tactile receptors in the skin are also characterized by exceptionally short response latency and a variety of adaptive behaviors producing transient (phasic) or sustained (tonic) signals (Lumpkin et al. 2010, Delmas et al. 2011), thereby allowing us to discern details of surface textures (Yau et al. 2009). Bacterial MS channels acting as fast osmolyte release valves reduce hydrostatic pressure in small cells subjected to abrupt osmotic shock within ~100 ms (Boer et al. 2011).

5.2 HOW THE STIMULUS REACHES THE GATE: CHANNEL LOCATION AND DISTRIBUTION OF FORCES

If we consider the general principles of channel gating, every class of ion channels responds to a specific stimulus, which inevitably generates local forces. Transitions in voltage-gated channels are driven by the electric field acting on a segment of the charged voltage-sensor helix spanning the apolar matrix of the membrane where the field is strongest (see Chapter 3 by Chanda). Ligand-gated channels are opened by local forces due to specific interactions between the ligand and groups forming a binding site that rearranges around the ligand, sending an allosteric signal to the gate (see Chapter 4 by Zagotta).

The complication with MS channels is that they gate in response to macroscopic forces (tensile, compression, or shear) usually applied to the entire tissue, and it is not always clear how these forces are distributed inside and where the point of application is on the transducer molecule. The situation is more tractable when a structurally characterized MS channels such as mechanosensitive channel of large conductance (MscL) or TWIK-related arachidonic acid–stimulated K^+ (TRAAK) channel are reconstituted in an artificial lipid bilayer of known composition (Moe and Blount 2005, Brohawn et al. 2014). In this case, applied tension can be calculated from patch curvature and applied pressure gradient according to the law of Laplace, whereas local distribution of forces at the molecular scale can be computed as the bilayer lateral pressure profile in MD simulations (Gullingsrud and Schulten 2004, Ollila et al. 2009). Recent studies of the insect MS channel *no mechanoreceptor potential C* (NompC) (Liang et al. 2013) strongly suggested that the long *N*-terminal domain comprises 29 ankyrin repeats acts as a *gating spring* connecting the first transmembrane helix of the channel with the supporting microtubule. These are the first data that specify the point of force delivery to this molecule. However, in most *in situ* experiments, pressure is typically applied to a patch containing the membrane and associated cytoskeleton (Zhang et al. 2000), or to the entire cell via a blunt glass probe (McCarter et al. 1999), or even to a piece of tissue such as skin (Lesniak et al. 2009), where distributions of stress and strain can only be approximated (Kim et al. 2012a).

Mechanical stimuli reaching MS channels may depend on the membrane inhomogeneity (lipid rafts, protein clusters), as well as on the presence of scaffolding elements such as the cell wall, cytoskeleton, and/or extracellular matrix (ECM), which can either attenuate or focus external forces to adjust the sensory *gain* to a desired level and meet the dynamic range of the intrinsic gating mechanism. In most somatic nonsensory cells, MS channels are believed to be shielded from mechanical membrane perturbations by the cortical cytoskeleton and ECM (Morris 2001, Suchyna et al. 2009). In contrast, in most specialized sensory organs and cells, MS channels are usually located in the most flexible apical processes or cilia and are free of shielding. Kidney epithelium, for instance, senses fluid flow utilizing channels from the transient receptor potential (TRP) family TRPP1–TRPP2 (aka PKD1-2) located at the base of the primary cilium (Nauli et al. 2003, Sharif-Naeini et al. 2009). In *Caenorhabditis elegans* (Bounoutas et al. 2009) and *Drosophila* (Liang et al. 2013), mechanotransduction channels are interconnected with the microtubular structures inside and the cuticle outside apparently to collect the stress from a larger *receptive field*. Interestingly, this design invokes current transients with both the onset and termination of a stepped mechanical stimulus. In the apical stereocilia of auditory hair cells, actomyosin maintains tension in special ECM elements called tip links (Assad et al. 1991), which interconnect stereocilia in one self-tuned mechanical system that collects stress arising from deflections of the hair bundle and conveys it to the auditory channels to maximize sensitivity (Vollrath et al. 2007).

5.3 TRANSIENT RESPONSES: ADAPTATION, DESENSITIZATION, AND INACTIVATION

Like neuronal voltage-gated channels, which produce conductance spikes with a step voltage and then abruptly inactivate (Hille 2001), most MS channels generate current transients in response to a stepped mechanical stimulation. The mechanism behind these transient responses may not always be the same. The simplest model implies a viscoelastic stress relaxation in the medium around the channel such that the stimulus reaching the gate decays with time due to passive rearrangement of soft stress-bearing elements (Sachs 2010). This type of adaptation in the cytoskeleton associated with the folded membrane could explain the transient activation of MS channels in many cells, including *Xenopus* oocytes (Zhang et al. 2000), where uncoupling of the cortical cytoskeleton from the patch membrane removes adaptation. Repetitive patch exercise also removes adaptation in Piezo1 (Gottlieb et al. 2012), a representative of the recently discovered class of MS ion channels (Coste et al. 2010). The fast (~1 ms) adaptation of the auditory channel currents in vertebrate hair cells was shown to depend on both membrane potential and the presence of Ca^{2+} on the apical side. This process was associated with a shift of the open probability–force curve apparently caused by Ca^{2+} entry into the cilia and binding to the transduction channel (Cheung and Corey 2006). This shift generates a returning *twitch* in the tip link—gating spring system that closes the channels. The slower components of hair bundle adaptation reset the activation curve according to the new resting position of the tip of the stereociliary bundle and are attributed to tension adjustments in the tip link through the slippage or pulling action of myosin motors (Gillespie and Cyr 2004, Vollrath et al. 2007).

Inactivation refers to an intrinsic process that renders the channel both nonconductive and completely insensitive to the stimulus. This can be viewed as uncoupling of the activation gate from the stimulus followed by gate reclosure, which some authors call *desensitization*. Alternatively, a return to a nonconductive state can be due to closure of a dedicated inactivation gate. Both mechanisms can produce a similar time course of channel activity and be described with the same kinetic scheme. The intrinsic mechanism of gate uncoupling in the bacterial tension-gated channel MscS will be described below. The 100 ms desensitization of TREK-1 also appears to be intrinsic to the channel complex (Honore et al. 2006) since it is unaffected by patch excision or cytoskeleton-depolymerizing agents; however, fast (~30 ms) adaptation of Piezo1 channels may in part be imposed by the environment (Gottlieb et al. 2012).

5.4 DIVERSITY OF MS CHANNELS: A BRIEF OVERVIEW

In contrast to voltage- and ligand-gated channels, MS channels do not represent a cohesive structural family with a characteristic membrane topology and conserved signature sequences exemplified by voltage-sensor domains of voltage-gated channels. MS channels belong to many structurally unrelated families. Some well-known representatives from ligand- and voltage-gated families also display some degree of mechanosensitivity (Paoletti and Ascher 1994, Morris 2011, Schmidt et al. 2012). In this section, we also intend to present the approximate chronology of key discoveries to reflect the conceptual and methodological development of the field.

5.4.1 AUDITORY CHANNEL

Hearing, the finest example of mechanosensitivity, attracted physiologists for many decades, in which experimenters focused on auditory hair cells and their environment (Lowenstein and Wersall 1959, Davis 1965). The ionic nature of the deflection-induced hair cell receptor potential was first reported by Hudspeth and Corey (1979), establishing that the primary auditory mechanotransducer must be an ion channel. This channel is located at the tips of the hair cell stereocilia near the points of the tip link insertion (Denk et al. 1995, Beurg et al. 2009). The cationic conductance ranges from 100 to 200 pS, and the active channel is large enough to pass the 611 Da FM1-43 dye into the tip of the stereocilium (Meyers et al. 2003). The conductance and kinetics of adaptation of transduction channels change with the position in the cochlea, according to the frequency they are most sensitive to (Ricci et al. 2003). Molecular identification of transduction channels has been the major challenge in the field, and up to now, the molecular entity of the vertebrate auditory channels remains unknown. The tetraspan membrane protein of hair cell stereocilia (TMHS) and transmembrane channel-like proteins (TMCs) 1 and 2 have been cloned through analysis of mutations causing deafness (Longo-Guess et al. 2005, Kawashima et al. 2011). They appear to be non–pore components of the auditory channel complex colocalizing and assembling with several other proteins at the tips of stereocilia (Xiong et al. 2012, Kim et al. 2013, Pan et al. 2013).

5.4.2 CATIONIC CHANNELS IN NONSENSORY CELLS

The first direct observation of activation of cationic channels in chicken myoblasts by stretching the patch membrane was published by Guharay and Sachs (1984), followed by numerous reports of MS channels' activities in various cells from yeast (Gustin et al. 1988) to vertebrates (Hamill and Martinac 2001) and plants (Cosgrove and Hedrich 1991). The following attempts to apply pharmacological tools to the characterization of mechanoactivated currents revealed that gadolinium (Gd^{3+}) ions (Yang and Sachs 1989), amiloride (Hamill et al. 1992), aminoglycoside antibiotics (Kroese et al. 1989), ruthenium red (Farris et al. 2004, Coste et al. 2012), and *Grammostola* spider peptide GsMTx4 (Suchyna et al. 2000) block different subsets of MS channels. However, no universal agent that could reliably block and *tag* the enigmatic MS channel molecules the way tetrodotoxin binds to neuronal Na_V channels has been reported (Hamill and McBride 1996).

5.4.3 DEG/ENaC/MEC FAMILY

This family originally comprised degenerins (DEGs) and epithelial sodium channels (ENaCs). The first molecular candidates for the role of MS channels came from genetic screens in the nematode *C. elegans* (Chalfie and Sulston 1981), which revealed 18 *mec* genes (*mechanosensory* abnormal) necessary for sensing light (~10 μN) touch (Chalfie and Au 1989) mediated by 6 specialized *touch* neurons characterized by unusual 15-protofilament microtubules (Chalfie and Thomson 1982). Out of them, 12 *mec* genes encode for components responsible not for neuron development, but specifically for mechanotransduction, which include tubulins (*mec-7, mec-12*), ECM elements (*mec-5, mec-9*), paraoxonase-like (*mec-6*) proteins, and cholesterol recruiting stomatin-like (*mec-2*) proteins (Goodman and Schwarz 2003, Arnadottir and Chalfie 2010). *mec-4* and *mec-10* encode channel-like proteins belonging to the DEG/ENaC channel family (Huang and Chalfie 1994, Lai et al. 1996). Direct electrophysiological recordings of mechanoactivated currents in these miniature neurons hidden under the worm cuticle were challenging (O'Hagan et al. 2005), but these studies have shown that out of these two genes, *mec-4* is absolutely indispensable for transduction (Arnadottir et al. 2011). Other members of the family, acid-sensitive ion channel (ASIC-1) and DEG-1 in particular, are essential for mechanotransduction in other mechanosensory neurons in *C. elegans* (Geffeney and Goodman 2012). In *Drosophila*, bristle receptors and multidendritic sensory neurons rely on the presence of several ENaC-like pickpocket (Ppk) proteins (Adams et al. 1998). In vertebrates, peripheral and somatosensory mechanotransduction by hair follicles, free nerve endings, Pacinian and Meissner's corpuscles, and Merkel cells involves α, β, γ ENaC (epithelial Na channel) and ASIC1-3 subunits (Drummond et al. 2000, Garcia-Anoveros et al. 2001, Delmas et al. 2011). ENaC β and γ subunits were also reported to be components of the aortic baroreceptor (Drummond et al. 2001).

5.4.4 BACTERIAL CHANNELS

Bacterial MS channels appeared more amenable to experiment in terms of abundance, biochemistry, reconstitution, and finally crystallographic analysis. Preparation of giant bacterial spheroplasts enabled exploration of prokaryotes as a source of MS channels. The first patch-clamp surveys of the *E. coli* cytoplasmic membrane and reconstituted proteoliposomes (Martinac et al. 1987, Sukharev et al. 1993, Berrier et al. 1996) revealed three major phenotypical classes of MS channels differing in conductance and activating pressure: large (MscL, 3 nS), small (MscS, 1 nS), and mini (MscM, 100–300 pS). The large-conductance MS channel MscL was identified and cloned through biochemical isolation and reconstitution (Sukharev et al. 1994). The MS channel of small conductance MscS was cloned by homology to the potassium efflux protein KefA (Levina et al. 1999), which turned out to be a K-dependent MS channel now called MscK (Li et al. 2002). Subsequently, analysis of the *E. coli* genome revealed four more MscS paralogs: YnaI, YbdG, YbiO, and YjeP. The mini-channel activities were attributed to the products of *ynaI*, *yjeP*, and *ybdG* genes (Schumann et al. 2010, Edwards et al. 2012). Together, these bacterial channels constitute an adaptive system that adjusts turgor pressure in the cell by releasing excessive osmolytes from the cytoplasm in the event of osmotic downshift. Remarkably, the two structurally unrelated channels, MscS and MscL, are functionally redundant and essentially equivalent in terms of rescuing bacteria from lysis in the osmotic downshock experiments. Only the deletion of both genes produces an osmotically fragile *E. coli* phenotype (Levina et al. 1999).

MscL and MscS were purified, functionally reconstituted with lipids, and shown to gate directly by tension in the lipid bilayer (Hase et al. 1995, Blount et al. 1996, Sukharev 2002). Both MscL (homolog from *Mycobacterium tuberculosis*) and *E. coli* MscS were crystallized and their structures successfully solved (Chang et al. 1998, Bass et al. 2002, Steinbacher et al. 2007). These two major bacterial osmolyte release valves became model systems for mechanistic studies of channel gating by tension. While MscL homologs are confined to the Prokarya domain, many MscS homologs have been found in essentially all organisms with walled cells including fission yeast, alga, and higher plants (Balleza and Gomez-Lagunas 2009). Finding MscS-like channels not only in the yeast plasma membrane but also in internal membranes (Nakayama et al. 2012) poses important questions of how the small cells subjected to environmental stresses handle abrupt changes in hydration through redistributing ions and osmolytes between internal compartments.

5.4.5 MS CHANNELS IN PLANTS

First patch-clamp surveys of plant protoplasts prepared from the guard, epidermal, and mesenchymal cells revealed a variety of currents activated with gentle suction or pressure and carried by Cl^-, K^+, or Ca^{2+} (Cosgrove and Hedrich 1991, Qi et al. 2004, Haswell 2007). Epidermal channels permeable to Ca^{2+} and Mg^{2+} were blocked by micromolar Gd^{3+}, the ion that abolished gravitropic responses in plant roots (Ding and Pickard 1993). Analysis of the *Arabidopsis thaliana* genome indicated the presence of ten homologs of bacterial MscS (MSL1–10) that may potentially be responsible for some of these conductances (Pivetti et al. 2003, Haswell and Meyerowitz 2006). Two of these channels, MSL2 and MSL3, were localized to plastids. Elimination of these two genes led to morphological abnormalities and swelling of plastids, which constantly experience internal hyperosmotic stress due to sugar production (Veley et al. 2012). A similar regulation of shape and division of chloroplasts by MSC1, a MscS homolog in the unicellular alga *Chlamydomonas reinhardtii* (Nakayama et al. 2012), confirmed the critical role of MscS-like channels in the maintenance of osmotic balance in these endosymbiotic organelles. The MSL9 and 10 were found in the plasma membrane of *Arabidopsis* root cells, whereas MSL8 is expressed specifically in pollen. Disruption of *MSL9* and *MSL10* along with *MSL4*, *MSL6*, and *MSL6* eliminated essentially all MS channel activities in root cell protoplasts without visible changes in root morphology or response to mechanical signals (Haswell et al. 2008). Individual properties of these channel subunits, the possibility of their heteromerization, and their roles in triggering mechanoactivated Ca^{2+} responses were discussed (Monshausen and Haswell 2013). The new MCA family of Ca^{2+}-permeable osmotically activated channels was identified in *Arabidopsis* (*mca-1* and *2*) using complementation cloning in yeast deficient for Ca^{2+} uptake (Nakagawa et al. 2007). MCA-1, which is directly sensitive to stretch, mediates osmotic influx of Ca^{2+} and confers sensitivity to touch and hardness of the substrate in growing *Arabidopsis roots*. In rice, the sole MCA homolog is involved in osmotic stress signaling that generates reactive oxygen species (Kurusu et al. 2012). The mechanosensory role of a single Piezo homolog in *Arabidopsis* has not yet been elucidated.

5.4.6 TWO-PORE POTASSIUM (K2P) CHANNEL

When genomic databases for eukaryotic organisms became more complete, not only were known channel families such as TRPs expanded with new members, but also the entire new classes of channels such as two-pore potassium (K2P) channels (Ketchum et al. 1995, Patel et al. 1998) were found. The *tandem of P domains in a weak inwardly rectifying* K^+ (TWIK) channel was described first (Lesage et al. 1996). Further cloning and characterization of several K2P channels revealed that these constitute a family of ubiquitous leakage channels setting the resting potential in excitable and nonexcitable cells. Among them, TREK-1 and TREK-2 (TWIK-related K^+) and TRAAK channels (Maingret et al. 1999, Lesage et al. 2000) were activated by stretch, producing robust fast-adapting currents (Honore et al. 2006). These channels are unique in a sense that they (TREK-1 in particular) respond to many different types of stimuli such as pH, temperature, polyunsaturated fatty acids, phosphoinositide lipids, and general anesthetics and for this reason are called multimodal (Honore 2007). The recently solved crystal structures of TRAAK (Brohawn et al. 2012, 2013) open broad opportunities for exploration of its own gating mechanism and mechanisms of its close relatives. The first attempt to purify TREK-1, reconstitute into liposomes, and record with patch clamp resulted in constitutively open channels that closed under applied tension (Berrier et al. 2013). The successful reconstitution of human TRAAK and zebrafish TREK-1 by the MacKinnon group (Brohawn et al. 2014) clearly demonstrated that adaptive gating of these channels is driven directly by tension in the lipid bilayer and does not require any additional force-transmitting elements. Some studies suggest that the gating of TREK-1 channel is not implemented through motions of a conventional gate formed by cytoplasmic ends of pore-lining helices, but rather uses a *C-type gate* working through rearrangements in the selectivity filter (Cohen et al. 2008, Bagriantsev et al. 2011). The structures of TRAAK, on the other hand, reveal an unusual connectivity of the pore interior with the aliphatic core of the membrane through *lateral openings* suggesting potential role of lipids in the gating (Brohawn et al. 2012). K2P channels are highly selective for K^+ and therefore they are hyperpolarizing and inhibitory. Some of them colocalize and work in conjunction with excitatory channels (TRPV1) and attenuate their effect (Alloui et al. 2006). Along with other types of K^+ channels, K2P channels appear to represent part of the *leakage* channel population, and their roles include a general regulation of excitability by adjusting the resting potential (Gonzalez et al. 2012), setting receptor sensitivity thresholds (Noel et al. 2009), and neuroprotection from adverse factors such as ischemia and anesthesia (Honore 2007).

5.4.7 TRP CHANNELS

The highly diverse TRP family of channels serves multiple functions such as heat and cold sensation, nociception and pain, taste and olfaction, regulation of intracellular Ca^{2+}, sensing the redox status of the cell, and osmo- and mechanosensation (Gees et al. 2012). Several representatives have been suggested to be MS; they belong to the subfamilies of TRPC (canonical), TRPA (ankyrin domain containing), TRPV (vanilloid), TRPN (*no mechanoreceptor potential*), and TRPP (related to polycystic disease). TRPN1 (NompC), initially identified in *Drosophila* (Walker et al. 2000), was directly implicated in sensing bristle deflections, in vibrations by the specialized ciliated cells in the chordotonal organ, and in proprioception. In zebrafish, TRPN is localized at the tips of stereocilia of lateral line and inner ear hair cells and its knockdown results in deafness and disoriented swimming (Sidi et al. 2003). TRP-4, a TRPN1 homolog in *C. elegans*, was demonstrated to be the transduction channel generating fast electrical responses in ciliated mechanosensory neurons (Kang et al. 2010). The function of TRPN1 in *Drosophila* depends on the presence of an intact *N*-terminal domain containing multiple ankyrin repeats and its association with microtubule (Liang et al. 2013). Sound and touch perception in *Drosophila* larvae, besides TRPN1, also relies on two other TRP channels, NANCHUNG and INACTIVE (Zhang et al. 2013).

In searches for a possible involvement of TRP channels in auditory transduction, TRPA1 was localized to the tips of stereocilia and proposed to be the component of the auditory channel in vertebrates (Corey et al. 2004), but not confirmed in genetic knockout studies (Bautista et al. 2006). The osmosensitive TRPV4 channel activated by cell swelling (Liedtke et al. 2000) has also been shown to directly activate in patches under applied tension (Loukin et al. 2010). The yeast vacuolar channel TRPY1, resembling the putative ancestor of the TRP family (D. van Rossum, personal commun.), was also shown to be directly sensitive to membrane tension (Zhou et al. 2003). Another interesting representative of the TRPV subfamily, TRP11, was identified in the unicellular alga *Chlamydomonas* and localized to the proximal segment of its swimming flagella (Fujiu et al. 2011). The primary function of the TRP11 channel in flagella-first swimming protists appears to signal for the obstacle avoidance reaction. In mammals, Ca^{2+}-permeable channels TRPP1 (also known as PC2, Pkd2, or polycystin 2) are located at the base of the primary cilium on the apical side of kidney epithelia. Coassembled with another membrane protein Pkd1, TRPP1 channels fulfill the function of flow sensors responding to cilia deflection with the influx of Ca^{2+} (Nauli et al. 2003, Sharif-Naeini et al. 2009). Like K2P channels, many TRPV channels exhibit multimodal activation by a variety of stimuli, thus making them *integrators* of mechanical stimuli with temperature, osmolarity, and the presence of anionic lipids, polyunsaturated fatty acids, or Ca^{2+}. These modulating factors seem to act on different sites and domains in the protein, but somehow the information converges on the gate.

5.4.8 PIEZO CHANNELS

The Piezo family is the most recent find in the field. Utilizing SiRNA knockdown technology combined with electrophysiology, the Patapoutian group has correlated the presence of transient mechanoactivated currents in neuroblastoma cells with the product of the Fam 38A gene (now called Piezo1) widely expressed in the bladder, colon, kidney, lung, and skin (Coste et al. 2010). While most animal and plant genomes contain single homologs of Piezo1, vertebrates have a second paralog, Piezo2, highly expressed in the sensory dorsal root ganglion (DRG) neurons. The channels generate robust fast-decaying cationic currents, which can be blocked by ruthenium red or GsMTx4 (Bae et al. 2011). Both proteins are ~2500 amino acids long with predicted 24–32 transmembrane helices and are shown to function as tetramers representing one of the largest (~1.5 MDa) membrane protein complexes. Piezo1 or Piezo2 proteins were isolated under nondenaturing conditions as homogenous complexes with no other stoichiometrically copurifying subunits. In planar bilayers, Piezo complexes formed constitutively open channels blockable by ruthenium red (Coste et al. 2012). Piezo knockout mice have not been reported yet, but tissue distributions suggest Piezo2 may be the primary receptor in mechanical somatosensation, whereas Piezo1 appears to be a stretch receptor in the organs that are naturally subjected to distension. The recently engineered mice deficient in Piezo2 specifically in the skin have shown that the slowly adapting Merkel-cell responses completely depend on Piezo2 (Woo et al., 2014). *Drosophila* Piezo (DmPiezo), which has a lower conductance than mammalian counterparts, is expressed in all neurons and many nonneural tissues, and knockout experiments have shown that it functions as a mechanical nociceptor in *Drosophila* larvae (Kim et al. 2012b).

5.5 ENERGETICS OF MECHANOSENSITIVE CHANNELS: GATING BY LINEAR FORCE, TENSION, AND PRESSURE

The basic mechanochemical principle is that in order to change equilibrium between two functional states, a molecule should *comply* with the external force and change its geometry in that specific dimension. This paradigm fully applies to MS channels (Corey and Hudspeth 1983, Sachs and Lecar 1991, Markin and Sachs 2004). The cartoon representation of three conformational transitions driven by a linear force, 2D tension in the membrane, or 3D (bulk) pressure is shown in Figure 5.1a through c. In all three cases, the sensor molecule occupies two states, the inactive (closed or resting) and active (open), separated by energy G_o, with probabilities P_c and P_o, obeying the Boltzmann distribution (Equation 5.1):

$$\frac{P_o}{P_c} = e^{-(G_o - W_c)/kT} \tag{5.1}$$

In the absence of an external stimulus, the intrinsic energy bias G_o shifts the equilibrium toward the resting state (P_c is high), while the open state is accessible with a low probability. As shown for the 1D and 2D cases, the applied external force makes

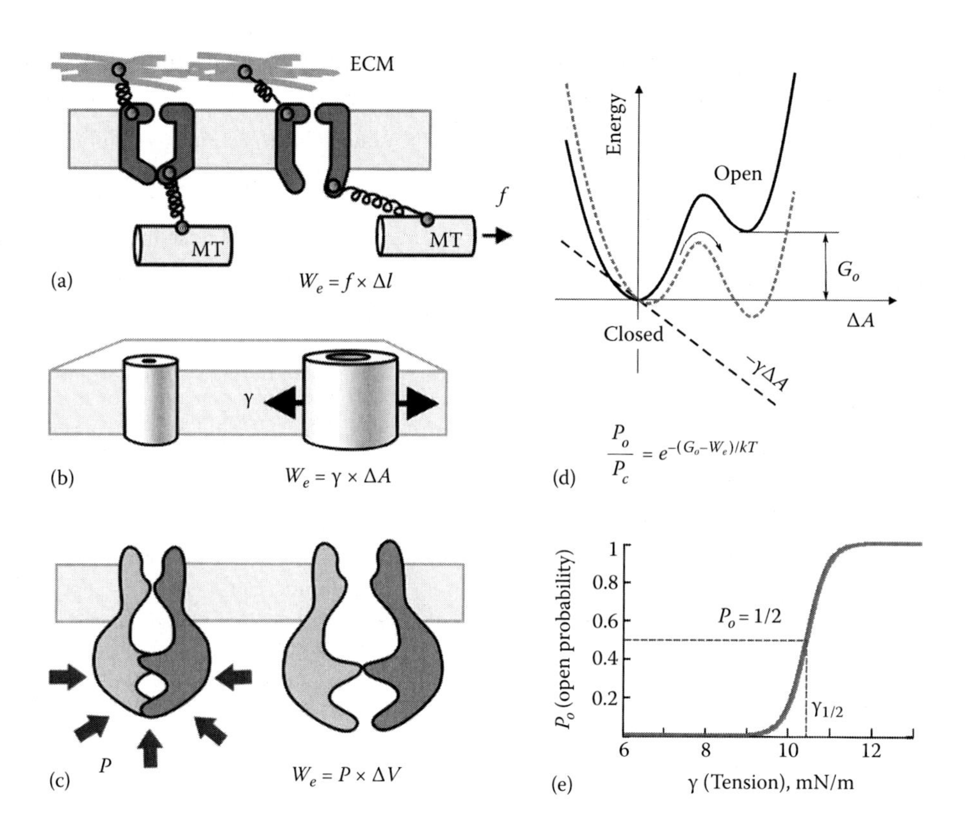

Figure 5.1 The three general ways a mechanical stimulus can reach the channel to produce a gating event. (a) Gating by linear force applied to the channel gate by movement of a microtubule through a gating spring; the spatial parameter Δl is the linear displacement of the channel gate in the direction of force and W_e is the work produced by the external force (f). (b) Gating of membrane-embedded channel by 2D tension (γ). The work in this case is equal to the product of tension and in-plane area change (ΔA) of the channel complex associated with opening. (c) In this hypothetical case, the opening of the channel is caused by an increase of cytoplasm hydration. The hollow sensor domain of the channel has openings allowing the passage of solvent and small molecules but is impermeable to large osmolytes. The domain changes its volume with the gating transition imparting sensitivity to osmotic (or crowding) pressure (P) in the cytoplasm, which is proportional to P and ΔV. (d) A two-well energetic diagram representing the energy of a tension-gated channel with expansion (ΔA) chosen as a transition coordinate. The open and closed states are separated by G_o, the energy gap between the two states in the absence of tension. Tension introduces a linear term, $\gamma\Delta A$, which changes the energy profile and lowers the energy of the open state. Obeying the Boltzmann distribution, the population of channels in the open state will increase. (e) A dose–response curve for a tension-activated channel plotted using Equation 5.2 with the experimental parameters G_o = 50 kT and ΔA = 20 nm² determined for MscL. (From Chiang, C.S. et al., *Biophys. J.*, 86(5), 2846, 2004.)

the expanded conformation more favorable, and the external energy input (W_e) stabilizing the active conformation is the product of *force* × *linear extension* or *tension* × *areal expansion*, respectively. For the 3D case, the cartoon depicts a hypothetical channel whose cytoplasmic domain is compacted by osmotic or crowding pressure, which keeps the pore closed. Dilution of the cytoplasm relieves pressure and the cytoplasmic domain expands. The product of *pressure* × *volume change* would provide the energy to increase the occupancy of the open state. All energies in the Boltzmann formalism are normalized to kT, the parameter characterizing the background thermal energy. It defines the threshold above, which the energy change will have a noticeable effect, and sets the scale for molecular driving forces and displacements. Changing the probability ratio e-fold (about 3 times) requires a change of the biasing energy by 1 kT = 4.1 × 10^{-21} J (k is the Boltzmann constant and T is the absolute temperature). This is equal to the work produced by a 4.1 pN (piconewton) force over a distance of 1 nm; similarly, a 1 kT work in 2D and 3D cases will amount to 4.1 mN/m × nm², or 4.1 atm × nm³, respectively.

As an example of a two-well transition energy profile, Figure 5.1d depicts energy as a function of a 2D molecule expansion driven by tension. As the magnitude of tension increases, it offsets the intrinsic bias more strongly, thus redistributing the population between the states. Rearrangement of Equation 5.1 with the two-state condition ($P_o = 1 - P_c$) leads to Equation 5.2, which produces an S-shaped dose–response curve for a simple tension-activated transition as depicted in Figure 5.1e. The maximal slope of the open probability (P_o) on force is determined solely by the spatial parameter of the transition, Δl, ΔA, or ΔV. The larger the expansion of the sensor molecule, the steeper is the activation curve. The midpoint of the dose–response curve, that is, the *equipartitioning* tension at which half of the channels are open ($P_c = P_o = 0.5$), as seen from Equation 5.1, would be achieved at the tension that compensates the intrinsic bias exactly ($G_o - f\Delta l = 0$). The actual response of the mechanotransducer, defined by the position and slope of the dose–response curve, is determined by both the intrinsic bias G_o and the spatial parameter of the transition:

$$P_o = \frac{1}{e^{(G_o - W_e)/kT} + 1} \tag{5.2}$$

A word of caution should be added to every consideration of dimensionality of the mechanical stimulus. Membranes and proteins behave like elastic bodies of finite dimensions, which interconvert and redirect forces arriving from the outside. For instance, the linear (1D) force exerted on the membrane at the tip of a stereocilium by the tip link causes *tenting* of the membrane (Kachar et al. 2000), which in turn creates 2D tension in the region near the insertion point of the tip link (Powers et al. 2012). This poses a dilemma as to what is the actual stimulus driving the transduction channel, a linear force or 2D tension. More detailed knowledge of the components constituting the system will permit a better approximation of force distribution and gauging of the stimulus acting on the channel.

5.6 EXPERIMENTAL PARAMETERS OF MS CHANNEL GATING

In this section, we will consider only three examples. The first is the auditory transduction channel, initially presumed to gate by linear force applied through a tip link. Despite the fact that the molecule forming the pore of this channel is yet to be identified, the mechanistic implications of the existing phenomenology are very rich. The two bacterial channels considered next, MscL and MscS, are both gated by tension in the surrounding lipid bilayer. The solved crystal structures of MscL from *M. tuberculosis* (Chang et al. 1998) and MscS from *E. coli* (Bass et al. 2002, Steinbacher et al. 2007) have an informed computational modeling and reconstruction of their main functional states based on experimental parameters (Sukharev et al. 2001, Akitake et al. 2007, Vasquez et al. 2008, Corry et al. 2010, Ward et al. 2014).

5.6.1 AUDITORY TRANSDUCTION CHANNEL

A basic theory for the gating for force-gated ion channels, which involves an elastic gating spring pulling on a movable channel gate (Figure 5.1a), came from two biophysical measurements: the kinetics of hair cell receptor currents evoked by the stereociliary bundle deflection (Corey and Hudspeth 1983) and concomitant changes of bundle compliance that matched the activation range of those receptor currents (Howard and Hudspeth 1988). This early model is reviewed by Howard et al. (1988) and Markin and Hudspeth (1995). Figure 5.2a represents the configuration of the stereociliary bundle at rest and upon deflection by force applied to the tip; panels b and c show two ways that a channel at the tip of a stereocilium could feel tension. In one, the elastic element is intracellular and connected directly to the channel; in the other, the tension is conveyed through the lipid membrane. When the bundle moves in the direction of the tallest stereocilium, all stereocilia pivot around their basal insertions, which tensions the tip links. The ratio of shear between adjacent stereocilia to the horizontal displacement of the tip of the stereociliary bundle, γ, varies with bundle geometry but is roughly 0.1–0.2. Panel d shows a simplified equivalent mechanical scheme of one interconnected pair made of model elements, including the gating spring (k_g) attached to the channel gate whose probability of being open (P_o) depends on tension and presence of Ca^{2+}. Parameter d is the gate *swing* associated with the channel opening, and Δx_g represents a calcium-induced relaxation of a hypothetical release element proposed as an additional contributor to the compliance increase on opening. The motor present in the system maintains a resting tension in the tip link. The parallel elastic (k_s) and viscous (Ξ_s) elements represent the pivoting stiffness of the stereocilium and fluid drag of the bundle. Panels e through g show data obtained by Cheung (Cheung and Corey 2006) using optical tweezers to simultaneously apply force and measure bundle movement. The receptor current (panel e) increases with deflection and then saturates, exhibiting a midpoint at a deflection near 50 nm. The force–deflection relationship (panel f) is linear at the left end of the graph, where all channels are closed, and at the right end, where they are open, indicating about the same stiffness of the bundle; however, the middle part, where the channels gate, shows flattening, and the stiffness in that region (panel g) has a minimum. The decrease of stiffness can be understood as a slackening of the gating spring associated with the opening or swing of the channel's gate in the direction of force (panel b). Fitting multiple experimental curves suggested the ranges of model parameters. In amphibian hair cells, the effective stiffness of the gating spring k_g was on the order of 700 μN/m, which collectively describes all the elastic elements that are in series with the transduction channel combined in a single element. To reach the midpoint of the activation curve (P_o = 0.5), the gating spring needs to be extended by about $50^*\gamma \approx 7$ nm under a force of about 5 pN. Channel opening is associated with an effective gate swing of ~4 nm (Cheung and Corey 2006).

More data (Peng et al. 2011, Hackney and Furness 2013) suggest that the scheme of Figure 5.2d might be an oversimplification. One should be careful interpreting the *swing* of the gate literally. It could be nearly as shown in Figure 5.2b if the channel gate is indeed directly attached to actin filaments through an elastic gating spring protein. The gating spring is not the tip link; however, crystal structures of protocadherin 15 and cadherin 23 forming the tip link and their analysis through molecular dynamics simulations have shown that these filamentous proteins are too stiff to satisfy the properties of the gating spring (Sotomayor et al. 2012). Hundreds of other proteins are in stereocilia, most uncharacterized, and one of them may have the requisite stiffness and extension.

In an alternative arrangement (Figure 5.2c), the channels may not be attached to the tip links or gating springs, but may reside in the membrane near the point of tip link insertion and may experience tension transmitted through lipids when the force applied to the tip link causes *tenting* of the membrane. The tip link with the tented membrane was shown to approximate the experimentally observed elasticity of the gating spring (Powers et al. 2012, 2014). The involvement of rigid cholesterol-containing lipid rafts around the tip link insertion zone has also been proposed (Zhao et al. 2012, Anishkin and Kung 2013) by analogy to cholesterol-rich multiprotein signaling complexes in the slit diaphragm of podocytes (Schermer and Benzing 2009).

5.6.2 BACTERIAL MS CHANNELS AS MODELS FOR GATING BY TENSION

The MS channels MscS and MscL are responsible for the bulk of osmotically induced permeability responses in prokaryotes (Levina et al. 1999). The large-conductance channel MscL is a high-threshold emergency osmolyte release valve, which opens a gigantic 3-nS pore permeable to ions, organic osmolytes, and even small proteins (van den Bogaart et al. 2007) in response to a near-lytic tension of 10–15 mN/m (Chiang et al. 2004). MscL stands out as an example of drastic and large-scale conformational rearrangement underlying its gating transition. MscS, a 1 nS

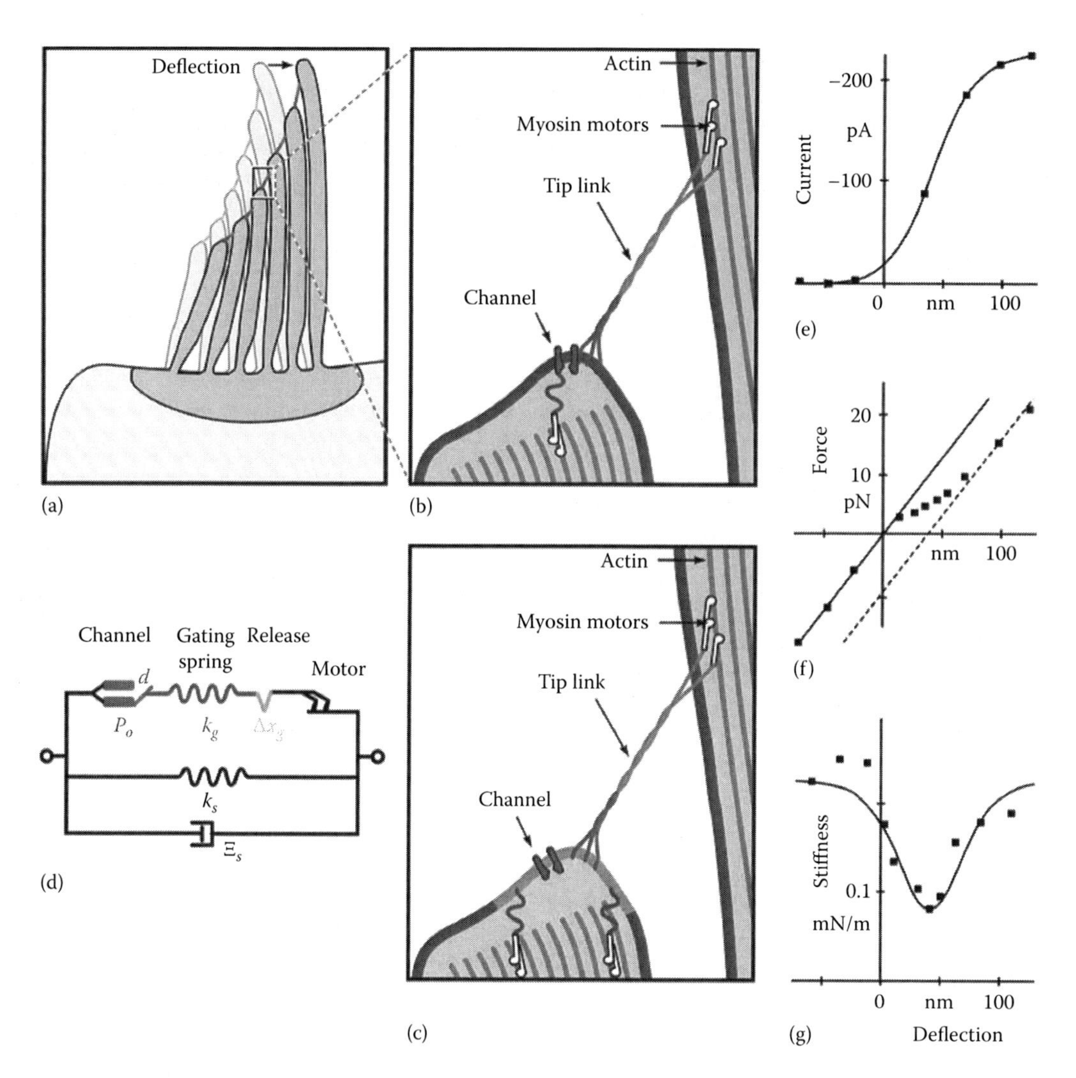

Figure 5.2 Gating of the inner ear transduction channel. (a) The arrangement of the stereociliary bundle at the apical side of a hair cell. The bundle is interconnected by the cadherin filaments called tip links. Deflection of the tip of the bundle produces stress of the tip links. (b) A hypothetical arrangement in which the transduction channel residing at the tip of the lower stereocilium is directly connected to the tip link (made of protocadherin 15 and cadherin 23) and the elastic element is attached to the actin below the channel. Tension in the tip link and the elastic element is maintained by the myosin motors. (c) The alternative arrangement in which the transduction channel resides in the membrane and experiences tension produced in the bilayer by the linear force exerted by the tip link near the point of insertion. (d) The equivalent mechanical scheme depicting the serial connection of the channel with its gate (P_o); the gating spring (k_g), which combines mechanical compliances of the tip link and all connected elements inside stereocilia; and the hypothetical *release* element that provides a Ca^{2+}-dependent extension, which can be associated with the channel and the motor. The two parallel elements k_s and Ξ_s represent the pivoting stiffness and drag, respectively. (e) The receptor–current–deflection curve for a bullfrog hair cell. (f) The force–deflection relationship showing linear regimes at both ends and a transition regime in the middle. (g) The stiffness–deflection curve with a minimum corresponding to the point of steepest current rise caused by transduction channel gating. (Panels d through g are redrawn from Cheung, E.L. and Corey, D.P., *Biophys. J.*, 90(1), 124, 2006.)

channel, opens at 5–7 mN/m, considerably below the lytic limit and attracts attention due to characteristic inactivating behavior that integrates tension in the membrane and crowding pressure in the cytoplasm.

5.6.3 LARGE-CONDUCTANCE CHANNEL MscL

Figure 5.3 shows the homology model of *E. coli* MscL (Sukharev et al. 2001) based on the 3.5A crystal structure of MscL from *M. tuberculosis* (Chang et al. 1998) in the closed (A) and open (B) states. MscL is a pentamer of 136 amino acid subunits spanning the membrane twice; the TM1 and TM2 helices are swapped between adjacent subunits such that periplasmic loops physically interconnect all subunits. The pore constriction, lined by aliphatic side chains in the closed state, is likely desolvated (Anishkin et al. 2010) and acts as a hydrophobic gate. Both the *N*- and *C*-terminal ends are in the cytoplasm and the *C*-terminal helices form a bundle that together with unstructured linkers forms a prefilter at the cytoplasmic entrance to the pore. The poorly resolved *N*-terminus was proposed to either form a short helical bundle acting as a second gate (Sukharev et al. 2001) or spread out along the inner surface of the membrane forming amphipathic *anchors* (Steinbacher et al. 2007). The iris-like opening transition was modeled as the tilting of TM1–TM2 helical pairs by 25°–30° and simultaneous ~12Å radial displacement that opens an aqueous pore 25–30 Å in diameter, satisfying the 3.2 nS unitary conductance. Disulfide cross-linking (Betanzos et al. 2002a), electron paramagnetic resonance (EPR) (Perozo et al. 2002a), and fluorescence resonance energy transfer (FRET) (Corry et al. 2005, 2010) experiments gave a strong support to this type of transition. This iris-like transition is also consistent with the

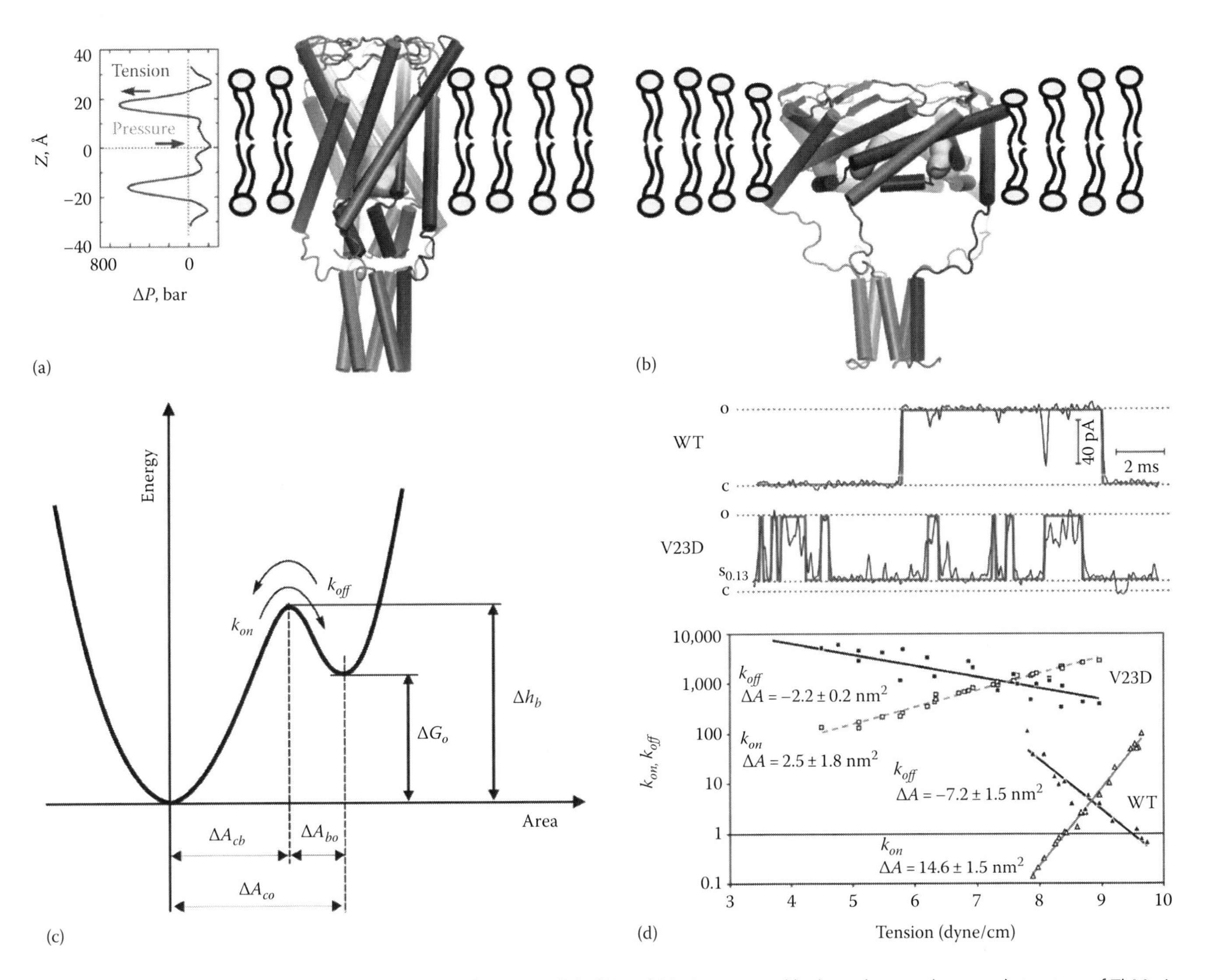

Figure 5.3 The gating mechanism of MscL. (a) The closed-state model of *E. coli* MscL generated by homology to the crystal structure of TbMscL. The lateral pressure profile in the surrounding bilayer shows two peaks of tension acting at the ends of transmembrane helices. (b) The model of the open state achieved through tilting of helical pairs. Flattening of the TM barrel leads to a distortion of the boundary lipids. (c) The spatial parameters defining the tension dependences for the opening and closing rates, k_{on} and k_{off} (see text). (d) The examples of single-channel traces for WT and V23D MscL and the logarithmic plot of k_{on} and k_{off} on tension. The slopes indicate area changes between the closed state and the transition barrier, ΔA_{cb}, and between the barrier and the open state, ΔA_{bo}. (From Anishkin, A. et al., *J. Gen. Physiol.*, 125(2), 155, 2005.)

computed lateral pressure profile of the lipid bilayer (Gullingsrud and Schulten 2004), where the peaks of tension correspond to the hydrophobic–hydrophilic boundaries inside the membrane (Figure 5.3a). Tension transmitted through these boundary layers is predicted to act on the ends of TM helices causing tilt. Since the lateral pressure profile is sensitive to the chemistry of both headgroups and aliphatic tails, channel gating is modulated by lipid composition (Moe and Blount 2005) and the presence of lipid-intercalating amphipathic substances. Asymmetrically inserted lysolipids act as potent activating agents (Perozo et al. 2002b).

Imaging experiments helped to visualize curvatures of pressure-stimulated patches and represent dose–response curves (P_o) as a function of membrane tension (γ). Thermodynamic analysis of $P_o(\gamma)$ dependences using the Boltzmann equation (Equation 5.1) estimated the intrinsic energy change associated with the opening transition as G_o = 50 kT (~125 kJ/mol) and the in-plane expansion of the MscL complex (ΔA) to be 20 nm² (Figure 5.3b), highly consistent with the models of iris-like transition predicting 20–23 nm² (Chiang et al. 2004, Corry et al. 2010). Generally, lateral expansion estimated for eukaryotic MS channels is smaller, on the order of 2–5 nm² (Honore et al. 2006, Maksaev et al. 2011).

The intrinsic energy difference between the states, G_o, appears to have three major components. One is the springlike action of the periplasmic loops (Ajouz et al. 2000). Another is the elastic energy stored in distorted lipids at the protein–lipid boundary. Helical tilting associated with pore expansion produces a substantial flattening of the channel complex along the pore axis. The thickness mismatch between the protein and the surrounding bilayer due to elastic deformation of the boundary lipids was calculated to provide a substantial component of the returning force contributing to G_o (Ursell et al. 2007, Phillips et al. 2009). The third contributor is unfavorable hydration of the hydrophobic gate that defines G_o and the barrier between the states (Anishkin et al. 2010). The energy and cooperativity of transitions critically depend on hydration properties of the pore as illustrated by the comparison of wild-type MscL with V23D mutant (Figure 5.3d). This mutation places charged side chains in the hydrophobic gate and produces easily opened (gain-of-function) channels with high occupancy of subconductive states (Anishkin et al. 2005). This type of gate modification led to the development of approaches to the chemical or light-induced MscL activation in reconstituted systems for the purposes of controlled drug delivery (Kocer et al. 2005).

Analysis of tension dependences for the separate opening (k_{on}) and closing (k_{off}) rates provides additional information about the shape of the energy profile along ΔA chosen as the reaction coordinate for the transition (Sukharev et al. 1999). Eyring-type equations for k_{on} and k_{off} take into account the height of the transition barrier as it is seen either from the bottom of the closed well or from the open-state well, respectively (Figure 5.3c):

$$k_{on} = k_0 \exp\left[\frac{-(\Delta h_b - \gamma \Delta A_{ob})}{kT}\right]$$

$$k_{off} = k_0 \exp\left[\frac{-(\Delta h_b - G_o - \gamma \Delta A_{cb})}{kT}\right]$$

Differentiation with respect to γ produces a simple relationship that equates the slope of $kT{\cdot}\ln(k_{on})$ on tension to ΔA_{cb}, the area change from the bottom of the closed-state well to the top of the transition barrier. By the same token, $kT \cdot d\ln(k_{off})/d\gamma = \Delta A_{ob}$, which is the area change from the center of the open-state well to the top of the barrier. Logarithmic plots of k_{on} and k_{off} on γ for WT and V23D MscL are shown in Figure 5.3d. The slopes assign the transition barrier position at about 0.65 of the total area change between the closed and open states for WT MscL. The location of the barrier closer to the open state makes the parabolic well for the closed state wider. The closed conformation is therefore predicted to be *softer* than the open conformation, and the channel should undergo a substantial *silent* expansion before it overcomes the transition barrier and opens. The ΔA_{cb} and ΔA_{ob} values for V23D MscL are lower and the entire transition is of smaller scale because the channel is preexpanded in the closed state due to an excessive pore hydration.

5.6.4 MscS CHANNEL AND ITS ADAPTIVE GATING MECHANISM

The MS channel of small conductance MscS opens a 1 nS pore at tensions of 5–7 mN/m with the slope of $P_o(\gamma)$ curve corresponding to ΔA of 15–18 nm^2 (Akitake et al. 2005, 2007). Steps of tension applied to patches excised from giant spheroplasts produce transient slowly decaying current responses, which represent a combination of two processes: closure due to stimulus adaptation within the patch membrane (Belyy et al. 2010b) and complete intrinsic inactivation. Analysis has shown that inactivation, like activation, is driven by tension from the closed state (Kamaraju et al. 2011). The inactivated channels are nonconductive and completely tension insensitive, and they remain in this state as long as tension persists. Upon tension release, the inactivated channels return back to the resting state within 2 s (Akitake et al. 2005). Importantly, the rate of inactivation critically depends not only on membrane tension but also on the presence of macromolecules or polymers (crowders) on the cytoplasmic side of the patch (Grajkowski et al. 2005). Here, we briefly present the hypothesis and key supporting data on how MscS integrates the two independent mechanical inputs, interfacial tension in the membrane and the bulk parameter of crowding pressure, to quickly disengage its gate as a result of their synergistic action.

The 3.9 Å crystal structure of *E. coli* MscS (Bass et al. 2002, Steinbacher et al. 2007) revealed the channel as a homoheptamer of 3TM subunits (PDB ID 2OAU, Figure 5.4). A subsequently solved structure of A106V mutant (PDB ID 2VV5) represents a partially open state (Wang et al. 2008). Today, there are six additional structures of *E. coli* MscS and its homologs from other bacteria solved in different conditions. All structures show splayed TM1–TM2 peripheral pairs are supposed to form the protein–lipid boundary, whereas TM3s line the central pore with a hydrophobic gate at the cytoplasmic end. Long C-termini from each subunit combine in a large hollow cytoplasmic domain with seven side portals, termed the *cage*. Other hallmarks of MscS structure (2OAU) are its sharply kinked TM3s just below the hydrophobic gate predicted to be dehydrated (Anishkin and Sukharev 2004), and a substantial separation of peripheral helices from the gate, suggesting an uncoupled state of the gate in the crystals. It is known that in the membrane, the channel resides in the closed *ready-to-fire* state; thus, it was inferred that the splayed state of helices in the crystal structure might be a result of lipid replacement with a detergent. The question of whether the splayed state of the peripheral TM1–TM2 helices can be changed by replacing detergent with membrane-mimic systems has recently been addressed using a pulsed EPR technique (Ward et al. 2014). The attempts to interpret several crystal structures of MscS have recently been reviewed by Naismith and Booth (2012).

Through modeling and simulations (Anishkin et al. 2008), the crystal structure has been reinterpreted as the inactivated state. The partially reconstructed functional cycle of MscS (Figure 5.4) now can explain the phenomenology of inactivation (Akitake et al. 2005, 2007, Rowe et al. 2014). Packing of the splayed TM1–TM2 pairs along TM3s was identified the step that restores physical contact between the gate and the lipid-facing helices. The reconstructed buried hydrophobic TM2–TM3 contact turned out to be the force transmission route from the lipid bilayer to the gate. The strength of apolar association was shown to be sufficient to convey external force to the gate and open the channel (Belyy et al. 2010a). The opening transition associated with the experimentally estimated 15 nm^2 expansion of the entire transmembrane domain produces a 16 Å wide pore well satisfying the 1.1–1.2 nS unitary conductance (Anishkin et al. 2008). The opening straightens up the TM3 helices, which have two conserved hinge points (glycines G113 and G121) and act as *collapsible struts* (Akitake et al. 2007). Importantly, hydrophilic substitutions at some positions in the buried TM2–TM3 zone led to a speedy channel inactivation without opening, indicating that the TM2–TM3 contact is labile and behaves as a dynamic *slip bond* conveying tension to the gate (Belyy et al. 2010a). Disruption of this bond leads to gate uncoupling and formation of the crystallographic kink at glycine 113, which stabilizes the inactivated state. Based on the tension dependences of the rates of inactivation and recovery (Kamaraju et al. 2011), it has been determined that gate uncoupling is associated with an

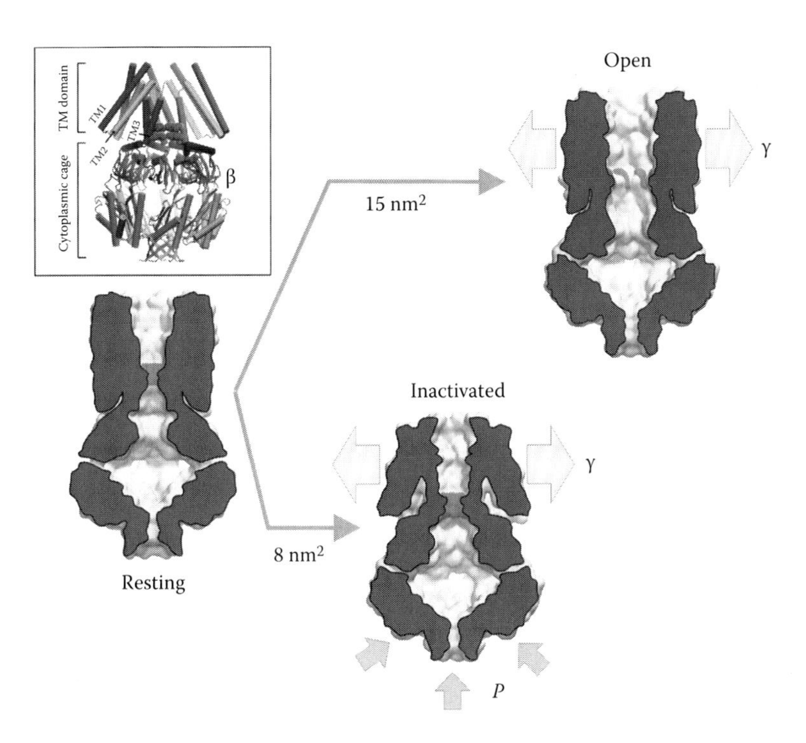

Figure 5.4 The crystal structure of *E. coli* MscS (framed) and the reconstructed functional cycle. The modeling and simulations produced the compact resting state by restoring the contact between splayed TM1–TM2 pairs and the central TM3s. The opening involves an ~15 nm² expansion of the TM barrel creating a 16Å-wide aqueous pore surrounded by kink-free TM3 helices; the gate is colored yellow. Entering the inactivated state involves the detachment of TM1–TM2 pairs from TM3s and reformation of the crystallographic kink at G113. Crowding pressure acting on the cage is predicted to cause an upward displacement of the cage producing a stronger splay of TM1–TM2 pairs thus assisting inactivation. (After Rowe, I. et al., *J. Gen. Physiol.*, 143(5), 543, 2014.)

approximately 8 nm² in-plane expansion, in good correspondence with the outward splaying motion of the TM1–TM2 pairs back to the crystal-like conformation. Thus, under tension, the channel has two alternative pathways, one into the open state and another into inactivation. How does crowding pressure make the inactivation path more preferable?

MD simulations of the closed and inactivated models revealed that the positions of the hollow cytoplasmic cage domain in the resting and inactivated states are different. In the latter case, the cage domain has a slightly smaller profile in the cytoplasm, making the inactivated state more compact and therefore more preferable in the presence of crowders (Rowe et al. 2014). In addition, the application of axial pressure to the cage domain in molecular dynamics simulations to mimic the effect of crowders produces a bigger splay of TM1–TM2 pairs, which additionally increases the *footprint* of the channel in the membrane. Thermodynamically, this means that crowding pressure will assist tension in separating the peripheral helices from the core and promote the uncoupling of the gate.

The biological meaning of the observed effect of crowders on MscS inactivation is that in the course of osmotically induced permeability response, the channel receives feedback on the state of the cytoplasm. The increased macromolecular excluded volume, perceived by the cage domain as a crowding pressure, disengages the gate, thus limiting the period of channel action. Because the fractional volume of free water in the cytoplasm should not decrease below a certain point (~30%), this feedback mechanism prevents cell *overdraining* when excessive cytoplasmic osmolytes are released during the turgor adjustment to a hypotonic medium (Rowe et al. 2014).

5.7 CONCLUSIONS, PERSPECTIVES, AND CRITICAL QUESTIONS

The common principle that the mechanotransducer molecule should comply with the stimulus and change its dimension in the direction of the applied force holds for the mechanically activated channels. Although the fraction of the external macroscopic stimulus reaching the channel is not always known, the accumulation of structural details about specific channels and the use of single-molecule techniques in the future will pinpoint the locus of force application, the magnitude required for gating, and the scales of structural rearrangements for each particular case. This will help us in understanding the basic mechanisms underlying the normal physiology and provide critical insights into acquired and inherited cases of deafness (Steel and Kros 2001), channel-related neurodegeneration (Goodman et al. 2002), progression of muscular dystrophy (Lansman and Franco 1991), triggering of cardiac arrhythmias and fibrillation (Kohl et al. 2011), neurogenic hypertension (Lu et al. 2009), red blood cell fragility (xerocytosis) (Zarychanski et al. 2012), and many other diseases.

The lipid bilayer is a highly anisotropic environment for membrane proteins, especially in the transverse direction. The lateral pressure profile, created jointly by hydrophobic interactions and the conformational dynamics of lipids, is the characteristic that defines the state of an embedded tension-gated channel. Variations of lipid composition and the presence of lipid-soluble messengers such as polyunsaturated fatty acids or general or local anesthetics change this profile and modulate polymodal channels that are activated both chemically and mechanically. This raises the question of what is the primary feature of the sensor channel and what is the by-product. Indeed, primary sensitivity to macroscopic tension may permit modulation by changes in the local distribution of lateral pressure. Conversely, intrinsic sensitivity to specific lipophilic messengers that change pressure/tension locally would also make the channel sensitive to macroscopic tension that may occur naturally (prestress) or is externally applied to the membrane. The solution to this problem will involve the development of probes for membrane tension that could monitor this parameter in live cells at rest or under applied external stress.

The occurrence of certain types of MS channels in different clades of organisms seems to follow an ancient separation according to the way the organisms handle ubiquitous osmotic forces. Prokaryotes covered by an elastic peptidoglycan wall evolved MscS to regulate internal turgor and hydration of the cytoplasm and MscL as an emergency valve to prevent osmotic lysis. Plants surrounded by a more rigid cell wall inherited and diversified MscS, which is now present in all organisms that use turgor pressure for maintenance of shape and rigidity. To take advantage of cellular contractility, evolving animals abandoned the external wall and developed a powerful and dynamic internal cytoskeleton that restrains osmotic swelling through attachments to a highly folded membrane, which now uses a large area excess to provide freedom for shape changes and volume adjustments. In the absence of strong perturbations, global tension in the animal cell membrane is presumed to be very low, yet cell contractility and attachments to the substrate can generate membrane stresses locally. Stretch-sensitive K2Ps, ENaCs, TRPs, and Piezos have evolved to function in the pleated and scaffolded plasma membrane of animal cells to monitor local changes in tension. When the cell needs to increase its sensitivity to external mechanical cues, it places MS channels in special processes and links them to the cytoskeleton to collect forces from larger areas. The channels from K2P, DEG/ENaC, or TRP families, shown or presumed to interact with the cytoskeleton, are not found in higher plants; however, a single Piezo homolog is present in their genomes. This suggests that Piezo channels are *platform independent* as they probably sense membrane tension directly through their gigantic protein–lipid interface and may not use stress-gathering attachments. This special *niche* in the sensory system may define physiological roles of Piezos.

Recordings made on individual MS channels in reconstituted or in heterologously expressed systems are most informative in terms of the mechanistic understanding of individual components. *In vivo* experiments, however, document responses of the native system as a whole. It is remarkable that genetic knockout or knockdown techniques rarely show an all-or-none functional effect attributable to the single protein of interest. Instead, experiments often reveal multiple components with overlapping ranges of sensitivity, partial or complete redundancy, or indicate hybrid assemblies made of different subunits producing new features. This suggests that the principles of evolvability (Kirschner and Gerhart 1998) fully apply to the evolution of sensory systems, which has picked and perfected several conserved designs of MS channels, diversified them, and mixed them in different contexts where a combination works better than any individual type. Matching various channels with different types of scaffolding may further change the activation threshold, gain, and adaptive capacity. Comparisons of functionally similar designs (hair cells vs. ciliated endothelial cells or sensory flagella in protozoa) may be beneficial in searches for similar components, adaptor domains, and common physical principles.

ACKNOWLEDGMENTS

The authors thank David Corey and Elizabeth Haswell for their critical comments.

REFERENCES

Abu Shah, E. and K. Keren (2013). Mechanical forces and feedbacks in cell motility. *Curr Opin Cell Biol* 25(5): 550–557.

Adams, C. M., M. G. Anderson, D. G. Motto, M. P. Price, W. A. Johnson, and M. J. Welsh (1998). Ripped pocket and pickpocket, novel Drosophila DEG/ENaC subunits expressed in early development and in mechanosensory neurons. *J Cell Biol* 140(1): 143–152.

Ajouz, B., C. Berrier, M. Besnard, B. Martinac, and A. Ghazi (2000). Contributions of the different extramembranous domains of the mechanosensitive ion channel MscL to its response to membrane tension. *J Biol Chem* 275(2): 1015–1022.

Akitake, B., A. Anishkin, N. Liu, and S. Sukharev (2007). Straightening and sequential buckling of the pore-lining helices define the gating cycle of MscS. *Nat Struct Mol Biol* 14(12): 1141–1149.

Akitake, B., A. Anishkin, and S. Sukharev (2005). The "Dashpot" mechanism of stretch-dependent gating in MscS. *J Gen Physiol* 125(2): 143–154.

Alloui, A., K. Zimmermann, J. Mamet, F. Duprat, J. Noel, J. Chemin, N. Guy et al. (2006). TREK-1, a K^+ channel involved in polymodal pain perception. *EMBO J* 25(11): 2368–2376.

Anishkin, A., B. Akitake, K. Kamaraju, C. S. Chiang, and S. Sukharev (2010). Hydration properties of mechanosensitive channel pores define the energetics of gating. *J Phys Condens Matter* 22(45): 454120.

Anishkin, A., B. Akitake, and S. Sukharev (2008). Characterization of the resting MscS: Modeling and analysis of the closed bacterial mechanosensitive channel of small conductance. *Biophys J* 94(4): 1252–1266.

Anishkin, A., C. S. Chiang, and S. Sukharev (2005). Gain-of-function mutations reveal expanded intermediate states and a sequential action of two gates in MscL. *J Gen Physiol* 125(2): 155–170.

Anishkin, A., K. Kamaraju, and S. Sukharev (2008). Mechanosensitive channel MscS in the open state: Modeling of the transition, explicit simulations, and experimental measurements of conductance. *J Gen Physiol* 132(1): 67–83.

Anishkin, A. and C. Kung (2013). Stiffened lipid platforms at molecular force foci. *Proc Natl Acad Sci USA* 110(13): 4886–4892.

Anishkin, A. and S. Sukharev (2004). Water dynamics and dewetting transitions in the small mechanosensitive channel MscS. *Biophys J* 86(5): 2883–2895.

Arnadottir, J. and M. Chalfie (2010). Eukaryotic mechanosensitive channels. *Annu Rev Biophys* 39: 111–137.

Arnadottir, J., R. O'Hagan, Y. Chen, M. B. Goodman, and M. Chalfie (2011). The DEG/ENaC protein MEC-10 regulates the transduction channel complex in *Caenorhabditis elegans* touch receptor neurons. *J Neurosci* 31(35): 12695–12704.

Assad, J. A., G. M. Shepherd, and D. P. Corey (1991). Tip-link integrity and mechanical transduction in vertebrate hair cells. *Neuron* 7(6): 985–994.

Bae, C., F. Sachs, and P. A. Gottlieb (2011). The mechanosensitive ion channel Piezo1 is inhibited by the peptide GsMTx4. *Biochemistry* 50(29): 6295–6300.

Bagriantsev, S. N., R. Peyronnet, K. A. Clark, E. Honore, and D. L. Minor, Jr. (2011). Multiple modalities converge on a common gate to control K2P channel function. *EMBO J* 30(17): 3594–3606.

Balleza, D. and F. Gomez-Lagunas (2009). Conserved motifs in mechanosensitive channels MscL and MscS. *Eur Biophys J* 38(7): 1013–1027.

Bass, R. B., P. Strop, M. Barclay, and D. C. Rees (2002). Crystal structure of Escherichia coli MscS, a voltage-modulated and mechanosensitive channel. *Science* 298(5598): 1582–1587.

Bautista, D. M., S. E. Jordt, T. Nikai, P. R. Tsuruda, A. J. Read, J. Poblete, E. N. Yamoah, A. I. Basbaum, and D. Julius (2006). TRPA1 mediates the inflammatory actions of environmental irritants and proalgesic agents. *Cell* 124(6): 1269–1282.

Belyy, V., A. Anishkin, K. Kamaraju, N. Liu, and S. Sukharev (2010a). The tension-transmitting 'clutch' in the mechanosensitive channel MscS. *Nat Struct Mol Biol* 17(4): 451-458.

Belyy, V., K. Kamaraju, B. Akitake, A. Anishkin, and S. Sukharev (2010b). Adaptive behavior of bacterial mechanosensitive channels is coupled to membrane mechanics. *J Gen Physiol* 135(6): 641–652.

Berrier, C., M. Besnard, B. Ajouz, A. Coulombe, and A. Ghazi (1996). Multiple mechanosensitive ion channels from *Escherichia coli*, activated at different thresholds of applied pressure. *J Membr Biol* 151(2): 175–187.

Berrier, C., A. Pozza, A. de Lacroix de Lavalette, S. Chardonnet, A. Mesneau, C. Jaxel, M. le Maire, and A. Ghazi (2013). The purified mechanosensitive channel TREK-1 is directly sensitive to membrane tension. *J Biol Chem* 288(38): 27307–27314.

Betanzos, M., C. S. Chiang, H. R. Guy, and S. Sukharev (2002). A large iris-like expansion of a mechanosensitive channel protein induced by membrane tension. *Nat Struct Biol* 9(9): 704–710.

Beurg, M., R. Fettiplace, J. H. Nam, and A. J. Ricci (2009). Localization of inner hair cell mechanotransducer channels using high-speed calcium imaging. *Nat Neurosci* 12(5): 553–558.

Blount, P., S. I. Sukharev, P. C. Moe, M. J. Schroeder, H. R. Guy, and C. Kung (1996). Membrane topology and multimeric structure of a mechanosensitive channel protein of *Escherichia coli. EMBO J* 15(18): 4798–4805.

Boer, M., A. Anishkin, and S. Sukharev (2011). Adaptive MscS gating in the osmotic permeability response in *E. coli*: The question of time. *Biochemistry* 50(19): 4087–4096.

Bounoutas, A., R. O'Hagan, and M. Chalfie (2009). The multipurpose 15-protofilament microtubules in *C. elegans* have specific roles in mechanosensation. *Curr Biol* 19(16): 1362–1367.

Brohawn, S. G., E. B. Campbell, and R. MacKinnon (2013). Domain-swapped chain connectivity and gated membrane access in a Fab-mediated crystal of the human TRAAK K^+ channel. *Proc Natl Acad Sci USA* 110(6): 2129–2134.

Brohawn, S. G., J. del Marmol, and R. MacKinnon (2012). Crystal structure of the human K2P TRAAK, a lipid- and mechano-sensitive K^+ ion channel. *Science* 335(6067): 436–441.

Brohawn, S. G., Z. Su, and R. MacKinnon (2014). Mechanosensitivity is mediated directly by the lipid membrane in TRAAK and TREK1 K^+ channels. *Proc Natl Acad Sci USA* 111(9): 3614–3619.

Chalfie, M. and M. Au (1989). Genetic control of differentiation of the *Caenorhabditis elegans* touch receptor neurons. *Science* 243(4894 Pt 1): 1027–1033.

Chalfie, M. and J. Sulston (1981). Developmental genetics of the mechanosensory neurons of *Caenorhabditis elegans. Dev Biol* 82(2): 358–370.

Chalfie, M. and J. N. Thomson (1982). Structural and functional diversity in the neuronal microtubules of *Caenorhabditis elegans. J Cell Biol* 93(1): 15–23.

Chang, G., R. H. Spencer, A. T. Lee, M. T. Barclay, and D. C. Rees (1998). Structure of the MscL homolog from *Mycobacterium tuberculosis*: A gated mechanosensitive ion channel. *Science* 282(5397): 2220–2226.

Cheung, E. L. and D. P. Corey (2006). Ca^{2+} changes the force sensitivity of the hair-cell transduction channel. *Biophys J* 90(1): 124–139.

Chiang, C. S., A. Anishkin, and S. Sukharev (2004). Gating of the large mechanosensitive channel in situ: Estimation of the spatial scale of the transition from channel population responses. *Biophys J* 86(5): 2846–2861.

Cohen, A., Y. Ben-Abu, S. Hen, and N. Zilberberg (2008). A novel mechanism for human K2P2.1 channel gating. Facilitation of C-type gating by protonation of extracellular histidine residues. *J Biol Chem* 283(28): 19448–19455.

Corey, D. P., J. Garcia-Anoveros, J. R. Holt, K. Y. Kwan, S. Y. Lin, M. A. Vollrath, A. Amalfitano et al. (2004). TRPA1 is a candidate for the mechanosensitive transduction channel of vertebrate hair cells. *Nature* 432(7018): 723–730.

Corey, D. P. and A. J. Hudspeth (1979). Ionic basis of the receptor potential in a vertebrate hair cell. *Nature* 281(5733): 675–677.

Corey, D. P. and A. J. Hudspeth (1983). Kinetics of the receptor current in bullfrog saccular hair cells. *J Neurosci* 3(5): 962–976.

Corry, B., A. C. Hurst, P. Pal, T. Nomura, P. Rigby, and B. Martinac (2010). An improved open-channel structure of MscL determined from FRET confocal microscopy and simulation. *J Gen Physiol* 136(4): 483–494.

Corry, B., P. Rigby, Z. W. Liu, and B. Martinac (2005). Conformational changes involved in MscL channel gating measured using FRET spectroscopy. *Biophys J* 89(6): L49–L51.

Cosgrove, D. J. and R. Hedrich (1991). Stretch-activated chloride, potassium, and calcium channels coexisting in plasma membranes of guard cells of *Vicia faba* L. *Planta* 186(1): 143–153.

Coste, B., J. Mathur, M. Schmidt, T. J. Earley, S. Ranade, M. J. Petrus, A. E. Dubin, and A. Patapoutian (2010). Piezo1 and Piezo2 are essential components of distinct mechanically activated cation channels. *Science* 330(6000): 55–60.

Coste, B., B. Xiao, J. S. Santos, R. Syeda, J. Grandl, K. S. Spencer, S. E. Kim et al. (2012). Piezo proteins are pore-forming subunits of mechanically activated channels. *Nature* 483(7388): 176–181.

Davis, H. (1965). A model for transducer action in the cochlea. *Cold Spring Harb Symp Quant Biol* 30: 181–190.

del Rio, A., R. Perez-Jimenez, R. Liu, P. Roca-Cusachs, J. M. Fernandez, and M. P. Sheetz (2009). Stretching single talin rod molecules activates vinculin binding. *Science* 323(5914): 638–641.

Delmas, P., J. Hao, and L. Rodat-Despoix (2011). Molecular mechanisms of mechanotransduction in mammalian sensory neurons. *Nat Rev Neurosci* 12(3): 139–153.

Denk, W., J. R. Holt, G. M. Shepherd, and D. P. Corey (1995). Calcium imaging of single stereocilia in hair cells: Localization of transduction channels at both ends of tip links. *Neuron* 15(6): 1311–1321.

Ding, J. P. and B. G. Pickard (1993). Mechanosensory calcium-selective cation channels in epidermal cells. *Plant J* 3(1): 83–110.

Drummond, H. A., F. M. Abboud, and M. J. Welsh (2000). Localization of beta and gamma subunits of ENaC in sensory nerve endings in the rat foot pad. *Brain Res* 884(1–2): 1–12.

Drummond, H. A., M. J. Welsh, and F. M. Abboud (2001). ENaC subunits are molecular components of the arterial baroreceptor complex. *Ann N Y Acad Sci* 940: 42–47.

Edwards, M. D., S. Black, T. Rasmussen, A. Rasmussen, N. R. Stokes, T. L. Stephen, S. Miller, and I. R. Booth (2012). Characterization of three novel mechanosensitive channel activities in *Escherichia coli. Channels (Austin)* 6(4): 272–281.

Farris, H. E., C. L. LeBlanc, J. Goswami, and A. J. Ricci (2004). Probing the pore of the auditory hair cell mechanotransducer channel in turtle. *J Physiol* 558(Pt 3): 769–792.

Fuentes, D. E. and P. J. Butler (2012). Coordinated mechanosensitivity of membrane rafts and focal adhesions. *Cell Mol Bioeng* 5(2): 143–154.

Fujiu, K., Y. Nakayama, H. Iida, M. Sokabe, and K. Yoshimura (2011). Mechanoreception in motile flagella of *Chlamydomonas. Nat Cell Biol* 13(5): 630–632.

Garcia-Anoveros, J., T. A. Samad, L. Zuvela-Jelaska, C. J. Woolf, and D. P. Corey (2001). Transport and localization of the DEG/ENaC ion channel BNaC1alpha to peripheral mechanosensory terminals of dorsal root ganglia neurons. *J Neurosci* 21(8): 2678–2686.
Gees, M., G. Owsianik, B. Nilius, and T. Voets (2012). TRP channels. *Compr Physiol* 2(1): 563–608.
Geffeney, S. L. and M. B. Goodman (2012). How we feel: Ion channel partnerships that detect mechanical inputs and give rise to touch and pain perception. *Neuron* 74(4): 609–619.
Gillespie, P. G. and J. L. Cyr (2004). Myosin-1c, the hair cell's adaptation motor. *Annu Rev Physiol* 66: 521–545.
Gonzalez, C., D. Baez-Nieto, I. Valencia, I. Oyarzun, P. Rojas, D. Naranjo, and R. Latorre (2012). K^+ channels: Function–structural overview. *Compr Physiol* 2(3): 2087–2149.
Goodman, M. B., G. G. Ernstrom, D. S. Chelur, R. O'Hagan, C. A. Yao, and M. Chalfie (2002). MEC-2 regulates *C. elegans* DEG/ENaC channels needed for mechanosensation. *Nature* 415(6875): 1039–1042.
Goodman, M. B. and E. M. Schwarz (2003). Transducing touch in *Caenorhabditis elegans*. *Annu Rev Physiol* 65: 429–452.
Gottlieb, P. A., C. Bae, and F. Sachs (2012). Gating the mechanical channel Piezo1: A comparison between whole-cell and patch recording. *Channels (Austin)* 6(4): 282–289.
Grajkowski, W., A. Kubalski, and P. Koprowski (2005). Surface changes of the mechanosensitive channel MscS upon its activation, inactivation, and closing. *Biophys J* 88(4): 3050–3059.
Guharay, F. and F. Sachs (1984). Stretch-activated single ion channel currents in tissue-cultured embryonic chick skeletal muscle. *J Physiol* 352: 685–701.
Gullingsrud, J. and K. Schulten (2004). Lipid bilayer pressure profiles and mechanosensitive channel gating. *Biophys J* 86(6): 3496–3509.
Gustin, M. C., X. L. Zhou, B. Martinac, and C. Kung (1988). A mechanosensitive ion channel in the yeast plasma membrane. *Science* 242(4879): 762–765.
Hackney, C. M. and D. N. Furness (2013). The composition and role of cross links in mechanoelectrical transduction in vertebrate sensory hair cells. *J Cell Sci* 126(Pt 8): 1721–1731.
Hamill, O. P., J. W. Lane, and D. W. McBride, Jr. (1992). Amiloride: A molecular probe for mechanosensitive channels. *Trends Pharmacol Sci* 13(10): 373–376.
Hamill, O. P. and B. Martinac (2001). Molecular basis of mechanotransduction in living cells. *Physiol Rev* 81(2): 685–740.
Hamill, O. P. and D. W. McBride, Jr. (1996). The pharmacology of mechanogated membrane ion channels. *Pharmacol Rev.* 48(2): 231–252.
Hase, C. C., A. C. Le Dain, and B. Martinac (1995). Purification and functional reconstitution of the recombinant large mechanosensitive ion channel (MscL) of *Escherichia coli*. *J Biol Chem* 270(31): 18329–18334.
Haswell, E. S. (2007). MscS-like proteins in plants. *Mechanosensitive Ion Channels, Part A* 58: 329–359.
Haswell, E. S. and E. M. Meyerowitz (2006). MscS-like proteins control plastid size and shape in Arabidopsis thaliana. *Curr Biol* 16(1): 1–11.
Haswell, E. S., R. Peyronnet, H. Barbier-Brygoo, E. M. Meyerowitz, and J. M. Frachisse (2008). Two MscS homologs provide mechanosensitive channel activities in the *Arabidopsis* root. *Curr Biol* 18(10): 730–734.
Hille, B. (2001). *Ion Channels of Excitable Membranes*. Sinauer, Sunderland, MA.
Honore, E. (2007). The neuronal background K2P channels: Focus on TREK1. *Nat Rev Neurosci* 8(4): 251–261.
Honore, E., A. J. Patel, J. Chemin, T. Suchyna, and F. Sachs (2006). Desensitization of mechano-gated K2P channels. *Proc Natl Acad Sci USA* 103(18): 6859–6864.
Howard, J. and A. J. Hudspeth (1988). Compliance of the hair bundle associated with gating of mechanoelectrical transduction channels in the bullfrog's saccular hair cell. *Neuron* 1(3): 189–199.
Howard, J., W. M. Roberts, and A. J. Hudspeth (1988). Mechanoelectrical transduction by hair cells. *Annu Rev Biophys Biophys Chem* 17: 99–124.
Huang, M. and M. Chalfie (1994). Gene interactions affecting mechanosensory transduction in *Caenorhabditis elegans*. *Nature* 367(6462): 467–470.
Kachar, B., M. Parakkal, M. Kurc, Y. Zhao, and P. G. Gillespie (2000). High-resolution structure of hair-cell tip links. *Proc Natl Acad Sci USA* 97(24): 13336–13341.
Kamaraju, K., V. Belyy, I. Rowe, A. Anishkin, and S. Sukharev (2011). The pathway and spatial scale for MscS inactivation. *J Gen Physiol* 138(1): 49–57.
Kang, L., J. Gao, W. R. Schafer, Z. Xie, and X. Z. Xu (2010). *C. elegans* TRP family protein TRP-4 is a pore-forming subunit of a native mechanotransduction channel. *Neuron* 67(3): 381–391.
Kawashima, Y., G. S. Geleoc, K. Kurima, V. Labay, A. Lelli, Y. Asai, T. Makishima et al. (2011). Mechanotransduction in mouse inner ear hair cells requires transmembrane channel-like genes. *J Clin Invest* 121(12): 4796–4809.
Ketchum, K. A., W. J. Joiner, A. J. Sellers, L. K. Kaczmarek, and S. A. Goldstein (1995). A new family of outwardly rectifying potassium channel proteins with two pore domains in tandem. *Nature* 376(6542): 690–695.
Kim, E. K., S. A. Wellnitz, S. M. Bourdon, E. A. Lumpkin, and G. J. Gerling (2012a). Force sensor in simulated skin and neural model mimic tactile SAI afferent spiking response to ramp and hold stimuli. *J Neuroeng Rehabil* 9: 45.
Kim, K. X., M. Beurg, C. M. Hackney, D. N. Furness, S. Mahendrasingam, and R. Fettiplace (2013). The role of transmembrane channel-like proteins in the operation of hair cell mechanotransducer channels. *J Gen Physiol* 142(5): 493–505.
Kim, S. E., B. Coste, A. Chadha, B. Cook, and A. Patapoutian (2012b). The role of *Drosophila* Piezo in mechanical nociception. *Nature* 483(7388): 209–212.
Kirschner, M. and J. Gerhart (1998). Evolvability. *Proc Natl Acad Sci USA* 95(15): 8420–8427.
Kocer, A., M. Walko, W. Meijberg, and B. L. Feringa (2005). A light-actuated nanovalve derived from a channel protein. *Science* 309(5735): 755–758.
Kohl, P., F. Sachs, M. R. Franz, and P. Kohl (2011). *Cardiac Mechano-Electric Coupling and Arrhythmias*. Oxford, U.K.: Oxford University Press.
Kroese, A. B., A. Das, and A. J. Hudspeth (1989). Blockage of the transduction channels of hair cells in the bullfrog's sacculus by aminoglycoside antibiotics. *Hear Res* 37(3): 203–217.
Kurusu, T., D. Nishikawa, Y. Yamazaki, M. Gotoh, M. Nakano, H. Hamada, T. Yamanaka et al. (2012). Plasma membrane protein OsMCA1 is involved in regulation of hypo-osmotic shock-induced Ca^{2+} influx and modulates generation of reactive oxygen species in cultured rice cells. *BMC Plant Biol* 12: 11.
Lai, C. C., K. Hong, M. Kinnell, M. Chalfie, and M. Driscoll (1996). Sequence and transmembrane topology of MEC-4, an ion channel subunit required for mechanotransduction in *Caenorhabditis elegans*. *J Cell Biol* 133(5): 1071–1081.
Lansman, J. B. and A. Franco, Jr. (1991). What does dystrophin do in normal muscle? *J Muscle Res Cell Motil* 12(5): 409–411.
Lesage, F., E. Guillemare, M. Fink, F. Duprat, M. Lazdunski, G. Romey, and J. Barhanin (1996). TWIK-1, a ubiquitous human weakly inward rectifying K^+ channel with a novel structure. *EMBO J* 15(5): 1004–1011.
Lesage, F., C. Terrenoire, G. Romey, and M. Lazdunski (2000). Human TREK2, a 2P domain mechano-sensitive K^+ channel with multiple regulations by polyunsaturated fatty acids, lysophospholipids, and Gs, Gi, and Gq protein-coupled receptors. *J Biol Chem* 275(37): 28398–28405.
Lesniak, D. R., S. A. Wellnitz, G. J. Gerling, and E. A. Lumpkin (2009). Statistical analysis and modeling of variance in the SA-I mechanoreceptor response to sustained indentation. *Conf Proc IEEE Eng Med Biol Soc* 2009: 6814–6817.

Levina, N., S. Totemeyer, N. R. Stokes, P. Louis, M. A. Jones, and I. R. Booth (1999). Protection of *Escherichia coli* cells against extreme turgor by activation of MscS and MscL mechanosensitive channels: Identification of genes required for MscS activity. *EMBO J* 18(7): 1730–1737.

Li, Y., P. C. Moe, S. Chandrasekaran, I. R. Booth, and P. Blount (2002). Ionic regulation of MscK, a mechanosensitive channel from *Escherichia coli*. *EMBO J* 21(20): 5323–5330.

Liang, X., J. Madrid, R. Gartner, J. M. Verbavatz, C. Schiklenk, M. Wilsch-Brauninger, A. Bogdanova, F. Stenger, A. Voigt, and J. Howard (2013). A NOMPC-dependent membrane-microtubule connector is a candidate for the gating spring in fly mechanoreceptors. *Curr Biol* 23(9): 755–763.

Liedtke, W., Y. Choe, M. A. Marti-Renom, A. M. Bell, C. S. Denis, A. Sali, A. J. Hudspeth, J. M. Friedman, and S. Heller (2000). Vanilloid receptor-related osmotically activated channel (VR-OAC), a candidate vertebrate osmoreceptor. *Cell* 103(3): 525–535.

Longo-Guess, C. M., L. H. Gagnon, S. A. Cook, J. Wu, Q. Y. Zheng, and K. R. Johnson (2005). A missense mutation in the previously undescribed gene Tmhs underlies deafness in hurry-scurry (hscy) mice. *Proc Natl Acad Sci USA* 102(22): 7894–7899.

Loukin, S., X. Zhou, Z. Su, Y. Saimi, and C. Kung (2010). Wild-type and brachyolmia-causing mutant TRPV4 channels respond directly to stretch force. *J Biol Chem* 285(35): 27176–27181.

Lowenstein, O. and J. Wersall (1959). Functional interpretation of the electron-microscopic structure of the sensory hairs in the cristae of the Elasmobranch Raja-Clavata in terms of directional sensitivity. *Nature* 184(4701): 1807–1808.

Lu, Y., X. Ma, R. Sabharwal, V. Snitsarev, D. Morgan, K. Rahmouni, H. A. Drummond et al. (2009). The ion channel ASIC2 is required for baroreceptor and autonomic control of the circulation. *Neuron* 64(6): 885–897.

Lumpkin, E. A., K. L. Marshall, and A. M. Nelson (2010). The cell biology of touch. *J Cell Biol.* 191(2): 237–248.

Maingret, F., M. Fosset, F. Lesage, M. Lazdunski, and E. Honore (1999). TRAAK is a mammalian neuronal mechano-gated K^+ channel. *J Biol Chem* 274(3): 1381–1387.

Maksaev, G., A. Milac, A. Anishkin, H. R. Guy, and S. Sukharev (2011). Analyses of gating thermodynamics and effects of deletions in the mechanosensitive channel TREK-1: Comparisons with structural models. *Channels (Austin)* 5(1): 34–42.

Markin, V. S. and A. J. Hudspeth (1995). Gating-spring models of mechanoelectrical transduction by hair cells of the internal ear. *Annu Rev Biophys Biomol Struct* 24: 59–83.

Markin, V. S. and F. Sachs (2004). Thermodynamics of mechanosensitivity. *Phys Biol* 1(1–2): 110–124.

Martinac, B., M. Buechner, A. H. Delcour, J. Adler, and C. Kung (1987). Pressure-sensitive ion channel in *Escherichia coli*. *Proc Natl Acad Sci USA* 84(8): 2297–2301.

McCarter, G. C., D. B. Reichling, and J. D. Levine (1999). Mechanical transduction by rat dorsal root ganglion neurons in vitro. *Neurosci Lett* 273(3): 179–182.

Meyers, J. R., R. B. MacDonald, A. Duggan, D. Lenzi, D. G. Standaert, J. T. Corwin, and D. P. Corey (2003). Lighting up the senses: FM1-43 loading of sensory cells through nonselective ion channels. *J Neurosci* 23(10): 4054–4065.

Moe, P. and P. Blount (2005). Assessment of potential stimuli for mechano-dependent gating of MscL: Effects of pressure, tension, and lipid headgroups. *Biochemistry* 44(36): 12239–12244.

Monshausen, G. B. and E. S. Haswell (2013). A force of nature: Molecular mechanisms of mechanoperception in plants. *J Exp Bot* 64(15): 4663–4680.

Morris, C. E. (2001). Mechanoprotection of the plasma membrane in neurons and other non-erythroid cells by the spectrin-based membrane skeleton. *Cell Mol Biol Lett* 6(3): 703–720.

Morris, C. E. (2011). Voltage-gated channel mechanosensitivity: Fact or friction? *Front Physiol* 2: 25.

Naismith, J. H. and I. R. Booth (2012). Bacterial mechanosensitive channels—MscS: Evolution's solution to creating sensitivity in function. *Annu Rev Biophys* 41: 157–177.

Nakagawa, Y., T. Katagiri, K. Shinozaki, Z. Qi, H. Tatsumi, T. Furuichi, A. Kishigami et al. (2007). *Arabidopsis* plasma membrane protein crucial for Ca^{2+} influx and touch sensing in roots. *Proc Natl Acad Sci USA* 104(9): 3639–3644.

Nakayama, Y., K. Yoshimura, and H. Iida (2012). Organellar mechanosensitive channels in fission yeast regulate the hypo-osmotic shock response. *Nat Commun* 3: 1020.

Nauli, S. M., F. J. Alenghat, Y. Luo, E. Williams, P. Vassilev, X. Li, A. E. Elia et al. (2003). Polycystins 1 and 2 mediate mechanosensation in the primary cilium of kidney cells. *Nat Genet* 33(2): 129–137.

Noel, J., K. Zimmermann, J. Busserolles, E. Deval, A. Alloui, S. Diochot, N. Guy et al. (2009). The mechano-activated K^+ channels TRAAK and TREK-1 control both warm and cold perception. *EMBO J* 28(9): 1308–1318.

O'Hagan, R., M. Chalfie, and M. B. Goodman (2005). The MEC-4 DEG/ENaC channel of *Caenorhabditis elegans* touch receptor neurons transduces mechanical signals. *Nat Neurosci* 8(1): 43–50.

Ollila, O. H., H. J. Risselada, M. Louhivuori, E. Lindahl, I. Vattulainen, and S. J. Marrink (2009). 3D pressure field in lipid membranes and membrane-protein complexes. *Phys Rev Lett* 102(7): 078101.

Pan, B., G. S. Geleoc, Y. Asai, G. C. Horwitz, K. Kurima, K. Ishikawa, Y. Kawashima, A. J. Griffith, and J. R. Holt (2013). TMC1 and TMC2 are components of the mechanotransduction channel in hair cells of the mammalian inner ear. *Neuron* 79(3): 504–515.

Paoletti, P. and P. Ascher (1994). Mechanosensitivity of NMDA receptors in cultured mouse central neurons. *Neuron* 13(3): 645–655.

Patel, A. J., E. Honore, F. Maingret, F. Lesage, M. Fink, F. Duprat, and M. Lazdunski (1998). A mammalian two pore domain mechano-gated S-like K^+ channel. *EMBO J* 17(15): 4283–4290.

Peng, A. W., F. T. Salles, B. Pan, and A. J. Ricci (2011). Integrating the biophysical and molecular mechanisms of auditory hair cell mechanotransduction. *Nat Commun* 2: 523.

Perozo, E., D. M. Cortes, P. Sompornpisut, A. Kloda, and B. Martinac (2002a). Open channel structure of MscL and the gating mechanism of mechanosensitive channels. *Nature* 418(6901): 942–948.

Perozo, E., A. Kloda, D. M. Cortes, and B. Martinac (2002b). Physical principles underlying the transduction of bilayer deformation forces during mechanosensitive channel gating. *Nat Struct Biol* 9(9): 696–703.

Phillips, R., T. Ursell, P. Wiggins, and P. Sens (2009). Emerging roles for lipids in shaping membrane-protein function. *Nature* 459(7245): 379–385.

Pivetti, C. D., M. R. Yen, S. Miller, W. Busch, Y. H. Tseng, I. R. Booth, and M. H. Saier, Jr. (2003). Two families of mechanosensitive channel proteins. *Microbiol Mol Biol Rev* 67(1): 66–85, table.

Powers, R. J., S. Kulason, E. Atilgan, W. E. Brownell, S. X. Sun, P. G. Barr-Gillespie, and A. A. Spector (2014). The local forces acting on the mechanotransduction channel in hair cell stereocilia. *Biophys J* 106(11): 2519–2528.

Powers, R. J., S. Roy, E. Atilgan, W. E. Brownell, S. X. Sun, P. G. Gillespie, and A. A. Spector (2012). Stereocilia membrane deformation: Implications for the gating spring and mechanotransduction channel. *Biophys J* 102(2): 201–210.

Qi, Z., A. Kishigami, Y. Nakagawa, H. Iida, and M. Sokabe (2004). A mechanosensitive anion channel in *Arabidopsis thaliana* mesophyll cells. *Plant Cell Physiol* 45(11): 1704–1708.

Ricci, A. J., A. C. Crawford, and R. Fettiplace (2003). Tonotopic variation in the conductance of the hair cell mechanotransducer channel. *Neuron* 40(5): 983–990.

Rowe, I., A. Anishkin, K. Kamaraju, K. Yoshimura, and S. Sukharev (2014). The cytoplasmic cage domain of the mechanosensitive channel MscS is a sensor of macromolecular crowding. *J Gen Physiol* 143(5): 543–557.

Sachs, F. (2010). Stretch-activated ion channels: What are they? *Physiology (Bethesda)* 25(1): 50–56.

Sachs, F. and H. Lecar (1991). Stochastic models for mechanical transduction. *Biophys J* 59(5): 1143–1145.

Schermer, B. and T. Benzing (2009). Lipid-protein interactions along the slit diaphragm of podocytes. *J Am Soc Nephrol* 20(3): 473–478.

Schmidt, D., J. del Marmol, and R. MacKinnon (2012). Mechanistic basis for low threshold mechanosensitivity in voltage-dependent K^+ channels. *Proc Natl Acad Sci USA* 109(26): 10352–10357.

Schumann, U., M. D. Edwards, T. Rasmussen, W. Bartlett, P. van West, and I. R. Booth (2010). YbdG in *Escherichia coli* is a threshold-setting mechanosensitive channel with MscM activity. *Proc Natl Acad Sci USA* 107(28): 12664–12669.

Sharif-Naeini, R., J. H. Folgering, D. Bichet, F. Duprat, P. Delmas, A. Patel, and E. Honore (2010). Sensing pressure in the cardiovascular system: Gq-coupled mechanoreceptors and TRP channels. *J Mol Cell Cardiol* 48(1): 83–89.

Sharif-Naeini, R., J. H. Folgering, D. Bichet, F. Duprat, I. Lauritzen, M. Arhatte, M. Jodar et al. (2009). Polycystin-1 and -2 dosage regulates pressure sensing. *Cell* 139(3): 587–596.

Sidi, S., R. W. Friedrich, and T. Nicolson (2003). NompC TRP channel required for vertebrate sensory hair cell mechanotransduction. *Science* 301(5629): 96–99.

Sotomayor, M., W. A. Weihofen, R. Gaudet, and D. P. Corey (2012). Structure of a force-conveying cadherin bond essential for inner-ear mechanotransduction. *Nature* 492(7427): 128–132.

Steel, K. P. and C. J. Kros (2001). A genetic approach to understanding auditory function. *Nat Genet* 27(2): 143–149.

Steinbacher, S., R. Bass, P. Strop, and D. C. Rees (2007). Structures of the prokaryotic mechanosensitive channels MscL and MscS. In S. Simon, O. Hamill, and D. Benos (eds.), *Mechanosensitive Ion Channels, Part A*, vol. 58, Academic Press, San Diego, CA, pp. 1–24.

Suchyna, T. M., J. H. Johnson, K. Hamer, J. F. Leykam, D. A. Gage, H. F. Clemo, C. M. Baumgarten, and F. Sachs (2000). Identification of a peptide toxin from *Grammostola spatulata* spider venom that blocks cation-selective stretch-activated channels. *J Gen Physiol* 115(5): 583–598.

Suchyna, T. M., V. S. Markin, and F. Sachs (2009). Biophysics and structure of the patch and the gigaseal. *Biophys J* 97(3): 738–747.

Sukharev, S. (2002). Purification of the small mechanosensitive channel of *Escherichia coli* (MscS): The subunit structure, conduction, and gating characteristics in liposomes. *Biophys J* 83(1): 290–298.

Sukharev, S., M. Betanzos, C. S. Chiang, and H. R. Guy (2001). The gating mechanism of the large mechanosensitive channel MscL. *Nature* 409(6821): 720–724.

Sukharev, S. I., P. Blount, B. Martinac, F. R. Blattner, and C. Kung (1994). A large-conductance mechanosensitive channel in *E. coli* encoded by mscL alone. *Nature* 368(6468): 265–268.

Sukharev, S. I., B. Martinac, V. Y. Arshavsky, and C. Kung (1993). Two types of mechanosensitive channels in the *Escherichia coli* cell envelope: Solubilization and functional reconstitution. *Biophys J* 65(1): 177–183.

Sukharev, S. I., W. J. Sigurdson, C. Kung, and F. Sachs (1999). Energetic and spatial parameters for gating of the bacterial large conductance mechanosensitive channel, MscL. *J Gen Physiol* 113(4): 525–540.

Ursell, T., K. C. Huang, E. Peterson, and R. Phillips (2007). Cooperative gating and spatial organization of membrane proteins through elastic interactions. *PLoS Comput Biol* 3(5): e81.

van den Bogaart, G., V. Krasnikov, and B. Poolman (2007). Dual-color fluorescence-burst analysis to probe protein efflux through the mechanosensitive channel MscL. *Biophys J* 92(4): 1233–1240.

Vasquez, V., M. Sotomayor, J. Cordero-Morales, K. Schulten, and E. Perozo (2008). A structural mechanism for MscS gating in lipid bilayers. *Science* 321(5893): 1210–1214.

Veley, K. M., S. Marshburn, C. E. Clure, and E. S. Haswell (2012). Mechanosensitive channels protect plastids from hypoosmotic stress during normal plant growth. *Curr Biol* 22(5): 408–413.

Vogel, V. (2006). Mechanotransduction involving multimodular proteins: Converting force into biochemical signals. *Annu Rev Biophys Biomol Struct* 35: 459–488.

Vollrath, M. A., K. Y. Kwan, and D. P. Corey (2007). The micromachinery of mechanotransduction in hair cells. *Annu Rev Neurosci* 30: 339–365.

von Wichert, G., D. Krndija, H. Schmid, G. von Wichert, G. Haerter, G. Adler, T. Seufferlein, and M. P. Sheetz (2008). Focal adhesion kinase mediates defects in the force-dependent reinforcement of initial integrin-cytoskeleton linkages in metastatic colon cancer cell lines. *Eur J Cell Biol* 87(1): 1–16.

Walker, R. G., A. T. Willingham, and C. S. Zuker (2000). A *Drosophila* mechanosensory transduction channel. *Science* 287(5461): 2229–2234.

Wang, L. C., L. K. Morgan, P. Godakumbura, L. J. Kenney, and G. S. Anand (2012). The inner membrane histidine kinase EnvZ senses osmolality via helix-coil transitions in the cytoplasm. *EMBO J* 31(11): 2648–2659.

Wang, W., S. S. Black, M. D. Edwards, S. Miller, E. L. Morrison, W. Bartlett, C. Dong, J. H. Naismith, and I. R. Booth (2008). The structure of an open form of an *E. coli* mechanosensitive channel at 3.45 A resolution. *Science* 321(5893): 1179–1183.

Ward, R., C. Pliotas, E. Branigan, C. Hacker, A. Rasmussen, G. Hagelueken, I. R. Boothet al. (2014). Probing the structure of the mechanosensitive channel of small conductance in lipid bilayers with pulsed electron-electron double resonance. *Biophys J* 106(4): 834–842.

Woo S.H., S. Ranade, A.D. Weyer, A.E. Dubin, Y. Baba, Z. Qiu, M. Petrus, T. Miyamoto, K. Reddy, E.A. Lumpkin, C.L. Stucky and A. Patapoutian (2014). Piezo2 is required for Merkel-cell mechanotransduction. *Nature* 509(7502): 622–626.

Xiong, W., N. Grillet, H. M. Elledge, T. F. Wagner, B. Zhao, K. R. Johnson, P. Kazmierczak, and U. Muller (2012). TMHS is an integral component of the mechanotransduction machinery of cochlear hair cells. *Cell* 151(6): 1283–1295.

Yang, X. C. and F. Sachs (1989). Block of stretch-activated ion channels in *Xenopus* oocytes by gadolinium and calcium ions. *Science* 243(4894 Pt 1): 1068–1071.

Yau, J. M., J. B. Olenczak, J. F. Dammann, and S. J. Bensmaia (2009). Temporal frequency channels are linked across audition and touch. *Curr Biol* 19(7): 561–566.

Zarychanski, R., V. P. Schulz, B. L. Houston, Y. Maksimova, D. S. Houston, B. Smith, J. Rinehart, and P. G. Gallagher (2012). Mutations in the mechanotransduction protein PIEZO1 are associated with hereditary xerocytosis. *Blood* 120(9): 1908–1915.

Zhang, W., Z. Yan, L. Y. Jan, and Y. N. Jan (2013). Sound response mediated by the TRP channels NOMPC, NANCHUNG, and INACTIVE in chordotonal organs of *Drosophila* larvae. *Proc Natl Acad Sci USA* 110(33): 13612–13617.

Zhang, Y., F. Gao, V. L. Popov, J. W. Wen, and O. P. Hamill (2000). Mechanically gated channel activity in cytoskeleton-deficient plasma membrane blebs and vesicles from *Xenopus* oocytes. *J Physiol* 523 Pt 1: 117–130.

Zhao, H., D. E. Williams, J. B. Shin, B. Brugger, and P. G. Gillespie (2012). Large membrane domains in hair bundles specify spatially constricted radixin activation. *J Neurosci* 32(13): 4600–4609.

Zhou, X. L., A. F. Batiza, S. H. Loukin, C. P. Palmer, C. Kung, and Y. Saimi (2003). The transient receptor potential channel on the yeast vacuole is mechanosensitive. *Proc Natl Acad Sci USA* 100(12): 7105–7110.

Ion channel methods

Patch clamping and single-channel analysis

León D. Islas

Contents

6.1 INTRODUCTION

The patch-clamp recording technique has allowed for the characterization of macroscopic ionic currents, single-channel currents, charge movement, and other electrical signals from diverse cell types and is now a commonplace experimental procedure in a multitude of laboratories (Sigworth, 1986; Colquhoun, 2007). One of the main uses of the patch-clamp technique has been the study of ion channels at the single-molecule level. Recording the activity of a single ion channel protein was achieved in the late 1960s and early 1970s in experiments from bacterial ion channels reconstituted in artificial lipid membranes. It was then recognized, as had been postulated on theoretical grounds, that ion channels behave as discrete two-conductance state systems with stochastic kinetics and that their conductance would be such that single-channel currents would be in the range of picoamperes (Mueller and Rudin, 1963; Ehrenstein et al., 1970, 1974).

Neher and Sakmann first reported single-channel patch-clamp recordings in the late 1970s (Neher and Sakmann, 1976; Neher et al., 1978). Since the publication of these and the seminal papers of Hamill et al. (1981) and Colquhoun and Hawkes (1981, 1982), our understanding of ion channels as molecular machines has increased dramatically, but, although automatic patch clamps and microfabricated and *polydimethylsiloxane* polymer (PDMS)–based pipette-free systems (Fertig et al., 2002; Klemic et al., 2002) are now available, the essential tools of single-channel recording and analysis have not changed dramatically.

As a technique with single-molecule and submillisecond temporal resolutions, the main purpose of single-channel recording is to measure the average kinetic parameters of a single ion channel and, through these, uncover molecular mechanisms and shed light on the biology of these remarkable proteins. The astonishing versatility and power of the patch clamp and single-channel electrophysiology is illustrated by the tens of thousands of papers published since initial development of the patch-clamp technique in the 1970s and 1980s.

Patch clamping originated from the suction electrode techniques that were in use in the early 1970s (Kostyuk et al., 1972, 1975). The crucial event in its development was the use of freshly prepared recording glass pipettes, very clean cell preparations, and the application of suction to the pipette (Sigworth and Neher, 1980). This allowed for the establishment of the gigaohm resistance seal or gigaseal between the pipette and the cell membrane, in which the leak conductance was no longer a determining factor of the noise in the recording, which instead was now limited by the electronics.

6.2 PATCH CLAMPING

6.2.1 ELECTRONICS

All patch-clamp amplifiers make use of an electronic circuit based on a current to voltage converter with very high feedback resistance (Hamill et al., 1981). Some patch clamps use a capacitor instead of a resistance for the feedback loop, with some advantages regarding noise in the electronics, since a capacitor has essentially infinite resistance (Horowitz and Hill, 1989). The main idea is to measure the very small current that flows into the pipette without changing the voltage at which the pipette is clamped, that is, under voltage-clamp conditions. This is accomplished by having an operational amplifier being connected as an inverting amplifier at the input stage (Figure 6.1). Several extra electronic stages are needed for signal conditioning and filtering or capacitance transient compensation and series resistance compensation, which can be used in experiments involving changes in voltage. A full account of patch-clamp electronics can be found in the article by Sigworth (1995). It is important to point out that, as with any experimental technique, artifacts can contaminate several aspects of patch-clamp recording. These are especially important in single-channel recording, since we are trying to record picoampere-magnitude signals in a system that may have a root mean square (RMS) current noise value of half a picoampere. Identifying the sources of these artifacts is as important as all other aspects of the experimental technique.

6.2.2 ESTABLISHING THE GIGASEAL

Obtaining a high resistance seal and recording from a patch-clamped membrane can be achieved as follows. The preparation must be as clean as possible. It is common to use cultured cells, be it in primary culture or cell lines heterologously expressing the channels of interest. Recordings can also be achieved from more complex preparation, such as *Xenopus laevis* oocytes, chronic slices, or even from tissues in vivo (Stuhmer et al., 1987; Edwards et al., 1989; Kitamura et al., 2008). In these cases, some degree of enzymatic treatment is still required in order to obtain clean membranes (Stuhmer, 1992). The patch pipette is then brought into contact with the membrane. It is advisable to allow for steady solution flow out of the pipette by application of gentle positive pressure to its back. By monitoring the electrode resistance, the pressure is released when resistance is approximately twice its initial value.

At this point, gentle suction is applied to the electrode and, in most cases, the electrical resistance, R_s, between the electrode and the cell membrane, will increase to gigaohm values within seconds.

6.2.3 CONFIGURATIONS

Once established, a seal with R_s in the gigaohm range is very stable. Recordings can be obtained in this configuration, which is commonly known as cell-attached or on-cell (Figure 6.2). The versatility of the patch-clamp technique is demonstrated by the ability to detach the piece of membrane attached to the glass electrode without disturbing the gigaseal (Horn and Patlak, 1980). These configurations are cell-free and allow precise control of the composition of the solutions bathing the membrane. If an on-cell patch electrode is slowly retrieved from the cell, the region of membrane under it detaches from the rest of the plasma membrane and remains part of the gigaseal. This configuration is known as inside-out and is useful when access to the intracellular part of the channel is needed, such as when studying intracellular acting blockers or the actions of modulatory substances or activators in the intracellular region of the channels. An example of this is the study of cyclic-nucleotide-gated (CNG) channels, in which the ligand is applied by perfusion methods to the exposed intracellular face of the patch (Benndorf et al., 1999; Sunderman and Zagotta, 1999).

After seal formation, one can apply a pulse of suction or a brief (microsecond duration) high voltage pulse, both of which

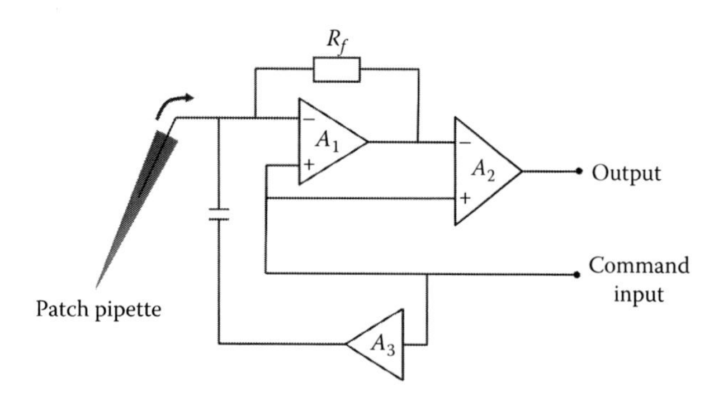

Figure 6.1 Simplified schematic of the main components of a patch-clamp amplifier. The first operational amplifier A_1 takes the pipette current and converts it to a voltage, which is then compared to a reference of command voltage by amplifier A_2. This signal can then be filtered and stored for analysis. The command input voltage sets the desired clamp voltage at the pipette electrode. Signal conditioning, such as capacitance compensation, is achieved by other stages, in this case, illustrated by amplifier A_3, which injects a fraction of the command voltage through a capacitor into the recording pipette.

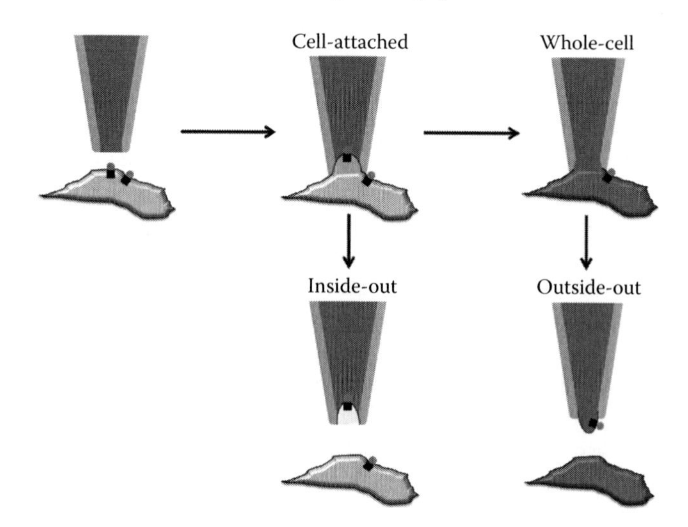

Figure 6.2 Cartoon representation of the recording configurations that are attainable in patch-clamp recording. The remarkable mechanical stability of the membrane and glass interaction accounts for the ability to maintain stable recording in the cell-free configurations. The glass-fabricated recording pipette is shown in light blue and the filling electrolyte in dark blue. A hypothetical channel is depicted with its extracellular domains represented by the circle and the intracellular domain by the square. Notice the opposite orientation of these domains in the inside-out and outside-out configurations.

will result in the rupture of the patch of membrane without disruption of the gigaseal (once again, demonstrating its high mechanical stability). In this configuration, the glass electrode has electrical access to the interior of the cell, making it possible to obtain current or voltage recordings arising from all the plasma membrane. This configuration is termed whole-cell mode. If the pipette is now slowly withdrawn, a membrane patch can be reformed in its tip, but this time, with an inverted orientation. In this new configuration, the patch is called an outside-out patch. With this type of patches, it is possible to very rapidly apply chemicals to the extracellular facing substructures of the channel and study ligand-gated ion channels with extracellular binding sites (Maconochie and Knight, 1989; Colquhoun et al., 1992) or the action of peptide blockers in potassium channels (Goldstein and Miller, 1993).

Patch-clamp recording is an incredibly versatile technique. For example, it can be used to control the intracellular composition of a cell by perfusing its interior in the whole-cell mode with the solution contained in the pipette (Horn and Marty, 1988). Using this configuration, it is possible to study the biophysics of macroscopic ion currents (Fox et al., 1987a), cell signaling and intercellular communication (Pfaffinger et al., 1988), and several other aspects of cellular dynamics, such as secretion or control of cell volume (Neher and Marty, 1982). Macroscopic currents can also be recorded in cell-attached, outside-out, and inside-out configurations, allowing for fast voltage-clamp recordings and providing some of the best characterizations of the kinetics of channels (Zagotta et al., 1994; Schoppa and Sigworth, 1998). However, the initial aim in the development of patch clamp was being able to record the activity of a single ion channel, and it is in this capability that its true uniqueness can be appreciated.

6.2.4 CONDITIONS FOR SINGLE-CHANNEL RECORDING

Single ion channel activity can be assayed when one (active) ion channel is confined within the seal established between a patch electrode and a small area of the plasma membrane of a cell. In order for this to happen, the seal must have an extremely high electrical resistance, $R_s \gg 1\ \text{G}\Omega$, such that a negligible amount of the current flowing through the channel is lost at the seal resistance and most is collected by the electronics. For an R_s of this magnitude, the interaction between the glass and the plasma membrane must occur at the molecular level. Although the exact details are not known, it is thought that strong electrostatic and Van der Waals interactions are responsible, since the pipette seal is mechanically very stable (Sokabe and Sachs, 1990; Suchyna et al., 2009).

Several factors are responsible for the *sealability* of a particular pipette and the quality and stability of the seals obtained. The type of glass used to fabricate patch pipettes is very important, with soft borosilicate glasses being the most widely used (Rae and Levis, 1984). Hard glass, although possessing better noise characteristics than borosilicate, tends to be more reticent to seal formation. The shape of the pipette is an important determinant, not only of the seal characteristics, but also of the electrical behavior of the patch. It is preferable to use short stubby pipettes to long pipettes, because the series resistance of the short pipette is smaller. Pipettes with longer shanks tend to allow more

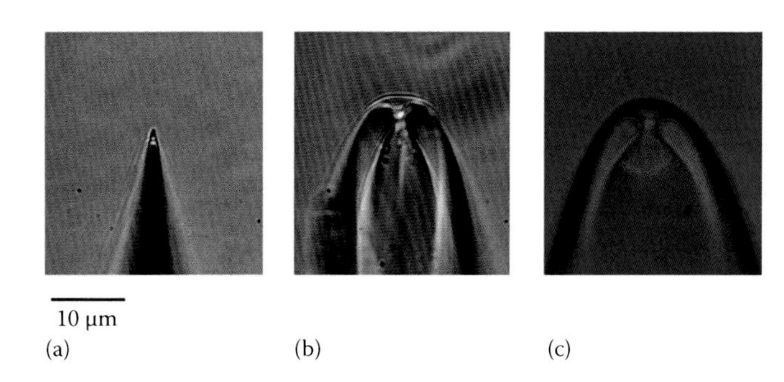

Figure 6.3 Shape of pipettes and shape of the membrane patch in an inside-out patch. (a) Small tip pipette appropriate for single-channel recording. Note the very steep approach to the tip. (b) Large tip pipette that can be used in larger cells such as *Xenopus* oocytes to obtain large membrane patches or macropatches. (c) A similar pipette to (b) with a membrane patch in the tip (inside-out configuration). The membrane contains fluorescently labeled ion channels, which allows direct visualization of the dome shape of the patch. The size scale is the same for all panels. 60× magnification.

membrane to be drawn into the tip. This can be a problem since areas of high series resistance and poor membrane attachment can generate membrane blebs with large voltage errors and the patch can get out of voltage-clamp (Bae et al., 2011). This is avoided in the short pipettes, where the seal is closer to the tip.

In general, the final shape of the pipette tip is determined by the fire-polishing step. After a pipette is pulled from capillary glass, it is brought in close proximity with a red-hot tungsten filament covered with glass. This process eliminates any roughness of the glass and burns off grease and other contaminants. Pipettes can be obtained with a large range of final tip diameters. For single-channel recordings of small patch sizes, small tips of 1 μm or less should be used (Figure 6.3a). To record larger currents and avoid problems with series resistance, pipettes with a large initial opening (10–20 μm) can be pulled and fire polished to a desired diameter between 1 and 10 μm (Figure 6.3b). The shape of the patch at the tip of the pipette is assumed to be that of a dome or the Greek letter Ω in cross section (Hamill et al., 1981; Sakmann and Neher, 1995). This can be readily appreciated in big tip pipettes expressing fluorescent channels where the shape of the patch can be easily observed (Figure 6.3c).

Given that glass is a dielectric, it contributes a large capacitance to the recording. This capacitance becomes a main noise source and is very relevant in recordings of voltage-gated ion channels, because a change in voltage has to charge the total system capacitance, of which the patch pipette can contribute the largest fraction. To reduce this contribution, patch electrodes are typically covered with a dielectric substance. This step also reduces the contribution of the pipette capacitance to current noise due to coupling to the head stage input. This has been achieved by the application to the region close to the pipette tip of polymers such as PDMS (Sylgard), dielectric varnishes (Q-dope), or even dental wax or bees' wax. These manipulations can and will leave traces of the coating substance at the pipette tip, so the fire-polishing step is also responsible for the most important characteristic of a new patch pipette. It is cleanliness. Any residue accumulated at the tip will be burned off during polishing, with the exception of Sylgard, which will only be cured and hardened by an elevated temperature.

6.3 ANALYSIS OF SINGLE-CHANNEL SIGNALS

Single-channel recordings are an electrophysiological report of the activity of the ion channel protein, as it stochastically gates its pore between a closed and an open conformation. This stochastic process is essentially a homogeneous Poisson process, where the rate constants are time independent. Single-channel analysis has thus two main purposes: to determine the average amplitude of events in the recording and the duration or lifetimes of these events. The lifetimes and their average values will be altered by several factors, such as transmembrane voltage, tension, temperature, or the presence of ligands and modulatory substances. It is in response to these variables where the interesting biology will be found (Hille, 1992; Hille et al., 1999).

Recordings can be achieved in two general cases. When the physical variable that controls the activity of the channel is held constant and the recording avoids any transient changes, the measurement is said to be in steady-state or being stationary. If the variable is instantaneously (or very rapidly) changed, the experiment is then a nonstationary measurement. The analysis of the channel response in each case has both common and particular methods, and each will be overviewed.

6.3.1 FILTERING THE DATA

Certain conditions are common to both steady-state and nonstationary recordings. Signal conditioning is achieved by low-pass filtering and is essential to impose the temporal resolution of the current recording and to improve the signal-to-noise ratio by eliminating high-frequency components in the recorded signal. This is most commonly done by passing the current being collected by the patch pipette through an analog filter, usually part of the amplifier's electronics. The best filter for this purpose is the multipole Bessel filter, because it produces less delay of the individual frequency components and, thus, less distortion of the shape of the signal. The corner frequency, f_c, of said filter corresponds to a 3 dB attenuation of the original signal or roughly a reduction of 50% amplitude. The f_c is a good parameter to characterize the temporal resolution of the recordings, as will be seen in Section 6.3.5. Since the analog current signal is to be digitally sampled, there is an advisable relationship between f_c and the sampling frequency, f_s. This is given by the Nyquist criterion, which states that $f_s \geq 2f_c$. This relationship ensures that aliasing, or the appearance of spurious frequency components in the spectrum of the signal, does not occur. In practice, sampling at $5f_c \leq f_s \leq 10f_c$ is necessary in order to record fast events and improve the detection of events, although at the expense of the signal-to-noise ratio.

6.3.2 NONSTATIONARY EXPERIMENTS

Examples of nonstationary single-channel measurements can be found in the study of ligand-activated or voltage-activated ion channels. In these cases, we are interested in the initial response of the channel to a sudden change in ligand concentration or in transmembrane voltage. Suppose we are dealing with a channel that is activated by voltage. In this case, we want to record the voltage-dependence of the dwell times and the patch-clamp apparatus can change the voltage within a millisecond or less. Although in general, there will be a delay between the moment of voltage change and the activation of the channel, the capacity transient produced by the changing voltage is slow enough that the initial opening of the channel will be distorted by the time course of the capacitive transient. In order to observe clear openings and closings, and to be able to study the time between voltage change and the first opening, we must subtract the capacitive transient and the leak current. In principle, a subtraction procedure like the *p*/*n* or −*p*/*n* protocols could be used (Armstrong and Bezanilla, 1974), but since the subtraction pulse must be scaled up by n or $-n$, respectively, this operation results in the problem of increased noise in the subtracted pulse (see Appendix 6.A), unless a very large number of subtraction pulses can be recorded and averaged. It is thus preferable to find and average several traces without openings or null traces. The null trace average can be used as a leak template to be subtracted from each trace. This procedure should not add significantly to the variance of the noise if enough null traces are accumulated in the average (Sigworth and Zhou, 1992).

6.3.3 STATIONARY RECORDINGS

Another modality of recording single-channel activity is under conditions in which the stimulus has reached a steady and constant value in time. In this case, channel gating has reached equilibrium, and openings and closings should only reflect the constant values of the underlying rate constants. Note that stationary recording can be applied to any type of ion channel, regardless of its activation modality. In this case, long stretches of channel activity are recorded (several seconds to minutes). Because dwell times are exponentially distributed, longer-lived events are less frequent, and may be entirely missed if only short pulses or short periods of time are recorded. Also, some channels may show different kinetic modes, switching between periods of high and low activity, and these phenomena can be captured only in longer stationary recordings (Auerbach and Lingle, 1986; Rothberg and Magleby, 1998).

Recordings in steady-state are an important tool for the characterization of ion channel behavior. For example, in the case of a ligand-gated ion channel, the probability of finding a channel in the open state is a function of the ligand concentration. At low agonist concentration, the probability might be in the order of less than 1%. In the case of voltage-gated channels, the open probability at low voltages can range from 0.01 to 10^{-8}. This implies that an extremely long time would be necessary to record sufficient events (openings) for statistical calculation of the probability and kinetics of the channel (Hirschberg et al., 1995; Islas and Sigworth, 1999). For example, if the open probability is 10^{-8} and the mean duration of an opening is 10 ms, one expects one opening in approximately 28 h.

6.3.4 DETECTION OF EVENTS

Data obtained from single-channel recordings can be stored directly into the computer memory. Once there, the next step is to analyze the recording and characterize the channel gating events. Two main methods of identifying events in a recording are currently in general use: direct fitting of the time course of the recording (Gibb and Colquhoun, 1991; Lape et al., 2008) and the threshold crossing method. This chapter will discuss the latter method. Direct fitting is a more specialized technique, and its advantages and drawbacks are discussed by Colquhoun and Sigworth (1995). The threshold crossing method is also applicable to both stationary and nonstationary recordings. Digital filtering previous to analysis is generally advisable. This last filtering step determines the final bandwidth of the recording. Since the analog signal has already passed through several serial filtering stages (the patch-clamp amplifier, any additional analog filter, any additional storage units apart from the computer hard disk), the digital filtering stage will set the bandwidth of the recorded data according to a cutoff frequency, f_c, which can be calculated by the formula

$$\frac{1}{f_c} = \frac{1}{f_1} + \frac{1}{f_2} + \cdots + \frac{1}{f_n}. \tag{6.1}$$

In this equation, n is the number of filtering stages, including the Gaussian digital filter, as mentioned in the last paragraph.

Digital filtering is generally carried out with a Gaussian filter algorithm, which more closely resembles the desirable characteristics of the multipole Bessel filter. Analytical expressions for the Gaussian filter are available, which can be used to obtain expression for other parameters such as the threshold for event detection (Colquhoun and Sigworth, 1995).

The first step is to find the average amplitude of the events in the recording. This can be done by first compiling an all-points histogram of several events and determining the average amplitude, $\langle I \rangle$. If the channel has a single, well-defined level of conductance, the all-points histogram is approximately distributed according to a Gaussian distribution, and a Gaussian fit can be used to determine the average amplitude, and thus the threshold, as $\langle I \rangle\theta$, where θ is a number between 0 and 1. This value is then fed back into the event detection algorithm. The all-points histogram is a useful tool, since it also permits one to detect the presence or absence of subconductance levels, which can introduce major errors in the analysis by the threshold method and that have to be dealt with in a special way.

Before events can be detected, the whole record must be interpolated to avoid distortion of the short-lived events as much as possible, and reduce the uncertainty associated with the determination of when exactly the threshold is crossed. A cubic spline is an effective interpolation, because it can be easily implemented in software and is not computationally very costly.

Defining the threshold, θ, is an important and complicated issue. The value of θ is critically related to the probability that random, Gaussian-distributed noise events in the recording will cross an arbitrary value defined by $\theta\langle I \rangle$. We want to avoid the detection of too many of these spurious events. If the value of θ is chosen to be too small, many false events will be detected. If it is too close to 1, too many short-lived real events will be missed and fluctuation events present in the noise when the channel is open will be counted as closing events. If the channels are being filtered with a Gaussian filter, there is a relatively simple rule of thumb that can be used to choose the value of θ. For small amplitude channel openings, heavy filtering has to be employed, perhaps less than 1 kHz and in this case, $\theta = 0.7\langle I \rangle$ can be used. In the case of openings with large amplitude, the threshold can be chosen as $\theta = 0.5\langle I \rangle$. This last case is often referred to as the 50% threshold crossing technique and is widely used as the standard method for event detection (Sachs et al., 1982). Once the threshold is selected, the duration of an opening is defined as the sum of all the points above the threshold times the sampling interval. Likewise, closings are recorded as all the events that remain below the threshold.

Nonstationary recordings present the opportunity to measure another important lifetime. This is the time between the change in voltage or agonist concentration and the first opening event, which is called the latency to first opening or first latency. This is a special parameter, since it cannot be obtained from steady-state recordings. If the channel has only one open state, the first latency represents all the closed states that the channel has to occupy before opening. Measuring first latencies poses the problem of deciding when the actual change in stimulus starts at the patch-clamped membrane. In the case of a voltage pulse, this is a parameter that depends on the particular instrumentation (amplifier and analog to digital/digital to analog, AD/DA converter) and filter settings used and should be measured in each case. An adequate way to estimate it is to measure an instantaneous change in current, for example, measuring the deactivation of a macroscopic tail current from a voltage-gated channel. The delay is then estimated as the time since the DA pulse and the midpoint of the current change. For voltage pulses, these delays are generally less than 1 μs and are not important except at extreme voltages where the intrinsic delay of channel activation might approach these values.

On the other hand, even the fastest changes in agonist concentration carried out in outside-out patches will incur in considerably longer delays, in the order of milliseconds.

The distribution of first latencies is not exponential when there are more than one gating transitions. It is common to display it as a cumulative distribution and to compare it with the probability that the channel is open after a time t after the change in voltage or ligand concentration. In this representation, it can be compared with a model like

$$C_0 \underset{k_{-1}}{\overset{k_1}{\rightleftarrows}} C_1 \ldots C_n \xrightarrow{\alpha} O.$$

In this model, the channel transitions between n closed states C_i with rates k_i and once it reaches the open state O, with rate α, it does not reopen, which is an analogous representation of the measurement of the cumulative distribution of first latencies (Sigworth and Zhou, 1992).

The end result of the event detection step is the generation of lists of event durations. Generally, one would obtain a list of open and closed state lifetimes (of which the first latency is the first closed time measured in each sweep), which will be used later for further analysis.

Even though programs of varying complexity can be written to carry out these analyses, it is crucial that the investigator visually inspects each detected event and validates them. It is important to understand that the selection of the threshold value is valid only for Gaussian-distributed (noise) events. If the recording contains fluctuations other than baseline noise and channel gating, such as changes of the average value of the baseline, noise can be *pushed* to cross the threshold, contaminating the events list with spurious events. If line-frequency interference is present in the recording, there is a real chance of periodically detecting spurious threshold crossings. Also, the seal resistance of the patch over long periods of recording can change, altering the baseline and the noise characteristics. Analysis under these circumstances can lead to detection of false events if the investigator does not validate each of them. Thus, the programs used for analysis should have a provision to remove false events from the lists of events.

6.3.5 RESOLUTION

As mentioned earlier, filtering the data imposes a temporal resolution. This can be calculated for a Gaussian filter as follows. If an instantaneous step input is passed through a Gaussian filter, the output is distorted in several ways. The output can be calculated as a function of time, *t*, with the step response *H*(*t*) of the filter, which is given by the equation

$$H(t) = \frac{1}{2}\left[1 + erf(5.336 f_c t)\right]. \quad (6.2)$$

In this equation, *erf* is the error function. Figure 6.4 illustrates the response of such a filter to several input pulses of equal amplitude but different durations. As can be gleaned from the figure, shorter pulses are reduced in amplitude by the filter until some become smaller than 50% of the amplitude of the original ones. These pulses do not cross the half-amplitude threshold and thus represent missed events. The rise time of the filter can be defined as the time it takes for the output to change from 10% to 90% of its maximum value, and it is given by

$$T_r = \frac{0.332}{f_c}. \quad (6.3)$$

For this type of filter, events that are roughly half the duration of T_r will be missed. The minimum duration of an event that we can expect to detect can be characterized by another metric called the filter death time T_d. The death time can be calculated as $T_d = 0.538T_r$ or $T_d = 0.179/f_c$.

Another consequence of the reduced amplitude of events caused by filtering is that two closely spaced events will be distorted to look as a single one. This will have the effect of missing short-lived events that will be counted as longer ones. Several procedures have been developed to cope with the distortions introduced by this and other types of missed events, but are too specialized to be discussed here. The interested reader can consult the works of Blatz and Magleby (1986), Crouzy and Sigworth (1990), and Qin et al. (1996).

6.3.6 DWELL TIME HISTOGRAMS AND FITTING OF DISTRIBUTIONS

In general, for a Poisson-like random process, such as channel gating, the dwell time in any given state visited by the process is an exponentially distributed random variable (Gardiner, 1994). The events lists are thus distributed as exponentials or sums of exponentials. The data contained in the events list can be compiled in histograms, and these can be plotted in a linear scale to graphically display the distribution of the data. This procedure has the drawback that, in the case of a channel with several states, the exponential components can be separated by several orders of magnitude. In a linear representation of such a complex histogram, it can be extremely difficult to make out the distinct components. A better visual representation can be obtained with the Sigworth–Sine transform (Sigworth and Sine, 1987). For a single exponential, if the intervals or bins of the histogram, T_i's, are transformed logarithmically according to

$$x_i = \ln(T_i),$$

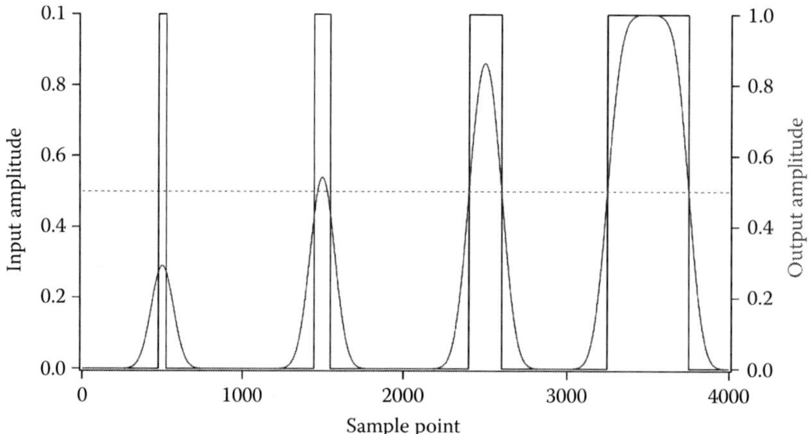

Figure 6.4 Response of the Gaussian filter to step inputs of varying duration. Four square pulses of 25, 50, 100, and 500 sample points (black trace) were filtered by a Gaussian filter, and the output is superimposed (gray trace). The rising and falling time courses of the filtered output response are distorted according to Equation 6.4 that describes the step response of the filter. Note that the shortest step does not reach the 50% value of the amplitude of the longer step and, as a consequence, will be undetected by the 50% threshold crossing technique. Also the second step barely crosses the 50% amplitude line and the measured width of the filtered pulse underestimates the real duration of the input step.

it can be shown that the probability density function (*pdf*) is given by

$$f(x) = a\tau^{-1}e^{\left(x-\tau^{-1}e^{x}\right)}, \tag{6.4}$$

where
τ is the mean dwell time
a is the amplitude of the exponential

This equation can be generalized as a sum of components for distributions with more than one exponential component.

In this display method, the individual dwell times t_i are transformed logarithmically and the ordinate is displayed as the square root of the number of events, because this transformation evens out the error across all the bin widths. The advantage of this transformation is that it very naturally provides a visual guide to the distribution of the dwell times, because as was shown for the *pdf*, an exponential appears as an approximately bell-shaped curve with its peak value occurring at a time equal to the mean of the exponential (Figure 6.5).

A crucial step in the analysis of single-channel records is the fitting of a theoretical distribution to the actual histograms of dwell times. This has the purpose of finding the parameters that describe the data, specially, the value of mean dwell time. In general, if the channel dwells in a single, discrete state, the distribution of lifetimes in that state will be described by a single exponential whose mean time is the inverse of the sum of all rate constants for leaving that state. Since the distribution of the data is known (it is exponential), the preferred method for fitting a distribution of single-channel dwell times is the use of a maximum likelihood algorithm, as opposed to least squares fitting or other methods. In this method, we are not interested in fitting exponential functions to the histogram of dwell times, but in actually using all the dwell times, t_i, in the events list having n entries, to calculate a likelihood function which, for a single exponential distribution with mean value τ, is defined as

$$Lik(\tau) = \prod^{n} \frac{1}{\tau} e^{(-t_i/\tau)}. \tag{6.5}$$

For actual computation of the likelihood, it is more useful to calculate the natural log of the function, also known as the log-likelihood, $L(\tau)$:

$$L(\tau) = \ln Lik(\tau). \tag{6.6}$$

A program can be written to maximize the value of L, which also maximizes *Lik*, by varying τ, which is the parameter that we are interested in finding out. It can be shown that the value of τ that maximizes L characterizes the exponential distribution that is more probable to describe the data (Horn and Lange, 1983; Colquhoun and Sigworth, 1995). Once τ has been found, an exponential distribution can be generated and superimposed in the actual histogram of the data (Figure 6.5).

For very rapid or very noisy, small amplitude data, there are methods to obtain the parameters describing the data that involve the use of hidden Markov models (HMMs). These specialized analysis technique (Qin et al., 2000; Venkataramanan and Sigworth, 2002) will be discussed in Chapter 8.

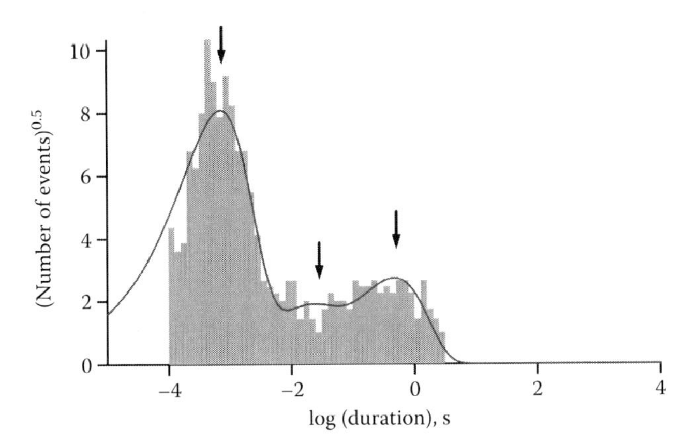

Figure 6.5 Distribution of closed dwell times. Data were obtained from a single-channel recording of a heat-activated TRPV1 channel. The closed times were measured with the 50% threshold crossing technique and are displayed according to the Sigworth–Sine transform. Note the presence of three bumps in the distribution, which correspond to three exponential components with fast (*f*), intermediate (*i*), and slow (*s*) mean times. The arrows approximately coincide with the peak of each component and represent the mean time, τ_f, τ_i, τ_s of each exponential. The smooth curve over the histogram is the sum of three exponentials estimated by maximum likelihood analysis of the closed times.

6.3.7 BURST ANALYSIS

Openings and closings often appear to be clustered in bursts separated by long visits to a closed state. Important parameters for the establishment and discrimination of kinetic mechanisms can be extracted by analyzing the burst characteristics, such as its duration, the number of openings and closings, and the mean duration of these events within the bursts (Colquhoun and Hawkes, 1982). To carry out burst analysis, one has first to identify a burst. Notice first that channel openings will be clustered in bursts when there are two or more closed states with widely varying mean dwell times. The channel openings will tend to be separated by visits to the short-lived closed state(s), and the bursts will be separated by sojourns to the long-lived closed states. In order to define the burst, one has to predetermine a criterion to set a maximal duration of the closed intervals within the burst, t_c. Any closed interval $t > t_c$ will be considered to be outside the burst and will signal the end of the identified burst. Several criteria have been proposed to define t_c. All of them try to define t_c as a value of t that lies between the fast and slow components of the closed time distribution. This implies that before burst analysis can be performed, the closed times distribution must be obtained as explained in the previous section. The method of Colquhoun and Sakmann (1985) may be less biased when the number of long and short closed events is very different. In this method, t_c is found by solving the equation

$$\exp\left(\frac{-t_c}{\tau_f}\right) = 1 - \exp\left(\frac{-t_c}{\tau_s}\right). \tag{6.7}$$

In this equation, τ_f and τ_s are the mean durations of the fast and slow components of the closed time distribution, respectively. Once t_c is defined, the program can use this value to identify bursts and determine their duration and other statistics. Generally, burst duration is also a random variable with exponential distribution, and this distribution, along with other parameters of the burst such as the total open time, etc., is important in the discrimination of kinetic mechanisms. A classic example of this can be found in the demonstration that blockers that only bind to the open state of a channel can be identified, because they produce an increase of the duration of bursts of openings (Neher and Steinbach, 1978).

6.4 SPECIAL PURPOSE TECHNIQUES

6.4.1 OTHER USES OF THE ALL-POINTS HISTOGRAM: BETA DISTRIBUTIONS

A phenomenon that can be observed at the single-channel level is the very rapid reaction of certain channel inhibitors that act by blocking or physically obstructing the passage of ions through the channel. If the kinetics of this process is faster than the death time of the filter, the transitions between the blocked and unblocked states (while the channel is open) will be entirely unresolved, and the observed effect will be a reduction of the channel current level (Moczydlowski, 1992). Under the assumption that the blocking process that occurs when the channel is open can be modeled as a two-state mechanism, the rate constants for blocker binding and unbinding can be estimated from the amplitude and shape of an all-points histogram of the current at different concentrations of the blocker. It has been shown that the kinetics of a rapid two-state process that is obscured by filtering can be estimated from the distribution of amplitudes of the resulting signal by fitting it to a beta distribution (Fitzhugh, 1983). This function is given by the following equations:

$$f(y) = \frac{y^{a-1}(1-y)^{b-1}}{B(a,b)} \quad \text{with } a = \alpha\tau b = \beta\tau$$

$$\text{and} \quad B(a,b) = \int_0^1 y^{a-1}(1-y)^{b-1}dy. \tag{6.8}$$

The parameters β and α of the beta distribution can be identified with the association and dissociation rates of the blocking process, respectively, and τ is the inverse of the corner frequency of the filter (Figure 6.6). This method was first used by Yellen to study rapid sodium block of the calcium-activated potassium channel when data was filtered by a Bessel filter (Yellen, 1984). The method has also been successfully applied to Gaussian-filtered data in the case of a study of the block by quaternary ammoniums of TRPV1 channels (Jara-Oseguera et al., 2008).

6.4.2 MEAN-VARIANCE ANALYSIS TO DETECT TRANSITIONS

It is not uncommon to observe one or more levels of conductance apart from the main open level, even when recording from a single channel. These openings to a subconductance level are

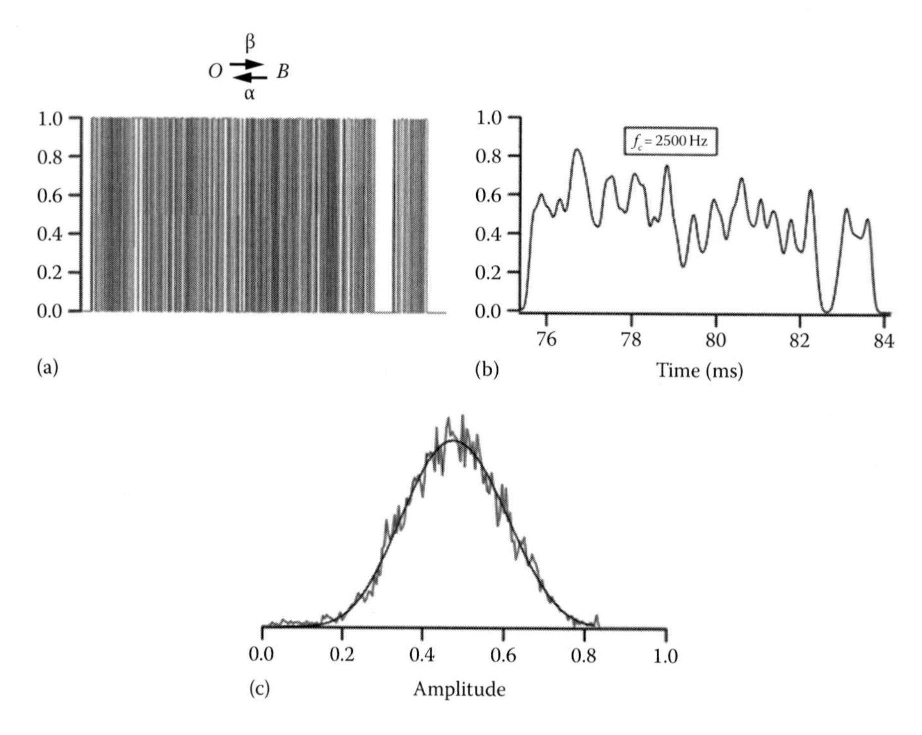

Figure 6.6 Use of the beta distribution to obtain the rate constants of a very fast simple blocking reaction. (a) Simulated blocking and unblocking events during two channel openings. (b) The data in (a) was filtered by a Gaussian digital filter having a corner frequency 2.5 kHz. Note that after filtering, individual blocking events are no longer discernible and the amplitude of the open channel current has been diminished. (c) All-points amplitude histogram from the data in (b). The dark curve is the fit to the beta distribution (see text) with blocking constant, β = 51,200 s^{-1} and unblocking constant, α = 50,950 s^{-1}. The actual constants used to simulate the data in (a), are β = 51,000 s^{-1} and unblocking constant α = 51,000 s^{-1}. (Reproduced with permission from Jara-Oseguera, A. et al., *J. Gen. Physiol.*, 132, 547, 2008.)

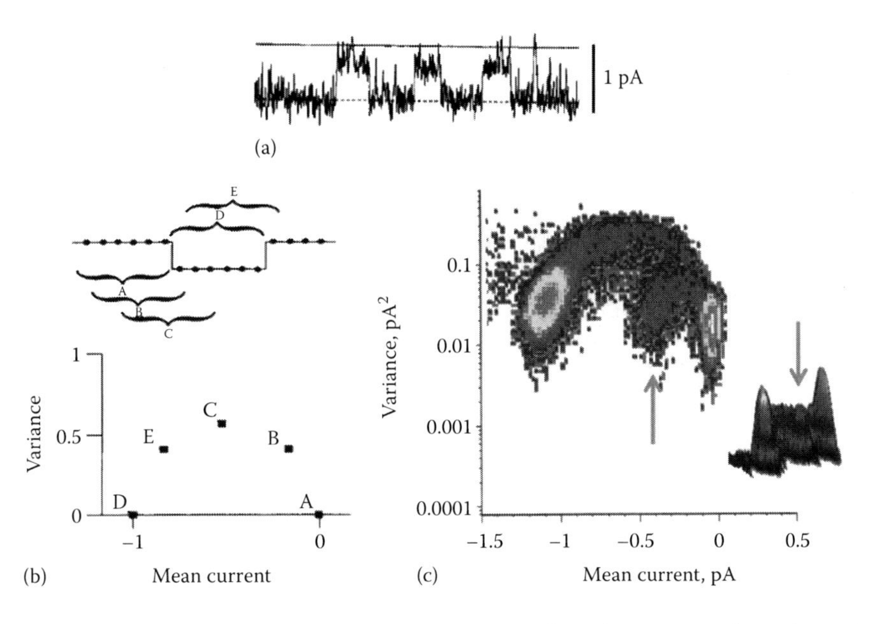

Figure 6.7 The mean-variance method applied to the detection of sublevels in small conductance ion channels. (a) Shows a single sweep from a voltage-dependent sodium channel. Notice the small amplitude of the currents and the presence of smaller open events or subconductances. (b) The method is based on the fact that the variance of the signal is larger at the point of a change of current (channel opening or closing) and the magnitude of the variance is directly proportional to the change in conductance, so that a larger change in current incurs in a larger variance. (c) Variance–mean plot obtained from several recordings as in (a). The coloring indicates the density of points having a particular mean–variance pair value. In these plots, it is easy to measure the amplitude of the subconductance state (indicated by the arrow) and the fully open state. (Modified from Patlak, J.B., *Biophys. J.*, 65, 29, 1993.)

difficult to detect when using a standard 50% threshold crossing technique. In order to analyze these kinds of records and detect dwell times in each conductance level, one could use a technique with multiple thresholds (Zheng and Sigworth, 1998) or use direct fitting of the current record (Lape et al., 2008). If the channel's main conductance is small, as is the case in voltage-gated sodium and calcium channels (Patlak and Horn, 1982; Fox et al., 1987b) and some potassium channels (Zagotta et al., 1989; Stocker, 2004), the subconductance states will be exceedingly small, making detection of their presence and estimations of durations very difficult with simple methods. In this case, one can use a technique called mean-variance analysis. In this technique, the current record is scanned with a running window of a few sample points. The mean and the variance are calculated in each window, producing a point in a variance vs. mean current plot. Every time there is a transition between two given conductance levels, the variance will increase and the mean will take the value of the new current level. The variance vs. mean plot will thus trace trajectories between discrete mean current levels. This technique has been successfully used to detect very small subconductance levels in voltage-gated sodium channels (Patlak, 1988) (Figure 6.7). Mean-variance analysis will not only effectively identify the presence and amplitude of subconductance levels, but has the added ability to provide kinetic information. Since the number of points in a given conductance level is proportional to the lifetime of that level, the relative number of points contributing to each conductance level in the variance vs. mean plot or volume is an index of the lifetime of each conductance. This can be estimated by measuring the volume of points for varying window widths (Patlak, 1993).

6.A APPENDIX

EFFECT OF *p/n* SUBTRACTION ON THE NOISE OF SINGLE-CHANNEL RECORDS

Let us call a single sweep obtained at voltage V_T, containing a capacitive transient, leak current and channel current, S_T. This sweep has a variance var(S_T). A subtraction sweep, without channel openings, S_L, obtained at the same gain but at a more negative voltage V_T/n will have a variance var(S_l) = var(S_T). Before subtraction, the sweep S_L is scaled by n to give a sweep S_{Ln}. This sweep will have the same size of the leak and capacitive transients as S_T but will have variance $\text{var}(S_{Ln}) = \text{var}(nS_l) = n^2\text{var}(S_T)$. If S_{Ln} is subtracted from S_T, the variance is increased by n^2, inevitably obscuring any channel openings present in S_T.

A simple property of the variance of uncorrelated variables can be used to see how to reduce the variance of the subtraction sweep. The variance of a sum of variables X_i is as follows:

$$\text{var}\left(\sum_{i=1}^{m} X_i\right) = \sum_{i=1}^{m}[\text{var}(X_i)]$$

if we average m of the scaled sweeps S_{Ln}:

$$\text{var}\left(\frac{1}{m}\sum_{i=1}^{m} S_{Ln}\right) = \sum_{i=1}^{m}\left[\text{var}\left(\frac{1}{m}S_{Ln}\right)\right] = \frac{1}{m^2}\sum_{i=1}^{m}\left[\text{var}\left(S_{Ln}\right)\right]$$

Here, the mean of the m sweeps S_{Ln} is

$$\frac{1}{m}\sum_{i=1}^{m} S_{Ln} = M_L,$$

so this implies that $\text{var}(M_L) = \text{var}(S_T)$ if $m = n^2$. So a minimum of n^2 subtraction sweeps must go into the average subtraction sweep M_L, before its subtraction from S_T does not significantly increase the noise of the original test pulses.

ACKNOWLEDGMENTS

Work in the laboratory of the author is supported by grants from CONACYT and DGAPA-PAPIIT-UNAM.

REFERENCES

Armstrong, C. M. and Bezanilla, F. 1974. Charge movement associated with the opening and closing of the activation gates of the Na channels. *J Gen Physiol*, 63, 533–552.

Auerbach, A. and Lingle, C. J. 1986. Heterogeneous kinetic properties of acetylcholine receptor channels in *Xenopus* myocytes. *J Physiol*, 378, 119–140.

Bae, C., Markin, V., Suchyna, T., and Sachs, F. 2011. Modeling ion channels in the gigaseal. *Biophys J*, 101, 2645–2651.

Benndorf, K., Koopmann, R., Eismann, E., and Kaupp, U. B. 1999. Gating by cyclic GMP and voltage in the alpha subunit of the cyclic GMP-gated channel from rod photoreceptors. *J Gen Physiol*, 114, 477–490.

Blatz, A. L. and Magleby, K. L. 1986. Correcting single channel data for missed events. *Biophys J*, 49, 967–980.

Colquhoun, D. 2007. What have we learned from single ion channels? *J Physiol*, 581, 425–427.

Colquhoun, D. and Hawkes, A. G. 1981. On the stochastic properties of single ion channels. *Proc R Soc Lond B Biol Sci*, 211, 205–235.

Colquhoun, D. and Hawkes, A. G. 1982. On the stochastic properties of bursts of single ion channel openings and of clusters of bursts. *Philos Trans R Soc Lond B Biol Sci*, 300, 1–59.

Colquhoun, D., Jonas, P., and Sakmann, B. 1992. Action of brief pulses of glutamate on AMPA/kainate receptors in patches from different neurones of rat hippocampal slices. *J Physiol*, 458, 261–287.

Colquhoun, D. and Sakmann, B. 1985. Fast events in single-channel currents activated by acetylcholine and its analogues at the frog muscle end-plate. *J Physiol*, 369, 501–557.

Colquhoun, D. and Sigworth, F. J. 1995. Fitting and statistical analysis of single-channel recors. In: Sakmann, B. and Neher, E. (eds.) *Single-Channel Recording*, 2nd edn. New York: Plenum.

Crouzy, S. C. and Sigworth, F. J. 1990. Yet another approach to the dwell-time omission problem of single-channel analysis. *Biophys J*, 58, 731–743.

Edwards, F. A., Konnerth, A., Sakmann, B., and Takahashi, T. 1989. A thin slice preparation for patch clamp recordings from neurones of the mammalian central nervous system. *Pflugers Arch*, 414, 600–612.

Ehrenstein, G., Blumenthal, R., Latorre, R., and Lecar, H. 1974. Kinetics of the opening and closing of individual excitability-inducing material channels in a lipid bilayer. *J Gen Physiol*, 63, 707–721.

Ehrenstein, G., Lecar, H., and Nossal, R. 1970. The nature of the negative resistance in bimolecular lipid membranes containing excitability-inducing material. *J Gen Physiol*, 55, 119–133.

Fertig, N., Blick, R. H., and Behrends, J. C. 2002. Whole cell patch clamp recording performed on a planar glass chip. *Biophys J*, 82, 3056–3062.

Fitzhugh, R. 1983. Statistical properties of the asymmetric random telegraph signal, with applications to single-channel analysis. *Math Biosci*, 64, 75–89.

Fox, A. P., Nowycky, M. C., and Tsien, R. W. 1987a. Kinetic and pharmacological properties distinguishing three types of calcium currents in chick sensory neurones. *J Physiol*, 394, 149–172.

Fox, A. P., Nowycky, M. C., and Tsien, R. W. 1987b. Single-channel recordings of three types of calcium channels in chick sensory neurones. *J Physiol*, 394, 173–200.

Gardiner, C. W. 1994. *Handbook of Stochastic Methods for Physics, Chemistry, and the Natural Sciences.* Berlin, Germany: Springer-Verlag.

Gibb, A. J. and Colquhoun, D. 1991. Glutamate activation of a single NMDA receptor-channel produces a cluster of channel openings. *Proc Biol Sci*, 243, 39–45.

Goldstein, S. A. and Miller, C. 1993. Mechanism of charybdotoxin block of a voltage-gated K^+ channel. *Biophys J*, 65, 1613–1619.

Hamill, O. P., Marty, A., Neher, E., Sakmann, B., and Sigworth, F. J. 1981. Improved patch-clamp techniques for high-resolution current recording from cells and cell-free membrane patches. *Pflugers Arch*, 391, 85–100.

Hille, B. 1992. *Ionic Channels of Excitable Membranes.* Sunderland, MA: Sinauer Associates.

Hille, B., Armstrong, C. M., and Mackinnon, R. 1999. Ion channels: From idea to reality. *Nat Med*, 5, 1105–1109.

Hirschberg, B., Rovner, A., Lieberman, M., and Patlak, J. 1995. Transfer of twelve charges is needed to open skeletal muscle Na^+ channels. *J Gen Physiol*, 106, 1053–1068.

Horn, R. and Lange, K. 1983. Estimating kinetic constants from single channel data. *Biophys J*, 43, 207–223.

Horn, R. and Marty, A. 1988. Muscarinic activation of ionic currents measured by a new whole-cell recording method. *J Gen Physiol*, 92, 145–159.

Horn, R. and Patlak, J. 1980. Single channel currents from excised patches of muscle membrane. *Proc Natl Acad Sci USA*, 77, 6930–6934.

Horowitz, P. and Hill, W. 1989. *The Art of Electronics.* Cambridge, U.K.: Cambridge University Press.

Islas, L. D. and Sigworth, F. J. 1999. Voltage sensitivity and gating charge in Shaker and Shab family potassium channels. *J Gen Physiol*, 114, 723–742.

Jara-Oseguera, A., Llorente, I., Rosenbaum, T., and Islas, L. D. 2008. Properties of the inner pore region of TRPV1 channels revealed by block with quaternary ammoniums. *J Gen Physiol*, 132, 547–562.

Kitamura, K., Judkewitz, B., Kano, M., Denk, W., and Hausser, M. 2008. Targeted patch-clamp recordings and single-cell electroporation of unlabeled neurons in vivo. *Nat Methods*, 5, 61–67.

Klemic, K. G., Klemic, J. F., Reed, M. A., and Sigworth, F. J. 2002. Micromolded PDMS planar electrode allows patch clamp electrical recordings from cells. *Biosens Bioelectron*, 17, 597–604.

Kostyuk, P. G., Krishtal, O. A., and Pidoplichko, V. I. 1972. Potential-dependent membrane current during the active transport of ions in snail neurones. *J Physiol*, 226, 373–392.

Kostyuk, P. G., Krishtal, O. A., and Pidoplichko, V. I. 1975. Effect of internal fluoride and phosphate on membrane currents during intracellular dialysis of nerve cells. *Nature*, 257, 691–693.

Lape, R., Colquhoun, D., and Sivilotti, L. G. 2008. On the nature of partial agonism in the nicotinic receptor superfamily. *Nature*, 454, 722–727.

Maconochie, D. J. and Knight, D. E. 1989. A method for making solution changes in the sub-millisecond range at the tip of a patch pipette. *Pflugers Arch*, 414, 589–596.

Moczydlowski, E. 1992. Analysis of drug action at single-channel level. In: Rudy, B. and Iverson, L. E. (eds.) *Methods in Enzymology: Ion Channels.* San Diego, CA: Academic Press.

Mueller, P. and Rudin, D. O. 1963. Induced excitability in reconstituted cell membrane structure. *J Theor Biol*, 4, 268–280.

Neher, E. and Marty, A. 1982. Discrete changes of cell membrane capacitance observed under conditions of enhanced secretion in bovine adrenal chromaffin cells. *Proc Natl Acad Sci USA*, 79, 6712–6716.

Neher, E. and Sakmann, B. 1976. Single-channel currents recorded from membrane of denervated frog muscle fibres. *Nature*, 260, 799–802.

Neher, E., Sakmann, B., and Steinbach, J. H. 1978. The extracellular patch clamp: A method for resolving currents through individual open channels in biological membranes. *Pflugers Arch*, 375, 219–228.

Neher, E. and Steinbach, J. H. 1978. Local anaesthetics transiently block currents through single acetylcholine-receptor channels. *J Physiol*, 277, 153–176.

Patlak, J. and Horn, R. 1982. Effect of N-bromoacetamide on single sodium channel currents in excised membrane patches. *J Gen Physiol*, 79, 333–351.

Patlak, J. B. 1988. Sodium channel subconductance levels measured with a new variance-mean analysis. *J Gen Physiol*, 92, 413–430.

Patlak, J. B. 1993. Measuring kinetics of complex single ion channel data using mean-variance histograms. *Biophys J*, 65, 29–42.

Pfaffinger, P. J., Leibowitz, M. D., Subers, E. M., Nathanson, N. M., Almers, W., and Hille, B. 1988. Agonists that suppress M-current elicit phosphoinositide turnover and Ca^{2+} transients, but these events do not explain M-current suppression. *Neuron*, 1, 477–484.

Qin, F., Auerbach, A., and Sachs, F. 1996. Estimating single-channel kinetic parameters from idealized patch-clamp data containing missed events. *Biophys J*, 70, 264–280.

Qin, F., Auerbach, A., and Sachs, F. 2000. Hidden Markov modeling for single channel kinetics with filtering and correlated noise. *Biophys J*, 79, 1928–1944.

Rae, J. L. and Levis, R. A. 1984. Patch clamp recordings from the epithelium of the lens obtained using glasses selected for low noise and improved sealing properties. *Biophys J*, 45, 144–146.

Rothberg, B. S. and Magleby, K. L. 1998. Kinetic structure of large-conductance Ca^{2+}-activated K^+ channels suggests that the gating includes transitions through intermediate or secondary states. A mechanism for flickers. *J Gen Physiol*, 111, 751–780.

Sachs, F., Neil, J., and Barkakati, N. 1982. The automated analysis of data from single ionic channels. *Pflugers Arch*, 395, 331–340.

Sakmann, B. and Neher, E. 1995. Geometric parameters of pipettes and membrane patches. In: Sakmann, B. and Neher, E. (eds.) *Single-Channel Recording*. New York: Plenum.

Schoppa, N. E. and Sigworth, F. J. 1998. Activation of Shaker potassium channels. III. An activation gating model for wild-type and V2 mutant channels. *J Gen Physiol*, 111, 313–342.

Sigworth, F. J. 1986. The patch clamp is more useful than anyone had expected. *Fed Proc*, 45, 2673–2677.

Sigworth, F. J. 1995. Electronic design of the patch clamp. In: Sakmann, B. and Neher, E. (eds.) *Single-Channel Recording*, 2nd edn. New York: Plenum.

Sigworth, F. J. and Neher, E. 1980. Single Na^+ channel currents observed in cultured rat muscle cells. *Nature*, 287, 447–449.

Sigworth, F. J. and Sine, S. M. 1987. Data transformations for improved display and fitting of single-channel dwell time histograms. *Biophys J*, 52, 1047–1054.

Sigworth, F. J. and Zhou, J. 1992. Analysis of nonstationary single-channel currents. In: Rudy, B. and Iverson, L. E. (eds.) *Methods in Enzymology: Ion Channels*. San Diego, CA: Academic Press.

Sokabe, M. and Sachs, F. 1990. The structure and dynamics of patch-clamped membranes: A study using differential interference contrast light microscopy. *J Cell Biol*, 111, 599–606.

Stocker, M. 2004. Ca(2+)-activated K^+ channels: Molecular determinants and function of the SK family. *Nat Rev Neurosci*, 5, 758–770.

Stuhmer, W. 1992. Electrophysiologycal recording from *Xenopus* oocytes. In: Iverson, B. R. A. L. E. (ed.) *Methods in Enzymology: Ion Channels*. San Diego, CA: Academic Press.

Stuhmer, W., Methfessel, C., Sakmann, B., Noda, M., and Numa, S. 1987. Patch clamp characterization of sodium channels expressed from rat brain cDNA. *Eur Biophys J*, 14, 131–138.

Suchyna, T. M., Markin, V. S., and Sachs, F. 2009. Biophysics and structure of the patch and the gigaseal. *Biophys J*, 97, 738–747.

Sunderman, E. R. and Zagotta, W. N. 1999. Sequence of events underlying the allosteric transition of rod cyclic nucleotide-gated channels. *J Gen Physiol*, 113, 621–640.

Venkataramanan, L. and Sigworth, F. J. 2002. Applying hidden Markov models to the analysis of single ion channel activity. *Biophys J*, 82, 1930–1942.

Yellen, G. 1984. Relief of Na^+ block of Ca^{2+}-activated K^+ channels by external cations. *J Gen Physiol*, 84, 187–199.

Zagotta, W. N., Hoshi, T., and Aldrich, R. W. 1989. Gating of single Shaker potassium channels in Drosophila muscle and in *Xenopus* oocytes injected with Shaker mRNA. *Proc Natl Acad Sci USA*, 86, 7243–7247.

Zagotta, W. N., Hoshi, T., and Aldrich, R. W. 1994. Shaker potassium channel gating. III: Evaluation of kinetic models for activation. *J Gen Physiol*, 103, 321–362.

Zheng, J. and Sigworth, F. J. 1998. Intermediate conductances during deactivation of heteromultimeric Shaker potassium channels. *J Gen Physiol*, 112, 457–474.

7 Models of ion channel gating

Frank T. Horrigan and Toshinori Hoshi

Contents

7.1 INTRODUCTION

Everything should be made as simple as possible, but no simpler.

Albert Einstein (attributed)

Models of ion channel gating are conceptual and/or mathematical constructs designed to *reproduce* and *predict* the kinetic and steady-state relationships between the opening/closing processes of a channel (gating, e.g., its open probability P_o; see the chapters by Armstrong, Chanda, and Zagotta in this volume) and the relevant stimuli, for instance, membrane potential, ligand concentration, mechanical deformation, and/or temperature. Successful models *reproduce* the known input–output (I–O) relationships or signal transduction properties of a channel. Such models also *predict* or simulate the properties of a channel not yet measured and sometimes practically impossible to do so. Consistent with a previously published view,[1] icons with multiple colors are not included as models in this chapter.

Gating models can take many forms, from data descriptor functions that provide a purely phenomenological description of I–O relationships without making any assumptions about the underlying mechanisms to all-atom molecular dynamics (MD) simulations based on explicit atomic representations of channel and membrane structure (although with some assumptions). However, more common than these extremes are gating schemes or semimechanistic models based on simplified representations of molecular structure and function such as kinetic states, functional domains, and/or *sensors* and *gates*. The strategies and procedures for developing models and level of detail incorporated depend in large part on the goal to be achieved as well as the availability of functional and structural information. Much of the *art* of model building lies in deciding what level of detail or simplification is required to address a goal and obtain a meaningful and useful answer.

Every model is a simplification, owing in part to incomplete information as well as philosophical tenants such as Occam's razor, which urges that hypotheses that can explain an effect by making the fewest assumptions are preferred. A simpler model is preferred over a complex one *provided* they cannot be distinguished based on the experimental observations. Parsimony is desirable as simpler models may be more intuitive and testable, and fits are better constrained when the number of free parameters is minimized. One caveat is that *simplest* does not have a unique definition, especially when the model must be consistent with multiple types of observations. For example, models with the fewest free parameters do not necessarily make the fewest mechanistic assumptions or have the fewest number of states, and vice versa.[2]

The purpose of this chapter is to serve as an instructional starting point for those who are curious about developing a gating model that reproduces and predicts properties of an ion channel. Based on the authors' own experience, the chapter will utilize voltage-gated K^+ channels and Ca^{2+}- and voltage-gated Slo1 K^+ (BK) channels as two exemplar systems to illustrate important concepts, principles, and rationales involved in model building. We regret that space constrain prevents us from presenting many excellent models of gating of ion channels, including various voltage-gated channels and ligand-gated channels. Readers interested in specific channels are referred to other chapters in this volume.

7.2 GOALS AND BENEFITS OF MODELING

Gating models provide quantitative predictions about ion channel function that can be used to understand the role of channels in cell signaling, to test hypotheses about molecular mechanisms of gating, or to infer information about elementary gating events based on function. These goals and the different requirements that they place on models are discussed next.

1. *To understand electrical excitability and the role of ion channels in cell signaling* has been a fundamental goal of modeling since the time of Hodgkin and Huxley[3] and of Katz.[4] Gating models provide insight into the functional role of ion channels because, once established, the number and/or properties of different channels can be adjusted to evaluate their contribution to cell signaling or to mimic effects of channel regulation in physiology, disease, and therapy. Electrical excitability involves feedback between ion channels and their effects on membrane potential (V_m; Figure 7.1a) that can, in some cases, be modeled using phenomenological descriptions of I–O relationships, without a detailed understanding of channel gating, but more frequently requires gating schemes or semimechanistic models to predict how channels respond and contribute to cell signaling, especially when multiple variables such as V_m and Ca^{2+} are involved.
2. *To predict channel behavior under conditions that are difficult to measure experimentally* is another goal of modeling. Measuring the response of a channel to all relevant stimuli that it could possibly encounter *in vivo* is often impractical, especially for

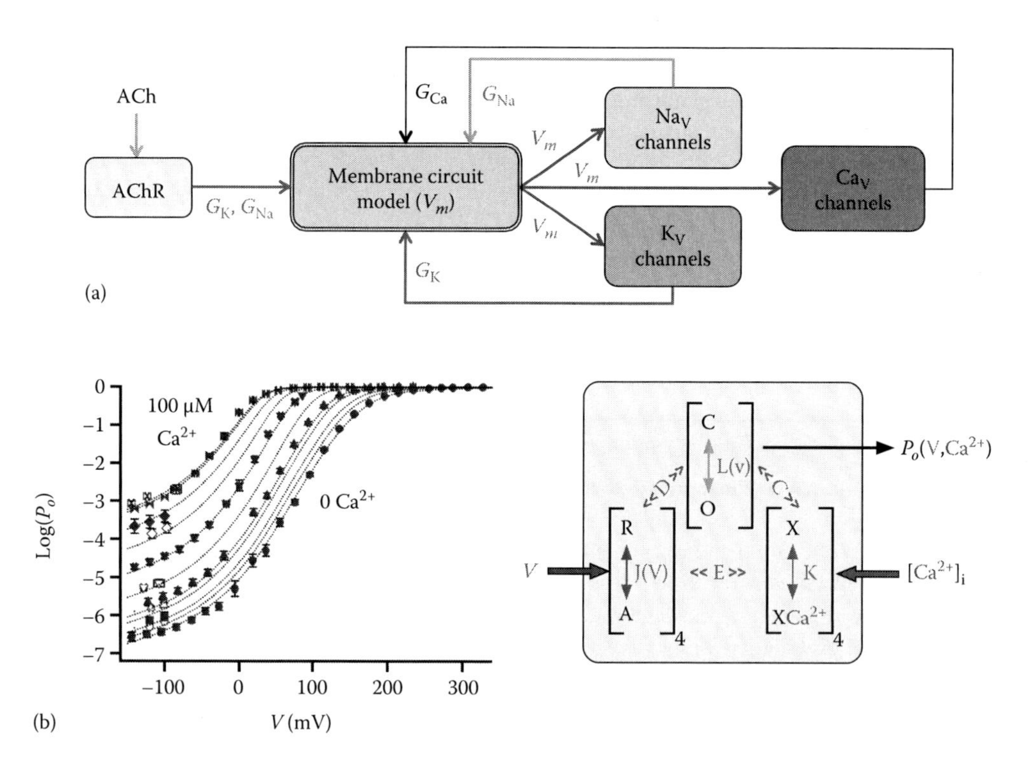

Figure 7.1 The goals of modeling. (a) A model of electrical excitability describing feedback between ion channels and their effects on membrane voltage (V_m). Models of different channel types, represented by boxes, including voltage-dependent K^+, Na^+, and Ca^{2+} channels and nicotinic acetylcholine receptors (AChRs), must reproduce time-dependent I–O relationships between their appropriate stimulus (e.g., V_m or ligand concentration) and open probability (P_o). The contribution of each channel to membrane conductance (G_x) for different ions is defined by their P_o, number and ion selectivity. A circuit model incorporating ionic conductances and the passive electrical properties of the cell is used to predict V_m. (b, left panel) Steady-state Log(P_o)–V relations for BK-type voltage- and Ca^{2+}-dependent K^+ channels measured at different $[Ca^{2+}]$ (in μM: 0 (●), 0.27 (■), 0.58 (Δ), 0.81 (▲), 1.8 (▽), 3.8 (▼), 8.2 (◇), 19 (◆), 68 (⧓), 99 (⋈)) are fit (curves) by the Horrigan–Aldrich (HA) model (b, right panel), which describes overall gating in terms of three gating processes: voltage-sensor activation [R–A], Ca^{2+} binding [X–XCa^{2+}] and opening of the activation gate [C–O], as well as interactions between these processes defined by allosteric factors C, D, and E. (From Horrigan, F.T. and Aldrich, R.W., *J. Gen. Physiol.*, 120, 267, 2002.) The model provides a basis for predicting how the channel integrates voltage and Ca^{2+} signaling and for extrapolating sparsely sampled data to different combinations of V_m and $[Ca^{2+}]_i$ when, as illustrated, the model is fit to a selection of data spanning a wide range of V_m and $[Ca^{2+}]_i$.

(Continued)

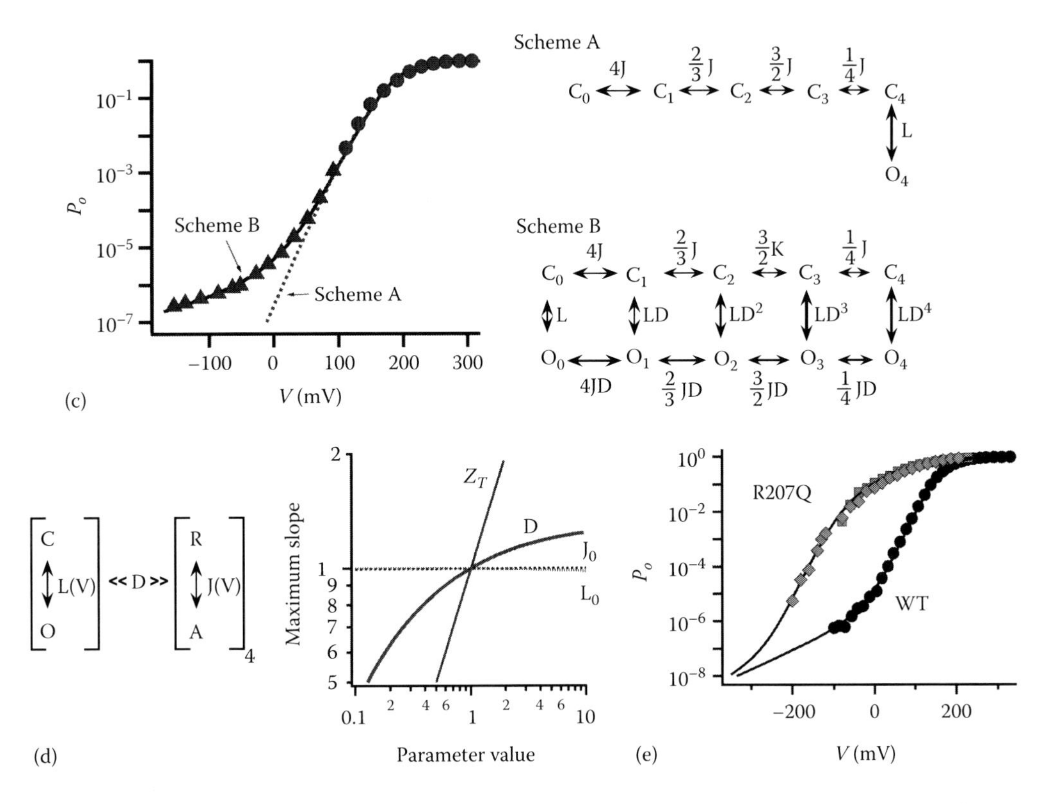

Figure 7.1 (Continued) The goals of modeling. (c) The mean P_o–V relation in 0 Ca^{2+} (right panel) is fit (curves) by two alternative models of voltage-dependent gating (left panel). (From Horrigan, F.T., *J. Gen. Physiol.*, 114, 277, 1999.) The model predictions are indistinguishable based on macroscopic data (●) but diverge at more negative voltages where scheme A achieves a maximum voltage dependence (i.e., limiting slope) while scheme B predicts a reduced slope. Data derived from unitary current recordings at low P_o (▲) rule out scheme A, but are well fit by scheme B, which is an allosteric model equivalent to the HA model in 0 Ca^{2+}. (d) Sensitivity analysis: the HA model in 0 Ca^{2+} (left panel) is defined by five parameters including the equilibrium constants for voltage-sensor (J_0) and gate (L_0) at $V = 0$, the gating charge associated with these transitions (z_J, z_L) and the allosteric factor D. To determine which parameters influence the voltage dependence of activation, the maximal logarithmic slope of the P_o–V relation predicted by the model was plotted as a function of the parameters J_0, L_0, D, or total gating charge $z_T = 4z_J + z_L$ (right panel), where both axes were normalized to the values in panel (c). This shows that the maximal slope cannot be taken as a direct measure of gating charge (z_T) because it is also sensitive to allosteric coupling (D). (From Ma, Z., *J. Gen. Physiol.*, 127, 309, 2006.) (e) Normalized P_o–V relations for wild-type (WT) and R207Q mutant BK channels were fit simultaneously by the HA model with only the voltage-sensor equilibrium constant J_0 altered,[70] suggesting that R207 in the S4 segment does not contribute to gating charge.

channels that are activated or regulated by multiple stimuli. In such cases, a gating model can help to predict a channel's response to numerous combinations of conditions poorly sampled by the measured results (Figure 7.1b, left panel). Similarly, a model may be used to predict channel activity under conditions that are difficult to achieve experimentally (e.g., high temperature). Different gating models vary considerably in their ability to predict a channel's response under conditions that were not measured experimentally. Extrapolating predictions *beyond* conditions that span the original data set typically requires semimechanistic models. An understanding of the physical mechanisms underlying voltage or ligand sensitivity provides a basis for predicting a channel's response over a broad range of conditions. Likewise, predicting how multimodal channels integrate combinations of stimuli is facilitated by understanding how the gating mechanisms are influenced by each stimulus and how those mechanisms interact with each other (Figure 7.1b, right panel).

3. *To translate a mechanistic hypothesis of ion channel function into a testable prediction* is another powerful function of gating models. Discrimination based on different mechanisms of gating specified by models provides mechanistic insight into many processes including stimulus integration and drug action. Models also facilitate the design of experiments to distinguish between competing hypotheses by revealing experimental conditions where the predictions of different hypotheses diverge (Figure 7.1c).
4. *To define the relationship between model parameters and channel response* is a goal of modeling for multiple reasons. First, *sensitivity analysis* may be performed on a model to determine the effect on channel function of systematic changes in parameters, and thereby facilitate identification of elementary processes (e.g., rate-limiting steps, charge movement, or ligand-binding events) that are critical for defining the channel's response (Figure 7.1d). Such analysis can help design experiments to detect perturbations in a gating process by revealing conditions under which the sensitivity of channel function to that process is most obvious (i.e., independent of other parameters). Models are also used for the so-called inverse problem: to derive estimates of important elementary parameters such as gating charge or ligand K_D by fitting the kinetic or steady-state properties of gating over a range of stimulus conditions.

5. *To understand relationships between channel structure and function* often involves gating models. Modeling allows the role of individual amino acid residues to be defined in terms of the gating mechanisms and parameters that are perturbed by mutation (e.g., Figure 7.1e); some examples encompassing voltage sensors and ligand sensors are discussed elsewhere in this chapter (also see the chapters by Chanda and Zagotta in this volume).

7.3 MODEL GENERATION

Figure 7.2 illustrates a typical workflow involved in development of a gating model of an ion channel. Here we will discuss some of the major steps in model building, starting with the first and the most important decision.

7.3.1 DO I NEED TO BUILD A (NEW) GATING MODEL?

The benefits of developing a *successful* gating model are numerous (see Section 7.2), but such development requires an intensive effort. Does the effort required justify *de novo* development? The answer to this question is probably *no*. This may come as a surprise in this chapter, but multiple lines of reasoning may be given.

1. The primary structures of many, if not most, ion channel proteins are now known. Thus, the channel of interest may have a better-studied homolog, for which a good gating model already exists. If so, the existing model may be modified to describe the channel under study. This approach requires less effort and the results will provide valuable comparative information about the original channel and the channel under study, and could potentially provide a nice basis for mutagenesis.
2. *De novo* development of a useful gating model typically requires an extensive and rigorous effort. Without such an effort, alternative models cannot be ruled out and the parameters of the resulting model will not be well constrained (see later text) and offer little insight and predictive power; using an ill-defined model may in fact be detrimental.

7.3.2 WHAT KIND OF A MODEL?

Clearly, the selection depends on the goal or the purpose of the model desired, and some of the possible model types are (1) phenomenological models, (2) statistically parsimonious models, and (3) semimechanistic models (Figure 7.3).

7.3.2.1 Phenomenological models

A phenomenological model is simply a set of equations that describes an aspect or aspects of channel gating. The equations may or may not have a realistic physical meaning attached. An extreme example is to use polynomial equations to describe ionic current kinetics. It is obvious that the polynomial approach fails to provide any physical or mechanistic interpretation; neither the coefficients nor the terms of polynomial equations correspond to any realistic physical aspect of ion channel protein function. Another probable phenomenological approach involves fractal mathematics.[5,6] The equations are not readily associated with a physical model of how channels function and the resulting models do not appear to offer any mechanistic insight. The utility and suitability of the fractal approach was once intensely debated[7,8] but now appears to be of limited utility in part evidenced by the observation that none of the currently prevailing gating models utilizes fractal mathematics.

While a phenomenological model may involve any mathematical function, some functions are more intimately associated with *physical realities*. One such example is the negative exponential function of the form $I(t) = A(1-e^{(-t/\tau)})$ or $I(t) = Ae^{(-t/\tau)}$, where $I(t)$ is the current size as a function of time t, A is a scaling factor, and τ is the time constant. The negative exponential function results from any system where the rate of change depends only on the present condition. A simple negative exponential function may be used to represent activation kinetics of an ionic current without any delay on presentation of a stimulus (e.g., depolarization or an agonist). A *product* of multiple negative exponential functions, for example, $I(t) = A(1-e^{(-t/\tau)})^n$,

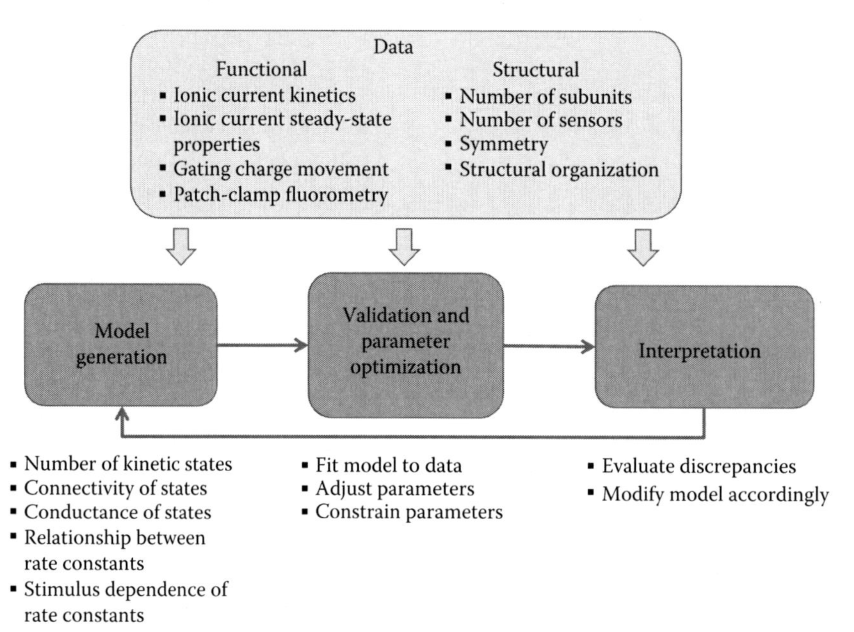

Figure 7.2 A typical model development workflow, including different types of data that may be taken into consideration during the iterative process of generating, testing, and modifying a model.

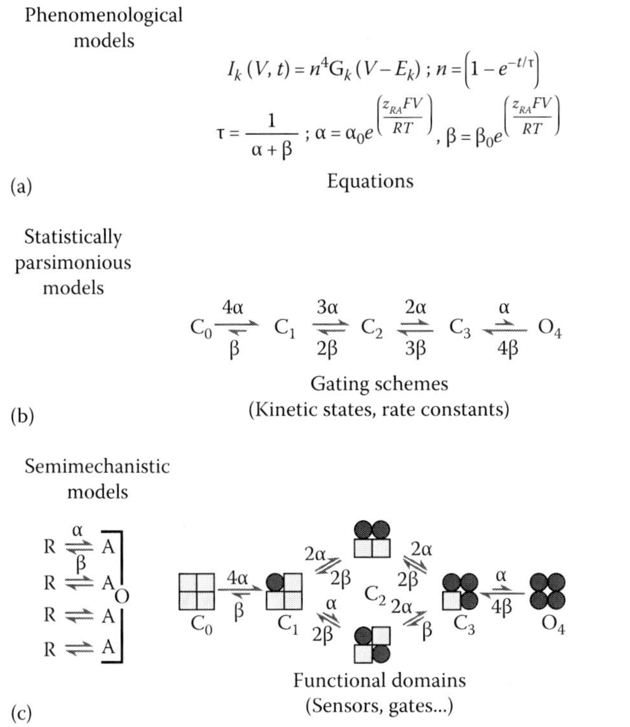

Phenomenological models

$$I_k(V,t) = n^4 G_k (V - E_k)\ ;\ n = \left(1 - e^{-t/\tau}\right)$$

$$\tau = \frac{1}{\alpha + \beta}\ ;\ \alpha = \alpha_0 e^{\left(\frac{z_{RA} FV}{RT}\right)},\ \beta = \beta_0 e^{\left(\frac{z_{RA} FV}{RT}\right)}$$

(a) Equations

Statistically parsimonious models

$$C_0 \underset{\beta}{\overset{4\alpha}{\rightleftharpoons}} C_1 \underset{2\beta}{\overset{3\alpha}{\rightleftharpoons}} C_2 \underset{3\beta}{\overset{2\alpha}{\rightleftharpoons}} C_3 \underset{4\beta}{\overset{\alpha}{\rightleftharpoons}} O_4$$

(b) Gating schemes (Kinetic states, rate constants)

Semimechanistic models

R ⇌ A (α, β); R ⇌ A; R ⇌ A; R ⇌ A — O

C_0, C_1, C_2, C_3, O_4 (4α, β; 2α, 2β; α, 2β; 2α, 2β; 2α, β; α, 4β)

(c) Functional domains (Sensors, gates...)

Figure 7.3 Three gating model types are illustrated in terms of the HH model of K_v channels. (a) Phenomenological models are equations that describe ion channel function. In this case, the relationship between K^+ current I_K, voltage V, and time t, is described by a negative exponential function to the fourth power, where the time constant τ is defined in terms of two voltage-dependent rate constants (α, β). For simplicity, the description of α_0 has been modified from the original description. (From Hodgkin, A.L. and Huxley, A.F., *J. Physiol. (Lond.)*, 117, 500, 1952.) (b) A statistically parsimonious gating scheme consistent with the HH model is a sequential scheme with four closed and one open state. (c) Two semimechanistic representations of the HH model assume that the channel contains four independent and identical subunits each with a voltage sensor that must activate for the channel to open. The left panel illustrates that each voltage sensor can undergo a transition between resting (R) and activated (A) conformations (with voltage-dependent rate constants α, β defined as in panel a) and that the channel is open (O) when all voltage sensors are in the A conformation. A cartoon (right panel) depicts the different states of the channel (C_0–C_3, O_4) defined by the conformation of each subunit (R: square; A: round). The forward rate constant for the C_0–C_1 transition (4α) is four times that for voltage sensor activation because the C_0 state contains four identical sensors, any one of which may activate to generate C_1 (assuming rotational symmetry about the central pore axis). Rate constants for the other forward and backward transitions are similarly dependent on the number of sensors that are available to activate from R to A or deactivate from A to R, respectively (i.e., statistical factors). Potential degeneracy in the C_2 state is illustrated by two conformations that are not equivalent by rotation. But if these conformations are assumed to be functionally indistinguishable, then they can be lumped into a single C_2 state, and the scheme is equivalent to that in panel b.

is often used to describe the *sigmoidal* time course or activation with a delay, as in the Hodgkin and Huxley model of voltage-gated Na^+ and K^+ channels (herein referred to as the HH model).[3] Multiplication of the equations implies that the underlying systems behave probabilistically in an independent manner. Sigmoidal activation kinetics can also be modeled as a *sum* of exponential functions, with fewer probabilistic implications.

Another function used frequently in phenomenological models is a Boltzmann function of the form

$$P_o(V) = \frac{1}{1 + e^{\left(\frac{V_{0.5} - V}{SF}\right)}}$$

where

$P_o(V)$ is P_o at the membrane potential V

$V_{0.5}$ is the membrane potential where $P_o = 0.5$ (half-activation voltage)

SF is a factor that determines the steepness of the curve (the smaller, the steeper)

This equation would be considered *mechanistic only for the simple two-state closed-open channel.* For every other channel type, the formulation is largely phenomenological; alterations in multiple aspects of channel gating can lead to changes in the equation parameters $V_{0.5}$ and SF. The aforementioned equation remains frequently utilized primarily for comparative reasons, but the parameter values should be considered as *phenomenological data descriptors* for most channel types.

7.3.2.2 Statistically parsimonious gating schemes

In developing a statistically parsimonious gating scheme, one relies on a statistical criterion (or, in some cases, including the work of Hodgkin and Huxley, visual inspection), to determine how many components such as kinetic states the scheme should contain and how they are arranged. A gating scheme developed in this manner is statistically parsimonious, balancing the overall goodness of fit (or the badness of fit[9] in some cases) and the cost of additional free parameters—an elegant and satisfying approach. However, this approach ignores infrequent transitions and infrequently visited states of the channel. Further, the model is parsimonious only for the data set analyzed and may not be for other conditions (see *semimechanistic models* later in this article); their general applicability or extensibility may be limited.

The simplest gating scheme to describe opening and closing of a channel is obviously the two-state model with one closed state and one open state. In the statistically parsimonious approach, additional states and transitions are added to describe the available observations only when a certain criterion is met. In this regard, this approach is an intensively data-driven method. As such, it is of utmost importance to have available high-quality experimental results to avoid the *garbage in, garbage out* problem. Some potential experimental pitfalls include but are not limited to voltage-clamp fidelity (e.g., space clamp, series resistance error, bandwidth, and junction potential), accumulation/depletion of ions, ligand-concentration change kinetics, and inhomogeneity of the channels under study (e.g., current separation and multiple channel types). Excellent experimental tips are found in Sakmann and Neher[10] and also in the chapter by Islas in this volume.

The gating scheme approach (and also the semimechanistic approach) almost always makes the following fundamental assumptions: (1) an ion channel has a relatively small number of *kinetically distinguishable states* and (2) the transitions among these discrete states can be described by specific time-invariant (often referred to as time-homogeneous) rate constants.

These two assumptions are generally accepted (with some reservations) in part because of the lack of any strong evidence otherwise (e.g., but see a commentary by Jones[11]). Regarding the first assumption about the number of states, an ion channel protein undoubtedly has numerous conformational states or free-energy minima, as suggested in part by recent all-atom MD simulations[12]; however, these conformations/free-energy minima are grouped into a smaller set of *functionally* (e.g., *electrophysiologically*) *distinguishable states*. Electrophysiological methods, for example, typically detect events lasting anywhere from ~20 μs to hundreds of seconds (depending on the exact experimental conditions such as the single-channel current size and the instrumentation/background noise level), and within this time range, up to a dozen or so *kinetically distinguishable states* have been reported using various criteria[13] (see later text; Figure 7.2).

The two fundamental assumptions reflect the idea that ion channel function is a time-homogeneous Markov process with finite states; the rate constants connecting discrete states *under a given condition* are *memoryless* and do not depend on how long the channel has stayed in one state. Consider, for example, the two-state scheme with one closed state (C) and one open state (O). The rate constant k_{CO} describes the mean number of transitions that this channel makes from C to O per unit time and k_{OC} is that for the reverse transition.

$$C \underset{k_{OC}}{\overset{k_{CO}}{\rightleftharpoons}} O \quad \text{(Scheme 7.1)}$$

According to the time-homogeneous Markov postulate, provided that *the experimental condition remains the same*, the value of k_{OC} remains invariant whether the channel has been open for 1 ms or 1 s. Similarly, the k_{CO} value is the same whether the channel has been closed for 10 ms or 100 s. Consequently, the dwell times within one state are distributed according to the negative exponential function. For example, in the two-state model, the closed durations are described by a single negative exponential function whose time constant is $1/k_{CO}$, and the open durations are described also by a single negative exponential function with a time constant of $1/k_{OC}$. Consistent with the Markov postulate, experimentally measured single-channel dwell times are generally well described by a negative exponential function or a sum of negative exponential functions[7] (for a comprehensive treatment of single-channel dwell time analysis, see various chapters in Sakmann and Neher[10]). However, it is vital to note that dwell times shorter than ~50–100 μs are not well resolved by the current patch-clamp method and *left-censor* correction procedures for these missed events may be imperfect. It is also uncertain theoretically whether rate constants remain truly *memoryless* on time scales approaching the molecular vibrational range (<1 ps).

Rate constants in gating schemes are often given molecular and physical connotations using the Arrhenius activation energy/transition state formulation $k = f'e^{(-(\Delta G/RT))}$, where k is the rate constant (1/s), ΔG is the Gibbs free energy difference between the state in which the channel resides and the transition state or the energy barrier (J/mol; herein referred to as the activation energy), and R (J/(mol × K)) and T(K) have their usual meanings. The term f', often called the frequency factor, vibrational/collisional factor or the pre-exponential factor, describes the mean number of transitions that the channel makes over the energy barrier to the next state as the barrier size for *that particular transition* diminishes to zero and is often quoted to be roughly 10^9–10^{12} s. However, *the value of f' is poorly defined and it has no direct experimental validation* in ion channel function.[14]

Experimentally observed single-channel parameters, such as P_o, of a single-channel molecule are not entirely stationary and they can change with time during the course of an experiment.[15] *Rundown* and *run-up* of ionic currents, observed in many channels especially after patch excision, also represent fluctuations in ion channel gating properties. Such changes are typically considered to reflect changes in channel function; the underlying rate constants may take on new values but are assumed to remain *memoryless* (if the conditions remain constant).

Overall, the Markov process postulate of ion channel function is considered reasonable and prevalent at the present day. Unless results that strongly reject the Markov postulate emerge, the most prudent path to develop a statistically parsimonious gating scheme is to adopt the Markov process assumption. The number of states, the allowed transitions, and the rate constant values are constrained by the experimental measurements and a criterion of choice.

How many states? At least two states of an ion channel are relatively easily recognized: closed and open. There may be additional kinetically distinct states that share very similar single-channel current amplitudes. In general, the following guidelines may be used to estimate the *minimum* number of states that a gating model must contain to be consistent with the experimental observations for a single class of channels.

1. If a channel visits N states during gating whether closed or open, there should be in theory $N - 1$ negative exponential components in ionic current kinetics. In practice, it may be difficult to detect all $N - 1$ components; the fractional amplitudes of some components may be very small and/or their time constants too similar to distinguish. Interested readers are referred to [16–18].
2. If a channel traverses m open states and n closed states, its open durations and closed durations are described, in theory, by sums of m and n negative exponential components, respectively. Again, not all the expected components may be experimentally detectable for various reasons. It is important to note that time constants do not correspond directly to the mean dwell time in any state, except for the simplest cases (closed-open model and closed-open-closed model). Even for the simple closed-closed-open model, the time constants of the two components in the closed durations do not simply correspond to the mean dwell times of the two closed states.[19] Detailed treatments of single-channel dwell time analysis are found in [16,17,20,21].

Additional information to determine the number of states and allowed transitions. Results obtained under different stimulus conditions may be analyzed according to the aforementioned general guidelines. Additional measurement types may also be incorporated to infer the number of states and how they are arranged, at least in select channels.

Gating currents. It is sometimes possible to measure the charge movements associated with gating of voltage-dependent channels.[22] Gating currents are useful for probing transitions among states with the same conductance; charged structural elements of ion channels may move while the ion conduction gate remains closed or in some cases while open.[2] Such measurements complement those of ionic currents and may eliminate certain models from consideration.[23–25]

Conditional probability analysis. In a typical analysis of single-channel dwell times, all noncensored open or closed events are lumped together and analyzed. Information regarding the sequential order in which individual events of different durations occur is discarded. Conditional probability analyses attempt to utilize some of this *sequence* information and may take on many different forms. For example, adjacent open and closed durations may be analyzed.[13,26–30] Such analysis has been used to determine whether voltage-gated Na^+ channels preferentially enter the inactivated state from closed states or the open state.[31]

The information content of single-channel recordings is even more efficiently utilized by the approach pioneered by Horn and colleagues.[20,32] The likelihood (≈probability) that an observed sequence of idealized single-channel closed and open events is generated by each model under consideration is calculated. The likelihood ratio test is then used to select the most parsimonious model among nested models and the Akaike information criterion (AIC) is typically used to rank order non-nested models.[20,33] This approach has been adapted and improved, incorporating the hidden Markov concept so that noisy and/or multichannel data can be analyzed.[34–38] Macroscopic ionic currents generated by a large number of homogeneous channels can be analyzed essentially in a similar manner so that different gating schemes and the associated rate constants can be quantitatively evaluated.[39]

Free software is available to implement various gating-scheme building processes and simulations on multiple operating system platforms. For example, interested readers are referred to http://www.qub.buffalo.edu/ and also http://www.ucl.ac.uk/Pharmacology/dc-bits/dcwinprogs.html.

Unresolved issues. The statistical gating scheme approach has a strong appeal in part because the available information appears objectively evaluated; data from multiple single-channel experiments of similar quality can be fit simultaneously,[40] and the method can also be applied to macroscopic data.[39,41] However, despite advances in the statistical gating scheme approach in part fueled by ever-increasing computational power, it has not been possible to address one particularly important issue—*how to weigh qualitatively different data types in the model evaluation process.* For example, a modern study of a voltage-gated ion channel may involve (1) macroscopic ionic currents, (2) macroscopic gating currents, (3) single-channel ionic currents, and (4) optical measurements (e.g., voltage-clamp fluorometry). The results from these diverse methods, each of which probes different aspects of gating, are clearly complementary. Yet, each has its own unique error and noise characteristics and the results from different types of measurements are difficult to compare in a straightforward manner. If a global fit analysis of macroscopic currents favors one candidate model and a hidden Markov-based single-channel analysis protocol favors another, what final criterion does one use to choose one over the other? The situation may be even more complex if voltage-clamp fluorometry results are to be somehow incorporated concurrently. No satisfactory solution to this important and nagging issue has been put forward to date.

Stimulus-dependent rate constants. Transitions among different states of an ion channel, beyond those induced by the inevitable prevailing thermal energy, are considered to reflect changes in one or more effective rate constant. For example, voltage-gated ion channels change their P_o because the values of select rate constants are influenced by changes in membrane potential (ΔV_m). To develop a successful gating scheme, the dependence of the rate constants on the relevant stimuli must be described. The stimulus dependence could be described by any function but some are more readily understood using well-known physical principles. Many of the formulations used in the gating scheme approach have ready physical interpretations and are also applicable in the semimechanistic approach.

Voltage-dependent rate constants. The voltage dependence of rate constant values has been described using a variety of functions. As noted earlier, the value of a rate constant is related to the activation energy of a transition although numerically in a poorly constrained way. Extending this concept, a voltage-dependent rate constant is thought to be determined by the underlying voltage-dependent activation energy $\Delta G_{total}(V) = \Delta G_i + \Delta G_V(V)$, where $\Delta G_{total}(V)$, representing the total activation energy at voltage V, is assumed to have two additive free energy components, the voltage-independent component ΔG_i and the voltage-dependent component $\Delta G_V(V)$. The latter represents the membrane potential energy transduced by the structural component responsible for this particular transition and it is expressed in many gating schemes and also in semimechanistic models as $\Delta G_V(V) = -zF\delta V$, where F is the Faraday constant (C/mol), V is the membrane potential (V = J/C), and δ represents the fraction of V *felt* by the structural element within the membrane electric field having a valence number of z (unitless). Because experimental determinations of z and δ are problematic, more commonly, these two terms are lumped together into the effective valence number z_e and expressed as $\Delta G_V(V) = -z_eFV$. For example, $z_e = 1$ may mean that a structural component with one elementary charge sensing the entire membrane electric field (V) or a structure with four charges, each sensing only 25% of the voltage, is responsible for the free energy difference between the state in which the channel resides and the transition state. Countless other combinations are of course possible. The value of z_e may be negative and/or noninteger.

Using the Arrhenius activation energy/transition state formulation presented earlier, the voltage-dependent rate constant $k(V)$ may be described as

$$k(V) = f'e^{\left(-\frac{\Delta G_i + \Delta G_V(V)}{RT}\right)} = f'e^{\left(-\frac{\Delta G_i + (-z_eFV)}{RT}\right)} = f'e^{\left(-\frac{\Delta G_i}{RT}\right)} e^{\left(-\frac{(-z_eFV)}{RT}\right)}$$

$$\text{let } k(0) = f'e^{\left(-\frac{\Delta G_i}{RT}\right)}$$

$$k(V) = k(0)e^{\left(-\frac{(-z_eFV)}{RT}\right)}$$

where

$k(0)$ is the rate constant value at 0 mV
$k(V)$ is the rate constant value at voltage V

At room temperature, F/RT is ~0.025 V (J/C), thus $k(V) = k(0)e^{\left(\frac{z_e V}{0.025}\right)}$, where V is the membrane potential in volts. Accordingly, the voltage dependence of a rate constant should be an exponential function of V_m. This is indeed observed in many cases, including β_n in the original HH model of K^+ current activation.[3] However, some rate constants appear to have nonexponential voltage dependence. The apparent nonexponential voltage dependence could be explained in multiple ways. First, the *rate constant* under study may not describe a single transition but it may be a lumped factor involving multiple rate constants with different voltage dependence.[24] Measurements of single transitions in isolation can be difficult to achieve experimentally. Second, the nonexponential voltage dependence has been explained using the concept of *induced-dipole*[42] and such dependence has been described as $k(V) \propto e^{(a + bV + cV^2)}$, where $k(V)$ is the rate constant value at the membrane potential V, and a, b, and c are constants.[42] For a more theoretical treatment of voltage-dependent rate constants, see Neher and Stevens.[42] Voltage-dependent characteristics of the two-state closed-open model are summarized in Figure 7.4.

Ligand-dependent rate constants. Ligand-dependent regulation of ion channels involves ligand-dependent effective rate constants, which are typically assumed to follow the pseudo first-order reaction as illustrated in the following simple hypothetical gating scheme:

$$\mathrm{C} \underset{k_d,\ \mathrm{Y}}{\overset{\mathrm{Y},\ k_a}{\rightleftharpoons}} \mathrm{C{\cdot}Y} \rightleftharpoons \mathrm{O{\cdot}Y} \qquad \text{(Scheme 7.2)}$$

where

Y represents a ligand Y

k_a is the pseudo first-order rate constant of binding (1/(s × M))

k_d (1/s) is the first-order rate constant of unbinding

C represents a nonconducting state without Y bound and in C·Y one molecule of Y is bound but the channel is closed

O·Y signifies the conductive state with one molecule of Y bound

The pseudo first-order binding rate constant k_a has units of 1/(s × M), and the effective rate constant of binding in this scheme is ligand dependent and given by the product [Y] × k_a (1/s), where [Y] represents the concentration (or activity) of the ligand Y *near the binding site.* The scheme is essentially the model of acetylcholine action on nicotinic ACh receptor channels proposed by del Castillo and Katz in 1957.[4] For those channels with multiple ligand-binding sites, multiple sequential binding steps are typically implemented, for example, in various gating schemes (flip/prime models) for many ligand-gated channels such as the nicotinic ACh receptor channel (see the chapters by Zagotta and Lester in this volume and also[43]). While in the scheme binding of Y to the channel is obligatory before the channel opens, some ligand-regulated channels including nicotinic ACh receptor channels and large-conductance Ca^{2+}- and voltage-dependent K^+ channels exhibit openings in the absence of their agonists, implying that in these channels, opening is not obligatory coupled to agonist binding. Mechanistic implications of such openings are discussed later in this chapter.

A direct blocking or inhibitory action of a ligand can be schematized in a similar manner, for example, as illustrated in the following scheme:

$$\mathrm{C} \underset{k_{\mathrm{OC}}}{\overset{k_{\mathrm{CO}}}{\rightleftharpoons}} \mathrm{O} \underset{k_{\mathrm{BO}},\ \mathrm{B}}{\overset{\mathrm{B},\ k_{\mathrm{OB}}}{\rightleftharpoons}} \mathrm{O{\cdot}B} \qquad \text{(Scheme 7.3)}$$

where

B is the blocking ligand B

k_{OB} represents the pseudo first-order rate constant of binding (1/(s × M))

k_{BO} is the first-order rate constant of unblocking (1/s)

In this scheme, binding of one blocking ligand (B) is capable of driving the channel into the nonconductive state (O·B) from the open state (O). In Markov gating schemes with time-homogeneous rate constants, the mean dwell time in one state is the reciprocal of the sum of all the *leaving* or *exiting* rate constants from that

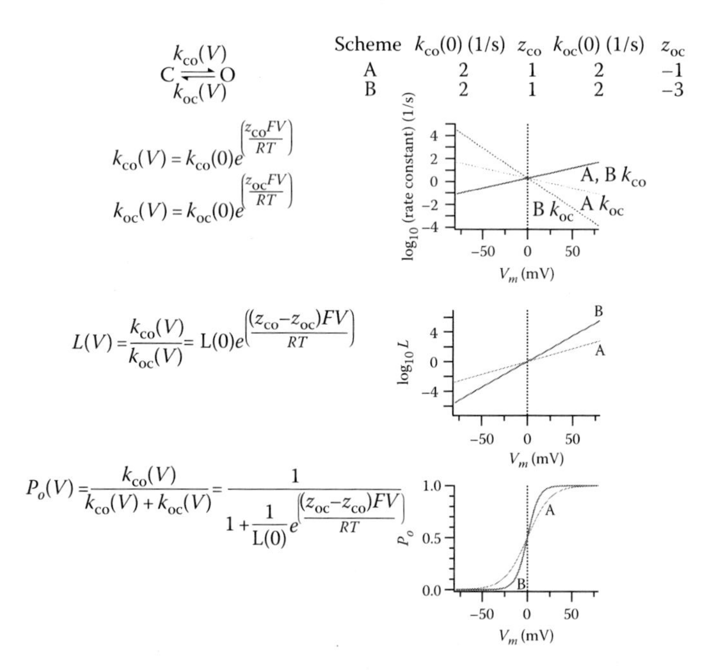

Figure 7.4 Properties of the voltage-dependent closed (C)–open (O) model. The equations describing the model behavior are shown below the scheme. L is the equilibrium constant between the C and O states. The rate constants values (*top*), L (*middle*), and P_o (*bottom*) are illustrated as a function of V_m using two data sets (A and B). Note that the rate constants (k_{CO} and k_{OC}) and the equilibrium constant (L) are exponentially voltage dependent while the voltage dependence of P_o is described by a Boltzmann equation with a slope factor $SF = RT/F(z_{OC}-z_{CO})$ that is inversely proportional to the *total* gating charge for the C–O transition.

particular state. Thus the mean open duration in the scheme is $1/(k_{OC} + [B] \times k_{OB})$, where [B] represents the concentration (activity) of the blocking ligand near the binding site; a graph relating the reciprocal of the mean open duration as a function of [B] should be linear with the slope of k_{OB} and the ordinate intercept of k_{OC}. In this particular simple example where the two nonconductive states (C and O · B) do not directly communicate with each other, the closed durations are described by a sum of two exponentials. The fast component with a time constant of $1/k_{CO}$ with a fractional amplitude of $k_{OC}/(k_{OC} + [B] \times k_{OB})$ reflects the sojourns in the state "C" and the slow component with a time constant of k_{BO} reflect those in the state "O · B."

Temperature-dependent rate constants. Changes in temperature have potential to influence the rate constant of any transition. The activation energy may be expressed as $\Delta G = \Delta H - T\Delta S$ at a given temperature, where ΔH and ΔS are the enthalpy and entropy differences, respectively, between the state and the transition state. Expanding the Arrhenius activation energy/transition state formulation and separating ΔG into the enthalpic and entropic components, the following expressions are obtained:

$$k = f'e^{\left(-\frac{\Delta G}{RT}\right)} = f'e^{\left(-\frac{\Delta H - T\Delta S}{RT}\right)} = f'e^{\left(\frac{\Delta S}{R}\right)}e^{\left(-\frac{\Delta H}{RT}\right)}$$

$$\text{let } f'' = f'e^{\left(\frac{\Delta S}{R}\right)}$$

$$\ln(k) = \ln(f'') - \frac{\Delta H}{RT}$$

A graph plotting $\ln(k)$ as a function of $1/T$ should be linear with a slope of $-\Delta H/R$ and an ordinate intercept of $\ln(f'')$. Transitions with greater ΔH values should be more temperature sensitive. All ion channels are temperature dependent to a degree but some, such as select TRP channels (see the chapter by McKemy in this volume), are exquisitely temperature sensitive, changing their P_o values steeply with small changes in temperature. Such temperature-dependent channels may have multiple transitions with large activation enthalpy levels (ΔH) and perhaps operate in a highly cooperative and allosteric manner.

Mechanosensitive rate constants. All ion channels are considered to be sensitive to mechanical stress,[44] but some are specialized for this purpose. Mechanosensitive ion channels in simple organisms have been particularly extensively studied (see the chapter by Sukharev and Anishkin in this volume). The mechanisms by which mechanical deformations of channel-lipid bilayer membrane complexes alter the rate constants of opening and closing are being intensively studied and interested readers are referred to the chapter by Sukharev and Anishkin in this volume and also to [44–46]. One of the critical issues in this field may be the experimental difficulty associated with stimulating mechanosensitive channels in a well-defined manner.[44]

7.3.2.3 Semimechanistic models

Truly mechanistic models of ion channel gating are built on descriptions of the elementary processes that give rise to functionally and kinetically distinct states. Mechanistic models ideally involve explicit simulation of protein conformational changes but, more commonly, are based on the widely held view that ion channels are modular proteins whose function can be described in terms of the collective activity of distinct functional domains such as sensors and gates. Thus, mechanistic models typically describe the movement or *activation* of functional domains as well as their interaction with each other, and combinatorial factors relating to the number of functional domains that can be activated within a channel. Such models are best described as *semimechanistic* in that they incorporate simplifying or parsimonious assumptions to describe the activation of individual functional domains. Nonetheless, even semimechanistic models offer several important advantages over phenomenological models and those built on purely statistical criteria as summarized next.

1. *Incorporation of diverse data types,* a priori *concepts, and nonquantifiable pieces of information.* Mechanistic and semimechanistic models incorporate a variety of information that cannot, or cannot easily, be integrated into phenomenological and statistically parsimonious gating scheme approaches. While some theoretical concepts, such as those defining stimulus dependence of rate constants, can be implemented in statistically parsimonious models, others describing how functional domains interact (e.g., allosteric factors; see Figure 7.5c through e) or the relationship between rate constants defined by independent functional domains (i.e., statistical factors; see Figure 7.3b and c), cannot be imposed on gating schemes without a mechanistic framework. Likewise, nonquantifiable structural information including the symmetry of homotetrameric channels and presence or absence of direct physical contact between functional domains can place important constraints on mechanistic models that cannot be implemented in a purely statistically parsimonious approach. Semimechanistic approaches capable of incorporating difficult-to-quantify structural information are therefore considered by many as conceptually elegant.
2. *Predictive power.* Semimechanistic models are more likely than phenomenological or purely statistically parsimonious models to be extensible or applicable beyond conditions where the model was originally tested. One reason is that semimechanistic models describe the effects of stimuli in terms of well-understood biophysical processes that are valid over a wide range of stimulus conditions. Statistically optimized gating schemes tend to exclude states that are rarely occupied under the testing condition and may therefore fail to provide accurate predictions when extended to conditions where those states are more frequently occupied.
3. *A framework for structure-function analysis.* Interpretations of the effect of mutagenesis in terms of their impact on functional domains and elementary processes typically require a semimechanistic model. Statistically optimized gating schemes often fail when perturbations are large enough that the *best* gating scheme for mutant and wild-type channels are different (e.g., due to differences in occupancy).
4. *Application to related channels.* Homologous channels are expected to contain the same number and type of functional domains. Therefore, a semimechanistic model for one channel should provide a reasonable framework for analyzing related channels. By contrast, statistically parsimonious gating schemes may not be applicable to homologous channels if, as in the case of mutation noted earlier, the distribution of occupied states is different.

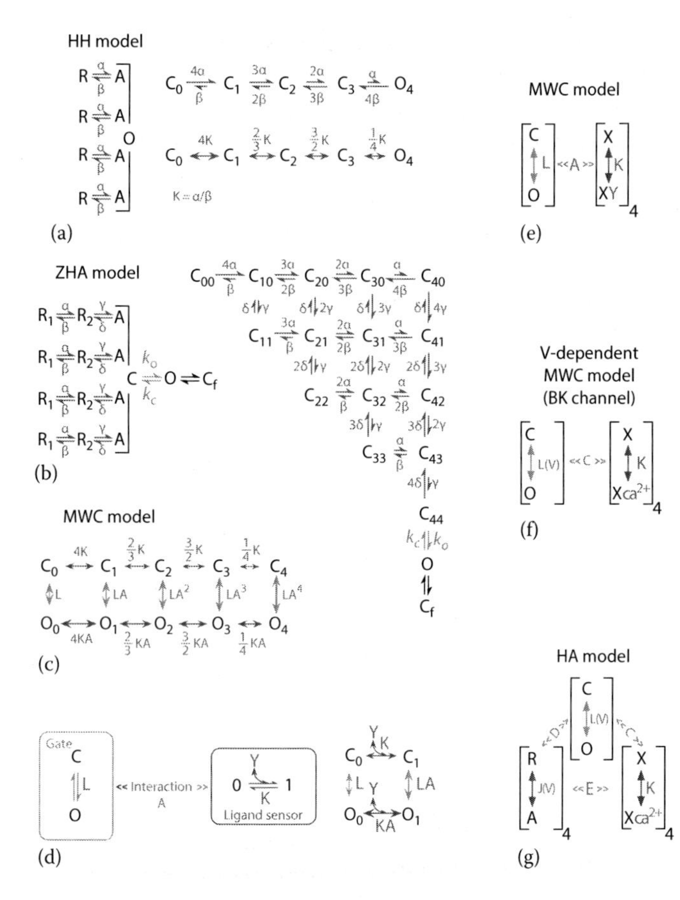

Figure 7.5 Semimechanistic models of voltage-gated K^+ (K_v) channels and allosteric Ca^{2+}- and voltage-gated Slo1 BK channels. (a) The HH model (left panel) generates a sequential gating scheme (right panel) with characteristic rate constants (upper scheme), as described in Figure 7.3. For comparison, the same scheme is shown with equilibrium constants (lower scheme) representing the ratio of forward to backward rate constants. (b) A modified ZHA model of K^+ channel activation (left panel) assumes that independent voltage sensors activate in two steps (R1 to R2 to A) and that a concerted opening transition (C–O) occurs after all voltage sensors are activated. The C_f state is needed to account for brief *flicker* closed events observed when single channels are maximally activated by voltage. The equivalent gating scheme (right panel) contains 15 closed states ($C_{m,n}$) in addition to C_f, where *m*, *n* represent the number of voltage sensors in R_2 and A conformations, respectively. (c) MWC allosteric model of ligand-dependent activation for a tetrameric channel contains five closed (C_0–C_4) and five open states (O_0–O_4), with subscripts indicating the number of activated ligand sensors. (From Hodgkin, A.L. and Huxley, A.F., *J. Physiol. (Lond.)*, 117, 500, 1952.) The channel can open in the absence of ligand (C_0–O_0) as defined by the equilibrium constant L. The equilibrium constant for activating a single ligand sensor is $K = [Y]/K_D(\text{closed})$ when channels are closed and KA when channels are open, where [Y] is the ligand concentration and the allosteric factor $A = K_D(\text{closed})/K_D(\text{open})$. (d) A simple MWC-type mechanism with a single-ligand sensor can be described by two processes, opening of a gate [C–O], binding to a ligand sensor [0–1], and an allosteric interaction between them (A). Ligand binding increases the equilibrium constant for opening A-fold and opening increases the equilibrium constant for ligand binding A-fold, as indicated in a four-state gating scheme (right panel). (e) The MWC model can be described in terms of interaction between one gate and four ligand sensors. (f) A voltage-dependent MWC model of BK channel activation assumes an allosteric interaction between one gate and four Ca^{2+} sensors described by an allosteric factor C, with the equilibrium constant for the C–O transition L(V) being voltage dependent. (From Cox, D.H., *J. Gen. Physiol.*, 110, 257, 1997.) (g) The HA model of BK channel activation describes gating in terms of allosteric interactions between one gate, four Ca^{2+} sensors, and four voltage-sensors (see also Figure 7.1b). (From Horrigan, F.T. and Aldrich, R.W., *J. Gen. Physiol.*, 120, 267, 2002.)

One *disadvantage* of semimechanistic models is that they *tend* to sacrifice some ability to fit functional data compared with statistically parsimonious models with an equivalent number of free parameters. One reason for this is that a semimechanistic model is constrained to *fit* observations that are not necessarily addressed by the statistically parsimonious model. Thus, a statistically parsimonious model may be preferred if the goal is to faithfully reproduce a particular functional response. But semimechanistic models may be preferred for a variety of other reasons as outlined earlier.

Semimechanistic and statistically parsimonious models are not necessarily mutually exclusive. In principle, statistically parsimonious approaches should/could converge on gating schemes that reflect underlying biophysical mechanisms and may be used to infer or test mechanistic assumptions. Likewise, constraints imposed on gating scheme architecture by semimechanistic models, such as the connectivity of states or relationships between rate constants, may be maintained in building and ranking gating schemes. However, in practice, low occupancy of some states and/or errors in the estimation of rate constants limit the utility of such approaches. Exceptions are the cases where the gating mechanism is simple enough and experimental conditions appropriate enough to adequately sample all *accessible* states and transitions. Typically this involves the use of stimulus conditions that limit the number of accessible states, such as saturating ligand concentrations,[47] voltages that fully activate voltage sensors (e.g., to study inactivation), or even the use of channels that have been mutated to reduce the number of functional sensors and simplify possible gating mechanisms.[48]

Stimulus-dependent rate constants in semimechanistic models. The rate constants in semimechanistic models are described essentially as presented earlier for statistically parsimonious gating schemes. In the semimechanistic and mechanistic approaches, one of the ultimate goals is to give structural correlates to the kinetic states and rate constants. Such an effort will be aided by eventual availability of physiologically relevant atomic structures of ion channel proteins. The currently available atomic structures have provided a wealth of information; however, the proteins (or protein fragments) are often in nonnative environments stabilized, in many cases, by nonphysiological crystal lattice interactions, and they are in unknown functional states. Despite these limitations, there is room to improve the existing semimechanistic models by incorporating the results obtained from structure-driven mutagenesis studies.

7.4 EXAMPLES OF SEMIMECHANISTIC MODELS AND ASSUMPTIONS

An advantage of phenomenological and statistically parsimonious models is that systematic approaches exist to build and refine them. For example, statistically parsimonious models may be built from a basic gating scheme by an iterative process of adding and subtracting states, optimizing parameters, and ranking the resulting models as discussed earlier. Similarly, phenomenological models can be chosen by comparing different equations or equations with different number of terms and testing their ability to fit data with parameters optimized through widely available nonlinear curve fitting routines. Indeed, software packages exist to facilitate the selection of a *best* phenomenological model by testing and ranking a large number of different equations (e.g., TableCurve). Semimechanistic models, by contrast, can be more difficult to refine because they are built upon mechanistic assumptions that may be difficult to validate directly or quantitatively. Furthermore, if a semimechanistic model fails to reproduce some features of the data, it cannot be *fixed* simply by extending the original hypothesis. Rather, alternative hypotheses must be formulated, which can be a complex process involving the synthesis of diverse information, identification of mechanisms are consistent with the data, and translation of biophysical principles into simplified models. *Therefore, it is useful to review some examples of semimechanistic models, both to illustrate the types of mechanistic assumptions that have been successful and the process of formulating new hypotheses.*

7.4.1 THE HODGKIN AND HUXLEY (HH) MODEL

The Hodgkin and Huxley (HH) model was developed originally as a phenomenological model but also given a mechanistic interpretation.[3] Although some of the mechanistic assumptions/implications have turned out to be too simple or incorrect, they nonetheless provided great insight at a time when the physical basis of membrane permeability was unknown, and have served as a foundation for many subsequent semimechanistic models. One accurate assumption/implication was (in modern terms) that voltage-dependent K^+ channels contain four identical voltage-sensor domains that can activate largely independently. This idea was based on the observation that the delayed activation of K^+ conductance in squid axon is well described by fourth-order kinetics, whereas deactivation exhibits a single exponential time course. A simple explanation for this observation is that channels are open only when four identical subunits each with a voltage sensor are activated, where activation of each voltage sensor-containing subunit is described by a two-state voltage-dependent transition between resting (R) and activated (A) states, as depicted in Figure 7.3c. The HH model predicts a sequential K^+ channel gating scheme (Figure 7.3b) with four closed states (C_0–C_3) and a single open state (O_4), defined by the number of activated voltage sensors (0–4), and also predicts a characteristic relationship between rate constants and equilibrium constants for different transitions determined by the number of voltage sensors that are available to activate or deactivate (i.e., statistical factors "4", "3", "2", and "1" in Figure 7.5a).

7.4.2 MODERN MODELS OF K_V CHANNEL GATING

Although Hodgkin and Huxley correctly predicted the number of voltage sensors in K^+ channels, this was in fact an estimate based on parsimony (including the computational limitations of the time—pencils and paper) rather than direct evidence, like most assumptions in their model. In subsequent decades, more complicated models including different numbers of independent or interacting voltage sensors were proposed to describe ionic and gating currents in squid axon. However, it was not until molecular biology and the patch-clamp technique were applied to K^+ channels that many mechanistic assumptions could be more properly tested. In the 1990s, models of the first cloned K_v-type channel, Shaker, were developed by Zagotta et al.[24] and by Schoppa and Sigworth[25] based on high-resolution patch-clamp recordings of single-channel and macroscopic K^+ current as well as gating currents and structural information. A critical piece of information was that K_v channels (but not Na_v or Ca_v channels) are indeed composed of four identical subunits, each of which contains a single voltage-sensor domain. This placed absolute constraints on the architecture and symmetry of any proposed model that together with more detailed functional measurements allowed other assumptions of the HH model to be tested and modified. The Zagotta, Hoshi Aldrich (ZHA) model of Shaker, as modified by Ledwell and Aldrich,[49] is depicted in Figure 7.5b and has been used to describe a variety of K_v channels. The key mechanistic assumptions are, briefly:

1. *The channel contains four independent and identical voltage sensors that activate in two steps (R1 to R2 to A).* That voltage sensors activate in *at least* two steps is required to describe multiexponential gating current kinetics; and more than two steps have been proposed in some cases[25] and a more recent study suggests that each K^+ channel voltage sensor activates, at least in theory, in up to four steps.[50] Some hints of cooperativity among the voltage sensors were also noted in the original description of the ZHA model.[24]
2. *Channel opening (C–O) represents a concerted conformational change that follows but is distinct from voltage-sensor activation.* The notion that subunits can undergo both independent transitions associated with voltage sensing and concerted transitions associated with opening was supported by a variety of evidence including the relationship between ionic and gating currents[24,51] and the ability of mutations to separate the voltage dependence of gating current and ionic current activation.[49] Again, the number of steps involved in opening may be subject to question.[25,50] But the fundamental concept that the voltage sensor and ion conduction gate represent structurally and functionally distinct domains whose activation in K_v channels is tightly coupled is now well supported by the atomic structure of K_v channels[52] as well as experiments designed to probe conformational changes in these different domains with chemical accessibility or voltage-clamp fluorometery.[53,54]

7.4.3 ALLOSTERIC MECHANISMS AND MODELS

Allostery, often defined broadly as *action at a distance*, should play a role in the function and regulation of most ion channels in principle simply because channels are modular proteins with structurally distinct functional domains (also see Colquhoun and Lape for more references[43]). Thus, the parts of the channel that detect stimuli (sensor domains) or those that interact with regulatory proteins or bind drugs are often distal from the pore-gate domain that regulates opening and closing. Allosteric mechanisms have been incorporated in many semimechanistic models of ion channel gating, predominantly to describe ligand-gated channels or drug action but also for channels that respond to other stimuli such as voltage as discussed later.

The term *allosteric* was originally coined to describe the idea that two different ligands, such as an enzyme substrate and inhibitor, can interact with a protein and influence each other's apparent binding affinity indirectly through a reversible protein conformational change, or *allosteric transition*, as opposed to directly competing via steric hindrance for a common binding site.[55,56] This concept was soon extended to describe mechanisms of cooperative binding in oligomeric proteins, whereby the interaction of a ligand with one subunit enhances or inhibits the binding of additional molecules of the same ligand to other subunits. To account for the cooperative binding of oxygen to hemoglobin, Monod, Wyman and Changeux devised the semimechanistic MWC allosteric model,[57] which has subsequently had a significant impact on the understanding of ligand action comparable to that of the HH model on membrane excitability, and has influenced or formed the basis for many ion channel models.

7.4.3.1 MWC model

Like the HH model, the MWC model was based on extensive experimental evidence including functional evidence concerning oxygen binding as well as structural and biochemical evidence that hemoglobin is a tetramer whose tertiary and quaternary conformation and subunit–subunit interactions are altered by ligand binding. The MWC model can be applied directly to describe the cooperative activation of tetrameric ligand-gated ion channels (see Zagotta in this volume) and is based on three plausible mechanistic assumptions, which stated in terms of ion channels are

1. The channel can undergo an allosteric transition between two conformations, closed (C) and open (O), involving a concerted change in all four subunits.
2. The affinity of ligand for its receptor in each subunit is the same, but is greater in the open than the closed conformation (i.e., K_D[closed] > K_D[open], where K_D[closed] and K_D[open] are the ligand dissociation constants when the channel is in C or O, respectively).
3. Opening can occur whether or not ligand is bound.

These assumptions define a two-tiered gating scheme (Figure 7.5c) with five closed and five open states, representing the two allosteric conformations (C, O) with different numbers of ligands bound (0–4). These assumptions are also sufficient to define the relationship between equilibrium constants indicated in the model and to explain why ligand binding induces channel opening and why ligand binding is cooperative. To understand these principles, it is useful to first consider a simpler MWC-type model with one ligand-binding site, which generates a four-state gating scheme with two closed and two open states (Figure 7.5d). First, if equilibrium constant between the closed states (C_0–C_1) is defined in terms of ligand concentration and dissociation constant as $K = [Y]/K_D[\text{closed}]$, then the equilibrium constant between the open states (O_0–O_1) can be defined as KA, where A is an allosteric factor representing the relative affinity of the ligand for the open and closed conformations ($A = K_D[\text{closed}]/K_D[\text{open}]$). Consequently, if the transition between closed and open in the absence of ligand (C_0–O_0) is assigned an equilibrium constant L, then the equilibrium constant for opening when ligand is bound (C_1–O_1) must be LA. That is, if opening favors ligand binding (A > 1), then ligand binding must favor opening (LA > L). This is required simply by thermodynamic principles that demand that the equilibrium constant between two states, which is the product of all the equilibrium constants for the intervening transitions, must be independent of path (e.g., $C_0 \Leftrightarrow C_1 \Leftrightarrow O_1$ vs. $C_0 \Leftrightarrow O_0 \Leftrightarrow O_1$).

Several additional features of Figure 7.5d are worth noting, both in terms of the MWC model and gating schemes in general. First, the mechanistic assumptions only constrain the equilibrium constants but not the rate constants in the model because ligand-binding affinity or K_D is an equilibrium concept that relates to the forward and backward rate constants for a ligand-binding step. Several additional assumptions are necessary to include rate constants in the model (see ligand-dependent rate constants given earlier). In fact, equilibrium models often serve a critical role in the development of kinetic and semimechanistic ion channel models because fewer mechanistic assumptions and free parameters are required to constrain the model and because equilibrium constants can be more directly linked to free energy than rate constants. That said, kinetic models are constrained by the same thermodynamic principles noted earlier. In particular, the principle of microscopic reversibility or detailed balance requires in a cyclical scheme like Figure 7.5d that the product of all the rate constants around a cycle in one direction (e.g., $C_0 \rightarrow C_1 \rightarrow O_1 \rightarrow O_0 \rightarrow C_0$) must equal the product of rate constants in the opposite direction ($C_0 \rightarrow O_0 \rightarrow O_1 \rightarrow C_1 \rightarrow C_0$). This constraint applies to any cyclical gating scheme, not just allosteric models, and must be incorporated in statistically parsimonious as well as semimechanistic approaches (otherwise one would have a *perpetual motion machine*). Finally, it is important to note that although allosteric mechanisms often result in a cyclic gating scheme, it is not true that all cyclic models are allosteric. For example, if the affinity of a ligand for closed and open conformation were identical, its interaction with a channel would be described by a 10-state model like MWC but with A = 1. Since A = 1 implies opening is independent of binding, the model is no longer allosteric. Indeed, the ZHA model of K_v channels generates a cyclical gating scheme owing to the assumption that independent voltage sensors can be in multiple states of activation (Figure 7.5b).

When the scheme in Figure 7.5d is extended to four ligand-binding sites (Figure 7.5c), it can be understood why the MWC model predicts cooperative binding. The 10-state MWC scheme exhibits a characteristic relationship between horizontal equilibrium constants, which is identical to that in the HH model (Figure 7.5a) reflecting that sensors, in this case ligand-binding sites, in different subunits are independent and identical. However, the equilibrium constants for transitions among open and closed conformations differ by the allosteric factor A. As a consequence, to satisfy detailed balance, the equilibrium constant for channel opening must increase by a factor A for each ligand bound. In other words, the effects of ligand binding on opening are energetically additive. This illustrates why binding of ligands to all four sites can produce a large increase (A^4-fold) in the equilibrium constant for opening, despite a much smaller A-fold allosteric affinity change for each site. Cooperative binding arises because binding promotes opening, and open channels in turn can bind additional ligand molecules with higher affinity than closed channels. The equilibrium properties of the MWC model can be described by a semimechanistic representation (Figure 7.5e), which includes a concerted conformational change to open the channel [C–O], ligand binding in each of four identical subunits $[X–XY]_4$, and the interaction between these processes described by an allosteric factor A.

The MWC model was innovative because it introduced the idea that the allosteric transition may be a *preexisting* conformational change that can occur in the absence of ligand binding and is *selected* by ligand binding[58] as opposed to concept of *induced fit* whereby ligand binding is required to produce the conformational change. Many ligand-gated ion channels have been shown to open in the absence of ligand although often at a low probability. Furthermore, the MWC model introduced the idea that the allosteric transition could involve a concerted quaternary conformational change. Alternative models of allostery and cooperative binding in hemoglobin and other proteins have been proposed such as the sequential KNF model,[59] which assumes individual subunits change conformation when ligand binds. However, the MWC model has been widely applied in ion channel models because the key allosteric transition in channels (ion conduction gate opening) is thought to be concerted or nearly concerted. Thus, this model provides an elegant and parsimonious mechanism to account for the cooperative response of many ion channels to ligands and other stimuli.

7.4.3.2 Allostery and voltage-dependent gating: Mechanisms of sensor/gate coupling

Before considering an example of an allosteric model of a voltage-dependent channel, it is useful to consider the definition of allostery in the context of voltage-dependent gating. While *action at a distance* may be a reasonable criterion for defining whether a ligand-dependent mechanism is allosteric, it is not directly applicable to voltage-dependent gating. Although the charged residues involved in voltage sensing are distal to the pore-gate domain, their action is not analogous to ligand binding; activation of a voltage sensor involves a conformational change, while ligand binding, at least in the MWC model, does not. A more useful distinction in the case of voltage-dependent gating relates to the difference between competitive and allosteric mechanisms or similarly obligatory vs. nonobligatory coupling. In the MWC model, ligand sensors and a gate show a nonobligatory coupling because opening does not require ligand binding; ligand binding and opening are distinct processes that can occur in isolation but can influence each other (Figure 7.5e). By contrast, the K_v channel models in Figure 7.5a and b, are obligatory because they assume that channels cannot open unless all four voltage sensors are activated. One possible explanation for such a mechanism is that the resting voltage sensor prevents gate opening by steric hindrance, analogous to direct (nonallosteric) competition between two ligands. By contrast, an allosteric mechanism, like the MWC model, allows channels to open whether or not voltage sensors are activated and could involve indirect interaction between the voltage sensor and gate.

The HA model of BK channels: BK channels (aka Slo1, Maxi K, K_{Ca}1.1, KCNMA1) are members of the voltage-dependent K^+ channel superfamily, and they are activated by both voltage and intracellular Ca^{2+}. Unlike K_v channels, BK channels were first studied in detail at the single-channel level, owing to their large conductance, and served as a model system for single-channel analysis as well as the application of statistically parsimonious model-building approach.[28,60–66] McManus and Magleby[65] showed through detailed single-channel analysis of native channels in different [Ca^{2+}] and constant voltage that BK channels can occupy at least three open and five closed states arranged in two (closed, open) tiers, as expected for a channel that can open and close with multiple Ca^{2+} ions bound and can bind Ca^{2+} whether open or closed. The authors noted the arrangement of states in their statistically parsimonious model resembled a subset of those predicted by the MWC model. Cox et al.[67] were the first to propose a voltage-dependent MWC model to describe the kinetic and steady-state properties of macroscopic K^+ current from heterologously expressed mouse Slo1 channels. To account for the effects of voltage, they assumed that the C–O transition in the MWC model was voltage dependent (Figure 7.5f). Consistent with the assumptions that Ca^{2+}-dependent activation can be described by an allosteric mechanism and that voltage- and Ca^{2+}-dependent transitions were distinct, they observed that BK channels could be fully activated by membrane depolarization in the absence of Ca^{2+} binding.[67] In addition, the simplifying assumption that voltage-dependent activation can be described by a two-state process appeared consistent with the near-exponential time course of K^+ current activation and deactivation. Nonetheless, a two-state model of voltage-dependent gating was difficult to reconcile with the tetrameric nature of BK channels and inconsistent with evidence that gating currents in BK channels activate much more rapidly than ionic currents.[68] Indeed, a detailed analysis of ionic and gating currents over a wide range of conditions by Horrigan et al.[69,70] indicated that voltage-dependent gating in BK channels, like Ca^{2+}-dependent gating, could best be described in terms of an allosteric mechanism as formulated in the Horrigan–Aldrich (HA) model[2] (Figure 7.5g). The HA model has been used to interpret the effects of many modulators and mutations on BK channels and has also been adapted successfully to the proton-sensitive BK channel homolog Slo3.[71] It has also served as a guide for developing semimechanistic models of other multimodal voltage- and ligand-gated channels (e.g., HCN, TRPV1; see other chapters in this volume). The key mechanistic assumptions of the HA model in terms of voltage-dependent activation are

1. *Channels contain four independent and identical voltage sensors whose activation can be described by a two-state process ($[R–A]_4$ in the model).* Two-state activation is not merely a simplifying assumption but is supported by the kinetics and steady-state activation of gating currents that are well described by single exponential and Boltzmann functions, respectively,[70] unlike in K_v channels. This might reflect that BK channels only have one voltage-sensing Arg residue in the S4 segment of the voltage sensor,[72] whereas K_v channels have four.[50] The assumption that voltage sensors can activate independently, as in K_v channels, is consistent with the relationship between the voltage dependence of ionic and gating current.[2] But this too may be considered a simplifying assumption because there is also some evidence for nonindependent activation of voltage sensors in the presence of Ca^{2+}.[2,48]
2. *Voltage-sensor activation and channel opening are distinct processes.* This is obvious in BK channels because the kinetics of gating currents and ionic currents differ by almost two orders of magnitude. Thus, the majority of gating charge moves before channels open (i.e., the ion conduction gate opens). This kinetic difference explains why the ionic current kinetics is nearly exponential following a brief delay.

3. *Voltage sensors promote channel opening through an allosteric interaction (D in the model).* This is one of the crucial differences between models of BK and K_v channel gating but is firmly supported by several lines of evidence. First, BK channels can activate in a nearly voltage-*in*dependent manner at extreme negative voltages where voltage sensors are in a resting state, but P_o is greatly increased when voltage sensors are activated. Second, there is a variety of direct and indirect evidence that voltage sensors can activate whether channels are open or closed, including gating current and *limiting-slope* analysis,[2] delays in both activation and deactivation of ionic current,[70] and single-channel analysis indicating that multiple open and closed states can be occupied in 0 Ca^{2+} or saturating Ca^{2+}.[19,30,73] Together, these observations support the notion that both voltage- and Ca^{2+}-sensor activation promote but are not required or obligatory for channel opening.
4. *Voltage- and Ca^{2+}-sensors interact (E in the model).* A weak allosteric interaction between the sensors is evidenced by the ability of Ca^{2+} to shift the voltage dependence of gating charge movement.[2] Consistent with such an interaction, a reciprocal effect of voltage on the apparent Ca^{2+}-affinity of channel activation has also been observed.[74]

7.4.4 SEMIMECHANISTIC MODELS REPRESENT A BALANCED APPROACH

The HH model, which started out as a phenomenological model in the middle part of the twentieth century, is still in use today, more than 60 years after its introduction. Despite some known shortcomings, the HH model has clearly stood the test of time and the reasons are manifold. For example, implementation of the HH model is relatively straightforward. Perhaps more significantly, the model has proven to be largely consistent with present knowledge about the modular structural nature of voltage-gated ion channels; the HH model has transformed itself from a phenomenological model to a semimechanistic model. The models of Shaker K^+ channels by Zagotta et al. (ZHA model) and Schoppa and Sigworth were developed from the start to balance statistical parsimony and structural information. A similar developmental approach was deployed for the HA model of BK channel gating and the resulting model has been applied to other channel types. The longevity and wide applicability of select models are due in part to their semimechanistic approach where the parameters were well constrained by the experimental observations and yet difficult-to-quantify structural information is incorporated.

7.5 SIMULATIONS

A gating scheme or a semimechanistic model is used to simulate state occupancy probabilities, ionic currents, and other functional properties during the model-building process (Figure 7.2) and also in later *application* phases (Figure 7.1).

State occupancy probabilities. With any Markov model, whether a statistically parsimonious model or a semimechanistic model, the most general approach to calculate state occupancy probabilities is the **Q** matrix approach and detailed descriptions of this approach are given in Colquhoun and Hawkes.[18] Suppose that a gating model contains n states (designated as 0 through $n-1$). The corresponding **Q** transition matrix is an n by n array filled with the underlying rate constants except for the diagonal cells from the top left to the bottom right. Those diagonal cells are set so that the sum of all entries in a given row is 0 (Figure 7.6).

$$\mathbf{Q}=\begin{array}{c} 0 \\ 1 \\ x \\ n-2 \\ n-1 \end{array}\overset{\begin{array}{ccccc} 0 & 1 & x & n-2 & n-1 \end{array}}{\begin{bmatrix} -\sum_{j=0,\,j\neq 0}^{n-1} k_{0,j} & k_{0,1} & k_{0,x} & k_{0,n-2} & k_{0,n-1} \\ k_{1,0} & -\sum_{j=0,\,j\neq 1}^{n-1} k_{1,j} & k_{1,x} & k_{1,n-2} & k_{1,n-1} \\ k_{x,0} & k_{x,1} & -\sum_{j=0,\,j\neq x}^{n-1} k_{x,j} & k_{x,n-2} & k_{x,n-1} \\ k_{n-2,0} & k_{n-2,1} & k_{n-2,x} & -\sum_{j=0,\,j\neq n-2}^{n-1} k_{n-2,j} & k_{n-2,n-1} \\ k_{n-1,0} & k_{n-1,1} & k_{n-1,x} & k_{n-1,n-2} & -\sum_{j=0,\,j\neq n-1}^{n-1} k_{n-1,j} \end{bmatrix}}$$

Figure 7.6 The **Q** transition matrix for a channel with n states, from 0 to $n-1$. The row number indicates the state the channel resides now and the column number indicates the state to which the channel transitions next. For example, the rate constant of moving from state "0" to "1" is indicated by $k_{0,1}$.

As an example, for the linear three-state model given next,

$$0 \underset{k_{10}}{\overset{k_{01}}{\rightleftharpoons}} 1 \underset{k_{21}}{\overset{k_{12}}{\rightleftharpoons}} 2 \qquad \text{(Scheme 7.4)}$$

the **Q** transition matrix is

$$\mathbf{Q}=\begin{array}{c} 0 \\ 1 \\ 2 \end{array}\overset{\begin{array}{ccc} 0 & 1 & 2 \end{array}}{\begin{bmatrix} -(k_{01}+k_{02}) & k_{01} & k_{02} \\ k_{10} & -(k_{10}+k_{12}) & k_{12} \\ k_{20} & k_{21} & -(k_{20}+k_{21}) \end{bmatrix}}$$

Further, because the state 0 and the state 2 do not directly communicate, $k_{02} = k_{20} = 0$.

The state occupancy probabilities as a function of time are evaluated using the following expression:

$$\frac{d\mathbf{p}(t)}{dt} = \mathbf{p}(t)\mathbf{Q}$$

$$\mathbf{p}(t) = \left[P_0(t) \;\; P_0(t) \;\; \ldots \;\; P_{n-2}(t) \;\; P_{n-1}(t)\right]$$

where

$\mathbf{p}(t)$ is a one-dimensional array

$P_x(t)$ is the probability of being in state x at time t

This equation has a general solution of the form:

$$\mathbf{p}(t) = \mathbf{p}(0)e^{\mathbf{Q}t}$$

For each state, $P_x(t)$, the probability of being state x at time t is given by

$$P_x(t) = P_x(\infty) + P_x(0)\left(w_0 e^{\frac{t}{\tau_0}} + w_1 e^{\frac{t}{\tau_1}} \cdots + w_{n-2} e^{\frac{t}{\tau_{n-2}}} \right)$$

where w_0, w_1, w_2, etc., are weighing factors for the exponential components with time constants τ_0, τ_1, τ_2, etc. This expression illustrates that there are $n - 1$ negative exponential components in state occupancy kinetics of a channel with n states. The matrix equation could be evaluated using a commercially available data analysis/mathematical package (for helpful tips on matrix operations, see Colquhoun and Hawkes[18]). While the matrix operations may be cumbersome, the approach is versatile and can simulate multiple functional properties of ion channels, including open and closed dwell time distributions.[18]

Alternatively, one may formulate the following expression for each state of the model and perform a straightforward one-dimensional numerical integration. This method is easy to implement in many commercially available data analysis/mathematical packages, and relatively robust if the integration time increment is small enough compared with the reciprocal of the largest rate constant:

$$\frac{dP_x(t)}{dt} = \sum_{\substack{i=0 \\ i \neq x}}^{n-1} P_i(t)k_{i,x} - \sum_{\substack{j=0 \\ j \neq x}}^{n-1} P_x(t)k_{x,j}$$

7.5.1 IONIC CURRENTS

Idealized (noise-free) single-channel ionic currents can be stochastically simulated by repeatedly applying the following three steps:

1. *Simulate one random dwell time event in the current state.* The mean dwell time is given by the reciprocal of the sum of all leaving rate constants out of the state. To simulate stochastic dwell times described by a negative exponential function with a mean of $E[x]$, the expression $x = -E[x] \times \ln$ (RND) may be used, where x is an event duration and RND is a number drawn from a uniform (pseudo)random distribution between 0 and 1.
2. *Set the current size during the determined dwell time to the current size associated with the state (e.g., 0 pA for nonconductive states).*
3. *Determine to which state the channel transitions.* The destination state is determined simply by comparing a number drawn from a uniform (pseudo)random distribution between 0 and 1 and the fractional amplitude of each of the leaving rate constants from the current state. Once the destination state is selected, the first step is repeated. The resulting one-dimensional array of idealized single-channel events may be processed through an analog (or a digital) filter and combined with a *noise* array that matches the instrumentation/background noise of the recording system. Such simulated records can then be analyzed in the same way as experimental records as a way to optimize the model parameter values (Figure 7.2). Assuming that the channels simulated work independently of each other, multichannel results, including macroscopic currents, could be simulated by simply summing the results of multiple one-channel trials.

Alternatively, macroscopic ionic currents through one homogeneous class of channels at a given voltage may be evaluated using the general formulation $I(t) = N\sum_{x=0}^{n-1} i_x P_x(t)$, where $I(t)$ is the ionic current size at time t, N is the number of functionally active channels present, i_x is the current size of state x out of n states, and P_x is the occupancy probability in state x. More specifically, ionic currents through voltage-gated ion channels as a function of the membrane potential (V) and time (t) are described by $I(V,t) = N\sum_{x=0}^{n-1} i_x(V)P_x(V,t)$. The occupancy probability term $P_x(V, t)$ is evaluated according to one of the methods described earlier. While it is convenient to assume that $i_x(V)$ is linear, this postulate must be verified experimentally. The single-channel current–voltage relation $i_k(V)$ should be experimentally determined using single-channel recordings (e.g., those obtained using fast voltage-ramp stimulations) and could be described using any mathematical function including polynomials. The preceding formulation is a convenient way to simulate macroscopic currents when N is large. For small N values, the currents simulated in this way are not realistic because the *noise* or stochastic fluctuations in current caused by discrete open events are absent. Especially for small N values, the stochastic single-channel current simulation method presented earlier provides much more realistic results.

Another potential shortcoming of the preceding expression is that the three terms, N, $i_x(V)$, and $P_x(V, t)$, are considered to be functionally independent. In reality, gating properties and ion permeation/conduction properties of some ion channels are intimately dependent on each other[75] and some cautions may be warranted. For instance, in many voltage-gated K^+ channels, the permeant ion occupancy of the selectivity filter region influences C-type inactivation.[76,77] Even a small, often unintended or unnoticed, increase in the extracellular K^+ concentration, such as that caused when a large number of K^+ channels open concurrently, can significantly alter the kinetics of C-type inactivation.[76,77] Additionally, opening and closing of the primary activation *gate* located at the cytoplasmic side are influenced by the presence of ions in the cavity/vestibule of the channel, which may influence the voltage-sensor function.[75,78,79]

7.5.2 GATING CURRENTS

Simulation of gating currents can be illustrated using the following example with two states “0” and “1”, where the transitions between them are specified by the rate constants k_{01} and k_{10} (1/s):

$$0 \underset{k_{10}}{\overset{k_{01}}{\rightleftharpoons}} 1 \qquad \text{(Scheme 7.5)}$$

If we assume that k_{01} and k_{10} are associated with charges characterized by two valence numbers (unitless), z_{01} and z_{10}, respectively, moving within the entire membrane electric field, the gating current at time t is given by the following expression $I_g(t) = NP_0(t)(z_{01}-z_{10})e_0k_{01}-NP_1(t)(z_{01}-z_{10})e_0k_{10}$, where N represents the total number of channels available, $P_0(t)$ and $P_1(t)$ are probabilities that the channel is in state 0 and state 1, respectively, and e_0 is the elementary charge (≈1.60 × 10^9 C). As one channel moves from state 0 to 1, $(z_{01}-z_{10})e_0$ C of charges move. Gating currents arising from multiple transitions are simply the sum of all the gating currents from the individual transitions. In most gating models, the valence numbers (e.g., z_{01} and z_{10}) associated with the rate constants describing opposing transitions (e.g., k_{01} vs. k_{10}) have different signs ("–" vs. "+").

7.6 VALIDATION AND PARAMETER OPTIMIZATION AND INTERPRETATION

As outlined in Figure 7.2, parameter optimization represents an important step in building statistically parsimonious models as well as semimechanistic models. The importance of this step is evidenced in part by the observation that those gating models where the parameters are well constrained by direct experimental observations (i.e., models with high degrees of *parameter identifiability*) tend to find wide use. For example, the HA model, while it generates a gating scheme with 70 states, has parameter values that can be identified and constrained if proper measurements are made, for example by manipulating critical variables such as V_m and Ca^{2+} over a wide range (Figure 7.1b) and by measuring ionic and gating currents.[2]

The importance of parameter optimization and constraint is twofold and cannot be overemphasized. First, during model development, validation and interpretation are not possible unless parameters can be constrained. *Validation* involves determining whether a model can fit the data, how well it fits, and possibly testing predictions that would help distinguish it from alternative models. Discrepancies between the model prediction and data must then be *interpreted* to construct improved models. But if parameters are poorly constrained, then a model is also poorly defined and difficult to test or improve. For example, the gating charge parameter (z) in a voltage-dependent model may be poorly constrained by fitting ionic current data alone and the valence number estimated in this manner represents zw, where z is the actual gating charge (as might be measured with gating current) and w is a *fudge factor* necessary to make the model fit the data. In this case, w is effectively an additional free parameter in the model with no clear physical interpretation. The model is not *valid* because the best fit gating charge parameter is inconsistent with a direct measure of gating charge. The model is also not *interpretable* because the factor w cannot be readily translated into a revised model that correctly describes the relationship between gating charge and ionic current data. Second, while we have emphasized the great effort required to develop and validate a new model, a considerable effort is also required even to extend a model that has been established/validated for wild-type (WT) channels for use with homologous channel, mutants, or modulators; and parameter optimization and constraint is critical to this process. Adapting a model in this way may not require that alternative models be tested and compared; however, many of the experimental efforts used to constrain the original model are still required to determine parameters in the modified model. For example, if a model has been shown to accurately describe the relationship between gating charge and macroscopic conductance–voltage (G–V) relationships based on WT ionic and gating current measurements, then one cannot assume that fitting the model to the G–V relations of a mutant channel will yield an accurate measure of gating charge. As for the WT, gating current or other measurements may be required. In general, when adapting a model for new use with limited data, it is important to perform a *reality check* to ensure the parameter values are feasible in comparison to parameters in the original model, and ideally to determine how well parameters are constrained as discussed later.

The parameter values of a model may be optimized to better describe the experimental observations in any number of ways. In some cases, this optimization may be relatively automatic if a single statistical criterion, such as the likelihood value, is used to evaluate different fits (see *Statistically parsimonious gating schemes*). In others, iterative heuristic comparisons of simulated and experimental results may be required. Point estimates of the model parameters are often reported, but such estimates alone do not indicate how *best* particular values are; one needs to know how well other values perform. Thus, interval estimates (i.e., error or confidence intervals associated with the estimates) should be considered.

Such error or interval information may be obtained in different ways. If the likelihood (see *Statistically parsimonious gating schemes*) is utilized as the optimization criterion, an m-dimensional likelihood surface could be constructed in theory for a model with m free parameters. Although not comprehensive, the results from different experiments or data sets may be analyzed separately and the parameter values may be simply presented and compared (e.g., [2,70]). Another useful analysis is to perform a sensitivity analysis. The simplest way is to vary one parameter at a time and assess how the perturbation alters the overall goodness/badness of fit although this method fails to account for potential interactions among multiple parameters. Resampling-/bootstrap[80]-based procedures can also provide useful information about the variability and error information associated with each parameter estimate.[33,81] If enough high-quality results are available, it is desirable to divide the data into two sets, one set is used to build the model and the other set is used to validate the model constructed (validation data set). In practice, this is rarely performed and all available results are used for the initial modeling process.

7.7 SUMMARIES, CONCLUSIONS, AND FUTURE DEVELOPMENTS

We presented here three classes of gating models—phenomenological models, statistically optimized gating schemes, and semimechanistic models.* Each has its unique

* The authors regret that space constraint forces us to limit our discussion to only a few select gating models. Our apologies extend to numerous authors whose gating models were not discussed.

strengths and weaknesses, crucially related to the goal of model building. Phenomenological models are relatively easy to develop but typically offer little mechanistic insight. Statistically optimized gating schemes provide detailed information about the channels in the data sets analyzed and some schemes may be given limited mechanistic and/or structural connotations. However, protein structural information is difficult to incorporate into purely statistically oriented gating schemes and they frequently suffer from the lack of extensibility or applicability to results other than the original data sets such as new stimulus conditions, mutants, and channel homologs. Semimechanistic models can incorporate difficult-to-quantify information including structural organizations of the multimeric channel complexes with multiple functional sensors and gates, which may interact in an allosteric manner. Semimechanistic models probably offer the best predictive power and extensibility, acting as elegant conceptual frameworks under a variety of conditions.

Even semimechanistic models of channel gating contain simplifying assumptions including the Markov postulate that ion channels contain discrete kinetically distinguishable states connected by memoryless rate constants. This premise may appear in contrast with the numerous but short-lived protein conformations observed in MD simulations (see the chapter by Allan in this volume). However, many of the conformations observed in the simulations are probably functionally indistinguishable. Undoubtedly, the fields of gating model development and structural simulations will start to converge eventually. In theory, structural simulations based on *physiologically relevant atomic structures* have great promise to provide insight into the conformational changes that underlie gating, and the mechanisms governing ion permeability. However, computational power has only recently advanced to the stage that such models can simulate single-channel behavior over timescales relevant to channel gating (i.e., microseconds to milliseconds[12]) and *the force field equations used in MD simulations are continuing to evolve*. The relative merits of various semimechanistic models and MD simulations are currently difficult to compare.

Gating models, especially semimechanistic models, will continue to be utilized not only to study ion channel structure-function issues but also to assess functional contributions of different channels to cell signaling. For example, the dynamic clamp method[82] to investigate functional roles of various ionic currents relies on simulated ionic currents. Simulations of cell function required for systems physiology and neuroscience will be greatly aided by further development of ion channel gating models.

REFERENCES

1. Colquhoun D. From shut to open: What can we learn from linear free energy relationships? *Biophys. J.* 89, 3673, 2005.
2. Horrigan FT and Aldrich RW. Coupling between voltage sensor activation, Ca^{2+} binding and channel opening in large conductance (BK) potassium channels. *J. Gen. Physiol.* 120, 267, 2002.
3. Hodgkin AL and Huxley AF. A quantitative description of membrane current and its application to conduction and excitation in nerve. *J. Physiol. (Lond.)* 117, 500, 1952.
4. del Castillo J and Katz B. Interaction at end-plate receptors between different choline derivatives. *Proc. R. Soc. Lond. B. Biol. Sci.* 146, 369, 1957.
5. Liebovitch LS and Sullilvan M. Fractal analysis of a voltage-dependent potassium channel from cultured mouse hippocampus neurons. *Biophys. J.* 52, 979, 1987.
6. Liebovitch LS. Testing fractal and Markov models of ion channel kinetics. *Biophys. J.* 55, 373, 1989.
7. McManus OB, Spivak CE, Blatz AL, Weiss DS, and Magleby KL. Fractal models, Markov models, and channel kinetics. *Biophys. J.* 55, 383, 1989.
8. Korn SJ and Horn R. Statistical discrimination of fractal and Markov models of single-channel gating. *Biophys. J.* 54, 871, 1988.
9. Kline RB. *Principles and Practice of Structural Equation Modeling*, third ed., Guilford Press, New York, p. xvi, 2011.
10. Sakmann B and Neher E. *Single-Channel Recording*, second ed., Plenum, New York, p. 700, 1995.
11. Jones SW. Are rate constants constant? *J. Physiol. (Lond.)* 571, 502, 2006.
12. Jensen MØ et al. Mechanism of voltage gating in potassium channels. *Science* 336, 229, 2012.
13. McManus OB, Blatz AL, and Magleby KL. Inverse relationship of the durations of adjacent open and shut intervals for Cl and K channels. *Nature* 317, 625, 1985.
14. Andersen OS. Graphic representation of the results of kinetic analyses. *J. Gen. Physiol.* 114, 589, 1999.
15. Silberberg SD, Lagrutta A, Adelman JP, and Magleby KL. Wanderlust kinetics and variable Ca^{2+}-sensitivity of *Drosophila*, a large conductance Ca^{2+}-activated K^+ channel, expressed in oocytes. *Biophys. J.* 70, 2640, 1996.
16. Magleby KL and Weiss DS. Estimating kinetic parameters for single channels with simulation. A general method that resolves the missed event problem and accounts for noise. *Biophys. J.* 58, 1411, 1990.
17. Colquhoun D and Hawkes AG. The principles of the stochastic interpretation of ion-channel mechanisms. *Single-Channel Recording*, Sakmann B and Neher E (Eds.), Plenum, New York, p. 397, 1995a.
18. Colquhoun D and Hawkes AG. A Q-matrix cookbook: How to write only one program to calculate the single-channel and macroscopic predictions for any kinetic mechanism. *Single-Channel Recording*, Sakmann B and Neher E (Eds.), Plenum, New York, p. 589, 1995b.
19. Shelley C and Magleby KL. Linking exponential components to kinetic states in Markov models for single-channel gating. *J. Gen. Physiol.* 132, 295, 2008.
20. Horn R and Lange K. Estimating kinetic constants from single channel data. *Biophys. J.* 43, 207, 1983.
21. Colquhoun D and Sigworth FJ. Fitting and statistical analysis of single-channel records. *Single-Channel Recording*, Sakmann B and Neher E (Eds.), Plenum, New York, p. 483, 1995.
22. Armstrong CM and Bezanilla F. Charge movement associated with the opening and closing of the activation gates of the Na channels. *J. Gen. Physiol.* 63, 533, 1974.
23. Armstrong CM and Bezanilla F. Inactivation of the sodium channel. II. Gating current experiments. *J. Gen. Physiol.* 70, 567, 1977.
24. Zagotta WN, Hoshi T, and Aldrich RW. Shaker potassium channel gating. III: Evaluation of kinetic models for activation. *J. Gen. Physiol.* 103, 321, 1994.
25. Schoppa NE and Sigworth FJ. Activation of Shaker potassium channels III—An activation gating model for wild-type and V2 mutant channels. *J. Gen. Physiol.* 111, 313, 1998.
26. Colquhoun D and Hawkes AG. A note on correlations in single ion channel records. *Proc. R. Soc. Lond. B Biol. Sci.* 230, 15, 1987.

27. Magleby KL and Song L. Dependency plots suggest the kinetic structure of ion channels. *Proc. Biol. Sci.* 249, 133, 1992.
28. Rothberg BS and Magleby KL. Kinetic structure of large-conductance Ca^{2+}-activated K^+ channels suggests that the gating includes transitions through intermediate or secondary states. A mechanism for flickers. *J. Gen. Physiol.* 111, 751, 1998.
29. Rothberg BS and Magleby KL. Gating kinetics of single large-conductance Ca^{2+}-activated K^+ channels in high Ca^{2+} suggest a two-tiered allosteric gating mechanism. *J. Gen. Physiol.* 114, 93, 1999.
30. Rothberg BS and Magleby KL. Voltage and Ca^{2+} activation of single large-conductance Ca^{2+}-activated K^+ channels described by a two-tiered allosteric gating mechanism. *J. Gen. Physiol.* 116, 75, 2000.
31. Aldrich RW and Stevens CF. Inactivation of open and closed sodium channels determined separately. *Cold Spring Harb. Symp. Quant. Biol.* 48, 147, 1983.
32. Horn R and Vandenberg CA. Statistical properties of single sodium channels. *J Gen. Physiol.* 84, 505, 1984.
33. Horn R. Statistical methods for model discrimination. Applications to gating kinetics and permeation of the acetylcholine receptor channel. *Biophys. J.* 51, 255, 1987.
34. Chung SH, Moore JB, Xia LG, Premkumar LS, and Gage PW. Characterization of single channel currents using digital signal processing techniques based on Hidden Markov Models. *Philos. Trans. R. Soc. Lond. B Biol. Sci.* 329, 265, 1990.
35. Klein S, Timmer J, and Honerkamp J. Analysis of multichannel patch clamp recordings by hidden Markov models. *Biometrics* 53, 870, 1997.
36. Qin F, Auerbach A, and Sachs F. Maximum likelihood estimation of aggregated Markov processes. *Proc. R. Soc. Lond. B Biol. Sci.* 264, 375, 1997.
37. Syed S, Mullner FE, Selvin PR, and Sigworth FJ. Improved hidden Markov models for molecular motors, part 2: Extensions and application to experimental data. *Biophys. J.* 99, 3696, 2010.
38. Mullner FE, Syed S, Selvin PR, and Sigworth FJ. Improved hidden Markov models for molecular motors, part 1: Basic theory. *Biophys. J.* 99, 3684, 2010.
39. Milescu LS, Akk G, and Sachs F. Maximum likelihood estimation of ion channel kinetics from macroscopic currents. *Biophys. J.* 88, 2494, 2005.
40. Colquhoun D, Hatton CJ, and Hawkes AG. The quality of maximum likelihood estimates of ion channel rate constants. *J. Physiol. (Lond.)* 547, 699, 2003.
41. Balser JR, Roden DM, and Bennett PB. Global parameter optimization for cardiac potassium channel gating models. *Biophys. J.* 57, 433, 1990.
42. Neher E and Stevens CF. Voltage-driven conformational changes in intrinsic membrane proteins. *The Neurosciences; Fourth Study Program*, Schmitt FO and Worden FG (Eds.), MIT Press, Cambridge, MA, p. 623, 1979.
43. Colquhoun D and Lape R. Perspectives on: Conformational coupling in ion channels: Allosteric coupling in ligand-gated ion channels. *J. Gen. Physiol.* 140, 599, 2012.
44. Sukharev S and Sachs F. Molecular force transduction by ion channels: Diversity and unifying principles. *J. Cell Sci.* 125, 3075, 2012.
45. Haselwandter CA and Phillips R. Connection between oligomeric state and gating characteristics of mechanosensitive ion channels. *PLoS Comput. Biol.* 9, e1003055, 2013.
46. Markin VS and Sachs F. Thermodynamics of mechanosensitivity. *Phys. Biol.* 1, 110, 2004.
47. Grosman C, Zhou M, and Auerbach A. Mapping the conformational wave of acetylcholine receptor channel gating. *Nature* 403, 773, 2000.
48. Shelley C, Niu X, Geng Y, and Magleby KL. Coupling and cooperativity in voltage activation of a limited-state BK channel gating in saturating Ca^{2+}. *J. Gen. Physiol.* 135, 461, 2010.
49. Ledwell JL and Aldrich RW. Mutations in the S4 region isolate the final voltage-dependent cooperative step in potassium channel activation. *J. Gen. Physiol.* 113, 389, 1999.
50. Tao X, Lee A, Limapichat W, Dougherty DA, and MacKinnon RA. Gating charge transfer center in voltage sensors. *Science* 328, 67, 2010.
51. Zagotta WN, Hoshi T, Dittman J, and Aldrich RW. Shaker potassium channel gating. II: Transitions in the activation pathway. *J. Gen. Physiol.* 103, 279, 1994.
52. Long SB, Tao X, Campbell EB, and MacKinnon R. Atomic structure of a voltage-dependent K^+ channel in a lipid membrane-like environment. *Nature* 450, 376, 2007.
53. Broomand A and Elinder F. Large-scale movement within the voltage-sensor paddle of a potassium channel-support for a helical-screw motion. *Neuron* 59, 770, 2008.
54. Mannuzzu LM, Moronne MM, and Isacoff EY. Direct physical measure of conformational rearrangement underlying potassium channel gating. *Science* 271, 213, 1996.
55. Monod J and Jacob F. Teleonomic mechanisms in cellular metabolism, growth, and differentiation. *Cold Spring Harb. Symp. Quant. Biol.* 26, 389, 1961.
56. Monod J, Changeux JP, and Jacob F. Allosteric proteins and cellular control systems. *J. Mol. Biol.* 6, 306, 1963.
57. Monod J, Wyman J, and Changeux J-P. On the nature of allosteric transitions: A plausible model. *J. Mol. Biol.* 12, 88, 1965.
58. Changeux JP. 50th Anniversary of the word "allosteric". *Protein Sci.* 20, 1119, 2011.
59. Koshland DE, Jr., Nemethy G, and Filmer D. Comparison of experimental binding data and theoretical models in proteins containing subunits. *Biochemistry* 5, 365, 1966.
60. Pallotta BS, Magleby KL, and Barrett JN. Single channel recordings of Ca^{2+}-activated K^+ currents in rat muscle cell culture. *Nature* 293, 471, 1981.
61. Barrett JN, Magleby KL, and Pallotta BS. Properties of single calcium-activated potassium channels in cultured rat muscle. *J. Physiol. (Lond.)* 1982, 211, 1982.
62. Methfessel C and Boheim G. The gating of single calcium-dependent potassium channels is described by an activation/blockade mechanism. *Biophys. Struct. Mech.* 9, 35, 1982.
63. Magleby KL and Pallotta BS. Burst kinetics of single calcium-activated potassium channels in cultured rat muscle. *J. Physiol. (Lond.)* 344, 605, 1983.
64. McManus OB and Magleby KL. Kinetic states and modes of single large-conductance calcium-activated potassium channels in cultured rat skeletal muscle. *J. Physiol. (Lond.)* 402, 79, 1988.
65. McManus OB and Magleby KL. Accounting for the Ca^{2+}-dependent kinetics of single large-conductance Ca^{2+}-activated K^+ channels in rat skeletal muscle. *J. Physiol. (Lond.)* 443, 739, 1991.
66. Moss BL, Silberberg SD, Nimigean CM, and Magleby KL. Ca^{2+}-dependent gating mechanisms for dSlo, a large-conductance Ca^{2+}-activated K^+ (BK) channel. *Biophys. J.* 76, 3099, 1999.
67. Cox DH, Cui J, and Aldrich RW. Allosteric gating of a large conductance Ca-activated K^+ channel. *J. Gen. Physiol.* 110, 257, 1997.
68. Stefani E et al. Voltage-controlled gating in a large conductance Ca^{2+}-sensitive K^+ channel (hslo). *Proc. Natl. Acad. Sci. U.S.A.* 94, 5427, 1997.
69. Horrigan FT and Aldrich RW. Allosteric voltage gating of potassium channels II. Mslo channel gating charge movement in the absence of Ca^{2+}. *J. Gen. Physiol.* 114, 305, 1999.
70. Horrigan FT, Cui J, and Aldrich RW. Allosteric voltage gating of potassium channels I. Mslo ionic currents in the absence of Ca^{2+}. *J. Gen. Physiol.* 114, 277, 1999.

71. Zhang X, Zeng X, and Lingle CJ. Slo3 K^+ channels: Voltage and pH dependence of macroscopic currents. *J. Gen. Physiol.* 128, 317, 2006.
72. Ma Z, Lou XJ, and Horrigan FT. Role of charged residues in the S1–S4 voltage sensor of BK channels. *J. Gen. Physiol.* 127, 309, 2006.
73. Talukder G and Aldrich RW. Complex voltage-dependent behavior of single unliganded calcium-sensitive potassium channels. *Biophys. J.* 78, 761, 2000.
74. Sweet TB and Cox DH. Measurements of the BK_{Ca} channel's high-affinity Ca^{2+} binding constants: Effects of membrane voltage. *J. Gen. Physiol.* 132, 491, 2008.
75. Yellen G. Single channel seeks permeant ion for brief but intimate relationship. *J. Gen. Physiol.* 110, 83, 1997.
76. Kurata HT and Fedida DA. Structural interpretation of voltage-gated potassium channel inactivation. *Prog. Biophys. Mol. Biol.* 92, 185, 2006.
77. Hoshi T and Armstrong CM. C-type inactivation of voltage-gated K^+ channels: Pore constriction or dilation? *J. Gen. Physiol.* 141, 151, 2013.
78. Chen FS, Steele D, and Fedida D. Allosteric effects of permeating cations on gating currents during K^+ channel deactivation. *J. Gen. Physiol.* 110, 87, 1997.
79. Pathak M, Kurtz L, Tombola F, and Isacoff E. The cooperative voltage sensor motion that gates a potassium channel. *J. Gen. Physiol.* 125, 57, 2005.
80. Simon JL. *Resampling: The New Statistics*, Duxbury Press, Belmont, CA, p. 290, 1992.
81. Hoshi T, Zagotta WN, and Aldrich RW. *Shaker* potassium channel gating. I: Transitions near the open state. *J. Gen. Physiol.* 103, 249, 1994.
82. Prinz AA, Abbott LF, and Marder E. The dynamic clamp comes of age. *Trends Neurosci.* 27, 218, 2004.

8 Utilizing Markov chains to model ion channel sequence variation and kinetics

Anthony Fodor

Contents

8.1 INTRODUCTION: SEQUENCE VARIATION DRIVES DIFFERENCES IN CHANNEL BIOPHYSICS, BUT LINKING THIS VARIATION TO HUMAN PHENOTYPE VARIATION REMAINS A CHALLENGE

In their Nobel Prize–winning formulation of the squid giant axon, Hudgkin and Hulxey [1] showed how quantitative models could be applied to describe how ion currents shaped electrical signaling. In the Hodgkin and Huxley models, parameters for activation and inactivation were fit from experimentally observed electrophysiological data. With the advent of molecular biology in the late 1980s, ion channels were cloned, and full-length sequences become known for many ion channels [2–5]. It was a natural question to ask how sequence variation in ion channels was linked to variation in parameters that described their electrical properties. With the introduction of site-directed mutagenesis and the ability to express ion channels in exogenous systems such as *Xenopus* oocytes [6], the effect of changes in sequence could be directly measured. This approach of introducing variants into a channel sequence and observing the results for channel function has had many successes in a wide variety of channels and has produced a detailed understanding of how sequence variation can constrain channel function. For example, in *Shaker* potassium channels, site-directed mutagenesis demonstrated that inactivation was tied to the amino-terminus region [7]. In the CFTR channel, distinct mutations associated with loss of channel function were determined to be causative of most types of cystic fibrosis [8].

While specific mutations in crucial regions can be linked to dramatic changes in channel function and disease, mutations throughout an entire ion channel's sequence can often affect ion-channel energetics in subtle ways. Studies that have utilized chimeras between related channels have found a modular design to channels in which different regions of the channel cooperate to influence channel energetics. For example, in cyclic-nucleotide-gated (CNG) channels, it has been demonstrated that a channel derived from olfactory epithelium is much more prone to favor an open state when compared with a homologous channel from rod cells. Chimeric channels that were intermediate in sequence between rod and olfactory channels had intermediate energetics with no one region of the channel sequence completely controlling the energetics of channel opening [9,10]. These results suggest that mutations that tune channel function can be spread throughout the channel sequence and that sequence variants found in sequence databases may cause subtle changes to channel energetics that may nonetheless be important in determining channel phenotype and function.

At the turn of the century, the human genome project was completed, initiating the era of genomics. While the first human genome project took decades of work with costs running into the billions of dollars, recent advances in next-generation sequencing have brought us to the edge of the *thousand dollar genome*. We can look forward to a near future in which the genomes of every subject in a clinical trial are known. A central challenge for all families of genes in the postgenomic era is to link changes in genome sequence to functional consequences in health and disease. Since mutations throughout the channel sequence can tune channel function in ways that presumably have profound consequences for host phenotype, it would be natural to think that whole-genome association studies, that link genotypic changes to phenotypes of interest, would yield a rich repository of important changes to ion channels that could explain diseases, such as heart disease and psychiatric disorders, that we know must involve ion channels. The expectation that whole-genome association studies would substantially explain complex diseases, however, has in general not been realized. In study after study, it has been found that variations described by whole-genome association studies have only been able to explain a small percentage of human phenotypic variation [11]. For example, a recent meta-analysis looked at genome-wide association studies (GWAS) that examined five neurological disorders: autism spectrum disorder, attention deficit hyperactivity disorder, bipolar disorder, major depressive disorder, and schizophrenia [12]. Intriguingly, single nucleotide polymorphisms (SNPs) within two L-type voltage-gated calcium channels were significantly associated with several of these disorders. Not only does a study of this sort suggest these calcium channels as a possible drug target but it also suggest that a single genetic mechanism might underlie a series of diseases that are currently categorized in the clinic in distinct ways. This raises the possibility that in the near future genetic approaches may allow for more clear and consistent diagnoses than are possible by only considering phenotype. However, as is the case in nearly all GWAS [13], while the contribution of the calcium channels to these disorders is reproducible across studies and statistically significant, the relative risk that carriers of mutant alleles of these channels have for the psychiatric disorders is modest. Put another way, the presence of these mutant channel alleles explains very little of who in the population does or does not present disease.

For most complex diseases, as in these psychiatric disorders, it is apparent that there is a great deal of human variation that cannot be modeled as an additive sum of independent changes in gene sequence from common alleles. There has been feverish interest in determining the source of the *missing heritability* that is generally not captured by GWAS studies [11,14]. Because most GWAS studies have focused on common alleles, one possible explanation for the low power of GWAS studies is that most human variation in complex diseases is in fact caused by low-frequency alleles. As SNP arrays become replaced by ever more affordable full-genome resequencing, it will become increasingly possible to measure rare variants. As more genomes are sequenced from more people allowing for more genomic detail for larger sample sizes, we will learn to what extent rare variants contain the missing variability. In addition, as technologies continue to develop that allow for measurement of genome structures that have not necessarily been captured by SNP arrays—such as copy number variation and epigenetic changes—we will learn to what extent genome variation not described by simple sequence explains the missing heritability. Gene–gene interactions [14] and gene–environment interactions [15,16] may also explain the low power of GWAS studies. If the missing heritability is caused by these sorts of complex interactions, then understanding the link between sequence variation and consequences for health in all genes families—including ion channels—will require new experimental paradigms that consider the context in which sequence variation within an individual gene occurs.

While whole-genome associations studies look for variation within a species to try and link sequence variation to function, it is also possible to look across species to track how evolutionary changes across long time spans constrain channel function. This approach is dependent upon finding conserved ion channels in organisms that may be distantly related. In this chapter, we examine fundamental algorithms that allow for detection of ion channels in large sequence databases. We note that some of these algorithms may already be familiar to students of ion channels as the Hidden Markov Model (HMM) formalism that describes the behavior of ion channels can easily be modified and applied to sequence alignments. We conclude with a discussion of how sequence variation is likely to be studied in the future within the context of gene–gene and gene–environment interactions.

8.2 THERE ARE HIGHLY EFFICIENT ALGORITHMS FOR FINDING ION CHANNELS IN LARGE DATABASES

As the cost of sequencing continues its exponential drop, the number of distinct proteins that have been identified has greatly increased. A dramatic example of this phenomenon was the Venter Institute's Sorcerer II global ocean survey, which generated 7.7 million sequencing reads from 41 ocean sites [17]. At the time of publication in 2007, this was the largest survey of mixed environmental microbial communities. Remarkably, this single sequencing effort nearly doubled the number of proteins known at the time [18]. With the increasing prevalence of next-generation sequencing, this tremendous increase in sequence diversity of proteins from across the phylogenetic tree has only continued.

Given a query sequence, say an ion channel or an ion-channel fragment, and a large database of sequences to search, the most basic bioinformatics task is to find sequences similar to the query within the database. The execution of such an algorithm allows us to discover new channels in a newly sequenced organism given knowledge of existing channels. These *homology detection algorithms* come in two flavors: searches based on *global alignments* attempt to find the entire query sequence in the database while searches based on *local alignments* look for a significant match of any part of the query sequence, for example, that of the potassium channel selectivity filter. Because sequence reads from Sanger sequencing or next-generation sequencing are often shorter than reads for an entire gene, especially for long genes like ion channels, and because evolution may have only conserved part of a protein across multiple species, algorithms based on local alignment are often more useful than those based on global alignments.

The most commonly used local alignment algorithm is basic local alignment search tool (BLAST) [19], developed over 25 years ago, but still indispensable for many bioinformatics applications. BLAST is an example of a *pairwise alignment* algorithm, meaning it compares the query sequence to each sequence in the database. Pairwise alignment algorithms often utilize a *substitution matrix*, which defines the similarity of symbols between sequences so that, for example, an alanine in the query sequence will be more likely to be matched to a glycine in the target rather than to a proline residue. Substitution matrices are empirically derived from sets of aligned sequences. Many implementations of BLAST default to the BLOSUM62 matrix [20], which was built on a training set of proteins with a threshold of a 62% identity. For all 20 possible pairs of amino acids, the frequency of each substitution within the training set is observed and converted to a log probability of a match divided by the background frequency of the residue pair. Like any data constructed with a training set, the choice of a substitution matrix captures or summarizes our expectations of what it means for two proteins to be similar. Since default training sets will be built from many different protein families, and not just membrane proteins or ion channels, these choices in principle can affect what is detected when using a membrane or ion-channel query sequence. Substitution matrices derived from membrane proteins may therefore be more sensitive for use in ion-channel homology detection than generic substitution matrices [21]. In practice, however, generic substitution matrices usually offer good enough performance in finding ion channels in databases of newly sequenced organisms.

Given a query sequence, a target sequence, a substitution matrix, and some strategy for how gaps in an alignment are scored, algorithms that exploit *dynamic programming* can generate an alignment between the target and the query that guarantees that no *better* alignment exists, where a *better* alignment would be one in which more similar residues as defined by the substitution matrix are aligned to each other (see [22] for an essential overview). These dynamic programming alignment algorithms (Needleman–Wunsch [23] for global alignments and Smith–Waterman [24] for local alignments) work by first building the best subalignment from the initial residues of the proteins and then recursively expanding on the subalignment. Crucial to this approach is the *assumption of independence*, which assumes that once a subalignment has been scored, no further residues that are later added to the subalignment can change the score of the subalignment. In terms of protein structure, this assumption could be interpreted as saying that no two regions of a protein interact. While this is clearly not a good assumption for ion channels, or any other complex protein, in practice, dynamic programming algorithms tend to give very reasonable alignments, although they will be insensitive to long-range interactions. So, for example, if a pore region of one ion channel is consistently linked to a ligand-gating region but these regions are separated by a long nonconserved region, the information that the two regions should only be found together cannot be used by dynamic programming algorithms to make a search more sensitive since the two subregions of the protein will be scored in an independent matter.

While dynamic programming algorithms guarantee the alignment with the *best* score, subject to the assumption of independence, for a given query and target sequence, they are generally too slow to be of much use in large-scale sequence analysis. The problem is that given a large database and a query sequence, the query sequence must be compared to every protein in the database, one at a time. We say that dynamic programming *scales* in terms of the size of the database, meaning that the larger the database gets, the longer it takes to search the database. Given that modern genomics databases can easily have terabases of sequence, this is far too slow to be practical. What is needed, instead, are algorithms that scale in terms of the size of the query sequence. As more sequences accrue, the databases become bigger every year, but the length of the genes themselves does not grow. It turns out that there are many data structures (including suffix trees and hash maps) that allow an *exact match* search to be performed in time proportional to the query rather than the database. That is, given a database of arbitrary size (subject to the amount of memory available in your system), it is possible to find an exact match from a query string in time that is proportional to the length of the query, not the size of the database. This always involves a *preprocessing* step in which some kind of dictionary is built from the database. This preprocessing step is often very slow, but only has to be done once. Once it is finished, queries against the database can be performed very quickly. This central trick in computer science is how search engines like Google are able to search the entire web and return results back from a query essentially instantly. The web is big (and growing), but the queries we type are small (and do not grow over time). Likewise, the number of sequences in databases is large (and growing), but the lengths of the queries (e.g., the length of ion-channel genes) do not increase over time. By using algorithms that scale on the length of the query and not the length of the database, our algorithms remain usable even as the universe of data continues to rapidly expand.

Algorithms like BLAST, therefore, use *heuristics* (time-saving approximations) that render them less sensitive to the size of the database. These heuristics mean that we give up the guarantee that our hit will be the best possible hit relative to a substitution matrix, but they allow the algorithms to execute in a reasonable amount of time. The most common heuristic is to constrain searches so that they will only succeed if there is some number of exact substring matches between the query and the target. This no longer guarantees the theoretically best hit and will ignore features such as chemical similarity since, for example, a search dependent on exact substrings cannot score a glycine-to-alanine substitution higher than glycine-to-proline. A type of search known as *k-mer* is dependent on the choice of a *word size* (e.g., three residues) and breaks the query sequence into all possible words of that length (in our example, every *word* of three residues within our protein). For each word, the database is searched, which can be done in *constant time* (i.e., in time independent of the size of the database). The top hit is the one that has the most k-mers (in our example 3 mers) in common between the query and the target. K-mer searches are fast but by themselves do not produce an alignment. Moreover, if the query and the protein target do not have many regions of identity that are equal in length to the word size, a k-mer search may miss proteins that are in actuality homologs. BLAST works by combining some of the features of a k-mer search with some of the features of dynamic programming. BLAST starts by selecting a word length (by default 3 for protein searches and 11 for nucleotide searches). Then for target sequences with matching k-mers, BLAST uses a dynamic programming like extension step to try and build an alignment from the k-mer seed. By changing the word length, a user can adjust speed vs. sensitivity. A longer word length will be less sensitive, requiring a longer region of exact match, but will also be faster as the extension step will be executed on fewer matches. A shorter word length will be more sensitive but slower as the extension step will be executed on more matches. Many derivatives and alternatives to BLAST have been developed, including the popular algorithm BLAT [25], which can be both faster and more memory efficient than BLAST.

While word size can have a profound effect on the results of a BLAST search, there are a number of other important parameters that can be set by the users. This includes which substitution matrix is used, how gaps are treated within the alignment, and whether searches are to be performed at the DNA or protein level. Because of the degeneracy of the genetic code, searches for distantly related ion channels can often fail at the nucleotide level but will succeed if both query and target sequences are translated into all six frames and the search is performed in protein space. These translated searches, however, can also take longer and can be more sensitive to low-complexity regions of the genome and other artifacts that can produce erroneous results. Using a filter to remove low-complexity regions of the genome can therefore be essential and is an option for many BLAST implementations.

Results of BLAST searches are typically ranked by their significance. BLAST results usually have an alignment score and an e-value, defined as the number of hits with at least that alignment score that one might expect by chance when searching a database of the same size with a query sequence of the same size. By this definition, an e-value >1 represents a hit that is worse than one would expect by chance while an e-score <1 could be considered to be significant. However, these e-values must be interpreted with due caution. The accuracy of the e-values is dependent on a model of the composition of a random sequence. The e-values can be thought of as representing how likely one is to find a protein of the same length as a query sequence madeup of random residues. However, the model that was used to define random residues will likely have been built on all proteins, not just membrane proteins or ion channels. E-values should therefore be considered only a rough guide to the significance of the results.

In general, the results of BLAST searches should be considered as an experiment dependent on a particular set of parameters and some understanding of how these parameters effect BLAST results can improve the chances of detecting homology across a wide phylogenetic space. There are excellent resources available discussing how these parameters can be tuned for particular searches (see, e.g., [26]). Blast searches can be run at many websites (e.g., at NCBI—http://blast.ncbi.nlm.nih.gov/Blast.cgi) or the BLAST software can be downloaded (http://blast.ncbi.nlm.nih.gov/Blast.cgi?CMD=Web&PAGE_TYPE=BlastDocs&DOC_TYPE=Download) and used to set up a database on almost any computer.

8.3 HIDDEN MARKOV MODELS ARE A FLEXIBLE FRAMEWORK THAT CAN BE USED TO MODEL MANY DIFFERENT PROBLEMS

Because of their speed and flexibility, pairwise alignment algorithms remain a workhorse in bioinformatics. However, it was realized some time ago that there may be information available within multiple sequence alignments that is not capturable by a pairwise approach, in particular, when the sequence similarity between the two sequence is low [19,27]. Rather than comparing a query sequence to every sequence in a database and returning the best hit, homology detection algorithms based on *profiles* build a model based on a *multiple sequence alignment* of sequences from the same family. It has been shown that by exploiting information from these alignments, searches conducted with these profiles can be more sensitive than pairwise searches [19,27].

The homology detection approach based on profiles is perhaps best exemplified by the popular PFAM database [28,29]. Rather than use a global substitution matrix, PFAM works by building a model in which each column within the multiple sequence alignment has its own expected distribution of amino acid residues. Using this model, one can ask how well a query sequence matches the protein family represented by the sequences within the multiple sequence alignment.

In order to build these probabilistic models that describe protein alignments, PFAM utilizes Markov chains. Markov chains, and the algorithms that manipulate them, are remarkably useful tools that can describe an extraordinary diversity of phenomena. Markov chains are utilized in every area of quantitative science, from speech recognition to statistical mechanics. As we will see later, Markov chains provide great flexibility in modeling and we will demonstrate their use in simple examples for both ion-channel sequence variation and ion-channel kinetics.

Markov chains can be used to model any set of data that is ordered. A Markov chain consists of a set of interconnected Markov states. Each observed piece of data is considered to be *emitted* from some state of the Markov chain. The data point at each interval is observed and known; however, which state the Markov model was in when the data was observed is not necessarily known. When the state the model is in corresponding to each emission is unknown, the Markov model is called hidden, hence the term *hidden Markov models* (*HMMs*). Each Markov state has an *emission probability*, which is the probability that data will be observed given that the model is in that state. States in a Markov chain are linked together by *transmission probabilities*, which is the probability that one moves from one Markov state to another.

8.4 MARKOV MODEL OF SINGLE ION-CHANNEL KINETICS

As an example of the flexibility of Markov chains, we will consider their application to ion-channel kinetics. We will here only consider a highly simplified model free from *real-world* problems such as baseline drift. The application of Markov models to real ion-channel data is discussed in detail elsewhere [30–32]. In our simplified model, we consider an ion channel that has only two states, closed and open (Figure 8.1 top panel). The rate parameter α defines the rate at which the channel moves from the closed state to the open state while the parameter β defines the rate at which the channel moves from the open state to the closed state. We say that the states of the channel display *independence* and do not have *memory*. In this simple two-state model, when the channel is in the closed state, it has no memory of how long it has been in the closed state or how many times the closed to open transition has occurred.

We can use a Markov chain to model the kinetics of our simple two-state channel (Figure 8.1, bottom panel). The Markov chain also has two states, closed and open. A Markov chain has a concept of an iteration, which in this case represents some short unit of time.* Rather than rate constants α and β,

* A convenient unit of time in a simulation of single-channel data might be the sampling time of the A/D converter used to acquire the single-channel data.

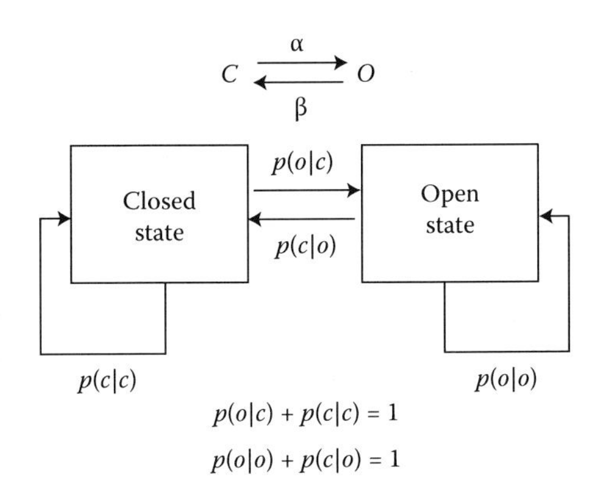

Figure 8.1 A simple two-state model of an ion channel capable of simulating single-channel recordings. Top panel: A chemical model with rate constants α and β. Bottom panel: The equivalent Markov chain with four transition parameters.

a Markov model defines *transition probabilities*. In our simple two-state model, there are four transition probabilities. If we are in the closed state, we can in the next iteration stay in the closed state. This transition probability is $p(C|C)$, which we read as *the probability that we stay in the closed state given that we are in the closed state*. Or alternatively, we can transition from the closed state to the open state, which is defined as $p(O|C)$ or *the probability that we move to the open state, given that we are in the closed state*. Obviously, in our simple model, $p(C|C) + p(O|C) = 1$, since in our two-state model, we must either stay in the closed state or move to the open state. We can likewise define transition probabilities from the open state, $p(C|O)$ and $p(O|O)$, which also must sum to one. This first-order Markov chain, like our chemical model, has no memory. The model only knows its current state, not its history. This feature also establishes independence as the behavior in a given state is independent of the history of how that state was achieved. These assumptions greatly simplify the implementations of algorithms that define and decode Markov chains.

In addition to transition probabilities, Markov chains also define *emission probabilities*. At each iteration, when the model is in a given state, it emits a value from some alphabet or distribution. As we will see, in Markov chains designed to model protein families, the emission symbol is one of the 20 protein residues or, possibly, a gap. For our simple ion-channel model, the emission probability describes the probability of observing a given current from each state. For each state, the emission probabilities over all possible emissions must sum to one. We can choose any distribution of probabilities, but a convenient choice is a simple Gaussian distribution with each state having a distinct mean and standard deviation for the emitted current (see [31] for a discussion of HMMs with more sophisticated noise models.) Given a set of emission parameters and transition parameters, it is straightforward to generate a simulated set of emissions. One simply chooses a start state* and samples an emission as defined by the emission probabilities of that state. Then, one either moves to the other state or stays in the same state as defined by the transition probabilities of that state. Then this process is repeated. The results of such a run for our two-state ion-channel model (current as a function of iteration/time) are shown in Figure 8.2. (The code used to generate this figure is available at https://github.com/afodor/IonChannelHMMs.)

We see that despite the great simplicity of our model, we can get surprisingly realistic single-channel data. By adding additional states to the model, we can increase the complexity of our simulated channels behavior. We can, for example, add states that represent long-lived inactivation and bursting (Figure 8.3).

In generating the simulated dataset, we know the state during each iteration, which forms a chain of states called a *path*, and use that information to produce a sequence of ordered emissions. The great strength of HMMs is that we can go in the opposite direction; that is, given just the emissions and a model with a given set of parameters, we can calculate the probability of being in each state for each iteration. This calculation is called *posterior decoding* (see [22]). For our ion-channel models, given the emissions that come from a *hidden* state in an HMM, we can efficiently calculate the probability that the hidden state was open or closed for each iteration. We say that given the data (and a Markov chain that is our model), we can calculate the $p(\text{open}|\text{data})$ and $p(\text{closed}|\text{data})$ for every emission. In the case of the model parameters shown in Figure 8.2, where there is no overlap between the open and closed states, this posterior decoding is perfect and the probability that the hidden state was the open state is either 1 or 0 (Figure 8.2, bottom panel). In a case where the magnitude of the difference between the open and closed states is not as large, our degree of confidence that the hidden state is open or closed is not as great and our posterior decoding therefore produces probabilities between 0 and 1 (Figure 8.4, bottom panel).

Of course, if one were modeling real data, one would wish to utilize a model that would produce the most insight into the nature of the data. Choosing models to apply to complex datasets can be as much art as science. In general, there is a trade-off between how well the model fits the data and model complexity. For Markov chains, the fit of the model to the data can be defined as how well the emission probabilities defined by the most likely path of states through the model (which is known as the Verterbi path; see [22]) match the data. For the high signal-noise data shown in Figure 8.2, there is essentially no model that could fit the data more closely than our two-state model. Our fit is essentially perfect in that in posterior decoding, we are certain we are either in the closed state (with a probability of 1) or the open state (with a probability of 1) and the actual amplitude and noise of each state are nearly identical to parameters that could be estimated from the large sample size represented by the many iterations shown in Figure 8.2. For the more difficult to model situation of low signal to noise shown in Figure 8.4, we are no longer always certain in posterior decoding which state we are in and therefore no longer have a perfect fit of model to data. If we were to introduce a model with more states, we could improve the fit to the data but at the cost of arbitrarily increasing model complexity. To take an extreme example, if we defined a separate Markov state for every observed distinct amplitude, we could perfectly fit our data to this model in that with each possible amplitude given its own state, we would always know with perfect certainty in posterior decoding which state we were in and the estimated parameters

* Technically, our model should also have initiation and termination with defined transition probabilities from these states to each other state in the model. See [22].

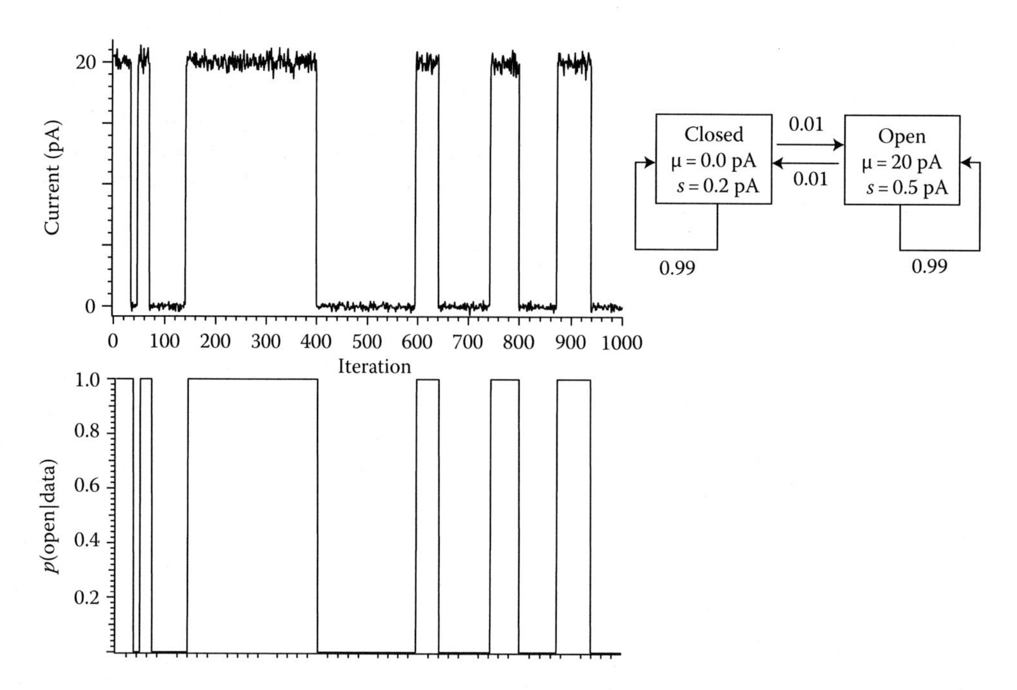

Figure 8.2 (top panel) Simulation of single-channel recordings from a two-state Markov chain; (bottom panel) given only the emissions and the underlying model, posterior decoding can determine the probability of being in the open or closed state for each iteration (time point). In this case, because the difference between the closed and the open state is large relative to the error, posterior decoding has no error or uncertainty and yields probabilities that are either zero or one. μ is the average and s is the standard deviation of the single ion channel current for each state as described by the Gaussian distribution.

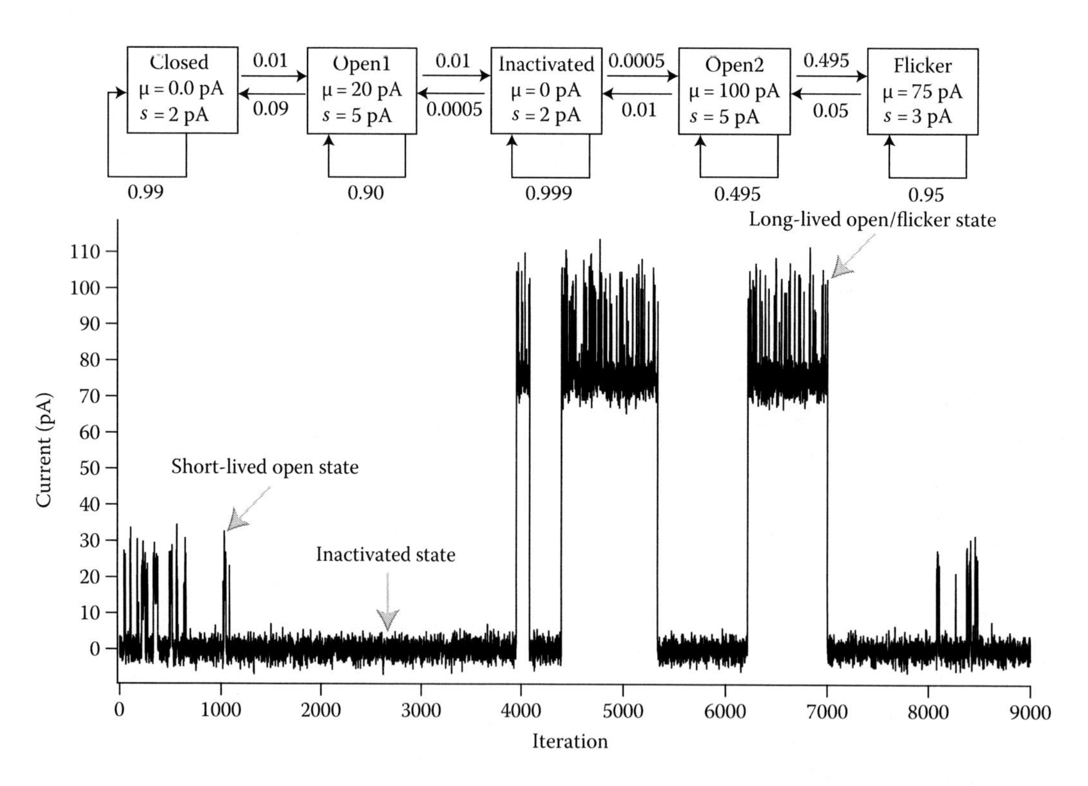

Figure 8.3 Simulations from a more complicated Markov chain that includes multiple open states, inactivation, and flicker.

from each state would perfectly match the observed amplitude at that iteration. This would, however, of course be an exercise in trivial overfitting that would not yield any insights into our data or any future predictive power. Our goal in finding the *best* model for a given dataset is to find the simplest model that explains most of our data. While there is an extensive literature devoted to this problem [33], there are no hard and fast rules that can determine when making a model more complex by adding a state produces a *better* model or simply leads to an overfit model. Ultimately, the most crucial test of an HMM or any other model constructed with statistical learning techniques is to be verified on a dataset that was in no way used for model construction.

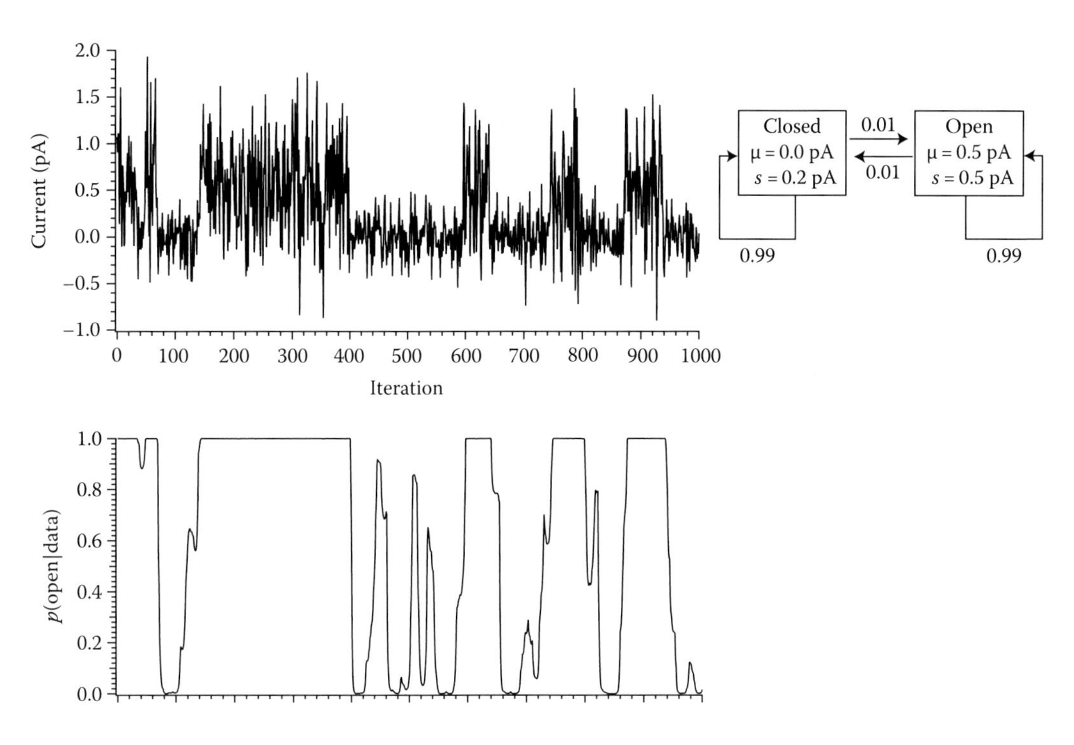

Figure 8.4 (top panel) Simulation of single-channel recordings from a two-state Markov chain with a smaller single-channel conductance than the simulations in Figure 8.2; (bottom panel) In this case, because the noise is as high as the single-channel amplitude, we cannot always tell when looking only at the emissions whether the underlying state was open or closed. Posterior decoding therefore produces probabilities other than zero and one.

8.5 MARKOV CHAIN FORMALISM APPLIED TO HOMOLOGY DETECTION

One remarkable property of Markov models is that the same formalism and algorithms that allow for simulation and decoding of single ion channels can be applied to a wide variety of problems in bioinformatics. As an illustrative example, let us say that we wanted to build a simple profile model with which to detect the conserved pore sequence in potassium channels. We start by building a Markov chain based on strongly conserved positions as seen in a published [34] alignment of the pore region of K^+ channels (covering the region TVGYGD in kcsa), which includes the more distantly related CNG channels (olCNG and rodCNG; see Figure 1 in [34]). Our highly simplified model of the six residue portion of the alignment could look something like Figure 8.5.

We have two kinds of states in our Markov chain. A *match* state emits residues with the frequency distribution derived from the residues in the column of the multiple sequence alignment. A *gap* state emits a gap in the query sequence relative to the alignment (with a frequency of 1). To calculate the probability of a particular sequence (say *T I G - - Y*) given our simple profile model, we start in the first state and record the probability of a T (14/16 since 14 of the residues in the alignment in [34] contain a *T* in the column corresponding to the first residue in our query sequence). We proceed to the second state (with a transition probability of 1) and record the probability of an I (4/16), then G in the next state (16/16). At this point, we make the transition to the gap state with a transition probability of (2/16). Once we are in the first gap state, our model requires a second gap (with a transition probability of 1) and finally in the last state, we recovered the probability of a Y (1/16). In general, the total probability of our path through the Markov chain is the product of all the emission and transition probabilities (see [22]). If we multiply all of these probabilities together, we achieve the *p*(sequence|model), which we read as the probability

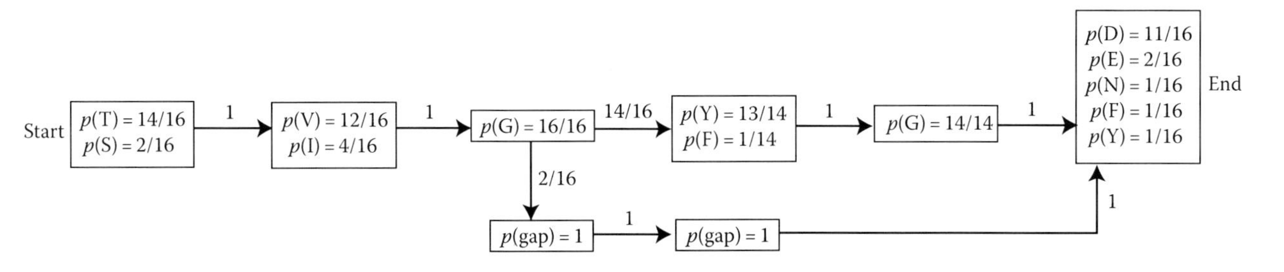

Figure 8.5 A highly simplified profile HMM capturing the alignment the *TVGYGD* region of the pore region of K^+ channels shown in Figure 1 from (Doyle, D.A. et al., *Science*, 280, 69, 1998). Each state represents one column of the alignment.

of the sequence given the underling model of the alignment. Of course, our simple model is not very flexible. For example, if only a single gap was observed instead of two in a row or our sequence were to start with any other residue other than T or S, the resulting probability of p(sequence|model) would be zero. A more sophisticated model would realize that not every possibility is captured by the alignment that serves as our training set. Such a model would add the possibility of a gap occurring anywhere along the alignment and pseudocounts of residues to each column (some small probability for each residue that was not observed in the alignment column) to avoid zero probabilities. With such modifications, our model could produce a small, but nonzero probability, for any sequence fragment. We could then report a *log-odds score* for the significance our query sequence:

$$\ln\left(p(\text{sequence}\,|\,\text{model})/p(\text{sequence}\,|\,\text{random model})\right),$$

where the random model is based on the properties of some large collection of randomly generated unaligned sequences. As is the case for BLAST e-values, the significance of this score will be based on how well the random model captures the possible variation of all proteins that we might search. So, a random model based on membrane sequences (which, e.g., would have emission probabilities higher in hydrophobic residues) might yield a different probability than a general model based on all proteins. In general, because our random model might not always be appropriate for our query sequence, the log-odd scores, like BLAST e-values, should be considered only rough guides to the significance of a hit. In fact, it can be shown that using ion channels as query sequences, proteins can be assigned to incorrect families even when the log-odds scores are highly significant [35].

The PFAM database (http://pfam.sanger.ac.uk/) attempts to build a profile HMM model that represents the alignment of every known protein family. The PFAM database takes an iterative approach to model building. First, a seed alignment is created from proteins known to share close homology. This alignment is used to construct a model, which is used to find more sequences, which in turn are used to build more refined models. The PFAM database provides aligned sequences for both seed and final alignments as well as representation of the HMMs that can be manipulated with the HMMer suite of software tools. Given a query sequence and a database of ion channels, represented as PFAM alignments, these tools can be used to ask whether the query sequence is represented by any of the models representing the ion channel.

Using the HMMer software and profiles built from the PFAM database, it has been demonstrated by comparing sequences to solved crystal structures that for a number of well-studied ion-channel families, the odds that prokaryotic and eukaryotic channels in fact have different protein folds are miniscule [35]. This validates a strategy of using prokaryotic structures to model eukaryotic channels. However, even within the context of a similar overall fold, small sequence variations can have enormous functional consequences. For example, although CLC channels have a high degree of sequence homology that all but guarantees that they share a common fold [35], there is increasing evidence that some members of this family function not as chloride channels but rather as proton-chloride antiporters [36]. These sorts of functional differences, while not currently predictable from just sequence information, presumably reflect small-scale differences in structure and energetics, which nonetheless must have enormous consequences for phenotype.

8.6 CONCLUSION

Electrophysiological recordings offer perhaps the most detailed measurements of real-time protein energetics and small conformational changes available for any protein. As described earlier, the advent of new sequencing technology combined with informatics tool is producing an avalanche of information on how sequence can vary within populations and across phylogenetic space. A central open question is whether we can use sequence variation to make a priori predictions about biophysical differences in ion-channel function or, instead, will determination of ion-channel function always be a strictly empirical matter in which the effects of sequence variation can be known only from experimental data.

In their influential review [11], Manolio et al. note that GWAS studies have focused on common alleles and have relied "almost exclusively on statistical evidence." They argue that if the missing heritability is to be found in rare variants then "the challenges of sifting through the millions of rare variants in which two individuals differ" may be substantial and may require a "return to biology if rare variants are to be grouped and analysed properly." If missing heritability is due to gene–gene [14] or gene–environment [15,16] interactions, the number of possible hypotheses that will need to be evaluated by future genomics studies likewise borders on the infinite. If the future progress of genomics indeed requires a *return to biology* to prioritize these many possible interactions, the study of ion channels will likely be front and center. Not only are ion channels crucially involved in most major human diseases, but electrophysiological techniques allows for exquisite linking of ion-channel structure and function. The merging of genomics techniques with tried-and-true experimental measurements points to a future in which ion-channel sequence variation captured from genomics is placed in a rich biological context encompassing both the biophysics of channel function and the important role that channels play in health and disease phenotypes.

ACKNOWLEDGMENTS

We thank Rick Aldrich and Zheng Chang Su for helpful comments on a previous draft of this chapter.

REFERENCES

1. Hodgkin, A.L. and A.F. Huxley, A quantitative description of membrane current and its application to conduction and excitation in nerve. *J Physiol*, 1952. **117**(4): 500–544.
2. Ho, K. et al., Cloning and expression of an inwardly rectifying ATP-regulated potassium channel. *Nature*, 1993. **362**(6415): 31–38.
3. Schrempf, H. et al., A prokaryotic potassium ion channel with two predicted transmembrane segments from Streptomyces lividans. *EMBO J*, 1995. **14**(21): 5170–5178.

4. Tamkun, M.M. et al., Molecular cloning and characterization of two voltage-gated K+ channel cDNAs from human ventricle. *FASEB J*, 1991. **5**(3): 331–337.
5. Tempel, B.L., Y.N. Jan, and L.Y. Jan, Cloning of a probable potassium channel gene from mouse brain. *Nature*, 1988. **332**(6167): 837–839.
6. Dascal, N. and I. Lotan, Expression of exogenous ion channels and neurotransmitter receptors in RNA-injected *Xenopus* Oocytes, in *Protocols in Molecular Neurobiology*, A. Longstaff and P. Revest (Eds.), 1993, Springer, New York. pp. 205–225.
7. Hoshi, T., W. Zagotta, and R. Aldrich, Biophysical and molecular mechanisms of Shaker potassium channel inactivation. *Science*, 1990. **250** (4980): 533–538.
8. Cheng, S.H. et al., Defective intracellular transport and processing of CFTR is the molecular basis of most cystic fibrosis. *Cell*, 1990. **63**(4): 827–834.
9. Gordan, S.E. and W.N. Zagotta, Localization of regions affecting an allosteric transition in cyclic nucleotide-activated channels. *Neuron*, 1995. **14**(4): 857–864.
10. Fodor, A.A., S.E. Gordon, and W.N. Zagotta, Mechanism of tetracaine block of cyclic nucleotide-gated channels. *J Gen Physiol*, 1997. **109**(1): 3–14.
11. Manolio, T.A. et al., Finding the missing heritability of complex diseases. *Nature*, 2009. **461**(7265): 747–753.
12. Consortium, C.-D.G.o.t.P.G., Identification of risk loci with shared effects on five major psychiatric disorders: A genome-wide analysis. *The Lancet*, 2013. **381**(9875): 1371–1379.
13. Hindorff, L.A. et al., Potential etiologic and functional implications of genome-wide association loci for human diseases and traits. *Proc Nat Acad Sci*, 2009. **106**(23): 9362–9367.
14. Zuk, O. et al., The mystery of missing heritability: Genetic interactions create phantom heritability. *Proc Nat Acad Sci*, 2012. http://www.pnas.org/content/early/2012/01/04/1119675109
15. Murcray, C.E., J.P. Lewinger, and W.J. Gauderman, Gene-environment interaction in genome-wide association studies. *Am J Epidemiol*, 2009. **169**(2): 219–226.
16. Parks, B.W. et al., Genetic control of obesity and gut microbiota composition in response to high-fat, high-sucrose diet in mice. *Cell Metab*, 2013. **17**(1): 141–152.
17. Rusch, D.B. et al., The Sorcerer II global ocean sampling expedition: Northwest Atlantic through Eastern Tropical Pacific. *PLoS Biol*, 2007. **5**(3): e77.
18. Yooseph, S. et al., The Sorcerer II Global Ocean sampling expedition: Expanding the universe of protein families. *PLoS Biol*, 2007. **5**(3): e16.
19. Altschul, S.F. et al., Gapped BLAST and PSI-BLAST: A new generation of protein database search programs. *Nucl Acids Res*, 1997. **25**(17): 3389–3402.
20. Eddy, S.R., Where did the BLOSUM62 alignment score matrix come from? *Nat Biotech*, 2004. **22**(8): 1035–1036.
21. Ng, P.C., J.G. Henikoff, and S. Henikoff, PHAT: A transmembrane-specific substitution matrix. *Bioinformatics*, 2000. **16**(9): 760–766.
22. Durbin, R. et al., *Biological Sequence Analysis: Probabilistic Models of Proteins and Nucleic Acids*, 1998, Cambridge University Press, Cambridge, U.K.
23. Needleman, S.B. and C.D. Wunsch, A general method applicable to the search for similarities in the amino acid sequence of two proteins. *J Mol Biol*, 1970. **48**(3): 443–453.
24. Smith, T.F. and M.S. Waterman, Identification of common molecular subsequences. *J Mol Biol*, 1981. **147**(1): 195–197.
25. Kent, W., BLAT—The BLAST-like alignment tool. *Genome Res*, 2002. **12**(4): 656–664.
26. Korf, I., M. Yandell, and J. Bedell, *BLAST*, 2003, O'Reilly Media, Sebastopol, CA.
27. Eddy, S.R., Profile hidden Markov models. *Bioinformatics*, 1998. **14**(9): 755–763.
28. Bateman, A. et al., The Pfam protein families database. *Nucl Acids Res*, 2002. **30**(1): 276–280.
29. Bateman, A. et al., The Pfam protein families database. *Nucl Acids Res*, 2004. **32**(suppl 1): D138–D141.
30. Qin, F., A. Auerbach, and F. Sachs, A direct optimization approach to hidden Markov modeling for single channel kinetics. *Biophys J*, 2000. **79**(4): 1915–1927.
31. Qin, F., A. Auerbach, and F. Sachs, Hidden Markov modeling for single channel kinetics with filtering and correlated noise. *Biophys J*, 2000. **79**(4): 1928–1944.
32. Venkataramanan, L. and F.J. Sigworth, Applying hidden Markov models to the analysis of single ion channel activity. *Biophys J*, 2002. **82**(4): 1930–1942.
33. Hastie, T., R. Tibshirani, and J. Friedman, *The Elements of Statistical Learning: Data Mining, Inference, and Prediction*, 2nd edn., 2009, Springer, Berlin, Germany.
34. Doyle, D.A. et al., The structure of the potassium channel: Molecular basis of K+ conduction and selectivity. *Science*, 1998. **280**(5360): 69–77.
35. Fodor, A.A. and R.W. Aldrich, Statistical limits to the identification of ion channel domains by sequence similarity. *J Gen Physiol*, 2006. **127**(6): 755–766.
36. Matulef, K. and M. Maduke, The CLC 'chloride channel' family: Revelations from prokaryotes (Review). *Mol Membr Biol*, 2007. **24**(5–6): 342–350.

9 Investigation of ion channel structure using fluorescence spectroscopy

Rikard Blunck

Contents

9.1 INTRODUCTION

Presently, an ever-increasing number of crystal structures of channels in different states are available. However, their structural information is restricted to "snapshots" in one conformation, and their validity has to be confirmed in a more native environment. Electrogenic transport proteins have the advantage that their function can be directly measured with electrophysiological techniques. What lacks is dynamic structural information that can be correlated with the functional electrophysiological data. Experimental structural dynamics can effectively be obtained by fluorescence techniques, which will be discussed in this chapter.

Historically, the first use of fluorescence to study ion channel structure was to monitor the intrinsic fluorescence of tryptophan and tyrosine (1,2). By exploiting intrinsic fluorescence, no additional labeling technique was required, making these experiments possible before the cloning of ion channels. However, using intrinsic fluorescence of aromatic amino acids only succeeds if they do not occur too frequently in the protein and if they are involved in the specific process investigated. Otherwise, labeling has to be directed to the region of interest, which was first reported with the synthesis of fluorescently labeled bungarotoxin and tetrodotoxin derivatives (3,4). Fluorescence recordings of channel activity picked up significantly with the development of fluorescent calcium indicators in the early 1980s (5,6). Finally, with the cloning and recombinant expression of the voltage-gated sodium (7–9), *Shaker* potassium (10–12), and other ion channels (13–17), the molecular biology tools as well as the structural information became available to specifically target important regions in the channels in order to follow their conformational changes. Accordingly, the first voltage-clamp fluorometry (VCF) measurements, simultaneous measurements of site-directed

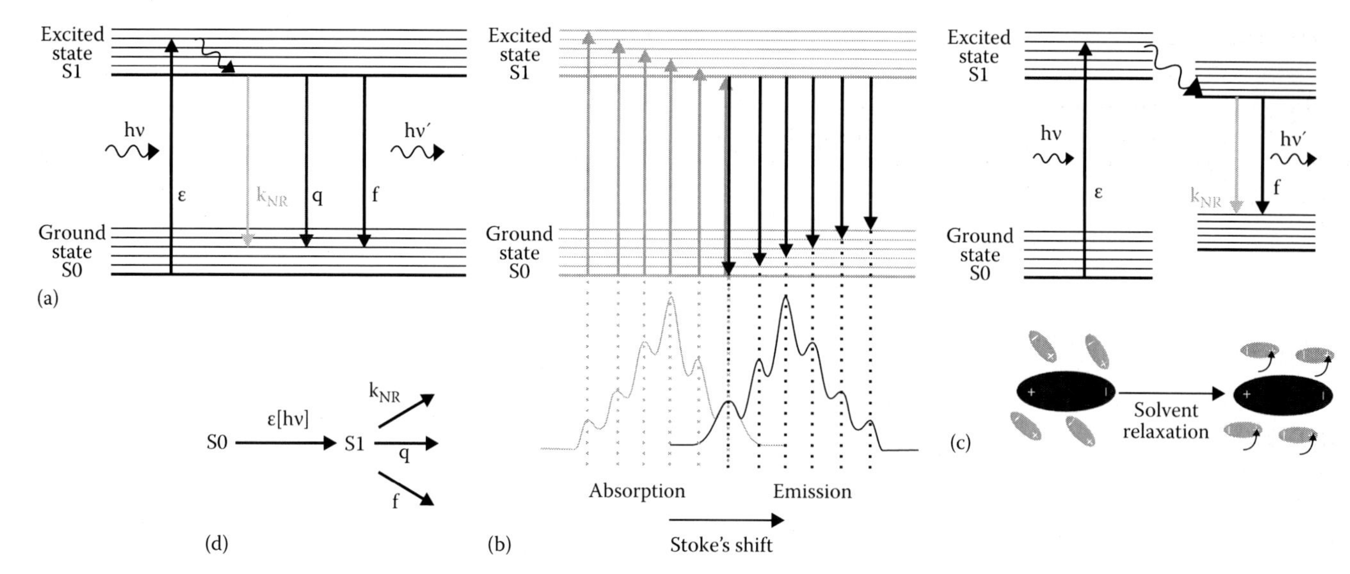

Figure 9.1 (a) Jablonski diagram of typical fluorescence process. An electron is lifted from the ground to one of the vibrational levels of the first excited state S1 by a photon of matching energy. The vibrational energy is quickly dissipated. The electron can return to the ground state either by emitting a photon (fluorescence, f), by nonradiative decay (k_{NR}), or by quenching (q) if a quencher is present (ε: extinction coefficient). (b) The electron may be excited to different vibrational levels of S1 with different energies, resulting in the excitation spectrum (*light gray, bottom*). The emission originates from the ground vibrational S1 state to different vibrational levels of S0, resulting in the emission spectrum. While during excitation, the vibrational energies are added to the energy difference between S0 and S1, it is subtracted during emission, leading to the Stoke's shift. (c) At the moment of excitation, the dipoles of the solvent are aligned to the S0 dipole moment. This leads to a higher energy of the S1 state, whose dipole moment is generally different. After the dipoles of the solvent realigned to the S1 state (solvent relaxation), the energy of S1 is reduced, whereas the energy of S0 is increased, as the dipoles are now misaligned to S0. As a consequence, the emission spectrum is shifted to longer wavelengths. (d) In a kinetic model of the fluorescence process, the rate constant into the S1 state is given by the extinction coefficient. The rate constants to leave the S1 state each describe a different process (f, fluorescence; q, quenching; k_{NR}, non-radiative decay).

fluorescence and electrophysiology, followed after just a few years (18,19) and were later extended to a wide variety of channels and other transport proteins. Single channel tracking in mammalian cells became possible by fusion with green fluorescent protein (GFP) or one of the derivatives (20,21), and investigation of ion channels by fluorescence spectroscopy was further improved by higher sensitivity down to the single molecule level and different labeling techniques.

Currently, fluorescence is used in three different aspects of ion channel research: first, as a readout for the conducting ion (i.e., Ca^{2+} imaging); second, to track ion channels in different expression systems or to determine their stoichiometry; third and finally, in spectroscopic measurements investigating the structure of ion channels, which may or may not be correlated directly with functional measurements (VCF). In this chapter, we will concentrate on the third point, namely, the use of fluorescence to obtain dynamic structural data.

9.2 MODULATION OF FLUORESCENCE

How information on conformational changes of proteins is detected by fluorophores can be derived from the general concept of fluorescence. Each fluorophore has characteristic excitation and emission spectra, which are related to its electronic energy levels and give the relative probability of absorbing and emitting a photon of a specific wavelength, respectively. When a fluorophore is excited, it absorbs a photon with the correct energy to lift an electron to a higher electronic state (typically from S0 to S1, Figure 9.1a). The excitation occurs into a higher vibrational level, but the energy is rapidly dissipated such that the electron will assume the S1 ground vibrational state. From the ground S1 state, the electron may return to the S0 state by emitting a photon corresponding to the energy difference (fluorescence, f). This transition again occurs to the higher vibrational states of S0 (Figure 9.1b). Thus, during excitation, the vibrational energy is added to the energy difference between S0 and S1, whereas it is subtracted from it during emission. Consequently, emission is shifted to wavelengths longer than excitation (Stoke's shift). Excitation and emission spectra, finally, result from a superposition of the transitions to each vibrational state (Figure 9.1b). The amplitude is given by the transition probabilities between the ground states and the higher vibrational states (Franck–Condon factors), which are symmetric for excitation and emission (22), that is, the probability for excitation from the S0 ground state to the n-th vibrational state of S1 equals the probability of emission from the ground S1 state to the n-th vibrational S0 state. In consequence, excitation and emission spectra are mirror images of one another (Figure 9.1b).

Not every photon is absorbed by the fluorophore reemitted and not every excited state results in emission of a photon. The absolute probability to absorb a photon from the ground state and emit a photon from the excited state is given by the *extinction coefficient* (ε)* and the *quantum yield* (QY), respectively. When no photon is emitted during return of the electron to its ground state, the energy may be dissipated

* More precisely, the extinction coefficient is the probability normalized to the concentration and pathway. The actual probability is given by the absorption cross section σ, which is related to the extinction coefficient by $\sigma = 1000 \cdot \ln(10) \cdot \varepsilon / N_A$.

into the molecule or surrounding (nonradiative decay, k_{NR}). If one considers the excitation/emission cycle in a kinetic model (Figure 9.1d), then the number of photons emitted per excitation cycle, the QY, is given by the rate constant for the emission of a photon (f) normalized to the sum of all pathways to the ground state:

$$QY = \frac{f}{f + k_{NR}}, \quad (9.1)$$

where k_{NR} is the rate constant for nonradiative decay. The *lifetime* of the fluorophore denotes the time constant τ of the exponential decay related to leaving the excited state. As in other kinetic models, τ equals the inverse of the sum of the rate constants, leaving the excited state

$$\tau = \frac{1}{f + k_{NR}}. \quad (9.2)$$

We will see in the following section that Stoke's shift, QY and lifetime of the fluorophore are the parameters typically observed when detecting modulation of fluorescence.

9.2.1 SOLVENT RELAXATION

In order for the fluorophore to interact with and emit electromagnetic radiation, it has to have a dipole moment. In the steady state, the dipoles of the surrounding medium (e.g., H_2O) will be aligned with inverse polarity to the fluorophore's dipole moment, as this would be energetically most favorable. The dipole moment alters during the transition from the S0 to the S1 state such that those of the surrounding solvent are no longer aligned to the fluorophore's dipole (Figure 9.1c). During the so-called solvent relaxation, the solvent dipoles realign with the new (S1) dipole moment, which lowers the energy of the excited state but at the same time increases the energy of the S0 ground state, since the dipoles are now aligned to S1 (Figure 9.1c). The lower energy gap between S1 and S0 leads to a redshift of the emission spectrum and to a larger shift between excitation and emission spectra.

The process of solvent relaxation occurs—dependent on the permittivity of the solvent—whenever a fluorophore is excited. In a hydrophilic environment, the shift is larger compared to a hydrophobic environment. Hence, if a fluorophore is moved from a hydrophilic to a hydrophobic environment for example, during protein movement, emission will be shifted to the blue. This shift might be directly measured (19) or may be translated into a change in fluorescence intensity using appropriate filters.

9.2.2 QUENCHING

More commonly, fluorescence is modulated by *quenching*. During quenching, the energy of the excited state is dissipated through interaction with a different molecule, similar to nonradiative decay mentioned earlier. Classically, one distinguishes between *static* and *dynamic* quenching depending on whether the quencher and the fluorophore form a stable nonfluorescent pair or whether the two partners *collide* and thus transfer the energy, respectively. Presence of a quencher offers an additional, very fast pathway back to the ground state (Figure 9.1a). If we assume that the rate constant for quenching is given by q (Figure 9.1d), the QY changes to

$$QY = \frac{f}{f + k_{NR} + q} \quad (9.3)$$

and the lifetime τ changes accordingly to

$$\tau = \frac{1}{f + k_{NR} + q}. \quad (9.4)$$

Both are reduced proportionally. For static quenchers, q is so large compared to f that the fluorophore immediately returns to the ground state without emitting a photon.

In proteins, one has to consider two different types of quenchers: those that are present in the surrounding solvent and those that are attached to the surrounding protein. The first group includes not only oxygen in the aqueous solution but also quenchers intentionally added to the solution to induce quenching. One possibility is the addition of potassium iodide in order to test accessibility of the fluorophore in one or several states (18,23,24). The second group of quenchers contains residues of the surrounding protein (25–29). Some residues, in particular, the aromatic ones, are especially well suited to quench fluorescence upon contact. Here again, one distinguishes between steady contact, where fluorophore and quenching residue form a complex that is no longer fluorescent (static quenching), and dynamic quenching, where the fluorophore is excited and the energy may be dissipated to the quencher with a certain rate constant or collision probability. Of all natural amino acids, tryptophan is the strongest quencher followed by tyrosine (25,27). Phenylalanine, in spite of featuring an aromatic ring, is much less effective as a quencher. The quenching abilities for different amino acid fluorophore pairs have been studied in detail by Marme et al. (25). For instance, tetramethylrhodamine (TMR) is quenched by 57%, 34%, and 11% in the presence of 30 mM tryptophan, tyrosine, and phenylalanine, respectively. Other amino acids such as methionine, histidine, glutamate, or aspartate also quench fluorescence, albeit at much higher concentrations (25–27). However, these measurements were mostly done by mixing fluorophores with the respective amino acids in solution. When both, fluorophore and the quenching amino acid, are attached to a protein, which keeps them in close proximity, concentrations are no longer defined. In order for quenching to occur, the van der Waals radii of fluorophore and amino acid have to overlap, be it by collision (concentration) or be it by constricting them to a small volume (protein). This is illustrated in the quenching of Cy5 by tryptophan. While Cy5 fluorescence is only reduced by 3% and 9% in the presence of 6 and 30 mM tryptophan, respectively, the quenching increases to 41% when both are attached to a small peptide (25).

The quenching efficiency also depends on the nature of the fluorophore. For typical organic fluorophores such as fluorescein, rhodamine, and bodipy, the efficacy of a specific quencher increases with increasing excitation and emission wavelengths, whereas carbocyanine dyes (Cy5, Alexa 647) are only slightly quenched even in the presence of tryptophan, and not quenched at all in the presence of tyrosine or methionine.

9.2.3 FÖRSTER RESONANCE ENERGY TRANSFER

Förster resonance energy transfer (FRET) is a near-field dipole–dipole interaction, which allows the transfer of an excited state from one fluorophore, the donor, to a second one, the acceptor. One may picture the donor as an emission antenna and the acceptor as a receiver antenna. If the dipole field of the donor reaches the acceptor, the donor induces the acceptor to oscillate. If this oscillation occurs with a resonance frequency of the acceptor, the acceptor may be excited, as if the oscillation was induced by a photon. The excitation of the acceptor leads in turn to a de-excitation of the donor, effectively transferring the excited state from donor to acceptor (Figure 9.2a). For this to occur, the energy difference between the excited state of the donor and the higher vibrational S0 states has to match the difference between the higher vibrational states of the acceptor and the ground S0 state. In other words, the emission spectrum of the donor fluorophore has to overlap with the excitation spectrum of the acceptor fluorophore (Figure 9.2b).

The energy transfer efficiency ET is the fraction by which the QY of the donor is decreased by the presence of the acceptor. Like the presence of a quencher discussed earlier, also the presence of an acceptor leads to a new pathway to leave the excited state with the rate constant k_{ET}, which may be determined from the properties of donor and acceptor and their environment (30):

$$k_{ET} = \frac{161.9 \cdot f \cdot \kappa^2}{\pi^5 N_A n^4} \cdot \frac{1}{R^6} \int_0^\infty Em_D(\lambda) \cdot \varepsilon_A(\lambda) \cdot \lambda^4 d\lambda, \qquad (9.5)$$

where
- R is the distance between donor and acceptor
- Em_D is the normalized emission spectrum of the donor
- ε_A is the absorption spectrum of the acceptor
- λ is the wavelength
- N_A is the Avogadro constant
- n is the refractive index of the surrounding
- κ is an orientation factor between donor and acceptor dipoles

The orientation factor κ is required as energy transfer is dependent on the arrangement of the two dipoles relative to one another. Full energy transfer efficiency is only possible if donor and acceptor are aligned in parallel, whereas a perpendicular orientation prevents energy transfer.

With the rate constant k_{ET}, the energy transfer efficiency becomes

$$ET = \frac{k_{ET}}{f + k_{NR} + k_{ET}} = 1 - \frac{f + k_{NR}}{f + k_{NR} + k_{ET}} = 1 - \frac{QY_{DA}}{QY_{DO}}, \qquad (9.6)$$

where QY_{DA} and QY_{DO} (Equation 9.1) are the QYs of the donor in the presence and absence of the acceptor, respectively:

$$QY_{DA} = \frac{f}{f + k_{NR} + k_{ET}}. \qquad (9.7)$$

Accordingly, also the fluorescence lifetime of the donor decreased:

$$\tau_{DA} = \frac{1}{f + k_{NR} + k_{ET}}, \qquad (9.8)$$

which simply reflects that the additional pathway to leave the excited state lowers the dwell time therein.

If k_{ET} equals the sum of the other rate constants ($f + k_{NR}$), then 50% of the energy is transferred to the acceptor ($k_{ET,0}$). With $k_{ET} \sim R^{-6}$, Equation 9.6 can be written as

$$ET = \frac{1}{1 + (R/R_0)^6}, \qquad (9.9)$$

with R_0 being the distance between donor and acceptor, where 50% of the energy is transferred to the acceptor. It can be calculated from Equation 9.5:

$$R_0 = \left[\frac{161.9 \cdot QY_D \cdot \kappa^2}{\pi^5 N_A n^4} \int_0^\infty Em_D(\lambda) \cdot \varepsilon_A(\lambda) \cdot \lambda^4 d\lambda \right]^{1/6}. \qquad (9.10)$$

R_0 typically ranges between 25 and 60 Å. The distance dependence of ET normalized to R_0 is shown in Figure 9.2c.

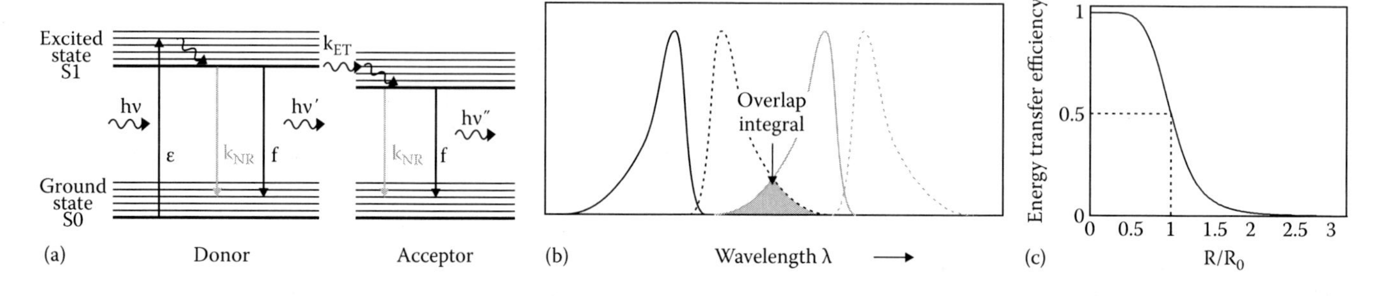

Figure 9.2 (a) Jablonski diagram for FRET. Excited electrons can leave the excited state by transferring their energy to an acceptor (k_{ET}) if a corresponding energy level exists. The electron itself remains with the donor and an electron of the acceptor is excited and returns to the ground state as usually. (b) In order for FRET to occur, the energy levels have to coincide. In terms of spectra, this means that the emission spectrum of the donor has to overlap with the excitation spectrum of the acceptor (donor: black, acceptor: gray, excitation spectrum: solid, emission spectrum: dashed). (c) Shown is the distance dependence of the energy transfer efficiency ET as a function of distance between donor and acceptor normalized to the R_0. Highest sensitivity is achieved in the range between $0.5R_0$ and $1.5R_0$.

9.3 OBTAINING STRUCTURAL INFORMATION FROM FLUORESCENCE MEASUREMENTS

Now that we have introduced the modulation of fluorescence by its surrounding, the question remains as to what kind of structural information these modulations can provide and how this information is obtained. This will be discussed in the following section.

9.3.1 KINETICS OF LOCAL STRUCTURAL REARRANGEMENTS

One of the most powerful uses of fluorescence is the detection of the kinetics of local structural rearrangements. This is particularly useful when the function of the channel is simultaneously monitored using electrophysiology, a technique called voltage-clamp fluorometry (18,19). In VCF, the fluorescence changes are measured simultaneously with the electrophysiological recordings and structural and functional data can be directly correlated to one another. VCF requires, in addition to electrophysiological, also optical access to the cells (Figure 9.3a). The electrophysiological setup is, therefore, mounted on a fluorescence microscope (either inverted or upright). VCF has been used in combination with two-electrode voltage-clamp (18,31), cut-open oocyte voltage-clamp (19), patch-clamp (24,26,32), planar lipid bilayer (33,34), and droplet interface bilayers (35,36). In order to monitor local conformational rearrangements, the channel is labeled at a specific site with a small fluorophore (see also Section 9.4) such that the properties of the fluorophore are determined by the immediate environment. If during stimulation of the channel, the fluorophore alters its position, that is, if this part of the channel undergoes a conformational change, the fluorescence will likely change due to the modulation mechanisms mentioned earlier. One possibility is that the fluorophore becomes accessible to the surrounding solution, which would lead to both quenching and a redshift of

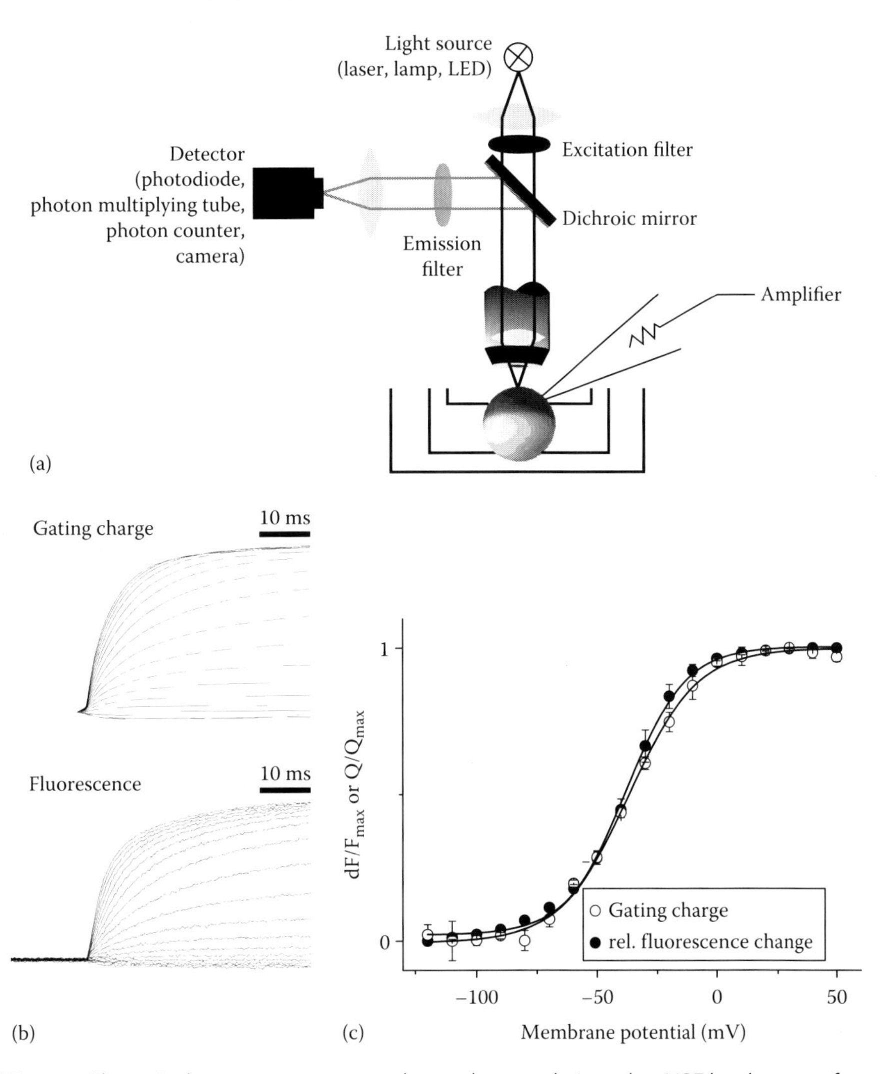

Figure 9.3 (a) Principal VCF setup. Shown is the cut-open oocyte voltage-clamp technique, but VCF has been performed also in two-electrode voltage clamp, patch-clamp and bilayer configurations. The electrophysiological setup is mounted on a fluorescence microscope. As light source, a low-noise lamp, laser, or LED source can be used. The emission is detected by a photodiode (PIN or APD), a photomultiplying tube, or a camera dependent on the configuration. For clarity, not all electrodes are shown. (b) Gating charge and fluorescence response of Shaker H4IR-W434F-A359C in response to depolarizing pulses from a holding potential of −90 mV. The channel was expressed in *Xenopus* oocytes and recordings performed in the cut-open oocyte voltage-clamp configuration. The fluorophore is linked to position A359C at the N-terminal end of the S4 and reports movements of the S4 during gating. (c) Gating charge–voltage and relative fluorescence change–voltage relations for the measurements in (b).

the emission spectrum. Another possibility is that the fluorophore is quenched by residues of the surrounding protein (26,29,37–39). One may check which mechanism is responsible for the fluorescence change either by addition of iodide as a quencher to the surrounding solution—to test whether accessibility changes—or by measuring the entire emission spectrum (18,19,40).

The resulting fluorescence signal will be an exponential or multiexponential intensity change, whose kinetics and amplitude will likely depend on the strength of the stimulus (e.g., voltage or ligand concentration, Figure 9.3b). It is important to realize that the time course of the fluorescence change reflects the kinetics of the transition that is monitored by the fluorescence and does not directly follow the conformational change of the protein itself. The movement of the protein is likely much faster than the observed changes. Let us assume that the observed transition can be described as a two-state process A ↔ B. The fluorophore will be in a different environment in each of the two positions and accordingly fluoresce with different intensities F_A and F_B. The total fluorescence intensity observed F_{total} will be

$$F_{total} = A \cdot F_A + B \cdot F_B, \tag{9.11}$$

where A and B are the occupancies of both states with A + B = 1. Let us define $\Delta F = F_B - F_A$. Then we can write

$$F_{total} = F_A + B \cdot \Delta F. \tag{9.12}$$

The fluorescence change will thus follow the occupancy of B and reflect the kinetics of the transition A ↔ B. For example, if the probe is attached to the top of the S4 in voltage-gated ion channels, the fluorescence signal will follow the gating charge equilibrium (Figure 9.3c, 18).

The fluorescence signal reports local rearrangements, which can be correlated with the functional data such as ionic current or gating currents. The local rearrangements can be mapped onto the channel by attaching the label at different positions throughout the protein (19,41,42) or scanning along a certain region (29,43,44). Also, the kinetics may be compared although here, one has to take care that the introduction of the label at different positions does not lead to intrinsically different kinetics. In this case, it would be advisable to use different labeling techniques to have two different fluorophores attached to the same channel at different sites and follow their kinetics simultaneously (41).

While quenching will provide precise information on kinetics of local conformational changes, it does not allow reconstructing the movement of the channel itself. Indirect estimations may be made based on the fact that the quenching requires the fluorophore to be within 5–10 Å from the quencher; however, this requires knowledge about the nature and location of the quencher. These are typically intrinsic amino acid residues, whose locations are not further defined. The exact location can be obtained if quenchers are specifically introduced in the proximity of the fluorophore. By comparing fluorescence changes in the presence and absence of the quenching residue, the relative movement between fluorophore and quencher can be probed (29,38,39). Fluorescence of bimane and coumarin, for instance, is quenched by tyrosines and tryptophans but not by other amino acids (27). The mechanism of quenching by tryptophan is thought to be an electron transfer process, which requires overlap of the van der Waals radii and decays rapidly with distance (25).

The first protein studied with directed fluorescence quenching was T4 lysozyme (28), but the technique has since been employed for the study of ion channels. In cyclic-nucleotide-gated (CNG) channels, it was used to study the conformational changes of the cyclic-nucleotide-binding domain (29). The authors observed a cGMP- or cAMP-dependent signal only in the presence of a tryptophan at specific positions, indicating proximity and relative movement between the positions of fluorophore and tryptophan. The concentration dependence of the response correlated with the cGMP- or cAMP-induced channel opening. In the large conductance calcium-activated potassium channels (BK_{Ca}) channels, structure and movements of the voltage-sensing domain S1–S4 has been studied. Besides the secondary structure of the S3–S4 linker (38), also proximity of the external ends of S2 and S4 and their relative movement during gating has been demonstrated (39).

9.3.1.1 Intra- and intermolecular distance measurements

While quenching allows investigating conformational changes, often distances between positions further apart within the protein and their movement relative to one another are of interest such as pore opening (45,46), movement of a position relative to the membrane (47,48), and the conformational changes related to ligand binding or other stimuli (40,48–53). Other processes that may be addressed by distance measurements are ligand binding or association of auxiliary subunits to the channels (54–57). Most of these processes include movements in the range of 5–15 Å and distances in the range of 20–70 Å. In this range FRET has proven extremely powerful. As described earlier, FRET efficiency is strongly dependent on the donor–acceptor distance R in the range between $0.5R_0$ and $1.5R_0$. Beyond that interval, the distance dependence flattens (Figure 9.2c). Thus, the fluorophore pair should be chosen such that the distances to be measured are similar to the R_0. Furthermore, the distance dependence of the FRET efficiency is normalized to R_0, meaning that the resolution, that is, the smallest distance changes that can be detected, is proportional to R_0—the higher the R_0, the lower the resolution.

If two positions move in relation to one another, the change in distance will lead to altered energy transfer and thus a fluorescence change (Figure 9.4a). FRET is observed either by exciting the donor and observing acceptor emission (sensitized emission), or better as the ratio of donor and acceptor emission upon donor excitation. As a control, one should ensure that no signal through quenching occurs if only donors or only acceptors are present. This method has first been employed by Berger et al. (58) to investigate drug binding to L-type Ca^{2+} channels. The authors used the intrinsic fluorescence of endogenous tryptophan residues in combination with bodipy-labeled dihydropyridines (DHPs), which are known to bind to a binding site close to the pore region of domain III. The time course of the distance change can be followed in the FRET signal after addition of labeled DHP. It has to be monitored as a reduction in the donor emission intensity, as the concentration of the acceptor is governed by diffusion and thus not necessarily constant.

The first time-resolved intramolecular rearrangements observed with FRET were obtained from the *Shaker* K^+ channel

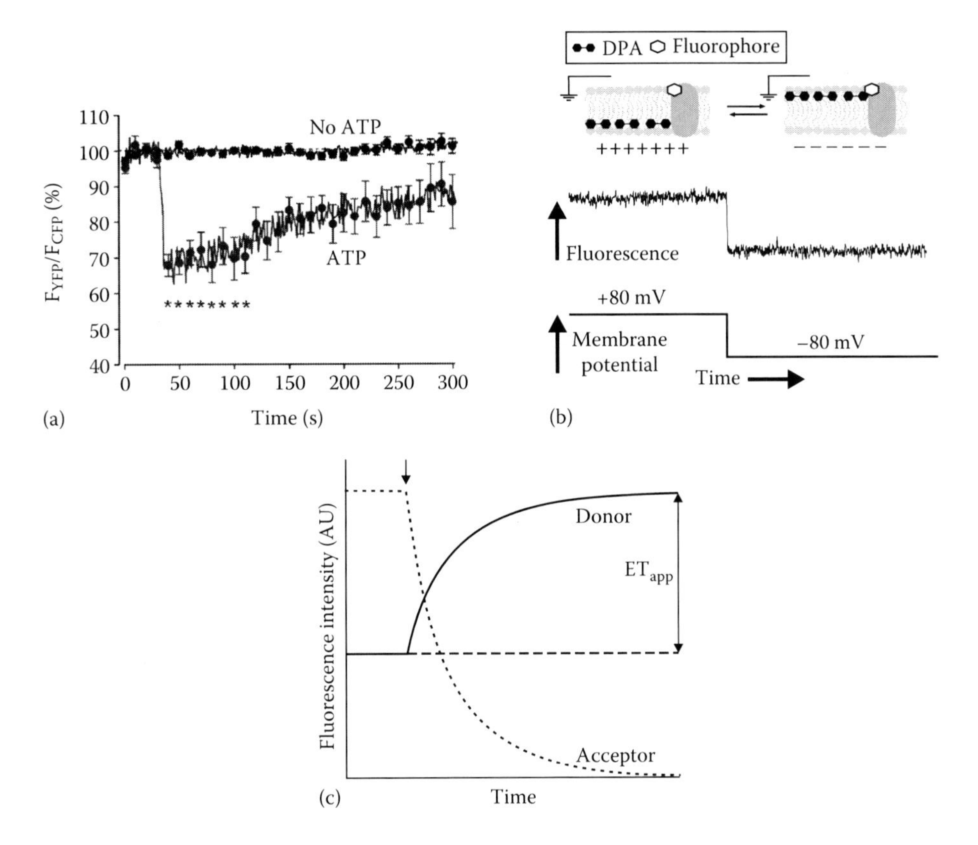

Figure 9.4 (a) The FRET acceptor is added to the membrane while the donor is linked to the protein of interest. At positive membrane potential, the negatively charged acceptor (dipicrylamine, DPA) is located at the inner leaflet, leading to high fluorescence. Upon reversal of the potential, the acceptor translocates to the outer leaflet, resulting in lower fluorescence due to energy transfer. (b) Conformational changes of P2X2 monitored with FRET between CFP and YFP: CFP and YFP were fused to the C-terminus of P2X2 and relative movement in response to ATP measured. (Reproduced from Fisher, J.A. et al., *J. Neurosci.*, 24, 10475, 2004.) (c) Apparent energy transfer efficiency can be determined by acceptor photobleaching. The arrow indicates the beginning of the photobleaching. As a consequence, acceptor fluorescence decreases. As photobleached acceptors no longer accept energy by FRET, the donor fluorescence increases. From the difference in donor fluorescence intensity before and after photobleaching, the efficiency is determined.

by Glauner et al. (59). In this work, corresponding sites in the S3–S4 linker on different subunits were stoichiometrically labeled (i.e., with different concentrations of donor and acceptor, resulting in binomial donor–acceptor distribution). Then the FRET efficiency was followed in response to a depolarizing pulse, leading to a relative displacement between the two sites.

When labeling identical positions on different subunits, any movement that does not change the distance to the central axis cannot be detected. For ion channels, this includes any synchronized movement vertical to the membrane surface. These may be detected either with respect to a different fix point in the channel (e.g., labeled toxin, 60–62) or relative to the membrane (Figure 9.4b). If the acceptor is added to the membrane, the mean distance between two acceptors has to fall in the range of the R_0 to achieve a sufficiently small distance between donor and acceptor. Therefore, a relatively high concentration of acceptors in the membrane is required. To avoid crosstalk from the many acceptors into donor fluorescence, a nonfluorescent acceptor can be used. The FRET mechanism only requires the acceptor to absorb in the correct range; it does not have to emit a photon during relaxation. For this purpose, for example, dipicrylamine (DPA) has been used (33,47,48). DPA is an amphipathic negatively charged compound with a maximum absorption in the blue (420 nm) that has been used with a wide variety of donors including fluorescein, TMR, Alexa488, Alexa588, Alexa647, oxonol, and different fluorescent proteins (FPs) (33,47,48,63,64). Due to its amphipathic character and net negative charge, DPA partitions into the membrane close to the lipid headgroups and distributes between the outer and the inner leaflets depending on the membrane potential (65). Thus, depolarization of the membrane from resting potential would move the acceptor from the outer to the inner leaflet of the membrane (Figure 9.4b). If the donor is located close to the outer membrane surface, such a redistribution of the acceptor would lead to a larger distance between donor and acceptor and thus increase donor fluorescence. A donor at the inner surface would result in an opposite change. The closer the donor is located to the center of the bilayer, the smaller the fluorescence change. This way, one can "calibrate" the position of the donor with respect to the membrane normal (47).

The most effective way to tag proteins is to fuse them with FPs. For FRET measurements, for instance, the FRET pair of cyan fluorescent protein (CFP) and yellow fluorescent protein (YFP) with an $R_0 = 49 \pm 1$ Å (66) has become very popular. It has been used to study the relative rearrangement between N- and C-termini of CNG channels depending on Ca/calmodulin (Ca/CaM) binding (50,67). When following the ratio of acceptor to donor fluorescence intensity over time, the FRET efficiency correlated with development of ionic current in response to transient application of calcium together with calmodulin, indicating that N- and C-terminal domains are in close proximity of one another but withdraw from one another upon Ca/CaM binding. Similarly, the conformational change in the cytosolic domains of P2X2 in response to increased ATP concentrations has been observed with CFP-/YFP-tagged pairs (Figure 9.4a, 68).

The displacements measured in both cases are relatively large and alter very slowly (seconds to minutes). Although high temporal resolution is possible (69), the size of the FPs makes it difficult to attach them tightly to the proteins, leading to low temporal and spatial resolution. However, often the time resolution does not play a significant role as the steady-state response upon a stimulus (voltage, ligand) and not the dynamic information is investigated (49,70). This has been done to measure rearrangements of the N- and C-termini in $K_v2.1$ channels (70), G-protein-coupled inwardly-rectifying potassium channels (GIRK) channels (71), and of the gating ring in BK_{Ca} channels (49).

9.3.1.2 Measuring static distances with FRET

The analysis of FRET efficiencies becomes more complex, if static distances and no stimulus-dependent changes are to be measured. In this case, the exact value of the energy transfer has to be obtained. If we assume that the energy transfer for all proteins remains constant, there are a number of techniques to obtain these values. Technically the easiest to accomplish are the so-called three-cube method (56) or the spectral FRET method (40). In the three-cube method, emission in the donor range upon donor excitation and emission in the acceptor range upon both donor and acceptor excitation are determined. In the spectral FRET method, the emission spectrum is recorded both upon donor and acceptor excitation. From these measurements, the apparent FRET efficiency can be determined from the ratio FR of sensitized emission (number of acceptors excited by FRET) to the direct acceptor excitation (number of acceptors present).* Before that the sensitized emission has to be corrected for donor fluorescence ("leak") detected in the sensitized emission and acceptor fluorescence excited directly by the light source. The required correction factors can be predetermined for the setup and filters used (40,56).

A different, more direct method to determine FRET efficiency is to observe the donor emission during photobleaching of the acceptor by high-intensity excitation at a wavelength that does not excite the donor. Once acceptors are photobleached, they no longer accept energy transfer from the donor, so that donor emission increases (Figure 9.4c, 68,72,73). FRET efficiency is then determined according to Equation 9.6.

The limitation of both techniques is that only an apparent energy transfer can be determined as seldom all donors are paired exactly with one acceptor. In the three-cube method and the acceptor photobleaching method, the number of acceptors and donors, respectively, may be overestimated. Both a lower FRET efficiency (longer distance) and more unpaired acceptors would lead to a lower apparent FRET efficiency. If, for instance, any unpaired donors were present, Equation 9.6 would transform to

$$ET = \frac{1}{1-a}\left(1 - \frac{QY_{DA}}{QY_{DO}}\right), \quad (9.13)$$

where a is the fraction of unpaired donors (30). The same problem occurs if two populations with different FRET efficiencies coexist. Occupancy of the populations may only be determined if the FRET efficiencies are known and vice versa. A detailed discussion about analysis of FRET signals can be found in Zal and Gascoigne (74).

In multimeric channels with arbitrary assembly between the subunits, the total apparent FRET efficiency can be calculated from a binomial distribution, and the intrinsic efficiency can be derived from it (75,171). This has, for instance, been used to investigate the relative arrangement of SUR1 and $K_{ir}6.2$ in K_{ATP} channels (57) or the assembly of transient receptor potential (TRP) channels (75).

9.3.1.3 FRET using fluorescence lifetime measurements

Strictly analyzing the fluorescence intensities in one condition will fail if the channels alter between two conformations or several populations coexist with distinct FRET efficiencies. However, the populations or conformations may be separated by determining the *fluorescence lifetimes* of donor and acceptor. The lifetime of a fluorophore is the time constant τ of the fluorescence decay given by Equation 9.2. It can experimentally be determined by exciting the fluorophore with a light pulse and observe the subsequent exponential decay (typically in the nanosecond range). Although fluorescence intensity and lifetime are proportional to one another, the addition of two intensities simply leads to a new value, whereas two decays with different time constants can be separated.

To determine the FRET efficiency, donor lifetime is measured in the absence (donor only, DO) and presence (donor in presence of acceptor, DA) of the acceptor, leading to two lifetimes τ_{DO} and τ_{DA}, respectively. As explained earlier, the lifetime of a fluorophore is not dependent on their number but is an intrinsic property of the fluorophore. The presence of an acceptor adds an additional pathway with a rate constant k_{ET} for the donor to leave the excited state (Figure 9.2a). Assuming several populations of donors i—each with a different distance to the acceptor—coexist, they will have different transfer rates k_{ETi}. According to Equation 9.8, each will also have a different time constant in the presence of the acceptor τ_{DAi} of

$$\tau_{DAi} = \frac{1}{f + k_{NR} + k_{ETi}}. \quad (9.14)$$

Upon an excitation pulse, fluorescence of each donor i will decay with the fluorescence intensity

$$I_{DAi}(t) = I_{DAi}(0) \cdot \exp\left(\frac{-t}{\tau_{DAi}}\right), \quad (9.15)$$

and since all donors are independent of one another, the final signal will be a superposition of all signals

$$I_{DA}(t) = \sum_i I_{DAi}(t) = \sum_i I_{DAi}(0) \cdot \exp\left(\frac{-t}{\tau_{DAi}}\right). \quad (9.16)$$

If they are not too similar, the time constants can be extracted from the fluorescence decay by fitting to a multiexponential decay. Please note that a donor with two acceptors in different

* To be exact, the ratio would have to be corrected by the ratio of the extinction coefficients of donor and acceptor, which determines the number of excited states in sensitized emission and direct excitation assuming equal excitation intensities.

distances will still decay with a single but distinct time constant. This is identical to the case of a single open state traversing into two different closed states, which will result in a single open dwell time. According to Equation 9.6, the energy transfer efficiency ET is given by

$$ET = 1 - \frac{QY_{DA}}{QY_{DO}}. \quad (9.17)$$

Comparison of Equations 9.1 and 9.2 gives QY = $f \cdot \tau$ so that

$$ET = 1 - \frac{\tau_{DA}}{\tau_{DO}}. \quad (9.18)$$

Thus, we can calculate the FRET efficiency directly from the experimentally obtained lifetimes. This has the additional advantage that lifetimes are independent of concentration and can be compared between different preparations. They also do not alter during photobleaching.* For each lifetime obtained from the multiexponential fit, we can determine one distance according to

$$R = R_0 \cdot \left(\frac{\tau_{DO}}{\tau_{DA}} - 1 \right)^{-1/6}. \quad (9.19)$$

Thus for each population or state that coexists, a distinct distance is calculated from the time constants of the multiexponential fit.

Biskup et al. have used lifetime measurements to investigate the association of α and $\beta 1$ subunits of the human cardiac sodium channel (76). The fluorescence decay of Na_V-α, tagged with CFP, was determined in the presence and absence of the $\beta 1$ subunit, fused to YFP, providing the lifetime of donor (CFP) in the absence (donor only, DO) and presence of the acceptor (DA), respectively. The fluorescence decay of DO could be fitted by a single exponential (τ_{DO}), whereas the decay of DA required two exponentials originating from α subunits with and without associated $\beta 1$ subunits. Accordingly, one of the time constants is the same as donor only (τ_{DO}) while the second one represents donors in the presence of acceptors (τ_{DA}). The two populations of sodium channels present in the cells ($\pm\beta 1$) were thus separated.

From the fit of the biexponential decay in the presence of acceptor, the distribution between both populations may be determined. Let us assume that the exponential decay I is described by

$$I(t) = I(0)\left(A_U \exp\left(-\frac{t}{\tau_U} \right) + A_B \exp\left(-\frac{t}{\tau_B} \right) \right), \quad (9.20)$$

where A_U and A_B and τ_U and τ_B are the amplitude and time constant of the un-/bound fraction, respectively. Then the fraction N_B of bound α subunits is

$$N_B = \left(1 + \frac{A_U \tau_B}{A_B \tau_U} \right)^{-1}. \quad (9.21)$$

* As long as each donor only transfers to a single acceptor.

The time constants are necessary to compensate for the reduced QY of the donor in presence of the acceptor. So, by determining fluorescence lifetimes, for each coexisting population, the exact FRET efficiencies as well as its occupancy can be evaluated.

Lifetime-based FRET measurements can also be performed spatially resolved by scanning a certain region of interest and determining the fluorescence decay for each pixel. This technique has been named fluorescence lifetime imaging (FLIM) (77,78).

9.3.1.4 Lanthanide-based resonance energy transfer (LRET)

Distance measurements using lifetime FRET pose two difficulties. First, when measuring lifetime FRET, the fast time constants, typically in the nanosecond range, require specialized detectors and recording equipment that can resolve such a fast decay. These might be time-correlated single-photon counting systems or streak cameras (76). Second, the fluorophores will probably be impaired in their movement, since they are attached to the proteins, and, as a result, the orientation factor κ in Equation 9.10 can only be approximated and might even change during a conformational change. While the first reason is more of technical nature, the second is a principal shortcoming of FRET using organic labels.

The preceding problems are solved, however, using a lanthanide-based donor. Lanthanides are rare earth metals of the f-block in the periodic table of elements. They can absorb light in the near UV range and emit luminescence in the visible wavelength range (79,80). Since they are single atoms, the emission occurs in distinct lines. The often employed terbium (Tb), for instance, emits at 489, 544, 583, and 621 nm with a typical peak width of 20 nm (Figure 9.5c). FRET thus occurs with a number of organic dyes, including TMR (60 Å), fluorescein (45 Å), Alexa-488 (43 Å), ATTO-465 (27 Å), Lucifer-yellow (23 Å), and ABD (15 Å) (46,60,80,81).

The terbium ion is protected in a chelate. The chelator can be a synthetic compound such as the polyaminocarboxylate-carbostyril (Figure 9.5a, 80) or a genetically encoded chelator based on the EF-hand motif (82,83). As a single atom, the emission of the donor is fully isotropic (84). The orientation between donor and acceptor is thus arbitrary, and angular integration will lead to an orientation factor $\kappa = 2/3$. By removing the uncertainty on the orientation factor, the distance between donor and acceptor can be calculated with higher accuracy.

Terbium luminesces with a lifetime in the range of 1–2 ms (Figure 9.5b). The long lifetime allows using standard recording equipment, and, more importantly, it allows to easily separate donor and acceptor lifetimes. One may think of the donor emission as the acceptor excitation in resonance energy transfer. Thus, acceptor fluorescence will be a convolution of donor and acceptor lifetimes, whereas donor emission will decay with a superposition of the DO and DA signals. Because lanthanide luminescence decays with a long lifetime, the acceptor lifetime in the nanosecond range can be neglected in comparison, and the acceptor decay will follow the slower donor (DA) decay. The sensitized emission of the acceptor will thus directly reflect the donor emission lifetime in the presence of the acceptor. The calculation of the distances from the time constants is identical to other FRET measurements (Equation 9.19).

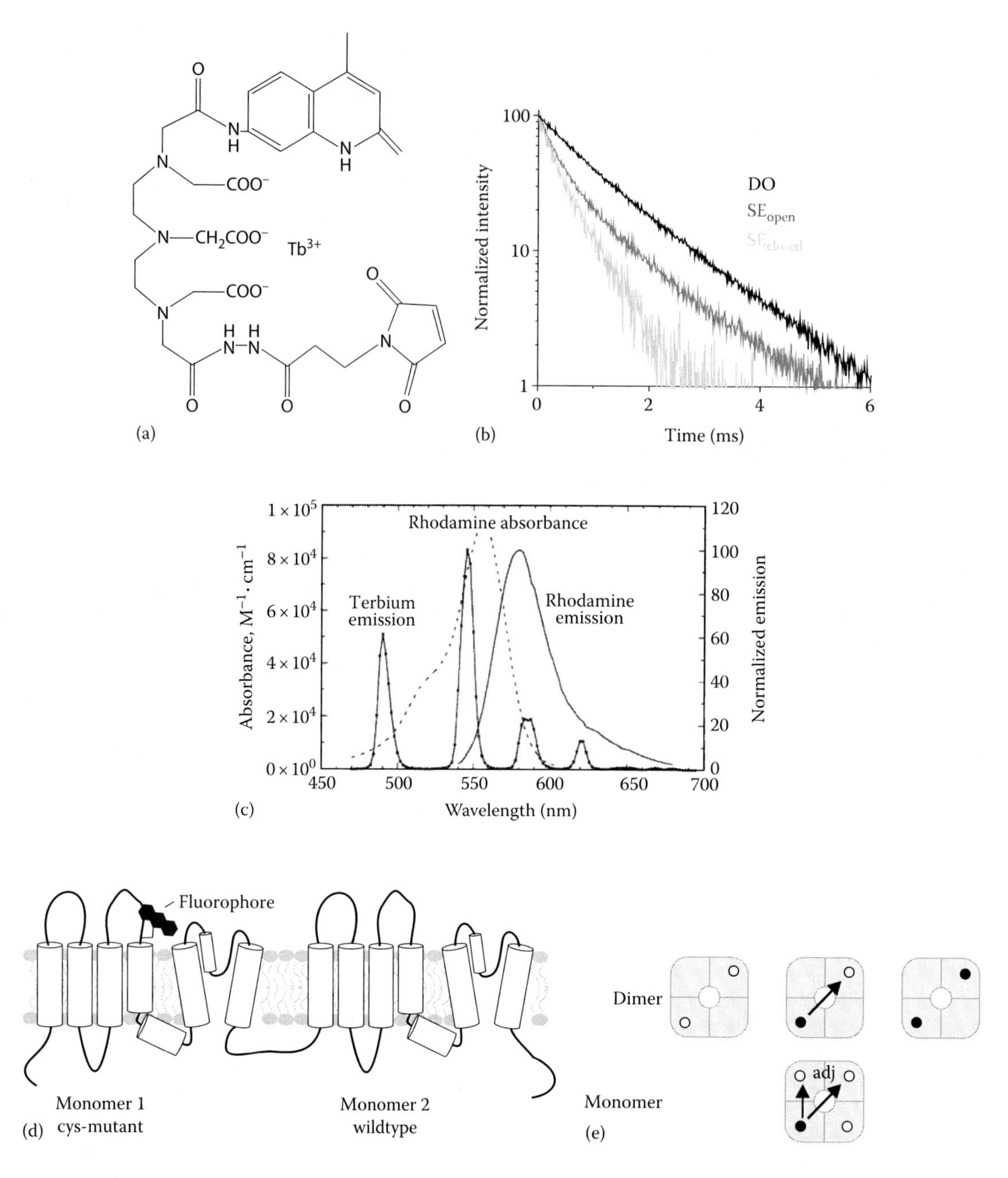

Figure 9.5 (a) Structure of terbium-chelate maleimide. (b) Example of LRET lifetime measurements using terbium chelate. Shown are the luminescent decays after excitation with a pulse at 337 nm for donor only (DO) and sensitized emission in the closed and open state. Shorter distance leads to higher LRET efficiency and faster decay times. (Reproduced from Faure, E. et al., *J. Biol. Chem.*, 287, 40091, 2012.) (c) Emission spectrum of terbium in comparison to excitation and emission spectra of rhodamine. (Reproduced from Selvin, P.R. and Hearst, J.E., *Proc. Natl. Acad. Sci. USA*, 91, 10024, 1994.) (d) Construction of a dimer to reduce the number of LRET distances demonstrated at an example of a 6TM ion channel. (e) Tetramers formed as dimers of dimers arrange with the identical subunits across the pore. (From Liu, Y.S. et al., *Nat. Struct. Biol.*, 8, 883, 2001.) Only dimers with each one acceptor and one donor will result in FRET/LRET.

But the long lifetime comes at a price; since each luminescent decay takes approximately 5 ms, temporal resolution in this range is lost. Conformational changes in the same time range would convolute with the lifetime decay. For this reason, all LRET measurements have been done in the steady state at different conformations (46,51,52,60–62,81,85–87).

The resolution of LRET distances is limited by the movement of the donor and acceptor. While linker length can be determined, its movement around the attachment site depends on the linker flexibility and the protein environment. This is a problem that does not only apply to LRET but also to FRET measurements. The diffusion of donor and acceptor can be estimated by statistical methods or simulations, increasing the accuracy of the distances determined by FRET/LRET (62,88–90).

9.3.1.5 Resonance energy transfer in multimeric channels

Similar to the FRET measurements, the information obtained from LRET is the distance between donor and acceptor. Thus, labeling of the channel proteins plays a crucial role. Often,

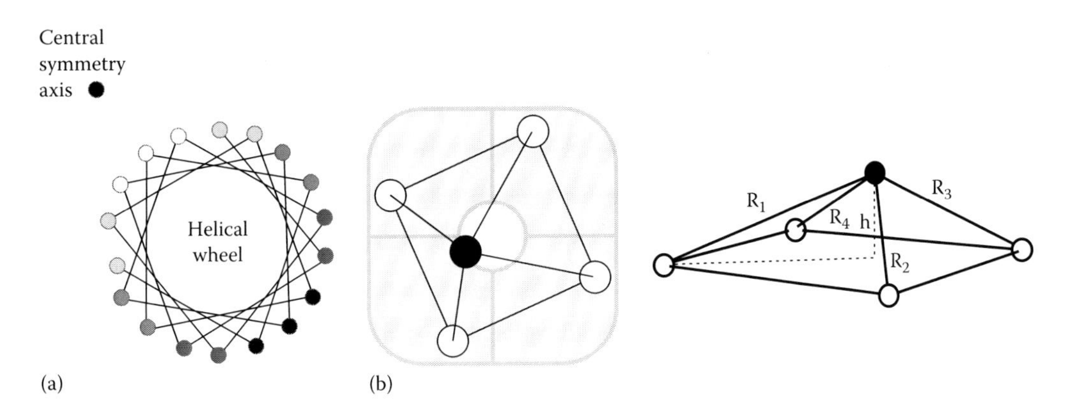

Figure 9.6 (a) The position of a helix in space can be reconstructed from the distance difference of the residues that are located closer (*lighter*) or further (*darker*) relative to the central symmetry axis. (b) For trilateration, donors are positioned at symmetric positions in the protein with an acceptor positioned off the central symmetry axis. From the four distances, the position in space of the donors can be reconstructed. (From Posson, D.J. and Selvin, P.R., *Neuron* 59, 98, 2008; Hyde, H.C. et al., *Structure*, 20, 1629, 2012.)

homomeric channels were investigated, and the identical positions on each subunit were labeled with a mixture of donors and acceptors.* Homomers are typically arranged symmetrically. For many homotetrameric ion channels, for instance, identical positions should be arranged in a square (Figure 9.6b, 91,92). Recent crystal structures, however, have shown that this is not always the case; for instance, the ligand-binding domains (LBDs) of ionotropic glutamate receptors are arranged as dimers of dimers and only show a twofold symmetry in spite of being homotetramers (93). Also the inactivated state of the prokaryotic sodium channel NavAB (94) is not fully symmetric.

Assuming a fourfold symmetric homotetrameric channel, labeled with three donors and one acceptor, each donor will be located adjacent or diagonal to its respective acceptor (Figure 9.5e, 81,86,87) and, consequently, two different time constants will be measured in the sensitized emission. The two distances are related by a Pythagorean so that also the time constants are interdependent.

However, Tb-chelate luminescence decays with two time constants. If one considers the two different distances, adjacent and diagonal, the sensitized emission will present four time constants, two of which are typically very close to each other (81). The interpretation of the results becomes thus quite complex. While it would be best to rely on the slowest time constant, it is not always clear where it originated from; theoretically, the slowest time constant originates from the diagonal distance, which is 1.4 times longer than the adjacent one. However, if the adjacent distance falls in the range of the R_0, the probability for energy transfer between diagonally arranged donor and acceptor becomes so small (ET = 0.5% if $R_{adj} = R_0$) that it would not be detected anymore. The *slowest* still detectable time constant would then originate from the adjacent distance (81). For this reason, several dyes with different R_0 have to be used to correctly assign the distances.

These problems can be avoided if only a single distance occurred between donor and acceptor. A single distance occurs naturally in homotrimeric channels such as P2X and acid-sensing channels (95–97). For the vast majority of ion channels, which form tetramers, concatemers of two channel proteins should be constructed with a single cysteine residue in only one of the two (Figure 9.5d, 46). Liu et al. (98) demonstrated that the pairwise identical subunits would be arranged diagonally in such a construct so that a single distance will be measured (Figure 9.5e). This also simplifies the labeling as resonance energy transfer only occurs in those concatemers that are labeled with each one donor and one acceptor.

In a monomeric channel or a multiple domain channel such as eukaryotic sodium or calcium channels, donor and acceptor position may be placed arbitrarily in the protein. This includes two positions within the same subunit in order to measure intrasubunit conformational changes. Two positions within the same subunit are principally also possible in multimeric channels (51,52); however, one should keep in mind that in the case of, for instance, tetrameric channels, eight labeling sites will be available. Thus, when choosing this option, special attention has to be paid that intra- and intersubunit distances are well separated; in other words, the intersubunit distances are well above the R_0 and do not contaminate the intrasubunit distance measurements. As a control that indeed intrasubunit distances are measured, LRET/sensitized emission measurements with a single labeling site are critical.

9.3.1.6 Reconstructing atomistic models from FRET/LRET distances

Due to its nature, FRET and LRET experiments provide exclusively distances between the donor and acceptor sites. This information is useful to verify distances of a known structure or to define or limit the extent of a known movement (e.g., enlarging the gating ring in BK_{Ca} channels (49) or movement of the S4 in K_V channels (47,59–61,87)). Nevertheless, in an era where research increasingly relies on atomistic models of protein movement,

* When labeling stoichiometrically, one should take care to label with an excess of donors. If more than one acceptor per donor is present, the lifetime would be altered. Each transfer to a donor would add a rate constant k_{T1} and k_{T2} to the luminescence decay. The lifetime would thus be: $\tau = 1/(f + k_{NR} + k_{T1} + k_{T2} + \ldots)$. Several donors per acceptor, on the other hand, would lead to a superposition of different lifetimes.

the question arises what structural information may be directly obtained from the distance measurements. One possibility is to scan several positions with known relation to one another; for instance, when scanning along an α-helix, the positions which point toward the central symmetry axis will generate shorter distances than the positions on the opposite side. By combining sufficient positions along a helix, its orientation in three-dimensional space can be reconstructed (Figure 9.6a, 46).

The distance and height* difference from a fixed position to a position symmetrically labeled in all subunits can also be obtained by *trilateration* (61,62). For trilateration, the symmetric positions in a homomer are labeled with the donor and a single acceptor is positioned off-center at a known position in the protein (Figure 9.6b). From the different distances between the donors and the acceptor combined with the geometric relation between them, the exact position with respect to the acceptor can be determined (62,90). This has been done by labeling a position atop the S4 in Shaker potassium channels with an acceptor bound to a pore blocker at a known position. If the acceptor is attached off-center, the fluorescence decay is a superposition of the four distances. Instead of directly extracting the time constants from the fluorescence decay with its known caveats (see the preceding text), these measurements can be combined with stochastic simulations of the diffusion of donor and acceptor such that the final position is extracted directly from the fluorescence decay (62,90). However, even this method still relies on homogeneity of the state of the channel and the symmetry of a multimeric channel.

More structural information can be obtained if the distances obtained from the FRET/LRET measurements are used as harmonic constraints for molecular dynamics (MD) simulations, which would require an initial model obtained by crystallography or simulations (46,47,49). When used as harmonic constraints in MD simulations, the link to the rest of the channel protein will also show movements invisible in symmetrically labeled FRET/LRET measurements such as rotations around the central axis or movements normal to the membrane surface.

9.3.2 STOICHIOMETRY OF MULTIMERIC CHANNELS

Most ion channels are assembled as multimeric proteins, including trimeric (P2X, acid sensing ion channels), tetrameric (K-channels, TRP channels, iGluR) and pentameric channels (*cys*-loop ion channel receptors). While most structure–function studies are performed on recombinantly expressed homomeric channels, in vivo often different family members assemble to form heteromers, and the composition of these heteromers regulates the function of the proteins. Other channels, such as the NMDA or GABA receptors, are intrinsically composed of different subunits (99,100). In addition, a multitude of auxiliary subunits exist that variably attach to the main channel proteins (101–104). To define the stoichiometry with which the subunits assemble, fluorescence techniques have proven very helpful.

* Height in this case is the shortest distance of the acceptor from the plane spanned by the donors.

9.3.2.1 Fluorescence intensity measurements

Assembly of two subunits can be studied effectively with FRET. As discussed in detail earlier, FRET requires a donor and acceptor to reside within approximately $1.5R_0$, which is typically within the protein diameter. Two subunits form a complex if labeling one each with either donor or acceptor results in occurrence of FRET. Expression levels should be kept sufficiently low to prevent accidental collision between donor and acceptor, leading to FRET. For example, FRET has been used to investigate the heteromeric subunit assembly of CNG, TRP, and K_v channels as well as kainate receptors (40,72,75,105,106).

While the occurrence of FRET will confirm an interaction between two subunits, it does not yet give information on the stoichiometry between the subunits. A fixed stoichiometry can be derived directly from the FRET efficiency (40,105). However, often multimeric channels are arbitrarily assembled, leading to a binomial distribution of donors and acceptors in the protein each with different FRET efficiencies (57,75). Although one may calculate the resulting apparent FRET efficiency, it would be difficult to distinguish different stoichiometries based on the apparent FRET efficiency, in particular when taking into account the low maturation rate of FPs in mammalian cells (107).

9.3.2.2 Single subunit counting

If populations with different stoichiometries coexist, the risk that their composition is averaged out in FRET measurements is relatively high. In this case, single molecule techniques are more promising. The most direct method to determine stoichiometry of ion channels and their auxiliary subunits is to label them with an organic dye or a FP (e.g., GFP) and count the number of fluorophores (Figure 9.7a). This can be done by determining the intensity per spot (108–111) or by counting the number of labels while photobleaching them (single subunit counting, 112). Single subunit counting makes use of the fact that each fluorophore, when exposed to high excitation intensity, is ultimately destroyed (photodestruction, photobleaching), leading to a stepwise decrease of the fluorescence intensity (Figure 9.7b). If each subunit of a specific kind is tagged, the number of fluorophores equals the number of subunits of that kind. In order to resolve single channel fluorescence, they are expressed in low copy number either in mammalian cells or *Xenopus* oocytes (Figure 9.7a) and imaged using total internal reflection fluorescence (TIRF, 107,112–118). The channels may also be purified from any expression system, reconstituted into lipid vesicles, and investigated in supported bilayers using TIRF imaging (36,119,120). TIRF excitation only reaches fluorophores within a distance of approximately half a wavelength (200–350 nm) from the glass surface, leading to effective reduction of background fluorescence and increasing the signal to noise ratio.

One should keep in mind that the number of detected photobleaching steps is not identical to the number of subunits present. Dependent on the labeling technique (see Section 9.4), each subunit has a certain probability of being labeled. If, for instance, thiol-reactive fluorophores are used, some of the thiol groups might be oxidized and thus not reactive. Also, a certain amount of background labeling might occur. This problem does not exist if the proteins are fused to one of the FPs as these are genetically encoded and FP and channel subunit will occur in a

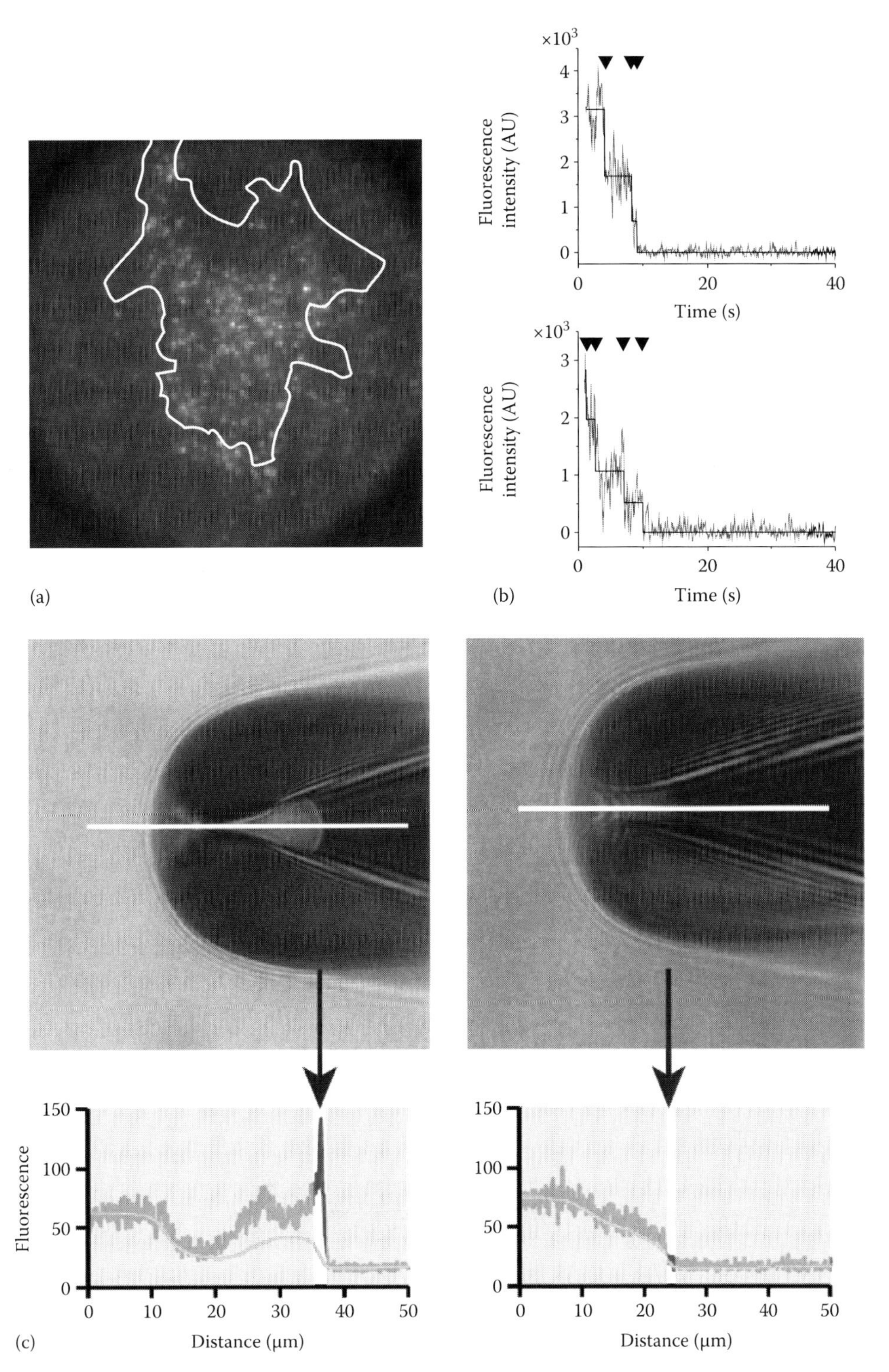

Figure 9.7 (a) Fluorescence image of a HEK293 cell expressing a fluorescently tagged kainate receptor 2 (GluK2) recorded in TIRF configuration. (b) Single subunit counting recordings of GluK2 as shown in (a). Upon bleaching of each subunit (*arrows*), fluorescence intensity decreases in discrete steps. The number of steps corresponds to the number of fluorophores attached to the protein. (c) Image of excised patch expressing CNGA2 channels (*left*) and control (*right*) labeled with 8-DY547-cGMP. In order to normalize to fluorescent background originating from fluorophores in solution, a *calibration* dye (DY647) was added to the solution and recorded using different filters. The graph below shows the increased fluorescence profiles along the white lines. 8-DY547-cGMP fluorescence (noisy traces) is significantly increased over the control dye (DY647, flat traces) at the position of the patch due to bound labeled cGMP, whereas in the absence of channels, both fluorophores show equal intensity profiles.

1:1 stoichiometry. However, also not every FP is visible as each FP has a certain probability to mature, which is dependent on different parameters: the derivative of the FP, the expression system, and the temperature (107,112,121–123). Certain FPs have a higher maturation rate than others, and the maturation rate generally decreases with increasing incubation temperature. Thus, FPs expressed in oocytes, incubated at 18°C, have a higher maturation rate than FPs expressed in mammalian cells. Independent of whether organic dyes or FPs are used, some photobleaching might already occur during the preparation of the samples. Thus, in any scenario, only a fraction (p_f) of the subunits are detected in fluorescence measurements. In consequence to the partial labeling, even a fixed number of subunits will lead to a binomial distribution of photobleaching steps:

$$P(k) = \frac{N!}{k! \cdot (N-k)!} \cdot p_f^k \cdot (1-p_f)^{N-k}, \qquad (9.22)$$

where
N is the number of subunits
k is the number of steps observed
P(k) is the probability to observe k photobleaching steps

These distributions are relatively easy to analyze as long as p_f is sufficiently high. In mammalian cells, incubated at 37°C, the maturation rate of FPs is relatively low and it would be prudent to use a version of *superfolder* GFP, which is optimized for expression at higher temperatures (122). In addition, it is important not to use FPs that dimerize, as that could potentially lead to biased results (107).

For the interpretation to be statistically unambiguous, often a high number of channel complexes need to be analyzed. For this reason, algorithms and software have been developed to aid in the analysis of the data (107,109,124). However, even then, coexistence of different stoichiometries will result in a superposition of the distributions. To avoid ambiguous interpretations, data from labeling different subunits and different expression ratios have to be combined.

9.3.3 LIGAND BINDING

One aspect of ligand-gated ion channels is the binding of the ligands to their respective binding sites and how this process is coupled to channel activation. In contrast to voltage-gated ion channels, where the activation of the voltage sensors can be detected as gating currents, no such direct readout exists for ligand binding. Therefore, fluorescence techniques have proven extremely helpful in this aspect. To address the problem, principally, two approaches are feasible: first, to monitor the conformational change of the binding site related to the ligand binding, and second, to synthesize fluorescently labeled ligand derivatives. The first approach has been discussed earlier. The region around the binding site is labeled either with a single dye for quenching assays (29), or with a FRET pair for distance measurements (49,51,125). The fluorescence change reports the local rearrangements related to ligand binding, whereas the current measurements record the functional state of the channel.

The second approach involves fluorescently labeled ligands. By simultaneously measuring current and fluorescence in excised patches (*patch-clamp fluorometry*), binding of the fluorescently labeled ligand to the channels in the patch leads to increased fluorescence intensity at the membrane. In principle, the *local concentration* of the ligand at the membrane is increased (Figure 9.7c; 32,50,54,55,126). This increase is not observed in the absence of channels, indicating that no unspecific labeling of the membrane occurred. As ligands remain in the solution, background fluorescence is removed by normalization to a second dye in the bulk solution (Figure 9.7c *bottom*). Using this technique, the relation between ligand binding and activation has been investigated for CNG channels using labeled cGMP (55) or calmodulin (50), HCN channels using labeled cGMP (32,54), and nAChR using labeled epibatidine (126,127).

9.3.4 SINGLE CHANNEL FLUORESCENCE

In electrophysiology, the step from macroscopic currents to single channel recordings by the patch-clamp technique led to a new dimension in the analysis of ion channel function, as now aspects of single channel behavior became accessible that were previously hidden in the ensemble average (128). We already reviewed single subunit counting as a method of determining the oligomeric state of a channel. Another important aspect is single channel tracking to investigate expression and trafficking of ion channels (172,173). Here, we want to examine how single channel fluorescence reports conformational changes of the ion channels (31,119,129–134). Certain aspects of structural changes are hidden in the ensemble fluorescence measurements such as the movement of single subunits relative to one another (119) or processes not occurring in a synchronized fashion, for example, in transporter proteins (129,130). For this reason, single channel fluorescence recordings entered the focus of interest.

When recording single molecule fluorescence, a high signal to noise ratio is required, as the absolute fluorescence changes will naturally be very small. Due to the quantum nature of light, however, noise cannot be reduced below Poisson noise, which is proportional to the square root of the light intensity. Therefore, background fluorescence has to be minimized and the signal maximized. In order to increase the signal, fluorophores with a high extinction coefficient are essential. To optimize the light collected, objectives with high numerical aperture as well as detectors with high quantum efficiency and low noise are required. Typically, electron-multiplied charged-coupled device (EMCCD) cameras or photon counters (avalanche photodiodes, [APD], or photon-multiplying tubes [PMT]) are used, as all of them feature a cascading amplification step for each photon before readout. For this reason, readout noise can be neglected. To further reduce background fluorescence, the excitation and detection volumes are optimized using TIRF or confocal imaging techniques. For more details, please refer to Roy et al. (135).

The signals are generated by quenching or (single pair, sp) FRET, as described earlier. As single proteins alternate between different conformations, the single channel fluorescence data will also follow the stochastic fluctuations just like single channel electrophysiology. Accordingly, similar techniques need to be used to analyze the data, including amplitude and dwell-time histograms as well as direct fits to Markov models (31,119,129,130,132). However, single molecule fluorescence has additional properties that need to be taken into account: photobleaching and photoblinking.

We already discussed photobleaching as a manner of distinguishing the number of subunits. When observing conformational changes, however, photobleaching limits the time available to observe the single channel. Thus, photostable dyes should be selected. Blinking is the more important problem; during photoblinking, the fluorophores temporarily enter a triplet state and do not fluoresce during this time. To reduce photobleaching and triplet state occupancy, triplet state quenchers and oxygen scavengers can be added to the solutions (135). Often blinking occurs in a similar time scale as the actual gating events. In this case, interpretation is easier using spFRET because donor and acceptor fluorescence have to change antisymmetrically (129,130,135).

The data also have to be correlated with known characteristics of the channel. While macroscopic properties may be called upon, it would be best to directly correlate single channel fluorescence and current. Currently, most single channel fluorescence results were obtained from purified proteins reconstituted in supported bilayers (119,129,130,132,134) so that no electrophysiological data was available. Sonnleitner et al. (31) recorded single channel fluorescence simultaneously to macroscopic currents in *Xenopus* oocytes. Only of the gramicidin toxin channels, spFRET and single channel recording was achieved simultaneously (34).

In a different approach, not conformational changes, but the ionic current itself is optically detected using Ca^{2+} dyes. Although in this chapter, the immense field of calcium imaging is not further discussed, it should be mentioned in this context. Using Ca^{2+} dyes, single channel current has been detected both electrophysiologically and optically (35,131,133). While imaging single Ca^{2+} currents does not provide any structural information, the spatial distribution of currents and their relation to one another can be investigated, as well as the diffusion characteristics of a single channel current (133). This is important information for simulating ion diffusion after channel opening in cells and synapses.

9.4 LABELING TECHNIQUES

One of the most important concerns for the study of ion channels using fluorescence is the labeling technique. Although mentioned in a different context earlier, we will summarize the various techniques in this section.

9.4.1 THIOL-REACTIVE CHEMISTRY

A commonly used technique is the reaction of the fluorophores with the sulfhydryl groups of cysteines. The fluorophores (or other compounds) are covalently bound to a thiol-reactive linker. These include maleimide (Mal), methanethiosulfonate (MTS), and iodoacetamide (IAA) linkers. Most fluorophores are commercially available with at least one of these linkers (Figure 9.8).

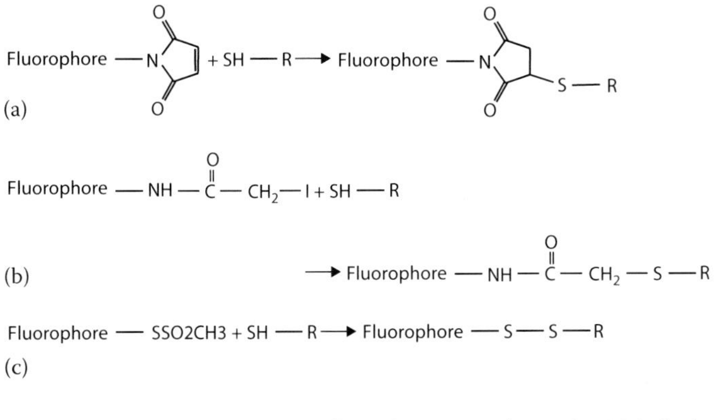

(a) Fluorophore —N(maleimide) + SH—R → Fluorophore —N(succinimide)—S—R

(b) Fluorophore —NH—C(=O)—CH_2—I + SH—R → Fluorophore —NH—C(=O)—CH_2—S—R

(c) Fluorophore — SSO2CH3 + SH —R → Fluorophore —S—S—R

Figure 9.8 Chemical reactions of a Mal- (a), IAA- (b), and MTS-linked (c) fluorophore with a cysteine residue.

IAA and Mal covalently bind to the thiol group of a cysteine, whereas the MTS linker can be separated by a reducing agent, for example, dithiothreitol (DTT). This is frequently used when modifying with chemical compound to vary properties like size or charge of a residue (136–138). However, when labeling with a fluorophore, a covalent, irreversible link will ensure permanent labeling. To improve labeling efficiency, cysteines should be reduced before labeling using DTT or TCEP (*tris*(2-carboxyethyl)phosphine).

One source of background fluorescence originates from labeled cysteines intrinsically available at the cell membrane. When labeling in oocytes, these may be blocked following an elegant protocol from Loots and Isacoff (139). Keeping *Xenopus* oocytes at 12°C allows for synthesis of the channel protein but blocks trafficking. Oocytes are thus injected and incubated at 12°C. Then all available (background) cysteines are blocked with tetraglycine-maleimide or maleimido propionic acid. Subsequent incubation at room temperature for 12–14 h exposes the previously synthesized channels at the surface.

Another possibility to block background labeling, named cysteine metal protection labeling, blocks cysteines temporarily with Cd^{2+} or Zn^{2+}, while background cysteines are reacted with a nonfluorescent compound. The increased affinity for Cd^{2+} or Zn^{2+} for the cysteines of interest is achieved by introducing a histidine in close proximity (140).

Other chemical linkers, for example, SNAP or HALO tags, are available but not commonly used for fluorescent labeling of ion channels.

9.4.2 FLUORESCENTLY LABELED LIGANDS OR TOXINS

Toxins and ligands have the immense advantage that they are highly specific for the ion channel in question. Typical affinity constants for toxins are in the range of 1–300 nM (141–143). Ligands bind to ion channels with affinities typically in the micromolar range. Therefore, very low concentrations of ligand or toxin allow in situ labeling. The labeled ligands or toxins can act as a donor or acceptor for FRET measurements (60–62) or act directly as optical readouts of ligand binding (32,50,54,55,127), as discussed in detail earlier.

9.4.3 GENETICALLY ENCODED FLUORESCENT LABELS

9.4.3.1 Fluorescent proteins

The most commonly used genetically encoded fluorophores are the FPs. Starting from GFP (20,21), a large number of derivatives are known today with a variety of spectral and chemical properties. FPs have been developed in all spectral colors, as chemical sensors, for example, for pH (144), photoswitchable FPs are used for high-resolution imaging (reviewed in 145) and split GFPs or two-component systems are used as membrane potential sensors (63,64,69,146,147). For the investigation of structural aspects of ion channels, FPs are used as labels for FRET and for single subunit counting experiments. For smaller conformational changes, the fluorescent labels would have to be inserted in the transmembrane region of ion channels, which is prevented by the size of the FPs compared to the ion channels (148).

A variation of FRET, namely, bioluminescence resonance energy transfer (BRET, 149), has been extensively used to study oligomerization of G-protein-coupled receptors (reviewed in 150). In BRET, *Renilla luciferase* (RLUC) is used as a donor. As RLUC is bioluminescent, that is, emits photons in response to a chemical reaction in the absence of excitation light, less background is produced, which is particularly useful in tissues. The emission spectrum of RLUC is similar to that of CFP, so that energy transfers well to YFP. In the jellyfish *Aequora*, BRET occurs naturally between aequorin and GFP to produce green light (151).

9.4.3.2 Ligand-binding domains

Another variation of the FRET system is the use of transition metals (Ni^{2+}, Cu^{2+}, and Co^{2+}) as acceptors. The transition metals can form a metal bridge between two or more cysteines or histidines. By introducing specific binding sites (double *cys* or *his*) into the protein, metals can be directed to specific sites (152,153), leading to an R_0 of typically, 10–20 Å. They are thus useful to detect short distances. Being differently colored, each of the transition metals has a characteristic absorption spectrum in the visible range. As the transition metals can be washed out, the acceptor can be removed and replaced during a single experiment. Alternatively, Ni-NTA (nitrilotriacetic acid) derivatives of fluorescent labels may be bound to a multi-*his*-motif (51).

A similar principle is used in the ligand-binding tags used for LRET experiments. A short motif (17aa) is introduced into the protein, which folds as a high-affinity lanthanide-binding site (LBT) similar to an EF-hand motif (82,83,153). By binding terbium ions to this *ligand-binding tag*, LRET experiments can be executed from the LBT to a chemically bound fluorophore. The LBTs are, however, large in comparison to transition metal bridges or cysteine replacements so that restrictions apply as to where in the protein, they can be introduced.

Both methods, the transition metals and ligand-binding tags, have the advantage that acceptor and donor, respectively, are linked to the protein with a different chemistry than the FRET partner. Thus, donor and acceptor can be more specifically directed to sites even within the same subunit without the need of stoichiometrical labeling.

9.4.3.3 Fluorescent unnatural amino acids

Genetically encoded fluorescent tags like the FPs have the immense advantage that they can be directed to any position in the protein and do not produce unspecific labeling. The use of FPs is limited, however, due to their large size and the protection of the chromophore by a β-barrel surrounding it, rendering it less susceptible to environmental changes. Therefore, fluorescent unnatural amino acids (fUAAs), small genetically encoded fluorescent tags substituting the endogenous amino acid, are more suitable. Due to their small size, just slightly larger than a tryptophan, they can be inserted almost anywhere in the protein. In addition, the site does not have to be accessible because they are incorporated during synthesis and do not require posttranslational modification. In spite of their size, fUAAs still have photochemical properties similar to organic dyes (154–157). To incorporate unnatural amino acids (UAAs) into the protein of interest, a nonsense codon is introduced at the position of interest and the corresponding tRNA–UAA complex provided to be inserted into the position in question. There are two principal manners of generating the aminoacylated tRNAs, pioneered by Peter G. Schultz (Scripps Research Institute, San Diego, CA). The first way is to directly synthesize the aminoacylated tRNA and coexpress it together with the protein RNA (154,158–162). In the second approach, the UAA–tRNA pair is directly synthesized in the expression system by coexpression with an orthogonal tRNA/tRNA synthetase pair providing only the UAA (41,163,164). While the use of aminoacylated tRNA allows more flexibility in the chemical structure of the UAA, using an orthogonal pair of tRNA/tRNA synthetase is more specific for the UAA and achieves higher yields.

fUAAs have been successfully incorporated into ion channels in an oocyte expression system using both an aminoacylated tRNA (154) and an orthogonal tRNA/tRNA synthetase pair (41). Like the FPs, using fUAAs allowed labeling of the cytosolic surface of ion channels. With fUAAs, it was possible to label positions close to the inner pore gate and at the C-terminus of the S4. Expression levels were comparable to channels not containing fUAAs such that VCF was possible, and gating and ionic current could be temporally correlated with the movement of the lower S4 and the opening of the inner pore gate (41).

9.5 CONCLUSIONS

With ever-improving imaging hardware and a plethora of fluorophores with various properties, fluorescence has become an indispensable tool for ion channel research, be it in structural biology, molecular neuroscience, or cardiac research. Fluorescence recordings are mainly limited by the ability to direct the label to a specific position. This aspect is being steadily improved, in particular, with the development of novel detection assays and increased availability of genetically encoded tags. The advantages of an optical readout include: first, fluorophores are sensitive to their environment and are thus modulated by a variety of factors, and second, optical recordings are spatially resolved. Furthermore, the optical recordings do not interfere with electrophysiology so that both can be recorded simultaneously. The spatial resolution permits parallelization of the systems, which plays a crucial role in the development of high-throughput assays or in recording from complex systems such as neuronal tissue. In this chapter, we concentrated on how to exploit fluorescence spectroscopy in order to obtain structural information. However, calcium, pH or voltage sensors are also common uses for optical readout that can be used in high-throughput systems. For instance, the temperature sensors in TRPV1 channels have been found by random mutagenesis and subsequent screening using a high-throughput calcium-influx assay on transiently transfected HEK293 cells (165).

Spatial resolution may also be exploited to optically excite ion channels in a specified region and thereby induce neuronal activity. By linking either a ligand or a blocker via a linker, whose length is photoswitchable (azo-benzene), to an ion channel, the channel can be turned on and off using light of different wavelengths (166–170).

REFERENCES

1. Feldman, I. and G. E. Norton. 1980. Effects of glucose and magnesium ion on the quenching of yeast hexokinase fluorescence by acrylamide. *Biochim Biophys Acta* 615:132–142.
2. Rao, A., P. Martin, R. A. Reithmeier, and L. C. Cantley. 1979. Location of the stilbenedisulfonate binding site of the human erythrocyte anion-exchange system by resonance energy transfer. *Biochemistry* 18:4505–4516.
3. Axelrod, D., P. Ravdin, D. E. Koppel, J. Schlessinger, W. W. Webb, E. L. Elson, and T. R. Podleski. 1976. Lateral motion of fluorescently labeled acetylcholine receptors in membranes of developing muscle fibers. *Proc Natl Acad Sci USA* 73:4594–4598.
4. Angelides, K. J. 1981. Fluorescent and photoactivatable fluorescent derivatives of tetrodotoxin to probe the sodium channel of excitable membranes. *Biochemistry* 20:4107–4118.
5. Tsien, R. Y. 1980. New calcium indicators and buffers with high selectivity against magnesium and protons: Design, synthesis, and properties of prototype structures. *Biochemistry* 19:2396–2404.
6. Grynkiewicz, G., M. Poenie, and R. Y. Tsien. 1985. A new generation of Ca^{2+} indicators with greatly improved fluorescence properties. *J Biol Chem* 260:3440–3450.
7. Noda, M., S. Shimizu, T. Tanabe, T. Takai, T. Kayano, T. Ikeda, H. Takahashi, H. Nakayama, Y. Kanaoka, and N. Minamino. 1984. Primary structure of electrophorus electricus sodium channel deduced from cDNA sequence. *Nature* 312:121–127.
8. Noda, M., T. Ikeda, H. Suzuki, H. Takeshima, T. Takahashi, M. Kuno, and S. Numa. 1986. Expression of functional sodium channels from cloned cDNA. *Nature* 322:826–828.
9. Goldin, A. L., T. Snutch, H. Lubbert, A. Dowsett, J. Marshall, V. Auld, W. Downey et al. 1986. Messenger RNA coding for only the alpha subunit of the rat brain Na channel is sufficient for expression of functional channels in *Xenopus* oocytes. *Proc Natl Acad Sci USA* 83:7503–7507.
10. Papazian, D. M., T. L. Schwarz, B. L. Tempel, Y. N. Jan, and L. Y. Jan. 1987. Cloning of genomic and complementary DNA from Shaker, a putative potassium channel gene from *Drosophila*. *Science* 237:749–753.
11. Tempel, B. L., D. M. Papazian, T. L. Schwarz, Y. N. Jan, and L. Y. Jan. 1987. Sequence of a probable potassium channel component encoded at Shaker locus of *Drosophila*. *Science* 237:770–775.
12. Iverson, L. E., M. A. Tanouye, H. A. Lester, N. Davidson, and B. Rudy. 1988. A-type potassium channels expressed from Shaker locus cDNA. *Proc Natl Acad Sci USA* 85:5723–5727.
13. Frech, G. C., A. M. VanDongen, G. Schuster, A. M. Brown, and R. H. Joho. 1989. A novel potassium channel with delayed rectifier properties isolated from rat brain by expression cloning. *Nature* 340:642–645.
14. Butler, A., S. Tsunoda, D. P. McCobb, A. Wei, and L. Salkoff. 1993. *mSlo*, a complex mouse gene encoding "maxi" calcium-activated potassium channels. *Science* 261:221–224.
15. Dhallan, R. S., K. W. Yau, K. A. Schrader, and R. R. Reed. 1990. Primary structure and functional expression of a cyclic nucleotide-activated channel from olfactory neurons. *Nature* 347:184–187.
16. Warmke, J., R. Drysdale, and B. Ganetzky. 1991. A distinct potassium channel polypeptide encoded by the *Drosophila* eag locus. *Science* 252:1560–1562.
17. Trudeau, M. C., J. W. Warmke, B. Ganetzky, and G. A. Robertson. 1995. HERG, a human inward rectifier in the voltage-gated potassium channel family. *Science* 269:92–95.
18. Mannuzzu, L. M., M. M. Moronne, and E. Y. Isacoff. 1996. Direct physical measure of conformational rearrangement underlying potassium channel gating. *Science* 271:213–216.
19. Cha, A. and F. Bezanilla. 1997. Characterizing voltage-dependent conformational changes in the Shaker K^+ channel with fluorescence. *Neuron* 19:1127–1140.
20. Prasher, D. C., V. K. Eckenrode, W. W. Ward, F. G. Prendergast, and M. J. Cormier. 1992. Primary structure of the *Aequorea victoria* green-fluorescent protein. *Gene* 111:229–233.
21. Heim, R., D. C. Prasher, and R. Y. Tsien. 1994. Wavelength mutations and posttranslational autoxidation of green fluorescent protein. *Proc Natl Acad Sci USA* 91:12501–12504.
22. Turro, N. J., V. Ramamurthy, and J. C. Scaiano. 2010. *Modern Molecular Photochemistry of Organic Molecules*. University Science Books, Sausalito, CA.
23. Cha, A. and F. Bezanilla. 1998. Structural implications of fluorescence quenching in the Shaker K^+ channel. *J Gen Physiol* 112:391–408.
24. Zheng, J. and W. N. Zagotta. 2000. Gating rearrangements in cyclic nucleotide-gated channels revealed by patch-clamp fluorometry. *Neuron* 28:369–374.
25. Marme, N., J. P. Knemeyer, M. Sauer, and J. Wolfrum. 2003. Inter- and intramolecular fluorescence quenching of organic dyes by tryptophan. *Bioconjug Chem* 14:1133–1139.
26. Blunck, R., D. M. Starace, A. M. Correa, and F. Bezanilla. 2004. Detecting rearrangements of shaker and NaChBac in real-time with fluorescence spectroscopy in patch-clamped mammalian cells. *Biophys J* 86:3966–3980.
27. Sato, E., M. Sakashita, Y. Kanaoka, and E. M. Kosower. 1988. Organic fluorescent reagents. XIV. Novel fluorogenic substrates for microdetermination of chymotrypsin and aminopeptidase—Bimane fluorescence appears after hydrolysis. *Bioorg Chem* 16:298–306.
28. Mansoor, S. E., H. S. McHaourab, and D. L. Farrens. 2002. Mapping proximity within proteins using fluorescence spectroscopy. A study of T4 lysozyme showing that tryptophan residues quench bimane fluorescence. *Biochemistry* 41:2475–2484.
29. Islas, L. D. and W. N. Zagotta. 2006. Short-range molecular rearrangements in ion channels detected by tryptophan quenching of bimane fluorescence. *J Gen Physiol* 128:337–346.
30. Lakowicz, J. R. 1999. *Principles of Fluorescence Spectroscopy*. Kluwer Academic, New York.
31. Sonnleitner, A., L. M. Mannuzzu, S. Terakawa, and E. Y. Isacoff. 2002. Structural rearrangements in single ion channels detected optically in living cells. *Proc Natl Acad Sci USA* 99:12759–12764.
32. Kusch, J., C. Biskup, S. Thon, E. Schulz, V. Nache, T. Zimmer, F. Schwede, and K. Benndorf. 2010. Interdependence of receptor activation and ligand binding in HCN2 pacemaker channels. *Neuron* 67:75–85.
33. Groulx, N., M. Juteau, and R. Blunck. 2010. Rapid topology probing using fluorescence spectroscopy in planar lipid bilayer: The pore-forming mechanism of the toxin Cry1Aa of *Bacillus thuringiensis*. *J Gen Physiol* 136:497–513.
34. Borisenko, V., T. Lougheed, J. Hesse, E. Fureder-Kitzmuller, N. Fertig, J. C. Behrends, G. A. Woolley, and G. J. Schutz. 2003. Simultaneous optical and electrical recording of single gramicidin channels. *Biophys J* 84:612–622.
35. Heron, A. J., J. R. Thompson, B. Cronin, H. Bayley, and M. I. Wallace. 2009. Simultaneous measurement of ionic current and fluorescence from single protein pores. *J Am Chem Soc* 131:1652–1653.
36. Thompson, J. R., B. Cronin, H. Bayley, and M. I. Wallace. 2011. Rapid assembly of a multimeric membrane protein pore. *Biophys J* 101:2679–2683.
37. Sorensen, J. B., A. Cha, R. Latorre, E. Rosenman, and F. Bezanilla. 2000. Deletion of the S3–S4 linker in the Shaker potassium channel reveals two quenching groups near the outside of S4. *J Gen Physiol* 115:209–222.
38. Semenova, N. P., K. Abarca-Heidemann, E. Loranc, and B. S. Rothberg. 2009. Bimane fluorescence scanning suggests secondary structure near the S3–S4 linker of BK channels. *J Biol Chem* 284:10684–10693.

39. Pantazis, A. and R. Olcese. 2012. Relative transmembrane segment rearrangements during BK channel activation resolved by structurally assigned fluorophore-quencher pairing. *J Gen Physiol* 140:207–218.
40. Zheng, J. and W. N. Zagotta. 2004. Stoichiometry and assembly of olfactory cyclic nucleotide-gated channels. *Neuron* 42:411–421.
41. Kalstrup, T. and R. Blunck. 2013. Dynamics of internal pore opening in K_V channels probed by a fluorescent unnatural amino acid. *Proc Natl Acad Sci USA* 110:8272–8277.
42. Chanda, B., O. K. Asamoah, and F. Bezanilla. 2004. Coupling interactions between voltage sensors of the sodium channel as revealed by site-specific measurements. *J Gen Physiol* 123:217–230.
43. Gandhi, C. S., E. Loots, and E. Y. Isacoff. 2000. Reconstructing voltage sensor-pore interaction from a fluorescence scan of a voltage-gated K^+ channel. *Neuron* 27:585–595.
44. Vaid, M., T. W. Claydon, S. Rezazadeh, and D. Fedida. 2008. Voltage clamp fluorimetry reveals a novel outer pore instability in a mammalian voltage-gated potassium channel. *J Gen Physiol* 132:209–222.
45. Wang, S., S. J. Lee, S. Heyman, D. Enkvetchakul, and C. G. Nichols. 2012. Structural rearrangements underlying ligand-gating in Kir channels. *Nat Commun* 3:617.
46. Faure, E., G. Starek, H. McGuire, S. Berneche, and R. Blunck. 2012. A limited 4 A radial displacement of the S4–S5 linker is sufficient for internal gate closing in Kv channels. *J Biol Chem* 287:40091–40098.
47. Chanda, B., O. K. Asamoah, R. Blunck, B. Roux, and F. Bezanilla. 2005. Gating charge displacement in voltage-gated ion channels involves limited transmembrane movement. *Nature* 436:852–856.
48. Taraska, J. W. and W. N. Zagotta. 2007. Structural dynamics in the gating ring of cyclic nucleotide-gated ion channels. *Nat Struct Mol Biol* 14:854–860.
49. Miranda, P., J. E. Contreras, A. J. Plested, F. J. Sigworth, M. Holmgren, and T. Giraldez. 2013. State-dependent FRET reports calcium- and voltage-dependent gating-ring motions in BK channels. *Proc Natl Acad Sci USA* 110(13):5217–5222.
50. Trudeau, M. C. and W. N. Zagotta. 2004. Dynamics of Ca^{2+}-calmodulin-dependent inhibition of rod cyclic nucleotide-gated channels measured by patch-clamp fluorometry. *J Gen Physiol* 124:211–223.
51. Gonzalez, J., A. Rambhadran, M. Du, and V. Jayaraman. 2008. LRET investigations of conformational changes in the ligand binding domain of a functional AMPA receptor. *Biochemistry* 47:10027–10032.
52. Du, M., A. Rambhadran, and V. Jayaraman. 2008. Luminescence resonance energy transfer investigation of conformational changes in the ligand binding domain of a kainate receptor. *J Biol Chem* 283:27074–27078.
53. Yang, F., Y. Cui, K. Wang, and J. Zheng. 2010. Thermosensitive TRP channel pore turret is part of the temperature activation pathway. *Proc Natl Acad Sci USA* 107:7083–7088.
54. Kusch, J., S. Thon, E. Schulz, C. Biskup, V. Nache, T. Zimmer, R. Seifert, F. Schwede, and K. Benndorf. 2012. How subunits cooperate in cAMP-induced activation of homotetrameric HCN2 channels. *Nat Chem Biol* 8:162–169.
55. Biskup, C., J. Kusch, E. Schulz, V. Nache, F. Schwede, F. Lehmann, V. Hagen, and K. Benndorf. 2007. Relating ligand binding to activation gating in CNGA2 channels. *Nature* 446:440–443.
56. Erickson, M. G., B. A. Alseikhan, B. Z. Peterson, and D. T. Yue. 2001. Preassociation of calmodulin with voltage-gated Ca(2+) channels revealed by FRET in single living cells. *Neuron* 31:973–985.
57. Wang, S., E. N. Makhina, R. Masia, K. L. Hyrc, M. L. Formanack, and C. G. Nichols. 2013. Domain organization of the ATP-sensitive potassium channel complex examined by fluorescence resonance energy transfer. *J Biol Chem* 288:4378–4388.
58. Berger, W., H. Prinz, J. Striessnig, H. C. Kang, R. Haugland, and H. Glossmann. 1994. Complex molecular mechanism for dihydropyridine binding to L-type Ca^{2+}-channels as revealed by fluorescence resonance energy transfer. *Biochemistry* 33:11875–11883.
59. Glauner, K. S., L. M. Mannuzzu, C. S. Gandhi, and E. Y. Isacoff. 1999. Spectroscopic mapping of voltage sensor movement in the Shaker potassium channel. *Nature* 402:813–817.
60. Posson, D. J., P. Ge, C. Miller, F. Bezanilla, and P. R. Selvin. 2005. Small vertical movement of a K^+ channel voltage sensor measured with luminescence energy transfer. *Nature* 436:848–851.
61. Posson, D. J. and P. R. Selvin. 2008. Extent of voltage sensor movement during gating of shaker K^+ channels. *Neuron* 59:98–109.
62. Hyde, H. C., W. Sandtner, E. Vargas, A. T. Dagcan, J. L. Robertson, B. Roux, A. M. Correa, and F. Bezanilla. 2012. Nano-positioning system for structural analysis of functional homomeric proteins in multiple conformations. *Structure* 20:1629–1640.
63. Chanda, B., R. Blunck, L. C. Faria, F. E. Schweizer, I. Mody, and F. Bezanilla. 2005. A hybrid approach to measuring electrical activity in genetically specified neurons. *Nat Neurosci* 8:1619–1626.
64. Wang, D., Z. Zhang, B. Chanda, and M. B. Jackson. 2010. Improved probes for hybrid voltage sensor imaging. *Biophys J* 99:2355–2365.
65. Fernandez, J. M., R. E. Taylor, and F. Bezanilla. 1983. Induced capacitance in the squid giant axon. Lipophilic ion displacement currents. *J Gen Physiol* 82:331–346.
66. Patterson, G. H., D. W. Piston, and B. G. Barisas. 2000. Forster distances between green fluorescent protein pairs. *Anal Biochem* 284:438–440.
67. Zheng, J., M. D. Varnum, and W. N. Zagotta. 2003. Disruption of an intersubunit interaction underlies Ca^{2+}-calmodulin modulation of cyclic nucleotide-gated channels. *J Neurosci* 23:8167–8175.
68. Fisher, J. A., G. Girdler, and B. S. Khakh. 2004. Time-resolved measurement of state-specific P2X2 ion channel cytosolic gating motions. *J Neurosci* 24:10475–10487.
69. Sakai, R., V. Repunte-Canonigo, C. D. Raj, and T. Knopfel. 2001. Design and characterization of a DNA-encoded, voltage-sensitive fluorescent protein. *Eur J Neurosci* 13:2314–2318.
70. Kobrinsky, E., L. Stevens, Y. Kazmi, D. Wray, and N. M. Soldatov. 2006. Molecular rearrangements of the Kv2.1 potassium channel termini associated with voltage gating. *J Biol Chem* 281:19233–19240.
71. Riven, I., E. Kalmanzon, L. Segev, and E. Reuveny. 2003. Conformational rearrangements associated with the gating of the G protein-coupled potassium channel revealed by fret microscopy. *Neuron* 38:225–235.
72. Kerschensteiner, D., F. Soto, and M. Stocker. 2005. Fluorescence measurements reveal stoichiometry of K^+ channels formed by modulatory and delayed rectifier alpha-subunits. *Proc Natl Acad Sci USA* 102:6160–6165.
73. Bastiaens, P. I. and T. M. Jovin. 1996. Microspectroscopic imaging tracks the intracellular processing of a signal transduction protein: Fluorescent-labeled protein kinase C beta I. *Proc Natl Acad Sci USA* 93:8407–8412.
74. Zal, T. and N. R. Gascoigne. 2004. Photobleaching-corrected FRET efficiency imaging of live cells. *Biophys J* 86:3923–3939.
75. Cheng, W., F. Yang, C. L. Takanishi, and J. Zheng. 2007. Thermosensitive TRPV channel subunits coassemble into heteromeric channels with intermediate conductance and gating properties. *J Gen Physiol* 129:191–207.
76. Biskup, C., T. Zimmer, and K. Benndorf. 2004. FRET between cardiac Na^+ channel subunits measured with a confocal microscope and a streak camera. *Nat Biotechnol* 22:220–224.

77. Becker, W., A. Bergmann, M. A. Hink, K. Konig, K. Benndorf, and C. Biskup. 2004. Fluorescence lifetime imaging by time-correlated single-photon counting. *Microsc Res Tech* 63:58–66.
78. Nomura, T., C. G. Cranfield, E. Deplazes, D. M. Owen, A. Macmillan, A. R. Battle, M. Constantine, M. Sokabe, and B. Martinac. 2012. Differential effects of lipids and lyso-lipids on the mechanosensitivity of the mechanosensitive channels MscL and MscS. *Proc Natl Acad Sci USA* 109:8770–8775.
79. Selvin, P. R. 2002. Principles and biophysical applications of lanthanide-based probes. *Annu Rev Biophys Biomol Struct* 31:275–302.
80. Selvin, P. R. and J. E. Hearst. 1994. Luminescence energy transfer using a terbium chelate: Improvements on fluorescence energy transfer. *Proc Natl Acad Sci USA* 91:10024–10028.
81. Richardson, J., R. Blunck, P. Ge, P. R. Selvin, F. Bezanilla, D. M. Papazian, and A. M. Correa. 2006. Distance measurements reveal a common topology of prokaryotic voltage-gated ion channels in the lipid bilayer. *Proc Natl Acad Sci USA* 103:15865–15870.
82. Nitz, M., M. Sherawat, K. J. Franz, E. Peisach, K. N. Allen, and B. Imperiali. 2004. Structural origin of the high affinity of a chemically evolved lanthanide-binding peptide. *Angew Chem Int Ed Engl* 43:3682–3685.
83. Franz, K. J., M. Nitz, and B. Imperiali. 2003. Lanthanide-binding tags as versatile protein coexpression probes. *ChemBioChem* 4:265–271.
84. Reifenberger, J. G., G. E. Snyder, G. Baym, and P. R. Selvin. 2003. Emission polarization of europium and terbium chelates. *J Phys Chem B* 107:12862–12873.
85. Rambhadran, A., J. Gonzalez, and V. Jayaraman. 2011. Conformational changes at the agonist binding domain of the N-methyl-D-aspartic acid receptor. *J Biol Chem* 286:16953–16957.
86. Gonzalez, J., M. Du, K. Parameshwaran, V. Suppiramaniam, and V. Jayaraman. 2010. Role of dimer interface in activation and desensitization in AMPA receptors. *Proc Natl Acad Sci USA* 107:9891–9896.
87. Cha, A., G. E. Snyder, P. R. Selvin, and F. Bezanilla. 1999. Atomic scale movement of the voltage-sensing region in a potassium channel measured via spectroscopy. *Nature* 402:809–813.
88. Sindbert, S., S. Kalinin, H. Nguyen, A. Kienzler, L. Clima, W. Bannwarth, B. Appel, S. Muller, and C. A. Seidel. 2011. Accurate distance determination of nucleic acids via Forster resonance energy transfer: Implications of dye linker length and rigidity. *J Am Chem Soc* 133:2463–2480.
89. Kalinin, S., T. Peulen, S. Sindbert, P. J. Rothwell, S. Berger, T. Restle, R. S. Goody, H. Gohlke, and C. A. Seidel. 2012. A toolkit and benchmark study for FRET-restrained high-precision structural modeling. *Nat Methods* 9:1218–1225.
90. Muschielok, A., J. Andrecka, A. Jawhari, F. Bruckner, P. Cramer, and J. Michaelis. 2008. A nano-positioning system for macromolecular structural analysis. *Nat Methods* 5:965–971.
91. Doyle, D. A., C. J. Morais, R. A. Pfuetzner, A. Kuo, J. M. Gulbis, S. L. Cohen, B. T. Chait, and R. MacKinnon. 1998. The structure of the potassium channel: Molecular basis of K^+ conduction and selectivity. *Science* 280:69–77.
92. Long, S. B., E. B. Campbell, and R. MacKinnon. 2005. Crystal structure of a mammalian voltage-dependent shaker family K^+ channel. *Science* 309:897–903.
93. Sobolevsky, A. I., M. P. Rosconi, and E. Gouaux. 2009. X-ray structure, symmetry and mechanism of an AMPA-subtype glutamate receptor. *Nature* 462:745–756.
94. Payandeh, J., T. M. Gamal El-Din, T. Scheuer, N. Zheng, and W. A. Catterall. 2012. Crystal structure of a voltage-gated sodium channel in two potentially inactivated states. *Nature* 486:135–139.
95. Gonzales, E. B., T. Kawate, and E. Gouaux. 2009. Pore architecture and ion sites in acid-sensing ion channels and P2X receptors. *Nature* 460:599–604.
96. Jasti, J., H. Furukawa, E. B. Gonzales, and E. Gouaux. 2007. Structure of acid-sensing ion channel 1 at 1.9 A resolution and low pH. *Nature* 449:316–323.
97. Kawate, T., J. C. Michel, W. T. Birdsong, and E. Gouaux. 2009. Crystal structure of the ATP-gated P2X(4) ion channel in the closed state. *Nature* 460:592–598.
98. Liu, Y. S., P. Sompornpisut, and E. Perozo. 2001. Structure of the KcsA channel intracellular gate in the open state. *Nat Struct Biol* 8:883–887.
99. Paoletti, P. and J. Neyton. 2007. NMDA receptor subunits: Function and pharmacology. *Curr Opin Pharmacol* 7:39–47.
100. Sigel, E. and M. E. Steinmann. 2012. Structure, function, and modulation of GABA(A) receptors. *J Biol Chem* 287:40224–40231.
101. Chahine, M. and M. E. O'Leary. 2011. Regulatory role of voltage-gated Na channel beta subunits in sensory neurons. *Front Pharmacol* 2:70.
102. Dolphin, A. C. 2012. Calcium channel auxiliary alpha(2)delta and beta subunits: Trafficking and one step beyond. *Nat Rev Neurosci* 13:542–555.
103. Tomita, S. 2010. Regulation of ionotropic glutamate receptors by their auxiliary subunits. *Physiology* 25:41–49.
104. Latorre, R., F. J. Morera, and C. Zaelzer. 2010. Allosteric interactions and the modular nature of the voltage- and Ca^{2+}-activated (BK) channel. *J Physiol* 588:3141–3148.
105. Zheng, J., M. C. Trudeau, and W. N. Zagotta. 2002. Rod cyclic nucleotide-gated channels have a stoichiometry of three CNGA1 subunits and one CNGB1 subunit. *Neuron* 36:891–896.
106. Ma-Hogemeier, Z. L., C. Korber, M. Werner, D. Racine, E. Muth-Kohne, D. Tapken, and M. Hollmann. 2010. Oligomerization in the endoplasmic reticulum and intracellular trafficking of kainate receptors are subunit-dependent but not editing-dependent. *J Neurochem* 113:1403–1415.
107. McGuire, H., M. R. Aurousseau, D. Bowie, and R. Blunck. 2012. Automating single subunit counting of membrane proteins in mammalian cells. *J Biol Chem* 287:35912–35921.
108. Swift, J. L., A. G. Godin, K. Dore, L. Freland, N. Bouchard, C. Nimmo, M. Sergeev, Y. De Koninck, P. W. Wiseman, and J. M. Beaulieu. 2011. Quantification of receptor tyrosine kinase transactivation through direct dimerization and surface density measurements in single cells. *Proc Natl Acad Sci USA* 108:7016–7021.
109. Godin, A. G., S. Costantino, L. E. Lorenzo, J. L. Swift, M. Sergeev, A. Ribeiro-da-Silva, Y. De Koninck, and P. W. Wiseman. 2011. Revealing protein oligomerization and densities in situ using spatial intensity distribution analysis. *Proc Natl Acad Sci USA* 108:7010–7015.
110. Kask, P., K. Palo, N. Fay, L. Brand, U. Mets, D. Ullmann, J. Jungmann, J. Pschorr, and K. Gall. 2000. Two-dimensional fluorescence intensity distribution analysis: Theory and applications. *Biophys J* 78:1703–1713.
111. Kask, P., K. Palo, D. Ullmann, and K. Gall. 1999. Fluorescence-intensity distribution analysis and its application in biomolecular detection technology. *Proc Natl Acad Sci USA* 96:13756–13761.
112. Ulbrich, M. H. and E. Y. Isacoff. 2007. Subunit counting in membrane-bound proteins. *Nat Methods* 4:319–321.
113. Hastie, P., M. H. Ulbrich, H. L. Wang, R. J. Arant, A. G. Lau, Z. Zhang, E. Y. Isacoff, and L. Chen. 2013. AMPA receptor/TARP stoichiometry visualized by single-molecule subunit counting. *Proc Natl Acad Sci USA* 110:5163–5168.
114. Yu, Y., M. H. Ulbrich, M. H. Li, S. Dobbins, W. K. Zhang, L. Tong, E. Y. Isacoff, and J. Yang. 2012. Molecular mechanism of the assembly of an acid-sensing receptor ion channel complex. *Nat Commun* 3:1252.
115. Reiner, A., R. J. Arant, and E. Y. Isacoff. 2012. Assembly stoichiometry of the GluK2/GluK5 kainate receptor complex. *Cell Rep* 1:234–240.

116. Tombola, F., M. H. Ulbrich, and E. Y. Isacoff. 2008. The voltage-gated proton channel Hv1 has two pores, each controlled by one voltage sensor. *Neuron* 58:546–556.
117. Plant, L. D., E. J. Dowdell, I. S. Dementieva, J. D. Marks, and S. A. Goldstein. 2011. SUMO modification of cell surface Kv2.1 potassium channels regulates the activity of rat hippocampal neurons. *J Gen Physiol* 137:441–454.
118. Plant, L. D., I. S. Dementieva, A. Kollewe, S. Olikara, J. D. Marks, and S. A. Goldstein. 2010. One SUMO is sufficient to silence the dimeric potassium channel K2P1. *Proc Natl Acad Sci USA* 107:10743–10748.
119. Blunck, R., H. McGuire, H. C. Hyde, and F. Bezanilla. 2008. Fluorescence detection of the movement of single KcsA subunits reveals cooperativity. *Proc Natl Acad Sci USA* 105:20263–20268.
120. Groulx, N., H. McGuire, R. Laprade, J. L. Schwartz, and R. Blunck. 2011. Single molecule fluorescence study of the *Bacillus thuringiensis* toxin Cry1Aa reveals tetramerization. *J Biol Chem* 286:42274–42282.
121. Garcia-Parajo, M. F., M. Koopman, E. M. van Dijk, V. Subramaniam, and N. F. van Hulst. 2001. The nature of fluorescence emission in the red fluorescent protein DsRed, revealed by single-molecule detection. *Proc Natl Acad Sci USA* 98:14392–14397.
122. Pedelacq, J. D., S. Cabantous, T. Tran, T. C. Terwilliger, and G. S. Waldo. 2006. Engineering and characterization of a superfolder green fluorescent protein. *Nat Biotechnol* 24:79–88.
123. Zacharias, D. A., J. D. Violin, A. C. Newton, and R. Y. Tsien. 2002. Partitioning of lipid-modified monomeric GFPs into membrane microdomains of live cells. *Science* 296:913–916.
124. Kerssemakers, J. W., E. L. Munteanu, L. Laan, T. L. Noetzel, M. E. Janson, and M. Dogterom. 2006. Assembly dynamics of microtubules at molecular resolution. *Nature* 442:709–712.
125. Du, M., S. A. Reid, and V. Jayaraman. 2005. Conformational changes in the ligand-binding domain of a functional ionotropic glutamate receptor. *J Biol Chem* 280:8633–8636.
126. Schmauder, R., D. Kosanic, R. Hovius, and H. Vogel. 2011. Correlated optical and electrical single-molecule measurements reveal conformational diffusion from ligand binding to channel gating in the nicotinic acetylcholine receptor. *ChemBioChem* 12:2431–2434.
127. Grandl, J., E. Sakr, F. Kotzyba-Hibert, F. Krieger, S. Bertrand, D. Bertrand, H. Vogel, M. Goeldner, and R. Hovius. 2007. Fluorescent epibatidine agonists for neuronal and muscle-type nicotinic acetylcholine receptors. *Angew Chem Int Ed* 46:3505–3508.
128. Hamill, O. P., A. Marty, E. Neher, B. Sakmann, and F. J. Sigworth. 1981. Improved patch-clamp techniques for high-resolution current recording from cells and cell-free membrane patches. *Pflugers Archiv: Eur J Physiol* 391:85–100.
129. Zhao, Y., D. Terry, L. Shi, H. Weinstein, S. C. Blanchard, and J. A. Javitch. 2010. Single-molecule dynamics of gating in a neurotransmitter transporter homologue. *Nature* 465:188–193.
130. Zhao, Y., D. S. Terry, L. Shi, M. Quick, H. Weinstein, S. C. Blanchard, and J. A. Javitch. 2011. Substrate-modulated gating dynamics in a Na^+-coupled neurotransmitter transporter homologue. *Nature* 474:109–113.
131. Demuro, A. and I. Parker. 2005. "Optical patch-clamping": Single-channel recording by imaging Ca^{2+} flux through individual muscle acetylcholine receptor channels. *J Gen Physiol* 126:179–192.
132. Wang, S. Z., S. Heyman, D. Enkvetchakul, and C. G. Nichols. 2013. Gating motions of KirBac1.1 cytoplasmic domain with respect to transmembrane domain revealed by FRET. *Biophys J* 104:129a.
133. Demuro, A. and I. Parker. 2006. Imaging single-channel calcium microdomains. *Cell Calcium* 40:413–422.
134. Ramaswamy, S., D. Cooper, N. Poddar, D. M. MacLean, A. Rambhadran, J. N. Taylor, H. Uhm, C. F. Landes, and V. Jayaraman. 2012. Role of conformational dynamics in alpha-amino-3-hydroxy-5-methylisoxazole-4-propionic acid (AMPA) receptor partial agonism. *J Biol Chem* 287:43557–43564.
135. Roy, R., S. Hohng, and T. Ha. 2008. A practical guide to single-molecule FRET. *Nat Methods* 5:507–516.
136. Holmgren, M., Y. Liu, Y. Xu, and G. Yellen. 1996. On the use of thiol-modifying agents to determine channel topology. *Neuropharmacology* 35:797–804.
137. Akabas, M. H., C. Kaufmann, P. Archdeacon, and A. Karlin. 1994. Identification of acetylcholine receptor channel-lining residues in the entire M2 segment of the alpha subunit. *Neuron* 13:919–927.
138. Roberts, D. D., S. D. Lewis, D. P. Ballou, S. T. Olson, and J. A. Shafer. 1986. Reactivity of small thiolate anions and cysteine-25 in papain toward methyl methanethiosulfonate. *Biochemistry* 25:5595–5601.
139. Loots, E. and E. Y. Isacoff. 2000. Molecular coupling of S4 to a K(+) channel's slow inactivation gate. *J Gen Physiol* 116:623–636.
140. Puljung, M. C. and W. N. Zagotta. 2011. Labeling of specific cysteines in proteins using reversible metal protection. *Biophys J* 100:2513–2521.
141. Gilchrist, J. and F. Bosmans. 2012. Animal toxins can alter the function of Nav1.8 and Nav1.9. *Toxins* 4:620–632.
142. Swartz, K. J. 2007. Tarantula toxins interacting with voltage sensors in potassium channels. *Toxicon: Off J Int Soc Toxinol* 49:213–230.
143. Restano-Cassulini, R., Y. V. Korolkova, S. Diochot, G. Gurrola, L. Guasti, L. D. Possani, M. Lazdunski, E. V. Grishin, A. Arcangeli, and E. Wanke. 2006. Species diversity and peptide toxins blocking selectivity of ether-a-go-go-related gene subfamily K^+ channels in the central nervous system. *Mol Pharmacol* 69:1673–1683.
144. Choi, W.-G., S. J. Swanson, and S. Gilroy. 2012. High-resolution imaging of Ca^{2+}, redox status, ROS and pH using GFP biosensors. *Plant J* 70:118–128.
145. Sengupta, P., S. Van Engelenburg, and J. Lippincott-Schwartz. 2012. Visualizing cell structure and function with point-localization superresolution imaging. *Dev Cell* 23:1092–1102.
146. Knopfel, T., K. Tomita, R. Shimazaki, and R. Sakai. 2003. Optical recordings of membrane potential using genetically targeted voltage-sensitive fluorescent proteins. *Methods* 30:42–48.
147. Guerrero, G., M. S. Siegel, B. Roska, E. Loots, and E. Y. Isacoff. 2002. Tuning flash: Redesign of the dynamics, voltage range, and color of the genetically encoded optical sensor of membrane potential. *Biophys J* 83:3607–3618.
148. Giraldez, T., T. E. Hughes, and F. J. Sigworth. 2005. Generation of functional fluorescent BK channels by random insertion of GFP variants. *J Gen Physiol* 126:429–438.
149. Xu, Y., D. W. Piston, and C. H. Johnson. 1999. A bioluminescence resonance energy transfer (BRET) system: Application to interacting circadian clock proteins. *Proc Natl Acad Sci USA* 96:151–156.
150. Lohse, M. J., S. Nuber, and C. Hoffmann. 2012. Fluorescence/bioluminescence resonance energy transfer techniques to study G-protein-coupled receptor activation and signaling. *Pharmacol Rev* 64:299–336.
151. Morin, J. G. and J. W. Hastings. 1971. Energy transfer in a bioluminescent system. *J Cell Physiol* 77:313–318.
152. Taraska, J. W., M. C. Puljung, N. B. Olivier, G. E. Flynn, and W. N. Zagotta. 2009. Mapping the structure and conformational movements of proteins with transition metal ion FRET. *Nat Methods* 6:532–537.
153. Sandtner, W., F. Bezanilla, and A. M. Correa. 2007. In vivo measurement of intramolecular distances using genetically encoded reporters. *Biophys J* 93:L45–L47.

154. Pantoja, R., E. A. Rodriguez, M. I. Dibas, D. A. Dougherty, and H. A. Lester. 2009. Single-molecule imaging of a fluorescent unnatural amino acid incorporated into nicotinic receptors. *Biophys J* 96:226–237.
155. Lee, H. S., J. Guo, E. A. Lemke, R. D. Dimla, and P. G. Schultz. 2009. Genetic incorporation of a small, environmentally sensitive, fluorescent probe into proteins in *Saccharomyces cerevisiae*. *J Am Chem Soc* 131:12921–12923.
156. Liu, C. C. and P. G. Schultz. 2010. Adding new chemistries to the genetic code. *Annu Rev Biochem* 79:413–444.
157. Summerer, D., S. Chen, N. Wu, A. Deiters, J. W. Chin, and P. G. Schultz. 2006. A genetically encoded fluorescent amino acid. *Proc Natl Acad Sci USA* 103:9785–9789.
158. Noren, C. J., S. J. Anthonycahill, M. C. Griffith, and P. G. Schultz. 1989. A general-method for site-specific incorporation of unnatural amino-acids into proteins. *Science* 244:182–188.
159. Ellman, J. A., D. Mendel, and P. G. Schultz. 1992. Site-specific incorporation of novel backbone structures into proteins. *Science* 255:197–200.
160. Chung, H. H., D. R. Benson, and P. G. Schultz. 1993. Probing the structure and mechanism of ras protein with an expanded genetic-code. *Science* 259:806–809.
161. Cornish, V. W. and P. G. Schultz. 1994. A new tool for studying protein-structure and function. *Curr Opin Struct Biol* 4:601–607.
162. Beene, D. L., D. A. Dougherty, and H. A. Lester. 2003. Unnatural amino acid mutagenesis in mapping ion channel function. *Curr Opin Neurobiol* 13:264–270.
163. Wang, L., A. Brock, B. Herberich, and P. G. Schultz. 2001. Expanding the genetic code of *Escherichia coli*. *Science* 292:498–500.
164. Chin, J. W., T. A. Cropp, J. C. Anderson, M. Mukherji, Z. W. Zhang, and P. G. Schultz. 2003. An expanded eukaryotic genetic code. *Science* 301:964–967.
165. Grandl, J., S. E. Kim, V. Uzzell, B. Bursulaya, M. Petrus, M. Bandell, and A. Patapoutian. 2010. Temperature-induced opening of TRPV1 ion channel is stabilized by the pore domain. *Nat Neurosci* 13:708–714.
166. Sandoz, G., J. Levitz, R. H. Kramer, and E. Y. Isacoff. 2012. Optical control of endogenous proteins with a photoswitchable conditional subunit reveals a role for TREK1 in GABA(B) signaling. *Neuron* 74:1005–1014.
167. Szobota, S. and E. Y. Isacoff. 2010. Optical control of neuronal activity. *Annu Rev Biophys* 39:329–348.
168. Janovjak, H., S. Szobota, C. Wyart, D. Trauner, and E. Y. Isacoff. 2010. A light-gated, potassium-selective glutamate receptor for the optical inhibition of neuronal firing. *Nat Neurosci* 13:1027–1032.
169. Volgraf, M., P. Gorostiza, R. Numano, R. H. Kramer, E. Y. Isacoff, and D. Trauner. 2006. Allosteric control of an ionotropic glutamate receptor with an optical switch. *Nat Chem Biol* 2:47–52.
170. Banghart, M., K. Borges, E. Isacoff, D. Trauner, and R. H. Kramer. 2004. Light-activated ion channels for remote control of neuronal firing. *Nat Neurosci* 7:1381–1386.
171. Bykova, E. A., X. D. Zhang, T. Y. Chen, and J. Zheng. 2006. Large movement in the C terminus of CLC-0 chloride channel during slow gating. *Nat Struct Mol Biol* 13(12):1115–1119.
172. Dahan, M., S. Lévi, C. Luccardini, P. Rostaing, B. Riveau, and A. Triller. 2003. Diffusion dynamics of glycine receptors revealed by single-quantum dot tracking. *Science* 302(5644):442–445.
173. Schwarzer, S., G. I. Mashanov, J. E. Molloy, and A. Tinker. 2013. Using total internal reflection fluorescence microscopy to observe ion channel trafficking and assembly. *Methods Mol Biol.* 998:201–208.

A practical guide to solving the structure of an ion channel protein

Tahmina Rahman and Declan A. Doyle

Contents

10.1 INTRODUCTION

Ion channels are some of the most structurally dynamic proteins in our bodies. This property is largely due to their function—to control the flow of ions (charged particles) across a biological membrane. As most ions are small charged particles, they can be transported over a short time frame (Hille 2001). This rapid transport produces a dilemma in that ion movement must be closely regulated, opening and closing the channel's main gate precisely and quickly in order to prevent dissipation of the ionic gradient and ultimately cell death. This has important consequences on the way we think of ion channels and how we may manipulate them.

Most ion channels can be defined as a multisubunit complex with the ion conduction pathway formed at the common interface, generally the center of the molecule (Doyle 2004). This applies to potassium channels, sodium channels, calcium channels, cationic channels, and anion channels and still holds true whether they are voltage-dependent, ligand-gated, or mechanically gated channels. The two groups that do not fall into this classification are the proton channels and some chloride channels such as the Cl^- (chloride) channels (CLC). This may be due to the special movement mechanism of protons in solution and the fact that CLC channels evolved from Cl^-/H^+ antiporters, respectively (Miller 2006; Ramsey et al. 2006; Sasaki et al. 2006; Lim et al. 2012). The main point is that all of the channels are multisubunit complexes, and therefore, there this additional complexity needs to be incorporated into the process of structure determination.

Even though there are many methods of determining the structure of an integral membrane protein (IMP), selecting the method that answers the specific question that you are interested in is key. Table 10.1 provides an overview of a selection of methods suitable for structural studies of an IMP. However, there are a number of guiding principles that apply across all membrane structural biology projects because of the common characteristics of IMPs. In this overview, we will discuss the issues that need to be addressed before reaching the ultimate goal of an atomic structure of an ion channel. These principles apply to any IMP.

The basic principles discussed here are relevant for many structural approaches used to study ion channels, but this review will focus on the use of x-ray protein crystallography. In particular, the early stages that deal with expression, purification, and stabilization of the target protein are still required for most of the methods listed in Table 10.1. Before proceeding to specific details, a brief outline of the general sequential steps involved in using x-ray protein crystallography to solve an ion channel structure, or indeed any IMP, is presented (Figure 10.1). They are as follows:

1. *P*rotein *E*xpression—Overproduction of the target protein to the milligram quantity level
2. *E*xtraction—Removing the newly formed protein from their membranes as a prelude to step 3
3. *P*urification—Isolation of the target protein from all other endogenous components to produce a near-homogeneous solution

Steps 1–3 constitute the PEEP procedure.

4. Crystallization—Formation of crystals suitable for structural studies
5. Phase determination—Generation of the electron density maps of sufficient quality that will allow step 6
6. Model building

Table 10.1 Provides a guide to some of the various techniques that produce a model of the target proteins ranging from boundary definition to atomistic details

METHOD	RESOLUTION	COMMENTS
X-ray crystallography	H	Excellent. Most IMPs solved using this method.
EM—single particles	H, I, L	Requires large proteins or complexes >200 kDa. Good for a structural overview and defining subunit boundaries and interactions. Recently, there has been an exciting advance in this field that has general implication for structural biology (Liao et al. 2013).
EM—projections	I, L	Good definition of boundaries; difficult to define helices.
EM—3D reconstruction	H	Excellent but lagging behind crystallography in technology and ease of use.
NMR—solid state	H	Excellent. 3D crystals required for largest protein structure, presently ~40 kDa (Tang et al. 2011).
NMR—solution state	H	Excellent. High temperatures are required (~50°C), therefore very stable IMPs are necessary; presently largest protein ~40 kDa (Van Horn et al. 2009).
SANS	L	Sensitive to aggregates obscuring data, requires additional structural information (Blakeley et al. 2008; Neylon 2008).
SAXS	L	Sensitive to aggregates obscuring data, requires additional structural information (Neylon 2008; Marasini et al. 2013).
AFM	L	Surface definition. Requires small amounts of protein, good for structural comparative analysis (Whited and Park 2013).
In silico	H	Excellent. Requires previous solved structure or close homologue.

Note: The general resolution limits presently obtainable for the techniques are categorized as high (H: >4 Å), intermediate (I: >15 <4 Å), and low (L: <15 Å).

Abbreviations: SANS, small angle neutron scattering; SAXS, small angle x-ray scattering; NMR, nuclear magnetic resonance; EM, electron microscopy; AFM, atomic force microscopy.

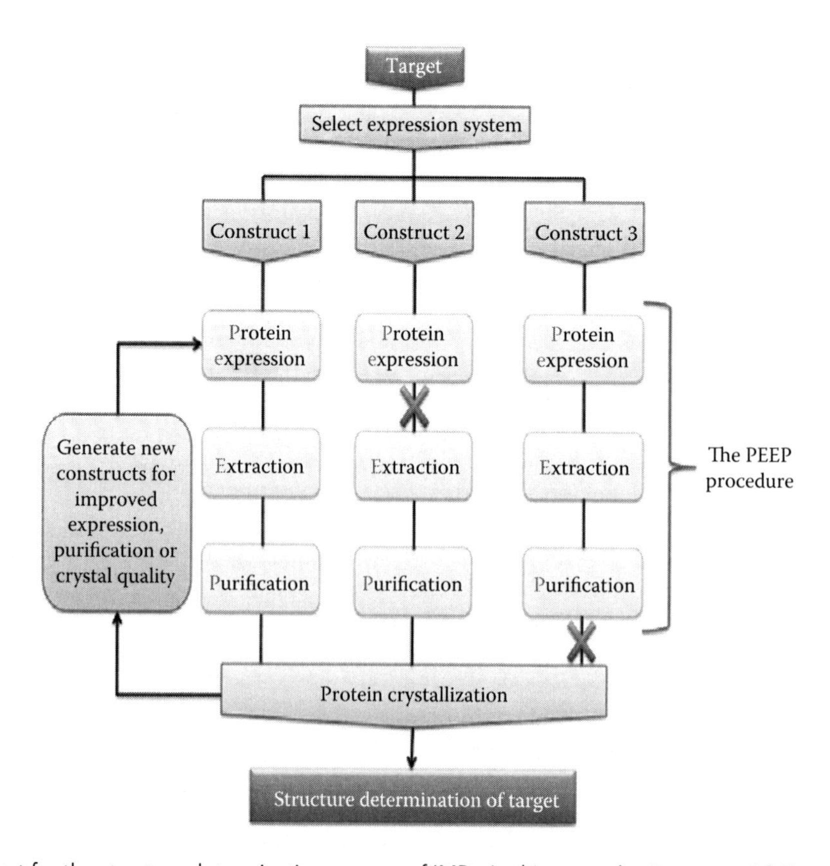

Figure 10.1 General flowchart for the structure determination process of IMPs. In this example, Construct 2 fails at the protein expression level while Construct 3 does not produce any suitable crystals. New constructs closely based on Construct 1 can be generated with the aim of improving any of the steps toward structure determination.

The first three steps are known as the PEEP procedure, which is the necessary groundwork required before the all important crystal formation stage.

10.2 BUILDING THE FOUNDATIONS

Before providing details on the various main steps outlined earlier, the reader needs to consider a number of important aspects in order to set up the correct approach. This has to be viewed as building the foundations for a successful outcome, that of solving the structure of an ion channel. As for any scientific research, adequate planning at an early stage should result in a successful outcome. In order to plan effectively, a detailed knowledge of the protein is required. This means that at each stage the researcher has to think about what the protein is doing, why it is doing it, and what it requires to keep it structurally stable and functionally active. Follow this simple idea, and the chances of success will improve, but bear in mind that it still requires careful work.

At the core of the protein x-ray crystallographic technique is the presence of a crystal. We will first think about what is required in order to generate a suitable protein crystal.

In order to visualize a protein, a radiation source with a wavelength in the order of a carbon–carbon bond is required. X-ray radiation is the most suitable, but due to their short wavelength, and therefore high intensity, x-rays will ultimately destroy any protein placed in their path. The crystallization process provides a means of protecting the protein with cryoprotection techniques while effectively providing a diffraction grating that amplifies the signal and allows a recordable signal in the form of a diffraction pattern.

A crystal is composed of a repeating identical unit arranged in a three-dimensional lattice. In all cases, the formation of a crystal will restrict the motions of the protein building blocks that make up this three-dimensional structure. Biological molecules add complexity for a number of reasons. Firstly, proteins are relatively delicate, needing an aqueous-based environment in order to maintain their integrity. Secondly, many proteins in their natural environment do not reside as simply static structures. They can be highly dynamic entities, performing their physiological roles by altering their conformations, linking up with other molecules, binding ligands, cofactors, ions, etc. An obvious example is the conformational changes associated with ion channels such as K^+ channels as they move between the closed and open states (Jiang et al. 2002; Kuo et al. 2003, 2005; Vargas et al. 2011; Whorton and MacKinnon 2013). This is an important consideration because the best crystals are formed from the most compact, rigid, identical components. Therefore, it is important to keep in mind the potential conformational changes that the target protein can adopt and the known methods of locking down a structure into a single conformation. The ultimate outcome of this restriction is that the final structure should be considered only as a snapshot of the protein which is fixed, generally, in one conformational state. Alternative conformations should be sought by coaxing the protein into a new conformational state prior to crystallization or by inducing relatively minor conformational changes within the crystal itself, for example, by adding a known binding small molecule ligand (Kuhn et al. 2002; Grey and Thompson 2010).

***PEEP*: Protein Expression, Extraction, and Purification**

Step 1: Protein Expression

From the outset, it should be noted that there is no single expression system that is suitable for all IMPs. Discovering which system is appropriate for your target can be a process of trial and error. Therefore, at the beginning, a strategy of screening, optimization, and selection should be adopted at each stage of the process. Once this defined procedure or approach has been optimized, it should be fixed before moving onto the next stage in which the same screening approach is repeated.

The most important question regarding expression is—how much protein is required? Presently, as there are no guaranteed means of making a crystal, especially one suitable for structural studies with diffraction limits around 3.5 Å or better, a screening approach must be implemented. Downscaling of the crystallization trials has dramatically reduced the total volume of the target protein solution, with 20 μL being easily sufficient to carry out a standard 96 crystallization conditions screen (Walter et al. 2005; Berry et al. 2006; Chayen 2009). Most of these screens have been optimized for a protein concentration within the 10–20 mg/mL range. From this simple outline, it appears that in total, microgram quantities of the target protein are required; however, this protein should be monodispersed and structurally homogeneous in order to provide the best chance of obtaining adequate crystals (Hitscherich et al. 2000; Gutmann et al. 2007; Proteau et al. 2010; De Marco 2012). In order to purify a membrane protein to this standard, excess amounts of overexpressed protein are required so that impurities can be completely removed. As a rough guide, this equates to an expression system that generates approximately 1 mg of protein per liter of culture broth.

There are a number of reasons why, as a rule of thumb, this amount of 1 mg/L of target protein is required as a good starting point. In general, in comparison to soluble proteins, more of the IMP will be expressed as incorrectly folded protein. This is amplified in the next part of the PEEP process, extraction, when the IMP is removed from its natural environment, the lipid bilayer. Ion channels tend to move through a number of intracellular compartments, the most direct being endoplasmic reticulum, through the Golgi to secretory vesicles and then to the cytoplasmic membrane. Overexpressed soluble proteins tend to be localized in the relatively homogeneous cytoplasmic space or are excreted. Hence, soluble proteins tend to experience less variation. In addition, depending on the expression system, posttranslational modifications can vary as the IMP moves through the various intracellular compartments (Hertel and Zhang 2013; Yang et al. 2013; Zhong and Wright 2013). Partially folded or unfolded IMPs can be moved through alternative pathways again altering their posttranslational modifications or binding partners (Wang et al. 2010; Lukacs and Verkman 2012). Staying with folding issues, many ion channels form oligomeric structures, adding complexity to the folding process and thus the structural diversity. Ideally, all of these structures should be driven toward a single conformation, but practically, this is not always possible, hence the need to produce a significantly higher amount of IMP protein than is ultimately required for crystallization. Finally, there are a large number of possible parameters and crystallization methods that can be tested again, requiring large amounts of protein.

10.2.1 EXPRESSION SYSTEMS

If we aim for the production of 1 mg/L of protein, then which expression systems are available? There are many options, but the best systems to use are those that have been already extensively characterized and are therefore relatively easiest to use. They are *Escherichia coli*, yeast, and the baculovirus expression systems (Hunte et al. 2003). As these three systems are present in most research institutes, the details of the methods will not be presented here. It should be said that these systems have been mostly set up for the expression of soluble proteins. This is significant as there are important differences in the use of each system depending on whether the target is a soluble protein or an IMP. As such, only the major differences will be highlighted. Other systems can, of course, be used, but care should be taken for the focus to remain on the ultimate goal of determining the structure of the target ion channel.

With all expression systems, the experiment needs to be carefully designed before starting. Maximizing ion channel expression levels is critical. With soluble proteins, expression levels ranging from 10 to 50 mg/L of culture are not unusual. In comparison, high-expression IMP targets generally produce 10-fold less protein than their soluble counterparts. This is related to the expression and folding issues already mentioned. Thus, optimization of expression levels for IMPs is critical. The issues to consider include the level of induction, the type and location of a purification tag, and the final orientation of the IMP in the membrane. Other variables that need to be addressed include the type of media to use, the induction method, the speed of culture growth, the cell density at induction or infection, induction temperature, and total induction time. These factors and more can be systematically screened using standard methods; however, the time and cost incurred to perform such a broad screen is considerable. To reduce the cost and associated time both to complete such screens, miniaturization of the process should be implemented. This requires the use of a readily detectable signal that correlates with expression levels. One such method is to tag the target ion channel with green fluorescent protein (GFP) and measure the whole cell fluorescence (Drew et al. 2005, 2006, 2008; D'Avanzo et al. 2010b). The advantage of this approach is that the protein does not have to be extracted prior to recording the fluorescence signal. All of the variables that were previously mentioned can be quickly tested, thereby rapidly allowing the identification of the best combination of conditions that provide the maximum expression level.

An important consideration even for membrane proteins is the exact construct to use. In this case, construct refers to the protein sequence of the target being overexpressed. Large, unstructured regions, defined as regions that on their own do not have any significant secondary structural elements, are generally not favorable for protein crystallization (Gileadi et al. 2007; Page 2008; Pan et al. 2010). This can refer to regions at the N- or C-termini or even internal loops. Solid, compact molecules are best at producing *good* quality crystals. Excessive flexibility should be avoided. Again, this links up with the idea of knowing your protein so that there is a good understanding of which regions can be altered. Even with this knowledge, it can be difficult to predict the exact start and end of the major structured domain(s) of your target protein. Therefore, multiple constructs should be generated that vary in the lengths of additional amino acids at either termini or internal loops (Figure 10.2). Keep in mind that some flexibility is necessary for the purification tag so that it is exposed and able to bind to the specific matrix. For this, it may be necessary to add a protease site so that the flexible tag can be removed prior to crystallization. Always try and crystallize the most compact form of your target protein. We shall see later how internal flexibility can be overcome with the use of protein binders.

Another issue when using multiple constructs is that in many instances, several constructs express to acceptable levels. So which construct should be selected for the more intensive scale-up process? Ideally, all of the constructs should be moved forward as it is unknown which of these will produce the magical protein–protein contacts that are found in good quality crystals. However, as another guiding principle, it is best to start off with the construct that expresses to the highest level. In general, this construct is likely to be the most compact and best folded; hence, the expression system is able to produce more with less issues related to proteolysis.

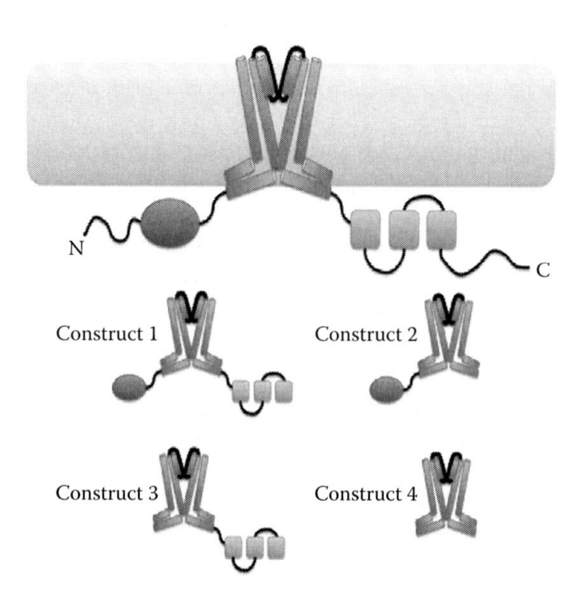

Figure 10.2 Schematic picture of a generic ion channel, embedded in a membrane (gray bar). The channel contains the transmembranous ion conducting pathway, a structured N-terminal domain (oval), and a repeating C-terminal domain (squares). The thick black lines indicate connecting loops. Four constructs can be generated from the full length channel by including or excluding various sections of the N- and C-termini while keeping the core of the protein.

Specific issues for the various expression systems include:

10.2.1.1 *E. coli*

1. Always use fresh transformations. All frozen stocks tend to reduce their expression levels over time. With low expressing IMP, this can have a detrimental effect further along the purification route as low-expression levels allow the purification of low affinity contaminants such as AcrB.
2. AcrB, which is an endogenous *E. coli* IMP that is stable, purifies as a contaminant during Ni^{2+}-IMAC (Immobilized Metal Affinity Chromatography) and crystallizes readily at low concentrations (Veesler et al. 2008; Psakis et al. 2009; Glover et al. 2011). Avoid by using a knockout strain or optimize the amount of Ni^{2+}-IMAC resin to be just enough to bind the amount of overexpressed protein. This allows the higher affinity target protein to displace any low affinity bound AcrB to the resin. Identification of contaminants that form crystals may be revealed early after initial data collection by comparing the unit cell of the novel crystal with those already deposited in the protein data bank (Berman et al. 2000). This can be achieved by using the program Nearest-cell which is freely available (Ramray et al. 2012).
3. In an ideal situation, more *E. coli* cells equal more protein. As for the use of fresh transformants, cells at their peak should be used, meaning induce protein expression at mid-log stage. For soluble proteins, it is recommended to induce when the cell density at 600 nm reaches values from 0.4 to 0.6. Membrane protein induction tends to slow down doubling rates and can, in many instances, be toxic. For IMPs, it is best to induce cells at a cell density at 600 nm from 1.0 to 1.2. Higher values than this usually produces poorer protein with more proteolysis while lower values result in less protein.
4. It is recommended you use a high-pressure cell disruption system to break cells. Sonication does work but fails to break open all the cells; thus, a significant loss of membranes with the membrane target protein incorporated is incurred when using sonication only.
5. Simple, small eukaryotic membrane proteins can be successfully overexpressed in *E. coli.*

10.2.1.2 Yeast

1. For speed, it is possible to use the vector system of *Saccharomyces cerevisiae* that allows for more rapid screening (D'Avanzo et al. 2010b). Once a suitable construct of the target has been identified and the expression levels still remain low, then the *Pichia pastoris* strain can be used in which the gene is inserted into the genome. The advantage of this yeast strain is that very high cell densities can be obtained. More cells equal more membrane, which equals more protein.
2. Cell breakage is critical for both strains of yeast mentioned. These yeast organisms have hard cell walls and as such require special mechanisms in order to break them open. Manufacturers test their cell breakage systems by following the release of a soluble protein marker; however, this is not appropriate for an IMP that is localized to the membranes as these remain inside the cell even after a cell wall is breached. Total breakage of the cells is required in order to release the cellular contents therefore a system that generates pressures of 40 kpsi or more such as a TS series homogenizer from Constant Systems Ltd. Another effective method is the BeadBeater (Biospec Products), but care should be taken to ensure that the sample does not overheat.

10.2.1.3 Baculovirus

1. Systematically screen for the best multiplicity of infection (MOI) values that produce the most protein. Try significantly lower MOI values than recommended for soluble proteins. This is an issue related to the cellular toxicity of overexpressing many IMPs.
2. The various cell strains, for example, Sf9, Sf21, Hi5, can have a significant effect on expression levels, so it is important to test as many as possible.
3. Consideration of the purification tag can be critical for this system, with the FLAG tag tending to produce more, cleaner protein than a hexahistidine tag. The drawback is the cost of the resin when using the FLAG system.

Noteworthy systems that are becoming more popular include the cell-free expression systems and the *Lactococcus lactis* system (Frelet-Barrand et al. 2010; Bernaudat et al. 2011; Abdine et al. 2012; Isaksson et al. 2012; Müller-Lucks et al. 2013). In using the cell-free system, care must be taken to ensure that the final product is functionally active (Jaehme and Michel 2013). This system clearly has an advantage in the use of specific amino acid labeling for NMR studies (Abdine et al. 2012).

PEEP: Protein Expression, Extraction, and Purification

Step 2: Extraction

Before proceeding directly to the extraction process, it is important to plan time carefully as it is usually best to complete this stage in one continuous time frame. This prevents loss of function as generally IMPs degrade rapidly once outside the membrane. In addition, each additional purification step decreases the amount of proteases present, thus increasing the quality of final protein produced. The best time to stop a preparation is either after overexpression with the cells still intact or after purification of the membrane that still contain the target IMP.

A major issue that needs to be considered before actually extracting the target ion channel from the membrane bilayer is the potential damage inflicted by endogenous proteases. If the target protein is small and deeply embedded in the lipid bilayer, then it may be protected and remain resistant to endogenous proteases once they are released from the membrane in the form of a protein/detergent complex. However, many target IMPs will become vulnerable to proteolytic attack and therefore need protection. In this instance, the best approach is first to purify the membranes containing the overexpressed target. The exact procedure varies depending on the expression system used, with much lower centrifugal forces being employed for insect cells. When using *E. coli* and yeast cells, the procedure can be as simple as performing a low-speed spin (30 min at ~25,000 g) to remove cell wall debris and unbroken cells, discarding the pellet. The supernatant containing the membranes is then subjected to a high-speed spin (1.5 h at ~100,000 g). The supernatant, containing many proteases, can then be removed. Then the pelleted membranes can be gently resuspended. An additional wash of the membranes using a standard buffer can be incorporated to ensure best results.

If all has gone well, then at least one construct has been overexpressed in sufficient quantities for further workup. The next step involves the extraction of the ion channel from the membrane either for functional analysis or further purification. This is necessary in order to remove nonspecific proteins, but it also has the immediate effect of destabilizing the target protein. The main reason for this is that the detergents used to break up the membrane and release the ion channels do not faithfully replicate the membrane bilayer environment. The process also has the effect of removing many lipids that may be essential for the integrity and activity of the target ion channel. As a result, the target IMP can unfold or, if oligomeric, fall apart, thus leading to protein loss by precipitation.

10.2.2 DETERGENTS

Detergents are the tools required for extracting an ion channel from their hydrophobic membrane. Detergent molecules can be visualized as occupying a cone-shaped space, with the detergent's hydrophilic head group at the widest end. Bringing together a large number of cone-shaped detergent molecules results in the formation of a ball-like structure, with the wedge-shaped acyl tails in the hydrophobic centre, protected from the solution by the hydrophilic head groups. These detergent aggregates are known as micelles. So how can a ball-shaped structure affect the integrity of the planar membrane bilayer? The key to this is the behavior of detergent molecules that coexist as monomers in solution. In fact, an equilibrium exists between the monomeric and oligomeric micellular states that is governed mainly by the degree of hydrophobicity of the acyl tail; longer and more hydrophobic tails favor micellular state, while the shorter and more hydrophilic tails require higher concentrations in order to form a micelle. The point at which the concentration of a particular detergent forms a micelle is known as the critical micelle concentration (cmc). This value varies depending on the character of the detergent but also the environment in which it exists, that is, salt concentration and temperature.

This simple, but important, concept is used to ensure that when the protein is extracted, there are sufficient detergent molecules to maintain stability of the ion channel as a detergent/protein complex. Returning to our question of how the detergents break apart the membrane bilayer, it is the detergent monomers that initially intercalate within the membrane. As more and more detergent molecules accumulate in the membrane, they overwhelm the lipids' ability to form a bilayer and disrupt the membrane. The membrane proteins that are subsequently released are more accurately thought of as detergent/lipid/protein complexes. We will come back to this triple complex in the "Purification" section.

From this description, the most important factor in the disruption of the cell membrane when using detergents is the quantity of detergent used. The lipid bilayer behaves as a sponge that can hold a considerable amount of detergent before it breaks down (De Foresta et al. 1989; Privé 2007; Lichtenberg et al. 2013). As a good starting point, the most commonly used detergent in the extraction process is decyl- or dodecyl-β-D-maltopyranoside (DM or DDM). These detergents tend to form a relatively large micelle in which the ion channel is deeply embedded. As such, an IMP can remain relatively stable even for oligomeric complexes such as ion channels (Doyle et al. 1998; Kuo et al. 2003; Bocquet et al. 2009; Sobolevsky et al. 2009; McCusker et al. 2012; Miller and Long 2012; Whorton and MacKinnon 2013). Another possibility is to carry on using the GFP-based screening approach to test a variety of detergents and select the detergent that yields the highest amount of fluorescence after solubilization (Backmark et al. 2013; Ellinger et al. 2013). Commonly used detergents that can be used from extraction to crystallization are *n*-decyl-β-D-maltopyranoside, DM, DDM, *N*,*N*-dimethyldodecylamine-*N*-oxide (LDAO), *n*-octyl-β-D-glucopyranoside (OG), and octyltetra-oxyethylene (C_8E_4).

From a practical point of view, a concentration of ~30 mM DM or DDM is sufficient to break down all of the membranes. Simply adding the detergent in solution along with membranes containing the target ion channel will not fully dissolve them unless the sample is agitated. When performing this, care should be taken not to form bubbles as this denatures the protein. Use a gently rocking motion and allow this to proceed for at least 2 h or longer at 4°C. Finally, a protease inhibitor cocktail should also be included in order to diminish unwanted proteolysis.

PEEP: Protein Expression, Extraction, and Purification

Step 3: Purification

Speed is essential in this step as the extracted ion channel is more likely to precipitate, aggregate, and lose activity the longer it spends outside a membrane. The maltoside-based detergents are an excellent starting point as they are capable of keeping many IMPs in a suitable biochemical state for at least two further purification steps: one usually involving an affinity-based tag while the other being size exclusion or ion exchange chromatography. The driving factor in terms of how far to carry on with the purification depends on the quality and quantity of the ion channel at each stage. A good value for the quality to aim for is ~95% purity. Lower purity can be used, but the possibility of generating false positives is increased during crystallization, that is, crystallization of a contaminant rather than the target ion channel.

At this point, the researcher needs to consider the method of crystallization (Figure 10.3) as this ultimately determines the purification protocol. There are three main methods for producing 3D crystals of a membrane that will be explained in greater detail later. They are

1. Detergent based
2. Lipidic cubic phase (LCP)
3. Bicelle crystallization

The LCP and bicelle methods both involve transferring the ion channel back into a membrane bilayer (Figure 10.3b); thus, the use of only one detergent is required as both effectively replace the detergent with phospholipid-like molecules. So, for both of these, the same detergent can be used, from extraction to the final purification step.

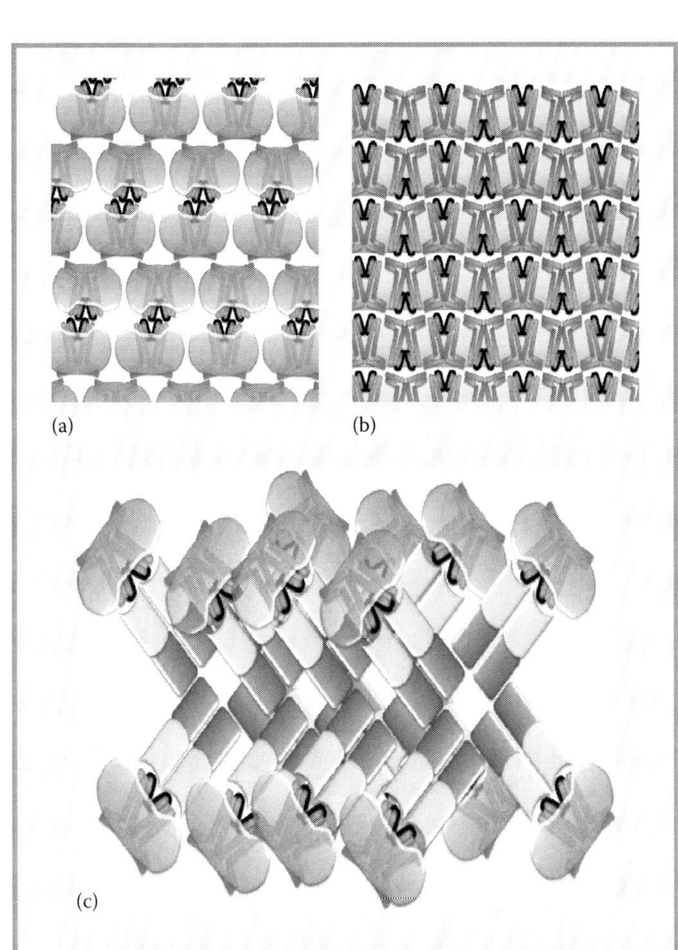

Figure 10.3 Schematic picture of the various protein packing arrangements as a result of the different IMP crystallization approaches. (a) Detergent-based crystallization. The donut arrangement of a detergent micelle surrounds the hydrophobic sections of the transmembrane ion channel. (b) LCP crystal packing. (c) Antibody fragment–assisted crystallization. The central bars represent the various domains of an antibody Fab fragment. This approach can be viewed as crystallizing the Fab fragment that happens to be holding an IMP.

In the detergent-based method, the target ion channel is transferred into a series of different detergents. Each one of the new detergent/protein complexes is set up in crystallization screens. Each new detergent can alter the characteristics of the protein by exposing residues, altering the oligomeric state, or inducing conformational changes. Any one of these changes could be the key event that allows a protein to form of a crystal. Unfortunately, it is impossible to predict which detergent/protein complex will be the winning combination. This is because, in general, the best contacts that contribute to a *good* crystal are protein–protein contacts, not protein–detergent or detergent–detergent contacts as shown in Figure 10.3a. With an ion channel of unknown structure, these connections are only known once the structure has been determined, and therefore, each detergent/protein complex must be screened in crystallization trials.

Hence, a detergent exchange step has to be accommodated within the purification protocol. This can be achieved at the initial IMAC when the target ion channel is still attached to the resin by washing with a buffer containing the new detergent. This can achieve near-complete detergent exchange. It can also be carried out during gel filtration in which the size exclusion column is equilibrated in a buffer containing a different detergent from the loaded protein sample. In this case, the final product is a mixture of the two detergents, with the new detergent dominating, as total detergent exchange generally does not occur. In a similar fashion to IMAC, near-complete detergent exchange can be accomplished for ion exchange chromatography.

An issue to consider during purification of the ion channel is the loss of lipids during the process. Not surprisingly, many IMPs require specific lipids in order to maintain their function (Zimmer and Doyle 2006; D'Avanzo et al. 2010a, 2013; Levitan et al. 2010; Hammond et al. 2012). Therefore, including lipids as part of the detergent buffer solutions during the purification process seems obvious. What is not so evident is the notion of maintaining a constant or at least a consistent amount of lipid during the purification process.

The amount of lipid associated with an ion channel is dependent on the total amount of membranes and detergent present. So long as these two factors are constant, then there should be a consistent amount of lipid. As a result of the detergent ousting lipids associated with the ion channel (delipidation), the more washing steps included or the greater the wash volume used, the less lipid will be present, if it is not replaced. In order to be consistent throughout the purification steps, the same volume of buffer and detergent concentration should be used. This is particularly critical so that the repeat preparations produce as near an identical final protein preparations as possible for the crystallization trials. As mentioned, the protein should be thought of as existing as a triple complex consisting of ion channel/detergent/lipids. The required approach is one in which it produces protein that is as homogeneous as possible in terms of the protein/lipid ratio on a prep-by-prep basis. What this means is that a purification protocol for an IMP should be built up with each step becoming fixed once the particular variables have been defined. As highlighted earlier, repeat purifications will be necessary due to the low expression levels and losses during each purification step. Screening using a GFP-tagged protein is once again a viable means of testing a broad range of conditions and optimizing the best findings.

10.3 WHICH CRITERIA ARE USEFUL CHECKS?

The assumption so far is that the protein being produced is fully functional and active. This assumption has to be tested, as it is possible that the overexpressed protein could be completely inactive. Protein misfolding, protein unfolding, proteolysis, missing cofactors, or loss of essential lipids are among the causes of such loss of activity. So what are the indicators that demonstrate that the purified protein is acceptable? This question is important and requires a thorough understanding of the protein's function.

Depending on the available resources, the following three investigative categories may be employed to help establish the

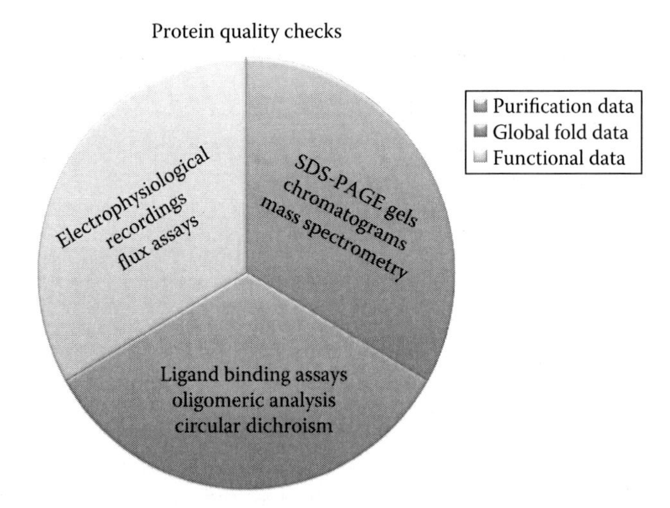

Figure 10.4 Various protein quality control checks can be grouped into three categories according to the information that they provide.

quality of the protein. They should also be used to guide the purification process so that the best quality protein is produced at each stage rather than the highest quantity of protein.

The three categories (Figure 10.4) are

1. Purification data
2. Global fold studies
3. Functional activity measurements

10.3.1 PURIFICATION DATA

One of the simplest data to obtain comes directly from the purification procedures. Measuring protein absorbance (280 nm) at the IMAC stage allows consistency and builds a deeper understanding of the preparation and the target protein's behavior. In judging quantity, a simple rule is that, after an initial affinity chromatography, the protein should be visible when run on sodium dodecyl-sulfate polyacrylamide gel electrophoresis (SDS-PAGE) and stained with Coomassie Brilliant Blue R-250. Unlike the standard approach to SDS-PAGE for soluble proteins, the ion channel sample should not be boiled prior to loading onto the gel. The detergent used, SDS, is central to the SDS-PAGE method. As the ion channel progresses through the gel, the detergent used during extraction is displaced with SDS. As such, many IMPs run as fully folded proteins. The drawback is that the membrane protein is not linearized and therefore does not run to their expected molecular weight. The ionic detergent SDS is generally thought of as being a *harsh* detergent that unfolds protein. With an oligomeric channel, it can break up the sample into the monomeric form so long as they are not covalently linked. When run on an SDS-PAGE gel, multiple bands can be observed of the different oligomers. When extracting KirBac1.1, we observed mostly monomers, but dimers, very weak trimers, and tetramers were observed. Boiling the sample will produce only monomers, as observed for KcsA, KirBac1.1, and KirBac3.1, but the monomeric bands tend to be smeared, probably due to the partial refolding of the hydrophobic helices in SDS. Overloading of the samples in which less of the purification detergent is displaced can help to identify which bands belong to the target protein and which are contaminants. It should be made clear that this is not a definitive test of the oligomeric state.

The best method to check if the protein is the correct target protein is to use a variety of mass spectrometry approaches. The first is to determine the total mass of the sample (Cadene and Chait 2000; Gabant and Cadene 2008; Berridge et al. 2011). The added advantage of this is that potential modifications to the protein can be identified. These include proteolysis, Arg to Lys substitutions, glycosylation, phosphorylation, the degree of protease cleavage, successful removal of the affinity tag, and so on. Mass spectrometry analysis can also be used to identify the proteins for each of the bands on an SDS-PAGE gel. In this case, each band is excised, tryptic digested, and the fragments sequenced by LS-MS/MS mass spectrometry (Lu and Zhu 2005; Liu et al. 2013). With IMPs, this method identifies soluble fragments such as loops or the N- or C-terminal sections. Particular attention should be given to any additional protein identified especially if they appear regularly. These may indicate overlapping targets in molecular weight or proteins that are copurifying.

Size exclusion chromatography, even if it is not used as a main purification step, should always be included. The main advantage of this method is that it indicates the potential oligomeric and aggregated state of the sample. Heavily aggregated protein rarely crystallizes even if the sample is fully active. The molecular weight of the sample can be estimated by calibrating the column with standards. Since the sample's molecular weight is a combination of the protein plus detergent, further complications arise if the protein is oligomeric as is the case for many ion channels.

A valuable addition to the general screening tools is fluorescence-detection size exclusion chromatography (FSEC). This method allows channels that are GFP tagged to be identified and monitored even in partially purified samples. Standard absorbance at 280 nm detects all proteins including contaminants; hence, the ability to detect a specific signal from the target protein allows downscaling the extraction and greater screening capabilities (Kawate and Gouaux 2006; Hattori et al. 2012; Parcej et al. 2013).

10.3.2 GLOBAL FOLD STUDIES

When a protein folds into its tertiary or quaternary structure, it possesses certain characteristics that can be examined before testing for functional activity. This can be an easier step for ion channels as it does not require the reconstitution of the channel back into a membrane bilayer as a prelude to electrophysiological recordings. The measurements can be as simple as predicting the secondary structure content of the ion channel using circular dichroism (CD). These predictions are best achieved by comparison with a database generated from known membrane protein structures (Whitmore et al. 2011). DichroMatch is a freely available web-based tool specifically designed for the analysis of CD spectra of IMPs (Klose et al. 2012).

Clearly, many ion channels go through dramatic conformational changes as they cycle between their closed and open states. There exists a wide range of ligands that can perturb the system, depending on the signals that drive channels from one state to another. Any of these can be used for testing using standard available laboratory procedures. Factors that may affect conformational changes are as follows: the ions themselves including protons, natural or synthetic ligands, proteins, toxins, and lipids. These conformational changes can be monitored by a number of general laboratory techniques such as fluorescence or absorption spectroscopy. More detailed techniques as for isothermal calorimetry or surface plasmon resonance can provide large amounts of detailed information on ligand affinities, but as for the spectroscopy approach, the focus should be on stabilizing or locking down the structure in a form suitable for crystallization.

Static light scattering and sedimentation equilibrium centrifugation can be applied in order to determine the oligomeric state of the channel/detergent complex (Hitscherich et al. 2000; Mogridge 2004; Gutmann et al. 2007). Another means of determining whether or not the purified protein is monodispersed is to use electron microscopy to reveal the physical nature of the protein sample. Well-behaved monodispersed protein appears as a grainy image while clear clumps are visible if the protein is severely aggregated (Horsefield et al. 2003).

10.3.3 FUNCTIONAL ACTIVITY MEASUREMENTS

Electrophysiological measurements are the gold standard for demonstrating that the purified channel functions correctly. There is a wealth of information describing how purified ion channels can be reconstituted and examined functionally. The book *Ion Channel Reconstitution* (Miller 1986) is an excellent starting point. One point that should be noted is that these sensitive methods usually only sample a small proportion of the purified channel. It is possible that 99% of the purified channel could be aggregated and not appropriate for structural studies, but the remaining 1% could provide excellent electrophysiological results. This could be misleading and waste valuable time in trying to use a purified channel sample that is not suitable for structural studies. To avoid such a situation, a flux assay is a better method of measuring ion movement through a large proportion of the purified channels (Lee et al. 2009). The focus should remain on using this technique to verify channel behavior but also in order to find ways in which all of the available channels can be driven into a small number of conformational states, preferably one, prior to crystallization.

Step 4: Crystallization

Armed with purified, functionally active ion channel that is locked into a single state, the next step is to generate a crystal that will produce a diffraction limit greater than 4 Å. Practically speaking, one crystal is not sufficient, with potentially hundreds being required. There are a number of precautions that can be taken in order to ensure reproducibility of the entire process in order to secure a good supply of crystals. As mentioned before, once each of the previous steps has been defined, these should be considered as fixed. This will ensure that the purified channel is as near to the same concentration with the same amounts of associated detergent, lipids, salts, pH, and cofactors for every crystal trial.

Despite using such a stringent approach, it may still be difficult to form reproducible crystals. Therefore, once a crystallization condition has been found, it is best to use a focused screen in which the crystallization conditions vary by 2% steps for the polyethylene glycol precipitants, 50 mM steps for salts, and steps of 0.25 pH units, and so on, depending on the components of the crystallization conditions.

Poor quality crystals are a reoccurring feature with IMPs especially when grown using the standard detergent-based approach. For many IMPs that do not have extensive extracellular or intracellular domains, little soluble protein surface is available to make contact with other proteins as they fall into the three-dimensional arrangement of a crystal. Added to this is the fact that the detergent is a highly dynamic, flexible component of the system; hence, it is little wonder that the crystals tend to be poorly diffracting, hard to handle, and difficult to reproduce. It was obvious early on that membrane proteins with large extramembranous domains produced better crystals with better resolution limits (Deisenhofer et al. 1985; Song et al. 1996; Xia et al. 1997; Iwata et al. 1998). Therefore, systems were set up to increase the soluble surface of the protein to aid the valuable protein–protein contacts. The first method involved adding another soluble protein to the target IMP. In order for this to work, the complex formed has to be of high affinity, stable, and rigid. Antibodies or antibody fragments were a clear candidate and have been successfully applied as an aid to crystallization and phasing (Figure 10.3c) (Ostermeier and Michel 1997; Ostermeier et al. 1997; Zhou et al. 2001). The drawbacks in using these are the high cost of production and the need for multiple antibody/target IMP complexes to be generated and tested. This is because there is no way of knowing which antibody will make the right contacts that result in the formation of high-resolution diffracting crystals. More recently, the overall cost has been lowered and speed of generating similar high affinity binding partners has been increased with the development of phage display screens. In these screens, a library of binding protein variants is used as the basis to select for low affinity binding partners. Once identified, further rounds of mutations and screening convert these low affinity possibilities into high affinity binding proteins (Tereshko et al. 2008; Koide 2009; Uysal et al. 2009, 2011; Bukowska and Grütter 2013). The high affinity binding partners function in the same manner as the antibody fragments. This is an exciting way forward, but the key to success is having the initial library sufficiently diverse so that several initial low affinity candidates can be obtained.

A simpler first-line approach would be to insert a protein into the ion channel that will perform a similar function to the antibody fragments in the preceding crystallization scenario. In this case, the idea is to have a protein that will self-interact to form protein–protein contacts, will not distort or interfere with the native structure, be stable, and have the added benefit of helping with phasing. Two excellent candidates (T4 lysozyme and apocytochrome b(562)RIL) have been used extensively in the recent explosion of G-protein coupled receptor (GPCR) structures (Cherezov et al. 2007; Rosenbaum et al. 2007; Rasmussen et al. 2011; Liu et al. 2012; Wang et al. 2013). For this approach to succeed, the selection of the internal loop into which the protein is to be inserted is critical. It should be a site that does not alter the structure of the channel, has sufficient freedom to allow proper folding of the transmembrane helices before and after the inserted site, and has little effect on the expression level.

Finding such a site may be difficult for ion channels because of their dynamic nature. Another approach, again taken from the experience of GPCR structural studies, is to lock the channel into a single conformational state by combining a series of stabilizing mutations (Warne et al. 2008; Lebon et al. 2011a,b). The guiding factor in this approach is the increase in thermal stability of the mutant GPCRs, with multiple mutations combining to further increase stability. This successful method comes back to the idea of generating a stable brick-like structure that is always best for producing good quality crystals.

Many of the previous crystallization methods described earlier use the channel/detergent mixture as the basis of the crystallization method. Ideally, the IMP should be crystallized from a lipid bilayer as the stability of the channel is expected to be highest when in its natural environment. There are two approaches that use this approach: LCP and the bicelle methods (Chiu et al. 2000; Nollert et al. 2002; Johansson et al. 2009; Ujwal and Bowie 2011; Agah and Faham 2012). Both methods start off with the purified IMP in detergent so that the sample is pure, but ultimately, both dilute the detergent so that it no longer dominates.

Returning the ion channel to a lipid bilayer after removing all contaminants appears to be the first logical step in a structural biology process. Unfortunately, this poses a problem related to the physical properties of a bilayer. The major components of a bilayer are phospholipids, and these lipids tend to form stable two-dimensional sheet-like structures. These restrict the ability of the protein to form a 3D crystal even though the IMP will have total freedom to move within the plane of the membrane. So, how have the LCP and bicelle methods overcome this problem?

Both methods use lipids that allow the formation of unusual bilayer structures. In the case of LCP, the lipid bilayers are highly curved, while for the bicelle lipids, the bilayers do not extend far beyond the IMP with the ends being capped. Because of this bilayer capping, the bicelle/IMP complex is not restricted in its movement in solution. It can be thought of as a direct replacement for the detergent micelle, hence the derivation of the name bicelle, a combination of a *bi*-layer and a mi-*celle*. Bicelle crystallization, therefore, can use the same approach and equipment as for the detergent-based crystal trials. Fewer IMP structures have been solved using this method. The reason for this remains unclear, but it may be simply due to the fact that it is a relatively recent addition to the membrane protein crystallization toolbox.

Steps 5 and 6: Phase Determination and Model Building

There are still numerous novel folds of IMPs to be discovered. This means that for phasing, de novo phasing methods are more likely to be required. These include multiple isomorphous replacement and multiple or single anomalous dispersion techniques. With dynamic structures, like ion channels, multiple conformational states that differ significantly from already solved structures will require one of these techniques. Weakly diffracting crystals and the requirement for multiple datasets requires many crystals. Crystals of a novel IMP structure that form part of a large antibody fragment complex may still be able to use the molecular replacement technique.

After all the hard work overexpressing, purifying, testing, crystallizing and determining the structure there should be one final thought; Will the protein adopt this conformation in the membrane? In other words, is it biologically correct? This is relevant considering the structural flexibility of many IMPs including ion channels. Freeing an IMPs from the constraints of the bilayer during a crystallization trial can have unforeseen consequences. In the best cases, these anomalies can drive a completely new direction of research, but all of the available biochemical and biophysical data should be employed to test and verify. This is also partly the reason for emphasizing that any protein target should be driven to one conformational stage prior to setting up crystallization trials.

In summary, ion channel structure determination requires a good understanding of the biochemical and biophysical properties of the protein. Manipulating these characteristics requires planning and flexibility, but as more and more channel structures are determined, this remains an exciting field of research.

REFERENCES

Abdine, A., K.-H. Park, and D.E. Warschawski. 2012. Cell-free membrane protein expression for solid-state NMR. *Methods in Molecular Biology (Clifton, N.J.)* 831: 85–109.

Agah, S. and S. Faham. 2012. Crystallization of membrane proteins in bicelles. *Methods in Molecular Biology (Clifton, N.J.)* 914: 3–16.

Backmark, A.E., N. Olivier, A. Snijder, E. Gordon, N. Dekker, and A.D. Ferguson. 2013. Fluorescent probe for high-throughput screening of membrane protein expression. *Protein Science: A Publication of the Protein Society* 22: 1124–1132.

Berman, M.H., J. Westbrook, Z. Feng, G. Gilliland, T.N. Bhat, H. Weissig, I.N. Shindyalov, and P.E. Bourne. 2000. The Protein Data Bank. *Nucleic Acids Research* 28 (1): 235–242.

Bernaudat, F., A. Frelet-Barrand, N. Pochon, S. Dementin, P. Hivin, S. Boutigny, J.-B. Rioux et al. 2011. Heterologous expression of membrane proteins: Choosing the appropriate host. *PloS One* 6 (12): e29191.

Berridge, G., R. Chalk, N. D'Avanzo, L. Dong, D. Doyle, J.-I. Kim, X. Xia et al. 2011. High-performance liquid chromatography separation and intact mass analysis of detergent-solubilized integral membrane proteins. *Analytical Biochemistry* 410 (2): 272–280.

Berry, I.M., O. Dym, R.M. Esnouf, K. Harlos, R. Meged, A. Perrakis, J.L. Sussman, T.S. Walter, J. Wilson, and A. Messerschmidt. 2006. SPINE high-throughput crystallization, crystal imaging and recognition techniques: Current state, performance analysis, new technologies and future aspects. *Acta Crystallographica. Section D, Biological Crystallography* 62 (Pt. 10): 1137–1149.

Blakeley, M.P., P. Langan, N. Niimura, and A. Podjarny. 2008. Neutron crystallography: Opportunities, challenges, and limitations. *Current Opinion in Structural Biology* 18 (5): 593–600.

Bocquet, N., H. Nury, M. Baaden, C.L. Poupon, J.-P. Changeux, M. Delarue, and P.-J. Corringer. 2009. X-ray structure of a pentameric ligand-gated ion channel in an apparently open conformation. *Nature* 457 (7225): 111–114.

Bukowska, M.A. and M.G. Grütter. 2013. New concepts and aids to facilitate crystallization. *Current Opinion in Structural Biology* 23 (3): 409–416.

Cadene, M. and B.T. Chait. 2000. A robust, detergent-friendly method for mass spectrometric analysis of integral membrane proteins. *Analytical Chemistry* 72 (22): 5655–5658.

Chayen, N.E. 2009. High-throughput protein crystallization. *Advances in Protein Chemistry and Structural Biology* 77: 1–22.

Cherezov, V., D.M. Rosenbaum, M.A. Hanson, S.G.F. Rasmussen, F.S. Thian, T.S. Kobilka, H.-J. Choi et al. 2007. High-resolution crystal structure of an engineered human beta2-adrenergic G protein-coupled receptor. *Science (New York, N.Y.)* 318 (5854): 1258–1265.

Chiu, M.L., P. Nollert, M.C. Loewen, H. Belrhali, E. Pebay-Peyroula, J.P. Rosenbusch, and E.M. Landau. 2000. Crystallization in cubo: General applicability to membrane proteins. *Acta Crystallographica. Section D, Biological Crystallography* 56 (Pt. 6): 781–784.

D'Avanzo, N., W.W.L. Cheng, D.A. Doyle, and C.G. Nichols. 2010a. Direct and specific activation of human inward rectifier K^+ channels by membrane phosphatidylinositol 4,5-bisphosphate. *The Journal of Biological Chemistry* 285 (48): 37129–37132.

D'Avanzo, N., W.W.L. Cheng, X. Xia, L. Dong, P. Savitsky, C.G. Nichols, and D.A. Doyle. 2010b. Expression and purification of recombinant human inward rectifier K^+ (KCNJ) channels in *Saccharomyces cerevisiae. Protein Expression and Purification* 71 (1): 115–121.

D'Avanzo, N., E.C. McCusker, A.M. Powl, A.J. Miles, C.G. Nichols, and B.A. Wallace. 2013. Differential lipid dependence of the function of bacterial sodium channels. *PloS One* 8 (4): e61216.

De Foresta, B., M. le Maire, S. Orlowski, P. Champeil, S. Lund, J.V. Møller, F. Michelangeli, and A.G. Lee. 1989. Membrane solubilization by detergent: Use of brominated phospholipids to evaluate the detergent-induced changes in Ca^{2+}-ATPase/lipid interaction. *Biochemistry* 28 (6): 2558–2567.

Deisenhofer, J., O. Epp, K. Miki, R. Huber, and H. Michel. 1985. Structure of the protein subunits in the photosynthetic reaction centre of *Rhodopseudomonas viridis* at 3Å resolution. *Nature* 318 (6047): 618–624.

De Marco, A. 2012. Optimization of purification protocols based on the step-by-step monitoring of the protein aggregates in soluble fractions. *Methods in Molecular Biology (Clifton, N.J.)* 824: 145–154.

Doyle, D.A. 2004. Structural changes during ion channel gating. *Trends in Neurosciences* 27 (6): 298–302.

Doyle, D.A., J.M. Cabral, R.A. Pfuetzner, A. Kuo, J.M. Gulbis, S.L. Cohen, B.T. Chait, and R. MacKinnon. 1998. The structure of the potassium channel: Molecular basis of K^+ conduction and selectivity. *Science (New York, N.Y.)* 280 (5360): 69–77.

Drew, D., M. Lerch, E. Kunji, D.-J. Slotboom, and J.-W. de Gier. 2006. Optimization of membrane protein overexpression and purification using GFP fusions. *Nature Methods* 3 (4): 303–313.

Drew, D., S. Newstead, Y. Sonoda, H. Kim, G. von Heijne, and S. Iwata. 2008. GFP-based optimization scheme for the overexpression and purification of eukaryotic membrane proteins in *Saccharomyces cerevisiae. Nature Protocols* 3 (5): 784–798.

Drew, D., D.-J. Slotboom, G. Friso, T. Reda, P. Genevaux, M. Rapp, N.M. Meindl-Beinker et al. 2005. A scalable, GFP-based pipeline for membrane protein overexpression screening and purification. *Protein Science: A Publication of the Protein Society* 14 (8): 2011–2017.

Ellinger, P., M. Kluth, J. Stindt, S.H.J. Smits, and L. Schmitt. 2013. Detergent screening and purification of the human liver ABC transporters BSEP (ABCB11) and MDR3 (ABCB4) expressed in the yeast *Pichia pastoris. PloS One* 8 (4): e60620.

Frelet-Barrand, A., S. Boutigny, E.R.S. Kunji, and N. Rolland. 2010. Membrane protein expression in *Lactococcus lactis. Methods in Molecular Biology (Clifton, N.J.)* 601: 67–85.

Gabant, G. and M. Cadene. 2008. Mass spectrometry of full-length integral membrane proteins to define functionally relevant structural features. *Methods (San Diego, California)* 46 (2): 54–61.

Gileadi, O., S. Knapp, W.H. Lee, B.D. Marsden, S. Müller, F.H. Niesen, K.L. Kavanagh et al. 2007. The scientific impact of the structural genomics consortium: A protein family and ligand-centered approach to medically-relevant human proteins. *Journal of Structural and Functional Genomics* 8 (2–3): 107–119.

Glover, C.A.P., V.L.G. Postis, K. Charalambous, S.B. Tzokov, W.I. Booth, S.E. Deacon, B.A. Wallace, S.A. Baldwin, and P.A. Bullough. 2011. AcrB contamination in 2-D crystallization of membrane proteins: Lessons from a sodium channel and a putative monovalent cation/proton antiporter. *Journal of Structural Biology* 176 (3): 419–424.

Grey, J.L. and D.H. Thompson. 2010. Challenges and opportunities for new protein crystallization strategies in structure-based drug design. *Expert Opinion on Drug Discovery* 5 (11): 1039–1045.

Gutmann, D.A., P.E. Mizohata, S. Newstead, S. Ferrandon, V. Postis, X. Xia, P.J.F. Henderson, H.W. van Veen, and B. Byrne. 2007. A high-throughput method for membrane protein solubility screening: The ultracentrifugation dispersity sedimentation assay. *Protein Science: A Publication of the Protein Society* 16 (7): 1422–1428.

Hammond, G.R.V., M.J. Fischer, K.E. Anderson, J. Holdich, A. Koteci, T. Balla, and R.F. Irvine. 2012. PI4P and PI(4,5)P2 are essential but independent lipid determinants of membrane identity. *Science (New York, N.Y.)* 337 (6095): 727–730.

Hattori, M., R.E. Hibbs, and E. Gouaux. 2012. A fluorescence-detection size-exclusion chromatography-based thermostability assay for membrane protein precrystallization screening. *Structure (London, England: 1993)* 20 (8): 1293–1299.

Hertel, F. and J. Zhang. 2014. Monitoring of post-translational modification dynamics with genetically encoded fluorescent reporters. *Biopolymers.* 101 (2): 180–187.

Hille, B. 2001. *Ion Channels of Excitable Membranes.* 3rd edn. Sunderland, MA: Sinauer.

Hitscherich, C., M. Allaman, J. Wiencek, J. Kaplan, and P.J. Loll. 2000. Static light scattering studies of OmpF porin: Implications for integral membrane protein crystallization. *Protein Science* 9 (8): 1559–1566.

Horsefield, R., V. Yankovskaya, S. Törnroth, C. Luna-Chavez, E. Stambouli, J. Barber, B. Byrne, G. Cecchini, and S. Iwata. 2003. Using rational screening and electron microscopy to optimize the crystallization of succinate: Ubiquinone oxidoreductase from *Escherichia Coli. Acta Crystallographica. Section D, Biological Crystallography* 59 (Pt. 3): 600–602.

Hunte, C., G. von Jagow, and H. Schagger. 2003. *Membrane Protein Purification and Crystallization: A Practical Guide.* 2nd edn. San Diego, CA: Elsevier Inc.

Isaksson, L., J. Enberg, R. Neutze, B.G. Karlsson, and A. Pedersen. 2012. Expression screening of membrane proteins with cell-free protein synthesis. *Protein Expression and Purification* 82 (1): 218–225.

Iwata, S., J.W. Lee, K. Okada, J.K. Lee, M. Iwata, B. Rasmussen, T.A. Link, S. Ramaswamy, and B.K. Jap. 1998. Complete structure of the 11-subunit bovine mitochondrial cytochrome Bc1 complex. *Science (New York, N.Y.)* 281 (5373): 64–71.

Jaehme, M. and H. Michel. 2013. Evaluation of cell-free protein synthesis for the crystallization of membrane proteins—Case study on a member of the glutamate transporter family from *Staphylothermus marinus*. *The FEBS Journal* 280 (4): 1112–1125.
Jiang, Y., A. Lee, J. Chen, M. Cadene, B.T. Chait, and R. MacKinnon. 2002. Crystal structure and mechanism of a calcium-gated potassium channel. *Nature* 417 (6888): 515–522.
Johansson, L.C., A.B. Wöhri, G. Katona, S. Engström, and R. Neutze. 2009. Membrane protein crystallization from lipidic phases. *Current Opinion in Structural Biology* 19 (4): 372–378.
Kawate, T. and E. Gouaux. 2006. Fluorescence-detection size-exclusion chromatography for precrystallization screening of integral membrane proteins. *Structure (London, England: 1993)* 14 (4): 673–681.
Klose, D.P., B.A. Wallace, and R.W. Janes. 2012. DichroMatch: A website for similarity searching of circular dichroism spectra. *Nucleic Acids Research* 40 (Web Server issue): W547–W552.
Koide, S. 2009. Engineering of recombinant crystallization chaperones. *Current Opinion in Structural Biology* 19 (4): 449–457.
Kuhn, P., K. Wilson, M.G. Patch, and R.C. Stevens. 2002. The genesis of high-throughput structure-based drug discovery using protein crystallography. *Current Opinion in Chemical Biology* 6 (5): 704–710.
Kuo, A., C. Domene, L.N. Johnson, D.A. Doyle, and C. Vénien-Bryan. 2005. Two different conformational states of the KirBac3.1 potassium channel revealed by electron crystallography. *Structure (London, England: 1993)* 13 (10): 1463–1472.
Kuo, A., J.M. Gulbis, J.F. Antcliff, T. Rahman, E.D. Lowe, J. Zimmer, J. Cuthbertson, F.M. Ashcroft, T. Ezaki, and D.A. Doyle. 2003. Crystal structure of the potassium channel KirBac1.1 in the closed state. *Science (New York, N.Y.)* 300 (5627): 1922–1926.
Lebon, G., K. Bennett, A. Jazayeri, and C.G. Tate. 2011a. Thermostabilisation of an agonist-bound conformation of the human adenosine A(2A) receptor. *Journal of Molecular Biology* 409 (3): 298–310.
Lebon, G., T. Warne, P.C. Edwards, K. Bennett, C.J. Langmead, A.G.W. Leslie, and C.G. Tate. 2011b. Agonist-bound adenosine A2A receptor structures reveal common features of GPCR activation. *Nature* 474 (7352): 521–525.
Lee, S.-Y., J.A. Letts, and R. MacKinnon. 2009. Functional reconstitution of purified human Hv1 H+ channels. *Journal of Molecular Biology* 387 (5): 1055–1060.
Levitan, I., Y. Fang, A. Rosenhouse-Dantsker, and V. Romanenko. 2010. Cholesterol and ion channels. *Sub-Cellular Biochemistry* 51: 509–549.
Liao, M., E. Cao, D. Julius, and Y. Cheng. 2013. Structure of the TRPV1 ion channel determined by electron cryo-microscopy. *Nature* 504 (7478): 107–112.
Lichtenberg, D., H. Ahyayauch, A. Alonso, and F.M. Goñi. 2013. Detergent solubilization of lipid bilayers: A balance of driving forces. *Trends in Biochemical Sciences* 38 (2): 85–93.
Lim, H.-H, T. Shane, and C. Miller. 2012. Intracellular proton access in a Cl(-)/H(+) antiporter. *PLoS Biology* 10 (12): e1001441.
Liu, F., M. Ye, C. Wang, Z. Hu, Y. Zhang, H. Qin, K. Cheng, and H. Zou. 2013. Polyacrylamide gel with switchable trypsin activity for analysis of proteins. *Analytical Chemistry* 85: 7024–7028.
Liu, W., E. Chun, A.A. Thompson, P. Chubukov, F. Xu, V. Katritch, G.W. Han et al. 2012. Structural basis for allosteric regulation of GPCRs by sodium ions. *Science (New York, N.Y.)* 337 (6091): 232–236.
Lu, X. and H. Zhu. 2005. Tube-gel digestion: A novel proteomic approach for high throughput analysis of membrane proteins. *Molecular & Cellular Proteomics: MCP* 4 (12): 1948–1958.
Lukacs, G.L. and A.S. Verkman. 2012. CFTR: Folding, misfolding and correcting the ΔF508 conformational defect. *Trends in Molecular Medicine* 18 (2): 81–91.
Marasini, C., L. Galeno, and O. Moran. 2013. A SAXS-based ensemble model of the native and phosphorylated regulatory domain of the CFTR. *Cellular and Molecular Life Sciences: CMLS* 70 (5): 923–933.
McCusker, E.C., C. Bagnéris, C.E. Naylor, A.R. Cole, N. D'Avanzo, C.G. Nichols, and B.A. Wallace. 2012. Structure of a bacterial voltage-gated sodium channel pore reveals mechanisms of opening and closing. *Nature Communications* 3: 1102.
Miller, A.N. and S.B. Long. 2012. Crystal structure of the human two-pore domain potassium channel K2P1. *Science (New York, N.Y.)* 335 (6067): 432–436.
Miller, C. 1986. *Ion Channel Reconstitution*. New York: Plenum Press.
Miller, C. 2006. ClC chloride channels viewed through a transporter lens. *Nature* 440 (7083): 484–489.
Mogridge, J. 2004. Using light scattering to determine the stoichiometry of protein complexes. *Methods in Molecular Biology (Clifton, N.J.)* 261: 113–118.
Müller-Lucks, A., P. Gena, D. Frascaria, N. Altamura, M. Svelto, E. Beitz, and G. Calamita. 2013. Preparative scale production and functional reconstitution of a human aquaglyceroporin (AQP3) using a cell free expression system. *New Biotechnology* 30 (5): 545–551.
Neylon, C. 2008. Small angle neutron and x-ray scattering in structural biology: Recent examples from the literature. *European Biophysics Journal: EBJ* 37 (5): 531–541.
Nollert, P., J. Navarro, and E.M. Landau. 2002. Crystallization of membrane proteins in cubo. *Methods in Enzymology* 343: 183–199.
Ostermeier, C., A. Harrenga, U. Ermler, and H. Michel. 1997. Structure at 2.7 Å resolution of the *Paracoccus denitrificans* two-subunit cytochrome c oxidase complexed with an antibody FV fragment. *Proceedings of the National Academy of Sciences of the United States of America* 94 (20): 10547–10553.
Ostermeier, C. and H. Michel. 1997. Crystallization of membrane proteins. *Current Opinion in Structural Biology* 7 (5): 697–701.
Page, R. 2008. Strategies for improving crystallization success rates. *Methods in Molecular Biology (Clifton, N.J.)* 426: 345–362.
Pan, X., C.A. Bingman, G.E. Wesenberg, Z. Sun, and G.N. Phillips Jr. 2010. Domain view: A web tool for protein domain visualization and analysis. *Journal of Structural and Functional Genomics* 11 (4): 241–245.
Parcej, D., R. Guntrum, S. Schmidt, A. Hinz, and R. Tampé. 2013. Multicolour fluorescence-detection size-exclusion chromatography for structural genomics of membrane multiprotein complexes. *PloS One* 8 (6): e67112.
Privé, G.G. 2007. Detergents for the stabilization and crystallization of membrane proteins. *Methods (San Diego, California)* 41 (4): 388–397.
Proteau, A., R. Shi, and M. Cygler. 2010. Application of dynamic light scattering in protein crystallization. *Current Protocols in Protein Science* Chapter 17 : Unit 17.10.
Psakis, G., J. Polaczek, and L.-O. Essen. 2009. AcrB et Al.: Obstinate contaminants in a picogram scale. One more bottleneck in the membrane protein structure pipeline. *Journal of Structural Biology* 166 (1): 107–111.
Ramraj, V., G. Evans, J.M. Diprose, and R.M. Esnouf. 2012. Nearest-cell: A fast and easy tool for locating crystal matches in the PDB. *Acta Crystallographica. Section D, Biological Crystallography* 68 (Pt. 12): 1697–1700.
Ramsey, I.S., M.M. Moran, J.A. Chong, and D.E. Clapham. 2006. A voltage-gated proton-selective channel lacking the pore domain. *Nature* 440 (7088): 1213–1216.
Rasmussen, S.G.F., B.T. DeVree, Y. Zou, A.C. Kruse, K.Y. Chung, T.S. Kobilka, F.S. Thian et al. 2011. Crystal structure of the B2 adrenergic receptor-Gs protein complex. *Nature* 477 (7366): 549–555.

Rosenbaum, D.M., V. Cherezov, M.A. Hanson, S.G.F. Rasmussen, F.S. Thian, T.S. Kobilka, H.-J. Choi et al. 2007. GPCR engineering yields high-resolution structural insights into beta2-adrenergic receptor function. *Science (New York, N.Y.)* 318 (5854): 1266–1273.

Sasaki, M., M. Takagi, and Y. Okamura. 2006. A voltage sensor-domain protein is a voltage-gated proton channel. *Science (New York, N.Y.)* 312 (5773): 589–592.

Sobolevsky, A.I., M.P. Rosconi, and E. Gouaux. 2009. X-ray structure, symmetry and mechanism of an AMPA-subtype glutamate receptor. *Nature* 462 (7274): 745–756.

Song, L., M.R. Hobaugh, C. Shustak, S. Cheley, H. Bayley, and J.E. Gouaux. 1996. Structure of staphylococcal alpha-hemolysin, a heptameric transmembrane pore. *Science (New York, N.Y.)* 274 (5294): 1859–1866.

Tang, M., L.J. Sperling, D.A. Berthold, C.D. Schwieters, A.E. Nesbitt, A.J. Nieuwkoop, R.B. Gennis, and C.M. Rienstra. 2011. High-resolution membrane protein structure by joint calculations with solid-state NMR and x-ray experimental data. *Journal of Biomolecular NMR* 51 (3): 227–233.

Tereshko, V., S. Uysal, A. Koide, K. Margalef, S. Koide, and A.A. Kossiakoff. 2008. Toward chaperone-assisted crystallography: Protein engineering enhancement of crystal packing and x-ray phasing capabilities of a camelid single-domain antibody (VHH) scaffold. *Protein Science: A Publication of the Protein Society* 17 (7): 1175–1187.

Ujwal, R. and J.U. Bowie. 2011. Crystallizing membrane proteins using lipidic bicelles. *Methods (San Diego, California)* 55 (4): 337–341.

Uysal, S., L.G. Cuello, D.M. Cortes, S. Koide, A.A. Kossiakoff, and E. Perozo. 2011. Mechanism of activation gating in the full-length KcsA K^+ channel. *Proceedings of the National Academy of Sciences of the United States of America* 108 (29): 11896–11899.

Uysal, S., V. Vásquez, V. Tereshko, K. Esaki, F.A. Fellouse, S.S. Sidhu, S. Koide, E. Perozo, and A. Kossiakoff. 2009. Crystal structure of full-length KcsA in its closed conformation. *Proceedings of the National Academy of Sciences of the United States of America* 106 (16): 6644–6649.

Van Horn, W.D., H.-J. Kim, C.D. Ellis, A. Hadziselimovic, E.S. Sulistijo, M.D. Karra, C. Tian, F.D. Sönnichsen, and C.R. Sanders. 2009. Solution nuclear magnetic resonance structure of membrane-integral diacylglycerol kinase. *Science (New York, N.Y.)* 324 (5935): 1726–1729.

Vargas, E., F. Bezanilla, and B. Roux. 2011. In search of a consensus model of the resting state of a voltage-sensing domain. *Neuron* 72 (5): 713–720.

Veesler, D., S. Blangy, C. Cambillau, and G. Sciara. 2008. There is a baby in the bath water: AcrB contamination is a major problem in membrane-protein crystallization. *Acta Crystallographica. Section F, Structural Biology and Crystallization Communications* 64 (Pt. 10): 880–885.

Walter, T.S., J.M. Diprose, C.J. Mayo, C. Siebold, M.G. Pickford, L. Carter, G.C. Sutton et al. 2005. A procedure for setting up high-throughput nanolitre crystallization experiments. Crystallization workflow for initial screening, automated storage, imaging and optimization. *Acta Crystallographica. Section D, Biological Crystallography* 61 (Pt. 6): 651–657.

Wang, C., I. Protasevich, Z. Yang, D. Seehausen, T. Skalak, X. Zhao, S. Atwell et al. 2010. Integrated biophysical studies implicate partial unfolding of NBD1 of CFTR in the molecular pathogenesis of F508del cystic fibrosis. *Protein Science: A Publication of the Protein Society* 19 (10): 1932–1947.

Wang, C., H. Wu, V. Katritch, G. Won Han, X.-P. Huang, W. Liu, F.Y. Siu, B.L. Roth, V. Cherezov, and R.C. Stevens. 2013. Structure of the human smoothened receptor bound to an antitumour agent. *Nature* 497 (7449): 338–343.

Warne, T., M.J. Serrano-Vega, J.G. Baker, R. Moukhametzianov, P.C. Edwards, R. Henderson, A.G.W. Leslie, C.G. Tate, and G.F.X. Schertler. 2008. Structure of a Beta1-adrenergic G-protein-coupled receptor. *Nature* 454 (7203): 486–491.

Whited, A.M. and P.S-H. Park. 2013. Atomic force microscopy: A multifaceted tool to study membrane proteins and their interactions with ligands. *Biochimica et Biophysica Acta* 1838: 56–68.

Whitmore, L., B. Woollett, A.J. Miles, D.P. Klose, R.W. Janes, and B.A. Wallace. 2011. PCDDB: The protein circular dichroism data bank, a repository for circular dichroism spectral and metadata. *Nucleic Acids Research* 39 (suppl. 1): D480–D486.

Whorton, M.R. and R. MacKinnon. 2013. X-ray structure of the mammalian GIRK2-βγ G-protein complex. *Nature* 498 (7453): 190–197.

Xia, D., C.A. Yu, H. Kim, J.Z. Xia, A.M. Kachurin, L. Zhang, L. Yu, and J. Deisenhofer. 1997. Crystal structure of the cytochrome Bc1 complex from bovine heart mitochondria. *Science (New York, N.Y.)* 277 (5322): 60–66.

Yang, Y., X. Jin, and C. Jiang. 2013. S-glutathionylation of ion channels: Insights into the regulation of channel functions, thiol modification crosstalk and mechanosensing. *Antioxidants and Redox Signaling* 20: 937–951.

Zhong, X. and J.F. Wright. 2013. Biological insights into therapeutic protein modifications throughout trafficking and their biopharmaceutical applications. *International Journal of Cell Biology* 2013: 273086.

Zhou, Y., J.H. Morais-Cabral, A. Kaufman, and R. MacKinnon. 2001. Chemistry of ion coordination and hydration revealed by a K^+ channel-Fab complex at 2.0 Å resolution. *Nature* 414 (6859): 43–48.

Zimmer, J. and D.A. Doyle. 2006. Phospholipid requirement and pH optimum for the in vitro enzymatic activity of the *E. coli* P-type ATPase ZntA. *Biochimica et Biophysica Acta* 1758 (5): 645–652.

Structural study of ion channels by cryo-electron microscopy

Qiu-Xing Jiang and Liang Shi

Contents

11.1 INTRODUCTION

Complementary to the structure determination by X-ray crystallography and nuclear magnetic resonance (NMR) spectroscopy, three-dimensional (3D) reconstruction from cryo-electron microscopic (cryoEM) images has become one of the major methods in structural biology (Frank 2002; Frank et al. 2002). It does not require large quantities of materials to start experiments, nor is high-quality crystalline arrangement of the target molecules a must. For some high-symmetry objects, 3D reconstructions at atomic resolutions have been achieved when high-quality specimens were imaged under more advanced microscopes (Yu et al. 2008; Zhang et al. 2008b, 2013; Zhou 2008, 2011). For a majority of the biological specimens, cryoEM analysis at the current state of the art is only able to deliver low- to intermediate-resolution structures, which are insufficient for the ab initio building of atomic models. Nevertheless, it does provide a venue to study the dynamics of biological complexes as well as the component organization of the target macromolecules in cells or even tissues (Frank 2002). For the study of some multidomain ion channels, the combination of all three structural techniques may be necessary in order to address the fundamental questions regarding the functional properties of these channels.

Ion channels are membrane proteins and fulfill their function in lipid bilayers. It is expected that the channels experience close interactions with lipids, and some of them are even strongly gated by lipids. To investigate the lipid effects on the structures of ion channels, electron crystallography of the channels in membrane would arguably be the best method of choice because it can deliver high-resolution structures of channels in membrane (Gonen et al. 2006). Alternatively, restrained spherical reconstruction of ion channels in small vesicles (Jiang et al. 2001; Wang and Sigworth 2009, 2010) will generate 3D reconstructions at better resolutions when enough molecules in vesicles can be imaged and analyzed.

To study the organization of ion channels in cells and tissues, conventional EM of ultrathin sections is usually a good option. But embedding, fixing, solvent substitution, and ultramicrotomy in the conventional EM have been known to introduce some changes to structures, and the resultant samples still are not able to deliver molecular resolutions. Freeze-etch, freeze-fracture, or cryo-sectioning of frozen samples in the conventional electron microscopy (EM) are difficult to implement in order to obtain high-resolution data (better than 5 nm) that would allow the reliable discrimination of ion channels (Heuser 2005, 2011). Nowadays, electron tomography and correlative light-electron microscopy (CLEM) are capable of generating 3D structures of ultrathin samples at 3–4 nm resolution (Steven and Aebi 2003; Gruska et al. 2008). With sufficient development in both specimen preparation and imaging technology using new phase plates and high-quality direct detectors, it will be feasible to recognize individual ion channels in the 3D structures of cells and tissues (Nagayama 2008; Nagayama and Danev 2008; Milazzo et al. 2011; Campbell et al. 2012).

In the next section, we will focus our discussions on three basic aspects of cryoEM: specimen preparation, mathematical principles of data analysis, and techniques for analyzing different specimens. As concrete examples of application, we will talk about the cryoEM studies of several types of ion channel preparations. Hopefully, these practice-oriented sections will provide enough information for the readers to understand the advantages and limitations of cryoEM in the structural studies of ion channels.

11.2 STRUCTURE DETERMINATION BY cryoEM

11.2.1 BASIC CONCEPTS

Three-dimensional cryoEM aims at building 3D models of target molecules from their transmission EM (TEM) images and using these models to gain new mechanistic insights on the function of these molecules (Frank 2001a,b). Images obtained in a TEM represent the projections of the 3D molecules along the direction of electron beam (Figure 11.1a). The procedure of 3D reconstruction is governed by the central section theorem (Figure 11.1b). The theorem states that a projection image from a 3D object in reciprocal space is equivalent to a central section of the Fourier transform of the object. It means that the Fourier transform of the projection image represents a sampling of the 3D object in the reciprocal space at a specific orientation. If the whole reciprocal space is sampled in as many directions as possible, the full structure of the 3D object can be calculated by an inverse transform. In other words, 3D reconstruction is a process of sampling the Fourier space by taking projection images at different directions and assigning to them correct orientations (Figure 11.1b). The general idea of central section theorem was first demonstrated in 2D objects and was found to be applicable in higher dimensions (Radermacher 1994). Its 3D version forges the foundation for 3D electron microscopy.

Depending on the specimens, the 3D electron microscopy falls into four different areas: electron crystallography (including helical reconstruction), single particle reconstruction (including the application to crystalline samples), electron tomography, and correlative light-electron microscopy (CLEM) (Jensen and Briegel 2007; Salje et al. 2009; Cortese et al. 2013; Mironov and Beznoussenko 2013; Polishchuk et al. 2013). The electron tomography and CLEM rely heavily on the proper preparation of ultrathin biological specimens. Either sections from embedded samples by ultramicrotomy or from frozen samples by cryo-sectioning are used for imaging. The room temperature electron tomogram from embedded samples may suffer from various complications introduced by embedding and sectioning. The cryo-sections retain the true structures of the biological samples but are relatively difficult to reproduce and easy to be damaged by electrons (Dubochet 2007; Dubochet et al. 2007). To protect these specimens from electron damage, the total tolerable dose of electrons is divided into 70–100 exposures that are usually collected from one specimen tilted at different angles. Such a dose-fractionation however limits the achievable resolution to 2–4 nm and makes it difficult to reveal structural details underlying the function of ion channels at molecular levels. We therefore will not discuss in detail the electron tomographic study of ion channels at the subcellular level. Instead, we will elaborate more on the technical details for both single particle reconstruction and electron crystallography because the applications of these two techniques are capable of revealing structural details of the ion channels at intermediate to high resolutions, and have been reported much more frequently in literature. Readers interested in more details about electron tomography and CLEM could refer to recent reviews (e.g., Polishchuk et al. 2013).

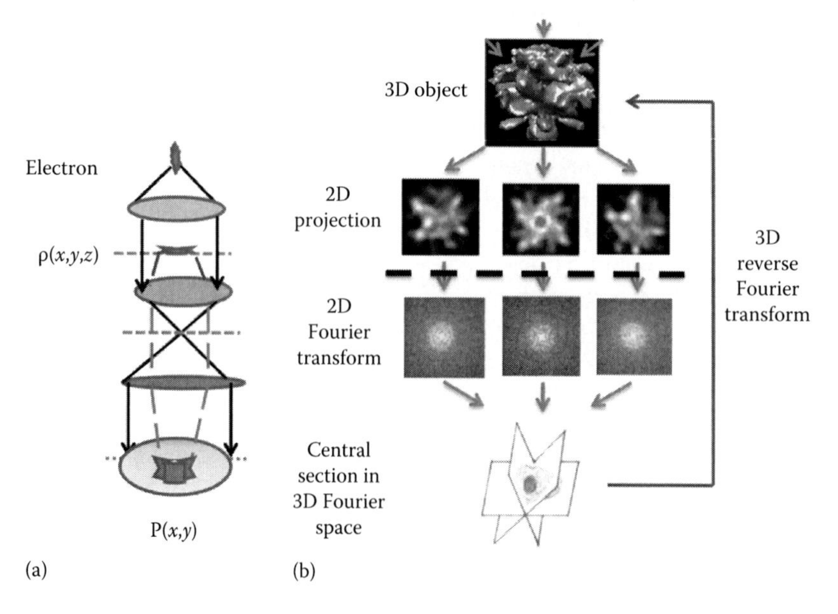

Figure 11.1 Central section theorem for 3D electron microscopy. (a) EM images are projections from 3D objects. (b) Fourier transform of a projection from a 3D object corresponds to a central section in the Fourier transform of the 3D object. The projections in different orientations lead to the sampling of the 3D Fourier space at different directions.

11.3 SAMPLE PREPARATION AND DATA COLLECTION BY cryoEM

11.3.1 BIOLOGICAL PREPARATIONS NEED TO BE OPTIMIZED BEFORE LOADING THEM TO EM GRIDS

EM grids are made of metals, such as copper, nickel, gold, molybdenum, etc. They are 3.05 mm in diameter, and are usually coated with carbon films that are about 3–5 nm thick. The high electrical conductivity of the carbon films is essential because it helps eliminate charges deposited on the specimens during electron exposure. The carbon films can be continuous or holey.

Even though sample preparation for cryoEM is not as demanding as for x-ray crystallography, good specimens are still the foremost requirement. For 2D crystals and helical filaments, the growth of these crystalline assemblies is fairly time-consuming and needs the exploration of a large parameter space. The parameters usually include protein concentration, lipid/protein ratio, lipid composition, salt and buffer (type and concentration), temperature, speed of detergent removal if the formation of lipid membranes is intended, and the type and concentration of precipitants that can facilitate crystallization (Walz and Grigorieff 1998). A screen usually starts with purified proteins at (sub)milligram levels. A protein at this level is suitable for screening both 2D and 3D crystals. When only micrograms of proteins are available, it is still possible to use carbon-coated grids or functionalized lipid monolayers to facilitate the growth of 2D crystals at the surfaces (Uzgiris and Kornberg 1983; Cyrklaff et al. 1995). Both of these small-scale methods make them inefficient in searching a broad parameter space.

In general, lack of high-throughput screening methods has limited the speed of identifying 2D crystals and their further refinement once the crystals show up. Recently, new progress has been made in designing and implementing high-throughput systems for screening 2D crystals (Stokes et al. 2013). The general availability of such systems will hopefully accelerate the speed of obtaining high-quality 2D crystals.

The helical filaments can be viewed as the linear stacking of the individual layers of molecules (Young et al. 1997). The screening of such an assembly is usually performed in the same solution phase for 3D crystallization. The facilitated helical crystallization on designed polymer structures was proposed to coax the molecules into interacting with each other in a confined space (Dang et al. 2005). But it remains a challenging and slow process, especially to obtain helical filaments that diffract to subnanometer resolutions.

The best example for the helical analysis of an ion channel is the nicotinic acetylcholine receptor nAchR structure (Miyazawa et al. 2003; Unwin 2003, 2005; Unwin and Fujiyoshi 2012). The purified nAchR in presynaptic membrane vesicles formed helical tubules after being stored at 4°C (Figure 11.2a). Here the large extracellular domains of the nAchR molecules are preferentially presented to the outside of the vesicles. During the growth of the helical tubules, these domains interact with each other and drive the crystalline packing. The same principle could be applicable to the growth of tubular crystals of other membrane proteins that protrude bulky ectodomains to one side of the membrane (Young et al. 1997).

For single particle specimens, the major demand is to make sure that the biological samples are as homogeneous as possible.

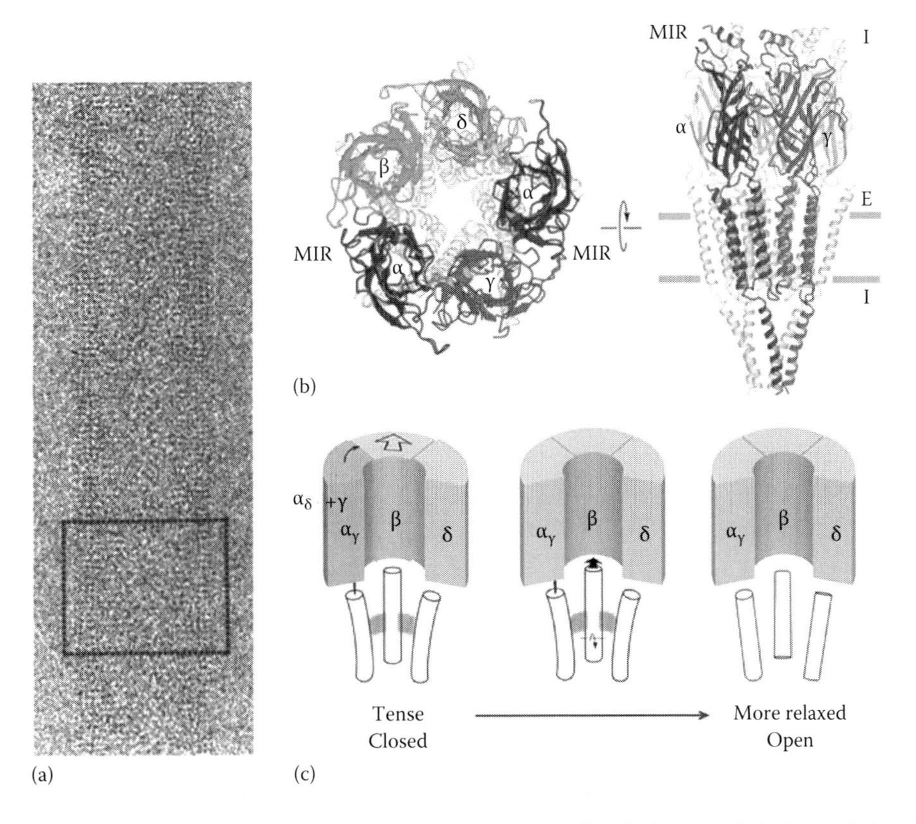

Figure 11.2 Helical reconstruction of nAchR. (a) cryoEM image of a short segment of a tubular crystal. (b) 3D model at 4.0 Å built from the cryoEM map. (c) Putative gating movement deduced from the two maps of nAchR in closed and open states at ~6.0 Å resolution.

For multicomponent biological complexes, the compositional and conformational heterogeneity is foreseeably inevitable, and will introduce structural discrepancies when different procedures are used to prepare the same complexes. For example, structural studies of the inositol 1,4,5-trisphosphate receptor (IP_3R) have generated contrastingly different 3D structures by four different research groups (Jiang et al. 2002; da Fonseca et al. 2003; Serysheva et al. 2003; Sato et al. 2004; Serysheva and Ludtke 2010; Ludtke et al. 2011). Two research groups reported very different structures when they varied the preparation of the protein from different sources using different protocols (da Fonseca et al. 2003; Serysheva et al. 2003; Wolfram et al. 2010; Ludtke et al. 2011). It is therefore worth reminding the readers that special attention should be paid to the preparation of biological specimens before loading them to EM grids. Once the specimens are suitable, the total amount of material for preparing EM grids is merely 1–20 µg, significantly lower than what is needed for crystallization.

11.3.2 NEGATIVE STAIN OF THE SPECIMENS

All three types of specimens can be loaded onto carbon-coated grids and negatively stained by heavy metal solutions for fast examination. For this purpose, the carbon films on the grids are rendered negatively or positively charged, or slightly hydrophobic. The sticking of the specimens to these carbon films will enrich them at the surface. The amount of material for negative stain EM can therefore be one or two orders of magnitude lower than that for cryo-preparation. The specimens in 1–3 µL solution are loaded onto carbon-coated grids. After a short period of incubation, usually 30–60 s, excess sample is blotted away by using the torn edge of a piece of absorptive filter paper. The specimens with some residual buffer are then washed with water before being incubated for a short time (~10–20 s) with the stain solutions. The stain solutions usually used in a lab are 1%–2% uranyl acetate or formate (pH 2.5–4), or 6%–10% ammonium molybdate (pH 6.3–7.0), or 1%–2% phosphotungstate (pH 7.5–8.0). After blotting, the stained specimens are air-dried before EM examination.

Preparation of 2D crystals sometimes requires special treatment of the carbon films because the crystals may not stick to the carbon films well. For example, it was reported that only after freshly evaporated carbon films were aged for a few days was it possible for good bacteriorhodopsin 2D crystals (purple membranes) to stick well (Baldwin et al. 1988). The other general issue encountered in negative staining is the crushing of the molecules during air-drying. For example, negatively stained ryanodine receptor (RyR) in detergents was found to have its transmembrane domain squashed, whose structural features were not as well recognized in the resultant 3D reconstruction as in the cryoEM map (Radermacher et al. 1992, 1994). Adding 0.5%–2% sugar in stain solutions in some cases can mitigate this problem.

11.3.3 PREPARATION OF CRYO SPECIMENS

In 1974, K. Taylor and R. Glaeser first reported that frozen hydrated thin catalase crystals in a cryoEM diffract electrons to 2.8 Å, suggesting that proper preparation of the cryo-specimens will allow the structure determination to atomic resolution (Taylor and Glaeser 1974). In an electron microscope, the specimens are under high vacuum. They cannot have high vapor pressure. To replace water molecules next to the specimen, cryo-protectants are usually introduced to preserve the hydrogen bonds at the surface of biological macromolecules. Glucose was first used to preserve (embed) the 2D crystals of bacteriorhodopsin. Tannin and trehalose were later found suitable for the same purpose. Back injection and carbon sandwiching methods were developed by Fujiyoshi's group to preserve 2D crystals for data collection (Gyobu et al. 2004).

In the 1980s, Dubochet's group found that fast-plunging of a thin layer of water (<2 µ) into liquid ethane freezes the water molecules so quickly that they do not have time to rearrange into ice crystals (Dubochet 1995, 2007). The resultant glass-state ice (called vitrified or vitreous ice) maintains the randomness of the water molecules as long as it is kept below its phase transition temperature. The vitrified specimens maintain the water layer next to the biological molecules and therefore keep them close to their native conditions. Since then, fast blotting and quick plunging has become the standard procedure to prepare biological samples for high-resolution cryoEM.

11.3.4 DATA COLLECTION IN cryoEM

Due to strong electron–specimen interactions, the highest dose of electron for biological samples is limited to about 20–30 per $Å^2$ (Unwin and Henderson 1975; Fujiyoshi 1998). Under this dose, the high-resolution information is preserved, but the signal-to-noise ratio (SNR) in the cryoEM images is very low (<0.1). The low tolerable dose also means that the samples can only be exposed to high dose once. A three-mode protocol was designed for cryoEM imaging. The samples are examined in a low magnification to find an area of interest (search mode). A small area, which is 2–3 µm away from the area of interest, is imaged at a high magnification to define zero defocus and set up the right underfocus level for imaging (focus mode). At last, the area of interest is exposed to 20–30 electrons/$Å^2$ to generate the final image. For 2D crystals, the imaging needs to be done from crystals tilted at different angles. The high-angle tilt usually makes it very difficult to collect high-resolution data, and slows down the progress significantly (Fujiyoshi 1998).

When a 2D crystal is sizeable enough (at least a few microns), its interaction with electron beam may allow the formation of strong electron diffraction pattern (Gonen 2013). The electron diffraction patterns have no convolution from the contrast transfer function (CTF) of the microscope, and thus represent more accurate measurements of the amplitudes of the structure factors. It is always preferred to collect electron diffraction data to high resolution and use them to refine the 3D map to a better resolution (Wisedchaisri and Gonen 2013). However, the limit in feasibility usually lies with the fact that it takes a lot of efforts to grow 2D crystals that are good enough for collecting diffraction data.

For helical filaments and single particle specimens, the three-mode data collection is suitable for high-resolution imaging. If the specimens are concentrated enough, holey grids can be used to suspend them in a thin layer of ice inside a hole, where no carbon background casts extra noise to the final image (Miyazawa et al. 1999). The helical tubules of nAchR trapped across holes were found to yield better data. If it is important to rely on the interaction between the specimens and the carbon films to enrich them, an ultrathin-layer carbon can be introduced to cover the holey grids and images can be taken from the covered holes. It is gradually becoming feasible to use two to three layers of graphene films to cover the holes for this purpose. Graphene has superior heat and electric conductivity, and will be superior in minimizing the charging effects (Pantelic et al. 2011). The carbon support also makes it possible to keep the ice thickness relative even across the exposed area. The attachment of the specimen on the carbon surface in a cryo-sample is likely retained through point contacts so that there is only a slim, if any, chance that the interaction would orient the specimens to a preferred direction. For example, the bacterial 70S ribosome usually takes two preferential orientations in negative-stain EM, but in cryoEM, the angular distribution in frozen samples is much broader and results in a complete coverage of all possible directions.

In summary, the practical operation of cryoEM requires that the biological samples are prepared to highest homogeneity, and are preserved by fast-freezing for high-resolution imaging. With high-quality cryo-specimens, a modern high-end electron microscope will be able to deliver the high-quality data, image, and/or diffraction data, for 3D reconstruction.

11.4 ANALYSIS OF cryoEM DATA

11.4.1 ELECTRON CRYSTALLOGRAPHY

Even though the term *crystallography* suggests that we need to generate good crystals of the ion channels in order to apply electron crystallography, it does not mean that we have to start with crystals diffracting to atomic resolutions. Electron crystallography can deal with crystals that diffract to intermediate resolutions. In fact, many published electron crystallographic studies are limited to 5–10 Å in resolution (Kuhlbrandt 2013). This is largely due to the high-quality phase information carried by the cryoEM images and the recursive real-space alignment procedures that can correct defects in crystalline packing and improve the SNR of the structural factors. However, the difficulty in obtaining well-ordered large crystals is still a significant roadblock. Largely due to the competition of x-ray crystallography and the lack of a reliable high-throughput screen for 2D crystals, only a handful of laboratories are still working on 2D crystals. For particular questions regarding the lipid–protein interactions, electron crystallography of channels in membranes made of different lipids may be arguably a better avenue. Because different techniques are utilized to analyze 2D planar crystals and the 1D helical ones, we will introduce them separately.

11.4.1.1 Analysis of helical assemblies (1D crystalline specimen)

Two different methods are available for analyzing the images of helical filaments. The first one was introduced by Aeron Klug, David DeRosier, and their colleagues in the 1960s (DeRosier and Klug 1968). It takes advantage of the helical arrangement and extracts the structure factors (amplitudes and phases) from the Fourier transform of individual filaments (DeRosier and Moore 1970). The 1D periodicity of the helical repeats leads to the regular layer lines in the reciprocal space. Because the Fourier transform of a helical object can be expressed as the sum of a series of Bessel functions in a polar cylindrical coordinate system, this technique is termed the Fourier–Bessel method. The structural factors from different filaments are compared and averaged when these filaments have the same symmetry and their individual units take the same structure. Within each helical filament, individual units in a helical specimen face different orientations and collectively make a good sampling of the Fourier space (the central section theorem). The assumption of simple symmetry allows the treatment of individual layer lines as the discrete sampling of the 3D lattice across specific directions. Because the Bessel functions represent the cylindrical harmonics and have their maximum at specific locations, it is possible to index the layer lines by examining the maxima along them. The indexing of higher Bessel orders can be further refined by examining the relationship among different lattice points on multiple layer lines. After the correction for out-of-plane tilt, the application of an unbending procedure to correct local distortions, and the correction of CTF, measurements from multiple filaments are put together by shifting them to common phase origins. The amplitudes of the layer lines could then be corrected by summed CTF weights before being combined with the phases. An inverse Fourier–Bessel transform of the layer line data over a range of the three cylindrical coordinates will give rise to the 3D map of the filament. The absolute handedness of the helical reconstructions can be determined by rotating the filaments along an equatorial axis, by imaging a rotationally shadowed filament, or by examining freeze-etched filaments (Diaz et al. 2010). It could become self-evident when the individual alpha helices of the helical subunits in a subnanometer reconstruction are clearly recognized and be compared with their x-ray structures. In practice, the Fourier–Bessel method has been successful in generating multiple 3D reconstructions of 6–30 Å resolutions, and two structures to ~4.0 Å that are good enough for building atomic models.

The second method for analyzing individual filaments is to treat short segments cut out of long ones as single particles. It was inspired by the *projection matching* method used in the real-space analysis of cryoEM images of individual particles (see Section 11.4.1.2) and was developed by Dr. Egelman's group as an alternative to the Fourier–Bessel method (Egelman 2000). It is named the iterative helical real-space refinement (IHRSR). The main idea of IHRSR is to express the helical symmetry in a general sense of rotational angle and axial rise without the need of integers to satisfy the helical selection rule. It starts with a good estimate of the symmetry parameters and then refines them through iterative optimization. With a set of images taken from the helical filaments, short segments are boxed out with successive boxes overlapping by ~90% in order to retrieve most structural information. These segments are then used for reference-based alignment and the refinement of helical symmetry. A 3D volume generated from back-projection is used for a least-squares search of symmetry and, after symmetry imposition, serves as new reference model in the next round of refinement. The convergence to the stable symmetry parameters from initial estimates that are a few degrees off in rotation or 1–2 angstroms away in axial rise is a good sign for robustness of the algorithm.

Starting with featureless cylinder and rough estimates of symmetry parameters, IHRSR can converge to a final solution that is representative of the original data and offers new insights to the target molecules. It thus has the potential to overcome the possible ambiguity in indexing and the related uncertainty in the helical symmetry encountered in the application of the Fourier–Bessel method. In some complicated cases, it may be able to deal with possible Bessel overlaps in different layer lines. More importantly, IHRSR can analyze filaments that are bent in various ways and are not suitable for the Fourier–Bessel method or exhibit variation in symmetry properties. The iterative refinement may allow objective classification of short filaments with different symmetry properties and the generation of different 3D reconstructions from subsets of data. This method has been mainly used by Dr. Egelman and his colleagues, and has recently been applied by other research labs to different problems (Egelman 2007; Pomfret et al. 2007; Sachse et al. 2007; Ge and Zhou 2011). Even though in many cases the resolutions of the final maps may be limited due to imperfect symmetry and local variation in helical filaments, IHRSR is a viable strategy in serving the best structural model for datasets that are not amendable to the Fourier–Bessel method. Because it is not often to see ion channels in helical filaments, this method has not found many applications in the channel field. In principle, the method is applicable to the tubular crystals of ion channels, such as nAchR (Figure 11.2a).

11.4.1.2 Analysis of planar 2D crystals

The analysis of proteins arranged in 2D arrays was pioneered by Drs. R. Henderson and N. Unwin in the 1970s (Henderson and Unwin 1975). It made history by generating the first 3D structure for a protein in membrane, the bacteriorhodopsin (Henderson et al. 1990).

The method takes consideration of the crystalline constraints that keep all the units in the same orientation. The constructive interference of the electrons scattered by these units makes it feasible to extract the structure factors of the target molecules out of the Fourier transforms of their projection images. Because the images contain both amplitude and phase information for the specimens, the analysis usually starts with unbending and merging of images from untilted crystals. After that, low-resolution phase and amplitude information of the images of the titled crystals (5°–30° initially) is introduced to build a reliable 3D map. Such a map can then be extended to subnanometer resolutions by including high-resolution information in images taken from crystals at both low and high tilts. If the crystals are large enough (>1 μ), accurate amplitude data from electron diffraction can be obtained and used to improve the resolution of the electron density map (Henderson et al. 1990).

The best studied channels by electron crystallography are water channels. Different aquaporins have been grown into 2D crystals. The best resolution (1.9 Å) was reached by Dr. T. Walz's group in their study of the aquaporin 0 in 1,2-dimyristoyl-sn-glycero-3-phosphocholine (DMPC) membranes (Gonen et al. 2004). The study for the first time revealed the arrangement of the annular lipid molecules at the protein–lipid interface and suggested that the phosphate groups from the lipids around the protein form a *phosphate belt*. The phosphate belt contributes to the intricate hydrogen-bonding network and stabilizes the protein in membrane (Hite et al. 2010). Other channels studied by electron crystallography will be reviewed in Section 11.5.1.

11.4.2 SINGLE PARTICLE RECONSTRUCTION FROM cryoEM IMAGES OF INDIVIDUAL ION CHANNELS

Single particle reconstruction assumes that all individual molecules take a similar structure so that images taken from the molecules at different orientations can be aligned, averaged, and put together to generate a genuine 3D structure (Verschoor et al. 1984). As we explained earlier, the EM images are projections of the 3D object from different orientations (or views). The reliability of individual views depends on how well the structural information is preserved and how much noise there is. Single particle analysis overcomes noise problem by putting together similar views and averaging them into class sums to enhance SNR. Large datasets are thus needed to improve the resolution of the final reconstruction.

For a small protein like a 300 kDa eukaryotic voltage-gated potassium channel, a million or so images of individual channels would be needed to approach atomic resolution if we assume that the channels are conformationally homogeneous. With the capability of high-resolution imaging, the availability of high-throughput data collection and analysis, and new algorithms for particle classification, alignment, and reconstruction as well as for variance analysis and objective separation of images belonging to different conformational states (Penczek et al. 2006a,b; Scheres et al. 2007a,b), single particle analysis of large biological complexes is seeing a broad spectrum of applications, and will provide important insights at molecular levels on the biological functions of these complexes. Atomic resolution structures have been achieved for the high-symmetry well-packed complexes, such as the viral particles or tight filamentous assemblies (Yonekura et al. 2003; Yu et al. 2008). Subnanometer resolution is becoming reachable for many of the low-symmetry complexes that do not suffer from sever intrinsic flexibility. Ion channels and their complexes that are at least 200 kDa in mass fall into this latter category. In addition, for ion channels that are smaller than 200 kDa, it may be feasible to introduce Fab fragments that bind to the channel with a restrained orientation (conformation-specific), increase the size of the complexes, and make them suitable for cryoEM analysis (Jiang et al. 2004).

We will describe the general flow of a typical project so that the readers will get a good idea about implementing it to a new protein, probably an ion channel. A project starts with a decent biochemical preparation of the target molecules. For ion channels and membrane proteins in general, significant efforts need to be invested in verifying that the biochemical conditions are suitable to keep the proteins in good monodispersity and in physiologically relevant states. The need of only a small amount of proteins to do single particle analysis does not relax the stringent requirements for well-behaving specimens.

Once the samples are biochemically ready, it is usually a good practice to look at them by negative-stain EM. Selection of the right stain solution and inclusion of small amount detergents in the stain solution may be critical in keeping the membrane proteins in good conditions. For proteins that do not have symmetry, it is a good idea to obtain the reference model by random conical tilt reconstruction or by electron tomography (Radermacher et al. 1987). For datasets that exhibit strong twofold or higher symmetry, common-line-based orientation determination (e.g., angular reconstitution implemented in IMAGIC) (van Heel 1987) can provide good starting reference models. The random conical tilt reconstruction starts with paired images from the same area on a grid at two different tilt angles, for example, 0° and 65°. Multiple pairs of particles are selected. The datasets from 0° images are classified (van Heel 1984). The images belonging to the same class are pooled together, and their corresponding images in the highly tilted images are selected to build a 3D volume. Different volumes can be calculated from different classes. These 3D volumes can be aligned and classified to generate one or a few discrete 3D models that will serve as the starting references.

Similarly, electron tomography can be performed on negatively stained samples to obtain low-resolution 3D volumes for individual particles. Images are taken from the same area with the grids tilted from −65° to +65°. The tilt angles smaller than 25° can be sampled in larger steps (4°–5°). Those above 25° are taken with small increments (1°–2°). The dose for each exposure is controlled so that the total dose will be tolerated by the specimen. Images are aligned and put together to generate the 3D volume. The 3D densities for individual particles are then taken out, and can be classified and averaged to provide a reliable first glimpse of the target molecule.

With good starting references, we can move onto the collection and analysis of cryoEM data. For an ion channel, the single particle data can come from channels in detergents or in small vesicles. The channels in vesicles need to be first separated from the densities of the vesicle membranes before they can be analyzed using the regular algorithm (Jiang et al. 2001; Wang and Sigworth 2009). When the signal from individual channel particles is strong, the datasets are subjected to reference-free classification and statistical analysis. Eigenimages out of the statistical analysis give a good estimate of symmetry. Common-line-based angular assignment can be utilized to build 3D models out of the high SNR class averages. The data can also be analyzed by using the reference models obtained from the negative-stain data.

From a Bayesian point of view, all available algorithms for single particle analysis search for the most probable model(s) to represent the 2D projection data (Sigworth 1998). Such a probability is enhanced when the 3D model is compared with the projection images by maximizing the cross-correlation coefficient, minimizing the weighted phase residual, decreasing weighted geometrical distance, etc. Starting with a reference model and a large dataset of individual images, we will refine the model by calculating its projections along all possible asymmetrical orientations, comparing every EM image with these projections to find the best matching projection and assign the orientation angle of the latter to the EM image, and selecting a good fraction of the matched EM images to calculate a new 3D model for the next refinement. The convergence of the model is reached when no further improvement can be seen.

But this convergence does not guarantee the correctness of the final 3D reconstruction. We usually need to analyze the map and evaluate if there is significant variability intrinsic to the raw data. During the refinement process, the distribution of the EM images into the classes represented by reference projections could be evaluated to recognize if certain orientations are preferred. The number of EM images at overpopulated orientations must be cut down in the final 3D reconstruction in order to avoid potential distortions in the final map. The variances among the particles in each orientational class are informative in recognizing significant structural variability in certain parts of the 3D map, and they can be used to calculate a 3D variance map. Similarly, a 3D covariance map can be used to evaluate the concerted motion between different mobile parts (Zhang et al. 2008a). Alternatively, the maximum-likelihood-based classification and multireference analysis may allow the original dataset to be divided into several subsets, each of which will yield the 3D reconstruction of a specific conformational state (Scheres et al. 2007a). The analysis of one macromolecular complex in distinct conformations is still a challenging problem for single particle analysis, but endows significant power to this method.

Once the maps are refined to stable solutions, the dataset can be divided into two separate sets for calculating the Fourier shell correlation. Based on the properties of the scattering power of the molecules and the setting in the microscope, the 3D map could be sharpened to enhance the high-resolution information. Depending on the resolution, the map interpretation needs to be done in conjunction with other information about the target molecules. Three-dimensional maps that do not reach 10 Å resolution can only give limited information about the domain organization of the individual subunits. When the map is at subnanometer resolution, some alpha helices will be visible. X-ray structures of individual domains of the protein are usually the best we can use to recognize specific parts of the molecules in these 3D maps. The interpretation of the low- or intermediate-resolution maps therefore relies on the available high-resolution structural models of individual domains by other methods. Additionally, bioinformatics analysis of different sequences, structural comparison and prediction based on the known structural folds, or distance measurements by electron paramagnetic resonance (EPR), double electron-electron resonance (DEER), fluorescence resonance energy transfer (FRET), or luminescence (or lanthanide-based) resonance energy transfer (LRET), mapping of neighboring sites by chemical cross-linking/mass spectrometry analysis, etc., can be combined with the 3D reconstructions in order to offer structural insights to the target molecules.

With the development of stable high-end cryoEM, high-sensitivity large-dimension direct electron detectors, automatic data collection and particle selection, and better algorithms to deal with conformational and/or compositional heterogeneity, the analysis of ion channels by single particle reconstruction will likely generate structural models in higher resolutions and allow the analysis of ion channels in different gating states.

11.4.3 ELECTRON TOMOGRAPHY

As explained earlier, the most difficult part of electron tomography lies in specimen preparation. For eukaryotic cells and tissues, cryo-sectioning is inevitable. The technical requirements for high-pressure freezing, free substitution, cryo-sectioning, etc., are achieved with special devices. We will not get into details here. Interested readers could refer to more recent reviews on these aspects (Gruska et al. 2008; Lucic et al. 2008).

11.5 TYPICAL RESULTS OF cryoEM STUDIES OF ION CHANNELS

With the introductory understanding of different cryoEM techniques, next we will examine some of the published cryoEM studies of ion channels. Due to technical limitations, most of the channel structures are in the low-resolution range (8–20 Å). Only a handful of structures by electron crystallography have provided atomic details.

11.5.1 ELECTRON CRYSTALLOGRAPHIC STUDIES OF ION CHANNELS

Following the structural studies of bacteriorhodopsin, light-harvesting complex, and water channels, a new frontier is evolved in understanding how lipids are packed against membrane proteins and affect their structures and functions (Jiang and Gonen 2012). For crystallographic studies, the limitation has always been the lack of high-quality crystals. We have seen crystals of nAchR (nicotinic acetylcholine receptor), gap junction channels (connexins), Kv1.2, KcsA, bacterial inward rectifying potassium channel (IRK), CFTR (cystic fibrosis transmembrane conductance regulator; in detergents), bacterial Nav, KvAP, MlotiK, and *Escherichia coli* ClC0 (Li et al. 1998; Unger et al. 1999; Mindell et al. 2001; Parcej and Eckhardt-Strelau 2003; Clayton et al. 2009; Ford et al. 2011; Shi et al. 2013). All these crystals diffract to better than 10 Å resolution. The structures of nAchR, connexin 26, and IRK have yielded 3D maps. Because of the x-ray structures resolved for Kv1.2, connexin 26, KcsA, bacterial Nav, IRK, and ClC0 at atomic resolutions, the electron crystallographic studies of these proteins were outcompeted. It is generally believed that if a protein can be crystallized in 2D, there is a high probability to grow 3D crystals. nAchR is the only one in the preceding list that has not been crystallized well in 3D, probably because of the tubular nature of its crystals. This trend in practice has made the 2D crystallography outpaced by X-ray crystallography. However, there is now mounting evidence suggesting that lipid–protein interaction may be a critical factor for the gating of multiple types of ion channels (Lee 2004, 2005; Powl et al. 2005; Jiang and Gonen 2012). Three-dimensional structures of these channels in membrane will be needed for us to understand the lipid effects.

The study of the nAchR tubular crystals by N. Unwin and colleagues revealed the structural differences between the ligand-free and ligand-bound receptors (Unwin and Fujiyoshi 2012). The fast spray of acetylcholine was introduced to trap the receptor in the open state before desensitization kicked in (Figure 11.2b and c). High-quality images from the liquid-helium cooled specimens were selected based on the quality of high-resolution layer lines. The unbending of the filaments and the correction of out-of-plane tilting were found to be essential. Lastly, the amplitude correction was facilitated by the 3D x-ray model of the pentameric acetylcholine-binding protein (Unwin 2005; Yonekura and Toyoshima 2007). These technical tricks led to the quantum leap that allows the calculations of the 3D maps at ~4.0 Å for the ligand-free receptor or at ~6Å for the ligand-bound open state of the receptor (Figure 11.2b and c). When the two structures were compared, it was found that the low-affinity acetylcholine-binding site at the interface between the alpha and gamma subunits is the critical point for driving the conformational changes in the extracellular domain (Unwin and Fujiyoshi 2012). The extracellular domain of the beta subunit moves outward by ~1 Å, which is coupled to the motion of its four transmembrane helices as well as the lateral motion of the M2 helices in the alpha(gamma) and delta subunits. The reorganization of the transmembrane helices widens the central part of the pore from ~6 to ~7 Å, shifts the narrowest part of the pore to the intracellular membrane surface, and increases the volume of the aqueous pore significantly. All these changes allow more water molecules to enter the pore and level off the energy barrier for hydrated cations to permeate through the pore. Even though an atomic structure for the open state is still not available, the structural insights from this study cannot be obtained from the x-ray structures of the bacterial homopentameric homologs.

The electron crystallographic analysis of the gap junction channels (Connexin, Cx26) was initiated in Mark Yeager's laboratory while he was at the Scripps Research Institute (Unger et al. 1999; Yeager and Harris 2007). The 3D map at 7.6 Å in-plane resolution revealed the arrangement of alpha helices in the transmembrane domains and the well-interdigitated extracellular domains that are responsible for forming the intercell connection and the long broad channel across the two apposed membranes (Figure 11.3a). Technically, the hexagonal lattice and the thickness of the sample due to double lipid bilayers was a challenge in the elucidation of the 3D structure. Cx26 was later resolved by X-ray crystallography to atomic resolution (Figure 11.3b). Comparison of the X-ray structure and the EM density showed almost the same arrangement, both of which were assigned to the open conformation (Maeda et al. 2009). The electron crystallographic structure of the Cx26 M34A mutant, a closed-pore mutant

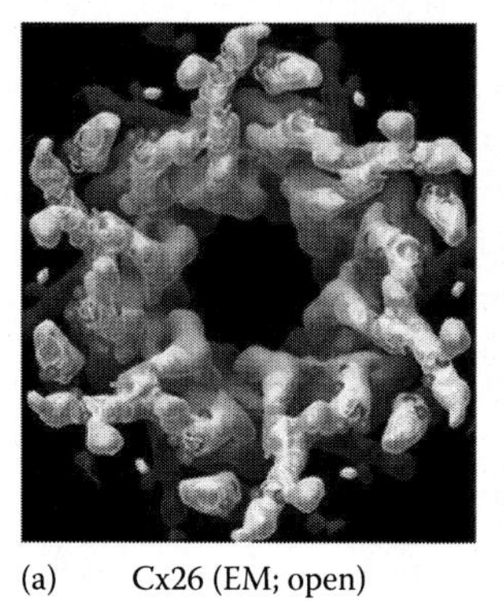
(a) Cx26 (EM; open)

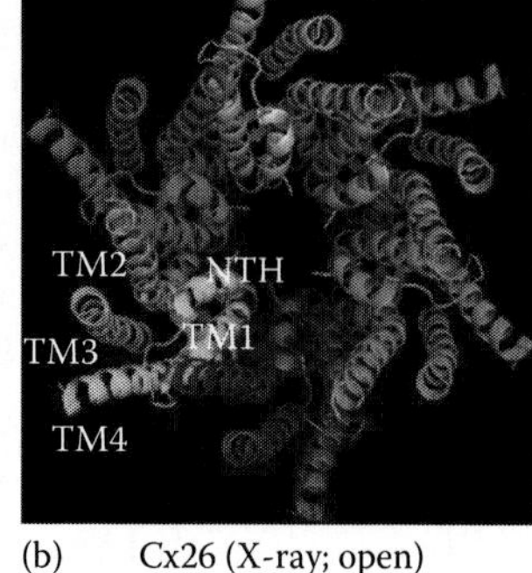

(b) Cx26 (X-ray; open)

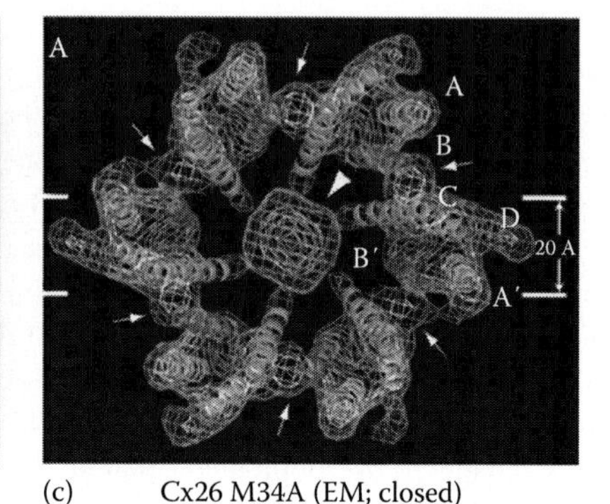

(c) Cx26 M34A (EM; closed)

Figure 11.3 Electron crystallographic study of gap junction channels. (a) Helical arrangement deduced from the 3D map of Cx26 at 7.6 Å in-plane resolution. (b) Helical arrangement determined by x-ray is very similar to that in (a). (c) Structure of the Cx26 M34A mutant in the closed state. The central density was proposed to be the gating plug.

(Figure 11.3c), suggested a plug formed by the N-terminal helices, which might serve as the gate (Oshima et al. 2007).

Electron crystallographic analysis of other ion channels will not be discussed here. It is almost certainly true that many questions related to the lipid effects on ion channels will benefit from their high-resolution structures in membranes. For example, to elucidate the structural basis of the lipid-dependent gating in voltage-gated ion channels, we are in great need of two structures of the same channel in different membranes.

11.5.2 SINGLE PARTICLE ANALYSIS OF ION CHANNELS

The first ion channel studied by single particle EM was type 1 RyR (Radermacher et al. 1992). The RyR is about 2.2 MDa in mass, clearly visible in cryoEM (Figure 11.4a). Its large cytosolic domain forms a square platform (the *foot* structure; Figure 11.4b), where other proteins, such as the L-type voltage-gated calcium channels, can interact and modulate the channel function. The recent cryoEM map at ~10 Å resolution had P4-symmetry imposed and was assumed to be in the closed state (Samso et al. 2005; Serysheva et al. 2008). It harbors a transmembrane domain that is much larger than the membrane integral part of a tetrameric voltage-gated potassium channel and can accommodate more than 10 transmembrane segments per subunit. The computational search identified five potential helices (Figure 11.4c), one of which (helix 1) was proposed to line the pore with a significant kink in the middle (Serysheva et al. 2008). The resolution of the transmembrane domain must be worse than the cytosolic domain, because most of the transmembrane alpha helices are not resolved. At the current resolution, it is unclear how the direct coupling controls the opening of the RyR, nor does it provide enough information to resolve the calcium-induced conformational changes in the foot structure or the pore domain. The study of RyR has evolved for more than 20 years by cryoEM. A lot of the questions about its pore structure and ligand-dependent gating remain unanswered. There is a strong need for whole-channel maps at a better resolution, maps of recombinant proteins containing landmark modifications, or high-resolution x-ray models of specific domains that can be recognized in the 3D maps. The main limits appear to be the intrinsic heterogeneity in the molecules. A high-yield recombinant expression system and the application of more advanced techniques in classifying a large dataset into different subsets may be able to improve the resolution to 5–6 Å.

As a different example of single particle analysis, we will discuss the studies of six-transmembrane segment Kv-like channels (Figure 11.5). The size of these proteins is 200–400 kDa, close to the detection limit in a modern cryoEM without a phase plate. The voltage-gated ion channels, the cyclic nucleotide-gated (CNG) and the transient receptor potential (TRP) channels, all fall into this category. Most of the published work was done by negative-stain EM. We will only talk about the three cryoEM studies in literature.

The best resolution so far achieved for such small channels in detergents was 10.5 Å for the KvAP/Fab complex (Figure 11.5a) imaged under cryo-negative EM condition (Jiang et al. 2004). Four Fab fragments were used to increase the mass of individual

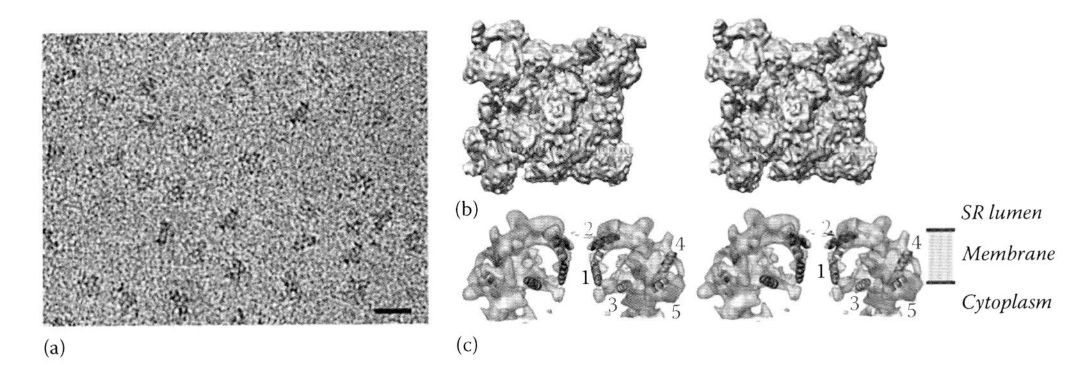

Figure 11.4 Single particle analysis of the massive RyR. (a) cryoEM image of RyRs. (b) 3D map in stereo view from an oblique angle. (c) Stereo view of the putative five helices in the transmembrane domain four.

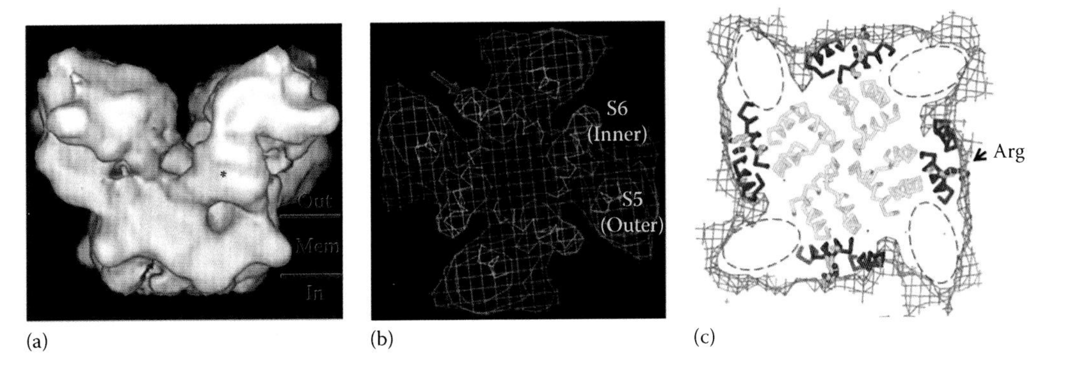

Figure 11.5 Single particle analysis of the small-sized Kv channel. (a) 3D map at 10.5 Å. (b) Features at the intracellular surface showing the docking of the four termini of the S6 helices into the densities. (c) Arrangement of the voltage sensor paddle (red) against the pore. The red oval represents the S1/S2 and the ball-model (arrow) is the second Arg residues.

particles, a scheme that was developed by others into a general method for analyzing small complexes. Interestingly, the flexibility between the two Ig domains of each Fab fragment did not prevent us from reaching good enough resolutions in the transmembrane domain. The density map showed the features on the intracellular surface of the pore domain that match the C-terminal ends of four S6 transmembrane helices (Figure 11.5b). The arrangement of the voltage sensors and the pore domain demonstrated that an S4 helix may have its first two Arg residues facing the lipid environment in the inactivated state (Figure 11.5c). This study opened the new direction for us to study the role of lipid–protein interaction in the gating of the voltage-gated ion channels.

The cryoEM maps of the voltage-gated Na channel from the *Torpedo* ray presynaptic membranes and the skeletal muscle L-type calcium channels (Cav1.1; dihydropyridin receptor DHPR) were resolved at nominal resolutions of 19 and 23 Å, respectively (Sato et al. 2001; Wang et al. 2002; Wolf et al. 2003). Many features in the maps are not well resolved such that the interpretation of the subunits and the assignment of different domains carry significant uncertainty. It is possible that both channels have multiple subunits and may suffer from significant compositional heterogeneity. Perhaps, a better preparation of these channels from a recombinant system will allow the cryoEM study to reach close to 10 Å resolutions in the future.

Last, we want to point out that many of the published single particle reconstructions of ion channels in detergents did not generate enough details or make well-grounded testable predictions. There is apparently a lot more to do in all these cases. With the technology advancement and the improvement in the biochemical preparation of these specimens, the structural studies of ion channels in detergents or in membrane will generate more and more useful information that will pave the way toward structural details and testable predictions.

11.6 GENERAL DISCUSSIONS ON THE APPLICATION OF cryoEM TO ION CHANNELS

Both electron crystallography and single particle cryoEM have made significant contributions to our understanding of various types of ion channels. However, until now, we only have very few cases where cryoEM studies have made conceptual advancement in specific directions. Considering the progress made in the past decade and the improvement of the resolution for high-symmetry objects from 7.0 Å in 1992 to 2.88 Å in 2011, we have a very good reason to be optimistic that cryoEM studies of ion channels will continue to advance so that eventually, it will be feasible to generate near-atomic resolution structures of ion channels in different gating conformations. A near-atomic resolution structure of TRPV1 channel was resolved by single particle cryoEM (Cao et al. 2013; Liao et al. 2013) while this chapter was in press.

ACKNOWLEDGMENTS

Research in Q-X Jiang's laboratory is supported by grants from NIH, CPRIT, AHA, and Welch Foundation.

REFERENCES

Baldwin, J.M., R. Henderson, E. Beckman, and F. Zemlin. 1988. Images of purple membrane at 2.8 Å resolution obtained by cryo-electron microscopy. *J Mol Biol* 202(3):585–591.

Campbell, M.G., A. Cheng, A.F. Brilot, A. Moeller, D. Lyumkis, D. Veesler, J. Pan et al. 2012. Movies of ice-embedded particles enhance resolution in electron cryo-microscopy. *Structure* 20(11):1823–1828.

Cao, E., M. Liao, Y. Cheng, and D. Julius. 2013. TRPV1 structures in distinct conformations reveal activation mechanisms. *Nature* 504(7478):113–118.

Clayton, G.M., S.G. Aller, J. Wang, V. Unger, and J.H. Morais-Cabral. 2009. Combining electron crystallography and x-ray crystallography to study the MlotiK1 cyclic nucleotide-regulated potassium channel. *J Struct Biol* 167(3):220–226.

Cortese, K., G. Vicidomini, M.C. Gagliani, P. Boccacci, A. Diaspro, and C. Tacchetti. 2013. High data output method for 3-D correlative light-electron microscopy using ultrathin cryosections. *Methods Mol Biol* 950:417–437.

Cyrklaff, M., M. Auer, W. Kuhlbrandt, and G.A. Scarborough. 1995. 2-D structure of the Neurospora crassa plasma membrane ATPase as determined by electron cryomicroscopy. *EMBO J* 14(9):1854–1857.

da Fonseca, P.C., S.A. Morris, E.P. Nerou, C.W. Taylor, and E.P. Morris. 2003. Domain organization of the type 1 inositol 1,4,5-trisphosphate receptor as revealed by single-particle analysis. *Proc Natl Acad Sci USA* 100(7):3936–3941.

Dang, T.X., R.A. Milligan, R.K. Tweten, and E.M. Wilson-Kubalek. 2005. Helical crystallization on nickel-lipid nanotubes: Perfringolysin O as a model protein. *J Struct Biol* 152(2):129–139.

DeRosier, D.J. and A. Klug. 1968. Reconstruction of three dimensional structures from electron micrographs. *Nature* 217(5124):130–134.

DeRosier, D.J. and P.B. Moore. 1970. Reconstruction of three-dimensional images from electron micrographs of structures with helical symmetry. *J Mol Biol* 52(2):355–369.

Diaz, R., W.J. Rice, and D.L. Stokes. 2010. Fourier-Bessel reconstruction of helical assemblies. *Methods Enzymol* 482:131–165.

Dubochet, J. 1995. High-pressure freezing for cryoelectron microscopy. *Trends Cell Biol* 5(9):366–368.

Dubochet, J. 2007. The physics of rapid cooling and its implications for cryoimmobilization of cells. *Methods Cell Biol* 79:7–21.

Dubochet, J., B. Zuber, M. Eltsov, C. Bouchet-Marquis, A. Al-Amoudi, and F. Livolant. 2007. How to "read" a vitreous section. *Methods Cell Biol* 79:385–406.

Egelman, E.H. 2000. A robust algorithm for the reconstruction of helical filaments using single-particle methods. *Ultramicroscopy* 85(4):225–234.

Egelman, E.H. 2007. Single-particle reconstruction from EM images of helical filaments. *Curr Opin Struct Biol* 17(5):556–561.

Ford, R.C., J. Birtley, M.F. Rosenberg, and L. Zhang. 2011. CFTR three-dimensional structure. *Methods Mol Biol* 741:329–346.

Frank, J. 2001a. Cryo-electron microscopy as an investigative tool: The ribosome as an example. *Bioessays* 23(8):725–732.

Frank, J. 2001b. Ribosomal dynamics explored by cryo-electron microscopy. *Methods* 25(3):309–315.

Frank, J. 2002. Single-particle imaging of macromolecules by cryo-electron microscopy. *Annu Rev Biophys Biomol Struct* 31:303–319.

Frank, J., T. Wagenknecht, B.F. McEwen, M. Marko, C.E. Hsieh, and C.A. Mannella. 2002. Three-dimensional imaging of biological complexity. *J Struct Biol* 138(1–2):85–91.

Fujiyoshi, Y. 1998. The structural study of membrane proteins by electron crystallography. *Adv Biophys* 35:25–80.

Ge, P. and Z.H. Zhou. 2011. Hydrogen-bonding networks and RNA bases revealed by cryo electron microscopy suggest a triggering mechanism for calcium switches. *Proc Natl Acad Sci USA* 108(23):9637–9642.

Gonen, T. 2013. The collection of high-resolution electron diffraction data. *Methods Mol Biol* 955:153–169.

Gonen, T., Y. Cheng, P. Sliz, Y. Hiroaki, Y. Fujiyoshi, S.C. Harrison, and T. Walz. 2006. Lipid-protein interactions in double-layered two-dimensional AQP0 crystals. *Nature* 438(7068):633–638.

Gonen, T., P. Sliz, J. Kistler, Y. Cheng, and T. Walz. 2004. Aquaporin-0 membrane junctions reveal the structure of a closed water pore. *Nature* 429(6988):193–197.

Gruska, M., O. Medalia, W. Baumeister, and A. Leis. 2008. Electron tomography of vitreous sections from cultured mammalian cells. *J Struct Biol* 161(3):384–392.

Gyobu, N., K. Tani, Y. Hiroaki, A. Kamegawa, K. Mitsuoka, and Y. Fujiyoshi. 2004. Improved specimen preparation for cryo-electron microscopy using a symmetric carbon sandwich technique. *J Struct Biol* 146(3):325–333.

Henderson, R., J.M. Baldwin, T.A. Ceska, F. Zemlin, E. Beckmann, and K.H. Downing. 1990. Model for the structure of bacteriorhodopsin based on high-resolution electron cryo-microscopy. *J Mol Biol* 213(4):899–929.

Henderson, R. and P.N. Unwin. 1975. Three-dimensional model of purple membrane obtained by electron microscopy. *Nature* 257(5521):28–32.

Heuser, J. 2005. Deep-etch EM reveals that the early poxvirus envelope is a single membrane bilayer stabilized by a geodetic "honeycomb" surface coat. *J Cell Biol* 169(2):269–283.

Heuser, J.E. 2011. The origins and evolution of freeze-etch electron microscopy. *J Electron Microsc* 60(1):S3–S29.

Hite, R.K., Z. Li, and T. Walz. 2010. Principles of membrane protein interactions with annular lipids deduced from aquaporin-0 2D crystals. *EMBO J* 29(10):1652–1658.

Jensen, G.J. and A. Briegel. 2007. How electron cryotomography is opening a new window onto prokaryotic ultrastructure. *Curr Opin Struct Biol* 17(2):260–267.

Jiang, Q.X., D.W. Chester, and F.J. Sigworth. 2001. Spherical reconstruction: A method for structure determination of membrane proteins from cryo-EM images. *J Struct Biol* 133(2–3):119–131.

Jiang, Q.X. and T. Gonen. 2012. The influence of lipids on voltage-gated ion channels. *Curr Opin Struct Biol* 22(4):529–536.

Jiang, Q.X., E.C. Thrower, D.W. Chester, B.E. Ehrlich, and F.J. Sigworth. 2002. Three-dimensional structure of the type 1 inositol 1,4,5-trisphosphate receptor at 24 Å resolution. *EMBO J* 21(14):3575–3581.

Jiang, Q.X., D.N. Wang, and R. MacKinnon. 2004. Electron microscopic analysis of KvAP voltage-dependent K^+ channels in an open conformation. *Nature* 430(7001):806–810.

Kuhlbrandt, W. 2013. Introduction to electron crystallography. *Methods Mol Biol* 955:1–16.

Lee, A.G. 2004. How lipids affect the activities of integral membrane proteins. *Biochim Biophys Acta* 1666(1–2):62–87.

Lee, A.G. 2005. Lipid sorting: Lipids do it on their own. *Curr Biol* 15(11):R421–R423.

Liao, M., E. Cao, D. Julius, and Y. Cheng. 2013. Structure of the TRPV1 ion channel determined by electron cryo-microscopy. *Nature* 504(7478):107–112.

Li, H.L., H.X. Sui, S. Ghanshani, S. Lee, P.J. Walian, C.I. Wu, K.G. Chandy, and B.K. Jap. 1998. Two-dimensional crystallization and projection structure of KcsA potassium channel. *J Mol Biol* 282(2):211–216.

Lucic, V., A. Leis, and W. Baumeister. 2008. Cryo-electron tomography of cells: Connecting structure and function. *Histochem Cell Biol* 130(2):185–196.

Ludtke, S.J., T.P. Tran, Q.T. Ngo, V.Y. Moiseenkova-Bell, W. Chiu, and I.I. Serysheva. 2011. Flexible architecture of IP3R1 by Cryo-EM. *Structure* 19(8):1192–1199.

Maeda, S., S. Nakagawa, M. Suga, E. Yamashita, A. Oshima, Y. Fujiyoshi, and T. Tsukihara. 2009. Structure of the connexin 26 gap junction channel at 3.5 Å resolution. *Nature* 458(7238):597–602.

Milazzo, A.C., A. Cheng, A. Moeller, D. Lyumkis, E. Jacovetty, J. Polukas, M.H. Ellisman, N.H. Xuong, B. Carragher, and C.S. Potter. 2011. Initial evaluation of a direct detection device detector for single particle cryo-electron microscopy. *J Struct Biol* 176(3):404–408.

Mindell, J.A., M. Maduke, C. Miller, and N. Grigorieff. 2001. Projection structure of a ClC-type chloride channel at 6.5 Å resolution. *Nature* 409(6817):219–223.

Mironov, A.A. and G.V. Beznoussenko. 2013. Correlative microscopy. *Methods Cell Biol* 113:209–255.

Miyazawa, A., Y. Fujiyoshi, M. Stowell, and N. Unwin. 1999. Nicotinic acetylcholine receptor at 4.6 Å resolution: Transverse tunnels in the channel wall. *J Mol Biol* 288(4):765–786.

Miyazawa, A., Y. Fujiyoshi, and N. Unwin. 2003. Structure and gating mechanism of the acetylcholine receptor pore. *Nature* 423 (6943):949–955.

Nagayama, K. 2008. Development of phase plates for electron microscopes and their biological application. *Eur Biophys J* 37(4):345–358.

Nagayama, K. and R. Danev. 2008. Phase contrast electron microscopy: Development of thin-film phase plates and biological applications. *Philos Trans R Soc Lond B Biol Sci* 363(1500):2153–2162.

Oshima, A., K. Tani, Y. Hiroaki, Y. Fujiyoshi, and G.E. Sosinsky. 2007. Three-dimensional structure of a human connexin26 gap junction channel reveals a plug in the vestibule. *Proc Natl Acad Sci USA* 104(24):10034–10039.

Pantelic, R.S., J.W. Suk, C.W. Magnuson, J.C. Meyer, P. Wachsmuth, U. Kaiser, R.S. Ruoff, and H. Stahlberg. 2011. Graphene: Substrate preparation and introduction. *J Struct Biol* 174(1):234–238.

Parcej, D.N. and L. Eckhardt-Strelau. 2003. Structural characterisation of neuronal voltage-sensitive K^+ channels heterologously expressed in *Pichia pastoris*. *J Mol Biol* 333(1):103–116.

Penczek, P.A., J. Frank, and C.M. Spahn. 2006a. A method of focused classification, based on the bootstrap 3D variance analysis, and its application to EF-G-dependent translocation. *J Struct Biol* 154(2):184–194.

Penczek, P.A., C. Yang, J. Frank, and C.M. Spahn. 2006b. Estimation of variance in single-particle reconstruction using the bootstrap technique. *J Struct Biol* 154(2):168–183.

Polishchuk, E.V., R.S. Polishchuk, and A. Luini. 2013. Correlative light-electron microscopy as a tool to study in vivo dynamics and ultrastructure of intracellular structures. *Methods Mol Biol* 931:413–422.

Pomfret, A.J., W.J. Rice, and D.L. Stokes. 2007. Application of the iterative helical real-space reconstruction method to large membranous tubular crystals of P-type ATPases. *J Struct Biol* 157(1):106–116.

Powl, A.M., J. Carney, P. Marius, J.M. East, and A.G. Lee. 2005. Lipid interactions with bacterial channels: Fluorescence studies. *Biochem Soc Trans* 33(Pt. 5):905–909.

Radermacher, M. 1994. Three-dimensional reconstruction from random projections: Orientational alignment via Radon transforms. *Ultramicroscopy* 53(2):121–136.

Radermacher, M., V. Rao, R. Grassucci, J. Frank, A.P. Timerman, S. Fleischer, and T. Wagenknecht. 1994. Cryo-electron microscopy and three-dimensional reconstruction of the calcium release channel/ryanodine receptor from skeletal muscle. *J Cell Biol* 127(2):411–423.

Radermacher, M., T. Wagenknecht, R. Grassucci, J. Frank, M. Inui, C. Chadwick, and S. Fleischer. 1992. Cryo-EM of the native structure of the calcium release channel/ryanodine receptor from sarcoplasmic reticulum. *Biophys J* 61(4):936–940.

Radermacher, M., T. Wagenknecht, A. Verschoor, and J. Frank. 1987. Three-dimensional reconstruction from a single-exposure, random conical tilt series applied to the 50S ribosomal subunit of *Escherichia coli*. *J Microsc* 146:113–136.

Sachse, C., J.Z. Chen, P.D. Coureux, M.E. Stroupe, M. Fandrich, and N. Grigorieff. 2007. High-resolution electron microscopy of helical specimens: A fresh look at tobacco mosaic virus. *J Mol Biol* 371(3):812–835.

Salje, J., B. Zuber, and J. Lowe. 2009. Electron cryomicroscopy of *E. coli* reveals filament bundles involved in plasmid DNA segregation. *Science* 323(5913):509–512.

Samso, M., T. Wagenknecht, and P.D. Allen. 2005. Internal structure and visualization of transmembrane domains of the RyR1 calcium release channel by cryo-EM. *Nat Struct Mol Biol* 12(6):539–544.

Sato, C., K. Hamada, T. Ogura, A. Miyazawa, K. Iwasaki, Y. Hiroaki, K. Tani, A. Terauchi, Y. Fujiyoshi, and K. Mikoshiba. 2004. Inositol 1,4,5-trisphosphate receptor contains multiple cavities and L-shaped ligand-binding domains. *J Mol Biol* 336(1):155–164.

Sato, C., Y. Ueno, K. Asai, K. Takahashi, M. Sato, A. Engel, and Y. Fujiyoshi. 2001. The voltage-sensitive sodium channel is a bell-shaped molecule with several cavities. *Nature* 409(6823):1047–1051.

Scheres, S.H., H. Gao, M. Valle, G.T. Herman, P.P. Eggermont, J. Frank, and J.M. Carazo. 2007a. Disentangling conformational states of macromolecules in 3D-EM through likelihood optimization. *Nat Methods* 4(1):27–29.

Scheres, S.H., R. Nunez-Ramirez, Y. Gomez-Llorente, C. San Martin, P.P. Eggermont, and J.M. Carazo. 2007b. Modeling experimental image formation for likelihood-based classification of electron microscopy data. *Structure* 15(10):1167–1177.

Serysheva, I.I, D.J. Bare, S.J. Ludtke, C.S. Kettlun, W. Chiu, and G.A. Mignery. 2003. Structure of the type 1 inositol 1,4,5-trisphosphate receptor revealed by electron cryomicroscopy. *J Biol Chem* 278(24):21319–21322.

Serysheva, I.I. and S.J. Ludtke. 2010. 3D Structure of IP(3) receptor. *Curr Top Membr* 66:171–189.

Serysheva, I.I, S.J. Ludtke, M.L. Baker, Y. Cong, M. Topf, D. Eramian, A. Sali, S.L. Hamilton, and W. Chiu. 2008. Subnanometer-resolution electron cryomicroscopy-based domain models for the cytoplasmic region of skeletal muscle RyR channel. *Proc Natl Acad Sci USA* 105(28):9610–9615.

Shi, L., H. Zheng, H. Zheng, B.A. Borkowski, D. Shi, T. Gonen, and Q.X. Jiang. 2013. Voltage sensor ring in a native structure of a membrane-embedded potassium channel. *Proc Natl Acad Sci USA* 110(9):3369–3374.

Sigworth, F.J. 1998. A maximum-likelihood approach to single-particle image refinement. *J Struct Biol* 122(3):328–339.

Steven, A.C. and U. Aebi. 2003. The next ice age: Cryo-electron tomography of intact cells. *Trends Cell Biol* 13(3):107–110.

Stokes, D.L., I. Ubarretxena-Belandia, T. Gonen, and A. Engel. 2013. High-throughput methods for electron crystallography. *Methods Mol Biol* 955:273–296.

Taylor, K.A. and R.M. Glaeser. 1974. Electron diffraction of frozen, hydrated protein crystals. *Science* 186(4168):1036–1037.

Unger, V.M., N.M. Kumar, N.B. Gilula, and M. Yeager. 1999. Three-dimensional structure of a recombinant gap junction membrane channel. *Science* 283(5405):1176–1180.

Unwin, N. 2003. Structure and action of the nicotinic acetylcholine receptor explored by electron microscopy. *FEBS Lett* 555(1):91–95.

Unwin, N. 2005. Refined structure of the nicotinic acetylcholine receptor at 4 Å resolution. *J Mol Biol* 346(4):967–989.

Unwin, N. and Y. Fujiyoshi. 2012. Gating movement of acetylcholine receptor caught by plunge-freezing. *J Mol Biol* 422(5):617–634.

Unwin, P.N. and R. Henderson. 1975. Molecular structure determination by electron microscopy of unstained crystalline specimens. *J Mol Biol* 94(3):425–440.

Uzgiris, E.E. and R.D. Kornberg. 1983. Two-dimensional crystallization technique for imaging macromolecules, with application to antigen–antibody–complement complexes. *Nature* 301(5896):125–129.

van Heel, M. 1984. Multivariate statistical classification of noisy images (randomly oriented biological macromolecules). *Ultramicroscopy* 13(1–2):165–183.

van Heel, M. 1987. Angular reconstitution: A posteriori assignment of projection directions for 3D reconstruction. *Ultramicroscopy* 21(2):111–123.

Verschoor, A., J. Frank, M. Radermacher, T. Wagenknecht, and M. Boublik. 1984. Three-dimensional reconstruction of the 30 S ribosomal subunit from randomly oriented particles. *J Mol Biol* 178(3):677–698.

Walz, T. and N. Grigorieff. 1998. Electron crystallography of two-dimensional crystals of membrane proteins. *J Struct Biol* 121(2):142–161.

Wang, L. and F.J. Sigworth. 2009. Structure of the BK potassium channel in a lipid membrane from electron cryomicroscopy. *Nature* 461(7261):292–295.

Wang, L. and F.J. Sigworth. 2010. Liposomes on a streptavidin crystal: A system to study membrane proteins by cryo-EM. *Methods Enzymol* 481:147–164.

Wang, M.C., G. Velarde, R.C. Ford, N.S. Berrow, A.C. Dolphin, and A. Kitmitto. 2002. 3D structure of the skeletal muscle dihydropyridine receptor. *J Mol Biol* 323(1):85–98.

Wisedchaisri, G. and T. Gonen. 2013. Phasing electron diffraction data by molecular replacement: Strategy for structure determination and refinement. *Methods Mol Biol* 955:243–272.

Wolf, M., A. Eberhart, H. Glossmann, J. Striessnig, and N. Grigorieff. 2003. Visualization of the domain structure of an L-type Ca^{2+} channel using electron cryo-microscopy. *J Mol Biol* 332(1):171–182.

Wolfram, F., E. Morris, and C.W. Taylor. 2010. Three-dimensional structure of recombinant type 1 inositol 1,4,5-trisphosphate receptor. *Biochem J* 428(3):483–489.

Yeager, M. and A.L. Harris. 2007. Gap junction channel structure in the early 21st century: Facts and fantasies. *Curr Opin Cell Biol* 19(5):521–528.

Yonekura, K., S. Maki-Yonekura, and K. Namba. 2003. Complete atomic model of the bacterial flagellar filament by electron cryomicroscopy. *Nature* 424(6949):643–650.

Yonekura, K. and C. Toyoshima. 2007. Structure determination of tubular crystals of membrane proteins. IV. Distortion correction and its combined application with real-space averaging and solvent flattening. *Ultramicroscopy* 107(12):1141–1158.

Young, H.S., J.L. Rigaud, J.J. Lacapere, L.G. Reddy, and D.L. Stokes. 1997. How to make tubular crystals by reconstitution of detergent-solubilized Ca2(+)-ATPase. *Biophys J* 72(6):2545–2558.

Yu, X., L. Jin, and Z.H. Zhou. 2008. 3.88 A structure of cytoplasmic polyhedrosis virus by cryo-electron microscopy. *Nature* 453(7193):415–419.

Zhang, W., M. Kimmel, C.M. Spahn, and P.A. Penczek. 2008a. Heterogeneity of large macromolecular complexes revealed by 3D cryo-EM variance analysis. *Structure* 16(12):1770–1776.

Zhang, X., P. Ge, X. Yu, J.M. Brannan, G. Bi, Q. Zhang, S. Schein, and Z.H. Zhou. 2013. Cryo-EM structure of the mature dengue virus at 3.5 Å resolution. *Nat Struct Mol Biol* 20(1):105–110.

Zhang, X., E. Settembre, C. Xu, P.R. Dormitzer, R. Bellamy, S.C. Harrison, and N. Grigorieff. 2008b. Near-atomic resolution using electron cryomicroscopy and single-particle reconstruction. *Proc Natl Acad Sci USA* 105(6):1867–1872.

Zhou, Z.H. 2008. Towards atomic resolution structural determination by single-particle cryo-electron microscopy. *Curr Opin Struct Biol* 18(2):218–228.

Zhou, Z.H. 2011. Atomic resolution cryo electron microscopy of macromolecular complexes. *Adv Protein Chem Struct Biol* 82:1–35.

12 Rosetta structural modeling

Vladimir Yarov-Yarovoy

Contents

12.1 INTRODUCTION

Ion channels are transmembrane proteins playing a key role in signal transduction in excitable cells. Mutations in some of human ion channel genes are the cause of various disorders. Furthermore, a number of human ion channels are primary targets of therapeutic drugs used for the treatment of diverse range of human diseases. Despite significant progress over the past 15 years in determining high-resolution structures of some representatives of ion channel family, it is still very difficult to obtain high-resolution structural information for a majority of mammalian ion channels. The ultimate goal of structural modeling methods is to use available structural and experimental data and increase power and accuracy of computational modeling to predict the structure of a channel protein at the atomic level of resolution. Successful pursuit of this goal will be useful for our understanding of ion channel function and modulation on structural level and rational design of drugs targeting ion channels.

De novo protein structure modeling predicts protein structure from its amino acid sequence when no homologous protein structure is available and can potentially generate membrane protein structure models at low resolution (2–4 Å root mean square deviation [RMSD] from the native structure) for proteins below ~150 residues (1). High-resolution structural models (below 2 Å RMSD from the native structure) have been generated de novo for soluble proteins below ~80–120 residues (2–4). Because of limitations in the modeling accuracy of de novo protein structure prediction methods, homology modeling is still the most useful approach for structure modeling of ion channels. Since determination of the first x-ray structure of KcsA channel by Roderick MacKinnon's group in 1998 (5), a number of new ion channel structures have become available. All of these ion channel structures provide valuable templates for modeling of homologous ion channels. However, we are still limited in a number of conformational states of ion channels captured at high resolution. Even when closed and/or open states of different homologous ion channels are available, they may not be representing structures of distantly homologous ion channels. For example, open states of MthK (6), KcsA (7), KvAP (8), and Kv1.2 (9,10) channels show a bend of the pore-lining helix at a different position. Therefore, experimentally derived constraints have to be used to explore multiple states simply of structurally flexible regions of a specific isoform of an ion channel.

There are a number of successful de novo and homology-based protein structure prediction methods, including MODELLER (11), TASSER (12), and I-TASSER (13). This chapter is focusing on Rosetta method for ion channel structure prediction that has been used for prediction of unknown ion channel conformational states and modeling of channel–peptide toxin interactions (14–22).

12.2 ROSETTA METHOD

The Rosetta de novo method is designed to predict tertiary structure of a protein directly from its amino acid sequence (Figure 12.1, Step 1). The method is based on the assumption that the native state of a protein is at the global free energy minimum and a large-scale search of conformational space for protein tertiary structures is carried out to select structures that are especially low in free energy for the given amino acid sequence. Extensive sampling of conformational space is a challenging task, and thousands to millions of independent protein-folding trajectories are explored by the method. The two key components of this method are the procedure for efficiently carrying out the conformational search and the free energy function used for evaluating possible conformations. The Rosetta method uses structural information for short protein fragments (extracted

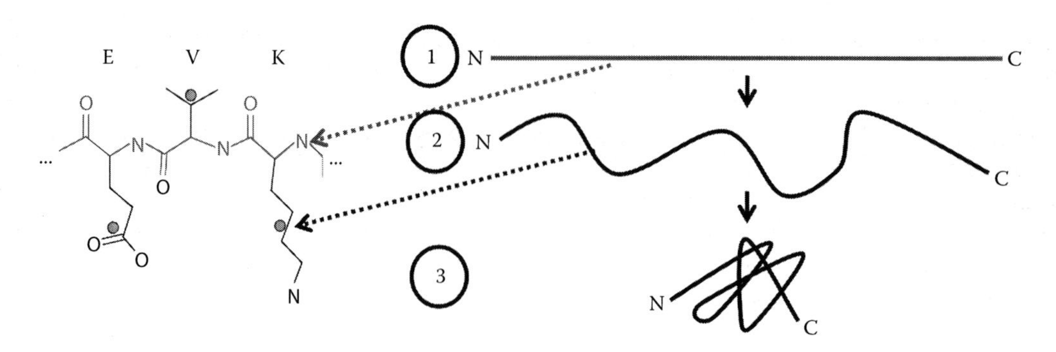

Figure 12.1 De novo protein folding using Rosetta method. Step 1, extended polypeptide chain. Step 2, initial three- and nine-residue segments from fragment library are inserted at random positions into the polypeptide chain. Step 3, folding of a protein using continuous random fragment insertion and Monte Carlo simulated annealing search of low-energy protein conformations.

from the protein data bank [PDB, http://www.rcsb.org/pdb/home/home.do]) to sample local conformations adopted by these fragments in a protein (24–26). Specifically, structural fragment libraries for each three- and nine-residue segment of protein sequence of interest are extracted from the PDB using a sequence profile comparison method that uses multiple sequence alignment of the homologous segments (Figure 12.1, Step 2). The use of fragments of known protein structures ensures that the energy of local interactions is close to optimal. In the Rosetta de novo method developed for modeling of water-soluble proteins, the conformational space defined by the fragment libraries is searched using a Monte Carlo procedure with an energy function that favors compact tertiary structures with buried hydrophobic residues and paired β-strands (Figure 12.1, Step 3). Finally, high-resolution full-atom refinement is performed to explore precise side-chain packing and modest backbone movements, and the best Rosetta models are the ones that have the lowest full-atom energy (2,4,24).

12.2.1 ROSETTA-MEMBRANE LOW-RESOLUTION ENERGY FUNCTION

The Rosetta-membrane de novo method has been developed for modeling of membrane proteins from amino acid sequence information alone (1) and uses similar procedures to the Rosetta de novo method for modeling of water-soluble proteins (24). However, the membrane environment–specific energy function favors burial of small hydrophobic residues and exposure of large hydrophobic residues within the hydrophobic layer of the membrane and minimizing hydrophobic residue exposure in the polar environment outside of the membrane (1,20,21). This method could be useful, for example, for de novo modeling of the voltage-sensing and voltage-sensing-like domains of ion channels or in combination with the Rosetta symmetry method (see the following text) can be used for modeling of pore-forming domains of ion channels.

The major steps in folding of an ion channel using the Rosetta-membrane de novo method are described later. Folding starts from completely extended polypeptide chain, and structural fragments of nine- and/or three-residue long (generated as described earlier) are inserted randomly into the polypeptide chain until these fragments completely cover the whole length of the protein. During this step, preliminary secondary structure elements are formed along the protein sequence according to the secondary structure of the inserted fragments. A low-resolution conformational search is then performed where the side chains of each residue are represented by a single point in space—centroid. The position of the centroid is calculated for each residue as the average position of the side-chain atoms in residues of the same amino acid identity and with φ and ψ angles in the same 10° × 10° bin and taken from known protein structures in the PDB (25). At this stage, the energy function disfavors residue–residue clashes using known Van der Waals radii of atoms and the minimal allowed distance between them. The membrane environment is simulated using an energy function derived from statistics generated from known high-resolution membrane protein structures (1). This membrane environment energy function includes residue environment, residue–residue interactions, and residue density terms. The residue environment term is based on amino acid propensities to be in each of four membrane layers (hydrophobic, interface, polar, and water (1)) and depends on the residue burial state—from being completely buried within a protein environment to being completely exposed either to the lipid or water environments. The residue–residue interaction term is based on the propensities of amino acid pairs to be in close proximity to each other within hydrophobic and polar layers. The polar layer for the purpose of this term includes the interface, polar, and water-exposed layer as defined for the residue environment term. The residue density term is based on the distribution of the number of residue neighbors within 6 and 12 Å spheres.

During the next step of protein folding, the membrane environment energy function is applied first only to two adjacent and randomly selected transmembrane helices along the protein sequence. The transmembrane regions location along the sequence are approximated from predictions by OCTOPUS transmembrane topology methods (27) or based on transmembrane topology in known homologous ion channel structures. Thousands of cycles of random fragment insertions into these two transmembrane helices and the loop connecting them are then carried out. Optimal embedding of transmembrane helices at this and all subsequent steps of folding is achieved using Monte Carlo simulated annealing search that favors insertion of structural fragments that reduce total membrane environment energy.

During the following steps of membrane protein folding, each new transmembrane region is *inserted* randomly into the simulated membrane environment on either side of the initially embedded transmembrane region until all transmembrane regions have been inserted into the membrane. A large number of independent simulations are carried out, which results in

10,000 different models. These models are then subjected to a clustering procedure to identify the broadest free energy minima (28) where all models are compared to each other using RMSD in backbone atom positions with 2–4 Å clustering threshold. The clustering procedure results in many clusters of models, and ten largest clusters represent protein conformations, which are most frequently generated by the Rosetta method and at least one of the clusters may represent near-native structure (1,20).

This method can be used to predict low-resolution structures of ion channels. In benchmark tests on a dozen of membrane protein structures, this method predicted up to 145 residues with RMSD 2–4 Å in backbone atom positions to the native structure (1).

12.2.2 ROSETTA-MEMBRANE FULL-ATOM ENERGY FUNCTION

The full-atom refinement of the low-resolution Rosetta-membrane models is performed to extensively sample backbone and side-chain conformations using membrane environment-specific full-atom free energy function (21,27). This energy function is similar to the energy function used for water-soluble protein design calculations (28), except that solvation and hydrogen bond potentials are functions of residue depth in the membrane. The membrane environment is described by two isotropic phases (water and hydrocarbon layers) and an anisotropic phase (interface layer) that interpolates the chemical properties of the two isotropic phases. An implicit atomic solvation potential for the hydrophobic layer is based on experimental transfer free energies of peptides from water to lipid bilayers (31). The atomic solvation energies in the water environment are based on a solvation model developed by Lazaridis and Karplus (32). The atomic solvation energies in the membrane interface layer were derived by interpolating the solvation properties from the two adjacent layers. The previously developed hydrogen-bond potential (33) was modified to model the effect of the environment on the strength of the hydrogen bonds involving solvent-exposed residues. At least 10,000 models are generated using Rosetta-membrane full-atom energy function and the lowest energy models are selected as the best models.

The Rosetta-membrane full-atom energy function successfully predicted side-chain conformations at 73% of buried residues in the native membrane proteins (29). The sequence recovery tests revealed that the Rosetta-membrane full-atom energy function performs at the level of accuracy of the water-soluble Rosetta full-atom energy function, with design tests predicting 34% of positions in the native membrane proteins (29,34). Furthermore, the Rosetta-membrane full-atom energy function is able to discriminate between native and nonnative structures of membrane proteins (29).

12.3 HOMOLOGY MODELING OF ION CHANNELS

The first step in homology modeling is selection of structural template(s). HHpred server (http://toolkit.tuebingen.mpg.de/hhpred) has proven to be one of the best homology detection servers (35). Sequence of ion channel of interest is used as an input to the HHpred server and at least several structural homologs are identified with corresponding pairwise sequence alignments between the protein of interest and the homologous ion channel(s) of known structure. The template with the highest percentage of sequence identity is used for homology modeling of the high-confidence sequence alignment regions. Poorly aligned or gapped regions are predicted de novo using Rosetta cyclic coordinate descent (CCD) and kinematic closure (KIC) loop modeling applications (36,37) and guided by membrane environment-specific energy function (21,29). Several rounds of CCD and KIC loop modeling are performed, and at least 10,000 models are generated during each round. All models after each round are ranked by total score and top 10% are clustered (28) using RMSD threshold that generates at least 150–200 models in the largest cluster. Models representing centers of the top 20 clusters (early rounds) and/or the best 10 models by total score (later rounds) are used as input for the next round of modeling. Rosetta full-atom relax application is used for exploring potential differences in backbone and side-chain conformation between ion channel of interest and structural template. Best 10 models by total score after the relax round are considered as the best models.

There are a number of successful homology-based protein structure prediction methods for soluble proteins, including MODELLER (11), TASSER (12), and I-TASSER (13). At present only GPCR-ITASSER (38) (http://zhanglab.ccmb.med.umich.edu/GPCRRD/GPCR_ITASSER.html) and Rosetta-membrane methods have been shown to be successful in homology modeling of membrane protein structures (15–18,20,21). However, only Rosetta-membrane method has been successfully applied to modeling of ion channel structures (15–18,20,21).

12.4 MODELING OF MULTIPLE STATES OF ION CHANNELS

A combination of computational methods and available experimental data can be used to model conformational states of ion channels that have not been captured by x-ray structure methods. For example, x-ray structure method has captured activated state conformations of the voltage-sensing domain of the voltage-gated ion channels (8–10,39–41). The Rosetta-membrane method has been used to explore the movement of voltage-sensing domain in the voltage-gated ion channels using disulfide cross-linking and other experimental data as constraints (15–17,21,42).

Preliminary template-based model is generated first that serves as a starting point for applying distance constraints between specific residues in the model (see Figure 12.2a). For disulfide bridge–based pairwise interaction observed experimentally, either disulfide bridge interaction is enforced by Rosetta method during modeling by specifying cysteine pair or ~6 Å distance constraint is applied between Cβ atoms of cysteine residues. In order to extend disulfide-locking data to modeling of interactions between native side chains, distance constraints between outermost carbon atoms are applied and based on average distances observed between corresponding residues in x-ray structures (21). At least 10,000 models are generated for each distance constraint and the 10 lowest energy models are used as input to full-atom relax with membrane-specific energy function and without applying experimental constraints. This step allows exploration of nearby deeper energy wells that are

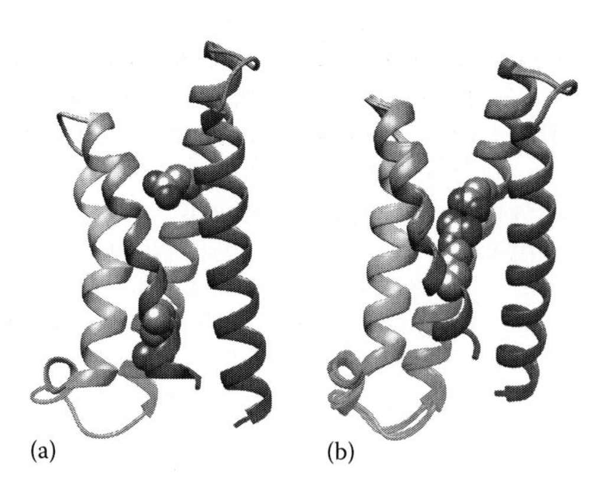

Figure 12.2 Distance constraints–based modeling of a conformational state of an ion channel. (a) Transmembrane view of a ribbon representation of a starting template model of a voltage-sensing domain of an ion channel. Side chains of a pair of residues that will be constrained (*panel* b) are shown in space-filling representation. (b) Transmembrane view of a ribbon representation of a constrained model of a voltage-sensing domain of an ion channel. Side chains of a pair of residues that were constrained during modeling are shown in space-filling representation.

not influenced by applying distance constraint between specific pair of residues in a channel. The 10 lowest energy models after this round of modeling represent the best models of a particular conformational state of a channel (see Figure 12.2b).

12.5 SYMMETRY MODELING OF ION CHANNELS

Pore-forming domains of ion channels can be modeled using a combination of Rosetta symmetry and homology or de novo methods. If homologous template is available, coordinates of pore-forming helices are kept rigid during modeling and variable region(s), such as a selectivity filter, could be modeled using Rosetta CCD and KCL loop modeling applications (36,37) and guided by membrane environment-specific energy function (21,29). Specific ion coordinating geometries can be also applied during modeling as constraints between specific residue atoms. For example, K^+ ion coordinating constraints could be based on ion coordination distances and angles from native KcsA (pdb: 1K4C) channel structure. Na^+ constraints could be based on ion coordination distances and angles from native NaK (pdb: 3E86) channel structure and Na^+-containing high-resolution membrane protein structures (pdbs: 2ITC, 352D, 3F3A, 2WGM). Rosetta symmetry-constraint protocol could search for a specific number of lowest energy ion coordination sites out of specified set of backbone and/or side-chain oxygen atoms within the loop lining the selectivity filter region. During the first round of modeling, Rosetta's CCD loop relax protocol (36) is used and top 20 cluster center models are passed to the second round. During the second and third rounds of modeling, kinematic loop relax protocol is used (37) and top 20 cluster center models are passed to the next round. During the fourth through seventh rounds of modeling, kinematic loop relax protocol is used (37) and top 20 models by score are passed to the next round. At least 10,000 models are generated in each round. The lowest energy models from the last round of iterative loop relax are selected as the best models (see Figure 12.3).

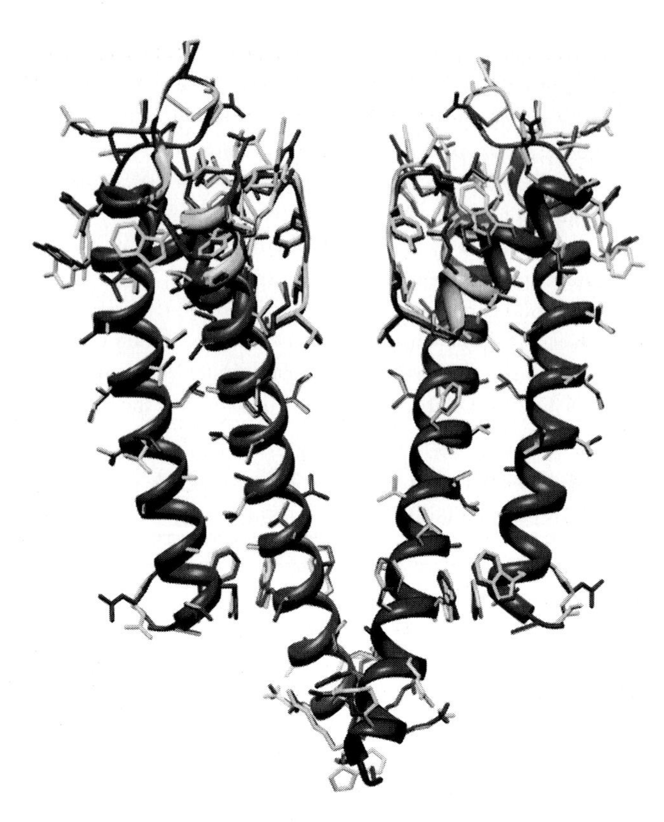

Figure 12.3 Symmetry modeling of a selectivity filter of an ion channel. Ribbon representation of the lowest energy Rosetta model (colored gray) and native KcsA structure (colored black) superimposed.

12.6 FUTURE DIRECTIONS

The Rosetta method has been successfully used to design new protein folds (30), redesign proteins to change their specificity or to perform new functions (43–46), and de novo design of high-affinity inhibitors (47). Rosetta can be potentially used in the future to redesign ion channel structures to perform new functions and design new high-affinity modulators of specific ion channel subtypes.

12.7 RELATED WEBSITES

Rosetta software suite—https://www.rosettacommons.org
Rosetta documentation—https://www.rosettacommons.org/manual_guide
Rosetta-based computer game for scientific research—http://fold.it/portal
David Baker group—http://depts.washington.edu/bakerpg/drupal
Vladimir Yarov-Yarovoy group—http://www.ucdmc.ucdavis.edu/physiology/faculty/yarovoy.html

REFERENCES

1. Yarov-Yarovoy, V., J. Schonbrun, and D. Baker. 2006. Multipass membrane protein structure prediction using Rosetta. *Proteins* 62:1010–1025.
2. Bradley, P., K. M. Misura, and D. Baker. 2005. Toward high-resolution de novo structure prediction for small proteins. *Science* 309:1868–1871.
3. Miller, A. N. and S. B. Long. 2012. Crystal structure of the human two-pore domain potassium channel K2P1. *Science* 335:432–436.

4. Qian, B., S. Raman, R. Das, P. Bradley, A. J. McCoy, R. J. Read, and D. Baker. 2007. High-resolution structure prediction and the crystallographic phase problem. *Nature* 450:259–264.
5. Doyle, D. A., J. Morais Cabral, R. A. Pfuetzner, A. Kuo, J. M. Gulbis, S. L. Cohen, B. T. Chait, and R. MacKinnon. 1998. The structure of the potassium channel: Molecular basis of K^+ conduction and selectivity. *Science* 280:69–77.
6. Jiang, Y., A. Lee, J. Chen, M. Cadene, B. T. Chait, and R. MacKinnon. 2002. The open pore conformation of potassium channels. *Nature* 417:523–526.
7. Cuello, L. G., V. Jogini, D. M. Cortes, and E. Perozo. 2010. Structural mechanism of C-type inactivation in K(+) channels. *Nature* 466:203–208.
8. Jiang, Y., A. Lee, J. Chen, V. Ruta, M. Cadene, B. T. Chait, and R. MacKinnon. 2003. X-ray structure of a voltage-dependent K^+ channel. *Nature* 423:33–41.
9. Long, S. B., E. B. Campbell, and R. Mackinnon. 2005. Crystal structure of a mammalian voltage-dependent Shaker family K^+ channel. *Science* 309:897–903.
10. Long, S. B., X. Tao, E. B. Campbell, and R. MacKinnon. 2007. Atomic structure of a voltage-dependent K^+ channel in a lipid membrane-like environment. *Nature* 450:376–382.
11. Eswar, N., B. Webb, M. A. Marti-Renom, M. S. Madhusudhan, D. Eramian, M. Y. Shen, U. Pieper, and A. Sali. 2006. Comparative protein structure modeling using modeller. *Current Protocols in Bioinformatics* Chapter 5:Unit 5.6.
12. Zhang, Y., A. K. Arakaki, and J. Skolnick. 2005. TASSER: An automated method for the prediction of protein tertiary structures in CASP6. *Proteins* 61(Suppl. 7):91–98.
13. Zhang, Y. 2008. I-TASSER server for protein 3D structure prediction. *BMC Bioinformatics* 9:40.
14. Cestele, S., V. Yarov-Yarovoy, Y. Qu, F. Sampieri, T. Scheuer, and W. A. Catterall. 2006. Structure and function of the voltage sensor of sodium channels probed by a beta-scorpion toxin. *Journal of Biological Chemistry* 281:21332–21344.
15. Decaen, P. G., V. Yarov-Yarovoy, T. Scheuer, and W. A. Catterall. 2011. Gating charge interactions with the S1 segment during activation of a Na^+ channel voltage sensor. *Proceedings of the National Academy of Sciences of the United States of America* 108:18825–18830.
16. DeCaen, P. G., V. Yarov-Yarovoy, E. M. Sharp, T. Scheuer, and W. A. Catterall. 2009. Sequential formation of ion pairs during activation of a sodium channel voltage sensor. *Proceedings of the National Academy of Sciences of the United States of America* 106:22498–22503.
17. DeCaen, P. G., V. Yarov-Yarovoy, Y. Zhao, T. Scheuer, and W. A. Catterall. 2008. Disulfide locking a sodium channel voltage sensor reveals ion pair formation during activation. *Proceedings of the National Academy of Sciences of the United States of America* 105:15142–15147.
18. Pathak, M. M., V. Yarov-Yarovoy, G. Agarwal, B. Roux, P. Barth, S. Kohout, F. Tombola, and E. Y. Isacoff. 2007. Closing in on the resting state of the shaker K(+) channel. *Neuron* 56:124–140.
19. Wang, J., V. Yarov-Yarovoy, R. Kahn, D. Gordon, M. Gurevitz, T. Scheuer, and W. A. Catterall. 2011. Mapping the receptor site for {alpha}-scorpion toxins on a Na^+ channel voltage sensor. *Proceedings of the National Academy of Sciences of the United States of America* 108:15426–15431.
20. Yarov-Yarovoy, V., D. Baker, and W. A. Catterall. 2006. Voltage sensor conformations in the open and closed states in ROSETTA structural models of K(+) channels. *Proceedings of the National Academy of Sciences of the United States of America* 103:7292–7297.
21. Yarov-Yarovoy, V., P. G. Decaen, R. E. Westenbroek, C. Y. Pan, T. Scheuer, D. Baker, and W. A. Catterall. 2012. Structural basis for gating charge movement in the voltage sensor of a sodium channel. *Proceedings of the National Academy of Sciences of the United States of America* 109:E93–E102.
22. Zhang, J. Z., V. Yarov-Yarovoy, T. Scheuer, I. Karbat, L. Cohen, D. Gordon, M. Gurevitz, and W. A. Catterall. 2011. Structure-function map of the receptor site for beta-scorpion toxins in domain II of voltage-gated sodium channels. *Journal of Biological Chemistry* 286:33641–33651.
23. Berman, H. M., J. Westbrook, Z. Feng, G. Gilliland, T. N. Bhat, H. Weissig, I. N. Shindyalov, and P. E. Bourne. 2000. The Protein Data Bank. *Nucleic Acids Research* 28:235–242.
24. Rohl, C. A., C. E. Strauss, K. M. Misura, and D. Baker. 2004. Protein structure prediction using Rosetta. *Methods in Enzymology* 383:66–93.
25. Simons, K. T., C. Kooperberg, E. Huang, and D. Baker. 1997. Assembly of protein tertiary structures from fragments with similar local sequences using simulated annealing and Bayesian scoring functions. *Journal of Molecular Biology* 268:209–225.
26. Simons, K. T., I. Ruczinski, C. Kooperberg, B. A. Fox, C. Bystroff, and D. Baker. 1999. Improved recognition of native-like protein structures using a combination of sequence-dependent and sequence-independent features of proteins. *Proteins* 34:82–95.
27. Viklund, H. and A. Elofsson. 2008. OCTOPUS: Improving topology prediction by two-track ANN-based preference scores and an extended topological grammar. *Bioinformatics* 24:1662–1668.
28. Bonneau, R., C. E. Strauss, C. A. Rohl, D. Chivian, P. Bradley, L. Malmstrom, T. Robertson, and D. Baker. 2002. De novo prediction of three-dimensional structures for major protein families. *Journal of Molecular Biology* 322:65–78.
29. Barth, P., J. Schonbrun, and D. Baker. 2007. Toward high-resolution prediction and design of transmembrane helical protein structures. *Proceedings of the National Academy of Sciences of the United States of America* 104:15682–15687.
30. Kuhlman, B., G. Dantas, G. C. Ireton, G. Varani, B. L. Stoddard, and D. Baker. 2003. Design of a novel globular protein fold with atomic-level accuracy. *Science* 302:1364–1368.
31. Hessa, T., H. Kim, K. Bihlmaier, C. Lundin, J. Boekel, H. Andersson, I. Nilsson, S. H. White, and G. von Heijne. 2005. Recognition of transmembrane helices by the endoplasmic reticulum translocon. *Nature* 433:377–381.
32. Lazaridis, T. and M. Karplus. 1999. Effective energy function for proteins in solution. *Proteins* 35:133–152.
33. Kortemme, T., A. V. Morozov, and D. Baker. 2003. An orientation-dependent hydrogen bonding potential improves prediction of specificity and structure for proteins and protein–protein complexes. *Journal of Molecular Biology* 326:1239–1259.
34. Kuhlman, B. and D. Baker. 2000. Native protein sequences are close to optimal for their structures. *Proceedings of the National Academy of Sciences of the United States of America* 97:10383–10388.
35. Soding, J., A. Biegert, and A. N. Lupas. 2005. The HHpred interactive server for protein homology detection and structure prediction. *Nucleic Acids Research* 33:W244–W248.
36. Wang, C., P. Bradley, and D. Baker. 2007. Protein–protein docking with backbone flexibility. *Journal of Molecular Biology* 373:503–519.
37. Mandell, D. J., E. A. Coutsias, and T. Kortemme. 2009. Sub-angstrom accuracy in protein loop reconstruction by robotics-inspired conformational sampling. *Nature Methods* 6:551–552.
38. Zhang, Y., M. E. Devries, and J. Skolnick. 2006. Structure modeling of all identified G protein-coupled receptors in the human genome. *PLoS Computational Biology* 2:e13.
39. Payandeh, J., T. Scheuer, N. Zheng, and W. A. Catterall. 2011. The crystal structure of a voltage-gated sodium channel. *Nature* 475:353–358.

40. Payandeh, J., T. M. Gamal El-Din, T. Scheuer, N. Zheng, and W. A. Catterall. 2012. Crystal structure of a voltage-gated sodium channel in two potentially inactivated states. *Nature* 486:135–139.
41. Zhang, X., W. Ren, P. DeCaen, C. Yan, X. Tao, L. Tang, J. Wang et al. 2012. Crystal structure of an orthologue of the NaChBac voltage-gated sodium channel. *Nature* 486:130–134.
42. Henrion, U., J. Renhorn, S. I. Borjesson, E. M. Nelson, C. S. Schwaiger, P. Bjelkmar, B. Wallner, E. Lindahl, and F. Elinder. 2012. Tracking a complete voltage-sensor cycle with metal-ion bridges. *Proceedings of the National Academy of Sciences of the United States of America* 109:8552–8557.
43. Ashworth, J., J. J. Havranek, C. M. Duarte, D. Sussman, R. J. Monnat, Jr., B. L. Stoddard, and D. Baker. 2006. Computational redesign of endonuclease DNA binding and cleavage specificity. *Nature* 441:656–659.
44. Jiang, L., E. A. Althoff, F. R. Clemente, L. Doyle, D. Rothlisberger, A. Zanghellini, J. L. Gallaher et al. 2008. De novo computational design of retro-aldol enzymes. *Science* 319:1387–1391.
45. Rothlisberger, D., O. Khersonsky, A. M. Wollacott, L. Jiang, J. DeChancie, J. Betker, J. L. Gallaher et al. 2008. Kemp elimination catalysts by computational enzyme design. *Nature* 453:190–195.
46. Siegel, J. B., A. Zanghellini, H. M. Lovick, G. Kiss, A. R. Lambert, J. L. St Clair, J. L. Gallaher et al. 2010. Computational design of an enzyme catalyst for a stereoselective bimolecular Diels-Alder reaction. *Science* 329:309–313.
47. Fleishman, S. J., T. A. Whitehead, D. C. Ekiert, C. Dreyfus, J. E. Corn, E. M. Strauch, I. A. Wilson, and D. Baker. 2011. Computational design of proteins targeting the conserved stem region of influenza hemagglutinin. *Science* 332:816–821.

13 Genetic methods for studying ion channel function in physiology and disease

Andrea L. Meredith

Contents

Genetic methods are a powerful tool for dissecting the function of ion channels in physiological contexts. Using molecular genetics, the function of ion channels can be decoded by phenotype, by genetic engineering in animal models, and through human channel variant-phenotype linkages. The evolution of our understanding of ion channel function has been stimulated by the development of genetically tractable model organisms (1970s), recombinant DNA technology and genetically modified mice (1980s), genome sequencing and annotation (late 1990s to early 2000s), and human SNP mapping (2000s). The resulting discoveries complemented biophysical studies of ion channel properties and revealed multifaceted interactions in vivo, including associations with more than 100 human disorders.

13.1 LINKING ION CHANNELS TO PHENOTYPES: FORWARD GENETICS

13.1.1 MODEL ORGANISMS

Beginning with Hodgkin and Huxley's biophysical studies of the giant squid axon, simple model organisms have revealed fundamental properties of the major ion channel classes. Moreover, given the ubiquity of ion channels in prokaryotes and eukaryotes, even the simplest single-celled organisms have facilitated formation of the primary connections between physiology and ion channel function. This function features in the rich diversity of behaviors exhibited across phyla, exemplified by modulation of membrane potential by light in the "giant unicell" *Acetabularia*, hyphal tip growth in the "nerve spore" *Neurospora*, escape response in the "swimming neuron" *Paramecium*, aggregation in the "social amoebae" *Dictyostelium*, and motility in the "plant-animal chimera" *Clamydomonas* (Martinac et al., 2008). Mediating growth, reproduction, and survival behaviors, these single-celled organisms possess specific channel-based ion transport and osmoregulation systems, intracellular Ca^{2+} signaling pathways, and membrane excitation networks (Hille, 2001), making them valuable models for understanding ion channel–related physiology.

Further up the phylogenetic tree, multicellular organisms from plants to mammals use ion channels with evolutionarily conserved functional mechanisms on a backdrop of increasing cellular and system specialization. Understanding the molecular basis for specialized electrical signaling requires knowledge of the molecular constituents of ion channels. Despite the rich diversity of electrical behaviors and ion transport mechanisms studied in simple model organisms, the confluence of experimental tractability and development of molecular genetic tools contributed to the rise of particular animal models, notably fly and mouse, for the study of ion channel function in vivo (Table 13.1).

Table 13.1 Comparison of standard model organisms

ORGANISM	GENOME	GENES SIMILAR TO HUMAN (%)	GENERATION TIME	EXPERIMENTAL ADVANTAGES	EXPERIMENTAL DISADVANTAGES
Yeast ***Saccharomyces cerevisiae***	6,352 genes	46	2 h	Inexpensive; stocks can be frozen; easy to clone genes; powerful modifier screening; possess all basic eukaryotic cellular components; *Saccharomyces* Genome Database. (http://yeastgenome.org/)	No distinct tissues; electrophysiology possible but not widely performed
Worm ***Caenorhabditis elegans***	21,187 genes	43	3 days	Inexpensive and tiny; can be stored frozen; only 959 cells; transparent development; hermaphrodite self-fertilization and sexual reproduction; complete morphological characterization by serial EM and lineage mapping; laser ablation of single identified cells; easy gene cloning (transposon tagging, SNP mapping, gene rescue, genome-wide deletion collection); comprehensive proteome (ORFeome) and promoter libraries; modifier screening; large collections of mutant, expression (GFP-tagged), and RNAi lines; reverse genetics by RNAi (can be fed to worms); Wormbase. (http://www.wormbase.org/#01-23-6)	Limited genetic, anatomical, and behavioral similarities to human; electrophysiology difficult but possible
Fly ***D. melanogaster***	15,431 genes	61	10 days	Inexpensive; oldest animal model with elegant forward genetic tools and easy gene cloning; modifier screening; precise temporal and spatial control of gene expression (Gal4-UAS); can make chimeric animals; large collections of mutant, expression, P-element lines; several reverse genetics techniques; more complex behaviors and similarities to human disease; amenable to drug discovery; electrophysiological access in larvae and adult muscle; Flybase. (http://flybase.org/)	Freezing protocols still under development; RNAi must be injected; electrophysiology limited in the adult brain
Zebrafish ***Danio rerio***	26,206 genes	76	10–12 weeks	Simplest vertebrate model with good forward and reverse genetics; external fertilization and development; large number of offspring per female; organ similarity to mammals (except lungs, mammary glands); transparent embryo optimal for imaging organogenesis; good transgenics for protein visualization; can make chimeric animals; reverse genetics by morpholino (antisense oligonucleotide knockdown) or genome editing with engineered nucleases (TALEN); in vivo and ex vivo electrophysiology recording capabilities; Zebrafish Model Organism Database. (http://zfin.org/)	Many duplicate genes; small but requires specialized housing and Institutional Animal Care and Use Committee (IACUC) protocols; homologous recombination limited by availability of embryonic stem (ES) cells; few cell lines available for cellular or biochemical studies

(*Continued*)

Table 13.1 (*Continued*) Comparison of standard model organisms

ORGANISM	GENOME	GENES SIMILAR TO HUMAN (%)	GENERATION TIME	EXPERIMENTAL ADVANTAGES	EXPERIMENTAL DISADVANTAGES
Mouse ***Mus musculus***	20,210 genes	99	6–8 weeks	Mammalian model possessing essentially all human tissues and cell types; inbred and outbred strains; cryopreservation; in vitro fertilization; reverse genetics by targeted homologous recombination (extensive resources available); can make chimeric animals; extensive molecular genetic tools (BAC libraries, gene trap libraries; expression arrays); well-characterized primary, ES, and stem cells for culture; superior in vivo and ex vivo electrophysiology recording capabilities; public and commercial mutant repositories; Mouse Genome Informatics. (http://www.informatics.jax.org/)	Expensive to house; longer generation time; incomplete genome coverage for conditional expression drivers (Cre lines); forward genetics laborious; requires IACUC protocols
Human ***Homo sapien***	19,040 genes		12+ years	>5000 genetically based diseases; familial pedigree analysis to link gene mutations and disease; detailed self-reporting of traits; genome-wide association studies (GWAS) linking gene haplotypes, traits, and disease risk; population studies (genome, transcriptome, proteome, epigenome, and microbiome); Online Mendelian Inheritance in Man. (http://www.ncbi.nlm.nih.gov/omim/)	Limited experimental access; requires extensive Institutional Review Board (IRB) protocols

Source: Barbazuk et al., 2000; Church et al., 2009; Howe et al., 2013; Jorgensen and Mango, 2002; Kile and Hilton, 2005; St Johnston, 2002; Woods et al., 2000. Database resources: NCBI Genome Database. Bethesda (MD): National Library of Medicine (US); [cited Nov. 5, 2013]. Available from: http://www.ncbi.nlm.nih.gov/genome/; NIH Model Organisms for Biomedical Research. Bethesda (MD): National Institutes of Health (US); [cited Nov. 5, 2013]. Available from: http://www.nih.gov/science/models/. Adapted from Bier and McGinnis, 2003.

13.1.2 CLONING

Early electrophysiological preparations such as squid axon and electric fish preceded the molecular biology and genetic revolutions. Biochemical purification enabled identification of the first ion channel proteins, isolated from the electric organs of the *Torpedo* ray (the nicotinic acetylcholine receptor; Noda et al., 1982) and the eel *Electrophorus electricus* (the voltage-gated Na^+ channel; Noda et al., 1984). Subsequent amino acid sequencing was used to generate DNA probes and screen cDNA libraries. The resulting isolation of coding sequences could be used to express channels in *Xenopus* oocytes or to identify other similar channel sequences by homology screening. While these methods identified a specific channel sequence, additional steps were required to discover roles for the channels in native physiological contexts.

Parallel to biochemical purification to identify channels, other researchers, notably Seymour Benzer at Caltech, were pioneering a different concept. Their approach centered around identifying genes based on inherent linkage to a functional role in animals, such as a behavioral alteration. In the early 1970s, Benzer's theory that mutations in single genes would produce a measurable effect on behavior was controversial, but quickly vetted using *Drosophila melanogaster* as a model organism. This approach for identifying genes, dubbed *forward genetics* (also referred to as classical genetics), denotes methods for identifying a *genotype* based on a *phenotype*. The genetic makeup of an organism, or more specifically the sequence of a particular gene of interest, is referred to as a *genotype*. *Phenotype* is the constellation of observable traits produced by a particular genotype.

With the rise of *Drosophila* as a model organism for the genetic dissection of behavior, ion channel genes not amenable to biochemical purification could be cloned, and importantly, linked to a specific phenotype. *Shaker* was the first ion channel sequence identified using a forward genetics approach (Papazian et al., 1987; Tempel et al., 1987). Like other *Drosophila* mutants named according to their phenotypic presentation, *Shaker* was so named for the leg shaking of mutant flies when exposed to ether. Electrophysiological analysis of independent *Shaker* lines revealed alterations in I_A, a fast transient K^+ current in muscle, suggesting the underlying mutations would be in a gene encoding structural component of I_A. A *positional cloning* strategy (using chromosomal location to identify genomic DNA sequence) was employed to locate the *Shaker* gene, which was shown to encode a K^+ channel with structural features similar

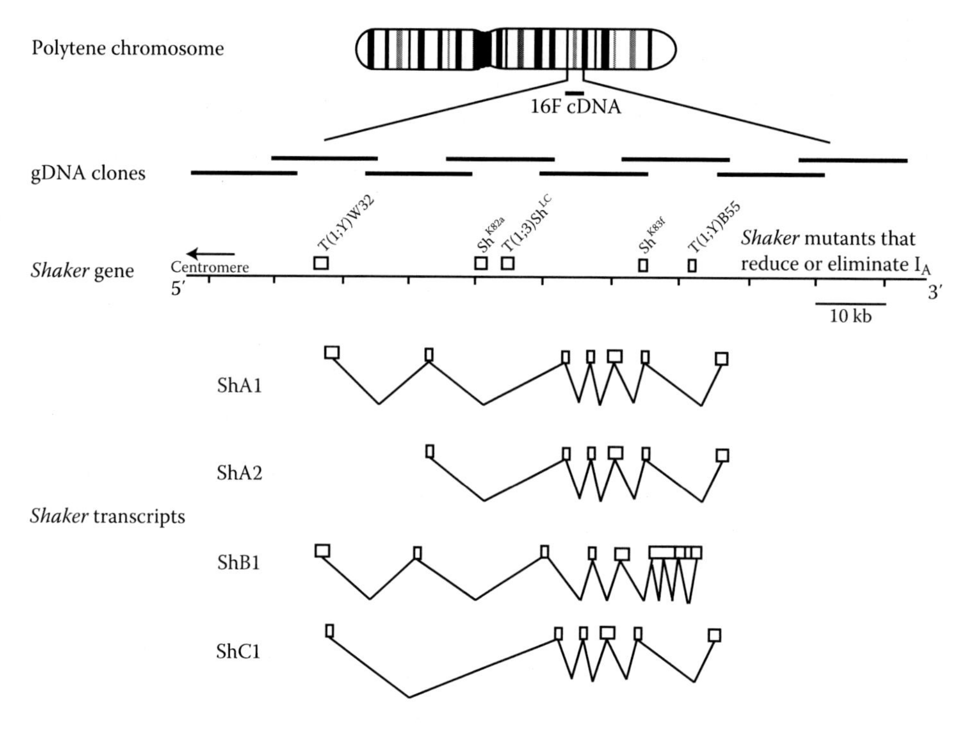

Figure 13.1 Positional cloning of *Shaker*. Abnormal banding patterns resulting from *Shaker* chromosomal aberrations were mapped to band 16F on *Drosophila* polytene chromosomes. Genomic DNA (gDNA) sequence encoding the *Shaker* locus was obtained by *chromosomal walking*, using a cDNA clone previously mapped to 16F to isolate a series of overlapping genomic clones from a library. Southern blot analysis was used to physically map the gDNA, revealing that independent *Shaker* mutations covered >65 kb. Restriction fragments from the *Shaker* gene were used to probe a cDNA library and identify the coding region (boxed regions indicate areas of hybridization between genomic and cDNA sequences). Four clones were isolated (ShA1, ShA2, ShB1, and ShC1) and shown to encode three distinct splice variants. (Adapted from Papazian, D.M. et al., *Science*, 237, 749, 1987.)

to the previously cloned Na^+ channel (Figure 13.1). Other *Drosophila* ion channel genes were identified by forward genetics, including K^+ channels (*slowpoke*, *ether a go-go*, and *seizure*) (Atkinson et al., 1991; Wang et al., 1997; Warmke et al., 1991) and the *transient receptor potential* cation channel (Montell and Rubin, 1989). Later, with the development of additional positional cloning resources in mouse, strains harboring spontaneous mutations such as *lurcher*, *tottering*, and *weaver* were used to independently identify the mammalian ion channel genes *GluRδ2*, *CACNA1a*, and *GIRK2*, respectively (Fletcher et al., 1996; Patil et al., 1995; Zuo et al., 1997).

Once an ion channel sequence was isolated, low stringency hybridization screens of cDNA or genomic libraries could be used to identify related channels (*homologs*). Related channels could stem from gene duplications (*paralogs*), such as within ion channel families, or across species (*orthologs*). After the cloning of *Shaker*, three additional K^+ channel subtypes in flies (*Shab*, *Shaw*, and *Shal*) and the mammalian voltage-gated K^+ channels were identified by homology screening (Butler et al., 1989; Wei et al., 1990).

13.1.3 PHENOTYPIC SCREENS

Mutant phenotype-driven forward genetic screens are used to identify genes and understand the physiological roles for gene products. In the case of *Shaker*, a specific ion channel subunit could now be linked to a deficit in I_A and prolonged transmitter release at neuromuscular junctions in flies. One advantage to forward genetic strategies was the purely genetic basis for linking genes with phenotypes, making large-scale screening feasible. Screening criteria did not necessarily require electrophysiology, which could be labor intensive. With some idea of the phenotypic characteristics that would involve ion channel function, high-throughput behavioral or morphological screens could be performed to isolate mutants. Early *Drosophila* screens were based on evaluating behavior to understand neural development, locomotor and flight behaviors, sensory responses, learning and memory, and sensitivity to toxins.

Large-scale mutagenesis screening made it possible to perform a relatively unbiased interrogation of the entire genome, without a priori knowledge of the genomic location or molecular composition of ion channel constituents. Small organisms, inexpensive to house, with a relatively short generation time were particularly amenable to high-throughput screens. In model organisms amenable to positional or molecular cloning techniques (Table 13.1), screening criteria could be designed to clone ion channels and their modulatory subunits, identify the genetic and physiological pathways in which they act, and reveal their importance in the function of the whole organism.

In a forward genetic screen, mutations are introduced into the genome by chemical exposure or irradiation (Figure 13.2). The recovery of mutants in subsequent *felial* (F) generations, progeny resulting from a genetically defined parental cross, is governed by Mendelian inheritance. F_1 progeny can be screened for dominant alleles, or recessive alleles in F_2. Dominant alleles are most often gain-of-function alterations in gene function, since they produce an effect in the presence of the *wild-type* (normal, unmutagenized) gene product. *Hypermorphs* increase normal gene function (such as caused by a duplication), *antimorphs* antagonize

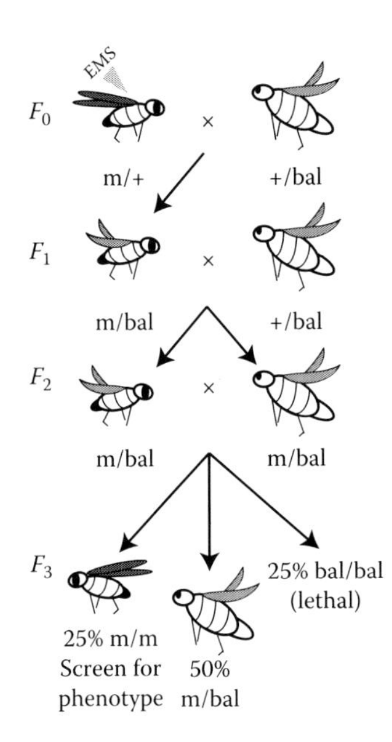

Figure 13.2 Mutagenesis Screens. Ethyl methane sulfonate (EMS), which generates mostly GC → AT base changes, is used to introduce random mutations into the *Drosophila* genome at an average of one mutation per 1000–5000 bp. In a typical screen for recessive *alleles* (variant forms of the gene), mutagenized male flies (m/+; F_0 generation) are crossed to females carrying chromosomal *balancers* (specialized chromosomes that do not undergo genetic recombination with the mutagenized chromosomes). F_1 progeny will have a random assortment of mutant alleles in *trans* to the balancers. Balancers carry visible mutations, such as wing shape or eye color, allowing selection of flies that harbor the mutant allele and the visible marker carried on the balancer. F_1 males carrying mutant alleles are backcrossed to the balancer stock to generate F_2 male and female progeny carrying the mutation over the balancer. Intercrossing F_2s produces a Mendelian ratio of homozygous mutant (m/m) progeny, which will express recessive phenotypes. Homozygous bal/bal progeny die due to inclusion of a recessive lethal mutation on the balancer chromosome.

normal gene function (such as a dominant negative), and *neomorphs* confer a new function to the gene (such as permitting new interactions or expression domains). More commonly, recessive phenotypes are desired, expressed in animals with a *homozygous* genotype (two copies of the allele). Recessive screens allow for recovery of deleterious loss-of-function mutations such as *hypomorphs* (partial loss of function) or *null* (complete loss of function) alleles. At some loci, several mutations mapping to the same gene may form an *allelic series*, a set of mutations with progressive severity. An allelic series is illustrated by the gene *CACNA1a* (P/Q Ca^{2+} channel) in mice with the *rocker*, *tottering*, and *rolling* alleles at the milder end, progressing to the more severe *leaner* allele (Zwingman et al., 2001).

Once a single gene is identified, additional screens can be designed to further reveal interacting gene products. In *modifier* or *sensitized* screens, mutagenesis is performed to introduce second-site mutations into a strain carrying the primary mutation. *Suppressor* screens identify second-site mutations that compensate or lessen the phenotype of the primary mutation, while *enhancer* screens identify mutations that act to amplify or worsen the primary mutation. Modifier screens can also be used to isolate second-site mutations within the same gene that interact with the primary mutated residues, providing an unbiased genetic screening method for double mutant cycle analysis. Modifier screens have been used to clone plant K^+ channels, investigate the structure and gating of mammalian K_{ir} channels, and identify pharmacological channel modulators (Minor, 2009).

13.2 MANIPULATING THE FUNCTION OF SPECIFIC ION CHANNELS: REVERSE GENETICS

Forward genetics drove the identification of many genes encoding ion channel subunits and relied on random mutagenesis in model organisms amenable to high-throughput phenotypic screening. In contrast, once a gene has been identified, *reverse genetics* allows the introduction of a specific, *targeted* alteration into the genome. The ability to engineer mutations into precise genetic loci was based on fundamental advances in molecular cloning techniques in the 1970s and 1980s. The discovery of restriction enzymes allowed DNA to be cut at specific sequences.* Ligases are then used to join together fragments of DNA with compatible ends.† Recombinant DNA can also be created and amplified using polymerase chain reaction (PCR).‡ This series of molecular biology techniques facilitated both the creation of ion channel cDNAs for introduction into plasmid-based expression vectors and the ability to manipulate sequences isolated from libraries, a prerequisite for reverse genetic alterations within a genome.

13.2.1 TYPES OF TRANSGENIC ALTERATIONS

In reverse genetics, *transgenic* colloquially refers to several types of animal models harboring an engineered genetic alteration. Mouse (*Mus musculus*) is a widely used laboratory animal model for reverse genetics based on its amenability to homologous recombination, shared genomic and physiological organization with humans, and accessibility to in vivo and ex vivo electrophysiological recordings. The following sections focus on transgenic manipulations in mouse (detailed methodology is covered in Nagy et al., 2003), although many of these approaches are routinely used in other organisms (Table 13.1).

Reverse genetic alterations are divided into targeted (*homologous*) and nontargeted (*random*) mutations. Homologous gene targeting involves mutation of a specific DNA sequence at its endogenous locus. This method is used to generate loss of function alleles through the deletion of essential genomic sequence (*knockout*) or gain of function alleles through the insertion of sequence that alters expression or function (*knock-in*). Detailed in the Section 13.2.2, these homologous recombination strategies require considerable time and resources to generate.

Simpler *nontargeted* mutagenesis (*conventional mutagenesis*; Palmiter et al., 1982) generates transgenic animals with a randomly inserted copy of a gene fragment into the genome

* For this work, the Nobel Prize in Physiology or Medicine was awarded to Werner Arber, Dan Nathans, and Hamilton Smith in 1978.

† Paul Berg was awarded the Nobel Prize in Chemistry in 1980 for generating the first recombinant DNA molecule, made from the synthetic ligation of several DNA fragments. Sharing the other half of the prize, Walter Gilbert and Fred Sanger developed methods for sequencing DNA.

‡ Kary B. Mullis and Michael Smith were awarded the Nobel Prize in Chemistry in 1993 for the development of PCR.

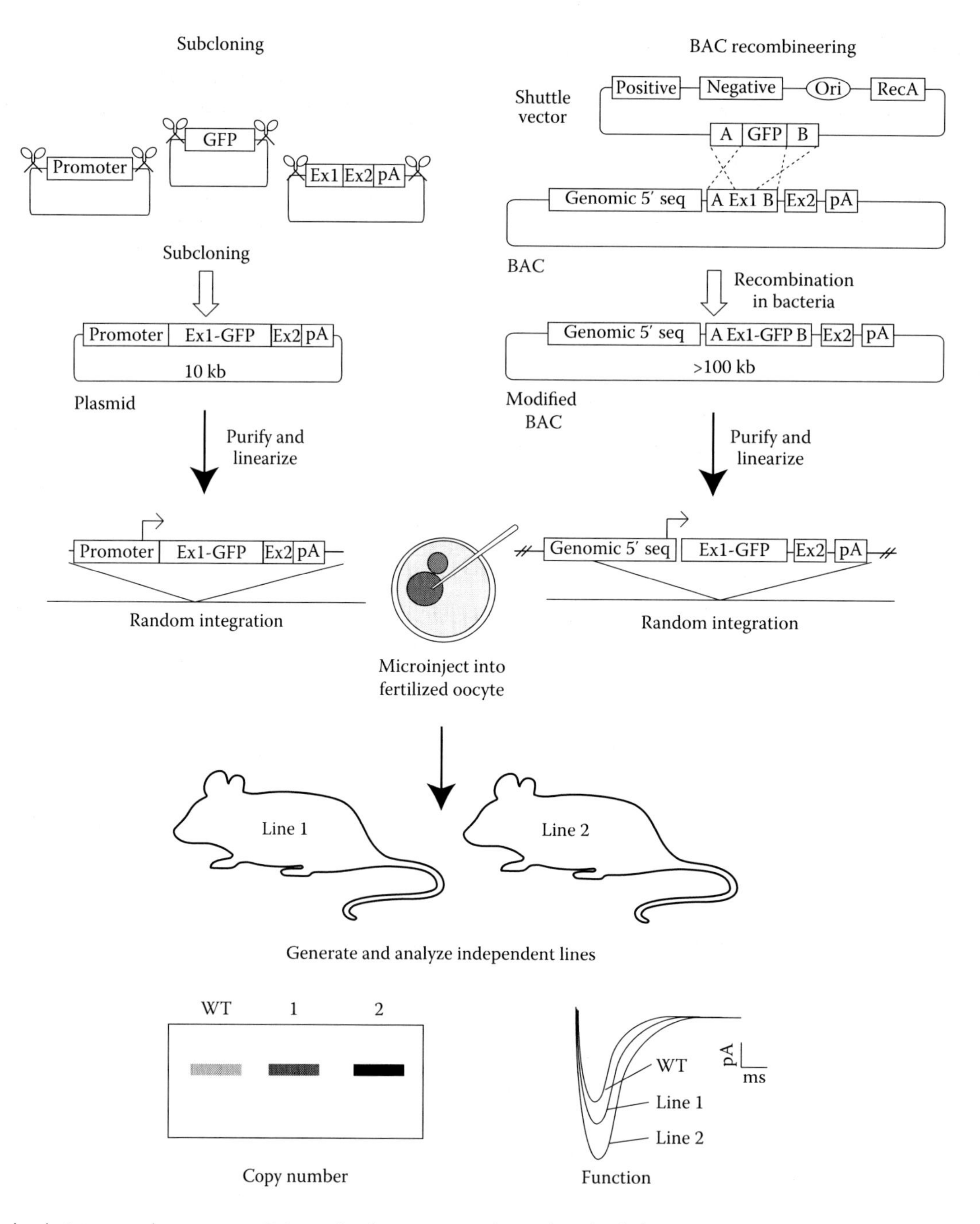

Figure 13.3 Randomly-integrated transgenes. Schematic of two commonly used methods for generating randomly integrated (nontargeted) transgenic mouse lines from an ion channel cDNA tagged with GFP. In a subcloning strategy (left), the three components of the transgene (promoter, cDNA, and GFP) are digested from their respective plasmid backbones and ligated together. The GFP marker is inserted in-frame with Exon1 of a cDNA, subcloned 3′ of the promoter fragment. This example construct is ~10 kb. A BAC-based strategy (right), joins the sequences by recombination in bacteria. A shuttle vector is constructed by subcloning short homology regions (A and B) flanking the GFP sequence. The shuttle vector is transformed into a recombination-competent host strain carrying a BAC with a large region of genomic sequence from the target gene (>100 kb). Recombination occurs across the homology regions, generating a mini-gene containing endogenous 5′ regulatory elements with GFP inserted (modified BAC). Both types of constructs are microinjected into fertilized embryos, where they randomly integrate into the genome, and lines are established from transgenic founders. Transgene copy number is analyzed by Southern blot, and the depicted ionic current magnitude is proportional to the transgene copy number in tissues that express the transgene.

(Figure 13.3). Based on the addition of a gene sequence, this type of transgenic is usually considered a gain-of-function. Minimally, the *transgene* is composed of promoter (sequences that direct transcription), a protein-coding sequence (cDNA), and a polyadenylation signal (sequences that terminate transcription and regulate mRNA stability). Basic transgenes are put together through sequential subcloning and/or PCR steps. The promoter and polyadenylation sequences may be derived from constitutively expressed housekeeping genes, such as β-actin, β-globin, or human growth hormone, or a strong viral promoter such as CMV or SV40. Other transcriptional regulatory elements may also be added to impart tissue-specific expression. The protein-coding sequence is usually composed of intron-less cDNA containing a translation start (ATG) and stop codon. The linearized construct is introduced into fertilized eggs by *pronuclear injection*, creating stable transgenic integrations in every cell that can be transmitted through the germline. This method is reasonably efficient, with 10%–40% of injected progeny carrying the transgene. The randomly integrated transgene incorporates into a single site, usually as a concatemer of several to hundreds of transgene copies. Due to copy number variation and the influence of surrounding sequence on transcription from the inserted transgene (*positional effects*), several lines may be characterized to choose the appropriate transgenic protein expression level (Figure 13.3). Routine genotyping is performed by PCR using primers that bind unique sequences within the transgene, distinguishing the transgenic gene product from wild-type.

Nontargeted strategies can be relatively straightforward, depending on the complexity of the transgenic construct. A major advantage over homologous recombination is the shorter time to generate transgenic mice and the higher rates of transgene integration, both of which significantly reduce cost. Nontargeted strategies are commonly used to express ion channels with protein tags, point mutations, or in new expression domains. Transgenic lines can be crossed onto knockout backgrounds to remove interactions with the wild-type gene product, or dominant functional effects can be characterized in the presence of the wild-type alleles.

A central drawback to nontargeted strategies is the relatively imprecise control of transgene expression due to both positional effects and incomplete regulatory elements. In practice, the number of synthetic DNA regulatory elements with defined expression patterns is small. Increasingly, more precise temporal and spatial control of expression is achieved by including transcriptional regulatory sequences derived from 5′ and 3′ untranslated regions (UTR), introns, and enhancer or repressor sequences from the genomic locus. Transgenes containing more comprehensive regulatory sequences can be generated from *bacterial artificial chromosomes* (BACs). BACs are large, stable plasmids containing inserts >100 kb, and libraries are available for most model organisms. Due to their size, BAC-based transgenes are less susceptible to positional effects, generally integrate in fewer copies, and allow delivery of large stretches of genomic regulatory sequence, even if the sequence has not yet been characterized. However, compared with conventional transgenic constructs, additional time and expertise is required for BAC-based strategies.

To generate a BAC transgene, restriction enzyme–based subcloning strategies are limited due to the very large size. Instead, modifications to the sequence, such as insertion of protein tags, are performed by *BAC recombineering* (recombination-mediated genetic engineering) (Testa et al., 2003). Recombination between a shuttle vector, containing modified coding sequence and selectable markers, and a BAC, containing the regulatory sequence needed flank the coding region, is used to generate a modified BAC transgene in recombination-competent bacterial strains. The modified BAC is microinjected and inserts randomly, similar to conventional constructs. BAC-based strategies can also be used for targeted homologous recombination (Valenzuela et al., 2003).

13.2.2 HOMOLOGOUS RECOMBINATION IN MOUSE

Targeted mutations in a gene can be generated in a number of ways, including deletions, replacements, or insertions of DNA (Joyner, 2000). A basic transgenic construct consists of two regions with homology to the target gene (*arms*) and positive and negative selection cassettes (Figure 13.4). The linearized targeting construct is electroporated into *embryonic stem* (*ES*) *cells*, an undifferentiated dividing cell line originally isolated from the pleuripotent inner cell mass of blastocyst stage embryos. At the blastocyst stage, chromosomal recombination occurs as a normal embryonic process yielding genetic variation. Exploiting this cellular recombination capability in ES cells, transgenes can be introduced into specific genomic sites defined by the homologous sequence arms on the targeting construct. The size and degree of match between the transgene homology arms and ES cell target locus determines the frequency of homologous recombination. Usually quite rare, the efficiency of intact targeting events can be enhanced by asymmetric homology arms (~5–10 kb on the long arm and 1–2 kb on the short arm), using the same strain DNA in the targeting construct and ES cell line, avoiding regions of repetitive DNA, and linearizing the targeting construct. Correct integration events are determined by PCR and Southern analysis, yielding an ES clone with a genomic modification introduced by homologous recombination.

ES cells have the capacity to differentiate into the three primary germ layers (ectoderm, endoderm, and mesoderm), including the mesoderm-derived gonads. Reintroduction of transgenic ES cells by injection into host blastocysts provides a principal mechanism for creating targeted germline modifications in animals (Figure 13.5). In 2007, the groundbreaking advance of targeted homologous recombination in mouse was recognized with the Nobel Prize, awarded to Martin Evans, Oliver Smithies, and Mario Capecchi.

CFTR was among the first gene loci to be targeted by a homologous recombination in mouse (Snouwaert et al., 1992). This fundamental first step was quickly followed by the targeted knock-in of $\Delta F508$, a deletion that accounts for ~70% of human cystic fibrosis cases (reviewed in Grubb and Boucher, 1999). In the following years, numerous mouse knockouts in voltage and ligand-gated ion channels followed. To date, almost all the ion channel pore-forming subunits have been knocked out in mouse (>100), many providing animal models for human channelopathies (Table 13.2). Recently, the International Knockout Mouse Consortium (IKMC) (https://www.mousephenotype.org/) has developed a high-throughput gene trap strategy to create a knockout of every gene in mouse.

A typical time line for generating a mouse line with a targeted genomic modification is about a year when the steps go smoothly and a core facility with routine expertise in ES cell injections is used. This estimate is prolonged by the number of steps required to make the targeting construct, efficiency of recombination in ES cells, and quality of the ES cells, which affects their ability to contribute the germline. The use of targeted ES cells generated by high-throughput screens can provide a shortcut, and ES cell repositories containing annotated ES cell clones are linked from the Trans-NIH Mouse Initiatives website (http://www.nih.gov/science/models/mouse/). The mutations carried out by these ES cells have generally not been verified as functional nulls and may not contain the mutation in a region of the gene that is desired. However, many knockout mouse lines have been successfully made, with less effort and cost, using ES cells from repositories.

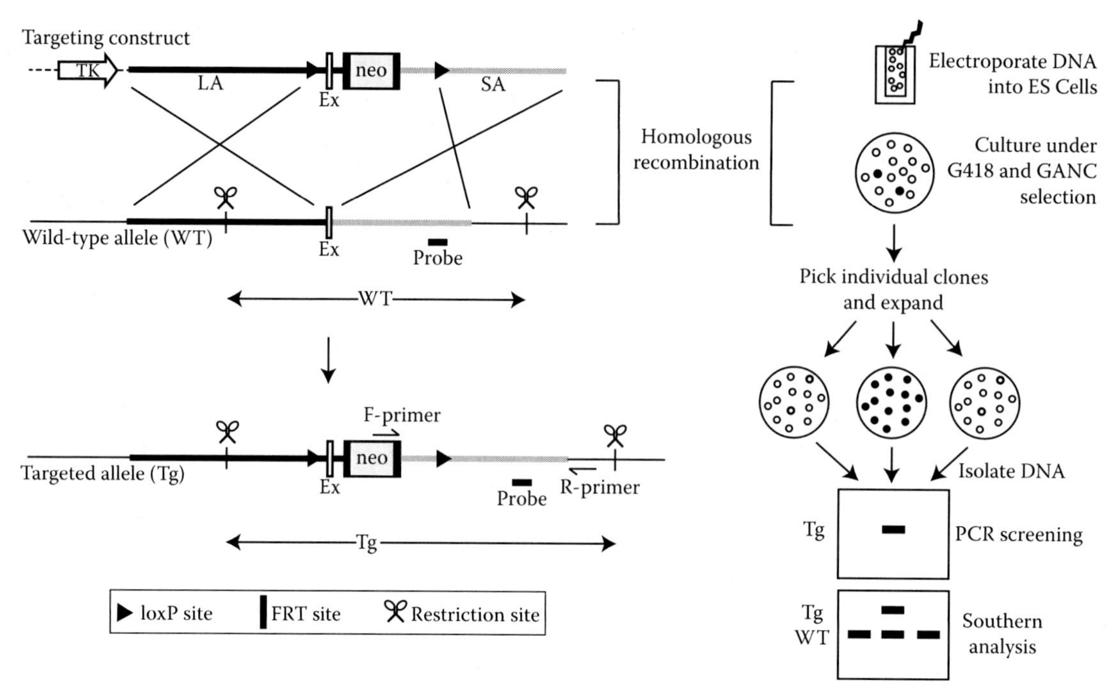

Figure 13.4 Homologous recombination in ES cells. Schematic of a homologous recombination strategy introducing *loxP* recombinase sites flanking a critical exon (Ex). A construct containing homology regions (LA, Long Arm and SA, Short Arm) to the target gene is electroporated into ES cells. Clones retaining Thymidine Kinase (TK), indicating random integration of the transgene, are eliminated by negative selection in gancyclovir (GANC). Homologous recombination with the wild-type allele causes loss of TK (survival in GANC). Positive selection in G418 is conferred by an antibiotic resistance gene (commonly *neomycin*, neo), which allows survival of clones that undergo a double crossover event resulting in intact integration of both homology arms. Distinct site-specific recombination sequences (e.g., Flippase recognition target, FRT) can be used to remove the *neo* positive selection cassette in ES cells or later in transgenic animals. Typically several hundred clones survive positive–negative selection. DNA from these clones is screened by PCR using primers that span the transgenic junction (F- and R-primers) in 96-well plates. Clones with correct PCR product sizes, usually fewer than 10, are further analyzed for complete integration of the transgene by Southern blot. DNA digested by restriction enzymes is hybridized with a common probe that produces a larger band due to transgene integration. (Adapted from Mortenson, R., Overview of gene targeting by homologous recombination, in: Ausubel, F.M. et al. (eds.), *Current Protocols in Molecular Biology*, John Wiley & Sons, Hoboken, NJ, 1999.)

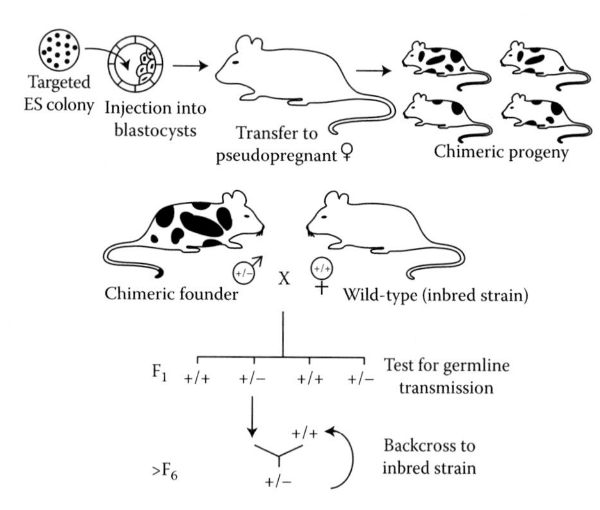

Figure 13.5 Generation of chimeric founders and a targeted mouse line. Correctly targeted ES cell clones are injected into blastocyst stage mouse embryos and implanted into pseudopregnant females. By using ES cells and host blastocysts with different coat colors, the relative ratio of ES cell-derived transgenic tissues can be estimated visually in the pups by coat color. Chimeric founders with a high contribution of ES cell–derived coat color are mated to wild-type mice. Progeny are genotyped to determine whether the ES cells have successfully populated the founder's gonads and resulted in germline transmission of the transgene. Routine genotyping is performed by PCR from tail DNA, usually with primers specific for the transgenic and wild-type alleles. To establish a transgenic mouse line, heterozygous progeny are backcrossed >6 generations to the desired inbred strain.

Although mouse methodology has been honed over several decades, going forward, targeted genetic manipulation in mammalian models is expected to become faster and less labor intensive by moving beyond first generation homologous recombination strategies. Two novel methods of genome engineering, the TALEN (transcription activator-like effector nuclease) and CRISPR (clustered regulatory interspaced short palindromic repeat) systems, have increased targeting efficiency and can be used in animals not amenable to homologous recombination, such as human cell lines (Gaj et al., 2013). With the genomic locations of most ion channel subunits known, the flexibility of newer genetic techniques will likely yield novel animal models, expanding the repertoire of physiological questions that can be addressed.

13.2.3 CONDITIONAL SITE-SPECIFIC RECOMBINATION

Conventional targeting strategies generate a permanently modified genomic locus in every cell. For genes with complex expression patterns, sorting out the pleiotropic effects in a conventional transgenic model can be complicated. In some cases, alteration of a gene throughout the animal can lead to lethality, developmental defects, or loss of phenotypes due to compensation for altered protein expression. Conditional modification of genes based on site-specific recombination of DNA was developed to alleviate these issues through more precise control over the tissues and temporal

Table 13.2 Ion channel genes, mouse models, and human diseases

TYPE	GENE ID (MGI)	TRANSGENIC ALLELES	MOUSE PHENOTYPE	HUMAN CHROMOSOME (OMIM)	HUMAN DISEASE
$Na_v1.1$	*SCN1A* (98246)	6	Seizures, behavioral deficits, postnatal lethality	2q24.3 (182389)	Dravet's Syndrome, generalized epilepsy, familial migraine
$Na_v1.2$	*SCN2A* (98248)	2	Neuronal apoptosis, severe hypoxia, neonatal lethality	2q24.3 (182390)	Epileptic encephalopathy, infantile seizures
$Na_v1.3$	*SCN3A* (98249)	2	Postnatal lethality; Het: improved glucose tolerance	2q24.3 (182391)	
$Na_v1.4$	*SCN4A* (98250)	8	Neonatal lethality, hind limb clasping; KI: K^+-sensitive myotonia, myofiber damage	17q23.3 (603967)	Hyperkalemic periodic paralysis, Myasthenic syndrome, Myotonia congenita, Paramyotonia congenita
$Na_v1.5$	*SCN5A* (98251)	11	Prenatal lethality, abnormal cardiovascular organogenesis; Het: abnormal cardiac conduction	3p22.2 (600163)	Atrial and ventricular fibrillation, Brugada syndrome, cardiomyopathy, long QT syndrome, heart block, sudden infant death susceptibility
$Na_v1.6$	*SCN8A* (103169)	4	Ataxia, dystonia, progressive paralysis, Purkinje cell loss, lethality; PM: deafness, tremor	12q13.13 (600702)	Cognitive impairment, cerebellar ataxia, infantile epileptic encephalopathy
$Na_v1.7$	*SCN9A* (107636)	4	Prenatal/neonatal lethality; KI: seizures	2q24.3 (603415)	Epilepsy, febrile seizures, pain disorders and small fiber neuropathy, Erythermalgia, Modifier of Dravet syndrome
$Na_v1.8$	*SCN10A* (108029)	4	Impaired pain response; PM: catalepsy, increased cold pain sensitivity	3p22.2 (604427)	Familial episodic pain syndrome
$Na_v1.9$	*SCN11A* (1345149)	6	Decreased pain and hyperalgesia responses	3p22.2 (604385)	Familial episodic pain syndrome, sensory and autonomic neuropathy
$Ca_v1.1$	*CACNA1S* (88294)	3	Neonatal lethality, muscle defects, secondary skeletal defects	1q32.1 (114208)	Hypokalemic periodic paralysis, malignant hyperthermia, thyrotoxic periodic paralysis
$Ca_v1.2$	*CACNA1C* (103013)	14	Prenatal lethality; beta cell deletion: impaired insulin secretion and glucose intolerance	12p13.33 (114205)	Brugada syndrome, Timothy syndrome
$Ca_v1.3$	*CACNA1D* (88293)	5	Growth defect, hypoinsulinemia, glucose intolerance, deafness, bradycardia, arrhythmia	3p14.3 (114206)	Sinoatrial node dysfunction and deafness
$Ca_v1.4$	*CACNA1F* (1859639)	5	Abnormal retinal morphology and electrophysiology	Xp11.23 (300110)	Aland Island eye disease, X-linked cone-rod dystrophy, congenital night blindness
$Ca_v2.1$	*CACNA1A* (109482)	20	Ataxia, episodic dyskinesia, cerebellar atrophy, and absence epilepsy	19p13.2 (601011)	Episodic ataxia, familial hemiplegic migraine with progressive cerebellar ataxia, spinocerebellar ataxia

(*Continued*)

Table 13.2 (*Continued*) Ion channel genes, mouse models, and human diseases

TYPE	GENE ID (MGI)	TRANSGENIC ALLELES	MOUSE PHENOTYPE	HUMAN CHROMOSOME (OMIM)	HUMAN DISEASE
$Ca_v2.2$	*CACNA1B* (88296)	6	Abnormal nociception, learning and memory, sleep–wake cycle, hyperactivity, and circulation	9q34.3 (601012)	
$Ca_v2.3$	*CACNA1E* (106217)	6	Increased timidity and body weight, impaired glucose tolerance, reduced locomotor activity	1q25.3 (601013)	
$Ca_v3.1$	*CACNA1G* (1201678)	7	Disrupted sleep, locomotor activity, neural activity, prolonged cardiac conduction	17q21.33 (604065)	
$Ca_v3.2$	*CACNA1H* (1928842)	3	Constitutive coronary arteriole contraction and focal myocardial fibrosis	16p13.3 (607904)	Childhood absence and idiopathic generalized epilepsy
$Ca_v3.3$	*CACNA1I* (2178051)	4		22q13.1 (608230)	
$K_v1.1$	*KCNA1* (96654)	3	Seizures, behavioral abnormalities, megencephaly, and prenatal lethality	12p13.32 (176260)	Episodic ataxia, myokymia syndrome
$K_v1.2$	*KCNA2* (96659)	4	Postnatal lethality, seizures, altered neuronal activity; PM: ataxia, impaired coordination, tremor	1p13 (176262)	
$K_v1.3$	*KCNA3* (96660)	3	Reduced body weight, increased metabolic rate and insulin sensitivity, resistance to obesity	1p13.3 (176263)	
$K_v1.4$	*KCNA4* (96661)	3	Subset of mutants exhibit spontaneous seizures	11q14.1 (176266)	
$K_v1.5$	*KCNA5* (96662)	3	Abnormal microglia and NO release after lesion; KI: impaired pulmonary vasoconstriction, resistance to drug-induced long QT	12p13.32 (176267)	Familial atrial fibrillation
$K_v1.6$	*KCNA6* (96663)	2	Increased thermal nociceptive threshold, increased circulating triglyceride levels	12p13.32 (176257)	
$K_v1.7$	*KCNA7* (96664)	3		19q13.33 (176268)	Progressive familial heart block
$K_v2.1$	*KCNB1* (96666)	3	Reduced fasting glucose levels, improved glucose tolerance, hyperinsulinemia	20q13.3 (600397)	
$K_v3.1$	*KCNC1* (96667)	3	Neurological, motor, and growth deficits	11p15.1 (176258)	Long QT syndrome
$K_v3.2$	*KCNC2* (96668)	3	Altered cortical excitability, seizures, increased anxiety, and abnormal sleep	12q21.1 (176256)	
$K_v3.3$	*KCNC3* (96669)	1	No overt phenotype; Kv3.1/Kv3.3: hyperactive, myoclonus, tremor, and ethanol hypersensitivity	19q13.33 (176264)	Spinocerebellar ataxia

(*Continued*)

Table 13.2 (*Continued*) Ion channel genes, mouse models, and human diseases

TYPE	GENE ID (MGI)	TRANSGENIC ALLELES	MOUSE PHENOTYPE	HUMAN CHROMOSOME (OMIM)	HUMAN DISEASE
$K_v3.4$	*KCNC4* (96670)	2		1p13.3 (176265)	
$K_v4.1$	*KCND1* (96671)	1		Xp11.23 (300281)	
$K_v4.2$	*KCND2* (102663)	3	Reduced I_A in spinal dorsal horn neurons, enhanced sensitivity to tactile and thermal stimuli	7q31.31-7q32 (605410)	
$K_v4.3$	*KCND3* (1928743)	2	No overt phenotype	1p13.2 (605411)	Spinocerebellar ataxia
$K_v7.1$	*KCNQ1* (108083)	8	Neurological defects and deafness	11p15.5-p15.4 (607542)	Atrial fibrillation, Jervell and Lange-Nielsen syndrome, Long and short QT syndrome
$K_v7.2$	*KCNQ2* (1309503)	7	Neonatal lethality, pulmonary atelectasis; Het: seizures; KI: seizures and lethality	20q13.33 (602235)	Epileptic encephalopathy, early infantile, myokymia, neonatal seizures
$K_v7.3$	*KCNQ3* (1336181)	3	Abnormal after hyperpolarization currents in granule cells; KI: seizures and lethality	8q24.22 (602232)	Neonatal seizures
$K_v7.4$	*KCNQ4* (1926803)	5	Progressive hearing loss, chronic depolarization and degeneration of cochlear outer hair cells	1p34.2 (603537)	Deafness
$K_v10.1$ EAG	*KCNH1* (1341721)	1	No overt phenotype, mild hyperactivity, decreased response during tail suspension test	1q32.2 (603305)	
$K_v11.1$ hERG	*KCNH2* (1341722)	4	Prenatal lethality, defects in cardiac development and function; 1a-only deletion: sinus bradycardia	7q36.1 (152427)	Long and short QT syndrome, schizophrenia
$K_v12.2$ ELK2	*KCNH3* (1341723)	7	Abnormal learning and memory, neuronal hyperexcitability, and seizures	12q13.12 (604527)	
$K_{ir}1.1$ ROMK	*KCNJ1* (1927248)	5	Postnatal lethality, impaired electrolyte/acid-base/fluid-volume homeostasis, Bartter's model	11q24.3 (600359)	Bartter's syndrome
$K_{ir}2.1$ IRK	*KCNJ2* (104744)	2	Neonatal lethality, cyanosis, loss of K^+-mediated vasodilatation in cerebral arteries	17q24.3 (600681)	Andersen syndrome, familial atrial fibrillation, short QT syndrome
$K_{ir}2.2$	*KCNJ12* (108495)	4	No overt phenotype	17p11.2 (602323)	
$K_{ir}3.1$ GIRK1	*KCNJ3* (104742)	3	Increased resting heart rate, altered vagal activation response	2q24.1 (601534)	
$K_{ir}3.2$ GIRK2	*KCNJ6* (104781)	1	Seizures, ataxia, hypotonia, increased mortality, Purkinje cell defects, male sterility	21q22.13 (600877)	
$K_{ir}3.3$ GIRK3	*KCNJ9* (108007)	3	No overt phenotype	1q23.2 (600932)	

(*Continued*)

Table 13.2 (*Continued*) Ion channel genes, mouse models, and human diseases

TYPE	GENE ID (MGI)	TRANSGENIC ALLELES	MOUSE PHENOTYPE	HUMAN CHROMOSOME (OMIM)	HUMAN DISEASE
$K_{ir}3.4$ GIRK4	*KCNJ5* (104755)	5	Tachycardia, reduced muscarinic-gated atrial potassium channel responses to stimulation	11q24.3 (600734)	Hyperaldosteronism, long QT syndrome
$K_{ir}4.1$	*KCNJ10* (1194504)	4	Abnormal Muller cell activity, endocochlear potential, and locomotion, spinal hypomyelination	1q23.2 (602208)	Enlarged vestibular aqueduct, SESAME syndrome
$K_{ir}4.2$	*KCNJ15* (1310000)	4	Impaired balance and coordination	21q22.13 (602106)	Diabetes mellitus, Down syndrome
$K_{ir}6.1$	*KCNJ8* (1100508)	5	Sudden cardiac death, dysregulation of coronary artery tone, Prinzmetal angina model	12p12.1 (600935)	
$K_{ir}6.2$ (K_{ATP})	*KCNJ11* (107501)	2	Impaired insulin secretion, hypoxia-induced seizure, arrhythmia and sudden death	11p15.1 (600937)	Diabetes mellitus, hyperinsulinemic hypoglycemia
$K_{ir}7.1$	*KCNJ13* (3781032)	3		2q37.1 (603208)	Leber congenital amaurosis, snowflake vitreoretinal degeneration
$K_{Ca}1.1$ BK	*KCNMA1* (99923)	4	Abnormal smooth muscle, locomotor, circadian rhythm, and hearing function, ataxia, tremor	10q22.3 (600150)	Generalized epilepsy and paroxysmal dyskinesia
$K_{Ca}2.1$ SK1	*KCNN1* (1933993)	5	No overt phenotype	19p13.1 (602982)	
$K_{Ca}2.2$ SK2	*KCNN2* (2153182)	3	Loss of apamin-sensitive after hyper polarization current; PM: tremor and gait abnormalities	5q22.3 (605879)	
$K_{Ca}2.3$ SK3	*KCNN3* (2153183)		No overt phenotype; KI: abnormal respiratory response to hypoxia, abnormal parturition and death	1q21.3 (602983)	Schizophrenia
$K_{Ca}3.1$ IK/ SK4	*KCNN4* (1277957)	4	Impaired erythrocytes, T lymphocyte volume regulation, and ACh-induced artery dilation, increased blood pressure	19q13.31 (602754)	
K_{Na}/Slo2.2 Slack	*KCNT1* (1924627)	3		9q34.3 (608167)	Frontal lobe epilepsy, infantile epileptic encephalopathy
K_{Na}/Slo2.1 Slick	*KCNT2* (3036273)	3		1q31.3 (610044)	
Slo3	*KCNU1* (1202300)	5	Male infertility with impaired sperm morphology, motility, and capacitation	8p11.23 (615215)	
$K_{2P}1.1$ TWIK1	*KCNK1* (109322)	3	Reduced urinary flow on a low phosphate diet, attenuated renal phosphate reabsorption	1q42.2 (601745)	
$K_{2P}2.1$ TREK	*KCNK2* (109366)	2	Increased sensitivity to pharmacologically induced seizures and ischemia	1q41 (603219)	

(*Continued*)

Table 13.2 (*Continued*) Ion channel genes, mouse models, and human diseases

TYPE	GENE ID (MGI)	TRANSGENIC ALLELES	MOUSE PHENOTYPE	HUMAN CHROMOSOME (OMIM)	HUMAN DISEASE
$K_{2P}3.1$ TASK1	*KCNK3* (1100509)	4	Impaired motor control and chemosensitivity in raphe neurons; KCNK3/9 deletion: hyperaldosteronism	2p23.3 (603220)	Pulmonary hypertension
$K_{2P}5.1$ TASK2	*KCNK5* (1336175)	2	Prenatal lethality, smaller size, abnormal respiratory physiology	6p21.2 (603493)	
$K_{2P}6.1$ TWIK2	*KCNK6* (1891291)	2	Vascular dysfunction and hypertension	19q13.2 (603939)	
$K_{2P}7.1$	*KCNK7* (1341841)	4	No overt phenotype, no significant alterations in volatile anesthetic sensitivity	11q13.1 (603940)	
HCN1	*HCN1* (1096392)	6	Forebrain deletion: deficits in motor learning and memory	5p12 (602780)	Infantile epileptic encephalopathy
HCN2	*HCN2* (1298210)	6	Altered resting membrane potential, cardiac sinus dysrhythmia, behavioral/ neurological abnormalities, tremors, absence seizures, altered neuropathic pain	19p13.3 (602781)	Febrile seizure, epilepsy
HCN3	*HCN3* (1298211)	3	Abnormal ventricular action potential waveform	1q22 (609973)	
HCN4	*HCN4* (1298209)	9	Prenatal lethality; cardiac deletion: sinus arrhythmia, severe bradycardia and death	15q24.1 (11)	Brugada syndrome, sick sinus syndrome
rodCNG	*CNGA1* (88436)	4		4p12 (123825)	Retinitis pigmentosa
olfCNG	*CNGA2* (108040)	7	Neonatal lethality, growth retardation, and abnormal olfactory neuron response to forskolin	Xq28 (300338)	
coneCNG	*CNGA3* (1341818)	2	Progressive loss of cone photoreceptor cells	2q11.2 (600053)	Achromatopsia-2
ANKTM1	*TRPA1* (3522699)	7	Deficits in inflammatory nociceptor excitation and pain hypersensitivity, decreased response to mustard oil, cold, and mechanical stimuli	8q13.3 (604775)	Episodic pain syndrome, familial
VR1 Capsaicin receptor	*TRPV1* (1341787)	5	Abnormal nociception, taste, and febrile response, increased sensitivity to renal damage, resistance to obesity, altered bladder function	17p13.2 (602076)	
VRL1	*TRPV2* (1341836)	4	Impaired macrophage migration, binding, and phagocytosis, increased mortality from infection	17p11.2 (606676)	
VRL3	*TRPV3* (2181407)	7	Deficits in response to innocuous and noxious heat but not in other sensory modalities	17p13.2 (607066)	Olmsted syndrome

(*Continued*)

Table 13.2 (*Continued*) Ion channel genes, mouse models, and human diseases

TYPE	GENE ID (MGI)	TRANSGENIC ALLELES	MOUSE PHENOTYPE	HUMAN CHROMOSOME (OMIM)	HUMAN DISEASE
VRL2	*TRPV4* (1926945)	5	Reduced responsiveness to pressure sensations and abnormal osmotic regulation	12q24.11 (605427)	Brachyolmia, familial digital arthropathy-brachydactyly, motor and sensory neuropathy, metatropic dysplasia, parastremmatic dwarfism, scapuloperoneal spinal muscular atrophy, spinal muscular atrophy, spondylometaphyseal dysplasia
ECaC1	*TRPV5* (2429764)	2	Increased calcium excretion and reduced bone thickness	7q34 (606679)	
ECaC2	*TRPV6* (1927259)	8	KI: impaired sperm motility	7q34 (606680)	Calcium oxalate nephrolithiasis
TRPM1	*TRPM1* (1330305)	3	Defects in rod and cone electrophysiology and impaired photoresponses	15q13.3 (603576)	Congenital night blindness
TRPM2	*TRPM2* (1351901)	5	Impaired ROS-induced chemokine production in monocytes, reduced neutrophil infiltration and ulceration in colitis inflammation model	21q22.3 (603749)	Bipolar disorder
TRPM3	*TRPM3* (2443101)	2	Impaired thermal and chemical nociception	9q21.12 (608961)	
TRPM4	*TRPM4* (1915917)	7	Altered Ca^{2+} responses, mast, and dendritic cell activation, increased vascular permeability and acute anaphylactic responses	19q13.33 (606936)	Progressive familial heart block, type IB
TRPM5	*TRPM5* (1861718)	7	Abnormal taste perception (respond to sour and salty, but not sweet or bitter stimuli)	11p15.5 (604600)	
TRPM6	*TRPM6* (2675603)	5	Postnatal lethality with exencephaly; Het: Some premature death, decreased serum Mg^{2+}	9q21.13 (607009)	Hypomagnesemia
TRPM7	*TRPM7* (1929996)	5	Prenatal lethality, thymocyte-specific deletion has blocked thymopoiesis	15q21.2 (605692)	Amyotrophic lateral sclerosis-parkinsonism dementia complex
TRPM8	*TRPM8* (2181435)	6	Decreased sensitivity to cold and reduced response to cold stimuli	2q37.1 (606678)	
TRPC1	*TRPC1* (109528)	4	Increased body weight, loss of salivary secretion from attenuation of store-operated Ca^{2+} currents	3q23 (602343)	
TRPC2	*TRPC2* (109527)	6	Altered sexual and social behavior, increased triglyceride and cholesterol levels	Pseudogene	
TRPC3	*TRPC3* (109526)	5	Abnormal gait, impaired mGluR transmission in Purkinje cells; PM: Purkinje cell death, lethality	4q27 (602345)	

(*Continued*)

Table 13.2 (*Continued*) Ion channel genes, mouse models, and human diseases

TYPE	GENE ID (MGI)	TRANSGENIC ALLELES	MOUSE PHENOTYPE	HUMAN CHROMOSOME (OMIM)	HUMAN DISEASE
TRPC4	*TRPC4* (109525)	3	Reduced agonist-induced Ca^{2+} entry and vasorelaxation of aortic rings	13q13.3 (603651)	
TRPC5	*TRPC5* (109524)	4	Diminished fear response with reduction in synaptic activation and amygdala	Xq23 (300334)	
TRPC6	*TRPC6* (109523)	3	No overt phenotype or increased thermal nociceptive response latency	11q22.1 (603652)	Glomerulosclerosis
TRPC7	*TRPC7* (1349470)	5	Persistence of light-evoked responses in melanopsin-ganglion cells		
CRAC	*ORAI1* (1925542)	4	Reduced size, defective mast cell degranulation, cytokine secretion and passive anaphalaxis	12q24.31 (610277)	Immunodeficiency due to impaired T-cell activation
CLC-1	*CLCN1* (88417)	2	PM: mild to severe spasms of the hind limbs and abnormal hind limb reflexes	7q34 (118425)	Myotonia congenita, Myotonia levior
CLC-2	*CLCN2* (105061)	5	Abnormal brain and eye morphology, male infertility	3q27.1 (600570)	Idiopathic and myoclonic epilepsy
CLC-K1	*CLCNKA* (1930643)	2	Increased urine, dehydration, loss of Cl^- gradient in thick ascending limb	1p36.13 (602024)	Bartter syndrome
TMEM16A	*ANO1* (2142149)	5	Postnatal lethal, slow weight gain, aerophagia, cyanosis, and tracheomalacia	11q13.3 (610108)	
TMEM16B	*ANO2* (2387214)	3	Absent Ca^{2+}-activated Cl^- currents in olfactory epithelium and VNO but normal olfaction in behavioral tasks	12p13.31 (610109)	
TMEM16E	*ANO5* (3576659)	1		11p14.3 (608662)	Gnathodiaphyseal dysplasia, Miyoshi muscular dystrophy, limb-girdle muscular dystrophy
TMEM16F	*ANO6* (2145890)	3	Impaired platelet coagulation, increased bleeding, decreased bone mineral deposition and skeletal defects	12q12 (608663)	Scott syndrome
TMEM16G	*ANO7* (3052714)			2q37.3 (605096)	
TMEM16K	*ANO10* (2143103)	2		3p22.1 (613726)	Spinocerebellar ataxia
Cx43	*GJA1* (95713)	23	Neonatal lethal with heart, eye lens, and male germ cell anomalies; ovary deletion affects follicular development; neuron/astrocyte deletion: behavioral and neural defects	6q22.31 (121014)	Atrioventricular septal defect, craniometaphyseal dysplasia, hypoplastic left heart syndrome, oculodentodigital dysplasia, syndactyly

(*Continued*)

Table 13.2 (*Continued*) Ion channel genes, mouse models, and human diseases

TYPE	GENE ID (MGI)	TRANSGENIC ALLELES	MOUSE PHENOTYPE	HUMAN CHROMOSOME (OMIM)	HUMAN DISEASE
Cx46	*GJA3* (95714)	23	Nuclear lens cataracts associated with the breakdown of gamma crystalline	13q12.11 (121015)	Cataract
Cx37	*GJA4* (95715)	4	Females are sterile, oocyte development arrest, lack of mature Graafian follicles, formation of premature corpora lutea	1p34.3 (121012)	Atherosclerosis, myocardial infarction
Cx32	*GJB1* (95719)	3	Decreased body weight, enhanced neuronal sensitivity to ischemic insults, and liver tumors	Xq13.1 (304040)	X-linked Charcot-Marie-Tooth neuropathy
Cx26	*GJB2* (95720)	10	Slow growth with impaired transplacental nutrient/glucose uptake, prenatal lethality; inner ear deletion: hearing impaired	13q12.11 (121011)	Bart-Pumphrey syndrome, deafness, hystrix-like and keratitis-ichthyosis with deafness, palmoplantar keratoderma, Vohwinkel syndrome
Cx31	*GJB3* (95721)	4	Partial lethality, transient placental dysmorphogenesis, no hearing or skin defect	1p34.3 (603324)	Deafness, erythrokeratodermia variabilis et progressiva
Cx30.3	*GJB4* (95722)	2	Reduced behavioral responses to a vanilla scent, suggesting impaired olfaction	1p34.3 (605425)	Erythrokeratodermia variabilis with erythema gyratum repens
Cx36	*GJD2* (1334209)	9	Loss of electrical synapses, impaired synchronous activity of olfactory/retinal networks, altered insulin secretion	15q14 (607058)	Familial epilepsy

Source: Mouse phenotypes are for homozygous null unless stated otherwise. Phenotypic descriptions may be condensed from more than one transgenic mouse line. Numbers refer to records in the MGI and OMIM databases. Transgenic alleles: number of targeted alleles in MGI (Eppig et al., 2012). Transgenic alleles are available for most ion channels not listed (see MGI). HET, heterozygous phenotype; KI, Knock-In allele phenotype; PM, point mutation phenotype; Online Mendelian Inheritance in Man, OMIM®, McKusick-Nathans Institute of Genetic Medicine, Johns Hopkins University, Baltimore, MD, 2013, World Wide Web URL: http://omim.org/; Mouse Genome Database (MGD) at the Mouse Genome Informatics website, The Jackson Laboratory, Bar Harbor, ME, World Wide Web (URL: http://www.informatics.jax.org), November, 2013.

aspects of expression from the transgenic locus. The first widely used conditional modification system in mouse was *Cre/loxP* (Gu et al., 1993; Orban et al., 1992). *Cre* (*c*auses *re*combination) is an integrase-type enzyme from bacteriophage that catalyzes a recombination event between two specific DNA sequences, termed *loxP* (*lo*cus of *x*-over in *P*1) sites. Depending on the orientation of the *loxP* sites, the intervening DNA may be deleted, translocated, or inverted (Figure 13.6). Cre/*loxP* recombination is highly specific, with little endogenous recombination in mouse, and efficient, nearing 100% recombination in some cases.

Typically, a transgene containing an entire gene sequence or an essential functional domain, such as the pore of an ion channel, is *floxed* (*fl*anked with *loxP* sites). The sites are introduced into the genomic locus by homologous recombination in ES cells, and a floxed mouse line is generated using standard transgenic methods. A separate mouse line expressing Cre recombinase must also be generated. Many Cre mouse lines have been generated as randomly integrated transgenes containing previously characterized regulatory sequence driving Cre expression in a specific tissue or temporal pattern. The two lines are crossed, yielding Cre$^+$, floxed progeny (double transgenics). In a straightforward conditional knockout strategy, double heterozygotes are intercrossed to generate F_2 animals with two copies of the floxed allele (homozygous) and at least one copy of the Cre transgene. Homozygous Cre is generally not required for efficient recombination. Following the tissue and temporal expression pattern of Cre, these animals will be chimeric for the conditionally modified floxed allele (Figure 13.7).

Gene inactivation is not the only outcome for conditional recombination. Gene expression can be turned on selectively by introducing a flox-stop-flox cassette. The presence of the stop sequence suppresses gene expression until Cre-mediated excision of the stop sequence. Similarly, expression from a cDNA can be silenced by inverting the sequence in an intron. Activation of expression occurs through Cre-mediated inversion, bringing the cDNA into frame for transcription (Figure 13.6). This technique can be used to conditionally express markers, such as fluorescent proteins, or exchange wild-type for mutant sequence, such as introducing a disease-linked point mutation.

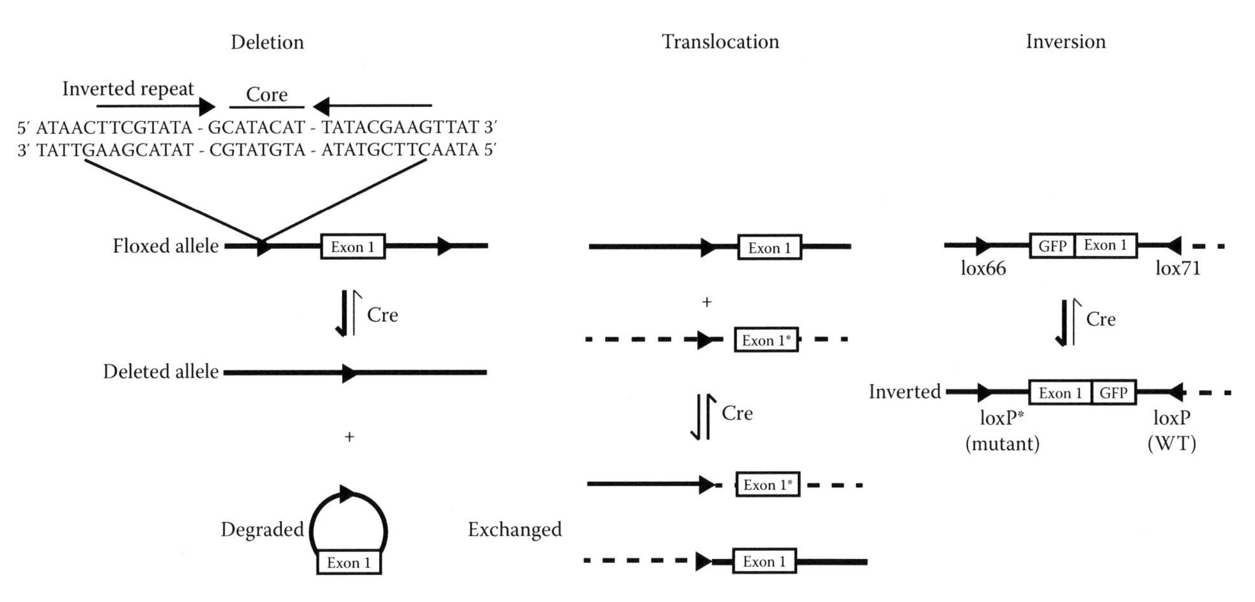

Figure 13.6 Cre-mediated recombination. Cre catalyzes several types of recombination events at *loxP* sites. A 34 bp *loxP* site consists of two 13 bp inverted repeats surrounding an 8 bp nonpalindromic core sequence. The orientation of the sites is determined by the direction of the core sequence. Placement of two direct repeat *loxP* sites in *cis* results in intramolecular recombination and deletion of the intervening DNA. This recombination event is generally limited to the forward reaction due to degradation of the excised sequence. *Trans* arrangement of direct *loxP* repeats results in intermolecular recombination and reciprocal exchange of DNA. The reaction may proceed in both directions. Inverted orientation *loxP* sites in *cis* will result in inversion of the intervening sequence. The reaction can be biased toward a single reaction product by using alternate *loxP* sites. Recombination between lox66 and lox71 regenerates one wild-type and one mutant *loxP* site, which inefficiently binds Cre and prevents the reverse reaction.

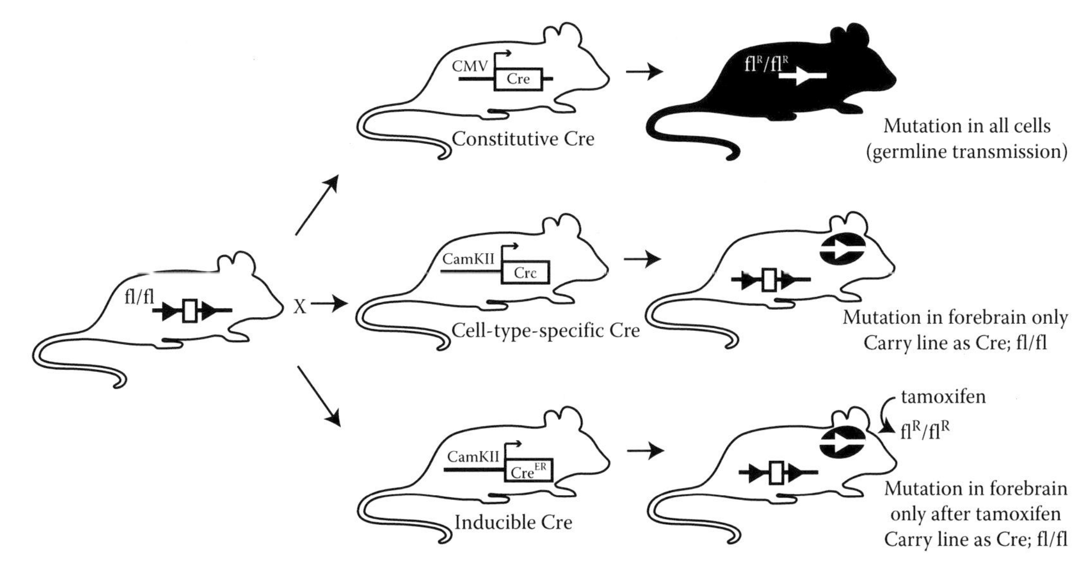

Figure 13.7 Tissue-specific gene modifications in Cre; floxed double transgenic mice. A mouse line carrying a floxed allele is mated to three representative Cre lines, which differ in their Cre expression patterns. (Top) A strong viral promoter (CMV) drives Cre expression in every tissue, including the germline. Once the recombination event occurs, the floxed allele (fl/fl) undergoes recombination everywhere (fl^R/fl^R). The mutation will be stably heritable through the germline. Tissue-specific expression is achieved using gene fragments from previously characterized promoter regions to drive Cre. (Middle) For example, an ~8 kb fragment of the calcium/calmodulin dependent protein kinase II (CamKII) gene drives forebrain-specific Cre expression. (From Tsien, J.Z. et al., *Cell*, 87, 1317, 1996.) (Bottom) Fusion of the Cre protein to a mutated estrogen receptor (Cre^{ER}) renders Cre translocation to the nucleus tamoxifen-dependent (From Feil, R. et al., *Proc. Natl. Acad. Sci. USA*., 93, 10887, 1996), combining tissue specificity with temporal control of Cre activity. Because the germline does not undergo recombination in a brain-specific Cre line, the line is maintained as a double transgenic.

A prerequisite for the conditional knockout strategy is that the unrecombined allele should produce normal gene expression until Cre is expressed. However, some selectable markers, such as the PGK*neo* cassette, have been shown to inadvertently disrupt expression from the endogenous gene or neighboring loci before Cre-mediated recombination occurs. The *neo* sequence contains a cryptic splice acceptor that can interfere with splicing and cause a hypomorphic allele in some loci. Alternately, the presence of a strong promoter required for negative selection in ES cells, such as phosphoglycerol kinase (PGK), can be disruptive to surrounding gene expression. For these reasons, it is routine to delete the selectable marker in ES cells before a mouse line is generated. A complex strategy involving Cre expression in ES cells and a target locus containing three loxP sites can be employed to achieve this.

In this case, ES clones that underwent a single recombination event to delete the selectable marker, but retained two *loxP* sites flanking the target sequence, are used to generate the floxed mouse line. Predictably, the correct recombination event is very rare. More commonly, a second recombinase system derived from yeast, Flp/FRT, is used to flank the selection cassette with distinct sequences (FRT sites; Liu et al., 2002). Flp recombinase is expressed in ES cells, specifically deleting the selection cassette. The use of both Cre/loxP and Flp/FRT is compatible with most targeting cassettes and makes the removal of a selectable marker formulaic (Figure 13.4).

The biggest consideration for precise control of conditional gene modification is the availability of mouse lines with the desired recombinase expression pattern. In 1996, there were ~10 published Cre mouse lines. By 2013, there were over 1400 Cre or Flp mouse lines, which along with floxed lines have been cataloged in the NIH-funded Jackson Laboratory Cre Repository (http://cre.jax.org/). Beyond lines where Cre is driven by tissue-specific promoters, there are several types of inducible lines available. Ligand activation of Cre can be conferred by fusion to a mutated human estrogen receptor (Cre^{ER}; Feil et al., 1996). Cre^{ER} has low affinity for endogenous estrogens and is activated by the antagonist tamoxifen. In the presence of tamoxifen, usually delivered to mice through their water bottle, Cre^{ER} is translocated to the nucleus where it can catalyze recombination at *loxP* sites. Another type of inducible system involves driving Cre expression via tet-ON or tet-OFF transgenes (Utomo et al., 1999). The strategy requires generation of a double or triple transgenic mouse line. In addition to the floxed allele, a second transgenic allele comprises the tetracycline-responsive promoter element (TRE; tetO) driving Cre expression. This allele is crossed with another mouse line expressing either reverse tetracycline-controlled transactivator protein (rtTA) or tetracycline-controlled transactivator protein (tTA) under the control of tissue-specific promoters. Alternately, the transactivators and Cre can be expressed from the same transgene, reducing the number of transgenes to just two. In the final transgenic line, Cre expression and *loxP* recombination in the appropriate tissues can be regulated with the tetracycline analog, doxycycline, given in the drinking water. Although inducible strategies provide an additional layer of specificity and control to conditional recombination, time and cost are also significant factors.

13.2.4 GENETICALLY ENCODED TOOLS FOR STUDYING ION CHANNEL FUNCTION

Beyond alterations in ion channel genes and expression, characterization of the wide array of physiological roles for ion channels is further accomplished through the expression of fluorescent markers or optical biosensors. The use of genetically encoded tools is performed in living tissues, reduces the invasiveness of physiological measurements, and can be combined with standard transgenic animal models. *Aequorea*-based GFP derivatives (EGFP, EYFP, EBFP2, Cerulean3, sfGFP, Citrine, and Venus) and red fluorescent proteins derived from *Discosoma* (DsRed, mCherry, and tdTomato) are frequently used for labeling proteins and visualizing cells. Fluorescent proteins are routinely incorporated into transgenes to mark cells that undergo DNA recombination or used to target specific cells for electrophysiological recordings (Livet et al., 2007).

Beyond passively marking cells, genetically encoded biosensors can be used to measure the spatial and temporal dynamics of cellular activity. Cameleon and GCamP belong to a family of GFP-based Ca^{2+} sensors that undergo changes in fluorescence upon Ca^{2+} binding (Miyawaki et al., 1997; Nakai et al., 2001). *Optical highlighters* have applications in channel trafficking and localization (Patterson, 2002). These proteins increase, decrease, or change their fluorescence after activation by a specific wavelength and come in three types: photoactivable, photoswitchable, and photoconvertible. Channel-based voltage sensors undergo changes in fluorescence based on fusion of a voltage sensor protein domain with GFP (FlaSh and ArcLight; Jin et al., 2012; Siegel and Isacoff, 1997). In practice, using genetically encoded sensors in tissues can be challenging due to low signal to noise ratios, kinetic limitations on temporal tracking, and short lifetimes compared with biological timescales.

Optically-gated ion channels are a newer class of genetically encoded tools (*optogenetics*). Light-driven modulation of channel gating has the advantage of tight spatial and temporal resolution. As such, these proteins are used to control, rather than simply monitor, cellular activity. The most well-known are Channelrhodopsin-1 and 2, both intrinsically light-sensitive ion channels identified from algae, neither of which requires expression of additional cofactors for activity. Photostimulation of ChR2 was shown to be effective at depolarizing mammalian neurons, driving action potential activity (Boyden et al., 2005). Other light-sensitive proteins mediate inhibition of activity, such as halorhodopsin (NpHR), a Cl^- pump with an excitation wavelength distinct from ChRs (Han and Boyden, 2007; Zhang et al., 2007). The precise control of membrane potential via optogenetics is subject to many factors stemming from the biophysical properties of the light-modulated channel and expression levels. Since the illumination field affords a high-level of spatial control, viral delivery has been successful for targeting optogenetic proteins to the desired location. Consequently, animal models not currently amenable to transgenic techniques, such as primates, are amenable to optogenetics (Han et al., 2009).

13.2.5 GENETIC INTEGRITY AND MAINTENANCE OF MOUSE LINES

Once transgenic founders have been obtained, the genetic background is "cleaned up," generating a genetically identical, inbred (*isogenic*) line. Several strategies exist for establishing mouse lines, but most require backcrossing to the desired inbred strain for >6 successive generations. A properly backcrossed transgenic line is *congenic* to the parent inbred strain, differing at only one locus (the transgenic allele). Generating a congenic line maximizes the ability to detect a phenotypic consequence stemming from the engineered mutation. A mixed genetic background can contribute to variable *penetrance* (the severity of a particular phenotype in animals carrying the transgene) due to the influence of modifier alleles and other extraneous mutations.

After a transgenic line is established, a typical mating scheme involves intercrossing transgene-positive heterozygote animals to allow the expression of recessive phenotypes in the progeny (Figure 13.8). Recessive mutations account for a major segment of the standard genetic manipulations. In some cases, the resulting

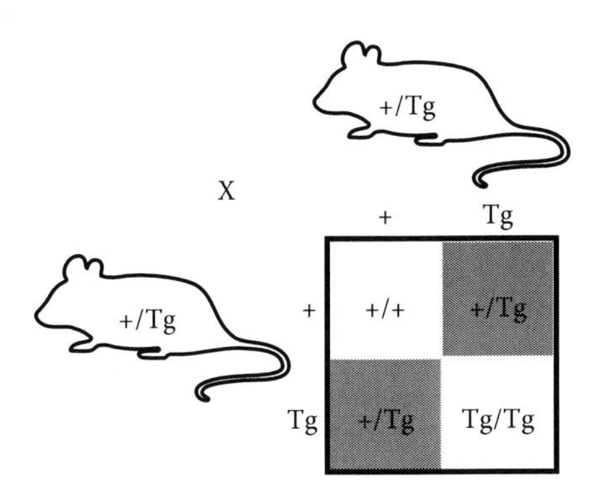

Figure 13.8 Mendelian Inheritance. To express recessive phenotypes, homozygous transgenic (Tg) animals are generated from heterozygous intercrosses. A Punnett square illustrates the allele segregation. The progeny from this cross display the expected Mendelian ratios: 25% wild-type (+/+), 50% heterozygous (+/Tg), and 25% homozygous (Tg/Tg).

homozygous progeny can be bred directly through intercrosses, if the transgenic manipulation does not affect reproduction. More commonly, colonies are maintained by heterozygous breeders that produce the homozygous transgenic animals and wild-type (nontransgenic) littermate controls. Careful consideration should be made when maintaining separate crosses to generate controls and transgenic experimental animals, due to the segregation of random alleles over successive generations (*genetic drift*). Periodic backcrossing (every 2–3 years) of transgenic animals to the inbred strain is essential, necessitating breeding colony records that tract filial generation number (F_1, F_2, etc). In the case of randomly integrated transgenics, choosing animals with one or two copies of the transgenic allele (*hemizygous* and homozygous, respectively) may also be an important consideration.

13.2.6 PHENOTYPIC CHARACTERIZATION

Some transgenic mouse models are generated to specifically recapitulate the critical features of a disease, disease risk, or pathophysiological condition. Other models are generated to identify the potential physiological roles for a particular ion channel. Either way, an initial, broad phenotypic characterization is essential to interpreting experimental data obtained from animal models. Outside of the primary phenotype of interest, transgenic lines are routinely assessed for basic developmental defects, overall health, neurological reflexes, motor and sensory function, learning and memory, and emotionality and social behaviors. Considerations for the optimal phenotypic tests, experimental design, and caveats to data interpretation are covered by an excellent comprehensive reference (Crawley, 2007). Phenotypic characterization is increasingly quantitative, with commercially available automated screening and data recording setups developed for a wide variety of behaviors. Many institutions support "phenotyping cores," recognizing the need for routine access to specialized expertise in the primary characterization of transgenic lines. Several commercial mouse vendors also offer breeding and phenotyping services. Existing information has been consolidated into several publicly accessible databases, including the Trans-NIH Mouse Initiatives (http://www.nih.gov/science/models/mouse/), Knockout Mouse Phenotyping project (http://commonfund.nih.gov/KOMP2/), and Mouse Genome Informatics (MGI) Database (http://www.informatics.jax.org/).

Because mice exhibit strain-dependent traits and disease susceptibility, transgenic alleles can be crossed onto sensitized backgrounds to enhance the phenotypic presentation or examine interactions with susceptibility loci. Traits between inbred mouse lines vary, generating strain-based differences in the presentation of seizure threshold, hearing loss, learning and memory tasks, and drug sensitivity, to name a few. The Mouse Phenome Database (http://phenome.jax.org/) is a valuable resource for selecting strains based on specific characteristics.

A principal consideration for phenotypic characterization is the selection of appropriate controls. Ideally, transgenic and control cohorts should be age- and sex-matched. Nontransgenic (wild-type) littermates are the optimal controls because in practice they often share a greater genetically similarity than inbred wild-type mice that have been maintained separately from the transgenic line (see previous section). The use of littermate controls also provides a more equivalent influence of environmental factors, such as animal colony housing and handling, on the experimental cohorts.

13.2.7 ANIMAL WELFARE

Use of live vertebrate animals for research requires compliance with the Animal Welfare Act. Funding agencies require compliance documentation in grant applications, generally following the federal guidelines enacted through the US Department of Health and Human Services Office of Laboratory Animal Welfare (OLAW) (http://grants.nih.gov/grants/olaw/olaw.htm). Of the common laboratory animal models, rodents and fish are subject to this policy, but as invertebrates, flies and worms are not (Table 13.1). In NIH grants, the "VertebrateAnimals" (http://grants.nih.gov/grants/olaw/VASfactsheet_v12.pdf) section includes specific details concerning the justification and proposed use of the animals (species, strains, ages, sex, and numbers), veterinary care, procedures for minimizing pain, and method and justification for euthanasia.

At the institutional level, Animal Care and Use Committees (IACUC) are responsible for oversight of the humane use guidelines in practice. IACUC protocols, which must be approved prior to commencing any research involving animals, contain some of the same information listed earlier, but with more comprehensive procedural and experimental protocols. IACUC protocols additionally require the identification of:

- General characteristics and necessary information to work with a particular strain, including breeding schemes and the number of animals produced with the desired genotype
- Any special needs or problems associated with a transgenic strain, including abnormal phenotypes and mortality
- Relevant humane end points for experiments
- The potential for duplication of specific experiments or lines of research

Generation of new transgenic lines, creating a new combination of transgenic alleles through breeding (double transgenics), or receipt of transgenic animals from collaborators or a commercial vendor requires IACUC prior approval. Creation of the recombinant DNA transgene itself also requires separate approval at most institutions, generally handled through the Environmental Health and Safety office.

13.3 ION CHANNELS IN HUMAN PHYSIOLOGY AND DISEASE

13.3.1 LINKAGE

Human allelic variation accounts for individual differences in physical traits, personality and cognitive function, and disease risk. The human genome contains three billion base pairs of DNA and >20,000 genes, creating a challenge in associating disease phenotypes with the specific sequences of a gene (*allele*). Progressive advancements in the technology available for making the connections between genotype and phenotype have resulted in an increased understanding of the mechanisms of human disease. Identifying specific genes associated with traits or disease in humans was previously limited by the low resolution of *genetic markers* used for linkage analysis. A *genetic marker* is a stretch of DNA of known location identifying a specific variation in size or sequence (*polymorphism*). Common genetic markers include restriction fragment length polymorphisms (RFLPs), microsatellite or variable number tandem repeats, and *single nucleotide polymorphisms* (SNPs, detailed in Section 13.3.3). The Human Genome Project (http://www.genome.gov/10001772), completed in 2003, vastly expanded the regions and density of the genome covered by various types of genetic markers, and combinations of these markers can now be used to identify individuals, such as for forensic analysis.

For disease association, co-segregation of a set of *linked* markers with the disease phenotype in a family pedigree or across a population is necessary. A minimal set of linked markers identifies a candidate genomic region that may contain the disease gene. When genetic markers are linked, indicated statistically by a high LOD (logarithm of odds) score, the frequency of recombination between them is low. Once linkage between a disease phenotype and genomic marker is resolved, the candidate genomic region can be sequenced, using PCR to isolate cDNAs from the interval between markers (*positional cloning*). For cDNAs containing a mutation in the coding region, subsequent mechanistic investigation is performed to shed light on the causal nature of the mutation in the disease phenotype.

13.3.2 CHANNELOPATHIES

Among the first human disease genes identified was the gene associated with cystic fibrosis, a Mendelian disease characterized by abnormal Cl^- transport and lung and pancreatic failure. Using a groundbreaking positional cloning strategy in 1989, the CFTR gene (cystic fibrosis transmembrane conductance regulator) was identified by John Riordan, Lap Chee Tsui, and Francis Collins (Riordan et al., 1989). Subsequent analysis revealed CFTR was a Cl^- channel in the ABC transporter family, with >1000 loss of function mutations linked to cystic fibrosis. Illustrating the complexity of the resulting deficits, mutations in CFTR fall into four functional classes: protein truncations, surface expression deficits, regulatory deficits (ATP and phosphorylation), and Cl^- conductance deficits.

Identification of other ion channel diseases (*channelopathies*) followed. Notably mutations in hERG were linked to prolonged QT interval in heart (Curran et al., 1995; Sanguinetti et al., 1995), mutations in the skeletal muscle Na^+ channel with hyperkalaemic periodic paralysis (HPP) (Fontaine et al., 1990), mutations in K^+ channels with episodic ataxia (Browne et al., 1994), and mutations in Ca^{2+} channels with migraine (Joutel et al., 1993; Steinlein et al., 1995). The number of known human ion channelopathies now includes hundreds of disease alleles associated with disorders of neuronal and muscle excitability, as well as disorders of renal, metabolic, neuroendocrine, skeletal, and immune systems (Table 13.2). As the number of channelopathies has grown, so too has the availability of genetic testing for ion channel mutations (see the NIH Genetic Testing Registry, http://www.ncbi.nlm.nih.gov/gtr/). Commercial tests are available for channelopathies, such as Long QT Syndrome caused by common mutations in KCNH2 (hERG), KCNQ1 (Kv7.1), and SCN5A (Nav1.5), and less common mutations in KCNJ2 (Kir2.1), KCNE1 (MinK), and KCNE2 (MIRP1). In other complex diseases such as epilepsy, a broader panel of several hundred ion channel mutations associated with increased risk can be screened. The results of such screening are used to confirm diagnosis or in some cases to select a specific pharmacological treatment (*genomic medicine*).

13.3.3 SINGLE NUCLEOTIDE POLYMORPHISMS AND THE ASSOCIATION OF COMPLEX TRAITS

Not all associations between a specific ion channel sequence and a disease phenotype are clear. Most complex traits and diseases are *polygenic*, attributed to the function of multiple gene products. Furthermore, detailed familial pedigrees are not widely available for linkage analysis. However, recent *genome-wide association studies* (*GWAS*) have been developed to harness the power of a systems approach for identifying genetic markers present in a population with a trait or disease. These studies rely on *SNPs*, genetic markers defined by sequence variation occurring at only one base. SNPs are found on average every 1200 base pairs and comprise the normal variation between the genes of individuals in a population. In contrast, a mutation is a rare polymorphism that differs from the sequences normally found in the population, such as the disease-causing mutations referred to as channelopathies.

While SNPs appear normally in the population, the linkage of a specific SNP, or several linked SNPs within a gene (haplotype), has been investigated for associations with particular phenotypes or disease susceptibility. The correlation of haplotypes and disease risk requires a comprehensive human SNP map, a central aim of the HapMap Project (http://hapmap.ncbi.nlm.nih.gov/thehapmap.html.en) which has genotyped 270 individuals and identified ten million SNPs (International HapMap Consortium et al., 2007). Linking ion channel SNPs and disease susceptibility is attractive because much of the basic function of ion channels is well-understood, suggesting that a functional alteration imparted by a SNP would manifest in the disease etiology. However, using idiopathic epilepsy as a model, it is not yet feasible to correlate an individual's SNP profile and the theoretical functional variations in ionic currents with their epilepsy phenotype (Klassen et al., 2011). At present, the lack of correlation between genetic markers and phenotype may be influenced by several factors, including an incomplete understanding of the interactions across an individual's complete genetic profile. In addition, a large number of SNPs are located in noncoding regions and have unknown function. Another distinct factor is *epigenetic* modification of DNA, which produces functional changes in a gene product by mechanisms distinct from sequence variation (genomic imprinting is an example). As a result of these unknowns, attribution of traits and diseases with particular SNPs continues to be a significant biological frontier.

New resources for querying publicly available SNP datasets have been developed to support this effort. The Phenotype-Genotype Integrator (http://www.ncbi.nlm.nih.gov/gap/phegeni) is a searchable integrated repository combining the National Human Genome Research Institute's catalog of GWAS data with several National Center for Biotechnology Information (NCBI) databases, including Gene, database of Genotypes and Phenotypes (dbGaP), Online Mendelian Inheritance in Man (OMIM), Genotype-Tissue Expression (GTEx), and The SNP database (dbSNP). PheGenI can be searched by phenotype, SNP, chromosomal interval, and most usefully, gene name.

13.4 SUMMARY

Ion channels are central to the physiology of cells and animals. Genetic methods in a diverse set of model organisms enabled each fundamental advance in our understanding of ion channel function in vivo, from cloning the genomic sequences, identifying the phenotypes stemming from ion channel dysfunction, to the linkage with human traits and disease. With centralized repositories for existing transgenic models, widespread access to transgenic techniques, and further optimization of targeted genomic manipulation, animal models can be engaged in the research programs of any lab.

To date, targeted mutations have been made in almost every ion channel gene, and several decades of research have been shaped by the analysis of single gene alterations on inbred backgrounds, facilitating identification of the principal physiological roles for ion channels and their mechanistic causality in human disease. Multifactorial genetic strategies will transform approaches to increasingly complex physiological questions. For ion channels, which exist not as singular entities but as components of membrane excitability and ion transport systems, the next frontier in physiology will integrate, rather than exclude, the issue of genetic variability. One new approach is the use of recombinant inbred strains, lines generated by crossing two inbred strains where each individual animal possesses a unique and random combination of inherited alleles (Complex Trait Consortium et al., 2004). Animal models developed using recombinant inbred strains are being used to address the relative contributions of genetic, developmental, and environmental factors in the expression of traits and disease, mimicking the genetic heterogeneity of human populations.

REFERENCES

Atkinson, N.S., Robertson, G.A., and Ganetzky, B. 1991. A component of calcium-activated potassium channels encoded by the *Drosophila slo* locus. *Science* 253, 551–555.

Barbazuk, W.B., Korf, I., Kadavi, C. et al. 2000. The syntenic relationship of the zebrafish and human genomes. *Genome Res* 10, 1351–1358.

Bier, E. and McGinnis, W. 2003. Model organisms in development and disease. In *Molecular Basis of Inborn Errors of Development*, eds. C.J. Epstein, E.P. Erickson, and A. Wynshaw-Boris. New York: Oxford University Press.

Boyden, E.S., Zhang, F., Bamberg, E. et al. 2005. Millisecond-timescale, genetically targeted optical control of neural activity. *Nat Neurosci* 8, 1263–1268.

Browne, D.L., Gancher, S.T., Nutt, J.G. et al. 1994. Episodic ataxia/myokymia syndrome is associated with point mutations in the human potassium channel gene, KCNA1. *Nat Genet* 8, 136–140.

Butler, A., Wei, A.G., Baker, K. et al. 1989. A family of putative potassium channel genes in *Drosophila*. *Science* 243, 943–947.

Church, D.M., Goodstadt, L., Hillier, L.W. et al. 2009. Lineage-specific biology revealed by a finished genome assembly of the mouse. *PLoS Biol* 7, e1000112.

Complex Trait Consortium, Churchill, G.A., Airey, D.C. et al. 2004. The collaborative cross, a community resource for the genetic analysis of complex traits. *Nature Genet* 35, 1133–1137.

Crawley, J.N. 2007. *What's Wrong with My Mouse? Behavioral Phenotyping of Transgenic and Knockout Mice*, 2nd edn. Hoboken, NJ: Wiley-Interscience.

Curran, M.E., Splawski, I., Timothy, K.W. et al. 1995. A molecular basis for cardiac arrhythmia: HERG mutations cause long QT syndrome. *Cell* 80, 795–803.

Eppig, J.T., Blake, J.A., Bult, C.J., Kadin, J.A., Richardson, J.E. and the Mouse Genome Database Group. 2012. *Nucleic Acids Res* 40, D881–D886.

Feil, R., Brocard, J., Mascrez, B. et al. 1996. Ligand-activated site-specific recombination in mice. *Proc Natl Acad Sci USA* 93, 10887–10890.

Fletcher, C.F., Lutz, C.M., O'Sullivan, T.N. et al. 1996. Absence epilepsy in tottering mutant mice is associated with calcium channel defects. *Cell* 87, 607–617.

Fontaine, B., Khurana, T.S., Hoffman, E.P. et al. 1990. Hyperkalemic periodic paralysis and the adult muscle sodium channel alpha-subunit gene. *Science* 250, 1000–1002.

Gaj, T., Gersbach, C.A., and Barbas, C.F. 2013. ZFN, TALEN, and CRISPR/Cas-based methods for genome engineering. *Trends Biotechnol* 31, 397–405.

Grubb, B.R. and Boucher, R.C. 1999. Pathophysiology of gene-targeted mouse models for cystic fibrosis. *Physiol Rev* 79, S193–S214.

Gu, H., Zou, Y.R., and Rajewsky, K. 1993. Independent control of immunoglobulin switch recombination at individual switch regions evidenced through Cre-loxP-mediated gene targeting. *Cell* 73, 1155–1164.

Han, X. and Boyden, E.S. 2007. Multiple-color optical activation, silencing, and desynchronization of neural activity, with single-spike temporal resolution. *PLoS One* 2, e299.

Han, X., Qian, X., Bernstein, J.G. et al. 2009. Millisecond-timescale optical control of neural dynamics in the nonhuman primate brain. *Neuron* 62, 191–198.

Hille, B. 2001. *Ion Channels of Excitable Membranes*, 3rd edn. Sunderland, MA: Sinauer Associates.

Howe, K., Clark, M.D., Torroja, C.F. et al. 2013. The zebrafish reference genome sequence and its relationship to the human genome. *Nature* 496, 498–503.

International HapMap Consortioum, Frazer, K.A., Ballinger, D.G. et al. 2007. A second generation human haplotype map of over 3.1 million SNPs. *Nature* 449, 851–861.

Jin, L., Han, Z., Platisa, J. et al. 2012. Single action potentials and subthreshold electrical events imaged in neurons with a fluorescent protein voltage probe. *Neuron* 75, 779–785.

Jorgensen, E.M. and Mango, S.E. 2002. The art and design of genetic screens: *Caenorhabditis elegans*. *Nat Rev Genet* 3, 356–369.

Joutel, A., Bousser, M.G., Biousse, V. et al. 1993. A gene for familial hemiplegic migraine maps to chromosome 19. *Nat Genet* 5, 40–45.

Joyner, A.L. 2000. *Gene Targeting: A Practical Approach*. New York: Oxford University Press.

Kile, B.T. and Hilton, D.J. 2005. The art and design of genetic screens: Mouse. *Nat Rev Genet* 6, 557–567.

Klassen, T., Davis, C., Goldman, A. et al. 2011. Exome sequencing of ion channel genes reveals complex profiles confounding personal risk assessment in epilepsy. *Cell* 145, 1036–1048.

Liu, P., Jenkins, N.A., and Copeland, N.G. 2002. Efficient Cre-loxP-induced mitotic recombination in mouse embryonic stem cells. *Nat Genet* 30, 66–72.

Livet, J., Weissman, T.A., Kang, H. et al. 2007. Transgenic strategies for combinatorial expression of fluorescent proteins in the nervous system. *Nature* 450, 56–62.

Martinac, B., Saimi, Y., and Kung, C. 2008. Ion channels in microbes. *Physiol Rev* 88, 1449–1490.

Minor, D.L., Jr. 2009. Searching for interesting channels: Pairing selection and molecular evolution methods to study ion channel structure and function. *Mol Biosyst* 5, 802–810.

Miyawaki, A., Llopis, J., Heim, R. et al. 1997. Fluorescent indicators for Ca^{2+} based on green fluorescent proteins and calmodulin. *Nature* 388, 882–887.

Montell, C. and Rubin, G.M. 1989. Molecular characterization of the *Drosophila* trp locus: A putative integral membrane protein required for phototransduction. *Neuron* 2, 1313–1323.

Mortenson, R. 1999. Overview of gene targeting by homologous recombination. In *Current Protocols in Molecular Biology*, eds. F.M. Ausubel, R. Brent, R.E. Kingston et al. Hoboken, NJ: John Wiley & Sons.

Nagy, A., Gertsenstein, M., Vintersten, K. et al. 2003. *Manipulating the Mouse Embryo: A Laboratory Manual.* Cold Spring Harbor, New York: Cold Spring Harbor Laboratory Press.

Nakai, J., Ohkura, M., and Imoto, K. 2001. A high signal-to-noise Ca^{2+} probe composed of a single green fluorescent protein. *Nat Biotechnol* 19, 137–141.

Noda, M., Shimizu, S., Tanabe, T. et al. 1984. Primary structure of *Electrophorus electricus* sodium channel deduced from cDNA sequence. *Nature* 312, 121–127.

Noda, M., Takahashi, H., Tanabe, T. et al. 1982. Primary structure of alpha-subunit precursor of *Torpedo californica* acetylcholine receptor deduced from cDNA sequence. *Nature* 299, 793–797.

Orban, P.C., Chui, D., and Marth, J.D. 1992. Tissue- and site-specific DNA recombination in transgenic mice. *Proc Natl Acad Sci USA* 89, 6861–6865.

Palmiter, R.D., Brinster, R.L., Hammer, R.E. et al. 1982. Dramatic growth of mice that develop from eggs microinjected with metallothionein-growth hormone fusion genes. *Nature* 300, 611–615.

Papazian, D.M., Schwarz, T.L., Tempel, B.L. et al. 1987. Cloning of genomic and complementary DNA from *Shaker*, a putative potassium channel gene from *Drosophila. Science* 237, 749–753.

Patil, N., Cox, D.R., Bhat, D. et al. 1995. A potassium channel mutation in weaver mice implicates membrane excitability in granule cell differentiation. *Nat Genet* 11, 126–129.

Patterson, G.H. 2002. Highlights of the optical highlighter fluorescent proteins. *J Microsc* 243, 1–7.

Riordan, J.R., Rommens, J.M., Kerem, B. et al. 1989. Identification of the cystic fibrosis gene: Cloning and characterization of complementary DNA. *Science* 245, 1066–1073.

Sanguinetti, M.C., Jiang, C., Curran, M.E. et al. 1995. A mechanistic link between an inherited and an acquired cardiac arrhythmia: *HERG* encodes the I_{Kr} potassium channel. *Cell* 81, 299–307.

Siegel, M.S. and Isacoff, E.Y. 1997. A genetically encoded optical probe of membrane voltage. *Neuron* 19, 735–741.

Snouwaert, J., Brigman, K.K., Latour, A.M., Malouf, N.N., Boucher, R.C., Smithies, O., and Koller, B.H. 1992. An animal model for cystic fibrosis made by gene targeting. *Science* 257, 1083–1088.

St. Johnston, D. 2002. The art and design of genetic screens: *Drosophila melanogaster. Nat Rev Genet* 3, 176–188.

Steinlein, O.K., Mulley, J.C., Propping, P. et al. 1995. A missense mutation in the neuronal nicotinic acetylcholine receptor alpha 4 subunit is associated with autosomal dominant nocturnal frontal lobe epilepsy. *Nat Genet* 11, 201–203.

Tempel, B.L., Papazian, D.M., Schwarz, T.L. et al. 1987. Sequence of a probable potassium channel component encoded at *Shaker* locus of *Drosophila. Science* 237, 770–775.

Testa, G., Zhang, Y., Vintersten, K. et al. 2003. Engineering the mouse genome with bacterial artificial chromosomes to create multipurpose alleles. *Nat Biotechnol* 21, 443–447.

Tsien, J.Z., Chen, D.F., Gerber, D. et al. 1996. Subregion- and cell type-restricted gene knockout in mouse brain. *Cell* 87, 1317–1326.

Utomo, A.R., Nikitin, A.Y., and Lee, W.H. 1999. Temporal, spatial, and cell type-specific control of Cre-mediated DNA recombination in transgenic mice. *Nat Biotechnol* 17, 1091–1096.

Valenzuela, D.M., Murphy, A.J., Frendewey, D. et al. 2003. High-throughput engineering of the mouse genome coupled with high-resolution expression analysis. *Nat Biotechnol* 21, 652–659.

Wang, X.J., Reynolds, E.R., Deak, P. et al. 1997. The *seizure* locus encodes the *Drosophila* homolog of the HERG potassium channel. *J Neurosci* 17, 882–890.

Warmke, J., Drysdale, R., and Ganetzky, B. 1991. A distinct potassium channel polypeptide encoded by the Drosophila eag locus. *Science* 252, 1560–1562.

Wei, A., Covarrubias, M., Butler, A. et al. 1990. K^+ current diversity is produced by an extended gene family conserved in *Drosophila* and mouse. *Science* 248, 599–603.

Woods, I.G., Kelly, P.D., Chu, F. et al. 2000. A comparative map of the zebrafish genome. *Genome Res* 10, 1903–1914.

Zhang, F., Wang, L.P., Brauner, M. et al. 2007. Multimodal fast optical interrogation of neural circuitry. *Nature* 446, 633–639.

Zuo, J., De Jager, P.L., Takahashi, K.A. et al. 1997. Neurodegeneration in Lurcher mice caused by mutation in delta2 glutamate receptor gene. *Nature* 388, 769–773.

Zwingman, T.A., Neumann, P.E., Noebels, J.L. et al. 2001. Rocker is a new variant of the voltage-dependent calcium channel gene *Cacna1a. J Neurosci* 21, 1169–1178.

14 Ion channel inhibitors

Jon Sack and Kenneth S. Eum

Contents

Nature, to be controlled, must be obeyed.

Sir Francis Bacon

14.1 MECHANISMS OF INHIBITION

Ion channel inhibitors are tools to control physiological function. The appropriate use of channel inhibitors can test a hypothesis or save a life. Blocker is the scientific vernacular for inhibitor. In the context of ion channels, this can conjure up the imagery of a plugged pore, and many channel inhibitors act in such a fashion. However, not all ion channel inhibitors block the pore; thus, the term lends itself to imprecise use. To avoid semantic confusion, the term *pore blocker* connotes a drug that itself occludes the ionic conduction pathway. The mechanistic alternative to pore blockade is allosteric inhibition. Allosteric means "other site" and in the context of ion channel inhibition refers to acting by a means other than blocking the pore. Allosteric inhibitors act by closing channels; they induce channel proteins to adopt nonconducting conformations. These two mechanisms, depicted in Figure 14.1, describe fundamental workings of all ion channel inhibitors. Many inhibitors work by a combination of both pore blockade and allosteric modulation. This chapter discusses the mechanisms by which ion channel inhibitors act. The physics of pore blockade and allosteric modulation are considered here, with particular attention paid to how the transmembrane voltage can influence inhibitor efficacy.

14.2 PORE BLOCKERS

The inhibitory mechanism of pore blockers makes intuitive sense; the ion channel blocker blocks ion flow. Pore block results from the inhibitor binding to a site in the pore, occluding the passage of permeant ions. Many ion channel inhibitors are pore blockers. A few pore-blocking inhibitors are especially widely used. Tetrodotoxin, produced by symbiotic bacteria in pufferfish, newts, and a variety of other poisonous creatures, is widely used in research to block the pore of sodium channels. Local anesthetics such as lidocaine are used clinically to block sodium channel pores. The anesthetics phencyclidine and ketamine are NMDA receptor pore blockers. The vasodilator verapamil is a calcium channel pore blocker. The expanding pharmacopeia of pore blockers is far too large to be discussed in a single chapter, so we concern ourselves here with mechanism of inhibition, which for all pore blockers is essentially the same.

In its simplest form, the dose–response of pore block is described by the physics of classical ligand–receptor interaction. To many physiologists, the relevant term is the fraction of channels that remain conducting, $f_{unblocked}$. For the simple process as depicted in Figure 14.2a, this fraction is determined by the product of a first-order association rate constant, α; inhibitor activity, $[I]$; and the inherent dissociation rate, β:

$$\frac{f_{unblocked}}{f_{blocked}} = \frac{\beta}{\alpha[I]} \tag{14.1}$$

The rate constants α and β differ for each inhibitor and in some cases can be measured directly. An inhibitor's chemical activity is usually similar to the concentration of the inhibitor. Generally a doubling of blocker concentration will lead to a doubling of chemical activity and hence doubling of binding

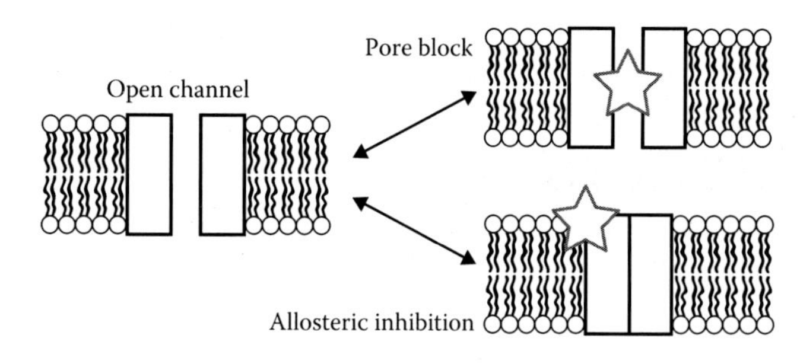

Figure 14.1 Types of inhibitors. Pore block results from inhibitors binding in the pore to occlude ion permeation. Allosteric inhibition results from inverse agonists that cause an open channel to close.

rate $\alpha[I]$. The ratio of the dissociation and association rate constants yields the dissociation constant, K_D:

$$K_D = \frac{\beta}{\alpha} \tag{14.2}$$

The K_D has units of concentration and is a useful value to develop intuition about dose dependence. In the case of a pore blocker, the K_D is the concentration where 50% of channels will be blocked, making it a straightforward measurable quantity. The K_D is sufficient to describe the dose response of the system at equilibrium. A function predicting the diminishment of conductance with increasing inhibitor concentration can then be easily derived:

$$\text{Conductance} = \frac{f_{unblocked}}{f_{unblocked} + f_{blocked}} = \frac{1}{1+([I]/K_D)} \tag{14.3}$$

A plot of Equation 14.3 demonstrates the dependence of channel block on the concentration of inhibitor (Figure 14.2b). Equation 14.3 is variously referred to as the law of mass action, a Langmuir binding isotherm, or Hill logistic with a coefficient of 1. Note that there is a broad concentration range of inhibitor action, with five orders of magnitude increase in inhibitor concentration to span the conductance decrement from 1% inhibition to 99% inhibition (Figure 14.2b). One can never fully inhibit a current, but only approach a saturating value. Thus, when determining what inhibitor concentration to use, consideration needs to be given to how much remaining current is tolerable.

14.3 PERTURBATION OF PORE BLOCK

Structural analysis of crystals of pore blockers inside K^+ channels has provided atomic level understanding of the pore blockade mechanism. A common theme in the configuration of the inhibitor within the channel is that the blocker displaces at least one permeant ion from the pore (Figure 14.3a through c). Thus, the binding of pore blockers is impacted by permeant ions. The interactions between permeant ions and pore blockers can be quite complex. Importantly, the effects of permeant ions are dynamic. Ion channels open and close, changing the access of permeant ions in solution to blocker sites and hence modulating their effects on inhibitors. Additionally, transmembrane voltage change impacts the localization of permeant ions and can exert force on the charged blockers directly. The effects of permeant ions and voltage change can dramatically alter the degree of inhibition by an ion channel blocker.

To quantitate the impact of a perturbation on channel inhibition, it is useful to calculate changes in binding energy.

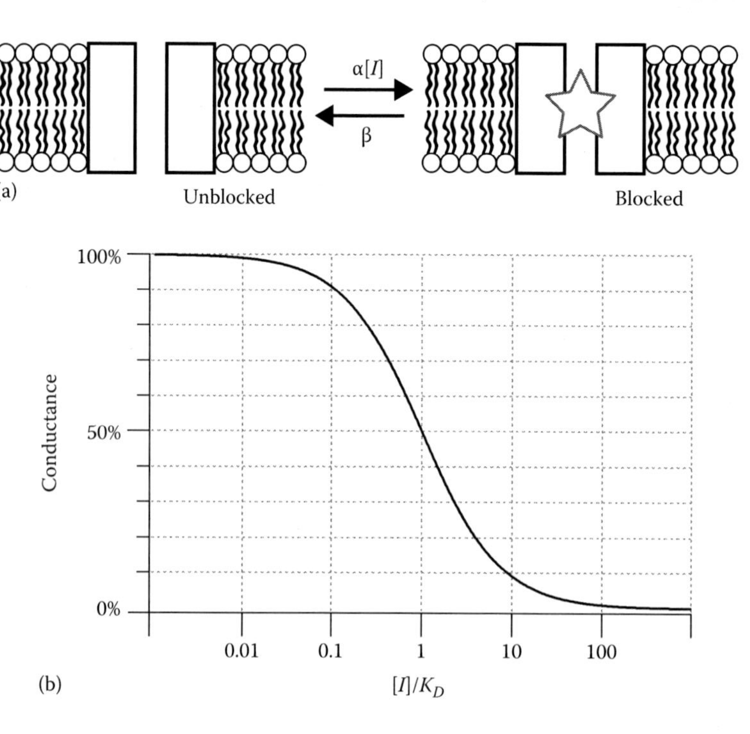

Figure 14.2 Pore blocker dose–response. (a) The fraction of channels blocked is determined by an association rate dependent on inhibitor concentration and an intrinsic dissociation rate. (b) When at equilibrium with a single binding site, a pore blocker will inhibit ionic conductance as a function of its concentration, $[I]$; and dissociation constant, K_D. Line is plot of Equation 14.3.

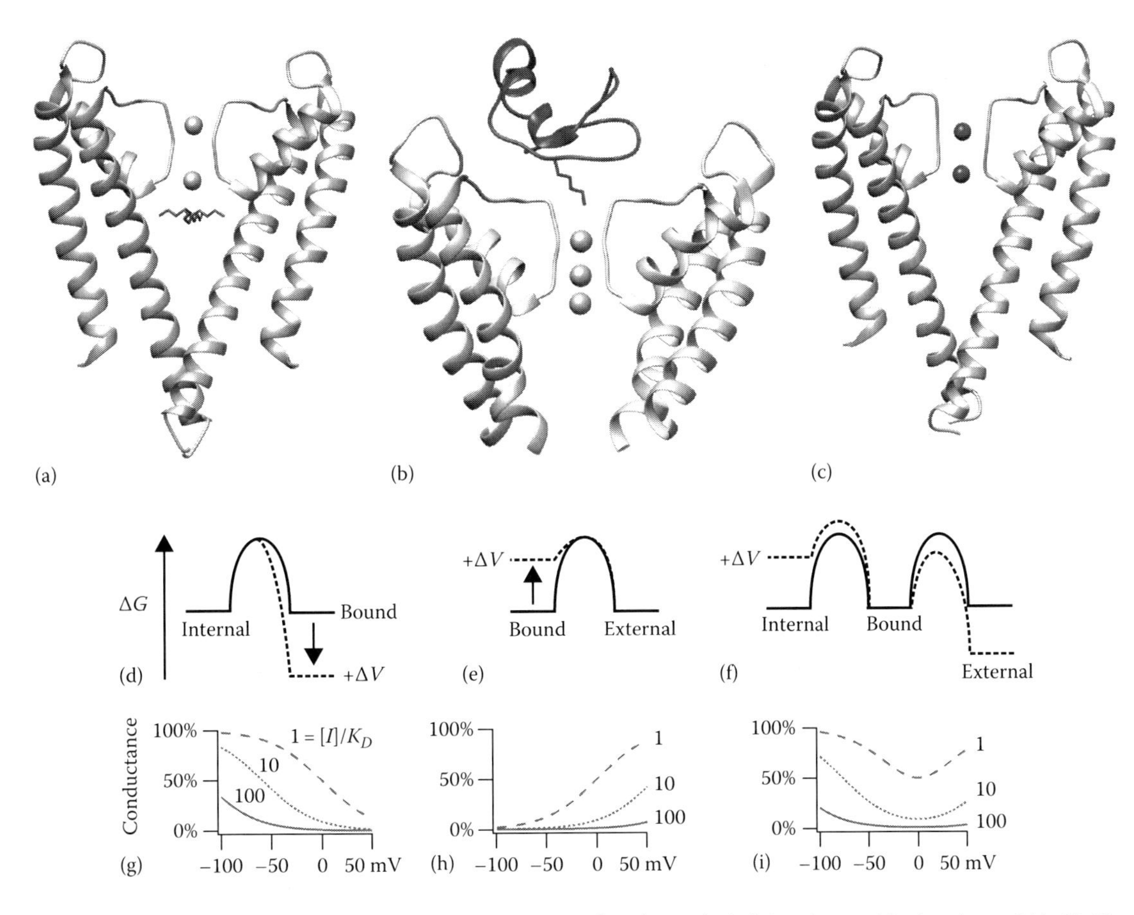

Figure 14.3 Pore blocker voltage dependence. (a–c) Crystal structures of K^+ channels (light) with pore blockers bound (dark). The extracellular side of channels are on top. Renderings generated by Drew Tilley, UC Davis. (a) Rendering of tetrabutylammonium blocking the pore of the KcsA channel, PDB ID 2HVJ. (b) Charybdotoxin blocking the pore of a Kv channel, PDB ID 4JTA. (c) Ba^{2+} blocking the pore of the KcsA channel, PDB ID 2ITD. (d–f) Reaction coordinate cartoons illustrating the voltage dependence of pore blocker binding. The ordinate represents increasing free energy. Dotted lines indicate energetic changes resulting from an intracellular voltage increase. (g–i) Voltage dependence of theoretical pore blockers. Lines are plots of Equation 14.3 with voltage dependence of K_D from Equation 14.8, with $z = 1$ (*G*), $z = -1$ (*H*), and from Equation 14.8 with $z_{in} = 1$, $z_{out} = -1$. (i) Dashed line indicates an inhibitor concentration equal to the K_D at 0 mV. Dotted line is 10 × K_D, solid line 100 × K_D.

Permeant ions and voltage change the energy of the blocked state to make it more or less favorable. The amount of binding energy for a simple pore block is described by Equation 14.4:

$$\Delta G_{blocked} = -k_B T \ln\left(1 + \frac{[I]}{K_D}\right) \tag{14.4}$$

where

$\Delta G_{blocked}$ is the Gibbs free energy for binding
k_B is the Boltzmann constant
T is the absolute temperature
ln is a logarithm

Note that binding energy does not saturate at any concentration. This is due to the binding rates of the blocker continuing to increase as concentration is increased.

When a blocker's binding is altered by a change of some kind, its binding energy will change. We can call ΔG_P a perturbation of inhibitor binding energy, and determine its impact on $\Delta G_{blocked}$:

$$\Delta G_{blocked} = \Delta G^0_{blocked} + \Delta G_P \tag{14.5}$$

$\Delta G^0_{blocked}$ is the energy of the inhibitor-bound state before perturbation. This energetic formulation can describe the effect of many perturbations, voltage changes, ionic concentration, effects of temperature, and other factors. Under physiological conditions, the two most relevant perturbations are likely to be voltage and ionic concentration. Concentration effects of competitive ligand are treated comprehensively elsewhere (1,2), but the effect of voltage on block of ion flow is a phenomenon unique to ion channels, and is considered further here.

The K_D of pore blockers will often have voltage dependence. The physical origins of voltage-dependent inhibition can be complex, especially when impacted by permeant ions. In the most simple, theoretical case, the voltage dependence of a pore blocker can be conveyed with a single parameter, z, the partial charge value that leads to voltage dependence. Measurements of the degree of inhibition at different voltages can empirically constrain z. From z, how the dose response of a pore blocker will be affected by voltage can be determined. With a simple dependence of pore block on z, the binding energy derived from a voltage change, ΔV, can be easily calculated with Faraday's constant, F:

$$\Delta G_V = \Delta V z F \tag{14.6}$$

Hence,

$$\Delta G_{blocked} = \Delta G^0_{blocked} + \Delta G_V \tag{14.7}$$

and the voltage dependence of the K_D is

$$K_D = K^0_{D}e^{-\Delta VzF} \tag{14.8}$$

The real mechanism of pore blockers binding to physiologically relevant channels is rarely (if ever!) as simple as the theoretical case discussed here, but the preceding equations can serve as reasonable approximations under many conditions. The exact nature of the voltage dependence of inhibition is determined by the geometry of the blocking interaction. A few specific examples are discussed later to elaborate some of the means by which voltage-dependent interactions can originate.

14.4 ONE-SIDED PORE BLOCKERS

Many pore blockers can reach their binding site from only one side of the cell membrane. For example, in Figure 14.3a, the inhibitor can reach its binding site only from the internal side of the pore; while in Figure 14.3b, the pore-blocking peptide can bind and dissociate only from the extracellular side. The sidedness of inhibitor binding determines the voltage dependence of its inhibition.

One of the best-studied types of pore blockers are the quaternary ammonium (QA) ions that inhibit K^+ channels. Recent crystal structures have revealed atomistic details of how this block occurs (3–5). The QA ion was found to fit nicely into the hydrophobic cavity of the K channels, just internal to the selectivity filter, where a permeant ion normally resides (Figure 14.3a). QA ions are too large to squeeze through the narrow selectivity filter of the channel. To dissociate, they must exit inward, and their voltage dependence arises from this dissociation to the intracellular side of the channel. As the voltage inside a cell is decreased, the force driving K^+ ions through the channel into the cell forces the pore blocker to be pushed into the intracellular solution by the permeant ions. The more rapid dissociation leads to a decreased affinity (larger K_D) for the pore blocker. This effect is schematized in the energy diagram of Figure 14.3d. When voltage increases, an internal blocker of a cation channel is liable to be more stable in the bound configuration. The kinetic interactions of QA ions with K^+ channels are thermodynamically complex and understood in precise detail (6–12). For the purposes of demonstration, however, it is useful to discuss an idealized simple scenario described by Equation 14.8 (Figure 14.3g). Note that within the physiologically relevant voltage range, of −100 to 50 mV, the percent conductance inhibited is altered dramatically. A concentration of pore blocker that inhibits only a small fraction of the conductance at a negative resting potential may inhibit the majority of the current if the cell is brought to a positive potential. The voltage dependence of an internal blocker can be far steeper than that depicted in Figure 14.3g. This type of voltage dependence forms the basis of inward rectification of the K_{ir} channels, where endogenous polyamines inhibit the outward flow of ions (13–15). Given the dramatic changes in inhibition with cell potential, it is worth carefully considering the voltage changes in any experimental preparation with voltage-dependent inhibitors.

Pore blockers that act from the external side of the pore will have the opposite voltage dependence to internal blockers. Many ion channel inhibitors are pore-blocking toxins that bind the extracellular side of the pore. One of these, the scorpion peptide charybdotoxin, has been crystallized bound to a K^+ channel (16). The resultant structure provides an example of how a peptide can physically occlude the conducting pore (Figure 14.3b). Akin to the internal pore blockers discussed earlier, the peptide mimics the chemistry of the channel's permeant ions to bind tightly to the extracellular side of the channel pore. Charybdotoxin displaces a K^+ ion from its binding site with the positively charged lysine residue. Due to its interactions with permeant ions, the dissociation of the toxin from the channel is voltage dependent (17–19). As the voltage inside a cell is increased, the force driving K^+ ions through the channel out of the cell leads to a decreased affinity for the toxin, which is pushed out to the external solution by the flow of K^+ ions (Figure 14.3e). Hence, the voltage dependence is opposite of blockers that exit to the intracellular solution (Figure 14.3g and h). Interestingly, this opposing voltage dependence of extracellular dissociation has been harnessed by NMDA receptors to allow conduction only when block by external Mg^{2+} has been relieved by positive cellular voltage (20,21). Hence, block of NMDA receptors by an endogenous ligand forms a voltage-dependent gate important for normal physiological function.

14.5 SLOWLY PERMEATING BLOCKING IONS

Some ion channel inhibitors pass all the way through the channel. These inhibitors block the pore by transiting through it at a rate slower than the ions that underlie its normal conductance. A common class of inhibitors used in electrophysiology experiments is small metal ions that block pores. In experimental preparations, Ca^{2+} channels can be blocked with Cd^{2+} or Co^{2+}, and K^+ channels with Cs^+ or Ba^{2+}. Ba^{2+} ion block of K^+ channels has been carefully investigated (22,23) and serves as an excellent case study of a slowly permeating blocker. Crystal structures have revealed the location of Ba^{2+} in K^+ channels (24,25), where it can be seen replacing K^+ in the selectivity filter (Figure 14.3c). The Ba^{2+} ion is nearly the same size as K^+, allowing it to fit snugly into these sites. Yet, due to its greater charge, Ba^{2+} dynamics differ and it remains in the channel pore for long periods of time, preventing the flux of other ions. Ba^{2+} eventually dissociates, and similar to K^+, it can exit from its binding site to either the internal or external solution.

The ability of permeant ions to enter and exit the pore from both sides of the channel leads to them often having biphasic voltage dependence. In addition to interactions with permeant ions, slowly permeating ions have a significant innate voltage dependence. As voltage is increased inside the cell, cationic blocking ions are driven into the channel from the internal solution and bound ions are pushed to exit the external side (Figure 14.3f). Thus, the rate of binding and dissociation of a slowly permeating blocker can have the same polarity. For this situation, voltage dependence has multiple components, and different steps in the slow permeation process can dominate at different voltages. A simple description of such behavior is as follows:

$$K_D = K_D^0\left(e^{-\Delta V z_{in} F} + e^{-\Delta V z_{out} F}\right) \tag{14.9}$$

where

z_{in} is the voltage dependence of binding/dissociation from the internal side of the pore

z_{out} is the voltage dependence of binding/dissociation from the external side of the pore

The multiple voltage components can lead to multiphasic voltage dependence. This is demonstrated in Figure 14.3i, where at negative voltages, inhibition decreases because ions exit to the internal side, and at positive voltages, the inhibition also decreases because ions exit to the external side. Biphasic voltage dependence is a hallmark of slowly permeating blockers that was first identified in careful studies of the proton block of sodium channels (26). The relief of block by blocker permeation is often referred to as punch-through, it is seen with many slowly permeating blockers (27).

As with one-sided poreblockers, the voltage dependence of slowly permeating blockers can result in different degrees of block at different voltages. This voltage-dependent inhibition can be problematic for researchers seeking to pharmacologically eliminate a current. As can be seen with the examples shown (Figure 14.3g through i), an inhibitor that produces near complete inhibition at some voltages can be ineffective at others. This variable inhibition is due to voltage dependence alone, and the efficacy of inhibitors becomes even more complicated when effects of channel conformation are considered.

14.6 STATE-DEPENDENT INHIBITION

The ability of most pore blockers to inhibit depends upon the conformational state of the channel. State-dependent blockers can require channel activity before any inhibition occurs. For example, the quaternary ammonium and Ba^{2+} blockers described earlier require that the channel be open to allow access to their binding sites. A wide variety of sodium channel inhibitors used to treat pain, arrhythmias, and epilepsy are also state-dependent pore blockers. Especially in the case of these sodium channel inhibitors, state-dependence is often referred to as use-dependence because the degree of inhibition of the channel increases when the channel is stimulated. Many manifestations of state-dependent inhibition exist, which in combination with voltage-dependence give inhibitors a dynamic specificity. State-dependent inhibitors can differentially inhibit structurally identical channels, if the stimulus provided to each channel differs.

Open channel blockers are drugs that preferentially bind when an ion channel is open. A classical example is the block of Kv channels by QA compounds. When presented to the intracellular face of a Kv channel, open channel blockers bind to a site inside of the channel that they cannot access when the channel is closed (Figure 14.4a). This can lead to a decay of current after an initial opening (Figure 14.4b). Upon repetitive stimulus, such as a train of action potentials, an open channel blocker or other use-dependent inhibitor will progressively inhibit their target (Figure 14.4c and d) (28). This is thought to be an important property of drugs that control excitotoxic pathologies, such as epilepsy (29).

14.7 ALLOSTERIC INHIBITORS

Not all ion channel inhibitors block the pore. Many act by inducing the channel to close. This mechanism is fundamentally distinct from pore blockade, as the inhibitory action is not due to the inhibitor obstructing the flow of ions, but to the inhibitor

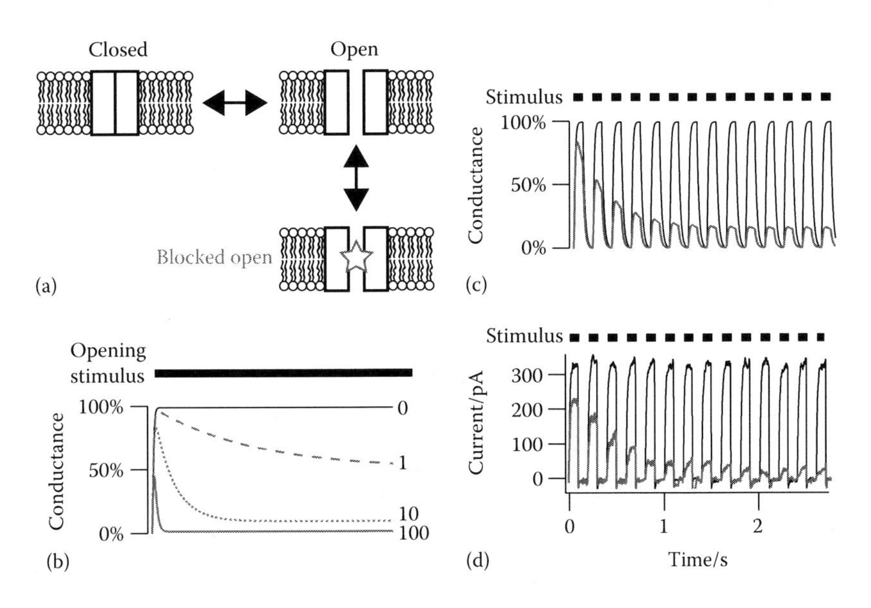

Figure 14.4 Use-dependent inhibition. (a) An open channel blocker is only able to occlude the pore when the channel opens. (b) Time dependence of pore blockade in a kinetic model of an open channel blocker. Dashed line indicates an inhibitor concentration equal to the K_D of the open state. Dotted line is 10 × K_D, solid line 100 × K_D. (c) Use dependence of block with repeated stimuli. (d) Use-dependent inhibition of Kv2.1 channels with 1 μM RY785. Inhibitor was applied for 10 min before first stimulus. Stimulus is to 50 mV from −80 mV. Whole-cell voltage clamp methods similar to Sack. (From Sack et al., *J. Gen. Physiol.*, 142, 315, 2013.)

stabilizing channels in nonconducting conformations. Despite these very different mechanisms, distinguishing pore blockade from allosteric inhibition can be difficult.

Allosteric inhibitors can act by a variety of physical means, such as membrane perturbation, surface charge screening, or direct binding to ion channels. The fundamental concept of allostery is that a modulatory chemical can selectively stabilize protein conformations, shifting the equilibrium between conformational states. A molecule that stabilizes nonconducting conformations is a negative allosteric modulator. By shifting the equilibrium of a channel to nonconducting, the channel will open less often and be inhibited.

Allosteric inhibitors act by energetically stabilizing the closed states relative to open ones. In the simplest case, depicted in Figure 14.5a, an inhibitor prevents a channel from opening. This type of allosteric inhibition is technically inverse agonism, as the channel cannot open with inhibitor bound. For an inverse agonist, the fraction of channel current inhibited is identical to the total fraction bound:

$$\Delta G_{closed\ bound} = -k_B T \ln\left(1 + \frac{[I]}{K_{D\ closed}}\right) \quad (14.10)$$

The thermodynamics of this type of inhibition make its macroscopic dose–response indistinguishable from pore block at any single voltage; the dose–response will have exactly the same shape as that in Figure 14.2b. The functional differences between a pore blocker and an allosteric inhibitor can be subtle or radical. In most cases, the binding of an allosteric inhibitor will be sensitive to the conformational changes associated with channel gating. Changing the stimulus that opens the channel will change the binding of the inhibitor. Figure 14.5c demonstrates the effect of an inverse agonist in response to increasing opening stimulus. The inverse agonist fights the opening stimulus, and the greater the concentration of inhibitor applied, more stimulus energy is needed to open the channel. The effects of a full inverse agonist (as depicted in Figure 14.5a) never saturate; more inhibitor produces more inhibition. The opening stimuli for most channels are ligands or voltage. As noted earlier, the interactions of inhibitors with ligand-stimulated proteins are discussed in great detail elsewhere (1,2); here, we discuss the particular interactions of allosteric inhibitors with voltage-gating processes.

In the case of most voltage-gated channels, voltage increase leads to more channel opening, and inverse agonists modulate these voltage-gating processes. Since the voltage alters the probability that a channel will be closed, the fraction of channels with inhibitor bound will change with voltage. The fact that voltage can also affect pore blocker binding can make these two types of inhibition even tougher to distinguish, and requires careful interpretation of inhibition data.

An example of an inverse agonist can be found in the defensive mucus of a marine snail. The gastropod *Calliostoma canaliculatum* secretes 6-bromo-2-mercatpotryptamine (BrMT) to deter predators. BrMT is an inverse agonist of voltage-gated K^+ (Kv) channels (30,31). BrMT acts by selectively binding to closed channels, as in Figure 14.5a, to prevent channels from opening. Its effects are consistent with full inverse agonism where the toxin must dissociate from the channel before it can open. In the presence of BrMT, the voltage required for activation is higher. Increasing concentrations require progressively higher voltages to open the channels (Figure 14.5d). The major effect of inverse agonists can be due to kinetic rather than equilibrium properties. It can be seen from a reaction coordinate diagram (Figure 14.5b) that an inverse agonist will increase the activation barrier, $\Delta G^{\ddagger}_{opening}$, by an amount equal to $\Delta G_{closed\ binding}$. Thus, if the inhibitor is preventing rate-limiting steps in channel opening, a full inverse agonist will slow the time course of channel opening by a factor of at least

$$\frac{\tau_{inhibited}}{\tau_{control}} = 1 + \frac{[I]}{K_{D\ closed}} \quad (14.11)$$

where τ represents the time constant of the rate-limiting kinetic process in activation. With increasing concentrations of an inhibitor, channel opening will be progressively slowed (Figure 14.5e). This progressive alteration of kinetics with increasing dose is a hallmark of a full inverse agonist.

14.8 PARTIAL INVERSE AGONISM

Some allosteric inhibitors will allow channels to open while the inhibitor remains bound. Inhibitors that act in such a way are partial inverse agonists (Figure 14.5f). A partial inverse agonist lowers the probability that a channel will be in a fully open state. The degree to which the binding of a partial inverse agonist leads to channel closing is the coupling between binding and inhibition. This can be quantified as a coupling energy, $\Delta G_{coupling}$, the amount of energy the inhibitor uses to close the channel. But where does $\Delta G_{coupling}$ arise from? The first law of thermodynamics demands that energy cannot just appear. It must be accounted for. It turns out that $\Delta G_{coupling}$ is derived from the binding energy of the inhibitor. When channels open with an inverse agonist bound, it weakens inhibitor binding (Figure 14.5f).

An energy diagram for this type of inhibition is given in Figure 14.5g. This depicts a general model of allosteric interaction where a portion of an inhibitor's binding energy is used to keep the channel from opening. How much coupling energy can arise from a change in binding can be calculated from the binding energies to the different channel states. The coupling energy comes from the difference in binding energy of the open and closed states:

$$\Delta G_{coupling} = \Delta G_{open\ binding} - \Delta G_{closed\ binding} \quad (14.12)$$

As binding affinities are related by dissociation constants, it can be seen that

$$\Delta G_{coupling} = -k_B T \ln\left(\frac{1 + \left([I]/K_{D\ open}\right)}{1 + \left([I]/K_{D\ closed}\right)}\right) \quad (14.13)$$

and at high concentrations of inhibitor, the effect of the inhibitor saturates:

$$\lim_{[I]\to\infty} \Delta G_{coupling} = -k_B T \ln\left(\frac{K_{D\ closed}}{K_{D\ open}}\right) \quad (14.14)$$

Therefore, the ratio of the dissociation constants for the open vs. closed conformations limits the potential of a partial inverse

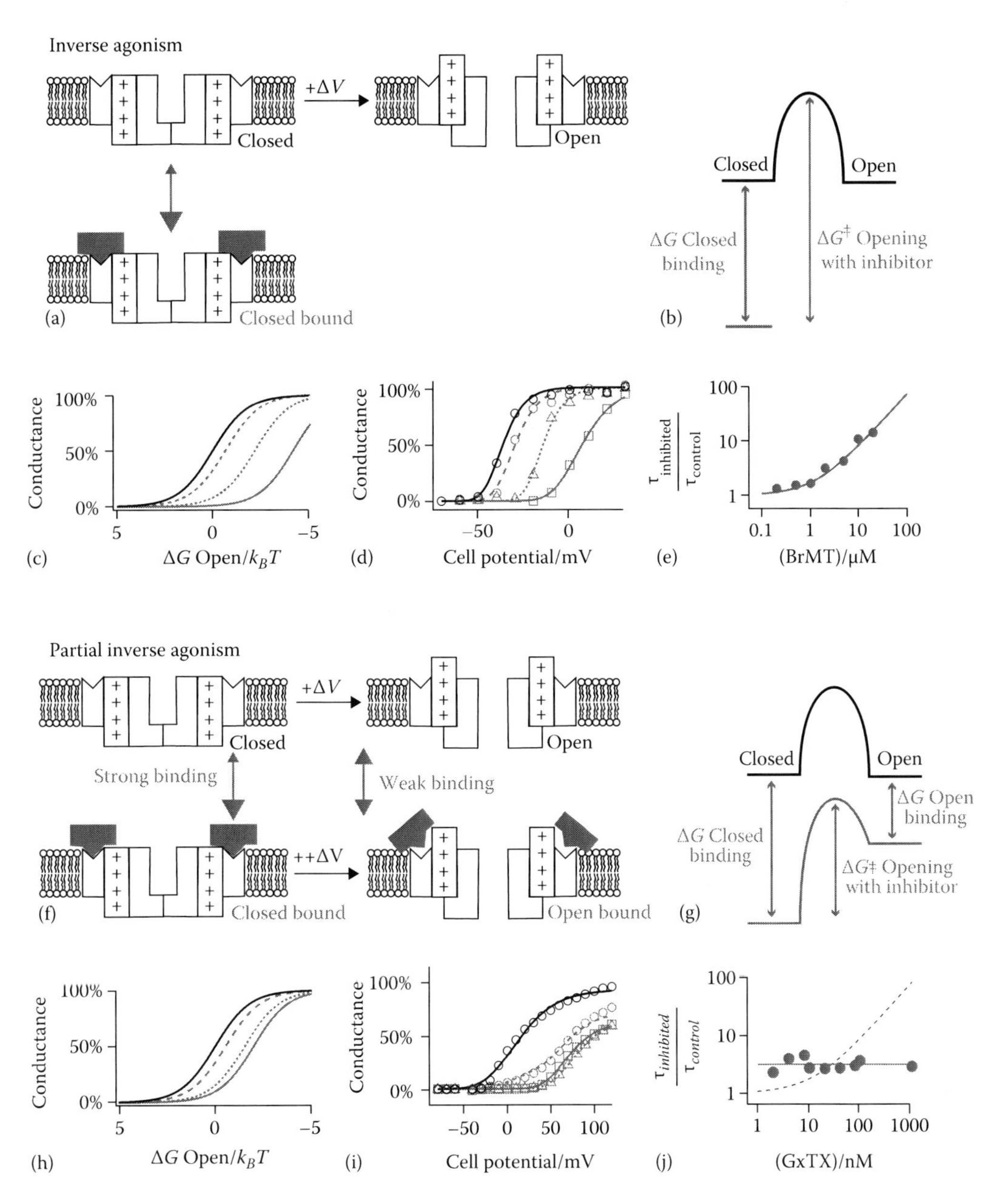

Figure 14.5 Allosteric inhibition. (a) A full inverse agonist needs to dissociate before the channel can open. Cartoon depicts a voltage activated ion channel that only binds inhibitor in its closed state. (b) The inverse agonist binding energetically stabilizes the closed state. (c) The dose–response of a full inverse agonist does not saturate. Lines are a Boltzmann distribution of conductance arising from increasing open probability. Black line is control condition. Dashed line indicates an inhibitor concentration equal to the K_D for the closed conformation. Dotted line is 10 × K_D, solid red line 100 × K_D. (d) Shift of conductance–voltage relation of Shaker K+ channels does not saturate with increasing concentrations of the inverse agonist BrMT. Black line is Boltzmann distribution fit to control data points. Dashed line is 1 μM BrMT, dotted line 5 μM, solid line 20 μM. Methods similar to Sack. (From Sack et al., *J. Gen. Physiol.*, 123, 685, 2004.) (e) Increasing inverse agonist concentration progressively slows channel opening. Data points are time constants of channel opening in indicated concentrations of BrMT relative to control. Line is Equation 14.11 with K_D set to 0.8 μM. (f) A partial inverse agonist can remain bound while the channel opens. Cartoon depicts a voltage-activated ion channel that strongly binds inhibitor in its closed state, and more weakly in its open conformation. (g) Open state binding of partial inverse agonists allow activation barrier for opening to be reduced. (h) The dose–response of a partial inverse agonist saturates. Lines are a Boltzmann distribution of conductance arising from increasing open probability. Black line is control condition. Dashed line indicates an inhibitor concentration equal to K_D closed where K_D open is 10 × K_D closed. Dotted line is 10 × K_D closed, solid red line 100 × K_D . (i) Shift of conductance–voltage relation of Kv2.1 channels saturates at high concentrations of the partial inverse agonist GxTX. Black line is Boltzmann distribution fit to control data points. Dashed line is 10 μM GxTX, dotted line 100 μM, solid line 1 μM. Methods similar to Sack. (From Sack et al., *J. Gen. Physiol.*, 142, 315, 2013.) (j) Rate of channel opening with partial reverse agonist is concentration independent. Data points are time constants of channel opening in indicated concentrations of GxTX relative to control. Solid line is 3.1-fold slowing of opening. Dashed line is Equation 14.11 with K_D set to 13 μM, K_D closed; poor fit indicates GxTX is not a full inverse agonist.

agonist to close a channel. This is demonstrated by the simulation in Figure 14.5h; the effects of partial inverse agonists saturate with increasing dose.

Tarantulas inject their prey with peptide toxins that are partial inverse agonists of Kv channels (32). One example is guangxitoxin-1E (GxTX) (33). In the presence of GxTX, Kv channels require more stimulus voltage to open, and its effects saturate (Figure 14.5i, (38)). This is due to channels opening with GxTX bound. As depicted in Figure 14.5g, a channel opening with inhibitor bound can have a $\Delta G^{\ddagger}_{opening}$ that is less than an opening pathway that first requires inhibitor dissociation. Because channels can open in the presence of inhibitor, the opening rate saturates and becomes insensitive to higher concentrations of toxin (Figure 14.5j). Thus, the mechanism of inverse agonist has a practical impact on experimental preparations. Partial inverse agonists have the advantage that a saturating amount of drug can be used, such that any experimental variation in inhibitor concentration has less impact on channel response. This limited efficacy of partial inverse agonists is advantageous in lowering the potential for error due to under- or overdose of an inhibitor.

14.9 INHIBITION BY LIPID BILAYER EFFECTS

The preceding treatments of inhibition mechanism assume that the inhibitor interacts directly with the ion channel. However, inhibitors can act without binding to channels. All ion channels are suspended in a sheet of the lipid bilayer that forms the cell membrane. The bilayer is intimately involved in channel function. Some inhibitors do not appear to bind the ion channels they affect at all, but rather act by perturbing the membrane that surrounds them (Figure 14.6).

A classically studied mechanism of inhibitor action is the surface charge effect. This is where multivalent metal ions adsorb to the surface of membranes. This perturbs the membrane electric field and hence the activity of channels in the membrane (34,35). Inhibitors that are hydrophobic or amphipathic in their physical chemistry can partition into the cell membrane. By changing the physical properties of the membrane itself, the equilibria between open and closed channels can be changed

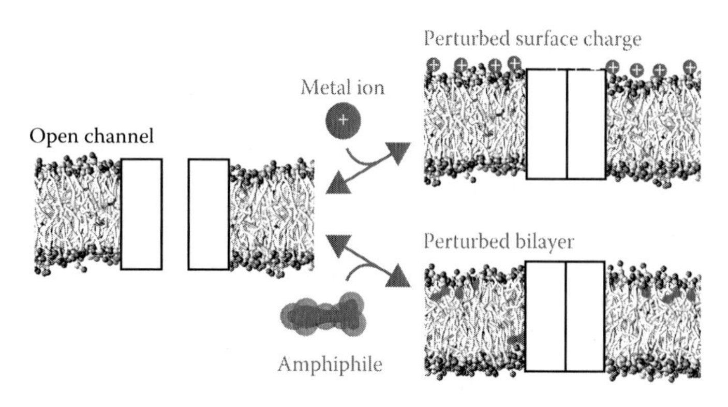

Figure 14.6 Membrane perturbation can alter channel function. Cartoons represent allosteric inhibition of channels by perturbing the surrounding membrane. Membrane images are molecular dynamics simulation snapshots of a phospholipid bilayer ±10 mol% resveratrol, an amphiphilic phytochemical. Simulation and rendering by Helgi I. Ingólfsson, University of Groningen, Groningen, the Netherlands.

(36,37). In cases of membrane perturbation, these effects can be similar to that of inverse agonists that bind the channel directly, but may have unusual dose response profiles. Membrane perturbing inhibitors have promiscuous effects on many different membrane proteins (39).

14.10 CONCLUDING REMARKS

We have discussed mechanisms by which inhibitors can act to decrease the conductance of ion channels. Inhibitors can bind in the pore and block ion flow, act allosterically by closing the channel, either by direct binding or perturbing the surrounding lipid membrane environment. Different types of inhibitors are well-suited for specific tasks. In attempting to dissect a channel's role in a complex physiological situation, a selective and complete inhibitor of a specific channel type is called for. In this case, a pore blocker may be preferable. While a partial inverse agonist is suboptimal for an experimentalist who wants to completely inhibit a current of interest, its mechanism may be valuable as a therapeutic that aims to partially inhibit a current at saturation. To inhibit channels only under hyperexcitable pathophysiological conditions, a use-dependent inhibitor is appropriate. To fine-tune the gating of a channel, an allosteric modulator is advantageous. In any case, understanding how inhibitor efficacy is affected by voltage changes, channel activity, and the membrane bilayer allows one to choose ion channel inhibitor doses and interpret results of experiments more wisely.

ACKNOWLEDGMENT

This chapter is dedicated to the memory of co-author Kenneth Eum (1987–2014). He was a bright soul and a talented physiologist.

REFERENCES

1. Wyman, J. and S. J. Gill. 1990. *Binding and Linkage: Functional Chemistry of Biological Macromolecules*. University Science Books, Mill Valley, CA.
2. Hilal-Dandan, R., L. L. Brunton, and L. S. Goodman. 2013. *Goodman and Gilman's Manual of Pharmacology and Therapeutics*. McGraw-Hill, New York.
3. Zhou, M., J. H. Morais-Cabral, S. Mann, and R. MacKinnon. 2001. Potassium channel receptor site for the inactivation gate and quaternary amine inhibitors. *Nature* 411:657–661.
4. Lenaeus, M. J., M. Vamvouka, P. J. Focia, and A. Gross. 2005. Structural basis of TEA blockade in a model potassium channel. *Nat Struct Mol Biol* 12:454–459.
5. Yohannan, S., Y. Hu, and Y. Zhou. 2007. Crystallographic study of the tetrabutylammonium block to the KcsA K^+ channel. *J Mol Biol* 366:806–814.
6. Posson, D. J., J. G. McCoy, and C. M. Nimigean. 2013. The voltage-dependent gate in MthK potassium channels is located at the selectivity filter. *Nat Struct Mol Biol* 20:159–166.
7. Li, W. and R. W. Aldrich. 2004. Unique inner pore properties of BK channels revealed by quaternary ammonium block. *J Gen Physiol* 124:43–57.
8. Holmgren, M., P. L. Smith, and G. Yellen. 1997. Trapping of organic blockers by closing of voltage-dependent K^+ channels: Evidence for a trap door mechanism of activation gating [see comments]. *J Gen Physiol* 109:527–535.

9. Wilkens, C. M. and R. W. Aldrich. 2006. State-independent block of BK channels by an intracellular quaternary ammonium. *J Gen Physiol* 128:347–364.
10. Choi, K. L., C. Mossman, J. Aube, and G. Yellen. 1993. The internal quaternary ammonium receptor site of Shaker potassium channels. *Neuron* 10:533–541.
11. Swenson, R. P., Jr. 1981. Inactivation of potassium current in squid axon by a variety of quaternary ammonium ions. *J Gen Physiol* 77:255–271.
12. Armstrong, C. M. 1969. Inactivation of the potassium conductance and related phenomena caused by quaternary ammonium ion injection in squid axons. *J Gen Physiol* 54:553–575.
13. Lopatin, A. N., E. N. Makhina, and C. G. Nichols. 1994. Potassium channel block by cytoplasmic polyamines as the mechanism of intrinsic rectification. *Nature* 372:366–369.
14. Xu, Y., H. G. Shin, S. Szep, and Z. Lu. 2009. Physical determinants of strong voltage sensitivity of K(+) channel block. *Nat Struct Mol Biol* 16:1252–1258.
15. Lu, Z. 2004. Mechanism of rectification in inward-rectifier K^+ channels. *Annu Rev Physiol* 66:103–129.
16. Banerjee, A., A. Lee, E. Campbell, and R. Mackinnon. 2013. Structure of a pore-blocking toxin in complex with a eukaryotic voltage-dependent K(+) channel. *eLife* 2:e00594.
17. MacKinnon, R. and C. Miller. 1988. Mechanism of charybdotoxin block of the high-conductance, Ca^{2+}-activated K^+ channel. *J Gen Physiol* 91:335–349.
18. Park, C. S. and C. Miller. 1992. Interaction of charybdotoxin with permeant ions inside the pore of a K^+ channel. *Neuron* 9:307–313.
19. Goldstein, S. A. and C. Miller. 1993. Mechanism of charybdotoxin block of a voltage-gated K^+ channel. *Biophys J* 65:1613–1619.
20. Nowak, L., P. Bregestovski, P. Ascher, A. Herbet, and A. Prochiantz. 1984. Magnesium gates glutamate-activated channels in mouse central neurones. *Nature* 307:462–465.
21. Mayer, M. L., G. L. Westbrook, and P. B. Guthrie. 1984. Voltage-dependent block by Mg^{2+} of NMDA responses in spinal cord neurones. *Nature* 309:261–263.
22. Neyton, J. and C. Miller. 1988. Discrete Ba^{2+} block as a probe of ion occupancy and pore structure in the high-conductance Ca^{2+}-activated K^+ channel. *J Gen Physiol* 92:569–586.
23. Neyton, J. and C. Miller. 1988. Potassium blocks barium permeation through a calcium-activated potassium channel. *J Gen Physiol* 92:549–567.
24. Jiang, Y. and R. MacKinnon. 2000. The barium site in a potassium channel by x-ray crystallography. *J Gen Physiol* 115:269–272.
25. Lockless, S. W., M. Zhou, and R. MacKinnon. 2007. Structural and thermodynamic properties of selective ion binding in a K^+ channel. *PLoS Biol* 5:e121.
26. Woodhull, A. M. 1973. Ionic blockage of sodium channels in nerve. *J Gen Physiol* 61:687–708.
27. Heginbotham, L. and E. Kutluay. 2004. Revisiting voltage-dependent relief of block in ion channels: A mechanism independent of punchthrough. *Biophys J* 86:3663–3670.
28. Courtney, K. R. 1975. Mechanism of frequency-dependent inhibition of sodium currents in frog myelinated nerve by the lidocaine derivative GEA. *J Pharmacol Exp Ther* 195:225–236.
29. Rogawski, M. A. and W. Loscher. 2004. The neurobiology of antiepileptic drugs. *Nat Rev Neurosci* 5:553–564.
30. Sack, J. T., R. W. Aldrich, and W. F. Gilly. 2004. A gastropod toxin selectively slows early transitions in the Shaker K channel's activation pathway. *J Gen Physiol* 123:685–696.
31. Kelley, W. P., A. M. Wolters, J. T. Sack, R. A. Jockusch, J. C. Jurchen, E. R. Williams, J. V. Sweedler, and W. F. Gilly. 2003. Characterization of a novel gastropod toxin (6-bromo-2-mercaptotryptamine) that inhibits shaker K channel activity. *J Biol Chem* 278:34934–34942.
32. Phillips, L. R., M. Milescu, Y. Li-Smerin, J. A. Mindell, J. I. Kim, and K. J. Swartz. 2005. Voltage-sensor activation with a tarantula toxin as cargo. *Nature* 436:857–860.
33. Herrington, J. et al. 2006. Blockers of the delayed-rectifier potassium current in pancreatic beta-cells enhance glucose-dependent insulin secretion. *Diabetes* 55:1034–1042.
34. Frankenhaeuser, B. and A. L. Hodgkin. 1957. The action of calcium on the electrical properties of squid axons. *J Physiol* 137:218–244.
35. Hille, B., A. M. Woodhull, and B. I. Shapiro. 1975. Negative surface charge near sodium channels of nerve: Divalent ions, monovalent ions, and pH. *Philos Trans R Soc Lond B Biol Sci* 270:301–318.
36. Andersen, O. S. and R. E. Koeppe, 2nd. 2007. Bilayer thickness and membrane protein function: An energetic perspective. *Annu Rev Biophys Biomol Struct* 36:107–130.
37. Lundbaek, J. A., S. A. Collingwood, H. I. Ingolfsson, R. Kapoor, and O. S. Andersen. 2010. Lipid bilayer regulation of membrane protein function: Gramicidin channels as molecular force probes. *J R Soc Interf* 7:373–395.
38. Sack, J. T., N. Stephanopoulos, D. C. Austin, M. B. Francis, and J. S. Trimmer. 2013. Antibody-guided photoablation of voltage-gated potassium currents. *J Gen Physiol* 142:315–324.
39. Ingolfsson, H. I., et al. 2014. Phytochemicals perturb membranes and promiscuously alter protein function. *ACS Chem Biol* 9:1788–1798.

15 High-throughput methods for ion channels

Haibo Yu and Min Li

Contents

15.1 INTRODUCTION

The common feature of ion channels is the movement of permeating inorganic ions across lipid membranes. Because charge is carried by the ion passage, the resultant minute ionic currents are the primary and the most direct signals to monitor ion channel activity. Several factors have made the endeavor of measuring these signals particularly difficult. *First*, ion channel activities in most cases, unlike second messengers in receptor signaling, do not involve any amplification. *Second*, electrical currents are the rate, not the accumulation, of ionic passage. In many cases, the net change of ionic concentration is extremely small. *Third*, most ion channels conduct ions transiently as part of their gating characteristics, for example, inactivation. These factors have posed considerable challenges in the development of high-throughput methods to monitor ion channel activities.

This chapter introduces the basic principles of different methods capable of achieving medium- or high-throughput measurements of ion channel activity. It is important to keep in mind that some methods are more effective or specific for certain channel classes, while other methods are more generally applicable. Historically, methods to study ion channels are illustrious for their exquisite sensitivity. In fact, patch clamp is the first, perhaps most commonly used, single-molecule analysis. However, it is almost universal that throughput increase tends to cause reduction in information content. Consideration and decision of necessary output therefore must guide the selection of methodology to be used.

Methodologies to study ion channels may be divided into functional and nonfunctional methods based on whether activity of the ion channel is measured. Electrophysiological methods afford the most direct functional means to measure ion channel activity and additionally allow flexibility in assay optimization for various channel types and gating preferences. The recent introduction of automated plate-based electrophysiology instruments has provided new opportunities in ion channel screening. However, cost, content acquisition, and throughput considerations limit high-throughput screen (HTS) application of electrophysiological methods to special conditions or medium-sized libraries. Improvements in chemical dye or protein sensors and instrumentation technologies have led to the widespread use of flux-based assays using the accumulation or depletion of ions as a functional readout. Detecting changes in ion concentration can be achieved by either atomic absorption spectroscopy or fluorescent indicator dyes. These assays have relatively low temporal resolution and information content, but enable robust and low-cost HTS. In contrast, nonfunctional assays such as ligand displacement are still used for ion channel screening, but are limited to cases in which a suitable labeled ligand is available. But these assays are conventional and usually amenable for high throughput in miniature volumes. A summary table for ion channel screening methods is shown in Table 15.1.

15.2 KEY ASSAY PARAMETERS

Before deciding the ideal method(s), it is important to consider the parameters to measure when comparing technologies and their applications. Among the many assay parameters, eight commonly considered parameters are sensitivity, specificity, throughput, temporal resolution, robustness, flexibility, cost, and of course, physiological relevance. While these are the most important parameters for ion channel assays, they apply to other bioassays as well.

Table 15.1 Screen methods for ion channel drug discovery

ASSAY TYPE	METHOD	PROPERTY	INSTRUMENT	CELL TYPE/TISSUE	MEASURED CHANNELS	REFERENCES AND EXAMPLES
Radio ligand binding assay	[^{3}H]dofetilide; [^{3}H]-astemizole	Easy and cost-effective; prone to false-negatives; cannot detect channel function and channel trafficking; underestimated compound potency	Liquid scintillation counter	hERG-HEK; SHSY5Y	hERG	(2–5)
Voltage-sensitive dyes	DiBAC4(3)	Slow response time (minutes); high false-positive rate; lack of voltage control	Fluorescent imaging plate reader (FLIPR)	hERG-CHO:SUR1 and Kir6.2 channels co-expressed in HEK-293 cells; HEK 293 cells expressing SK1, SK2, and SK3; cultured neonatal rat ventricular myocytes	hERG; KATP channel; Kir6.2/SUR1; SK1; SK2; SK3; KATP channel;	(86–89)
Voltage-sensitive dyes	FMP, MPAK	Response time in tens of seconds; high false-positive rate; lack of voltage control	FLIPR	HCN1-HEK 293 cells; Nav1.2-HEK 293 cells; GABA-HEK 293; Kv1.3-CHO cells	HCN1; NaV1.2; GABA; Kv1.3	(90–93)
Voltage-sensitive dyes	Fluorescence resonance energy transfer (FRET): DiSBAC2(3)/CC2-DMPE	Fast response time (in the range of seconds); high false-positive hit rate; lack of voltage control	VIPR; GENios Pro; FLiPRTetra	Nav1.5-HEK 293; Nav1.7-HEK 293; Nav1.8-HEK 293 cells	Nav1.5; Nav1.7; Nav1.8; Kir	(8,9,17,94)
	$^{86}Rb^+$ (radiolabeled)	Robust; less sensitive to disturbance; safety and environmental hazard due to the use of radioisotopes	Liquid scintillation counter	Rat blood vessel	K channels; Nonselective cation channels	(20,21)
Flux-based assays	$^{86}Rb^+$ (nonradiolabled)	Excellent tracer ion for K^+ channels; low background noise; lack of voltage control; low temporal resolution and low throughput	Atomic absorption spectroscopy	Cell lines expressing ion channels	Kv1.1; Kv1.3; Kv1.4; Kv1.5; hERG; KCNQ2; KCNQ2/3; Ca^{2+}-activated K^+ channels: SK and BK; Nonselective cation channels: TRPC6 and TRPA1,nAChR, P2X	(23–27,95,96)
Flux-based assays	Thallium influx (Tl^+)	Underestimate the potency of compound	FLiPR	Cell lines expressing ion channels	Potassium channels: KCNQ2, SK3, hERG; KCNQ4; Kir1.1; Kir2.1	(5,10–14,34–39, 72,97,98)

Flux-based assays	YFP/I^- influx	Noninvasive technique with the ability to measure fast responses; widely developed for chloride selective channels and receptors	FLiPR; FDSS 6000	FRT cell line expressing hTMEM16A and YFP-H148Q/I152L/F46L; YFP-H148Q/I152L and F508del-CFTR	Calcium-activated chloride channel (CaCC): TMEM16A; cystic fibrosis transmembrane conductance regulator (CFTR); GABA	(40,43–47,49)
Automated electrophysiology	Ionworks system (HT, Quattro, Barracuda)	Perforated patch; 384-well format; highest throughput unit; allows for the collection of 384 data points with ~60% success with IonWorks HT (single hole, 1 cell/well) and >95% success with Quattro (PPC, 64 cells/well); estimates of compound potency negatively influenced by propensity for lipophilic compounds to adhere to the plate; no intracellular perfusion.	Ionworks system (Molecular Devices, LLC)	Cell lines expressing ion channels	Voltage-gated channels and slowly desensitizing ligand-gated channels.	(13,26,38,39,59–61, 63,66–69,71, 72,96,99–102)
Automated electrophysiology	PatchXpress	Giga-ohm seals; 16-wells in parallel; 1 cell/well; desynchronized control of chambers; cumulative/multiple compound additions; ligand addition during recording; continuous voltage clamp; simultaneous visualization of all 16 channels; estimates of compound potency negatively influenced by propensity for lipophilic compounds to adhere to the plate; no simultaneous visualization of channels; no intracellular perfusion.	PatchXpress 7000A (Molecular Devices,LLC)	Cell lines expressing ion channels	Voltage-gated channels and ligand-gated channels.	(63,65,68,103,104)

(*Continued*)

Table 15.1 (*Continued*) Screen methods for ion channel drug discovery

ASSAY TYPE	METHOD	PROPERTY	INSTRUMENT	CELL TYPE/TISSUE	MEASURED CHANNELS	REFERENCES AND EXAMPLES
Automated electrophysiology	QPatch HT/ HTX	Giga-ohm seals; 16- or 48-wells in parallel; 1 cell/well (HT) or 10 cells/well (HTX); cumulative/ multiple compound additions; ligand addition during recording; continuous voltage clamp; estimates of compound potency negatively influenced by propensity for lipophilic compounds to adhere to the plate; no simultaneous visualization of channels; no intracellular perfusion.	Qpatch (Sophion Bioscience)	Cell lines expressing ion channels	hERG; Nav1.2; Nav1.5; Nav1.7; KCa2.3; TRPM8; Alpha7 nicotinic acetylcholine receptor	(35,68,73,74,104–110)
	Patchliner	Giga-ohm seals; 2-, 4-, 8- or 16-well recordings in parallel; primary cell recordings; heatable pipette—fast temperature jumps; automated current clamp recordings; perfusion of external and internal solution; ligand addition during recording; system performance information limited	Patchliner (Nanion Technologies)	KCa1.1/KCNMB1-HEK 293, P2X7-HEK 293, TRPC5-HEK 293 or TRPM3-HEK 293 cells, Rat type-I cortical astrocytes, human T lymphoblasts, human saphenous vein smooth muscle cells, human synoviocytes, human neutrophils, cultured Jurkat T cells, SH-SY5Y neuroblastoma cells	KCa1.1/KCNMB1, P2X7 receptors or TRPC5; TRPM3	(75,103)

Sensitivity describes the minimum change of signal that can be reliably detected. In the context of pharmaceutical screening, it describes the ability of an assay to detect weak stimuli over any noise signal. Sensitivity should be calibrated by the false-negative rate; a low false-negative rate corresponds to high sensitivity of an assay.

Specificity, also called selectivity, measures the ability of an assay to distinguish between different types of stimuli. The false-positive rate is a measurement of specificity; a low false-positive rate corresponds to a high specificity assay.

Throughput measures the speed of an assay to generate the data. It is usually expressed by the number of data points performed in unitary time (e.g., per day), by unitary resources (e.g., per person, or per instrument).

Temporal resolution is a frequency domain parameter measuring how fast the response data can be measured. It can be described by bandwidth. Assays with good temporal resolution show detailed kinetic response along the time axis. These data afford analyses at and between different time points.

Information content is a multidimensional attribute that describes the amount of data output from one test unit (e.g., a well of a microtiter plate) of an assay. It includes kinetic aspects (such as temporal resolution) as well as static or steady-state aspects of the measurements (such as fluorescence measurement at multiple wavelengths, simultaneous detection of fluorescence intensity, polarization and lifetime, multiple morphological attributes, and so on, or the combinations of these measurements). In some cases, for example, when testing voltage-gated ion channels, longer recording with different voltages would yield more information about the compound effects (content) but compromise the throughput.

Robustness measures the ability of the assay to tolerate interference and changing experimental conditions while still producing reasonable data. High reproducibility by the same protocol under different nontechnical conditions (instrumentation, plates, or researchers) corresponds to high robustness of an assay.

Flexibility relates directly to the ability of an assay to be adapted for different tests or different targets.

Cost measures the consumption of resources such as supplies, manpower, and machine time by an assay. Cost is usually measured by the cost per data point. Some assays require considerable effort to be devoted to analyses, consequently contributing to higher cost.

Finally, *physiological relevance* describes how relevant the result of an assay is to the physiological situation. The closer an assay setting is to physiological conditions, the more meaningful the data are regarded.

15.3 LIGAND BINDING ASSAYS

Generally, ligand binding (or displacement) assays are not considered as functional assays because they detect the binding of a compound to an ion channel rather than the compound's ability to alter channel function. Binding assays require previous knowledge of binding sites on the target and the generation of a labeled ligand specific for those binding sites by radioactive isotope or a fluorescent dye tag. A typical binding assay involves the labeling of a known ligand for the ion channel, such as a neurotoxin, commonly with a radioisotope. Activity of the test compound is indicated by the displacement of the labeled ligand. This method is similar to methods testing other protein classes, such as receptors and enzymes. Consequently, conventional instrumentation may be used, where throughput represents its major strength. This method is also prone to both false positives and negatives as undetected binding at the same site, or allosterically coupled sites, does not guarantee the compound has not bound to other sites on the target, and binding itself does not guarantee functional modulation of the ion channel (1–4). Binding assays do not provide information about the effect of novel agents on ion channel function. For example, an agonist cannot be distinguished from an antagonist in a binding assay. Secondary assays are necessary to determine whether the compound is an agonist, antagonist, or neither. The sensitivity of a binding assay is often determined by the affinity of the known, labeled ligand. A high-affinity ligand might not allow the detection of weak binders in displacement assay format. However, the selection of a low-affinity ligand could lead to increased detection of nonspecific binding. When the affinity of the ligand is within a certain range (e.g., from nano- to micromolar concentration), the IC_{50} values obtained from binding assays have a reasonable correlation with those obtained from patch clamping (5). Binding assays can be performed in 96-, 384-well, or even higher density formats with reasonable reagent cost that varies with respect to the ligand used. Furthermore, the scope of use of the binding assay is limited by the availability and affinity of radiolabeled ligands (2–4). In recent years, this assay has been used for evaluation of compound interaction with hERG channel to asscss cardiotoxicity (2–5).

15.4 FLUORESCENCE-BASED ASSAYS

Most ionic currents lead to changes in membrane potential. Fluorescence-based methods do not directly measure ionic current but, rather, measure membrane potential–dependent changes of fluorescence signals (from fluorescent dyes loaded into the cytosol or cell membrane) as a result of ionic flux. Because fluorescence-based methods give robust and homogeneous cell population measurement, these assays are similar to an assay for other protein classes. Therefore, more instrument choices and expertise are available. Consequently, these assays are relatively easy to implement and to be optimized to achieve high throughput.

Fluorescent voltage-sensor dyes are used to measure voltage changes across the cellular membrane through either the potential-dependent accumulation and redistribution (6) or the FRET mechanism (7). The lipophilic, negatively charged oxonol dyes, such as *bis*-(1,3-dibutylbarbituric acid) trimethine oxonol [DiBAC4(3)] are examples of redistribution dyes. DiBAC-based voltage-sensitive dyes have low fluorescence in an aqueous environment, but show increased quantum yield upon binding to hydrophobic intracellular molecules. Because the change of fluorescence occurs minutes after the change of membrane potential, these dyes are well suited to detect steady state instead of kinetic changes in membrane potential. Second and subsecond resolution of changes in membrane potential cannot be readily detected with these dyes. During HTS, compound-fluorescence and compound-dye interactions are major sources of artifacts that result in a relatively high false-positive rate. Molecular Devices (Sunnyvale, CA) and Hamamatsu Photonics (Hamamatsu, Japan) have recently introduced improved instruments and membrane potential kits that give faster response time (in tens of seconds) for use with its Fluorometric Imaging Plate Reader ($FLIPR^{Tetra}$, Molecular Devices) (8,9) or FDSS (Hamamatsu Photonics) (10–15) HTS platforms. These instruments use 384- or 1536-well microplate formats that give a throughput of up to 10 plates per hour depending on the particular assay format. In FRET-based voltage sensors, different negatively charged, membrane-soluble oxonol dyes are used as voltage-sensing FRET acceptors, *bis*-(1,3-dialkylthiobarbituric acid) trimethine oxonol [DiSBACn(3)]. The FRET donors are coumarin-tagged phospholipids (CC2-DMPE) that are integrated into the outer leaflet of the membrane when loaded into the cells. An increase or decrease of FRET in response to membrane hyperpolarization or depolarization produces fast, ratiometric changes. The ratiometric nature of the assay helps to eliminate many of the artifacts associated with DiBAC assays. Unlike DiBAC assays, the use of a phospholipid-anchored FRET donor restricts the location of FRET in the plasma membrane, which ensures the measurement of potential changes occurs at the cell membrane, rather than in other subcellular compartments such as the mitochondria. FRET-based voltage sensors also give subsecond temporal resolution, which allows for kinetic reading. Because the mobile oxonol molecules are charged, a dye-concentration-dependent dye-current might interfere with the change of membrane potential caused by the ionic current through the ion channels, especially when the current is less than a few hundred pico-amps. One way to reduce this effect is to use less dye, which might cause a decrease in the signal-to-noise ratio if the dye concentration is below a certain limit. Recently, Aurora Biosciences (San Diego, CA) has developed the Voltage Ion Probe Reader (VIPR™) as a high-throughput instrument for screening ion channel-related targets using these FRET-based voltage sensors (16,17). The newer VIPR II, like FLIPR, also adopts the 384-well format, capable of screening several plates per hour. The dye cost is about U.S. 10 cents–30 cents per data point plus a technology access fee. Like the DiBAC assays, VIPR assays are approximately one log value less sensitive during the screening of certain compounds compared with results obtained using patch clamp recordings (16,18,19).

15.5 FLUX-BASED ASSAYS

15.5.1 RADIOACTIVE PROBES

Radioactive ions permeable to targeted channels have been used to trace the cellular influx or efflux of specific ions. A commonly used assay format is $^{86}Rb^+$ efflux for most K^+ channels or nonselective cation channels. In this format, cells expressing the ion channel of interest are incubated with a buffer containing $^{86}Rb^+$ for several hours before they are washed and stimulated with an agonist to allow for $^{86}Rb^+$ efflux. The cells and supernatant are then collected for radioactive counting (20,21). However, radioactive-efflux assays also suffer from the inconvenience and cost associated with handling radioactive material, and, of course, the requirement for different radioisotopes for channels selective for different ions. Therefore a nonradioactive Rb^+ efflux assay was developed, which uses atomic absorption spectroscopy for the determination of rubidium (22). The flux assay was a format preferred by many screening laboratories because it measures ionic flux that has better correlation with the activity (23–26). This assay technology has found widespread application in the pharmaceutical industry for both drug discovery and hERG-related drug-safety screening for the identification of potential QT liabilities that might cause lethal arrhythmias (1,27). However, these assays tend to suffer from low temporal resolution (typically from seconds to minutes), uncontrolled membrane potential, less information content compared with voltage clamping and lower throughput compared with fluorescence-based assays. In fact, the assay generates a very weak signal for some ion channels, which thereby requires a high level of channel expression to achieve an acceptable signal-to-noise ratio.

15.5.2 ION-SPECIFIC FLUORESCENT PROBES

Assays that measure intracellular ionic concentrations are also widely used in research and pharmaceutical screening for ion channels. Among these dyes, calcium indicator dyes that alter the fluorescence emission upon calcium binding, are the most popular (28). Depending on the application, calcium dyes are available in a range of affinities to calcium ions (29, excitation and emission spectra, and chemical forms (membrane permeable or cytosolic). They show different temporal resolution (from milliseconds, like Fluo-3, to tens of seconds, like Fura-2), and different degrees of accuracy for each range of calcium concentration. For example, Indo-1 is preferable over Fluo-3 for measuring the large, relatively slow intracellular calcium transients associated with cellular contraction, whereas Fluo-3 is preferred for measuring small, fast transients associated with calcium "sparks" (30). These calcium dyes are usually used in conjunction with a FLIPR-type kinetic fluorescent reader to achieve high throughput, low-noise detection of both absolute levels and kinetic changes of cytosolic calcium concentration. Calcium dye-based ion channel assays suffer from the interference from other cellular processes that produce changes in intracellular calcium concentration. Overall, ion channel assays using ion indicator dyes are largely limited to the availability of high-performance ion-specific indicators. To date, only calcium indicators give a robust performance that can be used for HTS of calcium channels or nonselective cation channels (15,31–33).

It must be emphasized that temporal resolution in cell-based fluorescence ion channel assays (FLIPR, VIPR) is different from the ion channel characteristics that can be measured in patch-clamp analysis. In particular, the slower kinetics recorded with voltage-sensor dyes do not mirror the behavior of the ion channels under investigation (which open or close in the order of microseconds) but, rather, are related to the time of redistribution of the dye(s) upon membrane depolarization or hyperpolarization. What is measured is the convolution of the aggregated change in membrane potential over multiple cells as a result of ion channel gating and dye-charge movement and the dye response determined by the biophysical properties of the dye, the membrane fluidity, and so on. Therefore, fast dyes with response times in the micro- to millisecond range can provide more relevant information about the ion channels than slow dyes at the cell population and single cell, but not single channel, level. It should be noted, however, that temporal resolution is not necessarily a prerequisite to monitor the effects of compounds on ion channel activity using voltage-sensitive dyes.

More recently, a fluorescent Tl^+ assay was developed for the identification of potassium channel modulators (34). In this assay, Tl^+ is used as a tracer for K^+ and its influx into cells is measured with the thallium-sensitive fluorescent dye BTC. Although this assay is amenable to HTS, large concentrations (mM) of the highly toxic thallium and Cl^- free buffer conditions are necessary in order to generate a robust assay readout. Using a commercially available thallium assay kit (FluxOR), this assay has been developed for a number of potassium channels (10,12–14,34–39).

Another screening format for anion channels has been developed (40). Based on the halide-binding properties of yellow fluorescent variants (YFP) of the green fluorescent protein (GFP) (41), random mutation approaches have identified mutations of YFP, H148Q, and I152L, which further increase the YFP halide sensitivity (40,42). Upon ion channel activation, anions such as I^-, NO_3^-, Br^-, and Cl^- enter the cell, bind to YFP, and quench its fluorescence. Agonist-dependent quench of YFP fluorescence can then be measured with a fluorescent reader and used to determine channel activation, inhibition, and modulation. The YFP assay has the benefit of being a noninvasive technique with the ability to measure fast responses. The assay has been widely developed for chloride-selective channels and receptors including CFTR (43,44), calcium-activated Cl^- channels (CaCC) or TMEM16A (45–47), glycine receptor (48), and GABA receptor (48,49).

15.6 AUTOMATED ELECTROPHYSIOLOGY

Patch clamping is the gold standard for determining ion channel function (50–52). The patch clamp technique controls (or clamps) the electrical potential difference across a small patch of membrane or across the plasma membrane of an entire cell and directly assesses the current carried by ions crossing the membrane at that voltage through ionic channels. This technology provides high quality and physiologically relevant data of ion channel function at the single cell or single channel (within a small patch of membrane) level. Patch clamping is unparalleled in its microsecond to submillisecond temporal resolution for measuring the intrinsic biophysical properties of ion channels such as the time constants of transitions between closed, open, and inactivated states. For pharmacological testing of compounds, it provides a standard for measuring the potency of compound–channel interactions. Although the patch clamp technique is able to resolve ionic currents in the pico-amp to sub-pico-amp range, setting up patch clamping experiments is a complicated process requiring highly trained personnel to make the system less vulnerable to interference from vibration and electrical noise. Such low throughput and high labor cost is not most compatible with HTS purposes. Currently, manual patch clamping is used extensively for basic research. It is also a major method for secondary and safety screening. Patch clamping is a highly flexible technique that can be performed in either whole-cell configuration; cell attached, inside-out, or outside-out configurations in macropatch or single-channel mode (53); or even with membrane vesicles and reconstructed lipid bilayer configurations (54–57). Single-channel patch clamp recording provides event-based measurements such as unitary conductance, subconductance states, open and closed statistics, bursting behavior, and so on. No other existing technology can provide such direct, precise, and detailed measurements of the activity of an ion channel down to the single-molecule level.

Xenopus oocyte has become a popular ion channel recording and screening system for its faithful and high level expression, as well as relatively low endogenous background current. Its large size, together with its tolerance for being impaled by multiple microelectrodes, makes it easy to inject mRNA for expression cloning and drug screening. Because of the large size of oocytes, two-electrode voltage clamp (TEVC) is used for whole-cell recordings of *Xenopus* oocytes to enable one of the electrodes to be entirely dedicated to injecting sufficient current to clamp the voltage across its entire membrane. It is also impossible to control the intracellular ionic composition of oocytes because of their size. Oocyte recordings suffer from lower sensitivity during compound screening and pharmacological testing, largely because of the absorption of the compounds by egg yolk. In addition, seasonal variation of expression intrinsic to most oocyte preparations makes it difficult to perform screenings during the summer months (58).

Although conventional manual patch-clamp offers a direct, information-rich and real-time method to study channel function, it has very low throughput and a labor-intensive nature, requiring highly skilled and trained personnel. Over the last decade, the development of automated planar patch clamp has made a great breakthrough. Many automated electrophysiology platforms have been developed and made commercially available, including IonWorks™ HT, IonWorks Quattro™ and IonWorks Barracuda™, PatchXpress® 7000A (Molecular Devices, LLC); QPatch HT/HTX (Sophion, Copenhagen, Denmark); Dyna flow® HT (Cellectricon AB, Mölndal, Sweden); NPC-16 Patchliner® and SyncroPatch® 96 (Nanion Technologies GmbH, Munich, Germany); CytoPatch™ (Cytocentrics AG, Rostock, Germany); and IonFlux HT (Fluxion Bioscience Inc., San Francisco, CA). Depending on the specific system, they are providing giga-ohm seal quality of data comparable to manual electrophysiology recording. The use of perforated patch clamp technique has increased compound screening throughput with compromised data quality and pharmacology. IonWorks platform was the first of the commercially available automated electrophysiological screening platforms to gain widespread utility and validation in the field and is now available in the second generation IonWorks Quattro and the more recently launched IonWorks Barracuda. The system can operate in two modes, single-hole mode and population patch clamp mode (PPC) (59–61). IonWorks is able to record from 384 cells (1 cell/well of a 384-well plate) in single-hole mode. In PPC mode, the instrument reports average currents per well from recordings of up to 64 cells/well. The data recorded from single-hole mode show lower success rates and considerable well-to-well variability;

however, it is the preferred format for clonal screening and selection during cell line development (62). PPC mode provides improved data consistency and success rates in the measurement of ionic currents (59). One common limitation has been observed for lipophilic compounds with right-shifted compound potencies due to the use of plastic materials for the seal plates and potential nonspecific absorption of high-density cells in the recording wells (61,63). However, with optimization, an acceptable correlation between compound potencies derived from IonWorks assays and manual patch clamp electrophysiology can be achieved (14,61,64). For the IonWorks Quattro system, one important limitation is the inability of the system to place the recording and pipettor heads (E-Head) simultaneously into the recording well and thereby maintain continuous voltage clamp (due to a conflict of access between the robotic recording and perfusion heads). This largely restricts its utility for voltage-gated channels and is not optimal for fast-desensitizing ligand-gated channels. The advanced version IonWorks Barracuda has recently been introduced and demonstrates improved features related to these limitations (63,65). Despite the limitations of IonWorks Quattro, this has not restricted the IonWorks platform in practical use, where this caveat has been accepted as a trade-off for increased compound screening throughput without compromising data quality and pharmacology. Ion channel–targeted drug discovery efforts on a number of diverse ion channel subtypes have also benefited from the availability of the IonWorks platform with robust assays (60,65–72).

Several planar array-based automated electrophysiology systems were developed incorporating the precision and accuracy of manual electrophysiology recording. The platforms provide giga-ohm seal with compromised throughput but have gained prominence and acceptance as ion channel drug discovery platforms in the pharmaceutical industry, such as PatchXpress, QPatch, SynchroPatch, and IonFlux. The rapid solution exchange times (50–100 ms) and perfusion capability of these systems have also made them amenable for ligand-gated ion channel investigations. QPatch is the first commercial system to enable integrated microfluidic channels in the plate to allow fast solution exchanges. The system has evolved to increase throughput by threefold from a 16-channel to a 48-channel system available with HT (73,74) (a single-hole mode for compound profiling) and HTX modes (a multihole mode ideal for screening, similar to PPC mode of IonWorks). The system provides a precise voltage control and efficient contact time with the amplifier. One important feature of QPatch is the data analysis software compatible with all QPatch instruments. QPatch software fully automates the analysis of the data acquired from QPatch automated patch clamp systems and speeds up the data analysis process. The user can track experimental data reproducibility and build assays automatically. The built-in analysis offers various modules including IC_{50} and EC_{50} determination, IV plots, I–T plots, and Boltzmann fitting along with graphical illustrations and tables. Other systems are similar to QPatch system but with more or less similar features. Nanion's Patchliner operates in a single-hole mode and has microfluidic channels that allow raid perfusion and exchange of extracellular as well as intracellular solutions. In some recent reports (e.g., (75)), the system offers a capability for primary cell recording, both automated voltage clamp and current clamp recording modes. The pipette can be heated and provides a fast temperature jump that may be suitable for certain channel types.

Clearly high-throughput electrophysiology has many theoretical advantages and holds much promise. The continued evolution of existing and new platforms for automated ion channel screening by electrophysiology will keep up with the demand both for ion channel safety profiling and ion channel-targeted drug discovery.

15.7 EMERGING TECHNOLOGIES

In the human genome, approximately 30% of proteins are membrane proteins, among which many are ion channels. With increased recognition of the importance of ion channels in disease processes, one expects that in the near future such electrophysiological screening technologies will advance further to become more compatible with high-throughput screening. Several ion channel assay technologies are emerging with considerable advantages over current methodologies.

In recent years, there has been a significant expansion in the development of *label-free* biosensors for quantifying aspects of cell function (76). Initial applications included measuring cell proliferation and cytotoxicity, cell migration, and receptor activation. Two classes of label-free instruments, using either an electrical impedance–based or an optical biosensor, are now commercially available for investigating the effects of ligands on cellular targets. The two independent technologies have been developed with a fundamental feature of exquisite sensitivity, which should resolve cellular process previously undetectable. For example, impedance assays are capable of detecting actin-dependent cellular micromotion on the order of 1 nm or approximately one-third the thickness of the cell membrane. Similarly, optical instruments have the sensitivity to detect mass changes of ~300 Da. The availability of two independent technologies should be valuable for unraveling the cellular mechanism of GPCR activation detected in each assay. Studies of GPCR function have been especially prominent with these instruments due to the importance of this target class in drug discovery. Both impedance-based and optical biosensor instruments have been used to quantify GPCR pharmacology (77–80). Other applications for label-free detection remain under evaluation.

Compound screening against heterologously expressed ion channels, although intrinsic to the modern drug discovery process, will often uncover drug candidates whose high-order impact on networked neuronal systems are not necessarily inferable from their effects on recombinant expressed channel in cell lines. Thus, raising the throughput in such high-order systems has several unique challenges. The electrical properties of single neurons can be studied in great detail by electrophysiological methods in vitro and in vivo. However, for a simultaneous readout of electrical activity of different regions within one neuron or from large numbers of neurons in brain tissue, noninvasive optical imaging techniques are advantageous over other methods that require rupture of cell membranes. Optical potentiometric probes based on electrochromic organic dyes has nonspecific labeling of membranes and dye phototoxicity in mammalian tissue. The voltage-sensitive fluorescent proteins (VSFPs) can potentially overcome this problem. Now a different version of VESPs have been developed including VSFP 2.1, VSFP 2.3, VSFP 3.1, and VSFP Butterfly 1.2 etc., which were constructed as a fusion between the voltage-sensing domain of Ci-VSP (Ciona intestinalis voltage sensor containing phosphatase) and a fluorescence reporter composed of a pair of cyan and yellow fluorescent proteins (FPs) (81,82). While VSFPs afford genetic control of expression in target cells, they have the potential for further functional customization. Like voltage-gated ion channels, VSFPs activate with sigmoidal voltage-dependence as a consequence of the voltage-sensing mechanism that involves transition of the voltage sensor from a deactivated configuration with the sensing charge facing the intracellular environment, to an activated configuration with the charge facing the extracellular side. For the very same reason that nonlinearity of activation is essential for functional specialization and diversity of ion channels, VSFPs can in principle be tuned to customized profiles. It turns out that optical imaging of subthreshold fluctuations in membrane potential, allow for detection of voltage transients of genetically specified cell classes in a variety of experimental paradigms (82).

The emergence of a robot system to automatically conduct whole cell in vivo recording (83) and new voltage-sensitive fluorescence proteins (84) is providing new possibilities to develop and advance ion channel assays. Indeed, ion channel screening technologies are no longer limited to instrument development. Expression systems including cell lines and off-the-shelf assays are enabling new opportunities for the existing instrumentation. Recently, a few new products have emerged for cell line development, including MaxCyte STX electroporation instrument (49), BacMam system (10), Jump-In™ Cell Engineering Platform (85). The products should be able to enhance the process of stable cell line development and expedite screening for drug discovery, bioproduction, and cell-based therapy.

Data analysis and management is another important and critical aspect of the HTS process, especially for a large amount of data across different assays or targets. hERGCentral was developed as a website based on experimental data obtained from a primary screen by automated electrophysiology against more than 300,000 structurally diverse compounds (71). hERGCentral has annotated datasets of more than 300,000 compounds including structures and chemophysiological properties of compounds, raw traces, and biophysical properties. The system enables a variety of query formats, including searches for hERG effects according to either chemical structure or properties, and alternatively according to the specific biophysical properties of current changes caused by a compound. As an evolving platform, hERGCentral will be developed to both meet the needs and enhance the experience of end users, in which the major efforts include improved data and functionality enrichment. The website was designed for a fixed set of compound libraries, where pre-stored and annotated data allows users a rare chance to interact with the interface in the user end. There is an extensive need to have all-in-one, database-like software that users could not only use to manage the information for compounds and raw data, but also flexibly perform the data analysis for data from a single assay and across different assays or targets, trace view, data plot, hit selection, and compound search function etc. To meet this need, such a system is still under development.

ACKNOWLEDGMENTS

We thank members of the Li laboratory for valuable discussions and comments on the manuscript, Alison Neal for editorial assistance. The research in the Li laboratory has been supported by the National Institutes of Health (U54 MH084691) and Johns Hopkins University.

REFERENCES

1. Gill, S., Gill, R., Lee, S. S., Hesketh, J. C., Fedida, D., Rezazadeh, S., Stankovich, L., and Liang, D. (2003). Flux assays in high throughput screening of ion channels in drug discovery. *Assay Drug Dev Technol* 1, 709–717.
2. Finlayson, K., Pennington, A. J., and Kelly, J. S. (2001). [3H] dofetilide binding in SHSY5Y and HEK293 cells expressing a HERG-like K^+ channel? *Eur J Pharmacol* **412**, 203–212.
3. Diaz, G. J., Daniell, K., Leitza, S. T., Martin, R. L., Su, Z., McDermott, J. S., Cox, B. F., and Gintant, G. A. (2004). The [3H]dofetilide binding assay is a predictive screening tool for hERG blockade and proarrhythmia: Comparison of intact cell and membrane preparations and effects of altering $[K^+]_o$. *J Pharmacol Toxicol Meth* **50**, 187–199.
4. Chiu, P. J., Marcoe, K. F., Bounds, S. E., Lin, C. H., Feng, J. J., Lin, A., Cheng, F. C., Crumb, W. J., and Mitchell, R. (2004). Validation of a [^{3}H]astemizole binding assay in HEK293 cells expressing HERG K^+ channels. *J Pharmacol Sci* **95**, 311–319.
5. Huang, X. P., Mangano, T., Hufeisen, S., Setola, V., and Roth, B. L. (2010). Identification of human Ether-a-go-go related gene modulators by three screening platforms in an academic drug-discovery setting. *Assay Drug Dev Technol* **8**, 727–742.
6. Epps, D. E., Knechtel, T. J., Bacznskyj, O., Decker, D., Guido, D. M., Buxser, S. E., Mathews, W. R. et al. (1994). Tirilazad mesylate protects stored erythrocytes against osmotic fragility. *Chem Phys Lipids* **74**, 163–174.
7. Gonzalez, J. E. and Tsien, R. Y. (1995). Voltage sensing by fluorescence resonance energy transfer in single cells. *Biophys J* **69**, 1272–1280.
8. Solly, K., Cassaday, J., Felix, J. P., Garcia, M. L., Ferrer, M., Strulovici, B., and Kiss, L. (2008). Miniaturization and HTS of a FRET-based membrane potential assay for K(ir) channel inhibitors. *Assay Drug Dev Technol* **6**, 225–234.
9. Liu, C. J., Priest, B. T., Bugianesi, R. M., Dulski, P. M., Felix, J. P., Dick, I. E., Brochu, R. M. et al. (2006). A high-capacity membrane potential FRET-based assay for Nav1.8 channels. *Assay Drug Dev Technol* **4**, 37–48.
10. Titus, S. A., Beacham, D., Shahane, S. A., Southall, N., Xia, M., Huang, R., Hooten, E. et al. (2009). A new homogeneous high-throughput screening assay for profiling compound activity on the human ether-a-go-go-related gene channel. *Anal Biochem* **394**, 30–38.
11. Xia, M., Shahane, S. A., Huang, R., Titus, S. A., Shum, E., Zhao, Y., Southall, N. et al. (2011). Identification of quaternary ammonium compounds as potent inhibitors of hERG potassium channels. *Toxicol Appl Pharmacol* **252**, 250–258.
12. Wang, H. R., Wu, M., Yu, H., Long, S., Stevens, A., Engers, D. W., Sackin, H. et al. (2011). Selective inhibition of the K(ir)2 family of inward rectifier potassium channels by a small molecule probe: The discovery, SAR, and pharmacological characterization of ML133. *ACS Chem Biol* **6**, 845–856.

13. Yu, H., Wu, M., Townsend, S. D., Zou, B., Long, S., Daniels, J. S., McManus, O. B., Li, M., Lindsley, C. W., and Hopkins, C. R. (2011). Discovery, synthesis, and structure activity relationship of a series of N-aryl-bicyclo[2.2.1]heptane-2-carboxamides: Characterization of ML213 as a Novel KCNQ2 and KCNQ4 potassium channel opener. *ACS Chem Neurosci* **2**, 572–577.
14. Zou, B., Yu, H., Babcock, J. J., Chanda, P., Bader, J. S., McManus, O. B., and Li, M. (2010). Profiling diverse compounds by flux- and electrophysiology-based primary screens for inhibition of human Ether-a-go-go related gene potassium channels. *Assay Drug Dev Technol* **8**, 743–754.
15. Miller, M., Shi, J., Zhu, Y., Kustov, M., Tian, J. B., Stevens, A., Wu, M. et al. (2011). Identification of ML204, a novel potent antagonist that selectively modulates native TRPC4/C5 ion channels. *J Biol Chem* **286**, 33436–33446.
16. Gonzalez, J. E. and Maher, M. P. (2002). Cellular fluorescent indicators and voltage/ion probe reader (VIPR) tools for ion channel and receptor drug discovery. *Receptors Channels* **8**, 283–295.
17. Falconer, M., Smith, F., Surah-Narwal, S., Congrave, G., Liu, Z., Hayter, P., Ciaramella, G. et al. (2002). High-throughput screening for ion channel modulators. *J Biomol Screen* **7**, 460–465.
18. Huang, C. J., Harootunian, A., Maher, M. P., Quan, C., Raj, C. D., McCormack, K., Numann, R., Negulescu, P. A., and Gonzalez, J. E. (2006). Characterization of voltage-gated sodium-channel blockers by electrical stimulation and fluorescence detection of membrane potential. *Nat Biotechnol* **24**, 439–446.
19. Bugianesi, R. M., Augustine, P. R., Azer, K., Dufresne, C., Herrington, J., Kath, G. S., McManus, O. B. et al. (2006). A cell-sparing electric field stimulation technique for high-throughput screening of voltage-gated ion channels. *Assay Drug Dev Technol* **4**, 21–35.
20. Hamilton, T. C., Weir, S. W., and Weston, A. H. (1986). Comparison of the effects of BRL 34915 and verapamil on electrical and mechanical activity in rat portal vein. *Br J Pharmacol* **88**, 103–111.
21. Weir, S. W. and Weston, A. H. (1986). The effects of BRL 34915 and nicorandil on electrical and mechanical activity and on 86Rb efflux in rat blood vessels. *Br J Pharmacol* **88**, 121–128.
22. Terstappen, G. C. (1999). Functional analysis of native and recombinant ion channels using a high-capacity nonradioactive rubidium efflux assay. *Anal Biochem* **272**, 149–155.
23. Liu, K., Samuel, M., Harrison, R. K., and Paslay, J. W. (2010). Rb^+ efflux assay for assessment of non-selective cation channel activities. *Assay Drug Dev Technol* **8**, 380–388.
24. Chaudhary, K. W., O'Neal, J. M., Mo, Z. L., Fermini, B., Gallavan, R. H., and Bahinski, A. (2006). Evaluation of the rubidium efflux assay for preclinical identification of HERG blockade. *Assay Drug Dev Technol* **4**, 73–82.
25. Scott, C. W., Wilkins, D. E., Trivedi, S., and Crankshaw, D. J. (2003). A medium-throughput functional assay of KCNQ2 potassium channels using rubidium efflux and atomic absorption spectrometry. *Anal Biochem* **319**, 251–257.
26. Sorota, S., Zhang, X. S., Margulis, M., Tucker, K., and Priestley, T. (2005). Characterization of a hERG screen using the IonWorks HT: Comparison to a hERG rubidium efflux screen. *Assay Drug Dev Technol* **3**, 47–57.
27. Terstappen, G. C. (2004). Nonradioactive rubidium ion efflux assay and its applications in drug discovery and development. *Assay Drug Dev Technol* **2**, 553–559.
28. Hanson, G. T. and Hanson, B. J. (2008). Fluorescent probes for cellular assays. *Comb Chem High Throughput Screen* **11**, 505–513.
29. Paredes, R. M., Etzler, J. C., Watts, L. T., Zheng, W., and Lechleiter, J. D. (2008). Chemical calcium indicators. *Methods* **46**, 143–151.
30. Bailey, S. and Macardle, P. J. (2006). A flow cytometric comparison of Indo-1 to fluo-3 and Fura Red excited with low power lasers for detecting Ca^{2+} flux. *J Immunol Methods* **311**, 220–225.
31. Grynkiewicz, G., Poenie, M., and Tsien, R. Y. (1985). A new generation of Ca^{2+} indicators with greatly improved fluorescence properties. *J Biol Chem* **260**, 3440–3450.
32. Song, Y., Buelow, B., Perraud, A. L., and Scharenberg, A. M. (2008). Development and validation of a cell-based high-throughput screening assay for TRPM2 channel modulators. *J Biomol Screen* **13**, 54–61.
33. Dai, G., Haedo, R. J., Warren, V. A., Ratliff, K. S., Bugianesi, R. M., Rush, A., Williams, M. E. et al. (2008). A high-throughput assay for evaluating state dependence and subtype selectivity of Cav2 calcium channel inhibitors. *Assay Drug Dev Technol* **6**, 195–212.
34. Weaver, C. D., Harden, D., Dworetzky, S. I., Robertson, B., and Knox, R. J. (2004). A thallium-sensitive, fluorescence-based assay for detecting and characterizing potassium channel modulators in mammalian cells. *J Biomol Screen* **9**, 671–677.
35. Schmalhofer, W. A., Swensen, A. M., Thomas, B. S., Felix, J. P., Haedo, R. J., Solly, K., Kiss, L., Kaczorowski, G. J., and Garcia, M. L. (2010). A pharmacologically validated, high-capacity, functional thallium flux assay for the human Ether-a-go-go related gene potassium channel. *Assay Drug Dev Technol* **8**, 714–726.
36. Li, Q., Rottlander, M., Xu, M., Christoffersen, C. T., Frederiksen, K., Wang, M. W., and Jensen, H. S. (2011). Identification of novel KCNQ4 openers by a high-throughput fluorescence-based thallium flux assay. *Anal Biochem* **418**, 66–72.
37. Bridal, T. R., Margulis, M., Wang, X., Donio, M., and Sorota, S. (2010). Comparison of human Ether-a-go-go related gene screening assays based on IonWorks Quattro and thallium flux. *Assay Drug Dev Technol* **8**, 755–765.
38. Cheung, Y. Y., Yu, H., Xu, K., Zou, B., Wu, M., McManus, O. B., Li, M., Lindsley, C. W., and Hopkins, C. R. (2012). Discovery of a series of 2-phenyl-*N*-(2-(pyrrolidin-1-yl)phenyl)acetamides as novel molecular switches that modulate modes of K(v)7.2 (KCNQ2) channel pharmacology: Identification of (S)-2-phenyl-*N*-(2-(pyrrolidin-1-yl)phenyl)butanamide (ML252) as a potent, brain penetrant K(v)7.2 channel inhibitor. *J Med Chem* **55**, 6975–6979.
39. Mattmann, M. E., Yu, H., Lin, Z., Xu, K., Huang, X., Long, S., Wu, M. et al. (2012). Identification of (R)-N-(4-(4-methoxyphenyl) thiazol-2-yl)-1-tosylpiperidine-2-carboxamide, ML277, as a novel, potent and selective K(v)7.1 (KCNQ1) potassium channel activator. *Bioorg Med Chem Lett* **22**, 5936–5941.
40. Galietta, L. J., Haggie, P. M., and Verkman, A. S. (2001). Green fluorescent protein-based halide indicators with improved chloride and iodide affinities. *FEBS Lett* **499**, 220–224.
41. Wachter, R. M. and Remington, S. J. (1999). Sensitivity of the yellow variant of green fluorescent protein to halides and nitrate. *Curr Biol* **9**, R628–R629.
42. Jayaraman, S., Haggie, P., Wachter, R. M., Remington, S. J., and Verkman, A. S. (2000). Mechanism and cellular applications of a green fluorescent protein-based halide sensor. *J Biol Chem* **275**, 6047–6050.
43. Sui, J., Cotard, S., Andersen, J., Zhu, P., Staunton, J., Lee, M., and Lin, S. (2010). Optimization of a yellow fluorescent protein-based iodide influx high-throughput screening assay for cystic fibrosis transmembrane conductance regulator (CFTR) modulators. *Assay Drug Dev Technol* **8**, 656–668.
44. Galietta, L. J., Springsteel, M. F., Eda, M., Niedzinski, E. J., By, K., Haddadin, M. J., Kurth, M. J., Nantz, M. H., and Verkman, A. S. (2001). Novel CFTR chloride channel activators identified by screening of combinatorial libraries based on flavone and benzoquinolizinium lead compounds. *J Biol Chem* **276**, 19723–19728.

45. Namkung, W., Phuan, P. W., and Verkman, A. S. (2011). TMEM16A inhibitors reveal TMEM16A as a minor component of calcium-activated chloride channel conductance in airway and intestinal epithelial cells. *J Biol Chem* **286**, 2365–2374.
46. Namkung, W., Thiagarajah, J. R., Phuan, P. W., and Verkman, A. S. (2010). Inhibition of Ca^{2+}-activated Cl^- channels by gallotannins as a possible molecular basis for health benefits of red wine and green tea. *FASEB J* **24**, 4178–4186.
47. Namkung, W., Yao, Z., Finkbeiner, W. E., and Verkman, A. S. (2011). Small-molecule activators of TMEM16A, a calcium-activated chloride channel, stimulate epithelial chloride secretion and intestinal contraction. *FASEB J* **25**, 4048–4062.
48. Kruger, W., Gilbert, D., Hawthorne, R., Hryciw, D. H., Frings, S., Poronnik, P., and Lynch, J. W. (2005). A yellow fluorescent protein-based assay for high-throughput screening of glycine and GABAA receptor chloride channels. *Neurosci Lett* **380**, 340–345.
49. Johansson, T., Norris, T., and Peilot-Sjogren, H. (2013). Yellow fluorescent protein-based assay to measure GABAA channel activation and allosteric modulation in CHO-K1 cells. *PLoS One* **8**, e59429.
50. Hamill, O. P., Marty, A., Neher, E., Sakmann, B., and Sigworth, F. J. (1981). Improved patch-clamp techniques for high-resolution current recording from cells and cell-free membrane patches. *Pflugers Arch* **391**, 85–100.
51. Neher, E. and Sakmann, B. (1976). Single-channel currents recorded from membrane of denervated frog muscle fibres. *Nature* **260**, 799–802.
52. Dunlop, J., Bowlby, M., Peri, R., Vasilyev, D., and Arias, R. (2008). High-throughput electrophysiology: An emerging paradigm for ion-channel screening and physiology. *Nat Rev Drug Discov* **7**, 358–368.
53. Lippiat, J. D. and Wrighton, D. C. (2013). Conventional micropipette-based patch clamp techniques. *Methods Mol Biol* **998**, 91–107.
54. Poulos, J. L., Jeon, T. J., Damoiseaux, R., Gillespie, E. J., Bradley, K. A., and Schmidt, J. J. (2009). Ion channel and toxin measurement using a high throughput lipid membrane platform. *Biosens Bioelectron* **24**, 1806–1810.
55. Poulos, J. L., Jeon, T. J., and Schmidt, J. J. (2010). Automatable production of shippable bilayer chips by pin tool deposition for an ion channel measurement platform. *Biotechnol J* **5**, 511–514.
56. Poulos, J. L., Portonovo, S. A., Bang, H., and Schmidt, J. J. (2010). Automatable lipid bilayer formation and ion channel measurement using sessile droplets. *J Phys Condens Matter* **22**, 454105.
57. Thapliyal, T., Poulos, J. L., and Schmidt, J. J. (2011). Automated lipid bilayer and ion channel measurement platform. *Biosens Bioelectron* **26**, 2651–2654.
58. Kvist, T., Hansen, K. B., and Brauner-Osborne, H. (2011). The use of *Xenopus* oocytes in drug screening. *Expert Opin Drug Discov* **6**, 141–153.
59. Finkel, A., Wittel, A., Yang, N., Handran, S., Hughes, J., and Costantin, J. (2006). Population patch clamp improves data consistency and success rates in the measurement of ionic currents. *J Biomol Screen* **11**, 488–496.
60. Trivedi, S., Dekermendjian, K., Julien, R., Huang, J., Lund, P. E., Krupp, J., Kronqvist, R., Larsson, O., and Bostwick, R. (2008). Cellular HTS assays for pharmacological characterization of Na(v)1.7 modulators. *Assay Drug Dev Technol* **6**, 167–179.
61. Bridgland-Taylor, M. H., Hargreaves, A. C., Easter, A., Orme, A., Henthorn, D. C., Ding, M., Davis, A. M. et al. (2006). Optimisation and validation of a medium-throughput electrophysiology-based hERG assay using IonWorks HT. *J Pharmacol Toxicol Methods* **54**, 189–199.
62. Wible, B. A., Kuryshev, Y. A., Smith, S. S., Liu, Z., and Brown, A. M. (2008). An ion channel library for drug discovery and safety screening on automated platforms. *Assay Drug Dev Technol* **6**, 765–780.
63. Gillie, D. J., Novick, S. J., Donovan, B. T., Payne, L. A., and Townsend, C. (2013). Development of a high-throughput electrophysiological assay for the human ether-a-go-go related potassium channel hERG. *J Pharmacol Toxicol Methods* **67**, 33–44.
64. Terstappen, G. C., Roncarati, R., Dunlop, J., and Peri, R. (2010). Screening technologies for ion channel drug discovery. *Future Med Chem* **2**, 715–730.
65. Graef, J. D., Benson, L. C., Sidach, S. S., Wei, H., Lippiello, P. M., Bencherif, M., and Fedorov, N. B. (2013). Validation of a high-throughput, automated electrophysiology platform for the screening of nicotinic agonists and antagonists. *J Biomol Screen* **18**, 116–127.
66. Xie, X., Van Deusen, A. L., Vitko, I., Babu, D. A., Davies, L. A., Huynh, N., Cheng, H., Yang, N., Barrett, P. Q., and Perez-Reyes, E. (2007). Validation of high throughput screening assays against three subtypes of Ca(v)3 T-type channels using molecular and pharmacologic approaches. *Assay Drug Dev Technol* **5**, 191–203.
67. Lee, Y. T., Vasilyev, D. V., Shan, Q. J., Dunlop, J., Mayer, S., and Bowlby, M. R. (2008). Novel pharmacological activity of loperamide and CP-339,818 on human HCN channels characterized with an automated electrophysiology assay. *Eur J Pharmacol* **581**, 97–104.
68. Castle, N., Printzenhoff, D., Zellmer, S., Antonio, B., Wickenden, A., and Silvia, C. (2009). Sodium channel inhibitor drug discovery using automated high throughput electrophysiology platforms. *Comb Chem High Throughput Screen* **12**, 107–122.
69. Cao, X., Lee, Y. T., Holmqvist, M., Lin, Y., Ni, Y., Mikhailov, D., Zhang, H. et al. (2010). Cardiac ion channel safety profiling on the IonWorks Quattro automated patch clamp system. *Assay Drug Dev Technol* **8**, 766–780.
70. Ido, K., Ohwada, T., Yasutomi, E., Yoshinaga, T., Arai, T., Kato, M., and Sawada, K. (2013). Screening quality for Ca^{2+}-activated potassium channel in IonWorks Quattro is greatly improved by using BAPTA-AM and ionomycin. *J Pharmacol Toxicol Methods* **67**, 16–24.
71. Du, F., Yu, H., Zou, B., Babcock, J., Long, S., and Li, M. (2011). hERGCentral: A large database to store, retrieve, and analyze compound-human Ether-a-go-go related gene channel interactions to facilitate cardiotoxicity assessment in drug development. *Assay Drug Dev Technol* **9**, 580–588.
72. Zhang, H., Zou, B., Yu, H., Moretti, A., Wang, X., Yan, W., Babcock, J. J. et al. (2012). Modulation of hERG potassium channel gating normalizes action potential duration prolonged by dysfunctional KCNQ1 potassium channel. *Proc Natl Acad Sci USA* **109**, 11866–11871.
73. Beck, E. J., Hutchinson, T. L., Qin, N., Flores, C. M., and Liu, Y. (2010). Development and validation of a secondary screening assay for TRPM8 antagonists using QPatch HT. *Assay Drug Dev Technol* **8**, 63–72.
74. Liu, Y., Beck, E. J., and Flores, C. M. (2011). Validation of a patch clamp screening protocol that simultaneously measures compound activity in multiple states of the voltage-gated sodium channel Nav1.2. *Assay Drug Dev Technol* **9**, 628–634.
75. Milligan, C. J., Li, J., Sukumar, P., Majeed, Y., Dallas, M. L., English, A., Emery, P. et al. (2009). Robotic multiwell planar patch-clamp for native and primary mammalian cells. *Nat Protoc* **4**, 244–255.
76. Peters, M. F., Vaillancourt, F., Heroux, M., Valiquette, M., and Scott, C. W. (2010) Comparing label-free biosensors for pharmacological screening with cell-based functional assays. *Assay Drug Dev Technol* **8**, 219–227.
77. Lee, P. H., Gao, A., van Staden, C., Ly, J., Salon, J., Xu, A., Fang, Y., and Verkleeren, R. (2008). Evaluation of dynamic mass redistribution technology for pharmacological studies of recombinant and endogenously expressed G protein-coupled receptors. *Assay Drug Dev Technol* **6**, 83–94.

78. Yu, N., Atienza, J. M., Bernard, J., Blanc, S., Zhu, J., Wang, X., Xu, X., and Abassi, Y. A. (2006). Real-time monitoring of morphological changes in living cells by electronic cell sensor arrays: An approach to study G protein-coupled receptors. *Anal Chem* **78**, 35–43.
79. Peters, M. F., Knappenberger, K. S., Wilkins, D., Sygowski, L. A., Lazor, L. A., Liu, J., and Scott, C. W. (2007). Evaluation of cellular dielectric spectroscopy, a whole-cell, label-free technology for drug discovery on Gi-coupled GPCRs. *J Biomol Screen* **12**, 312–319.
80. Peters, M. F. and Scott, C. W. (2009). Evaluating cellular impedance assays for detection of GPCR pleiotropic signaling and functional selectivity. *J Biomol Screen* **14**, 246–255.
81. Akemann, W., Lundby, A., Mutoh, H., and Knopfel, T. (2009). Effect of voltage sensitive fluorescent proteins on neuronal excitability. *Biophys J* **96**, 3959–3976.
82. Akemann, W., Mutoh, H., Perron, A., Park, Y. K., Iwamoto, Y., and Knopfel, T. (2012). Imaging neural circuit dynamics with a voltage-sensitive fluorescent protein. *J Neurophysiol* **108**, 2323–2337.
83. Kodandaramaiah, S. B., Franzesi, G. T., Chow, B. Y., Boyden, E. S., and Forest, C. R. (2012). Automated whole-cell patch-clamp electrophysiology of neurons in vivo. *Nat Methods* **9**, 585–587.
84. Jin, L., Han, Z., Platisa, J., Wooltorton, J. R., Cohen, L. B., and Pieribone, V. A. (2012). Single action potentials and subthreshold electrical events imaged in neurons with a fluorescent protein voltage probe. *Neuron* **75**, 779–785.
85. Lieu, P. T., Machleidt, T., Thyagarajan, B., Fontes, A., Frey, E., Fuerstenau-Sharp, M., Thompson, D. V. et al. (2009). Generation of site-specific retargeting platform cell lines for drug discovery using phiC31 and R4 integrases. *J Biomol Screen* **14**, 1207–1215.
86. Dorn, A., Hermann, F., Ebneth, A., Bothmann, H., Trube, G., Christensen, K., and Apfel, C. (2005). Evaluation of a high-throughput fluorescence assay method for HERG potassium channel inhibition. *J Biomol Screen* **10**, 339–347.
87. Gopalakrishnan, M., Molinari, E. J., Shieh, C. C., Monteggia, L. M., Roch, J. M., Whiteaker, K. L., Scott, V. E., Sullivan, J. P., and Brioni, J. D. (2000). Pharmacology of human sulphonylurea receptor SUR1 and inward rectifier K^+ channel Kir6.2 combination expressed in HEK-293 cells. *Br J Pharmacol* **129**, 1323–1332.
88. Terstappen, G. C., Pellacani, A., Aldegheri, L., Graziani, F., Carignani, C., Pula, G., and Virginio, C. (2003). The antidepressant fluoxetine blocks the human small conductance calcium-activated potassium channels SK1, SK2 and SK3. *Neurosci Lett* **346**, 85–88.
89. Whiteaker, K. L., Davis-Taber, R., Scott, V. E., and Gopalakrishnan, M. (2001). Fluorescence-based functional assay for sarcolemmal ATP-sensitive potassium channel activation in cultured neonatal rat ventricular myocytes. *J Pharmacol Toxicol Methods* **46**, 45–50.
90. Vasilyev, D. V., Shan, Q. J., Lee, Y. T., Soloveva, V., Nawoschik, S. P., Kaftan, E. J., Dunlop, J., Mayer, S. C., and Bowlby, M. R. (2009). A novel high-throughput screening assay for HCN channel blocker using membrane potential-sensitive dye and FLIPR. *J Biomol Screen* **14**, 1119–1128.
91. Benjamin, E. R., Pruthi, F., Olanrewaju, S., Ilyin, V. I., Crumley, G., Kutlina, E., Valenzano, K. J., and Woodward, R. M. (2006). State-dependent compound inhibition of Nav1.2 sodium channels using the FLIPR Vm dye: On-target and off-target effects of diverse pharmacological agents. *J Biomol Screen* **11**, 29–39.
92. Joesch, C., Guevarra, E., Parel, S. P., Bergner, A., Zbinden, P., Konrad, D., and Albrecht, H. (2008). Use of FLIPR membrane potential dyes for validation of high-throughput screening with the FLIPR and microARCS technologies: Identification of ion channel modulators acting on the GABA(A) receptor. *J Biomol Screen* **13**, 218–228.
93. Slack, M., Kirchhoff, C., Moller, C., Winkler, D., and Netzer, R. (2006). Identification of novel Kv1.3 blockers using a fluorescent cell-based ion channel assay. *J Biomol Screen* **11**, 57–64.
94. Felix, J. P., Williams, B. S., Priest, B. T., Brochu, R. M., Dick, I. E., Warren, V. A., Yan, L. et al. (2004). Functional assay of voltage-gated sodium channels using membrane potential-sensitive dyes. *Assay Drug Dev Technol* **2**, 260–268.
95. Rezazadeh, S., Hesketh, J. C., and Fedida, D. (2004). Rb^+ flux through hERG channels affects the potency of channel blocking drugs: Correlation with data obtained using a high-throughput Rb^+ efflux assay. *J Biomol Screen* **9**, 588–597.
96. Jow, F., Shen, R., Chanda, P., Tseng, E., Zhang, H., Kennedy, J., Dunlop, J., and Bowlby, M. R. (2007). Validation of a medium-throughput electrophysiological assay for KCNQ2/3 channel enhancers using IonWorks HT. *J Biomol Screen* **12**, 1059–1067.
97. Raphemot, R., Weaver, C. D., and Denton, J. S. (2013). High-throughput screening for small-molecule modulators of inward rectifier potassium channels. *J Vis Exp.* (71). pii: 4209. doi: 10.3791/4209.
98. Felix, J. P., Priest, B. T., Solly, K., Bailey, T., Brochu, R. M., Liu, C. J., Kohler, M. G. et al. (2012). The inwardly rectifying potassium channel Kir1.1: Development of functional assays to identify and characterize channel inhibitors. *Assay Drug Dev Technol* **10**, 417–431.
99. Schroeder, K., Neagle, B., Trezise, D. J., and Worley, J. (2003). Ionworks HT: A new high-throughput electrophysiology measurement platform. *J Biomol Screen* **8**, 50–64.
100. John, V. H., Dale, T. J., Hollands, E. C., Chen, M. X., Partington, L., Downie, D. L., Meadows, H. J., and Trezise, D. J. (2007). Novel 384-well population patch clamp electrophysiology assays for Ca^{2+}-activated K^+ channels. *J Biomol Screen* **12**, 50–60.
101. Harmer, A. R., Abi-Gerges, N., Easter, A., Woods, A., Lawrence, C. L., Small, B. G., Valentin, J. P., and Pollard, C. E. (2008). Optimisation and validation of a medium-throughput electrophysiology-based hNav1.5 assay using IonWorks. *J Pharmacol Toxicol Methods* **57**, 30–41.
102. Ratliff, K. S., Petrov, A., Eiermann, G. J., Deng, Q., Green, M. D., Kaczorowski, G. J., McManus, O. B., and Herrington, J. (2008). An automated electrophysiology serum shift assay for K(v) channels. *Assay Drug Dev Technol* **6**, 243–253.
103. Milligan, C. J. and Moller, C. (2013). Automated planar patch-clamp. *Methods Mol Biol* **998**, 171–187.
104. Dunlop, J., Roncarati, R., Jow, B., Bothmann, H., Lock, T., Kowal, D., Bowlby, M., and Terstappen, G. C. (2007). In vitro screening strategies for nicotinic receptor ligands. *Biochem Pharmacol* **74**, 1172–1181.
105. Korsgaard, M. P., Strobaek, D., and Christophersen, P. (2009). Automated planar electrode electrophysiology in drug discovery: Examples of the use of QPatch in basic characterization and high content screening on Na(v), K(Ca)2.3, and K(v)11.1 channels. *Comb Chem High Throughput Screen* **12**, 51–63.
106. Friis, S., Mathes, C., Sunesen, M., Bowlby, M. R., and Dunlop, J. (2009). Characterization of compounds on nicotinic acetylcholine receptor alpha 7 channels using higher throughput electrophysiology. *J Neurosci Methods* **177**, 142–148.
107. Jones, K. A., Garbati, N., Zhang, H., and Large, C. H. (2009). Automated patch clamping using the QPatch. *Methods Mol Biol* **565**, 209–223.
108. Lenkey, N., Karoly, R., Lukacs, P., Vizi, E. S., Sunesen, M., Fodor, L., and Mike, A. (2010). Classification of drugs based on properties of sodium channel inhibition: A comparative automated patch-clamp study. *PLoS One* **5**, e15568.
109. Mathes, C. (2006). QPatch: The past, present and future of automated patch clamp. *Expert Opin Ther Targets* **10**, 319–327.
110. Mathes, C., Friis, S., Finley, M., and Liu, Y. (2009). QPatch: The missing link between HTS and ion channel drug discovery. *Comb Chem High Throughput Screen* **12**, 78–95.

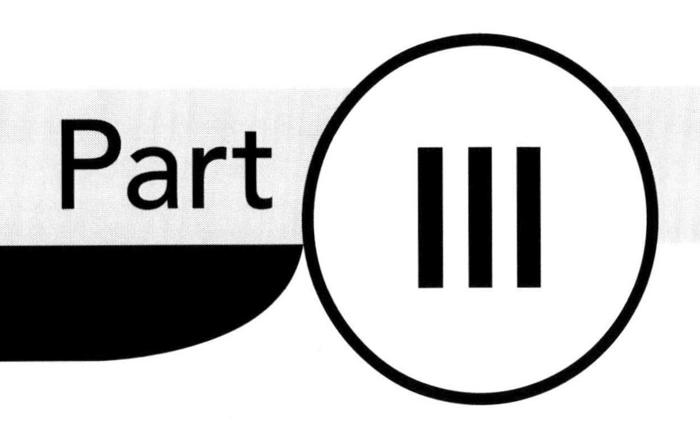

Ion channel families

16 Voltage-gated sodium channels

William A. Catterall

Contents

16.1 FUNCTIONAL ROLES OF VOLTAGE-GATED SODIUM CHANNELS

16.1.1 ACTION POTENTIAL GENERATION AND PROPAGATION

Electrical signaling in nerve, muscle, and endocrine cells depends on the initiation of action potentials by voltage-gated sodium channels, as described by Hodgkin and Huxley using the voltage-clamp technique (Hodgkin and Huxley, 1952a,b,c,d). By imposing a rapid depolarization upon the membrane of the giant nerve axon of the squid, they showed that sodium channels rapidly activate (within 1 ms) and then rapidly inactivate (within 5 ms). The brief pulse of inward sodium current produced by the activation of sodium channels is responsible for the rapidly rising (depolarizing) phase of the action potential and for its rapid conduction along nerve and muscle fibers. These studies first revealed the transient sodium currents as the mechanism of initiation and propagation of action potentials.

16.1.2 ACTIVATION, CONDUCTANCE, AND TWO PHASES OF INACTIVATION

The voltage-clamp studies of Hodgkin and Huxley also revealed the three essential functions of sodium channels: voltage-dependent activation, fast inactivation, and selective sodium conductance (Hodgkin and Huxley, 1952c,d). They showed that activation of the sodium current of squid giant axon is steeply dependent on transmembrane voltage and has sigmoid activation kinetics, consistent with the requirement for movement of three electrically charged *gating particles* across the membrane electric field during activation. In contrast, fast inactivation followed an exponential time course as if a single *gating particle* controlled this process. Their m^3h formulation of sodium channel gating kinetics has withstood the tests of six decades as a quantitative description of the sodium current.

The essence of voltage-dependent gating, as described by Hodgkin and Huxley (1952d), is the movement of gating charges that are part of the sodium channel protein across the membrane electric field. This essential gating charge movement was measured directly by Armstrong and Bezanilla (1973, 1974) and later by Keynes and Rojas (1974). In the absence of permeant ions, they measured a capacitative current in the squid giant axon that correlated closely with the activation of sodium channels. Subsequent studies with higher resolution methods showed that the gating charge movement that drives sodium channel activation amounts to transit of 12–16 positive charges across the full membrane electric field (Hirschberg et al., 1995; Kuzmenkin et al., 2004). This voltage-driven movement of gating charge is coupled to a conformational change that opens the pore of the sodium channel within 1 ms.

Once open, the sodium channel is highly selective for ions. Na^+ is approximately 10-fold more permeant than K^+ and 50-fold more permeant than Ca^{2+} (Hille, 1971, 1972). Detailed studies of permeation, saturation, and block by inorganic and organic monovalent cations led Hille to a model of the sodium channel selectivity filter as a high-field-strength site with one or more carboxyl groups poised to replace some or all of the waters of hydration of the Na^+ ion in a specific way and coordinate the permeating Na^+ ion as it moves over a series of four potential energy barriers and interacts with a series of three coordination sites on its way through the pore (Hille, 1975). This conceptual model of Na^+ permeation and selectivity agrees well with the emerging structural models of the ion selectivity filter of sodium channels (see the following text).

In addition to the fast inactivation process discovered by Hodgkin and Huxley, sodium channels have a second slow inactivation process that is engaged during prolonged single depolarizations and trains of repetitive depolarizations on the time scale of hundreds of milliseconds to seconds (Adelman and Palti, 1968, 1969; Rudy, 1978). In a physiological setting, this slow inactivation process is important to limit the frequency of firing and define the length of trains of action potentials in nerve and muscle and to protect cells against excitotoxic injury (Vilin and Ruben, 2001). In the central nervous system, the slow inactivation process contributes directly to encoding information in the frequency and duration of action potential trains (Carr et al., 2003).

16.1.3 PERSISTENT AND RESURGENT SODIUM CURRENTS

In addition to the classical transient sodium current, sodium channels have two additional well-described modes of activity. The transient sodium current inactivates incompletely in nerve and muscle cells, leaving a small, persistent sodium current in the range of 1% of the amplitude of peak sodium current (Crill, 1996). This persistent sodium current is important in neuronal dendrites and cell bodies where it boosts the size of excitatory postsynaptic potentials and helps to bring the cell to threshold during trains of action potentials. Many classes of neurons also generate resurgent sodium currents, which appear as a rebound inward current following a voltage-clamp pulse or an action potential (Raman and Bean, 1997, 1999). These currents are caused by reopening sodium channels, and therefore are distinct from persistent sodium currents that are caused by prolonged activity of previously open sodium channels. Resurgent sodium currents also contribute repetitive firing of action potentials by providing depolarizing drive following inactivation of the transient sodium current and repolarization of the cell by potassium currents. Persistent and resurgent sodium currents can work together with slow inactivation to generate complex patterns of firing of action potentials in central neurons (Do and Bean, 2003).

16.2 DISCOVERY AND BIOCHEMICAL PROPERTIES OF THE SODIUM CHANNEL PROTEIN

16.2.1 IDENTIFICATION, PURIFICATION, AND RECONSTITUTION OF SODIUM CHANNELS

Because of their essential role in generation of action potentials, sodium channel–directed neurotoxins have independently evolved in marine dinoflagellates found in plankton, sea anemones and other coelenterates, marine snails, fish, amphibians, spiders and scorpions. These toxins prevent nerve conduction and induce paralysis through actions at six distinct receptor sites on sodium channels (Table 16.1). Because of their specificity of interaction with sodium channels and their ability to bind to sodium channels with high affinity, these neurotoxins have been exploited as molecular probes of sodium channel structure and function.

Photoreactive derivatives of the polypeptide toxins of scorpion venom were covalently attached to Na^+ channels in intact nerve cell membranes from brain, allowing direct identification of the protein components of Na^+ channels (Beneski and Catterall, 1980). These experiments revealed large α subunits of 260 kDa and smaller β subunits of 30–40 kDa. Reversible binding of saxitoxin and tetrodotoxin to their common receptor site was used as a biochemical assay for the channel protein. Solubilization of brain membranes with nonionic detergents released the sodium channel, and the solubilized channel was purified by chromatographic techniques that separate glycoproteins by size, charge, and composition of covalently attached carbohydrate (Hartshorne and Catterall, 1981, 1984; Hartshorne et al., 1982; Figure 16.1). Sodium channel α subunits were also purified from eel electroplax, and a similar complex of α and β subunits was purified from skeletal muscle (Agnew et al., 1980; Barchi, 1983). An important step in the study of a purified membrane transport protein is to reconstitute its function in the pure state. This was accomplished by reconstitution of the sodium channel in phospholipid vesicles and in planar phospholipid bilayers (Tamkun et al., 1984; Hartshorne et al., 1985). Recordings of sodium currents from single purified brain sodium channels showed that they retained the voltage dependence, ion selectivity, and pharmacological properties of native channels, confirming that the correct sodium channel protein had been purified in functional form (Hartshorne et al., 1985; Figure 16.1).

16.2.2 PRIMARY STRUCTURES OF SODIUM CHANNEL SUBUNITS

The amino acid sequences of the Na^+ channel α, $\beta1$, and $\beta2$ subunits were determined by cloning DNA complementary to their mRNAs, using antibodies and oligonucleotide probes developed from work on purified Na^+ channels (Noda et al., 1984, 1986a,b; Goldin et al., 1986; Auld et al., 1988;

Table 16.1 Neurotoxin receptor sites on sodium channels

RECEPTOR SITE	TOXIN OR DRUG	LOCATION (DOMAIN/SEGMENT)
Neurotoxin receptor site 1	Tetrodotoxin	I/P, II/P, III/P, IV/P
	Saxitoxin	I/P, II/P, III/P, IV/P
	μ-Conotoxin	
Neurotoxin receptor site 2	Veratridine	
	Batrachotoxin	I/S6, IV/S6
	Grayanotoxin	
Neurotoxin receptor site 3	α-Scorpion toxins	I/S5–S6, IV/S1–S2, IV/S3–S4
	Sea anemone toxins	IV/S3–S4
Neurotoxin receptor site 4	β-Scorpion toxins	II/S1–S2, II/S3–S4, III/S5–S6
	Huwentoxin	IIS1–S2, IIS3–S4
Neurotoxin receptor site 5	Brevetoxins	I/S6, IV/S5
	Ciguatoxins	
Neurotoxin receptor site 6	δ-Conotoxins	IV/S3–S4

Sources: Reviewed in Catterall (1980), Rochat et al. (1984), Baden (1989), Lipkind and Fozzard (1994), Cestèle and Catterall (2000), Wang and Wang (2003), Terlau and Olivera (2004), Catterall et al. (2007), Moran et al. (2009), Zhang et al. (2009), and Tikhonov and Zhorov (2012), Minassianet et al. (2013), Xiao et al. (2014).

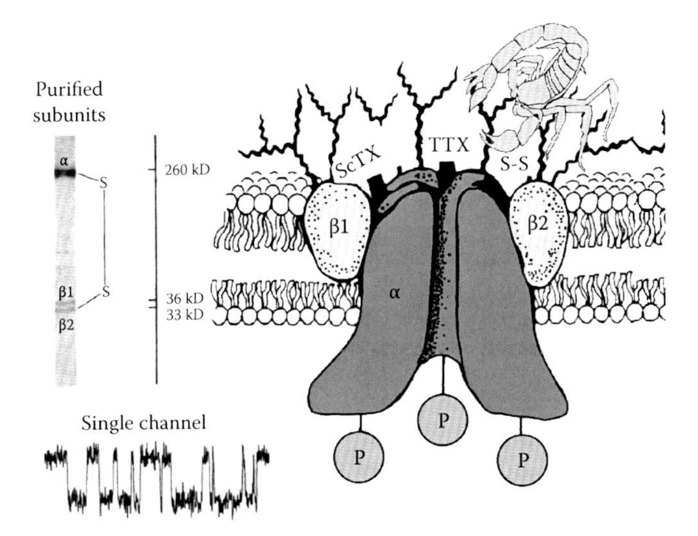

Figure 16.1 Subunit structure of voltage-gated sodium channels. (a) SDS polyacrylamide gel electrophoresis patterns illustrating the α and β subunits of the brain sodium channels. Sodium channel purified from rat brain showing the α, β1, and β2 subunits and their molecular weights. (From Hartshorne, R.P. et al., *J. Biol. Chem.*, 257, 13888, 1982.) As illustrated, the α and β2 subunits are linked by a disulfide bond. Tetrodotoxin and scorpion toxins bind to the α subunits of sodium channels as indicated and were used as molecular tags to identify and purify the sodium channel protein from brain. (From Beneski, D.A. and Catterall, W.A., *Proc. Natl. Acad. Sci. USA*, 77, 639, 1980; Hartshorne, R.P. et al., *J. Biol. Chem.*, 257, 13888, 1982; Hartshorne, R.P. and Catterall, W.A., *J. Biol. Chem.*, 259, 1667, 1984.) *Inset*: Single channel currents conducted by a single purified sodium channel incorporated into a planar bilayer. (From Hartshorne, R.P. et al., *Proc. Natl. Acad. Sci. USA*, 82, 240, 1985.)

Isom et al., 1992, 1995). The primary structures of these subunits are illustrated as transmembrane folding models in Figure 16.2. The large α subunits are composed of 1800–2000 amino acids and contain four repeated domains having greater than 50% internal sequence identity in their transmembrane regions. Each domain contains six segments that form transmembrane α-helices and an additional membrane reentrant loop that forms the outer mouth of the transmembrane pore (see the following text). In contrast, the smaller β1 and β2 subunits consist of a large extracellular N-terminal segment having a structure similar to antigen-binding regions of immunoglobulin (an Ig-fold), a single transmembrane segment, and a short intracellular segment (Isom et al., 1992, 1995; Figure 16.2). Only the principal α subunits of sodium channels are required for function, but the β subunits increase the level of cell surface expression and modify the voltage-dependent gating, conferring more physiologically correct functional properties on the expressed α subunits (Isom et al., 1992, 1995; Brackenbury and Isom, 2011). These results indicate that the principal α subunits of the voltage-gated sodium channels are functionally autonomous, but the auxiliary β subunits improve expression, help to determine subcellular localization, and modulate physiological properties.

16.2.3 MAPPING THE MOLECULAR COMPONENTS REQUIRED FOR SODIUM CHANNEL FUNCTION

Knowledge of the primary structures of the sodium channel subunits permitted detailed tests of their functional properties. cDNA clones encoding wild-type and mutant channels were expressed in recipient cells, and the resulting ion channels were studied by voltage-clamp methods. Much was learned about the molecular basis of sodium channel function by analyzing the effects of mutations, specific antibodies, drugs, and toxins on sodium channels expressed in recipient cells from their cDNAs. These insights are illustrated in the form of a color-coded molecular map of sodium channel functions in Figure 16.2. The voltage sensor is composed of the S1–S4 segments in each domain (Figure 16.2, white, yellow). The S4 segment contains four to eight repeated motifs having a positively

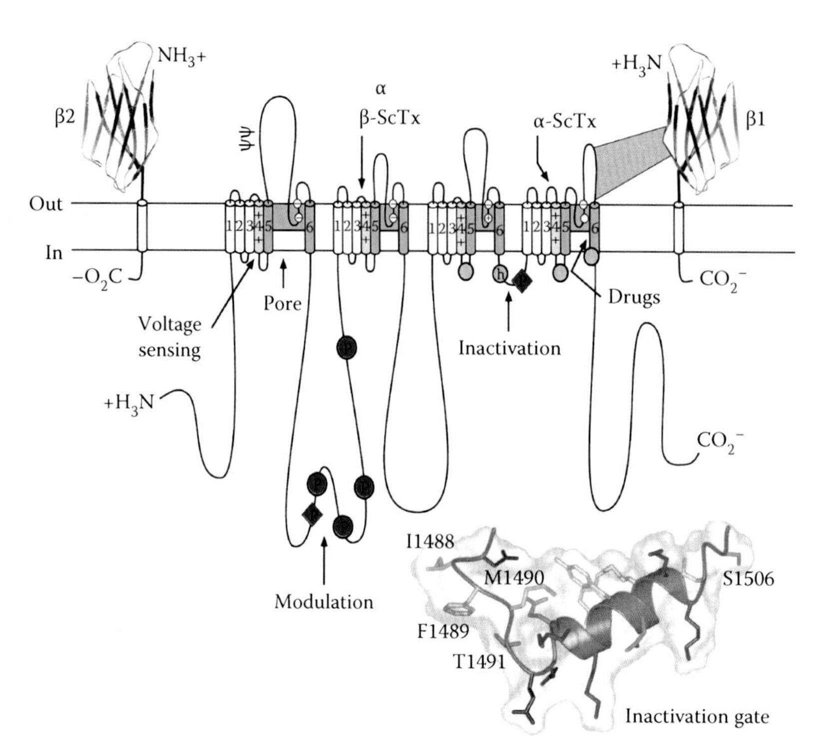

Figure 16.2 The primary structures of the subunits of the voltage-gated sodium channels. Cylinders represent probable alpha helical segments. Bold lines represent the polypeptide chains of each subunit with length approximately proportional to the number of amino acid residues in the brain sodium channel subtypes. The extracellular domains of the β1 and β2 subunits are shown as immunoglobulin-like folds. Ψ, sites of probable N-linked glycosylation; P in red circles, sites of demonstrated protein phosphorylation by PKA (circles) and PKC (diamonds); green, pore-lining segments; white circles, the outer and inner (DEKA) rings of amino residues that form the ion selectivity filter and the tetrodotoxin-binding site; yellow, S4 voltage sensors; h in blue circle, inactivation particle in the inactivation gate loop; blue circles, sites implicated in forming the inactivation gate receptor. Sites of binding of α- and β-scorpion toxins and a site of interaction between α and β1 subunits are also shown. Tetrodotoxin is a specific blocker of the pore of sodium channels, whereas the α- and β-scorpion toxins block fast inactivation and enhance activation, respectively, and thereby generate persistent sodium current that causes depolarization block of nerve conduction. Tetrodotoxin has been used as a tool to probe the pore of the sodium channel, whereas the scorpion toxins have been valuable as probes of voltage sensor function. *Inset*: Structure of the inactivation gate in solution determined by NMR.

charged amino acid residue (usually arginine) between two hydrophobic residues (Figure 16.2, yellow +). Mutation of these arginine residues shifts the voltage dependence of gating and reduces the steepness of its voltage dependence (Stuhmer et al., 1989; Kontis et al., 1997), showing that they are the gating charges that sense changes in the electric field across the cell membrane and initiate activation of sodium channels. Voltage sensors in all four domains contribute to the voltage-sensing process (Kontis et al., 1997; Cha et al., 1999), but the outward movements of the S4 segments are sequential, with domains I, II, and III moving rapidly, but in sequence, and then domain IV moving more slowly (Cha et al., 1999; Chanda and Bezanilla, 2002). The S5 and S6 segments and the P loop between them form the central pore (Figure 16.2, green). The pore motif was first identified by determination of the amino acid residues that are essential for binding the pore-blocking toxin tetrodotoxin (Noda et al., 1989; Terlau et al., 1991; Figure 16.2). Subsequent experiments showed that these amino acid residues also control ion selectivity of sodium channels (Heinemann et al., 1992; Favre et al., 1996; Sun et al., 1997). The fast inactivation gate, which inactivates sodium channels within a few milliseconds of opening, is formed by the intracellular loop connecting domains III and IV of the sodium channel α subunit (Figure 16.2). It was identified by antipeptide antibodies directed against it, which block the fast inactivation process (Vassilev et al., 1988, 1989), and further studied by site-directed mutagenesis (Stuhmer et al., 1989). This loop is thought to fold into the sodium channel structure and block the pore from the intracellular end during inactivation (Vassilev et al., 1988; Figure 16.2b). The structure of the fast inactivation gate peptide has been determined in solution by nuclear magnetic resonance (NMR) methods (Rohl et al., 1999; Figure 16.2, inset). It consists of a rigid α-helix preceded in the amino acid sequence by two loops that contain the conserved hydrophobic motif isoleucine–phenylalanine–methionine (IFM). The inactivation gate bends at hinge glycine residues, and the crucial motif binds and blocks the intracellular mouth of the pore and serves as a molecular latch to keep the inactivation gate closed (West et al., 1992; Eaholtz et al., 1994; McPhee et al., 1994, 1995, 1998; Kellenberger et al., 1996, 1997a,b; Smith and Goldin, 1997).

16.3 THREE-DIMENSIONAL STRUCTURE OF THE SODIUM CHANNEL

16.3.1 STRUCTURE OF BACTERIAL SODIUM CHANNELS

Analysis of the human genome revealed that there are 143 ion channel proteins whose pore-forming segments are related to Na^+ channels, and they are associated with at least 10 distinct families of auxiliary subunits (Yu and Catterall, 2004). The voltage-gated ion channels and their molecular relatives are one of the largest superfamilies of membrane signaling proteins and one of the most prominent targets for drugs used in the therapy of human diseases. The voltage-gated Na^+ channels were the founders of this large superfamily in terms of discovery of their function by Hodgkin and Huxley and the later discovery of the Na^+ channel protein itself. Surprisingly, the Na^+ channel family is also ancient in evolution (Ren et al., 2001). The bacterial Na^+ channel NaChBac and its prokaryotic relatives are composed of homotetramers of a single subunit whose structure resembles one of the domains of a vertebrate Na^+ channel (Ren et al., 2001). It is likely that these bacterial Na^+ channels are the evolutionary ancestors of the larger, four-domain Na^+ channels in eukaryotes, and a similar bacterial channel may have been the molecular ancestor of Ca^{2+} channels.

Sodium channel architecture has been revealed in three dimensions by determination of the crystal structure of the bacterial sodium channel NavAb from *Arcobacter butzleri* at high resolution (2.7 Å) (Payandeh et al., 2011; Figure 16.3). This ancient sodium channel has a very simple structure—four identical subunits that each is similar to one homologous domain of a mammalian sodium channel without the large intracellular and extracellular loops of the mammalian protein (Payandeh et al., 2011). This structure has revealed a wealth of new information about the structural basis for sodium selectivity and conductance, the mechanism for block of the channel by therapeutically important drugs, and the mechanism of voltage-dependent gating. As viewed from the top, NavAb has a central pore surrounded by four pore-forming modules composed of S5 and S6 segments and the intervening pore loop (Figure 16.3a). Four voltage-sensing modules composed of S1–S4 segments are symmetrically associated with the outer rim of the pore module (Figure 16.3a). The transmembrane architecture of NavAb shows that the adjacent subunits have swapped their functional domains such that each voltage-sensing module is most closely associated with the pore-forming module of its neighbor (Payandeh et al., 2011; Figure 16.3b and c). It is likely that this domain-swapped arrangement enforces concerted gating of the four subunits or domains of sodium channels.

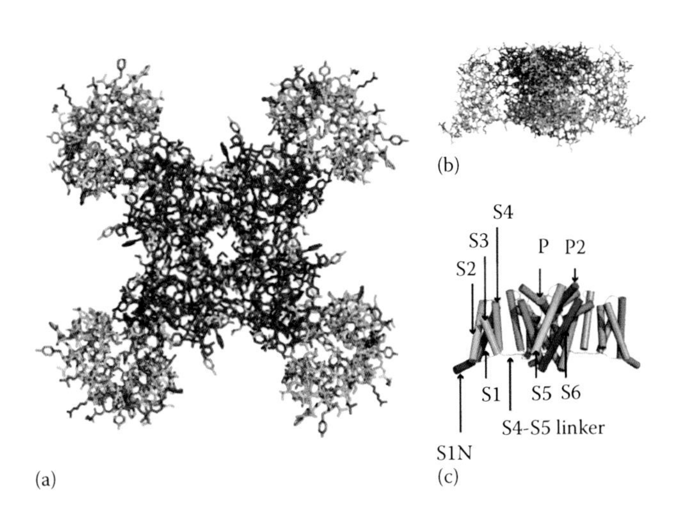

Figure 16.3 Structure of the bacterial sodium channel NavAb. (a) Top view of NavAb channels colored according to crystallographic temperature factors of the main chain (blue <50 Å² to red >150 Å²). The four pore modules in the center are rigid in the crystal structure and therefore are blue. The four voltage-sensing modules surround the pore and are more mobile, as illustrated by warmer colors. (b) Side view of NavAb. (c) Structural elements in NavAb. The structural components of one subunit are highlighted (1–6, transmembrane segments S1–S6).

16.3.2 PORE CONTAINS A HIGH-FIELD-STRENGTH CARBOXYL SITE AND TWO CARBONYL SITES

The overall pore architecture includes a large external vestibule, a narrow ion selectivity filter containing the amino acid residues shown to determine ion selectivity in vertebrate sodium and calcium channels, a large central cavity that is lined by the S6 segments and is water filled, and an intracellular activation gate formed at the crossing of the S6 segments at the intracellular surface of the membrane (Payandeh et al., 2011; Figure 16.4a). The activation gate is tightly closed in the NavAb structure (Figure 16.4b), and there is no space for ions or water to move through it. This general architecture resembles voltage-gated K^+ channels (see Chapter 1).

Although the overall pore architecture of sodium and potassium channels is similar, the structures of their ion selectivity filters and their mechanisms of ion selectivity and conductance are completely different. K^+ channels select K^+ by direct interaction with a series of four ion coordination sites formed by the backbone carbonyls of the amino acid residues that comprise the ion selectivity filter (Chapter 1). No charged amino acid residues are involved, and no water molecules intervene between K^+ and its interacting backbone carbonyls in the ion selectivity filter of potassium channels. In contrast, the NavAb ion selectivity filter has a high field strength site at its extracellular end (Payandeh et al., 2011; Figure 16.4c), which is formed by four glutamate residues in the positions of the key determinants of ion selectivity in vertebrate sodium and calcium channels (Figure 16.2). Considering its dimensions of approximately 4.6 Å², Na^+ with two planar waters of hydration could fit in this high-field-strength site. This outer site is followed by two ion coordination sites formed by backbone carbonyls (Figure 16.4d). These two carbonyl sites are perfectly designed to bind Na^+ with four planar waters of hydration but would be much too large to bind Na^+ directly. In fact, the NavAb selectivity filter is large enough to fit the backbone of the K^+ channel ion selectivity filter inside it. Thus, the chemistry of Na^+ selectivity and conductance is opposite to that of K^+: negatively charged residues interact with Na^+ to remove most (but not all) of its waters of hydration, and Na^+ is conducted as a hydrated ion interacting with the pore through its inner shell of bound waters. Theoretical considerations of sodium selectivity, saturation, and block predicted an outer high-field-strength site that would partially dehydrate the permeating ion and two inner sites that would conduct and rehydrate the permeant Na^+ ion in the four-barrier, three-site model of selectivity filter function (Hille, 1975). This congruence of theory and structure gives clear insight into the chemistry and biophysics of sodium permeation.

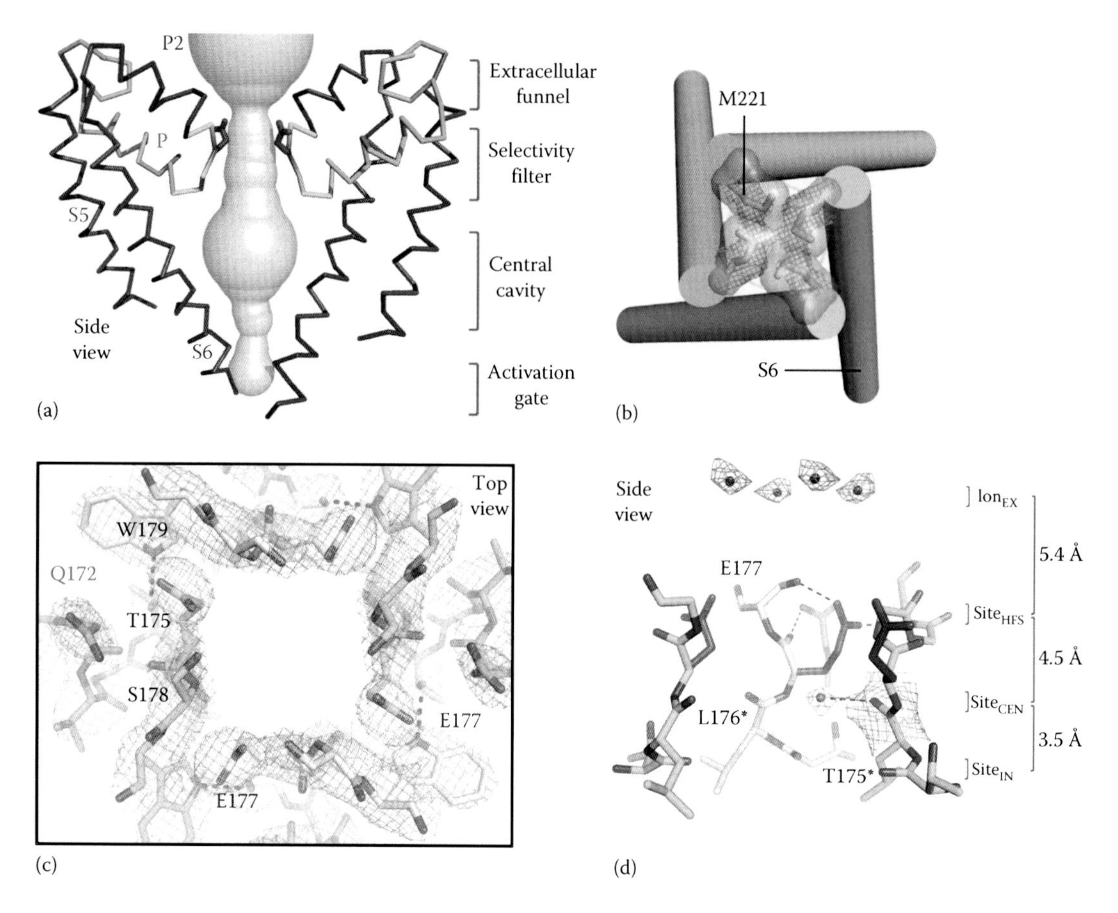

Figure 16.4 NavAb pore and selectivity filter. (a) Architecture of the NavAb pore. Glu177 side chains, purple; pore volume, gray. The P and P2 alpha helices that form the scaffold for the selectivity filter and outer vestibule are shown in green and red, respectively. (b) The closed activation gate at the intracellular end of the pore illustrating the close interaction of side chains Met221 residues in closing the pore. (c) Top view of the ion selectivity filter. Symmetry-related molecules are colored white and yellow; P-helix residues are colored green. Hydrogen bonds between Thr175 and Trp179 are indicated by gray dashes. (d) Side view of the selectivity filter. Glu177 (purple) interactions with Gln172, Ser178 and the backbone of Ser180 are shown in the far subunit; putative cations or water molecules (red spheres, Ion$_{EX}$). Electron-density around Leu176 (gray) and a bound water molecule are shown in gray. Na^+-coordination sites: Site$_{HFS}$, Site$_{CEN}$, and Site$_{IN}$.

16.3.3 VOLTAGE-DEPENDENT ACTIVATION INVOLVES A SLIDING-HELIX MOVEMENT OF GATING CHARGES

As pointed out by Hodgkin and Huxley (1952d), the steep voltage dependence of activation of sodium channels implies that *electrically charged particles* must move across the membrane in response to changes in membrane potential and thereby provide the driving force for opening of the Na^+ channel. The predicted transmembrane movement of these gating charges was detected as a small capacitative gating current in high-resolution voltage-clamp studies, first in the squid giant axon and later in other types of cells (Armstrong, 1981). The S4 segments in each of the homologous domains of Na^+ channels serve as voltage sensors, and the positively charged arginine and lysine residues at intervals of three in the S4 amino acid sequence serve as the gating charges (reviewed in Catterall, 2010). The x-ray crystal structure of the Na$_V$Ab channel provides a high-resolution model of an activated voltage sensor (Figure 16.5; Payandeh et al., 2011). The four transmembrane helices are organized in two helical hairpins composed of the S1–S2 and the S3–S4 transmembrane segments. The four gating-charge-carrying arginine residues in the S4 segment (R1–R4 in Na$_V$Ab) are arrayed in a sequence across the membrane (Figure 16.5a). Just below the center of the four-helix bundle, a cluster of hydrophobic residues, including a highly conserved phenylalanine residue (Phe56 in Na$_V$Ab), form the hydrophobic constriction site (HCS), which seals the voltage sensor to prevent transmembrane movement of water and ions

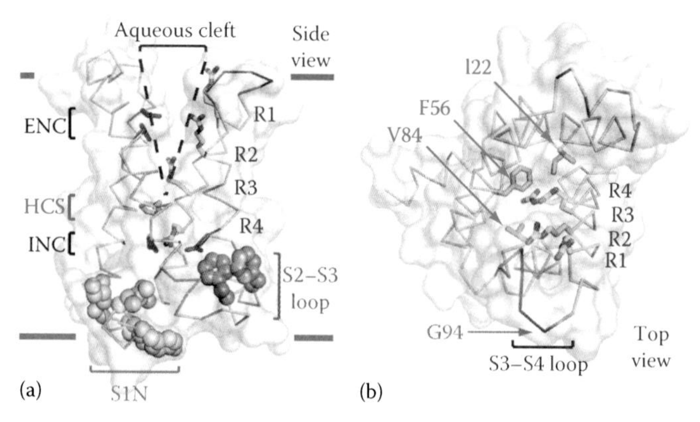

Figure 16.5 The structure of a voltage sensor of a sodium channel. Side and top views of the voltage-sensing module of NavAb illustrating the extracellular negative charge-cluster (red, ENC), the intracellular negative charge-cluster (red, INC), hydrophobic construction site (HCS) (green, HCS), residues of the S1N helix (cyan) and phenylalanines of the S2–S3 loop (purple). S4 segment and gating charges (R1–R4) are in yellow.

(HCS, Figure 16.5, green). An analogous Phe residue is crucial for voltage sensor function in K_V channels (Tao et al., 2010). Gating charges R1–R3 are located on the extracellular side of the HCS, and their arginine side chains interact with the negatively charged side chains of the extracellular negative cluster (ENC; Figure 16.5, red). The array of gating charges facing the aqueous cleft in the voltage sensor on the extracellular side is illustrated in the top view of Figure 16.5. Gating charge R4 is located on the intracellular side of the HCS and interacts with the intracellular negative cluster (INC; Figure 16.5, red). Overall, the structure of the voltage sensor seems designed to catalyze movement of the S4 gating charges through the HCS, exchanging ion-pair partners between the INC and the ENC. The voltage sensor in the Na_VAb structure has three of its gating charges on the extracellular side of the HCS (Figure 16.5; Payandeh et al., 2011). This conformation is nearly identical to the conformation of the voltage sensor in the structure of the $K_V1.2$ channel in its open state (Long et al., 2005a,b, 2007). Nevertheless, the activation gate of Na_VAb is tightly closed by interaction of the side chains of Met221 (Figure 16.4b). Therefore, it is likely that the Na_VAb structure has captured the preopen state, which is an expected intermediate in the activation process in which all four voltage sensors have been activated by depolarization and the intracellular activation gate is still closed, but poised to open rapidly in concerted conformational changes of all four subunits.

Much work has centered on the structure of the resting state of the sodium channel voltage sensor and how the gating charges in the S4 segments move outward upon depolarization. Unfortunately, crystal structures of voltage-gated ion channels in the resting state have not been achieved, probably because the resting state requires a negative resting membrane potential, which is not present in crystals. However, much has been learned from structure–function studies and structural modeling.

Toxin-binding studies show that the S3–S4 linkers of the voltage sensors in domains II and IV of sodium channels are available for binding of large, hydrophilic scorpion toxin polypeptides (60–70 residues) in both the resting and activated states (Catterall, 1979; Rogers et al., 1996; Cestèle et al., 1998, 2006; Wang et al., 2011; Zhang et al., 2011, 2012). These results place the S4 segment in a transmembrane position in both resting and activated states. Outward movement of the S4 segments in sodium channel voltage sensors has been detected in clever experiments that measure the movement of chemically reactive cysteine residues substituted for the native amino acids in S4 by analyzing functional effects of specific chemical reactions at those substituted cysteines (Yang and Horn, 1995; Yang et al., 1996, 1997). These results are consistent with the sliding-helix model of voltage sensing in which gating charges in the S4 segment move outward and exchange ion-pair partners to allow a low-energy pathway of gating charge movement (Catterall, 1986a,b; Guy and Seetharamulu, 1986; Yarov-Yarovoy et al., 2006, 2012; Shafrir et al., 2008a,b). This movement of the S4 helix is proposed to initiate a more general conformational change in each domain in the preopen state. After conformational changes have occurred in all four domains, the transmembrane pore can open in a concerted manner and conduct ions (Kuzmenkin et al., 2004; Zhao et al., 2004b; Yarov-Yarovoy et al., 2012; Movie 1*). This structural model shows that the S4 segment and its gating charges move through a narrow gating pore that focuses the transmembrane electric field to a distance of approximately 5 Å normal to the membrane (Starace and Bezanilla, 2004) and allows the gating charges to move from an intracellular aqueous vestibule to an extracellular aqueous vestibule with a short transit through the channel protein (Payandeh et al., 2011; Yarov-Yarovoy et al., 2012).

This sliding-helix mechanism of outward movement of the gating charges during the activation process illustrated in Movie 1 has been confirmed by extensive studies of the ion-pair interactions of gating charges with the ion-pair partners in the INC and ENC in sodium channels. In these experiments, cysteine residues were substituted for two amino acid residues in voltage sensor of the cysteine-free NaChBac channel, whose positions are predicted to be far apart in the resting state but close together in the activated state, or vice versa (DeCaen et al., 2008, 2009, 2011; Yarov-Yarovoy et al., 2012). For amino acid residues that are predicted to be far apart in the resting state but close together in the activated state, these experiments showed that NaChBac channels with paired cysteine substitutions were fully active during a first voltage-clamp depolarization, but were rapidly disulfide-locked and inactivated during depolarization and were unresponsive to further depolarizing stimuli until disulfide bonds were chemically reduced by β-mercaptoethanol or another disulfide-reducing agent. Conversely, for amino acid residues that are predicted to be close together in the resting state but far apart in the activated state, no sodium current is observed upon first depolarization of the cells, but sodium current appears upon reduction of disulfide bonds. The results from a portion of these extensive studies are summarized in Figure 16.6, which illustrates one of the resting states and one of the activated states of the voltage sensor. In the resting state (Figure 16.6), the R1 gating charge is in the HCS. This position has been confirmed by disulfide locking (Yarov-Yarovoy et al., 2012). In contrast, in the activated state model, the S4 segment has moved outward and the R3 gating charge has moved outside of the HCS. This position of the voltage sensor has also been confirmed by disulfide locking (DeCaen et al., 2009; Yarov-Yarovoy et al., 2012). These studies, and more extensive analyses of other amino acid residues in the S4 segment and its interacting partners, strongly support the sliding-helix movement of the S4 segment illustrated in Movie 1 (Yarov-Yarovoy et al., 2012). Moreover, an emerging consensus of laboratories working on this problem now supports this model for both sodium and potassium channels (Vargas et al., 2012).

From these structural models and movies, one can visualize the steps in the gating of a voltage-gated ion channel (Movie 1). In the closed state, the negative internal membrane potential of −70 to −90 mV pulls the S4 gating charges inward by electrostatic force. The inward position of the S4 segment exerts a force on the S4–S5 linker, straightens the S6 segment, and closes the pore at its inner mouth. When the cell is depolarized, the electrostatic force pulling the S4 segment inward is relieved. In response to the change in electrostatic force, the S4 segment moves outward with each positive gating charge interacting with

* See: http://www.pnas.org/content/suppl/2011/12/08/1118434109.DCSupplemental/sm01.mov.

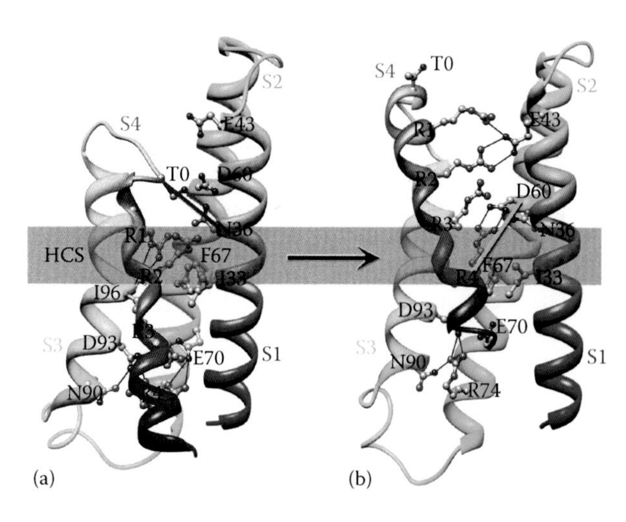

Figure 16.6 Sodium channel gating model. Transmembrane view of the lowest-energy Rosetta models of the voltage-sensing domain of NaChBac in resting state 1 (left) and activated state 3 (right). Side chains of the gating-charge-carrying arginines in S4 and key residues in S1, S2, and S3 segments are shown in stick representation and labeled. Gray, blue, and red atoms: C, N, and O. The HCS is highlighted by orange bars. R1 forms hydrogen bonds with the backbone carbonyl of I96 (in S3) at the extracellular edge of the HCS. On the intracellular side of the HCS, R3 makes ionic interactions with the amino acid residues of the intracellular negatively charged cluster, including E70 (in S2) and D93 (in S3), and R4 forms an ion pair with D93 (in S3). In addition, in the model shown, R1 forms an ion pair with E43 (in S1), R2 forms an ion pair with E43 (in S1), R3 forms hydrogen bond with Y156 (in S5) and makes ionic interactions with D60 (in S2) and E43 (in S1), and R4 forms an ion pair with D60 (in S2). These hydrophilic interactions stabilize the S4 gating charges in their transmembrane position and catalyze their transmembrane movement in response to changes in membrane potential.

charged amino acid side chains in turn to ease their movement through the voltage sensor module. When the R3 and R4 gating charges pass the HCS in the center of the voltage sensor module, the outward force on the S4–S5 linker is sufficient to open the pore.

16.3.4 IRIS-LIKE MOVEMENT OPENS THE PORE

The structure illustrated in Figure 16.4 likely represents the closed state because the α-helices lining the inner pore cross at their intracellular ends, apparently closing the pore (Figure 16.4a,b). In contrast, the structure of the open state of a structurally related K^+ channel has a bend and rotation of the S6 helix in comparison to the NavAb structure. This conformational change appears to open the pore at its intracellular end by splaying open the bundle of inner pore helices in an iris-like movement (Movie 1; Payandeh et al., 2011; Yarov-Yarovoy et al., 2012). Mutational studies of bacterial sodium channels support this pore-opening model because replacement of a conserved glycine with proline, which strongly favors a bend in the α-helix, also strongly favors pore opening and greatly slows closure (Zhao et al., 2004a).

16.3.5 PORE COLLAPSES DURING SLOW INACTIVATION

In addition to the fast inactivation process discovered by Hodgkin and Huxley in their classic work, a separate slow inactivation process operating on the timescale of 100 ms to seconds also terminates the Na^+ influx through Na^+ channels. This process is engaged during repetitive generation of action potentials in nerve and muscle cells and limits the length of trains of repetitive action potentials. Bacterial Na^+ channels whose structure has been determined have a slow inactivation process, even though their homotetrameric structure means that they do not have a structural component analogous to the intracellular loop connecting domains III and IV of vertebrate channels that mediates fast inactivation. The structure of the slow inactivated bacterial Na^+ channel reveals that the pore has partially collapsed by movement of two opposing S6 segments toward the central axis of the pore and corresponding movement of the two adjacent pairs of S6 segments away from the axis (Figure 16.7; Payandeh et al., 2012). This movement is observed at the selectivity filter at the extracellular end of the pore (Figure 16.7a), in the central cavity (Figure 16.7b), and at the activation gate at the intracellular end of the pore (Figure 16.7c). This asymmetric collapse of the pore is accompanied by subtle rotation of the voltage-sensing domain around the cylindrical exterior surface of the pore domain (Payandeh et al., 2012). It is likely that the pore collapse is important for stabilization of the sodium channel in the inactivated state, which requires strong, long-duration hyperpolarization for recovery to the resting state.

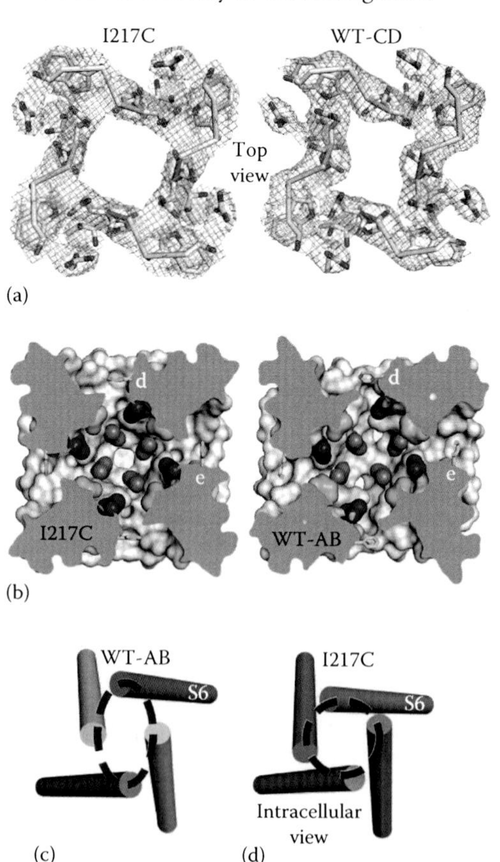

Figure 16.7 Collapse of the pore during slow inactivation of NavAb. During slow inactivation of NavAb, two S6 segments move inward to the central axis of the pore and two move outward to produce an asymmetric, partially collapsed conformation. (a) Top view of selectivity filter. Its structure has changed from nearly square to a partially collapsed parallelogram in inactivated state CD. (b) The central cavity is partially collapsed. (c) The activation gate is tightly closed, but collapsed into an asymmetric conformation.

16.4 DRUGS BLOCK THE PORE OF SODIUM CHANNELS

Many drugs used in therapy act on ion channels. Local anesthetics, which are used to block pain in dental and surgical procedures, bind in the inner pore of Na^+ channels in nerves and block them (Figure 16.8). In addition, related sodium channel–blocking drugs are used for treatment of epilepsy and cardiac arrhythmia. A combination of site-directed mutagenesis and molecular modeling reveals that these drugs bind to a receptor site formed by amino acid residues at specific positions in the S6 segments in domains I, III, and IV of the Na^+ channel, as illustrated in Figure 16.8a (Ragsdale et al., 1994, 1996; Wang et al., 1998; Nau et al., 1999, 2003; Yarov-Yarovoy et al., 2001, 2002; Liu et al., 2003). The aromatic and hydrophobic side chains of these amino acids contact the aromatic and substituted amino groups of the drug molecules and bind them in their receptor site, where they block ion movement through the pore. The structure of NavAb places this drug receptor site in three-dimensional context (Payandeh et al., 2011; Figure 16.8b and c). The amino acid residues that form the receptor sites for Na^+ channel blockers line the inner surface of the S6 segments and create a three-dimensional drug receptor site whose occupancy would block the pore (Figure 16.8b). Remarkably, fenestrations lead from the lipid phase of the membrane sideways into the drug receptor site, providing a hydrophobic access pathway for drug binding (Figure 16.8b). This form of drug binding from the membrane phase was predicted in early studies of the mechanism of block of Na^+ channels by different local anesthetics (Hille, 1977). Access to the drug receptor site in NavAb channels from the membrane phospholipid bilayer is limited by the side chain of a single amino acid residue (Phe203; Figure 16.8b). This amino acid residue is located in a position analogous to an amino acid residue in the IVS6 segment of brain sodium channels that controls egress of local anesthetics from their receptor and near amino acid residues that control access of extracellular sodium channel blockers to their receptor site in cardiac and brain sodium channels (Ragsdale et al., 1994; Qu et al., 1995; Sunami et al., 2000).

16.5 SODIUM CHANNEL DIVERSITY

In humans and other mammals, voltage-gated Na^+ channel α subunits are encoded by 10 genes, which are expressed in different excitable tissues (Table 16.2). $Na_V1.1$, 1.2, 1.3, and 1.6 are the primary Na^+ channels in the central nervous system. $Na_V1.7$, 1.8, and 1.9 are the primary Na^+ channels in the peripheral nervous system. $Na_V1.4$ is the primary Na^+ channel in skeletal muscle, whereas $Na_V1.5$ is primary in heart. Most of these Na^+ channels also have significant levels of expression outside of their primary tissues. The 10th Na^+ channel protein is not voltage-gated and is involved in salt-sensing (Watanabe et al., 2000; Hiyama et al., 2004). There is a small family of four $Na_V\beta$ subunits in total. β1 and β3 are associated noncovalently with α subunits and resemble each other most closely in amino acid sequence, whereas β2 and β4 form disulfide bonds with α subunits and also resemble each other closely (Isom et al., 1992, 1995; Morgan et al., 2000; Yu et al., 2003; Chen et al., 2012; Buffington and Rasband, 2013). $Na_V\beta$ subunits are promiscuous in their associations with α subunits in transfected cells, but restricted opportunities for the β2 and β4 to form disulfide bonds with different α subunits may add a level of subtype specificity in native cell environments (Chen et al., 2012; Buffington and Rasband, 2013).

16.6 SODIUM CHANNEL CELL BIOLOGY

In developing neurons in cell culture, sodium channels are synthesized, assembled into complexes with $Na_V\beta$ subunits, and held in an intracellular pool in the Golgi apparatus in preparation for insertion into the plasma membrane (Schmidt et al., 1985; Schmidt and Catterall, 1986). Sodium channels with β1 and β2 subunits bound are much more likely to reach the plasma membrane (Schmidt and Catterall, 1986). Glycosylation of sodium channels with both high-mannose and complex carbohydrate chains is also required for cell surface insertion (Schmidt and Catterall, 1986, 1987), and these carbohydrate chains include large amounts of polysialic acid (Zuber et al., 1992).

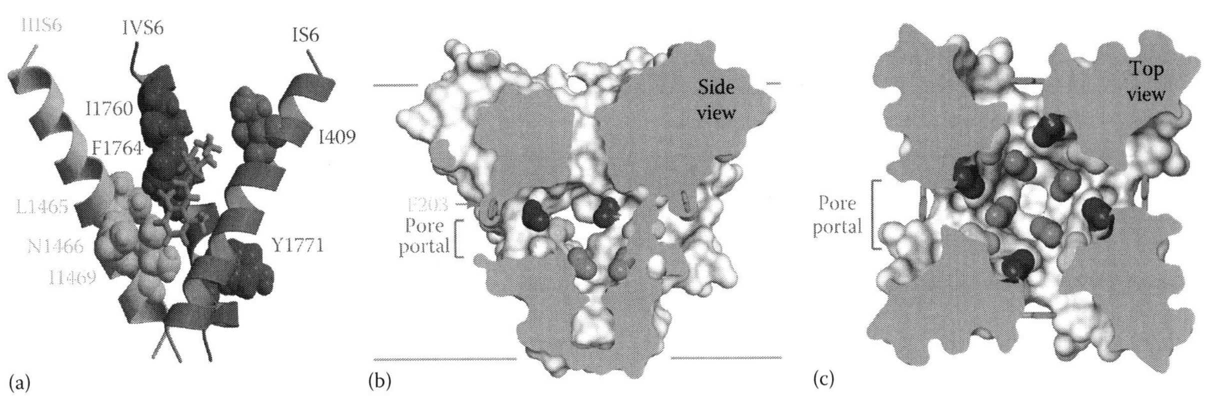

Figure 16.8 Drug-binding site and access pathways in the central cavity of sodium channels. (a) A model of the binding site for local anesthetics and related antiepileptic and antiarrhythmic drugs in the IS6, IIIS6, and IVS6 transmembrane segments of a mammalian sodium channel. Amino acid side chains involved in drug binding are shown in space-filling format. Yellow, bound etidocaine. (b) Side view through the pore module in the structure of NavAb illustrating fenestrations (portals) and hydrophobic access to central cavity. Phe203 side chains, yellow sticks. Surface representations of NavAb residues aligning with those implicated in drug binding and block, Thr206, blue; Met209, green; Val213, orange. Membrane boundaries, gray lines. (c) Top view sectioned below the selectivity filter, colored as in (a).

Table 16.2 The sodium channel protein family

CHANNEL	GENE	PRIMARY LOCATION	DISEASE
$Na_V1.1$	SCN1A	CNS	Generalized epilepsy with febrile seizures plus
			Dravet syndrome (Severe myoclonic epilepsy in infancy)
			Benign neonatal convulsions Familial hemiplegic migraine type III
$Na_V1.2$	SCN2A	CNS	Benign familial neonatal-infantile seizures
$Na_V1.3$	SCN3A	Embryonic CNS	
$Na_V1.4$	SCN4A	Skeletal Muscle	Hypokalemic periodic paralysis type II
			Normokalemic periodic paralysis
			Hyperkalemic periodic paralysis
			Paramyotonia congenita
$Na_V1.5$	SCN5A	Heart	Long QT syndrome type III
			Brugada syndrome
$Na_V1.6$	SCN7A	CNS	
$Na_V1.7$	SCN9A	PNS	Erythromelalgia
			Paroxysmal extreme pain disorder
			Congenital indifference to pain
$Na_V1.8$	SCN10A	DRG	Small fiber neuropathy
$Na_V1.9$	SCN11A	DRG	

Note: CNS, central nervous system; PNS, peripheral nervous system; DRG, sensory neurons in dorsal root ganglia.
Sources: Reviewed in Lehman-Horn and Jurkat-Rott (1999), Goldin et al. (2000), Keating and Sanguinetti (2001), Catterall et al. (2005, 2010), Moss and Kass (2005), Jurkat-Rott and Lehmann-Horn (2006), Venance et al. (2006), Catterall (2010), Dib-Hajj et al. (2013), and Waxman (2013).

Sodium channel subtypes are differentially expressed and localized within individual cells, in addition to their differential expression in different cell types. In brain neurons, $Na_V1.3$ channels are prominent in embryonic life, and they are replaced by $Na_V1.1$, $Na_V1.2$, and $Na_V1.6$ channels during postnatal development (Gordon et al., 1987; Beckh et al., 1989; García et al., 1998). Neurons have a high density of sodium channels in axon initial segments (Catterall, 1981; Wollner and Catterall, 1986; Jenkins and Bennett, 2001; Boiko et al., 2003), and skeletal muscle fibers have a high density of sodium channels in the peri-endplate region (Beam et al., 1985; Caldwell et al., 1986). Because action potentials are initiated in axon initial segments (Khaliq and Raman, 2006) and in the peri-endplate region of skeletal muscle fibers (Caldwell et al., 1986), these findings suggest that a high sodium channel density defines sites of action potential initiation. $Na_V1.1$ channels are localized in cell bodies and in axon initial segments (Westenbroek et al., 1989; Gong et al., 1999; Ogiwara et al., 2007; Vacher et al., 2008). $Na_V1.2$ channels are localized in unmyelinated axons and in dendrites (Westenbroek et al., 1989; Gong et al., 1999; Vacher et al., 2008). $Na_V1.6$ channels are highly localized in nodes of Ranvier in myelinated axons, where they replace $Na_V1.2$ channels during development and myelination (Boiko et al., 2001; Kaplan et al., 2001). They are also expressed prominently in the cell bodies of many excitatory neurons and in the Purkinje neurons of the cerebellum (Jenkins and Bennett, 2001; Vacher et al., 2008). Both axon initial segments and nodes of Ranvier have a highly specific localization of sodium channels, held in place by interactions with ankyrin G on the intracellular side and tenascin, neurofascin, and other cell adhesion molecules on the extracellular side (Srinivasan et al., 1998; Xiao et al., 1999; Malhotra et al., 2000; Kazarinova-Noyes et al., 2001; Ratcliffe et al., 2001; Susuki et al., 2013). Sodium channels also have a complex distribution in sensory neurons, with specific localizations of sodium channel subtypes in the peripheral nerve terminal, the peripheral axon, and the cell soma (Dib-Hajj et al., 2013). It is likely that this specialized distribution of sodium channel subtypes serves an important role in sensing and encoding pain information.

16.7 PATHOPHYSIOLOGY OF SODIUM CHANNEL MUTATIONS

A remarkable and unexpected facet of Na^+ channel biology that has evolved from the discovery of Na^+ channel genes, structure, and function has been the subsequent discovery of a large number of genetic diseases of Na^+ channels, including inherited forms of periodic paralysis, cardiac arrhythmia, epilepsy, and chronic pain (Table 16.2). In most cases, these are genetically dominant diseases in which the mutations cause a gain-of-function effect at both the molecular and cellular levels. In skeletal muscle, mutations in $Na_V1.4$ channels increase channel activity by impairing fast and/or slow inactivation in *paramyotonia congenita* and *hyperkalemic periodic paralysis* (Table 16.2). In the heart, mutations in $Na_V1.5$ channels cause *long QT syndrome* by impairing sodium channel inactivation. In sensory neurons, mutations in $Na_V1.7$ channels cause *erythromelalgia* by shifting the voltage dependence of activation to more negative membrane potentials and cause *paroxysmal extreme pain disorder* by impairing Na^+ channel inactivation. In the rare recessive pain disorder *congenital indifference to pain*, loss-of-function mutations in both alleles of the gene encoding $Na_V1.7$ channels cause complete loss of pain sensation.

In three of the autosomal dominant ion channelopathies, unexpected new mechanisms of increased cellular excitability have emerged from recent structure–function studies. In *Brugada syndrome*, loss-of-function mutations of $Na_V1.5$ channels create inhomogeneity of conduction across the ventricular wall and generate arrhythmias. In *Dravet syndrome*, and possibly also in *generalized epilepsy with febrile seizures plus*, loss-of-function mutations in $Na_V1.1$ channels selectively impair the excitability of GABAergic inhibitory neurons and thereby create hyperexcitability in neural circuits and cause epilepsy. In *hypokalemic periodic paralysis*, mutations of the gating charges in the voltage sensor of $Na_V1.4$ channels cause an ionic leak through the gating pore, resulting in excess Na^+ influx through the leaky voltage sensor, accumulation of intracellular Na^+, depolarization, and conduction block that lead to episodic paralysis. As these novel mechanisms of ion channelopathies illustrate, there is much more to do in analyzing the full extent of the pathophysiological implications of sodium channel dysfunction.

REFERENCES

Adelman, W.J., Jr. and Y. Palti. 1969. The effects of external potassium and long duration voltage conditioning on the amplitude of sodium currents in the giant axon of the squid, *Loligo pealei*. *J. Gen. Physiol.* 54:589–606.

Agnew, W.S., A.C. Moore, S.R. Levinson, and M.A. Raftery. 1980. Identification of a large molecular weight peptide associated with a tetrodotoxin binding proteins from the electroplax of *Electrophorus electricus*. *Biochem. Biophys. Res. Commun.* 92:860–866.

Armstrong, C.M. 1981. Sodium channels and gating currents. *Physiol. Rev.* 61:644–682.

Armstrong, C.M. and F. Bezanilla. 1973. Currents related to movement of the gating particles of the sodium channels. *Nature*. 242:459–461.

Armstrong, C.M. and F. Benzanilla. 1974. Charge movement associated with the opening and closing of the activation gates of the Na channels. *J. Gen. Physiol.* 63:533–552.

Auld, V.J., A.L. Goldin, D.S. Krafte et al. 1988. A rat brain sodium channel α subunit with novel gating properties. *Neuron*. 1:449–461.

Baden, D.G. 1989. Brevetoxins: Unique polyether dinoflagellate toxins. *FASEB J.* 3:1807–1817.

Barchi, R.L. 1983. Protein components of the purified sodium channel from rat skeletal sarcolemma. *J. Neurochem.* 36:1377–1385.

Beam, K.G., J.H. Caldwell, and D.T. Campbell. 1985. Na channels in skeletal muscle concentrated near the neuromuscular junction. *Nature*. 313:588–590.

Beckh, S., M. Noda, H. Lübbert, and S. Numa. 1989. Differential regulation of three sodium channel messenger RNAs in the rat central nervous system during development. *EMBO J.* 8:3611–3616.

Beneski, D.A. and W.A. Catterall. 1980. Covalent labeling of protein components of the sodium channel with a photoactivable derivative of scorpion toxin. *Proc. Natl. Acad. Sci. USA*. 77:639–643.

Boiko, T., M.N. Rasband, S.R. Levinson et al. 2001. Compact myelin dictates the differential targeting of two sodium channel isoforms in the same axon. *Neuron*. 30:91–104.

Boiko, T., A. Van Wart, J.H. Caldwell et al. 2003. Functional specialization of the axon initial segment by isoform-specific sodium channel targeting. *J. Neurosci.* 23:2306–2313.

Brackenbury, W.J. and L.L. Isom. 2011. Na channel β subunits: Overachievers of the ion channel family. *Front. Pharmacol.* 2:53.

Buffington, S.A. and M.N. Rasband. 2013. Na^+ channel-dependent recruitment of $Na_V\beta4$ to axon initial segments and nodes of Ranvier. *J. Neurosci.* 33:6191–6202.

Caldwell, J.H., D.T. Campbell, and K.G. Beam. 1986. Na channel distribution in vertebrate skeletal muscle. *J. Gen. Physiol.* 87:907–932.

Carr, D.B., M. Day, A.R. Cantrell et al. 2003. Transmitter modulation of slow, activity-dependent alterations in sodium channel availability endows neurons with a novel form of cellular plasticity. *Neuron*. 39:793–806.

Catterall, W.A. 1979. Binding of scorpion toxin to receptor sites associated with sodium channels in frog muscle. Correlation of voltage-dependent binding with activation. *J. Gen. Physiol.* 74:375–391.

Catterall, W.A. 1980. Neurotoxins that act on voltage-sensitive sodium channels in excitable membranes. *Annu. Rev. Pharmacol. Toxicol.* 20:15–43.

Catterall, W.A. 1981. Localization of sodium channels in cultured neural cells. *J. Neurosci.* 1:777–783.

Catterall, W.A. 1986a. Molecular properties of voltage-sensitive sodium channels. *Annu. Rev. Biochem.* 55:953–985.

Catterall, W.A. 1986b. Voltage-dependent gating of sodium channels: Correlating structure and function. *Trends Neurosci.* 9:7–10.

Catterall, W.A. 2010. Ion channel voltage sensors: Structure, function, and pathophysiology. *Neuron*. 67:915–928.

Catterall, W.A., S. Cestele, V. Yarov-Yarovoy et al. 2007. Voltage-gated ion channels and gating modifier toxins. *Toxicon*. 49:124–141.

Catterall, W.A., A.L. Goldin, and S.G. Waxman. 2005. International Union of Pharmacology. XLVII. Nomenclature and structure–function relationships of voltage-gated sodium channels. *Pharmacol. Rev.* 57:397–409.

Catterall, W.A., F. Kalume, and J.C. Oakley. 2010. $Na_V1.1$ channels and epilepsy. *J. Physiol.* 588:1849–1859.

Cestèle, S. and W.A. Catterall. 2000. Molecular mechanisms of neurotoxin action on voltage-gated sodium channels. *Biochimie*. 82:883–892.

Cestèle, S., Y. Qu, J.C. Rogers et al. 1998. Voltage sensor-trapping: Enhanced activation of sodium channels by β-scorpion toxin bound to the S3–S4 loop in domain II. *Neuron*. 21:919–931.

Cestèle, S., V. Yarov-Yarovoy, Y. Qu et al. 2006. Structure and function of the voltage sensor of sodium channels probed by a β-scorpion toxin. *J. Biol. Chem.* 281:21332–21344.

Cha, A., P.C. Ruben, A.L. George, E. Fujimoto, and F. Bezanilla. 1999. Voltage sensors in domains III and IV, but not I and II, are immobilized by Na^+ channel fast inactivation. *Neuron*. 22:73–87.

Chanda, B. and F. Bezanilla. 2002. Tracking voltage-dependent conformational changes in skeletal muscle sodium channel during activation. *J. Gen. Physiol.* 120:629–645.

Chen, C., J.D. Calhoun, Y. Zhang et al. 2012. Identification of the cysteine residue responsible for disulfide linkage of Na^+ channel α and β2 subunits. *J. Biol. Chem.* 287:39061–39069.

Crill, W.E. 1996. Persistent sodium current in mammalian central neurons. *Annu. Rev. Physiol.* 58:349–362.

DeCaen, P.G., V. Yarov-Yarovoy, T. Scheuer, and W.A. Catterall. 2011. Gating charge interactions with the S1 segment during activation of a Na^+ channel voltage sensor. *Proc. Natl. Acad. Sci. USA*. 108:18825–18830.

DeCaen, P.G., V. Yarov-Yarovoy, E.M. Sharp, T. Scheuer, and W.A. Catterall. 2009. Sequential formation of ion pairs during activation of a sodium channel voltage sensor. *Proc. Natl. Acad. Sci. USA*. 106:22498–22503.

DeCaen, P.G., V. Yarov-Yarovoy, Y. Zhao, T. Scheuer, and W.A. Catterall. 2008. Disulfide locking a sodium channel voltage sensor reveals ion pair formation during activation. *Proc. Natl. Acad. Sci. USA.* 105:15142–15147.

Dib-Hajj, S.D., Y. Yang, J.A. Black, and S.G. Waxman. 2013. The $Na_V1.7$ sodium channel: From molecule to man. *Nat. Rev. Neurosci.* 14:49–62.

Do, M.T. and B.P. Bean. 2003. Subthreshold sodium currents and pacemaking of subthalamic neurons: Modulation by slow inactivation. *Neuron.* 39:109–120.

Eaholtz, G., T. Scheuer, and W.A. Catterall. 1994. Restoration of inactivation and block of open sodium channels by an inactivation gate peptide. *Neuron.* 12:1041–1048.

Favre, I., E. Moczydlowski, and L. Schild. 1996. On the structural basis for ionic selectivity among Na^+, K^+, and Ca^{2+} in the voltage-gated sodium channel. *Biophys. J.* 71:3110–3125.

García, K.D., L.K. Sprunger, M.H. Meisler, and K.G. Beam. 1998. The sodium channel Scn8a is the major contributor to the postnatal developmental increase of sodium current density in spinal motoneurons. *J. Neurosci.* 18:5234–5239.

Goldin, A.L., R.L. Barchi, J.H. Caldwell et al. 2000. Nomenclature of voltage-gated sodium channels. *Neuron.* 28:365–368.

Goldin, A.L., T. Snutch, H. Lubbert et al. 1986. Messenger RNA coding for only the α subunit of the rat brain Na channel is sufficient for expression of functional channels in *Xenopus oocytes*. *Proc. Natl. Acad. Sci. USA.* 83:7503–7507.

Gong, B., K.J. Rhodes, Z. Bekele-Arcuri, and J.S. Trimmer. 1999. Type I and type II sodium channel α-subunit polypeptides exhibit distinct spatial and temporal patterning, and association with auxiliary subunits in rat brain. *J. Comp. Neurol.* 412:342–352.

Gordon, D., D. Merrick, V. Auld et al. 1987. Tissue-specific expression of the RI and RII sodium channel subtypes. *Proc. Natl. Acad. Sci. USA.* 84:8682–8686.

Guy, H.R. and P. Seetharamulu. 1986. Molecular model of the action potential sodium channel. *Proc. Natl. Acad. Sci. USA.* 508:508–512.

Hartshorne, R.P. and W.A. Catterall. 1981. Purification of the saxitoxin receptor of the sodium channel from rat brain. *Proc. Natl. Acad. Sci. USA.* 78:4620–4624.

Hartshorne, R.P. and W.A. Catterall. 1984. The sodium channel from rat brain. Purification and subunit composition. *J. Biol. Chem.* 259:1667–1675.

Hartshorne, R.P., B.U. Keller, J.A. Talvenheimo, W.A. Catterall, and M. Montal. 1985. Functional reconstitution of the purified brain sodium channel in planar lipid bilayers. *Proc. Natl. Acad. Sci. USA.* 82:240–244.

Hartshorne, R.P., D.J. Messner, J.C. Coppersmith, and W.A. Catterall. 1982. The saxitoxin receptor of the sodium channel from rat brain evidence for two nonidentical β subunits. *J. Biol. Chem.* 257:13888–13891.

Heinemann, S.H., H. Terlau, and K. Imoto. 1992. Molecular basis for pharmacological differences between brain and cardiac sodium channels. *Pflugers Arch.* 422:90–92.

Hille, B. 1971. The permeability of the sodium channel to organic cations in myelinated nerve. *J. Gen. Physiol.* 59:599–619.

Hille, B. 1972. The permeability of the sodium channel to metal cations in myelinated nerve. *J. Gen. Physiol.* 59:637–658.

Hille, B. 1975. Ionic selectivity, saturation, and block in sodium channels. A four-barrier model. *J. Gen. Physiol.* 66:535–560.

Hille, B. 1977. Local anesthetics: Hydrophilic and hydrophobic pathways for the drug-receptor reaction. *J. Gen. Physiol.* 69:497–515.

Hirschberg, B., A. Rovner, M. Lieberman, and J. Patlak. 1995. Transfer of twelve charges is needed to open skeletal muscle Na^+ channels. *J. Gen. Physiol.* 106:1053–1068.

Hiyama, T.Y., E. Watanabe, H. Okado, and M. Noda. 2004. The subfornical organ is the primary locus of sodium-level sensing by Na_X sodium channels for the control of salt-intake behavior. *J. Neurosci.* 24:9276–9281.

Hodgkin, A.L. and A.F. Huxley. 1952a. The components of membrane conductance in the giant axon of *Loligo*. *J. Physiol.* 116:473–496.

Hodgkin, A.L. and A.F. Huxley. 1952b. Currents carried by sodium and potassium ions through the membrane of the giant axon of *Loligo*. *J. Physiol.* 116:449–472.

Hodgkin, A.L. and A.F. Huxley. 1952c. The dual effect of membrane potential on sodium conductance in the giant axon of *Loligo*. *J. Physiol.* 116:497–506.

Hodgkin, A.L. and A.F. Huxley. 1952d. A quantitative description of membrane current and its application to conduction and excitation in nerve. *J. Physiol.* 117:500–544.

Isom, L.L., K.S. De Jongh, D.E. Patton et al. 1992. Primary structure and functional expression of the β-1 subunit of the rat brain sodium channel. *Science.* 256:839–842.

Isom, L.L., D.S. Ragsdale, K.S. De Jongh et al. 1995. Structure and function of the β-2 subunit of brain sodium channels, a transmembrane glycoprotein with a CAM-motif. *Cell.* 83:433–442.

Jenkins, S.M. and V. Bennett. 2001. Ankyrin-G coordinates assembly of the spectrin-based membrane skeleton, voltage-gated sodium channels, and L1 CAMs at Purkinje neuron initial segments. *J. Cell Biol.* 155:739–746.

Jurkat-Rott, K. and F. Lehmann-Horn. 2006. Paroxysmal muscle weakness: The familial periodic paralyses. *J. Neurol.* 253:1391–1398.

Kaplan, M.R., M.H. Cho, E.M. Ullian et al. 2001. Differential control of clustering of the sodium channels $Na_V1.2$ and $Na_V1.6$ at developing CNS nodes of Ranvier. *Neuron.* 30:105–119.

Kazarinova-Noyes, K., J.D. Malhotra, D.P. McEwen et al. 2001. Contactin associates with sodium channels and increases their functional expression. *J. Neurosci.* 21:7517–7525.

Keating, M.T. and M.C. Sanguinetti. 2001. Molecular and cellular mechanisms of cardiac arrhythmias. *Cell.* 104:569–580.

Kellenberger, S., T. Scheuer, and W.A. Catterall. 1996. Movement of the Na^+ channel inactivation gate during inactivation. *J. Biol. Chem.* 271:30971–30979.

Kellenberger, S., J.W. West, W.A. Catterall, and T. Scheuer. 1997a. Molecular analysis of potential hinge residues in the inactivation gate of brain type IIA Na^+ channels. *J. Gen. Physiol.* 19:607–617.

Kellenberger, S., J.W. West, T. Scheuer, and W.A. Catterall. 1997b. Molecular analysis of the putative inactivation particle in the inactivation gate of brain type IIA Na^+ channels. *J. Gen. Physiol.* 109:589–605.

Keynes, R.D. and E. Rojas. 1974. Kinetics and steady-state properties of the charged system controlling sodium conductance in the squid giant axon. *J. Physiol.* 239:393–434.

Khaliq, Z.M. and I.M. Raman. 2006. Relative contributions of axonal and somatic Na channels to action potential initiation in cerebellar Purkinje neurons. *J. Neurosci.* 26:1935–1944.

Kontis, K.J., A. Rounaghi, and A.L. Goldin. 1997. Sodium channel activation gating is affected by substitutions of voltage sensor positive charges in all four domains. *J. Gen. Physiol.* 110:391–401.

Kuzmenkin, A., F. Bezanilla, and A.M. Correa. 2004. Gating of the bacterial sodium channel, NaChBac: Voltage-dependent charge movement and gating currents. *J. Gen. Physiol.* 124:349–356.

Lehman-Horn, F. and K. Jurkat-Rott. 1999. Voltage-gated ion channels and hereditary disease. *Physiol. Rev.* 79:1317–1372.

Lipkind, G.M. and H.A. Fozzard. 1994. A structural model of the tetrodotoxin and saxitoxin binding site of the Na^+ channel. *Biophys. J.* 66:1–13.

Liu, G., V. Yarov-Yarovoy, M. Nobbs et al. 2003. Differential interactions of lamotrigine and related drugs with transmembrane segment IVS6 of voltage-gated sodium channels. *Neuropharmacology.* 44:413–422.

Long, S.B., E.B. Campbell, and R. Mackinnon. 2005a. Crystal structure of a mammalian voltage-dependent Shaker family K^+ channel. *Science.* 309:897–903.

Long, S.B., E.B. Campbell, and R. Mackinnon. 2005b. Voltage sensor of Kv1.2: Structural basis of electromechanical coupling. *Science.* 309:903–908.

Long, S.B., X. Tao, E.B. Campbell, and R. MacKinnon. 2007. Atomic structure of a voltage-dependent K^+ channel in a lipid membrane-like environment. *Nature.* 450:376–382.

Malhotra, J.D., K. Kazen-Gillespie, M. Hortsch, and L.L. Isom. 2000. Sodium channel β subunits mediate homophilic cell adhesion and recruit ankyrin to points of cell–cell contact. *J. Biol. Chem.* 275:11383–11388.

McPhee, J.C., D.S. Ragsdale, T. Scheuer, and W.A. Catterall. 1994. A mutation in segment IVS6 disrupts fast inactivation of sodium channels. *Proc. Natl. Acad. Sci. USA.* 91:12346–12350.

McPhee, J.C., D.S. Ragsdale, T. Scheuer, and W.A. Catterall. 1995. A critical role for transmembrane segment IVS6 of the sodium channel α subunit in fast inactivation. *J. Biol. Chem.* 270:12025–12034.

McPhee, J.C., D.S. Ragsdale, T. Scheuer, and W.A. Catterall. 1998. A critical role for the S4–S5 intracellular loop in domain IV of the sodium channel α subunit in fast inactivation. *J. Biol. Chem.* 273:1121–1129.

Minassian, N.A., A. Gibbs, A.Y. Shih et al. 2013. Analysis of the structural and molecular basis of voltage-sensitive sodium channel inhibition by the spider toxin huwentoxin-IV. *J. Biol. Chem.* 288:22707–22720.

Moran, Y., D. Gordon, and M. Gurevitz. 2009. Sea anemone toxins affecting voltage-gated sodium channels—Molecular and evolutionary features. *Toxicon.* 54:1089–1101.

Morgan, K., E.B. Stevens, B. Shah et al. 2000. β-3: An additional auxiliary subunit of the voltage-sensitive sodium channel that modulates channel gating with distinct kinetics. *Proc. Natl. Acad. Sci. USA.* 97:2308–2313.

Moss, A.J. and R.S. Kass. 2005. Long QT syndrome: From channels to cardiac arrhythmias. *J. Clin. Invest.* 115:2018–2024.

Nau, C., S.Y. Wang, G.R. Strichartz, and G.K. Wang. 1999. Point mutations at N434 in D1–S6 of μ1 Na^+ channels modulate binding affinity and stereoselectivity of local anesthetic enantiomers. *Mol. Pharmacol.* 56:404–413.

Nau, C., S.Y. Wang, and G.K. Wang. 2003. Point mutations at L1280 in Nav1.4 channel D3-S6 modulate binding affinity and stereoselectivity of bupivacaine enantiomers. *Mol. Pharmacol.* 63:1398–1406.

Noda, M., T. Ikeda, T. Kayano et al. 1986a. Existence of distinct sodium channel messenger RNAs in rat brain. *Nature.* 320:188–192.

Noda, M., T. Ikeda, T. Suzuki et al. 1986b. Expression of functional sodium channels from cloned cDNA. *Nature.* 322:826–828.

Noda, M., S. Shimizu, T. Tanabe et al. 1984. Primary structure of *Electrophorus electricus* sodium channel deduced from cDNA sequence. *Nature.* 312:121–127.

Noda, M., H. Suzuki, S. Numa, and W. Stühmer. 1989. A single point mutation confers tetrodotoxin and saxitoxin insensitivity on the sodium channel II. *FEBS Lett.* 259:213–216.

Ogiwara, I., H. Miyamoto, N. Morita et al. 2007. $Na_V1.1$ localizes to axons of parvalbumin-positive inhibitory interneurons: A circuit basis for epileptic seizures in mice carrying an Scn1a gene mutation. *J. Neurosci.* 27:5903–5914.

Payandeh, J., T.M. Gamal El-Din, T. Scheuer, N. Zheng, and W.A. Catterall. 2012. Crystal structure of a voltage-gated sodium channel in two potentially inactivated states. *Nature.* 486:135–139.

Payandeh, J., T. Scheuer, N. Zheng, and W.A. Catterall. 2011. The crystal structure of a voltage-gated sodium channel. *Nature.* 475:353–358.

Qu, Y., J. Rogers, T. Tanada, T. Scheuer, and W.A. Catterall. 1995. Molecular determinants of drug access to the receptor site for antiarrhythmic drugs in the cardiac Na^+ channel. *Proc. Natl. Acad. Sci. USA.* 270:25696–25701.

Ragsdale, D.S., J.C. McPhee, T. Scheuer, and W.A. Catterall. 1994. Molecular determinants of state-dependent block of sodium channels by local anesthetics. *Science.* 265:1724–1728.

Ragsdale, D.S., J.C. McPhee, T. Scheuer, and W.A. Catterall. 1996. Common molecular determinants of local anesthetic, antiarrhythmic, and anticonvulsant block of voltage-gated Na^+ channels. *Proc. Natl. Acad. Sci. USA.* 93:9270–9275.

Raman, I.M. and B.P. Bean. 1997. Resurgent sodium current and action potential formation in dissociated cerebellar Purkinje neurons. *J. Neurosci.* 17:4517–4526.

Raman, I.M. and B.P. Bean. 1999. Ionic currents underlying spontaneous action potentials in isolated cerebellar Purkinje neurons. *J. Neurosci.* 19:1664–1674.

Ratcliffe, C.F., R.E. Westenbroek, R. Curtis, and W.A. Catterall. 2001. Sodium channel β-1and β-3 subunits associate with neurofascin through their extracellular immunoglobulin-like domain. *J. Cell Biol.* 154:427–434.

Ren, D., B. Navarro, H. Xu et al. 2001. A prokaryotic voltage-gated sodium channel. *Science.* 294:2372–2375.

Rochat, H., H. Darbon, E. Jover et al. 1984. Interaction of scorpion toxins with the sodium channel. *J. Physiol.* (*Paris*). 79:334–337.

Rogers, J.C., Y. Qu, T.N. Tanada, T. Scheuer, and W.A. Catterall. 1996. Molecular determinants of high affinity binding of α-scorpion toxin and sea anemone toxin in the S3–S4 extracellular loop in domain IV of the Na^+ channel α subunit. *J. Biol. Chem.* 271:15950–15962.

Rohl, C.A., F.A. Boeckman, C. Baker et al. 1999. Solution structure of the sodium channel inactivation gate. *Biochemistry.* 38:855–861.

Rudy, B. 1978. Slow inactivation of the sodium conductance in squid giant axons. Pronase resistance. *J. Physiol.* (*Lond.*). 283:1–21.

Schmidt, J., S. Rossie, and W.A. Catterall. 1985. A large intracellular pool of inactive Na channel α subunits in developing rat brain. *Proc. Natl. Acad. Sci. USA.* 82:4847–4851.

Schmidt, J.W. and W.A. Catterall. 1986. Biosynthesis and processing of the α subunit of the voltage-sensitive sodium channel in rat brain neurons. *Cell.* 46:437–444.

Schmidt, J.W. and W.A. Catterall. 1987. Palmitylation, sulfation, and glycosylation of the α subunit of the sodium channel. Role of post-translational modifications in channel assembly. *J. Biol. Chem.* 262:13713–13723.

Shafrir, Y., S.R. Durell, and H.R. Guy. 2008a. Models of the structure and gating mechanisms of the pore domain of the NaChBac ion channel. *Biophys. J.* 95:3650–3662.

Shafrir, Y., S.R. Durell, and H.R. Guy. 2008b. Models of voltage-dependent conformational changes in NaChBac channels. *Biophys. J.* 95:3663–3676.

Smith, M.R. and A.L. Goldin. 1997. Interaction between the sodium channel inactivation linker and domain III S4–S5. *Biophys. J.* 73:1885–1895.

Srinivasan, J., M. Schachner, and W.A. Catterall. 1998. Interaction of voltage-gated sodium channels with the extracellular matrix molecules tenascin-C and tenascin-R. *Proc. Natl. Acad. Sci. USA.* 95:15753–15757.

Starace, D.M. and F. Bezanilla. 2004. A proton pore in a potassium channel voltage sensor reveals a focused electric field. *Nature.* 427:548–553.

Stuhmer, W., F. Conti, H. Suzuki et al. 1989. Structural parts involved in activation and inactivation of the sodium channel. *Nature.* 339:597–603.

Sun, Y.M., I. Favre, L. Schild, and E. Moczydlowski. 1997. On the structural basis for size-selective permeation of organic cations through the voltage-gated sodium channel—Effect of alanine mutations at the DEKA locus on selectivity, inhibition by Ca^{2+} and H^+, and molecular sieving. *J. Gen.Physiol.* 110:693–715.

Sunami, A., I.W. Glaaser, and H.A. Fozzard. 2000. A critical residue for isoform difference in tetrodotoxin affinity is a molecular determinant of the external access path for local anesthetics in the cardiac sodium channel. *Proc. Natl. Acad. Sci. USA.* 97:2326–2331.

Susuki, K., K.J. Chang, D.R. Zollinger et al. 2013. Three mechanisms assemble central nervous system nodes of Ranvier. *Neuron.* 78:469–482.

Tamkun, M.M., J.A. Talvenheimo, and W.A. Catterall. 1984. The sodium channel from rat brain. Reconstitution of neurotoxin-activated ion flux and scorpion toxin binding from purified components. *J. Biol. Chem.* 259:1676–1688.

Tao, X., A. Lee, W. Limapichat, D.A. Dougherty, and R. MacKinnon. 2010. A gating charge transfer center in voltage sensors. *Science.* 328:67–73.

Terlau, H., S.H. Heinemann, W. Stühmer et al. 1991. Mapping the site of block by tetrodotoxin and saxitoxin of sodium channel II. *FEBS Lett.* 293:93–96.

Terlau, H. and B.M. Olivera. 2004. Conus venoms: A rich source of novel ion channel-targeted peptides. *Physiol. Rev.* 84:41–68.

Tikhonov, D.B. and B.S. Zhorov. 2012. Architecture and pore block of eukaryotic voltage-gated sodium channels in view of NavAb bacterial sodium channel structure. *Mol. Pharmacol.* 82:97–104.

Vacher, H., D.P. Mohapatra, and J.S. Trimmer. 2008. Localization and targeting of voltage-dependent ion channels in mammalian central neurons. *Physiol. Rev.* 88:1407–1447.

Vargas, E., V. Yarov-Yarovoy, F. Khalili-Araghi et al. 2012. An emerging consensus on voltage-dependent gating from computational modeling and molecular dynamics simulations. *J. Gen. Physiol.* 140:587 594.

Vassilev, P., T. Scheuer, and W.A. Catterall. 1989. Inhibition of inactivation of single sodium channels by a site-directed antibody. *Proc. Natl. Acad. Sci. USA.* 86:8147–8151.

Vassilev, P.M., T. Scheuer, and W.A. Catterall. 1988. Identification of an intracellular peptide segment involved in sodium channel inactivation. *Science.* 241:1658–1661.

Venance, S.L., S.C. Cannon, D. Fialho et al. 2006. The primary periodic paralyses: Diagnosis, pathogenesis and treatment. *Brain.* 129:8–17.

Vilin, Y.Y. and P.C. Ruben. 2001. Slow inactivation in voltage-gated sodium channels: Molecular substrates and contributions to channelopathies. *Cell Biochem. Biophys.* 35:171–190.

Wang, G.K., C. Quan, and S. Wang. 1998. A common local anesthetic receptor for benzocaine and etidocaine in voltage-gated mu1 Na^+ channels. *Pflug. Archiv: Eur. J. Physiol.* 435:293–302.

Wang, J., V. Yarov-Yarovoy, R. Kahn et al. 2011. Mapping the receptor site for α-scorpion toxins on a Na^+ channel voltage sensor. *Proc. Natl. Acad. Sci. USA.* 108:15426–15431.

Wang, S.Y. and G.K. Wang. 2003. Voltage-gated sodium channels as primary targets of diverse lipid-soluble neurotoxins. *Cell. Signal.* 15:151–159.

Watanabe, E., A. Fujikawa, H. Matsunaga et al. 2000. Na_V2/NaG channel is involved in control of salt-intake behavior in the CNS. *J. Neurosci.* 20:7743–7751.

Waxman, S.G. 2013. Painful Na-channelopathies: An expanding universe. *Trends Mol. Med.* 19:406–409.

West, J.W., D.E. Patton, T. Scheuer et al. 1992. A cluster of hydrophobic amino acid residues required for fast Na^+ channel inactivation. *Proc. Natl. Acad. Sci. USA.* 89:10910–10914.

Westenbroek, R.E., D.K. Merrick, and W.A. Catterall. 1989. Differential subcellular localization of the RI and RII Na^+ channel subtypes in central neurons. *Neuron.* 3:695–704.

Wollner, D.A. and W.A. Catterall. 1986. Localization of sodium channels in axon hillocks and initial segments of retinal ganglion cells. *Proc. Natl. Acad. Sci. USA.* 83:8424–8428.

Xiao, Y., K. Blumenthal, and T.R. Cummins. 2014. Gating-pore currents demonstrate selective and specific modulation of individual sodium channel voltage-sensors by biological toxins. *Mol. Pharmacol.* 86(2):159–167.

Xiao, Z.C., D.S. Ragsdale, J.D. Malhotra et al. 1999. Tenascin-R is a functional modulator of sodium channel β subunits. *J. Biol. Chem.* 274:26511–26517.

Yang, N. and R. Horn. 1995. Evidence for voltage-dependent S4 movement in sodium channel. *Neuron.* 15:213–218.

Yang, N.B., A.L. George, Jr., and R. Horn. 1996. Molecular basis of charge movement in voltage-gated sodium channels. *Neuron.* 16:113–122.

Yang, N.B., A.L. George, and R. Horn. 1997. Probing the outer vestibule of a sodium channel voltage sensor. *Biophys. J.* 73:2260–2268.

Yarov-Yarovoy, V., D. Baker, and W.A. Catterall. 2006. Voltage sensor conformations in the open and closed states in ROSETTA structural models of K^+ channels. *Proc. Natl. Acad. Sci. USA.* 103:7292–7297.

Yarov-Yarovoy, V., J. Brown, E. Sharp et al. 2001. Molecular determinants of voltage-dependent gating and binding of pore-blocking drugs in transmembrane segment IIIS6 of the Na^+ channel α subunit. *J. Biol. Chem.* 276:20–27.

Yarov-Yarovoy, V., P.G. DeCaen, R.E. Westenbroek et al. 2012. Structural basis for gating charge movement in the voltage sensor of a sodium channel. *Proc. Natl. Acad. Sci. USA.* 109:E93–E102.

Yarov-Yarovoy, V., J.C. McPhee, D. Idsvoog et al. 2002. Role of amino acid residues in transmembrane segments IS6 and IIS6 of the Na^+ channel α subunit in voltage-dependent gating and drug block. *J. Biol. Chem.* 277:35393–35401.

Yu, F.H. and W.A. Catterall. 2004. The VGL-chanome: A protein superfamily specialized for electrical signaling and ionic homeostasis. *Sci STKE.* 2004:re15.

Yu, F.H., R.E. Westenbroek, I. Silos-Santiago et al. 2003. Sodium channel β-4, a new disulfide-linked auxiliary subunit with similarity to β-2. *J. Neurosci.* 23:7577–7585.

Zhang, J.Z., V. Yarov-Yarovoy, T. Scheuer et al. 2011. Structure–function map of the receptor site for β-scorpion toxins in domain II of voltage-gated sodium channels. *J. Biol. Chem.* 286:33641–33651.

Zhang, J.Z., V. Yarov-Yarovoy, T. Scheuer et al. 2012. Mapping the interaction site for a β-scorpion toxin in the pore module of domain III of voltage-gated Na^+ channels. *J. Biol. Chem.* 287:30719–30728.

Zhang, M.M., J.R. McArthur, L. Azam et al. 2009. Synergistic and antagonistic interactions between tetrodotoxin and mu-conotoxin in blocking voltage-gated sodium channels. *Channels (Austin).* 3:32–38.

Zhao, Y., T. Scheuer, and W.A. Catterall. 2004a. Reversed voltage-dependent gating of a bacterial sodium channel with proline substitutions in the S6 transmembrane segment. *Proc. Natl. Acad. Sci. USA.* 101:17873–17878.

Zhao, Y., V. Yarov-Yarovoy, T. Scheuer, and W.A. Catterall. 2004b. A gating hinge in Na^+ channels; a molecular switch for electrical signaling. *Neuron.* 41:859–865.

Zuber, C., P.M. Lackie, W.A. Catterall, and J. Roth. 1992. Polysialic acid is associated with sodium channels and the neural cell adhesion molecule N-CAM in adult rat brain. *J. Biol. Chem.* 267:9965–9971.

17 BK channels

Huanghe Yang and Jianmin Cui

Contents

17.1 FUNCTIONAL PROPERTIES OF BK CHANNELS

Large conductance, voltage, and Ca^{2+}-activated K^+ channels are known as BK or MaxiK channels to denote the large single channel conductance of 100–300 pS (Figure 17.1). BK channel currents were first discovered in the early 1980s when the newly invented patch clamp techniques were used to record various cell types, and their large single channel currents stood out from patch clamp recordings with intracellular solutions containing Ca^{2+} (1,2). A decade later, the gene that encodes the pore-forming α subunit of BK channels, *slo1* or *KCNMA1*, was first identified in *Drosophila* (3,4), and then in mouse (5) and human (6,7). BK channels activate in response to membrane depolarization and elevation of intracellular Ca^{2+} concentrations: the open probability of single BK channels increases with voltage and intracellular Ca^{2+} concentrations (Figure 17.1), while the macroscopic conductance (*G*) of many BK channels increases with voltage and the conductance–voltage (*G–V*) relation shifts toward more negative voltages with increasing intracellular Ca^{2+} concentrations (Figure 17.2). The dependence of BK channel activation on both voltage and Ca^{2+} makes the channel function versatile in two unique aspects. First, BK channel function is not defined by any single voltage or intracellular Ca^{2+} concentrations; the *G–V* relation of BK channels varies depending on the intracellular Ca^{2+} concentrations at which it is measured, while the dose–response curve of channel activation on intracellular Ca^{2+} concentrations also changes with voltage (8,9). Second, the channel function varies with voltages changing from −200 to +300 mV and intracellular Ca^{2+} concentrations variations from 0 to 10 mM (9) that cover all possible physiological and pathological conditions.

Besides Ca^{2+}, various physiologically important intracellular ions and molecules also alter BK channel activation. Intracellular Mg^{2+} around physiological concentrations (~mM) activates BK channels (10–12). The probability of single channel openings increases, and the *G–V* relation of macroscopic currents shifts to more negative voltage ranges with increasing $[Mg^{2+}]_i$ (11,13–16). Although the effects of Mg^{2+} on BK channel activation are similar to that of Ca^{2+}, the underlying mechanisms of Ca^{2+}- and Mg^{2+}-dependent activation are distinctive (see the following text). Other divalent cations also activate BK channels with the effectiveness $Cd^{2+} > Sr^{2+} > Mn^{2+} > Fe^{2+} > Co^{2+}$ (17). Ba^{2+}, an ion that effectively blocks BK channels, was recently found to also activate BK channels (18). Beside divalent ions, proton H^+ also activates BK channels (19,20). In addition, intracellular molecules other than ions such as alcohol (21,22), heme (23), and PIP_2 (24) also enhance BK channel activation.

In addition to activating BK channels, intracellular divalent cations also block BK channels. Single channel current–voltage relation of BK channels deviates from a straight line and the current becomes smaller at positive voltages (inwardly rectifying), which is more prominent in intracellular Ca^{2+} concentrations ≥ 100 μM (25,26). Likewise, Mg^{2+} also blocks the channel (16,27–29). These results suggest that Ca^{2+} and Mg^{2+} may reach a site in the channel pore or intracellular vestibule from the intracellular side; the blocking and unblocking events happen fast and cannot be resolved by the recording system (29).

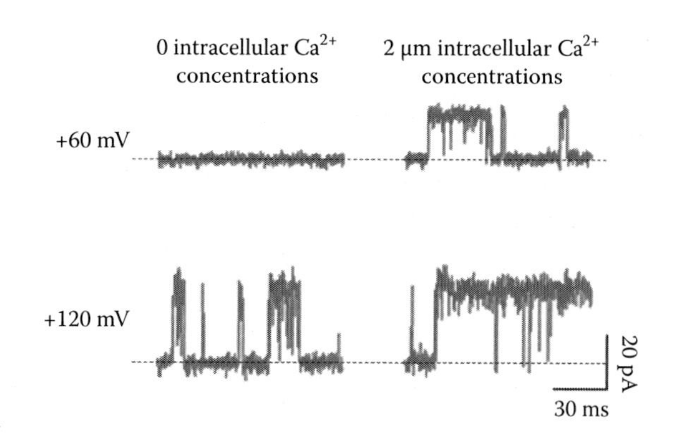

Figure 17.1 Single channel currents of a BK channel showing activation by intracellular Ca^{2+} and membrane depolarization. Dotted lines indicate the closed state of the channel. (Data were kindly provided by Dr. Frank Horrigan, personal communication.)

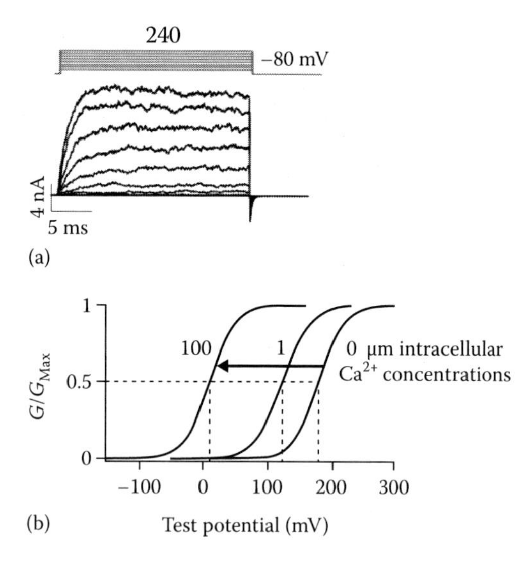

Figure 17.2 Macroscopic currents of BK channels. (a) Macroscopic currents of mSlo1 channels in response to the voltage pulses (top) in the absence of intracellular Ca^{2+} concentrations. (b) Increasing intracellular Ca^{2+} concentrations shifts the conductance–voltage (*G–V*) relation to more negative voltages.

On the other hand, Ba^{2+} blocks BK channels by entering the pore to cause blocking events longer than the close time of normal gating (30,31). Using Ba^{2+} blockade as a tool, Neyton and Miller elegantly predicted that there are four ion-binding sites in the ion conduction pathway of the BK channel: external K^+ lock-in site, Ba^{2+} block site, K^+ enhancement site, and internal K^+ lock-in site (32,33). Consistently, the crystal structure of KcsA (34), a bacterium K^+ channel, solved in Rb^+ solution shows four ion-binding sites in its selectivity filter (35,36) that correspond to the four ion-binding sites in the selectivity filter of BK channels, respectively, whereas the Ba^{2+} block site corresponds to the inner ion position of the internal sites of the selectivity filter (36,37). Since the blocking of BK channels by Ca^{2+} is prominent only at intracellular Ca^{2+} concentrations's higher than 100 μM and above +50 mV (25,38), at physiological conditions, Ca^{2+} increases BK channel current by activation. On the other hand, Mg^{2+} activates as well as blocks BK channels in physiological conditions so that the influence of Mg^{2+} on BK channel currents is more complex (26).

17.2 BK CHANNEL STRUCTURE

Four Slo1 subunits form a functional BK channel that can be activated by voltage and intracellular ions. A Slo1 subunit is composed of an N-terminal membrane-spanning domain (MSD) and a large cytosolic tail domain (CTD) at the C-terminus that consist of almost two-third of the entire sequence (Figure 17.3a). The MSD includes the transmembrane (TM) segments S0–S6, with S1–S4 being conserved as the voltage sensor domain (VSD) and S5–S6 as the pore-gate domain (PGD) (39). Unlike most of 6-TM voltage-gated K^+ (Kv) channels, the MSD of BK channels has an additional S0 segment that is linked to the VSD through a long intracellular S0–S1 linker, positioning the N-terminus of Slo1 to the extracellular side (40,41). Auxiliary β and γ subunits associate with the Slo1 channel to modify gating mechanisms (see the following text). The MSD of BK channels was modeled to have a similar structure as the MSD of other Kv channels. S5–S6 from all four subunits forms a central pore, which is surrounded by four S0-and-VSDs at the periphery (42). This model is consistent with the 3-D structure of the Slo1 BK channel solved using Cryo-electron microscopy (Cryo-EM) with a resolution of 17 Å (43). The structural homology with Kv channels suggests that a similar mechanism underlies voltage-dependent activation of BK and other Kv channels, that is, the voltage sensor changes conformation upon membrane depolarization, which promotes the opening of the activation gate (44).

The Cryo-EM structure of the BK channel also shows that the CTD forms a large structure that closely associates with the cytoplasmic side of the MSD (43). Later, x-ray crystallography

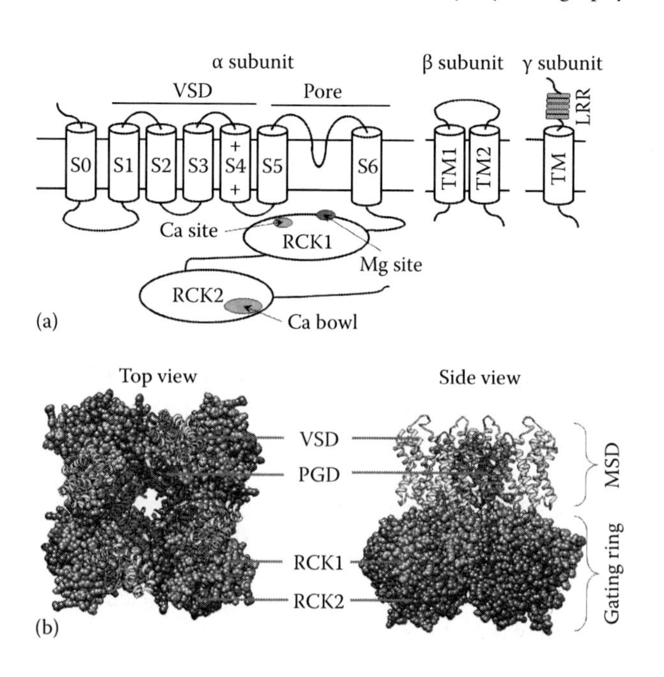

Figure 17.3 BK channel structure. (a) Diagram of the Slo1 β and γ subunits of BK channels. (b) Homology model of the Slo1 subunit was constructed by superimposing the crystal structures of the $K_V1.2$–$K_V2.1$ chimera (PDB ID: 2R9R) [174] and the MthK channel (PDB ID: 1LNQ) [175] at the selectivity filter, and then superimposing the BK channel gating ring (PDB ID: 3MT5) [45] onto that of the MthK channel using UCSF Chimera v1.4.1. MthK structure was then removed. Top view as seen from the extracellular side. RCK, regulators of K^+ conductance; LRR, leucine-rich-repeat; VSD, voltage sensor domain; PGD, pore-gate domain; MSD, membrane-spanning domain.

(45–47) reveals that the CTD of each Slo1 subunit contains two regulators of K^+ conductance (RCK1 and RCK2) domains, and the eight RCK domains from four subunits form a ring-like structure known as the gating ring (Figure 17.3b). In the gating ring, each RCK1 domain makes contacts with two RCK2 domains as well as with the cytoplasmic side of the MSD (43,48,49). A peptide linker (C linker) of 16 amino acids covalently connects the end of S6 in the MSD to the beginning of RCK1 in each Slo1 subunit (46,47,50). The RCK1 and RCK2 domains in each Slo1 subunit are connected by a peptide linker that is not conserved among BK channel orthologs. A cut was made in the linker between RCK1 and RCK2, and the resulting two pieces of the Slo1 subunit coexpressed to form functional channels (51), indicating that the free RCK2 can associate with RCK1 to endow correct structure and function.

The RCK domain structures adopt the arrangement of the Rossmann folds, with three parallel β sheets linked by two α helices in the topological order of β–α–β–α–β, which include αA-J and βA-J in RCK1 and αK-X and βK-T in RCK2, respectively (45–47). A Ca^{2+}-binding site, known as the Ca^{2+} bowl (52), is identified to reside in the RCK2 domain. The bound Ca^{2+} is coordinated by side-chain carboxylates from residues D898 and D900, as well as main chain carbonyl groups from residue Q892 and D895 (45) (residue numbers in mSlo1, GenBank accession number 347143). The site in the structure is consistent with that identified by mutagenesis studies in which neutralization of acidic residues reduced Ca^{2+} sensitivity (52–54). Another Ca^{2+}-binding site is formed by residues D367 and E535 in the RCK1 domain, identified by mutagenesis studies (55–57). No ion binding at this site was observed in the structural studies (45–47); however, the structure of RCK1 shows that these residues are located closely (46), consistent with the mutagenesis data that they form the Ca^{2+}-binding site (57).

Conformational differences were observed by comparing the gating ring structure of the BK channels in the presence and absence of Ca^{2+} (47,58). In the presence of 10 mM Ca^{2+}, RCK1 domains, especially their N-terminal lobes facing the plasma membrane, undergo large conformational changes by expanding from a diameter of 81–93 Å, while RCK2 domains only show very little change (47,58). These results suggest that in BK channels a conformational change of the cytosolic gating ring occurs upon Ca^{2+} binding that consequently opens the activation gate in the PGD. Interestingly, a recent fluorescence resonance energy transfer (FRET) study of functional BK channels suggested that the structural rearrangements of the gating ring upon Ca^{2+} binding are much larger than those predicted by these x-ray crystal structures of isolated gating rings (59).

This difference between isolated gating ring and the gating ring in functional channels may derive from the fact that only the Ca^{2+} bowl in the isolated gating ring could bind to Ca^{2+}, whereas no Ca^{2+} binding was observed in the RCK1 Ca^{2+} site (45,47). The conformational changes upon Ca^{2+} binding to the Ca^{2+} bowl involve primarily rigid body movements of the RCK domains (47), but Ca^{2+} binding to the RCK1 Ca^{2+} site may also alter the tertiary structure of the RCK1 domain, as suggested by mutagenesis studies (60). A mutational scan of the N-terminus of the RCK1 domain including βA-αC (the AC region) identified mutations of individual residues that alter Ca^{2+}-dependent activation via the RCK1 Ca^{2+} site, but none of the mutations affect Ca^{2+}-dependent activation via the Ca^{2+} bowl; in addition, the flexibility changes of the AC region as a result of mutation or alterations of intracellular osmolarity affect Ca^{2+}-dependent activation (60). Considering that the AC regions of the four RCK1 domains are directly in contact with the MSD, whereas RCK2 domains are separated from the MSD by RCK1 domains (Figure 17.3), these results suggest that Ca^{2+} binding to the two different sites causes different conformational changes and opens the activation gate through distinct mechanisms.

17.3 ALLOSTERIC MECHANISMS OF BK CHANNEL ACTIVATION

Although voltage-dependent activation of BK channels is affected by Ca^{2+} and vice versa, the mechanisms of voltage- and Ca^{2+}-dependent activation are distinct. Voltage can activate BK channels in the absence of Ca^{2+} binding (8), and likewise, Ca^{2+} can also activate BK channels when the voltage sensors are at the resting state (61,62). In the absence of Ca^{2+} binding, BK channels are activated only when voltage is extremely positive. For instance, the macroscopic currents of BK channels formed by homogeneous mSlo1 are measurable at about +110 mV, and the conductance–voltage (G–V) relation reaches half-maximum at the voltage ($V_{1/2}$) of ~180 mV (Figure 17.2) (8). On the other hand, intracellular Ca^{2+} concentrations increase from 0 to ~100 μM that is saturating for the two Ca^{2+}-binding sites in the gating ring can enhance BK channel open probability by four orders of magnitude from ~10^{-7} to 10^{-3} when the voltage sensors are at the resting state (60,63).

For *Shaker* type Kv channels, activation is strictly dependent on voltage; the logarithm of the probability of channel opening, $\log P_o$, depends on voltage with a straight line even at very negative voltages where P_o is ~10^{-7} (64). Such a relationship makes the coupling between the voltage sensor and the pore seems to be obligatory, that is, the channel opens when the voltage sensor is activated, and the channel closes when the voltage sensor stays at resting state. Contrary to *Shaker*, in BK channels, the coupling of voltage sensor or Ca^{2+} binding with channel opening is not obligatory. The BK channel has an intrinsic open probability of ~10^{-7} even in the absence of voltage sensor activation and Ca^{2+} binding. Because of this intrinsic open probability, at negative potentials, $\log P_o$ of BK channels deviates from a linear relation with voltage, becoming flat as P_o reaches 10^{-7} while voltage further decreases (61,62). Therefore, instead of being obligatory, the coupling of voltage or Ca^{2+} with the activation gate in BK channels is allosteric (62,63,65). Voltage sensor activation and Ca^{2+} binding promote opening of the channel, but the probability of channel opening is not bound to the state of the voltage sensor or the occupancy of Ca^{2+} binding. It is because of such an allosteric mechanism that either voltage or Ca^{2+} can activate BK channels independently. In a channel with obligatory coupling between the VSD and the activation gate, Ca^{2+} may not alter open probability at any given voltage unless it alters voltage-dependent activation of the VSD.

The allosteric mechanism of voltage- and Ca^{2+}-dependent activation of BK channels can be quantitatively described by Schemes I and II, respectively (Figure 17.4a and b). In Scheme I, voltage sensor activation occurs in both the closed (C) and open (O) states, and the channel can open when 0, 1, 2, 3, or

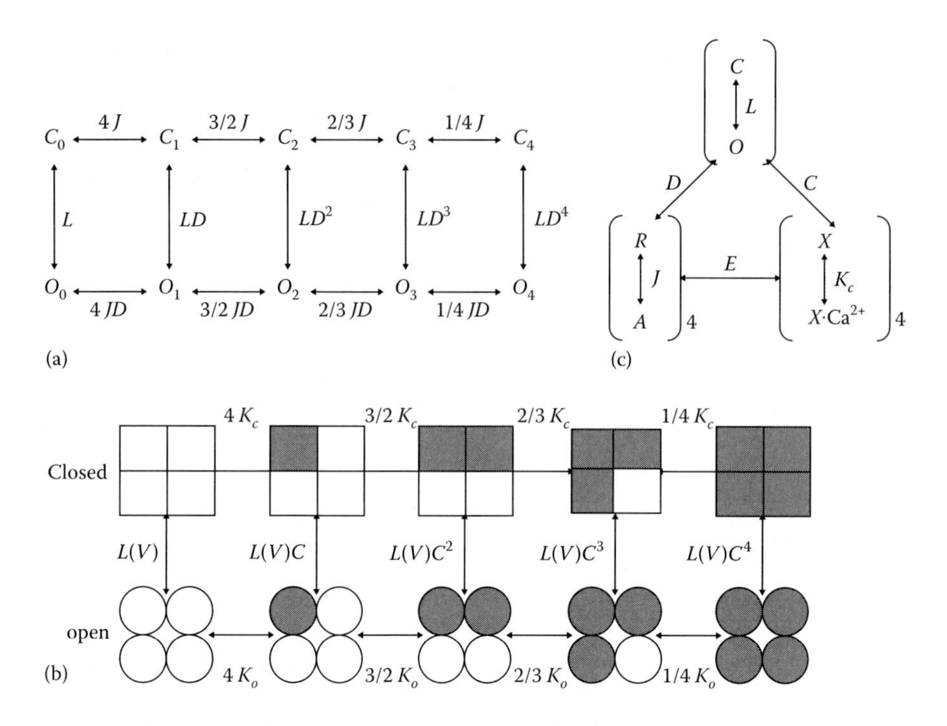

Figure 17.4 Allosteric gating mechanisms of BK channels. (a) An allosteric model of voltage-dependent gating in the absence of Ca^{2+} [62]. (b) A voltage-dependent Monod-Wyman-Changeux (MWC) model of Ca^{2+}-dependent gating [65]. (c) A general allosteric gating mechanism including allosteric interactions between Ca^{2+} binding, voltage sensor activation, and channel opening [63]. *C* and *O*: closed and open conformations of the channel, respectively; *L*: the equilibrium constant for the *C*–*O* transition; *J*: the equilibrium constant for voltage sensor activation; K_C: equilibrium constant for Ca^{2+} binding to closed channels; K_O: equilibrium constant for Ca^{2+} binding to open channels; *D*: allosteric factor describing interaction between channel opening and voltage sensor activation; *C*: allosteric factor describing interaction between channel opening and Ca^{2+} binding; *E*: allosteric factor describing interaction between voltage sensor activation and Ca^{2+} binding.

all 4 voltage sensors are activated. The voltage sensor activation, however, is more favored in the open state, with an equilibrium constant for activation larger than that in the closed state, and the ratio $D = J_o/J_c$ (J_o and J_c are equilibrium constants of voltage sensor movements at open and closed confirmation, respectively) is called the allosteric factor. Similarly, channel opening is favored by voltage sensor activation; with each additional voltage sensor activated, the equilibrium constant for channel opening is increased by a *D* factor. Such an allosteric mechanism can well fit the gating current data that measure voltage sensor activation as well as the ionic current data that measure channel opening in the voltage range of −450 ± 300 mV (66). Likewise, in Scheme II, Ca^{2+} can bind to both open and closed states with different affinities, and the ratio of this is called the allosteric factor *C*. Ca^{2+} binding favors channel opening by the factor *C*. This scheme in slightly different formats could fit the data of single channel and macroscopic currents with various intracellular Ca^{2+} concentrations's (65,67). More careful studies reveal that voltage sensor activation is influenced by Ca^{2+} binding, and vice versa, also with an allosteric mechanism (63,68,69). Thus, BK channel activation by voltage and Ca^{2+} is governed by all three sets of allosteric mechanisms integrated in Scheme III (Figure 17.4c) (63). Scheme III can fit the results of gating current and ionic current measurements at all intracellular Ca^{2+} concentrations's and voltages within the experimental limitations. It is noteworthy that the allosteric model of Ca^{2+}-dependent activation assumes only four identical Ca^{2+}-binding sites. The model can be more complex if two different Ca^{2+}-binding sites in each Slo1 subunit are considered. Nevertheless, the principle of allosteric mechanism should not change with increased complexities.

17.4 STRUCTURAL BASIS OF BK CHANNEL FUNCTION

17.4.1 LARGE SINGLE CHANNEL CONDUCTANCE

BK channels have the largest single channel conductance of all K^+ channels although the sequence of the selectivity filter is highly conserved within the K^+ channel superfamily (34). Part of such a large conductance is derived from two clusters of acidic residues that are located at the intracellular and extracellular vestibules within the K^+ permeation pathway. These clusters of negative charges thus serve as electrostatic traps to attract and concentrate local K^+ concentration, thereby increasing single channel conductance (70–72). At the intracellular vestibule, E321 and E324, which are conserved and specific to Slo1 located in the cytosolic end of the S6 segment, form a ring of eight negative charges. Reducing the negative charges and increasing positive charges by mutations gradually reduce single channel conductance by up to 50% (70,71). The effects of these charge changes on single channel conductance are diminished in the presence of high $[K^+]_i$ (~1 M), and these mutations only reduce the conductance of the outward current but not the inward current, supporting the mechanism that the negative charges attract K^+ to enhance the probability of K^+ entering the channel. Likewise, conserved D261/E264 in the extracellular vestibule of Slo1 also contributes to ~18% of BK channel single channel conductance, and the contribution of these negative charged residues to BK channel conductance in both intracellular and extracellular vestibules can be explained by their contributions to the electrostatic potentials at the intracellular and extracellular entrance of the pore, respectively (72). Nevertheless, these charges in the vestibules are not the sole determinant of the large single channel conductance as the single channel conductance of the mutant BK channel in the absence of these negative charges is still about six times larger than that of *Shaker* K^+ channels. This suggests that other structural aspects specific to BK channel pore may also be important for ion conduction (71,72).

17.4.2 VOLTAGE- AND Ca^{2+}-DEPENDENT ACTIVATION

Ion channels sense physiological stimuli to open and close, thereby regulating physiological processes. Channel activation generally involves three principal steps: sensor of stimuli changes conformation, the conformational change is propagated to the activation gate, and then gate opens. This section describes voltage sensor and Ca^{2+}-binding sites, the activation gate, and the structure underlying the coupling between the sensors and the activation gate of BK channels.

Similar to other Kv channels, the voltage sensor of BK channels contains S1–S4 TM segments. Basic residues R167 in S2, R213 in S4 and acidic residues D153 in S2 and D186 in S3 are gating charges (61,73,74). This differs from other Kv channels where the gating charges are primarily found in S4 (44). BK channels also contain an additional S0 that associates and interacts with the VSD (41,42). These features implicate unique movements of BK channel voltage sensors during activation (73,75). Ca^{2+} binding to either the RCK1 or RCK2 site can activate the channel; when mutations destroy one of the Ca^{2+}-binding sites the remaining site still activates the channel with a Ca^{2+} sensitivity of 40%–60% of the wildtype channel (54,55,60,68), measured by the voltage of *G–V* shifting in response to saturating intracellular Ca^{2+} concentrations (>100 μM), when mutations destroy either the RCK1 site or the Ca^{2+} bowl. However, the two sites may not be completely independent in activating the channel (76): the Ca^{2+} sensitivities contributed by each individual site do not add up to that of the wild-type channel, indicating that the two sites are cooperative in activating the channel (68).

Same as other Kv channels, the pore of BK channels is formed by S5–S6 and contains a highly conserved K^+ selectivity filter (34). For most of K^+ channels, the intracellular bundle crossing formed by the cytosolic side of four inner helices acts as the activation gate that restricts K^+ ion flux when the channels are closed (77). However, when the BK channel is closed, the entrance of the K^+ ion to the pore cavity from inside of the cell may not be restricted at the cytosolic side of S6, because even blockers and methanethiosulfonate (MTS) reagents with larger sizes can enter the cavity (78–83). Cysteine scanning and modification results (83) suggest that the S6 α-helix in BK channels may be disrupted at the unique diglycine motif G310/G311, resulting in a wider opening of the inner pore compared to other Kv channels that usually have a single glycine at the *glycine hinge* position (84) and narrower opening at the bundle crossing (85). Therefore, with the disrupted S6, the permeation gate for K^+ in BK channels may be only located at the selectivity filter. Both voltage sensor movement and Ca^{2+} binding regulate the opening and closing of this gate.

The linker between S6 and the cytosolic gating ring is important for coupling the activation gate with both voltage sensor movements and Ca^{2+} binding. Conformational changes of the gating ring upon Ca^{2+} binding are proposed to pull the gate open (47,50). In Kv channels, the S4–S5 linker interacts with the C linker that propagates the S4 movement to open the activation gate (86). A similar mechanism may also couple the voltage sensor to the activation gate in BK channels although no study of this mechanism has been reported. In addition, noncovalent interactions between the MSD and the cytosolic gating ring may be also involved in the coupling of the activation gate with voltage sensor movements and Ca^{2+} binding (69,87). The mechanism of coupling remains to be an active area of research.

17.4.3 ACTIVATION BY OTHER INTRACELLULAR IONS

A low-affinity metal-binding site was also identified in BK channels. Different from the two high-affinity Ca^{2+}-binding sites in the cytosolic RCK1 and RCK2 domains, this low-affinity cation-binding site is located in between the MSD and the cytosolic RCK1 domain, comprised of residues D99 and N172 from the MSD and D374 and E399 from the CTD (49,55,56). Both Mg^{2+} and Ca^{2+} at millimolar concentrations can bind to this site and activate BK channels (55,56,88), though at physiological conditions, this site is more likely occupied by Mg^{2+}. The bound Mg^{2+} is located closely with the gating charge R213 at the cytoplasmic side of S4 in the VSD (48,49,89). The electrostatic interaction between Mg^{2+} and R213 promotes the voltage sensor to the activated state. Thus, Mg^{2+} binding activates the channel primarily by promoting voltage sensor activation. Mg^{2+} also binds to the high-affinity Ca^{2+}-binding sites to compete with Ca^{2+}, thereby reducing the Ca^{2+}-dependent activation of the channel (11,12,55,56,88). However, the binding of Mg^{2+} to the Ca^{2+} sites per se does not seem to activate the channel.

Other divalent cations including Sr^{2+}, Cd^{2+}, Mn^{2+}, Co^{2+}, and Ni^{2+} can all bind to the Mg^{2+}-binding site to activate the channel (90). Sr^{2+} also binds to the two Ca^{2+}-binding sites to activate BK channels (90). On the other hand, Ba^{2+} activates BK channels by specifically binding to the Ca^{2+} bowl but not the RCK1 site or the Mg^{2+} site (18). Cd^{2+} is different from other divalent cations, which seems to activate the channel by binding to a unique site that differs from any of the Ca^{2+}- and Mg^{2+}-binding sites, although the site for Cd^{2+} may involve D367 in the RCK1 Ca^{2+} site (57,90). Proton activates the channel by binding H365 and H394 in the RCK1 domain and electrostatically interacting with D367 in the RCK1 high-affinity Ca^{2+}-binding site (91). Thus, both Cd^{2+} and H^+ may open the activation gate via the same mechanism as Ca^{2+} via the RCK1 Ca^{2+} site.

17.5 β AND γ SUBUNITS MODULATE BK CHANNEL FUNCTION

Four β (β1–β4) subunits (92) and four γ (γ1–γ4) subunits (93,94) of BK channels have been identified (Figure 17.3a). These auxiliary subunits show distinct tissue distribution and modulate BK channel function differently, thus providing a major mechanism for the diverse BK channel phenotypes observed in various tissues. All β (KCNMB) subunits share 20%–53% of sequence identity and a similar membrane topology with two TM segments (TM1 and TM2), a large extracellular loop (116–128 amino acid residues) and two cytosolic termini (Figure 17.3a). Up to four β subunits associate with four Slo1s to form a BK channel complex (95). TM1 and TM2 of β subunits are packed between the VSDs from two adjacent Slo1 subunits (96) and the extracellular loop extends to the central pore to cover extracellular vestibule of the channel (97,98). γ subunits are structurally and functionally distinct from β subunits (Figure 17.3a). All γ subunits belong to a large leucine-rich-repeat-containing (LRRC) protein family that contains a single TM segment with a large extracellular leucine-rich-repeat (LRR) motif, and a short cytosolic COOH tail (93,94). It is not known how γ subunits associate with Slo1. However, the γ1 subunit (also known as LRRC26) competes with the β1 subunit for association with Slo1 (93), suggesting that similar to other TM auxiliary subunits of Kv channels, the γ subunits also associate with the VSD at the cleft between adjacent Slo1 subunits (99,100).

All β subunits except for β3 alter voltage sensor movements by stabilizing the activated state; the β1 and β2 subunits shift the voltage dependence of gating charge movement, the *Q–V* relation, toward less positive voltages (101–105), while β4 reduces the slope of *Q–V* relation (105). The β1 and β4 subunits also shift the *G–V* relation toward more positive voltages (54,104,106). The β1, β2, and β4 subunits change Ca^{2+} sensitivity of BK channel activation. In the association of the β1 and β2 subunits, *G–V* relation shifts more toward negative voltages in response to Ca^{2+} concentration increase (102,107–109). In the absence of voltage sensor activation and Ca^{2+} binding, the β1, β2, and β4 subunits also reduce the intrinsic open probability of the channel by as much as an order of magnitude (102,103,108,110).

Apparently, the β1, β2, and β4 subunits, all alter BK channel gating with similar mechanisms. However, the effects of these β subunits on BK current differ significantly. For instance, neither the β1 nor β2 subunit shifts *G–V* relation by a large voltage at zero intracellular Ca^{2+} concentrations (102,105,109). Due to the increase of Ca^{2+} sensitivity, both the β1 and β2 subunits increase BK activation at physiological voltages and intracellular Ca^{2+} concentrations's. On the other hand, since the β4 subunit shifts the *G–V* relation to more positive voltages at 0 intracellular Ca^{2+} concentrations it inhibits BK currents at intracellular Ca^{2+} concentrations ≤ 10 μM, while enhances BK currents at intracellular Ca^{2+} concentrations ≥ 10 μM due to a larger Ca^{2+} sensitivity. In addition, β4 subunit slows down channel activation. Therefore, at physiological voltages and intracellular Ca^{2+} concentrations's, the β4 subunit reduces the function of BK channels (111). Besides enhancing Ca^{2+} sensitivity, the β2 subunit also inactivates BK channels with a *ball and chain* mechanism such that the N-terminus of β2 blocks the pore during channel opening to reduce BK currents (40,112).

All γ subunits alter voltage dependence by shifting *G–V* relation to more negative voltages. The amount of the shift caused by various γ subunits is γ1(LRRC26) > γ2 (LRRC52) > γ3 (LRRC55) > γ4 (LRRC38) (94). The shift of *G–V* relation caused by the association of the γ1 subunit is so large (approximately −140 mV) that the BK channel opens at the physiological voltages even when the intracellular intracellular Ca^{2+} concentrations is at the resting level (93). The γ subunits do not alter Ca^{2+} sensitivity of channel activation, and the overall effect of these auxiliary subunits is to enhance BK channel currents.

17.6 PHYSIOLOGICAL FUNCTIONS OF BK CHANNELS

BK channels are expressed throughout major tissues of the body and play important roles in controlling many physiological processes such as neuronal excitability (113–115), neurotransmitter release (116–118), muscle contraction (119–124), secretion (125), and endothelial functions (126). These diverse roles are originated from the unique large conductance and dual sensitivity to membrane potentials and intracellular Ca^{2+} concentration of BK channels. BK channels are located spatially proximal to various cell surface or intracellular Ca^{2+} channels including voltage-gated Ca^{2+} (Ca_V) channels, N-methyl-D-aspartate (NMDA) receptors (127), transient receptor potential (TRP) channels (128,129), inositol trisphosphate receptors (IP3R) (130,131), and ryanodine (RyR) receptors (132) (Figure 17.5). Thus, BK channels can quickly sense the elevation of intracellular Ca^{2+} level and membrane depolarization, and the large single

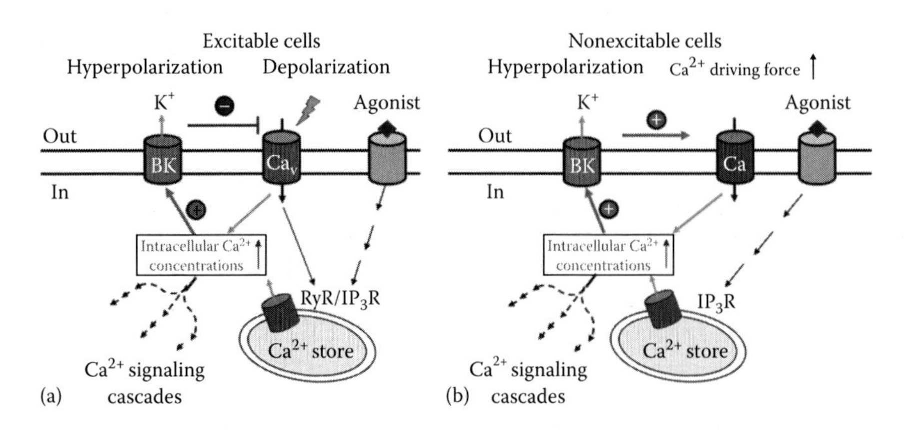

Figure 17.5 Physiological roles of BK channels in excitable (a) and nonexcitable cells (b). (a) In excitable cells, opening of BK channels in response to intracellular Ca^{2+} elevation hyperpolarizes the membrane and closes the voltage-gated Ca^{2+} (Ca_V) channels with a negative-feedback mechanism. (b) In nonexcitable cells, opening of BK channels in response to intracellular Ca^{2+} elevation hyperpolarizes the membrane and increases the driving force for Ca^{2+} influx through Ca^{2+}-permeable channels. Ca_V: voltage-gated Ca^{2+} channels; IP_3R: inositol 1,4,5-triphosphate receptor; RyR: ryanodine receptor; Ca: Ca^{2+}-permeable channels.

channel conductance of BK channels enables rapid K^+ efflux, thereby effectively hyperpolarizing the membrane.

The BK channel–induced membrane hyperpolarization has different physiological consequences in excitable and nonexcitable cells. In excitable cells, such as neurons, smooth muscles, secretary endocrine cells, and specialized sensory receptor cells, BK channels usually form protein complexes with Ca_V channels in which Ca_V and BK channels are about 10–15 nm apart (133,134). This spatial proximity ensures the formation of a negative-feedback regulation of Ca^{2+} entry in many cell types: Ca^{2+} entry through Ca_V channels activates proximal BK channels and results in membrane hyperpolarization, which in turn closes Ca_V channels and prevents further Ca^{2+} entry. This negative-feedback mechanism thereby controls membrane potential, intracellular Ca^{2+} dynamics, and the subsequent events such as neuronal excitability, neurotransmitter release, electrical secretion coupling, and tonic/phasic smooth muscle constriction.

In nonexcitable cells, including epithelial, endothelial, and glial cells, on the other hand, BK channels mainly serve as a positive-feedback regulator for Ca^{2+} entry through voltage-independent Ca^{2+} channels such as TRP channels, Ca^{2+}-release-activated (CRAC) channels, and P2X purinoreceptors (126). The BK-induced membrane hyperpolarization in response to the elevation of intracellular Ca^{2+} level can increase the Ca^{2+} electrochemical gradient across the plasma membrane, thereby promoting Ca^{2+} entry through these Ca^{2+} channels. With this arrangement, BK channels can play important roles in the control of Ca^{2+} homeostasis, thereby contributing to various physiological functions ranging from K^+ secretion, osmoregulation, cell proliferation to cell migration (126).

17.6.1 BK CHANNEL IN NEURONS

BK channels are broadly expressed throughout the nervous system with higher mRNA expression in neocortex, olfactory bulb, habenula, striatum, granule and pyramidal cell layer of the hippocampus, and Purkinje cells and deep nuclei in cerebellum (135). The primary role of BK channels is to link intracellular Ca^{2+} dynamics and the electrical activity of neurons. At the cellular level, BK channels show both pre- and postsynaptic localizations, including somas, dendrites, axons, and presynaptic terminals (135–140). In these subcellular compartments, BK channels contribute to the repolarization of the membrane and limiting further Ca^{2+} entry, thereby regulating both neuronal excitability and neurotransmitter release (138). At the soma, BK channels rapidly activate during the upstroke of an action potential when both membrane potential and intracellular Ca^{2+} concentration are raised, and quickly deactivate when the membrane potential drops to more negative range. In this way, BK channels help repolarize the action potential and contribute to the generation of the fast after-hyperpolarization (fAHP) that usually lasts 1–10 ms. Since AHP is a major determinant of the refractory period and therefore the firing frequency of a neuron, BK channels are important in controlling the shape and frequency of action potentials (138). At presynaptic terminals, BK channels play pivotal role in limiting Ca^{2+} influx and neurotransmitter release by reducing the duration of the presynaptic spikes and the opening of presynaptic Ca_V channels (118). Pharmacological blockade of BK channels, for instance, increases the synaptic efficacy of neurotransmitter release at synapses between hippocampal CA3 neurons (141). Compared to presynaptic BK channels, the functions of postsynaptic BK channels are less understood, despite their strong postsynaptic expression (137). In a recent attempt, the function of BK channels in the reciprocal feedback synapses between the rod bipolar cells and A17 amacrine cells of the retina was studied. BK channels in the postsynaptic A17 amacrine cells were found to modulate both excitatory and inhibitory synaptic transmission by suppressing postsynaptic depolarization, limiting Ca_V channel activation and GABA release from A17 cells, thereby reciprocally regulating the excitatory synaptic transmission of the glutamatergic ribbon synapses in the rod bipolar cell (142).

BK channels are also proposed to work together with Ca_V channels through negative-feedback mechanism to define the rhythmicity of spontaneous firing in suprachiasmatic nucleus (the master circadian pacemaker in mammals) (143), and cerebellar Purkinje cells (144) that generate the sole output from the cerebellar cortex to control motor coordination and motor learning. Consistent with their functional expression in these neuronal types, Slol channel–deficient mice exhibit abnormal circadian rhythm (143), and impaired motor behavior, including abnormal eye-blink reflex, abnormal locomotion, and cerebellar ataxia (144,145).

17.6.2 BK CHANNEL IN SMOOTH MUSCLE CELLS

In smooth muscle cells, BK channels are in close proximity to the Ca^{2+}-release sites of the sarcoplasmic reticulum (SR) (122,132). Calcium-induced calcium release (CICR) from SR through RyRs generates highly spatially and temporally localized *calcium sparks*, resulting in a localized and large (10–100 μM) increase in Ca^{2+} concentration (146). With the facilitation of the smooth muscle–selective expression of the accessory β1 subunit that enhances the open probability of Slol channels, smooth muscle BK channels effectively open in response to calcium sparks, generating spontaneous transient outward currents that hyperpolarize the membrane, shut down Ca_V channels that supply Ca^{2+} to initiate depolarization and contraction, ultimately causing smooth muscle relaxation (147). Indeed, targeted deletion of the β1 subunit in mouse decreases the Ca^{2+} sensitivity of BK channels and disrupts the functional coupling of calcium sparks to BK channel activation, resulting in increased arterial tone and blood pressure (148). Consistent with the results from this mouse model, a gain-of-function mutation (E65K) of the β1-subunit was reported to protect against human diastolic hypertension by increasing the apparent Ca^{2+} and voltage sensitivity of the BK channel (149,150). BK channels also play important roles in the nonvascular smooth muscle cells. In ureteral smooth muscle, the activation of BK channels by Ca^{2+} sparks limits the availability of Ca_V channels, resulting in a prolonged action potential refractory period (151). Consistently, *slo1* knockout mice exhibit increased urinary bladder contractility and elevation in urination frequency (152).

17.7 BK CHANNEL IN DISEASES

Consistent with their important roles in many physiological processes, BK channels have been proposed to involve in various pathological conditions. Based on recent studies of human patients and rodent models, BK channels are implicated in the pathogenesis of several diseases including epilepsy (153,154), cerebellar ataxia (145), autism and mental retardation (155), stroke (156), hypertension (148), asthma (157), obesity (158), and tumor progression (159,160).

The first human BK channelopathy due to a missense mutation of Asp-to-Gly (D434G) in the cytosolic RCK1 domain of Slo1 channel was identified from patients who have a syndrome of coexistent generalized epilepsy and paroxysmal dyskinesia (GEPD) (153). By reducing the flexibility of a key region in the RCK1 domain (the AC region) to enhance its apparent Ca^{2+} sensitivity (60), this gain-of-function mutation may lead to increased neuronal excitability by inducing rapid repolarization of action potential and fast firing frequency, ultimately resulting in the GEPD syndrome (153). Supporting this pathophysiological mechanism, mice null of BK channel accessory β4 subunit also exhibit spontaneous seizures and temporal lobe epilepsy (111). Lacking the inhibitory β4 subunit, neuronal BK channels are more ready to be activated, leading to shortening of the action potential duration and increasing of firing frequency. In addition to these gain-of-function effects of the BK channels in epilepsy, loss-of-function or downregulation of BK channels can also play important roles in the pathophysiology of epilepsy (154). For instance, studies from rodent models of the inherited generalized epilepsy and temporal lobe epilepsy demonstrate that BK current density is significantly reduced in various neuronal types such as cortical neurons, hippocampal CA3 neurons, and inferior colliculus neurons (161). In these cases, BK channel might behave as an *emergency brake* that protects against neuronal hyperexcitability by preventing excessive intracellular Ca^{2+} elevation and excitatory neurotransmitter release. The paradoxical roles of BK channels in epilepsy and seizure may depend on their anatomical and subcellular distribution, accessory subunit-specific regulation, posttranslational modification, and seizure types. Therefore, further understanding the molecular basis of how both gain-of-function and loss-of-function BK channels contribute to epileptogenesis and seizure generation is critical to develop rational therapies for some inherited and acquired epilepsy syndrome.

BK channels have been proposed to play a key neuroprotective role during and after brain ischemia, which causes Ca^{2+} accumulation, hyperexcitability, and energy deprivation of neurons that will ultimately lead to cell death through excitotoxicity (162). Opening of BK channels in response to ischemia can hyperpolarize membrane potential, limiting Ca^{2+} influx and neurotransmitter release in a negative feedback fashion, thereby promoting neuronal survival. In hippocampal slice cultures, BK channel blockers, paxilline and iberiotoxin, induced increased cell death in CA1 and CA3 neurons during and after oxygen and glucose deprivation (163). A recent *in vivo* study provided further support. Transient focal cerebral ischemia by middle cerebral artery occlusion produced larger infarct volume, more severe neurological deficits, and higher postischemic mortality in *slo1* knockout mice compared to WT littermates. Openers of BK channels, thus, have been pursued by pharmaceutical industry as neuroprotective agents to treat ischemic stroke (164). BMS-204352, a calcium-sensitive BK channel opener, potently activates BK channel only when intracellular Ca^{2+} concentration becomes higher than 1 μM (156). This ensures that only BK channels in ischemic neurons will be activated. This compound showed a promising neuroprotective effect against the focal ischemia in rat models of brain ischemia without affecting blood pressure or cerebral blood flow. Unfortunately, this compound failed to show superior efficacy in acute stroke patients compared to placebo in clinical trial phase III probably due to its dual activating effect on KCNQ channels (164).

Consistent with the functional importance of BK channels in vascular (165) and renal physiology (166), transgenic mice and rodent models under different pathological conditions suggest that BK channels are also critical in the pathogenesis of hypertension. Genetic ablation of Slo1 channels results in hypertension, extreme aldosteronism, and an absence of flow-induced K^+ secretion (167). On the other hand, deletion of the BK β1 subunit results in elevated mean arterial pressure (MAP) by 21 mm Hg (121,148), defect in K^+ secretion (168), and elevated production of aldosterone (166). β1 subunit was also found to be downregulated in hypertensive rats (169,170). Interestingly, a gain-of-function mutation (E65K) of the human β1 subunit was discovered to have protective effect in diastolic hypertension, especially for aging women, further supporting the importance of BK channels in hypertension (149,150).

In summary, BK channels are attracting drug targets to treat various human diseases based on their important physiological and pathophysiological roles. Nevertheless, it is challenging to develop effective therapeutics owing to the complexity of BK channel biology, which requires further understanding the structure–function, pharmacology, physiology, and pathophysiology of BK channels.

17.8 CONCLUDING REMARKS

Great progress in understanding of BK channels, from molecular architecture, working mechanism, cellular and *in vivo* functions, to human diseases, has been achieved since the initial discovery of BK currents three decades ago. BK channels have become one of the most studied ion channels owning to their large single channel conductance, dual activation by membrane depolarization and intracellular Ca^{2+}, and diverse physiological functions. We now know that the Slo1 subunit of BK channels contains three structural domains, the PGD, the voltage sensor domain (VSD), and the large cytosolic Ca^{2+}-binding domain (CTD) that are functionally coupled with allosteric mechanisms. The structure–function relations of each of these domains are quite well understood, and the biophysical principles of allosteric coupling among these domains have been shown. At present, understanding molecular mechanisms of the allosteric coupling among these functional domains is an emerging task (39), and the full-length BK channel structure with atomic resolution will provide important insights to this understanding.

BK channels play various roles in physiology mainly owing to their sensitivity to various physiological stimuli (39,171), posttranslational modifications (phosphorylation, methylation, heme) (172), different splicing isoforms (173), and tissue-specific expression of auxiliary β (92) and γ (94) subunits. All these factors fine-tune the properties of BK channel functions, thereby rendering BK channels tissue-specific functions in neurons, smooth muscle cells, secretary endocrine cells, and nonexcitable cells such as endothelial and glial cells (126,147,171). Mouse models of BK channels and congenital diseases (153) associated with BK channel mutations advanced our understanding of the physiological and pathophysiological roles of BK channels. Therefore, further understanding of the molecular architecture, allosteric gating mechanism, and biology of BK channels will greatly facilitate rational design of BK channel modulators to treat BK channel–related diseases, such as epilepsy (153,154), hypertension (148), asthma (157), stroke (156), and urinary incontinence (152).

REFERENCES

1. Marty, A. 1981. Ca-dependent K channels with large unitary conductance in chromaffin cell membranes. *Nature* 291:497–500.
2. Pallotta, B. S., K. L. Magleby, and J. N. Barrett. 1981. Single channel recordings of Ca^{2+}-activated K^+ currents in rat muscle cell culture. *Nature* 293:471–474.
3. Atkinson, N. S., G. A. Robertson, and B. Ganetzky. 1991. A component of calcium-activated potassium channels encoded by the *Drosophila* slo locus. *Science* 253:551–555.
4. Adelman, J. P., K. Z. Shen, M. P. Kavanaugh, R. A. Warren, Y. N. Wu, A. Lagrutta, C. T. Bond, and R. A. North. 1992. Calcium-activated potassium channels expressed from cloned complementary DNAs. *Neuron* 9:209–216.
5. Butler, A., S. Tsunoda, D. P. McCobb, A. Wei, and L. Salkoff. 1993. mSlo, a complex mouse gene encoding "maxi" calcium-activated potassium channels. *Science* 261:221–224.
6. Dworetzky, S. I., J. T. Trojnacki, and V. K. Gribkoff. 1994. Cloning and expression of a human large-conductance calcium-activated potassium channel. *Brain Res Mol Brain Res* 27:189–193.
7. McCobb, D. P., N. L. Fowler, T. Featherstone, C. J. Lingle, M. Saito, J. E. Krause, and L. Salkoff. 1995. A human calcium-activated potassium channel gene expressed in vascular smooth muscle. *Am J Physiol* 269:H767–H777.
8. Cui, J., D. H. Cox, and R. W. Aldrich. 1997. Intrinsic voltage dependence and Ca^{2+} regulation of mslo large conductance Ca-activated K^+ channels. *J Gen Physiol* 109:647–673.
9. Magleby, K. L. 2003. Gating mechanism of BK (Slo1) channels: So near, yet so far. *J Gen Physiol* 121:81–96.
10. Golowasch, J., A. Kirkwood, and C. Miller. 1986. Allosteric effects of Mg^{2+} on the gating of Ca^{2+}-activated K^+ channels from mammalian skeletal muscle. *J Exp Biol* 124:5–13.
11. Shi, J. and J. Cui. 2001. Intracellular Mg^{2+} enhances the function of BK-type Ca^{2+}-activated K^+ channels. *J Gen Physiol* 118:589–606.
12. Zhang, X., C. R. Solaro, and C. J. Lingle. 2001. Allosteric regulation of BK channel gating by Ca^{2+} and Mg^{2+} through a nonselective, low affinity divalent cation site. *J Gen Physiol* 118:607–636.
13. Squire, L. G. and O. H. Petersen. 1987. Modulation of Ca^{2+}- and voltage-activated K^+ channels by internal Mg^{2+} in salivary acinar cells. *Biochim Biophys Acta* 899:171–175.
14. Zamoyski, V. L., V. N. Serebryakov, and R. Schubert. 1989. Activation and blocking effects of divalent cations on the calcium—Dependent potassium channel of high conductance. *Biomed Biochim Acta* 48:S388–S392.
15. McLarnon, J. G. and D. Sawyer. 1993. Effects of divalent cations on the activation of a calcium-dependent potassium channel in hippocampal neurons. *Pflugers Arch* 424:1–8.
16. Zhang, X., E. Puil, and D. A. Mathers. 1995. Effects of intracellular Mg^{2+} on the properties of large-conductance, Ca^{2+}-dependent K^+ channels in rat cerebrovascular smooth muscle cells. *J Cereb Blood Flow Metab* 15:1066–1074.
17. Oberhauser, A., O. Alvarez, and R. Latorre. 1988. Activation by divalent cations of a Ca^{2+}-activated K^+ channel from skeletal muscle membrane. *J Gen Physiol* 92:67–86.
18. Zhou, Y., X. H. Zeng, and C. J. Lingle. 2012. Barium ions selectively activate BK channels via the Ca^{2+}-bowl site. *Proc Natl Acad Sci USA* 109:11413–11418.
19. Avdonin, V., X. D. Tang, and T. Hoshi. 2003. Stimulatory action of internal protons on Slo1 BK channels. *Biophys J* 84:2969–2980.
20. Hou, S., F. T. Horrigan, R. Xu, S. H. Heinemann, and T. Hoshi. 2009. Comparative effects of H^+ and Ca^{2+} on large-conductance Ca^{2+}- and voltage-gated Slo1 K^+ channels. *Channels (Austin)* 3:249–258.
21. Chu, B., A. M. Dopico, J. R. Lemos, and S. N. Treistman. 1998. Ethanol potentiation of calcium-activated potassium channels reconstituted into planar lipid bilayers. *Mol Pharmacol* 54:397–406.
22. Dopico, A. M., V. Anantharam, and S. N. Treistman. 1998. Ethanol increases the activity of Ca(++)-dependent K^+ (mslo) channels: Functional interaction with cytosolic Ca^{++}. *J Pharmacol Exp Ther* 284:258–268.
23. Tang, X. D., R. Xu, M. F. Reynolds, M. L. Garcia, S. H. Heinemann, and T. Hoshi. 2003. Haem can bind to and inhibit mammalian calcium-dependent Slo1 BK channels. *Nature* 425:531–535.
24. Vaithianathan, T., A. Bukiya, J. Liu, P. Liu, M. Asuncion-Chin, Z. Fan, and A. Dopico. 2008. Direct regulation of BK channels by phosphatidylinositol 4,5-bisphosphate as a novel signaling pathway. *J Gen Physiol* 132:13–28.
25. Cox, D. H., J. Cui, and R. W. Aldrich. 1997. Separation of gating properties from permeation and block in mslo large conductance Ca-activated K^+ channels. *J Gen Physiol* 109:633–646.
26. Geng, Y., X. Wang, and K. L. Magleby. 2013. Lack of negative slope in I-V plots for BK channels at positive potentials in the absence of intracellular blockers. *J Gen Physiol* 141:493–497.
27. Morales, E., W. C. Cole, C. V. Remillard, and N. Leblanc. 1996. Block of large conductance Ca^{2+}-activated K^+ channels in rabbit vascular myocytes by internal Mg^{2+} and Na^+. *J Physiol* 495:701–716.
28. Wachter, C. and K. Turnheim. 1996. Inhibition of high-conductance, calcium-activated potassium channels of rabbit colon epithelium by magnesium. *J Membr Biol* 150:275–282.
29. Zhang, Y., X. Niu, T. I. Brelidze, and K. L. Magleby. 2006. Ring of negative charge in BK channels facilitates block by intracellular Mg^{2+} and polyamines through electrostatics. *J Gen Physiol* 128:185–202.
30. Neyton, J. 1996. A Ba^{2+} chelator suppresses long shut events in fully activated high- conductance Ca^{2+}-dependent K^+ channels. *Biophys J* 71:220–226.
31. Miller, C. 1987. Trapping single ions inside single ion channels. *Biophys J* 52:123–126.
32. Neyton, J. and C. Miller. 1988. Potassium blocks barium permeation through a calcium-activated potassium channel. *J Gen Physiol* 92:549–567.
33. Neyton, J. and C. Miller. 1988. Discrete Ba^{2+} block as a probe of ion occupancy and pore structure in the high-conductance Ca^{2+} -activated K^+ channel. *J Gen Physiol* 92:569–586.
34. Doyle, D. A., J. Morais Cabral, R. A. Pfuetzner, A. Kuo, J. M. Gulbis, S. L. Cohen, B. T. Chait, and R. MacKinnon. 1998. The structure of the potassium channel: Molecular basis of K^+ conduction and selectivity. *Science* 280:69–77.
35. Morais-Cabral, J. H., Y. Zhou, and R. MacKinnon. 2001. Energetic optimization of ion conduction rate by the K^+ selectivity filter. *Nature* 414:37–42.

36. Jiang, Y. and R. MacKinnon. 2000. The barium site in a potassium channel by x-ray crystallography. *J Gen Physiol* 115:269–272.
37. Vergara, C., O. Alvarez, and R. Latorre. 1999. Localization of the K^+ lock-In and the Ba^{2+} binding sites in a voltage-gated calcium-modulated channel. Implications for survival of K^+ permeability. *J Gen Physiol* 114:365–376.
38. Schroeder, I., G. Thiel, and U. P. Hansen. 2013. Ca^{2+} block and flickering both contribute to the negative slope of the IV curve in BK channels. *J Gen Physiol* 141:499–505.
39. Lee, U. S. and J. Cui. 2010. BK channel activation: Structural and functional insights. *Trends Neurosci* 33:415–423.
40. Wallner, M., P. Meera, and L. Toro. 1999. Molecular basis of fast inactivation in voltage and Ca^{2+}-activated K^+ channels: A transmembrane beta-subunit homolog. *Proc Natl Acad Sci USA* 96:4137–4142.
41. Meera, P., M. Wallner, M. Song, and L. Toro. 1997. Large conductance voltage- and calcium-dependent K^+ channel, a distinct member of voltage-dependent ion channels with seven N-terminal transmembrane segments (S0–S6), an extracellular N terminus, and an intracellular (S9-S10) C terminus. *Proc Natl Acad Sci USA* 94:14066–14071.
42. Liu, G., S. I. Zakharov, L. Yang, S. X. Deng, D. W. Landry, A. Karlin, and S. O. Marx. 2008. Position and role of the BK channel alpha subunit S0 helix inferred from disulfide crosslinking. *J Gen Physiol* 131:537–548.
43. Wang, L. and F. J. Sigworth. 2009. Structure of the BK potassium channel in a lipid membrane from electron cryomicroscopy. *Nature* 461:292–295.
44. Bezanilla, F. 2000. The voltage sensor in voltage-dependent ion channels. *Physiol Rev* 80:555–592.
45. Yuan, P., M. D. Leonetti, A. R. Pico, Y. Hsiung, and R. MacKinnon. 2010. Structure of the human BK channel Ca^{2+}-activation apparatus at 3.0 A resolution. *Science* 329:182–186.
46. Wu, Y., Y. Yang, S. Ye, and Y. Jiang. 2010. Structure of the gating ring from the human large-conductance Ca(2+)-gated K(+) channel. *Nature* 466:393–397.
47. Yuan, P., M. D. Leonetti, Y. Hsiung, and R. MacKinnon. 2012. Open structure of the Ca^{2+} gating ring in the high-conductance Ca^{2+}-activated K^+ channel. *Nature* 481:94–97.
48. Yang, H., L. Hu, J. Shi, K. Delaloye, F. T. Horrigan, and J. Cui. 2007. Mg^{2+} mediates interaction between the voltage sensor and cytosolic domain to activate BK channels. *Proc Natl Acad Sci USA* 104:18270–18275.
49. Yang, H., J. Shi, G. Zhang, J. Yang, K. Delaloye, and J. Cui. 2008. Activation of Slo1 BK channels by Mg^{2+} coordinated between the voltage sensor and RCK1 domains. *Nat Struct Mol Biol* 15:1152–1159.
50. Niu, X., X. Qian, and K. L. Magleby. 2004. Linker-gating ring complex as passive spring and Ca^{2+}-dependent machine for a voltage- and Ca^{2+}-activated potassium channel. *Neuron* 42:745–756.
51. Schreiber, M., A. Yuan, and L. Salkoff. 1999. Transplantable sites confer calcium sensitivity to BK channels. *Nat Neurosci* 2:416–421.
52. Schreiber, M. and L. Salkoff. 1997. A novel calcium-sensing domain in the BK channel. *Biophys J* 73:1355–1363.
53. Bian, S., I. Favre, and E. Moczydlowski. 2001. Ca^{2+}-binding activity of a COOH-terminal fragment of the *Drosophila* BK channel involved in Ca^{2+}-dependent activation. *Proc Natl Acad Sci USA* 98:4776–4781.
54. Bao, L., C. Kaldany, E. C. Holmstrand, and D. H. Cox. 2004. Mapping the BKCa channel's "Ca^{2+} bowl": Side-chains essential for Ca^{2+} sensing. *J Gen Physiol* 123:475–489.
55. Xia, X. M., X. Zeng, and C. J. Lingle. 2002. Multiple regulatory sites in large-conductance calcium-activated potassium channels. *Nature* 418:880–884.
56. Shi, J., G. Krishnamoorthy, Y. Yang, L. Hu, N. Chaturvedi, D. Harilal, J. Qin, and J. Cui. 2002. Mechanism of magnesium activation of calcium-activated potassium channels. *Nature* 418:876–880.
57. Zhang, G., S. Y. Huang, J. Yang, J. Shi, X. Yang, A. Moller, X. Zou, and J. Cui. 2010. Ion sensing in the RCK1 domain of BK channels. *Proc Natl Acad Sci USA* 107:18700–18705.
58. Javaherian, A. D., T. Yusifov, A. Pantazis, S. Franklin, C. S. Gandhi, and R. Olcese. 2011. Metal-driven operation of the human large-conductance voltage- and Ca^{2+}-dependent potassium channel (BK) gating ring apparatus. *J Biol Chem* 286:20701–20709.
59. Miranda, P., J. E. Contreras, A. J. Plested, F. J. Sigworth, M. Holmgren, and T. Giraldez. 2013. State-dependent FRET reports calcium- and voltage-dependent gating-ring motions in BK channels. *Proc Natl Acad Sci USA* 110:5217–5222.
60. Yang, J., G. Krishnamoorthy, A. Saxena, G. Zhang, J. Shi, H. Yang, K. Delaloye, D. Sept, and J. Cui. 2010. An epilepsy/dyskinesia-associated mutation enhances BK channel activation by potentiating Ca^{2+} sensing. *Neuron* 66:871–883.
61. Cui, J. and R. W. Aldrich. 2000. Allosteric linkage between voltage and Ca^{2+}-dependent activation of BK-type mslo1 K^+ channels. *Biochemistry* 39:15612–15619.
62. Horrigan, F. T., J. Cui, and R. W. Aldrich. 1999. Allosteric voltage gating of potassium channels I. Mslo ionic currents in the absence of Ca^{2+}. *J Gen Physiol* 114:277–304.
63. Horrigan, F. T., and R. W. Aldrich. 2002. Coupling between voltage sensor activation, Ca^{2+} binding and channel opening in large conductance (BK) potassium channels. *J Gen Physiol* 120:267–305.
64. Islas, L. D., and F. J. Sigworth. 1999. Voltage sensitivity and gating charge in *Shaker* and *Shab* family potassium channels. *J Gen Physiol* 114:723–742.
65. Cox, D. H., J. Cui, and R. W. Aldrich. 1997. Allosteric gating of a large conductance Ca-activated K^+ channel. *J Gen Physiol* 110:257–281.
66. Horrigan, F. T. and R. W. Aldrich. 1999. Allosteric voltage gating of potassium channels II. Mslo channel gating charge movement in the absence of Ca^{2+}. *J Gen Physiol* 114:305–336.
67. McManus, O. B. and K. L. Magleby. 1991. Accounting for the Ca^{2+}-dependent kinetics of single large-conductance Ca^{2+}-activated K^+ channels in rat skeletal muscle. *J Physiol* 443:739–777.
68. Sweet, T. B. and D. H. Cox. 2008. Measurements of the BKCa channel's high-affinity Ca^{2+} binding constants: Effects of membrane voltage. *J Gen Physiol* 132:491–505.
69. Savalli, N., A. Pantazis, T. Yusifov, D. Sigg, and R. Olcese. 2012. The contribution of RCK domains to human BK channel allosteric activation. *J Biol Chem* 287:21741–21750.
70. Brelidze, T. I., X. Niu, and K. L. Magleby. 2003. A ring of eight conserved negatively charged amino acids doubles the conductance of BK channels and prevents inward rectification. *Proc Natl Acad Sci USA* 100:9017–9022.
71. Nimigean, C. M., J. S. Chappie, and C. Miller. 2003. Electrostatic tuning of ion conductance in potassium channels. *Biochemistry* 42:9263–9268.
72. Carvacho, I., W. Gonzalez, Y. P. Torres, S. Brauchi, O. Alvarez, F. D. Gonzalez-Nilo, and R. Latorre. 2008. Intrinsic electrostatic potential in the BK channel pore: Role in determining single channel conductance and block. *J Gen Physiol* 131:147–161.
73. Ma, Z., X. J. Lou, and F. T. Horrigan. 2006. Role of Charged Residues in the S1–S4 Voltage Sensor of BK Channels. *J Gen Physiol.* 127:309–328.
74. Diaz, L., P. Meera, J. Amigo, E. Stefani, O. Alvarez, L. Toro, and R. Latorre. 1998. Role of the S4 segment in a voltage-dependent calcium-sensitive potassium (hSlo) channel. *J Biol Chem* 273:32430–32436.

75. Pantazis, A., V. Gudzenko, N. Savalli, D. Sigg, and R. Olcese. 2010. Operation of the voltage sensor of a human voltage- and Ca^{2+}-activated K^+ channel. *Proc Natl Acad Sci USA* 107:4459–4464.
76. Qian, X., X. Niu, and K. L. Magleby. 2006. Intra- and intersubunit cooperativity in activation of BK channels by Ca^{2+}. *J Gen Physiol* 128:389–404.
77. Holmgren, M., K. S. Shin, and G. Yellen. 1998. The activation gate of a voltage-gated K^+ channel can be trapped in the open state by an intersubunit metal bridge. *Neuron* 21:617–621.
78. Li, W. and R. W. Aldrich. 2004. Unique inner pore properties of BK channels revealed by quaternary ammonium block. *J Gen Physiol* 124:43–57.
79. Li, W. and R. W. Aldrich. 2006. State-dependent block of BK channels by synthesized *Shaker* ball peptides. *J Gen Physiol* 128:423–441.
80. Wilkens, C. M. and R. W. Aldrich. 2006. State-independent block of BK channels by an intracellular quaternary ammonium. *J Gen Physiol* 128:347–364.
81. Brelidze, T. I. and K. L. Magleby. 2005. Probing the geometry of the inner vestibule of BK channels with sugars. *J Gen Physiol* 126:105–121.
82. Tang, Q. Y., X. H. Zeng, and C. J. Lingle. 2009. Closed-channel block of BK potassium channels by bbTBA requires partial activation. *J Gen Physiol* 134:409–436.
83. Zhou, Y., X. M. Xia, and C. J. Lingle. 2011. Cysteine scanning and modification reveal major differences between BK channels and Kv channels in the inner pore region. *Proc Natl Acad Sci USA* 108:12161–12166.
84. Jiang, Y., A. Lee, J. Chen, M. Cadene, B. T. Chait, and R. MacKinnon. 2002. The open pore conformation of potassium channels. *Nature* 417:523–526.
85. Webster, S. M., D. Del Camino, J. P. Dekker, and G. Yellen. 2004. Intracellular gate opening in *Shaker* K^+ channels defined by high-affinity metal bridges. *Nature* 428:864–868.
86. Long, S. B., E. B. Campbell, and R. Mackinnon. 2005. Voltage sensor of Kv1.2: Structural basis of electromechanical coupling. *Science* 309:903–908.
87. Yang, J., H. Yang, X. Sun, K. Delaloye, X. Yang, A. Moller, J. Shi, and J. Cui. 2013. Interaction between residues in the Mg^{2+}-binding site regulates BK channel activation. *J Gen Physiol* 141:217–228.
88. Hu, L., H. Yang, J. Shi, and J. Cui. 2006. Effects of multiple metal binding sites on calcium and magnesium-dependent activation of BK channels. *J Gen Physiol* 127:35–50.
89. Hu, L., J. Shi, Z. Ma, G. Krishnamoorthy, F. Sieling, G. Zhang, F. T. Horrigan, and J. Cui. 2003. Participation of the S4 voltage sensor in the Mg^{2+}-dependent activation of large conductance (BK) K^+ channels. *Proc Natl Acad Sci USA* 100:10488–10493.
90. Zeng, X. H., X. M. Xia, and C. J. Lingle. 2005. Divalent cation sensitivity of BK channel activation supports the existence of three distinct binding sites. *J Gen Physiol* 125:273–286.
91. Hou, S., R. Xu, S. H. Heinemann, and T. Hoshi. 2008. Reciprocal regulation of the Ca^{2+} and H^+ sensitivity in the SLO1 BK channel conferred by the RCK1 domain. *Nat Struct Mol Biol* 15:403–410.
92. Orio, P., P. Rojas, G. Ferreira, and R. Latorre. 2002. New disguises for an old channel: MaxiK channel beta-subunits. *News Physiol Sci* 17:156–161.
93. Yan, J. and R. W. Aldrich. 2010. LRRC26 auxiliary protein allows BK channel activation at resting voltage without calcium. *Nature* 466:513–516.
94. Yan, J. and R. W. Aldrich. 2012. BK potassium channel modulation by leucine-rich repeat-containing proteins. *Proc Natl Acad Sci USA* 109:7917–7922.
95. Wang, Y. W., J. P. Ding, X. M. Xia, and C. J. Lingle. 2002. Consequences of the stoichiometry of Slo1 alpha and auxiliary beta subunits on functional properties of large-conductance Ca^{2+}-activated K^+ channels. *J Neurosci* 22:1550–1561.
96. Liu, G., S. I. Zakharov, L. Yang, R. S. Wu, S. X. Deng, D. W. Landry, A. Karlin, and S. O. Marx. 2008. Locations of the beta1 transmembrane helices in the BK potassium channel. *Proc Natl Acad Sci USA* 105:10727–10732.
97. Zeng, X. H., X. M. Xia, and C. J. Lingle. 2003. Redox-sensitive extracellular gates formed by auxiliary beta subunits of calcium-activated potassium channels. *Nat Struct Biol* 10:448–454.
98. Meera, P., M. Wallner, and L. Toro. 2000. A neuronal beta subunit (KCNMB4) makes the large conductance, voltage- and Ca^{2+}-activated K^+ channel resistant to charybdotoxin and iberiotoxin [In Process Citation]. *Proc Natl Acad Sci USA* 97:5562–5567.
99. Pongs, O. and J. R. Schwarz. 2010. Ancillary subunits associated with voltage-dependent K^+ channels. *Physiol Rev* 90:755–796.
100. Sun, X., M. A. Zaydman, and J. Cui. 2012. Regulation of voltage-activated K(+) channel gating by transmembrane beta subunits. *Front Pharmacol* 3:63.
101. Uebele, V. N., A. Lagrutta, T. Wade, D. J. Figueroa, Y. Liu, E. McKenna, C. P. Austin, P. B. Bennett, and R. Swanson. 2000. Cloning and functional expression of two families of beta-subunits of the large conductance calcium-activated K^+ channel. *J Biol Chem* 275:23211–23218.
102. Bao, L. and D. H. Cox. 2005. Gating and ionic currents reveal how the BKCa channel's Ca^{2+} sensitivity is enhanced by its beta1 subunit. *J Gen Physiol* 126:393–412.
103. Wang, B. and R. Brenner. 2006. An S6 mutation in BK channels reveals {beta}1 subunit effects on intrinsic and voltage-dependent gating. *J Gen Physiol* 128:731–744.
104. Yang, H., G. Zhang, J. Shi, U. S. Lee, K. Delaloye, and J. Cui. 2008. Subunit-specific effect of the voltage sensor domain on Ca^{2+} sensitivity of BK channels. *Biophys J* 94:4678–4687.
105. Contreras, G. F., A. Neely, O. Alvarez, C. Gonzalez, and R. Latorre. 2012. Modulation of BK channel voltage gating by different auxiliary beta subunits. *Proc Natl Acad Sci USA* 109:18991–18996.
106. Brenner, R., T. J. Jegla, A. Wickenden, Y. Liu, and R. W. Aldrich. 2000. Cloning and functional characterization of novel large conductance calcium-activated potassium channel beta subunits, hKCNMB3 and hKCNMB4. *J Biol Chem* 275:6453–6461.
107. McManus, O. B., L. M. Helms, L. Pallanck, B. Ganetzky, R. Swanson, and R. J. Leonard. 1995. Functional role of the beta subunit of high conductance calcium-activated potassium channels. *Neuron* 14:645–650.
108. Orio, P. and R. Latorre. 2005. Differential effects of beta 1 and beta 2 subunits on BK channel activity. *J Gen Physiol* 125:395–411.
109. Lee, U. S., J. Shi, and J. Cui. 2010. Modulation of BK channel gating by the ss2 subunit involves both membrane-spanning and cytoplasmic domains of Slo1. *J Neurosci* 30:16170–16179.
110. Wang, B., B. S. Rothberg, and R. Brenner. 2006. Mechanism of beta4 subunit modulation of BK channels. *J Gen Physiol* 127:449–465.
111. Brenner, R., Q. H. Chen, A. Vilaythong, G. M. Toney, J. L. Noebels, and R. W. Aldrich. 2005. BK channel beta4 subunit reduces dentate gyrus excitability and protects against temporal lobe seizures. *Nat Neurosci* 8:1752–1759.
112. Xia, X. M., J. P. Ding, and C. J. Lingle. 2003. Inactivation of BK channels by the NH2 terminus of the beta2 auxiliary subunit: An essential role of a terminal peptide segment of three hydrophobic residues. *J Gen Physiol* 121:125–148.
113. Adams, P. R., A. Constanti, D. A. Brown, and R. B. Clark. 1982. Intracellular Ca^{2+} activates a fast voltage-sensitive K^+ current in vertebrate sympathetic neurones. *Nature* 296:746–749.
114. Lancaster, B. and R. A. Nicoll. 1987. Properties of two calcium-activated hyperpolarizations in rat hippocampal neurones. *J Physiol (Lond)* 389:187–203.

115. Storm, J. F. 1987. Action potential repolarization and a fast after-hyperpolarization in rat hippocampal pyramidal cells. *J Physiol (Lond)* 385:733–759.
116. Roberts, W. M., R. A. Jacobs, and A. J. Hudspeth. 1990. Colocalization of ion channels involved in frequency selectivity and synaptic transmission at presynaptic active zones of hair cells. *J Neurosci* 10:3664–3684.
117. Robitaille, R. and M. P. Charlton. 1992. Presynaptic calcium signals and transmitter release are modulated by calcium-activated potassium channels. *J Neurosci* 12:297–305.
118. Robitaille, R., M. L. Garcia, G. J. Kaczorowski, and M. P. Charlton. 1993. Functional colocalization of calcium and calcium-gated potassium channels in control of transmitter release. *Neuron* 11:645–655.
119. Brayden, J. E. and M. T. Nelson. 1992. Regulation of arterial tone by activation of calcium-dependent potassium channels. *Science* 256:532–535.
120. Wellman, G. C. and M. T. Nelson. 2003. Signaling between SR and plasmalemma in smooth muscle: Sparks and the activation of Ca^{2+}-sensitive ion channels. *Cell Calcium* 34:211–229.
121. Pluger, S., J. Faulhaber, M. Furstenau, M. Lohn, R. Waldschutz, M. Gollasch, H. Haller, F. C. Luft, H. Ehmke, and O. Pongs. 2000. Mice with disrupted BK channel beta1 subunit gene feature abnormal Ca^{2+} spark/STOC coupling and elevated blood pressure. *Circ Res* 87:E53–E60.
122. Nelson, M. T., H. Cheng, M. Rubart, L. F. Santana, A. D. Bonev, H. J. Knot, and W. J. Lederer. 1995. Relaxation of arterial smooth muscle by calcium sparks. *Science* 270:633–637.
123. Tanaka, Y., M. Aida, H. Tanaka, K. Shigenobu, and L. Toro. 1998. Involvement of maxi-K_{Ca} channel activation in atrial natriuretic peptide-induced vasorelaxation. *Naunyn Schmiedebergs Arch Pharmacol* 357:705–708.
124. Perez, G. J., A. D. Bonev, J. B. Patlak, and M. T. Nelson. 1999. Functional coupling of ryanodine receptors to K_{Ca} channels in smooth muscle cells from rat cerebral arteries. *J Gen Physiol* 113:229–238.
125. Ghatta, S., D. Nimmagadda, X. Xu, and S. T. O'Rourke. 2006. Large-conductance, calcium-activated potassium channels: Structural and functional implications. *Pharmacol Ther* 110:103–116.
126. Nilius, B. and G. Droogmans. 2001. Ion channels and their functional role in vascular endothelium. *Physiol Rev* 81:1415–1459.
127. Isaacson, J. S. and G. J. Murphy. 2001. Glutamate-mediated extrasynaptic inhibition: Direct coupling of NMDA receptors to Ca(2+)-activated K^+ channels. *Neuron* 31:1027–1034.
128. Fernandez-Fernandez, J. M., Y. N. Andrade, M. Arniges, J. Fernandes, C. Plata, F. Rubio-Moscardo, E. Vazquez, and M. A. Valverde. 2008. Functional coupling of TRPV4 cationic channel and large conductance, calcium-dependent potassium channel in human bronchial epithelial cell lines. *Pflugers Arch* 457:149–159.
129. Kim, E. Y., C. P. Alvarez-Baron, and S. E. Dryer. 2009. Canonical transient receptor potential channel (TRPC)3 and TRPC6 associate with large-conductance Ca^{2+}-activated K^+ (BKCa) channels: Role in BKCa trafficking to the surface of cultured podocytes. *Mol Pharmacol* 75:466–477.
130. Zhao, G., Z. P. Neeb, M. D. Leo, J. Pachuau, A. Adebiyi, K. Ouyang, J. Chen, and J. H. Jaggar. 2010. Type 1 IP3 receptors activate BKCa channels via local molecular coupling in arterial smooth muscle cells. *J Gen Physiol* 136:283–291.
131. Weaver, A. K., M. L. Olsen, M. B. McFerrin, and H. Sontheimer. 2007. BK channels are linked to inositol 1,4,5-triphosphate receptors via lipid rafts: A novel mechanism for coupling [Ca(2+)] (i) to ion channel activation. *J Biol Chem* 282:31558–31568.
132. Lifshitz, L. M., J. D. Carmichael, F. A. Lai, V. Sorrentino, K. Bellve, K. E. Fogarty, and R. ZhuGe. 2011. Spatial organization of RYRs and BK channels underlying the activation of STOCs by Ca(2+) sparks in airway myocytes. *J Gen Physiol* 138:195–209.
133. Berkefeld, H., C. A. Sailer, W. Bildl, V. Rohde, J. O. Thumfart, S. Eble, N. Klugbauer et al. 2006. BKCa-Cav channel complexes mediate rapid and localized Ca^{2+}-activated K^+ signaling. *Science* 314:615–620.
134. Marrion, N. V. and S. J. Tavalin. 1998. Selective activation of Ca^{2+}-activated K^+ channels by co-localized Ca^{2+} channels in hippocampal neurons. *Nature* 395:900–905.
135. Sausbier, U., M. Sausbier, C. A. Sailer, C. Arntz, H. G. Knaus, W. Neuhuber, and P. Ruth. 2006. Ca^{2+} -activated K^+ channels of the BK-type in the mouse brain. *Histochem Cell Biol* 125:725–741.
136. Knaus, H. G., C. Schwarzer, R. O. Koch, A. Eberhart, G. J. Kaczorowski, H. Glossmann, F. Wunder, O. Pongs, M. L. Garcia, and G. Sperk. 1996. Distribution of high-conductance Ca(2+)-activated K^+ channels in rat brain: Targeting to axons and nerve terminals. *J Neurosci* 16:955–963.
137. Sailer, C. A., W. A. Kaufmann, M. Kogler, L. Chen, U. Sausbier, O. P. Ottersen, P. Ruth, M. J. Shipston, and H. G. Knaus. 2006. Immunolocalization of BK channels in hippocampal pyramidal neurons. *Eur J Neurosci* 24:442–454.
138. Hu, H., L. R. Shao, S. Chavoshy, N. Gu, M. Trieb, R. Behrens, P. Laake et al. 2001. Presynaptic Ca^{2+}-activated K^+ channels in glutamatergic hippocampal terminals and their role in spike repolarization and regulation of transmitter release. *J Neurosci* 21:9585–9597.
139. Misonou, H., M. Menegola, L. Buchwalder, E. W. Park, A. Meredith, K. J. Rhodes, R. W. Aldrich, and J. S. Trimmer. 2006. Immunolocalization of the Ca^{2+}-activated K^+ channel Slo1 in axons and nerve terminals of mammalian brain and cultured neurons. *J Comp Neurol* 496:289–302.
140. Benhassine, N. and T. Berger. 2005. Homogeneous distribution of large-conductance calcium-dependent potassium channels on soma and apical dendrite of rat neocortical layer 5 pyramidal neurons. *Eur J Neurosci* 21:914–926.
141. Raffaelli, G., C. Saviane, M. H. Mohajerani, P. Pedarzani, and E. Cherubini. 2004. BK potassium channels control transmitter release at CA3-CA3 synapses in the rat hippocampus. *J Physiol* 557:147–157.
142. Grimes, W. N., W. Li, A. E. Chavez, and J. S. Diamond. 2009. BK channels modulate pre- and postsynaptic signaling at reciprocal synapses in retina. *Nat Neurosci* 12:585–592.
143. Meredith, A. L., S. W. Wiler, B. H. Miller, J. S. Takahashi, A. A. Fodor, N. F. Ruby, and R. W. Aldrich. 2006. BK calcium-activated potassium channels regulate circadian behavioral rhythms and pacemaker output. *Nat Neurosci* 9:1041–1049.
144. Chen, X., Y. Kovalchuk, H. Adelsberger, H. A. Henning, M. Sausbier, G. Wietzorrek, P. Ruth, Y. Yarom, and A. Konnerth. 2010. Disruption of the olivo-cerebellar circuit by Purkinje neuron-specific ablation of BK channels. *Proc Natl Acad Sci USA* 107:12323–12328.
145. Sausbier, M., H. Hu, C. Arntz, S. Feil, S. Kamm, H. Adelsberger, U. Sausbier et al. 2004. Cerebellar ataxia and Purkinje cell dysfunction caused by Ca^{2+}-activated K^+ channel deficiency. *Proc Natl Acad Sci USA* 101:9474–9478.
146. Nelson, M. T. and J. M. Quayle. 1995. Physiological roles and properties of potassium channels in arterial smooth muscle. *Am J Physiol* 268:C799–C822.
147. Ledoux, J., M. E. Werner, J. E. Brayden, and M. T. Nelson. 2006. Calcium-activated potassium channels and the regulation of vascular tone. *Physiology (Bethesda)* 21:69–78.

148. Brenner, R., G. J. Perez, A. D. Bonev, D. M. Eckman, J. C. Kosek, S. W. Wiler, A. J. Patterson, M. T. Nelson, and R. W. Aldrich. 2000. Vasoregulation by the beta1 subunit of the calcium-activated potassium channel. *Nature* 407:870–876.
149. Senti, M., J. M. Fernandez-Fernandez, M. Tomas, E. Vazquez, R. Elosua, J. Marrugat, and M. A. Valverde. 2005. Protective effect of the KCNMB1 E65K genetic polymorphism against diastolic hypertension in aging women and its relevance to cardiovascular risk. *Circ Res* 97:1360–1365.
150. Fernandez-Fernandez, J. M., M. Tomas, E. Vazquez, P. Orio, R. Latorre, M. Senti, J. Marrugat, and M. A. Valverde. 2004. Gain-of-function mutation in the KCNMB1 potassium channel subunit is associated with low prevalence of diastolic hypertension. *J Clin Invest* 113:1032–1039.
151. Burdyga, T. and S. Wray. 2005. Action potential refractory period in ureter smooth muscle is set by Ca sparks and BK channels. *Nature* 436:559–562.
152. Meredith, A. L., K. S. Thorneloe, M. E. Werner, M. T. Nelson, and R. W. Aldrich. 2004. Overactive bladder and incontinence in the absence of the BK large conductance Ca^{2+}-activated K^+ channel. *J Biol Chem* 279:36746–36752.
153. Du, W., J. F. Bautista, H. Yang, A. Diez-Sampedro, S. A. You, L. Wang, P. Kotagal et al. 2005. Calcium-sensitive potassium channelopathy in human epilepsy and paroxysmal movement disorder. *Nat Genet* 37:733–738.
154. N'Gouemo, P. 2011. Targeting BK (big potassium) channels in epilepsy. *Expert Opin Ther Targets* 15:1283–1295.
155. Laumonnier, F., S. Roger, P. Guerin, F. Molinari, R. M'Rad, D. Cahard, A. Belhadj et al. 2006. Association of a functional deficit of the BKCa channel, a synaptic regulator of neuronal excitability, with autism and mental retardation. *Am J Psychiatry* 163:1622–1629.
156. Gribkoff, V. K., J. E. Starrett, Jr., S. I. Dworetzky, P. Hewawasam, C. G. Boissard, D. A. Cook, S. W. Frantz et al. 2001. Targeting acute ischemic stroke with a calcium-sensitive opener of maxi-K potassium channels. *Nat Med* 7:471–477.
157. Seibold, M. A., B. Wang, C. Eng, G. Kumar, K. B. Beckman, S. Sen, S. Choudhry et al. 2008. An african-specific functional polymorphism in KCNMB1 shows sex-specific association with asthma severity. *Hum Mol Genet* 17:2681–2690.
158. Jiao, H., P. Arner, J. Hoffstedt, D. Brodin, B. Dubern, S. Czernichow, F. van't Hooft et al. 2011. Genome wide association study identifies KCNMA1 contributing to human obesity. *BMC Med Genomics* 4:51.
159. Sontheimer, H. 2008. An unexpected role for ion channels in brain tumor metastasis. *Exp Biol Med* (Maywood) 233:779–791.
160. Weaver, A. K., X. Liu, and H. Sontheimer. 2004. Role for calcium-activated potassium channels (BK) in growth control of human malignant glioma cells. *J Neurosci Res* 78:224–234.
161. Pacheco Otalora, L. F., E. F. Hernandez, M. F. Arshadmansab, S. Francisco, M. Willis, B. Ermolinsky, M. Zarei, H. G. Knaus, and E. R. Garrido-Sanabria. 2008. Down-regulation of BK channel expression in the pilocarpine model of temporal lobe epilepsy. *Brain Res* 1200:116–131.
162. Misonou, H. and J. S. Trimmer. 2009. The diverse roles of K^+ channels in brain Ischemia. In *New Strategies in Stroke Intervention*. L. Annunziato, ed. Humana Press, New York. pp. 211–224.
163. Runden-Pran, E., F. M. Haug, J. F. Storm, and O. P. Ottersen. 2002. BK channel activity determines the extent of cell degeneration after oxygen and glucose deprivation: A study in organotypical hippocampal slice cultures. *Neuroscience* 112:277–288.
164. Jensen, B. S. 2002. BMS-204352: A potassium channel opener developed for the treatment of stroke. *CNS Drug Rev* 8:353–360.
165. Carvalho-de-Souza, J. L., W. A. Varanda, R. C. Tostes, and A. Z. Chignalia. 2013. BK channels in cardiovascular diseases and aging. *Aging Dis* 4:38–49.
166. Holtzclaw, J. D., P. R. Grimm, and S. C. Sansom. 2011. Role of BK channels in hypertension and potassium secretion. *Curr Opin Nephrol Hypertens* 20:512–517.
167. Rieg, T., V. Vallon, M. Sausbier, U. Sausbier, B. Kaissling, P. Ruth, and H. Osswald. 2007. The role of the BK channel in potassium homeostasis and flow-induced renal potassium excretion. *Kidney Int* 72:566–573.
168. Grimm, P. R., D. L. Irsik, D. C. Settles, J. D. Holtzclaw, and S. C. Sansom. 2009. Hypertension of Kcnmb1-/- is linked to deficient K secretion and aldosteronism. *Proc Natl Acad Sci USA* 106:11800–11805.
169. Amberg, G. C., A. D. Bonev, C. F. Rossow, M. T. Nelson, and L. F. Santana. 2003. Modulation of the molecular composition of large conductance, Ca(2+) activated K(+) channels in vascular smooth muscle during hypertension. *J Clin Invest* 112:717–724.
170. Amberg, G. C. and L. F. Santana. 2003. Downregulation of the BK channel beta1 subunit in genetic hypertension. *Circ Res* 93:965–971.
171. Salkoff, L., A. Butler, G. Ferreira, C. Santi, and A. Wei. 2006. High-conductance potassium channels of the SLO family. *Nat Rev Neurosci* 7:921–931.
172. Weiger, T. M., A. Hermann, and I. B. Levitan. 2002. Modulation of calcium-activated potassium channels. *J Comp Physiol A Neuroethol Sens Neural Behav Physiol* 188:79–87.
173. Fury, M., S. O. Marx, and A. R. Marks. 2002. Molecular BKology: The study of splicing and dicing. *Sci STKE* 2002:PE12.
174. Long, S. B., X. Tao, E. B. Campbell, and R. MacKinnon. 2007. Atomic structure of a voltage-dependent K^+ channel in a lipid membrane-like environment. *Nature* 450:376–382.
175. Jiang, Y., A. Lee, J. Chen, M. Cadene, B. T. Chait, and R. MacKinnon. 2002. Crystal structure and mechanism of a calcium-gated potassium channel. *Nature* 417:515–522.

18 Inward rectifying potassium channels

Monica Sala-Rabanal and Colin G. Nichols

Contents

18.1 INTRODUCTION

Empirically, rectification means change of conductance with voltage, and most, if not all, ion channels show some degree of rectification. For potassium channels, inward rectification means that, given an equal but opposite electrochemical driving force, the inward flow of K^+ ions exceeds the outward flow. Strongly inwardly rectifying K^+ currents were first observed in frog skeletal muscle (Katz 1949) and termed *anomalous* to distinguish them from the *normal* or *delayed* outward rectification that is characteristic of K^+ currents in the squid axon, and expected for an electrodiffusive pathway in normal physiological $[K^+]$ gradients (Hodgkin et al. 1949). We now know that inward rectification is a consequence of voltage-dependent, asymmetric open channel pore block by cytoplasmic divalent cations, especially Mg^{2+}, and polyamines (Guo et al. 2003, Kurata et al. 2007, Lopatin et al. 1994, Lu and MacKinnon 1994, Matsuda et al. 1987, Tao et al. 2009, Vandenberg 1987, Yang et al. 1995). At hyperpolarizing membrane potentials, the blocking ions are cleared from the channel and K^+ currents flow, whereas at depolarizing voltages, the blockers are driven into the pore and the K^+ current is blocked (Figure 18.1a). Consequently, strongly rectifying Kir channels are conductive when excitable cells are at rest and nonconductive during excitation, and this property grants Kir channels a key role in the maintenance and modulation of cell membrane potential. The first Kir channel genes were isolated by expression cloning in 1993 (Ho et al. 1993, Kubo et al. 1993), and there are now at least seventeen eukaryotic Kir channel genes distributed in seven subfamilies (Kir1–7; Figure 18.1b), and multiple prokaryotic isoforms (KirBac1.1–9). Since their discovery, Kir channels have been found in many cell types, with widely varying rectification properties. For example, classical, strong inward rectifying K^+ currents are very prominent in cardiac myocytes, and in glial cells and neurons in the central nervous system (Brismar and Collins 1989, Hestrin 1987, Nakajima et al. 1988, Newman 1993, Vandenberg 1994). Rectification of these channels is sufficiently strong that very little current flows through them at potentials positive to about −40 mV; high conductance at negative voltages allows cells to maintain a stable resting potential, but the greatly reduced conductance at positive voltages avoids short-circuiting the action potential. ATP-sensitive K^+ (K_{ATP}) channels are present in all muscle cell types, in the brain, and in pancreatic cells (Ashcroft 1988). In contrast to classical inward rectifiers, K_{ATP} channels display only *weak* rectification and allow substantial outward current to flow at positive potentials (Nichols and Lederer 1991, Noma 1983). Between these extremes, the

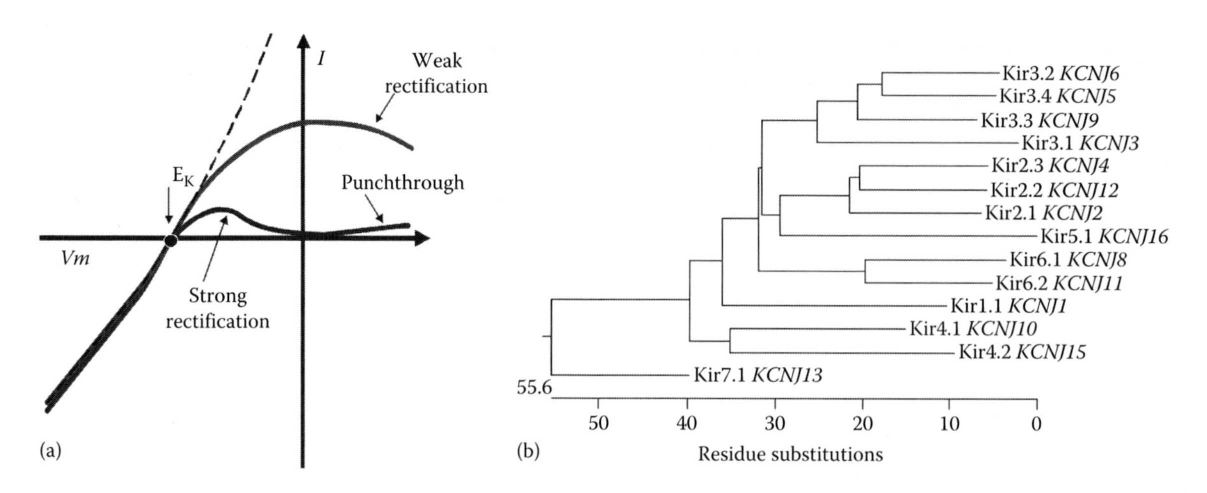

Figure 18.1 (a) Inward rectification. Under physiological conditions, conductance declines upon depolarization, due primarily to channel block by polyamines and Mg^{2+} ions. In weak inward rectifiers, this decline is very weakly voltage dependent. In strong inward rectifiers, the decline in conductance results in a marked *negative slope* region, which may be followed by a second region of positive conductance due to *punchthrough* of the blocking ions. (b) Phylogeny of Kir channels. There are seven major subfamilies of eukaryotic Kir genes, encoding subunits with ~60% amino acid identity within and ~40% identity between subfamilies, and generating various native inward rectifying channels as indicated.

brain is particularly well endowed with K^+ channels that show intermediate rectification properties, many of these channels being strongly dependent on ligand activation, often through G-proteins or other second messenger systems (Inanobe et al. 1999b, Lesage et al. 1995, Liao et al. 1996). Crystallization of the bacterial KcsA channel revolutionized our understanding of channel structure (Doyle et al. 1998), as it unveiled a basic architectural scaffold that is common to all major cation channel family members, including Kir channels, namely, two transmembrane helices bridged by an extracellular loop that generates the narrow portion of the pore and controls ion selectivity. Solving of the structure of full-length prokaryotic Kir channel homologs KirBac1.1 and KirBac3.1 (Kuo et al. 2003, 2005) revealed an extended cytoplasmic pore and the likely location of ligand binding sites, and the unique properties of eukaryotic Kir channels, as anticipated by decades of biochemical and biophysical experiments, were given a solid structural context with the crystallization of linked N- and C-terminal domains of Kir3.1 and Kir2.1 (Nishida and MacKinnon 2002, Pegan et al. 2005b) and, more recently, of the full-length Kir2.2 (Hansen et al. 2011, Tao et al. 2009) and Kir3.2 (Whorton and Mackinnon 2011).

18.2 KIR CHANNEL STRUCTURE AND GATING

Functional Kir channels are formed by the homomeric or heteromeric assembly of four subunits, each Kir subunit containing two membrane-spanning domains (TM1 and TM2), an extracellular pore-forming region (H5) that includes the selectivity filter, and cytosolic NH_2- and COOH-termini that associate with each other to form a cytoplasmic domain that controls gating (Figure 18.2) (Nishida et al. 2007, Whorton and MacKinnon 2011).

The Kir channel transmembrane domain is formed by outer (TM1) and inner (TM2) helices, the slide helix, and the pore helix (Figure 18.2). The selectivity filter, containing the signature sequence TXGY(F)G (where X is an aliphatic amino acid), is shared with other K^+-selective ion channels (Bichet et al. 2003) and forms a constriction in the conduction pathway that separates the central cavity from the extracellular solution. In eukaryotic Kir channels, the pore region is structurally constrained through covalent linkage by an absolutely conserved pair of Cys residues between the pore

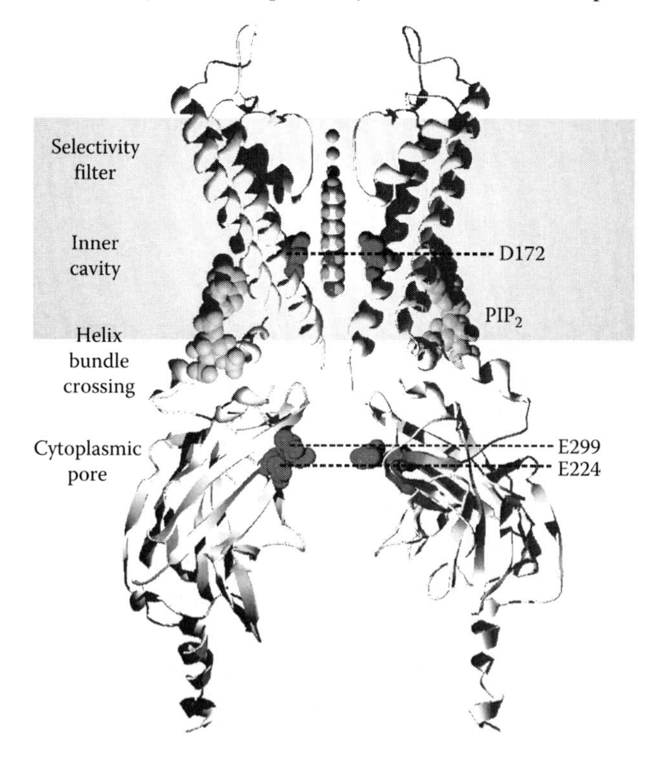

Figure 18.2 Kir channel structure. Ribbon diagram of two opposing Kir subunits. The channel is divided into two distinct domains, the transmembrane domain formed by the M1 and M2 helices and the pore loop (generating the inner cavity and selectivity filter, respectively), and the cytoplasmic domain formed by the N- and C-termini, lining the cytoplasmic extension of the pore. Conserved locations of pore-lining negatively charged residues (corresponding to D172, E224 and E299 in Kir2.1) are shown in ball-and-stick format. Also shown is a spermine molecule within the channel pore, at a location envisioned by the *fixed-tail* model of polyamine block, in which a polyamine is stabilized by interactions with charged residues in the inner cavity and the selectivity filter.

helix and the selectivity filter (e.g., C123 and C155 in Kir2.2). The central cavity is lined by the TM2 helix from each of the four Kir subunits, in an arrangement very similar to that of other K^+ channel families, including KcsA and bacterial and mammalian K_V channels (Doyle et al. 1998, Jiang et al. 2003, Long et al. 2005). Beneath the central cavity, large hydrophobic residues (such as I177 and M181 in Kir2.2) form two tight seals that close off the pore to the cytoplasm (Tao et al. 2009). Both rearrangements of the selectivity filter and motion of the TM2 helices have been proposed to occur, and may coexist, in Kir channel gating: mutations introduced in or around the selectivity filter modify the fast gating kinetics (Choe et al. 2001, Enkvetchakul et al. 2000, Proks et al. 2001, Yang et al. 1995), whereas mutations in the region where the bundle of inner TM2 helices cross at the bottom of the inner cavity have been shown to alter slow gating (Enkvetchakul and Nichols 2003, Enkvetchakul et al. 2000, Trapp et al. 1998, Yi et al. 2001); it has been hypothesized that bending of each TM2 around a hinge Gly residue halfway through the membrane would allow for the inner part of the helix to swing away from the permeation pathway (Jin et al. 2002). The TM1 domains are also implicated in the gating machinery of Kir channels: hydrogen bond linking between TM1 and TM2 at the bundle-crossing region modulates gating by intracellular protons and PIP_2 (Rapedius et al. 2007), and the slide helix, a transverse amphiphilic domain at the cytoplasmic end of TM1, serves as a mediator of coupling between the ligand-sensing, cytoplasmic interface, and the gating transmembrane domains (Doyle et al. 1998, Kuo et al. 2003).

The cytoplasmic domain of Kir channels, formed by the association of the NH_2- and COOH-termini of the four channel subunits (Figure 18.2), lines an extension of the channel pore into the cytoplasm and effectively doubles the conduction pathway length to over 60 Å. This domain also allows for the control of Kir channel gating by ATP, Na^+, nucleotides, G-proteins, and phosphatidylinositol-4,5-bisphosphate (PIP_2) (Nishida and MacKinnon 2002, Pegan et al. 2005a). For example, all Kir channels require PIP_2 for activity, and recent crystallographic and biochemical data suggest that the binding site for PIP_2 is formed predominantly by positively charged side chains at the interface between the transmembrane helices and the cytoplasmic domain (D'Avanzo et al. 2013, Hansen et al. 2011, Whorton and Mackinnon 2011). PIP_2 interactions determine the sensitivity of Kir channels to regulation by cytoplasmic factors such as ATP, phosphorylation, and pH, and differences among Kir channels in their specific regulation by a given modulator may reflect differences in their apparent affinity for PIP_2 (Du et al. 2004). In K_{ATP} channels, the ATP-binding pocket is formed at the cytoplasmic domain interfaces in Kir6 subunits (Antcliff et al. 2005), and in Kir1.1 and Kir4.1 channels, intracellular acidification disrupts an intersubunit cytoplasmic salt bridge (E302-R311 in Kir1.1; E288-R297 in Kir4.1), which contributes to channel closing (Rapedius et al. 2006, Sala-Rabanal et al. 2010).

18.3 MECHANISMS OF INWARD RECTIFICATION

Block by intracellular Mg^{2+} and polyamines underlies the Kir channel key functional property of preferential conduction of inward K^+ currents (Fakler et al. 1995, Ficker et al. 1994, Lopatin et al. 1994, 1995). As a rapid, voltage-dependent process, polyamine-mediated inward rectification provides a mechanism for K^+ current regulation in excitable tissues, which contributes to shaping both the action potential and the resting membrane potential in tissues such as the myocardium (Bianchi et al. 1996, Lopatin et al. 2000, Priori et al. 2005). Polyamine block entails an initial weakly voltage-dependent binding, probably in the cytoplasmic domain of the channel, followed by a steeply voltage-dependent step in which the blocker migrates to a stable binding site in the inner cavity (Guo and Lu 2000, Kurata et al. 2007, Shin and Lu 2005, Xie et al. 2002). Inward rectification is strongly voltage dependent, and the voltage at which Kir channels transition from a conductive state to a blocked state becomes more positive as the extracellular K^+ concentration is increased, reflecting competition between conducting ions, polyamines, and Mg^{2+} for sites in the pore (Guo et al. 2003, Pearson and Nichols 1998, Shin and Lu 2005, Tao et al. 2009). Residues critically involved in each step have been identified: mutations that affect the shallow binding step (e.g., E224 and E299 in the strong rectifier Kir2.1) cluster in the cytoplasmic domain, while the rectification controller residue (D172 in Kir2.1), critical for steep voltage-dependent block, lies at a pore-lining position in the channel inner cavity (Guo and Lu 2003, Kubo and Murata 2001, Kurata et al. 2007, Wible et al. 1994, Xie et al. 2003, Yang et al. 1995), and there is now substantial evidence that the polyamines bind between the rectification controller residue and the selectivity filter (Kurata et al. 2004, 2006, 2008, 2009, 2010).

Kir channel subtypes show differing strengths of inward rectification, and this is critically dependent on the presence or absence of a negatively charged residue at the *rectification controller* position (equivalent to D172 in Kir2.1) in the inner cavity: the Asp is conserved in strongly rectifying Kir3.x channels and only replaced by Glu in intermediate rectifiers such as Kir4.x, whereas in the weakly rectifying Kir1.1 and Kir6.2, this position is occupied by uncharged Asn residues (Kucheryavykh et al. 2007b, Kurata et al. 2004, Stanfield et al. 1994, Wible et al. 1994).

18.4 Kir CHANNEL FAMILIES

To date, seven subfamilies of Kir channel subunits have been identified, each sharing ~40% amino acid identity between one another, and ~60% identity between individual members within each subfamily (Figure 18.1b). They can be classified into four groups according to their functional and physiological properties: classical Kir channels (Kir2 subfamily members) are strong rectifiers highly expressed in the heart, where they are the key players in the cardiac inward rectifier current, I_{K1}; G-protein-gated Kir channels, or K_G channels (Kir3.x), underlie G-protein-coupled receptor-activated currents in the heart, brain, and endocrine tissues; ATP-sensitive K^+ channels, or K_{ATP} channels, are formed by the assembly of four pore-forming Kir6.x subunits and four regulatory sulfonylurea subunits (SURx), and are tightly coupled to cellular metabolism, and K^+ transport channels (Kir1.x, Kir4.x, Kir5.x, and Kir7.x) are intermediate or weak rectifiers predominantly expressed in barrier epithelia and in glial cells, where they contribute to K^+ reuptake and membrane repolarization.

18.4.1 CLASSIC INWARD RECTIFIERS: Kir2.x

Kir2 subunits generate channels that are constitutively active, strong inward rectifiers, and are robustly expressed in the heart and nervous system, where they are essential for the modulation of cell excitability. Five distinct mammalian Kir2 subfamily members have been identified: Kir2.1 (encoded by gene *KCNJ2*) was the first member of the subfamily to be cloned (Kubo et al. 1993), followed by Kir2.2 (*KCNJ12*) (Takahashi et al. 1994), Kir2.3 (*KCNJ4*) (Morishige et al. 1994), Kir2.4 (*KCNJ14*) (Topert et al. 1998), and, more recently, Kir2.6 (*KCNJ18*) (Ryan et al. 2010). Kir2.1 is the predominant isoform in skeletal muscle, and in atrial and ventricular myocytes where it associates with Kir2.2 or Kir2.3 to underlie the strong inwardly rectifying I_{K1} currents (Hibino et al. 2010a); the central role of I_{K1} currents in cardiac electrophysiology is manifested by severe arrhythmias linked to Kir2.x channel gain- or loss-of-function mutations, such as the Andersen–Tawil syndrome, certain forms of short QT syndrome (SQTS), tachycardia, and atrial fibrillation (AF) (Anumonwo and Lopatin 2010). In vascular endothelial cells, classical Kir channels, mostly made up of Kir2.2 subunits, have a prominent role in modulation of vascular tone (Adams and Hill 2004, Fang et al. 2005), and in arterial smooth muscle cells in the brain, K^+ reuptake by Kir2.1 contributes to vasodilation and the control of local blood flow (Bradley et al. 1999, Filosa et al. 2006). Kir2.x subunits are differentially expressed in the central nervous system, where classical Kir currents are critically involved in the regulation of neuronal excitability: Kir2.1 is expressed diffusely throughout the brain; Kir2.2 strongly in the cerebellum; Kir2.3 in spinal cord, forebrain, and olfactory bulb; and Kir2.4 in the midbrain, pons, and medulla (de Boer et al. 2010). Kir2.6 is transcriptionally regulated by thyroid hormone and controls skeletal muscle cell excitability; loss-of-function mutations in Kir2.6 channels have been associated with thyrotoxic hypokalemic periodic paralysis (Ryan et al. 2010).

18.4.1.1 I_{K1} channel assembly and regulation

I_{K1} currents dominate the background conductance and establish the negative resting potential of ventricular cardiomyocytes, including Purkinje fibers, and play a critical role in shaping the cardiac action potential; the lack of outward conductance at positive potentials prevents K^+ efflux and helps maintain the depolarization during the characteristic elongated plateau phase. Once repolarization is initiated by the activation of K_V channels, large outward currents pass through I_{K1} channels, which accelerate the final stages of repolarization (Anumonwo and Lopatin 2010, Hibino et al. 2010a).

While it is well established that members of the Kir2 subfamily underlie I_{K1} currents, subunit composition varies among species, cell types, and membrane domains. In the human heart, for example, Kir2.1 and Kir2.3 dominate over Kir2.2 in Purkinje cells; in the right ventricle, Kir2.1 is the most abundant subunit, whereas Kir2.3 is prominent in the right atrium; and Kir2.4 is not expressed (Gaborit et al. 2007). Kir2.2 appears to be absent in guinea pig and sheep cardiomyocytes (Dhamoon et al. 2004), and very little Kir2.3 is found in the mouse heart (Zaritsky et al. 2001). The molecular makeup of I_{K1} channels has been assessed in knock-out animal models, native cells, and heterologous expression systems based on the differential biophysical properties of individual Kir2 channel assemblies, but due to the tissue- and species-specific distribution, in many cases, the exact contribution of Kir2.x subunits to the conductance of heteromeric channels is still unknown. There is solid evidence, however, that in all species, Kir2.1 is the core subunit of cardiac I_{K1} channels; in mouse and rabbit heart, I_{K1} is generated by heteromeric Kir2.1/Kir2.2 channels (Zaritsky et al. 2001; Zobel et al. 2003); and in human cardiomyocytes, I_{K1} currents likely result from heteromultimerization of Kir2.1, Kir2.2, and Kir2.3 subunits (Schram et al. 2003).

I_{K1} currents are modulated by α and β adrenergic stimulation (Scherer et al. 2007, Zitron et al. 2008), intracellular Ca^{2+} (Zaza et al. 1998), and phosphorylation (Karle et al. 2002, Zitron et al. 2004). More notably, Kir2.x channels are selectively activated by PIP_2 and can be inhibited by other phosphoinositides (Cheng et al. 2011, D'Avanzo et al. 2010). Multiple positively charged residues in the cytoplasmic side of Kir2.1 have been found to control sensitivity to PIP_2 activation (Soom et al. 2001, Xie et al. 2008); recently, biochemical data and the atomic structures of Kir2.2 and PIP_2-bound Kir3.2 have revealed that some of these key residues cluster in one specific site just beyond the transmembrane segments, at the interface of the cytoplasmic NH_2- and COOH-termini (D'Avanzo et al. 2013, Tao et al. 2009, Whorton and Mackinnon 2011). Kir2 channels are constitutively closed in the absence of PIP_2; PIP_2 strengthens N- and C-termini interactions by tethering them to the cytoplasmic side of the membrane, thus mechanically opening the pore (D'Avanzo et al. 2013, Enkvetchakul and Nichols 2003, Xie et al. 2008). The cholesterol content of cell membranes has been shown to affect human Kir2.1 channel function, which could have pathophysiological implications (D'Avanzo et al. 2011).

18.4.1.2 Pathophysiology of I_{K1} channels

Atrial I_{K1} currents are upregulated in patients with chronic AF (Dobrev et al. 2002), and in congestive heart failure, atrial and ventricular I_{K1} current density may be moderately reduced (Anumonwo and Lopatin 2010). I_{K1} appears to be implicated in protective ischemic preconditioning after acute hypoxia or cyanide poisoning, by action potential shortening and early repolarization (Piao et al. 2007). More notably, four channelopathies with severe cardiovascular and musculoskeletal manifestations that originate from loss- or gain-of-function mutations in Kir2.1 have been identified: type 1 Andersen–Tawil syndrome (ATS1), catecholaminergic polymorphic ventricular tachycardia (CPTV), short QT3 syndrome, and hereditary AF. ATS1 is characterized by cardiac electrophysiological abnormalities, including prolongation of the QT interval that can result in biventricular tachycardia and arrhythmias, and may lead to AF and sudden cardiac death; additionally, ATS1 patients are dysmorphic, with short stature, scoliosis, and cleft palate, as well as skeletal muscle weakness with episodes of paralysis (Sansone and Tawil 2007). Since the causal relationship between mutant human Kir2.1 channels and ATS1 was first established (Plaster et al. 2001), over 30 *KCNJ2* dominant-negative mutations associated to the syndrome have been described, most located on the N-terminal slide helix and the C-terminal cytoplasmic domain, and all resulting in disruption of Kir2.1 channel function through mechanisms that include abnormal trafficking, altered interaction with second messengers or gating, and incorrect folding of the Kir2.1 protein (Tristani-Firouzi and Etheridge 2010). Several of these mutations are located within the cluster of residues involved in PIP_2 sensitivity (Lopes et al. 2002). CPTV, on the other hand, presents with ventricular arrhythmias and sudden cardiac death associated with physical activity, but no dysmorphic features (Lehnart et al. 2007); in most cases, the symptoms have been linked to mutations in the ryanodine receptors (CPTV1 or CPTV2), but in a subset of CPTV patients, several novel mutations in the *KCNJ2* gene, leading to loss of Kir2.1 function by mechanisms distinct from those of ATS1 mutations, have been identified (Eckhardt et al. 2007, Vega et al. 2009).

SQTS is an inherited condition that predisposes to atrial and ventricular fibrillation and sudden death; the SQT1 and SQT2 variants of the syndrome are a result of gain-of-function mutations in K_V channel genes *KCNH2* (HERG channel) and *KCNQ* (I_{Ks} channel), respectively, but a third form of the syndrome, SQT3, arises from overactive Kir2.1 channels (Priori et al. 2005). Finally, enhanced I_{K1} current density in the atrial myocardium, by overexpression of Kir2.1 protein or due to a gain-of-function mutation in the *KCNJ2* gene (Girmatsion et al. 2009, Li et al. 2004, Xia et al. 2005), has been linked to familial AF, a condition that often becomes chronic due to sustained action potential shortening and appearance of reentry circuits (Ravens and Cerbai 2008).

18.4.2 G-PROTEIN-GATED KIR CHANNELS: Kir3.x

Activation of G-protein-coupled receptors (GPCRs) by hormones or neurotransmitter ligands results in the release of two intracellular molecules, Gα and Gβγ, which modulate multiple cellular processes. Kir3 channels, or K_G channels, are targets of GPCRs (Pfaffinger et al. 1985): in atrial cardiac myocytes, for example, muscarinic K^+ channels (K_{ACh}), generated by Kir3.1/3.4 subunits, are activated by GPCRs in response to acetylcholine and adenosine, resulting in slowing of the heart rate (Anumonwo and Lopatin 2010). K_G channels are also prominent in the brain, where they have been implicated in synaptic plasticity and in mechanisms underlying drug addiction (Chung et al. 2009, Robbins and Everitt 1999), and in endocrine pancreas, where they help to regulate hormone secretion (Sharp 1996).

18.4.2.1 K_G channel assembly, regulation, and pharmacology

Functional K_G channels are homomeric or heteromeric tetramers of Kir3.x channel subunits. Presently, four Kir3 subfamily members are known, namely, Kir3.1, Kir3.2, Kir3.3, and Kir3.4, and various combinations of Kir3 subunits form K_G channels of unique properties, finely tuned by their localization, protein–protein interactions, and modulation by intracellular factors. Kir3.1/GIRK1 (encoded by gene *KCNJ3*) was the first Kir subunit shown to participate in the formation of K_G channels, and interestingly, Kir3.1 channel subunits do not generate K_G currents on their own (Corey and Clapham 1998, Wischmeyer et al. 1997) but are generally incorporated into heteromers with other Kir3.x subunits to form K_G channels in native cells and tissues (Duprat et al. 1995, Velimirovic et al. 1996); most notably, cardiac K_{ACh} channels are formed by the heteromeric assembly of Kir3.1 and Kir3.4 (GIRK4/*KCNJ5*) subunits (Krapivinsky et al. 1995, Wickman et al. 1998). There are four different isoforms of Kir3.2/GIRK2, generated by alternative splicing of the *KCNJ6* gene (Wickman et al. 2002), which exhibit differential expression patterns in various tissues, such as brain and pancreas, where they form functional homomeric channels or heteromerize with Kir3.1 or Kir3.3 subunits (Inanobe et al. 1999a,b, Jelacic et al. 2000); on the other hand, two isoforms of Kir3.3/GIRK3 (*KCNJ9*) have been cloned, but their expression profile and functional differences have not been fully established (Jelacic et al. 1999, Lesage et al. 1994).

K_G channel opening is triggered by membrane-bound G-proteins (Kurachi et al. 1986), and after a long controversy, it has been established that K_G channels are activated only by the Gβγ subunits (Kofuji et al. 1995, Wickman and Clapham 1995), although the Gα subunits may play important regulatory roles (Yamada et al. 1994). In the absence of agonists, K_G channels associate with GDP-bound Gα and Gβγ to make the *preformed* complex, which effectively reduces their basal activity; when the GPCR is activated, the Gα is released and Gβγ activates the associated K_G channel (Nobles et al. 2005, Riven et al. 2006). A recent Kir3.2-Gβγ cocrystal structure (Whorton and MacKinnon 2013) confirms an extensive interaction surface between Gβγ and the cytoplasmic domain of K_G channel subunits, and the key roles of specific residues identified functionally including L344 and G347 in Kir3.2 (Finley et al. 2004) and H64 and L262 in Kir3.4 (Ivanina et al. 2003). Several G-protein-mediated signal pathways are negatively modulated by regulator of G-protein signaling (RGS) proteins by accelerating intrinsic GTP hydrolysis in the Gα subunit (Ross and Wilkie 2000), and this has been implicated in the control of K_G channel activity (Fujita et al. 2000, Inanobe et al. 2001, Saitoh et al. 1997). Like other Kir channels, K_G channels require PIP_2 to maintain their activity (Huang et al. 1998), but unlike in most Kir channels, PIP_2-dependent activation in K_G channels is strengthened by interactions with Gβγ subunits and intracellular Na^+ (Ho and Murrell-Lagnado 1999, Rosenhouse-Dantsker et al. 2008, Zhang et al. 1999). Additionally, K_G channels are modulated by redox signaling, which may contribute to protecting cells from hypoxic or ischemic insults (Zeidner et al. 2001), and are inhibited by extracellular acidification, which may be involved in the control of cellular respiration and CO_2 chemoreception in tissues such as the brain (Mao et al. 2003). Ethanol, methanol, and other *n*-alcohols can activate K_G channels containing Kir3.1, Kir3.2, and Kir3.4 subunits (Lewohl et al. 1999), and K_G channels made up of various combinations of Kir3.x subunits can be modulated by antipsychotic drugs (Kobayashi et al. 2000), antidepressants (Kobayashi et al. 2004), and volatile and local anesthetics (Weigl and Schreibmayer 2001, Zhou et al. 2001), in many cases, by interaction with specific locations in the cytoplasmic domain.

18.4.2.2 Physiology and pathophysiology of K_G channels

Kir3 proteins are expressed throughout the brain (Koyrakh et al. 2005, Wickman et al. 2000) in various homomeric or heteromeric complexes. At least Kir3.1/Kir3.2, Kir3.2/Kir3.3, and Kir3.2/Kir3.4 assemblies have been identified in various regions of the brain (Jelacic et al. 2000, Liao et al. 1996). Kir3.1 and Kir3.2 are found in presynaptic and postsynaptic termini, whereas Kir3.3 and Kir3.4 localize to axons and dendrites (Hibino et al. 2010a). Postsynaptic Kir3.2 is critically involved in the generation of slow inhibitory postsynaptic potentials that suppress neuronal excitability, and Kir3.2-deficient mice develop spontaneous seizures and are more susceptible to seizure-inducing drugs (Signorini et al. 1997). A mutation in the signature sequence of the selectivity filter of Kir3.2 in weaver mice results in nonselective cation permeation, which leads to a severe phenotype including locomotor defects, male sterility, deficiencies in the midbrain dopaminergic system, and neuronal death (Navarro et al. 1996, Patil et al. 1995). Heteromeric Kir3.1/Kir3.4 channels have been identified in the rat pituitary gland, where the action of thyrotropin-releasing hormone enhances dopamine- and somatostatin-induced inhibitory K_G currents, and this constitutes an effective mechanism for control of plasma hormone levels (Morishige et al. 1999). In the heart, the acetylcholine released from vagal nerve endings activates K_{ACh} channels composed by Kir3.1 and Kir3.4 subunits (Krapivinsky et al. 1995), and this causes an increase in K^+ efflux and membrane hyperpolarization that results in a deceleration of the heartbeat. I_{KACh} has been found to play a critical role in AF, as electrical remodeling in AF patients causes K_{ACh} channels to be constitutively active, and this leads to persistent action potential shortening (Dobrev et al. 2005). Conversely, K_{ACh} channels could be a target of for AF treatment and prevention, as I_{KACh} can be inhibited by classical antiarrhythmic agents, such as quinidine (Kurachi et al. 1986).

18.4.3 ATP-SENSITIVE Kir CHANNELS: Kir6.x/SURx

ATP-sensitive potassium (K_{ATP}) channels are expressed in most excitable tissues and serve to couple intracellular energetics to electrical activity (Nichols 2006). K_{ATP} channels are uniquely generated by complexes of four pore-forming Kir6 channel subunits with four auxiliary sulfonylurea receptors (SURs). K_{ATP} channels are activated by Mg^{2+}-bound nucleotides in states of low metabolism and inhibited by cytosolic ATP in states of high metabolism. Since being first described in ventricular myocytes (Noma 1983, Trube and Hescheler 1984), K_{ATP} channels have been found in tissues throughout the body, including pancreatic β-cells (Cook and Hales 1984), skeletal muscle (Spruce et al. 1985), visceral and vascular smooth muscle (Standen et al. 1989), and brain (Bernardi et al. 1988, Gehlert et al. 1990).

The roles of K_{ATP} channels in glucose homeostasis and ischemic protection are well established (Flagg et al. 2010, Koster et al. 2005a, Rorsman et al. 2008), and novel functions continue to emerge: recognized as protective against neural apoptosis following a stroke (Wind et al. 1997), brain K_{ATP} channels have been implicated in memory (Betourne et al. 2009) and in the regulation of male reproductive behavior (McDevitt et al. 2009). Mutations leading to aberrant K_{ATP} channel function have been linked to a variety of diseases, from neonatal diabetes to congenital hyperinsulinism (Denton and Jacobson 2012, Flanagan et al. 2009) and, more recently, to the Cantu syndrome (Nichols et al. 2013). Classic K_{ATP} channel openers (KCOs), such as diazoxide and pinacidil, have been used to treat hypertension, angina, and hyperinsulinism of infancy, while antagonists such as sulfonylureas are established antidiabetic agents (Doyle and Egan 2003, Wickenden 2002). The drug industry continues to exploit the tissue-specific pharmacology of K_{ATP} channels in the design of novel therapeutic agents aimed at endocrine, vascular, neurological, urological, and even dermatological disorders (Jahangir and Terzic 2005).

18.4.3.1 K_{ATP} channel architecture and assembly

K_{ATP} channel activity in heterologous expression systems requires both Kir6 and SUR subunits to be coexpressed in a 4:4 stoichiometry to generate the functional K_{ATP} channel (Clement et al. 1997, Inagaki et al. 1995a, 1997, Shyng and Nichols 1997). Similarly, biochemical studies demonstrate that the SUR2 protein variants, SUR2A and SUR2B, can also coassemble with Kir6 subunits (Babenko et al. 1998, Inagaki et al. 1996, Okuyama et al. 1998, Yamada et al. 1997), presumably in a similar octameric arrangement. How the Kir6 and SUR subunits are physically connected is still unknown, but electron micrography and intersubunit FRET studies of complete K_{ATP} channel assemblies suggest the tight packing of four SUR and four Kir6.x subunits (Mikhailov et al. 2005, Wang et al. 2013). The genes for Kir6.2 and SUR1 lie sequentially on human chromosome 11p15.1 (Inagaki et al. 1995b), suggesting an as yet unconsidered coregulation at the gene level. In addition, the genes for Kir6.1 and SUR2 are also adjacent on chromosome 12p12.1 (Chutkow et al. 1996, Inagaki et al. 1995b), which implicates an evolutionary duplication event. Classic pancreatic K_{ATP} channels are generated by Kir6.2 and SUR1 subunits (Inagaki et al. 1995b), and in vascular smooth muscle (VSM), the predominant K_{ATP} current is represented by Kir6.1/SUR2B channels (Beech et al. 1993, Farzaneh and Tinker 2008, Zhang and Bolton 1996). However, the combination of subunits extends beyond the pairing of Kir6.2 with SUR1 and Kir6.1 with SUR2; for example, Kir6.2 associates with SUR2A in cardiac and skeletal muscle (Inagaki et al. 1996). Although it is possible to construct functional channels with more than one SUR isoform or Kir6 isoform in recombinant expression (Chan et al. 2008, Cheng et al. 2008, Cui et al. 2001, Kono et al. 2000, Wheeler et al. 2008), the presence of heteromultimeric K_{ATP} channels in native tissues remains controversial (Pountney et al. 2001, Seharaseyon et al. 2000). In addition, alternative splicing of the *ABCC9* gene gives rise to SUR2A or SUR2B, which confer distinct physiological and pharmacological properties on the channel complex (Chutkow et al. 1999, Shi et al. 2005).

18.4.3.2 Regulation and pharmacology of K_{ATP} channels

The key regulatory features of K_{ATP} channels are rapid and reversible closure by cytoplasmic ATP, and activation by nucleotide tri- and diphosphates (Nichols 2006). In the absence of Mg^{2+}, nucleotides inhibit K_{ATP} channel activity through interaction with the Kir6 subunit (Drain et al. 1998, Tucker et al. 1998), but in the presence of Mg^{2+}, both ATP and ADP stimulate channel activity (Dunne and Petersen 1986, Hopkins et al. 1992, Kakei et al. 1986, Lederer and Nichols 1989), through interaction with the SUR subunit (Gribble et al. 1997, Nichols et al. 1996, Shyng et al. 1997). Membrane phospholipids, in particular PIP_2, potently stimulate K_{ATP} activity by binding the Kir6.2 subunit (Baukrowitz et al. 1998, Fan and Makielski 1997, Ribalet et al. 2005, Shyng and Nichols 1998, Xie et al. 2008), and phospholipase C, which hydrolyzes PIP_2, reduces K_{ATP} channel activity (Baukrowitz et al. 1998, Fan and Makielski 1997, Hilgemann and Ball 1996,

Xie et al. 1999). Importantly, there exists a negative coupling between PIP_2 activation and ATP sensitivity of K_{ATP} channels such that as PIP_2 increases, channel open probability increases and ATP sensitivity decreases (Enkvetchakul and Nichols 2003). Residues involved in PIP_2 binding and activation overlap with the ATP-binding site on Kir6.2, consistent with their competitive effects observed in binding assays (Enkvetchakul and Nichols 2003, MacGregor et al. 2002, Ribalet et al. 2005). Collectively, these observations demonstrate that ATP sensitivity is not a fixed parameter and may change dynamically with changes in membrane composition. Positively charged amino acid residues in the slide helix region directly interact with PIP_2 in the membrane (Cukras et al. 2002, Schulze et al. 2003, Shyng et al. 2000), as in Kir2 channels. While less extensively studied, Kir6.1 channels are also activated by PIP_2 (Quinn et al. 2003), suggesting conservation of the fundamental determinants of phospholipid binding and gating. Long chain acyl-coA molecules (LC-CoA), intermediates of β-oxidation of fatty acids, have also been shown to modulate K_{ATP} channel activity in an analogous manner to membrane phosphoinositides (Liu et al. 2001a). The same residues on Kir6.2 that mediate PIP_2 activation also contribute to the stimulatory effect of LC-CoA (Manning Fox et al. 2004, Schulze et al. 2003), suggesting that Kir6.2 is the major site of LC-CoA action. Acidic intracellular pH stimulates K_{ATP} channel activity (Davies 1990, Davies et al. 1992, Xu et al. 2001), and intracellular acidification concomitant with anaerobic metabolism provides another potential physiological stimulus of K_{ATP} activity. Although the mechanistic basis of pH regulation is still not completely established, there is a general consensus that protons act to decrease sensitivity to inhibitory ATP. Agonist-dependent protein kinase A (PKA) phosphorylation regulates smooth muscle and pancreatic K_{ATP} channels: the Kir6.2 subunit has two consensus PKA phosphorylation sites which when phosphorylated increase channel open probability, and in human SUR1, a unique, constitutively phosphorylated PKA site acts to both increase the surface expression of the channel and decrease channel open probability (Beguin et al. 1999, Lin et al. 2000). Finally, protein kinase C (PKC) has mixed actions on native ventricular K_{ATP} channels, inhibiting at low micromolar ATP concentrations (Light et al. 1995), but activating at high ATP concentrations (Light et al. 1996), through phosphorylation of the highly conserved T180 residue in the Kir6.2 subunit (Light et al. 2000). Kir6.1/SUR2B channels are inhibited by acute PKC treatment due to phosphorylation of residues in Kir6.1 (Quinn et al. 2003), whereas Kir6.2/SUR2B channel activity is reportedly unaffected by PKC (Shi et al. 2008, Thorneloe et al. 2002), highlighting the specificity of K_{ATP} channel subunit combinations in physiological regulation.

The SUR subunit determines the sensitivity of the channel to a wide range of pharmacological KCOs and blockers (Babenko et al. 2000, D'Hahan et al. 1999, Hambrock et al. 2004). The sulfonylurea glibenclamide inhibits all SUR isoforms, with a potency that depends on the intracellular nucleotide concentration, and hence on the metabolic state (Koster et al. 1999, Reimann et al. 1999). The SUR subunits also confer sensitivity to KCOs such as diazoxide, cromakalim, and pinacidil (Kurata et al. 2006). Diazoxide is an effective activator of SUR1 and SUR2B, but not SUR2A, whereas pinacidil and cromakalim are effective activators of SUR2A and SUR2B, but not SUR1 (Flagg et al. 2008, Quayle et al. 1995), and this potential for tissue specificity remains an important pharmaceutical strategy (Ashcroft and Gribble 2000). These openers all require the presence of hydrolyzable ATP (Dickinson et al. 1997, Hambrock et al. 1998, Schwanstecher et al. 1998), which suggests that they act to stabilize or enhance ATP hydrolysis at the nucleotide-binding folds (NBFs).

18.4.3.3 K_{ATP} channel pathologies

In pancreatic β-cells, K_{ATP} channels are made up of Kir6.2 and SUR1, and link blood sugar levels to insulin secretion by modulating membrane excitability. Pancreatic K_{ATP} channels are constitutively active, due to the enhanced stimulatory effects conferred by SUR1 (vs. SUR2A) (Masia et al. 2005) as well as to the relatively low [ATP] to [ADP] ratio in β-cells during fasting, and help set a negative membrane potential. As blood glucose levels increase after a meal, glucose is taken up and metabolized by the β-cells, the [ATP] to [ADP] ratio is increased, and K_{ATP} channel activity is reduced. This results in cell membrane depolarization and activation of voltage-gated Ca^{2+} channels, and Ca^{2+}-dependent secretion of insulin granules (Ashcroft 2005). Loss-of-function mutations in pancreatic K_{ATP} channel subunits underlie congenital hyperinsulinism (Dunne et al. 2004), and gain-of-function K_{ATP} channel mutations lead to transient or permanent neonatal diabetes mellitus (Koster et al. 2005b). The most severe permanent form of the disease extends beyond the pancreas to neuronal or other tissues, such that patients experience motor and intellectual developmental delay, epilepsy, and neonatal diabetes, collectively known as the DEND syndrome (Hattersley and Ashcroft 2005); the extrapancreatic symptoms of DEND have been attributed to K_{ATP} overactivity in muscle and/or nerves and brain, which highlights the role of Kir6.2 and SUR1 in tissues outside the pancreas. Mouse models expressing overactive or inducible K_{ATP} channels established the K_{ATP}-dependent basis of neonatal diabetes (Girard et al. 2009, Koster et al. 2000, Remedi et al. 2009), and led to the realization that sulfonylureas could be used to treat the disease. Most patients with activating K_{ATP} mutations have now been successfully treated with oral sulfonylureas rather than insulin (Koster et al. 2008, Pearson et al. 2006, Wambach et al. 2010), and these have since become a preferred treatment option for neonatal diabetes (Karges et al. 2012).

Under normal metabolic conditions, cardiac sarcolemmal K_{ATP} channels are predominantly closed, and do not significantly contribute to cell excitability. However, these channels can open when exposed to a severe metabolic stress such as anoxia, metabolic inhibition, or ischemia, and this has been hypothesized to protect the cells against damage of Ca^{2+} overload (Flagg and Nichols 2005). In cardiac myocytes, K_{ATP} current activation leads to action potential shortening, which reduces Ca^{2+} entry and inhibits contractility, thereby reducing energy consumption and accelerating recovery after an ischemic event (Lederer et al. 1989). Kir6.2-null mice are intolerant of vigorous exercise, and show a predisposition to cardiac remodeling, heart failure, and death when exposed to hypertension, pressure overload, or ischemia (Kane et al. 2006, Liu et al. 2004, Suzuki et al. 2001, 2002, Yamada et al. 2006, Zingman et al. 2002). The function of cardiac K_{ATP} channels in less challenging physiological conditions is a matter of debate; for example, baseline cardiac action potential and contractility are unaffected in Kir6.2-null mice (Li et al. 2000, Suzuki et al. 2001, 2002), and no cardiovascular symptoms have been reported for individuals with gain-of-function or loss-of-function mutations in Kir6.2 or SUR1 (Flanagan et al. 2009, Nichols et al. 2007); also, Kir6.2 gain-of-function transgenic mice recapitulate the human condition of neonatal diabetes with no cardiac phenotype (Koster et al. 2001). Studies to ascertain the role of sarcolemmal K_{ATP} channels in cardiac rhythm have yielded conflicting results: treatment with KCOs has been shown to stabilize the resting membrane potential and reduce the frequency of arrhythmias in some studies (Grover and Garlid 2000), but has proven pro-arrhythmic in others (Wolleben et al. 1989). Similarly, both increased (Shigematsu et al. 1995) and decreased (Kantor et al. 1990, Wolleben et al. 1989) incidence of tachycardia and ventricular fibrillation following treatment with glibenclamide have been reported.

There have been reports of AF and heart failure associated with SUR2A loss-of-function mutations (Bienengraeber et al. 2004, Olson et al. 2007), and Kir6.1-null and SUR2-null mice recapitulate the clinical features of human Prinzmetal angina, including baseline hypertension, coronary artery vasospasm, and sudden cardiac death, presumably due to loss of K_{ATP} channel activity in either VSM or endothelium (Chutkow et al. 2002, Miki et al. 2002). One gain-of-function mutation in Kir6.1 (S422L) has been found in several individuals with early repolarization syndrome and Brugada syndrome (Barajas-Martinez et al. 2012, Delaney et al. 2012, Haissaguerre et al. 2009, Medeiros-Domingo et al. 2010), although a recent study suggests that S422L may be a benign variant (Veeramah et al. 2013), and mutations in Kir6.1 leading to reduced K_{ATP} channel activity have been reported in cases of sudden infant death syndrome (Tester et al. 2011). Recently, multiple different gain-of-function mutations in the gene encoding for SUR2 have been linked to the Cantu syndrome, a complex multiorgan disorder with a myriad of cardiovascular and musculoskeletal features, including hypertrichosis, osteochondrodysplasia, and edema (Harakalova et al. 2012, Nichols et al. 2013, van Bon et al. 2012).

Unlike cardiac K_{ATP} channels, pancreatic K_{ATP} channels are constitutively active, due to the enhanced stimulatory effects conferred by SUR1 (vs. SUR2A) (Masia et al. 2005). This results in cell membrane depolarization and activation of voltage-gated Ca^{2+} channels, and Ca^{2+}-dependent secretion of insulin granules (Ashcroft 2005). Due to the critical role that these channels play in glucose-stimulated insulin secretion, genetic modifications of Kir6.2 or SUR1 have the potential to cause insulin secretion disorders. Indeed, loss-of-function mutations in pancreatic K_{ATP} channel subunits underlie congenital hyperinsulinism (Dunne et al. 2004), and gain-of-function K_{ATP} channel mutations lead to transient or permanent neonatal diabetes mellitus (Koster et al. 2005a). The most severe permanent form of the disease extends beyond the pancreas to neuronal or other tissues, such that patients experience motor and intellectual developmental delay, epilepsy, and neonatal diabetes, collectively known as the DEND syndrome (Hattersley and Ashcroft 2005); the extrapancreatic symptoms of DEND have been attributed to K_{ATP} overactivity in muscle and/or nerves and brain, which highlights the role of Kir6.2 and SUR1 in tissues outside the pancreas. Mouse models expressing overactive or inducible K_{ATP} channels established the K_{ATP}-dependent basis of neonatal diabetes (Girard et al. 2009, Koster et al. 2000, Remedi et al. 2009), leading to the realization that sulfonylureas could be used to treat the disease. In early clinical studies, most patients with activating K_{ATP} mutations transitioned successfully from insulin to oral sulfonylureas (Koster et al. 2008, Pearson et al. 2006), and these have since become a preferred treatment option for neonatal diabetes (Karges et al. 2012).

18.4.3.4 Mitochondrial K_{ATP} channels

A K^+-selective, small-conductance channel was first identified in rat liver mitochondrial inner membrane, then reported to be reversibly inhibited by application of ATP and glibenclamide (Inoue et al. 1991), and finally activated by GTP, GDP, and diazoxide (Garlid et al. 1996, Paucek et al. 1996). The pharmacology of heterologously expressed SUR1/Kir6.1 complexes appears to most closely resemble such properties (Hu et al. 1999, Liu et al. 2001b), yet the function of these mitoK_{ATP} channels is apparently unaffected in both Kir6.1-null and Kir6.2-null mice (Miki et al. 2002, Suzuki et al. 2002), and whether specific SUR or Kir6 subunits are normally present in mitochondria remains unclear (Cuong et al. 2005, Foster et al. 2008, Hu et al. 1999, Singh et al. 2003, Suzuki et al. 1997, Zhou et al. 2005).

The lack of confirmed presence of canonical SUR or Kir6 subunits in mitochondria has led to alternative hypotheses regarding mitoK_{ATP} structure. In addition to opening K_{ATP} channels, diazoxide may inhibit succinate dehydrogenase (Hanley et al. 2002) and it has been suggested that this enzyme key to the Krebs cycle and the electron transport chain may be a component of the mitoK_{ATP} channel (Ardehali et al. 2004, 2005). Most recently, proteomic analysis and pharmacological studies on bovine mitochondrial inner membranes have implicated a short product of the KCNJ1 (Kir1.x) gene, in the formation of the mitoK_{ATP} channel (Foster et al. 2012).

18.4.4 K^+ TRANSPORT CHANNELS: Kir1.1, Kir4.X, Kir5.1, AND Kir7.1

18.4.4.1 Kir1.1

Kir1.1 (encoded by gene *KCNJ1*), initially described as *rat outer medullary K^+ channel* ROMK1, was the first member of the Kir family to be cloned (Ho et al. 1993). Six alternative splicing isoforms of *KCNJ1* are known, of which ROMK2, 4, 5, and 6 correspond to the same protein (Kir1.1b), respectively, 19 and 26 amino acids shorter in the NH_2-terminus than ROMK1 (Kir1.1a) and ROMK3 (Kir1.1c) (Boim et al. 1995, Kondo et al. 1996). Kir1.1 tetrameric assemblies generate very weak inward rectifiers, and this is attributed to the presence of an uncharged amino acid (N171) at the *rectification controller* site (Leng et al. 2006b). Kir1.1 channels are fundamental for blood and urine salt homeostasis, and mutations that disrupt Kir1.1 function result in an array of renal tubulopathy symptoms collectively known as Bartter's syndrome (Hebert et al. 2005, Peters et al. 2002).

18.4.4.1.1 Kir1.1 channel assembly and regulation

Intracellular pH (pH_i) modulates the activity of Kir1.1 channels; in particular, acidification closes the channels, with a pK_a of ~6.5 (Choe et al. 1997). There is evidence that pH_i sensitivity in Kir1.1 channels is controlled by three quartets of interactions, two within the same subunits (a salt bridge between R41 and E318, and a potential hydrogen bond between K80 and A177) and one between adjacent subunits (a salt bridge between E302-R311) (Fakler et al. 1996, Rapedius et al. 2006, Wang et al. 2005). Mutation of any of these six residues alters Kir1.1 pH_i sensitivity: for example, R311W and A177T, which underlie Bartter's syndrome, significantly increase the pK_a of ROMK channels and reduce open probability in the physiological pH_i range (Peters et al. 2003, Schulte et al. 1999). Mutations of K80 that are predicted to disrupt the H^+ bond lower pH_i sensitivity of Kir1.1, and also accelerate the recovery from inhibition induced by acidification. The K80-A177 bond has also been implicated in the control of PIP_2-mediated gating of Kir1.1 channels (Hibino et al. 2010a). Additionally, mutagenesis studies have suggested that the hydrophobic residue L160 may contribute to pH_i-dependent gating (Sackin et al. 2005). Finally, G167 and G176 in Kir1.1a and G148 and G157 in Kir1.1b are located close to the cytoplasmic apex of the TM2 helices, and mutation of either residue to Ala results in ROMK channels with increased pK_a, again consistent with a role in the mechanisms governing Kir1.1 pH_i sensitivity (Sackin et al. 2006).

Kir1.1 channel function is strongly regulated by surface expression. Thus, Kir1.1 has a carboxy-terminal endoplasmic reticulum (ER) retention signal, R368-A369-R370, which undergoes posttranslational modification by intracellular protein kinases and ultimately controls trafficking to the cell membrane (Ma et al. 2001). Kir1.1 channel activity requires PKA-dependent phosphorylation in three sites, namely, S44, S219, and S313, and phosphorylation in S44 by PKA or SGK has been found to increase Kir1.1 surface expression, by suppressing the ER retention signal (Hebert et al. 2005, McNicholas et al. 1994, O'Connell et al. 2005, Yoo et al. 2004). Other serine-threonine kinases, such as WNK1, WNK3, and WNK4, reduce Kir1.1 channel activity by decreasing surface expression (Kahle et al. 2003, Leng et al. 2006a, Wade et al. 2006), and mutations in WNK1 or WNK4 have been associated with Gordon's syndrome, a type of autosomal-dominant pseudo-hypoaldosteronism with hypertension and hyperkalemia than can be explained by decreased Kir1.1 function (Wilson et al. 2001). Finally, phosphorylation of S219 and S313 by PKA enhances channel activity by increasing sensitivity to PIP_2 (Liou et al. 1999), whereas PKC-dependent phosphorylation of S4 and S201 has been found to inhibit Kir1.1, possibly due to a reduction in membrane PIP_2 concentration (Lin et al. 2002, Wang and Giebisch 1991, Zeng et al. 2003).

18.4.4.1.2 Physiology and pathophysiology of Kir1.1 channels

Kir1.1 is expressed in the apical membrane of renal epithelial cells of the thick ascending (TAL) of the loop of Henle, the distal convoluted tubule (DCT), and the cortical collecting duct (CCD), where it plays a crucial role in the regulation not only of K^+ concentrations in blood and urine, but also of other ions such as Na^+ and Cl^-. Kidney TAL cells reabsorb ~25% of all the Na^+ in the renal filtrate, mainly through the apical membrane Na^+-K^+-$2Cl^-$ (NKCC) transporter, and Kir1.1 channels provide a route for K^+ efflux. Kir1.1 channels also facilitate reabsorption of Cl^- via NKCC, and Na^+ and Cl^- accumulated in the cytoplasm are returned to the blood by the Na^+-K^+-ATPase and the Cl^- channels at the basolateral membrane. Thus, functional coupling of Kir1.1 and NKCC results in unidirectional transport of Na^+ and Cl^-, sustains the activity of the ATPase, and establishes the transepithelial potential that drives paracellular reabsorption of Na^+, Ca^{2+}, and Mg^{2+} (Hebert et al. 2005). Additionally, the Cl^- channel CFTR colocalizes with Kir1.1 in the apical membrane of TAL cells, and may provide Kir1.1 channels with sensitivity to ATP and sulfonylureas through complex protein–protein interactions (Lu et al. 2006, Yoo et al. 2004).

Bartter's syndrome is an autosomal recessive renal tubulopathy characterized by hypokalemic metabolic alkalosis, renal salt wasting, hyperreninemia, and hyperaldosteronism (Bartter et al. 1962, Peters et al. 2002). Of the recognized types of the disease, genetic analysis revealed that type II Bartter's syndrome results specifically from mutations in the *KCNJ1* gene that affect PKA phosphorylation, pH sensing, channel gating, proteolytic processing, and sorting to the apical membrane. In TAL cells, loss of Kir1.1 function decreases K^+ supply to the NKCC and loss of the driving force for paracellular cation reabsorption (Hebert et al. 2005). *KCNJ1*-null mice recapitulate the symptoms of Bartter's syndrome, and present with hypokalemia and an excess of Na^+, Cl^-, and K^+ in their urine (Lorenz et al. 2002, Lu et al. 2002).

18.4.4.2 Kir4.x and Kir5.1

Kir4.1 (*KCNJ10*) was independently identified from a brain cDNA library by several groups, and variably referred to as BIR10, K_{AB}-2, BIRK-1, and Kir1.2 (Hibino et al. 2010a). Kir4.1 channels are essential for control of glial function and neuronal excitability, systemic K^+ homeostasis, and renal salt exchange (Butt and Kalsi 2006, Wagner 2010). Kir4.1 subunits form functional homotetramers, or coassemble with Kir5.1 (*KCNJ16*, initially termed BIR9) in heterotetramers with distinct physiological properties (Hibino et al. 2004a, Ishii et al. 2003, Tanemoto et al. 2000, Tucker et al. 2000). Kir4.2 (*KCNJ15*, also named Kir1.3) is found in kidney, liver, embryonic fibrocytes, and microvascular endothelial cells, where it may assemble in functional homotetramers or heterotetramers with Kir5.1 subunits (Pearson et al. 1999, Pessia et al. 2001). Linkage analysis has identified loss-of-function mutations in the *KCNJ10* gene that are responsible for an array of neural and somatic symptoms collectively referred to as the SeSAME (*s*eizures, *s*ensorineural deafness, *a*taxia, *m*ental retardation and *e*lectrolyte imbalance) or EAST (*e*pilepsy, *a*taxia, *s*ensorineural deafness, and *t*ubulopathy) syndrome (Bockenhauer et al. 2009, Sala-Rabanal et al. 2010, Scholl et al. 2009). Recently, antibodies against Kir4.1 have been found in a subset of multiple sclerosis (MS) patients (Srivastava et al. 2012), and thus is possible that some MS cases may also belong to an expanding number of autoimmune disorders caused by ion channel dysfunction (Ehling et al. 2011, Kleopa 2011).

18.4.4.2.1 Channel assembly, structure, and localization

Homomeric Kir4.1 and Kir4.2 channels display intermediate inward rectification, and Kir4.1 channels exhibit an incomplete rectification with a marked *punchthrough*, of polyamines through the channel pore (Kucheryavykh et al. 2007b). Homomeric Kir4.1 channels are inhibited by intracellular acidification with a $pK_a \sim 6$, whereas Kir4.1/Kir5.1 channels are suppressed by only a slight acidification and activated by alkalization, with a $pK_a \sim 7.5$ (Tanemoto et al. 2000, Yang et al. 2000). Homomeric Kir4.2 channels are more sensitive to pH_i than Kir4.1 homomers (Kir4.2, $pK_a \sim 7.1$), but heteromerization with Kir5.1 has only minor effects on pH_i sensitivity (Kir4.2/Kir5.1, $pK_a \sim 7.6$) (Pessia et al. 2001). Several residues in Kir4.1, including E158 in TM2, K67 in the cytoplasmic amino terminus, and H190 in the carboxy terminus, are involved in pH_i sensitivity of Kir4.1 and Kir4.1/Kir5.1 channels, but the mechanism remains elusive (Casamassima et al. 2003, Xu et al. 2000, Yang et al. 2000). Interestingly, mutation K66M in the NH_2 terminus of Kir4.2 renders the channel virtually insensitive to pH_i (Pessia et al. 2001), and this is consistent with the neutralization of the equivalent residue in Kir1.1, K80, which impairs pH_i sensing in these channels (Fakler et al. 1996, McNicholas et al. 1998). Homomeric Kir4.1 channels are found in the microvilli of gastric parietal cells (Fujita et al. 2002), and in the apical membrane of intermediate cells of the cochlear stria vascularis (Hibino and Kurachi 2006). Kir5.1 subunits are found in fibrocytes of the spiral ligament in the inner ear, where they are mostly confined to intracellular compartments (Hibino et al. 2004b). Both Kir4.1 and Kir5.1 subunits are expressed in brain astrocytes and in Müller cells of the retina (Hibino et al. 2010a): heteromeric channels are found in brain and retinal perisynaptic processes, whereas only Kir4.1 homomers are present at the end feet of the Müller cells facing the vitreous humor and the blood vessels (Ishii et al. 2003). Heteromeric Kir4.1/Kir5.1 channels constitute the majority of the basolateral K^+ conductance of the distal nephron, and renal epithelial cells do not express homomeric Kir4.1 channels (Lachheb et al. 2008, Lourdel et al. 2002, Wang et al. 2010). The cellular mechanisms governing Kir4.1/Kir5.1 heteromerization, or the preference for Kir4.1/Kir5.1 over homomeric Kir4.1 channels in the kidney, are still unclear. A 44-amino acid stretch [161–205] in the cytoplasmic region just distal to the Kir4.1 TM2 appears to be essential for association with Kir5.1 (Konstas et al. 2003), and PDZ-binding motifs implicated in oligomerization and plasma membrane localization have been identified in the COOH-terminal portion of Kir4.1 (Tanemoto et al. 2004, 2005, 2008).

18.4.4.2.2 Physiology and pathophysiology

In renal tubular epithelia, Kir4.1/Kir5.1 channels are responsible for the negative basolateral membrane potential that helps drive the electrogenic luminal processes of K^+ secretion and Na^+ reabsorption (Lachheb et al. 2008, Lourdel et al. 2002, Wang et al. 2010). In astrocytic glia, Kir4.1-containing channels account for the spatial buffering of K^+ released by neurons during action potential propagation (Butt and Kalsi 2006, Kucheryavykh et al. 2007a, Neusch et al. 2006). Kir4.1 is the predominant K^+ channel in retinal Müller glial cells (Bringmann et al. 2006), and in satellite glial cells of sensory ganglia (Tang et al. 2010). Kir4.1 channels in oligodendrocytes are critical for myelination (Kalsi et al. 2004, Neusch et al. 2001), and in the inner ear they play a role in K^+ regulation and generation of the endocochlear potential required for normal development of the cochlea and audition (Hibino et al. 2010b, Rozengurt et al. 2003). In the stomach, homomeric Kir4.1 channels have been implicated in acid secretion, as a mechanism of K^+ recycling for the H^+-K^+ pump (Kaufhold et al. 2008). On the other hand, Kir4.2 channels are involved in the migration of mouse embryonic fibrocytes and human microvascular endothelial cells (deHart et al. 2008), which suggests that these channels may play a crucial role in embryogenesis.

Missense and nonsense mutations in *KCNJ10* have been identified in SeSAME/EAST syndrome patients (Bockenhauer et al. 2009, Scholl et al. 2009). Individuals affected by this disorder present with an array of both neurological and renal symptoms, including seizures, ataxia, mental retardation, sensorineural deafness, and electrolyte imbalance. All of the SeSAME/EAST mutations have been found to disrupt channel function in Kir4.1 homomers and Kir4.1/Kir5.1 heteromers (Reichold et al. 2010, Sala-Rabanal et al. 2010). In contrast, potential gain-of-function mutations in Kir4.1 have been associated with autism with seizures and intellectual disability (Sicca et al. 2011), and Kir5.1-null mice present with hypokalemia and metabolic acidosis, that is, a renal phenotype essentially opposite to that of SeSAME/EAST, due to a compensatory expression of pH_i-insensitive, constitutively active homomeric Kir4.1 in the nephron, which leads to increased salt reabsorption (Paulais et al. 2011). Kir4.1 is highly expressed in glia, where it plays a major role in K^+ buffering and neuron repolarization (Haj-Yasein et al. 2011, Kucheryavykh et al. 2007a, Olsen and Sontheimer 2008). Loss of Kir4.1 function may cause reduced astrocyte-mediated K^+ clearance, neuronal hyperexcitability, and reduced seizure threshold; alternatively, an increase in Kir4.1 activity during intense neuronal activity may lead to increased and faster K^+ influx into glial cells, resulting in a larger and more sustained membrane depolarization that promotes local neuronal synchrony and epileptic activity. In the kidney, loss of Kir4.1/Kir5.1 activity results in a salt-wasting phenotype, presumably by secondarily reducing the activity of the Na^+-K^+-ATPase (Bandulik et al. 2011).

No pathology has yet been clearly linked to Kir4.2 deficiencies, but the *KCNJ15* gene is located close to the locus in chromosome 21 associated to Down syndrome (Gosset et al. 1997), and this suggests a possible linkage between some of the manifestations of the syndrome, which include dysmorphic features, hypotonia, and psychomotor delay, and Kir4.2 channel dysfunction.

18.4.4.3 Kir7.1

Kir7.1 (*KCNJ13*) is the sole representative of a seventh subfamily, most closely related to Kir4. The cDNA was independently isolated by three groups (Doring et al. 1998, Krapivinsky et al. 1998, Partiseti et al. 1998), and to date, only one isoform of the protein has been described. Kir7.1 is found in epithelial cells of the choroid plexus, thyroid follicular cells, enterocytes, renal epithelial cells of various regions along the nephron, and retinal pigmental epithelia (RPE); in all these cells, it colocalizes with the Na^+-K^+-ATPase, and thus may also serve as a mechanism for K^+ recycling for the establishment and maintenance of the membrane potential (Doring et al. 1998, Nakamura et al. 1999, Ookata et al. 2000). Loss-of-function mutations in *KCNJ13* have been found to underlie snowflake vitreoretinal degeneration, an autosomal-dominant, progressive eye disease that causes early-onset cataract, fibrillar degeneration of the vitreous humor, and retinal detachment (Hejtmancik et al. 2008, Pattnaik et al. 2013, Zhang et al. 2013), and this underscores the importance of Kir7.1 channels in normal retinal physiology.

Kir7.1 subunits assemble into functional homotetramers, and no evidence so far has been presented of heteromerization with other Kir family members. Kir7.1 channels are weak inward rectifiers, and this rectification appears to be independent of extracellular [K^+]; also, Kir7.1 channels have a much smaller single-channel conductance than other Kir channels and are about 10 times less sensitive to Ba^{2+} and Cs^+, and all these unique properties can be attributed to one residue in the pore region, M125, which is an Arg in all other Kir channel subunits (Doring et al. 1998, Hughes and Swaminathan 2008, Krapivinsky et al. 1998). Kir7.1 channels are activated by PIP_2, but their binding affinity is lower than in other Kir channels (Rohacs et al. 2003). More notably, Kir7.1 channels are sensitive to pH_i, but their response is bell shaped: thus, maximum activity is observed at $pH_i \sim 7.0$ and the current is attenuated at either higher or lower pH_i, and H26 in the cytoplasmic NH_2 terminus appears to control this behavior (Hughes and Swaminathan 2008).

REFERENCES

Adams DJ, Hill MA. 2004. Potassium channels and membrane potential in the modulation of intracellular calcium in vascular endothelial cells. *J. Cardiovasc. Electr.* 15: 598–610.

Antcliff JF, Haider S, Proks P, Sansom MS, Ashcroft FM. 2005. Functional analysis of a structural model of the ATP-binding site of the K_{ATP} channel Kir6.2 subunit. *EMBO J.* 24: 229–239.

Anumonwo J, Lopatin A. 2010. Cardiac strong inward rectifier potassium channels. *J. Mol. Cell. Cardiol.* 48: 45–54.

Ardehali H, Chen Z, Ko Y, Mejia-Alvarez R, Marban E. 2004. Multiprotein complex containing succinate dehydrogenase confers mitochondrial ATP-sensitive K^+ channel activity. *Proc. Natl. Acac. Sci. U.S.A.* 101: 11880–11885.

Ardehali H, O'Rourke B, Marban E. 2005. Cardioprotective role of the mitochondrial ATP-binding cassette protein 1. *Circ. Res.* 97: 740–742.

Ashcroft FM. 1988. Adenosine 5′-triphosphate-sensitive potassium channels. *Annu. Rev. Neurosci.* 11: 97–118.

Ashcroft FM. 2005. ATP-sensitive potassium channelopathies: Focus on insulin secretion. *J. Clin. Invest.* 115: 2047–2058.

Ashcroft FM, Gribble FM. 2000. New windows on the mechanism of action of K(ATP) channel openers. *Trends Pharmacol. Sci.* 21: 439–445.

Babenko AP, Gonzalez G, Aguilar-Bryan L, Bryan J. 1998. Reconstituted human cardiac K_{ATP} channels: Functional identity with the native channels from the sarcolemma of human ventricular cells. *Circ. Res.* 83: 1132–1143.

Babenko AP, Gonzalez G, Bryan J. 2000. Pharmaco-topology of sulfonylurea receptors. Separate domains of the regulatory subunits of K(ATP) channel isoforms are required for selective interaction with K(+) channel openers. *J. Biol. Chem.* 275: 717–720.

Bandulik S, Schmidt K, Bockenhauer D, Zdebik AA, Humberg E et al. 2011. The salt-wasting phenotype of EAST syndrome, a disease with multifaceted symptoms linked to the KCNJ10 K^+ channel. *Pflug. Arch. Eur. J. Physiol.* 461: 423–435.

Barajas-Martinez H, Hu D, Ferrer T, Onetti CG, Wu Y et al. 2012. Molecular genetic and functional association of Brugada and early repolarization syndromes with S422L missense mutation in KCNJ8. *Heart Rhythm* 9: 548–555.

Bartter FC, Pronove P, Gill JR, Jr., Maccardle RC. 1962. Hyperplasia of the juxtaglomerular complex with hyperaldosteronism and hypokalemic alkalosis. A new syndrome. *Am. J. Med.* 33: 811–828.

Baukrowitz T, Schulte U, Oliver D, Herlitze S, Krauter T et al. 1998. PIP_2 and PIP as determinants for ATP inhibition of KATP channels. *Science* 282: 1141–1144.

Beech DJ, Zhang H, Nakao K, Bolton TB. 1993. K channel activation by nucleotide diphosphates and its inhibition by glibenclamide in vascular smooth muscle cells. *Brit. J. Pharmacol.* 110: 573–582.

Beguin P, Nagashima K, Nishimura M, Gonoi T, Seino S. 1999. PKA-mediated phosphorylation of the human K(ATP) channel: Separate roles of Kir6.2 and SUR1 subunit phosphorylation. *EMBO J.* 18: 4722–4732.

Bernardi H, Fosset M, Lazdunski M. 1988. Characterization, purification, and affinity labeling of the brain [3H]glibenclamide-binding protein, a putative neuronal ATP-regulated K^+ channel. *Proc. Natl. Acac. Sci. U.S.A.* 85: 9816–9820.

Betourne A, Bertholet AM, Labroue E, Halley H, Sun HS et al. 2009. Involvement of hippocampal CA3 K(ATP) channels in contextual memory. *Neuropharmacology* 56: 615–625.

Bianchi L, Roy ML, Taglialatela M, Lundgren DW, Brown AM, Ficker E. 1996. Regulation by spermine of native inward rectifier K^+ channels in RBL-1 cells. *J. Biol. Chem.* 271: 6114–6121.

Bichet D, Haass FA, Jan LY. 2003. Merging functional studies with structures of inward-rectifier K(+) channels. *Nat. Rev. Neurosci.* 4: 957–967.

Bienengraeber M, Olson TM, Selivanov VA, Kathmann EC, O'Cochlain F et al. 2004. ABCC9 mutations identified in human dilated cardiomyopathy disrupt catalytic K_{ATP} channel gating. *Nat. Genet.* 36: 382–387.

Bockenhauer D, Feather S, Stanescu HC, Bandulik S, Zdebik AA et al. 2009. Epilepsy, ataxia, sensorineural deafness, tubulopathy, and KCNJ10 mutations. *N. Engl. J. Med.* 360: 1960–1970.

Boim MA, Ho K, Shuck ME, Bienkowski MJ, Block JH et al. 1995. ROMK inwardly rectifying ATP-sensitive K^+ channel. II. Cloning and distribution of alternative forms. *Am. J. Physiol. Renal Physiol.* 268: F1132–F1140.

Bradley KK, Jaggar JH, Bonev AD, Heppner TJ, Flynn ER et al. 1999. Kir2.1 encodes the inward rectifier potassium channel in rat arterial smooth muscle cells. *J. Physiol.* 515: 639–651.

Bringmann A, Pannicke T, Grosche J, Francke M, Wiedemann P et al. 2006. Muller cells in the healthy and diseased retina. *Prog. Retin. Eye Res.* 25: 397–424.

Brismar T, Collins VP. 1989. Inward rectifying potassium channels in human malignant glioma cells. *Brain Res.* 480: 249–258.

Butt AM, Kalsi A. 2006. Inwardly rectifying potassium channels (Kir) in central nervous system glia: A special role for Kir4.1 in glial functions. *J. Cell. Mol. Med.* 10: 33–44.

Casamassima M, D'Adamo MC, Pessia M, Tucker SJ. 2003. Identification of a heteromeric interaction that influences the rectification, gating, and pH sensitivity of Kir4.1/Kir5.1 potassium channels. *J. Biol. Chem.* 278: 43533–43540.

Chan KW, Wheeler A, Csanady L. 2008. Sulfonylurea receptors type 1and 2A randomly assemble to form heteromeric K_{ATP} channels of mixed subunit composition. *J. Gen. Physiol.* 131: 43–58.

Cheng WW, D'Avanzo N, Doyle DA, Nichols CG. 2011. Dual-mode phospholipid regulation of human inward rectifying potassium channels. *Biophys. J.* 100: 620–628.

Cheng WW, Tong A, Flagg TP, Nichols CG. 2008. Random assembly of SUR subunits in K(ATP) channel complexes. *Channels (Austin)* 2: 34–38.

Choe H, Sackin H, Palmer LG. 2001. Gating properties of inward-rectifier potassium channels: Effects of permeant ions. *J. Membrane. Biol.* 184: 81–89.

Choe H, Zhou H, Palmer LG, Sackin H. 1997. A conserved cytoplasmic region of ROMK modulates pH sensitivity, conductance, and gating. *Am. J. Physiol.* 273: F516–F529.

Chung HJ, Ge WP, Qian X, Wiser O, Jan YN, Jan LY. 2009. G protein-activated inwardly rectifying potassium channels mediate depotentiation of long-term potentiation. *Proc. Natl. Acac. Sci. U.S.A.* 106: 635–640.

Chutkow WA, Makielski JC, Nelson DJ, Burant CF, Fan Z. 1999. Alternative splicing of sur2 Exon 17 regulates nucleotide sensitivity of the ATP-sensitive potassium channel. *J. Biol. Chem.* 274: 13656–13665.

Chutkow WA, Pu J, Wheeler MT, Wada T, Makielski JC et al. 2002. Episodic coronary artery vasospasm and hypertension develop in the absence of Sur2 K(ATP) channels. *J. Clin. Invest.* 110: 203–208.

Chutkow WA, Simon MC, Le Beau MM, Burant CF. 1996. Cloning, tissue expression, and chromosomal localization of SUR2, the putative drug-binding subunit of cardiac, skeletal muscle, and vascular K_{ATP} channels. *Diabetes* 45: 1439–1445.

Clement JP, Kunjilwar K, Gonzalez G, Schwanstecher M, Panten U et al. 1997. Association and stoichiometry of K(ATP) channel subunits. *Neuron* 18: 827–838.

Cook DL, Hales CN. 1984. Intracellular ATP directly blocks K^+ channels in pancreatic B-cells. *Nature* 311: 271–273.

Corey S, Clapham DE. 1998. Identification of native atrial G-protein-regulated inwardly rectifying K^+ (GIRK4) channel homomultimers. *J. Biol. Chem.* 273: 27499–27504.

Cui Y, Giblin JP, Clapp LH, Tinker A. 2001. A mechanism for ATP-sensitive potassium channel diversity: Functional coassembly of two pore-forming subunits. *Proc. Natl. Acad. Sci. U.S.A.* 98: 729–734.

Cukras CA, Jeliazkova I, Nichols CG. 2002. Structural and functional determinants of conserved lipid interaction domains of inward rectifying Kir6.2 channels. *J. Gen. Physiol.* 119: 581–591.

Cuong DV, Kim N, Joo H, Youm JB, Chung JY et al. 2005. Subunit composition of ATP-sensitive potassium channels in mitochondria of rat hearts. *Mitochondrion* 5: 121–133.

D'Avanzo N, Cheng WW, Doyle DA, Nichols CG. 2010. Direct and specific activation of human inward rectifier K^+ channels by membrane phosphatidylinositol 4,5-bisphosphate. *J. Biol.Chem.* 285: 37129–37132.

D'Avanzo N, Hyrc K, Enkvetchakul D, Covey DF, Nichols CG. 2011. Enantioselective protein-sterol interactions mediate regulation of both prokaryotic and eukaryotic inward rectifier K^+ channels by cholesterol. *PloS one* 6: e19393.

D'Avanzo N, Lee SJ, Cheng WW, Nichols CG. 2013. Energetics and location of phosphoinositide binding in human Kir2.1 channels. *J. Biol. Chem.* 288: 16726–16737.

D'Hahan N, Moreau C, Prost AL, Jacquet H, Alekseev AE et al. 1999. Pharmacological plasticity of cardiac ATP-sensitive potassium channels toward diazoxide revealed by ADP. *Proc. Natl. Acad. Sci. U.S.A.* 96: 12162–12167.

Davies NW. 1990. Modulation of ATP-sensitive K^+ channels in skeletal muscle by intracellular protons. *Nature* 343: 375–377.

Davies NW, Standen NB, Stanfield PR. 1992. The effect of intracellular pH on ATP-dependent potassium channels of frog skeletal muscle. *J. Physiol.* 445: 549–568.

de Boer TP, Houtman MJ, Compier M, van der Heyden MA. 2010. The mammalian K(IR)2.x inward rectifier ion channel family: Expression pattern and pathophysiology. *Acta Physiol. (Oxf.)* 199: 243–256.

deHart GW, Jin T, McCloskey DE, Pegg AE, Sheppard D. 2008. The alpha9beta1 integrin enhances cell migration by polyamine-mediated modulation of an inward-rectifier potassium channel. *Proc. Natl. Acad. Sci. U.S.A.* 105: 7188–7193.

Delaney JT, Muhammad R, Blair MA, Kor K, Fish FA et al. 2012. A KCNJ8 mutation associated with early repolarization and atrial fibrillation. *Europace* 14: 1428–1432.

Denton JS, Jacobson DA. 2012. Channeling dysglycemia: Ion-channel variations perturbing glucose homeostasis. *Trends Endocrinol. Metab.* 23: 41–48.

Dhamoon AS, Pandit SV, Sarmast F, Parisian KR, Guha P et al. 2004. Unique Kir2.x properties determine regional and species differences in the cardiac inward rectifier K^+ current. *Circ. Res.* 94: 1332–1339.

Dickinson KE, Bryson CC, Cohen RB, Rogers L, Green DW, Atwal KS. 1997. Nucleotide regulation and characteristics of potassium channel opener binding to skeletal muscle membranes. *Mol. Pharmacol.* 52: 473–481.

Dobrev D, Friedrich A, Voigt N, Jost N, Wettwer E et al. 2005. The G protein-gated potassium current I(K,ACh) is constitutively active in patients with chronic atrial fibrillation. *Circulation* 112: 3697–3706.

Dobrev D, Wettwer E, Kortner A, Knaut M, Schuler S, Ravens U. 2002. Human inward rectifier potassium channels in chronic and postoperative atrial fibrillation. *Cardiovasc. Res.* 54: 397–404.

Doring F, Derst C, Wischmeyer E, Karschin C, Schneggenburger R et al. 1998. The epithelial inward rectifier channel Kir7.1 displays unusual K^+ permeation properties. *J. Neurosci.* 18: 8625–8636.

Doyle DA, Morais Cabral J, Pfuetzner RA, Kuo A, Gulbis JM et al. 1998. The structure of the potassium channel: Molecular basis of K^+ conduction and selectivity. *Science* 280: 69–77.

Doyle ME, Egan JM. 2003. Pharmacological agents that directly modulate insulin secretion. *Pharmacol. Rev.* 55: 105–131.

Drain P, Li L, Wang J. 1998. KATP channel inhibition by ATP requires distinct functional domains of the cytoplasmic C terminus of the pore-forming subunit. *Proc. Natl. Acac. Sci. U.S.A.* 95: 13953–13958.

Du X, Zhang H, Lopes C, Mirshahi T, Rohacs T, Logothetis DE. 2004. Characteristic interactions with phosphatidylinositol 4,5-bisphosphate determine regulation of Kir channels by diverse modulators. *J. Biol. Chem.* 279: 37271–37281.

Dunne MJ, Cosgrove KE, Shepherd RM, Aynsley-Green A, Lindley KJ. 2004. Hyperinsulinism in infancy: From basic science to clinical disease. *Physiol. Rev.* 84: 239–275.

Dunne MJ, Petersen OH. 1986. Intracellular ADP activates K^+ channels that are inhibited by ATP in an insulin-secreting cell line. *FEBS Lett.* 208: 59–62.

Duprat F, Lesage F, Guillemare E, Fink M, Hugnot JP et al. 1995. Heterologous multimeric assembly is essential for K^+ channel activity of neuronal and cardiac G-protein-activated inward rectifiers. *Biochem. Biophys. Res. Commun.* 212: 657–663.

Eckhardt LL, Farley AL, Rodriguez E, Ruwaldt K, Hammill D et al. 2007. KCNJ2 mutations in arrhythmia patients referred for LQT testing: A mutation T305A with novel effect on rectification properties. *Heart Rhythm* 4: 323–329.

Ehling P, Bittner S, Budde T, Wiendl H, Meuth SG. 2011. Ion channels in autoimmune neurodegeneration. *FEBS Lett.* 585: 3836–3842.

Enkvetchakul D, Loussouarn G, Makhina E, Shyng SL, Nichols CG. 2000. The kinetic and physical basis of K(ATP) channel gating: Toward a unified molecular understanding. *Biophys. J.* 78: 2334–2348.

Enkvetchakul D, Nichols CG. 2003. Gating mechanism of K_{ATP} channels: Function fits form. *J. Gen. Physiol.* 122: 471–480.

Fakler B, Brandle U, Glowatzki E, Weidemann S, Zenner HP, Ruppersberg JP. 1995. Strong voltage-dependent inward rectification of inward rectifier K^+ channels is caused by intracellular spermine. *Cell* 80: 149–154.
Fakler B, Schultz JH, Yang J, Schulte U, Brandle U et al. 1996. Identification of a titratable lysine residue that determines sensitivity of kidney potassium channels (ROMK) to intracellular pH. *EMBO J.* 15: 4093–4099.
Fan Z, Makielski JC. 1997. Anionic phospholipids activate ATP-sensitive potassium channels. *J. Biol. Chem.* 272: 5388–5395.
Fang Y, Schram G, Romanenko VG, Shi C, Conti L et al. 2005. Functional expression of Kir2.x in human aortic endothelial cells: The dominant role of Kir2.2. *Am. J. Physiol. Cell Physiol.* 289: C1134–C1144.
Farzaneh T, Tinker A. 2008. Differences in the mechanism of metabolic regulation of ATP-sensitive K^+ channels containing Kir6.1 and Kir6.2 subunits. *Cardiovasc. Res.* 79: 621–631.
Ficker E, Taglialatela M, Wible BA, Henley CM, Brown AM. 1994. Spermine and spermidine as gating molecules for inward rectifier K^+ channels. *Science* 266: 1068–1072.
Filosa JA, Bonev AD, Straub SV, Meredith AL, Wilkerson MK et al. 2006. Local potassium signaling couples neuronal activity to vasodilation in the brain. *Nat. Neurosci.* 9: 1397–1403.
Finley M, Arrabit C, Fowler C, Suen KF, Slesinger PA. 2004. betaL-betaM loop in the C-terminal domain of G protein-activated inwardly rectifying K(+) channels is important for G(betagamma) subunit activation. *J. Physiol.* 555: 643–657.
Flagg TP, Enkvetchakul D, Koster JC, Nichols CG. 2010. Muscle K_{ATP} channels: Recent insights to energy sensing and myoprotection. *Physiol. Rev.* 90: 799–829.
Flagg TP, Kurata HT, Masia R, Caputa G, Magnuson MA et al. 2008. Differential structure of atrial and ventricular K_{ATP}: Atrial K_{ATP} channels require SUR1. *Circ. Res.* 103: 1458–1465.
Flagg TP, Nichols CG. 2005. Sarcolemmal K(ATP) channels: What do we really know? *J. Mol. Cell. Cardiol.* 39: 61–70.
Flanagan SE, Clauin S, Bellanne-Chantelot C, de Lonlay P, Harries LW et al. 2009. Update of mutations in the genes encoding the pancreatic beta-cell K(ATP) channel subunits Kir6.2 (KCNJ11) and sulfonylurea receptor 1 (ABCC8) in diabetes mellitus and hyperinsulinism. *Hum. Mutat.* 30: 170–180.
Foster DB, Ho AS, Rucker JJ, Garlid AO, Chen L et al. 2012. Mitochondrial ROMK channel is a molecular component of MitoKATP. *Circ. Res.* 111: 446–454.
Foster DB, Rucker JJ, Marban E. 2008. Is Kir6.1 a subunit of mitoK(ATP)? *Biochem. Biophys. Res. Commun.* 366: 649–656.
Fujita A, Horio Y, Higashi K, Mouri T, Hata F et al. 2002. Specific localization of an inwardly rectifying K(+) channel, Kir4.1, at the apical membrane of rat gastric parietal cells; its possible involvement in K(+) recycling for the H(+)-K(+)-pump. *J. Physiol.* 540: 85–92.
Fujita S, Inanobe A, Chachin M, Aizawa Y, Kurachi Y. 2000. A regulator of G protein signalling (RGS) protein confers agonist- dependent relaxation gating to a G protein-gated K^+ channel. *J. Physiol.* 526(Part 2): 341–347.
Gaborit N, Le Bouter S, Szuts V, Varro A, Escande D et al. 2007. Regional and tissue specific transcript signatures of ion channel genes in the non-diseased human heart. *J. Physiol.* 582: 675–693.
Garlid KD, Paucek P, Yarov-Yarovoy V, Sun X, Schindler PA. 1996. The mitochondrial K_{ATP} channel as a receptor for potassium channel openers. *J. Biol. Chem.* 271: 8796–8799.
Gehlert DR, Mais DE, Gackenheimer SL, Krushinski JH, Robertson DW. 1990. Localization of ATP sensitive potassium channels in the rat brain using a novel radioligand, [125I]iodoglibenclamide. *Eur. J. Pharmacol.* 186: 373–375.
Girard CA, Wunderlich FT, Shimomura K, Collins S, Kaizik S et al. 2009. Expression of an activating mutation in the gene encoding the K_{ATP} channel subunit Kir6.2 in mouse pancreatic beta cells recapitulates neonatal diabetes. *J. Clin. Invest.* 119: 80–90.
Girmatsion Z, Biliczki P, Bonauer A, Wimmer-Greinecker G, Scherer M et al. 2009. Changes in microRNA-1 expression and IK1 up-regulation in human atrial fibrillation. *Heart Rhythm* 6: 1802–1809.
Gosset P, Ghezala GA, Korn B, Yaspo ML, Poutska A et al. 1997. A new inward rectifier potassium channel gene (KCNJ15) localized on chromosome 21 in the Down syndrome chromosome region 1 (DCR1). *Genomics* 44: 237–241.
Gribble FM, Tucker SJ, Ashcroft FM. 1997. The essential role of the walker A motifs of SUR1 in K-ATP channel activation by Mg-ADP and diazoxide. *EMBO J.* 16: 1145–1152.
Grover GJ, Garlid KD. 2000. ATP-Sensitive potassium channels: A review of their cardioprotective pharmacology. *J. Mol. Cell. Cardiol.* 32: 677–695.
Guo D, Lu Z. 2000. Mechanism of IRK1 channel block by intracellular polyamines. *J. Gen. Physiol.* 115: 799–814.
Guo D, Lu Z. 2003. Interaction mechanisms between polyamines and IRK1 inward rectifier K^+ channels. *J. Gen. Physiol.* 122: 485–500.
Guo D, Ramu Y, Klem AM, Lu Z. 2003. Mechanism of rectification in inward-rectifier K^+ channels. *J. Gen. Physiol.* 121: 261–276.
Haissaguerre M, Chatel S, Sacher F, Weerasooriya R, Probst V et al. 2009. Ventricular fibrillation with prominent early repolarization associated with a rare variant of KCNJ8/KATP channel. *J. Cardiovasc. Electrophysiol.* 20: 93–98.
Haj-Yasein NN, Jensen V, Vindedal GF, Gundersen GA, Klungland A et al. 2011. Evidence that compromised K^+ spatial buffering contributes to the epileptogenic effect of mutations in the human Kir4.1 gene (KCNJ10). *Glia* 59: 1635–1642.
Hambrock A, Kayar T, Stumpp D, Osswald H. 2004. Effect of two amino acids in TM17 of Sulfonylurea receptor SUR1 on the binding of ATP-sensitive K^+ channel modulators. *Diabetes* 53(Suppl 3): S128–S134.
Hambrock A, Loffler Walz C, Kurachi Y, Quast U. 1998. Mg^{2+} and ATP dependence of K(ATP) channel modulator binding to the recombinant sulphonylurea receptor, SUR2B. *Brit. J. Pharmacol.* 125: 577–583.
Hanley PJ, Mickel M, Loffler M, Brandt U, Daut J. 2002. K(ATP) channel-independent targets of diazoxide and 5-hydroxydecanoate in the heart. *J. Physiol.* 542: 735–741.
Hansen SB, Tao X, MacKinnon R. 2011. Structural basis of PIP_2 activation of the classical inward rectifier K^+ channel Kir2.2. *Nature* 477: 495–498.
Harakalova M, van Harssel JJ, Terhal PA, van Lieshout S, Duran K et al. 2012. Dominant missense mutations in ABCC9 cause Cantu syndrome. *Nat. Genet.* 44: 793–796.
Hattersley AT, Ashcroft FM. 2005. Activating mutations in Kir6.2 and neonatal diabetes: New clinical syndromes, new scientific insights, and new therapy. *Diabetes* 54: 2503–2513.
Hebert SC, Desir G, Giebisch G, Wang W. 2005. Molecular diversity and regulation of renal potassium channels. *Physiol. Rev.* 85: 319–371.
Hejtmancik JF, Jiao X, Li A, Sergeev YV, Ding X et al. 2008. Mutations in KCNJ13 cause autosomal-dominant snowflake vitreoretinal degeneration. *Am. J. Hum. Genet.* 82: 174–180.
Hestrin S. 1987. The properties and function of inward rectification in rod photoreceptors of the tiger salamander. *J. Physiol.* 390: 319–333.
Hibino H, Fujita A, Iwai K, Yamada M, Kurachi Y. 2004a. Differential assembly of inwardly rectifying K^+ channel subunits, Kir4.1 and Kir5.1, in brain astrocytes. *J. Biol. Chem.* 279: 44065–44073.

Hibino H, Higashi-Shingai K, Fujita A, Iwai K, Ishii M, Kurachi Y. 2004b. Expression of an inwardly rectifying K^+ channel, Kir5.1, in specific types of fibrocytes in the cochlear lateral wall suggests its functional importance in the establishment of endocochlear potential. *Eur. J. Neurosci.* 19: 76–84.

Hibino H, Inanobe A, Furutani K, Murakami S, Findlay I, Kurachi Y. 2010a. Inwardly rectifying potassium channels: Their structure, function, and physiological roles. *Physiol. Rev.* 90: 291–366.

Hibino H, Kurachi Y. 2006. Molecular and physiological bases of the K^+ circulation in the mammalian inner ear. *Physiology* 21: 336–345.

Hibino H, Nin F, Tsuzuki C, Kurachi Y. 2010b. How is the highly positive endocochlear potential formed? The specific architecture of the stria vascularis and the roles of the ion-transport apparatus. *Pflug. Arch. Eur. J. Physiol.* 459: 521–533.

Hilgemann DW, Ball R. 1996. Regulation of cardiac Na^+,Ca^{2+} exchange and K_{ATP} potassium channels by PIP2. *Science* 273: 956–959.

Ho IH, Murrell-Lagnado RD. 1999. Molecular determinants for sodium-dependent activation of G protein-gated K^+ channels. *J. Biol. Chem.* 274: 8639–8648.

Ho K, Nichols CG, Lederer WJ, Lytton J, Vassilev PM et al. 1993. Cloning and expression of an inwardly rectifying ATP-regulated potassium channel. *Nature* 362: 31–38.

Hodgkin AL, Huxley AF, Katz B. 1949. Ionic current underlying activity in the giant axon of the squid. *Arch. Sci. Physiol.* 3: 129–150.

Hopkins WF, Fatherazi S, Peter-Riesch B, Corkey BE, Cook DL. 1992. Two sites for adenine-nucleotide regulation of ATP-sensitive potassium channels in mouse pancreatic beta-cells and HIT cells. *J. Membr. Biol.* 129: 287–295.

Hu H, Sato T, Seharaseyon J, Liu Y, Johns DC et al. 1999. Pharmacological and histochemical distinctions between molecularly defined sarcolemmal K_{ATP} channels and native cardiac mitochondrial KATP channels. *Mol. Pharmacol.* 55: 1000–1005.

Huang CL, Feng S, Hilgemann DW. 1998. Direct activation of inward rectifier potassium channels by PIP_2 and its stabilization by Gbetagamma. *Nature* 391: 803–806.

Hughes BA, Swaminathan A. 2008. Modulation of the Kir7.1 potassium channel by extracellular and intracellular pH. *Am. J. Physiol. Cell Physiol.* 294: C423–C431.

Inagaki N, Gonoi T, Clement JP, Namba N, Inazawa J et al. 1995a. Reconstitution of IK_{ATP}: An inward rectifier subunit plus the sulfonylurea receptor. *Science* 270: 1166–1170.

Inagaki N, Gonoi T, Clement JP, Wang CZ, Aguilar-Bryan L et al. 1996. A family of sulfonylurea receptors determines the pharmacological properties of ATP-sensitive K^+ channels. *Neuron* 16: 1011–1017.

Inagaki N, Gonoi T, Clement JPt, Namba N, Inazawa J et al. 1995b. Reconstitution of IK_{ATP}: An inward rectifier subunit plus the sulfonylurea receptor. *Science* 270: 1166–1170.

Inagaki N, Gonoi T, Seino S. 1997. Subunit stoichiometry of the pancreatic beta-cell ATP-sensitive K^+ channel. *FEBS Lett.* 409: 232–236.

Inanobe A, Fujita S, Makino Y, Matsushita K, Ishii M et al. 2001. Interaction between the RGS domain of RGS4 with G protein alpha subunits mediates the voltage-dependent relaxation of the G protein-gated potassium channel. *J. Physiol.* 535: 133–143.

Inanobe A, Horio Y, Fujita A, Tanemoto M, Hibino H et al. 1999a. Molecular cloning and characterization of a novel splicing variant of the Kir3.2 subunit predominantly expressed in mouse testis. *J. Physiol.* 521: 19–30.

Inanobe A, Yoshimoto Y, Horio Y, Morishige KI, Hibino H et al. 1999b. Characterization of G-protein-gated K^+ channels composed of Kir3.2 subunits in dopaminergic neurons of the substantia nigra. *J. Neurosci.* 19: 1006–1017.

Inoue I, Nagase H, Kishi K, Higuti T. 1991. ATP-sensitive K^+ channel in the mitochondrial inner membrane. *Nature* 352: 244–247.

Ishii M, Fujita A, Iwai K, Kusaka S, Higashi K et al. 2003. Differential expression and distribution of Kir5.1 and Kir4.1 inwardly rectifying K^+ channels in retina. *Am. J. Physiol. Cell Physiol.* 285: C260–C267.

Ivanina T, Rishal I, Varon D, Mullner C, Frohnwieser-Steinecke B et al. 2003. Mapping the Gbetagamma-binding sites in GIRK1 and GIRK2 subunits of the G protein-activated K^+ channel. *J. Biol. Chem.* 278: 29174–29183.

Jahangir A, Terzic A. 2005. K(ATP) channel therapeutics at the bedside. *J. Mol. Cell. Cardiol.* 39: 99–112.

Jelacic TM, Kennedy ME, Wickman K, Clapham DE. 2000. Functional and biochemical evidence for G-protein-gated inwardly rectifying K^+ (GIRK) channels composed of GIRK2 and GIRK3. *J. Biol. Chem.* 275: 36211–36216.

Jelacic TM, Sims SM, Clapham DE. 1999. Functional expression and characterization of G-protein-gated inwardly rectifying K^+ channels containing GIRK3. *J. Membr. Biol.* 169: 123–129.

Jiang Y, Lee A, Chen J, Ruta V, Cadene M et al. 2003. X-ray structure of a voltage-dependent K^+ channel. *Nature* 423: 33–41.

Jin T, Peng L, Mirshahi T, Rohacs T, Chan KW et al. 2002. The betagamma subunits of G proteins gate a K(+) channel by pivoted bending of a transmembrane segment. *Mol. Cell* 10: 469–481.

Kahle KT, Wilson FH, Leng Q, Lalioti MD, O'Connell AD et al. 2003. WNK4 regulates the balance between renal NaCl reabsorption and K^+ secretion. *Nat. Genet.* 35: 372–376.

Kakei M, Kelly RP, Ashcroft SJ, Ashcroft FM. 1986. The ATP-sensitivity of K^+ channels in rat pancreatic B-cells is modulated by ADP. *FEBS Lett.* 208: 63–66.

Kalsi AS, Greenwood K, Wilkin G, Butt AM. 2004. Kir4.1 expression by astrocytes and oligodendrocytes in CNS white matter: A developmental study in the rat optic nerve. *J. Anat.* 204: 475–485.

Kane GC, Behfar A, Dyer RB, O'Cochlain DF, Liu XK et al. 2006. KCNJ11 gene knockout of the Kir6.2 K_{ATP} channel causes maladaptive remodeling and heart failure in hypertension. *Hum. Mol. Genet.* 15: 2285–2297.

Kantor PF, Coetzee WA, Carmeliet EE, Dennis SC, Opie LH. 1990. Reduction of ischemic K^+ loss and arrhythmias in rat hearts. Effect of glibenclamide, a sulfonylurea. *Circ. Res.* 66: 478–485.

Karges B, Meissner T, Icks A, Kapellen T, Holl RW. 2012. Management of diabetes mellitus in infants. *Nat. Rev. Endocrinol.* 8: 201–211.

Karle CA, Zitron E, Zhang W, Wendt-Nordahl G, Kathofer S et al. 2002. Human cardiac inwardly-rectifying K^+ channel Kir(2.1b) is inhibited by direct protein kinase C-dependent regulation in human isolated cardiomyocytes and in an expression system. *Circulation* 106: 1493–1499.

Katz B. 1949. Les constantes electriques de la membrane du muscle. *Arch. Sci. Physiol.* 2: 285–299.

Kaufhold MA, Krabbenhoft A, Song P, Engelhardt R, Riederer B et al. 2008. Localization, trafficking, and significance for acid secretion of parietal cell Kir4.1 and KCNQ1 K^+ channels. *Gastroenterology* 134: 1058–1069.

Kleopa KA. 2011. Autoimmune channelopathies of the nervous system. *Curr. Neuropharmacol.* 9: 458–467.

Kobayashi T, Ikeda K, Kumanishi T. 2000. Inhibition by various antipsychotic drugs of the G-protein-activated inwardly rectifying K(+) (GIRK) channels expressed in xenopus oocytes. *Brit. J. Pharmacol.* 129: 1716–1722.

Kobayashi T, Washiyama K, Ikeda K. 2004. Inhibition of G protein-activated inwardly rectifying K^+ channels by various antidepressant drugs. *Neuropsychopharmacology* 29: 1841–1851.

Kofuji P, Davidson N, Lester HA. 1995. Evidence that neuronal G-protein-gated inwardly rectifying K^+ channels are activated by G beta gamma subunits and function as heteromultimers. *Proc. Natl. Acad. Sci. U.S.A.* 92: 6542–6546.

Kondo C, Isomoto S, Matsumoto S, Yamada M, Horio Y et al. 1996. Cloning and functional expression of a novel isoform of ROMK inwardly rectifying ATP-dependent K+ channel, ROMK6 (Kir1.1f). *FEBS Lett.* 399: 122–126.

Kono Y, Horie M, Takano M, Otani H, Xie LH et al. 2000. The properties of the Kir6.1–6.2 tandem channel co-expressed with SUR2A. *Pflug. Arch. Eur. J. Physiol.* 440: 692–698.

Konstas AA, Korbmacher C, Tucker SJ. 2003. Identification of domains that control the heteromeric assembly of Kir5.1/Kir4.0 potassium channels. *Am. J. Physiol. Cell Physiol.* 284: C910–C917.

Koster JC, Cadario F, Peruzzi C, Colombo C, Nichols CG, Barbetti F. 2008. The G53D mutation in Kir6.2 (KCNJ11) is associated with neonatal diabetes and motor dysfunction in adulthood that is improved with sulfonylurea therapy. *J. Clin. Endocrinol. Metab.* 93: 1054–1061.

Koster JC, Knopp A, Flagg TP, Markova KP, Sha Q et al. 2001. Tolerance for ATP-insensitive K(ATP) channels in transgenic mice. *Circ. Res.* 89: 1022–1029.

Koster JC, Marshall BA, Ensor N, Corbett JA, Nichols CG. 2000. Targeted overactivity of beta cell K(ATP) channels induces profound neonatal diabetes. *Cell* 100: 645–654.

Koster JC, Permutt MA, Nichols CG. 2005a. Diabetes and insulin secretion: The ATP-sensitive K+ channel (K_{ATP}) connection. *Diabetes* 54: 3065–3072.

Koster JC, Remedi MS, Dao C, Nichols CG. 2005b. ATP and sulfonylurea sensitivity of mutant ATP-sensitive K+ channels in neonatal diabetes: Implications for pharmacogenomic therapy. *Diabetes* 54: 2645–2654.

Koster JC, Sha Q, Nichols CG. 1999. Sulfonylurea and K(+)-channel opener sensitivity of K(ATP) channels. Functional coupling of Kir6.2 and SUR1 subunits. *J. Gen. Physiol.* 114: 203–213.

Koyrakh L, Lujan R, Colon J, Karschin C, Kurachi Y et al. 2005. Molecular and cellular diversity of neuronal G-protein-gated potassium channels. *J. Neurosci.* 25: 11468–11478.

Krapivinsky G, Gordon EA, Wickman K, Velimirovic B, Krapivinsky L, Clapham DE. 1995. The G-protein-gated atrial K+ channel IKACh is a heteromultimer of two inwardly rectifying K(+)-channel proteins. *Nature* 374: 135–141.

Krapivinsky G, Medina I, Eng L, Krapivinsky L, Yang Y, Clapham DE. 1998. A novel inward rectifier K+ channel with unique pore properties. *Neuron* 20: 995–1005.

Kubo Y, Baldwin TJ, Jan YN, Jan LY. 1993. Primary structure and functional expression of a mouse inward rectifier potassium channel. *Nature* 362: 127–133.

Kubo Y, Murata Y. 2001. Control of rectification and permeation by two distinct sites after the second transmembrane region in Kir2.1 K+ channel. *J. Physiol.* 531: 645–660.

Kucheryavykh YV, Kucheryavykh LY, Nichols CG, Maldonado HM, Baksi K et al. 2007a. Downregulation of Kir4.1 inward rectifying potassium channel subunits by RNAi impairs potassium transfer and glutamate uptake by cultured cortical astrocytes. *Glia* 55: 274–281.

Kucheryavykh YV, Pearson WL, Kurata HT, Eaton MJ, Skatchkov SN, Nichols CG. 2007b. Polyamine permeation and rectification of Kir4.1 channels. *Channels* 1: 172–178.

Kuo A, Domene C, Johnson LN, Doyle DA, Venien-Bryan C. 2005. Two different conformational states of the KirBac3.1 potassium channel revealed by electron crystallography.[see comment]. *Structure* 13: 1463–1472.

Kuo A, Gulbis JM, Antcliff JF, Rahman T, Lowe ED et al. 2003. Crystal structure of the potassium channel KirBac1.1 in the closed state. *Science* 300: 1922–1926.

Kurachi Y, Nakajima T, Sugimoto T. 1986. On the mechanism of activation of muscarinic K+ channels by adenosine in isolated atrial cells: Involvement of GTP-binding proteins. *Pflug. Arch. Eur. J. Physiol.* 407: 264–274.

Kurata HT, Cheng WW, Arrabit C, Slesinger PA, Nichols CG. 2007. The role of the cytoplasmic pore in inward rectification of Kir2.1 channels. *J. Gen. Physiol.* 130: 145–155.

Kurata HT, Diraviyam K, Marton LJ, Nichols CG. 2008. Blocker protection by short spermine analogs: Refined mapping of the spermine binding site in a Kir channel. *Biophys. J.* 95: 3827–3839.

Kurata HT, Marton LJ, Nichols CG. 2006. The polyamine binding site in inward rectifier K+ channels. *J. Gen. Physiol.* 127: 467–480.

Kurata HT, Phillips LR, Rose T, Loussouarn G, Herlitze S et al. 2004. Molecular basis of inward rectification: Polyamine interaction sites located by combined channel and ligand mutagenesis. *J. Gen. Physiol.* 124: 541–554.

Kurata HT, Zhu EA, Nichols CG. 2010. Locale and chemistry of spermine binding in the archetypal inward rectifier Kir2.1. *J. Gen. Physiol.* 135: 495–508.

Lachheb S, Cluzeaud F, Bens M, Genete M, Hibino H et al. 2008. Kir4.1/Kir5.1 channel forms the major K+ channel in the basolateral membrane of mouse renal collecting duct principal cells. *Am. J. Physiol. Renal. Physiol.* 294: F1398–F1407.

Lederer WJ, Nichols CG. 1989. Nucleotide modulation of the activity of rat heart ATP- sensitive K+ channels in isolated membrane patches. *J. Physiol.* 419: 193–211.

Lederer WJ, Nichols CG, Smith GL. 1989. The mechanism of early contractile failure of isolated rat ventricular myocytes subjected to complete metabolic inhibition. *J. Physiol.* 413: 329–349.

Lehnart SE, Ackerman MJ, Benson DW, Jr., Brugada R, Clancy CE et al. 2007. Inherited arrhythmias: A National Heart, Lung, and Blood Institute and Office of Rare Diseases workshop consensus report about the diagnosis, phenotyping, molecular mechanisms, and therapeutic approaches for primary cardiomyopathies of gene mutations affecting ion channel function. *Circulation* 116: 2325–2345.

Leng Q, Kahle KT, Rinehart J, MacGregor GG, Wilson FH et al. 2006a. WNK3, a kinase related to genes mutated in hereditary hypertension with hyperkalaemia, regulates the K+ channel ROMK1 (Kir1.1). *J. Physiol.* 571: 275–286.

Leng Q, MacGregor GG, Dong K, Giebisch G, Hebert SC. 2006b. Subunit-subunit interactions are critical for proton sensitivity of ROMK: Evidence in support of an intermolecular gating mechanism. *Proc. Natl. Acad. Sci. U.S.A.* 103: 1982–1987.

Lesage F, Duprat F, Fink M, Guillemare E, Coppola T et al. 1994. Cloning provides evidence for a family of inward rectifier and G-protein coupled K+ channels in the brain. *FEBS Lett.* 353: 37–42.

Lesage F, Guillemare E, Fink M, Duprat F, Heurteaux C et al. 1995. Molecular properties of neuronal G-protein-activated inwardly rectifying K+ channels. *J. Biol. Chem.* 270: 28660–28667.

Lewohl JM, Wilson WR, Mayfield RD, Brozowski SJ, Morrisett RA, Harris RA. 1999. G-protein-coupled inwardly rectifying potassium channels are targets of alcohol action. *Nat. Neurosci.* 2: 1084–1090.

Li J, McLerie M, Lopatin AN. 2004. Transgenic upregulation of IK1 in the mouse heart leads to multiple abnormalities of cardiac excitability. *Am. J. Physiol. Heart Circ. Physiol.* 287: H2790–H2802.

Li RA, Leppo M, Miki T, Seino S, Marban E. 2000. Molecular basis of electrocardiographic ST-segment elevation. *Circ. Res.* 87: 837–839.

Liao YJ, Jan YN, Jan LY. 1996. Heteromultimerization of G-protein-gated inwardly rectifying K+ channel proteins GIRK1 and GIRK2 and their altered expression in weaver brain. *J. Neurosci.* 16: 7137–7150.

Light PE, Allen BG, Walsh MP, French RJ. 1995. Regulation of adenosine triphosphate-sensitive potassium channels from rabbit ventricular myocytes by protein kinase C and type 2A protein phosphatase. *Biochemistry* 34: 7252–7257.

Light PE, Bladen C, Winkfein RJ, Walsh MP, French RJ. 2000. Molecular basis of protein kinase C-induced activation of ATP-sensitive potassium channels. *Proc. Natl. Acad. Sci. U.S.A.* 97: 9058–9063.

Light PE, Sabir AA, Allen BG, Walsh MP, French RJ. 1996. Protein kinase C-induced changes in the stoichiometry of ATP binding activate cardiac ATP-sensitive K^+ channels. A possible mechanistic link to ischemic preconditioning. *Circ. Res.* 79: 399–406.

Lin D, Sterling H, Lerea KM, Giebisch G, Wang WH. 2002. Protein kinase C (PKC)-induced phosphorylation of ROMK1 is essential for the surface expression of ROMK1 channels. *J. Biol. Chem.* 277: 44278–44284.

Lin YF, Jan YN, Jan LY. 2000. Regulation of ATP-sensitive potassium channel function by protein kinase A-mediated phosphorylation in transfected HEK293 cells. *EMBO J.* 19: 942–955.

Liou HH, Zhou SS, Huang CL. 1999. Regulation of ROMK1 channel by protein kinase A via a phosphatidylinositol 4,5-bisphosphate-dependent mechanism. *Proc. Natl. Acad. Sci. U.S.A.* 96: 5820–5825.

Liu GX, Hanley PJ, Ray J, Daut J. 2001a. Long-chain acyl-coenzyme A esters and fatty acids directly link metabolism to K(ATP) channels in the heart. *Circ. Res.* 88: 918–924.

Liu XK, Yamada S, Kane GC, Alekseev AE, Hodgson DM et al. 2004. Genetic disruption of Kir6.2, the pore-forming subunit of ATP-sensitive K^+ channel, predisposes to catecholamine-induced ventricular dysrhythmia. *Diabetes* 53(Suppl 3): S165–S168.

Liu Y, Ren G, O'Rourke B, Marban E, Seharaseyon J. 2001b. Pharmacological comparison of native mitochondrial K(ATP) channels with molecularly defined surface K(ATP) channels. *Mol. Pharmacol.* 59: 225–230.

Long SB, Campbell EB, Mackinnon R. 2005. Crystal structure of a mammalian voltage-dependent Shaker family K^+ channel. *Science* 309: 897–903.

Lopatin AN, Makhina EN, Nichols CG. 1994. Potassium channel block by cytoplasmic polyamines as the mechanism of intrinsic rectification. *Nature* 372: 366–369.

Lopatin AN, Makhina EN, Nichols CG. 1995. The mechanism of inward rectification of potassium channels: "long-pore plugging" by cytoplasmic polyamines. *J. Gen. Physiol.* 106: 923–955.

Lopatin AN, Shantz LM, Mackintosh CA, Nichols CG, Pegg AE. 2000. Modulation of potassium channels in the hearts of transgenic and mutant mice with altered polyamine biosynthesis. *J. Mol. Cell. Cardiol.* 32: 2007–2024.

Lopes CM, Zhang H, Rohacs T, Jin T, Yang J, Logothetis DE. 2002. Alterations in conserved Kir channel-PIP2 interactions underlie channelopathies. *Neuron* 34: 933–944.

Lorenz JN, Baird NR, Judd LM, Noonan WT, Andringa A et al. 2002. Impaired renal NaCl absorption in mice lacking the ROMK potassium channel, a model for type II Bartter's syndrome. *J. Biol. Chem.* 277: 37871–37880.

Lourdel S, Paulais M, Cluzeaud F, Bens M, Tanemoto M et al. 2002. An inward rectifier K(+) channel at the basolateral membrane of the mouse distal convoluted tubule: Similarities with Kir4-Kir5.1 heteromeric channels. *J. Physiol.* 538: 391–404.

Lu M, Leng Q, Egan ME, Caplan MJ, Boulpaep EL et al. 2006. CFTR is required for PKA-regulated ATP sensitivity of Kir1.1 potassium channels in mouse kidney. *J. Clin. Invest.* 116: 797–807.

Lu M, Wang T, Yan Q, Yang X, Dong K et al. 2002. Absence of small conductance K^+ channel (SK) activity in apical membranes of thick ascending limb and cortical collecting duct in ROMK (Bartter's) knockout mice. *J. Biol. Chem.* 277: 37881–37887.

Lu Z, MacKinnon R. 1994. Electrostatic tuning of Mg^{2+} affinity in an inward-rectifier K^+ channel. *Nature* 371: 243–246.

Ma D, Zerangue N, Lin YF, Collins A, Yu M et al. 2001. Role of ER export signals in controlling surface potassium channel numbers. *Science* 291: 316–319.

MacGregor GG, Dong K, Vanoye CG, Tang L, Giebisch G, Hebert SC. 2002. Nucleotides and phospholipids compete for binding to the C terminus of K_{ATP} channels. *Proc. Natl. Acad. Sci. U.S.A.* 99: 2726–2731.

Manning Fox JE, Nichols CG, Light PE. 2004. Activation of adenosine triphosphate-sensitive potassium channels by acyl coenzyme A esters involves multiple phosphatidylinositol 4,5-bisphosphate-interacting residues. *Mol. Endocrinol.* 18: 679–686.

Mao J, Wu J, Chen F, Wang X, Jiang C. 2003. Inhibition of G-protein-coupled inward rectifying K^+ channels by intracellular acidosis. *J. Biol. Chem.* 278: 7091–7098.

Masia R, Enkvetchakul D, Nichols CG. 2005b. Differential nucleotide regulation of K_{ATP} channels by SUR1 and SUR2A. *J. Mol. Cell. Cardiol.* 39: 491–501.

Matsuda H, Saigusa A, Irisawa H. 1987. Ohmic conductance through the inwardly rectifying K channel and blocking by internal Mg^{2+}. *Nature* 325: 156–159.

McDevitt MA, Thorsness RJ, Levine JE. 2009. A role for ATP-sensitive potassium channels in male sexual behavior. *Horm. Behav.* 55: 366–374.

McNicholas CM, MacGregor GG, Islas LD, Yang Y, Hebert SC, Giebisch G. 1998. pH-dependent modulation of the cloned renal K^+ channel, ROMK. *Am. J. Physiol. Renal Physiol.* 275: F972–F981.

McNicholas CM, Wang W, Ho K, Hebert SC, Giebisch G. 1994. Regulation of ROMK1 K^+ channel activity involves phosphorylation processes. *Proc. Natl. Acad. Sci. USA* 91: 8077–8081.

Medeiros-Domingo A, Tan BH, Crotti L, Tester DJ, Eckhardt L et al. 2010. Gain-of-function mutation S422L in the KCNJ8-encoded cardiac K(ATP) channel Kir6.1 as a pathogenic substrate for J-wave syndromes. *Heart Rhythm* 7: 1466–1471.

Mikhailov MV, Campbell JD, de Wet H, Shimomura K, Zadek B et al. 2005. 3-D structural and functional characterization of the purified K_{ATP} channel complex Kir6.2-SUR1. *EMBO J.* 24: 4166–4175.

Miki T, Suzuki M, Shibasaki T, Uemura H, Sato T et al. 2002. Mouse model of Prinzmetal angina by disruption of the inward rectifier Kir6.1. *Nat. Med.* 8: 466–472.

Morishige K, Inanobe A, Yoshimoto Y, Kurachi H, Murata Y et al. 1999. Secretagogue-induced exocytosis recruits G protein-gated K^+ channels to plasma membrane in endocrine cells. *J. Biol. Chem.* 274: 7969–7974.

Morishige K, Takahashi N, Jahangir A, Yamada M, Koyama H et al. 1994. Molecular cloning and functional expression of a novel brain-specific inward rectifier potassium channel. *FEBS Lett.* 346: 251–256.

Nakajima Y, Nakajima S, Inoue M. 1988. Pertussis toxin-insensitive G protein mediates substance P-induced inhibition of potassium channels in brain neurons. *Proc. Natl. Acad. Sci. U.S.A.* 85: 3643–3647.

Nakamura N, Suzuki Y, Sakuta H, Ookata K, Kawahara K, Hirose S. 1999. Inwardly rectifying K^+ channel Kir7.1 is highly expressed in thyroid follicular cells, intestinal epithelial cells and choroid plexus epithelial cells: Implication for a functional coupling with Na^+,K^+-ATPase. *Biochem. J.* 342(Part 2): 329–336.

Navarro B, Kennedy ME, Velimirovic B, Bhat D, Peterson AS, Clapham DE. 1996. Nonselective and G(beta-gamma)-insensitive weaver K^+ channels. *Science* 272: 1950–1953.

Neusch C, Papadopoulos N, Muller M, Maletzki I, Winter SM et al. 2006. Lack of the Kir4.1 channel subunit abolishes K^+ buffering properties of astrocytes in the ventral respiratory group: Impact on extracellular K^+ regulation. *J. Neurophysiol.* 95: 1843–1852.

Neusch C, Rozengurt N, Jacobs RE, Lester HA, Kofuji P. 2001. Kir4.1 potassium channel subunit is crucial for oligodendrocyte development and in vivo myelination. *J. Neurosci.*21: 5429–5438.

Newman EA. 1993. Inward-rectifying potassium channels in retinal glial (Muller) cells. *J. Neurosci.* 13: 3333–3345.

Nichols CG. 2006. KATP channels as molecular sensors of cellular metabolism. *Nature* 440: 470–476.

Nichols CG, Koster JC, Remedi MS. 2007. Beta-cell hyperexcitability: From hyperinsulinism to diabetes. *Diabetes Obes. Metab.* 9: 81–88.

Nichols CG, Lederer WJ. 1991. Adenosine triphosphate-sensitive potassium channels in the cardiovascular system. *Am J. Physiol. Heart Circ. Physiol.* 261: H1675–H1686.

Nichols CG, Shyng SL, Nestorowicz A, Glaser B, Clement JP et al. 1996. Adenosine diphosphate as an intracellular regulator of insulin secretion. *Science* 272: 1785–1787.

Nichols CG, Singh GK, Grange DK. 2013. K_{ATP} channels and cardiovascular disease: Suddenly a syndrome. *Circ. Res.* 112: 1059–1072.

Nishida M, Cadene M, Chait BT, MacKinnon R. 2007. Crystal structure of a Kir3.1-prokaryotic Kir channel chimera. *EMBO J.* 26: 4005–4015.

Nishida M, MacKinnon R. 2002. Structural basis of inward rectification: Cytoplasmic pore of the G protein-gated inward rectifier GIRK1 at 1.8 A resolution. *Cell* 111: 957–965.

Noma A. 1983. ATP-regulated K^+ channels in cardiac muscle. *Nature* 305: 147–148.

O'Connell AD, Leng Q, Dong K, MacGregor GG, Giebisch G, Hebert SC. 2005. Phosphorylation-regulated endoplasmic reticulum retention signal in the renal outer-medullary K^+ channel (ROMK). *Proc. Natl. Acad. Sci. U.S.A.* 102: 9954–9959.

Okuyama Y, Yamada M, Kondo C, Satoh E, Isomoto S et al. 1998. The effects of nucleotides and potassium channel openers on the SUR2A/Kir6.2 complex K^+ channel expressed in a mammalian cell line, HEK293T cells. *Pflug. Arch. Eur. J. Physiol.* 435: 595–603.

Olsen ML, Sontheimer H. 2008. Functional implications for Kir4.1 channels in glial biology: From K^+ buffering to cell differentiation. *J. Neurochem.* 107: 589–601.

Olson TM, Alekseev AE, Moreau C, Liu XK, Zingman LV et al. 2007. K_{ATP} channel mutation confers risk for vein of Marshall adrenergic atrial fibrillation. *Nat. Clin. Pract. Cardiovasc. Med.* 4: 110–116.

Ookata K, Tojo A, Suzuki Y, Nakamura N, Kimura K et al. 2000. Localization of inward rectifier potassium channel Kir7.1 in the basolateral membrane of distal nephron and collecting duct. *J. Am. Soc. Nephrol.* 11: 1987–1994.

Partiseti M, Collura V, Agnel M, Culouscou JM, Graham D. 1998. Cloning and characterization of a novel human inwardly rectifying potassium channel predominantly expressed in small intestine. *FEBS Lett.* 434: 171–176.

Patil N, Cox DR, Bhat D, Faham M, Myers RM, Peterson AS. 1995. A potassium channel mutation in weaver mice implicates membrane excitability in granule cell differentiation. *Nat. Genet.* 11: 126–129.

Pattnaik BR, Tokarz S, Asuma MP, Schroeder T, Sharma A et al. 2013. Snowflake Vitreoretinal Degeneration (SVD) mutation R162W provides new insights into Kir7.1 ion channel structure and function. *PloS One* 8: e71744.

Paucek P, Yarov-Yarovoy V, Sun X, Garlid KD. 1996. Inhibition of the mitochondrial K_{ATP} channel by long-chain acyl-CoA esters and activation by guanine nucleotides. *J. Biol. Chem.* 271: 32084–32088.

Paulais M, Bloch-Faure M, Picard N, Jacques T, Ramakrishnan SK et al. 2011. Renal phenotype in mice lacking the Kir5.1 (Kcnj16) K^+ channel subunit contrasts with that observed in SeSAME/EAST syndrome. *Proc. Natl. Acad. Sci. U.S.A.* 108: 10361–10366.

Pearson ER, Flechtner I, Njolstad PR, Malecki MT, Flanagan SE et al. 2006. Switching from insulin to oral sulfonylureas in patients with diabetes due to Kir6.2 mutations. *N. Engl. J. Med.* 355: 467–477.

Pearson WL, Dourado M, Schreiber M, Salkoff L, Nichols CG. 1999. Expression of a functional Kir4 family inward rectifier K^+ channel from a gene cloned from mouse liver. *J. Physiol.* 514: 639–653.

Pearson WL, Nichols CG. 1998. Block of the Kir2.1 channel pore by alkylamine analogues of endogenous polyamines. *J. Gen. Physiol.* 112: 351–363.

Pegan S, Arrabit C, Zhou W, Kwiatkowski W, Collins A et al. 2005a. Cytoplasmic domain structures of Kir2.1 and Kir3.1 show sites for modulating gating and rectification. *Nat. Neurosci.* 8: 279–287.

Pegan S, Arrabit C, Zhou W, Kwiatkowski W, Collins A et al. 2005b. Cytoplasmic domain structures of Kir2.1 and Kir3.1 show sites for modulating gating and rectification. Erratum appears in *Nat. Neurosci.* 8:835, 279–287.

Pessia M, Imbrici P, D'Adamo MC, Salvatore L, Tucker SJ. 2001. Differential pH sensitivity of Kir4.1 and Kir4.2 potassium channels and their modulation by heteropolymerisation with Kir5.1. *J. Physiol.* 532: 359–367.

Peters M, Ermert S, Jeck N, Derst C, Pechmann U et al. 2003. Classification and rescue of ROMK mutations underlying hyperprostaglandin E syndrome/antenatal Bartter syndrome. *Kidney Int.* 64: 923–932.

Peters M, Jeck N, Reinalter S, Leonhardt A, Tonshoff B et al. 2002. Clinical presentation of genetically defined patients with hypokalemic salt-losing tubulopathies. *Am. J. Med.* 112: 183–190.

Pfaffinger PJ, Martin JM, Hunter DD, Nathanson NM, Hille B. 1985. GTP-binding proteins couple cardiac muscarinic receptors to a K channel. *Nature* 317: 536–538.

Piao L, Li J, McLerie M, Lopatin AN. 2007. Cardiac IK1 underlies early action potential shortening during hypoxia in the mouse heart. *J. Mol. Cell. Cardiol.* 43: 27–38.

Plaster NM, Tawil R, Tristani-Firouzi M, Canun S, Bendahhou S et al. 2001. Mutations in Kir2.1 cause the developmental and episodic electrical phenotypes of Andersen's syndrome. *Cell* 105: 511–519.

Pountney DJ, Sun ZQ, Porter LM, Nitabach MN, Nakamura TY et al. 2001. Is the molecular composition of K(ATP) channels more complex than originally thought? *J. Mol. Cell. Cardiol.* 33: 1541–1546.

Priori SG, Pandit SV, Rivolta I, Berenfeld O, Ronchetti E et al. 2005. A novel form of short QT syndrome (SQT3) is caused by a mutation in the KCNJ2 gene. *Circ. Res.* 96: 800 807.

Proks P, Capener CE, Jones P, Ashcroft FM. 2001. Mutations within the P-loop of Kir6.2 modulate the intraburst kinetics of the ATP-sensitive potassium channel. *J. Gen. Physiol.* 118: 341–353.

Quayle JM, Bonev AD, Brayden JE, Nelson MT. 1995. Pharmacology of ATP-sensitive K^+ currents in smooth muscle cells from rabbit mesenteric artery. *Am. J. Physiol. Cell Physiol.* 269: C1112–C1118.

Quinn KV, Cui Y, Giblin JP, Clapp LH, Tinker A. 2003. Do anionic phospholipids serve as cofactors or second messengers for the regulation of activity of cloned ATP-sensitive K^+ channels? *Circ. Res.* 93: 646–655.

Rapedius M, Haider S, Browne KF, Shang L, Sansom MS et al. 2006. Structural and functional analysis of the putative pH sensor in the Kir1.1 (ROMK) potassium channel. *EMBO Rep.* 7: 611–616.

Rapedius M, Paynter JJ, Fowler PW, Shang L, Sansom MS et al. 2007. Control of pH and PIP_2 gating in heteromeric Kir4.1/Kir5.1 channels by H-Bonding at the helix-bundle crossing. *Channels* 1: 327–330.

Ravens U, Cerbai E. 2008. Role of potassium currents in cardiac arrhythmias. *Europace* 10: 1133–1137.

Reichold M, Zdebik AA, Lieberer E, Rapedius M, Schmidt K et al. 2010. KCNJ10 gene mutations causing EAST syndrome (epilepsy, ataxia, sensorineural deafness, and tubulopathy) disrupt channel function. *Proc. Natl. Acad. Sci. U.S.A.* 107: 14490–14495.

Reimann F, Tucker SJ, Proks P, Ashcroft FM. 1999. Involvement of the n-terminus of Kir6.2 in coupling to the sulphonylurea receptor. *J. Physiol.* 518: 325–336.

Remedi MS, Kurata HT, Scott A, Wunderlich FT, Rother E et al. 2009. Secondary consequences of beta cell inexcitability: Identification and prevention in a murine model of K(ATP)-induced neonatal diabetes mellitus. *Cell Metab.* 9: 140–151.

Ribalet B, John SA, Xie LH, Weiss JN. 2005. Regulation of the ATP-sensitive K channel Kir6.2 by ATP and PIP(2). *J. Mol. Cell. Cardiol.* 39: 71–77.

Robbins TW, Everitt BJ. 1999. Drug addiction: Bad habits add up. *Nature* 398: 567–570.

Rohacs T, Lopes CM, Jin T, Ramdya PP, Molnar Z, Logothetis DE. 2003. Specificity of activation by phosphoinositides determines lipid regulation of Kir channels. *Proc. Natl. Acad. Sci. U.S.A.* 100: 745–750.

Rorsman P, Salehi SA, Abdulkader F, Braun M, MacDonald PE. 2008. K(ATP)-channels and glucose-regulated glucagon secretion. *Trends Endocrinol. Metab.* 19: 277–284.

Rosenhouse-Dantsker A, Sui JL, Zhao Q, Rusinova R, Rodriguez-Menchaca AA et al. 2008. A sodium-mediated structural switch that controls the sensitivity of Kir channels to PtdIns(4,5)P(2). *Nat. Chem. Biol.* 4: 624–631.

Ross EM, Wilkie TM. 2000. GTPase-activating proteins for heterotrimeric G proteins: Regulators of G protein signaling (RGS) and RGS-like proteins. *Annu. Rev. Biochem.* 69: 795–827.

Rozengurt N, Lopez I, Chiu CS, Kofuji P, Lester HA, Neusch C. 2003. Time course of inner ear degeneration and deafness in mice lacking the Kir4.1 potassium channel subunit. *Hearing Res.* 177: 71–80.

Ryan DP, da Silva MR, Soong TW, Fontaine B, Donaldson MR et al. 2010. Mutations in potassium channel Kir2.6 cause susceptibility to thyrotoxic hypokalemic periodic paralysis. *Cell* 140: 88–98.

Sackin H, Nanazashvili M, Palmer LG, Krambis M, Walters DE. 2005. Structural locus of the pH gate in the Kir1.1 inward rectifier channel. *Biophys. J.* 88: 2597–2606.

Sackin H, Nanazashvili M, Palmer LG, Li H. 2006. Role of conserved glycines in pH gating of Kir1.1 (ROMK). *Biophys. J.* 90: 3582–3589.

Saitoh O, Kubo Y, Miyatani Y, Asano T, Nakata H. 1997. RGS8 accelerates G-protein-mediated modulation of K^+ currents. *Nature* 390: 525–529.

Sala-Rabanal M, Kucheryavykh LY, Skatchkov SN, Eaton MJ, Nichols CG. 2010. Molecular mechanisms of EAST/SeSAME syndrome mutations in Kir4.1 (KCNJ10). *J. Biol. Chem.* 285: 36040–36048.

Sansone V, Tawil R. 2007. Management and treatment of Andersen-Tawil syndrome (ATS). *Neurotherapeutics* 4: 233–237.

Scherer D, Kiesecker C, Kulzer M, Gunth M, Scholz EP et al. 2007. Activation of inwardly rectifying Kir2.x potassium channels by beta 3-adrenoceptors is mediated via different signaling pathways with a predominant role of PKC for Kir2.1 and of PKA for Kir2.2. *Naunyn Schmiedebergs Arch. Pharmacol.* 375: 311–322.

Scholl UI, Choi M, Liu T, Ramaekers VT, Hausler MG et al. 2009. Seizures, sensorineural deafness, ataxia, mental retardation, and electrolyte imbalance (SeSAME syndrome) caused by mutations in KCNJ10. *Proc. Natl. Acad. Sci. U.S.A.* 106: 5842–5847.

Schram G, Pourrier M, Wang Z, White M, Nattel S. 2003. Barium block of Kir2 and human cardiac inward rectifier currents: Evidence for subunit-heteromeric contribution to native currents. *Cardiovasc. Res.* 59: 328–338.

Schulte U, Hahn H, Konrad M, Jeck N, Derst C et al. 1999. pH gating of ROMK (K(ir)1.1) channels: Control by an Arg-Lys-Arg triad disrupted in antenatal Bartter syndrome. *Proc. Natl. Acad. Sci. U.S.A.* 96: 15298–15303.

Schulze D, Krauter T, Fritzenschaft H, Soom M, Baukrowitz T. 2003. Phosphatidylinositol 4,5-bisphosphate (PIP2) modulation of ATP and pH sensitivity in Kir channels. A tale of an active and a silent PIP_2 site in the N terminus. *J. Biol. Chem.* 278: 10500–10505.

Schwanstecher M, Sieverding C, Dorschner H, Gross I, Aguilar-Bryan L et al. 1998. Potassium channel openers require ATP to bind to and act through sulfonylurea receptors. *EMBO J.* 17: 5529–5535.

Seharaseyon J, Sasaki N, Ohler A, Sato T, Fraser H et al. 2000. Evidence against functional heteromultimerization of the K_{ATP} channel subunits Kir6.1 and Kir6.2. *J. Biol. Chem.* 275: 17561–17565.

Sharp GW. 1996. Mechanisms of inhibition of insulin release. *Am. J. Physiol. Cell Physiol.* 271: C1781–C1799.

Shi NQ, Ye B, Makielski JC. 2005. Function and distribution of the SUR isoforms and splice variants. *J. Mol. Cell. Cardiol.* 39: 51–60.

Shi Y, Cui N, Shi W, Jiang C. 2008. A short motif in Kir6.1 consisting of four phosphorylation repeats underlies the vascular K_{ATP} channel inhibition by protein kinase C. *J. Biol. Chem.* 283: 2488–2494.

Shigematsu S, Sato T, Abe T, Saikawa T, Sakata T, Arita M. 1995. Pharmacological evidence for the persistent activation of ATP-sensitive K^+ channels in early phase of reperfusion and its protective role against myocardial stunning. *Circulation* 92: 2266–2275.

Shin HG, Lu Z. 2005. Mechanism of the voltage sensitivity of IRK1 inward-rectifier K^+ channel block by the polyamine spermine. *J. Gen. Physiol.* 125: 413–426.

Shyng S, Ferrigni T, Nichols CG. 1997. Regulation of K_{ATP} channel activity by diazoxide and MgADP. Distinct functions of the two nucleotide binding folds of the sulfonylurea receptor. *J. Gen. Physiol.* 110: 643–654.

Shyng S, Nichols CG. 1997. Octameric stoichiometry of the K_{ATP} channel complex. *J. Gen. Physiol.* 110: 655–664.

Shyng SL, Cukras CA, Harwood J, Nichols CG. 2000. Structural determinants of PIP(2) regulation of inward rectifier K(ATP) channels. *J. Gen. Physiol.* 116: 599–608.

Shyng SL, Nichols CG. 1998. Membrane phospholipid control of nucleotide sensitivity of K_{ATP} channels. *Science* 282: 1138–1141.

Sicca F, Imbrici P, D'Adamo MC, Moro F, Bonatti F et al. 2011. Autism with seizures and intellectual disability: Possible causative role of gain-of-function of the inwardly-rectifying K^+ channel Kir4.1. *Neurobiol. Dis.* 43: 239–247.

Signorini S, Liao YJ, Duncan SA, Jan LY, Stoffel M. 1997. Normal cerebellar development but susceptibility to seizures in mice lacking G protein-coupled, inwardly rectifying K^+ channel GIRK2. *Proc. Natl. Acad. Sci. U.S.A.* 94: 923–927.

Singh H, Hudman D, Lawrence CL, Rainbow RD, Lodwick D, Norman RI. 2003. Distribution of Kir6.0 and SUR2 ATP-sensitive potassium channel subunits in isolated ventricular myocytes. [see comment]. *J. Mol. Cell. Cardiol.* 35: 445–459.

Soom M, Schonherr R, Kubo Y, Kirsch C, Klinger R, Heinemann SH. 2001. Multiple PIP_2 binding sites in Kir2.1 inwardly rectifying potassium channels. *FEBS Lett.* 490: 49–53.

Spruce AE, Standen NB, Stanfield PR. 1985. Voltage-dependent ATP-sensitive potassium channels of skeletal muscle membrane. *Nature* 316: 736–738.

Srivastava R, Aslam M, Kalluri SR, Schirmer L, Buck D et al. 2012. Potassium channel KIR4.1 as an immune target in multiple sclerosis. *N. Engl. J. Med.* 367: 115–123.

Standen NB, Quayle JM, Davies NW, Brayden JE, Huang Y, Nelson MT. 1989. Hyperpolarizing vasodilators activate ATP-sensitive K^+ channels in arterial smooth muscle. *Science* 245: 177–180.

Stanfield PR, Davies NW, Shelton PA, Sutcliffe MJ, Khan IA et al. 1994. A single aspartate residue is involved in both intrinsic gating and blockage by Mg^{2+} of the inward rectifier, IRK1. *J. Physiol.* 478: 1–6.

Suzuki M, Kotake K, Fujikura K, Inagaki N, Suzuki T et al. 1997. Kir6.1: A possible subunit of ATP-sensitive K^+ channels in mitochondria. *Biochem. Biophys. Res. Commun.* 241: 693–697.

Suzuki M, Li RA, Miki T, Uemura H, Sakamoto N et al. 2001. Functional roles of cardiac and vascular ATP-sensitive potassium channels clarified by Kir6.2-knockout mice. *Circ. Res.* 88: 570–577.

Suzuki M, Sasaki N, Miki T, Sakamoto N, Ohmoto-Sekine Y et al. 2002. Role of sarcolemmal K(ATP) channels in cardioprotection against ischemia/reperfusion injury in mice. *J. Clin. Invest.* 109: 509–516.

Takahashi N, Morishige K, Jahangir A, Yamada M, Findlay I et al. 1994. Molecular cloning and functional expression of cDNA encoding a second class of inward rectifier potassium channels in the mouse brain. *J. Biol. Chem.* 269: 23274–23279.

Tanemoto M, Abe T, Ito S. 2005. PDZ-binding and di-hydrophobic motifs regulate distribution of Kir4.1 channels in renal cells. *J. Am. Soc. Nephrol.* 16: 2608–2614.

Tanemoto M, Abe T, Onogawa T, Ito S. 2004. PDZ binding motif-dependent localization of K^+ channel on the basolateral side in distal tubules. *Am. J. Physiol. Renal. Physiol.* 287: F1148–F1153.

Tanemoto M, Kittaka N, Inanobe A, Kurachi Y. 2000. In vivo formation of a proton-sensitive K^+ channel by heteromeric subunit assembly of Kir5.1 with Kir4.1. *J. Physiol.* 525(Pt 3): 587–592.

Tanemoto M, Toyohara T, Abe T, Ito S. 2008. MAGI-1a functions as a scaffolding protein for the distal renal tubular basolateral K^+ channels. *J. Biol. Chem.* 283: 12241–12247.

Tang X, Schmidt TM, Perez-Leighton CE, Kofuji P. 2010. Inwardly rectifying potassium channel Kir4.1 is responsible for the native inward potassium conductance of satellite glial cells in sensory ganglia. *Neuroscience* 166: 397–407.

Tao X, Avalos JL, Chen J, MacKinnon R. 2009. Crystal structure of the eukaryotic strong inward-rectifier K^+ channel Kir2.2 at 3.1 A resolution. *Science* 326: 1668–1674.

Tester DJ, Tan BH, Medeiros-Domingo A, Song C, Makielski JC, Ackerman MJ. 2011. Loss-of-function mutations in the KCNJ8-encoded Kir6.1 K(ATP) channel and sudden infant death syndrome. *Circulation* 4: 510–515.

Thorneloe KS, Maruyama Y, Malcolm AT, Light PE, Walsh MP, Cole WC. 2002. Protein kinase C modulation of recombinant ATP-sensitive K(+) channels composed of Kir6.1 and/or Kir6.2 expressed with SUR2B. *J. Physiol.* 541: 65–80.

Topert C, Doring F, Wischmeyer E, Karschin C, Brockhaus J et al. 1998. Kir2.4—A novel K^+ inward rectifier channel associated with motoneurons of cranial nerve nuclei. *J. Neurosci.*18: 4096–4105.

Trapp S, Proks P, Tucker SJ, Ashcroft FM. 1998. Molecular analysis of ATP-sensitive K channel gating and implications for channel inhibition by ATP. *J. Gen. Physiol.* 112: 333–349.

Tristani-Firouzi M, Etheridge SP. 2010. Kir 2.1 channelopathies: The Andersen-Tawil syndrome. *Pflug. Arch. Eur. J. Physiol.* 460: 289–294.

Trube G, Hescheler J. 1984. Inward-rectifying channels in isolated patches of the heart cell membrane: ATP-dependence and comparison with cell-attached patches. *Pflug. Arch. Eur. J. Physiol.* 401: 178–184.

Tucker SJ, Gribble FM, Proks P, Trapp S, Ryder TJ et al. 1998. Molecular determinants of K_{ATP} channel inhibition by ATP. *EMBO J.* 17: 3290–3296.

Tucker SJ, Imbrici P, Salvatore L, D'Adamo MC, Pessia M. 2000. pH dependence of the inwardly rectifying potassium channel, Kir5.1, and localization in renal tubular epithelia. *J. Biol. Chem.* 275: 16404–16407.

van Bon BW, Gilissen C, Grange DK, Hennekam RC, Kayserili H et al. 2012. Cantu syndrome is caused by mutations in ABCC9. *Am. J. Hum. Genet.* 90: 1094–1101.

Vandenberg CA. 1987. Inward rectification of a potassium channel in cardiac ventricular cells depends on internal magnesium ions. *Proc. Natl. Acad. Sci. U.S.A.* 84: 2560–2564.

Vandenberg CA. 1994. Cardiac inward rectifier potassium channel In *Ion Channels in the Cardiovascular System*, eds. PM Spooner, AM Brown. New York: Futura Publishing Co.

Veeramah KR, Karafet TM, Wolf D, Samson RA, Hammer MF. 2013. The KCNJ8-S422L variant previously associated with J-wave syndromes is found at an increased frequency in Ashkenazi Jews. *Eur. J. Hum. Genet.* doi.10.1038/ejhg.2013.78.

Vega AL, Tester DJ, Ackerman MJ, Makielski JC. 2009. Protein kinase A-dependent biophysical phenotype for V227F-KCNJ2 mutation in catecholaminergic polymorphic ventricular tachycardia. *Circul. Arrhythmia Electrophysiol.* 2: 540–547.

Velimirovic BM, Gordon EA, Lim NF, Navarro B, Clapham DE. 1996. The K^+ channel inward rectifier subunits form a channel similar to neuronal G protein-gated K^+ channel. *FEBS Lett.* 379: 31–37.

Wade JB, Fang L, Liu J, Li D, Yang CL et al. 2006. WNK1 kinase isoform switch regulates renal potassium excretion. *Proc. Natl. Acad. Sci. U.S.A.* 103: 8558–8563.

Wagner CA. 2010. New roles for renal potassium channels. *J. Nephrol.* 23: 5–8.

Wambach JA, Marshall BA, Koster JC, White NH, Nichols CG. 2010. Successful sulfonylurea treatment of an insulin-naive neonate with diabetes mellitus due to a KCNJ11 mutation. *Pediatr. Diabetes* 11: 286–288.

Wang R, Su J, Wang X, Piao H, Zhang X et al. 2005. Subunit stoichiometry of the Kir1.1 channel in proton-dependent gating. *J. Biol. Chem.* 280: 13433–13441.

Wang S, Makhina EN, Masia R, Hyrc KL, Formanack ML, Nichols CG. 2013. Domain organization of the ATP-sensitive potassium channel complex examined by fluorescence resonance energy transfer. *J. Biol. Chem.* 288: 4378–4388.

Wang WH, Giebisch G. 1991. Dual modulation of renal ATP-sensitive K^+ channel by protein kinases A and C. *Proc. Natl. Acad. Sci. U.S.A.* 88: 9722–9725.

Wang WH, Yue P, Sun P, Lin DH. 2010. Regulation and function of potassium channels in aldosterone-sensitive distal nephron. *Curr. Opin. Neprhol. Hypertens.* 19: 463–470.

Weigl LG, Schreibmayer W. 2001. G protein-gated inwardly rectifying potassium channels are targets for volatile anesthetics. *Mol. Pharmacol.* 60: 282–289.

Wheeler A, Wang C, Yang K, Fang K, Davis K et al. 2008. Coassembly of different sulfonylurea receptor subtypes extends the phenotypic diversity of ATP-sensitive potassium (KATP) channels. *Mol. Pharmacol.* 74: 1333–1344.

Whorton MR, MacKinnon R. 2011b. Crystal structure of the mammalian GIRK2 K^+ channel and gating regulation by G proteins, PIP_2, and sodium. *Cell* 147: 199–208.

Whorton MR, MacKinnon R. 2013. X-ray structure of the mammalian GIRK2-betagamma G-protein complex. *Nature* 498: 190–197.

Wible BA, Taglialatela M, Ficker E, Brown AM. 1994. Gating of inwardly rectifying K^+ channels localized to a single negatively charged residue. *Nature* 371: 246–249.

Wickenden A. 2002. K(+) channels as therapeutic drug targets. *Pharmacol. Therapeut.* 94: 157–182.

Wickman K, Clapham DE. 1995. Ion channel regulation by G proteins. *Physiol. Rev.* 75: 865–885.

Wickman K, Karschin C, Karschin A, Picciotto MR, Clapham DE. 2000. Brain localization and behavioral impact of the G-protein-gated K^+ channel subunit GIRK4. *J. Neurosci.*20: 5608–5615.

Wickman K, Nemec J, Gendler SJ, Clapham DE. 1998. Abnormal heart rate regulation in GIRK4 knockout mice. *Neuron* 20: 103–114.

Wickman K, Pu WT, Clapham DE. 2002. Structural characterization of the mouse Girk genes. *Gene* 284: 241–250.

Wilson FH, Disse-Nicodeme S, Choate KA, Ishikawa K, Nelson-Williams C et al. 2001. Human hypertension caused by mutations in WNK kinases. *Science* 293: 1107–1112.

Wind T, Prehn JH, Peruche B, Krieglstein J. 1997. Activation of ATP-sensitive potassium channels decreases neuronal injury caused by chemical hypoxia. *Brain Res.* 751: 295–299.

Wischmeyer E, Doring F, Wischmeyer E, Spauschus A, Thomzig A et al. 1997. Subunit interactions in the assembly of neuronal Kir3.0 inwardly rectifying K^+ channels. *Mol. Cell. Neurosci.* 9: 194–206.

Wolleben CD, Sanguinetti MC, Siegl PK. 1989. Influence of ATP-sensitive potassium channel modulators on ischemia-induced fibrillation in isolated rat hearts. *J. Mol. Cell. Cardiol.* 21: 783–788.

Xia M, Jin Q, Bendahhou S, He Y, Larroque MM et al. 2005. A Kir2.1 gain-of-function mutation underlies familial atrial fibrillation. *Biochem. Biophys. Res. Commun.* 332: 1012–1019.

Xie LH, Horie M, Takano M. 1999. Phospholipase C-linked receptors regulate the ATP-sensitive potassium channel by means of phosphatidylinositol 4,5-bisphosphate metabolism. *Proc. Natl. Acad. Sci. U.S.A.* 96: 15292–15297.

Xie LH, John SA, Ribalet B, Weiss JN. 2008. Phosphatidylinositol-4,5-bisphosphate (PIP2) regulation of strong inward rectifier Kir2.1 channels: Multilevel positive cooperativity. *J. Physiol.* 586: 1833–1848.

Xie LH, John SA, Weiss JN. 2002. Spermine block of the strong inward rectifier potassium channel Kir2.1: Dual roles of surface charge screening and pore block. *J. Gen. Physiol.* 120: 53–66.

Xie LH, John SA, Weiss JN. 2003. Inward rectification by polyamines in mouse Kir2.1 channels: Synergy between blocking components. *J. Physiol.* 550: 67–82.

Xu H, Wu J, Cui N, Abdulkadir L, Wang R et al. 2001. Distinct histidine residues control the acid-induced activation and inhibition of the cloned K_{ATP} channel. *J. Biol. Chem.* 276: 38690–38696.

Xu H, Yang Z, Cui N, Chanchevalap S, Valesky WW, Jiang C. 2000. A single residue contributes to the difference between Kir4.1 and Kir1.1 channels in pH sensitivity, rectification and single channel conductance. *J. Physiol.* 528(Pat 2): 267–277.

Yamada M, Ho YK, Lee RH, Kontanill K, Takahashill K et al. 1994. Muscarinic K^+ channels are activated by beta gamma subunits and inhibited by the GDP-bound form of alpha subunit of transducin. *Biochem. Biophys. Res. Commun.* 200: 1484–1490.

Yamada M, Isomoto S, Matsumoto S, Kondo C, Shindo T et al. 1997. Sulphonylurea receptor 2B and Kir6.1 form a sulphonylurea-sensitive but ATP-insensitive K^+ channel. *J. Physiol.* 499: 715–720.

Yamada S, Kane GC, Behfar A, Liu XK, Dyer RB et al. 2006. Protection conferred by myocardial ATP-sensitive K^+ channels in pressure overload-induced congestive heart failure revealed in KCNJ11 Kir6.2-null mutant. *J. Physiol.* 577: 1053–1065.

Yang J, Jan YN, Jan LY. 1995. Control of rectification and permeation by residues in two distinct domains in an inward rectifier K^+ channel. *Neuron* 14: 1047–1054.

Yang Z, Xu H, Cui N, Qu Z, Chanchevalap S et al. 2000. Biophysical and molecular mechanisms underlying the modulation of heteromeric Kir4.1-Kir5.1 channels by CO2 and pH. *J. Gen. Physiol.* 116: 33–45.

Yi BA, Lin YF, Jan YN, Jan LY. 2001. Yeast screen for constitutively active mutant G protein-activated potassium channels. *Neuron* 29: 657–667.

Yoo D, Flagg TP, Olsen O, Raghuram V, Foskett JK, Welling PA. 2004. Assembly and trafficking of a multiprotein ROMK (Kir 1.1) channel complex by PDZ interactions. *J. Biol. Chem.* 279: 6863–6873.

Zaritsky JJ, Redell JB, Tempel BL, Schwarz TL. 2001. The consequences of disrupting cardiac inwardly rectifying K(+) current (I(K1)) as revealed by the targeted deletion of the murine Kir2.1 and Kir2.2 genes. *J. Physiol.* 533: 697–710.

Zaza A, Rocchetti M, Brioschi A, Cantadori A, Ferroni A. 1998. Dynamic Ca^{2+}-induced inward rectification of K^+ current during the ventricular action potential. *Circ. Res.* 82: 947–956.

Zeidner G, Sadja R, Reuveny E. 2001. Redox-dependent gating of G protein-coupled inwardly rectifying K^+ channels. *J. Biol. Chem.* 276: 35564–35570.

Zeng WZ, Li XJ, Hilgemann DW, Huang CL. 2003. Protein kinase C inhibits ROMK1 channel activity via a phosphatidylinositol 4,5-bisphosphate-dependent mechanism. *J. Biol. Chem.* 278: 16852–16856.

Zhang H, He C, Yan X, Mirshahi T, Logothetis DE. 1999. Activation of inwardly rectifying K^+ channels by distinct PtdIns(4,5)P2 interactions. *Nat. Cell Biol.* 1: 183–188.

Zhang HL, Bolton TB. 1996. Two types of ATP-sensitive potassium channels in rat portal vein smooth muscle cells. *Br. J. Pharmacol.* 118: 105–114.

Zhang W, Zhang X, Wang H, Sharma AK, Edwards AO, Hughes BA. 2013. Characterization of the R162W Kir7.1 mutation associated with snowflake vitreoretinopathy. *Am. J. Physiol. Cell Physiol.* 304: C440–C449.

Zhou M, Tanaka O, Sekiguchi M, He HJ, Yasuoka Y et al. 2005. ATP-sensitive K^+-channel subunits on the mitochondria and endoplasmic reticulum of rat cardiomyocytes. *J. Histochem. Cytochem.* 53: 1491–1500.

Zhou W, Arrabit C, Choe S, Slesinger PA. 2001. Mechanism underlying bupivacaine inhibition of G protein-gated inwardly rectifying K^+ channels. *Proc. Natl. Acad. Sci. U.S.A.* 98: 6482–6487.

Zingman LV, Hodgson DM, Bast PH, Kane GC, Perez-Terzic C et al. 2002. Kir6.2 is required for adaptation to stress. *Proc. Natl. Acad. Sci. U.S.A.* 99: 13278–13283.

Zitron E, Gunth M, Scherer D, Kiesecker C, Kulzer M et al. 2008. Kir2.x inward rectifier potassium channels are differentially regulated by adrenergic alpha1A receptors. *J. Mol. Cell. Cardiol.* 44: 84–94.

Zitron E, Kiesecker C, Luck S, Kathofer S, Thomas D et al. 2004. Human cardiac inwardly rectifying current IKir2.2 is upregulated by activation of protein kinase A. *Cardiovasc. Res.* 63: 520–527.

Zobel C, Cho HC, Nguyen TT, Pekhletski R, Diaz RJ et al. 2003. Molecular dissection of the inward rectifier potassium current (IK1) in rabbit cardiomyocytes: Evidence for heteromeric co-assembly of Kir2.1 and Kir2.2. *J. Physiol.* 550: 365–372.

19 Two-pore domain potassium channels

Leigh D. Plant and Steve A.N. Goldstein

Contents

19.1 INTRODUCTION

The ability of the central nervous system (CNS) to receive, integrate, and process information is dependent on the operation of a finely tuned orchestra of receptors and ion channels that reside at the surface of excitable cells. Among the ensemble at the neuronal plasma membrane are many types of K^+ channels that act together to control some of the most fundamental determinants of excitable behavior, in particular, the frequency and duration of action potentials and the resting membrane potential (V_M). It is valuable therefore to understand the structural basis for K^+ channel function, the role of K^+ channels in physiology, and the factors that regulate K^+ channel activity in health and disease. In this chapter, we consider the most recently discovered family of mammalian K^+ channels, the predominant contributors to background (leak) K^+ currents, the two-pore domain K^+ (K2P) channels.

Encoded by 15 *kcnk* genes in humans (Figure 19.1a), K2P channels open and close (gate) in response to a variety of stimuli, including neurotransmitters, volatile anesthetics, and changes in pH. K2P channels are recognized by their distinct body plan of four transmembrane domains (TMD) and two-pore (P)-forming loops in each subunit (Figure 19.1b). Another distinction of the K2P channels is that they operate across the physiological voltage range (Figure 19.1c) and therefore influence both the shape of excitable events at depolarized voltages and the stability of V_M at hyperpolarized voltages below the firing threshold for action potentials.

The *kcnk* genes were identified in the late 1990s, advancing the study of background K^+ currents onto a molecular footing. Now, nearly two decades on, a wealth of research and discovery spanning the natural kingdom from yeast to man has propelled K^+ leak currents from insightful predictions by Hodgkin et al. required to model neuronal biophysics (Goldman 1943; Hodgkin et al. 1949) to K2P channels as identified mediators of the resting state from which neuronal action potentials rise and return.

Here, we present an overview of K2P channels and the regulatory pathways that modulate their roles in physiology; by necessity, this is an image in time, because our understanding of the mechanistic basis for their operation and the extent of their responsibilities in vivo are still rapidly expanding. As we consider individual K2P channels, it will become clear that a great deal of work has focused on the nervous system, but this should not mislead the reader; K2P channels are being discovered to be important to normal functions and in disease pathogenesis throughout the body, notably, in other excitable tissues like the heart and skeletal muscle, organs such as the kidney and pancreas, and even in circulating immune cells.

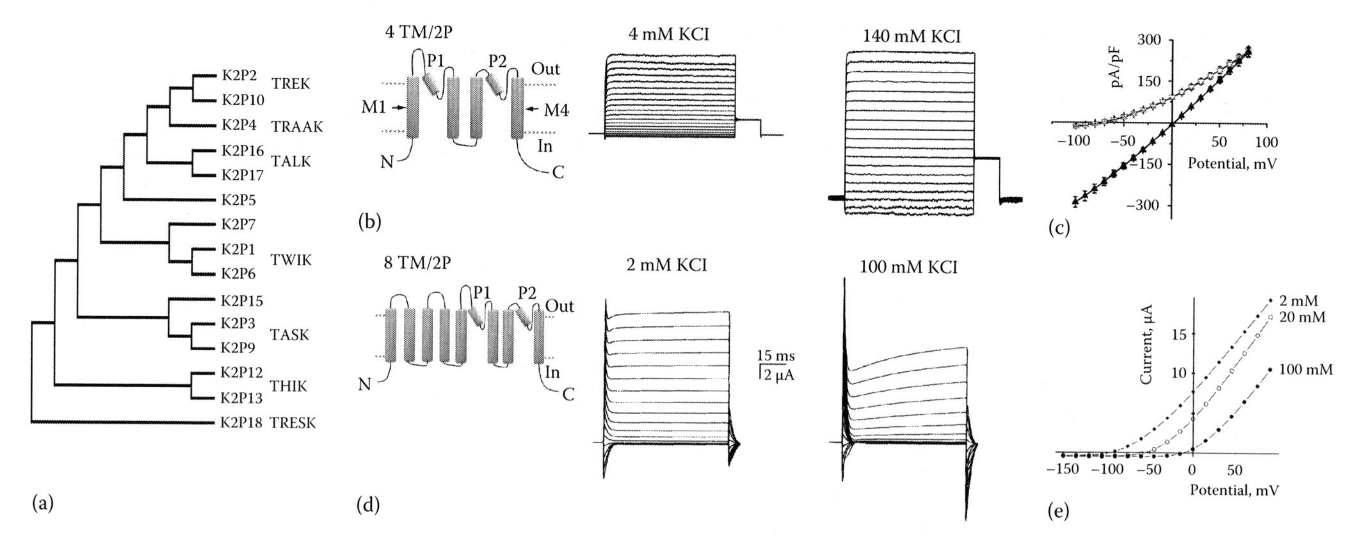

Figure 19.1 Classification and operation of two-P domain channels. (a) A phylogenetic tree calculated to show the relatedness of the 15 K2P subunits found in humans based on ClustalW alignments of the IUPHAR accession numbers for each clone (see http://www.iuphar-db.org/DATABASE/VoltageGatedSubunitListForward). To date, functional expression has not been observed for K2P7, K2P12, and K2P15 (b). *Left*: K2P subunits are integral membrane proteins with internal amino (N) and carboxy (C) termini, four TMDs (M1–M4), and two-pore-forming (P)-loops. *Middle*: Current recorded from a Chinese hamster ovary cell heterologously expressing active K2P1 channels (in whole cell mode with de-SUMOylating enzyme in the pipette). (Adapted from Plant, L.D., *Proc. Natl. Acad. Sci. U.S.A.*, 107, 10743, 2010.) The inside of the cell contains 140 mM K^+ and the external solution contains 4 mM K^+. *Right*: The same cell recorded with 140 mM K^+ on both sides of the membrane. (c). Mean current–voltage relationships for cells studied as in (b). Active K2P1 channels show open (GHK) rectification passing more outward current under quasiphysiologic conditions (Δ, 4 mM external K^+) and a linear current–voltage relationship with symmetrical 140 mM K^+ (▲). (d) *Left*: Fungal two-P domain subunits have eight TMDs. *Middle*: Current recorded from *Xenopus laevis* oocytes expressing TOK1 channels studied in 2 mM (*Middle*) or 100 mM external K^+ (*Right*). (Adapted from Ketchum, K.A. et al., *Nature*, 376, 690, 1995.) (e) Current–voltage relationships for the cell studied in panel D show outward rectification, that is, a shift of the potential where outward current is measured with changes in the external concentrations of K^+ that accord with the Nernst potential for K^+ (E_K, as described in Table 19.1).

19.2 RESTLESS MEMBRANE POTENTIAL

Even when neurons are electrically quiescent, the V_M is never truly at rest. Active transport proteins consume ATP to maintain ionic imbalance across the plasma membrane so that the concentration of Na^+ is low and the concentration of K^+ is high in the cytoplasm relative to the extracellular solution (Table 19.1). These ionic gradients are a repository of electrochemical energy that is harnessed when excitable cells respond to physicochemical stimuli. Thus, depolarization, temperature change, deformation of the surface membrane, and increases in the local concentrations of neurotransmitters are among the many stimuli that activate (or inactivate) ion channels and use the energy stored in the transmembrane ionic gradients to do the work of living cells.

When activation opens a channel conduction pore, large numbers of select ions diffuse down their electrochemical gradient to produce changes in cell behavior. Hence, in neurons, action potentials are triggered when V_M is depolarized to the firing threshold. At threshold, voltage-gated Na^+ (Na_V) channels are activated and this allows Na^+ to rush into the cell. The influx of Na^+ instigates greater depolarization, because additional Na_V channels become active. This rising phase of the action potential is regenerative and explosive. The phase is rapidly followed by fast inactivation of the channels, a process that stops Na^+ flow and allows the neuron to recover (repolarize) in preparation for the next firing event.

Outward flow of K^+ repolarizes the neuron. Thus, voltage-gated K^+ (K_V) channels activate in response to membrane depolarization, like Na_V channels, but with slower kinetics. K_V channels therefore allow the rapid rise of the action potential and then contribute to restoring V_M to resting levels until they inactivate or close with repolarization. Since K2P channels remain open during firing and recovery, net K^+ efflux continues through these pores until the magnitude of negative countercharge inside the cell is sufficient to balance the chemical and electrostatic driving force favoring K^+ outflow.

The V_M where there is no net flux of K^+ through open K^+-selective channels is called the Nernst potential for K^+ (E_K) and is usually around −95 mV in mammalian cells (Table 19.1). Typical Nernst potentials in human cells for Na^+ (E_{Na}), Ca^{2+} (E_{Ca}), and Cl^- (E_{Cl}) are 65, 120, and −90 mV, respectively, values that depend on the concentration of the ion on either side of the membrane and temperature in a manner described by the Nernst equation (Table 19.1). The V_M of cells at rest is determined by the sum of the Nernst potentials for the major ions scaled to reflect their relative permeability across the membrane, as described by the Goldman–Hodgkin–Katz (GHK) equation (Table 19.1). Typically, the resting V_M of mammalian cells is close to E_K, because K^+ channels populate most surface membranes densely and K^+ currents flow via K2P channels across the physiological voltage range.

Because different cells in the body express distinct types of ion channels and because regulators modify the level of expression and operation of ion channels, membrane permeability is dynamic. Thus, the resting V_M of neurons in dorsal root ganglia ranges from ~−45 to −75 mV, depending on the experimental condition (Baccaglini 1978; Puil et al. 1987; Wang et al. 1994; Baumann et al. 1996). In addition to the background K^+ currents

Table 19.1 Concentrations of free ions in mammalian skeletal muscle

ION	EXTRACELLULAR CONCENTRATION (mm)	INTRACELLULAR CONCENTRATION (mm)	$\frac{[\text{Ion}]_o}{[\text{Ion}]_i}$	EQUILIBRIUM POTENTIAL (mV)
Na^+	145	12	12	67
K^+	4	155	0.026	-98
Ca^{2+}	1.5	10^{-7} M	15,000	129
Cl^-	123	4.2	29	-90

Cellular membrane potential is dependent on the internal and external concentration of ions and the relative permeability of the membrane to each ion (Hille 2001). In many cells, resting V_M is strongly influenced by the magnitude of background K^+ currents. Equilibrium potentials for each ion (*Ex*) are calculated using the Nernst equation:

$$Ex = \frac{RT}{zF} \cdot \ln \frac{[X]_o}{[X]_i},$$

where

R is the Gas constant (1.987 cal K^{-1} mol^{-1})
T is temperature in Kelvin
z is ionic charge
F is Faraday's constant (9.648 × 10^4 C mol^{-1})
$[X]_o$ and $[X]_i$ are the external and internal concentrations of X, respectively

At physiological temperature (37°C), RT/zF is ~27 mV. The GHK equation shows that V_M is determined by the relative concentrations and permeabilities (P) of each ion, for example, X and Y:

$$V_M = \frac{RT}{F} \cdot ln \frac{Px[X]_o + Py[Y]_o}{Px[X]_i + Py[Y]_i}.$$

passed by K2P channels, V_M is also influenced by the activity of other channels that are open at resting potentials. These include K^+ currents passed by inwardly rectifying K^+ channels, for example, in guinea pig ventricular myocytes (Sakmann and Trube 1984), M-currents mediated by $K_V7.2$ and $K_V7.3$ subunits in central and peripheral neurons such as those in frog sympathetic neurons (Brown and Adams 1980) and rat dorsal root ganglia cells (Delmas and Brown 2005), and chloride currents observed in frog (Hodgkin and Horowicz 1959) and rat (Blatz and Magleby 1985) skeletal muscle.

19.3 K2P POTASSIUM CURRENTS SHAPE EXCITABILITY

Whereas the variety of Na_V and Ca_V channels in individual excitable cells is limited, the repertoire of K^+ channels that shape electrical activity is broad. Channels that pass background K^+ currents stabilize cells at hyperpolarized membrane voltages, below the firing threshold (Goldman 1943; Hodgkin et al. 1949). Regulatory stimuli that augment the magnitude of background K^+ currents decrease excitability, because V_M moves closer to E_K and larger depolarizations are required to initiate action potentials. In contrast, inhibition of background K^+ currents depolarizes V_M, allowing smaller stimuli to initiate action potentials. In this manner, background K^+ channels tune not only the magnitude of the stimulation required to initiate an action potential but modify the shape, frequency, and the magnitude of each spike (Hodgkin and Huxley 1952; Jones 1989).

Many distinct stimuli regulate the activity of background K^+ currents, including neurotransmitters (Siegelbaum et al. 1982; Shen et al. 1992; Watkins and Mathie 1996), changes in pO_2 (Buckler 1997; Plant et al. 2002), extracellular pH (Nattel et al. 1981), and pharmaceuticals such as volatile anesthetics (Nicoll and Madison 1982; Plant et al. 2012). Although the physiological significance of background K^+ currents had been appreciated since the 1940s (Goldman 1943; Hodgkin et al. 1949), the question of whether they were mediated by dedicated pathways or accumulated through the operation of other processes and membrane damage was not resolved until the mid-1990s.

19.4 FUNGAL TWO-PORE DOMAIN K+ CHANNELS

Two-P domain K^+ subunits were identified in the genomes of the budding yeast *Saccharomyces cerevisiae* and nematode *Caenorhabditis elegans* (Ketchum et al. 1995). Each subunit of the yeast channel, called TOK1 (for two-P domain, outwardly rectifying, K^+ channel 1), has two reentrant P-loops and eight TMDs, whereas subunits in the roundworm have two P-loops and four TMDs like their congeners in higher organisms (Figure 19.1). Subunits with two P-loops and four TMDs were described shortly thereafter in *Drosophila melanogaster* (Goldstein et al. 1996) and in mammals (Lesage et al. 1996; Plant et al. 2005).

To date, TOK-like channels with eight TMDs have been identified only in fungi—*S. cerevisiae* (Ketchum et al. 1995), the bread mold *Neurospora crassa* (Roberts 2003), and the opportunistic pathogen *Candida albicans* (Baev et al. 2003). In addition to their unique subunit architecture, TOK channels are novel for passing large outward K^+ currents when the membrane is depolarized above E_K but little or no inward currents when the V_M is below E_K (Ketchum et al. 1995) (Figure 19.1d).

Even though the V_M of *S. cerevisiae* is determined primarily by the transmembrane gradient of protons rather than K^+, an important role for TOK1 in fungal physiology was highlighted by a surprising view into biological competition between yeast populations. Yeast cells infected with RNA killer virus release K1 killer toxin, a peptide that activates TOK1 channels, causing fatal efflux of K^+ from virus-free neighbors yeast (Ahmed et al. 1999). In contrast, virus-positive cells are immune because internal K1 killer toxin blocks TOK1 channels, suppressing the action of external toxin (Sesti et al. 2001). Consistent with these observations, excess K^+ efflux is toxic for yeast cells engineered to overexpress the TOK1 channel (Loukin et al. 1997).

19.5 K2P CHANNELS IN INVERTEBRATES PASS BACKGROUND CURRENTS

Two-P domain K^+ subunits with four TMDs are now called K2P channel subunits. K2PØ, encoded by *kcnkØ*, was cloned from a *D. melanogaster* neuromuscular gene library and shown to exhibit the functional properties expected for background K^+ channels (Goldstein et al. 1996; Zilberberg et al. 2000, 2001; Ilan and Goldstein 2001). At first called dORK channels, to reflect their identification in *D. melanogaster* and operation as openly rectifying and K^+-selective leak pores, both the cloned channels and native background currents show GHK (or open) rectification—that is, the simple property of an ion-selective pore to pass permeant ions more readily across the membrane from the side of higher ion concentration to the side of lower concentration (Figure 19.1). Thus, intracellular K^+ is higher than extracellular and this favors K^+ efflux. Also like native background currents, cloned K2PØ channels expressed in experimental cells open and close with little or no dependence on voltage and time. *D. melanogaster* is now known to have 11 *kcnk* genes (Adams et al. 2000; Littleton and Ganetzky 2000) and K2PØ channels have been described to play a role in the circadian locomotor rhythms of fruit flies (Park and Griffith 2006) as well as in control of their cardiac rhythms by determining the slow diastolic depolarization phase of the heartbeat (Lalevee et al. 2006).

The nematode *C. elegans* has 47 genes that encode K2P subunits (Salkoff et al. 2005; Yook et al. 2012). This remarkable inventory is posited to represent a large number of rapidly evolving, nonconserved channels unique to *C. elegans*. This hypothesis is supported by the observation that there is poor conservation between the genes in *C. elegans* and the closely related roundworm *C. briggsae* (Salkoff et al. 2005). Genetic studies of *C. elegans* have determined that many of the channels are expressed in one neuron, or just a few, and that expression can be dynamic, allowing for fine, responsive control of excitability (Salkoff et al. 2001).

19.6 TWO-PORE DOMAIN K+ CHANNEL SUBUNITS IN MAMMALS

In the decade that followed the identification of two-P domain channels in yeast, worms, and fruit flies, 15 *kcnk* genes for K2P channel subunits were identified in the genomes of human, rat, and mouse (Figure 19.1a). The mammalian channels were initially named according to the biophysical, physiological, or pharmacological attributes of the currents that they passed (Table 19.2). However, discrepant observations in different species and experimental systems encouraged adoption of a formal nomenclature in 2005 (Plant et al. 2005). While proteins encoded by *kcnk* genes are now called K2P subunits (and numbered to match their encoding gene, thus *kcnk1* and K2P1), original descriptive names remain useful in ongoing efforts to correlate cloned channels and the background K^+ currents they mediate in native cells.

K2P channels fall into seven subfamilies based on sequence homology, and the channels in each subfamily are discussed later. To date, no currents have been reported to pass through three of the channels when they are expressed in experimental cells, K2P7, K2P12, and K2P15.

Consistent with studies of native leak currents, the operation of K2P channels is tightly controlled by a plethora of regulators (Goldstein et al. 2001). Regulatory pathways that decrease K2P channel activity by decreasing the density of channels at the cells surface, by blocking the conduction pore, or by reducing current magnitude due to lowered single channel conductance, open probability or altered ion selectivity, serve to increase cellular excitability. Reciprocally, those regulators that increase the activity of K2P channels increase the permeability of the membrane to K^+ and dampen excitability.

19.7 STRUCTURE OF K2P CHANNELS

Mammalian K2P channels have the membrane topology and subunit stoichiometry demonstrated for K2PØ channels: each subunit has two P-loops and four TMDs, the amino- and carboxy-termini reside in the cytoplasm, and channels operate as homodimers (Lopes et al. 2001; Kollewe et al. 2009; Plant et al. 2010) or heterodimers (Berg et al. 2004; Plant et al. 2012).

Initial insights into the 3D structure of K2P channels came from homology models that were based on K^+ channels with known x-ray structures and constraints defined by observations of K2P channels in action. Thus, a model of K2PØ was produced using the crystal structure of the $K_V1.2$ channel and 23 pairs of residues in K2PØ channels that were inferred by serial substitution analysis to interact directly based on second-site electrostatic compensation; the model predicted K2PØ channels to employ two subunits, arranged to form a single pore, with a K^+ selectivity filter that had fourfold symmetry (even though it was lined by the nonhomologous P1 and P2 loops from each subunit), and to have a channel corpus with bilateral symmetry, as expected for a channel complex formed by two subunits (Kollewe et al. 2009).

The strategy used to develop a homology model for K2P3 utilized the crystal structure of an aracheal K^+ channel, K_VAP, and the impact of serial alanine mutagenesis of K2P3 on pore blockade

Table 19.2 The nomenclature of mammalian K2P channels

IUPHAR CHANNEL NAME	HUGO GENE NAME	COMMON NAME	OTHER NAMES
K2P1	*kcnk1*	TWIK1	hOHO
K2P2	*kcnk2*	TREK1	TPKC1
K2P3	*kcnk3*	TASK1	TBAK1, OAT1
K2P4	*kcnk4*	TRAAK	KT4
K2P5	*kcnk5*	TASK2	
K2P6	*kcnk6*	TWIK2	TOSS
K2P7	*kcnk7*		*kcnk8*
K2P9	*kcnk9*	TASK3	
K2P10	*kcnk10*	TREK2	
K2P12	*kcnk12*	THIK2	
K2P13	*kcnk13*	THIK1	
K2P15	*kcnk15*	TASK5	*kcnk11, kcnk14*
K2P16	*kcnk16*	TALK1	
K2P17	*kcnk17*	TALK2	TASK4
K2P18	*kcnk18*	TRESK1	

K2P channels are named according to IUPHAR nomenclature (see http://www.iuphar-db.org/DATABASE/VoltageGatedSubunitListForward). *HUGO* (Human Genome Organization) designations for each gene are also listed. It is customary for gene names to be in capital letters for human genes (KCNK1), in lower case for rat genes (*kcnk1*), and in italics for mouse genes (*kcnk1*). Common names refer to a biophysical or pharmacological property from an early report.

to infer residues exposed in the ion conduction pathway (Streit et al. 2011). Supporting the predictions of the models of K2PØ and K2P3, the separate studies identified pore-lining residues in common even though the two channels do not share a high degree of sequence homology (Kollewe et al. 2009; Streit et al. 2011).

In 2012, x-ray structures were reported for human K2P1 (Miller and Long 2012) and K2P4 (Brohawn et al. 2012) at ~3.8 Å and appear to show the open state of the channels. As predicted by the homology models, both crystal structures revealed a fourfold symmetric ion conduction pore within a bilaterally symmetric channel corpus. Unexpectedly, the first external loop of each subunit was observed to extend ~35 Å beyond the outer leaflet of the plasma membrane to form a *cap-domain* above the outer mouth of the pore, bifurcating the entry to the K^+ conduction pathway. The cap-domain seen in both K2P1 and K2P4 has not been described in other channels and may explain failure of pore-blocking peptide neurotoxins to act on K2P channels due to limited access.

The crystal structure of K2P1 predicts a *C-helix* that follows directly from the final (M4) TMD and runs close to the intracellular lipid interface in K2P1 (Miller and Long 2012). In K2P4, an analogous helix is modeled as part of the second (M2) TMD (Brohawn et al. 2012). A role for these segments in gating was proposed based on their location below the conduction pore in the structure and previous reports that mutation of residues in the domains modified K2P channel activity.

In 2013, the structure of human K2P4 was obtained at higher resolution by crystallization of the channel in complex with a Fab antibody fragment (Figure 19.2). The new structure at 2.75 Å

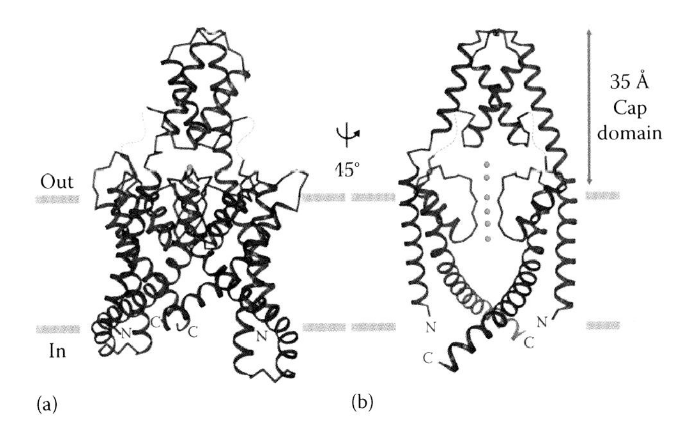

Figure 19.2 A 3D structure of K2P4 channels. (a) A ribbon representation of the x-ray structure of human K2P4 resolved at 2.75 Å. (Adapted from Brohawn, S.G. et al., *Proc. Natl. Acad. Sci. U.S.A.*, 110, 2129, 2013.) Fab fragments used to stabilize the protein have been removed for clarity. The channel is viewed from the membrane plane with one subunit in red and the other in blue, K^+ ions are shown in green, and the boundary of the membrane is in gray. Loops for which structure was not resolved are suggested by dashed lines. (b) A view of the channel rotated by ~45° demonstrating the unique domain swap wherein the outer pore helix interacts with the inner helix from the other subunit rather than its own. Note also the ~35 Å cap-domain above the outer mouth of the pore that bifurcates the entrance to the K^+ conduction pathway.

showed a novel arrangement not apparent at lower resolution: in contrast to other K^+ channels of known structure, each of the four outer pore helices of K2P4 was modeled to interact with the inner helix from the other subunit, an unexpected means for intersubunit communication (Brohawn et al. 2013). In addition, one of two lateral portals exposing the ion conduction pore to the hydrophobic interior of the lipid bilayer in the low resolution models of both K2P1 and K2P4 appeared in the Fab complex at higher resolution to be occluded.

19.8 K2P CHANNEL FAMILIES IN HUMANS: NORMAL OPERATION AND PATHOPHYSIOLOGY

19.8.1 K2P1, K2P6, AND K2P7 CHANNELS: SUMO REGULATION OF ION CHANNELS

Studies of K2P1 channels have been exciting, at times confusing, and ultimately very informative. Encoded by the *kcnk1* gene, and initially named TWIK1, for tandem of P-domains in a weak inward rectifying K^+ channel (Lesage et al. 1996), other groups were at first unable to observe any currents when the subunits were expressed in a variety of cell types (Goldstein et al. 1998; Pountney et al. 1999). Arguing that *kcnk1* was not a pseudogene but encoded a functional protein product, transcripts were demonstrated at first in the placenta, the lungs, the kidneys (Lesage et al. 1996; Talley et al. 2001), the CNS (Talley et al. 2001), and the cardiac conduction pathway (Gaborit et al. 2007), and later, K2P1 protein was shown in cerebellar granule neurons (Talley et al. 2001; Gaborit et al. 2007; Plant et al. 2012).

The paradox was resolved when a novel mechanism for regulating the electrical activity of ion channels was revealed—K2P1 channels are silenced by the small ubiquitin-like modifier protein (SUMO) pathway (Rajan et al. 2005). Mediated by an enzyme cascade, SUMOylation is the reversible, covalent modification of the ε-amino group of specific lysine(s) in target proteins by the 97 amino acid SUMO protein (Geiss-Friedlander and Melchior 2007). The SUMO pathway, present in all mammalian cells, was first identified to modulate the activity of transcription factors and nuclear import/export (Mahajan et al. 1997). Electrical silencing of K2P1 channels by SUMO was unexpected, because the pathway had not previously been found to operate beyond the nucleus (Wilson and Rosas-Acosta 2005).

Stoichiometric analysis using total internal reflection microscopy revealed that K2P1 channels in tissue culture cells carry two SUMO monomers, one on the Lys^{274} residue in each subunit of the channel and that SUMOylation of only one subunit was sufficient to suppress the function of channels in the plasma membrane (Plant et al. 2010). Thus, K2P1 channels can be activated when de-SUMOylated by SUMO-specific proteases or by mutation of Lys^{274} since only this residue can be SUMOylated.

As expected for a background K^+ current, and like K2PØ channels, K2P1 channels activated by de-SUMOylation pass openly rectifying, K^+ currents (Rajan et al. 2005). The channels are inhibited by external acidification. Thus, when active K2P1 channels are studied in expression systems (Rajan et al. 2005; Plant et al. 2010) or in neuronal cells (Plant et al. 2012), half-block is observed at physiological levels of pH and is sensitive to external K^+ concentration (~pH 6.7 at 4 mM external K^+). Block is due to protonation of a histidine residue in the outer portion of the first P-loop of each subunit, a mechanism described for K2P3 and K2P9 channels (Lopes et al. 2000, 2001). Reciprocally, K2P1, K2P3, and K2P9 channels appear to be less selective for K^+ when external pH is below normal (Ma et al. 2012).

Since 2005, SUMOylation has been shown to modulate the operation of a growing cadre of membrane proteins including transporters (Gibb et al. 2007), G-protein-coupled receptors (Tang et al. 2005), and other ion channels in the CNS (Martin et al. 2007; Plant et al. 2011), the cardiovascular system (Benson et al. 2007; Kruse et al. 2009), and the pancreas (Dai et al. 2009). The mechanism by which SUMOylation silences K2P1 channels remains unknown. The observation that silencing and activation are reversed by enzymes applied to the inside of off-cell patches argues the impact of SUMO, once covalently linked to the channel, does not require other proteins and suggests that SUMO may directly block the pore or interfere with channel gating. Both mechanisms are consistent with the x-ray structure of human K2P1, showing Lys^{274} to be exposed on the intracellular C-helix (Miller and Long 2012), and demonstrate that K2P channels appear to open and close at the K^+ selectivity filter (Zilberberg et al. 2001). What regulates SUMOylation of K2P1 channels is also as yet unknown.

K2P channels have been posited to pass *IKso*, the standing outward, background K^+ current that determines the V_M of cerebellar granule neurons and the response of these CNS cells to changes in pH (Millar et al. 2000), pO_2 (Plant et al. 2002) and volatile anesthetics. Recently, transcripts for K2P1 were observed in individual cerebellar granule neurons, along with mRNAs that encode for K2P3 and K2P9 (Plant et al. 2012). In vivo FRET analysis showed that K2P1, K2P3, and K2P9, but not K2P2, interacted with SUMO in cerebellar granule neurons, although only K2P1 was subject to SUMOylation when the cloned subunits were expressed in tissue culture cells. The findings were reconciled by the demonstration that K2P1 subunits assembled with K2P3 or K2P9 to form heterodimeric channels and that the mixed subunit channels were silenced by a single SUMO monomer on K2P1 Lys^{274} and reactivated by de-SUMOylation (Plant et al. 2012). This also explained why de-SUMOylation augmented *IKso* to decrease neuronal excitability. Because K2P1, K2P3, and K2P9 are expressed together throughout the brain, heart, and somatosensory systems, and the SUMO pathway is ubiquitous, regulation of K2P1 heterodimeric channels is expected to be commonplace.

SUMO regulation of K2P1 heterodimer channels was found to be important in the response of cerebellar granule neurons to volatile anesthetics, including halothane (Plant et al. 2012). Halogenated anesthetics act to hyperpolarize neurons via effects on multiple targets including the ligand-gated ion channel $GABA_A$ (Franks and Lieb 1994) and K^+ leak currents (Nicoll and Madison 1982; Sirois et al. 1998). Roles for K2P channels in the effects of volatile anesthetics implicate K2P3, K2P9 (Sirois et al. 2000; Talley and Bayliss 2002; Lazarenko et al. 2010b), K2P2, K2P10, (Patel et al. 1999), and K2P18 channels (Liu et al. 2004). Halogenated-volatile anesthetic agents such as halothane, isoflurane, and sevoflurane increase the activity of all these K2P channels, thereby decreasing neuronal excitability. In contrast, the activity of K2P13 (also known as THIK1 for two-P domain halothane-inhibited K^+ channel) is suppressed by volatile anesthetics, including halothane. While the effects of volatile anesthetics in humans are believed to be the most pronounced on K2P18 channels, marked species differences have been reported (Keshavaprasad et al. 2005). Of note, halothane has little impact on homomeric K2P1 channels and increases K2P9 channel currents by just ~30%, but it augments the current through heterodimeric channels formed of K2P1 and K2P9 by ~300% (Plant et al. 2012).

Volatile anesthetics appear to interact with a domain on the proximal C-terminus of K2P2, K2P3 (Patel et al. 1999), and K2P9 channels (Talley and Bayliss 2002). The same portion of these channels is implicated in binding of Gαq signaling molecules, leading to inhibition of channel activity. These observations suggest that volatile anesthetics augment the activity of K2P channels by suppressing the constitutive activity of Gαq proteins (Conway and Cotten 2011).

Recent studies have provided evidence that K2P1 channels contribute to the resting V_M of cardiac myocytes and mediate paradoxical depolarization, a pro-arrhythmic condition observed when the serum concentration of K^+ falls below 3 mM (hypokalemia) (Ma et al. 2011). Hypokalemic depolarization is paradoxical because low levels of serum K^+ are expected to hyperpolarize V_M (Table 19.1). The explanation appears to be that hypokalemia leads K2P1 channels to loosen their specificity for K^+ and to pass a sufficient amount of Na^+ to depolarize the cells. Still to be explained, the phenomenon is observed in cardiac myocytes but not in neurons even though both cell types express the channel.

Although the physiological role of K2P1 has been best explored in cerebellar granule neurons (Plant et al. 2012) and cardiac myocytes (Ma et al. 2011), the channel is also expressed in many other tissues and with reduced mRNA transcript levels in a variety of tumors, including melanoma, glioblastomas, and ovarian cancers (Beitzinger et al. 2008). Transcripts for *kcnk1* have also been detected throughout the renal nephron, leading to the suggestion that K2P1 is involved in K^+ recycling (Cluzeaud et al. 1998). Thus, knockout of the *kcnk1* gene in mice led to hyperpolarization of the principal cells of the cortical collecting duct and altered phosphate handling (Millar et al. 2006). This correlated with altered levels of the sodium–phosphate cotransporter in the proximal tubule and inappropriate internalization of aquaporin 2 in the collecting duct (Nie et al. 2005).

K2P6 and K2P7 show sequence similarity to K2P1 (Figure 19.1) and because they generate small currents on heterologous expression, clarity about their operation and roles has been slow to develop (Chavez et al. 1999; Salinas et al. 1999; Bockenhauer et al. 2000). K2P6 (previously TWIK2) was cloned from mouse where it is expressed in the eye, lung, and stomach (Salinas et al. 1999) and appears to play a role in regulating arterial blood pressure by determining the V_M of vascular smooth muscle cells (Lloyd et al. 2011). Transcript analysis shows that human and rat K2P6 are highly expressed in the aorta, esophagus, stomach, and spleen (Patel et al. 2000) as well as in isolated articular chondrocytes (Clark et al. 2011). Transcripts from aorta were posited to represent K2P6 channels in vascular smooth muscle, because blood vessels were denuded of endothelial cells (Patel et al. 2000).

Both human and rat K2P6 channels pass openly rectifying currents when heterologously expressed in tissue culture cells, although the rat isolate passes K^+-selective currents that are 15 times larger human K2P6 channels and therefore better characterized. Transcript analysis predicts that a splice variant of K2P6 lacking the first TMD is present in both human and rat with a similar expression pattern to the full-length channel (Patel et al. 2000).

K2P7 is expressed in the eye, lung, and stomach (Salinas et al. 1999; Bockenhauer et al. 2000) but does not appear to pass currents on heterologous expression. Sequence analysis predicts that K2P7 contains an EF-hand motif and four potential SH3-binding motifs in the C-terminal domain of each subunit. In common with K2P1, K2P3, K2P9, and K2P15, K2P7 has a histidine residue in the first P-domain (Gly-Tyr-Gly-His), that has been shown to mediate proton block of K2P1, K2P3, and K2P9 channels (Lopes et al. 2000; Rajan et al. 2000; Plant et al. 2010). In contrast, K2P7 is notable for an unconventional sequence in the second P-domain (Gly-Leu-Glu) not found in other human K2P subunits that carry instead Gly-Tyr-Gly, Gly-Leu-Gly, or Gly-Phe-Gly. The presence of a large, negatively charged glutamate residue in the second P-domain, rather than a small glycine, has led some to suggest that K2P7 may not function as a K^+ channel (Enyedi and Czirjak 2010).

19.8.2 K2P2, K2P4, AND K2P10 CHANNELS: POLYMODAL RHEOSTATS

K2P2 (or TREK1 for TWIK-related K^+ channel 1) and K2P4 (or TRAAK for TWIK-related arachidonic acid activated K^+ channel) are widely expressed throughout the CNS (Fink et al. 1996, 1998; Medhurst et al. 2001; Talley et al. 2001; Heurteaux et al. 2006) including in striatal neurons (Lauritzen et al. 2005), hippocampal neurons (Thomas et al. 2008) and cerebellar granule neurons (Plant et al. 2012). Both K2P2 and K2P4 have also been observed in the peripheral nervous system, trafficking along somatosensory neurons (Bearzatto et al. 2000). However, there is disagreement about the expression level of these channels in the periphery. This highlights that linking native background K^+ currents to the activity of specific K2P channels is challenging. In this case, discrepancies may reflect the expression of variants of K2P2 (Thomas et al. 2008) and K2P4 (Ozaita and Vega-Saenzde Miera 2002) that have altered function, native channels with similar phenotypes, such as K2P10 (Bang et al. 2000; Lesage et al. 2000; Gu et al. 2002; Simkin et al. 2008), and/or differences between species, experimental conditions and even among individual neurons.

In addition to wide expression in nervous tissue, transcript for K2P2 has also been observed in both atrial and ventricular myocytes isolated from rat heart (Terrenoire et al. 2001). Expression appears to increase with age (Liu and Saint 2004) and is stronger in subendocardium compared to subepicardium where the protein appears to be arranged in longitudinal stripes across cardiac myocytes (Tan et al. 2004).

The functional properties of neuronal K2P2 channels are subject to a novel type of regulation that leads altered excitability due to Na^+ permeation through the channels. Because the *kcnk2* gene has a weak Kozak initiation sequence, translation can also start at a second codon, to generate K2P2Δ, a channel variant lacking the first 56 residues (Thomas et al. 2008). Thus, alternative translation initiation (ATI) of *kcnk2* mRNA transcripts is a process that is regulated across brain regions over development. Although K2P2Δ channels have a truncated intracellular N-terminus, it is operation of the K^+ conduction pathway that is altered so that Na^+ ions pass under physiological conditions with a relative permeability of 0.18 compared to K^+, nearly an order of magnitude greater than for full-length channels (0.02) (Thomas et al. 2008). ATI has also been reported for rat *kcnk10* mRNA transcripts studied in experimental cells (Simkin et al. 2008).

The activities of K2P2, K2P4, and K2P10 channels are regulated by a diverse array of lipids including arachidonic acid, neurotransmitter-activated G-protein-coupled receptor pathways, anesthetics, and drugs (Honore 2007; Enyedi and Czirjak 2010). Several studies have reported on modulation of this family of K2P channels by changes in temperature (Maingret et al. 2000; Alloui et al. 2006), and by membrane stretch (Maingret et al. 1999a,b; Brohawn et al. 2012). Because a number of these stimuli often occur simultaneously, K2P2 channels have been posited to act as a polymodal signal integrators (Alloui et al. 2006; Honore 2007).

The temperature sensitivity of K2P2 and K2P4 channels has been reported to contribute to thermal regulation of nociceptive neurons. The activities of these channels increase as body temperature rises above normal (37°C) but falls again when temperature reaches the threshold for noxious heat (>~43°C) (Noel et al. 2009). Thus, increased activity of this group of channels is expected to dampen changes in the excitability of nociceptors up to the threshold for the detection of noxious heat, beyond which decreased background K^+ currents would facilitate excitability.

Background K^+ currents have similarly been implicated in the detection of cold. Temperature decrease to below ~18°C was observed to depolarize a subset of mouse trigeminal ganglion neurons (Viana et al. 2002) and rat dorsal root ganglion cells (Reid and Flonta 2001) by inhibiting a background K^+ conductance. Consistent with this observation, knockout mice engineered to lack the *kcnk2* and *kcnk4* genes are reported to have altered responses to temperature change (Noel et al. 2009). These findings suggest that K2P2 and K2P4 channels contribute as thermal rheostats, tuning the responsiveness of neurons to changes in temperatures. Consistent with this notion, K2P2 and K2P4 channels have been demonstrated to colocalize with various TRP channels in temperature-sensitive neurons (Yamamoto et al. 2009).

The Q_{10} for K2P2 and K2P4 channels increases by approximately sixfold for a 10°C rise. In contrast, the vanilloid family of transient receptor potential channels (TRPVs) has a Q_{10} close to 20. TRPV1 opens in response to noxious heat, while TRPV3 and TRPV4 channels are activated by more moderate, warm temperatures, >~30°C (Peier et al. 2002; Nilius et al. 2007). It remains unclear if temperature modulation of K2P2 and K2P4 channels is a direct effect or due to effects of temperature on second messenger pathways, phosphorylation cascades, or free fatty acid concentrations that act secondarily on the channels to modify their activity.

The operation of K2P2 channels is subject to regulation by protein kinase A–dependent phosphorylation of Ser^{348}. This regulatory change is rapid and notable because it transforms K2P2 channels from openly rectifying K^+ currents in hippocampal cells to currents that manifest voltage dependence (Bockenhauer et al. 2001). Furthermore, it appears that increased intracellular concentrations of phosphatidylinositol-4,5-bisphosphate (PIP2) shift the voltage dependence of K2P2 channels to hyperpolarized potentials, augmenting current magnitude across the physiological voltage range (Chemin et al. 2005; Lopes et al. 2005).

19.8.3 THE ACID-SENSITIVE CHANNELS, K2P3 AND K2P9: CONTROL OF SURFACE EXPRESSION

Originally called TASK1 and TASK3, for TWIK-related, acid-sensitive K^+ channel 1 and 3, K2P3 (Duprat et al. 1997) and K2P9 (Kim et al. 2000; Rajan et al. 2000) pass K^+-selective currents that are blocked by protonation of a histidine residue in the outer mouth of the pore in the P1 loop (Gly-Tyr-Gly-His) (Lopes et al. 2000, 2001), a mechanism of pH regulation shared with K2P1 (Rajan et al. 2005; Plant et al. 2010). Acid-sensitive background K^+ currents in tissues that express K2P1, K2P3, or K2P9 currents appear to be those identified as *IKso* in cerebellar granule neurons (Millar et al. 2000; Plant et al. 2002, 2012).

K2P3 and K2P9 channels are expressed throughout the central and peripheral nervous systems and have been proposed to mediate currents that depolarize neurons during acidification (Talley et al. 2000, 2001; Cooper et al. 2004; Rau et al. 2006; Plant et al. 2012). K2P3 is also expressed in the kidney, in adrenal glomerulosa cells, in the cardiac conduction pathway (Duprat et al. 1997; Gaborit et al. 2007), and in the carotid body; K2P3 channels are posited to depolarize type 1 glomus cells of the carotid body in response to acidosis and hypoxia (Peers 1990; Buckler 1997; Buckler et al. 2000). While the TASKs are not the only channels modulated by pH, changes in the magnitude of background K^+ currents correlate well with the slow kinetics of depolarization in somatosensory neurons in response to acidification. In contrast, other pH-modulated channels, such as the acid-sensing ion channels (ASICs), typically pass transient currents in response to acidification (Gold and Gebhart 2010).

K2P3 and K2P9 subunits are also notable because they form heterodimeric TASK channels (Czirjak and Enyedi 2002) with distinct sensitivities to acidification and pungent stimuli such as hydroxy-α-sanshool, the active ingredient in *Xanthoxylum* Szechuan peppercorns (Bautista et al. 2008). As noted earlier, K2P3 and K2P9 also coassemble with K2P1 subunits to form heterodimeric, SUMO-regulated channels in central neurons (Plant et al. 2012).

Surfaces levels of K2P3 and K2P9 are tightly regulated by the opposing action of signaling motifs that determine retention of the channels in the endoplasmic reticulum (ER) (O'Kelly et al. 2002). Thus, forward trafficking of K2P3 to the cell membrane requires the phosphorylation-dependent binding of the ubiquitous, soluble adapter protein, 14-3-3β. Binding of 14-3-3 suppresses the interaction between K2P3 and βCOP, a vesicular transport protein that promotes retention of the channel in the ER (O'Kelly et al. 2002). βCOP binding is mediated by separate, basic motifs on the N- and C-termini of K2P3 subunits and disruption of either site promotes forward trafficking of the channel to the cell membrane (O'Kelly and Goldstein 2008).

Competitive interaction between 14-3-3 and βCOP also regulates the forward trafficking of other membrane proteins including K2P9 (O'Kelly et al. 2002; Rajan et al. 2002; Zuzarte et al. 2009) and nicotinic acetylcholine receptors (O'Kelly et al. 2002). Forward trafficking of K2P9 was shown to require a di-acidic motif (EDE) located on the proximal C-terminus of the channel (Zuzarte et al. 2007).

Forward trafficking of K2P3 channels is also regulated by the annexin II subunit p11 (Girard et al. 2002; Renigunta et al. 2006;

O'Kelly and Goldstein 2008). The role of p11 in K2P3 channel trafficking appears to be via interaction with 14-3-3 and is specific to only some tissues, such as brain and lung where p11 is expressed prominently. In contrast, p11 is poorly expressed in the heart where expression of K2P3 is high (O'Kelly and Goldstein 2008).

K2P15 shows sequence similarity to K2P3 and K2P9, and *kcnk15* transcripts are identified in the pancreas, liver, lung, ovary, testis, and heart; nonetheless, the operation of the channel and its role in physiology remains unclear because it has thus far failed to pass currents in heterologous expression systems (Ashmole et al. 2001). Functional expression of K2P15 has been postulated to require an as-yet unidentified subunit or regulator.

19.8.4 K2P5, K2P16, AND K2P17: ALKALINE-ACTIVATED CHANNELS

K2P5 (previously TASK2) was initially considered to be a member of the TASK subfamily of K2P channels because it passed background K^+ currents regulated by changes in extracellular pH (Reyes et al. 1998). However, further characterization demonstrated that the channel passed currents only when the extracellular solution exceeded pH 7.5. K2P16 and K2P17 (also known as TALK1 and TALK2, respectively, for TWIK-related alkaline-activated K^+ channel) are also activated by extracellular alkalization above normal levels and, along with K2P5 are expressed at high levels in the pancreas (Decher et al. 2001; Girard et al. 2001). All three channels have been implicated in mediating the apical K^+ conductance that facilitates the secretion of bicarbonate from the epithelial cells in the tubular lumen of the exocrine pancreas (Fong et al. 2003).

K2P5 has been associated with bicarbonate handling in proximal tubule cells and the papillary collecting ducts of the kidney (Warth et al. 2004). Thus, K2P5 knockout mice show impaired bicarbonate reabsorption, metabolic acidosis, hyponatremia, and hypotension. Based on these findings, a role was suggested for K2P5 in renal acidosis syndromes.

Gating of K2P5, K2P16, and K2P17 by alkaline extracellular pH is proposed to be due to the neutralization of basic residues (arginine in K2P5 and K2P16, lysine in K2P17) that reside near the second P-loop and influence properties of the K^+ selectivity filter. A recent study employing concatemers of cloned K2P5 subunits demonstrated that both of these putative pH sensors must be neutralized for dimeric channels to conduct (Niemeyer et al. 2007).

19.8.5 K2P12 AND K2P13 CHANNELS

K2P12 and K2P13 (called THIK2 and THIK1, respectively, for TWIK-related halothane-inhibited K^+ channel 2 and 1) are expressed in the heart, skeletal muscle, and pancreas (Girard et al. 2001; Rajan et al. 2001). In situ hybridization suggests that both channels are expressed also in proximal tubules, thick ascending limbs and cortical collecting ducts of human kidney (Theilig et al. 2008), and in the CNS, including in the retrotrapezoid nucleus (Lazarenko et al. 2010a).

In common with other K2P clones, K2P13 channels pass K^+-selective leak currents in experimental cells. The currents are insensitive to physiological changes in pH, temperature, and free fatty acids. In contrast to many K2P channels, K2P13 currents are inhibited by the volatile halogenated anesthetics, such as halothane with a K_i of 2.8 mM (Rajan et al. 2001). Inhibition of a K2P13-like conductance has been proposed to activate central respiratory chemoreceptor neurons in the retrotrapezoid nucleus, and this mechanism is posited to preserve adequate respiratory motor activity and ventilation during anesthesia (Lazarenko et al. 2010a).

K2P12 clones do not pass currents in experimental systems despite being trafficked to the plasma membrane (Rajan et al. 2001), nor does coexpression of K2P12 appear to impact the magnitude of currents passed by K2P13, suggesting that these channel subunits do not form heterodimers (Rajan et al. 2001).

19.8.6 K2P18 CHANNELS

The final K2P channel identified in the human genome, K2P18 (also called TRESK for TWIK-related, spinal cord K^+ channel), has been found at the transcript level in human spinal cord (Sano et al. 2003), mouse cerebellum (Czirjak et al. 2004) and mouse testis (Kang et al. 2004). In common with other K2P channels, K2P18 channels pass K^+-selective currents and functions as an open rectifier (Keshavaprasad et al. 2005; Kang and Kim 2006; Dobler et al. 2007; Bautista et al. 2008).

Rodent clones of K2P18 are sensitive to external acidification because like K2P1, K2P3, and K2P9 channels, they are inhibited by protonation of a histidine in the first P-loop (pK_a ~6.8) (Dobler et al. 2007). In contrast to the rodent isoform, human K2P18 channels are insensitive to acidification, because the critical histidine residue is instead a tyrosine; the human channels can be endowed with sensitivity by mutation of the tyrosine to histidine (Dobler et al. 2007).

The activity of K2P18 channels is regulated by the intracellular concentration of Ca^{2+}. An increase in cytosolic Ca^{2+} activates the calmodulin-dependent protein phosphatase calcineurin and this leads to dephosphorylation of an intracellular serine (Ser^{264}) precluding the binding of 14-3-3 proteins that are observed to impact the activity of surface channels (Czirjak et al. 2004; Czirjak and Enyedi 2006). Channel activity is inhibited by the binding of the η and γ 14-3-3 but not by the β, ζ, ε, ς, or τ isoforms (Czirjak et al. 2008).

K2P18 is notable for its expression in the spinal cord, trigeminal and dorsal root ganglia, where it has been suggested to pass a significant portion of the background K^+ current that determines the excitability of somatosensory nociceptive fibers. K2P18 channels were further implicated in the normal operation of somatosensory nervous system by the observation that surgical denervation (axiotomy) of peripheral nociceptor neurons produced a decrease of ~50% in *kcnk18* mRNA (Tulleuda et al. 2011).

Formal correlation of K2P18 channels and pain was provided by linkage of a frame-shift mutation (F139WfsX24) in the *kcnk18* gene associated with migraine with aura in a large, multigenerational family (Lafreniere et al. 2010). The F139WfsX24 variant of K2P18 is truncated at 162 residues and appears to act as a dominant-negative subunit to suppress the function of wild type K2P18 channels.

19.9 FUTURE OF K2P CHANNEL RESEARCH

Over the last two decades, identification and study of the channels passing background K^+ currents has advanced our understanding of their roles in physiology and disease. K2P channels are now appreciated to operate in all phases of excitability, establishing resting V_M in many types of cells and influencing the rise to activation threshold, the shape of action potentials, the rate of recovery from firing (Plant 2012), and even the slow after-hyperpolarizations observed in developing starburst amacrine cells (Ford et al. 2013). The channels are now recognized to respond to anesthetics, changes in pH, temperature, membrane stretch, and regulatory pathways that include SUMO, kinases and phosphatases, GPCRs, lipids, and those that mediate trafficking (Plant et al. 2005; Honore 2007; Enyedi and Czirjak 2010). K2P channels are now recognized to have roles in clinical disorders including cardiac arrhythmia associated with hypokalemia (K2P1) (Ma et al. 2011), oncogenesis (K2P9) (Mu et al. 2003), and pain syndromes (K2P18) (Lafreniere et al. 2010). It is clear that we are just beginning to appreciate how and where K2P channels operate and how this varies at different times in different tissues. In some cases, we have yet to discover what the channels do even though they are synthesized and expressed at the plasma membrane. We have much to learn. Nonetheless, K2P channels are already understood to calm, control, and shape electrical activity, and this offers them, despite their recent addition to the family of K^+ channels, as attractive, tissue-specific targets for pharmaceutical therapy.

REFERENCES

Adams, M. D., S. E. Celniker, R. A. Holt, C. A. Evans, J. D. Gocayne, P. G. Amanatides, S. E. Scherer et al. 2000. The genome sequence of *Drosophila melanogaster*. *Science* 287 (5461):2185–2195.

Ahmed, A., F. Sesti, N. Ilan, T. M. Shih, S. L. Sturley, and S. A. Goldstein. 1999. A molecular target for viral killer toxin: TOK1 potassium channels. *Cell* 99 (3):283–291.

Alloui, A., K. Zimmermann, J. Mamet, F. Duprat, J. Noel, J. Chemin, N. Guy et al. 2006. TREK-1, a K^+ channel involved in polymodal pain perception. *EMBO J* 25 (11):2368–2376.

Ashmole, I., P. A. Goodwin, and P. R. Stanfield. 2001. TASK-5, a novel member of the tandem pore K^+ channel family. *Pflugers Arch* 442 (6):828–833.

Baccaglini, P. I. 1978. Action potentials of embryonic dorsal root ganglion neurones in *Xenopus tadpoles*. *J Physiol* 283:585–604.

Baev, D., A. Rivetta, X. S. Li, S. Vylkova, E. Bashi, C. L. Slayman, and M. Edgerton. 2003. Killing of *Candida albicans* by human salivary histatin 5 is modulated, but not determined, by the potassium channel TOK1. *Infect Immun* 71 (6):3251–3260.

Bang, H., Y. Kim, and D. Kim. 2000. TREK-2, a new member of the mechanosensitive tandem-pore K^+ channel family. *J Biol Chem* 275 (23):17412–17419.

Baumann, T. K., K. J. Burchiel, S. L. Ingram, and M. E. Martenson. 1996. Responses of adult human dorsal root ganglion neurons in culture to capsaicin and low pH. *Pain* 65 (1):31–38.

Bautista, D. M., Y. M. Sigal, A. D. Milstein, J. L. Garrison, J. A. Zorn, P. R. Tsuruda, R. A. Nicoll, and D. Julius. 2008. Pungent agents from Szechuan peppers excite sensory neurons by inhibiting two-pore potassium channels. *Nat Neurosci* 11 (7):772–779.

Bearzatto, B., F. Lesage, R. Reyes, M. Lazdunski, and P. M. Laduron. 2000. Axonal transport of TREK and TRAAK potassium channels in rat sciatic nerves. *Neuroreport* 11 (5):927–930.

Beitzinger, M., L. Hofmann, C. Oswald, R. Beinoraviciute-Kellner, M. Sauer, H. Griesmann, A. C. Bretz, C. Burek, A. Rosenwald, and T. Stiewe. 2008. p73 poses a barrier to malignant transformation by limiting anchorage-independent growth. *EMBO J* 27 (5):792–803.

Benson, M. D., Q. J. Li, K. Kieckhafer, D. Dudek, M. R. Whorton, R. K. Sunahara, J. A. Iniguez-Lluhi, and J. R. Martens. 2007. SUMO modification regulates inactivation of the voltage-gated potassium channel Kv1.5. *Proc Natl Acad Sci USA* 104 (6):1805–1810.

Berg, A. P., E. M. Talley, J. P. Manger, and D. A. Bayliss. 2004. Motoneurons express heteromeric TWIK-related acid-sensitive K^+ (TASK) channels containing TASK-1 (KCNK3) and TASK-3 (KCNK9) subunits. *J Neurosci* 24 (30):6693–6702.

Blatz, A. L. and K. L. Magleby. 1985. Single chloride-selective channels active at resting membrane potentials in cultured rat skeletal muscle. *Biophys J* 47 (1):119–123.

Bockenhauer, D., M. A. Nimmakayalu, D. C. Ward, S. A. Goldstein, and P. G. Gallagher. 2000. Genomic organization and chromosomal localization of the murine 2 P domain potassium channel gene Kcnk8: Conservation of gene structure in 2 P domain potassium channels. *Gene* 261 (2):365–372.

Bockenhauer, D., N. Zilberberg, and S. A. Goldstein. 2001. KCNK2: Reversible conversion of a hippocampal potassium leak into a voltage-dependent channel. *Nat Neurosci* 4 (5):486–491.

Brohawn, S. G., E. B. Campbell, and R. Mackinnon. 2013. Domain-swapped chain connectivity and gated membrane access in a Fab-mediated crystal of the human TRAAK K^+ channel. *Proc Natl Acad Sci USA* 110 (6):2129–2134.

Brohawn, S. G., J. del Marmol, and R. MacKinnon. 2012. Crystal structure of the human K2P TRAAK, a lipid- and mechano-sensitive K^+ ion channel. *Science* 335 (6067):436–441.

Brown, D. A. and P. R. Adams. 1980. Muscarinic suppression of a novel voltage-sensitive K^+ current in a vertebrate neurone. *Nature* 283 (5748):673–676.

Buckler, K. J. 1997. A novel oxygen-sensitive potassium current in rat carotid body type I cells. *J Physiol* 498 (Pt 3):649–662.

Buckler, K. J., B. A. Williams, and E. Honore. 2000. An oxygen-sensitive TASK-like potassium channel in rat carotid body type-1 cells. *J Physiol-Lond* 525:66pp.

Chavez, R. A., A. T. Gray, B. B. Zhao, C. H. Kindler, M. J. Mazurek, Y. Mehta, J. R. Forsayeth, and C. S. Yost. 1999. TWIK-2, a new weak inward rectifying member of the tandem pore domain potassium channel family. *J Biol Chem* 274 (12):7887–7892.

Chemin, J., A. J. Patel, F. Duprat, I. Lauritzen, M. Lazdunski, and E. Honore. 2005. A phospholipid sensor controls mechanogating of the K^+ channel TREK-1. *EMBO J* 24 (1):44–53.

Clark, R. B., C. Kondo, D. D. Belke, and W. R. Giles. 2011. Two-pore domain K(+) channels regulate membrane potential of isolated human articular chondrocytes. *J Physiol* 589 (Pt 21):5071–5089.

Cluzeaud, F., R. Reyes, B. Escoubet, M. Fay, M. Lazdunski, J. P. Bonvalet, F. Lesage, and N. Farman. 1998. Expression of TWIK-1, a novel weakly inward rectifying potassium channel in rat kidney. *Am J Physiol* 275 (6 Pt 1):C1602–C1609.

Conway, K. E. and J. F. Cotten. 2011. Covalent modification of a volatile anesthetic regulatory site activates TASK-3 (KCNK9) tandem pore potassium channels. *Mol Pharmacol* 81 (3):393–400.

Cooper, B. Y., R. D. Johnson, and K. K. Rau. 2004. Characterization and function of twik-related acid sensing K^+ channels in a rat nociceptive cell. *Neuroscience* 129 (1):209–224.

Czirjak, B., Z. E. Toth, and P. Enyedi. 2004. The two-pore domain K^+ channel, TRESK, is activated by the cytoplasmic calcium signal through calcineurin. *J Biol Chem* 279 (18):18550–18558.

Czirjak, G. and P. Enyedi. 2002. Formation of functional heterodimers between the TASK-1 and TASK-3 two-pore domain potassium channel subunits. *J Biol Chem* 277 (7):5426–5432.

Czirjak, G. and P. Enyedi. 2006. Targeting of calcineurin to an NFAT-like docking site is required for the calcium-dependent activation of the background K(+)channel, TRESK. *J Biol Chem* 281 (21):14677–14682.

Czirjak, G., D. Vuity, and P. Enyedi. 2008. Phosphorylation-dependent binding of 14-3-3 proteins controls TRESK regulation. *J Biol Chem* 283 (23):15672–15680.

Dai, X. Q., J. Kolic, P. Marchi, S. Sipione, and P. E. Macdonald. 2009. SUMOylation regulates Kv2.1 and modulates pancreatic beta-cell excitability. *J Cell Sci* 122 (Pt 6):775–779.

Decher, N., M. Maier, W. Dittrich, J. Gassenhuber, A. Bruggemann, A. E. Busch, and K. Steinmeyer. 2001. Characterization of TASK-4, a novel member of the pH-sensitive, two-pore domain potassium channel family. *FEBS Lett* 492 (1–2):84–89.

Delmas, P. and D. A. Brown. 2005. Pathways modulating neural KCNQ/M (Kv7) potassium channels. *Nat Rev Neurosci* 6 (11):850–862.

Dobler, T., A. Springauf, S. Tovornik, M. Weber, A. Schmitt, R. Sedlmeier, E. Wischmeyer, and F. Doring. 2007. TRESK two-pore-domain K(+) channels constitute a significant component of background potassium currents in murine dorsal root ganglion neurones. *J Physiol-Lond* 585 (3):867–879.

Duprat, F., F. Lesage, M. Fink, R. Reyes, C. Heurteaux, and M. Lazdunski. 1997. TASK, a human background K^+ channel to sense external pH variations near physiological pH. *EMBO J* 16 (17):5464–5471.

Enyedi, P. and G. Czirjak. 2010. Molecular background of leak K^+ currents: Two-pore domain potassium channels. *Physiol Rev* 90 (2):559–605.

Fink, M., F. Duprat, F. Lesage, R. Reyes, G. Romey, C. Heurteaux, and M. Lazdunski. 1996. Cloning, functional expression and brain localization of a novel unconventional outward rectifier K^+ channel. *EMBO J* 15 (24):6854–6862.

Fink, M., F. Lesage, F. Duprat, C. Heurteaux, R. Reyes, M. Fosset, and M. Lazdunski. 1998. A neuronal two P domain K^+ channel stimulated by arachidonic acid and polyunsaturated fatty acids. *EMBO J* 17 (12):3297–3308.

Fong, P., B. E. Argent, W. B. Guggino, and M. A. Gray. 2003. Characterization of vectorial chloride transport pathways in the human pancreatic duct adenocarcinoma cell line HPAF. *Am J Physiol Cell Physiol* 285 (2):C433–C445.

Ford, K. J., D. Arroyo, J. Kay, E. E. Lloyd, R. M. Bryan, Jr., J. Sanes, and M. B. Feller. 2013. A role for TREK1 in generating the slow after hyperpolarization in developing starburst amacrine cells. *J Neurophysiol* 109 (9):2250–2259.

Franks, N. P. and W. R. Lieb. 1994. Molecular and cellular mechanisms of general anaesthesia. *Nature* 367 (6464):607–614.

Gaborit, N., S. Le Bouter, V. Szuts, A. Varro, D. Escande, S. Nattel, and S. Demolombe. 2007. Regional and tissue specific transcript signatures of ion channel genes in the non-diseased human heart. *J Physiol* 582 (Pt 2):675–693.

Geiss-Friedlander, R. and F. Melchior. 2007. Concepts in sumoylation: A decade on. *Nat Rev Mol Cell Biol* 8 (12):947–956.

Gibb, S. L., W. Boston-Howes, Z. S. Lavina, S. Gustincich, R. H. Brown, Jr., P. Pasinelli, and D. Trotti. 2007. A caspase-3-cleaved fragment of the glial glutamate transporter EAAT2 is sumoylated and targeted to promyelocytic leukemia nuclear bodies in mutant SOD1-linked amyotrophic lateral sclerosis. *J Biol Chem* 282 (44):32480–32490.

Girard, C., F. Duprat, C. Terrenoire, N. Tinel, M. Fosset, G. Romey, M. Lazdunski, and F. Lesage. 2001. Genomic and functional characteristics of novel human pancreatic 2P domain K(+) channels. *Biochem Biophys Res Commun* 282 (1):249–256.

Girard, C., N. Tinel, C. Terrenoire, G. Romey, M. Lazdunski, and M. Borsotto. 2002. p11, an annexin II subunit, an auxiliary protein associated with the background K^+ channel, TASK-1. *EMBO J* 21 (17):4439–4448.

Gold, M. S. and G. F. Gebhart. 2010. Nociceptor sensitization in pain pathogenesis. *Nat Med* 16 (11):1248–1257.

Goldman, D. E. 1943. Potential, impedance, and rectification in membranes. *J Gen Physiol* 27 (1):37–60.

Goldstein, S. A., D. Bockenhauer, I. O'Kelly, and N. Zilberberg. 2001. Potassium leak channels and the KCNK family of two-P-domain subunits. *Nat Rev Neurosci* 2 (3):175–184.

Goldstein, S. A., L. A. Price, D. N. Rosenthal, and M. H. Pausch. 1996. ORK1, a potassium-selective leak channel with two pore domains cloned from *Drosophila melanogaster* by expression in *Saccharomyces cerevisiae*. *Proc Natl Acad Sci USA* 93 (23):13256–13261.

Goldstein, S. A., K. W. Wang, N. Ilan, and M. H. Pausch. 1998. Sequence and function of the two P domain potassium channels: Implications of an emerging superfamily. *J Mol Med (Berl)* 76 (1):13–20.

Gu, W., G. Schlichthorl, J. R. Hirsch, H. Engels, C. Karschin, A. Karschin, C. Derst, O. K. Steinlein, and J. Daut. 2002. Expression pattern and functional characteristics of two novel splice variants of the two-pore-domain potassium channel TREK-2. *J Physiol* 539 (Pt 3):657–668.

Heurteaux, C., G. Lucas, N. Guy, M. El Yacoubi, S. Thummler, X. D. Peng, F. Noble et al. 2006. Deletion of the background potassium channel TREK-1 results in a depression-resistant phenotype. *Nat Neurosci* 9 (9):1134–1141.

Hille, B. 2001. *Ion Channels of Excitable Membranes*. 3rd ed. Sunderland, MA: Sinauer.

Hodgkin, A. L. and P. Horowicz. 1959. The influence of potassium and chloride ions on the membrane potential of single muscle fibres. *J Physiol* 148:127–160.

Hodgkin, A. L. and A. F. Huxley. 1952. The components of membrane conductance in the giant axon of *Loligo*. *J Physiol-Lond* 116 (4):473–496.

Hodgkin, A. L., A. F. Huxley, and B. Katz. 1949. Ionic currents underlying activity in the giant axon of the *Squid*. *Arch Des Sci Physiologiq* 3 (2):129–150.

Honore, E. 2007. The neuronal background K2P channels: Focus on TREK1. *Nat Rev Neurosci* 8 (4):251–261.

Ilan, N. and S. A. Goldstein. 2001. Kcnko: Single, cloned potassium leak channels are multi-ion pores. *Biophys J* 80 (1):241–253.

Jones, S. W. 1989. On the resting potential of isolated frog sympathetic neurons. *Neuron* 3 (2):153–161.

Kang, D. and D. Kim. 2006. TREK-2 (K2P10.1) and TRESK (K2P18.1) are major background K^+ channels in dorsal root ganglion neurons. *Am J Physiol Cell Physiol* 291 (1):C138–C146.

Kang, D., E. Mariash, and D. Kim. 2004. Functional expression of TRESK-2, a new member of the tandem-pore K^+ channel family. *J Biol Chem* 279 (27):28063–28070.

Keshavaprasad, B., C. Liu, J. D. Au, C. H. Kindler, J. F. Cotten, and C. S. Yost. 2005. Species-specific differences in response to anesthetics and other modulators by the K2P channel TRESK. *Anesth Analg* 101 (4):1042–1049, table of contents.

Ketchum, K. A., W. J. Joiner, A. J. Sellers, L. K. Kaczmarek, and S. A. Goldstein. 1995. A new family of outwardly rectifying potassium channel proteins with two pore domains in tandem. *Nature* 376 (6542):690–695.

Kim, Y., H. Bang, and D. Kim. 2000. TASK-3, a new member of the tandem pore K(+) channel family. *J Biol Chem* 275 (13):9340–9347.

Kollewe, A., A. Y. Lau, A. Sullivan, B. Roux, and S. A. Goldstein. 2009. A structural model for K2P potassium channels based on 23 pairs of interacting sites and continuum electrostatics. *J Gen Physiol* 134 (1):53–68.

Kruse, M., E. Schulze-Bahr, V. Corfield, A. Beckmann, B. Stallmeyer, G. Kurtbay, I. Ohmert, P. Brink, and O. Pongs. 2009. Impaired endocytosis of the ion channel TRPM4 is associated with human progressive familial heart block type I. *J Clin Invest* 119 (9):2737–2744.

Lafreniere, R. G., M. Z. Cader, J. F. Poulin, I. Andres-Enguix, M. Simoneau, N. Gupta, K. Boisvert et al. 2010. A dominant-negative mutation in the TRESK potassium channel is linked to familial migraine with aura. *Nat Med* 16 (10):1157–U1501.

Lalevee, N., B. Monier, S. Senatore, L. Perrin, and M. Semeriva. 2006. Control of cardiac rhythm by ORK1, a *Drosophila* two-pore domain potassium channel. *Curr Biol* 16 (15):1502–1508.

Lauritzen, I., J. Chemin, E. Honore, M. Jodar, N. Guy, M. Lazdunski, and A. Jane Patel. 2005. Cross-talk between the mechano-gated K2P channel TREK-1 and the actin cytoskeleton. *EMBO Rep* 6 (7):642–648.

Lazarenko, R. M., M. G. Fortuna, Y. Shi, D. K. Mulkey, A. C. Takakura, T. S. Moreira, P. G. Guyenet, and D. A. Bayliss. 2010a. Anesthetic activation of central respiratory chemoreceptor neurons involves inhibition of a THIK-1-like background K(+) current. *J Neurosci* 30 (27):9324–9334.

Lazarenko, R. M., S. C. Willcox, S. Shu, A. P. Berg, V. Jevtovic-Todorovic, E. M. Talley, X. Chen, and D. A. Bayliss. 2010b. Motoneuronal TASK channels contribute to immobilizing effects of inhalational general anesthetics. *J Neurosci* 30 (22):7691–7704.

Lesage, F., E. Guillemare, M. Fink, F. Duprat, M. Lazdunski, G. Romey, and J. Barhanin. 1996. TWIK-1, a ubiquitous human weakly inward rectifying K^+ channel with a novel structure. *EMBO J* 15 (5):1004–1011.

Lesage, F., C. Terrenoire, G. Romey, and M. Lazdunski. 2000. Human TREK2, a 2P domain mechano-sensitive K^+ channel with multiple regulations by polyunsaturated fatty acids, lysophospholipids, and Gs, Gi, and Gq protein-coupled receptors. *J Biol Chem* 275 (37):28398–28405.

Littleton, J. T. and B. Ganetzky. 2000. Ion channels and synaptic organization: Analysis of the *Drosophila* genome. *Neuron* 26 (1):35–43.

Liu, C., J. D. Au, H. L. Zou, J. F. Cotten, and C. S. Yost. 2004. Potent activation of the human tandem pore domain K channel TRESK with clinical concentrations of volatile anesthetics. *Anesth Analg* 99 (6):1715–1722, table of contents.

Liu, W. and D. A. Saint. 2004. Heterogeneous expression of tandem-pore K^+ channel genes in adult and embryonic rat heart quantified by real-time polymerase chain reaction. *Clin Exp Pharmacol Physiol* 31 (3):174–178.

Lloyd, E. E., R. F. Crossland, S. C. Phillips, S. P. Marrelli, A. K. Reddy, G. E. Taffet, C. J. Hartley, and R. M. Bryan, Jr. 2011. Disruption of K(2P)6.1 produces vascular dysfunction and hypertension in mice. *Hypertension* 58 (4):672–678.

Lopes, C. M., P. G. Gallagher, M. E. Buck, M. H. Butler, and S. A. Goldstein. 2000. Proton block and voltage gating are potassium-dependent in the cardiac leak channel Kcnk3. *J Biol Chem* 275 (22):16969–16978.

Lopes, C. M., T. Rohacs, G. Czirjak, T. Balla, P. Enyedi, and D. E. Logothetis. 2005. PIP2 hydrolysis underlies agonist-induced inhibition and regulates voltage gating of two-pore domain K^+ channels. *J Physiol* 564 (Pt 1):117–129.

Lopes, C. M., N. Zilberberg, and S. A. Goldstein. 2001. Block of Kcnk3 by protons. Evidence that 2-P-domain potassium channel subunits function as homodimers. *J Biol Chem* 276 (27):24449–24452.

Loukin, S. H., B. Vaillant, X. L. Zhou, E. P. Spalding, C. Kung, and Y. Saimi. 1997. Random mutagenesis reveals a region important for gating of the yeast K^+ channel Ykc1. *EMBO J* 16 (16):4817–4825.

Ma, L., X. Zhang, and H. Chen. 2011. TWIK-1 two-pore domain potassium channels change ion selectivity and conduct inward leak sodium currents in hypokalemia. *Sci Signal* 4 (176):ra37.

Ma, L., X. Zhang, M. Zhou, and H. Chen. 2012. Acid-sensitive TWIK and TASK two-pore domain potassium channels change ion selectivity and become permeable to sodium in extracellular acidification. *J Biol Chem* 287 (44):37145–37153.

Mahajan, R., C. Delphin, T. Guan, L. Gerace, and F. Melchior. 1997. A small ubiquitin-related polypeptide involved in targeting RanGAP1 to nuclear pore complex protein RanBP2. *Cell* 88 (1):97–107.

Maingret, F., M. Fosset, F. Lesage, M. Lazdunski, and E. Honore. 1999a. TRAAK is a mammalian neuronal mechano-gated K^+ channel. *J Biol Chem* 274 (3):1381–1387.

Maingret, F., I. Lauritzen, A. J. Patel, C. Heurteaux, R. Reyes, F. Lesage, M. Lazdunski, and E. Honore. 2000. TREK-1 is a heat-activated background K(+) channel. *EMBO J* 19 (11):2483–2491.

Maingret, F., A. J. Patel, F. Lesage, M. Lazdunski, and E. Honore. 1999b. Mechano- or acid stimulation, two interactive modes of activation of the TREK-1 potassium channel. *J Biol Chem* 274 (38):26691–26696.

Martin, S., A. Nishimune, J. R. Mellor, and J. M. Henley. 2007. SUMOylation regulates kainate-receptor-mediated synaptic transmission. *Nature* 447 (7142):321–325.

Medhurst, A. D., G. Rennie, C. G. Chapman, H. Meadows, M. D. Duckworth, R. E. Kelsell, Gloger, II, and M. N. Pangalos. 2001. Distribution analysis of human two pore domain potassium channels in tissues of the central nervous system and periphery. *Brain Res Mol Brain Res* 86 (1–2):101–114.

Millar, J. A., L. Barratt, A. P. Southan, K. M. Page, R. E. Fyffe, B. Robertson, and A. Mathie. 2000. A functional role for the two-pore domain potassium channel TASK-1 in cerebellar granule neurons. *Proc Natl Acad Sci USA* 97 (7):3614–3618.

Millar, I. D., H. C. Taylor, G. J. Cooper, J. D. Kibble, J. Barhanin, and L. Robson. 2006. Adaptive downregulation of a quinidine-sensitive cation conductance in renal principal cells of TWIK-1 knockout mice. *Pflugers Arch* 453 (1):107–116.

Miller, A. N. and S. B. Long. 2012. Crystal structure of the human two-pore domain potassium channel K2P1. *Science* 335 (6067):432–436.

Mu, D., L. Chen, X. Zhang, L. H. See, C. M. Koch, C. Yen, J. J. Tong et al. 2003. Genomic amplification and oncogenic properties of the KCNK9 potassium channel gene. *Cancer Cell* 3 (3):297–302.

Nattel, S., V. Elharrar, D. P. Zipes, and J. C. Bailey. 1981. pH-dependent electrophysiological effects of quinidine and lidocaine on canine cardiac purkinje fibers. *Circ Res* 48 (1):55–61.

Nicoll, R. A. and D. V. Madison. 1982. General anesthetics hyperpolarize neurons in the vertebrate central nervous system. *Science* 217 (4564):1055–1057.

Nie, X., I. Arrighi, B. Kaissling, I. Pfaff, J. Mann, J. Barhanin, and V. Vallon. 2005. Expression and insights on function of potassium channel TWIK-1 in mouse kidney. *Pflugers Arch* 451 (3):479–488.

Niemeyer, M. I., F. D. Gonzalez-Nilo, L. Zuniga, W. Gonzalez, L. P. Cid, and F. V. Sepulveda. 2007. Neutralization of a single arginine residue gates open a two-pore domain, alkali-activated K^+ channel. *Proc Natl Acad Sci USA* 104 (2):666–671.

Nilius, B., G. Owsianik, T. Voets, and J. A. Peters. 2007. Transient receptor potential cation channels in disease. *Physiol Rev* 87 (1):165–217.

Noel, J., K. Zimmermann, J. Busserolles, E. Deval, A. Alloui, S. Diochot, N. Guy, M. Borsotto, P. Reeh, A. Eschalier, and M. Lazdunski, 2009. The mechano-activated K^+ channels TRAAK and TREK-1 control both warm and cold perception. *EMBO J* 28 (9):1308–1318.

O'Kelly, I., M. H. Butler, N. Zilberberg, and S. A. Goldstein. 2002. Forward transport. 14-3-3 binding overcomes retention in endoplasmic reticulum by dibasic signals. *Cell* 111 (4):577–588.

O'Kelly, I. and S. A. Goldstein. 2008. Forward transport of K2p3.1: Mediation by 14-3-3 and COPI, modulation by p11. *Traffic* 9 (1):72–78.

Ozaita, A. and E. Vega-Saenz de Miera. 2002. Cloning of two transcripts, HKT4.1a and HKT4.1b, from the human two-pore K^+ channel gene KCNK4. Chromosomal localization, tissue distribution and functional expression. *Brain Res Mol Brain Res* 102 (1–2):18–27.

Park, D. and L. C. Griffith. 2006. Electrophysiological and anatomical characterization of PDF-positive clock neurons in the intact adult *Drosophila* brain. *J Neurophysiol* 95 (6):3955–3960.
Patel, A. J., E. Honore, F. Lesage, M. Fink, G. Romey, and M. Lazdunski. 1999. Inhalational anesthetics activate two-pore-domain background K+ channels. *Nat Neurosci* 2 (5):422–426.
Patel, A. J., F. Maingret, V. Magnone, M. Fosset, M. Lazdunski, and E. Honore. 2000. TWIK-2, an inactivating 2P domain K+ channel. *J Biol Chem* 275 (37):28722–28730.
Peers, C. 1990. Hypoxic suppression of K+ currents in type I carotid body cells: Selective effect on the Ca2(+)-activated K+ current. *Neurosci Lett* 119 (2):253–256.
Peier, A. M., A. J. Reeve, D. A. Andersson, A. Moqrich, T. J. Earley, A. C. Hergarden, G. M. Story et al. 2002. A heat-sensitive TRP channel expressed in keratinocytes. *Science* 296 (5575):2046–2049.
Plant, L. D. 2012. A role for K2P channels in the operation of somatosensory nociceptors. *Front Mol Neurosci* 5 (21):1–5.
Plant, L. D., D. A. Bayliss, D. Kim, F. Lesage, and S. A. Goldstein. 2005. International Union of Pharmacology. LV. Nomenclature and molecular relationships of two-P potassium channels. *Pharmacol Rev* 57 (4):527–540.
Plant, L. D., I. S. Dementieva, A. Kollewe, S. Olikara, J. D. Marks, and S. A. Goldstein. 2010. One SUMO is sufficient to silence the dimeric potassium channel K2P1. *Proc Natl Acad Sci USA* 107 (23):10743–10748.
Plant, L. D., E. J. Dowdell, I. S. Dementieva, J. D. Marks, and S. A. Goldstein. 2011. SUMO modification of cell surface Kv2.1 potassium channels regulates the activity of rat hippocampal neurons. *J Gen Physiol* 137 (5):441–454.
Plant, L. D., P. J. Kemp, C. Peers, Z. Henderson, and H. A. Pearson. 2002. Hypoxic depolarization of cerebellar granule neurons by specific inhibition of TASK-1. *Stroke* 33 (9):2324–2328.
Plant, L. D., L. Zuniga, D. Araki, J. D. Marks, and S. A. Goldstein. 2012. SUMOylation silences heterodimeric TASK potassium channels containing K2P1 subunits in cerebellar granule neurons. *Sci Signal* 5 (251):ra84.
Pountney, D. J., I. Gulkarov, E. Vega-Saenz de Miera, D. Holmes, M. Saganich, B. Rudy, M. Artman, and W. A. Coetzee. 1999. Identification and cloning of TWIK-originated similarity sequence (TOSS): A novel human 2-pore K+ channel principal subunit. *FEBS Lett* 450 (3):191–196.
Puil, E., B. Gimbarzevsky, and R. M. Miura. 1987. Voltage dependence of membrane properties of trigeminal root ganglion neurons. *J Neurophysiol* 58 (1):66–86.
Rajan, S., L. D. Plant, M. L. Rabin, M. H. Butler, and S. A. Goldstein. 2005. Sumoylation silences the plasma membrane leak K+ channel K2P1. *Cell* 121 (1):37–47.
Rajan, S., R. Preisig-Muller, E. Wischmeyer, R. Nehring, P. J. Hanley, V. Renigunta, B. Musset et al. 2002. Interaction with 14-3-3 proteins promotes functional expression of the potassium channels TASK-1 and TASK-3. *J Physiol* 545 (Pt 1):13–26.
Rajan, S., E. Wischmeyer, C. Karschin, R. Preisig-Muller, K. H. Grzeschik, J. Daut, A. Karschin, and C. Derst. 2001. THIK-1 and THIK-2, a novel subfamily of tandem pore domain K+ channels. *J Biol Chem* 276 (10):7302–7311.
Rajan, S., E. Wischmeyer, G. Xin Liu, R. Preisig-Muller, J. Daut, A. Karschin, and C. Derst. 2000. TASK-3, a novel tandem pore domain acid-sensitive K+ channel. An extracellular histiding as pH sensor. *J Biol Chem* 275 (22):16650–16657.
Rau, K. K., B. Y. Cooper, and R. D. Johnson. 2006. Expression of TWIK-related acid sensitive K+ channels in capsaicin sensitive and insensitive cells of rat dorsal root ganglia. *Neuroscience* 141 (2):955–963.
Reid, G. and M. Flonta. 2001. Cold transduction by inhibition of a background potassium conductance in rat primary sensory neurones. *Neurosci Lett* 297 (3):171–174.
Renigunta, V., H. Yuan, M. Zuzarte, S. Rinne, A. Koch, E. Wischmeyer, G. Schlichthorl et al. 2006. The retention factor p11 confers an endoplasmic reticulum-localization signal to the potassium channel TASK-1. *Traffic* 7 (2):168–181.
Reyes, R., F. Duprat, F. Lesage, M. Fink, M. Salinas, N. Farman, and M. Lazdunski. 1998. Cloning and expression of a novel pH-sensitive two pore domain K+ channel from human kidney. *J Biol Chem* 273 (47):30863–30869.
Roberts, S. K. 2003. TOK homologue in *Neurospora crassa*: First cloning and functional characterization of an ion channel in a filamentous fungus. *Eukaryot Cell* 2 (1):181–190.
Sakmann, B. and G. Trube. 1984. Conductance properties of single inwardly rectifying potassium channels in ventricular cells from guinea-pig heart. *J Physiol* 347:641–657.
Salinas, M., R. Reyes, F. Lesage, M. Fosset, C. Heurteaux, G. Romey, and M. Lazdunski. 1999. Cloning of a new mouse two-P domain channel subunit and a human homologue with a unique pore structure. *J Biol Chem* 274 (17):11751–11760.
Salkoff, L., A. Butler, G. Fawcett, M. Kunkel, C. McArdle, G. Paz-y-Mino, M. Nonet et al. 2001. Evolution tunes the excitability of individual neurons. *Neuroscience* 103 (4):853–859.
Salkoff, L., A. D. Wei, B. Baban, A. Butler, G. Fawcett, G. Ferreira, and C. M. Santi. 2005. Potassium channels in *C. elegans*. *WormBook*:1–15.
Sano, Y., K. Inamura, A. Miyake, S. Mochizuki, C. Kitada, H. Yokoi, K. Nozawa, H. Okada, H. Matsushime, and K. Furuichi. 2003. A novel two-pore domain K(+) channel, TRESK, is localized in the spinal cord. *J Biol Chem* 278 (30):27406–27412.
Sesti, F., T. M. Shih, N. Nikolaeva, and S. A. Goldstein. 2001. Immunity to K1 killer toxin: Internal TOK1 blockade. *Cell* 105 (5):637–644.
Shen, K. Z., R. A. North, and A. Surprenant. 1992. Potassium channels opened by noradrenaline and other transmitters in excised membrane patches of guinea-pig submucosal neurones. *J Physiol* 445:581–599.
Siegelbaum, S. A., J. S. Camardo, and E. R. Kandel. 1982. Serotonin and cyclic AMP close single K+ channels in Aplysia sensory neurones. *Nature* 299 (5882):413–417.
Simkin, D., E. J. Cavanaugh, and D. Kim. 2008. Control of the single channel conductance of K2P10.1 (TREK-2) by the amino-terminus: Role of alternative translation initiation. *J Physiol* 586 (Pt 23):5651–5663.
Sirois, J. E., Q. Lei, E. M. Talley, C. Lynch, 3rd, and D. A. Bayliss. 2000. The TASK-1 two-pore domain K+ channel is a molecular substrate for neuronal effects of inhalation anesthetics. *J Neurosci* 20 (17):6347–6354.
Sirois, J. E., J. J. Pancrazio, C. L. Iii, and D. A. Bayliss. 1998. Multiple ionic mechanisms mediate inhibition of rat motoneurones by inhalation anaesthetics. *J Physiol* 512 (Pt 3):851–862.
Streit, A. K., M. F. Netter, F. Kempf, M. Walecki, S. Rinne, M. K. Bollepalli, R. Preisig-Muller et al. 2011. A specific two-pore domain potassium channel blocker defines the structure of the TASK-1 open pore. *J Biol Chem* 286 (16):13977–13984.
Talley, E. M. and D. A. Bayliss. 2002. Modulation of TASK-1 (Kcnk3) and TASK-3 (Kcnk9) potassium channels: Volatile anesthetics and neurotransmitters share a molecular site of action. *J Biol Chem* 277 (20):17733–17742.
Talley, E. M., Q. Lei, J. E. Sirois, and D. A. Bayliss. 2000. TASK-1, a two-pore domain K+ channel, is modulated by multiple neurotransmitters in motoneurons. *Neuron* 25 (2):399–410.
Talley, E. M., G. Solorzano, Q. B. Lei, D. Kim, and D. A. Bayliss. 2001. CNS distribution of members of the two-pore-domain (KCNK) potassium channel family. *J Neurosci* 21 (19):7491–7505.

Tan, J. H., W. Liu, and D. A. Saint. 2004. Differential expression of the mechanosensitive potassium channel TREK-1 in epicardial and endocardial myocytes in rat ventricle. *Exp Physiol* 89 (3):237–242.

Tang, Z., O. El Far, H. Betz, and A. Scheschonka. 2005. Pias1 interaction and sumoylation of metabotropic glutamate receptor 8. *J Biol Chem* 280 (46):38153–38159.

Terrenoire, C., I. Lauritzen, F. Lesage, G. Romey, and M. Lazdunski. 2001. A TREK-1-like potassium channel in atrial cells inhibited by beta-adrenergic stimulation and activated by volatile anesthetics. *Circ Res* 89 (4):336–342.

Theilig, F., I. Goranova, J. R. Hirsch, M. Wieske, S. Unsal, S. Bachmann, R. W. Veh, and C. Derst. 2008. Cellular localization of THIK-1 (K(2P)13.1) and THIK-2 (K(2P)12.1) K channels in the mammalian kidney. *Cell Physiol Biochem* 21 (1–3):63–74.

Thomas, D., L. D. Plant, C. M. Wilkens, Z. A. McCrossan, and S. A. Goldstein. 2008. Alternative translation initiation in rat brain yields K2P2.1 potassium channels permeable to sodium. *Neuron* 58 (6):859–870.

Tulleuda, A., B. Cokic, G. Callejo, B. Saiani, J. Serra, and X. Gasull. 2011. TRESK channel contribution to nociceptive sensory neurons excitability: Modulation by nerve injury. *Mol Pain* 7:30.

Viana, F., E. de la Pena, and C. Belmonte. 2002. Specificity of cold thermotransduction is determined by differential ionic channel expression. *Nat Neurosci* 5 (3):254–260.

Wang, Z., R. J. Van den Berg, and D. L. Ypey. 1994. Resting membrane potentials and excitability at different regions of rat dorsal root ganglion neurons in culture. *Neuroscience* 60 (1):245–254.

Warth, R., H. Barriere, P. Meneton, M. Bloch, J. Thomas, M. Tauc, D. Heitzmann et al. 2004. Proximal renal tubular acidosis in TASK2 K^+ channel-deficient mice reveals a mechanism for stabilizing bicarbonate transport. *Proc Natl Acad Sci USA* 101 (21):8215–8220.

Watkins, C. S. and A. Mathie. 1996. A non-inactivating K^+ current sensitive to muscarinic receptor activation in rat cultured cerebellar granule neurons. *J Physiol* 491 (Pt 2):401–412.

Wilson, V. G. and G. Rosas-Acosta. 2005. Wrestling with SUMO in a new arena. *Sci STKE* 2005 (290):pe32.

Yamamoto, Y., T. Hatakeyama, and K. Taniguchi. 2009. Immunohistochemical colocalization of TREK-1, TREK-2 and TRAAK with TRP channels in the trigeminal ganglion cells. *Neurosci Lett* 454 (2):129–133.

Yook, K., T. W. Harris, T. Bieri, A. Cabunoc, J. Chan, W. J. Chen, P. Davis et al. 2012. WormBase 2012: More genomes, more data, new website. *Nucleic Acids Res* 40 (Database issue):D735–D741.

Zilberberg, N., N. Ilan, and S. A. Goldstein. 2001. KCNKO: Opening and closing the 2-P-domain potassium leak channel entails "C-type" gating of the outer pore. *Neuron* 32 (4):635–648.

Zilberberg, N., N. Ilan, R. Gonzalez-Colaso, and S. A. Goldstein. 2000. Opening and closing of KCNKO potassium leak channels is tightly regulated. *J Gen Physiol* 116 (5):721–734.

Zuzarte, M., K. Heusser, V. Renigunta, G. Schlichthorl, S. Rinne, E. Wischmeyer, J. Daut, B. Schwappach, and R. Preisig-Muller. 2009. Intracellular traffic of the K^+ channels TASK-1 and TASK-3: Role of N- and C-terminal sorting signals and interaction with 14-3-3 proteins. *J Physiol* 587 (Pt 5):929–952.

Zuzarte, M., S. Rinne, G. Schlichthorl, A. Schubert, J. Daut, and R. Preisig-Muller. 2007. A di-acidic sequence motif enhances the surface expression of the potassium channel TASK-3. *Traffic* 8 (8):1093–1100.

20 KCNQ channels

Nikita Gamper and Mark S. Shapiro

Contents

20.1 INTRODUCTION

As for most of the ion channels discussed in this volume, the currents carried by KCNQ K^+ channels were characterized by pharmacological, kinetic, or functional features long before their gene identification, and so have established names reflecting those features. In the central nervous system (CNS) and the peripheral nervous system (PNS) neurons, this corresponds to the *M-current*, first described by David Brown and colleagues in sympathetic ganglia as a voltage-gated, noninactivating K^+ current depressed by the stimulation of muscarinic acetylcholine receptors (mAChRs) (1,2). These investigators were searching for the molecular basis of the slow excitatory postsynaptic potential (EPSP), the prolonged depolarization occurring with a delay after a synaptic EPSP seen after trains of action potentials (3). The slow EPSP proved to be due to the closure of the M-type K^+ current via mAChR stimulation and actions of $G_{q/11}$ G proteins, an effect also mediated by a variety of peptide neurotransmitters, such as Gonadotropin-releasing hormone (GnRH) (Luteinizing-hormone-releasing hormone (LHRH) in frog), substance P, angiotensin II, and others (4). The inhibition of M-current (I_M) generally increases excitability as the standing K^+ conductance at resting potentials is reduced (Figure 20.1). I_M is well poised to serve this role due to its lack of inactivation, threshold for activation near neuronal resting potentials, and slow kinetics (5). In the heart, a similar K^+ current with even slower kinetics, dubbed I_{Ks}, underlies much of the initial repolarization after the cardiac action potential and is sensitive to protein kinase A (PKA), making it partly responsible for the speeding of the heart rate upon adrenergic stimulation (6). Neither I_M nor I_{Ks} is particularly sensitive to the well-known blocker of most delayed rectifiers, tetraethylammonium (TEA) ions or various scorpion toxins. Much study has been spent answering two fundamental questions for these K^+ currents: the molecular correlates underlying them and the signal transduction mechanisms linking muscarinic or β-adrenergic stimulation to modulation of M-current, or I_{Ks}, respectively. As we see in the following, the major clues for both questions came from inherited diseases in people linked to specific gene loci.

20.2 GENE IDENTIFICATION

In 1996, an inherited form of human cardiac arrhythmia called *long QT* syndrome (LQTS) was localized to a novel K^+ channel gene, dubbed KvLQT1 (7–9). Similar to the tetrameric arrangement of four subunits of the prototypical *Shaker* K^+ channel, the KvLQT1 gene predicted each subunit to have six membrane-spanning domains (S1–S6), with the voltage sensor residing in S4, and a long carboxyl terminus. Careful study revealed native I_{Ks} to be composed of KvLQT1, as the pore-forming subunit, in conjunction with an auxiliary β-subunit, called minK/I_sK (this β-subunit was initially thought capable of itself being a channel; see the following text) (10–12). Concurrently, analysis of the genome of the worm, *C. elegans*, identified novel genes orthologous to KvLQT1, KQT1-2 (13). From this analysis, two other human KQT/KvLQT1-like channels were identified as the gene loci in which mutations underlie inherited epileptic syndromes in human newborns (14,15), mapped to chromosomes 20q1.3 and 8q24 (16,17). KvLQT1 was then renamed KCNQ1, and the two seizure-associated genes, *KCNQ2* and *KCNQ3*. From expression studies in oocytes, KCNQ2/3 heteromers were shown to display all the characteristics of native M-current (18), and expression in mammalian cell lines confirmed suppression of KCNQ2/3 current by M_1 mAChRs (19,20).

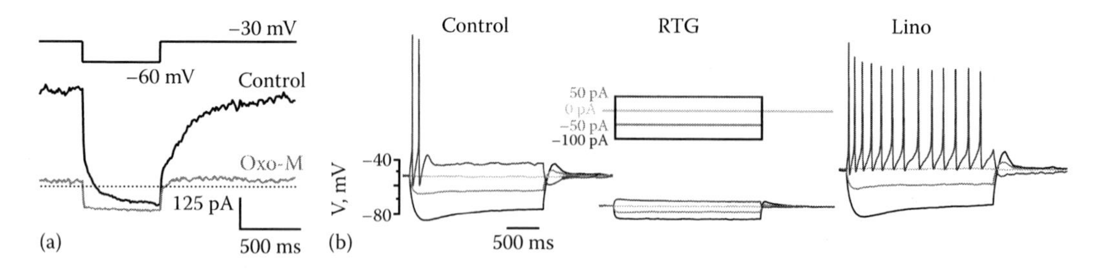

Figure 20.1 Hallmark features of neuronal M-current. (a) Whole-cell voltage clamp recording from a rat superior cervical ganglion (SCG) neuron, using the pulse protocol shown above. I_M is seen as the standing outward current at V_{hold} = −30 mV that slowly deactivates during the voltage step to −60 mV, and then slowly reactivates when the neuron is stepped back to −30 mV. Application of the muscarinic agonist, oxotremorine methiodide (oxo-M), causes a strong suppression of I_M within ~30 s. I_M is typically quantified in neurons as the amplitude of the relaxation at −60 mV. (b) A similar rat SCG neuron is studied under whole-cell current clamp, and depolarizing or hyperpolarizing currents applied, as indicated. Injection of the 50 pA current elicits only two action potentials (APs), although the current is maintained, typical of these *phasic* neurons. When the M-channel opener, retigabine(RTG) is applied, there is a hyperpolarization of V_{rest} by ~15 mV, an absence of any APs elicited by the 50 pA current, and a profound decrease in input resistance, as manifested by the greatly reduced hyperpolarization induced by the −50 or −100 pA current injections. All of these effects can be explained by the left-shift in the voltage dependence of M-channel activation, and increased open probability at all voltages, induced by retigabine (RTG). When the M-channel blocker, linopirdine (Lino), is applied, the phasic feature is abolished, as now the 50 pA injection causes an uninterrupted train of APs during the current pulse, indicating that the pronounced *spike-frequency adaptation* of these neurons is heavily M-current dependent. Similar phenomena due to M-currents are observed throughout the PNS and CNS (Unpublished data from the authors).

KCNQ2 is expressed as six to nine alternatively spliced forms in rodent (21) and human (22) brain, with the alternative splicing involving exons 7–15. With one or two exceptions, the isoforms seem functionally similar (23), although systematic investigation of this issue has not been performed.

Another subtype associated with inherited syndromes of deafness, KCNQ4, localizes to the inner ear (24) and to auditory nuclei in the brain (25–28). Finally, KCNQ5 contributes to neuronal M-channel composition in the brain, and sympathetic and sensory ganglia (29–31) and KCNQ5-associated epileptic syndromes have been suggested (32,33). KCNQ1, KCNQ4, and KCNQ5 have also been identified in various smooth muscle types, where they help control excitability and contractility (34–37). KCNQ1-containing channels are also expressed in various epithelia, where their function is almost certainly in K^+ transport (38).

20.3 BIOPHYSICAL, PHARMACOLOGICAL, AND STRUCTURAL FEATURES

20.3.1 OVERVIEW

As mentioned earlier, native M-current possesses ideal features for control over neuronal excitability with a threshold for activation around −60 mV (in some neurons, as negative as −80 mV (39)), no inactivation and kinetics of activation and deactivation of ~100–200 ms at 0 and −60 mV, respectively (Table 20.1). These characteristics endow M-channels with the ability to shape the somatic response to synaptic inputs and to be critical to spike-frequency adaptation and input/output relations. During bursts of action potentials, gradual M-channel activation increases the threshold for each subsequent action potential during the burst and, thus, underlies accommodation. As detailed in Table 20.1, the various KCNQ channels differ somewhat in their activation ranges, although all are well described by Boltzmann relations typical of voltage-gated channels (38). Structurally, their architecture is conserved with the other Kv families, possessing six transmembrane domains (S1–S6), a voltage sensor within S4, and an extended carboxy-terminus that is the locus of multiple modulatory signals. The single-channel conductance (γ) varies but is generally small (Table 20.1). The single-channel conductance of KCNQ1 homomers is too low to detect electrophysiologically (1–4 pS), but assembly with KCNE1 increases it considerably (40–42). Whereas KCNQ2-5 channels do not appreciably inactivate, KCNQ1 homomers do so modestly. Recovery from inactivation is faster than deactivation, resulting in a characteristic *hook* in the tail currents. Coexpression with KCNE1 removes inactivation (43), and boosts current amplitudes (44). The γ of other KCNQ channels is also generally low, with KCNQ3 having the largest and KCNQ4 and KCNQ5 the smallest (summarized in Table 20.1; (45–48)). At saturating voltages, the open probability (P_o) of KCNQ2-5 channels is strikingly divergent (46,48,49), related to their differential affinity for phosphatidylinositol 4,5-bisphosphate (PIP_2). The open probability of KCNQ3 is the highest, near unity; the maximal P_o of KCNQ2 is ~0.2; KCNQ4 has the lowest (<0.1), the open probability of KCNQ2/3 heteromers is intermediate between KCNQ2 and KCNQ3 (~0.3), and that of KCNQ1-containing channels remains indeterminate. M-channels display complex gating kinetics, including at least two open states and multiple closed states, and an unusually large number of transitions between states (47,48,50–52). Despite intense work, mostly by the Brown lab, their precise kinetic characteristics and number of state transitions, even for KCNQ2/3, remain unclear.

20.3.2 DETERMINANTS OF KCNQ CHANNEL ASSEMBLY

The extended carboxy-terminus of all KCNQ channels are thought to contain four helices, A–D, of which A and B, in the proximal part, are involved in channel regulation, whereas C and D, in the distal part, are involved in subunit-specific assembly and tetramerization (Figure 20.2; reviewed in (53)). As opposed to Kv1-Kv4 channels, in which homomeric vs. heteromeric assembly is controlled by the N-terminal *T1 domain* (54), this feature of KCNQ channels resides within the C and D helices. The Schwake and Lerche labs, using the *Xenopus* oocyte expression system, first defined a *subunit-interaction domain* (*sid*), encompassing the C and D helices (55–57), a finding later confirmed in hippocampal neurons (58). These domains dictate that KCNQ3 will coassemble with KCNQ2 (18, 20), KCNQ4 (59), and KCNQ5 (60), whereas KCNQ1 does not form heteromers with any of the neuronal KCNQ2-5 subunits (55,57,61). KCNQ2 is also discriminate, not forming heteromers with KCNQ4 (28) or KCNQ5 (29). Both C and D helices are thought to be coiled–coiled in nature, making the C-termini of KCNQ channels hard to purify biochemically. Nevertheless, the D helix alone has been purified (the B helix has been recently purified with calmodulin [CaM]; see the following text) and suggested to make critical intersubunit interactions that may control functional assembly (62).

The family of KCNQ channels are notable in their highly divergent current amplitudes when heterologously expressed (for a review, see (53)). Thus, KCNQ2/3 heteromers display >10-fold greater currents than KCNQ2 or KCNQ3 expressed alone, whereas KCNQ1, KCNQ4, and KCNQ5 express large homomeric currents. KCNQ4 homomers express so robustly that they are hard to isolate individually in patches (46,63). The situation for KCNQ3 homomers is especially noteworthy, since one is hard-pressed to record significant KCNQ3 homomers in oocytes or mammalian expression systems and most cell-attached patches are empty. Although there is evidence that the *sid* region plays a role in this (56,62), it is stunning that the replacement of a single alanine residue at the 315 position near the intracellular mouth of the pore (uniquely possessed by KCNQ3) with a hydrophilic serine or threonine results in ~15-fold greater currents, produced by ~25-fold greater expression of functional channels (53,64–66). That result, together with the fact that in expression systems the membrane abundance of KCNQ2/3 heteromers or KCNQ4 homomers is comparable with those of KCNQ3 homomers (65), has led to the hypothesis that most KCNQ3 homomers are *dormant* in the membrane and do not produce currents, with KCNQ2 thus acting as the *master switch* whose expression determines M-current levels in excitable cells (53). In addition, N-terminal domains have also been implicated

Table 20.1 Properties of KCNQ1-5 channels

KCNQ SUBTYPE	$V_{1/2}$ (mV)	τ_{ACT} (0 mV, ms)	τ_{DEACT} (−60 mV, ms)	γ (pS)	DRUG SENSITIVITY	β-SUBUNITS	TISSUE LOCALIZATION
KCNQ1	~15	150–200	200–450	<2	Blocked by lino, XE, chrom; RTG, NEM-insensitive	KCNE1-5	Heart, ear, epithelia, smooth muscle
KCNQ1/ KCNE1	2–8	>500	>200	3–6, varied	Blocked by lino, XE, chrom; RTG, NEM-insensitive	None	Heart, epithelia
KCNQ2	-15 to -30	~200	150–280	~6	Blocked by lino, XE; high TEA sensitivity; opened by RTG, FLU, NH29, ZnPy, NEM, H_2O_2	None	Brain, autonomic and sensory ganglia
KCNQ3	-35 to -50	100–150	100–200	~9	Blocked by lino, XE, TEA insensitive; opened by RTG, FLU, ZnPy, H_2O_2, NEM insensitive	None	Brain, autonomic and sensory ganglia, gut
KCNQ2/3	-20 to -27	130–170	75–100	9.0	Blocked by lino, XE, modest TEA sensitivity; opened by RTG, FLU, NH29, PPOs, NSAIDs, NEM, H_2O_2	None	Brain, autonomic and sensory ganglia
KCNQ4	-20 to -30	125–200	~100	~2	Blocked by lino, XE, slightly TEA sensitive, opened by RTG, FLU, NEM, BMS, ZnPy, NSAIDs, H_2O_2	KCNE1-4?	Cochlea, auditory nuclei, mechano-sensitive neurons, smooth muscle
KCNQ5	-25 to -44	~150	~125	~2	Blocked by lino, XE, slightly TEA sensitive, opened by RTG, FLU, NSAIDs, NEM, BMS, ZnPy, H_2O_2	ND	Brain, autonomic and sensory ganglia, smooth muscle
KCNQ4/5	-38	ND	ND	ND	Blocked by lino, XE, diclofenac, slightly TEA sensitive, opened by RTG, FLU, ZnPy, NSAIDs	ND	Smooth muscle

Note: Table 20.1 Shown are various properties of the KCNQ channels described in the text. For references, see the citations in the text in the corresponding sections.

Abbreviations: Lino, linopirdine; XE, XE991; TEA, tetraethylammonium; RTG, retigabine; FLU, flupirtine; NEM, *N*-ethylmaleimide; ZnPy, zinc pyrithione; PPOs, pyrazolo[1,5-a]pyrimidin-7(4*H*)-ones; NH29, *N*-phenylanthranilic acid-29; NSAIDs, nonsteroidal anti-inflammatory drugs; H_2O_2, hydrogen peroxide, BMS, BMS-204352, ND, not determined.

in these effects (64). Added to the mix is a postulated requirement of CaM for channel expression and functional density. Study continues on this topic.

20.3.3 PHARMACOLOGY

KCNQ channels have a unique, yet subtle, pharmacological profile (see Table 20.1), both presenting opportunities and challenges to study them experimentally, and to generate clinically relevant KCNQ-targeting drugs. With regard to TEA^+, the tyrosine at turret-position 284 in KCNQ2, analogous to position 449 in *Shaker* (67,68), uniquely confers high (<1 mM) sensitivity, whereas KCNQ3-5 and KCNQ1 are largely TEA^+ insensitive, and that of KCNQ2/3, as expected, displays moderate sensitivity (20,69,70). Given the importance of M-current to neuronal firing, a large effort has been made to develop KCNQ channel-specific drugs. Linopirdine, and its potent analogue, XE991, blocks all M-type channels at low μM concentrations (71–73), and XE991 has emerged as the *gold standard* for KCNQ channel identification. However, at the higher concentrations typically used by investigators,

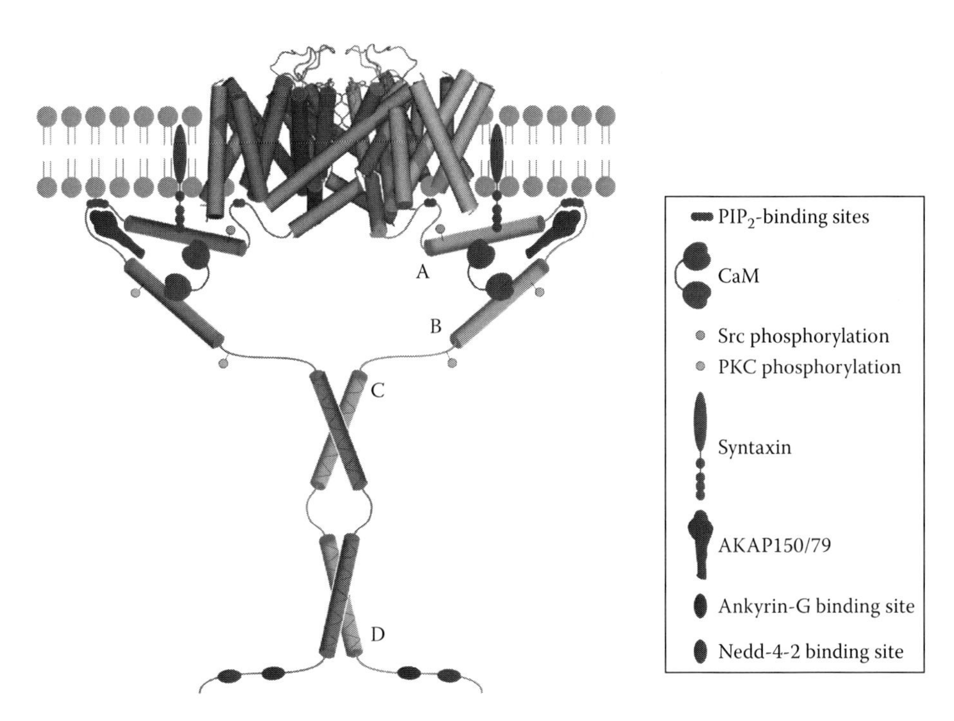

Figure 20.2 Hypothetical topology of the KCNQ channel complex. Two out of four subunits of the tetrameric KCNQ channel are shown. Helical segments A–D of the carboxy-terminus are featured; shown also are the phosphorylation sites and putative binding sites for regulatory and modulatory proteins and lipids (PIP_2, CaM, syntaxin, AKAP150/79, ankyrin-G, Nedd-4-2, as indicated in the graphical key). The coiled-coil nature of helices C and D is indicated with spirals. (The drawing of the KCNQ channel backbone is modified from Haitin, Y. and Attali, B., *J. Physiol.*, 586(7), 1803, 2008.)

both compounds will block many other K^+ currents ((74); unpublished observations by the authors); thus, care must be used in interpreting results from such doses. With the hope of developing novel anticonvulsants, a number of M-channel enhancers (openers) have appeared. The prototype, retigabine (RTG), potently augments opening of KCNQ2-5 and KCNQ2/3 channels (75–78), both by shifting the voltage dependence of activation toward more negative potentials, and by increasing their P_o at saturating voltages (49,79). KCNQ1-containing channels are RTG insensitive, due to its site of action on KCNQ2-5 at a tryptophan residue in S5 that KCNQ1 lacks (61,80). In line with expectations, animal models show RTG to be a potent anticonvulsant (81,82) and analgesic agent (31,83–85). RTG also induces a profound depression of neurotransmitter release (86). Novel lines of investigation have also showed RTG, and its close analogue, flupirtine (FLU), to be neuroprotective in models of stroke (87,88), and to regulate neuronal activity in various regions of the brain involved in the control of blood pressure (89), movement disorders, drug-seeking behavior, and emotional responses (90) (see the following text for further discussion on RTG actions in smooth muscle and on therapeutic potential of KCNQ channel modulators). Besides RTG/FLU, a number of other structurally distinct M-channel openers have recently been identified. These include N-phenylanthranilic acids (91,92), zinc pyrithione (93), and certain pyrazolo[1,5-a]pyrimidin-7(4*H*)-ones (94,95). In addition, hydrogen peroxide (85,87,96), *N*-ethylmaleimide (46,97), and cyclo-oxygenase (COX) inhibitors (92,98) have been shown to potently augment KCNQ channel activity, the latter's effects being wholly unrelated to their actions on COX enzymes (99).

20.3.4 RECEPTORS AND SIGNAL TRANSDUCTION

As alluded to in Section 20.1, M-current was discovered as a K^+ conductance depressed by muscarinic stimulation, prompting a long-lasting and intense search for the signaling mechanism linking mAChRs to M-channels. The mechanism was found to be via M_1 receptors (100–102), $G_{q/11}$ G proteins (103) acting via phospholipase C (PLC) (104,105) and requiring some intermediate second messenger (106). Although IP_3-mediated rises of intracellular Ca^{2+} and production of diacylglycerol (DAG) by activation of protein kinase C (PKC) are canonical second messengers in this pathway, neither was shown to be needed for muscarinic M-current depression (104,107,108) (see the following text for a more subtle role for PKC). The answer to the mystery came with the identification of the KCNQ genes underlying M-current and the discovery that M-channels require PIP_2 to activate ((Figure 20.3 (109–111), reviewed in (112)). This theme is elaborated next.

20.3.5 REGULATION OF KCNQ CHANNELS BY PIP_2

Although the pool of phosphoinositides, such as PIP_2, in mammalian membranes were thought to be nearly limitless, Hilgemann and colleagues were the first to show that significant depletions could occur, which would be sensed by membrane transport proteins that need the presence of PIP_2 to be activated (113,114). The field as a whole has a very robust literature, for which many inclusive reviews are available (115–120). The early work focused on the PIP_2 sensitivity of *inward-rectifier* (Kir) channels, soon followed by *T*ransient *R*eceptor *P*otential M7 (TRPM7) as a channel whose currents are depressed by depletion of membrane PIP_2 subsequent to stimulation of $G_{q/11}$-coupled

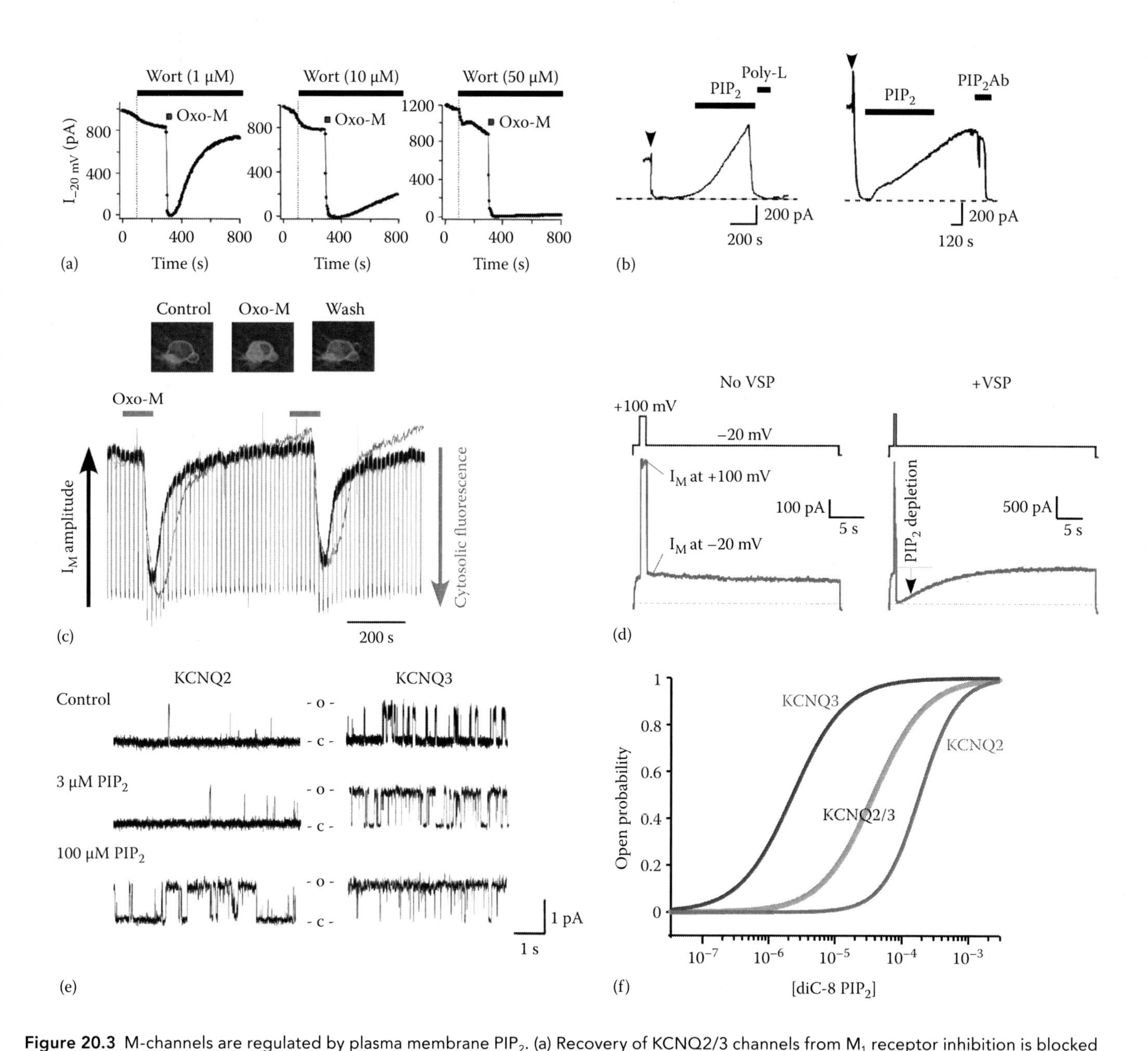

Figure 20.3 M-channels are regulated by plasma membrane PIP_2. (a) Recovery of KCNQ2/3 channels from M_1 receptor inhibition is blocked by the inhibitor of phosphatidylinositol kinases (PI-K), wortmannin (Wort). Plotted are the amplitudes of KCNQ2/3 currents at −20 mV from a tsA-201 cell that was heterologously transfected for KCNQ2 + KCNQ3 subunits, and M_1 receptors and studied under perforated-patch voltage clamp. With the cell exposed to Wort at 1 μM, a concentration sufficient to block PI 3-K, but not PI 4-K, application of the muscarinic agonist oxotremorine methiodide (oxo-M, 10 μM) nearly abolishes the KCNQ2/3 current within 30 s, and recovery is almost complete within ~7 min. However, at concentrations that block PI 4-K, I_M depression by oxo-M is unaltered, but recovery is increasingly abolished due to the inability of the cell to resynthesize PIP_2. (From Suh, B. and Hille, B., *Neuron*, 35, 507, 2002.) (b) *Run-down* of cloned KCNQ2/3 channels is rescued by PIP_2. Plotted is the current at 0 mV in a macropatch on an oocyte heterologously expressing KCNQ2 + KCNQ3 subunits. Upon excision into inside-out mode, the current rapidly runs down, is fully restored by bath application of PIP_2 (10 μM), and then abolished by bath application of a PIP_2 antibody or the PIP_2 scavenger, poly-L-lysine. (From Zhang, H. et al., *Neuron*, 37, 963, 2003.) (c) Translocation of the PIP_2/IP_3 probe, EGFP-PLCδ-PH, correlates with depression of I_M by stimulation of muscarinic receptors. Shown are I_M amplitudes and the cytosolic EGFP fluorescence from an SCG neuron transfected with EGFP-PLCδ-PH, and representative fluorescent micrographs of the neuron in the inset. Upon application of oxo-M, I_M is strongly depressed, and the EGFP-PLCδ-PH probe robustly translocates to the cytoplasm, with very similar time courses. (From Winks, J.S. et al., *J. Neurosci.*, 25, 3400, 2005.) (d) Dephosphorylation of PIP_2 suppresses M-channels. Shown are whole-cell current traces from a tsa-201 cell transfected with KCNQ2 + KCNQ3, alone (left) or together with a voltage-sensitive phosphatase from zebrafish (Dr-VSP) (right). In the latter, after a brief depolarization to 100 mV (voltage at which Dr-VSP is fully activated), the KCNQ2/3 current is strongly depressed, recovering within 60 s, reflecting the rate of recovery of PIP_2 levels by PI(4)P 5-kinase, and indicating that PI(4)P cannot maintain KCNQ2/3-channel activity. (From Falkenburger, B.H. et al., *J. Gen. Physiol.*, 135, 99, 2010.) (e) Single-channel recordings reveal KCNQ2 and KCNQ3 to have highly divergent apparent affinities for PIP_2. CHO cells were transfected with KCNQ2 or KCNQ3, alone and the channels studied at the single-channel level in patches. After excision to the inside-out mode, all the channels ran down, but could be reactivated by bath application of the water-soluble PIP_2 analogue, diC8-PIP_2, at divergent concentrations. *Left*: shown are current traces at 0 mV from an example of such an experiment for KCNQ2 and KCNQ3. (f) shown are pooled dose–response curves of P_o at 0 mV vs. [diC8-PIP_2] from experiments shown in (e), showing the highly divergent apparent affinities of KCNQ2, KCNQ3, and KCNQ2/3 heteromers. (From Li, Y. et al., *J. Neurosci.*, 25, 9825, 2005.)

M_1 mAChRs (121). Using blockade of PIP_2 resynthesis (Figure 20.3a; (109)) and analysis of inside-out *macropatches* (Figure 20.3b; (110)), M-channels were shown to be sensitive to PIP_2 abundance, and the muscarinic depression of I_M in rat sympathetic neuron (which continue to be a prime model neuron to study this topic) to be likely due mostly to PIP_2 depletion by PLC hydrolysis (Figure 20.3c). Similar results were found for KCNQ1-containing channels (122). Further support for this mechanism was found in frog ganglia (111,123), in mammalian cell lines heterologously expressing KCNQ2-4 channels (63), and by using membrane-permeant PIP_2 sequestering peptides (124). Li et al. (63), studying KCNQ channel activity in inside-out patches at the single-channel level, revealed the channels to have sharply divergent apparent affinities for PIP_2, with that of KCNQ3 being ~30 fold higher than for KCNQ2 or KCNQ4 (Figure 20.3e and f). As was predicted, the single-channel P_o of KCNQ isoforms is related to their apparent PIP_2 affinity, with the maximal P_o of KCNQ3 (at saturating voltages) nearing unity at tonic PIP_2 levels, whereas the maximal P_o of KCNQ2 and KCNQ4 is 5–10 times lower. Increasing tonic PIP_2 abundance by overexpression of a PI(4)P 5-kinase dramatically increases the KCNQ2 or KCNQ4 current amplitudes and their maximal P_o (63,125,126).

A very useful tool in the study of channel regulation by lipids has been the development of optical probes, allowing single-cell monitoring of PIP_2 metabolism in real time (127). The trick is to fuse short peptide domains that bind specific phosphoinositides to fluorescent proteins and then to monitor their translocation between plasma membrane and cytoplasm. For the case of PIP_2, the pleckstrin homology (PH) domain of PLCδ is most often used (128–130). At rest, the PLCδ-PH probe is localized to the membrane, bound to PIP_2. Upon stimulation of PLC-linked receptors (110,131), or depletion of PIP_2 by 5-phosphatases (125,132), the probe translocates to the IP_3 in the cytoplasm. Although the 10-fold higher affinity of this probe for IP_3 vs. PIP_2 (133) yields uncertainty in whether depletion of PIP_2 or production of IP_3 is being reported by the probe (134,135), the ability to correlate PIP_2 hydrolysis with KCNQ current regulation has proved invaluable. Indeed, Winks and colleagues exploited these properties of PLCδ-PH binding to strengthen the PIP_2 depletion hypothesis of muscarinic modulation of M-current using kinetic argument (Figure 20.3c) and careful titration of reactants (126). The Hille lab has embarked on a series of papers that dissect out the various steps in this signaling cascade, using fluorescently tagged components as Förster resonance energy transfer (FRET) probes, coupled with cellular modeling (134,136) of enzymes, reactants, and probes involved in the M_1R/$G_{q/11}$/PLC/PIP_2 cascade (137–141). All the modeling in these labs is consistent with M-channel modulation by loss of bound PIP_2, at least in small mammalian cells.

20.3.6 LIPID SPECIFICITY AND LOCATION OF THE PIP_2-BINDING SITE

Many phosphoinositide-sensitive proteins discriminate between poly-phosphorylated phosphoinositides on the basis of their phosphorylation at the 3′ position of the inositol ring (142). The 3′-phosphorylated species, such as $PI(3,4)P_2$ and $PI(3,4,5)P_3$, have very high affinity for Akt, Grb1, and other enzymes involved in the Ras/mitogen-activated protein kinase (MAPK) pathway of cell growth and proliferation. Indeed, the affinity of Akt or Grb1 for $PI(3,4)P_2$ over $PI(4,5)P_2$ (i.e., PIP_2) is several orders of magnitude, as is necessary since the plasma membrane abundance of the latter is nearly 100-fold greater than the former, even after stimulation of PI(3)-kinase. In contrast, the PLC isoforms activated by $G_{q/11}$-coupled receptors specifically hydrolyze PIP_2, which is most relevant for modulation of KCNQ channels. However, inside-out patch recordings have shown most KCNQ channels to be nearly as well activated by $PI(3,4)P_2$ or $PI(3,4,5)P_3$ as by PIP_2 (63,143), similar to certain Kir channels (144). The physiological relevance of these minor phosphoinositides species for KCNQ modulation is likely to be low, given the profound difference in their abundance. The use of invertebrate voltage-sensitive PIP_2 5′-phosphatases (VSPs) has shown the singly phosphorylated PI(4)P to minimally activate KCNQ2/3 channels (Figure 20.3d; (140) but cf. (143)), and the synthesis of PI(4)P by PI 4-kinase to be the rate-limiting step in recovery of M-current after its muscarinic depression. A similar conclusion was reached using the chemically induced dimerization method (132) of rapidly dephosphorylating PIP_2 into PI(4)P (125).

The study of the location of PIP_2 interactions with K_{ir} channels first showed their electrostatic interaction with negatively charged basic residues of channels (114,145). Zhang and colleagues identified a histidine just after S6 of KCNQ2 as being partly responsible (Figure 20.2; (110)), and a cluster of basic residues in the same region was identified for KCNQ1 (122,146). However, exploiting the dramatic difference in apparent affinity for PIP_2 among KCNQ2-4 channels, chimeras, and homology modeling within the proximal half of the C-terminus suggested the linker between the A and B helices of KCNQ2 and KCNQ3 (Figure 20.2) to be the primary interaction domain for PIP_2, and site-directed mutagenesis coupled with single-channel recording localized a cationic cluster of basic residues in this linker domain as being critical (45,147). Further support for this site comes from *PIP strip* immunoblots (143), although a more proximal site of interaction with PIP_2 has recently been suggested by this group (148). The recently solved crystal structure of a Kir channel bound by PIP_2 that cross-links the end of M2 (analogous to S6 in voltage-gated channels) to the beginning of the C-terminus (149) adds impetus to the clarification and further study of this issue.

20.3.7 RECEPTOR SPECIFICITY

Although numerous other $G_{q/11}$-coupled receptors also activate PLC and hydrolyze PIP_2 to produce IP_3 and DAG, and would thus be thought to act similarly toward M-channels, their mechanisms surprisingly diverge. Again, autonomic ganglia have served as best understood model neuron for these studies. Surprisingly, among the four $G_{q/11}$-coupled receptors most intensely studied for their modulation of M-current in sympathetic neurons: muscarinic M_1, bradykinin B_2, angiotensin AT_1, and purinergic P2Y receptors, stimulation of neither the M_1 nor AT_1 receptors provokes IP_3-mediated $[Ca^{2+}]_i$ rises, whereas the B_2 and P2Y receptors do. Nevertheless, stimulation of all four receptors induces robust PIP_2 hydrolysis, as reported by the PLCδ-PH probe (for a review, see (150)). This divergence has been ascribed to microdomains containing B_2 (and presumably P2Y) receptors together with IP_3 receptors (151), such that the IP_3 produced is sufficient to open reluctant IP_3 receptors (152).

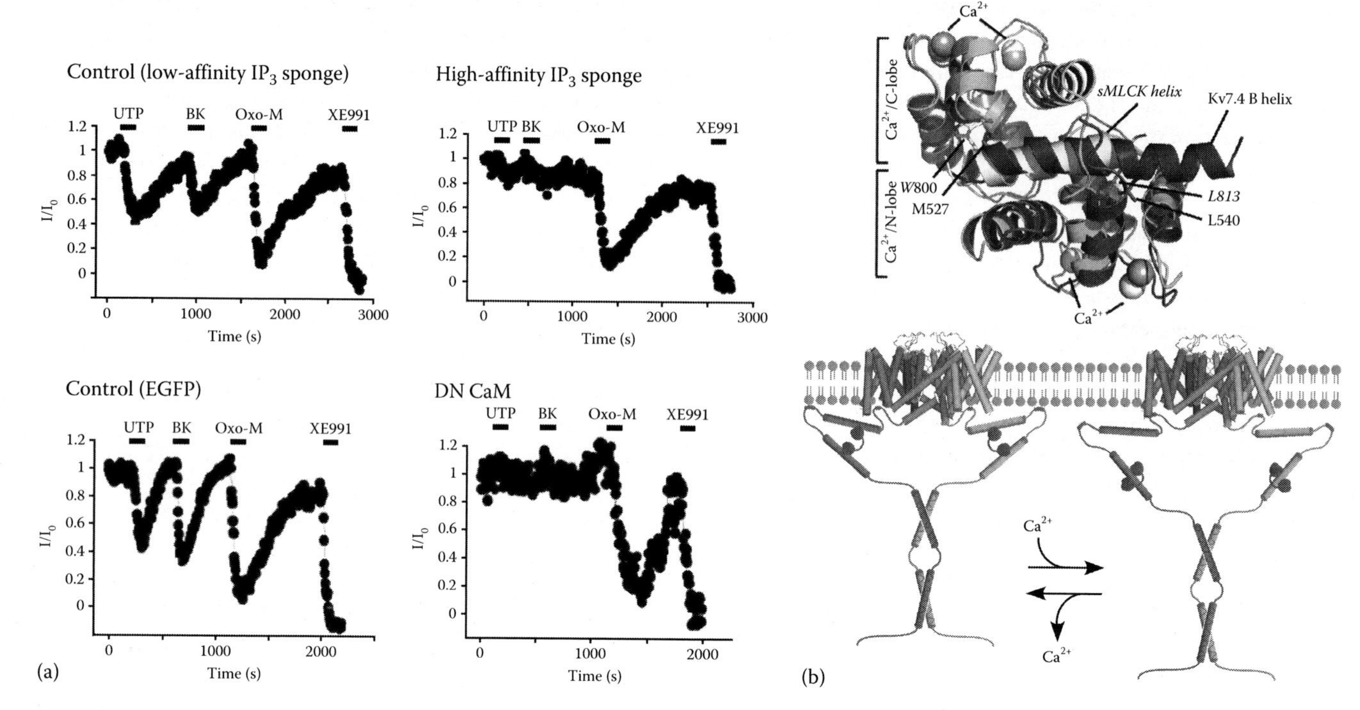

Figure 20.4 Role of CaM in regulation of neuronal KCNQ channels. (a) Receptor-specific involvement of IP_3-mediated Ca_i^{2+} signals and CaM. Plotted are normalized I_M amplitudes from a sympathetic neuron transfected with a low-affinity, or high-affinity, mutant of the IP_3-binding domain of the IP_3 receptor (the latter serving as an IP_3 sponge), only EGFP (control) or dominant-negative (no Ca^{2+}-binding) CaM. Expression of the high-affinity sponge or DN CaM prevents depression of I_M by stimulation of purinergic or bradykinin receptors, but does not perturb the muscarinic action. (From Zaika, O. et al., *J. Neurosci.*, 27, 8914, 2007.) (b) A structure suggests a model. Shown is the solved crystal structure of CaM bound to the KCNQ4 B-helix, and the model of Ca^{2+}-dependent CaM binding to KCNQ channels. In the model, the two lobes of apoCaM bridge the (a and b) helices; upon binding to Ca^{2+}, CaM changes its conformation, favoring both its lobes to bind to the B helix, which, in turn, results in disengagement of helices (a and b). (From Xu, Q. et al., *J. Mol. Biol.*, 425, 378, 2013.)

Furthermore, whereas blockade of IP_3 signals has no effect on muscarinic or angiotensin II depression of I_M, it severely blocks I_M depression by bradykinin or purinergic agonists (Figure 20.4a). The mechanism of the latter was shown to be via $[Ca^{2+}]_i$ signals and CaM (153–156), which binds directly to M-type channels (Figure 20.4b; (59,157–160)). More specific discussion on the possible mechanisms of receptor-specific modulation of KCNQ channels can be found in recent reviews (119,150). Alternative roles for CaM binding and regulation of M-channel function and expression are discussed in the next section.

Although several signaling intermediates of PLC cascade have been shown to contribute to receptor-mediated KCNQ channel inhibition, it is thought that many of these signals converge on PIP_2. Thus, PIP_2 unbinding from the channels can occur via cellular depletion of PIP_2, or by alteration of the affinity of a channel for PIP_2 by other signaling molecules (120). These modalities thus require the PIP_2 affinity of the channels to be only moderate for G protein-coupled receptor (GPCR) signaling to be effective, such that modest changes in PIP_2 abundance or PIP_2 affinity are physiologically sensed by the channels; for KCNQ subtypes with very high PIP_2 affinity (i.e., KCNQ3), their control by these pathways is probably not significant (119). This paradigm is similar to that postulated for control over the diverse family of Kir channels (161). In other peripheral ganglia, and in the CNS, the repertoire of $G_{q/11}$-coupled receptors differs from that of sympathetic ganglia, but the signal transduction mechanisms are very likely conserved. Thus, stimulation of B_2 (85,162), protease–activated receptor 2 (163) and MrgD (164) receptors inhibits M-channels in sensory neurons (*elaborated* in the following). $G_{q/11}$-coupled $5HT_3$ (165,166) and ghrelin (167) receptor action depress I_M in dopaminergic or serotonergic neurons; stimulation of P2Y (168), metabotropic glutamate (168), or muscarinic receptors (169,170) suppresses M-currents in the hippocampus and M_1 receptors do so in cholinergic striatum (171). All of these receptors depress I_M via some combination of Ca_i^{2+} signals, PKC and PIP_2 depletion, but their precise differences in intracellular signaling mechanism have yet to be elucidated.

20.4 AUXILIARY SUBUNITS AND REGULATORY PROTEINS OF KCNQ CHANNELS

20.4.1 CALMODULIN IN KCNQ CHANNEL FUNCTION

There are CaM-binding sites on both the A and B helices in the carboxy-terminus of all KCNQ subtypes (Figure 20.2; [153,157–159,172,173]), which have been suggested to have multiple functions. As introduced earlier, CaM is proposed to serve as the sensor in Ca_i^{2+}-dependent modulation of I_M, without the need of any CaM-binding kinase or phosphatase, similar to the role of CaM in activation of SK-type K^+ channels (174). As for SK channels (175), Ca^{2+} binding to the N-terminal, but not the C-terminal, lobe of CaM is required for the Ca^{2+}-dependent action (154). Given the obligate requirement for PIP_2 binding to M-channels, an obvious mechanism would be a CaM-mediated reduction in PIP_2 affinity (126,150), as has been proposed for TRP channels (176,177) and a variety of other PIP_2- and CaM-binding proteins (178) (but cf. (179)). Such a reduction in PIP_2 affinity could result from the Ca^{2+}/CaM-dependent disengagement of helixes A and B, which were recently suggested to be cross-bridged by CaM at resting conditions (173). We are currently testing this altered-affinity hypothesis. However, a second obligate role of CaM has been proposed in assembly (159,160), or trafficking (180–182) of KCNQ channels to the plasma membrane. Finally, CaM has been suggested to have a modifying role (172) in M-current regulation by syntaxin (183), consistent with M-channel function in control over neurotransmitter release (see the following text). At present, we cannot synchronize these diverse roles postulated for CaM action, although we expect such a global mechanism to emerge. Recently, the crystal structure of CaM bound to the B-helix of KCNQ4 was solved, suggesting that apoCaM binds to, and bridges, the A and B helices, whereas Ca^{2+}/CaM binds to the B-helix (173), perhaps indicating the structural mechanism of Ca^{2+}/CaM-mediated modulation of M-current.

20.4.2 A-KINASE ANCHORING PROTEIN (AKAP) 79/150 AND PKC ACTION

During the search for the *mystery* second messenger linking muscarinic receptors to M-channels, PKC was ruled out early on using pharmacological blockers (107), and indeed unbinding of PIP_2 is central to the muscarinic action. However, the Scott lab demonstrated PKC to phosphorylate KCNQ2, and AKAP79/150 was shown to have an important role in muscarinic, but not bradykinin, suppression of I_M in sympathetic neurons (184–186). As in its coordination of PKA to target proteins (187), AKAP79/150 (human/rodent) recruits PKC to KCNQ subunits, shown for KCNQ2 to phosphorylate S534 and S541 in the rat ortholog (184) (corresponding to S551 and S558 in the canonical human sequence), located in the distal B-helix. AKAP79/150 interacts with rat KCNQ2 at a site within residues 321–499 (184), which spans the A and B helices, including the linker between them. The action is to *sensitize* KCNQ2 channels (as well as endogenous M-channels in sympathetic neurons) to muscarinic suppression, such that the dose–response relationship of muscarinic agonist vs. current inhibition is shifted to lower concentrations (184,185,188). FRET experiments revealed AKAP79 to intimately associate also with KCNQ3-5 (likely within the corresponding domain), but not KCNQ1 (although a different AKAP, yotiao, interacts with KCNQ1, see the following text), subunits, and to sensitize KCNQ3-5 and KCNQ2/3 channels to muscarinic depression as well (188). It is presumed that AKAP79/150-recruited PKC also phosphorylates KCNQ3-5, at their threonine residues corresponding to S558 in KCNQ2 (KCNQ3-5 do not have a corresponding S/T analogous to S551), consistent with the T553A KCNQ4 mutant being insensitive to the presence of AKAP79 (188). In rodent sympathetic neurons, the suppression of I_M by AKAP150 RNAi (185) or expression of a dominant-negative (DN) AKAP79 (189) blunts I_M suppression at half-maximal agonist concentrations, similar to the responses in AKAP150 KO mice (190,191).

As mentioned earlier, the influence of AKAP79/150 is strikingly specific among the four $G_{q/11}$-coupled receptors in sympathetic neurons, sensitizing I_M to depression by M_1 and AT_1 receptors, but not by B_2 or P2Y receptors (Figure 20.5a; (185,189,190). We have suggested this specificity to arise from the involvement of Ca^{2+}/CaM in B_2 and P2Y, but not M_1 or AT_1, receptor modulation of I_M, and the overlapping site of AKAP79/150 interaction with M-channels, and of the PKC phosphorylation site, within the same B-helix that is critical to interactions with CaM (see earlier references). Thus, Ca^{2+}/CaM binding to the channels is suggested to block the interaction with AKAP79/150, preventing PKC B helix phosphorylation that sensitizes the channels to PLC-mediated inhibition. Consistent with that hypothesis, cotransfection of WT, but not DN, CaM, with KCNQ2/3 channels in CHO cells abrogated FRET between AKAP79 and the channels, and in cells coexpressing M_1 receptors, including WT CaM but not DN CaM, wholly blocked the sensitization of the KCNQ2/3 current by muscarinic stimulation (188). Furthermore, the expression of a DN AKAP79, or gene deletion of AKAP150 in rodent sympathetic neurons, reduced the depression of I_M by the stimulation of M_1 or AT_1 receptors but had no effect on I_M depression by B_2 or P2Y receptors (189). Interestingly, the earlier negative conclusions on PKC involvement in muscarinic depression of M-current, which were based on pharmacological tests (20,107), were shown due to the association of AKAP79/150 with PKC at its catalytic core (192), precisely the site where several PKC inhibitors act. Thus, those inhibitors have no effect on muscarinic depression of I_M, whereas those acting at the DAG binding site reduce the modulation at subsaturating doses of agonist (184). This AKAP79/150-dependent pharmacological profile was confirmed using a FRET-based reporter of PKC activity, and molecular modeling of the PKC/AKAP79 complex (179).

The AKAP, yotiao, is required for modulation of KCNQ1/KCNE1 (I_{Ks}) channels by PKA (Figure 20.5c; also see the following text). In addition, all KCNQ channels, except KCNQ1 and KCNQ2 homomers, have also been found to be regulated by Srk tyrosine kinases (Figure 20.5d); (193,194) and epidermal growth factor receptor tyrosine kinase (195). The physiological relevance of this pathway is yet to be determined.

20.4.3 REGULATION OF KCNQ CHANNELS BY KCNE SUBUNITS

One of the most studied examples of KCNQ channel auxiliary proteins is the family of small β subunits, KCNE1-5, also known as minK (KCNE1) and *minK-related peptides* (MiRP1-4) (196). KCNEs are small transmembrane proteins featuring a single transmembrane domain, an extracellular N-terminus, and cytosolic C-terminus (Figure 20.6a). Of the five KCNE isoforms, the most well studied is KCNE1 as, together with KCNQ1, this subunit forms the cardiac I_{Ks} channel. In addition to KCNQ, KCNEs interact with and regulate many other ion channels; however, these interactions are outside the scope of this review (the interested reader is directed to the recent review by Pongs et al. (197)).

20.4.3.1 KCNQ1/KCNE1

KCNE1 is abundantly expressed in tissues also known to express KCNQ1, that is, in the heart and epithelia. Several KCNE

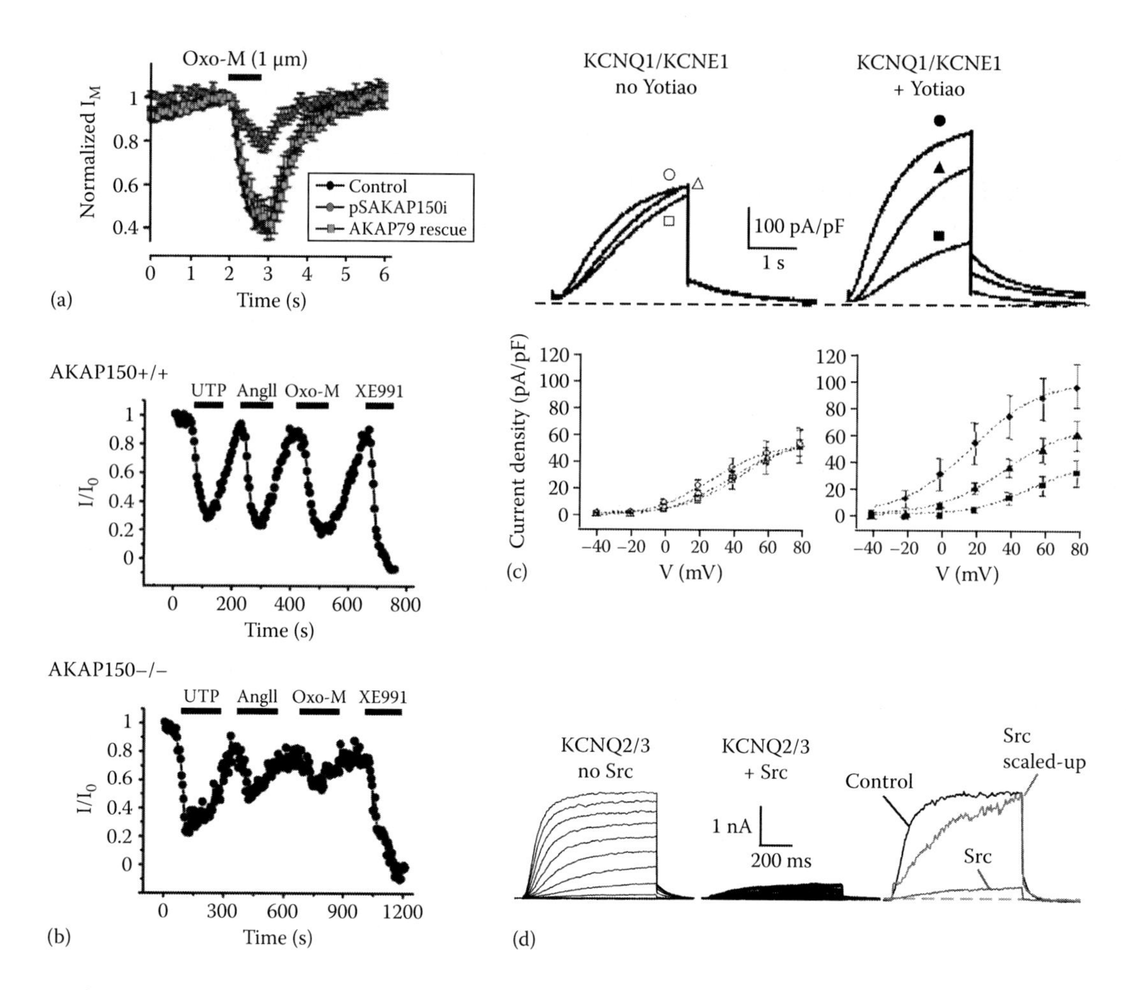

Figure 20.5 KCNQ channels are modulated by kinases, often recruited by AKAPs (a) AKAP79/150 sensitizes I_M to depression by muscarinic stimulation. Plotted are normalized I_M amplitudes under whole-cell voltage clamp, of a control sympathetic neuron, one injected with RNAi against AKAP150, or one injected with RNAi against AKAP150 and also heterologously expressing AKAP79 as *rescue*. The depression of I_M in response to the muscarinic agonist, oxotremorine methiodide (oxo-M), near its EC_{50} value for inhibition is much greater with functional AKAP79/150, than without. (From Hoshi, N. et al., *Nat. Cell. Biol.*, 7(11), 1066, 2005.) (b) AKAP79/150 sensitization of $G_{q/11}$-mediated suppression of I_M is receptor specific. Plotted are I_M amplitudes from sympathetic neurons studied under perforated-patch voltage clamp, from a WT (AKAP150+/+), or AKAP150−/− (KO), mouse. Inhibition of I_M was assayed in response to the P2Y, AT_1, or M_1 receptors agonists, uridine 5′-triphosphate (UTP), angiotensin II (AngII), or oxo-M, at concentrations near their EC_{50} values for I_M suppression. In the WT mouse, all three agonists elicit robust suppressions of I_M, but in the AKAP150 KO, that by AngII or oxo-M was markedly attenuated, whereas that by UTP was unaltered. (From Zaika, O. et al., *J. Neurosci.*, 27, 8914, 2007.) (c) The AKAP, yotiao, orchestrates PKA action on KCNQ1/KCNE1 (I_{Ks}) channels. CHO cells were transfected with KCNQ1 + KCNE1, either alone (*Left*) or together with yotiao (*Right*), and studied under whole-cell voltage clamp. Shown are the current traces and current/voltage relations with a control pipette solution, one with added cAMP (200 μM), or one with added cAMP (200 μM) and the phosphatase inhibitor, okadaic acid (OA). Open symbols, no yotiao; solid symbols, with yotiao. Squares, control; triangles, cAMP; circles, cAMP/OA. The increase in I_{Ks} is nearly absent in the absence of yotiao. (From Marx, S.O. et al., *Science*, 295, 496, 2002.) (d) Inhibition of KCNQ2/3 channels by Src tyrosine kinase. Shown are families of currents from CHO cells transfected with KCNQ2 + KCNQ3, either alone or together with c-Src. Besides suppression of current amplitudes (*Left*), Src slows the kinetics of activation (*Right*). (From Gamper, N. et al., *J. Neurosci.*, 23, 84, 2003.)

mutations are associated with LQTS. When coexpressed with KCNQ1 in expression systems, KCNE1 dramatically slows KCNQ1 activation without a marked effect on deactivation, removes inactivation, shifts the voltage dependence of activation by approximately 20 mV toward more positive potentials, and increases γ by threefold to sevenfold (Figure 20.6c; (10,40,41,43,198,199). In addition, KCNE1 dramatically reduces the sensitivity of KCNQ1 to XE991 (200) and increases its sensitivity to the I_{Ks} blockers, chromanol 293B and azimilide (201).

The stoichiometry of KCNQ1/KCNE1 assembly has been a matter of some debate as both fixed and variable stoichiometry models have been suggested (reviewed in (202)). According to the most common view, two KCNE molecules bind to KCNQ1 tetramers (203–206). Solution NMR, mutagenesis, and various simulations have resulted in several models of KCNQ1-KCNE1 interactions that share the idea of KCNE1 somehow aligning along the I_{Ks} channel pore, possibly by interacting with pore-forming S5-P-S6 residues of KCNQ1 (Figure 20.6b; (205,207,208)). It was also suggested that the single transmembrane domain of the KCNE1 protein is located in a cleft between voltage sensor domain of one KCNQ1 subunit and the pore helix of a neighboring subunit (205).

Various structural and kinetic models have been put forward to explain slowing of KCNQ1 gating by KCNE1 (some of these discussed in the recent review by Wrobel et al. (202)). Thus, it was suggested that KCNE1 might stabilize resting (closed)

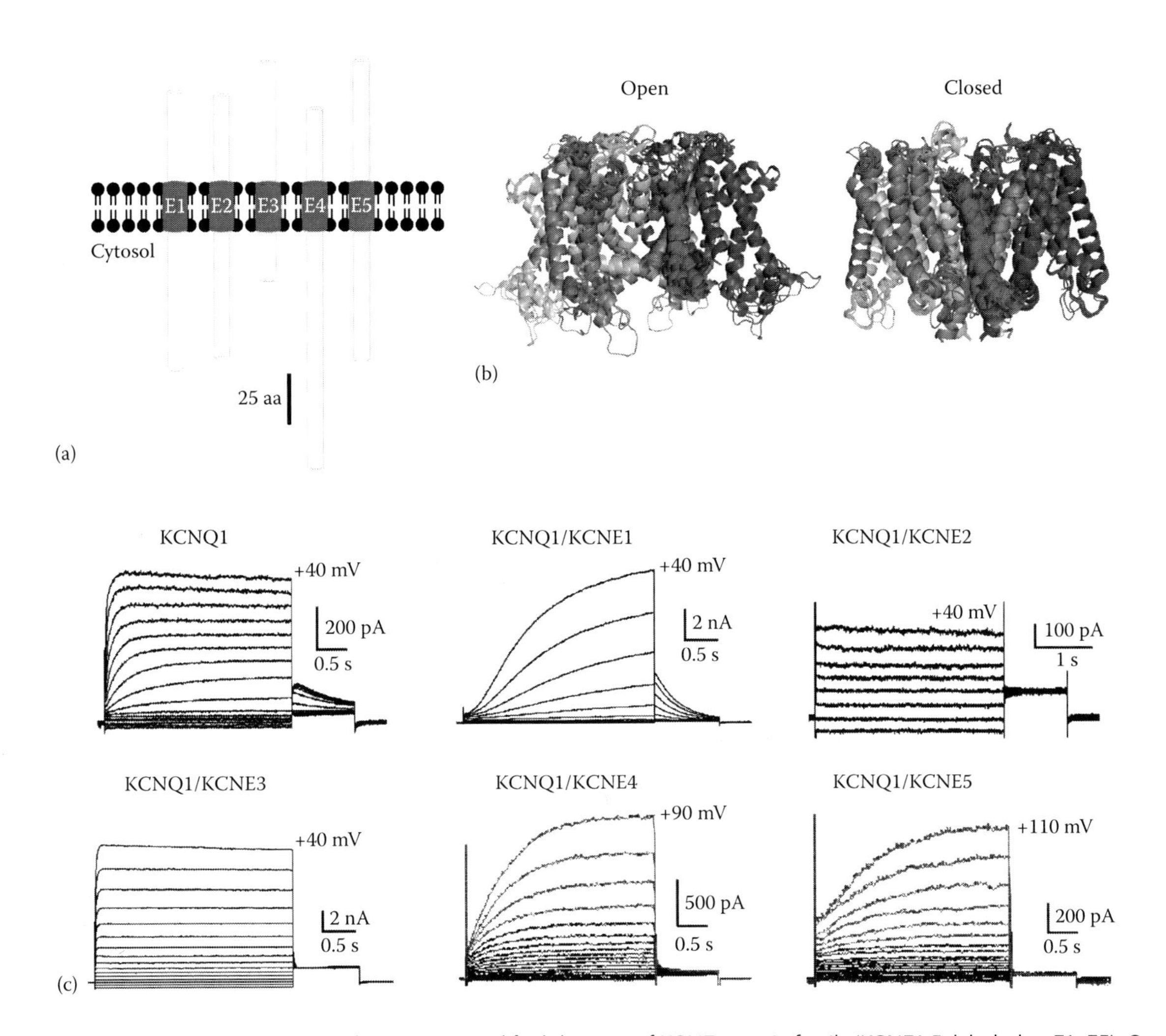

Figure 20.6 Regulation of KCNQ1 by KCNE subunits. (a), simplified diagram of KCNE protein family (KCNE1-5, labeled as E1–E5). Cytosolic domains are shown as gray boxes. (b) Structural model for KCNQ1-KCNE1 interaction (open and closed states, as labeled) obtained by a high-resolution docking simulation. The transmembrane domain of KCNE1 (residues 45–71) occupies a central position in the closed state model on the right and aligns to the left of the pore in the open state model on the left. (From Kang, C. et al., *Biochem.*, 47, 7999, 2008.) (c) Effects of KCNE subunits on the KCNQ1 currents studied by the whole-cell voltage clamp recordings from COS-7 cells transfected with KCNQ1 alone or with KCNQ1 and each of the 5 KCNE subunits (as indicated). Cells were held at −80 mV and currents elicited by train of voltage steps from −100 to + 40 mV, in 10 mV increment, for KCNE1 to KCNE3, and from −100 to + 90 or +110 mV for KCNE4 and KCNE5 respectively. (From Bendahhou, S. et al., *Cardiovasc. Res.*, 67, 529, 2005.)

states of KCNQ1 (209), retard pore opening (210), decelerate the movement of the S4 segment of the voltage-sensing domain of KCNQ1 (209,211), or disturb coupling between the voltage sensor movement and the pore opening (212). A recent study demonstrated that KCNE1 dramatically slows KCNQ1 gating currents (211), suggesting that slowing of S4 movement is indeed at the core of the KCNE1 action. At the single-channel level, the KCNE1 effect is manifested as an introduction of very long latencies to first opening to the first subconductance state and by increased occupancy of partially open subconductance states before channels reach the fully open state (42).

Importantly, although the association of KCNQ1 with KCNE1 reproduces kinetic properties of I_{Ks} fairly well, apparently for a full recapitulation of properties of native I_{Ks} the KCNQ1/KCNE1 complex needs to also include *yotiao* (213–217). Yotiao binds to a coiled-coil domain within the KCNQ1 C-terminus and acts as a scaffolding protein to recruit and position PKA and the phosphatase 1 (PP1) at the cytosolic domains of KCNQ1, thus controlling the phosphorylation/dephosphorylation of KCNQ1 channel at N-terminal Ser-27 (197,213,215,216). This control over KCNQ1 phosphorylation has an important physiological outcome as β-adrenergic stimulation results in PKA-dependent phosphorylation of KCNQ1 within the I_{Ks} complex which, in turn, leads to an increase in I_{Ks} amplitude and acceleration of action potential repolarization (Figure 20.5c; (197)). An LQTS-associated *KCNQ1* mutation (G589D), which disrupts assembly of the KCNQ1/KCNE1/yotiao complex, renders the KCNQ1 channel insensitive to PKA phosphorylation, thus emphasizing the importance of yotiao for normal I_{Ks} function and regulation by β-adrenergic signaling (197,213,215).

20.4.3.2 Other KCNE subunits

The association of KCNQ1 with KCNE2 or KCNE3 results in constitutively open, voltage-independent channels (Figure 20.6c), but with divergent effects on current densities; KCNE2 decreases it by over threefold (218,219), whereas KCNE3 increases it over 10-fold (218,220). The constitutively open gating of KCNQ1/KCNE3 or KCNQ1/KCNE2 channels resembles those of the native K+ currents in colonic crypt cells of small intestine and colon and, accordingly, KCNE2 knockout mice are deficient in gastric acid secretion (221). Mutations within *KCNE2* are also linked to LQTS although it is not clear whether it is due to its interaction with KCNQ1 as this subunit also interacts (at least in the expression systems) with a number of other K+ channels, including HERG (reviewed in (197)). KCNE2 moderately accelerates both activation and inactivation kinetics of KCNQ2 and KCNQ3 without much effect on their steady-state current amplitudes or voltage dependence (222). Interestingly, KCNE3 strongly suppressed KCNQ4 current in *Xenopus* oocytes (223) although the physiological significance of this phenomenon remains to be tested. KCNE4 and KCNE5 both strongly suppress KCNQ1 currents, resulting in little observable current at all if one uses a typical voltage family protocol that stops at +60 mV (Figure 20.6c; (218,224–227)). KCNE5 shifts the voltage dependence of activation of the heteromeric channel toward more positive voltages by about 140 mV as compared to KCNQ1 homomers (227). The physiological importance of interaction of KCNE4 or KCNE5 with KCNQ1 is not known. Other KCNQ channels are not reported to interact with KCNE4 or KCNE5.

20.4.3.3 Effects of KCNEs on KCNQ channel trafficking

The accumulating evidence suggests that coassembly of KCNQ with KCNE subunits affects the trafficking of the resulting heteromeric channels. Thus, it has been suggested that KCNQ1 and KCNE1 coassemble in the ER (228) and when overexpressed alone, most KCNE1 protein is localized to intracellular compartments (229). Moreover, disruption of KCNE1 glycosylation can impair KCNQ1/KCNE1 trafficking and plasma membrane insertion ((230), reviewed in (231)). The LQTS-associated KCNE1 mutation L51H disrupts KCNQ1 trafficking to the plasma membrane (228). On the other hand, KCNE1 has been also implemented in I_{Ks} complex internalization. Thus, Xu et al. (232) found that KCNE1 is responsible for the dynamin-dependent internalization of the I_{Ks} complex, whereas KCNQ1 homomers are not internalized in this way.

20.4.4 OTHER MOLECULAR PARTNERS OF KCNQ CHANNELS

Multiple other proteins have been reported to interact with KCNQ channels. Thus, KCNQ2, through its A helix, directly interacts with syntaxin 1A, which was suggested to target channels to presynaptic sites in neurons (172,183). KCNQ2 and KCNQ3 interact with the adaptor protein ankyrin-G, which targets the channels to the axon initial segment (AIS) and nodes of Ranvier. Both subunits have an ankyrin-G-binding site in the distal C-termini that is similar (but not identical) to that of nodal voltage-gated sodium channels (58,233,234). Degradation of KCNQ channels is controlled by the ubiquitin-protein ligase Nedd4-2 (44,235).

20.5 M-CHANNEL EXPRESSION AND FUNCTION IN MAMMALIAN TISSUES

M-channel subunits are abundantly expressed in several mammalian tissues, particularly (1) CNS and PNS neurons; (2) heart, smooth, and skeletal muscles; and (3) epithelia. In excitable cells, M-channels are mainly involved in control of excitability and synaptic transmission, whereas in epithelia, they recycle K+ ions and provide the driving force for epithelial transport.

20.5.1 FUNCTION OF KCNQ CHANNELS IN THE PERIPHERAL NERVOUS SYSTEM

20.5.1.1 Sympathetic system

M-current has been originally discovered in the peripheral sympathetic neurons and ever since, cultured sympathetic neurons (i.e., SCG) remain one of the most popular native cell preparation for studying M-channel properties and regulation (reviewed in (236,237)). Since much of the work discussed in this chapter is indeed obtained from such neurons, we will not specifically discuss the role of M-channels in these cells. Of note, however, is the fact that sympathetic neurons can be subdivided into *tonic* (firing continuously during depolarization) and *phasic* neurons (firing a single action potential or a short burst) and the relative abundance of M-current largely defines the pattern: tonic neurons display small tonic M-current amplitudes, whereas M-currents in phasic neurons are much larger (238,239). This exemplifies how tonic M-current density can define the firing properties of a neuron.

20.5.1.2 Peripheral somatosensory system

Another peripheral neuron type in which M-current is a major regulator of excitability is the peripheral somatosensory neuron and, in particular, painful stimuli-sensing (nociceptive) neuron. The recent literature suggests M-channels to be a major player in regulation of peripheral nociceptive transmission. Nociceptive and nonnociceptive somatosensory neurons express KCNQ2, KCNQ3, and KCNQ5 in various combinations (89,240–242); in addition, a subset of rapidly adapting mechano-sensitive neurons of both spinal and trigeminal systems abundantly express KCNQ4 (27,243). M-channels are expressed in cell bodies of dorsal root ganglion (DRG) (Figure 20.7a; (31,240,241)) and trigeminal ganglion (TG) (96,244) where they contribute to the slow delayed rectifier K+ current. Functional expression of M-channels is also confirmed in afferent fibers (240–242,245,246), as well as in the nociceptive nerve endings in the skin (Figure 20.7b; (242)). M-channel activity strongly contributes to the control of somatic (Figure 20.7c and d) and axonal (Figure 20.7e) firing in vitro (83,163,247–249). Moreover, growing evidence suggests that M-channels control peripheral fiber excitability in vivo. Thus, hind paw injection of the M-channel blocker XE991 induces moderate pain in rats (162,163), while similar injections of pharmacological M-channel enhancers such as RTG and FLU produce analgesic effect (162,240). Functional M-channels are also found in the central terminals of nociceptive fibers (250).

Among the voltage-gated K+ channels expressed in nociceptors, M-channels arguably have the most negative threshold for activation (negative to −60 mV). Given the resting membrane potential of these neurons in the range of −60 mV (85,96,251), M-current is probably the most abundant voltage-gated K+ current fraction present at rest. Indeed, in small DRG neurons, voltage clamped at −60 mV, there is a small outward current, which can be almost completely blocked by XE991; this current is also strongly augmented by M-channel openers (Figure 20.7c; (85)). Therefore, inhibition or downregulation of M-channel

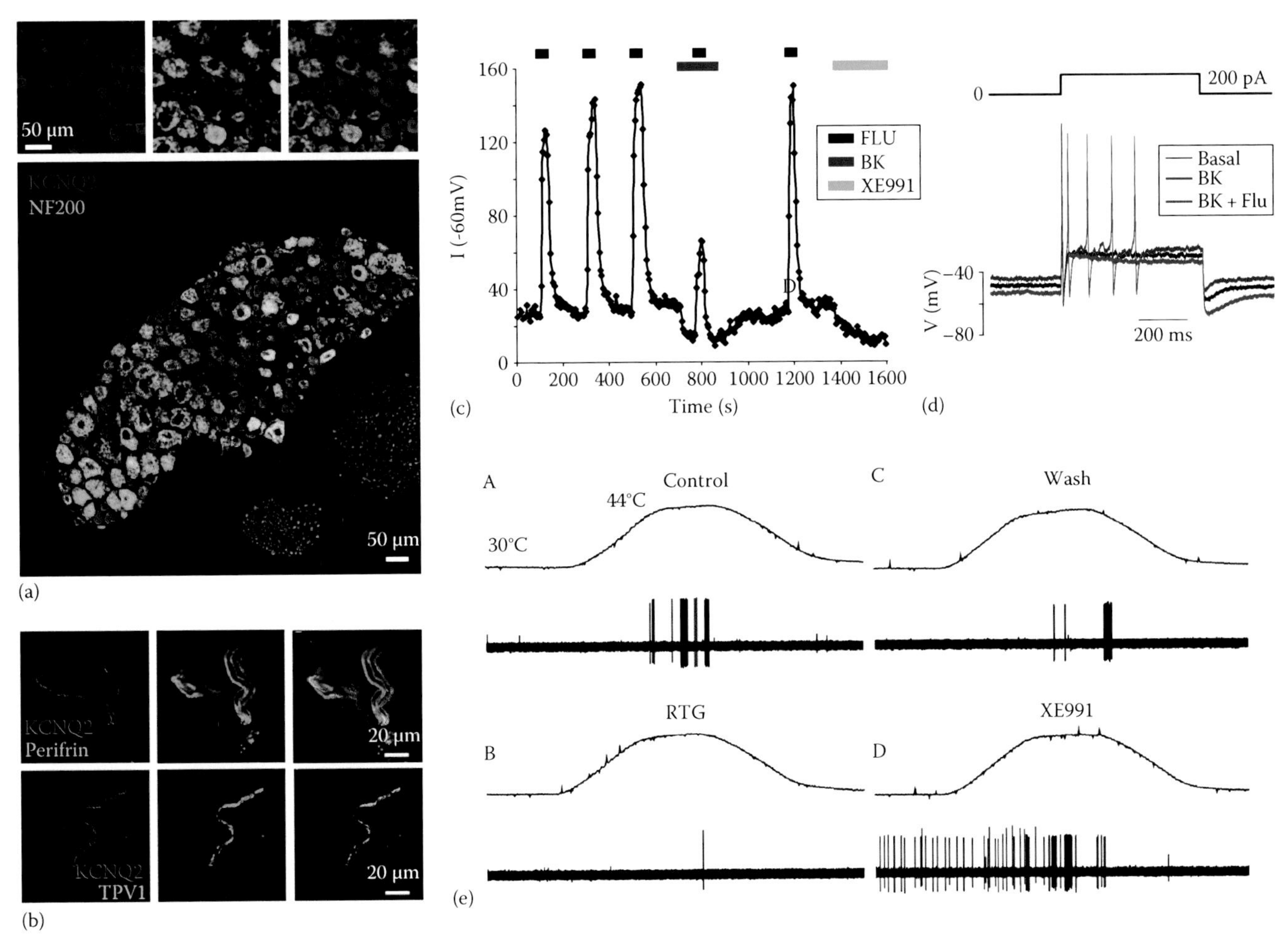

Figure 20.7 Functional expression of M-channels in nociceptive neurons and fibers. (a) Coimmunostaining of rat DRG sections with antibodies against a marker of large-diameter (nonnociceptive) neurons, neurofilament 200 (NF200, green), and against KCNQ2 (red). Most KCNQ2 immunoreactivity is displayed by small-diameter (presumed nociceptive) DRG neurons that express low levels of NF200. (From Rose, K. et al., *Pain*, 152, 742, 2011.) (b) Coimmunostaining of rat hairy skin with antibodies against KCNQ2, an unmyelinated fiber marker, peripherin (top, green), and a nociceptive fiber marker, TRPV1 (bottom, green; From Passmore, G.M. et al., *Front. Mol. Neurosci.*, 5, 63, 2012.) (c) Whole-cell voltage clamp recording from a cultured small-diameter DRG neuron showing the change in the steady-state current at −60 mV (elicited by voltage steps from holding potential of −30 mV) upon addition of FLU (FLU,10 µM), BK (1 µM) and XE991 (3 µM). (d) Current-clamp recording from a cultured DRG neuron showing an increase of action potential firing in response to BK application, which is reversed by the application of FLU. Firing was induced by injection of depolarizing current of 200 pA (c and d are From Linley, J.E. et al., *J. Physiol.*, 590, 793, 2012.) (e) Functional M-channels are present at nociceptive sensory nerve endings in the skin. An in vitro skin-saphenous nerve preparation has been used to examine responses to a noxious hit of an Aδ-fiber and how it is affected by RTG and XE991. Top panels show the heat ramps applied. The firing in response to a noxious heat ramp (A) was abolished by RTG (B), an effect that was reversed by wash (C). XE991 induced spontaneous firing at 30°C and enhanced the response to heat (D). (From Passmore, G.M. et al., *Front. Mol. Neurosci.*, 5, 63, 2012.)

activity/expression in nociceptive fibers is expected to produce hyperexcitability. The role of M-channels in pain will be discussed in more detail in Section 20.6.4.

20.5.2 FUNCTIONS OF KCNQ CHANNELS IN THE CENTRAL NERVOUS SYSTEM

20.5.2.1 KCNQ2/3

Neuronal KCNQ channel subunits (KCNQ2-5) are abundantly expressed throughout the mammalian CNS, with M-current identified in the hippocampus (169) shortly after its description in peripheral ganglia. KCNQ2 and KCNQ3 have very widespread CNS distribution, whereas KCNQ4 and KCNQ5 are also expressed in the CNS but in a more restricted fashion (*see* the following text). KCNQ2 and KCNQ3 subunits have been found in abundance throughout almost all major brain structures with particularly high expression at key sites controlling rhythmic neuronal activity and synchronization (252–255). In particular, Cooper et al. (252), using a immunohistochemical approach, found KCNQ2 to be most abundantly expressed in the following structures of the mouse brain: (1) In the basal ganglia, strong staining was observed in somata of dopaminergic and GABAergic cells of the substantia nigra, cholinergic large aspiny neurons of the striatum, and GABAergic and cholinergic neurons of the globus pallidus. (2) In the septum, somatic KCNQ2 labeling was found in GABAergic, purinergic, and cholinergic neurons that contribute to the septohippocampal and septohabenular pathways. (3) In the thalamus, anti-KCNQ2 antibodies labeled GABAergic nucleus reticularis neurons that regulate thalamocortical oscillations. (4) In the hippocampus, parvalbumin-positive and parvalbumin-negative interneurons were KCNQ2 positive. In another study, it was found that in hippocampus, KCNQ2 and KCNQ3 localize primarily to the somatic layers of dentate gyrus and CA1-3 (253).

In CNS neurons, KCNQ2 and KCNQ3 channels can be found in the both soma and axons (233,234,245,253,256) with almost no expression in dendrites (253,257–259). Growing evidence suggests that KCNQ subunit expression within the soma/axon is not uniform and that functional channels are localized in higher abundance at the sites of action potential generation and propagation: AIS and nodes of Ranvier (see the preceding text) (233,234,245,253,256). Numerous studies, mostly performed in the hippocampus, have investigated the neurophysiological significance of M-channels in the CNS. In general terms, activation of M-channels located presynaptically reduces neurotransmission, whereas the activation of M-channels expressed at postsynaptic sites (i.e., at the AIS) dampens excitability by increasing the threshold for action potential firing, shortening EPSPs, reducing synaptic integration, promoting accommodation within the action potential bursts, etc. (170,257–264).

The postsynaptic function of M-channels is relatively well understood. Thus, perisomatic M-channels have been shown to reduce somatic EPSP summation in CA1 pyramidal cells (Figure 20.8a; (258,259)). At the same time, due to the large distance between the soma and distal dendrites, these perisomatic M-channels have little impact on dendritic EPSP integration. Nevertheless, this perisomatic effect is sufficient for inhibition of spiking in response to dendritic EPSPs (258,259). As suggested by Hu and colleagues, the high density of M-channels at the AIS is probably an effective regulator of action potentials, as the pool of M-channels can effectively attenuate EPSPs before they reach the threshold for spike initiation; therefore, M-channels at the AIS may to some degree electrically isolate the somatodendritic and the axonal compartments of the neuron. M-channels also contribute to the medium component of afterhyperpolarization (mAHP) following an action potential (259,265–267), as well as possibly the slow AHP as well (268,269), phenomena that shunt action potential bursting. In addition, it has been suggested that in hippocampal neurons, M-channels, together with the persistent sodium current and I_h, generate subthreshold resonance in the theta frequency range (5–8 Hz) (259,265) and gamma oscillations (20–80 Hz) underlying synchronized network activity and seizure initiation (270,271). Theta oscillations in the hippocampal network are thought to be essential for hippocampal coding of navigation, learning, and memory (272) and, therefore, M-channel activity is likely to be involved in regulation of these processes. Axonally localized M-channels apparently do not contribute to EPSP integration but define action potential thresholds and resting membrane potentials of the axonal (nodal) membrane (257). In addition to the hippocampus, a functional role of KCNQ/M-channels in controlling neuronal excitability has also been demonstrated electrophysiologically in murine (273,274) and human (275) neocortex, cerebellar Golgi cells (276), striatum (171,277,278), and several other brain regions (e.g., (165,279)).

Growing evidence suggests that M-channels also have a presynaptic function (39,86,91,150,263,274,277). Thus, the Taglialatela group has found that KCNQ2 is localized presynaptically in cortical (274) and striatal (277) neurons; accordingly, M-channel activation with RTG reduced glutamate or dopamine release from cortical or striatal synaptosomes, respectively (274,277). Mechanistically, this effect can be explained by hyperpolarization of presynaptic terminals, which,

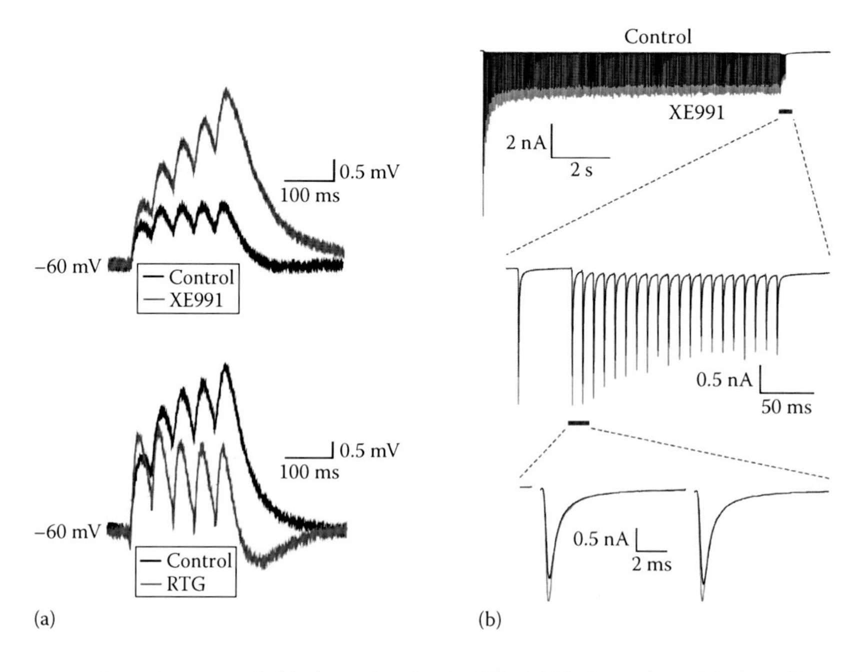

Figure 20.8 Examples of postsynaptic and presynaptic M-channel action in CNS. (a) M-channel modulation affects EPSP integration in CA1 pyramidal cells. At −60 mV, XE991 (3 μM) significantly enhanced and RTG (10 μM) significantly reduced the summation of 20 Hz trains of somatic 5 αEPSPs recorded using the whole-cell current clamp from hippocampal slices. (From Shah, M.M. et al., *J. Physiol.*, 589, 6029, 2011.) (b) Regulation of neurotransmitter release by KCNQ channels in the calyx of Held, a giant glutamatergic terminal in the MNTB. In a slice preparation, the postsynaptic cell was voltage clamped and presynaptic axons were stimulated with an extracellular electrode. The calyx was first preconditioned by 20 Hz stimuli for 10 s to mimic the spontaneous activity observed in the calyx of Held in vivo, followed immediately by a period of 100 Hz stimuli. XE991 (10 μM) increased all excitatory postsynaptic currents (EPSCs). (From Huang, H. and Trussell, L.O., *Nat. Neurosci.*, 14, 840, 2011.)

in turn, reduces Ca^{2+} influx into the terminals via voltage-gated Ca^{2+} channels and, thus, reduces the amount of neurotransmitter-containing vesicles released per arriving action potential. It should be noted that presynaptic localization of M-channels is somewhat inconsistent with the suggested AIS/nodal targeting of KCNQ2 and KCNQ3 and the lack of M-currents reported in dendrites (257–259). This matter will require further investigation; perhaps, M-channels in presynaptic terminals of some neurons are composed of different KCNQ subunit(s) (e.g., KCNQ5; see the following text).

20.5.2.2 KCNQ4

In contrast to KCNQ2 and KCNQ3, KCNQ4 has a much more restricted expression profile throughout the mammalian nervous system as it is specifically expressed in the auditory and vestibular pathways and structures (27,280,281). Thus, KCNQ4 is expressed in the cochlea where it is localized at the basal pole of outer hair cells (OHCs) (280). It is also found in postsynaptic calyx terminals innervating vestibular type I hair cells (282,283) and in auditory brainstem nuclei and tracts (27). KCNQ4 is important for hearing and vestibular function, and multiple mutations within KCNQ4 result in a dominant form of deafness, *DFNA2* (see the following text) (26,27,280). In some individuals, this disorder is also accompanied with vestibular abnormalities (284,285). In addition to auditory structures, KCNQ4 is also expressed in the peripheral nerve endings of cutaneous rapidly adapting hair follicle and Meissner corpuscle mechanoreceptors (as well as in the cell bodies of these neurons in DRG and TG). In these rapidly adapting mechanoreceptors, KCNQ4 shortens the duration of stimulus-induced bursts (243). Accordingly, human DFNA2 carriers are more sensitive to low-frequency tactile vibrations as compared to age-matched control subjects (243).

20.5.2.3 KCNQ5

KCNQ5 subunits are expressed throughout the CNS and PNS alongside KCNQ2 and KCNQ3, although it is believed that its contribution to perisomatic M-currents is less prominent. However, KCNQ5 has been found to express abundantly in synaptic terminals within the calyx of Held, a giant glutamatergic terminal in the medial nucleus of the trapezoid body (MNTB), where this subunit is likely to be the main constituent of M-channels (39). It has been demonstrated that at calyx terminals, KCNQ5 determines the presynaptic resting potential and controls synaptic strength (Figure 20.8b; (39)). Along with KCNQ4, KCNQ5 is also abundantly expressed throughout the auditory system where it localizes predominantly to synaptic endings (286), a finding that is compatible with the idea that KCNQ5 may represent a presynaptic KCNQ subunit. KCNQ5 has also been identified as underlying at least part of the slow AHP in the hippocampus (266).

20.5.2.4 KCNQ1

Although KCNQ1 is a cardiac KCNQ subunit, which is not considered to be widely expressed in the nervous system, one study reported abundant KCNQ1 expression in many brain regions, including forebrain and brainstem (287). Moreover, it was also shown that mice carrying LQT-associated mutations within KCNQ1 develop not only LQT but also the epilepsy. The coinheritance of LQT and epilepsy (in a form of KCNQ1 mutations) may explain sudden unexplained death in epilepsy (SUDEP) that can follow a seizure. This hypothesis is still under investigation.

20.5.3 HEART

Heteromeric association of KCNQ1 and KCNE1 results in a K^+ current with very slow kinetics that is abundantly expressed in the heart and underlies the slow component of the cardiac delayed rectifier current, I_{Ks}. The delayed rectifier K^+ current in the heart (I_K) was first described by Noble and Tsien (288,289), who also showed that the I_K is composed of two kinetically distinct components that were later identified as KCNQ1/KCNE1-mediated I_{Ks} (10,198) and HERG-mediated I_{Kr} (290–292). Cardiac I_K is responsible for the late repolarization phase of the cardiac action potential and regulates action potential duration in many species (see (293) for review). Particularly, the contribution of I_{Ks} is enhanced at high stimulation rates as its slow kinetics results in the incomplete deactivation at high pacing rates (294). In human heart, I_{Ks} probably plays little role in normal ventricular muscle action potential repolarization, but when the action potential duration is abnormally long, I_{Ks} is likely to provide an important safety mechanism that, when removed, increases arrhythmic risk (295). Accordingly, mutations within both KCNQ1 and HERG are associated with cardiac arrhythmias, particularly with a form of inherited LQTS, a cardiac disorder characterized by prolonged ventricular repolarization, resulting in episodic ventricular tachyarrhythmias that can lead to ventricular fibrillation and sudden death (296). Although the I_{Ks} has been known for decades, the elucidation of its molecular correlate required considerable study. I_{Ks} was originally attributed to the subunit that we now know as KCNE1 (see the preceding text) (297–299) as this subunit, when expressed in *Xenopus* oocytes, gives rise to a current with similar characteristics to mammalian I_{Ks}. This subunit was originally named *minK* (for *minimal K⁺ channel* (299)) although it was soon realized that such a small protein with only one transmembrane domain was unlikely to be able to assemble into functional voltage-gated K^+ channels on its own. Indeed, it was later discovered that minK, now called KCNE1, is indeed an auxiliary β-subunit of the KCNQ1 channel (10,198). *Xenopus* oocytes express an endogenous KCNQ1 orthologue (XK_vLQT1), which, when coassembled with exogenous KCNE1, recapitulates an I_{Ks}-like current (10). The biophysics of I_{Ks} is discussed in Section 20.4.3.1 and the I_{Ks} pathologies will be considered in Section 20.6.

20.5.4 OTHER TISSUES

Beside their roles in neurons and cardiac myocytes, KCNQ channels are expressed in epithelia, smooth and skeletal muscle, and also in some specialized cell types.

20.5.4.1 Epithelia

KCNQ1, in combination with KCNE1, KCNE2, or KCNE3, is expressed in the basolateral membranes of many epithelia including kidney, stomach, colon, and small intestine (220,300–304). Epithelial cells are polarized, and their apical (or luminal) and basolateral membranes serve distinct functions. One of the basic functions of the epithelia is transepithelial transport. The major driving force for this transport is the electrochemical gradient of Na^+ and K^+ across the epithelial membranes. KCNQ1 channels (together with other epithelial K^+ channels) are essential for both maintaining a negative membrane potential, providing a driving force for transepithelial transport, and for K^+ absorption or secretion. Interestingly, in many epithelia, voltage-gated KCNQ1 channels are converted into voltage-independent, constitutively open K^+ channels by coassembly with KCNE2 (219) or KCNE3 (220) (see the preceding text). The functional role of KCNQ1 in epithelia has been recently reviewed in references (44,305,306), so here we mention only some examples. Thus, basolateral KCNQ1-KCNE3 channels may be required for providing a driving force for cAMP-stimulated Cl^- secretion in the colon (220) and airway (307), whereas KCNQ1-KCNE2 channels mediating apical K^+ recycling in gastric parietal cells are necessary for gastric acid (H^+) secretion (308,309). KCNQ1/KCNE1 channels are expressed in the stria vascularis of the inner ear where they conduct a potassium current into the scala media to generate the K^+-rich endolymph (44,310).

20.5.4.2 Smooth and skeletal muscles

Smooth muscle is nonstriated muscle that mediates involuntary contraction and support organ dimensions. These muscles are found within the walls of blood vessels, as well as in urinary bladder, uterus, and reproductive, gastrointestinal, and respiratory tracts, and in some other more specified structures. In diverse smooth muscles, contractions are generated either mostly by the influx of Ca^{2+} through voltage-gated Ca^{2+} channels (bladder, gut), by a mixture of Ca^{2+} influx and Ca^{2+} release from intracellular stores (vascular, uterus), or mostly by release from intracellular stores (airway) (see (311,312) for reviews). Interestingly, M-type channels are localized to all of these smooth muscle types. The expression of KCNQ1, KCNQ4, and KCNQ5 has been found in several types of blood vessels, including portal vein, thoracic aorta, carotid artery, cerebral basilar artery, and femoral artery of mice (34,313–316) as well as in rat aorta and mesenteric artery (35,36). Although multiple KCNQ subunits were detected in these studies, the emerging pattern suggests that KCNQ4 homomers or KCNQ4/5 heteromers are probably the main KCNQ channels in smooth muscle. Functional expression of KCNQ4 in human arteries has also been demonstrated (317). A number of studies have described KCNQ expression in nonvascular smooth muscle. Thus, Ohya et al. (318) found KCNQ1, KCNQ3, KCNE1, and KCNE2 transcripts in rat gastric antral smooth muscle. Transcripts for KCNQ4, KCNQ5, and KCNE4, as well as functional M-like currents, were detected in murine gastrointestinal tract (319). Expression of KCNQ1, KCNQ5, and KCNE4 has been reported in murine myometrial smooth muscle throughout the estrus cycle (320). Recently, M-like currents have been reported in airway smooth muscle, composed of KCNQ1, KCNQ4, and KCNQ5 subunits (37), which has been suggested to play a role, in concert with BK channels, in control of resting potentials (321). It has been suggested that M-channel openers may represent novel therapeutic strategies for disorders of all of these systems (322). In addition to smooth muscle, expression of all five KCNQ subunits has been detected in skeletal muscle (29,30,323,324) where M-channels are suggested important for skeletal muscle proliferation, differentiation, and survival (323,324) as well as in their contractility (325). This latter topic is underexplored.

The functional significance of KCNQ channels in smooth muscle, as in the case of neurons, follows directly from their biophysical properties. Since KCNQ channels are open at rest, their activity contributes to setting resting membrane potentials. M-channel inhibition in muscle can result in depolarization and reduced repolarization following contraction-related depolarizations, which, in turn, can increase the influx of Ca^{2+} through voltage-gated Ca^{2+} channels in response to depolarizations (this simplified mechanism is more relevant to those smooth muscle types in which contractions are generated mostly by Ca^{2+} influx through voltage-gated Ca^{2+} channels). Therefore, in general terms, inhibition of myocyte M-channels should result in increased contractility, whereas M-channel activation (or pharmacological enhancement) should result in reduction of contractile responses. This general principle has been verified experimentally (36,313,315,325,326) (for a more in-depth discussion of these mechanisms, see (327)).

20.5.4.3 Other tissues/structures

Functional M-current is present in the neuroendocrine cells such as adrenal chromaffin cells (328) or prolactin-releasing cells (*lactotrophs*) of the pituitary gland (329); in these cells, M-current is believed to control neuroendocrine cell excitability and hormone release (i.e., the release of adrenaline and noradrenaline from chromaffin cells in response to histamine, which depresses I_M via $G_{q/11}$-coupled H_1 receptors). M-like currents were also recorded from rod photoreceptors (330) and retinal pigment epithelial cells (331). Recently, functional expression of KCNQ2 has been found in keratinocytes (332).

20.6 M-CHANNELOPATHIES AND THERAPEUTIC POTENTIAL OF KCNQ CHANNELS

As discussed earlier, genes encoding KCNQ subunits were discovered through the analysis of human diseases and, indeed, the KCNQ channel family is unique among other ion channel families in the variety of severe disorders that are associated with genetic mutations within KCNQ genes. In this section, we will consider channelopathies that are linked to KCNQ channels.

20.6.1 ARRHYTHMIAS

Mutations within genes encoding both subunits of I_{Ks}, KCNQ1 and KCNE1, underlie several arrhythmic heart disorders, including LQTS and short QT (SQT) syndrome, familial atrial fibrillation, and others (see the following text). The up-to-date database of LQTS-associated mutations (including over 200 LQTS1-associated mutations within KCNQ1 alone) can be found at http://www.fsm.it/cardmoc/.

20.6.1.1 Long QT syndrome

LQTS is an arrhythmic heart disorder characterized by the lengthening of the QT interval of the electrocardiogram, either due to the enhancement of inward depolarizing currents or the reduction in outward repolarizing currents (333,334). Most LQTS patients are asymptomatic until a triggering event (i.e., heavy exercise, stress, etc.) which initiates ventricular tachyarrhythmia (i.e., torsades de pointes) and may result in sudden cardiac death (306,333,334). Loss-of-function mutations within KCNQ1 often result in LQTS form 1 (LQTS1), with two major forms identified: (1) the autosomal dominant Romano–Ward syndrome (RW) and (2) the autosomal recessive Jervell and Lange-Nielsen syndrome (JLN). The latter is associated with bilateral deafness (335). A number of JLN KCNQ1 mutations have been identified ((336–343), reviewed in (293,344), many of them resulting in C-terminal *A-domain* truncations or missense sequences and, as discussed earlier, bring about defects in KCNQ1 subunit tetramerization. An additional reason for I_{Ks} loss of function due to JLN mutations could be in defective trafficking due to impaired interaction with yotiao (215,345). Generally, JLN mutations do not feature strong dominant-negative effects on wild-type KCNQ1 and, thus, heterozygous mutant gene carriers retain some functional I_{Ks} (293). In contrast, KCNQ1 mutations associated with RW often affect channel gating and RW KCNQ1 mutants confer strong dominant-negative effects when assembled with WT KCNQ1 (293,346). A number of RW mutations affecting KCNQ1 trafficking have also been identified (347–349). Mutations in KCNE1 are associated with another form of LQTS, LQT5 (350–352).

20.6.1.2 Short QT syndrome

Several gain-of-function KCNQ1 mutations are associated with a different form of arrhythmia that is characterized by a shortened QT interval, due to accelerated action potential repolarization (353–355). For example, an SQT-associated V307L KCNQ1 mutation was shown to shift the KCNQ1 voltage dependence toward more negative potentials and to accelerate activation kinetics, thus increasing I_{Ks} and shortening the QT interval (353). SQT is also associated with atrial fibrillation and sudden death (333).

20.6.2 DEAFNESS

Mutations within KCNQ4 are often associated with autosomal dominant type 2 deafness (DFNA2), a progressive form of sensory neuronal hearing loss, which starts with a loss of perception of the high sound frequencies but later in life develops into the progressive loss of hearing in the middle and low frequencies (26). KCNQ4 is expressed at high levels in the basolateral membrane of cochlear OHCs where it provides a major contribution to the K^+ conductance defined as $G_{K,n}$, (356,357). $G_{K,n}$ activates negative to −70 mV and functions to hold the OHC membrane potential near the equilibrium potential for K^+, thus maintaining a driving force for K^+ entry through the apical transducer channels (356). A similar K^+ conductance has also been found in the inner hair cells (306). KCNQ4 expression in the auditory hair cells has been verified by immunohistochemistry and RT-PCR (358), and accordingly, $G_{K,n}$ in cochlear hair cells is sensitive to linopirdine and XE991 (359–361). Genetic deletion of KCNQ4 in a mice model results in progressive hearing loss, which is paralleled by the loss of cochlear OHCs (280); therefore, it was hypothesized that loss of KCNQ4 currents may lead to chronic K^+ overload in these cells, causing the degeneration responsible for progressive hearing loss (280,357). There are however some noticeable biophysical and pharmacological differences between native K^+ currents in OHCs and heterologously expressed KCNQ4 (reviewed in (306,357)), which may result from heteromeric assembly of KCNQ4 with some other subunits (i.e., KCNE4 (223)) in cochlea. Nevertheless, the link between KCNQ4 and DFNA2 is firmly established by genetic analysis. A number of DFNA2-associated missense mutations and deletions of KCNQ4 have been identified (25,26,362–370). Five point mutations in the pore region (L274H, W276S, L281S, G285C, and G296S) as well as the C-terminal mutant G321S result in the endoplasmic reticulum retention of the mutant channels (371). Other mutations result in the pore collapse, defective assembly, and other loss-of-function defects (reviewed in (357)).

20.6.3 EPILEPSY/SEIZURES

As mentioned in Section 20.1, *KCNQ2* and *KCNQ3* were identified through positional cloning of chromosomal loci associated with an autosomal dominant form of epilepsy, benign familial neonatal seizures (BFNSs) (15–17). In this syndrome, seizures (generalized or focal tonic-clonic seizures) occur in otherwise healthy infants starting around 3 days of age. The seizures usually disappear spontaneously within the first 6 months of life, although 10–15% of infants with BFNC develop other forms of seizures later in life (372).

Sixty to seventy percent of families with BFNC carry mutations within *KCNQ2* or *KCNQ3;* NCBI's GeneReveiw ((372) http://www.ncbi.nlm.nih.gov/books/NBK32534/) lists over 70 *KCNQ2* mutations and 8 *KCNQ3* mutations leading to BFNC; thus, *KCNQ2* mutations are nearly 10 times more frequent than these in *KCNQ3*. All BFNC-associated *KCNQ3* mutations identified thus far are missense mutations; mutations within *KCNQ2* are more diverse and include various truncations, deletions, frame-shift, and missense mutations (306,372). Heterologous expression of these mutants in expression systems revealed a number of loss-of-function mechanisms: defective trafficking (373), CaM or other accessory protein binding (374,375), and gating abnormalities (376–378). Interestingly, it has been noted that heterologous expression of some BFNC-associated *KCNQ2* and *KCNQ3* mutants resulted in only rather mild (20–30%) reduction in I_M amplitude (379,380). This fact is usually interpreted as demonstrating the importance of M-current for control of neuronal excitability so that even a modest reduction in amplitude is sufficient to cause seizures (381). Another explanation has been suggested recently, in which some BFNC mutations may impair trafficking of KCNQ2/3 channels to axons (58,382) (an effect thats may not manifest itself in the large reduction of current amplitude when tested in the nonneuronal expression system); this would result in decreased axonal abundance of M-channels and, thus, abnormal excitability. The reasons why BFNC seizures disappear with age are not entirely clear, although it was suggested that it may have to do with the maturation of GABAergic system (382). Immature CNS neurons have high intracellular Cl^- concentration, which renders GABAergic transmission excitatory in early life, before becoming inhibitory (383). It was also reported that KCNQ channel expression in neonatal brain is low but increases with age (384). Thus, the developmental absence of one major inhibitory mechanism (GABA inhibition) in combination with low expression and genetic impairment in another (M-current) may result in the epileptic phenotype. With age, the GABAergic system engages fully and M-channel density in neurons increases, which may explain the disappearance of seizures.

20.6.4 PAIN

Pathological pain is a hyperexcitability disorder and, as such, has been referred to as a channelopathy (385). M-channels play a major role in controlling nociceptive neuron excitability (31,85,96,162,163,242). Accordingly, growing evidence suggests that inhibition or downregulation of M-channel activity/abundance in nociceptors may result in pain sensations. Thus, peripheral injection of M-channel blockers induces moderate pain (162,163) and hyperalgesia (242). Similarly, painful are the injections of inflammatory mediator bradykinin (96,162). Bradykinin inhibits M-current in a PLC- and Ca^{2+}-dependent manner (similarly to its action in sympathetic neurons described in detail earlier) and produces excitability of cultured nociceptive neurons, an action which is mimicked by XE991 and antagonized by FLU (85,162). Therefore, M-channel inhibition is likely to contribute significantly to the bradykinin-induced pain (85,162,386,387). Similarly, M-channels in small DRG neurons are inhibited by other PLC-coupled receptors such as PAR-2 (163) and MrgD (164). A recent study demonstrated that M-channels in trigeminal nociceptors can also be inhibited by nitric oxide (NO) via S-nitrosylation at the redox-sensitive module within the cytosolic loop between S2 and S3 (244). NO-mediated M-channel inhibition correlated with increased excitability and CGRP release in nociceptors and was suggested to contribute to excitatory effects of NO in trigeminal disorders such as headache and migraine (244).

In addition to inflammatory pain resulting from the acute inhibition of M-channels via GPCR cascades by the inflammatory mediators, longer-lasting downregulation of M channel expression may contribute to chronic pain. Thus, Rose et al. (240) reported strong downregulation of KCNQ2 expression in rat DRG following neuropathic injury. Similarly, a significant loss of KCNQ5 immunoreactivity was reported in DRG of rats after sciatic nerve transection (241). In accord with previous findings suggesting that augmentation of M-current in sensory fibers is anti-excitatory (83,248,249,388), thermal hyperalgesia produced by neuropathic injury in rats was significantly reduced by injection of FLU directly into the site of nerve injury, an effect completely prevented by XE991 (240). Similarly, a strong downregulation of both KCNQ2 and KCNQ3 in DRG neurons has been reported following the development of bone cancer pain in rats (389). This effect was accompanied by a marked decrease in I_M amplitude in nociceptive neurons. Application of RTG inhibited the bone cancer–induced hyperexcitability of DRG neurons and alleviated mechanical allodynia and thermal hyperalgesia in rats with bone cancer. These studies suggest that despite the decreased abundance of M-channels in nociceptive neurons affected by a chronic pain condition, the remaining M-channels can still be effectively targeted by the pharmacological enhancers in order to reduce nociceptor excitability (see the following text).

20.6.5 M-CHANNELS AS DRUG TARGETS

KCNQ channels are widely highlighted as prospective drug targets for the treatment of diverse hyperexcitability disorders. M-channel openers have been proposed to be used for the treatment of epilepsy, various types of pain, attention-deficit/hyperactivity disorder, stroke, hearing loss, bipolar disease, schizophrenia, mania, addiction to psychostimulants, and other disease states (390–392). Recently, RTG (Ezogabine™) was approved by the U.S. FDA for the adjunctive treatment of partial-onset seizures in adults (393). Since identification of RTG as an M-channel enhancer (76,79), hundreds if not thousands (e.g., screening by (394) alone reported over 600 unique M-channel enhancers) of new M-channel activators have been identified or synthesized in a number of large screens conducted by industrial and academic labs (87,93,394–400). A review of some of these efforts can be found in (390).

Although RTG has been introduced to the market only recently, its very close chemical analogue, FLU (Katadolon™, Awegal™, etc.), has been used as a nonopioid analgesic since the 1980s in Europe (although it was never certified in the United States) (401,402). Although FLU is not as selective as some of the latest-generation M-channel activators, it is believed that the analgesic efficacy of FLU arises mainly from its M-channel enhancer activity (403). FLU effectively reduces postoperative pain, chronic musculoskeletal pain, migraine, and neuralgia (401,403). As mentioned before, some nonsteroidal anti-inflammatory drugs such as diclofenac (92) and celecoxib (404,405) possess strong M-channel opener activity, which can be responsible for at least some of their analgesic efficacy.

Despite the facts that (1) the chemical structures of FLU and RTG are very similar and (2) RTG consistently reduced pain in a variety of animal models (83,406), a phase IIa proof-of-concept clinical trial of RTG for the treatment of postherpetic neuralgia pain was inconclusive (407). It is likely that this failure reflects the broad expression profile of M-channels within and outside of the nervous system (see the preceding text), and, therefore, a high risk for off-target effects of systemically applied broad-spectrum M-channel modulators. It is therefore hypothesized that modulators that cross the blood–brain barrier poorly and are more selective for KCNQ2 and KCNQ3 over KCNQ1 and KCNQ4 (which are highly expressed in the heart, epithelia and smooth muscle) should have a better safety profile while retaining analgesic activity (390).

20.7 REGULATION OF KCNQ GENE EXPRESSION

20.7.1 REGULATION BY REST

Until recently, little has been known about the transcriptional regulation of *KCNQ* genes. However, as much as the acute regulation of M-channel activity is important for regulation of neuronal excitability in the short term, long-term mechanisms of regulation of M-channel abundance are expected to contribute to mechanisms of plasticity within neuronal circuits. *KCNQ* genes have a highly conserved repressor element 1 (RE1, NRSE) binding site situated between the exons 1 and 2 (Figure 20.9a; (408)). Functional interactions of repressor element 1 silencing transcription factor (REST, NRSF) with *KCNQ2*, *KCNQ3*, *KCNQ5* (408), and *KCNQ4* (409) have been demonstrated. REST inhibits expression of genes through the recruitment of multiple chromatin-modifying enzymes (410). Accordingly, overexpression of REST in DRG neurons robustly suppressed native M-current

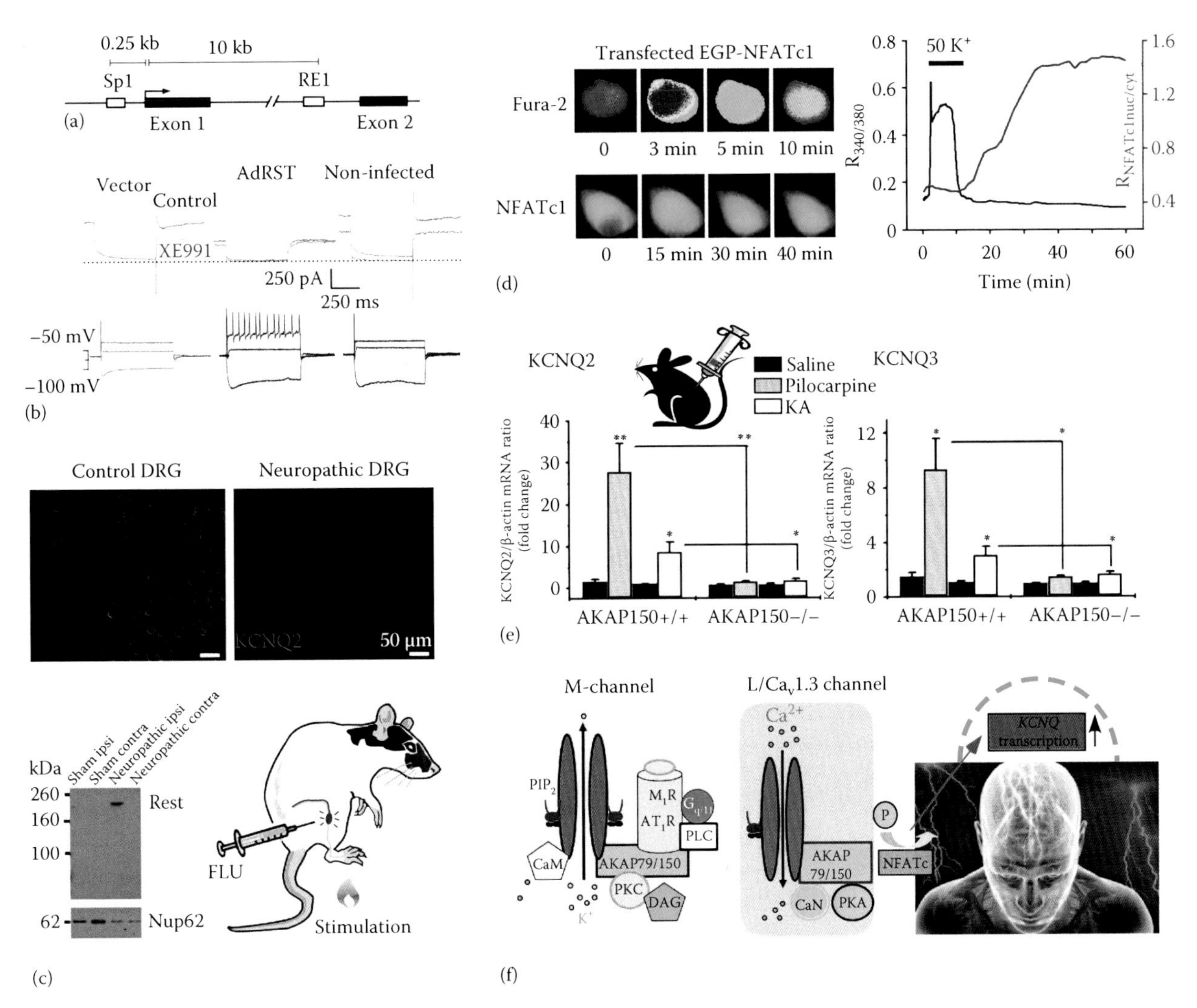

Figure 20.9 Emerging mechanisms of regulation of KCNQ2 gene transcription. (a–c) Regulation of *KCNQ* gene expression by REST. (a) Schematic representation of the 5′ of the *KCNQ2* gene. Black boxes represent exons, and the transcription start site is marked by an arrow. The location of regulatory elements, Sp1 and RE1, are shown as white boxes. (b) REST inhibits M-current and increases excitability of small-diameter DRG neurons. Top panel shows whole-cell voltage clamp recordings from cultured DRG neurons infected with an adenoviral construct expressing GFP only (Vector) or REST and GFP (AdREST). In cultures infected with REST and GFP, both green (AdREST) and nongreen (noninfected) neurons were tested. The XE991 (3 μM)-sensitive component of the whole-cell current elicited by voltage pulses from −30 to −60 mV was virtually absent in the REST-overexpressing neurons. Bottom panel shows whole-cell current-clamp traces in which the voltage was adjusted to −65 mV by current injection and 4 s square current pulses to different test currents applied. Consistent with the decrease in M-current density, REST-overexpressing neurons were overexcitable. (a, b; From Mucha, M. et al., *J. Neurosci.*, 30, 13235, 2010.) (c) Reciprocal changes in KCNQ2 and REST expression in DRG neurons following neuropathic injury in rat. Top panel shows KCNQ2 immunostaining of DRG sections from control animal (sham operated) and from rat with partial sciatic nerve ligation 30 days postsurgery (partial sciatic nerve ligation, PSNL) or control (sham) animals. Bottom left shows western blots of the nuclear-enriched fraction of DRG lysates from sham-operated and neuropathic animals using anti-REST antibody and nuclear protein anti-Nup 62. Ipsi and contra denotes samples from ipsilateral and contralateral DRGs, relative to the injury site. Despite strong decrease in KCNQ2 expression in neuropathic DRG, peripheral injection of FLU directly into the neuroma site of the neuropathic rats (as depicted in the bottom right cartoon) alleviated thermal hyperalgesia induced by the neuropathic injury. (From Rose, K. et al., *Pain*, 152, 742, 2011.) (d–f) Activity-dependent transcription of KCNQ2 and KCNQ3 channels is upregulated by NFAT transcription factors, orchestrated by AKAP79/150. In (d) is shown an experiment on a sympathetic neuron heterologously expressing EGFP-tagged NFATc1 and loaded with the Ca^{2+} reporter, fura-2. *Left*, pseudocolored 340/380 nm ratiometric images, and EGFP images in response to excitation by a high [K+] (50 mM) bath solution; *Right*, plotted are the fura-2 340/380 nm and the EGFP nucleus/cytoplasm localization ratios. Upon excitation by high [K+], there is a rapid elevation of $[Ca^{2+}]_i$ and a slower nuclear import of EGFP-NFATc1. (e) In response to chemoconvulsant seizures, mRNA for KCNQ2 and KCNQ3 in the hippocampus is profoundly increased, suggesting an *antiepileptogenic* protective role for activity-induced upregulation of M-channels in the brain. (f) AKAP79/150 model. Shown schematically are the signaling-protein complexes recruited to M-channels and L-type $Ca_V1.3$ (L-channels) by AKAP79/150 that underlie receptor-specific modulation of M-channels, PKA phosphorylation of L-channels, and upregulation of M-channel transcriptional expression. *Left*, arranged in a complex around KCNQ2/3 channels are CaM, AKAP79/150, PKC, and the $G_{q/11}$-coupled M_1 or AT_1 receptor types, and PIP_2 bound to both the channels, and AKAP. Right, clustered to L-channels is AKAP79/150, which recruits CaN and PKA, the latter mediating upregulation of L-channel currents. In response to activity, Ca^{2+} influx through L-channels calcifies CaN, dephosphorylating NFATs, revealing nuclear localization signals that direct NFATs into the nucleus, where they act on NFAT-binding regulatory domains of KCNQ2 and KCNQ3, increasing their transcription. (d–f; From Zhang, J. and Shapiro, M.S., *Neuron*, 76, 1133, 2012.)

density and increased tonic excitability of these neurons (Figure 20.9b; (408)). Tonic expression of REST in peripheral neurons is believed to be low, but it may increase greatly following inflammation (408) or after neuropathic injury (240,411). Interestingly, dramatic reduction in *KCNQ2* (240,389) and *KCNQ3* (389) transcripts and proteins following experimentally induced chronic pain development has been reported. It has therefore been suggested that M-channel downregulation upon the development of chronic pain can be mediated by the transcriptional suppression by REST. Indeed, increase of REST immunoreactivity and mRNA after nerve injury mirrored the reciprocal reduction in *KCNQ2* transcripts and KCNQ2 protein levels (Figure 20.9c; (240)). It is likely that transcriptional suppression of *KCNQ2* and *KCNQ3* expression in nociceptive neurons contributes to the long-lasting hyperexcitability of peripheral fibers observed in chronic pain states.

20.7.2 REGULATION BY NFAT

Recently, activity-dependent upregulation of M-channels at the transcriptional level has been reported in rodents, mediated by the Ca^{2+}-dependent phosphatase, calcineurin (CaN), nuclear-factor of activated T-cells (NFATs) transcription factors, and the same AKAP79/150 scaffold proteins mediating PKC phosphorylation (Figure 20.9d through f; (191)). Those authors found the reporter of neuronal activity to be $Ca_V1.3$ *L-type* Ca^{2+} channels, to which AKAP79/150 recruits CaN as the Ca^{2+} sensor, similar to the role of $Ca_V1.2$ channels (412). Upon binding Ca^{2+}, CaN dephosphorylates NFAT (413), which then translocates into the nucleus where it interacts with NFAT-binding regulatory elements of *KCNQ2* and *KCNQ3* genes, resulting in augmented mRNA levels and increased I_M amplitudes (191). If CaN activity is suppressed, L-type channels blocked, or AKAP150 knocked out in the animal, both effects were completely absent. Simultaneous imaging of GFP-tagged NFATs and $[Ca^{2+}]_i$ revealed selectivity for NFATc1 and NFATc2, and a time course of NFAT translocation to the nucleus much slower than that of $[Ca^{2+}]_i$ rises induced by high-$[K^+]_o$ or stimulation of nicotinic AChRs in sympathetic neurons, probably reflecting this NFAT pathway acting as a *working memory* of Ca_i^{2+} signals (414). The L-channel/CaN/NFAT pathway seen by Zhang and Shapiro (191) is similar to that first described in hippocampal neurons (415), suggesting a mechanism conserved in the CNS and PNS. Indeed, upon induction of a single chemoconvulsant seizure in mice, mRNA for *KCNQ2* and *KCNQ3* in the hippocampus was found to be increased by 3- to 30-fold, an effect likewise wholly absent in AKAP150 KO mice (191). Thus, transcriptional upregulation of I_M may represent an antiepileptogenic mechanism, acting to prevent an isolated seizure from progressing to full-blown epileptic disease.

Finally, increased transcription of *KCNQ2* and *KCNQ3* genes by the transcription factor Sp1 has also been described (408); however, the physiological relevance of this regulation remains to be determined.

20.8 CONCLUDING REMARKS

KCNQ/M-type channels have emerged as pivotal players in diverse excitable and nonexcitable tissues. In the former, they play powerful roles in control over electrical and chemical excitability, with corresponding effects on cognitive, autonomic, somatosensory, rhythmic activity, and visceral function. In the latter, M-type currents are crucial to normal K^+ flux and homeostasis. Not unexpectedly, dysfunction of KCNQ channels leads to severe dysfunction throughout the body and is seen as one of the *hottest* targets for novel modes of therapeutic intervention against a wide spectrum of human diseases. Equally logical is the employment of many modes of regulation of M-currents in cells in all of these tissues, mediated by G proteins, Ca^{2+}-binding proteins, kinases/phosphatases, lipid-signaling molecules, gasotransmitters, and regulators of transcription. Thus, not only are changes in KCNQ channels themselves exerting powerful actions on cellular states of excitability or activity, but subtle alterations among the plethora regulatory pathways and auxiliary subunits described in this chapter are expected to manifest critically in KCNQ current activity, with correspondingly powerful control over organismal function.

REFERENCES

1. Brown, D. A. and P. R. Adams. 1980. Muscarinic suppression of a novel voltage-sensitive K^+ current in a vertebrate neurone. *Nature* 283:673–676.
2. Constanti, A. and D. A. Brown. 1981. M-currents in voltage-clamped mammalian sympathetic neurones. *Neurosci Lett* 24:289–294.
3. Adams, P. R. and D. A. Brown. 1982. Synaptic inhibition of the M-current: Slow excitatory post-synaptic potential mechanism in bullfrog sympathetic neurones. *J Physiol* (*Lond*) 332:263–272.
4. Kandel, E. R. 2012. *Principles of Neural Science*. McGraw-Hill, New York.
5. Brown, D. A., F. C. Abogadie, T. G. Allen, N. J. Buckley, M. P. Caulfield, P. Delmas, J. E. Haley, J. A. Lamas, and A. A. Selyanko. 1997. Muscarinic mechanisms in nerve cells. *Life Sci* 60:1137–1144.
6. Walsh, K. B. and R. S. Kass. 1988. Regulation of a heart potassium channel by protein kinase A and C. *Science* 242:67–69.
7. Wang, Q., M. E. Curran, I. Splawski, T. C. Burn, J. M. Millholland, T. J. VanRaay, J. Shen et al. 1996. Positional cloning of a novel potassium channel gene: KVLQT1 mutations cause cardiac arrhythmias. *Nat Genet* 12:17–23.
8. Yang, W. P., P. C. Levesque, W. A. Little, M. L. Conder, F. Y. Shalaby, and M. A. Blanar. 1997. KvLQT1, a voltage-gated potassium channel responsible for human cardiac arrhythmias. *Proc Natl Acad Sci USA* 94:4017–4021.
9. Shalaby, F. Y., P. C. Levesque, W. P. Yang, W. A. Little, M. L. Conder, T. Jenkins-West, and M. A. Blanar. 1997. Dominant-negative KvLQT1 mutations underlie the LQT1 form of long QT syndrome. *Circulation* 96:1733–1736.
10. Sanguinetti, M. C., M. E. Curran, A. Zou, J. Shen, P. S. Spector, D. L. Atkinson, and M. T. Keating. 1996. Coassembly of K(V) LQT1 and minK (IsK) proteins to form cardiac I(Ks) potassium channel. *Nature* 384:80–83.
11. Romey, G., B. Attali, C. Chouabe, I. Abitbol, E. Guillemare, J. Barhanin, and M. Lazdunski. 1997. Molecular mechanism and functional significance of the MinK control of the KvLQT1 channel activity. *J Biol Chem* 272:16713–16716.
12. Kaczmarek, L. K., and E. M. Blumenthal. 1997. Properties and regulation of the minK potassium channel protein. *Physiol Rev* 77:627–641.
13. Wei, A., T. Jegla, and L. Salkoff. 1996. Eight potassium channel families revealed by the *C. elegans* genome project. *Neuropharmacology* 35:805–829.

14. Yang, W. P., P. C. Levesque, W. A. Little, M. L. Conder, P. Ramakrishnan, M. G. Neubauer, and M. A. Blanar. 1998. Functional expression of two KvLQT1-related potassium channels responsible for an inherited idiopathic epilepsy. *J Biol Chem* 273:19419–19423.
15. Charlier, C., N. A. Singh, S. G. Ryan, T. B. Lewis, B. E. Reus, R. J. Leach, and M. Leppert. 1998. A pore mutation in a novel KQT-like potassium channel gene in an idiopathic epilepsy family. *Nat Genet* 18:53–55.
16. Biervert, C., B. C. Schroeder, C. Kubisch, S. F. Berkovic, P. Propping, T. J. Jentsch, and O. K. Steinlein. 1998. A potassium channel mutation in neonatal human epilepsy. *Science* 279:403–406.
17. Singh, N. A., C. Charlier, D. Stauffer, B. R. DuPont, R. J. Leach, R. Melis, G. M. Ronen et al. 1998. A novel potassium channel gene, KCNQ2, is mutated in an inherited epilepsy of newborns. *Nat Genet* 18:25–29.
18. Wang, H. S., Z. Pan, W. Shi, B. S. Brown, R. S. Wymore, I. S. Cohen, J. E. Dixon, and D. McKinnon. 1998. KCNQ2 and KCNQ3 potassium channel subunits: Molecular correlates of the M-channel. *Science* 282:1890–1893.
19. Selyanko, A. A., J. K. Hadley, I. C. Wood, F. C. Abogadie, T. J. Jentsch, and D. A. Brown. 2000. Inhibition of KCNQ1-4 potassium channels expressed in mammalian cells via M1 muscarinic acetylcholine receptors. *J Physiol* (*Lond*) 522 Pt 3:349–355.
20. Shapiro, M. S., J. P. Roche, E. J. Kaftan, H. Cruzblanca, K. Mackie, and B. Hille. 2000. Reconstitution of muscarinic modulation of the KCNQ2/KCNQ3 K(+) channels that underlie the neuronal M current. *J Neurosci* 20:1710–1721.
21. Nakamura, M., H. Watanabe, Y. Kubo, M. Yokoyama, T. Matsumoto, H. Sasai, and Y. Nishi. 1998. KQT2, a new putative potassium channel family produced by alternative splicing. Isolation, genomic structure, and alternative splicing of the putative potassium channels. *Receptors Channels* 5:255–271.
22. Tinel, N., I. Lauritzen, C. Chouabe, M. Lazdunski, and M. Borsotto. 1998. The KCNQ2 potassium channel: Splice variants, functional and developmental expression. Brain localization and comparison with KCNQ3. *FEBS Lett* 438:171–176.
23. Pan, Z., A. A. Selyanko, J. K. Hadley, D. A. Brown, J. E. Dixon, and D. McKinnon. 2001. Alternative splicing of KCNQ2 potassium channel transcripts contributes to the functional diversity of M-currents. *J Physiol* 531:347–358.
24. Wangemann, P. 2002. K^+ cycling and the endocochlear potential. *Hear Res* 165:1–9.
25. Coucke, P. J., P. V. Hauwe, P. M. Kelley, H. Kunst, I. Schatteman, D. V. Velzen, J. Meyers et al. 1999. Mutations in the KCNQ4 gene are responsible for autosomal dominant deafness in four DFNA2 families. *Hum Mol Genet* 8:1321–1328.
26. Kubisch, C., B. C. Schroeder, T. Friedrich, B. Lutjohann, A. El-Amraoui, S. Marlin, C. Petit, and T. J. Jentsch. 1999. KCNQ4, a novel potassium channel expressed in sensory outer hair cells, is mutated in dominant deafness. *Cell* 96:437–446.
27. Kharkovets, T., J. P. Hardelin, S. Safieddine, M. Schweizer, A. El-Amraoui, C. Petit, and T. J. Jentsch. 2000. KCNQ4, a K^+ channel mutated in a form of dominant deafness, is expressed in the inner ear and the central auditory pathway. *Proc Natl Acad Sci USA* 97:4333–4338.
28. Sogaard, R., T. Ljungstrom, K. A. Pedersen, S. P. Olesen, and B. S. Jensen. 2001. KCNQ4 channels expressed in mammalian cells: Functional characteristics and pharmacology. *Am J Physiol Cell Physiol* 280:C859–C866.
29. Lerche, C., C. R. Scherer, G. Seebohm, C. Derst, A. D. Wei, A. E. Busch, and K. Steinmeyer. 2000. Molecular cloning and functional expression of KCNQ5, a potassium channel subunit that may contribute to neuronal M-current diversity. *J Biol Chem* 275:22395–22400.
30. Schroeder, B. C., M. Hechenberger, F. Weinreich, C. Kubisch, and T. J. Jentsch. 2000. KCNQ5, a novel potassium channel broadly expressed in brain, mediates M-type currents. *J Biol Chem* 275:24089–24095.
31. Passmore, G. M., A. A. Selyanko, M. Mistry, M. Al-Qatari, S. J. Marsh, E. A. Matthews, A. H. Dickenson et al. 2003. KCNQ/M currents in sensory neurons: Significance for pain therapy. *J Neurosci* 23:7227–7236.
32. Kananura, C., C. Biervert, M. Hechenberger, H. Engels, and O. K. Steinlein. 2000. The new voltage gated potassium channel KCNQ5 and neonatal convulsions. *Neuroreport* 11:2063–2067.
33. Yus-Najera, E., A. Munoz, N. Salvador, B. S. Jensen, H. B. Rasmussen, J. Defelipe, and A. Villarroel. 2003. Localization of KCNQ5 in the normal and epileptic human temporal neocortex and hippocampal formation. *Neuroscience* 120:353–364.
34. Ohya, S., G. P. Sergeant, I. A. Greenwood, and B. Horowitz. 2003. Molecular variants of KCNQ channels expressed in murine portal vein myocytes: A role in delayed rectifier current. *Circ Res* 92:1016–1023.
35. Brueggemann, L. I., C. J. Moran, J. A. Barakat, J. Z. Yeh, L. L. Cribbs, and K. L. Byron. 2007. Vasopressin stimulates action potential firing by protein kinase C-dependent inhibition of KCNQ5 in A7r5 rat aortic smooth muscle cells. *Am J Physiol Heart Circ Physiol* 292:H1352–H1363.
36. Mackie, A. R., L. I. Brueggemann, K. K. Henderson, A. J. Shiels, L. L. Cribbs, K. E. Scrogin, and K. L. Byron. 2008. Vascular KCNQ potassium channels as novel targets for the control of mesenteric artery constriction by vasopressin, based on studies in single cells, pressurized arteries, and in vivo measurements of mesenteric vascular resistance. *J Pharmacol Exp Ther* 325:475–483.
37. Brueggemann, L. I., P. P. Kakad, R. B. Love, J. Solway, M. L. Dowell, L. L. Cribbs, and K. L. Byron. 2012. Kv7 potassium channels in airway smooth muscle cells: Signal transduction intermediates and pharmacological targets for bronchodilator therapy. *Am J Physiol Lung Cell Mol Physiol* 302:L120–L132.
38. Robbins, J. 2001. KCNQ potassium channels: Physiology, pathophysiology, and pharmacology. *Pharmacol Ther* 90:1–19.
39. Huang, H. and L. O. Trussell. 2011. KCNQ5 channels control resting properties and release probability of a synapse. *Nat Neurosci* 14:840–847.
40. Pusch, M. 1998. Increase of the single-channel conductance of KvLQT1 potassium channels induced by the association with minK. *Pflugers Arch* 437:172–174.
41. Sesti, F. and S. A. Goldstein. 1998. Single-channel characteristics of wild-type IKs channels and channels formed with two minK mutants that cause long QT syndrome. *J Gen Physiol* 112:651–663.
42. Werry, D., J. Eldstrom, Z. Wang, and D. Fedida. 2013. Single-channel basis for the slow activation of the repolarizing cardiac potassium current, IKs. *Proc Natl Acad Sci USA* 110:E996–E1005.
43. Tristani-Firouzi, M. and M. C. Sanguinetti. 1998. Voltage-dependent inactivation of the human K^+ channel KvLQT1 is eliminated by association with minimal K^+ channel (minK) subunits. *J Physiol* (*Lond*) 510:37–45.
44. Jespersen, T., M. Grunnet, and S. P. Olesen. 2005. The KCNQ1 potassium channel: From gene to physiological function. *Physiology* (*Bethesda*) 20:408–416.
45. Hernandez, C. C., O. Zaika, and M. S. Shapiro. 2008. A carboxy-terminal inter-helix linker as the site of phosphatidylinositol 4,5-bisphosphate action on Kv7 (M-type) K^+ channels. *J Gen Physiol* 132:361–381.
46. Li, Y., N. Gamper, and M. S. Shapiro. 2004. Single-channel analysis of KCNQ K^+ channels reveals the mechanism of augmentation by a cysteine-modifying reagent. *J Neurosci* 24:5079–5090.

47. Selyanko, A. A. and D. A. Brown. 1999. M-channel gating and simulation. *Biophys J* 77:701–713.
48. Selyanko, A. A., J. K. Hadley, and D. A. Brown. 2001. Properties of single M-type KCNQ2/KCNQ3 potassium channels expressed in mammalian cells. *J Physiol* 534:15–24.
49. Tatulian, L. and D. A. Brown. 2003. Effect of the KCNQ potassium channel opener retigabine on single KCNQ2/3 channels expressed in CHO cells. *J Physiol* 549:57–63.
50. Owen, D. G., S. J. Marsh, and D. A. Brown. 1990. M-current noise and putative M-channels in cultured rat sympathetic ganglion cells. *J Physiol (Lond)* 431:269–290.
51. Stansfeld, C. E., S. J. Marsh, A. J. Gibb, and D. A. Brown. 1993. Identification of M-channels in outside-out patches excised from sympathetic ganglion cells. *Neuron* 10:639–654.
52. Telezhkin, V., D. A. Brown, and A. J. Gibb. 2012. Distinct subunit contributions to the activation of M-type potassium channels by PI(4,5)P2. *J Gen Physiol* 140:41–53.
53. Choveau, F. S. and M. S. Shapiro. 2012. Regions of KCNQ K(+) channels controlling functional expression. *Front Physiol* 3:397.
54. Li, M., Y. N. Jan, and L. Y. Jan. 1992. Specification of subunit assembly by the hydrophilic amino-terminal domain of the Shaker potassium channel. *Science* 257:1225–1230.
55. Maljevic, S., C. Lerche, G. Seebohm, A. K. Alekov, A. E. Busch, and H. Lerche. 2003. C-terminal interaction of KCNQ2 and KCNQ3 K^+ channels. *J Physiol* 548:353–360.
56. Schwake, M., D. Athanasiadu, C. Beimgraben, J. Blanz, C. Beck, T. J. Jentsch, P. Saftig, and T. Friedrich. 2006. Structural determinants of M-type KCNQ (Kv7) K^+ channel assembly. *J Neurosci* 26:3757–3766.
57. Schwake, M., T. J. Jentsch, and T. Friedrich. 2003. A carboxy-terminal domain determines the subunit specificity of KCNQ K(+) channel assembly. *EMBO Rep* 4:76–81.
58. Chung, H. J., Y. N. Jan, and L. Y. Jan. 2006. Polarized axonal surface expression of neuronal KCNQ channels is mediated by multiple signals in the KCNQ2 and KCNQ3 C-terminal domains. *Proc Natl Acad Sci USA* 103:8870–8875.
59. Bal, M., J. Zhang, O. Zaika, C. C. Hernandez, and M. S. Shapiro. 2008. Homomeric and heteromeric assembly of KCNQ (Kv7) K^+ channels assayed by total internal reflection fluorescence/fluorescence resonance energy transfer and patch clamp analysis. *J Biol Chem* 283:30668–30676.
60. Wickenden, A. D., A. Zou, P. K. Wagoner, and T. Jegla. 2001. Characterization of KCNQ5/Q3 potassium channels expressed in mammalian cells. *Br J Pharmacol* 132:381–384.
61. Schenzer, A., T. Friedrich, M. Pusch, P. Saftig, T. J. Jentsch, J. Grotzinger, and M. Schwake. 2005. Molecular determinants of KCNQ (Kv7) K^+ channel sensitivity to the anticonvulsant retigabine. *J Neurosci* 25:5051–5060.
62. Howard, R. J., K. A. Clark, J. M. Holton, and D. L. Minor, Jr. 2007. Structural insight into KCNQ (Kv7) channel assembly and channelopathy. *Neuron* 53:663–675.
63. Li, Y., N. Gamper, D. W. Hilgemann, and M. S. Shapiro. 2005. Regulation of Kv7 (KCNQ) K^+ channel open probability by phosphatidylinositol (4,5)-bisphosphate. *J Neurosci* 25:9825–9835.
64. Etxeberria, A., I. Santana-Castro, M. P. Regalado, P. Aivar, and A. Villarroel. 2004. Three mechanisms underlie KCNQ2/3 heteromeric potassium M-channel potentiation. *J Neurosci* 24:9146–9152.
65. Zaika, O., C. C. Hernandez, M. Bal, G. P. Tolstykh, and M. S. Shapiro. 2008. Determinants within the turret and pore-loop domains of KCNQ3 K^+ channels governing functional activity. *Biophys J* 95:5121–5137.
66. Choveau, F. S., S. M. Bierbower, and M. S. Shapiro. 2012. Pore helix-S6 interactions are critical in governing current amplitudes of KCNQ3 K^+ channels. *Biophys J* 102:2499–2509.
67. Heginbotham, L. and R. MacKinnon. 1992. The aromatic binding site for tetraethylammonium ion on potassium channels. *Neuron* 8:483–491.
68. MacKinnon, R. and G. Yellen. 1990. Mutations affecting TEA blockade and ion permeation in voltage-activated K^+ channels. *Science* 250:276–279.
69. Hadley, J. K., M. Noda, A. A. Selyanko, I. C. Wood, F. C. Abogadie, and D. A. Brown. 2000. Differential tetraethylammonium sensitivity of KCNQ1-4 potassium channels. *Br J Pharmacol* 129:413–415.
70. Hadley, J. K., G. M. Passmore, L. Tatulian, M. Al-Qatari, F. Ye, A. D. Wickenden, and D. A. Brown. 2003. Stoichiometry of expressed KCNQ2/KCNQ3 potassium channels and subunit composition of native ganglionic M channels deduced from block by tetraethylammonium. *J Neurosci* 23:5012–5019.
71. Aiken, S. P., B. J. Lampe, P. A. Murphy, and B. S. Brown. 1995. Reduction of spike frequency adaptation and blockade of M-current in rat CA1 pyramidal neurones by linopirdine (DuP 996), a neurotransmitter release enhancer. *Br J Pharmacol* 115:1163–1168.
72. Lamas, J. A., A. A. Selyanko, and D. A. Brown. 1997. Effects of a cognition-enhancer, linopirdine (DuP 996), on M-type potassium currents (IK(M)) and some other voltage- and ligand-gated membrane currents in rat sympathetic neurons. *Eur J Neurosci* 9:605–616.
73. Zaczek, R., R. J. Chorvat, J. A. Saye, M. E. Pierdomenico, C. M. Maciag, A. R. Logue, B. N. Fisher, D. H. Rominger, and R. A. Earl. 1998. Two new potent neurotransmitter release enhancers, 10,10-bis(4-pyridinylmethyl)-9(10H)-anthracenone and 10,10-bis(2-fluoro-4-pyridinylmethyl)-9(10H)-anthracenone: Comparison to linopirdine. *J Pharmacol Exp Ther* 285:724–730.
74. Elmedyb, P., K. Calloe, N. Schmitt, R. S. Hansen, M. Grunnet, and S. P. Olesen. 2007. Modulation of ERG channels by XE991. *Basic Clin Pharmacol Toxicol* 100:316–322.
75. Rundfeldt, C. 1997. The new anticonvulsant retigabine (D-23129) acts as an opener of K^+ channels in neuronal cells. *Eur J Pharmacol* 336:243–249.
76. Wickenden, A. D., W. Yu, A. Zou, T. Jegla, and P. K. Wagoner. 2000. Retigabine, a novel anti-convulsant, enhances activation of KCNQ2/Q3 potassium channels. *Mol Pharmacol* 58:591–600.
77. Main, M. J., J. E. Cryan, J. R. Dupere, B. Cox, J. J. Clare, and S. A. Burbidge. 2000. Modulation of KCNQ2/3 potassium channels by the novel anticonvulsant retigabine. *Mol Pharmacol* 58:253–262.
78. Rundfeldt, C. and R. Netzer. 2000. The novel anticonvulsant retigabine activates M-currents in Chinese hamster ovary-cells tranfected with human KCNQ2/3 subunits. *Neurosci Lett* 282:73–76.
79. Tatulian, L., P. Delmas, F. C. Abogadie, and D. A. Brown. 2001. Activation of expressed KCNQ potassium currents and native neuronal M-type potassium currents by the anti-convulsant drug retigabine. *J Neurosci* 21:5535–5545.
80. Wuttke, T. V., G. Seebohm, S. Bail, S. Maljevic, and H. Lerche. 2005. The new anticonvulsant retigabine favors voltage-dependent opening of the Kv7.2 (KCNQ2) channel by binding to its activation gate. *Mol Pharmacol* 67:1009–1017.
81. Lerche, H., K. Jurkat-Rott, and F. Lehmann-Horn. 2001. Ion channels and epilepsy. *Am J Med Genet* 106:146–159.
82. Otto, J. F., M. M. Kimball, and K. S. Wilcox. 2002. Effects of the anticonvulsant retigabine on cultured cortical neurons: Changes in electroresponsive properties and synaptic transmission. *Mol Pharmacol* 61:921–927.
83. Blackburn-Munro, G. and B. S. Jensen. 2003. The anticonvulsant retigabine attenuates nociceptive behaviours in rat models of persistent and neuropathic pain. *Eur J Pharmacol* 460:109–116.

84. Rivera-Arconada, I., J. Martinez-Gomez, and J. A. Lopez-Garcia. 2004. M-current modulators alter rat spinal nociceptive transmission: An electrophysiological study in vitro. *Neuropharmacology* 46:598–606.
85. Linley, J. E., L. Pettinger, D. Huang, and N. Gamper. 2012. M channel enhancers and physiological M channel block. *J Physiol* 590:793–807.
86. Martire, M., P. Castaldo, M. D'Amico, P. Preziosi, L. Annunziato, and M. Taglialatela. 2004. M channels containing KCNQ2 subunits modulate norepinephrine, aspartate, and GABA release from hippocampal nerve terminals. *J Neurosci* 24:592–597.
87. Gamper, N., O. Zaika, Y. Li, P. Martin, C. C. Hernandez, M. R. Perez, A. Y. Wang, D. B. Jaffe, and M. S. Shapiro. 2006. Oxidative modification of M-type K(+) channels as a mechanism of cytoprotective neuronal silencing. *EMBO J* 25:4996–5004.
88. Boscia, F., L. Annunziato, and M. Taglialatela. 2006. Retigabine and flupirtine exert neuroprotective actions in organotypic hippocampal cultures. *Neuropharmacology* 51:283–294.
89. Wladyka, C. L., B. Feng, P. A. Glazebrook, J. H. Schild, and D. L. Kunze. 2008. The KCNQ/M-current modulates arterial baroreceptor function at the sensory terminal in rats. *J Physiol* 586:795–802.
90. Maljevic, S., T. V. Wuttke, and H. Lerche. 2008. Nervous system KV7 disorders: Breakdown of a subthreshold brake. *J Physiol* 586:1791–1801.
91. Peretz, A., A. Sheinin, C. Yue, N. Degani-Katzav, G. Gibor, R. Nachman, A. Gopin, E. Tam, D. Shabat, Y. Yaari, and B. Attali. 2007. Pre- and postsynaptic activation of M-channels by a novel opener dampens neuronal firing and transmitter release. *J Neurophysiol* 97:283–295.
92. Peretz, A., N. Degani, R. Nachman, Y. Uziyel, G. Gibor, D. Shabat, and B. Attali. 2005. Meclofenamic acid and diclofenac, novel templates of KCNQ2/Q3 potassium channel openers, depress cortical neuron activity and exhibit anticonvulsant properties. *Mol Pharmacol* 67:1053–1066.
93. Xiong, Q., H. Sun, and M. Li. 2007. Zinc pyrithione-mediated activation of voltage-gated KCNQ potassium channels rescues epileptogenic mutants. *Nat Chem Biol* 3:287–296.
94. Qi, J., F. Zhang, Y. Mi, Y. Fu, W. Xu, D. Zhang, Y. Wu, X. Du, Q. Jia, K. Wang, and H. Zhang. 2011. Design, synthesis and biological activity of pyrazolo[1,5-a]pyrimidin-7(4H)-ones as novel Kv7/KCNQ potassium channel activators. *Eur J Med Chem* 46:934–943.
95. Jia, C., J. Qi, F. Zhang, Y. Mi, X. Zhang, X. Chen, L. Liu, X. Du, and H. Zhang. 2011. Activation of KCNQ2/3 potassium channels by novel pyrazolo[1,5-a]pyrimidin-7(4H)-one derivatives. *Pharmacology* 87:297–310.
96. Linley, J. E., L. Ooi, L. Pettinger, H. Kirton, J. P. Boyle, C. Peers, and N. Gamper. 2012. Reactive oxygen species are second messengers of neurokinin signaling in peripheral sensory neurons. *Proc Natl Acad Sci USA* 109:E1578–E1586.
97. Roche, J. P., R. Westenbroek, A. J. Sorom, B. Hille, K. Mackie, and M. S. Shapiro. 2002. Antibodies and a cysteine-modifying reagent show correspondence of M current in neurons to KCNQ2 and KCNQ3 K^+ channels. *Br J Pharmacol* 137:1173–1186.
98. Brueggemann, L. I., A. R. Mackie, J. L. Martin, L. L. Cribbs, and K. L. Byron. 2011. Diclofenac distinguishes among homomeric and heteromeric potassium channels composed of KCNQ4 and KCNQ5 subunits. *Mol Pharmacol* 79:10–23.
99. Brueggemann, L. I., B. K. Mani, A. R. Mackie, L. L. Cribbs, and K. L. Byron. 2010. Novel actions of nonsteroidal anti-inflammatory drugs on vascular ion channels: Accounting for cardiovascular side effects and identifying new therapeutic applications. *Mol Cell Pharmacol* 2:15–19.
100. Marrion, N. V., T. G. Smart, S. J. Marsh, and D. A. Brown. 1989. Muscarinic suppression of the M-current in the rat sympathetic ganglion is mediated by receptors of the M1-subtype. *Br J Pharmacol* 98:557–573.
101. Bernheim, L., A. Mathie, and B. Hille. 1992. Characterization of muscarinic receptor subtypes inhibiting Ca^{2+} current and M current in rat sympathetic neurons. *Proc Natl Acad Sci USA* 89:9544–9548.
102. Hamilton, S. E., M. D. Loose, M. Qi, A. I. Levey, B. Hille, G. S. McKnight, R. L. Idzerda, and N. M. Nathanson. 1997. Disruption of the m1 receptor gene ablates muscarinic receptor-dependent M current regulation and seizure activity in mice. *Proc Natl Acad Sci USA* 94:13311–13316.
103. Caulfield, M. P., S. Jones, Y. Vallis, N. J. Buckley, G. D. Kim, G. Milligan, and D. A. Brown. 1994. Muscarinic M-current inhibition via G alpha q/11 and alpha-adrenoceptor inhibition of Ca^{2+} current via G alpha o in rat sympathetic neurones. *J Physiol (Lond)* 477:415–422.
104. Pfaffinger, P. J., M. D. Leibowitz, E. M. Subers, N. M. Nathanson, W. Almers, and B. Hille. 1988. Agonists that suppress M-current elicit phosphoinositide turnover and Ca^{2+} transients, but these events do not explain M-current suppression. *Neuron* 1:477–484.
105. Brown, D. A., N. V. Marrion, and T. G. Smart. 1989. On the transduction mechanism for muscarine-induced inhibition of M-current in cultured rat sympathetic neurones. *J Physiol (Lond)* 413:469–488.
106. Selyanko, A. A., C. E. Stansfeld, and D. A. Brown. 1992. Closure of potassium M-channels by muscarinic acetylcholine-receptor stimulants requires a diffusible messenger. *Proc R Soc Lond B Biol Sci* 250:119–125.
107. Bosma, M. M., and B. Hille. 1989. Protein kinase C is not necessary for peptide-induced suppression of M current or for desensitization of the peptide receptors. *Proc Natl Acad Sci USA* 86:2943–2947.
108. Cruzblanca, H., D. S. Koh, and B. Hille. 1998. Bradykinin inhibits M current via phospholipase C and Ca^{2+} release from IP3-sensitive Ca^{2+} stores in rat sympathetic neurons. *Proc Natl Acad Sci USA* 95:7151–7156.
109. Suh, B. and B. Hille. 2002. Recovery from muscarinic modulation of M current channels requires phosphatidylinositol 4,5-bisphosphate synthesis. *Neuron* 35:507–520.
110. Zhang, H., L. C. Craciun, T. Mirshahi, T. Rohacs, C. M. Lopes, T. Jin, and D. E. Logothetis. 2003. PIP_2 activates KCNQ channels, and its hydrolysis underlies receptor-mediated inhibition of M currents. *Neuron* 37:963–975.
111. Ford, C. P., P. L. Stemkowski, P. E. Light, and P. A. Smith. 2003. Experiments to test the role of phosphatidylinositol 4,5-bisphosphate in neurotransmitter-induced M-channel closure in bullfrog sympathetic neurons. *J Neurosci* 23:4931–4941.
112. Brown, D. A., S. A. Hughes, S. J. Marsh, and A. Tinker. 2007. Regulation of M(Kv7.2/7.3) channels in neurons by PIP(2) and products of PIP(2) hydrolysis: Significance for receptor-mediated inhibition. *J Physiol* 582:917–925.
113. Hilgemann, D. W. and R. Ball. 1996. Regulation of cardiac Na^+,Ca^{2+} exchange and KATP potassium channels by PIP2. *Science* 273:956–959.
114. Huang, C. L., S. Feng, and D. W. Hilgemann. 1998. Direct activation of inward rectifier potassium channels by PIP_2 and its stabilization by $G\beta\gamma$. *Nature* 391:803–806.
115. Hilgemann, D. W., S. Feng, and C. Nasuhoglu. 2001. The complex and intriguing lives of PIP2 with ion channels and transporters. *Sci STKE* 2001:RE19.
116. Suh, B.-C. and B. Hille. 2005. Regulation of ion channels by phosphatidylinositol 4,5-bisphosphate. *Curr Opin Neurobiol* 15:370–378.

117. Suh, B. C. and B. Hille. 2008. PIP2 is a necessary cofactor for ion channel function: How and why? *Annu Rev Biophys* 37:175–195.
118. Gamper, N. and T. Rohacs. 2012. Phosphoinositide sensitivity of ion channels, a functional perspective. *Subcell Biochem* 59:289–333.
119. Gamper, N. and M. S. Shapiro. 2007. Target-specific PIP2 signalling: How might it work? *J Physiol* 582:967–975.
120. Gamper, N. and M. S. Shapiro. 2007. Regulation of ion transport proteins by membrane phosphoinositides. *Nat Rev Neurosci* 8:921–934.
121. Runnels, L. W., L. Yue, and D. E. Clapham. 2002. The TRPM7 channel is inactivated by PIP2 hydrolysis. *Nat Cell Biol* 4:329–336.
122. Loussouarn, G., K. H. Park, C. Bellocq, I. I. Baro, F. Charpentier, and D. Escande. 2003. Phosphatidylinositol-4,5-bisphosphate, PIP(2), controls KCNQ1/KCNE1 voltage-gated potassium channels: A functional homology between voltage-gated and inward rectifier K(+) channels. *EMBO J* 22:5412–5421.
123. Ford, C. P., P. L. Stemkowski, and P. A. Smith. 2004. Possible role of phosphatidylinositol 4,5 bisphosphate in luteinizing hormone releasing hormone-mediated M-current inhibition in bullfrog sympathetic neurons. *Eur J Neurosci* 20:2990–2998.
124. Robbins, J., S. J. Marsh, and D. A. Brown. 2006. Probing the regulation of M (Kv7) potassium channels in intact neurons with membrane-targeted peptides. *J Neurosci* 26:7950–7961.
125. Suh, B. C., T. Inoue, T. Meyer, and B. Hille. 2006. Rapid chemically induced changes of PtdIns(4,5)P2 gate KCNQ ion channels. *Science* 314:1454–1457.
126. Winks, J. S., S. Hughes, A. K. Filippov, L. Tatulian, F. C. Abogadie, D. A. Brown, and S. J. Marsh. 2005. Relationship between membrane phosphatidylinositol-4,5-bisphosphate and receptor-mediated inhibition of native neuronal M channels. *J Neurosci* 25:3400–3413.
127. Balla, T. 2007. Imaging and manipulating phosphoinositides in living cells. *J Physiol* 582:927–937.
128. Varnai, P. and T. Balla. 1998. Visualization of phosphoinositides that bind pleckstrin homology domains: Calcium- and agonist-induced dynamic changes and relationship to myo-[3H]inositol-labeled phosphoinositide pools. *J Cell Biol* 143:501–510.
129. Raucher, D., T. Stauffer, W. Chen, K. Shen, S. Guo, J. D. York, M. P. Sheetz, and T. Meyer. 2000. Phosphatidylinositol 4,5-bisphosphate functions as a second messenger that regulates cytoskeleton-plasma membrane adhesion. *Cell* 100:221–228.
130. Stauffer, T. P., S. Ahn, and T. Meyer. 1998. Receptor-induced transient reduction in plasma membrane PtdIns(4,5)P2 concentration monitored in living cells. *Curr Biol* 8:343–346.
131. Gamper, N., V. Reznikov, Y. Yamada, J. Yang, and M. S. Shapiro. 2004. Phosphatidylinositol 4,5-bisphosphate signals underlie receptor-specific Gq/11-mediated modulation of N-type Ca^{2+} channels. *J Neurosci* 24:10980–10992.
132. Varnai, P., B. Thyagarajan, T. Rohacs, and T. Balla. 2006. Rapidly inducible changes in phosphatidylinositol 4,5-bisphosphate levels influence multiple regulatory functions of the lipid in intact living cells. *J Cell Biol* 175:377–382.
133. Hirose, K., S. Kadowaki, M. Tanabe, H. Takeshima, and M. Iino. 1999. Spatiotemporal dynamics of inositol 1,4,5-trisphosphate that underlies complex Ca^{2+} mobilization patterns. *Science* 284:1527–1530.
134. Xu, C., J. Watras, and L. M. Loew. 2003. Kinetic analysis of receptor-activated phosphoinositide turnover. *J Cell Biol* 161:779–791.
135. Loew, L. M. 2007. Where does all the PIP2 come from? *J Physiol* 582:945–951.
136. Brown, S. A., F. Morgan, J. Watras, and L. M. Loew. 2008. Analysis of phosphatidylinositol-4,5-bisphosphate signaling in cerebellar Purkinje spines. *Biophys J* 95:1795–1812.
137. Suh, B. C., L. F. Horowitz, W. Hirdes, K. Mackie, and B. Hille. 2004. Regulation of KCNQ2/KCNQ3 current by G-protein cycling: The kinetics of receptor-mediated signaling by Gq. *J Gen Physiol* 123:663–683.
138. Horowitz, L. F., W. Hirdes, B. C. Suh, D. W. Hilgemann, K. Mackie, and B. Hille. 2005. Phospholipase C in living cells: Activation, inhibition, Ca^{2+} requirement, and regulation of M current. *J Gen Physiol* 126:243–262.
139. Jensen, J. B., J. S. Lyssand, C. Hague, and B. Hille. 2009. Fluorescence changes reveal kinetic steps of muscarinic receptor-mediated modulation of phosphoinositides and Kv7.2/7.3 K^+ channels. *J Gen Physiol* 133:347–359.
140. Falkenburger, B. H., J. B. Jensen, and B. Hille. 2010. Kinetics of PIP2 metabolism and KCNQ2/3 channel regulation studied with a voltage-sensitive phosphatase in living cells. *J Gen Physiol* 135:99–114.
141. Falkenburger, B. H., J. B. Jensen, and B. Hille. 2010. Kinetics of M1 muscarinic receptor and G protein signaling to phospholipase C in living cells. *J Gen Physiol* 135:81–97.
142. Lemmon, M. A. 2003. Phosphoinositide recognition domains. *Traffic* 4:201–213.
143. Telezhkin, V., J. M. Reilly, A. M. Thomas, A. Tinker, and D. A. Brown. 2012. Structural requirements of membrane phospholipids for M-type potassium channel activation and binding. *J Biol Chem* 287:10001–10012.
144. Rohacs, T., J. Chen, G. D. Prestwich, and D. E. Logothetis. 1999. Distinct specificities of inwardly rectifying K(+) channels for phosphoinositides. *J Biol Chem* 274:36065–36072.
145. Shyng, S. L., C. A. Cukras, J. Harwood, and C. G. Nichols. 2000. Structural determinants of PIP(2) regulation of inward rectifier K(ATP) channels. *J Gen Physiol* 116:599–608.
146. Thomas, A. M., S. C. Harmer, T. Khambra, and A. Tinker. 2011. Characterization of a binding site for anionic phospholipids on KCNQ1. *J Biol Chem* 286:2088–2100.
147. Hernandez, C. C., B. Falkenburger, and M. S. Shapiro. 2009. Affinity for phosphatidylinositol 4,5-bisphosphate determines muscarinic agonist sensitivity of Kv7 K^+ channels. *J Gen Physiol* 134:437–448.
148. Telezhkin, V., A. M. Thomas, S. C. Harmer, A. Tinker, and D. A. Brown. 2013. A basic residue in the proximal C-terminus is necessary for efficient activation of the M-channel subunit Kv7.2 by PI(4,5)P(2). *Pflugers Arch* 465:945–953.
149. Whorton, M. R. and R. MacKinnon. 2011. Crystal structure of the mammalian GIRK2 K^+ channel and gating regulation by G proteins, PIP2, and sodium. *Cell* 147:199–208.
150. Hernandez, C. C., O. Zaika, G. P. Tolstykh, and M. S. Shapiro. 2008. Regulation of neural KCNQ channels: Signalling pathways, structural motifs and functional implications. *J Physiol* 586:1811–1821.
151. Delmas, P. and D. A. Brown. 2002. Junctional signaling microdomains: Bridging the gap between the neuronal cell surface and Ca^{2+} stores. *Neuron* 36:787–790.
152. Zaika, O., J. Zhang, and M. S. Shapiro. 2011. Combined phosphoinositide and Ca^{2+} signals mediating receptor specificity toward neuronal Ca^{2+} channels. *J Biol Chem* 286:830–841.
153. Gamper, N. and M. S. Shapiro. 2003. Calmodulin mediates Ca^{2+}-dependent modulation of M-type K^+ channels. *J Gen Physiol* 122:17–31.
154. Gamper, N., Y. Li, and M. S. Shapiro. 2005. Structural requirements for differential sensitivity of KCNQ K^+ channels to modulation by Ca^{2+}/calmodulin. *Mol Biol Cell* 16:3538–3551.
155. Zaika, O., G. P. Tolstykh, D. B. Jaffe, and M. S. Shapiro. 2007. Inositol triphosphate-mediated Ca^{2+} signals direct purinergic P2Y receptor regulation of neuronal ion channels. *J Neurosci* 27:8914–8926.

156. Xu, T., L. Nie, Y. Zhang, J. Mo, W. Feng, D. Wei, E. Petrov et al. 2007. Roles of alternative splicing in the functional properties of inner ear-specific KCNQ4 channels. *J Biol Chem* 282:23899–23909.
157. Yus-Najera, E., I. Santana-Castro, and A. Villarroel. 2002. The identification and characterization of a non-continuous calmodulin binding site in non-inactivating voltage-dependent KCNQ potassium channels. *J Biol Chem* 24:24.
158. Wen, H. and I. B. Levitan. 2002. Calmodulin is an auxiliary subunit of KCNQ2/3 potassium channels. *J Neurosci* 22:7991–8001.
159. Shamgar, L., L. Ma, N. Schmitt, Y. Haitin, A. Peretz, R. Wiener, J. Hirsch, O. Pongs, and B. Attali. 2006. Calmodulin is essential for cardiac IKS channel gating and assembly. Impaired function in long-QT mutations. *Circ Res* 98(8):1055–1063.
160. Ghosh, S., D. A. Nunziato, and G. S. Pitt. 2006. KCNQ1 assembly and function is blocked by long-QT syndrome mutations that disrupt interaction with calmodulin. *Circ Res* 98:1048–1054.
161. Logothetis, D. E., D. Lupyan, and A. Rosenhouse-Dantsker. 2007. Diverse Kir modulators act in close proximity to residues implicated in phosphoinositide binding. *J Physiol* 582:953–965.
162. Liu, B., J. E. Linley, X. Du, X. Zhang, L. Ooi, H. Zhang, and N. Gamper. 2010. The acute nociceptive signals induced by bradykinin in rat sensory neurons are mediated by inhibition of M-type K^+ channels and activation of Ca^{2+}-activated Cl^- channels. *J Clin Invest* 120:1240–1252.
163. Linley, J. E., K. Rose, M. Patil, B. Robertson, A. N. Akopian, and N. Gamper. 2008. Inhibition of M current in sensory neurons by exogenous proteases: A signaling pathway mediating inflammatory nociception. *J Neurosci* 28:11240–11249.
164. Crozier, R. A., S. K. Ajit, E. J. Kaftan, and M. H. Pausch. 2007. MrgD activation inhibits KCNQ/M-currents and contributes to enhanced neuronal excitability. *J Neurosci* 27:4492–4496.
165. Hawryluk, J. M., T. S. Moreira, A. C. Takakura, I. C. Wenker, A. V. Tzingounis, and D. K. Mulkey. 2012. KCNQ channels determine serotonergic modulation of ventral surface chemoreceptors and respiratory drive. *J Neurosci* 32:16943–16952.
166. Roepke, T. A., A. W. Smith, O. K. Ronnekleiv, and M. J. Kelly. 2012. Serotonin 5-HT2C receptor-mediated inhibition of the M-current in hypothalamic POMC neurons. *Am J Physiol Endocrinol Metab* 302:E1399–E1406.
167. Shi, L., X. Bian, Z. Qu, Z. Ma, Y. Zhou, K. Wang, H. Jiang, and J. Xie. 2013. Peptide hormone ghrelin enhances neuronal excitability by inhibition of Kv7/KCNQ channels. *Nat Commun* 4:1435.
168. Filippov, A. K., R. C. Choi, J. Simon, E. A. Barnard, and D. A. Brown. 2006. Activation of P2Y1 nucleotide receptors induces inhibition of the M-type K^+ current in rat hippocampal pyramidal neurons. *J Neurosci* 26:9340–9348.
169. Halliwell, J. V. and P. R. Adams. 1982. Voltage-clamp analysis of muscarinic excitation in hippocampal neurons. *Brain Res* 250:71–92.
170. Lawrence, J. J., J. M. Statland, Z. M. Grinspan, and C. J. McBain. 2006. Cell type-specific dependence of muscarinic signalling in mouse hippocampal stratum oriens interneurones. *J Physiol* 570:595–610.
171. Shen, W., S. E. Hamilton, N. M. Nathanson, and D. J. Surmeier. 2005. Cholinergic suppression of KCNQ channel currents enhances excitability of striatal medium spiny neurons. *J Neurosci* 25:7449–7458.
172. Etzioni, A., S. Siloni, D. Chikvashvilli, R. Strulovich, D. Sachyani, N. Regev, D. Greitzer-Antes, J. A. Hirsch, and I. Lotan. 2011. Regulation of neuronal M-channel gating in an isoform-specific manner: Functional interplay between calmodulin and syntaxin 1A. *J Neurosci* 31:14158–14171.
173. Xu, Q., A. Chang, A. Tolia, and D. L. Minor, Jr. 2013. Structure of a Ca(2+)/CaM:Kv7.4 (KCNQ4) B-helix complex provides insight into M current modulation. *J Mol Biol* 425:378–394.
174. Xia, X. M., B. Fakler, A. Rivard, G. Wayman, T. Johnson-Pais, J. E. Keen, T. Ishii et al. 1998. Mechanism of calcium gating in small-conductance calcium-activated potassium channels. *Nature* 395:503–507.
175. Keen, J. E., R. Khawaled, D. L. Farrens, T. Neelands, A. Rivard, C. T. Bond, A. Janowsky, B. Fakler, J. P. Adelman, and J. Maylie. 1999. Domains responsible for constitutive and Ca(2+)-dependent interactions between calmodulin and small conductance Ca(2+)-activated potassium channels. *J Neurosci* 19:8830–8838.
176. Kwon, Y., T. Hofmann, and C. Montell. 2007. Integration of phosphoinositide- and calmodulin-mediated regulation of TRPC6. *Mol Cell* 25:491–503.
177. Cao, C., E. Zakharian, I. Borbiro, and T. Rohacs. 2013. Interplay between calmodulin and phosphatidylinositol 4,5-bisphosphate in Ca^{2+}-induced inactivation of transient receptor potential vanilloid 6 channels. *J Biol Chem* 288:5278–5290.
178. McLaughlin, S., G. Hangyas-Mihalyne, I. Zaitseva, and U. Golebiewska. 2005. Reversible–through calmodulin—Electrostatic interactions between basic residues on proteins and acidic lipids in the plasma membrane. *Biochem Soc Symp* 72:189–198.
179. Kosenko, A., S. Kang, I. M. Smith, D. L. Greene, L. K. Langeberg, J. D. Scott, and N. Hoshi. 2012. Coordinated signal integration at the M-type potassium channel upon muscarinic stimulation. *EMBO J* 31:3147–3156.
180. Etxeberria, A., P. Aivar, J. A. Rodriguez-Alfaro, A. Alaimo, P. Villace, J. C. Gomez-Posada, P. Areso, and A. Villarroel. 2008. Calmodulin regulates the trafficking of KCNQ2 potassium channels. *FASEB J* 22:1135–1143.
181. Alaimo, A., A. Alberdi, C. Gomis-Perez, J. Fernandez-Orth, J. C. Gomez-Posada, P. Areso, and A. Villarroel. 2013. Cooperativity between calmodulin-binding sites in Kv7.2 channels. *J Cell Sci* 126:244–253.
182. Alaimo, A., J. C. Gomez-Posada, P. Aivar, A. Etxeberria, J. A. Rodriguez-Alfaro, P. Areso, and A. Villarroel. 2009. Calmodulin activation limits the rate of KCNQ2 K^+ channel exit from the endoplasmic reticulum. *J Biol Chem* 284:20668–20675.
183. Regev, N., N. Degani-Katzav, A. Korngreen, A. Etzioni, S. Siloni, A. Alaimo, D. Chikvashvili, A. Villarroel, B. Attali, and I. Lotan. 2009. Selective interaction of syntaxin 1A with KCNQ2: Possible implications for specific modulation of presynaptic activity. *PLoS One* 4:e6586.
184. Hoshi, N., J. S. Zhang, M. Omaki, T. Takeuchi, S. Yokoyama, N. Wanaverbecq, L. K. Langeberg et al. 2003. AKAP150 signaling complex promotes suppression of the M-current by muscarinic agonists. *Nat Neurosci* 6:564–571.
185. Hoshi, N., L. K. Langeberg, and J. D. Scott. 2005. Distinct enzyme combinations in AKAP signalling complexes permit functional diversity. *Nat Cell Biol* 7(11):1066–1073.
186. Higashida, H., N. Hoshi, J. S. Zhang, S. Yokoyama, M. Hashii, D. Jin, M. Noda, and J. Robbins. 2005. Protein kinase C bound with A-kinase anchoring protein is involved in muscarinic receptor-activated modulation of M-type KCNQ potassium channels. *Neurosci Res* 51:231–234.
187. Wong, W. and J. D. Scott. 2004. AKAP signalling complexes: Focal points in space and time. *Nat Rev Mol Cell Biol* 5:959–970.
188. Bal, M., J. Zhang, C. C. Hernandez, O. Zaika, and M. S. Shapiro. 2010. Ca^{2+}/calmodulin disrupts AKAP79/150 interactions with KCNQ (M-Type) K^+ channels. *J Neurosci* 30:2311–2323.
189. Zhang, J., M. Bal, S. Bierbower, O. Zaika, and M. S. Shapiro. 2011. AKAP79/150 signal complexes in G-protein modulation of neuronal ion channels. *J Neurosci* 31:7199–7211.

190. Tunquist, B. J., N. Hoshi, E. S. Guire, F. Zhang, K. Mullendorff, L. K. Langeberg, J. Raber, and J. D. Scott. 2008. Loss of AKAP150 perturbs distinct neuronal processes in mice. *Proc Natl Acad Sci USA* 105:12557–12562.
191. Zhang, J. and M. S. Shapiro. 2012. Activity-dependent transcriptional regulation of M-Type (Kv7) K(+) channels by AKAP79/150-mediated NFAT actions. *Neuron* 76:1133–1146.
192. Faux, M. C., E. N. Rollins, A. S. Edwards, L. K. Langeberg, A. C. Newton, and J. D. Scott. 1999. Mechanism of A-kinase-anchoring protein 79 (AKAP79) and protein kinase C interaction. *Biochem J* 343(Pt 2):443–452.
193. Gamper, N., J. D. Stockand, and M. S. Shapiro. 2003. Subunit-specific modulation of KCNQ potassium channels by Src tyrosine kinase. *J Neurosci* 23:84–95.
194. Li, Y., P. Langlais, N. Gamper, F. Liu, and M. S. Shapiro. 2004. Dual phosphorylations underlie modulation of unitary KCNQ K(+) channels by Src tyrosine kinase. *J Biol Chem* 279:45399–45407.
195. Jia, Q., Z. Jia, Z. Zhao, B. Liu, H. Liang, and H. Zhang. 2007. Activation of epidermal growth factor receptor inhibits KCNQ2/3 current through two distinct pathways: Membrane PtdIns(4,5)P2 hydrolysis and channel phosphorylation. *J Neurosci* 27:2503–2512.
196. McCrossan, Z. A. and G. W. Abbott. 2004. The MinK-related peptides. *Neuropharmacology* 47:787–821.
197. Pongs, O. and J. R. Schwarz. 2010. Ancillary subunits associated with voltage-dependent K+ channels. *Physiol Rev* 90:755–796.
198. Barhanin, J., F. Lesage, E. Guillemare, M. Fink, M. Lazdunski, and G. Romey. 1996. K(V)LQT1 and lsK (minK) proteins associate to form the I(Ks) cardiac potassium current. *Nature* 384:78–80.
199. Yang, Y. and F. J. Sigworth. 1998. Single-channel properties of IKs potassium channels. *J Gen Physiol* 112:665–678.
200. Wang, H. S., B. S. Brown, D. McKinnon, and I. S. Cohen. 2000. Molecular basis for differential sensitivity of KCNQ and I(Ks) channels to the cognitive enhancer XE991. *Mol Pharmacol* 57:1218–1223.
201. Busch, A. E., G. L. Busch, E. Ford, H. Suessbrich, H. J. Lang, R. Greger, K. Kunzelmann, B. Attali, and W. Stuhmer. 1997. The role of the IsK protein in the specific pharmacological properties of the IKs channel complex. *Br J Pharmacol* 122:187–189.
202. Wrobel, E., D. Tapken, and G. Seebohm. 2012. The KCNE tango—How KCNE1 interacts with Kv7.1. *Front Pharmacol* 3:142.
203. Wang, K. W. and S. A. Goldstein. 1995. Subunit composition of minK potassium channels. *Neuron* 14:1303–1309.
204. Chen, H., L. A. Kim, S. Rajan, S. Xu, and S. A. Goldstein. 2003. Charybdotoxin binding in the I(Ks) pore demonstrates two MinK subunits in each channel complex. *Neuron* 40:15–23.
205. Kang, C., C. Tian, F. D. Sonnichsen, J. A. Smith, J. Meiler, A. L. George, Jr., C. G. Vanoye, H. J. Kim, and C. R. Sanders. 2008. Structure of KCNE1 and implications for how it modulates the KCNQ1 potassium channel. *Biochemistry* 47:7999–8006.
206. Morin, T. J. and W. R. Kobertz. 2008. Counting membrane-embedded KCNE beta-subunits in functioning K+ channel complexes. *Proc Natl Acad Sci USA* 105:1478–1482.
207. Tian, C., C. G. Vanoye, C. Kang, R. C. Welch, H. J. Kim, A. L. George, Jr., and C. R. Sanders. 2007. Preparation, functional characterization, and NMR studies of human KCNE1, a voltage-gated potassium channel accessory subunit associated with deafness and long QT syndrome. *Biochemistry* 46:11459–11472.
208. Strutz-Seebohm, N., M. Pusch, S. Wolf, R. Stoll, D. Tapken, K. Gerwert, B. Attali, and G. Seebohm. 2011. Structural basis of slow activation gating in the cardiac I Ks channel complex. *Cell Physiol Biochem* 27:443–452.
209. Nakajo, K. and Y. Kubo. 2007. KCNE1 and KCNE3 stabilize and/or slow voltage sensing S4 segment of KCNQ1 channel. *J Gen Physiol* 130:269–281.
210. Rocheleau, J. M. and W. R. Kobertz. 2008. KCNE peptides differently affect voltage sensor equilibrium and equilibration rates in KCNQ1 K+ channels. *J Gen Physiol* 131:59–68.
211. Ruscic, K. J., F. Miceli, C. A. Villalba-Galea, H. Dai, Y. Mishina, F. Bezanilla, and S. A. Goldstein. 2013. IKs channels open slowly because KCNE1 accessory subunits slow the movement of S4 voltage sensors in KCNQ1 pore-forming subunits. *Proc Natl Acad Sci USA*110:E559–E566.
212. Osteen, J. D., C. Gonzalez, K. J. Sampson, V. Iyer, S. Rebolledo, H. P. Larsson, and R. S. Kass. 2010. KCNE1 alters the voltage sensor movements necessary to open the KCNQ1 channel gate. *Proc Natl Acad Sci USA* 107:22710–22715.
213. Kurokawa, J., H. K. Motoike, J. Rao, and R. S. Kass. 2004. Regulatory actions of the A-kinase anchoring protein Yotiao on a heart potassium channel downstream of PKA phosphorylation. *Proc Natl Acad Sci USA* 101:16374–16378.
214. Marx, S. 2003. Ion channel macromolecular complexes in the heart. *J Mol Cell Cardiol* 35:37–44.
215. Marx, S. O., J. Kurokawa, S. Reiken, H. Motoike, J. D'Armiento, A. R. Marks, and R. S. Kass. 2002. Requirement of a macromolecular signaling complex for beta adrenergic receptor modulation of the KCNQ1-KCNE1 potassium channel. *Science* 295:496–499.
216. Nicolas, C. S., K. H. Park, A. El Harchi, J. Camonis, R. S. Kass, D. Escande, J. Merot, G. Loussouarn, F. Le Bouffant, and I. Baro. 2008. IKs response to protein kinase A-dependent KCNQ1 phosphorylation requires direct interaction with microtubules. *Cardiovasc Res* 79:427–435.
217. Haitin, Y. and B. Attali. 2008. The C-terminus of Kv7 channels: A multifunctional module. *J Physiol* 586(7):1803–1810.
218. Bendahhou, S., C. Marionneau, K. Haurogne, M. M. Larroque, R. Derand, V. Szuts, D. Escande, S. Demolombe, and J. Barhanin. 2005. In vitro molecular interactions and distribution of KCNE family with KCNQ1 in the human heart. *Cardiovasc Res* 67:529–538.
219. Tinel, N., S. Diochot, M. Borsotto, M. Lazdunski, and J. Barhanin. 2000. KCNE2 confers background current characteristics to the cardiac KCNQ1 potassium channel. *EMBO J* 19:6326–6330.
220. Schroeder, B. C., S. Waldegger, S. Fehr, M. Bleich, R. Warth, R. Greger, and T. J. Jentsch. 2000. A constitutively open potassium channel formed by KCNQ1 and KCNE3. *Nature* 403:196–199.
221. Roepke, T. K., A. Anantharam, P. Kirchhoff, S. M. Busque, J. B. Young, J. P. Geibel, D. J. Lerner, and G. W. Abbott. 2006. The KCNE2 potassium channel ancillary subunit is essential for gastric acid secretion. *J Biol Chem* 281:23740–23747.
222. Tinel, N., S. Diochot, I. Lauritzen, J. Barhanin, M. Lazdunski, and M. Borsotto. 2000. M-type KCNQ2-KCNQ3 potassium channels are modulated by the KCNE2 subunit. *FEBS Lett* 480:137–141.
223. Strutz-Seebohm, N., G. Seebohm, O. Fedorenko, R. Baltaev, J. Engel, M. Knirsch, and F. Lang. 2006. Functional coassembly of KCNQ4 with KCNE-beta- subunits in Xenopus oocytes. *Cell Physiol Biochem* 18:57–66.
224. Grunnet, M., T. Jespersen, N. MacAulay, N. K. Jorgensen, N. Schmitt, O. Pongs, S. P. Olesen, and D. A. Klaerke. 2003. KCNQ1 channels sense small changes in cell volume. *J Physiol* 549:419–427.
225. Grunnet, M., T. Jespersen, H. B. Rasmussen, T. Ljungstrom, N. K. Jorgensen, S. P. Olesen, and D. A. Klaerke. 2002. KCNE4 is an inhibitory subunit to the KCNQ1 channel. *J Physiol* 542:119–130.
226. Grunnet, M., S. P. Olesen, D. A. Klaerke, and T. Jespersen. 2005. hKCNE4 inhibits the hKCNQ1 potassium current without affecting the activation kinetics. *Biochem Biophys Res Commun* 328:1146–1153.
227. Angelo, K., T. Jespersen, M. Grunnet, M. S. Nielsen, D. A. Klaerke, and S. P. Olesen. 2002. KCNE5 induces time- and voltage-dependent modulation of the KCNQ1 current. *Biophys J* 83:1997–2006.

228. Krumerman, A., X. Gao, J. S. Bian, Y. F. Melman, A. Kagan, and T. V. McDonald. 2004. An LQT mutant minK alters KvLQT1 trafficking. *Am J Physiol Cell Physiol* 286:C1453–C1463.
229. Chandrasekhar, K. D., T. Bas, and W. R. Kobertz. 2006. KCNE1 subunits require co-assembly with K+ channels for efficient trafficking and cell surface expression. *J Biol Chem* 281:40015–40023.
230. Chandrasekhar, K. D., A. Lvov, C. Terrenoire, G. Y. Gao, R. S. Kass, and W. R. Kobertz. 2011. O-glycosylation of the cardiac I(Ks) complex. *J Physiol* 589:3721–3730.
231. Kanda, V. A. and G. W. Abbott. 2012. KCNE regulation of K(+) channel trafficking—A sisyphean task? *Front Physiol* 3:231.
232. Xu, X., V. A. Kanda, E. Choi, G. Panaghie, T. K. Roepke, S. A. Gaeta, D. J. Christini, D. J. Lerner, and G. W. Abbott. 2009. MinK-dependent internalization of the IKs potassium channel. *Cardiovasc Res* 82:430–438.
233. Cooper, E. C. 2011. Made for "anchorin": Kv7.2/7.3 (KCNQ2/KCNQ3) channels and the modulation of neuronal excitability in vertebrate axons. *Semin Cell Dev Biol* 22:185–192.
234. Pan, Z., T. Kao, Z. Horvath, J. Lemos, J. Y. Sul, S. D. Cranstoun, V. Bennett, S. S. Scherer, and E. C. Cooper. 2006. A common ankyrin-G-based mechanism retains KCNQ and NaV channels at electrically active domains of the axon. *J Neurosci* 26:2599–2613.
235. Ekberg, J., F. Schuetz, N. A. Boase, S. J. Conroy, J. Manning, S. Kumar, P. Poronnik, and D. J. Adams. 2007. Regulation of the voltage-gated K(+) channels KCNQ2/3 and KCNQ3/5 by ubiquitination. Novel role for Nedd4-2. *J Biol Chem* 282:12135–12142.
236. Brown, D. A. and G. M. Passmore. 2009. Neural KCNQ (Kv7) channels. *Br J Pharmacol* 156:1185–1195.
237. Shapiro, M. S., Gamper, N. 2009. Regulation of neuronal ion channels by g-protein-coupled receptors in sympathetic neurons. In *Structure, Function and Modulation of Neuronal Voltage-Gated Ion Channels*. V. K. Gribkoff and L. K. Kaczmarek, eds. Wiley, Hoboken, NJ. pp. 291–316.
238. Wang, H. S. and D. McKinnon. 1995. Potassium currents in rat prevertebral and paravertebral sympathetic neurones: Control of firing properties. *J Physiol (Lond)* 485:319–335.
239. Jia, Z., J. Bei, L. Rodat-Despoix, B. Liu, Q. Jia, P. Delmas, and H. Zhang. 2008. NGF inhibits M/KCNQ currents and selectively alters neuronal excitability in subsets of sympathetic neurons depending on their M/KCNQ current background. *J Gen Physiol* 131:575–587.
240. Rose, K., L. Ooi, C. Dalle, B. Robertson, I. C. Wood, and N. Gamper. 2011. Transcriptional repression of the M channel subunit Kv7.2 in chronic nerve injury. *Pain* 152:742–754.
241. King, C. H. and S. S. Scherer. 2012. Kv7.5 is the primary Kv7 subunit expressed in C-fibers. *J Comp Neurol* 520:1940–1950.
242. Passmore, G. M., J. M. Reilly, M. Thakur, V. N. Keasberry, S. J. Marsh, A. H. Dickenson, and D. A. Brown. 2012. Functional significance of M-type potassium channels in nociceptive cutaneous sensory endings. *Front Mol Neurosci* 5:63.
243. Heidenreich, M., S. G. Lechner, V. Vardanyan, C. Wetzel, C. W. Cremers, E. M. De Leenheer, G. Aranguez, M. A. Moreno-Pelayo, T. J. Jentsch, and G. R. Lewin. 2012. KCNQ4 K(+) channels tune mechanoreceptors for normal touch sensation in mouse and man. *Nat Neurosci* 15:138–145.
244. Ooi, L., Gigout, S., Pettinger, L., Gamper, N. 2013. Triple cysteine module within M-type K+ channels mediates reciprocal channel modulation by nitric oxide and reactive oxygen species. *J Neurosci* 33(14):6041–6046.
245. Devaux, J. J., K. A. Kleopa, E. C. Cooper, and S. S. Scherer. 2004. KCNQ2 is a nodal K+ channel. *J Neurosci* 24:1236–1244.
246. Roza, C., S. Castillejo, and J. A. Lopez-Garcia. 2011. Accumulation of Kv7.2 channels in putative ectopic transduction zones of mice nerve-end neuromas. *Mol Pain* 7:58.
247. Hirano, K., K. Kuratani, M. Fujiyoshi, N. Tashiro, E. Hayashi, and M. Kinoshita. 2007. K(v)7.2–7.5 voltage-gated potassium channel (KCNQ2-5) opener, retigabine, reduces capsaicin-induced visceral pain in mice. *Neurosci Lett* 413:159–162.
248. Lang, P. M., J. Fleckenstein, G. M. Passmore, D. A. Brown, and P. Grafe. 2008. Retigabine reduces the excitability of unmyelinated peripheral human axons. *Neuropharmacology* 54:1271–1278.
249. Roza, C. and J. A. Lopez-Garcia. 2008. Retigabine, the specific KCNQ channel opener, blocks ectopic discharges in axotomized sensory fibres. *Pain* 138:537–545.
250. Rivera-Arconada, I. and J. A. Lopez-Garcia. 2006. Retigabine-induced population primary afferent hyperpolarisation in vitro. *Neuropharmacology* 51:756–763.
251. Sapunar, D., M. Ljubkovic, P. Lirk, J. B. McCallum, and Q. H. Hogan. 2005. Distinct membrane effects of spinal nerve ligation on injured and adjacent dorsal root ganglion neurons in rats. *Anesthesiology* 103:360–376.
252. Cooper, E. C., E. Harrington, Y. N. Jan, and L. Y. Jan. 2001. M channel KCNQ2 subunits are localized to key sites for control of neuronal network oscillations and synchronization in mouse brain. *J Neurosci* 21:9529–9540.
253. Klinger, F., G. Gould, S. Boehm, and M. S. Shapiro. 2011. Distribution of M-channel subunits KCNQ2 and KCNQ3 in rat hippocampus. *Neuroimage* 58:761–769.
254. Geiger, J., Y. G. Weber, B. Landwehrmeyer, C. Sommer, and H. Lerche. 2006. Immunohistochemical analysis of KCNQ3 potassium channels in mouse brain. *Neurosci Lett* 400:101–104.
255. Kanaumi, T., S. Takashima, H. Iwasaki, M. Itoh, A. Mitsudome, and S. Hirose. 2008. Developmental changes in KCNQ2 and KCNQ3 expression in human brain: Possible contribution to the age-dependent etiology of benign familial neonatal convulsions. *Brain Dev* 30:362–369.
256. Rasmussen, H. B., C. Frokjaer-Jensen, C. S. Jensen, H. S. Jensen, N. K. Jorgensen, H. Misonou, J. S. Trimmer, S. P. Olesen, and N. Schmitt. 2007. Requirement of subunit co-assembly and ankyrin-G for M-channel localization at the axon initial segment. *J Cell Sci* 120:953–963.
257. Shah, M. M., M. Migliore, I. Valencia, E. C. Cooper, and D. A. Brown. 2008. Functional significance of axonal Kv7 channels in hippocampal pyramidal neurons. *Proc Natl Acad Sci USA* 105:7869–7874.
258. Shah, M. M., M. Migliore, and D. A. Brown. 2011. Differential effects of Kv7 (M-) channels on synaptic integration in distinct subcellular compartments of rat hippocampal pyramidal neurons. *J Physiol* 589:6029–6038.
259. Hu, H., K. Vervaeke, and J. F. Storm. 2007. M-channels (Kv7/KCNQ channels) that regulate synaptic integration, excitability, and spike pattern of CA1 pyramidal cells are located in the perisomatic region. *J Neurosci* 27:1853–1867.
260. Lawrence, J. J., Z. M. Grinspan, J. M. Statland, and C. J. McBain. 2006. Muscarinic receptor activation tunes mouse stratum oriens interneurones to amplify spike reliability. *J Physiol* 571:555–562.
261. Lawrence, J. J., F. Saraga, J. F. Churchill, J. M. Statland, K. E. Travis, F. K. Skinner, and C. J. McBain. 2006. Somatodendritic Kv7/KCNQ/M channels control interspike interval in hippocampal interneurons. *J Neurosci* 26:12325–12338.
262. Otto, J. F., Y. Yang, W. N. Frankel, H. S. White, and K. S. Wilcox. 2006. A spontaneous mutation involving Kcnq2 (Kv7.2) reduces M-current density and spike frequency adaptation in mouse CA1 neurons. *J Neurosci* 26:2053–2059.
263. Vervaeke, K., N. Gu, C. Agdestein, H. Hu, and J. F. Storm. 2006. Kv7/KCNQ/M-channels in rat glutamatergic hippocampal axons and their role in regulation of excitability and transmitter release. *J Physiol* 576:235–256.

264. Yue, C., and Y. Yaari. 2004. KCNQ/M channels control spike afterdepolarization and burst generation in hippocampal neurons. *J Neurosci* 24:4614–4624.
265. Gu, N., K. Vervaeke, H. Hu, and J. F. Storm. 2005. Kv7/KCNQ/M and HCN/h, but not KCa2/SK channels, contribute to the somatic medium after-hyperpolarization and excitability control in CA1 hippocampal pyramidal cells. *J Physiol* 566:689–715.
266. Tzingounis, A. V., M. Heidenreich, T. Kharkovets, G. Spitzmaul, H. S. Jensen, R. A. Nicoll, and T. J. Jentsch. 2010. The KCNQ5 potassium channel mediates a component of the afterhyperpolarization current in mouse hippocampus. *Proc Natl Acad Sci USA* 107:10232–10237.
267. Tzingounis, A. V., and R. A. Nicoll. 2008. Contribution of KCNQ2 and KCNQ3 to the medium and slow afterhyperpolarization currents. *Proc Natl Acad Sci USA* 105:19974–19979.
268. Kim, K. S., M. Kobayashi, K. Takamatsu, and A. V. Tzingounis. 2012. Hippocalcin and KCNQ channels contribute to the kinetics of the slow afterhyperpolarization. *Biophys J* 103:2446–2454.
269. Tzingounis, A. V., M. Kobayashi, K. Takamatsu, and R. A. Nicoll. 2007. Hippocalcin gates the calcium activation of the slow afterhyperpolarization in hippocampal pyramidal cells. *Neuron* 53:487–493.
270. Piccinin, S., A. D. Randall, and J. T. Brown. 2006. KCNQ/Kv7 channel regulation of hippocampal gamma-frequency firing in the absence of synaptic transmission. *J Neurophysiol* 95:3105–3112.
271. Leao, R. N., H. M. Tan, and A. Fisahn. 2009. Kv7/KCNQ channels control action potential phasing of pyramidal neurons during hippocampal gamma oscillations in vitro. *J Neurosci* 29:13353–13364.
272. Buzsaki, G. 2002. Theta oscillations in the hippocampus. *Neuron* 33:325–340.
273. Gigout, S., G. A. Jones, S. Wierschke, C. H. Davies, J. M. Watson, and R. A. Deisz. 2012. Distinct muscarinic acetylcholine receptor subtypes mediate pre- and postsynaptic effects in rat neocortex. *BMC Neurosci* 13:42.
274. Luisi, R., E. Panza, V. Barrese, F. A. Iannotti, D. Viggiano, A. Secondo, L. M. Canzoniero, M. Martire, L. Annunziato, and M. Taglialatela. 2009. Activation of pre-synaptic M-type K^+ channels inhibits [3H]D-aspartate release by reducing Ca^{2+} entry through P/Q-type voltage-gated Ca^{2+} channels. *J Neurochem* 109:168–181.
275. Gigout, S., S. Wierschke, T. N. Lehmann, P. Horn, C. Dehnicke, and R. A. Deisz. 2012. Muscarinic acetylcholine receptor-mediated effects in slices from human epileptogenic cortex. *Neuroscience* 223:399–411.
276. Forti, L., E. Cesana, J. Mapelli, and E. D'Angelo. 2006. Ionic mechanisms of autorhythmic firing in rat cerebellar Golgi cells. *J Physiol* 574:711–729.
277. Martire, M., M. D'Amico, E. Panza, F. Miceli, D. Viggiano, F. Lavergata, F. A. Iannotti et al. 2007. Involvement of KCNQ2 subunits in [3H]dopamine release triggered by depolarization and pre-synaptic muscarinic receptor activation from rat striatal synaptosomes. *J Neurochem* 102:179–193.
278. Jensen, M. M., S. C. Lange, M. S. Thomsen, H. H. Hansen, and J. D. Mikkelsen. 2011. The pharmacological effect of positive KCNQ (Kv7) modulators on dopamine release from striatal slices. *Basic Clin Pharmacol Toxicol* 109:339–342.
279. Koyama, S. and S. B. Appel. 2006. Characterization of M-current in ventral tegmental area dopamine neurons. *J Neurophysiol* 96:535–543.
280. Kharkovets, T., K. Dedek, H. Maier, M. Schweizer, D. Khimich, R. Nouvian, V. Vardanyan, R. Leuwer, T. Moser, and T. J. Jentsch. 2006. Mice with altered KCNQ4 K^+ channels implicate sensory outer hair cells in human progressive deafness. *EMBO J* 25:642–652.
281. Hurley, K. M., S. Gaboyard, M. Zhong, S. D. Price, J. R. Wooltorton, A. Lysakowski, and R. A. Eatock. 2006. M-like K^+ currents in type I hair cells and calyx afferent endings of the developing rat utricle. *J Neurosci* 26:10253–10269.
282. Sousa, A. D., L. R. Andrade, F. T. Salles, A. M. Pillai, E. D. Buttermore, M. A. Bhat, and B. Kachar. 2009. The septate junction protein caspr is required for structural support and retention of KCNQ4 at calyceal synapses of vestibular hair cells. *J Neurosci* 29:3103–3108.
283. Spitzmaul, G., L. Tolosa, B. H. Winkelman, M. Heidenreich, M. A. Frens, C. Chabbert, C. I. de Zeeuw, and T. J. Jentsch. 2013. Vestibular role of KCNQ4 and KCNQ5 K^+ channels revealed by mouse models. *J Biol Chem* 288(13):9334–9344.
284. Marres, H., M. van Ewijk, P. Huygen, H. Kunst, G. van Camp, P. Coucke, P. Willems, and C. Cremers. 1997. Inherited nonsyndromic hearing loss. An audiovestibular study in a large family with autosomal dominant progressive hearing loss related to DFNA2. *Arch Otolaryngol Head Neck Surg* 123:573–577.
285. De Leenheer, E. M., P. L. Huygen, P. J. Coucke, R. J. Admiraal, G. van Camp, and C. W. Cremers. 2002. Longitudinal and cross-sectional phenotype analysis in a new, large Dutch DFNA2/KCNQ4 family. *Ann Otol Rhinol Laryngol* 111:267–274.
286. Caminos, E., E. Garcia-Pino, J. R. Martinez-Galan, and J. M. Juiz. 2007. The potassium channel KCNQ5/Kv7.5 is localized in synaptic endings of auditory brainstem nuclei of the rat. *J Comp Neurol* 505:363–378.
287. Goldman, A. M., E. Glasscock, J. Yoo, T. T. Chen, T. L. Klassen, and J. L. Noebels. 2009. Arrhythmia in heart and brain: KCNQ1 mutations link epilepsy and sudden unexplained death. *Sci Transl Med* 1:2ra6.
288. Noble, D. and R. W. Tsien. 1968. The kinetics and rectifier properties of the slow potassium current in cardiac Purkinje fibres. *J Physiol* 195:185–214.
289. Noble, D. and R. W. Tsien. 1969. Outward membrane currents activated in the plateau range of potentials in cardiac Purkinje fibres. *J Physiol* 200:205–231.
290. Trudeau, M. C., J. W. Warmke, B. Ganetzky, and G. A. Robertson. 1995. HERG, a human inward rectifier in the voltage-gated potassium channel family. *Science* 269:92–95.
291. Curran, M. E., I. Splawski, K. W. Timothy, G. M. Vincent, E. D. Green, and M. T. Keating. 1995. A molecular basis for cardiac arrhythmia: HERG mutations cause long QT syndrome. *Cell* 80:795–803.
292. Sanguinetti, M. C., C. Jiang, M. E. Curran, and M. T. Keating. 1995. A mechanistic link between an inherited and an acquired cardiac arrhythmia: HERG encodes the IKr potassium channel. *Cell* 81:299–307.
293. Charpentier, F., J. Merot, G. Loussouarn, and I. Baro. 2010. Delayed rectifier K(+) currents and cardiac repolarization. *J Mol Cell Cardiol* 48:37–44.
294. Jurkiewicz, N. K., and M. C. Sanguinetti. 1993. Rate-dependent prolongation of cardiac action potentials by a methanesulfonanilide class III antiarrhythmic agent. Specific block of rapidly activating delayed rectifier K^+ current by dofetilide. *Circ Res* 72:75–83.
295. Jost, N., L. Virag, M. Bitay, J. Takacs, C. Lengyel, P. Biliczki, Z. Nagy et al. 2005. Restricting excessive cardiac action potential and QT prolongation: A vital role for IKs in human ventricular muscle. *Circulation* 112:1392–1399.
296. Vohra, J. 2007. The long QT syndrome. *Heart Lung Circ* 16(Suppl 3):S5–S12.
297. Takumi, T., H. Ohkubo, and S. Nakanishi. 1988. Cloning of a membrane protein that induces a slow voltage-gated potassium current. *Science* 242:1042–1045.

298. Goldstein, S. A. and C. Miller. 1991. Site-specific mutations in a minimal voltage-dependent K+ channel alter ion selectivity and open-channel block. *Neuron* 7:403–408.
299. Hausdorff, S. F., S. A. Goldstein, E. E. Rushin, and C. Miller. 1991. Functional characterization of a minimal K+ channel expressed from a synthetic gene. *Biochemistry* 30:3341–3346.
300. Vallon, V., F. Grahammer, K. Richter, M. Bleich, F. Lang, J. Barhanin, H. Volkl, and R. Warth. 2001. Role of KCNE1-dependent K+ fluxes in mouse proximal tubule. *J Am Soc Nephrol* 12:2003–2011.
301. Demolombe, S., D. Franco, P. de Boer, S. Kuperschmidt, D. Roden, Y. Pereon, A. Jarry, A. F. Moorman, and D. Escande. 2001. Differential expression of KvLQT1 and its regulator IsK in mouse epithelia. *Am J Physiol Cell Physiol* 280:C359–C372.
302. Dedek, K. and S. Waldegger. 2001. Colocalization of KCNQ1/KCNE channel subunits in the mouse gastrointestinal tract. *Pflugers Arch* 442:896–902.
303. Horikawa, N., T. Suzuki, T. Uchiumi, T. Minamimura, K. Tsukada, N. Takeguchi, and H. Sakai. 2005. Cyclic AMP-dependent Cl-secretion induced by thromboxane A2 in isolated human colon. *J Physiol* 562:885–897.
304. Kunzelmann, K., M. Hubner, R. Schreiber, R. Levy-Holzman, H. Garty, M. Bleich, R. Warth, M. Slavik, T. von Hahn, and R. Greger. 2001. Cloning and function of the rat colonic epithelial K+ channel KVLQT1. *J Membr Biol* 179:155–164.
305. Hamilton, K. L. and D. C. Devor. 2012. Basolateral membrane K+ channels in renal epithelial cells. *Am J Physiol Renal Physiol* 302:F1069–F1081.
306. Soldovieri, M. V., F. Miceli, and M. Taglialatela. 2011. Driving with no brakes: Molecular pathophysiology of Kv7 potassium channels. *Physiology (Bethesda)* 26:365–376.
307. Grahammer, F., R. Warth, J. Barhanin, M. Bleich, and M. J. Hug. 2001. The small conductance K+ channel, KCNQ1: Expression, function, and subunit composition in murine trachea. *J Biol Chem* 276:42268–42275.
308. Grahammer, F., A. W. Herling, H. J. Lang, A. Schmitt-Graff, O. H. Wittekindt, R. Nitschke, M. Bleich, J. Barhanin, and R. Warth. 2001. The cardiac K+ channel KCNQ1 is essential for gastric acid secretion. *Gastroenterology* 120:1363–1371.
309. Heitzmann, D., F. Grahammer, T. von Hahn, A. Schmitt-Graff, E. Romeo, R. Nitschke, U. Gerlach et al. 2004. Heteromeric KCNE2/KCNQ1 potassium channels in the luminal membrane of gastric parietal cells. *J Physiol* 561:547–557.
310. Maljevic, S., T. V. Wuttke, G. Seebohm, and H. Lerche. 2010. KV7 channelopathies. *Pflugers Arch* 460:277–288.
311. Thorneloe, K. S. and M. T. Nelson. 2005. Ion channels in smooth muscle: Regulators of intracellular calcium and contractility. *Can J Physiol Pharmacol* 83:215–242.
312. Hill-Eubanks, D. C., M. E. Werner, T. J. Heppner, and M. T. Nelson. 2011. Calcium signaling in smooth muscle. *Cold Spring Harbor Perspect Biol* 3:a004549.
313. Yeung, S., M. Schwake, V. Pucovsky, and I. Greenwood. 2008. Bimodal effects of the Kv7 channel activator retigabine on vascular K+ currents. *Br J Pharmacol* 155:62–72.
314. Yeung, S. Y., W. Lange, M. Schwake, and I. A. Greenwood. 2008. Expression profile and characterisation of a truncated KCNQ5 splice variant. *Biochem Biophys Res Commun* 371:741–746.
315. Yeung, S. Y., V. Pucovsky, J. D. Moffatt, L. Saldanha, M. Schwake, S. Ohya, and I. A. Greenwood. 2007. Molecular expression and pharmacological identification of a role for K(v)7 channels in murine vascular reactivity. *Br J Pharmacol* 151:758–770.
316. Mani, B. K., L. I. Brueggemann, L. L. Cribbs, and K. L. Byron. 2011. Activation of vascular KCNQ (Kv7) potassium channels reverses spasmogen-induced constrictor responses in rat basilar artery. *Br J Pharmacol* 164:237–249.
317. Ng, F. L., A. J. Davis, T. A. Jepps, M. I. Harhun, S. Y. Yeung, A. Wan, M. Reddy et al. 2011. Expression and function of the K+ channel KCNQ genes in human arteries. *Br J Pharmacol* 162:42–53.
318. Ohya, S., K. Asakura, K. Muraki, M. Watanabe, and Y. Imaizumi. 2002. Molecular and functional characterization of ERG, KCNQ, and KCNE subtypes in rat stomach smooth muscle. *Am J Physiol Gastrointest Liver Physiol* 282:G277–G287.
319. Jepps, T. A., I. A. Greenwood, J. D. Moffatt, K. M. Sanders, and S. Ohya. 2009. Molecular and functional characterization of Kv7 K+ channel in murine gastrointestinal smooth muscles. *Am J Physiol Gastrointest Liver Physiol* 297:G107–G115.
320. McCallum, L. A., I. A. Greenwood, and R. M. Tribe. 2009. Expression and function of K(v)7 channels in murine myometrium throughout oestrous cycle. *Pflugers Arch* 457:1111–1120.
321. Evseev, A., I. Semenov, J. L. Medina, C. R. Archer, P. H. Dube, M. S. Shapiro, and R. Brenner. 2013. Functional effects of KCNQ K+ channel modifiers in airway smooth muscle. *Front Physiol* 4:277.
322. Mackie, A. R. and K. L. Byron. 2008. Cardiovascular KCNQ (Kv7) potassium channels: Physiological regulators and new targets for therapeutic intervention. *Mol Pharmacol* 74:1171–1179.
323. Roura-Ferrer, M., L. Sole, R. Martinez-Marmol, N. Villalonga, and A. Felipe. 2008. Skeletal muscle Kv7 (KCNQ) channels in myoblast differentiation and proliferation. *Biochem Biophys Res Commun* 369:1094–1097.
324. Iannotti, F. A., E. Panza, V. Barrese, D. Viggiano, M. V. Soldovieri, and M. Taglialatela. 2010. Expression, localization, and pharmacological role of Kv7 potassium channels in skeletal muscle proliferation, differentiation, and survival after myotoxic insults. *J Pharmacol Exp Ther* 332:811–820.
325. Su, T. R., W. S. Zei, C. C. Su, G. Hsiao, and M. J. Lin. 2012. The effects of the KCNQ openers retigabine and flupirtine on myotonia in mammalian skeletal muscle induced by a chloride channel blocker. *Evid Based Complement Altern Med* 2012:803082.
326. Yeung, S. Y. and I. A. Greenwood. 2005. Electrophysiological and functional effects of the KCNQ channel blocker XE991 on murine portal vein smooth muscle cells. *Br J Pharmacol* 146:585–595.
327. Greenwood, I. A. and S. Ohya. 2009. New tricks for old dogs: KCNQ expression and role in smooth muscle. *Br J Pharmacol* 156:1196–1203.
328. Wallace, D. J., C. Chen, and P. D. Marley. 2002. Histamine promotes excitability in bovine adrenal chromaffin cells by inhibiting an M-current. *J Physiol* 540:921–939.
329. Sankaranarayanan, S. and S. M. Simasko. 1996. Characterization of an M-like current modulated by thyrotropin-releasing hormone in normal rat lactotrophs. *J Neurosci* 16:1668–1678.
330. Wollmuth, L. P. 1994. Mechanism of Ba^{2+} block of M-like K channels of rod photoreceptors of tiger salamanders. *J Gen Physiol* 103:45–66.
331. Takahira, M. and B. A. Hughes. 1997. Isolated bovine retinal pigment epithelial cells express delayed rectifier type and M-type K+ currents. *Am J Physiol* 273:C790–C803.
332. Reilly, J. M., G. M. Passmore, S. J. Marsh, and D. A. Brown. 2013. Kv7/M-type potassium channels in rat skin keratinocytes. *Pflugers Arch* 465(9):1371–1381.
333. Morita, H., J. Wu, and D. P. Zipes. 2008. The QT syndromes: Long and short. *Lancet* 372:750–763.
334. Giudicessi, J. R. and M. J. Ackerman. 2012. Potassium-channel mutations and cardiac arrhythmias—Diagnosis and therapy. *Nat Rev Cardiol* 9:319–332.
335. Neyroud, N., F. Tesson, I. Denjoy, M. Leibovici, C. Donger, J. Barhanin, S. Faure et al. 1997. A novel mutation in the potassium channel gene KVLQT1 causes the Jervell and Lange-Nielsen cardioauditory syndrome. *Nat Genet* 15:186–189.

336. Neyroud, N., P. Richard, N. Vignier, C. Donger, I. Denjoy, L. Demay, M. Shkolnikova et al. 1999. Genomic organization of the KCNQ1 K+ channel gene and identification of C-terminal mutations in the long-QT syndrome. *Circ Res* 84:290–297.
337. Huang, L., M. Bitner-Glindzicz, L. Tranebjaerg, and A. Tinker. 2001. A spectrum of functional effects for disease causing mutations in the Jervell and Lange-Nielsen syndrome. *Cardiovasc Res* 51:670–680.
338. Piippo, K., H. Swan, M. Pasternack, H. Chapman, K. Paavonen, M. Viitasalo, L. Toivonen, and K. Kontula. 2001. A founder mutation of the potassium channel KCNQ1 in long QT syndrome: Implications for estimation of disease prevalence and molecular diagnostics. *J Am Coll Cardiol* 37:562–568.
339. Wei, J., F. A. Fish, R. J. Myerburg, D. M. Roden, and A. L. George, Jr. 2000. Novel KCNQ1 mutations associated with recessive and dominant congenital long QT syndromes: Evidence for variable hearing phenotype associated with R518X. *Hum Mutat* 15:387–388.
340. Neyroud, N., I. Denjoy, C. Donger, F. Gary, E. Villain, A. Leenhardt, K. Benali, K. Schwartz, P. Coumel, and P. Guicheney. 1998. Heterozygous mutation in the pore of potassium channel gene KvLQT1 causes an apparently normal phenotype in long QT syndrome. *Eur J Hum Genet* 6:129–133.
341. Mohammad-Panah, R., S. Demolombe, N. Neyroud, P. Guicheney, F. Kyndt, M. van den Hoff, I. Baro, and D. Escande. 1999. Mutations in a dominant-negative isoform correlate with phenotype in inherited cardiac arrhythmias. *Am J Hum Genet* 64:1015–1023.
342. Tyson, J., L. Tranebjaerg, S. Bellman, C. Wren, J. F. Taylor, J. Bathen, B. Aslaksen et al. 1997. IsK and KvLQT1: Mutation in either of the two subunits of the slow component of the delayed rectifier potassium channel can cause Jervell and Lange-Nielsen syndrome. *Hum Mol Genet* 6:2179–2185.
343. Chen, Q., D. Zhang, R. L. Gingell, A. J. Moss, C. Napolitano, S. G. Priori, P. J. Schwartz et al. 1999. Homozygous deletion in KVLQT1 associated with Jervell and Lange-Nielsen syndrome. *Circulation* 99:1344–1347.
344. Loussouarn, G., I. Baro, and D. Escande. 2006. KCNQ1 K+ channel-mediated cardiac channelopathies. *Methods Mol Biol* 337:167–183.
345. Marx, S. O. and J. Kurokawa. 2006. AKAPs as antiarrhythmic targets? *Handb Exp Pharmacol* 171:221–233.
346. Chouabe, C., N. Neyroud, P. Guicheney, M. Lazdunski, G. Romey, and J. Barhanin. 1997. Properties of KvLQT1 K+ channel mutations in Romano-Ward and Jervell and Lange-Nielsen inherited cardiac arrhythmias. *EMBO J* 16:5472–5479.
347. Peroz, D., S. Dahimene, I. Baro, G. Loussouarn, and J. Merot. 2009. LQT1-associated mutations increase KCNQ1 proteasomal degradation independently of Derlin-1. *J Biol Chem* 284:5250–5256.
348. Seebohm, G., N. Strutz-Seebohm, R. Birkin, G. Dell, C. Bucci, M. R. Spinosa, R. Baltaev et al. 2007. Regulation of endocytic recycling of KCNQ1/KCNE1 potassium channels. *Circ Res* 100:686–692.
349. Seebohm, G., N. Strutz-Seebohm, O. N. Ureche, U. Henrion, R. Baltaev, A. F. Mack et al. 2008. Long QT syndrome-associated mutations in KCNQ1 and KCNE1 subunits disrupt normal endosomal recycling of IKs channels. *Circ Res* 103:1451–1457.
350. Splawski, I., J. Shen, K. W. Timothy, M. H. Lehmann, S. Priori, J. L. Robinson, A. J. Moss et al. 2000. Spectrum of mutations in long-QT syndrome genes. KVLQT1, HERG, SCN5A, KCNE1, and KCNE2. *Circulation* 102:1178–1185.
351. Tester, D. J., M. L. Will, C. M. Haglund, and M. J. Ackerman. 2005. Compendium of cardiac channel mutations in 541 consecutive unrelated patients referred for long QT syndrome genetic testing. *Heart Rhythm Offic J Heart Rhythm Soc* 2:507–517.
352. Splawski, I., M. Tristani-Firouzi, M. H. Lehmann, M. C. Sanguinetti, and M. T. Keating. 1997. Mutations in the hminK gene cause long QT syndrome and suppress IKs function. *Nat Genet* 17:338–340.
353. Bellocq, C., A. C. van Ginneken, C. R. Bezzina, M. Alders, D. Escande, M. M. Mannens, I. Baro, and A. A. Wilde. 2004. Mutation in the KCNQ1 gene leading to the short QT-interval syndrome. *Circulation* 109:2394–2397.
354. Chen, Y. H., S. J. Xu, S. Bendahhou, X. L. Wang, Y. Wang, W. Y. Xu, H. W. Jin et al. 2003. KCNQ1 gain-of-function mutation in familial atrial fibrillation. *Science* 299:251–254.
355. Hong, K., D. R. Piper, A. Diaz-Valdecantos, J. Brugada, A. Oliva, E. Burashnikov, J. Santos-de-Soto et al. 2005. De novo KCNQ1 mutation responsible for atrial fibrillation and short QT syndrome in utero. *Cardiovasc Res* 68:433–440.
356. Housley, G. D. and J. F. Ashmore. 1992. Ionic currents of outer hair cells isolated from the guinea-pig cochlea. *J Physiol* 448:73–98.
357. Nie, L. 2008. KCNQ4 mutations associated with nonsyndromic progressive sensorineural hearing loss. *Curr Opin Otolaryngol Head Neck Surg* 16:441–444.
358. Beisel, K. W., N. C. Nelson, D. C. Delimont, and B. Fritzsch. 2000. Longitudinal gradients of KCNQ4 expression in spiral ganglion and cochlear hair cells correlate with progressive hearing loss in DFNA2. *Brain Res Mol Brain Res* 82:137–149.
359. Marcotti, W. and C. J. Kros. 1999. Developmental expression of the potassium current IK,n contributes to maturation of mouse outer hair cells. *J Physiol* 520(Pt 3):653–660.
360. Oliver, D., M. Knipper, C. Derst, and B. Fakler. 2003. Resting potential and submembrane calcium concentration of inner hair cells in the isolated mouse cochlea are set by KCNQ-type potassium channels. *J Neurosci* 23:2141–2149.
361. Wong, W. H., K. M. Hurley, and R. A. Eatock. 2004. Differences between the negatively activating potassium conductances of Mammalian cochlear and vestibular hair cells. *J Assoc Res Otolaryngol* 5:270–284.
362. Talebizadeh, Z., P. M. Kelley, J. W. Askew, K. W. Beisel, and S. D. Smith. 1999. Novel mutation in the KCNQ4 gene in a large kindred with dominant progressive hearing loss. *Hum Mutat* 14:493–501.
363. Van Hauwe, P., P. J. Coucke, R. J. Ensink, P. Huygen, C. W. Cremers, and G. Van Camp. 2000. Mutations in the KCNQ4 K+ channel gene, responsible for autosomal dominant hearing loss, cluster in the channel pore region. *Am J Med Genet* 93:184–187.
364. Akita, J., S. Abe, H. Shinkawa, W. J. Kimberling, and S. Usami. 2001. Clinical and genetic features of nonsyndromic autosomal dominant sensorineural hearing loss: KCNQ4 is a gene responsible in Japanese. *J Hum Genet* 46:355–361.
365. Van Camp, G., P. J. Coucke, J. Akita, E. Fransen, S. Abe, E. M. De Leenheer, P. L. Huygen, C. W. Cremers, and S. Usami. 2002. A mutational hot spot in the KCNQ4 gene responsible for autosomal dominant hearing impairment. *Hum Mutat* 20:15–19.
366. Topsakal, V., R. J. Pennings, H. te Brinke, B. Hamel, P. L. Huygen, H. Kremer, and C. W. Cremers. 2005. Phenotype determination guides swift genotyping of a DFNA2/KCNQ4 family with a hot spot mutation (W276S). *Otol Neurotol* 26:52–58.
367. Kamada, F., S. Kure, T. Kudo, Y. Suzuki, T. Oshima, A. Ichinohe, K. Kojima et al. 2006. A novel KCNQ4 one-base deletion in a large pedigree with hearing loss: Implication for the genotype-phenotype correlation. *J Hum Genet* 51:455–460.
368. Mencia, A., D. Gonzalez-Nieto, S. Modamio-Hoybjor, A. Etxeberria, G. Aranguez, N. Salvador, I. Del Castillo et al. 2008. A novel KCNQ4 pore-region mutation (p.G296S) causes deafness by impairing cell-surface channel expression. *Hum Genet* 123:41–53.

369. Su, C. C., J. J. Yang, J. C. Shieh, M. C. Su, and S. Y. Li. 2007. Identification of novel mutations in the KCNQ4 gene of patients with nonsyndromic deafness from Taiwan. *Audiol Neuro-otol* 12:20–26.
370. Abdelfatah, N., D. A. McComiskey, L. Doucette, A. Griffin, S. J. Moore, C. Negrijn, K. A. Hodgkinson et al. 2013. Identification of a novel in-frame deletion in KCNQ4 (DFNA2A) and evidence of multiple phenocopies of unknown origin in a family with ADSNHL. *Eur J Hum Genet* 21(10):1112–1119.
371. Kim, H. J., P. Lv, C. R. Sihn, and E. N. Yamoah. 2011. Cellular and molecular mechanisms of autosomal dominant form of progressive hearing loss, *DFNA2. J Biol Chem* 286:1517–1527.
372. Bellini, G., F. Miceli, M. V. Soldovieri, E. Miraglia del Giudice, A. Pascotto, and M. Taglialatela. 2010. Benign familial neonatal seizures. In *Gene Reviews*. R. A. Pagon, T. D. Bird, C. R. Dolan, K. Stephens, and M. P. Adam, eds. University of Washington, Seattle, WA.
373. Soldovieri, M. V., P. Castaldo, L. Iodice, F. Miceli, V. Barrese, G. Bellini, E. Miraglia del Giudice et al. 2006. Decreased subunit stability as a novel mechanism for potassium current impairment by a KCNQ2 C terminus mutation causing benign familial neonatal convulsions. *J Biol Chem* 281:418–428.
374. Richards, M. C., S. E. Heron, H. E. Spendlove, I. E. Scheffer, B. Grinton, S. F. Berkovic, J. C. Mulley, and A. Davy. 2004. Novel mutations in the KCNQ2 gene link epilepsy to a dysfunction of the KCNQ2-calmodulin interaction. *J Med Genet* 41:e35.
375. Borgatti, R., C. Zucca, A. Cavallini, M. Ferrario, C. Panzeri, P. Castaldo, M. V. Soldovieri et al. 2004. A novel mutation in KCNQ2 associated with BFNC, drug resistant epilepsy, and mental retardation. *Neurology* 63:57–65.
376. Castaldo, P., E. M. del Giudice, G. Coppola, A. Pascotto, L. Annunziato, and M. Taglialatela. 2002. Benign familial neonatal convulsions caused by altered gating of KCNQ2/KCNQ3 potassium channels. *J Neurosci* 22:RC199.
377. Wuttke, T. V., J. Penzien, M. Fauler, G. Seebohm, F. Lehmann-Horn, H. Lerche, and K. Jurkat-Rott. 2008. Neutralization of a negative charge in the S1-S2 region of the KV7.2 (KCNQ2) channel affects voltage dependent activation in neonatal epilepsy. *J Physiol* 586:545–555.
378. Soldovieri, M. V., M. R. Cilio, F. Miceli, G. Bellini, E. Miraglia del Giudice, P. Castaldo, C. C. Hernandez et al. 2007. Atypical gating of M-type potassium channels conferred by mutations in uncharged residues in the S4 region of KCNQ2 causing benign familial neonatal convulsions. *J Neurosci* 27:4919–4928.
379. Singh, N. A., P. Westenskow, C. Charlier, C. Pappas, J. Leslie, J. Dillon, T. B. Consortium, V. E. Anderson, M. C. Sanguinetti, and M. F. Leppert. 2003. KCNQ2 and KCNQ3 potassium channel genes in benign familial neonatal convulsions: Expansion of the functional and mutation spectrum. *Brain* 126(Pt 12):2726–2737.
380. Schroeder, B. C., C. Kubisch, V. Stein, and T. J. Jentsch. 1998. Moderate loss of function of cyclic-AMP-modulated KCNQ2/KCNQ3 K^+ channels causes epilepsy. *Nature* 396:687–690.
381. Jentsch, T. J. 2000. Neuronal KCNQ potassium channels: Physiology and role in disease. *Nat Rev Neurosci* 1:21–30.
382. Cooper, E. C. 2012. Potassium channels (including KCNQ) and epilepsy. In *Jasper's Basic Mechanisms of the Epilepsies*. J. L. Noebels, M. Avoli, M. A. Rogawski, R. W. Olsen, and A. V. Delgado-Escueta, eds. National Center for Biotechnology Information, Bethesda, MD.
383. Ben-Ari, Y. 2002. Excitatory actions of gaba during development: The nature of the nurture. *Nat Rev Neurosci* 3:728–739.
384. Safiulina, V. F., P. Zacchi, M. Taglialatela, Y. Yaari, and E. Cherubini. 2008. Low expression of Kv7/M channels facilitates intrinsic and network bursting in the developing rat hippocampus. *J Physiol* 586:5437–5453.
385. Raouf, R., K. Quick, and J. N. Wood. 2010. Pain as a channelopathy. *J Clin Invest* 120:3745–3752.
386. Brown, D. A. and G. M. Passmore. 2010. Some new insights into the molecular mechanisms of pain perception. *J Clin Invest* 120:1380–1383.
387. Petho, G. and P. W. Reeh. 2012. Sensory and signaling mechanisms of bradykinin, eicosanoids, platelet-activating factor, and nitric oxide in peripheral nociceptors. *Physiol Rev* 92:1699–1775.
388. Hansen, H. H., C. Ebbesen, C. Mathiesen, P. Weikop, L. C. Ronn, O. Waroux, J. Scuvee-Moreau, V. Seutin, and J. D. Mikkelsen. 2006. The KCNQ channel opener retigabine inhibits the activity of mesencephalic dopaminergic systems of the rat. *J Pharmacol Exp Ther* 318:1006–1019.
389. Zheng, Q., D. Fang, M. Liu, J. Cai, Y. Wan, J. S. Han, and G. G. Xing. 2013. Suppression of KCNQ/M (Kv7) potassium channels in dorsal root ganglion neurons contributes to the development of bone cancer pain in a rat model. *Pain* 154:434–448.
390. Du, X. and N. Gamper. 2013. Potassium channels in peripheral pain pathways: Expression, function and therapeutic potential. *Curr Neuropharmacol* 11(6):621–640.
391. Weisenberg, J. L. and M. Wong. 2011. Profile of ezogabine (retigabine) and its potential as an adjunctive treatment for patients with partial-onset seizures. *Neuropsychiatr Dis Treat* 7:409–414.
392. Maljevic, S. and H. Lerche. 2012. Potassium channels: A review of broadening therapeutic possibilities for neurological diseases. *J Neurol* 260(9):2201–2211.
393. Stafstrom, C. E., S. Grippon, and P. Kirkpatrick. 2011. Ezogabine (retigabine). *Nat Rev Drug Discov* 10:729–730.
394. Zhang, D., R. Thimmapaya, X. F. Zhang, D. J. Anderson, J. L. Baranowski, M. Scanio, A. Perez-Medrano et al. 2011. KCNQ2/3 openers show differential selectivity and site of action across multiple KCNQ channels. *J Neurosci Methods* 200:54–62.
395. Hewawasam, P., W. Fan, D. A. Cook, K. S. Newberry, C. G. Boissard, V. K. Gribkoff, J. Starrett, and N. J. Lodge. 2004. 4-Aryl-3-(mercapto)quinolin-2-ones: Novel maxi-K channel opening relaxants of corporal smooth muscle. *Bioorg Med Chem Lett* 14:4479–4482.
396. L'Heureux, A., A. Martel, H. He, J. Chen, L. Q. Sun, J. E. Starrett, J. Natale et al. 2005. (S,E)-N-[1-(3-heteroarylphenyl) ethyl]-3-(2-fluorophenyl)acrylamides: Synthesis and KCNQ2 potassium channel opener activity. *Bioorg Med Chem Lett* 15:363–366.
397. Wu, Y. J., H. He, L. Q. Sun, A. L'Heureux, J. Chen, P. Dextraze, J. E. Starrett, Jr. et al. 2004. Synthesis and structure-activity relationship of acrylamides as KCNQ2 potassium channel openers. *J Med Chem* 47:2887–2896.
398. Dupuis, D. S., R. L. Schroder, T. Jespersen, J. K. Christensen, P. Christophersen, B. S. Jensen, and S. P. Olesen. 2002. Activation of KCNQ5 channels stably expressed in HEK293 cells by BMS-204352. *Eur J Pharmacol* 437:129–137.
399. Fritch, P. C., G. McNaughton-Smith, G. S. Amato, J. F. Burns, C. W. Eargle, R. Roeloffs, W. Harrison, L. Jones, and A. D. Wickenden. 2009. Novel KCNQ2/Q3 agonists as potential therapeutics for epilepsy and neuropathic pain. *J Med Chem* 53(2):887–896.
400. Gao, Z., T. Zhang, M. Wu, Q. Xiong, H. Sun, Y. Zhang, L. Zu, W. Wang, and M. Li. 2010. Isoform-specific prolongation of Kv7 (KCNQ) potassium channel opening mediated by new molecular determinants for drug-channel interactions. *J Biol Chem* 285:28322–28332.
401. Mastronardi, P., M. D'Onofrio, E. Scanni, M. Pinto, S. Frontespezi, M. G. Ceccarelli, F. Bianchi, and B. Mazzarella. 1988. Analgesic activity of flupirtine maleate: A controlled double-blind study with diclofenac sodium in orthopaedics. *J Int Med Res* 16:338–348.

402. Gribkoff, V. K. 2008. The therapeutic potential of neuronal K V 7 (KCNQ) channel modulators: An update. *Expert Opin Ther Targets* 12:565–581.
403. Devulder, J. 2010. Flupirtine in pain management: Pharmacological properties and clinical use. *CNS Drugs* 24:867–881.
404. Brueggemann, L. I., A. R. Mackie, B. K. Mani, L. L. Cribbs, and K. L. Byron. 2009. Differential effects of selective cyclooxygenase-2 inhibitors on vascular smooth muscle ion channels may account for differences in cardiovascular risk profiles. *Mol Pharmacol* 76:1053–1061.
405. Du, X. N., X. Zhang, J. L. Qi, H. L. An, J. W. Li, Y. M. Wan, Y. Fu et al. 2011. Characteristics and molecular basis of celecoxib modulation on K(v)7 potassium channels. *Br J Pharmacol* 164:1722–1737.
406. Munro, G., H. K. Erichsen, and N. R. Mirza. 2007. Pharmacological comparison of anticonvulsant drugs in animal models of persistent pain and anxiety. *Neuropharmacology* 53:609–618.
407. Passmore, G. and P. Delmas. 2011. Does cure for pain REST on Kv7 channels? *Pain* 152:709–710.
408. Mucha, M., L. Ooi, J. E. Linley, P. Mordaka, C. Dalle, B. Robertson, N. Gamper, and I. C. Wood. 2010. Transcriptional control of KCNQ channel genes and the regulation of neuronal excitability. *J Neurosci* 30:13235–13245.
409. Iannotti, F. A., V. Barrese, L. Formisano, F. Miceli, and M. Taglialatela. 2012. Specification of skeletal muscle differentiation by repressor element-1 silencing transcription factor (REST)-regulated Kv7.4 potassium channels. *Mol Biol Cell* 24(3):274–284.
410. Ooi, L. and I. C. Wood. 2007. Chromatin crosstalk in development and disease: Lessons from REST. *Nat Rev Genet* 8:544–554.
411. Uchida, H., L. Ma, and H. Ueda. 2010. Epigenetic gene silencing underlies C-fiber dysfunctions in neuropathic pain. *J Neurosci* 30:4806–4814.
412. Oliveria, S. F., M. L. Dell'Acqua, and W. A. Sather. 2007. AKAP79/150 anchoring of calcineurin controls neuronal L-type Ca^{2+} channel activity and nuclear signaling. *Neuron* 55:261–275.
413. Beals, C. R., N. A. Clipstone, S. N. Ho, and G. R. Crabtree. 1997. Nuclear localization of NF-ATc by a calcineurin-dependent, cyclosporin-sensitive intramolecular interaction. *Genes Dev* 11:824–834.
414. Tomida, T., K. Hirose, A. Takizawa, F. Shibasaki, and M. Iino. 2003. NFAT functions as a working memory of Ca^{2+} signals in decoding Ca^{2+} oscillation. *EMBO J* 22:3825–3832.
415. Graef, I. A., P. G. Mermelstein, K. Stankunas, J. R. Neilson, K. Deisseroth, R. W. Tsien, and G. R. Crabtree. 1999. L-type calcium channels and GSK-3 regulate the activity of NF-ATc4 in hippocampal neurons. *Nature* 401:703–708.

Ionotropic glutamate receptors

Andrew Plested

Contents

21.1 INTRODUCTION

Nine out of every 10 synapses in the brain are excitatory (Braitenberg and Schuz, 1998). The glutamate receptors that reside in these synapses are cornerstones of fast signaling in the brain and participate in some forms of activity-dependent plasticity thought to underlie learning and memory (Bredt and Nicoll, 2003; Whitlock et al., 2006).

21.2 SUBUNIT DIVERSITY, ALTERNATIVE SPLICING, AND EVOLUTIONARY RELATIONSHIPS

The varied responses of neurons to excitatory amino acids and their derivatives suggested that multiple receptors for these agents were found in brain (Watkins and Evans, 1981). This hypothesis was confirmed by molecular cloning (Hollmann and Heinemann, 1994), which delineated the AMPA, *N*-methyl-D-aspartate (NMDA), kainate (KA), and delta subtypes of glutamate receptor ion channels (iGluRs), and a separate family of metabotropic glutamate receptors that act on downstream targets via second messengers. These receptor ion channels are conserved down to worms, and glutamate-receptor-like genes are found in bacteria, plants, and algae. There are 18 mammalian genes. Some can form homomeric, functional channels, but in vivo, glutamate receptors are almost always found as heteromeric complexes. Receptors from each subfamily do not intermix. AMPA receptors, named for their semiselective agonist, alpha-amino-3-hydroxy-5-methylisoxazole-4-proprionic-acid (Honoré et al., 1981), are formed from the subunits designated GluA1-4 (with gene names GRIA1-4). NMDA receptors (NMDARs) are obligate heterotetramers, formed from GluN1 (GRIN 1) and any selection from the GluN2A-D (grin2A-D) or GluN3A-B (grin3A-B) subunits. The kainate receptor subunits GluK1-3 can form channels but in the brain are usually accompanied by GluK4 or K5 subunits (GRIK1-5). The GluD1 and D2 (GRID1,2) (Lomeli et al., 1993) belong to the delta family, which unless mutated as in the case of the Lurcher mouse (Kohda et al., 2000), are not functional ion channels.

In insects, the glutamate receptor template has extraordinarily wide usage. Not only are iGluRs involved in central synaptic transmission (Featherstone et al., 2005), they are also the receptor of the neuromuscular junction (Vesikansa et al., 2012). Further, a highly divergent family of about 60 subunits with the same domain architecture appear to be odorant receptors (Benton et al., 2009). At least some of these genes can generate agonist-activated ion channels with similar pores to AMPA and KA receptors (Abuin et al., 2011). However, it is conceivable that some are not channels, given the signaling roles of delta, NMDA, and kainate subtypes distinct from ion transport (see Section 21.7).

21.2.1 SPLICE VARIATION

In mammalian channels, considerable diversity is generated from alternative splicing (examples are shown in Figure 21.1a). In AMPA receptors (AMPARs), an alternatively spliced cassette called flip-flop (Sommer et al., 1990) controls kinetics and allosteric modulation (Mosbacher et al., 1994). The role of the flip-flop cassette in physiology remains somewhat opaque, but it has been hypothesized to participate in a homeostatic mechanism to ensure high-fidelity transmission (Penn et al., 2012). AMPARs also have long and short C-terminal variants (Köhler et al., 1994). In the case of the NMDA receptor, the fifth exon of GluN1, which is in the amino terminal domain (ATD) (Traynelis et al., 1995), controls allosteric modulation. In the GluN1 subunits, the cytoplasmic domains are assembled from two alternatively spliced exons (Sugihara et al., 1992). Stretches C0 and C1 (exons 21 and 22 respectively) contain motifs that associate with calmodulin, (Ataman et al., 2007) causing strong apparent inhibition of open probability (Ehlers et al., 1996). GluN3A also has a long C-terminal splice variant (Sasaki et al., 2002). The GluK1c isoform of kainate receptors (KARs) has an RXR motif and is retained in endoplasmic reticulum (ER; Jaskolski et al., 2004). Differential expression of this variant during development (Vesikansa et al., 2012) may impact synaptic function.

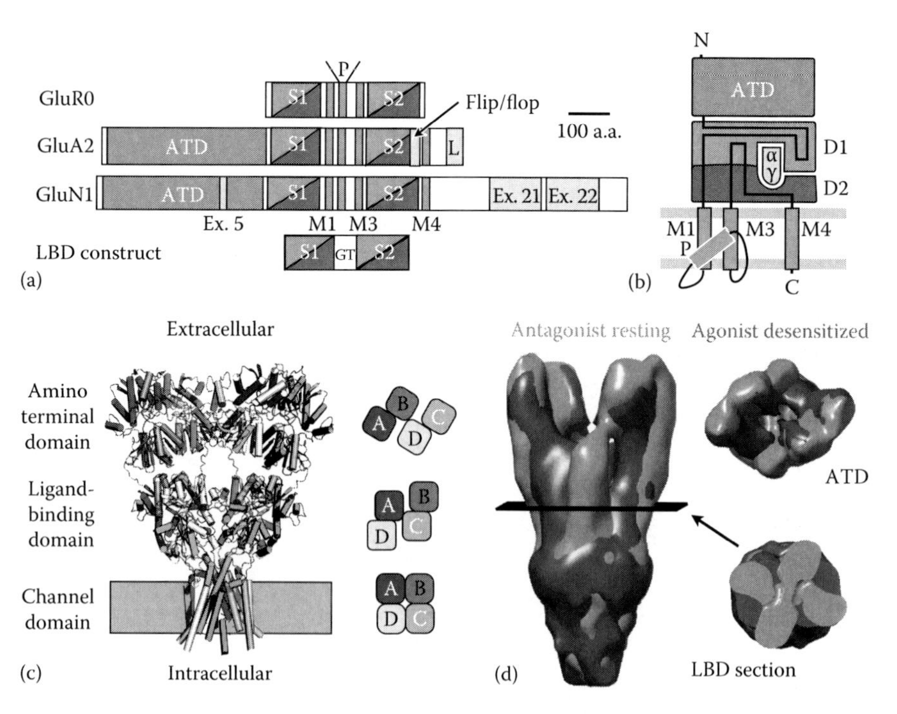

Figure 21.1 Receptor structure. (a) Primary structure of iGluRs, showing the increase in complexity from prokaryotic iGluRs to the mammalian NMDA receptor. Alternatively spliced exons are indicated in yellow. Membrane helical segments (M1, M3, M4) and the pore loop (P, or M2) are gray. The S1 and S2 segments are linked in the artificial isolated ligand binding domain (LBD) construct (used for crystallography and biophysics) by a Gly-Thr linker. ATD: amino terminal domain. L: long form of GluA2 C-terminus. (b) Topology of a single subunit in the membrane (orange bands). Coloring as in panel (a), alternate exons are omitted. D1 and D2: Upper and lower lobes of the LBD, respectively. Glutamate (with carboxyl groups indicated) is in yellow. (c) The crystal structure of GluA2 in the antagonist bound state, with the symmetry of the individual domain layers indicated in plan view. The cell membrane is shown as orange bar. (d) EM tomograms of kainate receptor GluK2 in antagonist (blue surface) and agonist bound states (red). Section through the LBD layer shows the switch to fourfold symmetry in the agonist bound, presumably desensitized, state, whereas the ATD layer is undisturbed.

21.3 STRUCTURE AND ORGANIZATION

At first, the topology of the glutamate receptor was imagined to be like that of the nicotinic receptor, because hydrophobicity plots indicated four membrane segments (M1–4). However, glycosylation analysis mandated that the loop between M3 and M4 was extracellular (Hollmann et al., 1994). This focused attention on ancient binding domains appropriated from prokaryotes (Stern-Bach et al., 1994) formed from two discontinuous polypeptide sequences. The organization of the transmembrane domain (TMD) containing the ion pore, which interrupts the two halves of the clamshell binding domain, is grossly the same as in any tetrameric ion channel, with a reentrant loop (P loop; M2) forming a narrow selectivity filter (Figure 21.1b) (Panchenko et al., 2001), but the orientation of the pore domain is reversed. Most tetrameric channels are gated by intracellular stimuli or voltage sensors through linkages at their intracellular faces, whereas iGluRs are gated from an external site, and with the re-entrant loop at the intracellular mouth of the channel.

Unparalleled insights into iGluRs have accrued over the past 15 years from crystallographic studies. Engineering of a soluble form of the ligand-binding domain (LBD) from rat GluA2 allowed crystallization (Armstrong et al., 1998), revealing the chemistry of neurotransmitter binding (Figure 21.2). A rich harvest of (to date) more than 200 crystal structures of iGluR domains has been amassed, elucidating the basis of neurotransmitter selectivity (Furukawa and Gouaux, 2003; Mayer, 2005; Naur et al., 2007). More recently, the ATDs have been crystallized (Jin et al., 2009; Kumar et al., 2009), opening the door to an understanding of subunit-specific assembly directed by this domain, and in NMDA receptors, allosteric modulation (Karakas et al., 2011).

These domains could be properly placed in the context of the tetrameric receptor for the first time following the solution of the crystal structure of the intact GluA2 tetramer in complex with an antagonist at 3.5 Å (Sobolevsky et al., 2009) (Figure 21.1c). This resting-like structure revealed highly unusual features, including a symmetry break and, thus, subunits with distinct structures. The extracellular domains have a delicate, extended arrangement that develops from hierarchical intersubunit contacts of varying

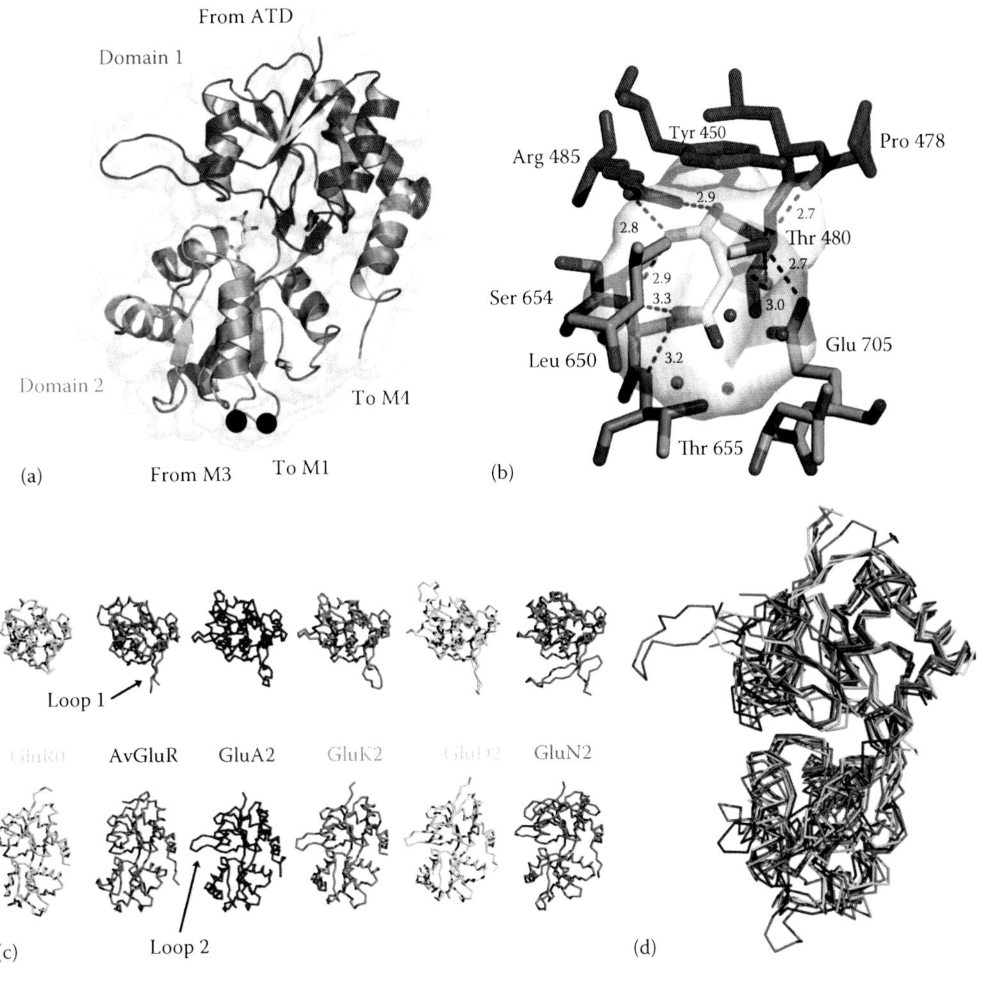

Figure 21.2 The glutamate-binding domain. (a) Glutamate (yellow)-bound LBD of GluA2 with upper lobe (purple) and lower lobe (green). Surface of the LBD shown in gray. GT linker site is shown as black spheres. (b) Close-up view of glutamate in the binding cavity between lobes. Distances are in angstroms. Water molecules are shown as red balls. (c) Top (upper row) and side views (lower row) of LBDs from prokaryotic, plant, and mammalian LBDs. Loop regions (Loop 1 and Loop 2) become increasingly ramified with evolutionary distance. All LBDs are glutamate bound, except for the GluD2, which does not bind glutamate and is shown in the D-serine-bound configuration. The accession numbers in the protein data bank are as follows: GluA2, 1FTJ; GluR0, 1LL5; GluN2, 2A5S; GluD2, 2V3U; GluK2, 3G3F; AvGluR, 4I02. (d) Overlay of the six LBDs shown in panel (c) reveals the remarkable conservation of LBD geometry from bacteria to mammalian channels.

strengths. In this structure, the amino terminal and LBDs are largely identical to their isolated colleagues, and form the same dimer interfaces. These similarities bolster the reliability of using previously determined structures of isolated domains to describe elements of the full-length receptor. However, some work remains to decide whether all features seen in the antagonist-bound, resting-like structure are observed in the wild-type receptor. The construct used to obtain the structure from the Gouaux lab exhibited near-normal function but appeared to be harder to activate. Further structures should reveal whether engineering of the construct to allow crystallization caused any consequent deviations from the native structure.

Reconciling the results of electron microscopic studies with the known structures of the ATD and LBD has not always been easy (Midgett and Madden, 2008; Tichelaar et al., 2004), but reconstitution suggests that the receptors used in these studies were functional (Baranovic et al., 2013). AMPA receptors purified from rat brain exhibited a large disruption of the ATD layer upon desensitization (Nakagawa et al., 2005). More recently, EM tomography of highly purified kainate receptor constructs with comparatively little modification has provided an intriguing comparison between resting and desensitized states (see Section 21.6; Schauder et al., 2013).

The AMPA, kainate, and delta subtypes lack extensive intracellular domains, but NMDA receptors have substantial intracellular C-termini, including sites for binding of calmodulin (Ataman et al., 2007). This region thus shows the greatest variation between subtypes, and probably lacks well-defined secondary structure. Bacterial orthologs lack the amino-terminal domain, the final transmembrane domain, and thus have an extracellular C-terminal (Chen et al., 1999), suggesting a minimal core of the receptor required for gating.

21.4 PHYSIOLOGICAL ROLES AND EXPRESSION PATTERN

The predominant form of AMPA receptor (AMPAR) in the brain contains GluA1 and GluA2 subunits. GluA2 mRNA is found almost uniformly in brain, although not in some glia (Keinänen et al., 1990). GluA3 and GluA4 can substitute GluA1 in calcium-impermeable AMPA receptors, and GluA3 may well associate with GluA1 to form calcium-permeable AMPARs, at least in the absence of GluA2 (Sans et al., 2003). However, the lack of a selective antibody between GluA2 and GluA3 makes unequivocal statements about GluA3 difficult. GluA4 is a *fast* AMPA subunit expressed strongly in cerebellum that appears in auditory neurons after the onset of hearing (Koike-Tani et al., 2005). It is also strongly expressed in Bergmann glia and astrocytes (Martin et al., 1993).

Kainate receptors subunits are somewhat sparsely distributed. GluK1 mRNA is found in various nuclei, hippocampus, and cortex. GluK2 was cloned from cerebellum and is expressed there, in the granule cell layer. It is also found in the hippocampus (Egebjerg et al., 1991). The GluK5 subunit, which only forms heteromeric receptors with GluK1-3, has extraordinarily wide expression in brain (Herb et al., 1992), given the limited extent to which kainate receptor currents are detectable. On the other hand, GluK4 seems to be exclusively present in the CA3 region of the hippocampus (Werner et al., 1991).

While the GluN1 subunit is expressed throughout the brain, the four GluN2 subunits show distinct spatial and developmental profiles (Monyer et al., 1994). GluN2B and GluN2D subunits are present at embryonic stages, whereas GluN2A and GluN2C appear later, the latter having very strong expression in the cerebellum. Antibody staining on wild-type mice indicates that N3A has a broadly similar distribution to N1, N2A, and N2B (Wong et al., 2002). N3A knockout mice have more dendritic spines and have larger NMDA currents than wild-type littermates (Das et al., 1998), but evidence for functional receptors incorporating N3 subunits in vivo is scant at best, partly because of technical difficulties in discerning the various NMDA receptor subtypes.

The delta 1 transcript has a peak during development but is very lowly expressed in adult rodents. Delta 2 has strong expression in the Purkinje cells of cerebellum (Lomeli et al., 1993).

21.4.1 BASAL SYNAPTIC TRANSMISSION

Glutamate receptors participate in synaptic currents that last from milliseconds to seconds (e.g., Logan et al., 2007; Silver et al., 1992) and efficient neural computation depends on this dynamic range (Attwell and Gibb, 2005). Steady-state desensitization of AMPA receptors is profound, even when the receptor is not saturated, whereas NMDA receptors retain activity at almost any glutamate concentration. These distinct biophysical properties have physiological impacts, allowing NMDA receptors to perform tasks that AMPA receptors cannot. Segregation of receptor subtypes on the subcellular level may be a simple compensation for distinct electrotonic and physicochemical neuronal environments. For example, the higher affinity of NMDA receptors for agonist could accommodate larger responses to spillover of glutamate (Scimemi et al., 2004). NMDA spikes and regenerative events occur in basal dendrites (Nevian et al., 2007). There is some evidence for functional segregation of receptors formed from different subunits to synaptic and perisynaptic sites (GluN2A vs. GluN2B; Stocca and Vicini, 1998). However, most of the functional data involves antagonists that are not particularly selective (such as NVP-AAM077 (Liu et al., 2004)) or noncompetitive antagonists that only inhibit at high glutamate concentrations (and may potentiate extrasynaptic receptor currents, because they boost glutamate binding) (Neyton and Paoletti, 2006). Given the abundance of triheteromeric NMDA receptors with distinct properties at hippocampal synapses, these data should be interpreted with caution (Hatton and Paoletti, 2005; Tovar et al., 2013).

Cells spike in the range 0.1–1000 Hz. Tuning of biophysical properties among the iGluR family allows parallel signaling (DeVries, 2000; DeVries and Schwartz, 1999), dendritic computation (Branco et al., 2010), and interplay between the subtypes (Geiger et al., 1995) in this frequency range. Sharpening of signaling with development has been linked to changes in the subunit composition of AMPA receptors (Joshi et al., 2004) and NMDA receptors, where the GluN2A subunit replaces GluN2B (Flint et al., 1997). Activity can drive the synaptic involvement of different subtypes (Liu and Cull-Candy, 2000, 2002) or expression of different splice variants (Penn et al., 2012), which in turn controls the kinetics of postsynaptic currents. Subcellular expression of distinct AMPAR complexes with differing kinetics (and influences on plasticity) has also been detected and may prove to be widespread (Sambandan et al., 2010).

21.4.2 CONTROL OF VESICLE RELEASE

The function of kainate receptors at presynaptic terminals (Chittajallu et al., 1996) has been refined in parallel with the development of more precise antagonists (such as ACET [(S)-1-(2-Amino-2-carboxyethyl)-3-(2-carboxy-5-phenylthiophene-3-yl-methyl)-5-methylpyrimidine-2,4-dione; Dargan et al., 2009]) and with the repeated examination of knockout mice (Breustedt and Schmitz, 2004). NMDA receptors are proposed to participate in presynaptic self-depression (Rodríguez-Moreno et al., 2013), and to enhance gamma-Aminobutyric acid (GABA) release in concert with calcium from intracellular stores in the cerebellum (Duguid and Smart, 2004). The latter idea is controversial because some investigators have failed to detect calcium influx following aspartate application to, for example, basket cell axons, where such NMDA receptors should be (Pugh and Jahr, 2011). In other cases, target cell–specific expression of NMDA receptors in boutons has been reported (Buchanan et al., 2012).

It is worth bearing in mind that "presynaptic effects" are only ultimately mediated through the presynaptic terminal, and don't require glutamate receptors that are localised in the terminal itself. There is little immunocytochemical evidence for presynaptic NMDARs in cortex, for example, although some axons of the hippocampus are stained (Petralia et al., 1994). The calcium required to enhance release may come from calcium channels activated by passive spread of depolarization from somatodendritic NMDA receptors (Christie and Jahr, 2008). Distinct glutamate receptor subtypes may be located throughout presynaptic cells, and thus, these mechanisms may overlap.

At least for hippocampal CA1 pyramidal cells, optical quantal analysis suggests that changes in release probability can account for almost all plasticity (Ahmed et al., 2011). In this context, it is worth noting that the affinity of AMPA receptors for glutamate is low (Patneau and Mayer, 1990), and that the peak of neurotransmitter concentration in the cleft probably hits the middle range of the AMPA receptor concentration response curve (around 1 mM glutamate; Clements et al., 1992). Thus, AMPA receptors might be tuned to respond in a linear way to presynaptic plasticity. At some synapses, receptor saturation is exploited to produce a maximal response (Foster et al., 2002). Recent data suggest that the amount of glutamate released may be optimized to maximize resources (Savtchenko et al., 2013).

Calcium entry through iGluRs can also trigger neurotransmitter release directly. At reciprocal synapses in the retina, glutamate release precipitates immediate GABAergic feedback inhibition. NMDA receptors produce slow feedback (Isaacson and Strowbridge, 1998; Vigh and von Gersdorff, 2005), whereas Ca-permeable AMPA receptors can provide much faster inhibition (Chávez et al., 2006).

21.4.3 SYNAPTOGENESIS

The original observation of silent synapses that lack AMPA receptors stimulated the idea that glutamate receptors play a definite role in synapse maturation (Isaac et al., 1995). Roles in synaptogenesis center on ATD interactions with trans-synaptic signaling systems. The interaction of Purkinje cell GluD2 with a protein secreted from cerebellar granule cells, Cbln1 (Matsuda et al., 2010), is essential for synapse formation. The ATD of GluD2 alone, expressed in nonneuronal cells, is enough to induce synaptogenesis (Panatier et al., 2006), which can support excitatory transmission, and an interaction with neurexin from parallel fibers can dynamically control axon morphology (Ito-Ishida et al., 2012). The ATD of GluA2 promotes spine formation in cultured hippocampal cells (Saglietti et al., 2007), and this activity depends on N-cadherins, which can associate selectively with the GluA2 ATD. In the case of GluA4, an interaction with pentraxins is also proposed to drive synaptogenesis (Sia et al., 2007).

21.4.4 SYNAPTIC PLASTICITY: LONG TERM POTENTIATION AND LONG TERM DEPRESSION

The plasticity of glutamatergic synapses has been repeatedly proposed to have a role in learning and other memory-related behaviors. The cornerstones in this edifice include the absolute NMDA receptor dependence of long term potentiation (LTP) in some pathways of the hippocampus (Collingridge et al., 1983) and the relatively mild impairments of certain forms of learning by the NMDA receptor antagonist, APV ((2*R*)-amino-5-phosphonovaleric acid; Morris, 1989). This hypothesis is highly attractive given that NMDARs are the molecular embodiment of Hebb's principle of coincidence detection in learning and memory (Hebb, 1949), and their high calcium permeability (Mayer and Westbrook, 1987).

Some of the first region-specific knockout mice apparently provided further evidence for the NMDA–hippocampal–LTP–spatial memory storage axis and cemented the role of glutamate receptors as memory molecules (Tsien et al., 1996b). However, Cre expression in these mice was not restricted to the hippocampus (Tsien et al., 1996a), but extended to the cortex in older mice, and probably led to complete loss of NMDA receptors also in cortex. Even in this extreme case, place fields (which were thought to rely on NMDA receptor–driven synaptic plasticity) were intact (McHugh et al., 1996), but less sharp. Genuinely specific Cre-CA3 mice (generated using the Ca^{2+}/calmodulin-dependent protein kinase-CaMKII-promoter) show no defects in spatial memory storage (Bannerman et al., 2012), but do have problems processing spatial information.

CaMKII and protein kinase C (PKC) are needed to induce LTP (Malinow et al., 1989), but not for its maintenance. CaMKII is activated at the level of individual dendritic spines, by binding to NMDA receptors via the GluN1 subunit (Krupp et al., 1999) and remains active for minutes (Grunwald and Kaplan, 2003). The specific activation of CaMKII by the NMDA receptor seems to be sufficient to greatly stabilize recently formed dendritic spines (Hill and Zito, 2013)—a form of Hebbian plasticity at the microanatomical level. Long term depression (LTD) in the cerebellum requires the GluA2 subunit, and phosphorylation of Ser880 on this subunit by PKC (Watanabe et al., 2002). This form of LTD also seems to involve the delta 2 subunit, in a signaling role that is not related to ion flux, but that may be controlled by D-serine (Kakegawa et al., 2011) Tyrosine phosphorylation at a neighboring site (Y876 in A2) is able to block LTD, and for this reason, Ptpmeg (megakaryocyte Protein Tyrosine Phosphatase) trapped by the C-terminus of GluD2, is needed to allow LTD (Kohda et al., 2013).

Dialyzing postsynaptic cells with calcium chelators having different kinetics has given weight to the suggestion that a short, sharp Ca influx produces LTP, whereas longer, weaker signals drive LTD (Sjöström et al., 2008). However, uncaging calcium (Neveu and Zucker, 1996) produced potentiation or depression independent of both the calcium concentration and other parameters. Further, despite the central role of Ca, unblock of the NMDA receptor and subsequent calcium influx is not the only possible plasticity signal. Voltage-gated calcium channel activation is probably part of spike-timing dependent plasticity (Magee and Johnston, 1997), among the most physiologically relevant induction protocol for LTP, and store release of calcium is observed in LTD (Miyata et al., 2000).

21.4.5 SHORT-TERM PLASTICITY

Short-term plasticity refers to intrinsically reversible changes that revert to the basal condition as soon as there is a lull in activity. Potentiation can be accounted for by accumulation of Ca^{2+} in terminals with low release probability. Vesicle depletion is responsible for depression at some synapses, but others seem indefatigable (Saviane and Silver, 2006). Here receptor desensitization underlies synaptic depression, allowing gain control and hence computation (Rothman et al., 2009). AMPA receptor subtypes with stable desensitization are enough to bring about short-term depression (Rozov et al., 2001), and activity dependence of polyamine block may permit short-term facilitation of Ca-permeable AMPARs (Rozov and Burnashev, 1999).

21.4.6 HOMEOSTASIS

Regulation of synaptic strength is achieved via changes in AMPA receptor abundance, perhaps triggered by various neurotrophic factors, (reviewed by Turrigiano, 2008). Given that LTP may purely depend on AMPA receptor abundance in extrasynaptic membranes (Granger et al., 2012), this process could be executed by synthesis and export of more or less receptors to the surface, which would tend to equilibrate by diffusion to distributed synapses. The timescale of hours to days for this kind of synaptic scaling seems consistent with the recovery of synaptic currents in cell culture following photoinactivation (Adesnik et al., 2005).

21.4.7 EXPRESSION OUTSIDE THE BRAIN

Expression of glutamate receptors outside the nervous system has not been a focus of intense investigation. AMPA receptor subtypes are expressed in beta and delta cells in the islets of Langerhans and can regulate somatostatin release (Muroyama et al., 2004). AMPA receptors are targets of CUX1 (cut-like homeobox 1 protein) in pancreatic cancer, and tumor cell survival in vitro can be altered by high concentrations of AMPA-R antagonists (Ripka et al., 2010). There is evidence that NMDA receptors are expressed in the stomach, and gastric epithelial cells may regulate expression (Watanabe et al., 2008) in response to tumors or infection by *Helicobacter pylori* (Sachs et al., 2011; Seo et al., 2011). Details of any agonist-inducted activation of these NMDA receptors are lacking. Kainate receptors are expressed in taste buds (Chaudhari et al., 1996; Chung et al., 2005).

21.5 PORE PROPERTIES (SELECTIVITY, PERMEATION, AND GATE)

Glutamate receptors depolarize cells by allowing sodium entry. AMPA and kainate receptors are nonselective cation channels with similar ion selectivity (Burnashev et al., 1996). Calcium permeation is controlled by mRNA editing of GluA2 (Sommer et al., 1991) and GluK1-3 (Köhler et al., 1993). Deamination of adenosine to imidine places a positively charged arginine at the tip of the pore loop, instead of the genomically encoded glutamine. Edited receptors are no longer blocked by polyamines, and are not permeable to calcium. Receptors that contain a full complement of pore arginines are permeant to fluoride, but their conductance is very small (Burnashev et al., 1996). A lack of editing by the responsible enzyme ADAR2 heightens neurodegeneration (Peng et al., 2006)—perhaps because all Gln-containing AMPA receptors allow uncontrolled calcium entry. Kainate receptors are also edited at sites in TM1 (Köhler et al., 1993). An open question, despite strenuous efforts (Mansour et al., 2001), is to what extent does RNA editing fix the conductance of channels by regulating stoichiometry. Work on AMPA receptor–TARP (transmembrane AMPA receptor–associated protein) fusion proteins in human embyronic kidney (HEK-293) cells suggests that except in extreme conditions, heteromers are 2:2 mixtures (Shi et al., 2009).

NMDA receptors allow glutamate to trigger intracellular calcium signaling in neurons (MacDermott et al., 1986). Most subtypes of NMDA receptors are tonically blocked (and thus silent) at the resting membrane potential. Channel block by Mg^{2+} is relieved at around −40 mV, leading to a region of negative conductance (Mayer et al., 1984; Nowak et al., 1984). Relief of magnesium block is not instantaneous (Kampa et al., 2004), narrowing the window for spike-timing dependent plasticity (Bi and Poo, 1998) but allowing rolling depolarization along dendrites to do nonlinear addition (Branco et al., 2010). Divalent ions also weakly block AMPA and kainate receptors. The N2C and N2D subunits reduce the strength of magnesium block (Kuner and Schoepfer, 1996). The GluN3 subunit also reduces Mg^{2+} block but reduces conductance as well (Chatterton et al., 2002). NMDA receptors also have subunit-dependent Ca^{2+} permeation and conductance properties (Kuner and Schoepfer, 1996), with a single residue (S632 in GluN2A and L657 in GluN2D) exhibiting dominant behavior (Siegler Retchless et al., 2012).

The NMDA pore region should be twofold symmetric, because it is an obligate heterotetramer (Furukawa et al., 2005) with 2 GluN1 and 2 GluN2 subunits diametrically opposed (in the LBD layer; Riou et al., 2012). The AMPA receptor clearly has a fourfold symmetry in the homomeric form (Panatier et al., 2006), and kainate receptor pores are likely to be similar. However, editing of GluA2 and GluK2 subunits to arginine at the tip of the pore loop and the preponderance of native receptors that are heteromeric and incorporate these subunits mean almost all native glutamate receptors will have at least some twofold symmetric sections in their pore regions.

21.6 PHARMACOLOGY AND BLOCKERS

21.6.1 COMPETITIVE ANTAGONISTS

The first separation between NMDA and non-NMDA currents was made possible by the NMDA *glutamate site* antagonist D-APV (Evans et al., 1982) that binds to GluN2. Later, antagonists of the non-NMDA families (AMPA and kainate receptors) including CNQX and NBQX (Honoré et al., 1988; Sheardown et al., 1990) complemented these observations. Further derivatization of quinoxalinedione has not been successful in distinguishing between AMPA and kainate receptors (Turski et al., 1998). Likewise, the binding sites of GluN1 and GluN3 receptors are too similar for strongly selective antagonists. For an exhaustive comparison of the binding of GluN1 vs. GluN3 antagonists, see Yao and Mayer (2006). In contrast, highly selective kainate receptor antagonists have been built on the scaffolds of known willardiine ligands (Dargan et al., 2009). This work was driven by the structures of the KAR LBDs that emerged in the middle of the last decade, which allowed rational exploration of the binding site (Chaudhry et al., 2009a; Mayer, 2005).

Depending on the presence or absence of auxiliary subunits, some competitive antagonists of AMPA receptors can be converted into weak partial agonists (Menuz et al., 2007).

21.6.2 NONCOMPETITIVE ANTAGONISTS AND ALLOSTERIC MODULATORS

GYKI 53666 and related drugs bind outside the LBD in the linker region (see Figure 21.3 and Balannik et al., 2005) and achieve selectivity for block of the AMPA receptor in this way. GYKI and its relatives were intensively derivatized, giving rise to the drug perampanel (see Section 21.9 and Krauss et al., 2012). Noncompetitive NMDA receptor antagonists selective for the GluN2C and GluN2D subunits can also bind at the linkers between LBDs and the TMD (e.g., QZN46 [Hansen and Traynelis, 2011]).

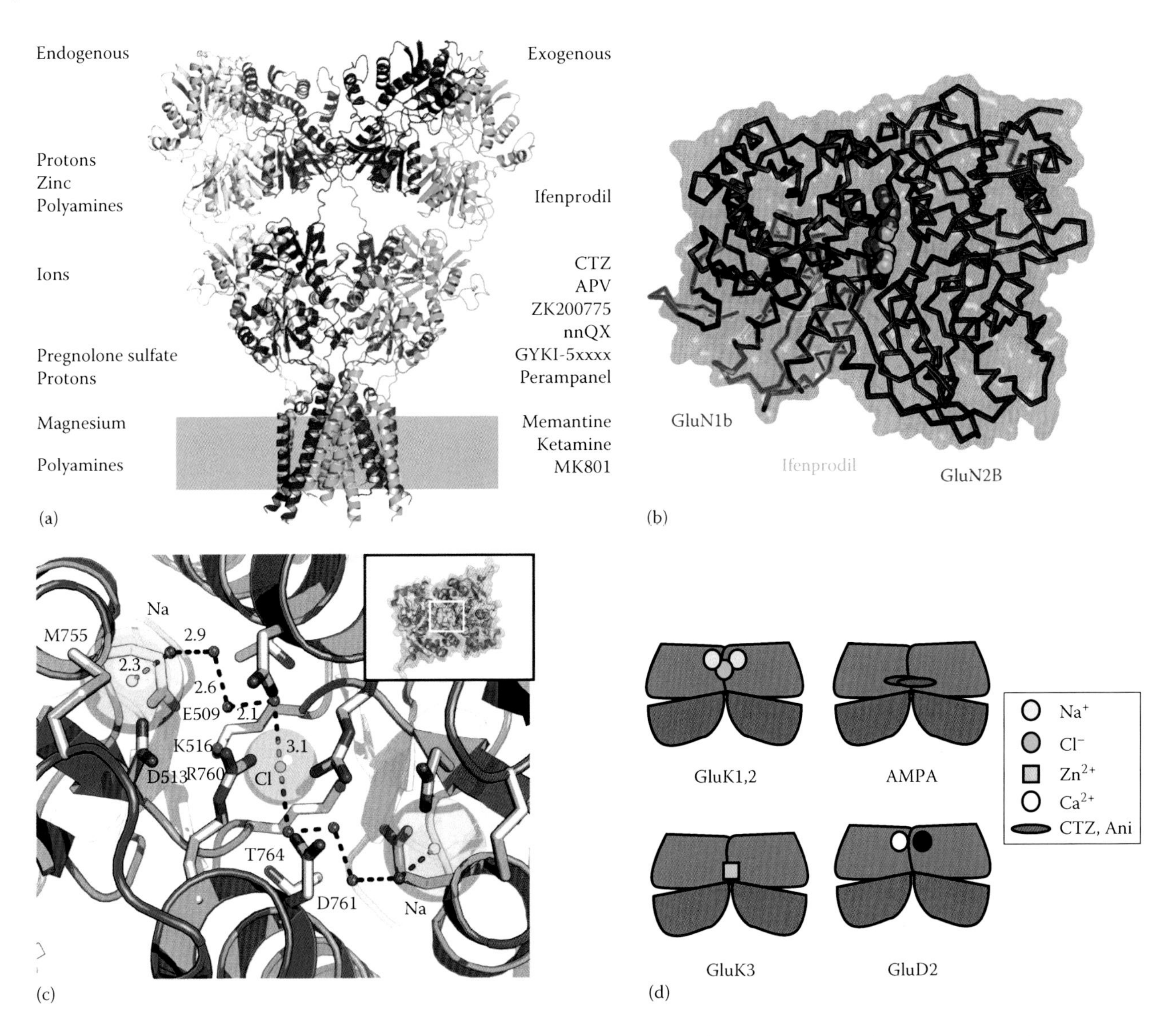

Figure 21.3 Modulator binding sites. (a) Endogenous and exogenous modulators and antagonists, with their approximate binding sites respect to the long axis of the receptor. Bold face indicates therapeutic drugs in clinical use. GYKI-5xxxx indicates the series of ligands that bind in the linker region between LBD and pore domain. nnQX: Quinoxaline dione family of competitive antagonists. CTZ: cyclothiazide. (b) Ifenprodil (green) binds in the NMDA receptor ATD heterodimer (GluN1: blue, GluN2: red). The upper lobes of both subunits and the lower lobe of GluN2 are involved in the binding site. (c) LBD dimers of GluK2 stabilized by adventitious binding of sodium (gold spheres) and chloride (green sphere) between subunits (orange and blue). Distances of hydrogen bonds in the water network (red spheres) are shown in angstroms. Inset shows the interdimer site in the context of one glutamate-bound LBD dimer. (d) Schematic of LBD dimer modulators in the glutamate receptor superfamily. Upper lobes: blue, lower lobes, purple.

The NMDA receptor ATDs contain several modulatory binding sites. The ATDs cannot interact directly with the channel, but exert their effects by changing the quaternary arrangement of the LBDs (Gielen et al., 2008), or by altering agonist affinity (Zhu et al., 2013). Zinc, which is present in synaptic vesicles at a subset of glutamatergic synapses (Tóth, 2011), inhibits GluN2A subunits with high affinity (Paoletti et al., 2000) by stabilizing a closed form of the GluN2A ATD. Ifenprodil very strongly promotes heterodimerization of isolated GluN1 and GluN2B ATDs (Karakas et al., 2011), binding at an interfacial site (Figure 21.3) nearby to that at which polyamines bind and potentiate the receptor (Mony et al., 2011). A selective potentiator of GluN2C and D might act through the ATD-LBD linker (Mullasseril et al., 2010).

AMPA receptors are relatively immune to pH changes, but protons strongly inhibit NMDA receptor gating in the physiological range (Traynelis and Cull-Candy, 1991). This inhibition may occur spontaneously in vivo upon glutamate release, because of the acidity of vesicles (Palmer et al., 2003). There is evidence that the relatively mild proton sensitivity of kainate receptors (and modulation by external polyamines) centers on a site proximal to the ATD-LBD linker (Mott et al., 2003). In contrast, the proton-binding site in NMDARs seems likely to sit between the channel and the LBD (Low et al., 2000).

A toxin from snails called con-ikot-ikot that binds very stably to AMPA receptors (but not kainate or NMDA receptors), and blocks desensitization, is highly damaging to neurons (Walker et al., 2009). These observations mirror the effects of a desensitization-blocking mutation in knock-in mice (Christie et al., 2010). Other toxins from *Conus* snails that lack disulfide bonds have unusual posttranslational modifications and represent a large family that has some selectivity between NMDA receptor subtypes (e.g., Teichert et al., 2007). Intriguingly, some glutamate receptor auxiliary proteins that affect desensitization have forms reminiscent of tethered toxins (von Engelhardt et al., 2010).

Cyclothiazide (CTZ) blocks desensitization by binding at the dimer interface between AMPA receptor subunits, if position 754 is a serine (as it is in flip isoforms; Partin et al., 1995). CTZ also potentiates responses, possibly because it reduces the angle between LBDs in each dimer (Sun et al., 2002), in turn promoting linker separation. Aniracetam and CX-614 block desensitization roughly in proportion to their ability to stabilize the LBD dimer in solution. Consistent with their binding modes at the hinge region of the LBDs (Sobolevsky et al., 2009), they might act by bracing the LBDs into closed conformations and in so doing slow glutamate unbinding and lengthen deactivation decays. NMDA receptor antagonists that bind at or near the dimer interface are also reported (Bettini et al., 2010). Concanavalin A partially prevents desensitization of kainate but not AMPA receptors (Partin et al., 1993). The mechanism of action remains unclear (Fay and Bowie, 2006) but must involve glycans.

Several general anesthetics inhibit glutamate receptors, but at clinically relevant doses, only nitrous oxide (Jevtovic-Todorović et al., 1998) and xenon (de Sousa et al., 2000) inhibit the NMDA receptor in a selective way. AMPA receptors are inhibited by volatile anesthetics, whereas kainate receptors are potentiated, but all effects occur at concentrations far too high to have clinical relevance (Minami et al., 1998; Plested et al., 2004).

21.6.3 PORE BLOCKERS

Phencyclidine and ketamine are NMDA receptor pore blockers (Anis et al., 1983) that show some subunit specificity (Dravid et al., 2007). MK801 is a very stable open channel blocker (Huettner and Bean, 1988) that, in common with 9-aminoacridine (Benveniste and Mayer, 1995), traps glutamate and glycine in their binding sites in a voltage-dependent manner. In contrast, ketamine and memantine allow the channel to close (Mealing et al., 1999), perhaps because they bind at more sites closer to the extracellular face of the channel (Kotermanski et al., 2009), explaining their lower affinity.

Species of AMPA and kainate receptors that do not include pore-edited subunits are blocked by endogenous cytosolic polyamines at positive potentials (Bowie and Mayer, 1995; Kamboj et al., 1995). Generally, activity relieves channel block by endogenous agents, either through depolarization by AMPAR relieving the block of NMDA by Mg, or inward ion flow competing with endogenous polyamines in AMPA or kainate receptors.

A toxin from the Joro spider was first identified as a blocker of crustacean glutamatergic neuromuscular transmission, and soon thereafter as an inhibitor of hippocampal glutamatergic synapses (Abe et al., 1983; Saito et al., 1985). Similar toxins based on polyamines are found in other spider and wasp venoms (Eldefrawi et al., 1988; Herlitze et al., 1993). These polyamine toxins and their derivatives depend on the editing of the pore Q/R site, with Arg-containing pores being spared (Blaschke et al., 1993). They have thus been presumed to isolate receptors containing the GluA2 subunit. This simple picture has been confused to some extent by the variable relief of channel block and changes in polyamine affinity induced by auxiliary proteins (see Section 21.8).

21.6.4 PHOTOACTIVE ANALOGUES OF GLUTAMATE

Several strategies for photochromic ligands, and ligands tethered to iGluRs by photoswitches, have been reported (Stawski et al., 2012). The prototype system, a GluK2 cysteine mutant, was successfully targeted by tethered agonists (Volgraf et al., 2006) and used to control behavior (Szobota et al., 2007). A derivative of the quinoxaline diones containing a photoreactive azido group has been developed (ANQX) (Chambers et al., 2004), and used to photoinactivate receptors in neurons (Adesnik et al., 2005). MNI-glutamate (Canepari et al., 2001) and RuBi glutamate (Fino et al., 2009) are extensively used for uncaging, with excess glutamate rapidly cleared by the basal uptake systems. NCM-D-aspartate (Huang et al., 2005) and other selective agonists with 4-methoxy-7-nitroindolinyl-(MNI) or 1-(2-nitrophenyl) ethoxycarbonyl (NPEC) cages are becoming available (Palma-Cerda et al., 2012). Multiphoton uncaging of such selective ligands may help to identify family-specific subcellular roles of iGluRs.

21.7 GATING MECHANISMS, AGONIST SELECTIVITY, AND SUBUNIT CONTRIBUTIONS AND INTERACTIONS

The gamma carboxylate is stabilized against domain 2 via waters and backbone oxygens (Armstrong and Gouaux, 2000). Glutamate analogues show relatively weak selectivity between receptor subtypes. Somewhat confusingly, AMPA receptors are activated by kainate (Patneau et al., 1993), although the activation is quite weak, because domain closure is limited (Armstrong and Gouaux, 2000) compared to that in GluK1 and GluK2 (Mayer, 2005). Heteromeric kainate receptors and GluK1 are activated by AMPA (Egebjerg et al., 1991). AMPA cannot bind in GluK2 subunit-binding sites because of steric clashes (Mayer, 2005), and so presumably activates heteromers by binding to GluK5. Dysiherbaine is an agonist with extraordinarily high affinity for the GluK1 subunit (Swanson et al., 2002). In GluN2 subunits, the glutamate residue from domain 2 (e.g., Glu705 in GluA2) that coordinates the amino group in AMPA and kainate-binding pockets is instead an aspartate, which allows NMDA to bind (Furukawa et al., 2005). Selecting between individual subunits within subfamilies with agonists appears almost impossible because the binding sites are so well conserved (see, e.g., Jensen et al., 2007). More ramified ligands to reach into divergent regions pay an entropic penalty and bind more weakly, or simply lose efficacy and eventually turn into antagonists.

The GluN1 subunits bind glycine but not glutamate (Furukawa and Gouaux, 2003), explaining the obligate coactivation of GluN1–GluN2 heteromers with glycine and glutamate. GluN3A and GluN3B subunits bind glycine with a strong preference over glutamate (Yao and Mayer, 2006). GluN1 and N3 subunits also bind D-serine. Ambient D-Ser, and that released by astrocytes (Panatier et al., 2006), rather than glycine, may be the physiological activator of native NMDA receptors (Schell et al., 1995).

21.7.1 LBD CLOSURE AS THE DRIVING FORCE OF CHANNEL ACTIVATION

The now-canonical view is that channel opening is driven by a simple closure of the clamshell binding domains (Armstrong and Gouaux, 2000), arranged in a back-to-back active dimer (Nanao et al., 2005; Sun et al., 2002). A series of cocrystal structures of the GluA2 LBD with willardiine agonists illustrates the general principle (Jin et al., 2003). Larger partial agonists induce less domain closure and are less efficacious. Other manipulations that alter the geometry of the LBD dimer to reduce linker separation also seem to reduce efficacy (e.g., cross-linking the dimer; Weston et al., 2006). These data could be interpreted to indicate partial agonism (at least in AMPA and kainate receptors, see the following text) that is primarily a question of geometry with distinct conformations corresponding to different levels of efficacy.

However, several strands of evidence suggest that the situation is more complicated. Firstly, the LBDs are highly mobile, particularly in the unliganded form (Plested and Mayer, 2009). Second, twisting of the LBD may also contribute to efficacy (Birdsey-Benson et al., 2010). Graded domain closure of GluA2 LBDs was observed in crystals while AMPA was supplanted with 5-bromo-willardiine (Br-Will; Jin and Gouaux, 2003). Average occupancy of the binding site could be tracked by the anomalous diffraction from the willardiine bromine atom and correlated with geometry. However, an individual binding domain can either be occupied by Br-Will or AMPA at a given point in time. Most willardiine agonists are resolved to cause similar domain closure in solution NMR experiments and crystal structures (Maltsev et al., 2008), meaning that crystal packing is not responsible. Slow interconversion between fully and partly closed LBDs bound by glutamate has been inferred from single molecule FRET studies on isolated LBDs (Landes et al., 2011). Further evidence for LBD plasticity comes from domain closure with cysteine at jaw sites and CNQX and kainate, but not DNQX (Ahmed et al., 2011). Taken together, it appears that a range of LBD conformations are available to each ligand, except to very bulky antagonists, and thus, partial agonism is a time-average over occupancy of states with different stabilities. Finally, the structures of NMDA receptor–binding domains in complex with partial and full agonists differ much less for GluN1 (Inanobe et al., 2005), GluN2 (Hansen et al., 2013) or GluN3 (Yao et al., 2008). However, molecular dynamics simulations serve as a reminder that there is a much better correlation between efficacy and the energy of binding, than with the simple parameter of binding domain closure (Lau and Roux, 2011).

21.7.2 KINETICS OF ACTIVATION

The glutamate transient is very brief (~1 ms) at most synapses (Clements et al., 1992), but clearance may be retarded in the case of terminals with specialized morphologies (Barbour et al., 1994). Although glutamate is cleared to a level of about 10 μM within 5 ms, this concentration is still high enough to preferentially desensitize AMPA type glutamate receptors (Colquhoun et al., 1992). However, for the fastest AMPA receptors, a time constant of recovery from desensitization in the range of 5 ms at 32°C (Crowley et al., 2007) allows prolonged high-frequency signaling where required.

Although selective antagonists revealed that the source of the slow component of an excitatory postsynaptic current derives from separate receptor class (the NMDA receptor), the mechanism of the slow current was subject of some debate, until fast perfusion (Clements et al., 1992; Colquhoun et al., 1992) discriminated between the speed of activation and the ligand concentration profile in time. NMDA receptors are activated very slowly even if the glutamate pulse is very short (Wyllie et al., 1998), consistent with multiple steps between ligand binding and gating. The mechanisms of this major difference between members of the superfamily remain unknown. Modal gating has been proposed to contribute to the slower, multicomponent decays of NMDA receptors (Zhang et al., 2008). Receptors that incorporate the GluN2D subunit have extremely long decays (Misra et al., 2000; Wyllie et al., 1998). Some of this effect derives from differences between subunits the hinge region of the LBD (Vance et al., 2011). Fast perfusion also allowed the proper discrimination between partial agonists and full agonists (Patneau et al., 1993), and differences in the stability of the desensitized state to be measured (Bowie and Lange, 2002). The kainate receptor GluK2 is much more stable than any AMPA receptor in the desensitized state, recovering over seconds. The lower lobe of the LBD is responsible for this difference (Carbone and Plested, 2012).

21.7.3 CHANNEL GATING

The upper part of M3 probably forms an activation gate in all members of the superfamily. The SYTANLAAF motif is universally conserved, and mutations in this region give rise to the spontaneously active Lurcher mutant in GluD2 (Kohda et al., 2000). The GluA2 crystal structure has a closed channel, and the constriction is long with three residues (Thr 617; T2 in SYTANLAAF, Ala 621; A6 and Thr 625) zipping up the top of the pore (Sobolevsky et al., 2009). The drastic taper to ~1 Å diameter here is far too narrow for any ion or water molecule to penetrate.

Cadmium labeling of substituted cysteines is suggestive of the A6 position adopting twofold symmetry in open AMPA channels (Sobolevsky et al., 2004), whereas modification was absent for closed channels. In NMDA receptors, the A7 position (Ala 652 in GluN1 and Ala 651 in GluN2B) was convincingly demonstrated to be the narrowest part of the channel in the closed state (Chang and Kuo, 2008). Accessibility of this site tracks channel activation and mutants at this site cannot trap MK-801 (Huettner and Bean, 1988), presumably due to constitutive activation. The motif DRPEER in GluN1, extracellular to the A7 gate and also to the membrane but possibly still part of an extended M3 helix, regulates calcium conductance (Watanabe et al., 2002).

21.7.4 INTERSUBUNIT INTERACTIONS

There is exceptionally strong evidence that stabilization of the back-to-back interaction between the upper lobes of LBDs blocks desensitization in AMPA and kainate receptors (Figures 21.3c and d). In NMDA receptors, the situation is more complicated, partly because the wild-type heterodimers are much more stable. Cross-linking of dimers in AMPA and kainate receptors illustrates that D1 breaking is the first trigger that ends activation before the receptor desensitizes (Weston et al., 2006).

The point mutant Leu to Tyr that was identified from nondesensitizing chimeras (Stern-Bach et al., 1998) maps to this interface. In kainate receptors, the same sites can also slow entry to desensitization (Chaudhry et al., 2009b), but the stable desensitized state of GluK2 masks the effect to some extent. Kinetic control was identified early in the R/G editing site (Lomeli et al., 1994) in AMPA receptors. The same residue is an arginine in kainate receptors and participates in a dimer interfacial chloride ion–binding site, which is flanked by two sodium ions (Plested and Mayer, 2007; Plested et al., 2008). Ca sites that affect GluD2 activation (Wollmuth et al., 2000) are located in the same dimer interface between LBDs (Hansen et al., 2009) and stabilize it (Chaudhry et al., 2009a), as does zinc in GluK3 (Veran et al., 2012).

Wild-type AMPA and kainate receptor LBDs do not strongly associate into dimers in solution (Sun et al., 2002; Weston et al., 2006), but the dimers of ATDs are extremely strong, suggesting that they do not dissociate during gating (Kumar et al., 2009; Zhao et al., 2012). Heterodimers of NMDA receptor ATDs are stabilized by ifenprodil (Karakas et al., 2011), and because the lower lobe is more mobile, these interactions might be much more plastic in general (Figure 21.3b). These dimers assemble with weak interactions, with the ATD dimers sitting in an extended N-shape, whereas the LBD dimers are tilted outward with interdimer contacts between the lower lobes. Perhaps surprisingly, these dimers are formed by different subunit pairs in the ATD and LBD layers—the so-called domain swap. These arrangements are conserved, at least across the kainate (Das et al., 2010) and NMDA receptors. In all cases, it appears that cross-linking can have some effect (Das et al., 2010), consistent with appreciable movements, including a rocking motion of the LBDs into a compact arrangement (see Figure 21.4 and Lau et al., 2013) during activation.

Recent work suggests that the stabilization of the desensitized state might involve novel interfaces in the receptor (Figure 21.1d).

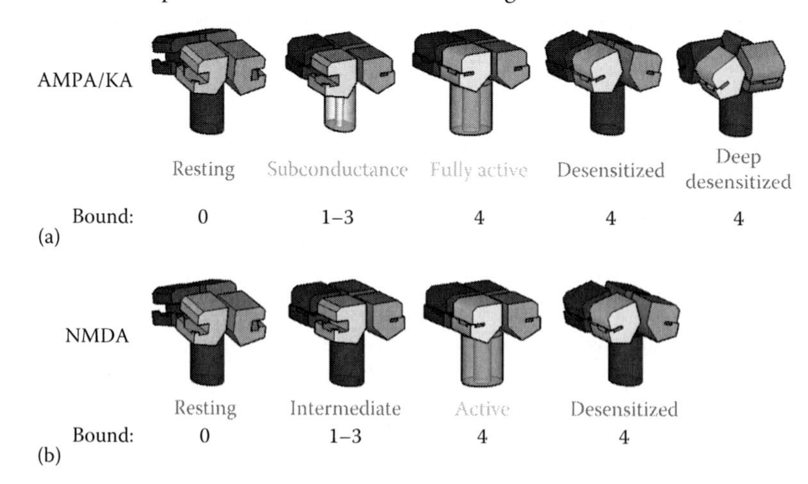

Figure 21.4 Activation gating and conformational change in the LBD layer. (a) Cartoon of AMPA and kainate receptor gating. In the resting state, the pore (cylinder) is closed and the four individual LBDs are in the open cleft conformation. Binding of glutamate induces LBD closure. Subconductance levels occur in partially bound receptors, in which the LBD dimers rock together, adopting a closed-angle arrangement for at least part of the time. The fully bound state has a maximally open pore. Desensitization is initiated by dimer relaxation through breaking of the interface between upper lobes. The LBD arrangement in fully bound states is only known for kainate receptor LBDs in the presence of agonist. This fourfold symmetric state (deep desensitized) matches the symmetry of the pore domain. (b) An analogous scheme, entirely hypothetical, for NMDA receptors, which are more poorly understood in terms of structural dynamics. The partially bound intermediate states of NMDA receptors do not generate conductance. NMDA receptors do not desensitize strongly, and so it is possible that the LBD dimers never dissociate into the fourfold desensitized arrangement of the kainate receptor.

Mutations in the lower lobe of the LBD can exert strong control over recovery rate in both AMPARs and KARs (Carbone and Plested, 2012). In the striking single-particle cryoelectron tomography of GluK2 in an agonist-bound (presumably desensitised) state, ATD dimers persist but the LBDs adopt a fourfold symmetric arrangement, with the lower lobes well packed (Schauder et al., 2013) (Figure 21.4). Although only a ~20 Å map, this novel conformation gains a great deal of credence because the same preparation with the antagonist yielded maps that closely match the published antagonist-bound GluA2 crystal structure. However, desensitized state complexes may differ within the superfamily, and the interfaces between the lower lobes, if they can be defined, will be tested to ensure that this conformation can be reversibly obtained in functional receptors. This rearrangement is likely to also involve the linkers, because point mutations here can reduce the tendency of AMPA and kainate receptors to desensitize (Yelshansky et al., 2004). Indeed, the upper reaches of the channel domain and the linkers have a markedly different configuration in resting and desensitized tomograms of GluK2 (Schauder et al., 2013).

21.7.5 SUBUNIT GATING

For AMPA receptors, it appears that the binding of two glutamate molecules is enough to generate a sublevel opening, and that subsequent binding of glutamate molecules increases conductance and open probability (Rosenmund et al., 1998). This relatively unusual property seems to be shared by cyclic nucleotide gated (CNG) channels (Kusch et al., 2004) but is quite distinct from the glycine and muscle nicotinic receptors, where partially bound openings are to the same amplitude as fully bound ones (Burzomato et al., 2004; Colquhoun and Sakmann, 1985). It is tempting to imagine that the similar architectures of CNG channels and iGluRs (related by inversion with respect to the membrane), with four LBDs driving opening of a transmembrane pore, are permissive of this activity. However, the tetrameric ligand-gated architecture does not mandate sublevel gating. Although sublevels are present in NMDA receptors, they have no relation to agonist concentration. Further, open-time distributions for NMDA receptors are not concentration dependent (Schorge et al., 2005), a strong suggestion that only receptors that are saturated by two glycine and two glutamate molecules are able to open. The basis of this differential coupling is unknown.

The twofold symmetry in the extracellular domains suggests that diametrically opposed subunits may have different influences on gating (Das et al., 2010), with the B and D subunits having greater potential to gate the channel because they sit further from the overall twofold axis. However, squaring this idea with the apparent independence of subunit gating seen in AMPA-KA chimeras (Rosenmund et al., 1998) might be difficult. The high affinity of kainate receptors for glutamate means that they may be tonically active at some synapses (Lauri et al., 2006). Intriguingly, compelling evidence shows that glutamate binding to only the GluK5 subunit can activate heteromeric kainate receptors, but desensitization only occurs at high concentrations that saturate GluK2 (Fisher and Mott, 2011). This functional asymmetry between subunits produces an inverse response to agonist concentration, which would allow receptors to be preferentially activated by low concentrations of spillover glutamate.

21.8 REGULATION (SECOND MESSENGERS, MECHANISMS, AND RELATED PHYSIOLOGY)

Glutamate receptors are subject to a myriad of posttranslational modifications and regulations, and in turn regulate the function of other neurotransmitters and enzymes. Most reports of iGluR modulation by metabotropic glutamate receptors (mGluRs) and muscarinic acetylcholine receptors (mAChRs) relate to kinase signaling (reviewed by Rojas and Dingledine, 2013), but direct effects are also seen (Rossi et al., 1996) with the most compelling being mGluR-dependent short-term plasticity that relies on slower glutamate diffusion and the postsynaptic protein Homer (Sylantyev et al., 2013).

21.8.1 PHOSPHORYLATION

As in many branches of biology, kinases and phosphatases oppose each other in continuous balance, regulating the activity of glutamate receptors (e.g., Src Kinase; Yu and Salter, 1999). In other cases, such as for homosynaptic LTD, persistent phosphatase activation is required (Mulkey et al., 1993) to maintain depression. Dephosphorylation of one intracellular tyrosine inhibits the NMDA receptor and happens independently of ion flux (Vissel et al., 2001). Serial induction of LTP and LTD and vice versa indicates that the biochemical changes underlying plasticity are reversible (Lee et al., 2000), but the changes in the degree of receptor phosphorylation in these experiments were small, mainly because only a subset of synapses underwent plasticity.

GluA1 is phosphorylated at positions 831 and 845 by PKC, PKA, and CaMKII (Mammen et al., 1997). Although control of gating of homomeric receptors in expression systems is appreciable (Banke et al., 2000; Derkach et al., 1999; Kristensen et al., 2011), acute reversal is not reported and unsurprisingly, the effects on native heteromers are less pronounced. Whether these modifications are the causal events in the potentiation of the synaptic response remains an open question. Knock-in mice with ablated phosphorylation sites in the GluA1 C-terminus (S831A S845A double mutant) lack hippocampal NMDA-receptor-dependent LTD but still exhibited (albeit reduced) LTP (Lee et al., 2003). Analysis of the individual mutants revealed no deficits at all in the S831A mutant, but a specific loss of LTD remained in mice with GluA1 S845A knocked in (Lee et al., 2010). Further, the idea that specific phosphorylation of AMPA receptor C-tails is essential for plasticity probably needs revising, because the identity of the C-tail (or even the glutamate receptor subtype) seems unimportant for LTP (Granger et al., 2012). In general, kinases that modify the AMPA receptor C-tails seem also to affect trafficking and localization through effects on auxiliary proteins (Opazo et al., 2010; Sumioka et al., 2010).

21.8.2 OTHER POSTTRANSLATIONAL MODIFICATIONS

Conjugation of Small Ubiquitin-like Modifier (SUMO) to GluK2, coincident with activation (e.g., by chronic application of kainate), drives endocytosis of kainate receptors (Martin et al., 2007). Sumoylation of synaptic proteins (but not necessarily AMPARs themselves) seems to underlie chemically induced LTP (Jaafari et al., 2013). GluN2 subunits have two palmitoylation clusters in their C-termini, indicating that multiple discontinuous parts of the large C-terminus of NMDA receptors are probably associated with the plasma membrane (Hayashi et al., 2009). Palmitoylation is also reported to alter plasma membrane accumulation of homomeric GluK2 receptors (Pickering et al., 1995).

NMDA receptors are sensitive to redox modulation (Aizenman et al., 1989), whereas AMPA and kainate receptors are not (Das et al., 2010; Plested and Mayer, 2009). S-Nitrosylation (due to free NO) of an extracellular site (Cys 399 of GluN1) appears to inhibit the receptor mildly (Choi et al., 2000). However, kinase modulation seems much more potent (Yu and Salter, 1999), because it increases open probability about fivefold (Yu and Salter, 1999).

To what extent do these myriad of modifications impact brain function? Determining the phosphorylation status of individual subunits in a complex is inconceivable with present methods. Such restrictions limit experimental investigation to population changes, which must be associated to hallmark plasticity events and/or behavior. Many studies have deployed short peptide inhibitors based on kinase substrates, either in brain slices or directly in the brain. However, some of these peptides lack specificity. For example, the zeta-inhibitory peptide (Pastalkova et al., 2006) can robustly reverse LTP in wild-type mice and those lacking the kinase, which is its presumed target (PKM-zeta) (Lee et al., 2013; Volk et al., 2013). The coherence of phosphorylation state of AMPA receptors and increased slope of field potentials (both correlates of LTP) with learned memories (Whitlock et al., 2006) is among the best evidence that glutamate receptor biochemistry has a direct relation to behavior. To gain more confidence in these insights, the same phenomenon should be studied at the single-cell (or even single synapse) resolution.

21.8.3 KAINATE RECEPTOR ACTION AT A DISTANCE

Kainate receptors have a *metabotropic function* that does not require current flow. Convincing experiments demonstrating that KARs have an action at a distance on other channels (Negrete-Díaz et al., 2006) are reminiscent of G-protein-coupled receptors. To date there is no mechanism, either from within the kainate receptor or its known partner subunits, by which to activate G-proteins. The GluK2 subunit seems to be essential for this action, but GluK4 and GluK5 subunits are dispensable (but see Fernandes et al., 2009).

Several strands of evidence links exogenously applied kainate to endocannabinoid signaling and alteration of inhibitory transmission (Lourenço et al., 2010, 2011). However, depression of inhibitory transmission can occur in the presence of CB1 blockers, and also if the potential of the presynaptic cell is controlled in paired recordings (Daw et al., 2010). Depolarization of axons or further metabotropic effects are not excluded in most cases, and complicate the situation.

21.9 CELL BIOLOGY (ASSEMBLY, TRAFFICKING, AND ASSOCIATED PROTEINS)

21.9.1 ASSEMBLY

There is some evidence that ATDs assemble into dimers first (Shanks et al., 2010). It may be that these domains fold rapidly (and even associate) while the rest of the receptor remains to be synthesized by the ribosome. The LBD cannot form as a monomer until the majority of the TMD is already synthesized, because of the interdigitated pore region. Mutations that greatly favor LBD dimer association seem to produce a large population of dead-end dimer intermediates (Shanks et al., 2010). The very tight associations of isolated ATDs (Zhao et al., 2012) raise the prospect that exchange into heteromeric forms may be a rate-limiting step in receptor assembly.

GluN1 assembles into stable intracellular dimers but only moves to the cell surface as part of heteromeric complexes (Riou et al., 2012). GluK5 does not go to the surface alone, which is due to a combination of poor matching in the ATD regions (Kumar et al., 2011) and ER retention motifs (Nasu-Nishimura et al., 2006; Vivithanaporn et al., 2006). Instead, a stable association between GluK2 and GluK5 allows this pair of subunits to form the predominant kainate receptor in neurons (Nasu-Nishimura et al., 2006). Pore editing of the GluA2 subunit is implicated in assembly and trafficking (Araki et al., 2010), with R-containing subunits being ER retained (Greger et al., 2002, 2003).

Mutant glutamate receptors can get trapped in the endoplasmic reticulum in worms (Grunwald and Kaplan, 2003), apparently connecting functional properties to trafficking. In the mammalian context, forward trafficking of kainate receptors (Mah et al., 2005), and later AMPA (Penn et al., 2008) and NMDA receptors (Kenny et al., 2009; She et al., 2012) can all be inhibited by mutants that disrupt binding. This general observation has evolved into the idea that a competent binding domain might be needed for proper surface expression of iGluRs. Intact desensitization has been proposed as a further quality control (Priel et al., 2006), but this apparent correlation may instead be a more general property of cysteine mutants, which generally express poorly (Das et al., 2010). However, it makes sense that manipulations that artificially link subunits can restrict expression, given the complex architecture of iGluRs. Whether receptors actually gate at intracellular sites remains an open question.

GluN1 and GluN3 can combine to make *excitatory glycine receptors* in oocytes and perhaps in neurons as well (Chatterton et al., 2002). Intriguingly, in mammalian cells, paired expression of GluN1 with GluN3A or GluN3B initially failed to produce currents, a riddle that was solved when GluN1-GluN3A-GluN3B receptors were robustly expressed in HEK cells (Smothers and Woodward, 2007). Subsequent work illustrated that the GluN1 splice variant critically controls current expression—the splice variants with the longer C-tails give the most current in single combinations with GluN3A or GluN3B—suggesting a link to impaired desensitization (Smothers and Woodward, 2009).

A bewildering complement of intracellular proteins act as adaptors or matchmakers between glutamate receptors and kinases and cytoskeletal elements (Henley et al., 2011). Most adaptors are suggested to determine the sign and the magnitude of plasticity, linking AMPA receptor abundance to stabilization and biochemical transformation at the hands of kinases and other modifiers. In some cases, it is difficult to ascertain precisely how these proteins associate with receptors. A good example is the protein interacting with C-kinase 1 (PICK1). Interactions between GluA2 and PICK1 seem important for LTD (Steinberg et al., 2006), but the involvement of PICK1 in LTP and LTD is controversial (Terashima et al., 2008). A further complication is that all trafficking that is related to plasticity is not created equal—intracellular pooling of AMPA receptors due to NMDA and mGluR activation and may involve the binding of different proteins. Elimination of PICK1 seems to destabilize intracellular stocks of AMPA receptors (Citri et al., 2010), but the timescales of trafficking (~15 min), combined with the lack of tools to attack intracellular proteins acutely, leave many unanswered questions. Although it is clear that adaptors such as NSF (N-ethylmaleimide-sensitive factor; Nishimune et al., 1998), GRIP1 (Glutamate Receptor Interacting Protein), and GRIP2 (also called ABP) (Takamiya et al., 2008) are involved in synaptic plasticity, the complexity of the cascades leaves many details in question.

Mass spectrometry of isolated, intact iGluR complexes from native tissue has proved invaluable in expanding the zoo of AMPA receptor auxiliary proteins (Schwenk et al., 2009, 2012; Shanks et al., 2012; von Engelhardt et al., 2010). Other approaches, such as sequence-based approaches, have yielded interactors that control synaptic transmission, for example, SynDIG1 (synapse differentiation-induced gene I; Kalashnikova et al., 2010). SynDIG1 itself has not been found to be an AMPA receptor interactor using mass spectrometry—but its homologs were found (Schwenk et al., 2012; Shanks et al., 2012). Isolation of complexes cannot reveal the functional relevance of particular subunits, or whether glutamate receptor subunits are associated internally or at surface sites or both, but related native gel techniques provide some information about stoichiometry. However, as the complement of auxiliary proteins may vary between cell types (Kim et al., 2010), such information cannot be treated as absolute.

21.9.2 TRANSMEMBRANE AMPA RECEPTOR-ASSOCIATED PROTEINS

On account of having been found first, TARPS, including gamma-2 or Stargazin (Stg), are by far the best investigated of all auxiliary subunits. What do they do? Two central effects on gating are an increase in single channel conductance and an apparent increase in the channel opening rate (Tomita et al., 2005). Later, it was discovered that TARPs reduce channel block by spermine and related ligands (Soto et al., 2007), by lowering the affinity of the receptor for intracellular polyamines. Paradoxically, the affinity for extracellular polyamine toxins increases in a TARP-dependent fashion (Jackson et al., 2011). Good evidence points to dissociation of the TARP–AMPA receptor complex under the action of glutamate (Morimoto-Tomita et al., 2009), which may or may not relate to the differential spatial dynamics of TARPs and AMPA receptors at synapses, as inferred from imaging (Tomita et al., 2004). The timescale of any dissociation is unclear however, and it seems just as likely that different conformational states may underlie the fast relaxations observed in HEK cells (Semenov et al., 2012). An intriguing property related to gating of TARP–AMPA receptor properties is the rebound of steady-state current during long (~1 s) applications of saturating glutamate, termed resensitization. This phenomenon has been seen in expression systems and neurons (Kato et al., 2010), although not in all cases (Herring et al., 2013).

A functional interaction with the ligand-binding core has been proposed (Tomita et al., 2007) but, although plausible, no evidence of a physical contact is available. The only specificity with relation to the AMPA receptor seems to be the flip-flop isoform (Semenov et al., 2012), which is largely buried in the D1 interface, while the receptor is in the active state. Given the wide range of actions of TARPs, multiple receptor–TARP interactions seem likely.

Transmembrane auxiliary subunits of glutamate receptors may hold the key to understanding the selective deployment of glutamate receptor complexes. For example, GluA1 interacts specifically with CNIH2 (cornichon homolog 2; Herring et al., 2013), but this interaction is outcompeted by gamma-8, which is the most abundant TARP in the hippocampus (Rouach et al., 2005). Similarly, the flip-flop region seems to determine some kinetic properties in GluA2-Stg complexes. Long- and short-form AMPA receptors have selective association with the auxiliary subunit gamma-5 (Soto et al., 2009). The spectrum of TARPs with variant functional properties and, presumably, different avidity for synaptic sites, likely, permits diversity in synaptic signaling by members of the AMPAR family.

21.9.3 NETO

Neto1 (Neuropilin- and Tolloid-like protein 1) has profound effects on brain function, which was first proposed to act via the NMDA receptor (Ng et al., 2009). Since this early report, the evidence that Neto1 is a kainate receptor interacting protein has become overwhelming. Both Neto1 and Neto2 profoundly change the biophysical properties of kainate receptors, lengthening activations (Straub et al., 2011; Zhang et al., 2009) and reducing polyamide block (Fisher and Mott, 2012), but to date, evidence for altered NMDA receptor function due to Neto1 is lacking. Neto1 may associate with NMDA receptors in the presence of the amyloid precursor protein, through interactions with the GluN2 subunit (Cousins et al., 2013). Neto2 seems to be promiscuous, also interacting with ion transporters (Ivakine et al., 2013).

The presence of Neto is enough to drive kainate receptor accumulation at synapses (Copits et al., 2011), suggesting Netos may be expression chaperones. Neto1 has a PDZ ligand and therefore seems equally likely to act as a synaptic anchor for kainate receptors. The hallmark summation of serial responses by kainate receptors is due at least in part to Neto1–kainate receptor complexes (Straub et al., 2011).

21.9.4 SYNAPTIC TRAPPING AND AUXILIARY PROTEINS

All membrane proteins that are not attached to cytomatrix diffuse laterally within plasma membranes (Figure 21.5). The relation between the physiology of excitatory synaptic transmission and this obligatory physical process has been extensively studied over the past decade (Borgdorff and Choquet, 2002). AMPA receptors reach synapses by diffusing laterally, and to some extent are trapped by the necks of dendritic spines (Ashby et al., 2006). Single particle tracking indicates that stargazin and AMPA receptor diffusion is correlated, suggesting complexation both within and outside synapses (Bats et al., 2007). However, subcellular exchange between different auxiliary partners is not yet excluded. It is conceivable that some glutamate receptors in the brain lack TARPs, because they are not an absolute requirement for synaptic clustering (Bats et al., 2012).

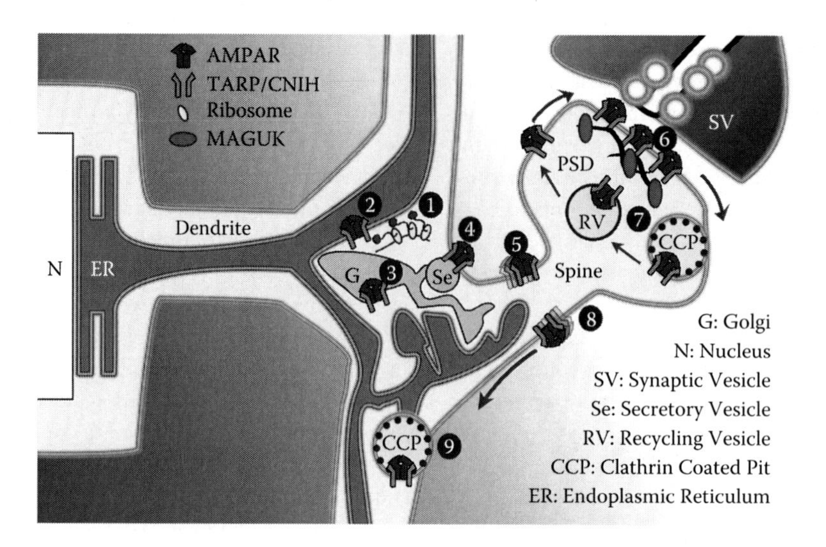

Figure 21.5 The glutamate receptor trafficking cycle. (1) Glutamate receptors (AMPAR) are synthesized from mRNA at or near to dendritic branch points. They pass through highly complex reticular structures near to spines, assembling with auxiliary subunits (2). Their glycans are pruned within Golgi outposts (3), and complexes are secreted to the plasma membrane (4). Diffusion through the spine neck (5) leads into a perisynaptic recycling loop. About half of these AMPA receptor complexes are trapped stably in nanodomains within the postsynaptic density (PSD; 6) but the rest diffuse rapidly within the spine and may be recycled by immediately adjacent clathrin pits (7), which are tethered to Homer and possibly other PSD components. Receptors may also escape the spine (8) to be reclaimed by clathrin-coated pits at dendritic or somatic sites (9), entering the endolysosomal pathway for degradation.

The PSD itself is a dense web that is connected to the glutamate receptors at the business end of the postsynaptic membrane synapse by surprisingly sparse interactions. The macromolecular architectures that underlie trapping of glutamate receptors at synapses remain somewhat opaque, due to both their complexity and the dynamic nature of the synapse (reviewed in MacGillavry et al., 2011). Membrane-associated guanylate kinases (MAGUKs) such as PSD95 are in fact not associated with the membrane, and directly bind NMDA receptors through the ramified C-tails of the GluN2 subunits. The binding to N2A and N2B subunits is distinct (Bard et al., 2010) and may contribute to their differential distributions at synapses. PSD95 also directly binds neuroligins, for example, but not AMPA receptors. Instead, the interaction between the PDZ domain of stargazin (and other TARPs) and PSD-95 seems to stabilize receptors at the postsynaptic density (PSD) (Bats et al., 2007). Other auxiliary proteins (such as Neto1) are likely to also interact with PSD components.

21.10 CHANNELOPATHIES AND DISEASE MECHANISMS

Cognitive disorders are associated with GluA3 mutations that have functional effects on desensitization and expression (Wu et al., 2007)—how these relatively minor functional effects play out in vivo is questionable, because most mutations have a dose-dependent effect (Robert et al., 2001) and, where measured, GluA3 seems to take a secondary role to GluA1 (Herring et al., 2013). A mutation in the GluK2 C-terminus has been linked to severe mental retardation (Motazacker et al., 2007). Genetic studies show that GluN2 subunits are associated with developmental defects (Endele et al., 2010). Given the essential roles for these subunits, the mechanism seems most likely to involve incomplete compensation, rather than biophysical alterations to channel function. The strength of these associations could be tested by reverse-engineering of NMDA receptors in mice, similar to what has been done for zinc sensitivity (Nozaki et al., 2011). In a broader sense, excitatory synapses are now a major focus of autism spectrum disorder research, because inherited mutations proteins of the PSD (e.g., Shank) cause cognitive defects (Peça et al., 2011). Presynaptic tethers like neurexin are associated with schizophrenia and autism (Reichelt et al., 2012), and can, for example, alter NMDA-receptor-dependent LTP (Aoto et al., 2013).

Due to their ubiquity, glutamate receptors have indirect involvement in a number of brain diseases and pathological states. They have been the object of intensive (but not particularly successful) research as targets for therapeutic agents. However, some FDA-approved agents are becoming available. Perampanel is a noncompetitive AMPA receptor antagonist approved for epilepsy treatment as an adjunct to existing therapies (Krauss et al., 2012). Perampanel was developed from the prototype noncompetitive AMPA antagonist GYKI-52466 (Donevan and Rogawski, 1993). Some animal models of Alzheimer's disease display altered LTP, implicating the NMDA receptor (Chapman et al., 1999). The interaction between Ephrin type-B receptor 2 and NMDA receptors is via the ATD, and Alzheimer's disease may impinge on this interaction (Cissé et al., 2011). Memantine is used as a therapeutic drug in Alzheimer's disease. It blocks the pore of NMDA-Rs, perhaps preferentially attacking extrasynaptic receptors, which may explain drug tolerance (Xia et al., 2010).

During brain injuries such as stroke, glutamate is released by reversed uptake (Rossi et al., 2000), because ion gradients driving glutamate transporters collapse. The lack of desensitization and higher glutamate affinity of NMDA receptors renders them especially susceptible in this situation. Targeting AMPA receptors as a second line of defense (with, e.g., ZK200755 [Turski et al., 1998]) to date proved to be far too damaging to brain function for clinical use (Walters et al., 2005).

Nitric oxide produced by neuronal nitric oxide synthase mediates the greater part of neurotoxicity due to the NMDA receptor overactivation (Dawson et al., 1991, 1996).

The identity of the C-terminal in GluN2A and N2B subunits influences this excitotoxicity (Martel et al., 2012). Consistent with this observation, disrupting the C-terminal interaction between NMDARs and PSD95 with a peptide is neuroprotective in ischemia (Aarts et al., 2002). Synaptic NMDA receptor activation may in fact be neuroprotective, on the basis that it stimulates separate antioxidant pathways (Papadia et al., 2008). Synaptic NMDA receptors are isolated in these experiments by replacing magnesium in the extracellular solution for MK801. This exchange preferentially removes synaptic receptors from action while sparing extrasynaptic receptors. The GluN3A subunit can be considered neuroprotective in its own right, because its deletion renders cells in the retina and other brain regions highly susceptible to NMDA (Nakanishi et al., 2009).

The regulation of NMDA receptors by zinc appears important in pain, because genetically targeted mice with a point mutation that abolishes the high-affinity zinc inhibition of NMDA receptors are subject to increased sensitivity to painful stimuli in a number of ways (Nozaki et al., 2011). Src kinase attaches to the NMDA receptor complex via ND2 (NADH dehydrogenase 2), a mitochondrial protein (Gingrich et al., 2004), and disrupting this interaction with a targeted peptide is effective in reducing neuropathic pain (Liu et al., 2008). Because painful stimuli correlate with the degree of NMDA receptor tyrosine phosphorylation, it is plausible that NMDA receptor phosphorylation (and the associated boosted receptor activity) regulates certain types of pain, though this has not yet been demonstrated directly.

REFERENCES

Aarts, M., Liu, Y., Liu, L., Besshoh, S., Arundine, M., Gurd, J.W., Wang, Y.T., Salter, M.W., and Tymianski, M. (2002). Treatment of ischemic brain damage by perturbing NMDA receptor- PSD-95 protein interactions. *Science 298*, 846–850.

Abe, T., Kawai, N., and Miwa, A. (1983). Effects of a spider toxin on the glutaminergic synapse of lobster muscle. *J Physiol 339*, 243–252.

Abuin, L., Bargeton, B., Ulbrich, M.H., Isacoff, E.Y., Kellenberger, S., and Benton, R. (2011). Functional architecture of olfactory ionotropic glutamate receptors. *Neuron 69*, 44–60.

Adesnik, H., Nicoll, R.A., and England, P.M. (2005). Photoinactivation of native AMPA receptors reveals their real-time trafficking. *Neuron 48*, 977–985.

Ahmed, A.H., Wang, S., Chuang, H.H., and Oswald, R.E. (2011). Mechanism of AMPA receptor activation by partial agonists: Disulfide trapping of closed lobe conformations. *J Biol Chem 286*, 35257–35266.

Aizenman, E., Lipton, S.A., and Loring, R.H. (1989). Selective modulation of NMDA responses by reduction and oxidation. *Neuron 2*, 1257–1263.

Anis, N.A., Berry, S.C., Burton, N.R., and Lodge, D. (1983). The dissociative anaesthetics, ketamine and phencyclidine, selectively reduce excitation of central mammalian neurones by N-methyl-aspartate. *Br J Pharmacol 79*, 565–575.

Aoto, J., Martinelli, D., Malenka, R., Tabuchi, K., and Südhof, T. (2013). Presynaptic neurexin-3 alternative splicing trans-synaptically controls postsynaptic (AMPA) receptor trafficking. *Cell 154*, 75–88.

Araki, Y., Lin, D.T., and Huganir, R.L. (2010). Plasma membrane insertion of the AMPA receptor GluA2 subunit is regulated by NSF binding and Q/R editing of the ion pore. *Proc Natl Acad Sci USA 107*, 11080–11085.

Armstrong, N. and Gouaux, E. (2000). Mechanisms for activation and antagonism of an AMPA-sensitive glutamate receptor: Crystal structures of the GluR2 ligand binding core. *Neuron 28*, 165–181.

Armstrong, N., Sun, Y., Chen, G.Q., and Gouaux, E. (1998). Structure of a glutamate-receptor ligand-binding core in complex with kainate. *Nature 395*, 913–917.

Ashby, M.C., Maier, S.R., Nishimune, A., and Henley, J.M. (2006). Lateral diffusion drives constitutive exchange of AMPA receptors at dendritic spines and is regulated by spine morphology. *J Neurosci 26*, 7046–7055.

Ataman, Z.A., Gakhar, L., Sorensen, B.R., Hell, J.W., and Shea, M.A. (2007). The NMDA receptor NR1 C1 region bound to calmodulin: Structural insights into functional differences between homologous domains. *Structure 15*, 1603–1617.

Attwell, D. and Gibb, A. (2005). Neuroenergetics and the kinetic design of excitatory synapses. *Nat Rev Neurosci 6*, 841–849.

Balannik, V., Menniti, F.S., Paternain, A.V., Lerma, J., and Stern-Bach, Y. (2005). Molecular mechanism of AMPA receptor noncompetitive antagonism. *Neuron 48*, 279–288.

Banke, T.G., Bowie, D., Lee, H., Huganir, R.L., Schousboe, A., and Traynelis, S.F. (2000). Control of GluR1 AMPA receptor function by cAMP-dependent protein kinase. *J Neurosci 20*, 89–102.

Bannerman, D.M., Bus, T., Taylor, A., Sanderson, D.J., Schwarz, I., Jensen, V., Hvalby, Ø., Rawlins, J.N., Seeburg, P.H., and Sprengel, R. (2012). Dissecting spatial knowledge from spatial choice by hippocampal NMDA receptor deletion. *Nat Neurosci 15*, 1153–1159.

Baranovic, J., Ramanujan, C.S., Kasai, N., Midgett, C.R., Madden, D.R., Torimitsu, K., and Ryan, J.F. (2013). Reconstitution of homomeric GluA2(flop) receptors in supported lipid membranes: functional and structural properties. *J Biol Chem 288*, 8647–8657.

Barbour, B., Keller, B.U., Llano, I., and Marty, A. (1994). Prolonged presence of glutamate during excitatory synaptic transmission to cerebellar Purkinje cells. *Neuron 12*, 1331–1343.

Bard, L., Sainlos, M., Bouchet, D., Cousins, S., Mikasova, L., Breillat, C., Stephenson, F.A., Imperiali, B., Choquet, D., and Groc, L. (2010). Dynamic and specific interaction between synaptic NR2-NMDA receptor and PDZ proteins. *Proc Natl Acad Sci USA 107*, 19561–19566.

Bats, C., Groc, L., and Choquet, D. (2007). The interaction between Stargazin and PSD-95 regulates AMPA receptor surface trafficking. *Neuron 53*, 719–734.

Bats, C., Soto, D., Studniarczyk, D., Farrant, M., and Cull-Candy, S.G. (2012). Channel properties reveal differential expression of TARPed and TARPless AMPARs in stargazer neurons. *Nat Neurosci 15*, 853–861.

Benton, R., Vannice, K.S., Gomez-Diaz, C., and Vosshall, L.B. (2009). Variant ionotropic glutamate receptors as chemosensory receptors in *Drosophila*. *Cell 136*, 149–162.

Benveniste, M. and Mayer, M.L. (1995). Trapping of glutamate and glycine during open channel block of rat hippocampal neuron NMDA receptors by 9-aminoacridine. *J Physiol 483(Pt 2)*, 367–384.

Bettini, E., Sava, A., Griffante, C., Carignani, C., Buson, A., Capelli, A.M., Negri, M. et al. (2010). Identification and characterization of novel NMDA receptor antagonists selective for NR2A- over NR2B-containing receptors. *J Pharmacol Exp Ther 335*, 636–644.

Bi, G.Q. and Poo, M.M. (1998). Synaptic modifications in cultured hippocampal neurons: Dependence on spike timing, synaptic strength, and postsynaptic cell type. *J Neurosci 18*, 10464–10472.

Birdsey-Benson, A., Gill, A., Henderson, L.P., and Madden, D.R. (2010). Enhanced efficacy without further cleft closure: Reevaluating twist as a source of agonist efficacy in AMPA receptors. *J Neurosci 30*, 1463–1470.

Blaschke, M., Keller, B.U., Rivosecchi, R., Hollmann, M., Heinemann, S., and Konnerth, A. (1993). A single amino acid determines the subunit-specific spider toxin block of alpha-amino-3-hydroxy-5-methylisoxazole-4-propionate/kainate receptor channels. *Proc Natl Acad Sci USA 90*, 6528–6532.

Borgdorff, A.J. and Choquet, D. (2002). Regulation of AMPA receptor lateral movements. *Nature 417*, 649–653.
Bowie, D. and Lange, G.D. (2002). Functional stoichiometry of glutamate receptor desensitization. *J Neurosci 22*, 3392–3403.
Bowie, D. and Mayer, M.L. (1995). Inward rectification of both AMPA and kainate subtype glutamate receptors generated by polyamine-mediated ion channel block. *Neuron 15*, 453–462.
Braitenberg, V. and Schuz, A. (1998). *Cortex: Statistics and Geometry of Neuronal Connectivity* (Berlin, Germany: Springer).
Branco, T., Clark, B.A., and Häusser, M. (2010). Dendritic discrimination of temporal input sequences in cortical neurons. *Science 329*, 1671–1675.
Bredt, D.S. and Nicoll, R.A. (2003). AMPA receptor trafficking at excitatory synapses. *Neuron 40*, 361–379.
Breustedt, J. and Schmitz, D. (2004). Assessing the role of GLUK5 and GLUK6 at hippocampal mossy fiber synapses. *J Neurosci 24*, 10093–10098.
Buchanan, K.A., Blackman, A.V., Moreau, A.W., Elgar, D., Costa, R.P., Lalanne, T., Tudor Jones, A.A., Oyrer, J., and Sjöström, P.J. (2012). Target-specific expression of presynaptic NMDA receptors in neocortical microcircuits. *Neuron 75*, 451–466.
Burnashev, N., Villarroel, A., and Sakmann, B. (1996). Dimensions and ion selectivity of recombinant AMPA and kainate receptor channels and their dependence on Q/R site residues. *J Physiol 496(Pt 1)*, 165–173.
Burzomato, V., Beato, M., Groot-Kormelink, P.J., Colquhoun, D., and Sivilotti, L.G. (2004). Single-channel behavior of heteromeric alpha1beta glycine receptors: An attempt to detect a conformational change before the channel opens. *J Neurosci 24*, 10924–10940.
Canepari, M., Nelson, L., Papageorgiou, G., Corrie, J.E., and Ogden, D. (2001). Photochemical and pharmacological evaluation of 7-nitroindolinyl-and 4-methoxy-7-nitroindolinyl-amino acids as novel, fast caged neurotransmitters. *J Neurosci Methods 112*, 29–42.
Carbone, A.L. and Plested, A.J. (2012). Coupled control of desensitization and gating by the ligand binding domain of glutamate receptors. *Neuron 74*, 845–857.
Chambers, J.J., Gouda, H., Young, D.M., Kuntz, I.D., and England, P.M. (2004). Photochemically knocking out glutamate receptors in vivo. *J Am Chem Soc 126*, 13886–13887.
Chang, H.R. and Kuo, C.C. (2008). The activation gate and gating mechanism of the NMDA receptor. *J Neurosci 28*, 1546–1556.
Chapman, P.F., White, G.L., Jones, M.W., Cooper-Blacketer, D., Marshall, V.J., Irizarry, M., Younkin, L. et al. (1999). Impaired synaptic plasticity and learning in aged amyloid precursor protein transgenic mice. *Nat Neurosci 2*, 271–276.
Chatterton, J.E., Awobuluyi, M., Premkumar, L.S., Takahashi, H., Talantova, M., Shin, Y., Cui, J. et al. (2002). Excitatory glycine receptors containing the NR3 family of NMDA receptor subunits. *Nature 415*, 793–798.
Chaudhari, N., Yang, H., Lamp, C., Delay, E., Cartford, C., Than, T., and Roper, S. (1996). The taste of monosodium glutamate: Membrane receptors in taste buds. *J Neurosci 16*, 3817–3826.
Chaudhry, C., Plested, A.J., Schuck, P., and Mayer, M.L. (2009a). Energetics of glutamate receptor ligand binding domain dimer assembly are modulated by allosteric ions. *Proc Natl Acad Sci USA 106*, 12329–12334.
Chaudhry, C., Weston, M.C., Schuck, P., Rosenmund, C., and Mayer, M.L. (2009b). Stability of ligand-binding domain dimer assembly controls kainate receptor desensitization. *EMBO J 28*, 1518–1530.
Chávez, A.E., Singer, J.H., and Diamond, J.S. (2006). Fast neurotransmitter release triggered by Ca influx through AMPA-type glutamate receptors. *Nature 443*, 705–708.
Chen, G.Q., Cui, C., Mayer, M.L., and Gouaux, E. (1999). Functional characterization of a potassium-selective prokaryotic glutamate receptor. *Nature 402*, 817–821.
Chittajallu, R., Vignes, M., Dev, K.K., Barnes, J.M., Collingridge, G.L., and Henley, J.M. (1996). Regulation of glutamate release by presynaptic kainate receptors in the hippocampus. *Nature 379*, 78–81.
Choi, Y.B., Tenneti, L., Le, D.A., Ortiz, J., Bai, G., Chen, H.S., and Lipton, S.A. (2000). Molecular basis of NMDA receptor-coupled ion channel modulation by S-nitrosylation. *Nat Neurosci 3*, 15–21.
Christie, J.M. and Jahr, C.E. (2008). Dendritic NMDA receptors activate axonal calcium channels. *Neuron 60*, 298–307.
Christie, L.A., Russell, T.A., Xu, J., Wood, L., Shepherd, G.M., and Contractor, A. (2010). AMPA receptor desensitization mutation results in severe developmental phenotypes and early postnatal lethality. *Proc Natl Acad Sci USA 107(20)*, 9412–9417.
Chung, K.M., Lee, S.B., Heur, R., Cho, Y.K., Lee, C.H., Jung, H.Y., Chung, S.H., Lee, S.P., and Kim, K.N. (2005). Glutamate-induced cobalt uptake elicited by kainate receptors in rat taste bud cells. *Chem Senses 30*, 137–143.
Cissé, M., Halabisky, B., Harris, J., Devidze, N., Dubal, D.B., Sun, B., Orr, A. et al. (2011). Reversing EphB2 depletion rescues cognitive functions in Alzheimer model. *Nature 469*, 47–52.
Citri, A., Bhattacharyya, S., Ma, C., Morishita, W., Fang, S., Rizo, J., and Malenka, R.C. (2010). Calcium binding to PICK1 is essential for the intracellular retention of AMPA receptors underlying long-term depression. *J Neurosci 30*, 16437–16452.
Clements, J.D., Lester, R.A., Tong, G., Jahr, C.E., and Westbrook, G.L. (1992). The time course of glutamate in the synaptic cleft. *Science 258*, 1498–1501.
Collingridge, G.L., Kehl, S.J., and McLennan, H. (1983). Excitatory amino acids in synaptic transmission in the Schaffer collateral-commissural pathway of the rat hippocampus. *J Physiol 334*, 33–46.
Colquhoun, D. and Sakmann, B. (1985). Fast events in single-channel currents activated by acetylcholine and its analogues at the frog muscle end-plate. *J Physiol 369*, 501–557.
Colquhoun, D., Jonas, P., and Sakmann, B. (1992). Action of brief pulses of glutamate on AMPA/kainate receptors in patches from different neurones of rat hippocampal slices. *J Physiol 458*, 261–287.
Copits, B.A., Robbins, J.S., Frausto, S., and Swanson, G.T. (2011). Synaptic targeting and functional modulation of GluK1 kainate receptors by the auxiliary neuropilin and tolloid-like (NETO) proteins. *J Neurosci 31*, 7334–7340.
Cousins, S.L., Innocent, N., and Stephenson, F.A. (2013). Neto1 associates with the Nmda receptor/amyloid precursor protein complex. *J Neurochem 126*, 554–564.
Crowley, J.J., Carter, A.G., and Regehr, W.G. (2007). Fast vesicle replenishment and rapid recovery from desensitization at a single synaptic release site. *J Neurosci 27*, 5448–5460.
Dargan, S.L., Clarke, V.R., Alushin, G.M., Sherwood, J.L., Nisticò, R., Bortolotto, Z.A., Ogden, A.M. et al. (2009). ACET is a highly potent and specific kainate receptor antagonist: Characterisation and effects on hippocampal mossy fibre function. *Neuropharmacology 56*, 121–130.
Das, S., Sasaki, Y.F., Rothe, T., Premkumar, L.S., Takasu, M., Crandall, J.E., Dikkes, P. et al. (1998). Increased NMDA current and spine density in mice lacking the NMDA receptor subunit NR3A. *Nature 393*, 377–381.
Das, U., Kumar, J., Mayer, M.L., and Plested, A.J. (2010). Domain organization and function in GluK2 subtype kainate receptors. *Proc Natl Acad Sci USA 107*, 8463–8468.
Daw, M.I., Pelkey, K.A., Chittajallu, R., and McBain, C.J. (2010). Presynaptic kainate receptor activation preserves asynchronous GABA release despite the reduction in synchronous release from hippocampal cholecystokinin interneurons. *J Neurosci 30*, 11202–11209.

Dawson, V.L., Dawson, T.M., London, E.D., Bredt, D.S., and Snyder, S.H. (1991). Nitric oxide mediates glutamate neurotoxicity in primary cortical cultures. *Proc Natl Acad Sci USA 88*, 6368–6371.

Dawson, V.L., Kizushi, V.M., Huang, P.L., Snyder, S.H., and Dawson, T.M. (1996). Resistance to neurotoxicity in cortical cultures from neuronal nitric oxide synthase-deficient mice. *J Neurosci 16*, 2479–2487.

Derkach, V., Barria, A., and Soderling, T.R. (1999). Ca^{2+}/calmodulin-kinase II enhances channel conductance of alpha-amino-3-hydroxy-5-methyl-4-isoxazolepropionate type glutamate receptors. *Proc Natl Acad Sci USA 96*, 3269–3274.

de Sousa, S.L., Dickinson, R., Lieb, W.R., and Franks, N.P. (2000). Contrasting synaptic actions of the inhalational general anesthetics isoflurane and xenon. *Anesthesiology 92*, 1055–1066.

DeVries, S.H. (2000). Bipolar cells use kainate and AMPA receptors to filter visual information into separate channels. *Neuron 28*, 847–856.

DeVries, S.H. and Schwartz, E.A. (1999). Kainate receptors mediate synaptic transmission between cones and "Off" bipolar cells in a mammalian retina. *Nature 397*, 157–160.

Donevan, S.D. and Rogawski, M.A. (1993). GYKI 52466, a 2,3-benzodiazepine, is a highly selective, noncompetitive antagonist of AMPA/kainate receptor responses. *Neuron 10*, 51–59.

Dravid, S.M., Erreger, K., Yuan, H., Nicholson, K., Le, P., Lyuboslavsky, P., Almonte, A. et al. (2007). Subunit-specific mechanisms and proton sensitivity of NMDA receptor channel block. *J Physiol 581*, 107–128.

Duguid, I.C. and Smart, T.G. (2004). Retrograde activation of presynaptic NMDA receptors enhances GABA release at cerebellar interneuron-Purkinje cell synapses. *Nat Neurosci 7*, 525–533.

Egebjerg, J., Bettler, B., Hermans-Borgmeyer, I., and Heinemann, S. (1991). Cloning of a cDNA for a glutamate receptor subunit activated by kainate but not AMPA. *Nature 351*, 745–748.

Ehlers, M.D., Zhang, S., Bernhadt, J.P., and Huganir, R.L. (1996). Inactivation of NMDA receptors by direct interaction of calmodulin with the NR1 subunit. *Cell 84*, 745–755.

Eldefrawi, A.T., Eldefrawi, M.E., Konno, K., Mansour, N.A., Nakanishi, K., Oltz, E., and Usherwood, P.N. (1988). Structure and synthesis of a potent glutamate receptor antagonist in wasp venom. *Proc Natl Acad Sci USA 85*, 4910–4913.

Endele, S., Rosenberger, G., Geider, K., Popp, B., Tamer, C., Stefanova, I., Milh, M. et al. (2010). Mutations in GRIN2A and GRIN2B encoding regulatory subunits of NMDA receptors cause variable neurodevelopmental phenotypes. *Nat Genet 42(11)*, 1021–1026.

Evans, R.H., Francis, A.A., Jones, A.W., Smith, D.A., and Watkins, J.C. (1982). The effects of a series of omega-phosphonic alpha-carboxylic amino acids on electrically evoked and excitant amino acid-induced responses in isolated spinal cord preparations. *Br J Pharmacol 75*, 65–75.

Fay, A.M. and Bowie, D. (2006). Concanavalin-A reports agonist-induced conformational changes in the intact GluR6 kainate receptor. *J Physiol 572*, 201–213.

Featherstone, D.E., Rushton, E., Rohrbough, J., Liebl, F., Karr, J., Sheng, Q., Rodesch, C.K., and Broadie, K. (2005). An essential *Drosophila* glutamate receptor subunit that functions in both central neuropil and neuromuscular junction. *J Neurosci 25*, 3199–3208.

Fernandes, H.B., Catches, J.S., Petralia, R.S., Copits, B.A., Xu, J., Russell, T.A., Swanson, G.T., and Contractor, A. (2009). High-affinity kainate receptor subunits are necessary for ionotropic but not metabotropic signaling. *Neuron 63*, 818–829.

Fino, E., Araya, R., Peterka, D.S., Salierno, M., Etchenique, R., and Yuste, R. (2009). RuBi-glutamate: Two-photon and visible-light photoactivation of neurons and dendritic spines. *Front Neural Circuits 3*, 2.

Fisher, J.L. and Mott, D.D. (2011). Distinct functional roles of subunits within the heteromeric kainate receptor. *J Neurosci 31*, 17113–17122.

Fisher, J.L. and Mott, D.D. (2012). The auxiliary subunits Neto1 and Neto2 reduce voltage-dependent inhibition of recombinant kainate receptors. *J Neurosci 32*, 12928–12933.

Flint, A.C., Maisch, U.S., Weishaupt, J.H., Kriegstein, A.R., and Monyer, H. (1997). NR2A subunit expression shortens NMDA receptor synaptic currents in developing neocortex. *J Neurosci 17*, 2469–2476.

Foster, K.A., Kreitzer, A.C., and Regehr, W.G. (2002). Interaction of postsynaptic receptor saturation with presynaptic mechanisms produces a reliable synapse. *Neuron 36*, 1115–1126.

Furukawa, H. and Gouaux, E. (2003). Mechanisms of activation, inhibition and specificity: Crystal structures of the NMDA receptor NR1 ligand-binding core. *EMBO J 22*, 2873–2885.

Furukawa, H., Singh, S.K., Mancusso, R., and Gouaux, E. (2005). Subunit arrangement and function in NMDA receptors. *Nature 438*, 185–192.

Geiger, J.R., Melcher, T., Koh, D.S., Sakmann, B., Seeburg, P.H., Jonas, P., and Monyer, H. (1995). Relative abundance of subunit mRNAs determines gating and Ca^{2+} permeability of AMPA receptors in principal neurons and interneurons in rat CNS. *Neuron 15*, 193–204.

Gielen, M., Le Goff, A., Stroebel, D., Johnson, J.W., Neyton, J., and Paoletti, P. (2008). Structural rearrangements of NR1/NR2A NMDA receptors during allosteric inhibition. *Neuron 57*, 80–93.

Gingrich, J.R., Pelkey, K.A., Fam, S.R., Huang, Y., Petralia, R.S., Wenthold, R.J., and Salter, M.W. (2004). Unique domain anchoring of Src to synaptic NMDA receptors via the mitochondrial protein NADH dehydrogenase subunit 2. *Proc Natl Acad Sci USA 101*, 6237–6242.

Granger, A.J., Shi, Y., Lu, W., Cerpas, M., and Nicoll, R.A. (2012). LTP requires a reserve pool of glutamate receptors independent of subunit type. *Nature 493(7433)*, 495–500.

Greger, I.H., Khatri, L., and Ziff, E.B. (2002). RNA editing at arg607 controls AMPA receptor exit from the endoplasmic reticulum. *Neuron 34*, 759–772.

Greger, I.H., Khatri, L., Kong, X., and Ziff, E.B. (2003). AMPA receptor tetramerization is mediated by Q/R editing. *Neuron 40*, 763–774.

Grunwald, M.E. and Kaplan, J.M. (2003). Mutations in the ligand-binding and pore domains control exit of glutamate receptors from the endoplasmic reticulum in *C. elegans*. *Neuropharmacology 45*, 768–776.

Hansen, K.B. and Traynelis, S.F. (2011). Structural and mechanistic determinants of a novel site for noncompetitive inhibition of GluN2D-containing NMDA receptors. *J Neurosci 31*, 3650–3661.

Hansen, K.B., Naur, P., Kurtkaya, N.L., Kristensen, A.S., Gajhede, M., Kastrup, J.S., and Traynelis, S.F. (2009). Modulation of the dimer interface at ionotropic glutamate-like receptor delta 2 by D-serine and extracellular calcium. *J Neurosci 29*, 907.

Hansen, K.B., Tajima, N., Risgaard, R., Perszyk, R.E., Jorgensen, L., Vance, K.M., Ogden, K.K., Clausen, R.P., Furukawa, H., and Traynelis, S.F. (2013). Structural determinants of agonist efficacy at the glutamate binding site of NMDA receptors. *Mol Pharmacol 84(1)*, 114–127.

Hatton, C.J. and Paoletti, P. (2005). Modulation of triheteromeric NMDA receptors by N-terminal domain ligands. *Neuron 46*, 261–274.

Hayashi, T., Thomas, G.M., and Huganir, R.L. (2009). Dual palmitoylation of NR2 subunits regulates NMDA receptor trafficking. *Neuron 64*, 213–226.

Hebb, D.O. (1949). *The Organization of Behavior* (New York: Wiley).

Henley, J.M., Barker, E.A., and Glebov, O.O. (2011). Routes, destinations and delays: Recent advances in AMPA receptor trafficking. *Trends Neurosci 34(5)*, 258–268.

Herb, A., Burnashev, N., Werner, P., Sakmann, B., Wisden, W., and Seeburg, P.H. (1992). The KA-2 subunit of excitatory amino acid receptors shows widespread expression in brain and forms ion channels with distantly related subunits. *Neuron 8*, 775–785.

Herlitze, S., Raditsch, M., Ruppersberg, J.P., Jahn, W., Monyer, H., Schoepfer, R., and Witzemann, V. (1993). Argiotoxin detects molecular differences in AMPA receptor channels. *Neuron 10*, 1131–1140.

Herring, B., Shi, Y., Suh, Y., Zheng, C.-Y., Blankenship, S., Roche, K., and Nicoll, R. (2013). Cornichon proteins determine the subunit composition of synaptic AMPA receptors. *Neuron 77*, 1083–1096.

Hill, T.C. and Zito, K. (2013). LTP-induced long-term stabilization of individual nascent dendritic spines. *J Neurosci 33*, 678–686.

Hollmann, M. and Heinemann, S. (1994). Cloned glutamate receptors. *Annu Rev Neurosci 17*, 31–108.

Hollmann, M., Maron, C., and Heinemann, S. (1994). N-glycosylation site tagging suggests a three transmembrane domain topology for the glutamate receptor GluR1. *Neuron 13*, 1331–1343.

Honoré, T., Davies, S.N., Drejer, J., Fletcher, E.J., Jacobsen, P., Lodge, D., and Nielsen, F.E. (1988). Quinoxalinediones: Potent competitive non-NMDA glutamate receptor antagonists. *Science 241*, 701–703.

Honoré, T., Krogsgaard-Larsen, P., Hansen, J.J., and Lauridsen, J. (1981). Glutamate and aspartate agonists structurally related to ibotenic acid. *Mol Cell Biochem 38 Spec No*, 123–128.

Huang, Y.H., Muralidharan, S., Sinha, S.R., Kao, J.P., and Bergles, D.E. (2005). Ncm-D-aspartate: A novel caged D-aspartate suitable for activation of glutamate transporters and N-methyl-D-aspartate (NMDA) receptors in brain tissue. *Neuropharmacology 49*, 831–842.

Huettner, J.E. and Bean, B.P. (1988). Block of N-methyl-D-aspartate-activated current by the anticonvulsant MK-801: Selective binding to open channels. *Proc Natl Acad Sci USA 85*, 1307–1311.

Inanobe, A., Furukawa, H., and Gouaux, E. (2005). Mechanism of partial agonist action at the NR1 subunit of NMDA receptors. *Neuron 47*, 71–84.

Isaac, J.T., Nicoll, R.A., and Malenka, R.C. (1995). Evidence for silent synapses: Implications for the expression of LTP. *Neuron 15*, 427–434.

Isaacson, J.S. and Strowbridge, B.W. (1998). Olfactory reciprocal synapses: Dendritic signaling in the CNS. *Neuron 20*, 749–761.

Ito-Ishida, A., Miyazaki, T., Miura, E., Matsuda, K., Watanabe, M., Yuzaki, M., and Okabe, S. (2012). Presynaptically Released Cbln1 induces dynamic axonal structural changes by interacting with GluD2 during cerebellar synapse formation. *Neuron 76*, 549–564.

Ivakine, E.A., Acton, B.A., Mahadevan, V., Ormond, J., Tang, M., Pressey, J.C., Huang, M.Y. et al. (2013). Neto2 is a KCC2 interacting protein required for neuronal Cl^- regulation in hippocampal neurons. *Proc Natl Acad Sci USA 110(9)*, 3561–3566.

Jaafari, N., Konopacki, F.A., Owen, T.F., Kantamneni, S., Rubin, P., Craig, T.J., Wilkinson, K.A., and Henley, J.M. (2013). SUMOylation is required for glycine-induced increases in AMPA receptor surface expression (ChemLTP) in hippocampal neurons. *PLoS One 8*, e52345.

Jackson, A.C., Milstein, A.D., Soto, D., Farrant, M., Cull-Candy, S.G., and Nicoll, R.A. (2011). Probing TARP modulation of AMPA receptor conductance with polyamine toxins. *J Neurosci 31*, 7511–7520.

Jaskolski, F., Coussen, F., Nagarajan, N., Normand, E., Rosenmund, C., and Mulle, C. (2004). Subunit composition and alternative splicing regulate membrane delivery of kainate receptors. *J Neurosci 24*, 2506–2515.

Jensen, A.A., Christesen, T., Bølcho, U., Greenwood, J.R., Postorino, G., Vogensen, S.B., Johansen, T.N., Egebjerg, J., Bräuner-Osborne, H., and Clausen, R.P. (2007). Functional characterization of Tet-AMPA [tetrazolyl-2-amino-3-(3-hydroxy-5-methyl-4-isoxazolyl) propionic acid] analogues at ionotropic glutamate receptors GluR1-GluR4. The molecular basis for the functional selectivity profile of 2-Bn-Tet-AMPA. *J Med Chem 50*, 4177–4185.

Jevtovic-Todorović, V., Todorović, S.M., Mennerick, S., Powell, S., Dikranian, K., Benshoff, N., and CF Zorumski, J.W. (1998). Nitrous oxide (laughing gas) is an NMDA antagonist, neuroprotectant and neurotoxin. *Nat Med 4*, 460–463.

Jin, R. and Gouaux, E. (2003). Probing the function, conformational plasticity, and dimer-dimer contacts of the GluR2 ligand-binding core: Studies of 5-substituted willardiines and GluR2 S1S2 in the crystal. *Biochemistry 42*, 5201–5213.

Jin, R., Banke, T.G., Mayer, M.L., Traynelis, S.F., and Gouaux, E. (2003). Structural basis for partial agonist action at ionotropic glutamate receptors. *Nat Neurosci 6*, 803–810.

Jin, R., Singh, S.K., Gu, S., Furukawa, H., Sobolevsky, A.I., Zhou, J., Jin, Y., and Gouaux, E. (2009). Crystal structure and association behaviour of the GluR2 amino-terminal domain. *EMBO J 28*, 1812–1823.

Joshi, I., Shokralla, S., Titis, P., and Wang, L.Y. (2004). The role of AMPA receptor gating in the development of high-fidelity neurotransmission at the calyx of Held synapse. *J Neurosci 24*, 183–196.

Kakegawa, W., Miyoshi, Y., Hamase, K., Matsuda, S., Matsuda, K., Kohda, K., Emi, K. et al. (2011). D-serine regulates cerebellar LTD and motor coordination through the δ2 glutamate receptor. *Nat Neurosci 14*, 603–611.

Kalashnikova, E., Lorca, R.A., Kaur, I., Barisone, G.A., Li, B., Ishimaru, T., Trimmer, J.S., Mohapatra, D.P., and Díaz, E. (2010). SynDIG1: An activity-regulated, AMPA- receptor-interacting transmembrane protein that regulates excitatory synapse development. *Neuron 65*, 80–93.

Kamboj, S.K., Swanson, G.T., and Cull-Candy, S.G. (1995). Intracellular spermine confers rectification on rat calcium-permeable AMPA and kainate receptors. *J Physiol 486(Pt 2)*, 297–303.

Kampa, B.M., Clements, J., Jonas, P., and Stuart, G.J. (2004). Kinetics of Mg^{2+} unblock of NMDA receptors: Implications for spike-timing dependent synaptic plasticity. *J Physiol 556*, 337–345.

Karakas, E., Simorowski, N., and Furukawa, H. (2011). Subunit arrangement and phenylethanolamine binding in GluN1/GluN2B NMDA receptors. *Nature 475*, 249–253.

Kato, A.S., Gill, M.B., Ho, M.T., Yu, H., Tu, Y., Siuda, E.R., Wang, H. et al. (2010). Hippocampal AMPA receptor gating controlled by both TARP and cornichon proteins. *Neuron 68*, 1082–1096.

Keinänen, K., Wisden, W., Sommer, B., Werner, P., Herb, A., Verdoorn, T.A., Sakmann, B., and Seeburg, P.H. (1990). A family of AMPA-selective glutamate receptors. *Science 249*, 556–560.

Kenny, A.V., Cousins, S.L., Pinho, L., and Stephenson, F.A. (2009). The integrity of the glycine co-agonist binding site of N-methyl-D-aspartate receptors is a functional quality control checkpoint for cell surface delivery. *J Biol Chem 284*, 324–333.

Kim, K.S., Yan, D., and Tomita, S. (2010). Assembly and stoichiometry of the AMPA receptor and transmembrane AMPA receptor regulatory protein complex. *J Neurosci 30*, 1064–1072.

Kohda, K., Kakegawa, W., Matsuda, S., Yamamoto, T., Hirano, H., and Yuzaki, M. (2013). The δ2 glutamate receptor gates long-term depression by coordinating interactions between two AMPA receptor phosphorylation sites. *Proc Natl Acad Sci USA 110*, E948–E957.

Kohda, K., Wang, Y., and Yuzaki, M. (2000). Mutation of a glutamate receptor motif reveals its role in gating and delta2 receptor channel properties. *Nat Neurosci 3*, 315–322.

Koike-Tani, M., Saitoh, N., and Takahashi, T. (2005). Mechanisms underlying developmental speeding in AMPA-EPSC decay time at the calyx of Held. *J Neurosci 25*, 199–207.

Kotermanski, S.E., Wood, J.T., and Johnson, J.W. (2009). Memantine binding to a superficial site on NMDA receptors contributes to partial trapping. *J Physiol 587*, 4589–4604.

Köhler, M., Burnashev, N., Sakmann, B., and Seeburg, P.H. (1993). Determinants of Ca^{2+} permeability in both TM1 and TM2 of high affinity kainate receptor channels: Diversity by RNA editing. *Neuron 10*, 491–500.

Köhler, M., Kornau, H.C., and Seeburg, P.H. (1994). The organization of the gene for the functionally dominant alpha-amino-3-hydroxy-5-methylisoxazole-4-propionic acid receptor subunit GluR-B. *J Biol Chem 269*, 17367–17370.

Krauss, G.L., Serratosa, J.M., Villanueva, V., Endziniene, M., Hong, Z., French, J., Yang, H. et al. (2012). Randomized phase III study 306: Adjunctive perampanel for refractory partial-onset seizures. *Neurology 78*, 1408–1415.

Kristensen, A.S., Jenkins, M.A., Banke, T.G., Schousboe, A., Makino, Y., Johnson, R.C., Huganir, R., and Traynelis, S.F. (2011). Mechanism of Ca^{2+}/calmodulin-dependent kinase II regulation of AMPA receptor gating. *Nat Neurosci 14*, 727–735.

Krupp, J.J., Vissel, B., Thomas, C.G., Heinemann, S.F., and Westbrook, G.L. (1999). Interactions of calmodulin and alpha-actinin with the NR1 subunit modulate Ca^{2+}-dependent inactivation of NMDA receptors. *J Neurosci 19*, 1165–1178.

Kumar, J., Schuck, P., and Mayer, M.L. (2011). Structure and assembly mechanism for heteromeric kainate receptors. *Neuron 71*, 319–331.

Kumar, J., Schuck, P., Jin, R., and Mayer, M.L. (2009). The N-terminal domain of GluR6-subtype glutamate receptor ion channels. *Nat Struct Mol Biol 16*, 631–638.

Kuner, T. and Schoepfer, R. (1996). Multiple structural elements determine subunit specificity of Mg^{2+} block in NMDA receptor channels. *J Neurosci 16*, 3549–3558.

Kusch, J., Nache, V., and Benndorf, K. (2004). Effects of permeating ions and cGMP on gating and conductance of rod-type cyclic nucleotide-gated (CNGA1) channels. *J Physiol 560*, 605–616.

Landes, C.F., Rambhadran, A., Taylor, J.N., Salatan, F., and Jayaraman, V. (2011). Structural landscape of isolated agonist-binding domains from single AMPA receptors. *Nat Chem Biol 7*, 168–173.

Lau, A.Y. and Roux, B. (2011). The hidden energetics of ligand binding and activation in a glutamate receptor. *Nat Struct Mol Biol 18*, 283–287.

Lau, A.Y., Salazar, H., Blachowicz, L., Ghisi, V., Plested, A.J., and Roux, B. (2013). A conformational intermediate in glutamate receptor activation. *Neuron 79(3)*, 492–503.

Lauri, S.E., Vesikansa, A., Segerstråle, M., Collingridge, G.L., Isaac, J.T., and Taira, T. (2006). Functional maturation of CA1 synapses involves activity-dependent loss of tonic kainate receptor-mediated inhibition of glutamate release. *Neuron 50*, 415–429.

Lee, A.M., Kanter, B.R., Wang, D., Lim, J.P., Zou, M.E., Qiu, C., McMahon, T., Dadgar, J., Fischbach-Weiss, S.C., and Messing, R.O. (2013). Prkcz null mice show normal learning and memory. *Nature 493*, 416–419.

Lee, H.K., Barbarosie, M., Kameyama, K., Bear, M.F., and Huganir, R.L. (2000). Regulation of distinct AMPA receptor phosphorylation sites during bidirectional synaptic plasticity. *Nature 405*, 955–959.

Lee, H.K., Takamiya, K., Han, J.S., Man, H., Kim, C.H., Rumbaugh, G., Yu, S. et al. (2003). Phosphorylation of the AMPA receptor GluR1 subunit is required for synaptic plasticity and retention of spatial memory. *Cell 112*, 631–643.

Lee, H.K., Takamiya, K., He, K., Song, L., and Huganir, R.L. (2010). Specific roles of AMPA receptor subunit GluR1 (GluA1) phosphorylation sites in regulating synaptic plasticity in the CA1 region of hippocampus. *J Neurophysiol 103*, 479–489.

Liu, L., Wong, T.P., Pozza, M.F., Lingenhoehl, K., Wang, Y., Sheng, M., Auberson, Y.P., and Wang, Y.T. (2004). Role of NMDA receptor subtypes in governing the direction of hippocampal synaptic plasticity. *Science 304*, 1021–1024.

Liu, S.Q.J. and Cull-Candy, S.G. (2000). Synaptic activity at calcium-permeable AMPA receptors induces a switch in receptor subtype. *Nature 405*, 454–458.

Liu, S.Q.J. and Cull-Candy, S.G. (2002). Activity-dependent change in AMPA receptor properties in cerebellar stellate cells. *J Neurosci 22*, 3881–3889.

Liu, X.J., Gingrich, J.R., Vargas-Caballero, M., Dong, Y.N., Sengar, A., Beggs, S., Wang, S.H., Ding, H.K., Frankland, P.W., and Salter, M.W. (2008). Treatment of inflammatory and neuropathic pain by uncoupling Src from the NMDA receptor complex. *Nat Med 14*, 1325–1332.

Logan, S.M., Partridge, J.G., Matta, J.A., Buonanno, A., and Vicini, S. (2007). Long-lasting NMDA receptor-mediated EPSCs in mouse striatal medium spiny neurons. *J Neurophysiol 98*, 2693–2704.

Lomeli, H., Mosbacher, J., Melcher, T., Höger, T., Geiger, J.R., Kuner, T., Monyer, H., Higuchi, M., Bach, A., and Seeburg, P.H. (1994). Control of kinetic properties of AMPA receptor channels by nuclear RNA editing. *Science 266*, 1709–1713.

Lomeli, H., Sprengel, R., Laurie, D.J., Kohr, G., Herb, A., Seeburg, P.H., and Wisden, W. (1993). The rat delta-1 and delta-2 subunits extend the excitatory amino acid receptor family. *FEBS Lett 315*, 318–322.

Lourenço, J., Cannich, A., Carta, M., Coussen, F., Mulle, C., and Marsicano, G. (2010). Synaptic activation of kainate receptors gates presynaptic CB(1) signaling at GABAergic synapses. *Nat Neurosci 13*, 197–204.

Lourenço, J., Matias, I., Marsicano, G., and Mulle, C. (2011). Pharmacological activation of kainate receptors drives endocannabinoid mobilization. *J Neurosci 31*, 3243–3248.

Low, C.M., Zheng, F., Lyuboslavsky, P., and Traynelis, S.F. (2000). Molecular determinants of coordinated proton and zinc inhibition of N-methyl-D-aspartate NR1/NR2A receptors. *Proc Natl Acad Sci USA 97*, 11062–11067.

MacDermott, A.B., Mayer, M.L., Westbrook, G.L., Smith, S.J., and Barker, J.L. (1986). NMDA-receptor activation increases cytoplasmic calcium concentration in cultured spinal cord neurones. *Nature 321*, 519–522.

MacGillavry, H.D., Kerr, J.M., and Blanpied, T.A. (2011). Lateral organization of the postsynaptic density. *Mol Cell Neurosci 48*, 321–331.

Magee, J.C. and Johnston, D. (1997). A synaptically controlled, associative signal for Hebbian plasticity in hippocampal neurons. *Science 275*, 209–213.

Mah, S.J., Cornell, E., Mitchell, N.A., and Fleck, M.W. (2005). Glutamate receptor trafficking: Endoplasmic reticulum quality control involves ligand binding and receptor function. *J Neurosci 25*, 2215–2225.

Malinow, R., Schulman, H., and Tsien, R.W. (1989). Inhibition of postsynaptic PKC or CaMKII blocks induction but not expression of LTP. *Science 245*, 862–866.

Maltsev, A.S., Ahmed, A.H., Fenwick, M.K., Jane, D.E., and Oswald, R.E. (2008). Mechanism of partial agonism at the GluR2 AMPA receptor: Measurements of lobe orientation in solution. *Biochemistry 47*, 10600–10610.

Mammen, A.L., Kameyama, K., Roche, K.W., and Huganir, R.L. (1997). Phosphorylation of the alpha-amino-3-hydroxy-5-methylisoxazole4-propionic acid receptor GluR1 subunit by calcium/calmodulin-dependent kinase II. *J Biol Chem 272*, 32528–32533.

Mansour, M., Nagarajan, N., Nehring, R.B., Clements, J.D., and Rosenmund, C. (2001). Heteromeric AMPA receptors assemble with a preferred subunit stoichiometry and spatial arrangement. *Neuron 32*, 841–853.

Martel, M.A., Ryan, T.J., Bell, K.F., Fowler, J.H., McMahon, A., Al-Mubarak, B., Komiyama, N.H. et al. (2012). The subtype of GluN2 C-terminal domain determines the response to excitotoxic insults. *Neuron 74*, 543–556.

Martin, L.J., Blackstone, C.D., Levey, A.I., Huganir, R.L., and Price, D.L. (1993). AMPA glutamate receptor subunits are differentially distributed in rat brain. *Neuroscience 53*, 327–358.

Martin, S., Nishimune, A., Mellor, J.R., and Henley, J.M. (2007). SUMOylation regulates kainate-receptor-mediated synaptic transmission. *Nature 447*, 321–325.

Matsuda, K., Miura, E., Miyazaki, T., Kakegawa, W., Emi, K., Narumi, S., Fukazawa, Y. et al. (2010). Cbln1 is a ligand for an orphan glutamate receptor delta2, a bidirectional synapse organizer. *Science 328*, 363–368.

Mayer, M.L. (2005). Crystal structures of the GluR5 and GluR6 ligand binding cores: Molecular mechanisms underlying kainate receptor selectivity. *Neuron 45*, 539–552.

Mayer, M.L. and Westbrook, G.L. (1987). Permeation and block of N-methyl-D-aspartic acid receptor channels by divalent cations in mouse cultured central neurones. *J Physiol 394*, 501–527.

Mayer, M.L., Westbrook, G.L., and Guthrie, P.B. (1984). Voltage-dependent block by Mg^{2+}; of NMDA responses in spinal cord neurones.

McHugh, T.J., Blum, K.I., Tsien, J.Z., Tonegawa, S., and Wilson, M.A. (1996). Impaired hippocampal representation of space in CA1-specific NMDAR1 knockout mice. *Cell 87*, 1339–1349.

Mealing, G.A., Lanthorn, T.H., Murray, C.L., Small, D.L., and Morley, P. (1999). Differences in degree of trapping of low-affinity uncompetitive N-methyl-D-aspartic acid receptor antagonists with similar kinetics of block. *J Pharmacol Exp Ther 288*, 204–210.

Menuz, K., Stroud, R.M., Nicoll, R.A., and Hays, F.A. (2007). TARP auxiliary subunits switch AMPA receptor antagonists into partial agonists. *Science 318*, 815–817.

Midgett, C.R. and Madden, D.R. (2008). The quaternary structure of a calcium-permeable AMPA receptor: Conservation of shape and symmetry across functionally distinct subunit assemblies. *J Mol Biol 382*, 578–584.

Minami, K., Wick, M.J., Stern-Bach, Y., Dildy-Mayfield, J.E., Brozowski, S.J., Gonzales, E.L., Trudell, J.R., and Harris, R.A. (1998). Sites of volatile anesthetic action on kainate (Glutamate receptor 6) receptors. *J Biol Chem 273*, 8248–8255.

Misra, C., Brickley, S.G., Wyllie, D.J., and Cull-Candy, S.G. (2000). Slow deactivation kinetics of NMDA receptors containing NR1 and NR2D subunits in rat cerebellar Purkinje cells. *J Physiol 525(Pt 2)*, 299–305.

Miyata, M., Finch, E.A., Khiroug, L., Hashimoto, K., Hayasaka, S., Oda, S.I., Inouye, M., Takagishi, Y., Augustine, G.J., and Kano, M. (2000). Local calcium release in dendritic spines required for long-term synaptic depression. *Neuron 28*, 233–244.

Mony, L., Zhu, S., Carvalho, S., and Paoletti, P. (2011). Molecular basis of positive allosteric modulation of GluN2B NMDA receptors by polyamines. *EMBO J 30*, 3134–3146.

Monyer, H., Burnashev, N., Laurie, D.J., Sakmann, B., and Seeburg, P.H. (1994). Developmental and regional expression in the rat brain and functional properties of four NMDA receptors. *Neuron 12*, 529–540.

Morimoto-Tomita, M., Zhang, W., Straub, C., Cho, C.-H., Kim, K.S., Howe, J.R., and Tomita, S. (2009). Autoinactivation of neuronal AMPA receptors via glutamate-regulated TARP interaction. *Neuron 61*, 101–112.

Morris, R.G. (1989). Synaptic plasticity and learning: Selective impairment of learning rats and blockade of long-term potentiation in vivo by the N-methyl-D-aspartate receptor antagonist AP5. *J Neurosci 9*, 3040–3057.

Mosbacher, J., Schoepfer, R., Monyer, H., Burnashev, N., Seeburg, P.H., and Ruppersberg, J.P. (1994). A molecular determinant for submillisecond desensitization in glutamate receptors. *Science 266*, 1059–1062.

Motazacker, M.M., Rost, B.R., Hucho, T., Garshasbi, M., Kahrizi, K., Ullmann, R., Abedini, S.S. et al. (2007). A defect in the ionotropic glutamate receptor 6 gene (GRIK2) is associated with autosomal recessive mental retardation. *Am J Hum Genet 81*, 792–798.

Mott, D.D., Washburn, M.S., Zhang, S., and Dingledine, R.J. (2003). Subunit-dependent modulation of kainate receptors by extracellular protons and polyamines. *J Neurosci 23*, 1179–1188.

Mulkey, R.M., Herron, C.E., and Malenka, R.C. (1993). An essential role for protein phosphatases in hippocampal long-term depression. *Science 261*, 1051–1055.

Mullasseril, P., Hansen, K.B., Vance, K.M., Ogden, K.K., Yuan, H., Kurtkaya, N.L., Santangelo, R., Orr, A.G., Le, P., Vellano, K.M., Liotta, D.C., and Traynelis, S.F. (2010). A subunit-selective potentiator of NR2C- and NR2D-containing NMDA receptors *Nat Commun 1*, 90.

Muroyama, A., Uehara, S., Yatsushiro, S., Echigo, N., Morimoto, R., Morita, M., Hayashi, M., Yamamoto, A., Koh, D.S., and Moriyama, Y. (2004). A novel variant of ionotropic glutamate receptor regulates somatostatin secretion from delta-cells of islets of *Langerhans. Diabetes 53*, 1743–1753.

Nakagawa, T., Cheng, Y., Ramm, E., Sheng, M., and Walz, T. (2005). Structure and different conformational states of native AMPA receptor complexes. *Nature 433*, 545–549.

Nakanishi, N., Tu, S., Shin, Y., Cui, J., Kurokawa, T., Zhang, D., Chen, H.S., Tong, G., and Lipton, S.A. (2009). Neuroprotection by the NR3A subunit of the NMDA receptor. *J Neurosci 29*, 5260–5265.

Nanao, M.H., Green, T., Stern-Bach, Y., Heinemann, S.F., and Choe, S. (2005). Structure of the kainate receptor subunit GluR6 agonist-binding domain complexed with domoic acid. *Proc Natl Acad Sci USA 102*, 1708–1713.

Nasu-Nishimura, Y., Hurtado, D., Braud, S., Tang, T.T., Isaac, J.T., and Roche, K.W. (2006). Identification of an endoplasmic reticulum-retention motif in an intracellular loop of the kainate receptor subunit KA2. *J Neurosci 26*, 7014–7021.

Naur, P., Hansen, K.B., Kristensen, A.S., Dravid, S.M., Pickering, D.S., Olsen, L., Vestergaard, B. et al. (2007). Ionotropic glutamate-like receptor delta2 binds D-serine and glycine. *Proc Natl Acad Sci USA 104*, 14116–14121.

Negrete-Díaz, J.V., Sihra, T.S., Delgado-García, J.M., and Rodríguez-Moreno, A. (2006). Kainate receptor-mediated inhibition of glutamate release involves protein kinase A in the mouse hippocampus. *J Neurophysiol 96*, 1829–1837.

Neveu, D. and Zucker, R.S. (1996). Postsynaptic levels of $[Ca^{2+}]_i$ needed to trigger LTD and LTP. *Neuron 16*, 619–629.

Nevian, T., Larkum, M.E., Polsky, A., and Schiller, J. (2007). Properties of basal dendrites of layer 5 pyramidal neurons: A direct patch-clamp recording study. *Nat Neurosci 10*, 206–214.

Neyton, J. and Paoletti, P. (2006). Relating NMDA receptor function to receptor subunit composition: Limitations of the pharmacological approach. *J Neurosci 26*, 1331–1333.

Ng, D., Pitcher, G.M., Szilard, R.K., Sertié, A., Kanisek, M., Clapcote, S.J., Lipina, T. et al. (2009). Neto1 is a novel CUB-domain NMDA receptor-interacting protein required for synaptic plasticity and learning. *PLoS Biol 7*, e41.

Nishimune, A., Isaac, J.T., Molnar, E., Noel, J., Nash, S.R., Tagaya, M., Collingridge, G.L., Nakanishi, S., and Henley, J.M. (1998). NSF binding to GluR2 regulates synaptic transmission. *Neuron 21*, 87.

Nowak, L., Bregestovski, P., Ascher, P., Herbet, A., and Prochiantz, A. (1984). Magnesium gates glutamate-activated channels in mouse central neurones. *Nature 307*, 462–465.

Nozaki, C., Vergnano, A.M., Filliol, D., Ouagazzal, A.M., Le Goff, A., Carvalho, S., Reiss, D. et al. (2011). Zinc alleviates pain through high-affinity binding to the NMDA receptor NR2A subunit. *Nat Neurosci 14*, 1017–1022.

Opazo, P., Labrecque, S., Tigaret, C.M., Frouin, A., Wiseman, P.W., De Koninck, P., and Choquet, D. (2010). CaMKII triggers the diffusional trapping of surface AMPARs through phosphorylation of stargazin. *Neuron 67*, 239–252.

Palma-Cerda, F., Auger, C., Crawford, D.J., Hodgson, A.C., Reynolds, S.J., Cowell, J.K., Swift, K.A. et al. (2012). New caged neurotransmitter analogs selective for glutamate receptor sub-types based on methoxynitroindoline and nitrophenylethoxycarbonyl caging groups. *Neuropharmacology 63*, 624–634.

Palmer, M.J., Hull, C., Vigh, J., and von Gersdorff, H. (2003). Synaptic cleft acidification and modulation of short-term depression by exocytosed protons in retinal bipolar cells. *J Neurosci 23*, 11332–11341.

Panatier, A., Theodosis, D.T., Mothet, J.P., Touquet, B., Pollegioni, L., Poulain, D.A., and Oliet, S.H. (2006). Glia-derived D-serine controls NMDA receptor activity and synaptic memory. *Cell 125*, 775–784.

Panchenko, V.A., Glasser, C.R., and Mayer, M.L. (2001). Structural similarities between glutamate receptor channels and K(+) channels examined by scanning mutagenesis. *J Gen Physiol 117*, 345–360.

Paoletti, P., Perin-Dureau, F., Fayyazuddin, A., Le Goff, A., Callebaut, I., and Neyton, J. (2000). Molecular organization of a zinc binding n-terminal modulatory domain in a NMDA receptor subunit. *Neuron 28*, 911–925.

Papadia, S., Soriano, F.X., Léveillé, F., Martel, M.A., Dakin, K.A., Hansen, H.H., Kaindl, A. et al. (2008). Synaptic NMDA receptor activity boosts intrinsic antioxidant defenses. *Nat Neurosci 11*, 476–487.

Partin, K.M., Bowie, D., and Mayer, M.L. (1995). Structural determinants of allosteric regulation in alternatively spliced AMPA receptors. *Neuron 14*, 833–843.

Partin, K.M., Patneau, D.K., Winters, C.A., Mayer, M.L., and Buonanno, A. (1993). Selective modulation of desensitization at AMPA versus kainate receptors by cyclothiazide and concanavalin A. *Neuron 11*, 1069–1082.

Pastalkova, E., Serrano, P., Pinkhasova, D., Wallace, E., Fenton, A.A., and Sacktor, T.C. (2006). Storage of spatial information by the maintenance mechanism of LTP. *Science 313*, 1141–1144.

Patneau, D.K. and Mayer, M.L. (1990). Structure-activity relationships for amino acid transmitter candidates acting at N-methyl-D-aspartate and quisqualate receptors. *J Neurosci 10*, 2385–2399.

Patneau, D.K., Vyklicky, L., and Mayer, M.L. (1993). Hippocampal neurons exhibit cyclothiazide-sensitive rapidly desensitizing responses to kainate. *J Neurosci 13*, 3496–3509.

Peça, J., Feliciano, C., Ting, J.T., Wang, W., Wells, M.F., Venkatraman, T.N., Lascola, C.D., Fu, Z., and Feng, G. (2011). Shank3 mutant mice display autistic-like behaviours and striatal dysfunction. *Nature 472(7344)*, 437–442.

Peng, P.L., Zhong, X., Tu, W., Soundarapandian, M.M., Molner, P., Zhu, D., Lau, L., Liu, S., Liu, F., and Lu, Y. (2006). ADAR2-dependent RNA editing of AMPA receptor subunit GluR2 determines vulnerability of neurons in forebrain ischemia. *Neuron 49*, 719–733.

Penn, A.C., Balik, A., Wozny, C., Cais, O., and Greger, I.H. (2012). Activity-mediated AMPA receptor remodeling, driven by alternative splicing in the ligand-binding domain. *Neuron 76*, 503–510.

Penn, A.C., Williams, S.R., and Greger, I.H. (2008). Gating motions underlie AMPA receptor secretion from the endoplasmic reticulum. *EMBO J 27*, 3056–3068.

Petralia, R.S., Yokotani, N., and Wenthold, R.J. (1994). Light and electron microscope distribution of the NMDA receptor subunit NMDAR1 in the rat nervous system using a selective anti-peptide antibody. *J Neurosci 14*, 667–696.

Pickering, D.S., Taverna, F.A., Salter, M.W., and Hampson, D.R. (1995). Palmitoylation of the GluR6 kainate receptor. *Proc Natl Acad Sci USA 92*, 12090–12094.

Plested, A.J. and Mayer, M.L. (2009). AMPA receptor ligand binding domain mobility revealed by functional cross linking. *J Neurosci 29*, 11912–11923.

Plested, A.J.R. and Mayer, M.L. (2007). Structure and mechanism of kainate receptor modulation by anions. *Neuron 53*, 829–841.

Plested, A.J.R., Vijayan, R., Biggin, P.C., and Mayer, M.L. (2008). Molecular basis of kainate receptor modulation by sodium. *Neuron 58*, 720–735.

Plested, A.J.R., Wildman, S.S., Lieb, W.R., and Franks, N.P. (2004). Determinants of the sensitivity of AMPA receptors to xenon. *Anesthesiology 100*, 347–358.

Priel, A., Selak, S., Lerma, J., and Stern-Bach, Y. (2006). Block of kainate receptor desensitization uncovers a key trafficking checkpoint. *Neuron 52*, 1037–1046.

Pugh, J.R. and Jahr, C.E. (2011). NMDA receptor agonists fail to alter release from cerebellar basket cells. *J Neurosci 31*, 16550–16555.

Reichelt, A.C., Rodgers, R.J., and Clapcote, S.J. (2012). The role of neurexins in schizophrenia and autistic spectrum disorder. *Neuropharmacology 62*, 1519–1526.

Riou, M., Stroebel, D., Edwardson, J.M., and Paoletti, P. (2012). An alternating GluN1-2-1-2 subunit arrangement in mature NMDA receptors. *PLoS One 7*, e35134.

Ripka, S., Riedel, J., Neesse, A., Griesmann, H., Buchholz, M., Ellenrieder, V., Moeller, F., Barth, P., Gress, T.M., and Michl, P. (2010). Glutamate receptor GRIA3--target of CUX1 and mediator of tumor progression in pancreatic cancer. *Neoplasia 12*, 659–667.

Robert, A., Irizarry, S.N., Hughes, T.E., and Howe, J.R. (2001). Subunit interactions and AMPA receptor desensitization. *J Neurosci 21*, 5574–5586.

Rodríguez-Moreno, A., González-Rueda, A., Banerjee, A., Upton, A.L., Craig, M.T., and Paulsen, O. (2013). Presynaptic self-depression at developing neocortical synapses. *Neuron 77*, 35–42.

Rojas, A. and Dingledine, R. (2013). Ionotropic glutamate receptors: Regulation by g-protein-coupled receptors. *Mol Pharmacol 83*, 746–752.

Rosenmund, C., Stern-Bach, Y., and Stevens, C.F. (1998). The tetrameric structure of a glutamate receptor channel. *Science 280*, 1596–1599.

Rossi, D.J., Oshima, T., and Attwell, D. (2000). Glutamate release in severe brain ischaemia is mainly by reversed uptake. *Nature 403*, 316–321.

Rossi, P., D'Angelo, E., and Taglietti, V. (1996). Differential long-lasting potentiation of the NMDA and non-NMDA synaptic currents induced by metabotropic and NMDA receptor coactivation in cerebellar granule cells. *Eur J Neurosci 8*, 1182–1189.

Rothman, J.S., Cathala, L., Steuber, V., and Silver, R.A. (2009). Synaptic depression enables neuronal gain control. *Nature 457*, 1015–1018.

Rouach, N., Byrd, K., Petralia, R.S., Elias, G.M., Adesnik, H., Tomita, S., Karimzadegan, S., Kealey, C., Bredt, D.S., and Nicoll, R.A. (2005). TARP gamma-8 controls hippocampal AMPA receptor number, distribution and synaptic plasticity. *Nat Neurosci 8*, 1525–1533.

Rozov, A. and Burnashev, N. (1999). Polyamine-dependent facilitation of postsynaptic AMPA receptors counteracts paired-pulse depression. *Nature 401*, 594–598.

Rozov, A., Jerecic, J., Sakmann, B., and Burnashev, N. (2001). AMPA receptor channels with long-lasting desensitization in bipolar interneurons contribute to synaptic depression in a novel feedback circuit in layer 2/3 of rat neocortex. *J Neurosci 21*, 8062–8071.

Sachs, G., Marcus, E.A., and Scott, D.R. (2011). The role of the NMDA receptor in *Helicobacter pylori*–induced gastric damage. *Gastroenterology 141*, 1967–1969.

Saglietti, L., Dequidt, C., Kamieniarz, K., Rousset, M.C., Valnegri, P., Thoumine, O., Beretta, F. et al. (2007). Extracellular interactions between GluR2 and N-cadherin in spine regulation. *Neuron 54*, 461–477.

Saito, M., Kawai, N., Miwa, A., Pan-Hou, H., and Yoshioka, M. (1985). Spider toxin (JSTX) blocks glutamate synapse in hippocampal pyramidal neurons. *Brain Res 346*, 397–399.

Sambandan, S., Sauer, J.F., Vida, I., and Bartos, M. (2010). Associative plasticity at excitatory synapses facilitates recruitment of fast-spiking interneurons in the dentate gyrus. *J Neurosci 30*, 11826–11837.

Sans, N., Vissel, B., Petralia, R.S., Wang, Y.X., Chang, K., Royle, G.A., Wang, C.Y., O'Gorman, S., Heinemann, S.F., and Wenthold, R.J. (2003). Aberrant formation of glutamate receptor complexes in hippocampal neurons of mice lacking the GluR2 AMPA receptor subunit. *J Neurosci 23*, 9367–9373.

Sasaki, Y.F., Rothe, T., Premkumar, L.S., Das, S., Cui, J., Talantova, M.V., Wong, H.K. et al. (2002). Characterization and comparison of the NR3A subunit of the NMDA receptor in recombinant systems and primary cortical neurons. *J Neurophysiol 87*, 2052–2063.

Saviane, C. and Silver, R.A. (2006). Fast vesicle reloading and a large pool sustain high bandwidth transmission at a central synapse. *Nature 439*, 983–987.

Savtchenko, L.P., Sylantyev, S., and Rusakov, D.A. (2013). Central synapses release a resource-efficient amount of glutamate. *Nat Neurosci 16*, 10–12.

Schauder, D.M., Kuybeda, O., Zhang, J., Klymko, K., Bartesaghi, A., Borgnia, M.J., Mayer, M.L., and Subramaniam, S. (2013). Glutamate receptor desensitization is mediated by changes in quaternary structure of the ligand binding domain. *Proc Natl Acad Sci USA 110*, 5921–5926.

Schell, M.J., Molliver, M.E., and Snyder, S.H. (1995). D-serine, an endogenous synaptic modulator: Localization to astrocytes and glutamate-stimulated release. *Proc Natl Acad Sci USA 92*, 3948–3952.

Schorge, S., Elenes, S., and Colquhoun, D. (2005). Maximum likelihood fitting of single channel NMDA activity with a mechanism composed of independent dimers of subunits. *J Physiol 569*, 395–418.

Schwenk, J., Harmel, N., Brechet, A., Zolles, G., Berkefeld, H., Müller, C.S., Bildl, W. et al. (2012). High-resolution proteomics unravel architecture and molecular diversity of native AMPA receptor complexes. *Neuron 74*, 621–633.

Schwenk, J., Harmel, N., Zolles, G., Bildl, W., Kulik, A., Heimrich, B., Chisaka, O. et al. (2009). Functional proteomics identify cornichon proteins as auxiliary subunits of AMPA receptors. *Science 323*, 1313–1319.

Scimemi, A., Fine, A., Kullmann, D.M., and Rusakov, D.A. (2004). NR2B-containing receptors mediate cross talk among hippocampal synapses. *J Neurosci 24*, 4767–4777.

Semenov, A., Möykkynen, T., Coleman, S.K., Korpi, E.R., and Keinänen, K. (2012). Autoinactivation of the stargazin-AMPA receptor complex: Subunit-dependency and independence from physical dissociation. *PLoS One 7*, e49282.

Seo, J.H., Fox, J.G., Peek, R.M., and Hagen, S.J. (2011). N-methyl D-aspartate channels link ammonia and epithelial cell death mechanisms in Helicobacter pylori Infection. *Gastroenterology 141*, 2064–2075.

Shanks, N.F., Maruo, T., Farina, A.N., Ellisman, M.H., and Nakagawa, T. (2010). Contribution of the global subunit structure and stargazin on the maturation of AMPA receptors. *J Neurosci 30*, 2728–2740.

Shanks, N.F., Savas, J.N., Maruo, T., Cais, O., Hirao, A., Oe, S., Ghosh, A. et al. (2012). Differences in AMPA and kainate receptor interactomes facilitate identification of AMPA receptor auxiliary subunit GSG1L. *Cell Rep 1*, 590–598.

She, K., Ferreira, J.S., Carvalho, A.L., and Craig, A.M. (2012). Glutamate binding to the GluN2B subunit controls surface trafficking of N-methyl-D-aspartate (NMDA) receptors. *J Biol Chem 287*, 27432–27445.

Sheardown, M.J., Nielsen, E.O., Hansen, A.J., Jacobsen, P., and Honoré, T. (1990). 2,3-Dihydroxy-6-nitro-7-sulfamoyl-benzo(F)quinoxaline: A neuroprotectant for cerebral ischemia. *Science 247*, 571–574.

Shi, Y., Lu, W., Milstein, A.D., and Nicoll, R.A. (2009). The stoichiometry of AMPA receptors and TARPs varies by neuronal cell type. *Neuron 62*, 633–640.

Sia, G.M., Béïque, J.C., Rumbaugh, G., Cho, R., Worley, P.F., and Huganir, R.L. (2007). Interaction of the N-terminal domain of the AMPA receptor GluR4 subunit with the neuronal pentraxin NP1 mediates GluR4 synaptic recruitment. *Neuron 55*, 87–102.

Siegler Retchless, B., Gao, W., and Johnson, J.W. (2012). A single GluN2 subunit residue controls NMDA receptor channel properties via intersubunit interaction. *Nat Neurosci 15*, 406–413, S1–S2.

Silver, R.A., Traynelis, S.F., and Cull-Candy, S.G. (1992). Rapid-time-course miniature and evoked excitatory currents at cerebellar synapses in situ. *Nature 355*, 163–166.

Sjöström, P.J., Rancz, E.A., Roth, A., and Häusser, M. (2008). Dendritic excitability and synaptic plasticity. *Physiol Rev 88*, 769–840.

Smothers, C.T. and Woodward, J.J. (2007). Pharmacological characterization of glycine-activated currents in HEK 293 cells expressing N-methyl-D-aspartate NR1 and NR3 subunits. *J Pharmacol Exp Ther 322*, 739–748.

Smothers, C.T. and Woodward, J.J. (2009). Expression of glycine-activated diheteromeric NR1/NR3 receptors in human embryonic kidney 293 cells Is NR1 splice variant-dependent. *J Pharmacol Exp Ther 331*, 975–984.

Sobolevsky, A.I., Rosconi, M.P., and Gouaux, E. (2009). X-ray structure, symmetry and mechanism of an AMPA-subtype glutamate receptor. *Nature 462*, 745–756.

Sobolevsky, A.I., Yelshansky, M.V., and Wollmuth, L.P. (2004). The outer pore of the glutamate receptor channel has 2-fold rotational symmetry. *Neuron 41*, 367–378.

Sommer, B., Keinanen, K., Verdoorn, T.A., Wisden, W., Burnashev, N., Herb, A., Kohler, M., Takagi, T., Sakmann, B., and Seeburg, P.H. (1990). Flip and flop: A cell-specific functional switch in glutamate-operated channels of the CNS. *Science 249*, 1580–1585.

Sommer, B., Köhler, M., Sprengel, R., and Seeburg, P.H. (1991). RNA editing in brain controls a determinant of ion flow in glutamate-gated channels. *Cell 67*, 11–19.

Soto, D., Coombs, I.D., Kelly, L., Farrant, M., and Cull-Candy, S.G. (2007). Stargazin attenuates intracellular polyamine block of calcium-permeable AMPA receptors. *Nat Neurosci 10*, 1260–1267.

Soto, D., Coombs, I.D., Renzi, M., Zonouzi, M., Farrant, M., and Cull-Candy, S.G. (2009). Selective regulation of long-form calcium-permeable AMPA receptors by an atypical TARP, gamma-5. *Nat Neurosci 12*, 277–285.

Stawski, P., Sumser, M., and Trauner, D. (2012). A photochromic agonist of AMPA receptors. *Angew Chem Int Ed Engl 51*, 5748–5751.

Steinberg, J.P., Takamiya, K., Shen, Y., Xia, J., Rubio, M.E., Yu, S., Jin, W., Thomas, G.M., Linden, D.J., and Huganir, R.L. (2006). Targeted in vivo mutations of the AMPA receptor subunit GluR2 and its interacting protein PICK1 eliminate cerebellar long-term depression. *Neuron 49*, 845–860.

Stern-Bach, Y., Bettler, B., Hartley, M., Sheppard, P.O., O'Hara, P.J., and Heinemann, S.F. (1994). Agonist selectivity of glutamate receptors is specified by two domains structurally related to bacterial amino acid-binding proteins. *Neuron 13*, 1345–1357.

Stern-Bach, Y., Russo, S., Neuman, M., and Rosenmund, C. (1998). A point mutation in the glutamate binding site blocks desensitization of AMPA receptors. *Neuron 21*, 907–918.

Stocca, G. and Vicini, S. (1998). Increased contribution of NR2A subunit to synaptic NMDA receptors in developing rat cortical neurons. *J Physiol 507(Pt 1)*, 13–24.

Straub, C., Hunt, D.L., Yamasaki, M., Kim, K.S., Watanabe, M., Castillo, P.E., and Tomita, S. (2011). Distinct functions of kainate receptors in the brain are determined by the auxiliary subunit Neto1. *Nat Neurosci 14*, 866–873.

Sugihara, H., Moriyoshi, K., Ishii, T., Masu, M., and Nakanishi, S. (1992). Structures and properties of seven isoforms of the NMDA receptor generated by alternative splicing. *Biochem Biophys Res Commun 185*, 826–832.

Sumioka, A., Yan, D., and Tomita, S. (2010). TARP phosphorylation regulates synaptic AMPA receptors through lipid bilayers. *Neuron 66*, 755–767.

Sun, Y., Olson, R., Horning, M., Armstrong, N., Mayer, M., and Gouaux, E. (2002). Mechanism of glutamate receptor desensitization. *Nature 417*, 245–253.

Swanson, G.T., Green, T., Sakai, R., Contractor, A., Che, W., Kamiya, H., and Heinemann, S.F. (2002). Differential activation of individual subunits in heteromeric kainate receptors. *Neuron 34*, 589–598.

Sylantyev, S., Savtchenko, L.P., Ermolyuk, Y., Michaluk, P., and Rusakov, D.A. (2013). Spike-driven glutamate electrodiffusion triggers synaptic potentiation via a homer-dependent mGluR-NMDAR link. *Neuron 77*, 528–541.

Szobota, S., Gorostiza, P., Del Bene, F., Wyart, C., Fortin, D.L., Kolstad, K.D., Tulyathan, O. et al. (2007). Remote control of neuronal activity with a light-gated glutamate receptor. *Neuron 54*, 535–545.

Takamiya, K., Mao, L., Huganir, R.L., and Linden, D.J. (2008). The glutamate receptor-interacting protein family of GluR2-binding proteins is required for long-term synaptic depression expression in cerebellar Purkinje cells. *J Neurosci 28*, 5752–5755.

Teichert, R.W., Jimenez, E.C., Twede, V., Watkins, M., Hollmann, M., Bulaj, G., and Olivera, B.M. (2007). Novel conantokins from *Conus parius* venom are specific antagonists of N-methyl-D-aspartate receptors. *J Biol Chem 282*, 36905–36913.

Terashima, A., Pelkey, K.A., Rah, J.C., Suh, Y.H., Roche, K.W., Collingridge, G.L., McBain, C.J., and Isaac, J.T. (2008). An essential role for PICK1 in NMDA receptor-dependent bidirectional synaptic plasticity. *Neuron 57*, 872–882.

Tichelaar, W., Safferling, M., Keinänen, K., Stark, H., and Madden, D.R. (2004). The three-dimensional structure of an ionotropic glutamate receptor reveals a dimer-of-dimers assembly. *J Mol Biol 344*, 435–442.

Tomita, S., Adesnik, H., Sekiguchi, M., Zhang, W., Wada, K., Howe, J.R., Nicoll, R.A., and Bredt, D.S. (2005). Stargazin modulates AMPA receptor gating and trafficking by distinct domains. *Nature 435*, 1052–1058.

Tomita, S., Fukata, M., Nicoll, R.A., and Bredt, D.S. (2004). Dynamic interaction of stargazin-like TARPs with cycling AMPA receptors at synapses. *Science 303*, 1508–1511.

Tomita, S., Shenoy, A., Fukata, Y., Nicoll, R.A., and Bredt, D.S. (2007). Stargazin interacts functionally with the AMPA receptor glutamate-binding module. *Neuropharmacology 52*, 87–91.

Tovar, K.R., McGinley, M.J., and Westbrook, G.L. (2013). Triheteromeric NMDA receptors at hippocampal synapses. *J Neurosci 33*, 9150–9160.

Tóth, K. (2011). Zinc in neurotransmission. *Annu Rev Nutr 31*, 139–153.

Traynelis, S.F. and Cull-Candy, S.G. (1991). Pharmacological properties and H^+ sensitivity of excitatory amino acid receptor channels in rat cerebellar granule neurones. *J Physiol 433*, 727–763.

Traynelis, S.F., Hartley, M., and Heinemann, S.F. (1995). Control of proton sensitivity of the NMDA receptor by RNA splicing and polyamines. *Science 268*, 873–876.

Tsien, J.Z., Chen, D.F., Gerber, D., Tom, C., Mercer, E.H., Anderson, D.J., Mayford, M., Kandel, E.R., and Tonegawa, S. (1996a). Subregion- and cell type-restricted gene knockout in mouse brain. *Cell 87*, 1317–1326.

Tsien, J.Z., Huerta, P.T., and Tonegawa, S. (1996b). The essential role of hippocampal CA1 NMDA receptor-dependent synaptic plasticity in spatial memory. *Cell 87*, 1327–1338.

Turrigiano, G. (2008). Homeostatic synaptic plasticity. In *Structural and Functional Organization of the Synapse* (Boston, MA: Springer), pp. 535–552.

Turski, L., Huth, A., Sheardown, M., McDonald, F., Neuhaus, R., Schneider, H.H., Dirnagl, U., Wiegand, F., Jacobsen, P., and Ottow, E. (1998). ZK200775: A phosphonate quinoxalinedione AMPA antagonist for neuroprotection in stroke and trauma. *Proc Natl Acad Sci USA 95*, 10960–10965.

Vance, K.M., Simorowski, N., Traynelis, S.F., and Furukawa, H. (2011). Ligand-specific deactivation time course of GluN1/GluN2D NMDA receptors. *Nat Commun 2*, 294.

Veran, J., Kumar, J., Pinheiro, P.S., Athané, A., Mayer, M.L., Perrais, D., and Mulle, C. (2012). Zinc potentiates GluK3 glutamate receptor function by stabilizing the ligand binding domain dimer interface. *Neuron 76*, 565–578.

Vesikansa, A., Sakha, P., Kuja-Panula, J., Molchanova, S., Rivera, C., Huttunen, H.J., Rauvala, H., Taira, T., and Lauri, S.E. (2012). Expression of GluK1c underlies the developmental switch in presynaptic kainate receptor function. *Sci Rep 2*, 310.

Vigh, J. and von Gersdorff, H. (2005). Prolonged reciprocal signaling via NMDA and GABA receptors at a retinal ribbon synapse. *J Neurosci 25*, 11412–11423.

Vissel, B., Krupp, J.J., Heinemann, S.F., and Westbrook, G.L. (2001). A use-dependent tyrosine dephosphorylation of NMDA receptors is independent of ion flux. *Nat Neurosci 4*, 587–596.

Vivithanaporn, P., Yan, S., and Swanson, G.T. (2006). Intracellular trafficking of KA2 kainate receptors mediated by interactions with coatomer protein complex I (COPI) and 14–3–3 chaperone systems. *J Biol Chem 281*, 15475–15484.

Volgraf, M., Gorostiza, P., Numano, R., Kramer, R.H., Isacoff, E.Y., and Trauner, D. (2006). Allosteric control of an ionotropic glutamate receptor with an optical switch. *Nat Chem Biol 2*, 47–52.

Volk, L.J., Bachman, J.L., Johnson, R., Yu, Y., and Huganir, R.L. (2013). PKM-ζ is not required for hippocampal synaptic plasticity, learning and memory. *Nature 493*, 420–423.

von Engelhardt, J., Mack, V., Sprengel, R., Kavenstock, N., Li, K.W., Stern-Bach, Y., Smit, A.B., Seeburg, P.H., and Monyer, H. (2010). CKAMP44: A brain-specific protein attenuating short-term synaptic plasticity in the dentate gyrus. *Science 327*, 1518–1522.

Walker, C.S., Jensen, S., Ellison, M., Matta, J.A., Lee, W.Y., Imperial, J.S., Duclos, N. et al. (2009). A novel *Conus snail* polypeptide causes excitotoxicity by blocking desensitization of AMPA receptors. *Curr Biol 19*, 900–908.

Walters, M.R., Kaste, M., Lees, K.R., Diener, H.C., Hommel, M., De Keyser, J., Steiner, H., and Versavel, M. (2005). The AMPA antagonist ZK 200775 in patients with acute ischaemic stroke: A double-blind, multicentre, placebo-controlled safety and tolerability study. *Cerebrovasc Dis 20*, 304–309.

Watanabe, J., Beck, C., Kuner, T., Premkumar, L.S., and Wollmuth, L.P. (2002). DRPEER: A motif in the extracellular vestibule conferring high Ca^{2+} flux rates in NMDA receptor channels. *J Neurosci 22*, 10209–10216.

Watanabe, K., Kanno, T., Oshima, T., Miwa, H., Tashiro, C., and Nishizaki, T. (2008). Vagotomy upregulates expression of the N-methyl-D-aspartate receptor NR2D subunit in the stomach. *J Gastroenterol 43*, 322–326.

Watkins, J.C. and Evans, R.H. (1981). Excitatory amino acid transmitters. *Annu Rev Pharmacol Toxicol 21*, 165–204.

Werner, P., Voigt, M., Keinänen, K., Wisden, W., and Seeburg, P.H. (1991). Cloning of a putative high-affinity kainate receptor expressed predominantly in hippocampal CA3 cells. *Nature 351*, 742–744.

Weston, M.C., Schuck, P., Ghosal, A., Rosenmund, C., and Mayer, M.L. (2006). Conformational restriction blocks glutamate receptor desensitization. *Nat Struct Mol Biol 13*, 1120–1127.

Whitlock, J.R., Heynen, A.J., Shuler, M.G., and Bear, M.F. (2006). Learning induces long-term potentiation in the hippocampus. *Science 313*, 1093–1097.

Wollmuth, L.P., Kuner, T., Jatzke, C., Seeburg, P.H., Heintz, N., and Zuo, J. (2000). The Lurcher mutation identifies delta 2 as an AMPA/kainate receptor-like channel that is potentiated by Ca(2+). *J Neurosci 20*, 5973–5980.

Wong, H.K., Liu, X.B., Matos, M.F., Chan, S.F., Pérez-Otaño, I., Boysen, M., Cui, J. et al. (2002). Temporal and regional expression of NMDA receptor subunit NR3A in the mammalian brain. *J Comp Neurol 450*, 303–317.

Wu, Y., Arai, A.C., Rumbaugh, G., Srivastava, A.K., Turner, G., Hayashi, T., Suzuki, E. et al. (2007). Mutations in ionotropic AMPA receptor 3 alter channel properties and are associated with moderate cognitive impairment in humans. *Proc Natl Acad Sci USA 104*, 18163–18168.

Wyllie, D.J., Béhé, P., and Colquhoun, D. (1998). Single-channel activations and concentration jumps: Comparison of recombinant NR1a/NR2A and NR1a/NR2D NMDA receptors. *J Physiol 510(Pt 1)*, 1–18.

Xia, P., Chen, H.S., Zhang, D., and Lipton, S.A. (2010). Memantine preferentially blocks extrasynaptic over synaptic NMDA receptor currents in hippocampal autapses. *J Neurosci 30*, 11246–11250.

Yao, Y. and Mayer, M.L. (2006). Characterization of a soluble ligand binding domain of the NMDA receptor regulatory subunit NR3A. *J Neurosci 26*, 4559–4566.

Yao, Y., Harrison, C.B., Freddolino, P.L., Schulten, K., and Mayer, M.L. (2008). Molecular mechanism of ligand recognition by NR3 subtype glutamate receptors. *EMBO J 27*, 2158–2170.

Yelshansky, M.V., Sobolevsky, A.I., Jatzke, C., and Wollmuth, L.P. (2004). Block of AMPA receptor desensitization by a point mutation outside the ligand-binding domain. *J Neurosci 24*, 4728–4736.

Yu, X.M. and Salter, M.W. (1999). Src, a molecular switch governing gain control of synaptic transmission mediated by N-methyl-D-aspartate receptors. *Proc Natl Acad Sci USA 96*, 7697–7704.

Zhang, W., Howe, J.R., and Popescu, G.K. (2008). Distinct gating modes determine the biphasic relaxation of NMDA receptor currents. *Nat Neurosci 11*, 1373–1375.

Zhang, W., St-Gelais, F., Grabner, C.P., Trinidad, J.C., Sumioka, A., Morimoto-Tomita, M., Kim, K.S. et al. (2009). A transmembrane accessory subunit that modulates kainate-type glutamate receptors. *Neuron 61*, 385–396.

Zhao, H., Berger, A.J., Brown, P.H., Kumar, J., Balbo, A., May, C.A., Casillas, E. et al. (2012). Analysis of high-affinity assembly for AMPA receptor amino-terminal domains. *J Gen Physiol 139*, 371–388.

Zhu, S., Stroebel, D., Yao, C.A., Taly, A., and Paoletti, P. (2013). Allosteric signaling and dynamics of the clamshell-like NMDA receptor GluN1 N-terminal domain. *Nat Struct Mol Biol 20(4)*, 477–485.

22 5-HT$_3$ receptors

Sarah C.R. Lummis

Contents

22.1 INTRODUCTION

The 5-HT$_3$ receptor is a Cys-loop ligand-gated ion channel and is structurally and functionally distinct from the other six classes of 5-HT receptors whose actions are mediated via G-proteins. 5-HT$_3$ receptors are pentamers, and five classes of subunit have been identified. These are widely distributed, both in the nervous system and in other tissues. 5-HT$_3$ receptor activation opens a cation-selective ion channel, and receptor function can be modulated by a wide range of compounds including anesthetics, opioids, and alcohols. 5-HT$_3$ receptors play a major role in the vomiting reflex; regulate gut motility, secretion, and peristalsis in the enteric nervous system; and are involved in information transfer in the gastrointestinal (GI) tract. Disturbances within the 5-HT$_3$ receptor system may contribute to the etiopathogenesis of a range of neurological, GI, and immunological disorders.

22.2 SUBUNIT DIVERSITY

22.2.1 5-HT3A SUBUNITS

The first cDNA clone encoding a 5-HT$_3$ receptor subunit, the mouse 5-HT3A receptor subunit, was isolated by functional screening of a mouse neuroblastoma (NCB20) cDNA library (Maricq et al., 1991). Subsequently, the full-length cDNAs for orthologous 5-HT3A receptor subunits have been cloned from a range of species including human (Belelli et al., 1995; Miyake et al., 1995), guinea pig (Lankiewicz et al., 1998), ferret (Mochizuki et al., 2000), and dog (Jensen et al., 2006). The homology between 5-HT3A receptor subunits and those from other Cys-loop receptors clearly indicates 5-HT$_3$ receptors are members of this family, although somewhat unusually, they can readily form functional homomeric receptors. This suggests they are perhaps evolutionarily ancient, but the lack of 5-HT$_3$ receptor homologues in invertebrates indicates they are more recently evolved than at least some other Cys-loop receptors, such as those for acetylcholine and GABA (Dent, 2006). 5-HT$_3$ receptors can also function as heteromeric proteins, and the presence of additional subunits was indicated from a range of studies some years before such subunits were identified, for example, when expressed in HEK-293 cells, 5-HT$_3$A receptors had a single channel conductance of <1 pS, while channel activity in rabbit nodose ganglion revealed a single-channel conductance of 19 pS (Gill et al., 1995). In 1999, a second subunit, the 5-HT3B subunit, was identified (Davies et al., 1999; Dubin et al., 1999). Coexpression of this subunit with the 5-HT3A subunit resulted in functional receptors with properties that more closely represented those found in some native receptors. Since then, three other subunits (5-HT3C, 5-HT3D, and 5-HT3E; see Figure 22.1) have been identified (Niesler et al., 2003).

The repertoire of 5-HT$_3$ receptor subunits is increased by a number of different isoforms and promoters, alternative splicing, single nucleotide polymorphisms (SNPs), and posttranslational modifications (Figure 22.2) (Holbrook et al., 2009; Niesler et al., 2003, 2008). Alternative splicing of the transcript encoding guinea pig, mouse, and rat (but not dog, ferret, or human), 5-HT3A subunits results in long (5-HT3A(a)) and short (5-HT3A(b)) isoforms, where the 5-HT3A(b) isoform lacks 5 or 6 amino acid residues within the M3–M4 intracellular loop, resulting in some subtle differences in receptor properties (Hope et al., 1993; Lankiewicz et al., 1998). In the human 5-HT3A subunit, three different splice variants have been described. Two of these (5-HT3AL, HT3Rext) would result in larger receptors, while the other (5-HT3AT) codes for a partial receptor containing only a single transmembrane domain (TMD). HT3Rext has not yet been functionally evaluated, but 5-HT3AT and 5-HT3AL, while not functional when expressed alone, form receptors with modified functional properties when coexpressed with canonical 5-HT3A subunits (Bruss et al., 2000).

22.2.2 5-HT3B, 5-HT3C, 5-HT3D, AND 5-HT3E SUBUNITS

In the human HT3B gene, alternative tissue-specific promoters have the potential to create truncated versions of the canonical 5-HT3B subunit (Figure 22.3) (Tzvetkov et al., 2007). One of these only has a few amino acids missing from the N-terminus, while the other is devoid of a large proportion of the extracellular domain (ECD). The functional significance of these isoforms has not yet been evaluated.

The 5-HT3C subunits appear to have the largest number of isoforms with at least five homologous genes reported in humans, compared to two for the 5-HT3D subunit and three for 5-HT3E (Holbrook et al., 2009; Niesler et al., 2003, 2007, 2008). As yet many of these subunits have not been demonstrated to contribute to physiological receptors, so their significance is yet to be established.

Studies have indicated that microRNAs could also play a role in regulation of 5-HT$_3$R subunit expression. A mutation in the 3′UTR of the 5-HT3E gene impairs binding of has-miR-510, resulting in enhanced expression (Kapeller et al., 2008). Further work is required to see if other genes are similarly influenced.

The stoichiometry of heteromeric receptors is still not clear, although it has been established that only 5-HT$_3$A subunits can form functional homomeric 5-HT$_3$ receptors, and the presence of at least one 5-HT$_3$A subunit appears to be obligatory in heteromeric receptors (Holbrook et al., 2009; Niesler et al., 2008). 5-HT$_3$AB receptors were originally suggested to possess a 3B/2A subunit ratio, and atomic force microscopy with tagged subunits indicated a BABBA arrangement (Barrera et al., 2005). However more recent data reveal a 3A/2B ration with an ABAAB arrangement (Miles et al., 2013; Lochner and Lummis, 2010; Thompson et al., 2011b), which is more consistent with the characteristics of these receptors (Thompson and Lummis, 2013). The arrangement and number of 5-HT3C, 5-HT3D, and 5-HT3E subunits in functional receptors have not yet been determined, although there are many possibilities (Figure 22.2).

22.3 STRUCTURE

5-HT$_3$ receptors, like other Cys-loop receptors, are pentameric assemblies of five identical or nonidentical subunits that pseudosymmetrically surround the ion pore (Boess et al., 1992; Green et al., 1995). Each subunit has a large ECD, a TMD consisting of four membrane-spanning α-helices (M1–M4), and an intracellular domain (ICD) between M3 and M4 (Figures 22.1 and 22.4). The structure of the homomeric 5-HT3A receptor only became available in 2014 (Hassaine et al.,2014), some years after the structures of related proteins were solved. Many of these related structures, which were derived using cryoelectron microscope and x-ray crystallography, are good structural representatives, and those that were used for templates before the structure became available include the nACh receptor, many acetylcholine-binding proteins (AChBPs), the invertebrate glutamate-gated chloride channel (GluCl), and the bacterial homologues *Erwinia* and *Gloeobacter* ligand-gated ion channels (ELIC and GLIC). Examples of such homology models can be found in recent publications (e.g., Reeves and Lummis, 2002; Thompson et al., 2010; 2011a; Verheij et al., 2012).

22.3.1 EXTRACELLULAR DOMAIN

Homology models of the ECD (Figure 22.4) are largely supported by experimental data, and thus we currently have a good understanding of many details of this region. The ECD contains the agonist-binding site, which is located at the interface of two adjacent subunits and is formed by three loops (A–C) from one (the principal) subunit and three β-strands (referred to as loops D–F) from the adjacent or complementary subunit; key residues that contribute to the binding pocket in these loops have been identified from a range of studies (see Barnes et al., 2009; Thompson and Lummis, 2006; Thompson et al., 2010 for reviews). In loop A, substitutions in the sequence ^{128}AsnGluPhe^{130} modify receptor function, although the only residue that has been extensively investigated is Glu129, which forms one or more hydrogen bonds critical for binding (Boess et al., 1997; Price et al., 2008; Steward et al., 2000; Sullivan et al., 2006). Loop B is very sensitive to modification, with substitution of many residues ablating function; this region is an obligate rigid structure with an extensive hydrogen bond network (Hassaine et al., 2014;

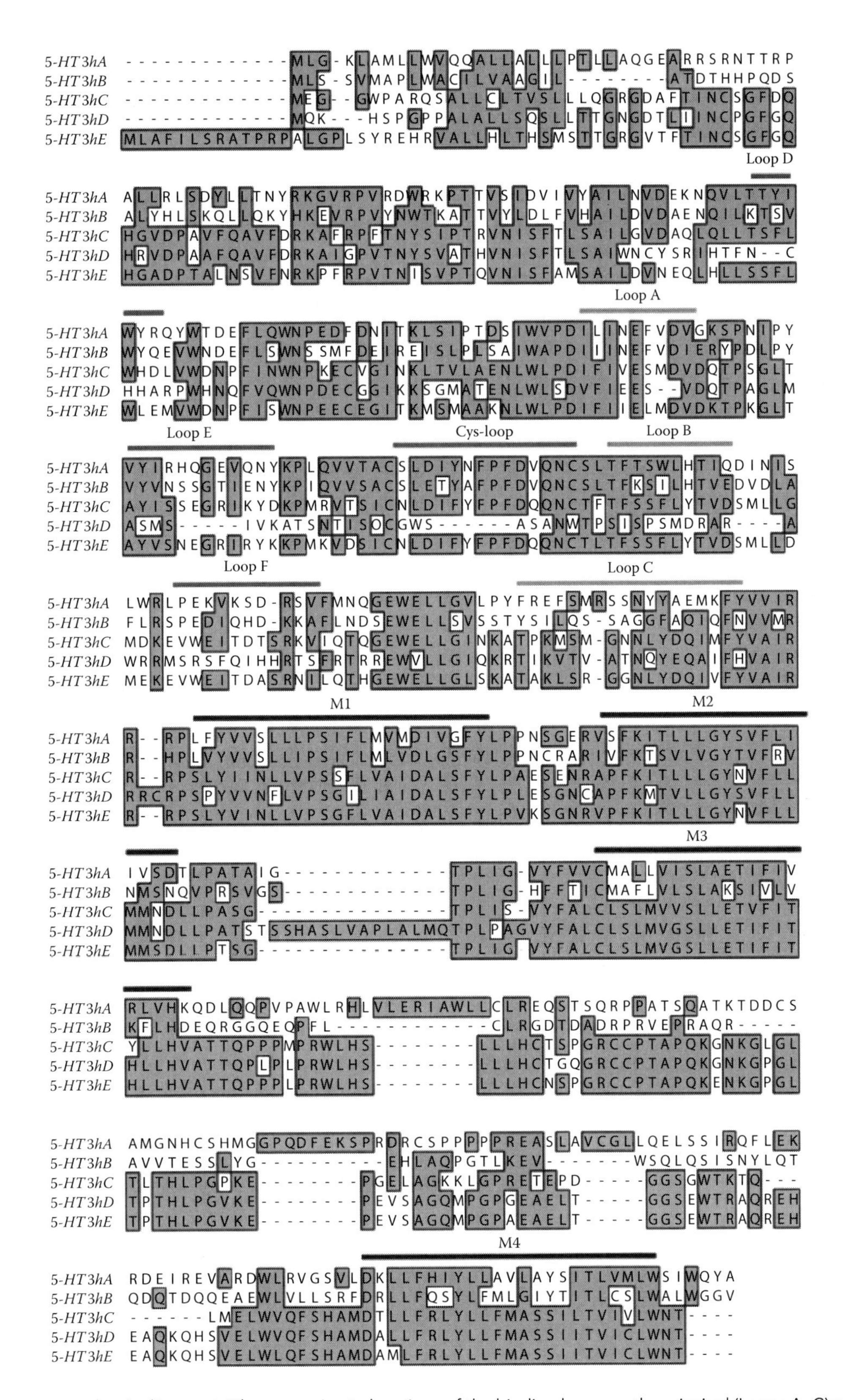

Figure 22.1 5-HT$_3$ receptor subunit alignment. The approximate locations of the binding loops on the principal (Loops A–C) and complementary (Loops D–F) faces, the Cys loop, and the transmembrane α-helices (M1–M4). Each subunit has multiple isoforms only one of which is shown here.

Thompson et al., 2008). The loop B Trp residue plays an especially critical role as it forms a cation–π interaction with the primary amine of 5-HT (Beene et al., 2002; Spier and Lummis, 2000). Loop C shows the largest variability and is important in determining the species specificity of various ligands, with multiple regions of the loop being important. Loops C and D both contribute a Trp to the *aromatic box* found in all Cys loop receptors, and Loop D also contributes an Arg to the binding site. Many Loop E residues, and at least three loop F residues, have been shown to be important for ligand binding, although a role in

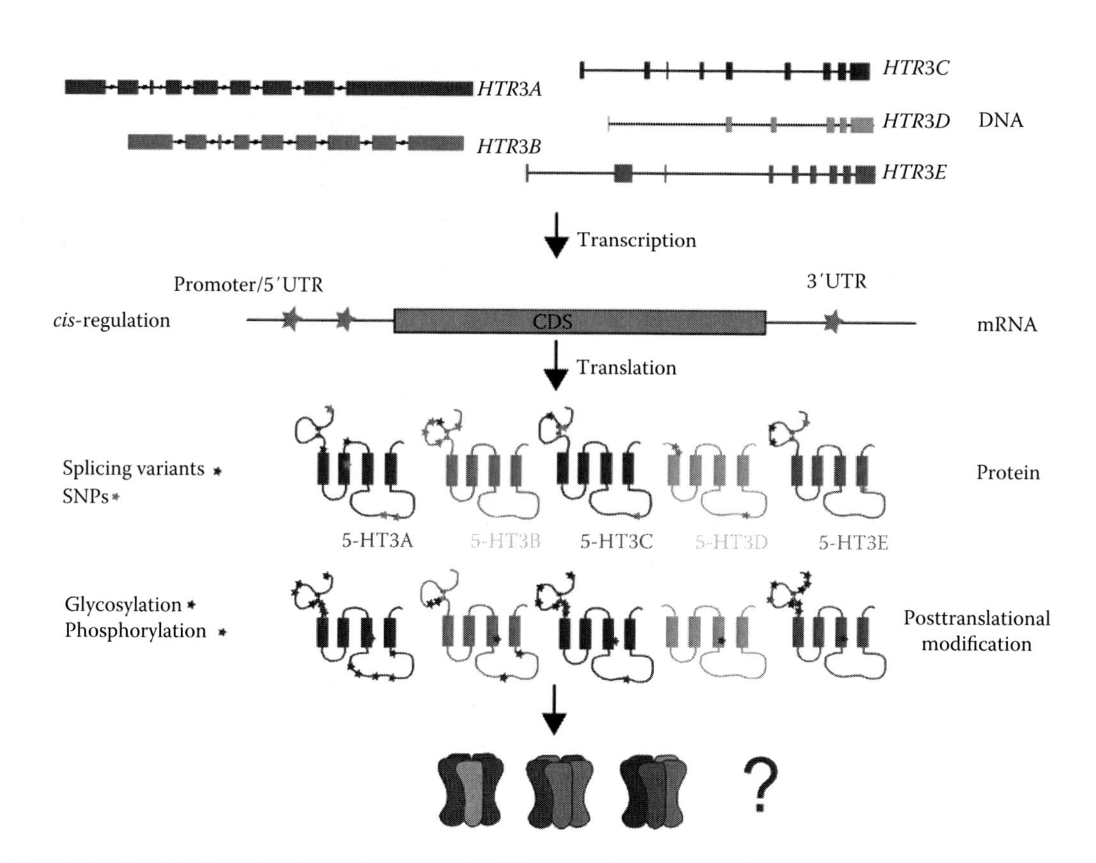

Figure 22.2 Molecular basis of the functional and pharmacological diversity of receptors in the 5-HT_3 receptor family. Receptor heterogeneity may be achieved at many levels in humans including (1) the existence of different subunits (5-HT3A-5-HT3E), (2) the use of alternative tissue-specific promoters driving differential expression, (3) alternative splicing (mostly SNPs), and (4) posttranslational modification (e.g., phosphorylation or glycosylation). The subunit composition of physiologically expressed heteromeric 5-HT_3 receptors is not yet known but there are many possibilities. (Reproduced from Walstab, J. et al., *Pharmacol. Ther.*, 128, 146, 2010. With permission.)

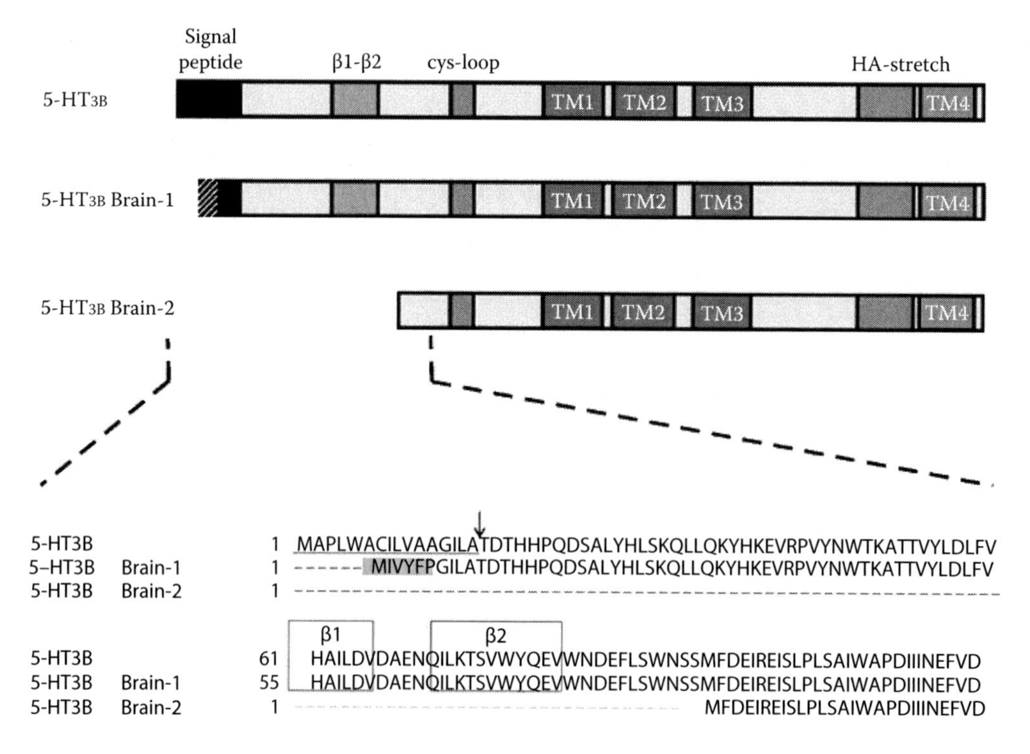

Figure 22.3 Protein sequences predicted from the alternative 5-HT3B transcripts. The N-terminal localization signal and the β1–β2 loop structure mediating channel gating are missing in the Brain-2 isoform. The part of the localization signal that differs between the intestinal and Brain-1 form is marked by hatching. The N-terminal sequences expected to differ between the isoforms are shown in the lower part of the figure. The signal sequence (underlined) and the cutting site of the signal peptidase (arrow) are shown for the canonical 5-HT_{3B} form. The six amino acids of the N-terminus of the Brain-1 isoform that differ from the canonical form are shaded. The two β-sheets are boxed and the potential N-glycosylation sites are shown in bold. (Reproduced from Tzvetkov, M.V. et al., *Gene*, 386, 52, 2007. With permission.)

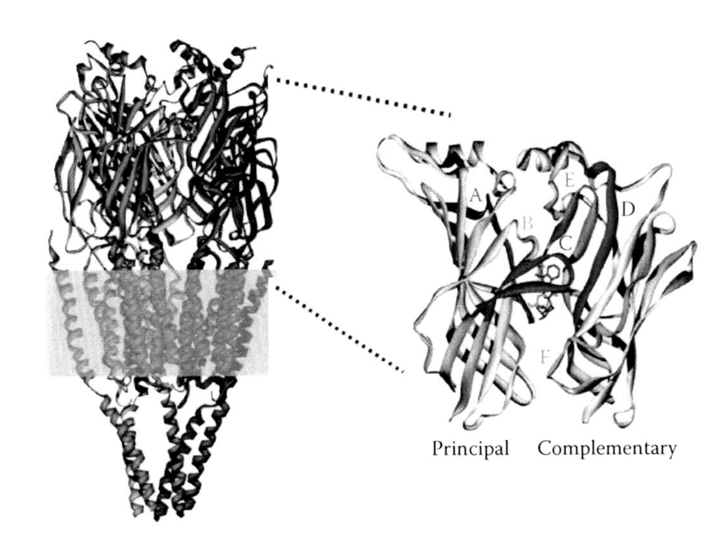

Figure 22.4 Homology model of the 5-HT$_3$ receptor. The five different subunits are shown in different colors, and the transmembrane region is in a grey box in the main image, which is based on data from the nACh receptor. The inset shows a homology model of the ECD based on AChBP, showing two adjacent subunits and the regions that contribute to the binding site (loops A–E). 5-HT (green) and granisetron (orange) are shown docked into the binding site.

conformational changes associated with receptor function, or that alters the structure of the binding pocket, cannot yet be ruled out (Price and Lummis, 2004; Thompson and Lummis, 2006, 2006a; Venkataraman et al., 2002).

22.3.2 TRANSMEMBRANE DOMAIN

The TMD of each of the 5-HT$_3$ receptor subunits is composed of four transmembrane α-helices (M1–M4), with short loops between M1 and M2 (intracellular) and M2 and M3 (extracellular). The M2 α-helices line the ion pore and extrapolation from nACh receptor data indicates that the α-helices of M1 and M2 project above the level of the membrane (Figure 22.5). M1, M3, and M4 protect M2 from the surrounding membrane lipids, although various mutations in these α-helices have been shown to modify channel function, suggesting that some residues or regions in these α-helices are involved in receptor activation. The conserved proline in M1, for example, is essential for activation, and the receptor is expressed but cannot function when this proline is replaced by alanine, glycine, or leucine (Dang et al., 2000). However, substitution with *trans*-3-methyl-proline, pipecolic acid, or leucic acid yields active channels similar to wild-type receptors. The commonality between these residues and proline is the lack of hydrogen bond donor activity; thus, the data suggest this is a key element in channel gating, possibly because of the resulting flexibility in secondary structure in this region of M1. Some regions, however, are probably purely structural; a minumum number of C-terminal residues in M4, for example, are essential for the expression of 5-HT$_3$ receptor on the cell surface (Butler et al., 1990).

22.3.3 INTRACELLULAR DOMAIN

The ICD is formed by the large M3–M4 intracellular loop, and is responsible for receptor modulation, and possibly also plays a role in trafficking. Deletion studies reveal the ICD is not essential, as the mouse 5-HT$_3$A receptor subunit ICD can be replaced by the heptapeptide M3–M4 linker of GLIC without loss of function (Jansen et al., 2008). Further evidence that the ICD can function as a separate domain comes from studies where it was added to the GLIC linker peptide, resulting in modification of GLIC function by the intracellular protein RIC-3 (Goyal et al., 2011).

ICD structural details are sparse, but each subunit is known to possess an α-helix that contributes to openings, known as portals, just below the level of the membrane. The residues that line these portals are important for ion conductance: when the 5-HT3A subunit residues are replaced with those found in the 5-HT3B subunit, the single-channel conductance, which is very low in the homomeric 5-HT$_3$A receptor, is increased to that of the heteromeric 5-HT$_3$AB receptor (Kelley et al., 2003; see Section 22.5 for more details).

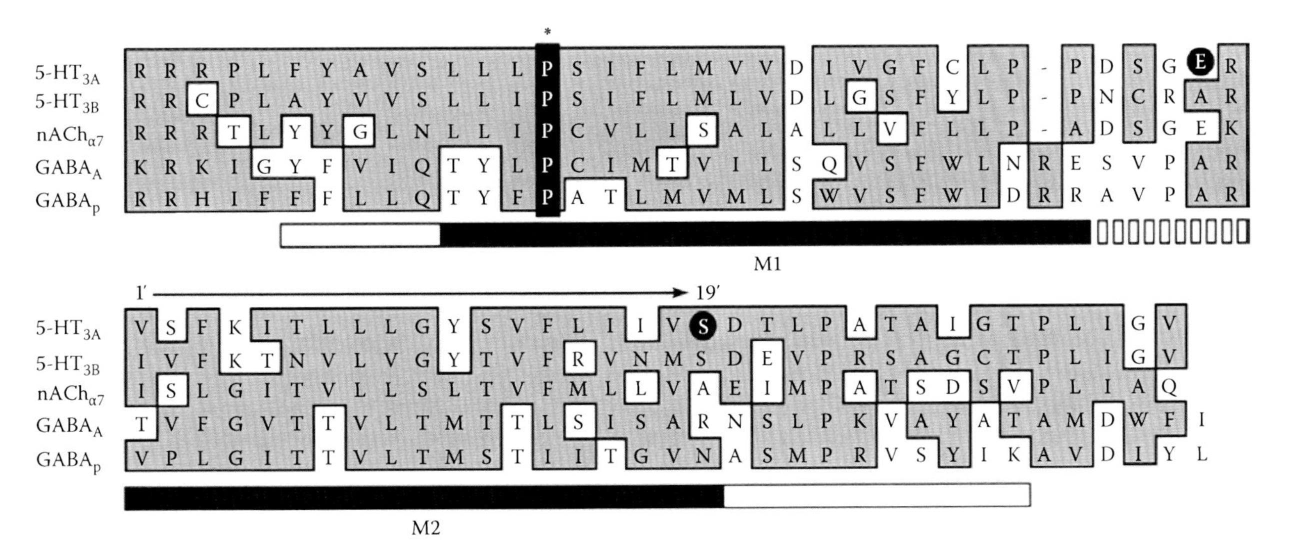

Figure 22.5 Sequence alignment of the M1, M2 (indicated by bars), and flanking regions of mouse 5-HT$_3$ and other representative Cys-loop receptor subunits. The white sections indicate regions of the structure that protrude above the membrane. The dashed line is the M1–M2 loop. The conserved proline residue in M1 is highlighted, as are the two residues in the 5-HT$_3$ receptor that result in a reversal of ion selectivity when mutated. The prime (′) labeling system in M2 is also shown. (Reproduced from Lummis, 2004. With permission.)

22.4 PHYSIOLOGICAL ROLES AND EXPRESSION

5-HT$_3$ receptors are located in many brain areas including the hippocampus, entorhinal cortex, frontal cortex, cingulated cortex, dorsal horn ganglia, amygdala, nucleus accumbens, substantia nigra, and ventral tegmental area (Barnes et al., 2009; Pratt et al., 1990). The dorsal vagal complex in the brainstem, which is key to the vomiting reflex and contains the area postrema and nucleus tractus solitarius, has the highest levels, consistent with the potent antiemetic properties of 5-HT$_3$ receptor antagonists (Pratt et al., 1990).

5-HT$_3$ receptor activation in the CNS can modulate the release of a variety of neurotransmitters, including dopamine, cholecystokinin, GABA, substance P, and acetylcholine (Barnes et al., 2009). In the rat brain, 5-HT$_3$ receptors are expressed by subsets of inhibitory interneurons in the CA1 area and dentate gyrus of the hippocampus and layer I of the cerebral cortex (Kawa, 1994; Ropert and Guy, 1991; Zhou and Hablitz, 1999). Presynaptic 5-HT$_3$ receptor activation at these interneurons likely results in sufficient Ca^{2+} entry to influence GABA release and cause an increase in the frequency of GABA$_A$ receptor-mediated spontaneous inhibitory postsynaptic currents (sIPSCs). Presynaptic 5-HT$_3$ receptors also facilitate the release of glutamate onto dorsal vagal preganglionic, nucleus tractus solitarius, and area postrema neurons (Glaum et al., 1992; Jeggo et al., 2005).

Postsynaptic 5-HT$_3$ receptor activation contributes to fast excitatory synaptic transmission in a range of locations including the lateral amygdala, nucleus tractus solitarius, visual cortex, and neocortical GABAergic interneurons that contain cholecystokinin and vasoactive intestinal peptide (Ferezou et al., 2002; Glaum et al., 1992; Koyama et al., 2000; Sugita et al., 1992). Indeed, 5-HT$_3$ receptors are often colocalized with cholecystokinin and also with central CB1 cannabinoid receptors: in neurons of the rat telencephalon, anterior olfactory nucleus, cortex, hippocampus, dentate gyrus, and amygdala, 37%–53% of all neurons expressing the 5-HT3A receptor subunit also expressed CB1 transcripts, and 5-HT3A/CB1-expressing neurons also contained GABA (Morales et al., 2004).

5-HT$_3$ receptors have long been known to play a role in the emetic response, and thus it is not surprising that they are involved in information transfer in the GI tract, and in the enteric nervous system they regulate gut motility and peristalsis (Galligan, 2002). They also play an important role in the urinary tract, and expression of constitutively active hypersensitive 5-HT$_3$ receptors in mice leads to excitotoxic neuronal cell death, resulting in fatal uropathy (Bhattacharya et al., 2004).

The functional studies are supported by expression studies, with 5-HT$_3$A receptor mRNA and protein being observed in regions known to have 5-HT$_3$ receptors. Such studies also indicate 5-HT$_3$ receptor expression in a wide range of tissues other than the CNS, including peripheral and sensory ganglia, and the GI tract (Barnes et al., 2009). In addition, expression of the 5-HT3A subunit has been reported in immune cells such as monocytes, chondrocytes, T cells, synovial tissue, and platelets (Fiebich et al., 2004; Stratz et al., 2008).

There was initial controversy as to the presence of 5-HT3B subunits in brain, but later studies show it is expressed here, with a preference for distinct *brain-type* isoforms. The longer canonical B subunit is broadly expressed in many tissues including the kidney, liver, and the GI tract, with relatively high levels in the spleen, colon, small intestine, and kidney (Holbrook et al., 2009; Tzvetkov et al., 2007).

5-HT$_3$C, 5-HT$_3$D, and 5-HT$_3$E receptor subunits were first identified in humans, and genes for these proteins have now been shown to exist in a range of species, although not in rodents (Holbrook et al., 2009; Niesler et al., 2008). Initial studies suggested that the 5-HT3D and 5-HT3E subunits had a very restricted expression in the GI tract, but more recent data suggest all these subunits have a widespread distribution. Studies examining protein levels have lagged behind the genetic work, but expression of 5-HT3C, 5-HT3D, and 5-HT3E subunits at the protein level in the GI tract has been confirmed (Kapeller et al., 2011).

22.5 BIOPHYSICAL PROPERTIES

22.5.1 RECEPTOR ACTIVATION

5-HT$_3$A receptors mediate rapidly activating and desensitizing inward currents, which show inward rectification. Concentration response studies from many groups indicate multiple agonist-binding sites, and occupation of at least two binding sites is needed for maximal activation (Jackson and Yakel, 1995). Detailed studies using a high conductance mutant of the 5-HT$_3$A receptor show that full activation arises from receptors with three agonist molecules bound (Corradi et al., 2009). The kinetic model derived from these data shows that a conformational change of the fully liganded receptor occurs while the channel is still closed, and the receptor subsequently enters an open–closed cycle involving three open states.

22.5.2 IONIC SELECTIVITY

The 5-HT$_3$ receptor pore is a relatively nonselective cation channel, constructed of five pseudosymmetrically arranged M2 α-helices (one from each of the five subunits). The residues that line the ion-accessible face of M2 are predominantly nonpolar and are the major controlling influence on ion flux (McKinnon et al., 2011; Panicker et al., 2002; Reeves et al., 2001). Currents are primarily carried by Na^+ and K^+ ions, although divalent and small organic cations are also permeable (Derkach et al., 1989; Maricq et al., 1991; Yang, 1990).

Ionic selectivity is predominantly mediated via residues in M2: a triple mutant receptor with a proline insertion at −1′ and the substitutions E−1′A and V13′T resulted in an anion permeable mouse 5-HT$_3$A receptor (Gunthorpe and Lummis, 2001), although subsequent studies showed that the replacement of only two residues (E−1′, S19′R) was needed to invert ion selectivity (Figure 22.3) (Thompson and Lummis, 2003). Changing −1′E alone to Ala resulted in nonselective channels indicating that the rings of charge at either end of M2 charge make the most critical contribution (Thompson and Lummis, 2003). Equivalent residues participate in the selectivity filters of other Cys-loop receptors (Thompson et al., 2010).

5-HT_3A receptors are almost equally permeable to monovalent and divalent cations (P_{Ca}/P_{Cs} = 1.0–1.4) (Brown et al., 1998; Davies et al., 1999; Livesey et al., 2011). However, human 5-HT_3AB receptors have lower Ca^{2+} permeability (P_{Ca}/P_{Cs} = 0.6 [Davies et al., 1999]), possibly the consequence of the 20′ residue being neutral (Asn and not Asp) in human 5-HT3B subunits. Consistent with this, a D20′A substitution in 5-HT3A subunits reduces Ca^{2+} permeability (P_{Ca}/P_{Cs} = 0.4) (Livesey et al., 2008). Studies suggest that the ICD may also play a role in Ca^{2+} permeability as substitutions of charged residues in this region can have a major effect on Ca^{2+} permeability (Livesey et al., 2011).

22.5.3 SINGLE-CHANNEL CONDUCTANCE

The single-channel conductance of the homomeric receptor is low: values of 0.4–0.76 pS have been reported (Davies et al., 1999; Gunthorpe et al., 2000; Kelley et al., 2003). The presence of Lys in the M2 region was originally considered a possible explanation, but substitutions here revealed this was not the case (Gunthorpe et al., 2000). Subsequent work revealed that the low conductance was due to residues located in the amphipathic helix of the M3–M4 loop, which forms portals through which ions can cross between the intracellular vestibule of the receptor and the cell interior (Kelley et al., 2003).

22.5.4 HETEROMERIC RECEPTOR PROPERTIES

Heteromeric 5-HT_3AB receptors show a range of distinct biophysical characteristics when compared to 5-HT3A receptors: their desensitization is more rapid, and concentration response curves reveal lower EC_{50}s and Hill slopes (Figure 22.6); their voltage dependence is linear, and their divalent cation permeability is much reduced; most noticeable, however, is their large single-channel conductance (13–16 pS) (Peters et al., 2005). This has been shown to be due to the substitution of Arg residues in the M3–M4 helical region of the 5-HT3A subunit by neutral or negatively charged residues in the 5-HT3B subunit (Kelley et al., 2003).

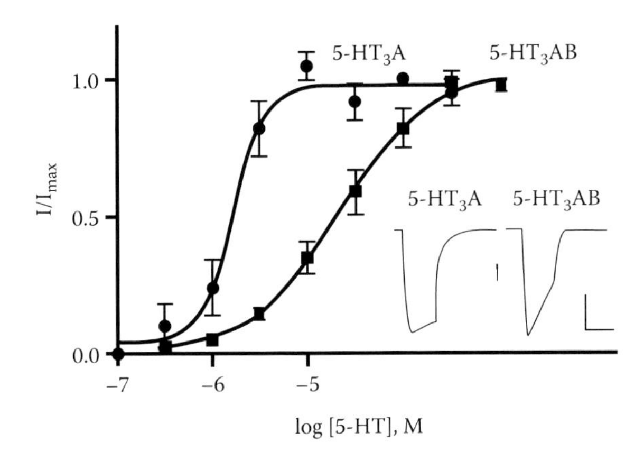

Figure 22.6 Typical whole-cell responses of 5-HT_3A and 5-HT_3AB receptors. Concentration response curves of 5-HT_3A and 5-HT_3AB receptors expressed in *Xenopus* oocytes and measured using two voltage electrode clamps reveal different EC_{50}s and Hill slopes, and (inset) typical current traces at 100 μM 5-HT reveal distinct kinetics of the responses. Scale bar represents I μA and 5 s.

Other heteromeric receptors have not yet been extensively investigated, but studies to date indicate that they are more similar to homomeric 5-HT_3A receptors.

22.5.5 CHANNEL GATE

The gate of the 5-HT_3A receptor channel is located centrally within M2 (Panicker et al., 2002), consistent with the *hydrophobic girdle* model of channel gating, the hydrophobic girdle being a region in the center of M2 that is less than 3.5 Å in diameter over a distance of ~8 Å; the residues that face the pore here are hydrophobic (13′Val and 9′Leu) making it effectively impermeable to ions in the closed conformation (Miyazawa et al., 2003). Data consistent with this hypothesis are substitutions of Val13′ residues in the 5-HT_3A receptor by threonine or serine, which causes an increase in agonist potency (Dang et al., 2000), or spontaneous channel openings (Bhattacharya et al., 2004), and substitutions of Leu9′ by a range of amino acids, which affects agonist potency and desensitization rates (Yakel et al., 1993).

22.6 PHARMACOLOGY

22.6.1 5-HT_3 RECEPTOR AGONISTS

There are currently many selective and potent compounds that act at the 5-HT_3 receptor (Figure 22.7). 5-HT_3 receptor agonists have in common a basic amine, an aromatic ring, a hydrophobic group, and two hydrogen bond acceptors, and active compounds include 2-methyl-5-HT, phenylbiguanide, and m-chlorophenylbiguanide (Cockcroft et al., 1995; Kilpatrick et al., 1990). Data from AChBP suggest that agonists are relatively small compounds that cause the C-loop to close over the binding site, initiating the gating process, and the structure of 5-HTBP, which is a modified version of AChBP that binds serotonin and granisetron, supports a similar mechanism of action for the 5-HT_3 receptor (Kesters et al., 2013; Tsetlin and Hucho, 2009).

22.6.2 5-HT_3 RECEPTOR ANTAGONISTS

5-HT_3 receptor competitive antagonists, which bind at the orthosteric (agonist) binding site, are usually larger than agonists; they require an aromatic part, a basic moiety, and an intervening hydrogen bond acceptor. For most antagonists, these are a rigid aromatic or heteroaromatic ring system, a basic amine, and a carbonyl group (or isosteric equivalent) that is coplanar to the aromatic system (Evans et al., 1991), and there are slightly longer distances between the aromatic and amine groups when compared to the agonist pharmacophore. Only small substituents, such as a methyl group, can be accommodated on the charged amine (Schmidt and Peroutka, 1989). Many potent antagonists of 5-HT_3 receptors have 6.5 heterocyclic rings, and the most potent compounds contain an aromatic 6-membered ring. Morphine and cocaine were the first antagonists used to characterize the 5-HT_3 receptor (Gaddum and Picarelli, 1957), with more selective 5-HT_3 antagonists being developed in the 1980s: MDL72222 or bemestron (Fozard, 1984) and ICS 205–930 or tropisetron (Donatsch et al., 1985). Later compounds that were developed include ondansetron, granisetron, and zacopride, which act at nanomolar concentrations, and there is now a wide range of similarly potent compounds, with many containing

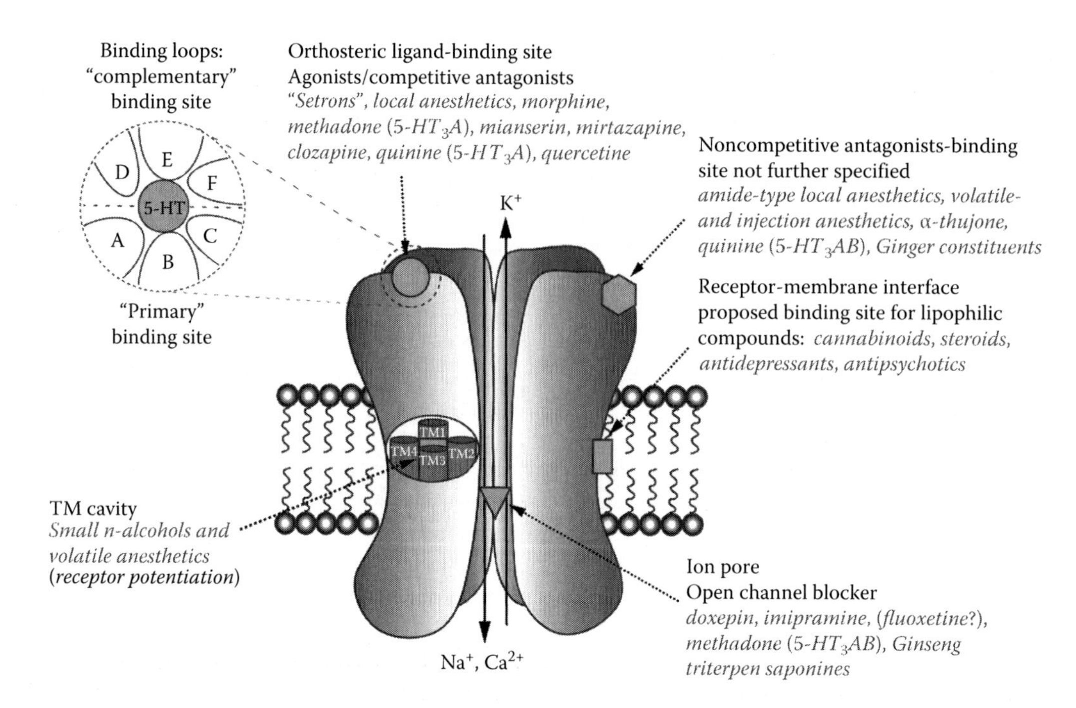

Figure 22.7 Schematic representation of the 5-HT$_3$ receptor. One subunit has been removed to reveal the cation-selective channel. The binding sites of agonist, antagonist, and modulators are shown. The *setrons* include ondansetron, granisetron, tropisetron, bemesetron, dolasetron, and palonosetron. (Reproduced from Walstab, J. et al., *Pharmacol. Ther.*, 128, 146, 2010. With permission.)

bicyclic heteroaromatic structures, that is, quinoxalines, quinazolines, or quinolines (Verheij et al., 2012). Data indicate that these compounds, and therefore possibly all ligands that bind to the orthosteric site, are stabilized in the binding pocket by interaction with water molecules (Figure 22.8).

Antagonists may also act in 5-HT$_3$ receptor channel. Picrotoxin, which was originally considered to be relatively specific as a GABA$_A$ receptor noncompetitive inhibitor, blocks the 5-HT$_3$ receptor channel; the binding of picrotoxinin (the active component of picrotoxin) has been localized to the 6′ position of M2 (Das and Dillon, 2003; Thompson et al., 2011a). Compounds structurally similar to picrotoxin, such as the ginkgolides and bilobalide, act similarly (Thompson et al., 2011a). Diltiazem, which blocks voltage-gated calcium channels, is also known to block the 5-HT$_3$ receptor pore and act close to the 7′ and/or 12′ residues in homomeric receptors (Gunthorpe and Lummis, 1999; Thompson et al., 2011a). Morphine and its analogue methadone and the antimalarial compounds quinine and mefloquine may also exert their inhibitory effects via binding to the pore, highlighting the common mechanisms that many of these drugs share, and also the promiscuity that many of these compounds display (Baptista-Hon et al., 2012; Brady et al., 2001; Deeb et al., 2009).

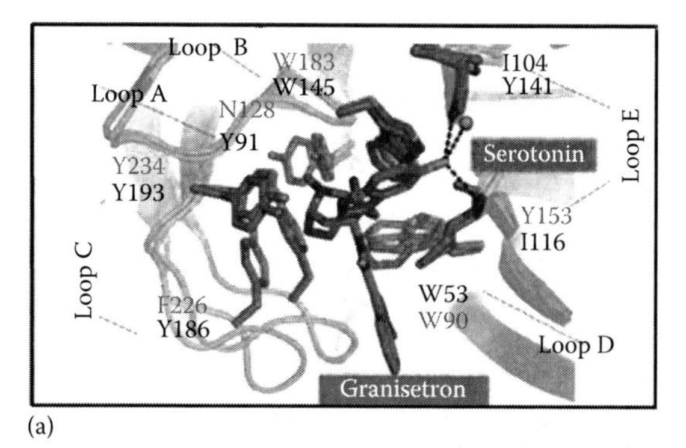

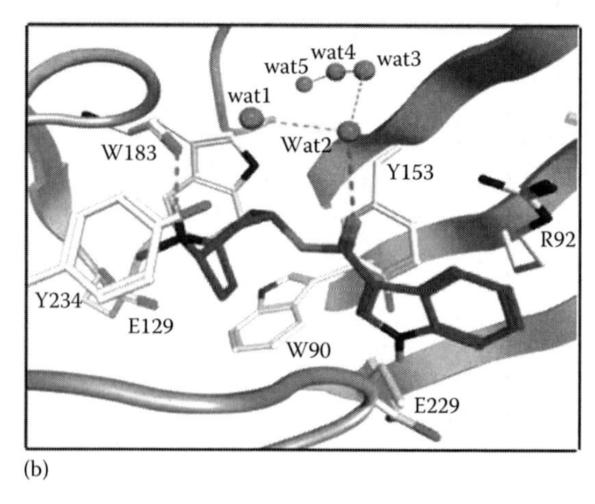

Figure 22.8 The 5-HT$_3$ receptor–binding pocket. (a) The orientation of 5-HT and granisetron has been revealed in 5-HTBP, a version of AChBP whose binding site has been modified to resemble that in the 5-HT$_3$ receptor. The equivalent residues found in the 5-HT3A subunit are shown in grey. (Reproduced from Kesters et al., 2013 with permission.) (b) A homology model of the 5-HT$_3$ receptor bound to the antagonist tropisetron reveals that water is integral to ligand binding. (Reproduced from Verheij, M.H. et al., *J. Med. Chem.*, 55, 8603, 2012. With permission.)

22.6.3 5-HT$_3$ RECEPTOR MODULATORS

There are a number of allosteric modulators that effect 5-HT$_3$ receptor function including n-alcohols, anesthetics, antidepressants, cannabinoids, opioids, steroids, and natural compounds; these can inhibit or enhance receptor activity and many also modulate other Cys-loop receptors, although not always in the same direction (see reviews by Davies, 2011; Machu, 2011; Walstab et al., 2010). Specific binding sites for these compounds have mostly not yet been confirmed, although some may bind in an intersubunit binding cavity at the top of the TMD (Trattnig et al., 2012). The effects of these compounds have mostly been studied on 5-HT$_3$A receptors to date, although alcohols and inhaled anesthetics have been shown to have reduced sensitivity at 5-HT$_3$AB receptors, while the effects of etomidate, propofol, and pentobarbital are similar at 5-HT$_3$A and 5-HT$_3$AB receptors (Rusch et al., 2007; Solt et al., 2005; Stevens et al., 2005).

22.7 GATING MECHANISMS AND SUBUNIT CONTRIBUTIONS

22.7.1 GATING MECHANISMS

The 5-HT$_3$ receptor is an allosteric protein, and extrapolation of data from nACh receptors suggests that binding of agonists causes a structural change in the ECD that is transduced through the protein to the TMD, opening the integral pore. Structural data suggest the ECD links to the TMD via the β1–β2 and β8–β9 loops and the β10 strand, which are close to residues in the M2–M3 linker, regions known to be important for receptor function (Thompson et al., 2010; Tsetlin and Hucho, 2009). In support of this mechanism, coupling of binding to gating in a chimeric AChBP (ECD)/5-HT$_3$ (ICD) receptor could only be achieved when three amino acid regions from the ECD of the 5-HT$_3$ receptor were substituted into the corresponding regions of AChBP (Bouzat et al., 2004). There is, however, a critical M1–M2 proline but no ECD/TMD salt bridge in the 5-HT$_3$ receptor in contrast to nACh and GABAC receptors (Lee and Sine, 2005; Lummis et al., 2005; Price et al., 2007; Wang et al., 2007), although it is clear that charged residues are important (Dougherty, 2008; Xiu et al., 2005). Thus, it may be that specific details of channel gating are subtly different in different Cys-loop receptors.

22.7.2 SUBUNIT CONTRIBUTIONS

5-HT$_3$B receptor subunits do not form functional homomeric receptors, but can coexpress with 5-HT3A subunits to yield heteromeric 5-HT$_3$AB receptors that differ from 5-HT$_3$A receptors in a range of properties (EC$_{50}$, Hill slope, desensitization kinetics, calcium permeability, shape of current–voltage relationship, and single-channel conductance) as described in Section 22.5. Despite these biophysical differences, the pharmacology of the orthosteric site of 5-HT$_3$A and 5-HT$_3$AB receptors is almost identical. This is consistent with the action of agonists and competitive antagonists being at an AA interface, as has been described for both human and mouse receptors (Lochner and Lummis, 2010; Thompson et al., 2011b), but conflicting with older data suggesting a BABBA arrangement determined using atomic force microscopy (Barrera et al., 2005).

There have been only two studies to date examining the functional effects of subunits 5HT3C, 5HT3D, and 5HT3E (Holbrook et al., 2009; Niesler et al., 2007). None of these subunits form functional homomers, but they do form functional receptors when coexpressed with 5-HT3A subunits, although there is no reported difference in radioligand-binding characteristics, current–voltage relationships, or kinetics of whole-cell currents of the heteromers compared to homomers. Thus, although we are aware of their sequences (Figure 22.9), the physiological roles of these more recently identified subunits are still unclear.

22.8 REGULATION

5-HT$_3$ receptors can undergo posttranslational modification through phosphorylation at kinase consensus sites that have been identified in the ICD. While phosphorylation has been observed on a putative protein kinase A (PKA) site at S409 in the guinea pig 5-HT$_3$ receptor (Lankiewicz et al., 1998), most studies have been indirect. Thus, activation of PKA substantially accelerates the desensitization kinetics of 5-HT$_3$ receptors in mouse neuroblastoma and HEK-293 cells, and activators of protein kinase C (PKC) increase the amplitude of 5-HT-activated currents (Hubbard et al., 2000; Shao et al., 1991; Yakel et al., 1991; Zhang et al., 1995). Unexpectedly, point mutations of putative PKC sites did not affect the sensitivity of the mutant receptors to PKC potentiation (Coultrap and Machu, 2002; Sun et al., 2003). Instead, the authors propose that enhancement of 5-HT-elicited responses results from increased cell surface expression of the receptor due to structural rearrangements of F-actin with which the receptor clusters (Sun et al., 2003). This suggests that PKC modulation of 5-HT$_3$A receptor function occurs via an F-actin-dependent mechanism, a conclusion that is consistent with observations that 5-HT$_3$A receptors colocalize with F-actin-rich membrane domains. As neurotransmitter release can be regulated through an actin-dependent mechanism, and 5-HT$_3$A receptors can modulate neurotransmitter release, it seems possible that PKC enhancement of 5-HT$_3$ receptor function may play a role in modulating the efficacy of 5-HT$_3$ receptor–elicited transmission.

5-HT$_3$A receptors can also specifically interact with the light chain of microtubule-associated protein 1B (MAP1B-LC1) altering receptor desensitization (Sun et al., 2008). This may play a role in shaping the efficacy of 5-HT$_3$ receptor-mediated synaptic transmission.

22.9 CELL BIOLOGY

5-HT$_3$ receptors have been followed from *birth* to *death* in elegant studies from Vogel's group (Ilegems et al., 2004). 5-HT$_3$A receptors formed in the endoplasmic reticulum (ER) and Golgi are trafficked in vesicle-like structures along microtubules to the plasma membrane. They aggregate in specific subcellular sites that can be disrupted by F-actin depolymerization (Emerit et al., 2002). These aggregations have been identified in a range of cells, including neurons, consistent with the precise anatomical location that is required to mediate fast synaptic neurotransmission (Emerit et al., 2002; Grailhe et al., 2004; Ilegems et al., 2004). Agonist interaction with cell surface 5-HT$_3$ receptors results in their internalization and destruction or recycling (Ilegems et al., 2004).

The other subunits have been less well studied but it has been shown that 5-HT3B receptor subunits, when expressed alone, fail to exit the ER, providing a possible explanation as to why these subunits cannot form functional homomeric receptors. ER retention is due, at least in part, to a CRAR retention motif that forms part of the M1–M2 intracellular loop (Boyd et al., 2003). Coexpression with the 5-HT3A subunit may shield this ER retention motif allowing heteromeric 5-HT$_3$AB receptors to reach the cell surface. There is some evidence that the 5-HT3B

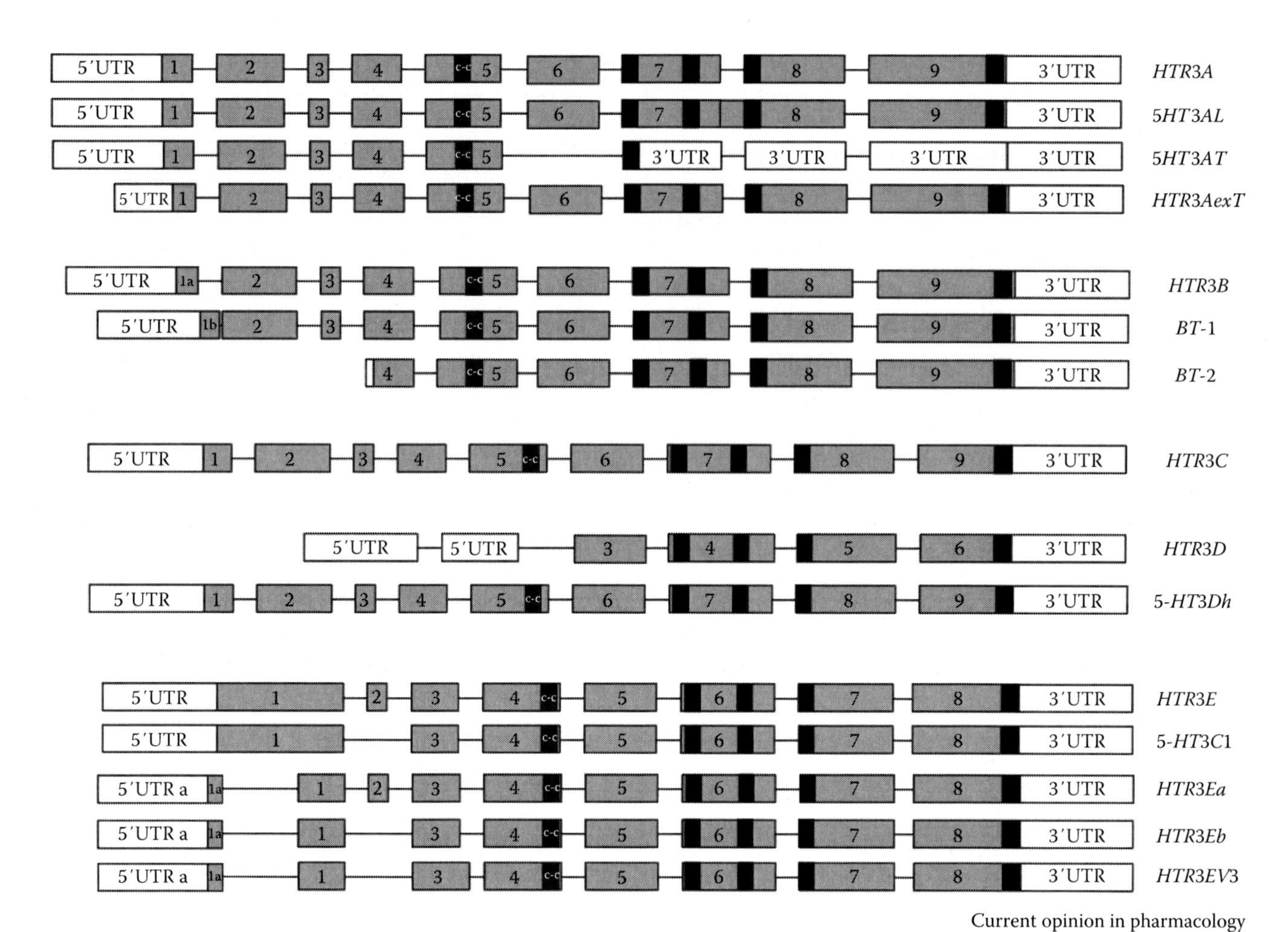

Figure 22.9 Schematic diagram of the different *HTR3* isoforms in their respective genomic contexts. Exons are indicated as light grey boxes and introns as black lines connecting the exons, and the untranslated regions (UTRs) are depicted as white boxes. Conserved domains of the respective 5-HT$_3$ subunits are indicated in black (transmembrane region) and dark grey (Cys-loop) boxes. The genes are not drawn to scale. (Reproduced from Niesler, 2011. With permission.)

subunit forces a preference for expression of the heteromeric 5-HT$_3$ receptor, as coexpression of the 5-HT3A and 5-HT3B subunits in tsA-201 cells did not indicate the presence of homomeric 5-HT3A receptors (Barrera et al., 2005), although the physiological relevance of this finding is not yet clear.

The human 5-HT3A subunit has four consensus sequence *N*-glycosylation sites in the N-terminal ECD domain and all can be *N*-glycosylated (Monk et al., 2004). *N*-glycosylation is essential for export from the ER, cell surface expression, and radioligand binding, although it is not necessary to preserve a ligand-binding site once the receptor has matured (Boyd et al., 2003; Green et al., 1995; Quirk et al., 2004). Three of the four *N*-glycosylation sites are conserved between a range of species (N104, N170, and N186) and appear to be critical, while the *N*-glycosylation site at residue 28 is less important and indeed absent in rodents.

BiP, calnexin, and RIC-3 have been identified as ER chaperone proteins that associate with the 5-HT$_3$ receptor and are likely to promote correct folding, oligomerization, posttranslational modification, and/or export from the ER (Boyd et al., 2003). RIC-3 has been the most widely studied but has different effects depending on the subunits, the species they originated from, and the expression system (Castillo et al., 2005; Cheng et al., 2005, 2007). Expression of human homomeric 5-HT$_3$A receptors in transfected mammalian cells, for example, is enhanced by RIC-3, but it causes inhibition of heteromeric (5-HT$_3$AB) receptor expression, and mouse 5-HT$_3$A receptor expression in oocytes is completely abolished (Cheng et al., 2005). This apparent discrepancy may be due to other proteins that could influence 5-HT$_3$ receptor expression. Cyclophilin A, for example, promotes 5-HT$_3$A receptor expression in the cell membrane via an integral peptidyl prolyl isomerase activity (Helekar and Patrick, 1997), and there may be a range of other proteins yet to be identified that can modify 5-HT$_3$ receptor expression.

22.10 CHANNELOPATHIES AND THERAPEUTIC POTENTIAL

Studies implicate the malfunction of 5-HT$_3$ receptors in a range of neurological and GI disorders (Niesler, 2011; Walstab et al., 2010). The identification of 5-HT$_3$ receptors in immune cells also suggests a possible role of 5-HT$_3$ receptors in immunological processes and inflammation and suggests that they may plausibly be involved in diseases like atherosclerosis, tendomyopathies, and fibromyalgia (Fiebich et al., 2004; Stratz et al., 2008).

Some SNPs have been identified in patients with bipolar affective disorder (BPAD) or schizophrenia, disorders that segregate with cytogenetic abnormalities involving a region on chromosome 11 that harbors the HTR3A gene (Weiss et al., 1995). Further studies are needed with two SNPs found in schizophrenic patients (R344H and P391R) to determine if they contribute to disease in these patients, but a significant association was found in BPAD with a P16S mutation in the 5-HT3A subunit, with reporter constructs indicating this mutant could modulate expression levels (Krzywkowski et al., 2007; Niesler et al., 2001; Thompson et al., 2006b). Additional SNPs in the HTR3A gene result in the 5-HT3A^{A33T} and 5-HT3A^{M257I} subunit variants, both of which are associated with reduced levels of cell surface expression, and the 5-HT3A^{S253N} variant that does not appear to compromise plasma membrane expression (Krzywkowski et al., 2007). The significance of these is yet to be determined.

In the 5-HT3B subunit, there has been an extensive investigation into a very common SNP, Y129S, which is linked both to BPAD and major depression in women (Hammer et al., 2012; Krzywkowski, 2006). Unusually, the Y129S variant is a gain of function mutation, as 5-HT$_3$AB^{Y129S} receptors have an increased maximal response to 5-HT, decreased desensitization and deactivation kinetics, and a sevenfold increase in mean-channel open time in comparison to heteromeric receptors containing the wild-type 5-HT3B subunit (Krzywkowski et al., 2008). An intermediate effect is apparent for receptors assembled from a mixture of wild-type 5-HT$_3$A, wild-type 5-HT$_3$B, and 5-HT$_3$B(Y129S) subunits, suggesting that signaling via the 5-HT$_3$AB receptor in heterozygous, as well as homozygous, individuals is altered by this SNP (Krzywkowski et al., 2008).

Studies in the more recently discovered HTR3C, HTR3D, and HTR3E genes also indicate possible involvement in disease. An SNP in the 5-HT3C gene (N163K) has been correlated with irritable bowel syndrome (IBS), and expression studies suggest it causes an increase in receptor density (Kapeller et al., 2011). Increased expression has also been associated with an SNP in the 3′UTR of the HTR3E gene, also associated with IBS, which inhibits the binding of a microRNA as described earlier (Kapeller et al., 2008).

IBS is one of the disorders in which 5-HT$_3$ receptor antagonists can be effective therapeutic agents, although their major use is to ameliorate nausea and vomiting in cancer patients receiving chemo- or radiation therapy, consistent with the role of 5-HT$_3$ receptors in the vomiting center. Studies suggest that a wide range of other diseases have the potential to be treated with 5-HT$_3$ receptor–selective drugs, including addiction, pruritus, emesis, migraine, chronic heart pain, bulimia, and neurological phenomena such as anxiety, psychosis, nociception, and cognitive function (Thompson and Lummis, 2007; Walstab et al., 2010).

REFERENCES

Baptista-Hon, D. T., Deeb, T. Z., Othman, N. A., Sharp, D., Hales, T. G., 2012. The 5-HT3B subunit affects high-potency inhibition of 5-HT$_3$ receptors by morphine. *Br J Pharmacol* 165, 693–704.

Barnes, N. M., Hales, T. G., Lummis, S. C., Peters, J. A., 2009. The 5-HT$_3$ receptor—The relationship between structure and function. *Neuropharmacology* 56, 273–284.

Barrera, N. P., Herbert, P., Henderson, R. M., Martin, I. L., Edwardson, J. M., 2005. Atomic force microscopy reveals the stoichiometry and subunit arrangement of 5-HT$_3$ receptors. *Proc Natl Acad Sci USA* 102, 12595–12600.

Beene, D. L., Brandt, G. S., Zhong, W., Zacharias, N. M., Lester, H. A., Dougherty, D. A., 2002. Cation-pi interactions in ligand recognition by serotonergic (5-HT$_3$A) and nicotinic acetylcholine receptors: The anomalous binding properties of nicotine. *Biochemistry* 41, 10262–10269.

Belelli, D., Balcarek, J. M., Hope, A. G., Peters, J. A., Lambert, J. J., Blackburn, T. P., 1995. Cloning and functional expression of a human 5-hydroxytryptamine type 3AS receptor subunit. *Mol Pharmacol* 48, 1054–1062.

Bhattacharya, A., Dang, H., Zhu, Q. M., Schnegelsberg, B., Rozengurt, N., Cain, G., Prantil, R. et al., 2004. Uropathic observations in mice expressing a constitutively active point mutation in the 5-HT$_3$A receptor subunit. *J Neurosci* 24, 5537–5548.

Boess, F. G., Lummis, S. C., Martin, I. L., 1992. Molecular properties of 5-hydroxytryptamine3 receptor-type binding sites purified from NG108-15 cells. *J Neurochem* 59, 1692–1701.

Boess, F. G., Steward, L. J., Steele, J. A., Liu, D., Reid, J., Glencorse, T. A., Martin, I. L., 1997. Analysis of the ligand binding site of the 5-HT$_3$ receptor using site directed mutagenesis: Importance of glutamate 106. *Neuropharmacology* 36, 637–647.

Bouzat, C., Gumilar, F., Spitzmaul, G., Wang, H. L., Rayes, D., Hansen, S. B., Taylor, P., Sine, S. M., 2004. Coupling of agonist binding to channel gating in an ACh-binding protein linked to an ion channel. *Nature* 430, 896–900.

Boyd, G. W., Doward, A. I., Kirkness, E. F., Millar, N. S., Connolly, C. N., 2003. Cell surface expression of 5-hydroxytryptamine type 3 receptors is controlled by an endoplasmic reticulum retention signal. *J Biol Chem* 278, 27681–27687.

Brady, C. A., Stanford, I. M., Ali, I., Lin, L., Williams, J. M., Dubin, A. E., Hope, A. G., Barnes, N. M., 2001. Pharmacological comparison of human homomeric 5-HT$_3$A receptors versus heteromeric 5-HT3A/3B receptors. *Neuropharmacology* 41, 282–284.

Brown, A. M., Hope, A. G., Lambert, J. J., Peters, J. A., 1998. Ion permeation and conduction in a human recombinant 5-HT$_3$ receptor subunit (h5-HT3A). *J Physiol* 507 (Part 3), 653–665.

Bruss, M., Barann, M., Hayer-Zillgen, M., Eucker, T., Gothert, M., Bonisch, H., 2000. Modified 5-HT3A receptor function by co-expression of alternatively spliced human 5-HT3A receptor isoforms. *Naunyn Schmiedebergs Arch Pharmacol* 362, 392–401.

Butler, A., Elswood, C. J., Burridge, J., Ireland, S. J., Bunce, K. T., Kilpatrick, G. J., Tyers, M. B., 1990. The pharmacological characterization of 5-HT$_3$ receptors in three isolated preparations derived from guinea-pig tissues. *Br J Pharmacol* 101, 591–598.

Castillo, M., Mulet, J., Gutierrez, L. M., Ortiz, J. A., Castelan, F., Gerber, S., Sala, S., Sala, F., Criado, M., 2005. Dual role of the RIC-3 protein in trafficking of serotonin and nicotinic acetylcholine receptors. *J Biol Chem* 280, 27062–27068.

Cheng, A., Bollan, K. A., Greenwood, S. M., Irving, A. J., Connolly, C. N., 2007. Differential subcellular localization of RIC-3 isoforms and their role in determining 5-HT$_3$ receptor composition. *J Biol Chem* 282, 26158–26166.

Cheng, A., McDonald, N. A., Connolly, C. N., 2005. Cell surface expression of 5-hydroxytryptamine type 3 receptors is promoted by RIC-3. *J Biol Chem* 280, 22502–22507.

Cockcroft, V., Ortells, M., Lunt, G., 1995. Ligands, receptor models, and evolution. *Ann N Y Acad Sci* 757, 40–47.

Corradi, J., Gumilar, F., Bouzat, C., 2009. Single-channel kinetic analysis for activation and desensitization of homomeric 5-HT$_3$A receptors. *Biophys J* 97, 1335–1345.

Coultrap, S. J., Machu, T. K., 2002. Enhancement of 5-hydroxytryptamine 3A receptor function by phorbol 12-myristate, 13-acetate is mediated by protein kinase C and tyrosine kinase activity. *Receptors Channels* 8, 63–70.

Dang, H., England, P. M., Farivar, S. S., Dougherty, D. A., Lester, H. A., 2000. Probing the role of a conserved M1 proline residue in 5-hydroxytryptamine$_3$ receptor gating. *Mol Pharmacol* 57, 1114–1122.

Das, P., Dillon, G. H., 2003. The 5-HT3B subunit confers reduced sensitivity to picrotoxin when co-expressed with the 5-HT3A receptor. *Brain Res Mol Brain Res* 119, 207–212.

Davies, P. A., 2011. Allosteric modulation of the 5-HT_3 receptor. *Curr Opin Pharmacol* 11, 75–80.

Davies, P. A., Pistis, M., Hanna, M. C., Peters, J. A., Lambert, J. J., Hales, T. G., Kirkness, E. F., 1999. The 5-HT3B subunit is a major determinant of serotonin-receptor function. *Nature* 397, 359–363.

Deeb, T. Z., Sharp, D., Hales, T. G., 2009. Direct subunit-dependent multimodal 5-hydroxytryptamine3 receptor antagonism by methadone. *Mol Pharmacol* 75, 908–917.

Dent, J. A., 2006. Evidence for a diverse Cys-loop ligand-gated ion channel superfamily in early bilateria. *J Mol Evol* 62, 523–535.

Derkach, V., Surprenant, A., North, R. A., 1989. 5-HT_3 receptors are membrane ion channels. *Nature* 339, 706–709.

Donatsch, P., Engel, G., Richardson, B. P., Stadler, P. A., 1985. ICS 205–930: A highly selective and potent antagonist at peripheral neuronal 5-hydroxytryptamine (5-HT) receptors. *Br J Pharmacol*, 81, 34P.

Dougherty, D. A., 2008. Cys-loop neuroreceptors: Structure to the rescue? *Chem Rev* 108, 1642–1653.

Dubin, A. E., Huvar, R., D'Andrea, M. R., Pyati, J., Zhu, J. Y., Joy, K. C., Wilson, S. J. et al., 1999. The pharmacological and functional characteristics of the serotonin 5-HT_3A receptor are specifically modified by a 5-HT(3B) receptor subunit. *J Biol Chem* 274, 30799–30810.

Emerit, M. B., Doucet, E., Darmon, M., Hamon, M., 2002. Native and cloned 5-HT_3A(S) receptors are anchored to F-actin in clonal cells and neurons. *Mol Cell Neurosci* 20, 110–124.

Evans, S. M., Galdes, A., Gall, M., 1991. Molecular modeling of 5-HT_3 receptor ligands. *Pharmacol Biochem Behav* 40, 1033–1040.

Ferezou, I., Cauli, B., Hill, E. L., Rossier, J., Hamel, E., Lambolez, B., 2002. 5-HT_3 receptors mediate serotonergic fast synaptic excitation of neocortical vasoactive intestinal peptide/cholecystokinin interneurons. *J Neurosci* 22, 7389–7397.

Fiebich, B. L., Akundi, R. S., Seidel, M., Geyer, V., Haus, U., Muller, W., Stratz, T., Candelario-Jalil, E., 2004. Expression of 5-HT_3A receptors in cells of the immune system. *Scand J Rheumatol Suppl* 119, 9–11.

Fozard, J. R., 1984. MDL 72222: A potent and highly selective antagonist at neuronal 5-hydroxytryptamine receptors. *Naunyn Schmiedebergs Arch Pharmacol* 326, 36–44.

Gaddum, J. H., Picarelli, Z. P., 1957. Two kinds of tryptamine receptor. *Br J Pharmacol Chemother* 12, 323–328.

Galligan, J. J., 2002. Ligand-gated ion channels in the enteric nervous system. *Neurogastroenterol Motil* 14, 611–623.

Gill, C. H., Peters, J. A., Lambert, J. J., 1995. An electrophysiological investigation of the properties of a murine recombinant 5-HT_3 receptor stably expressed in HEK 293 cells. *Br J Pharmacol* 114, 1211–1221.

Glaum, S. R., Brooks, P. A., Spyer, K. M., Miller, R. J., 1992. 5-Hydroxytryptamine-3 receptors modulate synaptic activity in the rat nucleus tractus solitarius in vitro. *Brain Res* 589, 62–68.

Goyal, R., Salahudeen, A. A., Jansen, M., 2011. Engineering a prokaryotic Cys-loop receptor with a third functional domain. *J Biol Chem* 286, 34635–34642.

Grailhe, R., de Carvalho, L. P., Paas, Y., Le Poupon, C., Soudant, M., Bregestovski, P., Changeux, J. P., Corringer, P. J., 2004. Distinct subcellular targeting of fluorescent nicotinic alpha 3 beta 4 and serotoninergic 5-HT3A receptors in hippocampal neurons. *Eur J Neurosci* 19, 855–862.

Green, T., Stauffer, K. A., Lummis, S. C., 1995. Expression of recombinant homo-oligomeric 5-hydroxytryptamine$_3$ receptors provides new insights into their maturation and structure. *J Biol Chem* 270, 6056–6061.

Gunthorpe, M. J., Lummis, S. C., 1999. Diltiazem causes open channel block of recombinant 5-HT_3 receptors. *J Physiol* 519(Part 3), 713–722.

Gunthorpe, M. J., Lummis, S. C., 2001. Conversion of the ion selectivity of the 5-HT_3a receptor from cationic to anionic reveals a conserved feature of the ligand-gated ion channel superfamily. *J Biol Chem* 276, 10977–10983.

Gunthorpe, M. J., Peters, J. A., Gill, C. H., Lambert, J. J., Lummis, S. C., 2000. The 4′ lysine in the putative channel lining domain affects desensitization but not the single-channel conductance of recombinant homomeric 5-HT_3A receptors. *J Physiol* 522(Part 2), 187–198.

Hammer, C., Cichon, S., Muhleisen, T. W., Haenisch, B., Degenhardt, F., Mattheisen, M., Breuer, R. et al., 2012. Replication of functional serotonin receptor type 3A and B variants in bipolar affective disorder: A European multicenter study. *Transl Psychiatry* 2, e103.

Hassaine, G., Deluz, C., Grasso, L., Wyss, R., Tol, M. B., Hovius, R., Graff, A., Stahlberg, H., Tomizaki, T., Desmyter, A., Moreau, C., Li, X. D., Poitevin, F., Vogel, H., Nury, H., 2014. X-ray structure of the mouse serotonin 5-HT_3 receptor, *Nature* 512, 276–281.

Helekar, S. A., Patrick, J., 1997. Peptidyl prolyl cis-trans isomerase activity of cyclophilin A in functional homo-oligomeric receptor expression. *Proc Natl Acad Sci USA* 94, 5432–5437.

Holbrook, J. D., Gill, C. H., Zebda, N., Spencer, J. P., Leyland, R., Rance, K. H., Trinh, H. et al., 2009. Characterisation of 5-HT3C, 5-HT3D and 5-HT3E receptor subunits: Evolution, distribution and function. *J Neurochem* 108, 384–396.

Hope, A. G., Downie, D. L., Sutherland, L., Lambert, J. J., Peters, J. A., Burchell, B., 1993. Cloning and functional expression of an apparent splice variant of the murine 5-HT_3 receptor A subunit. *Eur J Pharmacol* 245, 187–192.

Hubbard, P. C., Thompson, A. J., Lummis, S. C., 2000. Functional differences between splice variants of the murine 5-HT_3A receptor: Possible role for phosphorylation. *Brain Res Mol Brain Res* 81, 101–108.

Ilegems, E., Pick, H. M., Deluz, C., Kellenberger, S., Vogel, H., 2004. Noninvasive imaging of 5-HT_3 receptor trafficking in live cells: From biosynthesis to endocytosis. *J Biol Chem* 279, 53346–53352.

Jackson, M. B., Yakel, J. L., 1995. The 5-HT_3 receptor channel. *Annu Rev Physiol* 57, 447–468.

Jansen, M., Bali, M., Akabas, M. H., 2008. Modular design of Cys-loop ligand-gated ion channels: Functional 5-HT_3 and GABA rho1 receptors lacking the large cytoplasmic M3M4 loop. *J Gen Physiol* 131, 137–146.

Jeggo, R. D., Kellett, D. O., Wang, Y., Ramage, A. G., Jordan, D., 2005. The role of central 5-HT_3 receptors in vagal reflex inputs to neurones in the nucleus tractus solitarius of anaesthetized rats. *J Physiol* 566, 939–953.

Jensen, T. N., Nielsen, J., Frederiksen, K., Ebert, B., 2006. Molecular cloning and pharmacological characterization of serotonin 5-HT_3A receptor subtype in dog. *Eur J Pharmacol* 538, 23–31.

Kapeller, J., Houghton, L. A., Monnikes, H., Walstab, J., Moller, D., Bonisch, H., Burwinkel, B. et al., 2008. First evidence for an association of a functional variant in the microRNA-510 target site of the serotonin receptor-type 3E gene with diarrhea predominant irritable bowel syndrome. *Hum Mol Genet* 17, 2967–2977.

Kapeller, J., Moller, D., Lasitschka, F., Autschbach, F., Hovius, R., Rappold, G., Bruss, M., Gershon, M. D., Niesler, B., 2011. Serotonin receptor diversity in the human colon: Expression of serotonin type 3 receptor subunits 5-HT3C, 5-HT3D, and 5-HT3E. *J Comp Neurol* 519, 420–432.

Kawa, K., 1994. Distribution and functional properties of 5-HT_3 receptors in the rat hippocampal dentate gyrus: A patch-clamp study. *J Neurophysiol* 71, 1935–1947.

Kelley, S. P., Dunlop, J. I., Kirkness, E. F., Lambert, J. J., Peters, J. A., 2003. A cytoplasmic region determines single-channel conductance in 5-HT_3 receptors. *Nature* 424, 321–324.

Kesters, D., Thompson, A. D., Brams, M., van Elk, R., Spurny, R., Geitmann, M., Villalgordo, J. M. et al., 2013. Structural basis of ligand recognition in 5-HT_3 receptors. *EMBO J* 14, 49–56.

Kilpatrick, G. J., Butler, A., Burridge, J., Oxford, A. W., 1990. 1-(m-chlorophenyl)-biguanide, a potent high affinity 5-HT_3 receptor agonist. *Eur J Pharmacol* 182, 193–197.

Koyama, S., Matsumoto, N., Kubo, C., Akaike, N., 2000. Presynaptic 5-HT3 receptor-mediated modulation of synaptic GABA release in the mechanically dissociated rat amygdala neurons. *J Physiol* 529(Part 2), 373–383.

Krzywkowski, K., 2006. Do polymorphisms in the human 5-HT3 genes contribute to pathological phenotypes? *Biochem Soc Trans* 34, 872–876.

Krzywkowski, K., Davies, P. A., Feinberg-Zadek, P. L., Brauner-Osborne, H., Jensen, A. A., 2008. High-frequency HTR3B variant associated with major depression dramatically augments the signaling of the human 5-HT3AB receptor. *Proc Natl Acad Sci USA* 105, 722–727.

Krzywkowski, K., Jensen, A. A., Connolly, C. N., Brauner-Osborne, H., 2007. Naturally occurring variations in the human 5-HT3A gene profoundly impact 5-HT_3 receptor function and expression. *Pharmacogenet Genomics* 17, 255–266.

Lankiewicz, S., Lobitz, N., Wetzel, C. H., Rupprecht, R., Gisselmann, G., Hatt, H., 1998. Molecular cloning, functional expression, and pharmacological characterization of 5-hydroxytryptamine3 receptor cDNA and its splice variants from guinea pig. *Mol Pharmacol* 53, 202–212.

Lee, W. Y., Sine, S. M., 2005. Principal pathway coupling agonist binding to channel gating in nicotinic receptors. *Nature* 438, 243–247.

Livesey, M. R., Cooper, M. A., Deeb, T. Z., Carland, J. E., Kozuska, J., Hales, T. G., Lambert, J. J., Peters, J. A., 2008. Structural determinants of Ca^{2+} permeability and conduction in the human 5-hydroxytryptamine type 3A receptor. *J Biol Chem* 283, 19301–19313.

Livesey, M. R., Cooper, M. A., Lambert, J. J., Peters, J. A., 2011. Rings of charge within the extracellular vestibule influence ion permeation of the 5-HT_3A receptor. *J Biol Chem* 286, 16008–16017.

Lochner, M., Lummis, S. C., 2010. Agonists and antagonists bind to an A-A interface in the heteromeric 5-HT_3AB receptor. *Biophys J* 98, 1494–1502.

Lummis, S. C., 2004. The transmembrane domain of the 5-HT_3 receptor: its role in selectivity and gating, *Biochemical Society Transactions* 32, 535–539.

Lummis, S. C., Beene, D. L., Lee, L. W., Lester, H. A., Broadhurst, R. W., Dougherty, D. A., 2005. Cis-trans isomerization at a proline opens the pore of a neurotransmitter-gated ion channel, *Nature* 438, 248–252.

Machu, T. K., 2011. Therapeutics of 5-HT3 receptor antagonists: Current uses and future directions. *Pharmacol Ther* 130, 338–347.

Maricq, A. V., Peterson, A. S., Brake, A. J., Myers, R. M., Julius, D., 1991. Primary structure and functional expression of the 5-HT_3 receptor, a serotonin-gated ion channel. *Science* 254, 432–437.

McKinnon, N. K., Reeves, D. C., Akabas, M. H., 2011. 5-HT3 receptor ion size selectivity is a property of the transmembrane channel, not the cytoplasmic vestibule portals. *J Gen Physiol* 138, 453–466.

Miles, T. F., Dougherty, D. A., Lester, H. A., 2013. The 5-HT_3AB receptor shows an A3B2 stoichiometry at the plasma membrane, *Biophysical Journal* 105, 887–898.

Miyake, A., Mochizuki, S., Takemoto, Y., Akuzawa, S., 1995. Molecular cloning of human 5-hydroxytryptamine3 receptor: Heterogeneity in distribution and function among species. *Mol Pharmacol* 48, 407–416.

Miyazawa, A., Fujiyoshi, Y., Unwin, N., 2003. Structure and gating mechanism of the acetylcholine receptor pore. *Nature* 423, 949–955.

Mochizuki, S., Watanabe, T., Miyake, A., Saito, M., Furuichi, K., 2000. Cloning, expression, and characterization of ferret 5-HT_3 receptor subunit. *Eur J Pharmacol* 399, 97–106.

Monk, S. A., Williams, J. M., Hope, A. G., Barnes, N. M., 2004. Identification and importance of N-glycosylation of the human 5-hydroxytryptamine 3A receptor subunit. *Biochem Pharmacol* 68, 1787–1796.

Morales, M., Wang, S. D., Diaz-Ruiz, O., Jho, D. H., 2004. Cannabinoid CB1 receptor and serotonin 3 receptor subunit A (5-HT3A) are co-expressed in GABA neurons in the rat telencephalon. *J Comp Neurol* 468, 205–216.

Niesler, B., 2011. 5-HT(3) receptors: Potential of individual isoforms for personalised therapy. *Curr Opin Pharmacol* 11, 81–86.

Niesler, B., Flohr, T., Nothen, M. M., Fischer, C., Rietschel, M., Franzek, E., Albus, M., Propping, P., Rappold, G. A., 2001. Association between the 5′ UTR variant C178T of the serotonin receptor gene HTR3A and bipolar affective disorder. *Pharmacogenetics* 11, 471–475.

Niesler, B., Frank, B., Kapeller, J., Rappold, G. A., 2003. Cloning, physical mapping and expression analysis of the human 5-HT3 serotonin receptor-like genes HTR3C, HTR3D and HTR3E. *Gene* 310, 101–111.

Niesler, B., Kapeller, J., Hammer, C., Rappold, G., 2008. Serotonin type 3 receptor genes: HTR3A, B, C, D, E. *Pharmacogenomics* 9, 501–504.

Niesler, B., Walstab, J., Combrink, S., Moller, D., Kapeller, J., Rietdorf, J., Bonisch, H., Gothert, M., Rappold, G., Bruss, M., 2007. Characterization of the novel human serotonin receptor subunits 5-HT3C,5-HT3D, and 5-HT3E. *Mol Pharmacol* 72, 8–17.

Panicker, S., Cruz, H., Arrabit, C., Slesinger, P. A., 2002. Evidence for a centrally located gate in the pore of a serotonin-gated ion channel. *J Neurosci* 22, 1629–1639.

Peters, J. A., Hales, T. G., Lambert, J. J., 2005. Molecular determinants of single-channel conductance and ion selectivity in the Cys-loop family: Insights from the 5-HT_3 receptor. *Trends Pharmacol Sci* 26, 587–594.

Pratt, G. D., Bowery, N. G., Kilpatrick, G. J., Leslie, R. A., Barnes, N. M., Naylor, R. J., Jones, B. J. et al., 1990. Consensus meeting agrees distribution of 5-HT_3 receptors in mammalian hindbrain. *Trends Pharmacol Sci* 11, 135–137.

Price, K. L., Bower, K. S., Thompson, A. J., Lester, H. A., Dougherty, D. A., Lummis, S. C., 2008. A hydrogen bond in loop A is critical for the binding and function of the 5-HT_3 receptor. *Biochemistry* 47, 6370–6377.

Price, K. L., Lummis, S. C., 2004. The role of tyrosine residues in the extracellular domain of the 5-hydroxytryptamine3 receptor. *J Biol Chem* 279, 23294–23301.

Price, K. L., Millen, K. S., Lummis, S. C., 2007. Transducing agonist binding to channel gating involves different interactions in 5-HT_3 and $GABA_C$ receptors. *J Biol Chem* 282, 25623–25630.

Quirk, P. L., Rao, S., Roth, B. L., Siegel, R. E., 2004. Three putative N-glycosylation sites within the murine 5-HT_3A receptor sequence affect plasma membrane targeting, ligand binding, and calcium influx in heterologous mammalian cells. *J Neurosci Res* 77, 498–506.

Reeves, D. C., Goren, E. N., Akabas, M. H., Lummis, S. C., 2001. Structural and electrostatic properties of the 5-HT_3 receptor pore revealed by substituted cysteine accessibility mutagenesis. *J Biol Chem* 276, 42035–42042.

Reeves, D. C., Lummis, S. C., 2002. The molecular basis of the structure and function of the 5-HT_3 receptor: A model ligand-gated ion channel (review). *Mol Membr Biol* 19, 11–26.

Ropert, N., Guy, N., 1991. Serotonin facilitates GABAergic transmission in the CA1 region of rat hippocampus in vitro. *J Physiol* 441, 121–136.

Rusch, D., Braun, H. A., Wulf, H., Schuster, A., Raines, D. E., 2007. Inhibition of human 5-HT(3A) and 5-HT(3AB) receptors by etomidate, propofol and pentobarbital. *Eur J Pharmacol* 573, 60–64.

Schmidt, A. W., Peroutka, S. J., 1989. Three-dimensional steric molecular modeling of the 5-hydroxytryptamine$_3$ receptor pharmacophore. *Mol Pharmacol* 36, 505–511.

Shao, X. M., Yakel, J. L., Jackson, M. B., 1991. Differentiation of NG108–15 cells alters channel conductance and desensitization kinetics of the 5-HT_3 receptor. *J Neurophysiol* 65, 630–638.

Solt, K., Stevens, R. J., Davies, P. A., Raines, D. E., 2005. General anesthetic-induced channel gating enhancement of 5-hydroxytryptamine type 3 receptors depends on receptor subunit composition. *J Pharmacol Exp Ther* 315, 771–776.

Spier, A. D., Lummis, S. C., 2000. The role of tryptophan residues in the 5-HT_3 receptor ligand binding domain. *J Biol Chem* 275, 5620–5625.

Stevens, R., Rusch, D., Solt, K., Raines, D. E., Davies, P. A., 2005. Modulation of human 5-hydroxytryptamine type 3AB receptors by volatile anesthetics and n-alcohols. *J Pharmacol Exp Ther* 314, 338–345.

Steward, L. J., Boess, F. G., Steele, J. A., Liu, D., Wong, N., Martin, I. L., 2000. Importance of phenylalanine 107 in agonist recognition by the 5-hydroxytryptamine(3A) receptor. *Mol Pharmacol* 57, 1249–1255.

Stratz, C., Trenk, D., Bhatia, H. S., Valina, C., Neumann, F. J., Fiebich, B. L., 2008. Identification of 5-HT3 receptors on human platelets: Increased surface immunoreactivity after activation with adenosine diphosphate (ADP) and thrombin receptor-activating peptide (TRAP). *Thromb Haemost* 99, 784–786.

Sugita, S., Shen, K. Z., North, R. A., 1992. 5-hydroxytryptamine is a fast excitatory transmitter at 5-HT_3 receptors in rat amygdala. *Neuron* 8, 199–203.

Sullivan, N. L., Thompson, A. J., Price, K. L., Lummis, S. C., 2006. Defining the roles of Asn-128, Glu-129 and Phe-130 in loop A of the 5-HT_3 receptor. *Mol Membr Biol* 23, 442–451.

Sun, H., Hu, X. Q., Emerit, M. B., Schoenebeck, J. C., Kimmel, C. E., Peoples, R. W., Miko, A., Zhang, L., 2008. Modulation of 5-HT_3 receptor desensitization by the light chain of microtubule-associated protein 1B expressed in HEK 293 cells. *J Physiol* 586, 751–762.

Sun, H., Hu, X. Q., Moradel, E. M., Weight, F. F., Zhang, L., 2003. Modulation of 5-HT_3 receptor-mediated response and trafficking by activation of protein kinase C. *J Biol Chem* 278, 34150–34157.

Thompson, A. J., Duke, R. K., Lummis, S. C. R., 2011a. Binding sites for bilobalide, diltiazem, ginkgolide, and picrotoxinin at the 5-HT_3 receptor. *Mol Pharmacol* 80, 183–190.

Thompson, A. J., Lester, H. A., Lummis, S. C. R., 2010. The structural basis of function in Cys-loop receptors. *Q Rev Biophys* 43, 449–499.

Thompson, A. J., Lochner, M., Lummis, S. C. R., 2008. Loop B is a major structural component of the 5-HT_3 receptor. *Biophys J* 95, 5728–5736.

Thompson, A. J., Lummis, S. C. R., 2003. A single ring of charged amino acids at one end of the pore can control ion selectivity in the 5-HT_3 receptor. *Br J Pharmacol* 140, 359–365.

Thompson, A. J., Lummis, S. C. R., 2006. 5-HT_3 receptors. *Curr Pharm Des* 12, 3615–3630.

Thompson, A. J., Lummis, S. C. R., 2007. The 5-HT_3 receptor as a therapeutic target. *Expert Opin Ther Targets* 11, 527–540.

Thompson, A. J., Lummis, S. C., 2013. Discriminating between 5-HT_3A and 5-HT_3AB receptors, *British Journal of Pharmacology* 169, 736–747.

Thompson, A. J., Padgett, C. L., Lummis, S. C., 2006a. Mutagenesis and molecular modeling reveal the importance of the 5-HT_3 receptor F-loop. *J Biol Chem* 281, 16576–16582.

Thompson, A. J., Price, K. L., Lummis, S. C., 2011b. Cysteine modification reveals which subunits form the ligand binding site in human heteromeric 5-HT_3AB receptors. *J Physiol* 589, 4243–4257.

Thompson, A. J., Sullivan, N. L., Lummis, S. C. R., 2006b. Characterization of 5-HT_3 receptor mutations identified in schizophrenic patients. *J Mol Neurosci* 30, 273–281.

Trattnig, S. M., Harpsoe, K., Thygesen, S. B., Rahr, L. M., Ahring, P. K., Balle, T., Jensen, A. A., 2012. Discovery of a Novel Allosteric Modulator of 5-HT_3 Receptors: Inhibition and potentiation of Cys-loop receptor signaling through a conserved transmembrane intersubunit site. *J Biol Chem* 287, 25241–25254.

Tsetlin, V., Hucho, F., 2009. Nicotinic acetylcholine receptors at atomic resolution. *Curr Opin Pharmacol* 9, 306–310.

Tzvetkov, M. V., Meineke, C., Oetjen, E., Hirsch-Ernst, K., Brockmoller, J., 2007. Tissue-specific alternative promoters of the serotonin receptor gene HTR3B in human brain and intestine. *Gene* 386, 52–62.

Venkataraman, P., Venkatachalan, S. P., Joshi, P. R., Muthalagi, M., Schulte, M. K., 2002. Identification of critical residues in loop E in the 5-HT_3ASR binding site. *BMC Biochem* 3, 15.

Verheij, M. H., Thompson, A. J., van Muijlwijk-Koezen, J. E., Lummis, S. C., Leurs, R., de Esch, I. J., 2012. Design, synthesis, and structure-activity relationships of highly potent 5-HT_3 receptor ligands. *J Med Chem* 55, 8603–8614.

Walstab, J., Rappold, G., Niesler, B., 2010. 5-HT_3 receptors: Role in disease and target of drugs. *Pharmacol Ther* 128, 146–169.

Wang, J., Lester, H. A., Dougherty, D. A., 2007. Establishing an ion pair interaction in the homomeric rho1 gamma-aminobutyric acid type A receptor that contributes to the gating pathway. *J Biol Chem* 282, 26210–26216.

Weiss, B., Mertz, A., Schrock, E., Koenen, M., Rappold, G., 1995. Assignment of a human homolog of the mouse Htr3 receptor gene to chromosome 11q23.1-q23.2. *Genomics* 29, 304–305.

Xiu, X., Hanek, A. P., Wang, J., Lester, H. A., Dougherty, D. A., 2005. A unified view of the role of electrostatic interactions in modulating the gating of Cys loop receptors. *J Biol Chem* 280, 41655–41666.

Yakel, J. L., Lagrutta, A., Adelman, J. P., North, R. A., 1993. Single amino acid substitution affects desensitization of the 5-hydroxytryptamine type 3 receptor expressed in *Xenopus oocytes*. *Proc Natl Acad Sci USA* 90, 5030–5033.

Yakel, J. L., Shao, X. M., Jackson, M. B., 1991. Activation and desensitization of the 5-HT_3 receptor in a rat glioma x mouse neuroblastoma hybrid cell. *J Physiol* 436, 293–308.

Yang, J., 1990. Ion permeation through 5-hydroxytryptamine-gated channels in neuroblastoma N18 cells. *J Gen Physiol* 96, 1177–1198.

Zhang, L., Oz, M., Weight, F. F., 1995. Potentiation of 5-HT_3 receptor-mediated responses by protein kinase C activation. *Neuroreport* 6, 1464–1468.

Zhou, F. M., Hablitz, J. J., 1999. Activation of serotonin receptors modulates synaptic transmission in rat cerebral cortex. *J Neurophysiol* 82, 2989–2999.

23 GABA$_A$ receptors

Trevor G. Smart

Contents

23.1 INTRODUCTION

A functional central nervous system (CNS) relies, at a simplistic level, on the interplay between neuronal excitation that is punctuated, at appropriate times, by neuronal inhibition. For limiting the extent and duration of neuronal excitability, inhibition is important and this is mediated largely by inhibitory neurotransmitters such as GABA and glycine. These transmitters rapidly increase the membrane conductance to anions, effectively shunting excitatory activity. Overall, GABA is the ubiquitous and predominant inhibitory transmitter throughout the CNS, while glycine assumes greater significance in the brainstem and spinal cord (Lynch, 2004; Betz and Laube, 2006; Bowery and Smart, 2006). To cause inhibition, GABA activates specific ionotropic GABA$_A$ receptors that are Cl^--permeable ligand-gated channels and metabotropic receptors (classified as GABA$_B$ receptors that signal via G protein activation to numerous downstream effectors). In this chapter, the focus is on the GABA$_A$ receptor family (Rabow et al., 1996; Sieghart and Sperk, 2002; Olsen and Sieghart, 2009).

GABA$_A$ receptors are part of a family originally referred to as Cys-loop receptors, due to the presence of a highly conserved disulfide bridge located in receptor subunits' extracellular domain (ECD). Members of this family include nicotinic acetylcholine receptors (nAChRs), type 3 5-hydroxytryptamine receptors (5HT$_3$Rs), γ-aminobutyric acid receptors (types A and C, GABA$_{A/C}$R), and glycine receptors (GlyRs) (Grenningloh et al., 1987; Betz, 1990). Also included are the Zn^{2+}-activated cation channel (ZAC) (Davies et al., 2003) and invertebrate receptors activated either by glutamate or serotonin (anionic channels) or GABA (cationic channels) (Ortells and Lunt, 1995; Lester et al., 2004). However, recently, related bacterial homologues from *Gloeobacter violaceus* (GLIC) (Bocquet et al., 2009) and *Erwinia chrysanthemi* (ELIC) (Hilf and Dutzler, 2008) have been added to the family, but they lack the characteristic Cys-loop signature. As a consequence, Cys-loop receptors are now referred to as pentameric ligand-gated ion channels (pLGICs) to which we can also add the glutamate-activated Cl^- channel (GluCl) from *Caenorhabditis elegans* (Hibbs and Gouaux, 2011).

23.2 RECEPTOR SUBUNIT DIVERSITY AND STRUCTURE

For GABA$_A$ receptors, eight discrete subunit families have been identified: α1–6, β1–3, γ1–3, δ, ε, π, θ, and ρ1–3, contributing a total of 19 subunits to the mammalian repertoire (Barnard et al., 1987; Schofield et al., 1987; Pritchett et al., 1989; Shivers et al., 1989; Sieghart and Sperk, 2002; Sigel and Steinmann, 2012). The similarities and differences, and the extent of divergence between different receptor subunits, can be best appreciated from a dendrogram (Figure 23.1). Here, α subunits discretely subsegregate into α1, 2, 3, and 5 from another cluster

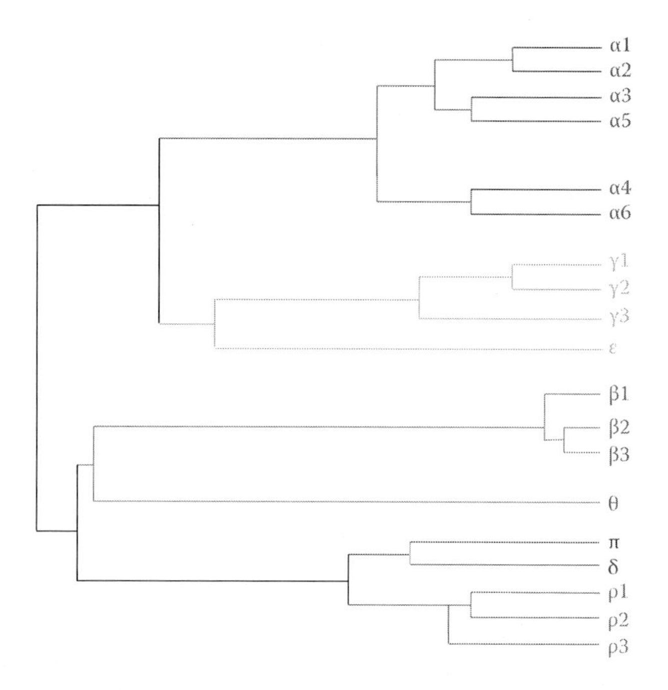

Figure 23.1 Dendrogram of GABA$_A$ receptor subunits. This shows the relative evolutionary relationship between the sequences of GABA$_A$ receptor subunit families and individual subunits. The length of the horizontal lines reflects the extent of amino acid sequence divergence between subunit families and individual subunits. (Analysis taken from Simon, J. et al., *J. Biol. Chem.* 279, 41422, 2004.)

incorporating α4 and α6. The late divergence of the β, γ, and ρ subunit families reflects their relative similarity, while the *rare* subunits, ε, π, and θ, segregate with γ, δ, and β subunits, respectively (Simon et al., 2004).

By taking 19 subunits and assuming a final pentameric assembly, the expression of many receptor subunit combinations seems possible in native neural tissue. However, in reality, naturally occurring GABA$_A$ receptor isoforms are likely to be far fewer in number (Whiting et al., 1995; Olsen and Sieghart, 2008). Preeminent amongst these are receptors containing αβγ and αβδ subunits, which populate synaptic and extrasynaptic domains, with more rare αβ isoforms (Olsen and Sieghart, 2009) and those containing ρ subunits, sometimes referred to as GABA$_C$ receptors.

The structure of GABA$_A$ receptors is yet to be precisely defined due to the paucity of crystal structures for these membrane proteins, apart from the recent crystal structure for the β3 homomer (Miller & Aricescu, 2014). Thus, most information on structure has been accrued from homology modeling with closely related receptor templates and/or crystals in conjunction with site-directed mutagenesis and functional studies (Corringer et al., 2012). Structures that are proving to be helpful in this regard include the *Torpedo marmorata* nAChR (Unwin, 1998, 2005), the acetylcholine-binding protein (AChBP) (Brejc et al., 2001), and GLIC (Bocquet et al., 2009), ELIC (Hilf and Dutzler, 2008), and GluCl (Hibbs and Gouaux, 2011).

These comparisons have resulted in a consensus structural view for GABA$_A$ receptors. Essentially, they are pseudosymmetrical rings containing five subunits, formed from an ECD with antiparallel arrays of inner and outer β-sheets, a four α-helical transmembrane domain (TMD), and an intracellular domain (ICD) of unknown structure (Figure 23.2). Other notable structural signatures of pentameric GABA$_A$ receptors include a central, contiguous ion-conducting pathway through the ECD and TMD, traversing an ion channel gate before reaching the ICDs (Figure 23.2), and the Cys-loop structure sited at the base of the ECDs, which, although not present in all pLGICs, plays a vital role in communicating neurotransmitter binding to ion channel activation.

23.3 RECEPTOR TRAFFICKING AND CLUSTERING

Synaptic GABA$_A$ receptors are clustered at inhibitory synapses, and this requires receptor-associated molecules to provide scaffold/anchorage support near the cell surface membrane that influences GABA$_A$ receptor trafficking. Receptor clusters are dynamic, and by posttranslational modification of GABA$_A$

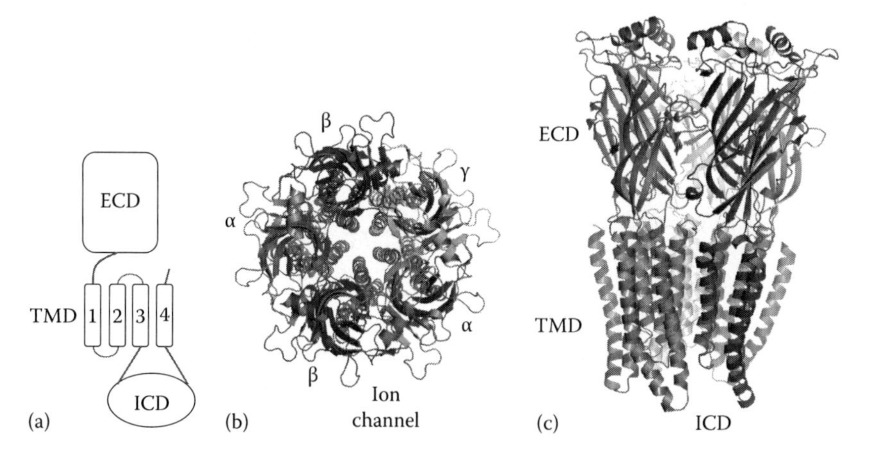

Figure 23.2 GABA$_A$ receptor structure. (a) Depicts a schematic domain structure for a typical GABA$_A$ receptor subunit showing the ECD, the TMD, and the ICD. The linking amino acids are shown as lines. (b) A view of a synaptic-type αβγ GABA$_A$ receptor pentamer from the perspective of a presynaptically released GABA molecule (plane view). The ion channel is the central pore lined by M2 from each subunit. (c) Side elevation of a GABA$_A$ receptor with the ECD and TMD shown. The structure of the ICD is unknown and not shown apart from the M1–M2 linker. The homology models in b and c are based on the crystal structure of GluCl. (Hibbs, R. E. and E. Gouaux, *Nature*, 474, 54, 2011.)

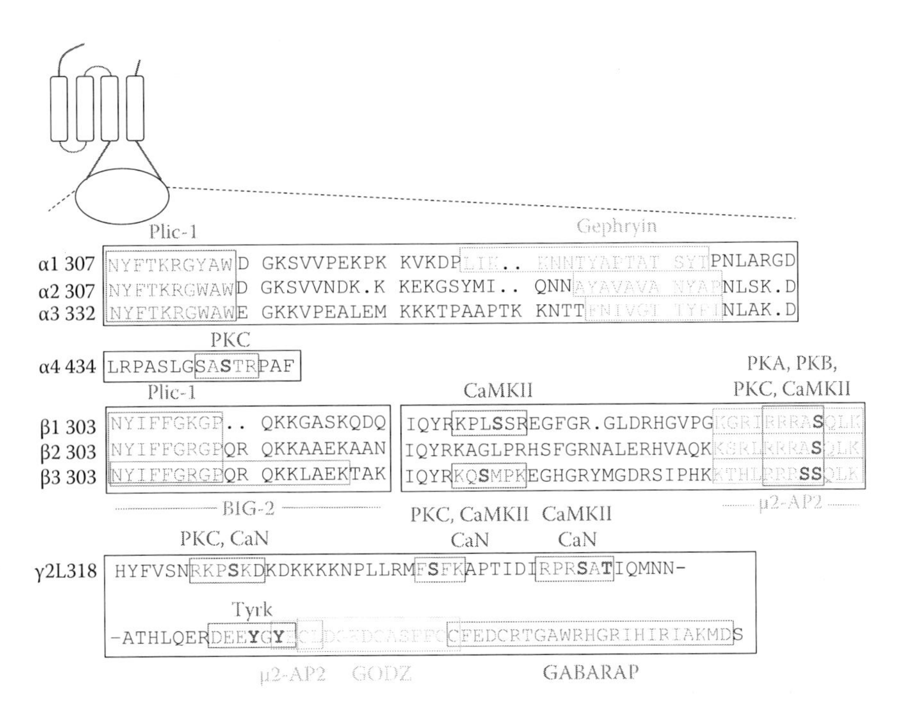

Figure 23.3 $GABA_A$ receptor ICD—mapping binding sites for interacting proteins. Selected portions of the ICD primary sequences for $GABA_A$ receptor α, β, and γ subunits are shown. These illustrate the consensus sequences identified for protein kinase phosphorylation of receptor subunits and the binding motifs that are critical for $GABA_A$ receptor-interacting proteins. *References:* Reviews (Chen and Olsen, 2007; Araud et al., 2010; Luscher et al., 2011); Gephyrin (Tretter et al., 2008; Mukherjee et al., 2011; Tretter et al., 2011); Protein kinase consensus sequences (Moss et al., 1995; Moss and Smart, 1996; Houston et al., 2009); BIG-2 (Charych et al., 2004); GABARAP (Nymann-Andersen et al., 2002); AP2 (Kittler et al., 2008; Smith et al., 2008); Plic-1 (Bedford et al., 2001); Calcineurin (CaN) (Wang et al., 2003; Muir et al., 2010); Golgi-specific DHHC (Asp-His-His-Cys) zinc finger protein (GODZ) (Fang et al., 2006).

receptors and/or their scaffolds, receptor expression levels, and thereby receptor function and plasticity, can all be regulated (Moss and Smart, 2001; Araud et al., 2010; Luscher et al., 2011).

Gephyrin is a vital component for securing $GABA_A$ receptors at inhibitory synapses. It is capable of forming a lattice structure to connect receptor subunits to the cytoskeleton (Feng et al., 1998; Kneussel and Betz, 2000; Sassoe-Pognetto and Fritschy, 2000; Luscher and Keller, 2004; Fritschy et al., 2008; Jacob et al., 2008; Tyagarajan and Fritschy, 2010). Gephyrin is able to bind directly to $GABA_A$ receptor α1, α2, and α3 subunits (Tretter et al., 2008; Mukherjee et al., 2011; Tretter et al., 2011). Given that these subunits can be found at inhibitory synapses, it provides a plausible explanation as to why some $GABA_A$ receptors are clustered at synapses (see Figure 23.3), but not others, for example, α4βδ.

However, recent evidence indicates that the ICDs of $GABA_A$ receptor β subunits may also directly bind to gephyrin (Kowalczyk et al., 2013). This interaction raises several interesting issues not least if gephyrin binds to β subunits, then in principle, receptors other than those considered to be typically synaptic could be clustered by gephyrin (e.g., α4βδ), but this is thought not to occur for receptors that are predominantly extrasynaptic. Additionally, gephyrin binding to the receptor might be more complex encompassing an interaction between receptor α and β subunits. Despite evidence that some $GABA_A$ receptors are able to cluster in the absence of gephyrin (Kneussel et al., 2001), the presence of this structural protein is considered a reliable indicator of the location of inhibitory synapses (Sassoe-Pognetto et al., 1995, 2000) though it can also be found at extrasynaptic sites. A number of other proteins can associate with $GABA_A$ receptors, including $GABA_A$R-associated protein (GABARAP) (Wang et al., 1999; Nymann-Andersen et al., 2002), Plic-1 (Bedford et al., 2001), and huntingtin-associated protein 1 (HAP-1) (Kittler et al., 2004). All these proteins regulate the surface stability and trafficking of $GABA_A$Rs (Luscher and Keller, 2004; Luscher et al., 2011) by interacting with the major ICD of $GABA_A$ receptor subunits (Figure 23.3).

With regard to extrasynaptic receptors, knowledge of their binding partners, if any, is sparse. Extrasynaptic receptors containing δ subunits are thought not to enter synapses, but extrasynaptic anchoring proteins remain, so far, unidentified. The one exception concerns receptors containing the α5 subunit. Here, radixin has been identified as a potential interactor to slow the diffusion of the receptor in the cell surface membrane (Loebrich et al., 2006), and its disruption can disperse α5 subunit-containing receptor clusters. Recently, the cell adhesion molecule, neuroplastin-65, has been found to colocalize with α1 and α2 subunits, but not α3 subunits, at inhibitory synapses, and also with α5 subunits in the extrasynaptic zone (Sarto-Jackson et al., 2012), suggesting it might anchor receptors in a subtype-specific manner.

Although $GABA_A$ receptors can be compartmentalized into synaptic and extrasynaptic zones in the cell surface membrane in a subunit-dependent manner (Figure 23.4), they are still capable of lateral mobility in the plane of the membrane, which serves as an important pathway for replenishing receptors at inhibitory synapses (Thomas et al., 2005; Triller and Choquet, 2005; Bogdanov et al., 2006; Choquet and Triller, 2013).

Many membrane protein interactions with the receptor are also regulated by phosphorylation of receptor subunits

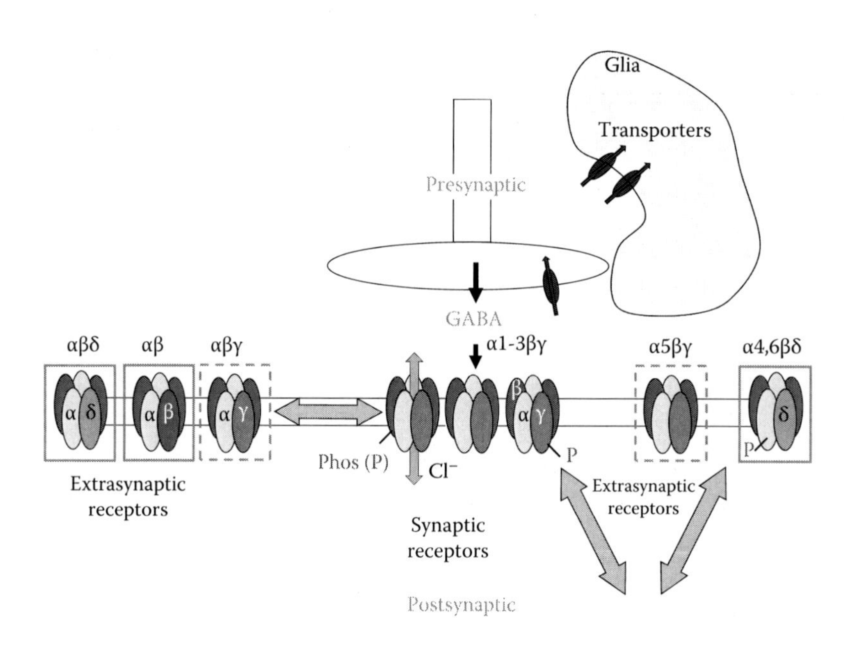

Figure 23.4 GABAergic inhibitory synapse. Schematic diagram showing the major elements of a GABAergic synaptic and extrasynaptic zone. Predominant GABA$_A$ receptors at synapses are composed of α1–3βγ subunits with extrasynaptic receptors comprising α4 or α6βδ and αβ subunit combinations (Moss, S.J. and Smart, T.G., *Nat. Rev. Neurosci.*, 2, 240, 2001; Fritschy, J.M. and Brunig, I., *Pharmacol. Ther.*, 98, 299, 2003; Mortensen, M. and Smart, T.G., *J. Physiol. (Lond.)*, 577, 841, 2006; Glykys, J. et al., *J. Neurosci.*, 28, 1421, 2008; Thomson, A.M. and Jovanovic, J.N., *Eur. J. Neurosci.*, 31, 2193, 2010.) The position of α5βγ receptors is less precise with evidence for an extrasynaptic location and also evidence for its presence at synapses (Thomson, A.M. and Jovanovic, J.N., *Eur. J. Neurosci.*, 31, 2193, 2010.) Sites for phosphorylation are indicated on β and γ subunits and also on the α4 subunit. The arrows indicate potential trafficking routes for the receptor into and within the membrane.

and their associated molecules, thereby modulating receptor trafficking, assembly, and stability in the surface membrane. Consensus sites for serine/threonine and tyrosine kinase phosphorylation of GABA$_A$ receptor subunits have been identified in the ICDs between M3 and M4 of α4, β, and γ subunits (Figure 23.3) (Moss and Smart, 1996; Brandon et al., 2002; Abramian et al., 2010). Phosphorylation-induced regulation of the trafficking of GABA$_A$ receptors is important for determining not only the steady-state numbers of cell surface receptors but, as a consequence, the macroscopic efficacy of synaptic and tonic inhibition. The broad consensus view is that GABA$_A$ receptors composed of α1–3βγ subunits will populate inhibitory synapses, while the extrasynaptic zone will be composed mainly of α4βδ or α6βδ, α5βγ, and αβ isoforms (Figure 23.4). However, there are exceptions, and following the discovery that functional GABA receptors are mobile in the cell surface membrane, it is now evident that αβγ receptors can also be found in the extrasynaptic zone (Thomas et al., 2005). Conversely, the less abundant α4βγ and α6βγ receptors may also be found within selected inhibitory synapses. Overall, while it is difficult to be precise about the relative weightings of GABA receptor isoforms in the synaptic and extrasynaptic zones, generally, αβγ receptors predominate at synaptic and αβδ at extrasynaptic sites.

In addition to varying expression of GABA$_A$ receptor isoforms both within and outside of inhibitory synapses, there are also differences in receptor subunit expression patterns across the CNS and during stages of development (Wisden et al., 1991, 1992; Fritschy et al., 1994; Fritschy and Mohler, 1995; Pirker et al., 2000; Fritschy and Brunig, 2003; Hortnagl et al., 2013). Although GABA$_A$ receptors have an almost ubiquitous distribution throughout the CNS, patterns of expression are noticeable, particularly for receptor α subunits (Figure 23.5) that can have a profound effect on the physiological and pharmacological properties of GABA$_A$ receptors.

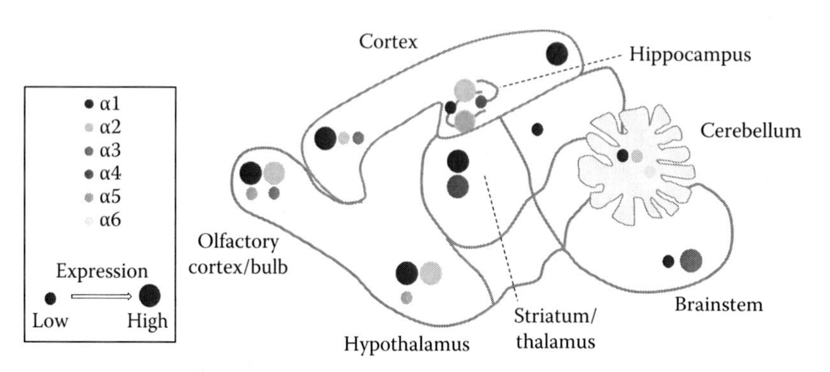

Figure 23.5 GABA$_A$ receptor α subunit distribution. Parasagittal section of the brain in diagrammatic form demonstrating major brain areas of GABA$_A$ receptor α subunit expression. Data representation is taken from immunocytochemical studies. See text for references.

23.4 RECEPTOR ACTIVATION

The main function of GABA-gated ion channels is to rapidly convert the process of GABA binding to ion channel opening, permitting transmembrane Cl^- flux. GABA binds at two locations on the receptor that are defined by the interfaces between neighboring β and α subunits. These binding sites involve ECD structures known as binding loops of which six have been identified (denoted A–F) with three forming the principal *face* (P or +) of the site (A–C, on the β subunit) and the remaining three (D–F) comprising the complementary face (C or –) provided by the adjacent α subunit (Figure 23.6; Lummis, 2009; Miller and Smart, 2010). These loops are important since mutating residues within the loops considerably affects the potency of GABA and the activation state of the receptor. For example, mutations in loops A, D, and E enable receptors to activate spontaneously in the absence of GABA (Boileau et al., 2002; Torres and Weiss, 2002; Sedelnikova et al., 2005). By contrast, loop F seems less directly involved in gating although some conformational movement in this region is thought to reflect the attachment of loop F to loop C that is relatively mobile after agonist binding tending to *close* over part of the β–α subunit interface (Khatri et al., 2009; Khatri and Weiss, 2010; Wang et al., 2010).

From structural homology models and cocrystallization of Cys-loop receptors with agonists or competitive antagonists bound to the neurotransmitter binding site, it is clear that loop C plays a prominent role in agonist binding by closing toward the receptor interface once the agonist molecule is bound (Celie et al., 2004; Hansen et al., 2005). For a bound antagonist, particularly of large volume and occupying the neurotransmitter binding site, loop C becomes displaced away from the receptor (Bourne et al., 2005; Celie et al., 2005; Hansen et al., 2005).

The coupling of these binding loop conformational changes to receptor activation and thus the opening of the ion channel involve a key interface between the receptor's ECD and the TMD. Here, loops 2, 7, and 9 of the ECD interact with the pre-M1 region and the M2–M3 linker to enable channel opening (Figure 23.7) (Bera et al., 2002; Newell and Czajkowski, 2003;

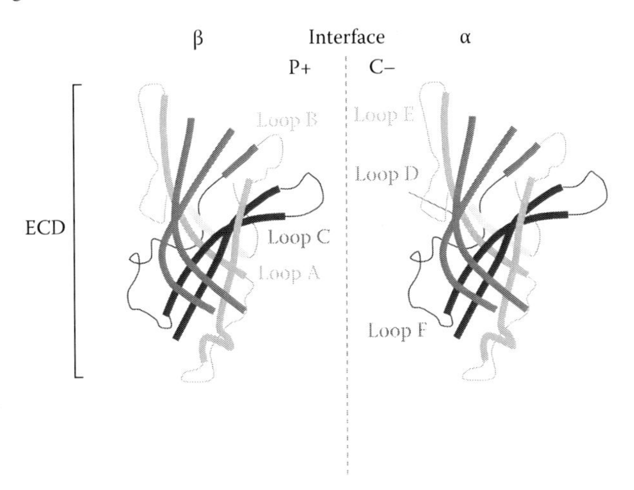

Figure 23.6 $GABA_A$ receptor-binding loops. Parts of the ECDs of two juxtaposed $GABA_A$ receptor subunits (β and α) are shown. The interface with its principal (P, +) and complementary (C, –) sides is also displayed together with the antiparallel β-sheet arrays and the six neurotransmitter-binding loops.

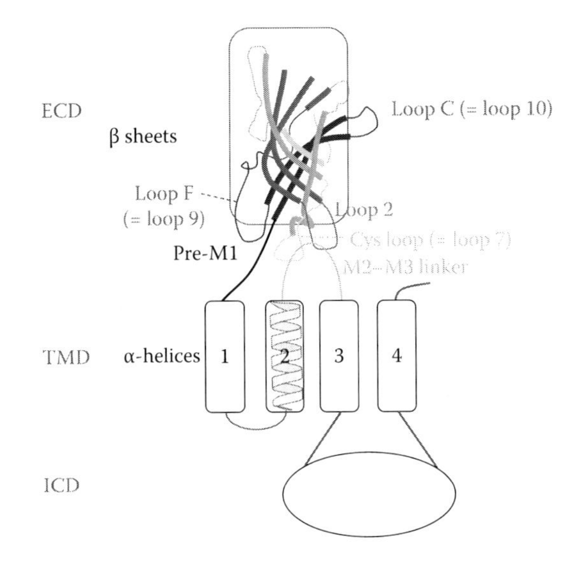

Figure 23.7 $GABA_A$ receptor activation—key structural domains. Diagram of the critical domains of a $GABA_A$ receptor that play a role in activation following ligand binding, including loops 2, 7, and 9 from the ECD, the pre-M1 linker, and the M2–M3 linker region.

Bouzat et al., 2004; Mercado and Czajkowski, 2006; Smart and Paoletti, 2012). This functional linkage between the ECD and the TMD relies on both hydrophobic and electrostatic interactions (Kash et al., 2003; Xiu et al., 2005).

For αβγ subunit $GABA_A$ receptors located at inhibitory synapses, these will be activated by presynaptically released GABA that reaches concentrations in the millimolar range causing synaptic inhibition (Farrant and Nusser, 2005). However, in the extrasynaptic zone, GABA receptors (e.g., αβδ subunits) will be exposed to much lower ambient and synaptic spillover GABA concentrations, possibly as low as nanomolar, which causes a persistent tonic inhibition (Glykys and Mody, 2007; Brickley and Mody, 2012).

23.5 ION CHANNEL DOMAIN

The GABA ion channel is formed by a pentameric ring of M2 α-helical segments drawn from all five individual receptor subunits. These α-helices have a discernible constriction in the middle of the channel before tapering outwards toward the extracellular end of the ion channel. Notably, and in keeping with other Cys-loop receptors, GABA ion channels possess concentric rings of residues with hydrophobic side chains (leucines and valines) dominating the mid-portion of the channel and residues with charged side chains (glutamates and arginines) predominating near the ends of the channel. For most Cys-loop receptors, the principal ion channel gate, which opens upon agonist activation to allow ion flux, is thought to be centered around the hydrophobic leucines and valine rings near the center of the channel (Miyazawa et al., 2003; Unwin, 2005; Miller and Smart, 2010). The GABA-induced channel opening is likely to involve a retraction of the M2 α-helices toward M1 and M3 facilitated by the aqueous cavity behind M2 (Bera et al., 2002; Bera and Akabas, 2005).

The charged ring of residues that are located at both ends of the GABA channel are net positively charged and can regulate channel conductance in Cys-loop receptors as well as anion selectivity in GABA channels (Jensen et al., 2002; Keramidas et al., 2004).

23.6 SUBUNIT COMPOSITION INFLUENCES RECEPTOR PROPERTIES

The properties of GABA$_A$ receptors depend upon their subunit composition. Physiologically, the sensitivity to GABA, the extent of receptor deactivation, and desensitization all vary with the type of α subunit assembled into a typical synaptic-type αβγ GABA$_A$ receptor pentameric complex (Figure 23.8). The potency of GABA in causing receptor activation distinguishes three receptor isoform groupings that differ over two orders of magnitude: receptors containing α6 subunits show the highest sensitivity to GABA followed by intermediate sensitivity for α1, α4, and α5 subunits, and receptors containing α2 and α3 subunits are the least sensitive (Sigel et al., 1990; Verdoorn et al., 1990; Picton and Fisher, 2007; Mortensen et al., 2011) (Figure 23.9).

Although β subunits display a high degree of homology, αβγ receptors containing β3 subunits are approximately fivefold more sensitive to GABA than either β1 or β2 subunit-containing receptors (Figure 23.9). For extrasynaptic GABA receptors containing δ subunits, or just αβ binary subunit combinations, a consistent trend is evident for GABA potency, dependent upon the identity of the α subunit, with α6 subunit receptors (α6βxδ and α6βx) exhibiting the highest sensitivity to GABA. This is followed by α4 and α1 subunit-containing receptors, and finally, α3 subunit receptors are least sensitive in this comparison.

The α subunit also influences the kinetic parameters for GABA$_A$ receptor activation, deactivation, and desensitization (Gingrich et al., 1995; Tia et al., 1996; Lavoie et al., 1997; Fisher, 2004; Picton and Fisher, 2007). With regard to the speed of receptor activation, α1 and α2 subunit receptors are generally submillisecond, with α4 and α6 subunit receptors requiring 1–2 ms and α3 and α5 receptors between 2 and 10 ms to activate (Figure 23.10). The rate of deactivation associated with GABA binding is invariably an order of magnitude slower than activation, with α4 and α5 subunit receptors deactivating fastest followed by an intermediate group containing α1 and α2 subunits and α3 and α6 showing the slowest rate of deactivation (Figure 23.10).

Entry into desensitized states is a familiar characteristic of most GABA$_A$ receptors. In measuring the rates of desensitization, many studies have detected more than one time constant. Here, we consider those studies where two rates have been detected, which represent the most common situation. On this basis, receptors containing α1, α2, or α4 subunits show the fastest rates of desensitization, followed by α3 and α6 subunit receptors with α5 subunit receptors showing the slowest rates of desensitization (Figure 23.10).

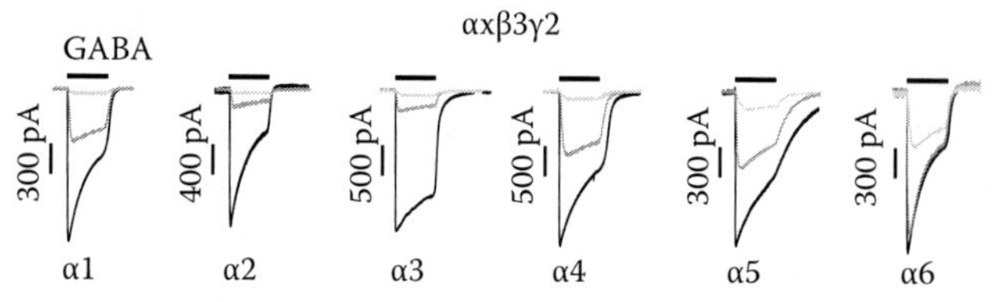

Figure 23.8 GABA current profiles. Whole-cell GABA current recordings evoked by 0.1, 1, and 100 μM GABA rapidly applied to HEK293 cells expressing αxβ3γ2 GABA$_A$ receptors. Note different profiles with regard to desensitization and deactivation. (Currents adapted from Mortensen, M. et al., *Front. Cell Neurosci.* 6, 1, 2011.)

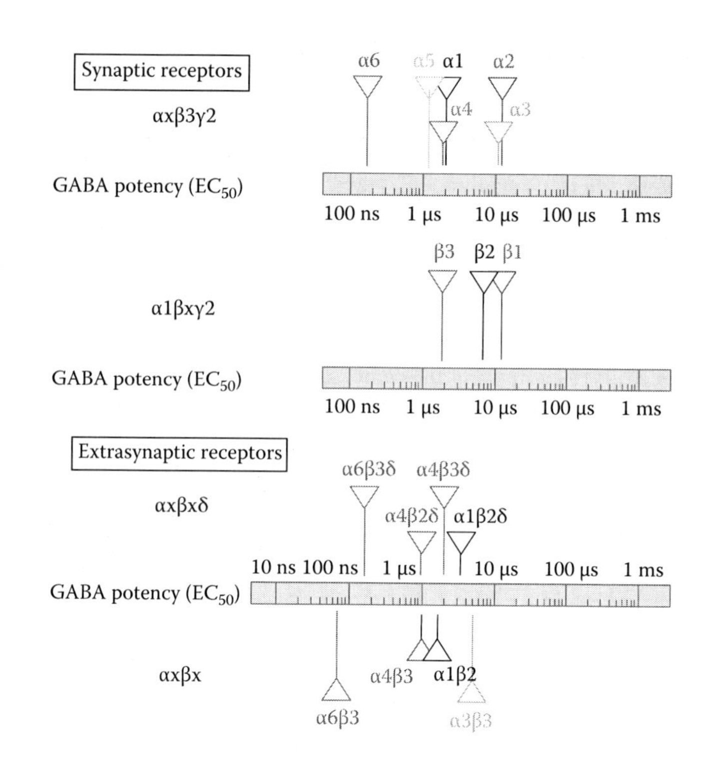

Figure 23.9 GABA potency at synaptic and extrasynaptic GABA$_A$ receptors. The potency of GABA (defined as the EC_{50}) in activating recombinant GABA$_A$ receptors expressed in HEK293 cells was determined from dose–response curve measurements for a selection of synaptic-type and extrasynaptic-type receptors. See text for references.

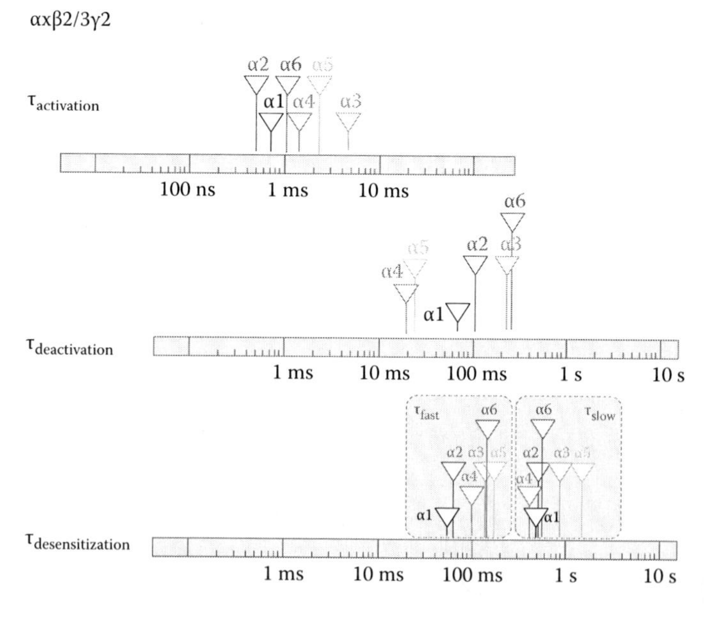

Figure 23.10 Kinetic profiles for GABA$_A$ receptors. Each measurement bar shows the time constants (τ) for GABA current activation (top), deactivation (middle), and desensitization (lower) for synaptic-type GABA$_A$ receptors comprising αβγ subunits. See text for references.

23.7 MODULATION OF GABA$_A$ RECEPTORS

GABA$_A$ receptors possess a large number of binding sites for ligands that can act directly at the neurotransmitter binding site (orthosteric) or in an allosteric capacity to regulate receptor function. Many of these sites have proven therapeutic potential and value. If we generalize, ligand-binding sites for many drugs at GABA$_A$ receptors can be broadly classified into one of the three major categories: ECD and interfacial, the TMDs, and the ion channel. These categories are not mutually exclusive and it is possible to find residues in the ion channel that affect drug action even though such drugs may bind elsewhere on the receptor complex.

23.8 INTERFACIAL-BINDING SITES

23.8.1 AGONISTS

The neurotransmitter-binding site is a classic example of an interfacial binding site being located at the interfaces between β and α subunits (Figure 23.11). Thus, two molecules of GABA can bind to each receptor. This is also the site of action for numerous GABA agonists (Ebert et al., 1994; Krogsgaard-Larsen et al., 2002; Mortensen et al., 2004; Storustovu and Ebert, 2006; Lummis, 2009). For synaptic-type receptors composed of αβγ subunits, GABA, muscimol, and isoguvacine are all classed as full agonists (with similar relative maximal efficacies determined from agonist dose–response curves) with potencies in the order of muscimol > GABA > isoguvacine. Other agonists can be classified into several categories of partial agonist with different potencies and relative maximal efficacies as measured at the whole-cell level. For example, 4,5,6,7-tetrahydro-isoxazolo[5,4-c]pyridine-3(2H)-one (THIP) and isonipecotic acid are less potent than GABA with relative maximal efficacies of ~0.8 compared to GABA.

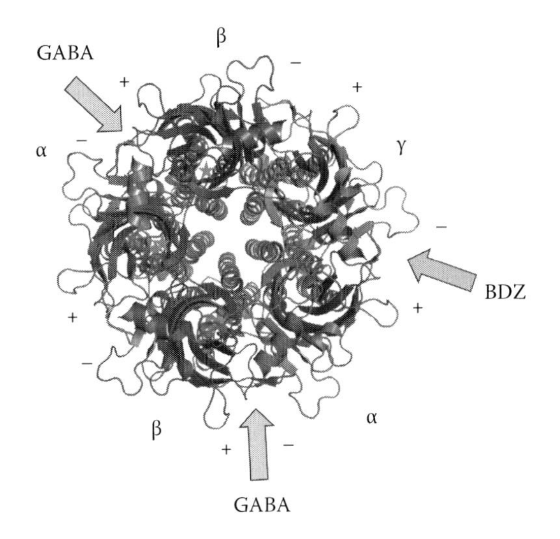

Figure 23.11 GABA$_A$ receptor-binding sites. A plan view of a typical synaptic GABA$_A$ receptor (αβγ subunits) showing the binding site locations for GABA and (BDZs), as well as the postulated aqueous pockets between the TMDs of individual subunits (two are shown for α and β subunits—openstars).

Piperidine-4-sulfonic acid (P4S) and imidazole acetic acid (IAA) attain only ~0.4 relative efficacy with similar or reduced potency to GABA, respectively. Very weak partial agonists are represented by Thio-4-PIOL and 4-PIOL (Frolund et al., 2000, 2002) with low potencies, and relative efficacies of not more than 0.01–0.05, compared to GABA (Mortensen et al., 2002, 2004).

23.8.2 ANTAGONISTS

Residues involved in GABA binding would also be accessible to antagonists, particularly competitive antagonists such as bicuculline (Johnston, 2013), and it has been largely assumed these would bind at the orthosteric binding site on the GABA$_A$ receptor. Indeed, *GABA structures* can be traced within the core structure of bicuculline implying commonality of binding sites (Steward et al., 1971; Bowery and Brown, 1974; Simmonds, 1980). Gabazine is another example of a competitive type antagonist (Heaulme et al., 1986). However, both bicuculline and gabazine are capable of inhibiting GABA$_A$ receptor activity initiated by either a steroid such as alphaxalone or a barbiturate such as pentobarbitone, from sites that are different from the GABA-binding site (Ueno et al., 1997). Since these antagonists do not compete for the steroid and barbiturate sites, it is assumed that they are capable of allosteric inhibition (Ueno et al., 1997), suggesting they are not simply competitive antagonists even if they do occupy the orthosteric-binding site.

23.8.3 POSITIVE ALLOSTERIC MODULATORS

The αβγ subunit GABA$_A$ receptor contains five subunit–subunit interfaces, of these four are distinct interfaces by virtue of the juxtaposed paired subunits in the pentamer (two β–α and one each for α–β, β–γ, and α–γ, Figure 23.11). The α–γ interface is significant because it is here that benzodiazepines (BDZs) will bind, potentiating receptor function, while being unable to directly activate the receptor (Mohler et al., 1980; Wieland et al., 1992; Sigel and Buhr, 1997; Boileau et al., 1998; Sigel et al., 1998; Kucken et al., 2000; Sigel, 2002; Tan et al., 2009; Richter et al., 2012). Given the interfacial (α+γ−) positioning of the binding site, it is unsurprising that the γ subunit is an absolute requirement for BDZ sensitivity of GABA$_A$ receptors (Pritchett et al., 1989). However, there is also a dependence on the type of α subunit present with α1–3 and α5 displaying BDZ sensitivity, while α4 and α6 subunit receptors are relatively insensitive. BDZ sensitivity depends on the presence of a crucial histidine residue in the ECD of α subunits 1–3 and 5 (H101 in α1) that is replaced by arginine in α4 and α6, which is sufficient to render the BDZs ineffective (Wieland et al., 1992; Kleingoor et al., 1993).

One of the bacterial homologues to GABA receptors, ELIC, is also activated by GABA and is modulated by BDZs (Spurny et al., 2012; Thompson et al., 2012). Detailed analysis of crystal structures has identified two binding sites for the BDZs (flurazepam) in crystal structures. One site, unexpectedly, is located within a subunit (intrasubunit site) close to the GABA-binding site and facing the ion channel vestibule, causing potentiation of ELIC function, while the other *inhibitory* site is located between subunits (intersubunit site) and partially overlaps the GABA-binding site (Spurny et al., 2012). It will be interesting to understand the importance of these sites for the modulation of mammalian GABA$_A$ receptors.

23.8.4 FORGOTTEN INTERFACE

Until relatively recently, the corresponding interface to β+α– (where GABA binds), which lies between α and β subunits (α+β–), was not generally appreciated as a site for GABA receptor modulation (Sieghart et al., 2012). This interface, with regard to the α subunit principal (+) side, is similar to the α+γ– interface where BDZs bind in GABA$_A$ receptors. A recent screen of numerous compounds has, however, revealed that an anxiolytic pyrazoloquinoline-based compound, CGS9895, that acts as an antagonist at the BDZ site is capable of enhancing GABA currents by binding to a site at the α+β– interface (Ramerstorfer et al., 2011).

Structural analogues of CGS9895 also act at this site (Varagic et al., 2013b), and given the unusual nature of the interface, it is predicted that modulation resulting from ligand binding at this site may depend on the identity of the β subunit, as well as the α subunit. Indeed, specific modulators have been reported for the α6β2/3γ2 receptor over α1–5β2/3γ2 receptors, by binding at the α6+β– interface (Varagic et al., 2013a). This discovery indicates that drugs acting at the α+β– interface will in principle be able to modulate αβ, αβγ, and αβδ receptors (i.e., the major synaptic- and extrasynaptic-type GABA$_A$ receptors) without causing direct receptor activation. Thus, they will achieve a broader therapeutic span, far in excess of that attributed to the BDZs that specifically require the inclusion of the γ subunit (along with α and β subunits) for any functional effect.

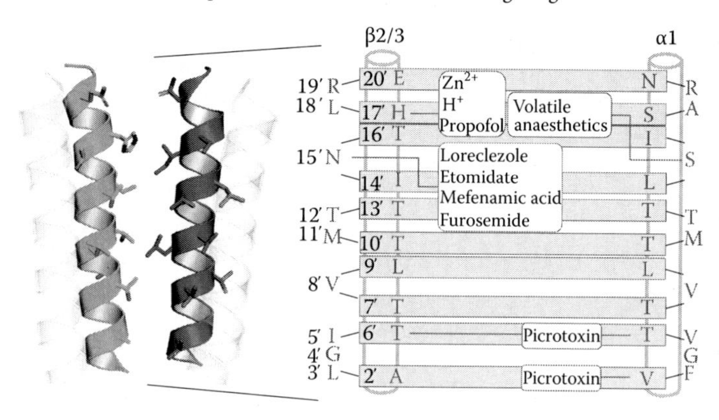

Figure 23.12 GABA$_A$ receptor transmembrane and ion channel residues. Left, homology model of two M2 segments for a GABA$_A$ receptor β and α subunits showing residues that are accessible from the ion channel. Right, schematic representation of accessible and nonaccessible residues in β and α M2 segments determined from cysteine-scanning mutagenesis. The listed drugs rely on the ion channel residues indicated for their functional effect and/or are considered to form part of the drug/ion binding sites. To facilitate a comparison between the amino acid residues of Cys-loop receptor ion channels, a prime number notation is employed that starts with the highly conserved positively charged residues at the cytoplasmic end of M2, which are defined as 0′, increasing in number to another ring of charged residues at the extracellular end that are defined as 20′.

23.9 TRANSMEMBRANE DOMAIN–BINDING SITES

23.9.1 INTRASUBUNIT

The second major domain that is a target for drugs on GABA$_A$ receptors involves the TMDs. Structural homology models suggest the existence of *aqueous pockets* between the TMDs of each individual receptor subunit (Ernst et al., 2005) (Figure 23.11), which are thought to house binding sites for several classes of ligand including the endogenous modulatory neurosteroids (Hosie et al., 2006, 2007) and exogenous inhalational and intravenous anesthetics (Mihic et al., 1997; Jenkins et al., 2001; Yamakura et al., 2001).

By using photolabeling and cysteine substitution experiments, it is becoming clear that some general anesthetics (etomidate) may bind within the TMDs but at interfacial sites between β and α subunits (β+α–) (Li et al., 2006, 2010; Stewart et al., 2013). Molecules binding in this vicinity of the TMDs differ from the BDZs in not just potentiating receptor function but also causing direct receptor activation in the absence of GABA. Traditionally, these two effects were thought to proceed via two distinct binding sites, but it is also apparent that a single class of binding site could mediate both effects (Rusch et al., 2004; Forman, 2012).

To obtain structural information on anesthetic-binding sites, GLIC, a bacterial homologue of the Cys-loop receptors, has been used (Bocquet et al., 2007; Corringer et al., 2012). Crystal structures of GLIC with the anesthetic propofol or desflurane bound *in situ* clearly indicate a binding site for anesthetics located within a single subunit (intrasubunit) (Nury et al., 2011). By using a diazirine-based photolabel attached to the anesthetic propofol, residues in the receptor β subunit that line the ion channel close to the ECD have also been identified, particularly H267 (Yip et al., 2013) (Figure 23.12). This is of interest since it apparently overlaps with neither the propofol binding site in GLIC nor the photolabeled binding site for azietomidate previously identified in GABA$_A$ receptors.

23.9.2 ION CHANNEL

The third major area for drug action on GABA$_A$ receptors involves the ion channel (Figure 23.12). By systematically converting amino acid residues thought to constitute the ion channel structure to cysteines and utilizing sulfhydryl-modifying reagents to covalently bind to these cysteines, it is possible to discover which are the most accessible residues facing the ion channel lumen (Xu and Akabas, 1993, 1996; Karlin and Akabas, 1998; Bera and Akabas, 2005) (Figure 23.12).

By progressively entering the ion channel from the extracellular end and considering those amino acids presented by the β subunit, a histidine residue has been identified as a major coordinating site for Zn^{2+}. This divalent cation acts as receptor subtype-selective inhibitor for GABA$_A$ receptors, potently inhibiting αβ, αβδ receptor function over αβγ isoforms (Draguhn et al., 1990; Smart and Constanti, 1990; Smart et al., 1991; Smart, 1992; Fisher and MacDonald, 1998; Nagaya and MacDonald, 2001). Mutation of this histidine causes a significant reduction in Zn^{2+} potency (Wooltorton et al., 1997; Horenstein and Akabas, 1998). The complete abolition of GABA$_A$ receptor sensitivity to Zn^{2+} requires the mutation of additional residues in the ECD at the α+β– subunit interface (E147 and H141 in α1 subunits and E182 in β subunits) (Dunne et al., 2002; Hosie et al., 2003).

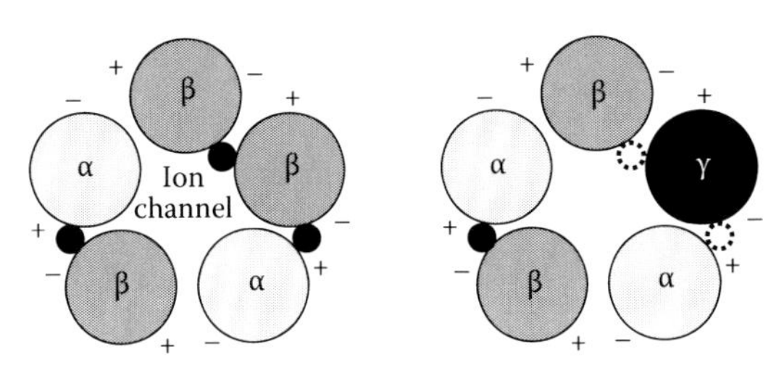

Figure 23.13 Zn^{2+} sites in $GABA_A$ receptors. Plan view of two $GABA_A$ receptors comprising αβ (left) or αβγ (right) subunits showing the locations for Zn^{2+} inhibitory coordination sites (black filled circles). The replacement of a β subunit by a γ subunit disrupts Zn^{2+} coordination in the ion channel and at one of the α+β− interface sites (dashed open circles), leaving only one weak Zn^{2+} coordination site intact.

The subtype-selective nature of Zn^{2+} inhibition at $GABA_A$ receptors could be explained by considering each of the Zn^{2+}-binding sites in αβγ and αβ $GABA_A$ receptors. The reduced sensitivity of αβγ receptors to Zn^{2+} most likely arose as a consequence of incorporating a γ subunit (lacking Zn^{2+}-binding residues) into the receptor pentamer and thus displacing a β subunit. This exchange of subunits would naturally disrupt the ion channel–binding site for Zn^{2+}, as well as one of the two sites located at the α+β− interfaces, since one interface would now be changed to α+γ− (Figure 23.13).

$GABA_A$ receptors are sensitive to modulation by a variety of cations (Smart and Constanti, 1982; Narahashi et al., 1994); amongst these, Cu^{2+} has recently been found to act as a potent inhibitor of tonic GABA currents in the cerebellum and striatum mediated by δ subunit $GABA_A$ receptors (McGee et al., 2013). Copper ions share many chemical similarities with Zn^{2+} with regard to the type of complexes formed with organic ligands. It is therefore unsurprising that the block of GABA currents by Cu^{2+} can be affected by some residues that also impact on Zn^{2+} inhibition (Fisher and MacDonald, 1998; Kim and MacDonald, 2003).

Protons are also modulators at $GABA_A$ receptors causing both potentiation and inhibition of currents depending on the receptor subunit composition (Robello et al., 1994; Krishek et al., 1996; Huang and Dillon, 1999; Feng and MacDonald, 2004; Huang et al., 2004). Histidine 267, a residue important for Zn^{2+} inhibition, also plays a key role in the proton sensitivity of $GABA_A$ receptors (Wilkins et al., 2002), though other residues, notably in the M2–M3 linker of β subunits (Wilkins et al., 2005) and ECD (Huang et al., 2004), are also important. Proton sensitivity of $GABA_A$ receptors prompts a cautionary note on the ubiquitous use of the pH buffer, HEPES, which, when protonated, can also cause inhibition of GABA currents (Hugel et al., 2012).

The recently identified interaction of propofol with H267 (Yip et al., 2013) suggests that general anesthesia induced by this agent may be affected by the presence of Zn^{2+} or H^+ that similarly interact with this ion channel–located histidine residue.

Further into the GABA ion channel, another residue of interest in receptor β2/3 subunits is an asparagine at 15′ (Figure 23.12). This is replaced by serine in β1 subunits. This change is sufficient to largely remove modulation of $GABA_A$ receptors by several therapeutically different drugs: loreclezole (anticonvulsant), methyl 6,7-dimethoxy-4-ethly-β-carboline (DMCM, a BDZ inverse agonist), etomidate (anesthetic), furosemide (diuretic), and mefenamic acid (anti-inflammatory). The importance of this 15′ residue is emphasized by the observation that loreclezole potentiates αβγ $GABA_A$ receptor function when the receptor contains β2 or β3 but not β1 subunits (Wingrove et al., 1994).

The asparagine at 15′ is not readily accessible from the lumen (Figure 23.12). Inserting the asparagine residue of β2/3 subunits into β1 subunits confers a sensitivity to loreclezole on αβ1γ receptors. This finding was subsequently extended to the BDZ inverse agonist β-carboline DMCM, which typically is thought to bind at the BDZ site α+γ− subunit interface, but clearly its modulation is affected by the β subunit and in particular N290 in β2/3 subunits (Stevenson et al., 1995).

The diuretic furosemide also shows β subunit selectivity dependent on N290, though there are residues elsewhere in the $GABA_A$ receptor (e.g., in M1) that are important for its modulatory action (Thompson et al., 1999). Another member of this quartet of drugs, mefenamic acid, is a modulator of $GABA_A$ receptors that is also dependent on receptor subunit composition causing potentiation at α1β2/3γ and α1β2/3 receptors, but inhibition at α1β1 receptors. A dependence on N290 was evident following its mutation (N290S) that markedly reduced potentiation at α1β2/3γ2 receptors (Halliwell et al., 1999). Finally, the anesthetic etomidate, acting as a potentiator at $GABA_A$ receptors, also shows a preference for β2/3 subunit-containing receptors (Hill-Venning et al., 1997), and substitution of N289 in β3 subunits reduced both direct etomidate-induced currents and its GABA-modulatory actions at αβ3γ2 receptors (Belelli et al., 1997; McGurk et al., 1998; Moody et al., 1998).

The natural question arising from these observations with drugs taken from four distinct therapeutic classes is whether the asparagine at 15′ in the β2/3 subunits constitutes part of a generic binding site or represents a crucial region for the propagation of signal transduction. Comparing the chemical structures of mefenamic acid, etomidate and loreclezole suggest their hydrophobic and electronegative moieties can be overlaid in minimum energy conformations (Halliwell et al., 1999), which would be in accord with them sharing a similar binding site. However, other evidence argues against the idea of a single binding site, not least the fact that photolabeling with etomidate indicates that it binds to a different interfacial site on $GABA_A$ receptors (Li et al., 2006) and that furosemide is critically dependent on other residues in M1 and shows a preference toward α6 subunit-containing receptors (Thompson et al., 1999). Taken together, the balance of evidence suggests the asparagine at 15′ is unlikely to form part of a binding site but more likely is an important part of an allosteric transduction pathway.

Travelling deeper into the GABA ion channel reveals other critical residues that are conserved across all $GABA_A$ receptor α subunits, namely, threonine at 6′ and a valine at 2′ (Figure 23.12). These residues play an important role in the antagonism of GABA receptors by the convulsant picrotoxin (Ffrench-Constant et al., 1993; Zhang et al., 1994; Gurley et al., 1995; Xu et al., 1995; Bali and Akabas, 2007) and quite likely by other similar convulsants such as t-butyl-bicyclophosphorothionate (TBPS).

The location of these residues at 2′ and 6′, deep within the ion channel, beyond the ion channel gate, seems ideal for an ion channel–blocking mechanism. Picrotoxin may well bind in this region, particularly as it can become *trapped* in what is assumed to be the ion channel when it closes (Bali and Akabas, 2007). However, for picrotoxin and the active component, picrotoxinin, the mode of inhibition is more complex than simple open-channel block with several studies promoting an allosteric inhibitory role that requires more than just open-channel block (Smart and Constanti, 1986; Newland and Cull-Candy, 1992; Goutman and Calvo, 2004; Qian et al., 2005; Korshoej et al., 2010). Future structural studies will no doubt cast light on this and other binding site debates.

23.10 CONCLUSIONS

The variety of GABA$_A$ receptor subunits and thus receptor isoforms, with their concomitant differences in physiological properties, provides a diverse signaling platform to influence the activity patterns of individual neurons and neural networks. Neurons can exert exquisite control over the excitability of the cell surface membrane by placement (clustering) of receptors at key points and by switching the receptor subunit composition to effect a change in physiology at other membrane locations. How these various receptor isoforms function within single cells, and their relative importance, remains a difficult task to address. Having a greater armamentarium of specific pharmacological agents would be helpful.

The GABA$_A$ receptor is sensitive to a myriad of drugs, and some of these do show receptor subtype selectivity, but nevertheless, more specific agents are required to deduce the underlying basis of receptor function. An alternative approach to the study of GABA receptors is to develop photochemical ligands, which could be valuable analytical tools. Recent studies suggest these can be used for analyzing receptors in membrane domains that are exposed to photons, thereby affecting receptor function in the presence of ligands whose conformation can be changed by light (Yue et al., 2012). In addition, inactive GABA molecules can be *uncaged* by light to map GABA$_A$ receptor density and clustering along neuronal processes (Kanemoto et al., 2011). Overall it is evident, given the therapeutic portfolio of the GABA$_A$ receptor, that the development of selective ligands to target specific GABA receptor isoforms is a prize worth pursuing.

REFERENCES

Abramian, A. M., E. Comenencia-Ortiz, M. Vithlani, E. V. Tretter, W. Sieghart, P. A. Davies, and S. J. Moss. 2010. Protein kinase C phosphorylation regulates membrane insertion of GABA$_A$ receptor subtypes that mediate tonic inhibition. *J. Biol. Chem.* 285:41795–41805.

Araud, T., S. Wonnacott, and D. Bertrand. 2010. Associated proteins: The universal toolbox controlling ligand gated ion channel function. *Biochem. Pharmacol.* 80:160–169.

Bali, M. and M. H. Akabas. 2007. The location of a closed channel gate in the GABA$_A$ receptor channel. *J. Gen. Physiol.* 129:145–159.

Barnard, E. A., M. G. Darlison, and P. Seeburg. 1987. Molecular biology of the GABA$_A$ receptor: The receptor/channel super family. *Trends Neurosci.* 10:502–509.

Bedford, F. K., J. T. Kittler, E. Muller, P. Thomas, J. M. Uren, D. Merlo, W. Wisden, A. Triller, T. G. Smart, and S. J. Moss. 2001. GABA$_A$ receptor cell surface number and subunit stability are regulated by the ubiquitin-like protein Plic-1. *Nat. Neurosci.* 4:908–916.

Belelli, D., J. J. Lambert, J. A. Peters, K. Wafford, and P. J. Whiting. 1997. The interaction of the general anesthetic etomidate with the γ-aminobutyric acid type A receptor is influenced by a single amino acid. *Proc. Natl. Acad. Sci. U.S.A.* 94:11031–11036.

Bera, A. K. and M. H. Akabas. 2005. Spontaneous thermal motion of the GABA$_A$ receptor M2 channel-lining segments. *J. Biol. Chem.* 280:35506–35512.

Bera, A. K., M. Chatav, and M. H. Akabas. 2002. GABA$_A$ receptor M2-M3 loop secondary structure and changes in accessibility during channel gating. *J. Biol. Chem.* 277:43002–43010.

Betz, H. 1990. Ligand-gated ion channels in the brain: The amino acid receptor superfamily. *Neuron* 5:383–392.

Betz, H. and B. Laube. 2006. Glycine receptors: Recent insights into their structural organization and functional diversity. *J. Neurochem.* 97:1600–1610.

Bocquet, N., H. Nury, M. Baaden, C. Le Poupon, J. P. Changeux, M. Delarue, and P. J. Corringer. 2009. X-ray structure of a pentameric ligand-gated ion channel in an apparently open conformation. *Nature* 457:111–114.

Bocquet, N., L. Prado de Carvalho, J. Cartaud, J. Neyton, C. Le Poupon, A. Taly, T. Grutter, J. P. Changeux, and P. J. Corringer. 2007. A prokaryotic proton-gated ion channel from the nicotinic acetylcholine receptor family. *Nature* 445:116–119.

Bogdanov, Y., G. Michels, C. Armstrong-Gold, P. G. Haydon, J. Lindstrom, M. Pangalos, and S. J. Moss. 2006. Synaptic GABA$_A$ receptors are directly recruited from their extrasynaptic counterparts. *EMBO J.* 25:4381–4389.

Boileau, A. J., A. M. Kucken, A. R. Evers, and C. Czajkowski. 1998. Molecular dissection of benzodiazepine binding and allosteric coupling using chimeric γ-aminobutyric acid$_A$ receptor subunits. *Mol. Pharmacol.* 53:295–303.

Boileau, A. J., J. G. Newell, and C. Czajkowski. 2002. GABA$_A$ receptor β2 Tyr97 and Leu99 line the GABA-binding site. Insights into mechanisms of agonist and antagonist actions. *J. Biol. Chem.* 277:2931–2937.

Bourne, Y., T. T. Talley, S. B. Hansen, P. Taylor, and P. Marchot. 2005. Crystal structure of a Cbtx-AChBP complex reveals essential interactions between snake alpha-neurotoxins and nicotinic receptors. *EMBO J.* 24:1512–1522.

Bouzat, C., F. Gumilar, G. Spitzmaul, H. L. Wang, D. Rayes, S. B. Hansen, P. Taylor, and S. M. Sine. 2004. Coupling of agonist binding to channel gating in an ACh-binding protein linked to an ion channel. *Nature* 430:896–900.

Bowery, N. G. and D. A. Brown. 1974. Depolarizing actions of gamma-aminobutyric acid and related compounds on rat superior cervical ganglia in vitro. *Br. J. Pharmacol.* 50:205–218.

Bowery, N. G. and T. G. Smart. 2006. GABA and glycine as neurotransmitters: A brief history. *Br. J. Pharmacol.* 147:S109–S119.

Brandon, N., J. Jovanovic, and S. Moss. 2002. Multiple roles of protein kinases in the modulation of γ-aminobutyric acid a receptor function and cell surface expression. *Pharmacol. Ther.* 94:113–122.

Brejc, K., W. J. van Dijk, R. V. Klaassen, M. Schuurmans, O. J. van Der, A. B. Smit, and T. K. Sixma. 2001. Crystal structure of an ACh-binding protein reveals the ligand-binding domain of nicotinic receptors. *Nature* 411:269–276.

Brickley, S. G. and I. Mody. 2012. Extrasynaptic GABA$_A$ receptors: Their function in the CNS and implications for disease. *Neuron* 73:23–34.

Celie, P. H., I. E. Kasheverov, D. Y. Mordvintsev, R. C. Hogg, P. van Nierop, R. van Elk, S. E. Rossum-Fikkert et al. 2005. Crystal structure of nicotinic acetylcholine receptor homolog AChBP in complex with an α-conotoxin PnIA variant. *Nat. Struct. Mol. Biol.* 12:582–588.

Celie, P. H., S. E. van Rossum-Fikkert, W. J. van Dijk, K. Brejc, A. B. Smit, and T. K. Sixma. 2004. Nicotine and carbamylcholine binding to nicotinic acetylcholine receptors as studied in AChBP crystal structures. *Neuron* 41:907–914.

Charych, E. I., W. Yu, C. P. Miralles, D. R. Serwanski, X. Li, M. Rubio, and A. L. De Blas. 2004. The brefeldin A-inhibited GDP/GTP exchange factor 2, a protein involved in vesicular trafficking, interacts with the beta subunits of the GABA receptors. *J. Neurochem.* 90:173–189.

Chen, Z. W. and R. W. Olsen. 2007. $GABA_A$ receptor associated proteins: A key factor regulating $GABA_A$ receptor function. *J. Neurochem.* 100:279–294.

Choquet, D. and A. Triller. 2013. The dynamic synapse. *Neuron* 80:691–703.

Corringer, P. J., F. Poitevin, M. S. Prevost, L. Sauguet, M. Delarue, and J. P. Changeux. 2012. Structure and pharmacology of pentameric receptor channels: From bacteria to brain. *Structure* 20:941–956.

Davies, P. A., W. Wang, T. G. Hales, and E. F. Kirkness. 2003. A novel class of ligand-gated ion channel is activated by Zn^{2+}. *J. Biol. Chem.* 278:712–717.

Draguhn, A., T. A. Verdoorn, M. Ewert, P. H. Seeburg, and B. Sakmann. 1990. Functional and molecular distinction between recombinant rat $GABA_A$ receptor subtypes by Zn^{2+}. *Neuron* 5:781–788.

Dunne, E. L., A. M. Hosie, J. R. Wooltorton, I. C. Duguid, K. Harvey, S. J. Moss, R. J. Harvey, and T. G. Smart. 2002. An N-terminal histidine regulates Zn^{2+} inhibition on the murine $GABA_A$ receptor β3 subunit. *Br. J. Pharmacol.* 137:29–38.

Ebert, B., K. A. Wafford, P. J. Whiting, P. Krogsgaard-Larsen, and J. A. Kemp. 1994. Molecular pharmacology of γ-aminobutyric acid type A receptor agonists and partial agonists in oocytes injected with different α, β, and γ receptor subunit combinations. *Mol. Pharmacol.* 46:957–963.

Ernst, M., S. Bruckner, S. Boresch, and W. Sieghart. 2005. Comparative models of $GABA_A$ receptor extracellular and transmembrane domains: Important insights in pharmacology and function. *Mol. Pharmacol.* 68:1291–1300.

Fang, C., L. Deng, C. A. Keller, M. Fukata, Y. Fukata, G. Chen, and B. Luscher. 2006. GODZ-mediated palmitoylation of $GABA_A$ receptors is required for normal assembly and function of GABAergic inhibitory synapses. *J. Neurosci.* 26:12758–12768.

Farrant, M. and Z. Nusser. 2005. Variations on an inhibitory theme: Phasic and tonic activation of $GABA_A$ receptors. *Nat. Rev. Neurosci.* 6:215–229.

Feng, G., H. Tintrup, J. Kirsch, M. C. Nichol, J. Kuhse, H. Betz, and J. R. Sanes. 1998. Dual requirement for gephyrin in glycine receptor clustering and molybdoenzyme activity. *Science* 282:1321–1324.

Feng, H. J. and R. L. MacDonald. 2004. Proton modulation of α1β3δ $GABA_A$ receptor channel gating and desensitization. *J. Neurophysiol.* 92:1577–1585.

Ffrench-Constant, R. H., T. A. Rocheleau, J. C. Steichen, and A. E. Chalmers. 1993. A point mutation in a *Drosophila* GABA receptor confers insecticide resistance. *Nature* 363:449–451.

Fisher, J. L. 2004. The α1 and α6 subunit subtypes of the mammalian $GABA_A$ receptor confer distinct channel gating kinetics. *J. Physiol.* 561:433–448.

Fisher, J. L. and R. L. MacDonald. 1998. The role of an α subtype M_2–M_3 His in regulating inhibition of $GABA_A$ receptor current by zinc and other divalent cations. *J. Neurosci.* 18:2944–2953.

Forman, S. A. 2012. Monod–Wyman–Changeux allosteric mechanisms of action and the pharmacology of etomidate. *Curr. Opin. Anaesthesiol.* 25:411–418.

Fritschy, J. M. and I. Brunig. 2003. Formation and plasticity of GABAergic synapses: Physiological mechanisms and pathophysiological implications. *Pharmacol. Ther.* 98:299–323.

Fritschy, J. M., J. Paysan, A. Enna, and H. Mohler. 1994. Switch in the expression of rat $GABA_A$-receptor subtypes during postnatal development: An immunohistochemical study. *J. Neurosci.* 14:5302–5324.

Fritschy, J.-M. and H. Mohler. 1995. $GABA_A$ receptor heterogeneity in the adult rat brain: Differential regional and cellular distribution of seven major subunits. *J. Comp. Neurol.* 359:154–194.

Fritschy, J. M., R. J. Harvey, and G. Schwarz. 2008. Gephyrin: Where do we stand, where do we go? *Trends Neurosci.* 31:257–264.

Frolund, B., A. T. Jorgensen, L. Tagmose, T. B. Stensbol, H. T. Vestergaard, C. Engblom, U. Kristiansen, C. Sanchez, P. Krogsgaard-Larsen, and T. Liljefors. 2002. Novel class of potent 4-arylalkyl substituted 3-isoxazolol $GABA_A$ antagonists: Synthesis, pharmacology, and molecular modeling. *J. Med. Chem.* 45:2454–2468.

Frolund, B., L. Tagmose, T. Liljefors, T. B. Stensbol, C. Engblom, U. Kristiansen, and P. Krogsgaard-Larsen. 2000. A novel class of potent 3-isoxazolol $GABA_A$ antagonists: Design, synthesis, and pharmacology. *J. Med. Chem.* 43:4930–4933.

Gingrich, K. J., W. A. Roberts, and R. S. Kass. 1995. Dependence of the $GABA_A$ receptor gating kinetics on the α-subunit isoform: Implications for structure-function relations and synaptic transmission. *J. Physiol. (Lond.)* 489:529–543.

Glykys, J., E. O. Mann, and I. Mody. 2008. Which $GABA_A$ receptor subunits are necessary for tonic inhibition in the hippocampus? *J. Neurosci.* 28:1421–1426.

Glykys, J. and I. Mody. 2007. Activation of $GABA_A$ receptors: Views from outside the synaptic cleft. *Neuron* 56:763–770.

Goutman, J. D. and D. J. Calvo. 2004. Studies on the mechanisms of action of picrotoxin, quercetin and pregnanolone at the GABAρ1 receptor. *Br. J. Pharmacol.* 141:717–727.

Grenningloh, G., E. D. Gundelfinger, B. Schmitt, H. Betz, M. G. Darlison, E. A. Barnard, P. R. Schofield, and P. H. Seeburg. 1987. Glycine vs GABA receptors. *Nature* 330:25–26.

Gurley, D., J. Amin, P. C. Ross, D. S. Weiss, and G. White. 1995. Point mutations in the M2 region of the α, β, or γ subunit of the $GABA_A$ channel that abolish block by picrotoxin. *Receptors Channels* 3:13–20.

Halliwell, R. F., P. Thomas, D. Patten, C. H. James, A. Martinez Torres, R. Miledi, and T. G. Smart. 1999. Subunit-selective modulation of $GABA_A$ receptors by the non-steroidal anti-inflammatory agent, mefenamic acid. *Eur. J. Neurosci.* 11:2897–2905.

Hansen, S. B., G. Sulzenbacher, T. Huxford, P. Marchot, P. Taylor, and Y. Bourne. 2005. Structures of Aplysia AChBP complexes with nicotinic agonists and antagonists reveal distinctive binding interfaces and conformations. *EMBO J.* 24:3635–3646.

Heaulme, M., J. P. Chambon, R. Leyris, J. C. Molimard, C. G. Wermuth, and K. Biziere. 1986. Biochemical characterization of the interaction of three pyridazinyl-GABA derivatives with the $GABA_A$ receptor site. *Brain Res.* 384:224–231.

Hibbs, R. E. and E. Gouaux. 2011. Principles of activation and permeation in an anion-selective Cys-loop receptor. *Nature* 474:54–60.

Hilf, R. J. C. and R. Dutzler. 2008. X-ray structure of a prokaryotic pentameric ligand-gated ion channel. *Nature* 452:375–379.

Hill-Venning, C., D. Belelli, J. A. Peters, and J. J. Lambert. 1997. Subunit-dependent interaction of the general anaesthetic etomidate with the γ-aminobutyric acid type A receptor. *Br. J. Pharmacol.* 120:749–756.

Horenstein, J. and M. H. Akabas. 1998. Location of a high affinity Zn^{2+} binding site in the channel of α1β1 γ-aminobutyric $acid_A$ receptors. *Mol. Pharmacol.* 53:870–877.

Hortnagl, H., R. O. Tasan, A. Wieselthaler, E. Kirchmair, W. Sieghart, and G. Sperk. 2013. Patterns of mRNA and protein expression for 12 $GABA_A$ receptor subunits in the mouse brain. *Neuroscience* 236:345–372.

Hosie, A. M., E. L. Dunne, R. J. Harvey, and T. G. Smart. 2003. Zinc-mediated inhibition of GABA$_A$ receptors: Discrete binding sites underlie subtype specificity. *Nat. Neurosci.* 6:362–369.
Hosie, A. M., M. E. Wilkins, H. M. A. da Silva, and T.G. Smart. 2006. Endogenous neurosteroids regulate GABA$_A$ receptors through two discrete transmembrane sites. *Nature* 444:486–489.
Hosie, A. M., M. E. Wilkins, and T. G. Smart. 2007. Neurosteroid binding sites on GABA$_A$ receptors. *Pharmacol. Ther.* 116:7–19.
Houston, C. M., Q. He, and T. G. Smart. 2009. CaMKII phosphorylation of the GABA$_A$ receptor: Receptor subtype- and synapse-specific modulation. *J. Physiol.* 587:2115–2125.
Huang, R. Q., Z. Chen, and G. H. Dillon. 2004. Molecular basis for modulation of recombinant $\alpha 1\beta 2\gamma 2$ GABA$_A$ receptors by protons. *J. Neurophysiol.* 92:883–894.
Huang, R. Q. and G. H. Dillon. 1999. Effect of extracellular pH on GABA-activated current in rat recombinant receptors and thin hypothalamic slices. *J. Neurophysiol.* 82:1233–1243.
Hugel, S., N. Kadiri, J. L. Rodeau, S. Gaillard, and R. Schlichter. 2012. pH-dependent inhibition of native GABA$_A$ receptors by HEPES. *Br. J. Pharmacol.* 166:2402–2416.
Jacob, T. C., S. J. Moss, and R. Jurd. 2008. GABA$_A$ receptor trafficking and its role in the dynamic modulation of neuronal inhibition. *Nat. Rev. Neurosci.* 9:331–343.
Jenkins, A., E. P. Greenblatt, H. J. Faulkner, E. Bertaccini, A. Light, A. Lin, A. Andreasen, A. Viner, J. R. Trudell, and N. L. Harrison. 2001. Evidence for a common binding cavity for three general anesthetics within the GABA$_A$ receptor. *J. Neurosci.* 21:RC136.
Jensen, M. L., D. B. Timmermann, T. H. Johansen, A. Schousboe, T. Varming, and P. K. Ahring. 2002. The β subunit determines the ion selectivity of the GABA$_A$ receptor. *J. Biol. Chem.* 277:41438–41447.
Johnston, G. A. 2013. Advantages of an antagonist: Bicuculline and other GABA antagonists. *Br. J. Pharmacol.* 169:328–336.
Kanemoto, Y., M. Matsuzaki, S. Morita, T. Hayama, J. Noguchi, N. Senda, A. Momotake, T. Arai, and H. Kasai. 2011. Spatial distributions of GABA receptors and local inhibition of Ca^{2+} transients studied with GABA uncaging in the dendrites of CA1 pyramidal neurons. *PLoS One* 6:e22652.
Karlin, A. and M. H. Akabas. 1998. Substituted-cysteine accessibility method. *Methods Enzymol.* 293:123–145.
Kash, T. L., A. Jenkins, J. C. Kelley, J. R. Trudell, and N. L. Harrison. 2003. Coupling of agonist binding to channel gating in the GABA$_A$ receptor. *Nature* 421:272–275.
Keramidas, A., A. J. Moorhouse, P. R. Schofield, and P. H. Barry. 2004. Ligand-gated ion channels: Mechanisms underlying ion selectivity. *Prog. Biophys. Mol. Biol.* 86:161–204.
Khatri, A., A. Sedelnikova, and D. S. Weiss. 2009. Structural rearrangements in loop F of the GABA receptor signal ligand binding, not channel activation. *Biophys. J.* 96:45–55.
Khatri, A. and D. S. Weiss. 2010. The role of loop F in the activation of the GABA receptor. *J. Physiol.* 588:59–66.
Kim, H. and R. L. MacDonald. 2003. An N-terminal histidine is the primary determinant of alpha subunit-dependent Cu^{2+} sensitivity of $\alpha\beta 3\gamma 2L$ GABA$_A$ receptors. *Mol. Pharmacol.* 64:1145–1152.
Kittler, J. T., G. Chen, V. Kukhtina, A. Vahedi-Faridi, Z. Gu, V. Tretter, K. R. Smith et al. 2008. Regulation of synaptic inhibition by phospho-dependent binding of the AP2 complex to a YECL motif in the GABA$_A$ receptor $\gamma 2$ subunit. *PNAS* 105:3616–3621.
Kittler, J. T., P. Thomas, V. Tretter, Y. D. Bogdanov, V. Haucke, T. G. Smart, and S. J. Moss. 2004. Huntingtin-associated protein 1 regulates inhibitory synaptic transmission by modulating γ-aminobutyric acid type A receptor membrane trafficking. *Proc. Natl. Acad. Sci. U.S.A.* 101:12736–12741.
Kleingoor, C., H. A. Wieland, E. R. Korpi, P. H. Seeburg, and H. Kettenmann. 1993. Current potentiation by diazepam but not GABA sensitivity is determined by a single histidine residue. *Neuro. Rep.* 4:187–190.
Kneussel, M. and H. Betz. 2000. Receptors, gephyrin and gephyrin-associated proteins: Novel insights into the assembly of inhibitory postsynaptic membrane specializations. *J. Physiol. (Lond.)* 525:1–9.
Kneussel, M., J. H. Brandstatter, B. Gasnier, G. Feng, J. R. Sanes, and H. Betz. 2001. Gephyrin-independent clustering of postsynaptic GABA$_A$ receptor subtypes. *Mol. Cell Neurosci.* 17:973–982.
Korshoej, A. R., M. M. Holm, K. Jensen, and J. D. Lambert. 2010. Kinetic analysis of evoked IPSCs discloses mechanism of antagonism of synaptic GABA receptors by picrotoxin. *Br. J. Pharmacol.* 159:636–649.
Kowalczyk, S., A. Winkelmann, B. Smolinsky, B. Förstera, I. Neundorf, G. Schwarz, and J. C. Meier. 2013. Direct binding of GABA$_A$ receptor $\beta 2$ and $\beta 3$ subunits to gephyrin. *Eur. J. Neurosci.* 37:544–554.
Krishek, B. J., A. Amato, C. N. Connolly, S. J. Moss, and T. G. Smart. 1996. Proton sensitivity of the GABA$_A$ receptor is associated with the receptor subunit composition. *J. Physiol.* 492:431–443.
Krogsgaard-Larsen, P., B. Frolund, and T. Liljefors. 2002. Specific GABA$_A$ agonists and partial agonists. *Chem. Rec.* 2:419–430.
Kucken, A. M., D. A. Wagner, P. R. Ward, J. A. Boileau, and C. Czajkowski. 2000. Identification of benzodiazepine binding site residues in the $\gamma 2$ subunit of the γ-aminobutyric acid$_A$ receptor. *Mol. Pharmacol.* 57:932–939.
Lavoie, A. M., J. J. Tingey, N. L. Harrison, D. B. Pritchett, and R. E. Twyman. 1997. Activation and deactivation rates of recombinant GABA$_A$ receptor channels are dependent on α-subunit isoform. *Biophys. J.* 73:2518–2526.
Lester, H. A., M. I. Dibas, D. S. Dahan, J. F. Leite, and D. A. Dougherty. 2004. Cys-loop receptors: New twists and turns. *Trends Neurosci.* 27:329–336.
Li, G. D., D. C. Chiara, J. B. Cohen, and R. W. Olsen. 2010. Numerous classes of general anesthetics inhibit etomidate binding to γ-aminobutyric acid type A (GABA$_A$) receptors. *J. Biol. Chem.* 285:8615–8620.
Li, G. D., D. C. Chiara, G. W. Sawyer, S. S. Husain, R. W. Olsen, and J. B. Cohen. 2006. Identification of a GABA$_A$ receptor anesthetic binding site at subunit interfaces by photolabeling with an etomidate analog. *J. Neurosci.* 26:11599–11605.
Loebrich, S., R. Bahring, T. Katsuno, S. Tsukita, and M. Kneussel. 2006. Activated radixin is essential for GABA$_A$ receptor $\alpha 5$ subunit anchoring at the actin cytoskeleton. *EMBO J.* 25:987–999.
Lummis, S. C. 2009. Locating GABA in GABA receptor binding sites. *Biochem. Soc. Trans.* 37:1343–1346.
Luscher, B., T. Fuchs, and C. Kilpatrick. 2011. GABA$_A$ receptor trafficking-mediated plasticity of inhibitory synapses. *Neuron* 70:385–409.
Luscher, B. and C. A. Keller. 2004. Regulation of GABA$_A$ receptor trafficking, channel activity, and functional plasticity of inhibitory synapses. *Pharmacol. Ther.* 102:195–221.
Lynch, J. W. 2004. Molecular structure and function of the glycine receptor chloride channel. *Physiol. Rev.* 84:1051–1095.
McGee, T. P., C. M. Houston, and S. G. Brickley. 2013. Copper block of extrasynaptic GABA$_A$ receptors in the mature cerebellum and striatum. *J. Neurosci.* 33:13431–13435.
McGurk, K. A., M. Pistis, D. Belelli, A. G. Hope, and J. J. Lambert. 1998. The effect of a transmembrane amino acid on etomidate sensitivity of an invertebrate GABA receptor. *Br. J. Pharmacol.* 124:13–20.
Mercado, J. and C. Czajkowski. 2006. Charged residues in the $\alpha 1$ and $\beta 2$ pre-M1 regions involved in GABA$_A$ receptor activation. *J. Neurosci.* 26:2031–2040.

Mihic, S. J., Q. Ye, M. J. Wick, V. V. Koltchine, M. D. Krasowski, S. E. Finn, M. P. Mascia et al. 1997. Sites of alcohol and volatile anaesthetic action on $GABA_A$ and glycine receptors. *Nature* 389:385–389.

Miller, P. S. and T. G. Smart. 2010. Binding, activation and modulation of Cys-loop receptors. *Trends Pharmacol. Sci.* 31:161–174.

Miller, P. S. and A. R. Aricescu. 2014. Crystal structure of a human GABA-A receptor. *Nature* 512:270–275.

Miyazawa, A., Y. Fujiyoshi, and N. Unwin. 2003. Structure and gating mechanism of the acetylcholine receptor pore. *Nature* 424:949–955.

Mohler, H., M. K. Battersby, and J. G. Richards. 1980. Benzodiazepine receptor protein identified and visualized in brain tissue by a photoaffinity label. *Proc. Natl. Acad. Sci. U.S.A.* 77:1666–1670.

Moody, E. J., C. S. Knauer, R. Granja, M. Strakhovaua, and P. Skolnick. 1998. Distinct structural requirements for the direct and indirect actions of the anaesthetic etomidate at $GABA_A$ receptors. *Toxicol. Lett.* 100–101:209–215.

Mortensen, M., B. Frolund, A. T. Jorgensen, T. Liljefors, P. Krogsgaard-Larsen, and B. Ebert. 2002. Activity of novel 4-PIOL analogues at human $\alpha 1\beta 2\gamma 2S$ $GABA_A$ receptors—correlation with hydrophobicity. *Eur. J. Pharmacol.* 451:125–132.

Mortensen, M., U. Kristiansen, B. Ebert, B. Frolund, P. Krogsgaard-Larsen, and T. G. Smart. 2004. Activation of single heteromeric $GABA_A$ receptor ion channels by full and partial agonists. *J. Physiol.* 557:389–413.

Mortensen, M., B. Patel, and T. G. Smart. 2011. GABA potency at $GABA_A$ receptors found in synaptic and extrasynaptic zones. *Front. Cell Neurosci.* 6:1–10.

Mortensen, M. and T. G. Smart. 2006. Extrasynaptic $\alpha\beta$ subunit $GABA_A$ receptors on rat hippocampal pyramidal neurons. *J. Physiol. (Lond.)* 577:841–856.

Moss, S. J., G. H. Gorrie, A. Amato, and T. G. Smart. 1995. Modulation of $GABA_A$ receptors by tyrosine phosphorylation. *Nature* 377:344–348.

Moss, S. J. and T. G. Smart. 1996. Modulation of amino acid-gated ion channels by protein phosphorylation. *Int. Rev. Neurobiol.* 39:1–52.

Moss, S. J. and T. G. Smart. 2001. Constructing inhibitory synapses. *Nat. Rev. Neurosci.* 2:240–250.

Muir, J., I. L. Arancibia-Carcamo, A. F. MacAskill, K. R. Smith, L. D. Griffin, and J. T. Kittler. 2010. NMDA receptors regulate $GABA_A$ receptor lateral mobility and clustering at inhibitory synapses through serine 327 on the $\gamma 2$ subunit. *Proc. Natl. Acad. Sci. U.S.A.* 107:16679–16684.

Mukherjee, J., K. Kretschmannova, G. Gouzer, H. M. Maric, S. Ramsden, V. Tretter, K. Harvey et al. 2011. The residence time of $GABA_ARs$ at inhibitory synapses is determined by direct binding of the receptor $\alpha 1$ subunit to gephyrin. *J. Neurosci.* 31:14677–14687.

Nagaya, N. and R. L. MacDonald. 2001. Two $\gamma 2L$ subunit domains confer low Zn^{2+} sensitivity to ternary $GABA_A$ receptors. *J. Physiol.* 532:17–30.

Narahashi, T., J. Y. Ma, O. Arakawa, E. Reuveny, and M. Nakahiro. 1994. GABA receptor-channel complex as a target site of mercury, copper, zinc, and lanthanides. *Cell Mol. Neurobiol.* 14:599–621.

Newell, J. G. and C. Czajkowski. 2003. The $GABA_A$ receptor $\alpha 1$ subunit Pro174-Asp191 segment is involved in GABA binding and channel gating. *J. Biol. Chem.* 278:13166–13172.

Newland, C. F. and S. G. Cull-Candy. 1992. On the mechanism of action of picrotoxin on GABA receptor channels in dissociated sympathetic neurones of the rat. *J. Physiol. (Lond.)* 447:191–213.

Nury, H., C. Van Renterghem, Y. Weng, A. Tran, M. Baaden, V. Dufresne, J. P. Changeux, J. M. Sonner, M. Delarue, and P. J. Corringer. 2011. X-ray structures of general anaesthetics bound to a pentameric ligand-gated ion channel. *Nature* 469:428–431.

Nymann-Andersen, J., H. Wang, L. Chen, J. T. Kittler, S. J. Moss, and R. W. Olsen. 2002. Subunit specificity and interaction domain between $GABA_A$ receptor-associated protein (GABARAP) and $GABA_A$ receptors. *J. Neurochem.* 80:815–823.

Olsen, R. W. and W. Sieghart. 2008. International Union of Pharmacology. LXX. Subtypes of gamma-aminobutyric $acid_A$ receptors: Classification on the basis of subunit composition, pharmacology, and function. *Pharmacol. Rev.* 60:243–260.

Olsen, R. W. and W. Sieghart. 2009. $GABA_A$ receptors: Subtypes provide diversity of function and pharmacology. *Neuropharmacology.* 56:141–148.

Ortells, M. O. and G. G. Lunt. 1995. Evolutionary history of the ligand-gated ion-channel superfamily of receptors. *Trends Neurosci.* 18:121–127.

Picton, A. J. and J. L. Fisher. 2007. Effect of the α subunit subtype on the macroscopic kinetic properties of recombinant $GABA_A$ receptors. *Brain Res.* 1165:40–49.

Pirker, S., C. Schwarzer, A. Wieselthaler, W. Sieghart, and G. Sperk. 2000. $GABA_A$ receptors: Immunocytochemical distribution of 13 subunits in the adult rat brain. *Neuroscience* 101:815–850.

Pritchett, D. B., H. Sontheimer, B. D. Shivers, S. Ymer, H. Kettenmann, P. R. Schofield, and P. H. Seeburg. 1989. Importance of a novel $GABA_A$ receptor subunit for benzodiazepine pharmacology. *Nature* 338:582–585.

Qian, H., Y. Pan, Y. Zhu, and P. Khalili. 2005. Picrotoxin accelerates relaxation of $GABA_C$ receptors. *Mol. Pharmacol.* 67:470–479.

Rabow, L. E., S. J. Russek, and D. H. Farb. 1996. From ion currents to genomic analysis: Recent advances in $GABA_A$ receptor research. *Synapse* 21:189–274.

Ramerstorfer, J., R. Furtmuller, I. Sarto-Jackson, Z. Varagic, W. Sieghart, and M. Ernst. 2011. The $GABA_A$ Receptor $\alpha+\beta-$ interface: A novel target for subtype selective drugs. *J. Neurosci.* 31:870–877.

Richter, L., C. de Graaf, W. Sieghart, Z. Varagic, M. Mörzinger, I. J. P. de Esch, G. F. Ecker, and M. Ernst. 2012. Diazepam-bound $GABA_A$ receptor models identify new benzodiazepine binding-site ligands. *Nat. Chem. Biol.* 8:455–464.

Robello, M., P. Baldelli, and A. Cupello. 1994. Modulation by extracellular pH of the activity of $GABA_A$ receptors on rat cerebellum granule cells. *Neuroscience* 61:833–837.

Rusch, D., H. Zhong, and S. A. Forman. 2004. Gating allosterism at a single class of etomidate sites on $\alpha 1\beta 2\gamma 2L$ $GABA_A$ receptors accounts for both direct activation and agonist modulation. *J. Biol. Chem.* 279:20982–20992.

Sarto-Jackson, I., I. Milenkovic, K. H. Smalla, E. D. Gundelfinger, T. Kaehne, R. Herrera-Molina, S. Thomas, M. A. Kiebler, and W. Sieghart. 2012. The cell adhesion molecule Neuroplastin-65 is a novel interaction partner of γ-aminobutyric acid type A receptors. *J. Biol. Chem.* 287:14201–14214.

Sassoe-Pognetto, M. and J. M. Fritschy. 2000. Mini-review: Gephyrin, a major postsynaptic protein of GABAergic synapses. *Eur. J. Neurosci.* 12:2205–2210.

Sassoe-Pognetto, M., J. Kirsch, U. Grunert, U. Greferath, J. M. Fritschy, H. Mohler, H. Betz, and H. Wassle. 1995. Colocalization of gephyrin and $GABA_A$—Receptor subunits in the rat retina. *J. Comp. Neurol.* 357:1–14.

Sassoe-Pognetto, M., P. Panzanelli, W. Sieghart, and J. M. Fritschy. 2000. Colocalization of multiple $GABA_A$ receptor subtypes with gephyrin at postsynaptic sites. *J. Comp. Neurol.* 420:481–498.

Schofield, P. R., M. G. Darlison, N. Fujita, D. R. Burt, F. A. Stephenson, H. Rodriguez, L. M. Rhee et al. 1987. Sequence and functional expression of the $GABA_A$ receptor shows a ligand-gated receptor super-family. *Nature* 328:221–227.

Sedelnikova, A., C. D. Smith, S. O. Zakharkin, D. Davis, D. S. Weiss, and Y. Chang. 2005. Mapping the $\rho 1$ $GABA_C$ receptor agonist binding pocket. Constructing a complete model. *J. Biol. Chem.* 280:1535–1542.

Shivers, B. D., I. Killisch, R. Sprengel, H. Sontheimer, M. Kohler, P. R. Schofield, and P. H. Seeburg. 1989. Two novel $GABA_A$ receptor subunits exist in distinct neuronal subpopulations. *Neuron* 3:327–337.

Sieghart, W. and G. Sperk. 2002. Subunit composition, distribution and function of GABA$_A$ receptor subtypes. *Curr. Top. Med. Chem.* 2:795–816.
Sieghart, W., J. Ramerstorfer, I. Sarto-Jackson, Z. Varagic, and M. Ernst. 2012. A novel GABA$_A$ receptor pharmacology: Drugs interacting with the $\alpha^+\beta^-$ interface. *Br. J. Pharmacol.* 166:476–485.
Sigel, E. 2002. Mapping of the benzodiazepine recognition site on GABA$_A$ receptors. *Curr. Top. Med. Chem.* 2:833–839.
Sigel, E. and A. Buhr. 1997. The benzodiazepine binding site of GABA$_A$ receptors. *Trends Pharmacol. Sci.* 18:425–429.
Sigel, E., R. Baur, G. Trube, H. Mohler, and P. Malherbe. 1990. The effect of subunit composition of rat brain GABA$_A$ receptors on channel function. *Neuron* 5:703–711.
Sigel, E., M. T. Schaerer, A. Buhr, and R. Baur. 1998. The benzodiazepine binding pocket of recombinant $\alpha 1\beta 2\gamma 2$ γ-aminobutyric acid$_A$ receptors: Relative orientation of ligands and amino acid side chains. *Mol. Pharmacol.* 54:1097–1105.
Sigel, E. and M. E. Steinmann. 2012. Structure, function, and modulation of GABA$_A$ receptors. *J. Biol. Chem.* 287:40224–40231.
Simmonds, M. A. 1980. Evidence that bicuculline and picrotoxin act at separate sites to antagonize gamma-aminobutyric acid in rat cuneate nucleus. *Neuropharmacology.* 19:39–45.
Simon, J., H. Wakimoto, N. Fujita, M. Lalande, and E. A. Barnard. 2004. Analysis of the set of GABA$_A$ receptor genes in the human genome. *J. Biol. Chem.* 279:41422–41435.
Smart, T. G. 1992. A novel modulatory binding site for zinc on the GABA$_A$ receptor complex in cultured rat neurones. *J. Physiol. (Lond.)* 447:587–625.
Smart, T. G. and A. Constanti. 1982. A novel effect of zinc on the lobster muscle GABA receptor. *Proc. R. Soc. Lond. Ser. B, Biol. Sci.* 215:327–341.
Smart, T. G. and A. Constanti. 1986. Studies on the mechanism of action of picrotoxinin and other convulsants at the crustacean muscle GABA receptor. *Proc. R. Soc. Lond. Ser. B, Biol. Sci.* 227:191–216.
Smart, T. G. and A. Constanti. 1990. Differential effect of zinc on the vertebrate GABA$_A$-receptor complex. *Br. J. Pharmacol.* 99:643–654.
Smart, T. G., S. J. Moss, X. Xie, and R. L. Huganir. 1991. GABA$_A$ receptors are differentially sensitive to zinc: Dependence on subunit composition. *Br. J. Pharmacol.* 103:1837–1839.
Smart, T. G. and P. Paoletti. 2012. Synaptic neurotransmitter-gated receptors. In *The Synapse*, Sheng, M., B. L. Sabatini, and T. C. Sudhof, (Eds.), Cold Spring Harbor Laboratory Press, New York, pp. 191–216.
Smith, K. R., K. McAinsh, G. Chen, I. L. Arancibia-Carcamo, V. Haucke, Z. Yan, S. J. Moss, and J. T. Kittler. 2008. Regulation of inhibitory synaptic transmission by a conserved atypical interaction of GABA$_A$ receptor β- and γ-subunits with the clathrin AP2 adaptor. *Neuropharmacology* 55:844–850.
Spurny, R., J. Ramerstorfer, K. Price, M. Brams, M. Ernst, H. Nury, M. Verheij et al. 2012. Pentameric ligand-gated ion channel ELIC is activated by GABA and modulated by benzodiazepines. *Proc. Natl. Acad. Sci. U.S.A.* 109:E3028–E3034.
Stevenson, A., P. B. Wingrove, P. J. Whiting, and K. A. Wafford. 1995. β-Carboline γ-aminobutyric acid$_A$ receptor inverse agonists modulate γ-aminobutyric acid via the loreclezole binding site as well as the benzodiazepine site. *Mol. Pharmacol.* 48:965–969.
Steward, E. G., R. Player, J. P. Quilliam, D. A. Brown, and M. J. Pringle. 1971. Molecular conformation of GABA. *Nat. New Biol.* 233:87–88.
Stewart, D. S., M. Hotta, G. D. Li, R. Desai, D. C. Chiara, R. W. Olsen, and S. A. Forman. 2013. Cysteine substitutions define etomidate binding and gating linkages in the α-M1 domain of γ-aminobutyric acid type A (GABA$_A$) receptors. *J. Biol. Chem.* 288:30373–30386.
Storustovu, S. and B. Ebert. 2006. Pharmacological characterization of agonists at δ-containing GABA$_A$ receptors: Functional selectivity for extrasynaptic receptors is dependent on the absence of γ2. *J. Pharmacol. Exp. Ther.* 316:1351–1359.
Tan, K. R., R. Baur, S. Charon, M. Goeldner, and E. Sigel. 2009. Relative positioning of diazepam in the benzodiazepine-binding-pocket of GABA receptors. *J. Neurochem.* 111:1264–1273.
Thomas, P., M. Mortensen, A. M. Hosie, and T. G. Smart. 2005. Dynamic mobility of functional GABA$_A$ receptors at inhibitory synapses. *Nat. Neurosci.* 8:889–897.
Thompson, A. J., M. Alqazzaz, C. Ulens, and S. C. R. Lummis. 2012. The pharmacological profile of ELIC, a prokaryotic GABA-gated receptor. *Neuropharmacology* 63:761–767.
Thompson, S. A., S. A. Arden, G. Marshall, P. B. Wingrove, P. J. Whiting, and K. A. Wafford. 1999. Residues in transmembrane domains I and II determine γ-aminobutyric acid type A$_A$ receptor subtype-selective antagonism by furosemide. *Mol. Pharmacol.* 55:993–999.
Thomson, A. M. and J. N. Jovanovic. 2010. Mechanisms underlying synapse-specific clustering of GABA$_A$ receptors. *Eur. J. Neurosci.* 31:2193–2203.
Tia, S., J. F. Wang, N. Kotchabhakdi, and S. Vicini. 1996. Distinct deactivation and desensitization kinetics of recombinant GABA$_A$ receptors. *Neuropharmacology* 35:1375–1382.
Torres, V. I. and D. S. Weiss. 2002. Identification of a tyrosine in the agonist binding site of the homomeric rho1 gamma-aminobutyric acid (GABA) receptor that, when mutated, produces spontaneous opening. *J. Biol. Chem.* 277:43741–43748.
Tretter, V., T. C. Jacob, J. Mukherjee, J. M. Fritschy, M. N. Pangalos, and S. J. Moss. 2008. The clustering of GABA$_A$ receptor subtypes at inhibitory synapses is facilitated via the direct binding of receptor α2 Subunits to gephyrin. *J. Neurosci.* 28:1356–1365.
Tretter, V., B. Kerschner, I. Milenkovic, S. L. Ramsden, J. Ramerstorfer, L. Saiepour, H. M. Maric et al. 2011. Molecular basis of the γ-aminobutyric acid a receptor α3 subunit interaction with the clustering protein gephyrin. *J. Biol. Chem.* 286:37702–37711.
Triller, A. and D. Choquet. 2005. Surface trafficking of receptors between synaptic and extrasynaptic membranes: And yet they do move! *Trends Neurosci.* 28:133–139.
Tyagarajan, S. K. and J. M. Fritschy. 2010. GABA$_A$ receptors, gephyrin and homeostatic synaptic plasticity. *J. Physiol.* 588:101–106.
Ueno, S., J. Bracamontes, C. Zorumski, D. S. Weiss, and J. H. Steinbach. 1997. Bicuculline and gabazine are allosteric inhibitors of channel opening of the GABA$_A$ receptor. *J. Neurosci.* 17:625–634.
Unwin, N. 1998. The nicotinic acetylcholine receptor of the Torpedo electric ray. *J. Struct. Biol.* 121:181–190.
Unwin, N. 2005. Refined structure of the nicotinic acetylcholine receptor at 4A resolution. *J. Mol. Biol.* 346:967–989.
Varagic, Z., J. Ramerstorfer, S. Huang, S. Rallapalli, I. Sarto-Jackson, J. Cook, W. Sieghart, and M. Ernst. 2013a. Subtype selectivity of α+β– site ligands of GABA$_A$ receptors: Identification of the first highly specific positive modulators at α6β2/3γ2 receptors. *Br. J. Pharmacol.* 169:384–399.
Varagic, Z., L. Wimmer, M. Schnurch, M. D. Mihovilovic, S. Huang, S. Rallapalli, J. M. Cook et al. 2013b. Identification of novel positive allosteric modulators and null modulators at the GABA$_A$ receptor α+β– interface. *Br. J. Pharmacol.* 169:371–383.
Verdoorn, T. A., A. Draguhn, S. Ymer, P. H. Seeburg, and B. Sakmann. 1990. Functional properties of recombinant rat GABA$_A$ receptors depend upon subunit composition. *Neuron* 4:919–928.
Wang, H. B., F. K. Bedford, N. J. Brandon, S. J. Moss, and R. W. Olsen. 1999. GABA$_A$-receptor-associated protein links GABA$_A$ receptors and the cytoskeleton. *Nature* 397:69–72.

Wang, J., S. Liu, U. Haditsch, W. Tu, K. Cochrane, G. Ahmadian, L. Tran et al. 2003. Interaction of calcineurin and type-A GABA receptor $\gamma 2$ subunits produces long-term depression at CA1 inhibitory synapses. *J. Neurosci.* 23:826–836.

Wang, Q., S. A. Pless, and J. W. Lynch. 2010. Ligand- and subunit-specific conformational changes in the ligand-binding domain and the TM2–TM3 linker of $\alpha 1\beta 2\gamma 2$ $GABA_A$ receptors. *J. Biol. Chem.* 285:40373–40386.

Whiting, P. J., R. M. McKernan, and K. A. Wafford. 1995. Structure and pharmacology of vertebrate $GABA_A$ receptor subtypes. *Int. Rev. Neurobiol.* 38:95–138.

Wieland, H. A., H. Luddens, and P. H. Seeburg. 1992. A single histidine in $GABA_A$ receptors is essential for benzodiazepine agonist binding. *J. Biol. Chem.* 267:1426–1429.

Wilkins, M. E., A. M. Hosie, and T. G. Smart. 2002. Identification of a β subunit TM2 residue mediating proton modulation of GABA type A receptors. *J. Neurosci.* 22:5328–5333.

Wilkins, M. E., A. M. Hosie, and T. G. Smart. 2005. Proton modulation of recombinant $GABA_A$ receptors: Influence of GABA concentration and the β subunit TM2–TM3 domain. *J. Physiol.* 567:365–377.

Wingrove, P. B., K. A. Wafford, C. Bain, and P. J. Whiting. 1994. The modulatory action of loreclezole at the γ-aminobutyric acid type A receptor is determined by a single amino acid in the β_2 and β_3 subunit. *Proc. Natl. Acad. Sci. U.S.A.* 91:4569–4573.

Wisden, W., A. L. Gundlach, E. A. Barnard, P. H. Seeburg, and S. P. Hunt. 1991. Distribution of $GABA_A$ receptor subunit mRNAs in rat lumbar spinal cord. *Brain Res. Mol. Brain Res.* 10:179–183.

Wisden, W., D. J. Laurie, H. Monyer, and P. H. Seeburg. 1992. The distribution of 13 $GABA_A$ receptor subunit mRNAs in the rat brain. I. Telencephalon, diencephalon, mesencephalon. *J. Neurosci.* 12:1040–1062.

Wooltorton, J. R., B. J. McDonald, S. J. Moss, and T. G. Smart. 1997. Identification of a Zn^{2+} binding site on the murine $GABA_A$ receptor complex: Dependence on the second transmembrane domain of β subunits. *J. Physiol.* 505 (Pt 3):633–640.

Xiu, X., A. P. Hanek, J. Wang, H. A. Lester, and D. A. Dougherty. 2005. A unified view of the role of electrostatic interactions in modulating the gating of Cys-loop receptors. *J. Biol. Chem.* 280:41655–41666.

Xu, M. and M. H. Akabas. 1993. Amino acids lining the channel of the gamma-aminobutyric acid type A receptor identified by cysteine substitution. *J. Biol. Chem.* 268:21505–21508.

Xu, M. and M. H. Akabas. 1996. Identification of channel-lining residues in the M2 membrane-spanning segment of the $GABA_A$ receptor $\alpha 1$ subunit. *J. Gen. Physiol.* 107:195–205.

Xu, M., D. F. Covey, and M. H. Akabas. 1995. Interaction of picrotoxin with $GABA_A$ receptor channel-lining residues probed in cysteine mutants. *Biophys. J.* 69:1858–1867.

Yamakura, T., E. Bertaccini, J. R. Trudell, and R. A. Harris. 2001. Anesthetics and ion channels: Molecular models and sites of action. *Annu. Rev. Pharmacol. Toxicol.* 41:23–51.

Yip, G. M. S., Z. W. Chen, C. J. Edge, E. H. Smith, R. Dickinson, E. Hohenester, R. R. Townsend et al. 2013. A propofol binding site on mammalian $GABA_A$ receptors identified by photolabeling. *Nat. Chem. Biol.* 9:715–720.

Yue, L., M. Pawlowski, S. S. Dellal, A. Xie, F. Feng, T. S. Otis, K. S. Bruzik, H. Qian, and D. R. Pepperberg. 2012. Robust photoregulation of $GABA_A$ receptors by allosteric modulation with a propofol analogue. *Nat. Commun.* 3:1095.

Zhang, H.-G., R. H. Ffrench-Constant, and M. B. Jackson. 1994. A unique amino acid of the *Drosophila* GABA receptor with influence of drug sensitivity by two mechanisms. *J. Physiol.* 479.1:65–75.

24 Cyclic nucleotide–gated channels

Michael D. Varnum and Gucan Dai

Contents

24.1 INTRODUCTION

Ion channels that are activated by the direct binding of intracellular cyclic nucleotides are best known for the essential role they play in sensory transduction, including vision and olfaction. In these sensory cells, they transduce a chemical signal produced by the stimulus input—a change in the intracellular concentration of guanosine 3′,5′-cyclic monophosphate (cGMP) or adenosine 3′,5′-cyclic monophosphates (cAMP)—into an electrical response via a change in cation conductance through the channel pore. Cyclic nucleotide–gated (CNG) channels are tetrameric complexes of homologous subunits, with each subunit contributing a cyclic nucleotide–binding domain (CNBD), part of the ion conduction pathway, and the fundamental machinery to covert cGMP or cAMP binding into channel opening. Here, we review their functional and structural properties and their physiological and pathophysiological roles.

24.2 PHYSIOLOGICAL ROLES OF CNG CHANNELS

The most prominent known role of CNG channels is in sensory cells, where they help to convert sensory inputs that alter cyclic nucleotide concentrations into electrical responses. CNG channels were first described by Fesenko and coworkers in the outer segment membrane of retinal rod photoreceptors (Fesenko et al., 1985). In rod and cone photoreceptors, these channels conduct a cation current (the *dark current*) in the absence of light; photoactivation of the light receptor opsin activates the G protein transducin, which in turn activates a cGMP-specific phosphodiesterase (PDE) leading to the hydrolysis of cGMP and closure of the CNG channel (Figure 24.1) (Stryer, 1987; Burns and Arshavsky, 2005). Decreased channel conductance produces hyperpolarization of the membrane and decreased neurotransmitter release onto second-order cells. In addition, CNG channels located at cone photoreceptor synapses onto bipolar cells have been shown to modulate synaptic transmission and mediate the effects of nitric oxide (Rieke and Schwartz, 1994; Savchenko et al., 1997).

CNG channels also play a critical role in olfactory transduction (Lancet, 1986; Nakamura and Gold, 1987; Menini, 1995). Olfactory sensory neurons (OSNs) in the olfactory epithelium detect odorants via diverse G protein–coupled olfactory receptors. Olfactory receptors are coupled to G_{olf} activation, which turns on adenylate cyclase (AC), thus leading to increased production of cAMP. Increased cAMP opens the olfactory CNG channels, which depolarize the olfactory neurons. Native olfactory CNG channels are highly permeable to Ca^{2+}, and Ca^{2+} amplifies olfactory signal transduction by opening Ca^{2+}-activated chloride channels. Chloride ions flowing out of the cell further depolarize the OSNs, which helps induce neurotransmitter release from olfactory receptor cells onto second-order neurons (Frings, 2001). A subset of olfactory receptor neurons utilize cGMP signaling components, including a specialized CNG channel, rather than the cAMP-based components found in the principal olfactory neurons (Meyer et al., 2000).

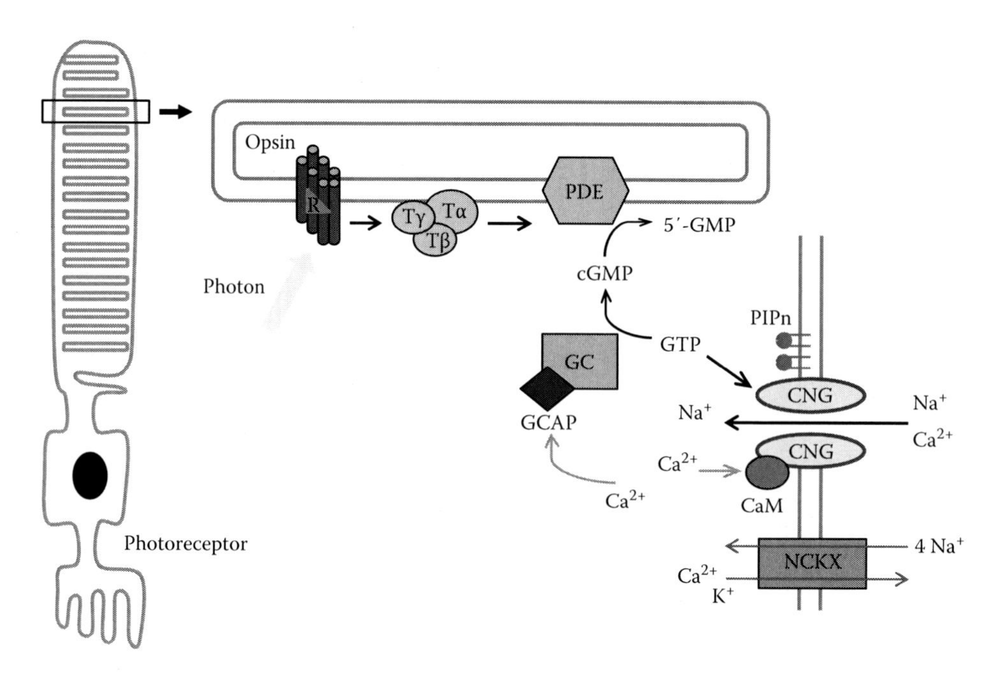

Figure 24.1 Phototransduction cascade in vertebrate photoreceptors. Shown is a magnified cartoon view of the phototransduction cascade within a single membrane disc as well as the plasma membrane of a rod photoreceptor outer segment (left). A photon can isomerize the 11-*cis* retinal (R) to all-trans retinal, which in turn activates the seven-transmembrane G protein–coupled receptor opsin, and the trimeric G protein transducin (T). The α subunit of activated transducin activates a cGMP-specific PDE, which hydrolyses cGMP to 5′-GMP. A cation-selective CNG channel is shown within the plasma membrane of the outer segment. Ca^{2+} entering through CNG channels can regulate guanylyl cyclase (GC) via binding to GC-activating protein (GCAP). In addition, CNG channel activity can be modulated by Ca^{2+}–CaM or membrane-bound phosphoinositides (PIP_n). The potassium-dependent, Na^+-Ca^{2+} exchanger (NCKX) extrudes Ca^{2+} from the photoreceptor, which helps to balance Ca^{2+} levels within the outer segment.

CNG channels are proposed to have several physiological functions outside of their canonical contributions to sensory signaling (vision and olfaction); they have been shown to be expressed in a wide variety of cell types (Distler et al., 1994) and to play diverse roles in multiple tissues. For example, CNG channels contribute to nerve growth cone guidance by Sema3A (Togashi et al., 2008), alveolar liquid reabsorption in the lung (Wilkinson et al., 2011), inhibition of inflammatory pain hypersensitivity in the spinal cord (Heine et al., 2011), and coolness-induced responses of Grueneberg ganglion neurons (Mamasuew et al., 2010). CNG channels are also present in vestibular and cochlear hair cells (Selvakumar et al., 2012, 2013) and in cerebellar granular cells (Lopez-Jimenez et al., 2012). Furthermore, CNG channels in the brain may contribute to hippocampal plasticity (Bradley et al., 1997; Kuzmiski and MacVicar, 2001; Michalakis et al., 2011a) and amygdala-dependent fear memory (Michalakis et al., 2011a). Gene expression for CNG channels has been described for several other cell types as well, where in most cases their precise physiological roles remain to be determined.

24.3 CNG CHANNEL SUBUNIT DIVERSITY AND BASIC STRUCTURAL ORGANIZATION

Six paralogous genes in mammals encode the fundamental building blocks (subunits) of CNG channels. The first CNG channel cDNA to be cloned and heterologously expressed encoded the rod photoreceptor *alpha subunit* (CNGA1) (Kaupp et al., 1989). Subsequently, cDNAs for CNGA2, CNGA3, CNGA4, CNGB1, and CNGB3 were isolated (Dhallan et al., 1990; Bönigk et al., 1993; Chen et al., 1993; Bradley et al., 1994; Liman and Buck, 1994; Gerstner et al., 2000). CNG channels are homo- or heteromeric assemblies of some combination of the six possible pore-forming subunits (Figure 24.2). The CNGA1–CNGA3 subunits can form functional channels when expressed alone as homomultimers, but the other subunit types cannot. Native rod photoreceptor channels are composed of three CNGA1 and one CNGB1a subunits (Weitz et al., 2002; Zheng et al., 2002; Zhong et al., 2002); cone CNG channels are thought to consist of two CNGA3 and two CNGB3 subunits (Peng et al., 2004; but also see Zhong et al., 2003; Ding et al., 2012); and the olfactory channels have two CNGA2, one CNGA4, and one CNGB1b subunits (Zheng and Zagotta, 2004). CNG channels are part of the pore loop, cation-selective superfamily of ion channels (Jan and Jan, 1990). The basic architecture of CNG channels is similar to that of voltage-dependent K^+ channels, characterized by tetrameric assembly of pore-forming subunits. Like K^+ channels, each CNG channel subunit has six transmembrane domains (S1–S6), with a vestigial voltage sensor-like domain (S1–S4), and an ion conduction pathway lined by part of the pore loop linking S5 and S6 along with the distal S6 region. Each subunit also presents cytoplasmic amino- (N) and carboxy- (C) terminal domains, with a CNBD located in the C-terminal region. A C-linker domain connects the CNBD to the pore-forming domain, helping to couple ligand binding to channel opening.

The nearest relatives of CNG channels are hyperpolarization-activated and cyclic nucleotide–regulated (HCN1–4) channels and the KCNH family of voltage-gated potassium channels, which includes ether-à-go-go (EAG, Kv10), EAG-related gene (ERG, Kv11), and EAG-like (ELK, Kv12) channels. In contrast

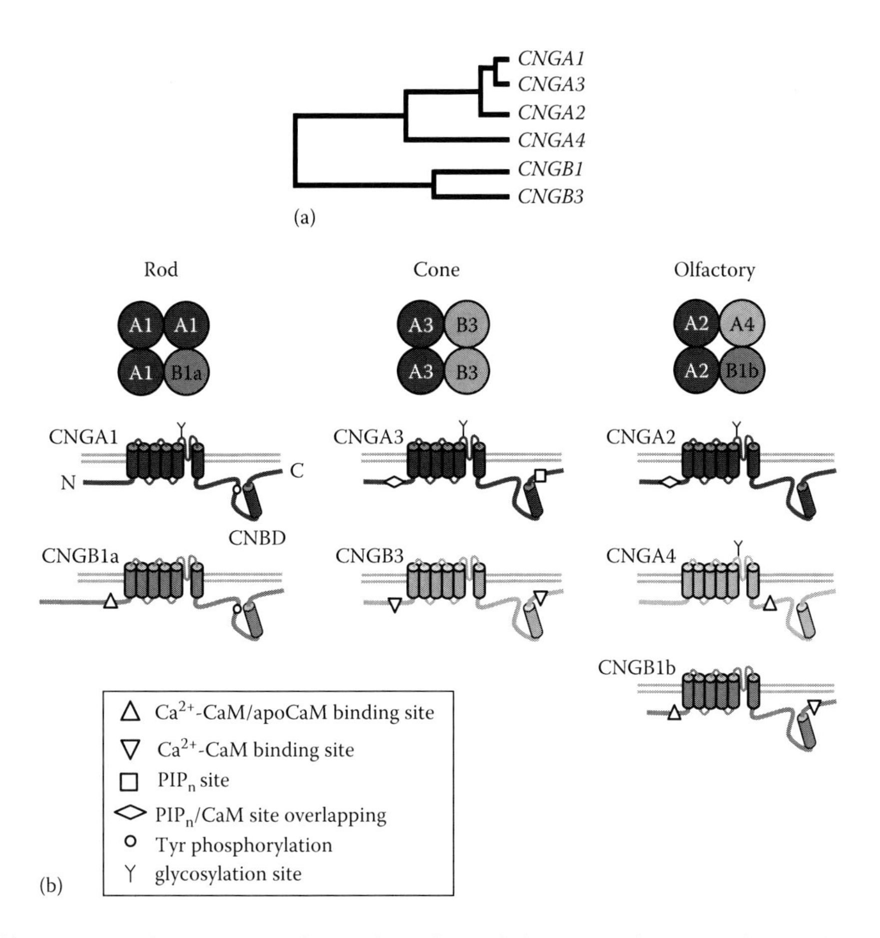

Figure 24.2 Diversity and basic structural organization of CNG channels. (a) Phylogenetic relationship showing the six genes encoding mammalian CNG channel subunits. (b) The respective stoichiometries for native rod, cone, and olfactory CNG channels, along with the basic topology of individual CNG channel subunits. Each subunit contains cytoplasmic N- and C-terminal regions, S1–S6 transmembrane domains, a pore loop region, and the CNBD within the proximal C-terminal region. The regions that represent CaM-binding sites, phosphoinositide regulation modules, tyrosine phosphorylation sites, and glycosylation sites are highlighted using different symbols (see the following key).

to CNG channels, HCN channels require membrane voltage changes for activation, but cyclic nucleotide binding can *regulate* their gating properties. KCNH channels also are voltage gated; the region homologous to the CNBD of CNG and HCN channels, however, does not bind cyclic nucleotides (Brelidze et al., 2009) but instead possesses an *intrinsic ligand* formed by part of the channel polypeptide that substitutes for the cyclic nucleotide within this cyclic nucleotide–binding *homology* domain (CNBHD) (Brelidze et al., 2012).

Diverse, related CNG channels exist in more distant parts of the phylogenetic landscape. Noted examples include the TAX2 and TAX4 subunits of the nematode *Caenorhabditis elegans*, forming channels that play roles in chemosensation in these organisms (Hellman and Shen, 2011) and the bacterial (Nimigean et al., 2004) and plant CNG channels (Kaplan et al., 2007). Some related channels have unique structural features that distinguish them from their mammalian relatives. For example, bacterial MloK CNG channels exhibit a dramatically shorter C-linker region proximal to the CNBD (Clayton et al., 2004). In addition, structure–function studies of MloK channels suggest that compared to mammalian CNG channels, this bacterial CNG channel exhibits less cooperativity during gating (Cukkemane et al., 2007), presumably because the CNBDs function independently (Chiu et al., 2007). More recently, a novel potassium-selective cyclic nucleotide-gated channel (CNGK) from *Arbacia punctulata* was discovered and characterized (Strünker et al., 2006; Bönigk et al., 2009). This channel comprises a single polypeptide chain with four pseudo-subunit domains, reflecting an architecture similar to that of voltage-dependent Na^+- and Ca^{2+}-selective channels. For CNGK channels, only the CNBD in pseudo-subunit repeat three appears to be necessary for channel activation; consistent with this observation, these channels exhibit little binding cooperativity (Bönigk et al., 2009).

In addition to the combinatorial assembly of differentially expressed subunits described earlier, CNG channel diversity also arises from alternative splicing of precursor mRNAs. The generation of CNG channel subunit variants via alternative splicing was first described for the *CNGB1* gene. A long form CNGB1 subunit (CNGB1a) is expressed in rod photoreceptor cells, while olfactory receptor neurons produce a shorter isoform (CNGB1b) that lacks much of the N-terminal cytoplasmic domain present in the long variant (Körschen et al., 1995; Sautter et al., 1998). In addition, alternative splicing can generate two soluble isoforms (long and short) representing only the N-terminal glutamic acid–rich protein (GARP) domain of CNGB1, without the channel-forming region (Figure 24.3c). The GARP domain of CNGB1a and the related soluble forms have been shown to be critical for protein–protein interactions within the photoreceptor outer segment (Körschen et al., 1999; Haber-Pohlmeier et al., 2007; Zhang et al., 2009;

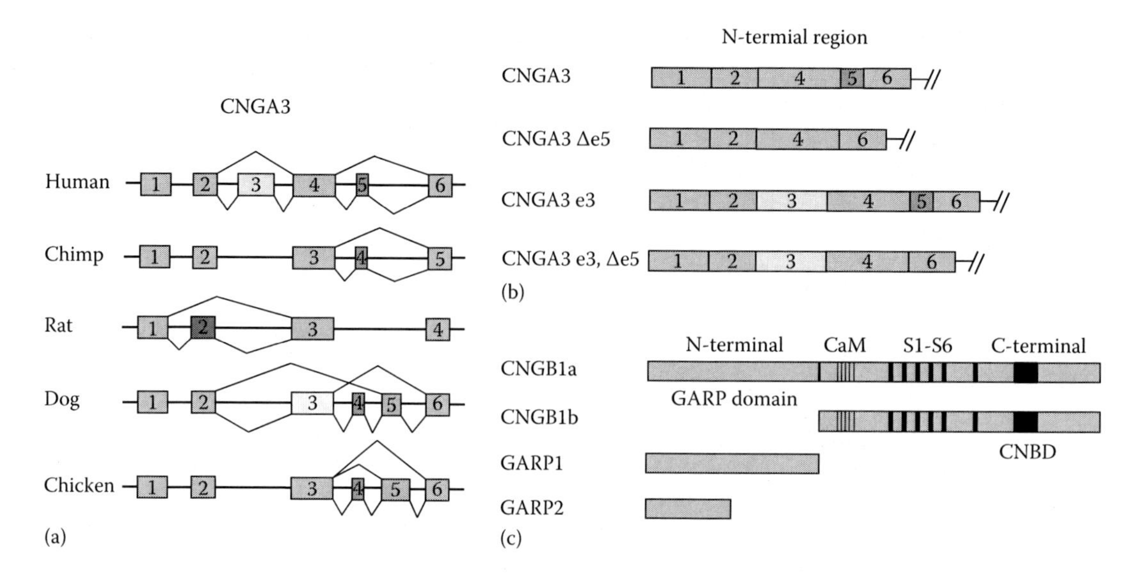

Figure 24.3 Alternative splicing generates CNG channel subunit diversity. (a) *CNGA3* splicing patterns for constitutive and alternative cassette exons across representative species. Exon (box) but not intron (line) length is presented to scale. (b) Four protein isoforms of human CNGA3 produced by inclusion or exclusion of regions encoded by optional exons 3 and 5. (c) Protein isoforms of CNGB1 produced by alternative splicing. The long CNGB1a isoform includes a GARP domain, while the short CNGB1b isoform does not. CaM-binding sites are located within the N-terminal region. GARP domains also are expressed as soluble proteins (GARP1 and GARP2) within rod photoreceptors.

Ritter et al., 2011). Recently, examination of a disease-associated mutation in CNGB1 has demonstrated that the GARP domain can act as a gating inhibitor in the context of heteromeric channels (Michalakis et al., 2011b). Second, *CNGA3* transcripts can produce subunit isoforms via the use of optional cassette exons that encode regions of the N-terminal cytoplasmic domain (Figure 24.3a and b). While alternative splicing of *CNGA3* is conserved across several species, some of the incorporated optional exons and the exact splicing patterns differ among *CNGA3* orthologs. One optional cassette exon found in human *CNGA3* transcripts (Wissinger et al., 2001; Cassar et al., 2004) appears to be unique to humans and functionally absent even from *CNGA3* of other primates. While *CNGA3* alternative splicing does not influence the fundamental gating properties of the channel, it appears to confer variations in sensitivity to modulation by second messengers (Bönigk et al., 1996; Dai et al. 2014). Finally, sequence evidence suggests that *CNGA1* can also give rise to protein variants via alternative splicing and/or alternative transcriptional start sites. Similar to CNGB1 and CNGA3 variants, these events lead to alterations within the N-terminal cytoplasmic region of the protein. To our knowledge, the functional significance of these changes has not yet been determined. Other CNG channel genes have not been extensively examined regarding possible protein variants.

While voltage sensor domains (VSDs) (S1–S4) are present in CNG channels, including positively charged residues in a repeated R/KXX pattern within S4, these do not support voltage-dependent channel activation. CNG channels instead exhibit only weak voltage dependence in their apparent affinity for cGMP, showing less than a twofold change $K_{1/2,cGMP}$ with a membrane voltage change from −100 to +100 mV. Tang and Papazian have shown that the S4 domain of CNGA2 can functionally replace S4 in voltage-gated EAG channels; they proposed that in CNG channels, the voltage sensor may exist in a constitutively *activated* position (Tang and Papazian, 1997). At the level of the transmembrane domains, CNG channels are thought to exhibit the *swapped-domain* configuration common to the voltage-gated channel superfamily (Long et al., 2005), with vestigial VSDs from each CNG channel subunit interacting with the pore-forming domains of adjacent subunits. Offset VSDs may position N-terminal cytoplasmic domains for interactions with C-terminal cytoplasmic domains. Indeed, intersubunit N-C interactions are a well-characterized feature of CNG channels, demonstrating intriguing roles in channel regulation (Gordon et al., 1997; Varnum and Zagotta, 1997; Rosenbaum and Gordon, 2002; Trudeau and Zagotta, 2002a; Zheng et al., 2003; Michalakis et al., 2011b; Dai and Varnum, 2013), although detailed structural information about these N–C interactions is currently missing.

The basic structural organization of CNG channels at the level of the C-linker and CNBD is thought to share fundamental features with the corresponding parts of homologous channels (Craven and Zagotta, 2006). Structural studies of the C-terminal region of HCN channels with bound ligand reveal a symmetrical tetramer arrangement, with contacts made between the C-linker regions of adjacent subunits (Zagotta et al., 2003; Flynn et al., 2007). Structural studies of related zELK and mEAG1 channels demonstrate dimeric assembly of C-terminal domains (Brelidze et al., 2012; Haitin et al., 2013); yet, the C-linker/CNBHD structure of agERG is monomeric (Brelidze et al., 2013). Another defined structural element in CNG channels is the post-CNBD region of CNGA subunits, which presents a carboxy-terminal, leucine-zipper (CLZ) domain that has been shown to play an important role in channel assembly (Trudeau and Zagotta, 2002a; Zhong et al., 2003). In addition, the CLZ domain is proposed to be a critical determinant for subunit stoichiometry and arrangement in heteromeric channels (Zhong et al., 2002, 2003; Shuart et al., 2011).

The exact arrangement of these functional units in CNG channels is unknown but may reflect many of the basic themes described earlier (Liu et al., 1998; Craven and

Zagotta, 2004; Zhou et al., 2004; Biskup et al., 2007), with possible state-dependent variations. There are likely to be as yet uncharacterized dynamic relationships between the C-linkers, CNBDs, post-CNBD regions, and/or N-terminal regions of CNG channels, involving changes in symmetry, orientation, and/or interactions among the domains (within and between subunits), which ultimately help govern channel gating and regulation.

24.4 GATING OF CNG CHANNELS

CNG channels are activated by binding of cyclic nucleotides to the CNBDs. Cyclic nucleotide binding induces allosteric conformational changes that are transduced via the C-linker, the inner helix (S6), the pore helix, and the selectivity filter, ultimately opening the CNG channel ion conduction pathway.

24.4.1 LIGAND BINDING TO CNG CHANNELS

Each subunit of CNG channels contains a CNBD within the cytoplasmic C-terminal region. Apparent ligand affinity and ligand efficacy for CNG channels vary depending on subunit composition, but generally cGMP is a better agonist than cAMP (Figure 24.4). The CNBD comprises of three α-helices, termed the A-, B-, and C-helix, and a β-roll domain with eight β strands located between the A-helix and the B-helix (Figure 24.5a). The CNBDs of CNG channels are homologous to other proteins with CNBDs, including the protein kinase A (PKA) regulatory subunit, protein kinase G (PKG), the catabolite gene-activator protein of *Escherichia coli* (CAP), HCN channels, bacterial cyclic nucleotide–regulated K^+ (MloK) channels, and the KCNH family of K^+ channels having CNBHDs. Although to date the crystal structure for a mammalian CNG channel CNBD has not been solved, the structures of many homologous CNBDs have been determined by x-ray crystallography or NMR (McKay and Steitz, 1981; Su et al., 1995; Zagotta et al., 2003; Clayton et al., 2004; Flynn et al., 2007; Das et al., 2009; Schünke et al., 2009; Brelidze et al., 2012).

There are two fundamental steps for ligand interaction with the CNBD of CNG channels. The first step is initial ligand docking to the CNBD; the second step is the conformational change in the CNBD that helps couple cyclic nucleotide binding to channel opening (Varnum et al., 1995; Flynn et al., 2007). The initial ligand-docking step is mediated mainly by interactions between the β-roll in the CNBD and the ribose and cyclic phosphate moiety of the cyclic nucleotide. There is a conserved short α-helical structure between the β-6 and β-7 strands of the CNBD for CNG channels called the phosphate-binding cassette (PBC). The PBC contains a conserved GE sequence; the glycine and glutamic acid are thought to interact with 2′-OH of the ribose for ligand docking. Furthermore, there is a conserved arginine within the β-7 strand that is mainly responsible for interacting with the cyclic phosphate of cyclic nucleotides; mutating this Arg to Glu dramatically decreases the ligand affinity but does not affect ligand efficacy (Tibbs et al., 1998). Therefore, this R-to-E mutation has been used to *cripple* the ligand-binding site of a CNG subunit in order to investigate subunit contributions to ligand binding and activation of CNG channels (Liu et al., 1998; Waldeck et al., 2009; Nache et al., 2012). Moreover, the threonine adjacent to this arginine within the β-7 strand also has been shown to be important for ligand docking as well as contributing to selectivity for cGMP over cAMP in CNG and HCN channels (Altenhofen et al., 1991; Varnum et al., 1995; Zagotta et al., 2003; Flynn et al., 2007; Zhou and Siegelbaum, 2007).

After the initial ligand-docking event, the C-helix of the CNBD moves to interact with the purine ring of cyclic nucleotides, and this interaction is coupled to the opening conformational change of CNG channels (Varnum et al., 1995; Sunderman and Zagotta, 1999a; Flynn et al., 2007). In addition, stabilization of the helical structure of the C-helix itself has been demonstrated to be linked to promoting channel opening after ligand binding (Taraska et al., 2009; Puljung and Zagotta, 2013). Based on the orientation of the purine ring relative to the ribose of cyclic nucleotides, cGMP and cAMP have two possible configurations: *syn* (with the pyrimidine ring closer to the ribose than the imidazole ring) and *anti* (180° rotation of the purine ring compared to the *syn* configuration) (Figure 24.5b). According to studies of HCN channels, cAMP binds to the CNBD in the *anti* configuration, while cGMP binds to the CNBD in the *syn* conformation (Zagotta et al., 2003; Flynn et al., 2007; Zhou and Siegelbaum, 2007); it is likely that cGMP and cAMP maintain

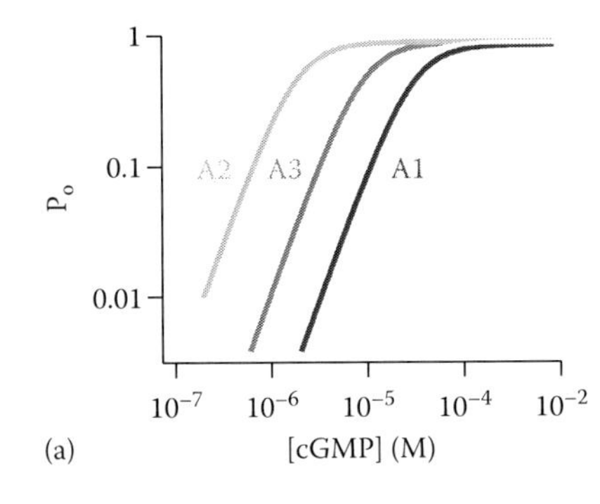

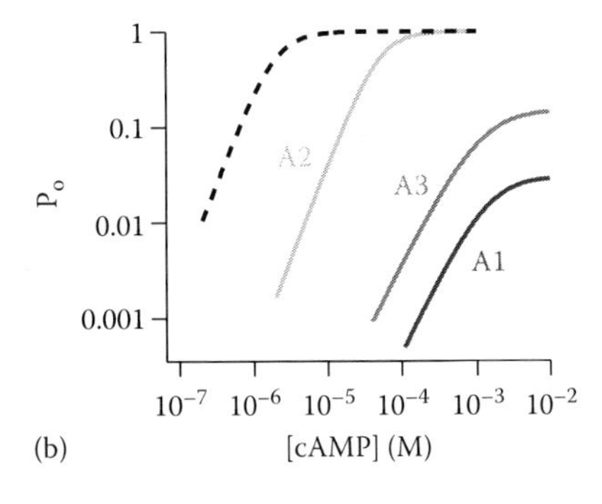

Figure 24.4 Diverse ligand sensitivities for CNG channels. (a) Representative cGMP dose–response relationships for the activation of homomeric CNGA1, CNGA2, and CNGA3 channels. Continuous curves represent fits using the Hill equation. Representative Hill parameters are as follows: for A1 channels, $K_{1/2,cGMP}$ = 33.0 μM, h = 2.0; for A2 channels, $K_{1/2,cGMP}$ = 1.9 μM, h = 2.0; and for A3 channels, $K_{1/2,cGMP}$ = 9.8 μM, h = 2.0. (b) Representative cAMP dose–response relationships for the activation of homomeric CNGA1, CNGA2, and CNGA3 channels. The Hill parameters are as follows: for A1 channels, $K_{1/2,cAMP}$ = 1.40 mM, h = 1.6; for A2 channels, $K_{1/2,cAMP}$ = 49.2 μM, h = 2.0; for A3 channels, $K_{1/2,cAMP}$ = 1.18 mM, h = 1.5. The dotted curve represents the cGMP dose–response relationships for A2 channels in panel A, emphasizing that cGMP is a more effective agonist compared to cAMP.

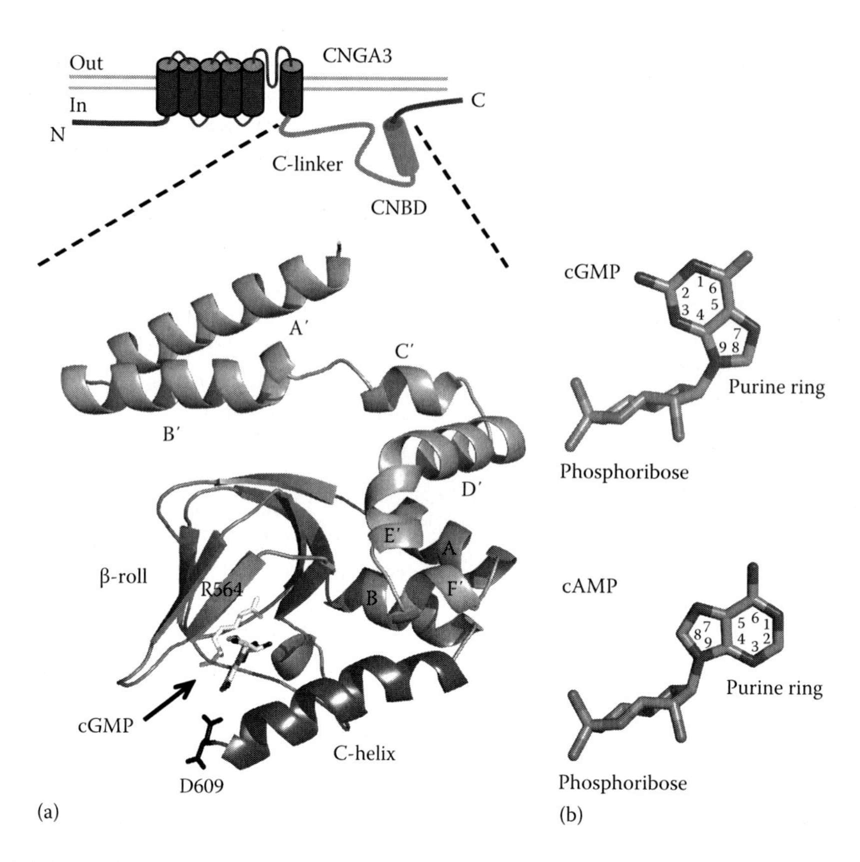

Figure 24.5 Model for the C-linker and CNBD region of CNGA3. (a) The gray portion of the CNGA3 subunit cartoon (preceding) is depicted as follows in complex with cGMP, based on homology modeling using the x-ray crystal structure of the C-terminal region of HCN2-I636D subunits. (From Flynn, G.E. et al., *Struct. 1993*, 15, 671, 2007.) The A′ to F′ helices of the C-linker region and the β-roll region and the A-, B-, and C-helices of the CNBD are indicated. Also highlighted are the arginine (R564 in stick form) within the β-roll that is important for initial ligand binding and the aspartic acid (D609 in stick form) within the C-helix of the CNBD that is critical for ligand discrimination. (b) Structures of cGMP in the *syn* configuration and cAMP in *anti* configuration.

the same configuration for binding to CNG channels. The conserved threonine in the β-7 strand of the β-roll is proposed to interact with the purine ring of cGMP when cGMP is in the *syn* configuration (Altenhofen et al., 1991). However, this interaction between the threonine in the β-roll and cGMP is not sufficient to explain high selectivity for cGMP over cAMP in CNG channels.

Instead, cGMP selectivity is largely dependent on a crucial ligand discrimination residue located near the C-terminal end of the C-helix (an aspartic acid in CNGA1 and CNGA3; a glutamic acid in CNGA2) (Varnum et al., 1995). For CNGA subunits, the negatively charged residue is thought to present a favorable electrostatic interaction with cGMP via two hydrogen bonds with N1 and N2 of the guanine ring; this interpretation is supported by recent structural studies of CNBDs in HCN channels (Flynn et al., 2007; Zhou and Siegelbaum, 2007). In contrast, the aspartic acid residue forms an unfavorable interaction with cAMP, possibly due to repulsion between the negative charge of the side chain and the unshared pair of electrons in the sp^2 orbital at N1 of the adenine. These differences make cGMP a nearly full agonist and cAMP a partial agonist for CNGA1 or CNGA3 channels (Varnum et al., 1995; Peng et al., 2004). Mutating this aspartic acid to methionine (D604M for CNGA1; D609M for CNGA3) reversed ligand selectivity, making cAMP a better agonist than cGMP (Varnum et al., 1995; Peng et al., 2004). Homomeric CNGA2 channels, which have a glutamic acid at this position, are still selective for cGMP over cAMP, showing an apparent affinity for cGMP about 25-fold higher than that for cAMP (Nache et al., 2012). For CNGA2 channels, cGMP and cAMP have a similar efficacy because the ligand-independent intrinsic tendency for opening of CNGA2 channels is very high. The equivalent residue within the CNBD of CNGB1 subunits is an asparagine; this difference accounts for the increase in cAMP efficacy for heteromeric rod CNG channels (Pagès et al., 2000; He and Karpen, 2001). For CNGB3 subunits, the equivalent residue is a positively charged lysine; thus, ligand discrimination for CNGB3 subunits may be mediated by a different mechanism compared to other types of CNG channel subunits. In CNGA4, the residue equivalent to D604 in CNGA1 instead is a methionine, which helps explain why CNGA4 confers high cAMP selectivity to heteromeric olfactory CNG channel (Bradley et al., 1994; Shapiro and Zagotta, 2000). However, other regions within the CNBD, including the β-roll, have been shown to contribute to the ligand selectivity for CNG channels (Young and Krougliak, 2004). In addition, it has been reported that an arginine in the C-helix of the CNBD helps couple ligand binding to channel activation for HCN and for bacterial cyclic nucleotide–regulated K^+ channels (Zagotta et al., 2003; Clayton et al., 2004).

Several analogues of natural cyclic nucleotides have been used to characterize the CNBD of CNG channels. It was found that chemical modifications of the ribose and phosphate moiety of cyclic nucleotide are less tolerated than those of the purine ring. Modification of the ribose and cyclic phosphate group of cGMP significantly impaired activation of photoreceptor CNG channels (Zimmerman et al., 1985; Tanaka et al., 1989). This led to the development of a *caged* 4,5-dimethoxy-2-nitrobenzyl ester of cGMP that is not capable of activating CNG channels unless altered by photolysis (Nerbonne et al., 1984; Karpen et al., 1988). Caged cGMP has been used to study the kinetics of CNG channel activation by fast jumps in cGMP concentration via laser flash photolysis. In contrast, the CNBD of CNG channels can accommodate large modifications of the purine ring of cyclic nucleotides, particularly at the C8 position. Generally, cGMP analogues with substitutions at the C8 position such as 8-Br-cGMP and 8-pCPT-cGMP have higher ligand affinity, are resistant to hydrolysis by PDE, and exhibit membrane permeability (Zimmerman et al., 1985; Wei et al., 1998). The photoaffinity analogue 8-pAPT-[^{32}P]-cGMP, which can be covalently attached to the CNBD after UV light activation, labels a region containing hydrophobic residues within the β-4 strand of the CNBD, indicating that this region is close to the C8 position of the purine ring (Brown et al., 1995). Moreover, a polymer-linked cGMP dimer has been engineered to enhance ligand affinity and to measure the distance between two ligand-binding sites in CNG channels (Kramer and Karpen, 1998).

24.4.2 C-LINKER OF CNG CHANNELS

The C-linker region connecting S6 and the CNBD has been demonstrated to be critical for coupling ligand binding to opening of rod, cone, and olfactory CNG channels (Gordon and Zagotta, 1995a,b; Brown et al., 1998; Zong et al., 1998; Zhou et al., 2004). Based on the crystal structure of this region in homologous HCN2 channels (Zagotta et al., 2003), the C-linker comprises six α-helices (A′–F′). The C-linker is thought to support several intersubunit interactions in the tetrameric channel, including an *elbow on the shoulder* contact with the A′–B′ helices of one subunit representing the *elbow* and C′–D′ helices of the adjacent subunit forming the *shoulder* (Zagotta et al., 2003; Craven and Zagotta, 2004). Ni^{2+} coordination sites within the A′ helix of the C-linker can potentiate or inhibit channel gating, depending on whether histidines in adjacent subunits coordinate Ni^{2+} better in open or closed conformations, respectively (Gordon and Zagotta, 1995a,b; Johnson and Zagotta, 2001). Disulfide-bond formation between introduced cysteines in this region of the A′ helix also can stabilize the open state (Hua and Gordon, 2005). In addition, a conserved tripeptide located in the A′ helix of the C-linker has been shown to be important for assembly and ligand-dependent gating of both CNG and HCN channels (Zhou et al., 2004). Craven and coworkers have characterized salt bridges that exist in parallel within the C-linker regions of CNG and HCN channels (Craven and Zagotta, 2004; Craven et al., 2008). In bovine CNGA1 channels, R431 in the B′ helix interacts via salt bridges with E462 in the D′ helix of the adjacent subunit and D502 in the CNBD β-roll of the same subunit (Craven and Zagotta, 2004; Craven et al., 2008). Disrupting these interactions enhanced channel opening, substantiating the view that these interactions occur in the closed state of the channel and that the C-linker region plays a critical role in CNG channel gating. The C-linker region of CNGA1 also was found to interact with the N-terminal region between and within subunits (Rosenbaum and Gordon, 2002). Considering the important roles of the C-linker and N-terminal regions for activation of CNG channels (Goulding et al., 1994; Gordon and Zagotta, 1995b; Tibbs et al., 1997; Möttig et al., 2001), N–C interactions may serve to adjust the gating properties of CNG channels.

24.4.3 CNG CHANNEL GATE

During CNG channel gating, the conformational change in the C-linker is transmitted to opening of the pore. For homologous voltage-gated potassium channels, the pore gate is well characterized as a *bundle crossing* formed by the inner helix (Hackos et al., 2002; Labro et al., 2003; Webster et al., 2004) with a hinge motif bending during channel opening (Jiang et al., 2002, 2003; Long et al., 2005). For CNG channels, the inner helix is unlikely to represent the gate, despite evidence for analogous conformational rearrangements here during channel activation (Flynn and Zagotta, 2001). Ag^+, Cd^{2+}, and some MTS reagents were able to enter the inner vestibule of CNGA1 channels in both the closed and open states (Flynn and Zagotta, 2001; Contreras and Holmgren, 2006; Nair et al., 2009). Furthermore, in contrast to intracellular quaternary ammonium (QA) block of voltage-gated K^+ channels (Armstrong and Hille, 1972), the extent of blockage of CNGA1 channels by QA is inversely proportional to the open probability of the channel, indicating that QA ions can bind to the inner vestibule of CNGA1 channels more readily during the closed state (Contreras and Holmgren, 2006). Together, the results indicate that gating of CNG channels involves a movement of the S6 helix, dilating or constricting the pathway to the inner vestibule during the open or closed state, respectively. However, because of the relative lack of state-dependent accessibility in this region compared to K^+ channels, the S6 helix probably does not represent the gate of CNG channels.

Other evidence favors the view that the gate of CNG channels is located at the selectivity filter (Fodor et al., 1997a; Contreras and Holmgren, 2006; Contreras et al., 2008; Martínez-François et al., 2009; Mazzolini et al., 2009). Using cysteine-scanning mutagenesis of the entire selectivity filter of CNGA1 channels, a recent study has found that the accessibility of high-affinity cysteine-binding agents, Cd^{2+} and Ag^+, to these mutant channels is state dependent (Contreras et al., 2008). Consistent with this finding, mutation of residues in the selectivity filter not only changed the conductance of the channel but also decreased the open probability of the channel (Bucossi et al., 1996; Becchetti and Gamel, 1999; Becchetti et al., 1999). Together, these results strongly imply that there is a reorientation of the pore region of the channel during channel gating. Furthermore, subconductance states that have been observed in CNG channels at subsaturating concentrations of cyclic nucleotides (Taylor and Baylor, 1995; Ruiz and Karpen, 1997) also are consistent with a conformational change in the channel pore during CNG channel gating.

If the selectivity filter acts as the gate of CNG channels, then how are the conformational changes within the C-linker and S6 helix transmitted to the pore of the channel? It has been demonstrated recently that S6 helix movement may be linked to the pore helix during the gating of CNG channels. F380 in the S6 helix has been shown to interact with L356 in the pore helix, stabilizing the open state of the channel via coupling of the S6 helix with the pore (Mazzolini et al., 2009). In addition, the pore helix has been demonstrated to present a rotational movement during channel gating (Liu and Siegelbaum, 2000), probably induced by movement of the C-linker and S6 helix. The gate at the selectivity filter of CNG channels is reminiscent of C-type inactivation of K^+ channels, where the selectivity filter acts as a secondary gate in addition to the *bundle-crossing* gate composed of S6 helices. Interactions between the selectivity filter and pore helix are critical for C-type inactivation of K^+ channels (Cordero-Morales et al., 2006, 2011; Cuello et al., 2010). Interestingly, this type of interaction is also present in CNGA1 channels. E363 in the selectivity filter of CNGA1 has been shown to interact with Thr 355 in the pore helix; disrupting the hydrogen bond between E363 and T355 impaired gating of the channel, producing desensitization during constant cGMP application (Mazzolini et al., 2009). Furthermore, mutating residues within the selectivity filter and pore helix of CNGA1 can render the channel voltage sensitive, producing strong outward rectification at saturating concentrations of cGMP (Martínez-François et al., 2009). Together, these results suggest that gating of CNG channels and C-type inactivation of K^+ channels share a similar mechanism at the selectivity filter.

24.4.4 CNG CHANNEL ACTIVATION SCHEMES AND SUBUNIT CONTRIBUTIONS

A number of different activation schemes have been employed to interpret the gating behavior of CNG channels. The Monod–Wyman–Changeux (MWC) concerted allosteric model (Monod et al., 1965) has been useful to describe ligand binding and channel activation of CNG channels (Goulding et al., 1994; Varnum and Zagotta, 1996; Tibbs et al., 1997) (Figure 24.6b). The MWC model for CNG channels assumes four equivalent binding sites; the cooperativity of ligand binding is dependent on the concerted conformational change and purely additive ligand-binding energies. One advantage of the MWC model compared to sequential models is that it can account for the well-defined spontaneous channel opening events in the absence of ligand binding, which reflect the ligand-independent intrinsic gating properties of the channel (Tibbs et al., 1997). For homomeric CNG channels, the spontaneous channel open probability (P_{sp}) has been estimated to be in the following order (from low to high): CNGA1 (P_{sp} = ~10^{-5}), CNGA3 (P_{sp} = ~10^{-4}), and CNGA2 (P_{sp} = ~10^{-3}) channels (Ruiz and Karpen, 1997; Tibbs et al., 1997; Gerstner et al., 2000). However, this concerted MWC model can be less than satisfactory in describing all properties integral to activation of CNG channels. Furthermore, multimerization of related but divergent subunit types is likely to make asymmetrical subunit contributions an essential feature of heteromeric channel gating. In order to study subunit contributions to channel gating, photoaffinity 8-pAPT-cGMP was used to covalently lock rod CNGA1 channels in one, two, three, or four ligand-bound states (Ruiz and Karpen, 1997). Utilizing this approach, it was found that (1) four ligands are required to fully activate CNGA1 channels, (2) the equilibrium constant for channel opening does not increase by a constant factor for each ligand-binding event, and (3) a single ligand cannot induce appreciable channel opening (Ruiz and Karpen, 1997). Other studies differ regarding the threshold number of bound cGMP molecules necessary for appreciable channel activation, indicating that channels with a single bound ligand can have a considerable open probability

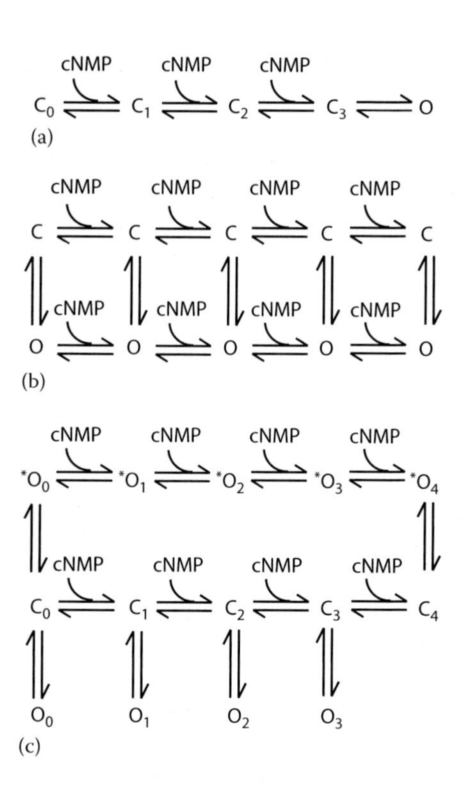

Figure 24.6 Gating schemes proposed for CNG channels. (a) A simple sequential model for activation of CNG channels. Cyclic nucleotides bind to channel subunits independently, followed by an allosteric transition from the closed state (C) to the open state (O). (b) MWC model for activation of CNG channels. (c) A C4L-*O4L sequential model proposed for activation and deactivation of homomeric CNGA2 channels. (From Nache, V. et al., *Sci. Signal.*, 5, ra48, 2013.) *O indicates a separate open state predominantly accessed in saturating concentrations of ligand; channel deactivation is thought to follow the *O to C_0 pathway.

(P_o > 0.01) (Liu et al., 1998; Biskup et al., 2007). These features may reflect in part the specific subunit compositions of the channels. Furthermore, partially liganded channels have been found to exhibit openings to one or more subconductance states (Ruiz and Karpen, 1997); this observation is not consistent with the MWC model, where there is a single transition from closed to open conformations. However, the existence of subconductance states for CNGA1 channel activation remains controversial (Sunderman and Zagotta, 1999b).

Sequential models involving two to four ligand-binding steps followed by an allosteric opening conformational change also have been proposed to describe activation of CNG channels (Karpen et al., 1988; Gordon and Zagotta, 1995a) (Figure 24.6a). One tool used in studying ligand binding and gating of rod and olfactory CNG channels is photolysis-induced jumps of *caged* cGMP or cAMP (Karpen et al., 1988; Nache et al., 2005). For rod CNG channels, activation was well described by sequential cGMP-binding steps and a rapid closed-open transition with three or four ligands bound; the rate-limiting step was the third cGMP-binding event (Karpen et al., 1988). For CNGA2 channels, it was found that the activation time course generated by cGMP jumps from zero to a saturating concentration was better fit with a double-exponential trace rather than a sigmoidal trace predicted by the MWC model (Nache et al., 2005). In addition, the time constant for CNGA2 activation was slowest when elicited by cGMP jumps from zero to an intermediate concentration, indicating that activation is highly cooperative (Nache et al., 2005). Moreover, ligand binding and channel activation can be monitored simultaneously via the patch-clamp fluorometry technique using a fluorescent cGMP analogue, 8-DY547-cGMP (Biskup et al., 2007). Using this approach, it was observed that under steady-state conditions, ligand binding was more favorable than channel opening at low concentrations of cGMP, whereas channel gating was more favorable at high concentrations of cGMP. Interestingly, it seems that only two ligands are sufficient to fully activate CNGA2 channels and the second ligand-binding event is the rate-limiting step for channel activation (Biskup et al., 2007). Furthermore, using the same fluorescent cGMP analogue, a recent study showed that the gating pathways for CNGA2 channels may depend on cyclic nucleotide concentration; fully liganded channels might adopt an open conformation different from that of partially liganded channels (Nache et al., 2013). Based on these results, an alternative (C4L-*O4L) sequential model was proposed to describe the activation and deactivation of CNGA2 channels (Nache et al., 2013) (Figure 24.6c). In addition, it has been proposed that for olfactory CNGA2+CNGA4+CNGB1b channels, the first cAMP binds to CNGA4 since CNGA4 has the highest cAMP affinity (Waldeck et al., 2009). The second cAMP then binds to one of the two CNGA2 subunits to fully activate the channel; heteromeric channels with crippled CNGA2 CNBDs were not able to produce detectable currents (Waldeck et al., 2009). However, cAMP still binds to all the four subunits including CNGB1b. In contrast to cAMP, it was found that CNGA2 subunits dominate cGMP-dependent gating and that CNGB1b subunits were not able to efficiently bind cGMP (Nache et al., 2012).

An alternative model that has been proposed to describe the activation of CNG channels is the coupled-dimer model (Liu et al., 1998). The coupled-dimer model arose from experiments using single-channel recordings of channels comprised of chimeric subunits having some number of crippled CNBDs—associated with a unique single-channel conductance. The behavior of these channels was best described by a model where two adjacent subunits of the channel could form a dimer unit for the activation of the tetrameric channel (Liu et al., 1998). Contrary to the view that CNG channels have fourfold symmetry at the level of the cytoplasmic domains, CNG channels might exist in a dimer-of-dimers arrangement (Higgins et al., 2002).

24.5 ION PERMEABILITY OF CNG CHANNELS

CNG channels are nonselective cation channels, conducting monovalent K^+, Na^+, Li^+, Rb^+, and Cs^+. CNG channels also are permeable to Ca^{2+} but can be blocked by Ca^{2+} from both the intracellular and extracellular sides of the membrane (Zimmerman and Baylor, 1992; Karpen et al., 1993; Root and MacKinnon, 1993). The presence of a glutamic acid within the selectivity filter of CNGA subunits has been shown to be critical for coordination of Ca^{2+} ions, conferring Ca^{2+} permeability and Ca^{2+} block to the channel (Root and MacKinnon, 1993; Eismann et al., 1994; Derebe et al., 2011b). Ion selectivity and the fraction of current carried by Ca^{2+} differ among homomeric CNGA subunits and for heteromeric channels containing CNGB subunits (Frings et al., 1995; Dzeja et al., 1999; Seifert et al., 1999). Moreover, Ca^{2+} selectivity is much greater for native cone CNG channels compared to rod CNG channels (Picones and Korenbrot, 1995). Instead of having a glutamic acid within the selectivity filter, CNGB1 and CNGB3 subunits contain a glycine at the respective position. The difference between E and G within the selectivity filter is thought to have two opposing effects on ion permeation, considering that Ca^{2+} is both a permeant ion and a blocker for CNG channels. First, this glycine residue decreases Ca^{2+} affinity, therefore reducing the dwell time of Ca^{2+} within the selectivity filter and concomitantly increasing Ca^{2+} influx. Second, the glycine residue relieves the Ca^{2+} block of monovalent cation permeation through the pore, thus enhancing the influx of Na^+ and K^+ ions.

The mechanisms underlying the nonselectivity among cations in CNG channels have been studied by engineering a NaK chimera mimicking the CNG channel pore, combined with x-ray crystallography (Derebe et al., 2011a). The sequence of the selectivity filter is *TIGET* for CNGA1, CNGA2, and CNGA3 subunits and *TIGGL* for CNGB1 and CNGB3; these differ from the *TVGYG* signature sequence found in K^+-selective channels. NaK channels are prokaryotic, nonselective cation channels with the *TVGDG* sequence in the selectivity filter. For NaK channels, the aspartic acid in the selectivity filter is equivalent to the glutamic acid of CNG channels. Ca^{2+} is also a permeable blocker for NaK channels, but the Ca^{2+} affinity is lower than that of CNG channels. In contrast to K^+-selective channels, there are only two contiguous ion-binding sites within the selectivity filter of NaK channels, corresponding to sites 3 and 4 for K^+-selective channels (Derebe et al., 2011a). In comparison, there are three contiguous ion-binding sites for the NaK–CNG chimera containing the CNG channel pore mimic, corresponding to sites 2, 3, and 4 for K^+-selective channels (Derebe et al., 2011a). These results suggest that the number of ion-binding sites within the selectivity filter helps determine the selectivity properties of CNG or NaK channels. If the number of ion-binding sites is less than 4, then the channel is relatively nonselective, conducting both Na^+ and K^+.

The molecular mechanism underlying Ca^{2+} block of CNG channels has been determined at the structural level by engineering a NaK channel chimera mimicking CNG channels (Derebe et al., 2011b). The NaK chimera mimicking the CNG channel pore was cocrystalized with bound Ca^{2+} ions. Contrary to the opinion that Ca^{2+} is coordinated by the negatively charged side chains of the aspartic or glutamic acid residues in the selectivity filter, it was found that Ca^{2+} is chelated by backbone carbonyl oxygens at sites 2 and 3 of the selectivity filter (Derebe et al., 2011b). In addition, the crystal structure of the NaK–CNG chimera showed that the D or E side chains within the selectivity filter form hydrogen bonds with tyrosine and threonine residues within the pore helix, corroborating the view that interactions occur between the selectivity filter and pore helix for CNG channels (Derebe et al., 2011b).

The selectivity filter of CNG channels is proposed to serve a dual function as both a channel gate and an ion selectivity feature. This view is supported by studies demonstrating that gating alters the ion permeability of CNG channels. The selectivity of native photoreceptor CNG channels for Ca^{2+} over Na^+ was found to increase as the open probability of the channel increased (Hackos and Korenbrot, 1999). In addition, the apparent cGMP affinity of CNGA1 channels is higher when K^+ is the permeant ion compared to when Na^+ is the permeant ion (Holmgren, 2003). K^+ prolonged the open time duration and increased the maximum open probability of the channel compared to Na^+ (Holmgren, 2003).

24.6 PHARMACOLOGY OF CNG CHANNELS

The best-known blocker for CNG channels is L-*cis*-diltiazem, an isomer of the clinically used drug Cardizem. L-*cis*-diltiazem has been shown to block CNG channels from rod and cone photoreceptors and OSNs (Koch and Kaupp, 1985; Frings et al., 1992; Haynes, 1992). The block of CNG channels by L-*cis*-diltiazem is voltage dependent, exhibiting an increase in current suppression with membrane depolarization (McLatchie and Matthews, 1992). Furthermore, L-*cis*-diltiazem block is a useful reporter to confirm the formation of heteromeric CNG channels when expressed as recombinant proteins; L-*cis*-diltiazem effectively blocks heteromeric CNG channels containing CNGB1 or CNGB3 subunits, while it has little effect on homomeric CNG channels (Chen et al., 1993; Gerstner et al., 2000). However, the mechanisms underlying L-*cis*-diltiazem block of CNG channels are not well understood; it may act by modifying channel gating instead of directly occluding the pore.

Several agents used to block other types of ion channels have been shown to block CNG channels. The local anesthetic tetracaine blocks CNG channels in a voltage-dependent and state-dependent manner (Fodor et al., 1997a,b); the profound state dependence of tetracaine block makes it a useful reporter for the equilibrium between closed and open CNG channels. Dichlorobenzamil, a derivative of amiloride and a well-known Na^+ channel blocker, was shown to block CNG channels (Nicol et al., 1987; Kolesnikov et al., 1990). Pimozide, D-600, and Nifedipine, which are Ca^{2+} channel blockers, have been reported to block CNG channels (Frings et al., 1992; Nicol, 1993; Zufall and Firestein, 1993). Dequalinium, an extracellular blocker of the small conductance Ca^{2+}-activated K^+ channel, blocks CNGA1 channels from the intracellular side (Rosenbaum et al., 2003). In addition, the polyamine spermine, which blocks inward-rectifier K^+ channels (Shin and Lu, 2005), has been reported not only to block CNG channels but also to permeate through the pore (Lu and Ding, 1999; Guo and Lu, 2000). Furthermore, the ball peptide mediating N-type inactivation of Shaker potassium channels (Hoshi et al., 1990; Zagotta et al., 1990) blocks CNG channels (Kramer et al., 1994). Finally, pseudechetoxin, a peptide toxin isolated from the venom of the Australian king brown snake, blocks CNG channels in a voltage-dependent manner (Brown et al., 1999). The extent of block by pseudechetoxin was demonstrated to be subunit specific, depending on sequence differences within the turret region of CNGA subunits; heteromeric channels containing CNGB1 subunits were insensitive to pseudechetoxin (Brown et al., 2003).

24.7 REGULATION OF CNG CHANNELS

Regulation of CNG channels is important for adaptation in retinal photoreceptors and in olfactory receptor neurons and for paracrine and circadian controls within photoreceptors. CNG channels can be regulated on a rapid time scale, ranging from milliseconds to seconds, for example, via Ca^{2+}-dependent feedback mechanisms, or on a relatively slow time scale, for example, for channel regulation by phosphorylation or by membrane phosphoinositides.

Ca^{2+}-dependent regulation of CNG channels, mediated by binding of calmodulin (CaM) or other calcium-sensing proteins, contributes to the rapid adaptive properties of olfactory neurons and photoreceptors. Sensory CNG channels are highly permeable to calcium, and Ca^{2+}–CaM serves to decrease channel apparent ligand affinity, thus providing a negative feedback signal for control of channel activity (Trudeau and Zagotta, 2003). OSNs desensitize dramatically after prolonged odorant stimulation (Reisert and Matthews, 1999). Ca^{2+}-mediated feedback regulation of olfactory CNG channels represents a central mechanism underlying odorant adaptation (Kurahashi and Menini, 1997). There are three CaM-binding sites located in olfactory CNG channel subunits: a *Baa*-type CaM-binding site within the N-terminal region of CNGA2, an *IQ*-type CaM-binding site within the C-linker region of CNGA4, and an *IQ*-type CaM-binding site within the N-terminal region of CNGB1b (Bradley et al., 2001). It was found that Ca^{2+}-free apocalmodulin (apoCaM) is permanently associated with the IQ-type CaM-binding sites within CNGA4 and CNGB1b subunits (Bradley et al., 2004). Homomeric CNGA2 channels also are sensitive to inhibition by Ca^{2+}–CaM (Liu et al., 1994); Ca^{2+}–CaM disrupts autoexcitatory intersubunit interactions between N- and C-terminal cytoplasmic regions of CNGA2 subunits (Varnum and Zagotta, 1997; Zheng et al., 2003). However, deleting the CaM-binding site in CNGA2 does not prevent regulation of CNGA2+CNGA4+CNGB1b channels, a result that suggested that the physiologically relevant CaM-binding sites are those present in CNGA4 and CNGB1b (Bradley et al., 2004). Moreover, a recent study has shown that the CNGB1b CaM-binding site is sufficient to mediate Ca^{2+}-dependent desensitization conferred by endogenous CaM (Waldeck et al., 2009). A mouse model where the CaM-binding domain within CNGB1b was genetically deleted has provided new insights into the physiological role of CaM regulation of olfactory CNG channels (Song et al., 2008). This study suggested that Ca^{2+}–CaM regulation of olfactory CNG channels is not responsible for adjusting sensitivity of olfactory neurons to repeated stimuli but instead contributes mainly to the rapid termination of the odorant response (within ~100 ms) (Song et al., 2008).

Rod CNG channels also have been shown to be directly regulated by Ca^{2+}–CaM, and this feature is thought to contribute to Ca^{2+}-dependent adjustment of phototransduction (Hsu and Molday, 1993; Gordon et al., 1995b; Haynes and Stotz, 1997). The Ca^{2+} concentration within the photoreceptor outer segments is mainly determined by the opening of CNG channels and also by the activity of the K^+-dependent Na^+–Ca^{2+} exchanger

(NCKX1 in rods; NCKX2 in cones) that extrudes Ca^{2+} from the outer segment (Szerencsei et al., 2002). The N-terminal region of the rod CNGB1a subunit contains an *IQ*-type Ca^{2+}/CaM-binding site (Figure 24.2) (Grunwald et al., 1998; Weitz et al., 1998). Ca^{2+}–CaM binding to this site disrupts an intersubunit interaction between the N-terminal region of CNGB1a and the C-terminal region of CNGA1, which decreases cGMP affinity of CNGA1+CNGB1a channels (Trudeau and Zagotta, 2002b, 2004). However, using a caged Ca^{2+} chelator to rapidly decrease Ca^{2+} concentration, Ca^{2+}-dependent modulation of CNG channels was detected only in cones but not rods in intact ground squirrel photoreceptors (Rebrik and Korenbrot, 2004). In addition, a recent paper utilizing a mouse model in which the CaM-binding domain within CNGB1a was genetically ablated demonstrated that CaM-dependent modulation of rod CNG channels is not a major mechanism mediating light adaptation for rod photoreceptors (Chen et al., 2010).

The native cone CNG channel exhibits greater sensitivity to calcium feedback regulation of apparent ligand affinity than the rod channel, and this modulation is thought to be of much greater importance for adaptation in cones (Korenbrot and Rebrik, 2002; Korenbrot, 2012). The N-terminal region of CNGA3 encompasses a *Baa*-type CaM-binding site, but for most species, homomeric CNGA3 channels are not sensitive to Ca^{2+}–CaM regulation. However, mutations downstream of this CaM-binding site within the CNGA3 N-terminal domain can unmask CaM sensitivity (Grunwald et al., 1999). Instead, functional CaM-binding sites within the N- and C-terminal regions of CNGB3 subunits support Ca^{2+}-dependent regulation of human cone CNG channels (Peng et al., 2003a). However, the magnitude of Ca^{2+}–CaM-dependent inhibition of native and recombinant cone CNG channels appears to be too small to completely account for Ca^{2+}-dependent desensitization in intact cones (Hackos and Korenbrot, 1997; Haynes and Stotz, 1997; Rebrik and Korenbrot, 1998, 2004). Furthermore, a recent study shows that CNG-modulin, a calcium-binding protein interacting with the N-terminal region of CNGB3, may be the authentic calcium sensor for Ca^{2+}-dependent control of cone CNG channels in striped bass (Rebrik et al., 2012).

Another type of CNG channel regulation operating on a relatively slow time is via phosphorylation of tyrosine and serine/threonine residues in channel subunits (Figure 24.2). Serine/threonine phosphorylation and dephosphorylation can adjust the apparent ligand affinity of native rod CNG channels (Gordon et al., 1992). In addition, tyrosine phosphorylation at positions Y498 of bovine CNGA1 and Y1097 of CNGB1a (within the respective CNBDs) inhibits the activity of CNGA1+CNGB1a channels, inducing an approximately twofold decrease in apparent cGMP affinity (Molokanova et al., 2003). Interestingly, there is crosstalk between regulations by tyrosine phosphorylation and CaM for rod CNG channels: CaM regulation was attenuated when the channel was thiophosphorylated; this effect was abolished in CNGA1-Y498F+CNGB1a channels but not in CNGA1+CNGB1a-Y1097F channels (Krajewski et al., 2003). In contrast to rod channels, the equivalent position in CNGA3 is a phenylalanine, while the equivalent tyrosine in CNGB3 (Y545) does not confer tyrosine phosphorylation-dependent regulation of cone CNG channels (Bright et al., 2007).

Signaling molecules downstream of phospholipase C (PLC) are associated with regulation of CNG channels. Diacylglycerol generated after PLC activation has been shown to inhibit rod CNG channels (Gordon et al., 1995a; Crary et al., 2000) and influence the ligand affinity of native cone CNG channels (Chen et al., 2007). Furthermore, protein kinase C (PKC) can modulate the activities of CNG channels in a subunit-specific way. It was reported that phosphorylation of olfactory CNGA2 subunits by PKC, at a serine residue within the N-terminal domain adjacent to the CaM-binding site, was able to increase apparent cGMP affinity (Müller et al., 1998). In addition, bovine CNGA3 channels were found to be regulated by PKC-mediated phosphorylation of serine residues within the CNBD, exhibiting a decrease in cGMP affinity for homomeric CNGA3 channels (Müller et al., 2001).

Membrane-bound phosphoinositides, which have been shown to be ubiquitous ion channel modulators (Gamper and Shapiro, 2007; Suh and Hille, 2008), also have been reported to regulate CNG channels. $PI(4,5)P_2$ and $PI(3,4,5)P_3$ produced inhibition of rod (Womack et al., 2000) and cone CNG channels (Bright et al., 2007). Furthermore, PI (3,4,5) P_3 has been reported to inhibit olfactory CNG channels in rat OSNs as well as recombinant CNGA2 channels expressed in heterologous expression systems (Spehr et al., 2002; Zhainazarov et al., 2004). Moreover, phosphoinositide regulation has been shown to interact with Ca^{2+}–CaM regulation of olfactory CNG channels (Brady et al., 2006). It was found that $PI(3,4,5)P_3$ inhibited the apparent ligand affinity of homomeric CNGA2 channels and heteromeric CNGA2+CNGA4+CNGB1b channels, an effect that was dependent on a putative N-terminal PIP_3-binding site within CNGA2 (Figure 24.2). PIP_3 suppressed Ca^{2+}–CaM regulation of heteromeric CNGA2+CNGA4+CNGB1b channels, presumably by preventing CaM binding to CNGA4 and CNGB1b subunits (Brady et al., 2006). In addition, the molecular mechanism for PIP_2 and PIP_3 regulation of cone CNG channels has recently been elucidated (Dai and Varnum, 2013; Dai et al., 2013). However, the physiological role for this regulation in photoreceptors is not well established. Phosphoinositide levels within photoreceptors are controlled by light, intracellular Ca^{2+}, paracrine signals, and circadian oscillators (Chen et al., 2007; Li et al., 2008; Ko et al., 2009). Thus, phosphoinositide regulation of CNG channels has the potential to contribute to several critical physiological processes necessary for normal photoreceptor function.

The ligand affinity of native cone photoreceptor CNG channels is modulated by retinal circadian oscillations, generating a low apparent cGMP affinity during the daytime and an almost twofold increase in cGMP affinity during the night (Ko et al., 2003, 2004; Chae et al., 2007; Chen et al., 2007). Tyrosine phosphorylation of chicken CNGB3 subunits has been proposed as one possible pathway responsible for circadian regulation of cone CNG channels (Chae et al., 2007), but other mechanisms such as activation of lipid kinases or lipid phosphatases, for example, PLC and/or PI3-kinase, might also contribute to changes in ligand affinity (Bright et al., 2007; Chen et al., 2007; Ko et al., 2009). Furthermore, circadian regulation of dopamine and somatostatin release may contribute to the circadian regulation of chicken cone CNG channels. Dopamine levels are higher during the day compared to the night; applying dopamine D2 receptor agonist decreased the cGMP affinity of CNG channels during the night but not during the daytime (Ko et al., 2003). Somatostatin release from retinal amacrine cells is high during the night and low during the day; activation of somatostatin receptors increased the ligand affinity of native cone CNG channels during the early part of the day, possibly through activation of PLC and PKC (Chen et al., 2007). Moreover, the cAMP–PKA signaling pathway, the small monomeric G protein, Ras, the mitogen-activated protein kinase (MARK) signaling pathway, and calcium–calmodulin kinase II (CaMKII) all have been reported to contribute to circadian regulation of cone CNG channels (Ko et al., 2001a, 2004).

Besides regulation of photoreceptor CNG channels by the intrinsic circadian clock, light is able to indirectly regulate CNG channels. Light-induced insulin receptor activation has been shown to inhibit rod CNG channels via phosphorylation of Y498 and Y503 in CNGA1 (Gupta et al., 2012). This finding adds a novel physiological link for the tyrosine phosphorylation of rod CNG channel subunits that was previously described by Kramer and coworkers (Molokanova et al., 1999, 2003). In addition, Grb 14, an insulin receptor-binding protein, has been shown to suppress rod CNG channel activity by directly binding to the C-terminal region of CNGA1 in a light-dependent manner (Gupta et al., 2010; Rajala et al., 2012). Direct phosphorylation of CNGA1 after insulin receptor activation and Grb 14 binding to CNGA1 together have an additive effect for inhibition of rod CNG channel activity (Gupta et al., 2012). Furthermore, PI3-kinase activity was enhanced after light-dependent phosphorylation of insulin receptors, which could potentially regulate photoreceptor CNG channels by manipulating membrane PIP_3 and PIP_2 levels in the outer segment of photoreceptors (Rajala et al., 2002). Consistent with this finding, light exposure has been demonstrated to induce a robust increase in PIP_3 levels within the photoreceptor outer segment (Li et al., 2008).

24.8 CNG CHANNEL CELL BIOLOGY: BIOGENESIS, TRAFFICKING, AND TURNOVER

Compared to the detailed biophysical characterization of CNG channels described earlier, much less is known about cell biology aspects of their function, including channel biogenesis and trafficking mechanisms. CNGA (but not CNGB) subunits are modified by N-glycosylation within the extracellular pore-turret region following the fifth transmembrane domain (TM5) (Figure 24.2). The precise site for N-glycosylation varies among subunit types and shows some species-specific differences. There is no obvious functional requirement for subunit glycosylation (Rho et al., 2000), but lack of glycan addition can report defects in channel folding or maturation (Faillace et al., 2004; Liu and Varnum, 2005; Duricka et al., 2012). In addition, CNGA subunit glycosylation at some positions can protect channels from MMP-dependent processing and subsequent potentiation of channel gating (Meighan et al., 2012, 2013). Furthermore, the N-terminal region of CNGA1 subunits (and chicken CNGA3 subunits) is subject to proteolytic processing in photoreceptors, with cleavage at amino acid 92 in bovine CNGA1 (Molday et al., 1991; Bönigk et al., 1993). The possible functional significance and tissue-type specificity of this subunit processing is not known.

There appear to be subunit-specific trafficking rules for CNG channels. In heterologous expression systems, CNGA1, CNGA2, and CNGA3 subunits exhibit efficient plasma-membrane localization in the absence of CNGB1 or CNGB3 (or CNGA4) subunits, but in the absence of CNGA1, CNGA2, or CNGA3 subunits, the CNGB or CNGA4 subunits are retained within intracellular compartments (Trudeau and Zagotta, 2002a; Peng et al., 2003b; Zheng and Zagotta, 2004; Nache et al., 2012). On the other hand, CNGB1 has been shown to be essential for ciliary trafficking of CNG channels (Hüttl et al., 2005; Michalakis et al., 2006). For the olfactory CNG channel, ciliary trafficking requires the kinesin motor protein Kif17, which interacts with the C-terminal region of CNGB1b (Jenkins et al., 2006). In addition, PACS-1 binding and serine/threonine protein kinase CK2–mediated phosphorylation of CNGB1b together regulate ciliary trafficking of the olfactory channel (Jenkins et al., 2009). The zebrafish KIF3A kinesin protein has been shown to be important for ciliary trafficking of cone CNG channels, as well as other phototransduction proteins localizing within the cone photoreceptor outer segment (Avasthi et al., 2009). Furthermore, targeting of rod CNG channels to the outer segment depends on interactions between the C-terminus of CNGB1a and the cytoskeletal protein ankyrin G (Kizhatil et al., 2009). For cone CNG channels, CNGB3 knockout in mice decreases CNGA3 protein levels and outer segment localization (Ding et al., 2009). However, the precise protein–protein interactions regulating ciliary targeting of cone CNG channels remain to be determined. There are other protein–protein interactions that are proposed to influence CNG channel localization. These include contacts made between channel subunits and protein components of the outer segment disc membranes (Körschen et al., 1999), including peripherin-2 (Poetsch et al., 2001). Also, rod and cone CNG channels assemble with the NCKX1 and NCKX2, respectively, presumably via interactions with the CNGA subunits (Bauer and Drechsler, 1992; Kim et al., 1998; Schnetkamp, 2004). Since NCKX1 and NCKX2 represent the primary extrusion pathway for outer segment calcium, channel–exchanger complexes are expected to create highly localized regulation of outer segment calcium levels (Prinsen et al., 2000; Schnetkamp, 1995).

Regarding channel stability, CNG channel subunits appear to be subject to degradation via the ubiquitin–proteasome system (UPS), particularly in terms of quality-control surveillance (Michalakis et al., 2006; Becirovic et al., 2010; Duricka and Varnum, unpublished data). The UPS may also exert a role in mediating normal turnover of the channels. The half-life of CNG channels in embryonic chick cone photoreceptors has been estimated to be less than 12 h (Ko et al., 2001b). This channel lifetime is much shorter than the expected turnover rate via outer segment phagocytosis mediated by the retinal pigment epithelium (RPE), which is thought to engulf about 10% of the photoreceptor outer segment per day (Kevany and Palczewski, 2010). Intriguingly, several enzymes mediating ubiquitination and deubiquitination are present in photoreceptor outer segments (Obin et al., 1996; Esteve-Rudd et al., 2010; Hajkova et al., 2010).

24.9 CNG CHANNELOPATHIES AND DISEASE MECHANISMS

The most common forms of CNG channelopathies involve disturbances of CNG channel function in the retina (Biel and Michalakis, 2007). Mutations in the genes encoding rod CNG channel subunits (*CNGA1* and *CNGB1*) are associated with autosomal recessive retinitis pigmentosa (arRP) (Dryja et al., 1995; Paquet-Durand et al., 2011). RP is characterized by progressive degeneration of rod photoreceptors followed by loss of cones. Mutations in *CNGA3* and *CNGB3* have been linked to complete or incomplete achromatopsia, progressive cone dystrophy, oligocone trichromacy, inherited macular degeneration, and/or macular malfunction (Kohl et al., 1998, 2000, 2005; Sundin et al., 2000; Wissinger et al., 2001; Rojas et al., 2002; Johnson et al., 2004; Michaelides et al., 2004; Nishiguchi et al., 2005; Khan et al., 2007; Wiszniewski et al., 2007; Thiadens et al., 2010; Vincent et al., 2011). These disorders are typically inherited in an autosomal recessive manner and are characterized by absent or limited cone function (but intact rod function), compromised visual acuity, nystagmus, photophobia, and in some cases cone degeneration.

Determining the pathophysiological mechanisms, at the molecular and cellular levels, arising from disease-associated mutations in CNG channel genes is an arena of keen ongoing interest. The most commonly occurring *CNGA3* mutations are missense mutations, and many have been shown to produce a loss-of-function phenotype at the molecular level—with decreased channel ligand sensitivity or absence of functional subunits, impaired folding, mislocalization, and/or increased subunit turnover (Tränkner et al., 2004; Liu and Varnum, 2005; Patel et al., 2005; Muraki-Oda et al., 2007; Reuter et al., 2008). Most *CNGA1* and *CNGB1* mutations also are thought to produce channel loss of function (Dryja et al., 1995; Trudeau and Zagotta, 2002a; Kizhatil et al., 2009). For some loss-of-function mutations, ER stress has been implicated in disease progression; improperly folded or misassembled channels arising from missense mutations (or null mutations in one subunit type) may accumulate in the endoplasmic reticulum and induce the unfolded protein response (Duricka et al., 2012; Thapa et al., 2012).

CNG channel gene knockouts in mice represent informative disease models that mimic pathological features arising from CNG channel null mutations in humans. Cone photoreceptors of *CNGA3* -/- mice demonstrate impaired photoresponses and progressive degeneration (Biel et al., 1999; Michalakis et al., 2005, 2011a). *CNGB1* knockout leads to functional and structural defects in both rod photoreceptors and olfactory receptor neurons (Hüttl et al., 2005; Michalakis et al., 2006). CNGA1 subunit knockdown using an antisense approach also leads to retinal degeneration (Leconte and Barnstable, 2000). In contrast, the effect of *CNGB3* absence appears to be less severe, with more subtle early changes, evidence of residual function, and slower progression overall (Ding et al., 2009). CNG channel subunit knockouts also have been shown to produce effects outside the retina, including impaired olfaction for CNGA2 -/- (Brunet et al., 1996) and CNGB1b -/- (Michalakis et al., 2006) and altered hippocampal LTP for CNGA2 -/- (Parent et al., 1998) and CNGA3 -/- (Michalakis et al., 2011a).

Disease-associated mutations can lead to gain-of-function changes in CNG channel activity. While the most commonly occurring *CNGB3* mutations lead to subunit truncation via frameshifts, splice-site defects, and/or premature stop codons, many *CNGB3* missense mutations (and some *CNGA3* mutations) appear to produce gain-of-function changes in channel gating (Peng et al., 2003b; Bright et al., 2005) (Figure 24.7). For example, the achromatopsia-associated CNGB3-F525N mutation produces an approximately threefold increase in cGMP sensitivity (Figure 24.7). Hyperactive CNG channels are expected to disturb calcium homeostasis and lead to photoreceptor dysfunction and cell death (Liu et al., 2013). In addition, disease-causing mutations in other photoreceptor proteins, particularly those producing elevated cGMP levels, can result in uncontrolled CNG channel activity. Consistent with this idea, CNG channel subunit knockout or knockdown via siRNA has been shown to have a profound rescue effect, slowing progression of retinal degeneration in mice having mutations in other critical phototransduction proteins such as PDE (Paquet-Durand et al., 2011; Tosi et al., 2011).

Other functional defects arising from CNG channel mutations include altered pore properties such as ion selectivity and/or single-channel conductance (Seifert et al., 1999; Peng et al., 2003b; Tränkner et al., 2004; Koeppen et al., 2010). Finally, disease-associated mutations in CNG channels can interfere with

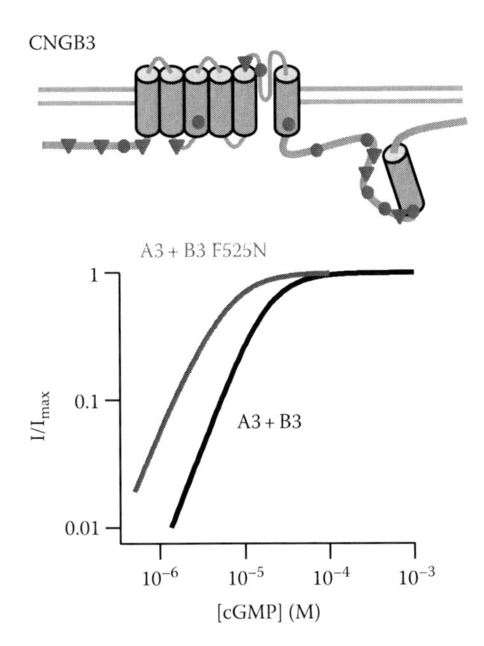

Figure 24.7 Disease-associated mutations in CNG channels can produce gain-of-function changes in channel gating. Cartoon showing the location of disease-linked mutations in CNGB3 (aforementioned), where circles indicate missense mutations and triangles represent truncations produced by frameshift mutations, defective splice sites, and/or premature stop codons. The achromatopsia-associated F525N mutation in CNGB3 enhances apparent cGMP affinity compared to wild-type CNGA3+CNGB3 channels. Representative cGMP dose–response relationships are shown with continuous curves representing fits using the Hill equation. The representative Hill parameters are as follows: for CNGA3+CNGB3 channels, $K_{1/2}$ = 17.6 μM, h = 1.8; for CNGA3+CNGB3-F525N channels, $K_{1/2}$ = 5.9 μM, h = 1.6.

intersubunit interactions that are critical for channel assembly, gating, and/or regulation (Trudeau and Zagotta, 2002a; Michalakis et al., 2011b; Dai and Varnum, 2013). In one example, an achromatopsia-associated mutation in CNGA3 was shown to enhance cone CNG channel sensitivity to regulation by phosphoinositides by altering intersubunit coupling (Dai and Varnum, 2013).

24.10 CONCLUSIONS

Since the discovery of CNG channels nearly 30 years ago, detailed physiological and biophysical studies have provided many insights into fundamental CNG channel mechanisms. Information revealed by structural studies of bacterial CNG channels and of related HCN and EAG family channels also provides a rich context for understanding mammalian CNG channel structure and function. Issues that appear to require further study include several cell biology aspects of CNG channels, including a deeper characterization of the features and protein–protein interactions that control channel assembly, trafficking, and stability/turnover, as these are likely to inform our thinking about both normal physiology and pathophysiology in humans. Additional structural information, including a closer view of the interactions between N- and C-terminal cytoplasmic domains, may uncover structural mechanisms influencing ligand-dependent gating and channel regulation by second messengers. Intriguing but incompletely characterized roles for CNG channels in nonsensory cells are another key area for additional investigation. Furthermore, to complement existing channel knockout models, the development and characterization of animal models representing CNG channel missense mutations is expected to provide important information about retinal disease mechanisms and potential treatment strategies.

REFERENCES

Altenhofen, W., J. Ludwig, E. Eismann, W. Kraus, W. Bönigk, and U.B. Kaupp. 1991. Control of ligand specificity in cyclic nucleotide-gated channels from rod photoreceptors and olfactory epithelium. *Proc. Natl. Acad. Sci. U.S.A.* 88:9868–9872.

Armstrong, C.M. and B. Hille. 1972. The inner quaternary ammonium ion receptor in potassium channels of the node of ranvier. *J. Gen. Physiol.* 59:388–400. doi:10.1085/jgp.59.4.388.

Avasthi, P., C.B. Watt, D.S. Williams, Y.Z. Le, S. Li, C.-K. Chen, R.E. Marc, J.M. Frederick, and W. Baehr. 2009. Trafficking of membrane proteins to cone but not rod outer segments is dependent on heterotrimeric kinesin-II. *J. Neurosci.* 29:14287–14298. doi:10.1523/JNEUROSCI.3976-09.2009.

Bauer, P.J. and M. Drechsler. 1992. Association of cyclic GMP-gated channels and Na(+)-Ca(2+)-K+ exchangers in bovine retinal rod outer segment plasma membranes. *J. Physiol.* 451:109–131.

Becchetti, A. and K. Gamel. 1999. The properties of cysteine mutants in the pore region of cyclic-nucleotide-gated channels. *Pflüg. Arch. Eur. J. Physiol.* 438:587–596.

Becchetti, A., K. Gamel, and V. Torre. 1999. Cyclic nucleotide-gated channels. Pore topology studied through the accessibility of reporter cysteines. *J. Gen. Physiol.* 114:377–392.

Becirovic, E., K. Nakova, V. Hammelmann, R. Hennel, M. Biel, and S. Michalakis. 2010. The retinitis pigmentosa mutation c.3444+1G>A in CNGB1 results in skipping of exon 32. *PloS One.* 5:e8969. doi:10.1371/journal.pone.0008969.

Biel, M. and S. Michalakis. 2007. Function and dysfunction of CNG channels: Insights from channelopathies and mouse models. *Mol. Neurobiol.* 35:266–277.

Biel, M., M. Seeliger, A. Pfeifer, K. Kohler, A. Gerstner, A. Ludwig, G. Jaissle, S. Fauser, E. Zrenner, and F. Hofmann. 1999. Selective loss of cone function in mice lacking the cyclic nucleotide-gated channel CNG3. *Proc. Natl. Acad. Sci. U.S.A.* 96:7553–7557.

Biskup, C., J. Kusch, E. Schulz, V. Nache, F. Schwede, F. Lehmann, V. Hagen, and K. Benndorf. 2007. Relating ligand binding to activation gating in CNGA2 channels. *Nature.* 446:440–443. doi:10.1038/nature05596.

Bönigk, W., W. Altenhofen, F. Müller, A. Dose, M. Illing, R.S. Molday, and U.B. Kaupp. 1993. Rod and cone photoreceptor cells express distinct genes for cGMP-gated channels. *Neuron.* 10:865–877.

Bönigk, W., A. Loogen, R. Seifert, N. Kashikar, C. Klemm, E. Krause, V. Hagen, E. Kremmer, T. Strünker, and U.B. Kaupp. 2009. An atypical CNG channel activated by a single cGMP molecule controls sperm chemotaxis. *Sci. Signal.* 2:ra68. doi:10.1126/scisignal.2000516.

Bönigk, W., F. Müller, R. Middendorff, I. Weyand, and U.B. Kaupp. 1996. Two alternatively spliced forms of the cGMP-gated channel alpha-subunit from cone photoreceptor are expressed in the chick pineal organ. *J. Neurosc.* 16:7458–7468.

Bradley, J., W. Bönigk, K.-W. Yau, and S. Frings. 2004. Calmodulin permanently associates with rat olfactory CNG channels under native conditions. *Nat. Neurosci.* 7:705–710. doi:10.1038/nn1266.

Bradley, J., J. Li, N. Davidson, H.A. Lester, and K. Zinn. 1994. Heteromeric olfactory cyclic nucleotide-gated channels: A subunit that confers increased sensitivity to cAMP. *Proc. Natl. Acad. Sci. U.S.A.* 91:8890–8894.

Bradley, J., D. Reuter, and S. Frings. 2001. Facilitation of calmodulin-mediated odor adaptation by cAMP-gated channel subunits. *Science.* 294:2176–2178. doi:10.1126/science.1063415.

Bradley, J., Y. Zhang, R. Bakin, H.A. Lester, G.V. Ronnett, and K. Zinn. 1997. Functional expression of the heteromeric "olfactory" cyclic nucleotide-gated channel in the hippocampus: A potential effector of synaptic plasticity in brain neurons. *J. Neurosci.* 17:1993–2005.

Brady, J.D., E.D. Rich, J.R. Martens, J.W. Karpen, M.D. Varnum, and R.L. Brown. 2006. Interplay between PIP3 and calmodulin regulation of olfactory cyclic nucleotide-gated channels. *Proc. Natl. Acad. Sci. U.S.A.* 103:15635–15640. doi:10.1073/pnas.0603344103.

Brelidze, T.I., A.E. Carlson, B. Sankaran, and W.N. Zagotta. 2012. Structure of the carboxy-terminal region of a KCNH channel. *Nature.* 481:530–533. doi:10.1038/nature10735.

Brelidze, T.I., A.E. Carlson, and W.N. Zagotta. 2009. Absence of direct cyclic nucleotide modulation of mEAG1 and hERG1 channels revealed with fluorescence and electrophysiological methods. *J. Biol. Chem.* 284:27989–27997. doi:10.1074/jbc.M109.016337.

Brelidze, T.I., E.C. Gianulis, F. DiMaio, M.C. Trudeau, and W.N. Zagotta. 2013. Structure of the C-terminal region of an ERG channel and functional implications. *Proc. Natl. Acad. Sci. U.S.A.* 110:11648–11653. doi:10.1073/pnas.1306887110.

Bright, S.R., T.E. Brown, and M.D. Varnum. 2005. Disease-associated mutations in CNGB3 produce gain of function alterations in cone cyclic nucleotide-gated channels. *Mol. Vis.* 11:1141–1150.

Bright, S.R., E.D. Rich, and M.D. Varnum. 2007. Regulation of human cone cyclic nucleotide-gated channels by endogenous phospholipids and exogenously applied phosphatidylinositol 3,4,5-trisphosphate. *Mol. Pharmacol.* 71:176–183. doi:10.1124/mol.106.026401.

Brown, R.L., R. Gramling, R.J. Bert, and J.W. Karpen. 1995. Cyclic GMP contact points within the 63-kDa subunit and a 240-kDa associated protein of retinal rod cGMP-activated channels. *Biochemistry (Mosc.).* 34:8365–8370.

Brown, R.L., T.L. Haley, K.A. West, and J.W. Crabb. 1999. Pseudechetoxin: A peptide blocker of cyclic nucleotide-gated ion channels. *Proc. Natl. Acad. Sci. U.S.A.* 96:754–759.

Brown, R.L., L.L. Lynch, T.L. Haley, and R. Arsanjani. 2003. Pseudechetoxin binds to the pore turret of cyclic nucleotide-gated ion channels. *J. Gen. Physiol.* 122:749–760. doi:10.1085/jgp.200308823.

Brown, R.L., S.D. Snow, and T.L. Haley. 1998. Movement of gating machinery during the activation of rod cyclic nucleotide-gated channels. *Biophys. J.* 75:825–833.

Brunet, L.J., G.H. Gold, and J. Ngai. 1996. General anosmia caused by a targeted disruption of the mouse olfactory cyclic nucleotide-gated cation channel. *Neuron.* 17:681–693.

Bucossi, G., E. Eismann, F. Sesti, M. Nizzari, M. Seri, U.B. Kaupp, and V. Torre. 1996. Time-dependent current decline in cyclic GMP-gated bovine channels caused by point mutations in the pore region expressed in *Xenopus oocytes*. *J. Physiol.* 493(Part 2):409–418.

Burns, M.E. and V.Y. Arshavsky. 2005. Beyond counting photons: Trials and trends in vertebrate visual transduction. *Neuron.* 48:387–401. doi:10.1016/j.neuron.2005.10.014.

Cassar, S.C., J. Chen, D. Zhang, and M. Gopalakrishnan. 2004. Tissue specific expression of alternative splice forms of human cyclic nucleotide gated channel subunit CNGA3. *Mol. Vis.* 10:808–813.

Chae, K.-S., G.Y.-P. Ko, and S.E. Dryer. 2007. Tyrosine phosphorylation of cGMP-gated ion channels is under circadian control in chick retina photoreceptors. *Invest. Ophthalmol. Vis. Sci.* 48:901–906. doi:10.1167/iovs.06–0824.

Chen, J., M.L. Woodruff, T. Wang, F.A. Concepcion, D. Tranchina, and G.L. Fain. 2010. Channel modulation and the mechanism of light adaptation in mouse rods. *J. Neurosci.* 30:16232–16240. doi:10.1523/JNEUROSCI.2868–10.2010.

Chen, S.-K., G.Y.-P. Ko, and S.E. Dryer. 2007. Somatostatin peptides produce multiple effects on gating properties of native cone photoreceptor cGMP-gated channels that depend on circadian phase and previous illumination. *J. Neurosci.* 27:12168–12175. doi:10.1523/JNEUROSCI.3541–07.2007.

Chen, T.Y., Y.W. Peng, R.S. Dhallan, B. Ahamed, R.R. Reed, and K.W. Yau. 1993. A new subunit of the cyclic nucleotide-gated cation channel in retinal rods. *Nature.* 362:764–767. doi:10.1038/362764a0.

Chiu, P.-L., M.D. Pagel, J. Evans, H.-T. Chou, X. Zeng, B. Gipson, H. Stahlberg, and C.M. Nimigean. 2007. The structure of the prokaryotic cyclic nucleotide-modulated potassium channel MloK1 at 16 A resolution. *Struct. Lond. Engl. 1993.* 15:1053–1064. doi:10.1016/j.str.2007.06.020.

Clayton, G.M., W.R. Silverman, L. Heginbotham, and J.H. Morais-Cabral. 2004. Structural basis of ligand activation in a cyclic nucleotide regulated potassium channel. *Cell.* 119:615–627. doi:10.1016/j.cell.2004.10.030.

Contreras, J.E. and M. Holmgren. 2006. Access of quaternary ammonium blockers to the internal pore of cyclic nucleotide-gated channels: Implications for the location of the gate. *J. Gen. Physiol.* 127:481–494. doi:10.1085/jgp.200509440.

Contreras, J.E., D. Srikumar, and M. Holmgren. 2008. Gating at the selectivity filter in cyclic nucleotide-gated channels. *Proc. Natl. Acad. Sci. U.S.A.* 105:3310–3314. doi:10.1073/pnas.0709809105.

Cordero-Morales, J.F., L.G. Cuello, Y. Zhao, V. Jogini, D.M. Cortes, B. Roux, and E. Perozo. 2006. Molecular determinants of gating at the potassium-channel selectivity filter. *Nat. Struct. Mol. Biol.* 13:311–318. doi:10.1038/nsmb1069.

Cordero-Morales, J.F., V. Jogini, S. Chakrapani, and E. Perozo. 2011. A multipoint hydrogen-bond network underlying KcsA C-type inactivation. *Biophys. J.* 100:2387–2393. doi:10.1016/j.bpj.2011.01.073.

Crary, J.I., D.M. Dean, W. Nguitragool, P.T. Kurshan, and A.L. Zimmerman. 2000. Mechanism of inhibition of cyclic nucleotide-gated ion channels by diacylglycerol. *J. Gen. Physiol.* 116:755–768.

Craven, K.B., N.B. Olivier, and W.N. Zagotta. 2008. C-terminal movement during gating in cyclic nucleotide-modulated channels. *J. Biol. Chem.* 283:14728–14738. doi:10.1074/jbc.M710463200.

Craven, K.B. and W.N. Zagotta. 2004. Salt bridges and gating in the COOH-terminal region of HCN2 and CNGA1 channels. *J. Gen. Physiol.* 124:663–677. doi:10.1085/jgp.200409178.

Craven, K.B. and W.N. Zagotta. 2006. CNG and HCN channels: Two peas, one pod. *Annu. Rev. Physiol.* 68:375–401. doi:10.1146/annurev.physiol.68.040104.134728.

Cuello, L.G., V. Jogini, D.M. Cortes, and E. Perozo. 2010. Structural mechanism of C-type inactivation in K(+) channels. *Nature.* 466:203–208. doi:10.1038/nature09153.

Cukkemane, A., B. Grüter, K. Novak, T. Gensch, W. Bönigk, T. Gerharz, U.B. Kaupp, and R. Seifert. 2007. Subunits act independently in a cyclic nucleotide-activated K(+) channel. *EMBO Rep.* 8:749–755. doi:10.1038/sj.embor.7401025.

Dai, G., T. Sherpa, and M.D. Varnum. 2014. *J Biol Chem.* May 9;289(19):13680–13690.

Dai, G., C. Peng, C. Liu, and M.D. Varnum. 2013. Two structural components in CNGA3 support regulation of cone CNG channels by phosphoinositides. *J. Gen. Physiol.* 141:413–430. doi:10.1085/jgp.201210944.

Dai, G. and M.D. Varnum. 2013. CNGA3 achromatopsia-associated mutation potentiates the phosphoinositide sensitivity of cone photoreceptor CNG channels by altering intersubunit interactions. *Am. J. Physiol. Cell Physiol.* doi:10.1152/ajpcell.00037.2013.

Das, R., S. Chowdhury, M.T. Mazhab-Jafari, S. Sildas, R. Selvaratnam, and G. Melacini. 2009. Dynamically driven ligand selectivity in cyclic nucleotide binding domains. *J. Biol. Chem.* 284:23682–23696. doi:10.1074/jbc.M109.011700.

Derebe, M.G., D.B. Sauer, W. Zeng, A. Alam, N. Shi, and Y. Jiang. 2011a. Tuning the ion selectivity of tetrameric cation channels by changing the number of ion binding sites. *Proc. Natl. Acad. Sci. U.S.A.* 108:598–602. doi:10.1073/pnas.1013636108.

Derebe, M.G., W. Zeng, Y. Li, A. Alam, and Y. Jiang. 2011b. Structural studies of ion permeation and Ca^{2+} blockage of a bacterial channel mimicking the cyclic nucleotide-gated channel pore. *Proc. Natl. Acad. Sci. U.S.A.* 108:592–597. doi:10.1073/pnas.1013643108.

Dhallan, R.S., K.W. Yau, K.A. Schrader, and R.R. Reed. 1990. Primary structure and functional expression of a cyclic nucleotide-activated channel from olfactory neurons. *Nature.* 347:184–187. doi:10.1038/347184a0.

Ding, X.-Q., C.S. Harry, Y. Umino, A.V. Matveev, S.J. Fliesler, and R.B. Barlow. 2009. Impaired cone function and cone degeneration resulting from CNGB3 deficiency: Down-regulation of CNGA3 biosynthesis as a potential mechanism. *Hum. Mol. Genet.* 18:4770–4780. doi:10.1093/hmg/ddp440.

Ding, X.-Q., A. Matveev, A. Singh, N. Komori, and H. Matsumoto. 2012. Biochemical characterization of cone cyclic nucleotide-gated (CNG) channel using the infrared fluorescence detection system. *Adv. Exp. Med. Biol.* 723:769–775. doi:10.1007/978-1-4614-0631-0_98.

Distler, M., M. Biel, V. Flockerzi, and F. Hofmann. 1994. Expression of cyclic nucleotide-gated cation channels in non-sensory tissues and cells. *Neuropharmacology.* 33:1275–1282.

Dryja, T.P., J.T. Finn, Y.W. Peng, T.L. McGee, E.L. Berson, and K.W. Yau. 1995. Mutations in the gene encoding the alpha subunit of the rod cGMP-gated channel in autosomal recessive retinitis pigmentosa. *Proc. Natl. Acad. Sci. U.S.A.* 92:10177–10181.

Duricka, D.L., R.L. Brown, and M.D. Varnum. 2012. Defective trafficking of cone photoreceptor CNG channels induces the unfolded protein response and ER-stress-associated cell death. *Biochem. J.* 441:685–696. doi:10.1042/BJ20111004.

Dzeja, C., V. Hagen, U.B. Kaupp, and S. Frings. 1999. Ca^{2+} permeation in cyclic nucleotide-gated channels. *EMBO J.* 18:131–144. doi:10.1093/emboj/18.1.131.
Eismann, E., F. Müller, S.H. Heinemann, and U.B. Kaupp. 1994. A single negative charge within the pore region of a cGMP-gated channel controls rectification, Ca^{2+} blockage, and ionic selectivity. *Proc. Natl. Acad. Sci. U.S.A.* 91:1109–1113.
Esteve-Rudd, J., L. Campello, M.-T. Herrero, N. Cuenca, and J. Martín-Nieto. 2010. Expression in the mammalian retina of parkin and UCH-L1, two components of the ubiquitin-proteasome system. *Brain Res.* 1352:70–82. doi:10.1016/j.brainres.2010.07.019.
Faillace, M.P., R.O. Bernabeu, and J.I. Korenbrot. 2004. Cellular processing of cone photoreceptor cyclic GMP-gated ion channels: A role for the S4 structural motif. *J. Biol. Chem.* 279:22643–22653. doi:10.1074/jbc.M400035200.
Fesenko, E.E., S.S. Kolesnikov, and A.L. Lyubarsky. 1985. Induction by cyclic GMP of cationic conductance in plasma membrane of retinal rod outer segment. *Nature.* 313:310–313.
Flynn, G.E., K.D. Black, L.D. Islas, B. Sankaran, and W.N. Zagotta. 2007. Structure and rearrangements in the carboxy-terminal region of SpIH channels. *Struct. Lond. Engl. 1993.* 15:671–682. doi:10.1016/j.str.2007.04.008.
Flynn, G.E. and W.N. Zagotta. 2001. Conformational changes in S6 coupled to the opening of cyclic nucleotide-gated channels. *Neuron.* 30:689–698.
Fodor, A.A., K.D. Black, and W.N. Zagotta. 1997a. Tetracaine reports a conformational change in the pore of cyclic nucleotide-gated channels. *J. Gen. Physiol.* 110:591–600.
Fodor, A.A., S.E. Gordon, and W.N. Zagotta. 1997b. Mechanism of tetracaine block of cyclic nucleotide-gated channels. *J. Gen. Physiol.* 109:3–14.
Frings, S. 2001. Chemoelectrical signal transduction in olfactory sensory neurons of air-breathing vertebrates. *Cell. Mol. Life Sci. CMLS.* 58:510–519.
Frings, S., J.W. Lynch, and B. Lindemann. 1992. Properties of cyclic nucleotide-gated channels mediating olfactory transduction. Activation, selectivity, and blockage. *J. Gen. Physiol.* 100:45–67.
Frings, S., R. Seifert, M. Godde, and U.B. Kaupp. 1995. Profoundly different calcium permeation and blockage determine the specific function of distinct cyclic nucleotide-gated channels. *Neuron.* 15:169–179.
Gamper, N. and M.S. Shapiro. 2007. Regulation of ion transport proteins by membrane phosphoinositides. *Nat. Rev. Neurosci.* 8:921–934. doi:10.1038/nrn2257.
Gerstner, A., X. Zong, F. Hofmann, and M. Biel. 2000. Molecular cloning and functional characterization of a new modulatory cyclic nucleotide-gated channel subunit from mouse retina. *J. Neurosci.* 20:1324–1332.
Gordon, S.E., D.L. Brautigan, and A.L. Zimmerman. 1992. Protein phosphatases modulate the apparent agonist affinity of the light-regulated ion channel in retinal rods. *Neuron.* 9:739–748.
Gordon, S.E., J. Downing-Park, B. Tam, and A.L. Zimmerman. 1995a. Diacylglycerol analogs inhibit the rod cGMP-gated channel by a phosphorylation-independent mechanism. *Biophys. J.* 69:409–417. doi:10.1016/S0006-3495(95)79913-1.
Gordon, S.E., J. Downing-Park, and A.L. Zimmerman. 1995b. Modulation of the cGMP-gated ion channel in frog rods by calmodulin and an endogenous inhibitory factor. *J. Physiol.* 486 (Part 3):533–546.
Gordon, S.E., M.D. Varnum, and W.N. Zagotta. 1997. Direct interaction between amino- and carboxyl-terminal domains of cyclic nucleotide-gated channels. *Neuron.* 19:431–441.
Gordon, S.E. and W.N. Zagotta. 1995a. A histidine residue associated with the gate of the cyclic nucleotide-activated channels in rod photoreceptors. *Neuron.* 14:177–183.
Gordon, S.E. and W.N. Zagotta. 1995b. Localization of regions affecting an allosteric transition in cyclic nucleotide-activated channels. *Neuron.* 14:857–864.
Goulding, E.H., G.R. Tibbs, and S.A. Siegelbaum. 1994. Molecular mechanism of cyclic-nucleotide-gated channel activation. *Nature.* 372:369–374. doi:10.1038/372369a0.
Grunwald, M.E., W.P. Yu, H.H. Yu, and K.W. Yau. 1998. Identification of a domain on the beta-subunit of the rod cGMP-gated cation channel that mediates inhibition by calcium-calmodulin. *J. Biol. Chem.* 273:9148–9157.
Grunwald, M.E., H. Zhong, J. Lai, and K.W. Yau. 1999. Molecular determinants of the modulation of cyclic nucleotide-activated channels by calmodulin. *Proc. Natl. Acad. Sci. U.S.A.* 96:13444–13449.
Guo, D. and Z. Lu. 2000. Mechanism of cGMP-gated channel block by intracellular polyamines. *J. Gen. Physiol.* 115:783–798.
Gupta, V.K., A. Rajala, R.J. Daly, and R.V.S. Rajala. 2010. Growth factor receptor-bound protein 14: A new modulator of photoreceptor-specific cyclic-nucleotide-gated channel. *EMBO Rep.* 11:861–867. doi:10.1038/embor.2010.142.
Gupta, V.K., A. Rajala, and R.V.S. Rajala. 2012. Insulin receptor regulates photoreceptor CNG channel activity. *Am. J. Physiol. Endocrinol. Metab.* 303:E1363–E1372. doi:10.1152/ajpendo.00199.2012.
Haber-Pohlmeier, S., K. Abarca-Heidemann, H.G. Körschen, H.K. Dhiman, J. Heberle, H. Schwalbe, J. Klein-Seetharaman, U.B. Kaupp, and A. Pohlmeier. 2007. Binding of Ca^{2+} to glutamic acid-rich polypeptides from the rod outer segment. *Biophys. J.* 92:3207–3214. doi:10.1529/biophysj.106.094847.
Hackos, D.H., T.-H. Chang, and K.J. Swartz. 2002. Scanning the intracellular S6 activation gate in the shaker K^+ channel. *J. Gen. Physiol.* 119:521–532.
Hackos, D.H. and J.I. Korenbrot. 1997. Calcium modulation of ligand affinity in the cyclic GMP-gated ion channels of cone photoreceptors. *J. Gen. Physiol.* 110:515–528.
Hackos, D.H. and J.I. Korenbrot. 1999. Divalent cation selectivity is a function of gating in native and recombinant cyclic nucleotide-gated ion channels from retinal photoreceptors. *J. Gen. Physiol.* 113:799–818.
Haitin, Y., A.E. Carlson, and W.N. Zagotta. 2013. The structural mechanism of KCNH-channel regulation by the eag domain. *Nature.* 501:444–448. doi:10.1038/nature12487.
Hajkova, D., Y. Imanishi, V. Palamalai, K.C.S. Rao, C. Yuan, Q. Sheng, H. Tang et al. 2010. Proteomic changes in the photoreceptor outer segment upon intense light exposure. *J. Proteome Res.* 9:1173–1181. doi:10.1021/pr900819k.
Haynes, L.W. 1992. Block of the cyclic GMP-gated channel of vertebrate rod and cone photoreceptors by l-cis-diltiazem. *J. Gen. Physiol.* 100:783–801.
Haynes, L.W. and S.C. Stotz. 1997. Modulation of rod, but not cone, cGMP-gated photoreceptor channels by calcium-calmodulin. *Vis. Neurosci.* 14:233–239.
He, Y. and J.W. Karpen. 2001. Probing the interactions between cAMP and cGMP in cyclic nucleotide-gated channels using covalently tethered ligands. *Biochemistry (Mosc.).* 40:286–295.
Heine, S., S. Michalakis, W. Kallenborn-Gerhardt, R. Lu, H.-Y. Lim, J. Weiland, D. Del Turco et al. 2011. CNGA3: A target of spinal nitric oxide/cGMP signaling and modulator of inflammatory pain hypersensitivity. *J. Neurosci.* 31:11184–11192. doi:10.1523/JNEUROSCI.6159-10.2011.
Hellman, A.B. and K. Shen. 2011. Sensory transduction channel subunits, tax-4 and tax-2, modify presynaptic molecular architecture in *C. elegans*. *PloS One.* 6:e24562. doi:10.1371/journal.pone.0024562.
Higgins, M.K., D. Weitz, T. Warne, G.F.X. Schertler, and U.B. Kaupp. 2002. Molecular architecture of a retinal cGMP-gated channel: The arrangement of the cytoplasmic domains. *EMBO J.* 21:2087–2094. doi:10.1093/emboj/21.9.2087.

Holmgren, M. 2003. Influence of permeant ions on gating in cyclic nucleotide-gated channels. *J. Gen. Physiol.* 121:61–72.

Hoshi, T., W.N. Zagotta, and R.W. Aldrich. 1990. Biophysical and molecular mechanisms of Shaker potassium channel inactivation. *Science.* 250:533–538.

Hsu, Y.T. and R.S. Molday. 1993. Modulation of the cGMP-gated channel of rod photoreceptor cells by calmodulin. *Nature.* 361:76–79. doi:10.1038/361076a0.

Hua, L. and S.E. Gordon. 2005. Functional interactions between A′ helices in the C-linker of open CNG channels. *J. Gen. Physiol.* 125:335–344. doi:10.1085/jgp.200409187.

Hüttl, S., S. Michalakis, M. Seeliger, D.-G. Luo, N. Acar, H. Geiger, K. Hudl et al. 2005. Impaired channel targeting and retinal degeneration in mice lacking the cyclic nucleotide-gated channel subunit CNGB1. *J. Neurosci.* 25:130–138. doi:10.1523/JNEUROSCI.3764-04.2005.

Jan, L.Y. and Y.N. Jan. 1990. A superfamily of ion channels. *Nature.* 345:672. doi:10.1038/345672a0.

Jenkins, P.M., T.W. Hurd, L. Zhang, D.P. McEwen, R.L. Brown, B. Margolis, K.J. Verhey, and J.R. Martens. 2006. Ciliary targeting of olfactory CNG channels requires the CNGB1b subunit and the kinesin-2 motor protein, KIF17. *Curr. Biol. CB.* 16:1211–1216. doi:10.1016/j.cub.2006.04.034.

Jenkins, P.M., L. Zhang, G. Thomas, and J.R. Martens. 2009. PACS-1 mediates phosphorylation-dependent ciliary trafficking of the cyclic-nucleotide-gated channel in olfactory sensory neurons. *J. Neurosci.* 29:10541–10551. doi:10.1523/JNEUROSCI.1590-09.2009.

Jiang, Y., A. Lee, J. Chen, M. Cadene, B.T. Chait, and R. MacKinnon. 2002. The open pore conformation of potassium channels. *Nature.* 417:523–526. doi:10.1038/417523a.

Jiang, Y., A. Lee, J. Chen, V. Ruta, M. Cadene, B.T. Chait, and R. MacKinnon. 2003. X-ray structure of a voltage-dependent K^+ channel. *Nature.* 423:33–41. doi:10.1038/nature01580.

Johnson, J.P., Jr. and W.N. Zagotta. 2001. Rotational movement during cyclic nucleotide-gated channel opening. *Nature.* 412:917–921. doi:10.1038/35091089.

Johnson, S., M. Michaelides, I.A. Aligianis, J.R. Ainsworth, J.D. Mollon, E.R. Maher, A.T. Moore, and D.M. Hunt. 2004. Achromatopsia caused by novel mutations in both CNGA3 and CNGB3. *J. Med. Genet.* 41:e20.

Kaplan, B., T. Sherman, and H. Fromm. 2007. Cyclic nucleotide-gated channels in plants. *FEBS Lett.* 581:2237–2246. doi:10.1016/j.febslet.2007.02.017.

Karpen, J.W., R.L. Brown, L. Stryer, and D.A. Baylor. 1993. Interactions between divalent cations and the gating machinery of cyclic GMP-activated channels in salamander retinal rods. *J. Gen. Physiol.* 101:1–25.

Karpen, J.W., A.L. Zimmerman, L. Stryer, and D.A. Baylor. 1988. Gating kinetics of the cyclic-GMP-activated channel of retinal rods: Flash photolysis and voltage-jump studies. *Proc. Natl. Acad. Sci. U.S.A.* 85:1287–1291.

Kaupp, U.B., T. Niidome, T. Tanabe, S. Terada, W. Bönigk, W. Stühmer, N.J. Cook, K. Kangawa, H. Matsuo, and T. Hirose. 1989. Primary structure and functional expression from complementary DNA of the rod photoreceptor cyclic GMP-gated channel. *Nature.* 342:762–766. doi:10.1038/342762a0.

Kevany, B.M. and K. Palczewski. 2010. Phagocytosis of retinal rod and cone photoreceptors. *Physiology (Bethesda).* 25:8–15. doi:10.1152/physiol.00038.2009.

Khan, N.W., B. Wissinger, S. Kohl, and P.A. Sieving. 2007. CNGB3 achromatopsia with progressive loss of residual cone function and impaired rod-mediated function. *Invest. Ophthalmol. Vis. Sci.* 48:3864–3871. doi:10.1167/iovs.06-1521.

Kim, T.S., D.M. Reid, and R.S. Molday. 1998. Structure-function relationships and localization of the Na/Ca-K exchanger in rod photoreceptors. *J. Biol. Chem.* 273:16561–16567.

Kizhatil, K., S.A. Baker, V.Y. Arshavsky, and V. Bennett. 2009. Ankyrin-G promotes cyclic nucleotide-gated channel transport to rod photoreceptor sensory cilia. *Science.* 323:1614–1617. doi:10.1126/science.1169789.

Ko, G.Y., M.L. Ko, and S.E. Dryer. 2001a. Circadian regulation of cGMP-gated cationic channels of chick retinal cones. Erk MAP kinase and Ca^{2+}/calmodulin-dependent protein kinase II. *Neuron.* 29:255–266.

Ko, G.Y., M.L. Ko, and S.E. Dryer. 2001b. Developmental expression of retinal cone cGMP-gated channels: Evidence for rapid turnover and trophic regulation. *J. Neurosci.* 21:221–229.

Ko, G.Y.-P., M.L. Ko, and S.E. Dryer. 2003. Circadian phase-dependent modulation of cGMP-gated channels of cone photoreceptors by dopamine and D2 agonist. *J. Neurosci.* 23:3145–3153.

Ko, G.Y.-P., M.L. Ko, and S.E. Dryer. 2004. Circadian regulation of cGMP-gated channels of vertebrate cone photoreceptors: Role of cAMP and Ras. *J. Neurosci.* 24:1296–1304. doi:10.1523/JNEUROSCI.3560-03.2004.

Ko, M.L., K. Jian, L. Shi, and G.Y.-P. Ko. 2009. Phosphatidylinositol 3 kinase-Akt signaling serves as a circadian output in the retina. *J. Neurochem.* 108:1607–1620. doi:10.1111/j.1471/-4159.2009.05931.x.

Koch, K.W. and U.B. Kaupp. 1985. Cyclic GMP directly regulates a cation conductance in membranes of bovine rods by a cooperative mechanism. *J. Biol. Chem.* 260:6788–6800.

Koeppen, K., P. Reuter, T. Ladewig, S. Kohl, B. Baumann, S.G. Jacobson, A.S. Plomp, C.P. Hamel, A.R. Janecke, and B. Wissinger. 2010. Dissecting the pathogenic mechanisms of mutations in the pore region of the human cone photoreceptor cyclic nucleotide-gated channel. *Hum. Mutat.* 31:830–839. doi:10.1002/humu.21283.

Kohl, S., B. Baumann, M. Broghammer, H. Jägle, P. Sieving, U. Kellner, R. Spegal et al. 2000. Mutations in the CNGB3 gene encoding the beta-subunit of the cone photoreceptor cGMP-gated channel are responsible for achromatopsia (ACHM3) linked to chromosome 8q21. *Hum. Mol. Genet.* 9:2107–2116.

Kohl, S., T. Marx, I. Giddings, H. Jägle, S.G. Jacobson, E. Apfelstedt-Sylla, E. Zrenner, L.T. Sharpe, and B. Wissinger. 1998. Total colourblindness is caused by mutations in the gene encoding the alpha-subunit of the cone photoreceptor cGMP-gated cation channel. *Nat. Genet.* 19:257–259. doi:10.1038/935.

Kohl, S., B. Varsanyi, G.A. Antunes, B. Baumann, C.B. Hoyng, H. Jägle, T. Rosenberg et al. 2005. CNGB3 mutations account for 50% of all cases with autosomal recessive achromatopsia. *Eur. J. Hum. Genet. EJHG.* 13:302–308. doi:10.1038/sj.ejhg.5201269.

Kolesnikov, S.S., A.B. Zhainazarov, and A.V. Kosolapov. 1990. Cyclic nucleotide-activated channels in the frog olfactory receptor plasma membrane. *FEBS Lett.* 266:96–98.

Korenbrot, J.I. 2012. Speed, adaptation, and stability of the response to light in cone photoreceptors: The functional role of Ca-dependent modulation of ligand sensitivity in cGMP-gated ion channels. *J. Gen. Physiol.* 139:31–56. doi:10.1085/jgp.201110654.

Korenbrot, J.I. and T.I. Rebrik. 2002. Tuning outer segment Ca^{2+} homeostasis to phototransduction in rods and cones. *Adv. Exp. Med. Biol.* 514:179–203.

Körschen, H.G., M. Beyermann, F. Müller, M. Heck, M. Vantler, K.W. Koch, R. Kellner et al. 1999. Interaction of glutamic-acid-rich proteins with the cGMP signalling pathway in rod photoreceptors. *Nature.* 400:761–766. doi:10.1038/23468.

Körschen, H.G., M. Illing, R. Seifert, F. Sesti, A. Williams, S. Gotzes, C. Colville, F. Müller, A. Dosé, and M. Godde. 1995. A 240 kDa protein represents the complete beta subunit of the cyclic nucleotide-gated channel from rod photoreceptor. *Neuron.* 15:627–636.

Krajewski, J.L., C.W. Luetje, and R.H. Kramer. 2003. Tyrosine phosphorylation of rod cyclic nucleotide-gated channels switches off Ca^{2+}/calmodulin inhibition. *J. Neurosci.* 23:10100–10106.

Kramer, R.H., E. Goulding, and S.A. Siegelbaum. 1994. Potassium channel inactivation peptide blocks cyclic nucleotide-gated channels by binding to the conserved pore domain. *Neuron.* 12:655–662.

Kramer, R.H. and J.W. Karpen. 1998. Spanning binding sites on allosteric proteins with polymer-linked ligand dimers. *Nature.* 395:710–713. doi:10.1038/27227.

Kurahashi, T. and A. Menini. 1997. Mechanism of odorant adaptation in the olfactory receptor cell. *Nature.* 385:725–729. doi:10.1038/385725a0.

Kuzmiski, J.B. and B.A. MacVicar. 2001. Cyclic nucleotide-gated channels contribute to the cholinergic plateau potential in hippocampal CA1 pyramidal neurons. *J. Neurosci.* 21:8707–8714.

Labro, A.J., A.L. Raes, I. Bellens, N. Ottschytsch, and D.J. Snyders. 2003. Gating of shaker-type channels requires the flexibility of S6 caused by prolines. *J. Biol. Chem.* 278:50724–50731. doi:10.1074/jbc.M306097200.

Lancet, D. 1986. Vertebrate olfactory reception. *Annu. Rev. Neurosci.* 9:329–355. doi:10.1146/annurev.ne.09.030186.001553.

Leconte, L. and C.J. Barnstable. 2000. Impairment of rod cGMP-gated channel alpha-subunit expression leads to photoreceptor and bipolar cell degeneration. *Invest. Ophthalmol. Vis. Sci.* 41:917–926.

Li, G., A. Rajala, A.F. Wiechmann, R.E. Anderson, and R.V.S. Rajala. 2008. Activation and membrane binding of retinal protein kinase Balpha/Akt1 is regulated through light-dependent generation of phosphoinositides. *J. Neurochem.* 107:1382–1397. doi:10.1111/j.1471-4159.2008.05707.x.

Liman, E.R. and L.B. Buck. 1994. A second subunit of the olfactory cyclic nucleotide-gated channel confers high sensitivity to cAMP. *Neuron.* 13:611–621.

Liu, C., T. Sherpa, and M.D. Varnum. 2013. Disease-associated mutations in CNGB3 promote cytotoxicity in photoreceptor-derived cells. *Mol. Vis.* 19:1268–1281.

Liu, C. and M.D. Varnum. 2005. Functional consequences of progressive cone dystrophy-associated mutations in the human cone photoreceptor cyclic nucleotide-gated channel CNGA3 subunit. *Am. J. Physiol. Cell Physiol.* 289:C187–C198. doi:10.1152/ajpcell.00490.2004.

Liu, D.T., G.R. Tibbs, P. Paoletti, and S.A. Siegelbaum. 1998. Constraining ligand-binding site stoichiometry suggests that a cyclic nucleotide-gated channel is composed of two functional dimers. *Neuron.* 21:235–248.

Liu, J. and S.A. Siegelbaum. 2000. Change of pore helix conformational state upon opening of cyclic nucleotide-gated channels. *Neuron.* 28:899–909.

Liu, M., T.Y. Chen, B. Ahamed, J. Li, and K.W. Yau. 1994. Calcium-calmodulin modulation of the olfactory cyclic nucleotide-gated cation channel. *Science.* 266:1348–1354.

Long, S.B., E.B. Campbell, and R. Mackinnon. 2005. Crystal structure of a mammalian voltage-dependent Shaker family K^+ channel. *Science.* 309:897–903. doi:10.1126/science.1116269.

Lopez-Jimenez, M.E., J.C. González, I. Lizasoain, J. Sánchez-Prieto, J.M. Hernández-Guijo, and M. Torres. 2012. Functional cGMP-gated channels in cerebellar granule cells. *J. Cell. Physiol.* 227:2252–2263. doi:10.1002/jcp.22964.

Lu, Z. and L. Ding. 1999. Blockade of a retinal cGMP-gated channel by polyamines. *J. Gen. Physiol.* 113:35–43.

Mamasuew, K., S. Michalakis, H. Breer, M. Biel, and J. Fleischer. 2010. The cyclic nucleotide-gated ion channel CNGA3 contributes to coolness-induced responses of Grueneberg ganglion neurons. *Cell. Mol. Life Sci. CMLS.* 67:1859–1869. doi:10.1007/s00018-010-0296-8.

Martínez-François, J.R., Y. Xu, and Z. Lu. 2009. Mutations reveal voltage gating of CNGA1 channels in saturating cGMP. *J. Gen. Physiol.* 134:151–164. doi:10.1085/jgp.200910240.

Mazzolini, M., C. Anselmi, and V. Torre. 2009. The analysis of desensitizing CNGA1 channels reveals molecular interactions essential for normal gating. *J. Gen. Physiol.* 133:375–386. doi:10.1085/jgp.200810157.

McKay, D.B. and T.A. Steitz. 1981. Structure of catabolite gene activator protein at 2.9 Å resolution suggests binding to left-handed B-DNA. *Nature.* 290:744–749.

McLatchie, L.M. and H.R. Matthews. 1992. Voltage-dependent block by L-cis-diltiazem of the cyclic GMP-activated conductance of salamander rods. *Proc. Biol. Sci.* 247:113–119. doi:10.1098/rspb.1992.0016.

Meighan, P.C., S.E. Meighan, E.D. Rich, R.L. Brown, and M.D. Varnum. 2012. Matrix metalloproteinase-9 and -2 enhance the ligand sensitivity of photoreceptor cyclic nucleotide-gated channels. *Channels Austin Tex.* 6:181–196. doi:10.4161/chan.20904.

Meighan, S.E., P.C. Meighan, E.D. Rich, R.L. Brown, and M.D. Varnum. 2013. Cyclic nucleotide-gated channel subunit glycosylation regulates matrix metalloproteinase-dependent changes in channel gating. *Biochemistry (Mosc.).* 52:8352–8362. doi:10.1021/bi400824x.

Menini, A. 1995. Cyclic nucleotide-gated channels in visual and olfactory transduction. *Biophys. Chem.* 55:185–196.

Meyer, M.R., A. Angele, E. Kremmer, U.B. Kaupp, and F. Muller. 2000. A cGMP-signaling pathway in a subset of olfactory sensory neurons. *Proc. Natl. Acad. Sci. U.S.A.* 97:10595–10600.

Michaelides, M., I.A. Aligianis, J.R. Ainsworth, P. Good, J.D. Mollon, E.R. Maher, A.T. Moore, and D.M. Hunt. 2004. Progressive cone dystrophy associated with mutation in CNGB3. *Invest. Ophthalmol. Vis. Sci.* 45:1975–1982.

Michalakis, S., H. Geiger, S. Haverkamp, F. Hofmann, A. Gerstner, and M. Biel. 2005. Impaired opsin targeting and cone photoreceptor migration in the retina of mice lacking the cyclic nucleotide-gated channel CNGA3. *Invest. Ophthalmol. Vis. Sci.* 46:1516–1524. doi:10.1167/iovs.04-1503.

Michalakis, S., T. Kleppisch, S.A. Polta, C.T. Wotjak, S. Koch, G. Rammes, L. Matt, E. Becirovic, and M. Biel. 2011a. Altered synaptic plasticity and behavioral abnormalities in CNGA3-deficient mice. *Genes Brain Behav.* 10:137–148. doi:10.1111/j.1601-183X.2010.00646.x.

Michalakis, S., J. Reisert, H. Geiger, C. Wetzel, X. Zong, J. Bradley, M. Spehr et al. 2006. Loss of CNGB1 protein leads to olfactory dysfunction and subciliary cyclic nucleotide-gated channel trapping. *J. Biol. Chem.* 281:35156–35166. doi:10.1074/jbc.M606409200.

Michalakis, S., X. Zong, E. Becirovic, V. Hammelmann, T. Wein, K.T. Wanner, and M. Biel. 2011b. The glutamic acid-rich protein is a gating inhibitor of cyclic nucleotide-gated channels. *J. Neurosci.* 31:133–141. doi:10.1523/JNEUROSCI.4735-10.2011.

Molday, R.S., L.L. Molday, A. Dosé, I. Clark-Lewis, M. Illing, N.J. Cook, E. Eismann, and U.B. Kaupp. 1991. The cGMP-gated channel of the rod photoreceptor cell characterization and orientation of the amino terminus. *J. Biol. Chem.* 266:21917–21922.

Molokanova, E., J.L. Krajewski, D. Satpaev, C.W. Luetje, and R.H. Kramer. 2003. Subunit contributions to phosphorylation-dependent modulation of bovine rod cyclic nucleotide-gated channels. *J. Physiol.* 552:345–356. doi:10.1113/jphysiol.2003.047167.

Molokanova, E., F. Maddox, C.W. Luetje, and R.H. Kramer. 1999. Activity-dependent modulation of rod photoreceptor cyclic nucleotide-gated channels mediated by phosphorylation of a specific tyrosine residue. *J. Neurosci.* 19:4786–4795.

Monod, J., J. Wyman, and J.P. Changeux. 1965. On the nature of allosteric transitions: A plausible model. *J. Mol. Biol.* 12:88–118.

Möttig, H., J. Kusch, T. Zimmer, A. Scholle, and K. Benndorf. 2001. Molecular regions controlling the activity of CNG channels. *J. Gen. Physiol.* 118:183–192.

Müller, F., W. Bönigk, F. Sesti, and S. Frings. 1998. Phosphorylation of mammalian olfactory cyclic nucleotide-gated channels increases ligand sensitivity. *J. Neurosci.* 18:164–173.

Müller, F., M. Vantler, D. Weitz, E. Eismann, M. Zoche, K.W. Koch, and U.B. Kaupp. 2001. Ligand sensitivity of the 2 subunit from the bovine cone cGMP-gated channel is modulated by protein kinase C but not by calmodulin. *J. Physiol.* 532:399–409.

Muraki-Oda, S., F. Toyoda, A. Okada, S. Tanabe, S. Yamade, H. Ueyama, H. Matsuura, and M. Ohji. 2007. Functional analysis of rod monochromacy-associated missense mutations in the CNGA3 subunit of the cone photoreceptor cGMP-gated channel. *Biochem. Biophys. Res. Commun.* 362:88–93. doi:10.1016/j.bbrc.2007.07.152.

Nache, V., T. Eick, E. Schulz, R. Schmauder, and K. Benndorf. 2013. Hysteresis of ligand binding in CNGA2 ion channels. *Nat. Commun.* 4:2864. doi:10.1038/ncomms3866.

Nache, V., E. Schulz, T. Zimmer, J. Kusch, C. Biskup, R. Koopmann, V. Hagen, and K. Benndorf. 2005. Activation of olfactory-type cyclic nucleotide-gated channels is highly cooperative. *J. Physiol.* 569:91–102. doi:10.1113/jphysiol.2005.092304.

Nache, V., T. Zimmer, N. Wongsamitkul, R. Schmauder, J. Kusch, L. Reinhardt, W. Bönigk et al. 2012. Differential regulation by cyclic nucleotides of the CNGA4 and CNGB1b subunits in olfactory cyclic nucleotide-gated channels. *Sci. Signal.* 5:ra48. doi:10.1126/scisignal.2003110.

Nair, A.V., C.H.H. Nguyen, and M. Mazzolini. 2009. Conformational rearrangements in the S6 domain and C-linker during gating in CNGA1 channels. *Eur. Biophys. J.* 38:993–1002. doi:10.1007/s00249-009-0491-4.

Nakamura, T. and G.H. Gold. 1987. A cyclic nucleotide-gated conductance in olfactory receptor cilia. *Nature.* 325:442–444. doi:10.1038/325442a0.

Nerbonne, J.M., S. Richard, J. Nargeot, and H.A. Lester. 1984. New photoactivatable cyclic nucleotides produce intracellular jumps in cyclic AMP and cyclic GMP concentrations. *Nature.* 310:74–76.

Nicol, G.D. 1993. The calcium channel antagonist, pimozide, blocks the cyclic GMP-activated current in rod photoreceptors. *J. Pharmacol. Exp. Ther.* 265:626–632.

Nicol, G.D., P.P. Schnetkamp, Y. Saimi, E.J. Cragoe Jr., and M.D. Bownds. 1987. A derivative of amiloride blocks both the light-regulated and cyclic GMP-regulated conductances in rod photoreceptors. *J. Gen. Physiol.* 90:651–669.

Nimigean, C.M., T. Shane, and C. Miller. 2004. A cyclic nucleotide modulated prokaryotic K^+ channel. *J. Gen. Physiol.* 124:203–210. doi:10.1085/jgp.200409133.

Nishiguchi, K.M., M.A. Sandberg, N. Gorji, E.L. Berson, and T.P. Dryja. 2005. Cone cGMP-gated channel mutations and clinical findings in patients with achromatopsia, macular degeneration, and other hereditary cone diseases. *Hum. Mutat.* 25:248–258. doi:10.1002/humu.20142.

Obin, M.S., J. Jahngen-Hodge, T. Nowell, and A. Taylor. 1996. Ubiquitinylation and ubiquitin-dependent proteolysis in vertebrate photoreceptors (rod outer segments). Evidence for ubiquitinylation of Gt and rhodopsin. *J. Biol. Chem.* 271:14473–14484.

Pagès, F., M. Ildefonse, M. Ragno, S. Crouzy, and N. Bennett. 2000. Coexpression of alpha and beta subunits of the rod cyclic GMP-gated channel restores native sensitivity to cyclic AMP: Role of D604/N1201. *Biophys. J.* 78:1227–1239. doi:10.1016/S0006-3495(00)76680-X.

Paquet-Durand, F., S. Beck, S. Michalakis, T. Goldmann, G. Huber, R. Mühlfriedel, D. Trifunović et al. 2011. A key role for cyclic nucleotide gated (CNG) channels in cGMP-related retinitis pigmentosa. *Hum. Mol. Genet.* 20:941–947. doi:10.1093/hmg/ddq539.

Parent, A., K. Schrader, S.D. Munger, R.R. Reed, D.J. Linden, and G.V. Ronnett. 1998. Synaptic transmission and hippocampal long-term potentiation in olfactory cyclic nucleotide-gated channel type 1 null mouse. *J. Neurophysiol.* 79:3295–3301.

Patel, K.A., K.M. Bartoli, R.A. Fandino, A.N. Ngatchou, G. Woch, J. Carey, and J.C. Tanaka. 2005. Transmembrane S1 mutations in CNGA3 from achromatopsia 2 patients cause loss of function and impaired cellular trafficking of the cone CNG channel. *Invest. Ophthalmol. Vis. Sci.* 46:2282–2290. doi:10.1167/iovs.05-0179.

Peng, C., E.D. Rich, C.A. Thor, and M.D. Varnum. 2003a. Functionally important calmodulin-binding sites in both NH_2- and COOH-terminal regions of the cone photoreceptor cyclic nucleotide-gated channel CNGB3 subunit. *J. Biol. Chem.* 278:24617–24623. doi:10.1074/jbc.M301699200.

Peng, C., E.D. Rich, and M.D. Varnum. 2003b. Achromatopsia-associated mutation in the human cone photoreceptor cyclic nucleotide-gated channel CNGB3 subunit alters the ligand sensitivity and pore properties of heteromeric channels. *J. Biol. Chem.* 278:34533–34540. doi:10.1074/jbc.M305102200.

Peng, C., E.D. Rich, and M.D. Varnum. 2004. Subunit configuration of heteromeric cone cyclic nucleotide-gated channels. *Neuron.* 42:401–410.

Picones, A. and J.I. Korenbrot. 1995. Permeability and interaction of Ca^{2+} with cGMP-gated ion channels differ in retinal rod and cone photoreceptors. *Biophys. J.* 69:120–127. doi:10.1016/S0006-3495(95)79881-2.

Poetsch, A., L.L. Molday, and R.S. Molday. 2001. The cGMP-gated channel and related glutamic acid-rich proteins interact with peripherin-2 at the rim region of rod photoreceptor disc membranes. *J. Biol. Chem.* 276:48009–48016. doi:10.1074/jbc.M108941200.

Prinsen, C.F.M., R.T. Szerencsei, and P.P.M. Schnetkamp. 2000. Molecular cloning and functional expression of the potassium-dependent sodium calcium exchanger from human and chicken retinal cone photoreceptors. *J. Neurosci.* 20:1424–1434.

Puljung, M.C. and W.N. Zagotta. 2013. A secondary structural transition in the C-helix promotes gating of cyclic nucleotide-regulated ion channels. *J. Biol. Chem.* doi:10.1074/jbc.M113.464123.

Rajala, R.V.S., M.E. McClellan, J.D. Ash, and R.E. Anderson. 2002. In vivo regulation of phosphoinositide 3-kinase in retina through light-induced tyrosine phosphorylation of the insulin receptor beta-subunit. *J. Biol. Chem.* 277:43319–43326. doi:10.1074/jbc.M206355200.

Rajala, R.V.S., A. Rajala, and V.K. Gupta. 2012. Conservation and divergence of Grb7 family of Ras-binding domains. *Protein Cell.* 3:60–70. doi:10.1007/s13238-012-2001-1.

Rebrik, T.I., I. Botchkina, V.Y. Arshavsky, C.M. Craft, and J.I. Korenbrot. 2012. CNG-modulin: A novel Ca-dependent modulator of ligand sensitivity in cone photoreceptor cGMP-gated ion channels. *J. Neurosci.* 32:3142–3153. doi:10.1523/JNEUROSCI.5518-11.2012.

Rebrik, T.I. and J.I. Korenbrot. 1998. In intact cone photoreceptors, a Ca^{2+}-dependent, diffusible factor modulates the cGMP-gated ion channels differently than in rods. *J. Gen. Physiol.* 112:537–548.

Rebrik, T.I. and J.I. Korenbrot. 2004. In intact mammalian photoreceptors, Ca^{2+}-dependent modulation of cGMP-gated ion channels is detectable in cones but not in rods. *J. Gen. Physiol.* 123:63–75. doi:10.1085/jgp.200308952.

Reisert, J. and H.R. Matthews. 1999. Adaptation of the odour-induced response in frog olfactory receptor cells. *J. Physiol.* 519(Part 3):801–813.

Reuter, P., K. Koeppen, T. Ladewig, S. Kohl, B. Baumann, and B. Wissinger. 2008. Mutations in CNGA3 impair trafficking or function of cone cyclic nucleotide-gated channels, resulting in achromatopsia. *Hum. Mutat.* 29:1228–1236. doi:10.1002/humu.20790.

Rho, S., H.M. Lee, K. Lee, and C. Park. 2000. Effects of mutation at a conserved N-glycosylation site in the bovine retinal cyclic nucleotide-gated ion channel. *FEBS Lett.* 478:246–252.

Rieke, F. and E.A. Schwartz. 1994. A cGMP-gated current can control exocytosis at cone synapses. *Neuron.* 13:863–873.

Ritter, L.M., N. Khattree, B. Tam, O.L. Moritz, F. Schmitz, and A.F.X. Goldberg. 2011. In situ visualization of protein interactions in sensory neurons: Glutamic acid-rich proteins (GARPs) play differential roles for photoreceptor outer segment scaffolding. *J. Neurosci.* 31:11231–11243. doi:10.1523/JNEUROSCI.2875-11.2011.

Rojas, C.V., L.S. María, J.L. Santos, F. Cortés, and M.A. Alliende. 2002. A frameshift insertion in the cone cyclic nucleotide gated cation channel causes complete achromatopsia in a consanguineous family from a rural isolate. *Eur. J. Hum. Genet. EJHG.* 10:638–642. doi:10.1038/sj.ejhg.5200856.

Root, M.J. and R. MacKinnon. 1993. Identification of an external divalent cation-binding site in the pore of a cGMP-activated channel. *Neuron.* 11:459–466.

Rosenbaum, T. and S.E. Gordon. 2002. Dissecting intersubunit contacts in cyclic nucleotide-gated ion channels. *Neuron.* 33:703–713.

Rosenbaum, T., L.D. Islas, A.E. Carlson, and S.E. Gordon. 2003. Dequalinium: A novel, high-affinity blocker of CNGA1 channels. *J. Gen. Physiol.* 121:37–47.

Ruiz, M.L. and J.W. Karpen. 1997. Single cyclic nucleotide-gated channels locked in different ligand-bound states. *Nature.* 389:389–392. doi:10.1038/38744.

Sautter, A., X. Zong, F. Hofmann, and M. Biel. 1998. An isoform of the rod photoreceptor cyclic nucleotide-gated channel beta subunit expressed in olfactory neurons. *Proc. Natl. Acad. Sci. U.S.A.* 95:4696–4701.

Savchenko, A., S. Barnes, and R.H. Kramer. 1997. Cyclic-nucleotide-gated channels mediate synaptic feedback by nitric oxide. *Nature.* 390:694–698. doi:10.1038/37803.

Schnetkamp, P.P. 1995. Calcium homeostasis in vertebrate retinal rod outer segments. *Cell Calcium.* 18:322–330.

Schnetkamp, P.P.M. 2004. The SLC24 Na^+/Ca^{2+}-K^+ exchanger family: Vision and beyond. *Pflüg. Arch. Eur. J. Physiol.* 447:683–688. doi:10.1007/s00424-003-1069-0.

Schünke, S., M. Stoldt, K. Novak, U.B. Kaupp, and D. Willbold. 2009. Solution structure of the Mesorhizobium loti K1 channel cyclic nucleotide-binding domain in complex with cAMP. *EMBO Rep.* 10:729–735. doi:10.1038/embor.2009.68.

Seifert, R., E. Eismann, J. Ludwig, A. Baumann, and U.B. Kaupp. 1999. Molecular determinants of a Ca^{2+}-binding site in the pore of cyclic nucleotide-gated channels: S5/S6 segments control affinity of intrapore glutamates. *EMBO J.* 18:119–130. doi:10.1093/emboj/18.1.119.

Selvakumar, D., M.J. Drescher, J.R. Dowdall, K.M. Khan, J.S. Hatfield, N.A. Ramakrishnan, and D.G. Drescher. 2012. CNGA3 is expressed in inner ear hair cells and binds to an intracellular C-terminus domain of EMILIN1. *Biochem. J.* 443:463–476. doi:10.1042/BJ20111255.

Selvakumar, D., M.J. Drescher, and D.G. Drescher. 2013. Cyclic nucleotide-gated channel α-3 (CNGA3) interacts with stereocilia tip-link cadherin 23 + exon 68 or alternatively with myosin VIIa, two proteins required for hair cell mechanotransduction. *J. Biol. Chem.* 288:7215–7229. doi:10.1074/jbc.M112.443226.

Shapiro, M.S. and W.N. Zagotta. 2000. Structural basis for ligand selectivity of heteromeric olfactory cyclic nucleotide-gated channels. *Biophys. J.* 78:2307–2320. doi:10.1016/S0006-3495(00)76777-4.

Shin, H.-G. and Z. Lu. 2005. Mechanism of the voltage sensitivity of IRK1 inward-rectifier K^+ channel block by the polyamine spermine. *J. Gen. Physiol.* 125:413–426. doi:10.1085/jgp.200409242.

Shuart, N.G., Y. Haitin, S.S. Camp, K.D. Black, and W.N. Zagotta. 2011. Molecular mechanism for 3:1 subunit stoichiometry of rod cyclic nucleotide-gated ion channels. *Nat. Commun.* 2:457. doi:10.1038/ncomms1466.

Song, Y., K.D. Cygnar, B. Sagdullaev, M. Valley, S. Hirsh, A. Stephan, J. Reisert, and H. Zhao. 2008. Olfactory CNG channel desensitization by Ca^{2+}/CaM via the B1b subunit affects response termination but not sensitivity to recurring stimulation. *Neuron.* 58:374–386. doi:10.1016/j.neuron.2008.02.029.

Spehr, M., C.H. Wetzel, H. Hatt, and B.W. Ache. 2002. 3-phosphoinositides modulate cyclic nucleotide signaling in olfactory receptor neurons. *Neuron.* 33:731–739.

Strünker, T., I. Weyand, W. Bönigk, Q. Van, A. Loogen, J.E. Brown, N. Kashikar, V. Hagen, E. Krause, and U.B. Kaupp. 2006. A K^+-selective cGMP-gated ion channel controls chemosensation of sperm. *Nat. Cell Biol.* 8:1149–1154. doi:10.1038/ncb1473.

Stryer, L. 1987. The molecules of visual excitation. *Sci. Am.* 257:42–50.

Su, Y., W.R. Dostmann, F.W. Herberg, K. Durick, N.H. Xuong, L. Ten Eyck, S.S. Taylor, and K.I. Varughese. 1995. Regulatory subunit of protein kinase A: Structure of deletion mutant with cAMP binding domains. *Science.* 269:807–813.

Suh, B.-C. and B. Hille. 2008. PIP2 is a necessary cofactor for ion channel function: How and why? *Annu. Rev. Biophys.* 37:175–195. doi:10.1146/annurev.biophys.37.032807.125859.

Sunderman, E.R. and W.N. Zagotta. 1999a. Sequence of events underlying the allosteric transition of rod cyclic nucleotide-gated channels. *J. Gen. Physiol.* 113:621–640.

Sunderman, E.R. and W.N. Zagotta. 1999b. Mechanism of allosteric modulation of rod cyclic nucleotide–gated channels. *J. Gen. Physiol.* 113:601–620. doi:10.1085/jgp.113.5.601.

Sundin, O.H., J.M. Yang, Y. Li, D. Zhu, J.N. Hurd, T.N. Mitchell, E.D. Silva, and I.H. Maumenee. 2000. Genetic basis of total colourblindness among the Pingelapese islanders. *Nat. Genet.* 25:289–293. doi:10.1038/77162.

Szerencsei, R.T., R.J. Winkfein, C.B. Cooper, C. Prinsen, T.G. Kinjo, K. Kang, and P.P.M. Schnetkamp. 2002. The Na/Ca-K exchanger gene family. *Ann. N. Y. Acad. Sci.* 976:41–52.

Tanaka, J.C., J.F. Eccleston, and R.E. Furman. 1989. Photoreceptor channel activation by nucleotide derivatives. *Biochemistry (Mosc.).* 28:2776–2784.

Tang, C.Y. and D.M. Papazian. 1997. Transfer of voltage independence from a rat olfactory channel to the *Drosophila* ether-à-go-go K^+ channel. *J. Gen. Physiol.* 109:301–311.

Taraska, J.W., M.C. Puljung, N.B. Olivier, G.E. Flynn, and W.N. Zagotta. 2009. Mapping the structure and conformational movements of proteins with transition metal ion FRET. *Nat. Methods.* 6:532–537. doi:10.1038/nmeth.1341.

Taylor, W.R. and D.A. Baylor. 1995. Conductance and kinetics of single cGMP-activated channels in salamander rod outer segments. *J. Physiol.* 483(Part 3):567–582.

Thapa, A., L. Morris, J. Xu, H. Ma, S. Michalakis, M. Biel, and X.-Q. Ding. 2012. Endoplasmic reticulum stress-associated cone photoreceptor degeneration in cyclic nucleotide-gated channel deficiency. *J. Biol. Chem.* 287:18018–18029. doi:10.1074/jbc.M112.342220.

Thiadens, A.A.H.J., S. Roosing, R.W.J. Collin, N. van Moll-Ramirez, J.J.C. van Lith-Verhoeven, M.J. van Schooneveld, A.I. den Hollander et al. 2010. Comprehensive analysis of the achromatopsia genes CNGA3 and CNGB3 in progressive cone dystrophy. *Ophthalmology.* 117:825–830.e1. doi:10.1016/j.ophtha.2009.09.008.

Tibbs, G.R., E.H. Goulding, and S.A. Siegelbaum. 1997. Allosteric activation and tuning of ligand efficacy in cyclic-nucleotide-gated channels. *Nature*. 386:612–615. doi:10.1038/386612a0.

Tibbs, G.R., D.T. Liu, B.G. Leypold, and S.A. Siegelbaum. 1998. A state-independent interaction between ligand and a conserved arginine residue in cyclic nucleotide-gated channels reveals a functional polarity of the cyclic nucleotide binding site. *J. Biol. Chem.* 273:4497–4505.

Togashi, K., M.J. von Schimmelmann, M. Nishiyama, C.-S. Lim, N. Yoshida, B. Yun, R.S. Molday, Y. Goshima, and K. Hong. 2008. Cyclic GMP-gated CNG channels function in Sema3A-induced growth cone repulsion. *Neuron*. 58:694–707. doi:10.1016/j.neuron.2008.03.017.

Tosi, J., R.J. Davis, N.-K. Wang, M. Naumann, C.-S. Lin, and S.H. Tsang. 2011. shRNA knockdown of guanylate cyclase 2e or cyclic nucleotide gated channel alpha 1 increases photoreceptor survival in a cGMP phosphodiesterase mouse model of retinitis pigmentosa. *J. Cell. Mol. Med.* 15:1778–1787. doi:10.1111/j.1582-4934.2010.01201.x.

Tränkner, D., H. Jägle, S. Kohl, E. Apfelstedt-Sylla, L.T. Sharpe, U.B. Kaupp, E. Zrenner, R. Seifert, and B. Wissinger. 2004. Molecular basis of an inherited form of incomplete achromatopsia. *J. Neurosci.* 24:138–147. doi:10.1523/JNEUROSCI.3883-03.2004.

Trudeau, M.C. and W.N. Zagotta. 2002a. An intersubunit interaction regulates trafficking of rod cyclic nucleotide-gated channels and is disrupted in an inherited form of blindness. *Neuron*. 34:197–207.

Trudeau, M.C. and W.N. Zagotta. 2002b. Mechanism of calcium/calmodulin inhibition of rod cyclic nucleotide-gated channels. *Proc. Natl. Acad. Sci. U.S.A.* 99:8424–8429. doi:10.1073/pnas.122015999.

Trudeau, M.C. and W.N. Zagotta. 2003. Calcium/calmodulin modulation of olfactory and rod cyclic nucleotide-gated ion channels. *J. Biol. Chem.* 278:18705–18708. doi:10.1074/jbc.R300001200.

Trudeau, M.C. and W.N. Zagotta. 2004. Dynamics of Ca^{2+}-calmodulin-dependent inhibition of rod cyclic nucleotide-gated channels measured by patch-clamp fluorometry. *J. Gen. Physiol.* 124:211–223. doi:10.1085/jgp.200409101.

Varnum, M.D., K.D. Black, and W.N. Zagotta. 1995. Molecular mechanism for ligand discrimination of cyclic nucleotide-gated channels. *Neuron*. 15:619–625.

Varnum, M.D. and W.N. Zagotta. 1996. Subunit interactions in the activation of cyclic nucleotide-gated ion channels. *Biophys. J.* 70:2667–2679. doi:10.1016/S0006-3495(96)79836-3.

Varnum, M.D. and W.N. Zagotta. 1997. Interdomain interactions underlying activation of cyclic nucleotide-gated channels. *Science*. 278:110–113.

Vincent, A., T. Wright, G. Billingsley, C. Westall, and E. Héon. 2011. Oligocone trichromacy is part of the spectrum of CNGA3-related cone system disorders. *Ophthalmic Genet.* 32:107–113. doi:10.3109/13816810.2010.544366.

Waldeck, C., K. Vocke, N. Ungerer, S. Frings, and F. Möhrlen. 2009. Activation and desensitization of the olfactory cAMP-gated transduction channel: Identification of functional modules. *J. Gen. Physiol.* 134:397–408. doi:10.1085/jgp.200910296.

Webster, S.M., D. Del Camino, J.P. Dekker, and G. Yellen. 2004. Intracellular gate opening in Shaker K^+ channels defined by high-affinity metal bridges. *Nature*. 428:864–868. doi:10.1038/nature02468.

Wei, J.Y., E.D. Cohen, H.G. Genieser, and C.J. Barnstable. 1998. Substituted cGMP analogs can act as selective agonists of the rod photoreceptor cGMP-gated cation channel. *J. Mol. Neurosci. MN*. 10:53–64. doi:10.1007/BF02737085.

Weitz, D., N. Ficek, E. Kremmer, P.J. Bauer, and U.B. Kaupp. 2002. *Neuron*. Dec 5;36(5):881–889.

Weitz, D., M. Zoche, F. Müller, M. Beyermann, H.G. Körschen, U.B. Kaupp, and K.W. Koch. 1998. Calmodulin controls the rod photoreceptor CNG channel through an unconventional binding site in the N-terminus of the beta-subunit. *EMBO J.* 17:2273–2284. doi:10.1093/emboj/17.8.2273.

Wilkinson, W.J., A.R. Benjamin, I. De Proost, M.C. Orogo-Wenn, Y. Yamazaki, O. Staub, T. Morita et al. 2011. Alveolar epithelial CNGA1 channels mediate cGMP-stimulated, amiloride-insensitive, lung liquid absorption. *Pflüg. Arch. Eur. J. Physiol.* 462:267–279. doi:10.1007/s00424-011-0971-0.

Wissinger, B., D. Gamer, H. Jägle, R. Giorda, T. Marx, S. Mayer, S. Tippmann et al. 2001. CNGA3 mutations in hereditary cone photoreceptor disorders. *Am. J. Hum. Genet.* 69:722–737. doi:10.1086/323613.

Wiszniewski, W., R.A. Lewis, and J.R. Lupski. 2007. Achromatopsia: The CNGB3 p.T383fsX mutation results from a founder effect and is responsible for the visual phenotype in the original report of uniparental disomy 14. *Hum. Genet.* 121:433–439. doi:10.1007/s00439-006-0314-y.

Womack, K.B., S.E. Gordon, F. He, T.G. Wensel, C.C. Lu, and D.W. Hilgemann. 2000. Do phosphatidylinositides modulate vertebrate phototransduction? *J. Neurosci.* 20:2792–2799.

Young, E.C. and N. Krougliak. 2004. Distinct structural determinants of efficacy and sensitivity in the ligand-binding domain of cyclic nucleotide-gated channels. *J. Biol. Chem.* 279:3553–3562. doi:10.1074/jbc.M310545200.

Zagotta, W.N., T. Hoshi, and R.W. Aldrich. 1990. Restoration of inactivation in mutants of Shaker potassium channels by a peptide derived from ShB. *Science*. 250:568–571.

Zagotta, W.N., N.B. Olivier, K.D. Black, E.C. Young, R. Olson, and E. Gouaux. 2003. Structural basis for modulation and agonist specificity of HCN pacemaker channels. *Nature*. 425:200–205. doi:10.1038/nature01922.

Zhainazarov, A.B., M. Spehr, C.H. Wetzel, H. Hatt, and B.W. Ache. 2004. Modulation of the olfactory CNG channel by Ptdlns(3,4,5)P3. *J. Membr. Biol.* 201:51–57.

Zhang, Y., L.L. Molday, R.S. Molday, S.S. Sarfare, M.L. Woodruff, G.L. Fain, T.W. Kraft, and S.J. Pittler. 2009. Knockout of GARPs and the β-subunit of the rod cGMP-gated channel disrupts disk morphogenesis and rod outer segment structural integrity. *J. Cell Sci.* 122:1192–1200. doi:10.1242/jcs.042531.

Zheng, J., M.C. Trudeau, and W.N. Zagotta. 2002. *Neuron*. Dec 5;36(5):891–896.

Zheng, J., M.D. Varnum, and W.N. Zagotta. 2003. Disruption of an intersubunit interaction underlies Ca^{2+}-calmodulin modulation of cyclic nucleotide-gated channels. *J. Neurosci.* 23:8167–8175.

Zheng, J. and W.N. Zagotta. 2004. Stoichiometry and assembly of olfactory cyclic nucleotide-gated channels. *Neuron*. 42:411–421.

Zhong, H., J. Lai, and K.-W. Yau. 2003. Selective heteromeric assembly of cyclic nucleotide-gated channels. *Proc. Natl. Acad. Sci. U.S.A.* 100:5509–5513. doi:10.1073/pnas.0931279100.

Zhong, H., L.L. Molday, R.S. Molday, and K.-W. Yau. 2002. The heteromeric cyclic nucleotide-gated channel adopts a 3A:1B stoichiometry. *Nature*. 420:193–198. doi:10.1038/nature01201.

Zhou, L., N.B. Olivier, H. Yao, E.C. Young, and S.A. Siegelbaum. 2004. A conserved tripeptide in CNG and HCN channels regulates ligand gating by controlling C-terminal oligomerization. *Neuron*. 44:823–834. doi:10.1016/j.neuron.2004.11.012.

Zhou, L. and S.A. Siegelbaum. 2007. Gating of HCN channels by cyclic nucleotides: Residue contacts that underlie ligand binding, selectivity, and efficacy. *Struct. 1993*. 15:655–670. doi:10.1016/j.str.2007.04.012.

Zimmerman, A.L. and D.A. Baylor. 1992. Cation interactions within the cyclic GMP-activated channel of retinal rods from the tiger salamander. *J. Physiol.* 449:759–783.
Zimmerman, A.L., G. Yamanaka, F. Eckstein, D.A. Baylor, and L. Stryer. 1985. Interaction of hydrolysis-resistant analogs of cyclic GMP with the phosphodiesterase and light-sensitive channel of retinal rod outer segments. *Proc. Natl. Acad. Sci. U.S.A.* 82:8813–8817.
Zong, X., H. Zucker, F. Hofmann, and M. Biel. 1998. Three amino acids in the C-linker are major determinants of gating in cyclic nucleotide-gated channels. *EMBO J.* 17:353–362. doi:10.1093/emboj/17.2.353.
Zufall, F. and S. Firestein. 1993. Divalent cations block the cyclic nucleotide-gated channel of olfactory receptor neurons. *J. Neurophysiol.* 69:1758–1768.

25 Acid sensing ion channels

Cecilia Canessa

Contents

25.1 INTRODUCTION

Acid sensing ion channels (ASICs) are proton-activated ionotropic receptors that belong to the Epithelial N_a Channel/Degenerins (ENaC/DEG) family of ion channels. All members of this class share a common protein structure consisting in three identical or homologous pore-forming subunits of approximately 550 amino acids in length. Each subunit has two transmembrane segments, a large extracellular domain with seven conserved cysteine bridges and short intracellular amino- and carboxy-termini. In spite of a highly similar structure, DEG/ENaC channels differ markedly in their functional roles and means of activation; some are constitutively active (Canessa et al., 1994), while others are gated by peptides (Lingueglia et al., 1995), hormones (Thistle et al., 2012), osmolality (Cameron et al., 2010), and mechanical forces (Zhang et al., 2002; O'Hagan et al., 2005), and in many more instances, the agonists are not yet known.

There are four ASIC genes in the mammalian genome, ACCN1–ACCN4 (Figure 25.1a). A fifth gene *ACCN5*—the protein is known as BLINaC for brain liver intestine Na channel (Sakai et al., 1999)—encodes a channel that bears 26% amino acid identity with ASIC1 but is not activated by protons; thus, it will not be discussed further. By differential splicing, these four genes, *ACCN1–ACCN4*, give rise to eight distinct proteins. ASIC1a and ASIC1b are spliced variants of the *ACCN2* gene; the proteins differ in the first 185 amino acids encompassing the cytosolic aminoterminus, the first transmembrane segment (TM1), and approximately the first third of the extracellular domain (Chen et al., 1998). Similarly, ASIC2a and ASIC2b are spliced forms of the *ACCN1* gene, and, as in the previous instance, the two ASIC2 proteins differ in the first 1/3 of the protein, while the distal 2/3 are identical (Figure 25.1b). Homotrimers of ASIC2a form proton-activated channels but those of ASIC2b do so only when ASIC2b associates with other subunits to form heteromeric channels (Lingueglia et al., 1997). Splicing of the *ACCN3* gene produces three isoforms that vary in the distal intracellular carboxyterminus; all three ASIC3 proteins produce channels with identical functional properties. The single product of ACCN4 is the subunit ASIC4 that expresses at low levels in the central nervous system (CNS) and is even less abundance in dorsal root ganglia. Protons do not gate homomeric ASIC4 channels, while ASIC4 assembled with other ASIC subunits decreases proton-evoked currents (Akopian et al., 2000; Gründer et al., 2000).

25.2 TISSUE DISTRIBUTION

Functional expression of proton-activated currents mediated by ASIC channels is for the most part limited to neurons and some glial cells (Berdiev et al., 2003; Feldman et al., 2008; Lin et al., 2010). Transcripts or proteins of some of the ASICs have been found in few additional tissues, but robust functional activity has not been detected outside the nervous system. Examples of expression outside the nervous system are ASIC3 in the lung (Su et al., 2006) and ASIC1, ASIC2, and ASIC3 in bones (Jahr et al., 2005). Instead, the CNS expresses all ASIC transcripts except for ASIC1b, which is present only in peripheral neurons. The most abundant subunit in the brain is ASIC1a that forms mainly homomeric channels in most areas of the brain but also heteromeric channels by association with ASIC2a. However, ASIC1a seems to be essential for the expression of H^+-mediated currents in the brain as they disappear when ASIC1a is inactivated in the mouse (Wemmie et al., 2002).

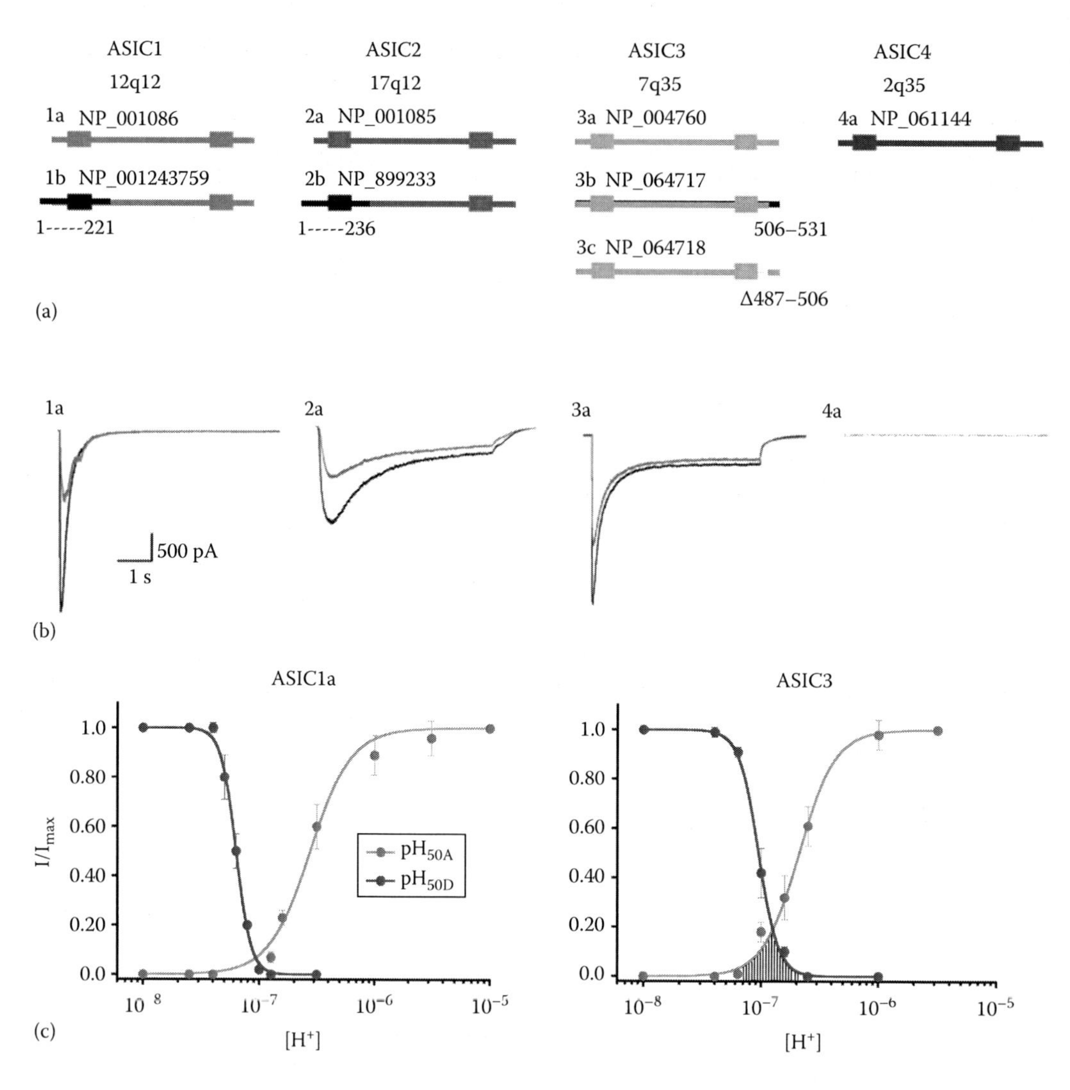

Figure 25.1 (a) *ACCN* genes and their localization in human chromosomes. NP numbers denote the NCBI reference sequence of the proteins. ASIC proteins and their spliced isoforms are represented with transmembrane domains shown as squares: TM1 and TM2. Numbers below are the amino acid positions of the spliced exons. (b) Representative examples of human ASIC currents evoked by changing the external pH from 7.4 to 6.5 (gray) and 5.0 (black), ASIC1 and ASIC3, and from 7.4 to 5.0 (gray) and 4.0 (black), ASIC2. (c) Plots of apparent pH_{50} of activation (pH_{50A}) and steady-state desensitization (pH_{50D}) of rat ASIC1 and ASIC3. Dashed area under the curves represents the pH range of sustained current, that is, *pH window* of nondesensitizing currents.

Peripheral neurons express all the ASIC isoforms in high abundance; any given neuron contains several types of ASIC subunits that associate in various combinations giving rise to a diversity of channels with different properties. The spliced variant ASIC1b is present almost exclusively in sensory neurons, and ASIC3 is highly abundant in peripheral ganglia that innervate the heart, where low pH evokes currents of large magnitude, ~8 nA (Benson et al., 1999). The single transcript of ACCN5 is expressed at low levels across the central and peripheral nervous systems and in peripheral tissues such as the liver and the intestine. The most comprehensive analysis of localization and expression levels of all ASIC transcripts, in human and mouse brain, can be found at the Allen Brain Atlas (http://www.brain-map.org).

The subcellular localization of ASICs, conducted with specific antibodies raised by several laboratories, agrees in demonstrating that ASICs reside at the plasma membrane and not in intracellular organelles. At the cell surface, ASICs are distributed over somata, axons, and dendrites of neurons including postsynaptic densities (Alvarez de la Rosa et al., 2003; Wemmie et al., 2003; Mercado et al., 2006). The ASIC2 protein also localizes to specialized nerve terminals that form mechanosensitive structures in the skin such as Meissner and Pacinian corpuscles (Cabo et al., 2012) and in hair follicles (García-Añoveros et al., 2001). Hence, the expression pattern of ASICs is broad across structures of the brain, and within neurons, it does not concentrate in a specific area of the cell but covers most of the plasma membrane.

25.3 FUNCTION

Almost every type of neuron examined to date, independent of its location (hippocampus, cerebellum, brain cortex, spinal cord, retina, or dorsal root ganglia) or specialization (excitatory glutamatergic, inhibitory GABAergic, sympathetic adrenergic, or sensory modality), expresses at least one type of ASIC-mediated current. The magnitude of the proton-evoked current, however, varies across structures of the nervous system being largest in some sensory neurons. The almost ubiquitous expression of ASIC in central and peripheral nervous systems and the high degree of sequence conservation across species spanning millions of years of evolutionary distance underscore the importance of these channels in the nervous system. However, despite significant progress made in recent years, the biological functions of many ASIC channels remain mostly unknown.

The present understanding of the physiological and pathological roles of ASICs has been derived chiefly from studying the phenotypes displayed by mice with gene inactivation of ASIC1a, ASIC2a/b, or ASIC3. In contrast, there is a dearth of information on the consequences of direct activation of ASIC in the CNS, in large part owing to technical difficulties of changing the external pH in selected areas of the brain. The other difficulty lies on obtaining accurate spatial and temporal measurements of external pH in microdomains of the CNS. This remains a challenge despite improvements in the sensitivity and time resolution of H^+-selective microelectrodes. The large size, invasiveness, and relative low temporal response in comparison to the high-frequency range of neuronal activity are all drawbacks of current microelectrodes. pH-sensitive fluorescent indicators might offer some advantages, though at present, they require large and complex detection apparatus not easily adapted for in vivo measurements in the mammalian brain.

All ASIC knockout mice exhibit normal gross morphology, fertility, and no aberrant behaviors unless examined under specific settings that uncover subtle defects. The first genetic inactivation of ASIC1a was made by replacement of exon 1 of the *ACCN2* gene by a cassette containing the antibiotic neomycin. Those animals express the variant ASIC1b because this isoform bypasses the first exon of ASIC1a by splicing of the mRNA of the same gene (Wemmie et al., 2002). However, since ASIC1b is expressed almost exclusively in sensory neurons, the ASIC1a knockout mouse ($ASIC1^{-/-}$) still is a valid model to study the consequences of eliminating ASIC1 from the CNS but not from sensory neurons. Proton-evoked currents are absent in all brain structures of the $ASIC1a^{-/-}$ mouse confirming that ASIC1a is the main subunit forming both homomeric and heteromeric ASIC channels in the brain. Loss of ASIC1a impairs long-term potentiation (LTP) in the mouse Schaffer collateral-CA1 synapses of hippocampus: the excitatory potentials and NMDA receptor activation during high-frequency stimulation are both reduced. Furthermore, $ASIC1^{-/-}$ mice display deficits in hippocampus-dependent spatial learning, cerebellum-dependent eye blinking (Wemmie et al., 2002), and amygdala-dependent fear conditioning (Wemmie et al., 2004; Ziemann et al., 2009). Some of those findings have been challenged by more recent results from another group that generated mice with brain-specific inactivation of ASIC1a (Wu et al., 2013). Wu et al. repeated the original experiments using their new strain of $ASIC1a^{-/-}$ mice, but in contrast to the previously published results, this group elicited normal LTP in hippocampal slices and normal spatial memory in conscious animals. However, they confirmed that ASIC1a has a potential role in regulating glutamate release (Cho et al., 2008) and in specific behavioral tasks such as contextual and cued fear conditioning (Wemmie et al., 2003).

The close structural similarity of ASICs with the degenerins, some of which are ion channels that participate in mechanotransduction in the nematode *Caenorhabditis elegans*, suggested that at least a few members of the ASICs might also transduce mechanical stimuli in vertebrates. Indeed, some early studies reported abnormal touch (Price et al., 2002) and baroreceptor reflexes (Lu et al., 2009) in $ASIC2^{-/-}$ mice, reinforcing the notion that ASIC2 might be a component of mechanosensors. However, subsequent studies reexamined the same knockout mice and additional $ASIC2^{-/-}$ lines generated by other laboratories (Drew et al., 2004; Roza et al., 2004) failing to demonstrate mechanosensory deficits after inactivation of ASIC2. These mice however are not completely normal; in early age, they exhibit enhanced electroretinograms and, as the mice age, they exhibit degeneration of the retina (Ettaiche et al., 2004), suggesting that ASIC2 and also ASIC3 (Ettaiche et al., 2009) are negative modulators of rod phototransduction. These results are consistent with ASIC2 and ASIC3 having a protective role in the eye that maintains integrity of the retina in normal subjects.

Studies of ASICs in peripheral neurons pose less challenges than in the CNS as sensory neuron terminals are accessible to the application of acid solutions, inflammatory agents, or mechanical stimuli. It is then not surprising that currently there is a better understanding of the ASIC's role in peripheral than in central neurons.

The finding that homomeric ASIC3 channels express at high levels in sensory neurons led to the proposal that ASIC3 was a pain receptor, specifically of noxious stimuli induced by acidification of peripheral tissues generated by ischemia such as myocardial pain (angina pectoris) (Benson et al., 1999; Sutherland et al., 2001) and skeletal muscle pain evoked by strenuous exercise (Birdsong et al., 2010). The basis for this assertion is that among all the ASIC proteins, only homomeric ASIC3 channels display a sustained noninactivating current necessary for transducing the long-lasting pain of protracted acidification associated with such conditions. The sustained current in the pH range of 6.8–7.0 originates from the overlap of the activation and the steady-state desensitization curves; thus, it is also known as a *window current* (Yagi et al., 2006). Although the sustained current is small—only a few pA compared to the peak current of several nA in magnitude—it persists without inactivation and is evoked by the range of pH drop induced by tissue ischemia or inflammation. Furthermore, it has been argued that lactic acid—a hallmark of ischemia—potentiates ASIC3 currents by chelating Ca^{2+} ions that compete with protons for the sites of activation in the extracellular domain of ASIC3. These experiments were conducted on recombinant rat and mouse ASIC3 channels expressed in heterologous systems or rodent dorsal root ganglia. However, the human ASIC3 does not exhibit the window of activation under mild acidification; thereby, no sustain current is evoked in the pH range of 6.8–7.0. The discrepancy arises from the steady-state desensitization curve of human ASIC3 being displaced to the left of that of rodent ASIC3. The consequence of this shift is that human ASIC3 is mostly desensitized at pH of 7.3–7.2 and the response to a further drop in pH is absent. Thus, the functional significance of the sustained current displayed by rat and mouse ASIC3 needs more investigation before it gains acceptance as a mechanism mediating nociception in humans. Furthermore, $ASIC3^{-/-}$ knockouts maintain normal nociception and mechanical responses, and mice bearing inactivation of three ASIC genes together—*ASIC1a*, *ASIC2a/b*, and *ASIC3*—also retain these sensory modalities. The only abnormality detected in the triple gene knockout is enhancement of sensitivity to mechanical stimuli (Borzan et al., 2010; Kang et al., 2012). The mechanism for this negative modulation remains speculative but might involve inhibition of molecules that mediate mechanosensitivity, perhaps by association with ASICs. Hence, currently, the evidence is insufficient to define unequivocally the functional roles of most of the ASICs in sensory modalities; it should be noticed, however, that most experiments have relied largely on the interpretation of phenotypes of knockout mice, which are compounded by compensations from other genes, and thereby might obscure subtle effects mediated by the ASICs.

25.4 AGONISTS, MODULATORS, AND INHIBITORS

25.4.1 AGONISTS AND MODULATORS

As the name indicates, ASICs are activated by external protons, but the question remains of whether protons are the sole agonists or additional endogenous stimuli can also gate ASICs. Several lines of evidence support such a possibility. First came the realization that some members of the ENaC/DEG superfamily are ionotropic receptors activated by neuropeptides. FMRFamide Na channel (FaNaCh) is a channel activated by the neuropeptide FMRFamide in neurons from several snails (*Helix aspersa* [Lingueglia et al., 1995], *Helisoma trivolvis* [Jeziorski et al., 2000], and *Aplysia californica* [Furukawa et al., 2006]). The other peptide-gated channel is hydra Na channel (HyNac) that was cloned from the freshwater polyp hydra and is activated by the neuropeptides hydra-RFamides I and II (Golubovic et al., 2007).

The most persuasive evidence directly relevant to the mammalian ASIC1 has come from the isolation of polypeptides from the venom of the Texas coral snake (*Micrurus tener tener*) whose bite produces excruciating pain (Bohlen et al., 2011). The toxin named MitTx-α/β consists of two separate polypeptides, a Kunitz-like protein (MitTx-α) and a phospholipase-A2-like protein (MitTx-β); when applied together at pH 7.4, they activate ASIC1a and ASIC1b with EC_{50} of 9.4 and 23 nM, respectively. ASIC3 is activated with EC_{50} of 830 nM and ASIC2a requires simultaneous lowering of the pH to <6.5. MitTxα/β markedly stabilizes the open conformation of ASIC1a since the toxin not only opens the channel at neutral pH but almost eliminates desensitization thereby the toxin-bound channel remains fully open—and thus the sustained pain—until the toxin is washed away. MitTxα/β illustrates the appeal of the notion of an endogenous peptide serving as agonist for ASIC gating. Albeit yet unproven, this idea is being pursued by laboratories and pharmaceutical companies.

In addition, searches of chemicals and small molecules have been conducted with the aim of finding alternative molecules to open ASIC channels. Screening of libraries containing compounds with basic groups that are able to bind to the negative charges of the proton sensor led to the isolation of molecules with a guanidinium group and a heterocyclic ring, exemplified by 2-guanidine-4-methylquinazoline (GMQ), which activate ASIC3 at neutral pH (Yu et al., 2010). GMQ alters the pH-dependent gating and amplifies the sustained current of ASIC3 and other ASIC channels (Alijevic and Kellenberger, 2012). The concentration of GMQ required for modifying the ASIC response is however very high, in the mM range, reducing its potential for use in vivo.

Enhancers of ASIC currents have been found also in the peptide category. The neuropeptide FMRFamide (Ph-Met-Arg-Phe-amide), although specific for the molluscan channel FANaCh, also binds ASIC1, ASIC2, and ASIC3 (EC_{50} ~30 μM) increasing the magnitude of the peak current, slowing the rate of channel desensitization, and inducing a sustained current in the continuous presence of both protons and the peptide (Askwith et al., 2000). Together, all these effects significantly increase charge transfer by the channel. Although FMRFamide is abundant in the nervous systems of invertebrates, where it serves as neurotransmitter and neuromodulator, it is not expressed in vertebrates. However, three FMRFamide-related peptides from mammals modulate the ASICs, although with low potency: neuropeptide FF (Phe-Leu-Phe-Gln-Pro-Gln-Arg-Phe-amide), neuropeptide AF (A18Famide), and neuropeptide SF (Ser-Leu-Ala-Ala-Pro-Gln-Arg-Phe-amide). Neuropeptide FF produces minimal effects in homomeric mammalian channels, but currents from the combination ASIC2a/ASIC3 are enhanced (EC_{50} ~2 μM) (Catarsi et al., 2001). Neuropeptide SF targets primarily ASIC3 channels with EC_{50} ~50 μM. In all cases, these neuropeptides bind to the closed state and slow down the kinetics of desensitization (Deval et al., 2003).

Externally applied divalent cations also modulate ASIC currents. Zn^{2+} increases the apparent affinity for H^+ of homomeric ASIC2a, heteromeric ASIC2a/ASIC1a, and ASIC2a/ASIC3 channels (Baron et al., 2001). The enhancement of ASIC2a currents by Zn^{2+} is likely to be relevant in glutamatergic synapses where neurotransmitter-containing vesicles also have high concentrations of both Zn^{2+} and H^+. The EC_{50} for the Zn^{2+} effect on ASIC2 has been estimated at ~120 μM, but the underlying mechanism has not been established. Corelease of Zn^{2+} and H^+ ions into the synaptic cleft raises Zn^{2+} concentration from 0.5 μM to approximately ~200 μM, which is sufficient to increase ASIC2a currents. Enhancement of proton affinity conferred by Zn^{2+} is potentially relevant since ASIC2a channels are abundant in the CNS but have intrinsic low sensitivity to protons.

25.4.2 INHIBITORS

The most successful source of potent inhibitors of ASICs has been found in venoms. Psalmotoxin 1 (PcTtx1), isolated from the venom of the tarantula *Psalmopoeus cambridgei*, was originally identified as a highly selective and potent inhibitor of ASIC1a homomeric channels (Escoubas et al., 2000). PcTtx1 is a 40 amino acid peptide with the cystine knot and triple-strand antiparallel β-sheet structure that defines a family of gating modifiers of ion channels (Norton and Pallaghy, 1998). PcTx1 increases the apparent affinity for H^+-mediated desensitization: it shifts the desensitization curve of mammalian ASIC1a from pH 7.19–7.46 leading to desensitization at neutral pH and consequently unresponsiveness to further pH drops (Chen et al., 2005). Later, it was found that a high concentration of PcTx1, in the μM range, opens and slows the desensitization rate of ASIC1a and also of ASIC1b when the toxin is applied together with a slight increase in the concentrations of protons, pH 7.1 (Chen et al., 2006). This observation was used to solve the structure of ASIC1

Table 25.1 Inhibitors of ASICs belong to various molecular classes and display a broad range of potency from a few nM to hundreds of μM. Some molecules inhibit all ASIC channels while others are subunit-specific

INHIBITOR	TARGET	IC_{50}	TYPE	REFERENCES
PcTx1	ASIC1a	2 nM	Peptide toxin	Escoubas et al. (2000)
APETx2	ASIC3	63 nM	Peptide toxin	Diochot et al. (2004)
Mambalgin 1 and 2	ASIC1a	55 nM	Peptide toxin	Diochot et al. (2012)
	ASIC1a+2a	246 nM		
	ASIC1a+2b	61 nM		
	ASIC1b	192 nM		
	ASIC1a+1b	72 nM		
Amiloride	ASIC1-to-3	100–200 μM		
A-317567	ASIC1	2 μM		Dubé et al. (2005)
	ASIC2	29 μM		
	ASIC3	9.5 μM		
Diminazen	ASIC1b	0.3 μM	Antiprotozoan	Chen et al. (2010)
Nafamostat mesilate	ASIC1a	13.5 μM		Ugawa et al. (2007)
	ASIC2a	70 μM		
	ASIC3	2.5 μM		
Flurbiprofen	ASIC1a	350 μM	NSAID	Voilley et al. (2001)
Salicylic acid	ASIC3	260 μM	NSAID	
Diclofenac	ASIC3	92 μM	NSAID	
Sevanol	ASIC3	350 μM	Lignan from thyme	Dubinnyi et al. (2012)

NSAID, nonsteroidal anti-inflammatory drug.

in the open state; cocrystallization of PcTx1 bound to the chicken ASIC1a led to the resolution of two different open conformations of the pore, one obtained at pH 7.25 and the other at pH 5.5 (Baconguis et al., 2012; Dawson et al., 2012). Since the discovery of PcTx1, it has been used extensively to inhibit ASIC1a both in in vitro and in vivo preparations. Intrathecal administration of PcTx1 produces analgesia in rodent models of acute pain suggesting that inhibition of ASIC1a in the CNS may serve as a new pathway to reduce pain (Mazzuca et al., 2007). However, in light of the evidence showing that PcTx1 also opens ASIC1a and ASIC1b channels, previous results using the toxin with the intension of inhibiting channels should be reinterpreted and allow for the possibility of activation of some of the channels in those preparations.

APETx2 is a toxin from the sea anemone *Anthopleura elegantissima* that inhibits homomeric (IC_{50} of 63 nM) and heteromeric ASIC3 channels (Diochot et al., 2004). APETx2 is a 42 amino acid peptide with three disulfide bridges and a structure similar to that of other inhibitory anemone toxins targeting voltage-gated Na^+ and K^+ channels. APETx2 has been used to induce analgesia in a rodent model of postoperative pain where protons produced by local inflammation presumably activate ASIC3 (Deval et al., 2011).

Mambalgins, mambalgin-1 and mambalgin-2, are toxins isolated from the venom of the black mamba snake that inhibit homomeric ASIC1a and ASIC1b channels and heteromeric channels formed by association of ASIC1 with either ASIC2a or ASIC2b (Diochot et al., 2012). These toxins, which differ in a single residue, are 58 amino acid peptides with a three-finger structure and four internal disulfide bridges. Mambalgins are reversible inhibitors of various ASIC channels that work as gate modifiers in both peripheral and central neurons with affinities that range from 55 to ~600 nM. ASIC inhibition produces potent analgesia against acute and chronic inflammatory pain whether the toxins are administered centrally or peripherally. However, as all the other toxins mentioned previously, the claim of inducing analgesia through inhibition of the ASICs lays mostly on effects obtained with the channels expressed in vitro. Other possible targets of the toxins in vivo should be kept in mind, in particular on the light that inactivation of ASIC1a, ASIC2, and ASIC3 together alters only mildly normal responses of sensory modalities, including pain (Tables 25.1 and 25.2).

25.5 BIOPHYSICAL PROPERTIES

25.5.1 ACTIVATION

The apparent affinity of external protons for activation of ASICs, pH_{50A}, varies widely according to the subunit composition of the channels. These differences are attributed not to the proton sensor, which is conserved across the ASICs, but to differences in the elements that transmit conformational changes from the sensor to the pore. Generally, homomeric ASIC1a channels are the most sensitive to protons; for instance, ASIC1a from human and the frog *Xenopus laevis* has apparent pH_{50A} of 6.8 and 7.0, respectively (Li et al., 2010c). The least sensitive channels are homomeric ASIC2a with pH_{50A} ~4.3 and human ASIC3, pH_{50A} of 5.0, while some isoforms (ASIC2b, ASIC4, and BLINaC) do not respond to protons at all.

Table 25.2 MitTx-α/β is the most potent H⁺-independent activator of ASIC1. PcTx1 functions as an inhibitor of ASIC1 at low concentrations (few nM) and as activator at high concentrations (100 nM). The other compounds synergize the effect of protons but do not activate by themselves any of the ASICs

ACTIVATOR/ MODULATOR	TARGET	IC_{50}	TYPE	REFERENCES
MitTx-α/β	ASIC1a	9.4 nM	Snake toxin	Bohlen et al. (2011)
	ASIC1b	230 nM		
	ASIC3	830 nM		
PcTx1	mASIC1b	140 nM	Tarantula toxin	Chen et al. (2006)
	mASIC1a	~600 nM		
	cASIC1	1 μM		Baconguis and Gouaux (2012)
GMQ	ASIC3	1 mM	2-Guanidine-4-methylquinazoline	Yu et al. (2010)
FMRFamide	ASIC3	50 μM	Neuropeptide	Askwith et al. (2000)
				Xie et al. (2003)
RFRP-1/2	ASIC1 ASIC3			
Zn^{2+}	ASIC2a	120 μM	Divalent cation	Baron et al. (2001)
	ASIC2a+1a	111 μM		
	ASIC2a+3			

Understanding of how external protons activate ASIC has been markedly advanced by the resolution of the crystal structure of chicken ASIC1a. The *proton sensor* consists of many negatively charged residues distributed in adjacent regions of the extracellular domain. The shape of the extracellular domain of ASIC has been compared to that of a hand; each subdomain has been assigned to an element of the hand illustrated in Figure 25.2 (Jasti et al., 2007). Residues in the thumb (Asp346, Asp350, Glu354) and the finger (Glu236, Glu238, Glu239, Glu243, Asp 260) of the same subunit and the palm of the adjacent subunit (Glu178, Glu220, Asp408) seem to be part of the proton sensor. Protonation of those residues leads to formation of H-bonds between the carboxylates of the pairs Asp238/Asp350, Glu239/Asp346, and Glu220/Asp408. Notwithstanding the importance of these interactions, many of them can be eliminated by substitutions with neutral amino acids, and the mutated channel still responds to protons, although with lower affinity for protons and with faster kinetics of desensitization (Li et al., 2009).

Lowering the concentration of external Ca^{2+} also opens ASIC channels. Initially, this observation suggested that displacement of a Ca^{2+} ion occluding the entrance of the pore was the mechanism whereby ASICs were gated (Immke and McCleskey, 2003). This notion was later dismissed by evidence provided from structural and functional studies that showed an allosteric mechanism wherein H^+ binding to the proton sensor, located far from the pore, leads to conformational changes of the extracellular domain that are transmitted to the transmembrane segments to open the pore (Zhang et al., 2006; Li et al., 2011). However, the notion that H^+ ions displacing Ca^{2+} does apply to several sites in the extracellular domain. In the resting state, pH 7.4, Ca^{2+} ions occupy many of the H^+

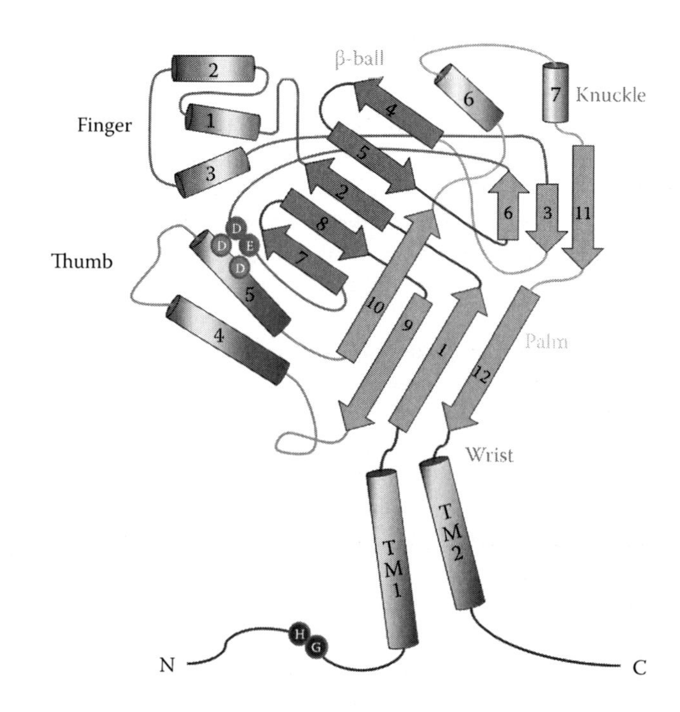

Figure 25.2 Cartoon representation of a single ASIC subunit. Segments belonging to the same protein domain are shown in the same shade of grey. α-helixes are shown as cylinders, β-strands as arrows, and coil structures as lines. Four residues in the putative proton sensor are shown in circles: D346 and D350 in the 5 α-helix of the thumb domain and D238 and E239 in a coil of the finger domain. The atomic structure of the N- and C-termini has not been solved in any of the available crystals from chicken ASIC1a; here, they are represented as coils with the conserved and functionally essential HG motif in the N-terminus.

binding sites stabilizing the close conformation. Upon lowering the external Ca^{2+} concentration, Ca^{2+} ions unbind leading to opening of the channels even at neutral pH. It is unlikely, however, that decreasing the concentration of external Ca^{2+} alone serves as a physiological means to opening ASIC because at pH ~7.4, the Ca^{2+} concentration must be reduced to the submicromolar range before channels open. On the other side, physiological concentrations of Ca^{2+} are essential for maintaining ASIC channels in a functional state since Ca^{2+} stabilizes the close conformation preventing channels entering steady-state desensitization (Alvarez de la Rosa et al., 2002). Thus, the action of external Ca^{2+} is to modulate rather than gating ASIC channels.

Correct characterization of ASIC currents evoked by protons requires rapid changes of external pH. Most of the variability in determining pH_{50A} values stems from inaccuracies of the measurements owing to slow perfusion systems. The use of excised patches in the outside-out configuration and a fast-exchange perfusion system eliminate most of the technical problems. The reason is that with slowly rising concentrations of protons, channel activity is highly desynchronized since some channels are opening or desensitizing, while others are still waiting to bind the agonist. Activation of channels with a low concentration of protons also tends to underestimate the magnitude of the peak current because a large fraction of channels enter steady-state desensitization owing to the much higher proton affinity for the desensitized (hASIC1a pH_{50D} 7.1) than for the active state (hASIC1a pH_{50A} 6.8). Representative examples of proton-evoked currents of human and rat ASICs are shown in Figure 25.1b.

25.5.2 ION SELECTIVITY

ASICs conduct monovalent cations with a selectivity sequence of $Na^+ > Li^+ > K^+$. Initially, high Ca^{2+} permeability of ASIC1a was also reported (Yermolaieva et al., 2004), but it has since been dismissed by several studies; for instance, the P_{Na}/P_{Ca} of human and rodent ASIC1a is ≥50 (Varming, 1999; Zhang and Canessa 2002; Lilley et al., 2004). The increase in intracellular Ca^{2+} observed after activation of ASIC1a is most likely indirect owing to membrane depolarization and secondary activation of Ca^{2+}-permeable pathways. On the other side, the small sustained currents carried by mammalian ASIC3 (Babinski et al., 1999) and shark ASIC1b (Springauf and Gründer, 2010) at low pH <5.0 seem to be less selective to monovalent cations implying that the open pore might adopt at least two distinct conformations that differ in ion selectivity. The structural counterpart to such observations may be found in the recently solved crystal structures of chicken ASIC1a in two different open conformations: one with a narrow pore and the other with a wide and nonselective pore, both with PcTx1 bound to the channel (Baconguis and Gouaux, 2012).

The most striking differences between the two open conformations of cASIC1a are at the level of the transmembrane domain. The structure of the channel solved at pH 5.5 has a narrow and asymmetric pore owing to the TM2 of one of the three subunits being shifted upward by one helix turn. The lumen is not cylindrical but has an elliptical shape and is lined by the TM2 of the three subunits and TM1 from only one subunit. The narrowest part of the pore is located halfway along the transmembrane domain and is delimited by the side chains of L440 of TM2; it has a diameter of ~5 Å × ~7 Å at the level of L440 from subunit C and of 4 Å × 10 Å at L440 from subunits A and B. The relatively large diameter of the pore compared to the Na^+ ionic radius of 1.02 Å suggests that permeating Na^+ ions may be partially hydrated in the selectivity filter. The structure of the selectivity filter solved at pH 5.5 is unexpected as it deviates substantially from the predicting factors governing Na^+ selectivity, that is, protein ligands with strong charge-donating ability such as carboxylate of side chains or backbone carbonyl groups, and coordination of the Na^+ ion by the protein rather than by hydrating water (Dudev and Lim, 2010).

In contrast, the structure solved at pH 7.25 with PcTX1 bound to the channel has a wide symmetric pore that is lined by both TM1 and TM2 from each of the three subunits. The region of the pore with the smallest diameter, ~10 Å, is located at the level of residue E433 in TM2. This residue forms the constriction that shuts the pore in the desensitized state of the channel (Gonzales et al., 2009) indicating that opening entails a major expansion of the pore achieved by straightening and rotating TM2 from each subunit and requires the disruption of many inter- and intrasubunit interactions between TM1 and TM2. The pore in this conformation is wide enough to accommodate not only monovalent cations but also divalent cations and even guanidinium ions.

A contentious point between these structures and the electrophysiological results is that the latter predict a highly Na^+-selective pore at pH >6.0 and, in certain ASIC channels, a less selective pore at pH <5.0; this is just opposite to the crystallographic data. However, it is possible that the discrepancy arises from an artifact induced by the toxin bound to the channel.

Most recently, another high-resolution structure of chicken ASIC1a was solved in the open state in complex with the snake toxin MitTx (Baconguis et al., 2014). The most salient feature of this structure lies in the pore that is substantially different from previous structures reported by the same group and is also uniquely distinctive among all other ion channels known to date. The TM2 is a discontinuous α-helix in which three highly conserved residues Gly443–Ala444–Ser445, in the midway of TM2, adopt an extended conformation swapping the cytoplasmic half of TM2 with an adjacent subunit (Figure 25.3c and d). These residues define the narrowest segment of the pore in the open conformation consistent with previous experimental results that identified this triad of amino acids as part of the selectivity filter. The Gly443 residues form a ring of three carbonyl oxygen atoms with a radius of ~3.6 Å, presenting an energetic barrier for hydrated ions. Together, the features of the latest structure coincide with predictions of numerous biophysical experimental and theoretical results wherein size of the pore is the primary determinant of ion selectivity of DEG/ENaC channels.

25.5.3 DESENSITIZATION

Desensitization influences the magnitude and time course of ASIC currents through several processes: (1) steady-state desensitization, (2) rate of decay of the peak current, and (3) kinetics of recovery from desensitization. ASIC1 channels in the resting state are highly sensitive to steady-state desensitization induced by small increases in the level of external protons. Since the pH_{50D} value is close to the baseline pH in many compartments of the CNS, small and slow acidification tends to silence rather than to activate ASIC1 channels. Other external

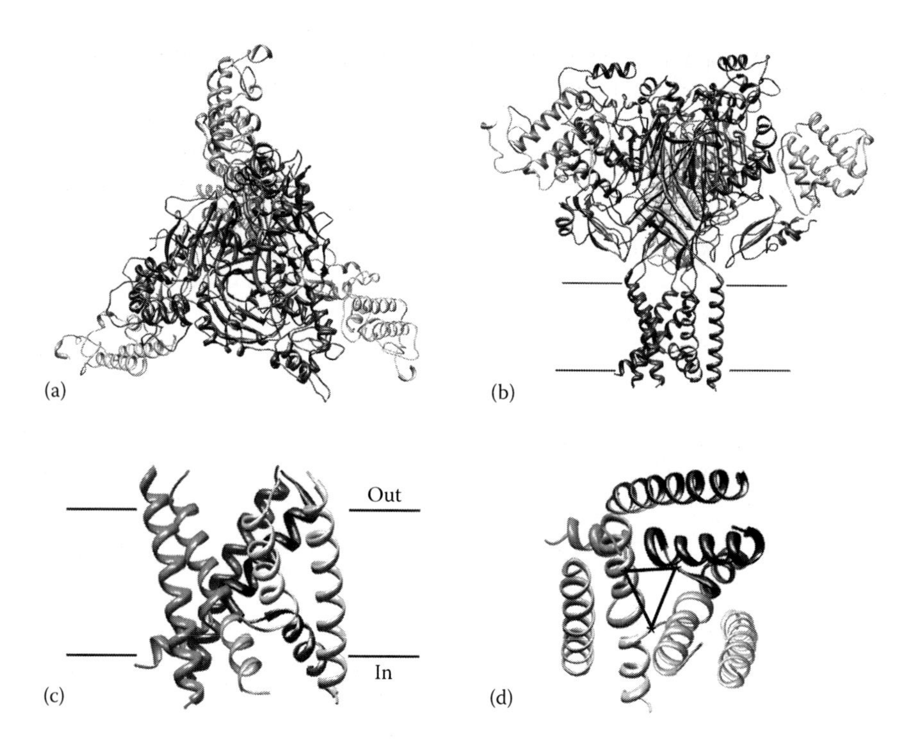

Figure 25.3 Crystal structure of open chicken ASIC1 in complex with MitTx1. Ribbon representation of cASIC1 bound to αMitTx and βMitTx viewed from the top (a) and side (b) derived from 4NTW.pbd using Chimera software. Each channel subunit binds one α- and one β-polypeptides of MitTx. The toxins make extensive interactions with the wrist, palm, and thumb domains. Close-up views of the transmembrane segments containing the channel pore viewed from the side (c) and from the top (d). Transmembrane domain swap of TM2 is mediated by an extended loop of the α-helix formed by the conserved GAS motif. The triangle indicates the carbonyl groups of G443 residues that line the selectivity filter.

factors also modulate the pH_{50D}; for instance, the concentration of Ca^{2+} and other divalent cations exerts an inverse effect on the pH_{50D}: a decrease in Ca^{2+} makes ASIC1 more sensitive to desensitization by shifting the pH_{50D} to the alkaline range. This shift reflects the competition between H^+ and Ca^{2+} ions for the same binding sites in the extracellular domain of the channel; therefore, when comparing values of pH_{50D}, care should be taken to prepare solutions with equal Ca^{2+} concentrations. Extracellular spermine and other polyamines affect the pH_{50D} in a way similar to that produced by divalent cations (Babini et al., 2002).

Elements in the channel protein that influence the value of the pH_{50D} are the β1–β2 (Li et al., 2010b) and β11–β12 linkers (Figure 25.2) (Li et al., 2010c). These short segments of few amino acids in length serve as flexible hinges that are crucial for transducing conformational changes of the extracellular domain to TM1 and TM2 in each subunit.

Putting together the available atomic structures of cASIC1a in various conformations, one can now attempt to reconstruct a likely sequence of events that recapitulates gating of ASIC1. Binding of H^+ to the proton sensor changes the conformation of various parts in the extracellular domain: the thumb and palm domains of adjacent subunits separate by 2–3 Å, the finger domains are displaced, and the β1–β2 and β11–β12 linkers flip to expand the cavity above the pore that is formed by β-strands at the base of the palm. The motions of the low palm domain converge at the wrist region where they are transmitted to the outer segments of TM1 and TM2. In the close/desensitized state, the side chains of residues D433 in the outer third of TM2 occlude the lumen of the pore. This constriction expands in the open state allowing ions to reach the selectivity filter (GAS) located about one third more distal along the length of TM2.

The aforementioned depictions of gating and desensitization are however a simplification of the actual events. Additional complexity arises from the fact that the mammalian ASIC1a has at least two functionally discernible desensitized states that differ markedly in stability; the time constants of the short-lived and long-lived desensitized states are <0.5 and 299 s (Li et al., 2012). These desensitized states are physiologically relevant since they determine whether channels respond or not to high-frequency trains of stimuli. A consequence of the long-lived desensitized state is that the currents of mammalian ASIC1a evoked by high-frequency stimuli become progressively smaller (Chen and Gründer, 2005). Mammalian ASIC1a channels however can escape silencing if the train of stimuli is made of short acid pulses, <100 ms; this is because channels desensitize slowly (τ_D ~ 770 ms); therefore, if the pulse is brief, channels close by inactivation (unbinding of H^+ from the open state) rather than by desensitization (closed state with H^+ bound). Together, these findings indicate that mammalian ASIC1a is silenced by prolonged stimuli, but they remain responsive to short stimuli even if they are applied at high frequency, which is indeed characteristic of neuronal activity in the CNS.

It is not clear yet what is the physiological meaning of the variability in rates of desensitization, rates of recovery from desensitization, and the values of the apparent pH_{50D} among ASICs of different species. They may represent evolutionary selection to life history and habitat, specifically baseline serum pH and temperature, which are factors that would impact the function of ASICs in vertebrates. Species-specific properties of ASICs have, however, served investigators identifying many of the structural determinants underlying the variability of those properties. For instance, the desensitization rate of mammalian ASIC1a is slow (τ_D ~ 770 ms) compared to that of other species such as chicken, fish, and lizard (τ_D ~ 100 ms). Residues that determine such variability are located in the β1–β2 linker (Coric et al., 2003). The β1–β2 linker also determines species variations of pH_{50D} as illustrated by the differences observed in rodent and human ASIC3 channels. The pH_{50D} of mouse ASIC3 is shifted to the acidic range in comparison to human ASIC3; thereby, only the mouse ASIC3 generates sustained currents at pH slightly below 7.4. This functional difference can be traced to a single amino acid: R83 in rodent and Q83 in human located in the β1–β2 linker (Delaunay et al., 2012).

25.6 PERSPECTIVES

Since the first description of proton-evoked currents in mammalian neurons by Krishtal and Pidoplichko (1980), almost two decades elapsed until the cloning of the molecules that mediate such currents (Waldmann et al., 1997a,b). Thereafter, a rapid pace of discoveries has led to great progress in understanding the biophysical properties and architecture of the ASIC molecules; in particular, a clear picture of the channel structure has emerged thanks to x-ray crystallography. Those developments stand in stark contrast to the scant progress made in elucidating the physiological roles of the ASICs. Chief among questions remaining to be answered is to find out whether pH fluctuations induced by neuronal activity can gate ASICs in the brain. Biophysical studies have shown that in order to activate efficiently mammalian ASIC1 channels, the external pH ought to drop by at least ≥0.4 units below 7.3, and the change must be fast to avoid desensitization and silencing of the channels. Advancements in this area will require developing novel technologies for delivering protons with high spatial and temporal precision in selected structures of the CNS while simultaneously monitoring the functional effects, ideally in conscious animals.

Equally important is the search for additional endogenous agonists to gate the ASICs. Finding such molecules has remained elusive despite the prediction that they should be ubiquitously expressed in the brain to match the broad distribution of ASICs across most structures of the CNS. Also intriguing is the meaning of the marked ASIC differences in biophysical properties exhibited by vertebrate species. Currently, it is not known whether they reflect particular physiological adaptations to species-specific internal conditions (temperature, baseline pH) or are mere curiosities of laboratory experiments without physiological importance.

Amidst the efforts to unravel ASIC functions, it is worth contemplating the possibility that some of the ASICs might work in ways distinct from ion channels. Without a doubt, protons activate ASICs in vitro, but gating of these channels by physiological fluctuations of pH in the brain has been only inferred and not directly proven. Furthermore, only in two instances among more than 60 members of the ENaC/DEG family there is definite proof that these molecules operate in vivo as ion channels: ENaC, a constitutively open channel from epithelial cells, is required to maintain blood volume of terrestrial vertebrates; FaNaCh is a peptide-gated ion channel from neurons that control cardiorespiratory in snails and a few degenerins involved in sensing light touch in *C. elegans*. For the large number of remaining channels, in both vertebrates (9 genes) and invertebrates (31 genes in *Drosophila* and 30 in *C. elegans*), the agonist is either unknown or the channels have been rendered active by the artificial introduction of gain-of-function mutations (García-Añoveros et al., 1998; Sakai et al., 1999; Li et al., 2010a).

Albeit it remains a great deal to be learned about the physiology of ASICs in the nervous system, the rapid pace of progress in recent years illustrated by the discovery of molecules that modify ASIC activity, and insightful structure-functional studies, is a clear indication that the field is moving toward the kind of detailed understanding that is necessary to enable the design of therapeutic interventions targeting the ASIC family of ion channels.

REFERENCES

Akopian AN, Chen CC, Ding Y, Cesare P, Wood JN. 2000. A new member of the acid-sensing ion channel family. *Neuroreport.* 11(10):2217–2222.

Alijevic O, Kellenberger S. 2012. Subtype-specific modulation of acid-sensing ion channel (ASIC) function by 2-guanidine-4-methylquinazoline. *J Biol Chem.* 287(43):36059–36070.

Alvarez de la Rosa D, Krueger SR, Kolar A, Shao D, Fitzsimonds RM, Canessa CM. 2003. Distribution, subcellular localization and ontogeny of ASIC1 in the mammalian central nervous system. *J Physiol.* 546(Pt. 1):77–87.

Alvarez de la Rosa D, Zhang P, Shao D, White F, Canessa CM. 2002. Functional implications of the localization and activity of acid-sensitive channels in rat peripheral nervous system. *Proc Natl Acad Sci USA.* 99(4):2326–2331.

Askwith CC, Cheng C, Ikuma M, Benson C, Price MP, Welsh MJ. 2000. Neuropeptide FF and FMRFamide potentiate acid-evoked currents from sensory neurons and proton-gated DEG/ENaC channels. *Neuron.* 26(1):133–141.

Babini E, Paukert M, Geisler HS, Grunder S. 2002. Alternative splicing and interaction with di- and polyvalent cations control the dynamic range of acid-sensing ion channel 1 (ASIC1). *J Biol Chem.* 277(44):41597–41603. Epub: Aug. 26, 2002.

Babinski K, Lê KT, Séguéla P. 1999. Molecular cloning and regional distribution of a human proton receptor subunit with biphasic functional properties. *J Neurochem.* 72(1):51–57.

Baconguis I, Bohlen CJ, Goehring A, Julius D, Gouaux E. 2014. X-ray structure of acid-sensing ion Channel-1-Snake toxin complex reveals open state of a Na^+-selective channel. *Cell.* 156:717–729.

Baconguis I, Gouaux E. 2012. Structural plasticity and dynamic selectivity of acid-sensing ion channel-spider toxin complexes. *Nature.* 489(7416):400–405.

Baron A, Schaefer L, Lingueglia E, Champigny G, Lazdunski M. 2001. Zn^{2+} and H^+ are coactivators of acid-sensing ion channels. *J Biol Chem.* 276(38):35361–35367.

Benson CJ, Eckert SP, McCleskey EW. 1999. Acid-evoked currents in cardiac sensory neurons: A possible mediator of myocardial ischemia sensation. *Circ Res.* 84:921–928.

Berdiev BK, Xia J, McLean LA, Markert JM, Gillespie GY, Mapstone TB, Naren AP et al. 2003. Acid-sensing ion channels in malignant gliomas. *J Biol Chem.* 278(17):15023–15034.

Birdsong WT, Fierro L, Williams FG, Spelta V, Naves LA, Knowles M, Marsh-Haffner J et al. 2010. Sensing muscle ischemia: Coincident detection of acid and ATP via interplay of two ion channels. *Neuron.* 68(4):739–749.

Bohlen CJ, Chesler AT, Sharif-Naeini R, Medzihradszky KF, Zhou S, King D, Sánchez EE, Burlingame AL, Basbaum AI, Julius D. 2011. A heteromeric Texas coral snake toxin targets acid-sensing ion channels to produce pain. *Nature.* 479(7373):410–414.

Borzan J, Zhao C, Meyer RA, Raja SN. 2010. A role for acid-sensing ion channel 3, but not acid-sensing ion channel 2, in sensing dynamic mechanical stimuli. *Anesthesiology.* 113:647–654.

Cabo R, Gálvez MA, San José I, Laurà R, López-Muñiz A, García-Suárez O, Cobo T, Insausti R, Vega JA. 2012. Immunohistochemical localization of acid-sensing ion channel 2 (ASIC2) in cutaneous Meissner and Pacinian corpuscles of *Macaca fascicularis. Neurosci Lett.* 516(2):197–201.

Cameron P, Hiroi M, Ngai J, Scott K. 2010. The molecular basis for water taste in *Drosophila. Nature.* 465(7294):91–95.

Canessa CM, Schild L, Buell G, Thorens B, Gautschi I, Horisberger JD, Rossier BC. 1994. Amiloride-sensitive epithelial Na^+ channel is made of three homologous subunits. *Nature.* 367(6462):463–467.

Catarsi S, Babinski K, Séguéla P. 2001. Selective modulation of heteromeric ASIC proton-gated channels by neuropeptide FF. *Neuropharmacology.* 41(5):592–600.

Chen CC, England S, Akopian AN, Wood JN. 1998. A sensory neuron-specific, proton-gated ion channel. *Proc Natl Acad Sci USA.* 95(17):10240–10245.

Chen, X, Grunder, S. 2005. Permeating protons contribute to tachyphylaxis of the acid-sensing ion channel (ASIC) 1a. *J Physiol (Lond).* 579:657–670.

Chen X, Kalbacher H, Gründer S. 2006. Interaction of acid-sensing ion channel (ASIC) 1 with the tarantula toxin psalmotoxin 1 is state dependent. *J Gen Physiol.* 127(3):267–276.

Chen X, Kalbacher H, Gründer SW. 2005. The tarantula toxin psalmotoxin 1 inhibits acid-sensing ion channel (ASIC) 1a by increasing its apparent H^+ affinity. *J Gen Physiol.* 126:71–79.

Chen X, Qiu L, Li M, Dürrnagel S, Orser BA, Xiong Z-G, MacDonald JF. 2010. Diarylamides: High potency inhibitors of acid-sensing ion channels. *Neuropharmacology* 58:1045–1053.

Cho JH, Askwith CC. 2008. Presynaptic release probability is increased in hippocampal neurons from ASIC1 knockout mice. *J Neurophysiol.* 99(2):426–441.

Coric T, Zhang P, Todorovic N, Canessa CM. 2003. The extracellular domain determines the kinetics of desensitization in acid-sensitive ion channel 1. *J Biol Chem.* 278(46):45240–45247.

Dawson RJ, Benz J, Stohler P, Tetaz T, Joseph C, Huber S, Schmid G et al. 2012. Structure of the acid-sensing ion channel 1 in complex with the gating modifier Psalmotoxin 1. *Nat Commun.* 3:936. doi: 10.1038/ncomms1917.

Delaunay A, Gasull X, Salinas M, Noël J, Friend V, Lingueglia E, Deval E. 2012. Human ASIC3 channel dynamically adapts its activity to sense the extracellular pH in both acidic and alkaline directions. *Proc Natl Acad Sci USA.* 109:13124–13129.

Deval E, Baron A, Lingueglia E, Mazarguil H, Zajac JM, Lazdunski M. 2003. Effects of neuropeptide SF and related peptides on acid sensing ion channel 3 and sensory neuron excitability. *Neuropharmacology.* 44(5):662–671.

Deval E, Noël J, Gasull X, Delaunay A, Alloui A, Friend V, Eschalier A, Lazdunski M, Lingueglia E. 2011. Acid-sensing ion channels in postoperative pain. *J Neurosci.* 31(16):6059–6066.

Diochot S, Baron A, Rash LD, Deval E, Escoubas P, Scarzello S, Salinas M Lazduncki M. 2004. A new sea anemone peptide, APETx2, inhibits ASIC3, a major acid-sensitive channel in sensory neurons. *EMBO J.* 23:1516–1525.

Diochot S, Baron A, Salinas M, Douguet D, Scarzello S, Dabert-Gay AS, Debayle D et al. 2012. Black mamba venom peptides target acid-sensing ion channels to abolish pain. *Nature.* 490(7421):552–555.

Drew LJ, Rohrer DK, Price MP, Blaver KE, Cockayne DA, Cesare P, Wood JN. 2004. Acid-sensing ion channels ASIC2 and ASIC3 do not contribute to mechanically activated currents in mammalian sensory neurons. *J Physiol.* 556(Pt. 3):691–710.

Dubé GR, Lehto SG, Breese NM, Baker SJ, Wang X, Matulenko MA, Honoré P, Stewart AO, Moreland RB, Brioni JD. 2005. Electrophysiological and in vivo characterization of A-317567, a novel blocker of acid sensing ion channels. *Pain.* 117:88–96.

Dubinnyi MA, Osmakov DI, Koshelev SG, Kozlov SA, Andreev YA, Zakaryan NA, Dyachenko IA, Bondarenko DA, Arseniev AS, Grishin EV. 2012. Lignan from thyme possesses inhibitory effect on ASIC3 channel current. *J Biol Chem.* 287(39):32993–3000. Epub: Aug. 1, 2012.

Dudev T, Lim C. 2010. Factors governing the Na^+ vs K^+ selectivity in sodium ion channels. *J Am Chem Soc.* 132:2321–2332.

Escoubas P, De Weille JR, Lecoq A, Diochot S, Waldmann R, Champigny G, Moinier D, Ménez A, Lazdunski M. 2000. Isolation of a tarantula toxin specific for a class of proton-gated Na^+ channels. *J Biol Chem.* 275(33):25116–25121.

Ettaiche M, Deval E, Pagnotta S, Lazdunski M, Lingueglia E. 2009. Acid-sensing ion channel 3 in retinal function and survival. *Invest Ophthalmol Vis Sci.* 50(5):2417–2426.

Ettaiche M, Guy N, Hofman P, Lazdunski M, Waldmann R. 2004. Acid-sensing ion channel 2 is important for retinal function and protects against light-induced retinal degeneration. *J Neurosci.* 24(5):1005–1012.

Feldman DH, Horiuchi M, Keachie K, Mccauley E, Bannerman P, Itoh A, Itoh T, Pleasure D. 2008. Characterization of acid-sensing ion channel expression in oligodendrocyte-lineage cells. *Glia.* 56(11):1238–1249.

Furukawa Y, Miyawaki Y, Abe G. 2006. Molecular cloning and functional characterization of the Aplysia FMRF-amide-gated Na^+ channel. *Pflugers Arch.* 451:646–656.

García-Añoveros J, García JA, Liu JD, Corey DP. 1998. The nematode degenerin UNC-105 forms ion channels that are activated by degeneration- or hypercontraction-causing mutations. *Neuron.* 20(6):1231–1241.

García-Añoveros J, Samad TA, Zuvela-Jelaska L, Woolf CJ, Corey DP. 2001. Transport and localization of the DEG/ENaC ion channel BNaC1alpha to peripheral mechanosensory terminals of dorsal root ganglia neurons. *J Neurosci.* 21(8):2678–2686.

Golubovic A, Kuhn A, Williamson M, Kalbacher H, Holstein TW, Grimmelikhuijzen CJP, Gründer S. 2007. A peptide-gated ion channel from the freshwater polyp hydra. *J Biol Chem.* 282:35098–35103.

Gonzales EB, Kawate T, Gouaux E. 2009. Pore architecture and ion sites in acid-sensing ion channels and P2X receptors. *Nature.* 460(7255):599–604.

Gründer S, Geissler HS, Bässler EL, Ruppersberg JP. 2000. A new member of acid-sensing ion channels from pituitary gland. *Neuroreport.* 11(8):1607–1611.

Immke DC, McCleskey EW. 2003. Protons open acid-sensing ion channels by catalyzing relief of Ca^{2+} blockade. *Neuron.* 37(1):75–84.

Jahr H, van Driel M, van Osch GJ, Weinans H, van Leeuwen JP. 2005. Identification of acid-sensing ion channels in bone. *Biochem Biophys Res Commun.* 337(1):349–354.

Jasti J, Furukawa H, Gonzales EB, Gouaux E. 2007. Structure of acid-sensing ion channel 1 at 1.9 Å resolution and low pH. *Nature.* 449(7160):316–323.

Jeziorski MC, Green KA, Sommerville J, Cottrell GA. 2000. Cloning and expression of FMRFamide-gated Na^+ channel from *Helisoma trivolvis* and comparison with the native neuronal channel. *J Physiol.* 526:13–25.

Kang S, Jang JH, Price MP, Gautam M, Benson CJ, Gong H, Welsh MJ, Brennan TJ. 2012. Simultaneous disruption of mouse ASIC1a, ASIC2 and ASIC3 genes enhances cutaneous mechanosensitivity. *PLoS One*. 7(4):e35225.

Krishtal OA, Pidoplichko VI. 1980. A receptor for protons in the nerve cell membrane. *Neuroscience*. 5:2325–2327.

Li T, Yang Y, Canessa CM. 2009. Interaction of the aromatics Tyr-72/Trp-288 in the interface of the extracellular and transmembrane domains is essential for proton gating of acid-sensing ion channels. *J Biol Chem*. 284(7):4689–4694.

Li T, Yang Y, Canessa CM. 2010a. Two residues in the extracellular domain convert a nonfunctional ASIC1 into a proton-activated channel. *Am J Physiol Cell Physiol*. 299(1):C66–C73.

Li T, Yang Y, Canessa CM. 2010b. Leu85 in the beta1–beta2 linker of ASIC1 slows activation and decreases the apparent proton affinity by stabilizing a closed conformation. *J Biol Chem*. 285(29):22706–22712.

Li T, Yang Y, Canessa CM. 2010c. Asn415 in the beta11–beta12 linker decreases proton-dependent desensitization of ASIC1. *J Biol Chem*. 285(41):31285–31291.

Li T, Yang Y, Canessa CM. 2011. Outlines of the pore in open and closed conformations describe the gating mechanism of ASIC1. *Nat Commun*. 2:399. doi: 10.1038/ncomms1409.

Li T, Yang Y, Canessa CM. 2012. Impact of recovery from desensitization on acid sensing ion channel-1a (ASIC1a) current and response to high-frequency stimulation. *J Biol Chem*. 287(48):40680–40689.

Lilley S, LeTissier P, Robbins J. 2004. The discovery and characterization of a proton-gated sodium current in rat retinal ganglion cells. *J Neurosci*. 24(5):1013–1022.

Lin YC, Liu YC, Huang YY, Lien CC. 2010. High-density expression of Ca^{2+}-permeable ASIC1a channels in NG2 glia of rat hippocampus. *PLoS One*. 5(9):e12665.

Lingueglia E, Champigny G, Lazdunski M, Barbry P. 1995. Cloning of the amiloride-sensitive FMRFamide peptide-gated sodium channel. *Nature*. 378:730–733.

Lingueglia E, de Weille JR, Bassilana F, Heurteaux C, Sakai H, Waldmann R, Lazdunski M. 1997. A modulatory subunit of acid sensing ion channels in brain and dorsal root ganglion cells. *J Biol Chem*. 272(47):29778–29783.

Lu Y, Ma X, Sabharwal R, Snitsarev V, Morgan D, Rahmouni K, Drummond HA et al. 2009. The ion channel ASIC2 is required for baroreceptor and autonomic control of the circulation. *Neuron*. 64(6):885–897.

Mazzuca M, Heurteaux C, Alloui A, Diochot S, Baron A, Voilley N, Blondeau N et al. 2007. A tarantula peptide against pain via ASIC1a channels and opioid mechanisms. *Nat Neurosci*. 10(8):943–945.

Mercado F, López IA, Acuna D, Vega R, Soto E. 2006. Acid-sensing ionic channels in the rat vestibular endorgans and ganglia. *J Neurophysiol*. 96(3):1615–1624.

Morris JA, Royall JJ, Bertagnolli D, Boe AF, Burnell JJ, Byrnes EJ, Copeland C, Desta T, Fischer SR, Goldy J, Glattfelder KJ, Kidney JM, Lemon T, Orta GJ, Parry SE, Pathak SD, Pearson OC, Reding M, Shapouri S, Smith KA, Soden C, Solan BM, Weller J, Takahashi JS, Overly CC, Lein ES, Hawrylycz MJ, Hohmann JG, Jones AR. 2010. Divergent and nonuniform gene expression patterns in mouse brain. *Proc Natl Acad Sci*. 107(44):1049–19054.

Norton RS, Pallaghy PK. 1998. The cystine knot structure of ion channel toxins and related polypeptides. *Toxicon* 36:1573–1583.

O'Hagan R, Chalfie M, Goodman MB. 2005. The MEC-4 DEG/ENaC channel of *Caenorhabditis elegans* touch receptor neurons transduce mechanical signals. *Nat Neurosci*. 8:43–50.

Price MP, Lewin GR, McIlwrath SL, Cheng C, Xie J, Heppenstall PA, Stucky CL et al. 2002. The mammalian sodium channel BNC1 is required for normal touch sensation. *Nature*. 407(6807):1007–1011.

Roza C, Puel JL, Kress M, Baron A, Diochot S, Lazdunski M, Waldmann R. 2004. Knockout of the ASIC2 channel in mice does not impair cutaneous mechanosensation, visceral mechanonociception and hearing. *J Physiol*. 558(Pt. 2):659–669.

Sakai H, Lingueglia E, Champigny G, Mattei MG, Lazdunski M. 1999. Cloning and functional expression of a novel degenerin-like Na^+ channel gene in mammals. *J Physiol* (*Lond*). 519(Pt. 2):323–333.

Springauf A, Gründer S. 2010. An acid-sensing ion channel from shark (*Squalus acanthias*) mediates transient and sustained responses to protons. *J Physiol*. 588(Pt. 5):809–820.

Su X, Li Q, Shrestha K, Cormet-Boyaka E, Chen L, Smith PR, Sorscher EJ, Benos DJ, Matalon S, Ji HL. 2006. Interregulation of proton-gated Na^+ channel 3 and cystic fibrosis transmembrane conductance regulator. *J Biol Chem*. 281(48):36960–36968.

Sutherland SP, Benson CJ, Adelman JP, McCleskey EW. 2001. Acid-sensing ion channel 3 matches the acid-gated current in cardiac ischemia-sensing neurons. *Proc Natl Acad Sci USA*. 98(2):711–716.

Thistle R, Cameron P, Ghorayshi A, Dennison L, Scott K. 2012. Contact chemoreceptors mediate male–male repulsion and male–female attraction during *Drosophila* courtship. *Cell*. 149(5):1140–1151.

Ugawa S, Ishida Y, Ueda T, Inoue K, Nagao M, Shimada S. 2007. Nafamostat mesilate reversibly blocks acid-sensing ion channel currents. *BBRC*. 363:203–208.

Varming T. 1999. Proton-gated ion channels in cultured mouse cortical neurons. *Neuropharmacology*. 38(12):1875–1881.

Voilley N, de Weille J, Mamet J, Lazdunski M. 2001. Nonsteroid anti-inflammatory drugs inhibit both the activity and the inflammation-induced expression of acid-sensing ion channels in nociceptors. *J Neurosci*. 21:8026–8033.

Waldmann M, Bassilana F, de Weille J, Champigny G, Heurteaux C, Lazdunski M. 1997a. Molecular cloning of a non-inactivating proton-gated Na^+ channel specific for sensory neurons. *J Biol Chem*. 272:20975–20978.

Waldmann R, Champigny G, Bassilana F, Heurteaux C, Lazdunski M. 1997b. A proton-gated cation channel involved in acid-sensing. *Nature*. 386(6621):173–177.

Wemmie JA, Askwith CC, Lamani E, Cassell MD, Freeman JH Jr, Welsh MJ. 2003. Acid-sensing ion channel 1 is localized in brain regions with high synaptic density and contributes to fear conditioning. *J Neurosci*. 23(13):5496–5502.

Wemmie JA, Chen J, Askwith CC, Hruska-Hageman AM, Price MP, Nolan BC, Yoder PG et al. 2002. The acid-activated ion channel ASIC contributes to synaptic plasticity, learning, and memory. *Neuron*. 34(3):463–477.

Wemmie JA, Coryell MW, Askwith CC, Lamani E, Leonard AS, Sigmund CD, Welsh MJ. 2004. Overexpression of acid-sensing ion channel 1a in transgenic mice increases acquired fear-related behavior. *Proc Natl Acad Sci USA*. 101(10):3621–3626.

Wu PY, Huang YY, Chen CC, Hsu TT, Lin YC, Weng JY, Chien TC, Cheng IH, Lien CC. 2013. Acid-sensing ion channel-1a is not required for normal hippocampal LTP and spatial memory. *J Neurosci*. 33(5):1828–1832.

Xie J, Price MP, Wemmie JA, Askwith CC, Welsh MJ. 2003. ASIC3 and ASIC1 mediate FMRFamide-related peptide enhancement of H^+-gated currents in cultured dorsal root ganglion neurons. *J Neurophysiol*. 89(5):2459–2465.

Yagi J, Wenk HN, Naves LA, McCleskey EW. 2006. Sustained currents through ASIC3 ion channels at the modest pH changes that occur during myocardial ischemia. *Circ Res*. 99(5):501–509.

Yermolaieva O, Leonard AS, Schnizler MK, Abboud FM, Welsh MJ. 2004. Extracellular acidosis increases neuronal cell calcium by activating acid-sensing ion channel 1a. *Proc Natl Acad Sci USA*. 101(17):6752–6757.
Yu Y, Chen Z, Li WG, Cao H, Feng EG, Yu F, Liu H, Jiang H, Xu TL. 2010. A nonproton ligand sensor in the acid-sensing ion channel. *Neuron*. 68(1):61–72.
Zhang P, Canessa CM. 2002. Single channel properties of rat acid-sensitive ion channel-1alpha, -2a, and -3 expressed in *Xenopus oocytes*. *J Gen Physiol*. 120(4):553–566.
Zhang P, Sigworth FJ, Canessa CM. 2006. Gating of acid-sensitive ion channel-1: Release of Ca^{2+} block vs. allosteric mechanism. *J Gen Physiol*. 127(2):109–117.
Zhang Y, Ma C, Delohery T, Nasipak B, Foat BC, Bounoutas A, Bussemaker HJ, Kim SK, Chalfie M. 2002. Identification of genes expressed in *C. elegans* touch receptor neurons. *Nature*. 418(6895):331–335.
Ziemann AE, Allen JE, Dahdaleh NS, Drebot II, Coryell MW, Wunsch AM, Lynch CM et al. 2009. The amygdala is a chemosensor that detects carbon dioxide and acidosis to elicit fear behavior. *Cell*. 139(5):1012–1021.

26 Degenerin/ENaC channels

James D. Stockand

Contents

26.1 INTRODUCTION

Degenerins (Deg) of *Caenorhabditis elegans* are the founding members of the Deg/ENaC ion channel family with the epithelial Na^+ channel (ENaC) being the first mammalian homolog cloned (Bianchi and Driscoll, 2002; Cottrell, 1997; Garty and Palmer, 1997; Kellenberger and Schild, 2002; Lingueglia, 2007). As shown in the dendrogram in Figure 26.1, in mammals, this ion channel family includes acid-sensing ion channels (ASICs); the brain, liver, and intestine Na^+ channel (BLINaC); and ENaC. The FMRFamide-gated Na^+ channel (FaNaC) and pickpockets (Ppks) along with Deg channels are found in invertebrates. Homologs in lamprey (lASIC; Coric et al., 2005) and lungfish (nENaC; Uchiyama et al., 2012) represent important transitional precursors of mammalian ASIC and ENaC, respectively. The HyNaCs expressed in the primitive animal hydra are representative of ancestral Deg/ENaC channels that existed prior to the radiation of bilateria early in evolution (Durrnagel et al., 2010; Golubovic et al., 2007). As listed in Table 26.1, humans express nine genes that encode Deg/ENaC proteins: four encoding the α, β, γ, and δ NaC subunits, four encoding ASIC1–4 subunits, and one encoding the related BLINaC protein, which sometimes is referred to as ASIC5.

The goals here are to provide an informative discussion about the expression pattern, function, structure, and regulation of Deg/ENaC channels and the role these channels play in human physiology and pathology. In addition, the molecular evolution of the Deg/ENaC channel family is discussed briefly with emphasis on the initial appearances of the genes encoding ASIC and ENaC and how their expression preadapted the forebears of modern terrestrial vertebrates to conquer the land.

26.2 EXPRESSION AND FUNCTION OF Deg/ENaC CHANNELS

Deg/ENaC channels serve diverse functions in physiology. In some invertebrate animals as typified by the nematode *C. elegans*, some members of the Deg/ENaC channel family function as ionotropic molecular sensory transducers in peripheral sensory neurons involved in perception of the external environment

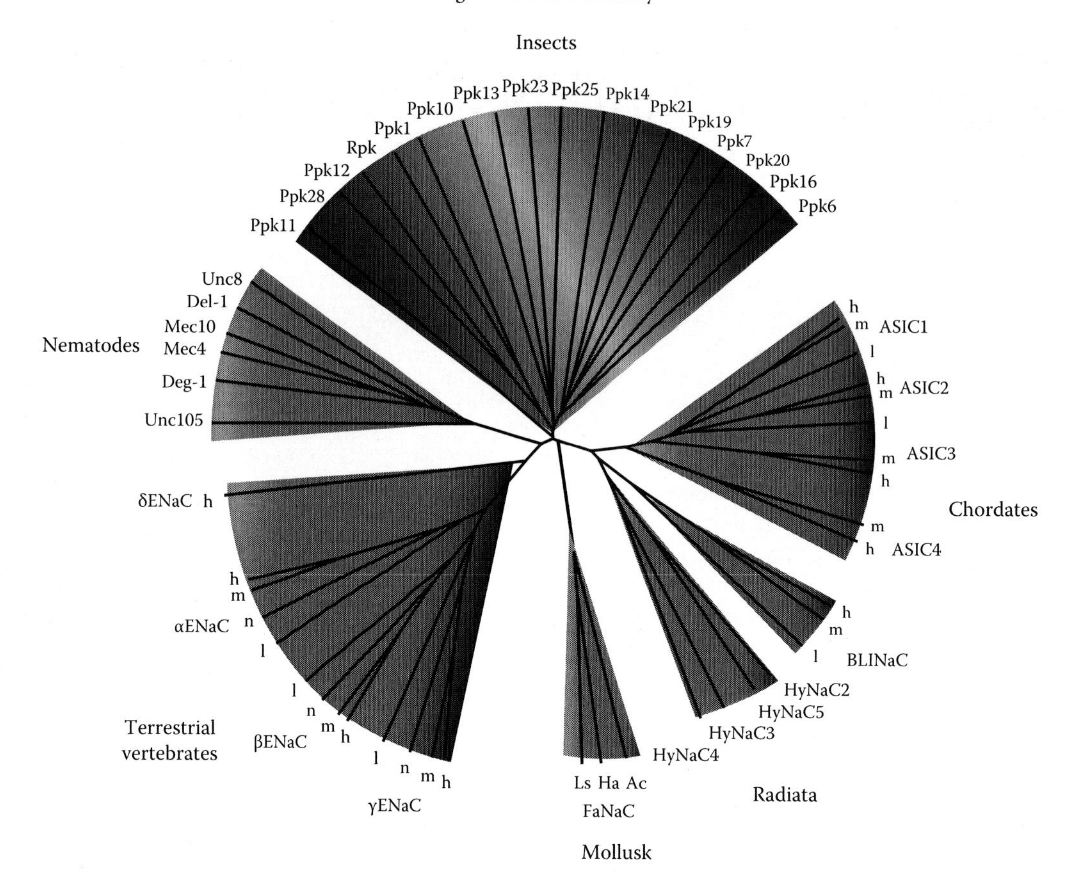

Figure 26.1 The Deg/ENaC ion channel family. In this dendrogram, lamprey, *N. forsteri*, mouse, and human are abbreviated as l, n, m, and h, respectively, and *H. aspersa*, *Lymnaea stagnalis*, and *Aplysia californica* are abbreviated as Ha, Ls, and Ac, respectively.

Table 26.1 Deg/ENaC channels expressed in humans

CHANNEL	ALTERNATIVE NAME	GENE	CHROMOSOME	LOCATION (Mb)
ASIC1	BNaC2	*ACCN2*	12	50.45–50.48
ASIC2	BNaC1, MDEG	*ACCN1*	17	31.34–32.5
ASIC3	DRASIC	*ACCN3*	7	150.75–150.75
ASIC4	BNaC4, SPASIC	*ACCN4*	2	220.38–220.4
BLINaC	hINaC, BASIC	*ACCN5*	4	156.75–156.79
αNaC	SCNEA	*SCNN1A*	12	6.46–6.49
βNaC	SCNEB	*SCNN1B*	16	23.29–23.39
γENaC	SCNEG	*SCNN1G*	16	23.19–23.23
δENaC	SCNED, dNaCH	*SCNN1D*	1	1.21–1.23

(Bianchi and Driscoll, 2002; Mano and Driscoll, 1999). In others, as typified by *Helix aspersa*, Deg/ENaC channels function as neuropeptide-gated channels in central neurons controlling excitation and synaptic transmission (Cottrell, 1997). In contrast, ENaC, which is expressed in epithelia of terrestrial vertebrates, functions during ion transport and consequently is involved in homeostatic control of fluid volumes and electrolyte content, making this channel a key end effector in the regulation of blood pressure (Garty and Palmer, 1997; Kellenberger and Schild, 2002). Mammalian ASIC, which are expressed in both peripheral and central neurons, more closely parallel the function of invertebrate Deg/ENaC channels as sensory receptors and during neurotransmission in the postsynaptic cell (Bianchi and Driscoll, 2002; Sherwood et al., 2012). Seven Deg/ENaC homologs, the invertebrate FaNaC, HyNaC, Deg and Ppk, and the mammalian ENaC, ASIC, and BLINaC are discussed here as proteins representative of the family and these various functions.

26.2.1 PHYSIOLOGICAL ROLE OF Deg/ENaC PROTEINS

Deg/ENaC proteins are ion channel subunits that contribute directly to a functional channel pore. Deg/ENaC channels preferentially conduct monovalent cations over divalent cations and anions with the majority of these channels being selective for Na^+. The degree of selectivity varies among family members. For instance, ENaC has a 100-fold preference for Na^+ over K^+ (Kellenberger and Schild, 2002). ASIC and many invertebrate Deg/ENaC channels in comparison have less preference for Na^+ (Cottrell, 1997; Kellenberger and Schild, 2002; Sherwood et al., 2012). Many of these less selective Deg/ENaC channels also conduct divalent cations. As a consequence, they contribute to the Ca^{++} permeability of the plasma membrane. Due to their selectivity, Deg/ENaC channels conduct depolarizing inward currents under physiological conditions and ionic gradients. Figure 26.2a and b shows typical macroscopic inward Na^+ currents conducted by ASIC1 activated in a rat CA1 hippocampal neuron and Ppk1 activated in a class IV multidendritic (md) peripheral sensory neuron from *Drosophila melanogaster*, both voltage clamped at typical resting membrane potentials (Boiko et al., 2013).

Deg/ENaC channels, like all channels, gate. While they share fundamental gating mechanisms as defined by structure, gating is differentially regulated among the distinct types of Deg/ENaC channels. For instance, ENaC is a noninactivating, constitutively active channel that gates independent of voltage or a ligand. In comparison, ASIC, BLINaC, FaNaC, and HyNaC are ligand-gated ion channels (Bianchi and Driscoll, 2002; Cottrell, 1997; Kellenberger and Schild, 2002; Sherwood et al., 2012). The prior is activated and subsequently desensitized by

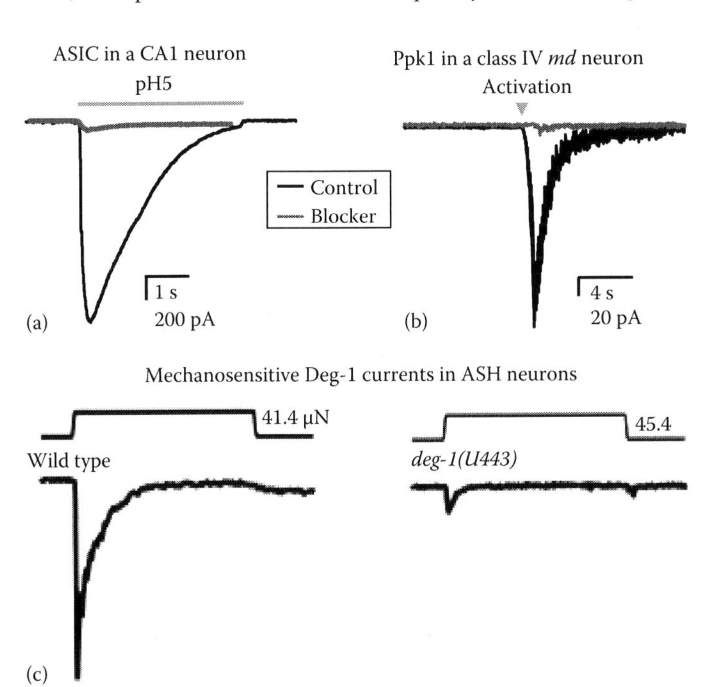

Figure 26.2 Activation of Deg/ENaC channels by stimuli. (a) Activation of the macroscopic ASIC current endogenous to rat CA1 hippocampal neurons by acidic pH. Activation in the absence and presence of the channel blocker indicated with black and gray, respectively. (Reprinted from Boiko, N. et al., *J. Biol. Chem.*, 288(13), 9418, 2013. With permission.) (b) Activation of the macroscopic Ppk1 current endogenous to class IV md neurons in *D. melanogaster.* (Reprinted from Boiko, N. et al., *J. Biol. Chem.*, 288(13), 9418, 2013. With permission.) (c) Activation by force of macroscopic Deg-1 currents endogenous to mechanosensitive ASH neurons in wild type and *C. elegans* harboring a loss of function Deg-1 mutation (*deg-1(u443)*). (Reprinted from Geffeney, S.L. et al., *Neuron*, 71, 845, 2011. With permission.)

acid with protons serving as ligands at extracellular binding sites. Bile acids function as extracellular ligands for BLINaC, and small neuropeptides, as typified by RFamides, serve as ligands for FaNaC and HyNaC. Ppk and Deg channels also gate in response to stimuli although certain of these channels respond to mechanical rather than chemical cues. For instance, as shown in Figure 26.2c, force directly activates mechanoreceptor currents mediated by Deg-1 in ASH neurons (Geffeney et al., 2011). ASH neurons are ciliated polymodal sensory neurons in *C. elegans* involved in mechanosensation.

Because they respond to stimuli, ASIC, BLINaC, Ppk, Deg, FaNaC, and HyNaC channels function as ionotropic receptors capable of affecting the electrical properties and excitability of a cell in response to an extracellular signal. They also may influence cell signaling by affecting the cell entry of the second messenger Ca^{++}. In comparison, ENaC does not respond directly to a ligand. Thus, ENaC does not function as an ionotropic receptor but rather as a regulated gateway allowing Na^+ to enter the cell in response to cell signaling. This enables ENaC to contribute to vectorial ion transport across epithelial barriers.

26.2.2 FaNaC: A PEPTIDE-GATED Na^+ CHANNEL

FaNaCs are widely expressed in ganglia neurons of Mollusca, such as the giant pedal neurons of the mollusk (Cottrell, 1997; Davey et al., 2001; Kellenberger and Schild, 2002; Lingueglia et al., 1995, 2006; Perry et al., 2001). The function of ganglia neurons in the simple rope-ladder-like nervous system of the mollusk is akin to that of central motor neurons in the more sophisticated nervous systems of complex animals. In ganglia neurons, FaNaC functions as a neuropeptide-gated channel that conducts a depolarizing inward Na^+ current upon activation. FaNaC activation is responsible for the fast excitatory action of neuropeptides in mollusks. As shown in Figure 26.3, FaNaC

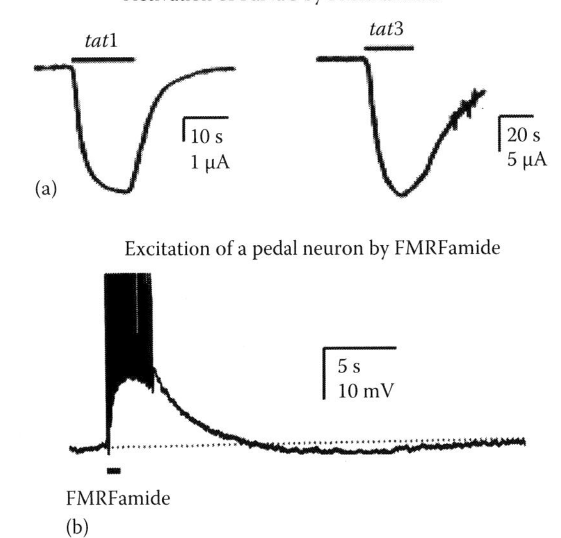

Figure 26.3 Activation of FaNaC excites neurons. (a) Activation of macroscopic FaNaC currents by FMRFamide neuropeptides *tat1* and *tat3*. (Reprinted from Cottrell, G.A. et al., *FEBS Lett.*, 489, 71, 2001. With permission.) (b) Stimulation of FaNaC with FMRFamide is able to excite a current-clamped pedal neuron to threshold, evoking a transient train of action potentials. (Reprinted from Perry, S.J. et al., *J. Neurosci.*, 21, 5559, 2001. With permission.)

rapidly activates and then desensitizes to FMRFamide (Cottrell et al., 2001). This effect of FMRFamide on FaNaC brings the neuron to threshold, evoking a transient train of action potentials (Perry et al., 2001). Such observations are consistent with FaNaC functioning as an excitatory, ionotropic receptor in the postsynaptic cell enabling peptidergic neurotransmission.

FaNaC is the first channel recognized to be peptide gated (Cottrell, 1997; Lingueglia et al., 1995). The fast excitatory response to FMRFamide as mediated by FaNaC is distinct from slower responses in these same neurons. Like the ionotropic and metabotropic glutamatergic and purinergic receptors of mammals, this ionotropic RFamide receptor is the complement to G-protein-coupled metabotropic RFamide receptors in invertebrates. Interestingly, while mammals express metabotropic receptors for small neuropeptides to include RFamides, an ionotropic counterpart has not been identified to date in these animals. Emerging evidence suggests that ASIC, BLINaC, or both perhaps serve such a function in mammals (Kellenberger and Schild, 2002; Lingueglia et al., 2006).

26.2.3 HyNaC: THE ANCESTRAL Deg/ENaC CHANNEL

HyNaC as expressed in *Hydra magnipapillata* is the closest to a primordial Deg/ENaC channel for which we currently have amino acid sequence and information about function (Durrnagel et al., 2010; Golubovic et al., 2007). Similar to FaNaC, HyNaC is directly gated by neuropeptides, in this case Hydra-RFamides I and II, and conducts a depolarizing inward Na^+ current upon activation as shown in Figure 26.4. HyNaCs are expressed in cells localized to the base of tentacles. These cells are likely to be epitheliomuscular cells and are adjacent to neurons that produce Hydra-RFamides, consistent with these neuropeptides being the natural ligands for these channels (Durrnagel et al., 2010; Golubovic et al., 2007). This organization is consistent with HyNaC functioning as a postsynaptic ionotropic receptor involved in fast peptidergic neurotransmission, possibly directing tentacle contractions during feeding.

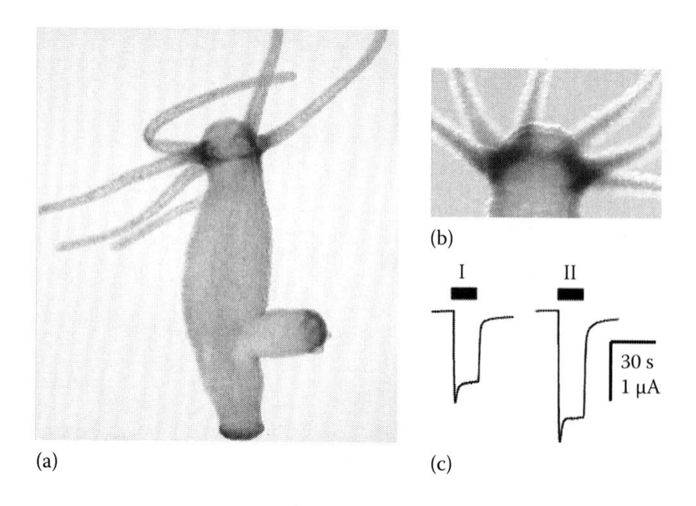

Figure 26.4 Stimulation of HyNaC by RFamides. (a and b) Expression of HyNaC in cells at the base of tentacles in hydra. (c) Activation of macroscopic HyNaC currents by RFamide I and II. (Reprinted from Golubovic, A. et al., *J. Biol. Chem.*, 282, 35098, 2007. With permission.)

The appearance of genes encoding these early Deg/ENaC channels in primitive animals parallels the appearance of the first nervous systems evolving in cnidarians, including hydra. That these ancient nervous systems extensively used peptides as neurotransmitters fits with Deg/ENaC channels serving a key role in the origins of synaptic transmission. Moreover, that at least three distinct HyNaC genes and one pseudogene are present in an organism in which the nervous system first evolved suggests that the expression of Deg/ENaC channels contributed to preadaptations that enabled the emergence of these early nervous systems (Durrnagel et al., 2010). The presence of peptide-gated Deg/ENaC channels in the primitive nervous systems of Cnidaria and Mollusca, in addition, indicates that peptide control of gating is an ancient feature of this channel family that has been preserved during evolution, particularly in Protostomia. It currently is unclear whether this feature is also conserved in Deuterostomia or whether neuropeptides have been completely replaced by other ligands in more complex animals.

26.2.4 Deg: MECHANOSENSITIVE CHANNELS OF THE NEMATODE

C. elegans express at least six different Deg/ENaC proteins involved in mechanosensation and possibly up to 21 Deg/ENaC homologs all together (Bianchi and Driscoll, 2002; Kellenberger and Schild, 2002; Mano and Driscoll, 1999). Those involved in mechanosensation, to include Mec-4 and Mec-10, form ion channels that are directly gated by mechanical force (Arnadottir et al., 2011; Driscoll and Chalfie, 1991; Hong and Driscoll, 1994; Hong et al., 2000; O'Hagan et al., 2005). Degs originally were identified because mutations in these proteins, for instance, the *mec-4(d)* mutant as shown in Figure 26.5a, caused degeneration of specific mechanosensitive neurons (Blum et al., 2008; Hong and Driscoll, 1994; Hong et al., 2000). This class of mutation resulted from the gain of function where hyperactivation of Degs

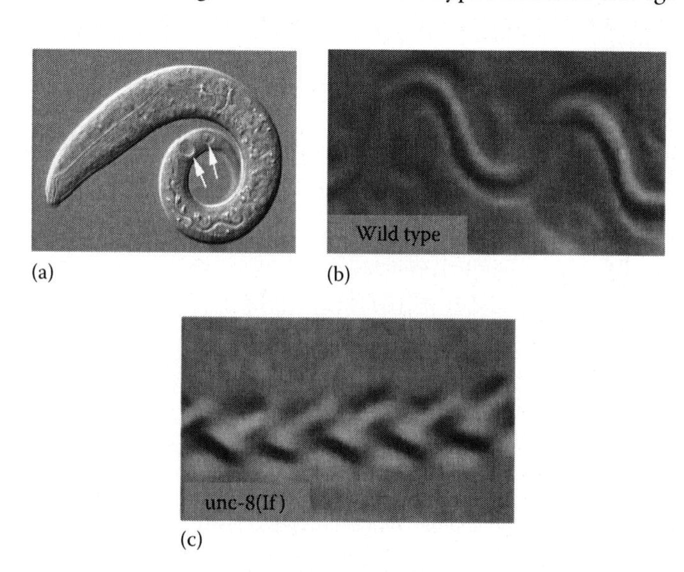

Figure 26.5 Degs are involved in mechanosensation. (a) Picture of a *C. elegans* that expresses a Deg mutation that causes swelling of neurons (noted by white arrows). (Reprinted from Blum, E.S. et al., *Cell Death Differ.*, 15, 1124, 2008. With permission.) (b) Crawling pattern of wild type and (c) *C. elegans* harboring the unc-8(f) loss of function mutation. (Reprinted from Tavernarakis, N. et al., *Neuron*, 18, 107, 1997. With permission.)

caused a constant influx of cations, leading to inappropriate membrane depolarization and ultimately necrotic cell death. In many respects, the cell death caused by hyperactivation of Degs mirrors the excitotoxic cell death that occurs in neurons of higher organisms in response to injury such as that in the brain during stroke (Bianchi and Driscoll, 2002; Sherwood et al., 2012).

Activation of Degs leads to an excitatory, depolarizing inward Na^+ current (see Figure 26.2c; Geffeney et al., 2011). Consequently, the Degs expressed in touch receptors of *C. elegans* function as molecular mechanoelectrical transduction machines capable of transforming force into a bioelectrical signal. Similarly, Degs expressed in the motor neurons that control nematode locomotion, such as Unc-8, also respond to mechanical cues. In this instance, a channel formed of Unc-8 and Del-1, which is expressed in specialized synapse-free processes, is believed to be activated by stretch (Mano and Driscoll, 1999; Tavernarakis et al., 1997). This is thought to contribute to nematode proprioception by providing feedback on body posture to fine-tune motor neuron activity. Consistent with this, disruption of Unc-8, as shown in Figure 26.5b and c, disrupts the normal sinusoidal body wave of the moving worm (Tavernarakis et al., 1997).

Yet another Deg, Unc-105, possibly activated by mechanical cues, also is thought to contribute to proprioception by monitoring muscle stretch (Kellenberger and Schild, 2002). In contrast to Unc-8 and Del-1 expressed in neurons, Unc-105 is expressed in muscle. Gain-of-function mutations in Unc-105 cause muscle hypercontraction because muscle cells are inappropriately depolarized by a sustained cation influx conducted by the activated channel.

Degs, such as Deg-1, which are directly activated by a mechanical stimulus (see Figure 26.2c), also are critical to nocifensive responses in the nematode (Geffeney et al., 2011). In this role, these nociceptors function as molecular ionotropic mechanoelectrical transducers capable of transforming applied force into a bioelectrical signal that influences sensory neuron excitation to ultimately provoke appropriate behavioral responses.

Although there is abundant experimental evidence documenting the function of Degs in mechanosensory transduction in *C. elegans*, it is less clear if their mammalian orthologs respond to identical modalities. Because of the potential importance of this possibility to how mammals may perceive touch, sound, and pain, this area is the focus of many contemporary studies.

26.2.5 Ppk: Deg/ENaC CHANNELS OF INSECTS

Insects express the largest number of distinct Deg/ENaC genes. For instance, *D. melanogaster* expresses at least 16 and possibly up to 30 different Deg/ENaC genes referred to as Ppk (Adams et al., 1998; Bianchi and Driscoll, 2002; Darboux et al., 1998; Kellenberger and Schild, 2002; Liu et al., 2003a; Mano and Driscoll, 1999). Correspondingly, the products of these genes have diverse expression profiles, activating stimuli and physiological functions to include involvement in mechanosensory transduction and control of locomotion, liquid clearance from the trachea, detection of pheromones and courtship behavior, egg-laying, and salt taste (Boiko et al., 2012; Liu et al., 2003b; Rezaval et al., 2012; Thistle et al., 2012; Zhong et al., 2010). The Ppk channels involved in sensory transduction are expressed in peripheral sensory neurons. The function of Ppk channels expressed in central neurons is obscure. The Ppk channels involved in transport are expressed in epithelial cells.

Akin to the function of ENaC in mammalian airways, Ppk4 and Ppk11 are necessary for liquid clearance from the trachea of the fly (Liu et al., 2003a). The proper clearance of liquid in airways is as critical to air breathing in insects as it is to mammals. This function of Ppk4 and Ppk11 is consistent with these channels contributing to Na^+ transport because such reabsorption provides the osmotic draw pulling water from the lumen of the trachea. As discussed in more detail later, involvement of Deg/ENaC channels in epithelial transport is retained in mammals, suggesting that this is a conserved function of the family.

The ability to detect salt is critical to the survival of terrestrial animals, including insects. In the fly, Ppk11 and Ppk19 are expressed in larval taste-sensing terminal organs and in adult taste bristles of the labelum, legs, and wing margins (Liu et al., 2003b). Disrupting these Ppk proteins decreases the ability of larvae to detect low concentrations of Na^+ and attenuates the electrophysiological response to small amounts of salt. Moreover, disrupting Ppk11 and Ppk19 changes the behavior of both larvae and adults relative to salt. Such findings argue that these Ppk proteins function as critical gateways facilitating cell entry of Na^+ into peripheral taste receptors. Activation of Ppk channels in taste receptors then is positioned to convey information about the presence of salt in the environment, ultimately driving appropriate behavioral responses.

Ppk channels, in addition, play a key role in reproduction in the fly. In females, Ppk channels are expressed in neurons that are part of a critical circuit that mediates postmating responses and egg-laying (Rezaval et al., 2012). In males, Ppk23 and Ppk29 are expressed in sensory neurons that influence through central circuits courtship behavior in response to sex-specific sensory cues (Thistle et al., 2012). Activation of these Ppk channels by pheromones in males is necessary and sufficient to promote courtship toward females. In this regard, Ppk23 and Ppk29 function as ionotropic sensory receptors responsive to chemical stimuli. It is reasonable to suggest that activation of these Ppk receptors excites sensory neurons to convey information about the presence of receptive females.

Ppk channels are also critical to normal fly locomotion and responses to noxious stimuli (Ainsley et al., 2003; Boiko et al., 2012; Zhong et al., 2010). Ppk1 is restrictively expressed in class IV md neurons of the fly. These are polymodal peripheral sensory neurons that form extensive dendritic networks that ramify beneath the epidermis and are required for normal proprioception and nociception. Class IV md neurons have much in common with polymodal neurons involved in sensing touch and noxious stimuli in the mammalian peripheral nervous system. As such, they have been used as a model to investigate mechanosensation. In class IV md neurons, Ppk1 conducts a transient depolarizing inward Na^+ current. As shown in Figure 26.6, targeted stimulation of Ppk1 excites class IV md neurons bringing them to threshold, evoking a transient train of action potentials as defined by the gating pattern of the channel (Boiko et al., 2012). Thus, Ppk1 functions in class IV md neurons as an ionotropic molecular sensory transducer capable of sensing and transforming a stimulus into a change in neural activity. Details about the specific physiological stimuli

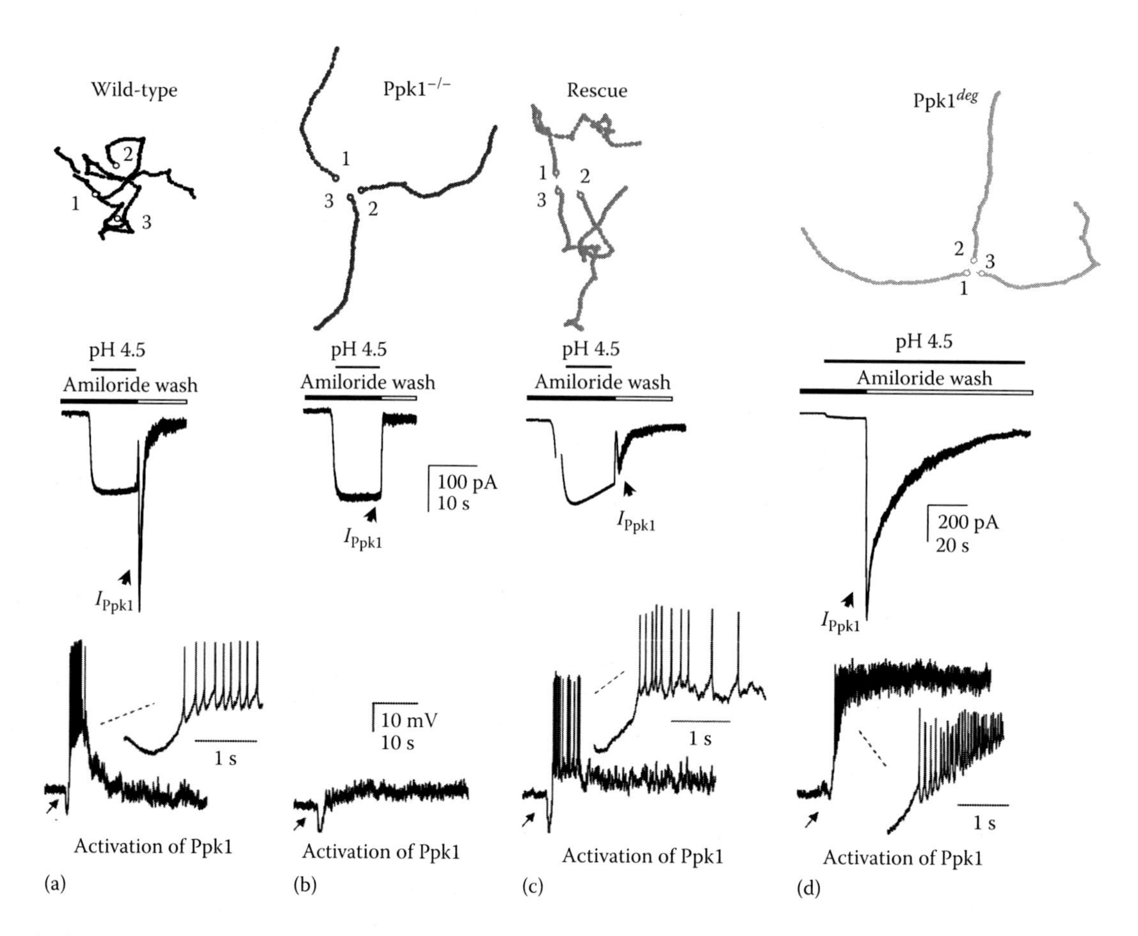

Figure 26.6 Ppk1 functions as an ionotropic molecular sensory transducer. (Top) Crawling pattern of three representative larvae, (middle) typical macroscopic Ppk1 currents in voltage-clamped md neurons, and (bottom) action potentials evoked by targeted activation of Ppk1 in current-clamped md neurons in (a) wild type, (b) $Ppk^{-/-}$, and (c) rescued flies and for (d) flies harboring the *Deg* gain-of-function mutation in Ppk1. (Reprinted from Boiko, N. et al., *J. Biol. Chem.*, 287, 39878, 2012. With permission.)

that activate Ppk1 currently are obscure, but indirect evidence suggests that akin to certain Degs, Ppk1 may be activated directly by mechanical cues. In any case, neural output from class IV md neurons as generated by activation of Ppk1 ultimately influences locomotive behavior. Cellular ablation of class IV md neurons like disruption of Ppk1 expression and function abrogates normal wandering of larvae in the foraging stage. Appropriate larval wandering is critical to normal food-seeking behavior. It is interesting that loss of function of Ppk1 phenocopies gain of function of this channel at the level of behavior (Boiko et al., 2012). This parallels the effects of loss and gain of function of Deg channels in mechanosensitive neurons in *C. elegans* (Gonzales et al., 2009; Mano and Driscoll, 1999). Loss of Ppk1 function abolishes the ability to sense a stimulus and respond appropriately. Gain of Ppk1 function abolishes the ability to transform the sensing of a stimulus properly and thus also abrogates an appropriate cellular response. Such findings demonstrate that appropriately graded Ppk1 activation is critical to proper signaling from mechanosensitive neurons of the fly. Perhaps mammalian orthologs serve a similar function. This, though, remains controversial.

26.2.6 ASIC: ACID-SENSING Na^+ CHANNELS EXPRESSED IN MAMMALIAN NEURONS

ASICs are also prominent members of the Deg/ENaC family. Although discussed here in brief, ASICs are also the focus of Chapter 25. Mammalian ASICs are widely expressed in neurons of both the central and peripheral nervous systems (Gonzales et al., 2009; Kellenberger and Schild, 2002; Lingueglia, 2007; Mano and Driscoll, 1999; Sherwood et al., 2012). Mammals express four ASIC genes with at least three of them, ASIC1–3, giving rise to several splice variants often referred to as a and b isoforms. ASIC1a, ASIC2a, and ASIC2b are richly expressed in the brain, chiefly in the hippocampus, cerebellum, cerebral cortex, striatum, and amygdala. They also are expressed in the spinal cord. The pituitary has strong ASIC4 expression. ASIC1a, ASIC2a, and ASIC3 are expressed in peripheral neurons with ASIC3 expressed primarily in small-diameter polymodal sensory neurons involved in sensation of pain and ASIC2a in medium- and large-diameter mechanosensory neurons, such as those in the dorsal root ganglia.

As their name implies, ASICs are activated by decreases in extracellular pH (see Figure 26.2a). Although it is clear that protons function as ligands that bind and activate ASIC, it is unclear if they are the only or even the primary physiological stimulus for every ASIC isoform. Active ASICs conduct excitatory, inward cation currents. Similar to FaNaC in postsynaptic cells and Ppk1 in sensory neurons, the magnitude of the depolarizing current conducted by ASIC is sufficient to evoke action potentials in mammalian neurons (Bianchi and Driscoll, 2002; Kellenberger and Schild, 2002; Lingueglia, 2007; Lingueglia et al., 2006; Sherwood et al., 2012).

The possibility that ASICs are involved in mechanosensation in mammals has generated much controversy. Currently, such a role is supported primarily by indirect and subtle evidence (Bianchi and Driscoll, 2002; Borzan et al., 2010; Lingueglia, 2007; Lingueglia et al., 2006; Lu et al., 2009; Staniland and McMahon, 2009). This hypothesis, however, is attractive and persists because of parallels with the function of Deg and Ppk homologs in invertebrates. The most compelling support for such a conserved function in touch and mechanosensation comes from the analysis of skin mechanoreceptor responses in ASIC2 and ASIC3 knockout mice. Moreover, ASIC2a expression in mechanosensory neurons is localized to specialized cutaneous nerve termini involved in sensing mechanical cues. Deletion of ASIC2 disrupts responses in two types of low-threshold, light touch-sensitive mechanosensitive fibers (Bianchi and Driscoll, 2002; Price et al., 2000). In ASIC2 knockout mice, rapidly adapting (RA) and to a lesser extent slowly adapting (SA) fibers are defective in increasing firing frequency in response to stronger displacement force stimuli. The threshold sensitivity and response frequency of AM fiber mechanonociceptors, which respond to high-threshold stimuli, including pinching, are significantly reduced in ASIC3 knockout mice (Price et al., 2001).

Although exciting, these two knockout models thus far have provided only subtle support for a conserved role for mammalian Deg/ENaC channels in mechanosensation because deletion of ASIC protein failed to eliminate completely either responses to light or harsh touch or responses to noxious mechanical stimuli. For instance, while defective, RA receptors in ASIC2 null mice maintain threshold sensitivity and other mechanoreceptors that express this channel appear normal (Bianchi and Driscoll, 2002; Price et al., 2000; Roza et al., 2004). Although such results could be explained by redundancy or overlapping function or that ASICs are modulatory but not necessary, additional research is needed before definitive conclusions can be made in this regard.

Expression of ASICs in neurons involved in sensing pain and activation by acidic pH is consistent with them playing a role in pain perception, particularly in response to tissue acidosis such as that arising from ischemia and inflammation (Bianchi and Driscoll, 2002; Chu and Xiong, 2013; Kellenberger and Schild, 2002; Leng and Xiong, 2012; Lingueglia, 2007; Sherwood et al., 2012). For instance, ASIC3 is expressed in sympathetic cardiac afferents, where compelling evidence supports their involvement in sensing nonadapting ischemic pain caused by acidosis (Chu and Xiong, 2013; Kellenberger and Schild, 2002; Leng and Xiong, 2012; Xiong et al., 2008). During cardiac ischemia, extracellular pH within affected areas decreases to values capable of activating ASIC3. It is reasonable to suggest that stimulation of ASIC3 excites these sensory neurons, evoking a train of action potentials that conveys the sensation of cardiac pain. Consistent with such a role, amiloride, the prototypical inhibitor of Deg/ENaC channels, has analgesic effects in a variety of animal pain models and inhibits activation of slowly conducting sensory fibers (C-type) by acid in the rat (Chu and Xiong, 2013; Kellenberger and Schild, 2002; Leng and Xiong, 2012; Staniland and McMahon, 2009; Xiong et al., 2008). Similarly, several different peptide toxins that inhibit ASICs, including venom from the black mamba and psalmotoxin 1 from the tarantula, have strong analgesic action (Diochot et al., 2012; Leng and Xiong, 2012; Mazzuca et al., 2007; Sherwood et al., 2012). Psalmotoxin 1 in particular has very potent analgesic properties against mechanical, chemical, inflammatory, and neuropathic pain in rodents. It is believed that these actions are mediated at least in part by inhibition of ASIC1a in nociceptors. Other peptidergic toxins, such as APETx2 from the sea anemone, inhibit ASIC3 and mitigate pain (Kellenberger and Schild, 2002; Lingueglia, 2007). Conversely, the coral snake toxin, MitTx, activates ASICs that induce pain (Bohlen et al., 2011). Thus, ASICs expressed in sensory neurons are targets for peptide toxins and other agents that inhibit or provoke pain. This is consistent with ASICs functioning as ionotropic molecular sensory receptors capable of exciting peripheral neurons involved in pain sensation.

Several of the toxins that inhibit ASIC and work as potent analgesics when applied locally in the PNS, including psalmotoxin 1, also mitigate pain when applied to the central nervous system (CNS) (Bianchi and Driscoll, 2002; Kellenberger and Schild, 2002; Sherwood et al., 2012). This suggests that ASICs also have a role in synaptic transmission as mediated by central neurons. Perhaps, they act as excitatory ionotropic receptors in postsynaptic neurons because they conduct depolarizing currents upon activation. Particulars about the degree they do this and in which neurons are currently obscure. Moreover, the physiological stimulus that would activate ASICs in this regard is unclear. Perhaps, like FaNaC and HyNaC, they respond to small neuropeptides. This, however, is speculation at this time. Alternatively, synaptic vesicles are acidic, and a transient drop in extracellular pH is associated with synaptic transmission (Bianchi and Driscoll, 2002). Perhaps, release of a neurotransmitter acidifies the synaptic cleft enough to activate postsynaptic ASICs that contribute to or modulate transmission in some regard. Although speculative, this is in agreement with the postsynaptic localization of these channels. Moreover, in some but not all studies, ASIC1 knockout mice have aberrant long-term potentiation as hippocampal neurons in null mice are defective in their response to high-frequency stimulation. This defect correlates with specific learning deficits in mutant mice (Gloor et al., 1993; Wemmie et al., 2002; Wu et al., 2013). Together, the bulk of this evidence is consistent with ASIC1 functioning at synapses in the CNS.

26.2.7 ENaC: THE MAMMALIAN EPITHELIAL Na^+ CHANNEL INVOLVED IN TRANSPORT

ENaC is expressed in the apical membrane of polarized epithelial cells involved in Na^+ transport (Canessa et al., 1993, 1994; Lingueglia et al., 1993). This includes epithelial cells lining the gastrointestinal tract and renal tubule as well as the lungs and airways to name a few (Garty and Palmer, 1997; Kellenberger and Schild, 2002). Figure 26.7a shows the polarized expression pattern of ENaC in the apical membranes of principal cells of the distal renal nephron (Mironova et al., 2012).

Sodium is maintained in disequilibrium across the plasma membranes of cells by the constant activity of the Na^+/K^+-ATPase pump combined with the barrier properties inherent to lipid bilayers. Such disequilibrium makes Na^+ a functional osmolyte capable of influencing the movement of water via osmosis. Activation of ENaC decreases the resistance of the apical membrane of an epithelial cell to Na^+, allowing this ion to

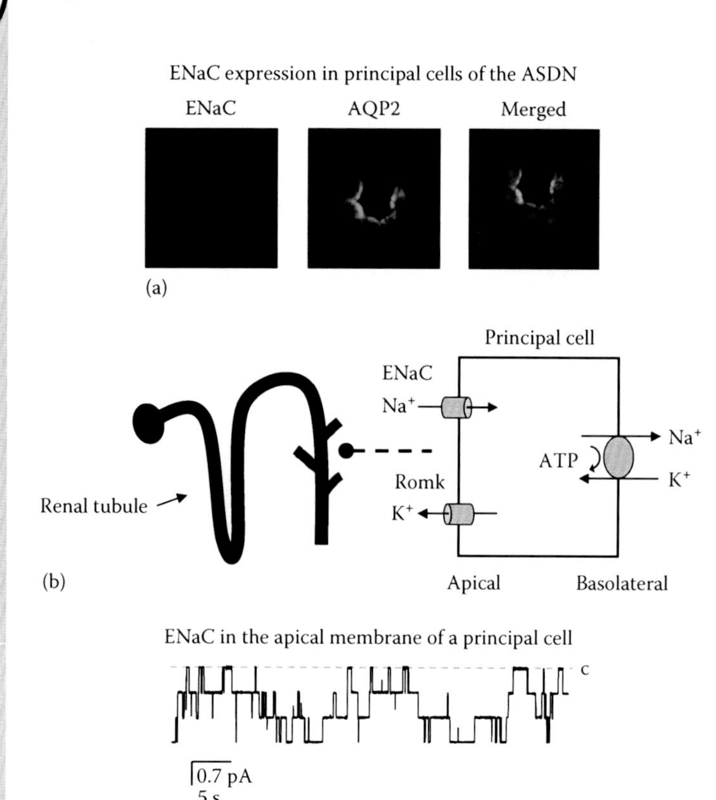

Figure 26.7 ENaC allows Na^+ transport in epithelial cells. (a) Fluorescence micrograph of ENaC expression (red) in aquaporin 2 (AQP2, green) positive cells (principal cells) of the ASDN. Nuclei labeled blue. (Reprinted from Mironova, E. et al., *Proc. Natl. Acad. Sci. USA*, 109, 10095, 2012. With permission.) (b) A drawing of a principal cell in the ASDN of the renal tubule with ion channels and transporters labeled. (c) A representative current trace of ENaC in a cell-attached patch formed on the apical membrane of a principal cell in an isolated, split-open murine ASDN.

enter the cell down its electrochemical gradient. This results in depolarization and an increase in the intracellular concentration of Na^+. In polarized epithelial cells, basolateral Na^+/K^+-ATPases then rapidly pump Na^+ out of the cell, extruding this ion across the membrane opposite from where it entered. Consequently, as depicted in Figure 26.7b, ENaC is an integral component of the transcellular pathway enabling vectorial transport (absorption) of Na^+ across polarized epithelial cells (Bonny and Hummler, 2000; Garty and Palmer, 1997; Hummler and Horisberger, 1999; Kellenberger and Schild, 2002; Rossier et al., 2002). Water follows via osmosis. Because ENaC is the major cell-entry pathway for Na^+ across the apical membrane in epithelial cells that express this channel, its activity is limiting for Na^+ and coupled water absorption across these cells.

As limiting for vectorial Na^+ absorption, ENaC plays a key role in setting the volume and electrolyte content of fluid compartments. This allows ENaC to serve several important physiological functions. In airways, it influences mucus hydration and consequently viscosity (Mall et al., 2004; Rauh et al., 2013). Proper fluidity of airway mucus is critical to the removal of inhaled foreign substances via mucociliary clearance. Deeper in the lungs, activation of ENaC facilitates water reabsorption, dehydrating alveolar spaces to decrease compliance (Bonny and Hummler, 2000; Hummler and Horisberger, 1999; Rossier et al., 2002). This is critical to the mechanics of air breathing. Thus, ENaC provides critical service in the lungs that allows air breathing. This in part enables terrestrial vertebrates to live out of water and inhabit the land.

In the gut and kidneys, ENaC enables (re)absorption of Na^+ and water into the body (Bonny and Hummler, 2000; Garty and Palmer, 1997; Kellenberger and Schild, 2002; Rossier et al., 2002). ENaC is also expressed in taste receptor cells of the tongue that ultimately influence NaCl-seeking behavior (Chandrashekar et al., 2010). Thus, in a broad sense, ENaC functions to bring in and retain sodium and water within an animal. This ENaC-dependent water conservation offers freedom from an immediate source of water and as such allows terrestrial life.

The Na^+ reabsorbed through ENaC in the kidneys ultimately influences the distribution of water within body fluid compartments and provides a mechanism for controlling blood pressure in mammals. The circulatory systems of higher animals are closed, meaning that vascular volume influences pressure in these systems. Na^+ and its conjugated bases, Cl^- and HCO_3^-, are the primary extracellular osmolytes in animals. Thus, terrestrial vertebrates control extracellular fluid volume, including plasma, by controlling serum Na^+. As is clear when considering the action of diuretics that suppress tubular Na^+ reabsorption, including those like amiloride that target ENaC, serum Na^+ and thus blood pressure are controlled in part by regulation of renal Na^+ excretion. Renal sodium excretion is fine-tuned in the aldosterone-sensitive distal nephron (ASDN). ENaC is expressed in the apical plasma membrane of principal cells of the ASDN (Bonny and Hummler, 2000; Garty and Palmer, 1997; Hummler and Horisberger, 1999; Kellenberger and Schild, 2002; Lifton et al., 2001). Here, ENaC serves as the primary cell-entry pathway for Na^+ reabsorption from the urine back into interstitial fluid and blood.

The activity of ENaC also influences the movement of Cl^- and K^+ through their respective channels and transporters in epithelial cells. This is a secondary effect stemming from the depolarizing actions of ENaC. Activation of ENaC increases the electrochemical forces driving K^+ efflux out of a cell. If an epithelial cell expresses active K^+ channels in its apical membrane along with ENaC, then activation of the latter channel will promote K^+ secretion from this cell. This explains in part why K^+ secretion is tied to Na^+ reabsorption in the distal nephron. Similarly, activation of ENaC decreases the electrochemical forces driving Cl^- from a cell. If an epithelial cell expresses a Cl^- channel along with ENaC in its apical membrane, then activation of ENaC suppresses secretion of Cl^-, and, conversely, activation of the Cl^- channel suppresses Na^+ absorption mediated by ENaC. This relation between Cl^- secretion and Na^+ (re)absorption also ultimately determines whether a specific epithelial tissue secretes or (re)absorbs water.

26.2.8 BLINaC: THE MAMMALIAN Deg/ENaC CHANNEL EXPRESSED IN BOTH NEURONS AND EPITHELIAL CELLS

The expression of BLINaC is more restrictive than ASIC and ENaC. As shown in Figure 26.8, BLINaC is expressed primarily in epithelial cells of the bile duct and a subset of interneurons in the granular layer of lobules IX and X of the cerebellum

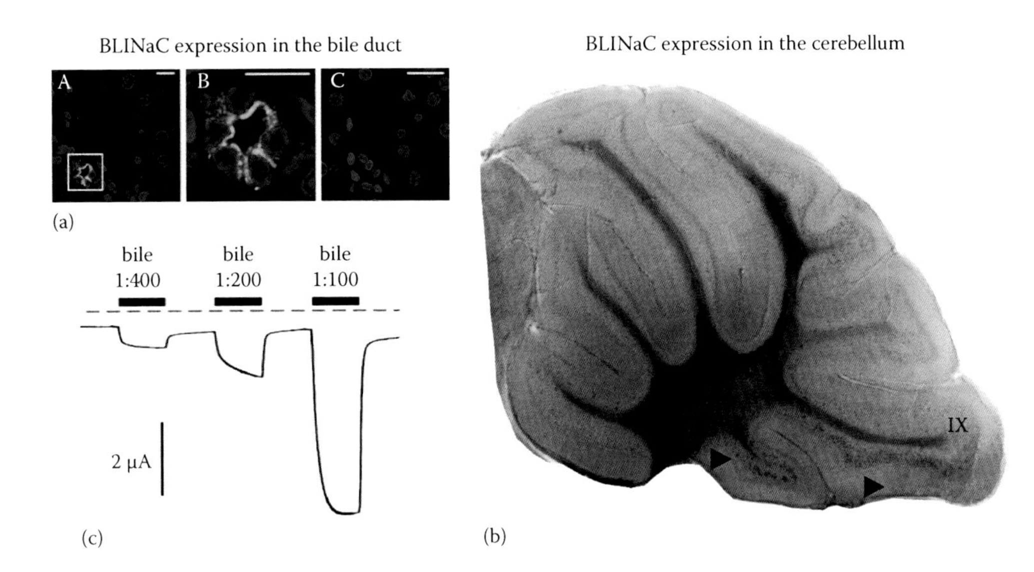

Figure 26.8 BLINaC is expressed in bile duct epithelial cells and neurons of the cerebellum. (a) Fluorescence micrograph of BLINaC expression in bile duct cells of the liver. (Reprinted from Wiemuth, D. et al., *FASEB J.*, 26, 4122, 2012. With permission.) (b) LacZ staining showing restrictive expression of BLINaC in interneurons of lobules IX and X of the cerebellum. Unpublished data from an ACCN5 null mouse, where exon 3 is disrupted with a neo-LacZ cassette created in the Stockand laboratory. (c) Activation of macroscopic BLINaC currents by bile acid. (Reprinted from Wiemuth, D. et al., *FASEB J.*, 26, 4122, 2012. With permission.)

(Wiemuth et al., 2012). BLINaC also is expressed in an as yet undefined cell type in the small intestine (Sakai et al., 1999; Schaefer et al., 2000).

The function of BLINaC in epithelia is unknown. However, inferences can be made considering its expression pattern and protein function. As a bile acid–gated Na^+ channel expressed in the apical membrane of cholangiocytes, BLINaC likely functions in these cells in a manner similar to the function of ENaC in renal and pulmonary epithelial cells (Wiemuth et al., 2012, 2013). BLINaC likely is a regulated gateway controlling cell entry of Na^+ from ductal fluid to influence the electrolyte content and volume of biliary secretions tuning these secretions to bile production.

The function of BLINaC in cerebellar interneurons also is unknown at this time. As a ligand-gated depolarizing ion channel (Sakai et al., 1999; Schaefer et al., 2000; Wiemuth and Grunder, 2011; Wiemuth et al., 2012), BLINaC is positioned to influence the excitability of these neurons perhaps in response to neuropeptides. Although not definitive, that the open probability of BLINaC approaches zero in the absence of a ligand is most consistent with this channel serving some role in the development or modulation of the action potential in the postsynaptic neuron in response to a specific stimulus.

26.3 EVOLUTION OF Deg/ENaC CHANNELS

Deg/ENaC channels represent an ancient ion channel family expressed by most if not all extant metazoan species (Golubovic et al., 2007). As shown in Figure 26.9, the genes encoding Deg/ENaC proteins are of primordial origin appearing first >600–700 million years ago, prior to the Radiata-Bilateria dichotomy. The rise of Deg/ENaC genes paralleled the emergences of neurons in the decentralized nerve net of primitive animals. The antiquity of this ion channel family is emphasized by the recent cloning of Deg/ENaC genes from the freshwater cnidaria *H. magnipapillata*, which has radial symmetry and a primitive nervous system (Durrnagel et al., 2010; Golubovic et al., 2007). Although the exact period and specific species in which Deg/ENaC channels first arose are obscure, the genes encoding these channels likely emerged subsequent to the appearance of multicellular animals because genes encoding Deg/ENaC proteins have not been identified to date in the genomes of bacteria, archaea, yeast, or any other unicellular eukaryotes.

Ancestral Deg/ENaC channels as represented by HyNaC served as rapid response ionotropic receptors for neuropeptides. These receptors complemented the function of more common but slower responding metabotropic neuropeptide receptors in the nervous systems of primitive animals. By the time bilaterian animals emerged, Deg/ENaC channels had diversified. Deg/ENaC channels began appearing in epithelial cells as well as neurons with corresponding expansion of channel function. As represented by FaNaC of *H. aspersa*, some family members retained their ability to sense neuropeptides and function as ionotropic RFamide receptors (Cottrell, 1997). Others, such as those appearing in ecdysozoans, evolved to respond to different stimuli enabling divergent cellular functions. For instance, certain Degs in *C. elegans* and Ppk channels in *D. melanogaster* specialized into ionotropic molecular sensory transducers in peripheral sensory neurons involved in perception of the external environment, while others specialized to function during epithelial transport and control of fluid volume and composition (Bianchi and Driscoll, 2002; Mano and Driscoll, 1999).

26.3.1 EMERGENCE OF ASIC

It is uncertain whether the complement of Deg/ENaC channels expressed in mammals retains the full scope of function and ligand sensitivity of their predecessors. For instance, none of the mammalian Deg/ENaC channels is as yet recognized to be sensitive to neuropeptides or definitively shown to be sensitive to

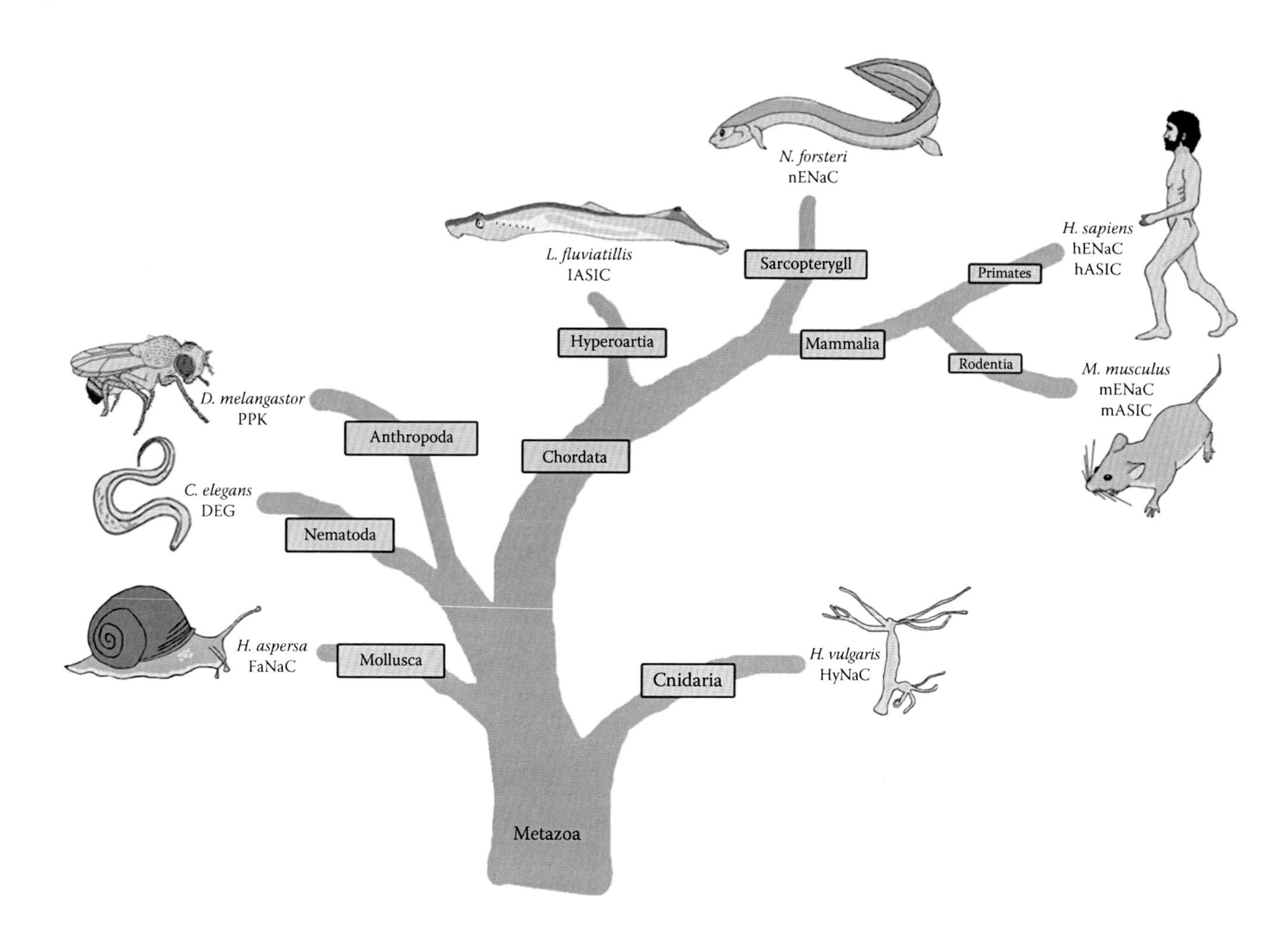

Figure 26.9 Tree of life showing the evolution of the Deg/ENaC ion channel family.

mechanical stimuli. Nevertheless, by the emergence of Chordates, in particular vertebrate fish, including the lamprey, a specialized branch of Deg/ENaC channels sensitive to extracellular pH had appeared (Coric et al., 2005). Such ASICs are common to the central and peripheral nervous systems of all extant vertebrate animals and play key roles in neuron function to include synaptic transmission and sensory perception (Sherwood et al., 2012).

26.3.2 EMERGENCE OF ENaC

ENaC is restrictively expressed in epithelial cells of all air-breathing sarcopterygians and terrestrial vertebrates (Cottrell, 1997; Garty and Palmer, 1997; Uchiyama et al., 2012). This specialized Deg/ENaC channel plays a critical role in Na^+ transport across epithelial barriers and thus influences hydration of mucus membranes and systemic electrolyte and water homeostasis. The genes encoding ENaC likely initially arose in the lobe-finned fish that served as the common ancestor of modern lungfish and terrestrial vertebrates. The sequences of the genes encoding ENaC proteins in the modern Dipnoi lungfish, *Neoceratodus forsteri*, and the Deg/ENaC proteins of living Coelacanth and Hyperoartia fish are the closest in existence to the ancestral ENaC progenitor (Uchiyama et al., 2012). Although the appearance of the genes encoding ENaC proteins is firmly established in the timeline of evolution, questions remain about whether the rise of these genes enabled animals to face better the selection pressures arising from the need for water conservation or air breathing using lungs necessary for terrestrial living. Regardless of which selection pressure was initially preeminent, proper expression and function of ENaC allows modern mammals to live on land. This is highlighted by the facts that loss of ENaC function in the lungs causes wet lungs, respiratory distress, and consequent laborious breathing in neonatal animals, which ultimately leads to their death, and loss of ENaC function in the kidneys causes inappropriate renal sodium and water wasting and dehydration that also lead to neonatal death (Bonny and Hummler, 2000; Hummler and Horisberger, 1999; Rossier et al., 2002).

26.4 STRUCTURE OF Deg/ENaC CHANNELS

The structure of Deg/ENaC channels is discussed here in brief. This also is covered in Chapter 57. The seminal work in which the structure of chicken ASIC1 recently was resolved at the atomic level has greatly informed understanding of this family of proteins (Jasti et al., 2007). Deg/ENaC channels are obligatory trimers where each of the three component subunits contributes to a central pore (Jasti et al., 2007; Staruschenko et al., 2005). Certain family members form obligatory heterotrimers composed of three related but distinct subunits as typified by ENaC, which comprises one α, one β, and one γ subunit (Canessa et al., 1994; Staruschenko et al., 2005). In some tissues, the α subunit is

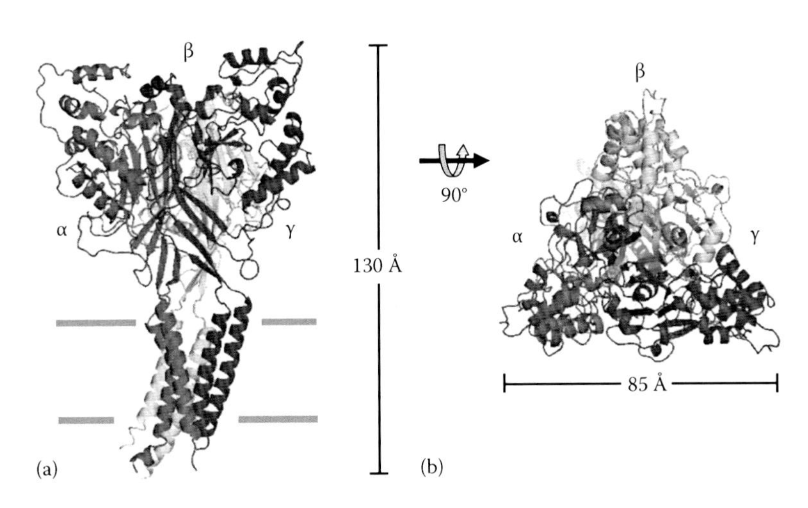

Figure 26.10 The predicted structure (a) perpendicular to the plasma membrane and (b) parallel to the plasma membrane of heterotrimeric ENaC. (Reprinted from Stockand, J.D. et al., *IUBMB Life*, 60, 620, 2008. With permission.)

replaced by the δ subunit. A predicted structure for αβγENaC as based on the crystal structure of cASIC1 is shown in Figure 26.10 (Stockand et al., 2008). Other family members form obligatory homotrimers, and others form both homo- and heterotrimers. ASICs fall into this latter category.

Deg/ENaC homologs share approximately 25%–35% sequence identity and have a conserved subunit topology, which is shown in Figure 26.11. All Deg/ENaC subunits contain a large extracellular domain bound by two transmembrane (TM) domains, TM1 and TM2, with intracellular NH_2- and COOH-termini (Gonzales et al., 2009; Jasti et al., 2007). The extracellular domain contains much secondary structure, with

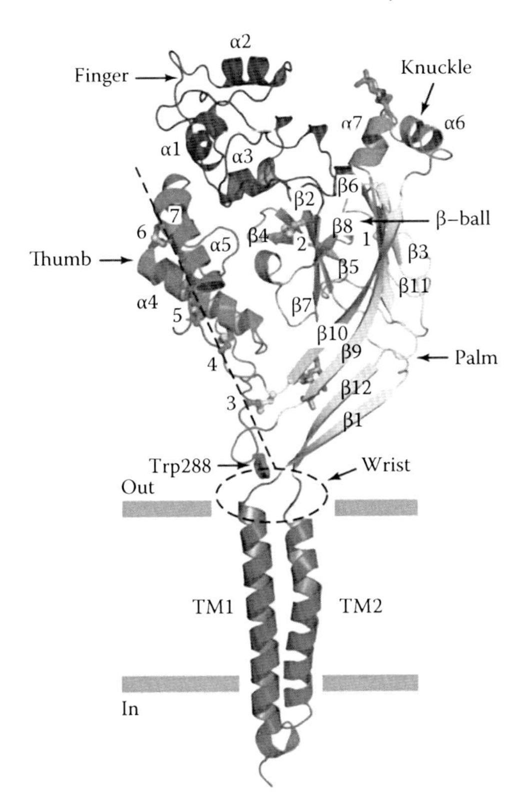

Figure 26.11 The 2° and 3° structure of a monomer within the cASIC1 homotrimer. (Reprinted from Jasti, J. et al., *Nature*, 449, 316, 2007. With permission.)

beta sheets and alpha helices from each of the three component subunits making extensive inter- and intrasubunit contacts. These secondary structures form five major extracellular domains in each subunit: the finger, knuckle, thumb, β-ball, and palm domains. Extracellular ligands bind in a pocket formed by the finger and thumb domains (Baconguis and Gouaux, 2012; Dawson et al., 2012). The pore of the channel is formed by the symmetry-related three helical TM2 domains, one from each subunit, as they run through the membrane in a linear manner.

The fully formed Deg/ENaC channel can be envisioned as a *chalice* where extracellular domains form a cup, TM domains a stem, and intracellular domains a base (Jasti et al., 2007). As shown in Figure 26.12, the pore is *hourglass* in shape with wide extracellular and intracellular vestibules and a narrowing in the middle. This pore is within the stem of the chalice. The extracellular mouth of the pore is joined to extracellular domains by short linker sequences referred to as the wrist.

26.4.1 PORE OF Deg/ENaC CHANNELS

The Deg/ENaC pore has threefold symmetry around its central axis perpendicular to the plane of the lipid bilayer. Ions enter and egress out of the extracellular and intracellular mouths of the pore proper through large vestibules that lay partially within the plane of the membrane. As shown in Figure 26.12, these vestibules have profound negative electrostatic potentials enabling them to act as cation reservoirs. This concentrating of cations around the mouths of the pore increases channel conductance (Gonzales et al., 2009).

The pore contains three Na^+ binding sites that are occupied during permeation. Permeant ions move through the pore in a single-file manner where adjacent sites are not occupied at the same time due to charge repulsion (Gonzales et al., 2009). The main-chain carbonyl oxygen atoms from the symmetry-related G432, G436, G439, and G443 residues in cASIC1 coordinate Na^+ at these binding sites. Ions in the pore are coordinated with trigonal antiprism geometry by three ligands on the upper triangular plane being staggered in comparison to those in the lower triangular plane. This arrangement provides the appropriate number of partial negative charges for coordination while perfectly accommodating the underlying threefold molecular

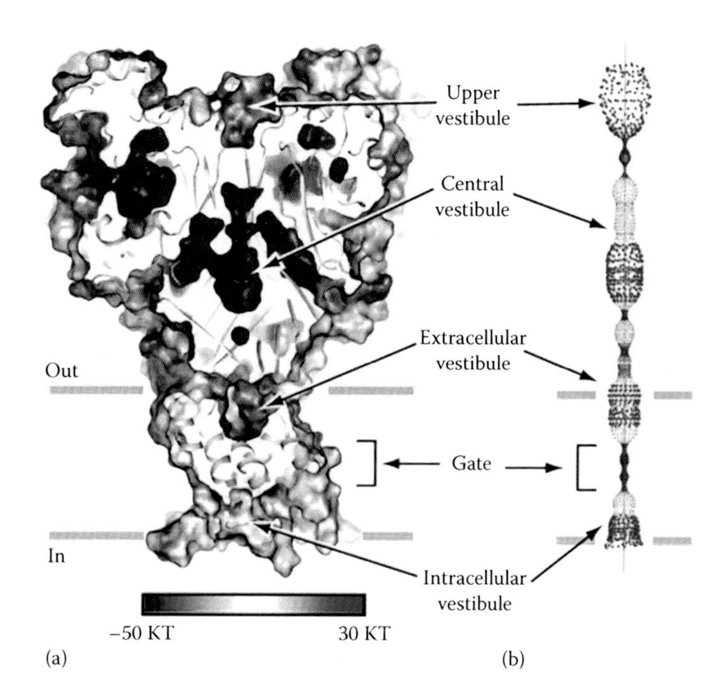

Figure 26.12 The pore of cASIC1. (a) A cartoon revealing the electrostatic surface potential and (b) radius of the conduction pathway (red < 1.4 Å < green < 2.3 Å < purple) for functional cASIC1 along the threefold axis of symmetry. (Reprinted from Gonzales, E.B. et al., *Nature*, 460, 599, 2009. With permission.)

symmetry of the trimeric Deg/ENaC channel. As such, trigonal antiprism coordination is believed to represent the archetype molecular basis of permeation through cation-selective ion channels containing three component subunits (Gonzales et al., 2009). Indeed, the P2X4 channel, which is trimeric, cation selective, and has 2°, 3°, and 4° structure resembling that of Deg/ENaC channels but shares no primary sequence with these channels, employs identical modes and means of permeation and coordination of permeant ions within the pore (Gonzales et al., 2009).

Several of the Gly residues involved in coordinating Na^+ within the pore of cASIC1 are replaced by Ser in ENaC. Interestingly, ENaC is about 10-fold more selective for Na^+ than is ASIC. Although structurally similar, Ser, which contains an extra carbon and hydroxyl moiety, occupies a larger volume than Gly. Perhaps, this introduces a steric constraint that better accommodates Na^+ compared to other cations in the permeation pathway of the more selective ENaC. Consistent with such a premise, mutations in and near the selectivity filter in ENaC reduce both the Na^+ conductance and Na^+ to K^+ selectivity (Kellenberger et al., 1999a,b; Snyder et al., 1999).

26.4.2 GATE OF Deg/ENaC CHANNELS

ASIC and most if not all ligand-gated Deg/ENaC channels assume three principal conformations: closed, open, and desensitized. ENaC, which gates independent of a ligand, does not inactivate or desensitize, having only closed and open conformations. It is assumed that all Deg/ENaC channels employ a common gating mechanism to move from the closed to open state with ligand-gated channels having an additional change in conformation leading to the desensitized state. In contrast to its inhibition of rat ASIC1, psalmotoxin 1 activates chicken ASIC1 (Grunder and Augustinowski, 2012). Understanding gleaned from the crystallization of psalmotoxin 1 with active cASIC1 has allowed rationalization of how Deg/ENaC channels may gate (Baconguis and Gouaux, 2012). The channel is thought to be in the open state following the untwisting of TM domains around the central axis of the pore perpendicular to the plane of the lipid bilayer. Specifically, the upper palm and knuckle domains of the three component subunits provide a fixed structural scaffold on which the lower palm domains shift to induce radial and rotational movements of the TM domains via the wrist linker (Baconguis and Gouaux, 2012; Grunder and Augustinowski, 2012). The finger and thumb domains, which make major contributions to the ligand-binding domain of Deg/ENaC channels, modulate movement of the lower palm, allowing the binding of a ligand to influence gating through an allosteric mechanism (Dawson et al., 2012; Grunder and Augustinowski, 2012).

Crystallization of psalmotoxin 1 with a variant of cASIC1 that can only assume the desensitized conformation revealed that three toxin molecules bind the trimeric channel making contact with thumb domains and extending deeply into ligand-binding pockets to contact palm domains (Dawson et al., 2012). This bimodal binding locks the relative arrangement between thumb and palm in the desensitized state for mammalian ASIC1 and in the open state for functional cASIC1 (Dawson et al., 2012; Grunder and Augustinowski, 2012).

Desensitization is thought to be the manifestation of a constriction formed by the crossing of the three TM2 domains at D433 (in cASIC1) occluding the pore (Gonzales et al., 2009; Jasti et al., 2007). This residue sits just below the large vestibule leading into the extracellular mouth of the pore. The *Deg* mutation that constitutively activates Deg/ENaC channels by locking them in long-lived open states is at 432 in cASIC1, one position upstream of D433 (Goodman et al., 2002; Waldmann et al., 1996). Placement of an amino acid larger than Gly at the Deg site would cause a steric clashing between the symmetry-related TM2 domains, providing a mechanism whereby mutations at this site perturb gating to lock the channel out of the desensitized conformation. It is interesting that Asn residues occupy the positions in ENaC subunits homologous to D433 in cASIC1. This is a signature feature of ENaC. Perhaps, this contributes to ENaC gating constitutively and being unable to enter a desensitized-like conformation.

26.5 Deg/ENaC CHANNELS IN HUMAN DISEASES

A pathological role for ASIC and ENaC in several different human diseases has been described. Not unexpectedly, these diseases contain a neuronal component for ASIC and an epithelial component for ENaC.

26.5.1 ROLE OF ASIC IN PATHOLOGY

Because of their sensitivity to pH, ASICs contribute to pathology resulting from ischemia, inflammation, and trauma that cause tissue acidosis (Bianchi and Driscoll, 2002; Leng and Xiong, 2012; Lingueglia, 2007; Staniland and McMahon, 2009). For instance, activation of ASIC1 by the metabolic acidosis induced by brain ischemia contributes to neuronal death associated with stroke. Abundant evidence now exists that Na^+ and Ca^{++} influx through ASICs activated by ischemic conditions contributes to anoxic depolarization, the rapid and pathological loss of membrane potential, which ultimately leads to cell death. Accordingly, inhibition of ASIC1 is an emerging therapy for stroke intervention and treatment of other neurological diseases associated with focal drops in pH and neuronal death (Chu and Xiong, 2013; Leng and Xiong, 2012).

Inappropriate activation of ASIC, moreover, contributes to the progression of multiple sclerosis (MS) (Friese et al., 2007; Vergo et al., 2011; Xiong et al., 2008). MS is an autoimmune neuroinflammatory disease of the CNS (Judge and Bever, 2006; Korenke et al., 2008; Solari et al., 2002). Hallmarks of this disease include CNS lesions marked by demyelination and axonal degeneration (Kornek et al., 2000; Lovas et al., 2000). Axonal degeneration in MS is caused by improper influx of Na^+ and Ca^{++} into neurons (Dutta and Trapp, 2007; Dutta et al., 2006; Lassmann, 2007; Nikolaeva et al., 2005; Petrescu et al., 2007; Stys, 2005; Waxman, 2006; Xiong et al., 2004). ASIC1 whose expression is increased in MS lesions by proinflammatory mediators serves as the gateway for this pathological influx of cations into diseased neurons (Vergo et al., 2011). Accordingly, inhibition of ASIC1 with amiloride and disruption of the ASIC1 gene provide a neuroprotective effect improving clinical symptoms by protecting both myelin and neurons from damage in an animal model of MS (Friese et al., 2007; Vergo et al., 2011).

26.5.2 REGULATION OF ENaC BY INTRACELLULAR SIGNALING

Mammalian Deg/ENaC channels are targets for diverse types of extracellular ligands, systemic hormones, and intracellular signaling cascades. Complete coverage of this subject is beyond the scope of this chapter. Because disruption of normal regulation of ENaC is causative for certain inheritable diseases in humans, key signaling pathways and domains within the channel involved in critical posttranslational regulation are discussed here in brief.

Corticosteroid hormones, including the mineralocorticoid aldosterone, are important positive regulators of ENaC. Upon binding its cognate receptor, aldosterone stimulates ENaC in renal epithelial cells by transactivating the gene-encoding serum and glucocorticoid-inducible kinase (Sgk1) (Pearce, 2003; Snyder, 2005; Staub et al., 2000). The activity of ENaC is controlled in part by regulation of its expression in the apical membrane. The ubiquitin ligase, Nedd4–2, associates with ENaC leading to ubiquitinylation of the channel. This tagging of the channel targets it for internalization. Sgk1 phosphorylates Nedd4–2 at a 14–3–3 binding site. Phosphorylation of this binding site allows 14–3–3 to sequester Nedd4–2 away from ENaC. Consequently, untagged ENaC is left in the apical membrane where it is active. Thus, aldosterone increases the activity of ENaC in part via a disinhibition mechanism lessening channel retrieval.

ENaC subunits contain a PY motif of the consensus PPPxY in their cytosolic COOH-termini (Kellenberger and Schild, 2002; Pearce, 2003; Snyder, 2005; Staub et al., 2000). This PY motif functions as a binding site for proteins containing the WW motif. For ENaC expressed in renal epithelia, Nedd4–2 is the primary binding partner to this site.

26.5.3 PATHOLOGY ARISING FROM ENaC DYSFUNCTION

Because it functions as a critical end effector of the renin–angiotensin–aldosterone system during feedback control of blood pressure, any mutation in ENaC or the upstream regulatory pathways governing channel activity that lead to hyperactivation in the ASDN elevates blood pressure. This elevation in blood pressure results from the disruption of normal renal sodium excretion. Inheritable diseases that fall into this category include Liddle's syndrome, apparent mineralocorticoid excess, glucocorticoid-remediable aldosteronism, and Geller's syndrome to name a few (Bonny and Hummler, 2000; Geller et al., 1998, 2000; Hummler and Horisberger, 1999; Lifton et al., 2001; Rossier et al., 2002). Certain dietary factors, such as glycyrrhetinic acid, also can lead to inappropriate activation of ENaC mimicking these inheritable hypertensive diseases. Because mutation of ENaC represents an end-organ defect where the channel is removed from normal feedback regulation, such channelopathy has hallmark elevated ENaC activity and concomitant elevations in blood pressure in the presence of decreased plasma renin activity, aldosterone levels, and serum potassium. Disease resulting from such gain of function of ENaC can be countered with amiloride and other drugs that inhibit the channel.

Most Liddle's mutations in ENaC that cause elevations in blood pressure result from frame shifts, early truncations that disrupt the cytosolic COOH terminus of one of its component subunits, or both. Specifically, it is the disruption of the Nedd4–2 binding PY motif that causes disease by impairing normal retrieval of the channel from the apical membrane leading to gain of channel function.

A disease-causing point mutation in ENaC outside of the PY motif that activates the channel and elevates blood pressure also has been described. The Liddle's mutation N530S in γ-hENaC increases channel activity but not membrane expression to cause disease (Hiltunen et al., 2002). N530, which corresponds to D433 in cASIC1, is one position downstream of the *Deg* position in the primary structure of ENaC at the extracellular apex of the pore-lining TM2. As discussed earlier, this region is important for the gating of Deg/ENaC channels. It is widely held that the *Deg* mutation locks open Deg/ENaC channels. Perhaps, the N530S Liddle's mutation does the same thing, leading to inappropriate ENaC hyperactivity.

Loss of function mutations in ENaC and its upstream regulatory pathways also result in inheritable forms of tubulopathy (Bonny and Hummler, 2000; Hummler and Horisberger, 1999; Lifton et al., 2001; Rossier et al., 2002). The majority of mutations resulting in loss of ENaC function completely disrupt normal expression of this channel in the ASDN where its absence compromises renal sodium reabsorption. Diseases caused by such mutations are termed pseudohypoaldosteronism (PHA) type I. Together, PHA type I represents a group of rare genetic diseases presenting with hallmark hyperkalemia and renal Na^+ wasting in the presence of high aldosterone (Chang et al., 1996; Riepe, 2009).

The missense mutation, G37S, in human βNaC also causes PHA type I (Chang et al., 1996; Grunder et al., 1997).

This Gly residue is in an HG motif conserved in the intracellular NH_2-terminal portions of all Deg/ENaC subunits. This motif is required for normal gating (Chang et al., 1996; Grunder et al., 1997, 1999). Substituting either the H or G residue in this motif decreases ENaC activity to where the channel is not active under physiological conditions (Kucher et al., 2011). A distinct missense mutation, S562P, in αNaC also causes familial PHA type I (Riepe et al., 2009). S562 occupies the third position in the selectivity-filter sequence, GSS, identifying the disruption of permeation as the most likely mechanism underlying loss of function (Kellenberger et al., 1999a,b; Sheng et al., 2001).

Because of its role in transport, ENaC also is involved in diseases of the lungs and airways. Loss of ENaC function in the lungs leads to fluid accumulation in alveolar spaces due to disruption of sodium reabsorption (Bonny and Hummler, 2000; Hummler and Horisberger, 1999; Kellenberger and Schild, 2002; Rossier et al., 2002). This increases the compliance of the lungs, which increases the energy cost of breathing, leading ultimately to exhaustion and death. Bacterial toxins and inflammatory mediators released in response to bacterial invasion of the lungs also decrease the activity of ENaC, causing a similar wet-lung phenotype. Through a related mechanism, abnormal ENaC activity is thought to explain certain instances of fluid accumulation in the lungs at high altitudes. Moreover, glucocorticoids modulate the expression of ENaC in the lungs. The surge of glucocorticoids and the forces and processes accompanying birthing activate ENaC to dehydrate the lungs to facilitate air breath upon parturition. Premature infants often are provided exogenous glucocorticoids in part to speed this process.

In airways, Na^+ reabsorbed via ENaC dehydrates mucus, countering the effects of Cl^- and coupled-water secretion as mediated by the cystic fibrosis transmembrane conductance regulator (CFTR). As such, gain of ENaC function in the lungs causes a cystic fibrosis–like phenotype with overly dry and sticky mucus similar to that resulting from loss of function of CFTR (Mall et al., 2004; Rauh et al., 2013).

The reverse of this also happens in the gut. Pathological hyperactivation of CFTR by bacterial toxins depolarizes the apical membrane to diminish the electrochemical forces favoring cell entry of Na^+ via ENaC, impeding the absorption of water. This underpins the conversion of intestinal epithelial cells from absorptive to secretory during secretory diarrhea such as that seen in cholera (Bonny and Hummler, 2000; Hummler and Horisberger, 1999; Kellenberger and Schild, 2002; Rossier et al., 2002).

REFERENCES

Adams, C. M., M. G. Anderson, D. G. Motto et al. 1998. Ripped pocket and pickpocket, novel *Drosophila* DEG/ENaC subunits expressed in early development and in mechanosensory neurons. *J. Cell. Biol.* 140:143–152.

Ainsley, J. A., J. M. Pettus, D. Bosenko et al. 2003. Enhanced locomotion caused by loss of the *Drosophila* DEG/ENaC protein pickpocket1. *Curr. Biol.* 13:1557–1563.

Arnadottir, J., R. O'Hagan, Y. Chen et al. 2011. The DEG/ENaC protein MEC-10 regulates the transduction channel complex in *Caenorhabditis elegans* touch receptor neurons. *J. Neurosci.* 31:12695–12704.

Baconguis, I. and E. Gouaux. 2012. Structural plasticity and dynamic selectivity of acid-sensing ion channel-spider toxin complexes. *Nature* 489:400–405.

Bianchi, L. and M. Driscoll. 2002. Protons at the gate: DEG/ENaC ion channels help us feel and remember. *Neuron* 34:337–340.

Blum, E. S., M. Driscoll, and S. Shaham. 2008. Noncanonical cell death programs in the nematode *Caenorhabditis elegans*. *Cell Death Differ.* 15:1124–1131.

Bohlen, C. J., A. T. Chesler, R. Sharif-Naeini et al. 2011. A heteromeric Texas coral snake toxin targets acid-sensing ion channels to produce pain. *Nature* 479:410–414.

Boiko, N., V. Kucher, B. A. Eaton et al. 2013. Inhibition of neuronal degenerin/epithelial Na^+ channels by the multiple sclerosis drug 4-aminopyridine. *J. Biol. Chem.* 288(13):9418–9427.

Boiko, N., V. Kucher, J. D. Stockand et al. 2012. Pickpocket1 is an ionotropic molecular sensory transducer. *J. Biol. Chem.* 287:39878–39886.

Bonny, O. and E. Hummler. 2000. Dysfunction of epithelial sodium transport: From human to mouse. *Kidney Int.* 57:1313–1318.

Borzan, J., C. Zhao, R. A. Meyer et al. 2010. A role for acid-sensing ion channel 3, but not acid-sensing ion channel 2, in sensing dynamic mechanical stimuli. *Anesthesiology* 113:647–654.

Canessa, C. M., J. D. Horisberger, and B. C. Rossier. 1993. Epithelial sodium channel related to proteins involved in neurodegeneration. *Nature* 361:467–470.

Canessa, C. M., L. Schild, G. Buell et al. 1994. Amiloride-sensitive epithelial Na channel is made of three homologous subunits. *Nature* 367:463–467.

Chandrashekar, J., C. Kuhn, Y. Oka et al. 2010. The cells and peripheral representation of sodium taste in mice. *Nature* 464:297–301.

Chang, S. S., S. Grunder, A. Hanukoglu et al. 1996. Mutations in subunits of the epithelial sodium channel causes salt wasting with hyperkalaemic acidosis, pseudohypoaldosteronism type 1. *Nat. Genet.* 12:248–253.

Chu, X. P. and Z. G. Xiong. 2013. Acid-sensing ion channels in pathological conditions. *Adv. Exp. Med. Biol.* 961:419–431.

Coric, T., D. Zheng, M. Gerstein, and C. M. Canessa. 2005. Proton sensitivity of ASIC1 appeared with the rise of fishes by changes of residues in the region that follows TM1 in the ectodomain of the channel. *J. Physiol.* 568:725–735.

Cottrell, G. A. 1997. The first peptide-gated ion channel. *J. Exp. Biol.* 200:2377–2386.

Cottrell, G. A., M. C. Jeziorski, and K. A. Green. 2001. Location of a ligand recognition site of FMRFamide-gated Na(+) channels. *FEBS Lett.* 489:71–74.

Darboux, I., E. Lingueglia, D. Pauron et al. 1998. A new member of the amiloride-sensitive sodium channel family in *Drosophila melanogaster* peripheral nervous system. *Biochem. Biophys. Res. Commun.* 246:210–216.

Davey, F., S. J. Harris, and G. A. Cottrell. 2001. Histochemical localisation of FMRFamide-gated Na^+ channels in *Helisoma trivolvis* and *Helix aspersa* neurones. *J. Neurocytol.* 30:877–884.

Dawson, R. J., J. Benz, P. Stohler et al. 2012. Structure of the acid-sensing ion channel 1 in complex with the gating modifier Psalmotoxin 1. *Nat. Commun.* 3:936.

Diochot, S., A. Baron, M. Salinas et al. 2012. Black mamba venom peptides target acid-sensing ion channels to abolish pain. *Nature* 490:552–555.

Driscoll, M. and M. Chalfie. 1991. The mec-4 gene is a member of a family of *Caenorhabditis elegans* genes that can mutate to induce neuronal degeneration. *Nature* 349:588–593.

Durrnagel, S., A. Kuhn, C. D. Tsiairis et al. 2010. Three homologous subunits form a high affinity peptide-gated ion channel in *Hydra*. *J. Biol. Chem.* 285:11958–11965.

Dutta, R., J. McDonough, X. Yin et al. 2006. Mitochondrial dysfunction as a cause of axonal degeneration in multiple sclerosis patients. *Ann. Neurol.* 59:478–489.

Dutta, R. and B. D. Trapp. 2007. Pathogenesis of axonal and neuronal damage in multiple sclerosis. *Neurology* 68:S22–S31.

Friese, M. A., M. J. Craner, R. Etzensperger et al. 2007. Acid-sensing ion channel-1 contributes to axonal degeneration in autoimmune inflammation of the central nervous system. *Nat. Med.* 13:1483–1489.

Garty, H. and L. G. Palmer. 1997. Epithelial sodium channels: Function, structure, and regulation. *Physiol. Rev.* 77:359–396.

Geffeney, S. L., J. G. Cueva, D. A. Glauser et al. 2011. DEG/ENaC but not TRP channels are the major mechanoelectrical transduction channels in a *C. elegans* nociceptor. *Neuron* 71:845–857.

Geller, D. S., A. Farhi, N. Pinkerton et al. 2000. Activating mineralocorticoid receptor mutation in hypertension exacerbated by pregnancy. *Science* 289:119–123.

Geller, D. S., J. Rodriguez-Soriano, B. A. Vallo et al. 1998. Mutations in the mineralocorticoid receptor gene cause autosomal dominant pseudohypoaldosteronism type I. *Nat. Genet.* 19:279–281.

Gloor, G. B., C. R. Preston, D. M. Johnson-Schlitz et al. 1993. Type I repressors of P element mobility. *Genetics* 135:81–95.

Golubovic, A., A. Kuhn, M. Williamson et al. 2007. A peptide-gated ion channel from the freshwater polyp *Hydra*. *J. Biol. Chem.* 282:35098–35103.

Gonzales, E. B., T. Kawate, and E. Gouaux. 2009. Pore architecture and ion sites in acid-sensing ion channels and P2X receptors. *Nature* 460:599–604.

Goodman, M. B., G. G. Ernstrom, D. S. Chelur et al. 2002. MEC-2 regulates *C. elegans* DEG/ENaC channels needed for mechanosensation. *Nature* 415:1039–1042.

Grunder, S. and K. Augustinowski. 2012. Toxin binding reveals two open state structures for one acid-sensing ion channel. *Channels (Austin)* 6:409–413.

Grunder, S., D. Firsov, S. S. Chang et al. 1997. A mutation causing pseudohypoaldosteronism type 1 identifies a conserved glycine that is involved in the gating of the epithelial sodium channel. *EMBO J.* 16:899–907.

Grunder, S., N. F. Jaeger, I. Gautschi et al. 1999. Identification of a highly conserved sequence at the N-terminus of the epithelial Na^+ channel alpha subunit involved in gating. *Pflugers Arch.* 438:709–715.

Hiltunen, T. P., T. Hannila-Handelberg, N. Petajaniemi et al. 2002. Liddle's syndrome associated with a point mutation in the extracellular domain of the epithelial sodium channel gamma subunit. *J. Hypertens.* 20:2383–2390.

Hong, K. and M. Driscoll. 1994. A transmembrane domain of the putative channel subunit MEC-4 influences mechanotransduction and neurodegeneration in *C. elegans*. *Nature* 367:470–473.

Hong, K., I. Mano, and M. Driscoll. 2000. In vivo structure-function analyses of *Caenorhabditis elegans* MEC-4, a candidate mechanosensory ion channel subunit. *J. Neurosci.* 20:2575–2588.

Hummler, E. and J. D. Horisberger. 1999. Genetic disorders of membrane transport. V. The epithelial sodium channel and its implication in human diseases. *Am. J. Physiol.* 276:G567–G571.

Jasti, J., H. Furukawa, E. B. Gonzales et al. 2007. Structure of acid-sensing ion channel 1 at 1.9 Å resolution and low pH. *Nature* 449:316–323.

Judge, S. I. and C. T. Bever, Jr. 2006. Potassium channel blockers in multiple sclerosis: Neuronal Kv channels and effects of symptomatic treatment. *Pharmacol. Ther.* 111:224–259.

Kellenberger, S., I. Gautschi, and L. Schild. 1999a. A single point mutation in the pore region of the epithelial Na^+ channel changes ion selectivity by modifying molecular sieving. *Proc. Natl. Acad. Sci. USA* 96:4170–4175.

Kellenberger, S., N. Hoffmann-Pochon, I. Gautschi et al. 1999b. On the molecular basis of ion permeation in the epithelial Na^+ channel. *J. Gen. Physiol.* 114:13–30.

Kellenberger, S. and L. Schild. 2002. Epithelial sodium channel/degenerin family of ion channels: A variety of functions for a shared structure. *Physiol. Rev.* 82:735–767.

Korenke, A. R., M. P. Rivey, and D. R. Allington. 2008. Sustained-release fampridine for symptomatic treatment of multiple sclerosis. *Ann. Pharmacother.* 42:1458–1465.

Kornek, B., M. K. Storch, R. Weissert et al. 2000. Multiple sclerosis and chronic autoimmune encephalomyelitis: A comparative quantitative study of axonal injury in active, inactive, and remyelinated lesions. *Am. J. Pathol.* 157:267–276.

Kucher, V., N. Boiko, O. Pochynyuk et al. 2011. Voltage-dependent gating underlies loss of ENaC function in pseudohypoaldosteronism type. *Biophys. J.* 100:1930–1939.

Lassmann, H. 2007. Multiple sclerosis: Is there neurodegeneration independent from inflammation? *J. Neurol. Sci.* 259:3–6.

Leng, T. D. and Z. G. Xiong. 2012. The pharmacology and therapeutic potential of small molecule inhibitors of acid-sensing ion channels in stroke intervention. *Acta Pharmacol. Sin.* 34:33–38.

Lifton, R. P., A. G. Gharavi, and D. S. Geller. 2001. Molecular mechanisms of human hypertension. *Cell* 104:545–556.

Lingueglia, E. 2007. Acid-sensing ion channels in sensory perception. *J. Biol. Chem.* 282:17325–17329.

Lingueglia, E., G. Champigny, M. Lazdunski et al. 1995. Cloning of the amiloride-sensitive FMRFamide peptide-gated sodium channel. *Nature* 378:730–733.

Lingueglia, E., E. Deval, and M. Lazdunski. 2006. FMRFamide-gated sodium channel and ASIC channels: A new class of ionotropic receptors for FMRFamide and related peptides. *Peptides* 27:1138–1152.

Lingueglia, E., N. Voilley, R. Waldmann et al. 1993. Expression cloning of an epithelial amiloride-sensitive Na^+ channel. A new channel type with homologies to *Caenorhabditis elegans* degenerins. *FEBS Lett.* 318:95–99.

Liu, L., W. A. Johnson, and M. J. Welsh. 2003a. *Drosophila* DEG/ENaC pickpocket genes are expressed in the tracheal system, where they may be involved in liquid clearance. *Proc. Natl. Acad. Sci. USA* 100:2128–2133.

Liu, L., A. S. Leonard, D. G. Motto et al. 2003b. Contribution of *Drosophila* DEG/ENaC genes to salt taste. *Neuron* 39:133–146.

Lovas, G., N. Szilagyi, K. Majtenyi et al. 2000. Axonal changes in chronic demyelinated cervical spinal cord plaques. *Brain* 123 (Pt. 2):308–317.

Lu, Y., X. Ma, R. Sabharwal et al. 2009. The ion channel ASIC2 is required for baroreceptor and autonomic control of the circulation. *Neuron* 64:885–897.

Mall, M., B. Grubb, J. Harkema et al. 2004. Increased airway epithelial Na(+) absorption produces cystic fibrosis-like lung disease in mice. *Nat. Med.* 10:487–493.

Mano, I. and M. Driscoll. 1999. DEG/ENaC channels: A touchy superfamily that watches its salt. *BioEssays* 21:568–578.

Mazzuca, M., C. Heurteaux, A. Alloui et al. 2007. A tarantula peptide against pain via ASIC1a channels and opioid mechanisms. *Nat. Neurosci.* 10:943–945.

Mironova, E., V. Bugaj, K. P. Roos et al. 2012. Aldosterone-independent regulation of the epithelial Na^+ channel (ENaC) by vasopressin in adrenalectomized mice. *Proc. Natl. Acad. Sci. USA* 109:10095–10100.

Nikolaeva, M. A., B. Mukherjee, and P. K. Stys. 2005. Na^+-dependent sources of intra-axonal Ca^{2+} release in rat optic nerve during in vitro chemical ischemia. *J. Neurosci.* 25:9960–9967.

O'Hagan, R., M. Chalfie, and M. B. Goodman. 2005. The MEC-4 DEG/ENaC channel of *Caenorhabditis elegans* touch receptor neurons transduces mechanical signals. *Nat. Neurosci.* 8:43–50.

Pearce, D. 2003. SGK1 regulation of epithelial sodium transport. *Cell. Physiol. Biochem.* 13:13–20.
Perry, S. J., V. A. Straub, M. G. Schofield et al. 2001. Neuronal expression of an FMRFamide-gated Na^+ channel and its modulation by acid pH. *J. Neurosci.* 21:5559–5567.
Petrescu, N., I. Micu, S. Malek et al. 2007. Sources of axonal calcium loading during in vitro ischemia of rat dorsal roots. *Muscle Nerve* 35:451–457.
Price, M. P., G. R. Lewin, S. L. McIlwrath et al. 2000. The mammalian sodium channel BNC1 is required for normal touch sensation. *Nature* 407:1007–1011.
Price, M. P., S. L. McIlwrath, J. Xie et al. 2001. The DRASIC cation channel contributes to the detection of cutaneous touch and acid stimuli in mice. *Neuron* 32:1071–1083.
Rauh, R., D. Soell, S. Haerteis et al. 2013. A mutation in the beta-subunit of ENaC identified in a patient with cystic fibrosis-like symptoms has a gain-of-function effect. *Am. J. Physiol. Lung Cell. Mol. Physiol.* 304:L43–L55.
Rezaval, C., H. J. Pavlou, A. J. Dornan et al. 2012. Neural circuitry underlying *Drosophila* female postmating behavioral responses. *Curr. Biol.* 22:1155–1165.
Riepe, F. G. 2009. Clinical and molecular features of type 1 pseudohypoaldosteronism. *Horm. Res.* 72:1–9.
Riepe, F. G., M. X. van Bemmelen, F. Cachat et al. 2009. Revealing a subclinical salt-losing phenotype in heterozygous carriers of the novel S562P mutation in the alpha subunit of the epithelial sodium channel. *Clin. Endocrinol.* (*Oxf*) 70:252–258.
Rossier, B. C., S. Pradervand, L. Schild et al. 2002. Epithelial sodium channel and the control of sodium balance: Interaction between genetic and environmental factors. *Annu. Rev. Physiol.* 64:877–897.
Roza, C., J. L. Puel, M. Kress et al. 2004. Knockout of the ASIC2 channel in mice does not impair cutaneous mechanosensation, visceral mechanonociception and hearing. *J. Physiol.* 558:659–669.
Sakai, H., E. Lingueglia, G. Champigny et al. 1999. Cloning and functional expression of a novel degenerin-like Na^+ channel gene in mammals. *J. Physiol.* 519 (Pt. 2):323–333.
Schaefer, L., H. Sakai, M. Mattei et al. 2000. Molecular cloning, functional expression and chromosomal localization of an amiloride-sensitive Na(+) channel from human small intestine. *FEBS Lett.* 471:205–210.
Sheng, S., K. A. McNulty, J. M. Harvey et al. 2001. Second transmembrane domains of ENaC subunits contribute to ion permeation and selectivity. *J. Biol. Chem.* 276:44091–44098.
Sherwood, T. W., E. N. Frey, and C. C. Askwith. 2012. Structure and activity of the acid-sensing ion channels. *Am. J. Physiol. Cell. Physiol.* 303:C699–C710.
Snyder, P. M. 2005. Regulation of epithelial Na^+ channel trafficking. *Endocrinology* 146:5079–5085.
Snyder, P. M., D. R. Olson, and D. B. Bucher. 1999. A pore segment in DEG/ENaC Na(+) channels. *J. Biol. Chem.* 274:28484–28490.
Solari, A., B. Uitdehaag, G. Giuliani et al. 2002. Aminopyridines for symptomatic treatment in multiple sclerosis. *Cochrane Database Syst. Rev.* 2002(4):CD1330–CD001330.
Staniland, A. A. and S. B. McMahon. 2009. Mice lacking acid-sensing ion channels (ASIC) 1 or 2, but not ASIC3, show increased pain behaviour in the formalin test. *Eur. J. Pain* 13:554–563.
Staruschenko, A., E. Adams, R. E. Booth et al. 2005. Epithelial Na^+ channel subunit stoichiometry. *Biophys. J.* 88:3966–3975.
Staub, O., H. Abriel, P. Plant et al. 2000. Regulation of the epithelial Na^+ channel by Nedd4 and ubiquitination. *Kidney Int.* 57:809–815.
Stockand, J. D., A. Staruschenko, O. Pochynyuk et al. 2008. Insight toward epithelial Na^+ channel mechanism revealed by the acid-sensing ion channel 1 structure. *IUBMB Life* 60:620–628.
Stys, P. K. 2005. General mechanisms of axonal damage and its prevention. *J. Neurol. Sci.* 233:3–13.
Tavernarakis, N., W. Shreffler, S. Wang et al. 1997. unc-8, a DEG/ENaC family member, encodes a subunit of a candidate mechanically gated channel that modulates *C. elegans* locomotion. *Neuron* 18:107–119.
Thistle, R., P. Cameron, A. Ghorayshi et al. 2012. Contact chemoreceptors mediate male-male repulsion and male-female attraction during *Drosophila* courtship. *Cell* 149:1140–1151.
Uchiyama, M., S. Maejima, S. Yoshie et al. 2012. The epithelial sodium channel in the Australian lungfish, *Neoceratodus forsteri* (Osteichthyes: Dipnoi). *Proc. Biol. Sci.* 279:4795–4802.
Vergo, S., M. J. Craner, R. Etzensperger et al. 2011. Acid-sensing ion channel 1 is involved in both axonal injury and demyelination in multiple sclerosis and its animal model. *Brain* 134:571–584.
Waldmann, R., G. Champigny, N. Voilley et al. 1996. The mammalian degenerin MDEG, an amiloride-sensitive cation channel activated by mutations causing neurodegeneration in *Caenorhabditis elegans*. *J. Biol. Chem.* 271:10433–10436.
Waxman, S. G. 2006. Axonal conduction and injury in multiple sclerosis: The role of sodium channels. *Nat. Rev. Neurosci.* 7:932–941.
Wemmie, J. A., J. Chen, C. C. Askwith et al. 2002. The acid-activated ion channel ASIC contributes to synaptic plasticity, learning, and memory. *Neuron* 34:463–477.
Wiemuth, D. and S. Grunder. 2011. The pharmacological profile of brain liver intestine Na^+ channel: Inhibition by diarylamidines and activation by fenamates. *Mol. Pharmacol.* 80:911–919.
Wiemuth, D., H. Sahin, B. H. Falkenburger et al. 2012. BASIC—A bile acid-sensitive ion channel highly expressed in bile ducts. *FASEB J.* 26:4122–4130.
Wiemuth, D., H. Sahin, C. M. Lefevre, H. E. Wasmuth, and S. Grunder. 2013. Strong activation of bile acid-sensitive ion channel (BASIC) by ursodeoxycholic acid. *Channels* (*Austin*) 7:38–42.
Wu, P. Y., Y. Y. Huang, C. C. Chen et al. 2013. Acid-sensing ion channel-1a is not required for normal hippocampal LTP and spatial memory. *J. Neurosci.* 33:1828–1832.
Xiong, Z. G., G. Pignataro, M. Li et al. 2008. Acid-sensing ion channels (ASICs) as pharmacological targets for neurodegenerative diseases. *Curr. Opin. Pharmacol.* 8:25–32.
Xiong, Z. G., X. M. Zhu, X. P. Chu et al. 2004. Neuroprotection in ischemia: Blocking calcium-permeable acid-sensing ion channels. *Cell* 118:687–698.
Zhong, L., R. Y. Hwang, and W. D. Tracey. 2010. Pickpocket is a DEG/ENaC protein required for mechanical nociception in *Drosophila* larvae. *Curr. Biol.* 20:429–434.

27 TRPC channels

Jin-Bin Tian, Dhananjay Thakur, Yungang Lu, and Michael X. Zhu

Contents

27.1 INTRODUCTION

The canonical subfamily of transient receptor potential (TRPC) channels was identified based on their sequence homology to the prototypical *Drosophila* TRP protein (Zhu et al., 1996; Montell et al., 2002), which when mutated caused a transient receptor potential (*trp*) phenotype, impairing phototransduction of the fruit fly (Cosens and Manning, 1969). Among the 28 TRP channels found in mammals, the 7 TRPC members have the closest homology with the *Drosophila* TRP. However, their functions are typically not associated with mammalian phototransduction. In fact, TRPC channels have been shown to play roles in many body systems. While early studies have concentrated on evaluating the contribution of TRPC channels in store-operated Ca^{2+} entry commonly present in nearly all cells, more recent investigations have focused on elucidating physiological functions of individual TRPC channels in different systems and cell types, resulting from not only Ca^{2+} but also Na^{+} influx mediated by the opened TRPC channels. This chapter will begin with an overview of the biophysical and pharmacological responses of, primarily, heterologously expressed TRPC channels and then provide a succinct discussion about main physiological functions of native TRPC channels, focusing mainly on the nervous system.

27.2 BIOPHYSICS AND PHARMACOLOGY OF TRPC CHANNELS

27.2.1 BASIC STRUCTURAL FEATURES

TRPC proteins are ubiquitously expressed in almost all mammalian tissues with varying expression levels and functional significance (Abramowitz and Birnbaumer, 2009). TRPC2 is a

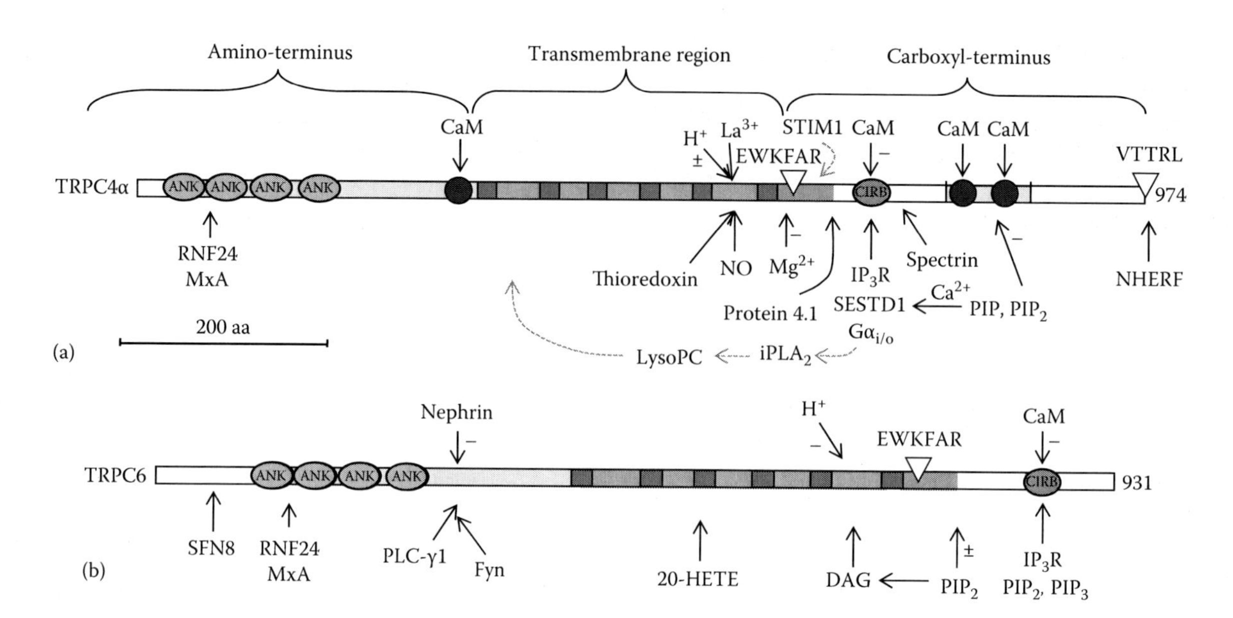

Figure 27.1 Structure features of TRPC4α (a) and TRPC6 (b). ANK-like repeats, transmembrane segments (*shaded small boxes*), and CaM binding sites (*dark circles*) are labeled. EWKFAR indicates the TRP motif. VTTRL indicates the PDZ-binding domain. Experimental evidence for some of the features indicated in TRPC4α has only been reported for TRPC5. −, experimental evidence exists for negative regulation; ±, experimental evidence exists for both positive and negative regulation. Binding motifs on the TRPCs have been defined for MxA (Lussier et al., 2005), nephrin and PLC-γ1 (Kanda et al., 2011), NHERF (Tang et al., 2000), PIP_3 (Kwon et al., 2007), protein 4.1 (Cioffi et al., 2005), RNF24 (Lussier et al., 2008), SFN8 (Carrasquillo et al., 2012), and STIM1 (Zeng et al., 2008), in addition to those discussed in the text.

pseudogene in humans, but in rodents it plays a central role in pheromone sensing (Freichel et al., 2004). All TRP channels are thought to be tetramers composed of either identical or related subunits. Each TRPC subunit is composed of six transmembrane segments (S1–S6) with both the amino (N)- and carboxyl (C)-termini located at the cytoplasmic side. A pore loop is found between S5 and S6, similar to the architecture of voltage-gated K^+ (Kv) channels, and determines ion selectivity. Based on sequence homology, TRPCs are subdivided into four groups, TRPC1, TRPC2, TRPC3/C6/C7, and TRPC4/C5. The latter two groups share about 75% and 65% amino acid identities among members within the groups, respectively. All TRPC members contain an absolutely conserved amino acid motif, EWKFAR, termed TRP domain, at their C-termini (Figure 27.1). They also have ankyrin (ANK)-like repeats at their N-termini and another conserved motif, namely, calmodulin (CaM) and IP_3 receptor-binding (CIRB) site at their C-termini (Tang et al., 2001; Zhang et al., 2001; Clapham, 2003). In addition, individual TRPC isoforms also interact with a diverse range of other proteins, for example, Homer, junctate, NHERF, and stromal interaction molecule 1 (STIM1), many of which have been covered extensively in recent reviews (Kiselyov et al., 2005; Eder et al., 2007b; Ong and Ambudkar, 2011).

27.2.2 ACTIVATION MECHANISMS

27.2.2.1 Phospholipase C signaling

TRPC channels are generally believed to be activated downstream from stimulation of phospholipase C (PLC). This typically occurs through ligand activation of G protein–coupled receptors (GPCRs) or receptor tyrosine kinases (RTKs). A subset of GPCRs are coupled to $G_{q/11}$ proteins, which stimulate PLCβ isoforms, while RTKs activate PLCγ isoforms, resulting in the hydrolysis of phosphatidylinositol 4,5-bisphosphate (PIP_2) and production of inositol 1,4,5-trisphosphate (IP_3) and diacylglycerol (DAG). DAGs can directly activate TRPC3, C6, and C7 independent of its effect on protein kinase C (PKC) (Hofmann et al., 1999). However, DAGs do not activate TRPC1, C4, and C5. It also remains questionable to what extent the DAG-induced activation can fully recapitulate the effect of stimulation of TRPC3/C6/C7 through receptor coupling (Albert and Large, 2003).

The other product of PIP_2 hydrolysis, IP_3, binds to IP_3 receptors, intracellular Ca^{2+} release channels located on the endoplasmic reticulum (ER) membrane. This action causes Ca^{2+} release from the ER, resulting in the depletion of Ca^{2+} from the ER store as well as the increase of intracellular Ca^{2+} concentration ($[Ca^{2+}]_i$). A large volume of early studies have provided evidence that some TRPC channels were activated by Ca^{2+} store depletion (Yuan et al., 2009); however, the mechanism(s) involved in coupling Ca^{2+} store depletion to TRPC channel activation was unclear. Both IP_3 receptors and its associated protein junctate have been considered as mediators capable of sensing Ca^{2+} levels in the ER lumen and communicating with TRPC channels (Treves et al., 2004, 2010). In fact, the C-terminal CIRB site that binds to a conserved N-terminal region of all IP_3 receptor isoforms is present in all TRPCs (Tang et al., 2001). More recent studies suggest that the ER-localized STIM proteins also interact and regulate the activation of TRPC channels. It is generally accepted that upon Ca^{2+} store depletion, STIMs form aggregates on the ER membrane and consequently trigger Ca^{2+} entry through interaction with channels formed by Orai proteins. While most investigators believe that Orai and TRPCs form separate channels regulated by STIMs, evidence exists that Orai and TRPC proteins may be present in the same channel complex. Detailed discussion on physical and functional links between ER Ca^{2+} store and TRPC channels can be found in a recent review and the references therein (Bavencoffe and Zhu, 2012).

Ca^{2+} plays an important part in TRPC gating, and its effects differ among different TRPC isoforms. Within the context of PLC signaling, IP_3-induced ER Ca^{2+} release probably provides the initial cytosolic Ca^{2+} signal to potentiate activation of TRPC4 and C5. Then the very Ca^{2+} that enters through the activated TRPC channel can further enhance its activity (positive feedback). Experiments supporting the Ca^{2+} requirement for TRPC4 and C5 activation have typically come from whole-cell patch clamp recording in which the intracellular solution contained either EGTA or BAPTA as the Ca^{2+} buffer and with the free $[Ca^{2+}]_i$ chelated at certain desired levels (Schaefer et al., 2000). Being a slower chelator, EGTA allows more $[Ca^{2+}]_i$ fluctuations than BAPTA and therefore helps reveal the effect of Ca^{2+} dynamics on the channel, including those due to the entrant Ca^{2+} from extracellular space. However, because TRPC4 and C5 activation is also strongly enhanced by extracellular Ca^{2+}, it can be difficult to distinguish the extracellular effect of Ca^{2+} from the intracellular one that results from Ca^{2+} entry via the open channel. In native cells, the distinction may be clearer because the Ca^{2+} influx and the resultant $[Ca^{2+}]_i$ rise that facilitate TRPC channel activation are often mediated by other Ca^{2+}-permeable channels, for example, voltage-gated Ca^{2+} channels (VGCCs) and Orai channels (see later in this chapter). Presently, direct activation by cytosolic Ca^{2+} has only been demonstrated for TRPC5 homomeric channels (Blair et al., 2009; Gross et al., 2009). Most likely, Ca^{2+} should only be considered a cofactor for activation of some TRPC channels. It is also important to note that although in general Ca^{2+} can sensitize TRPC4 and C5 and desensitize TRPC3, C6, and C7, the TRPC4 and C5 activities are also inhibited by high $[Ca^{2+}]_i$ (Ordaz et al., 2005).

The breakdown of PIP_2 by PLC also affects TRPC activities. However, conflicting results have been reported regarding the role of PIP_2 on TRPC activation. PIP_2 hydrolysis or inhibition of PI4P and PIP_2 synthesis induced TRPC5 activation, but dephosphorylation of PIP_2 to PI4P inhibited TRPC5 activity. Direct application of PIP_2 to the cytoplasmic side of inside-out patches, on the other hand, activated TRPC5 (Trebak et al., 2009). Whereas intracellular dialysis of PIP_2 prevented the activation by carbachol of TRPC4α, but not TRPC4β (Otsuguro et al., 2008), it slowed down the desensitization of TRPC4β and TRPC5 (Kim et al., 2008a, 2013). Moreover, intracellular PIP_2 dialysis rescued the inhibition of TRPC4β by coexpression of a constitutively active $G\alpha_q$ mutant (Jeon et al., 2012). Intriguingly, the effects of PIP_2 dialysis on carbachol-evoked native TRPC4-like currents in ileal myocytes contradicted each other, showing either enhancement (Otsuguro et al., 2008) or reduction of desensitization (Kim et al., 2008a). In vascular smooth muscle cells, the PIP_2-dependent activation of native TRPC1/C5 heteromeric channels has been attributed to TRPC1 (Shi et al., 2012). In inside-out patches, direct application of PIP_2 activated overexpressed TRPC6 and C7 (Lemonnier et al., 2008), but its depletion was shown to increase endogenous TRPC6-like activity in rabbit mesenteric artery smooth muscle cells (Albert et al., 2008). A more recent study using voltage-sensitive phosphatase to dephosphorylate PIP_2 in whole-cell recordings demonstrated the essential role of PIP_2 in supporting TRPC3, C6, and C7 function, suggesting a self-limiting role of PLC in the activation of mammalian TRPC channels via PIP_2 hydrolysis (Imai et al., 2012). Interestingly, in *Drosophila* photoreceptor cells, the activation of TRP and TRPL channels was dependent on the breakdown of phosphoinositides and a concomitant decrease of intracellular pH, yet direct application of PIP_2 to the cytoplasmic side strongly enhanced channel activity in inside-out patches excised from insect S2 cells that overexpressed TRPL (Huang et al., 2010). Therefore, the modulation of TRPC channels by phosphoinositides appears complex, including facilitation and inhibition. The relative levels and the number and positions of phosphates, that is, PI, PIP, or PIP_2, as well as the specific locations of these lipids on the plasma membrane in relation to the TRPC channels, are all important considerations that require further investigation.

27.2.2.2 $G_{i/o}$ protein signaling

A number of GPCRs that activate pertussis toxin (PTX)-sensitive $G_{i/o}$ proteins can induce TRPC4 and C5 channel activity. It has long been reported that the muscarinic cation current (mI_{CAT}) in intestinal smooth muscle cells, which was later shown to be mainly mediated by TRPC4-containing channels (Tsvilovskyy et al., 2009), was dependent on the coactivation of $G_{i/o}$-coupled M2 and $G_{q/11}$-coupled M3 receptors (Zholos and Bolton, 1997). GTPγS-evoked activation of expressed TRPC4 channels was indeed inhibited by treatment with PTX (Otsuguro et al., 2008). Other evidence supporting the involvement of $G_{i/o}$ proteins in TRPC channel activation includes findings that knocking down expression of $G\alpha_{i1}$, but not $G\alpha_{q/11}$, strongly reduced activity mediated by expressed TRPC5 in *Xenopus* oocytes (Tabata et al., 2002) and the TRPC5 activity stimulated by sphingosine-1-phosphate or oxidized phospholipids was inhibited by PTX (Xu et al., 2006; Al-Shawaf et al., 2010). Intriguingly, coexpression of constitutively active $G\alpha_{i2}$, $G\alpha_{i3}$, and $G\alpha_o$, but not $G\alpha_{i1}$, activated TRPC4 in the absence of any stimulus, with $G\alpha_{i2}$ being apparently more preferred. For TRPC5, this constitutive coupling seemed to prefer $G\alpha_{i3}$ (Jeon et al., 2012).

The critical site involved in interaction and functional coupling with $G\alpha_i$ proteins, interestingly, was found to overlap with the C-terminal CIRB site (Jeon et al., 2012), implicating a pivotal role of this region on overall gating of TRPC channels. Moreover, this region of TRPC4 and C5 also binds to SESTD1, a protein that contains a SEC14-like lipid binding domain and two spectrin domains and binds PIP and PIP_2 in the presence of Ca^{2+} (Miehe et al., 2010), implication of a modulation of the CIRB site by phosphoinositides and cytoskeleton (Figure 27.1a). Immediately downstream of the TRPC4 CIRB site is a region that binds to spectrins (Odell et al., 2008). Therefore, the interplay among CaM, IP_3 receptors, $G\alpha_i$'s, SESTD1, phosphoinositides, and cytoskeleton for functional regulation of TRPC4, C5, and perhaps other TRPC channels should be an interesting topic that warrants further exploration.

An alternative mechanism for $G_{i/o}$-mediated TRPC5 activation has also been shown using sphingosine-1-phosphate as the trigger. In this case, $G_{i/o}$ activation was thought to stimulate Ca^{2+}-independent group 6 (GVI) phospholipase A2 (iPLA2), generating lysophospholipids, which in turn activate TRPC5 (AL-Shawaf et al., 2011). Notably, the iPLA2 inhibitor used in this study, bromoenol lactone, was shown to inhibit TRPC5 in an iPLA2-independent manner by a different group (Chakraborty et al., 2011). The stimulatory effect of lysophospholipids on TRPC5, TRPC6, and native TRPC-like channels has been summarized in a recent review (Beech, 2012).

27.2.2.3 Other mechanisms

Several activation mechanisms unique to specific TRPC channels have been reported. These include redox states on TRPC3 (Balzer et al., 1999) and TRPC3/C4 heteromers (Poteser et al., 2006), arachidonic acid and its metabolite, 20-HETE, on TRPC6 (Basora et al., 2003; Inoue et al., 2009), nitric oxide on TRPC5 through S-nitrosylation of Cys-553 and Cys-558 accessible from the cytoplasmic side (Yoshida et al., 2006), mechanical stimulation on TRPC1 and C6 (Maroto et al., 2005; Spassova et al., 2006), and temperature cooling in the innocuous range on homomeric TRPC5 (Zimmermann et al., 2011). Paradoxically, the cysteine residues identified for nitric oxide modification are also critical for TRPC5 activation by extracellular thioredoxin or dithiothreitol (Figure 27.1a), which presumably breaks the disulfide bridge formed by the cysteines near the selectivity filter in the pore loop (Xu et al., 2008). Thus, both oxidizing and reducing agents activate TRPC5 (Takahashi and Mori, 2011). Questions have also been raised concerning mechanical activation of TRPC1 and C6 (Gottlieb et al., 2008), with some suggesting that the mechanosensitivity was at the level of GPCR (Mederos et al., 2008).

27.2.3 TRPC CHANNEL MODULATION

27.2.3.1 Phosphorylation

Phosphorylation sites and effects on TRPC channels identified through either site-direct mutagenesis or mass spectrometric analysis are summarized in Table 27.1. In general, TRPC3, C6, and C7 are inhibited by PKA, PKC, and PKG to varying degrees, and some of the identified sites are conserved among the three channels and can overlap between PKA and PKG. It was also shown that PKC could inhibit these channels through stimulating PKG (Kwan et al., 2006). For other TRPCs, the inhibition by PKC is also common (Venkatachalam et al., 2003). However, the native heteromeric TRPC1/C5 channels are activated by PKC (Shi et al., 2012). PKA and members of Src family of non-RTKs, for example, Fyn, have also been shown to regulate trafficking of some TRPC channels (Table 27.1).

27.2.3.2 Effects of cations

Ca^{2+}, Mg^{2+}, and H^+ can affect TRPC activities from either side of the plasma membrane. The effects of Ca^{2+}, some of which are included in the activation mechanisms discussed earlier, are complex and differ among different TRPCs (Schaefer et al., 2000; Shi et al., 2004). High concentrations of cytosolic Ca^{2+} are inhibitory to all TRPCs, even if the channel is dependent on cytosolic Ca^{2+} for activation (Ordaz et al., 2005). The inhibition of cytosolic Ca^{2+} may be mediated by CaM for some channels but not others (Zhang et al., 2001; Singh et al., 2002; Shi et al., 2004).

A few studies have explored the mechanism of pH regulation of TRPCs. While strong extracellular acidosis generally inhibits TRPC channels, mild acidosis potentiated TRPC4 and C5 activities. Glutamate residues E543, E595, and E598 were shown to modulate the pH sensitivity of TRPC5 (Semtner et al., 2007; Kim et al., 2008b). Interestingly, the potentiation effect of protons was not detected when extracellular Na^+ was replaced by Cs^+ (Kim et al., 2008b). In fact, extracellular Cs^+ seems to be important for the detection of TRPC4 and C5 currents under certain conditions when currents were too small or undetectable in normal Na^+-containing physiological bath solutions (Jeon et al., 2008). The reason for the facilitation by extracellular Cs^+, or inhibition by Na^+, however, is unclear.

The three glutamate residues identified for proton modulation of TRPC5 are also responsible for the potentiation of TRPC4 and C5 by lanthanides (La^{3+} and Gd^{3+}) (Jung et al., 2003). At about 10–300 μM, the lanthanides differentially modulate TRPCs with potentiation on TRPC4 and C5 and inhibition on TRPC1, C3, C6, and C7 and are thus commonly used to distinguish between the TRPC subgroups. Interestingly, lanthanides and protons both decreased unitary conductance but increased open probability of TRPC5, suggesting a similar mechanism (Semtner et al., 2007). Extracellular Ca^{2+} may mimic lanthanides to facilitate TRPC5 activation (Zeng et al., 2004).

Mercury compounds potently activate TRPC4 and C5 but not other TRPCs. The two cysteine residues in the TRPC5 pore loop, which were identified for nitric oxide and thioredoxin sensing, were shown to be critical for mercury stimulation. The mercury effect was not mimicked by Cd^{2+}, Ni^{2+}, and Zn^{2+}. It was proposed that TRPC4 and C5 may be responsible for the cytotoxic effect of mercurial compounds on neurodevelopment (Xu et al., 2012).

A Mg^{2+} block was found to account for the unusual *flat* segment or negative slope (see Figure 27.2) between +10 and +40 mV on the current–voltage (I–V) curves of TRPC5 currents. Asp-633 situated at the end of S6 transmembrane helix was shown to mediate the cytosolic Mg^{2+} block of outward current through TRPC5. This residue also affects unitary conductance of TRPC5 at negative potentials (Obukhov and Nowycky, 2005).

27.2.3.3 Regulated trafficking and translocation to plasma membrane

Ca^{2+} influx and/or receptor stimulation including both GPCRs and RTKs have been shown to promote vesicular trafficking to the plasma membrane, and as a result, TRPCs sequestered in vesicular membranes are brought to the plasma membrane, leading to increases in channel density and function (Bezzerides et al., 2004; Smyth et al., 2006). The Ca^{2+} influx that triggers TRPC trafficking may be mediated by a different channel, for example, Orai-mediated Ca^{2+} entry was shown to provide the local Ca^{2+} signal to enhance TRPC1 insertion into plasma membrane of cultured human salivary gland cells (Cheng et al., 2011). Additional mechanisms that enhance TRPC channel trafficking include pathways involving PKA, tyrosine kinases, PI3 kinase (see Table 27.1 for references), and cysteine oxidation (Graham et al., 2010).

27.2.4 BIOPHYSICAL PROPERTIES

27.2.4.1 Ionic selectivity and effects on membrane excitability

All homomeric TRPCs are nonselective cation channels that permeate Ca^{2+} and related divalent cations. In fact, the permeation to Sr^{2+}, Ba^{2+}, and Mn^{2+} had been used to assess TRPC channel activity in expression systems (Zhu et al., 1998; Venkatachalam et al., 2003). The reported ion selectivity values,

Table 27.1 Phosphorylation sites of TRPCs

KINASE	TRPC	PHOSPHO-RESIDUE	FUNCTION	CONSERVATION	REFERENCES
PKG	C3	Thr11 Ser263	Inhibition	Conserved in C6, C7	Kwan et al. (2004)
	C6	Thr69	Inhibition	Same as Thr11 of C3	Takahashi et al. (2008)
	C7	Thr15	Inhibition	Same as Thr11 of C3	Yuasa et al. (2011)
PKC	C3	Ser712	Inhibition	Conserved in TRPCs	Trebak et al. (2005)
	C6	Ser768	Inhibition	Same as Ser712 of C3	Kim and Saffen (2005)
		Ser448	Inhibition	Conserved in all TRPCs	Bousquet et al. (2010)
PKC-γ	C3	Thr573	Inhibition and calcineurin targeting	Conserved in C6, C7	Poteser et al. (2011)
PKA	C3	Ser251	Inhibition	Conserved in C6, C7	Nishioka et al. (2011)
		Undefined	Increasing translocation		Goel et al. (2010)
	C5	Ser794 Ser796	Partial inhibition	Unique to C5	Sung et al. (2005)
	C6	Ser28	Partial inhibition	Unique to C6	Horinouchi et al. (2011)
		Thr69	Inhibition	Conserved in C6, C7	Nishioka et al. (2011)
		Undefined	Increasing translocation		Fleming et al. (2007)
		Indirect	Activation via PI3K–PKB–MEK–ERK1/2 signaling pathway		Shen et al. (2011)
Src	C3	Tyr226	Required for activation		Vazquez et al. (2004); Kawasaki et al. (2006)
Src/Fyn	C4	Tyr959 Tyr972	EGF-induced membrane insertion by increasing NHREF binding	Unique to C4	Odell et al. (2005)
Fyn	C6	Tyr31 Tyr284	Activation, trafficking, PLC-γ1 binding		Hisatsune et al. (2004); Kanda et al. (2011)
WNK4	C3	Undefined	Inhibition of surface expression and activity		Park et al. (2011)
Undefined	C4β	Ser688[a] Thr691[a] Ser875[a] Thr879[a]	Not determined	Unique to C4	Lee et al. (2012)
Undefined	C6	Ser814[a]	No functional effect	Unique to C6	Bousquet et al. (2011)

PKA, PKC, PKG, protein kinase A, C, G, respectively; WNK4, serine-threonine kinase with no lysine 4.
[a] Phosphorylation residues identified by mass spectrometry.

based on reversal potential measurement, vary considerably. The PCa^{2+}/PNa^{+} (or Cs^{+}) ratios for TRPCs have been summarized to range from ~1 to 10, with a rank order of TRPC5 > C6 > C7 > C2 > C3 > C4 > C1 (Gees et al., 2010). So far there has been no report on fractional Ca^{2+} current measurement for TRPCs. Under physiological conditions, the main ion conducted by TRPCs is Na^{+} (Eder et al., 2005). The total cationic influx (INa^{+} + ICa^{2+}) mediated by TRPC leads to depolarization, which in excitable cells can activate voltage-gated Na^{+} and Ca^{2+} channels.

The Na^{+} influx mediated by TRPC3 has also been shown to drive Na^{+}/Ca^{2+} exchangers (NCX) in *reverse mode*, extruding Na^{+} in exchange for Ca^{2+} entry and thereby indirectly enhancing the Ca^{2+} signal (Eder et al., 2005, 2007a). It is possible that TRPC channels may facilitate or alter other Na^{+}-dependent transport activities via the increase in local $[Na^{+}]_i$.

The Ca^{2+} influx through TRPC1, however, has been shown to activate large-conductance Ca^{2+}-activated K^{+} (BK) channels in salivary gland and vascular smooth muscle cells (Liu et al., 2007; Kwan et al., 2009), leading to hyperpolarization instead of depolarization. Physical association between TRPC1 and BK was demonstrated, suggesting that the coupling is local and efficient (Kwan et al., 2009). Therefore, though the immediate local effect of TRPC activation is a depolarizing Na^{+} and Ca^{2+} influx, depending on association with other proteins, globally TRPC activation can lead to either depolarization or hyperpolarization. Clearly, TRPC channels function not only as mediators for $[Ca^{2+}]_i$ changes but also as key modulators of membrane potential.

In excitable cells, Ca^{2+} influx through TRPC channels may have special functional significance despite the notion that the

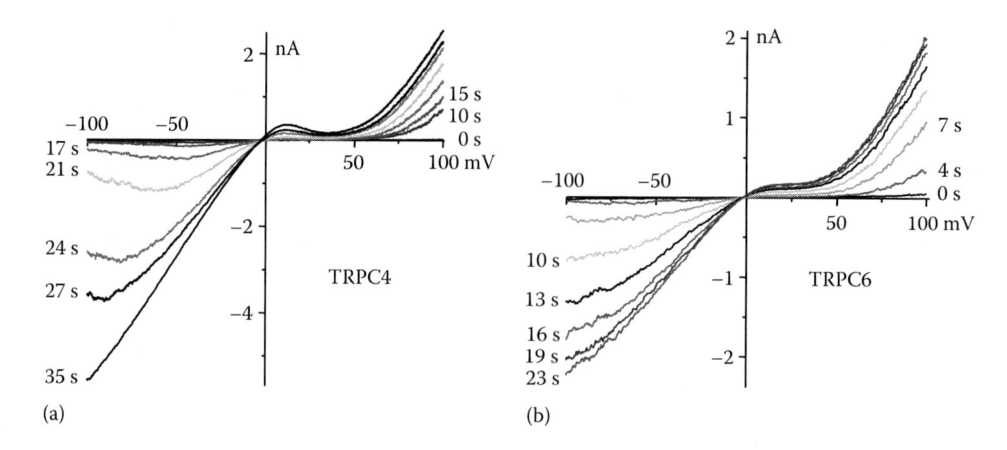

Figure 27.2 I–V relationships during activation of TRPC4 (a) and TRPC6 (b). (a) Whole-cell currents elicited by voltage ramps from +100 to −100 mV in a HEK293 cell that coexpressed TRPC4β and M2 muscarinic receptor during the course of stimulation by carbachol (5 μM). The time points (in seconds) following application of the agonists are indicated. Note the I–V curves at negative potentials changed from V-shaped to linear with time. (b) Similar to (a), but the cell coexpressed TRPC6 and M5 muscarinic receptor and the stimulation was carbachol (30 μM). The changes in the shapes of the I–V curves are less dramatic as compared to (a), but the trend is similar.

main Ca^{2+} entry pathway appears to rely on VGCCs, activated in response to TRPC-mediated membrane depolarization. For instance, a mutation at the TRPC3 selectivity filter, E633Q, disrupted its Ca^{2+} permeability without affecting the monovalent conductance or its functional coupling to L-type VGCCs in atrial myocytes but completely abolished TRPC3-dependent and calcineurin-mediated nuclear translocation of nuclear factor of activated T cells (NFAT) (Poteser et al., 2011).

27.2.4.2 Voltage dependence

The I–V relationships of TRPCs are not linear in the range of −120 to +100 mV, indicating voltage-dependent changes in conductance (Figure 27.2). There are at least three reasons for these changes. One is the block by divalent ions, Ca^{2+} and Mg^{2+}. As presented previously, Mg^{2+} blocks outward current of TRPC5 between +10 and +40 mV, giving rise to the *flat* segment on the I–V curve. The outward movement of Mg^{2+} appears to decrease unitary conductance in this voltage range (Obukhov and Nowycky, 2005). For the native TRPC-like current (mI_{CAT}) in ileal myocytes, the unitary current amplitude is reduced by extracellular Mg^{2+} and Ca^{2+} (Dresviannikov et al., 2006). However, the strong voltage dependence of expressed *Drosophila* TRPL channel was shown to result from open channel block by Ca^{2+}, which reduced channel open probability without affecting unitary conductance (Parnas et al., 2007).

Very few studies have examined the intrinsic voltage dependence of TRPC channels. Because the currents are quite unstable over extended time periods, they are typically detected using repetitive voltage ramps or gap-free recordings at a fix potential. Since the ramps are often quite fast (<0.5 s) and run from negative to positive potentials, the obtained I–V curves can be rather distorted from the true steady-state I–V relation. Nevertheless, the I–V relationships of TRPC4 and C5 often change from *U* or *V* shaped, or flat, to linear at negative potentials depending on the degree of activation (Figure 27.2) (Schaefer et al., 2000; Obukhov and Nowycky, 2008). Using slow (6 s) voltage ramps from positive to negative potential, a negative shift of the conductance–voltage curve can be clearly demonstrated during activation of native TRPC-like current in ileal myocytes by receptor agonist or intracellular dialysis of GTPγS (Dresviannikov et al., 2006), indicating that a negative shift of the voltage dependence to physiological potentials constitutes a part of TRPC activation.

Heteromultimerization can also affect voltage dependence. This has been shown for TRPC1–C4 and C1–C5 heteromeric channels, which when activated have stronger outward rectification and about sevenfold smaller unitary conductance at negative potentials than homomeric TRPC4 and C5 channels (Strübing et al., 2001). The measured unitary conductance of homomeric TRPCs varies significantly in the range of 25–130 pS. The values can vary at different voltages for the same channel (Obukhov and Nowycky, 2005, 2008). The unitary conductance of TRPC1–C5 heteromer, however, was determined to be ~5 pS between potentials −20 to −70 mV (Strübing et al., 2001).

27.2.5 PHARMACOLOGY

Until recently, only nonspecific blockers, such as SKF96365, 2-aminoethoxydiphenyl borate (2-APB), and flufenamic acid, were available to assess the involvement of TRPC channels in native systems. These drugs do not allow distinction among different TRPCs, and at the concentrations used, they also have many known non-TRPC targets. As such, studies are often complemented with approaches using isoform-specific blocking antibodies, siRNA, and/or knockout mouse models. KB-R7943, *N*-(*p*-amylcinnamoyl)anthranilic acid (ACA), ML-9, W-7, and BTP2 have also been shown to inhibit some expressed TRPC channels, but they again lack the specificity (Kraft, 2007; Harteneck and Gollasch, 2011). The pyrazole compounds, Pyr3 and its analog Pyr10, hold the promise to be TRPC3 selective inhibitors (Kiyonaka et al., 2009; Schleifer et al., 2012). More recent efforts on testing steroids and high-throughput screening of compound libraries have identified multiple new structures for selective TRPC inhibitors (Majeed et al., 2011; Miller et al., 2011; Miehe et al., 2012; Urban et al., 2012). It is anticipated that some of them or their derivatives will facilitate functional characterization and therapeutic development targeting against TRPC channels.

Direct activators are very scarce for TRPCs. 1-Oleoyl-2-acetyl-sn-glycerol (OAG) is commonly used to activate TRPC3, C6, and C7. Flufenamic acid enhances TRPC6 activity while inhibiting other TRPCs (Inoue et al., 2001). Hyperforin activates TRPC6 (Leuner et al., 2007).

27.3 PHYSIOLOGICAL FUNCTIONS OF TRPC CHANNELS

There is a rich literature on TRPC expression and function in different body systems. The physiological implications of TRPC channels are diverse and may differ from one cell type to another. Different TRPC isoforms may either have overlapping or different, and sometimes opposite, functions even within the same cell type. It will be exceedingly difficult to cover the physiological roles of TRPCs in all systems in the limited space. Here, we will use the nervous system as an example to illustrate the diverse functions of native TRPC channels and highlight the cell types and conditions when electric responses mediated by endogenous TRPCs have been successfully recorded.

27.3.1 TRPC CHANNELS IN NEURONAL DEVELOPMENT AND SURVIVAL

TRPCs are highly expressed in the central nervous system (Strübing et al., 2001, 2003; Sergeeva et al., 2003; Fowler et al., 2007), where they play roles in both neuronal development and synaptic transmission (Abramowitz and Birnbaumer, 2009; Birnbaumer, 2009; Bollimuntha et al., 2011). During neural development, a variety of growth-related environmental cues may act through TRPC channels and consequent Ca^{2+} signals to affect neural stem cell proliferation and differentiation, neuronal morphogenesis, axonal guidance, and synaptogenesis (Tai et al., 2009).

In rat hippocampus, TRPC6 expression peaks at P7–P14, a period critical for dendritic growth, to promote formation of dendrite spines and excitatory synapses through activating CaMKIV–CREB pathway (Tai et al., 2008; Zhou et al., 2008b). Brain-derived neurotrophic factor (BDNF) activates TRPC6 via the TrkB-PLCγ pathway in hippocampal pyramidal neurons. The Ca^{2+} influx through TRPC6 is linked to CaMKK stimulation, Akt phosphorylation, and subsequent mTOR activation, causing translocation of GluA1, a subunit of calcium-permeable AMPA receptors transiently expressed during early postnatal formation of glutamatergic synapses. This effect of TRPC6 on enhancing synaptic strength is shared with TRPC5 (Fortin et al., 2012). However, TRPC5 and TRPC6 have opposite effects on dendritic growth. While activation of TRPC6 by BDNF and neurotrophin-4 through TrKB-PLCγ promotes dendritic growth via CaMKIV, stimulation of TRPC5 by neurotrophin-3 through TrkC-PLCγ inhibits dendritic growth via CaMKIIα (He et al., 2012). Such antagonizing functions between TRPC5 and TRPC6 have also been reported for actin remodeling and cell motility in fibroblasts and kidney podocytes (Tian et al., 2010; Greka and Mundel, 2011).

For axonogenesis, semaphorin 3A induces growth cone collapses to deflect axons from inappropriate regions by causing a calpain-mediated cleavage of TRPC5 expressed in developing axonal growth cones (Kaczmarek et al., 2012). The highly localized Ca^{2+} influx via the truncated TRPC5 activates CaMKK and CaMKIγ, both are required for axon formation of cultured hippocampal neurons (Davare et al., 2009). By contrast, TRPC5 activated via cross-linking of GM1 ganglioside with α5β1 integrin promotes outgrowth of axon-like neurites in undifferentiated NG108-15 cells and primary cerebellar granule neurons (Wu et al., 2007).

In cultured cerebellar granule neurons, TRPC3 and TRPC6 are critical for BDNF-dependent growth cone guidance (Li et al., 2005), but growth cone turning of cultured *Xenopus* spinal neurons required TRPC1 (Shim et al., 2005; Wang and Poo, 2005). The BDNF-dependent survival of cerebellar granule neurons against serum starvation also required TRPC3 and TRPC6. This protective effect involves phosphorylation and activation of CREB signaling and is particularly important for granule neuron survival during P10–P12 in rodents (Jia et al., 2007).

Notably, most of the aforementioned studies were based on RNA interference, dominant negative TRPC constructs, and/or drugs that are unspecific. It is unclear to what extent the results can be reproduced in TRPC knockout mice. Global knockouts of individual *Trpc* genes do not appear to affect normal neuronal development; however, this might be explained by compensation by related channels. An exception was the marked reduction of seizure-induced neuronal death in dorsolateral septum and hippocampi of TRPC1/C4 double knockout mice and in hippocampi of $Trpc5^{-/-}$ mice, indicative of their role in excitotoxic neurodegeneration (Phelan et al., 2012). In this case, the TRPC channels were shown to be pro-death rather than pro-survival.

27.3.2 TRPC CHANNELS IN SYNAPTIC TRANSMISSION

In mature neurons, TRPC channels are best known to mediate membrane depolarization in response to stimulation of GPCRs (Plant and Schaefer, 2005; Bollimuntha et al., 2011) and/or neurotrophin receptors (Li et al., 1999, 2010; Amaral and Pozzo-Miller, 2007a). Comparing to the fast-gated ionotropic receptors (e.g., AMPA and NMDA receptors), TRPC-mediated depolarization and $[Ca^{2+}]_i$ elevation initiate more slowly but last much longer, which may be pivotal for neuromodulation, such as long-term potentiation (LTP), long-term depression (LTD), and synaptic plasticity related to learning and memory. Here, we highlight a few brain areas where TRPC activities in synaptic transmission have been demonstrated and their physiological roles illustrated.

27.3.2.1 Cerebellum

TRPC3 is highly expressed in cerebellar Purkinje cells of adult mice. These cells receive excitatory glutamatergic inputs from parallel fibers. Intensified stimulation (such as a train of electrical stimulations) of parallel fibers often causes spillover of glutamate from synaptic cleft, leading to activation of group I metabotropic glutamate receptors (mGluRs) located in perisynaptic regions of Purkinje cell dendrites. The response is the slow excitatory postsynaptic current (sEPSC), as opposed to the fast EPSC (fEPSC) mediated by AMPA receptors. Initially, it was reported that TRPC1 mediated the sEPSC (Kim et al., 2003), but a later study using multiple strains of TRPC knockout mice demonstrated that the sEPSC was dependent on the expression of TRPC3 but not any other tested TRPC isoforms (Hartmann et al., 2008).

The activation and/or postsynaptic localization of TRPC3 are regulated by GluRδ2, a glutamate insensitive homolog of AMPA receptors. TRPC3 is present in the same protein complex with type 1 mGluR (mGluR1), PKCγ and GluRδ2, and loss of GluRδ2 delayed sEPSC induction (Kato et al., 2012). Interestingly, TRPC1 is also physically associated with mGluR1 and is present in the perisynaptic regions of the parallel fiber Purkinje cell synapses (Kim et al., 2003), arguing that the TRPC1/C3 heteromeric channels may be responsible for sEPSC. However, the sEPSC of Purkinje neurons in *Trpc1*$^{-/-}$ appears normal (Hartmann et al., 2008). Notably, in addition to the $G_{q/11}$-PLCβ signaling, the coupling of mGluR1 stimulation to TRPC3 in Purkinje cells may also require Rho GTPase-mediated activation of phospholipase D (PLD) (Glitsch, 2010), and unlike the results obtained from heterologously expressed TRPC3 channels (Venkatachalam et al., 2003; Kwan et al., 2004; Trebak et al., 2005), the TRPC3-mediated current in Purkinje cells is not sensitive to inhibition by PKC or PKG (Nelson and Glitsch, 2012). Furthermore, supporting the critical role of the C-terminal CIRB site in functional regulation of TRPC channels (Tang et al., 2001; Zhang et al., 2001), an alternative splicing variant lacking about half of the TRPC3 CIRB site (TRPC3c) was found to express specifically in cerebellum and brainstem, and the channel displays enhanced activities, suggestive of reduced feedback inhibition by Ca^{2+} (Kim et al., 2012).

Blocking TRPC channels in rat Purkinje cells attenuated mGluR1-mediated $[Ca^{2+}]_i$ rise, PKC activation, and GluR2 internalization, and as a result, it abolished the induction of cerebellar LTD (Chae et al., 2012). Consistent with the role of TRPC3 in synaptic plasticity of Purkinje cells, *Trpc3*$^{-/-}$ mice exhibit an impaired walking behavior (Hartmann et al., 2008). The *moonwalker* mice harboring a gain-of-function mutation of TRPC3 exhibit cerebellar ataxia and Purkinje cell loss (Becker et al., 2009). Therefore, it is clear that TRPC3-containing channels in cerebellar Purkinje neurons underlie the mechanism of sEPSC evoked by mGluR1 activation and this pathway is critical for mGluR1-dependent cerebellar LTD and motor control and coordination.

27.3.2.2 Hippocampus

Multiple TRPC channels (TRPC1, C3, C4, C5, and C6) are expressed in hippocampal neurons. Similar to TRPC3 in Purkinje cells, the hippocampal TRPCs are also linked to mGluR activation. For example, synergistic activation of mGluR1 and mGluR5 on rat CA3 pyramidal neurons induced a TRPC-like inward current that was greatly enhanced by a $[Ca^{2+}]_i$ rise that was independent of G protein activation (Gee et al., 2003). In rat CA1 pyramidal neurons, group 1 mGluR activation and the accompanied Ca^{2+} waves elicited a sustained membrane depolarization that involved TRPC1, C4, and C5 (El-Hassar et al., 2011). This activity was suppressed by a rise in cAMP. In mouse, application of mGluR agonist DHPG or ACPD to brain slices directly elicited epileptiform discharge in both CA1 and CA3 pyramidal neurons (Wang et al., 2007; Phelan et al., 2013). The discharge in the CA1 neurons was absent in either *Trpc1*$^{-/-}$ or *Trpc1*$^{-/-}$/*Trpc4*$^{-/-}$ double-knockout mice but remained unaltered in *Trpc5*$^{-/-}$ mice (Phelan et al., 2013). By contrast, the high-frequency stimulus-induced LTP at the Schaffer collateral synapses in the hippocampal CA1 region was strongly reduced in TRPC5, but not TRPC1 or TRPC1/C4 double, knockout mice, suggesting distinct role of TRPC1/C4 and TRPC5 channels in this brain region (Phelan et al., 2013).

In addition to mGluRs, stimulation of muscarinic acetylcholine receptors also causes TRPC channel activation in rat CA1 pyramidal neurons. The activation required a brief stimulation of VGCCs via a depolarization pulse or current injection, but the TRPC-like inward current or membrane depolarization (referred to as plateau potential) lasted long after termination of the VGCC activity (Tai et al., 2011). Interestingly, the duration of the TRPC-like response was sensitive to intracellular ATP. The TRPC-like response was prevented by inhibition of CaM or PI3 kinase, treatments that also reduced the effect of carbachol on enhancing surface expression of TRPC5 in hippocampal cells (Tai et al., 2011).

TRPC3 underlies the slowly developing current activated by BDNF released by theta-burst stimulation of afferent inputs of CA1 or mossy fiber inputs of CA3 pyramidal neurons, which represents the most instant postsynaptic effect of BDNF (Amaral and Pozzo-Miller, 2007b; Li et al., 2010). BDNF also increased membrane expression of TRPC3 through PI3 kinase (Amaral and Pozzo-Miller, 2007b). BDNF may also regulate neurotransmitter release through activation of presynaptic TRPC channels as local BDNF stimulation to CA1 neurons increased the frequency of miniature EPSC (mEPSC) in a manner that required both Ca^{2+} release and influx as well as a channel(s) inhibited by SKF96365 (Amaral and Pozzo-Miller, 2012). Remarkably, a mutation in methyl-CpG-binding protein 2 (MeCP2) impaired both pre- and postsynaptic functions of BDNF in mouse CA3 neurons, including decreased BDNF-evoked current amplitude and dendritic Ca^{2+} signals as well as reduced presynaptic BDNF release. Since MeCP2 regulates TRPC3/C6 expression and its mutation is linked to Rett syndrome, increasing TRPC expression and function may be a potential therapeutic strategy to treat Rett syndrome (Li et al., 2012).

27.3.2.3 Hypothalamus

TRPC channels have been shown to mediate neurotransmission of several receptors in hypothalamic neurons, including those involved in energy metabolism, reproduction, and temperature regulation. TRPC channels modulate excitability of orexin/hypocretin and melanin-concentrating hormone (MCH) neurons in lateral hypothalamus (LH) and proopiomelanocortin (POMC) neurons in arcuate nucleus, implicating their involvement in energy metabolism. In the orexin/hypocretin neurons, a TRPC-like inward current and the accompanied Ca^{2+} influx are activated by cholecystokinin (CCK)-8S via type 1 CCK receptors (Tsujino et al., 2005). In several cell models, stimulation of orexin receptors by orexin A elicited TRPC-dependent Ca^{2+} influx and membrane current that were inhibited by PKC and extracellular Mg^{2+} (Larsson et al., 2005; Peltonen et al., 2009). The Ca^{2+} influx may be enhanced by the action of NCX (Louhivuori et al., 2010). Moreover, constitutive channel activity sensitive to block by a TRPC5 antibody was shown to contribute to the depolarized resting membrane potential of the active state of the orexin/hypocretin neurons (Cvetkovic-Lopes et al., 2010).

Thyrotropin-releasing hormone (TRH) inhibits MCH neurons through activation of GABAergic neurons in LH. This action, which may account for TRH's effect on decreasing food intake and sleep, appears to occur through the PLC–TRPC–NCX pathway (Zhang and vanden Pol, 2012).

TRPC channels also mediate the depolarization effect of leptin and serotonin (5-HT) on distinct populations of POMC neurons. Leptin acts at the long-form leptin receptor to activate TRPC1/C4/C5 via the Jak2–PI3K–PLCγ1 pathway (Qiu et al., 2010), while 5-HT acts at the 5-HT2c receptor to stimulate TRPC through the G_q–PLC pathway (Sohn et al., 2011). Leptin also excites kisspeptin neurons in arcuate nucleus via activation of TRPC1/C5 (Qiu et al., 2011).

Kisspeptin, on the other hand, induces depolarization of gonadotropin-releasing hormone (GnRH) neurons through activation of TRPCs (Zhang et al., 2008), implicating a role for these channels in reproduction. Furthermore, histamine acts at H1 histamine receptors to induce both transient and sustained activation of glutamatergic neurons in median preoptic nucleus (MnPO), an area involved in thermoregulation, osmoregulation, and sleep homeostasis. The sustained component, shown as persistent inward current and prolonged elevation of $[Ca^{2+}]_i$, had been attributed to TRPC1/C5 heteromers and was blocked by PKA (Tabarean, 2012).

27.3.2.4 Other brain areas

In several brain areas, such as prefrontal cortex, entorhinal cortex, and lateral septum, the TRPC-mediated response can be elicited by a depolarization pulse(s) in the presence of mGluR or muscarinic agonist, similar to that described earlier for hippocampal neurons. These relatively large (typically >10 mV) and prolonged (>1 s) depolarization responses have been referred to as delayed or slow after depolarization and/or plateau potential (Yan et al., 2009; Zhang et al., 2011) and in some cases thought to be involved in transient storage of memory traces for working memory (Sidiropoulou et al., 2009; Zhang et al., 2011). Depending on the protocol used and the degree of membrane depolarization, the response patterns can be quite variable, including desensitizing and nondesensitizing depolarizations as well as persistent firing and no firing on top of the plateau potential. A common theme is that the response was, typically, not elicited by the receptor agonist alone but required depolarizing current injection (Figure 27.3), which is thought to increase $[Ca^{2+}]_i$ through activation of VGCCs (Yan et al., 2009; Zhang et al., 2011). Thus, the activation of TRPCs through GPCRs in these neurons is greatly facilitated by a rise in $[Ca^{2+}]_i$, a well-described feature for TRPC4 and C5 (Schaefer et al., 2000; Gross et al., 2005; Blair et al., 2009). Supporting the role of TRPC4/C5, the carbachol-induced plateau potentials in entorhinal cortical neurons were inhibited by intracellular application of a peptide that represents the C-terminal PDZ-binding domain of these channels, EQVTTRL (Zhang et al., 2011).

TRPC5 or TRPC5-containing channels in layer III pyramidal neurons of rat entorhinal cortex and lateral and basolateral amygdala neurons are activated following stimulation of CCK2 receptors, causing membrane depolarization and an increase in firing frequency (Faber et al., 2006; Meis et al., 2007; Wang et al., 2011). Also in the amygdala neurons, TRPC-like current and depolarization responses were elicited by local tetanic stimulation in the presence of GABA and ionotropic glutamate receptor antagonists. The responses required group 1 mGluRs (Faber et al., 2006; Meis et al., 2007). Both the CCK2-mediated and tetanic stimulation-evoked responses were markedly reduced in neurons from *Trpc5*$^{-/-}$ mice, which, when combined with the finding that *Trpc5*$^{-/-}$ mice had impaired fear-related behaviors, demonstrated the critical contribution of TRPC5 in lateral amygdala neurons to fear conditioning (Riccio et al., 2009).

Studies using knockout mice have revealed the importance of TRPC4-containing channels in 5-HT-induced dendritic release

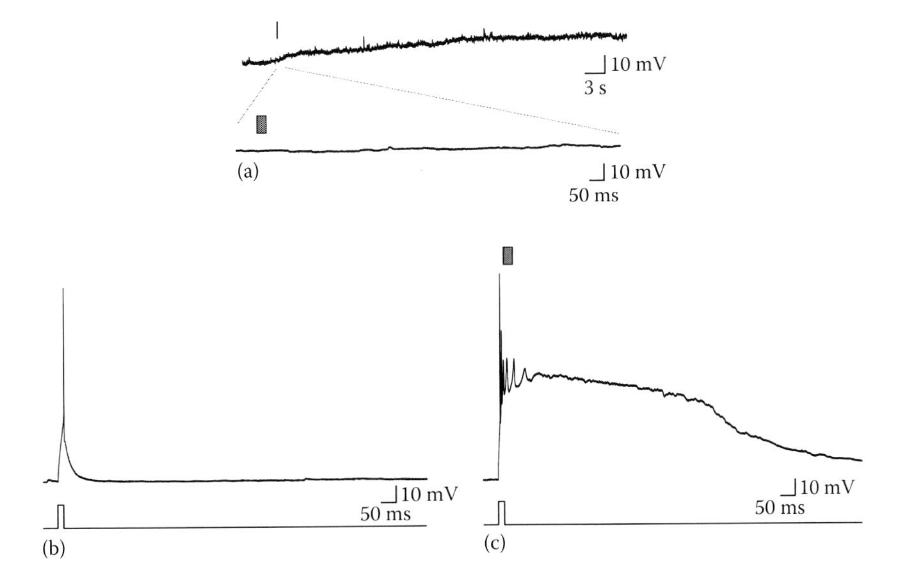

Figure 27.3 TRPC4-mediated plateau potential in lateral septal neurons. Current clamp traces show effects of pressure puffing of (S)-3,5-DHPG (30 μM) onto the soma of a lateral septal neuron in a brain slice from an adult C57BL/6 mouse. (a) The cell was held at −70 mV while DHPG was puffed (10 psi, 30 ms, vertical line or gray box). Lower trace shows expanded time scale. (b) The same cell was held at −80 mV, while a positive current (700 pA, 20 ms) was injected as indicated in the bottom trace. (c) Coapplication of current injection and DHPG puffing elicited a strong plateau depolarizing potential that lasted long after the cessation of the stimuli. No such response was detected in neurons from *Trpc4*$^{-/-}$ mice (*not shown*).

of GABA via 5-HT2 receptors from interneurons of thalamus (Munsch et al., 2003) and mGluR agonist-evoked plateau potential and epileptiform discharge in lateral septal neurons (Phelan et al., 2012). Likewise, the TRPC1/C4 channels are also critical for neurotransmission at dendritic–dendritic synapses between mitral/tufted cells (MTC) and granule cells in the olfactory bulb. Intriguingly, the long-lasting depolarization and sustained Ca^{2+} influx in the granule cells, which were attributed to TRPC1/C4 channels, were dependent on NMDA receptors but not mGluRs. Moreover, the activation of TRPCs in the granule cells also required a coincident Na^+ spike with the NMDA receptor activation (Stroh et al., 2012).

Besides TRPC1/C4/C5, TRPC3 has been implicated in a voltage-independent, TTX-insensitive *background* Na^+ conductance, responsible for the substantial depolarization of GABAergic neurons in the substantia nigra pars reticulata (SNr), a key basal ganglia output nucleus, characterized by depolarized membrane potential and rapid spontaneous firing that encode the basal ganglia output (Atherton and Bevan, 2005; Zhou et al., 2008a). The constitutively active TRPC3 is further enhanced by dopamine released from dendrites of substantia nigra pars compacta (SNc) dopaminergic neurons, which acts at dopamine D_1 and D_5 receptors in SNr GABA neurons. The tonic TRPC3 activity in SNr GABA neurons and its upregulation by dopamine from SNc neurons thus modulate the firing frequency and patterns of SNr GABA neurons and in turn control motor-related behaviors.

While most studies have focused on the functions of TRPC channels in neurons, a few have also explored their roles in glia. For instance, thrombin-induced TRPC3-dependent Ca^{2+} signaling in cortical astrocytes can initiate cellular processes that convert quiescent astrocytes to reactive astrocytes, leading to astrogliosis, implicating a contribution of TRPC3 to brain damage induced by thrombin leak into brain parenchyma, a condition when blood–brain barrier is disrupted during brain injury (Shirakawa et al., 2010). TRPC1 contributes to intracellular Ca^{2+} increase induced by GPCR agonists (e.g., DHPG and ATP) or mechanical stimulation and the subsequent Ca^{2+}-dependent glutamate release from astrocytes isolated from visual cortex (Malarkey et al., 2008).

27.3.3 TRPC CHANNEL FUNCTIONS IN OTHER BODY SYSTEMS

In other body systems, the functions of TRPC channels are also associated with Ca^{2+} and Na^+ influx. It is important to note that not only the Ca^{2+}/Na^+ entry can evoke membrane depolarization but also the Ca^{2+} entry, through preferential activation of Ca^{2+}-activated K^+ channels in certain cell types, can lead to membrane hyperpolarization (Liu et al., 2007; Kwan et al., 2009). Accumulating evidence has demonstrated the roles of TRPCs in the cardiovascular system and their contributions in pathological conditions, including cardiac contractility, arrhythmia, hypertrophy, endothelial cell permeability, vasodilator release, migration, angiogenesis, and atherogenesis, as well as vascular smooth muscle tone, hypoxic response, hypertension, and vascular smooth muscle proliferation (Vassort and Alvarez, 2009; Cioffi et al., 2010; Eder and Molkentin, 2011; Kurdi and Booz, 2011; Wang et al., 2013; Watanabe et al., 2013). In the renal system, mutations in TRPC6 are associated with certain familial forms of focal segmental glomerulosclerosis (FSGS), indicating a critical function for TRPC6 in glomerular filtration in the kidney. The altered TRPC6 activity due to the mutation disrupts slit diaphragm function of the podocytes (Dryer and Reiser, 2010). Evidence also exists for the role of TRPCs in principal cells of the collecting duct and glomerular mesangial cells (Goel and Schilling, 2010; Shen et al., 2011). In the digestive system, TRPC1 plays a critical role in saliva secretion (Liu et al., 2007); TRPC3 is involved in Ca^{2+}-dependent cytotoxicity of pancreatic and salivary gland acini (Kim et al., 2011); TRPC4 and C6 underlie the muscarinic agonist-induced cation current in gastrointestinal smooth muscle cells, responsible for cholinergic control of intestinal contractility and motility (Tsvilovskyy et al., 2009). In addition, expression and function of TRPC channels in hematopoietic cells, skeletal muscles, and bone cells have been documented (Abed et al., 2009; Tano et al., 2011; Gailly, 2012). Stem cell research and cancer studies have also revealed the involvement of these channels (Weick et al., 2009; Davis et al., 2012; Zhen et al., 2013). Therefore, the TRPC channels have been implicated in many physiological functions in almost all body systems and are potential drug targets of many diseases.

27.4 CONCLUDING REMARKS

Being the most closely related to the prototypical *Drosophila* TRP channel, mammalian TRPCs have been extensively studied for their involvements in store- and receptor-operated Ca^{2+} entry and membrane potential regulation downstream from activation of PLC pathways following ligand stimulation of GPCRs and/or RTKs. While the TRPC3/C6/C7 subgroup members are directly activated by DAG, the TRPC4/C5 subgroup members respond to PTX-sensitive $G_{i/o}$ proteins, especially the $G\alpha_{i/o}$ subunits, in a PLC-independent manner. The breakdown of PIP_2 and generation of IP_3 and Ca^{2+} signals also strongly influence the channel activity. TRPC channel functions are also regulated by redox conditions, temperature, mechanical force, phosphorylation/dephosphorylation, extracellular pH, Mg^{2+}, and membrane trafficking. All TRPC channels are Ca^{2+} permeable, but the inward currents at negative potentials are mostly carried by Na^+, eliciting membrane depolarization and/or facilitating Na^+-dependent transport activities. The voltage dependence of TRPC channels is influenced by permeation block by divalent cations, intrinsic voltage dependence, and heteromultimerization with another subunit, for example, TRPC1. The pharmacology for TRPCs is very limited, with most of the drugs used being nonspecific. Ongoing efforts are promising in generating subtype selective TRPC drugs. Despite these limitations, ample examples have revealed endogenous TRPC channel functions in native tissues using a combination of approaches including pharmacological inhibitors, functional blocking antibodies, siRNA, and *Trpc* gene knockout animals. In general, these functions are linked to Ca^{2+} entry and/or membrane potential changes mediated by TRPC channels activated downstream from PLC pathways. However, constitutive basal TRPC channel activities have also been described to exert homeostatic regulation, affecting response probability of neurons and other cell types. The presence of multiple factors that affect the relative response levels of TRPC channels suggests that these channels may serve as coincidence detectors of signaling by divergent inputs, through which they fine-tune the outputs of targeting cells. Continued efforts on elucidating the physiological and pathophysiological functions of TRPC channels and their delicate regulatory mechanisms will further enhance our understanding on cell signaling and health relevance of these fascinating channels.

ACKNOWLEDGMENT

The authors' work is supported in part by grants from U.S. National Institutes of Health (DK081654 and GM081658).

REFERENCES

Abed, E., Labelle, D., Martineau, C., Loghin, A., and Moreau, R. 2009. Expression of transient receptor potential (TRP) channels in human and murine osteoblast-like cells. *Mol. Membr. Biol.* 26:146–158.

Abramowitz, J. and Birnbaumer, L. 2009. Physiology and pathophysiology of canonical transient receptor potential channels. *FASEB J.* 23:297–328.

Albert, A.P. and Large, W.A. 2003. Synergism between inositol phosphates and diacylglycerol on native TRPC6-like channels in rabbit portal vein myocytes. *J. Physiol.* 552:789–795.

Albert, A.P., Saleh, S.N., and Large, W.A. 2008. Inhibition of native TRPC6 channel activity by phosphatidylinositol 4,5-bisphosphate in mesenteric artery myocytes. *J. Physiol.* 586:3087–3095.

Al-Shawaf, E., Naylor, J., Taylor, H., Riches, K., Milligan, C.J., O'Regan, D., Porter, K.E., Li, J., and Beech, D.J. 2010. Short-term stimulation of calcium-permeable transient receptor potential canonical 5-containing channels by oxidized phospholipids. *Arterioscler. Thromb. Vasc. Biol.* 30:1453–1459.

Al-Shawaf, E., Tumova, S., Naylor, J., Majeed, Y., Li, J., and Beech, D.J. 2011. GVI phospholipase A2 role in the stimulatory effect of sphingosine-1-phosphate on TRPC5 cationic channels. *Cell Calcium* 50:343–350.

Amaral, M.D. and Pozzo-Miller, L. 2007a. BDNF induces calcium elevations associated with IBDNF, a nonselective cationic current mediated by TRPC channels. *J. Neurophysiol.* 98:2476–2482.

Amaral, M.D. and Pozzo-Miller, L. 2007b. TRPC3 channels are necessary for brain-derived neurotrophic factor to activate a nonselective cationic current and to induce dendritic spine formation. *J. Neurosci.* 27:5179–5189.

Amaral, M.D. and Pozzo-Miller, L. 2012. Intracellular Ca^{2+} stores and Ca^{2+} influx are both required for BDNF to rapidly increase quantal vesicular transmitter release. *Neural Plast.* 2012:203536.

Atherton, J. and Bevan, M. 2005. Ionic mechanisms underlying autonomous action potential generation in the somata and dendrites of GABAergic substantia nigra pars reticulata neurons in vitro. *J. Neurosci.* 25:8272–8281.

Balzer, M., Lintschinger, B., and Groschner, K. 1999. Evidence for a role of Trp proteins in the oxidative stress-induced membrane conductances of porcine aortic endothelial cells. *Cardiovasc. Res.* 42:543–549.

Basora, N., Boulay, G., Bilodeau, L., Rousseau, E., and Payet, M.D. 2003. 20-Hydroxyeicosatetraenoic acid (20-HETE) activates mouse TRPC6 channels expressed in HEK293 cells. *J. Biol. Chem.* 278:31709–31716.

Bavencoffe, A. and Zhu, M.X. 2012. TRPC proteins as a link between plasma membrane ion transport and intracellular Ca^{2+} stores. In: Groschner, K., Graier, W.F., Romanin, C. (eds.) *Store-Operated Ca^{2+} Entry (SOCE) Pathways, Emerging Signaling Concepts in Human (PATHO) Physiology*. Vienna, Austria: Springer. Chapter 12.

Becker, E.B., Oliver, P.L., Glitsch, M.D., Banks, G.T., Achilli, F., Hardy, A., Nolan, P.M., Fisher, E.M., and Davies, K.E. 2009. A point mutation in TRPC3 causes abnormal Purkinje cell development and cerebellar ataxia in moonwalker mice. *Proc. Natl. Acad. Sci. USA* 106:6706–6711.

Beech, D.J. 2012. Integration of transient receptor potential canonical channels with lipids. *Acta Physiol. (Oxf.)* 204:227–237.

Bezzerides, V.J., Ramsey, I.S., Kotecha, S., Greka, A., and Clapham, D.E. 2004. Rapid vesicular translocation and insertion of TRP channels. *Nat. Cell Biol.* 6:709–720.

Birnbaumer, L. 2009. The TRPC class of ion channels: A critical review of their roles in slow, sustained increases in intracellular Ca^{2+} concentrations. *Annu. Rev. Pharmacol. Toxicol.* 49:395–426.

Blair, N.T., Kaczmarek, J.S., and Clapham, D.E. 2009. Intracellular calcium strongly potentiates agonist-activated TRPC5 channels. *J. Gen. Physiol.* 133:525–546.

Bollimuntha, S., Selvaraj, S., and Singh, B.B. 2011. Emerging roles of canonical TRP channels in neuronal function. *Adv. Exp. Med. Biol.* 704:573–593.

Bousquet, S.M., Monet, M., and Boulay, G. 2010. Protein kinase C-dependent phosphorylation of transient receptor potential canonical 6 (TRPC6) on serine 448 causes channel inhibition. *J. Biol. Chem.* 285:40534–40543.

Bousquet, S.M., Monet, M., and Boulay, G. 2011. The serine 814 of TRPC6 is phosphorylated under unstimulated conditions. *PLoS One* 6:e18121.

Carrasquillo, R., Tian, D., Krishna, S., Pollak, M.R., Greka, A., and Schlöndorff, J. 2012. SNF8, a member of the ESCRT-II complex, interacts with TRPC6 and enhances its channel activity. *BMC Cell Biol.* 13:33.

Chae, H.G., Ahn, S.J., Hong, Y.H., Chang, W.S., Kim, J., and Kim, S.J. 2012. Transient receptor potential canonical channels regulate the induction of cerebellar long-term depression. *J. Neurosci.* 32:12909–12914.

Chakraborty, S., Berwick, Z.C., Bartlett, P.J., Kumar, S., Thomas, A.P., Sturek, M., Tune, J.D., and Obukhov, A.G. 2011. Bromoenol lactone inhibits voltage-gated Ca^{2+} and transient receptor potential canonical channels. *J. Pharmacol. Exp. Ther.* 339:329–340.

Cheng, K.T., Liu, X., Ong, H.L., Swaim, W., and Ambudkar, I.S. 2011. Local Ca^{2+} entry via Orai1 regulates plasma membrane recruitment of TRPC1 and controls cytosolic Ca^{2+} signals required for specific cell functions. *PLoS Biol.* 9:e1001025.

Cioffi, D.L., Barry, C., and Stevens, T. 2010. Store-operated calcium entry channels in pulmonary endothelium: The emerging story of TRPCS and Orai1. *Adv. Exp. Med. Biol.* 661:137–154.

Cioffi, D.L., Wu, S., Alexeyev, M., Goodman, S.R., Zhu, M.X., and Stevens, T. 2005. Activation of the endothelial store-operated I_{SOC} Ca^{2+} channel requires interaction of protein 4.1 with TRPC4. *Circ. Res.* 97:1164–1172.

Clapham, D.E. 2003. TRP channels as cellular sensors. *Nature* 426:517–524.

Cosens, D.J. and Manning, A. 1969. Abnormal electroretinogram from a *Drosophila* mutant. *Nature* 224:285–287.

Cvetkovic-Lopes, V., Eggermann, E., Uschakov, A., Grivel, J., Bayer, L., Jones, B.E., Serafin, M., and Muhlethaler, M. 2010. Rat hypocretin/orexin neurons are maintained in a depolarized state by TRPC channels. *PLoS One* 5:e15673.

Davare, M., Fortin, D., Saneyoshi, T., Nygaard, S., Kaech, S., Banker, G., Soderling, T., and Wayman, G. 2009. Transient receptor potential canonical 5 channels activate Ca^{2+}/calmodulin kinase Ig to promote axon formation in hippocampal neurons. *J. Neurosci.* 29:9794–9808.

Davis, F.M., Peters, A.A., Grice, D.M., Cabot, P.J., Parat, M.O., Roberts-Thomson, S.J., and Monteith, G.R. 2012. Non-stimulated, agonist-stimulated and store-operated Ca^{2+} influx in MDA-MB-468 breast cancer cells and the effect of EGF-induced EMT on calcium entry. *PLoS One* 7:e36923.

Dresviannikov, A.V., Bolton, T.B., and Zholos, A.V. 2006. Muscarinic receptor-activated cationic channels in murine ileal myocytes. *Br. J. Pharmacol.* 149:179–187.

Dryer, S.E. and Reiser, J. 2010. TRPC6 channels and their binding partners in podocytes: Role in glomerular filtration and pathophysiology. *Am. J. Physiol. Renal Physiol.* 299:F689–F701.

Eder, P. and Molkentin, J.D. 2011. TRPC channels as effectors of cardiac hypertrophy. *Circ. Res.* 108:265–272.

Eder, P., Poteser, M., Romanin, C., and Groschner, K. 2005. Na^+ entry and modulation of Na^+/Ca^{2+} exchange as a key mechanism of TRPC signaling. *Pflugers Arch.* 451:99–104.

Eder, P., Probst, D., Rosker, C., Poteser, M., Wolinski, H., Kohlwein, S.D., Romanin, C., and Groschner, K. 2007a. Phospholipase C-dependent control of cardiac calcium homeostasis involves a TRPC3-NCX1 signaling complex. *Cardiovasc. Res.* 73:111–119.

Eder, P., Schindl, R., Romanin, C., and Groschner, K. 2007b. Protein–protein interactions in TRPC channel complexes. In: Liedtke, W.B., Heller, S. (eds.) *TRP Ion Channel Function in Sensory Transduction and Cellular Signaling Cascades*. Boca Raton, FL: CRC Press. Chapter 24.

El-Hassar, L., Hagenston, A.M., D'Angelo, L.B., and Yeckel, M.F. 2011. Metabotropic glutamate receptors regulate hippocampal CA1 pyramidal neuron excitability via Ca^{2+} wave-dependent activation of SK and TRPC channels. *J. Physiol.* 589:3211–3229.

Faber, E., Sedlak, P., Vidovic, M., and Sah, P. 2006. Synaptic activation of transient receptor potential channels by metabotropic glutamate receptors in the lateral amygdala. *Neuroscience* 137:781–794.

Fleming, I., Rueben, A., Popp, R., Fisslthaler, B., Schrodt, S., Sander, A., Haendeler, J. et al. 2007. Epoxyeicosatrienoic acids regulate Trp channel dependent Ca^{2+} signaling and hyperpolarization in endothelial cells. *Arterioscler. Thromb. Vasc. Biol.* 27:2612–2618.

Fortin, D., Srivastava, T., Dwarakanath, D., Pierre, P., Nygaard, S., Derkach, V., and Soderling, T. 2012. Brain-derived neurotrophic factor activation of CaM-kinase kinase via transient receptor potential canonical channels induces the translation and synaptic incorporation of GluA1-containing calcium-permeable AMPA receptors. *J. Neurosci.* 32:8127–8137.

Fowler, M.A., Sidiropoulou, K., Ozkan, E.D., Phillips, C.W., and Cooper, D.C. 2007. Corticolimbic expression of TRPC4 and TRPC5 channels in the rodent brain. *PLoS One* 2:e573.

Freichel, M., Vennekens, R., Olausson, J., Hoffmann, M., Müller, C., Stolz, S., Scheunemann, J., Weissgerber, P., and Flockerzi, V. 2004. Functional role of TRPC proteins in vivo: Lessons from TRPC-deficient mouse models. *Biochem. Biophys. Res. Commun.* 322:1352–1358.

Gailly, P. 2012. TRP channels in normal and dystrophic skeletal muscle. *Curr. Opin. Pharmacol.* 12:326–334.

Gee, C., Benquet, P., and Gerber, U. 2003. Group I metabotropic glutamate receptors activate a calcium-sensitive transient receptor potential-like conductance in rat hippocampus. *J. Physiol.* 546:655–664.

Gees, M., Colsoul, B., and Nilius, B. 2010. The role of transient receptor potential cation channels in Ca^{2+} signaling. *Cold Spring Harb. Perspect. Biol.* 2:a003962.

Glitsch, M.D. 2010. Activation of native TRPC3 cation channels by phospholipase D. *FASEB J.* 24:318–325.

Goel, M. and Schilling, W.P. 2010. Role of TRPC3 channels in ATP-induced Ca^{2+} signaling in principal cells of the inner medullary collecting duct. *Am. J. Physiol. Renal Physiol.* 299:F225–F233.

Goel, M., Zuo, C.D., and Schilling, W.P. 2010. Role of cAMP/PKA signaling cascade in vasopressin-induced trafficking of TRPC3 channels in principal cells of the collecting duct. *Am. J. Physiol. Renal Physiol.* 298:F988–F996.

Gottlieb, P., Folgering, J., Maroto, R., Raso, A., Wood, T.G., Kurosky, A., Bowman, C. et al. 2008. Revisiting TRPC1 and TRPC6 mechanosensitivity. *Pflugers Arch.* 455:1097–1103.

Graham, S., Ding, M., Ding, Y., Sours-Brothers, S., Luchowski, R., Gryczynski, Z., Yorio, T., Ma, H., and Ma, R. 2010. Canonical transient receptor potential 6 (TRPC6), a redox-regulated cation channel. *J. Biol. Chem.* 285:23466–23476.

Greka, A. and Mundel, P. 2011. Balancing calcium signals through TRPC5 and TRPC6 in podocytes. *J. Am. Soc. Nephrol.* 22:1969–1980.

Gross, S.A., Guzmán, G.A., Wissenbach, U., Philipp, S.E., Zhu, M.X., Bruns, D., and Cavalié, A. 2009. TRPC5 is a Ca^{2+}-activated channel functionally coupled to Ca^{2+}-selective ion channels. *J. Biol. Chem.* 284:34423–34432.

Harteneck, C. and Gollasch, M. 2011. Pharmacological modulation of diacylglycerol-sensitive TRPC3/6/7 channels. *Curr. Pharm. Biotechnol.* 12:35–41.

Hartmann, J., Dragicevic, E., Adelsberger, H., Henning, H., Sumser, M., Abramowitz, J., Blum, R. et al. 2008. TRPC3 channels are required for synaptic transmission and motor coordination. *Neuron* 59:392–398.

He, Z., Jia, C., Feng, S., Zhou, K., Tai, Y., Bai, X., and Wang, Y. 2012. TRPC5 channel is the mediator of neurotrophin-3 in regulating dendritic growth via CaMKIIα in rat hippocampal neurons. *J. Neurosci.* 32:9383–9395.

Hisatsune, C., Kuroda, Y., Nakamura, K., Inoue, T., Nakamura, T., Michikawa, T., Mizutani, A., and Mikoshiba, K. 2004. Regulation of TRPC6 channel activity by tyrosine phosphorylation. *J. Biol. Chem.* 279:18887–18894.

Hofmann, T., Obukhov, A.G., Schaefer, M., Harteneck, C., Gudermann, T., and Schultz, G. 1999. Direct activation of human TRPC6 and TRPC3 channels by diacylglycerol. *Nature* 397:259–263.

Horinouchi, T., Higa, T., Aoyagi, H., Nishiya, T., Terada, K., and Miwa, S. 2011. Adenylate cyclase/cAMP/protein kinase A signaling pathway inhibits endothelin type A receptor-operated Ca^{2+} entry mediated via transient receptor potential canonical 6 channels. *J. Pharmacol. Exp. Ther.* 340:143–151.

Huang, J., Liu, C.H., Hughes, S.A., Postma, M., Schwiening, C.J., and Hardie, R.C. 2010. Activation of TRP channels by protons and phosphoinositide depletion in *Drosophila* photoreceptors. *Curr. Biol.* 20:189–197.

Imai, Y., Itsuki, K., Okamura, Y., Inoue, R., and Mori, M.X. 2012. A self-limiting regulation of vasoconstrictor-activated TRPC3/C6/C7 channels coupled to $PI(4,5)P_2$-diacylglycerol signalling. *J. Physiol.* 590:1101–1119.

Inoue, R., Jensen, L.J., Jian, Z., Shi, J., Hai, L., Lurie, A.I., Henriksen, F.H. et al. 2009. Synergistic activation of vascular TRPC6 channel by receptor and mechanical stimulation via phospholipase C/diacylglycerol and phospholipase A2/omega-hydroxylase/20-HETE pathways. *Circ. Res.* 104:1399–1409.

Inoue, R., Okada, T., Onoue, H., Hara, Y., Shimizu, S., Naitoh, S., Ito, Y., and Mori, Y. 2001. The transient receptor potential protein homologue TRP6 is the essential component of vascular α1-adrenoceptor-activated Ca^{2+}-permeable cation channel. *Circ. Res.* 88:325–332.

Jeon, J.P., Hong, C., Park, E.J., Jeon, J.H., Cho, N.H., Kim, I.G., Choe, H., Muallem, S., Kim, H.J., and So, I. 2012. Selective Gai subunits as novel direct activators of transient receptor potential canonical (TRPC)4 and TRPC5 channels. *J. Biol. Chem.* 287:17029–17039.

Jeon, J.P., Lee, K.P., Park, E.J., Sung, T.S., Kim, B.J., Jeon, J.H., and So, I. 2008. The specific activation of TRPC4 by Gi protein subtype. *Biochem. Biophys. Res. Commun.* 377:538–543.

Jia, Y., Zhou, J., Tai, Y., and Wang, Y. 2007. TRPC channels promote cerebellar granule neuron survival. *Nat. Neurosci.* 10:559–567.

Jung, S., Mühle, A., Schaefer, M., Strotmann, R., Schultz, G., and Plant, T.D. 2003. Lanthanides potentiate TRPC5 currents by an action at extracellular sites close to the pore mouth. *J. Biol. Chem.* 278:3562–3571.

Kaczmarek, J., Riccio, A., and Clapham, D. 2012. Calpain cleaves and activates the TRPC5 channel to participate in semaphorin 3A-induced neuronal growth cone collapse. *Proc. Natl. Acad. Sci. USA* 109:7888–7892.

Kanda, S., Harita, Y., Shibagaki, Y., Sekine, T., Igarashi, T., Inoue, T., and Hattori, S. 2011. Tyrosine phosphorylation-dependent activation of TRPC6 regulated by PLC-γ1 and nephrin: Effect of mutations associated with focal segmental glomerulosclerosis. *Mol. Biol. Cell* 22:1824–1835.

Kato, A.S., Knierman, M.D., Siuda, E.R., Isaac, J.T., Nisenbaum, E.S., and Bredt, D.S. 2012. Glutamate receptor delta2 associates with metabotropic glutamate receptor 1 (mGluR1), protein kinase Cgamma, and canonical transient receptor potential 3 and regulates mGluR1-mediated synaptic transmission in cerebellar Purkinje neurons. *J. Neurosci.* 32:15296–15308.

Kawasaki, B.T., Liao, Y., and Birnbaumer, L. 2006. Role of Src in C3 transient receptor potential channel function and evidence for a heterogeneous makeup of receptor- and store-operated Ca^{2+} entry channels. *Proc. Natl. Acad. Sci. USA* 103:335–340.

Kim, B.J., Kim, M.T., Jeon, J.H., Kim, S.J., and So, I. 2008a. Involvement of phosphatidylinositol 4,5-bisphosphate in the desensitization of canonical transient receptor potential 5. *Biol. Pharm. Bull.* 31:1733–1738.

Kim, H., Jeon, J.P., Hong, C., Kim, J., Myeong, J., Jeon, J.H., and So, I. 2013. An essential role of $PI(4,5)P_2$ for maintaining the activity of the transient receptor potential canonical (TRPC)4β. *Pflugers Arch.* 465:1011–1021.

Kim, J.Y. and Saffen, D. 2005. Activation of M1 muscarinic acetylcholine receptors stimulates the formation of a multiprotein complex centered on TRPC6 channels. *J. Biol. Chem.* 280:32035–32047.

Kim, M.J., Jeon, J.P., Kim, H.J., Kim, B.J., Lee, Y.M., Choe, H., Jeon, J.H., Kim, S.J., and So I. 2008b. Molecular determinant of sensing extracellular pH in classical transient receptor potential channel 5. *Biochem. Biophys. Res. Commun.* 365:239–245.

Kim, M.S., Lee, K.P., Yang, D., Shin, D.M., Abramowitz, J., Kiyonaka, S., Birnbaumer, L., Mori, Y., and Muallem, S. 2011. Genetic and pharmacologic inhibition of the Ca^{2+} influx channel TRPC3 protects secretory epithelia from Ca^{2+}-dependent toxicity. *Gastroenterology* 140:2107–2115.

Kim, S.J., Kim, Y.S., Yuan, J.P., Petralia, R.S., Worley, P.F., and Linden, D.J. 2003. Activation of the TRPC1 cation channel by metabotropic glutamate receptor mGluR1. *Nature* 426:285–291.

Kim, Y., Wong, A., Power, J., Tadros, S., Klugmann, M., Moorhouse, A., Bertrand, P., and Housley, G. 2012. Alternative splicing of the TRPC3 ion channel calmodulin/IP3 receptor-binding domain in the hindbrain enhances cation flux. *J. Neurosci.* 32:11414–11423.

Kiselyov, K., Kim, J.Y., Zeng, W., and Muallem, S. 2005. Protein–protein interaction and functionTRPC channels. *Pflugers Arch.* 451:116–124.

Kiyonaka, S., Kato, K., Nishida, M., Mio, K., Numaga, T., Sawaguchi, Y., Yoshida, T. et al. 2009. Selective and direct inhibition of TRPC3 channels underlies biological activities of a pyrazole compound. *Proc. Natl. Acad. Sci. USA* 106:5400–5405.

Kraft, R. 2007. The Na^+/Ca^{2+} exchange inhibitor KB-R7943 potently blocks TRPC channels. *Biochem. Biophys. Res. Commun.* 361:230–236.

Kurdi, M. and Booz, G.W. 2011. Three 4-letter words of hypertension-related cardiac hypertrophy: TRPC, mTOR, and HDAC. *J. Mol. Cell. Cardiol.* 50:964–971.

Kwan, H.Y., Huang, Y., and Yao, X. 2004. Regulation of canonical transient receptor potential isoform 3 (TRPC3) channel by protein kinase G. *Proc. Natl. Acad. Sci. USA* 101:2625–2630.

Kwan, H.Y., Huang, Y., and Yao, X. 2006. Protein kinase C can inhibit TRPC3 channels indirectly via stimulating protein kinase G. *J. Cell Physiol.* 207:315–321.

Kwan, H.Y., Shen, B., Ma, X., Kwok, Y.C., Huang, Y., Man, Y.B., Yu, S., and Yao, X. 2009. TRPC1 associates with BK(Ca) channel to form a signal complex in vascular smooth muscle cells. *Circ. Res.* 104:670–678.

Kwon, Y., Hofmann, T., and Montell, C. 2007. Integration of phosphoinositide- and calmodulin-mediated regulation of TRPC6. *Mol. Cell* 25:491–503.

Larsson, K.P., Peltonen, H.M., Bart, G., Louhivuori, L.M., Penttonen, A., Antikainen, M., Kukkonen, J.P., and Akerman, K.E. 2005. Orexin-A-induced Ca^{2+} entry: Evidence for involvement of trpc channels and protein kinase C regulation. *J. Biol. Chem.* 280:1771–1781.

Lee, J.E., Song, M.Y., Shin, S.K., Bae, S.H., and Park, K.S. 2012. Mass spectrometric analysis of novel phosphorylation sites in the TRPC4β channel. *Rapid Commun. Mass Spectrom.* 26:1965–1970.

Lemonnier, L., Trebak, M., and Putney, J.W. Jr. 2008. Complex regulation of the TRPC3, 6 and 7 channel subfamily by diacylglycerol and phosphatidylinositol-4,5-bisphosphate. *Cell Calcium* 43:506–514.

Leuner, K., Kazanski, V., Müller, M., Essin, K., Henke, B., Gollasch, M., Harteneck, C., and Müller, W.E. 2007. Hyperforin—A key constituent of St. John's wort specifically activates TRPC6 channels. *FASEB J.* 21:4101–4111.

Li, H.S., Xu, X.Z., and Montell, C. 1999. Activation of a TRPC3-dependent cation current through the neurotrophin BDNF. *Neuron* 24:261–273.

Li, W., Calfa, G., Larimore, J., and Pozzo-Miller, L. 2012. Activity-dependent BDNF release and TRPC signaling is impaired in hippocampal neurons of Mecp2 mutant mice. *Proc. Natl. Acad. Sci. USA* 109:17087–17092.

Li, Y., Calfa, G., Inoue, T., Amaral, M., and Pozzo-Miller, L. 2010. Activity-dependent release of endogenous BDNF from mossy fibers evokes a TRPC3 current and Ca^{2+} elevations in CA3 pyramidal neurons. *J. Neurophysiol.* 103:2846–2856.

Li, Y., Jia, Y.C., Cui, K., Li, N., Zheng, Z.Y., Wang, Y.Z., and Yuan, X.B. 2005. Essential role of TRPC channels in the guidance of nerve growth cones by brain-derived neurotrophic factor. *Nature* 434:894–898.

Liu, X., Cheng, K.T., Bandyopadhyay, B.C., Pani, B., Dietrich, A., Paria, B.C., Swaim, W.D. et al. 2007. Attenuation of store-operated Ca^{2+} current impairs salivary gland fluid secretion in $TRPC1^{-/-}$ mice. *Proc. Natl. Acad. Sci. USA* 104:17542–17547.

Louhivuori, L.M., Jansson, L., Nordstrom, T., Bart, G., Nasman, J., and Akerman, K.E. 2010. Selective interference with TRPC3/6 channels disrupts OX1 receptor signalling via NCX and reveals a distinct calcium influx pathway. *Cell Calcium* 48:114–123.

Lussier, M.P., Cayouette, S., Lepage, P.K., Bernier, C.L., Francoeur, N., St-Hilaire, M., Pinard, M., and Boulay, G. 2005. MxA, a member of the dynamin superfamily, interacts with the ankyrin-like repeat domain of TRPC. *J. Biol. Chem.* 280:19393–19400.

Lussier, M.P., Lepage, P.K., Bousquet, S.M., and Boulay, G. 2008. RNF24, a new TRPC interacting protein, causes the intracellular retention of TRPC. *Cell Calcium* 43:432–443.

Majeed, Y., Amer, M.S., Agarwal, A.K., McKeown, L., Porter, K.E., O'Regan, D.J., Naylor, J., Fishwick, C.W., Muraki, K., and Beech, D.J. 2011. Stereo-selective inhibition of transient receptor potential TRPC5 cation channels by neuroactive steroids. *Br. J. Pharmacol.* 162:1509–1520.

Malarkey, E., Ni, Y., and Parpura, V. 2008. Ca^{2+} entry through TRPC1 channels contributes to intracellular Ca^{2+} dynamics and consequent glutamate release from rat astrocytes. *Glia* 56:821–835.

Maroto, R., Raso, A., Wood, T.G., Kurosky, A., Martinac, B., and Hamill, O.P. 2005. TRPC1 forms the stretch-activated cation channel in vertebrate cells. *Nat. Cell Biol.* 7:179–185.

Mederos y Schnitzler, M., Storch, U., Meibers, S., Nurwakagari, P., Breit, A., Essin, K., Gollasch, M., and Gudermann, T. 2008. Gq-coupled receptors as mechanosensors mediating myogenic vasoconstriction. *EMBO J.* 27:3092–3103.

Meis, S., Munsch, T., Sosulina, L., and Pape, H.C. 2007. Postsynaptic mechanisms underlying responsiveness of amygdaloid neurons to cholecystokinin are mediated by a transient receptor potential-like current. *Mol. Cell. Neurosci.* 35:356–367.

Miehe, S., Bieberstein, A., Arnould, I., Ihdene, O., Rütten, H., and Strübing, C. 2010. The phospholipid-binding protein SESTD1 is a novel regulator of the transient receptor potential channels TRPC4 and TRPC5. *J. Biol. Chem.* 285:12426–12434.

Miehe, S., Crause, P., Schmidt, T., Löhn, M., Kleemann, H.W., Licher, T., Dittrich, W., Rütten, H., and Strübing, C. 2012. Inhibition of diacylglycerol-sensitive TRPC channels by synthetic and natural steroids. *PLoS One* 7:e35393.

Miller, M., Shi, J., Zhu, Y., Kustov, M., Tian, J.B., Stevens, A., Wu, M., Xu, J. et al. 2011. Identification of ML204, a novel potent antagonist that selectively modulates native TRPC4/C5 ion channels. *J. Biol. Chem.* 286:33436–33446.

Montell, C., Birnbaumer, L., Flockerzi, V., Bindels, R.J., Bruford, E.A., Caterina, M.J., Clapham, D.E. et al. 2002. A unified nomenclature for the superfamily of TRP cation channels. *Mol. Cell* 9:229–231.

Munsch, T., Freichel, M., Flockerzi, V., and Pape, H.C. 2003. Contribution of transient receptor potential channels to the control of GABA release from dendrites. *Proc. Natl. Acad. Sci. USA* 100:16065–16070.

Nelson, C. and Glitsch, M.D. 2012. Lack of kinase regulation of canonical transient receptor potential 3 (TRPC3) channel-dependent currents in cerebellar Purkinje cells. *J. Biol. Chem.* 287:6326–6335.

Nishioka, K., Nishida, M., Ariyoshi, M., Jian, Z., Saiki, S., Hirano, M., Nakaya, M. et al. 2011. Cilostazol suppresses angiotensin II-induced vasoconstriction via protein kinase A-mediated phosphorylation of the transient receptor potential canonical 6 channel. *Arterioscler. Thromb. Vasc. Biol.* 31:2278–2286.

Obukhov, A.G. and Nowycky, M.C. 2005. A cytosolic residue mediates Mg^{2+} block and regulates inward current amplitude of a transient receptor potential channel. *J. Neurosci.* 25:1234–1239.

Obukhov, A.G. and Nowycky, M.C. 2008. TRPC5 channels undergo changes in gating properties during the activation-deactivation cycle. *J. Cell Physiol.* 216:162–171.

Odell, A.F., Scott, J.L., and Van Helden, D.F. 2005. Epidermal growth factor induces tyrosine phosphorylation, membrane insertion, and activation of transient receptor potential channel 4. *J. Biol. Chem.* 280:37974–37987.

Odell, A.F., Van Helden, D.F., and Scott, J.L. 2008. The spectrin cytoskeleton influences the surface expression and activation of human transient receptor potential channel 4 channels. *J. Biol. Chem.* 283:4395–4407.

Ong, H.L. and Ambudkar, I.S. 2011. The dynamic complexity of the TRPC1 channelosome. *Channels (Austin)* 5:424–431.

Ordaz, B., Tang, J., Xiao, R., Salgado, A., Sampieri, A., Zhu, M.X., and Vaca, L. 2005. Calmodulin and calcium interplay in the modulation of TRPC5 channel activity. Identification of a novel C-terminal domain for calcium/calmodulin-mediated facilitation. *J. Biol. Chem.* 280:30788–30796.

Otsuguro, K., Tang, J., Tang, Y., Xiao, R., Freichel, M., Tsvilovskyy, V., Ito, S., Flockerzi, V., Zhu, M.X., and Zholos, A.V. 2008. Isoform-specific inhibition of TRPC4 channel by phosphatidylinositol 4,5-bisphosphate. *J. Biol. Chem.* 283:10026–10036.

Park, H.W., Kim, J.Y., Choi, S.K., Lee, Y.H., Zeng, W., Kim, K.H., Muallem, S., and Lee, M.G. 2011. Serine-threonine kinase with-no-lysine 4 (WNK4) controls blood pressure via transient receptor potential canonical 3 (TRPC3) in the vasculature. *Proc. Natl. Acad. Sci. USA* 108:10750–10755.

Parnas, M., Katz, B., and Minke, B. 2007. Open channel block by Ca^{2+} underlies the voltage dependence of *Drosophila* TRPL channel. *J. Gen. Physiol.* 129:17–28.

Peltonen, H.M., Magga, J.M., Bart, G., Turunen, P.M., Antikainen, M.S., Kukkonen, J.P., and Akerman, K.E. 2009. Involvement of TRPC3 channels in calcium oscillations mediated by OX_1 orexin receptors. *Biochem. Biophys. Res. Commun.* 385:408–412.

Phelan, K., Shwe, U., Abramowitz, J., Wu, H., Rhee, S., Howell, M., Gottschall, P. et al. 2013. Canonical transient receptor channel 5 (TRPC5) and TRPC1/4 contribute to seizure and excitotoxicity by distinct cellular mechanisms. *Mol. Pharmacol.* 83:429–438.

Phelan, K.D., Mock, M.M., Kretz, O., Shwe, U.T., Kozhemyakin, M., Greenfield, L.J., Dietrich, A. et al. 2012. Heteromeric canonical transient receptor potential 1 and 4 channels play a critical role in epileptiform burst firing and seizure-induced neurodegeneration. *Mol. Pharmacol.* 81:384–392.

Plant, T.D. and Schaefer, M. 2005. Receptor-operated cation channels formed by TRPC4 and TRPC5. *Naunyn Schmiedebergs Arch. Pharmacol.* 371:266–276.

Poteser, M., Graziani, A., Rosker, C., Eder, P., Derler, I., Kahr, H., Zhu, M.X., Romanin, C., and Groschner, K. 2006. TRPC3 and TRPC4 associate to form a redox-sensitive cation channel. Evidence for expression of native TRPC3-TRPC4 heteromeric channels in endothelial cells. *J. Biol. Chem.* 281:13588–13595.

Poteser, M., Schleifer, H., Lichtenegger, M., Schernthaner, M., Stockner, T., Kappe, C.O., Glasnov, T.N., Romanin, C., and Groschner, K. 2011. PKC-dependent coupling of calcium permeation through transient receptor potential canonical 3 (TRPC3) to calcineurin signaling in HL-1 myocytes. *Proc. Natl. Acad. Sci. USA* 108:10556–10561.

Qiu, J., Fang, Y., Bosch, M., Rønnekleiv, O., and Kelly, M. 2011. Guinea pig kisspeptin neurons are depolarized by leptin via activation of TRPC channels. *Endocrinology* 152:1503–1514.

Qiu, J., Fang, Y., Ronnekleiv, O.K., and Kelly, M.J. 2010. Leptin excites proopiomelanocortin neurons via activation of TRPC channels. *J. Neurosci.* 30:1560–1565.

Riccio, A., Li, Y., Moon, J., Kim, K.S., Smith, K.S., Rudolph, U., Gapon, S. et al. 2009. Essential role for TRPC5 in amygdala function and fear-related behavior. *Cell* 137:761–772.

Schaefer, M., Plant, T.D., Obukhov, A.G., Hofmann, T., Gudermann, T., and Schultz, G. 2000. Receptor-mediated regulation of the nonselective cation channels TRPC4 and TRPC5. *J. Biol. Chem.* 275:17517–17526.

Schleifer, H., Doleschal, B., Lichtenegger, M., Oppenrieder, R., Derler, I., Frischauf, I., Glasnov, T.N., Kappe, C.O., Romanin, C., and Groschner, K.2012. Novel pyrazole compounds for pharmacological discrimination between receptor-operated and store-operated Ca^{2+} entry pathways. *Br. J. Pharmacol.* 167:1712–1722.

Semtner, M., Schaefer, M., Pinkenburg, O., and Plant, T.D. 2007. Potentiation of TRPC5 by protons. *J. Biol. Chem.* 282:33868–33878.

Sergeeva, O., Korotkova, T., Scherer, A., Brown, R., and Haas, H. 2003. Co-expression of non-selective cation channels of the transient receptor potential canonical family in central aminergic neurones. *J. Neurochem.* 85:1547–1552.

Shen, B., Kwan, H.Y., Ma, X., Wong, C.O., Du, J., Huang, Y., and Yao, X. 2011. cAMP activates TRPC6 channels via the phosphatidylinositol 3-kinase (PI3K)-protein kinase B (PKB)-mitogen-activated protein kinase kinase (MEK)-ERK1/2 signaling pathway. *J. Biol. Chem.* 286:19439–19445.

Shi, J., Ju, M., Abramowitz, J., Large, W.A., Birnbaumer, L., and Albert, A.P. 2012. TRPC1 proteins confer PKC and phosphoinositol activation on native heteromeric TRPC1/C5 channels in vascular smooth muscle: Comparative study of wild-type and $TRPC1^{-/-}$ mice. *FASEB J.* 26:409–419.

Shi, J., Mori, E., Mori, Y., Mori, M., Li, J., Ito, Y., and Inoue, R. 2004. Multiple regulation by calcium of murine homologues of transient receptor potential proteins TRPC6 and TRPC7 expressed in HEK293 cells. *J. Physiol.* 561:415–432.

Shim, S., Goh, E.L., Ge, S., Sailor, K., Yuan, J.P., Roderick, H.L., Bootman, M.D., Worley, P.F., Song, H., and Ming, G.L. 2005. XTRPC1-dependent chemotropic guidance of neuronal growth cones. *Nat. Neurosci.* 8:730–735.

Shirakawa, H., Sakimoto, S., Nakao, K., Sugishita, A., Konno, M., Iida, S., Kusano, A., Hashimoto, E., Nakagawa, T., and Kaneko, S. 2010. Transient receptor potential canonical 3 (TRPC3) mediates thrombin-induced astrocyte activation and upregulates its own expression in cortical astrocytes. *J. Neurosci.* 30:13116–13129.

Sidiropoulou, K., Lu, F.M., Fowler, M.A., Xiao, R., Phillips, C., Ozkan, E.D., Zhu, M.X., White, F.J., and Cooper, D.C. 2009. Dopamine modulates an mGluR5-mediated depolarization underlying prefrontal persistent activity. *Nat. Neurosci.* 12:190–199.

Singh, B.B., Liu, X., Tang, J., Zhu, M.X., and Ambudkar, I.S. 2002. Calmodulin regulates Ca^{2+}-dependent feedback inhibition of store-operated Ca^{2+} influx by interaction with a site in the C terminus of TrpC1. *Mol. Cell* 9:739–750.

Smyth, J.T., Lemonnier, L., Vazquez, G., Bird, G.S., and Putney, J.W. Jr. 2006. Dissociation of regulated trafficking of TRPC3 channels to the plasma membrane from their activation by phospholipase C. *J. Biol. Chem.* 281:11712–11720.

Sohn, J.-W., Xu, Y., Jones, J., Wickman, K., Williams, K., and Elmquist, J. 2011. Serotonin 2C receptor activates a distinct population of arcuate pro-opiomelanocortin neurons via TRPC channels. *Neuron* 71:488–497.

Spassova, M.A., Hewavitharana, T., Xu, W., Soboloff, J., and Gill, D.L. 2006. A common mechanism underlies stretch activation and receptor activation of TRPC6 channels. *Proc. Natl. Acad. Sci. USA* 103:16586–16591.

Stroh, O., Freichel, M., Kretz, O., Birnbaumer, L., Hartmann, J., and Egger, V. 2012. NMDA receptor-dependent synaptic activation of TRPC channels in olfactory bulb granule cells. *J. Neurosci.* 32:5737–5746.

Strübing, C., Krapivinsky, G., Krapivinsky, L., and Clapham, D.E. 2001. TRPC1 and TRPC5 form a novel cation channel in mammalian brain. *Neuron* 29:645–655.

Strübing, C., Krapivinsky, G., Krapivinsky, L., and Clapham, D.E. 2003. Formation of novel TRPC channels by complex subunit interactions in embryonic brain. *J. Biol. Chem.* 278:39014–39019.

Sung, T.S., Jeon, J.P., Kim, B.J., Hong, C., Kim, S.Y., Kim, J., Jeon, J.H. et al. 2005. Molecular determinants of PKA-dependent inhibition of TRPC5 channel. *Am. J. Physiol. Cell Physiol.* 301:C823–C832.

Tabarcan, I.V. 2012. Persistent histamine excitation of glutamatergic preoptic neurons. *PLoS One* 7:e47700.

Tabata, H., Tanaka, S., Sugimoto, Y., Kanki, H., Kaneko, S., and Ichikawa, A. 2002. Possible coupling of prostaglandin E receptor EP_1 to TRP5 expressed in *Xenopus laevis* oocytes. *Biochem. Biophys. Res. Commun.* 298:398–402.

Tai, C., Hines, D.J., Choi, H.B., and MacVicar, B.A. 2011. Plasma membrane insertion of TRPC5 channels contributes to the cholinergic plateau potential in hippocampal CA1 pyramidal neurons. *Hippocampus* 21:958–967.

Tai, Y., Feng, S., Du, W., and Wang, Y. 2009. Functional roles of TRPC channels in the developing brain. *Pflugers Arch.* 458:283–289.

Tai, Y., Feng, S., Ge, R., Du, W., Zhang, X., He, Z., and Wang, Y. 2008. TRPC6 channels promote dendritic growth via the CaMKIV-CREB pathway. *J. Cell Sci.* 121:2301–2307.

Takahashi, N. and Mori, Y. 2011. TRP channels as sensors and signal integrators of redox status changes. *Front. Pharmacol.* 2:58.

Tang, J., Lin, Y., Zhang, Z., Tikunova, S., Birnbaumer, L., and Zhu, M.X. 2001. Identification of common binding sites for calmodulin and inositol 1,4,5-trisphosphate receptors on the carboxyl termini of trp channels. *J. Biol. Chem.* 276:21303–21310.

Tang, Y., Tang, J., Chen, Z., Trost, C., Flockerzi, V., Li, M., Ramesh, V., and Zhu, M.X. 2000. Association of mammalian trp4 and phospholipase C isozymes with a PDZ domain-containing protein, NHERF. *J. Biol. Chem.* 275:37559–37564.

Tano, J.Y., Lee, R.H., and Vazquez, G. 2011. Macrophage function in atherosclerosis: Potential roles of TRP channels. *Channels (Austin)* 6:141–148.

Tian, D., Jacobo, S., Billing, D., Rozkalne, A., Gage, S., Anagnostou, T., Pavenstädt, H. et al. 2010. Antagonistic regulation of actin dynamics and cell motility by TRPC5 and TRPC6 channels. *Sci. Signal.* 3:ra77.

Trebak, M., Hempel, N., Wedel, B.J., Smyth, J.T., Bird, G.S., and Putney, J.W., Jr. 2005. Negative regulation of TRPC3 channels by protein kinase C-mediated phosphorylation of serine 712. *Mol. Pharmacol.* 67:558–563.

Trebak, M., Lemonnier, L., DeHaven, W.I., Wedel, B.J., Bird, G.S., and Putney, J.W. Jr. 2009. Complex functions of phosphatidylinositol 4,5-bisphosphate in regulation of TRPC5 cation channels. *Pflugers Arch.* 457:757–769.

Treves, S., Franzini-Armstrong, C., Moccagatta, L., Arnoult, C., Grasso, C., Schrum, A., Ducreux, S. et al. 2004. Junctate is a key element in calcium entry induced by activation of InsP3 receptors and/or calcium store depletion. *J. Cell Biol.* 166:537–548.

Treves, S., Vukcevic, M., Griesser, J., Armstrong, C.F., Zhu, M.X., and Zorzato, F. 2010. Agonist-activated Ca^{2+} influx occurs at stable plasma membrane and endoplasmic reticulum junctions. *J. Cell Sci.* 123:4170–4181.

Tsujino, N., Yamanaka, A., Ichiki, K., Muraki, Y., Kilduff, T.S., Yagami, K., Takahashi, S., Goto, K., and Sakurai, T. 2005. Cholecystokinin activates orexin/hypocretin neurons through the cholecystokinin A receptor. *J. Neurosci.* 25:7459–7469.

Tsvilovskyy, V.V., Zholos, A.V., Aberle, T., Philipp, S.E., Dietrich, A., Zhu, M.X., Birnbaumer, L., Freichel, M., and Flockerzi, V. 2009. Deletion of TRPC4 and TRPC6 in mice impairs smooth muscle contraction and intestinal motility in vivo. *Gastroenterology* 137:1415–1424.

Urban, N., Hill, K., Wang, L., Kuebler, W.M., and Schaefer, M. 2012. Novel pharmacological TRPC inhibitors block hypoxia-induced vasoconstriction. *Cell Calcium* 51:194–206.

Vassort, G. and Alvarez, J. 2009. Transient receptor potential: A large family of new channels of which several are involved in cardiac arrhythmia. *Can. J. Physiol. Pharmacol.* 87:100–107.

Vazquez, G., Wedel, B.J., Kawasaki, B.T., Bird, G.S., and Putney, J.W. Jr. 2004. Obligatory role of Src kinase in the signaling mechanism for TRPC3 cation channels. *J. Biol. Chem.* 279:40521–40528.

Venkatachalam, K., Zheng, F., and Gill, D.L. 2003. Regulation of canonical transient receptor potential (TRPC) channel function by diacylglycerol and protein kinase C. *J. Biol. Chem.* 278:29031–29040.

Wang, G.X. and Poo, M.M. 2005. Requirement of TRPC channels in netrin-1-induced chemotropic turning of nerve growth cones. *Nature* 434:898–904.

Wang, M., Bianchi, R., Chuang, S.-C., Zhao, W., and Wong, R. 2007. Group I metabotropic glutamate receptor-dependent TRPC channel trafficking in hippocampal neurons. *J. Neurochem.* 101:411–421.

Wang, P., Liu, D., Tepel, M., and Zhu, Z. 2013. Transient receptor potential canonical type 3 channels—Their evolving role in hypertension and its related complications. *J. Cardiovasc. Pharmacol.* 61:455–460.

Wang, S., Zhang, A.P., Kurada, L., Matsui, T., and Lei, S. 2011. Cholecystokinin facilitates neuronal excitability in the entorhinal cortex via activation of TRPC-like channels. *J. Neurophysiol.* 106:1515–1524.

Watanabe, H., Iino, K., Ohba, T., and Ito, H. 2013. Possible involvement of TRP channels in cardiac hypertrophy and arrhythmia. *Curr. Top. Med. Chem.* 13:283–294.

Weick, J.P., Austin Johnson, M., and Zhang, S.C. 2009. Developmental regulation of human embryonic stem cell-derived neurons by calcium entry via transient receptor potential channels. *Stem Cells* 27:2906–2916.

Wu, G., Lu, Z.H., Obukhov, A.G., Nowycky, M.C., and Ledeen, R.W. 2007. Induction of calcium influx through TRPC5 channels by cross-linking of GM1 ganglioside associated with a5b1 integrin initiates neurite outgrowth. *J. Neurosci.* 27:7447–7458.

Xu, S.Z., Muraki, K., Zeng, F., Li, J., Sukumar, P., Shah, S., Dedman, A.M. et al. 2006. A sphingosine-1-phosphate-activated calcium channel controlling vascular smooth muscle cell motility. *Circ. Res.* 98:1381–1389.

Xu, S.Z., Sukumar, P., Zeng, F., Li, J., Jairaman, A., English, A., Naylor, J. et al. 2008. TRPC channel activation by extracellular thioredoxin. *Nature* 451:69–72.

Xu, S.Z., Zeng, B., Daskoulidou, N., Chen, G.L., Atkin, S.L., and Lukhele, B. 2012. Activation of TRPC cationic channels by mercurial compounds confers the cytotoxicity of mercury exposure. *Toxicol. Sci.* 125:56–68.

Yan, H.D., Villalobos, C., and Andrade, R. 2009. TRPC channels mediate a muscarinic receptor-induced afterdepolarization in cerebral cortex. *J. Neurosci.* 29:10038–10046.

Yoshida, T., Inoue, R., Morii, T., Takahashi, N., Yamamoto, S., Hara, Y., Tominaga, M., Shimizu, S., Sato, Y., and Mori, Y. 2006. Nitric oxide activates TRP channels by cysteine S-nitrosylation. *Nat. Chem. Biol.* 2:596–607.

Yuan, J.P., Kim, M.S., Zeng, W., Shin, D.M., Huang, G., Worley, P.F., and Muallem, S. 2009. TRPC channels as STIM1-regulated SOCs. *Channels (Austin)* 3:221–225.

Zeng, B., Yuan, C., Yang, X., Atkin, S.L., and Xu, S.Z. 2013. TRPC channels and their splice variants are essential for promoting human ovarian cancer cell proliferation and tumorigenesis. *Curr. Cancer Drug Targets* 13:103–116.

Zeng, F., Xu, S.Z., Jackson, P.K., McHugh, D., Kumar, B., Fountain, S.J., and Beech, D.J. 2004. Human TRPC5 channel activated by a multiplicity of signals in a single cell. *J. Physiol.* 559:739–750.

Zeng, W., Yuan, J.P., Kim, M.S., Choi, Y.J., Huang, G.N., Worley, P.F., and Muallem, S. 2008. STIM1 gates TRPC channels, but not Orai1, by electrostatic interaction. *Mol. Cell* 32:439–448.

Zhang, C., Roepke, T.A., Kelly, M.J., and Ronnekleiv, O.K. 2008. Kisspeptin depolarizes gonadotropin-releasing hormone neurons through activation of TRPC-like cationic channels. *J. Neurosci.* 28:4423–4434.

Zhang, X. and van den Pol, A.N. 2012. Thyrotropin-releasing hormone (TRH) inhibits melanin-concentrating hormone neurons: Implications for TRH-mediated anorexic and arousal actions. *J. Neurosci.* 32:3032–3043.

Zhang, Z., Reboreda, A., Alonso, A., Barker, P., and Séguéla, P. 2011. TRPC channels underlie cholinergic plateau potentials and persistent activity in entorhinal cortex. *Hippocampus* 21:386–397.

Zhang, Z., Tang, J., Tikunova, S., Johnson, J., Chen, Z., Qin, N., Dietrich, A., Stefani, E., Birnbaumer, L., and Zhu, M.X. 2001. Activation of Trp3 by inositol 1,4,5-trisphosphate receptors through displacement of inhibitory calmodulin from a common binding domain. *Proc. Natl. Acad. Sci. USA* 98:3168–3173.

Zholos, A.V. and Bolton, T.B. 1997. Muscarinic receptor subtypes controlling the cationic current in guinea-pig ileal smooth muscle. *Br. J. Pharmacol.* 122:885–893.

Zhou, F.W., Matta, S., and Zhou, F.M. 2008a. Constitutively active TRPC3 channels regulate basal ganglia output neurons. *J. Neurosci.* 28:473–482.

Zhou, J., Du, W., Zhou, K., Tai, Y., Yao, H., Jia, Y., Ding, Y., and Wang, Y. 2008b. Critical role of TRPC6 channels in the formation of excitatory synapses. *Nat. Neurosci.* 11:741–743.

Zhu, X., Jiang, M., and Birnbaumer, L. 1998. Receptor-activated Ca^{2+} influx via human Trp3 stably expressed in human embryonic kidney (HEK)293 cells. Evidence for a non-capacitative Ca^{2+} entry. *J. Biol. Chem.* 273:133–142.

Zhu, X., Jiang, M., Peyton, M., Boulay, G., Hurst, R., Stefani, E., and Birnbaumer, L. 1996. trp, a novel mammalian gene family essential for agonist-activated capacitative Ca^{2+} entry. *Cell* 85:661–671.

Zimmermann, K., Lennerz, J.K., Hein, A., Link, A.S., Kaczmarek, J.S., Delling, M., Uysal, S., Pfeifer, J.D, Riccio, A., and Clapham, D.E. 2011. Transient receptor potential cation channel, subfamily C, member 5 (TRPC5) is a cold-transducer in the peripheral nervous system. *Proc. Natl. Acad. Sci. USA* 108:18114–18119.

28 TRPV channels

Sharona E. Gordon

Contents

Much can be learned from the structural and functional properties shared by the six members of the transient receptor potential vanilloid (TRPV) family of ion channels. In addition, the specializations of each for their physiological roles shed light on the workings of the entire family. This chapter will focus on *gating* of TRPV1–6 channels, with discussion of permeation, expression patterns, and physiological roles covered elsewhere.[1–5]

28.1 STRUCTURE OF TRPV CHANNELS

TRPV channels are members of the voltage-gated superfamily of ion channels with six-membrane-spanning domains, intracellular N- and C-termini, a (weak) voltage sensor domain in S1–S4, and a pore domain formed by S5–P–S6. Each functional TRPV channel is composed of four identical subunits.

Electron cryomicroscopy of TRPV1[6] and TRPV4[7] shows the structure of these channels to resemble that of Kv channels,[8] with a transmembrane structure of about 60–80 Å on a side with a hanging gondola about 100–110 Å tall. The N-termini of TRPV channels include an ankyrin repeat domain, a feature shared with TRPC channels. The x-ray crystal structures of the ankyrin repeat domains of TRPV1,[9] TRPV2,[10,11] TRPV4,[12] and TRPV6[13] channels have been solved in which six ankyrin repeats are observed. In the structures of TRPV1 and TRPV4, an ATP molecule binds to the *fingers* formed by the first and second ankyrin repeats. Interestingly, although the structure of the ankyrin repeats of TRPV2 is extremely similar to that of the ankyrin repeats of TRPV1, N-terminal fragments of TRPV1 bind ATP *in vitro* but fragments of TRPV2 do not. Although ATP has no effect on TRPV1 or TRPV2 in excised patches,[14] differences in the properties of the two channels may be useful in determining whether direct binding has a physiological role in regulating TRPV1.

28.2 TEMPERATURE DEPENDENCE OF GATING

TRPV1, TRPV2, TRPV3, and TRPV4 can all be activated by elevated temperature. Temperature dependence is typically discussed in terms of Q_{10}, the ratio of the rate of a reaction at one temperature to that at another. Although all enzymes are temperature sensitive to some extent, with typical Q_{10} values around 2, TRPV1–4 have Q_{10} values in the range of 10–40.

Whether a discrete temperature-sensing domain exists in TRP channels is currently the subject of active debate. Different regions of the N-terminal domain[15] as well as the C-terminal domain,[16] the pore turret,[17,18] and the pore/extracellular loop following the pore[19,20] have all been suggested to contain a temperature sensor.

An alternative mechanism to a discrete temperature-sensitive domain is based on first principles of thermodynamics.[21] Because proteins can have high heat capacity (ΔC_p), enthalpy (ΔH) and entropy (ΔS) cannot be thought of as independent of temperature. A relatively simple derivation leads to a surprising U-shaped dependence of the conformational change on temperature (Figure 28.1). Applied to TRP channels, this means that all that are activated by heat may also activated by cold, and *vice versa.* Whether we observe activation by heat or cold would depend on the temperature range for each turn of the U. Starting at the vertex (Figure 28.1, arrow), if the turn on the left is within the physiological range, then cold activation will be observed, whereas if the turn on the right is within the physiological range, then activation by heat is observed. This model-free framework predicts that, if it were practical to lower or raise the experimental temperature to any arbitrary value, then activation by both cold and heat would be observed for TRPV1–TRPV4. In this mechanism, a discrete temperature sensor would be simply

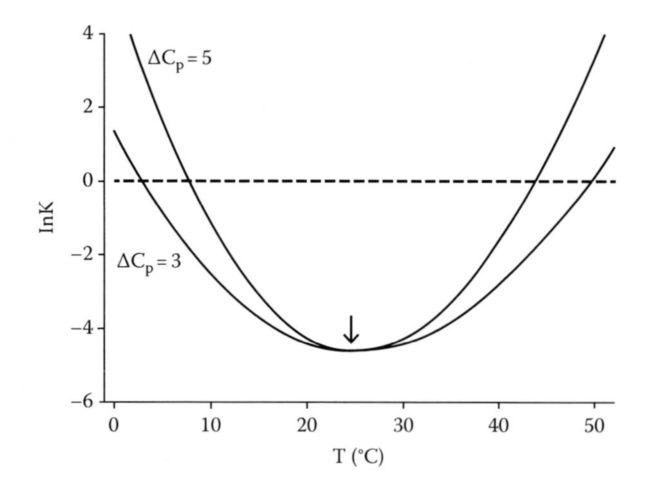

Figure 28.1 Temperature dependence of conformational equilibrium constant. (From Clapham, D.E. and Miller, C., *Proc. Natl. Acad. Sci. U.S.A.*, 108, 19492, 2011.)

a region of the protein with particularly high heat capacity. However, a discrete temperature sensor is not needed, as a change in heat capacity associated with the global conformational change would be sufficient to account for the steep temperature sensitivity of TRP channels.

28.3 REGULATION BY G-PROTEIN-COUPLED RECEPTORS

Increased sensitivity of pain-receptor neurons in the setting of inflammation is due, in part, to activation of G-protein-coupled receptors (GPCRs) and receptor tyrosine kinases. In particular, Gαq-coupled receptors, such as the bradykinin receptor, contribute to sensitization via their coupling to phospholipase C β (PLCβ). PLCβ hydrolyzes phosphatidylinositol 4,5-bisphosphate (PIP_2) to generate diacylglycerol and inositol trisphosphate. The diacylglycerol in turn activates protein kinase C. Phosphorylation of two serines, S502 and S800, potentiate activation of TRPV1 by capsaicin.[22] Association of TRPV1 (and TRPV2, TRPV3, and TRPV4) with A kinase anchoring protein 79/150 (AKAP79/150) appears to play a role in bradykinin-mediated sensitization of TRPV1 by positioning PKC appropriately to phosphorylate TRPV1.[23] Interestingly, AKAP79/150 may also play a role in desensitization.[23] The list of GPCRs that can regulate TRPV1 is extensive and growing; additional signaling pathways may well be revealed.

28.3.1 REGULATION BY PIP_2

As controversial as is the mechanism of heat activation, the mechanism and role of regulation by the signaling lipid PIP_2 are more so, with most of the work performed on TRPV1. Indeed, even whether PIP_2 is an activator or an inhibitor of TRPV1 is subject to intense debate. It is therefore worth expanding on the evidence for activation and inhibition.

28.3.2 PIP_2 AS AN INHIBITOR OF TRPV1

The first proposal that PIP_2 may inhibit TRPV1 was based on the sensitization of the channels by Gαq-coupled GPCRs, described earlier.[24] If Gαq activation of PLCβ were sufficiently strong, the level of PIP_2 in the inner leaflet of the plasma membrane might drop precipitously. Near-complete depletion of PIP_2 by Gαq-coupled receptors is believed to regulate a number of other ion channels, including K_{IR} and KCNQ channels.[25,26] If PIP_2 tonically inhibits TRPV1 in resting cells, and activation of Gαq-coupled GPCRs activates PLCβ to deplete PIP_2, then Gαq-coupled GPCRs could indeed relieve the tonic inhibition, giving sensitization.

Three types of evidence have been presented to support PIP_2 as a direct inhibitor of TRPV1. First, the application of an antibody against PIP_2 to excised patches from cells expressing TRPV1 transiently increases the capsaicin-activated current.[24] PIP_2 is not inherently antigenic; beads coated to present many PIP_2 head groups are used as the antigen. Furthermore, the specificity of the anti-PIP_2 antibody is unknown. Taken at face value, however, the increased TRPV1 activity observed in response to sequestration of PIP_2 by the anti-PIP_2 antibody in inside-out excised patches supports a role of PIP_2 as an inhibitor of TRPV1.

The second basis for PIP_2 as an inhibitor of TRPV1 is that the application of bacterial phosphatidylinositol (PI)-PLC to patches increased both basal currents and capsaicin-activated currents. However, biochemical and structural studies have shown that bacterial PI-PLC, in contrast to mammalian PI-PLC, does not hydrolyze phosphatidylinositol phosphate (PIP) or PIP_2.[27] It does cleave PI robustly, however, raising the question of whether PI may inhibit TRPV1.

The third line of evidence supporting PIP_2 as an inhibitor comes from experiments using purified TRPV1 reconstituted into synthetic liposomes of defined composition. The temperature and capsaicin dependence were measured in liposomes that included either no phosphoinositides, 4% PI, 4% phosphatidylinositol 4-phosphate (PI(4)P), 4% phosphatidylinositol 3-phosphate (PI(3)P), 4% PIP_2, or phosphatidylinositol 3,4,5-trisphosphate (PIP_3). With the exception of PIP_3, all the phosphoinositides significantly increased the threshold for activation by from about 27°C to about 38°C (Figure 28.2).[28] PI, PI(4)P, and PIP_2 also decreased the apparent affinity for capsaicin, whereas PIP_3 did not.

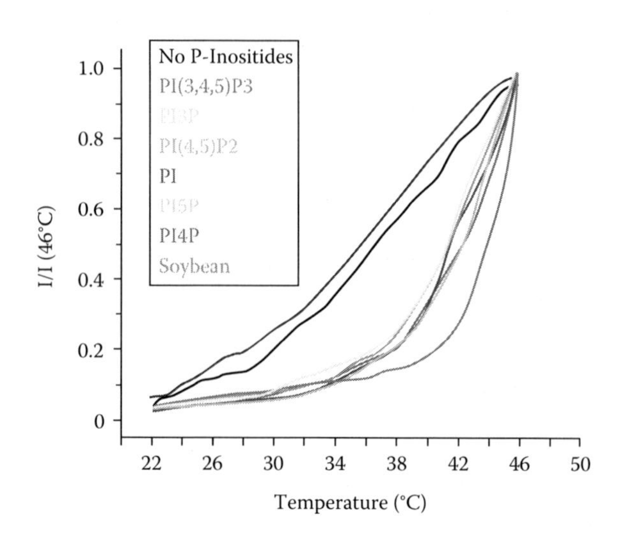

Figure 28.2 Change in the shape and threshold for activation of TRPV1 by heat in bilayers containing the phosphoinositides shown earlier. (From Cao, E. et al., *Neuron*, 77, 667, 2013.)

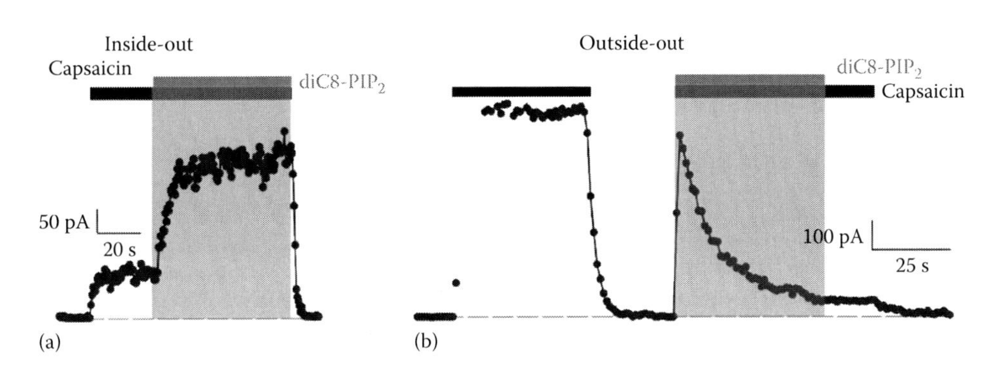

Figure 28.3 Leaflet-specific regulation of TRPV1 by PIP_2. (a) Application of PIP_2 to the intracellular leaflet, via addition to the bath with an inside-out patch, causes activation of TRPV1. (b) Application of PIP_2 to the extracellular leaflet, via addition to the bath with an outside-out patch, causes inhibition of TRPV1. (From Senning, E.N. et al., *J Biol Chem*. 289(16), 10999, 2014.)

Although the simplest explanation of these data is that phosphoinositides inhibit TRPV1, a number of experimental and conceptual concerns call this conclusion into question. The greatest of these concerns is the symmetrical nature of the liposome membrane, that is, what is the effect of having PIP_2 on both sides of the channels? When tested with outside-out patches from cells expressing TRPV1, adding PIP_2 to the extracellular leaflet robustly inhibited the capsaicin-activated currents (Figure 28.3).[29] Although reconstituting channels into asymmetric bilayers would be required to put the case to rest definitively, the observation that PIP_2 is activating when in the physiological leaflet of the plasma membrane only, and inhibits when present in both the inner and outer leaflets of both cell and synthetic membranes, suggests that the inhibition observed is not physiological.

A number of other caveats obfuscate the case for inhibition. These include the change in shape of the temperature dependence in TRPV1 reconstituted into phosphoinositides-free membranes compared to those including phosphoinositides (but not PIP_3; Figure 28.2). The relationship between the shape of the temperature vs. normalized current curve and the threshold for activation (Figure 28.2) is not understood, and the interpretation of changes in the shape of the temperature dependence of activation is unclear. In addition, no controls were described to show that the mole fraction of phosphoinositides in the TRPV1-containing liposomes reflected the mole fraction in the lipid mixture used to make the liposomes. This is an important point because charged phosphoinositides come out of the organic solvent (chloroform in this case) earlier than the other lipids used. Typically, this leaves separate patches of phosphoinositides and other lipids in the vessel, and a heterogeneous population is formed that does not fully mix even after many freeze–thaw cycles. The phosphoinositides with the highest charge, PIP_3, are the most likely to be affected by incorporation artifacts, interesting given that the behavior of TRPV1 in PIP_3-containing liposomes was distinct from that of TRPV1 in liposomes containing all other phosphoinositides.

Based, in part, on the experiments with reconstituted channels, a mechanism for inhibition of TRPV1 by phosphoinositides has been proposed. The distal C-terminal region contains a number of basic residues that are postulated to interact with the negatively charged phosphoinositides head groups. This interaction is proposed to stabilize an inhibited conformation of TRPV1. In the absence of phosphoinositides, the inhibited conformation is proposed to be less stable relative to the noninhibited conformation so that depletion of PIP_2 would be observed an increase in open probability. The interaction between the basic residues and the membrane must be more than just electrostatic in this mechanism, because PI inhibited reconstituted TRPV1 but phosphatidic acid, with the same charge as PI, did not. In addition, a purely electrostatic interaction would be expected to be the strongest for PIP_3, yet no effect of PIP_3 was observed. A protein that binds PI, PI(3)P, PI(4)P, and PIP_2, but not PIP_3 or phosphatidic acid, would indeed be very interesting.

28.3.3 PIP_2 AS AN ACTIVATOR OF TRPV1

PIP_2 has been proposed to play a role in Ca^{2+}-dependent desensitization of TRPV1, TRPV2, and TRPV6.[14,30–32] TRPV channels are highly permeable to Ca^{2+}, a convenient property that allows Ca^{2+} imaging to be used to assay their activity in cells. However, too much Ca^{2+} influx is poorly tolerated, and cells have a number of mechanisms to return calcium to resting levels. Among them is desensitization of TRPV channels, a self-limiting response in which the Ca^{2+} coming in through the channels also signals to reduce their activity. The desensitizing signaling pathway has been proposed to involve the stimulation of PLC by Ca^{2+} and PLC-mediated depletion of PIP_2. Depletion of the activator PIP_2 from TRPV1 would be observed as a decrease in TRPV1 activity, that is, desensitization.

For a signal to be identified as mediating an effect, a number of criteria must be met. The signal must be necessary, sufficient, and appropriately localized and occur either prior to or simultaneous with the observed effect. To answer whether depletion of PIP_2 is the signal that reduces TRPV1 activity in response to Ca^{2+}, we can examine each of these criteria in turn.

Is PIP_2 depletion necessary for desensitization? If so, then doping the plasma membrane's inner leaflet with extra PIP_2 should eliminate or reduce desensitization. Whole-cell voltage-clamp experiments in which PIP_2 was dialyzed into the cell via the patch pipette showed that desensitization was significantly reduced.[9] Is PIP_2 depletion sufficient for desensitization? This question has been addressed in both excised patches and in whole cells. In excised patches, application of a lysine polymer inhibited capsaicin-activated currents. The cationic lysine polymer is thought to bind to anionic lipids head groups, effectively reducing the free concentration of these lipids. Inhibition of TRPV1 by sequestering anionic lipids with the lysine polymer is consistent with an activating effect of PIP_2, but the nonselective nature of the reagent prevents a definitive interpretation.

More selective than a lysine polymer is the pleckstrin homology domain of PLCδ1 (PH-PLC). The isolated PH-PLC, lacking the enzymatic portion of the protein, binds to the headgroup of PIP_2 about 1000-fold better than to either PIP or PIP_3.[33,34] Recombinant PH-PLC can thus be used as a highly selective, reversible tool to reduce the free concentration of PIP_2 in the membrane. Recombinant, purified PH-PLC was applied to excised patches from cells expressing TRPV1.[30] PH-PLC produced a concentration-dependent, reversible inhibition of the capsaicin-activated currents. Neither boiled PH-PLC nor PH-PLC applied to outside-out patches altered the capsaicin-activated current. Moreover, a PIP_3-selective pleckstrin homology domain was ineffective.

The sufficiency of PIP_2 depletion in producing desensitization of TRPV1 was tested in whole cells by bypassing PLC to dephosphorylate PIP_2 with either a voltage-sensitive phosphatase or a chemically inducible lipid phosphatase.[30,32] In both cases, PIP_2 was dephosphorylated at the 5-position of the inositol ring, giving PI(4)P and avoiding generation of diacylglycerol and inositol trisphosphate. Activation of both the voltage-sensitive phosphatase and the chemically inducible phosphatase decreased the capsaicin-activated current[30] (but see also Lukacs et al.,[32] who found the effect of the chemically induced phosphatase to depend on the capsaicin concentration used).

Ca^{2+}-dependent desensitization in TRPV2 is indistinguishable from Ca^{2+}-dependent desensitization in TRPV1. In contrast to TRPV1, however, the ankyrin repeat domain of TRPV2 does not bind ATP or Ca^{2+}/calmodulin. These data suggest that Ca^{2+}-dependent desensitization does not require interactions among the ankyrin repeat domains and ATP and/or Ca^{2+}-calmodulin. However, Ca^{2+}-dependent desensitization of TRPV2 does appear to involve depletion of PIP_2. To determine whether depletion of PIP_2 and Ca^{2+}-dependent desensitization occurred with similar kinetics, simultaneous confocal imaging and whole-cell voltage clamp were performed on cells expressing both TRPV1 and the PH-PLC PIP_2-binding protein discussed earlier, genetically fused to GFP. The desensitization observed occurred with the same time course as depletion of PIP_2 (Figure 28.4).[14]

If PIP_2 is viewed as a cofactor required for TRPV1 activation, it is reasonable to ask whether, under physiological conditions, the PIP_2 concentration in the membrane ever falls low enough to create a pool of PIP_2-free TRPV1 channels. Recent work quantifying the partition coefficient of a water-soluble PIP_2, diC8-PIP_2, underscores the importance of this question. The EC_{50} value for activation of TRPV1 by diC8-PIP_2 has been reported to be in the range of 0.5–5 μM, depending on the cell type and the recording conditions. These concentrations are the concentrations in solution, but the important concentration is that of diC8-PIP_2 in the membrane. Using isothermal titration calorimetry with liposomes whose composition was modeled after the intracellular leaflet of neuronal plasma membranes, an EC_{50} of 1 μM translates into 0.002 mol% diC8-PIP_2 in the intracellular leaflet.[35] If the apparent affinity of TRPV1 for natural PIP_2 is within even an

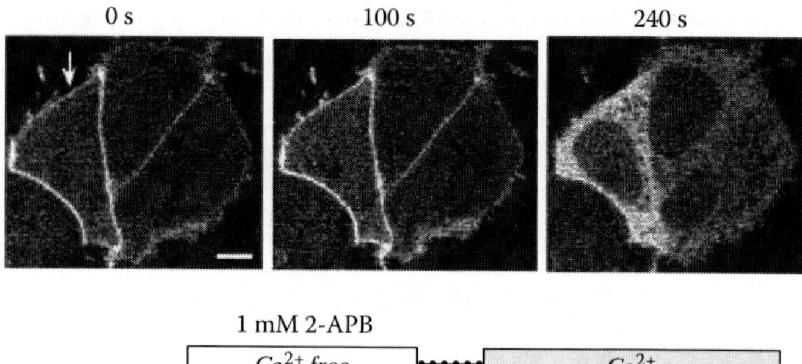

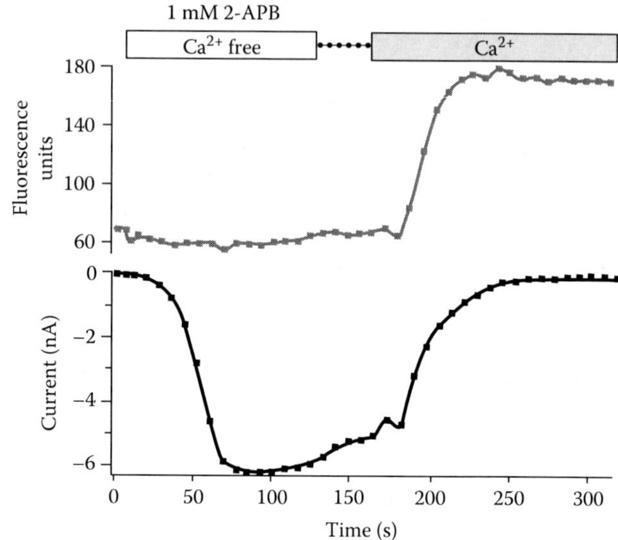

Figure 28.4 Ca^{2+}-dependent desensitization of TRPV2 coincides with depletion of PIP_2. (From Mercado, J. et al., *J. Neurosci.*, 30, 13338, 2010.)

order of magnitude of this value, then it is possible that the PIP_2 level in the cell membrane never falls sufficiently to remove PIP_2 from TRPV1 under physiological conditions.

PIP_2 appears to activate all the TRPV channels, TRPV1,[32,36] TRPV2,[14] TRPV4,[37] TRPV5,[38] and TRPV6,[31] with the exception of TRPV3. TRPV3 was found to be activated by PIP_2 depletion in whole cells and excised patches and to be inhibited by the direct application of diC8-PIP_2 to inside-out excised patches.[39] This fascinating inversion of regulation polarity, inhibition instead of activation, is reminiscent of the voltage dependence of HCN channels, compared to other voltage-gated channels.[40] Whether PIP_2 asserts a different conformational rearrangement in the PIP_2-binding site of TRPV3, compared to other TRPVs, or whether instead the coupling between the PIP_2-binding site and the pore gate is inverted, remains to be determined.

REFERENCES

1. Ramsey, I.S., Delling, M., and Clapham, D.E. An introduction to TRP channels. *Annu Rev Physiol* **68**, 619–647 (2006).
2. Gees, M., Owsianik, G., Nilius, B., and Voets, T. TRP Channels. *Compr Physiol* **2**, 563–608 (2012).
3. Nilius, B. and Owsianik, G. The transient receptor potential family of ion channels. *Genome Biol* **12**, 218 (2011).
4. Gees, M., Colsoul, B., and Nilius, B. The role of transient receptor potential cation channels in Ca^{2+} signaling. *Cold Spring Harb Perspect Biol* **2**, a003962 (2010).
5. Nilius, B. TRP channels in disease. *Biochim Biophys Acta* **1772**, 805–812 (2007).

6. Moiseenkova-Bell, V., Stanciu, L., Serysheva, I., Tobe, B., and Wensel, T. Structure of TRPV1 channel revealed by electron cryomicroscopy. *Proc Natl Acad Sci USA* **105**, 7451–7455 (2008).
7. Phelps, C.B., Wang, R.R., Choo, S.S., and Gaudet, R. Differential regulation of TRPV1, TRPV3, and TRPV4 sensitivity through a conserved binding site on the ankyrin repeat domain. *J Biol Chem* **285**, 731–740 (2010).
8. Long, S.B., Tao, X., Campbell, E.B., and MacKinnon, R. Atomic structure of a voltage-dependent K^+ channel in a lipid membrane-like environment. *Nature* **450**, 376–382 (2007).
9. Lishko, P.V., Procko, E., Jin, X., Phelps, C.B., and Gaudet, R. The ankyrin repeats of TRPV1 bind multiple ligands and modulate channel sensitivity. *Neuron* **54**, 905–918 (2007).
10. Jin, X., Touhey, J., and Gaudet, R. Structure of the N-terminal ankyrin repeat domain of the TRPV2 ion channel. *J Biol Chem* **281**, 25006–25010 (2006).
11. McCleverty, C.J., Koesema, E., Patapoutian, A., Lesley, S.A., and Kreusch, A. Crystal structure of the human TRPV2 channel ankyrin repeat domain. *Protein Sci* **15**, 2201–2206 (2006).
12. Inada, H., Procko, E., Sotomayor, M., and Gaudet, R. Structural and biochemical consequences of disease-causing mutations in the ankyrin repeat domain of the human TRPV4 channel. *Biochemistry* **51**, 6195–6206 (2012).
13. Phelps, C.B., Huang, R.J., Lishko, P.V., Wang, R.R., and Gaudet, R. Structural analyses of the ankyrin repeat domain of TRPV6 and related TRPV ion channels. *Biochemistry* **47**, 2476–2484 (2008).
14. Mercado, J., Gordon-Shaag, A., Zagotta, W., and Gordon, S. Ca^{2+}-dependent desensitization of TRPV2 channels is mediated by hydrolysis of phosphatidylinositol 4,5-bisphosphate. *J Neurosci* **30**, 13338–13347 (2010).
15. Yao, J., Liu, B., and Qin, F. Modular thermal sensors in temperature-gated transient receptor potential (TRP) channels. *Proc Natl Acad Sci USA* **108**, 11109–11114 (2011).
16. Brauchi, S., Orta, G., Salazar, M., Rosenmann, E., and Latorre, R. A hot-sensing cold receptor: C-terminal domain determines thermosensation in transient receptor potential channels. *J Neurosci* **26**, 4835–4840 (2006).
17. Yang, F., Cui, Y., Wang, K., and Zheng, J. Thermosensitive TRP channel pore turret is part of the temperature activation pathway. *Proc Natl Acad Sci USA* **107**, 7083–7088 (2010).
18. Cui, Y. et al. Selective disruption of high sensitivity heat activation but not capsaicin activation of TRPV1 channels by pore turret mutations. *J Gen Physiol* **139**, 273–283 (2012).
19. Kim, S.E., Patapoutian, A., and Grandl, J. Single residues in the outer pore of TRPV1 and TRPV3 have temperature-dependent conformations. *PLoS One* **8**, e59593 (2013).
20. Grandl, J. et al. Temperature-induced opening of TRPV1 ion channel is stabilized by the pore domain. *Nat Neurosci* **13**, 708–714 (2010).
21. Clapham, D.E. and Miller, C. A thermodynamic framework for understanding temperature sensing by transient receptor potential (TRP) channels. *Proc Natl Acad Sci USA* **108**, 19492–19497 (2011).
22. Bhave, G. et al. Protein kinase C phosphorylation sensitizes but does not activate the capsaicin receptor transient receptor potential vanilloid 1 (TRPV1). *Proc Natl Acad Sci USA* **100**, 12480–12485 (2003).
23. Zhang, X., Li, L., and McNaughton, P.A. Proinflammatory mediators modulate the heat-activated ion channel TRPV1 via the scaffolding protein AKAP79/150. *Neuron* **59**, 450–461 (2008).
24. Chuang, H. et al. Bradykinin and nerve growth factor release the capsaicin receptor from PtdIns(4,5)P2-mediated inhibition. *Nature* **411**, 957–962 (2001).
25. Logothetis, D.E., Petrou, V.I., Adney, S.K., and Mahajan, R. Channelopathies linked to plasma membrane phosphoinositides. *Pflugers Arch* **460**, 321–341 (2010).
26. Falkenburger, B.H., Jensen, J.B., Dickson, E.J., Suh, B.C., and Hille, B. Phosphoinositides: Lipid regulators of membrane proteins. *J Physiol* **588**, 3179–3185 (2010).
27. Heinz, D.W., Essen, L.O., and Williams, R.L. Structural and mechanistic comparison of prokaryotic and eukaryotic phosphoinositide-specific phospholipases C. *J Mol Biol* **275**, 635–650 (1998).
28. Cao, E., Cordero-Morales, J.F., Liu, B., Qin, F., and Julius, D. TRPV1 channels are intrinsically heat sensitive and negatively regulated by phosphoinositide lipids. *Neuron* **77**, 667–679 (2013).
29. Senning, E.N., Collins, M.D., and Stratiievska, A., Ufret-Vincenty, C.A., and Gordon, S.E. Regulation of TRPV1 by phosphoinositide (4,5)-bisphosphate: Role of membrane asymmetry. *J Biol Chem* **189**, 10999–1006 (2014).
30. Klein, R., Ufret-Vincenty, C., Hua, L., and Gordon, S. Determinants of molecular specificity in phosphoinositide regulation. Phosphatidylinositol (4,5)-bisphosphate (PI(4,5) P2) is the endogenous lipid regulating TRPV1. *J Biol Chem* **283**, 26208–26216 (2008).
31. Thyagarajan, B., Lukacs, V., and Rohacs, T. Hydrolysis of phosphatidylinositol 4,5-bisphosphate mediates calcium-induced inactivation of TRPV6 channels. *J Biol Chem* **283**, 14980–14987 (2008).
32. Lukacs, V. et al. Dual regulation of TRPV1 by phosphoinositides. *J Neurosci* **27**, 7070–7080 (2007).
33. Lemmon, M. Pleckstrin homology (PH) domains and phosphoinositides. *Biochem Soc Symp* **74**, 81–93 (2007).
34. Várnai, P. et al. Inositol lipid binding and membrane localization of isolated pleckstrin homology (PH) domains. Studies on the PH domains of phospholipase C delta 1 and p130. *J Biol Chem* **277**, 27412–27422 (2002).
35. Collins, M.D. and Gordon, S.E. Short-chain phosphoinositide partitioning into models of plasma membranes. *Biophys J* **105**, 2485–2494 (2013).
36. Stein, A., Ufret-Vincenty, C., Hua, L., Santana, L., and Gordon, S. Phosphoinositide 3-kinase binds to TRPV1 and mediates NGF-stimulated TRPV1 trafficking to the plasma membrane. *J Gen Physiol* **128**, 509–522 (2006).
37. Garcia-Elias, A. et al. Phosphatidylinositol-4,5-biphosphate-dependent rearrangement of TRPV4 cytosolic tails enables channel activation by physiological stimuli. *Proc Natl Acad Sci USA* **110**, 9553–9558 (2013).
38. Huang, Q. et al. Akt2 kinase suppresses glyceraldehyde-3-phosphate dehydrogenase (GAPDH)-mediated apoptosis in ovarian cancer cells via phosphorylating GAPDH at threonine 237 and decreasing its nuclear translocation. *J Biol Chem* **286**, 42211–42220 (2011).
39. Doerner, J.F., Hatt, H., and Ramsey, I.S. Voltage- and temperature-dependent activation of TRPV3 channels is potentiated by receptor-mediated PI(4,5)P2 hydrolysis. *J Gen Physiol* **137**, 271–288 (2011).
40. Baruscotti, M., Bottelli, G., Milanesi, R., DiFrancesco, J.C., and DiFrancesco, D. *Pflugers Arch*. Jul; **460**(2), 405–415 (2010). doi: 10.1007/s00424-010-0810-8.
41. Senning, E.N., Collins, M.D., Stratiievska, A., Ufret-Vincenty, C.A., and Gordon, S.E. Regulation of TRPV1 ion channel by phosphoinositide (4,5)-bisphosphate: the role of membrane asymmetry. *J Biol Chem*. Apr 18; **289**(16), 10999–1006 (2014). doi: 10.1074/jbc.M114.553180.

29 TRPM channels

David D. McKemy

Contents

29.1 INTRODUCTION

Transient receptor potential melastatin (TRPM) channels are a genetically and functionally diverse subfamily of the transient receptor potential family of ion channels and comprise eight members (Figure 29.1). Few characteristics are commonplace between these channels, but some share significant functional similarity. For example, TRPM2, TRPM6, and TRPM7 are unique among known ion channels in that they also contain enzymatically active protein domains in their C-termini. Moreover, TRPM6 and TRPM7 are functionally similar to each other, as are TRPM4 and TRPM5, yet a divergent expression profile between each member of these pairings leads to significantly different physiological roles. Each channel has been extensively studied and reviewed (see Refs. 1–8).

TRPM1 (otherwise known as melastatin) was the first TRPM identified and is considered a marker for metastasized melanomas. TRPM2 has a unique Nudix hydrolase domain that is homologous to the ADP pyrophosphatase NUDT9, which serves as a binding site for channel activation by adenine nucleotides, and the channel likely serves as a cellular sensor of oxidation. TRPM3 is an enigmatic channel in that it has several alternatively spliced variants that may serve a role in cellular responses to hypotonicity. TRPM4 and TRPM5 are the only members of the TRP channel family that are selective for only monovalent cations and do not permeate Ca^{2+}. However, both are activated by intracellular Ca^{2+}, serving as transduction channels downstream of increased cellular Ca^{2+} levels. Like TRPM2, TRPM6 and TRPM7 possess both ion channel and enzymatic domains, in this case atypical protein kinases that are structurally similar to protein kinase A (PKA). These channels appear to serve as sensors of the levels of intracellular Mg^{2+} and are inhibited under conditions of normal divalent cation

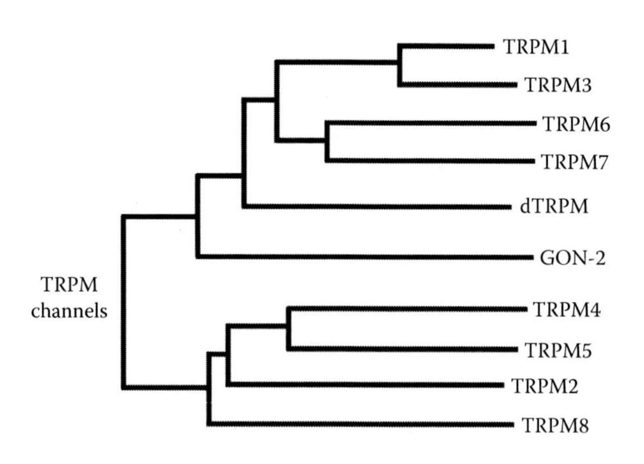

Figure 29.1 The phylogenetic tree of TRPM channels organized based on homology at the amino acid level. Nonmammalian TRPM channels dTRPM (*Drosophila*) and GON-2 (*Caenorhabditis elegans*) are placed in relation to their sequence similarity to mammalian channels.

concentrations suggesting they are important for general Mg^{2+} homeostasis. TRPM8 was first identified in prostate, in which it is proposed to be an androgen-responsive channel, but the channel chiefly serves as a sensor of cold temperatures in the peripheral nervous system. Thus, TRPM channels are a diverse array of ion channels key in a number of cellular processes. The following addresses salient points on each channel, putting them in the context of molecular and structural differences between each protein, as well as describes what is known of their basic cellular functions at this time.

29.2 TRPM1

TRPM1 is the founding member of the TRPM subfamily of TRP ion channels and was originally termed melastatin when first identified in a screen for genes involved in melanoma metastasis (9). Using two alternates of the B16 mouse melanoma cell line that vary by their capacity to metastasize, it was hypothesized that genes with differential expression would either serve as viable cancer markers or be relevant in metastasis. In this screen, TRPM1 was found to be robustly downregulated in the highly metastatic cell variants, consistent with subsequent analyses of expression patterns in melanocytic tumors where there is an inverse correlation with TRPM1 expression and melanoma metastasis, melanocytic tumor progression, and melanoma tumor thickness (10). Based on these findings, TRPM1 is suggested to be a tumor suppressor and that expression levels of TRPM1 transcripts may be a useful marker for the progression of melanoma metastasis (9,11,12).

In addition to melanocytes, TRPM1 transcripts were also found in the mouse eye (9), and the full-length channel was subsequently cloned from a human retinal cDNA library (13). Moreover, TRPM1 expression is almost absent in the retina of homozygous Appaloosa horses with congenital stationary night blindness (CSNB) (14), consistent with recent data linking mutations in TRPM1 to CSNB in humans (15–18). These results suggest that TRPM1 serves a key role in phototransduction, with recent evidence suggesting that it mediates a cationic current that modulates activity of retinal bipolar ON cells (19–21).

29.2.1 MOLECULAR STRUCTURE

It is notable that as the founding member of the TRPM channel subfamily, the transcript identified in mouse B16 cells only encodes a 542-amino-acid protein that lacks transmembrane domains and a putative pore region of the channel (9,19). However, the human clone subsequently identified in retina encoded a 1533-amino-acid protein with six predicted transmembrane domains and homology to other channels of TRP family (Figure 29.2) (9).

Recently, a full-length mouse cDNA that encodes a 1622-residue protein that, like the human clone, also contains transmembrane and the ion-permeating pore domains was identified, the results suggesting that the shorter melastatin mouse cDNA represents an amino (N)-terminal splice variant of a full-length isoform (19,22). Indeed, several splice variants of TRPM1 have been identified. For example, a short N-terminal variant directly interacts with the full-length channel and inhibits its translocation to the plasma membrane, thereby suppressing channel function (23).

In addition to shorter forms of the protein, a nucleotide polymorphism (C130 to T) in the 5′ untranslated region (UTR) of human TRPM1 results in a frameshift that generates an in-frame ATG initiation codon in exon 2 (normal start site is in exon 3). This viable transcript produces a novel 1603-residue protein, which is noted to be expressed higher in individuals of European decent compared to those of Asian or African heritage, suggesting a linkage between this variant and susceptibility to melanoma (1). Lastly, of the other splice forms identified, proteins with an additional 92, 102, and 109 amino acids on the N-terminus are also described (16,24), with functional proteins identified in melanocytes, brain, and retina.

In addition to melanocyte and retina expression, TRPM1 has also been detected in brain and heart tissues, albeit at very low levels (25,26). These analyses of TRPM1 channel expression were determined by transcript expression as the generation of viable TRPM1-specific antibodies has proved challenging. Those available have detected protein bands of sizes ranging from 120 to 250 kDa in lysates from melanocytes, suggesting poor antigenicity (27,28). However, TRPM1 immunoreactivity was recently observed in retina ON bipolar cells of wild type, but not $TRPM1^{-/-}$ mice (19,29). Cellular localization has proven problematic, with data suggesting that TRPM1 channels are largely localized intracellularly (24). However, these data were obtained with heterologously expressed fluorescently tagged proteins, and overexpression may hamper the ability to detect channels localized to the plasma membrane. This is particularly relevant as heterologous expression of these channels leads to outwardly rectifying, nonselective cation conductances (19,24), biophysical profiles consistent with TRP ion channels. Nonetheless, functional data of TRPM1 currents in bipolar cells show that at least in the visual system, TRPM1 channels are functional in the plasma membrane (19,20).

29.2.2 CELLULAR FUNCTION

Remarkably, despite year's research since the discovery of TRPM1 in melanocytes, the role of the channel in melanoma advancement is poorly understood. TRPM1 expression strongly

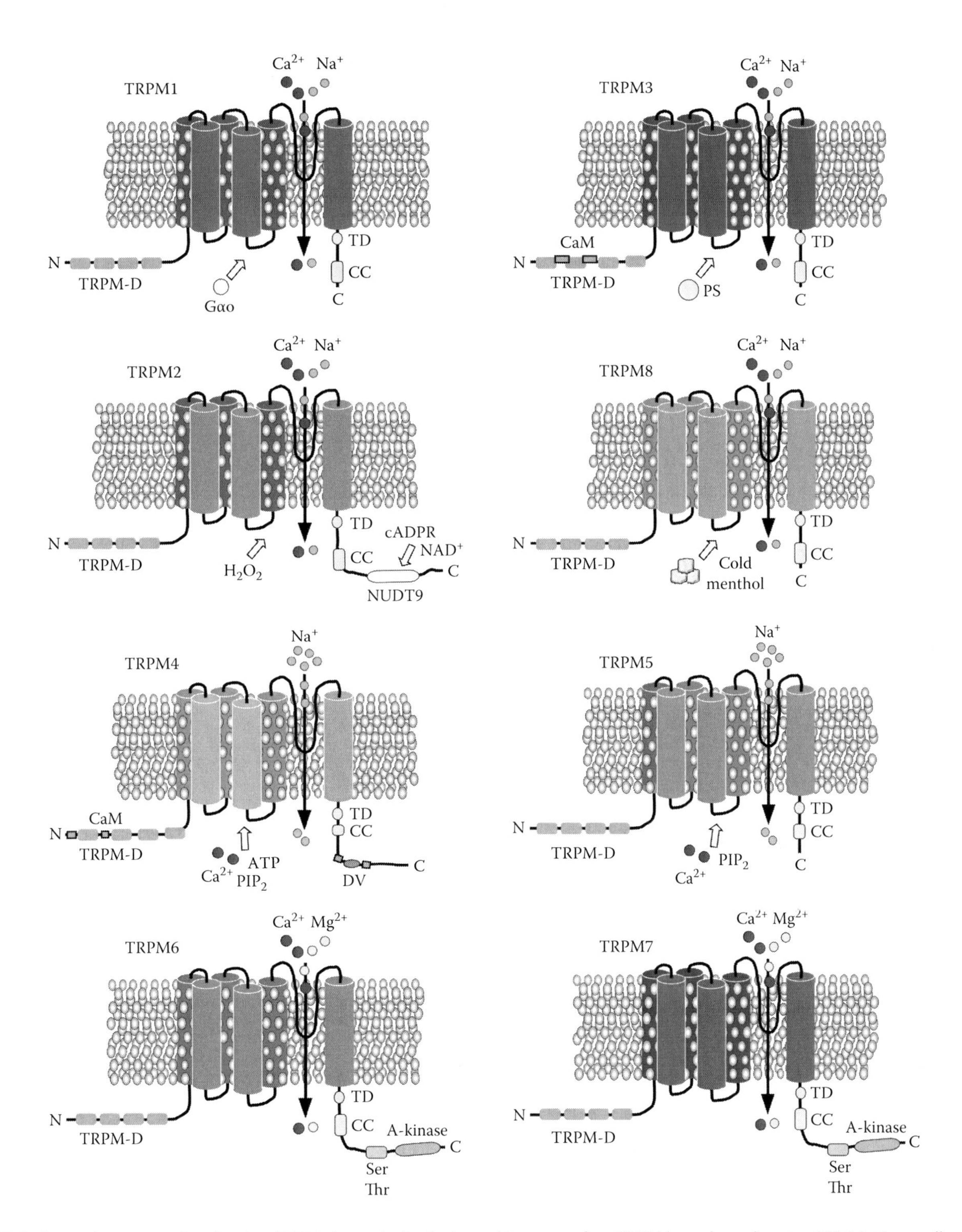

Figure 29.2 General structures localized to TRPM channels. Each channel has up to four TRPM homology domains (TRPM-D) as well as the TRP (TD) and coiled-coil domains (CC). TRPM1 channels are sensitive to the G-protein Gαo. TRPM2 channels contain an enzymatic Nudix-homology domain in the C-terminus (NUDT9), similar to the A-kinase domains in TRPM6 and TRPM7, with the former containing several ATP (not shown) and vanadate binding domains (DV). TRPM3 and TRPM4 channels contain Ca^{2+}/CaM-binding domains. Mostly all channels are sensitive to either PIP2 directly or as a consequence of PIP2 cleavage.

correlates with tumor thickness, the malignancy of both cutaneous and nodular melanomas, as well as survival rates in patients with malignant melanomas (10,12,30). Thus, there has been a significant effort to determine if TRPM1 functions as a tumor suppressor, but as of yet, this function has not been validated (1). Nonetheless, several lines of evidence support this hypothesis.

Within the TRPM1 promoter region, binding sites for the microphthalmia transcription factor (MITF), a helix–loop–helix leucine zipper transcription factor, were identified important as MITF has been shown to regulate expression of several melanocyte-specific genes, including TRPM1 (11,31). Moreover, in a screen of microRNA (miRNA) expression in melanoma cells and primary melanocytes identified miR-211 as the most differentially expressed, significant as it is encoded from the sixth intron of the TRPM1 gene (32,33). Expression of many predicted targets for miR-211 were reduced in melanoma cells lines ectopically expressing this miRNA and MITF is required for miR-211 expression, results suggesting a potential tumor suppressor mechanism (33). Similarly, mutations in, or regulation of, MITF alter coat color and eye pigmentation in mice (34,35), and Waardenburg syndrome, a form of deafness in humans that is due to the loss of melanocytes in the ear, results from mutations in MITF (35).

One significant confound in the study of TRPM1 in melanocyte function is the absence of robust animal models as rodents do not have melanocytes. In humans, melanocytes generate the pigment melanin in organelles termed melanosomes that are then transferred to keratinocytes where they protect cells from UV damage (36). Activation of the melanocortin-1 receptor (MC1R), a Gαs-protein-coupled receptor, regulates the synthesis of proteins needed to produce melanin, a process mediated by MITF expression, suggesting the involvement of TRPM1. TRPM1 expression correlates with melanin content (24,37), and signaling via metabotropic glutamate receptor subtype 6 (mGluR6) and changes in Ca^{2+} signaling in melanocytes regulated melanin content (28,38). Furthermore, inhibition of cellular TRPM1 expression reduces melanin content in melanocytes (24). Thus, TRPM1 may play a role in normal melanocyte function and regulate pigmentation, a hypothesis consistent with downregulation of TRPM1 in homozygous Appaloosa horses (14).

In the retina, rod and cone photoreceptors make synapses with ON and OFF bipolar cells, respectively, which transmit visual contrast to ganglion cells. ON cells detect increases in light intensity and are depolarized subsequent to photoreceptor hyperpolarization, which reduces the rate of glutamate release that occurs in response to light. The ON bipolar cell response is mediated by mGluR6 that, when activated, couples to the Gαo class of trimeric G proteins that regulate a depolarizing current in these cells. While this pathway was well appreciated, the identity of the channel responsible for the depolarizing current was unknown. As mentioned earlier, the coat spotting pattern in Appaloosa horses is due to a single autosomal-dominant gene termed leopard complex (LP) (39), and CSNB is associated with animals homozygous for this trait (LP/LP) (40). TRPM1 mapped to the LP locus, and transcript expression in the retina, was reduced by greater than 1800-fold in homozygous Appaloosa horses (14,41). Consistent with this mutation, TRPM1 expression was found in retinal cells isolated via mGluR6 expression (42), and functional evidence suggested the mGluR6-dependent current was likely to be a TRP channel (43). Lastly, several groups identified mutations in the TRPM1 gene in patients with CSNB (15–18). Thus, these data strongly supported a role for TRPM1 channels in light-evoked responses of ON bipolar cells.

This hypothesis was confirmed by analyses on TRPM1$^{-/-}$ mice (19,20). Electroretinogram analysis of TRPM1$^{-/-}$ mice found that b-waves associated with bipolar cell responses were abolished, yet these animals retained normal a-waves (20). Moreover, whole-cell patch-clamp recording from ON bipolar cells in retinal slices from TRPM1$^{-/-}$ mice found that light-evoked responses were abolished (19,20). Consistent with these functional data, TRPM1 transcript expression was found in the inner nucleus layer of the retina, and full-length TRPM1 channels were shown to be localized to the dendritic tips of ON bipolar cells and colocalized with mGluR6 (19,20,44). Using electron microscopy, TRPM1 immunoreactivity was found on the dendritic tips of ON bipolar cells; however, this expression was reduced in mGluR6$^{-/-}$ mice in the dendritic tips with the channel localized to the soma and primary dendrites, consistent with the lack of TRPM1 activity in these mice (45). These and other results identified TRPM1 as a visual transduction channel in mammalian retina.

As described, mGluR6 couples to Gαo trimeric proteins, yet how this leads to changes in TRPM1 channel gating is unclear. When mGluR6, TRPM1, and Gαo were heterologously expressed, a constitutively active nonselective cation current was recorded that was then absent if mGluR6 and Gαo were expressed alone (19). Furthermore, TRPM1 currents were suppressed if Gαo was replaced with a constitutively active mutant or if purified G-protein was added to the intracellular face of the membrane. However, in recordings from bipolar cells in retinal slices, dialysis of purified Gβγ dimers, but not activated Gαo, inhibited TRPM8 channels (21). Similarly, deletion of $G\beta_3$ subunit reduced light responses and either mislocalized or downregulated components of the signal transduction cascade (46). Thus, it is not clear how mGluR6 activation of Gαo leads to the regulation on TRPM1 channel function, but data suggest that direct interactions may be involved. Similarly, a proteomic search for proteins in the mouse retina associated with nyctalopin, a small leucine-rich repeat extracellular scaffolding protein, identified TRPM1 as the binding partner (47,48). Nyctalopin, which was shown to be required for glutamate-mediated bipolar cell responses (49), directly interacts with mGluR6, and disruption of mGluR6 prevented localization of TRPM1 channels to the postsynaptic compartment of ON bipolar neurons (48). Nyctalopin was also shown to interact with TRPM1, suggesting it acts as an accessory protein that is essential for appropriate channel localization and the organization of a macromolecular complex for precise phototransduction (47,48).

29.3 TRPM2

One of the distinguishing features of certain members of the TRPM subfamily is that in addition to having ion-permeating channel domains, they also have intracellular regions that have enzymatic function. TRPM2 is one such multifunctioning protein in that it is a nonselective cation channel but also has an enzymatic domain in its C-terminus that is important for gating of the channel (2). In addition, TRPM2 channels are important functionally in a variety of different cell types, including pancreatic β-cells and immune, endothelial, neuronal, and cardiac cells, supporting functions ranging from insulin release to the production of cytokines and cell motility to cell death (2,50). Unlike other TRPMs, TRPM2 channels show little voltage dependence with a linear current–voltage relationship that reverses at 0 mV (51). TRPM2 is critical for Ca^{2+} influx in these cells, as well as is localized intracellularly to the lysosome where it serves as a Ca^{2+} release channel (52). Lastly, TRPM2 is considered an intriguing target for diseases associated with oxidative stress, such as inflammation and neurodegeneration, as it is activated by reactive oxygen species (ROS) such as hydrogen peroxide.

29.3.1 MOLECULAR STRUCTURE

TRPM2 is unique among TRPM channels in that, like TRPM6 and TRPM7 (see Sections 29.7 and 29.8), it is considered a *chanzyme* in that it contains an NUDT9 adenosine diphosphate ribose (ADPR) pyrophosphatase domain (Nudix-like homology domain) in its C-terminus (Figure 29.2) (51,53). The human TRPM2 gene encodes a protein of 1503 amino acids consisting of six transmembrane domains with the >300 amino acids in the C-terminus comprising the NUDT9 domain (54). The mouse gene encodes for a slightly larger protein (1507 residues) but retains the same structural motifs as the human channel, including four TRPM homology domains (MHD) and a calmodulin (CaM)-binding IQ-like domain in the N-terminus and the characteristic TRP box and coiled–coiled region between the sixth transmembrane domain and the NUDT9 domain (54,55).

Several splice variants have been identified, and similar to TRPM1, a truncated form containing only residues corresponding to the N-terminus has been suggested to have a dominant negative role in TRPM2 channel activity in tissues such as brain, marrow, and the vasculature (2,56,57). Others include those lacking residues in both the N- and C-terminus, variants that have proven informative regarding channel function (58). Specifically, channel function is abolished in the N-terminal variant that is missing 20 amino acids corresponding to an IQ-like CaM-binding site as well as two PxxP protein–protein interaction motifs (58). However, mutation of individual residues within these regions had no effect on channel function, suggesting that individually they are dispensable for TRPM2 activation (59). The C-terminal splice variant lacks 34 residues that are within the NUDT9 domain and thus has been shown to be insensitive to ADPR but responds normally to H_2O_2 (58). However, as with the N-terminal variant, no specific residues were shown to be necessary for ADPR-induced gating, leading to speculation that the sequences deleted represent a spacer segment that stabilizes this region (60).

29.3.2 CELLULAR FUNCTION

TRPM2 channels are activated by ADPR, cyclic ADPR, nicotinamide dinucleotide (NAD^+), and ROS such as H_2O_2 and other substances that generate reactive oxygen and nitrogen species. The channel functions both in Ca^{2+} influx through the plasma membrane and as a Ca^{2+} release channel in lysosomes. ADP-ribose (ADPR) and cyclic ADP-ribose (cADPR) are produced by the ectoenzyme CD38 that has multiple enzymatic activities, including being a ADP ribosyl cyclase and cADPR hydrolase that uses NAD^+ and $NADP^+$ as substrates (61,62). Consistent with these products' activation of TRPM2, CD38 is coexpressed with the channel in many cell types (62).

Based on the profile of agonists activating TRPM2, the channel is thought to be important in cell death associated with oxidative stress. For example, expression of TRPM2 confers susceptibility of heterologous cells to H_2O_2, leading to cell death that is associated with an increase in intracellular Ca^{2+} that can be blocked by coexpression of the dominant negative TRPM2 splice variant (56,63). Moreover, downregulation of TRPM2 expression significantly reduced cell death induced by oxidative stress in several cell lines (63,64). In pancreatic β-cells, H_2O_2 and alloxan induce TRPM2-mediated Ca^{2+} release from lysosomes inducing apoptotic cell death (65–67). However, $TRPM2^{-/-}$ mice have higher levels of blood glucose that was found to associate with impaired glucose-stimulated insulin secretion both in the whole animal and at the cellular level, also suggesting a potential role for TRPM2 in diabetes (65,68,69).

In addition, oxidative stress is critical in neural degeneration, and TRPM2 channels are most robustly expressed in the brain where they have been linked to Alzheimer's and Parkinson's disease, as well as amyotrophic lateral sclerosis (ALS) (70–72). Ca^{2+} excitotoxicity plays a key role in the neuropathology associated with Alzheimer's, and it has been shown that H_2O_2-induced Ca^{2+} entry in cortical and basal ganglia neurons involves TRPM2 channels (71,73). However, a single amino acid substitution, P1018 to L, was identified in a variant of TRPM2 linked to pathogenesis of both Alzheimer's and Parkinson's disease, and mutant channels were shown to rapidly inactivate, thereby reducing Ca^{2+} influx (70). While this is counterintuitive in the context of Ca^{2+} excitotoxicity, it has been postulated that reduced Ca^{2+} entry through TRPM2 channels disrupts the coordination of critical Ca^{2+} signals or leads to depletion of intracellular stores, thereby leading to neurodegeneration (50). Similarly, another variant of TRPM2 has been linked to individuals with bipolar disorders (74). In line with the linkage of TRPM2 and oxidative stress, ADPR is also generated by the breakdown of ADPR polymers by poly-ADPR glycohydrolase (PARG), an enzyme whose activity increases as a result of DNA damage (75). Thus, while the exact physiological mechanism underlying TRPM2 and cell death is unclear, the channel plays a key role in these processes and will be a target of much needed research in the future.

29.4 TRPM3

TRPM3 is one of the least understood members of the TRPM subfamily and was the last member of this subfamily identified (3,76). One confound of this channel is that a large number of channel variants have been identified, thereby creating ambiguity in the study of channel function due to the uncertainty into the physiologically relevant isoforms (3). Nonetheless, the channel does show constitutive activity and likely regulates Ca^{2+} influx. TRPM3 is expressed in a limited number of tissues, with the most robust expression in humans localized to the kidney, although the channel is not found in mouse kidney (77,78). TRPM3 is also robustly expressed in the brain (dentate gyrus; choroid plexus), as well as in the pituitary, pancreas, ovaries, testis, sensory ganglia, and spinal cord (76,79). In the kidney, TRPM3 is localized to renal tubules and is proposed to play a role in osmoregulation and renal Ca^{2+} homeostasis (77,78). In the brain, TRPM3 channels are localized to epithelial cells of the choroid plexus, leading to the hypothesis that the channel regulates ion concentrations in cerebral spinal fluid in oligodendrocytes (3,80), suggesting that it plays a role in myelination (81) and in Purkinje neurons where it is implicated in glutamatergic neurotransmission in the developing nervous system (82). Lastly, recent evidence also implicates TRPM3 in thermosensation (79).

29.4.1 MOLECULAR STRUCTURE

The TRPM3 gene is remarkably large (850 kb in mouse) with introns separating the first few exons spanning over 250 kb of DNA sequence (80). Thus, it is not surprising that TRPM3 encodes for the largest number of splice variants of TRP channels, presenting a divergent population of channels with different functional domains and gating properties (Figure 29.2). In most cases, these variants are derived from alternative splicing mainly in the 5′ region of the transcript, as well as in the putative pore domain (see for review (76)).

For example, coding sequences corresponding to exon 2 are only found in one putative transcript in humans (83). Several splice variants alter sequences near the putative pore of the channel, and it has been found that cation permeability is affected in two mouse variants in exon 24 of the channel that differ only in the pore region (80). In most studies, TRPM3α2 is the variant of choice and lacks 12-amino-acid residues within the pore region but is highly permeable to divalent cations (80,84,85). Another defining feature of the TRPM3 gene is that, like TRPM1 (see Section 29.2), a miRNA, miR-204, is located in intron 8 of the mouse sequence (86,87). miR-204 is one of the most extensively studied miRNAs in the eye and is known to be physiological in human retinal pigmented epithelium, although its exact role is not understood (88).

As with other TRPM channels, TRPM3 has six transmembrane domains, a C-terminal coiled-coil region, as well as the TRPM homology domain located in the N-terminus of the channel (77,80). Moreover, four CaM-binding sites are also present in the N-terminus, and it has been shown that CaM can bind to heterologous proteins containing these sequences (89), suggesting regulation of the channel by Ca^{2+}.

Using both transcript and protein expression approaches, TRPM3 has been found localized to an array of cell types in the brain such as developing Purkinje cells (82), oligodendrocytes (81), and epithelial cells of the choroid plexus (78,80). As previously mentioned, TRPM3 channels are also in retinal tissues and were recently identified in primary sensory neurons. In nonneuronal tissues, TRPM3 is expressed in pancreatic islets and the pancreatic cell line INS-1 (25,90,91), human (but not mouse) kidney (77,78), vascular smooth muscle cells (92), and odontoblasts (93).

29.4.2 CELLULAR FUNCTION

To date, analyses of TRPM3 channel function, regardless of the variant tested, find that the channel is constitutively active when heterologously expressed (76) and is permeable to divalent cations such as Ca^{2+} and Mg^{2+} (80). TRPM3 channels are also blocked by gadolinium and lanthanum ions and by 2-aminoethoxydiphenyl borate (2-APB), an inhibitor of inositol trisphosphate (IP3) receptors, store-operated channels (SOCs), and several TRP channels (94,95). The pore is also permeable to manganese, and channel activity is blocked by hypertonic solutions and increased by hypotonicity (78), the results suggesting that TRPM3 channels in the kidney are involved in osmoregulation.

A number of studies have shown that TRPM3 channels are activated by the neurosteroid pregnenolone sulfate (PS) in pancreatic β-cells and the INS-1 cell line, two cells types shown to express a PS-sensitive channel, (90,91,96,97). TRPM3 is also activated by nifedipine, a dihydropyridine inhibitor of voltage-gated Ca^{2+} channels (91). Conversely, cholesterol and mefenamic acid block TRPM3 channels, as do rosiglitazone and troglitazone, two agonists for peroxisome proliferator-activated receptors (PPARg), and a number of citrus fruit flavanones (90,92,98,99). These agonists and antagonists have been instrumental in determining the potential roles of TRPM3 in physiology.

In pancreatic cells, PS potentiated glucose-induced insulin secretion from islets, effects that were blocked by mefenamic acid (90,91). Consistent with these in vivo data, PS increases intracellular Ca^{2+} in these cells, a cellular effect that is also blocked by mefenamic acid (90), which was ineffective in altering classical K_{ATP}-dependent pathways, thereby suggesting that TRPM3 channels work independent of glucose-mediated insulin secretion. It has been proposed that activation TRPM3 in the pancreas leads to sufficient depolarization that L-type Ca^{2+} channels open and change the levels of intracellular Ca^{2+} such that changes in gene expression are induced, likely through the Raf and ERK protein kinase pathways and increased production of the zinc finger transcription factor Egr-1 (96,100,101).

In neuronal tissues, TRPM3 is expressed in oligodendrocytes and activation of the channel leads to increased intracellular Ca^{2+}, the results suggesting involvement in myelination (81). Similarly, PS has been found to regulate the formation of glutamatergic synapses in the developing nervous system by enhancing glutamate release in synapses on Purkinje cells (102). Mefenamic acid blocks PS-induced glutamate release, the results suggesting that steroid-mediated modulation of glutamatergic neurotransmission during development uses TRPM3 channels to regulate intracellular Ca^{2+} (82). Lastly, TRPM3 was recently found in small-diameter neurons of the dorsal root and trigeminal ganglia and was shown to mediate aversive responses to PS in mice (79). This surprising result was coupled with evidence that TRPM3 channel activity increases with heating and that TRPM3$^{-/-}$ mice are deficient in heat responses and the development of inflammatory heat hyperalgesia.

29.5 TRPM4

Of the TRPM subfamily, TRPM4 (as well as TRPM5) is distinct in that the channel is not permeable to Ca^{2+}. However, Ca^{2+} is the primary regulator of channel function in that it can be gated by an increase in intracellular Ca^{2+}, establishing TRPM4 as a regulator of membrane excitability in a number of tissues. Another distinction of TRPM4 is that the channel appears to be expressed ubiquitously and that TRPM4 channels play important regulatory roles in Ca^{2+}-dependent processes, such as cardiac function and constriction of cerebral arteries, insulin secretion, activity of inspiratory neurons, and the immune response.

29.5.1 MOLECULAR STRUCTURE

TRPM4 channels are most closely related to TRPM5 in structure (~40% homologous at the amino acid level), but as with other TRPM channels described, the TRPM4 gene does encode for several splice variants, with TRPM4b considered the predominant channel variant and has been the primary transcript studied (Figure 29.2) (23,103–105). Human and mouse TRPM4b genes encode proteins of 1214 and 1213 amino acids, respectively, and as with other members of the TRPM subfamily, TRPM4 contains four TRPM homology domains in the N-terminus, a coiled-coil domain on the C-terminus, as well as putative CaM-binding sites in both the N- and C-termini (106). TRPM4 also contains motifs known to interact, including two ATP-binding cassette (ABC) transporter—like motifs, in addition to four Walker nucleotide binding domains (NBDs) that are localized in close proximity (107). Similarly, a region of positively charged amino acids in the C-terminus of TRPM4 binds decavanadate and increases channel activity (108). Lastly, TRPM4 channels have a phosphatidylinositol bisphosphate (PIP2) binding site, homologous to pleckstrin homology (PH) domains that confer the channel's sensitivity to phosphoinositides (4).

As described, TRPM4 is fairly ubiquitously expressed in a number of tissues, including expression of mRNA detected by northern blot analysis in pancreas, skeletal muscle, heart, liver, kidney, placenta, colon, lung, thymus, spleen, prostate, and small intestine from human tissues (23,103). Intracellular Ca^{2+} plays a significant role in most all physiological responses and thus is strongly regulated by many cellular elements. Thus, within these tissues, TRPM4 channels serve as a sensor of intracellular Ca^{2+}, comprising Ca^{2+}-activated nonselective cation channels (NSC_{Ca}). TRPM4 channel activation enables Na^+ entry into the cell, altering the driving forces for Ca^{2+} and other ions by changing the cellular membrane potential. Specifically, it has been shown that TRPM4 affects the activity of a variety of cellular ionic conductances such as voltage-gated and non-voltage-gated Ca^{2+} channels (109).

Channel function of both TRPM4 and TRPM5 is robustly sensitive to the phosphoinositide PIP2. Depletion of PIP2 in both excised membrane patches and by intracellular manipulation reduces channel activity, an effect that is reversed by the application of exogenous PIP2 (110–112). Both channels are activated by Ca^{2+} in excised patches, but their Ca^{2+} sensitivity decreases with time. This desensitization can be reversed by the application of PIP2 (110–112). TRPM4 is more sensitive to PIP2 in comparison to TRPM5 channels (112), and the levels of this phosphoinositide alter Ca^{2+} sensitivity and shift the voltage sensitivity of each channel.

29.5.2 CELLULAR FUNCTION

TRPM4 is expressed in several cell types in the heart, including sinoatrial node pacemaking cells and ventricular and atrial myocytes (113). In a rat model of hypertrophy, TRPM4 expression is increased in ventricular myocytes, consistent with the correlation of channel expression and cell capacitance, the latter a marker of hypertrophy (114). Arrhythmias related to hypertrophy involve a calcium-activated transient inward current (Iti) responsible for both delayed after depolarizations and early after depolarizations, and it has been suggested that TRPM4 channels may participate (109). Moreover, autosomal-dominant progressive familial heart block type 1 (PFHB1), a cardiac bundle branch disorder that affects electrical conduction of the heart leading to complete heart block, was linked to a single point mutation in TRPM4 (115,116). Surface membrane expression of this mutant is elevated due to impairment in endocytosis caused by altered SUMOylation of the channel, resulting in overall depolarization and progressive block of cardiac conduction. However, it is unclear if TRPM4's role is direct or indirect and how it supports conduction block. Lastly, to maintain constant blood flow, cerebral arteries constrict in response to increased internal pressure, a process termed the Bayliss effect (117). TRPM4 expression is detected at the transcript and functional levels, and decreased channel expression reduced the Bayliss effect (118,119).

In the immune system, TRPM4 appears to function to lessen the immune response by preventing Ca^{2+} overload in both T-lymphocytes and dendritic cells (120). Dendritic cells fail to migrate to lymphoid organs after being stimulated by a bacterial infection in TRPM4-deficient mice (120). In T-lymphocytes, reduction of TRPM4 expression increases Ca^{2+} responses, thereby elevating the production of immune factors, whereas activation of TRPM4 decreases factor release (121,122). TRPM4 is also implicated in inflammation and the allergic response due to increased activation of mast cells in TRPM4-deficient mice, as well as a decrease in mast cell migration during inflammation (123,124). Lastly, inflammatory diseases such as multiple sclerosis are a result, at least in part, of neuronal degeneration, and inhibition of TRPM4 channels was shown to reduce neuron loss, and TRPM4-null mice were protected against inflammatory stress (125). Thus, TRPM4 channels play a key role in immune responses associated with inflammation and other disorders.

The release of insulin from pancreatic β-cells in response to a glucose challenge is dependent on a rise in intracellular Ca^{2+}. TRPM4 channels are expressed in β-cells and it has been shown that glucose-stimulated insulin release from an immortalized β-cell line is reduced if TRPM4 channel activity is inhibited (126). However, in TRPM4-deficient mice, glucose and insulin homeostases were normal in vivo and in vitro, calling into questions the role of the channel in the pancreas (68). These, and other results, show that TRPM4 is involved in many cellular processes, a diversity in expression that has hampered interpretation of the role of this channel. Thus, future studies will hopefully shed light on the necessity of TRPM4 channels in various biological settings.

29.6 TRPM5

TRPM5, like TRPM4, is a Ca^{2+}-activated cation channel selective for monovalent ions and impermeable to divalents. However, unlike TRPM4, TRPM5 expression is restricted to only a few tissues, and therefore its role in physiology is similarly constrained. TRPM5 is found at high levels in the tongue, small intestine, and stomach, as well as in pancreatic β-cells where it appears to localize with insulin. TRPM5 channel activity is strongly regulated by voltage, plasma membrane phosphoinositides, acidity, and temperature and is required for transduction of bitter, sweet, and umami tastes. Moreover, TRPM5- deficient mice display impaired glucose homeostasis. Thus, like TRPM4, TRPM5 channels mediate depolarizing currents that regulate excitable cells, but in a more defined set of physiological responses.

29.6.1 MOLECULAR STRUCTURE

TRPM5 was originally identified in a screen for genes with homology to TRPM1 and encodes for a 1165-amino-acid protein in humans and a slightly smaller peptide of 1158 amino acids in the mouse (Figure 29.2) (127,128). To date, only a single functional transcript of TRPM5 has been identified and studied, and structurally TRPM5 does not contain many of the different motifs found in TRPM4. For example, TRPM5 lacks an ABC transporter—like motif and possesses only one Walker B site that is located in a region of the channel that is inaccessible to ATP (107). These sequence differences equate to function as TRPM4 is sensitive to block by decavanadate, presumably through the ABC motifs, whereas TRPM5 activity was unaffected (108,129).

For both TRPM4 and TRPM5, the site mediating Ca^{2+} sensitivity has yet to be identified. A direct comparison of Ca^{2+} sensitivity of each channel found that TRPM5 is activated at much lower Ca^{2+} concentrations than TRPM4 but that both channels strongly desensitize when studied in excised membrane patches (107,111). TRPM5 channels are not sensitive to regulation by the Ca^{2+}-binding protein CaM (130), but application of phosphoinositides such as PIP2 diminishes desensitization, suggesting coordination in the intracellular events regulating channel function (110,111). TRPM5 has been most extensively studied in taste cells where it is activated downstream of G-protein-coupled receptors and Phospholipase C (PLC)-mediated production of IP3, which is generated by the cleavage of PIP2, thereby implicating negative-feedback inhibitory pathway regulating channel function (5).

29.6.2 CELLULAR FUNCTIONS

TRPM5 is robustly expressed in taste receptor cells and is coexpressed with G-protein-coupled receptors for bitter, sweet, and umami tastants, as well as genes essential for taste transduction including the G-protein gustducin and PLCβ2 (131–133). The role of TRPM5 in taste was established when it was shown that TRPM5$^{-/-}$ mice were insensitive to these taste modalities but still detect sour and salty tastants (131). Mechanistically, TRPM5 serves as the downstream target of tastant receptor activation, which leads to the activation of PLC and the hydrolysis of PIP2 into diacylglycerol (DAG) and IP3, the latter inducing the liberation of Ca^{2+} from intracellular stores via IP3 receptors (5). As mentioned earlier, Ca^{2+} directly gates TRPM5 channels and tastant-evoked responses are blocked when intracellular Ca^{2+} is buffered or IP3 receptors are inhibited (111,130,134). TRPM5 channel function has also been shown to be temperature sensitive with heat leading to an increase in channel activity (135). However, heat does not directly gate TRPM5 as it does to TRPV1 channels, but likely serves to regulate the response of taste cells when temperature is altered. Similarly, pH also regulates TRPM5 channel function in taste cells, with channel activity reduced under acidic conditions (136), properties that might influence coding of different tastants under a variety of palatable conditions.

As stated earlier, in comparison to TRPM4, TRPM5 is less broadly expressed but, like TRPM4, is found in pancreatic islets and predominantly in cells containing insulin, suggesting expression in insulin-secreting β-cells (137–139). β-cells express a Ca^{2+}-sensitive nonselective cation channel, a current that is significantly reduced in TRPM5$^{-/-}$ mice (137), the results suggesting that TRPM5 channels contribute to electrical activity of β-cells under conditions of glucose-induced stimulation (140). Consistent with these cellular data, pancreatic islets obtained from TRPM5$^{-/-}$ mice exhibit deficits in glucose-induced insulin release, and these animals perform poorly in glucose tolerance tests and have lower plasma insulin levels (137,138).

Lastly, TRPM5 is also expressed in the gut in a chemosensory cell type that releases hormones such as glucagon-like peptide 1 (GLP-1) (141–144). The channel, as well as other molecules normally associated with taste cells, is found in a cohort of intestinal solitary brush cells and endocrine cells of the duodenal glands, suggesting that the channel may serve to detect chemicals in the gut lumen. In a model of type 2 diabetes, TRPM5 expression inversely correlated with blood glucose concentration (143), and enteroendocrine cells in mouse duodenum that express TRPM5 also express β-endorphin and other peptides that may have local roles in regulating intestinal function when under high glucose or hyperosmotic conditions (144).

29.7 TRPM6

Like TRPM4/TRPM5, the channels TRPM6 and TRPM7 are closely related with TRPM7 ubiquitously expressed, whereas TRPM6 expression is restricted to the intestine and kidney (6). These channels are also unique in that they possess kinase domains residing in their C-termini, making them, like TRPM2, multifunctional channels or chanzymes. Physiologically, TRPM6 and TRPM7 are critical for magnesium homeostasis, with a genetic defect in the former found to underlie hypomagnesemia with secondary hypocalcemia (HSH), an autosomal recessive disorder that is characterized by low serum magnesium (145,146). Thus, TRPM6 channels have a clear physiological role, but the mechanisms underlying how they regulate magnesium levels and if they form functional channels in the absence of TRPM7 are unclear.

29.7.1 MOLECULAR STRUCTURE

TRPM6 encodes for a protein of 2022 amino acids in humans and 2028 residues in mouse and shares 50% homology with TRPM7 (Figure 29.2) (145,146) and was originally cloned in a screen for transcripts with homology to elongation factor kinase 2 (147). Structurally, TRPM6 contains many similarities to other TRPM channels but as described earlier contains an A-kinase domain in its C-terminus and is reported to form both homotetrameric and heterotetrameric channels, the latter with TRPM7 (148,149). Indeed, the ubiquitous nature of TRPM7 expression, even in heterologous cells, suggests that TRPM6 may form functional channels when oligomerized with TRPM7, the latter suggested to be necessary for trafficking of the channel to the plasma membrane (148,150). However, the oligomerization of TRPM6 is still under debate (6,151), with evidence suggesting that there are different channel properties (i.e., conductance, ion permeation, and pharmacological sensitivity) when each channel is individually expressed, compared to when both TRPM6 and TRPM7 are coexpressed (149). TRPM6 currents are strongly outwardly rectifying with the small amount of current passed at negative membrane potentials made up primarily of Mg^{2+} or Ca^{2+}, but the channel also permeates a range of cations with little evidence that monovalents are conducted through the pore (149,151).

Alpha-kinases are unique in that they bear no resemblance to classical protein by phosphorylating residues within protein regions that are predicted to be alpha-helical (152). The alpha-kinase domain in TRPM6 does represent a functional kinase, although most of what is known of the enzymatic function comes from analysis on TRPM7 (see Section 29.8) and specific substrates have yet to be identified (150).

29.7.2 CELLULAR FUNCTIONS

Due to the channel's relatively exclusive expression in the kidney and intestine, as well as its link to HSH, the majority of what is known of TRPM6 function at the cellular level arises from analysis of Mg^{2+} homeostasis. Mg^{2+} is essential for energy metabolism and any defects in Mg^{2+} balance affects a number of physiological processes (153). Mg^{2+} concentration is regulated by dietary absorption in the intestine and kidney excretion, as well as from bone. Individuals with HSH have low levels of both Mg^{2+} and Ca^{2+}, a result of poor Mg^{2+} absorption in the intestine and altered renal excretion, but can be easily treated by supplemental Mg^{2+}. The gene locus for HSH was mapped to TRPM6 and shown that loss-of-function mutations of the channel were causative for this disease (145,146). A variety of mutations have been identified in the TRPM6 gene in HSH patients, including point and missense mutations, frameshifts, and premature stop codons (154).

Consistent with its role in Mg^{2+} homeostasis, TRPM6 was shown to be expressed on the apical membrane of the renal distal convoluted tubule as well as along the brush-border membrane of the small intestine (151). Channel function is regulated by the levels of intracellular Mg^{2+} (151) and enhanced by extracellular acidification (149). The A-kinase domain on TRPM6 has been shown to be nonessential for channel activity (155) but does appear to play a regulatory role in modulating channel function. The channel can be autophosphorylated, as well as increase phosphate incorporation in TRPM7 channels (150,155). However, the channel is inhibited by RACK1, a receptor for activated C-kinase 1 that interacts with the A-kinase domain of TRPM6 when autophosphorylated (155). Similarly, repressor of estrogen receptor activity (REA) also binds to this region and inhibits channel function in a kinase-dependent manner (155). Autophosphorylation regulates the ability of Mg^{2+} to inhibit TRPM6 channel function, thereby encompassing a feedback inhibitory mechanism to prevent Mg^{2+} overload.

Estrogen strongly regulates Mg^{2+} homeostasis and has been shown that expression of TRPM6 is downregulated in ovariectomized rats but is restored if these animals are given 17[beta]-estradiol (156). Similarly, epidermal growth factor (EGF) has been linked to Mg^{2+} homeostasis via activation of renal EGF receptors, which in turn increase the activity of TRPM6 channels (157). Thus, TRPM6 channel activity is critical for a number of convergent mechanisms regulating Mg^{2+} homeostasis.

29.8 TRPM7

As described earlier, TRPM7 channels are closely related to TRPM6, but its tissue distribution is much wider and has been implicated in a number of cellular functions, including Mg^{2+} homeostasis, cell death, proliferation, neuronal degeneration, and the cell cycle. Moreover, and likely due to this diversity in expression, considerably more research has been performed into TRPM7 channel function. Deletion of TRPM7 in mice results in embryonic lethality, a further evidence as to the general importance of this channel. TRPM7 channels are enhanced when intracellular Mg^{2+} concentration plummets and when placed in acidic environments, whereas channel function is diminished when intracellular levels of PIP2 are reduced in a PLC-dependent manner.

29.8.1 MOLECULAR STRUCTURE

The TRPM7 gene encodes for a 1865-amino-acid protein in humans and a slightly smaller 1863-residue protein in mouse (Figure 29.2) (158–160). As mentioned earlier, TRPM7 is closely homologous to TRPM6, 49%–52% sequence identity between these two proteins, and contains much the same molecular architecture as TRPM6, including the C-terminal A-kinase domain. TRPM7 is a ubiquitously expressed protein with expression found even in embryonic stem cells and is required for embryonic development (161,162). Moreover, native TRPM7 currents have been recorded in every cell type examined to date (7). The crystal structure of the kinase domain has been solved, the results suggesting that even though there is little shared amino acid similarity between this and other classes of kinases, this region forms a structure that is similar to cAMP-dependent PKA (163). This structure contains a zinc finger domain as well as a region involved in the catalysis of ATP. The A-kinase domain is a serine/threonine kinase that, as described of TRPM6, is unique in that it phosphorylates residues within alpha helices and has been shown to interact with various PLC proteins (164). When activated, PLC interacts with TRPM7 and inhibits TRPM7 channel activity in a PIP2-dependent manner (165,166).

TRPM7 channels are strongly outwardly rectifying and permeable to mainly divalent cations such as Zn^{2+}, Ni^{2+}, Mg^{2+}, Ca^{2+}, and Mn^{2+}, in that order (158,167), yet is normally inhibited under normal physiological conditions. However, when intracellular divalent concentrations are reduced, TRPM7 channel activity increases, mainly conduction K^+ and Na^+ currents (158,168,169). Moreover, physiological concentrations of Mg^{2+}-bound nucleotides, such as Mg–ATP, serve as a source of Mg^{2+} and are also reported to inhibit TRPM7 channels (169). As with TRPM6, TRPM7 channels are also sensitive pH; in this case, increased protons lead to an increase in inward current comprised mostly of monovalent cations as the extracellular environment acidifies (170). Lastly, TRPM7 forms heteromeric channels with TRPM6 and has been shown to facilitate trafficking of TRPM6 to the plasma membrane (150).

29.8.2 CELLULAR FUNCTION

Due to its ubiquitous expression, there is an overabundance of data examining TRPM7 function in various cell and tissue types, but these studies are hampered by the embryonic lethality of TRPM7-null mice (162). Moreover, mice in which only the kinase domain of TRPM7 is lacking are also embryonic lethal, but supplemental Mg^{2+} can restore the viability of embryonic stem cells from these mice (171), suggesting the importance of Mg^{2+} homeostasis as the underlying cause of death. However, deletion of TRPM7 from thymocytes does not alter Mg^{2+}

homeostasis, but does block development of these cells (162). Moreover, induced deletion of TRPM7 in adult mice resulted in no observable phenotype, suggesting that its role in development is critical for viability (161). Indeed TRPM7 function has been linked to cell survival in a number of tissues, including mast cells, vascular smooth muscle, hepatocytes, and various carcinomas.

It is still controversial as to whether or not the kinase domain is involved in channel gating, but TRPM7 has been shown to phosphorylate three distinct substrates, annexin-1, myosin IIA heavy chain, and calpain (172–174). Annexin I mediates the anti-inflammatory effects of glucocorticoids and has been shown to mediate in apoptosis and cell growth (172), suggesting that TRPM7-mediated phosphorylation of Annexin I regulates these processes. Similarly, the heavy chain of myosin II is also involved in cell growth and apoptosis, as well as cell migration and cytoskeletal structure (173). Calpain is critical for cell adhesion and motility, in addition to apoptosis, and directly interacts with TRPM7 (174). TRPM7 activation of calpain leads to degradation of peripheral cell adhesions and may play a role in neural injury by increasing the rate of cell death (175).

In the vasculature, TRPM7 expression in vascular smooth muscle cells is essential in regulating the influx of Mg^{2+}, as well as contraction, dilation, and proliferation (176–178). TRPM7 channels are translocated to the plasma membrane, which is a response to increased fluid flow and shear stress, resulting in increased membrane currents (179). These observations suggest that TRPM7 might act as a sensor of cell damage, particularly after damage to the endothelia. Consistent with this posit, angiotensin II and aldosterone both positively regulate expression of TRPM7, as does the proinflammatory agent bradykinin (176,177,180) (Yogi, 2009 #288).

In the heart, TRPM7 expression in cardiac fibroblasts has been linked to increased Ca^{2+} influx after upregulation of channel expression via stimulation with TGF-β1, a factor known to induce fibrosis (Du, 2010 #289). Ca^{2+} promotes fibroblasts to differentiate into myoblasts that then leads to atrial fibrillation and fibrosis, a model supported by the observation that atrial fibroblasts from arrhythmogenic patients show an increase in TRPM7 expression functionally.

Recent evidence suggests that TRPM7 also plays a critical role during ischemic events in the brain, underlying a nonselective cation current that is activated during oxygen–glucose deprivation and leads to enhanced Ca^{2+} uptake and anoxic cell death (181). Knockdown of TRPM7 expression in cultured primary neurons reduced this current that is activated by ROS, thereby reducing cell death. Similarly, during ischemic conditions, the levels of extracellular Ca^{2+} and Mg^{2+} are significantly reduced, leading to the activation of Ca^{2+}-sensing nonselective cation current that is likely mediated by TRPM7 (182). Indeed, reducing TRPM7 expression in hippocampal CA1 neurons made these cells impervious to cell death after an ischemic event (183). In these examples, Ca^{2+} overload likely underlies cell death, although zinc-induced neurotoxicity has also been suggested consistent with the permeability profile of TRPM7 (184).

A missense mutation in TRPM7 has been linked to the neurodegenerative diseases, Guamanian ALS, and parkinsonian dementia (PD) (185). This mutation, T1482 to I, is localized to a site between the channel and kinase domains and is proposed to be a site of autophosphorylation, thereby increasing the channel's Mg^{2+} sensitivity, although this is currently under debate (185–187). Moreover, TRPM7 channels are sensitive to regulation by phosphoinositides and proposed to be involved in familial Alzheimer's disease (188).

29.9 TRPM8

TRPM8 channels are most closely related to TRPM2 but lack the enzymatic domains located on the latter's C-terminus. TRPM8 was first identified in a screen for genes upregulated in cancerous prostate epithelial cells but then later found to be the receptor for menthol, the cooling ingredient in mint. Consistent with the psychophysical sensation of cold induced by menthol, TRPM8 channels are also temperature sensitive and directly gated when temperatures drop below 26°C. Thus, TRPM8, along with several TRPV channels, falls into the class of TRP channels that serve as temperature detectors in the peripheral nervous system but is the only TRPM channel that is a temperature sensor. TRPM8 mediates both innocuous cool and noxious cold perception, heightened responses to cold after injury or disease, as well as is critical for pain-relieving aspects of cooling in conditions of neuropathic and inflammatory pain. Moreover, TRPM8 serves an as yet undefined role in several malignant cell types, as well as is critical in controlling thermoregulatory responses.

29.9.1 MOLECULAR STRUCTURE

Similar to other TRPM channels, TRPM8 has six transmembrane domains as well as the characteristic TRP and coiled-coil domains in the C-terminal region (Figure 29.2). The coiled-coil domain in the C-terminus is reported to be necessary for tetramer assembly and channel trafficking (189,190). However, a later study found that TRPM8 channels with C-terminal deletions tetramerize and localize properly to the plasma membrane (191). These channels were inactive, as were mutants lacking a small region on the distal N-terminus, the results suggesting that these cytoplasmic domains are critical for proper channel function (191).

TRPM8 is activated by several cooling compounds including menthol and icilin, and regions in the channel important for each agonist have been identified. Specific residues in the second TM (Y745) and in C-terminus (Y1005 and L1009) have been shown to be important for menthol gating of TRPM8 (192), as in regions in the fourth TM and the linker region between TM5 and TM6 (193). Icilin activates TRPM8 in a manner that requires a coincident increase in intracellular Ca^{2+}, a requirement not needed for menthol or cold gating of the channel (194,195). Icilin's actions are localized to the linker region between the second and third TMs with residues N799, D802, and G805 shown to be critical (194). However, how Ca^{2+} and icilin interact to gate TRPM8 is unclear.

A key modulator of TRPM8 channel activity is the phosphoinositide PIP2, which has been shown to be obligatory for TRPM8 channel function. TRPM8 channels adapt to prolonged stimulation in a manner that requires the presence of external Ca^{2+} (195), and channel activity in excised membrane patches rapidly runs down after excision but can be recovered by the presence of exogenous PIP2 on the cytoplasmic face of the channel (196,197). Activation of PLC also leads to channel adaptation, suggesting that Ca^{2+} entry through the TRPM8 pore activates Ca^{2+}-sensitive PLCδ isoforms to cleave PIP2 and thereby adapt TRPM8 channels (198). Moreover, residues in the proximal region of the C-terminus near the TRP domain of TRPM8 have been proposed to interact with PIP2 (196).

The mechanisms underlying gating TRPM8 channels by cold are still unresolved, as is the case for all thermosensitive TRP channels. TRPM8 channels are weakly voltage sensitive and show characteristic outward rectification (195). This voltage sensitivity, along with the topological similarity between TRP channels and voltage-gated K^+ channels, suggested that activations by temperature and voltage are linked (193,199,200). Neutralization of positively charged residues in the fourth TM and the TM4–TM5 linker reduced the number of gating charges, suggesting these to be the site of a voltage sensor (193). However, the closest TRPM8 homologue TRPM2 has an identical pattern of charged residues in TM4 as TRPM8, but is not temperature or voltage sensitive (201). Moreover, there is evidence that temperature-, agonist-, and voltage-dependent gating mechanisms are independent processes since distinct activation domains for each have been identified, suggesting that the effect of one gating mechanism acts on another in an allosteric fashion (198,199,202–204). For example, PIP_2- and PLC-mediated adaptation leads to a change in the voltage dependence of the channel but does not alter thermal sensitivity of TRPM8 channels (198). Lastly, purified TRPM8 channels have been reconstituted into a planar lipid bilayer and shown to be activated by cold, the most convincing evidence that this channel is directly gated by temperature (205).

In addition to modulation by PIP2, a second messenger generated downstream of G-protein-coupled receptor activation, recent evidence suggests that Gαq itself directly binds to TRPM8 channels and inhibits activity (206,207). Gαq had no effect and no channel function, and a Gαq chimera that could not activate downstream signaling cascades was equally capable of inhibiting TRPM8. These data suggest that TRPM8 is inhibited under conditions of inflammation, thereby exacerbating heat responses. However, in vivo evidence suggests that inflammatory mediators such as the glial cell line–derived neurotrophic factor (GDNF) receptor ligand artemin and nerve growth factor (NGF) can potentiate cold responses in a TRPM8-dependent manner (208).

29.9.2 CELLULAR FUNCTION

In sensory ganglia, TRPM8 is expressed in <15% of small-diameter (~20 μm) sensory neurons, consistent with the proportion of neurons shown to be cold and menthol sensitive in neuronal cultures (195,209,210). TRPM8 is expressed in a heterogeneous population of neurons, and a subset of these also express nociceptive markers such as TRPV1 and calcitonin gene-related peptide (CGPR), as well as labels for both Aδ- and C-fibers, the receptor tyrosine kinase TrkA, and the artemin-receptor GFRα3 (210–212).

Biophysically, cold- and menthol-evoked TRPM8 currents have surprisingly similar properties to conductances recorded in native cells, including ion selectivity, menthol potency, and voltage dependence (195). More remarkably, TRPM8 currents are also evoked by temperature decreases with an activation temperature threshold of ~26°C, with activity increasing in magnitude down to 8°C. Interestingly, this broad range spans to what are considered both innocuous cool (~30°C to 15°C) and noxious cold temperatures (<15°C). Moreover, a number of cold-mimetic compounds activate TRPM8 channels, including Cool-actP, Cooling Agent 10, FrescolatMGA, FrescolatML, geraniol, hydroxycitronellal, linalool, PMD38, WS-3, and WS-23 (195,213,214). Many of these compounds, such as icilin, will induce characteristic shivering or *wet dog* shakes when given intravenously, a process that requires TRPM8 channels and neurons (215–217). Similarly, a number of antagonists have been identified, including PBMC, M8-B, BCTC, thio-BCTC, CTPC, and capsazepine, with PBMC and M8-B shown to induce a dramatic hypothermic response in mice, likely due to inhibition of cold-sensing afferents (214,218,219). Similarly, TRPM8 is functionally present in brown adipose tissue (BAT) and stimulating brown adipocytes with menthol, upregulated UCP1 expression, a regulator of thermogenesis (220).

TRPM8 is clearly involved in innocuous cool sensation, but a role for the channel in cold pain has been controversial. $TRPM8^{-/-}$ mice have severe deficits in innocuous cold sensation (215,216,221), with conflicting reports as to whether these mice are also insensitive to noxious cold. Two studies reported no difference between wild-type and $TRPM8^{-/-}$ mice using the cold plate tests at 10°C, 0°C, −1°C, −5°C, or −10°C (216,221), while others found a significant difference in withdrawal latency at 0°C, the results suggesting the animals were deficient in noxious cold sensation (215,222). Lightly restrained $TRPM8^{-/-}$ mice have significantly longer withdrawal latencies than wild type when their hindpaws were placed on a 10°C plate (223). Moreover, $TRPM8^{-/-}$ mice were completely unresponsive in the dynamic cold plate assay, suggesting that they do not perceive cold temperatures at all (224). Injury-evoked cold hypersensitivity is also reduced in $TRPM8^{-/-}$ mice, but not completely blocked, suggesting that other mechanisms also contribute (215,218,222).

Menthol and cooling are commonly used as topical analgesic (225,226), thereby suggesting that activation of TRPM8 may lead to pain relief. Topical application of cold or cooling compounds produces a temporary analgesic effect mediated by TRPM8-expressing afferents (215,227). In a rodent model of neuropathic pain, paw withdrawal latencies in response to mechanical or heat stimuli were significantly attenuated in animals first treated with cold or cooling compounds such as icilin. Analgesia persists for over 20–30 min after which the animals regained their hypersensitivity similar to before the cold stimuli were applied and at levels of those not pretreated with cold or cooling compounds. Only modest cooling and low doses of cooling compounds produced analgesia and were dependent on TRPM8 channels and neurons (215,227). Moreover, using formalin (a compound that evokes acute pain followed by inflammation) injections into wild-type mouse hindpaws, it was found that cooling to 17°C and 24°C produced a marked decrease in pain behaviors (licking and lifting hindpaws) during the acute pain phase (216). However, mice lacking TRPM8 did not behave similarly in that they continued to show nocifensive responses even after exposed to the cool surface and were indistinguishable from wild-type animals that were not exposed to cool temperatures. Together these data indicate that TRPM8 is mediating the analgesia provided by cool temperatures and cooling compounds, suggesting that modest activation of TRPM8 afferent nerves can serve as an endogenous mechanism to promote pain relief.

TRPM8 also serves other biological roles in addition to neuronal thermal sensing. Along with expression in normal prostate, TRPM8 is expressed in other nonprostatic tumors such as breast, colon, lung, and skin, as well as in the bladder and male genital tract (228–232). TRPM8 transcripts have been found in the gastric fundus, and cooling as a result of the consumption of cold foods induces contraction of gastrointestinal smooth muscles, resulting in a short-lived gastric voiding (233,234). These and other data strongly suggest that TRPM8 may have diverse biological functions outside of the peripheral nervous system.

ACKNOWLEDGMENT

This work was supported by U.S. National Institutes of Health grants (D.D.M.; NS071364 and NS078530).

REFERENCES

1. Oancea, E. and N. L. Wicks. 2011. TRPM1: New trends for an old TRP. *Advances in Experimental Medicine and Biology* 704:135–145.
2. Sumoza-Toledo, A. and R. Penner. 2011. TRPM2: A multifunctional ion channel for calcium signalling. *The Journal of Physiology* 589:1515–1525.
3. Oberwinkler, J. 2007. TRPM3, a biophysical enigma? *Biochemical Society Transactions* 35:89–90.
4. Vennekens, R. and B. Nilius. 2007. Insights into TRPM4 function, regulation and physiological role. *Handbook of Experimental Pharmacology* 179:269–285.
5. Liman, E. R. 2007. TRPM5 and taste transduction. *Handbook of Experimental Pharmacology* 179:287–298.
6. Runnels, L. W. 2011. TRPM6 and TRPM7: A Mul-TRP-PLIK-cation of channel functions. *Current Pharmaceutical Biotechnology* 12:42–53.
7. Bates-Withers, C., R. Sah, and D. E. Clapham. 2011. TRPM7, the Mg(2+) inhibited channel and kinase. *Advances in Experimental Medicine and Biology* 704:173–183.
8. McKemy, D. D. 2013. The molecular and cellular basis of cold sensation. *ACS Chemical Neuroscience* 4:238–247.
9. Duncan, L. M., J. Deeds, J. Hunter, J. Shao, L. M. Holmgren, E. A. Woolf, R. I. Tepper, and A. W. Shyjan. 1998. Down-regulation of the novel gene melastatin correlates with potential for melanoma metastasis. *Cancer Research* 58:1515–1520.
10. Deeds, J., F. Cronin, and L. M. Duncan. 2000. Patterns of melastatin mRNA expression in melanocytic tumors. *Human Pathology* 31:1346–1356.
11. Miller, A. J., J. Du, S. Rowan, C. L. Hershey, H. R. Widlund, and D. E. Fisher. 2004. Transcriptional regulation of the melanoma prognostic marker melastatin (TRPM1) by MITF in melanocytes and melanoma. *Cancer Research* 64:509–516.
12. Duncan, L. M., J. Deeds, F. E. Cronin, M. Donovan, A. J. Sober, M. Kauffman, and J. J. McCarthy. 2001. Melastatin expression and prognosis in cutaneous malignant melanoma. *Journal of Clinical Oncology: Official Journal of the American Society of Clinical Oncology* 19:568–576.
13. Hunter, J. J., J. Shao, J. S. Smutko, B. J. Dussault, D. L. Nagle, E. A. Woolf, L. M. Holmgren, K. J. Moore, and A. W. Shyjan. 1998. Chromosomal localization and genomic characterization of the mouse melastatin gene (Mlsn1). *Genomics* 54:116–123.
14. Bellone, R. R., S. A. Brooks, L. Sandmeyer, B. A. Murphy, G. Forsyth, S. Archer, E. Bailey, and B. Grahn. 2008. Differential gene expression of TRPM1, the potential cause of congenital stationary night blindness and coat spotting patterns (LP) in the Appaloosa horse (*Equus caballus*). *Genetics* 179:1861–1870.
15. Audo, I., S. Kohl, B. P. Leroy, F. L. Munier, X. Guillonneau, S. Mohand-Said, K. Bujakowska et al. 2009. TRPM1 is mutated in patients with autosomal-recessive complete congenital stationary night blindness. *American Journal of Human Genetics* 85:720–729.
16. Li, Z., P. I. Sergouniotis, M. Michaelides, D. S. Mackay, G. A. Wright, S. Devery, A. T. Moore, G. E. Holder, A. G. Robson, and A. R. Webster. 2009. Recessive mutations of the gene TRPM1 abrogate ON bipolar cell function and cause complete congenital stationary night blindness in humans. *American Journal of Human Genetics* 85:711–719.
17. Nakamura, M., R. Sanuki, T. R. Yasuma, A. Onishi, K. M. Nishiguchi, C. Koike, M. Kadowaki, M. Kondo, Y. Miyake, and T. Furukawa. 2010. TRPM1 mutations are associated with the complete form of congenital stationary night blindness. *Molecular Vision* 16:425–437.
18. van Genderen, M. M., M. M. Bijveld, Y. B. Claassen, R. J. Florijn, J. N. Pearring, F. M. Meire, M. A. McCall et al. 2009. Mutations in TRPM1 are a common cause of complete congenital stationary night blindness. *American Journal of Human Genetics* 85:730–736.
19. Koike, C., T. Obara, Y. Uriu, T. Numata, R. Sanuki, K. Miyata, T. Koyasu et al. 2010. TRPM1 is a component of the retinal ON bipolar cell transduction channel in the mGluR6 cascade. *Proceedings of the National Academy of Sciences of the United States of America* 107:332–337.
20. Morgans, C. W., J. Zhang, B. G. Jeffrey, S. M. Nelson, N. S. Burke, R. M. Duvoisin, and R. L. Brown. 2009. TRPM1 is required for the depolarizing light response in retinal ON-bipolar cells. *Proceedings of the National Academy of Sciences of the United States of America* 106:19174–19178.
21. Shen, Y., M. A. Rampino, R. C. Carroll, and S. Nawy. 2012. G-protein-mediated inhibition of the Trp channel TRPM1 requires the G beta gamma dimer. *Proceedings of the National Academy of Sciences of the United States of America* 109:8752–8757.
22. Koike, C., T. Numata, H. Ueda, Y. Mori, and T. Furukawa. 2010. TRPM1: A vertebrate TRP channel responsible for retinal ON bipolar function. *Cell Calcium* 48:95–101.
23. Xu, X. Z., F. Moebius, D. L. Gill, and C. Montell. 2001. Regulation of melastatin, a TRP-related protein, through interaction with a cytoplasmic isoform. *Proceedings of the National Academy of Sciences of the United States of America* 98:10692–10697.
24. Oancea, E., J. Vriens, S. Brauchi, J. Jun, I. Splawski, and D. E. Clapham. 2009. TRPM1 forms ion channels associated with melanin content in melanocytes. *Science Signaling* 2:ra21.
25. Fonfria, E., P. R. Murdock, F. S. Cusdin, C. D. Benham, R. E. Kelsell, and S. McNulty. 2006. Tissue distribution profiles of the human TRPM cation channel family. *Journal of Receptor and Signal Transduction Research* 26:159–178.
26. Kunert-Keil, C., F. Bisping, J. Kruger, and H. Brinkmeier. 2006. Tissue-specific expression of TRP channel genes in the mouse and its variation in three different mouse strains. *BMC Genomics* 7:159.
27. Zhiqi, S., M. H. Soltani, K. M. Bhat, N. Sangha, D. Fang, J. J. Hunter, and V. Setaluri. 2004. Human melastatin 1 (TRPM1) is regulated by MITF and produces multiple polypeptide isoforms in melanocytes and melanoma. *Melanoma Research* 14:509–516.
28. Devi, S., R. Kedlaya, N. Maddodi, K. M. Bhat, C. S. Weber, H. Valdivia, and V. Setaluri. 2009. Calcium homeostasis in human melanocytes: Role of transient receptor potential melastatin 1 (TRPM1) and its regulation by ultraviolet light. American journal of physiology. *Cell Physiology* 297:C679–687.

29. Morgans, C. W., R. L. Brown, and R. M. Duvoisin. 2010. TRPM1: The endpoint of the mGluR6 signal transduction cascade in retinal ON-bipolar cells. *BioEssays: News and Reviews in Molecular, Cellular and Developmental Biology* 32:609–614.
30. Hammock, L., C. Cohen, G. Carlson, D. Murray, J. S. Ross, C. Sheehan, T. M. Nazir, and J. A. Carlson. 2006. Chromogenic in situ hybridization analysis of melastatin mRNA expression in melanomas from American Joint Committee on Cancer stage I and II patients with recurrent melanoma. *Journal of Cutaneous Pathology* 33:599–607.
31. Levy, C., M. Khaled, and D. E. Fisher. 2006. MITF: Master regulator of melanocyte development and melanoma oncogene. *Trends in Molecular Medicine* 12:406–414.
32. Boyle, G. M., S. L. Woods, V. F. Bonazzi, M. S. Stark, E. Hacker, L. G. Aoude, K. Dutton-Regester, A. L. Cook, R. A. Sturm, and N. K. Hayward. 2011. Melanoma cell invasiveness is regulated by miR-211 suppression of the BRN2 transcription factor. *Pigment Cell & Melanoma Research* 24:525–537.
33. Mazar, J., K. DeYoung, D. Khaitan, E. Meister, A. Almodovar, J. Goydos, A. Ray, and R. J. Perera. 2010. The regulation of miRNA-211 expression and its role in melanoma cell invasiveness. *PloS One* 5:e13779.
34. Bauer, G. L., C. Praetorius, K. Bergsteinsdottir, J. H. Hallsson, B. K. Gisladottir, A. Schepsky, D. A. Swing et al. 2009. The role of MITF phosphorylation sites during coat color and eye development in mice analyzed by bacterial artificial chromosome transgene rescue. *Genetics* 183:581–594.
35. Steingrimsson, E., N. G. Copeland, and N. A. Jenkins. 2004. Melanocytes and the microphthalmia transcription factor network. *Annual Review of Genetics* 38:365–411.
36. Liu, J. J. and D. E. Fisher. 2010. Lighting a path to pigmentation: Mechanisms of MITF induction by UV. *Pigment Cell & Melanoma Research* 23:741–745.
37. Lu, S., A. Slominski, S. E. Yang, C. Sheehan, J. Ross, and J. A. Carlson. 2010. The correlation of TRPM1 (Melastatin) mRNA expression with microphthalmia-associated transcription factor (MITF) and other melanogenesis-related proteins in normal and pathological skin, hair follicles and melanocytic nevi. *Journal of Cutaneous Pathology* 37 Suppl 1:26–40.
38. Devi, S., Y. Markandeya, N. Maddodi, A. Dhingra, N. Vardi, R. C. Balijepalli, and V. Setaluri. 2013. Metabotropic glutamate receptor 6 signaling enhances TRPM1 calcium channel function and increases melanin content in human melanocytes. *Pigment Cell & Melanoma Research* 26:348–356.
39. Sponenberg, D. P., G. Carr, E. Simak, and K. Schwink. 1990. The inheritance of the leopard complex of spotting patterns in horses. *The Journal of Heredity* 81:323–331.
40. Sandmeyer, L. S., C. B. Breaux, S. Archer, and B. H. Grahn. 2007. Clinical and electroretinographic characteristics of congenital stationary night blindness in the Appaloosa and the association with the leopard complex. *Veterinary Ophthalmology* 10:368–375.
41. Bellone, R. R., G. Forsyth, T. Leeb, S. Archer, S. Sigurdsson, F. Imsland, E. Mauceli et al. 2010. Fine-mapping and mutation analysis of TRPM1: A candidate gene for leopard complex (LP) spotting and congenital stationary night blindness in horses. *Briefings in Functional Genomics* 9:193–207.
42. Nakajima, Y., M. Moriyama, M. Hattori, N. Minato, and S. Nakanishi. 2009. Isolation of ON bipolar cell genes via hrGFP-coupled cell enrichment using the mGluR6 promoter. *Journal of Biochemistry* 145:811–818.
43. Shen, Y., J. A. Heimel, M. Kamermans, N. S. Peachey, R. G. Gregg, and S. Nawy. 2009. A transient receptor potential-like channel mediates synaptic transmission in rod bipolar cells. *The Journal of Neuroscience: The Official Journal of the Society for Neuroscience* 29:6088–6093.
44. Klooster, J., J. Blokker, J. B. Ten Brink, U. Unmehopa, K. Fluiter, A. A. Bergen, and M. Kamermans. 2011. Ultrastructural localization and expression of TRPM1 in the human retina. *Investigative Ophthalmology & Visual Science* 52:8356–8362.
45. Xu, Y., A. Dhingra, M. E. Fina, C. Koike, T. Furukawa, and N. Vardi. 2012. mGluR6 deletion renders the TRPM1 channel in retina inactive. *Journal of Neurophysiology* 107:948–957.
46. Dhingra, A., H. Ramakrishnan, A. Neinstein, M. E. Fina, Y. Xu, J. Li, D. C. Chung, A. Lyubarsky, and N. Vardi. 2012. Gbeta3 is required for normal light ON responses and synaptic maintenance. *The Journal of Neuroscience: The Official Journal of the Society for Neuroscience* 32:11343–11355.
47. Pearring, J. N., P. Bojang, Jr., Y. Shen, C. Koike, T. Furukawa, S. Nawy, and R. G. Gregg. 2011. A role for nyctalopin, a small leucine-rich repeat protein, in localizing the TRP melastatin 1 channel to retinal depolarizing bipolar cell dendrites. *The Journal of Neuroscience: The Official Journal of the Society for Neuroscience* 31:10060–10066.
48. Cao, Y., E. Posokhova, and K. A. Martemyanov. 2011. TRPM1 forms complexes with nyctalopin in vivo and accumulates in postsynaptic compartment of ON-bipolar neurons in mGluR6-dependent manner. *The Journal of Neuroscience: The Official Journal of the Society for Neuroscience* 31:11521–11526.
49. Gregg, R. G., M. Kamermans, J. Klooster, P. D. Lukasiewicz, N. S. Peachey, K. A. Vessey, and M. A. McCall. 2007. Nyctalopin expression in retinal bipolar cells restores visual function in a mouse model of complete X-linked congenital stationary night blindness. *Journal of Neurophysiology* 98:3023–3033.
50. Takahashi, N., D. Kozai, R. Kobayashi, M. Ebert, and Y. Mori. 2011. Roles of TRPM2 in oxidative stress. *Cell Calcium* 50:279–287.
51. Perraud, A. L., A. Fleig, C. A. Dunn, L. A. Bagley, P. Launay, C. Schmitz, A. J. Stokes et al. 2001. ADP-ribose gating of the calcium-permeable LTRPC2 channel revealed by Nudix motif homology. *Nature* 411:595–599.
52. Lange, I., S. Yamamoto, S. Partida-Sanchez, Y. Mori, A. Fleig, and R. Penner. 2009. TRPM2 functions as a lysosomal Ca^{2+}-release channel in beta cells. *Science Signaling* 2:ra23.
53. Sano, Y., K. Inamura, A. Miyake, S. Mochizuki, H. Yokoi, H. Matsushime, and K. Furuichi. 2001. Immunocyte Ca^{2+} influx system mediated by LTRPC2. *Science* 293:1327–1330.
54. Nagamine, K., J. Kudoh, S. Minoshima, K. Kawasaki, S. Asakawa, F. Ito, and N. Shimizu. 1998. Molecular cloning of a novel putative Ca^{2+} channel protein (TRPC7) highly expressed in brain. *Genomics* 54:124–131.
55. Fleig, A. and R. Penner. 2004. The TRPM ion channel subfamily: Molecular, biophysical and functional features. *Trends in Pharmacological Sciences* 25:633–639.
56. Zhang, W., X. Chu, Q. Tong, J. Y. Cheung, K. Conrad, K. Masker, and B. A. Miller. 2003. A novel TRPM2 isoform inhibits calcium influx and susceptibility to cell death. *The Journal of Biological Chemistry* 278:16222–16229.
57. Vazquez, E. and M. A. Valverde. 2006. A review of TRP channels splicing. *Seminars in Cell & Developmental Biology* 17:607–617.
58. Wehage, E., J. Eisfeld, I. Heiner, E. Jungling, C. Zitt, and A. Luckhoff. 2002. Activation of the cation channel long transient receptor potential channel 2 (LTRPC2) by hydrogen peroxide. A splice variant reveals a mode of activation independent of ADP-ribose. *The Journal of Biological Chemistry* 277:23150–23156.
59. Kuhn, F. J., C. Kuhn, M. Naziroglu, and A. Luckhoff. 2009. Role of an N-terminal splice segment in the activation of the cation channel TRPM2 by ADP-ribose and hydrogen peroxide. *Neurochemical Research* 34:227–233.
60. Kuhn, F. J. and A. Luckhoff. 2004. Sites of the NUDT9-H domain critical for ADP-ribose activation of the cation channel TRPM2. *The Journal of Biological Chemistry* 279:46431–46437.

61. Schuber, F. and F. E. Lund. 2004. Structure and enzymology of ADP-ribosyl cyclases: Conserved enzymes that produce multiple calcium mobilizing metabolites. *Current Molecular Medicine* 4:249–261.
62. Malavasi, F., S. Deaglio, A. Funaro, E. Ferrero, A. L. Horenstein, E. Ortolan, T. Vaisitti, and S. Aydin. 2008. Evolution and function of the ADP ribosyl cyclase/CD38 gene family in physiology and pathology. *Physiological Reviews* 88:841–886.
63. Hara, Y., M. Wakamori, M. Ishii, E. Maeno, M. Nishida, T. Yoshida, H. Yamada et al. 2002. LTRPC2 Ca^{2+}-permeable channel activated by changes in redox status confers susceptibility to cell death. *Molecular Cell* 9:163–173.
64. Zhang, W., I. Hirschler-Laszkiewicz, Q. Tong, K. Conrad, S. C. Sun, L. Penn, D. L. Barber et al. 2006. TRPM2 is an ion channel that modulates hematopoietic cell death through activation of caspases and PARP cleavage. *American Journal of Physiology: Cell Physiology* 290:C1146–C1159.
65. Uchida, K., K. Dezaki, B. Damdindorj, H. Inada, T. Shiuchi, Y. Mori, T. Yada, Y. Minokoshi, and M. Tominaga. 2011. Lack of TRPM2 impaired insulin secretion and glucose metabolisms in mice. *Diabetes* 60:119–126.
66. Herson, P. S. and M. L. Ashford. 1997. Activation of a novel non-selective cation channel by alloxan and H_2O_2 in the rat insulin-secreting cell line CRI-G1. *The Journal of Physiology* 501(Pt 1):59–66.
67. Herson, P. S., K. Lee, R. D. Pinnock, J. Hughes, and M. L. Ashford. 1999. Hydrogen peroxide induces intracellular calcium overload by activation of a non-selective cation channel in an insulin-secreting cell line. *The Journal of Biological Chemistry* 274:833–841.
68. Uchida, K. and M. Tominaga. 2011. The role of thermosensitive TRP (transient receptor potential) channels in insulin secretion. *Endocrine Journal* 58:1021–1028.
69. Uchida, K. and M. Tominaga. 2011. TRPM2 modulates insulin secretion in pancreatic beta-cells. *Islets* 3:209–211.
70. Hermosura, M. C., A. M. Cui, R. C. Go, B. Davenport, C. M. Shetler, J. W. Heizer, C. Schmitz, G. Mocz, R. M. Garruto, and A. L. Perraud. 2008. Altered functional properties of a TRPM2 variant in Guamanian ALS and PD. *Proceedings of the National Academy of Sciences of the United States of America* 105:18029–18034.
71. Kaneko, S., S. Kawakami, Y. Hara, M. Wakamori, E. Itoh, T. Minami, Y. Takada et al. 2006. A critical role of TRPM2 in neuronal cell death by hydrogen peroxide. *Journal of Pharmacological Sciences* 101:66–76.
72. Fonfria, E., I. C. Marshall, I. Boyfield, S. D. Skaper, J. P. Hughes, D. E. Owen, W. Zhang, B. A. Miller, C. D. Benham, and S. McNulty. 2005. Amyloid beta-peptide(1–42) and hydrogen peroxide-induced toxicity are mediated by TRPM2 in rat primary striatal cultures. *Journal of Neurochemistry* 95:715–723.
73. Lee, C. R., R. P. Machold, P. Witkovsky, and M. E. Rice. 2013. TRPM2 channels are required for NMDA-induced burst firing and contribute to H(2)O(2)-dependent modulation in substantia nigra pars reticulata GABAergic neurons. *The Journal of Neuroscience: The Official Journal of the Society for Neuroscience* 33:1157–1168.
74. Xu, C., P. P. Li, R. G. Cooke, S. V. Parikh, K. Wang, J. L. Kennedy, and J. J. Warsh. 2009. TRPM2 variants and bipolar disorder risk: Confirmation in a family-based association study. *Bipolar Disorders* 11:1–10.
75. Giansanti, V., F. Dona, M. Tillhon, and A. I. Scovassi. 2010. PARP inhibitors: New tools to protect from inflammation. *Biochemical Pharmacology* 80:1869–1877.
76. Oberwinkler, J. and S. E. Phillipp. 2007. TRPM3. *Handbook of Experimental Pharmacology* 179:253–267.
77. Lee, N., J. Chen, L. Sun, S. Wu, K. R. Gray, A. Rich, M. Huang et al. 2003. Expression and characterization of human transient receptor potential melastatin 3 (hTRPM3). *The Journal of Biological Chemistry* 278:20890–20897.
78. Grimm, C., R. Kraft, S. Sauerbruch, G. Schultz, and C. Harteneck. 2003. Molecular and functional characterization of the melastatin-related cation channel TRPM3. *The Journal of Biological Chemistry* 278:21493–21501.
79. Vriens, J., G. Owsianik, T. Hofmann, S. E. Philipp, J. Stab, X. Chen, M. Benoit et al. 2011. TRPM3 is a nociceptor channel involved in the detection of noxious heat. *Neuron* 70:482–494.
80. Oberwinkler, J., A. Lis, K. M. Giehl, V. Flockerzi, and S. E. Philipp. 2005. Alternative splicing switches the divalent cation selectivity of TRPM3 channels. *The Journal of Biological Chemistry* 280:22540–22548.
81. Hoffmann, A., C. Grimm, R. Kraft, O. Goldbaum, A. Wrede, C. Nolte, U. K. Hanisch et al. 2010. TRPM3 is expressed in sphingosine-responsive myelinating oligodendrocytes. *Journal of Neurochemistry* 114:654–665.
82. Zamudio-Bulcock, P. A., J. Everett, C. Harteneck, and C. F. Valenzuela. 2011. Activation of steroid-sensitive TRPM3 channels potentiates glutamatergic transmission at cerebellar Purkinje neurons from developing rats. *Journal of Neurochemistry* 119:474–485.
83. Grimm, C., R. Kraft, G. Schultz, and C. Harteneck. 2005. Activation of the melastatin-related cation channel TRPM3 by D-erythro-sphingosine [corrected]. *Molecular Pharmacology* 67:798–805.
84. Wagner, T. F., A. Drews, S. Loch, F. Mohr, S. E. Philipp, S. Lambert, and J. Oberwinkler. 2010. TRPM3 channels provide a regulated influx pathway for zinc in pancreatic beta cells. *Pflugers Archiv: European Journal of Physiology* 460:755–765.
85. Colsoul, B., R. Vennekens, and B. Nilius. 2011. Transient receptor potential cation channels in pancreatic beta cells. *Reviews of Physiology, Biochemistry and Pharmacology* 161:87–110.
86. Weber, M. J. 2005. New human and mouse microRNA genes found by homology search. *The FEBS Journal* 272:59–73.
87. Rodriguez, A., S. Griffiths-Jones, J. L. Ashurst, and A. Bradley. 2004. Identification of mammalian microRNA host genes and transcription units. *Genome Research* 14:1902–1910.
88. Wang, F. E., C. Zhang, A. Maminishkis, L. Dong, C. Zhi, R. Li, J. Zhao et al. 2010. MicroRNA-204/211 alters epithelial physiology. *FASEB Journal: Official Publication of the Federation of American Societies for Experimental Biology* 24:1552–1571.
89. Holakovska, B., L. Grycova, M. Jirku, M. Sulc, L. Bumba, and J. Teisinger. 2012. Calmodulin and S100A1 protein interact with N terminus of TRPM3 channel. *The Journal of Biological Chemistry* 287:16645–16655.
90. Klose, C., I. Straub, M. Riehle, F. Ranta, D. Krautwurst, S. Ullrich, W. Meyerhof, and C. Harteneck. 2011. Fenamates as TRP channel blockers: Mefenamic acid selectively blocks TRPM3. *British Journal of Pharmacology* 162:1757–1769.
91. Wagner, T. F., S. Loch, S. Lambert, I. Straub, S. Mannebach, I. Mathar, M. Dufer et al. 2008. Transient receptor potential M3 channels are ionotropic steroid receptors in pancreatic beta cells. *Nature Cell Biology* 10:1421–1430.
92. Naylor, J., J. Li, C. J. Milligan, F. Zeng, P. Sukumar, B. Hou, A. Sedo et al. 2010. Pregnenolone sulphate- and cholesterol-regulated TRPM3 channels coupled to vascular smooth muscle secretion and contraction. *Circulation Research* 106:1507–1515.
93. Son, A. R., Y. M. Yang, J. H. Hong, S. I. Lee, Y. Shibukawa, and D. M. Shin. 2009. Odontoblast TRP channels and thermo/mechanical transmission. *Journal of Dental Research* 88:1014–1019.
94. Bootman, M. D., T. J. Collins, L. Mackenzie, H. L. Roderick, M. J. Berridge, and C. M. Peppiatt. 2002. 2-aminoethoxydiphenyl borate (2-APB) is a reliable blocker of store-operated Ca^{2+} entry but an inconsistent inhibitor of InsP3-induced Ca^{2+} release. *FASEB Journal: Official Publication of the Federation of American Societies for Experimental Biology* 16:1145–1150.

95. Xu, S. Z., F. Zeng, G. Boulay, C. Grimm, C. Harteneck, and D. J. Beech. 2005. Block of TRPC5 channels by 2-aminoethoxydiphenyl borate: A differential, extracellular and voltage-dependent effect. *British Journal of Pharmacology* 145:405–414.
96. Mayer, S. I., I. Muller, S. Mannebach, T. Endo, and G. Thiel. 2011. Signal transduction of pregnenolone sulfate in insulinoma cells: Activation of Egr-1 expression involving TRPM3, voltage-gated calcium channels, ERK, and ternary complex factors. *The Journal of Biological Chemistry* 286:10084–10096.
97. Muller, I., O. G. Rossler, and G. Thiel. 2011. Pregnenolone sulfate activates basic region leucine zipper transcription factors in insulinoma cells: Role of voltage-gated Ca^{2+} channels and transient receptor potential melastatin 3 channels. *Molecular Pharmacology* 80:1179–1189.
98. Straub, I., F. Mohr, J. Stab, M. Konrad, S. Philipp, J. Oberwinkler, and M. Schaefer. 2013. Citrus fruit and fabacea secondary metabolites potently and selectively block TRPM3. *British Journal of Pharmacology* 168:1835–1850.
99. Majeed, Y., S. Tumova, B. L. Green, V. A. Seymour, D. M. Woods, A. K. Agarwal, J. Naylor et al. 2012. Pregnenolone sulphate-independent inhibition of TRPM3 channels by progesterone. *Cell Calcium* 51:1–11.
100. Thiel, G., S. I. Mayer, I. Muller, L. Stefano, and O. G. Rossler. 2010. Egr-1-A Ca(2+)-regulated transcription factor. *Cell Calcium* 47:397–403.
101. Thiel, G., I. Muller, and O. G. Rossler. 2013. Signal transduction via TRPM3 channels in pancreatic beta-cells. *Journal of Molecular Endocrinology* 50:R75–83.
102. Zamudio-Bulcock, P. A. and C. F. Valenzuela. 2011. Pregnenolone sulfate increases glutamate release at neonatal climbing fiber-to-Purkinje cell synapses. *Neuroscience* 175:24–36.
103. Launay, P., A. Fleig, A. L. Perraud, A. M. Scharenberg, R. Penner, and J. P. Kinet. 2002. TRPM4 is a Ca^{2+}-activated nonselective cation channel mediating cell membrane depolarization. *Cell* 109:397–407.
104. Nilius, B., J. Prenen, G. Droogmans, T. Voets, R. Vennekens, M. Freichel, U. Wissenbach, and V. Flockerzi. 2003. Voltage dependence of the Ca^{2+}-activated cation channel TRPM4. *The Journal of Biological Chemistry* 278:30813–30820.
105. Murakami, M., F. Xu, I. Miyoshi, E. Sato, K. Ono, and T. Iijima. 2003. Identification and characterization of the murine TRPM4 channel. *Biochemical and Biophysical Research Communications* 307:522–528.
106. Nilius, B., J. Prenen, J. Tang, C. Wang, G. Owsianik, A. Janssens, T. Voets, and M. X. Zhu. 2005. Regulation of the Ca^{2+} sensitivity of the nonselective cation channel TRPM4. *The Journal of Biological Chemistry* 280:6423–6433.
107. Ullrich, N. D., T. Voets, J. Prenen, R. Vennekens, K. Talavera, G. Droogmans, and B. Nilius. 2005. Comparison of functional properties of the Ca^{2+}-activated cation channels TRPM4 and TRPM5 from mice. *Cell Calcium* 37:267–278.
108. Nilius, B., J. Prenen, A. Janssens, T. Voets, and G. Droogmans. 2004. Decavanadate modulates gating of TRPM4 cation channels. *The Journal of Physiology* 560:753–765.
109. Abriel, H., N. Syam, V. Sottas, M. Y. Amarouch, and J. S. Rougier. 2012. TRPM4 channels in the cardiovascular system: Physiology, pathophysiology, and pharmacology. *Biochemical Pharmacology* 84:873–881.
110. Zhang, Z., H. Okawa, Y. Wang, and E. R. Liman. 2005. Phosphatidylinositol 4,5-bisphosphate rescues TRPM4 channels from desensitization. *The Journal of Biological Chemistry* 280:39185–39192.
111. Liu, D. and E. R. Liman. 2003. Intracellular Ca^{2+} and the phospholipid PIP2 regulate the taste transduction ion channel TRPM5. *Proceedings of the National Academy of Sciences of the United States of America* 100:15160–15165.
112. Nilius, B., F. Mahieu, J. Prenen, A. Janssens, G. Owsianik, R. Vennekens, and T. Voets. 2006. The Ca^{2+}-activated cation channel TRPM4 is regulated by phosphatidylinositol 4,5-biphosphate. *The EMBO Journal* 25:467–478.
113. Guinamard, R., L. Salle, and C. Simard. 2011. The non-selective monovalent cationic channels TRPM4 and TRPM5. *Advances in Experimental Medicine and Biology* 704:147–171.
114. Guinamard, R., M. Demion, C. Magaud, D. Potreau, and P. Bois. 2006. Functional expression of the TRPM4 cationic current in ventricular cardiomyocytes from spontaneously hypertensive rats. *Hypertension* 48:587–594.
115. Kruse, M., E. Schulze-Bahr, V. Corfield, A. Beckmann, B. Stallmeyer, G. Kurtbay, I. Ohmert, E. Schulze-Bahr, P. Brink, and O. Pongs. 2009. Impaired endocytosis of the ion channel TRPM4 is associated with human progressive familial heart block type I. *The Journal of Clinical Investigation* 119:2737–2744.
116. Liu, H., L. El Zein, M. Kruse, R. Guinamard, A. Beckmann, A. Bozio, G. Kurtbay et al. 2010. Gain-of-function mutations in TRPM4 cause autosomal dominant isolated cardiac conduction disease. *Circulation: Cardiovascular Genetics* 3:374–385.
117. Voets, T. and B. Nilius. 2009. TRPCs, GPCRs and the Bayliss effect. *The EMBO Journal* 28:4–5.
118. Earley, S., B. J. Waldron, and J. E. Brayden. 2004. Critical role for transient receptor potential channel TRPM4 in myogenic constriction of cerebral arteries. *Circulation Research* 95:922–929.
119. Earley, S., S. V. Straub, and J. E. Brayden. 2007. Protein kinase C regulates vascular myogenic tone through activation of TRPM4. *American Journal of Physiology: Heart and Circulatory Physiology* 292:H2613–H2622.
120. Barbet, G., M. Demion, I. C. Moura, N. Serafini, T. Leger, F. Vrtovsnik, R. C. Monteiro, R. Guinamard, J. P. Kinet, and P. Launay. 2008. The calcium-activated nonselective cation channel TRPM4 is essential for the migration but not the maturation of dendritic cells. *Nature Immunology* 9:1148–1156.
121. Zitt, C., B. Strauss, E. C. Schwarz, N. Spaeth, G. Rast, A. Hatzelmann, and M. Hoth. 2004. Potent inhibition of Ca^{2+} release-activated Ca^{2+} channels and T-lymphocyte activation by the pyrazole derivative BTP2. *The Journal of Biological Chemistry* 279:12427–12437.
122. Launay, P., H. Cheng, S. Srivatsan, R. Penner, A. Fleig, and J. P. Kinet. 2004. TRPM4 regulates calcium oscillations after T cell activation. *Science* 306:1374–1377.
123. Vennekens, R., J. Olausson, M. Meissner, W. Bloch, I. Mathar, S. E. Philipp, F. Schmitz et al. 2007. Increased IgE-dependent mast cell activation and anaphylactic responses in mice lacking the calcium-activated nonselective cation channel TRPM4. *Nature Immunology* 8:312–320.
124. Shimizu, T., G. Owsianik, M. Freichel, V. Flockerzi, B. Nilius, and R. Vennekens. 2009. TRPM4 regulates migration of mast cells in mice. *Cell Calcium* 45:226–232.
125. Schattling, B., K. Steinbach, E. Thies, M. Kruse, A. Menigoz, F. Ufer, V. Flockerzi et al. 2012. TRPM4 cation channel mediates axonal and neuronal degeneration in experimental autoimmune encephalomyelitis and multiple sclerosis. *Nature Medicine* 18:1805–1811.
126. Cheng, H., A. Beck, P. Launay, S. A. Gross, A. J. Stokes, J. P. Kinet, A. Fleig, and R. Penner. 2007. TRPM4 controls insulin secretion in pancreatic beta-cells. *Cell Calcium* 41:51–61.
127. Enklaar, T., M. Esswein, M. Oswald, K. Hilbert, A. Winterpacht, M. Higgins, B. Zabel, and D. Prawitt. 2000. Mtr1, a novel biallelically expressed gene in the center of the mouse distal chromosome 7 imprinting cluster, is a member of the Trp gene family. *Genomics* 67:179–187.

128. Prawitt, D., T. Enklaar, G. Klemm, B. Gartner, C. Spangenberg, A. Winterpacht, M. Higgins, J. Pelletier, and B. Zabel. 2000. Identification and characterization of MTR1, a novel gene with homology to melastatin (MLSN1) and the trp gene family located in the BWS-WT2 critical region on chromosome 11p15.5 and showing allele-specific expression. *Human Molecular Genetics* 9:203–216.
129. Nilius, B., J. Prenen, T. Voets, and G. Droogmans. 2004. Intracellular nucleotides and polyamines inhibit the Ca^{2+}-activated cation channel TRPM4b. *Pflugers Archiv: European Journal of Physiology* 448:70–75.
130. Hofmann, T., V. Chubanov, T. Gudermann, and C. Montell. 2003. TRPM5 is a voltage-modulated and Ca(2+)-activated monovalent selective cation channel. *Current Biology* 13:1153–1158.
131. Zhang, Y., M. A. Hoon, J. Chandrashekar, K. L. Mueller, B. Cook, D. Wu, C. S. Zuker, and N. J. Ryba. 2003. Coding of sweet, bitter, and umami tastes: Different receptor cells sharing similar signaling pathways. *Cell* 112:293–301.
132. Perez, C. A., L. Huang, M. Rong, J. A. Kozak, A. K. Preuss, H. Zhang, M. Max, and R. F. Margolskee. 2002. A transient receptor potential channel expressed in taste receptor cells. *Nature Neuroscience* 5:1169–1176.
133. Margolskee, R. F. 2002. Molecular mechanisms of bitter and sweet taste transduction. *The Journal of Biological Chemistry* 277:1–4.
134. Prawitt, D., M. K. Monteilh-Zoller, L. Brixel, C. Spangenberg, B. Zabel, A. Fleig, and R. Penner. 2003. TRPM5 is a transient Ca^{2+}-activated cation channel responding to rapid changes in $[Ca^{2+}]_i$. *Proceedings of the National Academy of Sciences of the United States of America* 100:15166–15171.
135. Talavera, K., K. Yasumatsu, T. Voets, G. Droogmans, N. Shigemura, Y. Ninomiya, R. F. Margolskee, and B. Nilius. 2005. Heat activation of TRPM5 underlies thermal sensitivity of sweet taste. *Nature* 438:1022–1025.
136. Liu, D., Z. Zhang, and E. R. Liman. 2005. Extracellular acid block and acid-enhanced inactivation of the Ca^{2+}-activated cation channel TRPM5 involve residues in the S3-S4 and S5-S6 extracellular domains. *The Journal of Biological Chemistry* 280:20691–20699.
137. Colsoul, B., A. Schraenen, K. Lemaire, R. Quintens, L. Van Lommel, A. Segal, G. Owsianik et al. 2010. Loss of high-frequency glucose-induced Ca^{2+} oscillations in pancreatic islets correlates with impaired glucose tolerance in Trpm5-/- mice. *Proceedings of the National Academy of Sciences of the United States of America* 107:5208–5213.
138. Brixel, L. R., M. K. Monteilh-Zoller, C. S. Ingenbrandt, A. Fleig, R. Penner, T. Enklaar, B. U. Zabel, and D. Prawitt. 2010. TRPM5 regulates glucose-stimulated insulin secretion. *Pflugers Archiv: European Journal of Physiology* 460:69–76.
139. Enklaar, T., L. R. Brixel, B. U. Zabel, and D. Prawitt. 2010. Adding efficiency: The role of the CAN ion channels TRPM4 and TRPM5 in pancreatic islets. *Islets* 2:337–338.
140. Liman, E. R. 2010. A TRP channel contributes to insulin secretion by pancreatic beta-cells. *Islets* 2:331–333.
141. Bezencon, C., J. le Coutre, and S. Damak. 2007. Taste-signaling proteins are coexpressed in solitary intestinal epithelial cells. *Chemical Senses* 32:41–49.
142. Bezencon, C., A. Furholz, F. Raymond, R. Mansourian, S. Metairon, J. Le Coutre, and S. Damak. 2008. Murine intestinal cells expressing Trpm5 are mostly brush cells and express markers of neuronal and inflammatory cells. *The Journal of Comparative Neurology* 509:514–525.
143. Young, R. L., K. Sutherland, N. Pezos, S. M. Brierley, M. Horowitz, C. K. Rayner, and L. A. Blackshaw. 2009. Expression of taste molecules in the upper gastrointestinal tract in humans with and without type 2 diabetes. *Gut* 58:337–346.
144. Kokrashvili, Z., D. Rodriguez, V. Yevshayeva, H. Zhou, R. F. Margolskee, and B. Mosinger. 2009. Release of endogenous opioids from duodenal enteroendocrine cells requires TRPM5. *Gastroenterology* 137:598–606, 606 e591–e592.
145. Schlingmann, K. P., S. Weber, M. Peters, L. Niemann Nejsum, H. Vitzthum, K. Klingel, M. Kratz et al. 2002. Hypomagnesemia with secondary hypocalcemia is caused by mutations in TRPM6, a new member of the TRPM gene family. *Nature Genetics* 31:166–170.
146. Walder, R. Y., D. Landau, P. Meyer, H. Shalev, M. Tsolia, Z. Borochowitz, M. B. Boettger et al. 2002. Mutation of TRPM6 causes familial hypomagnesemia with secondary hypocalcemia. *Nature Genetics* 31:171–174.
147. Riazanova, L. V., K. S. Pavur, A. N. Petrov, M. V. Dorovkov, and A. G. Riazanov. 2001. Novel type of signaling molecules: Protein kinases covalently linked to ion channels. *Molekuliarnaia Biologiia* 35:321–332.
148. Chubanov, V., S. Waldegger, M. Mederos y Schnitzler, H. Vitzthum, M. C. Sassen, H. W. Seyberth, M. Konrad, and T. Gudermann. 2004. Disruption of TRPM6/TRPM7 complex formation by a mutation in the TRPM6 gene causes hypomagnesemia with secondary hypocalcemia. *Proceedings of the National Academy of Sciences of the United States of America* 101:2894–2899.
149. Li, M., J. Jiang, and L. Yue. 2006. Functional characterization of homo- and heteromeric channel kinases TRPM6 and TRPM7. *The Journal of General Physiology* 127:525–537.
150. Schmitz, C., M. V. Dorovkov, X. Zhao, B. J. Davenport, A. G. Ryazanov, and A. L. Perraud. 2005. The channel kinases TRPM6 and TRPM7 are functionally nonredundant. *The Journal of Biological Chemistry* 280:37763–37771.
151. Voets, T., B. Nilius, S. Hoefs, A. W. van der Kemp, G. Droogmans, R. J. Bindels, and J. G. Hoenderop. 2004. TRPM6 forms the Mg^{2+} influx channel involved in intestinal and renal Mg^{2+} absorption. *The Journal of Biological Chemistry* 279:19–25.
152. Drennan, D. and A. G. Ryazanov. 2004. Alpha-kinases: Analysis of the family and comparison with conventional protein kinases. *Progress in Biophysics and Molecular Biology* 85:1–32.
153. van der Wijst, J., J. G. Hoenderop, and R. J. Bindels. 2009. Epithelial Mg^{2+} channel TRPM6: Insight into the molecular regulation. *Magnesium Research: Official Organ of the International Society for the Development of Research on Magnesium* 22:127–132.
154. Woudenberg-Vrenken, T. E., R. J. Bindels, and J. G. Hoenderop. 2009. The role of transient receptor potential channels in kidney disease. *Nature Reviews Nephrology* 5:441–449.
155. Cao, G., S. Thebault, J. van der Wijst, A. van der Kemp, E. Lasonder, R. J. Bindels, and J. G. Hoenderop. 2008. RACK1 inhibits TRPM6 activity via phosphorylation of the fused alpha-kinase domain. *Current Biology* 18:168–176.
156. Groenestege, W. M., J. G. Hoenderop, L. van den Heuvel, N. Knoers, and R. J. Bindels. 2006. The epithelial Mg^{2+} channel transient receptor potential melastatin 6 is regulated by dietary Mg^{2+} content and estrogens. *Journal of the American Society of Nephrology* 17:1035–1043.
157. Groenestege, W. M., S. Thebault, J. van der Wijst, D. van den Berg, R. Janssen, S. Tejpar, L. P. van den Heuvel et al. 2007. Impaired basolateral sorting of pro-EGF causes isolated recessive renal hypomagnesemia. *The Journal of Clinical Investigation* 117:2260–2267.
158. Runnels, L. W., L. Yue, and D. E. Clapham. 2001. TRP-PLIK, a bifunctional protein with kinase and ion channel activities. *Science* 291:1043–1047.
159. Nadler, M. J., M. C. Hermosura, K. Inabe, A. L. Perraud, Q. Zhu, A. J. Stokes, T. Kurosaki et al. 2001. LTRPC7 is a Mg.ATP-regulated divalent cation channel required for cell viability. *Nature* 411:590–595.

160. Ryazanov, A. G. 2002. Elongation factor-2 kinase and its newly discovered relatives. *FEBS Letters* 514:26–29.
161. Jin, J., L. J. Wu, J. Jun, X. Cheng, H. Xu, N. C. Andrews, and D. E. Clapham. 2012. The channel kinase, TRPM7, is required for early embryonic development. *Proceedings of the National Academy of Sciences of the United States of America* 109:E225–E233.
162. Jin, J., B. N. Desai, B. Navarro, A. Donovan, N. C. Andrews, and D. E. Clapham. 2008. Deletion of TRPM7 disrupts embryonic development and thymopoiesis without altering Mg^{2+} homeostasis. *Science* 322:756–760.
163. Yamaguchi, H., M. Matsushita, A. C. Nairn, and J. Kuriyan. 2001. Crystal structure of the atypical protein kinase domain of a TRP channel with phosphotransferase activity. *Molecular Cell* 7:1047–1057.
164. Ryazanova, L. V., M. V. Dorovkov, A. Ansari, and A. G. Ryazanov. 2004. Characterization of the protein kinase activity of TRPM7/ChaK1, a protein kinase fused to the transient receptor potential ion channel. *The Journal of Biological Chemistry* 279:3708–3716.
165. Runnels, L. W., L. Yue, and D. E. Clapham. 2002. The TRPM7 channel is inactivated by PIP(2) hydrolysis. *Nature Cell Biology* 4:329–336.
166. Takezawa, R., C. Schmitz, P. Demeuse, A. M. Scharenberg, R. Penner, and A. Fleig. 2004. Receptor-mediated regulation of the TRPM7 channel through its endogenous protein kinase domain. *Proceedings of the National Academy of Sciences of the United States of America* 101:6009–6014.
167. Monteilh-Zoller, M. K., M. C. Hermosura, M. J. Nadler, A. M. Scharenberg, R. Penner, and A. Fleig. 2003. TRPM7 provides an ion channel mechanism for cellular entry of trace metal ions. *The Journal of General Physiology* 121:49–60.
168. Langeslag, M., K. Clark, W. H. Moolenaar, F. N. van Leeuwen, and K. Jalink. 2007. Activation of TRPM7 channels by phospholipase C-coupled receptor agonists. *The Journal of Biological Chemistry* 282:232–239.
169. Kozak, J. A. and M. D. Cahalan. 2003. MIC channels are inhibited by internal divalent cations but not ATP. *Biophysical Journal* 84:922–927.
170. Jiang, J., M. Li, and L. Yue. 2005. Potentiation of TRPM7 inward currents by protons. *The Journal of General Physiology* 126:137–150.
171. Ryazanova, L. V., L. J. Rondon, S. Zierler, Z. Hu, J. Galli, T. P. Yamaguchi, A. Mazur, A. Fleig, and A. G. Ryazanov. 2010. TRPM7 is essential for Mg(2+) homeostasis in mammals. *Nature Communications* 1:109.
172. Dorovkov, M. V. and A. G. Ryazanov. 2004. Phosphorylation of annexin I by TRPM7 channel-kinase. *The Journal of Biological Chemistry* 279:50643–50646.
173. Clark, K., M. Langeslag, B. van Leeuwen, L. Ran, A. G. Ryazanov, C. G. Figdor, W. H. Moolenaar, K. Jalink, and F. N. van Leeuwen. 2006. TRPM7, a novel regulator of actomyosin contractility and cell adhesion. *The EMBO Journal* 25:290–301.
174. Su, L. T., M. A. Agapito, M. Li, W. T. Simonson, A. Huttenlocher, R. Habas, L. Yue, and L. W. Runnels. 2006. TRPM7 regulates cell adhesion by controlling the calcium-dependent protease calpain. *The Journal of Biological Chemistry* 281:11260–11270.
175. Aarts, M. M. and M. Tymianski. 2005. TRPMs and neuronal cell death. *Pflugers Archiv: European Journal of Physiology* 451:243–249.
176. He, Y., G. Yao, C. Savoia, and R. M. Touyz. 2005. Transient receptor potential melastatin 7 ion channels regulate magnesium homeostasis in vascular smooth muscle cells: Role of angiotensin II. *Circulation Research* 96:207–215.
177. Touyz, R. M., Y. He, A. C. Montezano, G. Yao, V. Chubanov, T. Gudermann, and G. E. Callera. 2006. Differential regulation of transient receptor potential melastatin 6 and 7 cation channels by ANG II in vascular smooth muscle cells from spontaneously hypertensive rats. *American Journal of Physiology. Regulatory, Integrative and Comparative Physiology* 290:R73–R78.
178. Touyz, R. M. 2006. Magnesium and hypertension. *Current Opinion in Nephrology and Hypertension* 15:141–144.
179. Oancea, E., J. T. Wolfe, and D. E. Clapham. 2006. Functional TRPM7 channels accumulate at the plasma membrane in response to fluid flow. *Circulation Research* 98:245–253.
180. Sontia, B., A. C. Montezano, T. Paravicini, F. Tabet, and R. M. Touyz. 2008. Downregulation of renal TRPM7 and increased inflammation and fibrosis in aldosterone-infused mice: Effects of magnesium. *Hypertension* 51:915–921.
181. Yogi, A., G. E. Callera, R. Tostes, and R. M. Touyz. 2009. Bradykinin regulates calpain and proinflammatory signaling through TRPM7-sensitive pathways in vascular smooth muscle cells. *Am J Physiol Regul Integr Comp Physiol* 296:R201–207.
182. Du, J., J. Xie, Z. Zhang, H. Tsujikawa, D. Fusco, D. Silverman, B. Liang, and L. Yue. 2010. TRPM7-mediated Ca^{2+} signals confer fibrogenesis in human atrial fibrillation. *Circulation research* 106:992–1003.
183. Sun, H. S., M. F. Jackson, L. J. Martin, K. Jansen, L. Teves, H. Cui, S. Kiyonaka et al. 2009. Suppression of hippocampal TRPM7 protein prevents delayed neuronal death in brain ischemia. *Nature Neuroscience* 12:1300–1307.
184. Inoue, K., D. Branigan, and Z. G. Xiong. 2010. Zinc-induced neurotoxicity mediated by transient receptor potential melastatin 7 channels. *The Journal of Biological Chemistry* 285:7430–7439.
185. Hermosura, M. C., H. Nayakanti, M. V. Dorovkov, F. R. Calderon, A. G. Ryazanov, D. S. Haymer, and R. M. Garruto. 2005. A TRPM7 variant shows altered sensitivity to magnesium that may contribute to the pathogenesis of two Guamanian neurodegenerative disorders. *Proceedings of the National Academy of Sciences of the United States of America* 102:11510–11515.
186. Demeuse, P., R. Penner, and A. Fleig. 2006. TRPM7 channel is regulated by magnesium nucleotides via its kinase domain. *The Journal of General Physiology* 127:421–434.
187. Clark, K., J. Middelbeek, N. A. Morrice, C. G. Figdor, E. Lasonder, and F. N. van Leeuwen. 2008. Massive autophosphorylation of the Ser/Thr-rich domain controls protein kinase activity of TRPM6 and TRPM7. *PloS one* 3:e1876.
188. Landman, N., S. Y. Jeong, S. Y. Shin, S. V. Voronov, G. Serban, M. S. Kang, M. K. Park, G. Di Paolo, S. Chung, and T. W. Kim. 2006. Presenilin mutations linked to familial Alzheimer's disease cause an imbalance in phosphatidylinositol 4,5-bisphosphate metabolism. *Proceedings of the National Academy of Sciences of the United States of America* 103:19524–19529.
189. Tsuruda, P. R., D. Julius, and D. L. Minor, Jr. 2006. Coiled coils direct assembly of a cold-activated TRP channel. *Neuron* 51:201–212.
190. Erler, I., D. M. Al-Ansary, U. Wissenbach, T. F. Wagner, V. Flockerzi, and B. A. Niemeyer. 2006. Trafficking and assembly of the cold-sensitive TRPM8 channel. *The Journal of Biological Chemistry* 281:38396–38404.
191. Phelps, C. B. and R. Gaudet. 2007. The role of the N terminus and transmembrane domain of TRPM8 in channel localization and tetramerization. *The Journal of Biological Chemistry* 282:36474–36480.
192. Bandell, M., A. E. Dubin, M. J. Petrus, A. Orth, J. Mathur, S. W. Hwang, and A. Patapoutian. 2006. High-throughput random mutagenesis screen reveals TRPM8 residues specifically required for activation by menthol. *Nature Neuroscience* 9:493–500.
193. Voets, T., G. Owsianik, A. Janssens, K. Talavera, and B. Nilius. 2007. TRPM8 voltage sensor mutants reveal a mechanism for integrating thermal and chemical stimuli. *Nature Chemical Biology* 3:174–182.
194. Chuang, H. H., W. M. Neuhausser, and D. Julius. 2004. The super-cooling agent icilin reveals a mechanism of coincidence detection by a temperature-sensitive TRP channel. *Neuron* 43:859–869.

195. McKemy, D. D., W. M. Neuhausser, and D. Julius. 2002. Identification of a cold receptor reveals a general role for TRP channels in thermosensation. *Nature* 416:52–58.
196. Rohacs, T., C. M. Lopes, I. Michailidis, and D. E. Logothetis. 2005. PI(4,5)P2 regulates the activation and desensitization of TRPM8 channels through the TRP domain. *Nature Neuroscience* 8:626–634.
197. Liu, B. and F. Qin. 2005. Functional control of cold- and menthol-sensitive TRPM8 ion channels by phosphatidylinositol 4,5-bisphosphate. *The Journal of Neuroscience: The Official Journal of the Society for Neuroscience* 25:1674–1681.
198. Daniels, R. L., Y. Takashima, and D. D. McKemy. 2009. Activity of the neuronal cold sensor TRPM8 is regulated by phospholipase C via the phospholipid phosphoinositol 4,5-bisphosphate. *The Journal of Biological Chemistry* 284:1570–1582.
199. Brauchi, S., P. Orio, and R. Latorre. 2004. Clues to understanding cold sensation: Thermodynamics and electrophysiological analysis of the cold receptor TRPM8. *Proceedings of the National Academy of Sciences of the United States of America* 101:15494–15499.
200. Voets, T., G. Droogmans, U. Wissenbach, A. Janssens, V. Flockerzi, and B. Nilius. 2004. The principle of temperature-dependent gating in cold- and heat-sensitive TRP channels. *Nature* 430:748–754.
201. Kuhn, F. J., K. Witschas, C. Kuhn, and A. Luckhoff. 2010. Contribution of the S5-pore-S6 domain to the gating characteristics of the cation channels TRPM2 and TRPM8. *The Journal of Biological Chemistry* 285:26806–26814.
202. Brauchi, S., G. Orta, C. Mascayano, M. Salazar, N. Raddatz, H. Urbina, E. Rosenmann, F. Gonzalez-Nilo, and R. Latorre. 2007. Dissection of the components for PIP2 activation and thermosensation in TRP channels. *Proceedings of the National Academy of Sciences of the United States of America* 104:10246–10251.
203. Matta, J. A. and G. P. Ahern. 2007. Voltage is a partial activator of rat thermosensitive TRP channels. *The Journal of Physiology* 585:469–482.
204. Brauchi, S., G. Orta, M. Salazar, E. Rosenmann, and R. Latorre. 2006. A hot-sensing cold receptor: C-terminal domain determines thermosensation in transient receptor potential channels. *The Journal of Neuroscience: The Official Journal of the Society for Neuroscience* 26:4835–4840.
205. Zakharian, E., C. Cao, and T. Rohacs. 2010. Gating of transient receptor potential melastatin 8 (TRPM8) channels activated by cold and chemical agonists in planar lipid bilayers. *The Journal of Neuroscience: The Official Journal of the Society for Neuroscience* 30:12526–12534.
206. Zhang, X., S. Mak, L. Li, A. Parra, B. Denlinger, C. Belmonte, and P. A. McNaughton. 2012. Direct inhibition of the cold-activated TRPM8 ion channel by Galphaq. *Nature Cell Biology* 14:851–858.
207. Klasen, K., D. Hollatz, S. Zielke, G. Gisselmann, H. Hatt, and C. H. Wetzel. 2012. The TRPM8 ion channel comprises direct Gq protein-activating capacity. *Pflugers Archiv: European Journal of Physiology* 463:779–797.
208. Lippoldt, E. K., R. Elmes, D. D. McCoy, W. M. Knowlton, and D. D. McKemy. 2013. Artemin, a glial cell line-derived neurotrophic factor family member, induces TRPM8-dependent cold pain. *The Journal of Neuroscience: The Official Journal of the Society for Neuroscience* 33:12543–12552.
209. Peier, A. M., A. Moqrich, A. C. Hergarden, A. J. Reeve, D. A. Andersson, G. M. Story, T. J. Earley et al. 2002. A TRP channel that senses cold stimuli and menthol. *Cell* 108:705–715.
210. Takashima, Y., R. L. Daniels, W. Knowlton, J. Teng, E. R. Liman, and D. D. McKemy. 2007. Diversity in the neural circuitry of cold sensing revealed by genetic axonal labeling of transient receptor potential melastatin 8 neurons. *The Journal of Neuroscience: The Official Journal of the Society for Neuroscience* 27:14147–14157.
211. Takashima, Y., L. Ma, and D. D. McKemy. 2010. The development of peripheral cold neural circuits based on TRPM8 expression. *Neuroscience* 169:828–842.
212. Dhaka, A., T. J. Earley, J. Watson, and A. Patapoutian. 2008. Visualizing cold spots: TRPM8-expressing sensory neurons and their projections. *The Journal of Neuroscience: The Official Journal of the Society for Neuroscience* 28:566–575.
213. Weil, A., S. E. Moore, N. J. Waite, A. Randall, and M. J. Gunthorpe. 2005. Conservation of functional and pharmacological properties in the distantly related temperature sensors TRPV1 and TRPM8. *Molecular Pharmacology* 68:518–527.
214. Behrendt, H. J., T. Germann, C. Gillen, H. Hatt, and R. Jostock. 2004. Characterization of the mouse cold-menthol receptor TRPM8 and vanilloid receptor type-1 VR1 using a fluorometric imaging plate reader (FLIPR) assay. *British Journal of Pharmacology* 141:737–745.
215. Knowlton, W. M., R. Palkar, E. K. Lippoldt, D. D. McCoy, F. Baluch, J. Chen, and D. D. McKemy. 2013. A sensory-labeled line for cold: TRPM8-expressing sensory neurons define the cellular basis for cold, cold pain, and cooling-mediated analgesia. *The Journal of Neuroscience: The Official Journal of the Society for Neuroscience* 33:2837–2848.
216. Dhaka, A., A. N. Murray, J. Mathur, T. J. Earley, M. J. Petrus, and A. Patapoutian. 2007. TRPM8 is required for cold sensation in mice. *Neuron* 54:371–378.
217. Wei, E. T. and D. A. Seid. 1983. AG-3-5: A chemical producing sensations of cold. *Journal of Pharmacy and Pharmacology* 35:110–112.
218. Knowlton, W. M., R. L. Daniels, R. Palkar, D. D. McCoy, and D. D. McKemy. 2011. Pharmacological blockade of TRPM8 ion channels alters cold and cold pain responses in mice. *PloS One* 6:e25894.
219. Almeida, M. C., T. Hew-Butler, R. N. Soriano, S. Rao, W. Wang, J. Wang, N. Tamayo et al. 2012. Pharmacological blockade of the cold receptor TRPM8 attenuates autonomic and behavioral cold defenses and decreases deep body temperature. *The Journal of Neuroscience: The Official Journal of the Society for Neuroscience* 32:2086–2099.
220. Ma, S., H. Yu, Z. Zhao, Z. Luo, J. Chen, Y. Ni, R. Jin et al. 2012. Activation of the cold-sensing TRPM8 channel triggers UCP1-dependent thermogenesis and prevents obesity. *Journal of Molecular Cell Biology* 4:88–96.
221. Bautista, D. M., J. Siemens, J. M. Glazer, P. R. Tsuruda, A. I. Basbaum, C. L. Stucky, S. E. Jordt, and D. Julius. 2007. The menthol receptor TRPM8 is the principal detector of environmental cold. *Nature* 448:204–208.
222. Colburn, R. W., M. L. Lubin, D. J. Stone, Jr., Y. Wang, D. Lawrence, M. R. D'Andrea, M. R. Brandt, Y. Liu, C. M. Flores, and N. Qin. 2007. Attenuated cold sensitivity in TRPM8 null mice. *Neuron* 54:379–386.
223. Gentry, C., N. Stoakley, D. A. Andersson, and S. Bevan. 2010. The roles of iPLA2, TRPM8 and TRPA1 in chemically induced cold hypersensitivity. *Molecular Pain* 6:4.
224. Descoeur, J., V. Pereira, A. Pizzoccaro, A. Francois, B. Ling, V. Maffre, B. Couette et al. 2011. Oxaliplatin-induced cold hypersensitivity is due to remodelling of ion channel expression in nociceptors. *EMBO Molecular Medicine* 3:266–278.
225. Galeotti, N., L. Di Cesare Mannelli, G. Mazzanti, A. Bartolini, and C. Ghelardini. 2002. Menthol: A natural analgesic compound. *Neuroscience Letters* 322:145–148.
226. Green, B. G. 1992. The sensory effects of l-menthol on human skin. *Somatosensory & Motor Research* 9:235–244.

227. Proudfoot, C. J., E. M. Garry, D. F. Cottrell, R. Rosie, H. Anderson, D. C. Robertson, S. M. Fleetwood-Walker, and R. Mitchell. 2006. Analgesia mediated by the TRPM8 cold receptor in chronic neuropathic pain. *Current Biology* 16:1591–1605.
228. Stein, R. J., S. Santos, J. Nagatomi, Y. Hayashi, B. S. Minnery, M. Xavier, A. S. Patel et al. 2004. Cool (TRPM8) and hot (TRPV1) receptors in the bladder and male genital tract. *Journal of Urology* 172:1175–1178.
229. Tsavaler, L., M. H. Shapero, S. Morkowski, and R. Laus. 2001. Trp-p8, a novel prostate-specific gene, is up-regulated in prostate cancer and other malignancies and shares high homology with transient receptor potential calcium channel proteins. *Cancer Research* 61:3760–3769.
230. Bidaux, G., M. Roudbaraki, C. Merle, A. Crepin, P. Delcourt, C. Slomianny, S. Thebault et al. 2005. Evidence for specific TRPM8 expression in human prostate secretory epithelial cells: Functional androgen receptor requirement. *Endocrine Related Cancer* 12:367–382.
231. Thebault, S., L. Lemonnier, G. Bidaux, M. Flourakis, A. Bavencoffe, D. Gordienko, M. Roudbaraki et al. 2005. Novel role of cold/menthol-sensitive transient receptor potential melastatine family member 8 (TRPM8) in the activation of store-operated channels in LNCaP human prostate cancer epithelial cells. *The Journal of Biological Chemistry* 280:39423–39435.
232. Zhang, L. and G. J. Barritt. 2004. Evidence that TRPM8 is an androgen-dependent Ca^{2+} channel required for the survival of prostate cancer cells. *Cancer Research* 64:8365–8373.
233. Mustafa, S. and M. Oriowo. 2005. Cooling-induced contraction of the rat gastric fundus: Mediation via transient receptor potential (TRP) cation channel TRPM8 receptor and Rho-kinase activation. *Clinical and Experimental Pharmacology and Physiology* 32:832–838.
234. Mustafa, S. M. and O. Thulesius. 2001. Cooling-induced gastrointestinal smooth muscle contractions in the rat. *Fundamental & Clinical Pharmacology* 15:349–354.

30 TRPML channels

Qiong Gao, Xiaoli Zhang, and Haoxing Xu

Contents

Members of mammalian mucolipin transient receptor potential (TRPML) cation channels, TRPML1–3 or MCOLN1–3, are localized predominantly on the membranes of late endosomes and lysosomes (LELs) to mediate phosphoinositide-regulated release of Ca^{2+} and heavy metal Fe^{2+}/Zn^{2+} ions from the LEL lumen. Studies using genetic models and mammalian cell lines have revealed a broad spectrum of cellular functions for TRPMLs. These include the maintenance of LEL ion homeostasis and the regulation of membrane trafficking (fusion and fission) in the late endocytic pathways: autophagy, LEL-to-Golgi retrograde trafficking, and lysosomal exocytosis. At the whole-organism level, while loss-of-function mutations in human *TRPML1* cause mucolipidosis type IV (ML4), a childhood neurodegenerative lysosomal storage disease (LSD), gain-of-function mutations in mouse *TPRML3* result in the varitint-waddler (*Va*) phenotype with hearing and pigmentation defects. In this chapter, we will discuss the current understanding of TRPMLs regarding their tissue distribution, subcellular localization, channel properties, cellular functions, lysosome physiology, and disease relevance.

30.1 INTRODUCTION

The mucolipin subfamily of transient receptor potential (TRP) cation channels comprises three members in mammals, that is, TRPML1 (MCOLN1), TRPML2 (MCOLN2), and TRPML3 (MCOLN3). TRPML1 was the first member of the subfamily cloned during the search for the gene causing ML4, a childhood neurodegenerative lysosomal storage disease (LSD) manifested by psychomotor retardation and retinal degeneration (1–3). Subsequent homology cloning led to the identification of two additional members in the subfamily: TRPML2 and TRPML3 (4,5). Loss-of-function mutations of mouse *TRPML1* (6), *C. elegans* TRPML1 *(cup-5)* (7,8), and *Drosophila TRPML* (9) all result in cellular phenotypes and behavior defects that are reminiscent of ML4. No pathology has yet been associated with mammalian TRPML2. Gain-of-function mutations of mouse *TRPML3* cause the varitint-waddler (*Va*) phenotype with deafness, circling, and diluted coat color (4). Thus, TRPMLs represent an evolutionarily conserved ion channel family that is required for normal physiology in animals.

Unlike most TRP channels whose primary location is the plasma membrane (PM), TRPMLs are mainly localized in the intracellular LELs (10,11). Consistently, loss-of-function mutations in TRPMLs result in endosomal and lysosomal dysfunctions (10,11). The molecular mechanisms by which TRPMLs regulate endosomal and lysosomal functions, however, have been unclear. Fortunately, the development of the whole-endolysosome patch-clamp technique has allowed for detailed study of the channel properties and endogenous activation mechanisms of TRPMLs (12,13). Furthermore, heterologously overexpressed TRPMLs and/or their activating or surface-expressing trafficking mutations can be studied using whole-cell recordings (14–18). Together, these studies have characterized TRPMLs as inwardly rectifying Ca^{2+}-/Fe^{2+}-/Zn^{2+}-permeable cation channels that are activated by a LEL-specific phosphoinositide, $PI(3,5)P_2$ (12,13,19). Thus, TRPMLs are channels that can mediate the release of cations and heavy metal ions from the LEL lumen to the cytosol, in response to $PI(3,5)P_2$ elevation or unidentified trafficking cues.

Using genetic models of TRPML mutations, as well as the mammalian cell culture models in which the expression levels of TRPMLs are genetically manipulated, TRPMLs are found to regulate a variety of cellular processes, including endocytosis, endosomal and lysosomal membrane trafficking (fusion/fission), lysosome ion homeostasis, lysosomal exocytosis, and autophagy (10,20). Importantly, the recent identification of several small-molecule activators of TRPMLs has made it possible to further define the specific roles of TRPMLs in these cellular functions (18,20,21). Molecular and biochemical analyses have revealed several protein interaction partners for TRPMLs, defining the molecular context for TRPMLs' cellular functions. While lysosomes are essential for both catabolism and anabolism (22), TRPMLs may play important roles in many lysosome-associated physiological functions. TRPMLs represent a unique opportunity for studying the functions of ion channels in endosomes and lysosomes. Furthermore, TRPMLs may serve as a gateway to understand and treat ML4 and other lysosomal storage diseases (LSDs).

30.2 PUTATIVE MEMBRANE TOPOLOGY AND STRUCTURAL ASPECTS

Because crystallographic studies are still lacking for TRPMLs, the current understanding of TRPML structures is mostly obtained using bioinformatic analyses and biochemical characterization or extrapolation from studies of other related TRPs (23). TRPMLs putatively comprise six transmembrane (6TM)–spanning domains with the amino (NH_2)- and carboxyl (COOH)-terminal tails facing the cytosol (see Figure 30.1). The fifth and sixth transmembrane segments (S5 and S6) are predicted to form the channel gate (23). Indeed, several gain-of-function mutations that lock the channels in a nongated open state are mapped to the intracellular half of S5 (V432P or *Va* mutation as an example; Figure 30.1) (14,19). The S5–S6 link region is predicted to be the pore loop that constitutes the selectivity filter, and consistently, charge neutralization of acidic amino acid residues in the pore loop results in nonconducting pore-dead mutations (D^{471}D-KK in TRPML1; see Figure 30.1) (24). Several important regulatory sites in TRPMLs, especially TRPML1, are revealed

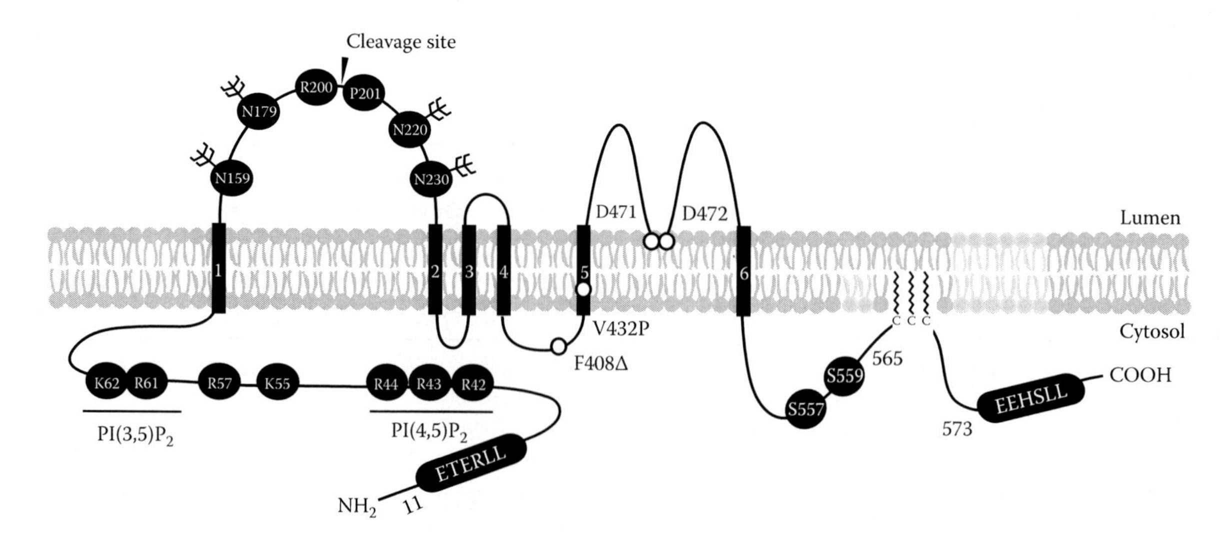

Figure 30.1 Putative membrane topology and structure of TRPML1. TRPML1 is predicted to have six transmembrane (6TM) domains with amino (NH_2)- and carboxyl (COOH)-terminal tails facing the cytosol. Two di-leucine (LL) lysosome-targeting motifs E^{11}TERLL and E^{573}EHSLL are located separately at each tail. Gain-of-function mutations are located at the bottom half of S5 (V432P is listed as an example), and loss-of-function ML4 mutations are located throughout the protein (V408Δ is listed as an example). Positively charged amino acid residues residing at the N-terminus of TRPML1 are potential phosphoinositide-interacting sties with Arg61 and Lys 62 for $PI(3,5)P_2$ and Arg42/Arg43/Arg44 for $PI(4,5)P_2$. Two potential PKA sites (S557 and S559) and three potential palmitoylation sites (C565–567) are localized in the C-terminus. The first luminal loop is highly N-glycosylated with a proteolytic cleavage site between the second and third glycosylation sites.

by bioinformatics sequence analyses. Considering that the three TRPMLs share high amino acid sequence homology (>75%), structural analyses of TRPML1 should provide important insights for TRPML2 and TRPML3 as well (Figure 30.1).

Two di-leucine (LL) motifs, found at both the N- and C-termini (Figure 30.1), mediate trafficking of TRPML1 to LELs (25) (also see the following text). Another potential structural determinant contributing to TRPML's LEL targeting is the palmitoylated site (C^{565}CC) at the C-terminal tail of TRPML1 (Figure 30.1), which may influence the efficiency of internalization of TRPML1 from the PM (25).

Similar to many lysosomal membrane proteins, TRPMLs are highly glycosylated (Figure 30.1) to confer the resistance to hydrolase-mediated degradation (26). There is a large extracellular loop between the first (S1) and the second (S2) transmembrane segments of TRPML1 (Figure 30.1), in which four consensus N-linked glycosylation sites are located (27,28). A proteolytic cleavage site has been identified between the second and the third glycosylation sites (27,28). Two protein kinase A (PKA) consensus motifs, Ser557 and Ser559, are located on the C-terminus of TRPML1 (29).

Several functional domains/sites that regulate the channel activity of TRPMLs have been identified. TRPML1 has a polybasic region comprising multiple positively charged amino acids in its N-terminus (13) (Figure 30.1). In vitro lipid-protein binding assays demonstrate that this polybasic domain binds directly with LEL-localized PI(3,5)P_2 and PM-specific phosphoinositide PI(4,5)P_2, which have been shown to activate and inhibit TRPML1, respectively (13,30). It was proposed that Arg41–43 residues mediate the inhibitory effect of PI(4,5)P_2 while Arg61 and Lys62 mediate the activation effect of PI(3,5)P_2 (30) (Figure 30.1). TRPML3 is regulated by extracellular pH and Na^+, and regulation is mapped to several histidine and acidic amino acid residues in the extracellular loop, respectively (20,24).

30.3 TISSUE DISTRIBUTION

TRPMLs are expressed in many tissues, which can be detected using nucleotide-based techniques, including RT-PCR, northern blot, and in situ hybridization (10,31). TRPML1 is ubiquitously expressed in all tissues with the highest level in the brain, kidney, spleen, liver, and heart. This expression pattern is consistent with the observation that the enlarged vacuole phenotype is seen in all cell types of ML4 patients and TRPML1 knockout mice (6,32). The expression spectra for TRPML2 and TRPML3 are much narrower. TRPML2 is highly expressed in immune tissues (33) and lymphoid and myeloid cell lines (34). However, it is also detectable in other tissues, including the lung and stomach (31). Likewise, TRPML3 is expressed in many tissues at low levels, but with the most robust expression in hair cells and melanocytes, which is consistent with the hearing and pigmentation defects associated with the *Va* phenotype in mice (4,15,17,35).

30.4 SUBCELLULAR LOCALIZATION

Expression studies using fluorescently tagged TRPMLs (5,27,36,37) indicate that TRPMLs are mainly localized in LELs, which is confirmed by Western blotting of the endogenous proteins in the lysosome fractionation (24,38). Heterologous expression studies demonstrate that the majority (>60%–80%) of TRPML1–3 proteins are localized in Lamp1- or Rab7-positive compartments that are often LysoTracker-positive, providing further support for the primary location of TRPMLs in LELs (10). TRPML1-GFP can also be detected in EEA1-positive early endosomes (EEs) (10), and TRPML1-specific channel activity can be measured in the inside-out patches excised from the PM of TRPML1-GFP-expressing HEK293T cells (30). Likewise, overexpressed TRPML2 is also detected at the PM, in addition to LELs and Arf-6-associated recycling endosomes (REs) (39) (Figure 30.2). Consistent with the fact that sizable whole-cell currents can be recorded from TRPML3-expressing HEK293T cells, a large portion of overexpressed TRPML3 is found at the PM and EEs (24). Thus, TRPMLs are primarily localized in LELs but are also present in other endocytic compartments (Figure 30.2).

Lysosomal membrane proteins can be transported to LELs from the Trans-Golgi Network (TGN) directly by an AP1/AP3-dependent mechanism (25). Alternatively, they can also be transported first to the PM and then to LELs via endocytosis (40). While the N-terminal LL mediates a direct AP1/3-dependent anterograde trafficking of TRPML1 from the TGN to LELs, the C-terminal LL interacts with AP2 to ensure rapid endocytosis of surface TRPML1 (25). Mutations in both LL motifs (L^{15}L/AA-L^{577}L/AA) result in a significant increase in the surface expression of TRPML1 and whole-cell TRPML1 currents (I_{TRPML1}) (18,30).

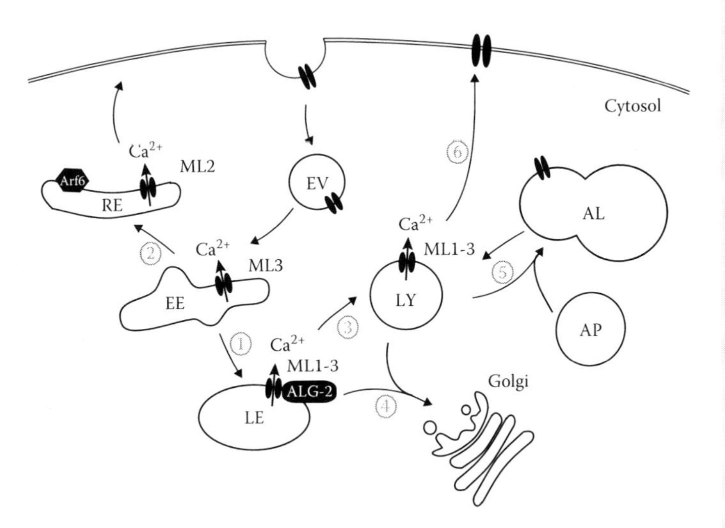

Figure 30.2 TRPMLs in the membrane trafficking pathways of endosomes and lysosomes. TRPML1–3 channels are predominantly localized in the LEs and lysosomes. TRPML3 is also found in the EEs and the PM; TRPML2 is also found in the REs. EEs undergo maturation to become LEs (trafficking step 1). REs are generated from the tubular structures of EEs (trafficking step 2). Further acidification of LEs leads to lysosomes (trafficking step 3). Transport vesicles derived from LEs and LYs mediate LEL-to-TGN retrograde trafficking (trafficking step 4). The autolysosomes are formed from the fusion of lysosomes with autophagosomes (trafficking step 5). Lysosomes can fuse with the PM to undergo exocytosis, in which Ca^{2+} release from TRPMLs might play an important role (trafficking step 6). Lysosomes could be reformed from autolysosomes and endolysosome hybrids. (EV, endocytic vesicle; EE, early endosome; LE, late endosome; LY, lysosome; AP, autophagosome; AL, autolysosome; RE, recycling endosome.)

30.5 PERMEATION PROPERTIES

The PM location of TRPML3 has allowed a characterization of TRPML3 using whole-cell patch clamp (14–17). Whole-cell currents from TRPML3 are larger in cells expressing a constitutively active mutant, TRPML3^{Va}, which is an alanine-to-proline mutation at the 419 position (A419P) that results in the *Va* phenotype in mice (15). TRPML3^{Va} displays a large inwardly rectifying current that is permeable to both monovalent (Na^+, K^+, Cs^+) and divalent (Ca^{2+}, Mg^{2+}) cations (14–17) (Table 30.1). Similarly, proline substitutions at the homologous positions in TRPML1 (V432P or TRPML1^{Va}; see Figure 30.1) and TRPML2 (A396P in the short splicing isoform or A424P in the long isoform; TRPML2^{Va}) lead to large inwardly rectifying whole-cell currents (15).

Endolysosome electrophysiology allows for the direct characterization of TRPMLs in their native intracellular compartments by mechanically isolating vacuolin-1-enlarged LELs (12). Vacuolin-1 is a lipid-soluble polycyclic triazine that can selectively enlarge endosomes and lysosomes, although the mechanism is not well understood (41). Under the whole-endolysosome configuration, most of the pore properties of vacuolar TRPMLs, such as the I–V relationship and ionic selectivity, resemble those of the PM-localized *Va* mutants (12,13). TRPMLs exhibit significant permeability to Ca^{2+}, providing a potential conduit for Ca^{2+} release from endolysosomes (10). Notably, detailed permeability analyses demonstrate a significant permeability of lysosomal TRPML1 and TRPML2, but not TRPML3, to divalent heavy trace metals, such as Fe^{2+}, Zn^{2+}, and Cu^{2+} (12) (Table 30.1). Finally, although TRPMLs appear to be important for LEL pH homeostasis (42,43), none of the TRPMLs are proton-permeable (10,15), suggesting that the effect of TRPMLs on the vacuolar pH is most likely secondary.

30.6 CHANNEL ACTIVATION AND GATING MECHANISMS

30.6.1 PHYSIOLOGICAL ACTIVATION MECHANISMS

Although predominantly residing on LEL membranes, TRPMLs undergo constant membrane trafficking and sorting. For example, TRPML1 can be temporarily located on the cell surface via lysosomal exocytosis (19). Likewise, TRPMLs may also appear in other intracellular compartments during autophagy and phagocytosis (24). As endolysosomal Ca^{2+}-permeable channels, TRPMLs are likely to be in the *driver's seat* during the Ca^{2+}-dependent endosomal and lysosomal trafficking, instead of cargos in the *passenger's seat*

Table 30.1 Summary of TRPML channels' characteristics

	TRPML1	TRPML2	TRPML3
Tissue distribution	Ubiquitously expressed with highest expression levels in brain, kidney, spleen, liver, and heart	Thymus, liver, kidney, heart, spleen, B cells, T cells, primary splenocytes, mastocytoma, and myeloma cell lines	Brain, thymus, lung, kidney, spleen, skin melanocytes, and hair cells
Subcellular localization	LEL	LEL, RE	LEL; EE; PM
Ion selectivity	Na^+, K^+, Ca^{2+}, Fe^{2+}, Zn^{2+}, etc.	Na^+, K^+, Cs^+, Ca^{2+}, Fe^{2+}	Ca^{2+}, Na^+, K^+, Mg^{2+}, Cs^+, etc.
I–V plot	Strong inwardly rectifying[a]	Strong inwardly rectifying[a]	Strong inwardly rectifying[a]
Single channel conductance (pS)	76 (from −140 to −100 mV) and 11 (from −80 to 40 mV)[b]	Not established	50 at negative voltages[c]
Activation mechanism	Voltage, extracellular or luminal low pH, PI(3,5)P_2	Voltage, extracellular or luminal low pH, PI(3,5)P_2	Voltage, PI(3,5)P_2
Activators	PI(3,5)P_2, ML-SA1, SF-22, SF-51	PI(3,5)P_2, ML-SA1, SF-21, SF-41, SF-81	PI(3,5)P_2, ML-SA1, SN-1, SN-2, SF-11 (21, 22, 23, 24, 31, 32, 33, 41, 51, 61, 71, and 81)
Pore blockers	La^{3+}	Not established	Gd^{3+}
Inhibitors	Sphingomyelins, PI(4,5)P_2, verapamil	Not established	Verapamil, H^+ (extracellular or luminal side), high (140 mM) sodium
Interacting proteins	TRPML2, TRPML3, Hsc70, HSP40, ALG-2, TPC2, LAMTMs	TRPML1, TRPML3, Hsc70?	TRPML1, TRPML2, TPC2, Hsc70?
Functions	Membrane trafficking in the late endocytic pathways, autophagy, lysosomal H^+/Fe^{2+}/Zn^{2+} homeostasis, gastric acid secretion	Membrane trafficking in the late endocytic pathways and Arf-6-regulated recycling pathways	Membrane trafficking along endolysosomal pathways, hair cell function, endolysosome pH homeostasis, autophagy
Human disease	ML4 and Niemann-Pick diseases	Not established	Not established
Mouse model	Knockout mice	Not established	Knockout mice, varitint-waddler mice

[a] In the whole-endolysosome configuration, *inward* currents indicate the cations flowing out of the lumen to the cytoplasm.
[b] Measured in the modified (pH 4.6) *Tyrode* solution (extracellular/luminal) using the activation mutation (V432P) of TRPML1.
[c] Measured using the activation mutation of TRPML3 (A419P or *Va* mutation).
[d] Direct interaction.

(11). Hence, the cellular functions of TRPMLs may depend on how TRPML's channel activity is regulated by various trafficking cues.

Luminal ions may serve as the trafficking cues to switch *on and off* the channel activity of TRPMLs. While TRPML3 (16,24) and TRPML1 (15) are sensitive to luminal pH, TRPML2 and TRPML3 are sensitive to the concentrations of luminal Na^+ (16,20,44). The pH and cation concentrations of intracellular compartments undergo drastic changes in the EEs and late endosomes (LEs) (45,46). Therefore, the regulation of TRPMLs by luminal ions may play an important role in the membrane trafficking of endosomes and lysosomes.

Endosomal and lysosomal phospholipids may regulate TRPMLs in a compartment-specific manner. Phosphoinositides, the signature lipids of diverse intracellular membranes, may act as permissive factors to assure the appropriate activity of intracellular channels in their *home* compartments (30). $PI(3,5)P_2$, but not any other phosphoinositide, directly and potently activates lysosomal recombinant TRPML1 at physiologically relevant concentrations (EC_{50} = 48 nM) (13). On the other hand, PM-enriched phosphoinositides, such as $PI(4,5)P_2$ and $PI(3,4,5)P_3$, inhibit TRPML1 in inside-out excised patches (30). Seven positively charged amino acid residues located on the N-terminus of TRPML1 (Figure 30.1) are required for the sensitivities of both $PI(3,5)P_2$ and $PI(4,5)P_2$ (30). Such distinct effects of two phosphoinositide isomers may be uniquely suited to meet the cellular functions of TRPMLs. While $PI(3,5)P_2$ may activate TRPMLs to induce endosomal and lysosomal Ca^{2+} release for the purpose of membrane trafficking, $PI(4,5)P_2$ may prevent the channels being active at the PM. Likewise, sphingomyelin, a lipid that is abundant at the PM and hydrolyzed in the lysosome, inhibits TRPML1 (18).

30.6.2 SYNTHETIC AGONISTS

To facilitate studies of TRPMLs in lysosome physiology, several synthetic TRPML activators have recently been identified (18,21). Using a high-throughput Ca^{2+} imaging-based screen assay, Grimm et al. identified 15 compounds that activate TRPML3 at high nanomolar to low micromolar concentrations (21) (Table 30.1). Several of these compounds including SF-51 also weakly activate whole-cell and whole-endolysosome TRPML1 currents (18). To search for more potent TRPML1 agonists, Shen et al. performed a low-throughput screening using SF-51 as the parent compound and identified ML-SA1 (mucolipin synthetic agonist 1), which can robustly activate TRPML1 (18). Interestingly, ML-SA1 was able to promote the LEL-to-TGN retrograde trafficking, which is defective in both ML4 and Niemann-Pick type C (NPC), a cholesterol-accumulating LSD (18). Although nonspecific TRP channel blockers, such as Gd^{3+} and La^{3+}, and the L-type Ca channel blocker, verapamil, were shown to inhibit TRPMLs (15,17), more selective and effective TRPML agonists and antagonists are needed for further studies to link the channel properties of TRPMLs to their cellular functions.

30.7 STOICHIOMETRY AND HETEROMERIZATION

Like other TRPs, homomeric TRPML channels are presumed to be tetrameric (10). Since TRPML1–3 channels are all localized in LELs, they may heteromultimerize with each other to form channels with novel properties that are different from homomeric ones (38). Consistently, dominant-negative pore-dead mutations in one TRPML can dramatically attenuate the cell death caused by the constitutively active mutations of another TRPML (20,44). Because TPCs are also localized in LELs, they may physically interact with TRPMLs to form TPC-TRPML heteromeric channels (5,38,47). However, although TPCs and TRPMLs co-immunoprecipitate with each other (20), they appear to function independently from each other (48). Like TRPMLs, TPCs are also activated by $PI(3,5)P_2$, but unlike TRPMLs, TPCs are Na^+ selective (46). It remains unknown whether TRPMLs and TPCs can form novel channels in LELs to mediate Ca^{2+} and Na^+ efflux.

30.8 PROTEIN INTERACTION PARTNERS

To investigate TRPMLs' cellular functions, it is important to identify proteins interacting with TRPMLs. A number of TRPML-interacting proteins have been identified using the yeast two-hybrid and protein pull-down assays (Table 30.1). While TPCs may form heteromeric channels with TRPMLs, some proteins such as ALG-2 can function downstream of TRPMLs, and others such as lysosome-associated protein transmembranes (LAPTMs) may function upstream to regulate TRPML channel activity (20).

Upon elevation of juxtaorganellar Ca^{2+}, Ca^{2+} sensor proteins may be recruited to translate Ca^{2+} changes into cellular responses. For example, synaptotagmin VII is the Ca^{2+} sensor in the lysosome that mediates lysosomal exocytosis (49). The Ca^{2+} sensors for the other steps of LEL membrane trafficking, however, are still unclear. An EF-hand-containing apoptosis-linked gene 2 protein (ALG-2) interacts directly with the N-terminus of TRPML1 in a Ca^{2+}-dependent manner (50). ALG-2 could serve as a Ca^{2+} sensor downstream of TRPML activation (Figure 30.2).

LAPTM proteins, whose transport substrates are still unknown, are found to interact with TRPML1 in a yeast two-hybrid screen (51). TRPML1 and LAPTMs are co-localized in LELs and can be co-immunoprecipitated with each other (51). siRNA-mediated knockdown of LAPTMs results in the accumulation of enlarged vacuoles and inclusions reminiscent of ML4 cells (51). Interestingly, lysosomal enlargement due to LAPTM4b overexpression could be suppressed by the expression of TRPML1 (51), suggesting that LAPTMs function upstream of TRPML1 to regulate lysosomal functions.

30.9 LYSOSOME ION HOMEOSTASIS

Lysosomes, first discovered by Christian de Duve 50 years ago, are dynamic organelles serving as terminal digestive compartments in the endocytic pathway. They receive substrates from endocytosis and autophagic pathways. One unique feature of the lysosome is its highly acidified intra-luminal environment (pH 4.5–5.0), which is optimal for hydrolytic enzymes to function properly (52). In addition, acidic pH is required to maintain the concentration gradients of other ions (52). Loss of TRPML1 was reported to result in increased (53) or decreased lysosomal pH (42). Likewise, TRPML3 overexpression causes an elevation of endosomal pH (43). However, TRPMLs do not exhibit measurable permeability to protons (15), suggesting that the role of TRPMLs in the regulation of luminal pH regulation is indirect.

TRPML1 can mediate the release of heavy metal ions such as Fe^{2+}/Zn^{2+} from lysosomes (12). Some ML4 patients exhibit anemia and Fe^{2+} deficiency (54). Using whole-cell and whole-endolysosome patch clamp, it was shown that Fe^{2+} conductance of ML4 mutations was much decreased compared with wild-type TRPML1 and TRPML1^{Va} (12). Consistently, high levels of intralysosomal Fe^{2+} and concurrent cytosolic Fe^{2+} deficiency were observed in ML4 cells, suggesting that TRPML1 is important for lysosome iron homeostasis (12). TRPML1 is also permeable to Zn^{2+} (12), and Zn^{2+} homeostasis is vital for brain function. Chelatable zinc accumulates in the enlarged endolysosomes in HEK293 cells treated with TRPML1 siRNA, and ML4 cells contain a high level of intralysosomal zinc (55). Hence, Zn^{2+} dyshomeostasis may also contribute to the symptoms/pathologies in ML4 patients (55).

30.10 LYSOSOMAL MEMBRANE TRAFFICKING

Intracellular membrane trafficking is highly regulated and connects many of the membrane-bound organelles in the cell. Endocytic trafficking involves multiple trafficking steps: EE formation, endocytic recycling, endosome maturation, endosome-to-TGN retrograde trafficking, autophagy, and lysosomal exocytosis (56). Accumulated evidence supports a role for Ca^{2+} release from endosomes and lysosomes in membrane trafficking (52). Juxtaorganellar Ca^{2+} has been proposed to be fusogenic because cell-free heterotypic fusion of LELs is inhibited by Ca^{2+} chelator BAPTA-AM (57). In the case of fusion, endosomal and lysosomal luminal Ca^{2+} is the primary source of the release, which is around 0.5 mM (~5000-fold higher than the cytosolic Ca^{2+}, ~100 nM) (58–60). However, the channels responsible for Ca^{2+} release in endosomes and lysosomes remain elusive. TRPMLs, Ca^{2+} channels localized on LELs, are natural candidates for release channels.

30.10.1 ENDOSOME MATURATION

The endocytic pathway starts with the endocytic fission from the PM to form endocytic vesicles and EEs (Figure 30.2), which then undergo a series of fusion/fission and maturation process to become LELs (61). Both TRPML1 and TRPLM3 may participate in endosome maturation. In ML4 cells, the delivery of cell surface receptors such as platelet-derived growth factor receptors to the lysosome is delayed (62). Consistently, the delivery of endocytosed molecules to the lysosome is delayed in cells treated with TRPML1-specific siRNA (37). Delayed delivery and degradation of endocytosed materials is also seen in cells overexpressing TRPML3 (24,43), suggesting that TRPML3 also plays a role in endosome maturation.

30.10.2 ENDOCYTIC RECYCLING

Cargos like glycosylphosphatidylinositol-anchored proteins (GPI-APs) are recycled back to the PM via the ADP-ribosylation factor (Arf-6)–regulated pathway. RFP-tagged TRPML2 co-localizes with CD59, one of the GPI-APs, in the tubular structures of the REs (39). Interestingly, recycling of CD59 is significantly delayed in cells expressing a dominant-negative form of TRPML2 (39), suggesting that TRPMLs may play a role in the endocytic recycling pathway (Figure 30.2).

30.10.3 LYSOSOME REFORMATION

In loss-of-function *Cup-5* (the *C. elegans* orthologue of TRPML1)-mutant worms, enlarged vacuoles are observed to contain both endosome and lysosome markers (7). Similarly, enlarged vacuoles are also observed in ML4 cells (10). These enlarged vacuoles are considered to be the hybrid organelles of endosomes and lysosomes (endolysosome hybrids). Based on the accumulation of endolysosome hybrid organelles, it was hypothesized that TRPML is required for the reformation of lysosomes, a process also referred to as lysosome biogenesis (63). Interestingly, lysosome biogenesis is also regulated by a master growth-regulated kinase mechanistic target of rapamycin (mTOR) (64) and a lysosome-associated transcription factor TFEB (65). While TFEB is shown to transcriptionally regulate TRPML1 (66), mTOR activation is significantly reduced in the *Drosophila trpml* mutant flies (67), suggesting that TRPML1 may interact with mTOR/TFEB signaling to regulate lysosome biogenesis.

30.10.4 LEL-TO-TGN RETROGRADE TRAFFICKING

Upon lysosomal degradation, digested products undergo *retrograde* trafficking to TGN (see Figure 30.2) for reutilization (61). This membrane fission process is also used by the membrane proteins that shuttle between LELs and TGN such as M6PR (61). M6PR is the key protein that transports hydrolytic enzymes from TGN to LELs (56). In human fibroblasts from ML4 patients, retrograde trafficking of LactosylCeramide-BODIPY (LacCer-BODIPY) from lysosomes to TGN is delayed or blocked (36,68), and this delay can be rescued by TPRML1 overexpression (36). Similar trafficking defects are also seen in $PI(3,5)P_2$-deficient cells, suggestive of a physiological relevance of the $PI(3,5)P_2$ regulation of TRPML1 (13). Interestingly, NPC cells also exhibit similar LEL-to-TGN trafficking defects, which can be attenuated by ML-SA1, a TRPML1 synthetic agonist (18).

30.10.5 AUTOPHAGY

Autophagy is a lysosome-mediated degradation process for damaged organelles and aged proteins. It starts with the formation of autophagosomes, which then undergo membrane fusion with lysosomes to form autolysosomes (Figure 30.2), in which autophagic substrates are degraded. Accumulation of autophagosomes is seen in ML4 cells, which is due to either increased autophagic flux or impaired fusion with lysosomes (62). Decreased degradation of autophagosomes has also been observed in TRPML-deficient *Drosophila* and mouse neurons (9,69,70). The impaired autophagy might account for neuronal cell death and neurodegeneration in ML4 (10). On the other hand, while siRNA-mediated knockdown of TRPML3 reduces the formation of autophagosomes, overexpression of TRPML3 results in a reduced level of autophagy by decreasing the formation of autolysosomes (24).

30.10.6 LYSOSOME EXOCYTOSIS

Lysosome exocytosis includes two steps: recruitment of lysosomes to proximity of cell surface and fusion of lysosomes with the PM upon transient Ca^{2+} elevation. Lysosome exocytosis is required for a variety of cellular processes, including membrane repair, transmitter release, neurite outgrowth, immune response, and phagocytosis (61). Lysosomal exocytosis is reduced in ML4 fibroblasts (71) but significantly increased in HEK293 cells expressing gain-of-function mutant TRPML1 channels (19). More interestingly, overexpression of TFEB has been shown to enhance lysosome exocytosis in a TRPML1-dependent manner (66).

30.11 ANIMAL MODELS AND HUMAN DISEASES

30.11.1 LOSS-OF-FUNCTION MUTATIONS

ML4 is caused by loss-of-function mutations of TPRML1, of which more than 15 are known (32). ML4 is an autosomal recessive lysosomal storage disease (LSD) characterized clinically by psychomotor retardation, retinal degeneration, corneal opacity, iron deficiency or anemia, and gastric abnormality (72). At the cellular level, membranous lipids are accumulated in enlarged vacuolar structures in most tissues. Similar phenotype is reported for TRPML1 knockout mice, *Drosophila TRPML* mutants, and *C. elegans CUP5* mutants (6,7,9,63).

Currently, there is no treatment for ML4. Small-molecule activators of TRPML1 might be able to boost the residual activity of hypofunctional (ML4) mutant TRPML1 channels. Lysosome-targeting iron or zinc chelators could reduce the lysosome stress associated with lysosomal Fe^{2+}/Zn^{2+} overload. In addition, modulation of TFEB-dependent autophagic pathway might prove promising for ML4 and other LSDs (66). In flies, high protein diet is able to rescue the defects associated with the loss of *Drosophila TRPML* (67), suggesting that amino acid supplementation and mTOR activation are beneficial for ML4. Although promising, the therapeutic potentials of these approaches remain to be tested.

30.11.2 GAIN-OF-FUNCTION MUTATIONS

A single missense amino acid substitution in the S5 of TRPML3 (A419P) results in *Va* mice with embryonic lethality, pigmentation defects, vestibular dysfunction (circling behavior, imbalance, head bobbing, and waddling), and deafness (73,74). The Va^J mutations that include an additional I326T substitution in the second extracellular loop exhibit similar but less severe phenotypes (4). Because the *Va* mutation may lock the channel at the open state, the sustained influx of Ca^{2+} results in early apoptosis in HEK293 cells (15). In melanocytes and hair cells, the *Va*- or Va^J-induced apoptosis might account for the pigmentation and vestibular/hearing defects in these mice (15). Because loss-of-function TRPML3 mutations do not display vestibular/hearing defects (75), *Va* is a purely gain-of-function phenotype.

30.11.3 OTHER LYSOSOME STORAGE DISEASES

The cellular phenotypes of ML4 exhibit many similarities to other LSDs, suggesting a role of TRPMLs in the housekeeping functions of lysosomes. In NPC cells, the accumulation of sphingomyelins on the luminal side of LEL inhibits TRPML1 and lysosomal Ca^{2+} release and Ca^{2+}-dependent lysosomal trafficking (18). Hence, TRPMLs may also play important roles in other LSDs and lysosome-associated diseases.

30.12 CONCLUDING REMARKS

After TRPML1 was cloned in 2000 as the genetic cause of ML4, rapid progress has been made toward the understanding of the molecular and cellular functions of TRPMLs. Studies in patient cells, model organisms, and cultured cells have established dual roles of TRPMLs in the regulation of membrane trafficking and lysosome ion homeostasis. However, many important questions still remain unanswered. We hope to see development in the following areas in the future:

- Development of effective TRPML antibodies to detect the tissue distribution and subcellular localization of endogenous TRPMLs
- Crystallographic studies of TRPMLs
- More potent TRPML agonists and antagonists to determine whether the cellular phenotypes caused by the genetic inactivation of TRPMLs are direct effects
- Using high-resolution live imaging methods to investigate the roles of TRPMLs in lysosome physiology
- Generation of animal models to study the in vivo function of TRPML2
- Testing potential therapeutic approaches in the animal models of ML4 and other TRPML-associated LSDs

ACKNOWLEDGMENTS

This work carried out in the author's laboratory is supported by NIH grants (NS062792, MH096595, and AR060837 to Haoxing Xu). We appreciate the encouragement and helpful comments from the other members of the Xu laboratory.

REFERENCES

1. Bargal, R., N. Avidan, E. Ben-Asher, Z. Olender, M. Zeigler, A. Frumkin, A. Raas-Rothschild, G. Glusman, D. Lancet, and G. Bach. 2000. Identification of the gene causing mucolipidosis type IV. *Nature Genetics* 26:118–123.
2. Bassi, M. T., M. Manzoni, E. Monti, M. T. Pizzo, A. Ballabio, and G. Borsani. 2000. Cloning of the gene encoding a novel integral membrane protein, mucolipidin-and identification of the two major founder mutations causing mucolipidosis type IV. *American Journal of Human Genetics* 67:1110–1120.
3. Sun, M., E. Goldin, S. Stahl, J. L. Falardeau, J. C. Kennedy, J. S. Acierno, Jr., C. Bove et al. 2000. Mucolipidosis type IV is caused by mutations in a gene encoding a novel transient receptor potential channel. *Human Molecular Genetics* 9:2471–2478.
4. Di Palma, F., I. A. Belyantseva, H. J. Kim, T. F. Vogt, B. Kachar, and K. Noben-Trauth. 2002. Mutations in Mcoln3 associated with deafness and pigmentation defects in varitint-waddler (Va) mice. *Proceedings of the National Academy of Sciences of the United States of America* 99:14994–14999.
5. Venkatachalam, K., T. Hofmann, and C. Montell. 2006. Lysosomal localization of TRPML3 depends on TRPML2 and the mucolipidosis-associated protein TRPML1. *The Journal of Biological Chemistry* 281:17517–17527.

6. Venugopal, B., M. F. Browning, C. Curcio-Morelli, A. Varro, N. Michaud, N. Nanthakumar, S. U. Walkley, J. Pickel, and S. A. Slaugenhaupt. 2007. Neurologic, gastric, and opthalmologic pathologies in a murine model of mucolipidosis type IV. *American Journal of Human Genetics* 81:1070–1083.
7. Fares, H. and I. Greenwald. 2001. Regulation of endocytosis by CUP-5, the *Caenorhabditis elegans* mucolipin-1 homolog. *Nature Genetics* 28:64–68.
8. Hersh, B. M., E. Hartwieg, and H. R. Horvitz. 2002. The *Caenorhabditis elegans* mucolipin-like gene cup-5 is essential for viability and regulates lysosomes in multiple cell types. *Proceedings of the National Academy of Sciences of the United States of America* 99:4355–4360.
9. Venkatachalam, K., A. A. Long, R. Elsaesser, D. Nikolaeva, K. Broadie, and C. Montell. 2008. Motor deficit in a *Drosophila* model of mucolipidosis type IV due to defective clearance of apoptotic cells. *Cell* 135:838–851.
10. Cheng, X., D. Shen, M. Samie, and H. Xu. 2010. Mucolipins: Intracellular TRPML1–3 channels. *FEBS Letters* 584:2013–2021.
11. Abe, K. and R. Puertollano. 2011. Role of TRP channels in the regulation of the endosomal pathway. *Physiology* 26:14–22.
12. Dong, X. P., X. Cheng, E. Mills, M. Delling, F. Wang, T. Kurz, and H. Xu. 2008. The type IV mucolipidosis-associated protein TRPML1 is an endolysosomal iron release channel. *Nature* 455:992–996.
13. Dong, X. P., D. Shen, X. Wang, T. Dawson, X. Li, Q. Zhang, X. Cheng, Y. Zhang, L. S. Weisman, M. Delling, and H. Xu. 2010. PI(3,5)P(2) controls membrane trafficking by direct activation of mucolipin Ca(2+) release channels in the endolysosome. *Nature Communications* 1:38.
14. Grimm, C., M. P. Cuajungco, A. F. van Aken, M. Schnee, S. Jors, C. J. Kros, A. J. Ricci, and S. Heller. 2007. A helix-breaking mutation in TRPML3 leads to constitutive activity underlying deafness in the varitint-waddler mouse. *Proceedings of the National Academy of Sciences of the United States of America* 104:19583–19588.
15. Xu, H., M. Delling, L. Li, X. Dong, and D. E. Clapham. 2007. Activating mutation in a mucolipin transient receptor potential channel leads to melanocyte loss in varitint-waddler mice. *Proceedings of the National Academy of Sciences of the United States of America* 104:18321–18326.
16. Kim, H. J., Q. Li, S. Tjon-Kon-Sang, I. So, K. Kiselyov, A. A. Soyombo, and S. Muallem. 2008. A novel mode of TRPML3 regulation by extracytosolic pH absent in the varitint-waddler phenotype. *The EMBO Journal* 27:1197–1205.
17. Nagata, K., L. Zheng, T. Madathany, A. J. Castiglioni, J. R. Bartles, and J. Garcia-Anoveros. 2008. The varitint-waddler (Va) deafness mutation in TRPML3 generates constitutive, inward rectifying currents and causes cell degeneration. *Proceedings of the National Academy of Sciences of the United States of America* 105:353–358.
18. Shen, D., X. Wang, X. Li, X. Zhang, Z. Yao, S. Dibble, X. P. Dong et al. 2012. Lipid storage disorders block lysosomal trafficking by inhibiting a TRP channel and lysosomal calcium release. *Nature Communications* 3:731.
19. Dong, X. P., X. Wang, D. Shen, S. Chen, M. Liu, Y. Wang, E. Mills, X. Cheng, M. Delling, and H. Xu. 2009. Activating mutations of the TRPML1 channel revealed by proline-scanning mutagenesis. *The Journal of Biological Chemistry* 284:32040–32052.
20. Grimm, C., S. Hassan, C. Wahl-Schott, and M. Biel. 2012. Role of TRPML and two-pore channels in endolysosomal cation homeostasis. *The Journal of Pharmacology and Experimental Therapeutics* 342:236–244.
21. Grimm, C., S. Jors, S. A. Saldanha, A. G. Obukhov, B. Pan, K. Oshima, M. P. Cuajungco, P. Chase, P. Hodder, and S. Heller. 2010. Small molecule activators of TRPML3. *Chemistry and Biology* 17:135–148.
22. Efeyan, A., R. Zoncu, and D. M. Sabatini. 2012. Amino acids and mTORC1: From lysosomes to disease. *Trends in Molecular Medicine*18:524–533.
23. Wu, L. J., T. B. Sweet, and D. E. Clapham. 2010. International union of basic and clinical pharmacology. LXXVI. Current progress in the mammalian TRP ion channel family. *Pharmacological Reviews* 62:381–404.
24. Kim, H. J., A. A. Soyombo, S. Tjon-Kon-Sang, I. So, and S. Muallem. 2009. The Ca(2+) channel TRPML3 regulates membrane trafficking and autophagy. *Traffic* 10:1157–1167.
25. Vergarajauregui, S. and R. Puertollano. 2006. Two di-leucine motifs regulate trafficking of mucolipin-1 to lysosomes. *Traffic* 7:337–353.
26. Fukuda, M. 1991. Lysosomal membrane glycoproteins. Structure, biosynthesis, and intracellular trafficking. *Journal of Biological Chemistry* 266:21327–21330.
27. Kiselyov, K., J. Chen, Y. Rbaibi, D. Oberdick, S. Tjon-Kon-Sang, N. Shcheynikov, S. Muallem, and A. Soyombo. 2005. TRP-ML1 is a lysosomal monovalent cation channel that undergoes proteolytic cleavage. *The Journal of Biological Chemistry* 280:43218–43223.
28. Miedel, M. T., K. M. Weixel, J. R. Bruns, L. M. Traub, and O. A. Weisz. 2006. Posttranslational cleavage and adaptor protein complex-dependent trafficking of mucolipin-1. *The Journal of Biological Chemistry* 281:12751–12759.
29. Vergarajauregui, S., R. Oberdick, K. Kiselyov, and R. Puertollano. 2008. Mucolipin 1 channel activity is regulated by protein kinase A-mediated phosphorylation. *The Biochemical Journal* 410:417–425.
30. Zhang, X., X. Li, and H. Xu. 2012. Phosphoinositide isoforms determine compartment-specific ion channel activity. *Proceedings of the National Academy of Science of the United States of America* 109:11384–11389.
31. Samie, M. A., C. Grimm, J. A. Evans, C. Curcio-Morelli, S. Heller, S. A. Slaugenhaupt, and M. P. Cuajungco. 2009. The tissue-specific expression of TRPML2 (MCOLN-2) gene is influenced by the presence of TRPML1. *Pflugers Archives: European Journal of Physiology* 459:79–91.
32. Slaugenhaupt, S. A. 2002. The molecular basis of mucolipidosis type IV. *Current Molecular Medicine* 2:445–450.
33. Song, Y., R. Dayalu, S. A. Matthews, and A. M. Scharenberg. 2006. TRPML cation channels regulate the specialized lysosomal compartment of vertebrate B-lymphocytes. *European Journal of Cell Biology* 85:1253–1264.
34. Lindvall, J. M., K. E. Blomberg, A. Wennborg, and C. I. Smith. 2005. Differential expression and molecular characterisation of Lmo7, Myo1e, Sash1, and Mcoln2 genes in Btk-defective B-cells. *Cellular Immunology* 235:46–55.
35. Cuajungco, M. P., and M. A. Samie. 2008. The varitint-waddler mouse phenotypes and the TRPML3 ion channel mutation: Cause and consequence. *Pflugers Archives: European Journal of Physiology* 457:463–473.
36. Pryor, P. R., F. Reimann, F. M. Gribble, and J. P. Luzio. 2006. Mucolipin-1 is a lysosomal membrane protein required for intracellular lactosylceramide traffic. *Traffic* 7:1388–1398.
37. Thompson, E. G., L. Schaheen, H. Dang, and H. Fares. 2007. Lysosomal trafficking functions of mucolipin-1 in murine macrophages. *BMC Cell Biology* 8:54.
38. Zeevi, D. A., S. Lev, A. Frumkin, B. Minke, and G. Bach. 2010. Heteromultimeric TRPML channel assemblies play a crucial role in the regulation of cell viability models and starvation-induced autophagy. *Journal of Cell Science* 123:3112–3124.

39. Karacsonyi, C., A. S. Miguel, and R. Puertollano. 2007. Mucolipin-2 localizes to the Arf6-associated pathway and regulates recycling of GPI-APs. *Traffic* 8:1404–1414.
40. Saftig, P. and J. Klumperman. 2009. Lysosome biogenesis and lysosomal membrane proteins: Trafficking meets function. *Nature Reviews Molecular Cell Biology* 10:623–635.
41. Huynh, C. and N. W. Andrews. 2005. The small chemical vacuolin-1 alters the morphology of lysosomes without inhibiting Ca^{2+}-regulated exocytosis. *EMBO Reports* 6:843–847.
42. Soyombo, A. A., S. Tjon-Kon-Sang, Y. Rbaibi, E. Bashllari, J. Bisceglia, S. Muallem, and K. Kiselyov. 2006. TRP-ML1 regulates lysosomal pH and acidic lysosomal lipid hydrolytic activity. *The Journal of Biological Chemistry* 281:7294–7301.
43. Martina, J. A., B. Lelouvier, and R. Puertollano. 2009. The calcium channel mucolipin-3 is a novel regulator of trafficking along the endosomal pathway. *Traffic* 10:1143–1156.
44. Grimm, C., S. Jors, Z. Guo, A. G. Obukhov, and S. Heller. 2012. Constitutive activity of TRPML2 and TRPML3 channels versus activation by low extracellular sodium and small molecules. *The Journal of Biological Chemistry* 287:22701–22708.
45. Gerasimenko, J. V., A. V. Tepikin, O. H. Petersen, and O. V. Gerasimenko. 1998. Calcium uptake via endocytosis with rapid release from acidifying endosomes. *Current Biology* 8:1335–1338.
46. Wang, X., X. Zhang, X. P. Dong, M. Samie, X. Li, X. Cheng, A. Goschka et al. 2012. TPC proteins are phosphoinositide- activated sodium-selective ion channels in endosomes and lysosomes. *Cell* 151:372–383.
47. Zeevi, D. A., A. Frumkin, V. Offen-Glasner, A. Kogot-Levin, and G. Bach. 2009. A potentially dynamic lysosomal role for the endogenous TRPML proteins. *The Journal of Pathology* 219:153–162.
48. Yamaguchi, S., A. Jha, Q. Li, A. A. Soyombo, G. D. Dickinson, D. Churamani, E. Brailoiu, S. Patel, and S. Muallem. 2011. Transient receptor potential mucolipin 1 (TRPML1) and two-pore channels are functionally independent organellar ion channels. *The Journal of Biological Chemistry* 286:22934–22942.
49. Czibener, C., N. M. Sherer, S. M. Becker, M. Pypaert, E. Hui, E. R. Chapman, W. Mothes, and N. W. Andrews. 2006. Ca^{2+} and synaptotagmin VII-dependent delivery of lysosomal membrane to nascent phagosomes. *Journal of Cell Biology* 174:997–1007.
50. Vergarajauregui, S., J. A. Martina, and R. Puertollano. 2009. Identification of the penta-EF-hand protein ALG-2 as a Ca^{2+}-dependent interactor of mucolipin-1. *The Journal of Biological Chemistry* 284:36357–36366.
51. Vergarajauregui, S., J. A. Martina, and R. Puertollano. 2011. LAPTMs regulate lysosomal function and interact with mucolipin 1: New clues for understanding mucolipidosis type IV. *Journal of Cell Science* 124:459–468.
52. Morgan, A. J., F. M. Platt, E. Lloyd-Evans, and A. Galione. 2011. Molecular mechanisms of endolysosomal Ca^{2+} signalling in health and disease. *The Biochemical Journal* 439:349–374.
53. Bach, G., C. S. Chen, and R. E. Pagano. 1999. Elevated lysosomal pH in Mucolipidosis type IV cells. *Clinica Chimica Acta: International Journal of Clinical Chemistry* 280:173–179.
54. Altarescu, G., M. Sun, D. F. Moore, J. A. Smith, E. A. Wiggs, B. I. Solomon, N. J. Patronas et al. 2002. The neurogenetics of mucolipidosis type IV. *Neurology* 59:306–313.
55. Eichelsdoerfer, J. L., J. A. Evans, S. A. Slaugenhaupt, and M. P. Cuajungco. 2010. Zinc dyshomeostasis is linked with the loss of mucolipidosis IV-associated TRPML1 ion channel. *The Journal of Biological Chemistry* 285:34304–34308.
56. Huotari, J. and A. Helenius. 2011. Endosome maturation. *EMBO Journal* 30:3481–3500.
57. Pryor, P. R., B. M. Mullock, N. A. Bright, S. R. Gray, and J. P. Luzio. 2000. The role of intraorganellar Ca(2+) in late endosome-lysosome heterotypic fusion and in the reformation of lysosomes from hybrid organelles. *Journal of Cell Biology* 149:1053–1062.
58. Michelangeli, F., O. A. Ogunbayo, and L. L. Wootton. 2005. A plethora of interacting organellar Ca^{2+} stores. *Current Opinion in Cell Biology* 17:135–140.
59. Samie, M. and H. Xu. 2011. Studying TRP channels in intracellular membranes. In *TRP Channels*. M. X. Zhu, ed., Boca Raton, FL, CRC Press.
60. Dong, X. P., X. Wang, and H. Xu. 2010. TRP channels of intracellular membranes. *Journal of Neurochemistry* 113:313–328.
61. Luzio, J. P., P. R. Pryor, and N. A. Bright. 2007. Lysosomes: Fusion and function. *Nature Reviews Molecular Cell Biology* 8:622–632.
62. Vergarajauregui, S., P. S. Connelly, M. P. Daniels, and R. Puertollano. 2008. Autophagic dysfunction in mucolipidosis type IV patients. *Human Molecular Genetics* 17:2723–2737.
63. Treusch, S., S. Knuth, S. A. Slaugenhaupt, E. Goldin, B. D. Grant, and H. Fares. 2004. *Caenorhabditis elegans* functional orthologue of human protein h-mucolipin-1 is required for lysosome biogenesis. *Proceedings of the National Academy of Sciences of the United States of America* 101:4483–4488.
64. Zoncu, R., L. Bar-Peled, A. Efeyan, S. Wang, Y. Sancak, and D. M. Sabatini. 2011. mTORC1 senses lysosomal amino acids through an inside-out mechanism that requires the vacuolar H-ATPase. *Science* 334:678–683.
65. Roczniak-Ferguson, A., C. S. Petit, F. Froehlich, S. Qian, J. Ky, B. Angarola, T. C. Walther, and S. M. Ferguson. 2012. The transcription factor TFEB links mTORC1 signaling to transcriptional control of lysosome homeostasis. *Science Signaling* 5:ra42.
66. Medina, D. L., A. Fraldi, V. Bouche, F. Annunziata, G. Mansueto, C. Spampanato, C. Puri et al. 2011. Transcriptional activation of lysosomal exocytosis promotes cellular clearance. *Developmental Cell* 21:421–430.
67. Wong, C. O., R. Li, C. Montell, and K. Venkatachalam. 2012. Drosophila TRPML is required for TORC1 activation. *Current Biology* 22:1616–1621.
68. Chen, C. S., G. Bach, and R. E. Pagano. 1998. Abnormal transport along the lysosomal pathway in mucolipidosis, type IV disease. *Proceedings of the National Academy of Sciences of the United States of America* 95:6373–6378.
69. Curcio-Morelli, C., F. A. Charles, M. C. Micsenyi, Y. Cao, B. Venugopal, M. F. Browning, K. Dobrenis, S. L. Cotman, S. U. Walkley, and S. A. Slaugenhaupt. 2010. Macroautophagy is defective in mucolipin-1-deficient mouse neurons. *Neurobiology of Disease* 40:370–377.
70. Micsenyi, M. C., K. Dobrenis, G. Stephney, J. Pickel, M. T. Vanier, S. A. Slaugenhaupt, and S. U. Walkley. 2009. Neuropathology of the Mcoln1(-/-) knockout mouse model of mucolipidosis type IV. *Journal of Neuropathology and Experimental Neurology* 68:125–135.
71. LaPlante, J. M., C. P. Ye, S. J. Quinn, E. Goldin, E. M. Brown, S. A. Slaugenhaupt, and P. M. Vassilev. 2004. Functional links between mucolipin-1 and Ca^{2+}-dependent membrane trafficking in mucolipidosis IV. *Biochemical and Biophysical Research Communications* 322:1384–1391.
72. Amir, N., J. Zlotogora, and G. Bach. 1987. Mucolipidosis type IV: Clinical spectrum and natural history. *Pediatrics* 79:953–959.
73. Sekimata, M., A. Murakami-Sekimata, and Y. Homma. 2011. CpG methylation prevents YY1-mediated transcriptional activation of the vimentin promoter. *Biochemical and Biophysical Research Communications* 414:767–772.
74. Zhao, R., T. Nakamura, Y. Fu, Z. Lazar, and D. L. Spector. 2011. Gene bookmarking accelerates the kinetics of post-mitotic transcriptional re-activation. *Nature Cell Biology* 13:1295–1304.
75. Jors, S., C. Grimm, L. Becker, and S. Heller. 2010. Genetic inactivation of Trpml3 does not lead to hearing and vestibular impairment in mice. *Plos One* 5:e14317.

31 CLC chloride channels and transporters

Giovanni Zifarelli and Michael Pusch

Contents

31.1 OVERVIEW OF CLC CHANNELS AND TRANSPORTERS

CLC proteins are a family comprising anion channels and transporters. The acronym "CLC" stands for "**Ch**L**o**ride **C**hannel," and CLC-0, the curious *double-barreled* Cl^- channel from the *Torpedo* electric organ, is the first cloned member of the family (Jentsch et al., 1990; Miller and White, 1980). Strictly speaking, the acronym is a misnomer because we now know that many CLC homologues are actually not passive Cl^- channels but secondary active Cl^-/H^+ antiporters (Accardi and Miller, 2004; Zifarelli and Pusch, 2007). Thus, we will be referring to these proteins as *CLC channels and transporters*, or just *CLC proteins*. All CLC proteins studied so far seem to share the same three-dimensional architecture: they are homodimeric proteins with physically separate, but identical ion translocation pathways in each subunit.

In the present chapter, we will provide a general overview of the CLC family focusing on the biophysical mechanisms underlying their function in straight connection to the knowledge obtained from x-ray structure analysis of bacterial and algal CLCs. Bretag and Ma will describe in more detail the physiological roles of mammalian CLC channels and transporters and in particular their involvement in various human genetic diseases.

31.1.1 DOUBLE-BARRELED *TORPEDO* CHANNEL

In their attempts to characterize the nicotinic acetylcholine receptor from the electric organ from *Torpedo*, Miller and colleagues discovered a novel Cl^- channel that showed a curious *double-barreled* behavior (Miller, 1982; Miller and Richard, 1990; Miller and White, 1980, 1984): Channel activity occurred in bursts that were characterized by three distinct current levels, as if the channel was composed of two identical protopores (Figure 31.1). Within a burst of activity, the occupation probabilities of the three conductance levels obeyed a simple binomial distribution, consistent with the assumption that the two protopores are gating independently from each other (Miller, 1982). Nevertheless, the presence of long silent periods of activity between successive bursts demonstrated that the two pores are not independent (Figure 31.1). Based on these and other results (Miller and White, 1984), Miller proposed the double-barreled shotgun model (Figure 31.1) according to which the channel is composed of two identical protopores, each bearing a gate, called fast gate or protopore gate. The slow gate, on the other hand, shuts off both pores simultaneously. The double-barreled architecture has been fully confirmed by the crystal structures of bacterial and algal CLC proteins (Dutzler et al., 2002; Feng et al., 2010), which will be described in Section 31.2.

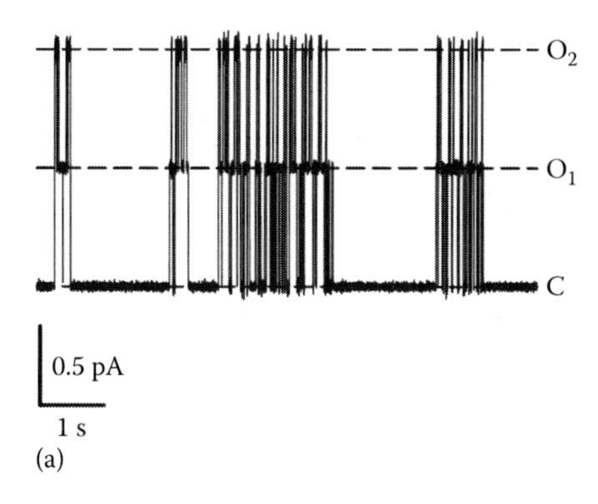

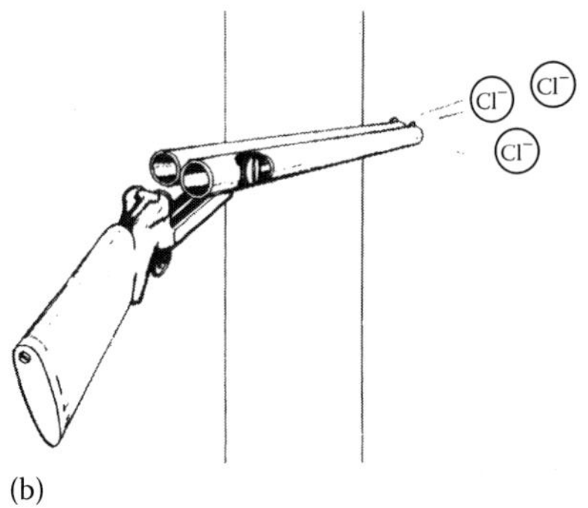

Figure 31.1 The double-barreled Cl⁻ channel from *Torpedo*. Panel (a) shows schematically the gating behavior of a single channel. Openings occur in bursts, which correspond to periods of activated slow gate, separated by long closures of the slow gate (conductance level C, for closed). During the bursts, the channel fluctuates between three equidistant conductance levels with binomial occupation probability of the states due to opening and closing of the fast gate of each protopore. These and other experiments have led Miller and colleagues to propose the double-barreled shotgun model shown in (b). (b: It was taken from Miller, C., *Philos. Trans. R. Soc. Lond. B. Biol. Sci.*, 299, 401, 1982.)

Initially, the *Torpedo* Cl⁻ channel remained somewhat a biophysical curiosity considered of scarce general relevance. This changed radically with the molecular cloning of CLC-0 by Jentsch et al. (1990), which led to the discovery of a new gene family present in all phyla. Having in hand the primary sequence allowed the application of genetic, biophysical, biochemical, and crystallographic methods. CLC-0 is an 89 kD protein with 805 amino acids (Jentsch et al., 1990). The CLC-0 peptide alone is sufficient to generate a fully functional double barreled channel in heterologous systems (Bauer et al., 1991). Experiments with concatameric channels and co-expression of mutant and WT proteins showed that the channel is a homodimer with two physically separate Cl⁻ conducting pathways (Ludewig et al., 1996; Middleton et al., 1996), in agreement with the double-barreled shotgun model.

31.1.2 OVERVIEW OF THE CLC FAMILY

CLC homologues are present in all phyla with nine distinct CLC genes in humans (Figure 31.2) (Jentsch, 2008; Jentsch et al., 1993, 1999, 2005; Mindell and Maduke, 2001). Among these, the muscle CLC-1, the ubiquitous CLC-2, and the kidney and inner ear epithelial CLC-Ka and CLC-Kb channels are most similar in primary sequence to the prototype *Torpedo* CLC-0. The other five mammalian CLCs (CLC-3-CLC-7)

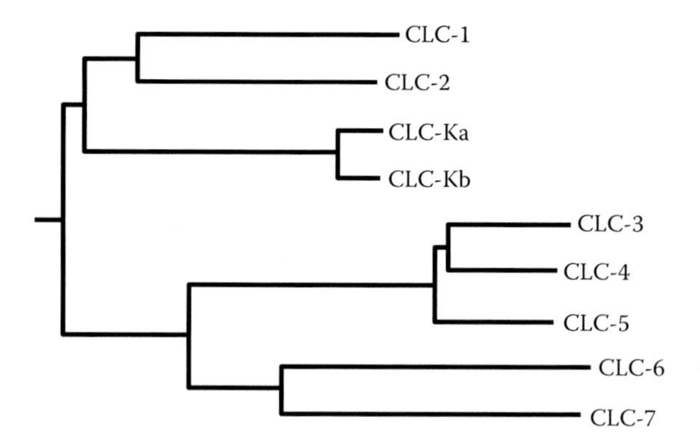

Figure 31.2 Dendrogram of the human CLC proteins. Length of lines reflects sequence similarity based on a standard ClustalW algorithm.

are quite distinct from these *muscle-type* CLC channels. Interestingly, all *muscle-type* CLC channels (CLC-1, CLC-2, CLC-Ka, CLC-Kb) are localized in the plasma membrane, while the CLC-3-CLC-7 proteins are predominantly found in endosomal and lysosomal membranes (Jentsch et al., 2005). Some CLC proteins associate with ancillary, small, β subunits: CLC-K channels (CLC-Ka and CLC-Kb) necessitate the two transmembrane domain protein barttin for proper targeting and function (Estévez et al., 2001); CLC-7 requires the highly glycosylated Ostm1 protein for proper functioning (Costa et al., 2012; Lange et al., 2006; Leisle et al., 2011). The interaction of CLC-K channels with barttin and that of CLC-7 with Ostm1 is obligatory. In contrast, the interaction of the rather ubiquitous CLC-2 channel with the cell-adhesion molecule GlialCAM is restricted to glia in the central nervous system (Jeworutzki et al., 2012).

Even though the physiological role of CLCs in bacteria is not well understood (with some exceptions [Iyer et al., 2002]), the study of a particular CLC from *E. coli*, CLC-ec1 (Maduke et al., 1999), has been of exceptional importance for the understanding of the mammalian CLCs. First, crystallization of CLC-ec1 has provided an extremely useful structural framework for the transmembrane part of CLC proteins (Dutzler et al., 2002). Second, functional studies of CLC-ec1 have demonstrated that this protein is not a passive Cl⁻ conducting anion channel, but a secondary active Cl^-/H^+ antiporter with a 2 Cl^-:1 H^+ stoichiometry (Accardi and Miller, 2004). This was a surprising finding from several perspectives. In fact, it is important to underline that passive diffusion (as in classical ion channels) and secondary active antiport are thermodynamically fundamentally different processes. Passive diffusion dissipates gradients and *destroys* free energy, whereas strictly coupled exchange of two substrates conserves free energy, transforming energy stored in one ion gradient into energy of the gradient of the coupled substrate. Another surprising aspect of the Cl^-/H^+ antiport is that the two substrates are chemically and electrically completely different. Many familiar antiporters exchange substrates that are similar to each other, for example, the Na^+/H^+ exchangers or Cl^-/HCO_3^- exchangers. For these classical antiporters, the *ping-pong-model* of transport (Figure 31.3) provides a natural mechanism of transport and simultaneously guarantees practically perfect stoichiometric coupling. Such a ping-pong transport mechanism is a priori

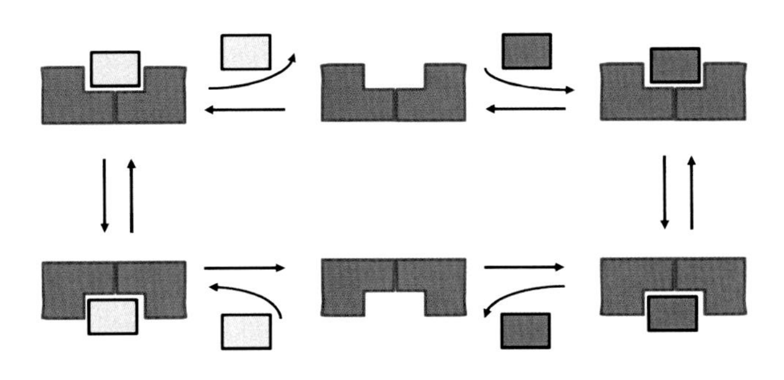

Figure 31.3 The classical ping-pong mechanism of alternate access for an antiporter with yellow and red substrates with a 1:1 stoichiometry. Forbidding that the transporter undergoes a substrate-free conformational change assures a perfect transport stoichiometry.

unlikely applicable to the exchange of such dissimilar substrates as protons and Cl^-. In fact, the available experimental evidence strongly disfavors a ping-pong mechanism (Picollo et al., 2012). An additional unusual feature of the 2 Cl^-/1 H^+ antiport is that it is highly electrogenic: A total of three elementary charges are translocated during each transport cycle. Thus, CLC exchange is intrinsically highly influenced by the membrane potential and, vice-versa, CLC activity directly affects the membrane potential. This feature is probably of significant physiological importance in endosomal and lysosomal CLC mediated Cl^-/H^+ exchange (Weinert et al., 2010).

Interestingly, among the mammalian CLC proteins, all plasma membrane localized CLCs (CLC-1, CLC-2, CLC-Ka, CLC-Kb) are passive Cl^- ion channels, whereas most if not all endosomal/lysosomal CLCs (CLC-3–7) are Cl^-/H^+ antiporters (Zifarelli and Pusch, 2007).

31.1.3 CLCs IN MODEL ORGANISMS

A single CLC is present in the yeast *Saccharomyces cerevisiae* (Greene et al., 1993), called Gef1. It is localized in intracellular membranes including Golgi (Hechenberger et al., 1996; Schwappach et al., 1998). *GEF1* mutant strains have deficits in iron and copper metabolism (Davis-Kaplan et al., 1998; Gaxiola et al., 1998; Greene et al., 1993; Schwappach et al., 1998). No direct functional data are available for Gef1. However, based on sequence similarity, in particular, the presence of the *proton glutamate* (see Section 31.2.2) and cell biological evidence Gef1 is likely a Cl^-/H^+ exchanger (Braun et al., 2010; Davis-Kaplan et al., 1998). This model organism may shed important light on the physiological role of intracellular CLC Cl^-/H^+ antiporters.

Among the six *Caenorhabditis elegans* CLC proteins (Schriever et al., 1999), the CLH-3b channel has been particularly well characterized (Strange, 2011). When CLH-3b is expressed in oocytes, it is activated by swelling and meiotic cell cycle progression (Rutledge et al., 2001). The channel is regulated by various kinases and phospatases that alter its gating properties (Denton et al., 2005; Falin et al., 2009; Rutledge et al., 2002). The channel is an interesting model for the dynamic interaction of the cytoplasmic intracellular domains with the membrane embedded part of the proteins (Strange, 2011).

The genome of the model plant *Arabidopsis thaliana* contains seven CLC genes (Hechenberger et al., 1996; Lurin et al., 1996; Lv et al., 2009) named AtCLC-a to AtCLC-g. A detailed overview of plant CLCs is provided in several reviews (De Angeli et al., 2009a; Ward et al., 2009; Zifarelli and Pusch, 2010).

CLCs from microorganisms have been useful for the structural and biochemical understanding of CLC exchanger function at a molecular level (Accardi and Picollo, 2010; Dutzler, 2004; Maduke et al., 2000; Miller, 2006; Mindell and Maduke, 2001). Relatively little is known about the physiological function of bacterial CLCs. For *E. coli*, it has been proposed that CLCs are important for acid resistance, helping the bacteria to transit the hostile acidic environment of the stomach (Iyer et al., 2002). An interesting recent addition to the repertoire of microbial CLCs is the CLC antiporter CLC-psy from the plant pathogen *Pseudomonas syringae* (Stockbridge et al., 2012). Transcription of the CLC-psy gene is activated by fluoride (F^-) and it functions as a 1 F^-/1 H^+ antiporter. Its activity has been proposed to confer F^- resistance to the bacteria (Stockbridge et al., 2012).

31.2 STRUCTURE AND FUNCTION OF CLC PROTEINS

31.2.1 STRUCTURES OF *E. COLI* AND ALGAE CLC TRANSPORTERS

Several crystal structures of prokaryotic and eukaryotic CLC antiporters (Dutzler et al., 2002, 2003; Feng et al., 2010) have provided a huge leap forward in our understanding of CLC protein's structure and function. The basic features emerging from these structures are very similar, but there are also important differences revealing key functional elements. As shown in Figure 31.4, the similarities mostly regard the *general morphology*

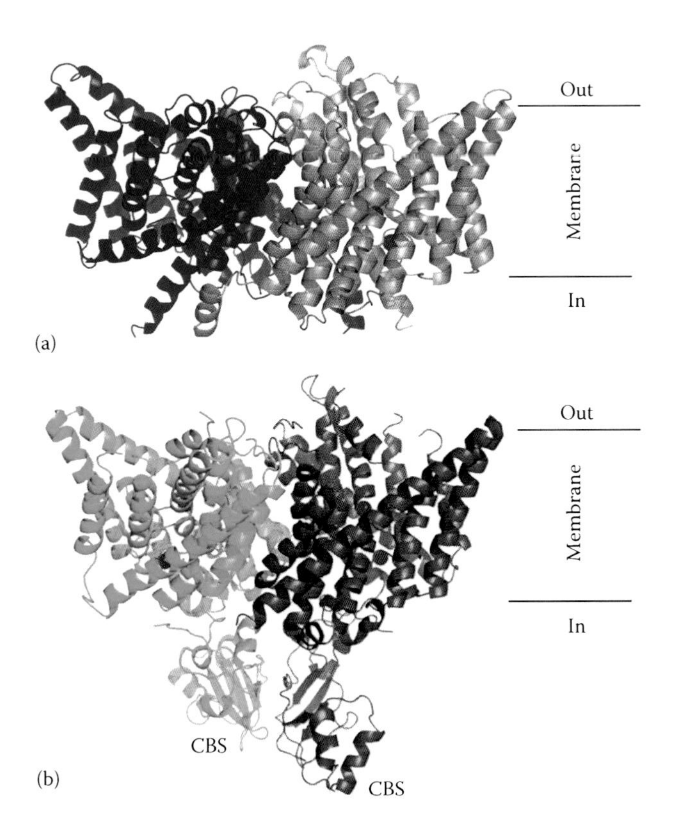

Figure 31.4 Crystallographic structures of CLC-ec1 (a) and CLC-cm (b). For each protein, the two subunits are indicated by different colors. The permeation pathway can be inferred from the position of two Cl^- ions in magenta.

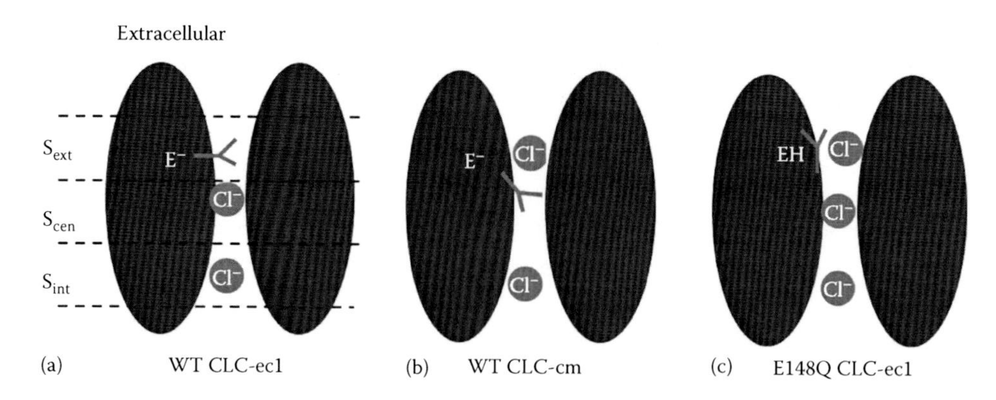

Figure 31.5 Different configurations of the selectivity filter in CLC antiporters. Cartoons representing the three different structures observed in crystallographic structures. Panels (a–c) represent respectively the structure of WT CLC-ec1 (From Dutzler, R. et al., *Nature*, 415, 287, 2002.), WT CLC-cm (From Feng, L. et al., *Science*, 330, 635, 2010.), and the mutant E148Q of CLC-ec1. (From Dutzler, et al., *Science*, 300, 108, 2003.) The side chain of the conserved gating glutamate (E148 in CLC-ec1 and E210 in CLC-cm) is shown. In C, the side chain of the glutamine residue is represented as a protonated glutamate side chain. The position of the three anion binding sites S_{ext}, S_{cen}, and S_{int} is indicated by dashed lines.

of the transmembrane domain. CLC proteins have a dimeric architecture with two identical subunits, each comprising an independent ion permeation pathway as suggested from the single channel behavior of the CLC-0 channel (Miller and White, 1980). The two subunits are related by a twofold axis of symmetry perpendicular to the membrane plane and have a triangular shape. Each subunit contains 18 α-helices arranged in a complex topology in which the N- and C-terminal parts of each subunit are organized in an antiparallel architecture that creates a pseudo twofold symmetry axis within each monomer (Dutzler et al., 2002). By virtue of pronounced tilt and partial unwinding of several helices, the two halves of each subunit wrap around the selectivity filter at a central position (Figure 31.4). It is interesting to note that the bound Cl^- ions do not make direct contact with a full positive charge from lysine or arginine residues. It has been hypothesized that a full positive charge would create a deep energy well and cause Cl^- to bind too tightly, compromising the efficiency of transport (Dutzler, 2004).

The ion permeation pathway has an hourglass shape: wide, water-filled vestibules present at both the internal and external extremities narrow to a small tunnel, 12 Å long, where three distinct anion binding sites were identified, S_{ext}, S_{cen}, and S_{int} (Figure 31.5). In fact, different crystal structures capture different configurations of this protein region. In the structure of WT CLC-ec1, two anions are bound to S_{int} and S_{cen}. The anion at S_{cen} is fully dehydrated and is coordinated by main chain amide nitrogen atoms and by side chain oxygen atoms from Ser-107 and Tyr-445 (Figure 31.6). The anion at S_{int} is still partially hydrated. In this structure, a conserved glutamate residue (Glu-148 in CLC-ec1 defined also gating glutamate) projects its side chain into the permeation pathway suggesting that this is a conformation in which transport does not occur (Figures 31.5a and 31.6). The structures of CLC-ec1 in which Glu-148 is mutated to alanine or glutamine is basically identical to the one of the WT except that an anion substitutes the side chain of the WT gating glutamate defining the third binding site, S_{ext} (Figure 31.5c). Interestingly, the side chain of the glutamine is oriented toward the extracellular space, leading to the suggestion that this might be the configuration assumed when the WT glutamate is protonated and transport is allowed. In the recent crystal structure of CLC-Cm, the gating glutamate side chain is oriented toward the intracellular side and occupies S_{cen} whereas anions are bound at S_{ext} and S_{int} (Figure 31.5b). Interestingly, this configuration of the gating glutamate was already suggested from molecular dynamics simulations (Bisset et al., 2005).

Several lines of evidence indicate that each subunit of the homodimer is independently able to carry out ion transport. The most direct proof of this concept comes from the investigation of a monomeric CLC-ec1 obtained from the disruption of the nonpolar interface between the dimers that showed a normal transport activity (Robertson et al., 2010). Is it possible that this property extends also to eukaryotic proteins that differ from the CLC-ec1 by the presence of a large cytoplasmic domain? Concatameric constructs of CLC-5 in which one of the subunits carried a mutation of the gating glutamate that rendered it a purely passive Cl^- conducting pore are still able to perform Cl^-/H^+ exchange indicating that this function is *contained* in a

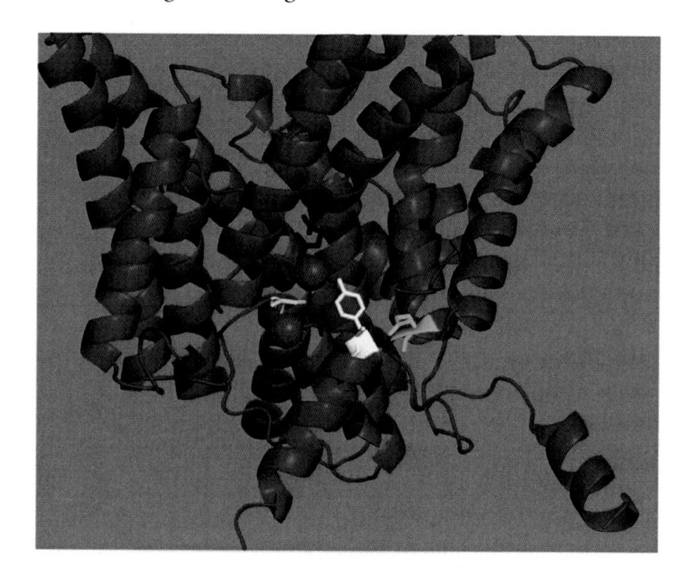

Figure 31.6 Structure of the selectivity filter of CLC-ec1. The gating glutamate E148 is indicated in red, the proton glutamate E203 is indicated in green. The two Cl^- ions at S_{cen} and S_{int} are represented as magenta spheres. The serine and the tyrosine residues coordinating the Cl^- ion at S_{cen} are indicated in yellow and white, respectively.

single subunit (Lísal and Maduke, 2008; Zdebik et al., 2008). However, this result does not strictly rule out the possibility that in CLC proteins carrying the cytoplasmic domain, unlike CLC-ec1, concerted conformational changes at the dimer interface might be required for transport. In this context, it is important to note that the slow gate mechanism of CLC-0 relies on the interdependence of the two subunits although the physical nature of the conformational changes associated with slow gate is still obscure. The evolutionary significance of the dimeric architecture of CLC proteins remains to date a mystery.

31.2.2 PROTON GLUTAMATE

The mechanism by which CLC proteins transport intracellular H^+ across the membrane has an obvious critical importance for CLC antiporters in which H^+ is one of the substrates. However, it has been recently proposed that intracellular H^+ have an unanticipated fundamental role also in the gating of CLC channels (Lísal and Maduke, 2008; Traverso et al., 2006).

In CLC antiporters, a key residue for the process of H^+ transport is the so-called proton glutamate, located toward the intracellular side, but still 10 Å away from the internal solution (Figure 31.6). Functional studies on CLC-ec1 (Accardi and Miller, 2004) and CLC-5 (Zdebik et al., 2008) suggested that this residue is essential for proton transport in all CLC antiporters, a notion supported by the conservation of this residue in CLC antiporters and its substitution by valine in CLC channels (Figure 31.7). However, it is remarkable that in CLC-ec1 substitutions of this residue with nonprotonatable residues turned the antiporter into a pure anionic conductance whereas in CLC-5 similar substitutions completely abrogated transport activity. This argues for important differences in the mechanism of coupled transport among CLC antiporters.

CLC-Ec1	FII**E**E
CLC-5	FSL**E**E
CLC-0	FSI**E**V
CLC-1	FSI**E**V
CLC-Cm	YSI**E**T
GEF1	FGL**E**E
CLC-Ck2	FIA**E**I

Figure 31.7 Sequence alignment for representative CLC transporters and channels showing the sequence stretch comprising the conserved glutamate at position 202 (bold characters) and the residue at position 203 (residues numbering for CLC-ec1). At position 203, residues in gray indicate that the corresponding protein function as an antiporter. For CLC-ec1 and CLC-5 the glutamate residue at position 203 is also defined as proton glutamate.

Interestingly, while substitution of the *proton glutamate* with alanine abrogated steady-state transport in CLC-5, this mutant displayed transient currents that originate from partial reactions in the transport cycle (Smith and Lippiat, 2010; Zifarelli et al., 2012).

However, it has more recently emerged that the eukaryotic antiporter CLC-Cm has a threonine residue substituting the proton glutamate (Feng et al., 2010), whereas a bacterial antiporter from *Citrobacter koseri*, CLC-Ck2, has an isoleucine residue at the corresponding position (Phillips et al., 2012) (Figure 31.7). Taken together, these results suggest that a protonatable residue at this position is not a universal requisite to support coupled anion/H^+ transport in CLC proteins. In this respect, a very recent investigation suggests that in CLC-ec1 another glutamate residue, Glu-202, immediately preceding the proton glutamate, Glu-203, is fundamental for H^+ delivery from the cytoplasm to the gating glutamate acting as a *water organizer* at the interface between the two subunits (Lim et al., 2012) even though the trajectory followed by H^+ from the intracellular space to the gating glutamate, 15 Å apart, remains to be elucidated. Interestingly, for the channel CLC-0, it has been suggested that H^+ delivery from the intracellular side is not mediated by protonatable residues but by the dissociation of an intrapore water molecule (Zifarelli et al., 2008).

31.2.3 MECHANISMS OF COUPLED TRANSPORT IN CLC EXCHANGERS: COUPLING MECHANISM AND SELECTIVITY

In all CLC antiporters in which the anion/H^+ coupling stoichiometry was investigated, CLC-ec1, AtCLC-a (that unlike the other transporters is a NO_3^-/H^+ antiporter), CLC-5, CLC-7, and CLC-Cm show a 2:1 stoichiometry (Accardi and Miller, 2004; De Angeli et al., 2006; Feng et al., 2010; Graves et al., 2008; Zifarelli and Pusch, 2009a). Polyatomic anions as NO_3^- and SCN^- are efficiently transported by the Cl^-/H^+ antiporters but they lead to a partial (in the case of NO_3^-) and complete (in the case of SCN^-) uncoupling of anion and proton coupling (Nguitragool and Miller, 2006; Zdebik et al., 2008).

The mechanism responsible for the coupling of anion and H^+ transport is still elusive. S_{cen} and its occupancy by anions is an important element for coupling. Crystallographic studies, in which S_{cen} was altered by mutating the conserved tyrosine residue, showed that the extent of anion occupancy at this site correlated with the ability of CLC-ec1 to carry out coupled Cl^-/H^+ exchange. Mutations that disturbed anion binding at this site displayed weaker coupling (Accardi et al., 2006). Reinforcing this notion, SCN^-, which is efficiently transported by CLC-ec1 but that abolishes H^+ coupling, does not bind to S_{cen} and occupies only S_{int} (Nguitragool and Miller, 2006).

Besides the conserved tyrosine residue, anions at S_{cen} are coordinated by a conserved serine that has been shown to be important for anion selectivity and anion/H^+ coupling. This serine is conserved in all Cl^-/H^+ exchangers, but it is substituted by a proline in the plant NO_3^-/H^+ antiporter AtCLC-a. Exchanging the WT serine with a proline turns CLC-5 from a Cl^-/H^+ to a NO_3^-/H^+ antiporter and confers NO_3^- over Cl^- selectivity to CLC-0 and CLC-ec1 (Bergsdorf et al., 2009; Picollo et al., 2009; Zifarelli and Pusch, 2009a). Recently, it has been suggested that also S_{ext} might be involved in anion selectivity as neutralization of a conserved positive charge in CLC-5 (Lys-210), located close to the gating glutamate Glu-211, led to an inversion of the canonical NO_3^- over Cl^- selectivity with no further effect on the WT anion/H^+ coupling ratio (De Stefano et al., 2011).

31.2.4 TRANSPORT MECHANISM

The transport mechanism of CLC antiporters remains a mystery. In particular the fact that the alanine mutant of the proton glutamate of CLC-ec1 is not able to transport H^+ and is transformed into a pure anionic conductance is difficult to reconcile with the classical transport model of alternate access. MacKinnon and coworkers suggested a transport mechanism in which the gating glutamate's side chain at S_{cen} can be protonated from the intracellular side, but, when it changes conformation and occupies S_{ext}, it can be deprotonated (Feng et al., 2010). These conformational changes of the gating glutamate between the two binding sites are accompanied by Cl^- transport. The model accounts for the 2:1 stoichiometry under the assumption that the access of Cl^-_{int} into the pore is slow compared with the deprotonation of the gating glutamate in the extracellular space. Interestingly, Picollo et al. found with isothermal calorimetry (ITC) measurements a Cl^-/H^+ binding stoichiometry of 2:1 suggesting a thermodynamic rather than kinetic coupling as the basis for the transport mechanism at least in CLC-ec1 (Picollo et al., 2012).

Independent of the detailed transport mechanism, several lines of evidence suggest how transport in CLC antiporters can be viewed as a competition between anions and the carboxylate side chain of the gating glutamate at the binding sites S_{ext} and S_{cen}. Feng et al. showed that in the absence of Cl^-, the gating glutamate is not able to transport H^+ in CLC-Cm and CLC-ec1 (Feng et al., 2012). This has been explained in terms of the high affinity of the glutamate carboxylic group for the anion binding in the pore such as, in the absence of Cl^-, the glutamate side chain would be permanently bound inside the transport pathway. Surprisingly, if the gating glutamate is mutated to alanine, H^+ influx can occur in the absence of Cl^- if a carboxylic group is provided by glutamate or gluconate in solution. This finding has been explained in terms of the ability of the carboxylic group in solution to reach S_{cen}, bind a H^+ from the intracellular side and transport it extracellularly (Feng et al., 2012). Glutamate/gluconate mediated H^+ transport does not occur in WT transporters suggesting that a carboxylate in solution does not effectively compete with the carboxylate of the gating glutamate for occupancy of the anion binding sites possibly due to weaker binding. Interestingly, Cl^- appears to inhibit H^+ transport mediated by carboxylic groups in solution. Aspartate, although very similar to glutamate, is not able to support transport activity consistent with its predicted inability to easily reach S_{cen} (Feng et al., 2012). Consistent with this overall view, on the basis of transient currents from WT and mutant CLC-5 (E268A), it was suggested that the first steps in the transport cycle are a conformational change of the gating glutamate followed by binding of Cl^-_{ext} (Zifarelli et al., 2012).

31.2.5 GATING OF CLC EXCHANGERS

The concept of gating is normally associated with ion channels to describe the process that controls their open probability and is much less common for transporters. However, it has already been shown for the Na^+/Ca^{2+} exchanger that transporters activity is not necessarily continuous and transporters can be in *active* and *inactive* states (Hilgemann, 1996). The presence of a gating mechanism regulating the transport activity of CLC exchanger is an important issue not only from the biophysical point of view but also from the physiological perspective. For example, CLC-5 is activated only at positive voltages and it is strongly rectifying when expressed in heterologous systems but this seems to conflict with its proposed role in endosomal physiology (Ishida et al., 2013; Lippiat and Smith, 2012; Stauber and Jentsch 2013). The concept that the activity of CLC transporters could be gated was originally proposed for CLC-5 on the basis of the characteristic noise spectrum of macroscopic currents (Zdebik et al., 2008). However, the first direct demonstration of the presence of a gating process for a CLC transporter has been obtained for the lysosomal antiporter CLC-7, which exhibits very slow activation and deactivation kinetics with time constants in the seconds range and similar to CLC-5 is strongly rectifying (Leisle et al., 2011). Very recently, a mutation of CLC-5 that significantly slowed down current kinetics, D76H, manifested inward tail currents at negative potentials and convincingly proved the presence of a gating mechanism that regulates transport activity (De Stefano et al., 2013). However, while the voltage dependence of the gating mechanism of CLC-7 is able to completely account for the voltage dependence of macroscopic currents and the rectifying behavior (Leisle et al., 2011), the situation is less clear for CLC-5 where the rectification may additionally reflect an intrinsic voltage dependence of the transport cycle (De Stefano et al., 2013).

31.2.6 FAST AND SLOW GATING IN MUSCLE-TYPE CLC CHANNELS

In muscle-type CLC channels that comprise CLC-0, CLC-1, and CLC-2, there are two different gating mechanisms defined as slow or common gate and fast or protopore gate. The slow gate acts simultaneously on both protopores of the dimeric channel whereas the fast gate acts on the single protopores. The attributes slow and fast are due to the fact that in CLC-0 they act on timescales that differ by orders of magnitude. Fast and common gating of CLC channels depend on different factors, like voltage, anion concentration, temperature, and pH. Most notably, the dependence on the proton concentration shows that CLC channels and transporters have strong mechanistic links.

31.2.6.1 Fast gate

Gating and permeation are strictly coupled in CLC channels and it has been proposed early on that part of the voltage dependence of fast gating arises indirectly from a voltage-dependent binding and/or translocation of a Cl^- ion entering the channel from the outside (Chen and Miller, 1996; Pusch et al., 1995). This hypothesis is in agreement with the fact that fast gate opening is favored by increasing $[Cl^-]_{ext}$, and positive voltages, with an apparent *gating valence* of ~1. However, the voltage dependence of the fast gate does not follow a simple Boltzmann distribution. In particular, the open probability does not approach zero at negative values but reaches a finite value (Chen and Miller, 1996; Ludewig et al., 1997). This has been ascribed to two possible pathways that lead to channel opening: one opening route with a rate constant that increases at positive voltages with a relatively large voltage dependence, and a less steeply voltage-dependent route whose rate increases at negative voltages (Chen and Miller, 1996). The closing rate constant shows a monotonic exponential voltage dependence (Chen and Miller, 1996; Zifarelli et al., 2008). The closing rate of the fast gate depends on $[Cl^-]_{int}$ (Chen and Miller, 1996) but not $[Cl^-]_{ext}$ (Chen et al., 2003). It has been more recently shown that permeation and gating are not always linked as some nonpermeant anions have a strong effect on gating (Chen et al., 2003; Rychkov et al., 1998).

The role of the conserved *gating glutamate* for the gating of CLC-0 mentioned earlier has been extensively studied. Neutralizing Glu-166 completely abolishes the normal voltage-dependent gating relaxations and locks the channel in an open state (Dutzler et al., 2003; Traverso et al., 2003). Thus, Glu-166 appears to be the major determinant of the fast protopore gate of the channel, even though additional conformational changes may accompany channel opening (Accardi and Pusch, 2003).

Extracellular acidification increases the open probability of the fast gate (Chen and Chen, 2001; Dutzler et al., 2003). Importantly, it seems that only the relatively voltage independent route of opening is affected by external pH, such that at acidic external pH, the open probability becomes effectively voltage-independent (Chen and Chen, 2001). In contrast, the opening rate constant for the steeply voltage-dependent route and the closing rate constant were found to be independent from pH_{ext} (Chen and Chen, 2001). The interpretation of these results is that extracellular protons can directly protonate Glu-166, leading to channel opening (Dutzler et al., 2003).

Hanke and Miller showed that intracellular pH has a dramatic effect on fast gating (Hanke and Miller, 1983). The effect is qualitatively different from that of pH_{ext}; increasing pH_{int} leads to a *shift* of the voltage-dependent activation curve to more positive membrane potentials. This issue was recently investigated in detail (Traverso et al., 2006; Zifarelli et al., 2008). It was shown that the mutant E166D could be strongly activated by acidic intracellular pH, with pH *shifting* the activation curve along the voltage axis. The pH_{int} effect was consistent with the idea that the relatively large voltage dependence of opening was caused by voltage-dependent protonation of the D166 residue from the intracellular side (Traverso et al., 2006). A similar idea was independently proposed by Miller (Miller, 2006). This proposal was quite speculative and contrasted with the previously established model ascribing the voltage dependence of fast gating opening to the voltage-dependent binding and/or translocation of a Cl^- ion (Chen and Miller, 1996; Pusch et al., 1995). The mechanism proposed to explain the effect of intracellular pH on fast gate is that the proton originates from the dissociation of a water molecule present in the pore (Zifarelli et al., 2008).

31.2.6.2 Slow gate

The slow gate controls opening of the two pores of the dimer simultaneously. In ClC-0, it is activated by hyperpolarized potentials (Richard and Miller, 1990) and does not deactivate completely at depolarized voltages (Pusch et al., 1997). An increase in both $[Cl^-_{ext}]$ and $[Cl^-_{int}]$ favor opening of the slow gate (Chen and Miller, 1996; Pusch et al., 1999).

Although the mechanism responsible for the slow gate is still unknown, several elements suggest that it is associated with large conformational change. First, the kinetics of the slow gate strongly depends on the temperature with a Q_{10} factor of ~40, more than 30 times higher compared with the expectations for a diffusion limited process (Pusch et al., 1997). Second, mutations in different regions of the channel affect the slow gate; in particular, the clustering of mutations at the dimer interface both at the level of the transmembrane domains (Duffield et al., 2003) and of the cytoplasmic domains (Estévez et al., 2004; Fong et al., 1998) further suggest that the slow gate relies on subunit interaction. Third, it has been suggested on the basis of FRET experiments that slow gate transitions are accompanied by a relative movement of the cytoplasmic termini of about 20 Å with closure of the gate associated with greater separation (Bykova et al., 2006).

In CLC-1, the slow gate has considerably lower temperature sensitivity (Q_{10} factor is ~4), an opposite voltage dependence to that of CLC-0, being activated by depolarization and its kinetics are orders of magnitude faster (Bennetts et al., 2001). Very recently, it has been suggested that in CLC-1 fast and slow gates are strongly coupled (Ma et al., 2011). As it has been already described, Richard and Miller observed that the gating of CLC-0 is not in thermodynamic equilibrium (Richard and Miller, 1990). The finding that this irreversibility is powered by the proton electrochemical gradient across the membrane (Lísal and Maduke, 2008) implicates that protons permeate the CLC-0 channel and, in doing so, favor slow gate opening. This has led to the provocative suggestion that CLC-0 is a broken Cl^-/H^+ antiporter.

The mechanistic relationship of the complex gating observed in other CLC channels like CLC-2 and CLC-Ka and -Kb with the fast and slow gate of CLC-0 is still unclear.

31.2.6.3 Gating of CLC-2

CLC-2 is an ubiquitously expressed Cl^- channel. It is not only activated by hyperpolarization, cell swelling, and intracellular Cl^- but is also modulated by pH_{ext} in a complex manner: Currents increase at moderate acidic pH but are reduced at more acidic pH. This observation was explained by postulating the presence of two independent proton binding sites with opposite effects (Arreola et al., 2002b). The activating site has been identified as E207, the *gating glutamate* of ClC-2 (Niemeyer et al., 2003). Neutralization of E207 abolished H^+ induced potentiation without influencing inhibition by more acidic pH. More recently, H532 has been identified as the H^+ sensor responsible for channel block at acidic pH (Niemeyer et al., 2009). Mutating H532 to phenylalanine selectively removed channel inhibition at acidic pH, leaving a pure H^+ induced activation. In spite of the clear participation of protons in the activation of CLC-2, the origin of the voltage dependence of the activation is still controversial. According to Sepùlveda and coworkers, the voltage dependence of activation is conferred by the voltage-dependent protonation of the gating glutamate (Niemeyer et al., 2009). As E207 is located approximately halfway across the conduction pathway, protons have to traverse some distance across the membrane electric field and therefore the rate of protonation is increased (and/or the rate of deprotonation is decreased) by negative voltages (Niemeyer et al., 2009). On the other hand, Arreola and coworkers propose that it is binding of Cl^-_{int} to the pore rather than protons that confers voltage sensitivity to CLC-2 gating (Sanchez-Rodriguez et al., 2010, 2012).

31.2.7 CBS DOMAINS AND REGULATION BY NUCLEOTIDES

The eukaryotic CLC proteins have a long cytoplasmic C-terminus (compare Figure 31.4a and b), which is physiologically very important as several disease-causing mutations cluster in this region (Estévez and Jentsch, 2002; Jentsch, 2002; Pusch, 2002). The C-terminal region contains two so-called CBS domains (from cystathionine-β-synthase). These structural domains normally occur in pairs and constitute nucleotide binding modules that regulate the activity of several unrelated protein families (Bateman, 1997; Ponting, 1997). Functional analysis and the presence of disease-causing mutations in this region in several CLCs indicate that this domain is important for correct assembly and gating properties (Jentsch, 2008). Nucleotide binding has been directly observed only in the isolated C-terminal part of CLC-5 (Meyer et al., 2007). Recently, the structure of the eukaryotic CLC-Cm showed for the first time the relative position of the transmembrane core and the C-terminal region with several contacts formed between them (Feng et al., 2010), providing a structural interpretation of how rearrangements of the C-terminal domain might transmit to the transmembrane core and influence pore and gating properties. This might explain the functional effects of nucleotides demonstrated for CLC-1, ClC-2, CLC-5, and AtCLC-a (Bennetts et al., 2007; De Angeli et al., 2009b; Scott et al., 2004; Zifarelli and Pusch, 2008, 2009b). Initial controversial results about the effect of ATP on CLC-1 (Bennetts et al., 2005; Zifarelli and Pusch, 2008) have been explained by the fact that this modulation is extremely sensitive to the presence of reducing/oxidizing agents (Zhang et al., 2008). Interestingly, while ATP potentiates the activity of CLC-5 (Zifarelli and Pusch, 2009b), it has an inhibitory effect on AtClC-a (De Angeli et al., 2009b) and CLC-1 (Bennetts et al., 2005). It has been recently shown that CLC-1 is modulated also by beta-nicotinamide adenine dinucleotide (Bennetts et al., 2012).

31.3 MODULATION AND PHARMACOLOGY OF CLC PROTEINS

Relatively few specific ligands are available for CLC proteins. The pharmacologically best characterized CLC proteins are the CLC-1 and the CLC-K channels (Gradogna and Pusch, 2010; Pusch et al., 2007). In experiments on both skeletal muscle, where it has been estimated that CLC-1 mediates around 80% of the total resting conductance, and in heterologous systems, CLC-1 is blocked with rather high affinity by 9-anthracene carboxylic acid (9-AC) (Figure 31.8a) (Astill et al., 1996;

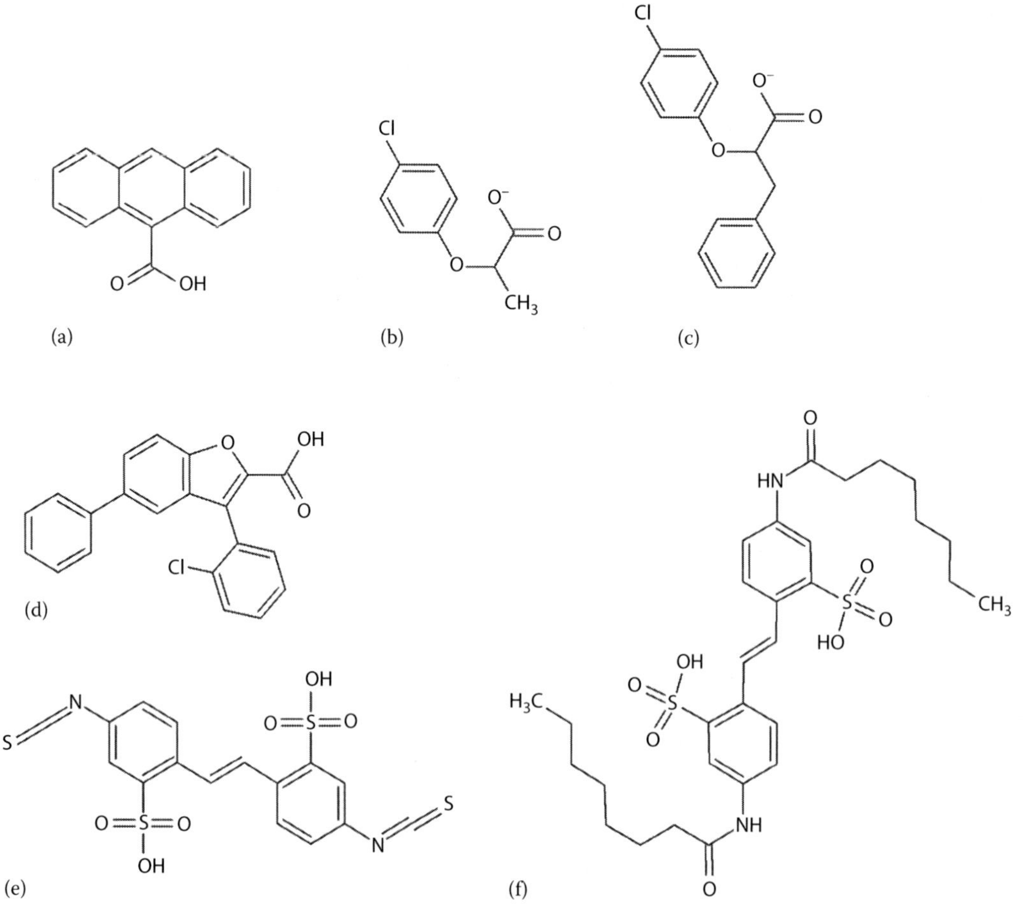

Figure 31.8 Compounds that are active on various CLC Cl^- channels and transporters. (a) 9-anthracene carboxylic acid (9-AC). (b) p-chlorophenoxy propionic acid (CPP). (c) 3-phenyl-CPP. (From Liantonio, A. et al., *J. Am. Soc. Nephrol.*, 15, 13, 2004.) (d) RT-93. (From Liantonio, A. et al., *Proc. Natl. Acad. Sci. U.S.A.*, 105, 1369, 2008.) (e) Diisothiocyano-2,2′-stilbenedisulfonic acid (DIDS). (f) OADS. (From Howery, A.E. et al., *Chem. Biol.*, 19, 1460, 2012.)

Bryant and Morales-Aguilera, 1971; Palade and Barchi, 1977; Steinmeyer et al., 1991) and by an unrelated class of compounds, clofibric acid (Figure 31.8b) and derivatives like p-chlorophenoxy-acetic acid (CPA) (Aromataris et al., 1999; De Luca et al., 1992; Pusch et al., 2000). Both types of compounds (clofibric acid and 9-AC) act from the intracellular side but, being rather hydrophobic, can reach their binding site also when applied from the extracellular side (Pusch et al., 2000, 2002). Inhibition by both types of compounds is also highly voltage-dependent with almost complete relief from block at positive voltages (Aromataris et al., 1999; Pusch et al., 2000, 2002). The mechanism of inhibition by clofibrates was studied in detail using the CLC-0 channel as a surrogate. Clofibrates inhibit the individual protopores of the double-barreled channel, compete with intracellular Cl^- ions, and have a much lower affinity for the open state of the fast gate compared with the closed state (Accardi and Pusch, 2003; Pusch et al., 2001; Traverso et al., 2003). Based on the differential sensitivity of CLC-1 and CLC-2 to 9-AC/CPA, the inhibitor binding site of 9-AC and CPA could be mapped on the CLC-1 channels in a putative binding pocket close to the putative ion permeation pathway (Estévez et al., 2003).

Zn^{2+} interacts with many proteins and inhibits several CLC-proteins (Arreola et al., 2002a; Chen, 1998; Clark et al., 1998; Duffield et al., 2005; Gradogna et al., 2012; Kürz et al., 1997, 1999; Salazar et al., 2004). The Zn^{2+} block has been studied in particular detail for CLC-0. Here, Zn^{2+} binding leads to an allosteric closure of the slow gate (Chen, 1998). Interestingly, the mutation of a single cysteine to serine, C212S, locks the slow gate completely open and abolishes the Zn^{2+} block (Lin et al., 1999). It is unclear, however, if C212 is directly involved in Zn^{2+} binding. Similar mechanisms seem to be at work also in CLC-1 (Duffield et al., 2005) and for Cd^{2+} block of CLC-2 (Zúñiga et al., 2004).

The pharmacological properties of CLC-K channels have been studied rather extensively because they are considered as potential targets for alternative diuretic drugs (Fong, 2004). While CPP (Figure 31.8b) itself is ineffective (from the outside), several derivatives of CPP (e.g., GF-100, and 3-phenyl-CPP, Figure 31.8c) were found to be relatively potent inhibitors of the CLC-K channels (K_D ~ 100 μM) (Liantonio et al., 2002, 2004) and recently benzofuran derivatives (benzofuran is a heterocyclic compound consisting of fused benzene and furan rings) such as RT-93 (Figure 31.8d) have been shown to block CLC-K channels with low micromolar affinity (Liantonio et al., 2008). Block of CLC-K channels by CPP derivatives is competitive with extracellular Cl^- (Liantonio et al., 2004) and based on the differential sensitivity of CLC-Ka and CLKC-Kb to DIDS (Figure 31.8e) and 3-phenyl-CPP the extracellular binding site could be mapped to the extracellular pore vestibule (Picollo et al., 2004). Surprisingly, niflumic acid potentiates human (but not rat) CLC-K channels at low concentrations and blocks at high concentrations (Liantonio et al., 2006; Picollo et al., 2007). However, the molecular basis for this activating effect is still unclear (Zifarelli et al., 2010).

CLC-K channels are also strongly modulated by extracellular pH, like most CLCs. However, while the pH dependence of most CLC channels is mediated (at least partially) by the protonation state of the gating glutamate (Chen and Chen, 2001; Dutzler et al., 2003; Niemeyer et al., 2009), CLC-K channels are unique in that they possess a valine at the place of the gating glutamate. Acidic pH blocks CLC-K channels (Estévez et al., 2001; Uchida et al., 1995; Waldegger et al., 2002) similar to CLC-2 (Arreola et al., 2002a). In both channels protonation of a histidine in the N-terminal part of helix Q underlies the inhibition by acidic pH (Gradogna et al, 2010; Niemeyer et al., 2009). In addition, CLC-K channels are inhibited at alkaline pH ≥ 9 by the deprotonation of the pore lysine K165 (Gradogna and Pusch, 2013).

Among CLCs, CLC-K channels are unique in that they are activated by extracellular Ca^{2+} and other divalent cations in the millimolar range (Estévez et al., 2001; Gradogna et al., 2010, 2012; Uchida et al., 1995; Waldegger et al., 2002). Ca^{2+} binds to two symmetrically related low affinity sites located on the extracellular surface of the channel. Each site is formed by four acidic residues from the loop connecting helices I and J (Figure 31.9) (Gradogna et al, 2010, 2012). Two of the residues are contributed by one subunit; the other two are from the neighboring subunit. The physiological relevance of Ca^{2+} and pH modulation of CLC-K channels remains to be demonstrated.

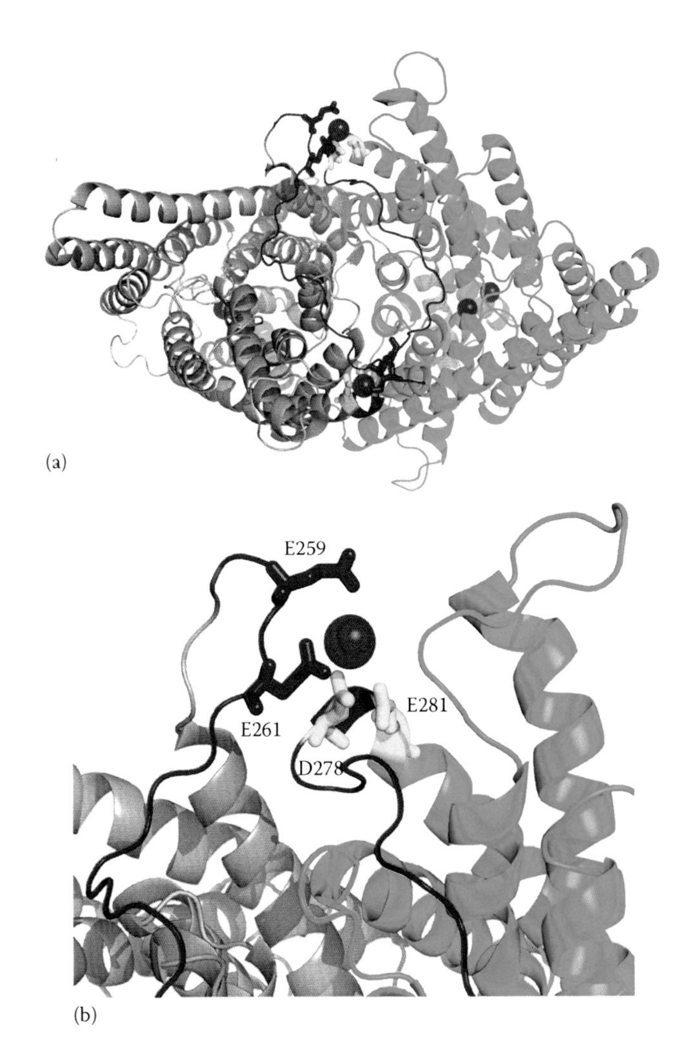

Figure 31.9 The two symmetrically localized, extracellular, inter-subunit Ca^{2+} binding sites modeled on a homology model of CLC-Ka (From Gradogna, A. et al., *J. Gen. Physiol.*, 140, 681, 2012.) The loop connecting helices I and J is colored in red, residues E259 and E261 from one subunit are shown in red and D278 and E281 from the neighboring subunit are shown in yellow. Panel (b) shows a blow up of one of the sites.

Lubiprostone is a compound used to treat constipation. It has been proposed to activate the CLC-2 channel in colonic epithelia resulting in enhanced fluid secretion (Cuppoletti et al., 2004). More recently, however, a direct action on CLC-2 and an involvement of CLC-2 in the pharmacological effects of lubiprostone could not be confirmed (Norimatsu et al., 2012).

Peptide toxins from scorpions, spiders, and other organisms have been identified as potent modulators for many ion channels (Hille, 2001). Interestingly, McCarty and colleagues described a high affinity scorpion toxin, called GaTx2, that inhibits CLC-2 from the extracellular side with subnanomolar affinity (Thompson et al., 2005, 2009). Apparently, the channel is blocked only in the closed state, making GaTx2 thus a gating modifier toxin (Thompson et al., 2009). So far, however, no follow-up studies have employed this highly promising tool.

Very few pharmacological tools exist for CLC Cl^-/H^+ antiporters, with most advances made for the model CLC transporter CLC-ec1 from *E. coli*. CLC-ec1 can be studied electrophysiologically in planar lipid bilayers (Accardi et al., 2004). However, interpretation of the bilayer results is complicated by the random insertion of the protein. In order to functionally silence the transporter inserted in a specific orientation, Maduke and colleagues tested several compounds that would inhibit CLC-ec1 specifically from one side of the membrane. They found that derivatives of DIDS (Figure 31.8e) which formed spontaneously from the hydrolysis of DIDS were potent inhibitors when applied from the intracellular side (Matulef et al., 2008; Matulef and Maduke, 2005). Recently, a rather high affinity inhibitor, called OADS (Figure 31.8f), of the *E. coli* CLC-ec1 transporter has been developed by Howery et al. (2012). OADS binds to the intracellular face of the transporter, but not within the ion permeation pathway, suggesting that it might interfere with conformational changes that occur during the transport cycle (Howery et al., 2012).

31.4 OUTLOOK

Many questions regarding CLC channels and transporters are unresolved. These include the mechanism of transport coupling in CLC exchangers, the mechanisms of slow gating, the precise origin of the voltage-dependence of CLC channels, the nature of the gating processes of CLC transporters, the precise molecular nature of the transient currents seen in CLC exchangers, the role of the cytoplasmic CBS domains, the physiological role of nucleotide regulation, the physiological role of Ca^{2+} and pH regulation of CLC-K channels, the stoichiometry and molecular nature of β-subunit interaction, and the substrates of many distantly related microbial CLC homologues. One of the most important questions for human physiology is the role of endosomal and lysosomal CLC Cl^-/H^+ exchangers. From this list, it can be seen that CLC proteins still harbor many secrets awaiting elucidation.

REFERENCES

Accardi A, Kolmakova-Partensky L, Williams C, Miller C (2004) Ionic currents mediated by a prokaryotic homologue of CLC Cl^- channels. *J Gen Physiol* **123**: 109–119.

Accardi A, Lobet S, Williams C, Miller C, Dutzler R (2006) Synergism between halide binding and proton transport in a CLC-type exchanger. *J Mol Biol* **362**: 691–699.

Accardi A, Miller C (2004) Secondary active transport mediated by a prokaryotic homologue of ClC Cl^- channels. *Nature* **427**: 803–807.

Accardi A, Picollo A (2010) CLC channels and transporters: Proteins with borderline personalities. *Biochim Biophys Acta* **1798**: 1457–1464.

Accardi A, Pusch M (2003) Conformational changes in the pore of CLC-0. *J Gen Physiol* **122**: 277–293.

Aromataris EC, Astill DS, Rychkov GY, Bryant SH, Bretag AH, Roberts ML (1999) Modulation of the gating of ClC-1 by S-(-) 2-(4-chlorophenoxy) propionic acid. *Br J Pharmacol* **126**: 1375–1382.

Arreola J, Begenisich T, Melvin JE (2002a) Conformation-dependent regulation of inward rectifier chloride channel gating by extracellular protons. *J Physiol* **541**: 103–112.

Arreola J, Begenisich T, Nehrke K, Nguyen HV, Park K, Richardson L, Yang B, Schutte BC, Lamb FS, Melvin JE (2002b) Secretion and cell volume regulation by salivary acinar cells from mice lacking expression of the Clcn3 Cl^- channel gene. *J Physiol* **545**: 207–216.

Astill DS, Rychkov G, Clarke JD, Hughes BP, Roberts ML, Bretag AH (1996) Characteristics of skeletal muscle chloride channel C1C-1 and point mutant R304E expressed in Sf-9 insect cells. *Biochim Biophys Acta* **1280**: 178–186.

Bateman A (1997) The structure of a domain common to archaebacteria and the homocystinuria disease protein. *Trends Biochem Sci* **22**: 12–13.

Bauer CK, Steinmeyer K, Schwarz JR, Jentsch TJ (1991) Completely functional double-barreled chloride channel expressed from a single *Torpedo* cDNA. *Proc Natl Acad Sci USA* **88**: 11052–11056.

Bennetts B, Parker MW, Cromer BA (2007) Inhibition of skeletal muscle CLC-1 chloride channels by low intracellular pH and ATP. *J Biol Chem* **282**: 32780–32791.

Bennetts B, Roberts ML, Bretag AH, Rychkov GY (2001) Temperature dependence of human muscle ClC-1 chloride channel. *J Physiol* **535**: 83–93.

Bennetts B, Rychkov GY, Ng H-L, Morton CJ, Stapleton D, Parker MW, Cromer BA (2005) Cytoplasmic ATP-sensing domains regulate gating of skeletal muscle ClC-1 chloride channels. *J Biol Chem* **280**: 32452–32458.

Bennetts B, Yu Y, Chen T-Y, Parker MW (2012) Intracellular beta-nicotinamide adenine dinucleotide inhibits the skeletal muscle ClC-1 chloride channel. *J Biol Chem* **287**: 25808–25820.

Bergsdorf EY, Zdebik AA, Jentsch TJ (2009) Residues important for nitrate/proton coupling in plant and mammalian CLC transporters. *J Biol Chem* **284**: 11184–11193.

Bisset D, Corry B, Chung SH (2005) The fast gating mechanism in ClC-0 channels. *Biophys J* **89**: 179–186.

Braun NA, Morgan B, Dick TP, Schwappach B (2010) The yeast CLC protein counteracts vesicular acidification during iron starvation. *J Cell Sci* **123**: 2342–2350.

Bryant SH, Morales-Aguilera A (1971) Chloride conductance in normal and myotonic muscle fibres and the action of monocarboxylic aromatic acids. *J Physiol* **219**: 367–383.

Bykova EA, Zhang XD, Chen TY, Zheng J (2006) Large movement in the C terminus of CLC-0 chloride channel during slow gating. *Nat Struct Mol Biol* **13**: 1115–1119.

Chen MF, Chen TY (2001) Different fast-gate regulation by external Cl(-) and H(+) of the muscle- type ClC chloride channels. *J Gen Physiol* **118**: 23–32.

Chen TY (1998) Extracellular zinc ion inhibits ClC-0 chloride channels by facilitating slow gating. *J Gen Physiol* **112**: 715–726.

Chen TY, Chen MF, Lin CW (2003) Electrostatic control and chloride regulation of the fast gating of ClC-0 chloride channels. *J Gen Physiol* **122**: 641–651.

Chen TY, Miller C (1996) Nonequilibrium gating and voltage dependence of the ClC-0 Cl^- channel. *J Gen Physiol* **108**: 237–250.

Clark S, Jordt SE, Jentsch TJ, Mathie A (1998) Characterization of the hyperpolarization-activated chloride current in dissociated rat sympathetic neurons. *J Physiol* **506**: 665–678.

Costa A, Gutla PV, Boccaccio A, Scholz-Starke J, Festa M, Basso B, Zanardi I, Pusch M, Schiavo FL, Gambale F, Carpaneto A (2012) The *Arabidopsis* central vacuole as an expression system for intracellular transporters: Functional characterization of the Cl^-/H^+ exchanger CLC-7. *J Physiol* **590**: 3421–3430.

Cuppoletti J, Malinowska DH, Tewari KP, Li QJ, Sherry AM, Patchen ML, Ueno R (2004) SPI-0211 activates T84 cell chloride transport and recombinant human ClC-2 chloride currents. *Am J Physiol Cell Physiol* **287**: C1173–C1183.

Davis-Kaplan SR, Askwith CC, Bengtzen AC, Radisky D, Kaplan J (1998) Chloride is an allosteric effector of copper assembly for the yeast multicopper oxidase Fet3p: an unexpected role for intracellular chloride channels. *Proc Natl Acad Sci USA* **95**: 13641–13645.

De Angeli A, Monachello D, Ephritikhine G, Frachisse JM, Thomine S, Gambale F, Barbier-Brygoo H (2006) The nitrate/proton antiporter AtCLCa mediates nitrate accumulation in plant vacuoles. *Nature* **442**: 939–942.

De Angeli A, Monachello D, Ephritikhine G, Frachisse JM, Thomine S, Gambale F, Barbier-Brygoo H (2009a) Review. CLC-mediated anion transport in plant cells. *Philos Trans R Soc Lond B Biol Sci* **364**: 195–201.

De Angeli A, Moran O, Wege S, Filleur S, Ephritikhine G, Thomine S, Barbier-Brygoo H, Gambale F (2009b) ATP binding to the C terminus of the *Arabidopsis thaliana* nitrate/proton antiporter, AtCLCa, regulates nitrate transport into plant vacuoles. *J Biol Chem* **284**: 26526–26532.

De Luca A, Tricarico D, Wagner R, Bryant SH, Tortorella V, Conte Camerino D (1992) Opposite effects of enantiomers of clofibric acid derivative on rat skeletal muscle chloride conductance: Antagonism studies and theoretical modeling of two different receptor site interactions. *J Pharmacol Exp Ther* **260**: 364–368.

De Stefano S, Pusch M, Zifarelli G (2011) Extracellular determinants of anion discrimination of the Cl^-/H^+ antiporter CLC-5. *J Biol Chem* **286**: 44134–44144.

De Stefano S, Pusch M, Zifarelli G (2013) A single point mutation reveals gating of the human CLC-5 Cl(-)/H(+) antiporter. *J Physiol* **591**: 5879–5893.

Denton J, Nehrke K, Yin X, Morrison R, Strange K (2005) GCK-3, a newly identified Ste20 kinase, binds to and regulates the activity of a cell cycle-dependent ClC anion channel. *J Gen Physiol* **125**: 113–125.

Duffield MD, Rychkov GY, Bretag AH, Roberts ML (2003) Involvement of helices at the dimer Interface in ClC-1 common gating. *J Gen Physiol* **121**: 149–161.

Duffield MD, Rychkov GY, Bretag AH, Roberts ML (2005) Zinc inhibits human ClC-1 muscle chloride channel by interacting with its common gating mechanism. *J Physiol* **568**: 5–12.

Dutzler R (2004) The structural basis of ClC chloride channel function. *Trends Neurosci* **27**: 315–320.

Dutzler R, Campbell EB, Cadene M, Chait BT, MacKinnon R (2002) X-ray structure of a ClC chloride channel at 3.0 Å reveals the molecular basis of anion selectivity. *Nature* **415**: 287–294.

Dutzler R, Campbell EB, MacKinnon R (2003) Gating the selectivity filter in ClC chloride channels. *Science* **300**: 108–112.

Estévez R, Boettger T, Stein V, Birkenhäger R, Otto E, Hildebrandt F, Jentsch TJ (2001) Barttin is a Cl^- channel beta-subunit crucial for renal Cl^- reabsorption and inner ear K^+ secretion. *Nature* **414**: 558–561.

Estévez R, Jentsch TJ (2002) CLC chloride channels: Correlating structure with function. *Curr Opin Struct Biol* **12**: 531–539.

Estévez R, Pusch M, Ferrer-Costa C, Orozco M, Jentsch TJ (2004) Functional and structural conservation of CBS domains from CLC channels. *J Physiol* **557**: 363–378.

Estévez R, Schroeder BC, Accardi A, Jentsch TJ, Pusch M (2003) Conservation of chloride channel structure revealed by an inhibitor binding site in ClC-1. *Neuron* **38**: 47–59.

Falin RA, Morrison R, Ham AJ, Strange K (2009) Identification of regulatory phosphorylation sites in a cell volume- and Ste20 kinase-dependent ClC anion channel. *J Gen Physiol* **133**: 29–42.

Feng L, Campbell EB, Hsiung Y, Mackinnon R (2010) Structure of a eukaryotic CLC transporter defines an intermediate state in the transport cycle. *Science* **330**: 635–641.

Feng L, Campbell EB, MacKinnon R (2012) Molecular mechanism of proton transport in CLC Cl^-/H^+ exchange transporters. *Proc Natl Acad Sci USA* **109**: 11699–11704.

Fong P (2004) CLC-K channels: If the drug fits, use it. *EMBO Rep* **5**: 565–566.

Fong P, Rehfeldt A, Jentsch TJ (1998) Determinants of slow gating in ClC-0, the voltage-gated chloride channel of *Torpedo marmorata*. *Am J Physiol* **274**: C966–C973.

Gaxiola RA, Yuan DS, Klausner RD, Fink GR (1998) The yeast CLC chloride channel functions in cation homeostasis. *Proc Natl Acad Sci USA* **95**: 4046–4050.

Gradogna A, Babini E, Picollo A, Pusch M (2010) A regulatory calcium-binding site at the subunit interface of CLC-K kidney chloride channels. *J Gen Physiol* **136**: 311–323.

Gradogna A, Fenollar-Ferrer C, Forrest LR, Pusch M (2012) Dissecting a regulatory calcium-binding site of CLC-K kidney chloride channels. *J Gen Physiol* **140**: 681–696.

Gradogna A, Pusch M (2010) Molecular pharmacology of kidney and inner ear CLC-K chloride channels. *Front Pharmacol* **1**: 130.

Gradogna A, Pusch M (2013) Alkaline pH block of CLC-K kidney chloride channels mediated by a pore lysine residue. *Biophys J* **105**: 80–90.

Graves AR, Curran PK, Smith CL, Mindell JA (2008) The Cl^-/H^+ antiporter ClC-7 is the primary chloride permeation pathway in lysosomes. *Nature* **453**: 788–792.

Greene JR, Brown NH, DiDomenico BJ, Kaplan J, Eide DJ (1993) The GEF1 gene of *Saccharomyces cerevisiae* encodes an integral membrane protein; mutations in which have effects on respiration and iron-limited growth. *Mol Gen Genet* **241**: 542–553.

Hanke W, Miller C (1983) Single chloride channels from *Torpedo* electroplax. Activation by protons. *J Gen Physiol* **82**: 25–45.

Hechenberger M, Schwappach B, Fischer WN, Frommer WB, Jentsch TJ, Steinmeyer K (1996) A family of putative chloride channels from Arabidopsis and functional complementation of a yeast strain with a CLC gene disruption. *J Biol Chem* **271**: 33632–33638.

Hilgemann DW (1996) Unitary cardiac Na^+, Ca^{2+} exchange current magnitudes determined from channel-like noise and charge movements of ion transport. *Biophys J* **71**: 759–768.

Hille B (2001) *Ion Channels of Excitable Membranes*, 3rd edn. Sunderland, MA: Sinauer.

Howery AE, Elvington S, Abraham SJ, Choi KH, Dworschak-Simpson S, Phillips S, Ryan CM et al. (2012) A designed inhibitor of a CLC antiporter blocks function through a unique binding mode. *Chem Biol* **19**: 1460–1470.

Ishida Y, Nayak S, Mindell JA, Grabe M (2013) A model of lysosomal pH regulation. *J Gen Physiol* 141: 705–720.

Iyer R, Iverson TM, Accardi A, Miller C (2002) A biological role for prokaryotic ClC chloride channels. *Nature* **419**: 715–718.

Jentsch TJ (2002) Chloride channels are different. *Nature* **415**: 276–277.

Jentsch TJ (2008) CLC chloride channels and transporters: From genes to protein structure, pathology and physiology. *Crit Rev Biochem Mol Biol* **43**: 3–36.

Jentsch TJ, Friedrich T, Schriever A, Yamada H (1999) The CLC chloride channel family. *Pflügers Arch* **437**: 783–795.

Jentsch TJ, Poët M, Fuhrmann JC, Zdebik AA (2005) Physiological functions of CLC Cl channels gleaned from human genetic disease and mouse models. *Ann Rev Physiol* **67**: 779–807.

Jentsch TJ, Pusch M, Rehfeldt A, Steinmeyer K (1993) The ClC family of voltage-gated chloride channels: Structure and function. *Ann N Y Acad Sci* **707**: 285–293.

Jentsch TJ, Steinmeyer K, Schwarz G (1990) Primary structure of *Torpedo marmorata* chloride channel isolated by expression cloning in *Xenopus* oocytes. *Nature* **348**: 510–514.

Jeworutzki E, López-Hernández T, Capdevila-Nortes X, Sirisi S, Bengtsson L, Montolio M, Zifarelli G et al. (2012) GlialCAM, a protein defective in a leukodystrophy, serves as a ClC-2 Cl^- channel auxiliary subunit. *Neuron* **73**: 951–961.

Kürz L, Wagner S, George AL, Jr., Rüdel R (1997) Probing the major skeletal muscle chloride channel with Zn^{2+} and other sulfhydryl-reactive compounds. *Pflügers Arch* **433**: 357–363.

Kürz LL, Klink H, Jakob I, Kuchenbecker M, Benz S, Lehmann-Horn F, Rüdel R (1999) Identification of three cysteines as targets for the Zn^{2+} blockade of the human skeletal muscle chloride channel. *J Biol Chem* **274**: 11687–11692.

Lange PF, Wartosch L, Jentsch TJ, Fuhrmann JC (2006) ClC-7 requires Ostm1 as a beta-subunit to support bone resorption and lysosomal function. *Nature* **440**: 220–223.

Leisle L, Ludwig CF, Wagner FA, Jentsch TJ, Stauber T (2011) ClC-7 is a slowly voltage-gated 2Cl(–)/1H(+)-exchanger and requires Ostm1 for transport activity. *EMBO J* **30**: 2140–2152.

Liantonio A, Accardi A, Carbonara G, Fracchiolla G, Loiodice F, Tortorella P, Traverso S et al. (2002) Molecular requisites for drug binding to muscle CLC-1 and renal CLC-K channel revealed by the use of phenoxy-alkyl derivatives of 2-(p-chlorophenoxy)propionic acid. *Mol Pharmacol* **62**: 265–271.

Liantonio A, Picollo A, Babini E, Carbonara G, Fracchiolla G, Loiodice F, Tortorella V, Pusch M, Camerino DC (2006) Activation and inhibition of kidney CLC-K chloride channels by fenamates. *Mol Pharmacol* **69**: 165–173.

Liantonio A, Picollo A, Carbonara G, Fracchiolla G, Tortorella P, Loiodice F, Laghezza A et al. (2008) Molecular switch for CLC-K Cl^- channel block/activation: Optimal pharmacophoric requirements towards high-affinity ligands. *Proc Natl Acad Sci USA* **105**: 1369–1373.

Liantonio A, Pusch M, Picollo A, Guida P, De Luca A, Pierno S, Fracchiolla G, Loiodice F, Tortorella P, Conte Camerino D (2004) Investigations of pharmacologic properties of the renal CLC-K1 chloride channel co-expressed with barttin by the use of 2-(p-Chlorophenoxy)propionic acid derivatives and other structurally unrelated chloride channels blockers. *J Am Soc Nephrol* **15**: 13–20.

Lim HH, Shane T, Miller C (2012) Intracellular proton access in a Cl(–)/H(+) antiporter. *PLoS Biol* **10**: e1001441.

Lin YW, Lin CW, Chen TY (1999) Elimination of the slow gating of ClC-0 chloride channel by a point mutation. *J Gen Physiol* **114**: 1–12.

Lippiat JD, Smith AJ (2012) The CLC-5 2Cl(–)/H(+) exchange transporter in endosomal function and Dent's disease. *Front Physiol* 3: 449.

Lísal J, Maduke M (2008) The ClC-0 chloride channel is a 'broken' Cl^-/H^+ antiporter. *Nat Struct Mol Biol* **15**: 805–810.

Ludewig U, Jentsch TJ, Pusch M (1997) Analysis of a protein region involved in permeation and gating of the voltage-gated *Torpedo* chloride channel ClC-0. *J Physiol* **498**: 691–702.

Ludewig U, Pusch M, Jentsch TJ (1996) Two physically distinct pores in the dimeric ClC-0 chloride channel. *Nature* **383**: 340–343.

Lurin C, Geelen D, Barbier-Brygoo H, Guern J, Maurel C (1996) Cloning and functional expression of a plant voltage-dependent chloride channel. *Plant Cell* **8**: 701–711.

Lv Q-d, Tang R-j, Liu H, Gao X-s, Li Y-z, Zheng H-q, Zhang H-x (2009) Cloning and molecular analyses of the *Arabidopsis thaliana* chloride channel gene family. *Plant Science* **176**: 650–661.

Ma L, Rychkov GY, Bykova EA, Zheng J, Bretag AH (2011) Movement of hClC-1 C-termini during common gating and limits on their cytoplasmic location. *Biochem J* **436**: 415–428.

Maduke M, Miller C, Mindell JA (2000) A decade of CLC chloride channels: Structure, mechanism, and many unsettled questions. *Annu Rev Biophys Biomol Struct* **29**: 411–438.

Maduke M, Pheasant DJ, Miller C (1999) High-level expression, functional reconstitution, and quaternary structure of a prokaryotic ClC-type chloride channel. *J Gen Physiol* **114**: 713–722.

Matulef K, Howery AE, Tan L, Kobertz WR, Du Bois J, Maduke M (2008) Discovery of potent CLC chloride channel inhibitors. *ACS Chem Biol* **3**: 419–428.

Matulef K, Maduke M (2005) Side-dependent inhibition of a prokaryotic ClC by DIDS. *Biophys J* **89**: 1721–1730.

Meyer S, Savaresi S, Forster IC, Dutzler R (2007) Nucleotide recognition by the cytoplasmic domain of the human chloride transporter ClC-5. *Nat Struct Mol Biol* **14**: 60–67.

Middleton RE, Pheasant DJ, Miller C (1996) Homodimeric architecture of a ClC-type chloride ion channel. *Nature* **383**: 337–340.

Miller C (1982) Open-state substructure of single chloride channels from *Torpedo electroplax. Philos Trans R Soc Lond B Biol Sci* **299**: 401–411.

Miller C (2006) ClC chloride channels viewed through a transporter lens. *Nature* **440**: 484–489.

Miller C, Richard EA (1990) The voltage-dependent chloride channel of *Torpedo Electroplax*. Intimations of molecular structure from quirks of single-channel function. In *Chloride Channels and Carriers in Nerve, Muscle and Glial Cells*, Alvarez-Leefmans FJ, Russell JM (eds.), pp 383–405. New York: Plenum.

Miller C, White MM (1980) A voltage-dependent chloride conductance channel from *Torpedo* electroplax membrane. *Ann NY Acad Sci* **341**: 534–551.

Miller C, White MM (1984) Dimeric structure of single chloride channels from *Torpedo* electroplax. *Proc Natl Acad Sci U S A* **81**: 2772–2775.

Mindell JA, Maduke M (2001) ClC chloride channels. *Genome Biol* **2**: REVIEWS3003.

Nguitragool W, Miller C (2006) Uncoupling of a CLC Cl(–)/H(+) exchange transporter by polyatomic anions. *J Mol Biol* **362**: 682–690.

Niemeyer MI, Cid LP, Yusef YR, Briones R, Sepúlveda FV (2009) Voltage-dependent and -independent titration of specific residues accounts for complex gating of a ClC chloride channel by extracellular protons. *J Physiol* **587**: 1387–1400.

Niemeyer MI, Cid LP, Zúñiga L, Catalán M, Sepúlveda FV (2003) A conserved pore-lining glutamate as a voltage- and chloride-dependent gate in the ClC-2 chloride channel. *J Physiol* **553**: 873–879.

Norimatsu Y, Moran AR, MacDonald KD (2012) Lubiprostone activates CFTR, but not ClC-2, via the prostaglandin receptor (EP(4)). *Biochem Biophys Res Commun* **426**: 374–379.

Palade P, Barchi R (1977) On the inhibition of muscle membrane chloride conductance by aromatic carboxylic acids. *J Gen Physiol* **69**: 879–896.

Phillips S, Brammer Ashley E, Rodriguez L, Lim H-H, Stary-Weinzinger A, Matulef K (2012) Surprises from an Unusual CLC Homolog. *Biophys J* **103**: L44–L46.

Picollo A, Liantonio A, Babini E, Camerino DC, Pusch M (2007) Mechanism of interaction of niflumic acid with heterologously expressed kidney CLC-K chloride channels. *J Membr Biol* **216**: 73–82.

Picollo A, Liantonio A, Didonna MP, Elia L, Camerino DC, Pusch M (2004) Molecular determinants of differential pore blocking of kidney CLC-K chloride channels. *EMBO Rep* **5**: 584–589.

Picollo A, Malvezzi M, Houtman JC, Accardi A (2009) Basis of substrate binding and conservation of selectivity in the CLC family of channels and transporters. *Nat Struct Mol Biol* **16**: 1294–1301.

Picollo A, Xu Y, Johner N, Bernèche S, Accardi A (2012) Synergistic substrate binding determines the stoichiometry of transport of a prokaryotic H(+)/Cl(-) exchanger. *Nat Struct Mol Biol* **19**: 525–531.

Ponting CP (1997) CBS domains in CIC chloride channels implicated in myotonia and nephrolithiasis (kidney stones). *J Mol Med* **75**: 160–163.

Pusch M (2002) Myotonia caused by mutations in the muscle chloride channel gene CLCN1. *Hum Mutat* **19**: 423–434.

Pusch M, Accardi A, Liantonio A, Ferrera L, De Luca A, Camerino DC, Conti F (2001) Mechanism of block of single protopores of the *Torpedo* chloride channel ClC-0 by 2-(p-chlorophenoxy)butyric acid (CPB). *J Gen Physiol* **118**: 45–62.

Pusch M, Accardi A, Liantonio A, Guida P, Traverso S, Camerino DC, Conti F (2002) Mechanisms of block of muscle type CLC chloride channels (Review). *Mol Membr Biol* **19**: 285–292.

Pusch M, Jordt SE, Stein V, Jentsch TJ (1999) Chloride dependence of hyperpolarization-activated chloride channel gates. *J Physiol* **515**: 341–353.

Pusch M, Liantonio A, Bertorello L, Accardi A, De Luca A, Pierno S, Tortorella V, Camerino DC (2000) Pharmacological characterization of chloride channels belonging to the ClC family by the use of chiral clofibric acid derivatives. *Mol Pharmacol* **58**: 498–507.

Pusch M, Liantonio A, De Luca A, Conte Camerino D (2007) Pharmacology of CLC chloride channels and transporters. In *Chloride Transport across Biological Membranes*, Pusch M (ed), Vol. 38, 4, pp 83–108. Amsterdam, the Netherlands: Elsevier.

Pusch M, Ludewig U, Jentsch TJ (1997) Temperature dependence of fast and slow gating relaxations of ClC-0 chloride channels. *J Gen Physiol* **109**: 105–116.

Pusch M, Ludewig U, Rehfeldt A, Jentsch TJ (1995) Gating of the voltage-dependent chloride channel CIC-0 by the permeant anion. *Nature* **373**: 527–531.

Richard EA, Miller C (1990) Steady-state coupling of ion-channel conformations to a transmembrane ion gradient. *Science* **247**: 1208–1210.

Robertson JL, Kolmakova-Partensky L, Miller C (2010) Design, function and structure of a monomeric ClC transporter. *Nature* **468**: 844–847.

Rutledge E, Bianchi L, Christensen M, Boehmer C, Morrison R, Broslat A, Beld AM, George AL, Greenstein D, Strange K (2001) CLH-3, a ClC-2 anion channel ortholog activated during meiotic maturation in *C. elegans* oocytes. *Curr Biol* **11**: 161–170.

Rutledge E, Denton J, Strange K (2002) Cell cycle- and swelling-induced activation of a *Caenorhabditis elegans* ClC channel is mediated by CeGLC-7alpha/beta phosphatases. *J Cell Biol* **158**: 435–444. Epub 2002 August 2005.

Rychkov GY, Pusch M, Roberts ML, Jentsch TJ, Bretag AH (1998) Permeation and block of the skeletal muscle chloride channel, ClC-1, by foreign anions. *J Gen Physiol* **111**: 653–665.

Salazar G, Love R, Styers ML, Werner E, Peden A, Rodriguez S, Gearing M, Wainer BH, Faundez V (2004) AP-3-dependent mechanisms control the targeting of a chloride channel (ClC-3) in neuronal and non-neuronal cells. *J Biol Chem* **279**: 25430–25439.

Sanchez-Rodriguez JE, De Santiago-Castillo JA, Arreola J (2010) Permeant anions contribute to voltage dependence of ClC-2 chloride channel by interacting with the protopore gate. *J Physiol* **588**: 2545–2556.

Sanchez-Rodriguez JE, De Santiago-Castillo JA, Contreras-Vite JA, Nieto-Delgado PG, Castro-Chong A, Arreola J (2012) Sequential interaction of chloride and proton ions with the fast gate steer the voltage-dependent gating in ClC-2 chloride channels. *J Physiol* **590**: 4239–4253.

Schriever AM, Friedrich T, Pusch M, Jentsch TJ (1999) CLC chloride channels in *Caenorhabditis elegans*. *J Biol Chem* **274**: 34238–34244.

Schwappach B, Stobrawa S, Hechenberger M, Steinmeyer K, Jentsch TJ (1998) Golgi localization and functionally important domains in the NH2 and COOH terminus of the yeast CLC putative chloride channel Gef1p. *J Biol Chem* **273**: 15110–15118.

Scott JW, Hawley SA, Green KA, Anis M, Stewart G, Scullion GA, Norman DG, Hardie DG (2004) CBS domains form energy-sensing modules whose binding of adenosine ligands is disrupted by disease mutations. *J Clin Invest* **113**: 274–284.

Smith AJ, Lippiat JD (2010) Voltage-dependent charge movement associated with activation of the CLC-5 $2Cl^-/1H^+$ exchanger. *FASEB J* **24**: 3696–3705.

Steinmeyer K, Ortland C, Jentsch TJ (1991) Primary structure and functional expression of a developmentally regulated skeletal muscle chloride channel. *Nature* **354**: 301–304.

Stockbridge RB, Lim HH, Otten R, Williams C, Shane T, Weinberg Z, Miller C (2012) Fluoride resistance and transport by riboswitch-controlled CLC antiporters. *Proc Natl Acad Sci USA* **109**: 15289–15294.

Strange K (2011) Putting the pieces together: A crystal clear window into CLC anion channel regulation. *Channels (Austin)* **5**: 101–105.

Thompson CH, Fields DM, Olivetti PR, Fuller MD, Zhang ZR, Kubanek J, McCarty NA (2005) Inhibition of ClC-2 chloride channels by a peptide component or components of scorpion venom. *J Membr Biol* **208**: 65–76.

Thompson CH, Olivetti PR, Fuller MD, Freeman CS, McMaster D, French RJ, Pohl J, Kubanek J, McCarty NA (2009) Isolation and characterization of a high affinity peptide inhibitor of ClC-2 chloride channels. *J Biol Chem* **284**: 26051–26062.

Traverso S, Elia L, Pusch M (2003) Gating competence of constitutively open CLC-0 mutants revealed by the interaction with a small organic Inhibitor. *J Gen Physiol* **122**: 295–306.

Traverso S, Zifarelli G, Aiello R, Pusch M (2006) Proton sensing of CLC-0 mutant E166D. *J Gen Physiol* **127**: 51–66.

Uchida S, Sasaki S, Nitta K, Uchida K, Horita S, Nihei H, Marumo F (1995) Localization and functional characterization of rat kidney-specific chloride channel, ClC-K1. *J Clin Invest* **95**: 104–113.

Waldegger S, Jeck N, Barth P, Peters M, Vitzthum H, Wolf K, Kurtz A, Konrad M, Seyberth HW (2002) Barttin increases surface expression and changes current properties of ClC-K channels. *Pflügers Arch* **444**: 411–418.

Ward JM, Maser P, Schroeder JI (2009) Plant ion channels: Gene families, physiology, and functional genomics analyses. *Annu Rev Physiol* **71**: 59–82.

Weinert S, Jabs S, Supanchart C, Schweizer M, Gimber N, Richter M, Rademann J, Stauber T, Kornak U, Jentsch TJ (2010) Lysosomal pathology and osteopetrosis upon loss of H^+-driven lysosomal Cl^- accumulation. *Science* **328**: 1401–1403.

Zdebik AA, Zifarelli G, Bergsdorf EY, Soliani P, Scheel O, Jentsch TJ, Pusch M (2008) Determinants of anion-proton coupling in mammalian endosomal CLC proteins. *J Biol Chem* **283**: 4219–4227.

Zhang XD, Tseng PY, Chen TY (2008) ATP inhibition of CLC-1 is controlled by oxidation and reduction. *J Gen Physiol* **132**: 421–428.

Zifarelli G, De Stefano S, Zanardi I, Pusch M (2012) On the mechanism of gating charge movement of ClC-5, a human Cl^-/H^+ antiporter. *Biophys J* **102**: 2060–2069.
Zifarelli G, Liantonio A, Gradogna A, Picollo A, Gramegna G, De Bellis M, Murgia AR, Babini E, Camerino DC, Pusch M (2010) Identification of sites responsible for the potentiating effect of niflumic acid on ClC-Ka kidney chloride channels. *Br J Pharmacol* **160**: 1652–1661.
Zifarelli G, Murgia AR, Soliani P, Pusch M (2008) Intracellular proton regulation of ClC-0. *J Gen Physiol* **132**: 185–198.
Zifarelli G, Pusch M (2007) CLC chloride channels and transporters: A biophysical and physiological perspective. *Rev Physiol Biochem Pharmacol* **158**: 23–76.
Zifarelli G, Pusch M (2008) The muscle chloride channel ClC-1 is not directly regulated by intracellular ATP. *J Gen Physiol* **131**: 109–116.
Zifarelli G, Pusch M (2009a) Conversion of the 2 Cl(–)/1 H(+) antiporter ClC-5 in a NO(3)(–)/H(+) antiporter by a single point mutation. *EMBO J* **28**: 175–182.
Zifarelli G, Pusch M (2009b) Intracellular regulation of human ClC-5 by adenine nucleotides. *EMBO Rep* **10**: 1111–1116.
Zifarelli G, Pusch M (2010) CLC transport proteins in plants. *FEBS Lett* **584**: 2122–2127.
Zúñiga L, Niemeyer MI, Varela D, Catalán M, Cid LP, Sepúlveda FV (2004) The voltage-dependent ClC-2 chloride channel has a dual gating mechanism. *J Physiol* **555**: 671–682.

32 Ca-activated chloride channels

Xiuming Wong and Lily Jan

Contents

32.1 INTRODUCTION

Eukaryotes have evolved sophisticated and ubiquitous calcium signaling pathways, and correspondingly, calcium-activated chloride channels (CaCCs) are implicated in physiological processes in a multitude of organisms and tissue types.

CaCCs are present in ancient life forms such as green algae, where they are involved in action potential generation in the genera *Chara* and *Nitella* (Fromm and Lautner, 2007; Shiina and Tazawa, 1987). Highly evolved and complex systems also express these channels. Amphibian physiology is a prominent example, having provided many of the early descriptions of CaCCs. Endogenous CaCCs were observed in *Xenopus laevis* (Miledi and Parker, 1984) and *Rana pipiens* oocytes (Cross, 1981), where they prevent polyspermy by activating upon postfertilization intracellular calcium release, depolarizing the oocyte to block additional sperm entry. Sensory tissues like salamander retina photoreceptors (Bader et al., 1982) and frog olfactory cilia (Kleene and Gesteland, 1991) have also been reported to express CaCCs.

CaCC expression in sensory tissue is also a feature of mammalian physiology. In the nervous system, CaCCs were observed in rat dorsal root ganglion (DRG) (Mayer, 1985) where they were presumed to regulate somatosensation. CaCCs were additionally reported to be upregulated in a subpopulation of DRG neurons upon axotomy (Andre et al., 2003). Secretory epithelia like rat lachrymal glands (Evans and Marty, 1986) and human airway epithelia (Anderson and Welsh, 1991) were reported to express CaCCs, which control chloride movement and thus the vectorial transport of electrolytes and fluid in these

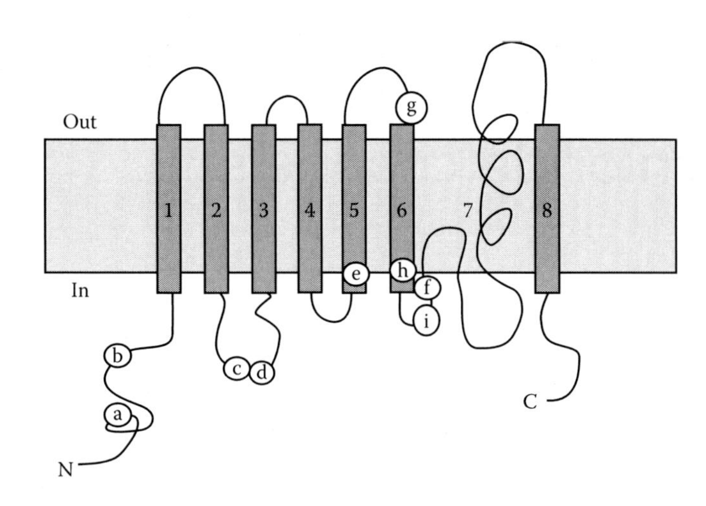

Figure 32.1 Topology of TMEM16A. (Adapted from Yu, K. et al., *Circ. Res.*, 110, 990, 2012.) with TMDs labeled 1–8. N and C termini of TMEM16A are predicted to be intracellular. (IN: intracellular space, OUT: extracellular space). (a) Dimerization domain described in Tien et al. (2012); (b) alternatively spliced exon affecting calcium sensitivity; (c) *calcium bowl* that affects voltage dependence. (From Ferrera, L. et al., *J. Biol. Chem.*, 284, 33360, 2009; Xiao, Q. et al., *Proc. Natl. Acad. Sci. U.S.A.*, 108, 8891, 2011.); (d) EAVK motif that affects voltage dependence; (e) lysine or glutamine residue that affects TMEM16A anionic and TMEM16F cationic selectivity respectively. (From Yang, H. et al., *Cell*, 151, 111, 2012.); (f) glutamate residues that affect calcium sensitivity in TMEM16A and TMEM16F. (From Yu, K. et al., *Circ. Res.*, 110, 990, 2012; Yang, H. et al., *Cell*, 151, 111, 2012.); (g–i) lysine residues reported to affect ionic selectivity of TMEM16A. (From Yang, Y.D. et al., *Nature*, 455, 1210, 2008.).

tissues. The upregulation of CaCCs in the airway epithelia of cystic fibrosis mouse models prompted interest in the role of these channels in airway disease (Grubb et al., 1994). CaCC expression in canine and guinea pig trachea smooth muscle (Janssen and Sims, 1992) and rat portal vein smooth muscle (Pacaud et al., 1989) suggests a role for these channels in smooth muscle tone regulation. Their expression in pulmonary arterial endothelial cells (Nilius et al., 1997), human mesenteric artery (Klockner, 1993), and rabbit ventricular myocytes (Zygmunt and Gibbons, 1991) implicates the channels in cardiovascular regulation.

In contrast to the long history and a hefty corpus of literature on the CaCC, molecular identification has been relatively recent. This was largely achieved by exploiting the known or predicted functions of these channels. Monitoring gene expression in a microarray of human bronchial epithelial cells treated with the inflammatory cytokines IL4/IL13 allowed identification of the upregulated CaCC-encoding gene (Caputo et al., 2008). In another report, fractionated *Xenopus* oocyte mRNA was injected into the oocytes of Axolotl, a polyspermic species lacking endogenous CaCCs, and the mRNA fraction giving rise to CaCCs in Axolotl oocytes allowed molecular identification of the CaCC (Schroeder et al., 2008). Together with a study prompted by a bioinformatics survey (Yang et al., 2008), three groups identified TMEM16A as the gene responsible for the classic CaCC. This has opened up the CaCC field to intensive molecular and functional dissection. A highly homologous family member, TMEM16B, was also found to be a CaCC, as will be discussed later.

Before the discovery of TMEM16A as a CaCC, the TMEM16 family was assigned to an eukaryotic protein superfamily known as DUF (domain of unknown function) 590, which corresponds to the series of eight predicted transmembrane domains (TMDs) and the short putative cytosolic loops between them (Figure 32.1). In general, there is less sequence conservation away from the TMDs. Pairwise alignment of various TMEM16 members from yeast to human on the PANDIT database gives an average of 28% identity in the DUF590 domain (http://www.ebi.ac.uk/research/goldman/software/pandit; Whelan et al., 2003). Mammals possess 10 TMEM16 members (Figure 32.2), and it remains to be determined whether the biophysical and functional properties of CaCCs are conserved throughout the TMEM16 family in mammals and beyond.

This chapter is a summary of literature on the TMEM16 family since TMEM16A was identified as a CaCC. Based on the anionic selectivity and eight putative TMDs of the founding member, the TMEM16 family has also been named the anoctamin family. Given uncertainties of the ionic selectivity and basic topology of this new family of proteins, this review will utilize the TMEM16 nomenclature. The bestrophins, a biophysically and functionally distinct major class of CaCCs, have been given excellent review elsewhere (Xiao et al., 2010).

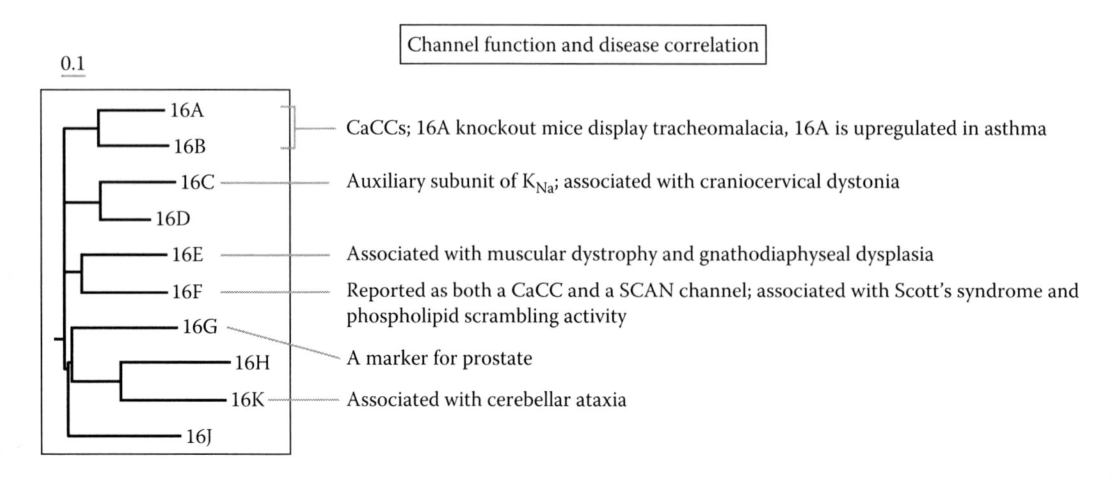

Figure 32.2 Phylogenetic diagram of the mammalian TMEM16 family. Alignment generated with ClustalW2 and phylogenetic tree generated with PHYLIP 3.67. Associations with channel activity and disease correlations are marked on the right. CaCC: calcium-activated chloride channel; SCAN channel: small-conductance calcium-activated nonselective cation channel.

32.2 TMEM16A ENCODES THE CLASSIC CaCC

Electrophysiological data from early studies has endowed us with a thorough biophysical description of the classic CaCC: currents arise from small-conductance channels activated by submicromolar concentrations of cytosolic calcium in a voltage- and time-dependent manner, and this voltage and time dependence is significantly reduced at higher concentrations of calcium. The classic CaCC was also observed to have modest ionic selectivity, with Na^+ only 10 times less permeable than Cl^-. Bulkier anions such as SCN^-, I^-, and Br^- have greater permeability than Cl^-.

TMEM16A recapitulates all of these properties. The EC_{50} of calcium-dependent activation for TMEM16A was shown to be 2.6 μM at −60 mV and 0.3 μM at +60 mV (Yang et al., 2008) (Figure 32.3). TMEM16A single-channel conductance was determined to be 8.6 pS in HEK 293 cells, similar to that reported for endogenous CaCCs in early studies (Yang et al., 2008). Reported sodium:chloride permeability ratios (P_{Na}/P_{Cl}) of TMEM16A range from 0.03 (Yang et al., 2008) to 0.14 (Yang et al., 2012). The preference for larger anions ($SCN^- > I^- > Br^- > Cl^-$) and the pharmacological profile of the classic CaCC is also generally recapitulated by TMEM16A expressed in HEK 293 cells (Yang et al., 2008) and Axolotl oocytes (Schroeder et al., 2008). These biophysical properties will be discussed in greater detail in later sections. Another property of classic CaCCs is the irrecoverable loss of currents over time in excised patches (Kuruma and Hartzell, 2000) or in whole-cell patch clamp (Ayon et al., 2009), suggesting the loss of a diffusible factor critical for channel activity. This run-down phenomenon was discussed in the characterization of TMEM16B, but the diffusible factor required for channel activity has yet to be identified (Pifferi et al., 2009).

32.3 BIOPHYSICAL PROPERTIES OF TMEM16A

32.3.1 TOPOLOGY

Bioinformatics hydropathy analysis predicts TMEM16A to have eight TMDs with cytosolic N and C termini. Early models of TMEM16A topology were based on a hemagglutinin (HA)-epitope insertion accessibility study of a different TMEM16 family member, TMEM16G (Das et al., 2008). Since then, HA-tag insertion and cysteine accessibility mutagenesis have been performed on TMEM16A itself, leading to the proposal of a different topological model (Figure 32.1). Relative to the older model, this model re-assigns the location of the sixth TMD, with ambiguity between the sixth and seventh TMDs (illustrated with free-form drawing in Figure 32.1) due to the lack of viable channels upon HA-tag insertion into this region (Yu et al., 2012). It remains to be determined how applicable these topological models are across the TMEM16 family.

32.3.2 CALCIUM AND VOLTAGE SENSING

At low calcium concentrations, TMEM16A and TMEM16B are preferentially activated at positive membrane potentials, whereas positive membrane potentials confer upon the channels higher calcium sensitivity (Figure 32.3). It is thus difficult to discuss calcium binding and voltage sensing separately, and it is likely that these two characteristics physically couple to activate the channel.

TMEM16A is rather sensitive to calcium, activating at submicromolar concentrations. The Hill coefficient of calcium binding for TMEM16A is greater than one, suggesting that multiple calcium ions are required to activate the channel

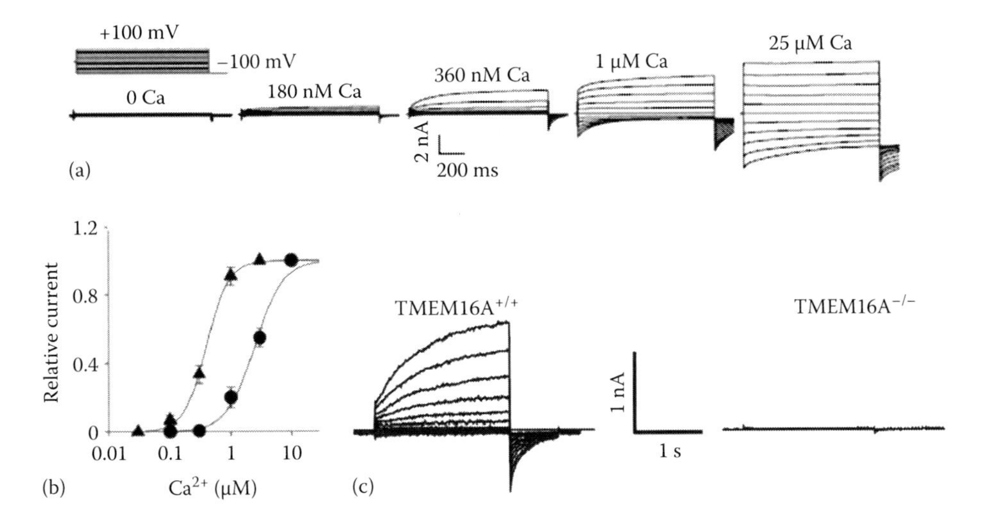

Figure 32.3 (a) TMEM16A is a calcium-dependent channel. In TMEM16A-transfected HEK293 cells, no currents are observed in whole-cell patch clamp experiments when zero calcium is added to the pipette solution. At low calcium concentrations, small time-dependent and outwardly rectifying currents are observed. Increasing calcium concentration in the pipette increases the amplitude of the current and abolishes rectification. High calcium concentrations also abolish time dependence, suggesting a constitutively open channel. (From Xiao, Q. et al., *Proc. Natl. Acad. Sci. U.S.A.*, 108, 8891, 2011.) (b) Calcium sensitivity is affected by voltage. Recordings from excised patches of TMEM16A-transfected HEK293 cells done at +60 mV (triangles) and −60 mV (circles) show that positive voltages increase calcium sensitivity. (From Yang, Y.D. et al., *Nature*, 455, 1210, 2008.) (c) TMEM16A-dependent currents in the salivary gland acinar cells. Recordings from the TMEM16A knockout mouse show little to no current, evidence that TMEM16A is indeed responsible for the CaCC in these cells. (From Romanenko, V.G. et al., *J. Biol. Chem.*, 285, 12990, 2010.) 250 nM calcium was used in these whole-cell patch clamp recordings. The observed currents in the wildtype mouse are similar to those obtained from heterologous expression of TMEM16A as seen in (a).

(Kuruma and Hartzell, 2000; Yang et al., 2008). However, common high affinity calcium-binding sites like EF hands and C2 domains cannot be identified from the primary protein sequence of TMEM16A or B.

A naturally occurring splice variant of TMEM16A lacking a 26 amino acid exon in the N terminus has been reported to be more calcium sensitive than channels containing that exon (Ferrera et al., 2009) (Figure 32.1b). The same group has also reported that a minimal isoform of TMEM16A lacking all four alternatively spliced exons is still a CaCC, suggesting that the calcium-sensing motif is present within this truncated form of the protein (Ferrera et al., 2011), although the voltage dependence of this channel is reduced.

The first cytosolic loop (between TMD2 and TMD3) has a short string of glutamate residues, which is speculated to contain a Ca^{2+}-binding site akin to the big conductance calcium-activated potassium (BK) channel calcium bowl (Figure 32.1c) (Ferrera et al., 2010; Schreiber and Salkoff, 1997). However, mutations of four consecutive glutamate residues in this motif only modestly lowered the apparent calcium sensitivity of the channel (Xiao et al., 2011). A similar study on TMEM16B, in which the corresponding glutamate string was deleted, reports that this region is critical for voltage but not calcium dependence of TMEM16B (Cenedese et al., 2012). An EAVK motif adjacent to the glutamate string has been implicated in calcium and voltage dependence (Figure 32.1d) (Ferrera et al., 2009; Xiao et al., 2011). These studies converge on the idea that the first intracellular loop plays a role in channel gating.

Different studies have shown that two glutamate residues within the proposed third intracellular loop are involved in calcium sensing. Mutation of either or both of these residues gives rise to a channel with an apparent decrease in calcium sensitivity (Figure 32.1f) (Yang et al., 2012; Yu et al., 2012).

It has also been observed that the classic CaCC calcium sensitivity increases upon exposure to more permeant anions, raising the possibility that anion–channel interactions couple to calcium binding and the gating process (Evans and Marty, 1986; Qu and Hartzell, 2000). In the absence of clear structural data on the channel, however, it remains a challenge to formulate a physical working model for calcium-dependent gating.

An alternative or parallel hypothesis posits that the binding of a calcium-sensitive auxiliary protein to TMEM16A mediates channel activation. Calmodulin has been postulated to carry out such a function. This small protein undergoes dramatic conformational change upon binding calcium, an event which occurs at submicromolar concentrations of calcium, recapitulating the observed calcium sensitivity of TMEM16A. Given calmodulin's well-known role as the calcium sensor for the small-conductance calcium-activated potassium channels (Xia et al., 1998), it is tempting to envision a similar scenario for the CaCC. Indeed, dominant negative mutants of calmodulin were shown to reduce the calcium sensitivity of CaCC in olfactory sensory neurons [OSNs] (Kaneko et al., 2006), which recent reports suggest is likely mediated by TMEM16B (Hengl et al., 2010; Rasche et al., 2010; Stephan et al., 2009). However, perfusing calmodulin over excised inside-out patches of heterologously expressed TMEM16B cannot rescue the loss of channel activity that occurs over time in these experiments (Pifferi et al., 2009).

It has been reported that calmodulin can be co-immunoprecipitated with TMEM16A and calmodulin inhibitors decrease whole-cell currents of TMEM16A (Tian et al., 2011). Given the ubiquity of calmodulin in numerous cell processes and its role in calcium-activated calcium release (Patel et al., 1999), it remains to be proven that calmodulin directly gates the channel. Another study reports that calmodulin affects the anionic selectivity of TMEM16A in a calcium-dependent fashion. This study finds that physical interaction of the channel with calmodulin increases the permeability of HCO_3^- relative to Cl^-, an observation with potential implications for TMEM16A function in epithelial secretion (Jung et al., 2013).

32.3.3 STOICHIOMETRY

TMEM16A has been observed as a homodimer in chemical cross-linking and native polyacrylamide gel electrophoresis experiments (Fallah et al., 2011; Sheridan et al., 2011). Förster resonance energy transfer (FRET) experiments show that TMEM16A multimerization occurs intracellularly before the channel is trafficked to the plasma membrane (PM) (Sheridan et al., 2011). The N-terminal cytosolic domain preceding the first TMD has been proposed to be the dimerization domain (Figure 32.1a) (Tien et al., 2013). It remains to be determined if the proposed dimer has a single pore or double-barrel pores like the voltage-gated chloride channels in the ClC family.

32.3.4 MOLECULAR DETERMINANTS OF IONIC SELECTIVITY

TMEM16A preferentially permeates bulkier anions like NO_3^-, I^-, and Br^- relative to the more physiologically abundant Cl^- (Schroeder et al., 2008; Yang et al., 2008). This is likely to be at least partially a consequence of hydration energies (Qu and Hartzell, 2000). There have also been reports that the anionic selectivity of TMEM16A and B shifts over the course of channel activation (Sagheddu et al., 2010; Schroeder et al., 2008). This could be due to a flexible pore that allows multiple open states of the channel to be sampled as changes in local calcium concentration occur, with higher calcium concentrations favoring permeation of larger anions relative to chloride (Sagheddu et al., 2010). The existence of multiple open states was reported in rabbit pulmonary artery smooth muscle CaCC (Piper and Large, 2003), and similar phenomenon has been observed (Boton et al., 1989) and computationally predicted in *Xenopus* CaCC (Kuruma and Hartzell, 2000).

Based on an older model of channel topology of TMEM16G (Das et al., 2008), a putative re-entrant loop between TM5 and 6 was proposed to form the pore of TMEM16A. Mutating conserved basic residues to acidic residues in this region was reported to increase the relative permeability of Na^+ over Cl^- (Figure 32.1g through i) (Yang et al., 2008). However, the more recently proposed topology calls the re-entrant loop model and the location of the pore into question, while determinants of ionic selectivity remain elusive (Yu et al., 2012).

32.3.5 PHARMACOLOGY

Cystic fibrosis, hypertension, asthma, and diarrhea are several medical conditions in which modulation of TMEM16A activity presents an opportunity for therapeutic intervention.

However, it remains an ongoing challenge to develop compounds that potently and specifically modulate the TMEM16 family of CaCCs. Traditional endogenous CaCC inhibitors include compounds such as niflumic acid (NFA), 4,4′-diisothiocyanatostilbene-2,2′-disulfonic acid (DIDS), and 5-nitro-2-(3-phenylpropylamino)benzoic acid (NPPB). These compounds have been shown to inhibit TMEM16A current with mid-micromolar potency, and are nonspecific compounds that target chloride channels in general, including the cystic fibrosis transmembrane conductance regulator (CFTR) channel and the ClC family of chloride channels. Other as yet unidentified channels have also been shown to be affected by broad-spectrum chloride channel inhibitors like NFA (Billig et al., 2011).

Since molecular elucidation, much effort has gone into developing TMEM16A- and TMEM16B -specific inhibitors, especially in the field of airway epithelial maintenance and asthma. One study found that siRNA of TMEM16A effectively reduces only an early transient current that arises from UTP stimulation of human bronchial epithelial cells taken from patients with cystic fibrosis. The authors went on to identify a small molecule, T16Ainh-A01, that specifically inhibits this early transient current, leading to the conclusion that T16Ainh-A01 is a TMEM16A-specific inhibitor (Namkung et al., 2011). T16Ainh-A01 inhibits TMEM16A with an IC_{50} of approximately 1 μM and does not inhibit forskolin-activated CFTR current. Another small molecule screen identified dichlorophen, benzbromarone, and hexachlorophen as TMEM16A and TMEM16B inhibitors with IC_{50} values ranging from 5 to 10 μM, with little effect on CFTR and epithelial sodium channel (ENaC) currents in normal human bronchial epithelial cells (Huang et al., 2012a). These pharmacological studies of airway CaCC show promise for TMEM16A modulation as a therapeutic handle in airway dysregulation, but a more thorough understanding of the contribution of TMEM16A activity to various disease states needs to be developed for TMEM16A based therapeutics to become a reality.

Another hurdle in developing TMEM16A inhibitors as therapeutic agents arises from discrepancies in endogenous CaCC pharmacological data. For example, tamoxifen has been reported to inhibit the endogenous CaCC in bovine pulmonary artery endothelial cells (Nilius et al., 1997) but not other types of tissue (Qu et al., 2003; Winpenny et al., 1998). The same compound inhibited TMEM16A current in HEK 293 cells (Yang et al., 2008) but poorly in *Axolotl oocytes* (Schroeder et al., 2008). In murine portal vein myocytes, CaCCs were shown to be sensitive to tamoxifen and paxillin, blockers of the BK_{Ca} channel. Pretreating the cells with methyl-β-cyclodextrin relieves this sensitivity, suggesting that BK/CaCC protein interactions within lipid microdomains affect the pharmacological profile of CaCC (Sones et al., 2010). In conclusion, tissue-specific factors might contribute to the observed disparities in the pharmacology of this channel.

32.3.6 POSTTRANSLATIONAL MODIFICATIONS AND CHANNEL FUNCTION

32.3.6.1 Phosphorylation

Endogenous CaCCs are regulated by kinase activity. Dialyzing AMP-PNP, a nonhydrolyzable ATP analog into rabbit arterial and portal vein smooth muscle cells reduces the loss of channel activity over time (Angermann et al., 2006). Phosphatase inhibitors can also antagonize recovery of CaCC after initial run-down in pulmonary arterial smooth muscle cells (PASMCs) (Ayon et al., 2009). Calmodulin-dependent kinase II (CaMKII) in particular has been associated with inhibition of CaCC in PASMCs (Greenwood et al., 2001). These observations are of particular interest because of reports suggesting that TMEM16A gives rise to the rat pulmonary arterial smooth muscle CaCC (Manoury et al., 2010). However, it has been suggested that TMEM16A activity in HEK 293 cells is not sensitive to staurosporine, a nonspecific kinase inhibitor, or the commonly used CaMKII inhibitor KN93 (Tian et al., 2011). This raises the possibility that in native tissue, tissue-specific factors mediate TMEM16A kinase sensitivity.

32.3.6.2 Glycosylation

TMEM16A is N-glycosidase F sensitive (Fallah et al., 2011; Yang et al., 2008), and PM-bound TMEM16A is Endo H insensitive, indicative of a mature, complex glycosylated ion channel (Fallah et al., 2011). Some glycosylation sites are conserved in the TMEM16 family (Das et al., 2008), but the significance of glycosylation on TMEM16A function remains unknown.

32.4 FUNCTIONAL IMPLICATIONS OF TMEM16A

32.4.1 DEVELOPMENT AND CANCER

The involvement of TMEM16A in development and cancer predates its molecular identification as a CaCC. A knockout mouse for TMEM16A was generated after it was observed that gene expression was enriched in the zone of polarizing activity (ZPA) at the murine distal limb (Rock et al., 2007). TMEM16A is expressed in the epithelia and smooth muscle of the developing trachea, and fetal knockout mice develop various defects in the trachea: an enlarged tracheal lumen, gaps in the tracheal cartilage rings, absence of the expected transverse orientation of the trachealis muscle, and a thin tracheal epithelium lacking stratification (Rock et al., 2008). These developmental defects cause tracheomalacia, a collapse of the upper airways. How TMEM16A is involved in developmental processes and whether or not its CaCC activity plays a role in development remain open questions.

In addition to its role in murine development, TMEM16A is widely implicated in oncology. Under the names GIST-1 (DOG-1) (West et al., 2004), FLJ10261 and oral cancer overexpressed 2 (ORAOV2) (Katoh, 2003), and tumor-amplified and overexpressed sequence 2 (TAOS2) (Huang et al., 2006), TMEM16A was early on noted for its high expression in various types of cancer and is frequently associated with poor prognosis. This channel has been shown to be tumorigenic in gastrointestinal stromal tumors (GISTs) (Simon et al., 2013), head and neck squamous cell carcinomas (HNSCC) (Ayoub et al., 2010; Ruiz et al., 2012), breast cancer (Britschgi et al., 2013), and metastatic prostate cancer (Liu et al., 2012). Various studies also suggest that TMEM16A promotes cancer metastasis and cell migration (Ayoub et al., 2010; Liu et al., 2012; Ruiz et al., 2012). In several reports, CaCC inhibitors have been shown to modulate tumor biology, but given the possibility of off-target effects, it remains an open question whether they exert their function via inhibition of TMEM16A channel activity.

32.4.2 SENSORY TRANSDUCTION

Immunostaining detects TMEM16A in DRG sensory neurons, with stronger signals in the small compared to large sensory neurons (Yang et al., 2008). Depolarizing CaCCs, observed in small DRG neurons upon treatment with the potent allogenic substance, BK, are inhibited by knockdown of TMEM16A, showing that this channel contributes to BK-induced DRG excitability and the perception of inflammatory pain (Liu et al., 2010). A separate biochemical study noted the physical interaction of TMEM16A, BK receptors, and the inositol 1,4,5-trisphosphate receptor 1 (IP3R1), forming a microdomain, which integrates TMEM16A activity downstream of specific calcium signaling pathways in nociceptive neurons (Jin et al., 2013).

Another study reports a role for TMEM16A in thermal nociception (Cho et al., 2012); CaCC activation occurs in DRG neurons in response to heat, and these currents are reduced upon intrathecal injection of TMEM16A-specific siRNA into wildtype mice. This study further reports that conditional knockout mice for TMEM16A display defects in heat-evoked behavioral responses, and small DRG neurons from these mice are deficient in thermally induced CaCCs.

In the normal mouse DRG, CaCCs are observed in a subset of medium-diameter (30–40 μm) sensory neurons. Sciatic nerve transection not only increases both CaCC amplitude and its prevalence in medium-diameter neurons, but also induces CaCC expression in large-diameter (40–50 μm) sensory neurons, generally thought to be low-threshold skin- and muscle-innervating neurons conveying touch and proprioceptive information. Expression of CaCC correlates with injury-induced neuron growth in vitro and in vivo (Andre et al., 2003). However, one study reported no detectable change of either TMEM16A or TMEM16B expression at the mRNA level in the DRG after axotomy (Boudes et al., 2009), calling into question the role of these channels in axotomy-induced CaCC upregulation.

32.4.3 SMOOTH MUSCLE TONE

TMEM16A expression can be detected by immunocytochemistry in smooth muscle cells in the murine airway, reproductive tract (Huang et al., 2009), portal vein, thoracic aorta, and carotid artery (Davis et al., 2010), but not in the gastrointestinal (GI) tract. Reduction of TMEM16A expression with siRNA in pulmonary artery smooth muscle cells (PASMCs) led to a significant loss of CaCC currents (Manoury et al., 2010). An increase in CaCC currents occurs in PASMCs upon hypertension, concurrent with an increase in TMEM16A expression under the same conditions (Forrest et al., 2012). The same study reported that a TMEM16A specific inhibitor, T16Ainh-A01, decreases smooth PASMC contraction in both hypertensive and nonhypertensive states.

In airway smooth muscle (ASM), CaCCs are regulated by calcium microdomains arising from ryanodine receptor activation, and are responsible for generating spontaneous transient inward currents (STICs) that are critical for smooth muscle contraction (Bao et al., 2008). ASM from TMEM16A knockout mice has weakened agonist-induced contraction (Zhang et al., 2013), as expected from CaCC-dependent positive feedback regulation in smooth muscle.

Cerebral (Thomas-Gatewood et al., 2011) and basilar artery smooth muscle cells (Wang et al., 2012) also display classic CaCC currents, which are decreased upon TMEM16A knockdown. There are reports that TMEM16A mediates swell or stretch-activated currents in arterial myocytes (Bulley et al., 2012) as well as epithelial cells (Almaca et al., 2009), but more biophysical characterization is necessary to prove that TMEM16A is indeed a pore-forming subunit of the swell or stretch-activated chloride channel.

Although TMEM16A mRNA cannot be detected in the smooth muscle cells of the mouse, nonhuman primate, and human GI tracts, it is expressed abundantly and specifically in the interstitial cells of Cajal (ICCs) (Huang et al., 2009). These are pacemaker cells that induce rhythmic slow waves in the electrically coupled smooth muscle cells of the GI tract, controlling smooth muscle phasic contractions that give rise to peristalsis (Sanders et al., 2006). TMEM16A plays a critical role in the pacemaker activity of ICCs. TMEM16A knockout mice fail to develop the slow wave and have defects in smooth muscle contractions in the GI tract (Gomez-Pinilla et al., 2009; Huang et al., 2009; Hwang et al., 2009). The channel also plays a similar role in the smooth muscle cells of the murine oviduct (Dixon et al., 2012). Alternative splice variants of TMEM16A in ICCs exhibit slower kinetics compared with the full-length TMEM16A in HEK 293 cells, and the authors speculate that this may directly contribute to the GI motility disorder in diabetic patients who express these splice variants (Mazzone et al., 2011).

32.4.4 EXOCRINE SECRETION AND EPITHELIAL MAINTENANCE

CaCCs have long been of interest for their role in regulating secretion. Chloride efflux from the apical surface of epithelia contributes to osmotic pressure for water efflux, driving fluid secretion (Melvin et al., 2005).

In TMEM16A knockout mice, tracheal epithelia have a 60% reduction in purinergic receptor–induced CaCC compared to wildtype mice (Rock et al., 2009), although both strains exhibit similar amounts of basal level CaCC. In another study, UTP treatment of human bronchial epithelial cells in a CFTR-deficient background stimulated biphasic currents—a transient and a prolonged CaCC; the former was blocked by a TMEM16A- and TMEM16B-specific inhibitor, but the latter was blocked only by broad-spectrum inhibitors (Namkung et al., 2011). This suggests firstly that channels other than TMEM16A or TMEM16B contribute to a basal level of airway epithelial CaCC, and secondly that TMEM16A is activated downstream of specific calcium signaling pathways in airway epithelia, similar to its regulation in the DRG and smooth muscle.

Apart from purinergic activation, TMEM16A expression is also upregulated in cultured airway epithelial cells upon treatment with inflammatory cytokines (Caputo et al., 2008) and is highly expressed in mucin-secreting cells in asthma (Huang et al., 2012a; Scudieri et al., 2012). TMEM16A expression has also been found in the porcine bronchial submucosal gland serous acinar cells, and may regulate agonist-induced fluid secretion in these cells (Lee and Foskett, 2010).

In addition to the lung and airway, TMEM16A expression is found in the apical membrane of acinar cells of the salivary glands and pancreas (Huang et al., 2009). Studies using TMEM16A knockout mice have demonstrated that TMEM16A is a critical component of submandibular gland acinar CaCCs (Figure 32.3c) (Romanenko et al., 2010). Local injection of siRNA to cause knockdown of TMEM16A has also been shown to affect murine saliva secretion (Yang et al., 2008).

32.5 TMEM16B

TMEM16B is another CaCC in the TMEM16 family. Relative to TMEM16A, it is less calcium sensitive, with an EC_{50} for calcium-dependent activation of 4.9 μM at −50 mV and 3.3 μM at +50 mV and a Hill coefficient of more than 2. Its single-channel conductance is 1.2 pS. TMEM16B permeation properties closely resemble those of TMEM16A: P_{Na}/P_{Cl} is 0.23, with preferential permeation of larger anions SCN^-, I^-, and Br^- over Cl^- (Pifferi et al., 2009). TMEM16A and TMEM16B have yet to be pharmacologically distinguished. In attempts to rescue the run-down observed in excised patches of TMEM16B, Pifferi et al. exposed patches to ATP, Na_3VO_4, DTT, calmodulin, cAMP, and PIP_3 but none of these compounds successfully restored TMEM16B current in the heterologous HEK 293 cell expression system.

Recent studies identify TMEM16B as a likely candidate for the CaCC observed in the presynaptic terminals of retinal photoreceptors (Lalonde et al., 2008) and in the cilia of OSNs (Hengl et al., 2010; Rasche et al., 2010; Stephan et al., 2009). Indeed, TMEM16B knockout mice have little to no CaCC currents in the olfactory epithelium and in the vomeronasal organ (Figure 32.4). However, the knockout mice showed no significant abnormality in olfactory behavioral tasks, demonstrating that cyclic nucleotide–gated cation channels do not need a boost by Cl^- channels to achieve near-physiological levels of olfaction (Billig et al., 2011). No detectable compensation occurred in the expression levels of other olfactory components in TMEM16B knockout mice, suggesting either subtler compensatory mechanisms for olfactory transduction in the TMEM16B knockout or a modest role, if any, for TMEM16B in olfaction.

The role of TMEM16B in the hippocampus has been studied by channel knockdown in hippocampal neurons (Huang et al., 2012b). This study reports that TMEM16B has an inhibitory influence on action potential duration, excitatory postsynaptic potential (EPSP) size, EPSP summation as well as EPSP-spike coupling. In these neurons, TMEM16B is activated downstream of calcium influx through *N*-methyl D-aspartate (NMDA) receptors and voltage-gated calcium channels. BAPTA (a fast calcium chelator), but not EGTA, blocked TMEM16B activation, suggesting that the channel is in close physical proximity to NMDA receptors and voltage-gated calcium channels.

32.6 OTHER MAMMALIAN TMEM16 FAMILY MEMBERS

32.6.1 TMEM16C, A REGULATORY SUBUNIT OF K_{Na}

Heterologous expression of TMEM16C has not yielded evidence that it functions as an ion channel, but nociceptive DRG neurons from TMEM16C knockout rats have greatly reduced sodium-activated potassium currents (K_{Na}) (Huang et al., 2013). A candidate K_{Na}, Slack (also known as Slo2.2), colocalizes with TMEM16C in small DRG neurons and its expression is reduced in the TMEM16C knockout. Super-resolution single-molecule imaging and co-immunoprecipitation assays reveal colocalization of Slack and TMEM16C. While TMEM16C expression alone does not yield measurable currents, co-expression of TMEM16C with Slack in HEK 293 cells alters Slack single-channel properties: the channels show higher sodium sensitivity and have a higher probability of accessing a full conductance state even in the absence of sodium. In agreement with its proposed role in potassium channel regulation in the DRG, TMEM16C knockout rats display hypersensitivity to noxious heat and mechanical stimulation.

Whole-exome sequencing has linked TMEM16C mutations to the pathogenesis of craniocervical dystonia (Charlesworth et al., 2012). The same study reported that a W490C mutant

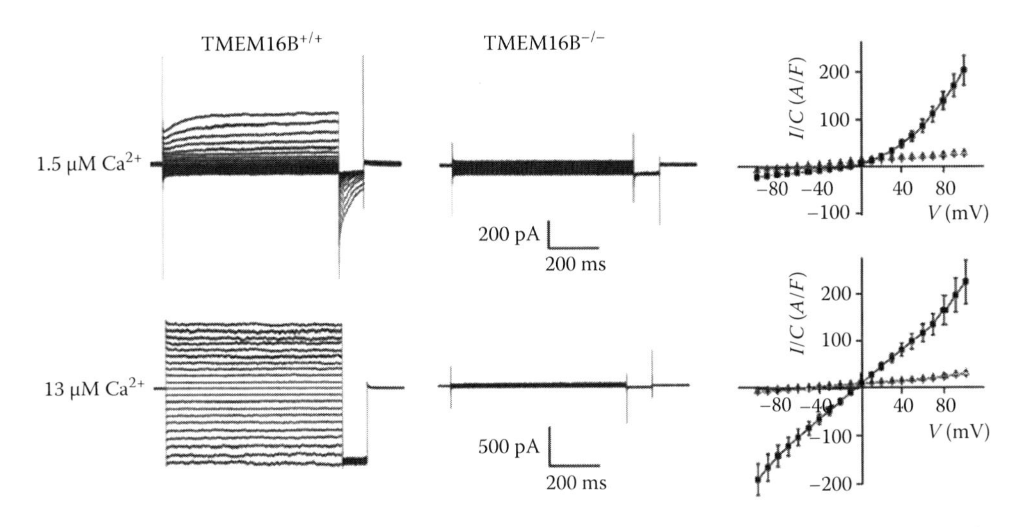

Figure 32.4 TMEM16B is the CaCC in OSNs. TMEM16B currents in wildtype mice are calcium dependent. At low calcium concentrations (top), currents from wildtype OSNs are time dependent and outwardly rectifying (top-left). OSNs from TMEM16B knockout mice have little to no current (top-center). I/V curves show the voltage dependence of the currents (top-right; wildtype: squares; knockout: triangles). Higher calcium concentrations (bottom) activate time-independent currents (bottom-left) absent in the knockout (bottom-center), which have a linear I/V relationship (bottom-right; wildtype: squares; knockout: triangles). (From Billig, G.M. et al., *Nat. Neurosci.*, 14, 763, 2011.)

of TMEM16C reduces calcium levels in the ER of mutation-carrying fibroblasts, suggesting a role for TMEM16C in intracellular calcium homeostasis.

32.6.2 TMEM16F, A CaCC OR A SMALL-CONDUCTANCE CALCIUM-ACTIVATED NONSELECTIVE CATION CHANNEL?

TMEM16F is responsible for calcium-dependent phospholipid scrambling, which may be enhanced by an aspartate to glycine mutation of TMEM16F in a B cell line (Suzuki et al., 2010). Putative loss-of-function mutations of TMEM16F have been shown to be associated with Scott syndrome, a bleeding disorder in which defective phosphatidylserine (PS) exposure on platelets results in attenuated blood coagulation and excessive bleeding (Castoldi et al., 2011; Suzuki et al., 2010). TMEM16F knockout mice also display reduced PS exposure and procoagulant activity (Yang et al., 2012). It has been reported that a fungal homologue of the TMEM16 family functions as a scramblase and a calcium-activated channel permeable to both anions and cations (Malvezzi et al., 2013), but TMEM16F has not yet been shown to display scramblase activity on its own, nor is it known if it forms a critical subunit of a larger scramblase complex in mammalian systems. In addition to TMEM16F, TMEM16C, TMEM16D, TMEM16G, and TMEM16J, but not the CaCCs TMEM16A and TMEM16B, have also been shown to be associated with phospholipid scrambling activity (Suzuki et al., 2013).

With regard to channel activity, TMEM16F has been described as a CaCC (Grubb et al., 2013; Shimizu et al., 2013), or a small-conductance calcium-activated nonselective cation (SCAN) channel (Yang et al., 2012). Grubb et al. report that heterologous expression of TMEM16F in HEK 293 cells gives rise to chloride currents with an independent and significant sodium conductance ($P_{Na}/P_{Cl} \sim 0.3$). These currents develop several minutes after establishing whole-cell mode. This delay could not be accounted for by the time taken for calcium to diffuse from the pipette into the cell, or calcium-dependent delivery of TMEM16F to the membrane. The anionic selectivity series of TMEM16F-dependent currents is altered by mutations in the channel, suggesting that TMEM16F is a pore-forming subunit of a CaCC with a modest selectivity for anions. This study reports an EC_{50} of calcium activation for TMEM16F as 100 μM at +70 mV.

In another study, Yang et al. report that SCAN currents ($P_{Na}/P_{Cl} \sim 7$) are present in inside-out excised patches from wildtype murine megakaryocytes but are absent in these cells from TMEM16F knockout mice. Heterologous expression of TMEM16F in *Axolotl oocytes* and HEK 293 cells also gives rise to SCAN currents in excised patches. The single-channel conductance for the TMEM16F SCAN channel is at the sub-picosiemen level, and the EC_{50} of calcium activation of SCAN currents in excised patches is 14 μM in HEK293 cells, and 5 μM in megakaryocytes (at +60 mV). To prove that TMEM16F is a pore-forming subunit of the observed SCAN currents, Yang et al. introduced various mutations into TMEM16F. An E667Q mutation caused a significant decrease in calcium sensitivity, corroborating a report that the corresponding residue in TMEM16A affects calcium sensitivity (Yu et al., 2012), further suggesting that calcium-sensing determinants are conserved within the TMEM16 family. Substitution of a TMD proximal and possibly pore-lining glutamine residue of TMEM16F with a lysine at the corresponding position in established CaCCs TMEM16A and TMEM16B results in a decrease in the cationic selectivity of TMEM16F, while the converse mutation decreases the anionic selectivity of TMEM16A. The authors also examined an aspartate to glycine mutation (D409G) on TMEM16F that allows it to mediate phospholipid scrambling in resting B cells (Suzuki et al., 2010), but this mutation did not alter either calcium permeability or calcium sensitivity of the SCAN channel.

CaCC inhibitors block TMEM16F-dependent currents recorded with whole-cell patch-clamp in some studies (Shimizu et al., 2013), but they have no effect on the TMEM16F-SCAN current recorded from excised inside-out patches (Yang et al., 2012). TMEM16F channel biophysics is thus far rife with controversy, likely arising in part from the broad expression of CaCC and related channels in various expression systems and other cell types. In whole-cell recordings, high calcium concentrations in the internal solution and the additional calcium permeability of TMEM16F might also lead to the activation of various endogenous calcium-dependent conductances over time, confounding electrophysiological analysis of TMEM16F.

32.6.3 TMEM16E

Recessive mutations of TMEM16E (also known as GDD1) have been identified as playing a causative role in proximal limb-girdle muscular dystrophy (LGMD2L) and distal nondysferlin Miyoshi myopathy (MMD3) (Bolduc et al., 2010). These include splice site and duplication aberrations, resulting in point mutations as well as premature termination of the protein-coding sequence. Cardiac myopathy has also been reported in patients carrying TMEM16E mutations (Wahbi et al., 2013).

Another disease attributed to TMEM16E dysfunction is gnathodiaphyseal dysplasia (GDD), a rare skeletal syndrome characterized by bone fragility, sclerosis of tubular bones, and cemento-osseous lesions of the jawbone (Tsutsumi et al., 2004). Mutations at the C356 position (to glycine or arginine) (Tsutsumi et al., 2004), as well as a T513I mutant (Marconi et al., 2013), have been linked to GDD.

Heterologous expression of TMEM16E is dogged by constitutive proteasomal degradation (Tran et al., 2013), and while the protein is predominantly localized intracellularly in immunohistochemical studies (Mizuta et al., 2007), it is not clear if a subpopulation of the protein resides on the PM, and it has yet to be described as an ion channel.

32.6.4 TMEM16G

TMEM16G (also known as new gene expressed in prostate [NGEP]) is considered a marker for prostate, though its function in this tissue remains unknown. An early model of TMEM16 family topology was based on TMEM16G (Das et al., 2008). It has yet to be described as an ion channel.

32.6.5 TMEM16K

TMEM16K has been linked to autosomal-recessive cerebellar ataxia (Vermeer et al., 2010), and interestingly, its *Drosophila* homologue, Aberrant X segregation (Axs), has been linked to X chromosome-nondisjunction in meiosis, and is localized to the meiotic spindle (Kramer and Hawley, 2003). Neither of these divergent homologues have been reported as ion channels.

32.6.6 FUNGAL TMEM16 FAMILY MEMBERS

Ist2 (increased sodium tolerance) is the single homologue of the TMEM16 family in *Saccharomyces cerevisiae*, drawing its name from a deletion strain reported to have increased tolerance to high salt growth conditions (Mannhaupt et al., 1994). IST2 mRNA has the unusual property of being transported by the actomyosin machinery to the yeast bud tip for local translation (Takizawa et al., 2000).

While earlier reports suggest that Ist2 is a PM)-bound protein found at PM-associated ER domains (Fischer et al., 2009), it is now more widely accepted to be an ER-bound protein, which has been reported to mediate ER-PM tethering (Wolf et al., 2012), as well as cortical ER formation (Lavieu et al., 2010). Ist2 has not yet been reported to demonstrate ion channel activity.

Another fungal TMEM16 homologue from *Aspergillus fumigatus* (afTMEM16) has been reconstituted in an in vitro lipid bilayer system and shown to be a nonselective calcium-activated and voltage-sensitive channel (P_K/P_{Cl} ~ 1.5), with a calcium dependence EC_{50} of approximately 0.4 μM. Reconstitution of purified protein from which calcium-activated currents can be recorded is also unequivocal proof that TMEM16 family members are the pore-forming subunit of channels. Amazingly, the reconstituted channel also functions as a calcium-dependent phospholipid scramblase, with EC_{50} ~ 0.5 μM. Point mutations reported to diminish mammalian TMEM16A and TMEM16F calcium sensitivity when introduced into the fungal channel also decrease afTMEM16-mediated ion flux and phospholipid scrambling activity. The authors thus propose a single calcium-regulated gate for both its channel and phospholipid scramblase activities. Interestingly, the channel activity of afTMEM16 is sensitive to lipid composition. This begs the questions of whether the lipid environment also affects mammalian channels, and if lipids are the diffusible factor that cause mammalian TMEM16 channel run-down. Unlike channel activity, the phospholipid scrambling activity of afTMEM16 is insensitive to lipid composition, suggesting physically separate pathways for ion permeation and phospholipid transport. In conclusion, afTMEM16 is an ancient channel with aggregate properties of its more diverged and specialized mammalian homologues (Malvezzi et al., 2013).

32.7 CONCLUSION

TMEM16 family CaCCs, TMEM16A and TMEM16B, recapitulate the properties of classic CaCCs that have been observed in various organisms and tissue types in nature. These channels are often elegantly integrated into specific calcium-mobilization pathways, where channel activation exerts a potent influence on physiology.

While TMEM16A and TMEM16B have been unambiguously described as CaCCs with a moderate permeability to cations, the ionic selectivity of TMEM16F is more controversial. Nonetheless, TMEM16A, TMEM16B, and TMEM16F are calcium- and voltage-dependent channels, suggesting that perhaps these properties are conserved within the TMEM16 family. The molecular bases of ionic selectivity, calcium and voltage dependence remain unclear in the absence of structural data. A physical mechanism for the phospholipid scrambling activity associated with mammalian TMEM16F has also not yet been described. In vitro reconstitution of a distant fungal homologue, which acts as a calcium- and voltage-regulated channel as well as a calcium-dependent phospholipid scramblase, promises to aid the molecular elucidation of these phenomena.

Unlike the earlier-mentioned TMEM16 family members, TMEM16C has not been observed to act as a channel by itself, but instead functions as an auxiliary subunit to a potassium channel. This raises the possibility that CaCCs in the TMEM16 family also moonlight as regulatory subunits of other channels, which may be a cause of disparity in the literature regarding the ionic selectivity, pharmacology, and other biophysical properties of TMEM16 family channels in different cell types.

Though not all have been established as ion channels, many of the mammalian TMEM16 family members are directly implicated in human disease and there is much interest in learning more about the function and properties of these putative channels. Future studies will undoubtedly uncover more exciting biology behind this new and enigmatic family of proteins.

REFERENCES

Almaca, J., Tian, Y., Aldehni, F., Ousingsawat, J., Kongsuphol, P., Rock, J.R., Harfe, B.D., Schreiber, R., and Kunzelmann, K. (2009). TMEM16 proteins produce volume-regulated chloride currents that are reduced in mice lacking TMEM16A. *J Biol Chem* **284**, 28571–28578.

Anderson, M.P. and Welsh, M.J. (1991). Calcium and cAMP activate different chloride channels in the apical membrane of normal and cystic fibrosis epithelia. *Proc Natl Acad Sci USA* **15**, 6003–6007.

Andre, S., Boukhaddaoui, H., Campo, B., Al-Jumaily, M., Mayeux, V., Greuet, D., Valmier, J., and Scamps, F. (2003). Axotomy-induced expression of calcium-activated chloride current in subpopulations of mouse dorsal root ganglion neurons. *J Neurophysiol* **90**, 3764–3773.

Angermann, J.E., Sanguinetti, A.R., Kenyon, J.L., Leblanc, N., and Greenwood, I.A. (2006). Mechanism of the inhibition of Ca^{2+}-activated Cl^- currents by phosphorylation in pulmonary arterial smooth muscle cells. *J Gen Physiol* **128**, 73–87.

Ayon, R., Sones, W., Forrest, A.S., Wiwchar, M., Valencik, M.L., Sanguinetti, A.R., Perrino, B.A., Greenwood, I.A., and Leblanc, N. (2009). Complex phosphatase regulation of Ca^{2+}-activated Cl^- currents in pulmonary arterial smooth muscle cells. *J Biol Chem* **284**, 32507–32521.

Ayoub, C., Wasylyk, C., Li, Y., Thomas, E., Marisa, L., Robe, A., Roux, M., Abecassis, J., de Reynies, A., and Wasylyk, B. (2010). ANO1 amplification and expression in HNSCC with a high propensity for future distant metastasis and its functions in HNSCC cell lines. *Br J Cancer* **103**, 715–726.

Bader, C.R., Bertrand, D., and Schwartz, E.A. (1982). Voltage-activated and calcium-activated currents studied in solitary rod inner segments from the salamander retina. *J Physiol* **331**, 253–284.

Bao, R., Lifshitz, L.M., Tuft, R.A., Bellve, K., Fogarty, K.E., and ZhuGe, R. (2008). A close association of RyRs with highly dense clusters of Ca^{2+}-activated Cl^- channels underlies the activation of STICs by Ca^{2+} sparks in mouse airway smooth muscle. *J Gen Physiol* **132**, 145–160.

Billig, G.M., Pal, B., Fidzinski, P., and Jentsch, T.J. (2011). Ca^{2+}-activated Cl^- currents are dispensable for olfaction. *Nat Neurosci* **14**, 763–769.

Bolduc, V., Marlow, G., Boycott, K.M., Saleki, K., Inoue, H., Kroon, J., Itakura, M. et al. (2010). Recessive mutations in the putative calcium-activated chloride channel Anoctamin 5 cause proximal LGMD2L and distal MMD3 muscular dystrophies. *Am J Hum Genet* **86**, 213–221.

Boton, R., Dascal, N., Gillo, B., and Lass, Y. (1989). Two calcium-activated chloride conductances in *Xenopus laevis* oocytes permeabilized with the ionophore A23187. *J Physiol* **408**, 511–534.

Boudes, M., Sar, C., Menigoz, A., Hilaire, C., Pequignot, M.O., Kozlenkov, A., Marmorstein, A., Carroll, P., Valmier, J., and Scamps, F. (2009). Best1 is a gene regulated by nerve injury and required for Ca^{2+}-activated Cl^- current expression in axotomized sensory neurons. *J Neurosci* **29**, 10063–10071.

Britschgi, A., Bill, A., Brinkhaus, H., Rothwell, C., Clay, I., Duss, S., Rebhan, M. et al. (2013). Calcium-activated chloride channel ANO1 promotes breast cancer progression by activating EGFR and CAMK signaling. *Proc Natl Acad Sci USA* **110**, E1026–E1034.

Bulley, S., Neeb, Z.P., Burris, S.K., Bannister, J.P., Thomas-Gatewood, C.M., Jangsangthong, W., and Jaggar, J.H. (2012). TMEM16A/ANO1 channels contribute to the myogenic response in cerebral arteries. *Circ Res* **111**, 1027–1036.

Caputo, A., Caci, E., Ferrera, L., Pedemonte, N., Barsanti, C., Sondo, E., Pfeffer, U., Ravazzolo, R., Zegarra-Moran, O., and Galietta, L.J. (2008). TMEM16A, a membrane protein associated with calcium-dependent chloride channel activity. *Science* **322**, 590–594.

Castoldi, E., Collins, P.W., Williamson, P.L., and Bevers, E.M. (2011). Compound heterozygosity for 2 novel TMEM16F mutations in a patient with Scott syndrome. *Blood* **117**, 4399–4400.

Cenedese, V., Betto, G., Celsi, F., Cherian, O.L., Pifferi, S., and Menini, A. (2012). The voltage dependence of the TMEM16B/anoctamin2 calcium-activated chloride channel is modified by mutations in the first putative intracellular loop. *J Gen Physiol* **139**, 285–294.

Charlesworth, G., Plagnol, V., Holmstrom, K.M., Bras, J., Sheerin, U.M., Preza, E., Rubio-Agusti, I. et al. (2012). Mutations in ANO3 cause dominant craniocervical dystonia: Ion channel implicated in pathogenesis. *Am J Hum Genet* **91**, 1041–1050.

Cho, H., Yang, Y.D., Lee, J., Lee, B., Kim, T., Jang, Y., Back, S.K. et al. (2012). The calcium-activated chloride channel anoctamin 1 acts as a heat sensor in nociceptive neurons. *Nat Neurosci* **15**, 1015–1021.

Cross, N.L. (1981). Initiation of the activation potential by an increase in intracellular calcium in eggs of the frog, *Rana pipiens*. *Dev Biol* **85**, 380–384.

Das, S., Hahn, Y., Walker, D.A., Nagata, S., Willingham, M.C., Peehl, D.M., Bera, T.K., Lee, B., and Pastan, I. (2008). Topology of NGEP, a prostate-specific cell: Cell junction protein widely expressed in many cancers of different grade level. *Cancer Res* **68**, 6306–6312.

Davis, A.J., Forrest, A.S., Jepps, T.A., Valencik, M.L., Wiwchar, M., Singer, C.A., Sones, W.R., Greenwood, I.A., and Leblanc, N. (2010). Expression profile and protein translation of TMEM16A in murine smooth muscle. *Am J Physiol Cell Physiol* **299**, C948–C959.

Dixon, R.E., Hennig, G.W., Baker, S.A., Britton, F.C., Harfe, B.D., Rock, J.R., Sanders, K.M., and Ward, S.M. (2012). Electrical slow waves in the mouse oviduct are dependent upon a calcium activated chloride conductance encoded by Tmem16a. *Biol Reprod* **86**, 1–7.

Evans, M.G. and Marty, A. (1986). Calcium-dependent chloride currents in isolated cells from rat lacrimal glands. *J Physiol* **378**, 437–460.

Fallah, G., Romer, T., Detro-Dassen, S., Braam, U., Markwardt, F., and Schmalzing, G. (2011). TMEM16A(a)/anoctamin-1 shares a homodimeric architecture with CLC chloride channels. *Mol Cell Proteomics* **10**, M110 004697.

Ferrera, L., Caputo, A., and Galietta, L.J. (2010). TMEM16A protein: A new identity for Ca(2+)-dependent Cl(−) channels. *Physiology (Bethesda)* **25**, 357–363.

Ferrera, L., Caputo, A., Ubby, I., Bussani, E., Zegarra-Moran, O., Ravazzolo, R., Pagani, F., and Galietta, L.J. (2009). Regulation of TMEM16A chloride channel properties by alternative splicing. *J Biol Chem* **284**, 33360–33368.

Ferrera, L., Scudieri, P., Sondo, E., Caputo, A., Caci, E., Zegarra-Moran, O., Ravazzolo, R., and Galietta, L.J. (2011). A minimal isoform of the TMEM16A protein associated with chloride channel activity. *Biochim Biophys Acta* **1808**, 2214–2223.

Fischer, M.A., Temmerman, K., Ercan, E., Nickel, W., and Seedorf, M. (2009). Binding of plasma membrane lipids recruits the yeast integral membrane protein Ist2 to the cortical ER. *Traffic* **10**, 1084–1097.

Forrest, A.S., Joyce, T.C., Huebner, M.L., Ayon, R.J., Wiwchar, M., Joyce, J., Freitas, N. et al. (2012). Increased TMEM16A-encoded calcium-activated chloride channel activity is associated with pulmonary hypertension. *Am J Physiol Cell Physiol* **303**, C1229–C1243.

Fromm, J. and Lautner, S. (2007). Electrical signals and their physiological significance in plants. *Plant Cell Environ* **30**, 249–257.

Gomez-Pinilla, P.J., Gibbons, S.J., Bardsley, M.R., Lorincz, A., Pozo, M.J., Pasricha, P.J., Van de Rijn, M. et al. (2009). Ano1 is a selective marker of interstitial cells of Cajal in the human and mouse gastrointestinal tract. *Am J Physiol Gastrointest Liver Physiol* **296**, G1370–G1381.

Greenwood, I.A., Ledoux, J., and Leblanc, N. (2001). Differential regulation of Ca(2+)-activated Cl(−) currents in rabbit arterial and portal vein smooth muscle cells by Ca(2+)-calmodulin-dependent kinase. *J Physiol* **534**, 395–408.

Grubb, B.R., Vick, R.N., and Boucher, R.C. (1994). Hyperabsorption of Na^+ and raised Ca(2+)-mediated Cl^- secretion in nasal epithelia of CF mice. *Am J Physiol* **266**, C1478–C1483.

Grubb, S., Poulsen, K.A., Juul, C.A., Kyed, T., Klausen, T.K., Larsen, E.H., and Hoffmann, E.K. (2013). TMEM16F (Anoctamin 6), an anion channel of delayed Ca(2+) activation. *J Gen Physiol* **141**, 585–600.

Hengl, T., Kaneko, H., Dauner, K., Vocke, K., Frings, S., and Mohrlen, F. (2010). Molecular components of signal amplification in olfactory sensory cilia. *Proc Natl Acad Sci USA* **107**, 6052–6057.

Huang, F., Rock, J.R., Harfe, B.D., Cheng, T., Huang, X., Jan, Y.N., and Jan, L.Y. (2009). Studies on expression and function of the TMEM16A calcium-activated chloride channel. *Proc Natl Acad Sci USA* **106**, 21413–21418.

Huang, F., Wang, X., Ostertag, E.M., Nuwal, T., Huang, B., Jan, Y.N., Basbaum, A.I., and Jan, L.Y. (2013). TMEM16C facilitates Na(+)-activated K(+) currents in rat sensory neurons and regulates pain processing. *Nat Neurosci* **16**, 1284–1290.

Huang, F., Zhang, H., Wu, M., Yang, H., Kudo, M., Peters, C.J., Woodruff, P.G. et al. (2012a). Calcium-activated chloride channel TMEM16A modulates mucin secretion and airway smooth muscle contraction. *Proc Natl Acad Sci USA* **109**, 16354–16359.

Huang, W.C., Xiao, S., Huang, F., Harfe, B.D., Jan, Y.N., and Jan, L.Y. (2012b). Calcium-activated chloride channels (CaCCs) regulate action potential and synaptic response in hippocampal neurons. *Neuron* **74**, 179–192.

Huang, X., Godfrey, T.E., Gooding, W.E., McCarty, K.S., Jr., and Gollin, S.M. (2006). Comprehensive genome and transcriptome analysis of the 11q13 amplicon in human oral cancer and synteny to the 7F5 amplicon in murine oral carcinoma. *Genes Chromosomes Cancer* **45**, 1058–1069.

Hwang, S.J., Blair, P.J., Britton, F.C., O'Driscoll, K.E., Hennig, G., Bayguinov, Y.R., Rock, J.R., Harfe, B.D., Sanders, K.M., and Ward, S.M. (2009). Expression of anoctamin 1/TMEM16A by interstitial cells of Cajal is fundamental for slow wave activity in gastrointestinal muscles. *J Physiol* **587**, 4887–4904.

Janssen, L.J. and Sims, S.M. (1992). Acetylcholine activates non-selective cation and chloride conductances in canine and guinea-pig tracheal myocytes. *J Physiol* **453**, 197–218.

Jin, X., Shah, S., Liu, Y., Zhang, H., Lees, M., Fu, Z., Lippiat, J.D. et al. (2013). Activation of the Cl^- channel ANO1 by localized calcium signals in nociceptive sensory neurons requires coupling with the IP3 receptor. *Sci Signal* **6**, ra73.

Jung, J., Nam, J.H., Park, H.W., Oh, U., Yoon, J.H., and Lee, M.G. (2013). Dynamic modulation of ANO1/TMEM16A HCO3(–) permeability by Ca^{2+}/calmodulin. *Proc Natl Acad Sci USA* **110**, 360–365.

Kaneko, H., Mohrlen, F., and Frings, S. (2006). Calmodulin contributes to gating control in olfactory calcium-activated chloride channels. *J Gen Physiol* **127**, 737–748.

Katoh, M. (2003). FLJ10261 gene, located within the CCND1-EMS1 locus on human chromosome 11q13, encodes the eight-transmembrane protein homologous to C12orf3, C11orf25 and FLJ34272 gene products. *Int J Oncol* **22**, 1375–1381.

Kleene, S.J. and Gesteland, R.C. (1991). Calcium-activated chloride conductance in frog olfactory cilia. *J Neurosci* **11**, 3624–3629.

Klockner, U. (1993). Intracellular calcium ions activate a low-conductance chloride channel in smooth-muscle cells isolated from human mesenteric artery. *Pflugers Arch* **424**, 231–237.

Kramer, J. and Hawley, R.S. (2003). The spindle-associated transmembrane protein Axs identifies a membranous structure ensheathing the meiotic spindle. *Nat Cell Biol* **5**, 261–263.

Kuruma, A. and Hartzell, H.C. (2000). Bimodal control of a Ca(2+)-activated Cl(–) channel by different Ca(2+) signals. *J Gen Physiol* **115**, 59–80.

Lalonde, M.R., Kelly, M.E., and Barnes, S. (2008). Calcium-activated chloride channels in the retina. *Channels (Austin)* **2**, 252–260.

Lavieu, G., Orci, L., Shi, L., Geiling, M., Ravazzola, M., Wieland, F., Cosson, P., and Rothman, J.E. (2010). Induction of cortical endoplasmic reticulum by dimerization of a coatomer-binding peptide anchored to endoplasmic reticulum membranes. *Proc Natl Acad Sci USA* **107**, 6876–6881.

Lee, R.J. and Foskett, J.K. (2010). Mechanisms of Ca^{2+}-stimulated fluid secretion by porcine bronchial submucosal gland serous acinar cells. *Am J Physiol Lung Cell Mol Physiol* **298**, L210–L231.

Liu, B., Linley, J.E., Du, X., Zhang, X., Ooi, L., Zhang, H., and Gamper, N. (2010). The acute nociceptive signals induced by bradykinin in rat sensory neurons are mediated by inhibition of M-type K^+ channels and activation of Ca^{2+}-activated Cl^- channels. *J Clin Invest* **120**, 1240–1252.

Liu, W., Lu, M., Liu, B., Huang, Y., and Wang, K. (2012). Inhibition of Ca(2+)-activated Cl(–) channel ANO1/TMEM16A expression suppresses tumor growth and invasiveness in human prostate carcinoma. *Cancer Lett* **326**, 41–51.

Malvezzi, M., Chalat, M., Janjusevic, R., Picollo, A., Terashima, H., Menon, A.K., and Accardi, A. (2013). Ca(2+)-dependent phospholipid scrambling by a reconstituted TMEM16 ion channel. *Nat Commun* **4**, 2367.

Mannhaupt, G., Stucka, R., Ehnle, S., Vetter, I., and Feldmann, H. (1994). Analysis of a 70 kb region on the right arm of yeast chromosome II. *Yeast* **10**, 1363–1381.

Manoury, B., Tamuleviciute, A., and Tammaro, P. (2010). TMEM16A/anoctamin 1 protein mediates calcium-activated chloride currents in pulmonary arterial smooth muscle cells. *J Physiol* **588**, 2305–2314.

Marconi, C., Brunamonti Binello, P., Badiali, G., Caci, E., Cusano, R., Garibaldi, J., Pippucci, T. et al. (2013). A novel missense mutation in ANO5/TMEM16E is causative for gnathodiaphyseal dyplasia in a large Italian pedigree. *Eur J Hum Genet* **21**, 613–619.

Mayer, M.L. (1985). A calcium-activated chloride current generates the after-depolarization of rat sensory neurones in culture. *J Physiol* **364**, 217–239.

Mazzone, A., Bernard, C.E., Strege, P.R., Beyder, A., Galietta, L.J., Pasricha, P.J., Rae, J.L. et al. (2011). Altered expression of Ano1 variants in human diabetic gastroparesis. *J Biol Chem* **286**, 13393–13403.

Melvin, J.E., Yule, D., Shuttleworth, T., and Begenisich, T. (2005). Regulation of fluid and electrolyte secretion in salivary gland acinar cells. *Annu Rev Physiol* **67**, 445–469.

Miledi, R. and Parker, I. (1984). Chloride current induced by injection of calcium into *Xenopus oocytes*. *J Physiol* **357**, 173–183.

Mizuta, K., Tsutsumi, S., Inoue, H., Sakamoto, Y., Miyatake, K., Miyawaki, K., Noji, S., Kamata, N., and Itakura, M. (2007). Molecular characterization of GDD1/TMEM16E, the gene product responsible for autosomal dominant gnathodiaphyseal dysplasia. *Biochem Biophys Res Commun* **357**, 126–132.

Namkung, W., Phuan, P.W., and Verkman, A.S. (2011). TMEM16A inhibitors reveal TMEM16A as a minor component of calcium-activated chloride channel conductance in airway and intestinal epithelial cells. *J Biol Chem* **286**, 2365–2374.

Nilius, B., Prenen, J., Voets, T., Van den Bremt, K., Eggermont, J., and Droogmans, G. (1997). Kinetic and pharmacological properties of the calcium-activated chloride-current in macrovascular endothelial cells. *Cell Calcium* **22**, 53–63.

Pacaud, P., Loirand, G., Mironneau, C., and Mironneau, J. (1989). Noradrenaline activates a calcium-activated chloride conductance and increases the voltage-dependent calcium current in cultured single cells of rat portal vein. *Br J Pharmacol* **97**, 139–146.

Patel, S., Joseph, S.K., and Thomas, A.P. (1999). Molecular properties of inositol 1,4,5-trisphosphate receptors. *Cell Calcium* **25**, 247–264.

Pifferi, S., Dibattista, M., and Menini, A. (2009). TMEM16B induces chloride currents activated by calcium in mammalian cells. *Pflugers Arch* **458**, 1023–1038.

Piper, A.S., and Large, W.A. (2003). Multiple conductance states of single Ca^{2+}-activated Cl^- channels in rabbit pulmonary artery smooth muscle cells. *J Physiol* **547**, 181–196.

Qu, Z. and Hartzell, H.C. (2000). Anion permeation in Ca(2+)-activated Cl(–) channels. *J Gen Physiol* **116**, 825–844.

Qu, Z., Wei, R.W., and Hartzell, H.C. (2003). Characterization of Ca^{2+}-activated Cl^- currents in mouse kidney inner medullary collecting duct cells. *Am J Physiol Renal Physiol* **285**, F326–F335.

Rasche, S., Toetter, B., Adler, J., Tschapek, A., Doerner, J.F., Kurtenbach, S., Hatt, H., Meyer, H., Warscheid, B., and Neuhaus, E.M. (2010). Tmem16b is specifically expressed in the cilia of olfactory sensory neurons. *Chem Senses* **35**, 239–245.

Rock, J.R., Futtner, C.R., and Harfe, B.D. (2008). The transmembrane protein TMEM16A is required for normal development of the murine trachea. *Dev Biol* **321**, 141–149.

Rock, J.R., Lopez, M.C., Baker, H.V., and Harfe, B.D. (2007). Identification of genes expressed in the mouse limb using a novel ZPA microarray approach. *Gene Expr Patterns* **8**, 19–26.

Rock, J.R., O'Neal, W.K., Gabriel, S.E., Randell, S.H., Harfe, B.D., Boucher, R.C., and Grubb, B.R. (2009). Transmembrane protein 16A (TMEM16A) is a Ca^{2+}-regulated Cl^- secretory channel in mouse airways. *J Biol Chem* **284**, 14875–14880.

Romanenko, V.G., Catalan, M.A., Brown, D.A., Putzier, I., Hartzell, H.C., Marmorstein, A.D., Gonzalez-Begne, M., Rock, J.R., Harfe, B.D., and Melvin, J.E. (2010). Tmem16A encodes the Ca^{2+}-activated Cl^- channel in mouse submandibular salivary gland acinar cells. *J Biol Chem* **285**, 12990–13001.

Ruiz, C., Martins, J.R., Rudin, F., Schneider, S., Dietsche, T., Fischer, C.A., Tornillo, L. et al. (2012). Enhanced expression of ANO1 in head and neck squamous cell carcinoma causes cell migration and correlates with poor prognosis. *PLoS One* **7**, e43265.

Sagheddu, C., Boccaccio, A., Dibattista, M., Montani, G., Tirindelli, R., and Menini, A. (2010). Calcium concentration jumps reveal dynamic ion selectivity of calcium-activated chloride currents in mouse olfactory sensory neurons and TMEM16b-transfected HEK 293T cells. *J Physiol* **588**, 4189–4204.

Sanders, K.M., Koh, S.D., and Ward, S.M. (2006). Interstitial cells of cajal as pacemakers in the gastrointestinal tract. *Annu Rev Physiol* **68**, 307–343.

Schreiber, M. and Salkoff, L. (1997). A novel calcium-sensing domain in the BK channel. *Biophys J* **73**, 1355–1363.

Schroeder, B.C., Cheng, T., Jan, Y.N., and Jan, L.Y. (2008). Expression cloning of TMEM16A as a calcium-activated chloride channel subunit. *Cell* **134**, 1019–1029.

Scudieri, P., Caci, E., Bruno, S., Ferrera, L., Schiavon, M., Sondo, E., Tomati, V. et al. (2012). Association of TMEM16A chloride channel overexpression with airway goblet cell metaplasia. *J Physiol* **590**, 6141–6155.

Sheridan, J.T., Worthington, E.N., Yu, K., Gabriel, S.E., Hartzell, H.C., and Tarran, R. (2011). Characterization of the oligomeric structure of the Ca(2+)-activated Cl^- channel Ano1/TMEM16A. *J Biol Chem* **286**, 1381–1388.

Shiina, T. and Tazawa, M. (1987). Ca^{2+}-activated Cl^- channel in plasmalemma of *Nitellopsis obtusa*. *J Membrane Biol* **99**, 137–146.

Shimizu, T., Iehara, T., Sato, K., Fujii, T., Sakai, H., and Okada, Y. (2013). TMEM16F is a component of a Ca^{2+}-activated Cl^- channel but not a volume-sensitive outwardly rectifying Cl^- channel. *Am J Physiol Cell Physiol* **304**, C748–C759.

Simon, S., Grabellus, F., Ferrera, L., Galietta, L., Schwindenhammer, B., Muhlenberg, T., Taeger, G. et al. (2013). DOG1 regulates growth and IGFBP5 in gastrointestinal stromal tumors. *Cancer Res* **73**, 3661–3670.

Sones, W.R., Davis, A.J., Leblanc, N., and Greenwood, I.A. (2010). Cholesterol depletion alters amplitude and pharmacology of vascular calcium-activated chloride channels. *Cardiovasc Res* **87**, 476–484.

Stephan, A.B., Shum, E.Y., Hirsh, S., Cygnar, K.D., Reisert, J., and Zhao, H. (2009). ANO2 is the cilial calcium-activated chloride channel that may mediate olfactory amplification. *Proc Natl Acad Sci USA* **106**, 11776–11781.

Suzuki, J., Umeda, M., Sims, P.J., and Nagata, S. (2010). Calcium-dependent phospholipid scrambling by TMEM16F. *Nature* **468**, 834–838.

Takizawa, P.A., DeRisi, J.L., Wilhelm, J.E., and Vale, R.D. (2000). Plasma membrane compartmentalization in yeast by messenger RNA transport and a septin diffusion barrier. *Science* **290**, 341–344.

Thomas-Gatewood, C., Neeb, Z.P., Bulley, S., Adebiyi, A., Bannister, J.P., Leo, M.D., and Jaggar, J.H. (2011). TMEM16A channels generate Ca(2+)-activated Cl(−) currents in cerebral artery smooth muscle cells. *Am J Physiol Heart Circ Physiol* **301**, H1819–H1827.

Tian, Y., Kongsuphol, P., Hug, M., Ousingsawat, J., Witzgall, R., Schreiber, R., and Kunzelmann, K. (2011). Calmodulin-dependent activation of the epithelial calcium-dependent chloride channel TMEM16A. *FASEB J* **25**, 1058–1068.

Tien, J., Lee, H.Y., Minor, D.L., Jr., Jan, Y.N., and Jan, L.Y. (2013). Identification of a dimerization domain in the TMEM16A calcium-activated chloride channel (CaCC). *Proc Natl Acad Sci USA* **110**, 6352–6357.

Tran, T.T., Tobiume, K., Hirono, C., Fujimoto, S., Mizuta, K., Kubozono, K., Inoue, H., Itakura, M., Sugita, M., and Kamata, N. (2013). TMEM16E (GDD1) exhibits protein instability and distinct characteristics in chloride channel/pore forming ability. *J Cell Physiol* **229**, 181–190.

Tsutsumi, S., Kamata, N., Vokes, T.J., Maruoka, Y., Nakakuki, K., Enomoto, S., Omura, K. et al. (2004). The novel gene encoding a putative transmembrane protein is mutated in gnathodiaphyseal dysplasia (GDD). *Am J Hum Genet* **74**, 1255–1261.

Vermeer, S., Hoischen, A., Meijer, R.P., Gilissen, C., Neveling, K., Wieskamp, N., de Brouwer, A. et al. (2010). Targeted next-generation sequencing of a 12.5 Mb homozygous region reveals ANO10 mutations in patients with autosomal-recessive cerebellar ataxia. *Am J Hum Genet* **87**, 813–819.

Wahbi, K., Behin, A., Becane, H.M., Leturcq, F., Cossee, M., Laforet, P., Stojkovic, T. et al. (2013). Dilated cardiomyopathy in patients with mutations in anoctamin 5. *Int J Cardiol* **168**, 76–79.

Wang, M., Yang, H., Zheng, L.Y., Zhang, Z., Tang, Y.B., Wang, G.L., Du, Y.H. et al. (2012). Downregulation of TMEM16A calcium-activated chloride channel contributes to cerebrovascular remodeling during hypertension by promoting basilar smooth muscle cell proliferation. *Circulation* **125**, 697–707.

West, R.B., Corless, C.L., Chen, X., Rubin, B.P., Subramanian, S., Montgomery, K., Zhu, S. et al. (2004). The novel marker, DOG1, is expressed ubiquitously in gastrointestinal stromal tumors irrespective of KIT or PDGFRA mutation status. *Am J Pathol* **165**, 107–113.

Whelan, S., de Bakker, P.I., and Goldman, N. (2003). Pandit: A database of protein and associated nucleotide domains with inferred trees. *Bioinformatics* **19**, 1556–1563.

Winpenny, J.P., Harris, A., Hollingsworth, M.A., Argent, B.E., and Gray, M.A. (1998). Calcium-activated chloride conductance in a pancreatic adenocarcinoma cell line of ductal origin (HPAF) and in freshly isolated human pancreatic duct cells. *Pflugers Arch* **435**, 796–803.

Wolf, W., Kilic, A., Schrul, B., Lorenz, H., Schwappach, B., and Seedorf, M. (2012). Yeast Ist2 recruits the endoplasmic reticulum to the plasma membrane and creates a ribosome-free membrane microcompartment. *PLoS One* **7**, e39703.

Xia, X.M., Fakler, B., Rivard, A., Wayman, G., Johnson-Pais, T., Keen, J.E., Ishii, T. et al. (1998). Mechanism of calcium gating in small-conductance calcium-activated potassium channels. *Nature* **395**, 503–507.

Xiao, Q., Hartzell, H.C., and Yu, K. (2010). Bestrophins and retinopathies. *Pflugers Arch* **460**, 559–569.

Xiao, Q., Yu, K., Perez-Cornejo, P., Cui, Y., Arreola, J., and Hartzell, H.C. (2011). Voltage- and calcium-dependent gating of TMEM16A/Ano1 chloride channels are physically coupled by the first intracellular loop. *Proc Natl Acad Sci USA* **108**, 8891–8896.

Yang, H., Kim, A., David, T., Palmer, D., Jin, T., Tien, J., Huang, F. et al. (2012). TMEM16F forms a Ca^{2+}-activated cation channel required for lipid scrambling in platelets during blood coagulation. *Cell* **151**, 111–122.

Yang, Y.D., Cho, H., Koo, J.Y., Tak, M.H., Cho, Y., Shim, W.S., Park, S.P. et al. (2008). TMEM16A confers receptor-activated calcium-dependent chloride conductance. *Nature* **455**, 1210–1215.

Yu, K., Duran, C., Qu, Z., Cui, Y.Y., and Hartzell, H.C. (2012). Explaining calcium-dependent gating of anoctamin-1 chloride channels requires a revised topology. *Circ Res* **110**, 990–999.

Zhang, C.H., Li, Y., Zhao, W., Lifshitz, L.M., Li, H., Harfe, B.D., Zhu, M.S., and ZhuGe, R. (2013). The transmembrane protein 16A Ca(2+)-activated Cl^- channel in airway smooth muscle contributes to airway hyperresponsiveness. *Am J Respir Crit Care Med* **187**, 374–381.

Zygmunt, A.C. and Gibbons, W.R. (1991). Calcium-activated chloride current in rabbit ventricular myocytes. *Circ Res* **68**, 424–437.

33 Store-operated CRAC channels

Murali Prakriya

Contents

In many animal cells, store-operated Ca^{2+} release-activated Ca^{2+} (CRAC) channels function as an essential route for Ca^{2+} entry. CRAC channels control fundamental cellular functions including gene expression, motility, and cell proliferation and are involved in the etiology of several disease processes including a severe combined immunodeficiency syndrome. With several distinguishing biophysical and molecular features, CRAC channels are constructed as a two-component system consisting of the stromal interaction molecule (STIM) proteins, which serve as the ER Ca^{2+} sensors, and the ORAI proteins, which form the channel pores. ER Ca^{2+} store depletion evokes direct interactions between the STIM sensors and the ORAI channels, driving their redistribution and accumulation into overlapping puncta at peripheral cellular sites to elicit localized elevations of $[Ca^{2+}]_i$ at clusters of CRAC channels. This chapter examines the molecular features of the STIM and ORAI proteins that regulate the operation of CRAC channels and highlights their physiological roles in select organ systems.

33.1 INTRODUCTION

Ca^{2+} is a ubiquitous intracellular signaling messenger involved in a wide range of cellular functions including enzyme activation, gene expression, chemotaxis, and neurotransmitter release. Cellular Ca^{2+} signals generally arise from the opening of Ca^{2+} permeable ion channels, a diverse family of membrane proteins that include voltage-gated Ca^{2+} channels, ligand-gated channels, and sensory transduction channels such as the TRP channels. In this large family, store-operated channels (SOCs) are recognized as a major mechanism for eliciting long-lasting Ca^{2+} signals in metazoans. Physiologically, SOCs are activated by the depletion of intracellular Ca^{2+} stores that occurs in response to stimulation of cell surface receptors coupled to the generation of ionositol trisphosphate (IP3). The ensuing store-operated calcium entry (SOCE) is implicated in diverse cellular functions that include gene expression, cell motility, and differentiation. Clinical studies have revealed that patients with mutations in SOCs suffer from a devastating immunodeficiency, muscle weakness, and abnormalities in the skin and teeth [1,2]. Moreover, animal studies have implicated a growing list of possible diseases from allergy [3], multiple sclerosis [4,5], cancer [6], thrombosis [7], and inflammatory bowel disease [8] to loss or gain of SOC activity, highlighting the potential importance of these channels for human health and disease.

Early studies of SOCE relied on cytosolic Ca^{2+} indicator dyes to examine Ca^{2+} entry through SOCs [9,10]. These investigations were strongly aided by the discovery of thapsigargin, a plant alkaloid that is a potent inhibitor of the SERCA pump in the ER membrane [11,12]. Thapsigargin allowed investigators to directly deplete intracellular Ca^{2+} stores without stimulating surface receptors thereby eliminating the myriad confounding effects of signaling that results from receptor stimulation. Electrophysiological studies that followed characterized the biophysical and functional features of store-operated currents in T lymphocytes and mast cells [13,14]. The SOCs in these cells, which were termed "calcium release-activated calcium (CRAC) channels," exhibit high Ca^{2+} selectivity and can be distinguished from other Ca^{2+}-selective channels based on their low unitary conductance and low permeability to large monovalent cations [15].

Despite early identification and characterization of CRAC currents, however, the molecular identity of CRAC channels and the mechanism linking ER Ca^{2+} store depletion to their activation remained unknown for several decades. This changed dramatically following the identification of STIM1 in 2005 as the ER Ca^{2+} sensor and ORAI1 in 2006 as a prototypic SOC subunit. These discoveries were a major catalyst for elucidation of the molecular mechanisms of CRAC channels and their physiological functions in many organ systems [16]. We now know that CRAC channels are activated through the binding of the ER Ca^{2+} sensors stromal interaction molecule 1 (STIM1) and STIM2 to the CRAC channel proteins ORAI1, ORAI2, and ORAI3 [17,18]. The STIM proteins bind to and directly activate ORAI channels, and these two families of molecules can fully reconstitute SOCE in heterologous expression systems, indicating that these proteins are both necessary and sufficient for SOCE. This chapter focuses on the molecular characteristics of STIM and ORAI proteins that regulate the activation of CRAC channels and their ion conduction mechanisms.

33.2 THE BIOPHYSICAL FINGERPRINT OF THE CRAC CHANNEL

Long before the identification of the STIM and ORAI proteins, several decades of electrophysiological studies had already revealed many important biophysical and pharmacological features of the CRAC channel. Overall, CRAC channels are widely noted for their exquisite Ca^{2+} selectivity ($P_{Ca}/P_{Na} > 1000$), which places them in a unique category of highly Ca^{2+} selective channels together with voltage-gated Ca^{2+} (Ca_v) channels [19]. However, its unitary conductance is >100-fold smaller than that of Ca_v channels; too small to measure from single-channel openings. It has been estimated from noise analysis to be 10–30 fS for Ca^{2+} and ~1 pS for Na^+ in the absence of extracellular divalent cations [14,20,21]. Interestingly, high Ca^{2+} selectivity is only manifested in Ca^{2+}-containing solutions, CRAC channels readily conduct a variety of small monovalent ions (Na^+, Li^+, and K^+) in divalent-free solutions [20,22,23], indicating that high Ca^{2+} selectivity is not an intrinsic feature of the CRAC channel pore but arises due to ion–ion and ion–pore interactions. This is clearly revealed by the blockade of monovalent currents by micromolar concentrations of Ca^{2+} (K_i ~ 20 μM at −100 mV) [19–24]. Occupancy by a single Ca^{2+} ion appears sufficient to block the large monovalent conductance, and, as expected for a binding site within the pore, Ca^{2+} block is voltage-dependent [21,25]. These characteristics are qualitatively reminiscent of the properties of L-type Ca_v channels, in which Ca^{2+} ions similarly bind tightly to a high-affinity binding site within the pore to occlude Na^+ flux [26,27]. In contrast to Ca_v channels, however, CRAC channels are virtually impermeable to the large monovalent cation, Cs^+ ($P_{Cs}/P_{Na} < 0.1$). This feature has been traced to a relatively narrow CRAC channel pore that likely limits the electrodiffusion of Cs^+ [21,25]. The channels are not sensitive to many of the common inhibitors of Ca_v channels, but are inhibited by imidazole antimycotics, submicromolar concentrations of lanthanides, and 2-aminoethyldiphenyl borate (2-APB), among others [28]. Overall, the high Ca^{2+} selectivity, low conductance, and high sensitivity to lanthanides emerged as stringent criteria by which to screen candidates for the CRAC channel gene, resulting in the exclusion of many candidates belonging to the TRP gene family, as reviewed in [29,30].

33.3 STIM1 IS THE ER Ca^{2+} SENSOR FOR SOCE

STIM (stromal interaction molecule) proteins were first discovered as cell surface proteins that can bind to pre-B lymphocytes and mature B lymphocytes [31]. Initial studies suggested a role for these proteins in tumor growth but subsequent studies using RNAi screens to inhibit SOCE found an unmistakable role for these proteins in regulating CRAC channel function. These screens used thapsigargin to deplete ER Ca^{2+} stores and examined suppression of SOCE in either *Drosophila* S2 [32] and mammalian cells [33]. The mammalian homologue identified from these studies, STIM1, is a 77 kDa single-pass ER membrane protein, with a luminal N-terminal domain containing the signal peptide and a large C-terminal domain in the cytosol [18]. Both regions exhibit several domains critical to its function including a sterile alpha motif (SAM) and two EF hands in the N-terminus (Figure 33.1a) and three coiled–coiled domains, a Ser/Pro-rich region, and a Lys-rich region in the C-terminus. Whereas *Drosophila* has a single STIM gene, mammals have two closely related genes, *STIM1* and *STIM2*, which differ significantly in their C-terminal region. In resting cells, STIM1 is largely localized in the bulk ER [33–38]. ER Ca^{2+} store depletion triggers unbinding of Ca^{2+} from the luminal EF-hand, which ultimately results in the redistribution of STIM1 from the bulk ER into puncta located in close apposition to the plasma membrane [33–38]. The EF-hand and SAM domains (EF-SAM) mediate critical roles in this process. EF-hand STIM1 mutants with impaired Ca^{2+} binding form puncta and activate CRAC channels independently of ER Ca^{2+} store depletion [33,34]; the absence of Ca^{2+} binding in these mutants essentially tricks the molecules into responding as if stores are depleted. Moreover, deletion of the SAM domain abrogates oligomerization and puncta formation in response to store depletion [35], indicating that this domain mediates a critical role in initiating the STIM1 conformational response to store depletion. The cytosolic C-terminal portion is essential for the redistribution of STIM1 oligomers to ER-PM junctions and subsequent CRAC channel activation occurs through a critical channel interaction domain encompassing the second coiled–coiled domain [35,39–42]. In this manner, changes in Ca^{2+}-binding at the N-terminus are coupled to SOCE initiation through protein–protein interactions in the STIM1 C-terminus. Thus, STIM1 fulfills two critical roles in the activation process of CRAC channels: sensing the depletion of ER Ca^{2+} stores, and communicating store depletion to CRAC channels located in the plasma membrane.

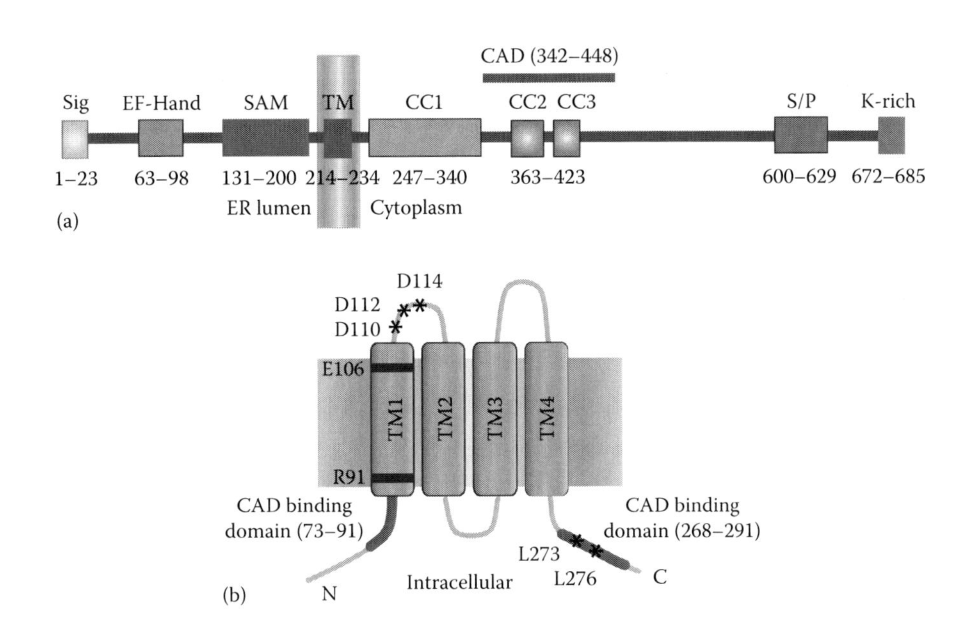

Figure 33.1 (a) A schematic representation of STIM1 and its key functional domains. These domains include Sig, signal peptide; SAM, sterile alpha motif; TM, transmembrane domain; CC, coiled-coil domain; CAD, CRAC activation domain; ID_{STIM1}, Inactivation domain of STIM1. (b) The predicted topology of ORAI1. The STIM1 binding domains are depicted as thick lines in the intracellular N- and C-termini. . Key residues (mentioned in the text) are labeled.

33.4 THE DISCOVERY OF ORAI PROTEINS

Over a a two-decade period from 1990, large scale efforts were undertaken to identify the molecular basis of the CRAC channel and how it is turned on by ER Ca^{2+} store depletion. These efforts culminated in 2006 with the identification of ORAI1 as the canonical CRAC channel protein [43–46]. In the intervening period, several candidate molecules including several TRP channels and voltage-gated Ca^{2+} channels were presented as possible candidates as CRAC channel pore [29,30], only to be discarded due to inconsistencies in their pore properties with those of native CRAC channels. Ultimately, efforts that led to the identification of the CRAC channel protein harnessed the power of high-throughput screening, linkage analysis, and the human genome sequencing project, tools that became available only in the new millennium. An important step in this discovery was the identification of human patients with a severe combined immunodeficiency lacking CRAC channel function in T-cells [47–49]. These patients exhibited a devastating immunodeficiency characterized by impaired T cell activation and effector gene expression [50], which confirmed earlier pharmacological and genetic evidence that CRAC channels orchestrate many aspects of lymphocyte development and function [51,52]. Feske et al., took advantage of a partial reduction in Ca^{2+} entry in the heterozygotes in the patient pedigree to localize the source of the defect to a small region in chromosome 12 with ~70 genes [43]. Simultaneously, genome-wide RNAi screens for genes involved in SOCE in *Drosophila* S2 cells carried out by three groups identified a novel gene as a critical mediator of *Drosophila* SOCE [43,53,54]. A human ortholog of this protein was mapped to the same region on chromosome 12 identified by linkage analysis [43]. This molecule, ORAI1, is a widely expressed 33 kDa cell surface protein with four predicted transmembrane domains, intracellular N- and C- termini (Figure 33.1b) and no significant sequence homology to other previously identified ion channels. The human SCID defect was found to arise from a missense mutation in ORAI1 (R91W) that abrogated CRAC channel activity [43].

33.5 ORAI1: A PROTOTYPIC CRAC CHANNEL PROTEIN

The conclusion that ORAI1 is a canonical CRAC channel pore-forming protein came quickly following its identification. One line of evidence came from studies showing that overexpression of ORAI1 together with STIM1 in HEK293 cells produced large currents [55–57] with characteristics consistent with native CRAC channels including high Ca^{2+} selectivity, low Cs^{+} permeability, and a narrow pore [25,55,56]. Other properties including Ca^{2+} block of Na^{+} current and pharmacological responses to 2-APB and La^{3+} reaffirmed similarities to native CRAC channels [25,45]. A second line of evidence came from mutation of a highly conserved acidic residue in ORAI1 (E106D), which significantly diminished the Ca^{2+} selectivity of the CRAC channels and altered a wide range of properties intimately associated with the pore, including La^{3+} block, the voltage-dependence of Ca^{2+} blockade, and Cs^{+} permeation [44–46]. In the ensuing years, cysteine-scanning studies that identified the pore-lining residues of the CRAC channel [58,59] and the crystal structure of *Drosophila* ORAI firmly established the ORAI proteins as the Ca^{2+}-selective pores of CRAC channels. Mammalian cells express two other closely related homologues, ORAI2 and ORAI3, that differ primarily in their C-terminal and the 3–4 loop sequence. All three isoforms function similarly in producing store-operated Ca^{2+} entry when co-expressed with STIM1 in HEK293 cells [60–62] and are widely expressed in most tissues [60,62,63]. However, ORAI1 remains the best-studied CRAC channel protein and appears to be the predominant isoform mediating SOCE in most cells. By contrast, there is no direct genetic evidence for a role of ORAI2 or ORAI3 channels in any cell type yet.

33.6 PHYSIOLOGICAL FUNCTIONS OF CRAC CHANNELS

Knockout studies and human patients with mutated *STIM/ORAI* genes have revealed many specific roles for CRAC channels in different organs systems. One well characterized physiological role is the generation of long-lasting $[Ca^{2+}]_i$ elevations essential for Ca^{2+}-dependent gene transcription and cytokine production in lymphocytes. Evidence for this role has come from several lines of study. For example, SKF96365, an imidazole compound that inhibits CRAC channels was also found to block IL-2 production in T-cells with similar efficacy [64]. Likewise, nanomolar concentrations of La^{3+} were found to block I_{CRAC}, the $[Ca^{2+}]_i$ rise, and the induction of T-cell activation markers such as CD25 and CD69 in response to CD3- or thapsigargin-stimulation [65]. Subsequent genetic studies provided compelling evidence that these pharmacological effects occur through the blockade of CRAC channels. For example, mutant Jurkat T cells lacking Ca^{2+} release-activated Ca^{2+} current, I_{CRAC}, displayed severely attenuated production of cytokines such as IL-2 [51]. More recently, several studies of human patients have shown that severe immunodeficiencies arise from mutations in CRAC channels that render them inactive [2,43,47–49,66]. The abrogation of CRAC channel function in these cells results in the elimination of Ca^{2+} elevations necessary to drive nuclear translocation of NFAT [48], an important and widely expressed transcription factor involved in cytokine gene expression [67].

In addition to T-cells, CRAC channels serve as the primary Ca^{2+} influx route for agonist-evoked Ca^{2+} entry in platelets. STIM1-deficient mice exhibit impaired agonist stimulated platelet SOCE and impaired thrombus formation resulting in a mild increase in bleeding time following injury [68,69]. STIM1–/– mice were significantly protected from arterial thrombosis and ischemic brain infarction. Likewise, ORAI1 deficient mice exhibited defective SOCE in platelets, and impaired thrombus formation and resistance to various measures of thrombus formation including pulmonary thromboembolism, arterial thrombosis, and ischemic brain infarction [70]. These findings indicate that CRAC channels in platelets are crucial mediators of cerebrovascular thrombus formation, raising the possibility that blockade of channel function might be beneficial for the treatment of this condition.

In addition to the immune phenotypes, genetic studies in humans and mice reveal defects in skeletal muscle function and development [71,72]. Mice lacking STIM1 show no SOCE in skeletal muscle and this phenotype is accompanied by marked propensity for rapid muscle fatigue during repeated stimulation [71]. Interestingly, unlike in nonexcitable cells, STIM1 and ORAI1 proteins appear to be pre-localized in proximity to each other within the triad junction in skeletal muscle under resting conditions, thus permitting extremely fast and efficient trans-sarcolemmal Ca^{2+} influx during store depletion [72].

STIM and ORAI proteins have also emerged as important regulators of several types of cancers, including breast, cervical, and glioblastomas [73]. One recent study found that ORAI1 and STIM1 were essential for breast tumor cell migration in vitro and tumor metastasis in mice [74]. Pharmacological inhibition of SOCE using SKF96365 or knockdown of ORAI1 or STIM1 RNA interference resulted in decreased tumor metastasis in mice. Another study showed that human breast cancer lines displayed increased levels of ORAI1, and microarray analysis 295 breast cancers found that women with a transcriptional profile characterized by STIM1high/STIM2low ratios had the poorest prognosis [75]. A different study has reported that breast cancer cell lines expressing the estrogen receptors exhibit SOCE that is mediated by ORAI3, in contrast to estrogen-receptor negative lines that reportedly depend on ORAI1 [73], raising the possibility that a switch in the machinery of ORAI channels mediates a causal role in the phenotypic switch to tumorogenesis [76]. The mechanisms by which Ca^{2+} signaling is altered in breast cancers remains uncertain, although one report indicates that ORAI1 in breast cancer cells functions independently of STIM1, through a mechanism that involves activation by the Golgi Ca^{2+}-ATPase, SPCA2 [77]. In the highly tumorigenic breast cancer line MCF-7, these authors found that SPCA2 redistributes from its normal localization in the Golgi to the plasma membrane, thereby positioning it to constitutively activate ORAI1 channels [77]. Although the specific mechanisms through which CRAC channels are modulated in the different disease states is unknown in most cases, these studies nevertheless establish a critical role for CRAC channels in different disease processes and provide strong impetus for understanding its molecular physiology at many levels.

33.7 OLIGOMERIZATION AND REDISTRIBUTION OF STIM1 TO THE ER–PLASMA MEMBRANE JUNCTIONS

The STIM proteins respond to alterations in ER Ca^{2+} store content through Ca^{2+} unbinding–binding from an EF-hand in the luminal domain. Studies of the isolated STIM1 EF-hand SAM domain in solution indicate that the dissociation constant of Ca^{2+} binding is ~500 μM [78,79], which is consistent with the range of Ca^{2+} concentrations known to exist in ER lumen [80,81]. Similar measurements for STIM2 indicate a lower affinity Ca^{2+} binding affinity [79], which is in agreement with the ability of STIM2 to form puncta more readily in resting cells [82]. Measurements of the dependence of $[Ca^{2+}]_{ER}$ on puncta formation indicate an apparent $K_{1/2}$ of 210 μM for STIM1 and 406 μM for STIM2 [82], which implies that that STIM2 needs significantly lower levels of store depletion than STIM1 for puncta formation [37]. This difference likely explains why STIM2, but not STIM1, promotes constitutive activation of SOCE when overexpressed with the ORAI proteins [82,83].

Structural studies of the isolated luminal domain fragments indicated that Ca^{2+} unbinding from the N-terminal EF-hand triggers the unfolding and aggregation of the luminal domain, resulting in the appearance of dimers and higher-order multimers [78]. This finding led to the idea that Ca^{2+} store depletion results in the formation of higher order oligomers of STIM1 [78], a hypothesis that was confirmed in follow-up studies using FRET with full-length STIM1 [39,84]. STIM1 oligomerization is an early step in the channel activation process, occurring well before STIM1 redistribution to the plasma membrane [39,80].

The critical role of STIM1 oligomerization for CRAC channel activation is underscored by the finding that artificially oligomerizing engineered STIM1 in which the luminal domain is deleted and replaced with the FRB-FKBP dimerizer leads to puncta formation and activation of SOCE independently of ER store depletion [80]. STIM1 oligomerization thus serves as a critical upstream activation switch that unfolds all subsequent steps of the channel activation process.

Perhaps the most striking feature of STIM1 behavior is its redistribution from the bulk ER in resting cells with full stores, to the plasma membrane where it accumulates into discrete puncta [33–35,85]. Accumulation of STIM1 near the plasma membrane causes ER tubules to move toward the plasma membrane and STIM1 appears to facilitate this process [36], indicating that store depletion rearranges the ER, with STIM1 facilitating this change. Consistent with the idea that changes in $(Ca^{2+})_{ER}$ (and not cytoplasmic (Ca^{2+})) directly regulate this process, depletion of ER Ca^{2+} stores with TPEN, a low affinity membrane permeant Ca^{2+} buffer, which is not expected to affect cytoplasmic (Ca^{2+}), also causes STIM1 and ORAI1 puncta formation [86]. Luik and colleagues further showed that the accumulation of Cherry-STIM1 at the plasma membrane exhibits the same dependence on ER Ca^{2+} concentration as activation of I_{CRAC} [80]. This is consistent with the notion that channel activation requires a local interaction between STIM1 and ORAI subunits, which can only occur following the redistribution of STIM1 from the bulk ER to the periphery. Moreover, the appearance of STIM1 near the surface of the cell precedes the development of I_{CRAC} by ~10 s [36], indicating that STIM1 translocation is required not only for CRAC activation, but that channel activation likely also requires additional steps.

Although STIM redistribution is triggered by the same initial conformational change that causes oligomerization—with both steps requiring unbinding of Ca^{2+} from its N-terminal EF-hand, these processes are distinct and fully separable. Redistribution occurs with a lag of tens of seconds following STIM1 oligomerization [37,39], indicating that the two steps are kinetically separable. Additionally, truncation of a basic region at the extreme C-terminus of STIM1 attenuates redistribution of the truncated STIM1 to peripheral sites without affecting STIM1 oligomerization [39], indicating that the molecular determinants of these processes are distinct. Interestingly, the deleterious effect of removing the K-rich C-terminal region on STIM1 redistribution is only seen in cells overexpressing STIM1 alone: when co-expressed with ORAI1, STIM1ΔK accumulates into puncta to the same extent as that seen in full-length STIM1 and supports the normal extent of puncta formation [40], though activation of I_{CRAC} is still delayed compared with full-length STIM1 [87]. Because the polybasic tail is not found in *Caenorhabditis elegans* or *Drosophila* STIM proteins, these results suggest that the polybasic domain is a vertebrate adaption that facilitates STIM1 migration to the plasma membrane, but is not essential for ORAI channel activation. It is also tempting to postulate in this context that the accumulation and stability of STIM1 at the ER-PM junctions is influenced by the strength of the STIM1 plasma membrane binding interactions, with weaker interactions diminishing the efficacy of STIM1 migration and binding to the plasma membrane.

In contrast to STIM1 oligomerization, relatively little is known about the mechanisms controlling STIM1 redistribution to peripheral puncta. Studies with fluorescently labeled STIM1 (CFP- or YFP-) indicate that STIM1 is at least partially associated with microtubules (MTs) and moves rapidly along tubulovesicular structures that overlap with MTs in resting cells with replete stores [35,88,89]. STIM1 also co-IPs with the MT-associated proteins, EB1 and EB3, indicating that STIM1 is closely associated with microtubules [88]. The functional relevance of this association, however, is nebulous. Nocodozole, which depolymerizes microtubules and therefore would be expected to severely impair STIM1 association with microtubules, does not affect puncta formation or even SOCE [35,88], although it does eliminate the tubulovesicular STIM1 movements. Likewise, depletion of cellular ATP eliminates the tubulovesicular movement, but does not impact puncta formation [90]. Thus, the relationship between MT association and CRAC channel activation currently remains mysterious. One possibility is that STIM1 exhibits two forms of movement, one along MTs that is powered by motors, and second diffusive mode of migration that is MT-independent. The available data suggests that SOCE is driven solely by the diffusive mode of STIM1 mobility, but this would be predicted to limit its movement to relatively short distances, a prediction that appears borne out by limited diffusional mobility of STIM1 [39,40].

33.8 CONFORMATIONAL CHANGES IN STIM1

STIM1 contains a variety of functional domains in its luminal and cytoplasmic regions (Figure 33.1). Current models of STIM1 activation indicate that interactions between these domains both within the same molecule as well as between neighboring STIM1 molecules regulate the activation state of STIM1. Some early reports suggested that internal electrostatic interactions between different regions of STIM1 are critical for the transition from resting STIM1 oligomers to their active state [91,92]. An acidic region in the CC1 domain was found to interact with a basic region in the CC2 domain to mask the active site of STIM1 that interacts with ORAI1 (termed the CAD domain). Oligomerization appears to favor the removal of this internal autoinhibition to reveal the CAD and polybasic domains, thereby permitting productive interactions between CAD and ORAI proteins leading to channel activation [91,92]. Similar conclusions were drawn by Muik et al., who made use of an intramolecular STIM1 FRET sensor to show that the cytoplasmic region of STIM1 that interacts with ORAI1 switches from a closed to an open configuration upon interaction with ORAI1 [93]. They suggested that the closed confirmation of STIM1 is stabilized by coiled-coil interactions within the C-terminal region of STIM1 and interaction with ORAI1 opens up STIM1 to expose its active ORAI1 binding site [93]. These findings, however, have recently been called into question by more recent evidence indicating that mutations that would be expected to disrupt the electrostatic interactions have no effects on STIM1 function [94]. Moreover, the recent structures of the CAD domain are not readily compatible with interactions between acidic and basic STIM1 regions [95]. Thus, the molecular mechanisms underlying dimer formation and stimulus-induced conformational changes in STIM1 remain to be clarified. In this context, a recent study employing lanthanide-acceptor energy transfer (LRET) to probe conformational changes in STIM1 following activation found that the CC1 domain of STIM1 directly interacts with the distal STIM1 region including CC3 in the resting state [96]. This interaction is thought to maintain STIM1 in the closed conformation wherein the CAD domain is hidden. Conversely, interactions between neighboring CC1 domains of the STIM1 dimer were suggested to extend the cytoplasmic STIM1 region, producing a conformational change that exposes CAD for association with its targets. Overall, this model is harmonious with previous findings indicating that the cytoplasmic region of STIM1 is folded back to hide CAD in the resting state [93].

33.9 STIM1 BINDS DIRECTLY TO ORAI1

CRAC channels are operationally defined as store-operated channels. The mechanistic basis for store-dependent activation of CRAC channel activation involves direct binding of STIM1 and ORAI1. Direct binding of the two proteins has been demonstrated by several methods using co-immunoprecipitation [45,46,97,98], FRET microscopy [37,38], and pull-downs [40]. ORAI1 and STIM1 associate even in a system of only purified components in solution [40,99], indicating that the interaction between these proteins is strong enough to persist in a variety of chemical environments and can occur even in detergent-solubilized extracts.

Deletion and serial truncations enabled identification of the region of STIM1 required for ORAI1 activation. The C-terminal domain consists of three putative coiled coils, but the rest of the sequence bears little similarity to any known motifs (Figure 33.1a). Huang et al., found that expressing just the cytoplasmic portion of STIM1 (STIM1-ct) was sufficient to activate SOCE, though not to the same extent as full-length STIM1 [41]. They also found that deletion of a large stretch of this domain that includes the coiled-coil motifs (amino acids 231–535) eliminated constitutive activity of this fragment [41]. Baba et al., found that deleting either 249–390 or 391-end in full-length STIM1 reduced SOCE, suggesting that these regions likely contribute to ORAI1 activation [35]. Subsequently, several groups identified a minimal region in STIM1 encompassing the second and the third CC domains as the critical element required for binding and activating ORAI [40,42,100,101]. This region is variously called the CRAC activation domain (CAD) [40], STIM1–ORAI1 activating region (SOAR) [42], or Ccb9 [101] and includes the amino acids 342–444. The structure of this domain has recently been solved, revealing a R-shaped dimeric module with the main functional domains (CC2 and CC3), forming a hair-pin motif [95]. Whether this dimeric module represents the active state structure of CAD and how this differs from its resting state configuration is not known yet, nevertheless, this high-resolution structure provides several testable hypotheses for dissecting the structural underpinnings of STIM1 function.

Interestingly, the regions involved in STIM1–ORAI1 binding overlap significantly with the domains found to be important for induced STIM1–STIM1 oligomerization. Muik et al. tested a series of STIM1 C-terminal fragments and found that Orai activating STIM fragment (OASF) fragments homomerized in situ and fragments shorter than OASF lacking key elements of the CAD/SOAR domain are monomeric on native gels, don't self-associate in vivo, and fail to activate ORAI1 [100]. Covington et al. directly investigated the regions of the STIM1 C-terminus important for store-depletion induced oligomerization and found that the critical region for oligomerization and puncta formation overlaps with the CAD/SOAR region [84]. One critical mutation identified by their analysis, A369K, exhibited enhanced resting-state oligomerization, co-localization with ORAI, and constitutive CRAC-channel activity, while at the same time nearly eliminating enhancement of oligomerization upon store-depletion [84]. A second mutation, A376K, caused STIM1 to constitutively self-associate and form puncta, but eliminated co-localization with ORAI and CRAC channel activity before or after store depletion. Their results predicted that both residues lie on a hydrophobic face of the alpha helix in CC2 [84], a prediction borne out in the crystal structure of the CAD domain. Collectively, these results suggest that residues important for ORAI1 binding are also important for STIM oligomerization.

33.10 ORAI1 DOMAINS INVOLVED IN STIM1 BINDING

It is now known that STIM1 interacts with both the intracellular C- and N-termini of ORAI subunits. Early studies found that both CRAC channel activation and STIM1–ORAI1 FRET following store-depletion was eliminated by deletions of the C-terminus, suggesting the presence of an essential STIM1 binding site in this region [37,40,87]. In vitro pull-down assays with isolated fragments of ORAI1 subsequently indicated that the ORAI1 C-terminal domain fragments encompassing the region between residues 254–301 associate with the cytosolic domains of STIM1 [40,99,100], demonstrating that the STIM1 C-terminus directly interacts with the ORAI1 C-terminus. As in STIM1, the critical binding motif in the ORAI1 C-terminus is a coiled-coil domain, and two point mutations in this region (L273S and L276D) completely disrupt the ORAI1–STIM1 interaction [37,38]. Although it has been suggested that the binding defects caused by these mutations may be traced to disruption of the tertiary structure of the ORAI1 CC domain [102], it is also plausible that the residues are located directly at the STIM1 binding interface itself; hence the mutations may affect STIM1 binding directly [103]. Both L273 and L276 are highly conserved and mutations in other ORAI isoforms at equivalent positions also impede STIM–ORAI binding [102,104]. Interestingly, the recent crystal structure of ORAI1 reveals that the C-terminal helices of neighboring subunits form an anti-parallel coiled-coil domain and directly interface with each other, with residues L273 and L276 of one subunit contacting L276 and L273 residues of the conjugate subunit [105]. The structural study suggested that the hydrophobic interface formed by the anti-parallel CC domains likely open to interact with the CAD domain during STIM1 binding [105–107]. However, an alternate possibility is that the anti-parallel domains stay together even when bound to STIM1, and that the STIM1 binding site on the CRAC channel is formed collectively by the combined surface of the anti-parallel coiled-coil domain. Indeed, a recent NMR study of a ORAI1 C-terminal fragment bound to a portion of the CC1–CC2 STIM1 domain suggests that the conformation alteration in the ORAI1 C-terminus required for STIM1 binding may be quite small [103], raising the possibility that the self-associated ORAI C-termini may not completely unravel. These differing conclusions highlight the need to better define the exact binding interfaces between STIM1 and ORAI1, which still remains unclear. Moreover, many prevailing models are based on results from small fragments of STIM1 and ORAI1. Whether results from protein fragments can be readily extrapolated to full-length molecules is unclear. More studies using full-length molecules are needed to test the validity of the proposed binding models.

In addition to the well-described interaction at the C-terminus of ORAI1, studies using systems of purified components have revealed a second interaction site on ORAI1, located at the N-terminus [40,99]. A purified peptide consisting of the cytosolic N-terminal region corresponding to the region 68–91 interacts with CAD in co-immunoprecipitation and split-ubiquitin assays [40]. Further, GST-pulldown assays have shown direct interaction of a purified n-terminal fragment 65–87 with purified STIM11-ct and the STIM1 c-terminal fragment 233–498 [99].

Functional studies have presented a confusing picture for the roles for the C- and N-terminal sites. In contrast to the complete elimination of STIM1 binding seen in the C-terminal mutations, some early studies found that N-terminal deletions retain significant levels of STIM1-ORAI1 binding, and yet, these deletions abrogated SOCE and I_{CRAC} [37,40,87,108]. A construct with only 73–84 deleted fails to support CRAC activity when co-expressed with STIM [40], yet, some N-terminal deletion mutants formed puncta [87] and supported increases in ORAI1–STIM1 FRET following store depletion [37]. These results led to a modular functional assignment for the roles of the two binding sites, with the C-terminal site thought to mediate STIM1 binding and the N-terminal site thought to strictly regulate only channel gating. However, a more recent study indicates that the N-terminus contributes significantly to the overall stability of STIM1–ORAI1 binding with deletions and mutations at this site strongly diminishing ORAI1 recruitment into puncta and STIM1–ORAI1 binding [109]. This study also found that completely deleting the C-terminal site resulted in nonfunctional channels even when CAD was directly tethered to the ORAI1 C-terminus [109], indicating that the C-terminal site has a role in gating beyond STIM1 binding. These revised results indicate that the C- and N-terminal STIM1 binding sites are both essential for multiple aspects of ORAI1 function including STIM1–ORAI1 association, ORAI1 trapping, and channel gating.

33.11 THE SUBUNIT STOICHIOMETRY OF CRAC CHANNELS

There is strong evidence that the ORAI subunits interact with each other to form a multi-subunit complex [38,46,60,61,97,99]. However, attempts to evaluate the stoichiometry of this interaction from biochemical and functional assays have proven controversial. Gwack et al. reported that purified ORAI1 co-migrates with STIM1 in glycerol-gradient centrifugation, and that this fraction runs as monomers and dimers on denaturing SDS-PAGE [60]. Maruyama et al. reported that purified ORAI1 is 3× larger than a tetramer [110], and likewise, Park et al. found that purified ORAI1 elutes in a 290 kDa complex [40]. While these studies reaffirmed that ORAI exists in a higher order oligomer, they did not provide an easily interpretable ORAI1 stoichiometry.

Two labs applied the subunit-counting approach wherein channel stoichiometry is evaluated by counting the number of photo-bleach steps of GFP fused to ORAI1 monomers [111]. With this approach, these studies concluded that ORAI1 channels stably bound to soluble STIM C-terminus fragment (STIM1-ct) have 4 ORAI1 copies per channel complex [97,112]. However, in the absence of STIM1, the two groups came to differing conclusions. Penna et al. reported that most GFP-ORAI1 complexes bleach in only two steps [97]. Likewise, ORAI3 channels gated directly by the small-molecule, 2-APB, were found to bleach mostly in two steps [113]. These results were interpreted in favor of the model in which ORAI exists as a dimer in the resting state, with STIM1 assembling the ORAI1 dimers to form functional, tetrameric channels. In contrast, Ji et al. found that co-expression of STIM1-ct did not affect the number of bleach steps, which occurred in three or four steps in both conditions [112]. In an alternate approach, Madl et al. used a combination of photobleaching and single molecule brightness analysis on the mobile fraction of ORAI1 in resting cells and concluded that ORAI1 predominantly diffuses as a tetramer [114]. They also showed that FRET between ORAI1 dimers was unaltered upon store-depletion suggesting that the stoichiometry of ORAI1 was independent of its association with STIM1 [59]. In yet another approach, Mignen et al. exploited the ability of pore mutants of ORAI1 (e.g., E106Q) to suppress CRAC channel activity through a dominant-negative effect [115]. They found that the ability of (monomeric) ORAI1 E106Q to inhibit I_{CRAC} is eliminated when co-expressed with tandem wt ORAI1 constructs containing four protomers [115]. Taken together, these studies concluded that the active channel is a tetramer, although they reached different conclusions on the resting state stoichiometry. All of these studies, however, suffered from the caveat that the underlying results did not rule out the possibility of the stoichmetry being more than four.

The most definitive evidence for the subunit stoichiometry of the CRAC channel has come from the recent crystal structure of the *Drosophila* ORAI1 channel [105]. This structure of the closed channel (crystallized in the absence of STIM1) revealed a multimeric complex composed of six ORAI subunits arranged in a threefold axis of symmetry (Figure 33.2a). The C-termini of neighboring subunits were found to be self-associated in an interesting anti-parallel coiled-coil configuration in this structure. Moreover, the cytoplasmic N-terminus was found to extend the pore considerably longer into previously suspected, resulting in an unusually long ion putative ion conduction pathway of ~55 Å (Figure 33.2b) [105]. The hexameric stoichiometry is consistent with the published evidence from size-exclusion and light scattering studies showing a complex of 290 kDa [40]. Thus, the structure of the purified ORAI complex has produced a model that differs from all previous studies of stoichiometry. It remains unclear, however, whether the purified

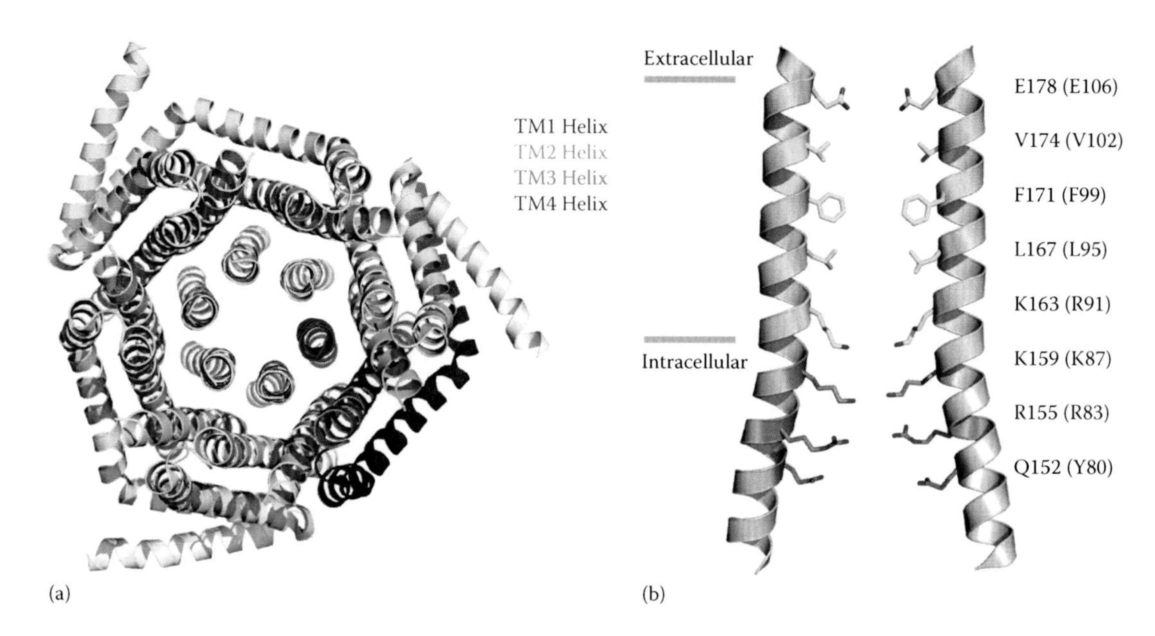

Figure 33.2 Crystal structure of *Drosophila* ORAI in the closed state. (a) Cross-sectional view of ORAI1 from the extracellular side showing hexameric ORAI complex arranged around a central axis. The transmembrane helices of a single subunit are highlighted in color. (b) Architecture of the ORAI pore showing two TM1 helices. The predicted pore-lining residues in TM1 are highlighted in yellow along with their side chains. The corresponding residues for human ORAI1 are shown in parentheses. Residues in the ORAI N-terminus including K159, R155, and Q152 (K87, R83, and Y80 in human ORAI1) that may be a part of the ion conduction pathway are highlighted in orange.

hexameric complex exhibits the canonical properties of CRAC channels including high Ca^{2+} selectivity and low permeability to Cs^{+}. Indeed, a recent study using concatenated ORAI1 monomers has contended that an overexpressed hexameric concatemer does not produce currents with the signature high Ca^{2+} of CRAC channels [116]. Thus, additional studies are clearly needed to evaluate findings predicted from the structure. Electrophysiological and molecular approaches promise to provide healthy debate on this issue.

33.12 HOW MANY STIM MOLECULES DOES IT TAKE TO ACTIVATE CRAC CHANNELS?

An important mechanistic attribute of CRAC channels with far-reaching consequences for CRAC channel operation is the functional stoichiometry of channel activation: How many STIM1 molecules does it take to drive the opening of CRAC channels? To tackle this issue, Li et al. used tandem constructs to determine the effect of STIM:ORAI1 ratio on CRAC channel activity [117]. In their approach, tandem constructs with varying number of ORAI1 protomers were fused to a STIM1 region containing the minimal activation domain (amino acids 336–485, called S). Their results indicated that for tandem constructs in which the S:ORAI1 ratio was 1:1 or 1:2, addition of a tandem S–S construct increased I_{CRAC} magnitude, but for complexes where the ratio is 2:1, the exogenous expression of S–S had no impact on current magnitude [117]. Moreover, (when expressed alone) I_{CRAC} was largest for constructs with a 2:1 S-ORAI1 ratio, and decreased as the STIM1–ORAI1 ratio decreased [117]. Their data suggested that a 2:1 STIM1:ORAI1 ratio gives optimal CRAC activity. If each complex contains four copies of ORAI1, then the active complex would have eight copies of STIM1. Crucially, their data showed that if the STIM1:ORAI1 ratio is less than optimal, I_{CRAC} is diminished but not eliminated.

In an alternate approach, Hoover and Lewis varied the relative expression of full-length STIM1 and ORAI1 proteins fused to mCherry and GFP to study the functional requirements of ORAI1 activation as well as trapping at the ER-PM junctions [118]. Their results confirmed that activation is graded with increasing STIM1 concentration, but in contrast to the findings using concatenated constructs, this study found that the dependence on STIM1 concentration is highly nonlinear [118]. Maximal CRAC current activation requires the binding of two STIM1 molecules per ORAI1 subunit, and declines sharply with diminishing STIM1 such that the minimal stoichiometry for trapping ORAI proteins into puncta fails to evoke significant activation [118]. These results were interpreted in terms of a model in which once an optimal STIM threshold is reached, individual channels open abruptly to their fully active state in an all-or-none manner due to high cooperativity of channel opening. At face value, this model is consistent with findings indicating that the slow increase in I_{CRAC} following store depletion occurs from the stepwise recruitment of closed channels to a very high P_o state [21]. However, whether this type of modal gating occurs due to STIM1 binding or represents an intrinsic property of ORAI proteins remains to be resolved.

Interestingly, there is strong evidence that the STIM1:ORAI1 ratio affects not only STIM1-dependent activation, but also calcium-dependent fast inactivation (CDI) [119,120] and as described further later, the ion selectivity of CRAC channels [121]. An acidic region in the C-terminus of STIM1 (amino acids 470–491) is critical for CDI although the precise mechanism by which this domain confers fast inactivation remains unclear [120,122,123]. STIM1/ORAI1 ratio also affects permeation to Ba^{2+} and Sr^{2+}, as well as inhibition by the compound, 2-APB [119]. These results suggest that in addition to serving as a ligand to promote channel activity, STIM1 likely serves as a mobile subunit, influencing many key functional attributes of the channel.

33.13 MECHANISMS OF Ca^{2+} SELECTIVITY AND PERMEATION

CRAC channels are widely noted for their exquisite Ca^{2+} selectivity ($P_{Ca}/P_{Na} > 1000$), which places them in a unique category of highly Ca^{2+} selective channels together with voltage-gated Ca^{2+} (Ca_v) channels [19]. As noted earlier, high Ca^{2+} selectivity is only manifested in Ca^{2+}-containing solutions: In divalent-free solutions, CRAC channels readily conduct a variety of small monovalent ions (Na^+, Li^+, and K^+) [20,22,23] illustrating that Ca^{2+} selectivity is not due to steric exclusion of monovalent ions. Supplementing divalent free solutions with micromolar concentrations of Ca^{2+} ($K_i \sim 20$ μM at −100 mV) [19–24] blocks monovalent currents, revealing that Ca^{2+} ions occlude monovalent flux through ion–ion and ion–pore interactions. As expected for a binding site within the pore, Ca^{2+} block is voltage-dependent [21,25]. These characteristics are qualitatively reminiscent of the properties of L-type Ca_v channels, in which Ca^{2+} ions similarly bind tightly to a high-affinity binding site within the pore to occlude Na^+ flux [26,27]. Unlike Ca_v channels, however, the affinity of Ca^{2+} blockade is significantly lower in CRAC channels (0.7 μM in L-type Ca^{2+} channels vs. 20–25 μM in CRAC channels) [22,23,25,26,124,125]. This key functional difference suggests that from a biophysical standpoint, the underlying mechanisms that confer Ca^{2+} selectivity are likely to be different between the two classes of highly Ca^{2+} selective channels. Indeed, a recent study has argued that in addition to Ca^{2+} binding at the selectivity filter, CRAC channels achieve high Ca^{2+} selectivity by restricting the flux of ions (both preferred and nonpreferred) by high energy barriers [125]. Enhancing Na^+ and Ca^{2+} flux rates by lowering the entry and exit barriers paradoxically reduces Ca^{2+} selectivity, as seen in ORAI3 channels gated directly by the small molecule, 2-APB [125]. Thus, both high affinity Ca^{2+} binding as well as kinetic factors seem to contribute to high Ca^{2+} selectivity in CRAC channels.

It is widely accepted that Ion permeation is governed by the chemistry and arrangement of pore-lining residues, prompting considerable interest in determining the identity of the pore-lining residues in CRAC channels. Toward this goal, one study applied the substituted cysteine accessibility method (SCAM) [58]. In this approach, residues in the pore-lining region are mutated individually to Cys and the sensitivity of the mutated channels to blockade by aqueous thiol-labeling reagents, such as MTS reagents is assessed [126]. This study indicated that residues in TM1 flank the pore, and ruled out TM3, and specifically, E190, as pore-lining residues [58]. Similar conclusions were reached in an independent report that examined the pattern of disulfide cross-linking of Cys residues introduced into ORAI1 [59]. The SCAM study also indicated that the TM1–TM2 loop segments interact tightly with both large (>8 Å) and small (<3 Å), and with positively charged and negatively charged probes, suggesting that these loops form an outer vestibule with sufficient flexibility to accommodate ions of different size and charge [58]. Strong Cd^{2+} reactivity of several residues in TM1 indicated that the centrally located TM1 helices are close to one another and therefore line a narrow pore, a feature that is likely to account for the low permeability of the CRAC channel to large cations (>3.8 Å) and its low unitary conductance. Moreover, differences in the accessibility of probes of different sizes showed that the pore narrows sharply at the base of the vestibule, near the Ca^{2+} binding site formed by E106. These results provided the first step toward building a structural model of the open pore, and were largely confirmed by the recent crystal structure of the *Drosophila* ORAI protein [105].

The crystal structure of *Drosophila* ORAI confirmed that TM1 flanks the ion conduction pathway, and, with the exception of F99, the pore-lining residues observed in the structure matched the residues found from cysteine accessibility analysis, including E106, V102, L95, and R91 (Figure 33.2b) [105]. The difference at F99 (G98 in the cysteine scan study) could indicate a possible structural alteration caused by introducing the Cys mutation, but may also reflect a difference between the structures of closed and open channels. The closed x-ray structure also revealed a Ca^{2+} ion density a few angstroms above the predicted selectivity filter formed by E106 rather than at the selectivity filter itself [105]. Why the Ca^{2+} ion density is not localized to the predicted center of the selectivity filter remains unclear, but this may be related to the presumed closed, nonconducting state of the crystallized channels. The cysteine accessibility studies and the x-ray crystal structure have provided us with a firm framework for gaining a better understanding of the mechanisms of ion conduction in CRAC channels, but clearly much more needs to be understood before the structure can be incorporated into plausible models for selectivity and permeation. Ultimately, the channel structures of both the closed and open states are required to illuminate the dynamics of the steps of CRAC channel ion transport cycle.

33.14 STIM1 REGULATES CRAC CHANNEL ION SELECTIVITY

A surprising functional aspect of CRAC channels revealed by analysis of the ion selectivity of a mutant ORAI1 channel is that STIM1 not only controls CRAC channel gating, but also bestows many fundamental features that have historically defined the fingerprint of the CRAC channel pore [121]. A variety of substitutions at the pore-lining residue, V102, including substitutions to Cys, Ala, Ser, and Thr, produce constitutively open channels that are open even in the absence of bound STIM1 [121]. The ion selectivities of the STIM1-bound and -free channels, however, are strikingly different. STIM1-free V102C mutant channels exhibit poor Ca^{2+} selectivity and allow permeation of Cs^+ and several other large cations that are normally impermeable through CRAC channels [121]. However, interaction of the mutant channels with STIM1 restores high Ca^{2+} selectivity to the poorly selective STIM1-free channels. In effect, the aberrant ion selectivity of the STIM1-free mutant channels is corrected following STIM1 binding [121]. These changes are accompanied by alterations in the pore geometry, specifically, significant narrowing of the pore to state that more closely resembles the dimensions seen for WT ORAI1 channels. The tuning of ORAI1 ion selectivity by STIM1 is not unique to the V102C mutant channels, but is also seen in wild-type ORAI1 channels as the amount of STIM1 bound to ORAI1 is increased [121], suggesting that the V102X mutations merely unmask a native intermediate channel activation state due to a leaky gate.

The regulation of Ca^{2+} selectivity of ORAI1 channels by STIM1 is surprising, for this feature implies that STIM1-free ORAI1 channels are intrinsically poorly Ca^{2+} selective. Instead, the distinguishing characteristics of CRAC channels including high Ca^{2+} selectivity, low Cs^+ permeability, and a narrow pore are bestowed to the otherwise poorly Ca^{2+} selective ORAI1 channels by STIM1. How does STIM1 modulate ORAI1 ion selectivity? Given that the ORAI1 N-terminus bears a STIM1 binding site in proximity to the pore-forming TM1 segment (Figure. 33.1b), it is not difficult to envision that STIM1 binding to the N-terminus could exert powerful effects on the energetic stability of the selectivity filter.

In addition to biophysical implications, the finding that STIM1 regulates ORAI1 selectivity has many important implications for the nature of Ca^{2+} signals arising from opening of ORAI1 channels under different conditions. The ability of CRAC channels to conduct Na^+ under certain conditions may expand their potential functions to include novel modes by which they encode and process cellular information. Because emerging evidence suggests that CRAC channels, aside from activation by STIM1, can also be activated in a STIM1-independent fashion by other ligands, including the small molecule, 2-APB [104,127–130], and the Golgi Ca^{2+}-ATPase, SPCA2 [77], these findings raise the possibility that ORAI1 channels may function either as highly Ca^{2+} selective channels or nonselective channels depending on the nature of the upstream activation signal. In addition, the tight coupling of permeation and gating found for CRAC channels provides an alternative perspective on ion channel gating that contradicts conventional ion channel postulates on the separation of gating and selectivity. The picture that emerges is of a hydrophobic gate (V102) located in proximity to the selectivity filter (E106); the proximity of the two structures likely results in a variety of conformational alterations in the selectivity filter during gating. These findings reinforce the emerging viewpoint that there is much more happening in the vicinity of the selectivity filter in ion channels than initially imagined [131,132].

33.15 REGULATION OF CRAC CHANNELS BY CALCIUM: CALCIUM-DEPENDENT INACTIVATION (CDI)

Ca^{2+}-dependent inactivation is a prominent hallmark of CRAC channels involving feedback inhibition of channel activity by the high local (Ca^{2+}) around individual CRAC channels, resulting in current decay over 100–300 ms steps during hyperpolarizing steps [19,133,134]. Multiple protein–protein interactions and motifs appear to be involved in this process, including an acidic region of the C-terminal region of STIM1 and calmodulin (CaM) binding to the N-terminus of ORAI1 [120,122,123]. An early indication of a role for STIM1 came from a study showing that increasing the STIM1:ORAI1 transfection ratio increases the extent, rate, and calcium dependence of fast inactivation [119]. This also suggested that multiple STIM1 must likely bind the CRAC channel to evoke fast inactivation. It light of this finding, much of the available data on fast inactivation is difficult to interpret because it is not clear in these studies whether the STIM1:ORAI1 ratio was controlled between the different conditions. For example, mutations of several ORAI1 regions (C-terminus, N-terminus, and the II–III loop are reported to affect inactivation, but it is difficult to know if these effects were really to the mutations or simply due to the mutants expressing at a different level than wt ORAI1.

CDI is not affected by mutations of STIM1 in the N-terminus, including the D76A mutation, which renders STIM1 constitutively active [122]. However, mutations or deletions of the region ~474–490 in the c-terminal domain of STIM1 significantly affect fast inactivation. In particular, neutralizing a set of negative charges in this region can enhance or inhibit fast inactivation [120,122,123]. However, while calcium binding to this region was found to affect CDI, it does not appear to be related to Ca^{2+} binding in a straightforward manner since some mutants exhibit reduced calcium binding affinity but increased CDI [120]. Thus, the precise allosteric mechanism by which this putative Ca^{2+} binding module within STIM1 regulates CDI remains to be clarified.

Interestingly, mutations in ORAI1 selectivity filter that diminish ion selectivity also strongly attenuate fast inactivation [25]. Diminished inactivation is not due to differences in channel expression or because of lower Ca^{2+} permeability of mutant channels [25]. The molecular basis of this effect remains unknown. One possibility is that the inactivation gating mechanism is closely coupled to ion permeation, such that mutations that alter permeation also have effects on inactivation gating [25]. A second possibility is that the mutations allosterically affect the gating mechanism, which is located elsewhere. Although the location of the inactivation gate is poorly understood, Srikanth et al. showed that mutations in the loop 2–3 region greatly decrease fast inactivation and enhance SOCE and I_{CRAC} amplitudes [135]. Overexpressing a 37 amino acid peptide encompassing the intracellular 2–3 loop or including this peptide directly in the patch pipette resulted in dramatically reduced CRAC currents. These results were interpreted in terms of a model in which the intracellular loop acts as a blocking peptide to produce open channel blockade at the intracellular mouth of the channel [135]. More tests are needed to elucidate if this peptide truly comprises the inactivation gate and determine how mutations in the selectivity filter might alter its function.

33.16 CONCLUSIONS

The identification of the STIM and ORAI protein families has produced dramatic advances in many aspects of CRAC channel function. In particular, we now have a firm mechanistic framework for understanding how CRAC channels are activated by the depletion of intracellular Ca^{2+} stores, and we know the key structural elements involved in ion permeation and selectivity. However, there remain many broad unresolved issues. A major unknown is the gating mechanism of the channel: How does STIM1 binding to ORAI1 open the pore? As described earlier, the structure of the closed channel is now available, many critical steps of STIM1 activation have been described in considerable detail, and recent studies are beginning to dissect conformational changes that occur during gating. We can therefore expect the coming years to provide steady progress on the possible structural mechanisms of CRAC channel gating. Another unknown relates to the mechanisms and functions of the "other" ORAI and STIM proteins, ORAI2, ORAI3, and STIM2. The study of these noncanonical CRAC channel proteins is currently constrained by the pressing lack of genetic evidence in both animals and humans for any function. However, recent evidence reveals fascinating differences in the behavior of these proteins from those of canonical CRAC channels, including but not limited to store-independent regulation and altered ion conduction properties [106,136]. The development of genetic tools such as specific knockouts and/or transgenic mice should further open the door for providing new information on the physiological roles of these proteins. Finally, although some work has occurred in the development of small molecular drugs against CRAC channels, the mechanism and targets of most CRAC channel inhibitors are unclear. Recent breakthroughs in solving the crystal structures of the CRAC channel proteins should influence structure-based drug design, and hopefully provide a new generation of CRAC channel blockers with therapeutic potential.

ACKNOWLEDGMENTS

I thank Rich Lewis (Stanford) and the members of my own laboratory for stimulating discussions, and Leidamarie Tirado-Lee for making Figure 33.2. This work was supported by grants from the NIH and American Heart Association.

REFERENCES

1. Feske, S. 2009. ORAI1 and STIM1 deficiency in human and mice: Roles of store-operated Ca^{2+} entry in the immune system and beyond. *Immunol Rev* 231:189–209.
2. Feske, S. 2010. CRAC channelopathies. *Pflugers Arch* 460:417–435.
3. Di Capite, J. L., G. J. Bates, and A. B. Parekh. 2011. Mast cell CRAC channel as a novel therapeutic target in allergy. *Curr Opin Allergy Clinc Immunol* 11:33–38.
4. Ma, J., C. A. McCarl, S. Khalil, K. Luthy, and S. Feske. 2010. T-cell-specific deletion of STIM1 and STIM2 protects mice from EAE by impairing the effector functions of Th1 and Th17 cells. *Eur J Immunol* 40:3028–3042.
5. Schuhmann, M. K., D. Stegner, A. Berna-Erro, S. Bittner, A. Braun, C. Kleinschnitz, G. Stoll, H. Wiendl, S. G. Meuth, and B. Nieswandt. 2010. Stromal interaction molecules 1 and 2 are key regulators of autoreactive T cell activation in murine autoimmune central nervous system inflammation. *J Immunol* 184:1536–1542.
6. Prevarskaya, N., R. Skryma, and Y. Shuba. 2011. Calcium in tumour metastasis: New roles for known actors. *Nat Rev Cancer* 11:609–618.
7. Varga-Szabo, D., A. Braun, and B. Nieswandt. 2011. STIM and Orai in platelet function. *Cell Calcium* 50:270–278.
8. McCarl, C. A., S. Khalil, J. Ma, M. Oh-hora, M. Yamashita, J. Roether, T. Kawasaki et al. 2010. Store-operated Ca^{2+} entry through ORAI1 is critical for T cell-mediated autoimmunity and allograft rejection. *J Immunol* 185:5845–5858.
9. Takemura, H., A. R. Hughes, O. Thastrup, and J. W. Putney, Jr. 1989. Activation of calcium entry by the tumor promoter thapsigargin in parotid acinar cells. Evidence that an intracellular calcium pool and not an inositol phosphate regulates calcium fluxes at the plasma membrane. *J Biol Chem* 264:12266–12271.
10. Mason, M. J., C. Garcia-Rodriguez, and S. Grinstein. 1991. Coupling between intracellular Ca^{2+} stores and the Ca^{2+} permeability of the plasma membrane. Comparison of the effects of thapsigargin, 2,5-di-(tert-butyl)-1,4-hydroquinone, and cyclopiazonic acid in rat thymic lymphocytes. *J Biol Chem* 266:20856–20862.
11. Thastrup, O., A. P. Dawson, O. Scharff, B. Foder, P. J. Cullen, B. K. Drobak, P. J. Bjerrum, S. B. Christensen, and M. R. Hanley. 1989. Thapsigargin, a novel molecular probe for studying intracellular calcium release and storage. *Agents Actions* 27:17–23.
12. Brayden, D. J., M. R. Hanley, O. Thastrup, and A. W. Cuthbert. 1989. Thapsigargin, a new calcium-dependent epithelial anion secretagogue. *Br J Pharmacol* 98:809–816.
13. Hoth, M. and R. Penner. 1992. Depletion of intracellular calcium stores activates a calcium current in mast cells. *Nature* 355:353–356.
14. Zweifach, A. and R. S. Lewis. 1993. Mitogen-regulated Ca^{2+} current of T lymphocytes is activated by depletion of intracellular Ca^{2+} stores. *Proc Natl Acad Sci USA* 90:6295–6299.
15. Prakriya, M. 2009. The molecular physiology of CRAC channels. *Immunol Rev* 231:88–98.
16. Lewis, R. S. 2007. The molecular choreography of a store-operated calcium channel. *Nature* 446:284–287.
17. Hogan, P. G., R. S. Lewis, and A. Rao. 2010. Molecular basis of calcium signaling in lymphocytes: STIM and ORAI. *Annu Rev Immunol* 28:491–533.
18. Lewis, R. S. 2011. Store-operated calcium channels: New perspectives on mechanism and function. *Cold Spring Harb Perspect Biol* 3:a003970.
19. Hoth, M. and R. Penner. 1993. Calcium release-activated calcium current in rat mast cells. *J Physiol* 465:359–386.
20. Prakriya, M. and R. S. Lewis. 2002. Separation and characterization of currents through store-operated CRAC channels and Mg^{2+}-inhibited cation (MIC) channels. *J Gen Physiol* 119:487–507.
21. Prakriya, M. and R. S. Lewis. 2006. Regulation of CRAC channel activity by recruitment of silent channels to a high open-probability gating mode. *J Gen Physiol* 128:373–386.
22. Lepple-Wienhues, A. and M. D. Cahalan. 1996. Conductance and permeation of monovalent cations through depletion-activated Ca^{2+} channels (I_{CRAC}) in Jurkat T cells. *Biophys J* 71:787–794.
23. Bakowski, D. and A. B. Parekh. 2002. Monovalent cation permeability and Ca^{2+} block of the store-operated Ca^{2+} current I_{CRAC} in rat basophilic leukemia cells. *Pflugers Arch* 443:892–902.
24. Su, Z., R. L. Shoemaker, R. B. Marchase, and J. E. Blalock. 2004. Ca^{2+} modulation of Ca^{2+} release-activated Ca^{2+} channels is responsible for the inactivation of its monovalent cation current. *Biophys J* 86:805–814.
25. Yamashita, M., L. Navarro-Borelly, B. A. McNally, and M. Prakriya. 2007. ORAI1 mutations alter ion permeation and Ca^{2+}-dependent inactivation of CRAC channels: Evidence for coupling of permeation and gating. *J Gen Physiol* 130:525–540.
26. Lansman, J. B., P. Hess, and R. W. Tsien. 1986. Blockade of current through single calcium channels by Cd^{2+}, Mg^{2+}, and Ca^{2+}. Voltage and concentration dependence of calcium entry into the pore. *J Gen Physiol* 88:321–347.
27. Sather, W. A. and E. W. McCleskey. 2003. Permeation and selectivity in calcium channels. *Annu Rev Physiol* 65:133–159.
28. Lewis, R. S. 1999. Store-operated calcium channels. *Adv Second Messenger Phosphoprotein Res* 33:279–307.
29. Prakriya, M. and R. S. Lewis. 2003. CRAC channels: Activation, permeation, and the search for a molecular identity. *Cell Calcium* 33:311–321.
30. Prakriya, M. and R. S. Lewis. 2004. Store-operated calcium channels: Properties, functions and the search for a molecular mechanism. *Adv Molec Cell Biol* 32:121–140.

31. Oritani, K. and P. W. Kincade. 1996. Identification of stromal cell products that interact with pre-B cells. *J Cell Biol* 134:771–782.
32. Roos, J., P. J. DiGregorio, A. V. Yeromin, K. Ohlsen, M. Lioudyno, S. Zhang, O. Safrina et al. 2005. STIM1, an essential and conserved component of store-operated Ca^{2+} channel function. *J Cell Biol* 169:435–445.
33. Liou, J., M. L. Kim, W. D. Heo, J. T. Jones, J. W. Myers, J. E. Ferrell, Jr., and T. Meyer. 2005. STIM is a Ca^{2+} sensor essential for Ca^{2+}-store-depletion-triggered Ca^{2+} influx. *Curr Biol* 15:1235–1241.
34. Zhang, S. L., Y. Yu, J. Roos, J. A. Kozak, T. J. Deerinck, M. H. Ellisman, K. A. Stauderman, and M. D. Cahalan. 2005. STIM1 is a Ca^{2+} sensor that activates CRAC channels and migrates from the Ca^{2+} store to the plasma membrane. *Nature* 437:902–905.
35. Baba, Y., K. Hayashi, Y. Fujii, A. Mizushima, H. Watarai, M. Wakamori, T. Numaga et al. 2006. Coupling of STIM1 to store-operated Ca^{2+} entry through its constitutive and inducible movement in the endoplasmic reticulum. *Proc Natl Acad Sci USA* 103:16704–16709.
36. Wu, M. M., J. Buchanan, R. M. Luik, and R. S. Lewis. 2006. Ca^{2+} store depletion causes STIM1 to accumulate in ER regions closely associated with the plasma membrane. *J Cell Biol* 174:803–813.
37. Muik, M., I. Frischauf, I. Derler, M. Fahrner, J. Bergsmann, P. Eder, R. Schindl et al. 2008. Dynamic coupling of the putative coiled-coil domain of ORAI1 with STIM1 mediates ORAI1 channel activation. *J Biol Chem* 283:8014–8022.
38. Navarro-Borelly, L., A. Somasundaram, M. Yamashita, D. Ren, R. J. Miller, and M. Prakriya. 2008. STIM1-ORAI1 interactions and ORAI1 conformational changes revealed by live-cell FRET microscopy. *J Physiol* 586:5383–5401.
39. Liou, J., M. Fivaz, T. Inoue, and T. Meyer. 2007. Live-cell imaging reveals sequential oligomerization and local plasma membrane targeting of stromal interaction molecule 1 after Ca^{2+} store depletion. *Proc Natl Acad Sci USA* 104:9301–9306.
40. Park, C. Y., P. J. Hoover, F. M. Mullins, P. Bachhawat, E. D. Covington, S. Raunser, T. Walz, K. C. Garcia, R. E. Dolmetsch, and R. S. Lewis. 2009. STIM1 clusters and activates CRAC channels via direct binding of a cytosolic domain to ORAI1. *Cell* 136:876–890.
41. Huang, G. N., W. Zeng, J. Y. Kim, J. P. Yuan, L. Han, S. Muallem, and P. F. Worley. 2006. STIM1 carboxyl-terminus activates native SOC, I(crac) and TRPC1 channels. *Nat Cell Biol* 8:1003–1010.
42. Yuan, J. P., W. Zeng, M. R. Dorwart, Y. J. Choi, P. F. Worley, and S. Muallem. 2009. SOAR and the polybasic STIM1 domains gate and regulate Orai channels. *Nat Cell Biol* 11:337–343.
43. Feske, S., Y. Gwack, M. Prakriya, S. Srikanth, S. H. Puppel, B. Tanasa, P. G. Hogan, R. S. Lewis, M. Daly, and A. Rao. 2006. A mutation in ORAI1 causes immune deficiency by abrogating CRAC channel function. *Nature* 441:179–185.
44. Prakriya, M., S. Feske, Y. Gwack, S. Srikanth, A. Rao, and P. G. Hogan. 2006. ORAI1 is an essential pore subunit of the CRAC channel. *Nature* 443:230–233.
45. Yeromin, A. V., S. L. Zhang, W. Jiang, Y. Yu, O. Safrina, and M. D. Cahalan. 2006. Molecular identification of the CRAC channel by altered ion selectivity in a mutant of Orai. *Nature* 443:226–229.
46. Vig, M., A. Beck, J. M. Billingsley, A. Lis, S. Parvez, C. Peinelt, D. L. Koomoa et al. 2006. CRACM1 multimers form the ion-selective pore of the CRAC channel. *Curr Biol* 16:2073–2079.
47. Feske, S., M. Prakriya, A. Rao, and R. S. Lewis. 2005. A severe defect in CRAC Ca^{2+} channel activation and altered K^+ channel gating in T cells from immunodeficient patients. *J Exp Med* 202:651–662.
48. Feske, S., J. Giltnane, R. Dolmetsch, L. M. Staudt, and A. Rao. 2001. Gene regulation mediated by calcium signals in T lymphocytes. *Nat Immunol* 2:316–324.
49. Partiseti, M., F. Le Deist, C. Hivroz, A. Fischer, H. Korn, and D. Choquet. 1994. The calcium current activated by T cell receptor and store depletion in human lymphocytes is absent in a primary immunodeficiency. *J Biol Chem* 269:32327–32335.
50. Feske, S., J. M. Muller, D. Graf, R. A. Kroczek, R. Drager, C. Niemeyer, P. A. Baeuerle, H. H. Peter, and M. Schlesier. 1996. Severe combined immunodeficiency due to defective binding of the nuclear factor of activated T cells in T lymphocytes of two male siblings. *Eur J Immunol* 26:2119–2126.
51. Fanger, C. M., M. Hoth, G. R. Crabtree, and R. S. Lewis. 1995. Characterization of T cell mutants with defects in capacitative calcium entry: Genetic evidence for the physiological roles of CRAC channels. *J Cell Biol* 131:655–667.
52. Lewis, R. S. 2001. Calcium signaling mechanisms in T lymphocytes. *Annu Rev Immunol* 19:497–521.
53. Zhang, S. L., A. V. Yeromin, X. H. Zhang, Y. Yu, O. Safrina, A. Penna, J. Roos, K. A. Stauderman, and M. D. Cahalan. 2006. Genome-wide RNAi screen of Ca^{2+} influx identifies genes that regulate Ca^{2+} release-activated Ca^{2+} channel activity. *Proc Natl Acad Sci USA* 103:9357–9362.
54. Vig, M., C. Peinelt, A. Beck, D. L. Koomoa, D. Rabah, M. Koblan-Huberson, S. Kraft et al. 2006. CRACM1 is a plasma membrane protein essential for store-operated Ca^{2+} entry. *Science* 312:1220–1223.
55. Mercer, J. C., W. I. Dehaven, J. T. Smyth, B. Wedel, R. R. Boyles, G. S. Bird, and J. W. Putney, Jr. 2006. Large store-operated calcium selective currents due to co-expression of ORAI1 or ORAI2 with the intracellular calcium sensor, STIM1. *J Biol Chem* 281:24979–24990.
56. Peinelt, C., M. Vig, D. L. Koomoa, A. Beck, M. J. Nadler, M. Koblan-Huberson, A. Lis, A. Fleig, R. Penner, and J. P. Kinet. 2006. Amplification of CRAC current by STIM1 and CRACM1 (ORAI1). *Nat Cell Biol* 8:771–773.
57. Soboloff, J., M. A. Spassova, X. D. Tang, T. Hewavitharana, W. Xu, and D. L. Gill. 2006. ORAI1 and STIM reconstitute store-operated calcium channel function. *J Biol Chem* 281:20661–20665.
58. McNally, B. A., M. Yamashita, A. Engh, and M. Prakriya. 2009. Structural determinants of ion permeation in CRAC channels. *Proc Natl Acad Sci USA* 106:22516–22521.
59. Zhou, Y., S. Ramachandran, M. Oh-Hora, A. Rao, and P. G. Hogan. 2010. Pore architecture of the ORAI1 store-operated calcium channel. *Proc Natl Acad Sci USA* 107:4896–4901.
60. Gwack, Y., S. Srikanth, S. Feske, F. Cruz-Guilloty, M. Oh-hora, D. S. Neems, P. G. Hogan, and A. Rao. 2007. Biochemical and functional characterization of Orai proteins. *J Biol Chem* 282:16232–16243.
61. Lis, A., C. Peinelt, A. Beck, S. Parvez, M. Monteilh-Zoller, A. Fleig, and R. Penner. 2007. CRACM1, CRACM2, and CRACM3 are store-operated Ca^{2+} channels with distinct functional properties. *Curr Biol* 17:794–800.
62. Gross, S. A., U. Wissenbach, S. E. Philipp, M. Freichel, A. Cavalie, and V. Flockerzi. 2007. Murine ORAI2 splice variants form functional Ca^{2+} release-activated Ca^{2+} (CRAC) channels. *J Biol Chem* 282:19375–19384.
63. Wissenbach, U., S. E. Philipp, S. A. Gross, A. Cavalie, and V. Flockerzi. 2007. Primary structure, chromosomal localization and expression in immune cells of the murine ORAI and STIM genes. *Cell Calcium* 42:439–446.
64. Chung, S. C., T. V. McDonald, and P. Gardner. 1994. Inhibition by SK&F 96365 of Ca^{2+} current, IL-2 production and activation in T lymphocytes. *Br J Pharmacol* 113:861–868.
65. Aussel, C., R. Marhaba, C. Pelassy, and J. P. Breittmayer. 1996. Submicromolar La^{3+} concentrations block the calcium release-activated channel, and impair CD69 and CD25 expression in CD3- or thapsigargin-activated Jurkat cells. *Biochem J* 313 (Pt 3):909–913.

66. Le Deist, F., C. Hivroz, M. Partiseti, C. Thomas, H. A. Buc, M. Oleastro, B. Belohradsky, D. Choquet, and A. Fischer. 1995. A primary T-cell immunodeficiency associated with defective transmembrane calcium influx. *Blood* 85:1053–1062.
67. Hogan, P. G., L. Chen, J. Nardone, and A. Rao. 2003. Transcriptional regulation by calcium, calcineurin, and NFAT. *Genes Dev* 17:2205–2232.
68. Varga-Szabo, D., K. S. Authi, A. Braun, M. Bender, A. Ambily, S. R. Hassock, T. Gudermann, A. Dietrich, and B. Nieswandt. 2008. Store-operated Ca(2+) entry in platelets occurs independently of transient receptor potential (TRP) C1. *Pflugers Arch* 457:377–387.
69. Varga-Szabo, D., A. Braun, C. Kleinschnitz, M. Bender, I. Pleines, M. Pham, T. Renne, G. Stoll, and B. Nieswandt. 2008. The calcium sensor STIM1 is an essential mediator of arterial thrombosis and ischemic brain infarction. *J Exp Med* 205:1583–1591.
70. Braun, A., J. E. Gessner, D. Varga-Szabo, S. N. Syed, S. Konrad, D. Stegner, T. Vogtle, R. E. Schmidt, and B. Nieswandt. 2009. STIM1 is essential for Fcgamma receptor activation and autoimmune inflammation. *Blood* 113:1097–1104.
71. Stiber, J., A. Hawkins, Z. S. Zhang, S. Wang, J. Burch, V. Graham, C. C. Ward et al. 2008. STIM1 signalling controls store-operated calcium entry required for development and contractile function in skeletal muscle. *Nat Cell Biol* 10:688–697.
72. Dirksen, R. T. 2009. Checking your SOCCs and feet: The molecular mechanisms of Ca^{2+} entry in skeletal muscle. *J Physiol* 587:3139–3147.
73. Motiani, R. K., I. F. Abdullaev, and M. Trebak. 2010. A novel native store-operated calcium channel encoded by ORAI3: Selective requirement of ORAI3 versus ORAI1 in estrogen receptor-positive versus estrogen receptor-negative breast cancer cells. *J Biol Chem* 285:19173–19183.
74. Yang, S., J. J. Zhang, and X. Y. Huang. 2009. ORAI1 and STIM1 are critical for breast tumor cell migration and metastasis. *Cancer Cell* 15:124–134.
75. McAndrew, D., D. M. Grice, A. A. Peters, F. M. Davis, T. Stewart, M. Rice, C. E. Smart, M. A. Brown, P. A. Kenny, S. J. Roberts-Thomson, and G. R. Monteith. 2011. ORAI1-mediated calcium influx in lactation and in breast cancer. *Mol Cancer Ther* 10:448–460.
76. Motiani, R. K., X. Zhang, K. E. Harmon, R. S. Keller, K. Matrougui, J. A. Bennett, and M. Trebak. 2013. ORAI3 is an estrogen receptor alpha-regulated Ca(2+) channel that promotes tumorigenesis. *FASEB J* 27:63–75.
77. Feng, M., D. M. Grice, H. M. Faddy, N. Nguyen, S. Leitch, Y. Wang, S. Muend et al. 2010. Store-independent activation of ORAI1 by SPCA2 in mammary tumors. *Cell* 143:84–98.
78. Stathopulos, P. B., G. Y. Li, M. J. Plevin, J. B. Ames, and M. Ikura. 2006. Stored Ca^{2+} depletion-induced oligomerization of stromal interaction molecule 1 (STIM1) via the EF-SAM region: An initiation mechanism for capacitive Ca^{2+} entry. *J Biol Chem* 281:35855–35862.
79. Zheng, L., P. B. Stathopulos, R. Schindl, G. Y. Li, C. Romanin, and M. Ikura. 2011. Auto-inhibitory role of the EF-SAM domain of STIM proteins in store-operated calcium entry. *Proc Natl Acad Sci USA* 108:1337–1342.
80. Luik, R. M., B. Wang, M. Prakriya, M. M. Wu, and R. S. Lewis. 2008. Oligomerization of STIM1 couples ER calcium depletion to CRAC channel activation. *Nature* 454:538–542.
81. Demaurex, N. and M. Frieden. 2003. Measurements of the free luminal ER Ca(2+) concentration with targeted "cameleon" fluorescent proteins. *Cell Calcium* 34:109–119.
82. Brandman, O., J. Liou, W. S. Park, and T. Meyer. 2007. STIM2 is a feedback regulator that stabilizes basal cytosolic and endoplasmic reticulum Ca^{2+} levels. *Cell* 131:1327–1339.
83. Bird, G. S., S. Y. Hwang, J. T. Smyth, M. Fukushima, R. R. Boyles, and J. W. Putney, Jr. 2009. STIM1 is a calcium sensor specialized for digital signaling. *Curr Biol* 19:1724–1729.
84. Covington, E. D., M. M. Wu, and R. S. Lewis. 2010. Essential role for the CRAC activation domain in store-dependent oligomerization of STIM1. *Mol Biol Cell* 21:1897–1907.
85. Luik, R. M., M. M. Wu, J. Buchanan, and R. S. Lewis. 2006. The elementary unit of store-operated Ca^{2+} entry: Local activation of CRAC channels by STIM1 at ER-plasma membrane junctions. *J Cell Biol* 174:815–825.
86. Gwozdz, T., J. Dutko-Gwozdz, V. Zarayskiy, K. Peter, and V. M. Bolotina. 2008. How strict is the correlation between STIM1 and ORAI1 expression, puncta formation, and ICRAC activation? *Am J Physiol Cell Physiol* 295:C1133–C1140.
87. Li, Z., J. Lu, P. Xu, X. Xie, L. Chen, and T. Xu. 2007. Mapping the interacting domains of STIM1 and ORAI1 in Ca^{2+} release-activated Ca^{2+} channel activation. *J Biol Chem* 282:29448–29456.
88. Grigoriev, I., S. M. Gouveia, B. van der Vaart, J. Demmers, J. T. Smyth, S. Honnappa, D. Splinter et al. 2008. STIM1 is a MT-plus-end-tracking protein involved in remodeling of the ER. *Curr Biol* 18:177–182.
89. Honnappa, S., S. M. Gouveia, A. Weisbrich, F. F. Damberger, N. S. Bhavesh, H. Jawhari, I. Grigoriev et al. 2009. An EB1-binding motif acts as a microtubule tip localization signal. *Cell* 138:366–376.
90. Chvanov, M., C. M. Walsh, L. P. Haynes, S. G. Voronina, G. Lur, O. V. Gerasimenko, R. Barraclough et al. 2008. ATP depletion induces translocation of STIM1 to puncta and formation of STIM1-ORAI1 clusters: Translocation and re-translocation of STIM1 does not require ATP. *Pflugers Arch* 457:505–517.
91. Calloway, N., D. Holowka, and B. Baird. 2010. A basic sequence in STIM1 promotes Ca^{2+} influx by interacting with the C-terminal acidic coiled coil of ORAI1. *Biochemistry* 49:1067–1071.
92. Korzeniowski, M. K., I. M. Manjarres, P. Varnai, and T. Balla. 2010. Activation of STIM1-ORAI1 involves an intramolecular switching mechanism. *Sci Signal* 3:ra82.
93. Muik, M., M. Fahrner, R. Schindl, P. Stathopulos, I. Frischauf, I. Derler, P. Plenk et al. 2011. STIM1 couples to ORAI1 via an intramolecular transition into an extended conformation. *EMBO J* 30:1678–1689.
94. Yu, F., L. Sun, S. Hubrack, S. Selvaraj, and K. Machaca. 2013. Intramolecular shielding maintains STIM1 in an inactive conformation. *J Cell Sci* 126:2401–2410.
95. Yang, X., H. Jin, X. Cai, S. Li, and Y. Shen. 2012. Structural and mechanistic insights into the activation of Stromal interaction molecule 1 (STIM1). *Proc Natl Acad Sci USA* 109:5657–5662.
96. Zhou, Y., P. Srinivasan, S. Razavi, S. Seymour, P. Meraner, A. Gudlur, P. Stathopulos, M. Ikura, A. Rao, and P. G. Hogan. 2013. Initial activation of STIM1, the regulator of store-operated calcium entry. *Nature Mol Struct Biol* 20:973–981.
97. Penna, A., A. Demuro, A. V. Yeromin, S. L. Zhang, O. Safrina, I. Parker, and M. D. Cahalan. 2008. The CRAC channel consists of a tetramer formed by STIM-induced dimerization of ORAI dimers. *Nature* 456:116–120.
98. Parvez, S., A. Beck, C. Peinelt, J. Soboloff, A. Lis, M. Monteilh-Zoller, D. L. Gill, A. Fleig, and R. Penner. 2008. STIM2 protein mediates distinct store-dependent and store-independent modes of CRAC channel activation. *FASEB J* 22:752–761.
99. Zhou, Y., P. Meraner, H. T. Kwon, D. Machnes, M. Oh-hora, J. Zimmer, Y. Huang, A. Stura, A. Rao, and P. G. Hogan. 2010. STIM1 gates the store-operated calcium channel ORAI1 in vitro. *Nat Struct Mol Biol* 17:112–116.
100. Muik, M., M. Fahrner, I. Derler, R. Schindl, J. Bergsmann, I. Frischauf, K. Groschner, and C. Romanin. 2009. A cytosolic homomerization and a modulatory domain within STIM1 C terminus determine coupling to ORAI1 channels. *J Biol Chem* 284:8421–8426.

101. Kawasaki, T., I. Lange, and S. Feske. 2009. A minimal regulatory domain in the C terminus of STIM1 binds to and activates ORAI1 CRAC channels. *Biochem Biophys Res Commun* 285:25720–25730.
102. Frischauf, I., M. Muik, I. Derler, J. Bergsmann, M. Fahrner, R. Schindl, K. Groschner, and C. Romanin. 2009. Molecular determinants of the coupling between STIM1 and ORAI channels: Differential activation of ORAI1–3 channels by a STIM1 coiled-coil mutant. *J Biol Chem* 284:21696–21706.
103. Stathopulos, P. B., R. Schindl, M. Fahrner, L. Zheng, G. M. Gasmi-Seabrook, M. Muik, C. Romanin, and M. Ikura. 2013. STIM1/ORAI1 coiled-coil interplay in the regulation of store-operated calcium entry. *Nat Commun* 4:2963.
104. Yamashita, M., A. Somasundaram, and M. Prakriya. 2011. Competitive modulation of CRAC channel gating by STIM1 and 2-aminoethyldiphenyl borate (2-APB). *J Biol Chem* 286:9429–9442.
105. Hou, X., L. Pedi, M. M. Diver, and S. B. Long. 2012. Crystal structure of the calcium release-activated calcium channel ORAI. *Science* 338:1308–1313.
106. Soboloff, J., B. S. Rothberg, M. Madesh, and D. L. Gill. 2012. STIM proteins: Dynamic calcium signal transducers. *Nature Rev Mol Cell Biol* 13:549–565.
107. Rothberg, B. S., Y. Wang, and D. L. Gill. 2013. ORAI channel pore properties and gating by STIM: Implications from the ORAI crystal structure. *Sci Signal* 6:pe9.
108. Lis, A., S. Zierler, C. Peinelt, A. Fleig, and R. Penner. 2010. A single lysine in the N-terminal region of store-operated channels is critical for STIM1-mediated gating. *J Gen Physiol* 136:673–686.
109. McNally, B. A., A. Somasundaram, A. Jairaman, M. Yamashita, and M. Prakriya. 2013. The C- and N-terminal STIM1 binding sites on ORAI1 are required for both trapping and gating CRAC channels. *J Physiol.*
110. Maruyama, Y., T. Ogura, K. Mio, K. Kato, T. Kaneko, S. Kiyonaka, Y. Mori, and C. Sato. 2009. Tetrameric ORAI1 is a teardrop-shaped molecule with a long, tapered cytoplasmic domain. *J Biol Chem* 284:13676–13685.
111. Ulbrich, M. H. and E. Y. Isacoff. 2007. Subunit counting in membrane-bound proteins. *Nat Methods* 4:319–321.
112. Ji, W., P. Xu, Z. Li, J. Lu, L. Liu, Y. Zhan, Y. Chen, B. Hille, T. Xu, and L. Chen. 2008. Functional stoichiometry of the unitary calcium-release-activated calcium channel. *Proc Natl Acad Sci USA* 105:13668–13673.
113. Demuro, A., A. Penna, O. Safrina, A. V. Yeromin, A. Amcheslavsky, M. D. Cahalan, and I. Parker. 2011. Subunit stoichiometry of human ORAI1 and ORAI3 channels in closed and open states. *Proc Natl Acad Sci USA* 108:17832–17837.
114. Madl, J., J. Weghuber, R. Fritsch, I. Derler, M. Fahrner, I. Frischauf, B. Lackner, C. Romanin, and G. J. Schutz. 2010. Resting-state ORAI1 diffuses as homotetramer in the plasma membrane of live mammalian cells. *J Biol Chem* 285:41135–41142.
115. Mignen, O., J. L. Thompson, and T. J. Shuttleworth. 2008. ORAI1 subunit stoichiometry of the mammalian CRAC channel pore. *J Physiol* 586:419–425.
116. Thompson, J. L. and T. J. Shuttleworth. 2013. How many ORAI's does it take to make a CRAC channel? *Sci Rep* 3:1961.
117. Li, Z., L. Liu, Y. Deng, W. Ji, W. Du, P. Xu, L. Chen, and T. Xu. 2010. Graded activation of CRAC channel by binding of different numbers of STIM1 to ORAI1 subunits. *Cell Res* 21:305–315.
118. Hoover, P. J. and R. S. Lewis. 2011. Stoichiometric requirements for trapping and gating of Ca^{2+} release-activated Ca^{2+} (CRAC) channels by stromal interaction molecule 1 (STIM1). *Proc Natl Acad Sci USA* 108:13299–13304.
119. Scrimgeour, N., T. Litjens, L. Ma, G. J. Barritt, and G. Y. Rychkov. 2009. Properties of ORAI1 mediated store-operated current depend on the expression levels of STIM1 and ORAI1 proteins. *J Physiol* 587:2903–2918.
120. Mullins, F. M., C. Y. Park, R. E. Dolmetsch, and R. S. Lewis. 2009. STIM1 and calmodulin interact with ORAI1 to induce Ca^{2+}-dependent inactivation of CRAC channels. *Proc Natl Acad Sci USA* 106:15495–15500.
121. McNally, B. A., A. Somasundaram, M. Yamashita, and M. Prakriya. 2012. Gated regulation of CRAC channel ion selectivity by STIM1. *Nature* 482:241–245.
122. Derler, I., M. Fahrner, M. Muik, B. Lackner, R. Schindl, K. Groschner, and C. Romanin. 2009. A Ca^{2+} release-activated Ca^{2+} (CRAC) modulatory domain (CMD) within STIM1 mediates fast Ca^{2+}-dependent inactivation of ORAI1 channels. *J Biol Chem* 284:24933–24938.
123. Lee, K. P., J. P. Yuan, W. Zeng, I. So, P. F. Worley, and S. Muallem. 2009. Molecular determinants of fast Ca^{2+}-dependent inactivation and gating of the Orai channels. *Proc Natl Acad Sci USA* 106:14687–14692.
124. Almers, W., E. W. McCleskey, and P. T. Palade. 1984. A non-selective cation conductance in frog muscle membrane blocked by micromolar external calcium ions. *J Physiol* 353:565–583.
125. Yamashita, M. and M. Prakriya. 2014. Divergence of Ca(2+) selectivity and equilibrium Ca(2+) blockade in a Ca(2+) release-activated Ca(2+) channel. *J Gen Physiol* 143:325–343.
126. Karlin, A. and M. H. Akabas. 1998. Substituted-cysteine accessibility method. *Methods Enzymol* 293:123–145.
127. Peinelt, C., A. Lis, A. Beck, A. Fleig, and R. Penner. 2008. 2-Aminoethoxydiphenyl borate directly facilitates and indirectly inhibits STIM1-dependent gating of CRAC channels. *J Physiol* 586:3061–3073.
128. Zhang, S. L., J. A. Kozak, W. Jiang, A. V. Yeromin, J. Chen, Y. Yu, A. Penna, W. Shen, V. Chi, and M. D. Cahalan. 2008. Store-dependent and -independent modes regulating Ca^{2+} release-activated Ca^{2+} channel activity of human ORAI1 and ORAI3. *J Biol Chem* 283:17662–17671.
129. Schindl, R., J. Bergsmann, I. Frischauf, I. Derler, M. Fahrner, M. Muik, R. Fritsch, K. Groschner, and C. Romanin. 2008. 2-aminoethoxydiphenyl borate alters selectivity of ORAI3 channels by increasing their pore size. *J Biol Chem* 283:20261–20267.
130. DeHaven, W. I., J. T. Smyth, R. R. Boyles, G. S. Bird, and J. W. Putney, Jr. 2008. Complex actions of 2-aminoethyldiphenyl borate on store-operated calcium entry. *J Biol Chem* 283:19265–19273.
131. Contreras, J. E., D. Srikumar, and M. Holmgren. 2008. Gating at the selectivity filter in cyclic nucleotide-gated channels. *Proc Natl Acad Sci USA* 105:3310–3314.
132. Thompson, J. and T. Begenisich. 2012. Selectivity filter gating in large-conductance Ca^{2+}-activated K^+ channels. *J Gen Physiol* 139:235–244.
133. Zweifach, A. and R. S. Lewis. 1995. Rapid inactivation of depletion-activated calcium current (I_{CRAC}) due to local calcium feedback. *J Gen Physiol* 105:209–226.
134. Fierro, L. and A. B. Parekh. 1999. Fast calcium-dependent inactivation of calcium release-activated calcium current (CRAC) in RBL-1 cells. *J Membr Biol* 168:9–17.
135. Srikanth, S., H. J. Jung, B. Ribalet, and Y. Gwack. 2010. The intracellular loop of ORAI1 plays a central role in fast inactivation of Ca^{2+} release-activated Ca^{2+} channels. *J Biol Chem* 285:5066–5075.
136. Shuttleworth, T. J. 2012. ORAI3—The 'exceptional' ORAI? *J Physiol* 590:241–257.

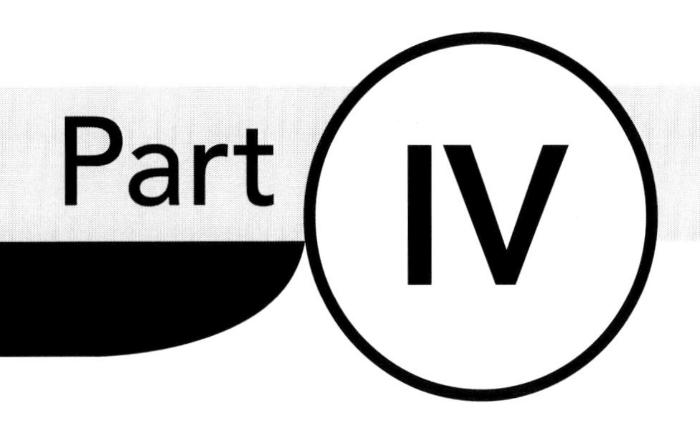

Part IV

Ion channel regulation

34 Mechanism of G-protein regulation of K+ channels

Rahul Mahajan and Diomedes E. Logothetis

Contents

The βγ subunits of heterotrimeric G-proteins (Gβγ) were first demonstrated to directly participate in signaling by their activation of K+ channels (I_{KACh}) underlying the acetylcholine (ACh)-induced decrease in heart rate. Outside of a membrane-targeting role, how Gβγ subunits specifically regulate the conformations of their effector proteins to alter activity is not understood at a molecular level. Several crystal structures of G-protein-gated inwardly rectifying K+ (GIRK) channels were published in the past decade, but attempts to cocrystallize them with Gβγ failed until 2013, when the GIRK2/Gβγ complex structure was reported. A parallel computational approach aimed to develop a multistage docking algorithm that combines several known methods in protein–protein docking. Application of the docking protocol to Gβγ and GIRK1 structures produced a clear signal of a favored binding mode. Analysis of this binding mode suggested a mechanism by which Gβγ promotes the open state of the channel. The channel–Gβγ interactions predicted by the model could be disrupted by mutation of one protein and rescued by additional mutation of reciprocal residues in the other protein. These interactions were found to extend to agonist-induced activation of the channels as well as to activation of the native heteromeric channels. The complex structures of Gβγ with GIRK1 (computational) and GIRK2 (crystallographic) show not only remarkable similarities but also interesting differences. Future challenges include determination of three-dimensional structures of additional members of the receptor/G-protein/channel macromolecular complex that will reveal the structural basis of agonist-independent and agonist-dependent channel activation.

34.1 VAGAL INHIBITION OF HEART RATE

GIRK channels (or Kir3 channels) are known to play diverse roles including important regulation of cardiac, neuronal, and endocrine physiology.[28] The first known effect of GIRK channels was their role in underlying I_{KACh}, the cardiac current largely responsible for the negative chronotropic effects of vagally released ACh. The inhibitory action of the vagus nerve on the heart has been demonstrated since at least the nineteenth century when the brothers Eduard and Ernst Weber communicated their results to an Italian congress of scientists in 1845.[20,38] They showed that heart rate in a frog preparation could be slowed and brought to a stop by stimulation of the vagus nerves using a rotary galvano-magnetic apparatus. The sensitivity of this vagal inhibition to atropine and its accompaniment by the hyperpolarization of the heart muscle were recognized as early as 1886 by Gaskell using extracellular recordings.[21,22] Furthermore, through careful chemical quantitation of extracellular fluids, Howell and Duke had demonstrated by 1908 that vagal inhibition of the heart is accompanied by a small release of potassium.[29]

Otto Loewi ushered in modern neuroscience by the direct demonstration of chemical transmission of nervous impulses.[51] Extract from a frog heart which had undergone vagal stimulation contained a substance deemed *Vagusstoff*, which could be applied to a second heart and cause its inhibition (Figure 34.1). This substance was later identified to be ACh.[52] The advent of intracellular microelectrodes and voltage-clamp techniques allowed for more rigorous exploration of these phenomena. Burgen and Terroux revisited the vagal-induced hyperpolarization of heart muscle reported by Gaskell. Using microelectrodes, they confirmed that hyperpolarization is induced by ACh application.[6] By measuring the effect of external K+ concentration on resting potential in the absence and presence of ACh, they also demonstrated that increased cell permeability to potassium may underlie hyperpolarization. Del Castillo and Katz used microelectrodes to directly show hyperpolarization of the sinus node upon vagal stimulation.[9] Voltage clamp allowed Trautwein and Dudel to directly confirm changes in cell potassium permeability by measuring K+ reversal potentials.[82] Noma and Trautwein studied activation kinetics of ACh-induced K+ currents and concluded that ACh binding activates a specific ion channel, K_{ACh}.[65] The introduction of the patch-clamp technique[62] led to the first single-channel recordings of K_{ACh} currents,[74] which

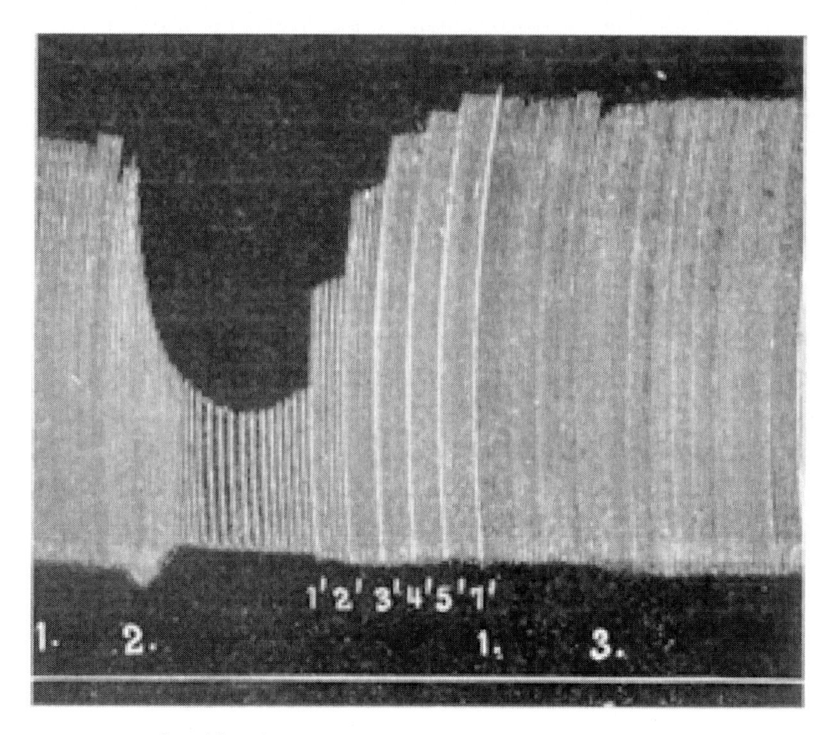

Figure 34.1 Frog heart contractions in a perfused frog heart, measured by suspension-lever. 1. Marks application of Ringer's solution. 2. Marks the application of *Vagusstoff* (extract from a separate heart after 15 min of vagal stimulation with esterases inactivated). 3. Marks the application of inactive *Vaguststoff* (just as in 2 but with the esterases active-not inactivated). Negative ionotropic (vertical amplitude) and chronotropic (horizontal frequency) effects can be seen. (1′–7′) represent increasing dilution of the applied heart extract in Ringer's solution demonstrating the concentration dependence of the effects. (Adapted with permission from Loewi, O. and Navratil, E., *Pflügers Archiv Eur. J. Physiol.*, 214(1), 678, 1926.)

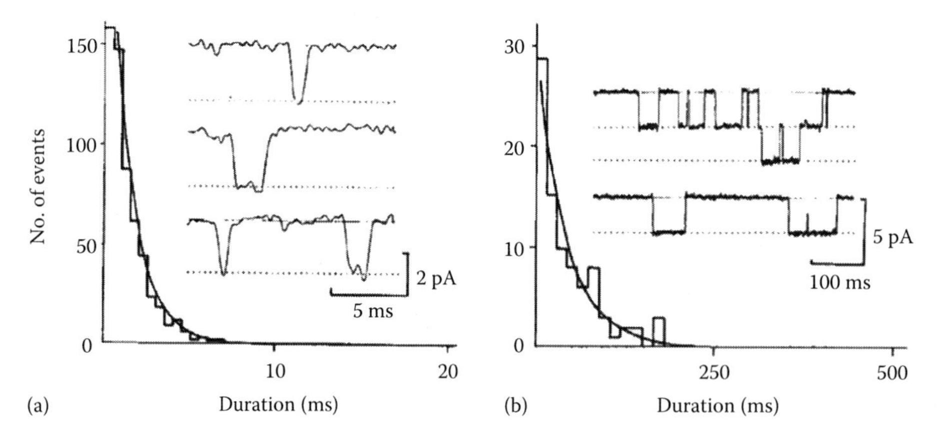

Figure 34.2 Single-channel recordings were performed on dispersed nonbeating AV nodal cells under identical conditions in (a) and (b). Frequency histograms of observed channel opening durations are displayed. Although the single-channel conductances were similar between (a) and (b) (see inset traces), the kinetics clearly differentiated the two classes of potassium channels. The channels in (a) were shown to be ACh sensitive and correspond to I_{KACh}, while those in (b) were ACh insensitive and correspond to background I_{K1} channels. (Adapted with permission from Sakmann, B. et al., *Nature*, 303(5914), 250, 1983.)

clearly demonstrated kinetic properties distinct from other background potassium channels (Figure 34.2).

Thus, the negative chronotropic effect of vagal stimulation is due to hyperpolarization of the sinus node due to the activation of a specific K+ channel in response to ACh released at the vagal termini. This I_{KACh} current has been shown to play important roles in cardiac physiology. In vivo loss of function of this current was shown to cause an almost complete loss of variability in the heart's beat-to-beat frequency and a large reduction in heart rate vagal response.[89] Loss of its function by mutations in its component subunits have also been associated with long QT syndrome in humans.[93] Excessive I_{KACh} activity is associated with atrial fibrillation in humans.[8,39]

34.2 MOLECULAR COMPONENTS OF K_{ACh} REGULATION

In order to examine the molecular regulators of I_{KACh}, a brief discussion of guanosine triphosphate (GTP)-binding proteins is necessary. Work from Earl Sutherland in the 1950s had demonstrated that several hormones lead to stimulation of adenylyl cyclase and the production of cyclic adenosine monophosphate (cAMP) within cells where it went on to act as a soluble second messenger.[4] At first the GTP dependence of this process was not known, because the process of purifying adenosine triphosphate used as a substrate for the generation of cAMP was imperfect and allowed for contamination by GTP (reviewed in [59]). Rodbell and colleagues showed that when supplying low concentrations of ATP, application of hormone or agonist became insufficient to stimulate cAMP production unless GTP was supplied.[72] This identified a GTP-dependent step in the signaling process, leading to stimulation of cAMP production, but its identity was not known.

Gill and Meren showed that cholera toxin leads to sustained elevation of cAMP levels by ADP-ribosylation of some unidentified cellular protein component.[23] Haga and colleagues in Al Gilman's laboratory generated a cell line, which failed to elevate cAMP levels in response to known agonists.[25] Assessment of direct adenylyl cyclase function showed that the activity of the cAMP-producing enzyme was intact in these cells, and thus, the deficiency arose from a component in the transduction mechanism. Using cholera toxin and a radioactive ADP substrate, they demonstrated radioisotope incorporation into a 45-kilodalton (kDa) polypeptide occurred in normal cells but not in their deficient cell line. These advances eventually allowed Gilman's laboratory to purify the 45 kDa polypeptide identified as the alpha subunit of the adenylyl cyclase–stimulating GTP-binding protein (Gs).[66] This polypeptide copurified with 35 kDa and 8–10 kDa proteins identified respectively as the beta (β) and gamma (γ) subunits of heterotrimeric GTP-binding proteins (G-proteins).

Today, the G-protein family has been extended to include three more classes of Gα subunits: Gq/11, G12/13, and the pertussis toxin-sensitive Gi/o. Various isoforms of twenty known Gα subunits heteromerize with one of five known Gβ isoforms and one of twelve known Gγ subunits to form the heterotimeric G-protein. Their coupling is promiscuous, but not every combination of isoforms can be found physiologically (reviewed in [59]). The heterotrimeric G-proteins couple to transmembrane G-protein-coupled receptors (GPCRs) and act as molecular switches, which help transduce extracellular signals to downstream effector proteins (Figure 34.3).

Several lines of evidence implicated G-proteins in the signal transduction mechanism that allowed ACh to activate K_{ACh}. Soejima and Noma showed in 1984[77] that the mechanism of K_{ACh} activation is membrane-delimited. Bath application of ACh did not activate K_{ACh} current in a cell-attached patch, but inclusion of ACh within the patch pipette solution did activate the channels.[77] Thus, it was concluded that ACh must be applied directly to the patch of membrane being recorded and no freely diffusible intracellular signaling mechanism could be responsible for activating the channel (discussed further below as part of Figure 34.5). Evidence from two groups implicated guanine nucleotide–binding proteins (G-proteins) in the signaling mechanism. Breitwieser and Szabo used intracellular application of nonhydrolyzable GTP analogs to maximally activate I_{KACh} such that it no longer responded to extracellular application of ACh[5]. This implied the channel was a distinct entity from the ACh receptor and that the activation mechanism likely involved G-proteins. Pfaffinger and colleagues demonstrated the sensitivity of the current to pertussis toxin and thus implicated G-proteins and specifically, the Gi/o family.[69] Logothetis et al demonstrated that it was the Gβγ subunits of heterotrimeric G-proteins that were responsible for channel activation.[53] Purified Gβγ but not Gα protein could be applied to the intracellular surface of an excised membrane patch to activate the channels (Figure 34.4). Although activation of I_{KACh} was the

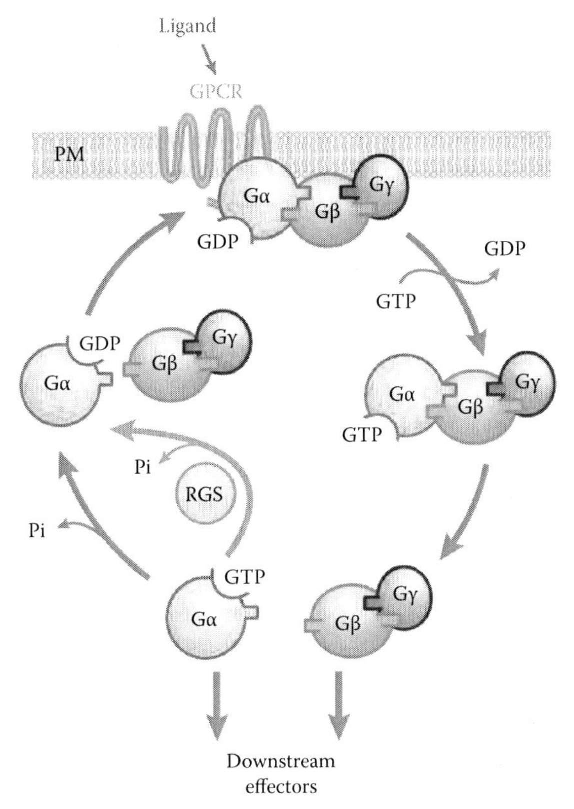

Figure 34.3 The G-protein cycle. In the inactive state, G-proteins exist as heterotrimers. Gα is bound to GDP and thus stabilized in a conformation with high affinity for Gβγ. Agonist binding to the GPCR elicits a conformational change in the receptor, allowing it to catalyze the exchange of GTP for GDP on the Gα subunit. Binding of GTP to Gα reduces its affinity for Gβγ, but complete dissociation of the two may or may not occur.[45] This represents the active form of both G-protein subunits and each of Gα and Gβγ may interact with various downstream effector proteins to modulate their activity. Gα has intrinsic GTPase activity, which causes it to hydrolyze the GTP back into GDP, releasing a pyrophosphate. Alternately, association of a regulator of G-protein signaling (RGS) molecule with Gα may accelerate its GTPase activity. The conversion of GTP to GDP returns alpha to a conformation with high affinity for Gβγ and the inactive heterotrimer is re-formed. (Adapted with permission from Li, L. et al., *Annu. Rev. Microbiol.*, 61(1), 423, 2007.)

first example of direct Gβγ signaling, many Gβγ effectors are now known and their numbers rival Gα effectors (reviewed in [7]).

Molecular cloning of the first component underlying I_{KACh}[16,43] led to the identification of the atrial heterotetrameric K_{ACh} channel comprised of GIRK1 and GIRK4 subunits.[41] A homotetramer of GIRK4 has also been reported in atrial myocytes, implying an unidentified role for homomeric GIRK channel in cardiac physiology.[15] Heterologous expression of GIRK1 alone yields no currents, and GIRK4 alone yields very small currents compared to the heterotetramer. Thus, functional GIRK1-containing channels can only exist in heterotetrameric form. Other known GIRK subunits are the neuronal GIRK2 and GIRK3 (reviewed in [28] and discussed in the following). Introduction of single point mutations in the reentrant pore helix region of these channels yields the GIRK1* (GIRK1-F137S), GIRK2* (GIRK2-E152D), and GIRK4* (GIRK4-S143T) channels, which yield robust currents and allow for the study of functional homomeric channels.[10,83,94] GIRK channels have now been found to be expressed in endocrine tissues, besides heart and brain, such as in pancreas and thyrotrophs of the rat pituitary gland (reviewed in [28]).

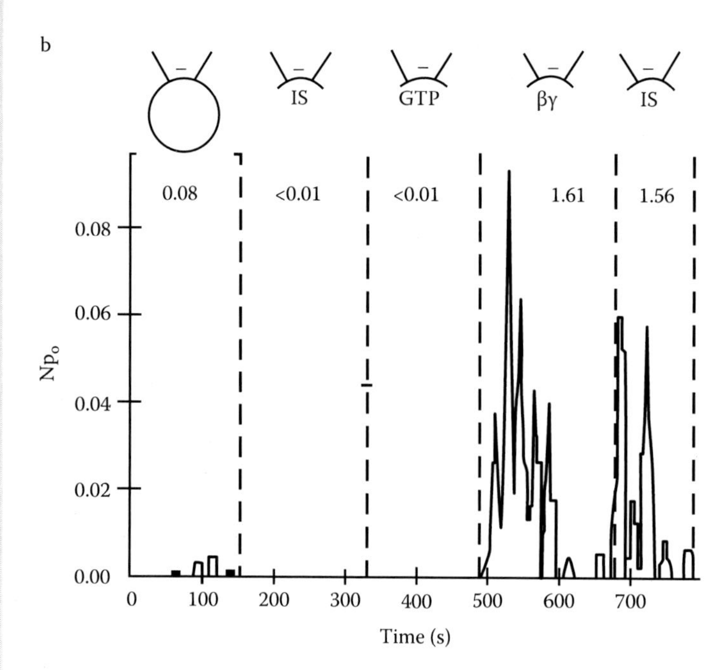

Figure 34.4 Np_o is plotted for observed potassium channel activity in a recording from cell-attached (first section) and inside-out patch (remaining sections) configurations of chick atrial myocytes. Application of purified Gβγ protein to the intracellular surface of the patch causes robust activation even in the absence of ACh in the pipette (–). The perfused Gβγ protein is not readily washed out. (Adapted with permission from Logothetis, D.E. et al., *Nature*, 325(6102), 321, 1987.)

Although it has been shown that Gβγ activates GIRK channels, there is some evidence that Gα plays a modulatory role. Receptor activation of native GIRK currents is selective to Gi-coupled receptors. Because a variety of Gβγ subunit combinations can stimulate GIRK currents,[88] the reason for this specificity is not known. It has been suggested that such specificity is achieved by colocalization of signaling components into preformed complexes. Chimeric analysis of different Gα subunits suggested that domains of Gαi may play a role in conferring this specificity.[73] Several studies have shown biochemically that Gα can bind to various domains of GIRK.[12,14,31,73] Fluorescent techniques also suggest that there is a basal interaction between Gα and GIRK channels even in the absence of receptor activation.[3,71] Reconstitution studies of pure components of a GIRK signaling system, namely of a GIRK1 chimera[64] with $G\beta_1\gamma_2$ and/or $G\alpha i_1$ subunits, but in the absence of GPCRs, suggested an active and required role of activated Gα subunits for Gβγ stimulation of channel activity.[47] Whether these results also hold true for GIRK1 channels whose origin is from only mammalian sources remains to be examined. In contrast, Gα subunits were not found to be required for Gβγ stimulation of GIRK2 channels both in a purified liposome assay as well as in lipid bilayers.[85,87] These results suggest that there may be differences in the way different GIRK isoforms couple to the G-protein signaling system.

Like all members of the Kir inward rectifying potassium channel family, GIRK channels require the presence of phosphatidylinositol 4,5-bisphosphate (PIP_2) for function.[32,80,97] PIP_2 is necessary for channel activation but is not sufficient.[97] An additional gating molecule such as Gβγ is required. Another gating molecule, which can activate GIRK2 and GIRK4 (but not GIRK1) channels, is intracellular Na^+. Both Gβγ and Na^+ appear to stabilize channel PIP_2 interactions.[32,97] Other activators include alcohols such as ethanol[2] and strongly reducing intracellular environments.[96] The multiple modulators of GIRK activity have been summarized in past comprehensive reviews.[28,79]

As mentioned earlier, in cell-attached recordings, Gi signaling stimulated outside the patch (agonist applied in the bath rather than in the patch pipette) does not result in channel, activity within the patch.[77] In contrast, Gq signaling does not appear to be restricted by the patch pipette. Stimulation of Gq signaling outside the patch consistently inhibits GIRK activity recorded from a cell-attached patch via a mechanism involving hydrolysis of PIP_2 in an intramembrane-diffusible (or membrane delimited) manner[96] (Figure 34.5). Thus isolation of the patch, as in the cell-attached mode of the patch-clamp technique, does prevent

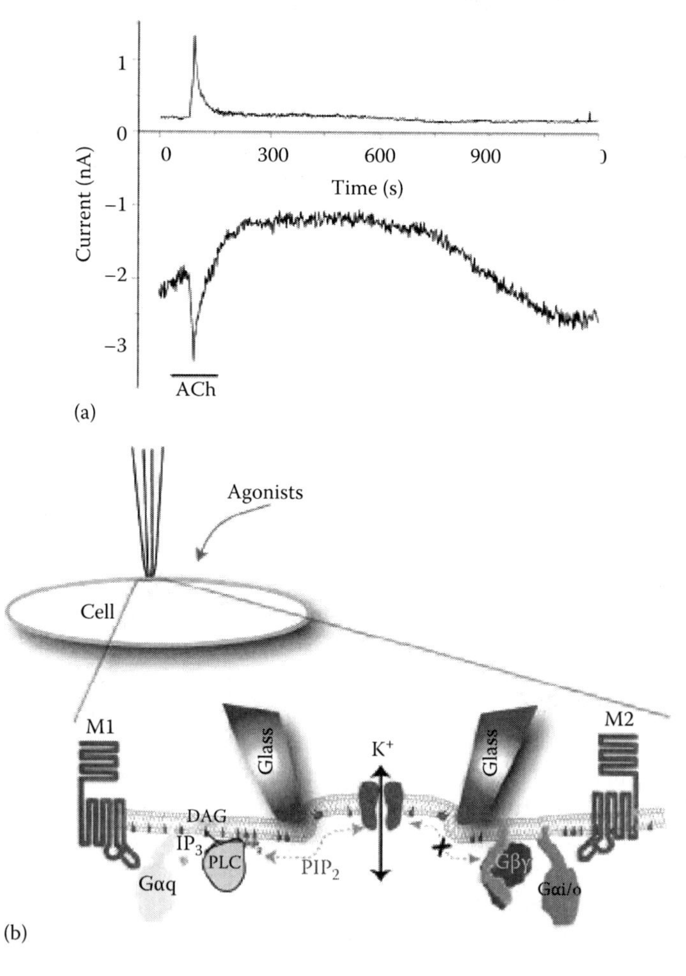

Figure 34.5 Signaling through a membrane-delimited diffusible second messenger. (a) Bath application of ACh outside the patch pipette activates M1 receptor and inhibits the active homomeric GIRK4(S143T) channel currents[83] recorded in a cell-attached patch from a *Xenopus* oocyte. M1 receptor activation also stimulates endogenous calcium-activated Cl^- currents (outward and inward spikes during ACh application) that are elicited by the increase in intracellular Ca^{2+} released by IP_3 receptors in the endoplasmic reticulum (IP_3 is generated by the M1-mediated activation of PLCβ1 and hydrolysis of PIP_2 to DAG and IP_3). Symmetrical high-K^+ solutions were used in the pipette as well as in the bath, 5 μM ACh was applied to the cell via the bathing solution. Representative record is from three similar experiments. (b) Cartoon depicting the experimental setup[77] showing that diffusion of Gβγ subunits across the patch is not possible, unlike diffusion of PIP_2. (Adapted with permission from Zhang, H. et al., *Neuron*, 37(6), 963, March 27, 2003.)

G-protein-signaling from outside the patch (Gi-mediated activation) but does not prevent the diffusion of phosphoinositides out of (Gq-mediated inhibition) and into (recovery via PIP_2 resynthesis) the patch. This restriction of G-protein diffusion is consistent with the idea of a G-protein macromolecular complex (receptor/G-protein subunits/channel) remaining intact within and outside the isolated patch.

34.3 NEURONAL GIRK CHANNELS

All four mammalian GIRK channel subunits are expressed in the nervous system. Neuronal GIRK channels play important roles in neuronal function including pain perception, reward-related behavior, mood, cognition, and memory modulation. Malfunctions in GIRK-mediated signaling in the brain have been linked to epilepsy, Down syndrome, Parkinson's disease, and drug addiction.[54,55] Knockout studies of neuronal GIRK channels have revealed their critical involvement in the formation of inhibitory postsynaptic potentials (IPSPs) in hippocampal and cerebral neurons.[55,75] GIRK knockout mice develop a number of defects that have been summarized in multiple comprehensive reviews.[28,54,55]

GIRK1–3 proteins are expressed throughout the brain, while GIRK4 is found in specific areas.[24,33,35,40,49,90] GIRK channels are involved in mediating mainly IPSPs but also presynaptic modulation of neuronal activity. There are four alternatively spliced isoforms of GIRK2 expressed as homomers (e.g., in dopaminergic neurons of the substantia nigra) or heteromers with GIRK1, GIRK3, or GIRK4 [reviewed in 28]. Some of the GIRK2 isoforms contain a PDZ domain that can interact with PDZ-binding proteins. Some GIRK2 isoforms may associate with proteins enriched in lipid rafts, such as the neural cell adhesion molecule. Such complexes can regulate the localization and function of neuronal GIRK channels.[54,55]

34.4 STRUCTURAL INSIGHTS

The first crystal structure of any potassium ion channel confirmed many predicted features of the potassium channel structure.[17] Among these were the presence of pore constrictions comprising the selectivity filter at the extracellular end of the pore and the helix bundle crossing (HBC) gate (also referred in the literature as *inner helix gate*) toward the intracellular end. The first structures of GIRK channels consisted of a fusion construct of the intracellular N- and C-termini of the channel. The transmembrane regions were deleted, and the termini were connected with a linker region.[63,68] These structures reveal the presence of a third putative gate at the apex of the intracellular region, the G-loop gate. A subsequent GIRK structure of Kir3.1 was a chimera between a mammalian and prokaryotic channel.[64] Substitution of the top three-fourths of the transmembrane region of GIRK1 with prokaryotic residues allowed the crystallization of a more complete channel, which places the intracellular termini in the proper context of a transmembrane region. The intracellular regions are organized through the cascading arrangement of secondary structure elements (Figure 34.6). Beginning most centrally near the pore is the G-loop, which comprises a gate in the channel structure. Moving outward and downward are the CD loop, the N-terminus of the adjacent subunit, the LM loop, and the DE loop of the adjacent subunit. Furthermore, this crystal structure captures the G-loop gate in two distinct conformations such that it is dilated *open* in one and constricted or *closed* in the other. The secondary structure elements also show some reorganization between the two conformations.

A previous paper reported molecular dynamics simulations of these conformations of the Kir3.1-chimera structure in the absence and presence of PIP_2 to study the interactions of the channel which allow for PIP_2 stabilization of the *open* conformation.[58] The conclusions of the detailed channel motions observed in this study can be summarized in terms of movements of the secondary structure elements. Transition to the *open* conformation of the channel stabilized by PIP_2 saw a dramatic upward movement of the LM loop, causing it to interact strongly with the N-terminus. The LM loop thus moves up and acts as a *sink* for the N-terminus so that the N-terminus switches its interactions from the CD to the LM loop. This frees the CD loop to interact with G-loop to stabilize its open state. PIP_2 interacts

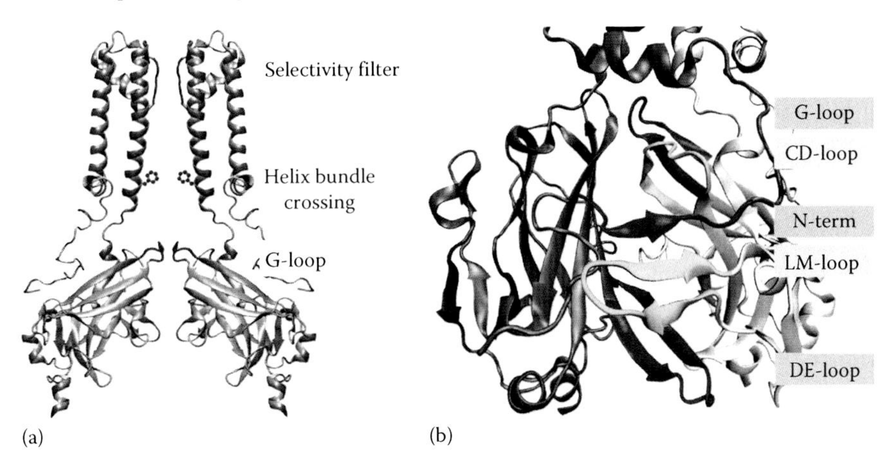

Figure 34.6 Summary of key structural features of GIRK1channels. The structure depicted is a Kir3.1 chimera.[64] (a) shows a cartoon depiction of two opposite subunits of the channel. Putative gates along the potassium permeation pathway are highlighted in red and labeled. (b) shows a close-up view of a cartoon depiction of the intracellular region of two adjacent subunits of the channel. Secondary structure elements that play important roles in gating are highlighted alternately in red and yellow and labeled. The configuration depicted corresponds to the *open* conformation of the crystal structure and thus shows the LM loop in the *raised* conformation interacting closely with the N-terminus, while the CD loop interacts closely with the G-loop.

directly with the CD loop and parts of the N-terminus to stabilize a conformation containing these interactions.

Several lines of evidence suggest that the movements of the putative channel gates are not independent but that the gates likely undergo correlated movements. Clarke and colleagues examined eleven crystal structures of bacterial Kir channels and concluded that changes in intracellular domain orientations were correlated with changes in the selectivity filter gate.[13] Xiao and colleagues examined the state-dependence of accessibility of intracellular cationic modifiers to a cysteine-modified pore residue.[92] They suggest that the HBC gate may not close completely to exclude their cationic modifiers, but its motions are correlated to changes in the selectivity filter gate. Finally, by determining multiple crystal structures of the full-length GIRK2 channel in the presence and absence of PIP_2, Whorton and MacKinnon suggest that PIP_2 acts to couple opening of the G-loop gate to movements of the transmembrane helices to cause opening of the HBC gate.[86]

Unlike channel structures, which have only been achieved within the last decade, G-protein structures have existed since the mid-1990s. Structures of the inactive GDP bound heterotrimer revealed that Gα consists of an upper GTPase domain and a helical domain. Its interaction surface with Gβγ consists of loops in the GTPase domain called the switch regions as well as its long N-terminal helix[46,84] (Figure 34.7). This N-terminal helix is disordered when Gα is not bound to Gβγ [reviewed in 67].

The Gβγ structure consists of a 7-blade beta-propeller structure. It has been crystallized alone[78] or together with regulatory proteins such as beta adrenergic receptor kinase (βARK), phosducin,[81,101] or of course, Gα. Comparisons of the various Gβγ structures do not reveal any major conformational changes, although small changes in interstrand loops and side-chain positions are observable. The exception is the cocrystal of Gβγ with phosducin where a separation of blades 6 and 7 is observed.[50,102] This conformation may be unique to the effect of phosducin. The farnesyl moiety, which normally anchors the C-terminus of $G\gamma_1$ to the membrane, is observed to occupy the cleft created between the propeller blades. Phosducin has the particular ability to dissociate Gβγ from the membrane and cause it to translocate to the cytoplasm. The opening of a cleft in the protein to bind the lipid anchor would be consistent with this function.

Numerous studies have addressed GIRK–Gβγ interactions. Biochemical binding studies employ different strategies for choosing fragments of the channel and testing their ability to bind Gβγ.[14,30,31,34,36,37,41,42,44] Other studies have focused on making and functionally characterizing chimeras between GIRK channels and the closely related but G-protein-insensitive IRK channels.[18,26,27,76] Some studies have used the chimeric analysis to suggest a functionally critical region and then created specific point mutants within the region. Such a previous study identified a residue in the LM loop at position L333 of GIRK1 to play a critical role in activation by Gβγ.[26] Mutation of GIRK1* L333 to the corresponding glutamate residue in IRK1 produced a phenotype such that the mutant channel showed intact basal activity but was not activated by Gβγ coexpression or agonist-induced receptor activation. While by no means exhaustive, results from many of these studies are summarized in Figure 34.8.

Studies have also attempted to identify important regions of Gβγ for channel interaction. Ford et al. performed an alanine scan of all residues, which comprised the Gα binding site on Gβγ.[19] These mutants were tested for their ability to regulate various Gβγ effectors, including their ability to activate GIRK channels upon coexpression. Albsoul-Younes and colleagues preformed a chimeric analysis between mammalian Gβ and a yeast Gβ deficient in activating GIRK channels.[1] They identified blades 1 and 2 as the critical regions for channel interaction.

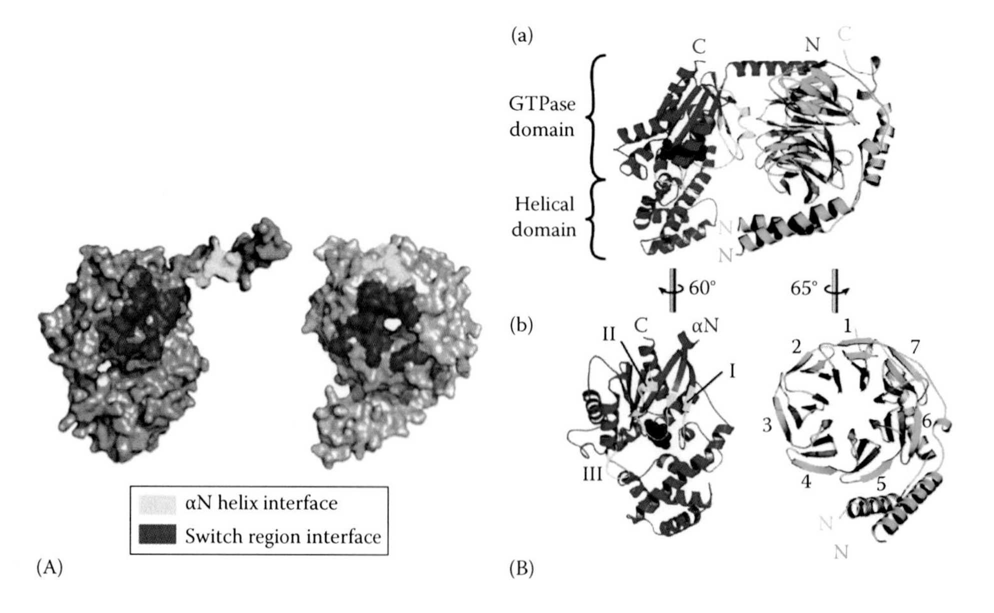

Figure 34.7 (A) Surface depiction of Gα (left) and Gβγ (right). The binding site of Gα on Gβγ can be separated into two regions: the regions contacting Gα switch regions (blue) or the Gα N-terminal helix (yellow). (Adapted with permission from Lambert, N.A., *Sci. Signal.*, 1(25), re5, June 2008.) (B) (a) The inactive heterotrimeric configuration is depicted as a cartoon (Gα, blue; Gβ, green; Gγ, yellow). The domains of Gα are labeled. (b) A separated view of Gα and Gβγ is shown in order to label the individual switch regions of Gα and the individual propeller blades of Gβγ. (Adapted with permission and with minor revisions from Oldham, W.M. and Hamm, H.E., *Q. Rev. Biophys.*, 39(2), 117, 2006.)

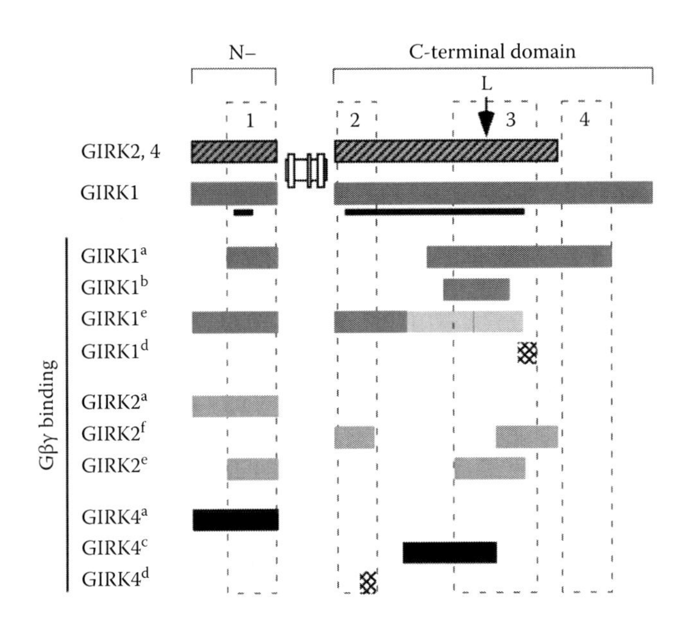

Figure 34.8 Critical regions of the GIRK channel for Gβγ binding as determined by eight different studies (designated by a–f superscript labels) of biochemical binding of fragments of channel protein to Gβγ. The position marked L at the top represents the critical leucine reside (position 333 in GIRK1) discovered by He et al. [26]. This residue was shown to be critical for both channel activation by Gβγ coexpression and by agonist-induced stimulation. While no obvious consensus region emerges from the data, the LM loop area, which contains the highlighted leucine residue overlaps with the critical regions identified by four of the six studies. (Studies depicted are in the following references: a[30,31], b[44], c[26,27], d[42], e[36], f[18]). (Adapted with permission from Finley, M. et al., *J. Physiol.*, 555(3), 643, 2004.)

Several other studies also identified mutants with reduced ability to activate GIRK channels within Gb blades 1 and 2.[60,61,99]

In 2013, two independent studies reported the long awaited complexes of Gβγ with GIRK1 [a computational study[56]] and GIRK2 [a crystallographic study[87]]. The two studies identified Gβγ to interact between two subunits making contacts within a cleft generated by the LM loop and the DE loop of two adjacent channel subunits. Other interaction contacts between Gβγ and the two channel subunits involved the βN strand and the βK-L loop. There was excellent agreement between the two complex structures with some interesting differences that will be discussed below.

Using computational modeling, Mahajan and colleagues generated models of the complex between the βγ subunits of G-proteins and the GIRK1 channel. The predictions of the models were tested experimentally using electrophysiological and biochemical techniques, providing a compelling picture of how Gβγ interacts with GIRK1 channels to stimulate their activity.[56] A multistage docking strategy was adopted (Figure 34.9). Two models stood out, the best scoring model (BSM) was found within a steep energy well, while the largest cluster model (LCM) represented a large group of favorable models that localized nearby in conformational space. A largely hydrophobic contact surface (~1800 Å) was seen between GIRK1 and Gβ in the BSM (Figure 34.10). The footprint of Gβγ onto the GIRK1 cytosolic domain predicted by these models was in excellent agreement with NMR data.[95] Gβ blades 7, 1, and 2 of Gβγ interacted with two adjacent subunits of the channel. Interestingly, several residues shown previously to be important[19,26,27,60,99,100] were identified in these structural studies.[56,86] In particular, key residues (GIRK1-L333,[26] and the Gα-interacting Gβ-L55, Gβ-K89[19]) in Gβγ stimulation of GIRK activity were found to interact with residues near the DE-LM cleft between adjacent channel subunits and stabilize the cleft in a *raised* conformation (the LM loop apart from the DE loop) (Figure 34.11). Electrophysiological evidence of paired interactions as predicted by the BSM of the GIRK1–Gβγ complex in *Xenopus laevis* oocytes showed that (1) steric defects in protein-protein interactions caused by mutations in one protein could be rescued by compensating mutations in the interacting protein; (2) introduced electrostatic repulsion between the LM and DE loops could stabilize the LM loop in the raised conformation; (3) disulfide cross-linking of Gβ(L55C) with GIRK1(L333C) caused channel activation; and (4) salt bridge stabilization of the LM loop (E334) could be achieved with the Gβ(K89)γ interaction.

These studies have given rise to a model for how the Gβγ–channel interactions comprising this binding mode promote the open state of the channel: by stabilizing the *raised* conformation of the LM loop to allow it to interact strongly with the N-terminus (Figure 34.6). The work by Mahajan and colleagues[56] extends the gating mechanism proposed in Meng et al. to include the role of the DE loop and the LM–DE loop cleft.[58] Meng and colleagues had proposed that in the closed state, the CD loop and N-terminus are closely interacting, while the LM loop has moved down and away sharply and the G-loop has shifted to its closed configuration (Figure 34.12). Introduction of PIP_2 stabilizes a different conformation of the secondary structure elements such that each element switches its close interactions from one adjacent element to its other adjacent element. Rather than constricting the pore, the G-loop interacts with the adjacent CD loop. CD loop interactions with N-terminus are in turn weakened, and the N-terminus switches to interacting with the adjacent LM loop.

The work by Mahajan and colleagues proposes that this cascade of switching adjacent element interactions continues down to the DE loop. In the closed state, the downward moved LM loop closely interacts with the adjacent DE loop, but this interaction switches in the open state, so the LM loop instead interacts closely with the adjacent N-terminus. Thus, this model proposes that PIP_2 and Gβγ modulate the same cascade of switching interactions, but their sites of action are distinct. PIP_2 acts close to the pore and gates by directly interacting with residues of the CD loop and N-terminus. Gβγ acts at the level of the LM and DE loops. Ethanol, which activates the channel, also interacts with the channel at the LM–DE loop cleft,[2] supporting the importance of this cleft in channel activation.

Furthermore, the Gβγ residues implicated in channel activation by the BSM of Mahajan and colleagues are part of the Gα-binding site on Gβγ. Specifically, these residues including L55 and K89 are among the residues that interact with the N-terminal helix of Gα.[19,84] Thus, we may speculate that although this model and experiments provide no information about the interactions of Gα, agonist-induced activation involves the unbinding of the Gα-N-terminal helix to reveal these important residues to allow them to interact with the channel. As it is known that the N-terminal helix adopts a disordered conformation upon Gβγ unbinding from Gα,[67] even a partial unbinding of the two proteins may be enough to remove the Gα N-terminal helix from these residues.

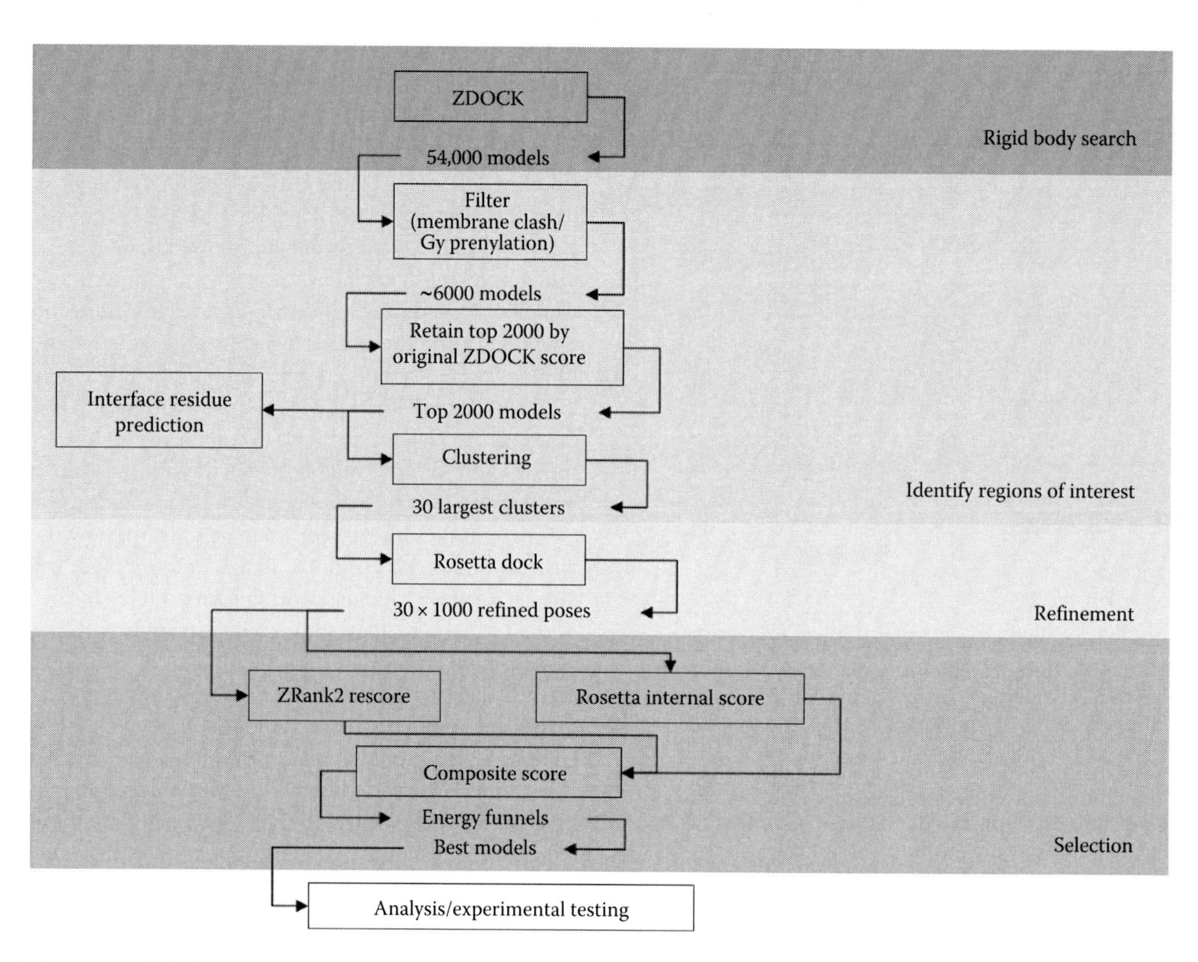

Figure 34.9 Summary of docking protocol used to predict the channel-Gβγ binding mode. ZDock was used as the global rigid-body docking program. Only the intracellular N and C termini of two adjacent subunits of the channel were included in the docking. 54,000 poses were retained, and these were then subject to a filter, which excluded any poses containing any Gβγ atoms protruding more than 8 Å above the expected plane of the membrane at the interfacial helix of the channel. Similarly, the filter also excluded any poses where the C-terminus of Gγ was more than 30 Å from the expected plane of the membrane. The first constraint reflects exclusion of the protein by the lipid bilayer and the second constraint reflects the expected prenylation (geranyl-geranylation) of the Gγ2 C-terminus, which would anchor it to the lipid bilayer. The top 2000 scored poses, which passed the filter, were subjected to the Cluspro 1.0 algorithm to sample the energy landscape and look for broad energy minima by simple hierarchical clustering. Clustering was done based on interface root mean square deviation (RMSD) and we employed the 9 Å clustering radius recommended for average. Structures representing the centers of the 30 largest clusters were retained for further analysis. Refinement via flexible docking was performed for each of these 30 structures using the local refinement module of RosettaDock. One thousand models were calculated to sample the energy landscape around each of the 30 starting structures. The scoring function used for selection employed a combination of the rigid and flexible docking algorithms and is detailed in Ref. [56].

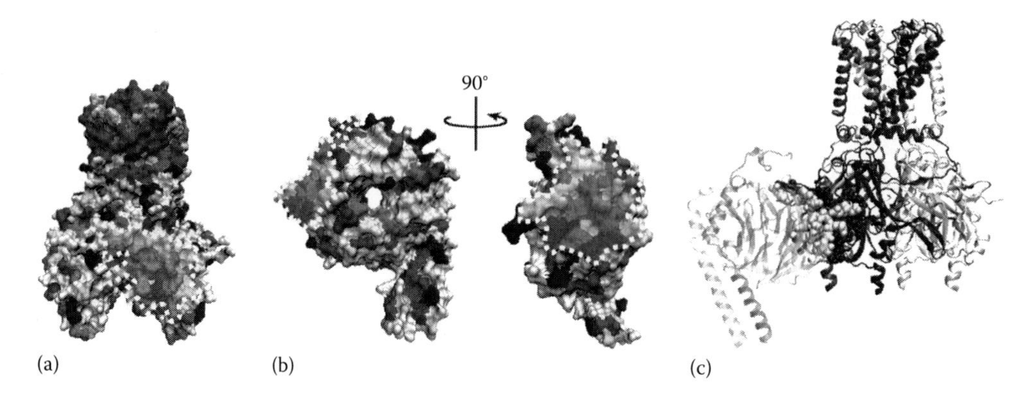

Figure 34.10 Surface representations of the channel and Gβγ (a and b respectively) are colored by residue hydrophobicity[91]: blue is most hydrophobic, white is intermediate, and red is least hydrophobic. Interface regions found in the BSM are highlighted in yellow. (c) Cartoon illustration of the two proteins together: two adjacent subunits of the channel are highlighted in red and gray, while the Gβ1 is yellow (transparent), Gγ2 is tan (transparent). Interface residues of the channel (red/gray) and the corresponding residues of Gβγ (yellow) are illustrated as spheres.

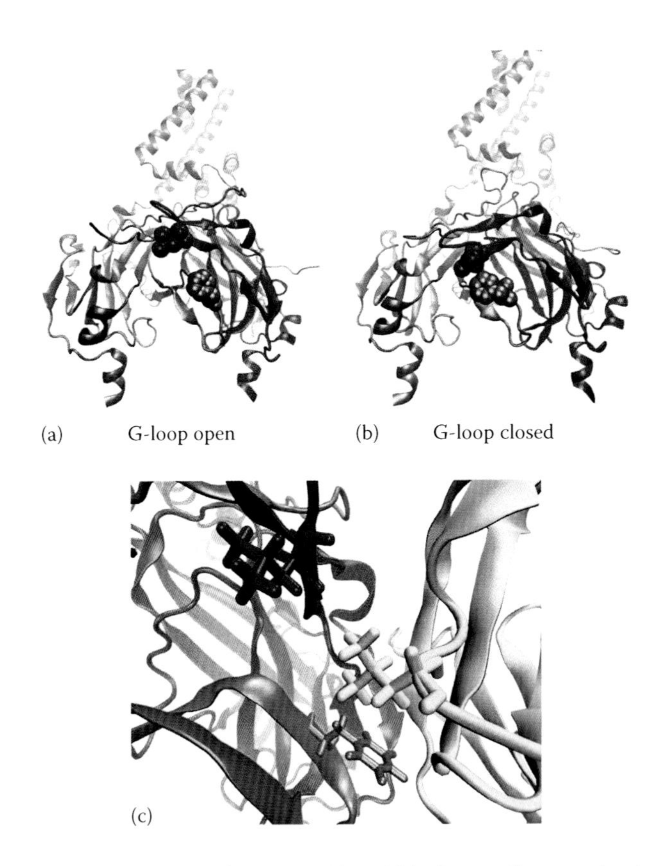

Figure 34.11 Cartoon depictions (a) and (b) of two adjacent subunits of the channel in the G-loop open and closed conformations respectively. GIRK1 residues F243 (gray) and L333 (red) are depicted as spheres in both panels. (c) Close-up view of the cleft between LM and DE loops in the BSM. Cartoon depictions of two adjacent channel subunits are in red and gray, while the $G\beta_F$ is yellow. Specific residues are highlighted in stick representation: GIRK1 L333 (red), GIRK1 F243 (gray), $G\beta_1$ L55 (yellow).

Whorton and MacKinnon[87] combined individually purified GIRK2 and Gβγ with diC8-PIP_2 (a soluble form of PIP_2 that has been used extensively in electrophysiological dose–response studies[98]) and incubated them at room temperature before crystal trials that were conducted in a high salt solution containing Na^+. Crystals of the complex diffracted at 3.5 Å resolution capturing the channel in a conformation that may represent an intermediate between the closed conformation and a partially open conformation previously determined by the same group.[86] The complex included a PIP_2 molecule and a Na^+ ion, revealing a smaller contact surface (~700 Å) (Figure 34.13) than that seen in the BSM of GIRK1 or other Gβγ effectors for which complex structures have been determined.[70] The GIRK2 secondary structure elements involved in contact with Gβ were the βK, βL, βM, and βN from one subunit with the βD and βE from an adjacent subunit. The Gβ secondary structure elements involved in contact with GIRK2 were the β-sheet elements forming blades 1 and 7 on one edge of the propeller. The GIRK2-binding site on Gβγ overlapped the Gα-binding site, consistent with the notion that Gβγ can interact with either the channel or Gα utilizing this interaction surface. Comparisons of the complex structure with prior structures in the absence of Gβγ suggested a 4° clockwise rotation (viewed from the inside) of the cytoplasmic domains along the central axis of the channel relative to the transmembrane domains. The F192 side chains were partially disordered (6–7 Å apart), although not enough to conduct hydrated K^+ ions (minimum of 10 Å apart). The authors concluded that the conformation captured by the crystal structure represents a *pre-open* state, consistent with the low open probability and burst kinetic behavior of unitary GIRK currents that show rapid flickering between open and closed (or pre-open) conformations.

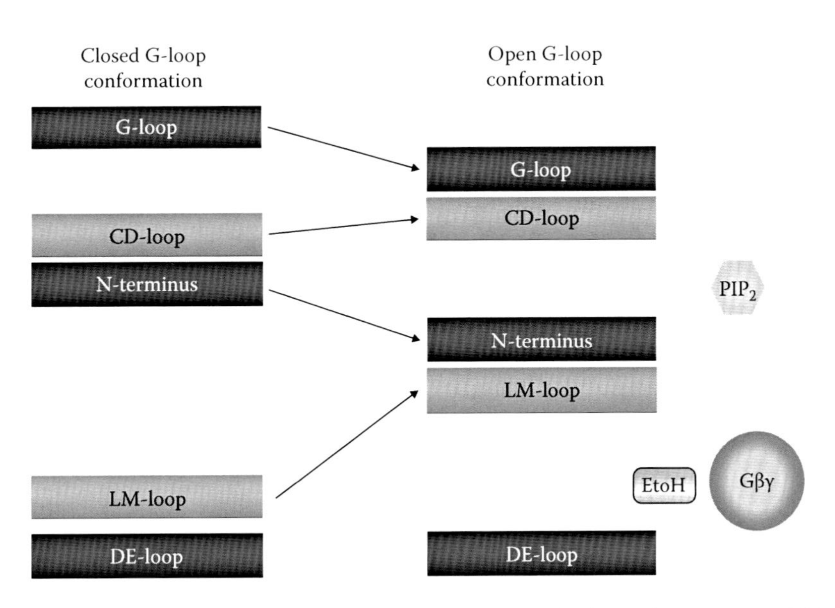

Figure 34.12 Summary of the major results of Meng et al. and their extension to include the DE loop and LM–DE loop cleft.[52] Transitioning from the closed to open, the secondary structure elements switch their close interactions from one adjacent element to the other. PIP_2 stabilizes the conformation on the right by direct interactions with the CD loop and N-terminus. Mahajan and colleagues have proposed that Gβγ works through a similar mechanism by stabilizing the same overall conformation but by direct interactions with a different part of the channel. Its proposed site of action at the DE–LM loop cleft is shared with the site of ethanol (EtOH) action. (Adapted with permission from Mahajan, R. et al., *Sci. Signal.*, 6(288), ra69, 2013.)

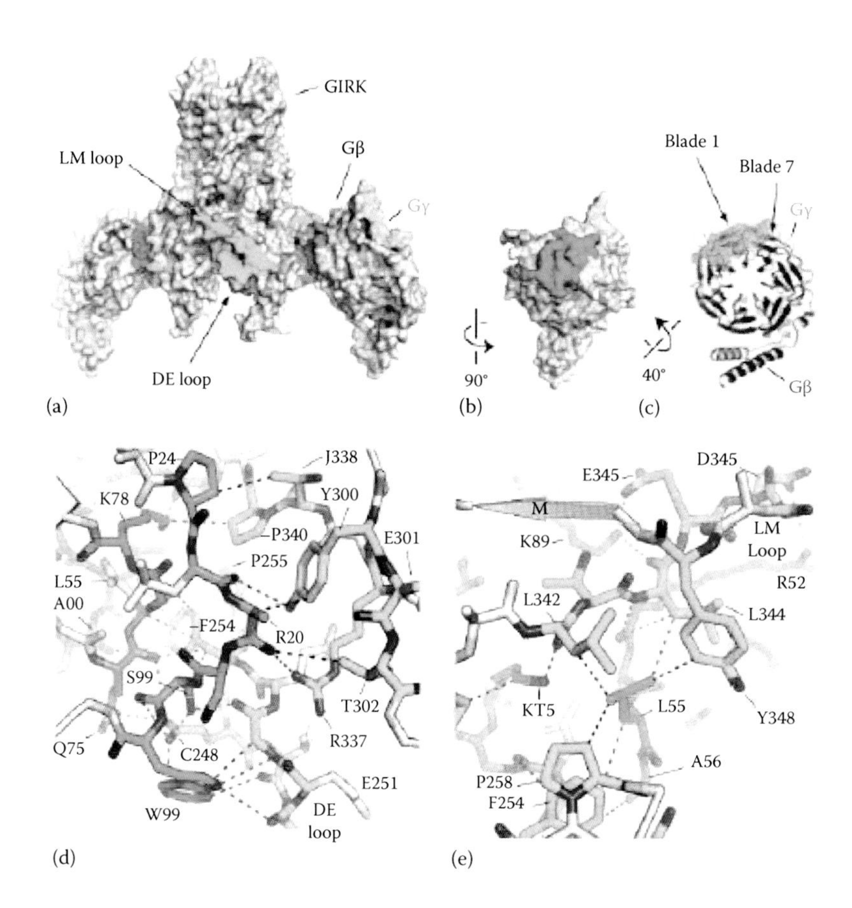

Figure 34.13 The GIRK–Gβγ binding interface. (a) Surface representation of the GIRK–Gβγ complex. The binding site on GIRK is colored yellow and the binding site on Gβγ is colored cyan. The front Gβγ dimer is removed for clarity. The overall orientation in a–c is similar as in Figure 34.10a and b. (b) A 90°-rotated view of a Gβγ dimer from panel a to more clearly show the binding interface. (c) The Gβγ dimer is rotated upward to orient the central axis of the β-propeller orthogonal to the page. (d, e) A close-up of the GIRK–Gβγ interaction, focused on the DE loop, βK and βN region (d) or the LM loop region (e) of GIRK2. Selected hydrogen bond and van der Waals interactions are shown as dashed lines as a visual aid. (Adapted with permission from Whorton, M.R. and MacKinnon, R., *Nature*, 498(7453), 190, 2013.)

The footprint of Gβγ on the GIRK1 and GIRK2 involved the same channel regions. The mGIRK2 shows only a 55.4% identity with the hGIRK1, and the specific interactions of the two channels with Gβγ show many similarities but also several interesting differences.

Comparison of the crystal structure model of the GIRK2–Gβγ complex with the two computational models showed that 64% of the GIRK2–Gβ interactions were the same in the LCM GIRK1–Gβ complex, while only 30% of the GIRK2–Gβγ interactions were the same in the BSM GIRK1–Gβγ complex. Gβγ in the LCM GIRK1 and GIRK2 structures is slightly rotated to include in its interaction surface residues absent from the BSM interaction surface, such as W99 and F335, on the *front* of the molecule closer to the center of the Gβγ propeller structure. In the BSM of GIRK1, Gβγ rotates slightly to instead uniquely engage residues in the *back* of the molecule, such as N88, E130, and N132.[56] The Gβ residues mentioned above (W99, F335, N88, E130, N132) reside in regions implicated by previous mutagenesis studies to be important in GIRK activation;[19,60,97] Binding site residue: in the front of the Gβ molecule are shared with Gα[19] and it is possible that interaction surfaces involving more of these residues underlie agonist-induced currents,[26,60,61] while residues towards the back of the molecule, not shared with Gα, may participate in stimulating basal (or agonist-independent) currents.[27,60] The pattern of interactions between Gβγ and either GIRK1 or GIRK2 suggests that common interactions between Gβγ with GIRK2 and GIRK1 may serve a fundamental role by which Gβγ activates these two channels, while differences could underlie distinct functional effects on the two channel subunits. For example, the unique C-terminus of the GIRK1 subunit confers robust receptor-dependent activity to GIRK heteromers.[11] It is possible that both the crystal structure of the GIRK2/Gβγ complex and the LCM of the GIRK1/Gβγ complex that show the largest similarity reside at a broad energy minimum (a pre-open state) near the final energy well, thus increasing the likelihood that the protein will *find* the most stable conformation represented by the BSM, which resides in a steep energy well representing one of the smaller clusters.

34.5 FUTURE QUESTIONS

The structural insights afforded by complexes of Gβγ with GIRK1 and GIRK2 have stimulated a number of questions, answers to which are likely not only to illuminate our understanding of homomeric vs. heteromeric GIRK channel gating by Gβγ but also to generally provide a structural understanding of G-protein signaling to effector proteins.

- How do the structures of different GIRK homomeric and heteromeric subunits with Gβγ underlie their functional differences?

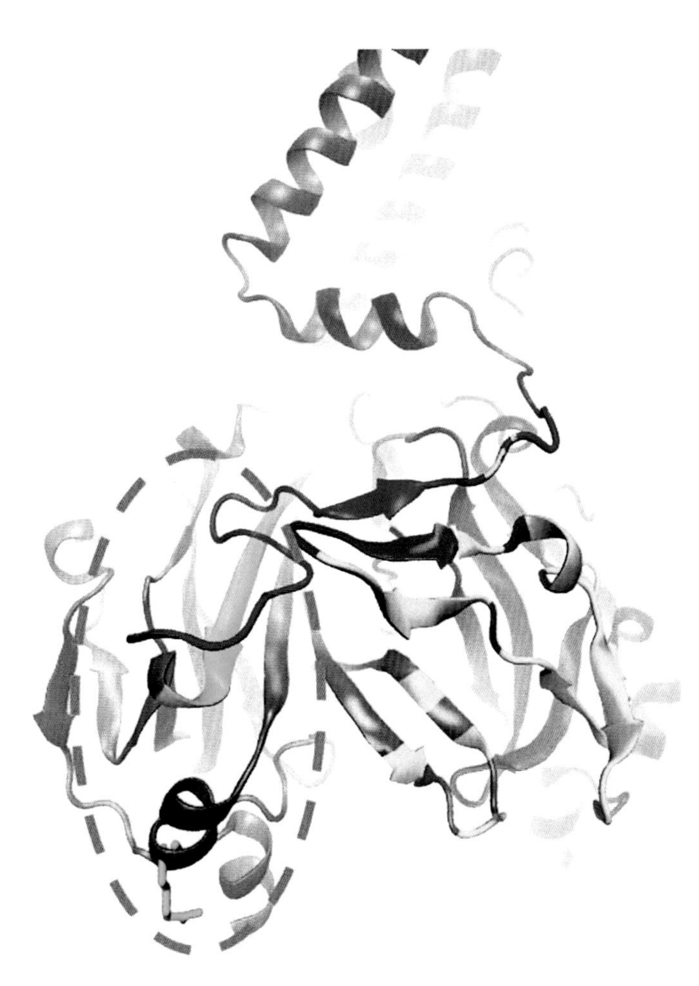

Figure 34.14 Regions of GIRK channels implicated in heterotrimer association. Cartoon depiction of two adjacent subunits of the channel IS shown in gray and red. Green represents the region implicated in Clancy and colleagues.[12] Blue represents the N-terminus implicated by multiple studies.[14,37] Green sticks represent the GIRK1* R286 residue that preliminary results have implicated as Gα-interacting.[57] Yellow highlights the predicted Gβγ interface from Mahajan and colleagues.[56] Dashed pink circle highlights a possible consensus region for heterotrimer association.

- How do Gα-GDP or Gα-GTP structures with different GIRK homomeric and heteromeric subunits explain inhibitory effects on Gβγ stimulation of activity or possibly permit stimulation of GIRK1 activity by Gβγ?
- How does the heterotrimeric G-protein (Gαβγ) interact with the GIRK channels to inhibit their function? Figure 34.14 suggests sites of interaction of Gβγ and Gα on the GIRK1 channel.
- How do structures of the entire macromolecular complex of GPCR/G-proteins/channel fit into producing agonist-independent and agonist-dependent channel activity?

ACKNOWLEDGMENTS

The authors are grateful to the many scientists who have intensely pursued the study of the mechanism of G-protein regulation of K^+ channels over the years. Special thanks goes to those of our colleagues who inspired this work: David Clapham, Yoshihisa Kurachi, and Eva Neer, and those who devoted their energy to this problem in the Logothetis lab: Scott Adney, Spiros Angelopoulos, Lia Baki, Kim Chan, Ciprian Craciun, Meng Cui, Wu Deng, Amr Ellaithy, Miguel Fribourg, Junghoon Ha, Candice Hatcher, Cheng He, Taihao Jin, Takeharu Kawano, Inna Keselman, Evgeny Kobrinsky, Noelle Langan, Edgar Leal-Pinto, George Liapakis, Coeli Lopes, Xuan-Yu Meng, Tooraj Mirshahi, Amanda Pabon, Gyu Park, Luying Peng, Jerome Petit-Jacques, Tibor Rohacs, Avia Rosenhouse-Dantsker, Radda Rusinova, Albert Shen, Jin-Liang Sui, Shobana Sundaram, Michel Vivaudou, Xixin Yan, Chul-Ho Yang, Jason Younkin, Hailin Zhang, Miao Zhang, and Qi Zhao. The authors are also grateful to the NIH (R01-HL054185, R01-HL090882, R01-HL059949, and F30-HL097582) and the AHA for their support of this work in the Logothetis lab over the past 20 years.

REFERENCES

1. Albsoul-Younes, A. M., P. M. Sternweis, P. Zhao, H. Nakata, S. Nakajima, Y. Nakajima, and T. Kozasa, Interaction sites of the G protein beta subunit with brain GIRK, *J. Biol. Chem.*, 276(16), 12712–12717, 2001. [73]
2. Aryal, P., H. Dvir, S. Choe, and P. A. Slesinger, A discrete alcohol pocket involved in GIRK channel activation, *Nat. Neurosci.*, 12(8), 988–995, 2009. [46]
3. Berlin, S., T. Keren-Raifman, R. Castel, M. Rubinstein, C. W. Dessauer, T. Ivanina, and N. Dascal, Gαi and Gβγ jointly regulate the conformations of a Gβγ effector, the neuronal G protein-activated K^+ channel (GIRK), *J. Biol. Chem.*, 285(9), 6179–6185, 2010. [42]
4. Berthet, J., T. W. Rall, and E. W. Sutherland, The relationship of epinephrine and glucagon to liver phosphorylase. IV. Effect of epinephrine and glucagon on the reactivation of phosphorylase in liver homogenates, *J. Biol. Chem.*, 224(1), 463–475, 1957. [19]
5. Breitwieser, G. E. and G. Szabo, Uncoupling of cardiac muscarinic and β-adrenergic receptors from ion channels by a guanine nucleotide analogue, *Nature*, 317(6037), 538–540, 1985. doi:10.1038/317538a0. [26]
6. Burgen, A. S. V. and K. G. Terroux, On the negative inotropic effect in the cat's auricle, *J. Physiol.*, 120(4), 449–464, 1953. [9]
7. Cabrera-Vera, T. M., J. Vanhauwe, T. O. Thomas, M. Medkova, A. Preininger, M. R. Mazzoni, and H. E. Hamm, Insights into G protein structure, function, and regulation, *Endocr. Rev.*, 24(6), 765–781, 2003. [29]
8. Carlsson, L., G. Duker, and I. Jacobson, New pharmacological targets and treatments for atrial fibrillation, *Trends Pharmacol. Sci.*, 31(8), 364–371, 2010. [17]
9. Castillo, J. D. and B. Katz, Production of membrane potential changes in the Frog's heart by inhibitory nerve impulses, *Nature*, 175(4467), 1035–1035, 1955. doi:10.1038/1751035a0. [10]
10. Chan, K. W., J.-L. Sui, M. Vivaudou, and D. E. Logothetis, Control of channel activity through a unique amino acid residue of a G protein-gated inwardly rectifying K^+ channel subunit, *Proc. Natl. Acad. Sci. U.S.A.*, 93(24), 14193–14198, 1996. [33]
11. Chan, K. W., J. L. Sui, M. Vivaudou, and D. E. Logothetis, Specific regions of heteromeric subunits involved in enhancement of G protein-gated K^+ channel activity. *J. Biol. Chem.*, 272(10), 6548–6555, 1997.
12. Clancy, S. M., C. E. Fowler, M. Finley, K. F. Suen, C. Arrabit, F. Berton, T. Kosaza, P. J. Casey, and P. A. Slesinger, Pertussis-toxin-sensitive Galpha subunits selectively bind to C-terminal domain of neuronal GIRK channels: Evidence for a heterotrimeric G-protein-channel complex, *Mol. Cell. Neurosci.*, 28(2), 375–389, 2005. [38]
13. Clarke, O. B., A. T. Caputo, A. P. Hill, J. I. Vandenberg, B. J. Smith, and J. M. Gulbis, Domain reorientation and rotation of an intracellular assembly regulate conduction in kir potassium channels, *Cell*, 141(6), 1018–1029, 2010. [53]

14. Cohen, N. A., Q. Sha, E. N. Makhina, A. N. Lopatin, M. E. Linder, S. H. Snyder, and C. G. Nichols, Inhibition of an inward rectifier potassium channel (Kir2.3) by G-protein betagamma subunits, *J. Biol. Chem.*, 271(50), 32301–32305, 1996. [40]
15. Corey, S. and D. E. Clapham, Identification of native atrial G-protein-regulated inwardly rectifying K+ (GIRK4) channel homomultimers. *J. Biol. Chem.*, 273(42), 27499–27504, 1998.
16. Dascal, N., W. Schreibmayer, N. F. Lim, W. Wang, C. Chavkin, L. DiMagno, C. Labarca, B. L. Kieffer, C. Gaveriaux-Ruff, and D. Trollinger, Atrial G protein-activated K+ channel: Expression cloning and molecular properties, *Proc. Natl. Acad. Sci. U.S.A.*, 90(21), 10235–10239, 1993. [31]
17. Doyle, D. A., J. Morais Cabral, R. A. Pfuetzner, A. Kuo, J. M. Gulbis, S. L. Cohen, B. T. Chait, and R. MacKinnon, The structure of the potassium channel: Molecular basis of K+ conduction and selectivity, *Science*, 280(5360), 69–77, 1998. [48]
18. Finley, M., C. Arrabit, C. Fowler, K. F. Suen, and P. A. Slesinger, βL–βM loop in the C-terminal domain of G protein-activated inwardly rectifying K+ channels is important for Gβγ subunit activation, *J. Physiol.*, 555(3), 643–657, 2004. [69]
19. Ford, C. E., N. P. Skiba, H. Bae, Y. Daaka, E. Reuveny, L. R. Shekter, R. Rosal et al., Molecular basis for interactions of G Protein βγ subunits with effectors, *Science*, 280(5367), 1271–1274, 1998. [72]
20. Fulton, J. F. and L. G. Wilson, *Selected Readings in the History of Physiology*. C.C. Thomas, Springfield, MA, 1966. [2]
21. Gaskell, W. H., The electrical changes in the quiescent cardiac muscle which accompany stimulation of the vagus nerve, *J. Physiol.*, 7(5–6), 451–452, 1886. [4]
22. Gaskell, W. H., On the action of muscarin upon the heart, and on the electrical changes in the non-beating cardiac muscle brought about by stimulation of the inhibitory and augmentor nerves, *J. Physiol.*, 8(6), 404–i8, 1887. [5]
23. Gill, D. M. and R. Meren, ADP-ribosylation of membrane proteins catalyzed by cholera toxin: Basis of the activation of adenylate cyclase, *PNAS*, 75(7), 3050–3054, 1978. [22]
24. Grosse, G., D. Eulitz, T. Thiele, I. Pahner, S. Schröter, S. Takamori, J. Grosse et al. Axonal sorting of Kir3.3 defines a GABA-containing neuron in the CA3 region of rodent hippocampus. *Mol. Cell. Neurosci.*, 24(3), 709–724, 2003.
25. Haga, T., E. M. Ross, H. J. Anderson, and A. G. Gilman, Adenylate cyclase permanently uncoupled from hormone receptors in a novel variant of S49 mouse lymphoma cells, *Proc. Natl. Acad. Sci. U.S.A.*, 74(5), 2016–2020, 1977. [23]
26. He, C., H. Zhang, T. Mirshahi, and D. E. Logothetis, Identification of a potassium channel site that interacts with G protein βγ subunits to mediate agonist-induced signaling, *J. Biol. Chem.*, 274(18), 12517–12524, 1999. [70]
27. He, C., X. Yan, H. Zhang, T. Mirshahi, T. Jin, A. Huang, and D. E. Logothetis, Identification of critical residues controlling G protein-gated Inwardly Rectifying K+ channel activity through interactions with the βγ subunits of G proteins, *J. Biol. Chem.*, 277(8), 6088–6096, 2002. [71]
28. Hibino, H., A. Inanobe, K. Furutani, S. Murakami, I. Findlay, and Y. Kurachi, Inwardly rectifying potassium channels: Their structure, function, and physiological roles, *Physiol. Rev.*, 90(1), 291–366, 2010. [1]
29. Howell, W. H. and W. W. Duke, The effect of vagus inhibition on the output of potassium from the heart, *Am. J. Physiol.*, 21(1), 51–63, 1908. [6]
30. Huang, C. L., P. A. Slesinger, P. J. Casey, Y. N. Jan, and L. Y. Jan, Evidence that direct binding of G beta gamma to the GIRK1 G protein-gated inwardly rectifying K+ channel is important for channel activation, *Neuron*, 15(5), 1133–1143, 1995. [62]
31. Huang, C. L., Y. N. Jan, and L. Y. Jan, Binding of the G protein betagamma subunit to multiple regions of G protein-gated inward-rectifying K+ channels, *FEBS Lett.*, 405(3), 291–298, 1997. [39]
32. Huang, C. L., S. Feng, and D. W. Hilgemann, Direct activation of inward rectifier potassium channels by PIP2 and its stabilization by Gbetagamma, *Nature*, 391(6669), 803–806, 1998. [45]
33. Iizuka, M., Tsunenari, I., Momota, Y., Akiba, I., and Kono, T. Localization of a G-protein-coupled inwardly rectifying K+ channel, CIR, in the rat brain. *Neuroscience*, 77(1), 1–13, 1997.
34. Inanobe, A., K. I. Morishige, N. Takahashi, H. Ito, M. Yamada, T. Takumi, H. Nishina et al., Gβγ directly binds to the carboxyl terminus of the G protein-gated muscarinic K+ channel, GIRK1, *Biochem. Biophys. Res. Commun.*, 212(3), 1022–1028, 1995. [63]
35. Inanobe, A., Y. Yoshimoto, Y. Horio, K. I. Morishige, H. Hibino, S. Matsumoto, Y. Tokunaga, Characterization of G-protein-gated K+ channels composed of Kir3.2 subunits in dopaminergic neurons of the substantia nigra. *J. Neurosci.*, 19(3), 1006–1017, 1999.
36. Ivanina, T., I. Rishal, D. Varon, C. Mullner, B. Frohnwieser-Steinecke, W. Schreibmayer, C. W. Dessauer, and N. Dascal, Mapping the Gbetagamma-binding sites in GIRK1 and GIRK2 subunits of the G protein-activated K+ channel, *J. Biol. Chem.*, 278(31), 29174–29183, 2003. [66]
37. Ivanina, T., D. Varon, S. Peleg, I. Rishal, Y. Porozov, C. W. Dessauer, T. Keren-Raifman, and N. Dascal, Galphai1 and Galphai3 differentially interact with, and regulate, the G protein-activated K+ channel, *J. Biol. Chem.*, 279(17), 17260–17268, 2004. [67]
38. Jones, J. F., Vagal control of the rat heart, *Exp. Physiol.*, 86(6), 797–801, 2001. [3]
39. Kovoor, P., K. Wickman, C. T. Maguire, W. Pu, J. Gehrmann, C. I. Berul, and D. E. Clapham, Evaluation of the role of I_{KACh} in atrial fibrillation using a mouse knockout model, *J. Am. Coll. Cardiol.*, 37(8), 2136–2143, 2001. [18]
40. Koyrakh, L., R. Luján, J. Colón, C. Karschin, Y. Kurachi, A. Karschin, and K. Wickman, Molecular and cellular diversity of neuronal G-protein-gated potassium channels. *J. Neurosci.*, 25(49), 11468–11478, 2005.
41. Krapivinsky, G., E. A. Gordon, K. Wickman, B. Velimirović, L. Krapivinsky, and D. E. Clapham, The G-protein-gated atrial K+ channel I_{KACh} is a heteromultimer of two inwardly rectifying K+-channel proteins, *Nature*, 374(6518), 135–141, 1995. [32]
42. Krapivinsky, G., I. Medina, L. Eng, L. Krapivinsky, Y. Yang, and D. E. Clapham, A novel inward rectifier K+ channel with unique pore properties, *Neuron*, 20(5), 995–1005, 1998. [65]
43. Kubo, Y., E. Reuveny, P. A. Slesinger, Y. N. Jan, and L. Y. Jan, Primary structure and functional expression of a rat G-protein-coupled muscarinic potassium channel, *Nature*, 364(6440), 802–806, 1993. [30]
44. Kunkel, M. T. and E. G. Peralta, Identification of domains conferring G protein regulation on inward rectifier potassium channels, *Cell*, 83(3), 443–449, 1995. [64]
45. Lambert, N. A., Dissociation of heterotrimeric G proteins in cells, *Sci. Signal.*, 1(25), re5, 2008. [78]
46. Lambright, D. G., J. Sondek, A. Bohm, N. P. Skiba, H. E. Hamm, and P. B. Sigler, The 2.0 A crystal structure of a heterotrimeric G protein, *Nature*, 379(6563), 311–319, 1996. [57]
47. Leal-Pinto, E., Y. Gómez-Llorente, S. Sundaram, Q. Y. Tang, T. Ivanova-Nikolova, R. Mahajan, L. Baki, Z. Zhang, J. Chavez, I. Ubarretxena-Belandia, and D. E. Logothetis. Gating of a G protein-sensitive mammalian Kir3.1 prokaryotic Kir channel chimera in planar lipid bilayers. *J. Biol. Chem.*, 2010 285(51), 39790–39800, 2010.
48. Li, L., S. J. Wright, S. Krystofova, G. Park, and K. A. Borkovich, Heterotrimeric G protein signaling in filamentous fungi, *Annu. Rev. Microbiol.*, 61(1), 423–452, 2007. [77]

49. Liao, Y. J., Y. N. Jan, and L. Y. Jan. Heteromultimerization of G-protein-gated inwardly rectifying K^+ channel proteins GIRK1 and GIRK2 and their altered expression in weaver brain. *J. Neurosci.*, 16(22):7137–7150, 1996.
50. Loew, A., Y. K. Ho, T. Blundell, and B. Bax, Phosducin induces a structural change in transducin beta gamma, *Structure*, 6(8), 1007–1019, 1998. [61]
51. Loewi, O., Über humorale übertragbarkeit der Herznervenwirkung, *Pflügers Archiv Eur. J. Physiol.*, 189(1), 239–242, 1921. [7]
52. Loewi, O. and E. Navratil, Über humorale Übertragbarkeit der Herznervenwirkung, *Pflügers Archiv Eur. J. Physiol.*, 214(1), 678–688, 1926. [8]
53. Logothetis, D. E., Y. Kurachi, J. Galper, E. J. Neer, and D. E. Clapham, The beta gamma subunits of GTP-binding proteins activate the muscarinic K^+ channel in heart, *Nature*, 325(6102), 321–326, 1987. [28]
54. Luján, R., E. Marron Fernandez de Velasco, C. Aguado, and K. Wickman, New insights into the therapeutic potential of GIRK channels, *Trends Neurosci.*, pii: S0166–S2236(13)00201–002014. Review, 2013.
55. Lüscher, C. and P. A. Slesinger, Emerging roles for G protein-gated inwardly rectifying potassium (GIRK) channels in health and disease, *Nat. Rev. Neurosci.*, 11(5), 301–315, 2010. [36]
56. Mahajan, R., J. Ha, M. Zhang, T. Kawano, T. Kozasa, D. E. Logothetis, A computational model predicts that $G\beta\gamma$ acts at a cleft between channel subunits to activate GIRK1 channels. *Sci. Signal.*, 6(288), ra69, 2013.
57. Mahajan, R. and D. E. Logothetis, Functional sites of interaction between G-protein $\beta\gamma$ subunits and GIRK channels. *Biophys. J.* 100(3), 358a–359a, 2011.
58. Meng, X.-Y., H.-X. Zhang, D. E. Logothetis, and M. Cui, The molecular mechanism by which PIP2 opens the intracellular G-loop gate of a Kir3.1 channel, *Biophys. J.*, 102(9), 2049–2059, 2012. [52]
59. Milligan, G. and E. Kostenis, Heterotrimeric G-proteins: A short history, *Br. J. Pharmacol.*, 147(Suppl 1), S46–S55, 2006. [20]
60. Mirshahi, T., L. Robillard, H. Zhang, T. E. Hébert, and D. E. Logothetis, $G\beta$ residues that do not interact with $G\alpha$ underlie agonist-independent activity of K^+ channels, *J. Biol. Chem.*, 277(9), 7348–7355, 2002. [75]
61. Mirshahi, T., V. Mittal, H. Zhang, M. E. Linder, and D. E. Logothetis, Distinct sites on G protein $\beta\gamma$ subunits regulate different effector functions, *J. Biol. Chem.*, 277(39), 36345–36350, 2002. [76]
62. Neher, E. and B. Sakmann, Single-channel currents recorded from membrane of denervated frog muscle fibres, *Nature*, 260(5554), 799–802, 1976. doi:10.1038/260799a0. [13]
63. Nishida, M. and R. MacKinnon, Structural basis of inward rectification: Cytoplasmic pore of the G protein-gated inward rectifier GIRK1 at 1.8 A resolution, *Cell*, 111(7), 957–965, 2002. [49]
64. Nishida, M., M. Cadene, B. T. Chait, and R. MacKinnon, Crystal structure of a Kir3.1-prokaryotic Kir channel chimera, *EMBO J.*, 26(17), 4005–4015, 2007. [51]
65. Noma, A. and W. Trautwein, Relaxation of the ACh-induced potassium current in the rabbit sinoatrial node cell, *Pflugers Archiv Eur. J. Physiol.*, 377(3), 193–200, 1978. [12]
66. Northup, J. K., P. C. Sternweis, M. D. Smigel, L. S. Schleifer, E. M. Ross, and A. G. Gilman, Purification of the regulatory component of adenylate cyclase, *Proc. Natl. Acad. Sci. U.S.A.*, 77(11), 6516–6520, 1980. [24]
67. Oldham, W. M. and H. E. Hamm, Structural basis of function in heterotrimeric G proteins, *Q. Rev. Biophys.*, 39(2), 117–166, 2006. [58]
68. Pegan, S., C. Arrabit, W. Zhou, W. Kwiatkowski, A. Collins, P. A. Slesinger, and S. Choe, Cytoplasmic domain structures of Kir2.1 and Kir3.1 show sites for modulating gating and rectification, *Nat. Neurosci.*, 8(3), 279–287, 2005. [50]
69. Pfaffinger, P. J., J. M. Martin, D. D. Hunter, N. M. Nathanson, and B. Hille, GTP-binding proteins couple cardiac muscarinic receptors to a K channel, *Nature*, 317(6037), 536–538, 1985. [27]
70. Reuveny, E. Structural biology: Ion channel twists to open. *Nature*, 498(7453), 182–183, 2013.
71. Riven, I., S. Iwanir, and E. Reuveny, GIRK channel activation involves a local rearrangement of a preformed G protein channel complex, *Neuron*, 51(5), 561–573, 2006. [41]
72. Rodbell, M., L. Birnbaumer, and S. L. Pohl, Characteristics of glucagon action on the hepatic adenylate cyclase system, *Biochem. J.*, 125(3), 58P–59P, 1971. [21]
73. Rusinova, R., T. Mirshahi, and D. E. Logothetis, Specificity of Gbetagamma signaling to Kir3 channels depends on the helical domain of pertussis toxin-sensitive Galpha subunits, *J. Biol. Chem.*, 282(47), 34019–34030, 2007. [37]
74. Sakmann, B., A. Noma, and W. Trautwein, Acetylcholine activation of single muscarinic K^+ channels in isolated pacemaker cells of the mammalian heart, *Nature*, 303(5914), 250–253, 1983. [14]
75. Signorini, S., Y. J. Liao, S. A. Duncan, L. Y. Jan, M. Stoffel, Normal cerebellar development but susceptibility to seizures in mice lacking G protein-coupled, inwardly rectifying K^+ channel GIRK2. *Proc. Natl. Acad. Sci. U.S.A.*, 94(3), 923–927, 1997.
76. Slesinger, P. A., E. Reuveny, Y. N. Jan, and L. Y. Jan, Identification of structural elements involved in G protein gating of the GIRK1 potassium channel, *Neuron*, 15(5), 1145–1156, 1995. [68]
77. Soejima, M. and A. Noma, Mode of regulation of the ACh-sensitive K-channel by the muscarinic receptor in rabbit atrial cells, *Pflügers Archiv Eur. J. Physiol.*, 400(4), 424–431, 1984. [25]
78. Sondek, J., A. Bohm, D. G. Lambright, H. E. Hamm, and P. B. Sigler, Crystal structure of a G-protein beta gamma dimer at 2.1 Å resolution, *Nature*, 379(6563), 369–374, 1996. [59]
79. Stanfield, P. R., S. Nakajima, and Y. Nakajima, Constitutively active and G-protein coupled inward rectifier K^+ channels: Kir2.0 and Kir3.0. *Rev. Physiol. Biochem. Pharmacol.*, 145, 47–179, 2002.
80. Sui, J. L., J. Petit-Jacques, and D. E. Logothetis, Activation of the atrial KACh channel by the betagamma subunits of G proteins or intracellular Na^+ ions depends on the presence of phosphatidylinositol phosphates, *Proc. Natl. Acad. Sci. U.S.A.*, 95(3), 1307–1312, 1998. [44]
81. Tesmer, J. J. G., V. M. Tesmer, D. T. Lodowski, H. Steinhagen, and J. Huber, Structure of human G protein-coupled receptor kinase 2 in complex with the kinase inhibitor balanol, *J. Med. Chem.*, 53(4), 1867–1870, 2010. [60]
82. Trautwein, W. and J. Dudel, Zum Mechanismus der Membranwirkung des Acetylcholin an der Herzmuskelfaser, *Pflügers Archiv Eur. J. Physiol.*, 266(3), 324–334, 1958. [11]
83. Vivaudou, M., K. W. Chan, J. L. Sui, L. Y. Jan, E. Reuveny, and D. E. Logothetis, Probing the G-protein regulation of GIRK1 and GIRK4, the two subunits of the KACh channel, using functional homomeric mutants, *J. Biol. Chem.*, 272(50), 31553–31560, 1997. [34]
84. Wall, M. A., D. E. Coleman, E. Lee, J. A. Iñiguez-Lluhi, B. A. Posner, A. G. Gilman, and S. R. Sprang, The structure of the G protein heterotrimer Gi[alpha]1[beta]1[gamma]2, *Cell*, 83(6), 1047–1058, 1995. [56]
85. Wang, W., M. R. Whorton, and R. Mackinnon, Quantitative analysis of mammalian GIRK2 channel regulation by G proteins, PIP_2 and Na^+ in a reconstituted system. eLife 10.7554/eLife 03671, 2014.
86. Whorton, M. R. and R. MacKinnon, Crystal structure of the mammalian GIRK2 K^+ channel and gating regulation by G proteins, PIP2, and sodium, *Cell*, 147(1), 199–208, 2011. [55]
87. Whorton, M. R. and R. MacKinnon, X-ray structure of the mammalian GIRK2-$\beta\gamma$ G-protein complex, *Nature*, 498(7453), 190–197, 2013.

88. Wickman, K. D., J. A. Iñiguez-Lhuhi, P. A. Daven Port, R. Taussig, G. B. Krapivinsky, M. E. Linder, A. G. Gilman, and D. E. Clapham, Recombinant G-protein $\beta\gamma$ subunits activate the muscarinic-gated atrial potassium channel. *Nature* 368(6468): 255–257, 1994.
89. Wickman, K., J. Nemec, S. J. Gendler, and D. E. Clapham, Abnormal heart rate regulation in GIRK4 knockout mice, *Neuron*, 20(1), 103–114, 1998. [15]
90. Wickman, K., C. Karschin, A. Karschin, M. R. Picciotto, D. E. Clapham, Brain localization and behavioral impact of the G-protein-gated K^+ channel subunit GIRK4. *J. Neurosci.*, 20(15), 5608–5015, 2000.
91. Wolfenden, R., L. Andersson, P. M. Cullis, and C. C. B. Southgate, Affinities of amino acid side chains for solvent water, *Biochemistry*, 20(4), 849–855, 1981. [110]
92. Xiao, J., X. Zhen, and J. Yang, Localization of PIP2 activation gate in inward rectifier K^+ channels, *Nat. Neurosci.*, 6(8) 811–818, 2003. [54]
93. Yang, Y., Y. Yang, B. Liang, J. Liu, J. Li, M. Grunnet, S.-P. Olesen et al., Identification of a Kir3.4 mutation in congenital long QT syndrome, *Am. J. Hum. Genet.*, 86(6), 872–880, 2010. [16]
94. Yi, B. A., Y. F. Lin, Y. N. Jan, and L. Y. Jan, Yeast screen for constitutively active mutant G protein-activated potassium channels, *Neuron*, 29(3), 657–667, 2001.
95. Yokogawa, M., M. Osawa, K. Takeuchi, Y. Mase, and I. Shimada, NMR analyses of the Gbetagamma binding and conformational rearrangements of the cytoplasmic pore of G protein-activated inwardly rectifying potassium channel 1 (GIRK1), *J. Biol. Chem.*, 286, 3, 2215–2223, 2011. [109]
96. Zeidner, G., R. Sadja, and E. Reuveny, Redox-dependent gating of G protein coupled inwardly rectifying K^+ channels, *J. Biol. Chem.*, 276(38), 35564–35570, 2001. [47]
97. Zhang, H., C. He, X. Yan, T. Mirshahi, and D. E. Logothetis, Activation of inwardly rectifying K^+ channels by distinct PtdIns(4,5)P2 interactions, *Nat. Cell Biol.*, 1(3), 183–188, 1999. [43]
98. Zhang, H., L. C. Craciun, T. Mirshahi, T. Rohács, C. M. Lopes, T. Jin, and D. E. Logothetis, PIP(2) activates KCNQ channels, and its hydrolysis underlies receptor-mediated inhibition of M currents, *Neuron*, 37(6), 963–975, 2003.
99. Zhao, Q., T. Kawano, H. Nakata, Y. Nakajima, S. Nakajima, and T. Kozasa, Interaction of G protein beta subunit with inward rectifier K^+ channel Kir3, *Mol. Pharmacol.*, 64(5), 1085–1091, 2003. [74]
100. Zhao, Q., A. M. Albsoul-Younes, P. Zhao, T. Kozasa, Y. Nakajima, and S. Nakajima, Dominant negative effects of a Gbeta mutant on G-protein coupled inward rectifier K+ channel. *FEBS Lett.*, 580(16), 3879–3882, 2006.
101. Gaudet, R., A. Bohm, P. B. Sigler, Crystal structure at 2.4 angstroms resolution of the complex of transducin betagamma and its regulator, phosducin, *Cell*, 87(3), 577–588, 1996.
102. Gaudet, R., J. R. Savage, J. N. McLaughlin, B. M. Willardson, and P. B. Sigler, A molecular mechanism for the phosphorylation-dependent regulation of heterotrimeric G proteins by phosducin, *Mol. Cell.*, 3(5), 649–660, 1999.

Calmodulin regulation of voltage-gated calcium channels and beyond

Manu Ben-Johny and David T. Yue

Contents

35.1 INTRODUCTION

In recent years, the ubiquitous Ca^{2+}-binding protein calmodulin (CaM) has emerged as a preeminent modulator of ion channel function (Saimi and Kung 2002), exhibiting exquisite Ca^{2+}-sensing capabilities (Chin and Means 2000; Tadross et al. 2008) and supporting vital Ca^{2+} feedback to many biological systems. An early example of such modulation was discovered via mutations in CaM that resulted in aberrant motile behavior of *Paramecium*. Organisms were either *underexcitable* or *overexcitable* to certain stimuli, reflecting the loss of a Ca^{2+}-dependent Na^+ current or K^+ current respectively (Kink et al. 1990). More recently, numerous ion channels have been found to be regulated by CaM, as reviewed elsewhere (Budde et al. 2002; Saimi and Kung 2002; Trudeau and Zagotta 2003; Halling et al. 2006; Gordon-Shaag et al. 2008; Minor and Findeisen 2010; Adelman et al. 2012; Van Petegem et al. 2012). This review will mainly focus on voltage-gated Ca^{2+} channels, but we briefly review the ion channel field in general, as follows. In small-conductance K^+ channels (SK channels), CaM initially preassociates in a Ca^{2+}-independent manner (Xia et al. 1998; Schumacher et al. 2004) and then activates these channels in response to submicromolar elevations in cytosolic Ca^{2+} (Xia et al. 1998), generating after hyperpolarizing current that shapes neuronal excitability (Stocker 2004; Adelman et al. 2012). The The KCNQ (K_V7) channels also constitutively bind Ca^{2+}-free CaM (apoCaM) (Ghosh et al. 2006), but intracellular Ca^{2+} may either inhibit (Gamper and Shapiro 2003; Gamper et al. 2005) or enhance these K^+ currents in an isoform specific manner (Ghosh et al. 2006; Shamgar et al. 2006). The cyclic nucleotide–gated (CNG) ion channels are also bestowed with apoCaM (Bradley et al. 2005), that upon Ca^{2+} binding induce a conformational change (Trudeau and Zagotta 2004) that inactivates channel activity (Bradley et al. 2004; Chen and Yau 1994; Trudeau and Zagotta 2003; Bradley et al. 2005). N-methyl-D-aspartate (NMDA) glutamate receptor activity is also downregulated by the direct interaction of Ca^{2+}/CaM with the carboxy terminus of the channel (Ehlers et al. 1996; Zhang et al. 1998). Ca^{2+}/CaM has been shown to bind a number of Transient Receptor Potential (TRP) channels, though the functional consequences of such binding remain controversial (Gordon-Shaag et al. 2008; Lau et al. 2012; Numazaki et al. 2003; Rosenbaum et al. 2004; Mercado et al. 2010). The Ca^{2+} release-activated Ca^{2+} channels (CRACs) undergo Ca^{2+}-dependent inactivation (CDI) orchestrated by CaM binding to Stromal interaction molecules (STIM) (Mullins et al. 2009). molecules (Mullins et al. 2009). Lastly, both the voltage-gated Na (Deschenes et al. 2002; Tan et al. 2002; Van Petegem et al. 2012; Ben-Johny et al. 2014; Biswas et al. 2008; Sarhan et al. 2009) and Ca^{2+} channels (Lee et al. 1999; Peterson et al. 1999; Zuhlke et al. 1999; DeMaria et al. 2001; Budde et al. 2002; Halling et al. 2006; Minor and Findeisen 2010) are under tight feedback regulation by Ca^{2+}/CaM, with far-reaching biological consequences (Alseikhan et al. 2002; Xu and Wu 2005; Adams et al. 2010). This review focuses on the rapid millisecond modulation of gating of voltage-gated Ca^{2+} channels, rich with biological impact, therapeutic possibilities, and mechanistic elegance.

35.2 CLASSIC HISTORICAL CONTEXT OF DISCOVERY

The earliest hints of CaM regulation of voltage-gated Ca^{2+} channels came not from a traditional biochemical approach wherein channels and CaM were copurified. Rather, functional indications emerged that elevated intracellular free Ca^{2+} concentration could accelerate the inactivation of voltage-gated Ca^{2+} currents in *Paramecium* (Brehm and Eckert 1978), invertebrate neurons (Tillotson 1979), and insect muscle (Ashcroft and Stanfield 1981). These data gave rise to the concept of CDI, a then revolutionary proposal that something besides transmembrane voltage could modulate the gating of ion channels (Eckert and Tillotson 1981).

Years elapsed before the discovery of corresponding behavior within vertebrate and mammalian systems, which awaited technological refinements enabling routine isolation and voltage clamp of Ca^{2+} currents in cells from these organisms. Among other systems, a flurry of reports then arose in cardiac muscle regarding the existence of CDI in L-type Ca^{2+} currents of the heart (carried by $Ca_V1.2$ channels); this story was revealed in preparations extending from the multicellular context (Kass and Sanguinetti 1984; Mentrard et al. 1984) to the level of isolated single cells (Lee et al. 1985). As an illustration, Figure 35.1a (Ca) shows Ca^{2+} current from frog myocytes, as evoked by a test voltage pulse to E_R + 80 mV (Mentrard et al. 1984). The actual extent of Ca^{2+} current (shaded area) was gauged by cobalt blockade of Ca^{2+} current (Co). To characterize CDI, Ca^{2+} entry during a prepulse of voltage to E_R + 80 mV (Figure 35.1b) can be seen to inactivate Ca^{2+} current in a closely ensuing test pulse (diminished area of shading). To distinguish this inactivation from that produced simply by voltage depolarization, a prepulse to E_R + 120 mV (Figure 35.1c) can be observed to produce far less inactivation in a subsequent test-pulse current. Because this prepulse entails strong depolarization, but decreased Ca^{2+} entry from attenuated Ca^{2+} entry driving force, the restoration of test Ca^{2+} current here argues that the potent inactivation in Figure 35.1b could be attributed to CDI. Data such as these reveal an inactivation process that depends upon voltage as a U-shaped function, now considered one hallmark of CDI (Eckert and Tillotson 1981).

Figures 35.1d through f exhibit the manifestation of CDI at the single-molecule level (Yue et al. 1990; Imredy and Yue 1992, 1994). Data such as these continue to be technically demanding, because of the small amplitude of unitary Ca^{2+} current (~0.3 pA), and the submillisecond gating timescale involved. Nonetheless, these data demonstrate the existence of U-shaped inactivation in the gating of a single cardiac L-type Ca^{2+} channel fluxing Ca^{2+}, as present in an adult rat ventricular myocyte (Imredy and Yue 1994). The prepulse voltage protocol is shown at the top; the dashed horizontal lines affiliated with exemplar traces (middle four rows) indicate the amplitude of open-channel elementary current; and the ensemble average current is shown at the bottom. These data proved that the Ca^{2+} influx of a single channel suffices

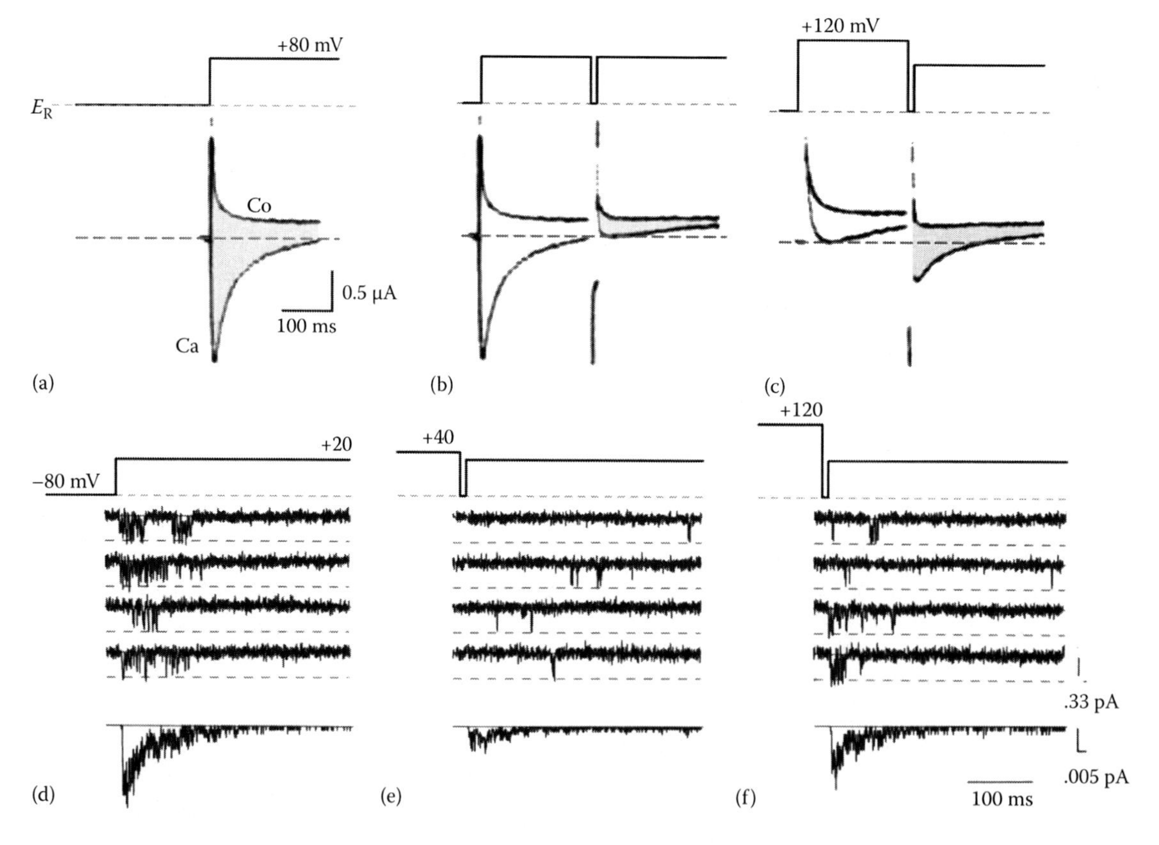

Figure 35.1 Classic U-shaped dependence of CDI. (a–c) Voltage-clamp experiments show CDI in macroscopic Ca^{2+} currents from multicellular recordings of frog atrial trabeculae cells. (Adapted from Mentrard, D. et al., *J. Gen. Physiol.*, 83, 105, 1984. With permission.) (a) Depolarizing voltage pulse to E_R + 80 mV elicits robust inactivating Ca^{2+} currents (shaded region). Fast capacitive current was measured by blocking these Ca^{2+} currents with 3 mM Co^{2+} solution. The vertical scale bar, 0.5 μA of current and horizontal scale bar, 100 ms. (b) When E_R + 80 mV conditioning pulse preceded the E_R + 80 mV test pulse, the Ca^{2+} currents in response to the test pulse were sharply reduced (shaded region), demonstrating robust inactivation of Ca_V channel. Format as in (a). (c) Further increase in the prepulse potential (E_R + 120 mV) led to a surprisingly weaker inactivation of Ca^{2+} current (compare to b). Format as in (a). (d–f) Single-channel recordings of L-type Ca^{2+} channels from adult rat ventricular myocytes show U-shaped dependence of Ca^{2+} channel inactivation. (Adapted Imredy, J.P. and Yue, D.T., *Neuron*, 12(6), 1301, 1994. With permission.) (d) Depolarizing pulse to +20 mV elicits representative elementary Ca^{2+} currents. Throughout, ensemble average currents are shown in bottom. (e) When depolarizing pulse was preceded by +40 mV prepulse, the elementary Ca^{2+} currents are sparser with the first opening delayed. This reduction in open probability reflects the sharp reduction of macroscopic Ca^{2+} currents seen in (b). Ensemble average, bottom. (f) Further increase in prepulse voltage to +120 mV resulted in a reversal of this pattern highlighting the U-shaped dependence of CDI. Ensemble average, bottom.

to trigger CDI in some contexts, a theme developed further in Section 35.4. More broadly, the striking correspondence of single-channel and multicellular behavior (Figure 35.1) undeniably established CDI as a legitimate molecular-level modulatory process. With its existence firmly in hand, there was the deep-seated belief that CDI would turn out to be of vital importance for biological Ca^{2+} homeostasis, a theme being continually corroborated with the passage of time (Section 35.6).

35.3 ADVENT OF CALCIUM CHANNEL CALMODULATION

The actual Ca^{2+} sensor of Ca^{2+} regulation of Ca^{2+} channels remained the subject of debate for some time. Proposed sensing mechanisms long included the direct binding of Ca^{2+} to the channel complex (Standen and Stanfield 1982; Plant et al. 1983; Eckert and Chad 1984), and Ca^{2+}-dependent phosphorylation/dephosphorylation of channels (Chad and Eckert 1986; Armstrong et al. 1988). Rapid progress toward identification of an actual Ca^{2+} sensor was enabled by the cloning and expression of recombinant Ca^{2+} channels, and the structure–function studies that then ensued (Snutch and Reiner 1992). Much of this wave of discovery was presaged by a bioinformatic gambit (Babitch 1990) that identified somewhat of a Ca^{2+}-binding EF-hand motif in the proximal carboxy termini of voltage-gated Na and Ca^{2+} channels. The alignment in Figure 35.2a explicitly documents the existence of a segment (EF1) with moderate resemblance to an EF-hand Ca^{2+}-binding motif in these channels, with the EF-hand scoring template displayed above the sequence in question. The overall position of this motif is represented within the channel cartoon shown in Figure 35.2b. Once sensitized to the existence of a first EF-hand analog, one could discern the existence of a second EF-hand-like motif in these channels (Figure 35.2a, EF2), just downstream of EF1. Alerted to the possibility that the carboxy terminus of Ca^{2+} channels might play an important role in CDI, chimeric channel analysis was pursued (de Leon et al. 1995), in which portions of the carboxy termini of L-type ($Ca_V1.2$) and R-type ($Ca_V2.3$) Ca^{2+} channels were swapped (Figure 35.2c). In particular, these types of channels, respectively, demonstrate strong and weak CDI under elevated buffering of intracellular Ca^{2+} buffering (Liang et al. 2003), rendering them useful for

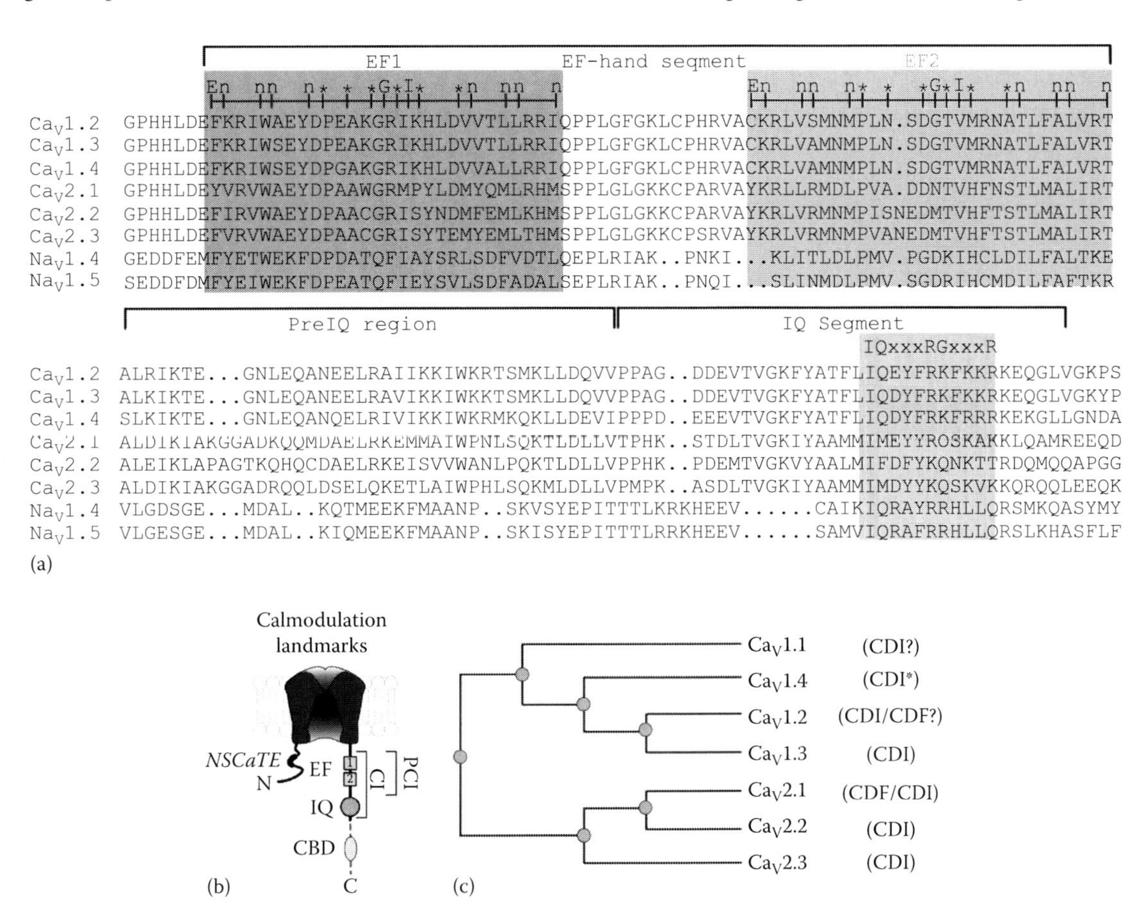

Figure 35.2 Sequence similarities across family of voltage-gated Ca channels foreshadow discovery of calmodulation across Ca_V1 and Ca_V2 channels. (a) Sequence alignment for the Ca^{2+}-inactivation (CI) regions of $Ca_V1.2$, $Ca_V1.3$, $Ca_V1.4$, $Na_V1.4$, and $Na_V1.5$ channels is shown. The CI region, in the proximal channel carboxy terminus (~160 aa), contains segments essential for CaM regulation. The dual vestigial EF-hand (EF) segments (shaded in top block of alignments) span the proximal ~100 aa of the CI region. Consensus sequence for EF-hand motif shown earlier (Tufty and Kretsinger 1975; de Leon et al. 1995). The IQ domain, long proposed as preeminent for CaM/channel binding, is a canonical CaM-binding motif. The consensus sequence for the IQ motif shaded in bottom block of alignments. (b) Channel cartoon depicts overall arrangement of calmodulatory landmarks. The NSCaTE (N-terminal spatial Ca^{2+} transforming element) segment on channel amino terminus of $Ca_V1.2$–1.3 channels is an N-lobe Ca^{2+}/CaM effector site. The CBD element is a CaM-binding segment of Ca_V2 channels touted to be critical for CDI (oval). (c) Phylogenetic tree depicts evolutionary relationships between the various Ca_V channel isoforms. The known functional manifestations of Ca^{2+}/CaM regulation for these channels are summarized to the right. $Ca_V1.1$ shows weak CDI. Latent CDI of $Ca_V1.4$ is revealed by truncation of distal carboxy terminus. CDF of $Ca_V1.2$ channels is unveiled by IQ/AA mutation in the IQ domain.

chimeric analysis. Results from these experiments demonstrated that the proximal third of the carboxy terminus (Figure 35.2b, CI region) was indeed important for CDI, and that the EF1 region of $Ca_V1.2$ channels was critical in particular for the strong CDI of these channels (de Leon et al. 1995). Thus, it was plausible that EF1 might be the Ca^{2+}-binding site for CDI postulated in earlier mechanisms of this process (Standen and Stanfield 1982; Plant et al. 1983; Eckert and Chad 1984). However, the sparing of CDI by point mutations within the Ca^{2+}-binding loop of EF1 rendered implausible this interpretation of the importance of EF1 (Zhou et al. 1997; Peterson et al. 2000).

A related intriguing observation was that CaM-binding targets themselves often resemble CaM (Jarrett and Madhavan 1991). Given the bilobed organization of CaM, each with two EF-hand binding sites, the dual vestigial EF hands in the carboxy terminus of channels (Figure 35.2a) could be viewed as resembling a lobe of CaM, thereby hinting that CaM itself might be the Ca^{2+} sensor underlying CDI. On the other hand, pharmacological inhibition of CaM did not eliminate CDI of L-type Ca^{2+} channels (Imredy and Yue 1994; Zuhlke and Reuter 1998). Nonetheless, systematic deletions of overlapping segments of the $Ca_V1.2$ channel CI region identified an IQ motif (Figure 35.2a and b) as another critical element in the CDI process (Zuhlke and Reuter 1998), and subsequent mutagenesis of the $Ca_V1.2$ IQ domain confirmed a critical role for CDI (Qin et al. 1999). As IQ motifs are frequently associated with the binding of Ca^{2+}-free CaM (apoCaM) (Jurado et al. 1999), the likelihood of CaM serving as CDI sensor arose anew. Definitive identification of CaM as the sensor for $Ca_V1.2$ channel came with the development of a mutant Ca^{2+}-insensitive mutant CaM, where point mutations are introduced within all four EF-hand Ca^{2+}-binding sites (CaM_{1234}) (Xia et al. 1998). If modulation of a target molecule were to require *preassociation* with apoCaM, followed by Ca^{2+} regulation induced via Ca^{2+} binding to this apoCaM, then CaM_{1234} could act as a dominant negative to eliminate regulatory function, just as seen for small-conductance Ca^{2+}-activated K channels (Xia et al. 1998). Indeed, when CaM_{1234} or its analogs was coexpressed with $Ca_V1.2$ channels, CDI was essentially eliminated (Peterson et al. 1999; Zuhlke et al. 1999). Furthermore, apoCaM was shown to bind to the $Ca_V1.2$ CI region, with strong dependence upon the IQ region (Erickson et al. 2001, 2003; Pitt et al. 2001). Interestingly, the preassociation of apoCaM with $Ca_V1.2$ channels would sterically protect CaM from pharmacological inhibition (Dasgupta et al. 1989), rationalizing the earlier insensitivity of CDI to such small-molecule perturbation. In all, these data firmly established CaM as the sensor of $Ca_V1.2$ channel Ca^{2+} regulation.

This scheme of CaM-mediated Ca^{2+} regulation was not only applicable to $Ca_V1.2$ channels, as perhaps first believed (Liang et al. 2003); this system of *calmodulation* was gradually recognized to pertain across most of the Ca_V1 and Ca_V2 (but not Ca_V3) branches of the Ca_V channel superfamily (Figure 35.2c). At first, P-type (Cav2.1) channels were shown to be Ca^{2+} regulated by CaM (Lee et al. 1999), but the role of CaM preassociation in this regulation was unclear, and a Ca^{2+}-/CaM-binding site downstream of the IQ element (CBD, Figure 35.2b) was argued to be important. The IQ domain was not identified as a structurally important element in this initial report. Moreover, Ca^{2+} regulation of $Ca_V2.1$ channels takes the form of a rapidly induced facilitation of current (Ca^{2+}-dependent facilitation [CDF]), followed by a slowly developing CDI. Thus, it was not initially clear whether the Ca^{2+} regulation of $Ca_V2.1$ channels followed a similar mechanistic scheme as found in $Ca_V1.2$ channels. Nonetheless, shortly thereafter, apoCaM was found to preassociate with $Ca_V2.1$ channels (Erickson et al. 2001); Ca^{2+}-insensitive mutant CaM molecules were found to eliminate Ca^{2+} regulation in $Ca_V2.1$ channels (DeMaria et al. 2001); and the IQ domain was determined to be structurally essential for this regulation (DeMaria et al. 2001). Curiously, deletion of the entire carboxy terminus downstream of the IQ element (including the CBD element) appears to spare completely the CDF and CDI of $Ca_V2.1$ channels (DeMaria et al. 2001; Chaudhuri et al. 2005), although there is ongoing debate on this point (Lee et al. 2002). Overall, however, $Ca_V2.1$ and $Ca_V1.2$ channels were seen to be Ca^{2+} regulated by a largely conserved CaM regulatory scheme.

In rapid succession, calmodulation was seen to extend to both $Ca_V2.2$ and $Ca_V2.3$ channels, leading to the realization that this modulatory system was likely to extend generally across large swaths of the Ca_V1–2 channel superfamily (Liang et al. 2003). This expectation has been largely fulfilled: strong CaM-mediated CDI was identified in $Ca_V1.3$ channels (Xu and Lipscombe 2001; Shen et al. 2006; Yang et al. 2006); the latent capacity for CaM-mediated CDI was revealed in $Ca_V1.4$ channels (Singh et al. 2006; Wahl-Schott et al. 2006); and there are potential indications of CaM-mediated regulation in $Ca_V1.1$ channels (Stroffekova 2008, 2011).

Certain nuances regarding the overall manifestation of calmodulation across Ca^{2+} channels merit some attention. First, some forms of calmodulation are insensitive to strong buffering of intracellular Ca^{2+} (e.g., $Ca_V1.2$ CDI), whereas others are not ($Ca_V2.3$ CDI). This differential sensitivity allowed for successful application of a chimeric channel approach to identify important structural determinants (de Leon et al. 1995), despite the existence of CaM regulation across the channel superfamily. We will explore the deeper implications of this contrast in Section 35.4. Second, the nature of CDF in $Ca_V1.2$ channels remains mysterious, in that its manifestation in native channels of the heart is rather weak to begin with, and curiously attenuated by blockade of Ca^{2+} release from neighboring ryanodine receptor channels (RYRs) (Wu et al. 2001). Moreover, such CDF was not observed in recombinant $Ca_V1.2$ channels expressed in mammalian cells lines (Peterson et al. 1999). Indeed, CDF is strongly manifest only in recombinant $Ca_V1.2$ channels bearing point mutations within the IQ domain, and expressed in frog oocytes (Zuhlke et al. 2000). By contrast, CDI of $Ca_V1.2$ channels is strong and universally observed across experimental platforms. These nuances aside, the overall scope and generality of calmodulation across the Ca_V1–2 channel superfamily is staggering (Figure 35.2c).

35.4 FUNCTIONAL BIPARTITION OF CALMODULIN AND SELECTIVITY FOR LOCAL/GLOBAL CALCIUM SOURCES

Apart from generality, the calmodulation of Ca^{2+} channels also turned out to exhibit an elegant and powerful form of Ca^{2+} decoding. This decoding echoes older findings of Ching Kung and colleagues, who discovered a *functional bipartition* of CaM in *Paramecium* (Kink et al. 1990; Saimi and Kung 2002). Here, *underexcitable* behavioral mutants had mutations only in the N-lobe of CaM, whereas *overexcitable* mutants had mutations only in the C-lobe. This led to the realization that one lobe of CaM can be important for signaling to one set of functions, whereas the other lobe signals to an alternative set of functions. This discovery gave reason to wonder whether an analog of this bipartition pertains to the behavior of higher-order animals (Wei et al. 2003), still an intriguing and open question. However, the large-scale generalization of this phenomenon to mammalian systems awaited the discovery of CaM regulation of mammalian Ca^{2+} channels. The requirement that regulation requires apoCaM to preassociate with these mammalian channels readily permitted exploration of the bipartition hypothesis. By coexpressing channels with mutant CaM constructs in which only C- or N-terminal lobes can bind Ca^{2+} (CaM_{12} and CaM_{34}, respectively), the ability of individual lobes to trigger distinct components of channel regulation was discovered. In most Ca_V1 channels, the C-lobe was found to trigger a kinetically rapid phase of CDI (Figure 35.4d), whereas the N-lobe induced a slower component (Peterson et al. 1999; Yang et al. 2006; Liu et al. 2010) (see Section 35.5). Most strikingly, in $Ca_V2.1$ channels, the lobes of CaM produce opposing polarities of regulation: the C-lobe of CaM triggers a kinetically rapid CDF, whereas the N-lobe evokes a slower CDI (DeMaria et al. 2001). $Ca_V2.2$ and $Ca_V2.3$ channels were found to manifest CDI triggered mainly by the N-lobe of CaM (Liang et al. 2003). This manner of functional bipartition across the Ca_V1–2 channel family is summarized in Figure 35.3a.

Beyond simple bipartition, however, the calmodulation of mammalian Ca^{2+} channels revealed still greater power in the functional capabilities of the individual lobes of CaM. Hints as to this enormous capability came from the differential sensitivity of calmodulation in various channels to Ca^{2+} buffering. Processes triggered by the C-lobe of CaM are invariably insensitive to Ca^{2+} buffering, whereas N-lobe triggered processes in Ca_V2 channels can be eliminated by the same manipulation. Such buffering would only spare Ca^{2+} elevations near the cytoplasmic mouth of channels where strong point-source Ca^{2+} influx would overwhelm the capabilities of buffers (Neher 1986; Stern 1992). Thus, it could be argued that local Ca^{2+} influx through individual channels suffices to trigger C-lobe signaling of affiliated CaM; in other words, the C-lobe of CaM exhibited a *local Ca^{2+} selectivity*. By contrast, a spatially global elevation of Ca^{2+} (present in the absence of strong Ca^{2+} buffering) is required for N-lobe signaling of Ca_V2 channels; the N-lobe in this context manifests a *global selectivity*. The existence of global selectivity is remarkable, given that local Ca^{2+} influx yields far larger Ca^{2+} elevations near a channel than does globally sourced Ca^{2+} (Tay et al. 2012). The exception to this pattern is the local Ca^{2+} selectivity of the N-lobe component of CDI in $Ca_V1.2/3$ channels, which we will explain later. Further corroboration of the existence of spatial Ca^{2+} selectivity comes from the ability of single $Ca_V1.2$ channels to undergo CDI (Figure 35.1d through f). By definition, only a local Ca^{2+} source is present in the single-channel configuration; thus, the presence of CDI indicates local Ca^{2+} selectivity. Additionally, single-channel records of $Ca_V2.1$ channels only exhibit CDF (driven by C-lobe), but not CDI (triggered by N-lobe) (Chaudhuri et al. 2007), entirely consistent with the proposed differential selectivities for this channel. Figure 35.3a summarizes the remarkable arrangement of spatial Ca^{2+} selectivities according to the lobes of CaM, believed broadly important for the Ca^{2+} signaling repertoire in many contexts.

The mechanisms underlying these contrasting spatial Ca^{2+} selectivities long remained mysterious. These behaviors have recently been explained in terms of emergent behaviors of a

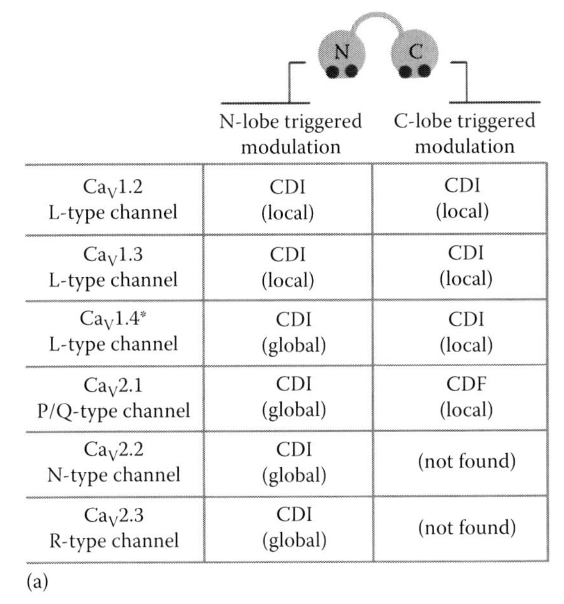

	N-lobe triggered modulation	C-lobe triggered modulation
$Ca_V1.2$ L-type channel	CDI (local)	CDI (local)
$Ca_V1.3$ L-type channel	CDI (local)	CDI (local)
$Ca_V1.4$* L-type channel	CDI (global)	CDI (local)
$Ca_V2.1$ P/Q-type channel	CDI (global)	CDF (local)
$Ca_V2.2$ N-type channel	CDI (global)	(not found)
$Ca_V2.3$ R-type channel	CDI (global)	(not found)

(a)

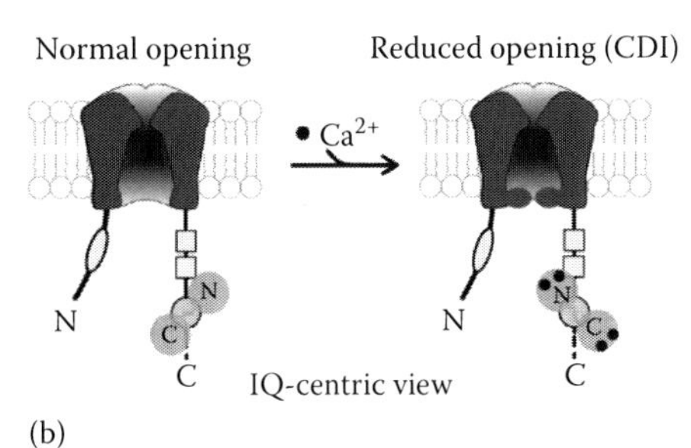

(b)

Figure 35.3 Functional bipartition of CaM regulation. (a) Table outlines functional bipartition of CaM in Ca_V channels and the corresponding spatial Ca^{2+} selectivities. In $Ca_V1.2$ and $Ca_V1.3$, both the C-lobe and the N-lobe of CaM support fast CDI with local Ca^{2+} selectivity. Latent CDI of $Ca_V1.4$ channels revealed by deletion of a distal carboxy terminal autoinhibitory domain. In Ca_V2 channels, Ca^{2+} binding to N-lobe of CaM elicits slow CDI with global Ca^{2+} spatial selectivity. In $Ca_V2.1$ channels, C-lobe of CaM supports a fast CDF. No C-lobe triggered modulation of $Ca_V2.2$ and $Ca_V2.3$ has been reported. (b) Historic IQ-centric view for CaM regulation of Ca_V channels. ApoCaM starts off preassociated to the IQ domain. Following Ca^{2+} binding, CaM rebinds the same IQ domain with an even higher affinity. *Indicates latent capability for CaM regulation.

system where a lobe of apoCaM must transiently detach from a channel preassociation site to bind Ca^{2+}, and where a Ca^{2+}-bound lobe of CaM must bind a channel effector site to induce regulation (Tadross et al. 2008). When the slow Ca^{2+} unbinding kinetics of the C-lobe of CaM are imposed on this system, local Ca^{2+} sensitivity results. By contrast, when the rapid Ca^{2+} unbinding kinetics of the N-lobe are interfaced with this same architecture, global Ca^{2+} selectivity arises when the channel preferentially binds apoCaM, as in the case of Ca_V2 channels (Figure 35.3a). By contrast, when the channel preferentially interacts with Ca^{2+}/CaM, a local Ca^{2+} selectivity emerges, as in the context of Ca_V1 channels (Figure 35.3a). Indeed, an NSCaTE Ca^{2+}-/CaM-binding site present only on the amino terminus of $Ca_V1.2/3$ channels (Figure 35.2b) accounts for channel preference for the N-lobe of Ca^{2+}/CaM over apoCaM (Ivanina et al. 2000; Dick et al. 2008), explaining the local Ca^{2+} selectivity of N-lobe CDI in these channels. Elimination of the NSCaTE site in these channels switches their N-lobe CDI to a global profile; donation of NSCaTE to Ca_V2 channels endows their N-lobe CDI with local Ca^{2+} selectivity (Dick et al. 2008). Thus, nature itself utilizes the NSCaTE module to fine-tune the spatial Ca^{2+} selectivity of N-lobe CDI.

35.5 MOLECULAR BASIS OF CaM REGULATION

The actual arrangement of apoCaM and Ca^{2+}/CaM on Ca^{2+} channels long remained a standing challenge, given the >2000 amino acids comprising the main pore-forming α_1 subunit alone, the capability of multiple peptide segments of the channel to bind CaM in vitro, and the challenging nature of obtaining atomic level structures for these channels.

The stoichiometry between channel and CaM has been debated. Based on crystal structures of Ca^{2+}/CaM complexed with portions of the carboxy tail CI region of $Ca_V1.2$ channels (Fallon et al. 2009; Kim et al. 2010), it has been argued that multiple CaM molecules interact with a channel to produce the full tapestry of Ca^{2+} regulatory functions. By contrast, covalent fusion of single CaM molecules to $Ca_V1.2$ channels strongly suggests that only one CaM per channel is involved in functional regulation (Mori et al. 2004). Moreover, live-cell fluorescence resonance energy transfer (FRET) studies between CaM and holochannels ($Ca_V1.2$) also advocate strongly for a 1:1 CaM:channel ratio (Ben-Johny et al. 2012). In all, while peptide fragments of Ca^{2+} channels may interact with multiple CaM molecules in some contexts, we favor the idea that only one CaM interacts with intact holochannels.

The IQ domain has long been known to figure importantly in the calmodulation of various channels. Mutations within the IQ domain strongly modulate the Ca^{2+} regulation of $Ca_V1.2$ channels (Zuhlke and Reuter 1998; Qin et al. 1999; Zuhlke et al. 1999, 2000; Erickson et al. 2003), $Ca_V1.3$ channels (Yang et al. 2006; Ben-Johny et al. 2013), $Ca_V2.1$ channels (DeMaria et al. 2001; Mori et al. 2008), $Ca_V2.2$, and $Ca_V2.3$ channels (Liang et al. 2003). ApoCaM preassociation with $Ca_V1.2/1.3$ channels depends strongly on the IQ domain (Erickson et al. 2001, 2003; Pitt et al. 2001; Ben-Johny et al. 2013). Moreover, Ca^{2+}/CaM binds well to IQ-domain peptides of many Ca_V channels (Peterson et al. 1999; Zuhlke et al. 1999; DeMaria et al. 2001; Pitt et al. 2001; Liang et al. 2003; Ben-Johny et al. 2013). These findings have led to the long prevailing IQ-centric hypothesis shown in Figure 35.3b. Here, the IQ domain is important not only as an apoCaM preassociation site (left cartoon) but also as the Ca^{2+}/CaM effector site (right cartoon). The predominance of the IQ-centric paradigm has prompted resolution of several crystal structures of Ca^{2+}/CaM complexed with IQ-domain peptides of various Ca_V1–2 channels (Fallon et al. 2005; Van Petegem et al. 2005; Kim et al. 2008; Mori et al. 2008).

However, this viewpoint remained problematic in three regards: (1) The atomic structures of Ca^{2+}/CaM bound to wild-type and mutant IQ peptides of $Ca_V1.2$ show that a central isoleucine in the IQ element is deeply buried within the C-lobe of Ca^{2+}/CaM, and that alanine substitution at this site hardly changes structure (Fallon et al. 2005). Additionally, Ca^{2+}/CaM dissociation constants for corresponding wild-type and mutant IQ peptides are about the same (Zuhlke et al. 2000). It is then puzzling that alanine substitution at this well-encapsulated locus can influence the remainder of the channel to intensely disrupt regulation (Fallon et al. 2005). (2) For $Ca_V1.2/1.3$ channels, the N-lobe of Ca^{2+}/CaM effector site appears to be an NSCaTE element of the channel amino terminus (Dick et al. 2008; Tadross et al. 2008; Liu and Vogel 2012), separate from the IQ element. (3) In-depth study of the crystal structure of Ca^{2+}/CaM complexed with the IQ peptide of $Ca_V2.1$ channels argues that the C-lobe effector site also resides at a site beyond the IQ module (Mori et al. 2008).

A major concern with older IQ-domain analyses is that the calmodulatory system was not considered conceptually as a whole, as diagrammed in Figure 35.4a (shown with specific reference to $Ca_V1.3$ channels for didactic clarity) (Ben-Johny et al. 2013). Configuration E portrays channels lacking apoCaM. Such channels can open normally but do not manifest CDI, because Ca^{2+}/CaM from bulk solution cannot efficiently access a channel in configuration E to induce CDI (Mori et al. 2004; Yang et al. 2007; Liu et al. 2010; Findeisen et al. 2011). ApoCaM binding with configuration E gives rise to configuration A, where opening can also occur normally, but CDI can now transpire. For CDI, Ca^{2+} binding to both lobes of CaM yields configuration I_{CN}, corresponding to a fully inactivated channel with strongly reduced opening. As for intermediate configurations, Ca^{2+} binding only to the C-lobe induces configuration I_C, synonymous with a C-lobe inactivated channel; Ca^{2+} binding only to the N-lobe yields the N-lobe-inactivated configuration (I_N). Importantly, subsequent entry into I_{CN} likely involves positively cooperative interactions indicated by a λ symbol.

This framework underscores one challenge for older analyses of the IQ domain, where function was mostly characterized with only endogenous CaM present. This regime is ambiguous because IQ-domain mutations could alter calmodulation via changes at multiple steps within Figure 35.4a, while interpretations mainly ascribed effects to perturbed Ca^{2+}/CaM binding with an IQ effector site. For example, IQ mutations could weaken apoCaM preassociation and reduce CDI by favoring configuration E. Moreover, mutations that do weaken interaction with one lobe of Ca^{2+}/CaM may have their functional effects largely masked by positively cooperative steps (λ in Figure 35.4a).

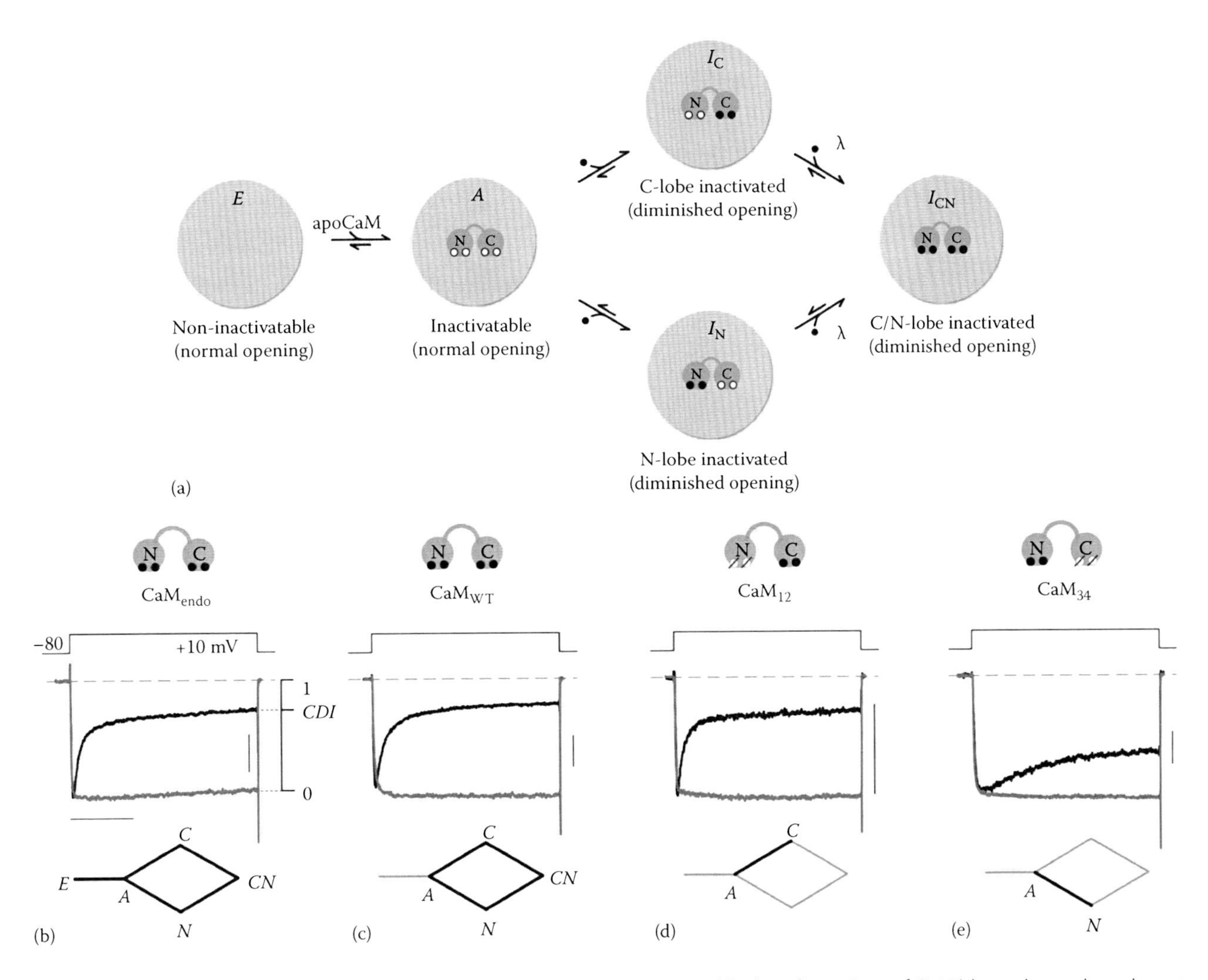

Figure 35.4 General schema for CaM regulation of L-type $Ca_V1.3$ channels. (a) Simplified configurations of CaM/channel complex relevant for CaM regulation (*E*, *A*, I_C, I_N, and I_{CN}). In configuration *E*, channels lack a resident CaM and are unable to undergo CDI. Configuration *A* corresponds to channels fully charged with CaM, capable of undergoing robust CDI. Ca^{2+} binding to the C-lobe and N-lobe of CaM results in configurations I_C and I_N. Ca^{2+} binding to both lobes of CaM yields configuration I_{CN} with both C and N lobes of CaM engaged toward CDI. (b) Whole-cell $Ca_V1.3$ currents in HEK293 cells demonstrate robust CDI with only endogenous CaM. CDI observed here reflects properties of the entire system (bottom, stick-figure) diagrammed in (a). Here and throughout, the vertical scale bar pertains to 0.2 nA of Ca^{2+} current (black), and the Ba^{2+} current (gray) has been scaled approximately threefold downward to aid comparison of decay kinetics. Horizontal scale bar, 100 ms. (c) Currents during overexpression of CaM_{WT}, isolating the behavior of the diamond-shaped subsystem at bottom. CDI unchanged from (b), since CaM is plentiful in HEK293 cells. (d) Currents during overexpression of CaM_{12} isolates C-lobe form of CDI. (e) Currents during overexpression of CaM_{34} isolates N-lobe form of CDI. (a–e: Adapted from Ben-Johny, M. et al., *Nat. Commun.*, 4, 1717, 2013. With permission.)

To minimize these errors in a recent study (Ben-Johny et al. 2013), CDI of $Ca_V1.3$ channels was characterized during strong coexpression with various mutant CaM molecules, as diagrammed in Figure 35.4b through e. For orientation, CDI of channels expressed with only endogenous CaM present is shown in Figure 35.4b. Strong CDI is evident from the rapid decay of whole-cell Ca^{2+} current (black trace), compared with the nearly absent decline of Ba^{2+} current (gray trace). Because Ba^{2+} binds little with CaM (Chao et al. 1984), the decline of Ca^{2+} vs. Ba^{2+} current after 300 ms depolarization quantifies CDI (Figure 35.4b, right, *CDI* parameter). We can isolate the diamond-shaped subsystem lacking configuration *E* (Figure 35.4c), by using mass action and strong coexpression of wild-type CaM (CaM_{WT}). Full simplification of CDI arises upon strong coexpression of channels with CaM_{12} (Figure 35.4d), which depopulates configuration *E* by mass action, and excludes configurations I_N and I_{CN}. Thus, the isolated C-lobe component of CDI can be resolved, with signature rapid timecourse of current decay. Critically, this regime avoids interplay with cooperative λ steps in Figure 35.4a. Likewise, strongly coexpressing CaM_{34} isolates the slower N-lobe form of CDI, with affiliated simplifications (Figure 35.4e). By judicious characterization of these subsystems (Figure 35.4b through e), combined with binding and electrophysiological characterization of an alanine scan covering the entire carboxyl tail of $Ca_V1.3$ channels, the proposed model of CaM/channel configurations underlying CDI shown in Figure 35.5 was achieved (Ben-Johny et al. 2013).

Figure 35.5a displays a proposed configuration *A* for apoCaM interaction with the channel. This includes a homology model of the apoCaM C-lobe complexed with the IQ domain (blue), based on an analogous atomic structure from Na_V channels (Chagot and Chazin 2011; Feldkamp et al. 2011). The portrayal of the apoCaM N-lobe incorporates ab initio structural prediction of the CI domain, containing a PCI domain (green) with

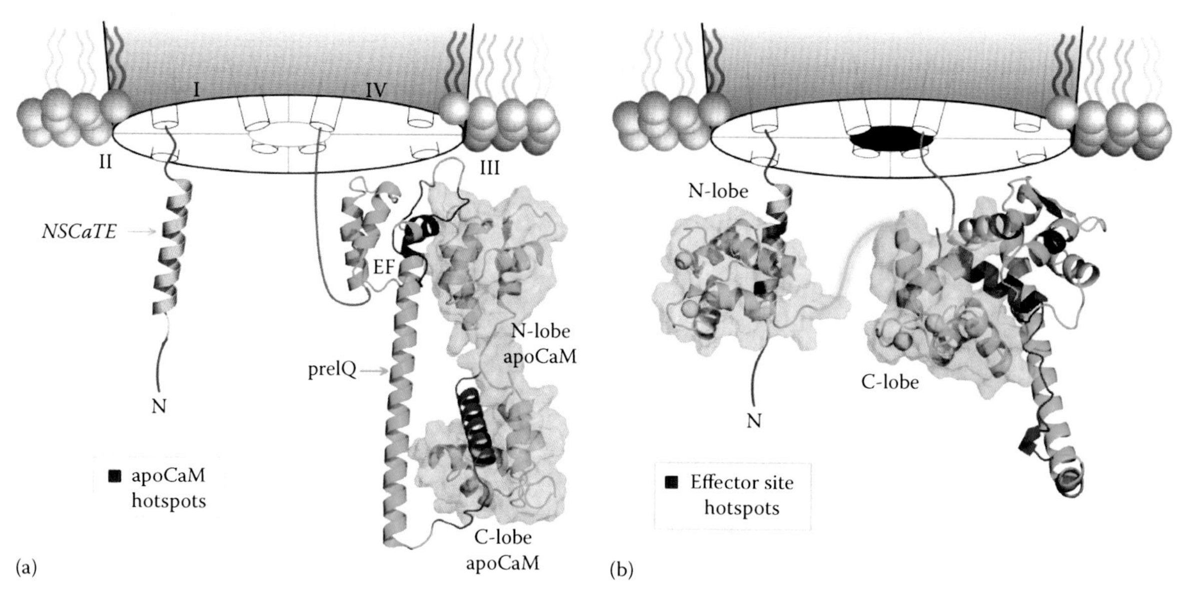

Figure 35.5 Next-generation view of CaM regulatory configurations of Ca_V channels. (a) De novo molecular model of $Ca_V1.3$ CI region docked to apoCaM (PCI region: green; IQ domain: blue). apoCaM hotspots shaded red. C-lobe of apoCaM contacts IQ domain, while N-lobe the EF-hand segment. (b) Model for Ca^{2+} inactivated state of channel. The amino terminus, NMR structure of NSCaTE bound to N-lobe of Ca^{2+}/CaM (2LQC). NSCaTE, tan; N-lobe of Ca^{2+}/CaM, cyan. N-lobe CDI hotspots on NSCaTE, red. The carboxy terminus, De novo model of tripartite IQ-PCI-Ca^{2+}/CaM complex (PCI region, green; IQ domain, blue). C-lobe CDI hotspots in red for PCI and IQ domains. (a and b: Adapted from Ben-Johny, M. et al., *Nat. Commun.*, 4, 1717, 2013. With permission.)

two vestigial EF hands (EF), and a protruding helix (pre-IQ subelement). The EF-hand module (EF) resembles the structure of a homologous segment of Na_V channels (Chagot et al. 2009; Wang et al. 2012), and a helical segment has been seen in crystal structures of analogous $Ca_V1.2$ peptides (Fallon et al. 2009; Kim et al. 2010). The atomic structure of the apoCaM N-lobe (1CFD) was interfaced with shape-complementarity docking algorithms.

Figure 35.5b displays a proposed configuration of configuration I_{CN}. The N-lobe bound to NSCaTE is an NMR (nuclear magnetic resonance) structure (Liu and Vogel 2012). An alternative ab initio model of the PCI is computationally docked with the C-lobe of Ca^{2+}/CaM (3BXL) and IQ module, which together form a ternary complex. Intriguingly, the C-lobe configuration resembles a rather canonical CaM/peptide complex, where the channel EF module contributes a surrogate lobe of CaM. Overall, this framework may aid future structural-biology and structure–function work. As well, this scheme of CaM exchange with its target molecule may generalize beyond the Ca^{2+} channel family.

35.6 BIOLOGICAL CONSEQUENCES AND PROSPECTS FOR NEW DISEASE THERAPIES

The biological consequences of Ca^{2+} regulation by CaM promise to be wide ranging and immense. In heart, elimination of $Ca_V1.2$ CDI via dominant-negative CaM yields several-fold prolongation of ventricular action potential duration (APD), implicating CDI as a dominant control factor in specifying APD (Alseikhan et al. 2002). As APD is one of the main determinants of electrical stability and arrhythmias in the heart, pharmacological manipulation of such regulation looms as a future antiarrhythmic strategy (Mahajan et al. 2008; Anderson and Mohler 2009). Whole-exome sequencing of patients with unexplained long QT syndrome and arrhythmias has recently identified point mutations within CaM as causative mutations (Crotti et al. 2013), acting via disruption of $Ca_V1.2/1.3$ channel CDI (Limpitikul et al. 2014).

In brain, $Ca_V1.3$ channels constitute a prominent Ca^{2+} entry portal into pacemaking and oscillatory neurons (Bean 2007), owing to the more negative voltages required to open these ion channels. These channels are subject to extensive alternative splicing (Hui et al. 1991; Xu and Lipscombe 2001; Shen et al. 2006; Bock et al. 2011; Tan et al. 2011) and RNA editing (Huang et al. 2012) in the carboxy tail, in ways that strongly modulate the strength of CDI (Shen et al. 2006; Liu et al. 2010; Huang et al. 2012). This fine-tuning of CDI appears to be important for circadian rhythms (Huang et al. 2012). From the specific perspective of disease, $Ca_V1.3$ channels contribute an unusually substantial portion of Ca^{2+} entry into substantia nigral neurons (Bean 2007; Chan et al. 2007; Puopolo et al. 2007; Guzman et al. 2009), which exhibit high-frequency pacemaking (Chan et al. 2007) that drives dopamine release important for movement control. Importantly, loss of these neurons is intimately related to Parkinson's disease (PD), and Ca^{2+} disturbances and overload are critical to this neurodegeneration (Bezprovanny 2009; Surmeier and Sulzer 2013). Accordingly, a highly attractive therapeutic possibility for PD now involves the discovery of small molecules that selectively downregulate the opening of $Ca_V1.3$ vs. other closely related Ca^{2+} channels (Kang et al. 2012). Thus, understanding the mechanisms underlying the modulation of $Ca_V1.3$ CDI is crucial, particularly to furnish specific molecular interfaces as targets of rational screens for small-molecule modulators.

More broadly, recent state-of-the-art analysis genome-wide single-nucleotide polymorphisms (SNPs) identify Ca_V channels as a major risk factor for several forms of psychiatric disorders (Cross-Disorder Group of the Psychiatric Genomics Consortium 2013). This finding multiplies further the expectations that calmodulation of Ca^{2+} channels will have extensive implications for novel pharmacology and genetic treatments of diverse diseases.

35.7 CALMODULATION REFLECTS AN ANCIENT AND WIDELY SHARED DESIGN

The unmistakable sequence conservation in the CI region across the $Ca_V1/2$ channel superfamily (Figure 35.2a) is impressive. Given the key role of this module in supporting richly contrasting versions of Ca^{2+} regulation (Figure 35.3a), this extensive sequence conservation speaks to calmodulation as a system of ancient and widely shared design. In this regard, the shared homology with numerous voltage-gated Na_V channels in this very region (Figure 35.2a) is even more intriguing, raising the possibility of further generality. However, the indications for analogous CaM-mediated Ca^{2+} regulation of these channels has been recently reviewed (Van Petegem et al. 2012), and the functional signature of Ca^{2+} regulation in these channels appears more subtle and variable, and the involvement of CaM less than universally accepted. Nonetheless, given the shared homology in the CI region between Ca^{2+} and Na channels, it is satisfying that greater mechanistic commonalities are now emerging in full measure between these two superfamilies (Ben-Johny et al. 2014). If so, calmodulation may represent in the Ca^{2+} signaling field something akin to a unified field theory of physics.

ACKNOWLEDGMENTS

This work was supported by grants from the NHLBI (R37HL076795), NIMH (R01MH065531), and NINDS (R01NS073874) to David T. Yue, and the NIMH (F31MH088109) to Manu Ben-Johny.

REFERENCES

Adams, P. J., R. L. Rungta, E. Garcia, A. M. van den Maagdenberg, B. A. MacVicar, and T. P. Snutch. 2010. Contribution of calcium-dependent facilitation to synaptic plasticity revealed by migraine mutations in the P/Q-type calcium channel. *Proc Natl Acad Sci USA* 107 (43):18694–18699.

Adelman, J. P., J. Maylie, and P. Sah. 2012. Small-conductance Ca^{2+}-activated K^+ channels: Form and function. *Annu Rev Physiol* 74:245–269.

Alseikhan, B. A., C. D. DeMaria, H. M. Colecraft, and D. T. Yue. 2002. Engineered calmodulins reveal the unexpected eminence of Ca^{2+} channel inactivation in controlling heart excitation. *Proc Natl Acad Sci U S A* 99 (26):17185–17190.

Anderson, M. E. and P. J. Mohler. 2009. Rescuing a failing heart: Think globally, treat locally. *Nat Med* 15 (1):25–26.

Armstrong, D., C. Erxleben, D. Kalman, Y. Lai, A. Nairn, and P. Greengard. 1988. Intracellular calcium controls the activity of dihydropyridine-sensitive calcium channels through protein phosphorylation and its removal. *J Gen Physiol* 92:10a.

Ashcroft, F. M. and P. R. Stanfield. 1981. Calcium dependence of the inactivation of calcium currents in skeletal muscle fibers of an insect. *Science* 213 (4504):224–226.

Babitch, J. 1990. Channel hands. *Nature* 346 (6282):321–322.

Bean, B. P. 2007. Neurophysiology: Stressful pacemaking. *Nature* 447 (7148):1059–1060.

Ben-Johny, M., P. S. Yang, H. X. Bazzazi, and D. T. Yue. 2013. Dynamic switching of calmodulin interactions underlies Ca^{2+} regulation of $Ca_V1.3$ channels. *Nat Commun* 4:1717.

Ben-Johny, M., D. N. Yue, and D. T. Yue. 2012. A novel FRET-based assay reveals 1:1 stoichiometry of apocalmodulin binding across Ca_V and Na_V ion channels (abstr.). *Biophys J* 102:125a–126a.

Ben-Johny, M., P. S. Yang., J. Niu, W. Yang, R. Joshi-Mukherjee, and D. T. Yue. 2014. Conservation of Ca^{2+}/calmodulin regulation across Na and Ca^{2+} channels. *Cell* 157:1657–1670.

Bezprovanny, I. 2009. Calcium signaling and neurodegenerative diseases. *Trends Mol Med* 15:89–100.

Biswas, S., I. Deschenes, D. Disilvestre, Y. Tian, V. L. Halperin, and G. F. Tomaselli. 2008. Calmodulin regulation of Nav1.4 current: Role of binding to the carboxyl terminus. *J Gen Physiol* 131 (3):197–209.

Bock, G., M. Gebhart, A. Scharinger et al. 2011. Functional properties of a newly identified C-terminal splice variant of Cav1.3 L-type Ca^{2+} channels. *J Biol Chem* 286 (49):42736–42748.

Bradley, J., W. Bonigk, K. W. Yau, and S. Frings. 2004. Calmodulin permanently associates with rat olfactory CNG channels under native conditions. *Nat Neurosci* 7 (7):705–710.

Bradley, J., J. Reisert, and S. Frings. 2005. Regulation of cyclic nucleotide-gated channels. *Curr Opin Neurobiol* 15:343–349.

Brehm, P. and R. Eckert. 1978. Calcium entry leads to inactivation of calcium channel in *Paramecium*. *Science* 202:1203–1206.

Budde, T., S. Meuth, and H. C. Pape. 2002. Calcium-dependent inactivation of neuronal calcium channels. *Nat Rev Neurosci* 3 (11):873–883.

Chad, J. E. and R. Eckert. 1986. An enzymatic mechanism for calcium current inactivation in dialysed Helix neurones. *J Physiol* 378:31–51.

Chagot, B. and W. J. Chazin. 2011. Solution NMR structure of Apo-calmodulin in complex with the IQ motif of human cardiac sodium channel NaV1.5. *J Mol Biol* 406 (1):106–119.

Chagot, B., F. Potet, J. R. Balser, and W. J. Chazin. 2009. Solution NMR structure of the C terminal EF-hand domain of human cardiac sodium channel NaV1.5. *J Biol Chem* 284 (10):6436–6445.

Chan, C. S., J. N. Guzman, E. Ilijic et al. 2007. 'Rejuvenation' protects neurons in mouse models of Parkinson's disease. *Nature* 447 (7148):1081–1086.

Chao, S. H., Y. Suzuki, J. R. Zysk, and W. Y. Cheung. 1984. Activation of calmodulin by various metal cations as a function of ionic radius. *Mol Pharmacol* 26 (1):75–82.

Chaudhuri, D., B. A. Alseikhan, S. Y. Chang, T. W. Soong, and D. T. Yue. 2005. Developmental activation of calmodulin-dependent facilitation of cerebellar P-type Ca^{2+} current. *J Neurosci* 25 (36):8282–8294.

Chaudhuri, D., J. B. Issa, and D. T. Yue. 2007. Elementary mechanisms producing facilitation of Cav2.1 (P/Q-type) channels. *J Gen Physiol* 129 (5):385–401.

Chin, D. and A. R. Means. 2000. Calmodulin: A prototypical calcium sensor. *Trends Cell Biol* 10 (8):322–328.

Chen, T. Y. and K. W. Yau. 1994. Direct modulation by Ca(2+)-calmodulin of cyclic nucleotide-activated channel of rat olfactory receptor neurons. *Nature* 368 (6471):545–548.

Cross-Disorder Group of the Psychiatric Genomics, Consortium. 2013. Identification of risk loci with shared effects on five major psychiatric disorders: A genome-wide analysis. *Lancet* 381:1371–1379.

Crotti, L., C. N. Johnson, E. Graf et al. 2013. Calmodulin mutations associated with recurrent cardiac arrest in infants. *Circulation* 127 (9):1009–1017.

Dasgupta, M., T. Honeycutt, and D. K. Blumenthal. 1989. The gamma-subunit of skeletal muscle phosphorylase kinase contains two noncontiguous domains that act in concert to bind calmodulin. *J Biol Chem* 264:17156–17163.

de Leon, M., Y. Wang, L. Jones et al. 1995. Essential Ca(2+)-binding motif for Ca(2+)-sensitive inactivation of L-type Ca^{2+} channels. *Science* 270 (5241):1502–1506.

DeMaria, C. D., T. W. Soong, B. A. Alseikhan, R. S. Alvania, and D. T. Yue. 2001. Calmodulin bifurcates the local Ca^{2+} signal that modulates P/Q-type Ca^{2+} channels. *Nature* 411 (6836):484–489.

Deschenes, I., N. Neyroud, D. DiSilvestre, E. Marban, D. T. Yue, and G. F. Tomaselli. 2002. Isoform-specific modulation of voltage-gated Na(+) channels by calmodulin. *Circ Res* 90 (4):E49–E57.

Dick, I. E., M. R. Tadross, H. Liang, L. H. Tay, W. Yang, and D. T. Yue. 2008. A modular switch for spatial Ca^{2+} selectivity in the calmodulin regulation of Ca_V channels. *Nature* 451:830–834.

Eckert, R. and J. Chad. 1984. Inactivation of Ca channels. *Prog Biophys Mol Biol (Lond)* 44:215–267.

Eckert, R. and D. Tillotson. 1981. Calcium-mediated inactivation of the calcium conductance in caesium-loaded giant neurones of *Aplysia californica*. *J Physiol* 314:265–280.

Ehlers, M. D., S. Zhang, J. P. Bernhadt, and R. L. Huganir. 1996. Inactivation of NMDA receptors by direct interaction of calmodulin with the NR1 subunit. *Cell* 84 (5):745–755.

Erickson, M. G., B. A. Alseikhan, B. Z. Peterson, and D. T. Yue. 2001. Preassociation of calmodulin with voltage-gated Ca(2+) channels revealed by FRET in single living cells. *Neuron* 31 (6):973–985.

Erickson, M. G., H. Liang, M. X. Mori, and D. T. Yue. 2003. FRET two-hybrid mapping reveals function and location of L-type Ca^{2+} channel CaM preassociation. *Neuron* 39 (1):97–107.

Fallon, J. L., M. R. Baker, L. Xiong et al. 2009. Crystal structure of dimeric cardiac L-type calcium channel regulatory domains bridged by Ca^{2+} calmodulins. *Proc Natl Acad Sci U S A* 106 (13):5135–5140.

Fallon, J. L., D. B. Halling, S. L. Hamilton, and F. A. Quiocho. 2005. Structure of calmodulin bound to the hydrophobic IQ domain of the cardiac Ca(v)1.2 calcium channel. *Structure* 13 (12):1881–1886.

Feldkamp, M. D., L. Yu, and M. A. Shea. 2011. Structural and energetic determinants of apo calmodulin binding to the IQ motif of the Na(V)1.2 voltage-dependent sodium channel. *Structure* 19 (5):733–747.

Findeisen, F., A. Tolia, R. Arant, E. Y. Kim, E. Isacoff, and D. L. Minor, Jr. 2011. Calmodulin overexpression does not alter Cav1.2 function or oligomerization state. *Channels (Austin)* 5 (4):320–324.

Gamper, N. and M. S. Shapiro. 2003. Calmodulin mediates Ca^{2+}-dependent modulation of M-type K^+ channels. *J Gen Physiol* 122 (1):17–31.

Gamper, N., Y. Li, and M. S. Shapiro. 2005. Structural requirements for differential sensitivity of KCNQ K^+ channels to modulation by Ca^{2+}/calmodulin. *Mol Biol Cell* 16 (8):3538–3551.

Ghosh, S., D. A. Nunziato, and G. S. Pitt. 2006. KCNQ1 assembly and function is blocked by long-QT syndrome mutations that disrupt interaction with calmodulin. *Circ Res* 98 (8):1048–1054.

Gordon-Shaag, A., W. N. Zagotta, and S. E. Gordon. 2008. Mechanism of Ca(2+)-dependent desensitization in TRP channels. *Channels (Austin)* 2 (2):125–129.

Guzman, J. N., J. Sanchez-Padilla, C. S. Chan, and D. J. Surmeier. 2009. Robust pacemaking in substantia nigra dopaminergic neurons. *J Neurosci* 29 (35):11011–11019.

Halling, D. B., P. Aracena-Parks, and S. L. Hamilton. 2006. Regulation of voltage-gated Ca^{2+} channels by calmodulin. *Sci STKE* 2006 (318):er1.

Huang, H., B. Z. Tan, Y. Shen et al. 2012. RNA editing of the IQ domain of $Ca_V1.3$ channels modulates their Ca^{2+}-dependent inactivation. *Neuron* 73:304–316.

Hui, A., P. T. Ellinor, O. Krizanova, J. J. Wang, R. J. Diebold, and A. Schwartz. 1991. Molecular cloning of multiple subtypes of a novel rat brain isoform of the alpha 1 subunit of the voltage-dependent calcium channel. *Neuron* 7 (1):35–44.

Imredy, J. P. and D. T. Yue. 1992. Submicroscopic Ca^{2+} diffusion mediates inhibitory coupling between individual Ca^{2+} channels. *Neuron* 9 (2):197–207.

Imredy, J. P. and D. T. Yue. 1994. Mechanism of Ca^{2+}-sensitive inactivation of L-type Ca^{2+} channels. *Neuron* 12 (6):1301–1318.

Ivanina, T., Y. Blumenstein, E. Shistik, R. Barzilai, and N. Dascal. 2000. Modulation of L-type Ca^{2+} channels by $G_{\beta\gamma}$ and calmodulin via interactions with N and C termini of α_{1C}. *J Biol Chem* 275 (51):39846–39854.

Jarrett, H. W. and R. Madhavan. 1991. Calmodulin-binding proteins also have a calmodulin-like binding site within their structure. The flip-flop model. *J Biol Chem* 266 (1):362–371.

Jurado, L. A., P. S. Chockalingam, and H. W. Jarrett. 1999. Apocalmodulin. *Physiol Rev* 79 (3):661–682.

Kang, S., G. Cooper, S. F. Dunne et al. 2012. CaV1.3-selective L-type calcium channel antagonists as potential new therapeutics for Parkinson's disease. *Nat Commun* 3:1146.

Kass, R. S. and M. Sanguinetti. 1984. Inactivation of calcium channel current in the calf cardiac Purkinje fiber. Evidence for voltage- and calcium-mediated mechanisms. *J Gen Physiol* 84:705–726.

Kim, E. Y., C. H. Rumpf, Y. Fujiwara, E. S. Cooley, F. Van Petegem, and D. L. Minor, Jr. 2008. Structures of CaV2 Ca^{2+}/CaM-IQ domain complexes reveal binding modes that underlie calcium-dependent inactivation and facilitation. *Structure* 16 (10):1455–1467.

Kim, E. Y., C. H. Rumpf, F. Van Petegem et al. 2010. Multiple C-terminal tail Ca(2+)/CaMs regulate Ca(V)1.2 function but do not mediate channel dimerization. *Embo J* 29 (23):3924–3938.

Kink, J. A., M. E. Maley, R. R. Preston et al. 1990. Mutations in paramecium calmodulin indicate functional differences between the C-terminal and N-terminal lobes in vivo. *Cell* 62 (1):165–174.

Lau, S. Y., E. Procko, and R. Gaudet. 2012. Distinct properties of Ca^{2+}-calmodulin binding to N- and C-terminal regulatory regions of the TRPV1 channel. *J Gen Physiol* 140 (5):541–555.

Lee, A., R. E. Westenbroek, F. Haeseleer, K. Palczewski, T. Scheuer, and W. A. Catterall. 2002. Differential modulation of Ca(v)2.1 channels by calmodulin and Ca^{2+}-binding protein 1. *Nat Neurosci* 5 (3):210–217.

Lee, A., S. T. Wong, D. Gallagher et al. 1999. Ca^{2+}/calmodulin binds to and modulates P/Q-type calcium channels. *Nature* 399 (6732):155–159.

Lee, K., E. Marban, and R. W. Tsien. 1985. Inactivation of calcium channels in mammalian heart cells: Joint dependence on membrane potential and intracellular calcium. *J Physiol* 364:395–411.

Liang, H., C. D. DeMaria, M. G. Erickson, M. X. Mori, B. A. Alseikhan, and D. T. Yue. 2003. Unified mechanisms of Ca^{2+} regulation across the Ca^{2+} channel family. *Neuron* 39 (6):951–960.

Limpitikul, W. B., I. E. Dick, R. Joshi-Mukherjee, M. T. Overgaard, A. L. George, and D. T. Yue. 2014. Calmodulin mutations associated with long QT syndrome prevent inactivation of cardiac L-type Ca^{2+} currents and promote proarrhythmic behavior in ventricular myocytes. *J Mol Cell Cardiol* 74:115–124.

Liu, X., P. S. Yang, W. Yang, and D. T. Yue. 2010. Enzyme-inhibitor-like tuning of Ca^{2+} channel connectivity with calmodulin. *Nature* 463 (7283):968–972.

Liu, Z. and H. J. Vogel. 2012. Structural basis for the regulation of L-type voltage-gated calcium channels: Interactions between the N-terminal cytoplasmic domain and Ca(2+)-calmodulin. *Front Mol Neurosci* 5:38.

Mahajan, A., D. Sato, Y. Shiferaw et al. 2008. Modifying L-type calcium current kinetics: Consequences for cardiac excitation and arrhythmia dynamics. *Biophys J* 94 (2):411–423.

Mentrard, D., G. Vassort, and R. Fischmeister. 1984. Calcium-mediated inactivation of the calcium conductance in cesium-loaded frog heart cells. *J Gen Physiol* 83:105–131.

Mercado, J., A. Gordon-Shaag, W. N. Zagotta, and S. E. Gordon. 2010. Ca^{2+}-dependent desensitization of TRPV2 channels is mediated by hydrolysis of phosphatidylinositol 4,5-bisphosphate. *J Neurosci* 30 (40):13328–13347.

Minor, D. L., Jr. and F. Findeisen. 2010. Progress in the structural understanding of voltage-gated calcium channel (Ca_V) function and modulation. *Channels (Austin)* 4 (6):459–474.

Mori, M. X., M. G. Erickson, and D. T. Yue. 2004. Functional stoichiometry and local enrichment of calmodulin interacting with Ca^{2+} channels. *Science* 304 (5669):432–435.

Mori, M. X., C. W. Vander Kooi, D. J. Leahy, and D. T. Yue. 2008. Crystal structure of the Ca_V2 IQ domain in complex with Ca^{2+}/calmodulin: High-resolution mechanistic implications for channel regulation by Ca^{2+}. *Structure* 16 (4):607–620.

Mullins, F. M., C. Y. Park, R. E. Dolmetsch, and R. S. Lewis. 2009. STIM1 and calmodulin interact with Orai1 to induce Ca^{2+}-dependent inactivation of CRAC channels. *Proc Natl Acad Sci USA* 106 (36):15495–15500.

Neher, E. 1986. Concentration profiles of intracellular calcium in the presence of a diffusible chelator. *Exp Brain Res* 14:80–96.

Numazaki, M., T. Tominaga, K. Takeuchi, N. Murayama, H. Toyooka, and M. Tominaga. 2003. Structural determinant of TRPV1 desensitization interacts with calmodulin. *Proc Natl Acad Sci USA* 100 (13):8002–8006.

Peterson, B. Z., C. D. DeMaria, J. P. Adelman, and D. T. Yue. 1999. Calmodulin is the Ca^{2+} sensor for Ca^{2+}-dependent inactivation of L-type calcium channels. *Neuron* 22 (3):549–558.

Peterson, B. Z., J. S. Lee, J. G. Mulle, Y. Wang, M. DeLeon, and D. T. Yue. 2000. Critical determinants of Ca^{2+}-dependent inactivation within an EF-hand motif of L-type Ca^{2+} channels. *Biophys J* 78:1906–1920.

Pitt, G. S., R. D. Zuhlke, A. Hudmon, H. Schulman, H. Reuter, and R. W. Tsien. 2001. Molecular basis of calmodulin tethering and Ca^{2+}-dependent inactivation of L-type Ca^{2+} channels. *J Biol Chem* 276 (33):30794–30802.

Plant, T., N. Standen, and T. Ward. 1983. The Effects of injection of calcium ions and calcium chelators on calcium channel inactivation in helix neurones. *J Physiol* 334:189–212.

Puopolo, M., E. Raviola, and B. P. Bean. 2007. Roles of subthreshold calcium current and sodium current in spontaneous firing of mouse midbrain dopamine neurons. *J Neurosci* 27 (3):645–656.

Qin, N., R. Olcese, M. Bransby, T. Lin, and L. Birnbaumer. 1999. Ca^{2+}-induced inhibition of the cardiac Ca^{2+} channel depends on calmodulin. *Proc Natl Acad Sci U S A* 96 (5):2435–2438.

Rosenbaum, T., A. Gordon-Shaag, M. Munari, and S. E. Gordon. 2004. Ca^{2+}/calmodulin modulates TRPV1 activation by capsaicin. *J Gen Physiol* 123 (1):53–62.

Saimi, Y. and C. Kung. 2002. Calmodulin as an ion channel subunit. *Annu Rev Physiol* 64:289–311.

Sarhan, M. F., F. Van Petegem, and C. A. Ahern. 2009. A double tyrosine motif in the cardiac sodium channel domain III-IV linker couples calcium-dependent calmodulin binding to inactivation gating. *J Biol Chem* 284 (48):33265–33274.

Schumacher, M. A., M. Crum, and M. C. Miller. 2004. Crystal structures of apocalmodulin and an apocalmodulin/SK potassium channel gating domain complex. *Structure* 12 (5):849–860.

Shamgar, L., L. Ma, N. Schmitt, Y. Haitin, A. Peretz, R. Wiener, J. Hirsch, O. Pongs, and B. Attali. 2006. Calmodulin is essential for cardiac IKS channel gating and assembly: Impaired function in long-QT mutations. *Circ Res* 98:1055–1063.

Shen, Y., D. Yu, H. Hiel et al. 2006. Alternative splicing of the Ca(v)1.3 channel IQ domain, a molecular switch for Ca^{2+}-dependent inactivation within auditory hair cells. *J Neurosci* 26 (42):10690–10699.

Singh, A., D. Hamedinger, J. C. Hoda et al. 2006. C-terminal modulator controls Ca^{2+}-dependent gating of $Ca_V1.4$ L-type Ca^{2+} channels. *Nat Neurosci* 9 (9):1108–1116.

Snutch, T. P. and P. B. Reiner. 1992. Ca^{2+} channels: Diversity of form and function. *Curr Opin Neurobiol* 2 (3):247–253.

Standen, N. and P. Stanfield. 1982. A binding-site model for calcium channel inactivation that depends on calcium entry. *Proc R Soc (Lond)* 217:101–110.

Stern, M. D. 1992. Buffering of calcium in the vicinity of a channel pore. *Cell Calcium* 13 (3):183–192.

Stocker, M. 2004. Ca(2+)-activated K^+ channels: Molecular determinants and function of the SK family. *Nat. Rev. Neurosci.* 5:758–770.

Stroffekova, K. 2008. Ca^{2+}/CaM-dependent inactivation of the skeletal muscle L-type Ca^{2+} channel (Cav1.1). *Pflugers Arch* 455 (5):873–884.

Stroffekova, K. 2011. The IQ motif is crucial for Cav1.1 function. *J Biomed Biotechnol* 2011:504649.

Surmeier, D. J. and D. Sulzer. 2013. The pathology roadmap in Parkinson disease. *Prion* 7 (1):85–91.

Tadross, M. R., I. E. Dick, and D. T. Yue. 2008. Mechanism of local and global Ca^{2+} sensing by calmodulin in complex with a Ca^{2+} channel. *Cell* 133 (7):1228–1240.

Tan, B. Z., F. Jiang, M. Y. Tan et al. 2011. Functional characterization of alternative splicing in the C terminus of L-type CaV1.3 channels. *J Biol Chem* 286 (49):42725–42735.

Tan, H. L., S. Kupershmidt, R. Zhang et al. 2002. A calcium sensor in the sodium channel modulates cardiac excitability. *Nature* 415 (6870):442–447.

Tay, L. H., I. E. Dick, W. Yang, M. Mank, O. Griesbeck, and D. T. Yue. 2012. Nanodomain Ca(2)(+) of Ca(2)(+) channels detected by a tethered genetically encoded Ca(2)(+) sensor. *Nat Commun* 3:778.

Tillotson, D. 1979. Inactivation of Ca conductance dependent on entry of Ca ions in molluscan neurons. *Proc Natl Acad Sci U S A* 76 (3):1497–1500.

Trudeau, M. C. and W. N. Zagotta. 2003. Calcium/calmodulin modulation of olfactory and rod cyclic nucleotide-gated ion channels. *J Biol Chem* 278 (21):18705–18708.

Trudeau, M. C. and W. N. Zagotta. 2004. Dynamics of Ca^{2+}-calmodulin-dependent inhibition of rod cyclic nucleotide-gated channels measured by patch-clamp fluorometry. *J Gen Physiol* 124 (3):211–223.

Tufty, R. M. and R. H. Kretsinger. 1975. Troponin and parvalbumin calcium binding regions predicted in myosin light chain and T4 lysozyme. *Science* 187 (4172):167–169.

Van Petegem, F., F. C. Chatelain, and D. L. Minor, Jr. 2005. Insights into voltage-gated calcium channel regulation from the structure of the CaV1.2 IQ domain-Ca^{2+}/calmodulin complex. *Nat Struct Mol Biol* 12 (12):1108–1115.

Van Petegem, F., P. A. Lobo, and C. A. Ahern. 2012. Seeing the forest through the trees: Towards a unified view on physiological calcium regulation of voltage-gated sodium channels. *Biophys J* 103 (11):2243–2251.

Wahl-Schott, C., L. Baumann, H. Cuny, C. Eckert, K. Griessmeier, and M. Biel. 2006. Switching off calcium-dependent inactivation in L-type calcium channels by an autoinhibitory domain. *Proc Natl Acad Sci U S A* 103 (42):15657–15662.
Wang, C., B. C. Chung, H. Yan, S. Y. Lee, and G. S. Pitt. 2012. Crystal structure of the ternary complex of a NaV C-terminal domain, a fibroblast growth factor homologous factor, and calmodulin. *Structure* 20 (7):1167–1176.
Wei, F., X. M. Xia, J. Tang et al. 2003. Calmodulin regulates synaptic plasticity in the anterior cingulate cortex and behavioral responses: A microelectroporation study in adult rodents. *J Neurosci* 23 (23):8402–8409.
Wu, Y., I. Dzhura, R. J. Colbran, and M. E. Anderson. 2001. Calmodulin kinase and a calmodulin-binding 'IQ' domain facilitate L-type Ca^{2+} current in rabbit ventricular myocytes by a common mechanism. *J Physiol* 535 (Part 3):679–687.
Xia, X. M., B. Fakler, A. Rivard et al. 1998. Mechanism of calcium gating in small-conductance calcium-activated potassium channels. *Nature* 395 (6701):503–507.
Xu, J. and L. G. Wu. 2005. The decrease in the presynaptic calcium current is a major cause of short-term depression at a calyx-type synapse. *Neuron* 46 (4):633–645.
Xu, W. and D. Lipscombe. 2001. Neuronal Ca(V)1.3alpha(1) L-type channels activate at relatively hyperpolarized membrane potentials and are incompletely inhibited by dihydropyridines. *J Neurosci* 21 (16):5944–5951.
Yang, P. S., B. A. Alseikhan, H. Hiel et al. 2006. Switching of Ca^{2+}-dependent inactivation of $Ca_V1.3$ channels by calcium binding proteins of auditory hair cells. *J Neurosci* 26 (42):10677–10689.
Yang, P. S., M. X. Mori, E. A. Antony, M. R. Tadross, and D. T. Yue. 2007. A single calmodulin imparts distinct N- and C-lobe regulatory processes to individual $Ca_V1.3$ channels (abstr.). *Biophys J* 92:354a.
Yue, D. T., P. H. Backx, and J. P. Imredy. 1990. Calcium-sensitive inactivation in the gating of single calcium channels. *Science* 250 (4988):1735–1738.
Zhang, S., M. D. Ehlers, J. P. Bernhardt, C. T. Su, and R. L. Huganir. 1998. Calmodulin mediates calcium-dependent inactivation of N-methyl-D-aspartate receptors. *Neuron* 21 (2):443–453.
Zhou, J., R. Olcese, N. Qin, F. Noceti, L. Birnbaumer, and E. Stefani. 1997. Feedback inhibition of Ca^{2+} channels by Ca^{2+} depends on a short sequence of the C terminus that does not include the Ca^{2+}—Binding function of a motif with similarity to Ca^{2+}-binding domains. *Proc Natl Acad Sci U S A* 94 (6):2301–2305.
Zuhlke, R. D., G. S. Pitt, K. Deisseroth, R. W. Tsien, and H. Reuter. 1999. Calmodulin supports both inactivation and facilitation of L-type calcium channels. *Nature* 399 (6732):159–162.
Zuhlke, R. D., G. S. Pitt, R. W. Tsien, and H. Reuter. 2000. Ca^{2+}-sensitive inactivation and facilitation of L-type Ca^{2+} channels both depend on specific amino acid residues in a consensus calmodulin-binding motif in the(alpha)1C subunit. *J Biol Chem* 275 (28):21121–21129.
Zuhlke, R. D. and H. Reuter. 1998. Ca^{2+}-sensitive inactivation of L-type Ca^{2+} channels depends on multiple cytoplasmic amino acid sequences of the α_{1c} subunit. *Proc Natl Acad Sci U S A* 95:3287–3294.

36 Phosphorylation of voltage-gated ion channels

James S. Trimmer and Hiroaki Misonou

Contents

Phosphorylation is a fundamental biological process conserved in prokaryotes and eukaryotes and controls the function of a wide variety of cellular proteins (1). Ion channels are no exception, and studies over the past few decades have provided striking examples of the fundamental role of phosphorylation as determinants of ion channel expression, localization and function, and in important physiological events mediated by dynamic and reversible changes in this phosphorylation. In this chapter, we will cover known phosphorylation on voltage-gated ion channel proteins and how they affect diverse aspects of the function of these proteins, as well as recent advances in the methodology to study ion channel phosphorylation.

36.1 INTRODUCTION

Protein phosphorylation is the covalent addition of a single phosphate group to the side chain of specific residues in a polypeptide. In the vast majority of cases, the linkage occurs to the hydroxyl group of Serine (Ser or S), threonine (Thr or T), and tyrosine (Tyr or Y) residues, although aspartate (Asp), histidine, cysteine, arginine, and lysine residues can also be phosphorylated in rare cases, the most notable case may be the phosphorylation of an Asp residue in the Na^+/K^+ ATPase during its transport cycle (2). The covalent addition of a phosphate group to a protein substrate is catalyzed by protein kinases, which the human genome encodes >500 (3). The predominant form of phosphorylation in mammalian cells is when these enzymes transfer the gamma phosphate group from ATP to the Ser, Thr, or Tyr side chain of a polypeptide. The covalent addition of a phosphate group itself is irreversible as the free energy of phosphorylation is very large. However, the phosphate group can be enzymatically removed by protein phosphatases, of which the human genome encodes approximately 150 (4,5), thereby making protein phosphorylation reversible. Phosphorylation has evolved as a mechanism to achieve rapid reversible posttranslational changes in protein structure and function not possible with primary structure alone (6).

Although phosphorylation occurs on a protein surface where enzymes can have an access to side chain substrates, its impact on the protein structure can be substantial due to two main biophysical characteristics of phosphate (7). First, the addition of a phosphate group adds two negative charges to the side chain of an otherwise uncharged amino acid (Ser, Thr, Tyr) on a polypeptide, which could dramatically change the electrostatic landscape of the protein surface. Second, a phosphate group can form three or more hydrogen bonds in the environment, and the tetrahedral geometry of the phosphate group makes these hydrogen bonds highly directional, thereby influencing the positions of surrounding structures of a protein. These changes would affect the conformation of a target protein and therefore its function.

Another prominent effect of phosphorylation is to change the nature and the strength of protein–protein interactions. Tyrosine phosphorylation in a specific peptide sequence can dramatically increase the affinity to a SH2 domain from immeasurable K_D to 10^{-9} M, thereby *creating* a de novo SH2 binding site (8). Other examples are the MH2 domain of SMAD and 14-3-3 proteins, which both bind to phosphorylated Ser/Thr in specific peptide sequences (9,10).

The activity of many proteins is regulated by reversible changes in phosphorylation altering these intramolecular structural alterations and/or changes in these classes of intermolecular interactions. Ion channels are no exception, such that the expression, localization, and function of ion channels can also be regulated through these mechanisms, although in many cases the exact molecular mechanisms and structural basis of how phosphorylation mediates these aspects of ion channel biology are not yet elucidated. Nevertheless, recent advances in biochemical, molecular, and mass spectrometric techniques summarized in the next section have yielded a large number of in vivo phosphorylation sites on ion channel proteins in many mammalian tissues, which are listed in the following section. This information represents an important resource for structure–function analyses of the role of these sites in mediating the diverse modes of ion channel regulation by phosphorylation.

36.2 TECHNIQUES TO STUDY ION CHANNEL PHOSPHORYLATION

Although we have accumulated vast knowledge regarding phosphorylation of ion channel proteins as described in the following section, we still do not have a complete picture of ion channel phosphoproteome, its regulation, and functional consequences. Here, we summarize current techniques to detect protein phosphorylation and to identify phosphorylation sites on ion channels, which might help to fill the gap in our knowledge base.

36.2.1 TECHNICAL CONSIDERATIONS

Phosphorylation is a reversible process, which can change in the time scale of seconds. Therefore, a cautious approach must be taken toward sample preparation. This is particularly relevant to tissue preparations that are difficult to access (e.g., mammalian brain) and that may require a certain amount of time (for anesthesia, dissection and homogenization) until target proteins are protected from unregulated post vivo kinase and phosphatase activity. Cultured cells have an advantage in this respect as they can be rapidly cooled, metabolically fixed, or lysed to stop enzymatic reactions. Nevertheless, one must be cautious in interpreting the results obtained by the techniques currently available as to whether they accurately reflect the nature and extent of in vivo phosphorylation under a given condition.

36.2.2 RADIOACTIVE ASSAYS

The classical way of detecting protein phosphorylation is the use of the phosphate donor, ATP, labeled at the gamma position with ^{32}P. This method provides a sensitive and reliable way to detect any protein phosphorylation in cultured cells grown in the presence of ^{32}P-labeled inorganic phosphate that is incorporated in vivo into ^{32}P-γ-ATP, or in in vitro kinase reactions employing ^{2}P-γ-ATP itself. Radiosequencing can also provide a way to identify target amino acid residues. These techniques were used in early studies of ion channel phosphorylation. However, more recently a number of nonradioactive methods have been developed to study protein phosphorylation as discussed later.

36.2.3 PHOSPHOPROTEIN STAINS

There are some chemicals reagents that bind to phosphorylated amino acid residues regardless of the surrounding peptide sequences. These reagents allow detection of phosphorylation on a specific protein when used in combination with immunoreagents. One of these reagents widely used is Pro-Q Diamond (11). Affinity-purified proteins can be probed with this fluorescent reagent, whose fluorescence intensity increases upon binding to phosphorylated amino acid residues (unfortunately, the basis of its specific binding to phospho-amino acid is proprietary). However, the detection sensitivity of Pro-Q Diamond is similar to that of coomassie brilliant blue and in the range of nanograms of protein, thereby somewhat limiting the usefulness of this reagent for low abundance proteins, such as ion channels in native tissue samples. Another phosphate-binding reagent is Phos-tag (12). This chemical binds to phosphorylated Ser, Thr, or Tyr residues through bivalent metal ions in the structure. As the biotin-tagged version of the chemical is available, it might provide a more sensitive way of detecting protein phosphorylation.

36.2.4 PHOSPHO-DEPENDENT ANTIBODIES

There are two classes of phospho-specific antibodies, residue-specific antibodies and sequence-dependent antibodies. Antibodies against phosphorylated Tyr are a good example of the former category (13). These antibodies have been shown to be specific to phospho-Tyr irrespective of the surrounding sequences. A typical approach is to affinity-purify a target protein, for example, by immunoprecipitation with a target-specific antibody, and test whether or not any Tyr residues on the protein are phosphorylated by performing an immunoblot using an anti-phospho-Tyr antibody. Alternatively, one can perform the converse experiment (i.e., immunoprecipitate with anti-phospho-Tyr antibody and immunoblot for the target protein). Antibodies against phospho-Ser and phospho-Thr are also available, although their specificity to phosphorylated residues in the context of different surrounding sequences have not been fully validated.

More widely used antibodies for the detection of phosphorylation are those antibodies that are both sequence- and phospho-specific, which are normally developed and selected using phospho-peptides corresponding to individual phosphorylation sites on a specific target protein (14,15). These have the dual requirement of being molecule-specific (they only recognize that protein) and phospho-specific (and they only recognize it when it is phosphorylated at that particular site).

This is a high bar, but that is necessary if ones wishes to use these antibodies, especially for immunohistochemistry (16), where the molecular nature of the observed signal cannot be directly discerned. In the case of immunoblotting, one can make the argument that absolute molecular specificity is not required, and that as long as one can identify the molecular target, for example, based on electrophoretic mobility on SDS gels or other criteria, the antibody can be used to study the regulation of phosphorylation of that site on the phosphoprotein target. As these types of antibodies have to be generated and validated for each phosphorylation site/phosphoprotein combination, their application is limited to the study of that site on that protein. However, these represent extremely powerful tools for studying phosphorylation in native cells and tissues.

36.2.5 GEL ANALYSIS

Given the changes in the charge on amino acids side chains that occurs upon phosphorylation, it is not surprising that the electrophoretic mobility of proteins is altered by changes in phosphorylation state. The addition of phosphate groups to a protein changes its isoelectric point, which makes each of the phosphorylated forms of a protein distinct from one another, and from the nonphosphorylated form, in isoelectric focusing analyses such as two-dimensional electrophoresis. However, these techniques are not easily applied to ion channels due to their hydrophobicity. In some cases, phosphorylation can also be detected in the standard one-dimensional SDS–PAGE as a shift in electrophoretic mobility due to both the direct effects of changing charge and the indirect effects that this has on SDS binding. This approach, especially when combined with alkaline phosphatase digestion to provide a comparison sample lacking phosphorylation, was used as an easy method to determine the extent of phosphorylation state of a given protein and was applied extensively to ion channels. In many cases, the choice of SDS can impact the nature and extent of the electrophoretic mobility shift, with less pure preparations yielding accentuated shifts (17–19). The Phos-tag reagents (see Section 36.2.3) is also available as an acrylamide conjugate, such that its binding to phosphorylated proteins changes their electrophoretic mobility.

36.2.6 IN VITRO PHOSPHORYLATION ASSAYS

Phosphorylation of a protein can also be tested in vitro using a peptide from the target protein or the recombinant protein with a panel of purified protein kinases. Conventionally, solution containing a peptide is incubated with a purified kinase and the radioactive ^{32}P-γ-ATP for detection of phosphorylation. More recently, mass spectrometry (see later section) was used to analyze the products of nonradioactive in vitro phosphorylation reactions (20). Finally, unnatural fluorescent amino acids have been developed as a general protein kinase substrate in commercially available kinase assay kits for use with a peptide of interest (21,22). One must be cautious, however, that these methods only report whether or not the target protein is capable of being phosphorylated (or dephosphorylated) with the enzymes, but not whether it is really subjected to modifications from these enzymes in its native environment, as subcellular compartmentalization of these enzymes relative to their substrates is a major mechanism or regulation in vivo.

36.2.7 SITE-DIRECTED MUTAGENESIS

If there are only a handful of Ser, Thr, or Tyr residues in a target protein, one can mutate these residues to test whether they are phosphorylated in cultured cells. Typically, this is done in combination with other phosphorylation assays, such as the radioactive assays, to determine the exact phosphorylated residues in the protein.

Another common way of assessing phosphorylation using mutagenesis is to test whether mutating a putative phosphorylation site changes the expression, localization, or function of the target protein. For ion channels, this is particularly useful because phosphorylation often affects their gating properties, which can be directly assessed with electrophysiological methods. Typically, this type of experiment begins with a finding that stimulating a signaling pathway known to increase the activity of a protein kinase (or phosphatase) changes a specific functional characteristic (e.g., the voltage-dependence of activation gating) of an ion channel, which indicates phosphorylation of the ion channel by that particular kinase. If mutation of a specific Ser, Thr, or Tyr residue eliminates this kinase-dependent change in function, then that position might be the site phosphorylated by that protein kinase. Some of the putative phosphorylation sites of ion channels have actually been determined using these methods. However, as any mutation can cause conformational changes that may change gating in a manner similar to phosphorylation, or indirectly affect phosphorylation at the critical sites, direct validation that this amino acid is chemically modified with phosphate, using one or more of the other methods listed in this section, is desirable. For this reason, in the data summarized later, we do not include phosphorylation sites of ion channels that have been inferred by site-directed mutagenesis alone and not having validated as being chemically modified by phosphate by at least one other method.

36.2.8 MASS SPECTROMETRY

Mass spectrometry (MS)-based techniques have recently emerged as the primary tool for the identification of phosphorylation on proteins (23), including voltage-gated ion channels, and have revealed unanticipated extent of multisite phosphorylation on many ion channel subunits (24,25). Typically, target proteins are enriched by affinity chromatography, separated in SDS–PAGE, and digested with a protease into peptide fragments. These peptides are then separated by high-resolution liquid chromatography (LC) and injected into a mass spectrometer via electrospray, which converts peptides to ions in the gas phase. These ionized peptides are identified in MS using a peptide database and sequenced in a secondary MS after fragmentation (MS/MS). This allows the assignment of phosphorylation to a specific amino acid residue. Recent advances in MS allow detection of phosphopeptides derived from ion channels present in a sample at femtomole levels (26,27). Further enrichment of phosphopeptides by simple methods such as immobilized metal affinity chromatography (IMAC) can lead to identification of additional phosphosites not detected in unenriched samples (26).

Given the high complexity of the brain proteome, and highly variable levels of expression of different ion channels, enrichment for specific channel subunits can greatly benefit attempts at a comprehensive analysis of the extent and nature of their phosphorylation. Antibody-based approaches represent a powerful tool to directly isolate, as target antigens, specific ion channel from samples with a high degree of proteomic complexity. Ion channel subunits can then be subjected to MS analysis in a sample with a limited complexity. The antibody-dependent approach was used in a number of recent studies employing MS-based approaches to identify phosphosites on brain Nav (26) and Kv (27–30) channel α subunits. This approach has also been applied to identification of phosphosites on copurifying interacting proteins, such as in the MS-based identification of phosphosites on the auxiliary Kvβ2 subunit, following its copurification from brain with antibodies directed against the Kv1.2 α subunit (31).

An alternative antibody-based approach is to use antibodies that recognize an entire family of ion channel proteins. A particularly useful set of such *pan* antibodies, targeting Nav channel α subunits, was raised by immunizing animals with the cytoplasmic linker region between domains III and IV (i.e., the ID III–IV linker) that is absolutely conserved among all vertebrate Nav channels. Polyclonal (32) and monoclonal (33) antibodies raised against this segment have proven extremely useful in labeling Nav channel α subunits on immunoblots and by immunocytochemistry, and have recently been used in concert with isoform-specific antibodies in immunopurification of Nav channels from mammalian brain for MS-based studies of in vivo phosphorylation sites (26).

36.2.9 PHOSPHOPROTEOME-BASED ANALYSES

Because of the robust ability of modern LC–MS/MS instruments to separate peptides from very complex mixtures, large-scale and global proteomic identification of all phosphorylation sites (i.e., the phosphoproteome) in a given sample (cultured cells or native tissues) is now approachable (34–37). These studies have led to the identification of a large and rapidly expanding set of identified in vivo phosphosites of ion channel proteins.

A broader antibody-based approach uses antiphosphotyrosine (anti-pTyr) antibodies to isolate and identify pTyr-containing proteins, or from tryptic digests, pTyr-containing phosphopeptides (38). Enrichment can be performed on material that has already been enriched by other means (e.g., subcellular fractionation, immunopurification with specific antibodies, etc.), on membrane fractions, or on whole brain extracts in large-scale proteomic studies. The latter approach was used to begin to define the extent and nature of mouse brain Tyr phosphorylation and led to the identification of 414 unique pTyr phosphosites, including sites on ion channel polypeptides (38).

High-throughput, antibody-independent approaches represent a powerful approach to obtain important information on ion channel phosphorylation in the absence of suitable antibodies, or when one aims to obtain a more global view of phosphorylation that does not focus on individual channel subunits. A number of recent studies have effectively used antibody-independent approaches that incorporate various fractionation procedures, as well as phosphoprotein/phosphopeptide enrichment methods, to reduce the overall complexity of the sample. This has allowed for detection of relatively low abundant ion channel phosphopeptides.

A number of key recent studies have provided important new information on the ion channel phosphoproteome, without having this as a specific focus per se. Global phosphoproteomic analyses of mouse brain proteins present in synaptic membranes and/or synaptosomes yielded 1367 unique in vivo phosphosites, including a number of ion channels (35). A global analysis of the phosphoproteome of a mouse brain extract yielded over 12,000 phosphosites on ≈4,600 brain phosphoproteins, using an antibody-independent approach for pSer and pThr sites, and an antibody-dependent approach for less abundant pTyr sites (36,37). The complexity of whole brain proteins and peptides was reduced by multiple fractionation steps, including strong anion exchange (SAX), strong cation exchange (SCX), size exclusion chromatography (SEC), and immobilized metal affinity chromatography (IMAC) and TiO_2 phospho-peptide enrichment. A similar study performed on a variety of mouse tissues identified ≈36,000 phosphosites on over 12,000 phosphoproteins (36). The largest number of sites (≈15,000) was from brain and included many ion channel proteins. Additional ion channel phosphosites were found on ion channels in heart, brown fat, and other tissue types. Finally, a recent analysis of mouse brain synaptosomal preparations yielded 16,500 phosphorylation sites on 3,135 proteins (39). One item of note from these studies is that the overall ratio obtained for pSer:pThr:pTyr phosphosites in these mouse brain samples was 85:14:1 (35) and 83:15:2 (37), and 83:15:2 across all tissues analyzed by Huttlin et al. (36), comparable to those (86:12:2) obtained from previous studies on a human cell line (40).

36.3 PHOSPHORYLATION OF ION CHANNELS

Here, we describe our current knowledge of ion channel phosphorylation, specifically for voltage-gated sodium (Nav), calcium (Cav), and potassium (Kv) channels. Initial attempts to identify ion channel phosphorylation relied mainly on radioactive assays, in vitro phosphorylation, and site-directed mutagenesis. These studies provided limited but valuable information regarding ion channel phosphorylation. However, MS-based studies of individual ion channels purified from tissue samples, and of global phosphoproteomic analyses of specific tissues, have provided a number of novel in vivo phosphorylation sites from ion channels present in native tissue, which may not have been otherwise discovered. Most of the phosphorylation sites listed in this section are those identified by these global phosphoproteomic studies. It should also be noted that we specifically avoided phosphorylation sites that have been only identified by site-directed mutagenesis but not validated as being sites chemically modified by phosphate for the reasons described in the previous section. Much of the information listed later can also be found on the PhosphoSitePlus website (41).

36.3.1 VOLTAGE-GATED SODIUM (NAV) CHANNEL α SUBUNITS

Each pore-forming α subunit of Nav channels consists of four homologous domains, each containing the S1–S6 transmembrane domains that form the voltage-sensing and pore-forming modules. These four internally repeated domains, each of

Table 36.1 Phosphorylation sites on Nav channel α subunits

SUBTYPE	SITES	REFERENCES
Nav1.1/SCN1A	T464, T465, S467, S470, S474, S537, S550, S551, S555, S558, S565, S573, S576, S586, S607, S620, S694, S696, S700, S707, T710, T712, T721, T723, S730, S1885, S1928, S1939 (Mouse) S470, S551, S607 (Rat)	(26,34–37,39)
Nav1.2/SCN2A	Y110, T113, S468, S471, S475, S484, S486, S488, S526, S528, S531, S540, S553, S554, S558, S561, S568, S573, S576, S579, S589, T590, S599, T600, S606, S610, S623, S626, Y689, S692, S707, S710, T713, T715, S722, S1052, S1055, S1887, T1944, S1959, T1960, T1964, T1967, S1972 (Mouse) S4, Y66, S468, S471, S484, S528, S554, S579, S610, S623, S687, S688, S721, Y730, S1112, S1124, S1126, S1497, S1498, S1506, Y1893, S1930, T1966, S1968, T1969, T1970, S1971/1975, Y1975 (Rat)	(26,34–37,39,44–50)
Nav1.3/SCN3A	T105, S106, Y109, S554, S558, S568, S579, S710, S711, S1024, S1833 (Mouse) Y1613, S1616 (Human)	(36,37,39,51)
Nav1.4/SCN4A	S487 (Rat)	(52)
Nav1.5/SCN5A	S39, S42, S457, S459, S460, S483, S484, S497, S510, S516, S524, S525, S539, S571, S664, S667, S1927 (Mouse) T109, S116, S529 (Rat) T594, Y1494, Y1495 (Human)	(36,46,50,53–55)
Nav1.6/SCN8A	S43, S51, S504, S518, S520, S522, S579, S600, S678, S682, S688, S1083, Y1881 (Mouse)	(36,37,39,56–58)
Nav1.7/SCN7A	S4, T57, S443, S446, T449, S790, S793, Y806, S840, T1509, S1511, T1617, T1619 (Mouse)	(36,37,39,59)
Nav1.8/SCN10A	S551, S556, S1452 (Rat)	(60,61)
Nav1.9/SCN11A	S502, S504, S1062, S1064 (Mouse) T471, S475, S477, S484 (Rat)	(46,59,62)

which resembles a single Kv channel α subunit, are connected by cytoplasmic segments called the interdomain or ID loops, such that there are ID I–II, ID II–III, and ID III–IV loops. Biochemical studies using purified brain Nav channel α subunits, of which the predominant isoform is Nav1.2 (26,42), in in vitro phosphorylation assays demonstrated that ID I–II is a particularly rich substrate for phosphorylation by protein kinases A and C (PKA and PKC) (43). Peptide mapping, sequencing, and site-directed mutagenesis later provided additional phosphorylation sites in this region. Recent MS-based approaches have provided confirmation of many of these sites, as well as numerous additional sites in this region, and in other cytoplasmic domains (Table 36.1). These analyses were accomplished by antibody-based approaches targeting Nav channel α subunits (26) and through antibody-independent global phosphoproteomics studies (34–37,39). Among the mammalian brain ion channel α subunits analyzed to date, Nav1.2 is among the most highly phosphorylated, with over 60 phosphorylation sites identified on Nav1.2 in mouse and rat brain.

36.3.2 VOLTAGE-GATED CALCIUM (CAV) CHANNEL α SUBUNITS

Cav channel α subunits share some common structural features with those of Nav channels, particularly the membrane topology, with four internally repeated homologous domains each containing six transmembrane segments and connected by intracellular loops. Cav channel α subunits have been known as rich substrates for PKA, PKC, and calmodulin-dependent kinase II (CaMKII) (63,64). As in the case of Nav channels, a number of sites phosphorylated by these protein kinases were initially identified by biochemical studies using Cav channel α subunits purified from various chicken, rabbit, and rat tissues (65). These studies have been extended by recent MS-based analyses of in vivo Cav channel α subunit phosphorylation sites in mouse and rat tissues. Most of the phosphorylation sites listed in Table 36.2 were obtain in antibody-independent global phosphoproteome studies (34–37,39,44,46,66). Many phosphorylation sites are located in the N-terminus, the I–II loop, the II–III loop, and the C-terminus of Cav channel α subunits.

36.3.3 VOLTAGE-GATED POTASSIUM (Kv) CHANNEL α SUBUNITS

Kv channels contain four independent α subunits, each of which resembles an internally repeated domain of a Nav or Cav channel α subunit. The diversity of Kv channel α subunits is large as compared with Nav and Cav channels, with ≈40 Kv channel genes in the human genome. Phosphoproteomic analysis of Kv channel α subunits was accomplished using antibody-dependent (80–83) for Kv1.2, Kv2.1, Kv4.2, Kv7.2, and Kv7.3 α subunits. Phosphorylation sites of these and many other Kv channels have been identified through antibody-independent global phosphoproteomic approaches (34–37,39,66). The majority of the phosphorylation sites identified are in the cytoplasmic N- and C-termini of

Table 36.2 Phosphorylation sites on Cav channel α subunits

SUBTYPE	SITES	REFERENCES
Cav1.1	S393, T395, S397, S1575, T1579 (Mouse) S687, 1575, T1579, S1617, S1757, S1854 (Rabbit) Y971, Y975 (Human)	(36,39,67–70)
Cav1.2	S77, S469, T471, S473, T476, S783, S808, S815, S1650, S1670, S1691, S1714, S1848, S1897, S2125 (Mouse) S528, S533, S1517, S1575, S1674, S1700, T1704, S1829, S1842, S1843, S1928 (Rabbit) S1927, Y2148 (Rat)	(36,46,64,71–75)
Cav1.3	S81, S1940, S1679, S1958, T2001, T2005, S2075 (Mouse) S45, S46, T49, S52, S81, S121, T443, T504, S517, S519, S858, T863, S923, S929, S1703, S1743, S1788, T1795, S1816, S1944, S1964, S2064, Y2067, T2068, S2074, Y2075, S2082, T2147, S2165, S2152 (Rat)	(36,39,46,64,71–75)
Cav1.4	T1538, S1539 (Mouse) T1696, S1699 (Rat)	(46,76)
Cav2.1	T411, S450, S453, S468, S752, S755, S792, S867, S1038, S1042, S1046, S1051, S1491, T1935, S1981, S1998, S2016, T2024, Y2027, S2028, S2030, S2068, S2071, S2078, S2091, S2200, S2202, S2220, S2252, S2273, S2303, S2318, S2329, Y2360, S2361, S2363 (Mouse) S792 (Rat)	(34–37,39,44,77,78)
Cav2.2	S48, S411, S424, S446, S745, S748, S753, S783, S892, S915, T920, S1058, S1929, S1951, S2007, S2014, S2056, S2062, S2197, S2212, S2221, S2244 (Mouse) S774, S784, S802, S896, S898, S2016, S2126 (Rat) S411, S447 (Rabbit) Y804, Y815 (Chicken)	(34–37,39,44,66,79)
Cav2.3	S15, S20, S23, T29, T421, T425, S428, S429, S438, T441, S737, S746, S794, S816, S856, S873, S876, S944, S948, S954, S988, S1049, S1051, T1094, S1099, S2017, S2054, T2067, S2073, S2097 (Mouse)	(34–37,39,44,59,66)
Cav3.1	S467, S715, S1096, S1140, S1146, S1147, S1942, S1972, S1980, S2036, S2051, S2083, S2086, S2089, S2143, S2164, S2168, S2174, S2252, S2273, S2361 (Mouse)	(35–37,39)
Cav3.2	Y440, S445, S541, S1171, T1172, T1196, S2201, S2350, S2360 (Mouse)	(36,39,76)
Cav3.3	S492, S911, T469, S914, S948, S951, S958, S1023, S1017, S1745, S1850 (Mouse)	(36,37,39)

Kv channel α subunits. Among these, Kv2.1 and Kv7.2 are the most heavily modified by phosphorylation. Note that antibody-dependent (27) and -independent (34–37,39,66) studies of phosphorylation of voltage- and calcium-dependent large-conductance potassium or BK channel α subunits have also yielded a wealth of data on in vivo phosphorylation sites on this important ion channel. However, the complexity of BK channel isoforms arising from alternative splicing, species differences between these isoforms, and the complexity in mapping known phosphorylation sites onto these isoforms precludes their inclusion here (84,85) (Table 36.3).

36.3.4 HYPERPOLARIZATION AND CYCLIC NUCLEOTIDE-DEPENDENT HCN CHANNEL α SUBUNITS

HCN channel α subunits are topologically very similar to Kv channel α subunits in their overall structures, although they are activated by hyperpolarization instead of depolarization of membrane potentials. These channels mediate the pacemaker I_H current in the heart and in certain neurons. The I_H current is modulated by signaling cascades involving Src and p38 MAP kinases (107). In vitro analysis revealed Tyr residues phosphorylated in HCN2 and HCN4 (108,109), including a recent mass spectrometric-based analysis of PKA-dependent in vitro phosphorylation of HCN4 (110). Recent global phosphoproteome studies have also provided a number of phosphorylation sites in all four HCN α subunits (34–37,39,66) (Table 36.4).

36.4 FUNCTIONAL CONSEQUENCES OF ION CHANNEL PHOSPHORYLATION AT IDENTIFIED SITES

A number of studies over the last few decades have provided compelling examples of the crucial role of phosphorylation in regulating the function of ion channels, and the signaling cascades, protein kinases, and protein phosphatases involved (113). Identification of phosphorylation sites on individual ion channel polypeptides has provided important information as to the molecular basis for such regulation. Although we cannot list the known effects of each phosphorylation sites on ion channels within the space provided, we will describe some of the notable examples later in each type of phospho-dependent regulation of ion channels.

Table 36.3 Phosphorylation sites on Kv channel α subunits

SUBTYPE ACCESSION	SITES	REFERENCES
Kv1.1/KCNA1	S5, S13, T14, S23 (Mouse)	(37,39)
Kv1.2/KCNA2	T22, T421, Y429, T433, S434, S440, S441, S447, S468 (Mouse) S434, S440, S441, S449 (Rat)	(29,35–39,44,66)
Kv1.3/KCNA3 IPI00133732 mouse	S29, 452, S457, S464 (Mouse)	(36,39)
Kv1.4/KCNA4	S82, S122, S349 (Mouse) S229 (Rat)	(36,37,39,66,86)
Kv1.5/KCNA5	S535, S564 (Mouse)	(37,39)
Kv1.6/KCNA6	S3, S6, T8, T486, S511 (Mouse)	(36,37,39,66)
Kv1.7/KCNA7	ND	
Kv1.8/KCNA8	ND	
Kv2.1/KCNB1	S12, T13, S15, S444, S447, S457, S484, S517, S518, S519, S520, S541, S564, S567, S607, S655, S719, S782, S799, T803, S804 (Mouse) S15, S457, S484, S496, S503, S520, S541, S567, S590, S607, S655, S719, S771, S799, S804, T836 (Rat)	(35–37,39,44,66,80)
Kv2.2/KCNB2	S19, S448, S451, S461, S481, S488, S520, S602, S605, T805 (Mouse) T170 (Rat)	(36,37,39,44,87)
Kv3.1/KCNC1	S44, S130, S142, S158, S160, S468, Y471, S474, S478, T483, S485, T487, T527, S570 (Mouse) S503 (Rat)	
Kv3.2/KCNC2	S509, S526, S549, S557, S604, S619, S634, S636 (Mouse) S57 (Human)	(36,37,39,88)
Kv3.3/KCNC3	S687, S692, T695, S698, S717, S731, S734, S742, T750, S754 (Mouse) Y574 (Human)	(36,37,39,44,70)
Kv3.4/KCNC4	S21, S508, S555 (Mouse)	(37,39)
Kv4.1/KCND1	Y317, T318, S458, S460, S550, S553, S555, S584 (Mouse)	(36,37,39)
Kv4.2/KCND2	T38, T154, Y315, T316, S473, S546, S548, S552, S572, S574, S575, T602, T606, T607, S616, S620 (Mouse) S438, S447, S459, S537 (Rat) Y134 (Human)	(36,37,39,66,89–95)
Kv4.3/KCND3	S153, Y312, T313, T459, S569, S585 (Mouse) Y108, Y136, T435, Y441 (Human)	(36,37,39,96,97)
Kv5.1	S444, S470, S472 (Mouse) Y373, Y377 (Human)	(37,44,70)
Kv6.1	ND	
Kv6.2	ND	
Kv7.1/KCNQ1	S91, S94, S406, S408, S456, S462, S462, S463 (Mouse)	
Kv6.3/KCNG3	ND	(77)
Kv6.4/KCNG4	S214 (Mouse)	(36,70,98)
Kv7.1/KCNQ1	S91, S94, S406, S408, S456, S463 (Mouse) S27, Y184 (Human)	(36,37,39,66,70,82)
Kv7.2/KCNQ2	S52, S302, S352, S406, S429, S438, S440, S444, S448, S450, S457, S479, S483, S592, S655, S657, S664, Y667, Y679, S681, T695, S697, S736, S785, S787, S799, S801, S808, Y831, S837, T839, S841 (Mouse) Y74, T217, Y372 (Human)	(36,37,39,66,70,82,99,100)

(Continued)

Table 36.3 (*Continued*) Phosphorylation sites on Kv channel α subunits

SUBTYPE ACCESSION	SITES	REFERENCES
Kv7.3/KCNQ3	S31, T82, S454, S457, S485, T490, T493, T574, S579, S579, T580, S596, S599, S785, T787 (Mouse) Y105, Y349 (Rat) Y125, T246, Y502 (Human)	(36,101)
Kv7.5/KCNQ5	S89, S448, S450, S458, T460, S467, S660, S832 (Mouse)	(36,101,102)
Kv8.1/KCNV1	ND	
Kv8.2/KCNV2	ND	
Kv9.1/KCNS1	ND	
Kv9.2/KCNS2	S463, S471 (Mouse) Y324 (Human)	(39,70)
Kv9.3/KCNS3	ND	
Kv10.1/KCNH1	T897, S899, S904, S966, S974, S978, S981 (Mouse) T219, T228, Y237, S240 (Rat)	(36,37,39,46)
Kv10.2/KCNH5	S883, S885 (Mouse)	(36,37)
Kv11.1/KCNH2	S241, S245, S265, S268, S285, S286, S322, S353, S356, S873, S876, S881, S1032, S1140 (Mouse) S613 (Rat) S322 (Human)	(34,36,37,39,46,103–105)
Kv11.2/KCNH6	ND	
Kv11.3/KCNH7	S174, S238, T318, S891, S1007, S1135, S1169, S1188 (Mouse) S269 (Human)	
Kv12.1/KCNH8	ND	
Kv12.2/KCNH3	ND	
Kv12.3/KCNH4	S105 (Human)	(106)

Table 36.4 Phosphorylation sites on HCN channel α subunits

SUBTYPE ACCESSION	SITES	REFERENCES
HCN1	T39, S69, T474, S478, S588 (Mouse)	(34,36,37,39,44,66,103)
HCN2	S64, S67, S70, S90, S101, S119, S131, S134, Y476, S641, S726, S743, S750, S756, S757, S764, S771, S795, S834, S840, S842, S847, S860 (Mouse) S816, S817 (Rat)	(34,36,37,39,44,46,66,111,112)
HCN3	S33, S633, S636, S648, S651, S654, S687, S720 (Mouse)	(34,36,37,39,44,66)
HCN4	S14, S99, S110, S117, S125, S139, Y554, S704, S816, S903, S921, S990, S996, S1011, S1036, T1056, T1060, S1090, S1093, S1094, S1097, S1113, T1138, S1139, S1140, S1142 (Mouse)	(34,36,37,39,66,110)

36.4.1 EFFECTS OF PHOSPHORYLATION ON ION CHANNEL CURRENT DENSITY

Phosphorylation can affect the biology of ion channels in many different ways. The most common way of ion channel regulation is changing the ion channel macroscopic current amplitude. This can be accomplished either by changing the conductance of ions through individual channels or by changing the number of channels, the latter being much more common. The current amplitude of Cav1.2 L-type channels is upregulated by beta-adrenergic signaling and PKA, which increases the heart rate and contractility (114). Single-channel recording studies have shown that the beta-adrenergic stimulation can change the gating mode of the channel and thereby increase the number of active Cav1.2 channels at a given membrane potential (115). A recent study suggests that phosphorylation of S439 in the ID I–II loop (identified only by mutagenesis) or S1517 in the C-terminal tail of Cav1.2 can cause the same effect on Cav1.2 gating (116), although whether these are the sites phosphorylated upon beta-adrenergic stimulation has not been directly determined.

A more common mechanism to alter current density is through modulating the density of ion channel proteins in the plasma membrane. The current density of Nav1.6, a prominent

Nav channel α subunit in central and peripheral nervous system (117), is reduced by the activation of p38 MAP kinase (58). In vitro phosphorylation assays (verified by radioactive assays and mutagenesis) showed that S553 in the ID I–II loop is phosphorylated by this protein kinase (58). Pharmacological block of endocytosis by Dynasore completely abolishes the p38-mediated current reduction, suggesting that channel endocytosis underlies the current reduction by p38 MAP kinase (57).

The Kv1.2 delayed rectifier Kv channel is also regulated by endocytosis in response to metabotropic glutamate signaling. Mutagenesis studies revealed that phosphorylation of Kv1.2 Y415 and Y417 causes disruption of Kv1.2 interaction with the cytoskeletal protein cortactin, and loss of interaction with cortactin diminishes the total Kv1.2 current in heterologous expression systems (118). Tyrosine phosphorylation of Kv1.2 at Y132, Y466, and Y482 leads to enhanced endocytosis of Kv1.2 channels (119,120). Phosphorylation of Kv1.2 at Y466 and Y482 partially releases Kv1.2 from its interaction with cortactin (120), triggering endocytosis, which is then dependent on phosphorylation at Y132 (121). The endocytosis of Kv1.2 may also be regulated by PKA through phosphorylation of T46, although phosphorylation at this site was shown only by mutagenesis (121).

The Kv4.2 transient A-type channel on CA1 pyramidal cell dendrites is subjected to clathrin-mediated endocytosis in response to glutamate receptor activation (122). Pharmacological activation of PKA leads to Kv4.2 internalization from dendritic spines, whereas PKA inhibition blocks AMPA receptor-mediated internalization. This process is inhibited by the S552A mutation, suggesting that activity-dependent internalization of Kv4.2 is dependent on PKA phosphorylation on S552 (123), which has been identified as an in vivo Kv4.2 phosphorylation site (Table 36.3).

In contrast to phosphorylation-dependent downregulation of Kv1.2 and Kv4.2, the surface density of Kv2.1 delayed rectifier channel is increased by phosphorylation. Seminal work by Choi and colleagues revealed that delayed rectifier current was enhanced in mouse cortical neurons during apoptosis, and that the K^+ efflux associated with the increased channel Kv activity was required for induction of apoptosis (124). Later studies revealed that Kv2.1-containing channels mediated these proapoptotic increases in delayed rectifier K^+ current (125). Increased phosphorylation of Kv2.1 on S800, which has been identified as an in vivo Kv2.1 phosphorylation site (Table 36.3), via p38 MAP kinase activation leads to enhanced K^+ current levels (126) through enhanced SNARE-mediated membrane insertion (127). A number of recent studies have suggested that a wide variety of proapoptotic stimuli trigger increased Kv2.1 currents in diverse types of mammalian neurons (128–130).

Although phosphorylation generally serves as a mean to dynamically change the surface density of ion channels as described earlier, it may also be required for their constitutive trafficking. Site-directed mutagenesis studies showed that Ser-to-Ala mutations at either S440 or S441, which have been identified as in vivo Kv1.2 phosphorylation sites (Table 36.3), diminished Kv1.2 surface expression levels and increased the levels of intracellular Kv1.2, resulting in decreased ionic current (29). Moreover, these mutations decreased the population of Kv1.2 carrying a processed oligosaccharide chain and increased the population with a high mannose chain, suggesting that phosphorylation of these Ser residues is required for the biosynthetic delivery of Kv1.2 to the plasma membrane. However, as S440 was identified as a substrate for PKA (83), Kv1.2 trafficking to the surface can also be dynamically regulated by PKA-dependent signaling cascades, as recently shown in secretin receptor-mediated regulation of Kv1.2 in cerebellum (131).

36.4.2 EFFECTS OF PHOSPHORYLATION ON VOLTAGE-DEPENDENT GATING

Phosphorylation can also affect ion channel functioning via changing their voltage-dependent gating. Reversible changes in phosphorylation state can change in the voltage-dependence of channel activation, as well as activation gating kinetics, and inactivation.

Early studies revealed that Kv2.1 is extensively phosphorylated in mammalian central neurons (132), but rapidly dephosphorylated in response to treatments that increase neuronal cytosolic Ca^{2+}, such as the activation of ionotropic glutamate receptors (133) and G-protein signaling (134). This dephosphorylation causes a large (~20 mV) hyperpolarizing shift in the voltage-dependent activation and an increase in activation kinetics at the activation midpoint voltage of the Kv2.1 current (80). This effect appears to be due to the removal of phosphate because intracellular dialysis of alkaline phosphatase causes the same hyperpolarizing shift (80,132). However, the sites mediating this modulation remained elusive for some time, in part due to almost 100 potential phospho-acceptor residues in the Kv2.1 C-terminus. An antibody-based mass spectrometric analysis led to identification of 16 Ser and Thr phosphorylation sites, 15 of which were located in the C-terminus (80). To specifically identify residues regulated by treatments that modulate Kv2.1 function, samples from control cultures and cultures stimulated to activate calcineurin were subjected to SILAC labeling (135), allowing for quantitative mass spectrometric identification of eight sites regulated by calcineurin (80). Individually mutating these sites to alanine partially mimicked the effects of stimuli that induced dephosphorylation in yielding hyperpolarizing shifts in the voltage-dependence of activation. Mutations at more than one site gave larger magnitude effects, suggesting that multisite phosphorylation could yield graded regulation of Kv2.1 function (80). Kv2.1 phosphorylation exhibits bidirectional regulation in vivo in response to treatments that alter neuronal activity (136).

Similar to Kv2.1, the voltage-dependent activation of Kv3.1 channel is modulated by phosphorylation state, in that AP treatment causes a hyperpolarizing shift in the voltage-dependent activation of Kv3.1 currents in MNTB neurons (137). Treatment with CK2 inhibitors mimicked the effects of AP on the voltage-dependent activation of the Kv3.1 channel in auditory neurons (137). Phosphorylation of the Kv3.1 delayed rectifier channel at S503 (verified with a sequence-specific phospho-dependent antibody) in the C-terminus by PKC also decreases the probability of opening of single channels, thereby suppressing the current density of Kv3.1 in auditory neurons (138). Kv3.1 is rapidly dephosphorylated in response to high-frequency auditory input or synaptic stimulation, leading to increased Kv3.1 current that facilitates high-frequency spiking (139).

Voltage-dependent inactivation can also be affected by phosphorylation. Nav1.2 channel abundant in the brain and heart is regulated by two type of inactivation, fast and slow inactivation. Phosphorylation of Tyr residues in the loop I–II and III–IV by Fyn kinase accelerates fast inactivation of Nav1.2, thereby reducing the current density (45,140). Conversely, dephosphorylation of these Tyr residues by receptor protein Tyr phosphatase can restore the inactivation kinetics and the current density (141). Slow inactivation, which occurs in the time scale of second as compared with the millisecond time scale of the fast inactivation, can also be modulated by phosphorylation. PKA and PKC enhance slow inactivation and reduce the availability of Nav1.2 (142,143). Site-directed mutagenesis identified S1466 as the candidate phosphorylation site for this phospho-dependent modulation (143), although this site has not been observed in the analyses of in vivo Nav1.2 phosphorylation detailed earlier.

Slow inactivation (or steady-state inactivation) of Cav2.2 N-type channel is also regulated by phosphorylation. The binding of a SNARE protein, syntaxin 1A, was shown to cause a hyperpolarizing shift of Cav2.2 slow inactivation (144), which reduces the overall availability of Cav2.2 channels. In contrast, activation of PKC was shown to increase the availability of Cav2.2 channels (145,146). Biochemical studies show that phosphorylation of S896 and 898 disrupts the binding of syntaxin 1A to the synprint site in loop II–III of Cav2.2 (147,148), which then abolishes the hyperpolarizing shift in slow inactivation (149).

A recent review highlights the instances where disease-causing mutations in ion channels either create or destroy phosphorylation sites (150). This raises questions as to the extent that changes in ion channel phosphorylation, whether due to mutations leading to changes in phosphorylation site, alterations in the normal regulation of protein kinases and/or phosphatases, or other mechanisms acting through ion phosphorylation, contribute to disease phenotypes, and lead to *ion channel phosphorylopathies* (150).

36.5 CONCLUDING REMARKS

The recent advances in MS-based identification of in vivo phosphorylation sites on ion channel α subunits has initiated a new era of studies of the impact of phosphorylation on ion channel expression, localization, and function, and how this can be dynamically modulated to impact the physiology of excitable and nonexcitable cells. We now appreciate that ion channel phosphorylation is more extensive than previously appreciated, and that new avenues of research aimed at relating the patterns of in vivo phosphorylation that have been recently revealed to the physiologically relevant modulation of ion channels in normal tissue, how this is altered in disease, and how ion channel phosphorylation could be targeted for therapeutic intervention.

REFERENCES

1. Cohen, P. 2001. The role of protein phosphorylation in human health and disease. The Sir Hans Krebs Medal Lecture. *Eur J Biochem.* 268: 5001–5010.
2. Post, R. L. and Kume, S. 1973. Evidence for an aspartyl phosphate residue at the active site of sodium and potassium ion transport adenosine triphosphatase. *J Biol Chem.* 248: 6993–7000.
3. Manning, G., Whyte, D. B., Martinez, R., Hunter, T., and Sudarsanam, S. 2002. The protein kinase complement of the human genome. *Science.* 298: 1912–1934.
4. Alonso, A., Sasin, J., Bottini, N., Friedberg, I., Friedberg, I., Osterman, A., Godzik, A., Hunter, T., Dixon, J., and Mustelin, T. 2004. Protein tyrosine phosphatases in the human genome. *Cell.* 117: 699–711.
5. Mustelin, T. 2007. A brief introduction to the protein phosphatase families. *Methods Mol Biol.* 365: 9–22.
6. Pearlman, S. M., Serber, Z., and Ferrell, J. E. J. 2011. A mechanism for the evolution of phosphorylation sites. *Cell.* 147: 934–946.
7. Groban, E. S., Narayanan, A., and Jacobson, M. P. 2006. Conformational changes in protein loops and helices induced by post-translational phosphorylation. *PLoS Comput Biol.* 2: e32.
8. Felder, S., Zhou, M., Hu, P., Urena, J., Ullrich, A., Chaudhuri, M., White, M., Shoelson, S. E., and Schlessinger, J. 1993. SH2 domains exhibit high-affinity binding to tyrosine-phosphorylated peptides yet also exhibit rapid dissociation and exchange. *Mol Cell Biol.* 13: 1449–1455.
9. Pawson, T. and Nash, P. 2000. Protein–protein interactions define specificity in signal transduction. *Genes Dev.* 14: 1027–1047.
10. Obsil, T. and Obsilova, V. 2011. Structural basis of 14-3-3 protein functions. *Semin Cell Dev Biol.* 22: 663–672.
11. Martin, K., Steinberg, T. H., Cooley, L. A., Gee, K. R., Beechem, J. M., and Patton, W. F. 2003. Quantitative analysis of protein phosphorylation status and protein kinase activity on microarrays using a novel fluorescent phosphorylation sensor dye. *Proteomics.* 3: 1244–1255.
12. Kinoshita, E., Kinoshita-Kikuta, E., Takiyama, K., and Koike, T. 2006. Phosphate-binding tag, a new tool to visualize phosphorylated proteins. *Mol Cell Proteomics.* 5: 749–757.
13. Ross, A. H., Baltimore, D., and Eisen, H. N. 1981. Phosphotyrosine-containing proteins isolated by affinity chromatography with antibodies to a synthetic hapten. *Nature.* 294: 654–656.
14. Sternberger, L. A. and Sternberger, N. H. 1983. Monoclonal antibodies distinguish phosphorylated and nonphosphorylated forms of neurofilaments in situ. *Proc Natl Acad Sci USA.* 80: 6126–6130.
15. Archuleta, A. J., Stutzke, C. A., Nixon, K. M., and Browning, M. D. 2011. Optimized protocol to make phospho-specific antibodies that work. *Methods Mol Biol.* 717: 69–88.
16. Mandell, J. W. 2008. Immunohistochemical assessment of protein phosphorylation state: The dream and the reality. *Histochem Cell Biol.* 130: 465–471.
17. Best, D., Warr, P. J., and Gull, K. 1981. Influence of the composition of commercial sodium dodecyl sulfate preparations on the separation of alpha- and beta-tubulin during polyacrylamide gel electrophoresis. *Anal Biochem.* 114: 281–284.
18. Margulies, M. M. and Tiffany, H. L. 1984. Importance of sodium dodecyl sulfate source to electrophoretic separations of thylakoid polypeptides. *Anal Biochem.* 136: 309–313.
19. Shi, G., Kleinklaus, A. K., Marrion, N. V., and Trimmer, J. S. 1994. Properties of Kv2.1 K^+ channels expressed in transfected mammalian cells. *J Biol Chem.* 269: 23204–23211.
20. Hattori, S., Iida, N., and Kosako, H. 2008. Identification of protein kinase substrates by proteomic approaches. *Expert Rev Proteomics.* 5: 497–505.
21. Rothman, D. M., Shults, M. D., and Imperiali, B. 2005. Chemical approaches for investigating phosphorylation in signal transduction networks. *Trends Cell Biol.* 15: 502–510.
22. Tarrant, M. K. and Cole, P. A. 2009. The chemical biology of protein phosphorylation. *Annu Rev Biochem.* 78: 797–825.
23. Lemeer, S. and Heck, A. J. 2009. The phosphoproteomics data explosion. *Curr Opin Chem Biol.* 13: 414–420.

24. Baek, J. H., Cerda, O., and Trimmer, J. S. 2011. Mass spectrometry-based phosphoproteomics reveals multisite phosphorylation on mammalian brain voltage-gated sodium and potassium channels. *Semin Cell Dev Biol.* 22: 153–159.
25. Cerda, O., Baek, J. H., and Trimmer, J. S. 2011. Mining recent brain proteomic databases for ion channel phosphosite nuggets. *J Gen Physiol.* 137: 3–16.
26. Berendt, F. J., Park, K. S., and Trimmer, J. S. 2010. Multisite phosphorylation of voltage-gated sodium channel alpha subunits from rat brain. *J Proteome Res.* 9: 1976–1984.
27. Yan, J., Olsen, J. V., Park, K. S., Li, W., Bildl, W., Schulte, U., Aldrich, R. W., Fakler, B., and Trimmer, J. S. 2008. Profiling the phospho-status of the BKCa channel alpha subunit in rat brain reveals unexpected patterns and complexity. *Mol Cell Proteomics.* 7: 2188–2198.
28. Park, K. S., Yang, J. W., Seikel, E., and Trimmer, J. S. 2008. Potassium channel phosphorylation in excitable cells: Providing dynamic functional variability to a diverse family of ion channels. *Physiology (Bethesda).* 23: 49–57.
29. Yang, J. W., Vacher, H., Park, K. S., Clark, E., and Trimmer, J. S. 2007. Trafficking-dependent phosphorylation of Kv1.2 regulates voltage-gated potassium channel cell surface expression. *Proc Natl Acad Sci USA.* 104: 20055–20060.
30. Seikel, E. and Trimmer, J. S. 2009. Convergent modulation of Kv4.2 channel alpha subunits by structurally distinct DPPX and KChIP auxiliary subunits. *Biochemistry.* 48: 5721–5730.
31. Vacher, H., Yang, J. W., Cerda, O., Autillo-Touati, A., Dargent, B., and Trimmer, J. S. 2011. Cdk-mediated phosphorylation of the Kv{beta}2 auxiliary subunit regulates Kv1 channel axonal targeting. *J Cell Biol.* 192: 813–824.
32. Dugandzija-Novakovic, S., Koszowski, A. G., Levinson, S. R., and Shrager, P. 1995. Clustering of Na channels and node of Ranvier formation in remyelinating axons. *J Neurosci.* 15: 492–502.
33. Rasband, M. N., Peles, E., Trimmer, J. S., Levinson, S. R., Lux, S. E., and Shrager, P. 1999. Dependence of nodal sodium channel clustering on paranodal axoglial contact in the developing CNS. *J Neurosci.* 19: 7516–7528.
34. Trinidad, J. C., Thalhammer, A., Specht, C. G., Lynn, A. J., Baker, P. R., Schoepfer, R., and Burlingame, A. L. 2008. Quantitative analysis of synaptic phosphorylation and protein expression. *Mol Cell Proteomics.* 7: 684–696.
35. Tweedie-Cullen, R. Y., Reck, J. M., and Mansuy, I. M. 2009. Comprehensive mapping of post-translational modifications on synaptic, nuclear, and histone proteins in the adult mouse brain. *J Proteome Res.* 8: 4966–4982.
36. Huttlin, E. L., Jedrychowski, M. P., Elias, J. E., Goswami, T., Rad, R., Beausoleil, S. A., Villen, J., Haas, W., Sowa, M. E., and Gygi, S. P. 2010. A tissue-specific atlas of mouse protein phosphorylation and expression. *Cell.* 143: 1174–1189.
37. Wisniewski, J. R., Nagaraj, N., Zougman, A., Gnad, F., and Mann, M. 2010. Brain phosphoproteome obtained by a FASP-based method reveals plasma membrane protein topology. *J Proteome Res.* 9: 3280–3289.
38. Ballif, B. A., Carey, G. R., Sunyaev, S. R., and Gygi, S. P. 2008. Large-scale identification and evolution indexing of tyrosine phosphorylation sites from murine brain. *J Proteome Res.* 7: 311–318.
39. Trinidad, J. C., Barkan, D. T., Gulledge, B. F., Thalhammer, A., Sali, A., Schoepfer, R., and Burlingame, A. L. 2012. Global identification and characterization of both O-GlcNAcylation and phosphorylation at the murine synapse. *Mol Cell Proteomics.* 11: 215–229.
40. Olsen, J. V., Blagoev, B., Gnad, F., Macek, B., Kumar, C., Mortensen, P., and Mann, M. 2006. Global, in vivo, and site-specific phosphorylation dynamics in signaling networks. *Cell.* 127: 635–648.
41. Hornbeck, P. V., Kornhauser, J. M., Tkachev, S., Zhang, B., Skrzypek, E., Murray, B., Latham, V., and Sullivan, M. 2012. PhosphoSitePlus: A comprehensive resource for investigating the structure and function of experimentally determined post-translational modifications in man and mouse. *Nucleic Acids Res.* 40: D261–D270.
42. Gordon, D., Merrick, D., Auld, V., Dunn, R., Goldin, A. L., Davidson, N., and Catterall, W. A. 1987. Tissue-specific expression of the RI and RII sodium channel subtypes. *Proc Natl Acad Sci USA.* 84: 8682–8686.
43. Scheuer, T. 2011. Regulation of sodium channel activity by phosphorylation. *Semin Cell Dev Biol.* 22: 160–165.
44. Goswami, T., Li, X., Smith, A. M., Luderowski, E. M., Vincent, J. J., Rush, J., and Ballif, B. A. 2012. Comparative phosphoproteomic analysis of neonatal and adult murine brain. *Proteomics.* 12: 2185–2189.
45. Beacham, D., Ahn, M., Catterall, W. A., and Scheuer, T. 2007. Sites and molecular mechanisms of modulation of Na(v)1.2 channels by Fyn tyrosine kinase. *J Neurosci.* 27: 11543–11551.
46. Hoffert, J. D., Pisitkun, T., Wang, G., Shen, R. F., and Knepper, M. A. 2006. Quantitative phosphoproteomics of vasopressin-sensitive renal cells: Regulation of aquaporin-2 phosphorylation at two sites. *Proc Natl Acad Sci USA.* 103: 7159–7164.
47. Brechet, A., Fache, M. P., Brachet, A., Ferracci, G., Baude, A., Irondelle, M., Pereira, S., Leterrier, C., and Dargent, B. 2008. Protein kinase CK2 contributes to the organization of sodium channels in axonal membranes by regulating their interactions with ankyrin G. *J Cell Biol.* 183: 1101–1114.
48. Cantrell, A. R., Tibbs, V. C., Yu, F. H., Murphy, B. J., Sharp, E. M., Qu, Y., Catterall, W. A., and Scheuer, T. 2002. Molecular mechanism of convergent regulation of brain Na(+) channels by protein kinase C and protein kinase A anchored to AKAP-15. *Mol Cell Neurosci.* 21: 63–80.
49. Smith, R. D. and Goldin, A. L. 1997. Phosphorylation at a single site in the rat brain sodium channel is necessary and sufficient for current reduction by protein kinase A. *J Neurosci.* 17: 6086–6093.
50. Murphy, B. J., Rossie, S., De Jongh, K. S., and Catterall, W. A. 1993. Identification of the sites of selective phosphorylation and dephosphorylation of the rat brain Na^+ channel alpha subunit by cAMP-dependent protein kinase and phosphoprotein phosphatases. *J Biol Chem.* 268: 27355–27362.
51. Brill, L. M., Xiong, W., Lee, K. B., Ficarro, S. B., Crain, A., Xu, Y., Terskikh, A., Snyder, E. Y., and Ding, S. 2009. Phosphoproteomic analysis of human embryonic stem cells. *Cell Stem Cell.* 5: 204–213.
52. Kraner, S. D., Novak, K. R., Wang, Q., Peng, J., and Rich, M. M. 2012. Altered sodium channel-protein associations in critical illness myopathy. *Skelet Muscle.* 2: 17.
53. Marionneau, C., Lichti, C. F., Lindenbaum, P., Charpentier, F., Nerbonne, J. M., Townsend, R. R., and Merot, J. 2012. Mass spectrometry-based identification of native cardiac Nav1.5 channel alpha subunit phosphorylation sites. *J Proteome Res.* 11: 5994–6007.
54. Ashpole, N. M., Herren, A. W., Ginsburg, K. S., Brogan, J. D., Johnson, D. E., Cummins, T. R., Bers, D. M., and Hudmon, A. 2012. Ca^{2+}/calmodulin-dependent protein kinase II (CaMKII) regulates cardiac sodium channel NaV1.5 gating by multiple phosphorylation sites. *J Biol Chem.* 287: 19856–19869.
55. Ahern, C. A., Zhang, J. F., Wookalis, M. J., and Horn, R. 2005. Modulation of the cardiac sodium channel NaV1.5 by Fyn, a Src family tyrosine kinase. *Circ Res.* 96: 991–998.
56. Li, H., Xing, X., Ding, G., Li, Q., Wang, C., Xie, L., Zeng, R., and Li, Y. 2009. SysPTM: A systematic resource for proteomic research on post-translational modifications. *Mol Cell Proteomics.* 8: 1839–1849.

57. Gasser, A., Cheng, X., Gilmore, E. S., Tyrrell, L., Waxman, S. G., and Dib-Hajj, S. D. 2010. Two Nedd4-binding motifs underlie modulation of sodium channel Nav1.6 by p38 MAPK. *J Biol Chem.* 285: 26149–26161.
58. Wittmack, E. K., Rush, A. M., Hudmon, A., Waxman, S. G., and Dib-Hajj, S. D. 2005. Voltage-gated sodium channel Nav1.6 is modulated by p38 mitogen-activated protein kinase. *J Neurosci.* 25: 6621–6630.
59. Rinschen, M. M., Yu, M. J., Wang, G., Boja, E. S., Hoffert, J. D., Pisitkun, T., and Knepper, M. A. 2010. Quantitative phosphoproteomic analysis reveals vasopressin V2-receptor-dependent signaling pathways in renal collecting duct cells. *Proc Natl Acad Sci USA.* 107: 3882–3887.
60. Hudmon, A., Choi, J. S., Tyrrell, L., Black, J. A., Rush, A. M., Waxman, S. G., and Dib-Hajj, S. D. 2008. Phosphorylation of sodium channel Na(v)1.8 by p38 mitogen-activated protein kinase increases current density in dorsal root ganglion neurons. *J Neurosci.* 28: 3190–3201.
61. Wu, D. F., Chandra, D., McMahon, T., Wang, D., Dadgar, J., Kharazia, V. N., Liang, Y. J., Waxman, S. G., Dib-Hajj, S. D., and Messing, R. O. 2012. PKCepsilon phosphorylation of the sodium channel NaV1.8 increases channel function and produces mechanical hyperalgesia in mice. *J Clin Invest.* 122: 1306–1315.
62. Dai, J., Jin, W. H., Sheng, Q. H., Shieh, C. H., Wu, J. R., and Zeng, R. 2007. Protein phosphorylation and expression profiling by Yin-yang multidimensional liquid chromatography (Yin-yang MDLC) mass spectrometry. *J Proteome Res.* 6: 250–262.
63. Dai, S., Hall, D. D., and Hell, J. W. 2009. Supramolecular assemblies and localized regulation of voltage-gated ion channels. *Physiol Rev.* 89: 411–452.
64. Fuller, M. D., Emrick, M. A., Sadilek, M., Scheuer, T., and Catterall, W. A. 2010. Molecular mechanism of calcium channel regulation in the fight-or-flight response. *Sci Signal.* 3: ra70.
65. Catterall, W. A. 2000. Structure and regulation of voltage-gated Ca^{2+} channels. *Annu Rev Cell Dev Biol.* 16: 521–555.
66. Munton, R. P., Tweedie-Cullen, R., Livingstone-Zatchej, M., Weinandy, F., Waidelich, M., Longo, D., Gehrig, P. et al. 2007. Qualitative and quantitative analyses of protein phosphorylation in naive and stimulated mouse synaptosomal preparations. *Mol Cell Proteomics.* 6: 283–293.
67. Rohrkasten, A., Meyer, H. E., Nastainczyk, W., Sieber, M., and Hofmann, F. 1988. cAMP-dependent protein kinase rapidly phosphorylates serine-687 of the skeletal muscle receptor for calcium channel blockers. *J Biol Chem.* 263: 15325–15329.
68. Emrick, M. A., Sadilek, M., Konoki, K., and Catterall, W. A. 2010. Beta-adrenergic-regulated phosphorylation of the skeletal muscle Ca(V)1.1 channel in the fight-or-flight response. *Proc Natl Acad Sci USA.* 107: 18712–18717.
69. Mitterdorfer, J., Froschmayr, M., Grabner, M., Moebius, F. F., Glossmann, H., and Striessnig, J. 1996. Identification of PK-A phosphorylation sites in the carboxyl terminus of L-type calcium channel alpha 1 subunits. *Biochemistry.* 35: 9400–9406.
70. Rikova, K., Guo, A., Zeng, Q., Possemato, A., Yu, J., Haack, H., Nardone, J. et al. 2007. Global survey of phosphotyrosine signaling identifies oncogenic kinases in lung cancer. *Cell.* 131: 1190–1203.
71. Yang, L., Doshi, D., Morrow, J., Katchman, A., Chen, X., and Marx, S. O. 2009. Protein kinase C isoforms differentially phosphorylate Ca(v)1.2 alpha(1c). *Biochemistry.* 48: 6674–6683.
72. Lee, T. S., Karl, R., Moosmang, S., Lenhardt, P., Klugbauer, N., Hofmann, F., Kleppisch, T., and Welling, A. 2006. Calmodulin kinase II is involved in voltage-dependent facilitation of the L-type Cav1.2 calcium channel: Identification of the phosphorylation sites. *J Biol Chem.* 281: 25560–25567.
73. Takahashi, E., Fukuda, K., Miyoshi, S., Murata, M., Kato, T., Ita, M., Tanabe, T., and Ogawa, S. 2004. Leukemia inhibitory factor activates cardiac L-Type Ca^{2+} channels via phosphorylation of serine 1829 in the rabbit Cav1.2 subunit. *Circ Res.* 94: 1242–1248.
74. Hulme, J. T., Westenbroek, R. E., Scheuer, T., and Catterall, W. A. 2006. Phosphorylation of serine 1928 in the distal C-terminal domain of cardiac CaV1.2 channels during beta1-adrenergic regulation. *Proc Natl Acad Sci USA.* 103: 16574–16579.
75. Davare, M. A. and Hell, J. W. 2003. Increased phosphorylation of the neuronal L-type Ca(2+) channel Ca(v)1.2 during aging. *Proc Natl Acad Sci USA.* 100: 16018–16023.
76. Kim, B. G., Lee, J. H., Ahn, J. M., Park, S. K., Cho, J. H., Hwang, D., Yoo, J. S., Yates, J. R. R., Ryoo, H. M., and Cho, J. Y. 2009. Two-stage double-technique hybrid (TSDTH) identification strategy for the analysis of BMP2-induced transdifferentiation of premyoblast C2C12 cells to osteoblast. *J Proteome Res.* 8: 4441–4454.
77. Hsu, P. P., Kang, S. A., Rameseder, J., Zhang, Y., Ottina, K. A., Lim, D., Peterson, T. R. et al. 2011. The mTOR-regulated phosphoproteome reveals a mechanism of mTORC1-mediated inhibition of growth factor signaling. *Science.* 332: 1317–1322.
78. Wu, C. C., MacCoss, M. J., Howell, K. E., and Yates, J. R. R. 2003. A method for the comprehensive proteomic analysis of membrane proteins. *Nat Biotechnol.* 21: 532–538.
79. Richman, R. W., Tombler, E., Lau, K. K., Anantharam, A., Rodriguez, J., O'Bryan, J. P., and Diverse-Pierluissi, M. A. 2004. N-type Ca^{2+} channels as scaffold proteins in the assembly of signaling molecules for GABAB receptor effects. *J Biol Chem.* 279: 24649–24658.
80. Park, K.-S., Mohapatra, D. P., Misonou, H., and Trimmer, J. S. 2006. Graded regulation of the Kv2.1 potassium channel by variable phosphorylation. *Science.* 313: 976–979.
81. Yang, E. K., Alvira, M. R., Levitan, E. S., and Takimoto, K. 2001. Kv beta subunits increase expression of Kv4.3 channels by interacting with their C termini. *J Biol Chem.* 276: 4839–4844.
82. Surti, T. S., Huang, L., Jan, Y. N., Jan, L. Y., and Cooper, E. C. 2005. Identification by mass spectrometry and functional characterization of two phosphorylation sites of KCNQ2/KCNQ3 channels. *Proc Natl Acad Sci USA.* 102: 17828–17833.
83. Johnson, R. P., El-Yazbi, A. F., Hughes, M. F., Schriemer, D. C., Walsh, E. J., Walsh, M. P., and Cole, W. C. 2009. Identification and functional characterization of protein kinase A-catalyzed phosphorylation of potassium channel Kv1.2 at serine 449. *J Biol Chem.* 284: 16562–16574.
84. Fury, M., Marx, S. O., and Marks, A. R. 2002. Molecular BKology: The study of splicing and dicing. *Sci STKE.* 2002: PE12.
85. Fodor, A. A. and Aldrich, R. W. 2009. Convergent evolution of alternative splices at domain boundaries of the BK channel. *Annu Rev Physiol.* 71: 19–36.
86. Tao, Y., Zeng, R., Shen, B., Jia, J., and Wang, Y. 2005. Neuronal transmission stimulates the phosphorylation of Kv1.4 channel at Ser229 through protein kinase A1. *J Neurochem.* 94: 1512–1522.
87. Deng, W. J., Nie, S., Dai, J., Wu, J. R., and Zeng, R. 2010. Proteome, phosphoproteome, and hydroxyproteome of liver mitochondria in diabetic rats at early pathogenic stages. *Mol Cell Proteomics.* 9: 100–116.
88. Herskowitz, J. H., Seyfried, N. T., Duong, D. M., Xia, Q., Rees, H. D., Gearing, M., Peng, J., Lah, J. J., and Levey, A. I. 2010. Phosphoproteomic analysis reveals site-specific changes in GFAP and NDRG2 phosphorylation in front temporal lobar degeneration. *J Proteome Res.* 9: 6368–6379.
89. Anderson, A. E., Adams, J. P., Qian, Y., Cook, R. G., Pfaffinger, P. J., and Sweatt, J. D. 2000. Kv4.2 phosphorylation by cyclic AMP-dependent protein kinase. *J Biol Chem.* 275: 5337–5346.

90. Schrader, L. A., Birnbaum, S. G., Nadin, B. M., Ren, Y., Bui, D., Anderson, A. E., and Sweatt, J. D. 2006. ERK/MAPK regulates the Kv4.2 potassium channel by direct phosphorylation of the pore-forming subunit. *Am J Physiol Cell Physiol.* 290: C852–C861.
91. Hu, H. J., Carrasquillo, Y., Karim, F., Jung, W. E., Nerbonne, J. M., Schwarz, T. L., and Gereau, R. W. 2006. The kv4.2 potassium channel subunit is required for pain plasticity. *Neuron.* 50: 89–100.
92. Hu, H. J., Alter, B. J., Carrasquillo, Y., Qiu, C. S., and Gereau, R. W. T. 2007. Metabotropic glutamate receptor 5 modulates nociceptive plasticity via extracellular signal-regulated kinase-Kv4.2 signaling in spinal cord dorsal horn neurons. *J Neurosci.* 27: 13181–13191.
93. Varga, A. W., Yuan, L. L., Anderson, A. E., Schrader, L. A., Wu, G. Y., Gatchel, J. R., Johnston, D., and Sweatt, J. D. 2004. Calcium-calmodulin-dependent kinase II modulates Kv4.2 channel expression and upregulates neuronal A-type potassium currents. *J Neurosci.* 24: 3643–3654.
94. Schrader, L. A., Ren, Y., Cheng, F., Bui, D., Sweatt, J. D., and Anderson, A. E. 2009. Kv4.2 is a locus for PKC and ERK/MAPK cross-talk. *Biochem J.* 417: 705–715.
95. Moritz, A., Li, Y., Guo, A., Villen, J., Wang, Y., MacNeill, J., Kornhauser, J. et al. 2010. Akt-RSK-S6 kinase signaling networks activated by oncogenic receptor tyrosine kinases. *Sci Signal.* 3: ra64.
96. Molina, H., Horn, D. M., Tang, N., Mathivanan, S., and Pandey, A. 2007. Global proteomic profiling of phosphopeptides using electron transfer dissociation tandem mass spectrometry. *Proc Natl Acad Sci USA.* 104: 2199–2204.
97. Zhang, Y. H., Wu, W., Sun, H. Y., Deng, X. L., Cheng, L. C., Li, X., Tse, H. F., Lau, C. P., and Li, G. R. 2012. Modulation of human cardiac transient outward potassium current by EGFR tyrosine kinase and Src-family kinases. *Cardiovasc Res.* 93: 424–433.
98. Heijman, J., Spatjens, R. L., Seyen, S. R., Lentink, V., Kuijpers, H. J., Boulet, I. R., de Windt, L. J., David, M., and Volders, P. G. 2012. Dominant-negative control of cAMP-dependent IKs upregulation in human long-QT syndrome type 1. *Circ Res.* 110: 211–219.
99. Li, Y., Langlais, P., Gamper, N., Liu, F., and Shapiro, M. S. 2004. Dual phosphorylations underlie modulation of unitary KCNQ K(+) channels by Src tyrosine kinase. *J Biol Chem.* 279: 45399–45407.
100. Raijmakers, R., Kraiczek, K., de Jong, A. P., Mohammed, S., and Heck, A. J. 2010. Exploring the human leukocyte phosphoproteome using a microfluidic reversed-phase-TiO_2-reversed-phase high-performance liquid chromatography phosphochip coupled to a quadrupole time-of-flight mass spectrometer. *Anal Chem.* 82: 824–832.
101. Choudhary, C., Olsen, J. V., Brandts, C., Cox, J., Reddy, P. N., Bohmer, F. D., Gerke, V. et al. 2009. Mislocalized activation of oncogenic RTKs switches downstream signaling outcomes. *Mol Cell.* 36: 326–339.
102. Shu, H., Chen, S., Bi, Q., Mumby, M., and Brekken, D. L. 2004. Identification of phosphoproteins and their phosphorylation sites in the WEHI-231 B lymphoma cell line. *Mol Cell Proteomics.* 3: 279–286.
103. Zanivan, S., Gnad, F., Wickstrom, S. A., Geiger, T., Macek, B., Cox, J., Fassler, R., and Mann, M. 2008. Solid tumor proteome and phosphoproteome analysis by high resolution mass spectrometry. *J Proteome Res.* 7: 5314–5326.
104. Yu, Y., Yoon, S. O., Poulogiannis, G., Yang, Q., Ma, X. M., Villen, J., Kubica, N. et al. 2011. Phosphoproteomic analysis identifies Grb10 as an mTORC1 substrate that negatively regulates insulin signaling. *Science.* 332: 1322–1326.
105. Mayya, V., Lundgren, D. H., Hwang, S. I., Rezaul, K., Wu, L., Eng, J. K., Rodionov, V., and Han, D. K. 2009. Quantitative phosphoproteomic analysis of T cell receptor signaling reveals system-wide modulation of protein–protein interactions. *Sci Signal.* 2: ra46.
106. Iliuk, A. B., Martin, V. A., Alicie, B. M., Geahlen, R. L., and Tao, W. A. 2010. In-depth analyses of kinase-dependent tyrosine phosphoproteomes based on metal ion-functionalized soluble nanopolymers. *Mol Cell Proteomics.* 9: 2162–2172.
107. Poolos, N. P., Bullis, J. B., and Roth, M. K. 2006. Modulation of h-channels in hippocampal pyramidal neurons by p38 mitogen-activated protein kinase. *J Neurosci.* 26: 7995–8003.
108. Arinsburg, S. S., Cohen, I. S., and Yu, H. G. 2006. Constitutively active Src tyrosine kinase changes gating of HCN4 channels through direct binding to the channel proteins. *J Cardiovasc Pharmacol.* 47: 578–586.
109. Huang, J., Huang, A., Zhang, Q., Lin, Y. C., and Yu, H. G. 2008. Novel mechanism for suppression of hyperpolarization-activated cyclic nucleotide-gated pacemaker channels by receptor-like tyrosine phosphatase-alpha. *J Biol Chem.* 283: 29912–29919.
110. Liao, Z., Lockhead, D., Larson, E. D., and Proenza, C. 2010. Phosphorylation and modulation of hyperpolarization-activated HCN4 channels by protein kinase A in the mouse sinoatrial node. *J Gen Physiol.* 136: 247–258.
111. Zong, X., Eckert, C., Yuan, H., Wahl-Schott, C., Abicht, H., Fang, L., Li, R. et al. 2005. A novel mechanism of modulation of hyperpolarization-activated cyclic nucleotide-gated channels by Src kinase. *J Biol Chem.* 280: 34224–34232.
112. Hammelmann, V., Zong, X., Hofmann, F., Michalakis, S., and Biel, M. 2011. The cGMP-dependent protein kinase II Is an inhibitory modulator of the hyperpolarization-activated HCN2 channel. *PLoS One.* 6: e17078.
113. Levitan, I. B. 1994. Modulation of ion channels by protein phosphorylation and dephosphorylation. *Annu Rev Physiol.* 56: 193–212.
114. Harvey, R. D. and Hell, J. W. 2013. CaV1.2 signaling complexes in the heart. *J Mol Cell Cardiol.* 58: 143–152.
115. Bean, B. P., Nowycky, M. C., and Tsien, R. W. 1984. Beta-adrenergic modulation of calcium channels in frog ventricular heart cells. *Nature.* 307: 371–375.
116. Erxleben, C., Liao, Y., Gentile, S., Chin, D., Gomez-Alegria, C., Mori, Y., Birnbaumer, L., and Armstrong, D. L. 2006. Cyclosporin and Timothy syndrome increase mode 2 gating of CaV1.2 calcium channels through aberrant phosphorylation of S6 helices. *Proc Natl Acad Sci USA.* 103: 3932–3937.
117. Vacher, H., Mohapatra, D. P., and Trimmer, J. S. 2008. Localization and targeting of voltage-dependent ion channels in mammalian central neurons. *Physiol Rev.* 88: 1407–1447.
118. Hattan, D., Nesti, E., Cachero, T. G., and Morielli, A. D. 2002. Tyrosine phosphorylation of Kv1.2 modulates its interaction with the actin-binding protein cortactin. *J Biol Chem.* 277: 38596–38606.
119. Nesti, E., Everill, B., and Morielli, A. D. 2004. Endocytosis as a mechanism for tyrosine kinase-dependent suppression of a voltage-gated potassium channel. *Mol Biol Cell.* 15: 4073–4088.
120. Williams, M. R., Markey, J. C., Doczi, M. A., and Morielli, A. D. 2007. An essential role for cortactin in the modulation of the potassium channel Kv1.2. *Proc Natl Acad Sci USA.* 104: 17412–17417.
121. Connors, E. C., Ballif, B. A., and Morielli, A. D. 2008. Homeostatic regulation of Kv1.2 potassium channel trafficking by cyclic AMP. *J Biol Chem.* 283: 3445–3453.
122. Kim, J., Jung, S. C., Clemens, A. M., Petralia, R. S., and Hoffman, D. A. 2007. Regulation of dendritic excitability by activity-dependent trafficking of the A-type K^+ channel subunit Kv4.2 in hippocampal neurons. *Neuron.* 54: 933–947.
123. Hammond, R. S., Lin, L., Sidorov, M. S., Wikenheiser, A. M., and Hoffman, D. A. 2008. Protein kinase a mediates activity-dependent Kv4.2 channel trafficking. *J Neurosci.* 28: 7513–7519.

124. Yu, S. P., Yeh, C. H., Sensi, S. L., Gwag, B. J., Canzoniero, L. M., Farhangrazi, Z. S., Ying, H. S., Tian, M., Dugan, L. L., and Choi, D. W. 1997. Mediation of neuronal apoptosis by enhancement of outward potassium current. *Science.* 278: 114–117.
125. Pal, S., Hartnett, K. A., Nerbonne, J. M., Levitan, E. S., and Aizenman, E. 2003. Mediation of neuronal apoptosis by Kv2.1-encoded potassium channels. *J Neurosci.* 23: 4798–4802.
126. Redman, P. T., He, K., Hartnett, K. A., Jefferson, B. S., Hu, L., Rosenberg, P. A., Levitan, E. S., and Aizenman, E. 2007. Apoptotic surge of potassium currents is mediated by p38 phosphorylation of Kv2.1. *Proc Natl Acad Sci USA.* 104: 3568–3573.
127. Pal, S. K., Takimoto, K., Aizenman, E., and Levitan, E. S. 2006. Apoptotic surface delivery of K^+ channels. *Cell Death Differ.* 13: 661–667.
128. Shen, Q. J., Zhao, Y. M., Cao, D. X., and Wang, X. L. 2009. Contribution of Kv channel subunits to glutamate-induced apoptosis in cultured rat hippocampal neurons. *J Neurosci Res.* 87: 3153–3160.
129. Jiao, S., Liu, Z., Ren, W. H., Ding, Y., Zhang, Y. Q., Zhang, Z. H., and Mei, Y. A. 2007. cAMP/protein kinase A signalling pathway protects against neuronal apoptosis and is associated with modulation of Kv2.1 in cerebellar granule cells. *J Neurochem.* 100: 979–991.
130. Yao, H., Zhou, K., Yan, D., Li, M., and Wang, Y. 2009. The Kv2.1 channels mediate neuronal apoptosis induced by excitotoxicity. *J Neurochem.* 108: 909–919.
131. Williams, M. R., Fuchs, J. R., Green, J. T., and Morielli, A. D. 2012. Cellular mechanisms and behavioral consequences of Kv1.2 regulation in the rat cerebellum. *J Neurosci.* 32: 9228–9237.
132. Murakoshi, H., Shi, G., Scannevin, R. H., and Trimmer, J. S. 1997. Phosphorylation of the Kv2.1 K^+ channel alters voltage-dependent activation. *Mol Pharmacol.* 52: 821–828.
133. Misonou, H., Mohapatra, D. P., Park, E. W., Leung, V., Zhen, D., Misonou, K., Anderson, A. E., and Trimmer, J. S. 2004. Regulation of ion channel localization and phosphorylation by neuronal activity. *Nat Neurosci.* 7: 711–718.
134. Mohapatra, D. P. and Trimmer, J. S. 2006. The Kv2.1 C terminus can autonomously transfer Kv2.1-like phosphorylation-dependent localization, voltage-dependent gating, and muscarinic modulation to diverse Kv channels. *J Neurosci.* 26: 685–695.
135. Ong, S. E., Blagoev, B., Kratchmarova, I., Kristensen, D. B., Steen, H., Pandey, A., and Mann, M. 2002. Stable isotope labeling by amino acids in cell culture, SILAC, as a simple and accurate approach to expression proteomics. *Mol Cell Proteomics.* 1: 376–386.
136. Misonou, H., Menegola, M., Mohapatra, D. P., Guy, L. K., Park, K. S., and Trimmer, J. S. 2006. Bidirectional activity-dependent regulation of neuronal ion channel phosphorylation. *J Neurosci.* 26: 13505–13514.
137. Macica, C. M. and Kaczmarek, L. K. 2001. Casein kinase 2 determines the voltage dependence of the Kv3.1 channel in auditory neurons and transfected cells. *J Neurosci.* 21: 1160–1168.
138. Song, P. and Kaczmarek, L. K. 2006. Modulation of Kv3.1b potassium channel phosphorylation in auditory neurons by conventional and novel protein kinase C isozymes. *J Biol Chem.* 281: 15582–15591.
139. Song, P., Yang, Y., Barnes-Davies, M., Bhattacharjee, A., Hamann, M., Forsythe, I. D., Oliver, D. L., and Kaczmarek, L. K. 2005. Acoustic environment determines phosphorylation state of the Kv3.1 potassium channel in auditory neurons. *Nat Neurosci.* 8: 1335–1342.
140. Ahn, M., Beacham, D., Westenbroek, R. E., Scheuer, T., and Catterall, W. A. 2007. Regulation of Na(v)1.2 channels by brain-derived neurotrophic factor, TrkB, and associated Fyn kinase. *J Neurosci.* 27: 11533–11542.
141. Ratcliffe, C. F., Qu, Y., McCormick, K. A., Tibbs, V. C., Dixon, J. E., Scheuer, T., and Catterall, W. A. 2000. A sodium channel signaling complex: Modulation by associated receptor protein tyrosine phosphatase beta. *Nat Neurosci.* 3: 437–444.
142. Carr, D. B., Day, M., Cantrell, A. R., Held, J., Scheuer, T., Catterall, W. A., and Surmeier, D. J. 2003. Transmitter modulation of slow, activity-dependent alterations in sodium channel availability endows neurons with a novel form of cellular plasticity. *Neuron.* 39: 793–806.
143. Chen, Y., Yu, F. H., Surmeier, D. J., Scheuer, T., and Catterall, W. A. 2006. Neuromodulation of Na^+ channel slow inactivation via cAMP-dependent protein kinase and protein kinase C. *Neuron.* 49: 409–420.
144. Bezprozvanny, I., Scheller, R. H., and Tsien, R. W. 1995. Functional impact of syntaxin on gating of N-type and Q-type calcium channels. *Nature.* 378: 623–626.
145. Yang, J. and Tsien, R. W. 1993. Enhancement of N- and L-type calcium channel currents by protein kinase C in frog sympathetic neurons. *Neuron.* 10: 127–136.
146. Stea, A., Soong, T. W., and Snutch, T. P. 1995. Determinants of PKC-dependent modulation of a family of neuronal calcium channels. *Neuron.* 15: 929–940.
147. Yokoyama, C. T., Sheng, Z. H., and Catterall, W. A. 1997. Phosphorylation of the synaptic protein interaction site on N-type calcium channels inhibits interactions with SNARE proteins. *J Neurosci.* 17: 6929–6938.
148. Yokoyama, C. T., Myers, S. J., Fu, J., Mockus, S. M., Scheuer, T., and Catterall, W. A. 2005. Mechanism of SNARE protein binding and regulation of Cav2 channels by phosphorylation of the synaptic protein interaction site. *Mol Cell Neurosci.* 28: 1–17.
149. Jarvis, S. E. and Zamponi, G. W. 2001. Distinct molecular determinants govern syntaxin 1A-mediated inactivation and G-protein inhibition of N-type calcium channels. *J Neurosci.* 21: 2939–2948.
150. Gentile, S. 2012. Ion channel phosphorylopathy: A link between genomic variation and human disease. *ChemMedChem.* 7: 1757–1761.

37 Alternative splicing

Andrea L. Meredith

Contents

Genomes contain a finite number of genes. Yet this limited genetic code must produce a vast diversity of protein function that spans developmental programs to tissue-specific tuning of physiological function. A fundamental and nearly ubiquitous genetic mechanism for creating this diversity is alternative splicing of mRNA, producing a multitude of unique transcripts encoded by a single gene. Alternative splicing results in addition, deletion, or alteration of functional protein modules, arising from independent regulation of discrete functional domains (Raingo et al., 2007) or coordinate regulation of functionally interconnected domains (Glauser et al., 2011; Johnson et al., 2011). In ion channels, exon variation can lead to incremental transitions in current properties or create *on-/off*-type regulatory switching (Lipscombe et al., 2009). Electrophysiological recordings of ionic currents have proven to be a particularly sensitive detection method for the functional consequences of alternative splicing. Ion channel splice variants exhibit different gating properties, allosteric modulation, permeability, regulation by signaling pathways, and expression and localization. Alternative splicing is particularly prevalent in the brain, where the complexities of ion channel function are most elaborate. Tailoring of ion channel properties by alternative splicing enables intricate fine-tuning of the ionic networks underlying excitability and homeostasis, vastly expanding the parameter space for physiological solutions when amplified across the repertoire of ion channels expressed within a cell or tissue.

The prevalence of alternative splicing in mammalian genomes is very high, estimated to occur in 95% of transcripts (Kornblihtt et al., 2013). Missplicing is associated with disease, and emerging therapeutic strategies selectively target critical splicing events. The number of known splice variants among ion channels is large, with alternative splicing occurring in the majority of multiexon ion channel gene products. Much of this information is already curated on annotated genomic sequence, openly accessible via the Entrez Search and Retrieval System at the National Center for Biotechnology Information.* This chapter addresses the basic mechanisms of splicing, alternative splicing of ion channels in physiological function and disease, and techniques for identification of splice variants from tissues or genomes.

37.1 GENE STRUCTURE, EXPRESSION, AND PROTEIN SYNTHESIS

A gene is a heritable unit of DNA sequence that is expressed through a process of transcription and translation. Eukaryotic genes contain both protein coding sequence (exons) and noncoding sequence (cis-regulatory regions and introns). Mammalian genes encoding ion channels range from intronless that span 3–40 kb of genomic sequence (e.g., K_V1 channels) to transcripts comprised of >50 alternative and constitutive exons that span >500 kb of genomic sequence (Ca^{2+} channels).* Exon sequences are phylogenetically conserved, stemming

* National Center for Biotechnology Information Database [internet]. Available from: http://www.ncbi.nlm.nih.gov/gene and http://www.ncbi.nlm.nih.gov/unigene.

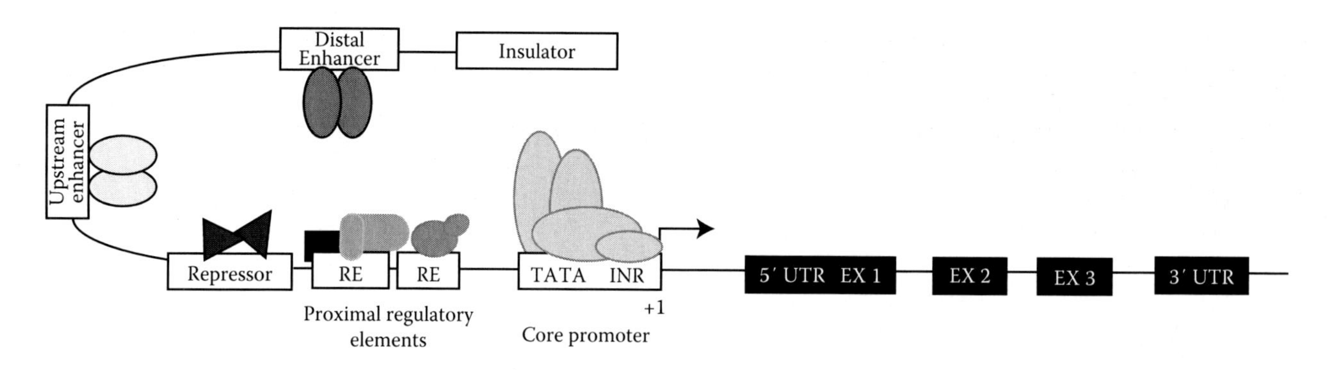

Figure 37.1 Transcription initiation. Enhancers, repressors, and the RNA polymerase II preinitiation complex bind DNA at distinct sequences within a gene. Enhancers and repressors exert short- and long-range effects on the core promoter complex through proximal regulatory elements (RE), which serve to tether the interactions. Tissue specificity is achieved through multiple enhancer and repressor sequences that recruit transcription factors and direct transcription in distinct cells or expression domains. These regulatory elements may cover >100 kb in genes with complex expression patterns. Insulator sequences prevent regulatory elements from one gene from affecting transcription at adjacent loci. (Adapted from Levine, M. and Tijan, R., *Nature*, 424, 147, 2003.)

from evolutionary selection for particular functionality, while intronic sequences diverge widely due to tolerance of base changes in noncoding regions. In many genes, exons code for functional modules of a protein, with the number of exons increasing with protein length and additional functionality (Lewin, 2007).

Genome-scale transcriptional regulatory mechanisms determine temporal and tissue-specific gene expression. Transcription of genes into an mRNA template is regulated by specific DNA sequences that bind transcription factor complexes (Figure 37.1). Basal promoter elements initiate formation of the RNA transcript, while transcription factor binding sites in cis-regulatory sequence mediate the specificity of gene expression. The combinatorial actions of enhancers and repressors provide for distinct programs of ion channel expression in particular cell and tissue types. This specificity is achieved through both stimulation of transcription in excitable tissues, such as neurons and muscle, and active repression of transcription in other tissues. Many muscle and neuronally expressed ion channels contain E-box sequences (CANNTG)

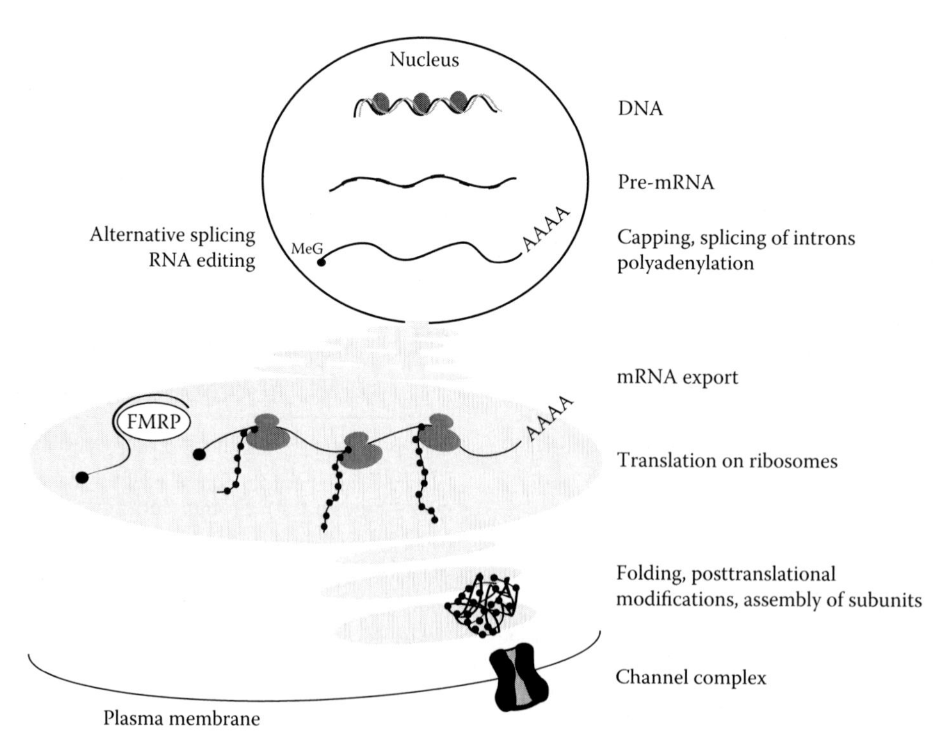

Figure 37.2 Pre-mRNA processing and protein synthesis. A gene is transcribed with introns (pre-mRNA). Pre-mRNA are modified with the addition of the 5′ cap, removal of introns by splicing, and addition of the polyA tail within the nucleus. Alternative splicing, resulting in removal or insertion of exonic sequences, and mRNA editing also occur at this step. Mature mRNA contain signal sequences that direct protein synthesis in the endoplasmic reticulum. FMRP (Fragile X mental retardation protein) inhibits translation in part by directly binding to some mRNA transcripts. The translated polypeptides fold, undergo posttranslational modifications such as glycosylation, associate with accessory subunits, and are sorted through the golgi for targeted delivery to subcellular locations.

in their promoters, conferring binding by canonical basic helix-loop-helix (bHLH) transcription factors. For example, the enhancers MyoD, Myogenin, and MRF4 have been shown to positively regulate the expression of the skeletal muscle $Na_V1.4$ channel (Kraner et al., 1998, 1999). Conversely, gene expression can be inhibited by factors such as REST (Repressor Element Silencing Transcription or Neuron-Restrictive Silencer Factor), a zinc-finger protein. During development and in nonneuronal tissues, $Na_V1.2$ is repressed through REST binding at RE1 sequences (Chong et al., 1995). RE1 sequences are also found in the cis-regulatory regions of several other voltage and ligand-gated channels (Hille, 2001). In addition to bHLH and REST factors, which shape the early spatiotemporal patterns of gene expression, other transcription factors confer regulation through cellular second messenger cascades linked to G-protein coupled receptors or activity-dependent signaling. For example, the factors CREB, CBP, CaRF, NFAT, DREAM, and FosB are involved in activity dependent regulation of transcription through Ca^{2+}-dependent pathways (West et al., 2002).

After assembly of the enhancers, repressors, and core promoter complexes, transcription is initiated by RNA polymerase II, generating a full-length copy of the DNA gene sequence (pre-mRNA). The pre-mRNA is rapidly modified for recognition by the ribosome (5′ cap), removal of introns by splicing, and addition of a polyadenylated (polyA) tail for transcriptional termination (Figure 37.2). Some pre-mRNAs are alternatively spliced or edited at this step, involving alteration of the nucleotide sequence in the mature mRNA transcript (see Section 37.2).

Mature mRNAs are transported out of the nucleus to the cytoplasm. Most mRNAs are translated on ribosomes of the endoplasmic reticulum, but in neurons, local protein synthesis has also been demonstrated from transcripts distally translocated to cellular compartments such as dendrites. mRNA localized to dendrites includes >60 voltage- and ligand-gated ion channels (Cajigas et al., 2012). On the ribosome, translation commences with formation of the ternary initiation complex, generating the polypeptide protein. Fragile X mental retardation protein (FMRP), which alters neuronal activity when mutated, has been shown to inhibit translation of a subset of mRNAs, including $K_V3.1$ and $K_V4.2$ channels (Strumbos et al., 2010; Lee et al., 2011).

37.2 Pre-mRNA PROCESSING

37.2.1 SPLICEOSOME

Splicing of the pre-mRNA to remove introns generally occurs co-transcriptionally, catalyzed by the spliceosome RNA-protein complex (comprehensively reviewed in Lewin, 2007; Li et al., 2007; and Kornblihtt et al., 2013). The consensus RNA sequences that define a splice site are found in the intron and typically consist of the bases GU at the 5′ end of the splice site, the branchpoint sequence with a conserved A residue, a polypyrimidine-tract near the 3′ end of the intron, and the bases AG at the 3′ end of the splice site. These sequences recruit the canonical five small nuclear riboparticles (snRNPs) of the major spliceosome and several accessory proteins to the splice site, resulting in intron excision and joining of exons (Figure 37.3). These steps comprise the basic events of constitutive splicing to remove introns.

Alternative splicing can arise from transition of a constitutive to an alternative exon, but acquisition of adaptive functional diversity is also linked to exon shuffling and exonization of genomic sequence (Keren et al., 2010). Using the same basic spliceosome machinery as constitutive splicing, the major types of alternative splicing are (1) exon skipping, (2) mutually exclusive exons, (3) alternative 5′ or 3′ splicing sites, and (4) intron retention (Figure 37.4). One key determinant of alternative splicing events is the *strength* of the splice site sequences (Keren et al., 2010). Constitutive splicing is mediated by strong consensus splice site sequences, while alternative splice sites deviate from these consensus sequences (weak). Sites of alternative splicing are subject to further regulation by additional exonic and intronic splicing enhancers and silencers. Members of the Ser/Arg-rich proteins (SRs) bind to enhancer sequences in exons and activate adjacent splice sites. Heterogeneous nuclear ribonucleoproteins (hnRNPs) factors suppress splicing via intronic silencers. Regulation of splicing may also occur through interactions with the transcriptional machinery, where the two processes may be functionally coupled (Kornblihtt et al., 2013). Transcriptional kinetics during RNA polymerase-mediated elongation is another determinant of alternative splicing. Sequences that cause pausing of the polymerase are associated with increased exon retention, while faster elongation is associated with exon skipping. These effects are context-dependent, contingent upon the strength of the 5′ and 3′ splice sequences. Similar to the transcription factors that initiate RNA synthesis in a tissue-specific manner, these additional control elements differ by tissue and cause deviation from the default constitutive pattern of splicing in a regulated manner.

Several factors have been implicated in ion channel alternative splicing (Figure 37.5). Nova1 and Nova2 are neuron-specific splicing factors in the KH-type RNA-binding family, related to hnRNPs (Li et al., 2007). Alternative exons are included when Nova binds to the downstream intron and excluded when Nova binds within exons or to the upstream intron. Nova targets include GlyRα2, $GABA_AR\gamma2$, $Ca_V2.1$, and $Ca_V2.2$ (Buckanovich and Darnell, 1997; Dredge and Darnell, 2003; Ule et al., 2006; Allen et al., 2010). The splicing factors, rbFox and PTB, regulate additional sites of alternative splicing in Ca_V1, Ca_V2, and $Na_V1.6$ genes (Tang et al., 2009, 2011; Gehman et al., 2011; O'Brien et al., 2012). The actions of these proteins families may be coordinated as suggested by the ability of rbFox2 to regulate Nova1 expression (Zhang et al., 2010).

37.2.2 RNA EDITING

Splicing of pre-mRNA is not the only mechanism for altering the transcripts produced by a single gene. Alterations in transcript sequence by RNA editing also mediate functional variation in proteins. The main form of RNA editing in mammals is an adenosine to inosine base change, an irreversible deamination of

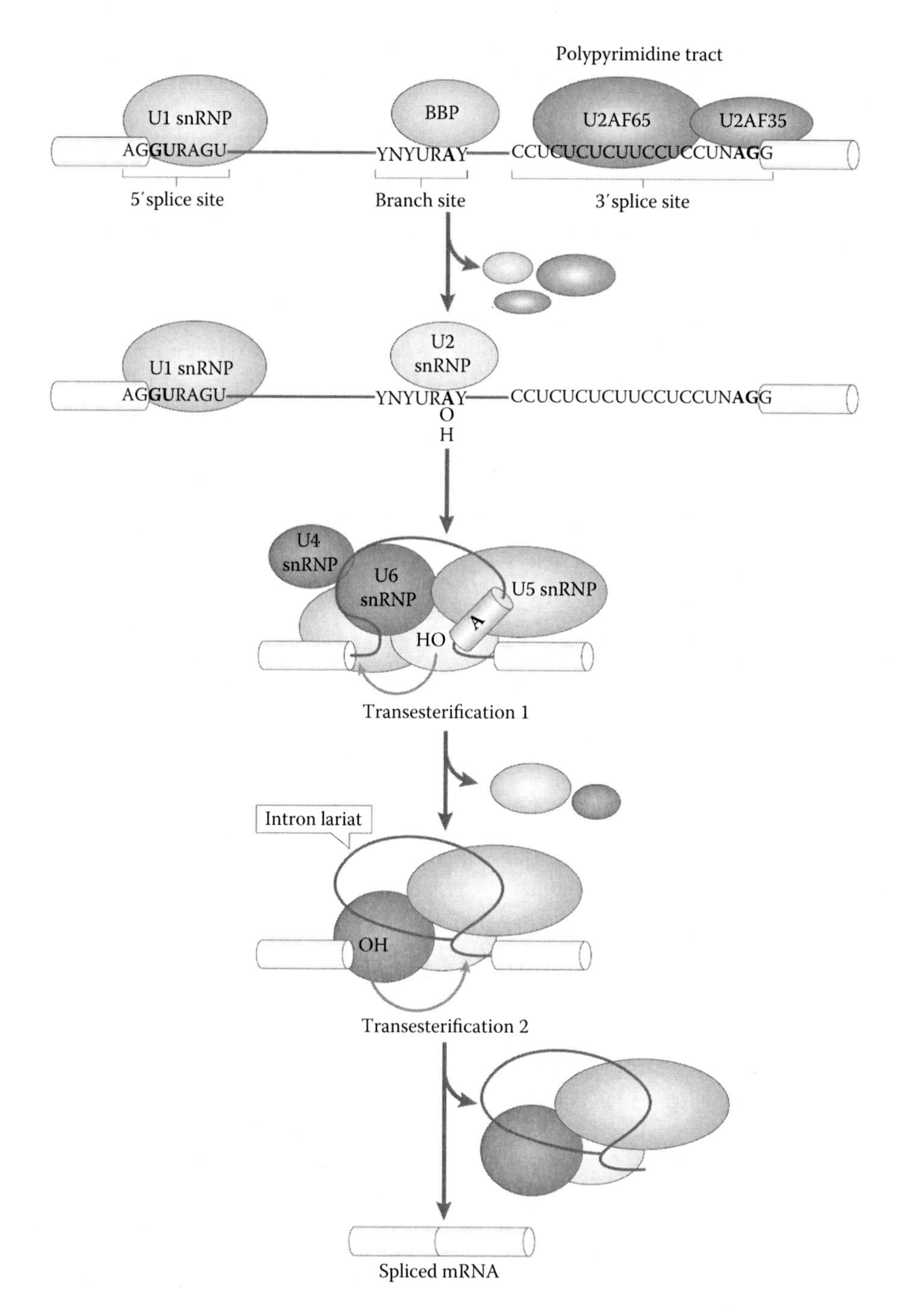

Figure 37.3 The splicing reaction. The spliceosome is formed by the recruitment of the snRNPs U1, U2, U4, U5, U6 and the factors U2AF65, U2AF35, and the branchpoint-binding protein (BBP). Activation of the spliceosome occurs through a series of protein rearrangements (Li et al., 2007; Kornblihtt et al., 2013), bringing the U6 snRNP into contact with the 5′ splice site and catalyzing a two-step transesterification reaction. In the first reaction, a specific branchpoint nucleotide makes a nucleophilic attack on the first nucleotide of the intron at the 5′ splice site, forming the lariat intermediate. In the second reaction, the 3′ OH of the freed 5′ exon makes a nucleophilic attack at the last nucleotide of the 3′ splice site in the intron. This second step joins the exons and releases the intron lariat. (Reproduced from Kornblihtt, A.R. et al., *Nat. Rev. Mol. Cell. Biol.*, 14, 153, 2013. With permission.)

adenosine catalyzed by ADARs (adenosine deaminases acting on RNA). Two ADAR enzymes are widely expressed, but ADAR3 appears to be restricted to the nervous system (Nishikura, 2010). After editing, inosine is read as guanosine during translation, creating an alteration of the protein sequence. However, the base changes introduced by editing have also been proposed to affect RNA secondary structure, localization, alternative splicing, and RNAi efficacy.

The first examples of RNA editing in mammals were in the GluR2, GluR5, and GluR6 subunits in brain, where A to I editing changes a glutamine to an arginine (Q/R site) in the pore-loop of the receptors (Table 37.1; Sommer et al., 1991). The Q/R site essentially 100% edited in GluR2 subunits, while other subunits show variable levels of edited transcripts. The presence of the charged arginine restricts Ca^{2+} permeability, fundamentally altering the properties of channels resulting from the edited transcripts

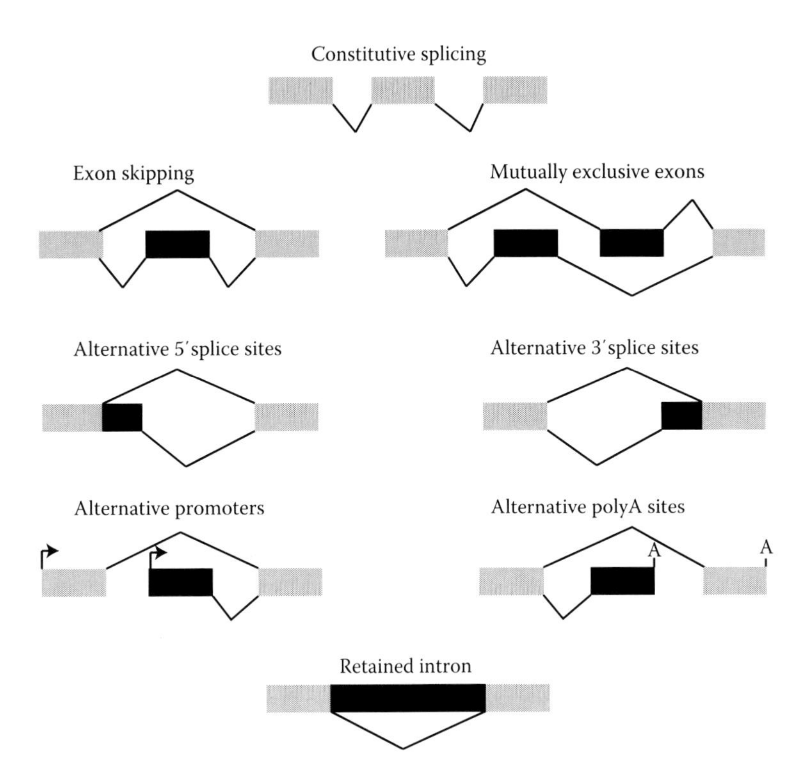

Figure 37.4 Types of alternative splicing. The most common type of alternative splicing event is exon skipping, occurring in about 40% of the splicing events in higher eukaryotic transcripts. (From Keren, H. et al., *Nat. Revs. Genet.*, 11, 345, 2010). Most alternatively spliced ion channels undergo this type of event. In exon-skipping events, a cassette exon is spliced out of the pre-mRNA along with its flanking introns. Less common are mutually exclusive exons, exemplified by inclusion of e37a or e37b in Ca_V2 transcripts (Gray et al., 2007), and the recognition of two or more 5′ or 3′ splice sites within an exon, resulting in short and long versions of the alternative exon. Other less frequent events include intron retention, found less than 5% of spliced transcripts including the BK channel (Bell et al., 2008), and use of alternative promoter or polyadenylation sites. (From Emerick, M.C. et al., *Proteins*, 64, 320, 2006; Brown, M.R. et al., *J. Physiol.*, 586, 5161, 2008; Gong, Q. et al., *J. Biol. Chem.*, 285, 32233, 2010.) Trans-splicing, not typically found in mammalian transcriptomes, is a spliceosome-mediated event between two transcripts. Rescue of genetic diseases by trans-splicing is under development for mutations in CFTR. (From Liu, X. et al., *Nat. Biotechnol.*, 20, 47, 2002.)

(Lomeli et al., 1994). Furthermore, a mechanistic relationship between RNA editing and downstream alternative splice site choice has been established in GluR2 (Schoft et al., 2007). RNA editing has also been discovered in other ion channel transcripts such as *Drosophila* Na^+ (*para*), eag, and Ca_V2 (*cacophony*) channels (Smith et al., 1996; Reenan et al., 2000; Hoopengardner et al., 2003) and mammalian $K_V1.1$, $Ca_V1.3$, and $GABA_A$ channels (Bhalla et al., 2004; Ohlson et al., 2007; Huang et al., 2012).

37.3 PHYSIOLOGICAL THEMES INVOLVING ALTERNATIVE SPLICING

Expression of specific ion channel variants is a determining factor in the physiology of both static (e.g., functional differences between species or tissues) and dynamic (development, homeostasis, and activity dependent) processes. Several of the following systems serve as exemplars, where identified ion channel variants have a quantifiable association with physiological function.

37.3.1 SPECIES OR TISSUE-SPECIFIC ALTERNATIVE SPLICING

The types of alternative splicing events within genomes and the patterns of alternative exon expression within genes differ between species (Keren et al., 2010). Constitutive exons are more conserved across species than alternative exons, exemplified by the extensively alternatively spliced BK K^+ channel (Figure 37.6; Fodor and Aldrich, 2009). Despite the lack of conservation of alternative splicing between species, sites of alternative splicing are generally conserved at functionally recognized domain boundaries. Sequence alterations at these positions may be more tolerated, allowing functional integrity to be preserved (Keren et al., 2010). As the genomes evolved across species, the sequence variation introduced by alternative splicing probably yielded variants with little influence on function at first. As a result, negligible functional consequences and a low prevalence of the splicing event would not promote negative selection. Over time, as these alternative sequences acquired mutations, their presence in transcripts could be enhanced if they conferred beneficial function, leading to stable maintenance of the splicing event.

In vampire bats, adjusted thermal activation of the capsaicin receptor, the ion channel TRPV1, for infrared detection of prey provides an elegant example of species-specific alternative splicing. While snakes and other blood feeders evolved an infrared detector from a non-heat-sensitive channel (TRPA1), vampire bats produce an alternatively spliced TRPV1 channel with a truncated C-terminus. TRPV1-short variants have a lowered thermal activation threshold compared to the long variant, which provides classical noxious heat detection (Gracheva et al., 2011). In addition to the species-specific acquisition of this splicing event, it also occurs in a tissue-specific pattern. TRPV1-short variants are

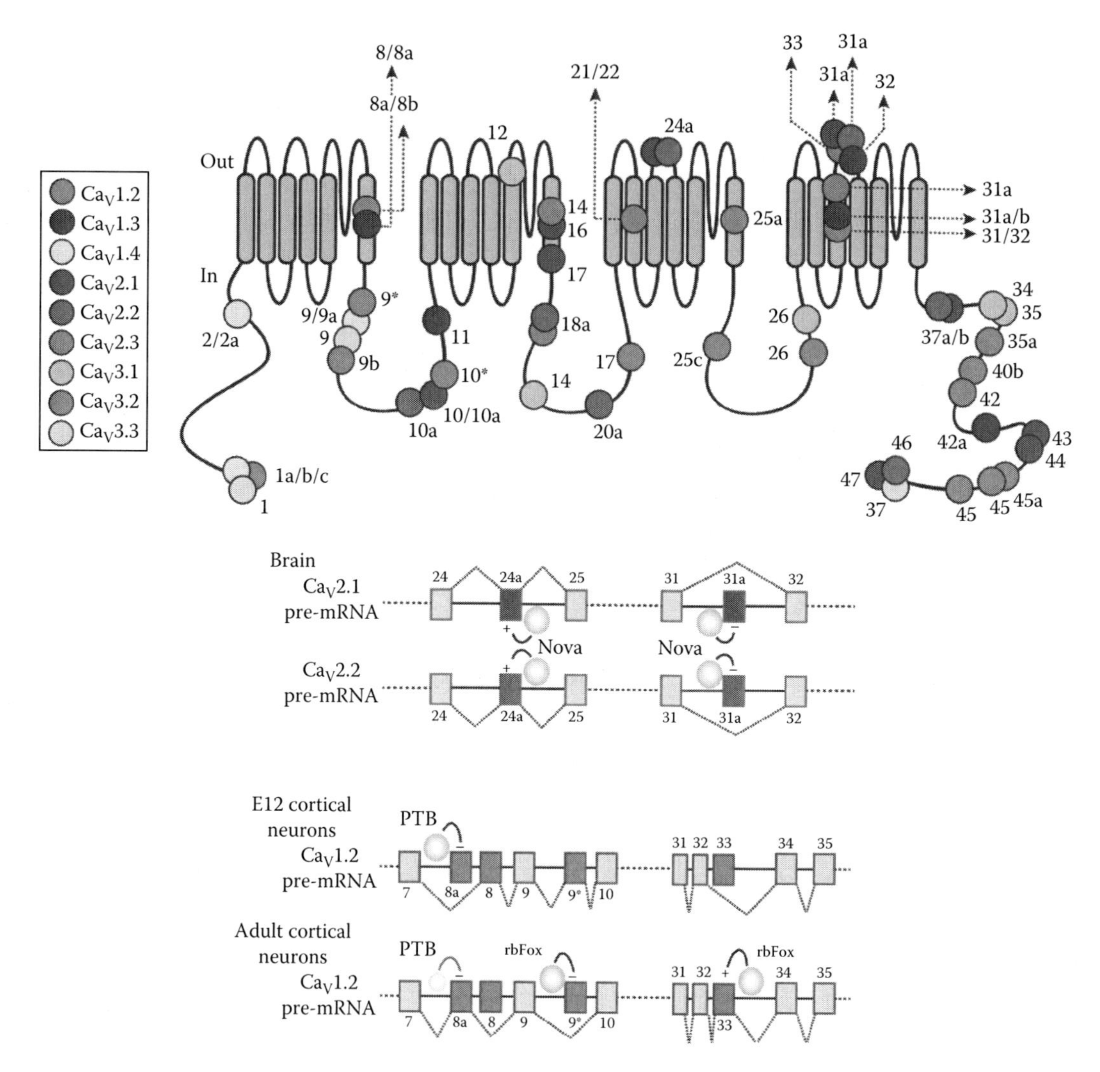

Figure 37.5 Ca_V alternative splicing. Sites of alternative splicing mapped onto Ca_V exons (top; comprehensively reviewed in Lipscombe, D. et al., Alternative splicing of neuronal Ca_V2 calcium channels, in: Gribkoff, V.K. and Kaczmarek, L.K. (eds.), *Structure, Function, and Modulation of Neuronal Voltage-Gated Ion Channels*, John Wiley & Sons, Hoboken, NJ, 2009; Lipscombe, D. et al., *Trends Neurosci.*, 36, 598, 2013.) (Bottom) Nova splicing factors are expressed in brain, where they enhance e24a, and suppress e31a, inclusion in $Ca_V2.1$ and $Ca_V2.2$ pre-mRNA. Developmental expression of PTB (polypyrimidine tract-binding protein) suppresses e8a in $Ca_V1.2$ pre-mRNA. PTB levels decline in adult neurons resulting in e8a inclusion. rbFox also mediates developmental regulation of $Ca_V1.2$ splicing, suppressing e9* and enhancing e33 inclusion. (Reproduced from Lipscombe, D. et al., *Trends Neurosci.*, 36, 598, 2013. With permission.)

Table 37.1 Ion channels with RNA editing sites in the coding region

TRANSCRIPT	AA CHANGE	FUNCTIONAL CONSEQUENCE OF EDIT	REFERENCES
GluR2, 5	Q/R	Decreased Ca^{2+} permeability and single channel conductance, linear I–V Severe seizures and early lethality[a]	Sommer et al. (1991); Rosenthal and Seeberg (2012); Brusa et al. (1995)
GluR2, 3, 4	R/G	Faster recovery from desensitization, larger steady-state currents	Lomeli et al. (1994)
GluR6	Q/R	Increased Ca^{2+} permeability NMDA-independent LTP and increased susceptibility to seizures[a]	Kohler et al. (1993); Vissel et al. (2001)
	I/V, Y/C	Q/R context-dependent changes in Ca^{2+} permeability	Kohler et al. (1993)
$GABA_A$-α3	I/M	Reduced peak current, slower activation, and faster deactivation	Ohlson et al. (2007); Rula et al. (2008)
$K_V1.1$ channel	I/V	Faster recovery from inactivation, altered lipid modulation and open channel block	Bhalla et al. (2004); Decher et al. (2010)
$Ca_V1.3$ channel	Q/R, I/M, Y/C	Reduced Ca^{2+}-dependent inactivation Disrupted circadian rhythms in SCN neural activity[a]	Huang et al. (2012)

[a] Phenotype associated with loss of RNA editing.

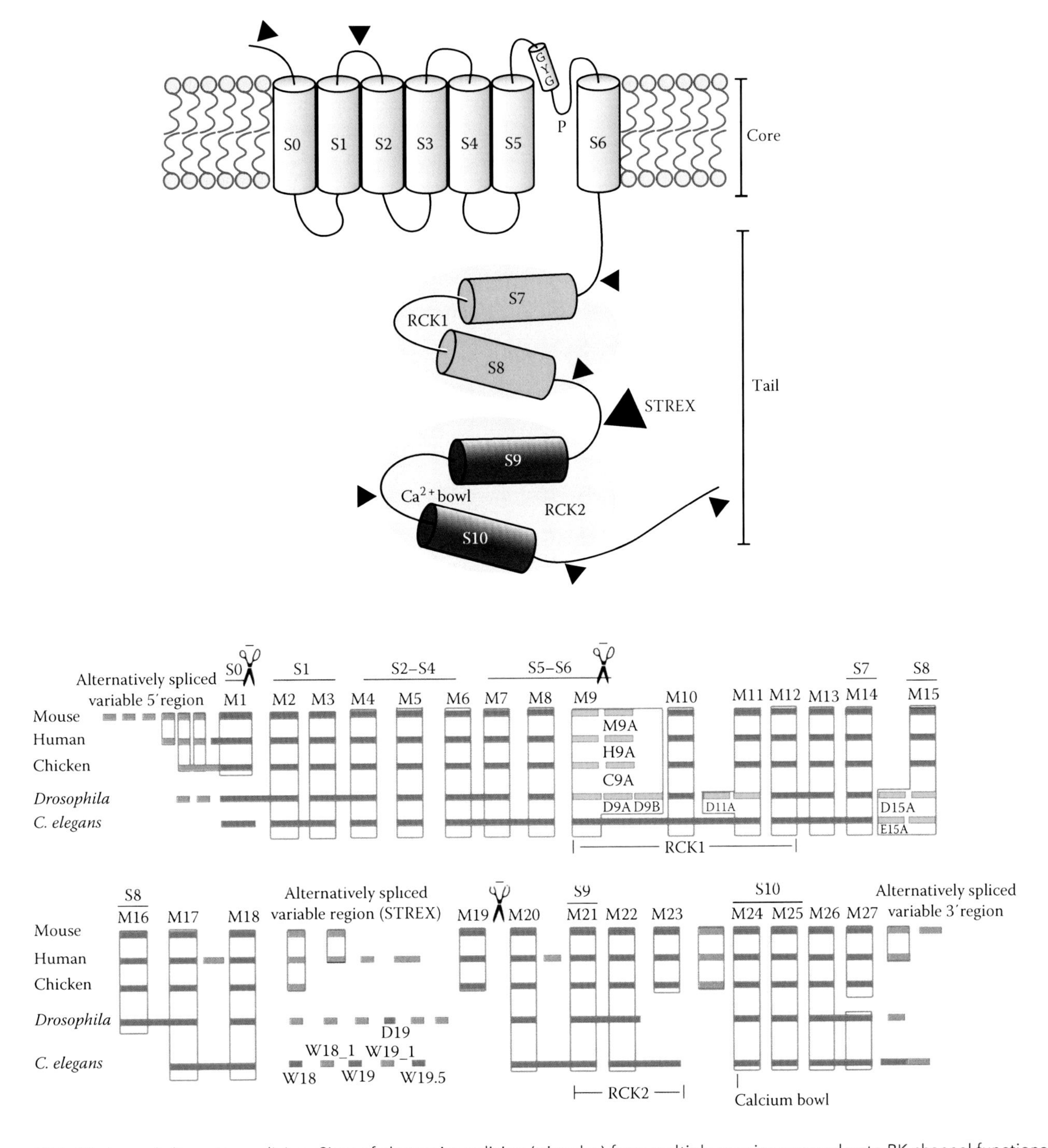

Figure 37.6 BK channel alternative splicing. Sites of alternative splicing (triangles) from multiple species mapped onto BK channel functional domains (top). Bottom: pattern of constitutive (blue), tandem duplicate (green), and alternative (orange) exons in BK channel transcripts from several species. Scissors indicate domain boundary regions where separate transcripts can be coinjected to produce functional channels. Exons encoding less than four amino acids or present in only one transcript are not depicted. (Bottom panel reproduced from Fodor, A.A. and Aldrich, R.W., *Annu. Rev. Physiol.*, 71, 19, 2009. With permission.)

only expressed in trigeminal ganglia where the infrared detection capabilities are housed, preserving the role for TRPV1-long variants as the noxious heat detector in dorsal root ganglia.

Most alternative splicing events exhibit some degree of tissue-specificity, associated with the overall tissue-specificity of transcriptional control. Alternative splicing of ion channel transcripts underlies the overt differences in electrical properties between tissues, muscle vs. neuron, for example, but also provides more precise fine-tuning of membrane properties. The expression of BK K^+ channel splice variants in the nonmammalian cochlea is a classic example (Fettiplace and Fuchs, 1999). In hair cells, electrical tuning in response to sound is the product of membrane potential oscillations that vary across the tonotopic axis of the cochlea. The basic oscillation is driven by depolarization-induced Ca^{2+} influx, which activates the Ca^{2+}- and voltage-sensitive BK channels, hyperpolarizing the membrane. The continuum of resonant frequencies across the tonotopic axis is correlated with spatial alteration of BK gating kinetics, where cells responding to higher frequencies exhibit faster BK kinetics. These differences in gating kinetics are in part driven by alternative splicing at six sites in the BK channel (Navaratnam et al., 1997; Rosenblatt et al., 1997; Jones et al., 1998, 1999; Ramanathan et al., 1999).

37.3.2 DEVELOPMENT

Development is a highly dynamic process, genetically programmed by concerted transcriptional regulation. These temporal patterns of splicing are tied to transcriptional control, changing with lineage commitment, differentiation, and maintenance of the mature phenotype. The expression of particular classes of ion channels plays a major role in the transitions that occur during the acquisition of a cellular identity. For example, *Xenopus* BK channel variants (*xSlo*) are expressed during embryogenesis prior to differentiation of excitability. However, in spinal motor neurons, the neural-specific *xSlo59* variant is absent during early development and becomes upregulated during functional maturation, when the neurons transition to firing short action potentials (Kukuljan et al., 2003). BK currents produced by the *xSlo59* variant are activated at lower voltages, correlated with the faster repolarization phase and the generation of afterhyperpolarizations observed during spinal neuron maturation. Another example is the developmental expression of *flip* or *flop* variants in GluR1–4 transcripts, affecting the rate of receptor desensitization (Sommer et al., 1990; Mosbacher et al., 1994). Prenatal expression of *flip* variants implicate enhanced glutamate-induced current during circuit formation, while distinct postnatal onset of *flop* expression correlates with reduced excitability, circuit consolidation, and maintenance of different neuronal types (Monyer et al., 1991).

Developmental patterns of alternative splicing have been detected for almost every class of ion channel. Ca^{2+} channels provide several additional examples. In $Ca_V1.2$ (L-type) transcripts, the developmental abundance of alternative exons 8/8a, 9*, and 33 have been linked to expression of Fox and PTB splicing factors (Figure 37.5; Tang et al., 2009, 2011). In rat superior cervical ganglion neurons, which exclusively use $Ca_V2.2$ (N-type) channels for transmitter release, the expression of e18a increases during development, coinciding with synaptogenesis onto target neurons (Gray et al., 2007). Interestingly, in mature brain, the corresponding e18a alternative exon in $Ca_V2.3$ is reciprocally downregulated, suggesting a developmentally coordinated regulatory mechanism for alternative splicing in $Ca_V2.2$ and $Ca_V2.3$ transcripts. In $Ca_V2.1$ (P/Q type) channels, coordinated expression of alternative exons 37 and 47 result in three variants of an EF hand domain with the potential to confer a developmental shift in Ca^{2+}-dependent facilitation (Chaudhuri et al., 2004).

37.3.3 ACTIVITY AND HORMONE-DEPENDENT SPLICING

Physiology and behavior are dynamically driven, and postdevelopmentally, patterns of alternative splicing in ion channels respond to changes in cellular activity and environmental conditions. Alternative splicing of two C-terminal cassettes (C2 and C2′) in the NMDA receptor subunit NR1 is linked to synaptic activity and plasticity in cortical neurons (Mu et al., 2003). Increasing activity with bicuculline decreased C2′ inclusion in transcripts. Conversely, blocking activity with tetrodotoxin favored C2′ inclusion, enhancing ER export and channel trafficking to postsynaptic structures. In GluR1 and GluR2 AMPA receptor transcripts, chronic reduction of synaptic activity in CA1, but not CA3 hippocampal neurons, increased inclusion of the *flop* exon in the ligand binding domain. Flop variants produced altered channel kinetics and dimerization properties, correlated with increased synaptic fidelity, proposed to underlie the homeostatic response to activity deprivation (Penn et al., 2012).

Outside of the synapse, ion channels involved in intrinsic cellular excitability are also subject to activity-dependent regulation of alternative splicing. Depolarization leads to decreased stress-regulated exon (STREX) inclusion in BK channel transcripts and decreased C1 cassette inclusion in NR1 (Xie and Black, 2001; An and Grabowski, 2007; Lee et al., 2007). These splicing events are downstream of calcium-/calmodulin-dependent kinase (CamK) activation, which represses STREX or C1 inclusion via CamKIV-responsive RNA elements (CaRRE) in the alternative exons. Depolarization and CaRRE motifs are found in proteins involved in calcium signaling (Li et al., 2007).

Other physiological stimuli, such as hormones, growth factors, cytokines, nutrients, and ethanol, can influence alternative splicing. For example, the expression of STREX-containing BK channel variants is tied to the systemic hypothalamus-pituitary-adrenal (HPA) stress response. Release of adrenocorticotropic hormone (ACTH) from the pituitary promotes STREX inclusion in BK transcripts expressed in adrenal chromaffin cells (Xie and McCobb, 1998). STREX variants deactivate more slowly and may influence the repetitive firing linked to epinephrine-secreting cells. Ethanol has also been shown to modulate alternative splicing of NR1, $GABA_A$, $Ca_V1.2$, and $Ca_V2.2$ channels (Hardy et al., 1999; Walter et al., 2000; Petrie et al., 2001; Nagy et al., 2003; Newton et al., 2005; Raeder et al., 2008). In BK channel transcripts, ethanol exposure in the brain was found to downregulate a subset of variants containing either the STREX or ALCOREX alternative exons. However, the mechanism relies on ethanol-dependent increase in miR-9 micro-RNA and selective degradation of the variants, rather than spliceosome-mediated regulation of exon inclusion (Pietrzykowski et al., 2008).

37.3.4 CIRCADIAN REGULATION OF SPLICING

Accumulating evidence has linked alternative splicing to other physiological control mechanisms. One ubiquitous intrinsic control system is the daily circadian (~24 h) rhythm, a dynamic, transcriptionally regulated process that drives a major component of predictive homeostasis in living organisms. Circadian entrainment allows appropriate phasing of cyclic fluctuations in physiology and behavior to the day-night cycle. Autonomous rhythms in pacemaker tissues, such as the suprachiasmatic nucleus of the hypothalamus in mammals and the ~150 clock neurons in *Drosophila*, are mediated by an evolutionarily conserved transcriptional feedback loop involving *clock genes* (King and Takahashi, 2000). Up to 10% of the genome is subject to circadian control (Panda et al., 2002), including factors involved in alternative splicing and RNA processing (Sanchez et al., 2010; Hughes et al., 2012).

The *Drosophila* clock gene *Period* undergoes alternative splicing (Majercak et al., 1999; Sanchez et al., 2010), and conversely, mutating *Period* alters the pattern of alternative splicing and RNA editing in a broad set of transcripts (Hughes et al., 2012) suggesting RNA processing is a central feature of the core clock mechanism. This core clock mechanism in turn drives circadian variation in membrane excitability via regulation of ion channel expression and function (Meredith et al., 2006; Ko et al., 2009; Colwell, 2011). In the suprachiasmatic nucleus, day-night differences in delayed rectifier K^+ currents, the BK Ca^{2+}-activated K^+ current, L-type Ca^{2+} current, and TEA-sensitive K^+ currents are involved in the circadian patterning of action potentials. Outside of the pacemaker, other brain areas and peripheral tissues also exhibit circadian differences in excitability (or ion homeostasis), notably in long-term potentiation in the hippocampus (Chaudhury et al., 2005), the Q-T interval of cardiac conduction (Jeyaraj et al., 2012), and renal Na^+ reabsorption (Zuber et al., 2009).

Ion channel transcripts have been identified in screens for alternative splicing events and RNA editing in *Drosophila* and mammalian circadian pacemakers (Sanchez et al., 2010; Hughes et al., 2012). In the suprachiasmatic nucleus, a splice variant of the BK channel that exhibits reduced current in response to action potential stimuli is more abundant during the day, suggesting alternative splicing facilitates high-resolution control of cyclic alterations in membrane excitability (Shelley et al., 2013), and RNA editing of $Ca_V1.3$ transcripts is linked to circadian rhythms in firing of suprachiasmatic nucleus neurons (Huang et al., 2012).

37.3.5 ALTERNATIVE SPLICING AND DISEASE

Missplicing can result in disease, either by altering the function or expression of proteins. Genetic mutations associated with missplicing are proposed to be a factor in >15% of human diseases (Wang and Cooper, 2007). In addition, single nucleotide polymorphisms (SNPs) also play a role in splicing-based disease susceptibility. Deleterious (due to mutation) or subtle (due to SNPs) changes in transcript sequences cause disruption of splice sites, creation of new cryptic splice sites, or alterations in RNA secondary structure that affect patterns of alternative splicing. Furthermore, disruption of trans factors, the proteins that regulate splicing, also cause splicing-based disease, such as spinal muscular atrophy, retinitis pigmentosa, and Prader–Willi syndrome.

Several ion channel-based diseases are associated with altered expression of particular splice variants or missplicing events. For example, a SNP in *hERG* (*KCNH2*) has been linked to schizophrenia, associated with altered expression of a primate- and brain-specific *hERG* splice variant with unique properties (Huffaker et al., 2009). Altered expression or function of several Ca^{2+} channels are linked to missplicing, including $Ca_V1.2$ with Timothy Syndrome and heart failure, $Ca_V2.1$ with spinocerebellar ataxia, epilepsy, and migraine, and $Ca_V2.2$ with neuropathic pain (Liao et al., 2009). Mutations and SNPs in CFTR are also associated with altered splicing patterns. Exon 9 skipping is associated with both classical cystic fibrosis and the symptomatic penetrance in individuals with atypical cystic fibrosis due to less severe CFTR mutations (Chu et al., 1993; Steiner et al., 2004). R553X, the third most common CFTR mutation, has been proposed to cause skipping of exon 11 by creating a silencer near a 5′ splice site (Aznarez et al., 2007). Interestingly, even synonymous SNPs, which are generally assumed to be functionally neutral, cause altered splicing with deleterious consequences on CFTR function (Pagani et al., 2005).

37.4 SUMMARY

Given the prevalence of alternative splicing among ion channels, identification of full-length variants and quantification of their relative abundance in tissues remains a major aim (Table 37.2). In transcripts with several sites of alternative splicing, such as BK and Ca^{2+} channels, the theoretical number of transcripts is in the thousands. In tissues, identification of the full repertoire

Table 37.2 Techniques for identification and quantitation of splice variants

TECHNIQUE	REFERENCES
Library screening	Moriyoshi et al. (1991); McCobb et al. (1995); Regan et al. (2000)
Semi-quantitative, real-time RT-PCR	Glauser et al. (2011); Johnson et al. (2011); Chen and Shipston (2008); Kodama et al. (2012)
5′ or 3′ RACE	Brown et al. (2008); Gong et al. (2010)
In situ hybridization	Monyer et al. (1991); Rosenblatt et al. (1997)
Exon tiling or exon–exon junction microarrays	Johnson et al. (2003); Kwan et al. (2008)
Transcript (EST) alignment	NCBI Database: http://www.ncbi.nlm.nih.gov/gene, http://www.ncbi.nlm.nih.gov/unigene
Whole transcriptome sequencing (RNAseq)	Wang et al. (2008); Pan et al. (2008); Mortazavi et al. (2008)
Computational genomic sequence analysis	Barash et al. (2010); Zhang et al. (2010); Mudge et al. (2013) Encode database: http://www.genome.gov/10005107
Splice variant specific antibodies	Pereverzev et al. (1998); Bannister et al. (2011); Tan et al. (2011)
Mass spectrometric analyses	Yan et al. (2008); Singh et al. (2013)
Transgenic alternative splicing reporters	Mottes and Iverson (1995); Kuroyanagi et al. (2006)

of actual alternative exon combinations has been challenging (Emerick et al., 2006; Glauser et al., 2011; Johnson et al., 2011). For example, $Ca_V3.1$, comprised of 38 exons, has 15 sites of transcript variation, including two alternative promoters (5′ UTR), two alternative polyadenylation sites (3′ UTR), and 11 sites of alternative splicing (Emerick et al., 2006). Nevertheless, traditional molecular biological approaches involving isolation of RNA from tissues and cloning splice variants by RT-PCR are the mainstay of most experimental investigations in this area, providing a more direct link to the physiology of the parent tissue than computational approaches. Comprehensive collections of ion channel alternative variants expressed in particular cell and tissue types will reveal both the transcriptional solutions to coordinate regulation of functional modules in proteins as well as the linkage to physiological activities.

REFERENCES

Allen, S.E., Darnell, R.B., and Lipscombe, D. 2010. The neuronal splicing factor Nova controls alternative splicing in N-type and P-type Ca_V2 calcium channels. *Channels* 4, 483–489.

An, P. and Grabowski, P.J. 2007. Exon silencing by UAGG motifs in response to neuronal excitation. *PLoS Biol* 5, e36.

Aznarez, I., Zielenski, J., Rommens, J.M. et al. 2007. Exon skipping through the creation of a putative exonic splicing silencer as a consequence of the cystic fibrosis mutation R553X. *J Med Genet* 44, 341–346.

Bannister, J.P., Thomas-Gatewood, C.M., Neeb, Z.P. et al. 2011. $Ca_V1.2$ channel N-terminal splice variants modulate functional surface expression in resistance size artery smooth muscle cells. *J Biol Chem* 286, 15058–15066.

Barash, Y., Calarco, J.A., Gao, W. et al. 2010. Deciphering the splicing code. *Nature* 465, 53–59.

Bell, T.J., Miyashiro, K.Y., Sul, J.Y. et al. 2008. Cytoplasmic BK_{Ca} channel intron-containing mRNAs contribute to the intrinsic excitability of hippocampal neurons. *Proc Natl Acad Sci USA* 105, 1901–1906.

Bhalla, T., Rosenthal, J.J., Holmgren, M. et al. 2004. Control of human potassium channel inactivation by editing of a small mRNA hairpin. *Nat Struct Mol Biol* 11, 950–956.

Brown, M.R., Kronengold, J., Gazula, V. et al. 2008. Amino-termini isoforms of the Slack K^+ channel, regulated by alternative promoters, differentially modulate rhythmic firing and adaptation. *J Physiol* 586, 5161–5179.

Brusa, R., Zimmermann, F., Koh, D.S. et al. 1995. Early-onset epilepsy and postnatal lethality associated with an editing-deficient GluR-B allele in mice. *Science* 270, 1677–1680.

Buckanovich, R.J. and Darnell, R.B. 1997. The neuronal RNA binding protein Nova-1 recognizes specific RNA targets in vitro and in vivo. *Mol Cell Biol* 17, 3194–3201.

Cajigas, I.J., Tushev, G., Will, T.J. et al. 2012. The local transcriptome in the synaptic neuropil revealed by deep sequencing and high-resolution imaging. *Neuron* 74, 453–466.

Chaudhuri, D., Chang, S.Y., DeMaria, C.D. et al. 2004. Alternative splicing as a molecular switch for Ca^{2+}/calmodulin-dependent facilitation of P/Q-type Ca^{2+} channels. *J Neurosci* 24, 6334–6342.

Chaudhury, D., Wang, L.M., and Colwell, C.S. 2005. Circadian regulation of hippocampal long-term potentiation. *J Biol Rhythms* 20, 225–236.

Chen, L. and Shipston, M.J. 2008. Cloning of potassium channel splice variants from tissues and cells. *Methods Mol Biol* 491, 35–60.

Chong, J.A., Tapia-Ramírez, J., Kim, S. et al. 1995. REST: A mammalian silencer protein that restricts sodium channel gene expression to neurons. *Cell* 80, 949–957.

Chu, C.S., Trapnell, B.C., Curristin, S. et al. 1993. Genetic basis of variable exon 9 skipping in cystic fibrosis transmembrane conductance regulator mRNA. *Nat Genet* 3, 151–156.

Colwell, C.S. 2011. Linking neural activity and molecular oscillations in the SCN. *Nat Rev Neurosci* 12, 553–969.

Decher, N., Streit, A.K., Rapedius, M. et al. 2010. RNA editing modulates the binding of drugs and highly unsaturated fatty acids to the open pore of Kv potassium channels. *EMBO J* 29, 2101–2113.

Dredge, B.K. and Darnell, R.B. 2003. Nova regulates $GABA_A$ receptor γ2 alternative splicing via a distal downstream UCAU-rich intronic splicing enhancer. *Mol Cell Biol* 23, 4687–4700.

Emerick, M.C., Stein, R., Kunze, R. et al. 2006. Profiling the array of $Ca_V3.1$ variants from the human T-type calcium channel gene *CACNA1G*: Alternative structures, developmental expression, and biophysical variations. *Proteins* 64, 320–342.

Fettiplace, R. and Fuchs, P.A. 1999. Mechanisms of hair cell tuning. *Annu Rev Physiol* 61, 809–834.

Fodor, A.A. and Aldrich, R.W. 2009. Convergent evolution of alternative splices at domain boundaries of the BK channel. *Annu Rev Physiol* 71, 19–36.

Gehman, L.T., Stoilov, P., Maguire, J. et al. 2011. The splicing regulator Rbfox1 (A2BP1) controls neuronal excitation in the mammalian brain. *Nat Genet* 43, 706–711.

Glauser, D.A., Johnson, B.E., Aldrich, R.W. et al. 2011. Intragenic alternative splicing coordination is essential for *Caenorhabditis elegans slo-1* gene function. *Proc Natl Acad Sci USA* 108, 20790–20795.

Gong, Q., Stump, M.R., Dunn, A.R. et al. 2010. Alternative splicing and polyadenylation contribute to the generation of hERG1 C-terminal isoforms. *J Biol Chem* 285, 32233–32241.

Gracheva, E.O., Cordero-Morales, J.F., González-Carcacía, J.A. et al. 2011. Ganglion-specific splicing of TRPV1 underlies infrared sensation in vampire bats. *Nature* 476, 88–91.

Gray, A.C., Raingo, J., and Lipscombe, D. 2007. Neuronal calcium channels: Splicing for optimal performance. *Cell Calcium* 42, 409–417.

Hardy, P.A., Chen, W., and Wilce, P.A. 1999. Chronic ethanol exposure and withdrawal influence NMDA receptor subunit and splice variant mRNA expression in the rat cerebral cortex. *Brain Res* 819, 33–39.

Hille, B. 2001. *Ion Channels of Excitable Membranes*, 3rd edn. Sunderland, MA: Sinauer Associates.

Hoopengardner, B., Bhalla, T., Staber, C. et al. 2003. Nervous system targets of RNA editing identified by comparative genomics. *Science* 301, 832–836.

Huang, H., Tan, B.Z., Shen, Y. et al. 2012. RNA editing of the IQ domain in $Ca_V1.3$ channels modulates their Ca^{2+}-dependent inactivation. *Neuron* 73, 304–316.

Huffaker, S.J., Chen, J., Nicodemus, K.K. et al. 2009. A primate-specific, brain isoform of *KCNH2* affects cortical physiology, cognition, neuronal repolarization and risk of schizophrenia. *Nat Med* 15, 509–518.

Hughes, M.E., Grant, G.R., Paquin, C. et al. 2012. Deep sequencing the circadian and diurnal transcriptome of *Drosophila* brain. *Genome* 22, 1266–1281.

Jeyaraj, D., Haldar, S.M., Wan, X. et al. 2012. Circadian rhythms govern cardiac repolarization and arrhythmogenesis. *Nature* 483, 96–99.

Johnson, J.M., Castle, J., Garrett-Engele, P. et al. 2003. Genome-wide survey of human alternative pre-mRNA splicing with exon junction microarrays. *Science* 302, 2141–2144.

Johnson, B.E., Glauser, D.A., Dan-Glauser, E.S. et al. 2011. Alternatively spliced domains interact to regulate BK potassium channel gating. *Proc Natl Acad Sci USA* 108, 20784–20789.

Jones, E.M., Gray-Keller, M., Fettiplace, R. et al. 1999. The role of Ca^{2+}-activated K^+ channel spliced variants in the tonotopic organization of the turtle cochlea. *J Physiol* 518, 653–665.

Jones, E.M.C., Laus, C., and Fettiplace, R. 1998. Identification of Ca^{2+}-activated K^+ channel splice variants and their distribution in the turtle cochlea. *Proc R Soc London* 265, 685–692.

Keren, H., Lev-Maor, G., and Ast, G. 2010. Alternative splicing and evolution: Diversification, exon definition, and function. *Nat Revs Genet* 11, 345–355.

King, D.P. and Takahashi J.S. 2000. Molecular genetics of circadian rhythms in mammals. *Annu Rev Neurosci* 23, 713–742.

Ko, G.Y., Shi, L., and Ko, M.L. 2009. Circadian regulation of ion channels and their functions. *J Neurochem* 110, 1150–1169.

Kodama, T., Guerrero, S., Shin, M. et al. 2012. Neuronal classification and marker gene identification via single-cell expression profiling of brainstem vestibular neurons subserving cerebellar learning. *J Neurosci* 32, 7819–7831.

Kohler, M., Burnashev, N., Sakmann, B. et al. 1993. Determinants of Ca^{2+} permeability in both TM1 and TM2 of high affinity kainate receptor channels: Diversity by RNA editing. *Neuron* 10, 491–500.

Kornblihtt, A.R., Schor, I.E., Alló, M. et al. 2013. Alternative splicing: A pivotal step between eukaryotic transcription and translation. *Nat Rev Mol Cell Biol* 14, 153–165.

Kraner, S.D., Rich, M.M., Kallen, R.G. et al. 1998. Two E-boxes are the focal point of muscle-specific skeletal muscle type 1 Na^+ channel gene expression. *J Biol Chem* 273, 11327–11334.

Kraner, S.D., Rich, M.M., Sholl, M.A. et al. 1999. Interaction between the skeletal muscle type 1 Na^+ channel promoter E-box and an upstream repressor element. *J Biol Chem* 274, 8129–8136.

Kukuljan, M., Taylor, A., Chouinard, H. et al. 2003. Selective regulation of *xSlo* splice variants during *Xenopus* embryogenesis. *J Neurophysiol* 90, 3352–3360.

Kuroyanagi, H., Kobayashi, T., Mitani, S. et al. 2006. Transgenic alternative-splicing reporters reveal tissue-specific expression profiles and regulation mechanisms in vivo. *Nat Methods* 3, 909–915.

Kwan, T., Benovoy, D., Dias, C. et al. 2008. Genome-wide analysis of transcript isoform variation in humans. *Nat Genet* 40, 225–231.

Lee, H.Y., Ge, W.P., Huang, W. et al. 2011. Bidirectional regulation of dendritic voltage-gated potassium channels by the fragile X mental retardation protein. *Neuron* 72, 630–642.

Lee, J.A., Xing, Y., Nguyen, D. et al. 2007. Depolarization and CaM kinase IV modulate NMDA receptor splicing through two essential RNA elements. *PLoS Biol* 5, e40.

Levine, M. and Tjian, R. 2003. Transcription regulation and animal diversity. *Nature* 424, 147–151.

Lewin, B. 2007. *Genes IX.* Boston, MA: Jones and Bartlett Publishers.

Li, Q., Lee, J., and Black, D.L. 2007. Neuronal regulation of alternative pre-mRNA splicing. *Nature Rev Neurosci* 8, 819–831.

Liao, P., Zhang, H.Y., and Soong, T.W. 2009. Alternative splicing of voltage-gated calcium channels: From molecular biology to disease. *Pflugers Arch* 458, 481–487.

Lipscombe, D., Allen, S.E., Gray, A.C. et al. 2009. Alternative splicing of neuronal Ca_V2 calcium channels. In *Structure, Function, and Modulation of Neuronal Voltage-Gated Ion Channels*, eds. V.K. Gribkoff and L.K. Kaczmarek. Hoboken, NJ: John Wiley & Sons.

Lipscombe, D., Allen, S.E., and Toro, C.P. 2013. Control of neuronal voltage-gated calcium ion channels from RNA to protein. *Trends Neurosci* 36, 598–608.

Liu, X., Jiang, Q., Mansfield, S.G. et al. 2002. Partial correction of endogenous ΔF508 CFTR in human cystic fibrosis airway epithelia by spliceosome-mediated RNA trans-splicing. *Nat Biotechnol* 20, 47–52.

Lomeli, H., Mosbacher, J., Melcher, T. et al. 1994. Control of kinetic properties of AMPA receptor channels by nuclear RNA editing. *Science* 266, 1709–1713.

Majercak, J., Sidote, D., Hardin, P.E. et al. 1999. How a circadian clock adapts to seasonal decreases in temperature and day length. *Neuron* 24, 219–230.

McCobb, D.P., Fowler, N.L., Featherstone, T. et al. 1995. A human calcium-activated potassium channel gene expressed in vascular smooth muscle. *Am J Physiol* 269, 767–777.

Meredith, A.L., Wiler, S.W., Miller, B.H. et al. 2006. BK calcium-activated potassium channels regulate circadian behavioral rhythms and pacemaker output. *Nat Neurosci* 9, 1041–1049.

Monyer, H., Seeburg, P.H., and Wisden, W. 1991. Glutamate-operated channels: Developmentally early and mature forms arise by alternative splicing. *Neuron* 6, 799–810.

Moriyoshi, K., Masu, M., Ishii, T. et al. 1991. Molecular cloning and characterization of the rat NMDA receptor. *Nature* 354, 31–37.

Mortazavi, A., Williams, B.A., McCue, K. et al. 2008. Mapping and quantifying mammalian transcriptomes by RNA-Seq. *Nat Methods* 5, 621–628.

Mosbacher, J., Schoepfer, R., Monyer, H. et al. 1994. A molecular determinant for submillisecond desensitization in glutamate receptors. *Science* 266, 1059–1062.

Mottes, J.R. and Iverson, L.E. 1995. Tissue-specific alternative splicing of hybrid *Shaker*/lacZ genes correlates with kinetic differences in *Shaker* K^+ currents *in vivo*. *Neuron* 14, 613–623.

Mu, Y., Otsuka, T., Horton, A.C. et al. 2003. Activity-dependent mRNA splicing controls ER export and synaptic delivery of NMDA receptors. *Neuron* 40, 5815–5894.

Mudge, J.M., Frankish, A., and Harrow, J. 2013. Functional transcriptomics in the post-ENCODE era. *Genome Res* 23, 1961–1973.

Nagy, J., Kolok, S., Dezso, P. et al. 2003. Differential alterations in the expression of NMDA receptor subunits following chronic ethanol treatment in primary cultures of rat cortical and hippocampal neurones. *Neurochem Int* 42, 35–43.

Navaratnam, D.S., Bell, T.J., Tu, T.D. et al. 1997. Differential distribution of Ca^{2+}-activated K^+ channel splice variants among hair cells along the tonotopic axis of the chick cochlea. *Neuron* 19, 1077–1085.

Newton, P.M., Tully, K., McMahon, T. et al. 2005. Chronic ethanol exposure induces an N-type calcium channel splice variant with altered channel kinetics. *FEBS Lett* 579, 671–676.

Nishikura, K. 2010. Functions and regulation of RNA editing by ADAR deaminases. *Ann Rev Biochem* 79, 321–349.

O'Brien, J.E., Drews, V.L., Jones, J.M. et al. 2012. Rbfox proteins regulate alternative splicing of neuronal sodium channel SCN8A. *Mol Cell Neurosci* 49, 120–126.

Ohlson, J., Pedersen, J.S., Haussler, D. et al. 2007. Editing modifies the $GABA_A$ receptor subunit α3. *RNA* 13, 698–703.

Pagani, F., Raponi, M., and Baralle, F.E. 2005. Synonymous mutations in CFTR exon 12 affect splicing and are not neutral in evolution. *Proc Natl Acad Sci USA* 102, 6368–6372.

Pan, Q., Shai, O., Lee, L.J. et al. 2008. Deep surveying of alternative splicing complexity in the human transcriptome by high-throughput sequencing. *Nature Genet* 40, 1413–1415.

Panda, S., Antoch, M.P., Miller, B.H. et al. 2002. Coordinated transcription of key pathways in the mouse by the circadian clock. *Cell* 109, 307–320.

Penn, A.C., Balik, A., Wozny, C. et al. 2012. Activity-mediated AMPA receptor remodeling, driven by alternative splicing in the ligand-binding domain. *Neuron* 76, 503–510.

Pereverzev, A., Klöckner, U., Henry, M. et al. 1998. Structural diversity of the voltage-dependent Ca^{2+} channel α1E-subunit. *Eur J Neurosci* 10, 916–925.

Petrie, J., Sapp, D.W., Tyndale, R.F. et al. 2001. Altered $GABA_A$ receptor subunit and splice variant expression in rats treated with chronic intermittent ethanol. *Alcohol Clin Exp Res* 25, 819–828.

Pietrzykowski, A.Z., Friesen, R.M., Martin, G.E. et al. 2008. Posttranscriptional regulation of BK channel splice variant stability by miR-9 underlies neuroadaptation to alcohol. *Neuron* 59, 274–287.
Raeder, H., Holter, S.M., Hartmann, A.M. et al. 2008. Expression of N-methyl-D-aspartate (NMDA) receptor subunits and splice variants in an animal model of long-term voluntary alcohol self-administration. *Drug Alcohol Depend* 96, 16–21.
Raingo, J., Castiglioni, A.J., and Lipscombe, D. 2007. Alternative splicing controls G protein-dependent inhibition of N-type calcium channels in nociceptors. *Nat Neurosci* 10, 285–292.
Ramanathan, K., Michael, T.H., Jiang, G.J. et al. 1999. A molecular mechanism for electrical tuning of cochlear hair cells. *Science* 283, 215–217.
Reenan, R.A., Hanrahan, C.J., and Ganetzky, B. 2000. The mle^{napts} RNA helicase mutation in *Drosophila* results in a splicing catastrophe of the *para* Na^+ channel transcript in a region of RNA editing. *Neuron* 25, 139–149.
Regan, M.R., Emerick, M.C., and Agnew, W.S. 2000. Full-length single-gene cDNA libraries: Applications in splice variant analysis. *Anal Biochem* 286, 265–276.
Rosenblatt, K.P., Sun, Z.P., Heller, S. et al. 1997. Distribution of Ca^{2+}-activated K^+ channel isoforms along the tonotopic gradient of the chicken's cochlea. *Neuron* 19, 1061–1075.
Rosenthal, J.J. and Seeburg, P.H. 2012. A-to-I RNA editing: Effects on proteins key to neural excitability. *Neuron* 74, 432–439.
Rula, E.Y., Lagrange, A.H., Jacobs, M.M. et al. 2008. Developmental modulation of $GABA_A$ receptor function by RNA editing. *J Neurosci* 28, 6196–6201.
Sanchez, S.E., Petrillo, E., Beckwith, E.J. et al. 2010. A methyl transferase links the circadian clock to the regulation of alternative splicing. *Nature* 468, 112–116.
Schoft, V.K., Schopoff, S., and Jantsch, M.F. 2007. Regulation of glutamate receptor B pre-mRNA splicing by RNA editing. *Nucleic Acids Res* 35, 3723–3732.
Shelley, C., Whitt, J.P., Montgomery, J.R., and Meredith, A.L. 2013. Phosphorylation of a constitutive serine inhibits BK channel variants containing the alternate exon "SRKR". *J Gen Physiol* 142, 585–598.
Singh, H., Lu, R., Bopassa, J.C. et al. 2013. MitoBK$_{Ca}$ is encoded by the *Kcnma1* gene, and a splicing sequence defines its mitochondrial location. *Proc Natl Acad Sci USA* 110, 10836–10841.
Smith, L.A., Wang, X., Peixoto, A.A. et al. 1996. A *Drosophila* calcium channel α1 subunit gene maps to a genetic locus associated with behavioral and visual defects. *J Neurosci* 16, 7868–7879.
Sommer, B., Keinänen, K., Verdoorn, T.A. et al. 1990. Flip and flop: A cell-specific functional switch in glutamate-operated channels of the CNS. *Science* 249, 1580–1585.
Sommer, B., Köhler, M., Sprengel, R. et al. 1991. RNA editing in brain controls a determinant of ion flow in glutamate-gated channels. *Cell* 67, 11–19.
Steiner, B., Truninger, K., Sanz, J. et al. 2004. The role of common single-nucleotide polymorphisms on exon 9 and exon 12 skipping in nonmutated CFTR alleles. *Hum Mutat* 24, 120–129.
Strumbos, J.G., Brown, M.R., Kronengold, J. et al. 2010. Fragile X mental retardation protein is required for rapid experience-dependent regulation of the potassium channel $K_V3.1b$. *J Neurosci* 30, 10263–10271.
Tan, B.Z., Jiang, F., Tan, M.Y. et al. 2011. Functional characterization of alternative splicing in the C terminus of L-type $Ca_V1.3$ channels. *J Biol Chem* 286, 42725–42735.
Tang, Z.Z., Sharma, S., Zheng, S. et al. 2011. Regulation of the mutually exclusive exons 8a and 8 in the $Ca_V1.2$ calcium channel transcript by polypyrimidine tract-binding protein. *J Biol Chem* 286, 10007–10016.
Tang, Z.Z., Zheng, S., Nikolic, J. et al. 2009. Developmental control of $Ca_V1.2$ L-type calcium channel splicing by Fox proteins. *Mol Cell Biol* 29, 4757–4765.
Ule, J., Stefani, G., Mele, A. et al. 2006. An RNA map predicting Nova-dependent splicing regulation. *Nature* 444, 580–586.
Vissel, B., Royle, G.A., Christie, B.R. et al. 2001. The role of RNA editing of kainate receptors in synaptic plasticity and seizures. *Neuron* 29, 217–227.
Walter, H.J., McMahon, T., Dadgar, J. et al. 2000. Ethanol regulates calcium channel subunits by protein kinase C delta-dependent and -independent mechanisms. *J Biol Chem* 18, 25717–25722.
Wang, E.T., Sandberg, R., Luo, S.J. et al. 2008. Alternative isoform regulation in human tissue transcriptomes. *Nature* 456, 470–476.
Wang, G.S. and Cooper, T.A. 2007. Splicing in disease: Disruption of the splicing code and the decoding machinery. *Nat Rev Genet* 8, 749–761.
West, A.E., Griffith, E.C., and Greenberg, M.E. 2002. Regulation of transcription factors by neuronal activity. *Nature Rev Neurosci* 3, 921–931.
Xie, J. and Black, D.L. 2001. A CaMK IV responsive RNA element mediates depolarization-induced alternative splicing of ion channels. *Nature* 410, 936–939.
Xie, J. and McCobb, D.P. 1998. Control of alternative splicing of potassium channels by stress hormones. *Science* 280, 443–446.
Yan, J., Olsen, J.V., Park, K.S. et al. 2008. Profiling the phospho-status of the BK$_{Ca}$ channel alpha subunit in rat brain reveals unexpected patterns and complexity. *Mol Cell Proteomics* 7, 2188–2198.
Zhang, C., Frias, M.A., Mele, A. et al. 2010. Integrative modeling defines the Nova splicing-regulatory network and its combinatorial controls. *Science* 329, 439–443.
Zuber, A.M., Centeno, G., Pradervand, S. et al. 2009. Molecular clock is involved in predictive circadian adjustment of renal function. *Proc Natl Acad Sci USA* 106, 16523–16528.

Single transmembrane regulatory subunits of voltage-gated potassium channels

Anatoli Lvov and William R. Kobertz

Contents

38.1 INTRODUCTION

The human genome encodes over 40 voltage-gated potassium (K^+) channels that have unique unitary conductances, voltage sensitivities, and activation/deactivation/inactivation gating kinetics. Despite this diverse portfolio of K^+ channels, most voltage-gated K^+ channels must co-assemble with water-soluble and membrane regulatory subunits to meet the potassium conduction and voltage-gating requirements for a wide variety of tissues. Maligned in the literature as ancillary or auxiliary, these K^+ channel subunits are essential for proper physiological function, and mutations in them cause several human diseases. In this chapter, we examine the structure, biogenesis, trafficking, modulation properties, and physiology of three different families of single-pass transmembrane (TM) K^+ channel regulatory subunits: KCNE peptides, dipeptidyl peptidase–like proteins (DPPs), and leucine-rich repeat-containing proteins (LRRCs). Although much smaller than the channel-forming α-subunits, these proverbial Davids can force the Goliath-sized channels to the membrane, alter their single channel conductance, and impair or disable their voltage sensing, enabling a voltage-gated channel to function in both excitable and nonexcitable cells.

38.2 KCNE TYPE I TRANSMEMBRANE PEPTIDES*

The KCNE genes encode type I TM (Takumi et al. 1988) MinK-related peptides (MiRPs) that interact with pore-forming subunits of K^+ channel and modulate their biophysical properties. The first isoform identified was KCNE1—a 129 amino acid (aa) protein that was originally named MinK, which stood for *Min*imal K^+ channel because it generated slowly activating, cardiac I_{Ks} current when its mRNA was injected into *Xenopus* oocytes (Takumi et al. 1988; Kaczmarek 1991). Upon the realization that MinK was forming membrane-embedded complexes with xKCNQ1 in oocytes (Barhanin et al. 1996; Sanguinetti et al. 1996), five human genes were identified (Abbott and Goldstein 1998; Piccini et al. 1999) and were denoted KCNE1–KCNE5 since they all encode regulatory subunits of the voltage-gated K^+ channel (Kv) superfamily potassium (K) channels nomenclature (KCN). Because the MiRPs are regulatory subunits and not channels as first thought, most in the field have adopted potassium voltage-gated channel (KCN) subfamily E member to describe both the

* KCNE1 aliases: MinK and IsK; KCNE2 alias: MiRP1; KCNE3 alias: MiRP2; KCNE4 alias: MiRP3; KCNE5 aliases: MiRP4 and KCNE1-L.

gene and protein. KCNE peptides have been shown to interact with various voltage-gated K+ channels, including different members of Kv1 (Abbott et al. 1999), Kv2 (Gordon et al. 2006), Kv3 (Abbott et al. 2001), Kv4 (Liu et al. 2008), Kv7 (Barhanin et al. 1996), and Kv11 (Bell Et al. 2005) families. In addition, mRNA profiling indicates that multiple KCNE peptides are expressed in the same tissue, raising the possibility that multiple KCNE family members can assemble with the same K+ channel (Alders and Mannens 1993; Lundquist et al. 2006). Given the large number of K+ channels that KCNE peptides have been shown to modulate, for this section, we will focus on KCNE1–5 modulation of KCNQ1 (Kv7) K+ channels, because it illustrates the diversity of potassium currents that can be generated by co-assembly with the different KCNE family members (Barhanin et al. 1996; Wallner et al. 1999; Abbott et al. 2001b; Grunnet et al. 2002; Abbott et al. 2008; Alders et al. 2009).

38.2.1 SEQUENCE AND STRUCTURE

KCNE1–5 are type I membrane proteins with a single TM domain flanked by extracellular N- and intracellular C-termini. They are tiny proteins (100–180 aa), contain multiple N-glycosylation sites 28, 29, and share modest sequence similarity in their TM domain and C-terminus that is predicted to be juxtaposed to the inner leaflet of the membrane. KCNE proteins are essentially found throughout the human body. *KCNE1* is expressed in heart (Barhanin et al. 1996), brain (Ohya et al. 2002), lung (Grahammer et al. 2001b), kidney (Barriere et al. 2003), testis (Tsevi et al. 2005), small intestine (Dedek and Waldegger 2001), exocrine pancreas (Thevenod 2002), inner ear (Vetter et al. 1996), and peripheral blood leukocytes (Chouabe et al. 1997). According to Lundquist, there are two KCNE1 splice variants: *KCNE1a*, which is expressed more ubiquitously, and *KCNE1b*, which is predominantly found in heart (Lundquist et al. 2006). *KCNE2* is widely expressed in brain (Tinel et al. 2000), eye (Warth and Barhanin 2002), heart (Tinel et al. 2000), skeletal muscle (Abbott et al. 1999), stomach (Grahammer et al. 2001b), lung (Warth and Barhanin 2002), heart (Tinel et al. 2000a), bladder, and kidney (Ohya et al. 2002). A small but significant expression of *KCNE2* was found in liver, ovary, testis, prostate, small intestine, and leukocytes. *KCNE3* is highly expressed in kidney (Warth and Barhanin 2002), skeletal muscle (Abbott et al. 2001a), brain (Abbott et al. 2001a), and heart (Abbott et al. 2001a), with moderate levels in small intestine (Dedek and Waldegger 2001). *KCNE4* is predominantly expressed in reproductive system (embryo and adult uterus), with low expression found in kidney, small intestine, lung, and heart (Grunnet et al. 2002). The last family member, *KCNE5*, is expressed in heart, skeletal muscle, brain, spinal cord, and placenta (Piccini et al. 1999). In sum, at least at mRNA level, KCNEs are ubiquitously expressed.

The solution nuclear magnetic resonance (NMR) (spectroscopy) structure (Kang et al. 2008) of bacterially expressed KCNE1 demonstrated that the KCNE TM domain is a curved α-helix. α-Helices were also identified in the N-terminus and the proximal half of the C-terminal domains. Complementary mutagenesis studies also indicate that the TM domains are helical (Goldstein and Miller 1991; Takumi et al. 1991; Wilson et al. 1994); alanine-scanning suggests that the C-terminal α-helix is split in half at a conserved proline (e.g., proline 77 in KCNE1) (Rocheleau et al. 2006). Several models of the KCNQ1-KCNE1 complex have been generated (Kang et al. 2008; Xu et al. 2008; Chung et al. 2009; Lvov et al. 2010) by mapping the data from many KCNE structure-function studies onto the high resolution structures of the *Shaker*-type K+ channel (Long et al. 2005) and the NMR structure of the isolated KCNE1 peptide (Kang et al. 2008).

38.2.2 BIOGENESIS/ASSEMBLY/TRAFFICKING

After the start of KCNE protein translation, the next step in biogenesis is recognition by the signal recognition particle (SRP) (Figure 38.1, *type I*). Because KCNE peptides do not possess a cleavable ER signal sequence, the SRP binds to the hydrophobic KCNE TM domain and brings the ribosomally elongating peptide to the ER membrane for insertion into the translocational tunnel (*Sec61*). Upon docking, the entire TM domain is inserted en masse into the translocation tunnel, placing the N-terminus into the ER lumen. The N-termini of KCNE peptides have 1–3 N-linked glycosylation sites: N-X-T/S-Y, where X and Y are any residue other than proline. *N*-glycans are added to KCNE peptides during protein translation (co-translationally) as well as after protein synthesis is complete (posttranslationally) (Chandrasekhar et al. 2006, 2011; Bas et al. 2011). Mutations that reduce co-translational N-glycosylation result in unglycosylated KCNE peptides that do not reach the cell surface (Bas et al. 2011). In addition to N-glycosylation, KCNE1 is O-glycosylated in the Golgi of cardiomyocytes (Chandrasekhar et al. 2011). Presently, no native disulfide bonds have been detected for KCNE subunits; however, KCNE3 and KCNE5 do possess one cysteine residue in their extracellular N-termini (Figure 38.2a).

Once inserted into the ER membrane, KCNE peptides are primed to assemble with voltage-gated K+ channels. KCNE's widespread tissue distribution and proclivity to assemble with any ion-conducting α-subunit has resulted in KCNEs being dubbed the *promiscuous subunits*. Therefore, many in the field have begun to question how many of the KCNE ion channel complexes overexpressed in cells are physiologically relevant. However, most of the assembly studies have focused on the well-documented complex between KCNE1 and KCNQ1 (cardiac I_{Ks} current). The stoichiometry between KCNE and pore-forming α-subunits in K+ channel complex is still a subject to controversy: several studies have shown that only two KCNE (β) subunits are in the channel tetramer (4α:2β) (Wang and Goldstein 1995; Chen et al. 2003a; Morin and Kobertz 2008a), but experiments with concatemers and green fluorescent protein (GFP) proteins tagged with KCNE subunits indicate that the ratio could be variable with up to four β-subunits per channel (4α:2β to 4α:4β) (Wang et al. 1998; Nakajo et al. 2010).

Similarly, the exact site of KCNE1-KCNQ1 co-assembly in the secretory pathway is still debated. Initially, it was proposed that the membrane-embedded complex formed (or transiently interacted (Poulsen and Klaerke 2007)) at the plasma membrane (Romey et al. 1997; Grunnet et al. 2002; Poulsen and Klaerke 2007). Later, it was shown that co-assembly takes place early in the secretory pathway (ER or *cis*-Golgi), because immaturely N-glycosylated KCNE1 co-immunoprecipitates with KCNQ1 (Chandrasekhar et al. 2006). In addition, another study trapped KCNQ1 in the ER using a trafficking-deficient KCNE1 point

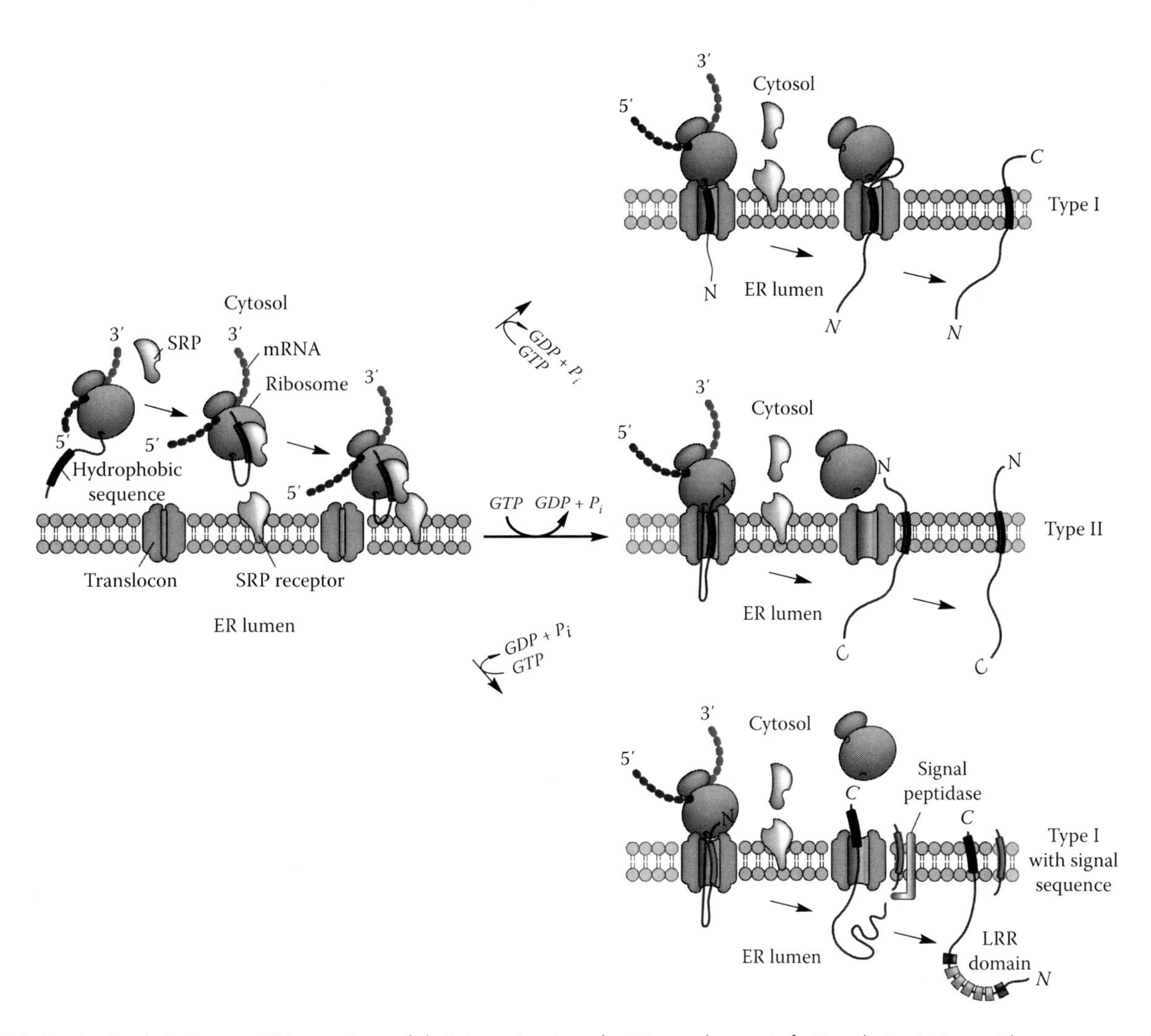

Figure 38.1 Synthesis of single-pass TM proteins and their insertion into the ER membrane. Left: Translating TM peptides are targeted to the ER by the SRP, which recognizes and binds to hydrophobic sequences that emerge from the ribosome. Upon binding to the SRP receptor, the ribosome–nascent chain complex becomes associated with the ER translocon. Right: For *type I* TM peptides, the N-terminus of the nascent chain is inserted into the translocon with the N-terminus leading (*head-first*). After protein synthesis is complete, the ribosome is released and the hydrophobic TM domain slides in the ER membrane. In contrast, *type II* peptides are inserted *tail-first* (C-terminus leads) and the growing polypeptide is secreted into the lumen of the ER. The ribosome detaches after protein synthesis, and the TM peptide exits the translocon. For *type I TM peptides with signal sequence*, the insertion process is similar to type II peptides in that the C-terminus of the signal sequence is inserted into the ER lumen. Signal peptidase cleaves the signal sequence and the TM segment is threaded into the translocon with the N-terminus leading, giving rise to a type I TM peptide. Both the cleaved signal sequence and the type I peptide slide out of the translocon and enter the ER membrane.

mutation (L51H), supporting the notion of assembly in the ER (Krumerman et al. 2004). Exactly how early does co-assembly take place? While the association of pore-forming subunits with their water-soluble accessory subunits has been shown to occur co-translationally (Shi et al. 1996), assembly of KCNE1 with KCNQ1 can occur independent of protein translation, as revealed by experiments with bacterially expressed KCNE1 protein injected into oocytes expressing KCNQ1 (Vanoye et al. 2010). Recently, KCNE1 has been shown to independently traffic to the *trans*-Golgi in kidney cells, suggesting that assembly in these polarized cells may occur later in the secretory pathway (David et al. 2013). Until a reliable kinetic assay is developed to follow KCNQ1-KCNE co-assembly in living cells, the assembly site(s) and trafficking (*vide infra*) for the KCNQ1-KCNE1 complex will remain controversial (Kanda and Abbott 2012).

The anterograde (forward) trafficking of KCNE1 to the plasma membrane strongly depends on the presence of a K^+ channel subunit (Chandrasekhar et al. 2006). In addition, both N- and O-glycosylation influence anterograde trafficking, as an arrhythmogenic mutation that disrupts both forms of glycosylation (T7I) severely hampers the forward progression of KCNE1 subunits in most cells (Bas et al. 2011). Interestingly, the different KCNE subunits may traffic to the cell surface at different speeds (e.g., KCNE2 travels faster than KCNE1) via distinct secretory pathways, including exosomes and multivesicular endosomes (MVEs) that bypass the Golgi (Um and McDonald 2007). Thus, the preferential assembly with K^+ channel subunits in native cells may also depend on the different rates of protein processing, anterograde trafficking, and retention of various KCNE family members (Um and McDonald 2007).

KCNE peptides not only regulate potassium currents by promotion of the surface expression of KCNQ channels, but may also serve as an endocytic chaperone for channels internalization. Xu and colleagues have described a KCNE1-independent

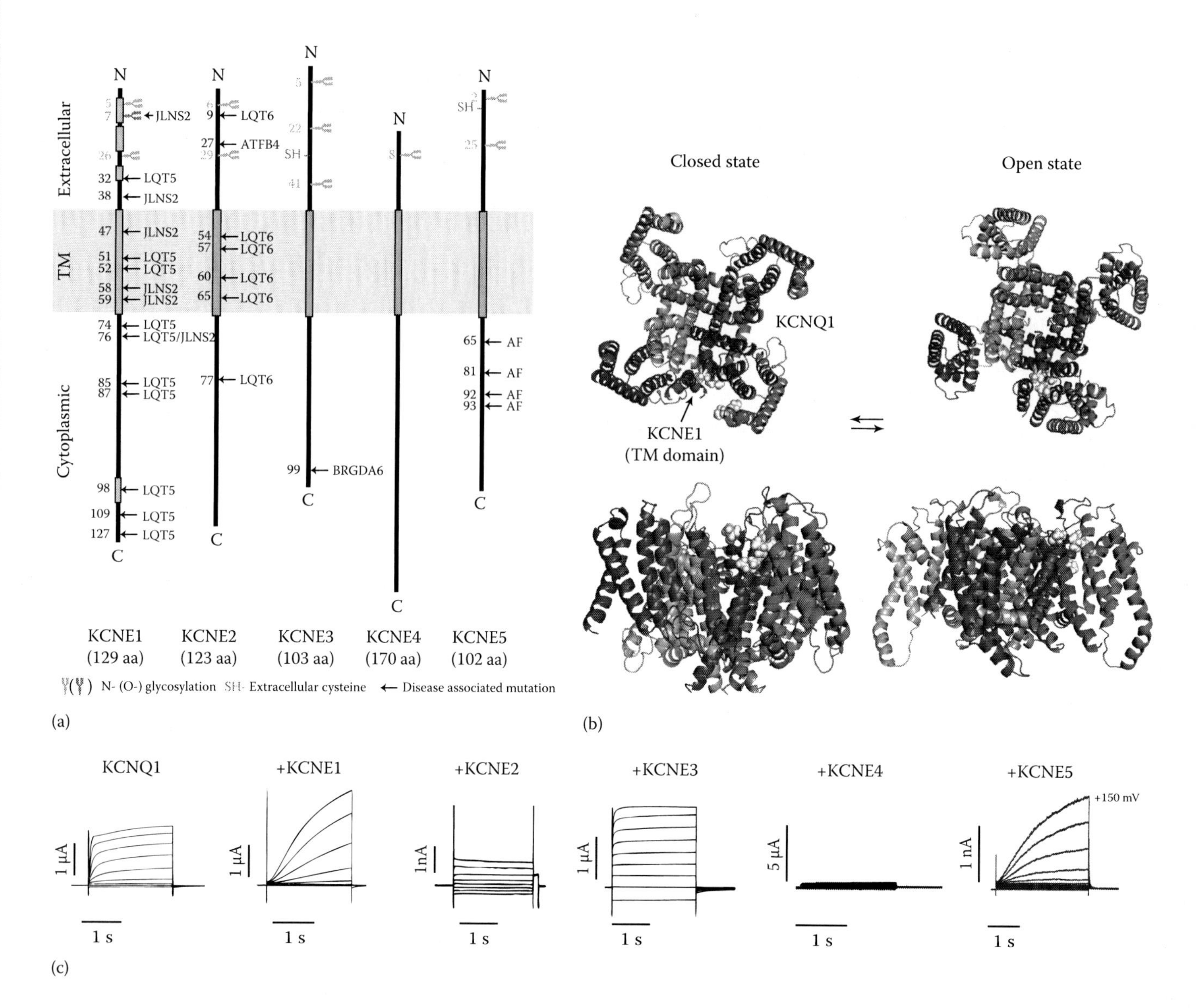

Figure 38.2 The KCNE family of K^+ channel subunits. (a) Membrane topology of KCNE1–5 proteins. NMR-determined (KCNE1) and sequence-predicted (KCNE2–5) α-helices are indicated as rectangular boxes. Approximate positions of known glycosylation sites, extracellular cysteines, and disease-associated mutations are depicted. (b) Putative location of KCNE1 within the cardiac I_{Ks} channel complex. Ribbon diagrams of experimentally restrained docking model for a KCNQ1 tetramer in the closed and open states with superimposed KCNE1-TM. (This model is adopted from Kang, C. et al., *Biochemistry*, 47(31), 7999, 2008.) (c) Representative current recordings of KCNQ1/KCNE complexes. Whole-cell voltage-clamp recordings were obtained from *Xenopus* oocytes, except for KCNE2 (*HEK-293*) and KCNE5 (*CHO-K1*); KCNE1–KCNE4: −100 to 40 mV, step 20 mV; KCNE5: −80 to 150 mV, step 10 mV. (Traces are adopted from Dedek, K. et al., *Pflugers Arch.*, 442(6), 896, 2001 [*+KCNE2*], Grunnet, M. et al., *J. Physiol.*, 542, 119, 2002 [*+KCNE4*], and Angelo, K. et al., *Biophys. J.*, 83(4), 1997, 2002 [*+KCNE5*].)

mechanism of KCNQ1 internalization, which leads to surface expression of KCNQ1 complexes lacking KCNE1 (Xu et al. 2009). Another important conclusion drawn from this study is that channel may travel to and from the plasma membrane by alternative, KCNE-independent vesicular pathway, most likely with other yet unidentified protein partners. A possible candidate is Nedd4-/Nedd4-like protein, an ubiquitin ligase that is involved in regulated internalization of a number of ion channels, including sodium channels ENaC (Weber et al. 1995) and $Na_v1.5$ (van Bemmelen et al. 2004), a chloride transporter ClC-5 (Garcia-Patrone and Tandecarz 1995) and the cardiac K^+ channel, hERG1 (Albesa et al. 2011). Nedd4 catalyzes posttranslational ubiquitylation: a covalent attachment of ubiquitin moieties to lysine residues of ion channels via PY motifs (P-P-X-Y-X-X-φ, where P is a proline, X any amino acid, and φ a hydrophobic amino acid) in their cytoplasmic C-termini (Kanelis, et al. 2001). Upon attachment, ubiquitin may initiate internalization and rapid ER-associated degradation of target proteins by lysosomes or proteasomes (Pickart 2001). Recently, ubiquitylation of KCNQ1 by Nedd4 ligase has been shown to reduce cell surface protein levels of both KCNQ1 channels and KCNQ1/KCNE1 complexes (Jespersen et al. 2007), providing a mechanism for the long-term modulation of cardiac I_{Ks} current.

38.2.3 K^+ CHANNEL MODULATION

Biochemical and electrophysiological studies have demonstrated that KCNE proteins interact with and modulate a wide variety of voltage-gated K^+ channels. For a comprehensive list, see the review by McCrossan and Abbott (2004). In this chapter, we will focus on the membrane-embedded complexes of the five KCNE family members with KCNQ1 (Kv7.1) channels.

KCNE1: When heterologously expressed, the homotetrameric KCNQ1 channel generates a rapidly activating and slowly deactivating voltage-dependent (Kv) current that shows incomplete voltage-dependent inactivation (Tristani-Firouzi and Sanguinetti 1998; Seebohm et al. 2003; Levy et al. 2008); co-expression with KCNE1 removes inactivation, shifts the voltage dependence of activation toward more positive potentials, increases single channel conductance, increases sensitivity to PIP_2, and provides the complex with sensitivity to PKA phosphorylation and pharmacological agents (Sanguinetti et al. 1996; Sesti and Goldstein 1998; Tristani-Firouzi and Sanguinetti 1998; Loussouarn et al. 2003; Nicolas et al. 2008). Moreover, co-expression of KCNQ1 with KCNE1 produces a slowly activating current strikingly reminiscent to cardiac I_{Ks} (Barhanin et al. 1996; Wallner et al. 1999) (Figure 38.2c). In the inner ear, the noninactivating KCNQ1/KCNE1 complexes allow potassium secretion through the apical membrane of marginal cells into the potassium-rich endolymph.

KCNE2: In contrast to KCNE1, association with KCNE2 (and KCNE3) results in currents that lack voltage dependency and are, at least, partially open at resting membrane potential (therefore, might be considered as *leak* potassium current) (Tinel et al. 2000a; Abbott et al. 2008) (Figure 38.2c). In vivo, the KCNQ1/KCNE2 complex is responsible for coupling of potassium to gastric acid secretion via co-localization of the channel with the gastric H-K-ATPase (Roepke et al. 2006). In addition, KCNQ1 and KCNE2 form a thyroid-stimulating hormone (TSH)-stimulated constitutively active K^+ channels in thyrocytes, which are required for normal thyroid hormone biosynthesis (Roepke et al. 2009).

KCNE3: KCNE3 also converts KCNQ1 into a voltage-independent, constitutively conducting channel, which serves as a potassium recycling pathway in epithelial cells at the intestinal crypt base (Schroeder et al. 2000). Additionally, colonic KCNQ1-KCNE3 channels provide a driving force for cAMP-stimulated intestinal Cl^- secretion (Schroeder et al. 2000). In contrast to KCNE2, KCNQ1–KCNE3 complexes are not inhibited by a reduction of external pH (Grahammer et al. 2001).

KCNE4: KCNE4 strongly inhibits KCNQ1 channel function (2002; Manderfield et al. 2009; Grunnet et al. 2002) (Figure 38.2c). KCNE4-induced inhibition of KCNQ1 requires a Ca^{2+}-dependent interaction between a tetraleucine motif in the juxtamembrane region of KCNE4 and calmodulin (Ciampa et al. 2011). KCNE4 was also shown to co-assemble, together with KCNE1, in the same KCNQ1 to form a *triple subunit* (KCNE1-KCNQ1-KCNE4) complex (Manderfield et al. 2009). The importance of this complex is underscored by existence of a single nucleotide polymorphism in KCNE4 gene (KCNE4–145E/D), which exerts the effect of *gain of function* on the KCNQ1 channel, leading to atrial fibrillation. KCNE4 inhibition of KCNQ1 channel function is not dominant, because heteromeric KCNQ1/KCNE complexes (KCNQ1/KCNE4/KCNE1 or KCNQ1/KCNE4/KCNE3) are functional, affording currents reminiscent of KCNQ1/KCNE1 and KCNQ1/KCNE3 complexes, respectively (Morin and Kobertz 2008b). The hierarchy for KCNE modulation of KCNQ1 is: KCNE3 > KCNE1 >> KCNE4 (Morin and Kobertz 2008b).

KCNE5: KCNE5 or KCNE1-like co-assembles with KCNQ1 to generate currents with activation kinetics similar to cardiac I_{Ks} (KCNQ1/KCNE1) (Figure 38.2c); however, the midpoint of activation is shifted to an apparently unphysiologically reachable voltage (~120 mV) (Angelo et al. 2002). Correspondingly, co-expression of KCNE5 with KCNQ1/KCNE1 (cardiac I_{Ks}) complex has been shown to reduce I_{Ks} current (Ravn et al. 2008). In the same study, a KCNE5 mutation, leading to *gain of function* of I_{Ks} current and potentially associated with atrial fibrillation, has been described.

The remarkable diversification of KCNQ1 voltage-gating by the KCNE peptides is in stark contrast to the other single-pass TM regulatory subunits of K^+ channels (see DPP and LRRC peptides). Thus, many investigators have focused on determining (1) how KCNE peptides manipulate KCNQ function and (2) which regions of the KCNE peptides are responsible for the modulatory effects. Although the KCNE peptides are relatively small in size, it has become clear that these type I TM peptides utilize their TM, extracellular, and intracellular domains to manipulate the biophysical properties of the KCNQ1 channel.

Because the operational core of the voltage-gated K^+ channel—voltage sensor, conducting pore, and its activation gate—is mostly immersed in the plasma membrane, the first structure-function studies investigated the membrane-embedded portions of KCNQ1/KCNE complexes. It was initially proposed that the TM segment of KCNE lined the conductivity pathway (Wang et al. 1996; Tai and Goldstein 1998; Chen et al. 2003b). However, this hypothesis was overturned by subsequent function-structure analyses (Kurokawa et al. 2001; Tapper and George 2001) and the high resolution structures (Doyle et al. 1998; Jiang et al. 2003) of the tightly packed pore domain of voltage-gated K^+ channel. Although not lining the ion conducting pathway, the TM segment of KCNE was shown to reside in close proximity to the S6 segment of the pore domain (Tapper and George 2001) and regulate activation gating (Melman et al. 2001, 2002; Gage and Kobertz 2004) by either physical or allosteric interaction with specific residues in pore forming S6 (Melman et al. 2004; Panaghie et al. 2006). In addition, the interaction between TM segments of KCNE and KCNQ proteins may not be limited by pore-forming domains, but also include voltage sensor (S4) (Nakajo and Kubo 2007; Rocheleau and Kobertz 2008; Shamgar et al. 2008) and its cytoplasmic S4-S5 linker (Chouabe et al. 2000).

There is a growing body of evidence that KCNE1 may affect KCNQ1 voltage sensors' conformation and activation movement trajectory. Initially, state-dependent movements of the voltage sensors in KCNQ1/KCNE channel complexes were probed using cysteine accessibility experiments with cysteine-reactive, methanethiolsulfonate (MTS) reagents (Nakajo and Kubo 2007; Rocheleau and Kobertz 2008). The common denominator in these studies was that KCNE3 peptides shift the equilibrium of the voltage sensor such that it favored the active state at resting potentials, allowing KCNQ1 to be open at very negative voltages. Subsequently, Shamgar and colleagues demonstrated that the KCNE1 and KCNE3 gating phenotypes could be mimicked by mutations of specific residues in KCNQ1 voltage sensor (Shamgar et al. 2008). Further, MTS modification experiments were used to demonstrate that KCNE1 alters both the conformation (Wu et al. 2010a) of and the electrostatic environment (Wu et al. 2010b) around the KCNQ1 voltage sensor. Finally, measurements of voltage-sensor movements using voltage-clamp fluorometry not only confirmed that voltage-sensor movement is altered in the presence of KCNE1, but also revealed uncoupling between the voltage sensor and the channel gate in KCNQ1/KCNE1 complex (Osteen et al. 2010; Ruscic et al. 2013).

The discovery of arrhythmogenic mutations in the S1–S2 loop of KCNQ1 (Chen et al. 2003c; Hong et al. 2005; Lundby et al. 2007) has shifted the attention toward the extracellular domains of the KCNQ1/KCNE1 complex. Engineered disulfide cross-linking experiments have identified protein–protein interactions between the N-terminus of KCNE1 and the extracellular tops of the S1 and S4 segments, positioning the KCNE1 TM domain into a putative *binding pocket* between the S1, S4, and S6 segments of three separate Q1 subunits (Xu et al. 2008; Chung et al. 2009) (Figure 38.2b). Conversely, a swath of mutations in the KCNE1 C-terminus linked to cardiac arrhythmias helped identify interactions between KCNQ1 and the cytoplasmic portion of KCNE1 that abuts the membrane (Takumi et al. 1991; Romey et al. 1997; Splawski et al. 2000; Schulze-Bahr et al. 2001; Tapper and George 2001; Ma et al. 2003; Gage and Kobertz 2004; Lai et al. 2005; Napolitano et al. 2005). Recently, we identified a protein–protein interaction in the cytoplasmic parts of KCNQ1/KCNE1 complex, which included S6 activation gate and S4–S5 linker (Lvov et al. 2010). The KCNE1 cytoplasmic domain also binds to the KCNQ1 channel *tetramerization domain*, providing KCNE1 with an anchoring point at the channel's C-terminus (Haitin et al. 2009).

38.2.4 PHYSIOLOGY AND DISEASE

KCNE1: Mutations in either KCNQ1 or KCNE1, which disrupt complex assembly, anterograde trafficking, or channel function (Figure 38.2a) have been linked to inherited long QT syndrome (LQT1 or LQT5, respectively), a condition leading to cardiac arrhythmia (Neyroud et al. 1997; Tyson et al. 1997; Duggal et al. 1998). The most common autosomal dominant form of this inherited potentially life-threatening condition is *Romano–Ward syndrome*, an electrophysiological disorder characterized by a prolonged Q-T interval and abnormalities in T-wave on the ECG, and the ventricular tachycardia (Alders and Mannens 1993). Homozygous recessive mutations in either KCNQ1 or KCNE1 underlie the *Jervell and Lange-Nielsen syndrome* with severe phenotypes, including congenital profound bilateral sensorineural hearing loss and long (usually greater than 500 ms) Q-T interval, which is associated with ventricular tachyarrhythmias and fibrillation (Tranebjaerg et al. 1993).

KCNE2: A number of autosomal dominant mutations in the KCNE2 gene have been linked to long QT syndrome type 6 (LQT6) leading to Romano–Ward syndrome and familial atrial fibrillation (Lu et al. 2003) (ATFB4) (Figure 38.2a). Because KCNE2 appears to have multiple cardiac K^+ channel partners, the molecular mechanisms underlying the cause of these disease mutations have been difficult to nail down. Targeted knock-down of KCNE2 genes indicates that KCNE2 regulates ventricular I_{Kslow1} and $I_{to,f}$ currents, impairing Kv1.5 trafficking and diminishing Kv4.2 current density (Roepke et al. 2008). These channels also co-immunoprecipitated with KCNE2 protein from membrane fractions of murine ventricular myocytes. In addition to cardiac K^+ channels, KCNE2 also affects the cell surface expression and biophysical properties of HCN channels, which give rise to the *funny* cardiac pacemaker current ($I_{(f)}$) (Brandt et al. 2009). Outside of the heart, constitutively active KCNQ1/KCNE2 channels are required for normal thyroid hormone biosynthesis (Roepke et al. 2009). Since thyroid hormones are crucial for proper cardiac functions, this finding suggests that mutations in KCNE2 gene may also lead to cardiac arrhythmias in hypothyroid patients. KCNE2 downregulation was also linked to gastric neoplasia (Roepke et al. 2010), where KCNE2 also forms a complex with KCNQ1. Overexpression of KCNE2 has antitumorigenic effects, slowing down cell proliferation and stomach tumorigenesis (Yanglin et al. 2007).

KCNE3: A missense mutation (R99H) in KCNE3 gene may underlie the development of Brugada syndrome (BRGDA6, Figure 38.2a) (Delpon et al. 2008), a condition that is characterized by uncoordinated electrical activity in ventricles and associated with high risk of sudden cardiac death (Brugada and Brugada 1992). This mutation was shown to affect Kv4.3 channel modulation by KCNE3, resulting in a significant increase in the I_{to} current intensity, without any effect on the I_{Ks} current (Delpon et al. 2008). Since KCNQ1-KCNE3 channels provide a driving force for cAMP-stimulated intestinal Cl^- secretion, they may also be involved in secretory diarrhea and cystic fibrosis (Schroeder et al. 2000).

KCNE4: Although it was shown by in vitro experiments that KCNE4 may regulate cardiac KCNQ1 and Kv4.3 channels (Grunnet et al. 2003; Teng et al. 2003), a physiological role for KCNE4 has not been clearly elucidated.

KCNE5: KCNE5 was first discovered as a gene deletion involved in the development of the cardiac and neurological abnormalities observed in patients with alport syndrome with mental retardation, midface hypoplasia and elliptocytosis (AMME) syndrome (Piccini et al. 1999). Similar to KCNE3, a number of novel KCNE5 mutations were shown to be responsible for gain-of-function effects on I_{to} current density in patients with Brugada syndrome (Ravn et al. 2008; Ohno et al. 2011) (AF, Figure 38.2a).

38.3 DPP6/10: DIPEPTIDYL PEPTIDASE-LIKE PROTEINS*

DPP6 and DPP10 are two enzymatically inactive members of dipeptidyl peptidase (DPP) IV family of serine proteases. Together with cytoplasmic Kv channel-interacting proteins (KChIPs), these type II TM glycoproteins co-assemble with members of the Kv4 *family* of voltage-gated K^+ channels to form macromolecular complexes that are responsible for the transient outward current (I_{to}) in cardiac myocytes and somatodendritic sub-threshold *A-type* potassium *current* (I_{SA}) in neurons (Jerng et al. 2005; Zagha et al. 2005; Amarillo et al. 2008;

* DPP6 aliases: dipeptidyl peptidase 6, dipeptidyl peptidase like protein 1 (DPL1), BSPL, DPPX; DPP10 aliases: dipeptidyl peptidase 10, dipeptidyl peptidase like protein 2 (DPL2), dipeptidyl peptidase IV–related protein 3 (DPRP3), DPPY.

Soh and Goldstein 2008). As single TM proteins with a large extracellular domain, DPP6 and DPP10 are well positioned to both detect extracellular signals and modulate Kv4 gating.

38.3.1 SEQUENCE AND STRUCTURE

Both DPP6 and DPP10* were first discovered in a mammalian brain. DPP6 was isolated from bovine and rat neurons (as a part of the Kv4 channel complex) in 1992 by Wada et al. (1992), and DPP10 was initially cloned from human hypothalamus in 2000 by Nagase et al. (2000), then fully described as a new member of the DPP IV serine proteases family by Qi et al. (2003). Human DPP6 is an 865 aa glycoprotein encoded by the homonymous gene located in chromosome 7 and expressed predominantly in the central nervous system (hippocampus, thalamus, hypothalamus, and striatum). Significant level of DPP6 expression was also found in pancreas and testes, and with lower levels—in the heart ventricles, lung, kidney, and ovary (Radicke et al. 2005). In contrast, the DPP10 gene is located in chromosome 2 and encodes a slightly shorter protein (796 aa) found mainly in brain and spinal cord, pancreas, and adrenal glands. DPP10 protein has also been detected in liver, trachea, and placenta (Allen et al. 2003; Qi et al. 2003). DPP6 and DPP10 bear the highest amino acid identity (up to 51%) among other members of DPPIV (or S9B) family of ectopeptidases (Qi et al. 2003), with the most variability in their N-termini. There are five isoforms of human DPP6 (DPP6a/E, DPP6K, DPP6-S, and DPP6-L) and 4 isoforms of DPP10 (DPP10a, DPP10b, DPP10c, and DPP610). In addition, there are also N-terminal variants of the isoforms, which are the result of alternative mRNA initiation (Wada et al. 1992).

DPP6 and DPP10 are single-pass type II membrane proteins, with a short and divergent cytoplasmic N-terminus attached to the TM domain by a highly conserved juxtamembrane sequence, and a long extracellular C-terminus (Kin et al. 2001) (Figure 38.3a). The extracellular domain of DPP-like proteins can be structurally divided into a β-propeller domain and an α/β hydrolase domain (Rasmussen et al. 2003; Strop et al. 2004) (Figure 38.3b). Enzymatically active family members cleave the N-terminal bond of proline-containing polypeptides. This selectivity arises from a cooperation between the β-propeller domain, which forms a *tunnel* leading to the active site, and the α/β hydrolase domain, which houses the active site (the *catalytic triad*: a nucleophilic *Ser*; an acidic *Asp*; and an absolutely conserved *His*) (Rasmussen et al. 2003). The inactive α/β hydrolase domains of DPP6 and DPP10 contain modified catalytic triad, in which the first nucleophilic Ser is replaced with Asp (DPP6) or Gly (DPP10). Although DPP6 and DPP10 lack this key nucleophilic serine residue, it is impossible to restore the proteolytic activity by simply mutating these residues back to Ser, indicating other structural differences in the DPP6/10 catalytic domains (Kin et al. 2001). Indeed, a crystal structure of the DPP6 extracellular domain revealed that the arrangement of the active site residues is inconsistent with the catalytic mechanism of serine proteases (Strop et al. 2004).

38.3.2 BIOGENESIS/ASSEMBLY/TRAFFICKING

As type II TM glycoproteins, the biogenesis and ER insertion of DPP6 and DPP10 is inverted compared to the KCNE type I peptides; thus, the N-terminus becomes anchored in the cytoplasm when the TM domain is inserted into translocation tunnel while the C-terminus is co-translationally threaded into the ER lumen (Figure 38.1, *type II*). Once in the ER lumen, the C-terminus of both DPP6 and DPP10 become heavily N-glycosylated—DPP6 contains seven consensus sites (173, 319, 404, 471, 535, 566, and 813); (Kin et al. 2001) DPP10 has eight sites (90, 111, 119, 257, 342, and 748) (Chen et al. 2006). Although DPP10 has more putative N-glycosylation sites, it appears to be less glycosylated than DPP6 (Jerng et al. 2004), which is consistent with a recent mutagenesis study showing only six of the eight consensus sites in DPP10 are used (Cotella et al. 2012). In addition to N-glycosylation, the

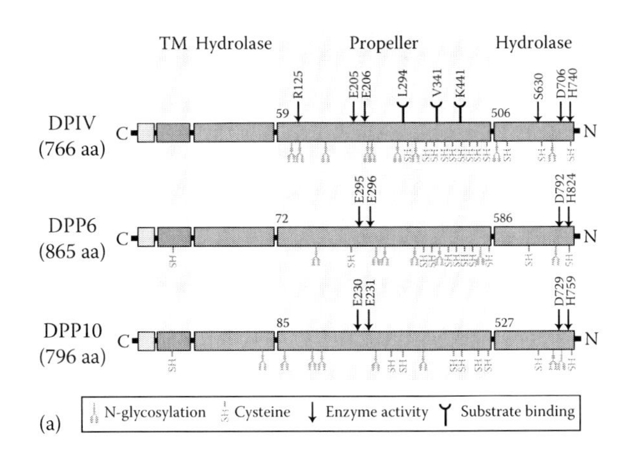

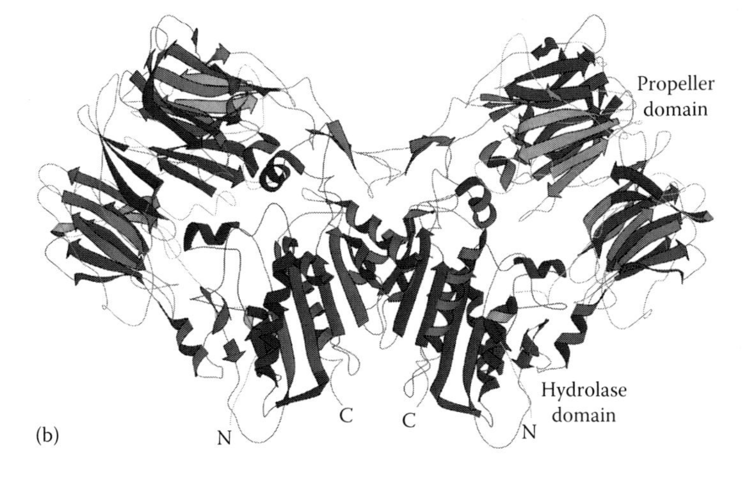

Figure 38.3 Dipeptidyl peptidase–like proteins DPP6/10. (a) Membrane and structural topology of proteins of the DPPIV family. Approximate positions of N-glycosylation sites, cysteine residues, and residues required for substrate binding and enzyme activity are depicted. (Schematic presentation is adopted from Gorrell, M.D., *Clin. Sci. (Lond.)*, 108(4), 277, 2005. (b) DPP6 homodimer depicted as a ribbon diagram. The figure was created using the atomic coordinates from the protein data bank (PDB) code 1XFD with KiNG display software ver. 2.21 (Duke University).
(Continued)

* See footnote in page 562.

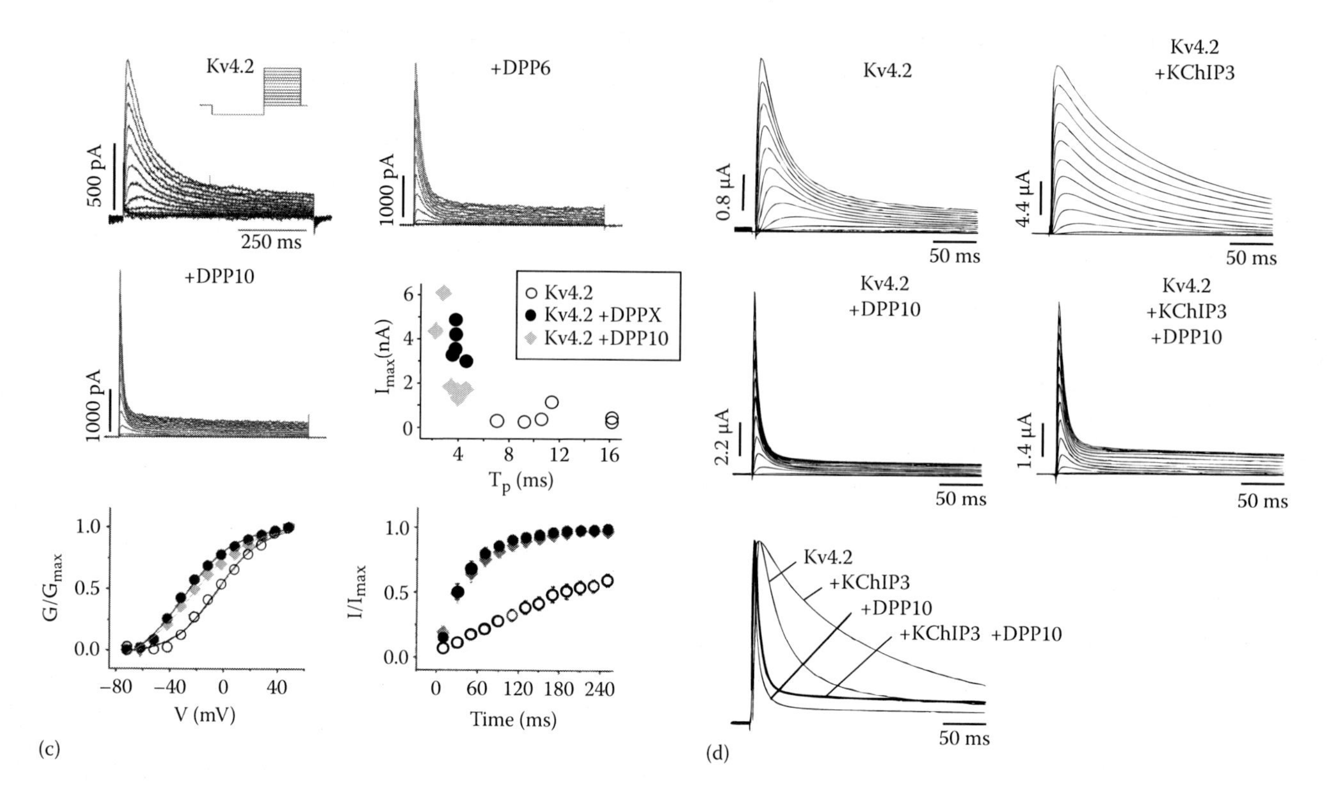

Figure 38.3 (Continued) Dipeptidyl peptidase–like proteins DPP6/10. (c) Modulation of A-type potassium current (Kv4.2) by DPP6 and DPP10 proteins. Whole-cell current traces elicited by the stimulation protocol shown in the inset, in the absence (Kv4.2) or presence of the indicated modulatory subunits (+DPP6, +DPP10), are depicted. CHO cells were voltage-clamped at a holding potential of −82 mV and hyperpolarized for 1 s to −112 mV to remove channel inactivation, followed by depolarizing test pulses from −82 to 48 mV in increments of 10 mV. Both DPPX and DPP10 considerably increase current magnitude (I_{max}) (note scale bar). In addition, the time to peak (t_p) is decreased, the voltage dependence of Kv4.2 activation is shifted to hyperpolarizing potentials, and the rate of recovery from inactivation is increased. (Illustrations adopted from Zagha, E. et al., *J. Biol. Chem.*, 280(19), 18853, 2005. (d) A ternary Kv4.2/KChIP3/DPP10 complex conducts current that inactivates differently from those of Kv4.2 alone or binary Kv4.2/DPP10 and Kv4.2/KChIP3 complexes. Representative families of traces as recorded by 1 s step depolarizations from −100 to +50 mV (10 mV increments). Normalized and overlapped traces at +50 mV are also shown on the bottom. (Illustrations adopted from Jerng, H.H. et al., *J. Physiol.*, 568, 767, 2005.)

C-terminus of DPP6 has four disulfide bonds: three are in the β-propeller domain, where they form stabilizing cross-links in blades 5, 7, and 8, and one is in the α/β-hydrolase domain near the C-terminus of the molecule (Strop et al. 2004). Although there is no experimental evidence for disulfide bonds in the C-terminal domain of DPP10, their presence in other DPP IV proteins (Strop et al. 2004; Gorrell 2005; Thielitz et al. 2008) and their roles in protein folding and stabilization of the mature protein (reviewed in Fass 2012) portend that the DPP10 C-terminus will also contain several disulfide bonds. Lastly, previous studies on the archetypical DPP IV proteins indicate three potential O-glycosylation sites; however, the specificity of trifluoromethane-sulfonic acid (TFMS) *O*-glycan cleavage (Naim et al. 1999), and GalNac-α-*O*-benzyl (BGN), a drug commonly used to block O-glycosylation (Huet et al. 1998), are controversial.

Co-immunoprecipitation (Jerng et al. 2005) and functional reconstitution experiments (Jerng et al. 2005; Soh and Goldstein 2008) have demonstrated that four DPP6/10 β-subunits assemble with four pore-forming Kv4 α-subunits. Since the transcript and protein expression patterns of Kv4, KChIPs, and DPPs overlap in all brain regions (Kunjilwar et al. 2004), the entire macromolecular complex that is responsible for the native I_{to} and I_{SA} currents is a dodecamer with the four cytoplasmic KChiPs attached to the Kv4 subunits' dangling T1 domain (Shibata et al. 2003; Jerng et al. 2005). Although it is possible that the dodecameric complex retains the fourfold symmetry of the Kv4 channel, the propensity of the DPP proteins to dimerize via the α/β-hydrolase and the β-propeller domains (McNicholas et al. 2009) and biogenesis studies of the homologous *Shaker* K^+ channel (Xu et al. 1998) indicate that the complex will either be twofold symmetric or asymmetric.

Co-assembly with DPP6 and DPP10 affects Kv4 channel anterograde (forward) trafficking and cell surface expression. DPP6/10 co-expression results in a redistribution of Kv4 channels away from the ER and Golgi apparatus to the cell surface (Nadal et al. 2003; Jerng et al. 2004; Foeger et al. 2012), resulting in increase in Kv4 current density (Jerng et al. 2004). Tunicamycin treatment (to block all N-glycosylation) greatly reduces the Kv4 current (Jerng et al. 2004), hinting that N-glycosylation is important for cell surface expression of the complex. Recently, Cotella et al. have shown that the *N*-glycans on the β-subunit (DPP10: N90, N119, *N257*, and N342) are required for the proper cell surface sorting and interaction with the Kv4.3/KChIP2a channel complex (Kaulin et al. 2009; Cotella et al. 2012). Although DPPs influence Kv4 channel trafficking, the converse does not appear to be true, because solitary DPP6 and DPP10 subunits traffic to the cell surface in the absence of Kv4 α subunits (Foeger et al. 2012).

38.3.3 K+ CHANNEL MODULATION

The most evident biophysical consequence of DPP6/10 on Kv4 channel function is the increase in current magnitude (Nadal et al. 2003; Jerng et al. 2004, 2005; Zagha et al. 2005; Amarillo et al. 2008); however, DPPs also modify Kv4 channel gating (Figure 38.3c). DPP6 and DPP10 accelerate both inactivation and recovery from inactivation; they also shift the voltage dependence of both activation and inactivation to more negative voltages (Jerng et al. 2004; Witzel et al. 2012). DPP subunits' effects on Kv4 gating are more prominent at room temperature and become somewhat tempered at physiological temperatures (37°C) (Radicke et al. 2013).

Based on a number of studies, the TM segment of DPP6 and DPP10 β-subunits operates as an integral voltage-sensor interacting protein, associating with the S1–S2 TM segments of the Kv4 channel's voltage-sensing domain (Ren et al. 2005; Zagha et al. 2005) and shifting the voltage dependence of channel activation and inactivation (Covarrubias et al. 2008). Gating current measurements indicate that DPP6-S remodels two aspects of gating charge dynamics: (1) it accelerates the outward movements of gating charge and its return upon repolarization; (2) it shifts the voltage dependence of gating charge movement toward more negative membrane potentials (Dougherty and Covarrubias 2006). Thus, the DPP6 TM segment (1) reduces the energy barrier for activation/deactivation and (2) destabilizes the resting conformation of the voltage sensor and/or uncouples voltage-sensor movement from channel opening (Jerng et al. 2004; Dougherty and Covarrubias 2006).

The divergent cytoplasmic N-terminal sequences of the different isoforms and splice variants also modulate Kv4 channel gating. For example, DPP6K shifts the steady-state inactivation curve to more hyperpolarizing potentials and slows down the recovery from inactivation by modulating KChIP3a effects (Jerng and Pfaffinger 2012) (Figure 38.3c). Furthermore, the proximal N-terminus of two splice variants, DPP6a and DPP10a, have a conserved N-terminal sequence (MNQTA) that directly induces ultrafast voltage-independent inactivation of Kv4 channel through an N-type pore-blocking mechanism (Jerng et al. 2007, 2009).

In contrast, the extracellular β-propeller domain of DPP6 and DPP10 does not appear to play a direct role in modulating Kv4 channel gating. Unglycosylated DPPs (tunicamycin treatment) (Jerng et al. 2004) that reach the plasma membrane with Kv4 channels have gating kinetics similar to fully N-glycosylated complexes. Moreover, replacing the entire DPP10 extracellular domain with an antibody epitope tag (Myc) affords a mutant protein that recapitulates wild type DPP10 modulation of Kv4.3 (Zagha et al. 2005). Similarly, investigations using DPP chimeric proteins indicate that the extracellular β-propeller domain is not involved in Kv4 channel gating (Ren et al. 2005). Deletion of the DPP10 extracellular domain does reduce Kv4 current density, presumably due to compromised sorting and cell surface trafficking of the resultant mutant complex (Zagha et al. 2005). Although the extracellular domain is dispensable for Kv4 channel gating, the β-propeller domain may stabilize the complex at the plasma membrane and serve as a receptor for signal transduction (McNicholas et al. 2009).

38.3.4 PHYSIOLOGY AND DISEASE

As β-subunits of voltage-gated K+ channels that operate at membrane potentials below the threshold for action potential generation, DPP6 and DPP10 are important regulators of cardiac and neuronal membrane excitability. Recently, a flurry of discoveries have linked expression and traffic aberrations of DPP6/10 subunits to various inherited and acquired disorders.

Neuronal diseases: There is a growing body of evidence that DPP6 and DPP10 are involved in neurodegenerative diseases. First, single nucleotide polymorphism in intron 3 of the DPP6 gene was implicated in amyotrophic lateral sclerosis (ALS), a disease of the central nervous system that affects voluntary muscle movement (Cronin et al. 2008; Del Bo et al. 2008; Garber 2008; van Es et al. 2008). Overexpression of DPP6 was also detected in patients suffering from the progressive forms of multiple sclerosis (PrMS), which shares some features with ALS (Brambilla et al. 2012). Moreover, Abbott and coworkers have detected truncated forms of DPP10 (789 aa) in Alzheimer's brain plaques and tangles, which are the anomalous clusters of mis-trafficked and aggregated proteins associated with the disease (Chen et al. 2008). Second, both DPP6 and DPP10 were implicated in inflammation and allergic responses in the brain. DPP6 was found to be robustly expressed in particular forms of encephalitis, triggered by the autoimmune reaction to the DPP6 antigen (Boronat et al. 2012). There are also intriguing findings that DPP10 gene polymorphisms may play a role in predisposition to asthma and bronchial hyper-responsiveness (Allen et al. 2003; Grammatikos 2008; Schade et al. 2008). Third, de novo and inherited copy number variations (CNVs) of both DPP6 and DPP10 were implicated in susceptibility to autism spectrum disorders (Marshall et al. 2008). Synaptic plasticity, which is the biological basis for the cognitive brain functions, critically depends on protein composition of glutamatergic synapses. DPP6 and DPP10, together with Kv4.2 channel and numerous scaffolding proteins (e.g., neurexins and neuroligins), are involved in the protein conglomerate, regulating glutamate release. Therefore, even minor deviances of the protein composition in these synapses could affect the balance in the autistic brain.

Cardiac diseases: Downregulation of transient outward potassium current (I_{to}) contributes to action potential profile alteration in numerous heart diseases, including congestive heart failure, myocardial infarction, and atrial fibrillations (reviewed in Nattel et al. 2007). Expression of both DPP6 and DPP10, which were shown to accelerate I_{to} current inactivation, are upregulated in human failing myocardium, potentially contributing to the I_{to} reduction (Radicke et al. 2005; Alders et al. 2009; Schmidt et al. 2010).

Cancer: Protein components of different K+ channel complexes have been correlated with tumor cell proliferation, growth, and survival. Recently, DPP6/DPP10 has been also put forward as possible prognostic marker for cancer progression and patient survival. RT-PCR analysis of DPP10 expression in the human primary malignant pleural mesothelioma (MPM) found that patients with tumors expressing *DPP10* transcripts had statistically significant better overall survival compared with patients whose tumors lacked any *DPP10* expression (Bueno et al. 2010). In addition, DPP6/10 is enzymatically silent structural homologues of DPP-IV, which is considered a cell adhesion molecule due to its ability to interact with extracellular matrix (ECM) proteins (e.g., collagen and fibronectin) (Hanski et al. 1985, 1988; Loster et al. 1995). Therefore, similarly to DPP-IV (Garin-Chesa et al. 1990; Sulda et al. 2006), DPP6, and DDP10 may also play a role in cancer biology.

38.4 LRRC26: LEUCINE-RICH REPEAT-CONTAINING PROTEIN 26 AND ITS PARALOGUES*

LRRC26 and LRRC52 are two members of a very large family of LRRC proteins (Bjorklund et al. 2006) that have recently been recognized as regulatory proteins of large-conductance, voltage, and calcium-activated K^+ channel (Yan and Aldrich 2010; Yang et al. 2011). LRRC26 and LRRC52 are essential tissue-specific activators of BK (Slo1) and KSper (Slo3) K^+ channels, respectively (Yan and Aldrich 2010, 2012; Yang et al. 2011; Leonetti et al. 2012). They were shown to co-assemble with ion-conducting channel subunits in nonexcitable tissues and shift activation gating of the resultant channel complexes to the physiologically relevant range of voltages and pH values (KSper channel only). Therefore, co-assembly with LRRC proteins is vital for defining specific native potassium currents in these tissues. Since the ubiquitously expressed LRRC proteins provide a versatile structural framework for achieving various protein–ligand interactions in many biologically important processes, including but not limited to regulation of gene expression and cellular trafficking, the major roles for these proteins in ion channel biology are just beginning to be discovered.

38.4.1 SEQUENCE AND STRUCTURE

The LRRC proteins are an astonishing large family of structurally dissimilar, functionally unrelated, and evolutionally unconnected proteins (the *LRRfinder.com*, a web-based tool for the identification of LRRC proteins, now uses a database of over 4000 unique, naturally occurring sequences). Sequence analyses have revealed at least seven distinct subfamilies of these proteins, with slightly different lengths and consensus sequences of individual leucine-rich repeats (LRRs) (Kobe and Kajava 2001; Dolan et al. 2007). Thus, the current number of LRRC proteins (4) that are known K^+ channel regulatory subunits is very small.

The typical LRR is 19–29 amino acids long with a highly conserved N-terminal sequence and a somewhat variable C-terminus. The water-exposed N-terminus (9–12 amino acids) is unusually rich in precisely positioned hydrophobic residues (typically leucines arranged in *LxxLxLxxN/CxL* sequence, can be also valines, isoleucines, and phenylalanines). This sequence corresponds to the β-strand and adjacent loop connected to the C-terminal stretch (10–19 amino acids), which forms the α-helix that is flanked by half-turns at both ends (Kobe and Deisenhofer 1993; Kajava 1998). These β-strand–turn–α-helix structural modules are further arrayed in tandems of 2–45 repeats, with each repeat forming one helical turn of an assembled solenoid domain. The first x-ray crystal structure of a LRR protein revealed that these multiple repeats adopt an arc, or horseshoe shape with an interior parallel β-sheet and an exterior array of α-helices (Kobe and Deisenhofer 1993). Thus, the concave side of this structure is a solvent-exposed β-sheet that forms a potentially large ligand-binding surface (Kobe and Kajava 2001). In this type of structure, hydrophobic core residues are buried between the α-helical array and β-sheet. Sequence and secondary structure variation of individual LRRs affects curvature of the domain (Bella et al. 2008) and contributes to the enormous diversity of potential substrates, which include proteins, nucleic acids, lipids, and small molecule hormones (Helft et al. 2011).

Based on localization of interacting domain, all LRRC proteins belong to either intracellular (iLRR) or extracellular (eLRR) proteins superfamily. Each superfamily is further categorized into classes, based on the architecture of extracellular domain and clustering with other proteins, but not on overall protein topology and localization; thus, each class may include secreted, GPI-anchored, single- and multiple-membrane-spanning proteins (Kobe and Kajava 2001; Dolan et al. 2007). LRRC26 and LRRC52 belong to the *e*xtracellular *L*eucine-*R*ich-*R*epeat–*on*ly (Elron) family, which also includes four other LRRC proteins: LRRC38, LRRC55, LRTM1, and LRTM2. These proteins are all predicted to be type I membrane proteins with a short (17–42 amino acids) cytoplasmic C-terminal tail, a single-pass TM segment, an extracellular LRRC domain made up 5 of LRRs flanked at the N- and C-terminal regions by cysteine-rich capping structures (known as LRR-NT and LRR-CT, respectively), and a cleavable N-terminal signal peptide for targeting to the ER (Figure 38.4a). The full-length isoform (isoform 1 or L-CAPC) of LRRC26 is predominantly expressed in prostate and salivary gland, and weakly in stomach, pancreas, intestine, colon, testis, and fetal brain. Isoform 1 is also expressed highly in many cancer cell lines, including breast, pancreatic, and colon carcinomas (Egland et al. 2006; Anaganti et al. 2009). There is also very short isoform (isoform 2 or S-CAPC), which is produced by alternative splicing and expressed in cancer cell lines only (Egland 2006; Anaganti et al. 2009). In contrast, LRRC52 is mainly expressed in testis and skeletal muscle (Yan and Aldrich 2012).

38.4.2 BIOGENESIS/ASSEMBLY/TRAFFICKING

Based on primary amino acid sequence, LRR-containing proteins of the *Elron* subfamily were predicted to be type I TM proteins with a cleavable N-terminal signal peptide, which is recognized by the SRP to target the protein to the ER translocational tunnel for translation of the LRR domain into the ER lumen (Figure 38.1, *type I with signal sequence*). Indeed, the N-terminal signal peptide of LRRC26/38/52/55 proteins is cleaved as if the proteins were properly inserted into the ER (Yan and Aldrich 2012); however, initial cellular localization studies showed that the mature proteins were not associated with the plasma membrane or any other intracellular membranes, but were predominantly located in the cytoplasm, and partially co-localized with cytoskeleton proteins (Egland et al. 2006). This befuddling result was recently resolved by the Aldrich group, which used immunoblotting and immunopurification combined

* Aliases: γ1–γ4.

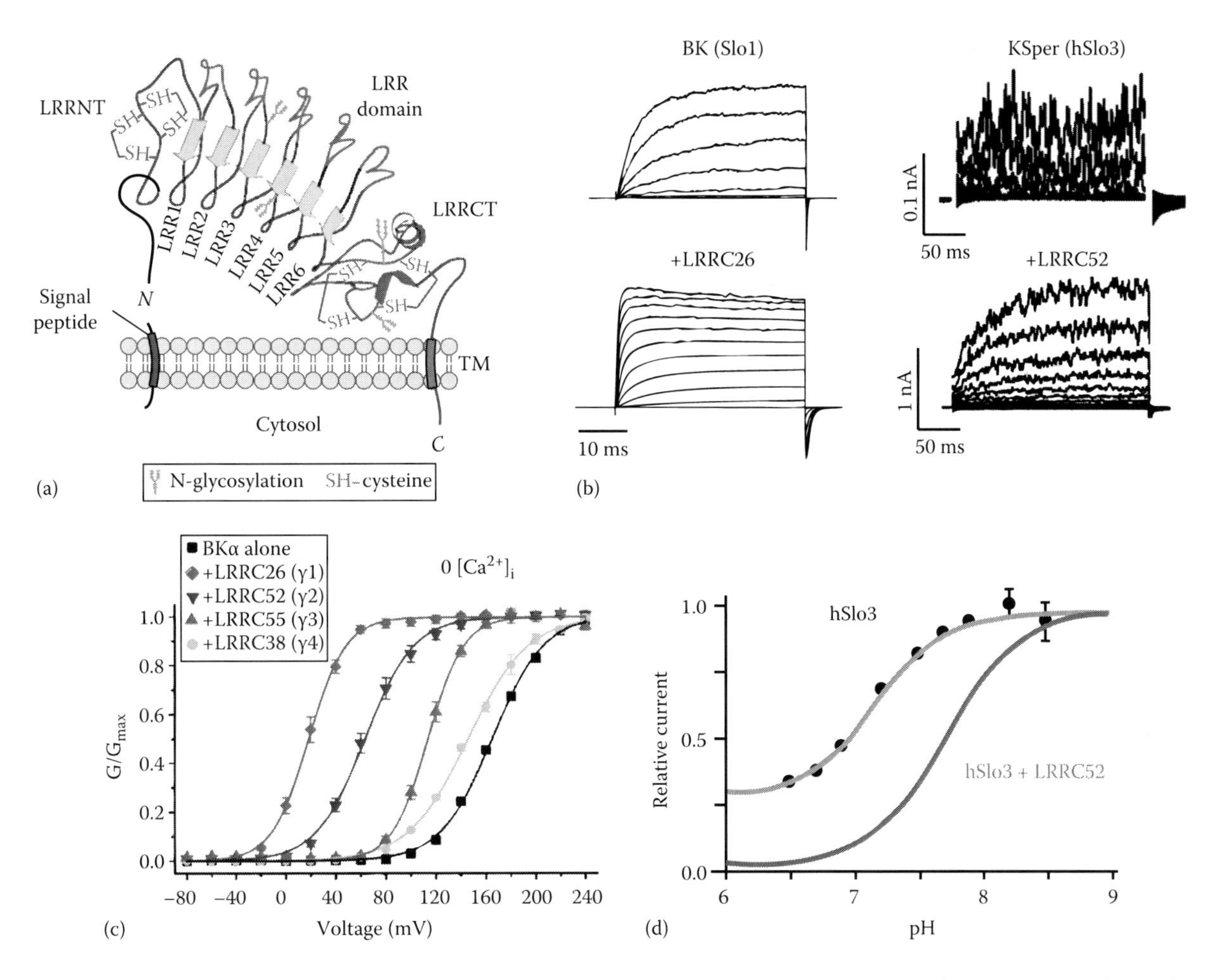

Figure 38.4 Slo potassium channels are modulated by LRRC proteins. (a) Predicted membrane topology of LRRC proteins of Elron family. LRRC domain was created using the atomic coordinates of homologous domain from hagfish VLR-B59 protein (PDB 2O6S). Approximate positions of N-glycosylation sites and cysteine residues for potential disulfide formations are depicted. (b) Families of BK (left) and KSper (right) current traces recorded from cells expressing channels alone, or co-expressed with LRRC26 or LRRC52, respectively. Command voltage steps for families of traces are 20 mV; BK(Slo1): −80 to +200 mV; KSper(hSlo3): −140 to +180 mV. (Illustration adopted from Yan, J. and Aldrich, R.W., *Proc. Natl. Acad. Sci. U.S.A.*, 109(20), 7917, 2012; Leonetti, M.D. et al., *Proc. Natl. Acad. Sci. U.S.A.*, 109(47), 19274, 2012.) (c) LRRC proteins shift the voltage dependence of BK channel activation toward more hyperpolarizing potentials in the absence of intracellular Ca^{2+}. (Illustration adopted from Yan, J. and Aldrich, R.W., *Proc. Natl. Acad. Sci. U.S.A.*, 109(20), 7917, 2012.) (d) pH dependence of hSlo3 currents is altered by hLRRC52 co-expression. (Illustration adopted from Leonetti, M.D. et al., *Proc. Natl. Acad. Sci. U.S.A.*, 109(47), 19274, 2012.)

with liquid chromatography and tandem mass spectrometry to show that membrane-embedded LRRC26 associates with α-subunits of BK (Slo1) channel (Yan and Aldrich 2010). Although the interaction between LRRC26/52 proteins with Slo1/3 channels is well documented (Yan and Aldrich 2010, 2012; Yang et al. 2011), this membrane-embedded protein–protein interaction is fickle. Yang and colleagues were unable to co-immunoprecipitate LRRC52 with KSper (Slo3) channel from native tissue lysate but were able to detect co-assembly between the heterologously expressed proteins (Yang et al. 2011). Similarly, a fusion construct between BK and LRRC that self-cleaves in the ER is needed to reproducibly observe the modulated currents in HEK 293 cells, indicating that co-translation of the subunits may be required in native tissues.

As with most of regulatory subunits, the assembly of LRRC proteins with pore-forming subunits appears to occur early during biogenesis (Yang et al. 2011). First, quantitative analysis of KSper and Elron LRRC proteins mRNA expression in postpartum mice revealed similar time course and level of abundance for KSper and LRRC52 messengers. Second, expression of both KSper and LRRC52 proteins was restricted to the similar areas in testis. Finally, biotinylation experiments in oocytes demonstrated that the KSper is required for the LRRC52 proteins to appear in surface membrane. Currently, there is no information about the stoichiometry of a LRRC-Slo channel complex. However, another membrane-embedded subunit of the Slo family (β1) functionally competes with LRRC26, suggesting that the LRRC proteins and the betas (β1-β4) may have similar or overlapping interaction sites on BK (Slo1) channels (Yan and Aldrich 2010). Thus, the channel complex may consist of up to four LRRC subunits, similar to the 1:1 stoichiometry of α1:β1 subunits (Wang et al. 2002). However, it remains unclear whether steric or allosteric factors prevent simultaneous binding of β1 and LRRC26 to the BK channel.

N-linked glycosylation has been shown to play essential roles in the trafficking and localization of various LRRC proteins (Weber et al. 2004; Sun et al. 2006; Yan et al. 2008). Both N-linked glycosylation and disulfide bonds are also crucial for structure, stability, and function of various LRRC proteins

(Meng et al. 2008; Yan et al. 2008). The horseshoe-shaped LRRC domain is decorated with glycosyl modifications, as demonstrated by several resolved crystal X-ray structures (Yan, Scott et al. 2008; Bell, Botos et al. 2005). In addition to N-linked glycosylation, both human LRRC26 and LRRC52 contain four pairs of conserved cysteines in their N- and C-terminal LRR-capping domains (see Q2I0M4 and Q8N7C0 entries in *UniProtKB/Swiss-Prot*), which are predicted to form disulfide bonds in the oxidizing environment of the ER and are thought to enhance the solubility of LRR domain and protect it from proteolysis when it reaches the plasma membrane (Hogg 2003). Based on the protein sequence, five N-glycosylation sites are predicted in human LRRC52 (Yang et al. 2011); however, the rapid folding of and disulfide formation in LRR domains has the potential to prevent posttranslational N-glycosylation in the ER (Ruiz-Canada et al. 2009), thereby reducing the number of consensus sites modified by the oligosaccharyltransferase (OST). In any case, deglycosylation with PNGase F (a glycosidase that removes most mature *N*-glycans) validated that multiple *N*-glycans are attached to LRRC52 (Yang et al. 2011) and at least one *N*-glycan at Asn-147 is present in the middle of LRR domain in LRRC26 (Yan and Aldrich 2012).

38.4.3 K^+ CHANNEL MODULATION

LRRC26 and LRC52 proteins represent a novel type of K^+ channel regulatory subunits that associate with the Big-conductance voltage- and Ca^{2+}-activated K^+ channels (BK, K_{Ca}) of the Slo family, and modulate the channel's surface expression and activation gating (Yan and Aldrich 2010, 2012; Yang et al. 2011), as well as sensitivity to pH (Yang et al. 2011) and pharmacological activation (Almassy and Begenisich 2012) (Figure 38.4b through d). BK channels are both voltage and ligand gated: they can be activated by either strong (to +100 mV or more) depolarizations or by elevation in free cytosolic Ca^{2+} (or Mg^{2+}) concentration. Usually, both a rise in cytosolic Ca^{2+} and membrane depolarization are required to elicit physiologically meaningful response in excitable cells, where the BK channel contributes to repolarization of the membrane potential. However, BK channels are ubiquitously expressed in the body and are also active in nonexcitable cells, which is due to their interactions with the different regulatory subunits that greatly modify the BK channel's gating kinetics and voltage/Ca^{2+} sensitivities.

LRRC proteins are recent, but most likely not the last addition to the list of BK channel regulatory subunits, which also include the membrane-embedded β-subunits (β1–β4) (Knaus et al. 1994; Wallner et al. 1999; Brenner et al. 2000; Weiger et al. 2000) and KCNE4 (Levy et al. 2008) peptide in mammals, Slob (Zhou Et al. 1999) and dSLIP1 (Xia et al. 1998) in *Drosophila*, and BKIP-1 (Chen et al. 2010) in roundworms. To differentiate the LRRC family from the membrane-embedded β-subunits, LRRC26/52/55/38 have been recently coined gamma-subunits (γ1–γ4), respectively. The main functional consequence of interaction with BK channels is a leftward (hyperpolarized) shift in the voltage dependence of activation (Figure 38.4c). This shift to lower voltages enables the BK channel to operate in nonexcitable tissues because the BK/LRRC complex is activated at the negative resting potential of most cells (and, for pH sensitive KSper, at lower pH, Figure 38.4d) without requiring an elevation of free cytosolic Ca^{2+}. Of the four members of Elron subfamily of LRRC proteins that have been found to modulate the BK channel's voltage dependence of activation, LRRC26 (γ1) is the most potent activator, producing a negative shift of almost ~140 mV (Yan and Aldrich 2012). In addition to shifting the voltage dependence of BK channels, interaction with LRRC proteins significantly modifies BK channel's current kinetics, accelerating channel activation and decelerating its deactivation.

To determine the regions of the LRRC proteins that are responsible for assembly with and modulation of BK channels, Yan and Aldrich were guided by the x-ray structures of the variable lymphocyte receptor B from hagfish (Kim et al. 2007a) for the LRRNT and LRR regions, and mouse TLR4 (Kim et al. 2007b) for the LRRCT. Subsequent functional analysis of deletion mutants demonstrated that the TM segment is needed for co-assembly, whereas removal of any of the LRR domains greatly diminishes LRRC26's (γ1) effect on shifting the voltage dependence of the BK channel (Yan and Aldrich 2010). The proposed mechanism underlying BK channel modulation by LRRC proteins is different than the β subunits (β1-β4) (Liu et al. 2008, 2010; Wu et al. 2009; Morera et al. 2012): LRRC26-induced effects on the channel gating are mainly through facilitation of the coupling between voltage-sensor movement and activation gate opening (Yan and Aldrich 2010); thus, the interaction with LRRC26 apparently lowers the energy barrier for the BK channel activation, stabilizing its open state at more negative membrane potentials (Braun 2010).

38.4.4 PHYSIOLOGY AND DISEASE

The discovery that LRRC proteins co-assemble and modulate Slo channels is relatively recent, so the physiological roles of the complexes are just starting to emerge. Ironically, both LRRC26 (γ1) and LRRC52 (γ2) were first associated with malignant neoplastic diseases—liver, pancreatic, colon, prostate, breast cancer (Egland et al. 2006; Wang et al. 2006; Anaganti et al. 2009; Liu et al. 2012), and t-cell leukemia (Cheung et al. 2009)—before they were identified as Slo channel regulatory subunits. A potential role for the LRRC26/BK complex in cancer is to hyperpolarize the membrane of nonexcitable cells, which may facilitate Ca^{2+} entry, and thus, uncontrolled cell growth and division (Braun 2010). If this scenario is true, disruption of LRRC26/BK channel interaction would be a viable growth suppressing strategy in the treatment of cancers expressing LRRC26.

REFERENCES

Abbott, G. W., M. H. Butler et al. (2001). MiRP2 forms potassium channels in skeletal muscle with Kv3.4 and is associated with periodic paralysis. *Cell* **104**(2): 217–231.

Abbott, G. W. and S. A. Goldstein (1998). A superfamily of small potassium channel subunits: Form and function of the MinK-related peptides (MiRPs). *Q Rev Biophys* **31**(4): 357–398.

Abbott, G. W., S. A. Goldstein et al. (2001). Do all voltage-gated potassium channels use MiRPs? *Circ Res* **88**(10): 981–983.

Abbott, G. W., B. Ramesh et al. (2008). Secondary structure of the MiRP1 (KCNE2) potassium channel ancillary subunit. *Protein Pept Lett* **15**(1): 63–75.

Abbott, G. W., F. Sesti et al. (1999). MiRP1 forms IKr potassium channels with HERG and is associated with cardiac arrhythmia. *Cell* **97**(2): 175–187.

Albesa, M., L. S. Grilo et al. (2011). Nedd4–2-dependent ubiquitylation and regulation of the cardiac potassium channel hERG1. *J Mol Cell Cardiol* **51**(1): 90–98.

Alders, M., T. T. Koopmann et al. (2009). Haplotype-sharing analysis implicates chromosome 7q36 harboring DPP6 in familial idiopathic ventricular fibrillation. *Am J Hum Genet* **84**(4): 468–476.

Alders, M. and M. Mannens (1993). Romano-ward syndrome. *Gene Reviews*. R. A. Pagon, T. D. Bird, C. R. Dolan, K. Stephens and M. P. Adam. (eds.), Seattle, WA.

Allen, M., A. Heinzmann et al. (2003). Positional cloning of a novel gene influencing asthma from chromosome 2q14. *Nat Genet* **35**(3): 258–263.

Almassy, J. and T. Begenisich (2012). The LRRC26 protein selectively alters the efficacy of BK channel activators. *Mol Pharmacol* **81**(1): 21–30.

Amarillo, Y., J. A. De Santiago-Castillo et al. (2008). Ternary Kv4.2 channels recapitulate voltage-dependent inactivation kinetics of A-type K+ channels in cerebellar granule neurons. *J Physiol* **586**(8): 2093–2106.

Anaganti, S., J. K. Hansen et al. (2009). Non-AUG translational initiation of a short CAPC transcript generating protein isoform. *Biochem Biophys Res Commun* **380**(3): 508–513.

Angelo, K., T. Jespersen et al. (2002). KCNE5 induces time- and voltage-dependent modulation of the KCNQ1 current. *Biophys J* **83**(4): 1997–2006.

Barhanin, J., F. Lesage et al. (1996). K(V)LQT1 and lsK (minK) proteins associate to form the I(Ks) cardiac potassium current. *Nature* **384**(6604): 78–80.

Barriere, H., I. Rubera et al. (2003). Swelling-activated chloride and potassium conductance in primary cultures of mouse proximal tubules. Implication of KCNE1 protein. *J Membr Biol* **193**(3): 153–170.

Bas, T., G. Y. Gao et al. (2011). Post-translational N-glycosylation of type I transmembrane KCNE1 peptides: Implications for membrane protein biogenesis and disease. *J Biol Chem* **286**(32): 28150–28159.

Bell, J. K., I. Botos et al. (2005). The molecular structure of the Toll-like receptor 3 ligand-binding domain. *Proc Natl Acad Sci USA* **102**(31): 10976–10980.

Bella, J., K. L. Hindle et al. (2008). The leucine-rich repeat structure. *Cell Mol Life Sci* **65**(15): 2307–2333.

Bjorklund, A. K., D. Ekman et al. (2006). Expansion of protein domain repeats. *PLoS Comput Biol* **2**(8): e114.

Boronat, A., J. M. Gelfand et al. (2012). Encephalitis and antibodies to dipeptidyl-peptidase-like protein-6, a subunit of Kv4.2 potassium channels. *Ann Neurol* **73**: 120–128.

Brambilla, P., F. Esposito et al. (2012). Association between DPP6 polymorphism and the risk of progressive multiple sclerosis in Northern and Southern Europeans. *Neurosci Lett* **530**(2): 155–160.

Brandt, M. C., J. Endres-Becker et al. (2009). Effects of KCNE2 on HCN isoforms: Distinct modulation of membrane expression and single channel properties. *Am J Physiol Heart Circ Physiol* **297**(1): H355–H363.

Braun, A. P. (2010). A new "opening" act on the BK channel stage: Identification of LRRC26 as a novel BK channel accessory subunit that enhances voltage-dependent gating. *Channels (Austin)* **4**(4): 249–250.

Brenner, R., T. J. Jegla et al. (2000). Cloning and functional characterization of novel large conductance calcium-activated potassium channel beta subunits, hKCNMB3 and hKCNMB4. *J Biol Chem* **275**(9): 6453–6461.

Brugada, P. and J. Brugada (1992). Right bundle branch block, persistent ST segment elevation and sudden cardiac death: A distinct clinical and electrocardiographic syndrome. A multicenter report. *J Am Coll Cardiol* **20**(6): 1391–1396.

Bueno, R., A. De Rienzo et al. (2010). Second generation sequencing of the mesothelioma tumor genome. *PLoS One* **5**(5): e10612.

Chandrasekhar, K. D., T. Bas et al. (2006). KCNE1 subunits require co-assembly with K+ channels for efficient trafficking and cell surface expression. *J Biol Chem* **281**(52): 40015–40023.

Chandrasekhar, K. D., A. Lvov et al. (2011). O-glycosylation of the cardiac I(Ks) complex. *J Physiol* **589**(Part 15): 3721–3730.

Chen, B., Q. Ge et al. (2010). A novel auxiliary subunit critical to BK channel function in *Caenorhabditis elegans*. *J Neurosci* **30**(49): 16651–16661.

Chen, H., L. A. Kim et al. (2003a). Charybdotoxin binding in the I(Ks) pore demonstrates two MinK subunits in each channel complex. *Neuron* **40**(1): 15–23.

Chen, H., F. Sesti et al. (2003b). Pore- and state-dependent cadmium block of I(Ks) channels formed with MinK-55C and wild-type KCNQ1 subunits. *Biophys J* **84**(6): 3679–3689.

Chen, T., K. Ajami et al. (2006). Molecular characterization of a novel dipeptidyl peptidase like 2-short form (DPL2-s) that is highly expressed in the brain and lacks dipeptidyl peptidase activity. *Biochim Biophys Acta* **1764**(1): 33–43.

Chen, T., X. Shen et al. (2008). Molecular characterisation of a novel dipeptidyl peptidase like protein: Its pathological link to Alzheimer's disease. *Clin Chem Lab Med* **46**(4): A13.

Chen, Y. H., S. J. Xu et al. (2003c). KCNQ1 gain-of-function mutation in familial atrial fibrillation. *Science* **299**(5604): 251–254.

Cheung, C. L., B. Y. Chan et al. (2009). Pre-B-cell leukemia homeobox 1 (PBX1) shows functional and possible genetic association with bone mineral density variation. *Hum Mol Genet* **18**(4): 679–687.

Chouabe, C., N. Neyroud et al. (1997). Properties of KvLQT1 K+ channel mutations in Romano-Ward and Jervell and Lange-Nielsen inherited cardiac arrhythmias. *EMBO J* **16**(17): 5472–5479.

Chouabe, C., N. Neyroud et al. (2000). Novel mutations in KvLQT1 that affect Iks activation through interactions with Isk. *Cardiovasc Res* **45**(4): 971–980.

Chung, D. Y., P. J. Chan et al. (2009). Location of KCNE1 relative to KCNQ1 in the I(KS) potassium channel by disulfide cross-linking of substituted cysteines. *Proc Natl Acad Sci USA* **106**(3): 743–748.

Ciampa, E. J., R. C. Welch et al. (2011). KCNE4 juxtamembrane region is required for interaction with calmodulin and for functional suppression of KCNQ1. *J Biol Chem* **286**(6): 4141–4149.

Cotella, D., S. Radicke et al. (2012). N-glycosylation of the mammalian dipeptidyl aminopeptidase-like protein 10 (DPP10) regulates trafficking and interaction with Kv4 channels. *Int J Biochem Cell Biol* **44**(6): 876–885.

Covarrubias, M., A. Bhattacharji et al. (2008). The neuronal Kv4 channel complex. *Neurochem Res* **33**(8): 1558–1567.

Cronin, S., S. Berger et al. (2008). A genome-wide association study of sporadic ALS in a homogenous Irish population. *Hum Mol Genet* **17**(5): 768–774.

David, J. P., M. N. Andersen et al. (2013). Trafficking of the I(Ks)-complex in MDCK cells: Site of subunit assembly and determinants of polarized localization. *Traffic* **14**: 399–411.

Dedek, K. and S. Waldegger (2001). Colocalization of KCNQ1/KCNE channel subunits in the mouse gastrointestinal tract. *Pflugers Arch* **442**(6): 896–902.

Del Bo, R., S. Ghezzi et al. (2008). DPP6 gene variability confers increased risk of developing sporadic amyotrophic lateral sclerosis in Italian patients. *J Neurol Neurosurg Psychiatry* **79**(9): 1085.

Delpon, E., J. M. Cordeiro et al. (2008). Functional effects of KCNE3 mutation and its role in the development of Brugada syndrome. *Circ Arrhythm Electrophysiol* **1**(3): 209–218.

Dolan, J., K. Walshe et al. (2007). The extracellular leucine-rich repeat superfamily; a comparative survey and analysis of evolutionary relationships and expression patterns. *BMC Genomics* **8**: 320.

Dougherty, K. and M. Covarrubias (2006). A dipeptidyl aminopeptidase-like protein remodels gating charge dynamics in Kv4.2 channels. *J Gen Physiol* **128**(6): 745–753.

Doyle, D. A., J. Morais Cabral et al. (1998). The structure of the potassium channel: Molecular basis of K^+ conduction and selectivity. *Science* **280**(5360): 69–77.

Duggal, P., M. R. Vesely et al. (1998). Mutation of the gene for IsK associated with both Jervell and Lange-Nielsen and Romano-Ward forms of Long-QT syndrome. *Circulation* **97**(2): 142–146.

Egland, K. A., X. F. Liu et al. (2006). High expression of a cytokeratin-associated protein in many cancers. *Proc Natl Acad Sci USA* **103**(15): 5929–5934.

Fass, D. (2012). Disulfide bonding in protein biophysics. *Annu Rev Biophys* **41**: 63–79.

Foeger, N. C., A. J. Norris et al. (2012). Augmentation of Kv4.2-encoded currents by accessory dipeptidyl peptidase 6 and 10 subunits reflects selective cell surface Kv4.2 protein stabilization. *J Biol Chem* **287**(12): 9640–9650.

Gage, S. D. and W. R. Kobertz (2004). KCNE3 truncation mutants reveal a bipartite modulation of KCNQ1 K^+ channels. *J Gen Physiol* **124**(6): 759–771.

Garber, K. (2008). Genetics. The elusive ALS genes. *Science* **319**(5859): 20.

Garcia-Patrone, M. and J. S. Tandecarz (1995). A glycoprotein multimer from Bacillus thuringiensis sporangia: Dissociation into subunits and sugar composition. *Mol Cell Biochem* **145**(1): 29–37.

Garin-Chesa, P., L. J. Old et al. (1990). Cell surface glycoprotein of reactive stromal fibroblasts as a potential antibody target in human epithelial cancers. *Proc Natl Acad Sci USA* **87**(18): 7235–7239.

Goldstein, S. A. and C. Miller (1991). Site-specific mutations in a minimal voltage-dependent K^+ channel alter ion selectivity and open-channel block. *Neuron* **7**(3): 403–408.

Gordon, E., T. K. Roepke et al. (2006). Endogenous KCNE subunits govern Kv2.1 K^+ channel activation kinetics in *Xenopus* oocyte studies. *Biophys J* **90**(4): 1223–1231.

Gorrell, M. D. (2005). Dipeptidyl peptidase IV and related enzymes in cell biology and liver disorders. *Clin Sci (Lond)* **108**(4): 277–292.

Grahammer, F., A. Herling et al. (2001). The cardiac K^+ channel KCNQ1 Is essential for gastric acid secretion. *Gastroenterology* **120**(6): 1363–1371.

Grahammer, F., R. Warth et al. (2001). The small conductance K^+ channel, KCNQ1: Expression, function, and subunit composition in murine trachea. *J Biol Chem* **276**(45): 42268–42275.

Grammatikos, A. P. (2008). The genetic and environmental basis of atopic diseases. *Ann Med* **40**(7): 482–495.

Grunnet, M., T. Jespersen et al. (2002). KCNE4 is an inhibitory subunit to the KCNQ1 channel. *J Physiol* **542**(Part 1): 119–130.

Grunnet, M., T. Jespersen et al. (2003). KCNQ1 channels sense small changes in cell volume. *J Physiol* **549**(Part 2): 419–427.

Haitin, Y., R. Wiener et al. (2009). Intracellular domains interactions and gated motions of I(KS) potassium channel subunits. *EMBO J* **28**(14): 1994–2005.

Hanski, C., T. Huhle et al. (1985). Involvement of plasma membrane dipeptidyl peptidase IV in fibronectin-mediated adhesion of cells on collagen. *Biol Chem Hoppe Seyler* **366**(12): 1169–1176.

Hanski, C., T. Huhle et al. (1988). Direct evidence for the binding of rat liver DPP IV to collagen in vitro. *Exp Cell Res* **178**(1): 64–72.

Helft, L., V. Reddy et al. (2011). LRR conservation mapping to predict functional sites within protein leucine-rich repeat domains. *PLoS One* **6**(7): e21614.

Hogg, P. J. (2003). Disulfide bonds as switches for protein function. *Trends Biochem Sci* **28**(4): 210–214.

Hong, K., D. R. Piper et al. (2005). De novo KCNQ1 mutation responsible for atrial fibrillation and short QT syndrome in utero. *Cardiovasc Res* **68**(3): 433–440.

Huet, G., S. Hennebicq-Reig et al. (1998). GalNAc-alpha-O-benzyl inhibits NeuAcalpha2–3 glycosylation and blocks the intracellular transport of apical glycoproteins and mucus in differentiated HT-29 cells. *J Cell Biol* **141**(6): 1311–1322.

Jerng, H. H., K. Dougherty et al. (2009). A novel N-terminal motif of dipeptidyl peptidase-like proteins produces rapid inactivation of KV4.2 channels by a pore-blocking mechanism. *Channels (Austin)* **3**(6): 448–461.

Jerng, H. H., K. Kunjilwar et al. (2005). Multiprotein assembly of Kv4.2, KChIP3 and DPP10 produces ternary channel complexes with ISA-like properties. *J Physiol* **568**(Part 3): 767–788.

Jerng, H. H., A. D. Lauver et al. (2007). DPP10 splice variants are localized in distinct neuronal populations and act to differentially regulate the inactivation properties of Kv4-based ion channels. *Mol Cell Neurosci* **35**(4): 604–624.

Jerng, H. H. and P. J. Pfaffinger (2012). Incorporation of DPP6a and DPP6K variants in ternary Kv4 channel complex reconstitutes properties of A-type K current in rat cerebellar granule cells. *PLoS One* **7**(6): e38205.

Jerng, H. H., Y. Qian et al. (2004). Modulation of Kv4.2 channel expression and gating by dipeptidyl peptidase 10 (DPP10). *Biophys J* **87**(4): 2380–2396.

Jespersen, T., M. Membrez et al. (2007). The KCNQ1 potassium channel is down-regulated by ubiquitylating enzymes of the Nedd4/Nedd4-like family. *Cardiovasc Res* **74**(1): 64–74.

Jiang, Y., A. Lee et al. (2003). X-ray structure of a voltage-dependent K^+ channel. *Nature* **423**(6935): 33–41.

Kaczmarek, L. K. (1991). Voltage-dependent potassium channels: minK and *Shaker* families. *New Biol* **3**(4): 315–323.

Kajava, A. V. (1998). Structural diversity of leucine-rich repeat proteins. *J Mol Biol* **277**(3): 519–527.

Kanda, V. A. and G. W. Abbott (2012). KCNE regulation of K(+) channel trafficking—A Sisyphean task? *Front Physiol* **3**: 231.

Kanelis, V., D. Rotin et al. (2001). Solution structure of a Nedd4 WW domain-ENaC peptide complex. *Nat Struct Biol* **8**(5): 407–412.

Kang, C., C. Tian et al. (2008). Structure of KCNE1 and implications for how it modulates the KCNQ1 potassium channel. *Biochemistry* **47**(31): 7999–8006.

Kaulin, Y. A., J. A. De Santiago-Castillo et al. (2009). The dipeptidyl-peptidase-like protein DPP6 determines the unitary conductance of neuronal Kv4.2 channels. *J Neurosci* **29**(10): 3242–3251.

Kim, H. M., S. C. Oh et al. (2007). Structural diversity of the hagfish variable lymphocyte receptors. *J Biol Chem* **282**(9): 6726–6732.

Kim, H. M., B. S. Park et al. (2007). Crystal structure of the TLR4-MD-2 complex with bound endotoxin antagonist Eritoran. *Cell* **130**(5): 906–917.

Kin, Y., Y. Misumi et al. (2001). Biosynthesis and characterization of the brain-specific membrane protein DPPX, a dipeptidyl peptidase IV-related protein. *J Biochem* **129**(2): 289–295.

Knaus, H. G., K. Folander et al. (1994). Primary sequence and immunological characterization of beta-subunit of high conductance Ca(2+)-activated K^+ channel from smooth muscle. *J Biol Chem* **269**(25): 17274–17278.

Kobe, B. and J. Deisenhofer (1993). Crystal structure of porcine ribonuclease inhibitor, a protein with leucine-rich repeats. *Nature* **366**(6457): 751–756.

Kobe, B. and A. V. Kajava (2001). The leucine-rich repeat as a protein recognition motif. *Curr Opin Struct Biol* **11**(6): 725–732.

Krumerman, A., X. Gao et al. (2004). An LQT mutant minK alters KvLQT1 trafficking. *Am J Physiol Cell Physiol* **286**(6): C1453–C1463.

Kunjilwar, K., C. Strang et al. (2004). KChIP3 rescues the functional expression of Shal channel tetramerization mutants. *J Biol Chem* **279**(52): 54542–54551.

Kurokawa, J., H. K. Motoike et al. (2001). TEA(+)-sensitive KCNQ1 constructs reveal pore-independent access to KCNE1 in assembled I(Ks) channels. *J Gen Physiol* **117**(1): 43–52.

Lai, L. P., Y. N. Su et al. (2005). Denaturing high-performance liquid chromatography screening of the long QT syndrome-related cardiac sodium and potassium channel genes and identification of novel mutations and single nucleotide polymorphisms. *J Hum Genet* **50**(9): 490–496.

Leonetti, M. D., P. Yuan et al. (2012). Functional and structural analysis of the human SLO3 pH- and voltage-gated K^+ channel. *Proc Natl Acad Sci USA* **109**(47): 19274–19279.

Levy, D. I., S. Wanderling et al. (2008). MiRP3 acts as an accessory subunit with the BK potassium channel. *Am J Physiol Renal Physiol* **295**(2): F380–F387.

Liu, G., X. Niu et al. (2010). Location of modulatory beta subunits in BK potassium channels. *J Gen Physiol* **135**(5): 449–459.

Liu, G., S. I. Zakharov et al. (2008a). Locations of the beta1 transmembrane helices in the BK potassium channel. *Proc Natl Acad Sci USA* **105**(31): 10727–10732.

Liu, W. J., H. T. Wang et al. (2008b). Co-expression of KCNE2 and KChIP2c modulates the electrophysiological properties of Kv4.2 current in COS-7 cells. *Acta Pharmacol Sin* **29**(6): 653–660.

Liu, X. F., L. Xiang et al. (2012). CAPC negatively regulates NF-kappaB activation and suppresses tumor growth and metastasis. *Oncogene* **31**(13): 1673–1682.

Long, S. B., E. B. Campbell et al. (2005). Crystal structure of a mammalian voltage-dependent *Shaker* family K^+ channel. *Science* **309**(5736): 897–903.

Loster, K., K. Zeilinger et al. (1995). The cysteine-rich region of dipeptidyl peptidase IV (CD 26) is the collagen-binding site. *Biochem Biophys Res Commun* **217**(1): 341–348.

Loussouarn, G., K. H. Park et al. (2003). Phosphatidylinositol-4,5-bisphosphate, PIP2, controls KCNQ1/KCNE1 voltage-gated potassium channels: A functional homology between voltage-gated and inward rectifier K^+ channels. *EMBO J* **22**(20): 5412–5421.

Lu, Y., M. P. Mahaut-Smith et al. (2003). Mutant MiRP1 subunits modulate HERG K^+ channel gating: A mechanism for pro-arrhythmia in long QT syndrome type 6. *J Physiol* **551**(Part 1): 253–262.

Lundby, A., L. S. Ravn et al. (2007). KCNQ1 mutation Q147R is associated with atrial fibrillation and prolonged QT interval. *Heart Rhythm* **4**(12): 1532–1541.

Lundquist, A. L., C. L. Turner et al. (2006). Expression and transcriptional control of human KCNE genes. *Genomics* **87**(1): 119–128.

Lvov, A., S. D. Gage et al. (2010). Identification of a protein-protein interaction between KCNE1 and the activation gate machinery of KCNQ1. *J Gen Physiol* **135**(6): 607–618.

Ma, L., C. Lin et al. (2003). Characterization of a novel Long QT syndrome mutation G52R-KCNE1 in a Chinese family. *Cardiovasc Res* **59**(3): 612–619.

Manderfield, L. J., M. A. Daniels et al. (2009). KCNE4 domains required for inhibition of KCNQ1. *J Physiol* **587**(Part 2): 303–314.

Marshall, C. R., A. Noor et al. (2008). Structural variation of chromosomes in autism spectrum disorder. *Am J Hum Genet* **82**(2): 477–488.

McCrossan, Z. A. and G. W. Abbott (2004). The MinK-related peptides. *Neuropharmacology* **47**(6): 787–821.

McNicholas, K., T. Chen et al. (2009). Dipeptidyl peptidase (DP) 6 and DP10: Novel brain proteins implicated in human health and disease. *Clin Chem Lab Med* **47**(3): 262–267.

Melman, Y. F., A. Domenech et al. (2001). Structural determinants of KvLQT1 control by the KCNE family of proteins. *J Biol Chem* **276**(9): 6439–6444.

Melman, Y. F., A. Krumerman et al. (2002). A single transmembrane site in the KCNE-encoded proteins controls the specificity of KvLQT1 channel gating. *J Biol Chem* **277**(28): 25187–25194.

Melman, Y. F., S. Y. Um et al. (2004). KCNE1 binds to the KCNQ1 pore to regulate potassium channel activity. *Neuron* **42**(6): 927–937.

Meng, J., P. Parroche et al. (2008). The differential impact of disulfide bonds and N-linked glycosylation on the stability and function of CD14. *J Biol Chem* **283**(6): 3376–3384.

Morera, F. J., A. Alioua et al. (2012). The first transmembrane domain (TM1) of beta2-subunit binds to the transmembrane domain S1 of alpha-subunit in BK potassium channels. *FEBS Lett* **586**(16): 2287–2293.

Morin, T. J. and W. R. Kobertz (2008a). Counting membrane-embedded KCNE beta-subunits in functioning K^+ channel complexes. *Proc Natl Acad Sci USA* **105**(5): 1478–1482.

Morin, T. J. and W. R. Kobertz (2008b). Tethering chemistry and K^+ channels. *J Biol Chem* **283**(37): 25105–25109.

Nadal, M. S., A. Ozaita et al. (2003). The CD26-related dipeptidyl aminopeptidase-like protein DPPX is a critical component of neuronal A-type K^+ channels. *Neuron* **37**(3): 449–461.

Nagase, T., R. Kikuno et al. (2000). Prediction of the coding sequences of unidentified human genes. XVII. The complete sequences of 100 new cDNA clones from brain which code for large proteins in vitro. *DNA Res* **7**(2): 143–150.

Naim, H. Y., G. Joberty et al. (1999). Temporal association of the N- and O-linked glycosylation events and their implication in the polarized sorting of intestinal brush border sucrase-isomaltase, aminopeptidase N, and dipeptidyl peptidase IV. *J Biol Chem* **274**(25): 17961–17967.

Nakajo, K. and Y. Kubo (2007). KCNE1 and KCNE3 stabilize and/or slow voltage sensing S4 segment of KCNQ1 channel. *J Gen Physiol* **130**(3): 269–281.

Nakajo, K., M. H. Ulbrich et al. (2010). Stoichiometry of the KCNQ1-KCNE1 ion channel complex. *Proc Natl Acad Sci USA* **107**(44): 18862–18867.

Napolitano, C., S. G. Priori et al. (2005). Genetic testing in the long QT syndrome: Development and validation of an efficient approach to genotyping in clinical practice. *JAMA* **294**(23): 2975–2980.

Nattel, S., A. Maguy et al. (2007). Arrhythmogenic ion-channel remodeling in the heart: Heart failure, myocardial infarction, and atrial fibrillation. *Physiol Rev* **87**(2): 425–456.

Neyroud, N., F. Tesson et al. (1997). A novel mutation in the potassium channel gene KVLQT1 causes the Jervell and Lange-Nielsen cardioauditory syndrome. *Nat Genet* **15**(2): 186–189.

Nicolas, C. S., K. H. Park et al. (2008). IKs response to protein kinase A-dependent KCNQ1 phosphorylation requires direct interaction with microtubules. *Cardiovasc Res* **79**(3): 427–435.

Ohno, S., D. P. Zankov et al. (2011). KCNE5 (KCNE1L) variants are novel modulators of Brugada syndrome and idiopathic ventricular fibrillation. *Circ Arrhythm Electrophysiol* **4**(3): 352–361.

Ohya, S., K. Asakura et al. (2002). Molecular and functional characterization of ERG, KCNQ, and KCNE subtypes in rat stomach smooth muscle. *Am J Physiol Gastrointest Liver Physiol* **282**(2): G277–G287.

Osteen, J. D., C. Gonzalez et al. (2010). KCNE1 alters the voltage sensor movements necessary to open the KCNQ1 channel gate. *Proc Natl Acad Sci USA* **107**(52): 22710–22715.

Panaghie, G., K. K. Tai et al. (2006). Interaction of KCNE subunits with the KCNQ1 K^+ channel pore. *J Physiol* **570**(Part 3): 455–467.

Piccini, M., F. Vitelli et al. (1999). KCNE1-like gene is deleted in AMME contiguous gene syndrome: Identification and characterization of the human and mouse homologs. *Genomics* **60**(3): 251–257.

Pickart, C. M. (2001). Mechanisms underlying ubiquitination. *Annu Rev Biochem* **70**: 503–533.
Poulsen, A. N. and D. A. Klaerke (2007). The KCNE1 beta-subunit exerts a transient effect on the KCNQ1 K+ channel. *Biochem Biophys Res Commun* **363**(1): 133–139.
Qi, S. Y., P. J. Riviere et al. (2003). Cloning and characterization of dipeptidyl peptidase 10, a new member of an emerging subgroup of serine proteases. *Biochem J* **373**(Part 1): 179–189.
Radicke, S., D. Cotella et al. (2005). Expression and function of dipeptidyl-aminopeptidase-like protein 6 as a putative beta-subunit of human cardiac transient outward current encoded by Kv4.3. *J Physiol* **565**(Part 3): 751–756.
Radicke, S., T. Riedel et al. (2013). Accessory subunits alter the temperature sensitivity of Kv4.3 channel complexes. *J Mol Cell Cardiol* **56**: 8–18.
Rasmussen, H. B., S. Branner et al. (2003). Crystal structure of human dipeptidyl peptidase IV/CD26 in complex with a substrate analog. *Nat Struct Biol* **10**(1): 19–25.
Ravn, L. S., Y. Aizawa et al. (2008). Gain of function in IKs secondary to a mutation in KCNE5 associated with atrial fibrillation. *Heart Rhythm* **5**(3): 427–435.
Ren, X., Y. Hayashi et al. (2005). Transmembrane interaction mediates complex formation between peptidase homologues and Kv4 channels. *Mol Cell Neurosci* **29**(2): 320–332.
Rocheleau, J. M., S. D. Gage et al. (2006). Secondary structure of a KCNE cytoplasmic domain. *J Gen Physiol* **128**(6): 721–729.
Rocheleau, J. M. and W. R. Kobertz (2008). KCNE peptides differently affect voltage sensor equilibrium and equilibration rates in KCNQ1 K+ channels. *J Gen Physiol* **131**(1): 59–68.
Roepke, T. K., A. Anantharam et al. (2006). The KCNE2 potassium channel ancillary subunit is essential for gastric acid secretion. *J Biol Chem* **281**(33): 23740–23747.
Roepke, T. K., E. C. King et al. (2009). Kcne2 deletion uncovers its crucial role in thyroid hormone biosynthesis. *Nat Med* **15**(10): 1186–1194.
Roepke, T. K., A. Kontogeorgis et al. (2008). Targeted deletion of kcne2 impairs ventricular repolarization via disruption of I(K,slow1) and I(to,f). *FASEB J* **22**(10): 3648–3660.
Roepke, T. K., K. Purtell et al. (2010). Targeted deletion of Kcne2 causes gastritis cystica profunda and gastric neoplasia. *PLoS One* **5**(7): e11451.
Romey, G., B. Attali et al. (1997). Molecular mechanism and functional significance of the MinK control of the KvLQT1 channel activity. *J Biol Chem* **272**(27): 16713–16716.
Ruiz-Canada, C., D. J. Kelleher et al. (2009). Cotranslational and posttranslational N-glycosylation of polypeptides by distinct mammalian OST isoforms. *Cell* **136**(2): 272–283.
Ruscic, K. J., F. Miceli et al. (2013). IKs channels open slowly because KCNE1 accessory subunits slow the movement of S4 voltage sensors in KCNQ1 pore-forming subunits. *Proc Natl Acad Sci USA* **110**(7): E559–E566.
Sanguinetti, M. C., M. E. Curran et al. (1996). Coassembly of K(V) LQT1 and minK (IsK) proteins to form cardiac I(Ks) potassium channel. *Nature* **384**(6604): 80–83.
Schade, J., M. Stephan et al. (2008). Regulation of expression and function of dipeptidyl peptidase 4 (DP4), DP8/9, and DP10 in allergic responses of the lung in rats. *J Histochem Cytochem* **56**(2): 147–155.
Schmidt, K., S. Radicke et al. (2010). The new beta-subunit DPP10 may contribute to Ito regulation in heart disease. *Cardiovasc Res* **87**(Suppl 1): S-371.
Schroeder, B. C., S. Waldegger et al. (2000). A constitutively open potassium channel formed by KCNQ1 and KCNE3. *Nature* **403**(6766): 196–199.
Schulze-Bahr, E., M. Schwarz et al. (2001). A novel long-QT 5 gene mutation in the C-terminus (V109I) is associated with a mild phenotype. *J Mol Med (Berl)* **79**(9): 504–509.
Seebohm, G., M. C. Sanguinetti et al. (2003). Tight coupling of rubidium conductance and inactivation in human KCNQ1 potassium channels. *J Physiol* **552**(Part 2): 369–378.
Sesti, F. and S. A. Goldstein (1998). Single-channel characteristics of wild-type IKs channels and channels formed with two minK mutants that cause long QT syndrome. *J Gen Physiol* **112**(6): 651–663.
Shamgar, L., Y. Haitin et al. (2008). KCNE1 constrains the voltage sensor of Kv7.1 K+ channels. *PLoS One* **3**(4): e1943.
Shi, G., K. Nakahira et al. (1996). Beta subunits promote K+ channel surface expression through effects early in biosynthesis. *Neuron* **16**(4): 843–852.
Shibata, R., H. Misonou et al. (2003). A fundamental role for KChIPs in determining the molecular properties and trafficking of Kv4.2 potassium channels. *J Biol Chem* **278**(September 19): 36445–36454.
Soh, H. and S. A. Goldstein (2008). I SA channel complexes include four subunits each of DPP6 and Kv4.2. *J Biol Chem* **283**(22): 15072–15077.
Splawski, I., J. Shen et al. (2000). Spectrum of mutations in long-QT syndrome genes. KVLQT1, HERG, SCN5A, KCNE1, and KCNE2. *Circulation* **102**(10): 1178–1185.
Strop, P., A. J. Bankovich et al. (2004). Structure of a human A-type potassium channel interacting protein DPPX, a member of the dipeptidyl aminopeptidase family. *J Mol Biol* **343**(4): 1055–1065.
Sulda, M. L., C. A. Abbott et al. (2006). DPIV/CD26 and FAP in cancer: A tale of contradictions. *Adv Exp Med Biol* **575**: 197–206.
Sun, J., K. E. Duffy et al. (2006). Structural and functional analyses of the human Toll-like receptor 3. Role of glycosylation. *J Biol Chem* **281**(16): 11144–11151.
Tai, K. K. and S. A. Goldstein (1998). The conduction pore of a cardiac potassium channel. *Nature* **391**(6667): 605–608.
Takumi, T., K. Moriyoshi et al. (1991). Alteration of channel activities and gating by mutations of slow ISK potassium channel. *J Biol Chem* **266**(33): 22192–22198.
Takumi, T., H. Ohkubo et al. (1988). Cloning of a membrane protein that induces a slow voltage-gated potassium current. *Science* **242**(4881): 1042–1045.
Tapper, A. R. and A. L. George, Jr. (2001). Location and orientation of minK within the I(Ks) potassium channel complex. *J Biol Chem* **276**(41): 38249–38254.
Teng, S., L. Ma et al. (2003). Novel gene hKCNE4 slows the activation of the KCNQ1 channel. *Biochem Biophys Res Commun* **303**(3): 808–813.
Thevenod, F. (2002). Ion channels in secretory granules of the pancreas and their role in exocytosis and release of secretory proteins. *Am J Physiol Cell Physiol* **283**(3): C651–C672.
Thielitz, A., S. Ansorge et al. (2008). The ectopeptidases dipeptidyl peptidase IV (DP IV) and aminopeptidase N (APN) and their related enzymes as possible targets in the treatment of skin diseases. *Front Biosci* **13**: 2364–2375.
Tinel, N., S. Diochot et al. (2000a). KCNE2 confers background current characteristics to the cardiac KCNQ1 potassium channel. *EMBO J* **19**(23): 6326–6330.
Tinel, N., S. Diochot et al. (2000b). M-type KCNQ2-KCNQ3 potassium channels are modulated by the KCNE2 subunit. *FEBS Lett* **480**(2–3): 137–141.
Tranebjaerg, L., R. A. Samson et al. (1993). Jervell and Lange-Nielsen syndrome. *Gene Reviews*. R. A. Pagon, T. D. Bird, C. R. Dolan, K. Stephens and M. P. Adam (eds.). Seattle, WA: University of Washington.
Tristani-Firouzi, M. and M. C. Sanguinetti (1998). Voltage-dependent inactivation of the human K+ channel KvLQT1 is eliminated by association with minimal K+ channel (minK) subunits. *J Physiol* **510** (Part 1): 37–45.

Tsevi, I., R. Vicente et al. (2005). KCNQ1/KCNE1 channels during germ-cell differentiation in the rat: Expression associated with testis pathologies. *J Cell Physiol* **202**(2): 400–410.

Tyson, J., L. Tranebjaerg et al. (1997). IsK and KvLQT1: Mutation in either of the two subunits of the slow component of the delayed rectifier potassium channel can cause Jervell and Lange-Nielsen syndrome. *Hum Mol Genet* **6**(12): 2179–2185.

Um, S. Y. and T. V. McDonald (2007). Differential association between HERG and KCNE1 or KCNE2. *PLoS One* **2**(9): e933.

van Bemmelen, M. X., J. S. Rougier et al. (2004). Cardiac voltage-gated sodium channel Nav1.5 is regulated by Nedd4–2 mediated ubiquitination. *Circ Res* **95**(3): 284–291.

van Es, M. A., P. W. van Vught et al. (2008). Genetic variation in DPP6 is associated with susceptibility to amyotrophic lateral sclerosis. *Nat Genet* **40**(1): 29–31.

Vanoye, C. G., R. C. Welch et al. (2010). KCNQ1/KCNE1 assembly, co-translation not required. *Channels (Austin)* **4**(2): 108–114.

Vetter, D. E., J. R. Mann et al. (1996). Inner ear defects induced by null mutation of the isk gene. *Neuron* **17**(6): 1251–1264.

Wada, K., N. Yokotani et al. (1992). Differential expression of two distinct forms of mRNA encoding members of a dipeptidyl aminopeptidase family. *Proc Natl Acad Sci USA* **89**(1): 197–201.

Wallner, M., P. Meera et al. (1999). Molecular basis of fast inactivation in voltage and Ca^{2+}-activated K^+ channels: A transmembrane beta-subunit homolog. *Proc Natl Acad Sci USA* **96**(7): 4137–4142.

Wang, A. G., S. Y. Yoon et al. (2006). Identification of intrahepatic cholangiocarcinoma related genes by comparison with normal liver tissues using expressed sequence tags. *Biochem Biophys Res Commun* **345**(3): 1022–1032.

Wang, K. W. and S. A. Goldstein (1995). Subunit composition of minK potassium channels. *Neuron* **14**(6): 1303–1309.

Wang, K. W., K. K. Tai et al. (1996). MinK residues line a potassium channel pore. *Neuron* **16**(3): 571–577.

Wang, W., J. Xia et al. (1998). MinK-KvLQT1 fusion proteins, evidence for multiple stoichiometries of the assembled IsK channel. *J Biol Chem* **273**(51): 34069–34074.

Wang, Y. W., J. P. Ding et al. (2002). Consequences of the stoichiometry of Slo1 alpha and auxiliary beta subunits on functional properties of large-conductance Ca^{2+}-activated K^+ channels. *J Neurosci* **22**(5): 1550–1561.

Warth, R. and J. Barhanin (2002). The multifaceted phenotype of the knockout mouse for the KCNE1 potassium channel gene. *Am J Physiol Regul Integr Comp Physiol* **282**(3): R639–R648.

Weber, A. N., M. A. Morse et al. (2004). Four N-linked glycosylation sites in human toll-like receptor 2 cooperate to direct efficient biosynthesis and secretion. *J Biol Chem* **279**(33): 34589–34594.

Weber, R. E., H. Malte et al. (1995). Mass spectrometric composition, molecular mass and oxygen binding of Macrobdella decora hemoglobin and its tetramer and monomer subunits. *J Mol Biol* **251**(5): 703–720.

Weiger, T. M., M. H. Holmqvist et al. (2000). A novel nervous system beta subunit that downregulates human large conductance calcium-dependent potassium channels. *J Neurosci* **20**(10): 3563–3570.

Wilson, G. G., A. Sivaprasadarao et al. (1994). Changes in activation gating of IsK potassium currents brought about by mutations in the transmembrane sequence. *FEBS Lett* **353**(3): 251–254.

Witzel, K., P. Fischer et al. (2012). Hippocampal A-type current and Kv4.2 channel modulation by the sulfonylurea compound NS5806. *Neuropharmacology* **63**(8): 1389–1403.

Wu, D., K. Delaloye et al. (2010a). State-dependent electrostatic interactions of S4 arginines with E1 in S2 during Kv7.1 activation. *J Gen Physiol* **135**(6): 595–606.

Wu, D., H. Pan et al. (2010b). KCNE1 remodels the voltage sensor of Kv7.1 to modulate channel function. *Biophys J* **99**(11): 3599–3608.

Wu, R. S., N. Chudasama et al. (2009). Location of the beta 4 transmembrane helices in the BK potassium channel. *J Neurosci* **29**(26): 8321–8328.

Xia, X., B. Hirschberg et al. (1998). dSLo interacting protein 1, a novel protein that interacts with large-conductance calcium-activated potassium channels. *J Neurosci* **18**(7): 2360–2369.

Xu, J., W. Yu et al. (1998). Distinct functional stoichiometry of potassium channel beta subunits. *Proc Natl Acad Sci USA* **95**(4): 1846–1851.

Xu, X., M. Jiang et al. (2008). KCNQ1 and KCNE1 in the IKs channel complex make state-dependent contacts in their extracellular domains. *J Gen Physiol* **131**(6): 589–603.

Xu, X., V. A. Kanda et al. (2009). MinK-dependent internalization of the IKs potassium channel. *Cardiovasc Res* **82**(3): 430–438.

Yan, J. and R. W. Aldrich (2010). LRRC26 auxiliary protein allows BK channel activation at resting voltage without calcium. *Nature* **466**(7305): 513–516.

Yan, J. and R. W. Aldrich (2012). BK potassium channel modulation by leucine-rich repeat-containing proteins. *Proc Natl Acad Sci USA* **109**(20): 7917–7922.

Yan, Y., D. J. Scott et al. (2008). Identification of the N-linked glycosylation sites of the human relaxin receptor and effect of glycosylation on receptor function. *Biochemistry* **47**(26): 6953–6968.

Yang, C., X. H. Zeng et al. (2011). LRRC52 (leucine-rich-repeat-containing protein 52), a testis-specific auxiliary subunit of the alkalization-activated Slo3 channel. *Proc Natl Acad Sci USA* **108**(48): 19419–19424.

Yanglin, P., Z. Lina et al. (2007). KCNE2, a down-regulated gene identified by in silico analysis, suppressed proliferation of gastric cancer cells. *Cancer Lett* **246**(1–2): 129–138.

Zagha, E., A. Ozaita et al. (2005). DPP10 modulates Kv4-mediated A-type potassium channels. *J Biol Chem* **280**(19): 18853–18861.

Zhou, Y., W. M. Schopperle et al. (1999). A dynamically regulated 14–3–3, Slob, and Slowpoke potassium channel complex in *Drosophila* presynaptic nerve terminals. *Neuron* **22**(4): 809–818.

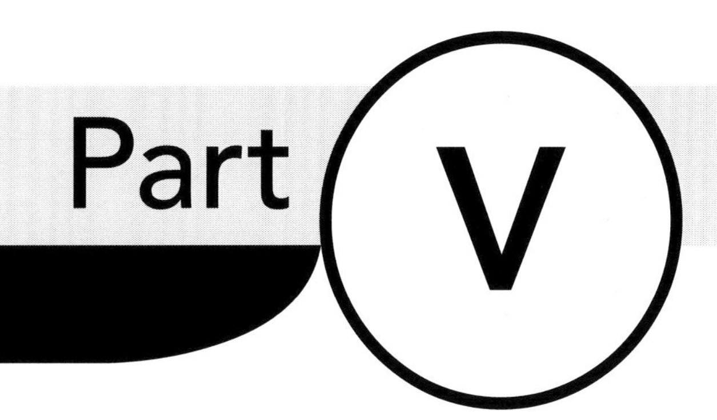

Ion channel physiology and diseases

39 Ion channels of the heart

Donald M. Bers and Eleonora Grandi

Contents

39.1 GENERATION OF THE CARDIAC ACTION POTENTIAL AND EXCITATION–CONTRACTION COUPLING

The action potential (AP) is a transient cell membrane depolarization emerging from the dynamic behavior of a diverse population of membrane ion channels. A prototypical ventricular myocyte AP, simulated with our human ventricular cell model (Grandi et al. 2010), is shown in Figure 39.1a and exhibits a steep upstroke, followed by a sustained slowly decaying plateau phase, which eventually gives way to rapid repolarization. Below the AP are shown the associated depolarizing currents, which are carried by inward Na^+ and Ca^{2+} currents. Under physiological conditions, the Na^+ current (I_{Na}) activates rapidly, producing the AP upstroke, and then inactivates nearly completely (Figure 39.1b). The L-type Ca^{2+} current (I_{CaL}) inactivates more slowly (Figure 39.1b), and less completely, allowing for the inward Ca^{2+} current to maintain the AP plateau phase.

The influx of Ca^{2+} via I_{CaL} triggers the release of Ca^{2+} from the sarcoplasmic reticulum (SR), the subcellular organelle that stores and releases the majority of Ca^{2+} during each heartbeat (Bers 2001). This event is known as Ca^{2+}-induced Ca^{2+} release and generates a rise in cytosolic [Ca^{2+}] (Figure 39.1a) that ultimately leads to cell contraction. A small amount of Ca^{2+} influx during depolarization is also contributed by the Na^+/Ca^{2+} exchanger (NCX, in reverse mode, outward current). During the largest fraction of the cardiac cycle, however, NCX mostly operates in the Ca^{2+} extrusion (forward) mode to maintain the transarcolemmal Ca^{2+} balance, by extruding Ca^{2+} that entered via I_{CaL} and (outward) reverse NCX mode. This yields a net inward charge movement when 1 Ca^{2+} ion is exchanged for 3 Na^+ ions (Figure 39.1d). The SR actively resequesters Ca^{2+} via the SR Ca^{2+} ATPase (SERCA), which is the primary mechanism removing Ca^{2+} from the cytosol to allow relaxation

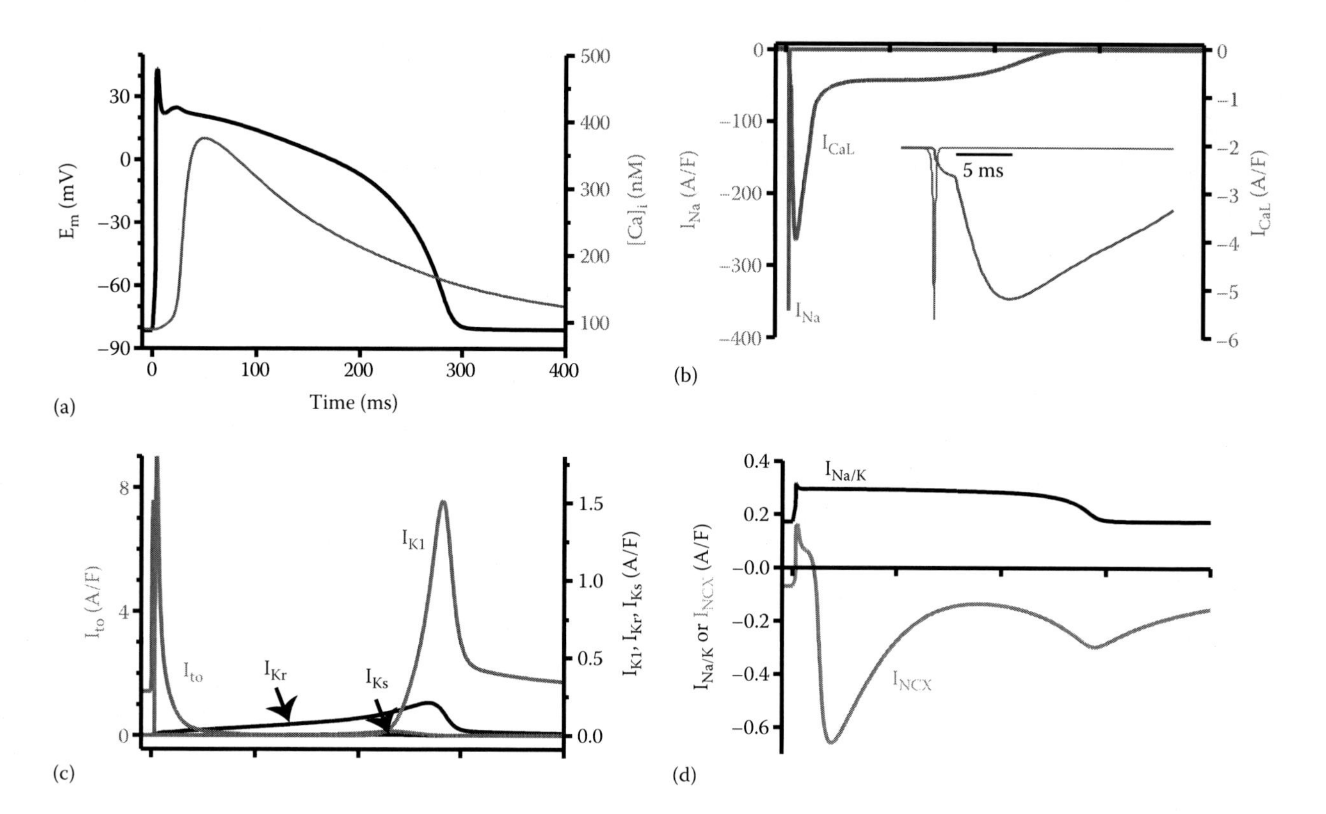

Figure 39.1 Cardiac AP, Ca^{2+} transient and ionic currents. (a) Human ventricular AP and Ca^{2+} transient simulated with the Grandi–Pasqualini–Bers model (Grandi et al. 2010), (b) Na^+ and Ca^{2+} currents (inset shows the different activation and decay times), (c) K^+ currents, (d) Na^+/K^+ pump and Na^+/Ca^{2+} exchange currents during an AP. (Modified from Grandi, E. and Bers, D.M., Models of the ventricular action potential in health and disease, in *Cardiac Electrophysiology: From Cell to Bedside*, (eds.) D.P. Zipes and J. Jalife, Saunders, Philadelphia, PA, 2013.)

in between heartbeats (Bassani et al. 1994). The plasma membrane Ca^{2+} ATPase (PMCA) also extrudes Ca^{2+} from the cell and may finely regulate cytosolic Ca^{2+} levels. Throughout this process, Ca^{2+} is buffered by Ca^{2+}-binding proteins such as calmodulin (CaM) and troponin. The intracellular Ca^{2+} signal also feeds back on the L-type Ca^{2+} channel (LTCC), mediating inactivation of the current, and therefore plays a role in influencing the AP shape. It also regulates a variety of processes including mitochondrial ATP production, intracellular signaling, and gene expression.

Various types of K^+ channels drive cell membrane repolarization. The transient outward currents (I_{to}, Figure 39.1c, and $I_{Cl(Ca)}$), carried by K^+ and Cl^-, respectively, determine the early AP repolarization (notch) that follows the upstroke, whereas the components of the delayed rectifier K^+ current (I_{Kr} and I_{Ks}, Figure 39.1c) contribute to AP repolarization. The inward rectifier current (I_{K1}, Figure 39.1c) maintains and stabilizes the resting potential. Another important player in shaping cardiac AP properties is the Na^+/K^+ pump (NKA), which generates an outward current (Figure 39.1d) by extruding three Na^+ ions and importing two K^+ ions on each cycle and contributes to AP repolarization.

While the biophysical properties of ion channels are examined in detail in various chapters, here we describe the fundamental aspects of the major cardiac ionic currents involved in generating and shaping the normal cardiac AP. Special emphasis is placed on the functional consequences of their alterations in arrhythmia-associated ion channel mutations and acquired diseases (Table 39.1).

39.2 CARDIAC Na^+ CHANNEL PHYSIOLOGY AND PATHOPHYSIOLOGY

39.2.1 CHANNEL STRUCTURE AND FUNCTION

The cardiac Na^+ channel (see recent reviews by Abriel (2010), Bezzina et al. (2001), and Herren et al. (2013)) is responsible for the generation of the rapid upstroke of the cardiac AP and thus plays a central role in cardiac cell excitability and in the propagation of electrical impulses in the heart. It belongs to the voltage-dependent family of Na^+ channels and has an α pore-forming subunit (Figure 39.2a) associated with one or two ancillary β subunits (β1–β4: one transmembrane domain proteins involved in different aspects of channel function). The SCN5A gene encodes the cardiac Na^+ channel α-subunit $Na_V1.5$, which is the principal Na^+ channel isoform expressed in cardiomyocytes and consists of four internally homologous domains (I–IV) each made up of six transmembrane segments (S1–S6). The S5–S6 linker includes the P-loops (pore region). The fourth transmembrane segment in each domain contains positively charged residues (voltage sensor) that are responsible for channel activation upon depolarization of the membrane potential (E_m). If depolarization is maintained, the channel enters a nonpermissive inactivated state (a cluster of three hydrophobic amino acids in the III–IV linker, IFM motif, are involved in fast inactivation gating). Upon repolarization, the channel recovers to a closed conformation that allows reopening.

Table 39.1 Cardiac ion currents and their molecular identities (α subunits)

CURRENT	PROTEIN	GENE	NOTES
Voltage-gated			
I_{Na}, I_{NaL}	Nav1.5	SCN5A	Na^+ channel isoforms (other than cardiac) are expressed in the heart. They account for less than 10% of peak I_{Na} and may underlie a greater fraction of I_{NaL}
	Nav1.1?	SCN1A	
	Nav1.2?	SCN2A	
	Nav1.4?	SCN4A	
	Nav1.6?	SCN8A	
	Nav1.8?	SCN10A	
I_{CaL}	Cav1.2	CACNA1C	
	Cav1.3	CACNA1D	Atrium, SAN, and AVN
I_{CaT}	Cav3.1	CACNA1G	
	Cav3.2	CACNA1H	Purkinje fibers
I_{Kr}	Kv11.1	hERG (KCNH2)	
I_{Ks}	KvLQT1	KCNQ1	
$I_{to,fast}$	Kv4.2/4.3	KCND2/3	Significant differences among species in expression profiles and possible role(s) of these two clones in the generation of $I_{to,fast}$
$I_{to,slow}$	Kv1.4	KCNA4	
I_{Kur}	Kv1.5	KCNA5	Mostly expressed in atrial cells and in ventricles of some species (e.g., rodents)
$I_{K,slow}$	Kv1.2	KCNA2	
I_{K1}	Kir2.1	KCNJ2	
I_f	HCN4	HCN4	Pacemaker cells
Ligand-gated			
$I_{K(Ach)}$	Kir3.1/3.4	GIRK1/GIRK4	
$I_{K(ATP)}$	Kir6.2	KCNJ11	
$I_{K(Ca)}$	SK2	KCNN2	
$I_{Cl(Ca)}$	CLCA-1	CLCA-1	
	Bestrophin	BEST	
	TMEM16	TMEM16	
$I_{Cl(CAMP)}$	CFTR	CFTR	
$I_{Cl,acid}$	?	?	
Mechanosensitive			
$I_{Cl(swell)}$	ClC-2, ClC-3	ClC-2, ClC-3	
$I_{NS(stretch)}$	?	?	

Immunocytochemical experiments revealed that in addition to the expected localization of Na^+ channels at the surface and t-tubular membranes, Na^+ channels localize to terminal intercalated disks, also the predominant sites of gap junctional coupling between cardiomyocytes (Cohen 1996). However, the precise localization of $Na_V1.5$ in cardiac cells is somewhat controversial. An intriguing hypothesis is that different pools of $Na_V1.5$ may be targeted to different cell regions (i.e., t-tubules vs. intercalated disks), serve different cellular functions, and undergo differential modulation (Lin et al. 2011).

39.2.2 INHERITED AND ACQUIRED CARDIAC DISEASES

Mutations in the gene SCN5A have been linked to cardiac disorders (Figure 39.2 and Table 39.2), including long QT syndrome (LQTS), Brugada syndrome (BrS), conduction defects, sinus dysfunction, and familial atrial fibrillation (fAF) (Wilde and Bezzina 2005), suggesting that the cardiac sodium channel is not limited to AP initiation and conduction but may have more important effects on shaping the AP and triggering arrhythmias. In cardiac cells, $Na_V1.5$ forms a macromolecular complex with

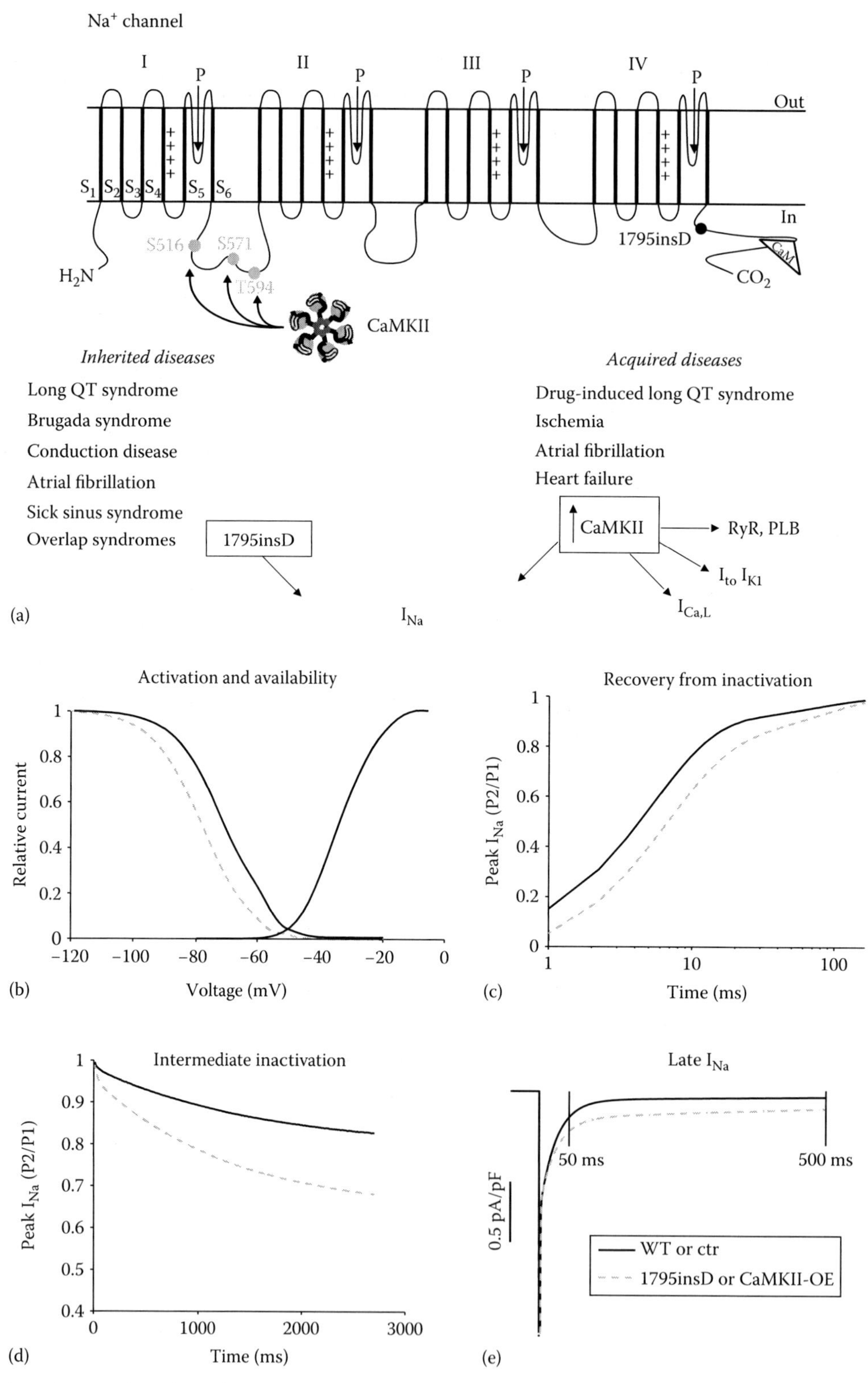

Figure 39.2 Modulation of cardiac Na⁺ channel gating. (a) Na⁺ channel α-subunit structure and the inherited and acquired disease linked to Na⁺ channel malfunction. CaMKII overexpression (CaMKII-OE) and hyperactivity and 1795InsD mutation in SCN5A gene cause (b) leftward shift in Na⁺ channel availability, with no changes in channel activation, (c) slower recovery from inactivation, (d) enhancement of intermediate inactivation, and (e) increased I_{NaL}. (Modified from Grandi, E. et al., Computer simulation of altered sodium channel gating in rabbit and human ventricular myocytes, in *Lecture Notes in Computer Science*, (eds.) F.B. Sachse and G. Seemann, Springer-Verlag GmbH, 2007a.)

partner proteins that may be anchoring/adaptor proteins, interact with and modify the channel, and modulate the localization and biophysical properties of $Na_V1.5$ (reviewed in Abriel 2010). Indeed, mutations disrupting Na⁺ channel protein interactions have also been associated with LQT3 and BrS (Table 39.2).

SCN5A mutations also lead to combinations of these phenotypes, known as overlap syndromes. Intriguingly, a single human mutation at this site (1795InsD) in SCN5A is linked to simultaneous LQT3 and BrS features (Bezzina et al. 1999; Veldkamp et al. 2000). Mutation-induced increase

Table 39.2 Inherited channelopathies

CHANNEL	SYNDROME	GENE	AFFECTED CURRENT	NOTES
Na^+ and Na^+-related	LQT3	SCN5A	I_{Na}	
	LQT9	Cav3	I_{Na}	
	LQT10	SCN4B	I_{Na}	β_4 subunit
	LQT12	SNTA1	I_{Na}	
	BrS1	SCN5A	I_{Na}	
	BrS2	GDP1-L	I_{Na}	
	BrS5	SCN1B	I_{Na}	β_1 subunit
	BrS7	SCN3B	I_{Na}	β_3 subunit
	Progressive cardiac conduction defect	SCN5A?	I_{Na}	
	SSS	SCN5A	I_{Na}	
	fAF	SCN5A	I_{Na}	
K^+	LQT1	KCNQ1	I_{Ks}	Autosomal dominant and recessive transmissions
	LQT2	KCNH2	I_{Kr}	
	LQT5	HERG (KCNH2)	I_{Ks}	β subunit
	LQT6	KCNE2/MiRP1	I_{Kr}	β subunit
	LQT7	KCNJ2	I_{K1}	
	LQT11	AKAP9	I_{Ks}	Yotiao forms complex with I_{Ks}
	LQT13	KCNJ5	$I_{K(Ach)}$	
	SQT1	KCNH2	I_{Kr}	
	SQT2	KCNQ1	I_{Ks}	
	SQT3	KCNJ2	I_{K1}	
	BrS6	KCNE3/MiRP2	I_{to}	β subunit
	BrS8	HCN4	I_f	
	CPVT3	KCNJ2	I_{K1}	
	fAF	KCNQ1	I_{Ks}	KCNQ1–KCNE2 current
		KCNA5	I_{Kur}	
		KCNE3/MiRP2	I_{to}	β subunit
		KCNE2/MiRP1	I_{Ks}	KCNQ1–KCNE2 current
		KCNJ2	I_{K1}	
Ca^{2+} and Ca^{2+} related	BrS3 and SQT4	CACNA1C	I_{CaL}	
	BrS4 and SQT5	CACNB2b	I_{CaL}	β_2 subunit
	Timothy syndrome/LQT8	CACNA1C	I_{CaL}	
	SQT6	CACNA2D1	I_{CaL}	$\alpha_2\delta$ subunit
	CPVT1	RyR2	I_{rel}	
	CPVT2	CASQ2	I_{rel}	
	Ankyrin-B syndrome/LQT4	ANK2	I_{NCX}, I_{NaK}	
	CPVT	CALM1	I_{rel}?	
	LQTS	CALM1, CALM2	I_{CaL}?, I_{Ks}?, I_{rel}?	

LQT, long QT syndrome; BrS, Brugada syndrome; fAF, familial atrial fibrillation; SSS, sick sinus syndrome; SQT, short QT syndrome; CPVT: catecholaminergic polymorphic ventricular tachycardia.

in intermediate I_{Na} inactivation, slowing of recovery from inactivation, and hyperpolarizing shift in channel availability (Figure 39.2b–d; Veldkamp et al. 2000) reduce I_{Na} (loss of function) and underlie the Brugada-like symptoms (slow conduction) of patients at higher heart rates (Clancy and Rudy 2002). In addition, this 1795InsD mutation causes an increase in the late component of I_{Na} (I_{NaL} [Veldkamp et al. 2000], Figure 39.2e) that remains active during the AP plateau. At a slow heart rates (where I_{Na} recovery from inactivation may be more complete and APD is intrinsically longer), the late current–dependent AP prolongation is responsible for LQTS (Clancy and Rudy 2002).

These inherited channelopathies have tremendously deepened our understanding of normal $Na_V1.5$ function and arrhythmia mechanisms. Emerging evidence also links Na^+ channel gating alterations to acquired diseases (Figure 39.2), for example, drug-induced LQTS, cardiac ischemia, heart failure (HF), and atrial fibrillation (AF). Altered Na^+ channel regulation may also occur in HF, causing a widespread form of acquired Na^+ channel dysfunction. A pathological increase in I_{NaL} has been linked to disease manifestation in inherited and acquired cardiac diseases, including LQT3, HF, and AF. Although small in magnitude as compared with the peak Na^+ current (~1%), I_{NaL} enhancement leads to AP prolongation, disruption of normal cellular repolarization, development of arrhythmia triggers, and propensity to ventricular arrhythmia. It has been proposed that I_{NaL} is modulated by the channel β subunits (Maltsev et al. 2009) and that other Na channel isoforms are responsible for I_{NaL} in normal and failing hearts (Biet et al. 2012; Xi et al. 2009), and β subunits may be targeting these channels, although their contribution to cardiac physiology is still poorly understood.

Posttranslational modifications also alter Na^+ channel function and exacerbate the effects of inherited or acquired diseases. For example, Ca^{2+}/CaM-dependent protein kinase II (CaMKII), which is upregulated in HF and more active, has been shown to regulate Na^+ channel gating. Wagner et al. (2006) showed that the upregulation of CaMKII in cardiac myocytes causes an extremely similar spectrum of gating changes to those seen for the combined LQT/Brugada phenotype seen with 1795InsD (see Figure 39.2b–e). We showed how less Na^+ channel availability at high heart rates and more inward I_{Na} during long APs at low heart rates could contribute to a propensity to arrhythmias (Grandi et al. 2007b).

Based on the similar effects of CaMKII and SCN5A C-terminus mutation, one might look for but (Aiba et al. 2010) provided evidence that the I–II loop might be a major CaMKII phosphorylation target. Hund et al. (2010) identified S571 as a potential CaMKII target, demonstrated that CaMKII phosphorylates S571 in vitro and affects channel inactivation. These effects were abolished when S571 was mutated to a nonphosphorylatable alanine and mimicked when S571 was mutated to a phosphomimetic glutamine reside. Our group (Ashpole et al. 2012) found that only the I–II loop of $hNa_V1.5$ was substantially phosphorylated by CaMKII and showed that S516 and T594 were the main in vitro CaMKII phosphorylation sites. In patch-clamp analysis, we found that alanine substitution of S516, S571, and T594 could all inhibit the CaMKII-dependent negative shift in I_{Na} availability and accumulation of intermediate inactivation observed in myocytes. However, only S516E and T594E phosphomimetic mutants could recapitulate CaMKII effects on I_{Na} availability. Thus, there may be three sites in this stretch of the I–II loop that participate in CaMKII-dependent regulation of cardiac I_{Na} gating (Figure 39.2).

Oxidative stress may play a crucial role in arrhythmogenesis. Reactive oxygen species (ROS) enhance I_{NaL}, and this could be due to a direct effect of channel oxidation (Kassmann et al. 2008) or to the indirect effect of ROS to activate CaMKII (Wagner et al. 2011). Both ROS and CaMKII are increased in diseases such as ischemia, HF, and AF. Hund et al. (2008) showed that CaMKII is more expressed and active during myocardial infarction and also implicated oxidized CaMKII in the impairment of cardiac conduction following myocardial infarction due to reduced I_{Na} availability (Christensen et al. 2009).

39.3 CARDIAC L-TYPE Ca^{2+} CHANNEL PHYSIOLOGY AND PATHOPHYSIOLOGY

39.3.1 CHANNEL FUNCTION AND REGULATION

Voltage-gated cardiac Ca^{2+} channels critically regulate excitation–contraction coupling (ECC) (Bers 2001) by triggering SR Ca^{2+} release and modulating AP shape and duration (APD), that is, maintaining the long AP plateau. Two main types of Ca^{2+} channels are expressed in cardiac myocytes, L-type (LTCCs) and T-type (TTCCs) Ca^{2+} channels(), due mainly to α_{1C} (and α_{1D} to a lesser extent) and α_{1G-H}, respectively and characterized by long-lasting openings at larger depolarization vs. more transient openings and activation at more negative potentials. Functional I_{CaL} is present in all cardiac myocytes, whereas TTCC current (I_{CaT}) is more prominent in atrial and Purkinje cells.

LTCCs comprise α_1, α_2/δ, β, and γ subunits (Figure 39.3a) that allow depolarization-induced Ca^{2+} influx into the cytosol. α_{1C} subunit consists of four homologous motifs (I–IV), each composed of six membrane-spanning α-helices (S1–S6) linked by variable cytoplasmic loops (linkers), the ion-selective pore (P-loops), voltage sensor, and gating machinery. The C-terminal domain of the α_{1C} subunit contains several regulatory sites, such as a protein kinase A (PKA) site, an EF-hand region, and an IQ motif at which CaM can bind. Cardiac I_{CaL} is rapidly activated by membrane depolarization, whereas LTCC inactivation is regulated by both Ca^{2+}- and voltage-dependent inactivation (CDI and VDI) (Lee et al. 1985). It has been proposed that CDI is due to binding of Ca^{2+} to CaM (Peterson et al. 1999; Zuhlke et al. 1999), which causes a channel conformational change that prevents the EF-hand in the C-terminus from interacting with the cytosolic I–II linker, which then occludes the channel pore, thus accelerating inactivation (Cens et al. 2006). I_{CaL} inactivates with a biexponential time course when Ca^{2+} is the charge carrier, and it has been widely assumed that the initial fast decay represents CDI and the slower phase reflects VDI. During ECC, with normal SR Ca^{2+} release and Ca^{2+} transients, local $[Ca^{2+}]_i$ near the LTCC is elevated and inactivation is rapid (Figure 39.3b and c, top I_{CaL} trace). The effect of SR Ca^{2+} release on the inactivation of I_{CaL} provides a unique bioassay for local cleft $[Ca^{2+}]_i$ (which may exceed 50 μM during ECC). That is, the rapid inactivation

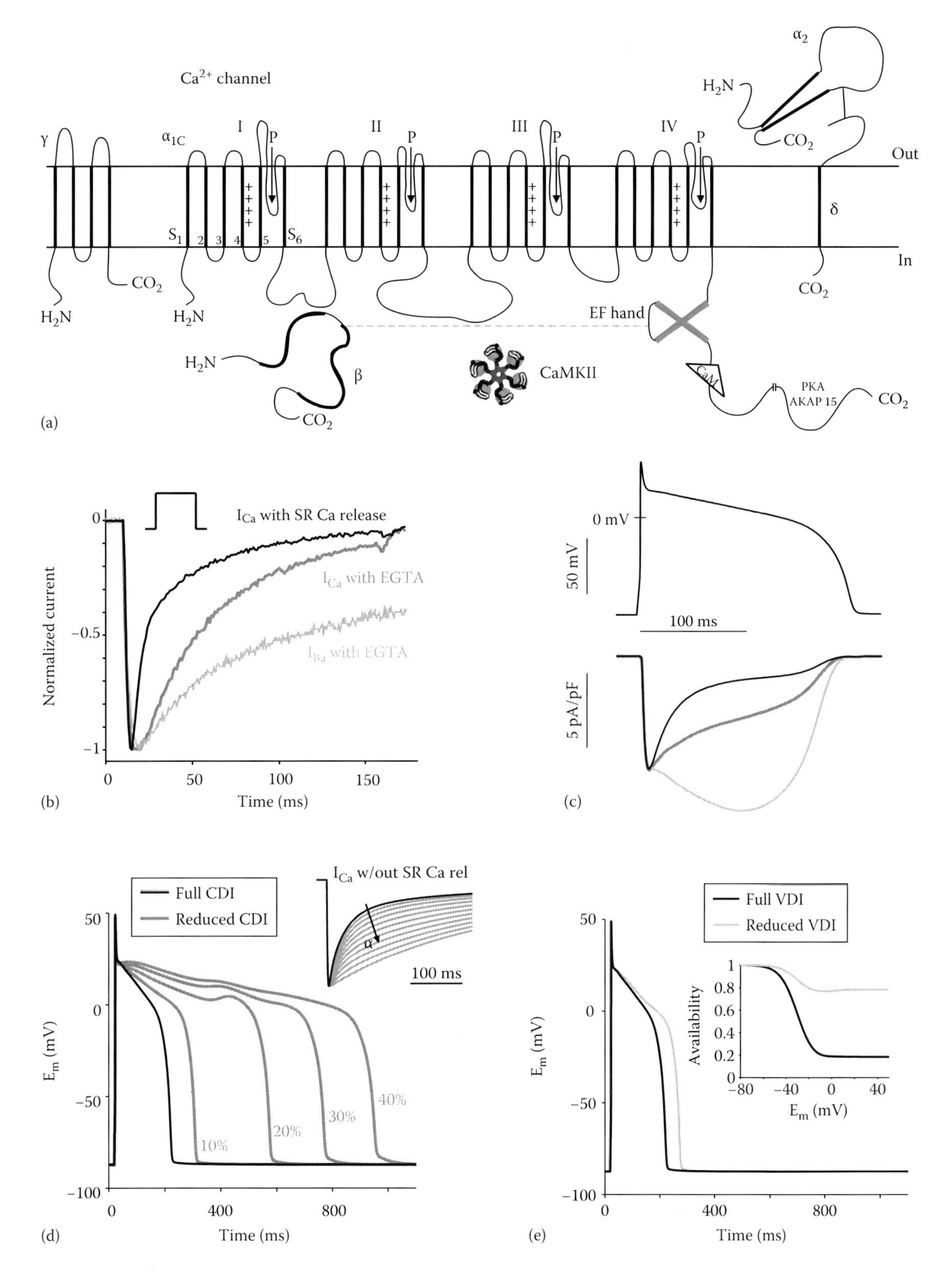

Figure 39.3 Cardiac Ca^{2+} channel inactivation. (a) Ca^{2+} channel structure. Current through LTCCs elicited by (b) a voltage step, from Bers (2001), and (c) an AP (simulated) under various ionic homeostasis conditions. (d) Effect of progressive impairment of LTCC CDI (inset) on the cardiac AP. (e) Effect of impaired VDI (inset) on the cardiac AP. (c–e: Modified from Morotti, S. et al., *J. Physiol.*, 590(Pt 18), 4465, 2012.)

of I_{CaL} allows measurement of the SR Ca^{2+} release time course (which peaks in 2–3 ms at 35°C). When Ca^{2+} transients are abolished (e.g., by ryanodine or very strong $[Ca^{2+}]_i$ buffering), I_{CaL} inactivation is slower (Figure 39.3b and c) and reflects a small rise in local $[Ca^{2+}]_i$ near the mouths of Ca^{2+} channels due to Ca^{2+} entering via the channels themselves. In addition to Ca^{2+}, LTCCs allow permeation by other ions including Na^+, K^+, Sr^{2+}, Cs^+, and Ba^{2+}, depending on ionic conditions, but inactivation is slower than with Ca^{2+} and monoexponential when Ba^{2+} (Figure 39.3b and c), Sr^{2+}, or Na^+ are the charge carrier.

39.3.2 INHERITED AND ACQUIRED CARDIAC DISEASES

CDI is an important feedback mechanism that limits the amount of Ca^{2+} entry during each Ca^{2+} transient, regulates SR Ca^{2+} load, and modulates APD (Bers 2001). The need for a faithful characterization of the roles of VDI and CDI has been underlined by experimental studies showing the deleterious effects of inhibiting one of the two mechanisms. Indeed, prevention of CDI by the expression of a mutant Ca^{2+}-insensitive CaM in adult cardiomyocytes induces a dramatic (four- to fivefold) prolongation of the cardiac AP (Alseikhan et al. 2002). A theoretical study also identified CDI as the key player in shortening the AP when external $[Ca^{2+}]$ is increased (Grandi et al. 2009). The physiological significance of VDI is illustrated by the dramatic consequences of Timothy syndrome (Splawski et al. 2004, 2005), which impairs VDI and causes severe ventricular arrhythmias (and dysfunction in other tissues, e.g., brain).

The effects of Timothy syndrome have been investigated in numerous simulation studies (Faber et al. 2007; Thiel et al. 2008). Our computational analysis (Morotti et al. 2012) confirmed that removing VDI, while retaining CDI, caused AP prolongation (Figure 39.3e). Notably, when simulating the expression of CaM_{1234}, which removes CDI without affecting VDI (Alseikhan et al. 2002), we predicted a much more marked APD prolongation (Figure 39.3d) and early afterdepolarizations (EADs, see Section 39.9), confirming previously observed experimental (Alseikhan et al. 2002) and simulation (Mahajan et al. 2008) results. The notion that gene mutations linked to disease may interfere with both VDI and CDI is intriguing (Yarotskyy et al. 2009), although not yet fully understood. Also, shifts in VDI could alter intracellular Ca^{2+} handling and signaling, which could in turn contribute to the Timothy syndrome phenotype (Thiel et al. 2008).

At physiologic holding E_m (−80 mV), a pulse-dependent progressive increase in I_{CaL} amplitude and a prominent slowing of inactivation that accumulates over 5–10 pulses at 1 Hz are characteristic. This positive I_{CaL} staircase is Ca^{2+} dependent (does not occur with Ba^{2+} current), occurs even without SR Ca^{2+} release, and is mediated by CaMKII-dependent phosphorylation (Anderson et al. 1994; Xiao et al. 1994; Yuan and Bers 1994). This facilitatory effect of Ca^{2+} entry on subsequent I_{CaL} is distinct from but coexists with the intercellular Ca^{2+}-dependent inactivation described earlier. The physiologic role of this Ca^{2+}-dependent facilitation is not entirely clear, but it may partly offset direct CDI and create LTCC memory on a timescale of a few seconds. The exact molecular mechanism underlying I_{CaL} facilitation is not fully resolved, but the CaM activating CaMKII in this process is likely to be distinct from the CaM involved in CDI. Activated CaMKII can associate with multiple CaMKII interaction sites on the C-terminal tail of the α1C subunit adjacent to the CaM-binding IQ motif (Hudmon et al. 2005; Pitt 2007). CaMKII phosphorylation sites mediating I_{CaL} facilitation have been described on the N and C termini of the α1C subunit (Hudmon et al. 2005) as well as on the accessory β2 subunit (Grueter et al. 2006).

39.4 CARDIAC RYANODINE RECEPTORS

39.4.1 CHANNEL FUNCTION AND REGULATION

Ryanodine receptors (RyRs) are large (~2.2 MDa) channels forming homotetrameric assemblies in a mushroom-like shape with a large cytoplasmic head. RyR2 channels directly control SR Ca^{2+} release in cardiac muscles, activating contraction during ECC. ECC requires Ca^{2+} influx (via LTCCs) that raises cleft $[Ca^{2+}]_i$ to activate RyR opening. Once a single RyR channel opens, it can recruit more RyR within the cleft. The coordinated activation of several RyRs in a cluster generates a local increase in cytoplasmic Ca^{2+} concentration (detected as a Ca^{2+} spark by a Ca^{2+} indicator). Within the tiny cleft space, local $[Ca^{2+}]$ probably rises to a peak of ~50–100 μM. However, as Ca^{2+} diffuses away, the local $[Ca^{2+}]_i$ at sites outside the cleft is much lower (peak $[Ca^{2+}]_i$ ~ 0.5–1 μM) and declines as one gets further away, due to three-dimensional diffusion. These local release events are synchronized by the AP and almost simultaneous activation of I_{CaL} at each junction throughout the myocyte and the heart. This synchronization is critical for functionally effective ECC (Bers 2008).

Ca^{2+} released from RyR is critical in controlling LTCC CDI (as discussed in Section 39.3) and drives inward NCX current (Section 39.6), with consequences on the AP configuration. There are several other channels whose gating is influenced by $[Ca^{2+}]_i$ and can influence AP configuration as well. These include Ca^{2+}-activated Cl^- current ($I_{Cl(Ca)}$), Ca^{2+}-activated K^+ current ($I_{K(Ca)}$), and the delayed rectifier (I_{Ks}). $I_{Cl(Ca)}$ (point 4 in Section 39.8) is more prominent at positive E_m and contributes to the early repolarization phase of the AP, where submembrane $[Ca^{2+}]$ is especially high. $I_{K(Ca)}$ (Section 39.7.3) is very small or nonexistent in ventricular myocytes but is more prominent in atrial myocytes and may play a role in AP repolarization. I_{Ks} (Section 39.7.1) is known to increase with higher $[Ca^{2+}]_i$, but the kinetics of this effect and how it interfaces with the β-adrenergic effects on I_{Ks} are not well understood (Xie et al. 2013). All of these currents will be sensitive to changes in Ca^{2+} transients, including the large increases in Ca^{2+} transient associated with physiological β-adrenergic receptor (β-AR) activation.

39.4.2 INHERITED AND ACQUIRED CARDIAC DISEASES

Spontaneous SR Ca^{2+} release events (Ca^{2+} sparks, Figure 39.4a and b) can also occur in the absence of Ca^{2+} current. This is because RyR opening is stochastic and influenced by $[Ca^{2+}]_i$, $[Ca^{2+}]_{SR}$, and RyR modulation (e.g., by phosphorylation, oxidation, or disease-related mutations, as seen in mutations in RyR2 or calsequestrin with catecholaminergic polymorphic ventricular tachycardia, CPVT). Ca^{2+} activation of the cardiac RyR begins at submicromolar $[Ca^{2+}]$, reaches a broad maximum (at very high P_o) near 100 μM Ca^{2+}, and decreases at very high $[Ca^{2+}]$ (5–10 mM; Rousseau and Meissner 1989; Xu et al. 1998) (Figure 39.4d). There is compelling evidence that cardiac RyR gating is influenced strongly by luminal $[Ca^{2+}]_{SR}$ such that RyR opening is favored at high $[Ca^{2+}]_{SR}$ ([Figure 39.4d] enhancing Ca^{2+} spark initiation)

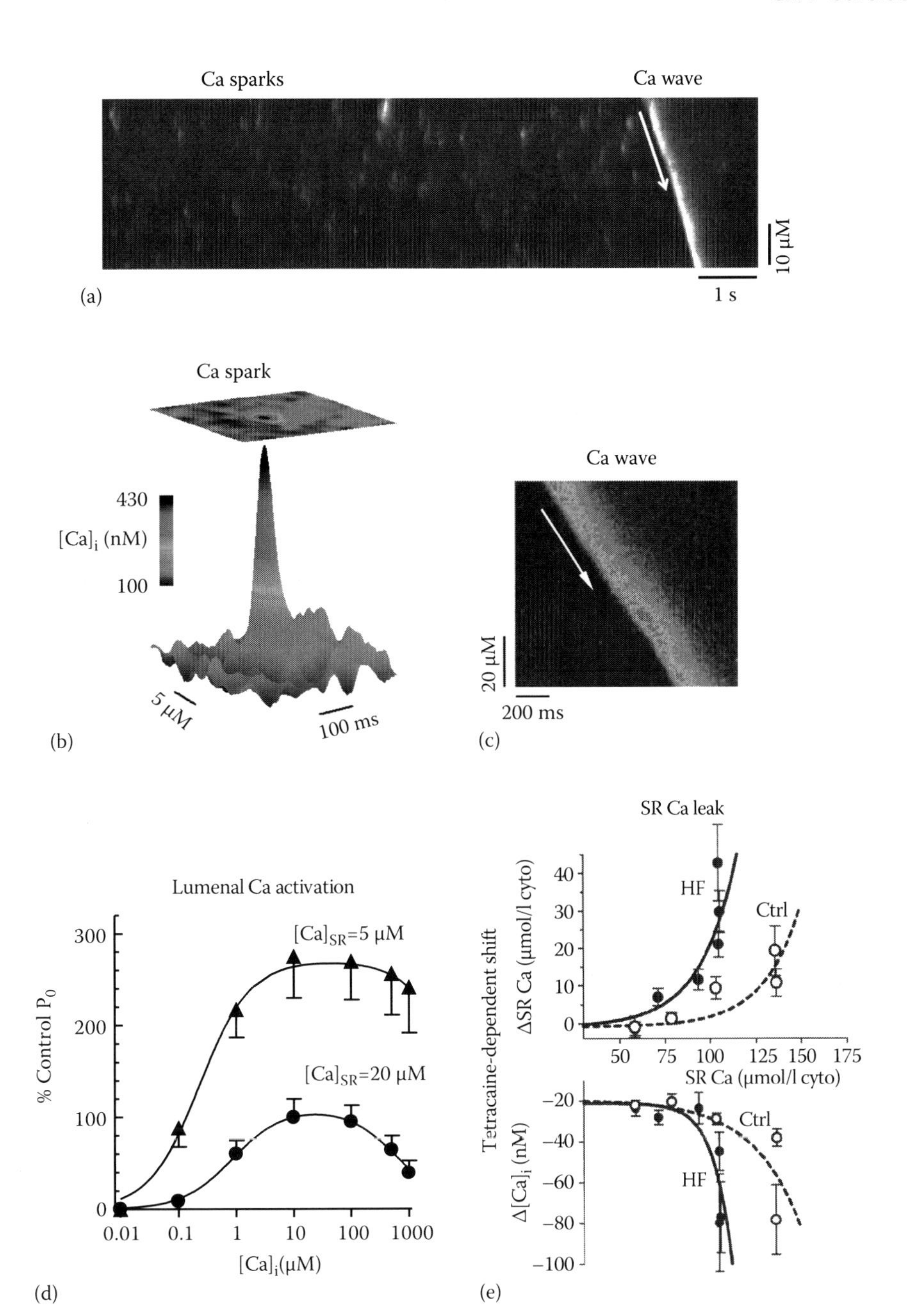

Figure 39.4 SR Ca^{2+} release. (a–c) Ca^{2+} sparks and wave recorded in intact (mouse) ventricular myocytes. All plots are longitudinal line scan records (length vs. time), except a surface plot (length vs. time vs. $[Ca^{2+}]_i$) in panel (b) (bottom). (Modified from Bers, D.M., *Nature*, 415(6868), 198, 2002.) (d) Dependence of RyR P_o on intracellular and SR $[Ca^{2+}]$. (Modified from Bers, D.M., *Excitation-Contraction Coupling and Cardiac Contractile Force*, Kluwer Academic Press, Dordrecht, The Netherlands, 2001.) (e) Dependence of SR Ca^{2+} leak on SR Ca^{2+} load in nonfailing and failing myocytes. (Modified from Shannon, T.R. et al., *Circ. Res.*, 93(7), 592, 2003.)

and RyR closing is favored at lower $[Ca^{2+}]_{SR}$. Experiments show that release terminates at ~400 μM $[Ca^{2+}]_{SR}$ during both Ca^{2+} sparks and ECC and when there is little spatial gradient of $[Ca^{2+}]_{SR}$ (Picht et al. 2011). Thus, there seems to be an internal brake to prevent this positive feedback from going to completion.

When RyR2 open probability is higher than normal during diastole, the frequency of sparks increases, and each can in turn activate neighboring RyR2 clusters to produce a Ca^{2+} wave that is propagated throughout the cell (Figure 39.4a and c; Bers 2008). Ca^{2+} waves can increase the cytoplasmic Ca^{2+} concentration above normal diastolic levels, activating NCX and other surface membrane ion channels, leading to the generation of delayed afterdepolarizations (DADs) and certain types of EADs (see Section 39.9). Furthermore, because Ca^{2+} diffuses out of the SR at the same time as it is being accumulated by SERCA, the SR does not load to its optimal diastolic capacity, and less Ca^{2+} is available for release during the subsequent systole.

Increased Ca^{2+} leak from the SR mediated by the defective regulation of RyRs has emerged as an important mechanistic paradigm in Ca^{2+} handling dysfunction in HF. It has been shown that diastolic Ca^{2+} leak is steeply $[Ca^{2+}]_{SR}$-dependent (Figure 39.4e) (Shannon et al. 2003), perhaps because of increased $[Ca^{2+}]_i$ sensitivity of the RyR at higher $[Ca^{2+}]_{SR}$. The leak-$[Ca^{2+}]_{SR}$ relationship is also shifted leftward in failing vs. nonfailing myocytes (Figure 39.4e), indicating increased SR Ca^{2+} leak at a given $[Ca^{2+}]_{SR}$ (Shannon et al. 2003). This may be due to hypersensitive RyR. Both PKA and CaMKII phosphorylate the RyRs, and RyR2 phosphorylation is probably increased in HF, possibly at both the S2809 (S2808) PKA and S2814 CaMKII sites (and maybe also at S2031, although it has been studied in less detail), as summarized recently (Bers 2012). Reduction in local phosphatase activity in HF may also contribute to increase phosphorylation levels. There is more consistent (but not unanimous) observation among groups that RyR2 phosphorylation is increased in HF at the S2814 CaMKII site. Much more controversial is the effect at the PKA site S2808 (or S2809 depending on species), and whether and how PKA phosphorylation alters RyR2 (Bers 2012). CaMKII-dependent RyR2 phosphorylation increases RyR2 opening, and CaMKII inhibition can inhibit SR Ca^{2+} leak in HF (Ai et al. 2005). ROS can also directly activate RyR, and that may be part of the mechanism of enhanced SR Ca^{2+} leak in HF. CaMKII activation state can also be elevated chronically by both autophosphorylation and ROS, both of which may reinforce this SR Ca^{2+} leak phenotype in HF.

Increased diastolic SR Ca^{2+} leak associated with HF is very likely to be arrhythmogenic but may not contribute to systolic dysfunction, in a manner analogous to CPVTmutants. Indeed, although the roles and mechanisms of RyRs in HF are still under debate, it has become clear that mishandling of Ca^{2+} in the cytoplasm due to mutations in the RyR2 genes that sensitize the channel to Ca^{2+}-dependent activation is associated with the human arrhythmia CPVT. CPVT is an inherited disease characterized by life-threatening arrhythmias elicited by stress and emotion and is caused by dominant mutations in the RYR2 gene and by recessive mutations in the CASQ2 gene, encoding the cardiac calsequestrin isoform (Table 39.2, reviewed in Priori and Chen 2011). CPVT RyR2 mutations alter the sensitivity of the channel to luminal and/or cytosolic Ca^{2+} activation, leading to enhanced spontaneous Ca^{2+} release during SR Ca^{2+} overload. Despite this mechanism, these mutations have little or no impact on ECC, as patients with CPVT RyR2 mutations do not show arrhythmias in nonstimulated conditions. Reduced SR Ca^{2+} buffering capacity is a common consequence of CPVT CASQ2 mutations. This results in a fast recovery of SR-free Ca^{2+} after each Ca^{2+} release and a potentially higher level of SR-free Ca^{2+} during a sudden increase in SR Ca^{2+} loading, both of which increase the propensity for spontaneous SR Ca^{2+} release and thus DADs and triggered activity. Some CASQ2 mutations may alter the interactions between CASQ2 and the RyR2 channel complex, thus affecting the response of RyR2 to elevating luminal Ca^{2+}.

39.5 CARDIAC IP_3 RECEPTORS

The IP_3 receptor (IP_3R) is closely related to the RyR, and these combine to form the superfamily of intracellular Ca^{2+} release channels. In whole heart, all three types of IP_3Rs are expressed, but in isolated ventricular myocytes, only the type-2 IP_3R is expressed (Perez et al. 1997). Despite their sequence and domain similarity, the three types of IP_3R are differentially regulated by IP_3 and Ca^{2+}. The type-2 IP_3R in ventricular myocytes is both more sensitive to IP_3 and almost insensitive to $[Ca^{2+}]$ over the physiological range of $[Ca^{2+}]_i$ (0.1–10 μM).

The number of IP_3Rs in ventricular myocytes is probably 2%–10% of the number of RyRs (Perez et al. 1997), raising questions about their function in ventricular myocytes. IP_3Rs are six times more highly expressed in atrial than in ventricular myocytes (Lipp et al. 2000). Atrial myocyte IP_3Rs were found to be subsarcolemmal and apparently colocalized with surface RyRs. IP_3 could produce SR Ca^{2+} release in skinned atrial cells, and a membrane permeant IP_3 analogue enhanced Ca^{2+} spark frequency and twitch Ca^{2+} transient in intact atrial myocytes. In cat atrial myocytes, the application of IP_3 increased SR Ca^{2+} release events through IP_3R (Ca^{2+} puffs) and facilitates SR Ca^{2+} release through RyR clusters (Ca^{2+} sparks) (Zima and Blatter 2004).

Immunolocalization studies with type-2-specific IP_3R antibodies show perinuclear localization in ventricular myocytes. Such perinuclear IP_3R have been shown to activate CaMKII and contribute to transcriptional regulation (Wu et al. 2006).

Similar to effects observed in atrial myocytes, the application of IP_3 to permeabilized ventricular myocytes produced an increase in cytosolic Ca^{2+} spark frequency and an associated decrease in SR Ca^{2+} content. This was a consequence of IP_3R-dependent facilitation of Ca^{2+} release from clusters of RyRs. In intact myocytes, endothelin-1 was used to stimulate IP_3 production and caused an increase in the amplitude of AP-induced Ca^{2+} transients (reversed by IP_3R inhibition) (Domeier et al. 2008). IP3R inhibition did not prevent RyR-independent Ca^{2+} leak (Zima et al. 2010), suggesting that IP_3Rs are not contributing to diastolic Ca^{2+} leak (at least in the absence of IP_3). However, Zima et al. (2010) found that activation of IP_3Rs by IP_3 application nearly doubled RyR-independent SR Ca^{2+} leak.

39.6 Na^+/Ca^{2+} EXCHANGER

The cardiac sarcolemmal Na^+/Ca^{2+} exchanger is a reversible transporter that exchanges three Na^+ for one Ca^{2+} and produces an ionic current (I_{NCX}). I_{NCX} exhibits a reversal potential that is analogous to those of ion channels ($E_{NCX} = 3E_{Na}–2E_{Ca}$, where E_{Na} and E_{Ca} are equilibrium potentials for Na^+ and Ca^{2+}). As $[Ca^{2+}]_i$, $[Na]_i$, and E_m change during the AP, net I_{NCX} changes dynamically. Outward I_{NCX} (or Ca^{2+} influx) is electrochemically favored by low $[Ca^{2+}]_i$, high $[Na]_i$, and depolarized E_m. Inward I_{NCX} (or Ca^{2+} efflux) is favored by the opposite conditions (Bers 2001). In addition to the electrochemical potential, which sets the direction of I_{NCX}, NCX is allosterically regulated by $[Ca^{2+}]_i$ such that when $[Ca^{2+}]_i$ is low, I_{NCX} deactivates (Miura and Kimura 1989). Increasing $[Ca^{2+}]_i$ can activate NCX, allowing I_{NCX} to flow. Allosteric regulation in intact myocytes at physiological $[Ca^{2+}]_i$, and under conditions where the NCX is able to dynamically control $[Ca^{2+}]_I$, has been shown by Weber et al. (2001) in both nonfailing and failing myocytes (Weber et al. 2003b).

39.6.1 SENSING LOCAL $[Ca^{2+}]$ AND $[Na^+]$ DURING PHYSIOLOGICAL Ca^{2+} RELEASE AND AP

In diastole, $E_m < E_{NCX}$, so Ca^{2+} extrusion is favored (inward I_{NCX}, Figure 39.1d). Early in the AP, the E_m exceeds E_{NCX}, which tends to drive Ca^{2+} entry by outward I_{NCX} (until $E_m = E_{NCX}$). Note that E_{NCX} changes because $[Ca^{2+}]_i$ (and thus E_{Ca}) changes. On AP repolarization, the negative E_m and high $[Ca^{2+}]_i$ drive a large inward I_{NCX}, and this reflects Ca^{2+} extrusion from the cell (Bers 2008). This is complicated by elevations in local submembrane $[Ca^{2+}]$ and $[Na^+]$ ($[Ca^{2+}]_{sm}$ and $[Na^+]_{sm}$), which are caused by rapid Na^+ and Ca^{2+} fluxes (through I_{Na}, I_{Ca}, and SR Ca^{2+} release). Although $[Ca^{2+}]_{sm}$ may not get as high as cleft $[Ca^{2+}]_i$ during I_{Ca} and SR Ca^{2+} release, the high $[Ca^{2+}]_{sm}$ causes I_{NCX} to become inward very early in the rise of the AP (Weber et al. 2003a), such that very little Ca^{2+} enters through I_{NCX} (≪1 μM). An I_{Na}-induced rise in $[Na^+]_{sm}$ during the AP might delay the reversal of I_{NCX} to inward (Weber et al. 2003), but outward I_{NCX} would still last for only 4 ms or less (with Ca^{2+} entry still less than 1 μM). Thus, under physiological conditions, NCX works mainly in the Ca^{2+} extrusion mode, driven mostly by the Ca^{2+} transient. The positive E_m during the AP plateau can, however, limit Ca^{2+} extrusion. This emphasizes again the importance of considering local vs. bulk ion concentration (as for inactivation of I_{Ca}).

39.6.2 NCX IN PATHOPHYSIOLOGY AND ARRHYTHMIA

NCX can contribute to arrhythmogenesis in two ways: (1) by affecting the Ca^{2+} balance in the cell and (2) by carrying an arrhythmogenic current. Inward NCX has been implicated in the generation of EADs and DADs. These arise from spontaneous release of Ca^{2+} from the SR, due to excessive Ca^{2+} load and/or alterations in the RyR gating properties, which causes $[Ca^{2+}]_i$ to rise and an inward current to be generated via NCX working in the Ca^{2+} extrusion (forward) mode. Whether this depolarizing current is sufficient to trigger an AP depends on the state of the other ion channels. For example, I_{K1} downregulation in HF lowers the threshold for DADs; that is, less inward current is required to depolarize the E_m to activate Na^+ channels (Pogwizd et al. 2001).

Although NCX normally works mainly in the Ca^{2+} efflux mode, an exception can occur in certain pathophysiologic conditions, like HF, where $[Na^+]_i$ is elevated, Ca^{2+} transients are small, and APD is prolonged. All of these factors shift NCX more in favor of Ca^{2+} influx, and so in HF, Ca^{2+} entry via NCX can persist for most of the AP plateau and significantly contribute to the Ca^{2+} transient. In a sense, this limits the extent of systolic dysfunction in HF, where NCX is enhanced, by bringing Ca^{2+} in and by indirectly helping load the SR with Ca^{2+}, but could also slow relaxation and worsen diastolic dysfunction, with increased spontaneous SR Ca^{2+} release followed by arrhythmogenic transient inward currents.

39.7 CARDIAC K+ CHANNELS

Cardiac K^+ channels produce outward currents that drive E_m repolarization toward E_K. Thus, they play a crucial role in maintaining the resting E_m and determining the duration of cardiac repolarization. Indeed, malfunction of K^+ channels, due to either gene mutations or acquired diseases, can alter cardiomyocyte excitability and the electrical balance of depolarization (Table 39.2) and repolarization, and underlies different types of cardiac arrhythmias. In the heart, K^+ channels are structurally categorized as voltage-gated and inward rectifiers.

39.7.1 VOLTAGE-GATED K+ CHANNELS

Voltage-gated K^+ channels containing six transmembrane regions (S1–S6) with a single pore loop formed by the S5 and S6 segments and S5–S6 linker. The S4 segment is rich in positively charged residues and serves as the voltage sensor (e.g., I_{to}, I_{Kur}, I_{Kr}, and I_{Ks}). Some E_m-dependent K^+ channels exhibit outward or delayed rectification (e.g., see I_{Ks} in Figure 39.5a), characterized by an increasingly positive slope at more positive E_m. Figure 39.5a shows that without rectification K^+ current would be a linear function of E_m with a reversal potential at E_K. That is, the slope or conductance (g_K) would be constant. The increasing slope upon depolarization constitutes apparent outward rectification. However, this conductance increase is attributable to the E_m-dependent gating of the channels rather than an intrinsic permeation property. That is, the conductance increases with depolarization because more channels are open, not because individual channels really exhibit rectification. It is referred to as delayed rectification because these channels take a finite time to be activated (e.g., I_{Ks} activation is very slow). Some K^+ channels also exhibit negative slope regions at positive potentials due to inactivation. While the term is somewhat imprecise, these E_m-dependent K^+ channels (I_{Kr}, I_{Ks}, and I_{Kur}) are referred to as outward or delayed rectifier channels.

39.7.1.1 Transient outward K+ current

The transient outward K^+ current (I_{to}, reviewed by Niwa and Nerbonne 2010) activates and inactivates rapidly during E_m depolarization and carries a prominent repolarizing current. Biophysical characterization revealed the presence of two components of I_{to}: the fast component ($I_{to,fast}$) recovers and inactivates with time constants of <100 ms, whereas the slow component ($I_{to,slow}$) recovers in hundreds of milliseconds up to several seconds and inactivates in ~200 ms. $I_{to,fast}$ and $I_{to,slow}$ are generated by different K^+ channel isoforms, whereby Kv1.4 encodes the α subunit of $I_{to,slow}$ and Kv4.2/3 encodes the α subunit of $I_{to,fast}$ (Kv4.3 is predominant in human and canine; both Kv4.2 and 4.3 are expressed in rodents and ferrets). These have been proposed to interact with a number of β subunits. For example, KChIP2 plays a critical role in current expression and seems to be a primary determinant of transmural gradient of $I_{to,fast}$. Interestingly, a mutation in KCNE3, coding MiRP2, has been associated with altered I_{to} and BrS (Table 39.2).

In rodents, I_{to} is very large and dominates AP repolarization, accounting for markedly short APD and lack of a plateau phase. In larger mammals, including canine and human, I_{to} is responsible for the early repolarization phase and influences the activation of LTCCs (Wang et al. 2006), thus affecting AP plateau amplitude and duration. In many species, $I_{to,fast}$ and $I_{to,slow}$ are variably expressed in different regions of the heart and contribute to regional AP heterogeneity. For example, higher I_{to} density (along with the expression of I_{Kur}) accelerates early repolarization and shortens APD in atrial vs. ventricular myocytes (Grandi et al. 2011). In human ventricles, I_{to} is larger in epicardial and midmyocardial vs. endocardial cells, generating a prominent *spike and dome* morphology (absent in endocardium and in guinea pig and porcine myocardium, where I_{to} is poorly expressed). I_{to} is mostly $I_{to,fast}$ in human atrial cells and in ventricular cells from

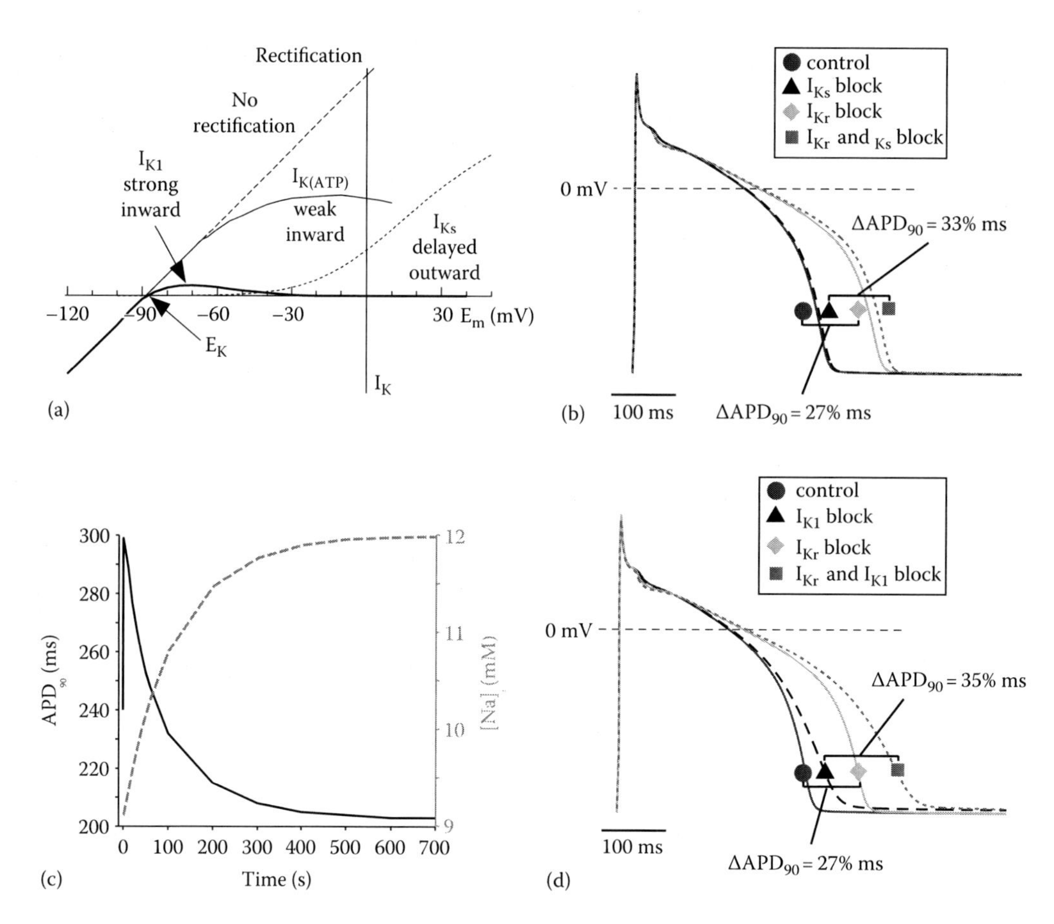

Figure 39.5 Cardiac repolarizing currents. (a) Rectification of K^+ currents. (Modified from Bers, D.M., *Excitation-Contraction Coupling and Cardiac Contractile Force*, Kluwer Academic Press, Dordrecht, The Netherlands, 2001.) (b) Simulated effect of I_{Ks} and I_{Kr} block on human ventricular AP. (c) Time course of APD and $[Na^+]_i$ changes following partial (50%) blockade of NKA. (d) Simulated effect of I_{K1} and I_{Kr} block on human ventricular AP. (b–d: Modified from Grandi, E. et al., *J. Mol. Cell Cardiol.*, 48(1), 112, 2010.)

the epicardium, whereas $I_{to,slow}$ is dominant in endocardial cells. In mice, $I_{to,slow}$ is only detected in the septum, and a left-to-right gradient is observed in $I_{to,fast}$ in ventricles. In rabbit, on the other hand, $I_{to,slow}$ is the primary component of I_{to} (75% vs. 25% $I_{to,fast}$). Notably, the slow recovery of $I_{to,slow}$ means that it becomes less available and thus may limit APD shortening at faster heart rates.

Downregulation of I_{to} has been described in animal models of hypertrophy and human HF, is associated with APD prolongation, and predisposes to EADs. Markedly reduced I_{to} density is also reported in chronic AF, and mutations are associated with fAF (Table 39.2). Alterations in I_{to} have also been linked to BrS (Table 39.2).

The gating properties and kinetics of cardiac I_{to} are modulated by various regulatory and signaling pathways, including serine/threonine and tyrosine phosphorylation. Multiple phosphorylation sites for PKA, protein kinase C (PKC), extracellular signal-regulated kinase (ERK), and CaMKII have been identified in Kv4.2/3 and Kv1.4. Tessier and colleagues (Tessier et al. 1999) provided evidence that CaMKII regulates (slows) the inactivation of I_{to} in human atrial myocytes, thus contributing to AF. It was suggested that CaMKII acts on Kv4.3 by a direct effect at Ser550, thereby prolonging open-state inactivation and accelerating the rate of recovery from inactivation (Sergeant et al. 2005). Recent evidence suggests that chronic inhibition of CaMKII in murine hearts results in APD shortening and prevents remodeling after myocardial infarction and excessive β-AR stimulation (Li et al. 2006) and suggests that CaMKII may be involved in the regulation of repolarization in HF. Another recent report from Xiao et al. (2008) has shown that I_{to} is downregulated in canine ventricular myocytes paced at 3 Hz and suggested that CaMKII is involved in the I_{to} remodeling during sustained tachycardia. Wagner et al. (2009) studied how CaMKII alters I_{to}, both acutely by adenoviral CaMKII overexpression in rabbit myocytes and chronically in CaMKII overexpressing transgenic mice with HF. CaMKII activation accelerates I_{to} recovery from inactivation for both $I_{to,fast}$ and $I_{to,slow}$, and this appears to be an acute regulatory effect of phosphorylation by CaMKII, as it is acutely reversed by CaMKII inhibitors. There are also longer-term alterations in $I_{to,fast}$ and $I_{to,slow}$, which may result from CaMKII-dependent changes in -subunit expression. Kv1.4, underlying $I_{to,slow}$, was enhanced by both acute and transgenic CaMKIIδ_C overexpression in rabbit and mouse myocytes. Conversely, KV4.2/4.3 (the molecular correlates of cardiac $I_{to,fast}$) was decreased during chronic CaMKIIδ_C overexpression in mice. In a modeling study (Grandi et al. 2007b), we showed that CaMKII may prolong APDs in the endocardium (via I_{Na} and I_{CaL} effects) and shorten APDs in the epicardium (via I_{to} effects). This could amplify transmural dispersion of repolarization, thus predisposing to the development of reentrant arrhythmias. An interesting finding by Wang et al. (2006) suggests that the I_{to} inhibitor 4-AP binds to the I_{to} channel and facilitates I_{CaL} via the activation of CaMKII. They postulated that a significant amount of inactive CaMKII forms molecular complex with Kv4.3, and 4-AP-induced dissociation from Kv4.3 facilitates CaMKII activation (Keskanokwong et al. 2011).

39.7.1.2 Delayed rectifier K^+ current

The delayed rectifier K^+ current is responsible for AP repolarization and controls APD in several species, including human (reviewed by Charpentier et al. 2010). It consists of slowly (I_{Ks}) and rapidly (I_{Kr}) activating components carried by KCNQ1 and hERG α subunits, respectively. An even faster component, carried by Kv1.5, has been described in atrial myocytes from various species and in rodent ventricular myocytes (I_{Kur}).

I_{Kr} exhibits a fast VDI (and rapid recovery there from) that limits K^+ flow at positive voltages and confers the current apparent inward rectification (with large tail currents at positive E_m). The critical role of I_{Kr} for cardiac repolarization is witnessed by its involvement in human cardiac channelopathies (Table 39.2). Gain-of-function mutations account for SQT1. Several mutations in KCNH2 (coding for hERG), virtually all causing loss-of-function and prolongation of repolarization due to altered channel gating or trafficking defects, are associated to the second most common form of LQTS (type 2, LQT2). LQT6 is caused by mutations in MiRP1, the channel modulatory β subunit. Interestingly, hERG is highly susceptible to blockade by a number of drugs. These drugs lengthen cardiac repolarization (acquired LQTS) and can predispose patients to lethal arrhythmias. Indeed, I_{Kr} blockade significantly prolongs APD in human ventricular myocytes, which was also recapitulated by our human ventricular model (Figure 39.5b) showing a more prominent AP prolongation in our endocardial cells. There is some evidence that hERG channels are modulated by various G protein–coupled receptors, including α- and β-ARs, which may be involved in I_{Kr} downregulation in HF.

I_{Ks} is carried by KCNQ1 (formerly KvLQT1), whose behavior and trafficking are regulated by the β-subunit minK (KCNE1). It is a very slowly activating current (100 ms to 1 s depending on the species) that shows no appreciable inactivation. I_{Ks} is regulated by α-AR via PKC and β-AR activation through the PKA pathway (and regulated by PKA anchoring proteins, AKAPs) (Marx et al. 2002). I_{Ks} is also enhanced by intracellular Ca^{2+} at a physiological range via a CaM-dependent pathway, which does not involve phosphorylation (Shamgar et al. 2006). Valuable information on the function of the KCNQ1–KCNE1 complex and its modulation comes from the study of numerous mutations related to human familial diseases (Table 39.2). I_{Ks} loss-of-function mutations lead to the autosomal dominant Romano–Ward and the autosomal recessive Jervell and Lange-Nielsen syndromes (LQT1), whereas defects in the β-subunit KCNE1 have been associated with LQT5. An interesting mutation in KCNQ1 is simultaneously associated with AF (gain of function in atria due to association with MiRP1) and LQTS (loss of function in ventricles). Yotiao is a KCNQ1-bound AKAP that is important in activating I_{Ks} during sympathetic stimulation. Mutations disrupting KCNQ1–Yotiao interaction and mutations in Yotiao itself are also associated with LQTS. I_{Ks} contribution to AP repolarization varies among species, and the role of this current in human ventricle has been debated. Experimental results in isolated ventricular myocytes showed no changes in AP duration or morphology when I_{Ks} is silenced, as also predicted by our human ventricular model (Figure 39.5b). This does not mean that I_{Ks} is irrelevant, but is consistent with the idea that in normal human ventricle it is only functionally important upon β-AR stimulation to decrease APD in the fight or flight response. Also, when the repolarization reserve is attenuated (e.g., by drugs or diseases that reduce other repolarizing currents), I_{Ks} may become increasingly important in limiting AP prolongation. Indeed, similar to experimental findings, our simulations (Grandi et al. 2010) predict that I_{Kr} block increases APD by 27% when I_{Ks} is present (Figure 39.5, ♦ vs. ●) vs. 33% with I_{Ks} block (Figure 39.5, ■ vs. ▲), which is an increase of 22% attributable to the loss of I_{Ks} contribution to repolarization reserve. Sex differences in I_{Ks} expression may underlie increased APD, QT interval, and incidence of torsades des pointes in females (Zhu et al. 2012). I_{Ks} contribution is enhanced at high stimulation rates, due to the incomplete slow deactivation of the current and accumulation in the open state (in guinea pigs), or to accumulation in closed states that are kinetically proximal to the open state (in larger mammals where deactivation in faster) (Silva and Rudy 2005). Although in guinea pig this may partly account for APD shortening at faster pacing rates, theoretical studies have shown that in large mammals, including humans, APD rate adaptation is mostly driven by the accumulation of intracellular Na^+ that drives Na^+ extrusion via NKA, thus increasing an outward repolarizing current (Grandi et al. 2010). In fact, partial block of NKA is predicted to instantaneously shorten APD, followed by a slow APD shortening as Na^+ increases as a result of reduced extrusion (Figure 39.5c). It is worth noting that I_{NKA} is comparable to I_K during the AP plateau and repolarization (Figure 39.1).

I_{Kur} activates more rapidly than I_{Ks} and I_{Kr} and inactivates slowly (in seconds). I_{Kur} is detected in atrial myocytes (and in ventricular myocytes in some species), where its rapid activation in the positive potential range following the AP upstroke may offset depolarizing I_{CaL} and hence lead to the less positive plateau phase in atrial than ventricular cells (Grandi et al. 2011). I_{Kur} is upregulated during β-AR activation, and we provided evidence that I_{Kur} in the atrium may serve the same function as I_{Ks} in the ventricle, which is opposing AP prolongation expected from larger inward I_{CaL} and I_{NCX} during β-adrenergic stress. Indeed, our simulations showed that block of I_{Kur} (to mimic Kv1.5 mutation that leads to nonfunctional current, and AF) in the presence of adrenergic challenge causes EADs (Grandi et al. 2011). While diminished expression of Kv1.5 has been reported in patients with chronic AF, others have reported no changes in I_{Kur} density (Grandi et al. 2012). Inconsistent results about I_{Kur} function have been commented on previously by Christ et al. (2008) and attributed to different strategies for the identification of I_{Kur} (e.g., pharmacological or with I_{to}-inactivating prepulse) and to a fraction of I_{Kur} that is not accounted for by Kv1.5 (Christ et al. 2008). Experimental evidence suggests that block of I_{Kur} enhances the force of contraction of isolated human atrial trabeculae both in patients in sinus rhythm and AF (Schotten et al. 2007), suggesting that I_{Kur} might be a potentially useful atrial-specific target to counteract hypocontractility associated with cAF. A slight AP prolongation associated with I_{Kur} blockade may also be beneficial (Grandi et al. 2011).

39.7.2 INWARDLY RECTIFYING K+ CHANNEL

Inward rectifier K^+ channels contain two transmembrane domains (M1–M2) with intracellular N- and C-termini and a single pore loop formed by the M1 and M2 domains (e.g., I_{K1}, I_{KACh}, and I_{KATP}). Inward rectification means that the channel conducts inward current better than outward current. With weak inward rectification (as shown for $I_{K(ATP)}$ in Figure 39.5a), the outward current at $E_m > E_K$ is less than expected. For strong inward rectification (as for I_{K1} and I_{KACh}, reviewed by Anumonwo and Lopatin 2010), outward current can be completely blocked at more positive E_m.

I_{K1} is responsible for stabilizing the resting E_m near E_K, determining the excitation threshold, and controlling the late repolarization phase (Figure 39.5d). Rectification of I_{K1} is also modulated by external [K^+], as when this increases the region of rectification shifts in parallel with E_K. I_{K1} is differentially expressed in various regions of the heart, conferring the AP different morphologies. It is almost nonexistent in sinoatrial node (SAN) cells, allowing for a relatively depolarized (–50 mV) resting E_m with respect to the ventricles (–80 mV), which have robust I_{K1} expression. I_{K1} current density is 6–10 times larger in the ventricles than in the atria, resulting in a more hyperpolarized resting E_m and faster repolarization (phase 3) in the ventricles (Grandi et al. 2011). Evidence now implicates KCNJ2 gene mutations and remodeling of the I_{K1} channel in cardiac diseases, causing the functional upregulation or downregulation of I_{K1}, which may have profound effects on cardiac excitability and increase the risk of life-threatening arrhythmia. To date, four channelopathies associated with inward rectifier channels have been identified (Table 39.2), all originating from loss-of-function or gain-of-function mutations in KCNJ2 (Kir2.1): type 1 Andersen syndrome (LQT7), CPVT, fAF, and SQT3. I_{K1} heterogeneity in the right vs. left ventricle has been shown to contribute to the generation and stability of reentrant activity (Samie et al. 2001). I_{K1} is downregulated in HF (Pogwizd et al. 2001), contributing to the prolongation of APs. Reduced I_{K1} contributes to the development of DADs and ventricular arrhythmias (by lowering the threshold for triggered activity). In contrast, I_{K1} is greatly increased during cAF and may constitute a compensatory response against the depolarizing effect of the high-rate electrical activity of the fibrillating atria. I_{K1} upregulation contributes markedly to atrial AP shortening, which promotes the initiation and maintenance of multiple reentrant wavelets and thus contributes to the self-perpetuation of AF (Pandit et al. 2005).

I_{KACh} is activated by muscarinic agonists such as acetylcholine and is prominently expressed in atrial, SAN, and atrioventricular node (AVN) cells, where vagal innervation in the heart is the highest but is sparse in ventricles. Activation of I_{KACh} causes an inward rectifier K^+ current that hyperpolarizes the E_m, shortens cardiac APD, slows the spontaneous firing rate of pacemaker cells in SAN and AVN, and slows atrioventricular conduction. The heterogeneity of I_{KACh} expression within and between the left and right atria correlates with potentially proarrhythmic ability of vagal nerve stimulation. Importantly, the atrial-specific localization and the functional upregulation (increased constitutive activity) of I_{KACh} during AF make it a possible antiarrhythmic target devoid of ventricular side effects (Dobrev and Nattel 2010).

I_{KATP} is carried by ATP-sensitive K^+ channels, composed by Kir6.2 (α-) and SUR2 (β subunit), which activate with a decrease in intracellular ATP concentration. I_{KATP} plays a pivotal role in maintaining cardiac homeostasis under stress conditions like myocardial ischemia/reperfusion and hypoxia and mediates the ischemia-induced electrophysiological changes and the cardioprotective effect of preconditioning. Activation of I_{KATP} shortens cardiac APD, reduces Ca^{2+} influx (and prevents cardiac Ca^{2+} overload), preserves ATP levels, and increases cell survival during myocardial ischemia. On the other hand, activation of I_{KATP} may also promote reentrant ventricular arrhythmias (Li and Dong 2010).

39.7.3 Ca2+-ACTIVATED K+ CHANNELS

Ca^{2+}-activated K^+ channels have been detected in cardiac myocytes. Small conductance SK2 channels are more abundant in the atria compared with the ventricles and have been suggested to play a crucial role in human and mouse cardiac repolarization, especially during the late phase of the cardiac AP (Xu et al. 2003), as their inhibition causes AP prolongation in human and mouse. However, another study showed that full blockade of these channels did not cause measurable electrophysiological changes in rat and dog atrial and ventricular multicellular preparations (Nagy et al. 2009). Nevertheless, knockout of SK2 channels prolongs AP and causes AF in mouse (Li et al. 2009).

39.7.4 FUNNY CURRENT I_f

I_f was so named because it was a funny channel activated upon hyperpolarization. It is carried by K^+ and Na^+, with a reversal potential between –10 and –20 mV. Funny channels are encoded by hyperpolarization-activated cyclic nucleotide-gated (HCN) isoforms, whereby HCN4 is the most highly expressed in the heart. It is expressed in SAN, where it is relevant to pacemaker activity, as witnessed by loss-of-function mutations in humans that cause heart rate disturbances. Albeit to a minor extent, I_f is also expressed in the AV node and Purkinje system and at very low levels in atrial and ventricular myocytes, where it does not play a significant physiological role but may constitute an arrhythmogenic source in certain cardiac diseases (see Baruscotti et al. 2010 for review).

39.8 CARDIAC Cl− CHANNELS

In comparison with cation (K^+, Na^+, and Ca^{2+}) channels, much less is known about the functional role of anion (Cl^-) channels in cardiovascular physiology and pathophysiology. The reversal potential for Cl^- is within a range (usually –65 to –40 mV) that is more positive than the resting E_m and can be either negative or positive to the actual E_m during the normal cardiac cycle. Thus, cardiac Cl^- channels can generate both inward and outward currents (and cause both depolarization and repolarization) and produce significant effects on cardiac AP characteristics and pacemaker activity (causing diastolic depolarization). Over the past decades, various types of Cl^- currents have been recorded in cardiac cells from different regions of the heart and in different species, including humans; and their biophysical, pharmacological, and molecular properties have been characterized (Duan 2009; Hume et al. 2000). At the molecular level, all cardiac Cl^- channels described so far may fall into the following Cl^- channel gene families (Duan 2009):

1. The cystic fibrosis transmembrane conductance regulator (CFTR) is a member of the adenosine triphosphate–binding cassette transporter superfamily and may be responsible for the Cl^- currents activated by PKA ($I_{Cl,PKA}$), PKC ($I_{Cl,PKC}$), and extracellular ATP ($I_{Cl,ATP}$) in the heart. A major physiological role of activation of CFTR may be to prevent excessive APD prolongation and protect the heart against the development of EADs and triggered activity caused by β-adrenergic stimulation. However, when background K^+ conductance is reduced in the case of myocardial hypokalemia, activation of CFTR channels will cause significant membrane depolarization and induce abnormal automaticity. It has been shown that activation of CFTR channels contributes to hypoxia-induced shortening in APD. Activation of CFTR channels may accelerate the development of reentry due to shortening of APD and refractoriness, plus a decrease in conduction velocity caused by a slight depolarization of diastolic potential, leading to Na^+ channel inactivation. Remodeling of CFTR channels (reduced expression and heterogeneity) has been observed in myocardial hypertrophy and HF, although the functional consequences are not fully known. Also, CFTR may play a role in the protective effects of ischemic preconditioning and postconditioning on cardiac function and myocardium injury against sustained ischemia.
2. ClC-2 is a member of the ClC voltage-gated Cl^- channel superfamily and may be responsible for the hyperpolarization- and cell-swelling-activated inwardly rectifying Cl^- current ($I_{Cl,ir}$). $I_{Cl,ir}$ is small under basal or isotonic conditions but may be of pathological importance during hypoxia- or ischemia-induced acidosis or cell swelling.
3. ClC-3 is also a member of the ClC voltage-gated Cl^- channel superfamily and may be responsible for the volume-regulated outwardly rectifying Cl^- current ($I_{Cl,vol}$), including the basally activated ($I_{Cl,b}$) and swelling activated ($I_{Cl,swell}$) components. ClC-3 channels may contribute to hypoxia-, ischemia-, and reperfusion-induced shortening of APD and arrhythmias and limit APD prolongation in myocardial hypertrophy and HF.
4. CLCA-1, which was thought to be responsible for the Ca^{2+}-activated Cl^- current ($I_{Cl(Ca)}$).
5. Bestrophin, a candidate also for $I_{Cl(Ca)}$.
6. TMEM16, a novel candidate for $I_{Cl(Ca)}$.

The kinetic behavior of $I_{Cl(Ca)}$ is significantly determined by the time course of the Ca^{2+} transient, being small during resting $[Ca^{2+}]_i$ and carrying a significant amount of transient outward current when $[Ca^{2+}]_i$ rises. $I_{Cl(Ca)}$ activates early during the AP in response to an increase in $I_{Cl(Ca)}$ associated with Ca^{2+}-induced Ca^{2+} release, whereas the time course of the decline of the Ca^{2+} transient will determine the extent to which $I_{Cl(Ca)}$ contributes to early repolarization (phase 1). Thus, in Ca^{2+}-overloaded cardiac preparations, $I_{Cl(Ca)}$ could contribute to arrhythmogenic phenomena. However, the role of $I_{Cl(Ca)}$ in phase 1 repolarization and the generation of EAD and DAD is potentially species dependent, and the impact on either normal or failing human heart seems very limited.

A novel Cl^- current activated by extracellular acidosis ($I_{Cl,acid}$) has also been observed in cardiac myocytes, but the molecular identity for $I_{Cl,acid}$ is currently not known.

39.9 TRIGGERED ARRHYTHMIA AND CARDIAC AUTOMATICITY

Normal APs remain at the resting potential after repolarization, maintained by high resting K^+ permeability through I_{K1}. EADs and DADs are proarrhythmic aberrations of the AP. DADs take off after completion of AP repolarization and are more commonly observed in conditions of increased cellular and SR Ca^{2+} loading. If after AP repolarization the conditions favor Ca^{2+} sparks and waves (i.e., Ca^{2+} overload and/or enhanced RyR sensitivity), more of these events will occur and will generate a Ca^{2+}-activated transient inward current and DAD, which is carried almost exclusively via NCX current. Notably, the negative diastolic E_m favors inward NCX. If the inward NCX current is sufficient to bring the myocyte to the threshold of Na^+ channel activation, then an AP can be triggered (Figure 39.6, lower right).

EADs that take off before repolarization is complete are mediated by two possible mechanisms. One mechanism is fundamentally the same as that for DADs (i.e., spontaneous SR Ca^{2+} release and inward NCX current), as in Figure 39.6 (EAD2, lower left). EADs are typically observed under conditions where APD is long (e.g., I_{Kr} block or LQTS), and this tends to load cells with Ca^{2+} because of the prolonged Ca^{2+} entry and reduced diastolic interval for Ca^{2+} efflux. However, the depolarized E_m reduces inward I_{NCX} at any given $[Ca^{2+}]_i$, making this more impactful as repolarization proceeds than at positive plateau potentials. Another mechanism is that some LTCCs can become reavailable during long APD and reactivate, creating an inward current surge and net depolarization (Figure 39.6, upper

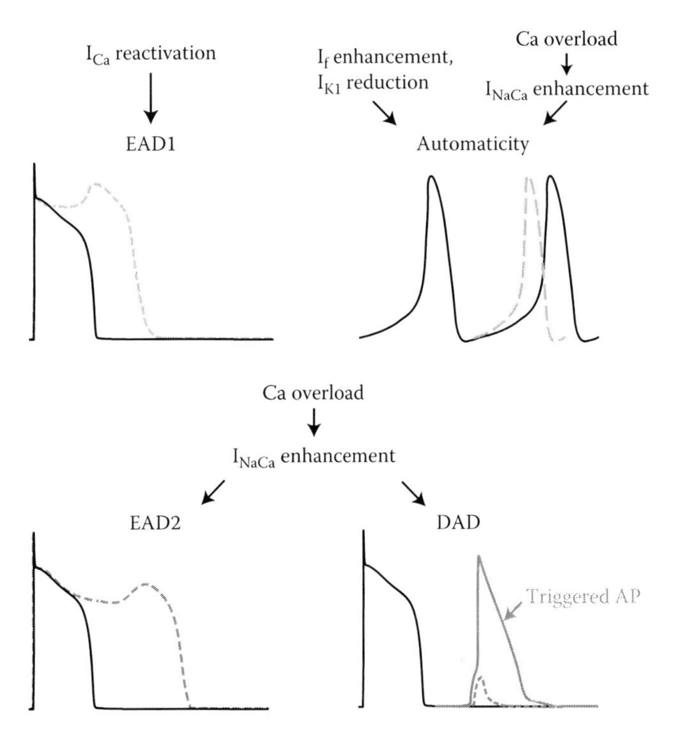

Figure 39.6 Mechanisms of EADs and DADs and cardiac cell automaticity.

Ion channel physiology and diseases

left, EAD1). There is good experimental and theoretical evidence for both of these mechanisms (Morotti et al. 2012), and they are not mutually exclusive. For example, if an SR Ca^{2+} release event occurs and produces only a very small plateau I_{NCX} and depolarization, it may be amplified by the recruitment of I_{Ca} to cause a larger EAD.

SAN and AVN myocytes show spontaneous Ca^{2+} transients and APs. Enhanced automaticity is caused by changes in this balance resulting from decreased I_{K1} and/or enhanced I_f. It is worth noting that the SR Ca^{2+} release and I_{NCX} mechanism described earlier for triggered DAD activity in atrial and ventricular myocytes is an important part of normal pacemaker activity in the SA node (Figure 39.6, upper right). These cells probably have a relatively Ca^{2+}-loaded physiological state, ancillary I_{CaL} that activates at more negative E_m ($Ca_v1.3$), lower stabilizing I_{K1} current, and a favorable source–sink relationship compared to ventricular myocytes (Bers 2001).

REFERENCES

Abriel, H. 2010. Cardiac sodium channel Na(v)1.5 and interacting proteins: Physiology and pathophysiology. *J Mol Cell Cardiol* 48(1): 2–11.

Ai, X., J. W. Curran, T. R. Shannon, D. M. Bers, and S. M. Pogwizd. 2005. Ca^{2+}/calmodulin-dependent protein kinase modulates cardiac ryanodine receptor phosphorylation and sarcoplasmic reticulum Ca^{2+} leak in heart failure. *Circ Res* 97(12): 1314–1322.

Aiba, T., G. G. Hesketh, T. Liu et al. 2010. Na^+ channel regulation by Ca^{2+}/calmodulin and Ca^{2+}/calmodulin-dependent protein kinase II in guinea-pig ventricular myocytes. *Cardiovasc Res* 85(3): 454–463.

Alseikhan, B. A., C. D. DeMaria, H. M. Colecraft, and D. T. Yue. 2002. Engineered calmodulins reveal the unexpected eminence of Ca^{2+} channel inactivation in controlling heart excitation. *Proc Natl Acad Sci USA* 99(26): 17185–17190.

Anderson, M. E., A. P. Braun, H. Schulman, and B. A. Premack. 1994. Multifunctional Ca^{2+}/calmodulin-dependent protein kinase mediates Ca^{2+}-induced enhancement of the L-type Ca^{2+} current in rabbit ventricular myocytes. *Circ Res* 75(5): 854–861.

Anumonwo, J. M. and A. N. Lopatin. 2010. Cardiac strong inward rectifier potassium channels. *J Mol Cell Cardiol* 48(1): 45–54.

Ashpole, N. M., A. W. Herren, K. S. Ginsburg et al. 2012. Ca^{2+}/calmodulin-dependent protein kinase II (CaMKII) regulates cardiac sodium channel $Na_v1.5$ gating by multiple phosphorylation sites. *J Biol Chem* 287(24): 19856–19869.

Baruscotti, M., A. Barbuti, and A. Bucchi. 2010. The cardiac pacemaker current. *J Mol Cell Cardiol* 48(1): 55–64.

Bassani, J. W., R. A. Bassani, and D. M. Bers. 1994. Relaxation in rabbit and rat cardiac cells: Species-dependent differences in cellular mechanisms. *J Physiol* 476(2): 279–293.

Bers, D. M. 2001. *Excitation-Contraction Coupling and Cardiac Contractile Force*. Dordrecht, The Netherlands: Kluwer Academic Press.

Bers, D. M. 2002. Cardiac excitation-contraction coupling. *Nature* 415(6868): 198–205.

Bers, D. M. 2008. Calcium cycling and signaling in cardiac myocytes. *Annu Rev Physiol* 70: 23–49.

Bers, D. M. 2012. Ryanodine receptor S2808 phosphorylation in heart failure: Smoking gun or red herring. *Circ Res* 110(6): 796–799.

Bezzina, C., M. W. Veldkamp, M. P. van Den Berg et al. 1999. A single Na^+ channel mutation causing both long-QT and Brugada syndromes. *Circ Res* 85(12): 1206–1213.

Bezzina, C. R., M. B. Rook, and A. A. Wilde. 2001. Cardiac sodium channel and inherited arrhythmia syndromes. *Cardiovasc Res* 49(2): 257–271.

Biet, M., H. Barajas-Martinez, A. T. Ton, J. F. Delabre, N. Morin, and R. Dumaine. 2012. About half of the late sodium current in cardiac myocytes from dog ventricle is due to non-cardiac-type Na^+ channels. *J Mol Cell Cardiol* 53(5): 593–598.

Cens, T., M. Rousset, J. P. Leyris, P. Fesquet, and P. Charnet. 2006. Voltage- and calcium-dependent inactivation in high voltage-gated Ca^{2+} channels. *Prog Biophys Mol Biol* 90(1–3): 104–117.

Charpentier, F., J. Merot, G. Loussouarn, and I. Baro. 2010. Delayed rectifier K^+ currents and cardiac repolarization. *J Mol Cell Cardiol* 48(1): 37–44.

Christ, T., E. Wettwer, N. Voigt et al. 2008. Pathology-specific effects of the $I_{Kur}/I_{to}/I_{K,ACh}$ blocker AVE0118 on ion channels in human chronic atrial fibrillation. *Br J Pharmacol* 154(8): 1619–1630.

Christensen, M. D., W. Dun, P. A. Boyden, M. E. Anderson, P. J. Mohler, and T. J. Hund. 2009. Oxidized calmodulin kinase II regulates conduction following myocardial infarction: A computational analysis. *PLoS Comput Biol* 5(12): e1000583.

Clancy, C. E. and Y. Rudy. 2002. Na^+ channel mutation that causes both Brugada and long-QT syndrome phenotypes: A simulation study of mechanism. *Circulation* 105(10): 1208–1213.

Cohen, S. A. 1996. Immunocytochemical localization of rH1 sodium channel in adult rat heart atria and ventricle. Presence in terminal intercalated disks. *Circulation* 94(12): 3083–3086.

Dobrev, D. and S. Nattel. 2010. New antiarrhythmic drugs for treatment of atrial fibrillation. *Lancet* 375(9721): 1212–1223.

Domeier, T. L., A. V. Zima, J. T. Maxwell, S. Huke, G. A. Mignery, and L. A. Blatter. 2008. IP3 receptor-dependent Ca^{2+} release modulates excitation-contraction coupling in rabbit ventricular myocytes. *Am J Physiol Heart Circ Physiol* 294(2): H596–H604.

Duan, D. 2009. Phenomics of cardiac chloride channels: The systematic study of chloride channel function in the heart. *J Physiol* 587(Pt 10): 2163–2177.

Faber, G. M., J. Silva, L. Livshitz, and Y. Rudy. 2007. Kinetic properties of the cardiac L-type Ca^{2+} channel and its role in myocyte electrophysiology: A theoretical investigation. *Biophys J* 92(5): 1522–1543.

Grandi, E. and D. M. Bers. 2013. Models of the ventricular action potential in health and disease. In *Cardiac Electrophysiology: From Cell to Bedside*, (eds.) D. P. Zipes and J. Jalife. Philadelphia, PA: Saunders.

Grandi, E., S. V. Pandit, N. Voigt et al. 2011. Human atrial action potential and Ca^{2+} model: Sinus rhythm and chronic atrial fibrillation. *Circ Res* 109(9): 1055–1066.

Grandi, E., F. S. Pasqualini, and D. M. Bers. 2010. A novel computational model of the human ventricular action potential and Ca transient. *J Mol Cell Cardiol* 48(1): 112–121.

Grandi, E., F. S. Pasqualini, C. Pes, C. Corsi, A. Zaza, and S. Severi. 2009. Theoretical investigation of action potential duration dependence on extracellular Ca^{2+} in human cardiomyocytes. *J Mol Cell Cardiol* 46(3): 332–342.

Grandi, E., J. L. Puglisi, S. Severi, and D. M. Bers. 2007a. Computer simulation of altered sodium channel gating in rabbit and human ventricular myocytes. In *Lecture Notes in Computer Science*, (eds.) F. B. Sachse and G. Seemann. Springer-Verlag GmbH, Berlin and Heidelberg.

Grandi, E., J. L. Puglisi, S. Wagner, L. S. Maier, S. Severi, and D. M. Bers. 2007b. Simulation of Ca-calmodulin-dependent protein kinase II on rabbit ventricular myocyte ion currents and action potentials. *Biophys J* 93(11): 3835–3847.

Grandi, E., A. J. Workman, and S. V. Pandit. 2012. Altered excitation contraction coupling in human chronic atrial fibrillation. *J Atr Fibrillation* 4(6): 37–53.

Grueter, C. E., S. A. Abiria, I. Dzhura et al. 2006. L-type Ca^{2+} channel facilitation mediated by phosphorylation of the beta subunit by CaMKII. *Mol Cell* 23(5): 641–650.

Herren, A., D. M. Bers, and E. Grandi. 2013. Post-translational modifications of the cardiac Na channel: contribution of CaMKII-dependent phosphorylation to acquired arrhythmias. *Am J Physiol Heart Circ Physiol.* 305(4): H431–H445.

Hudmon, A., H. Schulman, J. Kim, J. M. Maltez, R. W. Tsien, and G. S. Pitt. 2005. CaMKII tethers to L-type Ca^{2+} channels, establishing a local and dedicated integrator of Ca^{2+} signals for facilitation. *J Cell Biol* 171(3): 537–547.

Hume, J. R., D. Duan, M. L. Collier, J. Yamazaki, and B. Horowitz. 2000. Anion transport in heart. *Physiol Rev* 80(1): 31–81.

Hund, T. J., K. F. Decker, E. Kanter et al. 2008. Role of activated CaMKII in abnormal calcium homeostasis and I(Na) remodeling after myocardial infarction: Insights from mathematical modeling. *J Mol Cell Cardiol* 45(3): 420–428.

Hund, T. J., O. M. Koval, J. Li et al. 2010. A beta(IV)-spectrin/CaMKII signaling complex is essential for membrane excitability in mice. *J Clin Invest* 120(10): 3508–3519.

Kassmann, M., A. Hansel, E. Leipold et al. 2008. Oxidation of multiple methionine residues impairs rapid sodium channel inactivation. *Pflugers Arch* 456(6): 1085–1095.

Keskanokwong, T., H. J. Lim, P. Zhang et al. 2011. Dynamic Kv4.3-CaMKII unit in heart: An intrinsic negative regulator for CaMKII activation. *Eur Heart J* 32(3): 305–315.

Lee, K. S., E. Marban, and R. W. Tsien. 1985. Inactivation of calcium channels in mammalian heart cells: Joint dependence on membrane potential and intracellular calcium. *J Physiol* 364(1): 395–411.

Li, G. R. and M. Q. Dong. 2010. Pharmacology of cardiac potassium channels. *Adv Pharmacol* 59: 93–134.

Li, J., C. Marionneau, R. Zhang et al. 2006. Calmodulin kinase II inhibition shortens action potential duration by upregulation of K^+ currents. *Circ Res* 99(10): 1092–1099.

Li, N., V. Timofeyev, D. Tuteja et al. 2009. Ablation of a Ca^{2+}-activated K^+ channel (SK2 channel) results in action potential prolongation in atrial myocytes and atrial fibrillation. *J Physiol* 587(Pt 5): 1087–1100.

Lin, X., N. Liu, J. Lu et al. 2011. Subcellular heterogeneity of sodium current properties in adult cardiac ventricular myocytes. *Heart Rhythm* 8(12): 1923–1930.

Lipp, P., M. Laine, S. C. Tovey et al. 2000. Functional $InsP_3$ receptors that may modulate excitation-contraction coupling in the heart. *Curr Biol* 10(15): 939–942.

Mahajan, A., D. Sato, Y. Shiferaw et al. 2008. Modifying L-type calcium current kinetics: Consequences for cardiac excitation and arrhythmia dynamics. *Biophys J* 94(2): 411–423.

Maltsev, V. A., J. W. Kyle, and A. Undrovinas. 2009. Late Na^+ current produced by human cardiac Na^+ channel isoform Nav1.5 is modulated by its beta1 subunit. *J Physiol Sci* 59(3): 217–225.

Marx, S. O., J. Kurokawa, S. Reiken et al. 2002. Requirement of a macromolecular signaling complex for beta adrenergic receptor modulation of the KCNQ1-KCNE1 potassium channel. *Science* 295(5554): 496–499.

Miura, Y. and J. Kimura. 1989. Sodium–calcium exchange current. Dependence on internal Ca and Na and competitive binding of external Na and Ca. *J Gen Physiol* 93(6): 1129–1145.

Morotti, S., E. Grandi, A. Summa, K. S. Ginsburg, and D. M. Bers. 2012. Theoretical study of L-type Ca^{2+} current inactivation kinetics during action potential repolarization and early afterdepolarizations. *J Physiol* 590(Pt 18): 4465–4481.

Nagy, N., V. Szuts, Z. Horvath et al. 2009. Does small-conductance calcium-activated potassium channel contribute to cardiac repolarization? *J Mol Cell Cardiol* 47(5): 656–663.

Niwa, N. and J. M. Nerbonne. 2010. Molecular determinants of cardiac transient outward potassium current (I(to)) expression and regulation. *J Mol Cell Cardiol* 48(1): 12–25.

Pandit, S. V., O. Berenfeld, J. M. Anumonwo et al. 2005. Ionic determinants of functional reentry in a 2-D model of human atrial cells during simulated chronic atrial fibrillation. *Biophys J* 88(6): 3806–3821.

Perez, P. J., J. Ramos-Franco, M. Fill, and G. A. Mignery. 1997. Identification and functional reconstitution of the type 2 inositol 1,4,5-trisphosphate receptor from ventricular cardiac myocytes. *J Biol Chem* 272(38): 23961–23969.

Peterson, B. Z., C. D. DeMaria, J. P. Adelman, and D. T. Yue. 1999. Calmodulin is the Ca^{2+} sensor for Ca^{2+}-dependent inactivation of L-type calcium channels. *Neuron* 22(3): 549–558.

Picht, E., A. V. Zima, T. R. Shannon, A. M. Duncan, L. A. Blatter, and D. M. Bers. 2011. Dynamic calcium movement inside cardiac sarcoplasmic reticulum during release. *Circ Res* 108(7): 847–856.

Pitt, G. S. 2007. Calmodulin and CaMKII as molecular switches for cardiac ion channels. *Cardiovasc Res* 73(4): 641–647.

Pogwizd, S. M., K. Schlotthauer, L. Li, W. Yuan, and D. M. Bers. 2001. Arrhythmogenesis and contractile dysfunction in heart failure: Roles of sodium–calcium exchange, inward rectifier potassium current, and residual beta-adrenergic responsiveness. *Circ Res* 88(11): 1159–1167.

Priori, S. G. and S. R. Chen. 2011. Inherited dysfunction of sarcoplasmic reticulum Ca^{2+} handling and arrhythmogenesis. *Circ Res* 108(7): 871–883.

Rousseau, E. and G. Meissner. 1989. Single cardiac sarcoplasmic reticulum Ca^{2+}-release channel: Activation by caffeine. *Am J Physiol* 256(2 Pt 2): H328–H333.

Samie, F. H., O. Berenfeld, J. Anumonwo et al. 2001. Rectification of the background potassium current: A determinant of rotor dynamics in ventricular fibrillation. *Circ Res* 89(12): 1216–1223.

Schotten, U., S. de Haan, S. Verheule et al. 2007. Blockade of atrial-specific K^+-currents increases atrial but not ventricular contractility by enhancing reverse mode Na^+/Ca^{2+}-exchange. *Cardiovasc Res* 73(1): 37–47.

Sergeant, G. P., S. Ohya, J. A. Reihill et al. 2005. Regulation of Kv4.3 currents by Ca^{2+}/calmodulin-dependent protein kinase II. *Am J Physiol Cell Physiol* 288(2): C304–C313.

Shamgar, L., L. Ma, N. Schmitt et al. 2006. Calmodulin is essential for cardiac I_{Ks} channel gating and assembly: Impaired function in long-QT mutations. *Circ Res* 98(8): 1055–1063.

Shannon, T. R., S. M. Pogwizd, and D. M. Bers. 2003. Elevated sarcoplasmic reticulum Ca^{2+} leak in intact ventricular myocytes from rabbits in heart failure. *Circ Res* 93(7): 592–594.

Silva, J. and Y. Rudy. 2005. Subunit interaction determines I_{Ks} participation in cardiac repolarization and repolarization reserve. *Circulation* 112(10): 1384–1391.

Splawski, I., K. W. Timothy, N. Decher et al. 2005. Severe arrhythmia disorder caused by cardiac L-type calcium channel mutations. *Proc Natl Acad Sci USA* 102(23): 8089–8096.

Splawski, I., K. W. Timothy, L. M. Sharpe et al. 2004. $Ca_v1.2$ calcium channel dysfunction causes a multisystem disorder including arrhythmia and autism. *Cell* 119(1): 19.

Tessier, S., P. Karczewski, E. G. Krause et al. 1999. Regulation of the transient outward K^+ current by Ca^{2+}/calmodulin-dependent protein kinases II in human atrial myocytes. *Circ Res* 85(9): 810–819.

Thiel, W. H., B. Chen, T. J. Hund et al. 2008. Proarrhythmic defects in timothy syndrome require calmodulin kinase II. *Circulation* 118(22): 2225–2234.

Veldkamp, M. W., P. C. Viswanathan, C. Bezzina, A. Baartscheer, A. A. Wilde, and J. R. Balser. 2000. Two distinct congenital arrhythmias evoked by a multidysfunctional Na^+ channel. *Circ Res* 86(9): E91–E97.

Wagner, S., N. Dybkova, E. C. Rasenack et al. 2006. Ca^{2+}/calmodulin-dependent protein kinase II regulates cardiac Na^+ channels. *J Clin Invest* 116(12): 3127–3138.
Wagner, S., E. Hacker, E. Grandi et al. 2009. Ca/calmodulin kinase II differentially modulates potassium currents. *Circ Arrhythm Electrophysiol* 2(3): 285–294.
Wagner, S., H. M. Ruff, S. L. Weber et al. 2011. Reactive oxygen species-activated Ca/calmodulin kinase IIdelta is required for late I(Na) augmentation leading to cellular Na and Ca overload. *Circ Res* 108(5): 555–565.
Wang, Y., J. Cheng, S. Tandan, M. Jiang, D. T. McCloskey, and J. A. Hill. 2006. Transient-outward K^+ channel inhibition facilitates L-type Ca^{2+} current in heart. *J Cardiovasc Electrophysiol* 17(3): 298–304.
Weber, C. R., K. S. Ginsburg, and D. M. Bers. 2003a. Cardiac submembrane $[Na^+]$ transients sensed by Na^+-Ca^{2+} exchange current. *Circ Res* 92(9): 950–952.
Weber, C. R., K. S. Ginsburg, K. D. Philipson, T. R. Shannon, and D. M. Bers. 2001. Allosteric regulation of Na/Ca exchange current by cytosolic Ca in intact cardiac myocytes. *J Gen Physiol* 117(2): 119–131.
Weber, C. R., V. Piacentino, III, K. S. Ginsburg, S. R. Houser, and D. M. Bers. 2002. Na^+-Ca^{2+} exchange current and submembrane $[Ca^{2+}]$ during the cardiac action potential. *Circ Res* 90(2): 182–189.
Weber, C. R., V. Piacentino, III, S. R. Houser, and D. M. Bers. 2003b. Dynamic regulation of sodium/calcium exchange function in human heart failure. *Circulation* 108(18): 2224–2229.
Wilde, A. A. and C. R. Bezzina. 2005. Genetics of cardiac arrhythmias. *Heart* 91(10): 1352–1358.
Wu, X., T. Zhang, J. Bossuyt et al. 2006. Local InsP3-dependent perinuclear Ca^{2+} signaling in cardiac myocyte excitation-transcription coupling. *J Clin Invest* 116(3): 675–682.
Xi, Y., G. Wu, L. Yang et al. 2009. Increased late sodium currents are related to transcription of neuronal isoforms in a pressure-overload model. *Eur J Heart Fail* 11(8): 749–757.
Xiao, L., P. Coutu, L. R. Villeneuve et al. 2008. Mechanisms underlying rate-dependent remodeling of transient outward potassium current in canine ventricular myocytes. *Circ Res* 103(7): 733–742.
Xiao, R. P., H. Cheng, W. J. Lederer, T. Suzuki, and E. G. Lakatta. 1994. Dual regulation of Ca^{2+}/calmodulin-dependent kinase II activity by membrane voltage and by calcium influx. *Proc Natl Acad Sci USA* 91(20): 9659–9663.
Xie, Y., E. Grandi, J. L. Puglisi, D. Sato, and D. M. Bers. 2013. β-adrenergic stimulation activates early afterdepolarizations transiently via kinetic mismatch of PKA targets. *J Mol Cell Cardiol* 58: 153–161.
Xu, L., A. Tripathy, D. A. Pasek, and G. Meissner. 1998. Potential for pharmacology of ryanodine receptor/calcium release channels. *Ann N Y Acad Sci* 853: 130–148.
Xu, Y., D. Tuteja, Z. Zhang et al. 2003. Molecular identification and functional roles of a Ca^{2+}-activated K^+ channel in human and mouse hearts. *J Biol Chem* 278(49): 49085–49094.
Yarotskyy, V., G. Gao, B. Z. Peterson, and K. S. Elmslie. 2009. The Timothy syndrome mutation of cardiac Cav1.2 (L-type) channels: Multiple altered gating mechanisms and pharmacological restoration of inactivation. *J Physiol* 587(Pt 3): 551–565.
Yuan, W. and D. M. Bers. 1994. Ca-dependent facilitation of cardiac Ca current is due to Ca-calmodulin-dependent protein kinase. *Am J Physiol* 267(3 Pt 2): H982–H993.
Zhu, Y., X. Ai, R. A. Oster, D. M. Bers, and S. M. Pogwizd. 2012. Sex differences in repolarization and slow delayed rectifier potassium current and their regulation by sympathetic stimulation in rabbits. *Pflugers Arch.* 465(6): 805–818.
Zima, A. V. and L. A. Blatter. 2004. Inositol-1,4,5-trisphosphate-dependent Ca^{2+} signalling in cat atrial excitation-contraction coupling and arrhythmias. *J Physiol* 555(Pt 3): 607–615.
Zima, A. V., E. Bovo, D. M. Bers, and L. A. Blatter. 2010. Ca^{2+} spark-dependent and -independent sarcoplasmic reticulum Ca^{2+} leak in normal and failing rabbit ventricular myocytes. *J Physiol* 588(Pt 23): 4743–4757.
Zuhlke, R. D., G. S. Pitt, K. Deisseroth, R. W. Tsien, and H. Reuter. 1999. Calmodulin supports both inactivation and facilitation of L-type calcium channels. *Nature* 399(6732): 159–162.

Ion channels in pain

J.P. Johnson, Jr.

Contents

40.1 INTRODUCTION

Pain signaling is largely accomplished by neurons, and ion channels are critical for the function and signaling of all neurons. Neuronal signaling is a complex synthesis of the conductances of many channels and transporters, and all of the channels in the neurons of the pain pathway play some role in pain signaling. This chapter will introduce the basic pathways of pain signaling and discuss a few ion channels that have been most studied and most closely linked to pain.

Pain is a critical protective mechanism that allows healthy animals to avoid tissue damage and to prevent further damage to injured tissue. It allows people and animals to learn the physical constraints of their bodies and to identify dangerous stimuli. People who lack normal pain sensation have great difficulty in establishing their physical boundaries and as a result often inadvertently injure themselves. For example, mutations that prevent proper expression of the nerve growth factor receptor Trk-A (1) or the voltage-gated sodium (Na_V) channel $Na_V1.7$ confer complete congenital insensitivity to pain in humans (2). These patients invariably injure themselves by biting their lips, tongues, and fingers in infancy and suffer frequent self-injury throughout early life until they learn how to avoid harm. Even with vigilance, a mild injury like a foot blister, which would be quickly noticed and remedied by most people, can go unnoticed in the absence of pain and progress to a serious skin lesion before it is discovered.

Despite the great survival advantages that pain confers, many conditions exist where pain outlives its usefulness. Once the dangerous insult is removed or avoided and tissue has healed, pain should abate. Nonetheless, there are many cases where pain persists and becomes pathologic. In such cases, pain can have serious negative impacts on quality of life. Pain can be broadly categorized as acute, inflammatory, or neuropathic (caused by nerve injury). Acute pain is normal, protective pain. Inflammatory pain can also be useful as it induces the sufferer to protect damaged tissue, but it can become chronic and pathologic. Over 14 million people in the United States suffer from neuropathic pain, and it is most prevalent in women and the elderly. Current therapies have poor efficacy and/or a high risk of adverse events. Nearly 15,000 annual deaths in the United States are caused by opioid pain drugs alone. This review will focus on the channels involved in pain. The limited treatment options for pain, combined with a growing awareness of the risks of the currently available pain drugs, especially opioids, illustrate a need for new pain control therapies. Because of their role in the physiology of pain-sensing neurons, ion channels represent important potential targets for current and novel pain therapeutics.

40.1.1 PAIN PATHWAYS

Ultimately, pain is in the brain, because it is in the brain that pain signals are synthesized and integrated with the overall emotional outlook of the animal. But pain signaling can be broadly broken down into three compartments. The primary peripheral neurons of the dorsal root ganglia (DRG) detect painful stimuli and carry the signal to the spinal cord. The secondary wide dynamic range neurons of the spinal cord integrate the peripheral signals and propagate them to the brain. Tertiary neurons distributed in multiple brain regions ultimately assemble the perception of pain. Ion channels with signaling roles specific for pain in the brain have not yet been well identified. The current knowledge of pain in the brain is primarily a description of the brain regions responsible for pain perception and the interactions between these regions. In the relatively simpler circuitry of the peripheral and spinal neurons, the molecular components of pain signaling are better established, and these will be the focus of this chapter.

The first step in the pain pathway is the generation of an action potential at the sensory nerve endings of the peripheral afferent neurons. The somas of these neurons reside in the DRG, a small bundle of nerve tissue that lies just outside the spine in humans (Figure 40.1). The neurons of the DRG have one process that splits into two branches soon after leaving the soma. One branch leads to the peripheral sensory terminals in the skin or other organs where the sensation of pain normally originates. The other branch reaches into the spine to synapse onto the neurons of lamina I and II of the spinal cord. Once the electrical action potential reaches the spinal cord, the terminals of DRG neurons release neurotransmitters to excite the secondary spinal neurons (wide dynamic range neurons), and the action potential continues toward the brain. Once in the brain, multiple brain areas contribute to the synthesis of pain perception.

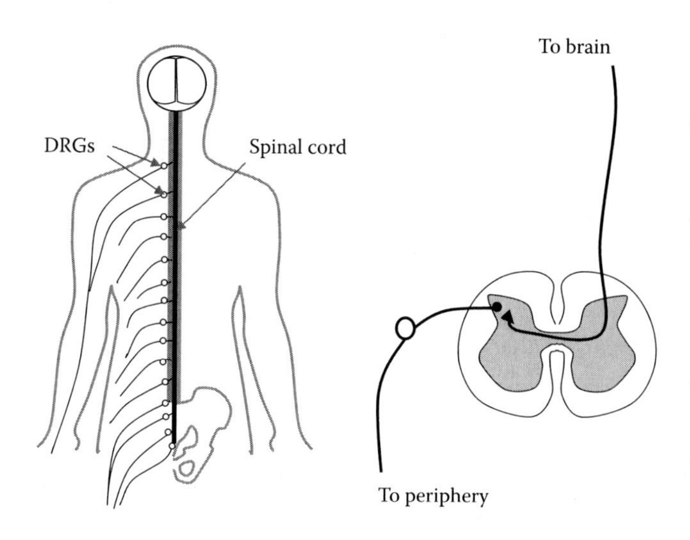

Figure 40.1 Schematic of the sensory pain pathway.

Along with the afferent signaling pathways that carry sensory signaling information to the brain, there are also inhibitory efferent neurons that descend from the brain to synapse on the spinal afferent neurons. These descending neurons normally dampen the excitability of the system and keep normal innocuous sensations from inducing pain.

40.1.2 NOCICEPTIVE NEURONS

The neurons of the DRG carry all the sensory information from the periphery, and they must encode a great diversity of different types of information. This likely explains why the ganglia are composed of such a heterogeneous group of neurons.

The axons of nociceptive neurons give rise to C- or Aδ-nerve fibers. These nerve fibers are classified empirically by their conduction speed. C-fibers are small and unmyelinated. Due to the lack of myelin insulation, they carry signals much more slowly than myelinated neurons. Aδ-fibers are small but myelinated and carry signals much more rapidly. Aδ-fibers carry the fast component of pain felt instantly upon injury, while C-fibers carry the slower aching or burning phase of pain sensation. Large myelinated Aβ-fibers carry mechanical signals and are not normally associated with pain, but they may be recruited into nociceptive roles in the case of chronic pain scenarios.

The primary channels discussed here will be the Na_V channels, the TRPV1 channel, and voltage-gated calcium channel, $Ca_V2.2$ (Figure 40.2). TRPV1 channels serve as primary sensors of painful stimuli in the peripheral nerve terminals. Na_V channels serve to sense and propagate the electrical action potentials of pain signaling along the primary nociceptor. $Ca_V2.2$ plays a critical role at the spinal terminal of the nociceptor, controlling the calcium influx that ultimately leads to neurotransmitter release and stimulation of the secondary neurons.

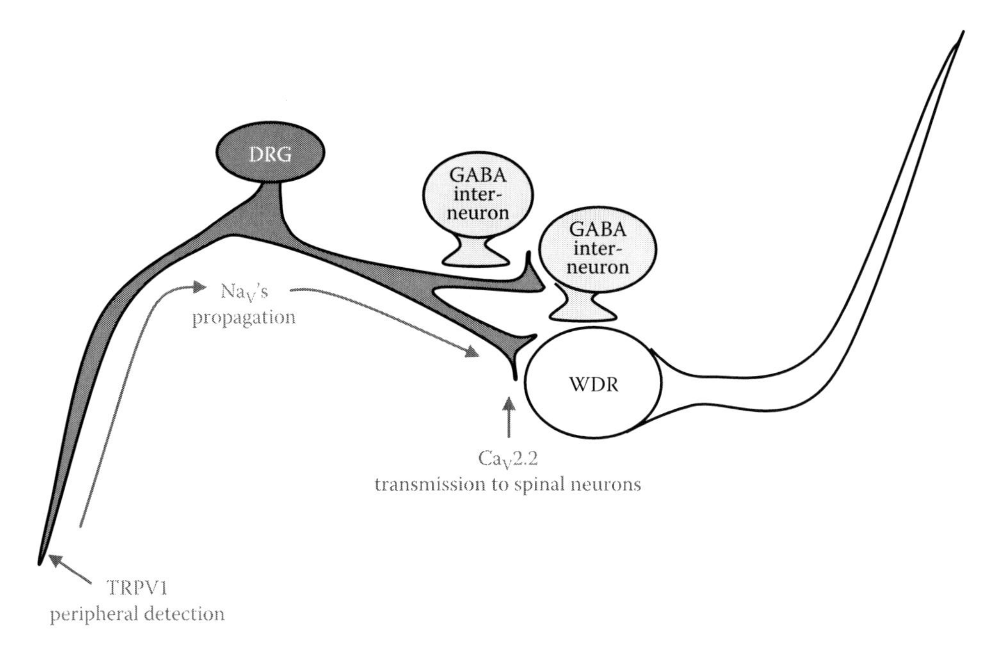

Figure 40.2 DRG neurons carry pain signals to the wide dynamic range neurons of the spinal cord.

40.2 VOLTAGE-GATED SODIUM CHANNELS

Na$_V$ channels play a critical role in pain signaling. The evidence for the role of these channels in normal physiology, the pathological states arising from mutations in sodium channel genes, preclinical work in animal models, and the clinical pharmacology of known sodium channel modulating agents all point to a critical role of Na$_V$s in pain sensation (3–6).

Na$_V$s are key biological mediators of electrical signaling as they are the primary mediators of the rapid upstroke of the action potential of many excitable cell types (e.g., neurons, skeletal myocytes, cardiac myocytes) (7). The peripheral neurons of the DRG express many isoforms of voltage-gated ion channel, including the tetrodotoxin (TTX)-sensitive isoforms Na$_V$1.1, Na$_V$1.2, Na$_V$1.6, and Na$_V$1.7 and the TTX-resistant isoforms Na$_V$1.8 and Na$_V$1.9 (8).

Because of the role Na$_V$s play in the initiation and propagation of neuronal signals, antagonists that reduce Na$_V$ currents can prevent or reduce neural signaling, and Na$_V$ channels have long been considered likely targets to reduce pain in conditions where hyperexcitability is observed (9). Several clinically useful analgesics have been identified as inhibitors of Na$_V$ channels (10). The local anesthetic drugs, for example, lidocaine, block pain by inhibiting Na$_V$ channels (Figure 40.3). Other compounds that have proven effective at reducing pain, like carbamazepine, lamotrigine, and tricyclic antidepressants, have also been suggested to act by sodium channel inhibition (11,12). When locally applied at high concentrations, lidocaine is effective at blocking pain, but it does so by unselectively shutting down all sensory nerve transmission. This results in the numbness and paralysis that can be experienced after a visit to a dentist. TTX is a deadly poison when administered systemically, but an intrathecal formulation of TTX is currently being developed as an analgesic for intractable cancer pain (13–15).

Lidocaine

Tetrodotoxin

Carbamazepine

Lamotrigine

Figure 40.3 Na$_V$ antagonists used clinically for pain.

When systemically administered, nonselective Na$_V$ blockers must be used at much lower concentrations than can be tolerated locally. The concentrations where these blockers begin to be effective to reduce pain are nearly the same as the concentrations where side effects become dangerous or intolerable for most patients. Thus, these compounds have a poor therapeutic window that is hypothesized to be due to their lack of isoform selectivity. This lack of selectivity leads to adverse events since the compounds block Na$_V$s in tissues not useful for pain reduction, such as the central nervous system (CNS) (Na$_V$1.1, Na$_V$1.2, Na$_V$1.3, and Na$_V$1.6), heart (Na$_V$1.5), and skeletal muscle (Na$_V$1.4). An obvious solution to this problem would be the development of compounds that block only the Na$_V$ isoforms critical to pain sensation while sparing the Na$_V$s that lead to adverse events. This approach has long been, and continues to be, a focus of pain research and drug development. The isoforms most closely linked with pain are $Na_V1.3$, $Na_V1.7$, $Na_V1.8$, and $Na_V1.9$, each of which will be discussed individually in the following sections.

40.3 $Na_V1.3$

40.3.1 $Na_V1.3$ LOCALIZATION AND ROLE IN NOCICEPTOR SIGNALING

The most compelling link between $Na_V1.3$ and pain pathways is the fact that it is upregulated in some painful conditions. $Na_V1.3$ is widely expressed in the neurons of the central, sympathetic, and peripheral sensory nervous systems of embryonic and neonatal animals, but expression quickly drops after birth and very little $Na_V1.3$ is found in adult brain, spinal cord, or peripheral sensory neurons. $Na_V1.3$ remains at high levels in the sympathetic neurons of adults (16). After nerve injury in rodent models of neuropathic pain, $Na_V1.3$ is dramatically upregulated (17–20). In contrast, all the other Na_V channels have reduced expression in injured DRG neurons (16,19,21,22). $Na_V1.3$ is also upregulated in injured human nerves (23–28). This unique regulation in response to injury suggests that $Na_V1.3$ might be an important contributor to the hyperexcitability that leads to hyperalgesia and spontaneous pain in patients or animals with neuropathic nerve injury pain. Intrathecal administration of sodium channel blockers can prevent both the upregulation of $Na_V1.3$ and the development of the hyperalgesia characteristic of animal models of neuropathic nerve injury pain (29,30).

The upregulation of $Na_V1.3$ is not limited to the peripheral sensory neurons but has also been observed in the neurons of the dorsal horn of the spinal cord (31,32), and even in the thalamus (33). Thus, the pathologic upregulation of $Na_V1.3$ appears to extend throughout the entire pain signaling pathway from the sensory terminals of the primary nociceptors to the tertiary neurons of the brain.

40.3.2 $Na_V1.3$ BIOPHYSICS

The function of $Na_V1.3$ is well suited to a role in pain signaling. The activation and inactivation kinetics of the channel, combined with a low propensity for closed-state inactivation, leads to large ramp currents evoked in response to slow depolarizations, as when a neuron is gradually depolarizing toward firing threshold (34–37). Recovery from inactivation is also quite fast, enabling the sort of rapid cycling through channel states that is necessary to support the rapid action potential firing characteristic of painful neurons.

40.3.3 $Na_V1.3$ IN SYMPATHETIC NEURONS

$Na_V1.3$ and $Na_V1.7$ are the primary sodium channels of the sympathetic neurons of the superior cervical ganglia. This may contribute to their role in pain pathways since sympathetic input can support neuropathic pain states. Increasing sympathetic innervation of the DRG has been shown to follow nerve injury (38), and sympatholytic treatments, interrupting sympathetic signaling either surgically or pharmacologically, are effective in many experimental animal pain models (39–42). They can also be effective for some pain patients in the clinic, particularly patients with complex regional pain syndrome, but sympatholytic treatments are used sparingly because of the potential for unwanted side effects (43).

40.3.4 GENETIC MANIPULATION OF $Na_V1.3$ IN ANIMAL PAIN MODELS

Despite the evidence of $Na_V1.3$ expression plasticity, genetic interventions have had mixed results. Antisense oligonucleotide knockdown of $Na_V1.3$ in the intrathecal space has been reported to be effective in spinal cord injury models (31,44) and ineffective in the spared nerve injury model (45). Systemic genetic postnatal knockout of $Na_V1.3$ leads to mice developing neuropathic hyperalgesia just as normal mice indicating that $Na_V1.3$ is not strictly necessary for neuropathic pain (46). It is possible that other Na_Vs might be upregulated in $Na_V1.3$ knockout mice. $Na_V1.3$ is upregulated in some neurons of $Na_V1.1$ knockout mice (47). A similar compensatory regulation of other channels might also occur in $Na_V1.3$ knockouts, rescuing the neuropathic pain phenotype and obscuring its role in neuropathic pain pathways.

40.4 $Na_V1.7$

40.4.1 $NA_V1.7$ LOCALIZATION AND ROLE IN NOCICEPTOR SIGNALING

$Na_V1.7$ is highly expressed constitutively in the peripheral and sympathetic nervous system. The primary sensory neurons of the DRG and trigeminal ganglia, the neurons of the myenteric plexus innervating the gut, and the sympathetic neurons of the superior cervical ganglia are all rich with $Na_V1.7$ expressing neurons. $Na_V1.7$ expression increases in the pulp of painful human teeth (48). Expression decreases after nerve injury, but the decrease is relatively small and significant $Na_V1.7$ expression remains.

40.4.2 $Na_V1.7$ BIOPHYSICS

$Na_V1.7$ inactivates at relatively negative potentials (midpoint of inactivation approximately −70 mV), and it has a very stable slow-inactivated state. As a result, a large fraction of the $Na_V1.7$ channels in most pain-sensing neurons, with resting membrane potentials around −55 mV, are likely inactivated. The stable slow-inactivated state of the channel also makes it poorly suited to sustain the high-frequency firing associated with excited nociceptors. Nonetheless, $Na_V1.7$ plays a major role in pain signaling, and it likely does this by setting the threshold, or the *gain* of nociceptor signaling (2). $Na_V1.7$ contributes to the subthreshold oscillations that trigger full-blown nociceptor activation and thus controls the initiation phase of pain signaling in the periphery (49). The high level of expression in the sympathetic neurons suggests that it might likewise control the threshold of those neurons as well.

40.4.3 $Na_V1.7$ HUMAN GENETICS

A large and growing number of human mutations in $Na_V1.7$ have been linked to human pain pathologies. These mutations confer either gain- or loss-of-function phenotypes on channel gating and in turn cause gain- or loss-of-pain phenotypes in the patients that harbor them. Patients with $Na_V1.7$ null mutations have congenital insensitivity to pain and are completely unable to feel pain even after breaking bones or during childbirth (2,50). These patients are prone to injury due to a lack of the normal caution that pain inspires. Despite the dramatic changes in pain perception, $Na_V1.7$ null patients otherwise seem largely normal. The one notable exception is that they have little or no sense of smell.

Three categories of $Na_V1.7$ gain-of-function mutations have been described in human patients, and biophysical characterization of the mutant channels has arrived at three distinct categories of phenotypes (51–53). Inherited erythromelalgia (IEM) patients experience spontaneously painful episodes. Mutations that cause this condition have been found to result in negative shifts of the voltage dependence of channel activation, such that channels open at more negative potentials than normal. Paroxysmal extreme pain disorder (PEPD) is a distinct syndrome associated with bouts of pain that is spontaneous or triggered by innocuous stimuli (53,54). These patients bear mutations that result in positive shifts in the voltage dependence of inactivation gating, resulting in channels that remain open longer than they should. One mutation has been identified that has an intermediate phenotype between IEM and PEPD, and this family bears a mutant channel with both activation and inactivation abnormalities (51).

A third gain-of-function phenotype has been described as a polymorphism present in up to 30% of American Caucasians (55). This variant, R1150W, has not been shown to directly cause any frank pain phenotypes. It results in a small positive shift in the voltage dependence of activation when the channel is expressed in human embryonic kidney (HEK) cells. In the simplest scenario, this would be expected to be a loss-of-function mutation, but expression in DRG neurons results in enhanced excitability (55). The R1150W polymorphism has been suggested to enhance susceptibility to pain syndromes like interstitial cystitis pain, postoperative pain, chronic widespread pain, and complex regional pain (56). The association of $Na_V1.7$ polymorphisms with a propensity for chronic pain remains controversial as some authors have been unable to replicate such a link (57).

40.4.4 $Na_V1.7$ IN ANIMAL PAIN MODELS

The overwhelming evidence of the human genetic data left little doubt that $Na_V1.7$ plays a role in human pain signaling, but genetic knockout in rodents was initially more confusing than helpful. Global $Na_V1.7$ knockout in mice results in viable pups that fail to feed and die on the first day after birth (58). It has not been clearly demonstrated why $Na_V1.7$ null mice die neonatally, but it is likely because they, like their human counterparts, have a compromised sense of smell (3 refs). Since humans are less dependent on smell to drive feeding, and they have a greater degree of parental involvement, human children without a sense of smell develop normally. Targeted knockout of $Na_V1.7$ in nociceptors driven by coupling to the $Na_V1.8$ promoter resulted in mice with increased thermal and mechanical pain thresholds and also with markedly reduced pain behaviors in response to inflammatory insults. Still, these mice clearly do respond to painful stimuli. They do not have the insensitivity to pain that the human null patients do (58), and they develop neuropathic pain after nerve injury just as do their wild-type littermates.

When mice were created with the knockout of $Na_V1.7$ in all DRG neurons (not just the $Na_V1.8$ expressing neurons), their resistance to pain became greater, but it required elimination of $Na_V1.7$ in all the sensory and sympathetic neurons to recreate the congenital insensitivity to pain observed in CIP humans. The subcellular localization of $Na_V1.7$ extends throughout DRG neurons, from the peripheral terminals to the terminals that synapse to the second-order neurons in the spinal cord (59). In parallel to observations in the DRG, the point where $Na_V1.7$ is most critical in transducing olfactory signaling is in facilitating the depolarization of the final nerve terminal to stimulate the release of neurotransmitters to the second-order neurons (60).

40.4.5 $Na_V1.7$ PHARMACOLOGY

The clear role of $Na_V1.7$ in pain signaling has led to great interest in $Na_V1.7$ as a therapeutic target for relieving human pain. Considerable work in many pharmaceutical companies has led to a flurry of activity in the patent literature, and $Na_V1.7$ antagonists may well be available in the clinic in the not too distant future. There remains the possibility that the impressive impact of $Na_V1.7$ genetic knockouts to prevent pain may not be able to be replicated by acute pharmacologic intervention, where the patient has developed and matured with an intact pain signaling system. However, $Na_V1.7$ gain-of-function PEPD patients do respond to block by nonselective sodium channel blockers like carbamazepine (61,62). Recent studies by Xenon Pharmaceuticals also demonstrated that IEM patients with $Na_V1.7$ gain-of-function mutations found pain relief after administration of the sodium channel blocker XEN402 (63). These studies suggest that pharmacologic block of $Na_V1.7$ will be an effective pain therapy, at least for some patients.

40.5 $Na_V1.8$

40.5.1 $Na_V1.8$ LOCALIZATION AND ROLE IN NOCICEPTOR SIGNALING

Long before the SCN10A gene was cloned, currents from $Na_V1.8$ were functionally identified as the slowly inactivating TTX-resistant sodium currents of DRG neurons (64,65). Biophysically, $Na_V1.8$ is quite distinct from other neuronal Na_Vs, activating and inactivating at much more positive potentials. The only other neuronal TTX-resistant Na_V is $Na_V1.9$, and $Na_V1.9$ produces low-voltage activated persistent currents very different from those of $Na_V1.8$. These unique properties, along with its high expression level in DRGs, make $Na_V1.8$ the neuronal Na_V channel most readily isolated and studied in native neurons.

$Na_V1.8$ expression is primarily restricted to peripheral sensory neurons, including the neurons of the DRG, trigeminal ganglia, and myenteric plexus but not the brain. $Na_V1.8$ is abundant in the DRG, and in fact, $Na_V1.8$ has been widely used as a marker for nociceptive neurons (58,66). The neurons of the DRG give rise to the axons that make up C-fibers and Aδ-fibers that carry pain signals from the nociceptive terminals to the central nervous system, and $Na_V1.8$ is highly expressed in these small nociceptors (67). Despite this concentration in the small neurons of the DRG, $Na_V1.8$ can also be found in larger DRG neurons, including a portion of the myelinated A-type mechanoreceptor neurons (68). In total, $Na_V1.8$ is expressed in about 80% of the neurons of the DRG and nodose ganglia and about 70% of the neurons of the trigeminal ganglia (69). In nociceptors of the DRG, that fraction is over 90% (68).

$Na_V1.8$ is the primary channel that mediates large amplitude action potentials in small neurons of the DRG (70). It is necessary for rapid repetitive action potentials in nociceptors and for spontaneous activity of damaged neurons (49,71,72). In depolarized or damaged DRG neurons, $Na_V1.8$ appears to be the primary driver of hyperexcitablility (73). In some animal pain models, $Na_V1.8$ mRNA expression levels have been shown to increase in the DRG (74–76). In rat nerve injury models, $Na_V1.8$ decreases in the DRG cell bodies, but this comes as a result of protein migrating into the peripheral axons near the site of injury. The relocalization of $Na_V1.8$ to the injured region of axons has been proposed to underlie the neuronal hyperexcitability that follows nerve injury (28). $Na_V1.8$ is seen in all portions of peripheral sensory nerves from the peripheral nerve terminals to the final projections into the spinal cord, and for this reason, $Na_V1.8$ is likely to play a role in nociceptor signaling throughout the entire nociceptor neuron, from the sensory terminal to the final synapse on the spinal neurons (23,77).

Outside of the sensory neurons, the expression of $Na_V1.8$ is limited, but several recent genome-wide analyses have linked human cardiac phenotypes to $Na_V1.8$ single-nucleotide polymorphisms (78–82). Subsequent studies identified signs of functional $Na_V1.8$ channels in cardiac tissue (83–85), but whether $Na_V1.8$ plays a significant role in cardiac function remains uncertain. $Na_V1.8$ is not normally expressed in the brain, but it has also been shown to appear in the brains of people suffering from multiple sclerosis (MS), and similarly in animal models of MS (86). This aberrant expression of $Na_V1.8$ in MS may contribute to the disease phenotype or its progression (87,88).

40.5.2 $Na_V1.8$ IN ANIMAL PAIN MODELS

Several approaches have been used to inhibit $Na_V1.8$ in preclinical animal pain models in order to better understand its role in pain signaling. These approaches fall into three categories; constitutive systemic genetic knockouts where the gene is absent throughout embryonic development, genetic knockdown, or pharmacologic inhibition in adult animals.

Systemic genetic knockout of $Na_V1.8$ in mice or specific destruction of $Na_V1.8$-expressing neurons with a toxin whose expression is driven by the $Na_V1.8$ promoter greatly reduces perception of acute mechanical, inflammatory, visceral, and cold pain (66,89–91). Knockout mice appear to develop neuropathic pain behaviors normally after nerve injury (66,89,90,92). In contrast, inhibition of $Na_V1.8$ in adult animals by intrathecal injection of antisense oligodeoxynucleotides has been shown to reduce neuropathic pain behaviors while leaving acute pain sensation intact (92,93). Selective pharmacologic inhibition of $Na_V1.8$ with A-803467 in rodents has been shown to reduce pain behaviors associated with neuropathic nerve injuries (94,95). A-803467 has also been shown to be effective in rat capsaicin pain models (94).

The different pattern of efficacy observed across pain models in knockdown animals vs. constitutive knockout animals indicates there may be compensatory changes during development in $Na_V1.8$ knockout animals that obscure the normal role of $Na_V1.8$ in neuropathic pain models. Consistent with this idea, it has been shown that TTX-sensitive Na_V channels are upregulated in $Na_V1.8$ knockout animals (89).

$Na_V1.8$ has been specifically linked to the sensation of pain in response to noxious cold temperatures. In the case of cold pain, both $Na_V1.8$ genetic knockout and knockdown rodents are insensitive to noxious cold (66,96–98). Mice with a gain-of-function mutation that leads to increased $Na_V1.8$ current density are hypersensitive to cold pain (99). $Na_V1.8$'s critical role in cold pain is due to its relative resistance to cold. At cold temperatures, the TTX-sensitive Na_Vs of sensory neurons become chronically inactivated, unable to pass Na^+, and thus unable to initiate or propagate action potentials. $Na_V1.8$ continues to work at cold temperatures and actually begins to open in response to smaller voltage stimuli (91). Thus, at noxious cold temperatures, pain seems to be dependent on $Na_V1.8$.

40.5.3 $Na_V1.8$ HUMAN DATA

As in rodents, $Na_V1.8$ mRNA is highly expressed in human and cynomolgus monkey DRG neurons (8,100). $Na_V1.8$ mRNA is also highly expressed in human trigeminal ganglia, and this expression is maintained in trigeminal neuralgia patients (101). $Na_V1.8$ protein expression has been demonstrated in damaged human dorsal root nerves as well as in the peripheral nerve terminals in the spinal cord (26). $Na_V1.8$ accumulates at the site of painful human nerve injuries in both adults and neonates (26,102). Painful human neuromas resulting from amputation or surgical interventions have elevated levels of $Na_V1.8$ (26), and $Na_V1.8$ is also upregulated in the pulp of painful human teeth (103).

Recently, $Na_V1.8$ mutations have been identified in patients with a debilitating chronic pain syndrome, idiopathic painful small-fiber neuropathy (IPSFN). Some of these mutations have been demonstrated to result in gain-of-function phenotypes, suggesting that, in approximately 10% of cases, IPSFN may be an $Na_V1.8$ channelopathy and that increased activity of $Na_V1.8$ may directly cause human pain syndromes (104). Thus, the existing human data are consistent with a crucial role for $Na_V1.8$ in nociceptor excitability similar to that demonstrated in lower species.

40.5.4 $Na_V1.8$ PHARMACOLOGY

Existing pharmaceutical inhibitors of Na_Vs like lidocaine are dose limited by the central nervous system side effects when administered systemically. Despite being generally nonselective, these compounds have been shown to inhibit $Na_V1.8$ with somewhat lower potency than the TTX-sensitive neuronal Na_Vs ($Na_V1.1$, $Na_V1.2$, $Na_V1.6$, and $Na_V1.7$) (105–107). This is due to the state-dependent mechanism of these drugs. They target inactivated states of the channel, and a large fraction of the TTX-sensitive Na_Vs are inactivated and available for block in resting sensory neurons. In contrast, $Na_V1.8$ inactivates at much more positive potentials, and few channels are inactivated and available for block in resting neurons. The poor potency of the local anesthetics on $Na_V1.8$ has been suggested to limit the clinical efficacy of those compounds (108). In agreement with this idea, systemic lidocaine is a more effective analgesic in $Na_V1.8$ knockout mice than in normal mice that have $Na_V1.8$ (89). These data lead to the hypothesis that tonic inhibitors of $Na_V1.8$ might be the most effective way to target this channel's role in pain (109).

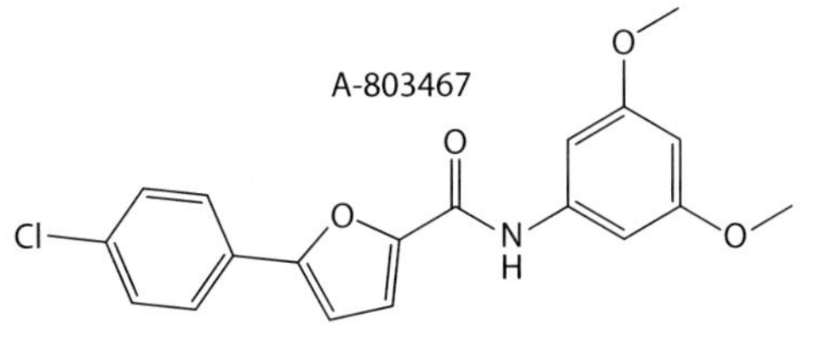

Figure 40.4 Selective Na$_V$1.8 antagonist.

Since Na$_V$1.8 is primarily restricted to the neurons that sense pain, selective Na$_V$1.8 blockers, like A-803467, are unlikely to induce the adverse events common to nonselective Na$_V$ blockers (Figure 40.4). Despite many years of effort by the pharmaceutical industry, no selective Na$_V$1.8 inhibitors have yet been tested in human pain patients.

40.6 Na$_V$1.9

40.6.1 Na$_V$1.9 LOCALIZATION AND ROLE IN NOCICEPTOR SIGNALING

Na$_V$1.9 is perhaps the most unusual and elusive of the Na$_V$ channels associated with pain. Despite some early hints that there might be multiple TTX-resistant Na$_V$s in DRG neurons (107), a clear demonstration of the behavior of Na$_V$1.9 was lacking until the advent of Na$_V$1.8 knockout mice (110). Even then, the link between the current phenotype of Na$_V$1.9 and the actual gene was largely by process of elimination. A TTX-resistant, persistent Na$_V$ current that activated at very negative potentials and inactivated (albeit slowly) at relatively negative potentials remained in mouse DRG neurons after Na$_V$1.8 knockout. All the other Na$_V$ isoforms in DRGs had been successfully overexpressed in heterologous cell expression systems and found to inactivate relatively rapidly; thus, the remaining persistent TTX-R current was likely due to the Na$_V$ clone that could not be expressed heterologously, Na$_V$1.9.

Na$_V$1.9, like Na$_V$1.8, is found primarily in the peripheral sensory neurons of the DRG, no dose ganglia, trigeminal ganglia, and the enteric nervous system. It is most prevalent in the small neurons that give rise to C-fibers. Most reports indicate very low expression in the central nervous system (111), though there are a few reports indicating that Na$_V$1.9 may have some role there (112), particularly in the retinal ganglion cells (113).

The concentrated expression of Na$_V$1.9 in the small nociceptive neurons of the DRG was the first indication that it might be involved in pain signaling. There have also been reports that Na$_V$1.9 expression in the DRGs increases after experimental injuries, particularly inflammatory injuries. Na$_V$1.9 expression is upregulated in the pulp of painful human teeth (114), in rat DRGs after insult with complete Freund's adjuvant (75,115), and in rat models of bone cancer pain (76). Na$_V$1.9 channel density, like Na$_V$1.8, decreases in the soma of DRG neurons after nerve injury in rat models of neuropathic pain (116–118). Unlike Na$_V$1.8, there have been no reports that the decrease of Na$_V$1.9 channels at the cell soma are associated with a concomitant increase in the axons at the site of injury.

40.6.2 Na$_V$1.9 BIOPHYSICS

Recently, Na$_V$1.9 has been successfully heterologously expressed in neurons from Na$_V$1.9 knockout mice (119) and more recently still in ND 7/23 cells, finally allowing confirmation of the functional behavior of the isolated channel (120). As in native neurons, recombinant Na$_V$1.9 activates at much more hyperpolarized voltages than the other neuronal Na$_V$s. It also has considerable overlap between the voltage dependence of activation and inactivation. This, along with its slow inactivation kinetics, indicates that in a sensory neuron with a normal resting membrane potential (–45 to –65 mV), a small fraction of the Na$_V$1.9 channels will be constitutively open and provide a small sodium leak current. This unusual profile allows Na$_V$1.9 to influence the resting membrane potential and also the subthreshold oscillations that ultimately trigger action potential firing (121,122).

40.6.3 Na$_V$1.9 IN ANIMAL PAIN MODELS

Na$_V$1.9 appears to have a special role in mediating inflammatory pain. Inflammatory insults like PGE2 acutely increase Na$_V$1.9 currents in DRG neurons by altering the voltage dependence of the channel via a G protein–dependent pathway (123). Intracellular GTPδS stimulation increases the channel open probability and mean open time (120).

The application of GTPδS enhances nociceptor excitability, and this effect is dependent on Na$_V$1.9, as DRG neurons of Na$_V$1.9 knockout mice fail to respond to GTPδS (119). Na$_V$1.9 null mice have normal acute thermal and mechanical pain sensitivity in the absence of injury. They also develop enhanced mechanical and thermal sensitivity after nerve injuries just as do wild-type mice. Thus, the potential for neuropathic pain appears preserved in these mice (122,124). Antisense knockdown of Na$_V$1.9 similarly failed to prevent the development of neuropathic hypersensitivity after nerve injury (93).

In contrast, pain behaviors are reduced in Na$_V$1.9 null mice after insult by a host of inflammatory mediators like carrageenan, CFA, or PGE2, bradykinin, and P2X agonists. Mechanical hypersensitivity after such injuries remains intact, comparable to wild-type mice, but thermal sensitivity is greatly reduced (122,124–126). Antisense knockdown of Na$_V$1.9 in rats likewise can reduce inflammatory pain behaviors (127), but at least one group found normal CFA responses in Na$_V$1.9 knockdown rats (115).

40.7 TRPV1

40.7.1 TRPV1 LOCALIZATION AND ROLE IN NOCICEPTOR SIGNALING

The link between the transient receptor potential channel (TRPV1) and pain is obvious to anyone who has ever eaten a hot chili pepper. Capsaicin, the active ingredient in chili peppers that gives them their spice, does so by directly binding to and opening the TRPV1 ion channel. Once the channel is open, sodium and calcium ions flood the nerve ending, depolarizing nociceptor neurons, initiating the action potential, and causing the burning pain characteristic of capsaicin.

TRPV1 channels are activated by a host of stimuli including voltage, noxious heat, acid, ethanol, mustard oil, vinegar, endocannabinoids, lipids, and inflammatory mediators (128–131). Phosphorylation by protein kinase A (PKA) (132,133), protein kinase C (PKC) (134), mitogen-activated protein kinase (MAPK)

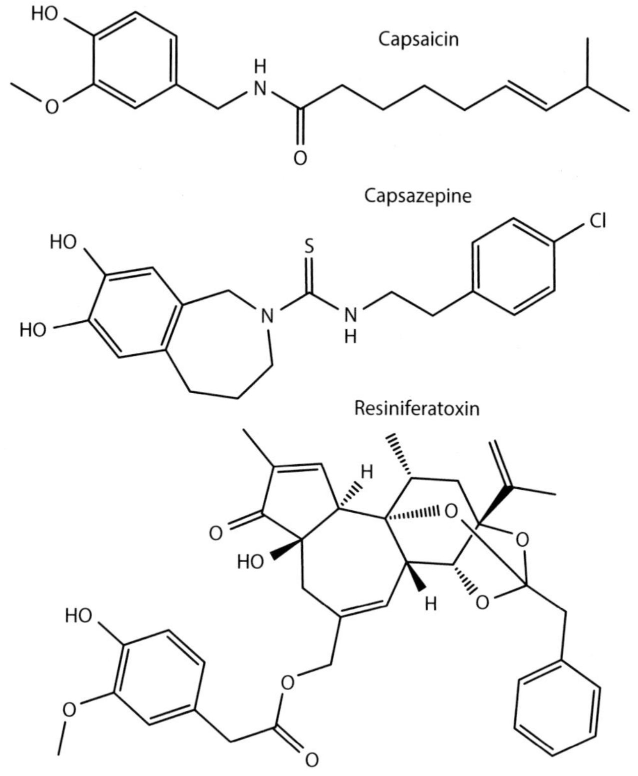

Figure 40.5 TRPV1 modulators.

(135), and Ca^{2+}/calmodulin-dependent kinase II have all been shown to sensitize TRPV1. Even light can activate the channel (136). TRPV1 can be synergistically activated by multiple types of stimuli, meaning that it can act as a coincidence detector for noxious insults.

TRPV1 is also known as the vanilloid receptor 1 (VR1). It was originally cloned from and thought restricted to the primary terminals of the sensory neurons of the DRG, and nearly 80% of DRG neurons respond to capsaicin (137,138). TRPV1 continues to be used as a functional and histological marker to identify nociceptor neurons among the subpopulations of DRG neurons. Cloning of the channel gene allowed detailed molecular expression studies, and it was soon realized that TRPV1 is expressed in many other diverse tissues like the trigeminal ganglia, nodose ganglia (134), the spinal cord, the brain (139), the bladder (140), the enteric nervous system, the innervation of the heart, vasculature (141), skin (142), and hair follicles (143). The availability of specific small molecule agonists for the channel, like capsaicin, as well as specific antagonists, like capsazepine, make the role of TRPV1 in pain signaling somewhat more tractable to query than those of the voltage-gated channels (Figure 40.5) (144).

40.7.2 TRPV1 IN ANIMAL PAIN MODELS

Genetic knockout of TRPV1 channels results in relatively normal appearing animals but impacts many measures of pain behavior in experimental animal models. Visceral pain (145); acute, thermal, and mechanical pain; inflammatory pain (146,147), and acid pain (148) have all been shown to be sensitive to genetic knockout of TRPV1.

Antisense knockdown of TRPV1 in mice has been reported to reduce neuropathic pain in some studies (149,150), but studies of mouse models of diabetic or chemotherapy-induced polyneuropathy actually behaved more sensitively to mechanical stimuli after genetic knockout of TRPV1 (146).

Capsaicin-stimulated pain responses in rats and mice are a commonly used pain model for testing the effects of analgesic compounds of many diverse mechanisms. Intradermal capsaicin administration has also been used as an experimental human pain model, both as a means to understand the basic mechanisms of pain signaling and as a way to test the efficacy and therapeutic potential for candidate analgesic treatments (151–154). The human capsaicin pain model provides a simple way to evaluate analgesics without the heterogeneity of pain-causing mechanisms present in real pain patients. Despite its simplicity, the utility of the human or rat capsaicin pain models for making actual predictions for the efficacy of novel analgesics remains unclear.

TRPV1 knockout mice also gave insight into the rich and widespread role of TRPV1 in physiologic processes outside of pain. TRPV1 null mice have changes in body temperature regulation (155), vasodilation (156), postischemic cardiac recovery (157), bladder function, digestive peristalsis (158), inflammation (159), pruritus, and hair growth.

40.7.3 TRPV1 PHARMACOLOGY

The clear connection of TRPV1 to pain signaling made it a nearly irresistible target for pharmaceutical companies intent on developing novel analgesic drugs. Unlike in the voltage-gated channels, TRPV1 channels have natural agonists (like capsaicin and resiniferatoxin) to use as chemical starting points for the design of small molecule antagonists. In fact, selective TRPV1 antagonists were developed by multiple groups quite quickly, and within 10 years of the cloning of the gene, clinical candidate compounds were undergoing trials in humans (160). At the time when these drug candidates appeared, the evidence for TRPV1's roles outside of pain was beginning to amass, but the general consensus was that these molecules would be well-tolerated analgesics. They are in fact quite good analgesics, and selective TRPV1 antagonists have proven efficacious in a host of rat pain models (161–163). They have also proven effective at reducing human pain (164,165).

Unfortunately, for both pharmaceutical companies and pain patients, these compounds were marked with intolerable side effects in the clinic. One problem was that they worked too well at relieving pain from noxious heat. The antagonists raised the temperature threshold for pain higher, and patients could not recognize high temperatures as painful. This predisposed them to burn injuries from everyday activities like sipping tea that was too hot or bathing in scalding water. These are just the sort of injuries that the patients who are congenitally insensitive to pain (as discussed in Section 40.4) must be ever vigilant to avoid.

Another critical problem was that the compounds raised the resting body temperature of patients, making them feverish. Similar hyperthermia was also seen in rodents. There were hints in the literature that TRPV1 might be linked to thermoregulation (166), but the very high level of TRPV1 channel expression in the primary nociceptors leads most to believe that analgesia would be the dominant effect of the drugs.

In the end, many drug discovery groups have abandoned TRPV1 antagonism as a mechanism for analgesia, though there remains some hope of separating the useful components of antagonism from the problematic aspects (167). For example, it might be possible to find molecules that spare the thermal sensitivity of TRPV1 while still blocking other means of activating the channel. Recently, it has been suggested that blocking TRPV1 in the ascending afferent pain pathways may actually be the wrong approach, and that activating TRPV1 in descending, inhibitory efferents that normally dampen pain signals may be a more practical approach (168). It may also be that other indications where TRPV1 plays a role may respond to these compounds at lower doses than are needed to impair thermoregulation and acute thermal pain.

Paradoxically, the best TRPV1-based therapeutics so far are agonists like capsaicin and resiniferatoxin that activate the channel and initially cause pain. Applied chronically, these agonists presumably work by chronically depolarizing the neurons, inactivating the voltage-gated channels needed to support action potential propagation. Prolonged exposure to high concentrations of TRPV1 agonists can cause the nerve terminals to retreat from the area of application, and high concentrations of resiniferatoxin are lethal to TRPV1-expressing neurons. A capsaicin patch applied directly to the skin is now used in the clinic to treat neuropathic pain indications and has shown some efficacy (169,170).

40.8 $Ca_V2.2$

40.8.1 $Ca_V2.2$ LOCALIZATION AND ROLE IN NOCICEPTOR SIGNALING

The $Ca_V2.2$ voltage-gated calcium channel encodes the N-type calcium channel currents of neurons. They are restricted to the nervous system, but are widely expressed throughout the peripheral and central nervous system, including the DRG, multiple brain regions, and the spinal cord (171–173). They are primarily associated with neurons but have also been described in astrocytes (174). $Ca_V2.2$ is localized to the presynaptic terminal of neurons. When an electrical action potential arrives at the presynaptic terminal, $Ca_V2.2$ is activated and permits the calcium influx that mobilizes neurotransmitter release. The neurotransmitters released then carry the signal on to the next neuron in the signaling pathway. $Ca_V2.2$ is not restricted to the pain signaling circuitry and so is critical for many types of neuronal signaling.

40.8.2 $Ca_V2.2$ IN ANIMAL PAIN MODELS

Genetic knockout of $Ca_V2.2$ results in a reduced propensity for acute, inflammatory, and neuropathic pain (175,176). These mice also have many other behavioral differences from wild-type mice, including changes in sympathetic nervous signaling, memory formation, decreased anxiety, and increased aggression. The significant differences in the behavior of wild-type and knockout mice make interpreting the pain responses of these animals somewhat challenging.

40.8.3 $Ca_V2.2$ PHARMACOLOGY

With such a critical role and such widespread expression, $Ca_V2.2$ would appear to be an unlikely candidate for analgesic intervention, but antagonists of $Ca_V2.2$ have in fact proven to be quite successful in the pain clinic. The critical pharmacologic tool used to recognize $Ca_V2.2$ in native neurons is the peptide toxin ω-conotoxin (177). This toxin was isolated from the venom of the cone snail *Conus magus*. The toxin selectively binds to the channel with high affinity and blocks it very effectively. Prior to cloning of the channel gene, ω-conotoxin sensitivity defined the N-type channel current. In recent years, a commercialized form of the peptide, Prialt, has been approved for the treatment of opioid intractable pain. Ziconotide is used only to manage severe chronic pain because of the difficulties in dosing patients with this peptide. Ziconotide must be dosed intrathecally via a pump because systemic exposure is not tolerated. Oral or intravenous administration is not possible since blocking $Ca_V2.2$ in the sympathetic nervous system has profound effects on blood pressure and cardiovascular function (178). Intrathecal ω-conotoxin is effective in relieving pain, but the therapeutic window is narrow. The side effects that limit the dose levels of ziconotide are likely due to the critical role of $Ca_V2.2$ in non-pain-related neurons and include confusion, dizziness, abnormal gate, anxiety, and hallucinations (179,180).

The antiepileptic drugs gabapentin and pregabalin also target $Ca_V2.2$. Gabapentin is a close chemical analog of the inhibitory neurotransmitter GABA and was designed to modulate GABA signaling pathways. Despite the intention, the effectivness of the drugs actually appears to occur through decreases in the protein trafficking of the $Ca_V2.2$ channel. The N-type calcium channel current is reduced, leading to relief for patients with chronic pain. Gabapentin was also recently reported to decrease the expression of $Na_V1.7$, so the complete story of the mechanism of gabapentin's analgesic properties is still being elucidated (181). Gabapentin is approved by the FDA for treatment of epilepsy and postherpetic neuralgia. Pregabalin is chemically and pharmacologically similar to gabapentin and is used to treat fibromyalgia, diabetic neuralgia, postherpetic neuralgia, epilepsy, and generalized anxiety disorder.

Pharmaceutical companies have invested heavily in an attempt to identify use-dependent blockers for $Ca_V2.2$. The rationale, as for Na_V blockers, is that a compound that targets the inactivated state will preferentially block channels in overactive, pain-causing, neurons (182). Several use-dependent selective blockers of $Ca_V2.2$ have recently been identified, and they are effective in animal behavioral models of pain (183–185). In the future, these compounds may provide a much-needed new class of pain relievers in the clinic.

40.9 OTHER CHANNELS LINKED TO PAIN

While the channels discussed in this chapter have the most comprehensive data linking them to pain physiology, many other channels participate in the initiation and maintenance of pain signaling. The hyperpolarization-activated cyclic nucleotide–gated (HCN) channels are likely important in determining the threshold and frequency of nociceptor firing (186,187). TRPA1 channels work synergistically with TRPV1 and plays a role in inflammatory pain (188–190). Acid-sensing ion channels (ASIC's) are linked to the pain of tissue acidosis (191). Virtually, all DRG neurons are sensitive to ATP stimulation, and ATP-activated P2X

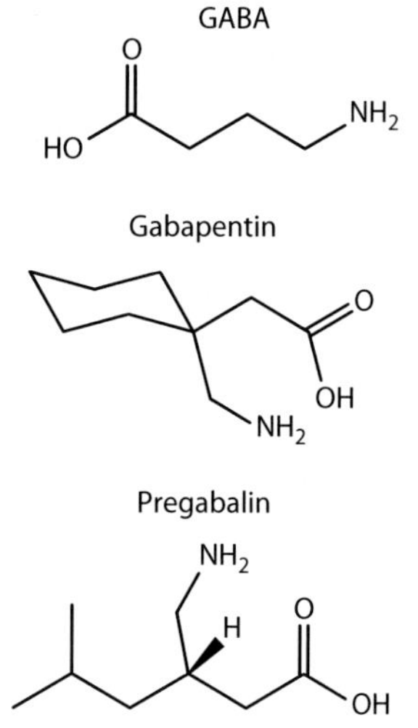

GABA

Gabapentin

Pregabalin

Figure 40.6 GABA and structurally similar $Ca_V2.2$ modulators used clinically for pain.

receptors have been considered as pain targets (192). Voltage-gated potassium channels are an important determinant of the resting membrane potential for all neurons and thereby impact nociceptor excitability (193). N-methyl-D-aspartate (NMDA) receptors and AMPA receptors open in response to the excitatory neurotransmitter glutamate and antagonists have been used to treat neuropathic pain (194–196). $GABA_B$ channels are critical in the descending inhibition from the CNS that dampens the excitability of the secondary spinal neurons, but GABA is excitatory when applied directly to adult neurons of the DRG due to the distinct physiology of those neurons. Thus $GABA_B$ channels may play a role in both pain induction and pain suppression (197).

40.10 CONCLUSIONS

Because of the close interdependence of ion channels and other conductances in neurons, teasing out the most critical channels relating to any given function is a challenge. The impact of other mental and physiologic states on pain signaling and perception only adds to this difficulty. The Na_V channels, TRPV1, and $Ca_V2.2$ are channel subtypes that stand out as clear modulators of pain pathways (Figure 40.6). There are currently multiple drugs that target these channels as analgesic agents, but their therapeutic windows and hence their usefulness are limited. There is a great need for new therapeutic interventions that target pain. More selective or effective methods for modulating these channels remain among the most promising ways to improve outcomes for people with pathological pain.

REFERENCES

1. Toscano, E., R. della Casa, S. Mardy, L. Gaetaniello, F. Sadile, Y. Indo, C. Pignata, and G. Andria. 2000. Multisystem involvement in congenital insensitivity to pain with anhidrosis (CIPA), a nerve growth factor receptor(Trk A)-related disorder. *Neuropediatrics* 31:39–41.
2. Cox, J. J., F. Reimann, A. K. Nicholas, G. Thornton, E. Roberts, K. Springell, G. Karbani, H. et al. 2006. An SCN9A channelopathy causes congenital inability to experience pain. *Nature* 444:894–898.
3. Cummins, T. R. and A. M. Rush. 2007. Voltage-gated sodium channel blockers for the treatment of neuropathic pain. *Expert Rev Neurother* 7:1597–1612.
4. Cummins, T. R., P. L. Sheets, and S. G. Waxman. 2007. The roles of sodium channels in nociception: Implications for mechanisms of pain. *Pain* 131:243–257.
5. England, S. 2008. Voltage-gated sodium channels: The search for subtype-selective analgesics. *Expert Opin Investig Drugs* 17:1849–1864.
6. Krafte, D. S. and A. W. Bannon. 2008. Sodium channels and nociception: Recent concepts and therapeutic opportunities. *Curr Opin Pharmacol* 8:50–56.
7. Hille, B. 2001. *Ion Channels of Excitable Membranes*. Sinauer Associates, Inc., Sunderland, MA.
8. Raymond, C. K., J. Castle, P. Garrett-Engele, C. D. Armour, Z. Kan, N. Tsinoremas, and J. M. Johnson. 2004. Expression of alternatively spliced sodium channel alpha-subunit genes. Unique splicing patterns are observed in dorsal root ganglia. *J Biol Chem* 279:46234–46241.
9. Chahine, M., A. Chatelier, O. Babich, and J. J. Krupp. 2008. Voltage-gated sodium channels in neurological disorders. *CNS Neurol Disord Drug Targets* 7:144–158.
10. Rush, A. M. and T. R. Cummins. 2007. Painful research: Identification of a small-molecule inhibitor that selectively targets $Na_v1.8$ sodium channels. *Mol Interv* 7:192–195.
11. Soderpalm, B. 2002. Anticonvulsants: Aspects of their mechanisms of action. *Eur J Pain* 6 Suppl A:3–9.
12. Wang, G. K., J. Mitchell, and S. Y. Wang. 2008. Block of persistent late Na^+ currents by antidepressant sertraline and paroxetine. *J Membr Biol* 222:79–90.
13. Hagen, N. A., P. du Souich, B. Lapointe, M. Ong-Lam, B. Dubuc, D. Walde, R. Love, and A. H. Ngoc. 2008. Tetrodotoxin for moderate to severe cancer pain: A randomized, double blind, parallel design multicenter study. *J Pain Symptom Manage* 35:420–429.
14. Hagen, N. A., K. M. Fisher, B. Lapointe, P. du Souich, S. Chary, D. Moulin, E. Sellers, and A. H. Ngoc. 2007. An open-label, multi-dose efficacy and safety study of intramuscular tetrodotoxin in patients with severe cancer-related pain. *J Pain Symptom Manage* 34:171–182.
15. Hagen, N. A., B. Lapointe, M. Ong-Lam, B. Dubuc, D. Walde, B. Gagnon, R. Love, R. et al. 2011. A multicentre open-label safety and efficacy study of tetrodotoxin for cancer pain. *Curr Oncol* 18:e109–e116.
16. Waxman, S. G., J. D. Kocsis, and J. A. Black. 1994. Type III sodium channel mRNA is expressed in embryonic but not adult spinal sensory neurons, and is reexpressed following axotomy. *J Neurophysiol* 72:466–470.
17. Dib-Hajj, S. D., J. Fjell, T. R. Cummins, Z. Zheng, K. Fried, R. LaMotte, J. A. Black, and S. G. Waxman. 1999. Plasticity of sodium channel expression in DRG neurons in the chronic constriction injury model of neuropathic pain. *Pain* 83:591–600.
18. Okuse, K., S. R. Chaplan, S. B. McMahon, Z. D. Luo, N. A. Calcutt, B. P. Scott, A. N. Akopian, and J. N. Wood. 1997. Regulation of expression of the sensory neuron-specific sodium channel SNS in inflammatory and neuropathic pain. *Mol Cell Neurosci* 10:196–207.
19. Kim, C. H., Y. Oh, J. M. Chung, and K. Chung. 2001. The changes in expression of three subtypes of TTX sensitive sodium channels in sensory neurons after spinal nerve ligation. *Brain Res Mol Brain Res* 95:153–161.
20. Kim, C. H., Y. Oh, J. M. Chung, and K. Chung. 2002. Changes in three subtypes of tetrodotoxin sensitive sodium channel expression in the axotomized dorsal root ganglion in the rat. *Neurosci Lett* 323:125–128.

21. Decosterd, I., R. R. Ji, S. Abdi, S. Tate, and C. J. Woolf. 2002. The pattern of expression of the voltage-gated sodium channels Na(v)1.8 and Na(v)1.9 does not change in uninjured primary sensory neurons in experimental neuropathic pain models. *Pain* 96:269–277.
22. Iwahashi, Y., T. Furuyama, S. Inagaki, Y. Morita, and H. Takagi. 1994. Distinct regulation of sodium channel types I, II and III following nerve transection. *Brain Res Mol Brain Res* 22:341–345.
23. Bucknill, A. T., K. Coward, C. Plumpton, S. Tate, C. Bountra, R. Birch, A. Sandison, S. P. Hughes, and P. Anand. 2002. Nerve fibers in lumbar spine structures and injured spinal roots express the sensory neuron-specific sodium channels SNS/PN3 and NaN/SNS2. *Spine (Phila Pa 1976)* 27:135–140.
24. Coward, K., A. Aitken, A. Powell, C. Plumpton, R. Birch, S. Tate, C. Bountra, and P. Anand. 2001. Plasticity of TTX-sensitive sodium channels PN1 and brain III in injured human nerves. *Neuroreport* 12:495–500.
25. Coward, K., A. Jowett, C. Plumpton, A. Powell, R. Birch, S. Tate, C. Bountra, and P. Anand. 2001. Sodium channel beta1 and beta2 subunits parallel SNS/PN3 alpha-subunit changes in injured human sensory neurons. *Neuroreport* 12:483–488.
26. Coward, K., C. Plumpton, P. Facer, R. Birch, T. Carlstedt, S. Tate, C. Bountra, and P. Anand. 2000. Immunolocalization of SNS/PN3 and NaN/SNS2 sodium channels in human pain states. *Pain* 85:41–50.
27. Black, J. A., L. Nikolajsen, K. Kroner, T. S. Jensen, and S. G. Waxman. 2008. Multiple sodium channel isoforms and mitogen-activated protein kinases are present in painful human neuromas. *Ann Neurol* 64:644–653.
28. Huang, H. L., C. M. Cendan, C. Roza, K. Okuse, R. Cramer, J. F. Timms, and J. N. Wood. 2008. Proteomic profiling of neuromas reveals alterations in protein composition and local protein synthesis in hyper-excitable nerves. *Mol Pain* 4:33.
29. Cheng, K. I., H. C. Wang, C. S. Lai, H. P. Tsai, A. L. Kwan, S. T. Ho, J. J. Wang, and L. L. Chang. 2011. Pre-emptive intrathecal quinidine alleviates spinal nerve ligation-induced peripheral neuropathic pain. *J Pharm Pharmacol* 63:1063–1069.
30. Cheng, K. I., C. S. Lai, F. Y. Wang, H. C. Wang, L. L. Chang, S. T. Ho, H. P. Tsai, and A. L. Kwan. 2011. Intrathecal lidocaine pretreatment attenuates immediate neuropathic pain by modulating $Na_v1.3$ expression and decreasing spinal microglial activation. *BMC Neurol* 11:71.
31. Hains, B. C., J. P. Klein, C. Y. Saab, M. J. Craner, J. A. Black, and S. G. Waxman. 2003. Upregulation of sodium channel $Na_v1.3$ and functional involvement in neuronal hyperexcitability associated with central neuropathic pain after spinal cord injury. *J Neurosci* 23:8881–8892.
32. Hains, B. C., C. Y. Saab, J. P. Klein, M. J. Craner, and S. G. Waxman. 2004. Altered sodium channel expression in second-order spinal sensory neurons contributes to pain after peripheral nerve injury. *J Neurosci* 24:4832–4839.
33. Hains, B. C., C. Y. Saab, and S. G. Waxman. 2005. Changes in electrophysiological properties and sodium channel $Na_v1.3$ expression in thalamic neurons after spinal cord injury. *Brain* 128:2359–2371.
34. Cummins, T. R., F. Aglieco, M. Renganathan, R. I. Herzog, S. D. Dib-Hajj, and S. G. Waxman. 2001. $Na_v1.3$ sodium channels: Rapid repriming and slow closed-state inactivation display quantitative differences after expression in a mammalian cell line and in spinal sensory neurons. *J Neurosci* 21:5952–5961.
35. Lampert, A., B. C. Hains, and S. G. Waxman. 2006. Upregulation of persistent and ramp sodium current in dorsal horn neurons after spinal cord injury. *Exp Brain Res* 174:660–666.
36. Estacion, M., C. Han, J. S. Choi, J. G. Hoeijmakers, G. Lauria, J. P. Drenth, M. M. Gerrits, S. D. Dib-Hajj, C. G. Faber, I. S. Merkies, and S. G. Waxman. 2011. Intra- and interfamily phenotypic diversity in pain syndromes associated with a gain-of-function variant of $Na_v1.7$. *Mol Pain* 7:92.
37. Estacion, M. and S. G. Waxman. 2012. The response of $Na_v1.3$ sodium channels to ramp stimuli: Multiple components and mechanisms. *J Neurophysiol* 109:306–314.
38. Zhang, J. M., H. Li, and M. A. Munir. 2004. Decreasing sympathetic sprouting in pathologic sensory ganglia: A new mechanism for treating neuropathic pain using lidocaine. *Pain* 109:143–149.
39. Xanthos, T., K. A. Ekmektzoglou, I. S. Vlachos, D. Dimitroulis, S. Tsitsilonis, T. Karatzas, and D. N. Perrea. 2008. A prognostic index for the successful use of adenosine in patients with paroxysmal supraventricular tachycardia in emergency settings: A retrospective study. *Am J Emerg Med* 26:304–309.
40. Chung, K., H. J. Kim, H. S. Na, M. J. Park, and J. M. Chung. 1993. Abnormalities of sympathetic innervation in the area of an injured peripheral nerve in a rat model of neuropathic pain. *Neurosci Lett* 162:85–88.
41. Kim, S. H., H. S. Na, K. Sheen, and J. M. Chung. 1993. Effects of sympathectomy on a rat model of peripheral neuropathy. *Pain* 55:85–92.
42. Sekiguchi, M., H. Kobayashi, Y. Sekiguchi, S. Konno, and S. Kikuchi. 2008. Sympathectomy reduces mechanical allodynia, tumor necrosis factor-alpha expression, and dorsal root ganglion apoptosis following nerve root crush injury. *Spine (Phila Pa 1976)* 33:1163–1169.
43. Happak, W., S. Sator-Katzenschlager, and L. K. Kriechbaumer. 2012. Surgical treatment of complex regional pain syndrome type II with regional subcutaneous venous sympathectomy. *J Trauma Acute Care Surg* 72:1647–1653.
44. Hains, B. C., C. Y. Saab, and S. G. Waxman. 2006. Alterations in burst firing of thalamic VPL neurons and reversal by Na(v)1.3 antisense after spinal cord injury. *J Neurophysiol* 95:3343–3352.
45. Lindia, J. A., M. G. Kohler, W. J. Martin, and C. Abbadie. 2005. Relationship between sodium channel $Na_v1.3$ expression and neuropathic pain behavior in rats. *Pain* 117:145–153.
46. Nassar, M. A., M. D. Baker, A. Levato, R. Ingram, G. Mallucci, S. B. McMahon, and J. N. Wood. 2006. Nerve injury induces robust allodynia and ectopic discharges in $Na_v1.3$ null mutant mice. *Mol Pain* 2:33.
47. Yu, F. H., M. Mantegazza, R. E. Westenbroek, C. A. Robbins, F. Kalume, K. A. Burton, W. J. Spain, G. S. McKnight, T. Scheuer, and W. A. Catterall. 2006. Reduced sodium current in GABAergic interneurons in a mouse model of severe myoclonic epilepsy in infancy. *Nat Neurosci* 9:1142–1149.
48. Luo, S., G. M. Perry, S. R. Levinson, and M. A. Henry. 2008. $Na_v1.7$ expression is increased in painful human dental pulp. *Mol Pain* 4:16.
49. Choi, J. S. and S. G. Waxman. 2011. Physiological interactions between $Na_v1.7$ and $Na_v1.8$ sodium channels: A computer simulation study. *J Neurophysiol* 106:3173–3184.
50. Goldberg, Y. P., J. MacFarlane, M. L. MacDonald, J. Thompson, M. P. Dube, M. Mattice, R. Fraser et al. 2007. Loss-of-function mutations in the $Na_v1.7$ gene underlie congenital indifference to pain in multiple human populations. *Clin Genet* 71:311–319.
51. Estacion, M., S. D. Dib-Hajj, P. J. Benke, R. H. Te Morsche, E. M. Eastman, L. J. Macala, J. P. Drenth, and S. G. Waxman. 2008. NaV1.7 gain-of-function mutations as a continuum: A1632E displays physiological changes associated with erythromelalgia and paroxysmal extreme pain disorder mutations and produces symptoms of both disorders. *J Neurosci* 28:11079–11088.

52. Lampert, A., A. O. O'Reilly, P. Reeh, and A. Leffler. 2010. Sodium channelopathies and pain. *Pflugers Arch* 460(2):249–263.
53. Fertleman, C. R., M. D. Baker, K. A. Parker, S. Moffatt, F. V. Elmslie, B. Abrahamsen, J. Ostman et al. 2006. SCN9A mutations in paroxysmal extreme pain disorder: Allelic variants underlie distinct channel defects and phenotypes. *Neuron* 52:767–774.
54. Fertleman, C. R., C. D. Ferrie, J. Aicardi, N. A. Bednarek, O. Eeg-Olofsson, F. V. Elmslie, D. A. Griesemer et al. 2007. Paroxysmal extreme pain disorder (previously familial rectal pain syndrome). *Neurology* 69:586–595.
55. Estacion, M., T. P. Harty, J. S. Choi, L. Tyrrell, S. D. Dib-Hajj, and S. G. Waxman. 2009. A sodium channel gene SCN9A polymorphism that increases nociceptor excitability. *Ann Neurol* 66:862–866.
56. Duan, G., G. Xiang, X. Zhang, R. Yuan, H. Zhan, and D. Qi. 2013. A single-nucleotide polymorphism in SCN9A may decrease postoperative pain sensitivity in the general population. *Anesthesiology* 118:436–442.
57. Holliday, K. L., W. Thomson, T. Neogi, D. T. Felson, K. Wang, F. C. Wu, I. T. Huhtaniemi et al. 2012. The non-synonymous SNP, R1150W, in SCN9A is not associated with chronic widespread pain susceptibility. *Mol Pain* 8:72.
58. Nassar, M. A., L. C. Stirling, G. Forlani, M. D. Baker, E. A. Matthews, A. H. Dickenson, and J. N. Wood. 2004. Nociceptor-specific gene deletion reveals a major role for $Na_v1.7$ (PN1) in acute and inflammatory pain. *Proc Natl Acad Sci USA* 101:12706–12711.
59. Black, J. A., N. Frezel, S. D. Dib-Hajj, and S. G. Waxman. 2012. Expression of Nav1.7 in DRG neurons extends from peripheral terminals in the skin to central preterminal branches and terminals in the dorsal horn. *Mol Pain* 8:82.
60. Weiss, J., M. Pyrski, E. Jacobi, B. Bufe, V. Willnecker, B. Schick, P. Zizzari et al. 2011. Loss-of-function mutations in sodium channel Nav1.7 cause anosmia. *Nature* 472:186–190.
61. Fischer, T. Z., E. S. Gilmore, M. Estacion, E. Eastman, S. Taylor, M. Melanson, S. D. Dib-Hajj, and S. G. Waxman. 2009. A novel $Na_v1.7$ mutation producing carbamazepine-responsive erythromelalgia. *Ann Neurol* 65:733–741.
62. Natkunarajah, J., D. Atherton, F. Elmslie, S. Mansour, and P. Mortimer. 2009. Treatment with carbamazepine and gabapentin of a patient with primary erythermalgia (erythromelalgia) identified to have a mutation in the SCN9A gene, encoding a voltage-gated sodium channel. *Clin Exp Dermatol* 34:e640–e642.
63. Goldberg, Y. P., N. Price, R. Namdari, C. J. Cohen, M. H. Lamers, C. Winters, J. Price et al. 2011. Treatment of $Na_v1.7$-mediated pain in inherited erythromelalgia using a novel sodium channel blocker. *Pain* 153:80–85.
64. Matsuda, Y., S. Yoshida, and T. Yonezawa. 1978. Tetrodotoxin sensitivity and Ca component of action potentials of mouse dorsal root ganglion cells cultured in vitro. *Brain Res* 154:69–82.
65. Yoshida, S., Y. Matsuda, and A. Samejima. 1978. Tetrodotoxin-resistant sodium and calcium components of action potentials in dorsal root ganglion cells of the adult mouse. *J Neurophysiol* 41:1096–1106.
66. Abrahamsen, B., J. Zhao, C. O. Asante, C. M. Cendan, S. Marsh, J. P. Martinez-Barbera, M. A. Nassar, A. H. Dickenson, and J. N. Wood. 2008. The cell and molecular basis of mechanical, cold, and inflammatory pain. *Science* 321:702–705.
67. Djouhri, L., X. Fang, K. Okuse, J. N. Wood, C. M. Berry, and S. N. Lawson. 2003. The TTX-resistant sodium channel Nav1.8 (SNS/PN3): Expression and correlation with membrane properties in rat nociceptive primary afferent neurons. *J Physiol* 550:739–752.
68. Shields, S. D., H. S. Ahn, Y. Yang, C. Han, R. P. Seal, J. N. Wood, S. G. Waxman, and S. D. Dib-Hajj. 2012. $Na_v1.8$ expression is not restricted to nociceptors in mouse peripheral nervous system. *Pain* 153:2017–2030.
69. Gautron, L., I. Sakata, S. Udit, J. M. Zigman, J. N. Wood, and J. K. Elmquist. 2011. Genetic tracing of $Na_v1.8$-expressing vagal afferents in the mouse. *J Comp Neurol* 519:3085–3101.
70. Blair, N. T. and B. P. Bean. 2002. Roles of tetrodotoxin (TTX)-sensitive Na^+ current, TTX-resistant Na^+ current, and Ca^{2+} current in the action potentials of nociceptive sensory neurons. *J Neurosci* 22:10277–10290.
71. Renganathan, M., T. R. Cummins, and S. G. Waxman. 2001. Contribution of $Na_v1.8$ sodium channels to action potential electrogenesis in DRG neurons. *J Neurophysiol* 86:629–640.
72. Roza, C., J. M. Laird, V. Souslova, J. N. Wood, and F. Cervero. 2003. The tetrodotoxin-resistant Na^+ channel $Na_v1.8$ is essential for the expression of spontaneous activity in damaged sensory axons of mice. *J Physiol* 550:921–926.
73. Rush, A. M., S. D. Dib-Hajj, S. Liu, T. R. Cummins, J. A. Black, and S. G. Waxman. 2006. A single sodium channel mutation produces hyper- or hypoexcitability in different types of neurons. *Proc Natl Acad Sci USA* 103:8245–8250.
74. Sun, W., B. Miao, X. C. Wang, J. H. Duan, W. T. Wang, F. Kuang, R. G. Xie et al. 2012. Reduced conduction failure of the main axon of polymodal nociceptive C-fibres contributes to painful diabetic neuropathy in rats. *Brain* 135:359–375.
75. Strickland, I. T., J. C. Martindale, P. L. Woodhams, A. J. Reeve, I. P. Chessell, and D. S. McQueen. 2008. Changes in the expression of $Na_v1.7$, $Na_v1.8$ and $Na_v1.9$ in a distinct population of dorsal root ganglia innervating the rat knee joint in a model of chronic inflammatory joint pain. *Eur J Pain* 12:564–572.
76. Qiu, F., Y. Jiang, H. Zhang, Y. Liu, and W. Mi. 2012. Increased expression of tetrodotoxin-resistant sodium channels Nav1.8 and Nav1.9 within dorsal root ganglia in a rat model of bone cancer pain. *Neurosci Lett* 512:61–66.
77. McGaraughty, S., K. L. Chu, M. J. Scanio, M. E. Kort, C. R. Faltynek, and M. F. Jarvis. 2008. A selective Nav1.8 sodium channel blocker, A-803467 [5-(4-chlorophenyl-N-(3,5-dimethoxyphenyl)furan-2-carboxamide], attenuates spinal neuronal activity in neuropathic rats. *J Pharmacol Exp Ther* 324:1204–1211.
78. Chambers, J. C., J. Zhao, C. M. Terracciano, C. R. Bezzina, W. Zhang, R. Kaba, M. Navaratnarajah et al. 2010. Genetic variation in SCN10A influences cardiac conduction. *Nat Genet* 42:149–152.
79. Pfeufer, A., C. van Noord, K. D. Marciante, D. E. Arking, M. G. Larson, A. V. Smith, K. V. Tarasov et al. 2010. Genome-wide association study of PR interval. *Nat Genet* 42:153–159.
80. Holm, H., D. F. Gudbjartsson, D. O. Arnar, G. Thorleifsson, G. Thorgeirsson, H. Stefansdottir, S. A. Gudjonsson et al. 2010. Several common variants modulate heart rate, PR interval and QRS duration. *Nat Genet* 42:117–122.
81. Smith, J. G., J. W. Magnani, C. Palmer, Y. A. Meng, E. Z. Soliman, S. K. Musani, K. F. Kerr et al. 2011. Genome-wide association studies of the PR interval in African Americans. *PLoS Genet* 7:e1001304.
82. Sotoodehnia, N., A. Isaacs, P. I. de Bakker, M. Dorr, C. Newton-Cheh, I. M. Nolte, P. van der Harst et al. 2010. Common variants in 22 loci are associated with QRS duration and cardiac ventricular conduction. *Nat Genet* 42:1068–1076.
83. Facer, P., P. P. Punjabi, A. Abrari, R. A. Kaba, N. J. Severs, J. Chambers, J. S. Kooner, and P. Anand. 2011. Localisation of SCN10A gene product $Na_v1.8$ and novel pain-related ion channels in human heart. *Int Heart J* 52:146–152.
84. Verkerk, A. O., C. A. Remme, C. A. Schumacher, B. P. Scicluna, R. Wolswinkel, B. de Jonge, C. R. Bezzina, and M. W. Veldkamp. 2012. Functional $Na_v1.8$ channels in intracardiac neurons: The link between SCN10A and cardiac electrophysiology. *Circ Res* 111(3):333–343.

85. Yang, T., T. C. Atack, D. M. Stroud, W. Zhang, L. Hall, and D. M. Roden. 2012. Blocking SCN10A channels in heart reduces late sodium current and is antiarrhythmic. *Circ Res* 111(3):322–332.
86. Black, J. A., S. Dib-Hajj, D. Baker, J. Newcombe, M. L. Cuzner, and S. G. Waxman. 2000. Sensory neuron-specific sodium channel SNS is abnormally expressed in the brains of mice with experimental allergic encephalomyelitis and humans with multiple sclerosis. *Proc Natl Acad Sci USA* 97:11598–11602.
87. Renganathan, M., M. Gelderblom, J. A. Black, and S. G. Waxman. 2003. Expression of $Na_v1.8$ sodium channels perturbs the firing patterns of cerebellar Purkinje cells. *Brain Res* 959:235–242.
88. Shields, S. D., X. Cheng, A. Gasser, C. Y. Saab, L. Tyrrell, E. M. Eastman, M. Iwata et al. 2012. A channelopathy contributes to cerebellar dysfunction in a model of multiple sclerosis. *Ann Neurol* 71:186–194.
89. Akopian, A. N., V. Souslova, S. England, K. Okuse, N. Ogata, J. Ure, A. Smith et al. 1999. The tetrodotoxin-resistant sodium channel SNS has a specialized function in pain pathways. *Nat Neurosci* 2:541–548.
90. Laird, J. M., V. Souslova, J. N. Wood, and F. Cervero. 2002. Deficits in visceral pain and referred hyperalgesia in $Na_v1.8$ (SNS/PN3)-null mice. *J Neurosci* 22:8352–8356.
91. Zimmermann, K., A. Leffler, A. Babes, C. M. Cendan, R. W. Carr, J. Kobayashi, C. Nau, J. N. Wood, and P. W. Reeh. 2007. Sensory neuron sodium channel $Na_v1.8$ is essential for pain at low temperatures. *Nature* 447:855–858.
92. Lai, J., M. S. Gold, C. S. Kim, D. Bian, M. H. Ossipov, J. C. Hunter, and F. Porreca. 2002. Inhibition of neuropathic pain by decreased expression of the tetrodotoxin-resistant sodium channel, $Na_v1.8$. *Pain* 95:143–152.
93. Porreca, F., J. Lai, D. Bian, S. Wegert, M. H. Ossipov, R. M. Eglen, L. Kassotakis et al. 1999. A comparison of the potential role of the tetrodotoxin-insensitive sodium channels, PN3/SNS and NaN/SNS2, in rat models of chronic pain. *Proc Natl Acad Sci USA* 96:7640–7644.
94. Jarvis, M. F., P. Honore, C. C. Shieh, M. Chapman, S. Joshi, X. F. Zhang, M. Kort et al. 2007. A-803467, a potent and selective Nav1.8 sodium channel blocker, attenuates neuropathic and inflammatory pain in the rat. *Proc Natl Acad Sci USA* 104:8520–8525.
95. Joshi, S. K., P. Honore, G. Hernandez, R. Schmidt, A. Gomtsyan, M. Scanio, M. Kort, and M. F. Jarvis. 2009. Additive antinociceptive effects of the selective $Na_v1.8$ blocker A-803467 and selective TRPV1 antagonists in rat inflammatory and neuropathic pain models. *J Pain* 10:306–315.
96. Foulkes, T. and J. N. Wood. 2007. Mechanisms of cold pain. *Channels (Austin)* 1:154–160.
97. Leo, S., R. D'Hooge, and T. Meert. 2009. Exploring the role of nociceptor-specific sodium channels in pain transmission using Nav1.8 and Nav1.9 knockout mice. *Behav Brain Res* 208:149–157.
98. Sturzebecher, A. S., J. Hu, E. S. Smith, S. Frahm, J. Santos-Torres, B. Kampfrath, S. Auer, G. R. Lewin, and I. Ibanez-Tallon. 2010. An in vivo tethered toxin approach for the cell-autonomous inactivation of voltage-gated sodium channel currents in nociceptors. *J Physiol* 588:1695–1707.
99. Blasius, A. L., A. E. Dubin, M. J. Petrus, B. K. Lim, A. Narezkina, J. R. Criado, D. N. Wills et al. 2011. Hypermorphic mutation of the voltage-gated sodium channel encoding gene Scn10a causes a dramatic stimulus-dependent neurobehavioral phenotype. *Proc Natl Acad Sci USA* 108:19413–19418.
100. Dib-Hajj, S. D., L. Tyrrell, T. R. Cummins, J. A. Black, P. M. Wood, and S. G. Waxman. 1999. Two tetrodotoxin-resistant sodium channels in human dorsal root ganglion neurons. *FEBS Lett* 462:117–120.
101. Siqueira, S. R., B. Alves, H. M. Malpartida, M. J. Teixeira, and J. T. Siqueira. 2009. Abnormal expression of voltage-gated sodium channels $Na_v1.7$, $Na_v1.3$ and $Na_v1.8$ in trigeminal neuralgia. *Neuroscience* 164:573–577.
102. Yiangou, Y., R. Birch, L. Sangameswaran, R. Eglen, and P. Anand. 2000. SNS/PN3 and SNS2/NaN sodium channel-like immunoreactivity in human adult and neonate injured sensory nerves. *FEBS Lett* 467:249–252.
103. Renton, T., Y. Yiangou, C. Plumpton, S. Tate, C. Bountra, and P. Anand. 2005. Sodium channel Nav1.8 immunoreactivity in painful human dental pulp. *BMC Oral Health* 5:5.
104. Faber, C. G., G. Lauria, I. S. Merkies, X. Cheng, C. Han, H. S. Ahn, A. K. Persson et al. 2012. Gain-of-function Nav1.8 mutations in painful neuropathy. *Proc Natl Acad Sci USA* 109(47):19444–19449.
105. Roy, M. L. and T. Narahashi. 1992. Differential properties of tetrodotoxin-sensitive and tetrodotoxin-resistant sodium channels in rat dorsal root ganglion neurons. *J Neurosci* 12:2104–2111.
106. Leffler, A., R. I. Herzog, S. D. Dib-Hajj, S. G. Waxman, and T. R. Cummins. 2005. Pharmacological properties of neuronal TTX-resistant sodium channels and the role of a critical serine pore residue. *Pflugers Arch* 451:454–463.
107. Rush, A. M. and J. R. Elliott. 1997. Phenytoin and carbamazepine: Differential inhibition of sodium currents in small cells from adult rat dorsal root ganglia. *Neurosci Lett* 226:95–98.
108. Kistner, K., K. Zimmermann, C. Ehnert, P. W. Reeh, and A. Leffler. 2010. The tetrodotoxin-resistant Na^+ channel $Na_v1.8$ reduces the potency of local anesthetics in blocking C-fiber nociceptors. *Pflugers Arch* 459:751–763.
109. Cohen, C. J. 2011. Targeting voltage-gated sodium channels for treating neuropathic and inflammatory pain. *Curr Pharm Biotechnol* 12:1715–1719.
110. Cummins, T. R., S. D. Dib-Hajj, J. A. Black, A. N. Akopian, J. N. Wood, and S. G. Waxman. 1999. A novel persistent tetrodotoxin-resistant sodium current in SNS-null and wild-type small primary sensory neurons. *J Neurosci* 19:RC43.
111. Dib-Hajj, S. D., J. A. Black, T. R. Cummins, A. M. Kenney, J. D. Kocsis, and S. G. Waxman. 1998. Rescue of alpha-SNS sodium channel expression in small dorsal root ganglion neurons after axotomy by nerve growth factor in vivo. *J Neurophysiol* 79:2668–2676.
112. Blum, R., K. W. Kafitz, and A. Konnerth. 2002. Neurotrophin-evoked depolarization requires the sodium channel $Na_v1.9$. *Nature* 419:687–693.
113. O'Brien, B. J., J. H. Caldwell, G. R. Ehring, K. M. Bumsted O'Brien, S. Luo, and S. R. Levinson. 2008. Tetrodotoxin-resistant voltage-gated sodium channels $Na_v1.8$ and $Na_v1.9$ are expressed in the retina. *J Comp Neurol* 508:940–951.
114. Wells, J. E., V. Bingham, K. C. Rowland, and J. Hatton. 2007. Expression of $Na_v1.9$ channels in human dental pulp and trigeminal ganglion. *J Endod* 33:1172–1176.
115. Yu, Y. Q., F. Zhao, S. M. Guan, and J. Chen. 2011. Antisense-mediated knockdown of $Na_v1.8$, but not $Na_v1.9$, generates inhibitory effects on complete Freund's adjuvant-induced inflammatory pain in rat. *PLoS One* 6:e19865.
116. Persson, A. K., M. Gebauer, S. Jordan, C. Metz-Weidmann, A. M. Schulte, H. C. Schneider, D. Ding-Pfennigdorff et al. 2009. Correlational analysis for identifying genes whose regulation contributes to chronic neuropathic pain. *Mol Pain* 5:7.
117. Cummins, T. R. and S. G. Waxman. 1997. Downregulation of tetrodotoxin-resistant sodium currents and upregulation of a rapidly repriming tetrodotoxin-sensitive sodium current in small spinal sensory neurons after nerve injury. *J Neurosci* 17:3503–3514.

118. Dib-Hajj, S. D., L. Tyrrell, J. A. Black, and S. G. Waxman. 1998. NaN, a novel voltage-gated Na channel, is expressed preferentially in peripheral sensory neurons and down-regulated after axotomy. *Proc Natl Acad Sci USA* 95:8963–8968.
119. Ostman, J. A., M. A. Nassar, J. N. Wood, and M. D. Baker. 2008. GTP up-regulated persistent Na^+ current and enhanced nociceptor excitability require $Na_v1.9$. *J Physiol* 586:1077–1087.
120. Vanoye, C. G., J. D. Kunic, G. R. Ehring, and A. L. George, Jr. 2013. Mechanism of sodium channel $Na_v1.9$ potentiation by G-protein signaling. *J Gen Physiol* 141:193–202.
121. Herzog, R. I., T. R. Cummins, and S. G. Waxman. 2001. Persistent TTX-resistant Na^+ current affects resting potential and response to depolarization in simulated spinal sensory neurons. *J Neurophysiol* 86:1351–1364.
122. Priest, B. T., B. A. Murphy, J. A. Lindia, C. Diaz, C. Abbadie, A. M. Ritter, P. Liberator et al. 2005. Contribution of the tetrodotoxin-resistant voltage-gated sodium channel $Na_v1.9$ to sensory transmission and nociceptive behavior. *Proc Natl Acad Sci USA* 102:9382–9387.
123. Rush, A. M. and S. G. Waxman. 2004. PGE2 increases the tetrodotoxin-resistant $Na_v1.9$ sodium current in mouse DRG neurons via G-proteins. *Brain Res* 1023:264–271.
124. Amaya, F., H. Wang, M. Costigan, A. J. Allchorne, J. P. Hatcher, J. Egerton, T. Stean et al. 2006. The voltage-gated sodium channel $Na_v1.9$ is an effector of peripheral inflammatory pain hypersensitivity. *J Neurosci* 26:12852–12860.
125. Maingret, F., B. Coste, F. Padilla, N. Clerc, M. Crest, S. M. Korogod, and P. Delmas. 2008. Inflammatory mediators increase $Na_v1.9$ current and excitability in nociceptors through a coincident detection mechanism. *J Gen Physiol* 131:211–225.
126. Ritter, A. M., W. J. Martin, and K. S. Thorneloe. 2009. The voltage-gated sodium channel $Na_v1.9$ is required for inflammation-based urinary bladder dysfunction. *Neurosci Lett* 452:28–32.
127. Lolignier, S., M. Amsalem, F. Maingret, F. Padilla, M. Gabriac, E. Chapuy, A. Eschalier, P. Delmas, and J. Busserolles. 2011. $Na_v1.9$ channel contributes to mechanical and heat pain hypersensitivity induced by subacute and chronic inflammation. *PLoS One* 6:e23083.
128. Everaerts, W., M. Gees, Y. A. Alpizar, R. Farre, C. Leten, A. Apetrei, I. Dewachter et al. 2011. The capsaicin receptor TRPV1 is a crucial mediator of the noxious effects of mustard oil. *Curr Biol* 21:316–321.
129. Ross, R. A. 2003. Anandamide and vanilloid TRPV1 receptors. *Br J Pharmacol* 140:790–801.
130. Clapham, D. E. 2003. TRP channels as cellular sensors. *Nature* 426:517–524.
131. Cao, E., J. F. Cordero-Morales, B. Liu, F. Qin, and D. Julius. 2013. TRPV1 channels are intrinsically heat sensitive and negatively regulated by phosphoinositide lipids. *Neuron* 77:667–679.
132. Mohapatra, D. P. and C. Nau. 2003. Desensitization of capsaicin-activated currents in the vanilloid receptor TRPV1 is decreased by the cyclic AMP-dependent protein kinase pathway. *J Biol Chem* 278:50080–50090.
133. De Petrocellis, L., S. Harrison, T. Bisogno, M. Tognetto, I. Brandi, G. D. Smith, C. Creminon, J. B. Davis, P. Geppetti, and V. Di Marzo. 2001. The vanilloid receptor (VR1)-mediated effects of anandamide are potently enhanced by the cAMP-dependent protein kinase. *J Neurochem* 77:1660–1663.
134. Tominaga, M., M. J. Caterina, A. B. Malmberg, T. A. Rosen, H. Gilbert, K. Skinner, B. E. Raumann, A. I. Basbaum, and D. Julius. 1998. The cloned capsaicin receptor integrates multiple pain-producing stimuli. *Neuron* 21:531–543.
135. Ji, R. R., T. A. Samad, S. X. Jin, R. Schmoll, and C. J. Woolf. 2002. p38 MAPK activation by NGF in primary sensory neurons after inflammation increases TRPV1 levels and maintains heat hyperalgesia. *Neuron* 36:57–68.
136. Gu, Q., L. Wang, F. Huang, and W. Schwarz. 2013. Stimulation of TRPV1 by green laser light. *Evid Based Complement Alternat Med* 2012:857123.
137. Caterina, M. J., M. A. Schumacher, M. Tominaga, T. A. Rosen, J. D. Levine, and D. Julius. 1997. The capsaicin receptor: A heat-activated ion channel in the pain pathway. *Nature* 389:816–824.
138. Hoffman, E. M., R. Schechter, and K. E. Miller. 2009. Fixative composition alters distributions of immunoreactivity for glutaminase and two markers of nociceptive neurons, Nav1.8 and TRPV1, in the rat dorsal root ganglion. *J Histochem Cytochem* 58:329–344.
139. Roberts, J. C., J. B. Davis, and C. D. Benham. 2004. [3H] Resiniferatoxin autoradiography in the CNS of wild-type and TRPV1 null mice defines TRPV1 (VR-1) protein distribution. *Brain Res* 995:176–183.
140. Birder, L. A., Y. Nakamura, S. Kiss, M. L. Nealen, S. Barrick, A. J. Kanai, E. Wang et al. 2002. Altered urinary bladder function in mice lacking the vanilloid receptor TRPV1. *Nat Neurosci* 5:856–860.
141. Scotland, R. S., S. Chauhan, C. Davis, C. De Felipe, S. Hunt, J. Kabir, P. Kotsonis, U. Oh, and A. Ahluwalia. 2004. Vanilloid receptor TRPV1, sensory C-fibers, and vascular autoregulation: A novel mechanism involved in myogenic constriction. *Circ Res* 95:1027–1034.
142. Bodo, E., I. Kovacs, A. Telek, A. Varga, R. Paus, L. Kovacs, and T. Biro. 2004. Vanilloid receptor-1 (VR1) is widely expressed on various epithelial and mesenchymal cell types of human skin. *J Invest Dermatol* 123:410–413.
143. Bodo, E., T. Biro, A. Telek, G. Czifra, Z. Griger, B. I. Toth, A. Mescalchin et al. 2005. A hot new twist to hair biology: Involvement of vanilloid receptor-1 (VR1/TRPV1) signaling in human hair growth control. *Am J Pathol* 166:985–998.
144. Walpole, C. S., S. Bevan, G. Bovermann, J. J. Boelsterli, R. Breckenridge, J. W. Davies, G. A. Hughes et al. 1994. The discovery of capsazepine, the first competitive antagonist of the sensory neuron excitants capsaicin and resiniferatoxin. *J Med Chem* 37:1942–1954.
145. Wang, X., R. L. Miyares, and G. P. Ahern. 2005. Oleoylethanolamide excites vagal sensory neurones, induces visceral pain and reduces short-term food intake in mice via capsaicin receptor TRPV1. *J Physiol* 564:541–547.
146. Bolcskei, K., Z. Helyes, A. Szabo, K. Sandor, K. Elekes, J. Nemeth, R. Almasi, E. Pinter, G. Petho, and J. Szolcsanyi. 2005. Investigation of the role of TRPV1 receptors in acute and chronic nociceptive processes using gene-deficient mice. *Pain* 117:368–376.
147. Negri, L., R. Lattanzi, E. Giannini, M. Colucci, F. Margheriti, P. Melchiorri, V. Vellani, H. Tian, M. De Felice, and F. Porreca. 2006. Impaired nociception and inflammatory pain sensation in mice lacking the prokineticin receptor PKR1: Focus on interaction between PKR1 and the capsaicin receptor TRPV1 in pain behavior. *J Neurosci* 26:6716–6727.
148. Rong, W., K. Hillsley, J. B. Davis, G. Hicks, W. J. Winchester, and D. Grundy. 2004. Jejunal afferent nerve sensitivity in wild-type and TRPV1 knockout mice. *J Physiol* 560:867–881.
149. Christoph, T., C. Gillen, J. Mika, A. Grunweller, M. K. Schafer, K. Schiene, R. Frank et al. 2007. Antinociceptive effect of antisense oligonucleotides against the vanilloid receptor VR1/TRPV1. *Neurochem Int* 50:281–290.
150. Christoph, T., A. Grunweller, J. Mika, M. K. Schafer, E. J. Wade, E. Weihe, V. A. Erdmann, R. Frank, C. Gillen, and J. Kurreck. 2006. Silencing of vanilloid receptor TRPV1 by RNAi reduces neuropathic and visceral pain in vivo. *Biochem Biophys Res Commun* 350:238–243.

151. Ando, K., M. S. Wallace, J. Braun, and G. Schulteis. 2000. Effect of oral mexiletine on capsaicin-induced allodynia and hyperalgesia: A double-blind, placebo-controlled, crossover study. *Reg Anesth Pain Med* 25:468–474.
152. Lam, V. Y., M. Wallace, and G. Schulteis. 2010. Effects of lidocaine patch on intradermal capsaicin-induced pain: A double-blind, controlled trial. *J Pain* 12:323–330.
153. Wallace, M. S., S. Quessy, and G. Schulteis. 2004. Lack of effect of two oral sodium channel antagonists, lamotrigine and 4030W92, on intradermal capsaicin-induced hyperalgesia model. *Pharmacol Biochem Behav* 78:349–355.
154. Scanlon, G. C., M. S. Wallace, J. S. Ispirescu, and G. Schulteis. 2006. Intradermal capsaicin causes dose-dependent pain, allodynia, and hyperalgesia in humans. *J Investig Med* 54:238–244.
155. Szelenyi, Z., Z. Hummel, J. Szolcsanyi, and J. B. Davis. 2004. Daily body temperature rhythm and heat tolerance in TRPV1 knockout and capsaicin pretreated mice. *Eur J Neurosci* 19:1421–1424.
156. Wang, L. H., M. Luo, Y. Wang, J. J. Galligan, and D. H. Wang. 2006. Impaired vasodilation in response to perivascular nerve stimulation in mesenteric arteries of TRPV1-null mutant mice. *J Hypertens* 24:2399–2408.
157. Wang, L. and D. H. Wang. 2005. TRPV1 gene knockout impairs postischemic recovery in isolated perfused heart in mice. *Circulation* 112:3617–3623.
158. Rahmati, R. 2012. The transient receptor potential vanilloid receptor 1, TRPV1 (VR1) inhibits peristalsis in the mouse jejunum. *Arch Iran Med* 15:433–438.
159. Fernandes, E. S., C. T. Vong, S. Quek, J. Cheong, S. Awal, C. Gentry, A. A. Aubdool et al. 2012. Superoxide generation and leukocyte accumulation: Key elements in the mediation of leukotriene B4-induced itch by transient receptor potential ankyrin 1 and transient receptor potential vanilloid 1. *FASEB J* 27(4):1664–1673.
160. Szallasi, A., D. N. Cortright, C. A. Blum, and S. R. Eid. 2007. The vanilloid receptor TRPV1: 10 years from channel cloning to antagonist proof-of-concept. *Nat Rev Drug Discov* 6:357–372.
161. Kitagawa, Y., A. Miyai, K. Usui, Y. Hamada, K. Deai, M. Wada, Y. Koga et al. 2012. Pharmacological characterization of (3S)-3-(hydroxymethyl)-4-(5-methylpyridin-2-yl)-N-[6-(2,2,2-trifluoroethoxy)pyrid in-3-yl]-3,4-dihydro-2H-benzo[b][1,4] oxazine-8-carboxamide (JTS-653), a novel transient receptor potential vanilloid 1 antagonist. *J Pharmacol Exp Ther* 342:520–528.
162. Honore, P., P. Chandran, G. Hernandez, D. M. Gauvin, J. P. Mikusa, C. Zhong, S. K. Joshi et al. 2009. Repeated dosing of ABT-102, a potent and selective TRPV1 antagonist, enhances TRPV1-mediated analgesic activity in rodents, but attenuates antagonist-induced hyperthermia. *Pain* 142:27–35.
163. Lim, K. M. and Y. H. Park. 2012. Development of PAC-14028, a novel transient receptor potential vanilloid type 1 (TRPV1) channel antagonist as a new drug for refractory skin diseases. *Arch Pharm Res* 35:393–396.
164. Krarup, A. L., L. Ny, M. Astrand, A. Bajor, F. Hvid-Jensen, M. B. Hansen, M. Simren, P. Funch-Jensen, and A. M. Drewes. 2011. Randomised clinical trial: The efficacy of a transient receptor potential vanilloid 1 antagonist AZD1386 in human oesophageal pain. *Aliment Pharmacol Ther* 33:1113–1122.
165. Chizh, B. A., M. B. O'Donnell, A. Napolitano, J. Wang, A. C. Brooke, M. C. Aylott, J. N. Bullman et al. 2007. The effects of the TRPV1 antagonist SB-705498 on TRPV1 receptor-mediated activity and inflammatory hyperalgesia in humans. *Pain* 132:132–141.
166. Hori, T. 1984. Capsaicin and central control of thermoregulation. *Pharmacol Ther* 26:389–416.
167. Nash, M. S., P. McIntyre, A. Groarke, E. Lilley, A. Culshaw, A. Hallett, M. Panesar, A. Fox, and S. Bevan. 2012. 7-tert-Butyl-6-(4-chloro-phenyl)-2-thioxo-2,3-dihydro-1H-pyrido[2,3-d]pyrimidin-4-one, a classic polymodal inhibitor of transient receptor potential vanilloid type 1 with a reduced liability for hyperthermia, is analgesic and ameliorates visceral hypersensitivity. *J Pharmacol Exp Ther* 342:389–398.
168. Palazzo, E., L. Luongo, V. de Novellis, L. Berrino, F. Rossi, and S. Maione. 2010. Moving towards supraspinal TRPV1 receptors for chronic pain relief. *Mol Pain* 6:66.
169. Simpson, D. M., S. Brown, J. K. Tobias, and G. F. Vanhove. 2013. NGX-4010, a capsaicin 8% dermal patch, for the treatment of painful HIV-associated distal sensory polyneuropathy: Results of a 52-week open-label study. *Clin J Pain* 30(2):134–142.
170. Derry, S., A. Sven-Rice, P. Cole, T. Tan, and R. A. Moore. 2013. Topical capsaicin (high concentration) for chronic neuropathic pain in adults. *Cochrane Database Syst Rev* 2:CD007393.
171. Westenbroek, R. E., J. W. Hell, C. Warner, S. J. Dubel, T. P. Snutch, and W. A. Catterall. 1992. Biochemical properties and subcellular distribution of an N-type calcium channel alpha 1 subunit. *Neuron* 9:1099–1115.
172. Dubel, S. J., T. V. Starr, J. Hell, M. K. Ahlijanian, J. J. Enyeart, W. A. Catterall, and T. P. Snutch. 1992. Molecular cloning of the alpha-1 subunit of an omega-conotoxin-sensitive calcium channel. *Proc Natl Acad Sci USA* 89:5058–5062.
173. Bell, T. J., C. Thaler, A. J. Castiglioni, T. D. Helton, and D. Lipscombe. 2004. Cell-specific alternative splicing increases calcium channel current density in the pain pathway. *Neuron* 41:127–138.
174. D'Ascenzo, M., M. Vairano, C. Andreassi, P. Navarra, G. B. Azzena, and C. Grassi. 2004. Electrophysiological and molecular evidence of L-(Ca_v1), N- ($Ca_v2.2$), and R- ($Ca_v2.3$) type Ca^{2+} channels in rat cortical astrocytes. *Glia* 45:354–363.
175. Tsunemi, T., H. Saegusa, K. Ishikawa, S. Nagayama, T. Murakoshi, H. Mizusawa, and T. Tanabe. 2002. Novel $Ca_v2.1$ splice variants isolated from Purkinje cells do not generate P-type Ca^{2+} current. *J Biol Chem* 277:7214–7221.
176. Kim, C., K. Jun, T. Lee, S. S. Kim, M. W. McEnery, H. Chin, H. L. Kim et al. 2001. Altered nociceptive response in mice deficient in the alpha(1B) subunit of the voltage-dependent calcium channel. *Mol Cell Neurosci* 18:235–245.
177. Reynolds, I. J., J. A. Wagner, S. H. Snyder, S. A. Thayer, B. M. Olivera, and R. J. Miller. 1986. Brain voltage-sensitive calcium channel subtypes differentiated by omega-conotoxin fraction GVIA. *Proc Natl Acad Sci USA* 83:8804–8807.
178. Bowersox, S. S., T. Singh, L. Nadasdi, Z. Zukowska-Grojec, K. Valentino, and B. B. Hoffman. 1992. Cardiovascular effects of omega-conopeptides in conscious rats: Mechanisms of action. *J Cardiovasc Pharmacol* 20:756–764.
179. Wallace, M. S., P. S. Kosek, P. Staats, R. Fisher, D. M. Schultz, and M. Leong. 2008. Phase II, open-label, multicenter study of combined intrathecal morphine and ziconotide: Addition of ziconotide in patients receiving intrathecal morphine for severe chronic pain. *Pain Med* 9:271–281.
180. Ellis, D. J., S. Dissanayake, D. McGuire, S. G. Charapata, P. S. Staats, M. S. Wallace, G. W. Grove, and P. Vercruysse. 2008. Continuous intrathecal infusion of ziconotide for treatment of chronic malignant and nonmalignant pain over 12 months: A prospective, open-label study. *Neuromodulation* 11:40–49.
181. Zhang, J. L., J. P. Yang, J. R. Zhang, R. Q. Li, J. Wang, J. J. Jan, and Q. Zhuang. 2012. Gabapentin reduces allodynia and hyperalgesia in painful diabetic neuropathy rats by decreasing expression level of $Na_v1.7$ and p-ERK1/2 in DRG neurons. *Brain Res* 1493:13–18.
182. Winquist, R. J., J. Q. Pan, and V. K. Gribkoff. 2005. Use-dependent blockade of $Ca_v2.2$ voltage-gated calcium channels for neuropathic pain. *Biochem Pharmacol* 70:489–499.

183. Tyagarajan, S., P. K. Chakravarty, M. Park, B. Zhou, J. B. Herrington, K. Ratliff, R. M. Bugianesi et al. 2011. A potent and selective indole N-type calcium channel ($Ca_v2.2$) blocker for the treatment of pain. *Bioorg Med Chem Lett* 21:869–873.
184. Abbadie, C., O. B. McManus, S. Y. Sun, R. M. Bugianesi, G. Dai, R. J. Haedo, J. B. Herrington et al. 2010. Analgesic effects of a substituted N-triazole oxindole (TROX-1), a state-dependent, voltage-gated calcium channel 2 blocker. *J Pharmacol Exp Ther* 334:545–555.
185. Shao, P. P., F. Ye, P. K. Chakravarty, D. J. Varughese, J. B. Herrington, G. Dai, R. M. Bugianesi et al. 2012. Aminopiperidine sulfonamide $Ca_v2.2$ channel inhibitors for the treatment of chronic pain. *J Med Chem* 55:9847–9855.
186. Emery, E. C., G. T. Young, and P. A. McNaughton. 2012. HCN2 ion channels: An emerging role as the pacemakers of pain. *Trends Pharmacol Sci* 33:456–463.
187. Wickenden, A. D., M. P. Maher, and S. R. Chaplan. 2009. HCN pacemaker channels and pain: A drug discovery perspective. *Curr Pharm Des* 15:2149–2168.
188. Lapointe, T. K. and C. Altier. 2011. The role of TRPA1 in visceral inflammation and pain. *Channels (Austin)* 5:525–529.
189. Nilius, B., G. Appendino, and G. Owsianik. 2012. The transient receptor potential channel TRPA1: From gene to pathophysiology. *Pflugers Arch* 464:425–458.
190. Andrade, E. L., F. C. Meotti, and J. B. Calixto. 2011. TRPA1 antagonists as potential analgesic drugs. *Pharmacol Ther* 133:189–204.
191. Wu, W. L., C. F. Cheng, W. H. Sun, C. W. Wong, and C. C. Chen. 2012. Targeting ASIC3 for pain, anxiety, and insulin resistance. *Pharmacol Ther* 134:127–138.
192. Burnstock, G. 2011. Targeting the visceral purinergic system for pain control. *Curr Opin Pharmacol* 12:80–86.
193. Takeda, M., Y. Tsuboi, J. Kitagawa, K. Nakagawa, K. Iwata, and S. Matsumoto. 2011. Potassium channels as a potential therapeutic target for trigeminal neuropathic and inflammatory pain. *Mol Pain* 7:5.
194. Collins, S., M. J. Sigtermans, A. Dahan, W. W. Zuurmond, and R. S. Perez. 2010. NMDA receptor antagonists for the treatment of neuropathic pain. *Pain Med* 11:1726–1742.
195. Mannino, R., P. Coyne, C. Swainey, L. A. Hansen, and L. Lyckholm. 2006. Methadone for cancer-related neuropathic pain: A review of the literature. *J Opioid Manag* 2:269–276.
196. Gormsen, L., N. B. Finnerup, P. M. Almqvist, and T. S. Jensen. 2009. The efficacy of the AMPA receptor antagonist NS1209 and lidocaine in nerve injury pain: A randomized, double-blind, placebo-controlled, three-way crossover study. *Anesth Analg* 108:1311–1319.
197. Benarroch, E. E. 2012. GABAB receptors: Structure, functions, and clinical implications. *Neurology* 78:578–584.

CLC-related proteins in diseases

Allan H. Bretag and Linlin Ma

Contents

41.1 INTRODUCTION

Proteins of the voltage-gated chloride (Cl^-) channel and transporter (CLC) family have proven to be unique in their general structure and fascinating in their differing functions and purposes. This knowledge has been painstakingly gained over more than a century largely through what has been able to be deduced from the diseases that occur in the absence of these proteins or when they malfunction. If we exclude the heterogeneous epilepsies and cardiac arrhythmias, which were known since ancient times, myotonia congenita (MC) (Thomsen disease) was the first ion channel disease to be described in detail as a clinical entity (Thomsen, 1876).

Importantly, for MC, a century of hypotheses on the nature of membrane pores and their contribution to cellular electrical activity led eventually to the accidental discovery of muscle-type Cl^- channels in the electroplaque of *Torpedo californica* (White and Miller, 1979). Rapid advances took place once the electroplaque Cl^- channel gene had been cloned (Jentsch et al., 1990). Almost immediately, a whole group of related mammalian (and human)

Table 41.1 An overview of those CLC proteins, their subunits and their genes known to be associated with disease

PROTEIN, (GENE)	TISSUE DISTRIBUTION	MEMBRANE (PROTEIN FUNCTION)	DYSFUNCTION WHEN MUTATED	NATURALLY OCCURRING ANIMAL MODELS, (KNOCKOUT MOUSE)	HUMAN DISEASE
ClC-1[a] (*CLCN1*)	Skeletal muscle	Sarcolemma, t-tubule membrane? (Cl^- ion channel)	Membrane voltage instability	Myotonic *adr* mouse; myotonic goat, dog, horse, buffalo, etc.	Dominant and recessive myotonia, myotonic dystrophy
	Brain	Plasma membrane?	?	—	Epilepsy?
ClC-2 (*CLCN2*)	Widespread	Basolateral in epithelia, junctional membranes in glia (Cl^- ion channel)	Disruption to interstitial fluid homeostasis and to secretion/reabsorption in epithelia	(*Clcn2*$^{-/-}$ retinal and testicular degeneration, leukoencephal-opathy)	Megalencephalic leukoencephalopathy with subcortical cysts? Epilepsy?
GlialCAM[a] (*HEPACAM*)		Junctional membranes in glia (auxiliary subunit to ClC-2)	Failure of ClC-2 trafficking to cell-cell junctions	—	Megalencephalic leukoencephalopathy with subcortical cysts
ClC-Ka[a] (*CLCNKA*)	Kidney, inner ear	Basolateral in epithelia (Cl^- ion channel)	Reduced epithelial Cl^- transport, can be compensated by ClC-Kb in humans	(*Clcnk1*$^{-/-}$ similar to human renal diabetes insipidus)	Bartter syndrome IVB with sensorineural deafness (only when digenic with *CLCNKB*)
ClC-Kb[a] (*CLCNKB*)	Kidney, inner ear	Basolateral in epithelia (Cl^- ion channel)	Reduced epithelial Cl^- transport, unable to be compensated by ClC-Ka		Bartter syndrome III, Bartter syndrome IVB as above, Gitelman syndrome
Barttin[a] (*BSND*)	Kidney, inner ear	Basolateral in epithelia (essential auxiliary subunit to ClC-Ka and ClC-Kb)	Simultaneous dysfunction of both ClC-Ka and ClC-Kb in kidney and inner ear		Bartter syndrome IVA with sensorineural deafness, non-syndromic deafness
ClC-3 (*CLCN3*)	Widespread	Endosomes, synaptic vesicles (Cl^-/H^+ antiporter, Cl^- ion channel?)	Inadequate acidification of small membrane-bound subcellular compartments?	(*Clcn3*$^{-/-}$ CNS degeneration similar to human NCL, blindness, degeneration of the hippocampus)	Neuronal ceroid lipofuscinosis (NCL)? Cardiovascular cerebrovascular diseases? Glioma invasion/ migration?
ClC-4 (*CLCN4*)	Widespread	Endosomes (Cl^-/H^+ antiporter)	?	(*Clcn4*$^{-/-}$ no known pathology)	
ClC-5[a] (*CLCN5*)	Kidney, intestine	Endosomes, apical membranes in epithelia (Cl^-/H^+ antiporter)	Reduced acidifica-tion of sub-apical endosomes, disruption of endocytosis	(*Clcn5*$^{-/-}$ similar to human Dent disease)	Dent disease
ClC-6 (*CLCN6*)	Widespread	Endosomes (Cl^-/H^+ antiporter)	Inadequate acidification of endosomes?	(*Clcn6*$^{-/-}$ similar to mild late onset human NCL but without the motor deficits or brain and retinal degeneration)	Cardiac disease? Neuronal ceroid lipofuscinosis?

(Continued)

Table 41.1 (*Continued*) An overview of those CLC proteins, their subunits and their genes known to be associated with disease

PROTEIN, (GENE)	TISSUE DISTRIBUTION	MEMBRANE (PROTEIN FUNCTION)	DYSFUNCTION WHEN MUTATED	NATURALLY OCCURRING ANIMAL MODELS, (KNOCKOUT MOUSE)	HUMAN DISEASE
ClC-7[a] (*CLCN7*)	Brain, kidney, liver, osteoclasts, quite widespread	Late endosomes, lysosomes (Cl^-/H^+ antiporter)	Failed acidification of osteoclast resorption lacunae, disturbed lysosomal Cl^- homeostasis	(*Clcn7*$^{-/-}$ severe osteopetrosis, etc., as in humans, neurodegeneration resembling severe early onset human thalamo-cortical NCL)	Dominant and recessive osteopetrosis, no CNS deficits through to severe neuro-degeneration resembling NCL, vision, hearing, bone marrow function affected by bone encroachment
Ostm1[a] (*OSTM1*)	Brain, kidney, liver, osteoclasts, quite widespread	Late endosomes, lysosomes (essential auxiliarysubunit to ClC-7)	ClC-7 destabilized and degraded with consequences upon bone resorption and lysosomal function	(*Grey lethal* severe osteopetrosis, etc., as in humans, neurodegeneration resembling severe early onset human thalamo-cortical NCL)	Severe malignant infantile osteopetrosis, retinal degeneration, severe neurodegeneration resembling NCL, vision, hearing, bone marrow function affected by bone encroachment

[a] CLC-related proteins known to be associated with human disease.

Cl^- channel genes, later named the *CLCN* family, was discovered by homology screening. Links between the *CLCN* genes and a variety of diseases soon followed (Table 41.1). Particularly important to the understanding of the relationship with disease has been the enormous effort put into observations on animal models, into cellular electrophysiological and membrane biophysical studies of the CLC proteins, and into their purification, crystallization, and structural determination by X-ray crystallography (Pusch and Zifarelli elsewhere in this volume). Some CLC proteins have proven to be transporters and not channels, while some are confined to internal membranes rather than cellular surfaces; nevertheless, much remains unknown and controversial with respect to the association between the CLC transporters/channels and disease.

41.2 CLC-1 (202* VARIATIONS IN THE *CLCN1* GENE ASSOCIATED WITH DISEASE)

41.2.1 MYOTONIA CONGENITA, THOMSEN DISEASE (MIM 160800)

Thomsen's MC is characterized by mild-to-severe, cramp-like, but usually painless, muscle stiffness on the first voluntary movement after rest, or within one or two repeats of that movement, with normal freedom to move returning only after a number of further repetitions (the *warm-up* phenomenon). A feature seldom discussed is that sudden changes in movement, and even in the rate of a previously warmed-up movement (as in speeding up from walking to running), may also be hindered by the reemergence of stiffness (Becker, 1977). This peculiarity has also been demonstrated in isolated rat skeletal muscle in vitro treated with Cl^- channel blockers (Figure 41.1). To be startled or tripped is considered serious by those affected as they find that their musculature becomes rigid and, if off balance, they topple in a wooden fashion, like felled trees, unable to protect themselves from the fall. Without training, they often have a strong and defined musculature and, if warmed up, can perform capably as athletes. Their well-developed muscles may result from the added involuntary exercise caused by their prolonged myotonic contractions (Becker, 1977). As implied by the descriptor *congenita*, it can be obvious in the stiff limbs of infants from birth. In other cases, however, MC can be so mild in its symptoms, late in its onset, and without unusually developed musculature (Lehmann-Horn et al., 1995; Burgunder et al., 2008; Kumar et al., 2010) that it has been termed myotonia levior. A worsening of the myotonic stiffness with cold has frequently been remarked upon as a symptom in some families with MC (Becker, 1977; Koty et al., 1996) although this could be a subjective opinion of patients in response to leading questions from the examining clinician rather than anything measurable (Ricker et al., 1977). Maintained cold suppresses myotonia in isolated rat muscle treated as in Figure 41.1, whereas cold-shock induces a myotonic contraction in the absence of any other stimulus (Bretag, 1987). In MC, a negative association with cold could result from myotonia initiated by rapid cooling of exposed musculature or by abrupt shivering, general muscle function not being impaired.

* HGMD 2013 (Human Gene Mutation Database).

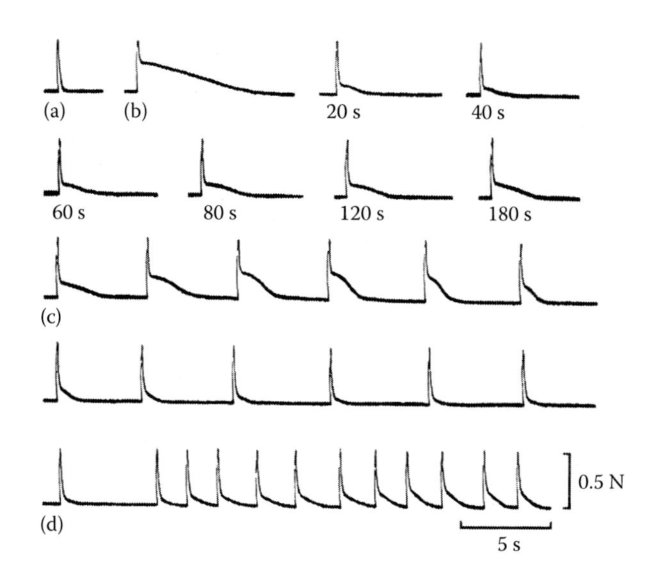

Figure 41.1 A demonstration of warm-up and of paradoxical myotonia in the isolated rat soleus muscle. (Reproduced from Bretag, A.H., *Proc. Aus. Physiol. Pharmacol. Soc.*, 14, 170, 1983.) The muscle was bathed in synthetic interstitial fluid (Bretag, 1969) at 37°C. Each individual isometric muscle contraction resulted from direct stimulation with a train of three, just-supra-maximal, square-wave, pulses each of one millisecond duration at 225 Hz. (a) Control contraction recorded following equilibration. (b) After 30 min of rest, during which the muscle was exposed to anthracene-9-carboxylate (2.25×10^{-5} mol·L^{-1}), it was stimulated at 20 s intervals with representative records shown to 180 s. A typical, long-lasting, myotonic contraction results from the first stimulus. At 20 and 40 s, the second and third contractions, respectively, show substantial warm-up that was partially maintained up to the 10th sequential contraction at 180 s. (c) Here, the 10th contraction record at 180 s is reproduced and followed in a continuous sequence by contractions initiated at 5 s intervals. Over several subsequent contractions, myotonia now increases in severity (paradoxical myotonia) and then warms up almost completely by the 11th contraction, at the end of the 4th line of records. The 11th contraction record is reproduced at the beginning of the fifth line. (d) Contractions are now initiated at approximately 2 s intervals showing that, even though almost complete warm-up had been achieved, a sudden increase in the frequency of contraction again results in paradoxical myotonia.

41.2.2 RECESSIVE GENERALIZED MYOTONIA, BECKER DISEASE (OMIM 255700)

Typically, Becker's recessive generalized myotonia (RGM) has similar features to MC but is a more severe condition and the musculature shows more pronounced hypertrophy. Initial movements after rest are not only stiff as in MC, they may also be extremely hesitant and slow or even completely arrested with a concomitant and subsequent period of significant weakness (Becker, 1977; Rüdel et al., 1988). Despite their athletic appearance, individuals may, therefore, be quite seriously debilitated. Unexpectedly, in RGM as in MC, there are reports of relatively mild cases without unusually hypertrophic muscles (Burgunder et al., 2008; Gurgel-Gianetti et al., 2012). A significant gender imbalance has been observed in the RGM population with a preponderance of males over females of 2:1 (Ulzi et al., 2012), males being more severely affected and male heterozygotic carriers more frequently having subclinical signs of myotonia, *latent myotonia* (Becker, 1977; Mailänder et al., 1996).

Notwithstanding some observations to the contrary (e.g., Kornblum et al., 2010), distal muscular weakness and/or atrophy reminiscent of myotonic dystrophy type 1 (DM1) (discussed later) have been observed in what has been presumed to be RGM (Becker, 1977) but also in cases of genetically defined RGM, especially at older ages (Nagamitsu et al., 2000; Hilbert et al., 2011; Weinberger et al., 2012; Morrow et al., 2013). Muscle ultrasound echo intensities are also increased inversely in relation to range of movement in RGM (Trip et al., 2009). No mechanisms for these muscle pathologies are known.

Skeletal muscle fibers in MC and RGM have a lower threshold for electrical and mechanical stimulation due to reduced membrane Cl^- conductance although they are not spontaneously active (Lipicky et al., 1971). Myotonic contractions are due to abnormal after-discharges of action potentials triggered because the repolarisation phase of each action potential of a voluntary, mechanically or electrically stimulated volley causes incremental accumulation of K^+ in t-tubules (Adrian and Bryant 1974). This build up is usually opposed by Cl^- influx, but in myotonic muscle its excitatory, depolarising influence, re-activates sufficient Na^+ channels that are recovering from inactivation during an action potential's relatively refractory period to initiate a new action potential. Inflow of positive charge through open Na^+ channels is also unopposed by Cl^- influx, allowing successive action potentials to continue until increasing depolarisation and a rising percentage of inactivated Na^+ channels terminates the process.

After years of speculation about putative sarcolemmal Cl^- channels that might be involved (Bretag, 1987), eventual success in cloning of the rat *CLCN* gene and characterizing its protein product, ClC-1, provided the answer (Steinmeyer et al., 1991a). Variations in the human *CLCN1* gene were quickly discovered associated with RGM (Koch et al., 1992) and with MC (George et al., 1993). Lists of disease-associated *CLCN1* variations compiled since, along with how they affect the function of heterologously expressed ClC-1, can be found in previous reviews (Pusch, 2002; Colding-Jørgensen, 2005; Lossin and George, 2008). Many more recent studies have added substantially to these lists (e.g., Burgunder et al., 2008;

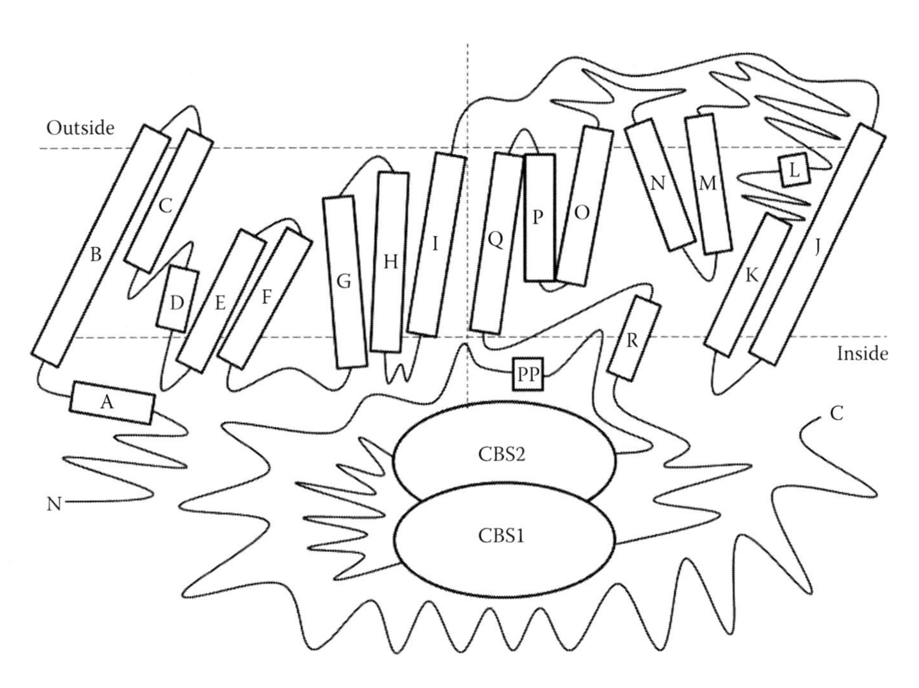

Figure 41.2 A two-dimensional representation of the three-dimensional structure of a ClC-1 monomer as predicted from the X-ray crystal structure of CmCLC (Feng et al., 2010) and from functional studies (Ma et al., 2011). Basically, the two halves of the membrane resident region of the monomer (helix B to helix I and helix J to helix Q) are structurally similar but inverted in the membrane with respect to each other. Helices B to G and J to O wrap around the two sides of the channel pore such that helices B and J lie antiparallel on opposite sides of the triangular prism-shaped monomer. Helices Q and P fold back behind O, while H and I fold back behind G to make up the QPHI surface (most of the dimer interface) when viewed from the base of the prism. Crystal positions of the helix–helix linkers are uncertain for ClC-1 and are not drawn to scale but are generally indicative of their respective loop lengths. Importantly, CBS2 lies close to the cytoplasmic helix–helix loops and the linker between CBS2 and the poly-proline (PP) helix lies close to helix R and its linker to CBS1. The structures of the segments linking CBS1 to CBS2 and linking helix PP to the C terminus are quite unknown, as is the position of helix PP, although structural and functional studies (Macías et al., 2007; Feng et al., 2010; Ma et al., 2011) suggest that helix PP and the C terminus lie close to the cytoplasmic face of the membrane resident region of the monomer.

Dupré et al., 2009; Moon et al., 2009; Gao et al., 2010; Kornblum et al., 2010; Modoni et al., 2011; Ivanova et al., 2012; Mazón et al., 2012; Ulzi et al., 2012) so that well over 200 *CLCN1* variants have now been associated with MC and RGM (HGMD, 2013). Although they are topologically distributed throughout the gene, there is a somewhat greater preponderance of variants (especially for MC) in the helix-H, H-I linker and helix-I (exon 8) region (Fialho et al., 2007) and an absence of any, so far, in helix-R (Figure 41.2).

In the majority of cases, the identified genetic variation adequately accounts for the myotonic symptoms of MC and RGM. Truncations in the long cytoplasmic carboxyl tail of ClC-1 that partially or completely delete the cystathionine β-synthase (CBS) regulatory domains (Figures 41.2 and 41.3) also compromise or eliminate channel function (Estévez et al., 2004; Hebeisen et al., 2004; Wu et al., 2006), explainable by a direct influence of the CBS domains and post CBS tail through a binding interface at the cytoplasmic surface of the membrane resident portion of the channel (Ma et al., 2008, 2009, 2011; Feng et al., 2010). Nonsense mutations in the distal part of the cytoplasmic carboxyl tail have also been associated with MC and RGM (George et al., 1994; Modoni et al., 2011), probably by reducing surface expression of the mutant ClC-1 due to trafficking failure or reduced protein stability (Macías et al., 2007; Papponen et al., 2008). Homozygous or compound heterozygous nonsense mutations in RGM can, therefore, limit sarcolemmal Cl^- conductance to a small percentage of that

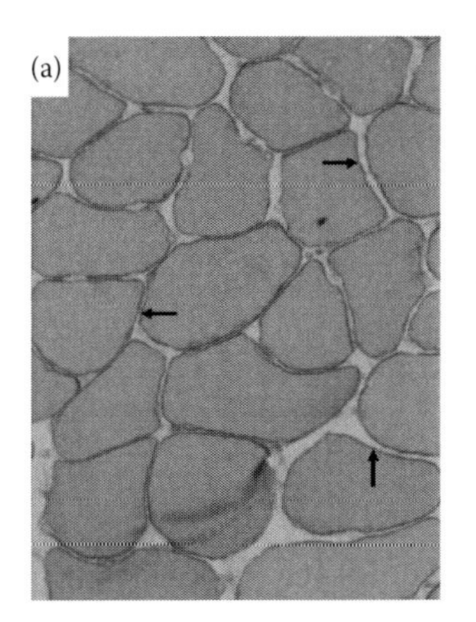

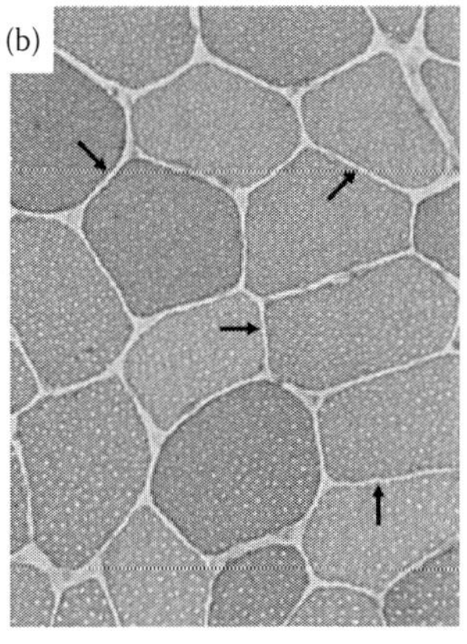

Figure 41.3 Transverse sections of muscle biopsies from a control (a) and a genetically confirmed RGM patient (b) showing presence and relative absence, respectively, of sarcolemmal staining (arrows) using a double staining with two specific ClC-1 antibodies. (Images courtesy of Bjarne Udd, Neuromuscular Research Center, University of Tampere, Finland, 2013, unpublished; for methods, see Raheem et al., 2012). There is distinct, continuous sarcolemmal staining in the control but lesser, and minimal residual staining in a minority of fibers in the RGM case indicating almost total loss of ClC-1 from the sarcolemma.

normally present due to fully functional ClC-1 and, therefore, explain the associated myotonia.

Many disease-associated *CLCN1* variations occur not as nonsense but as missense mutations (nonsynonymous SNPs) and as small insertions or deletions within exons, or as copy number variations (exon deletions, duplications, and inversions), as well as intronic mutations predicted to affect splicing (Lossin and George, 2008; Raja Rayan et al., 2012). Surprisingly, intraexonic synonymous SNPs (maintaining the same amino acid and, therefore, not missense mutations) have been associated with both human and buffalo myotonia where they also adversely influence ClC-1 splicing (Raheem et al., 2012; Borges et al., 2013).

By contrast with WT ClC-1, when some missense mutants are expressed in rat cell lines and myofibers, they fail to be trafficked to the sarcolemma or are rapidly degraded (Papponen et al., 2008). This has been confirmed directly (Figure 41.3) in human RGM muscle biopsies (Raheem et al., 2012). Other missense mutants have been found to be correctly localized when heterologously expressed but malfunction or lack function (Pusch, 2002; Colding-Jørgensen, 2005; Lossin and George, 2008; Ulzi et al., 2012). Types of malfunction include dramatic alterations in voltage dependence and ion selectivity and decreased single channel conductance. Function that appears to be normal in some heterologously expressed, disease-associated ClC-1 mutants (Colding-Jørgensen, 2005) might suggest that its associations, modulation, or processing are different in native skeletal muscle cells.

One of the most inexplicable actions of many missense mutations is their exertion of dominance in some instances, as in MC, but not in others, as in the heterozygous parents of RGM cases who usually have no disease symptoms. This is partially explainable in terms of the dimeric structure of ClC-1 and the presumed equal expression of its alleles. When expressed in heterologous systems, there is an adverse dominant effect of mutant MC monomers (mt_{MC}) on channel function. How this occurs is not yet fully understood but the unique properties of fast and common gating in the two pores of CLC channels (Push and Zifarelli, elsewhere in this volume) have been invoked in attempts to explain it. Frequently associated with it, a drastic effect on the voltage dependence of common gating in mt_{MC}-mt_{MC} and mt_{MC}-WT mutant dimers hinders the opening of the common gate at physiological membrane potentials (Pusch et al., 1995; Kubish et al., 1998; Duffield et al., 2003). While fast gating occurs independently in each individual pore of the double barreled channel with movement of the carboxyl side chain of its E232 residue into and out of the conduction pathway, common gating switches the dimer between two states, one in which both fast gates are simultaneously latched closed and the other in which they are both unlatched to resume their independent activity. The two fast gates are, therefore, also the final effectors of common gating, being linked through the intramembrane and cytoplasmic dimerization interfaces (Cederholm et al., 2010; Ma et al., 2011). Via this linkage, particular mt_{MC} subunits might adversely affect mt_{MC}-WT heterodimers by acting to bias the voltage dependence of common gating by varying amounts toward that of mt_{MC}-mt_{MC} homodimers and away from that of WT-WT homodimers (Kubish et al., 1998). Depending on the ClC-1 variant, different levels of bias could alter sarcolemmal Cl^- conductance from a minimum of 25% to 50% of normal, thereby explaining the clinical expression of myotonic symptoms from typical MC to the absence of symptoms in RGM carriers.

Severity of symptoms, however, does not necessarily correlate with what may be deduced from the genetic defect or from heterologous expression of ClC-1 constructs engineered to contain the mutation. Quite mild RGM has been associated with the K248X mutant (Gurgel-Gianetti et al., 2012) and in complete contrast, a dominant mutation, G233S (Richman et al., 2012), increases rather than reduces open probability. Its dominant negative effect could result from its location (adjacent to the fast gate, E322) affecting pore to pore communication of common gating information in its heterodimers (Cederholm et al., 2010).

There has been much speculation about the involvement of ClC-1 in muscle fatigue (Allen et al., 2008) and both NAD^+ (Figure 41.4) and adenosine nucleotides are able to bind in the cleft between the CBS1 and CBS2 domains of each subunit making gating especially sensitive to the pH changes predicted to occur during prolonged exercise (Bennetts et al., 2012). Any major influence of ClC-1 function on muscle fatigue, however, remains to be determined because neither unusual endurance nor abnormal fatigability occurs as a noteworthy feature of MC or of RGM (Becker, 1977) where ClC-1 is poorly functional or absent, respectively.

As with other genetically determined diseases, variable clinical expression of symptoms and unpredictable penetrance confound our understanding of dominance in MC and recessivity in RGM. Many factors might contribute to this including posttranscriptional regulation of ClC-1 expression (Chen et al., 1997), gene dosage (Bernard et al., 2008), copy number variation (Raja Rayan et al., 2012), seemingly benign polymorphisms (synonymous SNPs) that modulate splicing (Raheem et al., 2012; Borges et al., 2013), redundant codons (Li et al., 2012), other factors that might skew allelic expression

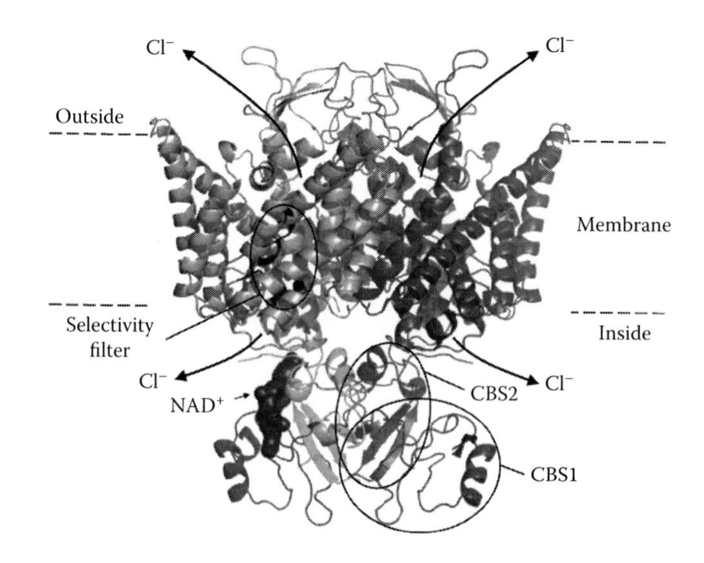

Figure 41.4 A ClC-1 homology model based on the structural coordinates of CmCLC (PDB ID: 3ORG, Feng et al., 2010), depicted in cartoon format with ribbons for both helices and beta sheets. (Image courtesy of Brett Bennetts, Biota Structural Biology Centre, St. Vincent's Institute of Medical Research, Melbourne, Victoria, Australia, 2013, unpublished). Subunits are arranged around the dimerisation interface at the vertical midline. In the left subunit, elements of the selectivity filter are outlined and shown in black, CBS1 and CBS2 are outlined in the right subunit. A docked NAD^+ molecule is depicted in space-fill. For most of the residues linking CBS1 to CBS2 (specifically, A678–V814), there are no coordinates on which to model ClC-1 and so these are excluded from the figure. Arrows indicate the two chloride permeation pathways of the dimer.

(Plassart-Scheiss et al., 1998; Dunø et al., 2004), possible epigenetic influences, and the expression of compensating channels or of channel regulatory proteins. Some of these issues might be resolved by analysis of mRNA from muscle biopsies (Nagamitsu et al., 2000; Mankodi et al., 2002; Raheem et al., 2012; Ursu et al., 2012; Wijnberg et al., 2012; Borges et al., 2013).

Hereditary myotonic diseases in a number of animal species, from the original goat studies (Bryant, 1962), to mice (Heller et al., 1982; Gronemeier et al., 1994), dogs (Bhalerao et al., 2002; Finnigan et al., 2007), the Murrah water buffalo (Borges et al., 2013), and the New Forest pony (Wijnberg et al., 2012), have frequently proven more insightful than human studies. Research on the association between Cl^- conductance and myotonia in the goat explained human MC and RGM, in the ADR mouse it led to the chromosomal localization of human ClC-1 (Steinmeyer et al., 1991b) and in buffalo myotonia it supports other evidence that even intraexonic synonymous SNPs in *CLCN1* seem able to have a potent effect on mRNA splicing (Borges et al., 2013).

41.2.3 MYOTONIC DYSTROPHY 1, STEINERT DISEASE (MIM 160900) AND MYOTONIC DYSTROPHY 2, PROXIMAL MYOTONIC MYOPATHY (PROMM), RICKER SYNDROME (MIM 602668)

Myotonic dystrophies are syndromic diseases with muscle wasting and weakness, myotonia, insulin resistance, cataracts, hypersomnia, frontal balding, testicular atrophy, cardiac conduction abnormalities, and hypogammaglobulinemia. In DM types 1 and 2, the genetic defects are abnormally expanded CTG and CCTG repeat sequences in the *DMPK* and *ZNF9* (*CNBP*) genes, respectively. Despite the different genetic etiology, clinically, DM1 and myotonic dystrophy 2 (DM2) are similar in most features already mentioned. Differences include a severe congenital form that occurs in DM1 and is absent in DM2, strong evidence of anticipation in DM1, a pronounced proximal muscle weakness and wasting (hence, proximal myotonic myopathy, PROMM) that presents initially in many DM2 cases compared to the distal muscular atrophy at onset in DM1 (occurring later in DM2) and gastrointestinal dysfunction that is mainly restricted to DM1 (Cho and Tapscott, 2007; Mankodi, 2008). In these diseases, it is generally the weakness and falling as a result of muscle wasting that causes patients to seek medical advice as the myotonia is considered less of a problem. Age at onset is very variable, some individuals only being diagnosed with mild symptoms late in life (Suominen et al., 2011) but debilitating symptoms tend to worsen progressively. When present, central nervous system deficits, including cognitive slowing and hypersomnia, also progress. In both DM1 and DM2, pre-mRNA containing the respective genetic repeat sequences (CUG and CCUG) binds certain nuclear proteins, such as MBNL1 and CUGBP1, that are essential to the regulation of normal mRNA splicing. Aberrant splicing of various pre-mRNAs, including those of the cardiac troponin and insulin receptor genes, then occurs with a substantially reduced production of normal ClC-1 mRNA (Cho and Tapscott, 2007; Mankodi, 2008). Because of the aberrant pre-mRNA splicing in both DM1 and DM2, expression of functional ClC-1 channels is diminished, as is sarcolemmal Cl^- conductance, with consequent myotonic symptoms (Mankodi et al., 2002). In complementary experiments, enhancement of functional ClC-1 expression by exon skipping (Wheeler et al., 2007) or by adenoviral transfection of WT ClC-1 (Lueck et al., 2010) was found to attenuate myotonia in mouse models of DM. An unexpected involvement of ClC-1 arises from what appears to be an unusual preponderance of DM2 in Finland and Germany. This has been explained by the co-segregation of *CLCN1* mutations that exaggerate the myotonic symptoms and hence bias diagnosis of the condition (Suominen et al., 2008).

41.2.4 EPILEPSY?

A possible association between ClC-1 and epilepsy has very recently been raised by the discovery of a novel nonsense variant in a case of idiopathic epilepsy (Chen et al., 2013) while screening a large idiopathic epilepsy cohort for *CLCN* variations. Reassessment of ClC-1 expression in the brain, because it was not obvious in the original study of mouse brain (Steinmeyer et al., 1991a), now found widespread distribution in human autopsy specimens as well as in mouse brain (Chen et al., 2013). It will be interesting to discover whether this brain ClC-1 could be involved in epilepsy, where, in neurons or glia, it is located and whether or not it utilizes an auxiliary subunit, as does ClC-2 (Jeworutzki et al., 2012).

41.3 CLC-2 (23* VARIATIONS IN THE *CLCN2* GENE—DISEASE ASSOCIATION NOT CONFIRMED)

There have been a number of reports implicating ClC-2 mutations in epilepsy (Combi et al., 2009; Kleefuß-Lie et al., 2009); however, none of the described mutations have definitely been shown to have effects on ClC-2 function that could be causative of epilepsy (Niemeyer et al., 2010). Especially, young ClC-2 knockout (*Clcn2*$^{-/-}$) mice did not show any increased tendency to spontaneous seizures nor were they more susceptible to pro-convulsant agents (Bösl et al., 2001; Blanz et al., 2007); nevertheless, there is evidence of neuronal instability when these mice reach older ages (Cortez et al., 2010). Some authors propose that the ClC-2 channel in neurons constitutes a part of the background conductance regulating input resistance and providing an efflux pathway to prevent excessive Cl^- accumulation during inhibitory synaptic activity (Földy et al., 2010; Rinke et al., 2010). Overall, possible ClC-2 involvement in neuronal excitability remains a contentious issue (Ratté and Prescott, 2011).

By contrast, there has been no disagreement about the pathologies that result from ClC-2 knockout (*Clcn2*$^{-/-}$) in mice. In the absence of ClC-2, these mice developed progressive retinal degeneration and infertility in males due to testicular degeneration (Bösl et al., 2001). It has been proposed that ClC-2 is essential in the close relationships of photoreceptors and spermatocytes with supporting cells of the blood–brain and blood–testis barrier, maintaining their ionic environment. In a later study, older ClC-2 knockout mice were found to develop

* HGMD 2013 (Human Gene Mutation Database).

a leukoencephalopathy with vacuolated myelin in brain and spinal cord white matter tracts (Blanz et al., 2007). As this resembled the subcortical cysts found in some megalencephalic leukoencephalopathies, *CLCN2* was screened in patients where causative gene mutations had not been ascertained (Scheper et al., 2010). While variations in *CLCN2* were found, none of these could be identified as disease causing.

Early observations found ClC-2 to be present in stomach and intestines where it has been proposed to be involved in acid secretion (Sherry et al., 2001) and fluid secretion (Cuppoletti et al., 2004), respectively, with an ongoing disagreement about its exact cellular localization and function (Cátalan et al., 2012). In ClC-2 knockout (*Clcn2*$^{-/-}$) mice, however, there was no significant reduction in gastric acid secretion (Bösl et al., 2001). Recent studies appear to resolve many of the other issues, indicating that ClC-2 is present in the basolateral rather than apical membranes of both small intestine and colon epithelial cells where its major function is in Cl^- reabsorption rather than secretion (Catalán et al., 2012; Jakab et al., 2012). A role for ClC-2 in maintenance of the integrity of intestinal mucosa seems to depend on its ability to modulate trafficking of the protein occludin to tight junctions and not on its Cl^- channel function, which might be incidental (Nighot and Blikslager, 2012).

Lubiprostone, a drug used to relieve constipation, had been proposed to activate ClC-2 directly and thereby increase intestinal Cl^- and fluid secretion (Cuppoletti et al., 2004). There is now substantial evidence that it acts, instead, via E type prostaglandin (EP) receptors to initiate internalization of basolateral ClC-2. This results in reduced Cl^- and fluid reabsorption; however, concurrent modulation of the trafficking and operation of many other membrane transporters also simultaneously increases secretion (Jakab et al., 2012).

41.4 GLIALCAM (HEPACAM) (15* VARIATIONS IN THE *HEPACAM* GENE ASSOCIATED WITH DISEASE)

41.4.1 MEGALENCEPHALIC LEUKOENCEPHALOPATHY WITH SUBCORTICAL CYSTS 2A AND 2B (MLC2A AND MLC2B) (MIM 613925 AND 613926)

Although *CLCN2* variations have not definitely been directly associated with disease, despite earlier clues (Blanz et al., 2007; Scheper et al., 2010), it is indirectly implicated by the discovery of variations in the *GLIALCAM* (*HEPACAM*) gene as a second genetic cause of megalencephalic leukoencephalopathy with subcortical cysts (MLC) (López-Hernández et al., 2011; Jeworutski et al., 2012). Most commonly, MLC is caused by variations in the *MLC1* gene.

These autosomal recessive diseases cause macroencephaly from infancy and diffuse disruption of the cerebral white matter with the characteristic presence of subcortical cysts. Seizures occur from an early age, followed by steady deterioration in motor function with ataxia and spasticity, and mild, later-onset, cognitive decline becoming terminal anywhere from the second to the fifth decade.

Reasons for the common phenotype associated with the two genotypes have now become apparent. Firstly, mutated MLC1 protein disturbs Cl^- transport and cell volume regulation (Ridder et al., 2011). Secondly, GlialCAM normally accompanies both ClC-2 and the MLC1 protein on their way to cell junctions (van der Knaap et al., 2012), especially where astrocyte end-feet contact each other as they surround blood vessels and form the blood–brain barrier, while disease-associated GlialCAM mutations eliminate this targeting. Thirdly, ClC-2, under the influence of MLC1 and GlialCAM, becomes a good candidate for the anion channel involved. It has, therefore, been concluded that GlialCAM must be an obligate auxiliary subunit of ClC-2 in glial cells. In fact, GlialCAM and ClC-2 co-localize in several different types of glial cell, including oligodendrocytes with their critical importance in myelination (Jeworutski et al., 2012). The interdependence of GlialCAM, ClC-2, and MLC1 begs the question of which of these proteins escorts the other(s) to cell junctions. Is it GlialCAM (van der Knaap, 2012), or is it ClC-2, as in intestinal mucosa (Nighot and Blikslager, 2012)?

41.5 CLC-Ka (2† VARIATIONS IN THE *CLCNKA* GENE ASSOCIATED WITH DISEASE)

So far, nothing is known of the effect of a discrete functional absence of human ClC-Ka (*CLCNKA* gene product) because no monogenic disease has been able to be attributed to mutations in it alone (but see Bartter Syndrome 4B).

Absence of ClC-K1 (the rodent *Clcnk1* gene product orthologous to ClC-Ka) in double knockout (*Clcnk1*$^{-/-}$) mice causes renal diabetes insipidus through failure of Cl^- transport in the thin ascending limb of the loop of Henle (Matsumura et al., 1999). This salt transport defect results in a decreased ability to generate and maintain a cortico-medullary interstitial fluid osmotic gradient and a consequent failure of kidney concentrating ability in the distal nephron. Possibly, in humans, ClC-Kb can compensate for the absence of functional ClC-Ka.

41.6 CLC-Kb (74‡ VARIATIONS IN THE *CLCNKB* GENE ASSOCIATED WITH DISEASE)

41.6.1 BARTTER SYNDROME TYPE 3, CLASSICAL BARTTER SYNDROME (MIM 607364)

As originally described, Bartter syndrome includes hypochloremia, hypokalemia, and alkalosis in the presence of normal, or even low, blood pressure, despite hyperaldosteronemia and hyperreninemia (Bartter et al., 1962). Clinically, the typical features of the disease are failure to thrive in infancy and childhood, vomiting, polyuria, and polydipsia. Cardiac, muscular, and nervous system consequences of the electrolyte imbalances include dysrhythmias, tetany, and convulsions. The common features of Bartter syndrome have been linked to functional deficiencies in any of several different proteins although it is homozygous or compound

* HGMD 2013 (Human Gene Mutation Database).

† HGMD 2013 (Human Gene Mutation Database).

‡ HGMD 2013 (Human Gene Mutation Database).

heterozygous mutations in the Cl^- channel, ClC-Kb, that are associated with the classical Bartter type 3 symptoms consequent upon mild salt wasting (Fremont and Chan, 2012).

Dependent upon the specific *CLCNKB* variation, symptoms that differ from those of classical Bartter type 3 sometimes occur, although strict genotype–phenotype correlations have not been found. For example, the classical symptoms have, in some cases, been observed to be coincident with hypocalciuria (Fukuyama et al., 2004) and, in others, antenatal symptoms of fetal polyuria, maternal polyhydramnios, fetal growth limitation, and premature birth have been present (Nozu et al., 2010; Lee et al., 2012). A number of examples of Bartter type 3 symptoms coincident with hypocalciuria have been reported (Fukuyama et al., 2004). One *CLCNKB* variation that produces gain of function in its ClC-Kb product (T481S) has been described and found to be associated with predisposition to a salt-sensitive hypertension (Jeck et al., 2004), although this relationship was not confirmed by later studies (Kokubo et al., 2005; Speirs et al., 2005).

Localized in the basolateral membrane of the epithelial cells of the thick ascending limb of the loop of Henle, the distal convoluted tubule and the collecting duct (Peters et al., 2002; Jeck and Seyberth, 2011; Eladari et al., 2012), ClC-Kb normally operates in conjunction with apical Cl^- transporters to facilitate salt reabsorption from the luminal fluid. Compromised function in this distal part of the nephron then results in salt loss. It has been suggested that ClC-Kb absence or malfunction might secondarily (via intracellular Cl^- accumulation) affect the normal operation of other transporters in the distal nephron such as the apically located, thiazide-sensitive, Na^+-Cl^- co-transporter, NCCT (Fukuyama et al., 2004). Activity of this co-transporter is known to influence urinary Na^+ and Ca^{2+} excretion in a reciprocal fashion in Gitelman syndrome and in long-term treatment with thiazide diuretics. In both cases, Na^+ is lost and Ca^{2+} retained, which might explain those instances of salt loss with Ca^{2+} retention (hypocalciuria) when the defect is in ClC-Kb.

41.6.2 GITELMAN SYNDROME (MIM #263800)

Typically, Gitelman syndrome is milder and of later onset than Bartter syndrome, having its initial presentation occur in late childhood or early adulthood. Metabolic alkalosis and hypokalemia, as in Bartter syndrome, are here accompanied by hypomagnesemia and abnormally low urinary Ca^{2+} excretion. Its clinical manifestations include tetany and other neuromuscular symptoms including muscle weakness, cramps, dizziness, and fatigue (Fremont and Chan, 2012) exacerbated by bouts of gastroenteritis when additional K^+ and Mg^{2+} are lost. Gitelman syndrome is usually caused by mutations in the thiazide-sensitive, Na^+-Cl^- co-transporter, NCCT; however, some cases of mixed Bartter–Gitelman phenotype are associated with *CLCNKB* variations (Lee et al., 2012).

41.6.3 BARTTER SYNDROME 4B, INFANTILE BARTTER SYNDROME WITH SENSORINEURAL DEAFNESS (MIM 613090)

On those extremely rare occasions where there is combined loss of functional ClC-Ka and ClC-Kb caused by variations in all four alleles of the *CLCNKA* and *CLCNKB* genes, there is, typically, extremely severe salt wasting and congenital sensorineural deafness. Antenatal complications of fetal polyuria and consequent maternal polyhydramnios and premature birth are followed postnatally by excessive excretion of prostaglandins in the urine, salt wasting, vomiting, volume loss, and severe growth retardation. One case in a consanguineous family (Haas et al., 2003; Schlingmann et al., 2004) and one isolated case (Nozu et al., 2008) have been described.

Since monogenic variation *CLCNKB* does not cause deafness and deafness due to monogenic variation in *CLCNKA* is unknown, it is likely that loss of function of one of either ClC-Ka or ClC-Kb is compensated by the presence of the other in the stria vascularis of the cochlea (Estévez et al., 2001; Krämer et al., 2008). Secretion of K^+ into the endolymph of the scala media by the marginal cells of the stria vascularis maintains the hair cells of the organ of Corti in the high K^+ environment that is essential for cochlear function (Wangeman, 2006). Secretion of K^+ fails if the marginal cells cannot simultaneously recycle Cl^- through ClC-Ka and ClC-Kb channels into surrounding intrastrial fluid spaces. A reduction in endolymph K^+ then adversely affects entry into the hair cells of the K^+ and Ca^{2+} necessary for mechano-electric transduction (Jagger et al., 2010; Kazmierczak and Müller, 2012).

41.7 BARTTIN (17* VARIATIONS IN THE *BSND* GENE ASSOCIATED WITH DISEASE)

41.7.1 BARTTER SYNDROME 4A (MIM 602522)

Bartter syndrome 4A is a serious perinatal disease with maternal polyhydramnios, premature birth, postnatal life-threatening salt wasting and volume loss, renal failure during infancy, and sensorineural deafness. Although its symptoms typically recapitulate those of Bartter syndrome 4B, in this case the disease results from certain variations in the *BSND* gene and its protein product, barttin, an essential auxiliary subunit of the ClC-K channels (Brennan et al., 1998; Birkenhäger et al., 2001).

Mutated versions of barttin modulate function and trafficking of human and rodent ClC-K channels (Scholl et al., 2006; Nomura et al., 2011), therefore, having the potential to compromise function mediated by ClC-Ka and ClC-Kb in both the distal nephron and cochlea resulting in the typical severity of the disease symptoms.

41.7.2 RARELY, DEAFNESS OCCURS WITH ONLY MILD RENAL DYSFUNCTION (RIAZUDDIN ET AL., 2009)

This non-syndromic barttin-associated disease results from certain homozygous and compound heterozygous mutation combinations. Barttin-ClC-Ka or barttin-ClC-Kb presence in the surface membrane of modèl cells does not appear to be affected but the distribution between the endoplasmic reticulum and surface is disturbed. It has been proposed, therefore, that the defect is in the final stages of trafficking of the barttin-bound ClC-K channels.

* HGMD 2013 (Human Gene Mutation Database).

41.8 CLC-3 (*NO *CLCN3* VARIATIONS HAVE BEEN ASSOCIATED WITH ANY HUMAN DISEASE)

41.8.1 NEURONAL CEROID LIPOFUSCINOSIS? CARDIOVASCULAR DISEASE? GLIOMA?

Expression of ClC-3 has been demonstrated in many rodent and human tissues, especially kidney and brain, but also in skeletal muscle, heart, vasculature, adrenals, liver, and pancreas (Kawasaki et al., 1994; Borsani et al., 1995), suggesting that ClC-3 mutations could result in many different diseases. Subsequently, double knockout (*Clcn3*$^{-/-}$) mice were found to be severely affected with impaired growth, reduced lifespan, motor abnormalities, degeneration of the ileal mucosa, and, especially, retinal and hippocampal neurodegeneration (Stobrawa et al., 2001, Yoshikawa et al., 2002). Their central nervous degeneration was likened to that of human neuronal ceroid lipofuscinosis (NCL) (Yoshikawa et al., 2002) where lipofuscin, a yellow brown, autofluorescent, granular pigment representing the lipid containing residues of lysosomal digestion, is abnormally accumulated by neurons. In NCL, this is accompanied by retinal, cerebral, and cerebellar neurodegeneration manifested as a progressive decline in motor and cognitive function, deteriorating vision, ataxia, myoclonus, and seizures eventually leading to a vegetative state (Goebel, 1995; Mole et al., 2005). About 90% of NCL cases have been linked to variations in twelve different genes (Santorelli et al., 2013) but none to ClC-3.

In the hippocampus, it is becoming well established that ClC-3 is involved in inhibitory synaptic transmission as it is co-localized with the vesicular GABA transporter in inhibitory synaptic vesicles (Riazanski et al., 2012) where it appears to be a co-requisite of GABA loading and hence of effective inhibitory transmission. While this, the hippocampal degeneration in the *Clcn3*$^{-/-}$ mouse and in epilepsy of temporal lobe origin, might suggest a possible link, nothing definite has been reported despite extensive screening (Chen et al., 2013).

Meanwhile, debate continues about whether ClC-3 can be identified with the volume regulated Cl^- channel (VRCC) and the associated current, I_{VRCC}, of the plasma membrane that has been found in many tissues and in heterologous systems expressing ClC-3. Strong evidence suggests that ClC-3 does not usually act as a Cl^- channel of the plasma membrane but is, rather, a Cl^-/H^+ antiporter because of its close homology with the known CLC antiporters, particularly at E281 (Jentsch, 2008; Lísal and Maduke, 2009) and its amino terminal sorting motif that directs it to endosomal-lysosomal and synaptic vesicles (Stauber and Jentsch, 2010). Nevertheless, there have been many proposals for the involvement of ClC-3, as a cell volume regulator, firstly, in maintenance and disease of the cardiovascular and cerebrovascular systems (Duan, 2010, 2011) and, secondly, in cell proliferation and migration, especially in cancers such as gliomas (Cuddapah and Sontheimer 2010). Although chlorotoxin, a blocker of small Cl^- channels, inhibits glioma cell motility and invasiveness, its mechanism of action through direct block of ClC-3 is disputed. It seems to act, instead, by binding to surface metalloproteinase 2 to initiate internalization of a complex group of surface proteins, including ClC-3, metalloproteinases, annexin A2, and possibly other proteins (Mcferrin and Sontheimer, 2006; Kesavan et al., 2010; Kasai et al., 2012). This could inhibit not only volume regulation by ClC-3 but also dissolution of the extracellular matrix by surface proteinases, as well as numerous motility and other cellular operations, all of which might normally be facilitators of tissue invasion.

* HGMD 2013 (Human Gene Mutation Database).

41.9 CLC-4 (†NO *CLCN4* VARIATIONS HAVE BEEN ASSOCIATED WITH ANY HUMAN DISEASE)

As for ClC-3, there is widespread expression of ClC-4 in mice; however, no associated morbidity or pathology has been reported in double knockout (*Clcn4*$^{-/-}$) mice (Jentsch, 2008; Mohammad-Pannah et al., 2009) and there has been nothing to implicate *CLCN4* variations in human disease. Its close identity with the ClC-3 and ClC-5 Cl^-/H^+ antiporters, however, suggests that an involvement should not be discounted.

41.10 CLC-5 (140‡ VARIATIONS IN THE *CLCN5* GENE HAVE BEEN ASSOCIATED WITH DISEASE)

41.10.1 DENT DISEASE 1 (MIM 300009) AND RELATED ALLELIC DISORDERS

Dent disease 1 is a hereditary x-linked condition, associated with variations in the *CLCN5* gene, typically characterized by the loss of low molecular weight (LMW) proteins, such as albumin, microglobulins, and vitamin- and hormone-binding proteins, and of excessive amounts of Ca^{2+} and amino acids in the urine (Devuyst and Thakker, 2010). *Dent (Japan) disease* has been reported to be a less severe variant (Lloyd et al., 1997), although this might simply reflect a younger age at ascertainment (Igarashi et al., 2000). Dent disease 2 has similar phenotypical features to Dent disease 1 but is genetically distinct involving variations in the *OCRL* gene (Devuyst and Thakker, 2010).

Accompanying the other aberrant processing of urinary filtrate, phosphate is not recovered resulting in lower than normal plasma phosphate levels and a tendency for Ca^{2+} salts to be precipitated calcifying the kidney tissue (nephrocalcinosis) and forming kidney stones (nephrolithiasis). This combination of kidney pathophysiologies can result in bone softening (rickets in children, osteomalacia in adults) and progressive renal failure. In around 50% of affected males, end-stage renal failure is reached between 20 and 50 years of age, some features of the disease having already begun during childhood. Carrier females can display aspects of the disease that range up to severe (Devuyst and Thakker, 2010),

† HGMD 2013 (Human Gene Mutation Database).
‡ HGMD 2013 (Human Gene Mutation Database).

perhaps depending on the random skewing of x-chromosome inactivation (but see also discussion of a possible effect on common gating given later).

A Bartter-like disease with hypokalemic metabolic alkalosis associated with hyperreninemic hyperaldosteronism has occasionally been attributed to variations in *CLCN5* (Bogdanović et al., 2010). It was suggested that the coexistence of LMW proteinuria with features of a Bartter-like syndrome might predict a *CLCN5* variation.

Endocytotic reuptake of LMW proteins from the glomerular filtrate is essential in the normal function of the kidney's proximal convoluted tubule (PCT). Mutated versions of ClC-5 have been classified functionally into those having an innate low electrical activity, those causing retention of the immature protein in the endoplasmic reticulum, and those that reduce stability of the mature protein. These mutations cluster at the dimer interface (Lourdel et al., 2012) and like in ClC-1 and ClC-7 (although there should be no dominant negative effect in this x-linked condition), some equivalent of common gating in ClC-5 could be disrupted.

While there is some correlation between the specific *CLCN5* variation and the cellular functional defect, there is no known equivalent genotype–phenotype relationship (Smith et al., 2009). Exactly how ClC-5 is normally utilized in the protein receptor complex of the apical membrane of PCT cells (Hryciw et al., 2012), in endocytosis and the endosome acidification process (Lippiat and Smith, 2012), remains controversial. It is certain, however, that beyond the requirement for the structural presence of ClC-5, it is not simply its Cl^- transporting ability but rather its Cl^-/H^+ exchange that is critical (Lippiat and Smith, 2012). There is little current understanding of the relationship between ClC-5 malfunction and nephrocalcinosis and nephrolithiasis (Devuyst and Thakker, 2010).

41.11 CLC-6 (*NO *CLCN6* VARIATIONS HAVE BEEN ASSOCIATED WITH ANY HUMAN DISEASE)

41.11.1 NEURONAL CEROID LIPOFUSCINOSIS? CARDIAC DISEASE?

During development, ClC-6 double knockout (*Clcn6*$^{-/-}$) mice showed increasing neuronal autofluorescence (Poët et al., 2006), eventually indicating the presence of lipofuscin throughout the brain, as described earlier under ClC-3. Motor deficits and brain and retinal degeneration as found in human NCL and in ClC-3 double knockout (*Clcn3*$^{-/-}$) mice were, however, absent. Variations in *CLCN6* were, nevertheless, sought in human cases of NCL and were found but could not be proven to be causative (Poët et al., 2006). An association of *CLCN6* variants with a marker of cardiac dysfunction, high blood serum levels of the N-terminal cleavage product of the B-type natriuretic peptide (NT-proBNP), has also been noted (Del Greco et al., 2011). Proposed hypothetical mechanisms by which this might arise remain untested.

* HGMD 2013 (Human Gene Mutation Database).

41.12 CLC-7 (65† VARIATIONS IN THE *CLCN7* GENE ASSOCIATED WITH DISEASE)

41.12.1 ALBERS-SCHÖNBERG DISEASE, AD OSTEOPETROSIS 2 (MIM 166600); INFANTILE MALIGNANT OSTEOPETROSIS 2, AR OSTEOPETROSIS 4 (MIM 611490)

Osteopetrosis, literally, "petrification, or, conversion-to-stone, of the bones," occurs in mild, intermediate, and severe forms most frequently as autosomal dominant osteopetrosis (ADO), less commonly as autosomal recessive osteopetrosis (ARO), and, much more rarely, with an x-linked inheritance pattern. Symptoms of osteopetrosis include a tendency to bone fragility, bone marrow failure with pan-hemocyopenia and increased infection frequency, growth impairment, deafness, and blindness (Stark and Savarirayan, 2009). Neuropathologies due to nerve compression or cranial confinement and hematological abnormalities due to bone marrow loss result from the abnormal density and encroachment of bone in the cranium and long bones. Mutations in *CLCN7* (Souraty et al., 2007; Pangrazio et al., 2009) are responsible for the majority of the ADO cases and many of the ARO cases. A number of other genes have also been associated with osteopetrosis and will not be discussed here (Stark and Savarirayan, 2009).

Severe or malignant infantile ARO is almost always accompanied by cranial nerve palsy with blindness, deafness, and facial paralysis, as well as neurodegeneration (Pangrazio et al., 2009; Al-Aama et al., 2012) but unlike the situation in the myotonias, few of these cases result from predicted ClC-7 truncations. But again, like the Cl^- channel myotonias, heterozygous parents of children with ARO are typically free of symptoms. Some of the less severe ARO and more severe ADO cases have been described as *intermediate* (Pangrazio et al., 2009), but only rarely has the ARO condition been reported to be compatible with life up to the third decade (Kantaputra et al., 2012). Severe ARO mimics that seen in double ClC-7 knockout (*Clcn7*$^{-/-}$) mice (Kornak et al., 2001) including the severe primary neurodegeneration that resembles NCL (described under ClC-3), the neurological deficits and the lethality.

By contrast, ADO (classical Albers-Schönberg Disease) has variable symptomatology ranging from relatively mild to occasionally quite severe but is usually compatible with a normal life expectancy (Stark and Savarirayan, 2009). From genotype–phenotype correlations and by analogy with the role of *CLCN1* variations in dominant and recessive myotonias, it seems likely that ADO due to a *CLCN7* variation is not a result of haploinsufficiency (Pangrazio et al., 2009).

Early treatment with hematopoietic stem cell transplantation has been found to ameliorate the osteopetrosis and anemia in infantile ARO (Dreissen et al., 2003; Pangrazio et al., 2009) but is ineffective in those rare instances associated with *CLCN7* where there is primary neurodegeneration rather than neurological deficit due to nerve compression (Pangrazio et al., 2009).

† HGMD 2013 (Human Gene Mutation Database).

Again, this recapitulates observations in ClC-7 knockout (*Clcn7*$^{-/-}$) mice where deficient ClC-7 function in osteoclasts has been ameliorated but neurodegeneration remains (Kasper et al., 2005; Rajan et al., 2011).

Osteopetrosis occurs as a result of the failure of basic bone homeostasis. Normally a dynamic balance is achieved by osteoblasts that synthesize the proteins, polysaccharides, and enzymes necessary for the deposition of organic and mineral bone matrix and osteoclasts that secrete HCl and a mixture of proteases for bone resorption (matrix breakdown). When secretion of HCl fails at the osteoclast resorption lacuna, due to malfunction of the ruffled membrane proton pump or a fault in the parallel Cl^- transporting pathway through ClC-7, bone resorption is compromised, resulting in the increased bone density of osteopetrosis.

Recently developed expression systems that allow the study of ClC-7 at the plasma membrane (Leisle et al., 2011; Costa et al., 2012) have proven ClC-7 to be a Cl^-/H^+ antiporter and promise the possibility of determining whether ADO is due to a dominant negative effect equivalent to that outlined for the ClC-1 channel earlier (e.g., Ma et al., 2011). It is already apparent that Cl^-/H^+ exchange in ClC-7 is voltage gated (Leisle et al., 2011) and the dimerization interfaces of its membrane resident and cytoplasmic CBS regions contain a high concentration of ADO mutations (Pangzrazio et al., 2009) as in MC (Duffield et al., 2003).

41.13 OSTM1 (4* VARIATIONS IN THE *OSTM1* GENE ASSOCIATED WITH DISEASE)

41.13.1 INFANTILE MALIGNANT OSTEOPETROSIS 3, AR OSTEOPETROSIS 5 (MIM 259720)

First observed as the *grey lethal* mouse mutant, variations in the human *OSTM1* gene disrupt the expression, localization or function of its product, Ostm1, an essential auxiliary subunit of ClC-7 in mammalian cells (Lange et al., 2006). In turn, localization or function of ClC-7 would be disrupted (Leisle et al., 2011). This explains why Ostm1 mutations result in clinical manifestations equivalent to those of malignant infantile ARO due to *CLCN7* variations (Souraty et al., 2007), including the NCL-like neurodegeneration and neurological deficits.

41.14 CONCLUSIONS

Close to 500 variations in five different CLC proteins and approaching 40 in three different CLC auxiliary subunits (HGMD 2013)† have been associated with some 20 distinct diseases, as described in this review. While great advances have been made in understanding CLC-related proteins in disease, in some cases, especially those involving the CLC transporters, the mechanisms by which their functional failure causes disease, especially neurodegeneration, in humans and in mouse models remain unknown.

Many more variations exist as synonymous SNPs encoding apparently benign genetic polymorphisms (see, e.g., LOVD, 2013, for *CLCN1*)‡, although as discussed earlier, these could be responsible for subtle effects on individual allele expression or on splicing or have profound effects in heterozygous combination with classical mutants.

Considering the ubiquity of CLC channels/transporters and their apparently essential functions, it is astonishing that most of the diseases associated with their genetic variations are relatively rare and often quite mild.

* HGMD 2013 (Human Gene Mutation Database).

† For the most recent information on these variations, please consult the latest HGMD database.

‡ For the most recent information on these variations, please consult the latest LOVD database.

REFERENCES

Adrian, R. H. and S. H. Bryant. 1974. On the repetitive discharge in myotonic muscle fibres. *J. Physiol.* 240:505–515.

Al-Aama, J. Y., A. A. Dabbagh, and A. Y. Edrees. 2012. A newly described mutation of the CLCN7 gene causes neuropathic autosomal recessive osteopetrosis in an Arab family. *Clin. Dysmorphol.* 21:1–7.

Allen, D. G., G. D. Lamb, and H. Westerblad. 2008. Skeletal muscle fatigue: Cellular mechanisms. *Physiol. Rev.* 88:287–332.

Bartter, F. C., P. Pronove, J. R. Gill, Jr., and R. C. MacCardle. 1962. Hyperplasia of the juxtaglomerular complex with hyperaldosteronism and hypokalemic alkalosis. A new syndrome. *Am. J. Med.* 33:811–828.

Becker, P. E. 1977. Myotonia congenita and syndromes associated with myotonia: Clinical-genetic studies of the nondystrophic myotonias. In *Topics in Human Genetics*, eds. P. E. Becker, W. Lenz, F. Vogel, and G. G. Wendt. Vol. 3. Stuttgart, Germany: Georg Thieme.

Bennetts, B., Y. Yu, T.-Y. Chen, and M. W. Parker. 2012. Intracellular β-nicotinamide adenine dinucleotide inhibits the skeletal muscle ClC-1 chloride channel. *J. Biol. Chem.* 287:25808–25820.

Bernard, G., C. Poulin, J. Puymirat, D. Sternberg, and M. Shevell. 2008. Dosage effect of a dominant CLCN1 mutation: A novel syndrome. *J. Child Neurol.* 23:163–166.

Bhalerao, D. P., Y. Rajpurohit, C. H. Vite, and U. Giger. 2002. Detection of a genetic mutation for myotonia congenita among Miniature Schnauzers and identification of a common carrier ancestor. *Am. J. Vet. Res.* 63:1443–1447.

Birkenhäger, R., E. Otto, M. J. Schürmann et al. 2001. Mutation of *BSND* causes Bartter syndrome with sensorineural deafness and kidney failure. *Nat. Genet.* 29:310–314.

Blanz, J., M. Schweizer, M. Auberson et al. 2007. Leukoencephalopathy upon disruption of the chloride channel ClC-2. *J. Neurosci.* 27:6581–6589.

Bogdanović, R., M. Draaken, A. Toromanović, M. Đorđević, N. Stajić, and M. Ludwig. 2010. A novel *CLCN5* mutation in a boy with Bartter-like syndrome and partial growth hormone deficiency. *Pediatr. Nephrol.* 25:2363–2368.

Borges, A. S., J. D. Barbosa, L. A. Resende et al. 2013. Clinical and molecular study of a new form of hereditary myotonia in Murrah water buffalo. *Neuromuscul. Disord.* 23:206–213.

Borsani, G., E. I. Rugarli, M. Taglialatela, C. Wong, and A. Ballabio. 1995. Characterization of a human and murine gene (CLCN3) sharing similarities to voltage-gated chloride channels and to a yeast integral membrane protein. *Genomics* 27:131–141.

Bösl, M. R., V. Stein, C. Hübner et al. 2001. Male germ cells and photoreceptors, both dependent on close cell-cell interactions, degenerate upon ClC-2 Cl^- channel disruption. *EMBO J.* 20:1289–1299.

Brennan, T. M. H., D. Landau, H. Shalev et al. 1998. Linkage of infantile Bartter syndrome with sensorineural deafness to chromosome 1p. *Am. J. Hum. Genet.* 62:355–361.

Bretag, A. H. 1969. Synthetic interstitial fluid for isolated mammalian tissue. *Life Sci.* 8:319–329.

Bretag, A. H. 1983. Antimyotonic agents and myotonia. *Proc. Aus. Physiol. Pharmacol. Soc.* 14:170–191.

Bretag, A. H. 1987. Muscle chloride channels. *Physiol. Rev.* 67:618–724.

Bryant, S. H. 1962. Muscle membrane of normal and myotonic goats in normal and low external chloride. *Fed. Proc.* 21:312.

Burgunder, J. M., S. Huifang, P. Beguin et al. 2008. Novel chloride channel mutations leading to mild myotonia among Chinese. *Neuromuscul. Disord.* 18:633–640.

Catalán, M. A., C. A. Flores, M. González-Begne, Y. Zhang, F. V. Sepúlveda, and J. E. Melvin. 2012. Severe defects in absorptive ion transport in distal colons of mice that lack ClC-2 channels. *Gastroenterology* 142:346–354.

Cederholm, J. M., G. Y. Rychkov, C. J. Bagley, and A. H. Bretag. 2010. Inter-subunit communication and fast gate integrity are important for common gating in hClC-1. *Int. J. Biochem. Cell Biol.* 42:1182–1188.

Chen, M.-f., R. Niggeweg, P. A. Iaizzo, F. Lehmann-Horn, and H. Jockusch. 1997. Chloride conductance in mouse muscle is subject to post-transcriptional compensation of the functional Cl^- channel 1 gene dosage. *J. Physiol.* 504:75–81.

Chen, T. T., T. L. Klassen, A. M. Goldman, C. Marini, R. Guerrini, and J. L. Noebels. 2013. Novel brain expression of ClC-1 chloride channels and enrichment of *CLCN1* variants in epilepsy. *Neurology* 80:1078–1085.

Cho, D. H. and S. J. Tapscott. 2007. Myotonic dystrophy: Emerging mechanisms for DM1 and DM2. *Biochim. Biophys. Acta* 1772:195–204.

Colding-Jørgensen, E. 2005. Phenotypic variability in myotonia congenita. *Muscle Nerve* 32:19–34.

Combi, R., D. Grioni, M. Contri et al. 2009. Clinical and genetic familial study of a large cohort of Italian children with idiopathic epilepsy. *Brain. Res. Bull.* 79:89–96.

Cortez, M. A., C. Li, S. N. Whitehead et al. 2010. Disruption of ClC-2 expression is associated with progressive neurodegeneration in aging mice. *Neuroscience* 167:154–162.

Costa, A., P. V. K. Gutla, A. Boccaccio et al. 2012. The *Arabidopsis* central vacuole as an expression system for intracellular transporters: functional characterization of the Cl^-/H^+ exchanger CLC-7. *J. Physiol.* 590:3421–3430.

Cuddapah, V. A. and H. Sontheimer. 2010. Molecular interaction and functional regulation of ClC-3 by Ca^{2+}/calmodulin-dependent protein kinase II (CaMKII) in human malignant glioma. *J. Biol. Chem.* 285:11188–11196.

Cuppoletti, J., D. H. Malinowska, K. P. Tewari et al. 2004. SPI-0211 activates T84 cell chloride transport and recombinant human ClC-2 chloride currents. *Am. J. Physiol. Cell. Physiol.* 287:C1173–C1183.

Del Greco, M. F., C. Pattaro, A. Leuchner et al. 2011. Genome-wide association analysis and fine mapping of NT-proBNP level provide novel insight into the role of the *MTHFR-CLCN6-NPPA-NPPB* gene cluster. *Hum. Mol. Genet.* 20:1660–1671.

Devuyst, O. and R. V. Thakker. 2010. Dent's disease. *Orphanet J. Rare Dis.* 5:28.

Dreissen, G. J., E. J. Gerritsen, A. Fischer et al. 2003. Long-term outcome of haematopoietic stem cell transplantation in autosomal recessive osteopetrosis: an EBMT report. *Bone Marrow Transplant.* 32:657–663.

Duan, D. D. 2010. Volume matters: Novel roles of the volume-regulated ClC-3 channels in hypertension-induced cerebrovascular remodeling. *Hypertension.* 56:346–348.

Duan, D. D. 2011. The ClC-3 chloride channels in cardiovascular disease. *Acta Pharmacol. Sin.* 32:675–684.

Duffield, M., G. Rychkov, A. Bretag, and M. Roberts. 2003. Involvement of helices at the dimer interface in ClC-1 common gating. *J. Gen. Physiol.* 121:149–161.

Dunø, M., E. Colding-Jørgensen, M. Grunnet, T. Jespersen, J. Vissing, and M. Schwartz. 2004. Difference in allelic expression of the *CLCN1* gene and the possible influence on the myotonia congenita phenotype. *Eur. J. Hum. Genet.* 12:738–743.

Dupré, N., N. Chrestian, J.-P. Bouchard et al. 2009. Clinical, electrophysiologic, and genetic study of non-dystrophic myotonia in French-Canadians. *Neuromuscul. Disord.* 19:330–334.

Eladari, D., R. Chambrey, and J. Peti-Peterdi. 2012. A new look at electrolyte transport in the distal tubule. *Annu. Rev. Physiol.* 74:325–349.

Estévez, R., T. Boettger, V. Stein et al. 2001. Barttin is a Cl^- channel β-subunit crucial for renal Cl^- reabsorption and inner ear secretion. *Nature* 414:558–561.

Estévez, R., M. Pusch, C. Ferrer-Costa, M. Orozco, and T. J. Jentsch. 2004. Functional and structural conservation of CBS domains from CLC chloride channels. *J. Physiol.* 557:363–378.

Feng, L., E. B. Campbell, Y. Hsiung, and R. MacKinnon. 2010. Structure of a eukaryotic CLC transporter defines an intermediate state in the transport cycle. *Science* 330:635–641.

Fialho, D., S. Schorge, U. Pucovska et al. 2007. Chloride channel myotonia: exon 8 hot-spot for dominant-negative interactions. *Brain* 130:3265–3274.

Finnigan, D. F., W. J. B. Hanna, R. Poma, and A. J. Bendall. 2007. A novel mutation of the *CLCN1* gene associated with myotonia hereditaria in an Australian cattle dog. *J. Vet. Intern. Med.* 21:458–463.

Földy, C., S.-H. Lee, R. J. Morgan, and I. Soltesz. 2010. Regulation of fast-spiking basket cell synapses by the chloride channel ClC-2. *Nat. Neurosci.* 13:1047–1049.

Fremont, O. T. and J. C. M. Chan. 2012. Understanding Bartter syndrome and Gitelman syndrome. *World J. Pediatr.* 8:25–30.

Fukuyama, S., M. Hiramatsu, M. Akagi, M. Higa, and T. Ohta. 2004. Novel mutations of the chloride channel Kb gene in two Japanese patients clinically diagnosed as Bartter syndrome with hypocalciuria. *J. Clin. Endocrin. Metab.* 89:5847–5850.

Gao, F., F. C. Ma, Z. F. Yuan et al. 2010. Novel chloride channel gene mutations in two unrelated Chinese families with myotonia congenita. *Neurol. India* 58:743–746.

George, Jr., A. L., M. A. Crackower, J. A. Abdalla et al. 1993. Molecular basis of Thomsen's disease (autosomal dominant myotonia congenita). *Nat. Genet.* 3:305–310.

George, Jr., A. L., K. Sloan-Brown, G. M. Fenichel et al. 1994. Nonsense and missense mutations of the muscle chloride channel gene in patients with myotonia congenita. *Hum. Mol. Genet.* 3:2071–2072.

Goebel, H. H. 1995. The neuronal ceroid-lipofuscinoses. *J. Child Neurol.* 10:424–437.

Gronemeier, M., A. Condie, J. Prosser, K. Steinmeyer, T. J. Jentsch, and H. Jockusch. 1994. Nonsense and missense mutations in the muscular chloride channel gene *Clc-1* of myotonic mice. *J. Biol. Chem.* 269:5963–5967.

Gurgel-Gianetti, J., A. S, Senkevics, D. Zilbersztajn-Gotlieb et al. 2012. Thomsen or Becker myotonia? A novel autosomal recessive nonsense mutation in the *CLCN1* gene associated with a mild phenotype. *Muscle Nerve* 45:279–283.

Haas, N. A., R. Nossal, C. H. Schneider et al. 2003. Successful management of an extreme example of neonatal hyperprostaglandin-E syndrome (Bartter's syndrome) with the new cyclooxygenase-2 inhibitor rofecoxib. *Pediatr. Crit. Care Med.* 4:249–251.

Hebeisen, S., A. Biela, B. Giese, G. Müller-Newen, P. Hidalgo, and C. Fahlke. 2004. The role of the carboxyl terminus in ClC chloride channel function. *J. Biol. Chem.* 279:13140–13147.

Heller, A. H., E. A. Eicher, M. Hallett, and R. L. Sidman. 1982. Myotonia, a new inherited muscle disease in mice. *J. Neurosci.* 2:924–933.

HGMD. 2013. Human gene mutation database. Accessed March 20, 2013. http://www.hgmd.cf.ac.uk

Hilbert, P., S. Frank, O. Raheem et al. 2011. Normal muscle MRI does not preclude increased connective tissue in muscle of recessive myotonia congenita. *Acta Neurol. Scand.* 124:146–147.

Hryciw, D. H., K. A. Jenkin, A. C. Simcocks, E. Grinfeld, A. J. McAinch, and P. Poronnik. 2012. The interaction between megalin and ClC-5 is scaffolded by the Na^+-H^+ exchanger regulatory factor 2 (NHERF2) in proximal tubule cells. *Int. J. Biochem. Cell Biol.* 44:815–823.

Igarashi, T., J. Inatomi, T. Ohara, T. Kuwahara, M. Shimadzu, and R. V. Thakker. 2000. Clinical and genetic studies of *CLCN5* mutations in Japanese families with Dent's disease. *Kidney Int.* 58:520–527.

Ivanova, E. A., E. L. Dadali, V. P. Fedotov et al. 2012. The spectrum of *CLCN1* gene mutations in patients with nondystrophic Thomsen's and Becker's myotonias. *Russ. J. Genet.* 48:952–961.

Jagger, D. J., G. Nevill, and A. Forge. 2010. The membrane properties of cochlear root cells are consistent with the roles in potassium recirculation and spatial buffering. *J. Assoc. Res. Otolaryngol.* 11:435–448.

Jakab, R. L., A. M. Collaco, and N. A. Ameen. 2012. Lubiprostone targets prostanoid signaling and promotes ion transporter trafficking, mucus exocytosis and contractility. *Dig. Dis. Sci.* 57:2826–2845.

Jeck, N. and H. W. Seyberth. 2011. Loop disorders: insights derived from defined genotypes. *Nephron. Physiol.* 118:7–14.

Jeck, N., S. Waldegger, A. Lampert et al. 2004. Activating mutation of the renal epithelial chloride channel ClC-Kb predisposing to hypertension. *Hypertension* 43:1175–1181.

Jentsch, T. J. 2008. CLC chloride channels and transporters: From genes to protein structure, pathology and physiology. *Crit. Rev. Biochem. Mol. Biol.* 43:3–36.

Jentsch, T. J., K. Steinmeyer, and G. Schwarz. 1990. Primary structure of *Torpedo marmorata* chloride channel isolated by expression cloning in *Xenopus* oocytes. *Nature* 348:510–514.

Jeworutzki, E., T. López-Hernández, X. Capdevila-Nortes et al. 2012. GlialCAM, a protein defective in a leukodystrophy serves as a ClC-2 Cl^- channel auxiliary subunit. *Neuron* 73:951–961.

Kantaputra, P. N., S. Thawanaphong, W. Issarangporn et al. 2012. Long-term survival in infantile malignant autosomal recessive osteopetrosis secondary to homozygous p.Arg526Gln mutation in CLCN7. *Am. J. Med. Genet. A.* 158A:909–916.

Kasai, T., K. Nakamura, A. Vaidyanath et al. 2012. Chlorotoxin fused to IgG-Fc inhibits glioblastoma cell motility via receptor-mediated endocytosis. *J. Drug Deliv.* 2012:975763.

Kasper, D., R. Planells-Cases, J. C. Fuhrmann et al. 2005. Loss of the chloride channel ClC-7 leads to lysosomal storage disease and neurodegeneration. *EMBO J.* 24:1079–1091.

Kawasaki, M., S. Uchida, T. Monkawa et al. 1994. Cloning and expression of a protein kinase C-regulated chloride channel abundantly expressed in rat brain neuronal cells. *Neuron* 12:597–604.

Kazmierczak, P. and U. Müller. 2012. Sensing sound: Molecules that orchestrate mechanotransduction by hair cells. *Trends Neurosci.* 35:220–229.

Kesavan, K., J. Ratliff, E W. Johnson et al. 2010. Annexin A2 is a molecular target for TM 601, a peptide with tumor-targeting and anti-angiogenic effects. *J. Biol. Chem.* 285:4366–4374.

Kleefuß-Lie, A., W. Friedl, S. Cichon et al. 2009. *CLCN2* variants in idiopathic generalized epilepsy. *Nat. Genet.* 41:954–955.

Koch, M. C., K. Steinmeyer, C. Lorenz et al. 1992. The skeletal muscle chloride channel in dominant and recessive human myotonia. *Science* 257:797–800.

Kokubo, Y., N. Iwai, N. Tago et al. 2005. Association analysis between hypertension and *CYBA*, *CLCNKB*, and *KCNMB1* functional polymorphisms in the Japanese population: The Suita study. *Circ. J.* 69:138–142.

Kornak, U., D. Kasper, M. R. Bösl et al. 2001. Loss of the ClC-7 chloride channel leads to osteopetrosis in mice and man. *Cell* 104:205–215.

Kornblum, C., G. G. Lutterbey, B. Czermin et al. 2010. Whole-body high-field MRI shows no skeletal muscle degeneration in young patients with recessive myotonia congenita. *Acta Neurol. Scand.* 121:131–135.

Koty, P. P., E. Pegoraro, G. Hobson et al. 1996. Myotonia and the muscle chloride channel: Dominant mutations show variable penetrance and founder effect. *Neurology* 47:963–968.

Krämer, B. K., T. Bergler, B. Stoelcker, and S. Waldegger. 2008. Mechanisms of disease: the kidney-specific chloride channels ClCKA and ClCKB, the Barttin subunit, and their clinical relevance. *Nat. Clin. Pract. Nephrol.* 4:38–46.

Kubisch, C., T. Schmidt-Rose, B. Fontaine, A. H. Bretag, and T. J. Jentsch. 1998. ClC-1 chloride channel mutations in myotonia congenita: variable penetrance of mutations shifting the voltage dependence. *Hum. Mol. Genet.* 7:1753–1760.

Kumar, K. R., K. Ng, H. Vandebona, M. R. Davis, and C. M. Sue. 2010. A novel *CLCN1* mutation (G1652A) causing a mild phenotype of thomsen disease. *Muscle Nerve* 41:412–415.

Lange, P. F., L. Wartosch, T. J. Jentsch, and J. C. Fuhrmann. 2006. ClC-7 requires Ostm1 as a β-subunit to support bone resorption and lysosomal function. *Nature* 440:220–223.

Lee, B. H., H. Y. Cho, H. K. Lee et al. 2012. Genetic basis of Bartter syndrome in Korea. *Nephrol. Dial. Transplant.* 27:1516–1521.

Lehmann-Horn, F., V. Mailänder, R. Heine, and A. L. George. 1995. Myotonia levior is a chloride channel disorder. *Hum. Mol. Genet.* 4:1397–1402.

Leisle, L., C. F. Ludwig, F. A. Wagner, T. J. Jentsch, and T. Stauber. 2011. ClC-7 is a slowly voltage-gated $2Cl^-/1H^+$-exchanger and requires Ostm1 for transport activity. *EMBO J.* 30:2140–2152.

Li, G.-W., E. Oh, and J. S. Weissman. 2012. The anti-Shine-Dalgarno sequence drives translational pausing and codon choice in bacteria. *Nature* 484:538–541.

Lipicky, R. J., S. H. Bryant, and J. H. Salmon. 1971. Cable parameters, sodium, potassium, chloride and water content, and potassium efflux in isolated external intercostals muscle of normal volunteers and patients with myotonia congenita. *J. Clin. Invest.* 50:2091–2103.

Lippiat, J. D. and A. J. Smith. 2012. The ClC-5 $2Cl^-/H^+$ exchange transporter in endosomal function and Dent's disease. *Front. Physiol.* 3:449.

Lísal, J. and M. Maduke. 2009. Proton-coupled gating in chloride channels. *Phil. Trans. R. Soc. B* 364:181–187.

Lloyd, S. E., S. H. S. Pearce, and W. Günther et al. 1997. Idiopathic low molecular weight proteinuria associated with hypercalciuric nephrocalcinosis in Japanese children is due to mutations of the renal chloride channel (CLCN5). *J. Clin. Invest.* 99:967–974.

López-Hernández, T., M. C. Ridder, M. Montolio et al. 2011. Mutant GlialCAM causes megalencephalic leukoencephalopathy with subcortical cysts, benign familial macrocephaly, and macrocephaly in retardation and autism. *Am. J. Hum. Genet.* 88:422–432.

Lossin, C. and A. L. George, Jr. 2008. Myotonia congenita. *Adv. Genet.* 63:25–55.

Lourdel, S., T. Grand, J. Burgos, W. González, F. V. Sepúlveda, and J. Teulon. 2012. ClC-5 mutations associated with Dent's disease: A major role of the dimer interface. *Pflugers Arch.* 463:247–256.

LOVD. 2013. Leiden open variation database: Chloride channel 1, skeletal muscle *CLCN1*. Accessed March 20, 2013. http://chromium.liacs.nl/LOVD2/home.php?select_db=CLCN1

Lueck, J. D., A. E. Rossi, C. A. Thornton, K. P. Campbell, and R. T. Dirksen. 2010. Sarcolemmal-restricted localization of functional ClC-1 channels in mouse skeletal muscle. *J. Gen. Physiol.* 136:597–613.

Ma, L., G. Y. Rychkov, and A. H. Bretag. 2009. Functional study of cytoplasmic loops of human skeletal muscle chloride channel, hClC-1. *Int. J. Biochem. Cell Biol.* 41:1402–1409.

Ma, L., G. Y. Rychkov, E. A. Bykova, J. Zheng, and A. H. Bretag. 2011. Movement of hClC-1 C-termini during common gating and limits on their cytoplasmic location. *Biochem. J.* 436:415–428.

Ma, L., G. Y. Rychkov, B. P. Hughes, and A. H. Bretag. 2008. Analysis of carboxyl tail function in the skeletal muscle Cl^- channel hClC-1. *Biochem. J.* 413:61–69.

Macías, M. J., O. Teijido, G. Zifarelli et al. 2007. Myotonia-related mutations in the distal C-terminus of ClC-1 and ClC-0 chloride channels affect the structure of a poly-proline helix. *Biochem. J.* 403:79–87.

Mailänder, V., R. Heine, F. Deymeer, and F. Lehmann-Horn. 1996. Novel muscle chloride channel mutations and their effects on heterozygous carriers. *Am. J. Hum. Genet.* 58:317–324.

Mankodi, A. 2008. Myotonic disorders. *Neurol. India* 56:298–304.

Mankodi, A., M. P. Takahashi, H. Jiang et al. 2002. Expanded CUG repeats trigger aberrant splicing of ClC-1 chloride channel pre-mRNA and hyperexcitability of skeletal muscle in myotonic dystrophy. *Mol. Cell* 10:35–44.

Matsumura, Y., S. Uchida, Y. Kondo et al. 1999. Overt nephrogenic diabetes insipidus in mice lacking the CLC-K1 chloride channel. *Nat. Genet.* 21:95–98.

Mazón, M. J., F. Barros, P. De la Peña et al. 2012. Screening for mutations in Spanish families with myotonia. Functional analysis of novel mutations in *CLCN1* gene. *Neuromuscul. Disord.* 22:231–243.

Mcferrin, M. B. and H. Sontheimer. 2006. A role for ion channels in glioma cell invasion. *Neuron Glia Biol.* 2:39–49.

Modoni, A., A. D'Amico, B. Dallapiccola et al. 2011. Low-rate repetitive nerve stimulation protocol in an Italian cohort of patients affected by recessive myotonia congenita. *J. Clin. Neurophysiol.* 28:39–44.

Mohammad-Panah, R., L. Wellhauser, B. E. Steinberg et al. 2008. An essential role for ClC-4 in transferrin receptor function revealed in studies of fibroblasts derived from *Clcn4*-null mice. *J. Cell Sci.* 122:1229–1237.

Mole, S. E., R. E. Williams, and H. H. Goebel. 2005. Correlations between genotype, ultrastructural morphology and clinical phenotype in the neural ceroid lipofuscinoses. *Neurogenetics* 6:107–126.

Moon, I. S., H. S. Kim, J. H. Shin et al. 2009. Novel *CLCN1* mutations and clinical features of Korean patients with myotonia congenita. *J. Korean Med. Sci.* 24:1038–1044.

Morrow, J. M., E. Matthews, D. L. Raja Rayan et al. 2013. Muscle MRI reveals distinct abnormalities in genetically proven non-dystrophic myotonias. *Neuromuscul. Disord.* 23:637–646.

Nagamitsu, S., T. Matsuura, M. Khajavi et al. 2000. A "dystrophic" variant of autosomal recessive myotonia congenita caused by novel mutations in the CLCN1 gene. *Neurology* 55:1697–1703.

Niemeyer, M. I., L. P. Cid, F. V. Sepúlveda, J. Blanz, M. Auberson, and T. J. Jentsch. 2010. No evidence for a role of *CLCN2* variants in idiopathic generalized epilepsy. *Nat. Genet.* 42:3.

Nighot, P. K. and A. T. Blikslager. 2012. Chloride channel ClC-2 modulates tight junction barrier function via intracellular trafficking of occludin. *Am. J. Physiol. Cell Physiol.* 302:C178–C187.

Nomura, N., M. Tajima, N. Sugawara et al. 2011. Generation and analysis of R8L barttin knockin mouse. *Am. J. Physiol. Renal Physiol.* 301:F297–F307.

Nozu, K., T. Inagaki, X. J. Fu et al. 2008. Molecular analysis of digenic inheritance in Bartter syndrome with sensorineural deafness. *J. Med. Genet.* 45:182–186.

Nozu, K., K. Iijima, K. Kanda et al. 2010. The pharmacological characteristics of molecular-based inherited salt-losing tubulopathies. *J. Clin. Endocrinol. Metab.* 95:E511–E518.

Pangrazio, A., M. Pusch, E. Caldana et al. 2010. Molecular and clinical heterogeneity in CLCN7-dependent osteopetrosis: Report of 20 novel mutations. *Hum. Mutat.* 31:E1071–E1080.

Papponen, H., M. Nissinen, T. Kaisto, V. V. Myllylä, R. Myllylä, and K. Metsikkö. 2008. F413C and A531V but not R894X myotonia congenita mutations cause defective endoplasmic reticulum export of the muscle-specific chloride channel CLC-1. *Muscle Nerve* 37:317–325.

Peters, M., N. Jeck, S. Reinalter et al. 2002. Clinical presentation of genetically defined patients with hypokalemic salt-losing tubulopathies. *Am. J. Med.* 112:183–190.

Plassart-Schiess, E., A. Gervais, B. Eymard et al. 1998. Novel muscle chloride channel (CLCN1) mutations in myotonia congenita with various modes of inheritance including incomplete dominance and penetrance. *Neurology* 50:1176–1179.

Poët, M., U. Kornak, M. Schweizer et al. 2006. Lysosomal storage disease upon disruption of the neuronal chloride transport protein ClC-6. *Proc. Natl. Acad. Sci. U.S.A.* 103:13854–13859.

Pusch, M. 2002. Myotonia caused by mutations in the muscle chloride channel gene CLCN1. *Hum. Mutat.* 19:423–434.

Pusch, M., K. Steinmeyer, M. C. Koch, and T. J. Jentsch. 1995. Mutations in dominant human myotonia congenita drastically alter the voltage dependence of the ClC-1 chloride channel. *Neuron* 15:1455–1463.

Raheem, O., S. Penttilä, T. Suominen et al. 2012. New immunohistochemical method for improved myotonia and chloride channel mutation diagnosis. *Neurology* 79:2194–2200.

Rajan, I., R. Read, D. L. Small, J. Perrard, and P. Vogel. 2011. An alternative splicing variant in *Clcn7*$^{-/-}$ mice prevents osteopetrosis but not neural and retinal degeneration. *Vet. Path.* 48:663–675.

Raja Rayan, D. L., A. Haworth, R. Sud et al. 2012. A new explanation for recessive myotonia congenita: Exon deletions and duplications in *CLCN1*. *Neurology* 78:1953–1958.

Ratté, S. and S. A. Prescott. 2011. ClC-2 channels regulate neuronal excitability, not intracellular chloride levels. *J. Neurosci.* 31:15838–15843.

Riazanski, V., L. V. Deriy, P. D. Shevchenko, B. Le, E. A. Gomez, and D. J. Nelson. 2011. Presynaptic CLC-3 determines quantal size of inhibitory transmission in the hippocampus. *Nat. Neurosci.* 14:487–494.

Riazuddin, S., S. Anwar, M. Fischer et al. 2009. Molecular basis of DFNB73: Mutations of *BSND* can cause nonsyndromic deafness or Bartter syndrome. *Am. J. Hum. Genet.* 85:273–280.

Richman, D. P., Y. Yu, T. T. Lee et al. 2012. Dominantly inherited myotonia congenita resulting from a mutation that increases open probability of the muscle chloride channel CLC-1. *Neuromol. Med.* 14:326–337.

Ricker, K., G. Hertel, K. Langscheid, and G. Stodieck. 1977. Myotonia not aggravated by cooling. Force and relaxation of the adductor pollicis in normal subjects and in myotonia as compared to paramyotonia. *J. Neurol.* 216:9–20.

Ridder, M. C., I. Boor, J. C. Lodder et al. 2011. Megalencephalic leucoencephalopathy with cysts: defect in chloride currents and cell volume regulation. *Brain* 134:3342–3354.

Rinke, I., J. Artmann, and V. Stein. 2010. ClC-2 voltage-gated channels constitute part of the background conductance and assist chloride extrusion. *J. Neurosci.* 30:4776–4786.

Rüdel, R., K. Ricker, and F. Lehmann-Horn. 1988. Transient weakness and altered membrane characteristic in recessive generalized myotonia (Becker). *Muscle Nerve* 11:202–211.

Santorelli, F. M., B. Garavaglia, F. Cardona et al. 2013. Molecular epidemiology of childhood neuronal ceroid-lipofuscinosis in Italy. *Orphanet J. Rare Dis.* 8:19.

Scheper, G. C., C. G. van Berkel, L. Leisle et al. 2010. Analysis of CLCN2 as candidate gene for megalencephalic leukoencephalopathy with subcortical cysts. *Genet. Test Mol. Biomarkers* 14:255–257.

Schlingmann, K. P., M. Konrad, N. Jeck et al. 2004. Salt wasting and deafness resulting from mutations in two chloride channels. *N. Engl. J. Med.* 350:1314–1319.

Scholl, U., S. Hebeisen, A. G. H. Janssen, G. Müller-Newen, A. Alekov, and C. Fahlke. 2006. Barttin modulates trafficking and function of ClC-K channels. *Proc. Natl. Acad. Sci. U.S.A.* 103:11411–11416.

Sherry, A. M., D. H. Malinowska, R. E. Morris, G. M. Ciraolo, and J. Cuppoletti. 2001. Localization of ClC-2 Cl^- channels in rabbit gastric mucosa. *Am. J. Physiol. Cell Physiol.* 280:C1599–C1606.

Smith, A. J., A. A. C. Reed, N. Y. Loh, R. V. Thakker, and J. D. Lippiat. 2009. Characterization of Dent's disease mutations of CLC-5 reveals a correlation between functional and cell biological consequences and protein structure. *Am. J. Physiol. Renal Physiol.* 296:F390–F397.

Souraty, N., P. Noun, C. Djambas-Khayat et al. 2007. Molecular study of six families originating from the Middle-East and presenting with autosomal recessive osteopetrosis. *Eur. J. Med. Genet.* 50:188–199.

Speirs, H. J. L., W. Y. S. Wang, A. V. Benjafield, and B. J. Morris. 2005. No association with hypertension of *CLCNKB* and *TNFRSF1B* polymorphisms at a hypertension locus on chromosome 1p36. *J. Hypertens.* 23:1491–1496.

Stark, Z. and R. Savarirayan. 2009. Osteopetrosis. *Orphanet J. Rare Dis.* 4:5.

Stauber, T. and T. J. Jentsch. 2010. Sorting motifs of the endosomal/lysosomal CLC chloride transporters. *J. Biol. Chem.* 285:34537–34548.

Steinmeyer, K., R. Klocke, C. Ortland et al. 1991b. Inactivation of muscle chloride channel by transposon insertion in myotonic mice. *Nature* 354: 304–308.

Steinmeyer, K., C. Ortland, and T. J. Jentsch. 1991a. Primary structure and functional expression of a developmentally regulated skeletal muscle chloride channel. *Nature* 354:301–304.

Stobrawa, S. M., T. Breiderhoff, S. Takamori et al. 2001. Disruption of ClC-3, a chloride channel expressed on synaptic vesicles, leads to a loss of the hippocampus. *Neuron* 29:185–196.

Suominen, T., L. L. Bachinski, S. Auvinen et al. 2011. Population frequency of myotonic dystrophy: higher than expected frequency of myotonic dystrophy type 2 (DM2) mutation in Finland. *Eur. J. Hum. Genet.* 19:776–782.

Suominen, T., B. Schoser, O. Raheem et al. 2008. High frequency of co-segregating *CLCN1* mutations among myotonic dystrophy type 2 patients from Finland and Germany. *J. Neurol.* 255:1731–1736.

Thomsen, J. 1876. Tonische Krämpfe in willkürlich beweglichen Muskeln infolge von ererbter psychischer Disposition (Ataxia muscularis?). *Eur. Arch. Psychiatry Clin. Neurosci.* 6:702–718.

Trip, J., S. Pillen, C. G. Faber, B. G. M. van Engelen, M. J. Zwarts, and G. Drost. 2009. Muscle ultrasound measurements and functional muscle parameters in non-dystrophic myotonias suggest structural muscle changes. *Neuromuscul. Disord.* 19:462–467.

Ulzi, G., M. Lecchi, V. Sansone et al. 2012. Myotonia congenita: Novel mutations in *CLCN1* gene and functional characterizations in Italian patients. *J. Neurol. Sci.* 318:65–71.

Ursu, S.-F., A. Alekov, N.-H. Mao, and K. Jurkat-Rott. 2012. ClC1 chloride channel in myotonic dystrophy type 2 and ClC1 splicing in vitro. *Acta Myol.* 31:144–153.

van der Knaap, M. S., I. Boor, and R. Estévez. 2012. Megalencephalic leukoencephalopathy with subcortical cysts: chronic white matter oedema due to a defect in brain ion and water homeostasis. *Lancet Neurol.* 11:973–985.

Wangemann, P. 2006. Supporting sensory transduction: cochlear fluid homeostasis and the endocochlear potential. *J. Physiol.* 576:11–21.

Weinberger, S., D. Wojciechowski, D. Sternberg et al. 2012. Disease-causing mutations C277R and C277Y modify gating of human ClC-1 chloride channels in myotonia congenita. *J. Physiol.* 590:3449–3464.

Wheeler, T. M., J. D. Lueck, M. S. Swanson, R. T. Dirksen, and C. A. Thornton. 2007. Correction of ClC-1 splicing eliminates chloride channelopathy and myotonia in mouse models of myotonic dystrophy. *J. Clin. Invest.* 117:3952–3957.

White, M. M. and C. Miller. 1979. A voltage-gated anion channel from the electric organ of *Torpedo californica*. *J. Biol. Chem.* 254:10161–10166.

Wijnberg, I. D., M. Owczarek-Lipska, R. Sacchetto et al. 2012. A missense mutation in the skeletal muscle chloride channel 1 (CLCN1) as candidate causal mutation for congenital myotonia in a New Forest pony. *Neuromuscul. Disord.* 22:361–367.

Wu, W., G. Y. Rychkov, B. P. Hughes, and A. H. Bretag. 2006. Functional complementation of truncated human skeletal-muscle chloride channel (hClC-1) using carboxyl tail fragments. *Biochem. J.* 395:89–97.

Yoshikawa, M., S. Uchida, J. Ezaki et al. 2002. CLC-3 deficiency leads to phenotypes similar to human neuronal ceroid lipofuscinosis. *Genes Cells* 7:597–605.

42 Cystic fibrosis and the CFTR anion channel

Yoshiro Sohma and Tzyh-Chang Hwang

Contents

42.1 INTRODUCTION

Cystic fibrosis transmembrane conductance regulator (CFTR) is no doubt a medicinally important channel protein. This chloride channel not only plays a critical role in the pathogenesis of cystic fibrosis (CF), a life-shortening hereditary disorder afflicting primarily the Caucasian population, its hyperactivity also constitutes the root cause of secretory diarrhea, which incapacitate millions of people mainly in developing countries. Since the major emphasis of this book is on the biophysical mechanism of ion channel function, the main focus of this chapter will be placed on discussing the mechanisms underlying regulation, gating, and ion permeation of the CFTR chloride channel. However, we feel obliged to include a brief history of CF, elaborating how a disease thought to be an anomaly of mucus rheology was eventually revealed to be an ion transport disease. Because of the obvious clinical relevance of CFTR, we also end this chapter by including sections glancing over how mutations of the CFTR gene cause various dysfunction of the protein as well as how recent development of small molecules to rectify these abnormalities impact basic sciences and clinical medicine. As these areas are likely covered in a cursory manner, more extensive reviews (Amaral, 2011; Becq et al., 2011; Cohen and Prince, 2012; Lukacs and Verkman, 2012) and Cystic Fibrosis Foundation website (http://www.cff.org/) are recommended for interested readers.

42.2 BRIEF HISTORY OF CYSTIC FIBROSIS

CF is an autosomal recessive genetic disorder affecting many epithelium-lining organ systems such as the lung, pancreas, liver, intestine, and sweat duct (Quinton 1999). Although CF is found in all ethnicities, it is most prevalent in people of Caucasian heritage (Bobadilla et al., 2002). Humans have died from CF since thousands of years ago. European folklore from the medieval time warned "woe is the child who tastes salty from a kiss on the brow, for he is cursed, and soon must die" as if infants with a salty skin were *hexed* or *bewitched* (Quinton, 1999). Although sporadic case reports for CF-like childhood diseases can be found in the literature before 1938, the first systematic description of CF as a clinical entity was often credited to a landmark paper by Dorothy Andersen (Andersen, 1938), who reported the clinicopathological findings in 49 children with the characteristic fibrocystic changes of the pancreas and associated this pancreatic pathology with neonatal meconium ileus, and respiratory complications.

While the prevailing view of CF in those early days was that CF is a disease mainly affecting the exocrine system of the pancreas, Sydney Farber in 1943 introduced the term *mucoviscidosis* and described CF as a generalized disorder also impairing the function of organs other than the pancreas. He was the first to accurately summarize the secondary consequences of the CF defect as "the respiratory tract damage therefore depends on primary obstruction by thick mucus, failure of proper lubrication of ciliated epithelium and secondary staphylococcal infection" (Farber, 1943). Indeed, we now know that nearly all organ systems with the only exception of sweat ducts afflicted by CF secrete mucus (Quinton, 2008). It is perhaps ironic that it is the work done with sweat ducts that eventually leads to the theory of chloride transport abnormality in CF (Quinton, 2007). It is also interesting to note how this long-recognized rheological abnormality of the mucus is linked to the molecular defect in CF remains a mystery (Quinton, 2010).

The first clue pointing to abnormal sweat electrolyte emerged out of an incident happened in 1948 in New York City. The extreme summer heat caused heat prostration especially in infants with a prior diagnosis of CF (Kessler and Andersen, 1951). This led to the recognition of the increased salt content of the sweat in patients with CF (diSant'Agnese et al., 1953). Measurements of sweat salt content hence became an important protocol for the diagnosis of CF, but collecting sufficient sweat for chemical analysis was a mounting challenge. Fortunately, diagnosis by sweat analysis became more practicable, accurate, safer, and more generally available when sweating was stimulated by the pilocarpine iontophoresis rather than various potentially dangerous methods of heating the patients (Gibson and Cooke, 1959).

Progress toward a mechanistic understanding of the basic defects in CF was held back for the ensuing 20 years despite the establishment of sweat chloride tests as the gold standard for diagnosing CF. However, steady improvements in the survival of CF patients were achieved mainly due to development of better nutritional support (Sinaasappel et al., 2002), application of nebulized and intravenous antibiotics (O'Sullivan and Freedman, 2009; Bals et al., 2011), and implementation of vigorous physical therapy (Dodd and Prasad, 2005). Of note, none of these therapeutical advancements targeted the root cause of the disease, which remained in the dark.

Then, in 1981, Michael Knowles and colleagues demonstrated an abnormal potential difference in the nasal mucosa of patients with CF, thus providing more direct evidence of a primary dysfunction in transepithelial electrolyte transport (Knowles et al., 1981). This was further supported by subsequent demonstration of the abnormality being already present in newborns with CF (Gowen et al., 1986), indicating the abnormality was primary rather than any secondary circulating substances. While this groundbreaking finding of abnormal transepithelial potential difference was attributed mostly to an increased sodium absorption, in 1983, Paul Quinton showed that it is chloride impermeability in CF sweat ducts that accounts for the abnormally high transepithelial voltage difference (Quinton, 1983). His results provided a satisfactory explanation for the elevated sweat chloride concentration in patients with CF. Sato and Sato (1984) also demonstrated a defect in CF sweat glands by showing that while the cholinergic stimulation of sweating is preserved, adrenergic agonists failed to stimulate sweating in CF patients. These landmark discoveries in the human sweat gland/duct system—equivalent to squid giant axons for fundamental electrophysiology of the nerve cell—found satisfying molecular basis once the gene responsible for CF was identified and the gene product CFTR can then be characterized in vitro with exquisite molecular details (see Section 42.3).

Following the great advances in the molecular biology and genetics, various groups attempted to identify the CF gene from the early 1980s. In 1985, three laboratories (Tsui et al., 1985; Wainwright et al., 1985; White et al., 1985) independently showed that the location of CF gene was on chromosome 7 using different DNA makers. The year 1989 is considered the annus mirabilis in the history of CF as the CF gene was cloned by research teams headed by Lap-Chee Tsui, Francis Collins, and Jack Riordan (Kerem et al., 1989; Riordan et al., 1989; Rommens et al., 1989). The CF gene product was named the CFTR. Hydropathy analysis of the amino acid sequence of CFTR (Figure 42.1a) placed this integral membrane protein in one of the largest protein families: ATP binding cassette (ABC) transporter superfamily (Riordan et al., 1989). Successes in cloning the CF gene not only established the genetic basis of CF, but also opened the door to comprehensive studies of the CFTR protein described in Section 42.3.

42.3 CFTR

42.3.1 OVERVIEW

The ABC transporter superfamily is one of the largest protein superfamilies widely distributed from prokaryotes to humans (Rees et al., 2009). The basic structural unit of an ABC protein consists of a membrane-spanning domain (MSD) and a cytoplasmic nucleotide-binding domain (NBD). In some members, one functional unit is formed by dimerization of two MSD–NBD complexes (half transporter). In others, two tandem MSD–NBD complexes are conjoined into a single gene product (full transporter). Most ABC proteins function as active transporters that enact an uphill transport of a wide variety of substrates using the energy of ATP hydrolysis.

Although CFTR shows a characteristic topology of an ABC exporter (Figure 42.1a), that is, two tandem repeats of the MSD–NBD complex with a 6 × 6 fold in its MSDs, unique to CFTR, a regulatory domain (RD) with multiple consensus sequences for protein kinase A (PKA)-dependent phosphorylation is

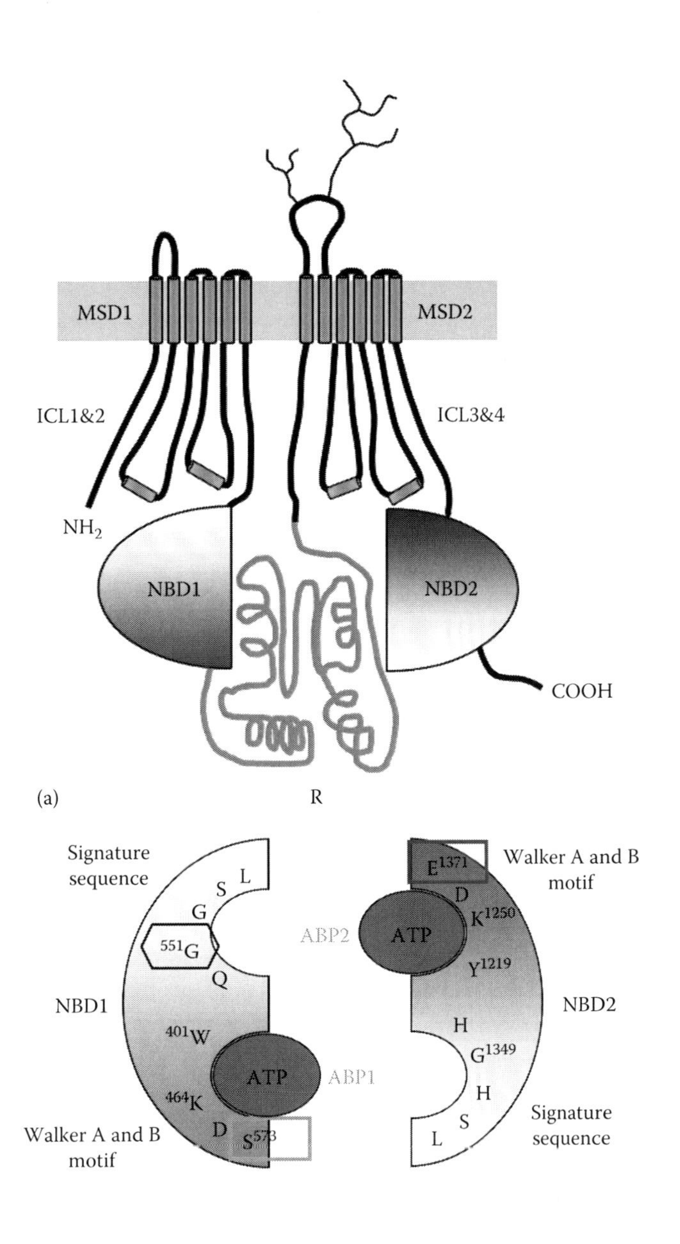

Figure 42.1 Schematic representation of CFTR structure. (a) A topological model of CFTR. MSD: membrane-spanning domain; NBD: nucleotide-binding domain; RD: regulatory domain. See text for more detail. (b) A cartoon depicting the head and tail subdomains of the NBDs of CFTR. Exact amino acid sequences for the Walker A, Walker B, and ABC signature sequence in NBD1 and NBD2 are shown. Note that the ABC signature in NBD2 is degenerate whereas that in NBD1 is conserved. Also note the glutamate residue following the Walker B motif in NBD1 is changed to serine, ^{573}S (circle). Two ATP-binding pockets (ABP1 and ABP2) are formed upon NBD dimerization in a head-to-tail configuration. The glycine residue at position 551 is highlighted by a square as the pathogenic mutation of this residue to aspartate results in a channel that fails to respond to ATP. (Adapted from Sohma, Y. et al., *Curr. Pharm. Des.*, 19, 3521, 2013. With permission.)

inserted between the two pseudo-symmetrical MSD–NBD repeats (Riordan et al., 1989). Despite the topological similarities between CFTR and a typical ABC exporter (Procko et al., 2009), CFTR is an ATP-gated ion channel (Anderson et al., 1991a; Nagel et al., 1992), mediating passive diffusion of anions (Bear et al., 1992). But PKA-dependent phosphorylation of the RD is a prerequisite for CFTR to function as an ion channel (see the following text for details).

Channels and transporters are thought to belong to two distinct classes of transport proteins. Active transporters use some form of free energy input to drive protein conformational changes, allowing the substrate binding site in the MSDs to become alternately accessible to the internal and external sides of the membrane. Figuratively, this can be achieved by having two *gates*, opening and closing alternately during a transport cycle (i.e., alternating access mechanism, see the following text). In contrast, a single gate is sufficient for the function of an ion channel as long as opening the gate affords a continuous ion diffusive pathway. Furthermore, kinetic transitions between the open and closed states of an ion channel are usually considered to be in equilibrium. In contrast, because of the imposed nonequilibrium condition either by a concentration gradient or by a prevailing concentration of the energy currency ATP, conformational changes in a transport cycle are driven in a preferred direction. This long-held boundary between channels and transporters was broken when phylogenic analysis revealed two families of integral membrane proteins containing both channels and transporters as their members: CLC protein family and ABC transporter superfamily (reviewed in [Chen and Hwang, 2008]). Recent biophysical studies of CFTR's MSDs provide data supporting the degraded transporter hypothesis: CFTR evolves from a primordial ABC exporter by degenerating its cytoplasmic gate (Bai et al., 2010, 2011; Gao et al., 2013). Thus, in addition to the obvious implications in CFTR-associated diseases, sorting out the fundamental mechanism of CFTR function could bear a broader ramification on ABC transporters in general.

42.3.2 NUCLEOTIDE-BINDING DOMAIN: GATING MECHANISM

42.3.2.1 Structures of NBDs and NBD dimer

Whereas the field still awaits high-resolution structure of the whole CFTR protein (Mio et al., 2008; Zhang et al., 2009, 2011; Rosenberg et al., 2011), the crystal structures of CFTR's two NBDs were published (Lewis et al., 2004, 2005; Atwell et al., 2010) (PDB code: 1XMI for NBD1; 3GD7 for NBD2). Like those in ABC proteins, each NBD of CFTR consists of two subdomains: an F1-type ATP-binding core *head* subdomain including Walker A and B motifs and an α-helical (ABCα) *tail* subdomain harboring the ABC signature sequence (LSGGQ) that defines the ABC transporter family. The remarkable resemblance of the crystal structures of CFTR's NBDs to those of ABC proteins indicates that the biochemical mechanisms for ATP binding and hydrolysis are well conserved. For example, (1) the aromatic side chain of tryptophan 401 (W401 in NBD1) and tyrosine 1219 (Y1219 in NBD2) in the *head* subdomain forms a π-electron stacking interaction with the adenine ring of ATP, (2) the positively charged Walker A lysine (K464 in NBD1 and K1250 in NBD2) electrostatically interacts with the negative charge of the γ-phosphate, whereas Walker B aspartate (D572 in NBD1 and D1370 in NBD2) coordinates the Mg^{2+} ion that is essential for ATP hydrolysis, and (3) once ATP is bound stably and Mg^{2+} ions and H_2O molecules are properly coordinated, the side chain of glutamate 1371 (E1371) in NBD2 functions as the catalytic base for hydrolyzing ATP (cf. Ernst et al., 2006; Hanekop et al., 2006). Of note, the equivalent residue of this *catalytic glutamate* in NBD1 is a serine.

Although CFTR's two NBDs show strikingly similar structural folds, sequence homology between NBD1 and NBD2 is very low except in several conserved regions, for example, Walker A, B motifs, signature sequence, D-loop, Q-loop, and H-loop. Even in these conserved regions, some differences were noticed. In addition to the catalytic glutamate residue mentioned earlier, a critical histidine in the H-loop is also replaced by a serine in NBD1. Most noticeably is the deviation of NBD1's signature sequence LSHGH from the consensus sequence of LSGGQ. These differences may explain why NBD2 but not NBD1 exhibits an appreciable ATP hydrolysis rate (Aleksandrov et al., 2002; Basso et al., 2003; Stratford et al., 2007). Interestingly, about half of human ABC proteins show this salient NBD asymmetry (Procko et al., 2009), corroborating the idea that hydrolysis of one ATP molecule is sufficient to drive a complete transport cycle ((Hou et al., 2000) MRP1; (Perria et al., 2006) TAP1/2; (Zhang et al., 2006), ABCG5 and ABCG8). For unknown reasons Csanady et al., 2005a; Aleksandrov et al., 2010; Jih et al., 2011), upstream from its Walker A motif, CFTR's NBD1 contains extra 30 amino acids known as regulatory insertion (RI).

The high homology of NBD structures among the ABC transporters also suggests that the two NBDs of CFTR are expected to form a prototypical *head-to-tail* dimer with two ATP-binding pockets (ABPs) sandwiched at the dimer interface (Figure 42.1b). Although definitive biochemical/structural evidence for the dimerization of CFTR's two NBDs is still missing, some functional studies elaborated in the following text do implicate just that (Vergani et al., 2005; Mense et al., 2006). If NBDs are viewed as the energy-harvesting machine for ABC transporters, it seems reasonable to hypothesize that CFTR adopts the same machinery to harness the energy of ATP hydrolysis to drive its gating conformational changes. This idea implies that unlike most other ion channel proteins, gating transitions of CFTR are not in equilibrium due to an input of free energy from ATP hydrolysis.

42.3.2.2 ATP-dependent gating driven by NBD dimerization

Since the cloning of CFTR in 1989, a plethora of functional studies have established firmly that CFTR, once phosphorylated by PKA, is an ATP-gated chloride channel (reviewed in Gadsby et al., 2006; Chen and Hwang, 2008; Hwang and Sheppard, 2009), However, unlike classical ligand-gated channels, CFTR's ligand ATP is hydrolyzed during the gating process, an idea supported by numerous reports. First, in the presence of ATP alone, wild-type (WT) channels open for hundreds of milliseconds, but the application of nonhydrolyzable ATP analogs such as AMP-PNP or pyrophosphate (PPi) in addition to ATP locks the channel in a stable open state for tens of seconds (Gunderson and Kopito, 1994; Hwang et al., 1994). Second, once ATP hydrolysis in NBD2 is abolished by mutating Walker A lysine (K1250) or the catalytic glutamate (E1371) (Figure 42.1), ATP alone can lock open the CFTR channels (Gunderson and Kopito, 1995; Ramjeesingh et al., 1999; Zeltwanger et al., 1999; Powe et al., 2002a; Bompadre et al., 2005b). Third, by scrutinizing single-channel recordings of WT-CFTR, Gunderson and Kopito (1995) reported two open states with distinct current amplitudes (O1 and O2). The preferential ordered transition between these states indicates a violation of microscopic reversibility and thus demands an input of the free energy likely from ATP hydrolysis. Fourth, microscopic kinetic analysis of single-channel traces reveals a paucity of the short-lived open events, an indication of irreversible steps in channel closure (Csanady et al., 2010). These gating data could be explained by a three-state, nonequilibrium gating scheme depicting a strict coupling between ATP hydrolysis and the gating cycle (Figure 42.2a). Although a role of ATP hydrolysis in channel closing was suggested way before any structural data became available, coupling of channel opening to NBD dimerization was not established until the landmark paper by Vergani et al. (2005).

Since the early 2000s, breakthroughs in solving the high-resolution structures of several ABC exporters have provided more concrete ideas regarding how these proteins function at a molecular level (Locher et al., 2002; Dawson and Locher, 2006; Aller et al., 2009; Hohl et al., 2012). The crystal structures of ABC exporters obtained under a nucleotide-free condition showed an outward-facing MSD with separated NBDs, for example (Locher et al., 2002; Aller et al., 2009; Hohl et al., 2012); in contrast, the crystal structure obtained with a nucleotide showed an inward-facing MSD but dimerized NBDs (Dawson and Locher, 2006). These crystallographic data support the long-held alternating access model depicting a coupled rigid body movement of each MSD/NBD complex.

Taking advantage of these structural insights as well as a wealth of sequence information, Vergani et al. (2005) showed

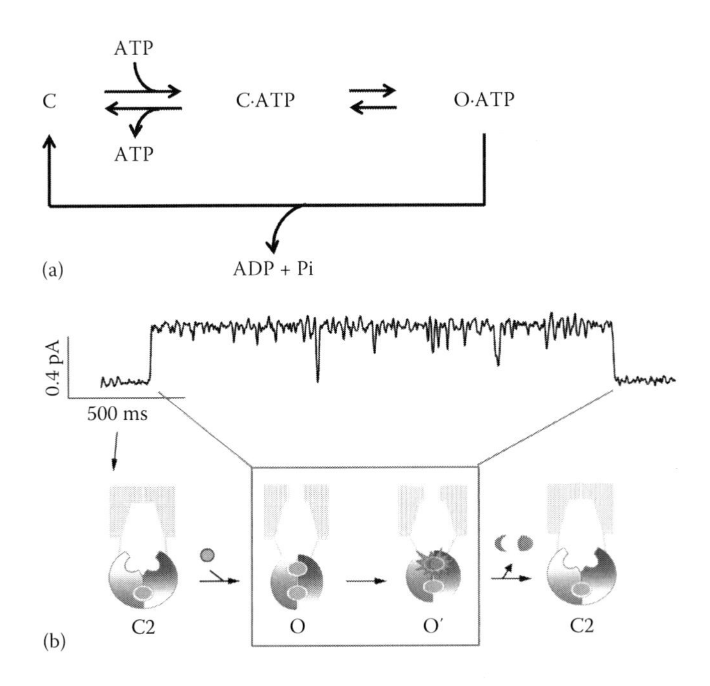

Figure 42.2 A classical model for ATP-dependent gating of the CFTR channel. (a) A simplified three-state kinetic scheme for ATP-dependent gating. C: a closed state, CATP: the closed state with ATP bound; OATP: an open state with ATP bound. Note that the transition from OATP to CATP is an irreversible step associated with ATP hydrolysis. (b) A proposed relationship between an opening/closing cycle in a single-channel recording (the trace) and the ATP-dependent conformational changes in NBDs. C2: the closed state with a partial NBD dimer; O: the open state prior to ATP hydrolysis; O′: the posthydrolytic open state with a full NBD dimer. This strict coupling model depicts opening of the gate is accompanied by NBD dimerization following ATP binding to ABP2, whereas closing of the gate is associated with the release of ATP hydrolysis products, ADP and Pi, which requires a partial separation of the NBD dimer at ABP2.

that an evolutionarily conserved hydrogen-bond pair (R555 in NBD1 and T1246 in NBD2) only interacts in the transition state upon channel opening and/or in the open state, supporting the idea that NBD dimerization does occur in CFTR and is associated with channel opening. Cross-linking studies also (Mense et al., 2006) confirmed that the interface of two NBDs should be close enough to make the cys-cys cross-linking during the ATP-dependent gating. That association and dissociation of NBDs accompanying gate opening and closing, respectively, not only bring the distant cousin CFTR one step closer to the ABC protein superfamily, but this idea also opens the door for kinetic studies that reveal intriguing features on the molecular mechanism of CFTR gating.

42.3.2.3 Hidden states during ATP-dependent gating cycle

Patch-clamp electrophysiological techniques bestow exquisite temporal resolution of the protein conformational changes during gating transitions. Through quantitative analysis of the currents, we can infer the transitions between conducting and nonconducting states, which report in real time the status of CFTR's gate. But the demonstration that opening of CFTR's gate is associated with NBD dimerization allows experimental protocols tackling the status of NBDs as the bound ligands can only be substituted by a new ligand after two NBDs have separated and hence opened a space that is large enough to accommodate the entry of a new ligand following the exit of the old ones. Thus, by applying a nucleotide analog at different times for different durations, one can fish out normally *invisible* states in CFTR gating.

By applying nonhydrolyzable PPi or AMP-PNP to the cytoplasmic side of the channel at different times following the removal of ATP, Tsai et al. (2009) identified two different closed states by their distinct response to these ligands: the robust-responding C2 emerges right after channel closing and subsequently dissipates into the poor-responding C1 state. As the lifetime of the C2 state can be modulated by using high-affinity ATP analog, N^6-phenylethyl-ATP (Zhou et al., 2005), or mutations that destabilize ATP binding in ABP1, it was concluded that *upon channel closing*, CFTR sojourns first to state C2, wherein ABP2 is vacant and ready for another ligand, but ABP1 remains occupied by ATP. Since the C2 state can exist for tens of seconds whereas each opening/closing cycle of CFTR only lasts for ~1 s, these results also suggest that each observed opening and closing of CFTR are associated with NBD dimerization and hydrolysis-triggered partial separation of the NBD dimer (Figure 42.2b).

The theory depicted in Figure 42.3(a) and (c) was further supported by a ligand-exchange protocol developed by Tsai et al. (2010). When abruptly changing the ligand from ATP to P-ATP, the single-channel open probability (P_o) increases in two distinct steps: an immediate rise by an increase of the opening rate and a delayed one with a decreased closing rate. These results were interpreted to mean that upon switching the ligand, ATP in ABP2 was substituted immediately while ATP in ABP1 was replaced with a long delay. As the second phase of ligand exchange can be manipulated by mutations at the head subdomain of NBD1 and the tail subdomain of NBD2, Tsai et al. (2010) concluded that the long residence time of ATP in ABP1

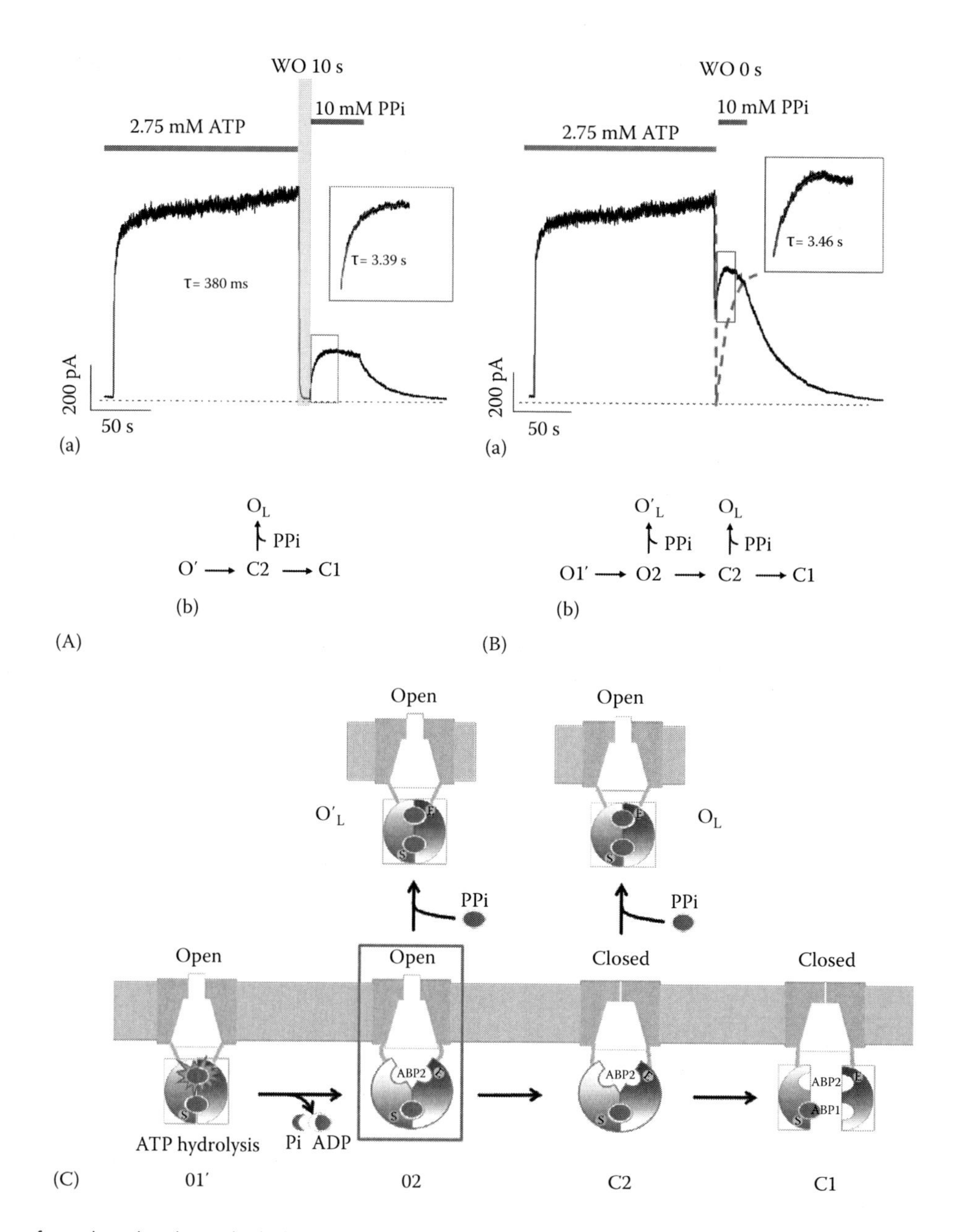

Figure 42.3 Evidence for a short-lived, posthydrolytic open state. (A and B) (a) PPi-induced reopening of WT-CFTR channels with different ligand-switch protocols: (A) washout of ATP for 10 s before applying PPi and (B) direct switch from ATP to PPi. (b) Simple gating models explaining the PPi-induced current changes upon respective ligand switches. (C) A cartoon depicting a revised CFTR gating model based on PPi-induced reopening of CFTR. O′, a posthydrolytic open state with a fully dimerized NBD; O2, the open state with a partially separated NBDs; C2, the closed state with a partially separated NBD; O'_L and O_L, locked open state following PPi binding to ABP2 in CFTR in C2 and O2 states, respectively. Note that O'_L and O_L are assumed to be structurally identical. (Adapted from Jih, K.Y. et al., *J. Gen. Physiol.*, 139, 359, 2012b. With permission.)

is due to trapping of the ligand at the NBD dimer interface. Of note, this idea was further supported by an independent approach in CFTR. Szollosi et al. (2011) employed mutant cycle analysis to identify any state-dependent interactions between residues in the head subdomain of NBD1 and those in the tail subdomain of NBD2. The lack of such interaction was interpreted as a continuously engaged ABP1 during CFTR's gating cycle. Interestingly, a recently published crystal structure of a bacterial ABC exporter, TM287/288 (Hohl et al., 2012), shows an inward-facing conformation of the MSDs (presumably equivalent to the closed state of CFTR) accompanied by partially engaged NBDs.

While by applying a nonhydrolyzable analog such as PPi to the closed channels fished out different closed states (e.g., Figure 42.3A), directly switching ATP to PPi identified a potential transient posthydrolytic open state (Jih et al., 2012b). As shown in Figure 42.3B(a), PPi (or AMP-PNP), applied immediately after the removal of ATP, could keep the gate open, suggesting that PPi induced a direct transition from one open state to a locked open state without closing. As PPi only locks open the channel by binding to ABP2 after the original ligand ATP is hydrolyzed and the hydrolytic products are released, these surprising results suggest that the gate can remain open despite a partial separation of NBDs following ATP hydrolysis. Thus, following gate opening by ATP-induced NBD dimerization (i.e., open state O1(′) in Figure 42.3C), hydrolysis of ATP at ABP2 and subsequent partial separation of NBDs and dissociation of

ADP and Pi does not close the gate immediately (i.e., open state O2 in Figure 42.3C). In theory, when the channel resides in state O2, ATP should be able to bind and re-dimerize NBDs to initiate another hydrolysis cycle without an obligatory closure of the gate (see the following text). Then the gating cycle and the ATP hydrolysis cycle do not necessarily follow a strict one-to-one relationship!

42.3.2.4 Coupling between NBDs and MSDs

For nearly two decades, coupling between CFTR gating and ATP hydrolysis is thought to bear a one-to-one stoichiometry (Figure 42.2). This theory is not only supported by numerous data in the literature (reviewed in Jih and Hwang, 2012), it also fits nicely with the proposed flip-flop motion of the MSD/MSD complexes that exposes the substrate binding site alternately to either side of the membrane by ATP binding–induced association and subsequent hydrolysis-triggered dissociation of NBDs for an ABC transporter (Rees et al., 2009). However, the observation that an open CFTR channel can accept a new ligand contradicts the strict coupling hypothesis (Jih et al., 2012b), demanding the existence of a posthydrolytic open state with ABP2 already vacated.

The most compelling evidence for a more plastic coupling mechanism came from the fortuitous discovery of a CFTR mutant R352C, which exhibits a change of the single-channel conductance following ATP hydrolysis during an open burst (Jih et al., 2012a). It was found, in single-channel recordings of R352C-CFTR, that a majority (>50%) of the open bursts show a preferred transition (C → O1 → O2 → C in Figure 42.4a), whereas abolishing ATP hydrolysis eliminates the O1 → O2 transition (Jih et al., 2012a). This suggests that the low conductance state O1 corresponds to a dimerized NBD state with two ATP molecules sandwiched at the dimer interface, and the high conductance state O2 represents a posthydrolytic state. Importantly, a significant population of the observed events (10%–20%) shows multiple repeats of the O1 → O2 transition within a single open burst (Figure 42.4a) (Jih et al., 2012a), indicating that multiple rounds of ATP hydrolysis could take place before gate closure. Thus, the coupling between the ATP hydrolysis cycle in NBD and the open/closed gating cycle in MSD violates the one-to-one stoichiometry (Figure 42.4b). Of note, the fraction of open bursts containing multiple cycles of the O1 → O2 transition can be modulated by mutations that alter the interaction of ATP with ABP1, further supporting the plastic nature of the coupling mechanism.

A more generalized gating model (Figure 42.4b) was proposed based on recent data supporting a nonstrict coupling mechanism (Jih et al., 2012a,b). This gating scheme abides by the classical Monod–Wyman–Changeux model for allosteric modulation (Changeux, 2012). All major kinetic transitions, ligand binding/unbinding (C ↔ CATP, O2 ↔ O2ATP), NBD association/dissociation (CATP ↔ CAD, O2ATP ↔ O1), and gate opening/closing (C ↔ O2, CATP ↔ O2ATP, CAD ↔ O1), are microscopically reversible steps. It was proposed that ATP hydrolysis, by providing the energy to allow the O1 state to escape from an otherwise exceedingly stable state, constitutes the irreversible kinetic step (see Jih et al., 2012a, for details). Thus, the model portrays MSDs and NBDs as two autonomous molecular entities but energetically coupled: motions of one domain promote the motion of the other. Although the structural nature of this new coupling mechanism remains unclear (Jih and Hwang, 2012), the idea of probabilistic connection between association/dissociation of NBDs and opening/closing of the gate in MSDs does not require a concurrent movement of two MSD/NBD complexes as a rigid body, a well-promulgated theory in the ABC transporter field (Procko et al., 2009; Rees et al., 2009). As more and more crystal structures of ABC transporters are solved, it will be interesting to see if this new idea of MSD/NBD coupling may find direct structural supports.

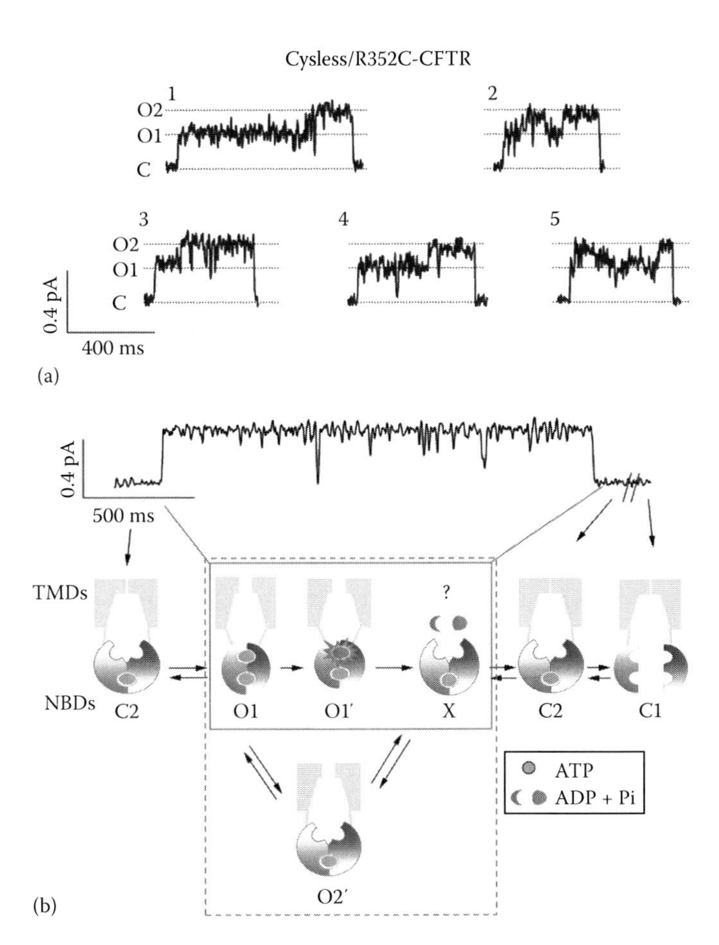

Figure 42.4 Nonstrict coupling between ATP hydrolysis cycle and the gating cycle. (a) Opening events of cys-less/R352C-CFTR containing one (# 1, 3, and 4) or more (# 2 and 5) O1 → O2 transitions. (b) An updated model illustrating the relationship between an opening/closing cycle of the gate and ATP consumption in CFTR's NBDs. The trace represents an opening burst of WT-CFTR. The cartoon in the following text shows transitions of gating states throughout the gating cycle. Note that the C2 state is extremely stable and can last for ~30 s before returning to the C1 state. (From Tsai, M.F. et al., *J. Gen. Physiol.*, 135, 399, 2010.) C1: the closed state with two separated NBDs; C2: the closed state with a partially separated NBDs; O1: the prehydrolytic open state; O1′: posthydrolytic open state with a full NBD dimer; O2: the open state with partially separated NBDs (Jih et al., 2012b); the dashed box encompasses the reentry pathway that may occur within each opening burst. (Adapted from Jih, K.Y. et al., *J. Gen. Physiol.*, 139, 359, 2012b. With permission.)

42.3.2.5 Future perspectives

One immediate consequence out of the nonstrict coupling mechanism for CFTR gating is a nonintegral stoichiometry between the gating cycle and the ATP hydrolysis cycle. When the same mechanism is extrapolated to explain the transport cycle of an ABC transporter, an inevitable conclusion of inefficient energy utilization emerges: instead of guaranteed translocation of a substrate across the membrane per molecule of ATP hydrolyzed, futile ATP hydrolysis cycles exist. Nevertheless, based on the results with the R352C mutant, one can conclude that at least 90% of the observed events comprise one single O1 → O2 transition, indicating the coupling does not deviate very far away from one-to-one stoichiometry. Indeed, a significant deviation will predict an increase of the mean open burst time with increasing [ATP] for WT-CFTR, a phenomenon very rarely reported in the literature (Zeltwanger et al., 1999). In theory, this tight and yet slightly imperfect coupling can be accomplished by simply proposing that the rate of O2 → C is much faster than that of O2 → O2ATP (Sohma and Hwang, unpublished computer simulations). Indeed, an [ATP]-dependent prolongation of the open time is seen experimentally with a conserved mutation W401F presumably because of an increased rate of O2 → O2ATP (Jih et al., 2012b). Future studies will likely identify other ways to manipulate different transitions in the gating scheme with verifiable predictions on the kinetics of CFTR gating.

As discussed in more detail in the following text in Sections 42.4.2 and 42.4.3, this newly proposed gating model may also explain how clinically applicable drugs work in improving the function of CFTR. Recently U.S. Food and Drug Administration (FDA) approved a CFTR potentiator, VX-770 (Kalydeco or ivacaftor), for treating CF patients carrying the G551D mutation, which impairs the ATP-dependent gating in spite of the normal expression in the plasma membrane (Van Goor et al., 2009; Accurso et al., 2010; Ramsey et al., 2011). Jih and Hwang (2013) showed that VX-770-induced effects on WT-CFTR can be accounted for by a simple mechanism of stabilizing the O2 state and consequently promoting decoupling between the gating cycle and ATP hydrolysis cycle. Such a clear understanding of drug actions on CFTR gating is expected to be very helpful for future development of the novel CF medications.

On the basic science front, many questions pertaining to the conformational changes involved in CFTR gating remain unanswered. For example, since during most of the gating cycle, ABP1 remains engaged, how does partial separation of NBDs initiate conformational changes in both MSDs? Although not depicted in Figure 42.5, a closed state with NBDs completely disengaged does exist (i.e., C1 closed state in Tsai et al., 2010), what is the consequence in gating kinetics once ATP in ABP1 is dissociated? As many of the ABC exporters possess only one catalysis-competent site in their NBDs like CFTR, these issues are readily relevant to the fundamental mechanism of these asymmetrical ABC transporters. Even for ABC proteins harboring two catalysis-competent sites, it would be interesting to ask if the stochastic nature of the biological energy metabolism revealed from CFTR gating studies may also be applicable.

42.3.3 REGULATORY DOMAIN

PKA-dependent phosphorylation of the R domain regulates the activity of CFTR in a very complex manner. First, there seems to be an all-or-none type of regulation. Even for mutants that completely lose the responsiveness to ATP, the presence of channel activity requires a prior phosphorylation (Bompadre et al., 2007; Cui et al., 2007; Wang et al., 2007). Second, since application of ATP alone before PKA-dependent phosphorylation fails to open CFTR, phosphorylation of the R domain may also regulate the function of NBDs (Anderson et al., 1991a; Nagel et al., 1992). Third, as different levels of channel activity were observed with different degrees of phosphorylation, a mechanism involving incremental regulation of CFTR was also proposed (Hwang et al., 1993, 1994; Mathews et al., 1998; Csanady et al., 2000; Wang et al., 2000).

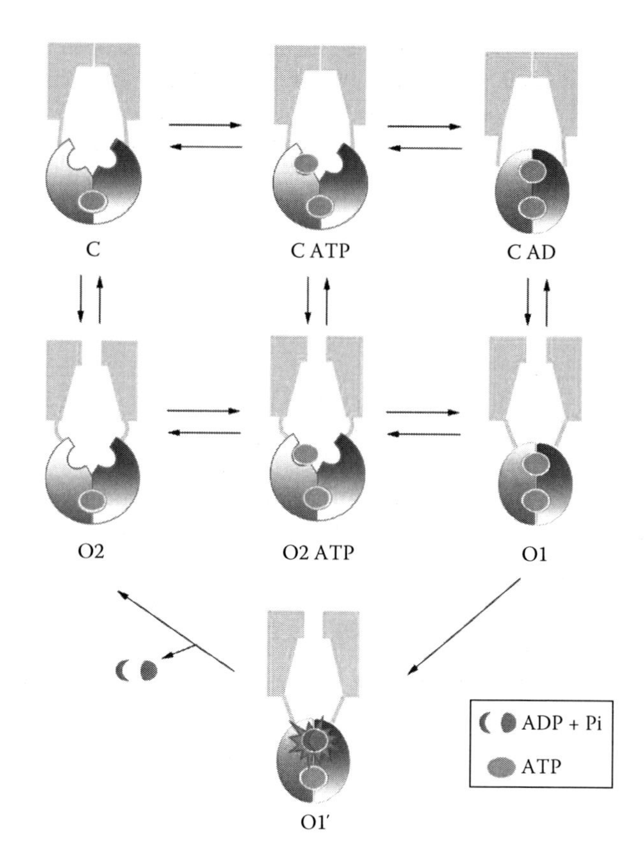

Figure 42.5 Energetic coupling model for CFTR gating. ATP hydrolysis provides a shortcut from the O1 to O2 state. The CATP state is added to represent ATP binding before NBD dimerization. (Adapted from Jih, K.Y. et al., *J. Gen. Physiol.*, 140, 347, 2012a. With permission.)

Nearly all the PKA consensus sites important for CFTR regulation are located within the R domain, a notion also supported by biochemical data (Seibert et al., 1995). These include both classical dibasic PKA sites and a number of monobasic sites. NBD1 does contain a dibasic site, S422, in its RI region, but its physiological role is unclear, since removal of RI does not seem to affect CFTR function ((Csanady et al., 2005b); cf. (Chang et al., 1993)). Biochemical experiments have shown phosphorylation of at least six dibasic PKA sites in the R domain (S660, S700, S737, S768, S795, and S813) in vivo (Cheng et al., 1991; Hegedus et al., 2009). However, converting all dibasic sites in the R domain to alanine significantly decreases the P_o but fails to completely abolish PKA-dependent activation of CFTR (Chang et al., 1993; Rich et al., 1993), suggesting the involvement of monobasic sites in CFTR regulation. Indeed, fifteen PKA sites must be eliminated to render CFTR channels completely PKA insensitive (Seibert et al., 1999; Hegedus et al., 2009). While most PKA sites play a stimulatory role in CFTR regulation, phosphorylation of two consensus serines (S737 and S768) has been implicated to inhibit CFTR channel function (Wilkinson et al., 1997; Csanady et al., 2005b, but cf. (Hegedus et al., 2009)). This unique inhibitory role may account for the downregulation by AMP kinase (King et al., 2009). Interestingly, a similar bimodal regulation of CFTR by protein kinase C (PKC) has been reported (Chappe et al., 2004), but the molecular details have yet to be worked out.

42.3.3.1 Mechanism of CFTR regulation by the R domain

A number of functional studies have been performed to tackle the molecular mechanism of CFTR regulation by PKA-dependent phosphorylation of the R domain. First, it was shown that deletion of a large portion of the R domain allows the channel to respond to ATP without prior phosphorylation (Rich et al., 1991). In fact, a construct with the whole R domain removed (ΔR-CFTR, missing residues 634–836) can assume a channel activity similar to that of fully phosphorylated WT-CFTR (Csanady et al., 2000; Bompadre et al., 2005a). These findings suggest that an unphosphorylated R domain inhibits channel opening, whereas phosphorylation of the R domain relieves this inhibition. Since replacing serines in PKA sites with acidic residues, that is, aspartate or glutamate, produces phenotypes very similar to that of ΔR-CFTR (Chang et al., 1993; Rich et al., 1993), part of the mechanism for phosphorylation-dependent activation of CFTR may involve an accumulation of negative charges in the R domain. However, different structural alterations likely occur with phosphorylation as the CD spectrum of the phosphorylated R domain differs significantly from that of the R domain with serine-to-glutamate mutations (Dulhanty and Riordan, 1994; Dulhanty et al., 1995). Furthermore, this theory of accumulation of negative charges may not easily explain the observation that phosphorylation of serine 737 or 768 may exert an inhibitory effect (Wilkinson et al., 1997; Csanady et al., 2005b).

As gating of CFTR is controlled by NBDs, one possible mechanism for phosphorylation-dependent activation of CFTR is for the R domain to modulate NBD function. Indeed, purified CFTR proteins exhibit significant ATPase activity only after phosphorylation by PKA (Li et al., 1996); cross-linking experiments show that phosphorylation of the R domain enhances NBD dimerization (Mense et al., 2006; He et al., 2008). However, since, as discussed earlier, NBDs and gating are functionally coupled, these results do not necessarily mean a direct action of the R domain on NBDs. Recent NMR experiments with individual fragments of CFTR (Baker et al., 2007) did provide direct evidence for a state-dependent interaction between the isolated R domain and NBD1. Since the contact points are located at or near the PKA phosphorylation sites, it was proposed that unphosphorylated R domain inhibits CFTR function by preventing dimerization of NBDs and phosphorylation relieves this inhibition (Figure 42.6).

It is worthy to note that several experimental data cannot be accounted for with this *via NBD dimerization* hypothesis. For example: (1) The activity of an NBD2-deleted CFTR mutant depends strictly on PKA phosphorylation (Cui et al., 2007), although this mutant does not respond to ATP (Wang et al., 2007, 2010b). (2) Consistent with an earlier computation study (Hegedus et al., 2008), electromicroscopic single-particle data suggest that the R domain may directly interact with the

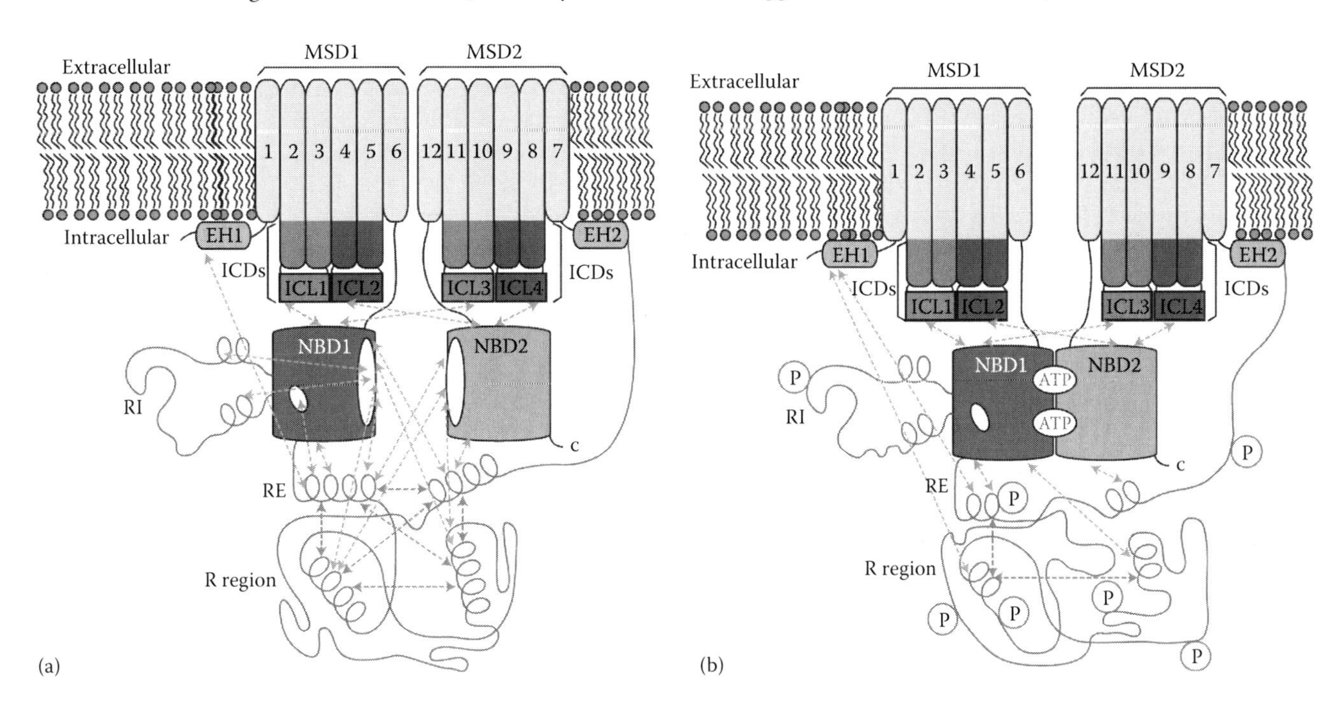

Figure 42.6 Schematic representation for the domain–domain interactions of CFTR in (a) nonphosphorylated and (b) phosphorylated states, based on NMR data. MSD, membrane-spanning domain; RI, regulatory insertion; RE, regulatory extension; EH, elbow helix; ICD, intracellular domain. Arrows indicate the identified and putative interactions between different parts of CFTR. The R region and RI in NBD1 are shown with curved lines to reflect their disordered states. Binding sites for the RI and R region on NBD1 and NBD2 are shown as white ovals in (a). Note that many of the multiple interactions within helical segments of the R region and interactions of the RI and the R region with NBD1 are disrupted upon phosphorylation, whereas interactions with HE1 are enhanced. See (Baker et al., 2007; Kanelis et al., 2010; Chong et al., 2013) for more detail. (Adapted from Chong, P.A. et al., *Cold Spring Harb. Perspect. Quant. Med.*, 3, a009522, 2013. With permission.)

cytoplasmic loops or MSDs in a phosphorylation-dependent manner (Zhang et al., 2011). (3) NMR studies with CFTR fragments also reveal phosphorylation-dependent interactions between CFTR's NBD1 and a peptide corresponding to the hypothetical coupling helix in the first intracellular loop (Kanelis et al., 2010). Thus, PKA phosphorylation might control CFTR channel gating at multiple levels. It is conceivable that parts of the R domain, by controlling the motion of MSDs, may decide if the gate of CFTR can be opened or not—thus an all-or-none control. On the other hand, other part(s) of the R domain may interact with NBDs to fine-tune the activity. More studies are needed to tackle this mysterious and yet important issue unique to CFTR (Hwang and Kirk, 2013).

42.3.4 MEMBRANE-SPANNING DOMAINS: CHANNEL PORE AND GATING MOTIONS

As an ion channel, CFTR must possess an ion-conducting pore in its MSDs. The presumed evolutionary relationship between CFTR and ABC exporters also predicts that CFTR may employ the conserved structural framework of MSDs found in all ABC exporters to enact the function of a gated pore. Interestingly, while CFTR is an anion channel, many of its close relatives (members of the ATP Binding Cassette transporter C (ABCC) subfamily) transport organic anions (e.g., Kool et al., 1999). However, unlike a transporter, CFTR's ion-conducting pathway must remain open to both sides of the membrane to permit anion diffusion down an electrochemical potential gradient. Thus, one expects similarities as well as differences in the structure and function of MSDs between CFTR and other members of the ABC exporters.

42.3.4.1 Biophysical characteristics of the CFTR channel pore

Chloride is the most prevailing permeant anions for the CFTR channel in tissues that mediate salt and fluid transport such as exocrine glands (Wine and Joo, 2004). Generally, chloride channels are known to be less selective among different anions in comparison to K^+ or Na^+ channels (Hille, 1973) probably because there is really no evolutionary pressure for an anion channel to develop strict selective mechanism for the most abundant anions in nearly all organisms. Thus, it is perhaps not surprising that the CFTR pore also affords a significant HCO_3^- conduction, which may play an important physiological role in the exocrine pancreas (Gray et al., 1990; Ishiguro et al., 2009) as well as in the airways (Quinton, 2001, 2008; Wang et al., 2003).

The selectivity of an ion conduction pore can be biophysically characterized by measuring the relative conductance, an index for the *throughput* rate of a particular ion, or the relative permeability, an index usually reflecting the easiness of an ion to enter the pore. The relative conductance sequence of CFTR follows $Cl^- > NO_3^- > Br^-$ acetate $> I^- > SCN^-$ (McCarty and Zhang, 2001), but the permeability sequence significantly differs: acetate $SCN^- > NO_3^- > Br^- > Cl^- > I^- >$ acetate (Linsdell et al., 2000). The observation that the permeability sequence follows a lyotropic sequence of small anions suggests that the dehydration of the anion from the bulk water constitutes the rate-limiting step for the pore entry (Smith et al., 1999). Based on the relationship between the permeability sequence and the hydration energies of the permeant anions, CFTR pore was modeled as a *dielectric or polarizable tunnel* with no specific interactions between the permeant anions and the pore walls (Smith et al., 1999; Liu et al., 2003). However, recent biophysical studies targeting various TMs of CFTR raised the possibility that permeant anions may physically interact with the pore-lining residues (see the following text).

As discussed earlier, physiologically, chloride and bicarbonate are two most important anions passing though the CFTR channel pore. The permeability ratio of bicarbonate over chloride ($P_{HCO_3^-}/P_{Cl^-}$) is ~0.3 to 0.5 and the conductance ratio ($G_{HCO_3^-}/G_{Cl^-}$) is also ~0.3 (Gray et al., 1990; O'Reilly et al., 2000). A recent study suggests that this relative selectivity of the CFTR pore between chloride and bicarbonate might be dynamically regulated by the intracellular signaling system, for example, WNK1-OSR1/SPAK pathway, in vivo (Park et al., 2010), although the molecular details have yet to be worked out.

42.3.4.2 Molecular structure of the CFTR channel pore

A number of site-directed mutagenesis and substituted-cysteine accessibility method (SCAM) experiments on CFTR's TMs have been performed to study the molecular structure of the permeation pore. Soon after the CFTR gene was cloned, site-directed mutagenesis was employed to test the idea that CFTR by itself encodes the cAMP-dependent chloride conductance. Anderson et al. (1991b) reported that mutating two lysines (K95 and K335) in predicted TMs 1 and 6 respectively to acidic residues altered the relative halide permeability sequence of the cAMP-activated current in CFTR-expressing cells (Anderson et al., 1991b), suggesting that these two TMs play a functional role in forming the chloride permeation pathway.

TM6 has since been most intensively studied. For example, mutations at residues F337 and T338 in TM6 also affected the halide permeability sequence (Linsdell et al., 2000). Moreover, the membrane-impermeant large methanethiosulfonate (MTS) reagents applied from the external side could react with cysteines introduced at several positions from 331 to 338, whereas the small-channel-permeant thiol-reactive pseudohalides ($Ag(CN)_2$ and $Au(CN)_2$) reached positions along the entire length of TM6 (to 353) (Alexander et al., 2009) (Figure 42.7A) (cf. (Fatehi and Linsdell, 2008)). These results support the idea that TM6 residues line the pore and strongly supported the proposed narrowing near position 338 from the extracellular end of the pore (Linsdell et al., 2000; McCarty and Zhang, 2001). On the other hand, applications of bulky MTS reagents from the cytoplasmic side of the channel identified position 341 as the accessibility limit (Bai et al., 2010) (Figure 42.7A) (cf. (El Hiani and Linsdell, 2010)). Thus, it appears that the pore is constructed with two fairly accessible internal and external vestibules with a *bottleneck* that traverses only one helical turn along TM6 (Norimatsu et al., 2012a). The twofold symmetry of MSDs seen in other ABC exporters (Dawson and Locher, 2006; Ward et al., 2007; Aller et al., 2009) (Figure 42.7A) suggests that TM12 should also contribute to the CFTR channel pore. Indeed, the SCAM studies revealed that the cytoplasmic half of TM12 likely assumes an α-helix structure and contributes to the pore formation (Bai et al., 2011; Qian et al., 2011). Besides TM6 and TM12, SCAM studies also implicate TM1, TM3, and TM9 as pore-forming segments (Wang et al., 2011; Norimatsu et al., 2012a; Gao et al., 2013).

TM 6

	Linsdell[37] (outside)	Dawson (outside)	Hwang[35] (inside)	Linsdell[36] (inside)
G330		–	–	
I331		MTS/Ag	–	
I332		–	–	
L333		MTS/Ag	–	
R334	MTS	MTS/Ag	–	–
K335	MTS	MTS/Ag	–	–
I336		MTS/Ag	–	–
F337	MTS	† MTS/Ag	–	MTS
T338	MTS	MTS/Ag	–	MTS
T339		Ag	–	–
I340		Ag	–	–
S341	MTS	Ag	MTS	MTS
F342		Ag	–	–
C343		–	–	–
I344		Ag	MTS	MTS
V345		Ag	MTS	MTS
L346		–	–	–
R347		–	–	–
M348		Ag	MTS	MTS
A349		Ag	–	MTS
V350		–	–	–
T351		–	–	–
R352		Ag	MTS	MTS
Q353		Ag	MTS	MTS
F354		–	–	–
P355		Ag		–
W356		Ag		
A357		–		
V358		–		

TM 12

	Linsdell[38] (outside)	Dawson (outside)	Hwang[9] (inside)	Linsdell[10] (inside)
G1125	–	–		
E1126	–	–		
G1127	MTS	MTS/Ag		
R1128	–	–		
V1129	MTS	Ag	–	
G1130	–	–	–	
I1131	MTS	MTS/Ag	–	–
I1132	MTS	–	–	–
L1133	–	–	–	–
T1134	–	Ag	–	–
L1135	–	Ag	–	–
A1136	–	–	–	–
M1137		Ag	–	–
N1138		Ag	–	MTS
I1139		Ag	–	–
M1140		–	MTS	MTS
S1141		Ag	MTS	MTS
T1142		Ag	MTS	MTS
L1143		–	–	–
Q1144		Ag	MTS	MTS
W1145		Ag	MTS	MTS
A1146		–	–	–
V1147		Ag	MTS	MTS
N1148		Ag	MTS	MTS
S1149		Ag	–	MTS
S1150		–	MTS	
I1151		–		

(A)

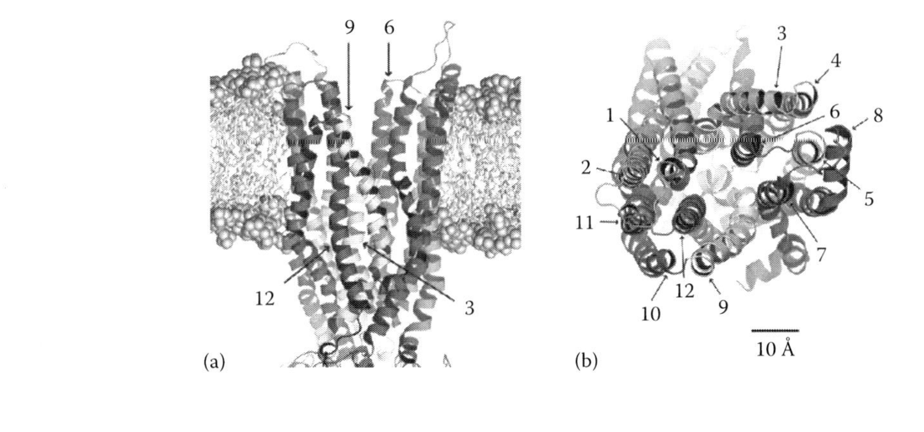

(B)

Figure 42.7 Putative structure of the CFTR pore. (A) Comparison of cysteine scanning results for TM6 and TM12 published from three laboratories. Each column shows SCAM results reported by the indicated laboratories using either externally applied or internally applied MTS reagents and/or $[Ag(CN)_2]^-$. Bold dash: no reactivity; Blank: not tested. (B) A homology model of CFTR's MSDs based on the crystal structure of Sav1866. Side (a) and top (b) views of the modeled CFTR protein inserted into a 1,2-dimyristoyl-*sn*-glycero-3-phosphocholine (DMPC) bilayer. Individual transmembrane segments are numbered. See Norimatsu et al. (2012a) for detail. (Adapted with permission from Norimatsu, Y., Ivetac, A., Alexander, C., Kirkham, J., O'Donnell, N., Dawson, D.C. and Sansom, M.S., Cystic fibrosis transmembrane conductance regulator: A molecular model defines the architecture of the anion conduction path and locates a *bottleneck* in the pore, *Biochemistry*, 51, 2199–2212. Copyright 2012a American Chemical Society.)

In addition to testing TMs as potential pore-lining segments, SCAM studies could also offer insights into possible gating conformational changes. For example, cysteines engineered into the cytoplasmic side of TM6 or TM12 can be modified readily by internal MTS reagents with or without ATP, indicating that this part of the pore is accessible in both open state and closed state (Bai et al., 2010, 2011; but cf. (El Hiani and Linsdell, 2010)). However, two types of gating motion were implicated: First, the alternate pattern of changes in the modification rate by ATP is consistent with the idea that TM12 may undergo a rotational movement upon gating (Bai et al., 2011). Second, since cysteines introduced in TM6 or TM12 can be modified by a 13-Å MTS reagent (MTSEA-Texas Red) in the closed state, but not in the open state, Bai et al. (2011) proposed a translational movement of TM6 and TM12 so that the internal entrance of the CFTR pore becomes smaller upon channel opening. Of note, these two proposed gating motions for CFTR find parallel propositions for ABC exporters (Procko et al., 2009; Gutmann et al., 2010).

Recent SCAM experiments on TM1 lend further support to the hypothesis that CFTR and other ABC exporters may share significant structural similarities in MSDs. Not only was the idea of a *bottleneck* region in the pore confirmed in TM1 (L102–I106), the periodicity of reactive positions also implicates a secondary structure of an α-helix (Gao et al., 2013). However, unlike SCAM results with TM6 and TM12, a strict open state-dependent modification was observed for TM1. This intriguing but puzzling result is actually consistent with the crystal structures of ABC exporters solved in different states: while TM1 forms part of the substrate translocation pathway in the *outward-facing* (equivalent to the open state of CFTR) configuration of a bacterial ABC exporter, SAV1866 (Dawson and Locher, 2006), several *inward-facing* (equivalent to the closed state of CFTR) structures of ABC exporters show that TM1, located on the periphery of the whole MSD complex, does not line the substrate translocation pathway (Ward et al., 2007; Aller et al., 2009; Hohl et al., 2012).

A latest SCAM study performed on TM5, TM7, and TM11 (Wang et al., 2014) also showed that several residues in TM11 were accessible to extracellular and/or intracellular cysteine reactive reagents, whereas no reactive cysteines were identified in TM5 or TM7. However, the reported state dependence of modification at 1115C and 1118C (Wang et al., 2014) contradicts the pattern reported in Bai et al. (2011) for TM12. Nevertheless, the surprising absence of any positive hit in TM7 of MSD2, which is equivalent to TM1 of MSD1, suggests asymmetric contributions of TMs to the open channel pore structure.

Taken together, these results lead to a rough picture of the CFTR pore: (1) At least 6 TMs (TM1, TM3, TM6, TM9, TM11, and TM12) line the chloride permeation pathway of CFTR. (2) The pore consists of a narrow region flanked by fairly sizable internal and external vestibules. (3) Opening and closing of the CFTR channel entail not only translational movements of two MSDs depicted in many cartoons describing alternating access of the substrate translocation pathway of ABC transporters, but also significant rearrangements of TMs as shown in different crystal structures of ABC exporters.

42.3.4.3 CFTR pore blockers and vestibules

A variety of agents with diverse chemical structures, for example, arylaminobenzoates and sulfonylureas, inhibit CFTR. These CFTR blockers share several characteristics; that is, all are anions, most are lipophilic, and many are large in size. Diphenylamine-2-carboxylate (DPC) (Di Stefano et al., 1985) and 5-nitro-2-(3-phenylpropylamino)-benzoic acid (NPPB) (Wangemann et al., 1986) are representative CFTR blockers of arylaminobenzoates. In the early stage, these blockers provided structural insights into the CFTR channel pore. The DPC inhibition of CFTR was dependent on the membrane potential and the external Cl^- concentration, suggesting that DPC inhibited CFTR by occluding the channel pore (McDonough et al., 1994). Site-directed mutagenesis studies suggested that both S341 in TM6 and S1141 in TM12 contributed to DPC binding (McDonough et al., 1994), inferring that both TM6 and TM12 line the CFTR pore.

Sulfonylureas, such as glibenclamide and tolbutamide, have been used to treat non-insulin-dependent diabetes mellitus (Panten et al., 1996) because of their inhibitory effects on the ATP-sensitive K^+ channels (K_{ATP} channels). Since the molecular target of sulfonylureas, SUR, is a member of the ABC transporter family, these chemicals were tested on CFTR by several investigators (Schultz et al., 1996; Venglarik et al., 1996; Sheppard and Robinson, 1997). Kinetic analysis indicates that glibenclamide and tolbutamide are categorized respectively into *intermediate* and *fast* open channel blockers of CFTR (Venglarik et al., 1996; Sheppard and Welsh, 1999; Zhou et al., 2002; Cui et al., 2012).

In addition to arylaminobenzoates and sulfonylureas, the CFTR channel can be blocked by a chemically diverse group of large organic anions including some common anion channel inhibitors, for example, the disulfonic stilbenes, 4,49-diisothiocyanostilbene-2,29-disulfonic acid (DIDS), and 4,49-dinitrostilbene-2,29-disulfonic acid (DNDS) (Linsdell and Hanrahan, 1996a), gluconate and glutamate (Linsdell and Hanrahan, 1996b) and 9-AC (Ai et al., 2004). Again, these blockers work when applied to the cytosolic side, but not to the extracellular side of the membrane. Furthermore, these CFTR blockers induced a voltage-dependent block as if their binding sites were located 15%–60% of the way through the transmembrane electric field from the cytoplasmic end. Because of their enormous size, the existence of a large intracellular vestibule that accommodates these sizeable molecules was proposed. As the high-affinity blockade is usually associated with more hydrophobic blocker, it was also speculated that the intracellular vestibule should include some hydrophobic residues (Zhou et al., 2002).

The internal vestibule might also contain some positively charged residues (e.g., K95, R303, and R352), as suggested from SCAM results on TM1 and TM6 (Bai et al., 2010; El Hiani and Linsdell, 2010; Wang et al., 2011; Gao et al., 2013) as well as mutagenesis studies that show an effect on the potency of the blocker by neutralizing these positively charged residues (Linsdell, 2005; Aubin and Linsdell, 2006; Zhou et al., 2010). Although it was proposed that some of these positive charges may actually form part of the blocker binding site (Linsdell, 2005), in light of the large caliber of the internal vestibule and significant voltage dependence of the blockade, it is perhaps more likely that these positive charges in the internal vestibule serve to increase the *capture radius* of the pore (Smith et al., 1999) and attract anions to the pore entrance.

For the extracellular vestibule, a similar role of a positively charged residue, R334, has been proposed to facilitate anion entry from the extracellular side (Smith et al., 2001). Recently, it was reported that GlyH-101, a CFTR-specific inhibitor discovered by the high-throughput screening (Muanprasat et al., 2004), blocked the CFTR currents with an opposite polarity of the voltage dependence to that of the internal blockers. Mutational studies as well as molecular dynamics (MD) simulations lead the authors to propose that GlyH-101 blocked the channel pore by entering from the extracellular side and binding to a site within the pore (Norimatsu et al., 2012b).

In summary, the open CFTR pore looks like an hourglass with a narrow region made of just a few amino acids that may serve as a low-grade anion selectivity filter. The two vestibules sandwiching this *bottleneck* region show significant asymmetry in physical dimension as well as chemical properties. The existence of a binding site for large hydrophobic anions in the internal vestibule may reflect the evolutionary relationship between CFTR and its ancestors, which transport organic anions out of the cell using ATP hydrolysis as the source of energy (Chen and Hwang, 2008).

42.4 MOLECULAR PATHOPHYSIOLOGY AND PHARMACOLOGY OF CFTR

42.4.1 CFTR BIOSYNTHESIS AND MEMBRANE TRAFFICKING

Like all other membrane proteins, before the CFTR can carry out its function in the cell membrane, numerous cell biological processes need to proceed uneventfully. The CFTR gene is first transcribed into mRNA in the nucleus. The mRNA is then translated to CFTR proteins in the endoplasmic reticulum (ER). In the CFTR folding process, it is suggested that whereas NBD1 folds largely cotranslationally, the folding of the full channel including NBD2 is completed posttranslationally (Du et al., 2005; Kleizen et al., 2005; Du and Lukacs, 2009).

Luminal chaperons such as calnexin and calreticulin support the CFTR protein to fold into the right conformation (Harada et al., 2006; Rosser et al., 2008). The folding of the cytosolic domains of CFTR is supported by both Hsc70/Hsp70 and Hsp90 with a subset of co-chaperones (e.g., Hdj2 (DNAJ1), HsBp1, Hop, and p23) and small heat shock proteins (Wang et al., 2003; Alberti et al., 2004; Sun et al., 2008; Glozman et al., 2009). In the Golgi apparatus, the CFTR protein is fully matured by the attachment of oligosaccharide side chains at two consensus Asn-linked *N*-glycosylation sites (asparagine residues) in the 4th extracellular loop within MSD2 (Lukacs et al., 1994; Ward and Kopito, 1994), which may also stabilize the CFTR fold directly (Glozman et al., 2009).

Finally, the mature, fully glycosylated, CFTR complex enters the secretory system via coat protein complex II (COPII) transport vesicles from the Golgi apparatus (Wang et al., 2006; Younger et al., 2006). Once CFTR reaches the cell membrane, the half-life of WT-CFTR is approximately 12–24 h (Lukacs et al., 1993; Ward and Kopito, 1994). Subsequently, plasma membrane CFTR is internalized through clathrin-coated endosomes and degraded in lysosomes, but some populations of CFTR are recycled back to the plasma membrane (Lukacs et al., 1997).

The quality control of CFTR is performed in the ER membrane and the cytoplasmic compartment (Meacham et al., 2001; Younger et al., 2004, 2006). The folding process of heterogeneously transfected CFTR proteins in ER is not so efficient that the misfolding rate is higher than 50% in WT CFTR (Lukacs et al., 1994; Ward and Kopito, 1994), whereas the maturation efficiency of CFTR endogenously expressed in two epithelial cell lines, Calu-3 and T84, was reported to be nearly 100% (Varga et al., 2004). In ER, misfolded CFTR proteins are ubiquitin (Ub) tagged by the protein quality control system and degraded by the ubiquitin proteasome system (UPS) (ER-associated degradation [ERAD]) (Lukacs et al., 1994; Ward and Kopito, 1994; Ward et al., 1995). Several E3 Ub ligases, for example CHIP, RMA1, Gp78, Nedd4–2, and Fbs1, and E4 Ub ligase, gp79, have been reported to recognize misfolded CFTR in a chaperone-dependent (CHIP, RMA1) or chaperone-independent (Gp78, Ndee4–2, Fbs1) manner and catalyze the cotranslational and posttranslational ubiquitination of nonnative nascent CFTR chains (Meacham et al., 2001; Yoshida et al., 2002; Younger et al., 2006; Morito et al., 2008; Caohuy et al., 2009). On the other hand, the ER-anchored USP19 (ubiquitin specific protease 19) and soluble UCH-L1 (ubiquitin C-terminal hydrolyse-L1) may deubiquitinate CFTR and hence modulate ERAD (Hassink et al., 2009; Henderson et al., 2010).

In addition to the ER quality control system, senescent or poorly functioning CFTR in the plasma membrane is also recognized and ubiquitinated in a chaperone (Hsc70, Hsp90)-dependent (CHIP, gp78) or chaperone-independent (Ndee4–2, cCbl) manner (Okiyoneda et al., 2010; Ye et al., 2010), rapidly internalized, and removed from the plasma membrane. Ubiquitinated CFTRs are subsequently recognized by the ESCRT0-I components (e.g., Hrs, STAM-1 and Tsg101) at early endosomes and rerouted from the recycling pathway toward the lysosomal degradation (Sharma et al., 2004; Okiyoneda et al., 2010).

42.4.2 MUTATIONS THAT LEAD TO CF

More than 1900 disease-associated CFTR mutations have been identified (http://www.genet.sickkids.on.ca/cftr) since the cloning of CFTR in 1989. These CF-associated mutations are categorized into five classes I–V according to the mechanisms underlying the CFTR dysfunctions (Welsh and Smith, 1993; Zielenski and Tsui, 1995).

1. Class I: defective protein production (e.g., W1282X)
   Class I mutations fail to produce full-length CFTR proteins because of nonsense mutations introducing stop codons into the CFTR gene.
2. Class II: defective protein processing (e.g., ΔF508)
   Class II mutations cause misfolding of the CFTR protein in the ER. The misfolded proteins then fail to be transported to the Golgi apparatus for maturation and thus lead to low expression of the CFTR protein in the cell membrane. The proteins retained in the ER are subsequently degraded by the proteasome (ERAD).
3. Class III: defective channel regulation (e.g., G551D)
   Class III mutations cause a loss of CFTR channel function by disrupting gating, although the mutant protein can process normally to the plasma membrane.
4. Class IV: defective channel conduction (e.g., R347P)
   Class III mutations impair the anion conduction through the open channel pore, whereas their PKA-dependent regulation and ATP-dependent gating appear to be intact.
5. Class V: reduced synthesis
   Class V mutations reduce CFTR gene transcription by altering the promoter sequence or inducing alternative splicing variants.

42.4.3 MECHANISM OF CFTR DYSFUNCTION

In this section, we will focus our discussion on two CF-associated mutations, ΔF508 (class II) and G551D (class III), not only because of their prevalence (first and third most common respectively), but also because of extensive literatures over the molecular mechanism underlying CFTR dysfunction caused by these mutations.

A deletion of phenylalanine at position 508 (ΔF508) is the most common CF-associated mutation present in nearly 90% of CF patients (Welsh and Smith, 1993). The F508 residue is located in NBD1 close to its interface with the fourth cytoplasmic loop (CL4). This mutation likely destabilizes NBD1 and its interaction with other domains (Serohijos et al., 2008; Rabeh et al., 2012), resulting in both trafficking and gating defects (Cui et al., 2006). In addition, when ΔF508-CFTR does reach the cell membrane, it exhibits a shorter half-life (~4 h) than WT channels because of either more rapid ubiquitin-dependent degradation or, in some cases, defective recycling (Lukacs et al., 1993; Swiatecka-Urban et al., 2005; Chang et al., 2008; Okiyoneda et al., 2010).

Early biochemical studies of isolated NBD1 with the ΔF508 mutation showed that deletion of F508 did not affect the backbone structure and thermodynamic stability significantly whereas its refolding is impaired (Qu and Thomas, 1996). Indeed, both crystallographic data (Lewis et al., 2005; Thibodeau et al., 2005; Atwell et al., 2010) and mass spectral measurements (Lewis et al., 2010) suggested minimal structural differences between WT- and ΔF508-NBD1 that is restricted to the flexible surface loop at residues 509–511, an experimental result corroborated by molecular dynamic simulations (Wieczorek and Zielenkiewicz, 2008; Bisignano and Moran, 2010; Lewis et al., 2010). On the other hand, recent isothermal calorimetric studies suggested that the ΔF508 mutation results in both kinetic and thermodynamic folding defects in the isolated NBD1 domain (Protasevich et al., 2010; Wang et al., 2010a).

Although the exact mechanism by which a local perturbation by the ΔF508 mutation in NBD1 leads to global misfolding of the ΔF508-CFTR protein is unknown, one possibility is that individual domains in CFTR are loosely folded and assembled cotranslationally, but a cooperative posttranslational folding involving proper assembly of the NBDs–MSDs interfaces is needed for acquiring the final native fold of the whole protein (Du and Lukacs, 2009). That is, the ΔF508-induced impaired assembly at the interface between NBD1 and MSDs might destabilize the conformations of MSD1, MSD2, and NBD2 (Du et al., 2005; Thibodeau et al., 2005; Cui et al., 2007; Rosser et al., 2008; Du and Lukacs, 2009). A subsequent failed cooperative assembly of the first four CFTR domains, MSD1, MBD1, RD, and MSD2, causes misfolding of NBD2 (Du et al., 2005; Cui et al., 2007) and other structural defects at the interfaces between MSD1 and MSD2 and between NBD1 and NBD2 (Loo et al., 2009; He et al., 2010; Jih et al., 2011) (see Figure 42.8). Once the ΔF508-CFTR protein is misfolded, it is degraded by the quality control system in the ER.

ΔF508-CFTR is also known to exhibit grave gating defects in addition to the trafficking defect described earlier (Dalemans et al., 1991; Hwang et al., 1997; Ostedgaard et al., 2007). The P_o of ΔF508-CFTR is ~15-fold less than that of WT-CFTR; however, a high-affinity ATP analog, N^6-(2-phenylethyl)-2′-deoxy-ATP (P-dATP), can completely rectify this gating defect by acting on CFTR's two ATP-binding sites (Miki et al., 2010). In addition to a low steady-state P_o, several interesting functional abnormalities were associated with the ΔF508 mutation. These include reduced lock-open time in cases when ATP hydrolysis is abolished and a destabilized C2 closed state (Jih et al., 2011), suggesting that the ΔF508 mutation may affect the stability of the partial and full NBD dimeric states (see Figure 42.4b in Section 42.3.2).

The glycine-to-aspartate missense mutation at position 551 (G551D) is the most well-known CF-associated mutation that cause primarily gating defects (Welsh and Smith, 1993). The G551 residue is located in the ABC transporter signature sequence in NBD1, which forms ABP2 with the Walker A/B motif of NBD2. G551D-CFTR exhibits a very low P_o resulting from a loss of ATP-dependent gating (Welsh and Liedtke, 1986; Li et al., 1996; Cai et al., 2006; Bompadre et al., 2007), whereas the mutant protein can be trafficked normally to the plasma membrane and is phosphorylated by cAMP-dependent protein kinase (Cutting et al., 1990; Chang et al., 1993; Welsh and Smith, 1993; Li et al., 1996). The idea that ATP binding to ABP2 plays a critical role in channel opening (Zhou et al., 2006) via NBD dimerization (Vergani et al., 2005) satisfactorily explains why the G551D mutation abolishes channel's response to ATP as the negatively charged side chain of aspartate at 551 may exert an electrostatic repulsion with the phosphate group in ATP upon NBD dimerization (Bompadre et al., 2007).

42.4.4 CFTR PHARMACOLOGY

For a more comprehensive coverage of CFTR pharmacology, readers are referred to several reviews focused on this topic (Hwang and Sheppard, 1999; Amaral and Kunzelmann, 2007; Cai et al., 2011; Hanrahan et al., 2013). In this section,

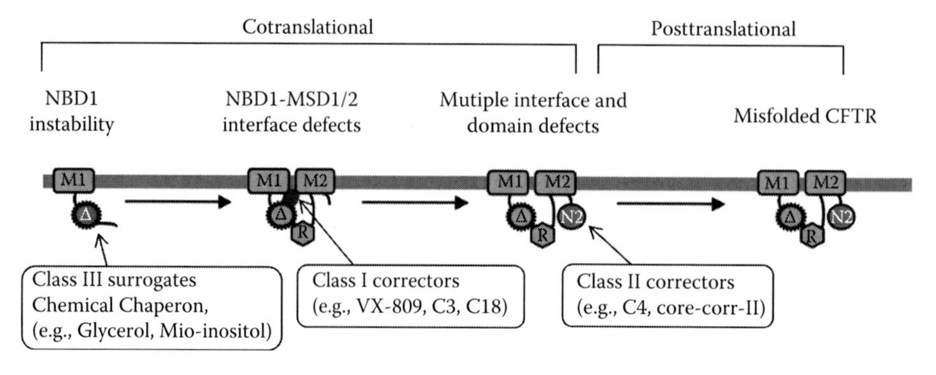

Figure 42.8 Working model for misfolding of ΔF508-CFTR and mechanism-based classification of CFTR correctors. Chemical chaperones such as glycerol and myo-inositol are considered as surrogates of class III correctors. See Okiyoneda et al. (2013) for detail. (Adapted from Okiyoneda, T. et al., *Nat. Chem. Biol.*, 9, 444, 2013. With permission.)

we will emphasize two recent directions of pharmaceutical development that are aimed to correct the folding defect for class II mutants especially the ΔF508 mutation (i.e., correctors), or to potentiate the activity of class III mutants such as G551D (i.e., potentiators).

Ever since the cloning of the CFTR gene makes it possible to characterize CFTR function in vitro, tremendous efforts have been placed in finding compounds that can increase the P_o of CFTR (Hwang and Sheppard, 1999). Despite a plethora of papers reporting such reagents over a wide spectrum of chemical scaffolds, success in translating these findings to clinical applications was limited until the discovery of VX-770 by high-throughput drug screening (Van Goor et al., 2009). Soon after the discovery of VX-770 on the bench, encouraging results were reported from clinical trials (Accurso et al., 2010; Ramsey et al., 2011), followed without delay by the approval of VX-770 for CF treatment by the FDA.

On the basic science front, a latest study unravels the unique mechanism by which VX-770 increases the P_o of CFTR (Jih and Hwang, 2013). What intrigued these investigators was the observation that VX-770 increases the activity of G551D-CFTR, a mutant that does not respond to ATP (Bompadre et al., 2007), and prolongs the open time of WT-CFTR, which is mainly controlled by ATP hydrolysis (Hwang et al., 1994; Carson et al., 1995; Powe et al., 2002b; Vergani et al., 2003; Bompadre et al., 2005b). It is known that the classical methods to increase the open time of WT-CFTR by disabling ATP hydrolysis with either mutations or nonhydrolyzable ATP analogs fail to work on the G551D channels (Cai et al., 2006; Bompadre et al., 2007). However, the newly proposed gating model in Jihand Hwang (2013) bequeaths a testable hypothesis that VX-770, by promoting gate opening (vertical transitions in Figure 42.5), can increase the activity of G551D-CFTR even though this mutation eliminates the transitions leading to NBD dimerization, and at the same time prolongs the open time of WT-CFTR by increasing [ATP]-dependent *reentry* events (see Section 42.3.2.4 for more detail). Indeed, several predictions based on this hypothesis were verified experimentally (Jih and Hwang, 2013). Importantly, the mechanism revealed for the action of VX-770 also predicts that this compound can be considered as a universal potentiator as it shifts the equilibrium of gating transitions toward the open channel conformation irrespective of the status of the gating machinery, NBDs. Indeed, VX-770 was reported to enhance the activity of several class III mutants (Yu et al., 2012). By understanding how VX-770 works mechanistically, one can also start to entertain the idea of how to complement the action of this drug by developing reagents targeting different domains of CFTR (Jih and Hwang, 2013).

High-throughput screening assays can be effectively used to find CFTR correctors. Indeed, VX-809 was developed and confirmed to increase the cell surface expression of functional ΔF508-CFTR in vitro (Van Goor et al., 2011). While the mechanism of its action is unclear, studies along this direction start to appear in the literature (He et al., 2013) and are expected to bloom in the foreseeable future. Nonetheless, clinical trials on CF patients with the ΔF508 mutation were not highly satisfactory (Clancy et al., 2012) probably due to the fact that the gating defect associated with the ΔF508 mutation is not rectified by VX-809. Ongoing clinical trials with a combination of VX-809 and VX-770 (NCT01225211) are clearly the right direction for this issue, and results may soon emerge.

Recently, Okiyoneda et al. (2013) reported combined effects of different CFTR correctors on ΔF508-CFTR and subsequently proposed a mechanism-based classification of CFTR correctors: class I correctors (VX-809, C3, C18) target the primary conformational defect at the NBD1-MSD1 or NBD1-MSD2 interfaces; class II correctors (e.g., C4, core-corr II) rectify NBD2 misassembly; class III correctors (chemical chaperons, e.g., glycerol, myo-inositol) promote thermodynamic stability of the ΔF508-NBD1. A combination of class I and/or class II with class III surrogates showed robust synergistic effects on the processing defects of ΔF508-CFTR (Okiyoneda et al., 2013) (see Figure 42.8). This new study thus documents the importance of a fundamental understanding about the molecular mechanism of disease-associated CFTR mutations in future therapeutical design for CF.

42.5 CONCLUDING REMARKS

CFTR is the only member of the ABC transporter superfamily that functions as an anion channel and hence has been most extensively studied at a functional level. More than two decades of investigations have advanced enormously our understanding of the structure/function relationship of CFTR. While these studies will go on, the field will benefit greatly once the high-resolution structure of CFTR is solved. From the basic science point of view, this kind of breakthrough, together with already available functional data, will illuminate the evolutionary path of a transporter-to-channel metamorphosis. On the other hand, the high-resolution crystal structure of CFTR is expected to establish the foundation for rational drug design for CF and other CFTR-associated diseases.

ACKNOWLEDGMENT

We thank our laboratory colleagues for valuable discussions. During the preparation of this review, Y. Sohma was supported by JSPS KAKENHI Grant Number 22590212, 25293049, MEXT KAKENHI Grant Number 23118714, and Keio Gijuku Academic Development Funds, and T. C. Hwang by the National Institutes of Health (grant R01DK55835) and a grant (Hwang11P0) from the Cystic Fibrosis Foundation.

REFERENCES

Accurso, F.J., Rowe, S.M., Clancy, J.P., Boyle, M.P., Dunitz, J.M., Durie, P.R., Sagel, S.D et al. (2010) Effect of VX-770 in persons with cystic fibrosis and the G551D-CFTR mutation. *The New England Journal of Medicine*, **363**, 1991–2003.

Ai, T., Bompadre, S.G., Sohma, Y., Wang, X., Li, M., and Hwang, T.C. (2004) Direct effects of 9-anthracene compounds on cystic fibrosis transmembrane conductance regulator gating. *Pflugers Archiv: European Journal of Physiology*, **449**, 88–95.

Alberti, S., Bohse, K., Arndt, V., Schmitz, A., and Hohfeld, J. (2004) The cochaperone HspBP1 inhibits the CHIP ubiquitin ligase and stimulates the maturation of the cystic fibrosis transmembrane conductance regulator. *Molecular Biology of the Cell*, **15**, 4003–4010.

Aleksandrov, A.A., Kota, P., Aleksandrov, L.A., He, L., Jensen, T., Cui, L., Gentzsch, M., Dokholyan, N.V., and Riordan, J.R. (2010) Regulatory insertion removal restores maturation, stability and function of DeltaF508 CFTR. *Journal of Molecular Biology*, **401**, 194–210.

Aleksandrov, L., Aleksandrov, A.A., Chang, X.B., and Riordan, J.R. (2002) The first nucleotide binding domain of cystic fibrosis transmembrane conductance regulator is a site of stable nucleotide interaction, whereas the second is a site of rapid turnover. *The Journal of Biological Chemistry*, **277**, 15419–15425.

Alexander, C., Ivetac, A., Liu, X., Norimatsu, Y., Serrano, J.R., Landstrom, A., Sansom, M., and Dawson, D.C. (2009) Cystic fibrosis transmembrane conductance regulator: Using differential reactivity toward channel-permeant and channel-impermeant thiol-reactive probes to test a molecular model for the pore. *Biochemistry*, **48**, 10078–10088.

Aller, S.G., Yu, J., Ward, A., Weng, Y., Chittaboina, S., Zhuo, R., Harrell, P.M. et al. (2009) Structure of P-glycoprotein reveals a molecular basis for poly-specific drug binding. *Science*, **323**, 1718–1722.

Amaral, M.D. (2011) Targeting CFTR: How to treat cystic fibrosis by CFTR-repairing therapies. *Current Drug Targets*, **12**, 683–693.

Amaral, M.D. and Kunzelmann, K. (2007) Molecular targeting of CFTR as a therapeutic approach to cystic fibrosis. *Trends in Pharmacological Sciences*, **28**, 334–341.

Andersen, D.H. (1938) Cystic fibrosis of the pancreas and its relation to celiac disease: A clinical and pathological study. *The American Journal of Disease of Children*, **56**, 344–399.

Anderson, M.P., Berger, H.A., Rich, D.P., Gregory, R.J., Smith, A.E., and Welsh, M.J. (1991a) Nucleoside triphosphates are required to open the CFTR chloride channel. *Cell*, **67**, 775–784.

Anderson, M.P., Gregory, R.J., Thompson, S., Souza, D.W., Paul, S., Mulligan, R.C., Smith, A.E., and Welsh, M.J. (1991b) Demonstration that CFTR is a chloride channel by alteration of its anion selectivity. *Science*, **253**, 202–205.

Atwell, S., Brouillette, C.G., Conners, K., Emtage, S., Gheyi, T., Guggino, W.B., Hendle, J. et al. (2010) Structures of a minimal human CFTR first nucleotide-binding domain as a monomer, head-to-tail homodimer, and pathogenic mutant. *Protein Engineering, Design & Selection*, **23**, 375–384.

Aubin, C.N. and Linsdell, P. (2006) Positive charges at the intracellular mouth of the pore regulate anion conduction in the CFTR chloride channel. *The Journal of General Physiology*, **128**, 535–545.

Bai, Y., Li, M., and Hwang, T.C. (2010) Dual roles of the sixth transmembrane segment of the CFTR chloride channel in gating and permeation. *The Journal of General Physiology*, **136**, 293–309.

Bai, Y., Li, M., and Hwang, T.C. (2011) Structural basis for the channel function of a degraded ABC transporter, CFTR (ABCC7). *The Journal of General Physiology*, **138**, 495–507.

Baker, J.M., Hudson, R.P., Kanelis, V., Choy, W.Y., Thibodeau, P.H., Thomas, P.J., and Forman-Kay, J.D. (2007) CFTR regulatory region interacts with NBD1 predominantly via multiple transient helices. *Nature Structural & Molecular Biology*, **14**, 738–745.

Bals, R., Hubert, D., and Tummler, B. (2011) Antibiotic treatment of CF lung disease: From bench to bedside. *Journal of Cystic Fibrosis: Official Journal of the European Cystic Fibrosis Society*, **10**(Suppl 2), S146–S151.

Basso, C., Vergani, P., Nairn, A.C., and Gadsby, D.C. (2003) Prolonged nonhydrolytic interaction of nucleotide with CFTR's NH2-terminal nucleotide binding domain and its role in channel gating. *The Journal of General Physiology*, **122**, 333–348.

Bear, C.E., Li, C.H., Kartner, N., Bridges, R.J., Jensen, T.J., Ramjeesingh, M., and Riordan, J.R. (1992) Purification and functional reconstitution of the cystic fibrosis transmembrane conductance regulator (CFTR). *Cell*, **68**, 809–818.

Becq, F., Mall, M.A., Sheppard, D.N., Conese, M., and Zegarra-Moran, O. (2011) Pharmacological therapy for cystic fibrosis: From bench to bedside. *Journal of Cystic Fibrosis: Official Journal of the European Cystic Fibrosis Society*, **10**(Suppl 2), S129–S145.

Bisignano, P. and Moran, O. (2010) Molecular dynamics analysis of the wild type and dF508 mutant structures of the human CFTR-nucleotide binding domain 1. *Biochimie*, **92**, 51–57.

Bobadilla, J.L., Macek, M., Jr., Fine, J.P., and Farrell, P.M. (2002) Cystic fibrosis: A worldwide analysis of CFTR mutations—Correlation with incidence data and application to screening. *Human Mutation*, **19**, 575–606.

Bompadre, S.G., Ai, T., Cho, J.H., Wang, X., Sohma, Y., Li, M., and Hwang, T.C. (2005a) CFTR gating I: Characterization of the ATP-dependent gating of a phosphorylation-independent CFTR channel (DeltaR-CFTR). *The Journal of General Physiology*, **125**, 361–375.

Bompadre, S.G., Cho, J.H., Wang, X., Zou, X., Sohma, Y., Li, M., and Hwang, T.C. (2005b) CFTR gating II: Effects of nucleotide binding on the stability of open states. *The Journal of General Physiology*, **125**, 377–394.

Bompadre, S.G., Sohma, Y., Li, M., and Hwang, T.C. (2007) G551D and G1349D, two CF-associated mutations in the signature sequences of CFTR, exhibit distinct gating defects. *The Journal of General Physiology*, **129**, 285–298.

Cai, Z., Taddei, A., and Sheppard, D.N. (2006) Differential sensitivity of the cystic fibrosis (CF)-associated mutants G551D and G1349D to potentiators of the cystic fibrosis transmembrane conductance regulator (CFTR) Cl^- channel. *The Journal of Biological Chemistry*, **281**, 1970–1977.

Cai, Z.W., Liu, J., Li, H.Y., and Sheppard, D.N. (2011) Targeting F508del-CFTR to develop rational new therapies for cystic fibrosis. *Acta pharmacologica Sinica*, **32**, 693–701.

Caohuy, H., Jozwik, C., and Pollard, H.B. (2009) Rescue of DeltaF508-CFTR by the SGK1/Nedd4–2 signaling pathway. *The Journal of Biological Chemistry*, **284**, 25241–25253.

Carson, M.R., Travis, S.M., and Welsh, M.J. (1995) The two nucleotide-binding domains of cystic fibrosis transmembrane conductance regulator (CFTR) have distinct functions in controlling channel activity. *The Journal of Biological Chemistry*, **270**, 1711–1717.

Chang, X.B., Mengos, A., Hou, Y.X., Cui, L., Jensen, T.J., Aleksandrov, A., Riordan, J.R., and Gentzsch, M. (2008) Role of N-linked oligosaccharides in the biosynthetic processing of the cystic fibrosis membrane conductance regulator. *Journal of Cell Science*, **121**, 2814–2823.

Chang, X.B., Tabcharani, J.A., Hou, Y.X., Jensen, T.J., Kartner, N., Alon, N., Hanrahan, J.W., and Riordan, J.R. (1993) Protein kinase A (PKA) still activates CFTR chloride channel after mutagenesis of all 10 PKA consensus phosphorylation sites. *The Journal of Biological Chemistry*, **268**, 11304–11311.

Changeux, J.P. (2012) Allostery and the Monod-Wyman-Changeux model after 50 years. *Annual Review of Biophysics*, **41**, 103–133.

Chappe, V., Hinkson, D.A., Howell, L.D., Evagelidis, A., Liao, J., Chang, X.B., Riordan, J.R., and Hanrahan, J.W. (2004) Stimulatory and inhibitory protein kinase C consensus sequences regulate the cystic fibrosis transmembrane conductance regulator. *Proceedings of the National Academy of Sciences of the United States of America*, **101**, 390–395.

Chen, T.Y. and Hwang, T.C. (2008) CLC-0 and CFTR: Chloride channels evolved from transporters. *Physiological Reviews*, **88**, 351–387.

Cheng, S.H., Rich, D.P., Marshall, J., Gregory, R.J., Welsh, M.J., and Smith, A.E. (1991) Phosphorylation of the R domain by cAMP-dependent protein kinase regulates the CFTR chloride channel. *Cell*, **66**, 1027–1036.

Chong, P.A., Kota, P., Dokholyan, N.V., and Forman-Kay, J.D. (2013) Dynamics intrinsic to cystic fibrosis transmembrane conductance regulator function and stability. *Cold Spring Harbor Perspectives in Medicine*, **3**, a009522.

Clancy, J.P., Rowe, S.M., Accurso, F.J., Aitken, M.L., Amin, R.S., Ashlock, M.A., Ballmann, M. et al. (2012) Results of a phase IIa study of VX-809, an investigational CFTR corrector compound, in subjects with cystic fibrosis homozygous for the F508del-CFTR mutation. *Thorax*, **67**, 12–18.

Cohen, T.S. and Prince, A. (2012) Cystic fibrosis: A mucosal immunodeficiency syndrome. *Nature Medicine*, **18**, 509–519.

Csanady, L., Chan, K.W., Nairn, A.C., and Gadsby, D.C. (2005a) Functional roles of nonconserved structural segments in CFTR's NH2-terminal nucleotide binding domain. *The Journal of General Physiology*, **125**, 43–55.

Csanady, L., Chan, K.W., Seto-Young, D., Kopsco, D.C., Nairn, A.C., and Gadsby, D.C. (2000) Severed channels probe regulation of gating of cystic fibrosis transmembrane conductance regulator by its cytoplasmic domains. *The Journal of General Physiology*, **116**, 477–500.

Csanady, L., Seto-Young, D., Chan, K.W., Cenciarelli, C., Angel, B.B., Qin, J., McLachlin, D.T. et al. (2005b) Preferential phosphorylation of R-domain Serine 768 dampens activation of CFTR channels by PKA. *The Journal of General Physiology*, **125**, 171–186.

Csanady, L., Vergani, P., and Gadsby, D.C. (2010) Strict coupling between CFTR's catalytic cycle and gating of its Cl^- ion pore revealed by distributions of open channel burst durations. *Proceedings of the National Academy of Sciences of the United States of America*, **107**, 1241–1246.

Cui, G., Song, B., Turki, H.W., and McCarty, N.A. (2012) Differential contribution of TM6 and TM12 to the pore of CFTR identified by three sulfonylurea-based blockers. *Pflugers Archiv: European Journal of Physiology*, **463**, 405–418.

Cui, L., Aleksandrov, L., Chang, X.B., Hou, Y.X., He, L., Hegedus, T., Gentzsch, M., Aleksandrov, A., Balch, W.E., and Riordan, J.R. (2007) Domain interdependence in the biosynthetic assembly of CFTR. *Journal of Molecular Biology*, **365**, 981–994.

Cui, L., Aleksandrov, L., Hou, Y.X., Gentzsch, M., Chen, J.H., Riordan, J.R., and Aleksandrov, A.A. (2006) The role of cystic fibrosis transmembrane conductance regulator phenylalanine 508 side chain in ion channel gating. *The Journal of Physiology*, **572**, 347–358.

Cutting, G.R., Kasch, L.M., Rosenstein, B.J., Zielenski, J., Tsui, L.C., Antonarakis, S.E., and Kazazian, H.H., Jr. (1990) A cluster of cystic fibrosis mutations in the first nucleotide-binding fold of the cystic fibrosis conductance regulator protein. *Nature*, **346**, 366–369.

Dalemans, W., Barbry, P., Champigny, G., Jallat, S., Dott, K., Dreyer, D., Crystal, R.G., Pavirani, A., Lecocq, J.P., and Lazdunski, M. (1991) Altered chloride ion channel kinetics associated with the delta F508 cystic fibrosis mutation. *Nature*, **354**, 526–528.

Dawson, R.J. and Locher, K.P. (2006) Structure of a bacterial multidrug ABC transporter. *Nature*, **443**, 180–185.

di Sant' Agnese, P.A., Darling, R.C., Perera, G.A., and Shea, E. (1953) Abnormal electrolyte composition of sweat in cystic fibrosis of the pancreas: Clinical implications and relationship to the disease. *Pediatrics*, **12**, 549–563.

Di Stefano, A., Wittner, M., Schlatter, E., Lang, H.J., Englert, H., and Greger, R. (1985) Diphenylamine-2-carboxylate, a blocker of the Cl(-)-conductive pathway in Cl(-)-transporting epithelia. *Pflugers Archiv: European Journal of Physiology*, **405**(Suppl 1), S95–S100.

Dodd, M.E. and Prasad, S.A. (2005) Physiotherapy management of cystic fibrosis. *Chronic Respiratory Disease*, **2**, 139–149.

Du, K. and Lukacs, G.L. (2009) Cooperative assembly and misfolding of CFTR domains in vivo. *Molecular Biology of the Cell*, **20**, 1903–1915.

Du, K., Sharma, M., and Lukacs, G.L. (2005) The DeltaF508 cystic fibrosis mutation impairs domain-domain interactions and arrests post-translational folding of CFTR. *Nature Structural & Molecular Biology*, **12**, 17–25.

Dulhanty, A.M., Chang, X.B., and Riordan, J.R. (1995) Mutation of potential phosphorylation sites in the recombinant R domain of the cystic fibrosis transmembrane conductance regulator has significant effects on domain conformation. *Biochemical and Biophysical Research Communications*, **206**, 207–214.

Dulhanty, A.M. and Riordan, J.R. (1994) Phosphorylation by cAMP-dependent protein kinase causes a conformational change in the R domain of the cystic fibrosis transmembrane conductance regulator. *Biochemistry*, **33**, 4072–4079.

El Hiani, Y. and Linsdell, P. (2010) Changes in accessibility of cytoplasmic substances to the pore associated with activation of the cystic fibrosis transmembrane conductance regulator chloride channel. *The Journal of Biological Chemistry*, **285**, 32126–32140.

Ernst, R., Koch, J., Horn, C., Tampe, R., and Schmitt, L. (2006) Engineering ATPase activity in the isolated ABC cassette of human TAP1. *The Journal of Biological Chemistry*, **281**, 27471–27480.

Farber, S. (1943) Pancreatic insufficiency and the celiac syndrome. *The New England Journal of Medicine*, **229**, 653–682.

Fatehi, M. and Linsdell, P. (2008) State-dependent access of anions to the cystic fibrosis transmembrane conductance regulator chloride channel pore. *The Journal of Biological Chemistry*, **283**, 6102–6109.

Gadsby, D.C., Vergani, P., and Csanady, L. (2006) The ABC protein turned chloride channel whose failure causes cystic fibrosis. *Nature*, **440**, 477–483.

Gao, X., Bai, Y., and Hwang, T.C. (2013) Cysteine scanning of CFTR's first transmembrane segment reveals its plausible roles in gating and permeation. *Biophysical Journal*, **104**, 786–797.

Gibson, L.E. and Cooke, R.E. (1959) A test for concentration of electrolytes in sweat in cystic fibrosis of the pancreas utilizing pilocarpine by iontophoresis. *Pediatrics*, **23**, 545–549.

Glozman, R., Okiyoneda, T., Mulvihill, C.M., Rini, J.M., Barriere, H., and Lukacs, G.L. (2009) N-glycans are direct determinants of CFTR folding and stability in secretory and endocytic membrane traffic. *The Journal of Cell Biology*, **184**, 847–862.

Gowen, C.W., Lawson, E.E., Gingras-Leatherman, J., Gatzy, J.T., Boucher, R.C., and Knowles, M.R. (1986) Increased nasal potential difference and amiloride sensitivity in neonates with cystic fibrosis. *The Journal of Pediatrics*, **108**, 517–521.

Gray, M.A., Pollard, C.E., Harris, A., Coleman, L., Greenwell, J.R., and Argent, B.E. (1990) Anion selectivity and block of the small-conductance chloride channel on pancreatic duct cells. *The American Journal of Physiology*, **259**, C752–C761.

Gunderson, K.L. and Kopito, R.R. (1994) Effects of pyrophosphate and nucleotide analogs suggest a role for ATP hydrolysis in cystic fibrosis transmembrane regulator channel gating. *The Journal of Biological Chemistry*, **269**, 19349–19353.

Gunderson, K.L. and Kopito, R.R. (1995) Conformational states of CFTR associated with channel gating: The role ATP binding and hydrolysis. *Cell*, **82**, 231–239.

Gutmann, D.A., Ward, A., Urbatsch, I.L., Chang, G., and van Veen, H.W. (2010) Understanding polyspecificity of multidrug ABC transporters: Closing in on the gaps in ABCB1. *Trends in Biochemical Sciences*, **35**, 36–42.

Hanekop, N., Zaitseva, J., Jenewein, S., Holland, I.B., and Schmitt, L. (2006) Molecular insights into the mechanism of ATP-hydrolysis by the NBD of the ABC-transporter HlyB. *FEBS Letters*, **580**, 1036–1041.

Hanrahan, J.W., Sampson, H.M., and Thomas, D.Y. (2013) Novel pharmacological strategies to treat cystic fibrosis. *Trends in Pharmacological Sciences*, **34**, 119–125.

Harada, K., Okiyoneda, T., Hashimoto, Y., Ueno, K., Nakamura, K., Yamahira, K., Sugahara, T. et al. (2006) Calreticulin negatively regulates the cell surface expression of cystic fibrosis transmembrane conductance regulator. *The Journal of Biological Chemistry*, **281**, 12841–12848.

Hassink, G.C., Zhao, B., Sompallae, R., Altun, M., Gastaldello, S., Zinin, N.V., Masucci, M.G., and Lindsten, K. (2009) The ER-resident ubiquitin-specific protease 19 participates in the UPR and rescues ERAD substrates. *EMBO Reports*, **10**, 755–761.

He, L., Aleksandrov, A.A., Serohijos, A.W., Hegedus, T., Aleksandrov, L.A., Cui, L., Dokholyan, N.V., and Riordan, J.R. (2008) Multiple membrane-cytoplasmic domain contacts in the cystic fibrosis transmembrane conductance regulator (CFTR) mediate regulation of channel gating. *The Journal of Biological Chemistry*, **283**, 26383–26390.

He, L., Aleksandrov, L.A., Cui, L., Jensen, T.J., Nesbitt, K.L., and Riordan, J.R. (2010) Restoration of domain folding and interdomain assembly by second-site suppressors of the DeltaF508 mutation in CFTR. *FASEB Journal: Official Publication of the Federation of American Societies for Experimental Biology*, **24**, 3103–3112.

He, L., Kota, P., Aleksandrov, A.A., Cui, L., Jensen, T., Dokholyan, N.V., and Riordan, J.R. (2013) Correctors of DeltaF508 CFTR restore global conformational maturation without thermally stabilizing the mutant protein. *FASEB Journal: Official Publication of the Federation of American Societies for Experimental Biology*, **27**, 536–545.

Hegedus, T., Aleksandrov, A., Mengos, A., Cui, L., Jensen, T.J., and Riordan, J.R. (2009) Role of individual R domain phosphorylation sites in CFTR regulation by protein kinase A. *Biochimica et Biophysica Acta*, **1788**, 1341–1349.

Hegedus, T., Serohijos, A.W., Dokholyan, N.V., He, L., and Riordan, J.R. (2008) Computational studies reveal phosphorylation-dependent changes in the unstructured R domain of CFTR. *Journal of Molecular Biology*, **378**, 1052–1063.

Henderson, M.J., Vij, N., and Zeitlin, P.L. (2010) Ubiquitin C-terminal hydrolase-L1 protects cystic fibrosis transmembrane conductance regulator from early stages of proteasomal degradation. *The Journal of Biological Chemistry*, **285**, 11314–11325.

Hille, B. (1973) Potassium channels in myelinated nerve. Selective permeability to small cations. *The Journal of General Physiology*, **61**, 669–686.

Hohl, M., Briand, C., Grutter, M.G., and Seeger, M.A. (2012) Crystal structure of a heterodimeric ABC transporter in its inward-facing conformation. *Nature Structural & Molecular Biology*, **19**, 395–402.

Hou, Y., Cui, L., Riordan, J.R., and Chang, X. (2000) Allosteric interactions between the two non-equivalent nucleotide binding domains of multidrug resistance protein MRP1. *The Journal of Biological Chemistry*, **275**, 20280–20287.

Hwang, T.C., Horie, M., and Gadsby, D.C. (1993) Functionally distinct phospho-forms underlie incremental activation of protein kinase-regulated Cl- conductance in mammalian heart. *The Journal of General Physiology*, **101**, 629–650.

Hwang, T.C. and Kirk, K.L. (2013) The CFTR ion channel: Gating, regulation, and anion permeation. *Cold Spring Harbor Perspectives in Medicine*, **3**, a009498.

Hwang, T.C., Nagel, G., Nairn, A.C., and Gadsby, D.C. (1994) Regulation of the gating of cystic fibrosis transmembrane conductance regulator C1 channels by phosphorylation and ATP hydrolysis. *Proceedings of the National Academy of Sciences of the United States of America*, **91**, 4698–4702.

Hwang, T.C. and Sheppard, D.N. (1999) Molecular pharmacology of the CFTR Cl- channel. *Trends in Pharmacological Sciences*, **20**, 448–453.

Hwang, T.C. and Sheppard, D.N. (2009) Gating of the CFTR Cl-channel by ATP-driven nucleotide-binding domain dimerisation. *The Journal of Physiology*, **587**, 2151–2161.

Hwang, T.C., Wang, F., Yang, I.C., and Reenstra, W.W. (1997) Genistein potentiates wild-type and delta F508-CFTR channel activity. *The American Journal of Physiology*, **273**, C988–C998.

Ishiguro, H., Steward, M.C., Naruse, S., Ko, S.B., Goto, H., Case, R.M., Kondo, T., and Yamamoto, A. (2009) CFTR functions as a bicarbonate channel in pancreatic duct cells. *The Journal of General Physiology*, **133**, 315–326.

Jih, K.Y. and Hwang, T.C. (2012) Nonequilibrium gating of CFTR on an equilibrium theme. *Physiology (Bethesda)*, **27**, 351–361.

Jih, K.Y. and Hwang, T.C. (2013) Vx-770 potentiates CFTR function by promoting decoupling between the gating cycle and ATP hydrolysis cycle. *Proceedings of the National Academy of Sciences of the United States of America*, **110**, 4404–4409.

Jih, K.Y., Li, M., Hwang, T.C., and Bompadre, S.G. (2011) The most common cystic fibrosis-associated mutation destabilizes the dimeric state of the nucleotide-binding domains of CFTR. *The Journal of Physiology*, **589**, 2719–2731.

Jih, K.Y., Sohma, Y., and Hwang, T.C. (2012a) Nonintegral stoichiometry in CFTR gating revealed by a pore-lining mutation. *The Journal of General Physiology*, **140**, 347–359.

Jih, K.Y., Sohma, Y., Li, M., and Hwang, T.C. (2012b) Identification of a novel post-hydrolytic state in CFTR gating. *The Journal of General Physiology*, **139**, 359–370.

Kanelis, V., Hudson, R.P., Thibodeau, P.H., Thomas, P.J., and Forman-Kay, J.D. (2010) NMR evidence for differential phosphorylation-dependent interactions in WT and DeltaF508 CFTR. *The EMBO Journal*, **29**, 263–277.

Kerem, B., Rommens, J.M., Buchanan, J.A., Markiewicz, D., Cox, T.K., Chakravarti, A., Buchwald, M., and Tsui, L.C. (1989) Identification of the cystic fibrosis gene: Genetic analysis. *Science*, **245**, 1073–1080.

Kessler, W.R. and Andersen, D.H. (1951) Heat prostration in fibrocystic disease of the pancreas and other condition. *Pediatrics*, **8**, 648.

King, J.D., Jr., Fitch, A.C., Lee, J.K., McCane, J.E., Mak, D.O., Foskett, J.K., and Hallows, K.R. (2009) AMP-activated protein kinase phosphorylation of the R domain inhibits PKA stimulation of CFTR. *American Journal of Physiology. Cell Physiology*, **297**, C94–C101.

Kleizen, B., van Vlijmen, T., de Jonge, H.R., and Braakman, I. (2005) Folding of CFTR is predominantly cotranslational. *Molecular Cell*, **20**, 277–287.

Knowles, M., Gatzy, J., and Boucher, R. (1981) Increased bioelectric potential difference across respiratory epithelia in cystic fibrosis. *The New England Journal of Medicine*, **305**, 1489–1495.

Kool, M., van der Linden, M., de Haas, M., Scheffer, G.L., de Vree, J.M., Smith, A.J., Jansen, G. et al. (1999) MRP3, an organic anion transporter able to transport anti-cancer drugs. *Proceedings of the National Academy of Sciences of the United States of America*, **96**, 6914–6919.

Lewis, H.A., Buchanan, S.G., Burley, S.K., Conners, K., Dickey, M., Dorwart, M., Fowler, R. et al. (2004) Structure of nucleotide-binding domain 1 of the cystic fibrosis transmembrane conductance regulator. *The EMBO Journal*, **23**, 282–293.

Lewis, H.A., Wang, C., Zhao, X., Hamuro, Y., Conners, K., Kearins, M.C., Lu, F. et al. (2010) Structure and dynamics of NBD1 from CFTR characterized using crystallography and hydrogen/deuterium exchange mass spectrometry. *Journal of Molecular Biology*, **396**, 406–430.

Lewis, H.A., Zhao, X., Wang, C., Sauder, J.M., Rooney, I., Noland, B.W., Lorimer, D. et al. (2005) Impact of the deltaF508 mutation in first nucleotide-binding domain of human cystic fibrosis transmembrane conductance regulator on domain folding and structure. *The Journal of Biological Chemistry*, **280**, 1346–1353.

Li, C., Ramjeesingh, M., Wang, W., Garami, E., Hewryk, M., Lee, D., Rommens, J.M., Galley, K. and Bear, C.E. (1996) ATPase activity of the cystic fibrosis transmembrane conductance regulator. *The Journal of Biological Chemistry*, **271**, 28463–28468.

Linsdell, P. (2005) Location of a common inhibitor binding site in the cytoplasmic vestibule of the cystic fibrosis transmembrane conductance regulator chloride channel pore. *The Journal of Biological Chemistry*, **280**, 8945–8950.

Linsdell, P., Evagelidis, A., and Hanrahan, J.W. (2000) Molecular determinants of anion selectivity in the cystic fibrosis transmembrane conductance regulator chloride channel pore. *Biophysical Journal*, **78**, 2973–2982.

Linsdell, P. and Hanrahan, J.W. (1996a) Disulphonic stilbene block of cystic fibrosis transmembrane conductance regulator Cl- channels expressed in a mammalian cell line and its regulation by a critical pore residue. *The Journal of Physiology*, **496**(Part 3), 687–693.

Linsdell, P. and Hanrahan, J.W. (1996b) Flickery block of single CFTR chloride channels by intracellular anions and osmolytes. *The American Journal of Physiology*, **271**, C628–C634.

Liu, X., Smith, S.S., and Dawson, D.C. (2003) CFTR: What's it like inside the pore? *Journal of Experimental Zoology. Part A, Comparative Experimental Biology*, **300**, 69–75.

Locher, K.P., Lee, A.T., and Rees, D.C. (2002) The E. coli BtuCD structure: A framework for ABC transporter architecture and mechanism. *Science*, **296**, 1091–1098.

Loo, T.W., Bartlett, M.C., and Clarke, D.M. (2009) Correctors enhance maturation of DeltaF508 CFTR by promoting interactions between the two halves of the molecule. *Biochemistry*, **48**, 9882–9890.

Lukacs, G.L., Chang, X.B., Bear, C., Kartner, N., Mohamed, A., Riordan, J.R., and Grinstein, S. (1993) The delta F508 mutation decreases the stability of cystic fibrosis transmembrane conductance regulator in the plasma membrane. Determination of functional half-lives on transfected cells. *The Journal of Biological Chemistry*, **268**, 21592–21598.

Lukacs, G.L., Mohamed, A., Kartner, N., Chang, X.B., Riordan, J.R., and Grinstein, S. (1994) Conformational maturation of CFTR but not its mutant counterpart (delta F508) occurs in the endoplasmic reticulum and requires ATP. *The EMBO Journal*, **13**, 6076–6086.

Lukacs, G.L., Segal, G., Kartner, N., Grinstein, S., and Zhang, F. (1997) Constitutive internalization of cystic fibrosis transmembrane conductance regulator occurs via clathrin-dependent endocytosis and is regulated by protein phosphorylation. *The Biochemical Journal*, **328**(Part 2), 353–361.

Lukacs, G.L. and Verkman, A.S. (2012) CFTR: Folding, misfolding and correcting the DeltaF508 conformational defect. *Trends in Molecular Medicine*, **18**, 81–91.

Mathews, C.J., Tabcharani, J.A., Chang, X.B., Jensen, T.J., Riordan, J.R., and Hanrahan, J.W. (1998) Dibasic protein kinase A sites regulate bursting rate and nucleotide sensitivity of the cystic fibrosis transmembrane conductance regulator chloride channel. *The Journal of Physiology*, **508**(Part 2), 365–377.

McCarty, N.A. and Zhang, Z.R. (2001) Identification of a region of strong discrimination in the pore of CFTR. *American Journal of Physiology. Lung Cellular and Molecular Physiology*, **281**, L852–L867.

McDonough, S., Davidson, N., Lester, H.A., and McCarty, N.A. (1994) Novel pore-lining residues in CFTR that govern permeation and open-channel block. *Neuron*, **13**, 623–634.

Meacham, G.C., Patterson, C., Zhang, W., Younger, J.M., and Cyr, D.M. (2001) The Hsc70 co-chaperone CHIP targets immature CFTR for proteasomal degradation. *Nature Cell Biology*, **3**, 100–105.

Mense, M., Vergani, P., White, D.M., Altberg, G., Nairn, A.C., and Gadsby, D.C. (2006) in vivo phosphorylation of CFTR promotes formation of a nucleotide-binding domain heterodimer. *The EMBO Journal*, **25**, 4728–4739.

Miki, H., Zhou, Z., Li, M., Hwang, T.C., and Bompadre, S.G. (2010) Potentiation of disease-associated cystic fibrosis transmembrane conductance regulator mutants by hydrolyzable ATP analogs. *The Journal of Biological Chemistry*, **285**, 19967–19975.

Mio, K., Ogura, T., Mio, M., Shimizu, H., Hwang, T.C., Sato, C., and Sohma, Y. (2008) Three-dimensional reconstruction of human cystic fibrosis transmembrane conductance regulator chloride channel revealed an ellipsoidal structure with orifices beneath the putative transmembrane domain. *The Journal of Biological Chemistry*, **283**, 30300–30310.

Morito, D., Hirao, K., Oda, Y., Hosokawa, N., Tokunaga, F., Cyr, D.M., Tanaka, K., Iwai, K., and Nagata, K. (2008) Gp78 cooperates with RMA1 in endoplasmic reticulum-associated degradation of CFTRDeltaF508. *Molecular Biology of the Cell*, **19**, 1328–1336.

Muanprasat, C., Sonawane, N.D., Salinas, D., Taddei, A., Galietta, L.J., and Verkman, A.S. (2004) Discovery of glycine hydrazide pore-occluding CFTR inhibitors: Mechanism, structure-activity analysis, and in vivo efficacy. *The Journal of General Physiology*, **124**, 125–137.

Nagel, G., Hwang, T.C., Nastiuk, K.L., Nairn, A.C., and Gadsby, D.C. (1992) The protein kinase A-regulated cardiac Cl- channel resembles the cystic fibrosis transmembrane conductance regulator. *Nature*, **360**, 81–84.

Norimatsu, Y., Ivetac, A., Alexander, C., Kirkham, J., O'Donnell, N., Dawson, D.C. and Sansom, M.S. (2012a) Cystic fibrosis transmembrane conductance regulator: A molecular model defines the architecture of the anion conduction path and locates a *bottleneck* in the pore. *Biochemistry*, **51**, 2199–2212.

Norimatsu, Y., Ivetac, A., Alexander, C., O'Donnell, N., Frye, L., Sansom, M.S., and Dawson, D.C. (2012b) Locating a plausible binding site for an open-channel blocker, GlyH-101, in the pore of the cystic fibrosis transmembrane conductance regulator. *Molecular Pharmacology*, **82**, 1042–1055.

Okiyoneda, T., Barriere, H., Bagdany, M., Rabeh, W.M., Du, K., Hohfeld, J., Young, J.C., and Lukacs, G.L. (2010) Peripheral protein quality control removes unfolded CFTR from the plasma membrane. *Science*, **329**, 805–810.

Okiyoneda, T., Veit, G., Dekkers, J.F., Bagdany, M., Soya, N., Xu, H., Roldan, A. et al. (2013) Mechanism-based corrector combination restores DeltaF508-CFTR folding and function. *Nature Chemical Biology*, **9**, 444–454.

O'Reilly, C.M., Winpenny, J.P., Argent, B.E., and Gray, M.A. (2000) Cystic fibrosis transmembrane conductance regulator currents in guinea pig pancreatic duct cells: Inhibition by bicarbonate ions. *Gastroenterology*, **118**, 1187–1196.

Ostedgaard, L.S., Rogers, C.S., Dong, Q., Randak, C.O., Vermeer, D.W., Rokhlina, T., Karp, P.H., and Welsh, M.J. (2007) Processing and function of CFTR-DeltaF508 are species-dependent. *Proceedings of the National Academy of Sciences of the United States of America*, **104**, 15370–15375.

O'Sullivan, B.P. and Freedman, S.D. (2009) Cystic fibrosis. *Lancet*, **373**, 1891–1904.

Panten, U., Schwanstecher, M., and Schwanstecher, C. (1996) Sulfonylurea receptors and mechanism of sulfonylurea action. *Experimental and Clinical Endocrinology & Diabetes: Official Journal, German Society of Endocrinology [and] German Diabetes Association*, **104**, 1–9.

Park, H.W., Nam, J.H., Kim, J.Y., Namkung, W., Yoon, J.S., Lee, J.S., Kim, K.S. et al. (2010) Dynamic regulation of CFTR bicarbonate permeability by [Cl-]i and its role in pancreatic bicarbonate secretion. *Gastroenterology*, **139**, 620–631.

Perria, C.L., Rajamanickam, V., Lapinski, P.E., and Raghavan, M. (2006) Catalytic site modifications of TAP1 and TAP2 and their functional consequences. *The Journal of Biological Chemistry*, **281**, 39839–39851.

Powe, A., Zhou, Z., Hwang, T.C., and Nagel, G. (2002a) Quantitative analysis of ATP-dependent gating of CFTR. *Methods in Molecular Medicine*, **70**, 67–98.
Powe, A.C., Jr., Al-Nakkash, L., Li, M., and Hwang, T.C. (2002b) Mutation of Walker-A lysine 464 in cystic fibrosis transmembrane conductance regulator reveals functional interaction between its nucleotide-binding domains. *The Journal of Physiology*, **539**, 333–346.
Procko, E., O'Mara, M.L., Bennett, W.F., Tieleman, D.P., and Gaudet, R. (2009) The mechanism of ABC transporters: General lessons from structural and functional studies of an antigenic peptide transporter. *FASEB Journal: Official Publication of the Federation of American Societies for Experimental Biology*, **23**, 1287–1302.
Protasevich, I., Yang, Z., Wang, C., Atwell, S., Zhao, X., Emtage, S., Wetmore, D., Hunt, J.F., and Brouillette, C.G. (2010) Thermal unfolding studies show the disease causing F508del mutation in CFTR thermodynamically destabilizes nucleotide-binding domain 1. *Protein Science: A Publication of the Protein Society*, **19**, 1917–1931.
Qian, F., El Hiani, Y., and Linsdell, P. (2011) Functional arrangement of the 12th transmembrane region in the CFTR chloride channel pore based on functional investigation of a cysteine-less CFTR variant. *Pflugers Archiv: European Journal of Physiology*, **462**, 559–571.
Qu, B.H. and Thomas, P.J. (1996) Alteration of the cystic fibrosis transmembrane conductance regulator folding pathway. *The Journal of Biological Chemistry*, **271**, 7261–7264.
Quinton, P.M. (1983) Chloride impermeability in cystic fibrosis. *Nature*, **301**, 421–422.
Quinton, P.M. (1999) Physiological basis of cystic fibrosis: A historical perspective. *Physiological Reviews*, **79**, S3–S22.
Quinton, P.M. (2001) The neglected ion: HCO3. *Nature Medicine*, **7**, 292–293.
Quinton, P.M. (2007) Cystic fibrosis: Lessons from the sweat gland. *Physiology (Bethesda)*, **22**, 212–225.
Quinton, P.M. (2008) Cystic fibrosis: Impaired bicarbonate secretion and mucoviscidosis. *Lancet*, **372**, 415–417.
Quinton, P.M. (2010) Role of epithelial HCO3(-) transport in mucin secretion: Lessons from cystic fibrosis. *American Journal of Physiology. Cell Physiology*, **299**, C1222–C1233.
Rabeh, W.M., Bossard, F., Xu, H., Okiyoneda, T., Bagdany, M., Mulvihill, C.M., Du, K. et al. (2012) Correction of both NBD1 energetics and domain interface is required to restore DeltaF508 CFTR folding and function. *Cell*, **148**, 150–163.
Ramjeesingh, M., Li, C., Garami, E., Huan, L.J., Galley, K., Wang, Y., and Bear, C.E. (1999) Walker mutations reveal loose relationship between catalytic and channel-gating activities of purified CFTR (cystic fibrosis transmembrane conductance regulator). *Biochemistry*, **38**, 1463–1468.
Ramsey, B.W., Davies, J., McElvaney, N.G., Tullis, E., Bell, S.C., Drevinek, P., Griese, M. et al. (2011) A CFTR potentiator in patients with cystic fibrosis and the G551D mutation. *The New England Journal of Medicine*, **365**, 1663–1672.
Rees, D.C., Johnson, E., and Lewinson, O. (2009) ABC transporters: The power to change. *Nature Reviews. Molecular Cell Biology*, **10**, 218–227.
Rich, D.P., Gregory, R.J., Anderson, M.P., Manavalan, P., Smith, A.E., and Welsh, M.J. (1991) Effect of deleting the R domain on CFTR-generated chloride channels. *Science*, **253**, 205–207.
Rich, D.P., Gregory, R.J., Cheng, S.H., Smith, A.E., and Welsh, M.J. (1993) Effect of deletion mutations on the function of CFTR chloride channels. *Receptors & Channels*, **1**, 221–232.
Riordan, J.R., Rommens, J.M., Kerem, B., Alon, N., Rozmahel, R., Grzelczak, Z., Zielenski, J. et al. (1989) Identification of the cystic fibrosis gene: Cloning and characterization of complementary DNA. *Science*, **245**, 1066–1073.
Rommens, J.M., Iannuzzi, M.C., Kerem, B., Drumm, M.L., Melmer, G., Dean, M., Rozmahel, R. et al. (1989) Identification of the cystic fibrosis gene: Chromosome walking and jumping. *Science*, **245**, 1059–1065.
Rosenberg, M.F., O'Ryan, L.P., Hughes, G., Zhao, Z., Aleksandrov, L.A., Riordan, J.R., and Ford, R.C. (2011) The cystic fibrosis transmembrane conductance regulator (CFTR): Three-dimensional structure and localization of a channel gate. *The Journal of Biological Chemistry*, **286**, 42647–42654.
Rosser, M.F., Grove, D.E., Chen, L., and Cyr, D.M. (2008) Assembly and misassembly of cystic fibrosis transmembrane conductance regulator: Folding defects caused by deletion of F508 occur before and after the calnexin-dependent association of membrane spanning domain (MSD) 1 and MSD2. *Molecular Biology of the Cell*, **19**, 4570–4579.
Sato, K. and Sato, F. (1984) Defective beta adrenergic response of cystic fibrosis sweat glands in vivo and in vitro. *The Journal of Clinical Investigation*, **73**, 1763–1771.
Schultz, B.D., DeRoos, A.D., Venglarik, C.J., Singh, A.K., Frizzell, R.A., and Bridges, R.J. (1996) Glibenclamide blockade of CFTR chloride channels. *The American Journal of Physiology*, **271**, L192–L200.
Seibert, F.S., Chang, X.B., Aleksandrov, A.A., Clarke, D.M., Hanrahan, J.W., and Riordan, J.R. (1999) Influence of phosphorylation by protein kinase A on CFTR at the cell surface and endoplasmic reticulum. *Biochimica et Biophysica Acta*, **1461**, 275–283.
Seibert, F.S., Tabcharani, J.A., Chang, X.B., Dulhanty, A.M., Mathews, C., Hanrahan, J.W. and Riordan, J.R. (1995) cAMP-dependent protein kinase-mediated phosphorylation of cystic fibrosis transmembrane conductance regulator residue Ser-753 and its role in channel activation. *The Journal of Biological Chemistry*, **270**, 2158–2162.
Serohijos, A.W., Hegedus, T., Aleksandrov, A.A., He, L., Cui, L., Dokholyan, N.V., and Riordan, J.R. (2008) Phenylalanine-508 mediates a cytoplasmic-membrane domain contact in the CFTR 3D structure crucial to assembly and channel function. *Proceedings of the National Academy of Sciences of the United States of America*, **105**, 3256–3261.
Sharma, M., Pampinella, F., Nemes, C., Benharouga, M., So, J., Du, K., Bache, K.G. et al. (2004) Misfolding diverts CFTR from recycling to degradation: Quality control at early endosomes. *The Journal of Cell Biology*, **164**, 923–933.
Sheppard, D.N. and Robinson, K.A. (1997) Mechanism of glibenclamide inhibition of cystic fibrosis transmembrane conductance regulator Cl- channels expressed in a murine cell line. *The Journal of Physiology*, **503**(Part 2), 333–346.
Sheppard, D.N. and Welsh, M.J. (1999) Structure and function of the CFTR chloride channel. *Physiological Reviews*, **79**, S23–S45.
Sinaasappel, M., Stern, M., Littlewood, J., Wolfe, S., Steinkamp, G., Heijerman, H.G., Robberecht, E., and Doring, G. (2002) Nutrition in patients with cystic fibrosis: A European consensus. *Journal of Cystic Fibrosis*, **1**, 51–75.
Smith, S.S., Liu, X., Zhang, Z.R., Sun, F., Kriewall, T.E., McCarty, N.A., and Dawson, D.C. (2001) CFTR: Covalent and noncovalent modification suggests a role for fixed charges in anion conduction. *The Journal of General Physiology*, **118**, 407–431.
Smith, S.S., Steinle, E.D., Meyerhoff, M.E., and Dawson, D.C. (1999) Cystic fibrosis transmembrane conductance regulator. Physical basis for lyotropic anion selectivity patterns. *The Journal of General Physiology*, **114**, 799–818.
Sohma, Y., Yu, Y.C., and Hwang, T.C. (2013) Curcumin and genistein: The combined effects on disease-associated CFTR mutants and their clinical implications. *Current Pharmaceutical Design*, **19**, 3521–3528.

Stratford, F.L., Ramjeesingh, M., Cheung, J.C., Huan, L.J., and Bear, C.E. (2007) The Walker B motif of the second nucleotide-binding domain (NBD2) of CFTR plays a key role in ATPase activity by the NBD1-NBD2 heterodimer. *The Biochemical Journal*, **401**, 581–586.

Sun, F., Mi, Z., Condliffe, S.B., Bertrand, C.A., Gong, X., Lu, X., Zhang, R. et al. (2008) Chaperone displacement from mutant cystic fibrosis transmembrane conductance regulator restores its function in human airway epithelia. *FASEB Journal*, **22**, 3255–3263.

Swiatecka-Urban, A., Brown, A., Moreau-Marquis, S., Renuka, J., Coutermarsh, B., Barnaby, R., Karlson, K.H. et al. (2005) The short apical membrane half-life of rescued {Delta}F508-cystic fibrosis transmembrane conductance regulator (CFTR) results from accelerated endocytosis of {Delta}F508-CFTR in polarized human airway epithelial cells. *The Journal of Biological Chemistry*, **280**, 36762–36772.

Szollosi, A., Muallem, D.R., Csanady, L., and Vergani, P. (2011) Mutant cycles at CFTR's non-canonical ATP-binding site support little interface separation during gating. *The Journal of General Physiology*, **137**, 549–562.

Thibodeau, P.H., Brautigam, C.A., Machius, M., and Thomas, P.J. (2005) Side chain and backbone contributions of Phe508 to CFTR folding. *Nature Structural & Molecular Biology*, **12**, 10–16.

Tsai, M.F., Li, M., and Hwang, T.C. (2010) Stable ATP binding mediated by a partial NBD dimer of the CFTR chloride channel. *The Journal of General Physiology*, **135**, 399–414.

Tsai, M.F., Shimizu, H., Sohma, Y., Li, M., and Hwang, T.C. (2009) State-dependent modulation of CFTR gating by pyrophosphate. *The Journal of General Physiology*, **133**, 405–419.

Tsui, L.C., Buchwald, M., Barker, D., Braman, J.C., Knowlton, R., Schumm, J.W., Eiberg, H. et al. (1985) Cystic fibrosis locus defined by a genetically linked polymorphic DNA marker. *Science*, **230**, 1054–1057.

Van Goor, F., Hadida, S., Grootenhuis, P.D., Burton, B., Cao, D., Neuberger, T., Turnbull, A. et al. (2009) Rescue of CF airway epithelial cell function in vitro by a CFTR potentiator, VX-770. *Proceedings of the National Academy of Sciences of the United States of America*, **106**, 18825–18830.

Van Goor, F., Hadida, S., Grootenhuis, P.D., Burton, B., Stack, J.H., Straley, K.S., Decker, C.J. et al. (2011) Correction of the F508del-CFTR protein processing defect in vitro by the investigational drug VX-809. *Proceedings of the National Academy of Sciences of the United States of America*, **108**, 18843–18848.

Varga, K., Jurkuvenaite, A., Wakefield, J., Hong, J.S., Guimbellot, J.S., Venglarik, C.J., Niraj, A. et al. (2004) Efficient intracellular processing of the endogenous cystic fibrosis transmembrane conductance regulator in epithelial cell lines. *The Journal of Biological Chemistry*, **279**, 22578–22584.

Venglarik, C.J., Schultz, B.D., DeRoos, A.D., Singh, A.K., and Bridges, R.J. (1996) Tolbutamide causes open channel blockade of cystic fibrosis transmembrane conductance regulator Cl- channels. *Biophysical Journal*, **70**, 2696–2703.

Vergani, P., Lockless, S.W., Nairn, A.C., and Gadsby, D.C. (2005) CFTR channel opening by ATP-driven tight dimerization of its nucleotide-binding domains. *Nature*, **433**, 876–880.

Vergani, P., Nairn, A.C., and Gadsby, D.C. (2003) On the mechanism of MgATP-dependent gating of CFTR Cl- channels. *The Journal of General Physiology*, **121**, 17–36.

Wainwright, B.J., Scambler, P.J., Schmidtke, J., Watson, E.A., Law, H.Y., Farrall, M., Cooke, H.J., Eiberg, H., and Williamson, R. (1985) Localization of cystic fibrosis locus to human chromosome 7cen-q22. *Nature*, **318**, 384–385.

Wang, C., Protasevich, I., Yang, Z., Seehausen, D., Skalak, T., Zhao, X., Atwell, S. et al. (2010a) Integrated biophysical studies implicate partial unfolding of NBD1 of CFTR in the molecular pathogenesis of F508del cystic fibrosis. *Protein Science: A Publication of the Protein Society*, **19**, 1932–1947.

Wang, F., Zeltwanger, S., Hu, S., and Hwang, T.C. (2000) Deletion of phenylalanine 508 causes attenuated phosphorylation-dependent activation of CFTR chloride channels. *The Journal of Physiology*, **524**(Part 3), 637–648.

Wang, W., Bernard, K., Li, G., and Kirk, K.L. (2007) Curcumin opens cystic fibrosis transmembrane conductance regulator channels by a novel mechanism that requires neither ATP binding nor dimerization of the nucleotide-binding domains. *The Journal of Biological Chemistry*, **282**, 4533–4544.

Wang, W., El Hiani, Y., and Linsdell, P. (2011) Alignment of transmembrane regions in the cystic fibrosis transmembrane conductance regulator chloride channel pore. *The Journal of General Physiology*, **138**, 165–178.

Wang, W., El Hiani, Y., Rubaiy, H.N., and Linsdell, P. (2014) Relative contribution of different transmembrane segments to the CFTR chloride channel pore. *Pflugers Archiv: European Journal of Physiology*, **466**, 477–490.

Wang, W., Wu, J., Bernard, K., Li, G., Wang, G., Bevensee, M.O., and Kirk, K.L. (2010b) ATP-independent CFTR channel gating and allosteric modulation by phosphorylation. *Proceedings of the National Academy of Sciences of the United States of America*, **107**, 3888–3893.

Wang, X., Venable, J., LaPointe, P., Hutt, D.M., Koulov, A.V., Coppinger, J., Gurkan, C., Kellner, W., Matteson, J., Plutner, H., Riordan, J.R., Kelly, J.W., Yates, J.R., 3rd and Balch, W.E. (2006) Hsp90 cochaperone Aha1 downregulation rescues misfolding of CFTR in cystic fibrosis. *Cell*, **127**: 803–815.

Wang, X.F., Zhou, C.X., Shi, Q.X., Yuan, Y.Y., Yu, M.K., Ajonuma, L.C., Ho, L.S. et al. (2003) Involvement of CFTR in uterine bicarbonate secretion and the fertilizing capacity of sperm. *Nature Cell Biology*, **5**, 902–906.

Wangemann, P., Wittner, M., Di Stefano, A., Englert, H.C., Lang, H.J., Schlatter, E., and Greger, R. (1986) Cl(-)-channel blockers in the thick ascending limb of the loop of Henle. Structure activity relationship. *Pflugers Archiv: European Journal of Physiology*, **407**(Suppl 2), S128–S141.

Ward, A., Reyes, C.L., Yu, J., Roth, C.B., and Chang, G. (2007) Flexibility in the ABC transporter MsbA: Alternating access with a twist. *Proceedings of the National Academy of Sciences of the United States of America*, **104**, 19005–19010.

Ward, C.L. and Kopito, R.R. (1994) Intracellular turnover of cystic fibrosis transmembrane conductance regulator. Inefficient processing and rapid degradation of wild-type and mutant proteins. *The Journal of Biological Chemistry*, **269**, 25710–25718.

Ward, C.L., Omura, S., and Kopito, R.R. (1995) Degradation of CFTR by the ubiquitin-proteasome pathway. *Cell*, **83**, 121–127.

Welsh, M.J. and Liedtke, C.M. (1986) Chloride and potassium channels in cystic fibrosis airway epithelia. *Nature*, **322**, 467–470.

Welsh, M.J. and Smith, A.E. (1993) Molecular mechanisms of CFTR chloride channel dysfunction in cystic fibrosis. *Cell*, **73**, 1251–1254.

White, R., Woodward, S., Leppert, M., O'Connell, P., Hoff, M., Herbst, J., Lalouel, J.M., Dean, M., and Vande Woude, G. (1985) A closely linked genetic marker for cystic fibrosis. *Nature*, **318**, 382–384.

Wieczorek, G. and Zielenkiewicz, P. (2008) DeltaF508 mutation increases conformational flexibility of CFTR protein. *Journal of Cystic Fibrosis: Official Journal of the European Cystic Fibrosis Society*, **7**, 295–300.

Wilkinson, D.J., Strong, T.V., Mansoura, M.K., Wood, D.L., Smith, S.S., Collins, F.S., and Dawson, D.C. (1997) CFTR activation: Additive effects of stimulatory and inhibitory phosphorylation sites in the R domain. *The American Journal of Physiology*, **273**, L127–L133.

Wine, J.J. and Joo, N.S. (2004) Submucosal glands and airway defense. *Proceedings of the American Thoracic Society*, **1**, 47–53.

Ye, S., Cihil, K., Stolz, D.B., Pilewski, J.M., Stanton, B.A., and Swiatecka-Urban, A. (2010) c-Cbl facilitates endocytosis and lysosomal degradation of cystic fibrosis transmembrane conductance regulator in human airway epithelial cells. *The Journal of Biological Chemistry*, **285**, 27008–27018.

Yoshida, Y., Chiba, T., Tokunaga, F., Kawasaki, H., Iwai, K., Suzuki, T., Ito, Y. et al. (2002) E3 ubiquitin ligase that recognizes sugar chains. *Nature*, **418**, 438–442.

Younger, J.M., Chen, L., Ren, H.Y., Rosser, M.F., Turnbull, E.L., Fan, C.Y., Patterson, C., and Cyr, D.M. (2006) Sequential quality-control checkpoints triage misfolded cystic fibrosis transmembrane conductance regulator. *Cell*, **126**, 571–582.

Younger, J.M., Ren, H.Y., Chen, L., Fan, C.Y., Fields, A., Patterson, C., and Cyr, D.M. (2004) A foldable CFTR{Delta}F508 biogenic intermediate accumulates upon inhibition of the Hsc70-CHIP E3 ubiquitin ligase. *The Journal of Cell Biology*, **167**, 1075–1085.

Yu, H., Burton, B., Huang, C.J., Worley, J., Cao, D., Johnson, J.P., Jr., Urrutia, A. et al. (2012) Ivacaftor potentiation of multiple CFTR channels with gating mutations. *Journal of Cystic Fibrosis: Official Journal of the European Cystic Fibrosis Society*, **11**, 237–245.

Zeltwanger, S., Wang, F., Wang, G.T., Gillis, K.D., and Hwang, T.C. (1999) Gating of cystic fibrosis transmembrane conductance regulator chloride channels by adenosine triphosphate hydrolysis. Quantitative analysis of a cyclic gating scheme. *The Journal of General Physiology*, **113**, 541–554.

Zhang, D.W., Graf, G.A., Gerard, R.D., Cohen, J.C., and Hobbs, H.H. (2006) Functional asymmetry of nucleotide-binding domains in ABCG5 and ABCG8. *The Journal of Biological Chemistry*, **281**, 4507–4516.

Zhang, L., Aleksandrov, L.A., Riordan, J.R., and Ford, R.C. (2011) Domain location within the cystic fibrosis transmembrane conductance regulator protein investigated by electron microscopy and gold labelling. *Biochimica et Biophysica Acta*, **1808**, 399–404.

Zhang, L., Aleksandrov, L.A., Zhao, Z., Birtley, J.R., Riordan, J.R., and Ford, R.C. (2009) Architecture of the cystic fibrosis transmembrane conductance regulator protein and structural changes associated with phosphorylation and nucleotide binding. *Journal of Structural Biology*, **167**, 242–251.

Zhou, J.J., Li, M.S., Qi, J., and Linsdell, P. (2010) Regulation of conductance by the number of fixed positive charges in the intracellular vestibule of the CFTR chloride channel pore. *The Journal of General Physiology*, **135**, 229–245.

Zhou, Z., Hu, S., and Hwang, T.C. (2002) Probing an open CFTR pore with organic anion blockers. *The Journal of General Physiology*, **120**, 647–662.

Zhou, Z., Wang, X., Li, M., Sohma, Y., Zou, X., and Hwang, T.C. (2005) High affinity ATP/ADP analogues as new tools for studying CFTR gating. *The Journal of Physiology*, **569**, 447–457.

Zhou, Z., Wang, X., Liu, H.Y., Zou, X., Li, M., and Hwang, T.C. (2006) The two ATP binding sites of cystic fibrosis transmembrane conductance regulator (CFTR) play distinct roles in gating kinetics and energetics. *The Journal of General Physiology*, **128**, 413–422.

Zielenski, J. and Tsui, L.C. (1995) Cystic fibrosis: Genotypic and phenotypic variations. *Annual Review of Genetics*, **29**, 777–807.

Drugs targeting ion channels

KeWei Wang

Contents

Ion channels represent the second largest class among 435 proven effect-mediating drug targets after G-protein-coupled receptors (GPCRs) (Rask-Andersen et al., 2011). They are primary targets for therapeutic areas of neuropsychiatric disorders (such as pain, epilepsy, stroke, Alzheimer's disease, anxiety, schizophrenia, etc.), cardiovascular and metabolic diseases (such as hypertension, arrhythmias, and diabetes), immunodiseases (such as asthma), nephrology (such as urinary retention and incontinence), irritable bowel syndrome, and pulmonary/respiratory diseases (such as chronic obstructive pulmonary airway disease). From 1939 to 2012, US Food and Drug Administration (FDA) approved a total of 2265 small molecule drugs (including the same molecular entity in different formulation) for all therapeutic areas. Among these approved drugs, there are about 730 drugs (more than one-third of the total) that have been classified as ion channel–targeting drugs although the exact mechanisms for most of them are not defined or unknown, in particular for drugs that were approved before the year of 2000. This chapter summarizes some of the representative drugs whose mechanisms of action are better studied or known. The description is organized based on target subclasses.

43.1 DRUGS ACTING ON VOLTAGE-GATED POTASSIUM CHANNELS

43.1.1 EZOGABINE OR RETIGABINE

Ezogabine (United States Adopted Name [USAN]) or *Retigabine* (International Nonproprietary Name [INN]) is an anticonvulsant drug used as a treatment for partial-onset seizures (Tatulian et al., 2001). This drug contains new molecular entity (NME) of *N*-(2-amino-4-[fluorobenzylamino]-phenyl) carbamic acid, codeveloped by Valeant and GlaxoSmithKline. It was approved by the European Medicines Agency (EMA) under the trade name *Trobalt* in March 2011 and by the US FDA under the trade name *Potiga* in June 2011.

Mechanism of action: Retigabine acts as a neuronal Kv7/KCNQ/M-channel opener, which is markedly different from that of other current anticonvulsants. Most anticonvulsants (such as Phenytoin, Zonisamide, and Valproate) target sodium/calcium channel and GABAa channel activity. Kv7/KCNQ channels are involved in setting the resting membrane potential and regulating neuronal activity (Jentsch et al., 2000; Wulff et al., 2009). Upregulating Kv7/KCNQ activity by retigabine repolarizes the membrane potential and inhibits repetitive firing that underlies epileptic activity. Retigabine activates four subtypes of Kv7/KCNQ channels including Kv7.2, Kv7.3, Kv7.2/7.3 and Kv7.4/7.5 channels, with an effective concentration for half maximum response (EC_{50}) of 1.9 μM at −30 mV for Kv7.2 (Wickenden et al., 2000; Tatulian et al., 2001; Xiong et al., 2008; Wulff et al., 2009). Its potential clinical applications include treatment of epilepsy, anxiety, neuropathic pain, and other neuropsychiatric disorders. Retigabine reaches maximum plasma concentrations between half an hour and 2 h after a single oral dose. It has a moderately high oral bioavailability (50%–60%), a high volume of distribution (6.2 L/kg), and a terminal half-life of 8–11 h (Luszczki, 2009). Retigabine is quickly absorbed and metabolized in the liver by N-glucuronidation and acetylation. Retigabine and its metabolites are excreted almost completely (84%) by kidney (Luszczki, 2009). Retigabine appears to be free of drug interactions with most commonly used anticonvulsants.

Retigabine also serves as an important chemical tool for studies of biological and pharmacological function as well as therapeutic potential of Kv7/KCNQ channel modulation.

43.1.2 DALFAMPRIDINE OR FAMPRIDINE

Dalfampridine (USAN) or *fampridine* (INN) is commonly known as 4-aminopyridine (4-AP) and is widely used as a research tool compound, in characterizing subtypes of potassium channel (see Figure 43.1). It is a small molecule with the chemical formula $C_5H_4N–NH_2$. In January 2010, FDA approved dalfampridine, under the trade name *Ampyra*, to be used to manage some of the symptoms of multiple sclerosis (MS), and Lambert–Eaton myasthenic syndrome (LEMS).

MS is an inflammatory disease in which the fatty myelin sheaths around the axons of the brain and spinal cord are damaged, leading to demyelination and scarring as well as a broad spectrum of signs and symptoms. LEMS (sometimes named Lambert–Eaton syndrome or Eaton–Lambert syndrome) is a rare autoimmune disorder characterized by muscle weakness of the limbs. The FDA approval of Ampyra was based on two clinical trials with a total of 510 patients with MS for a period of up to 21 weeks. During the double-blind treatment period, a significantly greater proportion of patients taking Ampyra 10 mg twice daily had increases in walking speed of as much as 30% from baseline, a substantial improvement that was absent in patients taking the placebo. Spinal cord injury patients have also seen improvement with 4-AP therapy. These improvements include sensory, motor, and pulmonary functions, with a decrease in spasticity and pain.

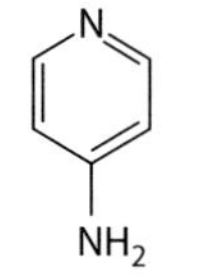

Mechanism of action: 4-AP is a relatively selective blocker of members of the Kv1 family of voltage-activated potassium channels (*KCNA1* gene for human homolog or *Shaker* for *Drosophila*). At concentration of 1 mM, 4-AP selectively and reversibly inhibits *Shaker* channels without significant effect on other sodium, calcium, and potassium conductances. Electrophysiological studies of demyelinated axons show that augmented potassium currents increase extracellular potassium ion concentration, which decreases action potential duration and amplitude, thus likely causing conduction failure. Potassium channel blockade reverses effects mediated by the increased potassium current seen in diseased neurons. Interestingly, 4-AP has also been shown as a potent calcium channel activator and it can improve synaptic and neuromuscular function by directly acting on the calcium channel beta subunit.

43.1.3 DRONEDARONE

Dronedarone is an antiarrhythmia drug developed by Sanofi-Aventis (see Figure 43.2). In July 2009, it was approved by the FDA under the trade name *Multaq*. It was recommended as an alternative to amiodarone for the treatment of atrial fibrillation (AF) and atrial flutter (AFL) in people whose heart has either returned to normal rhythm or who undergoes drug therapy or electric shock treatment to maintain normal rhythm.

The FDA approval of Multaq was based on results of three studies with a total of 5865 enrolled subjects in sinus rhythm with a prior episode of AF or AFL (Hohnloser et al., 2009). The subjects were at least 75 years old, or at least 70 years old with another risk factor. They were treated for up to 30 months with either Multaq 400 mg twice daily or placebo. The time to first hospitalization for cardiovascular reasons or death from any cause was significantly decreased in the Multaq group by 24%, and the risk of cardiovascular death was reduced by 30% (Singh et al., 2007).

Mechanism of action: Dronedarone is considered to block the *KCNH2* gene that codes for the alpha subunit of hERG (*h*uman *ether-à-go-go-*related *g*ene) potassium channel, also known as Kv11.1. hERG activity mediates the rapid delayed rectifier current (I_{Kr}) in cardiac myocytes, responsible for the rapid phase

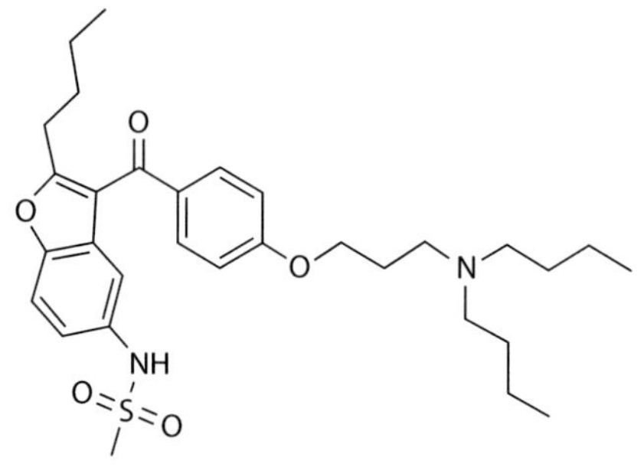

N
NH2

Figure 43.1 Molecular structure of dalfampridine or fampridine, which is commonly known as 4-aminopyridine (4-AP).

O
O
O
N
NH
O=S=O

Figure 43.2 Molecular structure of dronedarone, an antiarrhythmia drug.

of repolarization of ventricular action potential. Dronedarone is a benzofuran derivative of amiodarone, an effective yet toxic antiarrhythmic drug. Unlike amiodarone, dronedarone does not contain iodine atoms and hence retains the efficacy of amiodarone without its unique toxicity profile. Dronedarone also has a much smaller volume of distribution, and has an elimination half-life of 24 h due to its less lipophilic nature, in contrast to amiodarone's half-life of several weeks. As a result of these preferable pharmacokinetic characteristics, dronedarone dosing may be less complicated than amiodarone.

Studies also suggest that dronedarone is a multichannel blocker. It inhibits several inward potassium currents, including rapid delayed rectifier, slow delayed rectifier, and ACh-activated inward rectifier. It is also believed to reduce inward rapid Na^+ current and current from L-type Ca^{2+} channels. The reduction in K^+ current in some studies was shown to be due to the inhibition of K-ACh channel or associated GTP-binding proteins. Reduction of K^+ current by 69% led to increased action potential duration and increased effective refractory period, thus has been shown to suppress pacemaker potential of the sinoatrial (SA) node and return patients to a normal heart rhythm.

43.1.4 DOFETILIDE

Dofetilide is a potent and selective class III antiarrhythmic agent for the maintenance of normal sinus rhythm in patients with AF and AFL (see Figure 43.3). Dofetilide was approved by FDA in October 1999, and is marketed under the trade name *Tikosyn* by Pfizer. Dofetilide is for patients with AF/AFL of greater than 1-week duration. After taking the drug, patients have been converted to normal sinus rhythm. Dofetilide is also indicated for the conversion of AF and AFL to normal sinus rhythm.

Mechanism of action: Dofetilide works by selectively blocking the rapid component of the delayed rectifier outward potassium current (I_{Kr}) encoded primarily by hERG and therefore increasing the effective refractory period and action potential duration without affecting the fast inward sodium current. Dofetilide has also been shown to block potassium currents such as $K_{2P}2.1$ encoded by *KCNK2* gene and Kir2.2 encoded by *KCNJ12* gene. $K_{2P}2.1$ is a member of the two-pore-domain background potassium channel family, and Kir2.2 is an ATP-sensitive inward rectifier potassium channel. The pharmacokinetic profile of dofetilide in both healthy volunteers and patients includes a linear dose–plasma concentration relationship. The terminal plasma elimination half-life is approximately 9–10 h, and systemic bioavailability is in the region of 100% (Rasmussen et al., 1992).

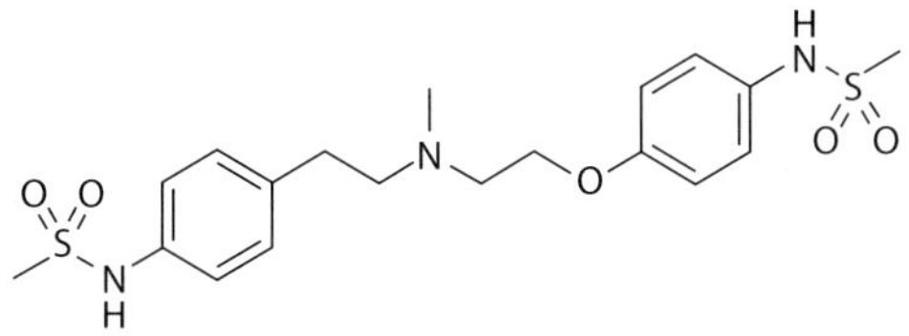

Figure 43.3 Molecular structure of dofetilide, a class III antiarrhythmic agent for the maintenance of normal sinus rhythm in patients with AF and AFL.

43.2 DRUGS ACTING ON ATP-SENSITIVE INWARD RECTIFIER K^+ CHANNELS

43.2.1 NATEGLINIDE

Nateglinide (INN) is an antidiabetic drug for treatment of type 2 diabetes mellitus (T2DM) (see Figure 43.4). Developed by Ajinomoto company, Nateglinide was approved by FDA in 2000 and marketed under the trade name *Starlix* by Novartis.

Type 2 diabetes occurs due to impaired insulin secretion that becomes chronic and progressive, resulting initially in impaired glucose tolerance and eventually in type 2 diabetes. As most patients with type 2 diabetes have both insulin resistance and insulin deficiency, therapy for T2DM is aimed at controlling not only fasting, but also postprandial plasma glucose levels (Tentolouris et al., 2007).

Mechanism of action: Nateglinide, an amino acid d-phenylalanine derivative, belongs to the meglitinide class of blood glucose–lowering medications. It acts by inhibiting ATP-sensitive K^+ channel (K_{ATP} channel) potassium channels in the membrane of β-cells. Inhibition of K_{ATP} channel activity depolarizes β-cells and causes voltage-gated calcium channels to open. The resulting calcium influx induces fusion of insulin-containing vesicles with the cell membrane, and insulin secretion occurs from the pancreas. Nateglinide restores postprandial early phase insulin secretion in a transient and glucose-sensitive manner without affecting the basal insulin level by directly acting on the pancreatic beta-cells to stimulate insulin secretion.

43.2.2 REPAGLINIDE

Repaglinide is an orally administered antidiabetic drug in the class of medications known as meglitinides that can be used to manage meal-related glucose loads (see Figure 43.5). It was approved by FDA in December 1997. Repaglinide is marketed by Novo Nordisk under the trade name *Prandin* in the United States, *GlucoNorm* in Canada, *Surepost* in Japan, and *NovoNorm* in the rest of the world.

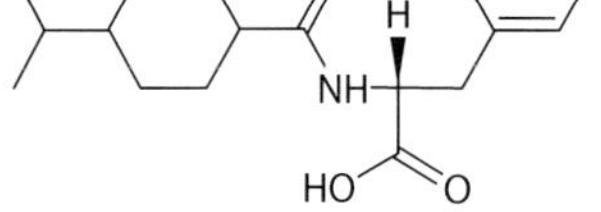

Figure 43.4 Molecular structure of nateglinide, an antidiabetic drug for treatment of T2DM.

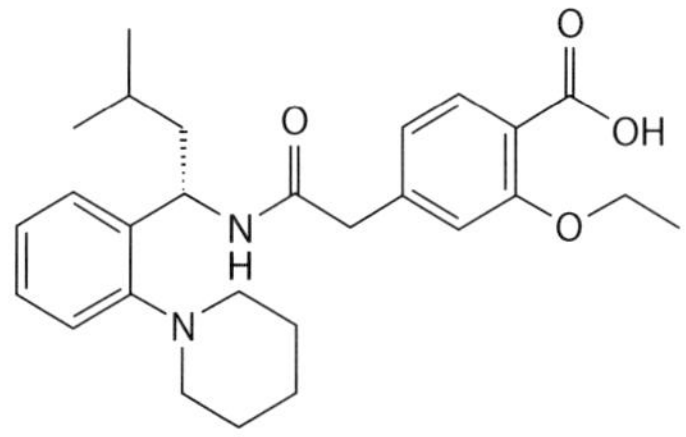

Figure 43.5 Molecular structure of repaglinide, an orally administered antidiabetic drug, among the meglitinides which are used to manage meal-related glucose loads.

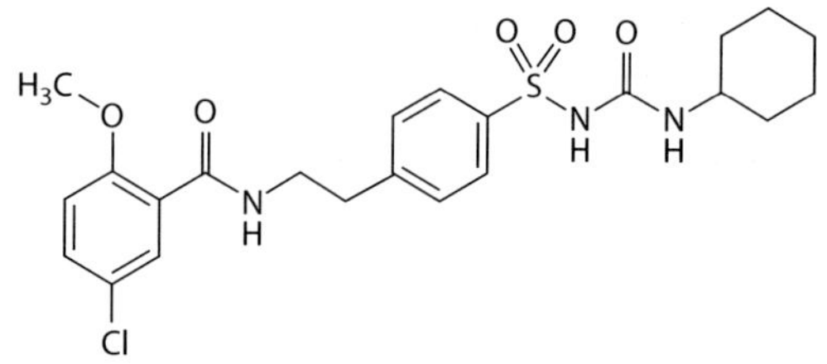

Figure 43.6 Molecular structure of glibenclamide, also known as glyburide (USAN), an antidiabetic blood glucose–lowering agent for patients with non-insulin-dependent T2DM.

Prandin's quick onset and short duration of action concentrates its effect around meal time glucose load, which is important to the treatment of type 2 diabetes. Prandin is minimally excreted by the kidney, which may be an advantage for patients who suffer from decreased kidney function. Prandin posts a low risk of hypoglycemia and potentially low risk of significant weight gain.

Mechanism of action: Prandin lowers blood glucose by stimulating the release of insulin from the pancreas. This is achieved by closing ATP-dependent potassium channels in the membrane of the β-cells. Closing the channel depolarizes β cell membrane, leading to opening of voltage-dependent calcium channels. The resulting calcium influx induces insulin secretion.

43.2.3 GLIBENCLAMIDE

Glibenclamide (INN), also known as *glyburide* (USAN), is an antidiabetic drug in a class of medications known as sulfonylureas that are closely related to sulfa drugs (see Figure 43.6). It is a blood glucose–lowering agent used in the treatment of patients with non-insulin-dependent diabetes mellitus (Type II), whose hyperglycemia cannot be controlled by diet alone. Glyburide was developed by Roche and approved by FDA in 1997.

Mechanism of action: Glibenclamide works by inhibiting the sulfonylureareceptorl (SUR1), which is a regulatory subunit of the ATP-sensitive inward rectifier potassium channels (KATP) in pancreatic β-cells (Chen et al., 2003; Davies et al., 2005; Zunkler, 2006). As is the case for Prandin, reduction of potassium current causes cell membrane depolarization and activation of voltage-dependent calcium channel. Activation of voltage-dependent calcium channel results in an increase in intracellular calcium in the β-cells and subsequent stimulation of insulin release.

43.3 DRUGS ACTING ON CALCIUM-ACTIVATED K+ CHANNELS

43.3.1 CHLOROTHIAZIDE

Chlorothiazide (Diuril) is a thiazide diuretic and antihypertensive agent, approved by FDA before 1982 (see Figure 43.7). As a diuretic Chlorothiazide helps prevent body from absorbing too much salt, which can cause fluid retention. Chlorothiazide treats fluid retention (edema) in people with congestive heart failure, cirrhosis of the liver, or kidney disorders, as well as edema caused by taking steroids

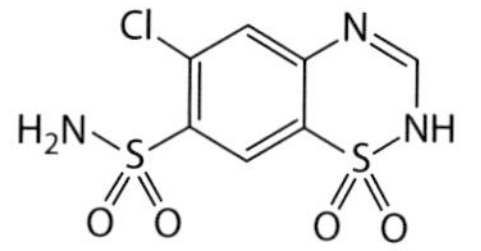

Figure 43.7 Molecular structure of chlorothiazide (Diuril), a thiazide diuretic and antihypertensive agent.

or estrogen. Its chemical name is 6-chloro-2 H-1,2,4-benzothiadiazine7-sulfonamide 1,1-dioxide, with an empirical formula $C_7H_6ClN_3O_4S_2$.

Mechanism of drug action: Chlorothiazide works by blocking potassium large conductance calcium-activated (or big potassium, BK channels) channel alpha subunit 1 encoded by *KCNMA1* gene, previously known as SLO1 or Maxi-K. BK channels are activated by changes of membrane potential and/or increases of intracellular calcium. BK channels, as an attractive drug target, play a pivotal and specific role in many pathophysiological conditions including the regulation of smooth muscle tone and neuronal excitability. Chlorothiazide is not metabolized but is eliminated rapidly by the kidney with plasma half-life about 45–120 min. Chlorothiazide crosses the placental but not the blood–brain barrier.

43.3.2 CHLORZOXAZONE

Chlorzoxazone is approved by FDA also before 1982 and is sold under the trade name *Muscol* (see Figure 43.8). Chlorzoxazone is a centrally acting muscle relaxant. It is used to treat muscle spasm and the resulting pain or discomfort.

Mechanism of drug action: As an activator of Ca^{2+}-dependent K^+ channels (I_{KCa}), chlorzoxazone reversibly increases the channel current in a concentration-dependent manner with an EC_{50} value of 30 μM (Cao et al., 2001; Liu et al., 2003). The chlorzoxazone-stimulated I_{KCa} was inhibited by iberitoxin (200 nM) or clotrimazole (10 μM), but not by glibenclamide (10 μM) or apamin (200 nM). Chlorzoxazone (30 μM) suppressed voltage-dependent L-type Ca^{2+} current. In the inside-out configuration, chlorzoxazone applied to the intracellular side of the patch did not modify single-channel conductance of large conductance Ca^{2+}-activated K^+ channels (BK), but did increase channel activity by increasing the mean open time and decreasing the mean closed time. Chlorzoxazone also caused a left shift in the activation curve of BK channels, which promotes channel opening under physiological conditions. The Ca^{2+} sensitivity of these channels was unaffected by chlorzoxazone. Under current-clamp condition, chlorzoxazone (10 μM) reduced the firing rate of action potentials.

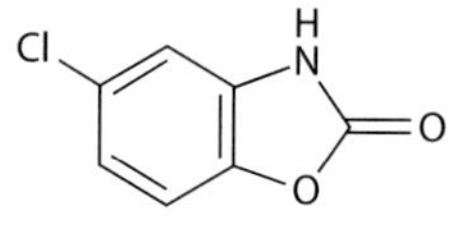

Figure 43.8 Molecular structure of chlorzoxazone, a centrally acting muscle relaxant used to treat muscle spasm and the resulting pain or discomfort.

43.4 DRUGS ACTING ON VOLTAGE-GATED Na+ CHANNELS

43.4.1 RANOLAZINE

Ranolazine (Ranexa) is an oral anti-ischemic/anti-anginal drug, designed to act without reducing heart rate or blood pressure (see Figure 43.9). It was developed by CV Therapeutics and approved under the trade name *Ranexa* by FDA in January 2006.

Ranexa is specifically a useful new option for patients with myocardial ischemia and chronic stable angina whose symptoms are not controlled with first-line anti-anginal therapy or who do not tolerate first-line anti-anginal agents. Approval of Ranexa was based on a pair of clinical trials, dubbed ERICA (Efficacy of Ranolazine In Chronic Angina) and CARISA (Combination Assessment of Ranolazine In Stable Angina). In general, myocardial ischemia is associated with reduced adenosine triphosphate fluxes and decreased energy supply, resulting in severe disturbances of intracellular ion homeostasis in cardiac myocytes. Experimental and clinical studies have shown that ranolazine is effective in reducing manifestations of ischemia and angina, and it also holds potential promise to be effective in the management of left ventricular dysfunction, particularly diastolic dysfunction, and arrhythmias (Stone, 2008; Maier, 2009; Reffelmann and Kloner, 2010). In the ventricles, ranolazine can suppress arrhythmias associated with acute coronary syndrome, long QT syndrome, heart failure, ischemia, and reperfusion. In atria, ranolazine effectively suppresses atrial tachyarrhythmias and AF.

Mechanism of drug action: Ranolazine's mechanism of action has not been fully characterized. The drug has been shown to exert its anti-anginal and anti-ischemic effects without reducing heart rate or blood pressure. Ranolazine, a specific inhibitor of late I(Na), reduces Na^+ influx and hence ameliorates disturbed Na^+ and Ca^{2+} homeostasis. The principal mechanism underlying ranolazine's antiarrhythmic actions is thought to be primarily via inhibition of the persistent or late I(Na) in the ventricles and via use-dependent inhibition of peak I(Na) and I(Kr) in the atria. Short- and long-term safety of ranolazine has been demonstrated in the clinic, even in patients with structural heart disease (Antzelevitch et al., 2011). Ranolazine also affects the sodium-dependent calcium channels during myocardial ischemia in rabbits by altering the intracellular sodium level. Thus, ranolazine indirectly prevents the calcium overload that causes cardiac ischemia in rats. The effects of ranolazine on the Na_v 1.7 and Na_v 1.8 sodium channels also make it potentially useful in the treatment of neuropathic pain (Casey et al., 2010).

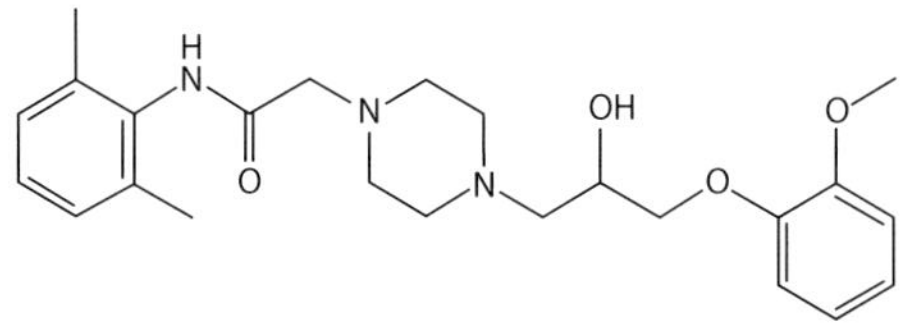

Figure 43.9 Molecular structure of ranolazine (Ranexa), an oral anti-ischemic/anti-anginal drug designed to act without reducing heart rate or blood pressure.

43.4.2 OXCARBAZEPINE

Oxcarbazepine is an anticonvulsant or antiepileptic drug (AED), used primarily in the treatment of partial seizures in adults with epilepsy and for the adjunctive treatment of partial seizures in children, aged 4–16, with epilepsy (see Figure 43.10). Oxcarbazepine is marketed as *Trileptal* by Novartis and approved by FDA in January 2000.

Mechanism of drug action: The precise mechanism by which oxcarbazepine exerts antiseizure effect is unknown; however, in vitro electrophysiological studies indicate that they produce blockade of voltage-sensitive sodium channels, resulting in the stabilization of hyperexcited neural membranes, inhibition of repetitive neuronal firing, and diminution of propagation of synaptic impulses. In addition, increased potassium conduction and modulation of high-voltage-activated calcium channels may contribute to the anticonvulsive effects of the drug.

43.5 DRUGS ACTING ON VOLTAGE-GATED CALCIUM CHANNELS

43.5.1 CLEVIDIPINE

Clevidipine (INN) was developed by *The Medicines Company* for the treatment of hypertension and was approved under the trade name *Cleviprex* by FDA in August 2008 (see Figure 43.11). Cleviprex is indicated for the reduction of blood pressure, acting by selectively relaxing smooth muscle cells that line small arteries. This results in widening of the arterial lumen and reduction of blood pressure since the small arterioles are the primary resistance vessel within the vasculature.

Mechanism of action: Cleviprex is an intravenous short-acting dihydropyridine L-type calcium channel blocker, highly selective for vascular, as opposed to myocardial, smooth muscle and, therefore, has little or no effect on myocardial contractility or cardiac conduction. It reduces mean arterial blood pressure by decreasing systemic vascular resistance. Cleviprex does not reduce cardiac filling pressure (pre-load), confirming lack of effects on

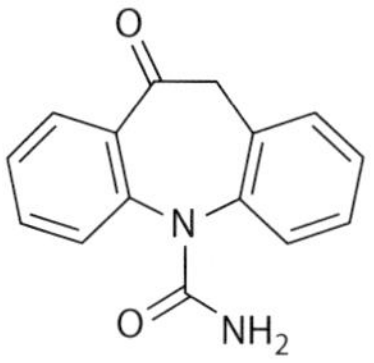

Figure 43.10 Molecular structure of oxcarbazepine, an anticonvulsant or AED used primarily in the treatment of partial seizures.

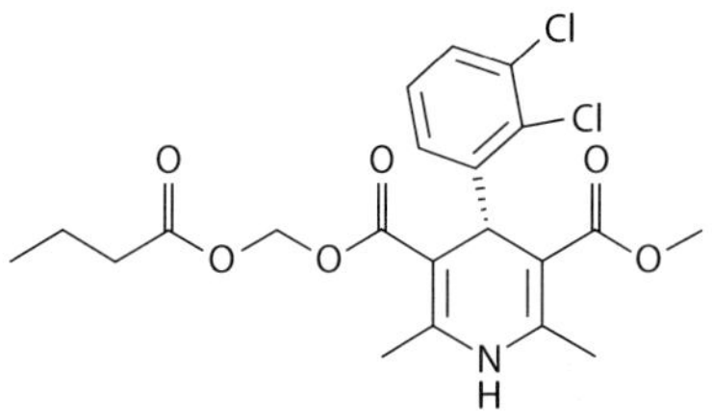

Figure 43.11 Molecular structure of clevidipine, a drug used to treat hypertension.

the venous capacitance vessels. No increase in myocardial lactate production in coronary sinus blood has been seen, confirming the absence of myocardial ischemia due to coronary steal.

43.5.2 PREGABALIN

Pregabalin (INN) (Lyrica) is an anticonvulsant drug used for neuropathic pain that was approved by FDA in December 2004 (see Figure 43.12). Pregabalin was marketed by Pfizer under the trade name *Lyrica*. It has also been found effective for generalized anxiety disorder and was approved for this use in the European Union (EU) in 2007. In addition, Pregabalin is also used off-label for the treatment of chronic pain, post-herpetic neuralgia (PHN), diabetic peripheral neuropathy and fibromyalgia, perioperative pain, and migraine. Pregabalin (Lyrica) was approved by FDA in June 2012 for the treatment of neuropathic pain associated with spinal cord injury.

Mechanism of action: Lyrica (pregabalin) is a modulator of voltage-gated calcium channels, designed to affect neurological transmission in multiple systems. The exact mechanism of Lyrica's action has not been fully characterized. Lyrica binds to the alpha2-delta auxiliary subunit of voltage-gated calcium channels. Blockade of these channels has been shown to inhibit the calcium-dependent release of a number of neurotransmitters. The drug is a structural derivative of the inhibitory neurotransmitter γ-Aminobutyric acid (GABA), though it does not bind directly to GABAa, GABAb, or benzodiazepine receptors, does not augment GABAa responses in cultured neurons, does not alter rat brain GABA concentration, or have acute effects on GABA uptake or degradation. In cultured neurons, prolonged application of pregabalin increased the density of GABA transporter protein and increased the rate of functional GABA transport. The drug does not achieve its antinociceptive or antiseizure activity through blockade of sodium channels, activation of opioid receptors, alteration of cyclooxygenase enzyme activity, through activity at serotonin and dopamine receptors, or through effect on dopamine, serotonin, or noradrenaline reuptake.

43.5.3 GABAPENTIN

Gabapentin is an anticonvulsant previously approved by FDA in March 2000 and marketed by Pfizer under the trade name *Neurotin* as an adjunct treatment for partial epileptic seizures in adults and children (see Figure 43.13). In October 2000, Neurotin in oral solution was also approved as an adjunctive therapy in the treatment of partial seizures in pediatric patients 3 years of age and older. In May 2002, FDA approved Neurotin again for the management of PHN. Neurontin is the first oral medication approved by the FDA for this indication.

Gabapentin enacarbil, as an extended-release formulation of gabapentin, was approved for the treatment of restless leg syndrome by FDA in April 2011 under the trade name *Horizant* developed by GSK. Gabapentin enacarbil is also approved by FDA in June 2012 for the treatment of PHN under the trade name *Horizant* developed by GSK. The FDA approval of Horizant for PHN was based on the results of one 12-week clinical trial in total of 279 adult subjects that are divided in three dosing groups. Treatment with Horizant statistically significantly improved the mean pain score and increased the proportion of subjects with at least a 50% reduction in pain score from baseline at all doses tested. A benefit over placebo was observed for all 3 doses of Horizant (1200, 2400, and 3600 mg/day) as early as week 1 and maintained to the end of treatment.

Mechanism of action: How Gabapentin works remains elusive, but it was considered to block voltage-gated calcium channels. Gabapentin was designed as a lipophilic GABA analog and was first synthesized as a potential anticonvulsant and was launched in 1994 as an add-on therapy for the treatment of epilepsy (Bryans and Wustrow, 1999). More recent studies suggest α2δ1 auxiliary subunit of voltage-gated calcium channels serves as the target for this drug's actions (Maneuf et al., 2006; Thorpe and Offord, 2010).

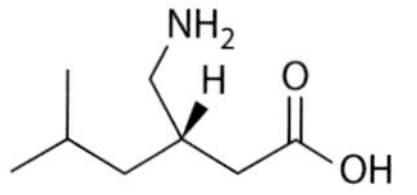

Figure 43.12 Molecular structure of pregabalin, an anticonvulsant drug used for neuropathic pain.

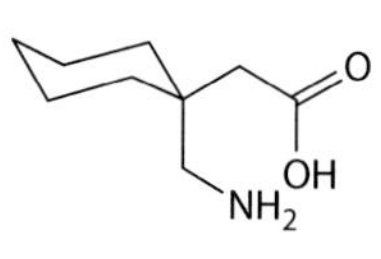

Figure 43.13 Molecular structure of gabapentin, an anticonvulsant.

43.6 DRUGS ACTING ON TRANSIENT RECEPTOR POTENTIAL (TRP) CHANNELS

43.6.1 CAPSAICIN

Capsaicin is the active component of chili peppers and produces a sensation of burning in tissues with which it comes into contact (see Figure 43.14). Capsaicin ($C_{18}H_{27}NO_3$) is currently used in a high-dose dermal patch that was developed by a company called NeurogesX and was approved by FDA in 2009 under the trade name *Qutenza* for neuropathic pain associated with PHN caused by shingles (Backonja et al., 2008; Babbar et al., 2009).

Qutenza (capsaicin) 8% is a transdermal patch containing capsaicin in a localized dermal delivery system. The capsaicin in Qutenza is a synthetic equivalent of the naturally occurring compound found in chili peppers. Qutenza works by targeting certain pain nerves in the area of skin where pain is being experienced.

Mechanism of action: Capsaicin selectively binds to TRPV1 that resides on the membranes of nociceptive and heat-sensing neurons, permitting cations to pass through the cell membrane and into the cell. The resulting depolarization of the neuron signals to the brain. TRPV1 is a heat-activated calcium-permeable

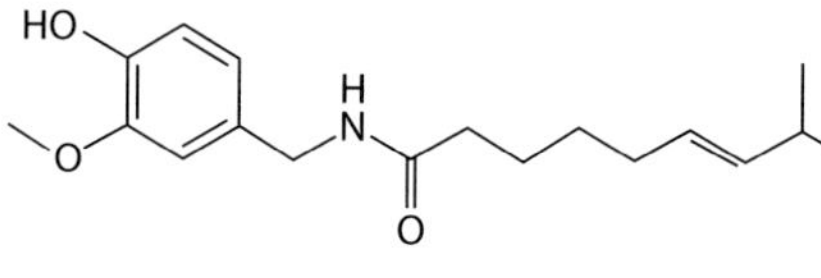

Figure 43.14 Molecular structure of capsaicin, the active component of chili peppers.

nonselective cation channel that was cloned in 1997 (Caterina et al., 1997). TRPV1 opens when temperature reaches at or above 43°C. When capsaicin binds to TRPV1, it activates the channel to open below 37°C, causing the sensation of heat. Prolonged activation of these neurons by capsaicin depletes presynaptic substance P, one of the body's neurotransmitters for pain and heat.

43.6.2 MENTHOL

Menthol is a natural compound synthesized or extracted from cornmint and peppermint with chemical molecular formula of $C_{10}H_{20}O$ (see Figure 43.15). Menthol, a monoterpene widely used in cosmetics and as a flavoring agent, was approved as antipruritics, also known as ant-itch drugs, by FDA before 1982 for use of inhibiting itching that is often associated with sunburns, allergic reactions, eczema, psoriasis, chickenpox, fungal infections, insect bites and stings, and contact dermatitis.

Mechanism of action: Menthol acts upon TRP channels by rapidly increasing intracellular calcium and mobilizing calcium flux through the channels. Menthol's characteristic well-known cooling sensation is due to the activation of TRP melastatin family member 8 (TRPM8) channels expressed in sensory neurons and the skin when inhaled, eaten, or applied to the skin. TRPM8 is activated when temperatures drop at or below 28°C. Application of menthol to dorsal root ganglia (DRG) neurons in culture results in a rapid increase in intracellular calcium at the presynaptic terminals from intracellular calcium store (Tsuzuki et al., 2004). The calcium influx caused by menthol activity within the presynaptic sites acts as a mediator for release of glutamate for enhanced glutamatergic and glycinergic neurotransmission at sensory synapses, resulting in cold sensation and analgesia (Tsuzuki et al., 2004).

It has been proposed that the specific interaction with two cysteine residues (C929 and C940) between transmembrane segments 5 and 6 may point to a menthol-specific interaction with the receptor, whereas binding to the transmembrane segment 2 and the carboxyl terminus refers to a preserved general binding site for modulators of the receptor. This would explain the differences in responses to the cold-sensitizing agent ilicin and menthol (Farco and Grundmann, 2013).

Menthol also activates TRP subfamily A member 1 (TRPA1) and modulates the channel to increase intracellular calcium. However, the modulation appears to be bimodal with activation of mouse TRPA1 receptor at low concentrations but inhibition of the receptor at high concentrations (Xiao et al., 2008). The pore region and transmembrane segments 5 and 6 of TRPA1 are critical for menthol responsiveness (Xiao et al., 2008). Aside from its cold-inducing sensation capabilities, menthol exhibits cytotoxic effects in cancer cells, induces reduction in malignant cell growth, engages in synergistic excitation of GABA receptors, and inhibits $Na_v1.8$ and $Na_v1.9$ as well as nicotinic acetylcholine receptors.

OH

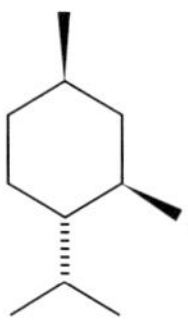

Figure 43.15 Molecular structure of menthol, a natural compound synthesized or extracted from cornmint and peppermint and used as anti-itch drugs.

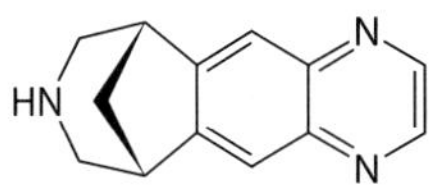

HN N N

Figure 43.16 Molecular structure of varenicline, a drug used for smoking cessation.

43.7 DRUGS ACTING ON NEURONAL ACETYLCHOLINE RECEPTOR SUBUNIT ALPHA-7

Varenicline ($C_{13}H_{13}N_3$) was discovered and marketed by Pfizer under the trade name *Chantix* in the United States and *Champix* in Canada, Europe, and other countries (see Figure 43.16). Varenicline was approved by FDA and EU in 2006 for smoking cessation. Varenicline functions as nicotinic receptor partial agonist as it stimulates nicotine receptors more weakly than nicotine itself does. As a partial agonist, it reduces cravings for and decreases the pleasurable effects of cigarettes and other tobacco products, the mechanisms by which varenicline can assist some patients to quit smoking (Jorenby et al., 2006).

Mechanism of drug action: Varenicline is a full agonist on α7 receptors, but it is a partial agonist of the α4β2 subtype of nicotinic acetylcholine receptor (Mihalak et al., 2006). Acting as a partial agonist, varenicline binds to α4β2 and partially stimulates the receptor without producing a full effect like nicotine. Varenicline also acts as an agonist at 5-HT3 receptors, which may contribute to mood-altering effects of varenicline. In addition, it acts on α3β4- and weakly on α3β2- and α6-containing receptors (Mihalak et al., 2006).

REFERENCES

Abbott, G. W., M. H. Butler et al. (2001). MiRP2 forms potassium channels in skeletal muscle with Kv3.4 and is associated with periodic paralysis. *Cell* **104**(2): 217–231.

Abbott, G. W. and S. A. Goldstein (1998). A superfamily of small potassium channel subunits: Form and function of the MinK-related peptides (MiRPs). *Q Rev Biophys* **31**(4): 357–398.

Abbott, G. W., S. A. Goldstein et al. (2001). Do all voltage-gated potassium channels use MiRPs? *Circ Res* **88**(10): 981–983.

Abbott, G. W., B. Ramesh et al. (2008). Secondary structure of the MiRP1 (KCNE2) potassium channel ancillary subunit. *Protein Pept Lett* **15**(1): 63–75.

Abbott, G. W., F. Sesti et al. (1999). MiRP1 forms IKr potassium channels with HERG and is associated with cardiac arrhythmia. *Cell* **97**(2): 175–187.

Albesa, M., L. S. Grilo et al. (2011). Nedd4–2-dependent ubiquitylation and regulation of the cardiac potassium channel hERG1. *J Mol Cell Cardiol* **51**(1): 90–98.

Alders, M., T. T. Koopmann et al. (2009). Haplotype-sharing analysis implicates chromosome 7q36 harboring DPP6 in familial idiopathic ventricular fibrillation. *Am J Hum Genet* **84**(4): 468–476.

Alders, M. and M. Mannens (1993). Romano-ward syndrome. *Gene Reviews.* R. A. Pagon, T. D. Bird, C. R. Dolan, K. Stephens and M. P. Adam. (eds.), University of Washington, Seattle, WA.

Allen, M., A. Heinzmann et al. (2003). Positional cloning of a novel gene influencing asthma from chromosome 2q14. *Nat Genet* **35**(3): 258–263.

Almassy, J. and T. Begenisich (2012). The LRRC26 protein selectively alters the efficacy of BK channel activators. *Mol Pharmacol* **81**(1): 21–30.

Amarillo, Y., J. A. De Santiago-Castillo et al. (2008). Ternary Kv4.2 channels recapitulate voltage-dependent inactivation kinetics of A-type K+ channels in cerebellar granule neurons. *J Physiol* **586**(8): 2093–2106.

Anaganti, S., J. K. Hansen et al. (2009). Non-AUG translational initiation of a short CAPC transcript generating protein isoform. *Biochem Biophys Res Commun* **380**(3): 508–513.

Angelo, K., T. Jespersen et al. (2002). KCNE5 induces time- and voltage-dependent modulation of the KCNQ1 current. *Biophys J* **83**(4): 1997–2006.

Barhanin, J., F. Lesage et al. (1996). K(V)LQT1 and lsK (minK) proteins associate to form the I(Ks) cardiac potassium current. *Nature* **384**(6604): 78–80.

Barriere, H., I. Rubera et al. (2003). Swelling-activated chloride and potassium conductance in primary cultures of mouse proximal tubules. Implication of KCNE1 protein. *J Membr Biol* **193**(3): 153–170.

Bas, T., G. Y. Gao et al. (2011). Post-translational N-glycosylation of type I transmembrane KCNE1 peptides: Implications for membrane protein biogenesis and disease. *J Biol Chem* **286**(32): 28150–28159.

Bell, J. K., I. Botos et al. (2005). The molecular structure of the Toll-like receptor 3 ligand-binding domain. *Proc Natl Acad Sci USA* **102**(31): 10976–10980.

Bella, J., K. L. Hindle et al. (2008). The leucine-rich repeat structure. *Cell Mol Life Sci* **65**(15): 2307–2333.

Bjorklund, A. K., D. Ekman et al. (2006). Expansion of protein domain repeats. *PLoS Comput Biol* **2**(8): e114.

Boronat, A., J. M. Gelfand et al. (2012). Encephalitis and antibodies to dipeptidyl-peptidase-like protein-6, a subunit of Kv4.2 potassium channels. *Ann Neurol* **73**: 120–128.

Brambilla, P., F. Esposito et al. (2012). Association between DPP6 polymorphism and the risk of progressive multiple sclerosis in Northern and Southern Europeans. *Neurosci Lett* **530**(2): 155–160.

Brandt, M. C., J. Endres-Becker et al. (2009). Effects of KCNE2 on HCN isoforms: Distinct modulation of membrane expression and single channel properties. *Am J Physiol Heart Circ Physiol* **297**(1): H355–H363.

Braun, A. P. (2010). A new "opening" act on the BK channel stage: Identification of LRRC26 as a novel BK channel accessory subunit that enhances voltage-dependent gating. *Channels (Austin)* **4**(4): 249–250.

Brenner, R., T. J. Jegla et al. (2000). Cloning and functional characterization of novel large conductance calcium-activated potassium channel beta subunits, hKCNMB3 and hKCNMB4. *J Biol Chem* **275**(9): 6453–6461.

Brugada, P. and J. Brugada (1992). Right bundle branch block, persistent ST segment elevation and sudden cardiac death: A distinct clinical and electrocardiographic syndrome. A multicenter report. *J Am Coll Cardiol* **20**(6): 1391–1396.

Bueno, R., A. De Rienzo et al. (2010). Second generation sequencing of the mesothelioma tumor genome. *PLoS One* **5**(5): e10612.

Chandrasekhar, K. D., T. Bas et al. (2006). KCNE1 subunits require co-assembly with K+ channels for efficient trafficking and cell surface expression. *J Biol Chem* **281**(52): 40015–40023.

Chandrasekhar, K. D., A. Lvov et al. (2011). O-glycosylation of the cardiac I(Ks) complex. *J Physiol* **589**(Part 15): 3721–3730.

Chen, B., Q. Ge et al. (2010). A novel auxiliary subunit critical to BK channel function in *Caenorhabditis elegans. J Neurosci* **30**(49): 16651–16661.

Chen, H., L. A. Kim et al. (2003a). Charybdotoxin binding in the I(Ks) pore demonstrates two MinK subunits in each channel complex. *Neuron* **40**(1): 15–23.

Chen, H., F. Sesti et al. (2003b). Pore- and state-dependent cadmium block of I(Ks) channels formed with MinK-55C and wild-type KCNQ1 subunits. *Biophys J* **84**(6): 3679–3689.

Chen, T., K. Ajami et al. (2006). Molecular characterization of a novel dipeptidyl peptidase like 2-short form (DPL2-s) that is highly expressed in the brain and lacks dipeptidyl peptidase activity. *Biochim Biophys Acta* **1764**(1): 33–43.

Chen, T., X. Shen et al. (2008). Molecular characterisation of a novel dipeptidyl peptidase like protein: Its pathological link to Alzheimer's disease. *Clin Chem Lab Med* **46**(4): A13.

Chen, Y. H., S. J. Xu et al. (2003c). KCNQ1 gain-of-function mutation in familial atrial fibrillation. *Science* **299**(5604): 251–254.

Cheung, C. L., B. Y. Chan et al. (2009). Pre-B-cell leukemia homeobox 1 (PBX1) shows functional and possible genetic association with bone mineral density variation. *Hum Mol Genet* **18**(4): 679–687.

Chouabe, C., N. Neyroud et al. (1997). Properties of KvLQT1 K+ channel mutations in Romano-Ward and Jervell and Lange-Nielsen inherited cardiac arrhythmias. *EMBO J* **16**(17): 5472–5479.

Chouabe, C., N. Neyroud et al. (2000). Novel mutations in KvLQT1 that affect Iks activation through interactions with Isk. *Cardiovasc Res* **45**(4): 971–980.

Chung, D. Y., P. J. Chan et al. (2009). Location of KCNE1 relative to KCNQ1 in the I(KS) potassium channel by disulfide cross-linking of substituted cysteines. *Proc Natl Acad Sci USA* **106**(3): 743–748.

Ciampa, E. J., R. C. Welch et al. (2011). KCNE4 juxtamembrane region is required for interaction with calmodulin and for functional suppression of KCNQ1. *J Biol Chem* **286**(6): 4141–4149.

Cotella, D., S. Radicke et al. (2012). N-glycosylation of the mammalian dipeptidyl aminopeptidase-like protein 10 (DPP10) regulates trafficking and interaction with Kv4 channels. *Int J Biochem Cell Biol* **44**(6): 876–885.

Covarrubias, M., A. Bhattacharji et al. (2008). The neuronal Kv4 channel complex. *Neurochem Res* **33**(8): 1558–1567.

Cronin, S., S. Berger et al. (2008). A genome-wide association study of sporadic ALS in a homogenous Irish population. *Hum Mol Genet* **17**(5): 768–774.

David, J. P., M. N. Andersen et al. (2013). Trafficking of the I(Ks)-complex in MDCK cells: Site of subunit assembly and determinants of polarized localization. *Traffic* **14**: 399–411.

Dedek, K. and S. Waldegger (2001). Colocalization of KCNQ1/KCNE channel subunits in the mouse gastrointestinal tract. *Pflugers Arch* **442**(6): 896–902.

Del Bo, R., S. Ghezzi et al. (2008). DPP6 gene variability confers increased risk of developing sporadic amyotrophic lateral sclerosis in Italian patients. *J Neurol Neurosurg Psychiatry* **79**(9): 1085.

Delpon, E., J. M. Cordeiro et al. (2008). Functional effects of KCNE3 mutation and its role in the development of Brugada syndrome. *Circ Arrhythm Electrophysiol* **1**(3): 209–218.

Dolan, J., K. Walshe et al. (2007). The extracellular leucine-rich repeat superfamily; a comparative survey and analysis of evolutionary relationships and expression patterns. *BMC Genomics* **8**: 320.

Dougherty, K. and M. Covarrubias (2006). A dipeptidyl aminopeptidase-like protein remodels gating charge dynamics in Kv4.2 channels. *J Gen Physiol* **128**(6): 745–753.

Doyle, D. A., J. Morais Cabral et al. (1998). The structure of the potassium channel: Molecular basis of K+ conduction and selectivity. *Science* **280**(5360): 69–77.

Duggal, P., M. R. Vesely et al. (1998). Mutation of the gene for IsK associated with both Jervell and Lange-Nielsen and Romano-Ward forms of Long-QT syndrome. *Circulation* **97**(2): 142–146.

Egland, K. A., X. F. Liu et al. (2006). High expression of a cytokeratin-associated protein in many cancers. *Proc Natl Acad Sci USA* **103**(15): 5929–5934.

Farco, J. A. and O. Grundmann (2013). Menthol–pharmacology of an important naturally medicinal "cool". Mini reviews in medicinal chemistry **13**: 124–131.

Fass, D. (2012). Disulfide bonding in protein biophysics. *Annu Rev Biophys* **41**: 63–79.

Foeger, N. C., A. J. Norris et al. (2012). Augmentation of Kv4.2-encoded currents by accessory dipeptidyl peptidase 6 and 10 subunits reflects selective cell surface Kv4.2 protein stabilization. *J Biol Chem* **287**(12): 9640–9650.

Gage, S. D. and W. R. Kobertz (2004). KCNE3 truncation mutants reveal a bipartite modulation of KCNQ1 K^+ channels. *J Gen Physiol* **124**(6): 759–771.

Garber, K. (2008). Genetics. The elusive ALS genes. *Science* **319**(5859): 20.

Garcia-Patrone, M. and J. S. Tandecarz (1995). A glycoprotein multimer from Bacillus thuringiensis sporangia: Dissociation into subunits and sugar composition. *Mol Cell Biochem* **145**(1): 29–37.

Garin-Chesa, P., L. J. Old et al. (1990). Cell surface glycoprotein of reactive stromal fibroblasts as a potential antibody target in human epithelial cancers. *Proc Natl Acad Sci USA* **87**(18): 7235–7239.

Goldstein, S. A. and C. Miller (1991). Site-specific mutations in a minimal voltage-dependent K+ channel alter ion selectivity and open-channel block. *Neuron* **7**(3): 403–408.

Gordon, E., T. K. Roepke et al. (2006). Endogenous KCNE subunits govern Kv2.1 K+ channel activation kinetics in *Xenopus* oocyte studies. *Biophys J* **90**(4): 1223–1231.

Gorrell, M. D. (2005). Dipeptidyl peptidase IV and related enzymes in cell biology and liver disorders. *Clin Sci (Lond)* **108**(4): 277–292.

Grahammer, F., A. Herling et al. (2001). The cardiac K+ channel KCNQ1 Is essential for gastric acid secretion. *Gastroenterology* **120**(6): 1363–1371.

Grahammer, F., R. Warth et al. (2001). The small conductance K+ channel, KCNQ1: Expression, function, and subunit composition in murine trachea. *J Biol Chem* **276**(45): 42268–42275.

Grammatikos, A. P. (2008). The genetic and environmental basis of atopic diseases. *Ann Med* **40**(7): 482–495.

Grunnet, M., T. Jespersen et al. (2002). KCNE4 is an inhibitory subunit to the KCNQ1 channel. *J Physiol* **542**(Part 1): 119–130.

Grunnet, M., T. Jespersen et al. (2003). KCNQ1 channels sense small changes in cell volume. *J Physiol* **549**(Part 2): 419–427.

Haitin, Y., R. Wiener et al. (2009). Intracellular domains interactions and gated motions of I(KS) potassium channel subunits. *EMBO J* **28**(14): 1994–2005.

Hanski, C., T. Huhle et al. (1985). Involvement of plasma membrane dipeptidyl peptidase IV in fibronectin-mediated adhesion of cells on collagen. *Biol Chem Hoppe Seyler* **366**(12): 1169–1176.

Hanski, C., T. Huhle et al. (1988). Direct evidence for the binding of rat liver DPP IV to collagen in vitro. *Exp Cell Res* **178**(1): 64–72.

Helft, L., V. Reddy et al. (2011). LRR conservation mapping to predict functional sites within protein leucine-rich repeat domains. *PLoS One* **6**(7): e21614.

Hogg, P. J. (2003). Disulfide bonds as switches for protein function. *Trends Biochem Sci* **28**(4): 210–214.

Hong, K., D. R. Piper et al. (2005). De novo KCNQ1 mutation responsible for atrial fibrillation and short QT syndrome in utero. *Cardiovasc Res* **68**(3): 433–440.

Huet, G., S. Hennebicq-Reig et al. (1998). GalNAc-alpha-O-benzyl inhibits NeuAcalpha2–3 glycosylation and blocks the intracellular transport of apical glycoproteins and mucus in differentiated HT-29 cells. *J Cell Biol* **141**(6): 1311–1322.

Jerng, H. H., K. Dougherty et al. (2009). A novel N-terminal motif of dipeptidyl peptidase-like proteins produces rapid inactivation of KV4.2 channels by a pore-blocking mechanism. *Channels (Austin)* **3**(6): 448–461.

Jerng, H. H., K. Kunjilwar et al. (2005). Multiprotein assembly of Kv4.2, KChIP3 and DPP10 produces ternary channel complexes with ISA-like properties. *J Physiol* **568**(Part 3): 767–788.

Jerng, H. H., A. D. Lauver et al. (2007). DPP10 splice variants are localized in distinct neuronal populations and act to differentially regulate the inactivation properties of Kv4-based ion channels. *Mol Cell Neurosci* **35**(4): 604–624.

Jerng, H. H. and P. J. Pfaffinger (2012). Incorporation of DPP6a and DPP6K variants in ternary Kv4 channel complex reconstitutes properties of A-type K current in rat cerebellar granule cells. *PLoS One* **7**(6): e38205.

Jerng, H. H., Y. Qian et al. (2004). Modulation of Kv4.2 channel expression and gating by dipeptidyl peptidase 10 (DPP10). *Biophys J* **87**(4): 2380–2396.

Jespersen, T., M. Membrez et al. (2007). The KCNQ1 potassium channel is down-regulated by ubiquitylating enzymes of the Nedd4/Nedd4-like family. *Cardiovasc Res* **74**(1): 64–74.

Jiang, Y., A. Lee et al. (2003). X-ray structure of a voltage-dependent K+ channel. *Nature* **423**(6935): 33–41.

Kaczmarek, L. K. (1991). Voltage-dependent potassium channels: minK and *Shaker* families. *New Biol* **3**(4): 315–323.

Kajava, A. V. (1998). Structural diversity of leucine-rich repeat proteins. *J Mol Biol* **277**(3): 519–527.

Kanda, V. A. and G. W. Abbott (2012). KCNE regulation of K(+) channel trafficking—A Sisyphean task? *Front Physiol* **3**: 231.

Kanelis, V., D. Rotin et al. (2001). Solution structure of a Nedd4 WW domain-ENaC peptide complex. *Nat Struct Biol* **8**(5): 407–412.

Kang, C., C. Tian et al. (2008). Structure of KCNE1 and implications for how it modulates the KCNQ1 potassium channel. *Biochemistry* **47**(31): 7999–8006.

Kaulin, Y. A., J. A. De Santiago-Castillo et al. (2009). The dipeptidyl-peptidase-like protein DPP6 determines the unitary conductance of neuronal Kv4.2 channels. *J Neurosci* **29**(10): 3242–3251.

Kim, H. M., S. C. Oh et al. (2007). Structural diversity of the hagfish variable lymphocyte receptors. *J Biol Chem* **282**(9): 6726–6732.

Kim, H. M., B. S. Park et al. (2007). Crystal structure of the TLR4-MD-2 complex with bound endotoxin antagonist Eritoran. *Cell* **130**(5): 906–917.

Kin, Y., Y. Misumi et al. (2001). Biosynthesis and characterization of the brain-specific membrane protein DPPX, a dipeptidyl peptidase IV-related protein. *J Biochem* **129**(2): 289–295.

Knaus, H. G., K. Folander et al. (1994). Primary sequence and immunological characterization of beta-subunit of high conductance Ca(2+)-activated K+ channel from smooth muscle. *J Biol Chem* **269**(25): 17274–17278.

Kobe, B. and J. Deisenhofer (1993). Crystal structure of porcine ribonuclease inhibitor, a protein with leucine-rich repeats. *Nature* **366**(6457): 751–756.

Kobe, B. and A. V. Kajava (2001). The leucine-rich repeat as a protein recognition motif. *Curr Opin Struct Biol* **11**(6): 725–732.

Krumerman, A., X. Gao et al. (2004). An LQT mutant minK alters KvLQT1 trafficking. *Am J Physiol Cell Physiol* **286**(6): C1453–C1463.

Kunjilwar, K., C. Strang et al. (2004). KChIP3 rescues the functional expression of Shal channel tetramerization mutants. *J Biol Chem* **279**(52): 54542–54551.

Kurokawa, J., H. K. Motoike et al. (2001). TEA(+)-sensitive KCNQ1 constructs reveal pore-independent access to KCNE1 in assembled I(Ks) channels. *J Gen Physiol* **117**(1): 43–52.

Lai, L. P., Y. N. Su et al. (2005). Denaturing high-performance liquid chromatography screening of the long QT syndrome-related cardiac sodium and potassium channel genes and identification of novel mutations and single nucleotide polymorphisms. *J Hum Genet* **50**(9): 490–496.

Leonetti, M. D., P. Yuan et al. (2012). Functional and structural analysis of the human SLO3 pH- and voltage-gated K+ channel. *Proc Natl Acad Sci USA* **109**(47): 19274–19279.

Levy, D. I., S. Wanderling et al. (2008). MiRP3 acts as an accessory subunit with the BK potassium channel. *Am J Physiol Renal Physiol* **295**(2): F380–F387.

Liu, G., X. Niu et al. (2010). Location of modulatory beta subunits in BK potassium channels. *J Gen Physiol* **135**(5): 449–459.

Liu, G., S. I. Zakharov et al. (2008a). Locations of the beta1 transmembrane helices in the BK potassium channel. *Proc Natl Acad Sci USA* **105**(31): 10727–10732.

Liu, W. J., H. T. Wang et al. (2008b). Co-expression of KCNE2 and KChIP2c modulates the electrophysiological properties of Kv4.2 current in COS-7 cells. *Acta Pharmacol Sin* **29**(6): 653–660.

Liu, X. F., L. Xiang et al. (2012). CAPC negatively regulates NF-kappaB activation and suppresses tumor growth and metastasis. *Oncogene* **31**(13): 1673–1682.

Long, S. B., E. B. Campbell et al. (2005). Crystal structure of a mammalian voltage-dependent *Shaker* family K+ channel. *Science* **309**(5736): 897–903.

Loster, K., K. Zeilinger et al. (1995). The cysteine-rich region of dipeptidyl peptidase IV (CD 26) is the collagen-binding site. *Biochem Biophys Res Commun* **217**(1): 341–348.

Loussouarn, G., K. H. Park et al. (2003). Phosphatidylinositol-4,5-bisphosphate, PIP2, controls KCNQ1/KCNE1 voltage-gated potassium channels: A functional homology between voltage-gated and inward rectifier K+ channels. *EMBO J* **22**(20): 5412–5421.

Lu, Y., M. P. Mahaut-Smith et al. (2003). Mutant MiRP1 subunits modulate HERG K+ channel gating: A mechanism for pro-arrhythmia in long QT syndrome type 6. *J Physiol* **551**(Part 1): 253–262.

Lundby, A., L. S. Ravn et al. (2007). KCNQ1 mutation Q147R is associated with atrial fibrillation and prolonged QT interval. *Heart Rhythm* **4**(12): 1532–1541.

Lundquist, A. L., C. L. Turner et al. (2006). Expression and transcriptional control of human KCNE genes. *Genomics* **87**(1): 119–128.

Lvov, A., S. D. Gage et al. (2010). Identification of a protein-protein interaction between KCNE1 and the activation gate machinery of KCNQ1. *J Gen Physiol* **135**(6): 607–618.

Ma, L., C. Lin et al. (2003). Characterization of a novel Long QT syndrome mutation G52R-KCNE1 in a Chinese family. *Cardiovasc Res* **59**(3): 612–619.

Manderfield, L. J., M. A. Daniels et al. (2009). KCNE4 domains required for inhibition of KCNQ1. *J Physiol* **587**(Part 2): 303–314.

Marshall, C. R., A. Noor et al. (2008). Structural variation of chromosomes in autism spectrum disorder. *Am J Hum Genet* **82**(2): 477–488.

McCrossan, Z. A. and G. W. Abbott (2004). The MinK-related peptides. *Neuropharmacology* **47**(6): 787–821.

McNicholas, K., T. Chen et al. (2009). Dipeptidyl peptidase (DP) 6 and DP10: Novel brain proteins implicated in human health and disease. *Clin Chem Lab Med* **47**(3): 262–267.

Melman, Y. F., A. Domenech et al. (2001). Structural determinants of KvLQT1 control by the KCNE family of proteins. *J Biol Chem* **276**(9): 6439–6444.

Melman, Y. F., A. Krumerman et al. (2002). A single transmembrane site in the KCNE-encoded proteins controls the specificity of KvLQT1 channel gating. *J Biol Chem* **277**(28): 25187–25194.

Melman, Y. F., S. Y. Um et al. (2004). KCNE1 binds to the KCNQ1 pore to regulate potassium channel activity. *Neuron* **42**(6): 927–937.

Meng, J., P. Parroche et al. (2008). The differential impact of disulfide bonds and N-linked glycosylation on the stability and function of CD14. *J Biol Chem* **283**(6): 3376–3384.

Morera, F. J., A. Alioua et al. (2012). The first transmembrane domain (TM1) of beta2-subunit binds to the transmembrane domain S1 of alpha-subunit in BK potassium channels. *FEBS Lett* **586**(16): 2287–2293.

Morin, T. J. and W. R. Kobertz (2008a). Counting membrane-embedded KCNE beta-subunits in functioning K+ channel complexes. *Proc Natl Acad Sci USA* **105**(5): 1478–1482.

Morin, T. J. and W. R. Kobertz (2008b). Tethering chemistry and K+ channels. *J Biol Chem* **283**(37): 25105–25109.

Nadal, M. S., A. Ozaita et al. (2003). The CD26-related dipeptidyl aminopeptidase-like protein DPPX is a critical component of neuronal A-type K+ channels. *Neuron* **37**(3): 449–461.

Nagase, T., R. Kikuno et al. (2000). Prediction of the coding sequences of unidentified human genes. XVII. The complete sequences of 100 new cDNA clones from brain which code for large proteins in vitro. *DNA Res* **7**(2): 143–150.

Naim, H. Y., G. Joberty et al. (1999). Temporal association of the N- and O-linked glycosylation events and their implication in the polarized sorting of intestinal brush border sucrase-isomaltase, aminopeptidase N, and dipeptidyl peptidase IV. *J Biol Chem* **274**(25): 17961–17967.

Nakajo, K. and Y. Kubo (2007). KCNE1 and KCNE3 stabilize and/or slow voltage sensing S4 segment of KCNQ1 channel. *J Gen Physiol* **130**(3): 269–281.

Nakajo, K., M. H. Ulbrich et al. (2010). Stoichiometry of the KCNQ1-KCNE1 ion channel complex. *Proc Natl Acad Sci USA* **107**(44): 18862–18867.

Napolitano, C., S. G. Priori et al. (2005). Genetic testing in the long QT syndrome: Development and validation of an efficient approach to genotyping in clinical practice. *JAMA* **294**(23): 2975–2980.

Nattel, S., A. Maguy et al. (2007). Arrhythmogenic ion-channel remodeling in the heart: Heart failure, myocardial infarction, and atrial fibrillation. *Physiol Rev* **87**(2): 425–456.

Neyroud, N., F. Tesson et al. (1997). A novel mutation in the potassium channel gene KVLQT1 causes the Jervell and Lange-Nielsen cardioauditory syndrome. *Nat Genet* **15**(2): 186–189.

Nicolas, C. S., K. H. Park et al. (2008). IKs response to protein kinase A-dependent KCNQ1 phosphorylation requires direct interaction with microtubules. *Cardiovasc Res* **79**(3): 427–435.

Ohno, S., D. P. Zankov et al. (2011). KCNE5 (KCNE1L) variants are novel modulators of Brugada syndrome and idiopathic ventricular fibrillation. *Circ Arrhythm Electrophysiol* **4**(3): 352–361.

Ohya, S., K. Asakura et al. (2002). Molecular and functional characterization of ERG, KCNQ, and KCNE subtypes in rat stomach smooth muscle. *Am J Physiol Gastrointest Liver Physiol* **282**(2): G277–G287.

Osteen, J. D., C. Gonzalez et al. (2010). KCNE1 alters the voltage sensor movements necessary to open the KCNQ1 channel gate. *Proc Natl Acad Sci USA* **107**(52): 22710–22715.

Panaghie, G., K. K. Tai et al. (2006). Interaction of KCNE subunits with the KCNQ1 K+ channel pore. *J Physiol* **570**(Part 3): 455–467.

Piccini, M., F. Vitelli et al. (1999). KCNE1-like gene is deleted in AMME contiguous gene syndrome: Identification and characterization of the human and mouse homologs. *Genomics* **60**(3): 251–257.

Pickart, C. M. (2001). Mechanisms underlying ubiquitination. *Annu Rev Biochem* **70**: 503–533.

Poulsen, A. N. and D. A. Klaerke (2007). The KCNE1 beta-subunit exerts a transient effect on the KCNQ1 K+ channel. *Biochem Biophys Res Commun* **363**(1): 133–139.

Qi, S. Y., P. J. Riviere et al. (2003). Cloning and characterization of dipeptidyl peptidase 10, a new member of an emerging subgroup of serine proteases. *Biochem J* **373**(Part 1): 179–189.

Radicke, S., D. Cotella et al. (2005). Expression and function of dipeptidyl-aminopeptidase-like protein 6 as a putative beta-subunit of human cardiac transient outward current encoded by Kv4.3. *J Physiol* **565**(Part 3): 751–756.

Radicke, S., T. Riedel et al. (2013). Accessory subunits alter the temperature sensitivity of Kv4.3 channel complexes. *J Mol Cell Cardiol* **56**: 8–18.

Rasmussen, H. B., S. Branner et al. (2003). Crystal structure of human dipeptidyl peptidase IV/CD26 in complex with a substrate analog. *Nat Struct Biol* **10**(1): 19–25.

Ravn, L. S., Y. Aizawa et al. (2008). Gain of function in IKs secondary to a mutation in KCNE5 associated with atrial fibrillation. *Heart Rhythm* **5**(3): 427–435.

Ren, X., Y. Hayashi et al. (2005). Transmembrane interaction mediates complex formation between peptidase homologues and Kv4 channels. *Mol Cell Neurosci* **29**(2): 320–332.

Rocheleau, J. M., S. D. Gage et al. (2006). Secondary structure of a KCNE cytoplasmic domain. *J Gen Physiol* **128**(6): 721–729.

Rocheleau, J. M. and W. R. Kobertz (2008). KCNE peptides differently affect voltage sensor equilibrium and equilibration rates in KCNQ1 K+ channels. *J Gen Physiol* **131**(1): 59–68.

Roepke, T. K., A. Anantharam et al. (2006). The KCNE2 potassium channel ancillary subunit is essential for gastric acid secretion. *J Biol Chem* **281**(33): 23740–23747.

Roepke, T. K., E. C. King et al. (2009). Kcne2 deletion uncovers its crucial role in thyroid hormone biosynthesis. *Nat Med* **15**(10): 1186–1194.

Roepke, T. K., A. Kontogeorgis et al. (2008). Targeted deletion of kcne2 impairs ventricular repolarization via disruption of I(K,slow1) and I(to,f). *FASEB J* **22**(10): 3648–3660.

Roepke, T. K., K. Purtell et al. (2010). Targeted deletion of Kcne2 causes gastritis cystica profunda and gastric neoplasia. *PLoS One* **5**(7): e11451.

Romey, G., B. Attali et al. (1997). Molecular mechanism and functional significance of the MinK control of the KvLQT1 channel activity. *J Biol Chem* **272**(27): 16713–16716.

Ruiz-Canada, C., D. J. Kelleher et al. (2009). Cotranslational and posttranslational N-glycosylation of polypeptides by distinct mammalian OST isoforms. *Cell* **136**(2): 272–283.

Ruscic, K. J., F. Miceli et al. (2013). IKs channels open slowly because KCNE1 accessory subunits slow the movement of S4 voltage sensors in KCNQ1 pore-forming subunits. *Proc Natl Acad Sci USA* **110**(7): E559–E566.

Sanguinetti, M. C., M. E. Curran et al. (1996). Coassembly of K(V) LQT1 and minK (IsK) proteins to form cardiac I(Ks) potassium channel. *Nature* **384**(6604): 80–83.

Schade, J., M. Stephan et al. (2008). Regulation of expression and function of dipeptidyl peptidase 4 (DP4), DP8/9, and DP10 in allergic responses of the lung in rats. *J Histochem Cytochem* **56**(2): 147–155.

Schmidt, K., S. Radicke et al. (2010). The new beta-subunit DPP10 may contribute to Ito regulation in heart disease. *Cardiovasc Res* **87**(Suppl 1): S-371.

Schroeder, B. C., S. Waldegger et al. (2000). A constitutively open potassium channel formed by KCNQ1 and KCNE3. *Nature* **403**(6766): 196–199.

Schulze-Bahr, E., M. Schwarz et al. (2001). A novel long-QT 5 gene mutation in the C-terminus (V109I) is associated with a mild phenotype. *J Mol Med (Berl)* **79**(9): 504–509.

Seebohm, G., M. C. Sanguinetti et al. (2003). Tight coupling of rubidium conductance and inactivation in human KCNQ1 potassium channels. *J Physiol* **552**(Part 2): 369–378.

Sesti, F. and S. A. Goldstein (1998). Single-channel characteristics of wild-type IKs channels and channels formed with two minK mutants that cause long QT syndrome. *J Gen Physiol* **112**(6): 651–663.

Shamgar, L., Y. Haitin et al. (2008). KCNE1 constrains the voltage sensor of Kv7.1 K+ channels. *PLoS One* **3**(4): e1943.

Shi, G., K. Nakahira et al. (1996). Beta subunits promote K+ channel surface expression through effects early in biosynthesis. *Neuron* **16**(4): 843–852.

Shibata, R., H. Misonou et al. (2003). A fundamental role for KChIPs in determining the molecular properties and trafficking of Kv4.2 potassium channels. *J Biol Chem* **278**(September 19): 36445–36454.

Soh, H. and S. A. Goldstein (2008). I SA channel complexes include four subunits each of DPP6 and Kv4.2. *J Biol Chem* **283**(22): 15072–15077.

Splawski, I., J. Shen et al. (2000). Spectrum of mutations in long-QT syndrome genes. KVLQT1, HERG, SCN5A, KCNE1, and KCNE2. *Circulation* **102**(10): 1178–1185.

Strop, P., A. J. Bankovich et al. (2004). Structure of a human A-type potassium channel interacting protein DPPX, a member of the dipeptidyl aminopeptidase family. *J Mol Biol* **343**(4): 1055–1065.

Sulda, M. L., C. A. Abbott et al. (2006). DPIV/CD26 and FAP in cancer: A tale of contradictions. *Adv Exp Med Biol* **575**: 197–206.

Sun, J., K. E. Duffy et al. (2006). Structural and functional analyses of the human Toll-like receptor 3. Role of glycosylation. *J Biol Chem* **281**(16): 11144–11151.

Tai, K. K. and S. A. Goldstein (1998). The conduction pore of a cardiac potassium channel. *Nature* **391**(6667): 605–608.

Takumi, T., K. Moriyoshi et al. (1991). Alteration of channel activities and gating by mutations of slow ISK potassium channel. *J Biol Chem* **266**(33): 22192–22198.

Takumi, T., H. Ohkubo et al. (1988). Cloning of a membrane protein that induces a slow voltage-gated potassium current. *Science* **242**(4881): 1042–1045.

Tapper, A. R. and A. L. George, Jr. (2001). Location and orientation of minK within the I(Ks) potassium channel complex. *J Biol Chem* **276**(41): 38249–38254.

Teng, S., L. Ma et al. (2003). Novel gene hKCNE4 slows the activation of the KCNQ1 channel. *Biochem Biophys Res Commun* **303**(3): 808–813.

Thevenod, F. (2002). Ion channels in secretory granules of the pancreas and their role in exocytosis and release of secretory proteins. *Am J Physiol Cell Physiol* **283**(3): C651–C672.

Thielitz, A., S. Ansorge et al. (2008). The ectopeptidases dipeptidyl peptidase IV (DP IV) and aminopeptidase N (APN) and their related enzymes as possible targets in the treatment of skin diseases. *Front Biosci* **13**: 2364–2375.

Tinel, N., S. Diochot et al. (2000a). KCNE2 confers background current characteristics to the cardiac KCNQ1 potassium channel. *EMBO J* **19**(23): 6326–6330.

Tinel, N., S. Diochot et al. (2000b). M-type KCNQ2-KCNQ3 potassium channels are modulated by the KCNE2 subunit. *FEBS Lett* **480**(2–3): 137–141.

Tranebjaerg, L., R. A. Samson et al. (1993). Jervell and Lange-Nielsen syndrome. *Gene Reviews*. R. A. Pagon, T. D. Bird, C. R. Dolan, K. Stephens and M. P. Adam (eds.). University of Washington, Seattle, WA.

Tristani-Firouzi, M. and M. C. Sanguinetti (1998). Voltage-dependent inactivation of the human K+ channel KvLQT1 is eliminated by association with minimal K+ channel (minK) subunits. *J Physiol* **510** (Part 1): 37–45.

Tsevi, I., R. Vicente et al. (2005). KCNQ1/KCNE1 channels during germ-cell differentiation in the rat: Expression associated with testis pathologies. *J Cell Physiol* **202**(2): 400–410.

Tyson, J., L. Tranebjaerg et al. (1997). IsK and KvLQT1: Mutation in either of the two subunits of the slow component of the delayed rectifier potassium channel can cause Jervell and Lange-Nielsen syndrome. *Hum Mol Genet* **6**(12): 2179–2185.

Um, S. Y. and T. V. McDonald (2007). Differential association between HERG and KCNE1 or KCNE2. *PLoS One* **2**(9): e933.

van Bemmelen, M. X., J. S. Rougier et al. (2004). Cardiac voltage-gated sodium channel Nav1.5 is regulated by Nedd4–2 mediated ubiquitination. *Circ Res* **95**(3): 284–291.

van Es, M. A., P. W. van Vught et al. (2008). Genetic variation in DPP6 is associated with susceptibility to amyotrophic lateral sclerosis. *Nat Genet* **40**(1): 29–31.

Vanoye, C. G., R. C. Welch et al. (2010). KCNQ1/KCNE1 assembly, co-translation not required. *Channels (Austin)* **4**(2): 108–114.

Vetter, D. E., J. R. Mann et al. (1996). Inner ear defects induced by null mutation of the isk gene. *Neuron* **17**(6): 1251–1264.

Wada, K., N. Yokotani et al. (1992). Differential expression of two distinct forms of mRNA encoding members of a dipeptidyl aminopeptidase family. *Proc Natl Acad Sci USA* **89**(1): 197–201.

Wallner, M., P. Meera et al. (1999). Molecular basis of fast inactivation in voltage and Ca2+-activated K+ channels: A transmembrane beta-subunit homolog. *Proc Natl Acad Sci USA* **96**(7): 4137–4142.

Wang, A. G., S. Y. Yoon et al. (2006). Identification of intrahepatic cholangiocarcinoma related genes by comparison with normal liver tissues using expressed sequence tags. *Biochem Biophys Res Commun* **345**(3): 1022–1032.

Wang, K. W. and S. A. Goldstein (1995). Subunit composition of minK potassium channels. *Neuron* **14**(6): 1303–1309.

Wang, K. W., K. K. Tai et al. (1996). MinK residues line a potassium channel pore. *Neuron* **16**(3): 571–577.

Wang, W., J. Xia et al. (1998). MinK-KvLQT1 fusion proteins, evidence for multiple stoichiometries of the assembled IsK channel. *J Biol Chem* **273**(51): 34069–34074.

Wang, Y. W., J. P. Ding et al. (2002). Consequences of the stoichiometry of Slo1 alpha and auxiliary beta subunits on functional properties of large-conductance Ca2+-activated K+ channels. *J Neurosci* **22**(5): 1550–1561.

Warth, R. and J. Barhanin (2002). The multifaceted phenotype of the knockout mouse for the KCNE1 potassium channel gene. *Am J Physiol Regul Integr Comp Physiol* **282**(3): R639–R648.

Weber, A. N., M. A. Morse et al. (2004). Four N-linked glycosylation sites in human toll-like receptor 2 cooperate to direct efficient biosynthesis and secretion. *J Biol Chem* **279**(33): 34589–34594.

Weber, R. E., H. Malte et al. (1995). Mass spectrometric composition, molecular mass and oxygen binding of Macrobdella decora hemoglobin and its tetramer and monomer subunits. *J Mol Biol* **251**(5): 703–720.

Weiger, T. M., M. H. Holmqvist et al. (2000). A novel nervous system beta subunit that downregulates human large conductance calcium-dependent potassium channels. *J Neurosci* **20**(10): 3563–3570.

Wilson, G. G., A. Sivaprasadarao et al. (1994). Changes in activation gating of IsK potassium currents brought about by mutations in the transmembrane sequence. *FEBS Lett* **353**(3): 251–254.

Witzel, K., P. Fischer et al. (2012). Hippocampal A-type current and Kv4.2 channel modulation by the sulfonylurea compound NS5806. *Neuropharmacology* **63**(8): 1389–1403.

Wu, D., K. Delaloye et al. (2010a). State-dependent electrostatic interactions of S4 arginines with E1 in S2 during Kv7.1 activation. *J Gen Physiol* **135**(6): 595–606.

Wu, D., H. Pan et al. (2010b). KCNE1 remodels the voltage sensor of Kv7.1 to modulate channel function. *Biophys J* **99**(11): 3599–3608.

Wu, R. S., N. Chudasama et al. (2009). Location of the beta 4 transmembrane helices in the BK potassium channel. *J Neurosci* **29**(26): 8321–8328.

Xia, X., B. Hirschberg et al. (1998). dSLo interacting protein 1, a novel protein that interacts with large-conductance calcium-activated potassium channels. *J Neurosci* **18**(7): 2360–2369.

Xu, J., W. Yu et al. (1998). Distinct functional stoichiometry of potassium channel beta subunits. *Proc Natl Acad Sci USA* **95**(4): 1846–1851.

Xu, X., M. Jiang et al. (2008). KCNQ1 and KCNE1 in the IKs channel complex make state-dependent contacts in their extracellular domains. *J Gen Physiol* **131**(6): 589–603.

Xu, X., V. A. Kanda et al. (2009). MinK-dependent internalization of the IKs potassium channel. *Cardiovasc Res* **82**(3): 430–438.

Yan, J. and R. W. Aldrich (2010). LRRC26 auxiliary protein allows BK channel activation at resting voltage without calcium. *Nature* **466**(7305): 513–516.

Yan, J. and R. W. Aldrich (2012). BK potassium channel modulation by leucine-rich repeat-containing proteins. *Proc Natl Acad Sci USA* **109**(20): 7917–7922.

Yan, Y., D. J. Scott et al. (2008). Identification of the N-linked glycosylation sites of the human relaxin receptor and effect of glycosylation on receptor function. *Biochemistry* **47**(26): 6953–6968.

Yang, C., X. H. Zeng et al. (2011). LRRC52 (leucine-rich-repeat-containing protein 52), a testis-specific auxiliary subunit of the alkalization-activated Slo3 channel. *Proc Natl Acad Sci USA* **108**(48): 19419–19424.

Yanglin, P., Z. Lina et al. (2007). KCNE2, a down-regulated gene identified by in silico analysis, suppressed proliferation of gastric cancer cells. *Cancer Lett* **246**(1–2): 129–138.

Zagha, E., A. Ozaita et al. (2005). DPP10 modulates Kv4-mediated A-type potassium channels. *J Biol Chem* **280**(19): 18853–18861.

Zhou, Y., W. M. Schopperle et al. (1999). A dynamically regulated 14–3–3, Slob, and Slowpoke potassium channel complex in *Drosophila* presynaptic nerve terminals. *Neuron* **22**(4): 809–818.

Index